(10)	1B (11)	2B (12)	3A (13)	4A (14)	5A (15)	6A (16)	7A (17)	8A (18)
							Hydrogen 1 **H** 1.0079	Helium 2 **He** 4.0026
			Boron 5 **B** 10.811	Carbon 6 **C** 12.011	Nitrogen 7 **N** 14.0067	Oxygen 8 **O** 15.9994	Fluorine 9 **F** 18.9984	Neon 10 **Ne** 20.1797
			Aluminum 13 **Al** 26.9815	Silicon 14 **Si** 28.0855	Phosphorus 15 **P** 30.9738	Sulfur 16 **S** 32.066	Chlorine 17 **Cl** 35.4527	Argon 18 **Ar** 39.948
Nickel 28 **Ni** 58.693	Copper 29 **Cu** 63.546	Zinc 30 **Zn** 65.39	Gallium 31 **Ga** 69.723	Germanium 32 **Ge** 72.61	Arsenic 33 **As** 74.9216	Selenium 34 **Se** 78.96	Bromine 35 **Br** 79.904	Krypton 36 **Kr** 83.80
Palladium 46 **Pd** 106.42	Silver 47 **Ag** 107.8682	Cadmium 48 **Cd** 112.411	Indium 49 **In** 114.82	Tin 50 **Sn** 118.710	Antimony 51 **Sb** 121.757	Tellurium 52 **Te** 127.60	Iodine 53 **I** 126.9045	Xenon 54 **Xe** 131.29
Platinum 78 **Pt** 195.08	Gold 79 **Au** 196.9665	Mercury 80 **Hg** 200.59	Thallium 81 **Tl** 204.3833	Lead 82 **Pb** 207.2	Bismuth 83 **Bi** 208.9804	Polonium 84 **Po** (209)	Astatine 85 **At** (210)	Radon 86 **Rn** (222)
110 Discovered Nov. 1994	111 Discovered Dec. 1994							

Europium 63 **Eu** 151.965	Gadolinium 64 **Gd** 157.25	Terbium 65 **Tb** 158.9253	Dysprosium 66 **Dy** 162.50	Holmium 67 **Ho** 164.9303	Erbium 68 **Er** 167.26	Thulium 69 **Tm** 168.9342	Ytterbium 70 **Yb** 173.04	Lutetium 71 **Lu** 174.967
Americium 95 **Am** (243)	Curium 96 **Cm** (247)	Berkelium 97 **Bk** (247)	Californium 98 **Cf** (251)	Einsteinium 99 **Es** (252)	Fermium 100 **Fm** (257)	Mendelevium 101 **Md** (258)	Nobelium 102 **No** (259)	Lawrencium 103 **Lr** (260)

P555

Chemistry & Chemical Reactivity

Third Edition

John C. Kotz

SUNY Distinguished Teaching Professor
State University of New York
College at Oneonta

Paul Treichel, Jr.

Professor of Chemistry
University of Wisconsin–Madison

Saunders Golden Sunburst Series

SAUNDERS COLLEGE PUBLISHING

Harcourt Brace College Publishers

Fort Worth Philadelphia San Diego New York Orlando Austin
San Antonio Toronto Montreal London Sydney Tokyo

Text Typeface: New Baskerville
Compositor: York Graphic Services
Publisher: John Vondeling
Developmental Editor: Elizabeth C. Rosato
Managing Editor and Project Editor: Carol Field
Copy Editor: Linda Davoli
Manager of Art & Design: Carol Bleistine
Art Directors: Anne Muldrow and Carol Bleistine
Text Design: Rebecca Lloyd Lemna
Cover Design: Lawrence R. Didona
Text Artwork: George Kelvin and JAK Studios
Layout Artist: Claudia Durrell
Photo Editor: Kathrine Kotz
Director of EDP: Tim Frelick
Production Manager: Charlene Squibb
Marketing Managers: Marjorie Waldron and Angus McDonald

Cover: Beaker boiling over by Jook Leung, © FPG International

Printed in the United States of America

CHEMISTRY & CHEMICAL REACTIVITY

ISBN 0-03-001291-0

Library of Congress Catalog Card Number: 95-069433

678 032 9876543

Preface

This is the third edition of *CHEMISTRY & CHEMICAL REACTIVITY.* Although some years have passed since the book was first conceived, the principal theme has remained the same: to provide a broad overview of the principles of chemistry and the reactivity of chemical elements and compounds. We also have hoped to convey that chemistry, a field with a lively history, is also dynamic, with important new developments on the horizon. In addition, we want to provide some insight into the chemical aspects of the world around us. For example, what materials are important to our economy, how does chemistry contribute to health care, and what role do chemists play in protecting the environment? By tackling the principles leading to answers to these questions, you can come to a better general understanding of nature and to an appreciation for some of the consumer products coming from the chemical industry. Indeed, one of the objectives of this book is to provide the tools and background information for you to function as an informed citizen in a technologically complex world. Learning something of the chemical world is as important as understanding some basic mathematics and biology and having an appreciation for fine music and literature.

We are also very excited by the fact that this is the first chemistry textbook to be offered on a CD-ROM (compact disc-read-only-memory). Computers have the capability to organize and convey information, and our CD-ROM is the first attempt to make this resource available to students. Not only is the actual textbook—with all the photos and figures—available on the CD-ROM, but the material in each chapter is presented in an interactive manner. In addition, there are mathematical and molecular modeling tools and an illustrated database of compounds and their properties. The disc is meant to be an individual learning tool. Therefore, it is available for purchase with the textbook or as a stand-alone product. The contents of the disc are outlined in more detail below.

The authors of this book became chemists because, simply put, it is exciting to discover new compounds and find new ways to apply chemical principles. We hope to have conveyed our enjoyment of chemistry in this book as well as our awe at what is known about chemistry and, just as important, what is not known!

AUDIENCE FOR THE BOOK AND CD-ROM

CHEMISTRY & CHEMICAL REACTIVITY is a textbook for introductory courses in chemistry for students interested in further study in science, whether that science is biology, chemistry, engineering, geology, physics, or related subjects. Our assumption is that students beginning this course have had some preparation in algebra and in general science. Although undeniably helpful, a previous exposure to chemistry is neither assumed nor required.

PHILOSOPHY AND APPROACH OF THE BOOK

When the first edition of this book was planned, we had two major, but not independent, goals. This edition shares these same goals. The first was to con-

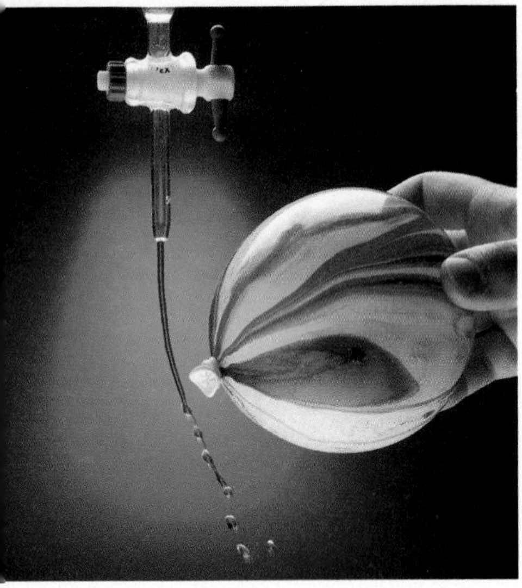

struct a book that students would enjoy reading and that would offer, at a reasonable level of rigor, chemistry and chemical principles in a format and organization typical of college and university courses today. Second, we wanted to convey the utility and importance of chemistry by introducing the properties of the elements, their compounds, and their reactions as early as possible and by focusing the discussion as much as possible on these subjects.

A glance at the introductory chemistry texts currently available shows that there is a generally common order of treatment of chemical principles used by educators. With a few minor changes, we followed that order. That is not to say that the chapters cannot be used in some other order. For example, although the behavior of gases is often studied early in a chemistry course, the chapter on this topic (Chapter 12) has been placed with chapters on liquids, solids, and solutions because it logically fits with these other topics. It can easily be read and understood, however, after covering only the first four or five chapters of the book.

The discussion of organic chemistry (Chapter 11) is typically left to one of the final chapters in chemistry textbooks. We believe, however, that the importance of organic compounds in biochemistry and in the chemical industry means that we should present that material earlier in the sequence of chapters. Therefore, we placed it in the middle of the book, following the chapters on structure and bonding. We chose this position because the principles of structure and bonding are particularly well illustrated by organic compounds. In any event, this chapter may be covered at almost any point in the book.

In addition, one of the authors of this text sometimes teaches the material on equilibria involving insoluble solids (Chapter 19) before acid-base equilibria (Chapters 17 and 18), and introduces kinetics (Chapter 15) and thermodynamics (Chapter 20) as a unit, after all of the material on equilibria. Although chapters are loosely organized into groups with common themes, *every attempt has been made to make the chapters as independent as possible.*

The order of topics in the text was also devised to introduce as early as possible the background required for the laboratory experiments usually done in General Chemistry. For this reason, chapters on common reaction types (especially acid-base and oxidation-reduction reactions, Chapter 4) and stoichiometry (Chapter 5) begin the book. In addition, because an understanding of energy is so important in the study of chemistry, thermochemistry is introduced in Chapter 6.

The American Chemical Society has been urging educators to put "chemistry" back into introductory chemistry courses, and we agree wholeheartedly. Therefore, we describe the elements, their compounds, and their reactions as early and as often as possible in three ways. First, there are numerous color photographs of reactions occurring, of the elements and common compounds, and of common laboratory operations and industrial processes. Second, we bring material on the properties of elements and compounds as early as possible into the Exercises and Study Questions and introduce new principles using realistic chemical situations. Third, we include sections called *Current Issues in Chemistry* that discuss such topics as the effects of chlorine-containing compounds in the environment (Chapter 2), buckyball chemistry (Chapter 3), the chemistry of black smokers (Chapter 4), the disposal of industrial wastes (Chapter 5), the use of hydrogen as a fuel (Chapter 6), the chemistry of NO (Chapter 9), surviving at sea (Chapter 14), and the destruction of the earth's ozone layer (Chapter 15).

PHILOSOPHY AND APPROACH OF THE CD-ROM

The CD-ROM was co-authored by William Vining and produced by Archipelago Productions. It is designed to take advantage of what computers do best: allow the user to *interact* with information. Therefore, our goal was to produce "an interactive movie about the book." The material in each chapter is presented in a series of "screens," each presenting an idea or concept and allowing the user to interact with the information in some manner—by seeing video of a reaction in progress, by changing a variable in a chemical experiment and watching what happens to the system, or by listening to important tips and ideas about ways to understand a concept or solve a problem. In addition, you will see practicing chemists describe how the topic of the chapter applies to their work.

To make the CD-ROM even more useful, the text of the actual book is also present, with all the photos and figures. Therefore, you will be able to study the material using the interactive presentation, and you can find the most pertinent section of the book with the click of a mouse. If you need information on a compound or need to plot a graph or view a molecule, the other tools on the disc are readily available. A workbook to accompany the CD-ROM will help you use the interactive presentation of the book.

The combination of an interactive presentation of chemistry, coupled with the text of the actual book and other tools for doing chemistry, means that the CD-ROM provides all the resources needed to learn chemistry. Depending on your needs, and the availability of computer facilities, you may purchase only the book, only the CD-ROM with its workbook, or both the book and the CD-ROM.

THE AUTHORS

John Kotz has been the principal author of the first two editions of this book, working with Keith Purcell. With this edition we welcome a new author to our team, Professor Paul Treichel of the University of Wisconsin–Madison. Paul has over 30 years of experience teaching General Chemistry to majors and nonmajors. His background in inorganic chemistry has contributed to writing this text, particularly with respect to the chemistry of the elements and to aspects of molecular structure.

John C. Kotz received his Ph.D. from *Cornell University* in 1964 and currently teaches chemistry at the *SUNY College at Oneonta,* where he was promoted to University Distinguished Teaching Professor in 1986. In 1979, he was a Fulbright Lecturer and Research Scholar in Lisbon, Portugal. He has received the National Catalyst Award in Chemical Education from the Chemical Manufacturers Association. Kotz is an editor of *Chem Matters* magazine and is on the board of editors of the *Journal of Chemical Education: Software.* He is also the co-author of two inorganic chemistry textbooks and another Saunders introductory general chemistry text, *The Chemical World: Concepts and Applications.*

Paul Treichel, Jr., received a B.S. from the *University of Wisconsin–Madison* in 1958 and a Ph.D. from *Harvard University* in 1962. After a year of postdoctoral study at *Queen Mary College* in London, he assumed a faculty position at the *University of Wisconsin–Madison,* where he has taught general and inorganic chemistry for 32 years and served as Department Chair from 1986 to 1995. Treichel's research in organometallic chemistry, aided by 75 graduate and undergraduate students, has resulted in the publication of more than 160 articles in scientific journals.

ORGANIZATION OF THE BOOK

CHEMISTRY & CHEMICAL REACTIVITY is organized in two ways. First, there are chapters on the *Principles of Reactivity,* and others on *Bonding and Molecular Structure* that are especially important in carrying the themes of the book.

The chapters on *Principles of Reactivity* introduce you to the factors involved in chemical reactions that lead to the successful production of products. Thus, under this topic you will study common types of reactions, the energy involved in reactions, and the factors that affect the speed of a reaction.

The principles of *Bonding and Molecular Structure* are particularly important. If you page through the book, you will notice the abundance of molecular models, some drawn by George Kelvin, our principal artist for this edition, and others drawn with a computer. As described in several places in the book (An Introduction—The Nature of Chemistry; Chapter 3: *Buckyballs, AIDS, and Chemistry;* and Chapter 9: *Computer Molecular Modeling*) an understanding of molecular structures is one cornerstone of modern chemistry. Using the latest laboratory techniques for uncovering molecular structures and computer programs that generate revealing portraits of structures, chemists have enormous insight into the ways molecules react.

Second, the book is also divided roughly into five sections, each with a grouping of chapters with a common theme.

Part 1: The Basic Tools of Chemistry

Certain ideas and methods form the fabric of chemistry, and these basic tools are introduced in Part 1. Chapter 1 defines some important terms and is a review of units and mathematical methods. Chapters 2 and 3 introduce atoms and molecules, and Chapter 2 introduces the periodic table, one of the most important resources available to chemists. In Chapters 4 and 5 we begin to discuss some principles of chemical reactivity and to introduce the numerical methods used by chemists to extract quantitative information from chemical reactions. Chapter 6 is the first introduction to the energy involved in chemical processes.

Part 2: The Structure of Atoms and Molecules

One major goal of this section is to outline (in Chapters 7 and 8) the current theories of the arrangement of electrons in atoms and some of the historical developments that led to these ideas. With this background information, we can understand better why atoms and their ions have different chemical and physical properties. This discussion is tied closely to the periodic table so that these properties can be recalled and predictions made. In Chapter 9 we discuss how the electrons of atoms in a molecule may lead to chemical bonding and the properties of these bonds. In addition, we show how to derive the three-dimen-

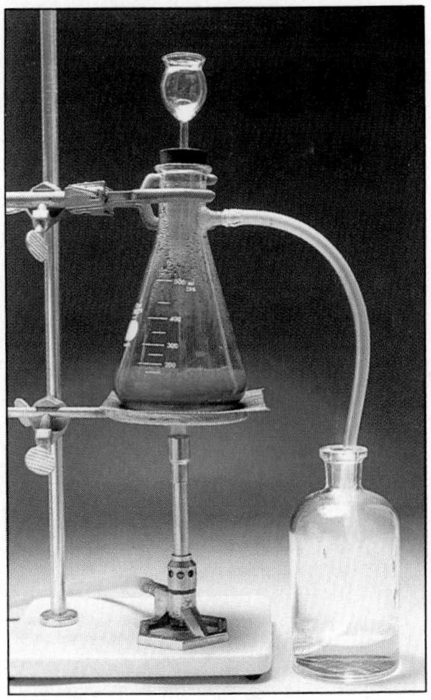

sional structure of simple molecules. Finally, Chapter 10 considers two of the major theories of chemical bonding in more detail.

This part of the book is completed with a discussion of organic chemistry in Chapter 11. Organic chemistry is such an enormous area of chemistry that we cannot hope to cover it in detail in this book. We chose to focus on the structures of compounds and on compounds of particular importance, such as synthetic polymers.

Part 3: States of Matter

The behavior of the three states of matter—gases, liquids, and solids—is described in that order in Chapters 12 and 13. The discussion of liquids and solids is tied to gases through the description of intermolecular forces, with particular attention given to liquid and solid water. Chapter 13 also considers the solid state, an area of chemistry currently undergoing a renaissance. In Chapter 14 we talk about the properties of solutions, intimate mixtures of gases, liquids, and solids.

Part 4: The Principles of Reactivity

This section is wholly concerned with the *Principles of Reactivity*. Chapter 15 examines the important question of the rates of chemical processes and the factors controlling these rates. With this in mind, we move to Chapters 16 through 19, to consider chemical reactions at equilibrium. After an introduction to equilibrium in Chapter 16, we highlight the reactions involving acids and bases in water (Chapters 17 and 18) and reactions leading to insoluble salts (Chapter 19). To tie together the discussion of chemical equilibria, we again explore thermodynamics in Chapter 20. As a final topic in this section, we describe in Chapter 21 a major class of chemical reactions—those involving the transfer of electrons—and the use of these reactions in electrochemical cells.

Part 5: The Chemistry of the Elements and Their Compounds

Although the chemistry of the various elements is described throughout the book, Part 5 considers this topic in a more systematic way. Chapter 22 presents the chemistry of the representative elements, and Chapter 23 is a discussion of the transition elements and their compounds. Finally, Chapter 24 is a brief discussion of nuclear chemistry.

ORGANIZATION OF THE CD-ROM

The organizational metaphor for the CD-ROM is that of a chemistry building. There is the "lecture room" where you will find the interactive version of the book. This is linked to the "library" in which you will find the full version of the book in computer-readable form, as well as a database of information on several hundred compounds. Finally, there is a "laboratory" with various tools that chemists use: graphing and mathematical tools and a program for viewing molecular structures.

To make the CD-ROM as useful as possible, we have also prepared a "workbook" with exercises and problems to be done as you work through a chapter. The workbook also contains the Appendices to the book.

NEW TO THIS EDITION

There are several new and exciting features to the third edition of *CHEMISTRY & CHEMICAL REACTIVITY.* For example

- The Saunders Interactive General Chemistry CD-ROM accompanies the text and provides students with the entire text of the book on the disc, extensive interactive capabilities, mathematical and molecular modeling tools, and an illustrated database of compounds. (See separate discussions on the CD-ROM in this Preface.)
- A more cohesive organization of chemical reactions and stoichiometry (Chapters 4 and 5, respectively) gives students the necessary background required for laboratory experimentation.
- An early discussion of organic chemistry (Chapter 11) stresses the importance of organic compounds in biochemistry and the chemical industry. Immediately following the chapters on structure and bonding, Chapter 11 applies the principles of those chapters. For those who wish to cover this material elsewhere in their course, this is a completely transportable chapter.
- A greater emphasis on conceptual problem solving is provided with "Problem-Solving Tips and Ideas," which target common errors and difficulties; revised examples integrated throughout the text; flow diagrams to immerse students in essential critical-thinking activities; and a dedicated section of conceptual problems at the end of each chapter.
- A refined art program features color-coded molecular models using both computer and artwork by George Kelvin, pedagogical use of color in periodic tables, and more than 600 full-color photos, many of which are completely new.
- High-interest "Portrait of a Scientist," "Current Issues in Chemistry," and "Interviews with Chemists" sections emphasize diversity and direct attention to the contributions of women and minorities in the field of chemistry.
- A new Introductory chapter presents the dynamic, current, applicable nature of chemistry, giving students insight into numerous areas of study.

FEATURES AND LEARNING AIDS IN THE BOOK

Problem Solving

Worked Examples and Solved Exercises

Several hundred worked-out examples serve as models for solving end-of-chapter problems. We developed the detailed solutions using the technique of dimensional analysis and highlighted the answers. Exercises follow all examples, with solutions given in Appendix L.

Problem-Solving Flow Diagrams

Scattered throughout the book and in the examples are problem-solving flow diagrams. Use these diagrams to help organize the information contained in the problems.

Problem-Solving Tips and Ideas

Based on our years of teaching chemistry, we have found that students make certain errors in solving problems and have very specific difficulties. These "Tips and Ideas" pass on our experience to you.

End-of-Chapter Study Questions

The end-of-chapter questions, some of which are illustrated with photographs or art, include review questions, questions classified by type, and general questions. *The classified problems are in matched pairs.* The first member of each pair is numbered with a bold-faced number and the answer is given in Appendix M.

Conceptual Questions

These questions ask you to think through the solution to a question or problem. Mathematics generally is not involved.

Summary Questions

Summary questions link the concepts discussed in the current chapter with those in previous chapters.

Essays on History and Current Issues

Portrait of a Scientist

These essays are about the lives of important scientists and provide some insight into the historical background of chemistry.

Current Issues in Chemistry

Scattered about the book are a number of essays that describe the applications of chemistry in the world today, from the current controversy over banning chlorine to the uses of buckyballs in medicine.

Interviews with Chemists

Interviews with three eminent chemists from different fields of chemistry appear in the book. These prominent teachers and researchers talk about how they became involved in science, about their research and their work in teaching or industry, and their views of the frontier areas of science, the importance of science in our society, and environmental concerns.

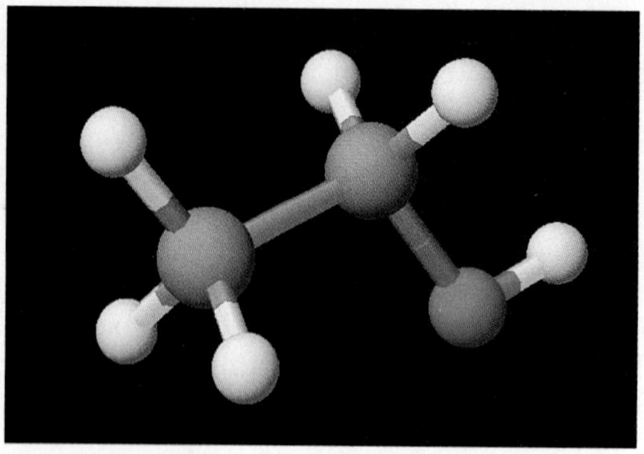

A computer-generated model of ethanol, CH_3CH_2OH. The C atoms are gray, the H atoms are white, and the O atom is red.

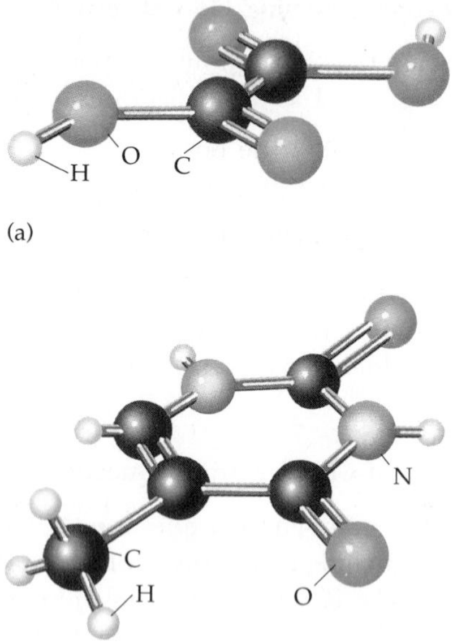

(a)

(b)

Two molecular formulas. (a) Oxalic acid, a compound found in many plants. (b) Thymine, one of the bases in DNA.

Illustrations

Full-Color Photography

Over 600 full-color photos, chosen specifically for this book, illustrate common elements, compounds, and minerals, as well as reactions and other processes in progress.

Use of Color in Art

Our pedagogical use of color makes the diagrams as attractive and meaningful as possible. For example,

- In the periodic tables in the text metals are shown in blue, metalloids in green, and nonmetals in yellow.
- A common color scheme is used in the illustration of molecular models.

Other Features

- Each chapter ends with Chapter Highlights, a summary of the important concepts, equations, and key terms.
- Boxes entitled "A Closer Look" delve deeper into topics related to the subject being discussed. They provide a more detailed discussion of a subject or a look into chemical research.

- Appendices at the back of the book include a review of mathematical methods, a table of conversion factors, important constants, and a glossary of terms in the combined index/glossary. Inside the back cover are short tables of useful constants and a listing of all of the data tables in the book.
- A Chemical Puzzler opens each chapter by asking thought-provoking questions. After reading a chapter, the students should be able to solve the Chemical Puzzler. We also provide brief answers to the puzzlers in Appendix L.

SUPPORTING MATERIALS

Written Materials for Student and Instructor

Pocket Guide by John DeKorte, Glendale Community College, contains useful summaries of each text section, as well as helpful problem-solving reminders and tips.

Study Guide by Harry Pence, SUNY–Oneonta, accompanies the text and has been designed around key objectives of the book. Each chapter includes a list of the main concepts, important terms, questions testing mastery of each objective, a test evaluating overall mastery of the chapter, and a set of comprehensive questions.

Student Solutions Manual by Alton Banks, North Carolina State University, contains detailed solutions to designated, end-of-chapter Study Questions.

The Use of Estimates in Solving Chemistry Problems by Michael Green and Denise Garland, both of The City College of the City University, New York, is designed to help students gain a feel for chemistry by first solving problems approximately and getting past the memorization of formulas.

Test Bank by Karen Eichstadt, Ohio University, contains over 1100 multiple choice questions and numerous fill-in printed questions for each chapter.

Instructor's Resource Manual by Jack Kotz, Susan Young of Roanoke College, Lynn Hunsberger of the University of Louisville, and Linda Zarzana of American River College, suggests alternative organizations for the course, classroom demonstrations, and worked-out solutions to questions not designated by a boldfaced number in the text.

Overhead Transparency Acetates provide a set of 150 full-color transparencies with large labels for viewing even in a large lecture hall. In addition, these transparencies as well as those from other Saunders College Publishing chemistry textbooks are available on the *Chemistry of Life* videodisc, available now, and the *Chemistry in Perspective* videodisc, due to publish January 1996. The transparencies and videodiscs are available at no charge to adopters of this book.

Chemical Principles in the Laboratory, 6th edition, by Emil Slowinski and Wayne Wolsey, both of Macalester College, and William Masterton, University of Connecticut, provides detailed directions and advance study assignments. This manual contains 42 experiments that have been selected with regard to cost and safety and that have been thoroughly class-tested.

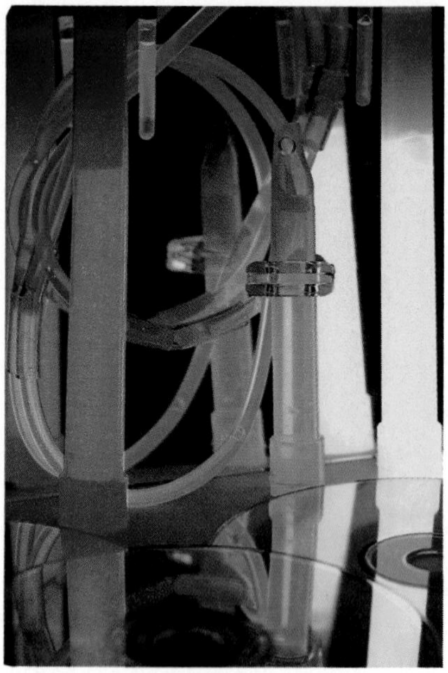

Instructor's Manual to accompany **Chemical Principles in the Laboratory,** 6th edition, provides lists of equipment and chemicals needed for each experiment.

Multimedia Materials

Saunders Interactive General Chemistry CD-ROM is a revolutionary interactive tool. This multimedia presentation serves as a companion to the text, or it can be used as an alternative to the text, because *CHEMISTRY & CHEMICAL REACTIVITY, 3e,* appears in its entirety on the CD-ROM. Divided into chapters, the CD-ROM presents ideas and concepts with which the user can interact in several different ways, for example, by watching a reaction in progress, changing a variable in an experiment and observing the results, and listening to tips and suggestions for understanding concepts or solving problems. Students navigate through the CD-ROM using original animation and graphics, interactive tools, pop-up definitions, over 100 video clips of chemical experiments, which are enhanced by sound effects and narration, and over 100 molecular models and animations.

Cambridge Scientific ChemDraw and Chem3D are available shrink-wrapped with the text for a nominal fee. These software packages enable students to draw molecular structures using ChemDraw. Users draw with ChemDraw and then can transfer their structure into Chem3D, which allows them to create, manipulate, and view three-dimensional color models for a clearer image of a molecule's shape and reaction sites. Cambridge Scientific provides an accompanying User's Guide and Quick Reference Card, written exclusively for Saunders College Publishing.

f(g) Scholar–Spreadsheet/Graphing Calculator/Graphing Software f(g) Scholar is a powerful scientific/engineering spreadsheet software program with over 300 built-in math functions, developed by Future Graph, Inc. It uniquely integrates graphing calculator, spreadsheet, and graphing applications into one, and allows for quick and easy movement between the applications. Students will find many uses for f(g) Scholar across their science, math, and engineering courses, including working through their laboratories from start to finished reports. Other features include a programming language for defining math functions, curve fitting, three-dimensional graphing, and equation displaying. When bookstores order f(g) Scholar through Saunders College Publishing, they can pass on our exclusive low price to the student.

CalTech Chemistry Animation Project (CAP) is a set of six video units of unmatched quality and clarity that cover the chemical topics of Atomic Orbitals, Valence Shell Electron Pair Repulsion Theory, Crystals and Unit Cells, Molecular Orbitals in Diatomic Molecules, Periodic Trends, and Hybridization and Resonance.

The Chemistry of Life Videodisc contains over 100 molecular model animations, chemical reaction videos, and approximately 2500 still images from a variety of Saunders College Publishing chemistry textbooks.

The Chemistry in Perspective Videodisc contains over 110 minutes of motion footage, including molecular model animations, chemical reaction videos, animated principles of chemistry, and videos demonstrating chemical principles at work in every day life, as well as 2000 still images from Saunders College Publishing 1996 chemistry titles.

LectureActive™ Software for Macintosh and IBM formats accompanies The Chemistry of Life Videodisc and contains references to all video clips and still images. Instructors can create custom lectures quickly and easily. Lectures can be read directly from the computer screen or printed with barcodes that contain videodisc instructions.

Barcode Manual for The Chemistry of Life Videodisc contains complete descriptions, barcode labels, and reference numbers for every video clip and still image. The Manual also provides practical advice about The Chemistry of Life Videodisc and set-up instructions for first-time users.

ExaMaster+™ Computerized Test Bank is the software version of the printed Test Bank. Instructors can create thousands of questions in a multiple-choice format. A command reformats the multiple-choice question into a short-answer question. Adding or modifying existing problems, as well as incorporating graphics, can be done. ExaMaster has gradebook capabilities for recording and graphing students' grades.

KC?Discoverer Software (JCE:Software), developed by a team of chemists, is an extensive database allowing users to explore 48 different properties of the chemical elements, such as atomic radii, density, ionization energy, color, reactivity with air, water, and acids and bases. The HELP menu provides a reference for the source of data for each of the properties and the database. KC?Discoverer Software has the capability to correlate with the Periodic Table Videodisc. Chosen as the "official" database for this book, references appear in the Annotated Instructor's Edition of this text, and all the data tables in this book reflect information from KC?Discoverer. This program is available to qualified adopters or it may be purchased directly from JCE: Software, Department of Chemistry, University of Madison, Wisconsin, 1101 University Avenue, Madison, WI 53706.

Periodic Table Videodisc: Reactions of the Elements (JCE: Software), by Alton Banks, North Carolina State University, features still and live footage of the elements, their uses, and their reactions with air, water, acids, and bases. Users operate the videodisc from a videodisc player with a hand-controlled keypad, a barcode reader, or an interface to a computer running KC?Discoverer Software. It is particularly useful as a way to demonstrate chemical reactions in a large lecture room. Available to qualified adopters.

Shakhashiri Demonstration Videotapes feature well-known instructor Bassam Shakhashiri, University of Wisconsin, performing 50 three- to five-minute chemical demonstrations. An accompanying Instructor's Manual describes each demonstration and includes discussion questions.

World of Chemistry Videotapes, taken from the popular PBS television series and hosted by Nobel laureate Roald Hoffmann, are two 40-minute videos highlighting topics such as the mole, bonding, and acid-base chemistry and their applications. Order through the Annenberg Foundation at 1-800-LEARNER.

Acknowledgments

Preparing the third edition of *CHEMISTRY & CHEMICAL REACTIVITY* took more than eighteen months of continuous effort. However, as in our work on the first two editions, we have had the support and encouragement of family and of some wonderful friends, colleagues, and students.

We begin by expressing our great appreciation to Mel Joesten, John Moore, and Jim Wood, co-authors with JCK on *The Chemical World.* This book set the tone for a new style of introductory chemistry books, and some of the ideas and materials developed during its writing have been used in this edition of *CHEMISTRY & CHEMICAL REACTIVITY.* In particular, we acknowledge John's contribution to the material on thermodynamics and kinetics and Mel's contribution to the section on polymer chemistry.

SAUNDERS COLLEGE PUBLISHING

The editorial staff of Saunders College Publishing has once again been extraordinarily helpful. The project has benefited from their good humor, friendship, and dedication. Much of the credit goes to our Publisher, John Vondeling. We have worked with John for many years and have become fast friends. His support and confidence are greatly appreciated—and he wears better ties than in the past. He has promised a long fishing trip when we retire.

The Developmental Editor for this edition was Beth Rosato. In addition to being a very pleasant colleague, she is a good chemist, is well organized, and has insight into the proper balance in a book. She has been a trusted friend and confidant, and we have appreciated her great efforts to make this a successful book.

Carol Field has once again served as both our Managing Editor and Project Editor. In addition, she has served as the Developmental Editor for CD-ROM. We have now worked with Carol for over eight years, and she has become part of our family. Her close attention to detail has helped make past editions of this book successful.

No book can be successful without proper marketing. Margie Waldron, Director of Marketing, and her staff have again been very helpful. Angus McDonald, Product Manager, joined the marketing team at Saunders specifically to market chemistry texts. We are always happy to work with Margie and Angus, and would even be willing to be part of a magic act again.

Our team at Saunders College Publishing is completed with Anne Muldrow, Carol Bleistine, Charlene Squibb, and Tim Frelick, who kept the art program and production of the book organized. Jay Freedman again did a wonderful job of creating the index/glossary.

PHOTOGRAPHY AND ART

Most of the color photographs for this edition have again been beautifully done by Charles D. Winters of Oneonta. He produced hundreds of photos for the

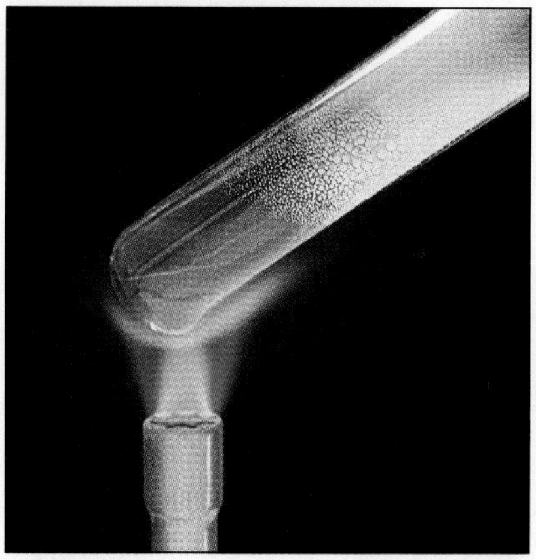

book and the CD-ROM, as well as the videos for the CD-ROM, often under great deadline pressure. We have worked with Charlie for some years and have become close friends. Charlie is still willing to eat the fried egg sandwiches one of us makes for him for lunch, and we listen to his old jokes—and always forget them.

Many of the illustrations in the book are by George Kelvin. George is simply the best scientific illustrator working today. He took our sketches and rendered them accurately and with an eye to a perspective most useful to students. We have thoroughly enjoyed working with him.

CD-ROM

William Vining of Hartwick College is largely responsible for authoring the CD-ROM version of the book. One of the authors (JCK) is proud that Bill is one of his former students and has become a close colleague. When we needed someone to translate the book into an interactive CD-ROM, Bill was our first choice. He is energetic, creative, pleasant to work with, and he has done a magnificent job.

The CD-ROM was produced by Archipelago Productions of Lake Tahoe, San Francisco, San Diego, and points in between. The vision was provided by Gary Lopez, an oceanographer who became interested in producing environmental films and educational software. It is Gary who has guided us through the very difficult, expensive, and complex task of creating the CD-ROM version of the text. He is a good friend.

Creating multimedia software cannot be done by a single person. We have had a large team of very talented people working on this project. Pat Harman designed the CD-ROM and played a principal role in translating our ideas into images on the computer screen. He has been a wonderful and creative colleague and we thank him for his patience. Bill Gudmundson created the computer art and animation sequences, and Brian Rowlett designed the software tools. Bruce Hoffman wrote the script for the spoken material and edited the written text. Nicole Taylor managed the project and kept track of the hundreds of photos and hours of video that we needed. Brian Griffith, Charles Hamper, Mark Keller, and Birgit Maddox completed the team of graphic artists on the project. We simply cannot thank them enough. They worked very hard against severe deadlines with good humor, and their creativity never waned. We all hope we have time to relax together someday.

Some of the original planning for the CD-ROM took place in an informal meeting at the University of Wisconsin–Madison. We wish to thank John and Betty Moore for hosting that gathering and for their comments, and Kathy Christoph, Jon Holmes, and Paul Schatz for their valuable insights.

CAChe SCIENTIFIC

Several years ago CAChe Scientific, Inc., made a grant of a Molecular Modeling Worksystem to JCK. This software has been used heavily by students in general and inorganic chemistry courses. It was also used to prepare molecular models for this book, the CD-ROM, and the new laserdisc from Saunders College Publishing. The people at CAChe—especially George Fabel, George Purvis, Evelyn Brosnan, and Rick DeHoff—have been extremely helpful, and we wish to acknowledge their support with gratitude.

OTHERS

Publishing a book and a CD-ROM is a complicated process. A large team of people is needed to carry out the task. Two members of that team not yet mentioned—but who have been the keys to its success—are Susan Young and Katie Kotz.

Susan Young is a postdoctoral associate of one of the authors (JCK). She has had a wide range of responsibilities on the book and CD-ROM. All of the computer-generated images were her creation, and she worked closely with Charlie Winters on the photography and video program. She checked galley proofs, worked out answers to problems, and prepared the Annotated Instructor's Edition. In addition, she taught General and Inorganic Chemistry and helped students at SUNY–Oneonta, and she did it all with energy and good humor. We shall miss her greatly when she moves on to a faculty position at Roanoke College.

Katie Kotz kept the work in Oneonta—photography, video production, text preparation, and photo research—organized. Her organizational skills and her expertise in maintaining a large database of information have been invaluable. In addition, she has been the wonderful wife of one of the authors for 34 years.

John C. Kotz
Paul Treichel, Jr.
July 1995

Reviewers

We believe the success of any book is due in no small way to the quality of the reviewers of the manuscript. The reviewers of the first and second editions made important contributions that are still part of this book. Reviewers of the third edition continued that tradition. We wish to acknowledge with gratitude the efforts of all those listed below. Several should be noted in particular, however. Gary Riley and Steve Landers worked all of the problems in the book. Steve Albrecht read the proofs for accuracy. Geoff Davies, Conrad Stanitski, Chris Willis, and Don Kleinfelter were very helpful in working on the early portions of the book. Finally, we want to add a special thanks to John DeKorte, who played a major role in the final development of the manuscript.

Lester Andrews
University of Virginia

Jeffrey R. Appling
Clemson University

Caroline L. Ayers
East Carolina University

Cindy A. Burkhardt
Radford University

Michael P. Castellani
Marshall University

Geoffrey Davies
Northeastern University

Randall Davy
Liberty College

Dan Decious
California State University, Sacramento

John DeKorte
Glendale Community College

Karen E. Eichstadt
Ohio University

Kevin Grundy
Dalhousie University

Mary Gurnee
Embry-Riddle Aeronautical University

Suzanne Harris
University of Wyoming

Jerry P. Jasinski
Keene State College

Martin Kellerman
California Polytechnical State University, San Luis Obispo

Christine S. Kerr
Montgomery College, Rockville Campus

D. Whitney King
Colby College

Donald Kleinfelter
The University of Tennessee Knoxville

Robert M. Kren
The University of Michigan–Flint

Joan Lebsack
Fullerton College

David E. Marx
University of Scranton

William H. Myers
University of Richmond

Frank A. Palocsay
James Madison University

Pete Poston
Western Oregon State College

Robert A. Pribush
Butler University

Nancy C. Reitz
American River College

Vic Shanbhag
Mississippi State University

Alka Shukla
Houston Community College–Southeast Branch

Saul I. Shupack
Villanova University

William E. Stanclift
Northern Virginia Community College

Conrad Stanitski
The University of Central Arkansas

Juan F. Villa
Herbert H. Lehman College of The City University of New York

Wayne Wesolowski
Illinois Benedictine College

Christopher J. Willis
The University of Western Ontario

Laura Yeakel
Henry Ford Community College

Reviewers of the Second Edition

Tom Baer
University of North Carolina at Chapel Hill

Muriel Bishop
Clemson University

Edward Booker
Texas Southern University

Donald Clemens
East Carolina University

Michael Davis
University of Texas, El Paso

Carl Ewig
Vanderbilt University

Russell Grimes
University of Virginia

Anthony W. Harmon
University of Tennessee–Martin

Alan S. Heyn
Montgomery College

Lisa Hibbard
Spelman College

Mary Hickey
Henry Ford Community College

William Jensen
South Dakota State University

Ronald Johnson
Emory University

Stanley Johnson
Orange Coast College

Lenore Kelly
Louisiana Tech University

Paul Loeffler
Sam Houston State University

Brian McGuire
Northeast Missouri State University

Jerry Mills
Texas Tech University

Mark Noble
University of Louisville

L.G. Pederson
University of North Carolina, Chapel Hill

Chester Pinkham
Tri-State University

Steve Ruis
American River College

Jerry Sarquis
Miami University

Steven Strauss
Colorado State University

Larry C. Thompson
University of Minnesota, Duluth

Milt Wieder
Metropolitan State College

Reviewers of the First Edition

Bruce Ault
University of Cincinnati

Alton Banks
Southwest Texas State University

O.T. Beachley
SUNY–Buffalo

Jon M. Bellama
University of Maryland

James M. Burlitch
Cornell University

Geoffrey Davies
Northeastern University

Glen Dirreen
University of Wisconsin

John M. DeKorte
Northern Arizona University

Darrell Eyman
University of Iowa

Lawrence Hall
Vanderbilt University

James D. Heinrich
Southwestern College

Forrest C. Hentz
North Carolina State

Marc Kasner
Montclair State College

Philip Keller
University of Arizona

Herbert C. Moser
Kansas State University

John Parson
Ohio State University

Lee G. Pedersen
University of North Carolina

Harry E. Pence
SUNY–Oneonta

Charles Perrino
California State University (Hayward)

Elroy Post
University of Wisconsin–Oshkosh

Ronald Ragsdale
University of Utah

Eugene Rochow
Harvard University

Steven Russo
Indiana University

Charles W.J. Scaife
Union College

George H. Schenk
Wayne State University

Peter Sheridan
Colgate University

Kenneth Spitzer
Washington State University

Donald D. Titus
Temple University

Charles A. Trapp
University of Louisville

Trina Valencich
California State University (Los Angeles)

Saunders College Publishing wishes to take this opportunity to thank the following respondents to surveys and focus groups conducted to provide feedback from users of the second edition to the authors.

Vernon Archer
University of Wyoming

Joe Brundage
Cuesta College

Juliette Bryson
Las Positas College

Robert Buckley
Hillsborough Community College

W. Centobene
Cypress College

Dale Chatfield
San Diego State University

Thomas Clark
Westark Community College

Daniel Decious
California State University, Sacramento

Howard DeVoe
University of Maryland

Jeffrey Dial
Olympic College

Dennis Drolet
Wartburg College

Joseph Gandler
California State University–Fresno

Suzanne Harris
University of Wyoming

Milo Johnson
College of the Redwoods

Doris Kimbrough
University of Colorado at Denver

Richard M. Kren
University of Michigan–Flint

Joan Lebsack
Fullerton College

William H. Myers
University of Richmond

R.L. Nemen
East Central University

Paul O'Brien
West Valley College

Nancy Reitz
American River College

John Rund
University of Arizona

David Sonnenberger
Illinois Benedictine College

David Stehly
Muhlenberg College

Catherine Travaglini
Parkland College

Dan Vinicor
Fresno City College

John Woodson
San Diego State University

A Note to Faculty and Students Who Will Use This Book

There are almost as many ways to teach chemistry properly as there are faculty members in the chemistry departments in North America. We make no pretense that we have found the best way of organizing chemistry, of explaining a principle, or of doing a problem. Further, even though we have put our best efforts into creating an error-free book, errors will surely be found. Therefore, if you find a way to explain something more clearly, a better way to demonstrate a reaction or a principle, or if you find errors in our discussions, we hope you will feel free to write to us or to our editors at Saunders College Publishing; they will see that we get your letter. Many specific things can be corrected in subsequent printings of the book, and more general ideas can be incorporated in any editions of this text that may follow. And don't be surprised if we call to find out more about your ideas.

A final word to students using the book. We believe chemistry is a challenging, exciting, and worthwhile area of study, and we hope you agree. However, like anything worthwhile, it does take some work to understand the subject. As you work through the book, just remember the encouraging words of Dr. Seuss from his book *Oh, the Places You'll Go* (Theodor S. Geisel and Audrey S. Geisel, Random House, New York, 1990).

> "*Onward up many*
> *a frightening creek*
> *though your arms may get sore*
> *and your sneakers may leak.*"

John C. Kotz
KOTZJC@oneonta.edu

Paul M. Treichel
Treichel@chem.wisc.edu

Contents Overview

Contents

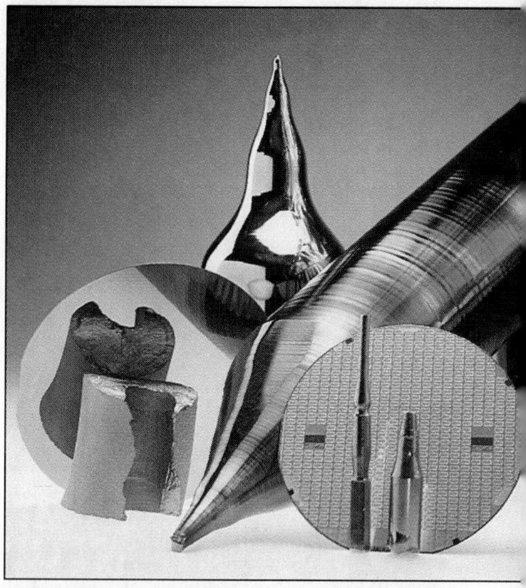

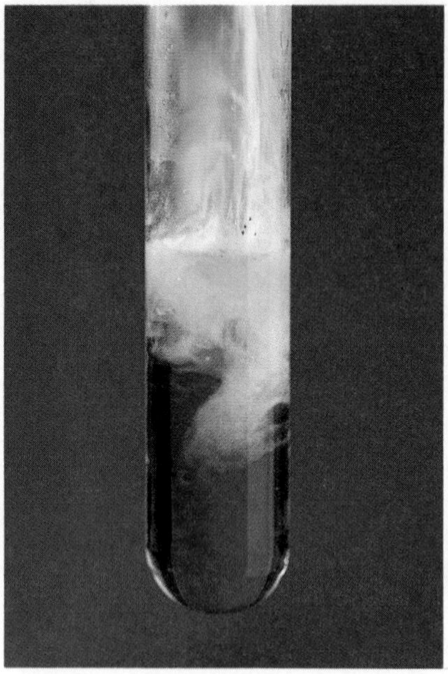

Part 2

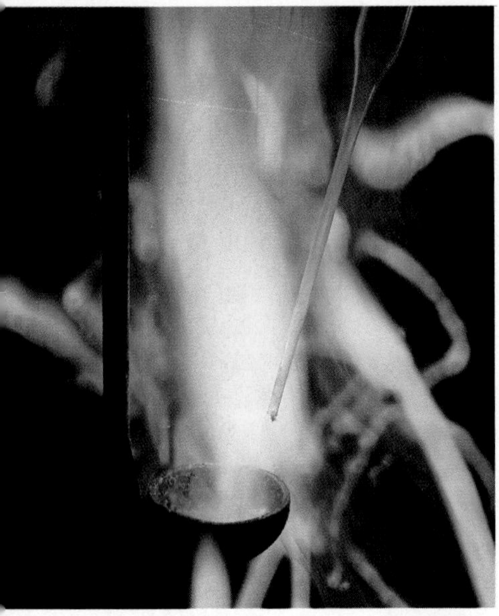

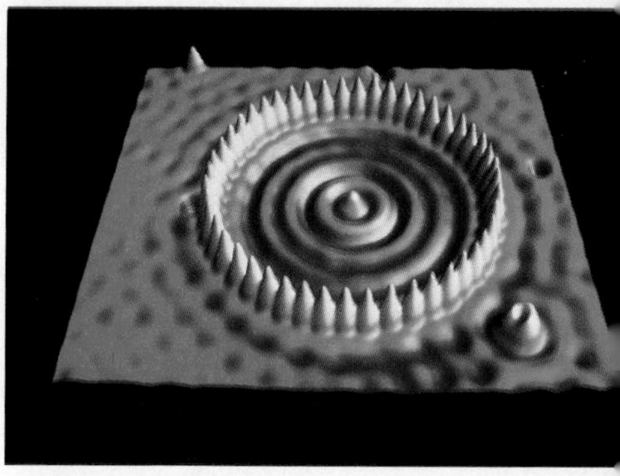

Part 5

I N T R O D U C T I O N

The Nature of Chemistry

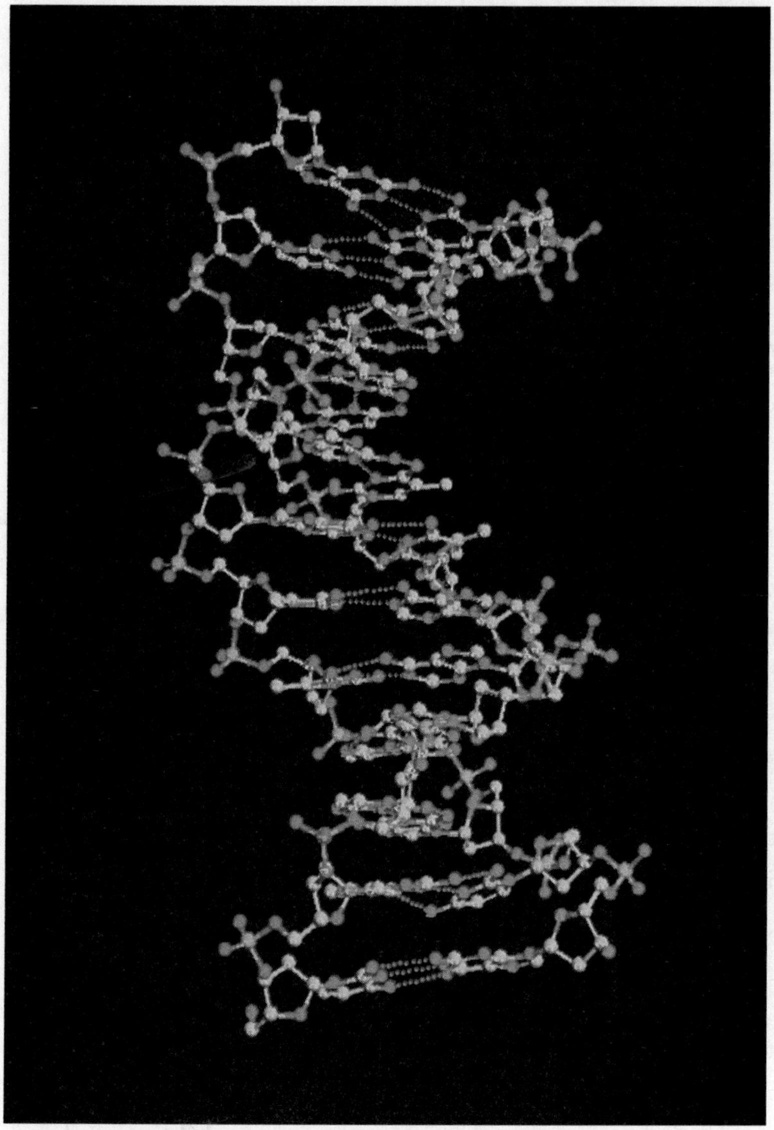

Molecular model of DNA structure.

Several years ago a research group in Michigan made an important announcement: The gene responsible for the often-fatal disease cystic fibrosis had been found. Now prospective parents could be screened to see if their offspring might inherit the disease. Or, even better, this information could eventually be used to find a cure for the disease.

Although developing tests for genetic diseases such as cystic fibrosis seems a worthy goal, difficulties abound. A report from the National Academy of Sciences has warned that screening to identify the genetic predisposition to a disease may lead to discrimination. For example, a person could be refused insurance coverage or a job because the insurance company or a potential employer fear large claims for compensation in the future.

In spite of the difficulties and unintended social consequences, research in genetics, genetic diseases, and the human gene is pursued by hundreds of scientists around the world. Companies have grown up almost overnight using the techniques of genetics to produce new medicines and even hormones to help cows produce more milk. This field of research has had, and will continue to have, an important influence on our society.

But what does genetics have to do with chemistry? A great deal! A gene is a piece of the chemical compound deoxyribonucleic acid, abbreviated DNA, which has been the subject of intense study for the past four decades because genetic information is encoded in the molecule. Although we know what a gene does, to understand how it functions we need to know about its chemical composition and structure. This crucial piece of information was uncovered by James D. Watson, Francis Crick, and Maurice Wilkins, who shared the Nobel Prize in medicine and physiology for their work in 1962 (Figure 1). Uncovering DNA's structure was one of the most important scientific discoveries of this century, and it has opened the way to rapid progress in molecular biology.

The tale of the discovery of DNA's structure was told by Watson in his book *The Double Helix.** As an undergraduate at the University of Chicago he said he had "managed to avoid taking any chemistry or physics courses which looked of even medium difficulty." Later, when he was a graduate student at Indiana University, Watson had an interest in the gene and said he hoped that its biological role might be solved "without my learning any chemistry." However, he and Crick used their imaginations and a knowledge of the kind of chemistry that is described in the book you are just beginning, together with the experiments of Maurice Wilkins and Rosalind Franklin, to make one of the major discoveries of this century.

*J. D. Watson: *The Double Helix, A Personal Account of the Discovery of the Structure of DNA.* New York, Atheneum, 1968.

Figure 1 The recipients of 1962 Nobel Prizes. In the foreground receiving their prize for their work on the structure of DNA are James D. Watson and Francis Crick. (Chemical Heritage Foundation)

Solving important problems requires teamwork among scientists of many kinds, so chemists often work with colleagues around the world. Therefore, after earning his Ph.D. in the United States, Watson went to study in Europe. After a brief stay in Copenhagen, Denmark, working on a problem that did not interest him, he went to Cambridge University in England in the fall of 1951. There he first met Francis Crick, who, Watson said, talked louder and faster than anyone else. Crick shared Watson's interest in DNA and the belief that it was fundamentally important. They soon learned that Maurice Wilkins and his colleague, Rosalind Franklin, at King's College in London (Figure 2) were already doing experiments on DNA. Wilkins and Franklin were using the technique of x-ray crystallography in the hope of learning more of its structure.

Watson and Crick believed that understanding DNA's structure was crucial to understanding genetics. To solve the problem, they needed experimental data, and the type of information they needed could come from the experiments at King's College. The problem was that Franklin had experimental data but did not seem willing to share it. To compound the problem, the experiments were difficult to perform, and Watson and Crick had neither the expertise nor the equipment to repeat them. And finally, it seemed to Watson and Crick that the King's College scientists did not really appreciate the significance of their work. So, Watson and Crick were faced with a dilemma: How could they convince the King's College group to share their

Figure 2 Rosalind Franklin was a colleague of Maurice Wilkins at King's College in London. She was an x-ray crystallographer and was studying the structure of DNA before Watson and Crick took up the problem. Watson says in *The Double Helix* that his "initial impressions of her, both scientific and personal, were often wrong." However, in the end he stated that "the x-ray work she did at King's was increasingly regarded as superb." Although the relationship between Franklin and Watson and Crick was strained initially, Watson has said that "we both came to appreciate greatly her personal honesty and generosity, realizing years too late the struggles that the intelligent woman faces to be accepted by a scientific world which often regards women as mere diversions from serious thinking." Rosalind Franklin died at the age of 37 in 1958. She did not receive the Nobel Prize for her contributions to the study of DNA because prizes are never awarded posthumously. (J. D. Watson: *The Double Helix, A Personal Account of the Discovery of the Structure of DNA,* p. 71. New York, Atheneum, 1968.) (Chemical Heritage Foundation)

data, and, more generally, was it ethical to work on a problem that others had claimed as theirs? "The English sense of fair play would not allow Francis to move in on Maurice's problem," said Watson.

Watson and Crick knew from the beginning that the overall structure of DNA was a helix (Figure 3), that is, that the molecule twisted in space like the threads of a screw. The experimental data they had were clear on this

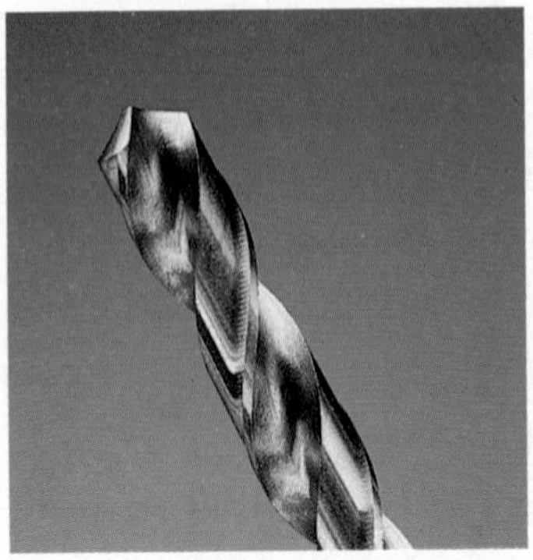

(a)

(b)

Figure 3 The helix. (a) The threads of a drill twist along the axis in the form of a helix. (b) Some plants also climb around a support or send out tendrils in a helical pattern. (C. D. Winters)

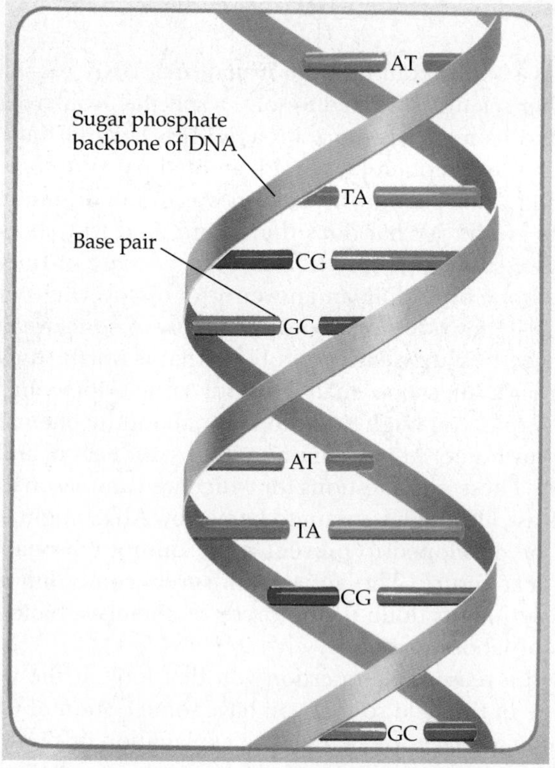

Figure 4 The overall structure of DNA. Two chains consisting of oxygen and phosphorus atoms, together with sugar molecules, twist together into a double helix. The chains are joined by molecules (here abbreviated by the letters A, T, G, and C) consisting of carbon, nitrogen, oxygen, and hydrogen.

point. They also knew what chemical elements it contained and roughly how they were grouped together. What they did not know was the detailed structure of the helix. Exactly how were the atoms arranged? How does the arrangement influence the functioning of the DNA? What role does it play in heredity? By the spring of 1953 they had the answer. DNA is a *double* helix of chains consisting of oxygen and phosphorus atoms, along with sugar molecules. Attached to each chain are a series of molecules called nucleotide bases, which are made up of atoms of carbon, nitrogen, hydrogen, and oxygen. The structure of DNA is a *double* helix because the nucleotides of one chain line up with the nucleotides of the second chain in a particular way. The two chains are held together by the forces of attraction between matched pairs of bases (Figure 4).

The double-helical structure of DNA is beautiful, and the underlying details are interesting and significant. Because an understanding of the structure of DNA and other molecules is so important, the structure of molecules is a major theme of this text.

SCIENCE AND ITS METHODS

Gould's definition of science is found in his "Essay on a Pig Roast" in *Bully for the Brontosaurus,* New York, W.W. Norton, 1991. The book also contains essays on textbooks ("The Case of the Creeping Fox Terrier"), on statistics ("The Median Isn't the Message"), and on the great chemist Lavoisier ("The Passion of Antoine Lavoisier").

The Double Helix is a book about how the structure of DNA was solved, about how chemists and other scientists worked to solve a specific problem. But how can we define science and its methods more broadly? It is hard to find a better definition than one given by Stephen Jay Gould, a noted paleontologist. "Science is a method for testing claims about the natural world, not an immutable compendium of absolute truths." What does this mean? And why should you care?

As scientists we study questions of our own choosing or ones that someone else poses in the hope of finding an answer or of discovering some useful information. Although it may not seem so now, it is easy to conceive an idea to study. The goal should be to state clearly a problem that is worth studying and that is also narrow enough in scope that a useful conclusion can realistically be reached. For example, you might want to know about the chemical composition of the water in your home. Are any unwanted substances in it, and, if so, how can they be removed? These are questions for which we can reasonably hope to find answers. In contrast, although we want to know how AIDS might be cured or how a vaccine could be developed to prevent it, we cannot reasonably expect these answers in the near future. The answer will surely come only as the result of thousands of experiments done by hundreds of chemists, biologists, and physicians in dozens of laboratories.

Having posed a reasonable question you first look at the work that others have already done in the field so that you have some notion of what direction to take. After forming a *hypothesis*—a tentative explanation or prediction of experimental observations—you perform experiments designed to give results that may confirm your hypothesis. In chemistry this surely requires that you collect both quantitative and qualitative information. *Quantitative* information usually means numerical data, such as the temperature at which a chemical substance melts. *Qualitative* information, in contrast, consists of nonnumerical observations, such as the color of a substance or its physical appearance.

Preliminary experimental data often require you to revise and extend the original hypothesis. This in turn creates the need for new experiments. After you have done a number of experiments and have continually checked to ensure your results are truly *reproducible,* a pattern of results may begin to emerge. At this point, you may be able to summarize your observations in the form of a general rule. Finally, after numerous experiments by many scientists over an extended time, the original hypothesis may have become a *law*—a concise verbal or mathematical statement of a relation that seems always to be the same under the same conditions.

We base much of what we do in science on laws because they help us predict what may occur under a new set of circumstances. For example, we know from experience that if we allow the chemical element sodium to come in contact with the chemical compound water, a very violent reaction occurs and several new chemical compounds are formed (Figure 5). But the result of an experiment might be different from what is expected based on a general rule. Then a chemist gets excited because experiments that do not follow the general rules of chemistry are the most interesting. We know that understanding the exceptions almost invariably gives new insights.

Once enough reproducible experiments have been done, and scientists have formulated a general rule or a law, it may be possible to conceive a theory to explain the observation. A *theory* is a unifying principle that explains a body of facts and the laws based on them. It is capable of suggesting new hypotheses. Theories abound, not only in the physical and biological sciences but also in economics and sociology. In chemistry, excellent examples of theories are those developed to account for chemical bonding (Chapters 9–10). It is a fact that atoms are held together, or bonded to one another, as in a molecule such as DNA. But how and why? Several theories are currently used in chemistry to answer these questions, but are they correct? What are their limits? Can the theories be improved, or are completely new theories necessary? Laws summarize the facts of nature and rarely change. Theories are inventions of the human mind. *Theories can and do change* as new facts are uncovered.

People who work outside science usually have the idea that science is an intensely logical field. They picture white-coated chemists moving logically from hypothesis to experiment and then to laws and theories without human emotion or foibles. Nothing could be further from the truth! Watson and Crick worked many months and made embarrassing errors before they understood DNA's structure.

Figure 5 The metallic element sodium reacts very vigorously with water. (C. D. Winters)

Often, scientific results and understanding arise quite by accident, otherwise known as serendipity. Creativity and insight are needed to transform a fortunate accident into useful and exciting results, and a wonderful example is the discovery of a cancer drug by Barnett Rosenberg (Figure 6). He and his collaborators discovered that a chemical compound that had been known for a century or more, a compound composed of the elements platinum, chlorine, nitrogen, and hydrogen, was effective in the treatment of certain types of cancers. The discovery of the utility of this compound, now commonly known as *cisplatin,* in cancer therapy is an example of serendipity in science.

In the case of cisplatin Rosenberg had set out to study a problem that had interested him for some time—the effect of electrical fields on living cells—but the results of the experiment were quite different from his expectations. He and his students had passed a suspension of a live culture of *Escherichia coli* bacteria in

Figure 6 Dr. Barnett Rosenberg moved from New York City, where he had studied physics, to Michigan State University in East Lansing, Michigan, to establish a department of biophysics in 1961. He is now the head of the Barros Research Institute near East Lansing. Dr. Rosenberg has received many awards for his work, including the Galileo Medal from the University of Padua (Italy) and the Charles F. Kettering Prize from the General Motors Cancer Foundation. (Doug Elbinger)

water through an electric field between supposedly inert platinum plates. Much to their surprise they found that cell growth was affected; cell division had stopped! After careful experimentation, the effect on cell division was traced to tiny amounts of cisplatin, which was produced by reaction of the platinum with electrically charged chemical species in the water.

Much to the benefit of cancer chemotherapy, Rosenberg recognized that these laboratory results had wider implications, and subsequent experiments led to compounds now used by cancer patients. He recently said that the use of cisplatin has meant that "Testicular cancer went from a disease that normally killed about 80% of the patients, to one which is close to 95% curable. This is probably the most exciting development in the treatment of cancers that we have had in the past 20 years. It is now the treatment of first choice in ovarian, bladder, and osteogenic sarcoma [bone] cancers as well."

SIMPLE BUT ELEGANT EXPERIMENTS

The remarks by Richard Feynman (1919–1988) and the story of his role in the Challenger disaster are found in his book *What Do You Care What Other People Think?* (New York, W.W. Norton and Company, 1988). See also his autobiography *Surely You're Joking Mr. Feynman* (New York, W.W. Norton and Company, 1985), in which he tells about his life in physics and his passion for the bongo drums.

Another person who understood very well how science works was Richard Feynman, the recipient of the Nobel Prize in physics in 1965 and one of the most original thinkers of this century. Feynman said that "Scientific knowledge is a body of statements of varying degrees of certainty—some most unsure, some nearly sure, but none *absolutely* certain. . . . Now, we scientists are used to this, and we take it for granted that it is perfectly consistent to be unsure, that it is possible to live and *not* know." It is well to remember Feynman's statement as you read statements about our environment, about health care and medicine, and about nutrition.

Richard Feynman gave the United States a lesson in how science is done when he used a simple experiment to uncover the reason for the disastrous explosion of the Space Shuttle, *Challenger.* The *Challenger* was launched on Tuesday, January 28, 1986. The day was unusually cold for Florida—the temperature at the time of launch was 29 °F. The world watched in horror when, after a minute or so of flight, the shuttle and its rockets exploded in a fireball, killing everyone on board.

To understand the reason for the explosion and the importance of Feynman's experiment, recall something of the design of the shuttle. The main engine is fueled by liquid hydrogen and oxygen, which are contained in the large tank strapped onto the belly of the shuttle. The hydrogen and oxygen combine to give water, and the energy this reaction produces provides the thrust to boost the shuttle into orbit. To provide additional thrust at the time of launch, however, solid-fuel booster rockets are strapped on each side of the shuttle. The solid-fuel rockets are made in sections, and the sections are shipped from the factory to the Kennedy Space Center where they are joined to make the completed booster rocket. The joint between the sections was designed so that hot gases from the burning solid fuel would not leak through the walls of the rocket. Part of the design to close the joint, and yet make it somewhat flexible, included a thin O-ring made of a special rubber (Figure 7). From the beginning Feynman and others thought that a possible cause of the *Challenger* explosion was that the solid fuel had burned through the wall of the booster rocket and then burned

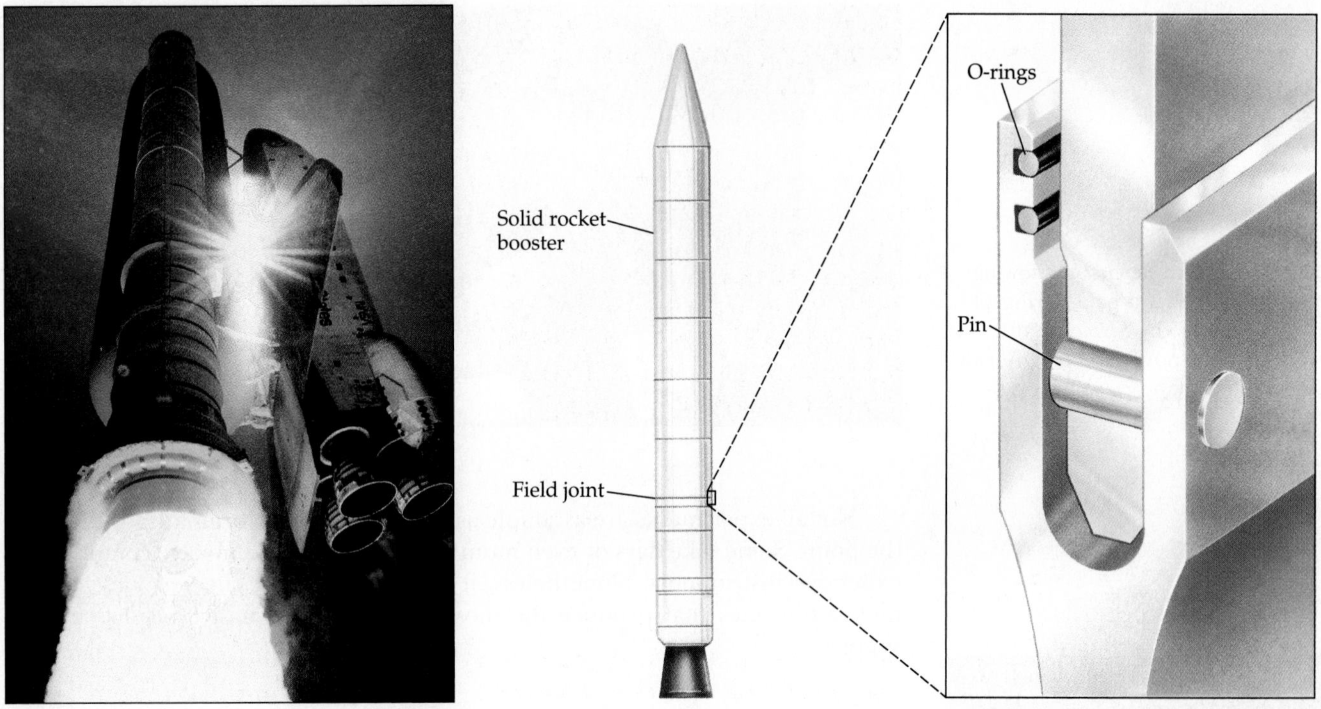

(a) (b)

Figure 7 (a) The Space Shuttle, *Atlantis,* lifting off on August 2, 1991. The large orange tank on the belly of the orbiter contains hydrogen and oxygen, which are burned in the orbiter's main engines. The long tubes on either side of the hydrogen-oxygen tank (only one is seen here) are the solid-fuel boosters. (b) Each solid-fuel booster rocket is made of a number of sections, bolted together. The joints between sections are complicated and involve, among other things, O-rings made of a special rubber. The function of the O-rings is to seal the joints and prevent hot gases from leaking out of the sides of the rocket. (a, NASA)

into the tank holding the liquid hydrogen, thus exploding the hydrogen. But how did this happen?

When the shuttle is launched and the solid fuel begins to burn, the walls of the rocket casing move slightly outward. If this movement caused a joint between sections to open, the fuel would burn through the joint. But the O-rings were supposed to prevent this. Based on information from engineers involved in the design of the solid rocket boosters, one hypothesis to explain the accident was that, due to the unusually cold weather, the rubber O-rings had not expanded properly, and flame burned through the joint. To prove this point, Feynman did a dramatic—but very simple—experiment. During a public hearing Feynman took a sample of the rubber O-ring, held it tightly in a C-clamp, and put it into a glass of ice water (Figure 8). Everyone could make the qualitative observation that the rubber did not spring back to its original shape! The poor resilience of the rubber at low temperatures doomed the *Challenger.* Feynman's hypothesis was supported by his elegantly simple experiment.

Figure 8 Richard Feynman showing experimentally that a cold O-ring of the type used in the Space Shuttle, *Challenger,* does not snap back quickly to its original shape. (Marilynn K. Yee/NYT Pictures)

Not all experiments are as simple as Feynman's nor so dramatically illustrate the point. Some take days or even months to complete and involve complex and expensive instruments. Nonetheless, it is often true that the very best experiments, the ones that produce the most useful and persuasive results, are the simplest.

Figure 9 Francis H. C. Crick (b. 1916) (*right*) and James D. Watson (b. 1928) (*left*) working in the Cavendish Laboratory at Cambridge University (England). Watson and Crick built scale models of the double-helical structure of DNA based on the x-ray data of Rosalind Franklin (1920–1958) and Maurice H. F. Wilkins (b. 1916). Knowing the distances and angles between atoms, they compared the task to solving a three-dimensional jigsaw puzzle. Watson, Crick, and Wilkins received the Nobel Prize for physiology and medicine in 1962 for their work on the structure of DNA (see Figure 1). (Peter Arnold, Inc.)

MODEL BUILDING, SCIENCE, AND MOLECULAR STRUCTURES

Another way to solve scientific problems is to build a model, something that we think mimics reality. This can be a mathematical model for the production and income of a small business. It can be the model an architect constructs to see how the rooms in a house fit together. Or a chemist might make a model of a molecule out of paper and plastic or construct and visualize one by computer.

To solve the DNA structural problem, Watson and Crick built physical models of the pieces of DNA. They then used their chemical knowledge and intuition to see how these pieces best fit together to create a molecule with properties that matched experiment (Figure 9). Watson said that he and Crick really did think that "model building represented a serious approach to science, not the easy resort of slackers who wanted to avoid the hard work necessitated by an honest scientific career." Their work was a truly successful use of model building, even though not everyone shared their enthusiasm at the time.

The powerful x-ray techniques that provided the experimental proof of the Watson-Crick model have been used extensively over the past forty years to explore the shapes, or *structures,* of thousands of molecules. Scientists now have well-defined ideas of molecular structure, and, just as importantly, our understanding of the relation between molecular structure and chemical behavior is increasing. Indeed, the importance of molecular structure cannot be understated. DNA is a helix because of the geometry of the linkages between the chemical elements in its backbone. It is a *double* helix because of the way that molecules from one helical chain fit together with those from the other chain (Figure 10). And it is the molecule involved in heredity precisely because of the geometries of the molecules involved.

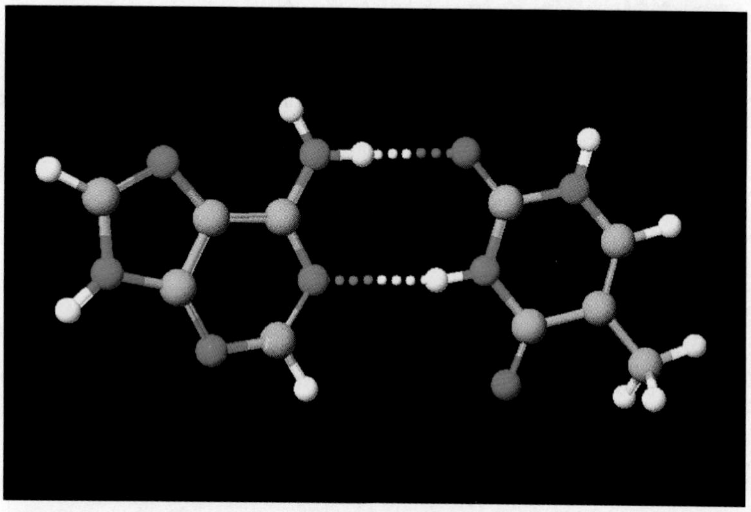

Figure 10 Four different molecules bridge the two strands of the "backbone" of DNA (see Figure 4). Two of them (left) (adenine) and (right) (thymine) are shown here. Because of their geometry, these two molecules fit tightly together, providing one of the many linkages that hold together the two strands of the DNA backbone. (The atoms in these computer-generated molecular structures are labeled by color: hydrogen = white; carbon = gray; oxygen = red; and nitrogen = blue.)

A knowledge of molecular structure is now regarded as fundamental to progress in chemistry. Indeed, it is so important that pharmaceutical and chemical companies, for example, model structures and chemical reactions using computer-generated structures before doing laboratory work (much as those in Figure 10). To ensure that you become familiar with these ideas, pictures of molecular models occur throughout this book. In addition, if you have access to a computer with a CD-ROM player, you will be able to use the library of molecular structures on the CD version of this book. Furthermore, the disk contains software from CAChe Scientific, Inc., that enables you to build and view the structure of any molecule.

WHY CARE ABOUT CHEMISTRY?

The structure of DNA, the *Challenger* disaster, and Feynman's experiment bring us to the reasons you should care about science and the way it is done. Many in the United States were saddened by the *Challenger* explosion, not only because of the deaths of the astronauts but also because, unless solved, it could clearly affect the future of our space program and its role in our national goals. Furthermore, such engineering failures can lead uninformed citizens to lose faith in all technology, to the detriment of our economy and society. Professor Feynman and his common-sense approach uncovered the reasons for the *Challenger* explosion, and the space program is moving cautiously ahead. Hundreds of problems confronting our society, however, depend in some way on science for their resolution. Almost every day we read or hear statements in newspapers or magazines or on television like the following:

- We could develop an AIDS vaccine if we just moved more quickly.
- Asbestos must be removed from all public buildings.
- All pesticides and herbicides should be banned.
- The Antarctic ozone hole is becoming larger.
- Burning tropical forests is an ecological disaster.

- Global warming is a reality.
- Global warming is not a reality.
- Homeopathic "medicines" are a sure cure for whatever ails us.
- You should not eat butter but substitute margarine.
- Health risks are associated with eating margarine.
- You should check the basement of your house for dangerous levels of radon.
- Eating foods with high levels of selenium prevents cancer.
- Incineration of wastes is highly damaging to the local ecology.

The list goes on and on. What are you to believe? How do you analyze the problem? We believe that some knowledge of chemistry and the methods used by chemists and other scientists can be of great benefit. This can help you at least begin to analyze the risks and benefits of government policies.

RISKS AND BENEFITS

Problems like those listed above, all of which are related directly to chemistry and chemicals, will be with us for many years, and others will surely be added. No matter where you live or what you do for a living, you are surrounded by chemicals. Many are necessary to life—proteins, carbohydrates, oxygen, and water, for example. Others are not necessary or are even toxic. You have decided you can tolerate some of the less desirable chemicals, however, because the risk they pose seems to be outweighed by the benefits they offer. Examples include having an occasional beer, which contains toxic alcohol, or using an herbicide on your garden or lawn. In other cases, however, it is difficult for us to weigh the risks and benefits. This means that we will hear more about the assessment and management of risks.

Figure 11 We voluntarily assume risks in our everyday lives, including smoking, eating charcoal-grilled food, sitting in the sun, and biking. Are there others in this photograph? (C. D. Winters)

"Risk assessment" is a process that brings together the scientific disciplines of chemistry, biology, toxicology, epidemiology, and statistics and that attempts to establish the severity of a health risk associated with exposure to a particular chemical. The first step in identifying a hazard is performing animal tests or examining data on human exposure. The second step is finding the relation between the amount of exposure and the chances of experiencing an adverse health effect. Also involved is determining how people are exposed and to how much. Once these things are known, we can try to estimate the overall risk.

In recent years a new discipline called "risk communication" developed to promote the accurate transfer of information between the scientists who assess risk, the government agencies that manage risk, and the public. Those who study risk communication have found that the response of people to risks depends on a number of interesting factors. For example, people accept voluntary risks, such as those posed by smoking or playing in a high school football game, much more readily than involuntary ones, such as being in a building that contains asbestos (Figure 11). Viewed objectively, however, the annual death rate per million people is 10 for playing high school football and 40 for long-term smoking, but the death rate from sitting in a classroom with an asbestos ceiling is estimated to be in the range 0.005 to 0.093.

It is also interesting that people often conclude that anything artificially made is "bad," but anything natural is "good." This is not always confirmed by risk assessment, however, as you can see in Table 1.

Risk management involves ethics, equity, economics, and other matters that are part of government and politics. The removal of asbestos from public buildings is an example of the management of a risk that many scientists believe to be

TABLE 1 Estimates of Risk: Activities that Produce One Additional Death per 1 Million People Exposed to the Risk[1]

Activity	Cause of Death
Smoking 1.4 cigarettes	Cancer, lung disease
Living 2 months with a cigarette smoker	Cancer, lung disease
Eating 40 tbsp of peanut butter	Liver cancer caused by the natural carcinogen aflatoxin B.
Drinking 40 cans of saccharin-sweetened soda	Cancer
Eating 100 charcoal-broiled steaks	Cancer
Traveling 6 min by canoe	Accident
Traveling 10 min by bicycle	Accident
Traveling 300 miles by car	Accident
Traveling 1000 miles by jet aircraft	Accident
Drinking water from the Miami water system for 1 year	Cancer from chloroform
Living 2 months in Denver	Cancer caused by cosmic radiation
1 Chest x-ray in a good hospital	Cancer
Living 5 years at the boundary of a typical nuclear power plant	Cancer

[1]L. Gough and M. Gough: "Risky Business," *Chem Matters,* pp. 10–12, December, 1993.

quite low for the general public. The public has not been willing to tolerate the risk, however, and so the political system has responded.

Another example of risk management is the ban on chlorofluorocarbons (CFCs) in many nations because of the damage CFCs are thought to cause to earth's ozone layer (see Section 15.7 for a complete discussion). Here morality and economics clearly clash. After many years of development, modern refrigeration equipment that uses CFCs is efficient, relatively inexpensive, and widespread. Its use is spreading to Third World countries, such as the nations of Africa and many in South America, and its availability has a profound effect on their economies. Now, just as these countries are beginning to develop, the countries of the so-called developed world tell them that CFC-based refrigeration equipment can no longer be used. What is better for the greater good: a ban on equipment and processes using CFCs, which will apparently have a long-term effect on the ozone layer, or some limited, reasonable use of CFCs in the refrigeration equipment that is vital to the development of Third World countries? If alternatives to CFCs can be found, are they economical and themselves without damaging environmental effects? Can the technology be transferred readily and efficiently to Third World countries?

Why study chemistry? The reasons are clear. You will be called on to make many decisions in your life for your own good and for the good of those in your community—your local community or the global community. An understanding of science in general and chemistry in particular can only help in these decisions.

BIBLIOGRAPHY

If you wish to learn more about the subjects introduced here about the importance of chemistry in our world and the thoughts of some scientists who have contemplated the role of science in our society, consider some of the following books and articles.

- J. Bronowski: "The creative process." *Scientific American,* September, 1958.
- Rachel Carson: *Silent Spring.* New York, Houghton Mifflin, 1962.
- J. D. Watson: *The Double Helix, A Personal Account of the Discovery of the Structure of DNA.* New York, Atheneum, 1968.
- Richard Feynman: *What Do You Care What Other People Think?* New York, W.W. Norton and Company, 1988.
- Richard Feynman: *Surely You're Joking Mr. Feynman.* New York, W.W. Norton and Company, 1985.
- Thomas S. Kuhn: *The Structure of Scientific Revolutions.* Chicago, The University of Chicago Press, 1970.
- Primo Levi: *The Periodic Table.* New York, Schocken Books, 1984.
- Lewis Thomas: *The Lives of a Cell.* New York, Penguin Books, 1978.
- Sharon B. McGrayne: *Nobel Prize Women in Science,* New York, Birch Lane Press, 1993.

CHAPTER 1

Matter and Measurement

(C. D. Winters)

A Chemical Puzzler

Processed foods often contain many additives. For example, iron is added to some foods, including breakfast cereal. But did you know that some breakfast cereals contain metallic iron? What experiment could you do to prove or disprove this? If you find a cereal that contains metallic iron, how would you separate the iron from the cereal? Finally—a question to think about much later in the book—why is metallic iron used and not some iron-containing chemical compound?

You put about half a pound of sugar (whose chemical name is sucrose) along with half a cup of water and half a cup of corn syrup (a glucose solution) in a pan and heat the mixture while stirring steadily. You continue to boil the syrup and watch as it slowly turns brown; a vapor rises above the bubbling mixture. When the temperature reaches 140 °C, you toss in a handful of peanuts and a pinch of sodium bicarbonate and pour the liquid mixture quickly onto a piece of aluminum foil. When it cools, you smash the solid translucent slab into small pieces and pop some into your mouth! You have just made peanut brittle, during the course of which you have done many of the things that chemists pay attention to every day. You have seen chemical and physical changes and different states of matter, have made qualitative and quantitative observations, and have made measurements. That is what this chapter is all about—making sense of the material world through quantitative and qualitative observations and using the knowledge so gained to adapt the material world to our needs.

1.1 PHYSICAL PROPERTIES

Your friends recognize you by your physical appearance: your height and weight and the color of your eyes and hair. The same is true of chemical substances. While making peanut brittle you can distinguish sugar from water because you know that sugar consists of small white solid particles, whereas water is a colorless liquid. Corn syrup is also a liquid, but it comes in light and dark colors and is much more viscous (pours more slowly) than water. Properties such as these, which can be observed and measured without changing the composition of a substance, are called **physical properties.** As you can see in Figure 1.1, the chemical elements iron and sulfur, both solids at room temperature, clearly differ in color, and liquid bromine and solid iodine differ in their physical states as well as their colors. Physical properties allow us to classify and identify substances of the material world. Table 1.1 lists some physical properties of matter that chemists commonly use. A few of these are discussed in more detail in this chapter, and others are taken up later in the text.

Making peanut brittle. For more on the chemistry of the process see E. Catelli: *Chem Matters,* December, p. 4, 1991. (C. D. Winters)

EXERCISE 1.1 ***Physical Properties***

Identify as many physical properties in Table 1.1 as you can for the following common substances: (a) iron; (b) water; (c) table salt (whose chemical name is sodium chloride); and (d) oxygen. ■

States of Matter

An easily observed and very useful property of matter is its physical state, or phase. Is it a solid, liquid, or gas at room temperature or at some other temperature? A **solid** can be recognized because it has a rigid shape and a fixed volume that changes very little as temperature and pressure change. Like a solid, a **liquid** has a fixed volume, but a liquid is fluid—it takes on the shape of its container

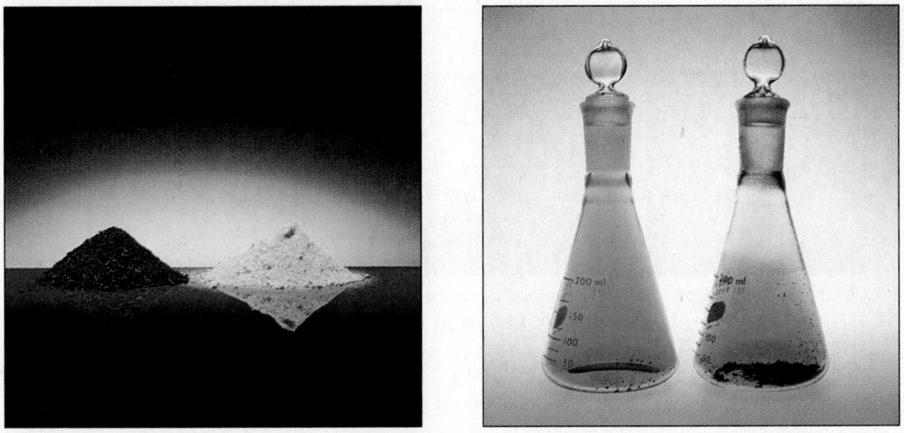

(a) (b)

Figure 1.1 The chemical elements differ from one another in their physical properties. Here are four elements, three of them solid and one a liquid. (a) Solid iron (*left*) and sulfur (*right*) are clearly different in color. (b) Bromine (*left*) and iodine (*right*) differ both in color and physical state. Notice that bromine is a liquid, but it is volatile (easily vaporized) as indicated by the red-orange color of bromine vapor in the flask. Iodine is a purple solid, but it too is volatile; iodine vapor fills the flask. (C. D. Winters)

and has no definite form of its own. **Gases** are fluid also, but gases expand to fill whatever container they occupy, and their volume varies considerably with temperature and pressure. For most substances, the volume of the solid is slightly less than the volume of the same mass of liquid, but the volume of the same mass of gas is much, much larger. Virtually all matter is found in the solid state at very

TABLE 1.1 Some Physical Properties of Matter

Property	Comment
Color	
State of matter	Is it a solid, liquid, or gas?
Melting point	The temperature at which a solid melts
Boiling point	The temperature at which a liquid boils
Heat of vaporization	The heat required to change a liquid to a vapor
Heat of fusion	The heat required to change a solid to a liquid
Density	Usually expressed in units of grams per milliliter or grams per cubic centimeter
Solubility	The amount of a substance that can dissolve in a given mass or volume of water or other solvent
Metallic character	
Electrical conductivity	
Conductivity of heat	
Magnetic properties	
Shape of the crystals of a solid	
Malleability	The ease with which a solid can be deformed
Ductility	The ease with which a solid can be drawn into a wire
Viscosity	The susceptibility of a liquid to flow

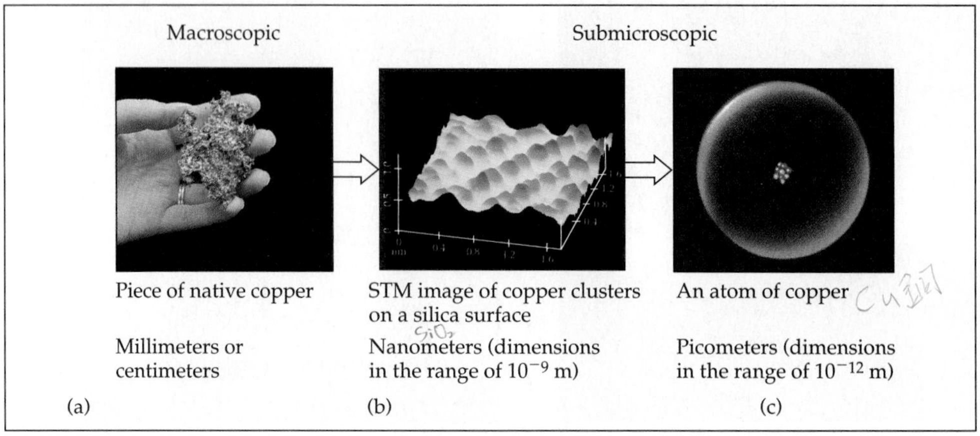

Figure 1.2 The macroscopic and submicroscopic world of chemistry. (a) A macroscopic piece of copper metal. This is native copper, a piece of elemental copper as it existed in the earth. (b) Scanning tunneling (STM) and atomic force microscopy (AFM) are powerful methods of probing the surface of solids with atomic resolution. They can reveal the submicroscopic world of atoms. Here the figure shows an AFM topographic image of the copper atoms on the surface of silica. The image is 170 nm square, and the rows of atoms are separated by about 0.44 nm. (c) Representation of an atom of copper. [(a) C. D. Winters. Part (b) from X. Xu, S. M. Vesecky, and D. W. Goodman, *Science,* Vol. 258, p. 788, 1992.]

Scientists have recently developed several new methods that are capable of "seeing" individual atoms. One of these is STM, which stands for "scanning tunneling microscopy."

low temperatures. As the temperature is raised, though, solids generally melt to form liquids. Eventually, if the temperature is raised high enough, liquids can evaporate to form gases.*

All the physical properties just described can be observed by the unaided human senses and refer to samples of matter large enough to be seen, measured, and handled. Such samples are called **macroscopic,** in contrast to **microscopic** samples, such as biological cells and microorganisms, which are so small that they can only be seen with a microscope. The structure of matter that really interests chemists, however, is at the **submicroscopic** scale of atoms and molecules (Figure 1.2).

A fundamental idea of chemistry is that matter exists as it does because of the nature of its parts, and those parts are very, very tiny. Therefore, imagination is required to discover useful ideas that connect the behavior of those tiny parts to the behavior of chemical substances in the macroscopic world.

Kinetic-Molecular Theory

One theory that helps us interpret the physical properties of solids, liquids, and gases is the **kinetic-molecular theory of matter.** According to this theory, all matter consists of extremely tiny particles (atoms and molecules), which are in constant motion. In a solid these particles are packed closely together in a regular array (Figure 1.3). The particles vibrate back and forth about their average

*A fourth state of matter, plasmas, are gases composed of charged particles. Plasmas exist naturally in the outer portion of the earth's atmosphere, in the atmosphere of stars, and in the beautiful aurora borealis or "northern lights."

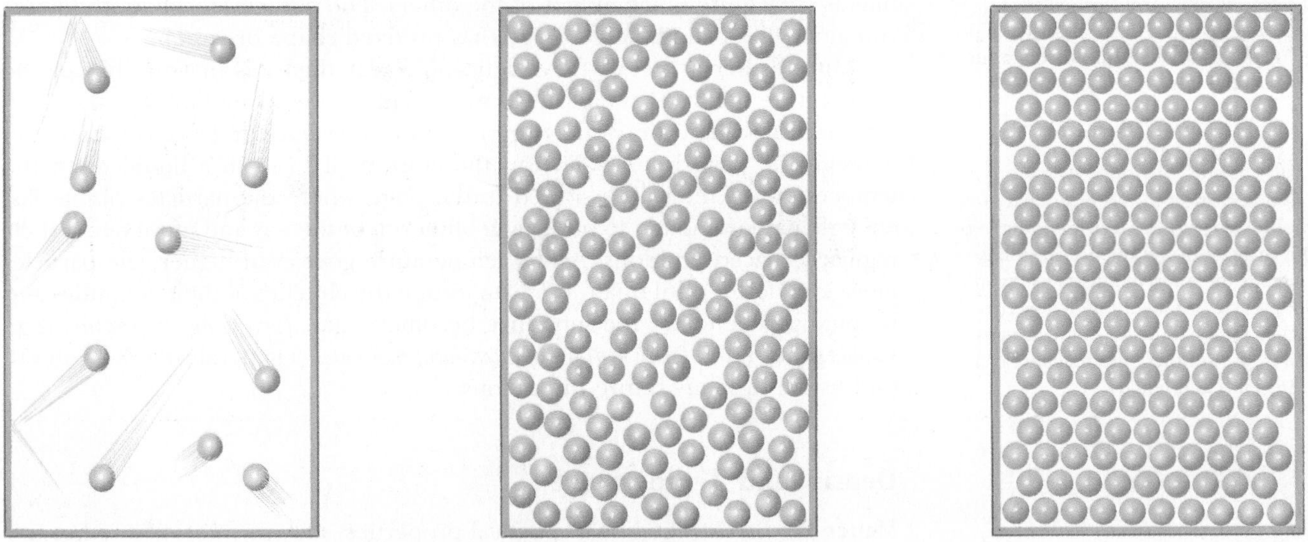

Figure 1.3 The three states of matter. In the gas phase, atoms or molecules move rapidly over distances larger than the sizes of the atoms or molecules themselves. There is little interaction between them. Cooling, increasing the pressure, or both, converts gases to liquids. The atoms or molecules are now much closer together, and they interact with one another. Motion of the particles is still very evident, although the particles move over only very small distances. Further cooling converts a liquid to a solid. The particles are even closer together and almost totally restricted to specific locations. They are arranged with a high degree of regularity.

positions, but seldom does a particle in a solid squeeze past its immediate neighbors to come into contact with a new set of particles. Because the particles are packed so tightly and in such a regular arrangement, a solid is rigid, its volume is fixed.

The external shape of a solid often reflects the internal arrangement of its particles. For example, a piece of "fool's gold," or iron pyrite (Figure 1.4), is made up of iron and sulfur, and the cubic solid you can hold in your hand is a vastly enlarged version of the cubic arrangement of individual iron and sulfur atoms in the solid crystal. This relation between the observable structure of the solid and the arrangement of the particles from which it is made is one reason scientists have long been fascinated by the shapes of crystals and minerals.

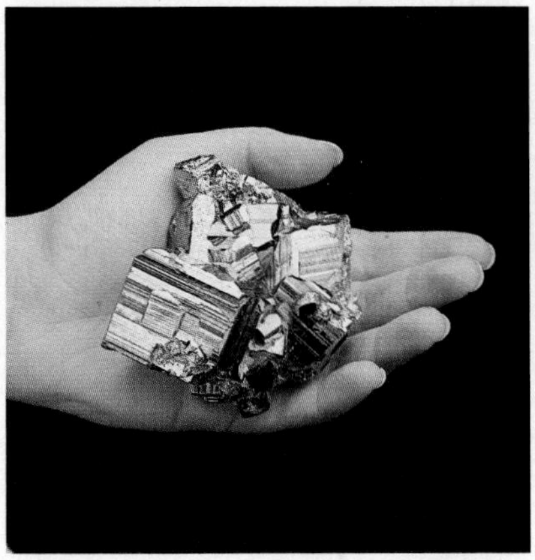

Figure 1.4 Iron pyrite, which is composed of atoms of iron and sulfur, can form large cubic crystals that reflect the arrangement of the atoms deep inside the crystal. (C. D. Winters)

The kinetic-molecular theory of matter can also be used to interpret the properties of liquids and gases, as shown in Figure 1.3. Liquids and gases are fluid because the atoms or molecules are arranged at random, rather than in the regular patterns found in solids. They are not confined to specific locations but rather can move past one another. Because the particles are a little farther apart in a liquid than in the corresponding solid, the volume is a little bigger. No particle goes very far without bumping into another—the particles in a liquid interact with one another constantly. In a gas the particles are very far from one another and are moving quite rapidly. (In air at room temperature, for example, the average molecule is going faster than 450 meters per second or over 1000 miles per hour.) A particle hits another particle every so often, but most of the

In the Introduction you learned about models used by chemists to describe molecules. The kinetic molecular theory is another example of a scientific model.

time each is quite independent of the others. The particles fly about to fill any container they are in, and so a gas has no fixed shape or volume.

One further aspect of the kinetic-molecular theory is that the higher the temperature the faster the particles move. The energy of motion that the particles possess (called the *kinetic energy*) acts to overcome the forces of attraction between the particles. A solid can therefore melt to form a liquid when the temperature of the solid is raised to the point where the particles vibrate fast enough and far enough to push each other out of the way and move out of their regularly spaced positions. As the temperature goes even higher, the particles move even faster until finally they can escape the clutches of their comrades and become independent; the substance becomes a gas. *Increasing temperature corresponds to faster and faster motions of atoms and molecules,* a general rule that you will find useful in many future discussions.

It is also important to recognize that attractive forces exist between the submicroscopic particles of a substance. In solids and liquids these forces must be substantial as the particles are held tightly together. Attractive forces also exist between particles in a gas, but these are much weaker than in solids and liquids.

Density and Temperature

Extensive properties (mass, volume) depend on the amount of matter present. Intensive properties (density, temperature, color) do not depend on the amount of sample.

Matter can be classified by its physical properties, such as color or whether it is a solid, liquid, or gas under normal conditions. Various minerals and gems, for example, are partly identified by their color (Figure 1.5). In addition, important quantitative measures, such as the density of the solid, liquid, or gas or the temperature at which it melts or boils, characterize a substance.

Density, the ratio of the mass of an object to its volume, is a physical property that is useful for identifying substances.

$$\text{Density} = \frac{\text{mass}}{\text{volume}}$$

D= g/cm³

Your brain unconsciously utilizes the density of an object you want to pick up by estimating volume visually and preparing your muscles to lift the expected mass. For example, the person shown in Figure 1.6 is holding objects of similar size but of widely different densities. By looking at the appearance of the solids—their physical properties—can you tell which one is denser?

Figure 1.5 A collection of minerals and gems. These can be differentiated by their physical properties of color and outward appearance. From left to right, pictured are galena, aquamarine, ruby, sulfur, silver, elbaite, sapphire, fluorite, copper, azurite, and malachite. (C. D. Winters)

Figure 1.6 A student is holding two different solids. From your experience and the appearance of the solids, can you tell which one is denser? The answer is given in Appendix L. (C. D. Winters)

Density can be determined more precisely by measuring the mass of an object and its volume. Nickel, the element that gave its name to the U.S. 5-cent coin, has a density of 8.90 g/cm^3 (grams per cubic centimeter), for example. That is, 1.00 cm^3 of nickel has a mass of 8.90 g. In contrast, titanium, a metal used for its resistance to corrosion, has a much lower density, only 4.5 g/cm^3. Pieces of both of these metals are shiny and gray, but you could tell the difference between them by measuring their densities or by selecting pieces of each metal of nearly identical volume and then weighing them.

If any two of three quantities—mass, volume, and density—are known for a sample of matter, they can be used to calculate the third. Density equals mass divided by volume; therefore

$$\text{Volume} \times \text{density} = \text{volume}(\cancel{\text{cm}^3}) \cdot \frac{\text{mass (g)}}{\text{volume }(\cancel{\text{cm}^3})} = \text{mass (g)}$$

You can use this approach to find the mass of 25 cm^3 [or 25 mL (milliliters)] of mercury (in the graduated cylinder in the photo). A handbook of information for chemistry lists the density of mercury as 13.534 g/cm^3 (at 20 °C).

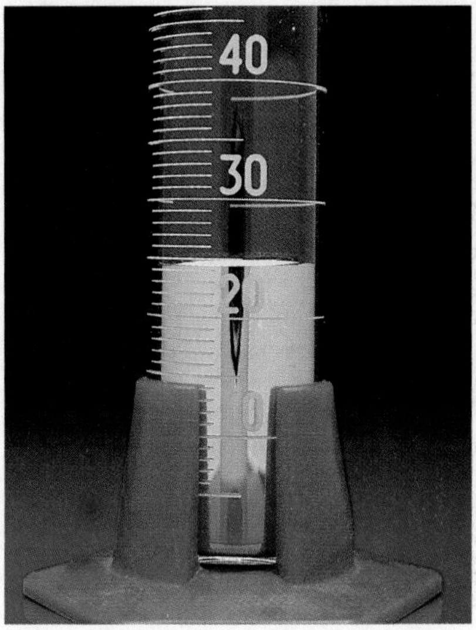

Graduated cylinder containing 25 mL of mercury. See text for a calculation using this information. (C. D. Winters)

$$25\,\cancel{\text{cm}^3} \cdot \frac{13.534\text{ g}}{1\,\cancel{\text{cm}^3}} = 340\text{ g}$$

Therefore, this very small volume of mercury has a large mass (about $\frac{3}{4}$ lb, using the English system of measurement).

EXAMPLE 1.1 *Density*

Suppose you have 225 cm^3 (that is, 225 mL) of ethanol (the common "alcohol" of alcoholic beverages; often called ethyl alcohol). If the density of ethanol is 0.789 g/cm^3 (at 20 °C), what mass, in grams, of alcohol does this represent?

The CD-ROM version of this book has an extensive database of properties of elements and compounds, including density.

Solution You know the density and volume of the sample. Because density is the ratio of the mass of a sample to its volume (density = mass/volume), then mass = volume × density. Therefore, to find the sample's mass, multiply the volume by the density; units of cm^3 cancel.

cm³ = mL

$$\text{Mass} = 225\,\cancel{\text{cm}^3} \cdot \frac{0.789\text{ g ethanol}}{1\,\cancel{\text{cm}^3}\text{ ethanol}} = \boxed{178\text{ g ethanol}}$$

Answers to all the exercises in the book are given in Appendix L.

EXERCISE 1.2 ***Density***

The density of dry air is 1.12×10^{-3} g/cm^3. What volume of air, in cubic centimeters, has a mass of 100. g? ■

PROBLEM-SOLVING TIPS AND IDEAS

1.1 Finding Data

All the information you may need to solve a problem in this book may not be presented in the problem. For example, although the density of alcohol was given in Example 1.1, we could have left it out and assumed you would (a) recognize that you needed density to convert a volume to a mass and (b) know where to find the information. The appendices of this book also contain a wealth of information, and even more is available in the CD-ROM version of the book. Finally, various handbooks of information are available in most libraries; among the best are the *Handbook of Chemistry and Physics* (CRC Press) and *Lange's Handbook of Chemistry* (McGraw-Hill). ■

Another very useful physical property of pure elements and compounds is the temperature at which the solid melts (its **melting point**) or the liquid boils (its **boiling point**). **Temperature** is the property of matter that determines whether heat (energy) can be transferred from one body to another *and* the direction of that transfer: heat energy transfers spontaneously *only* from a hotter object to a cooler one. The number that represents an object's temperature depends on the unit chosen for the measurement.

The number representing the distance between two points depends on its unit of measurement in the same way. The length of your arm can be expressed as 2.5 feet or 76 centimeters, for example. Expressing the length by different numbers does not mean your arm changes its length.

Three scales for temperature measurement are in common use today: Fahrenheit, Celsius, and Kelvin (Figure 1.7). The Celsius scale is generally used for measurements in the laboratory. When calculations incorporate temperature data, however, the data generally must be expressed in kelvins.

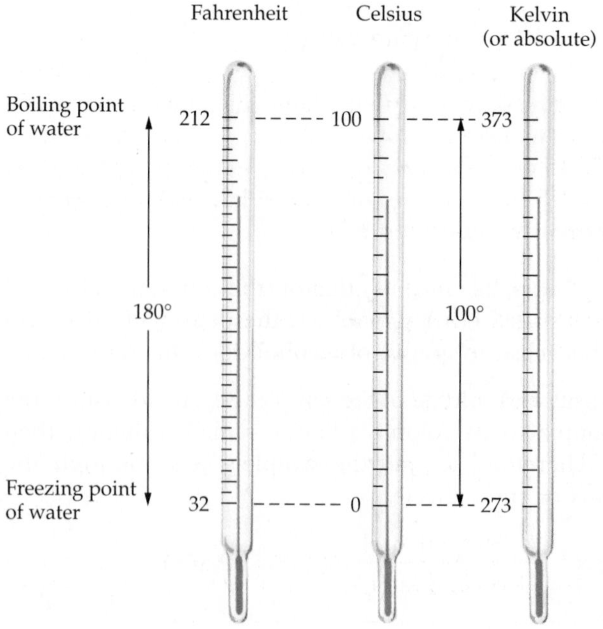

Figure 1.7 A comparison of Fahrenheit, Celsius, and Kelvin temperature scales. The reference, or starting point, for the Kelvin scale is absolute zero (0 K = −273.15 °C), which has been shown theoretically to be the lowest possible temperature. Note that the abbreviation K for the kelvin unit is used *without* the degree sign (°). Also note that 1 °C = 1 K = ($\frac{9}{5}$) °F.

The Celsius Temperature Scale In the United States everyday temperatures are reported using the Fahrenheit scale, but the Celsius scale is used in most other countries and in scientific notation. The latter scale was first suggested by Anders Celsius (1701–1744), a Swedish astronomer, and like the Fahrenheit scale it is based on the properties of water. The size of the Celsius degree is defined by assigning zero as the freezing point of pure water (0 °C) and 100 as its boiling point (100 °C).*

The size of the Fahrenheit degree is equally arbitrary. Gabriel Fahrenheit (1686–1736), a German physicist, defined 0 °F as the freezing point of a solution in which he had dissolved the maximum amount of salt (because this was the lowest temperature he could reproduce reliably), and he intended 100 °F to be the normal human body temperature (but this turned out to be 98.6 °F). Today, the reference points are set at 32 °F and 212 °F, the freezing and boiling points of pure water, respectively. The number of units between these points is 180 Fahrenheit degrees (see Figure 1.7). Comparing the two units, the Celsius degree is almost twice as large as the Fahrenheit degree. It takes only 5 Celsius degrees to cover the same temperature range as 9 Fahrenheit degrees, and this relationship can be used to convert a temperature on one scale to a temperature on the other.

See Table 1.5 for the method of converting between Fahrenheit and Celsius temperatures.

To help you think in terms of the Celsius scale, it is useful to know that water freezes at 0 °C and boils at 100 °C, a comfortable room temperature is about 22 °C (72 °F), your average body temperature is 37 °C (98.6 °F), and the hottest water you could put your hand into without serious burns is about 60 °C (140 °F).

The Kelvin Unit Winter temperatures in many places can easily drop below 0 °C, that is, to temperatures given by negative numbers on the Celsius scale. In the laboratory, even colder temperatures can be achieved easily, and the temperatures are given by even larger negative numbers. There is a limit to how low the temperature can go, however, and hundreds of experiments have found that that limiting temperature is −273.15 °C (or −459.67 °F).

0K = -273.15°C = -459.67°F

William Thomson, known as Lord Kelvin (1824–1907), first suggested a temperature scale that does not use negative numbers. Kelvin's scale, now adopted as the international standard for science, uses the same size degree as the Celsius scale, but it takes the lowest possible temperature as its zero, a point called **absolute zero.** Because Kelvin and Celsius units are the same size, the freezing point of water is reached 273.15 degrees *above* the starting point; that is, 0 °C is the same as 273.15 kelvins, or 273.15 K. Temperatures in Celsius units are readily converted to kelvins, and vice versa, using the relation

$$t(\mathrm{K}) = \frac{1\ \mathrm{K}}{1\ ^\circ\mathrm{C}}(t\,^\circ\mathrm{C} + 273.15\ ^\circ\mathrm{C})$$

Thus, a common room temperature of 23.5 °C is

$$t(\mathrm{K}) = \frac{1\ \mathrm{K}}{1\ ^\circ\mathrm{C}}(23.5\ ^\circ\mathrm{C} + 273.15) = 296.7\ \mathrm{K}.$$

*To be entirely correct, we must specify that water freezes at 0 °C and boils at 100 °C only when the pressure of the surrounding atmosphere is 1 standard atmosphere. We discuss pressure and its effect on boiling point in Chapter 13.

Figure 1.8 Lead melts—changes from the solid to the liquid state—at 327.5 °C, or 600.7 K (*left*). The molten lead is poured into an antique mold where it solidifies into a musket ball (*right*). These are examples of a physical change. (C. D. Winters)

Finally, be sure to notice that the degree symbol (°) is not used with Kelvin temperatures. The name of the unit on this scale is the *kelvin* (not capitalized), and such temperatures are designated with a capital K.

Chemistry studies changes in matter. Changes involving physical properties are called **physical changes.** In a physical change the chemical identity of a substance is preserved even though it may have changed its physical state or the gross size and shape of its pieces. An example of a physical change is the melting of a solid, and the temperature at which this occurs is often so characteristic of the solid that it can be used to identify the substance. For example, tin and lead resemble each other in outward appearance, but tin melts at 231.8 °C, and lead melts about 100 °C (or kelvins) higher (327.5 °C) (Figure 1.8).

$$\text{Melting point of tin (K)} = \frac{1\ \text{K}}{1\ °\text{C}}(231.8\ °\text{C} + 273.15\ °\text{C}) = 505.0\ \text{K}$$

$$\text{Melting point of lead (K)} = \frac{1\ \text{K}}{1\ °\text{C}}(327.5\ °\text{C} + 273.15\ °\text{C}) = 600.7\ \text{K}$$

EXERCISE 1.3 ***Temperature Conversions***

Liquefied nitrogen boils at 77 K. What is this temperature in Celsius degrees? ■

1.2 ELEMENTS AND ATOMS

When heated, pure table sugar (sucrose) decomposes in a complex series of chemical changes (caramelization) that produce the brown color and flavor of peanut brittle or caramel candy. If heated for a longer time at a high enough temperature, however, sucrose can be converted completely to two other pure substances: carbon and water (Figure 1.9). Furthermore, if the water is collected, it can be decomposed still further to pure hydrogen and oxygen by passing an electric current through it. Pure substances like carbon, hydrogen, and oxygen that are composed of only one type of atom are classified as **elements.** Only 110

Figure 1.9 Sucrose can be decomposed using sulfuric acid to form carbon (the black solid) and water (seen as steam emerging from the beaker). (C. D. Winters)

elements are known at this time, and of these only about 90 are found in nature; the remainder have been created by scientists. Each element has a *name* and a *symbol,* which are listed in the table at the front of the book. Carbon (C), sulfur (S), iron (Fe), copper (Cu), silver (Ag), tin (Sn), gold (Au), mercury (Hg), and lead (Pb) were known in relatively pure form to the early Greeks and Romans and to the alchemists of ancient China, the Arab world, and medieval Europe. Many others, however, such as aluminum (Al), silicon (Si), iodine (I), and helium (He), were not discovered in the minerals of the earth or the atmosphere until the 18th and 19th centuries. Finally, artificial elements, such as technetium (Tc), plutonium (Pu), and americium (Am), were not made until the 20th century when the techniques of modern physics became available.

An atom of element 110 was first detected on November 9, 1994 by a team of scientists at the Heavy Ion Research Center in Darmstadt, Germany. The atom existed for only 393 microseconds.

Many elements have names and symbols with Latin or Greek origins, but more recently discovered elements have been named for their place of discovery or for a person or place of significance. (Table 1.2). The table at the front of the book, in which the symbol and other information for each element are enclosed in a box, is called the **periodic table.** We describe this important tool of chemistry in more detail beginning in Chapter 2. For the moment, however, be sure to notice that only the first letter of an element's symbol is capitalized. For exam-

TABLE 1.2 The Names of Some Chemical Elements

Element	Symbol	Date of Discovery	Discoverer	Derivation of Name or Symbol
Berkelium	Bk	1950	G. T. Seaborg S. G. Thompson A. Ghioroso (U.S.)	Berkeley, California, was the site of Seaborg's laboratory.
Copper	Cu	Ancient		Latin, *cuprum,* copper, or *cyprium,* from Cyprus.
Lead	Pb	Ancient		Latin, *plumbum,* lead, meaning heavy.
Oxygen	O	1774	J. Priestley (Great Britain) K. W. Scheele (Sweden)	French, *oxygene,* generator of acid, derived from the Greek, *oxy* and *genes* meaning acid forming (because oxygen was thought to be part of all acids).

ple, cobalt is Co and not CO. The notation CO represents the combination of carbon (C) and oxygen (O) in the chemical compound carbon monoxide.

An **atom** is the smallest particle of an element that retains the chemical properties (see Section 1.4) of that element. Modern chemistry is based on an understanding and exploration of nature at the atomic level, and we have much more to say about atoms and atomic properties in Chapters 2, 7, and 8, in particular.

EXERCISE 1.4 ***Elements***

Using the periodic table in the front of the book,

a. Find the names of the elements with symbols Na, Cl, and Cr.

b. Find the symbols for the elements zinc, nickel, and potassium. ■

1.3 COMPOUNDS AND MOLECULES

Pure substances like sucrose and water that are composed of two or more different elements are referred to as **chemical compounds.** Even though only 110 elements are known, there appears to be no practical limit to the number of compounds that can be made from those elements. At the moment more than 12 million compounds are known, with about a half million added to the list each year.

In a compound, atoms of more than one element are combined in a manner distinctive to that compound. When elements become part of a compound, their original characteristic properties, such as color, hardness, and melting point, are replaced by the characteristic properties of the compound. Consider ordinary table sugar, sucrose, which is composed of three elements:

- Carbon, which is usually a black powder, but is also found in other forms such as diamond

CURRENT ISSUES IN CHEMISTRY

Gold Made by Bacteria?

From old westerns shown on television, your vision of a gold miner may be of a grizzled old prospector panning for gold in a stream. He scooped up some sand and gravel from the bottom of the stream and swirled it in a shallow pan. Because gold is much denser than gravel, the gold settled to the bottom of the pan and could be picked out as tiny flecks and even larger nuggets. This "placer" gold is material long thought to have been eroded from an ore bed by wind and water and washed into streams.

John R. Watterson of the U.S. Geological Survey in Denver has a new theory about the way this gold found its way to the streams. He has found evidence that stream-borne gold particles in Alaska, and presumably elsewhere, were formed by bacteria that produced a thin skin of pure gold! The photograph here is of a mass of bacteria (of the genus *Pedomicrobium*) that accumulated gold around themselves; the piece is about 0.1 mm (millimeter) wide. These coatings are extraordinary in that they are almost pure, 24-karat gold. Apparently the bacteria attract gold compounds selectively from ore beds and convert them into a metallic gold coating. When the bacteria die, the coating survives and can be washed into streams as a fleck of gold.

Watterson and others are far from understanding these "gold bugs," which are also known to form coatings of other minerals around them. When asked why the bacteria use gold, he replied, "That's still an absolute mystery to me." (See *Scientific American,* Vol. 267, p. 27, September 1992.)

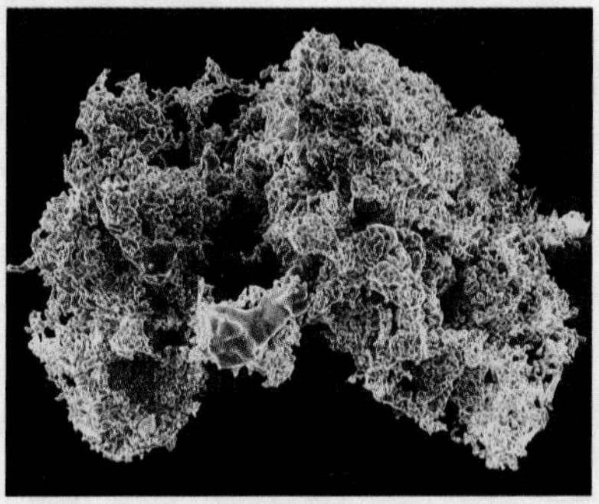

The shell from a microbe that accumulates gold. (John Watterson, U.S. Geological Survey, *Geology,* Vol. 20, pp. 315–318, 1992)

- Hydrogen, the lightest gas known
- Oxygen, a gas necessary for human and animal respiration

Sucrose, a white, crystalline solid, has properties completely unlike any of these three elements.

It is important to make a careful distinction between a mixture of elements and a *compound* of two or more elements. Iron and sulfur can be mixed in varying proportions (see Figure 1.1). In the chemical compound known as iron pyrite (see Figure 1.4), however, this kind of variation cannot occur. Not only does iron pyrite exhibit properties peculiar to itself and different from those of iron and sulfur, or a mixture of these elements, but it has a definite percentage composition by weight (46.55% Fe and 53.45% S, or 46.55 g of Fe and 53.45 g of S in 100.0 g of sample). Thus, two major differences exist between mixtures and pure compounds: compounds have distinctly different properties from their parent elements, and *compounds have a definite percentage composition (by mass) of their combining elements.*

Some compounds—such as salt, NaCl—are composed of **ions,** electrically charged atoms or groups of atoms. (We describe such compounds beginning in Chapter 3.) Other compounds—such as water and sugar—consist of **molecules,** which are the smallest discrete units that retain the chemical characteristics of the compound. The composition of any compound can be represented by its **formula.** You already know that the formula for water is H_2O. The symbol for

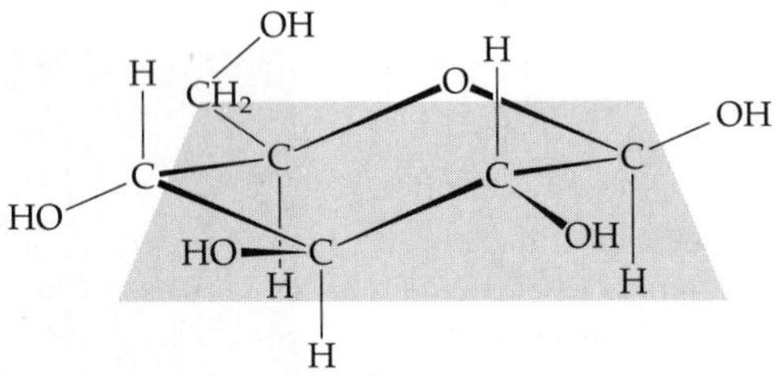

(a)

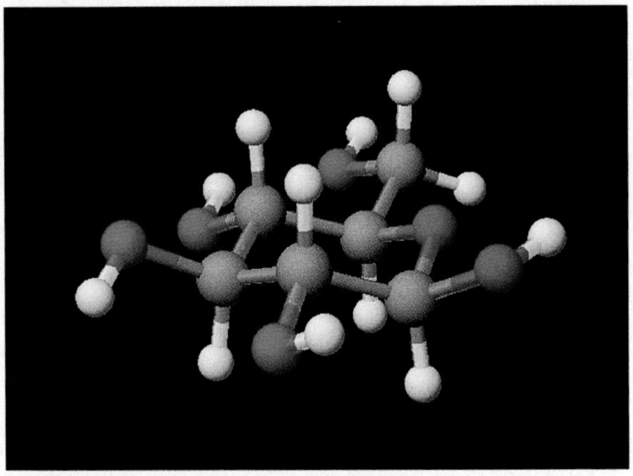

(b)

(c)

Figure 1.10 Some examples of models of a chemical compound. Here glucose, $C_6H_{12}O_6$, is illustrated as (a) a line drawing, (b) a model constructed using a computer program, and (c) a plastic model that can be held in your hand. ((c) C. D. Winters)

hydrogen, H, is followed by a subscript 2 to indicate that two atoms of this type occur in the water molecule. Similarly, the formula shows one O atom, because the appearance of the symbol for oxygen without a subscript means a single atom of that type occurs in the molecule.

The compound sucrose has a molecular formula of $C_{12}H_{22}O_{11}$. A single molecule is composed of 12 C atoms, 22 H atoms, and 11 O atoms. Similarly, the ratio of atoms in a spoonful of sugar is still 12 C atoms to 22 H atoms to 11 O atoms.

Many physical and chemical properties depend on *molecular structure.* Two general features define what is meant by this term. The first is how atoms fit together within a molecule; the second is how molecules fit together to form the liquid or solid state.

An example of a molecular structure is presented by the simple sugar, glucose, $C_6H_{12}O_6$ (Figure 1.10). This complicated molecule has 24 atoms. The molecule is based on a six-membered ring of five carbon atoms and one oxygen atom. Each carbon atom in the molecule is attached to four other atoms, and the hydrogen and oxygen atoms are attached to only one or two other atoms, respectively. As you can see in Figure 1.10, the molecule is three-dimensional, with specific distances and angles between atoms.

Visualizing molecular structures is one of the important aspects of chemistry. Structures may be drawn on paper, a computer "molecular-modeling" program can be used to construct a model, or we can use wood or plastic to make a model we can hold and handle (see Figure 1.10). As described in the Introduction, models of molecules are an example of modeling in science and are extremely useful. For this reason we use them extensively in this book.

We should not overlook the second aspect of structure: how molecules fit together in a solid or liquid. Physical properties, such as melting point and boiling point, relate to the forces of attraction between molecules. The melting point of glucose, 146 °C, is determined by the forces of attraction between glucose molecules in the solid.

1.4 CHEMICAL AND PHYSICAL CHANGE

Hydrogen and oxygen gas are mixed in a balloon. If a lighted match or candle is touched to the balloon, the mixture explodes, producing water. This is an example of a **chemical change,** or **chemical reaction,** because one or more substances (the **reactants**) are transformed into one or more different substances (the **products**). We say that hydrogen and oxygen *react* to form water, which can be represented as

$$\underbrace{2 \text{ molecules } H_2 \text{ gas} + 1 \text{ molecule } O_2 \text{ gas}}_{\text{reactants}} \xrightarrow{\text{produces}} \underbrace{2 \text{ molecules } H_2O \text{ vapor}}_{\text{product}}$$

At the molecular level a *chemical change* produces a new arrangement of atoms without a gain or loss in the number of atoms (Figure 1.11). The molecules present after the reaction are different from those present before the reaction. A *physical change* does not result in a new chemical substance. Instead, the molecules present before and after the change are the same, but their arrangement relative to one another (farther apart in a gas, closer together in a solid) is different.

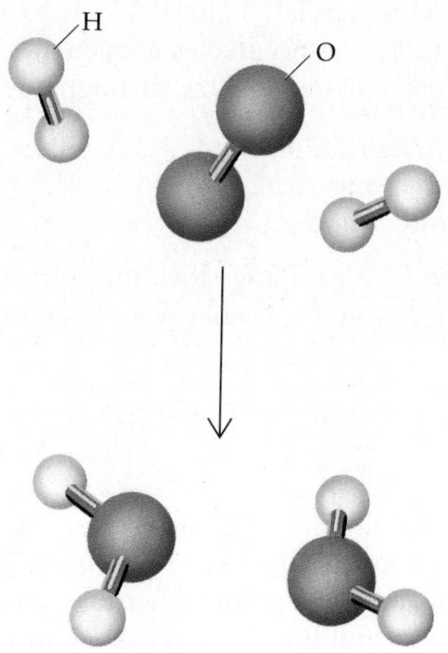

Figure 1.11 Hydrogen and oxygen molecules react to produce water molecules. The connections between hydrogen atoms and those between oxygen atoms have been broken, and new ones are formed between oxygen and hydrogen. At the molecular level the chemical change has produced a new arrangement of atoms without a gain or loss in the number of atoms.

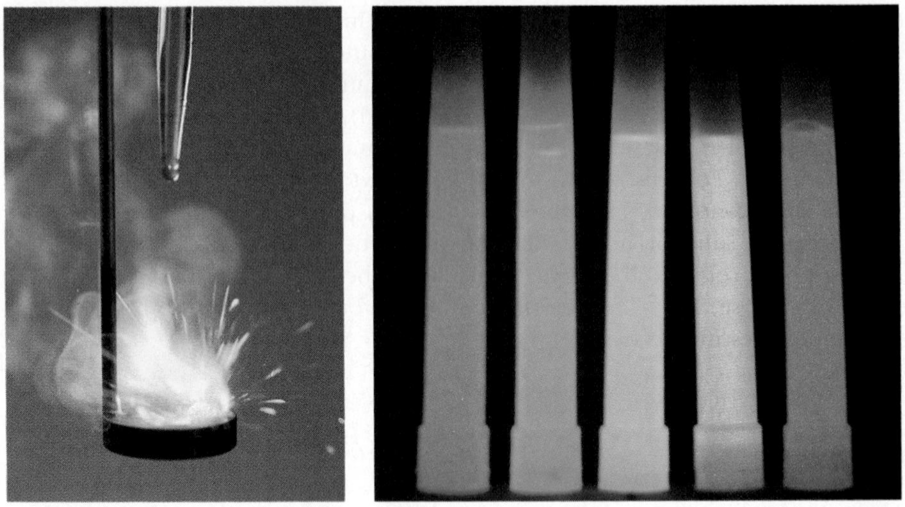

(a) (b)

Figure 1.12 Chemical changes. (a) The very rapid chemical change that occurs when water is dropped onto the element potassium. (b) A "light stick" produces light by a chemical change. (C. D. Winters)

The **chemical properties** of a substance are those properties that involve chemical changes. The following are some of the more common examples of chemical properties:

- Burn in air. (This is called *combustion.* See Chapter 4.)
- React with water
- React with an acid (see Chapter 4)
- React with a base (see Chapter 4)
- Undergo a change when subjected to an electric current. (This process is called *electrolysis;* see Chapter 21.)

In each case one or more new substances are produced.

Physical changes, and to a greater extent chemical changes, are accompanied by transfers of energy. The reaction of the element potassium with water (Figure 1.12a) transfers a tremendous amount of energy (in the form of heat and light) to its surroundings. The reaction in a commercial "light stick" evolves light energy and a little heat (Figure 1.12b), and a battery makes a calculator work because a chemical reaction forces electric current to flow through the circuits.

A pot of water heated on a propane-burning camping stove. See Exercise 1.5 (C. D. Winters)

EXERCISE 1.5 ***Chemical Reactions and Physical Changes***

In the photo shown here, you see water heated by a camping stove that burns propane. Name the chemical and physical changes you see. Is energy involved in the process? If so, how? ■

1.5 MIXTURES AND PURE SUBSTANCES

Most natural samples of matter consist of two or more substances; that is, they are mixtures (Figure 1.13). Peanut brittle is obviously a mixture, because one can see that the peanuts are different from the surrounding material. A mixture

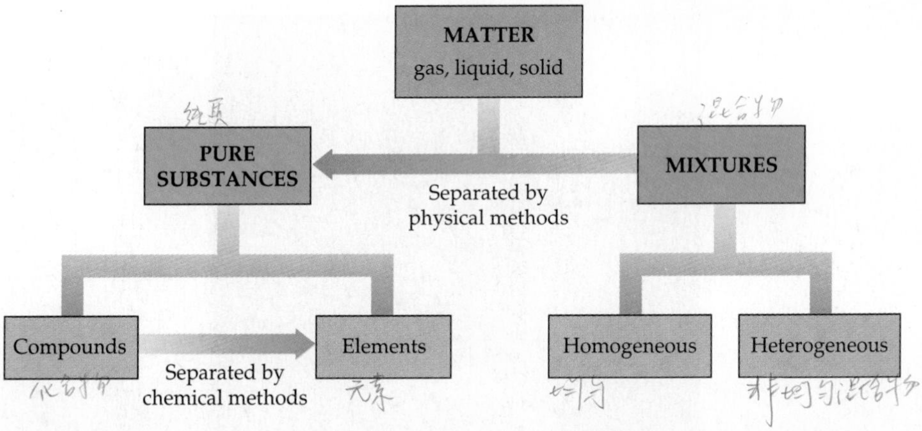

Figure 1.13 The components of matter and the relation between mixtures and pure substances.

Figure 1.14 The air in a forest seems to be homogeneous, but it is really a heterogeneous mixture. That is, in the proper light, particles of dust can be seen floating in the air. (C. D. Winters)

in which the uneven texture of the material is visible is called a **heterogeneous mixture.** Other heterogeneous mixtures may appear completely uniform but on closer examination are not. For example, the air in a room may look like a homogeneous gas until a beam of light enters the room, revealing floating dust particles (Figure 1.14). Milk appears smooth in texture to the unaided eye, but magnification reveals fat and protein globules within the liquid. In a heterogeneous mixture the properties in one region are different from those in another region.

A **homogeneous mixture,** or **solution,** is completely uniform at the macroscopic level and consists of two or more substances in the same phase (Figure 1.15). No amount of optical magnification reveals a solution to have different properties in one region than in another, because heterogeneity exists in a solution only at the atomic or molecular level, where the individual particles are too small to be seen with ordinary light. Examples are pure air (mostly a mixture of nitrogen and oxygen gases), sugar dissolved in water, and brass alloy (which is a homogeneous solid mixture of copper and zinc). The properties of a homoge-

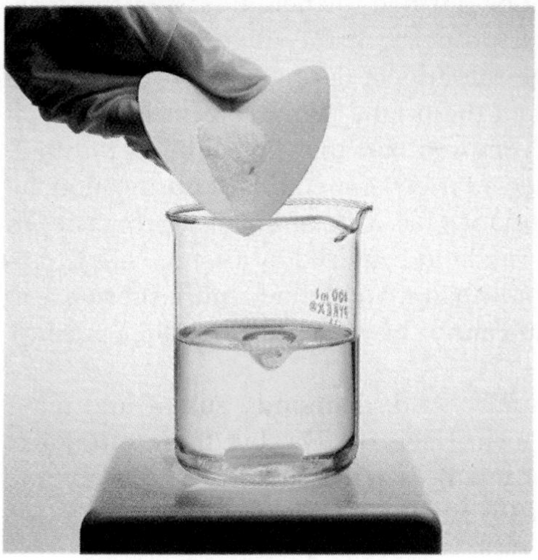

Figure 1.15 A homogeneous mixture or solution. A solid yellow chemical compound called potassium chromate is stirred into water, in which it dissolves to form an aqueous solution. (C. D. Winters)

Figure 1.16 The iron chips in a mixture of iron and sulfur may be removed by stirring the heterogeneous mixture with a magnet. (C. D. Winters)

neous mixture are the same everywhere in a sample, but they can vary from one sample to another depending on how much of one component is present relative to another component.

When a mixture is separated into its pure components, the components are said to be *purified* (see Figure 1.13). Most efforts at separation are not complete in a single step, however, and repetition is almost always necessary to give an increasingly purer substance. For example, the iron can be separated from a heterogeneous mixture of iron and sulfur by repeatedly stirring the mixture with a magnet (Figure 1.16). When the mixture is stirred the first time and the magnet is removed, much of the iron is removed with it, leaving the sulfur in a higher state of purity. After one stirring, however, the sulfur may still have a dirty appearance due to a small amount of iron that remains. Repeated stirrings with the magnet, or perhaps the use of a very strong magnet, finally leave a bright yellow sample of sulfur that apparently cannot be purified further, at least by this technique. This purification process uses a property of the mixture, its color, to measure the extent of purification. (The color depends on the relative quantities of iron and sulfur in the mixture.) When the bright yellow color is obtained, it is assumed that all the iron has been removed and that the sulfur is purified.

Using a magnet to separate the iron and sulfur in iron pyrite does not work. The elements are combined into a chemical compound in pyrite.

It is safe to call sulfur pure only when a variety of methods of purification fail to change its properties. Each pure substance has a set of physical properties by which it can be recognized, just as you can be recognized by a set of characteristics such as your hair or eye color or your height. We define a **pure substance** as a sample of matter with properties that cannot be changed by further purification.

Many pure substances occur naturally. Gold, diamonds, sulfur, and many minerals occur naturally in very pure form (Figure 1.17), but these substances are special cases. We live in a world of mixtures—all living things, the air and food we depend on, and many products of technology are mixtures.

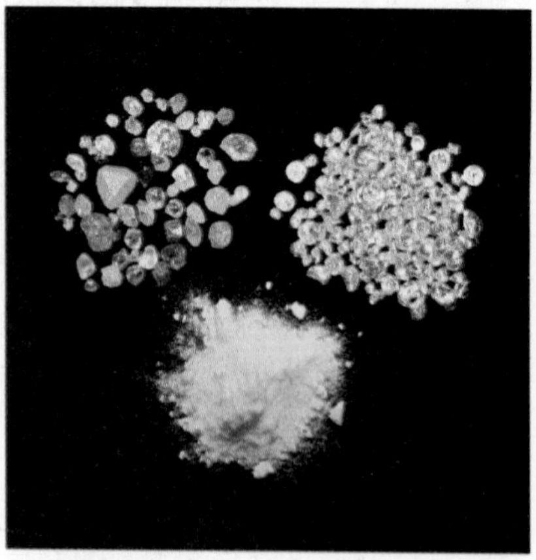

Figure 1.17 Pure substances from the earth. Clockwise from the bottom the substances are sulfur, diamonds, and gold. (C. D. Winters)

EXERCISE 1.6 ***Mixtures and Pure Substances***

The photo shown here is of two mixtures. Which one is homogeneous and which is heterogeneous? Which one is a solution? ■

Homogeneous and heterogeneous mixtures. (C. D. Winters)

Separation of Mixtures into Pure Substances

The separation of mixtures is usually more difficult than the magnetic separation of iron and sulfur described earlier (see Figure 1.16). Many trained chemists would find it difficult to separate some of the metal-containing minerals in Figure 1.5 into pure substances. Nevertheless, because each pure substance has a set of properties unlike those of any other pure substance, it should be possible to use these properties to separate the pure substances in these minerals, just as we described in the separation of iron and sulfur.

Many different methods have been developed to separate mixtures into pure substances. Different *physical* properties of the pure substances—such as solubility in water, density, and melting point or boiling point—are often exploited to make the separation. A practical example is the separation and purification of a metal-bearing ore. In an **ore,** a mineral (a compound containing the metal combined chemically with other elements) is usually mixed with dirt and other, unwanted minerals. The first step in obtaining the metal is separation of the desired mineral from the ore. The production of copper, the metal used in electrical wiring, illustrates the process. Copper-bearing ore is separated from unwanted material by a process called *froth flotation.* Here the powdered ore is mixed with oil and agitated with soapy water in a large tank (Figure 1.18). Compressed air is forced through the mixture, and lightweight, oil-covered particles of a nearly pure copper-containing compound are carried to the top and float on the froth. The other, heavier material settles to the bottom of the tank, and the froth with the copper-containing mineral is skimmed off.

Figure 1.18 "Winning" copper from its ores. The copper-containing ore (a copper sulfide) is enriched in the flotation process. The lighter particles containing the copper compound are trapped in soap bubbles and float on the water. The heavier "gangue," or waste material, settles to the bottom.

Nitrogen, N_2, and oxygen, O_2, are annually among the top five chemicals produced in the United States. Both are separated from air using a difference in physical properties. Liquid nitrogen boils at 77 K, whereas liquid oxygen boils at 90 K. Although this is a small difference, it is sufficient to allow for efficient separation.

Pure substances have a fixed composition and cannot be purified further by methods such as those we have mentioned. By using the *chemical properties* of the substance, however, it may be possible to decompose a pure substance into other

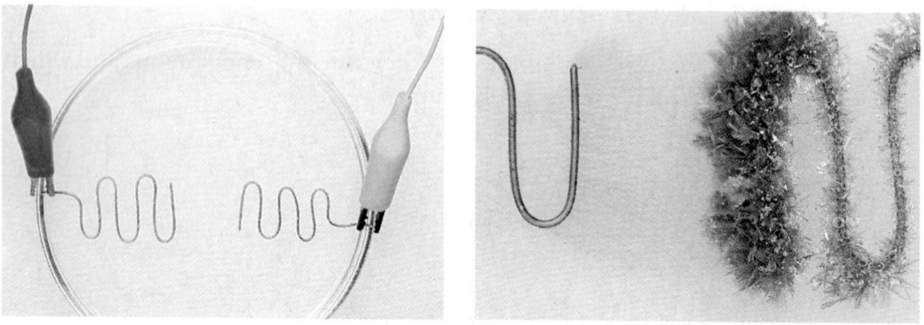

Chlorine, Cl_2, can be obtained by passing electricity through a solution in which a chlorine-containing compound has been dissolved. In this case tin(II) chloride was dissolved in water and then electricity passed through the solution. The yellow color of dissolved chlorine appears near the electrode on the left. Surrounding the electrode at the right are crystals of tin metal. (C. D. Winters)

pure substances, which may be chemical elements or other compounds. Once again the production of a commonly used chemical illustrates this technique. Chlorine, Cl_2, is tenth on the list of chemicals produced in the United States, with 24.06 billion lb produced in 1993. The element is obtained from salt, NaCl, which is usually obtained by mining salt beds deep in the earth or by forcing water into the beds and then pumping out the dissolved salt. Electricity is passed through the solution of NaCl in water, and Cl_2 is formed.

Figure 1.19 Nanotechnology: A scanning electron microscopic view of a micromotor. The rotor of the motor is only 150 μm across. This dimension (150 micrometers) is equal to 150×10^{-6} m, or 0.015 cm, or 0.0059 in. (From H. Guckel *Science,* Vol. 254, p. 1340, 1991. Copyright AAAS.)

1.6 UNITS OF MEASUREMENT

Doing chemistry requires observing chemical reactions and physical changes. When making peanut brittle you see the color turn brown and the evolution of steam as water escapes when the mixture is heated. These are **qualitative observations.** No measurements and numbers are involved, but something obviously happened, probably a chemical reaction, and it can be described in words.

To understand a chemical reaction more completely, chemists usually make **quantitative measurements.** For example, if two compounds react with each other, is some of one of them left over? How much product forms? Do we have to heat the reactants to make the change occur, and if so, how much heat is required and for how long? If light is given off, is it red, blue, green, or is it not in the visible range at all (infrared or ultraviolet)? If a new molecule is formed, what is its identity?

Such questions require us to think in terms of physical measurements, such as length, volume, temperature, and time. Because each measurement consists of a number and a unit of measurement, making a measurement consists of counting the number of units (how many centimeters, for example) that correspond to the measured item.

A frontier area of chemistry where an understanding of measurements and their units is crucial is *nanotechnology,* the construction of devices built from molecules or atoms, that have dimensions between 1 and 100 nanometers (Figure 1.19). A nanometer (nm) is 10^{-9} m, a very small dimension in the world of

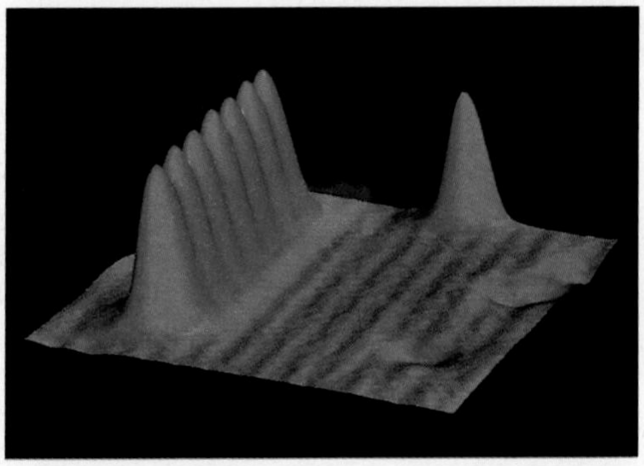

Figure 1.20 The first "hand-built" atomic structure. Scientists at IBM's Almaden Research Center in San Jose, California, were able to move individual atoms across a nickel surface and position them at will. Here seven xenon (Xe) atoms (each represented by a peak) form a linear chain. The image is 5 nm on a side. This is 5×10^{-9} m, or 5×10^{-7} cm (2×10^{-7} in.). The atoms are 6.4 billionths of an inch (0.16 nm) high and are spaced 20 billionths of an inch (0.5 nm) apart. See *Science,* Vol. 254, p. 1324, 1991. (IBM Corporation, Research Division, Almaden Research Center)

engineers, a large one to chemists, but a common one to biologists—a bacterium is about 1000 nm in length, whereas a typical molecule is only about 0.1 nm across. Figure 1.20 shows a nanostructure of xenon atoms on a nickel surface. This object is 5 nm on a side or only about 2×10^{-7} in.^{2}, and so the xenon atoms are only about 0.5 nm apart.

To know more about nanostructures—or about much of science and engineering—it is essential to be familiar with the dimensions of the objects being studied, the units being used, and the relation between them. The scientific community has chosen a modified version of the **metric system** as the standard system for recording and reporting measurements. This is a decimal system, in which all of the units are expressed as powers of 10 times some basic unit. The resulting system, as applied internationally in science, is called the *Système International d'Unités* (International System of Units), abbreviated **SI.**

SI Units

In SI, all units are derived from seven base units (listed in Table 1.3). Larger and smaller quantities are expressed by using an appropriate prefix with the base

TABLE 1.3 SI Base Units[1]

Measurement	Name of Unit	Abbreviation
mass	kilogram	kg
length	meter	m
time	second	s
temperature	kelvin	K
amount of substance	mole	mol
electrical current	ampere	A

[1]A seventh base unit, the candela, is used for luminous intensity.

TABLE 1.4 Selected Prefixes Used in the Metric System

Prefix	Abbreviation	Meaning	Example
mega-	M	10^6	1 megaton = 1×10^6 tons
kilo-	k	10^3	1 kilogram (kg) = 1×10^3 g
deci-	d	10^{-1}	1 decimeter (dm) = 1×10^{-1} m
centi-	c	10^{-2}	1 centimeter (cm) = 1×10^{-2} m
milli-	m	10^{-3}	1 millimeter (mm) = 1×10^{-3} m
micro-	μ[1]	10^{-6}	1 micrometer (μm) = 1×10^{-6} m
nano-	n	10^{-9}	1 nanometer (nm) = 1×10^{-9} m
pico-[2]	p	10^{-12}	1 picometer (pm) = 1×10^{-12} m

[1]This is the Greek letter mu (pronounced "mew").
[2]This prefix is pronounced "peako."

TABLE 1.5 Some Common Conversion Factors

LENGTH: SI Unit = meter (m)		
1 kilometer	=	1000 m
	=	0.62137 mile
1 meter	=	100 cm
1 centimeter	=	10 mm
1 nanometer	=	1×10^{-9} m
1 picometer	=	1×10^{-12} m
1 inch	=	2.54 cm (exactly)
1 Ångstrom	=	1×10^{-10} m
VOLUME: SI Unit = cubic meter (m^3)		
1 liter (L)	=	1×10^{-3} m^3
	=	1 dm^3
	=	1000 cm^3
	=	1.056710 qt
1 gallon	=	4.00 qt
MASS: SI Unit = kilogram (kg)		
1 kilogram	=	1000 g
1 gram	=	1000 mg
1 pound	=	453.59237 g = 16 oz
1 ton	=	2000 lb
TEMPERATURE: SI Unit = kelvin (K)		
0 K	=	−273.15 °C
K	=	(1 K/1 °C) (°C + 273.15 °C)
? °C	=	$\frac{5\ °C}{9\ °F}$(°F − 32 °F)
? °F	=	$\frac{9\ °F}{5\ °C}$(°C) + 32 °F

unit (Table 1.4). For instance, highway distances are given in *kilo*meters, in which 1 km (kilometer) is exactly 1000 or 10^3 m (meters). In chemistry, length is most often given in subdivisions of the meter, such as *centi*meters (cm) or *milli*meters (mm). The prefix *centi-* means 1/100, so 1 centimeter is 1/100 of a meter ($1 \text{ cm} = 1 \times 10^{-2}$ m); 1 millimeter is 1/1000 of a meter ($1 \text{ mm} = 1 \times 10^{-3}$ m). On the atomic scale, dimensions are often given in nanometers (nm) ($1 \text{ nm} = 1 \times 10^{-9}$ m) or picometers (pm) ($1 \text{ pm} = 1 \times 10^{-12}$ m). Several **conversion factors** that allow you to convert between SI and non-SI units are given in Table 1.5 and in Appendix C.

PROBLEM-SOLVING TIPS AND IDEAS

1.2 Using Scientific Notation

The number 0.001, or 1/1000, is written as 1×10^{-3}. This notation—called scientific, or exponential, notation—is used throughout the book and is explained further in Appendix A. Scientific notation makes it much simpler to handle very large or very small numbers.

Make sure you know how to use your calculator with exponential numbers. When entering a number such as 1.23×10^{-4} into your calculator, you first enter 1.23 and then press a key marked EE or EXP (or something very similar). This enters the "$\times$ 10" portion of the notation for you. You then complete the entry by keying in the exponent of the number, −4. (To change the exponent from +4 to −4, you need to press the "±" key.)

A common error made by students is to enter 1.23, then press the multiply key (×) and then key in 10 before finishing by pressing EE or EXP followed by −4. This gives an entry that is 10 times too large. ■

1.7 USING NUMERICAL INFORMATION

As part of your course in chemistry you will prepare materials in the laboratory that will require you to make some calculations. You will collect numerical data and use those data to calculate a result or you will look for correlations among the pieces of data. All these activities mean you have to be comfortable handling numerical information. This section describes some common calculations and proper ways to handle quantitative information. For more information see Appendix A.

Suppose you are given a rectangular piece of aluminum and are asked to find its density in units of grams per cubic centimeters. Because density is the ratio of mass to volume, you need to measure the mass and determine the volume of the piece. The data in the table were collected in the laboratory.

Measurement	Data Collected
Mass of aluminum	13.56 g
Length	6.45 cm
Width	2.50 cm
Thickness	3.1 mm

To find the volume of the aluminum in cubic centimeters, you multiply its length by its width and its thickness. First, however, all the measurements must be in the same unit, meaning that the thickness must be converted to centimeters.

$$3.1\ \text{mm} \cdot \frac{1\ \text{cm}}{10\ \text{mm}} = 0.31\ \text{cm}$$

You will find that **dimensional analysis,** an approach explored further in Appendix A, is very useful in making this conversion. This means you should multiply the number you wish to convert (3.1 mm) by a *conversion factor* (1 cm/10 mm) to produce a result in the desired unit (0.31 cm). Units are like numbers, and because the unit "mm" was both in the numerator and denominator, it canceled out and left the answer in centimeters, the desired unit.

The conversion factor expresses the equivalence of the distance in the two different units (here 1 cm = 10 mm). Because the numerator and denominator describe the same distance, the factor is equivalent to the number 1. Therefore, multiplication of the original distance by this factor does not change the measured distance, only its units.

A conversion factor is always written so that it has the form (new units divided by units to be converted).

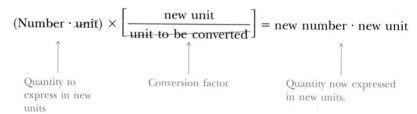

$$(\text{Number} \cdot \text{unit}) \times \left[\frac{\text{new unit}}{\text{unit to be converted}}\right] = \text{new number} \cdot \text{new unit}$$

↑ Quantity to express in new units

↑ Conversion factor

↑ Quantity now expressed in new units.

This way the units in the denominator cancel the units of the original data, leaving the desired units. In the previous problem, units of millimeters canceled, leaving the result in centimeters.

With all the dimensions of the piece of aluminum now in the same unit, the volume can be found,

$$\text{Length} \times \text{width} \times \text{thickness} = \text{volume}$$
$$6.48\ \text{cm} \cdot 2.50\ \text{cm} \cdot 0.31\ \text{cm} = 5.0\ \text{cm}^3$$

and the density can be calculated.

$$\text{Density} = \frac{13.56\ \text{g}}{5.0\ \text{cm}^3} = 2.7\ \text{g/cm}^3$$

EXAMPLE 1.2 *Distances on the Molecular Level*

A hydrogen atom is connected to each of five of the oxygen atoms in glucose (Figure 1.10). On average the distance between O and H is 95.7 pm. What is this distance in meters? In nanometers (nm)?

Solution You can solve this problem by knowing the relation between the units of the information you are given (picometers) and the units of the desired outcome (meters or nanometers). There is no direct conversion from nanometers to picometers given in our tables, but relationships are listed between meters and picometers and between meters and nanometers (Table 1.4; 1 pm = 1×10^{-12} m and 1 nm = 1×10^{-9} m). Therefore, we first convert picometers to meters and then convert meters to nanometers.

$$\text{Picometers} \xrightarrow{\times \frac{\text{m}}{\text{pm}}} \text{Meters} \xrightarrow{\times \frac{\text{nm}}{\text{m}}} \text{Nanometers}$$

$$95.7\ \cancel{\text{pm}} \cdot \frac{1 \times 10^{-12}\ \text{m}}{1\ \cancel{\text{pm}}} = 95.7 \times 10^{-12}\ \text{m} = 9.57 \times 10^{-11}\ \text{m}$$

$$9.57 \times 10^{-11}\ \cancel{\text{m}} \cdot \frac{1\ \text{nm}}{1 \times 10^{-9}\ \cancel{\text{m}}} = 9.57 \times 10^{-2}\ \text{nm}\ \ (\text{or } 0.0957\ \text{nm})$$

Notice how units cancel to leave an answer whose unit is that of the numerator of the conversion factor.

EXERCISE 1.7 ***Interconverting Units of Length***

The pages of a typical textbook are 10.0 in. long and 8.00 in. wide. If 1 inch = 2.54 cm, what is each length in centimeters? In meters? In millimeters? What is the area of a page in square centimeters? ■

EXERCISE 1.8 ***Density***

A platinum sheet is 2.50 cm square and has a mass of 1.656 g. The density of platinum is 21.45 g/cm^3. What is the thickness of the platinum sheet in millimeters? ■

PROBLEM-SOLVING TIPS AND IDEAS

1.3 Problem-Solving Strategies

Now that you have seen solutions for several numerical problems, it is an appropriate time to comment on the strategy of problem solving.

Step 1. *Define the problem.* Read the question carefully. What key principles are involved? What information is necessary, and what is only there to place the question in the context of chemistry? Organize the information to see what is necessary and see relationships among the data given to you. Try writing the information down in a table form. If it is numerical information, be sure to include units.

One of the greatest difficulties for a student in introductory chemistry is picturing what is being asked for. Try sketching a picture of situation involved. For example, sketch a picture of the piece of aluminum whose density we wanted to calculate, and put the dimensions on the drawing.

Step 2. *Develop a plan.* Have you done a problem of this type before? If not, perhaps the problem is really just a combination of several easier ones you have seen before. Break it down into those simpler components. Try reasoning backward from the units of the answer. What data do you need to find an answer in those units?

Step 3. *Execute the plan.* Carefully write down each step of the problem, being sure to keep track of the units on numbers. (Do the units cancel out to give you the answer in the desired units?) Don't skip steps. Don't do anything but the simplest steps in your head. Students often say they got a problem wrong because they "made a stupid mistake." Your instructor—and book authors—make them, too, and it is usually because they don't take the time to write down the steps of the problem clearly.

Step 4. *Is the answer reasonable?* As a final check when doing any calculation, ask yourself if the answer is reasonable. Let us say you are asked to convert 100. yd to a distance in meters. Using dimensional analysis, and some well known factors for converting from the English system to the metric, we have

$$100.\ \cancel{\text{yards}} \cdot \frac{3\ \text{ft}}{1\ \cancel{\text{yard}}} \cdot \frac{12\ \cancel{\text{in.}}}{1\ \text{ft}} \cdot \frac{2.54\ \cancel{\text{cm}}}{1\ \cancel{\text{in.}}} \cdot \frac{1\ \text{m}}{100\ \cancel{\text{cm}}} = 91.4\ \text{m}$$

You should recognize that a distance of 91.4 m is about right. Because a meter is a little more than 3 ft, the distance should be a little less than 100 m. In the first step, if you divided instead of multiplied by 3, the final answer would be a little more than 10 m. This is equivalent to only about 30 ft, and you know a 100-yd football field is longer than that. ■

Figure 1.21 Some common laboratory glassware. (C. D. Winters)

Chemists often handle chemicals in glassware, such as beakers, flasks, pipets, graduated cylinders, and burets, which are marked in volume units (Figure 1.21). The SI unit of volume is the cubic meter (m^3), which is too large for everyday laboratory use. For example, if you used cubic meters, the volume of a common laboratory beaker would be 0.0006 m^3 and you would often work with volumes of chemicals in the range of 0.001 m^3 or less in the laboratory. These are inconvenient numbers for routine use, so more often we use the unit called the **liter,** symbolized by **L.**

A cube with sides equal to 10 cm (0.1 m) has a volume of 10 cm × 10 cm × 10 cm = 1000 cm^3 (or 0.001 m^3). This is defined as 1 liter. Thus,

$$1\ \text{liter}\ (1\ \text{L}) = 1000\ \text{cm}^3$$

The liter is a convenient unit to use in the laboratory, as is the **milliliter, mL.** Because 1000 mL and 1000 cm^3 are in a liter, this means that

$$1\ \text{cm}^3 = 0.001\ \text{L} = 1\ \text{milliliter}\ (1\ \text{mL})$$

Because the liter is exactly equal to 1000 cm^3, the terms milliliter and cubic centimeter (or "cc") are interchangeable. Therefore, a flask that contains exactly 125 mL has a volume of 125 cm^3 or one eighth of a liter.

$$125\ \cancel{\text{cm}^3} \cdot \frac{1\ \text{L}}{1000\ \cancel{\text{cm}^3}} = 0.125\ \text{L}$$

The cubic decimeter (dm^3) is not widely used in the United States, but it is common in the rest of the world. A length of 10 cm is called a decimeter (dm) because it is $\frac{1}{10}$ of a meter. Therefore, because a cube 10 cm on a side defines a volume of 1 liter, a liter is equivalent to a cubic decimeter: 1 L = 1 dm^3. Products in Europe are often sold by the cubic decimeter.

EXAMPLE 1.3 *Units of Volume*

A laboratory beaker has a volume of 0.6 L. What is its volume in cubic centimeters and milliliters?

Solution The relation between liters and cubic centimeters is 1 L = 1000 cm^3. Therefore, you should multiply 0.6 L by the conversion factor (1000 cm^3/L) so that units of L cancel to leave an answer with units of cm^3.

$$0.6\,\cancel{L} \cdot \frac{1000\ cm^3}{1\,\cancel{L}} = 600\ cm^3$$

Because cubic centimeter and mL are equivalent, we can say the volume of the beaker is also 600 mL.

EXERCISE 1.9 ***Volume***

a. A standard wine bottle has a volume of 750 mL. How many liters does this represent?

b. One U.S. gallon is equivalent to 3.7865 L. How many liters are there in a 2.0-qt carton of milk? How many cubic decimeters? ■

To determine the density of a piece of aluminum, we needed to know its mass. The **mass** of a body is the fundamental measure of the amount of matter in that body, and the SI unit of mass is the **kilogram (kg).** Smaller masses are expressed in **grams (g)** or **milligrams (mg)** (see Table 1.4).

$$1\ kg = 1000\ g$$
$$1\ g = 1000\ mg$$

In the United States the English system of mass measurement is most commonly used, so most masses are given in pounds; *1 pound is equivalent to 453.59237 g.* This means that a mass of 1.00 kg is equivalent to 2.20 lb, or about the mass of a quart of milk. But let us say you are living in Portugal and you have bought 750 g of strawberries. How many kilograms does this represent and how many pounds?

$$750\,\cancel{g} \cdot \frac{1\ kg}{1000\,\cancel{g}} = 0.75\ kg$$

$$750\,\cancel{g} \cdot \frac{1\ pound}{454\,\cancel{g}} = 1.7\ pounds$$

EXAMPLE 1.4 ***Mass in Kilograms and Grams***

A new U.S. penny has a mass of 2.49 g. Express this mass in kilograms and milligrams.

Solution Here the relation between the unit of the desired answer and the unit of the information given is 1 kg = 1000 g and 1000 mg = 1 g. Therefore, multiply the mass in grams by a factor that has the form "units for answer divided by units of information given."

$$2.49\ \text{g} \cdot \frac{1\ \text{kg}}{1000\ \text{g}} = 2.49 \times 10^{-3}\ \text{kg}\ \ (\text{or } 0.00249\ \text{kg})$$

$$2.49\ \text{g} \cdot \frac{1000\ \text{mg}}{1\ \text{g}} = 2.49 \times 10^{3}\ \text{mg}$$

EXAMPLE 1.5 ***Density in Different Units***

The density of some substances may be given more conveniently in units of kilograms per cubic meter. The density of sea water is 1.025 g/cm^3 at 15 °C. What is this density in kilograms per cubic meter?

Solution To simplify this problem, let us break it into two steps. We shall first change grams to kilograms and then convert cubic centimeters to cubic meters. The final density is the ratio of the mass in kilograms to the volume in cubic meters.

$$1.025\ \text{g} \cdot \frac{1\ \text{kg}}{1000\ \text{g}} = 1.025 \times 10^{-3}\ \text{kg}$$

No direct conversion factor is available in one of our tables for changing units of cubic centimeters to cubic meters. We can find one, however, by cubing (raising to the third power) the relation between the meter and centimeter.

$$1\ \text{cm}^3 \cdot \left(\frac{1\ \text{m}}{100\ \text{cm}}\right)^3 = 1\ \text{cm}^3 \cdot \left(\frac{1\ \text{m}^3}{1 \times 10^6\ \text{cm}^3}\right) = 1 \times 10^{-6}\ \text{m}^3$$

Therefore, the density of sea water is

$$\text{Density} = \frac{1.025 \times 10^{-3}\ \text{kg}}{1 \times 10^{-6}\ \text{m}^3} = 1.025 \times 10^{3}\ \text{kg/m}^3$$

This answer makes sense because a cubic meter is a *large* volume. The mass of this volume of sea water is thus expected to be very large.

EXERCISE 1.10 ***Mass and Density***

a. The amount of vitamin C in a tablet is 500. mg. How many grams is this? How many kilograms?

b. The density of iron is 7.874 g/cm^3. How many kilograms of iron are there in a slab that is 4.5 m long, 1.5 m wide, and 25 cm thick? ■

Significant Figures

The **precision** of a measurement indicates how well several determinations of the same quantity agree. Precision is illustrated by the results of throwing darts at a bull's eye (Figure 1.22). In part (a) the darts are scattered all over the board; the dart thrower was apparently not very skillful (or threw the darts from a long distance away from the board), and the precision of their placement on the board is low. In Figure 1.22b the darts are all clustered together, indicating much better reproducibility on the part of the thrower, that is, greater precision. In addition, each dart has come very close to the bull's eye. This is described by saying that the thrower has been quite **accurate**—the average of all throws is very close to the accepted position, namely the bull's eye. Figure 1.22c illustrates that it is possible to be precise without being accurate—the dart thrower has consistently missed the bull's eye, although all the darts are clustered very precisely around a different point on the board. This third case is like an experiment with some flaw (either in its design or in a measuring device) that causes all results to differ from the correct value by the same amount.

In the laboratory we attempt to set up experiments so that the greatest possible accuracy can be obtained. As a further check on accuracy, results are usually compared among different laboratories so that any flaw in experimental design or measurement can be detected. For each individual experiment, several measurements are usually made and their precision determined. Usually better precision is taken as an indication of better experimental work, and it is necessary to know the experimenters' precision in order to compare results among them. If two different experimenters both had results like Figure 1.22a, their average values could differ quite a lot before they would say that their results did not agree.

In most experiments several different kinds of measurements must be made, and some can be made more precisely than others. It is common sense that a calculated result *can be no more precise than the least precise piece of information* that went into the calculation. This is where the rules for "significant figures" come in. Consider the example in which you collected information on a piece of aluminum to determine its density. Laboratory balances readily allow you to find the mass to four digits (13.56 g). To measure the length and width of the piece you had a good metric ruler and found dimensions of 6.45 cm and 2.50 cm. The thickness was found to be 0.31 cm. All these numbers have two digits to the right of the decimal, but they have different numbers of significant figures.

The best guideline for significant figures is to use your common sense!

The quantity 0.31 cm has two significant figures. This means that the three in 0.31 is exactly right, but the one is not known precisely. Unless indicated otherwise, we assume the final digit in a number is uncertain to the extent of ±1.

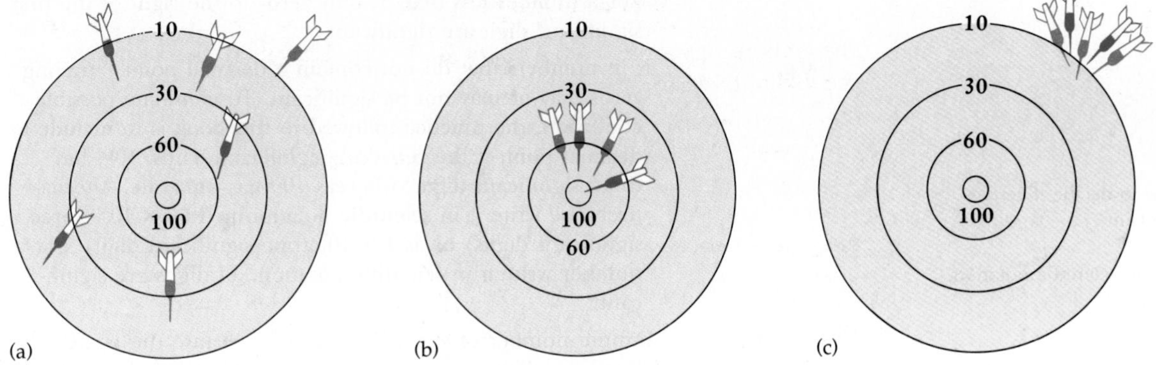

Figure 1.22 Precision and accuracy. (a) Poor precision and poor accuracy. (b) Good precision and good accuracy. (c) Good precision and poor accuracy.

Thus, the thickness could have been as small as 0.30 or as large as 0.32. In general, in a number that represents a scientific measurement, the last digit to the right is taken to be inexact, but all digits farther to the left are assumed to be exact. Therefore, the data collected for the density experiment are different in the number of significant figures they have.

Measurement	**Data Collected**	**Significant Figures**
Mass of aluminum	13.56 g	4
Length	6.45 cm	3
Width	2.50 cm	3
Thickness	0.31 cm	2

When these data were combined, the calculated density was 7.2 g/cm^3, a number with two significant figures. This follows from the previous statement that the calculated result can be no more precise than the least precise piece of information.

When doing calculations using measured quantities, you must follow some simple rules so that the results reflect the precision of all the measurements that go into the calculations.

Guidelines for Determining Significant Figures

Rule 1. To determine the number of significant figures in a measurement, read the number from left to right and count all digits, starting with the first digit that is *not* zero.

Example	**Number of significant figures**
1.23 g	3
0.00123 g	3; the zeros to the *left* of the 1 simply locate the decimal point. To avoid confusion, write numbers of this type in scientific notation; thus, $0.00123 = 1.23 \times 10^{-3}$.
2.0 g and 0.020 g	2; both have two significant digits. When a number is greater than 1, *all zeros to the right of the decimal point are significant.* For a number less than 1, only zeros to the right of the first significant digit are significant.
100 g	1; in numbers that do not contain a decimal point, "trailing" zeros may or may not be significant. To eliminate possible confusion, the practice followed in this book is to include a decimal point if the zeros are significant. Thus, 100. has three significant digits, whereas 100 has only one. Alternatively, we write it in scientific notation as 1.00×10^2 (three significant digits) or as 1×10^2 (one significant digit). For a number written in scientific notation, all digits are significant.
100 cm/m	Infinite number of significant figures, because this is a defined quantity
$\pi = 3.1415926\ldots.$	The value of π is known to a greater number of significant figures than any data you will ever use in a calculation.

Be sure to notice how to decide if zeroes in numbers are significant.

For a number written in scientific notation, all digits are significant.

The number π is now known to 1,011,196,691 digits. It is doubtful that you will need this accuracy in this course—or ever.

Rule 2. When adding or subtracting, the number of decimal places in the answer should be equal to the number of decimal places in the number with the *fewest* places.

0.12	2 significant figures	2 decimal places
1.6	2 significant figures	1 decimal place
10.976	5 significant figures	3 decimal places
12.696		

This answer should be reported as 12.7, a number with one decimal place, because 1.6 has only one decimal place.

Rule 3. In multiplication or division, the number of significant figures in the answer should be the same as that in the quantity with the *fewest* significant figures.

$$\frac{0.01208}{0.0236} = 0.512 \text{ or, in scientific notation, } 5.12 \times 10^{-1}$$

Because 0.0236 has only three significant figures and 0.01208 has four, the answer is limited to three significant figures.

Rule 4. When a number is rounded off (the number of significant figures is reduced), the last digit retained is increased by 1 only if the following digit is 5 or greater.*

Full Number	Number Rounded to Three Significant Figures
12.696	12.7
16.249	16.2
18.35	18.4
18.351	18.4

One last word regarding significant figures and calculations. In working problems on a pocket calculator, you should do the calculation using all the digits allowed by the calculator and round off only at the end of the problem. Rounding off in the middle can introduce errors. If your answers do not quite agree with those in the back of the book, this may be the source of the disagreement.

When working a problem involving many steps, we round off only at the end of the calculation.

*A modification of this rule is sometimes used to reduce the accumulation of roundoff errors. If the digit following the last permitted significant figure is *exactly* 5 (with no following digits or all following digits being zero only), then (a) increase the last significant figure by 1 if it is *odd* or (b) leave the last significant figure unchanged if it is *even*. Thus, both 18.35 and 18.45 are rounded to 18.4.

EXAMPLE 1.6 *Significant Figures*

An example of a calculation you will do later in the book (Chapter 12) is

$$\text{Volume of a gas} = \frac{(0.120)\,(0.08206)\,(273.15 + 23)}{(230/760.0)}$$

Calculate the final answer with the correct number of significant figures.

Solution As a first step, we analyze each number in the equation.

Number	Number of Significant Figures	Comments
0.120	3	The trailing 0 is significant. To see that the 0 to the left of the decimal point is not significant, write in scientific notation as 1.20×10^{-1}. See Rule 1.
0.08206	4	The 0 just to the right of the decimal point is not significant. Write in scientific notation as 8.206×10^{-2}. See Rule 1.
273.15 + 23 = 296	3	23 has no decimal places, so the answer can have none. See Rule 2.
230/760.0 = 0.30	2	230 has two significant figures because the last zero is not significant; there is no decimal point in the number. In scientific notation, it is 2.3×10^2. In contrast, there is a decimal point in 760.0, so there are four significant figures. The result of the division must have two significant figures. See Rules 1 and 3.

The analysis shows that one of the pieces of information is known with only two significant figures. Therefore, the answer must be volume of gas = 9.6, a number with two significant digits. (In this case, the answer is in units of liters, but this is not important for now.)

EXERCISE 1.11 *Significant Figures*

a. What are the sum and product of 11.19 and 0.054?

b. What is the result of the calculation

$$x = \frac{110.2 - 67}{0.021 + 0.00115}$$ ■

The Concept of "Percent"

Chemists often express the composition of matter in terms of **percent.** For example, we know that 88.81% of a given mass of water is oxygen, that sucrose is 42.11% carbon, or that a 5¢ coin, a nickel, is only about 25% nickel (the rest is copper). Because "percent" is a widely used concept in chemistry, it is worth thinking about it while we are discussing units.

You are familiar with "percent" from looking for a bargain at the local mall or from paying sales taxes. For example, paying a sales tax of 6.00% means that you pay 6.00 dollars per 100 dollars of things purchased. Therefore, if you buy a new shirt for $31.50, the tax is $1.89.

$$31.50\ \cancel{\$\text{ spent}} \cdot \frac{6.00\ \$\text{ tax}}{100\ \cancel{\$\text{ spent}}} = 1.89\ \$\text{ tax}$$

Two important points must be made here. First, notice that the problem—one that you have probably done in your head many times—was solved using units. We used the conversion factor "$ tax/$100 spent," and the unit "$ spent" cancels out and leaves units of "$ tax." Second, the word "percent" tells us what the conversion factor must be, because *per cent* is a shortened version of the Latin phrase *per centum,* meaning *in 100.* The conversion factor in a percent calculation is always the value of the percent divided by 100.

EXAMPLE 1.7 *Using Percent*

Battery plates in lead storage batteries (the type used in automobiles) are made from a mixture of two chemical elements: lead (Pb, 94.0%) and antimony (Sb, 6.0%). If a piece of a battery plate has a mass of 25.0 g, what masses of lead and antimony (in grams) are present?

Solution Let us first solve for the mass of lead, using the known percentage of lead. The plate is 94.0% lead, which means that 94.0 g of lead is present in every 100. g of plate.

$$25.0\text{ g}\ \cancel{\text{battery plate}} \cdot \frac{94.0\text{ g lead}}{100.\text{ g}\ \cancel{\text{battery plate}}} = 23.5\text{ g lead}$$

We know that the plate contains only lead and antimony, so

$$25.0\text{ g plate} = 23.5\text{ g lead} + x\text{ g antimony}$$

Solving for x, we find that the mass of antimony is 25.0 g − 23.5 g = 1.5 g of antimony. Of course, we could obtain the same result from the calculation

$$25.0\text{ g}\ \cancel{\text{battery plate}} \cdot \frac{6.0\text{ g antimony}}{100.\text{ g}\ \cancel{\text{battery plate}}} = 1.5\text{ g antimony}$$

Notice that the answer has only two significant figures, because there are only two in 6.0 g of antimony.

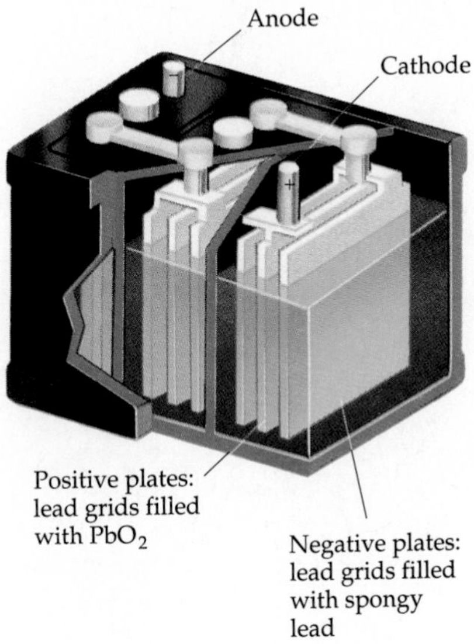

Lead plates in a lead storage battery.
(C. D. Winters)

EXERCISE 1.12 ***Percent***

What is the mass of gold in a 15.0-g earring if it is made of 14-karat gold? Fourteen-karat gold is 58% gold, the remainder being copper and silver. ■

CHAPTER HIGHLIGHTS

Having studied this chapter, you should be able to

- Define physical properties of matter and give some examples (Section 1.1)
- Recognize the different states of matter (solids, liquids, and gases) and give their characteristics (Section 1.1)
- Understand the basic ideas of the kinetic molecular theory (Section 1.1)
- Convert between temperatures on the Celsius and Kelvin scales (Section 1.1)
- Use density as a way to connect the volume and mass of a substance (Sections 1.1 and 1.7)

  The density of a substance is defined as the ratio of its mass to its volume.

  $$\text{Density} = \frac{\text{mass}}{\text{volume}}$$

  In chemistry we generally use density in units of grams per cubic centimeter.

- Identify the name or symbol for an element, given its symbol or name (Section 1.2)
- Be able to use the terms *atom, element, molecule,* and *compound* correctly (Sections 1.2 and 1.3)
- Explain the difference between chemical and physical change (Sections 1.1 and 1.4)
- Recognize the difference between homogeneous and heterogeneous mixtures (Section 1.5)
- Recognize and know how to use the prefixes that modify the sizes of metric units (Section 1.6)
- Begin using dimensional analysis to carry out unit conversions and other calculations (Section 1.7)
- Know the difference between precision and accuracy and understand the rules for using significant figures (Section 1.7).
- Use the concept of percent in chemistry (Section 1.7)

STUDY QUESTIONS

Answers to boldface-numbered questions appear in Appendix M.

Review Questions

1. Fluorite is a mineral that contains the elements calcium and fluorine. What are the symbols of these elements? How would you describe the shape of the crystals in the photo? What does this tell you about the arrangement of the atoms inside the crystal?

The mineral fluorite. (C. D. Winters)

2. What are the states of matter and how do they differ from one another?
3. The photo shows some copper balls, immersed in water, floating on top of mercury. What are the symbols of the elements copper and mercury? What are the liquids and solids in this photo? What does this photo tell you about the relative densities of these materials?

Water, copper balls, and mercury. (C. D. Winters)

4. Small chips of iron are mixed with sand (see photo). Is this a homogeneous or heterogeneous mixture? Suggest a way to separate the iron and sand from each other.

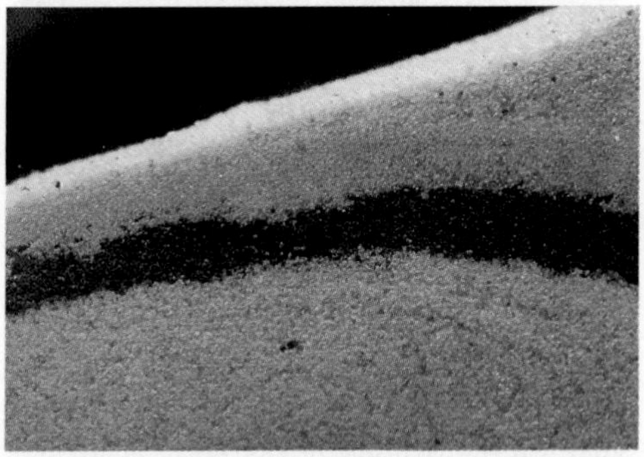

Chips of iron mixed with sand. (C. D. Winters)

5. The photo shows the label from a bottle of cranberry juice. Is this a homogeneous or heterogeneous mixture? Is the juice a pure substance or not?

Cranberry juice. (C. D. Winters)

6. What is the difference between the terms "compound" and "molecule"? Use these words in a sentence.
7. In the photo on the next page, you see a crystal of the mineral calcite surrounded by piles of calcium and carbon, two of the elements that combine to make the mineral. (The other element combined in calcite is oxygen.)

Calcite is the clear crystal, calcium chips are at the bottom, and graphite is at the right. (Calcium is normally a shiny metal, but these chips are covered with a thin film of white calcium oxide.) (C. D. Winters)

(a) What are the symbols of the three elements that combine to make calcite?
(b) Based on the photo, describe some of the physical properties of the elements and the mineral. Are any the same? Are any properties different?

8. In each case, decide if each underlined property is a physical or chemical property:
   (a) The normal <u>color</u> of bromine is red-orange.
   (b) Iron is <u>transformed into rust</u> in the presence of air and water.
   (c) Dynamite can <u>explode</u>.
   (d) The <u>density</u> of uranium metal is 19.07 g/cm^3.
   (e) Aluminum metal, the "foil" you use in the kitchen, <u>melts</u> at 933 K.
9. In each case, decide if the change is a chemical or physical change.
   (a) A cup of household bleach changes the color of your favorite T-shirt from purple to pink.
   (b) The fuels in the Space Shuttle (hydrogen and oxygen) combine to give water and provide the energy to lift the shuttle into space.
   (c) An ice cube in your glass of lemonade melts.
10. While camping in the mountains you build a small fire out of tree limbs you find on the ground near your campsite. The dry wood crackles and burns brightly and warms you. Before slipping into your sleeping bag for the night, you put the fire out by dousing it with cold water from a nearby stream. Steam rises when the water hits the hot coals. Describe the physical and chemical changes in this scene.

A sample of quartz called a Herkimer "diamond." (C. D. Winters)

11. In the photo you see a quartz crystal. If you studied the crystal, you might observe that it is colorless and clear, and it has a mass of 2.5 g and a length of 4.6 cm. Which of these observations are qualitative and which are quantitative?
12. The density of pure water is given below at various temperatures.

t, °C	*d* (g/cm^3)
5	0.99999
15	0.99913
25	0.99707
35	0.99406

Suppose your laboratory partner tells you the density of water at 20 °C is 0.99910 g/cm^3. Is this a reasonable number? Why or why not?

13. Give the number of significant figures in each of the following numbers:
   (a) 9.87
   (b) 0.00823
   (c) 1050
   (d) 1.67×10^{-6}

NUMERICAL QUESTIONS

Density

14. Ethylene glycol, $C_2H_6O_2$, is a liquid that is the base of the antifreeze you use in the radiator of your car. It has a density of 1.1135 g/cm^3 at 20 °C. If you need 500. mL of this liquid, what mass of the compound, in grams, is required?

15. The piece of silver metal in the photo has a mass of 2.365 g. If the density of silver is 10.5 g/cm^3, what is the volume of the silver?

Silver nuggets and a silver belt buckle. (C. D. Winters)

16. Water has a density at 25 °C of 0.997 g/cm^3. If you have 500. mL of water, what is its mass in grams? In kilograms?

17. A chemist needs 2.00 g of a liquid compound. (a) What volume of the compound is necessary if the density of the liquid is 0.718 g/cm^3? (b) If the compound costs \$2.41 per milliliter, what is the cost of 2.00 g?

18. The "cup" is a volume widely used by cooks in the United States. One cup is equivalent to 225 mL. If 1 cup of olive oil has a mass of 205 g, what is the density of the oil (in grams per cubic centimeter)?

19. Peanut oil has a density of 0.92 g/cm^3. If a recipe calls for 1 cup of peanut oil (1 cup = 225 mL), what mass of peanut oil (in grams) are you using?

20. A sample of unknown metal is placed in a graduated cylinder containing water. If the mass of the sample is 37.5 g, and the water levels before and after adding the sample to the cylinder are as shown in the figure, which metal listed below is most likely the sample? (*d* is the density of the metal)

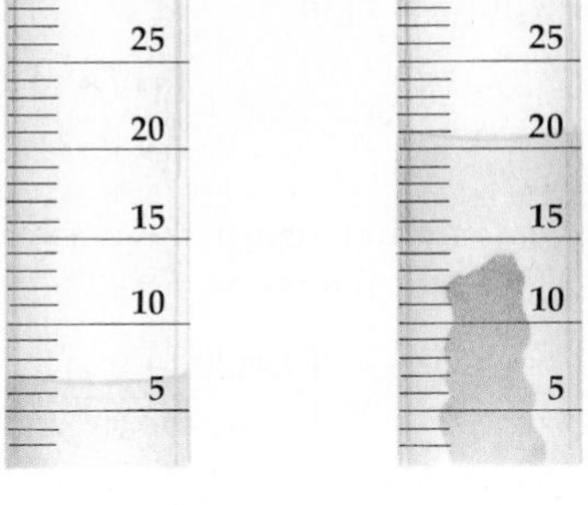

Graduated cylinders.

(a) Mg, $d = 1.74$ g/cm^3
(b) Fe, $d = 7.87$ g/cm^3
(c) Ag, $d = 10.5$ g/cm^3
(d) Al, $d = 2.70$ g/cm^3
(e) Cu, $d = 8.96$ g/cm^3
(f) Pb, $d = 11.3$ g/cm^3

21. Fool's gold, sometimes called iron pyrite, can look very much like gold (see Figure 1.4). Assume that you have a piece of a solid that looks as though it may be gold, but you believe it to be fool's gold. The sample has a mass of 23.5 g. When placed in some water in a graduated cylinder (see Study Question 20), the water level increases from 47.5 mL to 52.2 mL. Is the sample fool's gold ($d = 5.00$ g/cm^3) or real gold ($d = 19.3$ g/cm^3)?

Temperature

22. Many laboratories use 25 °C as a standard temperature. What is this temperature in kelvins?

23. The temperature on the surface of the sun is 5.50×10^3 °C. What is this temperature in kelvins?

24. Make the following temperature conversions:

	°C	K
(a)	16	_____
(b)	_____	370
(c)	−40	_____

25. Make the following temperature conversions:

	°C	K
(a)	_____	77
(b)	60	_____
(c)	_____	1450

26. Solid gallium has a melting point of 29.8 °C. If you hold this metal in your hand, what is its physical state? That is, is it a solid or a liquid? Explain briefly.

27. Neon, a gaseous element used in signs, has a melting point of −248.6 °C and a boiling point of −246.1 °C. Express these temperatures in kelvins.

Elements and Atoms

28. Give the name of each of the following elements:

(a) C	(c) Cl	(e) Mg
(b) Na	(d) P	(f) Ca

29. Give the name of each of the following elements:
(a) Mn (c) K (e) As (g) Cu
(b) F (d) Fe (f) Kr (h) V
30. Give the symbol for each of the following elements:
(a) Lithium (d) Silicon
(b) Titanium (e) Cobalt
(c) Iron (f) Zinc
31. Give the symbol for each of the following elements:
(a) Silver (e) Tin
(b) Aluminum (f) Barium
(c) Plutonium (g) Krypton
(d) Cadmium (h) Palladium

Units and Unit Conversions

32. The average lead pencil, new and unused, is 19 cm long. What is its length in millimeters? In meters?
33. A race covers a distance of 1500 m. What is this distance in kilometers? In centimeters?
34. A standard U.S. postage stamp is 2.5 cm long and 2.1 cm wide. What is the area of the stamp in square centimeters? In square meters?
35. A compact disk has a diameter of 11.8 cm. What is the surface area of the disk in square centimeters? In square meters?
36. A typical laboratory beaker has a volume of 800. mL. What is its volume in cubic centimeters? In liters? In cubic meters?
37. A large bottle of wine has a volume of 1.5 L. What is this volume in milliliters? In cubic centimeters?
38. A new U.S. quarter has a mass of 5.63 g. What is its mass in kilograms? In milligrams?
39. The tourmaline crystal (which contains the elements aluminum, oxygen, and silicon among others) in the photo has a density of 3.26 g/cm^3. What is the mass in milligrams of a crystal with a volume of 1.00 cm^3? What is the density of the crystal in kilograms per cubic meter? What is the volume (in cubic centimeters) of a crystal with a mass of 4.56 g?

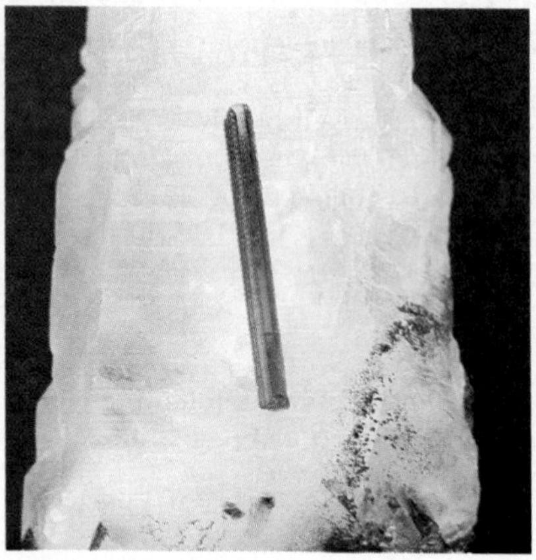

Tourmaline crystal. (C. D. Winters)

40. Complete the following table of masses:

Milligrams	Grams	Kilograms
______	0.693	______
156	______	______
______	______	2.23

41. Complete the following table of masses.

Milligrams	Grams	Kilograms
10.2	______	______
______	16.56	______
______	______	0.545

42. A standard sheet of notebook paper is $8\frac{1}{2} \times 11$ in. What are these dimensions in centimeters? What is the area of the paper in square centimeters?
43. Suppose your bedroom is 18 ft long, 15 ft wide, and the distance from floor to ceiling is 8 ft, 6 in. You need to know the volume of the room in metric units for some scientific calculations. What is the room's volume in cubic meters? In liters?

Significant Figures

44. What is the average mass of three objects whose individual masses are 10.3 g, 9.234 g, and 8.35 g?
45. What is the volume, in cubic centimeters, of a backpack whose dimensions are 22.86 cm × 38.0 cm × 76 cm?
46. Solve the equation, and report the answer in the correct number of significant numbers: 0.000523 × 0.0263 × 263.28
47. Solve the equation, and report the answer in the correct number of significant figures.

$$(1.68)(7.847)\left(\frac{1.0000}{55.85}\right)$$

48. Solve the equation, and report the answer in the correct number of significant figures.

$$(0.0345)\left[\frac{(25.35 - 2.4)}{1.678 \times 10^3}\right]$$

49. Solve the equation, and report the answer in the correct number of significant figures.

$$\left[\frac{(3.23 + 14.26 - 0.0025)}{13.24}\right](0.001734)$$

50. Solve the equation for n, and report the answer to the correct number of significant figures.

$$\left(\frac{36.3}{760.0}\right)(75.0) = n(0.0821)(298.3)$$

51. Solve the equation for V, and report the answer to the correct number of significant figures.

$$\left(\frac{234}{760.0}\right)V = (0.000214)(0.0821)(273.15 + 21.5)$$

Percent

52. Silver jewelry is actually a mixture of silver and copper. If a bracelet with a mass of 17.6 g contains 14.1 g of silver, what is the percentage of silver? Of copper?

53. The solder once used by plumbers to fasten copper pipes together consists of 67% lead and 33% tin. What is the mass of lead (in grams) in a 1.00-lb block of solder? What is the mass of tin? (1 lb = 453.59 g)

54. Automobile batteries are filled with sulfuric acid. What is the mass of the acid (in grams) in 500. mL of the battery acid solution if the density of the solution is 1.285 g/cm^3 and if the solution is 38.08% sulfuric acid by mass?

55. The density of a solution of sulfuric acid is 1.285 g/cm^3, and it is 38.08% acid by mass. What volume of the acid solution (in milliliters) do you need to supply 125 g of sulfuric acid?

GENERAL QUESTIONS

56. Molecular distances are usually given in nanometers (1 nm = 1×10^{-9} m) or in picometers (1 pm = 1×10^{-12} m). However, a commonly used unit has been the Ångstrom, where 1 Å = 1×10^{-10} m. (The Ångstrom unit is not an SI unit.) If the distance between the Pt atom and the N atom in the cancer chemotherapy drug cisplatin is 1.97 Å, what is this distance in nanometers? In picometers?

57. The separation between carbon atoms in diamond is 0.154 nm. (a) What is their separation in meters? (b) What is the carbon atom separation in Ångstrom units (where 1 Å = 10^{-10} m)?

58. The smallest repeating unit of a crystal of common salt is a cube with an edge length of 0.563 nm. What is the volume of this cube in cubic nanometers? In cubic centimeters?

59. The mass of a gemstone is often measured in "carats" where 1 carat = 0.200 g. If the annual worldwide production of diamonds is 12.5 million carats, how many grams does this represent?

60. Metals such as gold and platinum are sold in units called "troy ounces," where 1 troy ounce has a mass of 31.103 g. Is this larger or smaller than an ounce in the usual scale used in the United States? [Recall that there are 16 ounces in 1 pound (called an "avoirdupois" pound) and that 1 pound has a mass of 453.59237 g.]

61. As discussed in Chapter 7, light and other forms of radiation can be described as waves, and the distance between adjacent crests of a wave is called the wavelength. If a radio wave has a wavelength of 13 cm, what is its wavelength in meters? In feet?

62. The platinum-containing cancer drug cisplatin contains 65.0% platinum. If you have 1.53 g of the compound, how many grams of platinum can be recovered from this sample?

63. At 25 °C the density of water is 0.997 g/cm^3, whereas the density of ice at −10 °C is 0.917 g/cm^3. (a) If a soft-drink can (volume = 250. mL) is filled completely with pure water and then frozen at −10 °C, what volume does the solid occupy? (b) Can the ice be contained within the can?

64. When you heat popcorn, it pops because it loses water explosively. Assume a kernel of corn, weighing 0.125 g, weighs only 0.106 g after popping. What percent of its mass did the kernel lose on popping? Popcorn is sold by the pound. Using 0.125 g as the average mass of a popcorn kernel, how many kernels are there in a pound of popcorn?

65. An ancient gold coin is 2.2 cm in diameter and 3.0 mm thick. It is a cylinder for which volume = (π) (radius)^2 (thickness). If the density of gold is 19.3 g/cm^3, what is the mass of the coin in grams? Assume a price of gold of \$410 per troy ounce. How much is the coin worth? (1 troy ounce = 31.10 g)

66. You have a 100.0 mL graduated cylinder containing 50.0 mL of water. You drop a 154-g piece of brass (density = 8.56 g/cm^3) into the water. How high does the water rise in the graduated cylinder?

(a) (b)

(a) A graduated cylinder with 50.0 mL of water. (b) A piece of brass is added to the cylinder. (C. D. Winters)

67. The aluminum in a package containing 75 ft^2 of kitchen foil weighs approximately 12 oz. Aluminum has a density of 2.70 g/cm^3. What is the approximate thickness of the aluminum foil in millimeters? (1 ounce = 28.4 g.)

A package of aluminum foil. (C. D. Winters)

68. The fluoridation of city water supplies has been practiced in the United States for several decades because it is believed that fluoride prevents tooth decay, especially in young children. This is done by continuously adding sodium fluoride to water as it comes from a reservoir. Assume you live in a medium-sized city of 150,000 people and that each person uses 175 gal of water per day. How many tons of sodium fluoride must be added to the water supply each year (365 days) to have the required fluoride concentration of 1 ppm (part per million, that is, 1 ton of fluoride per million tons of water)? (Sodium fluoride is 45.0% fluoride, and 1 U.S. gallon of water has a mass of 8.34 lb.)

69. Measure the length of your foot in inches. What is the length in centimeters? In meters?

70. Copper has a density of 8.94 g/cm^3. If a factory has an ingot of copper that has a mass of 125 lb, and the ingot is drawn into wire with a diameter of 9.50 mm, how many feet of wire can be produced?

71. Which occupies a larger volume, 600 g of water (with a density of 0.995 g/cm^3) or 600 g of lead (with a density of 11.34 g/cm^3)?

72. You can identify a metal by carefully determining its density. An unknown piece of metal, with a mass of 29.454 g, is 2.35 cm long, 1.34 cm wide, and 1.05 cm thick. Which of the following is the element?
(a) Nickel, 8.91 g/cm^3
(b) Titanium, 4.50 g/cm^3
(c) Zinc, 7.14 g/cm^3
(d) Tin, 7.23 g/cm^3

73. About two centuries ago, Benjamin Franklin showed that 1 tsp (teaspoon) of oil covers about 0.5 acre of still water. If you know that 1.0×10^4 m^2 = 2.47 acres, and that there are approximately 5 cm^3 in a teaspoon, what is the thickness of the layer of oil? How might this thickness be related to the sizes of molecules?

74. It has been proposed that dinosaurs and many other organisms became extinct 65 million years ago because the earth was struck by a large asteroid. (See *Science,* Vol. 257, pp. 878–880, August 14, 1992.) The idea is that dust from the impact was lifted into the upper atmosphere all around the globe and blocked the sunlight reaching earth's surface. On the dark, cold earth that temporarily resulted, many forms of life became extinct. Available evidence suggests that about 20% of the asteroid's mass ended up as dust spread uniformly over the earth after eventually settling out of the upper atmosphere. This dust amounted to about 0.02 g/cm^2 of earth's surface. The asteroid likely had a density of about 2 g/cm^3. What was the mass of the asteroid? If the asteroid is considered to have been spherical, what was its diameter? (The earth has a surface area of 5.1×10^{14} m^2.) (From J. Harte: *Consider a Spherical Cow—A Course in Environmental Problem Solving.* Mill Valley, CA, University Science Books, 1988. Used with permission.)

75. What SI base unit, accompanied by what prefix, is most convenient for describing each of the quantities that follows?
(a) The mass of the earth
(b) The distance from New York City to London, England
(c) The mass of a single atom
(d) The mass of an orange
(e) The thickness of a human hair

76. The picture is an STM (scanning tunneling microscope) image of 48 iron atoms arranged on a copper surface. Physicists call this a "quantum corral." If the diameter of the circle of iron atoms is 143 Å, what is its diameter in nanometers? (1 Å = 1×10^{-10} m)

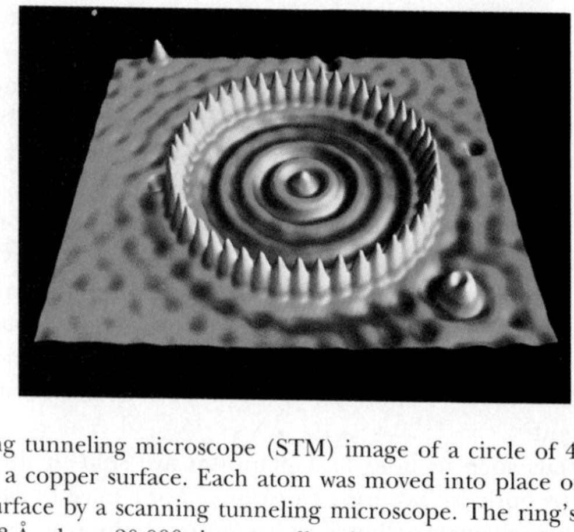

A scanning tunneling microscope (STM) image of a circle of 48 iron atoms on a copper surface. Each atom was moved into place on the copper surface by a scanning tunneling microscope. The ring's diameter is 143 Å, about 20,000 times smaller than the diameter of a human hair. (From M. F. Crommie, C. P. Lutz, and D. M. Eigler, *Science,* Vol. 262, p. 219, October 8, 1993.) (Photo courtesy of IBM Almaden Research Center.)

77. The density of pure water is given below at various temperatures:

t, °C	*d* (g/cm^3)
5	0.99999
15	0.99913
25	0.99707
35	0.99406

(a) Use these data to predict the density of water at 40 °C.
(b) Write an equation (of the form $y = mx + c$) that relates temperature and density and that allows you to estimate the density at any temperature.
(c) Refer back to Study Question 12. Is 0.99910 g/cm^3 a reasonable density for water at 20 °C? Why or why not? What is your estimate of the density at 20 °C?

CONCEPTUAL QUESTIONS

78. Lead is sold as tiny spherical beads called "shot." What is the average diameter (in millimeters) of a piece of lead shot if 25 beads have a total mass of 2.31 g? (Recall that the volume of a sphere is given by the formula $(\frac{4}{3})(\pi r^3)$ where r is the radius of the sphere. The density of lead is 11.3 g/cm^3.) Outline another, simple way to find the diameter.

79. A dry black powder is placed in a clean, dry glass tube. Pure hydrogen gas [$H_2(g)$] is passed down the tube in contact with the powder and drives out all the air. With the $H_2(g)$ still flowing, the powder is heated. The powder turns red, and water vapor can be detected coming out the end of the tube. Was the original black powder an element or a compound? Explain your reasoning.

80. Make a drawing, based on the kinetic-molecular theory and the ideas about atoms and molecules presented in this chapter, of a submicroscopic model of the arrangement of particles in each of the cases listed below. For each case draw ten submicroscopic particles of each substance. It is acceptable for your diagram to be two-dimensional; it need not be fancy. Represent each atom as a circle and distinguish each different kind of atom by shading its circle.
(a) A sample of solid aluminum (which consists of aluminum atoms)
(b) A sample of liquid water (which consists of H_2O molecules)
(c) A sample of pure water vapor
(d) A homogeneous mixture of water vapor and helium gas (which consists of helium atoms)
(e) A heterogeneous mixture consisting of liquid water and solid aluminum; show a region of the sample that includes both substances.
(f) A sample of brass (which is a homogeneous mixture of copper and zinc)

81. You are given a sample of a silvery metal. What information do you need to prove that the metal is silver?

82. Suggest a way to determine the density of solid salt. (Remember that salt is soluble in water.)

83. Describe an experimental method that can be used to determine the density of an irregularly shaped piece of metal.

84. Suggest a procedure to determine whether a clear liquid is pure water or a solution of salt dissolved in water.

85. Describe an experimental procedure to determine the density of a liquid. Identify the equipment you would use. Estimate the accuracy of each measurement, and indicate the number of significant figures in the answer.

86. A copper-colored metal is found to conduct an electrical current. Can you say with certainty that it is copper? Why or why not? Suggest additional information that could provide unequivocal confirmation that the metal is copper.

87. What experiment can you use to
(a) Separate salt from water?
(b) Separate iron filings from small pieces of lead?
(c) Separate elemental sulfur from sugar?

CHAPTER 2

Atoms and Elements

A Chemical Puzzler

How many jelly beans are in the jar? You could dump them out and count them, one by one. Alternatively, *if* all the beans had the same mass, and if you knew that mass and the total mass of the beans in the jar, you could calculate the number of beans. Chemistry involves "bean counting," too. We often want to know how many atoms are involved in a sample or in a chemical reaction. How do we find out how many atoms of mercury are in the antique flask in this photograph? What kind of "chemical counting unit" do we use? But first—what is an atom?

Hydrogen is the most abundant element in the universe, and an enormous amount of it occurs at the surface of the earth, mostly in water in the oceans. Nevertheless, it was not until 1660 that the British scientist Robert Boyle (1627–1691) prepared hydrogen by the reaction of iron with oil of vitriol (sulfuric acid), and not until 1766 that the French-born British scientist Henry Cavendish (1731–1810) prepared a pure sample and distinguished it from other gases, calling it "inflammable air." Cavendish found that hydrogen did not dissolve in water, had a smaller density than any other gas, formed an explosive mixture with normal air, and produced water when burned. Cavendish did not claim to have discovered a new element. Shortly after he reported his experiments, however, hydrogen was recognized as an element and given its name, which means "water former."

An early use of hydrogen depended on its low density. In 1783, Jacques Charles (1746–1823) filled a large balloon with hydrogen and flew over the French countryside in a basket suspended from the balloon. Hydrogen-filled balloons were used during World War I to take observers aloft to report on troop movements at the battlefront. In 1928, Germany built the *Graf Zeppelin,* a rigid airship that was lighter than air because it was filled with hydrogen. It carried more than 13,000 people between Germany and the United States until 1937, when it was replaced by the *Hindenburg.* The *Hindenburg* was designed to be filled with another gaseous element, helium, which does not react with air and presents no explosion hazard. But because Germany had no source of helium, the Hindenburg was filled with hydrogen. Unfortunately, while landing at Lakehurst, New Jersey, in May 1937, the airship exploded and burned (Figure 2.1), and hydrogen gas earned a reputation as a very dangerous substance.

Figure 2.1 The disastrous end of the hydrogen-filled dirigible, *Hindenburg,* in May 1937. The accident occurred on landing in Lakehurst, New Jersey. (The Bettmann Archive)

Actually, with proper techniques, hydrogen can be handled safely in large quantities. It is the principal fuel in the Space Shuttle (Figure 7 in the Introduction), for example, and until about 1950, when pipelines brought natural gas to homes throughout the United States, hydrogen was a major component of the gas used in kitchen ranges. Currently the most important use of hydrogen is to make ammonia (NH_3), which is used as fertilizer. Approximately 3 million tons of hydrogen are combined with nitrogen each year in the United States to make this valuable chemical. Hydrogen is also used to manufacture methanol, which is used as a gasoline additive and in windshield washer antifreeze, and to make "partially hydrogenated oils" such as those found in peanut butter.

Molecules such as hydrogen (H_2), water (H_2O), ammonia (NH_3), and methanol (CH_4O) are composed of atoms. The physical and chemical properties of all molecules depend ultimately on the atoms from which they are built and how these atoms are arranged in the molecule. Therefore, our discussion of chemistry begins with a discussion of atoms and their structure.

2.1 ORIGINS OF ATOMIC THEORY

Chapter 1 presented a picture of atoms that was not very detailed. Tiny spheres in constant motion were adequate to explain the macroscopic properties (such as differences among solids, liquids, and gases) that were described. This simple picture of atoms goes back a long way in history—to the Greek philosopher Leucippus and his student Democritus (460–370 BC). Democritus reasoned that if a piece of matter, such as gold, were divided into smaller and smaller pieces, one would ultimately arrive at a tiny particle of gold that could not be divided further but would still retain the properties of gold. He used the word **"atom,"** which literally means "uncuttable," to describe this undividable, ultimate particle of matter. According to Aristotle (384–322 BC), Democritus taught that atoms are hard, move spontaneously, and link to one another by some "hook-and-eye" connection. Epicurus (341–270 BC) attributed mass to atoms, and about 100 BC Asklepiades introduced the idea of clusters of atoms, corresponding to what we now call molecules.

Democritus used his concept of atoms to explain the properties of substances. For example, the high density and softness of lead can be interpreted if lead atoms are packed very closely together like marbles in a box and moved easily past one another. Iron, on the other hand, is less dense and harder than lead, and Democritus argued that iron atoms might be shaped like corkscrews so that they entangle in a rigid but relatively lightweight structure. Democritus was able to explain in a simple way other common observations, such as drying clothes, the appearance of moisture on the outside of a vessel of cold water, how an odor moves through a room, and how crystals grow from a solution. He imagined atoms scattering or collecting together as needed to explain the macroscopic events he saw. All atomic theory has been built on the assumptions of

Leucippus and Democritus: the properties of matter that we can see are explained by the properties and behavior of atoms that we cannot see.

Plato (427–347 BC) and Aristotle argued against the existence of atoms, and for centuries their ideas prevailed. Most of those in the mainstream of enlightened thought rejected or remained ignorant of the atomic theory proposed by Democritus, though a few well-known scientists did refer to atoms. Galileo Galilei (1564–1642) reasoned that the appearance of a new substance through chemical change involved a rearrangement of parts too small to be seen, and Francis Bacon (1561–1626) speculated that heat might be a form of motion of small particles. Robert Boyle (1627–1691) and Sir Isaac Newton (1642–1727) used atomic concepts to interpret physical phenomena. None of these people, however, provided detailed, quantitative explanations of physical and chemical facts in terms of atomic theory.

Jacob Bronowski, in a television series and book titled *The Ascent of Man,* had this to say about the importance of imagination, "There are many gifts that are unique in man; but at the center of them all, the root from which all knowledge grows, lies the ability to draw conclusions from what we see to what we do not see."

EXERCISE 2.1 ***Atomic Theory***

Use the idea that matter consists of atoms or molecules to interpret each of the following observations. Describe what the atoms or molecules are doing and how that explains what happens.

a. Wet clothes hung on a line eventually become dry.

b. Moisture appears on the outside of a glass of ice water.

c. Crystals of solid sugar dissolve in water.

d. Sugar dissolves faster in hot water than in cold water. ■

John Dalton and His Atomic Theory

In 1803, John Dalton (1766–1844) forcefully revived the idea of atoms. Dalton linked the existence of elements, which cannot be decomposed chemically, to the idea of atoms, which are indivisible. Compounds, which can be broken down into two or more new substances, must contain two or more different kinds of atoms. Dalton went further to say that each kind of atom must have its own properties—in particular a characteristic mass. This idea allowed his theory to account quantitatively for the masses of different elements that combine chemically to form compounds. Thus, Dalton's ideas could be used to interpret known chemical facts, and to do so quantitatively. The postulates of his atomic theory are

- All matter is made of atoms. These indivisible and indestructible objects are the ultimate chemical particles.
- All atoms of a given element are identical, both in mass and in properties. Atoms of different elements have different masses and different properties.
- Compounds are formed by combination of two or more different kinds of atoms. Atoms combine in the ratio of small whole numbers, for example, one atom of A with one atom of B, or two atoms of A with one atom of B.
- Atoms are the units of chemical change. A chemical reaction involves only combination, separation, or rearrangement of atoms, but atoms are not created, destroyed, divided into parts, or converted into other kinds of atoms during a chemical reaction.

John Dalton's ideas were accepted by the scientific community because they could be used to explain several general rules or scientific laws that were already known when he proposed the atomic theory. Some years earlier Antoine Lavoisier (1743–1794; see Chapter 4) had carried out a series of experiments in which the reactants were carefully weighed before a chemical reaction and the products were carefully weighed afterward. He found no change in mass when a reaction occurred. Lavoisier proposed this was true for every reaction and called his proposal the **law of conservation of matter.** Others verified his results, and the law became accepted. Dalton's second and fourth postulates imply the same thing. If each kind of atom has a particular characteristic mass, and if exactly the same number of each kind of atom exists before and after a reaction, the masses before and after must also be the same. Consequently, Dalton's theory is able to explain the quantitative chemical fact of the conservation of matter.

Another chemical law known in Dalton's time had been proposed by the French chemist Joseph Louis Proust (1754–1826), as a result of his analyses of minerals. Proust found that a particular compound, once purified, always contained the same elements in the same ratio by mass. This was called the **law of constant composition,** or the **law of definite proportions.** As an example, one of the compounds formed from carbon and oxygen always had one third greater mass of oxygen than carbon. Dalton therefore suggested that the carbon-oxygen compound contained one carbon atom for each oxygen atom, giving the formula CO, and that the mass of one oxygen atom was one third greater than the mass of a carbon atom. Using this kind of reasoning, Dalton was able to determine which atoms were heavier than others and how much of one element could be expected to combine with another element in a compound.

Dalton's theory was valuable because it explained existing facts, but Dalton went further and proposed a new law on the basis of his theory and a few experiments. To see how his reasoning worked, suppose that another compound can form in which two oxygen atoms occur for each carbon atom; that is, the mole-

John Dalton (Oesper Collection in the History of Chemistry/University of Cincinnati)

PORTRAIT OF A SCIENTIST

John Dalton (1766–1844)

John Dalton was born about the fifth of September, 1766, in the village of Eaglesfield in Cumberland, England. His family was quite poor, and his formal schooling ended at age 11. He was clearly a bright young man, however, and with the help of influential patrons, he began a teaching career at the age of 12. Shortly thereafter he made his first attempts at scientific investigation: observations of the weather. This was a study that was to last his lifetime. In fact, he made more than 200,000 observations of weather conditions during his lifetime.

In 1793, Dalton moved to Manchester, England, where he took up a post as tutor at the New College, but he left there in 1799 to pursue scientific inquiry full-time. Not long afterward, on October 21, 1803, he read his paper introducing the "Chemical Atomic Theory" to the Literary and Philosophical Society of Manchester. His presentation was followed by lectures in London and in other cities in England and Scotland, and his reputation as a scientist rapidly increased. He became a member of the top scientific society in Britain, the Royal Society, in 1810, and many other honors followed over his lifetime.

cules have the formula CO_2. This compound has twice as great a mass of oxygen per gram of carbon as does CO, because CO_2 has twice as many oxygen atoms per carbon atom. If an oxygen atom weighs one third more than a carbon atom, then in CO the ratio (mass of O)/(mass of C) is $1.3333/1 = 1.3333$; in the compound CO_2, with 2 O atoms per C atom, this ratio is $2 \times 1.3333/1 = 2.6666$, which is what the proportions turned out to be when they were measured. Dalton stated that when two elements form two different compounds, the mass ratio in one compound is a small whole number times the mass ratio in the other. (In the case of CO and CO_2, the small whole number is 2.) This law is called the **law of multiple proportions.**

Dalton's atomic theory was important because it suggested a new law and stimulated Dalton and his contemporaries to do a great deal more research, thereby contributing to scientific progress. A good theory not only accounts for existing knowledge but also stimulates the search for new knowledge. Though it was not until the 1860s that a consistent set of relative masses of the atoms was agreed on, Dalton's idea that the masses of atoms are crucial to quantitative chemistry was accepted from the early 1800s on.

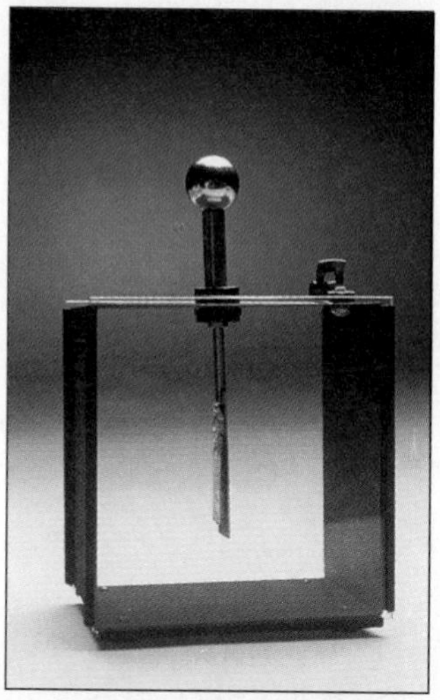

(a)

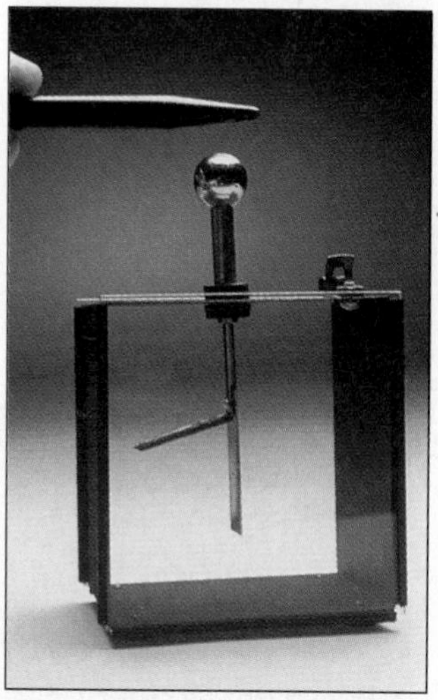

(b)

Figure 2.2 An electroscope demonstrates electric charge. (a) With no electrical influence, the foil leaf hangs straight down. (b) A rubber rod that had been rubbed with fur was touched to the bulb of the electroscope. (The rod was withdrawn slightly before this photograph was made.) The electric charge that had built up on the rod flowed onto the electroscope, and the movable leaf diverged from the stationary leaf. The reason for this observation is that the same charge has flowed onto both leaves. Because like charges repel, the leaves repel each other. (C. D. Winters)

2.2 ATOMIC STRUCTURE

Dalton's atomic theory said nothing about Democritus's ideas that atoms have structure. We know now that atoms do have a structure, and this knowledge is important because it gives us insights into how and why atoms stick together to form molecules. Though we now disagree with Democritus's idea that atoms are held together by "hooks and eyes," his fundamental notion that we can better understand how elements behave if we know about the structures of atoms turned out to be correct. The next few sections describe how scientists arrived at our current understanding of the structure of atoms.

Electricity

Electricity is involved in many of the experiments from which the theory of atomic structure was derived. Electric charge was first observed and recorded by the ancient Egyptians, who noted that amber, when rubbed with wool or silk, attracted small objects. You can observe the same thing when you comb your hair on a dry day—your hair is attracted to the comb. A bolt of lightning or the shock you get when touching a doorknob result when an electric charge moves from one place to another. Two types of electric charge had been discovered by the time of Benjamin Franklin (1706–1790), the American statesman and inventor. He named them positive (+) and negative (−) because they appear as opposites and can neutralize each other. Experiments with an electroscope (Figure 2.2) show that *like charges repel* one another and *unlike charges attract* one another. Franklin also concluded that charge is conserved: if a negative charge appears somewhere, a positive charge of the same size must appear somewhere else. The fact that a charge builds up when one substance is rubbed over another implies that the rubbing separates positive and negative charges. Apparently positive and negative charges are somehow associated with matter—perhaps with atoms.

Radioactivity

In 1896, French physicist Henri Becquerel (1852–1908) discovered that a uranium ore emitted rays that could expose a photographic plate, even though the plate was covered by black paper to protect it from being exposed by light rays. In 1898, Marie Curie (1867–1934) and coworkers isolated polonium and radium, which also emitted the same kind of rays, and in 1899 she suggested that atoms of radioactive substances disintegrate when they emit these unusual rays. She named this phenomenon **radioactivity.** About 25 elements have been found to be radioactive.

Radioactive elements spontaneously emit three kinds of radiation: alpha (α), beta (β), and gamma rays (γ). These behave differently when passed between electrically charged plates, as shown in Figure 2.3. Alpha and β rays are deflected, but γ rays pass straight through. This implies that α and β rays are electrically charged particles, because charges are attracted or repelled by the charged plates. Even though an α particle was found to have an electric charge (+2) twice as large as that of a β particle (−1), α particles are deflected less; hence, α particles must be heavier than β particles. Gamma rays have no detectable charge or mass; they behave like light rays.

Marie Curie's suggestion that atoms disintegrate contradicts Dalton's idea that atoms are indivisible, and requires an extension of Dalton's theory. If atoms can break apart, there must be something smaller than an atom; that is, atomic structure must involve subatomic particles.

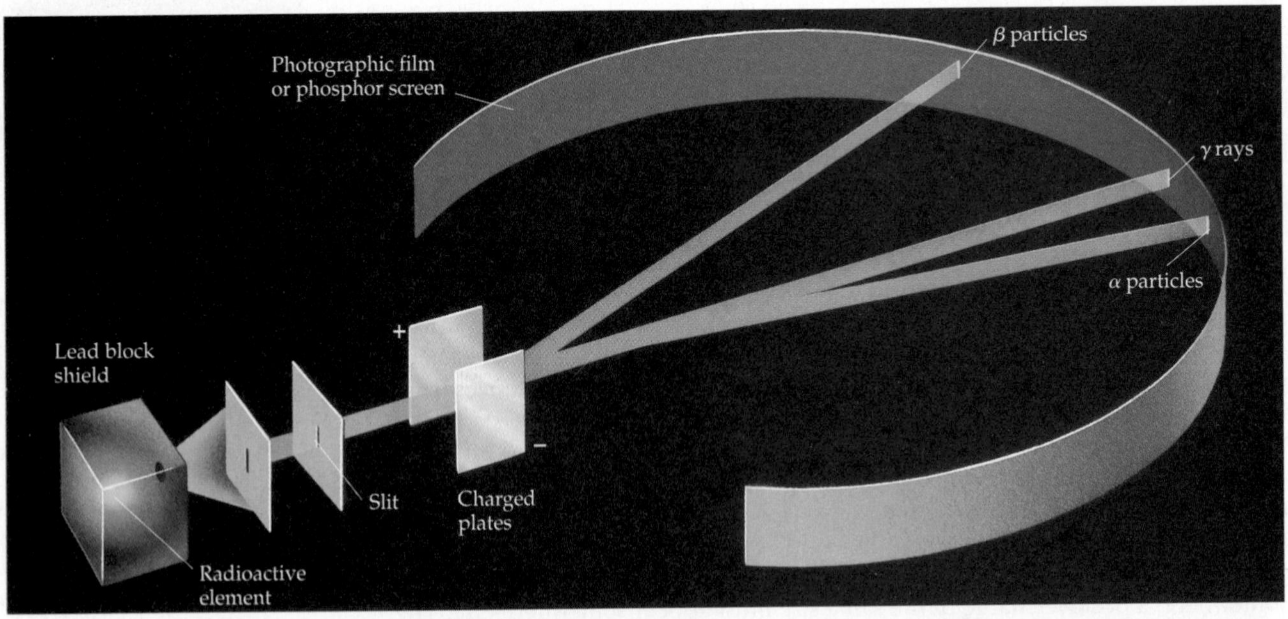

Figure 2.3 Alpha, β, and γ rays from a radioactive element are separated by passing them through electrically charged plates. Positively charged α particles are attracted to the negative plate, and the negative β particles are attracted to the positive plate. (Note that the heavier α particles are deflected less than the β particles.) Gamma rays have no electric charge and pass undeflected between the charged plates.

PORTRAIT OF A SCIENTIST

Marie Curie (1867–1934)

She was born Marya Sklodovska in Poland, but her family called her Manya. When she later lived in France she was Marie, and today she is usually referred to as Madame Curie.

She was the daughter of school teachers and grew up in a household full of books; she absorbed book after book. When she graduated from high school with honors, she wanted to attend the University of Warsaw—but they did not accept women—and neither she nor her family had enough money to allow her to go to Paris to study. She and her sister Bronya, however, worked out a plan to help each other. Marya would work to support Bronya while Bronya studied medicine in Paris, and then Bronya would work to support Marya. Their plan worked, and Marya enrolled at the Sorbonne in Paris in November 1891. She lived extremely frugally, spending almost nothing on luxuries or even on food. On several occasions she fainted because she worked constantly and lived only on a few cherries and radishes. But she successfully completed her degree, finishing first in her class. Almost exactly 15 years later, on November 5, 1906, Marie Curie became the first woman ever hired to teach at that same university.

By 1896, several physicists, especially Henri Becquerel, had been studying the curious radiation emanating from uranium. Marie Curie was thinking about possible topics for her doctoral thesis, and she had read about this work. Because her French physicist husband, Pierre, was working at the Ecole Supérieure de Physique et de Chimie de Paris, she was given a laboratory there—but she had to supply her own materials. As a chemist she was interested in whether other elements could emit such radiation, that is, whether they could be radioactive. After Pierre Curie had developed a highly sensitive way of detecting the radiation from a radioactive source, Marie Curie tested every substance she could find to answer this question. One of her first findings was to confirm Becquerel's observation that uranium metal itself was radioactive and that the degree to which a uranium-bearing sample was radioactive depended on the percentage of uranium present. Thus, she was astonished when she tested pitchblende, a common ore containing uranium and other metals (such as lead, bismuth, and copper), and found that it was even more radioactive than pure uranium. There was only one explanation: pitchblende contained an element more radioactive than uranium.

In 1898, the Curies published their work on the separation of pitchblende into the various metals that it contained. Using techniques of qualitative analysis—techniques that are so effective that they are still in use today in the introductory chemistry laboratory—they separated a new element from the pitchblende. The Curies stated that "We therefore think that the substance that we have extracted from pitchblende contains a metal previously unknown, a neighbor of bismuth in its analytical properties. If the existence of the new metal is confirmed, we propose to call it *polonium,* after the native land of one of us." They had discovered the next element after bismuth in the periodic table. And further investigation of the pitchblende by the Curies uncovered another new, highly radioactive element, radium.

To study radioactivity further, the Curies needed to isolate larger samples of radium, but they needed much larger amounts of pitchblende. The Austrian government soon sent them a ton of the mineral and, after months of back-breaking labor, they isolated 100 mg of radium chloride!

In 1903, Becquerel and Marie and Pierre Curie shared the Nobel Prize in physics for their discovery of "spontaneous radioactivity."

Marie Curie with her daughter Irène and husband Pierre (Mutter Museum, Philadelphia College of Physicians).

Electrons

Passing an electric current through a solution of a compound—a technique called **electrolysis**—can cause a chemical reaction, such as plating gold or silver onto another metal or the production of chlorine from sodium chloride. In 1833, the British scientist Michael Faraday (1791–1867) showed that the same current caused different quantities of different metals to deposit, and that those

The photo on page 34 demonstrates the electrolysis of tin(II) chloride to give tin and chlorine.

quantities were related to the relative masses of the atoms of those elements. Such experiments were interpreted to mean that, just as an atom is the fundamental particle of an element, a fundamental particle of electricity must exist. This "atom" of electricity was given the name **electron.**

For a very brief account of the discovery of the electron, proton, and neutron, and the origins of the names, see B. M. Peake: *Journal of Chemical Education,* Vol. 66, p. 738, 1989.

Further evidence that atoms are composed of smaller particles came from experiments with glass tubes from which most of the air had been removed and which had a piece of metal called an electrode sealed into each end (Figure 2.4). When a sufficiently high voltage was applied to the electrodes, a beam of particles called a "cathode ray" flowed from the negatively charged electrode (cathode) to the positive electrode (anode). Cathode rays travel in straight lines, cast sharp shadows, cause gases and fluorescent materials to glow, can heat metal objects red hot, can be deflected by a magnetic field, and are attracted toward positively charged plates. When cathode rays strike a fluorescent screen, light is given off in a series of tiny flashes. Thus a cathode ray is a beam of negatively charged particles, each one of which produces a flash of light when it hits a fluorescent screen. You are familiar with cathode rays because the deflection of a cathode ray by electrically charged plates is used to form the picture on a television picture tube or a computer's CRT (cathode-ray tube) screen. Neon and fluorescent lights also involve cathode rays.

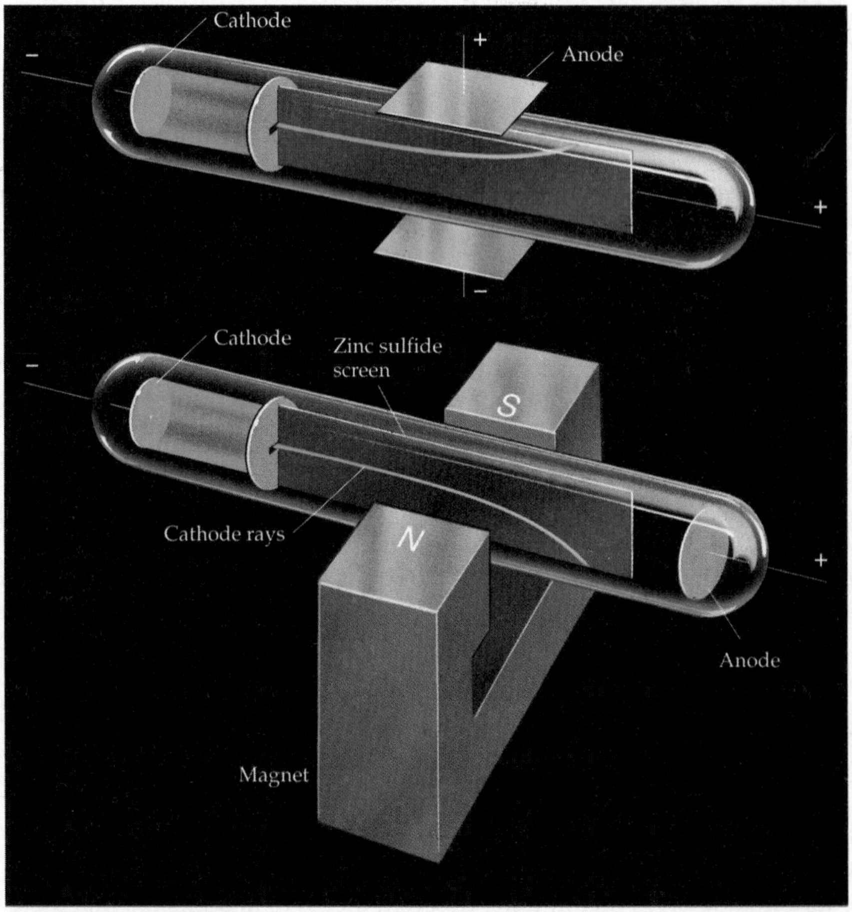

Figure 2.4 Deflection of a cathode ray by an electric field (top) and by a magnetic field (bottom). When an external electric field is applied, the cathode ray is deflected toward the positive pole. When a magnetic field is applied, the cathode ray is deflected from its normal straight path into a curved path. In both cases, the curvature is related to the mass and velocity of the particles of the cathode rays and the magnitude of the field.

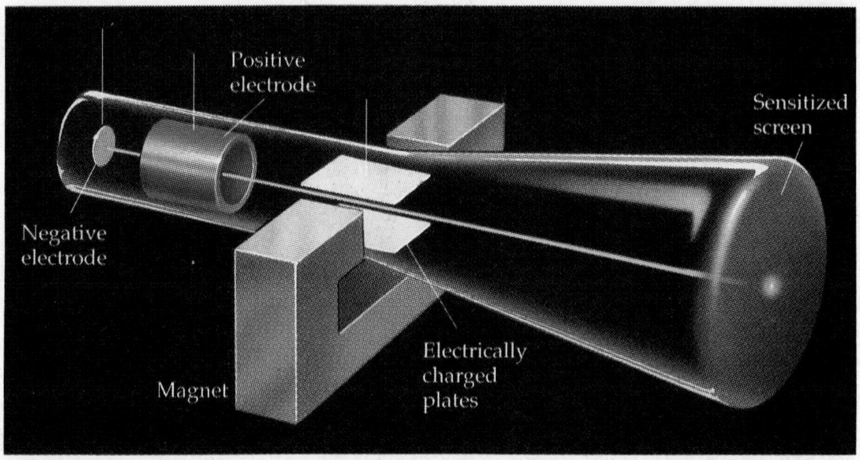

Figure 2.5 Thomson's experiment to measure the charge-to-mass ratio of the electron. A beam of electrons (cathode rays) passes through an electric field and a magnetic field. The experiment is arranged so that the electric field causes the beam to be deflected in one direction, and the magnetic field deflects the beam in the opposite direction. By balancing the effects of these fields, the charge-to-mass ratio of the electron can be determined.

In 1897, Sir Joseph John Thomson (1856–1940) used a specially designed cathode-ray tube (Figure 2.5) to apply electric and magnetic fields simultaneously to a beam of cathode rays. By balancing the effect of the electric field against that of the magnetic field and using basic laws of electricity and magnetism, Thomson was able to calculate the ratio of charge to mass for the particles in the beam. He was not able to determine either charge or mass independently, however. On the basis of the fact that cathode rays were a beam of negatively charged particles, Thomson suggested that they were the same as the electrons associated with Faraday's experiments. He obtained the same charge-to-mass ratio in experiments with 20 different metals and several different gases in the cathode-ray tube. These results suggested that electrons are present in all kinds of matter and that they presumably exist in atoms of all elements.

It remained for the American physicist Robert Andrews Millikan (1868–1953) to measure the charge on an electron and thereby enable calculation of its mass. His apparatus is shown schematically in Figure 2.6. Tiny droplets of oil were sprayed into a chamber from an atomizer. As they settled slowly through the air, they were exposed to x rays, which caused air molecules to transfer electrons to them. Millikan used a small telescope to observe individual droplets. He adjusted the electric charge on plates above and below the droplets until the electrostatic attraction pulling a droplet upward just balanced the force of gravity pulling the droplet downward. From the equations describing these forces, Millikan calculated the charge on the droplet. Different droplets had different charges, but Millikan found that each was a whole-number multiple of the same smaller charge. That smaller charge was 1.60×10^{-19} C (where C represents the coulomb, the SI unit of electric charge; Appendix D). Millikan assumed this to be the fundamental unit of charge, the charge on an electron. Because the charge-to-mass ratio of the electron was known, the mass of an electron could be calculated. The currently accepted value for the electron mass is 9.109389×10^{-28} g, and the currently accepted value of the electronic charge is

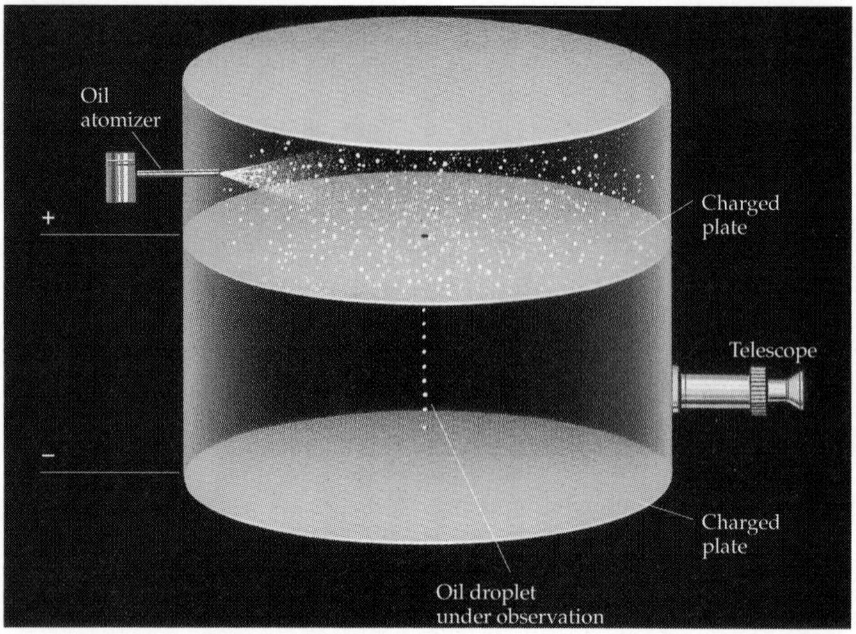

Figure 2.6 Millikan oil drop experiment. A fine mist of oil droplets is introduced into the chamber. The gas molecules in the chamber are ionized (split into electrons and positive ions) by a beam of x-rays. The electrons adhere to the oil droplets, some droplets having one electron, some two electrons, and so forth. These negatively charged oil droplets fall under the force of gravity into the region between the electrically charged plates. By carefully adjusting the voltage on the plates, the force of gravity is exactly counterbalanced by the attraction of the negative oil drop to the upper, positively charged plate. Analysis of these forces leads to a value for the charge on the electron.

$-1.60217733 \times 10^{-19}$ C. When talking about the properties of fundamental particles, we always express charge in terms of the charge on the electron, which is given the value of -1.

Additional experiments showed that cathode rays had the same properties as the β particles emitted by radioactive elements, providing further evidence that the electron is a fundamental particle of matter.

Protons

The first experimental evidence of a fundamental positive particle came from the study of "canal rays" (Figure 2.7), which were observed in a special cathode-ray tube with a perforated cathode. When high voltage is applied to the tube, cathode rays can be observed as in any cathode-ray tube. On the other side of the perforated cathode, however, a different kind of ray is observed. Because these rays are attracted toward a negatively charged plate, they must be composed of positively charged particles.

Each gas used in the tube gives a different charge-to-mass ratio for the positively charged particles (unlike the cathode rays, which are the same no matter what gas is used). When hydrogen gas is used, the largest charge-to-mass ratio is obtained, suggesting that hydrogen provides positive particles with the smallest

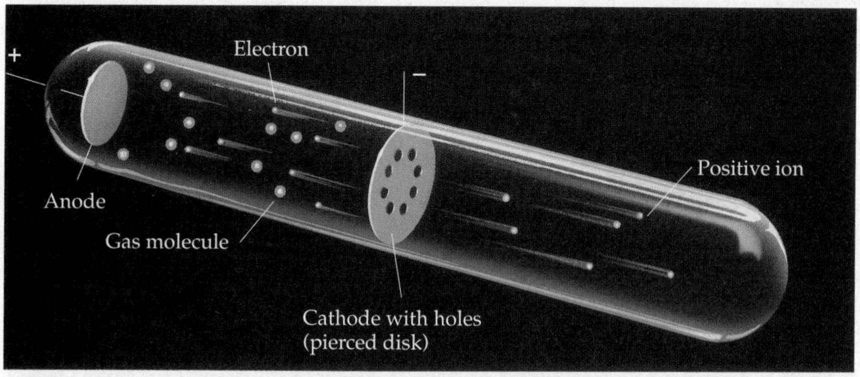

Figure 2.7 Cathode ray tube with perforated cathode. Electrons collide with gas molecules and produce positive ions, which are attracted to the negative cathode. Some of the positive ions pass through the holes and form a positive ray. Like cathode rays, positive rays (or "canal rays") are deflected by electric and magnetic fields but much less so for a given value of the field, because positive particles are much heavier.

mass. These were considered to be the fundamental, positively charged particles of atomic structure and were later called **protons** by Ernest Rutherford (from a Greek word meaning "the primary one").

The mass of a proton is known from experiment to be 1.672623×10^{-24} g. The charge on the proton, equal in size but opposite in sign to the charge on the electron, is +1.

Neutrons

Because atoms normally have no charge, *the number of protons and electrons in an atom must be equal.* Most atoms have masses greater than would be predicted on the basis of only protons and electrons, indicating that uncharged particles must be present in the atom. Because this third type of particle has no charge, the usual methods of detecting particles could not be used. Nonetheless, in 1932, many years after the discovery of the proton, the British physicist James Chadwick (1891–1974) devised a clever experiment that produced these expected neutral particles and then detected them by having them knock hydrogen ions, a detectable species, out of paraffin. It is now known that the fundamental particle called a **neutron** has no electric charge and a mass of $1.6749286 \times 10^{-24}$ g, nearly the same mass as a proton.

The Nucleus of the Atom

J. J. Thomson had supposed that an atom was a uniform sphere of positively charged matter within which thousands of electrons circulated in coplanar rings. But how many electrons were circulating within this sphere? To try to find an answer, Thomson and his students directed a beam of electrons at a very thin metal foil. They theorized that as the beam passed through the foil, the electrons of the beam would encounter the very large number of electrons within the atoms, and the negative charges would repel. A tiny deflection of the beam from its straight path should be observed at each encounter, with the size of the total deflection related to the number of electrons in the atom. Thomson did indeed observe a deflection, but it was much smaller than expected, and he and his students were forced to revise their estimate of the number of electrons, but not their model of the atom.

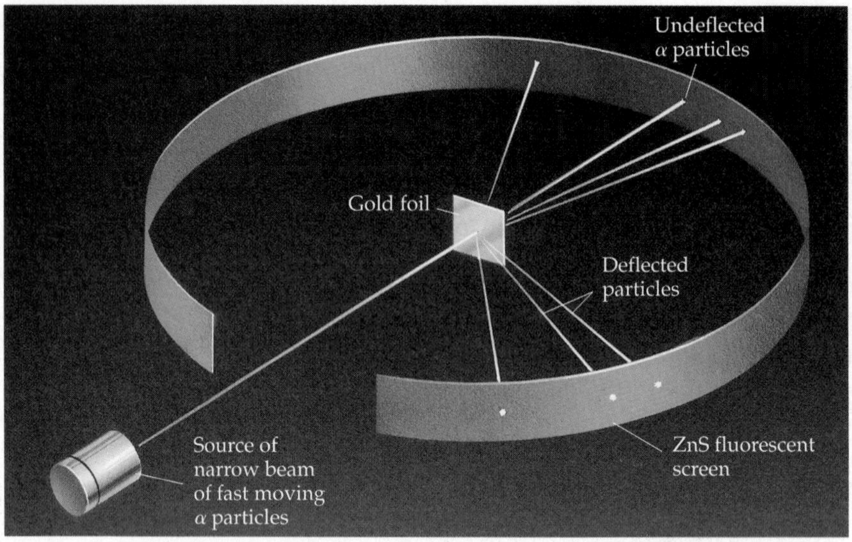

Figure 2.8 The experimental arrangement of the Rutherford experiment. A beam of positively charged α particles was directed at a very thin piece of gold foil. A luminescent screen coated with zinc sulfide (ZnS) was used to detect particles passing through or deflected by the foil. Most particles passed straight through. Some were deflected to some extent, and a few were even deflected backward. (Note that a circular luminescent screen is shown for simplicity; actually a smaller, movable screen was used.)

In about 1910, Ernest Rutherford (1871–1937) decided to test Thomson's model further. Rutherford had earlier discovered that α rays consisted of positively charged particles having the same mass as helium atoms. He reasoned that, if Thomson's atomic model were correct, a beam of such massive particles would be deflected very little as they passed through the atoms in a very thin sheet of gold foil. Rutherford's associate, Hans Geiger, and a young student, Ernst Marsden, set up the apparatus diagrammed in Figure 2.8 and observed what happened when α particles hit the foil. Most passed almost straight through, but Geiger and Marsden were amazed to find that a *very few* α particles were deflected through large angles, and some came almost straight back! Rutherford later described this unexpected result by saying, "It was about as credible as if you had fired a 15-in. [artillery] shell at a piece of paper and it came back and hit you."

The only way to account for this was to discard Thomson's model and to conclude that all of the positive charge and most of the mass of the atom is concentrated in a very small volume. Rutherford called this tiny core of the atom the **nucleus** (Figure 2.9). The electrons occupy the rest of the space in the atom. From their results Rutherford, Geiger, and Marsden calculated that the positive charge on the gold nucleus is in the range of 100 ± 20 and that the nucleus has a radius of about 10^{-12} cm. The currently accepted values for these results are 79 for the charge and about 10^{-13} cm for the radius, which makes the nucleus about 100,000 times smaller than that of the atom.

The experiments just described can be interpreted in terms of three primary constituents of atoms: protons, electrons, and neutrons. The nucleus, or core, of

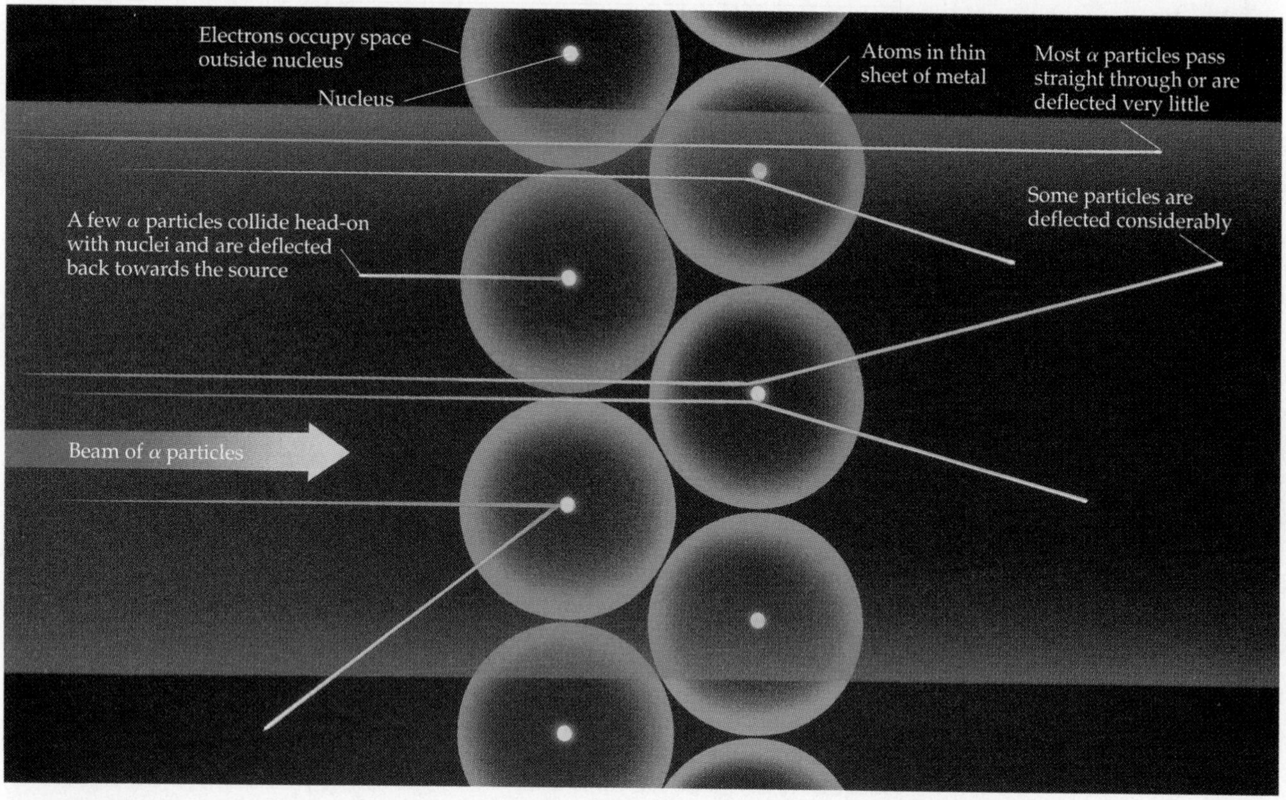

Figure 2.9 Rutherford's interpretation of the results of the experiment done by Geiger and Marsden.

the atom contains most of the mass and all of the positive charge; its radius is about 100,000 times smaller than for the atom itself. Protons and neutrons make up the nucleus. Negatively charged electrons occupy most of the volume of an atom but contribute very little mass. Because an atom has no net electric charge, *the number of electrons outside the nucleus must equal the number of protons inside the nucleus.*

To chemists the electrons are the most important part of the atom, because they are the first part of an atom that contacts another atom, should the two approach each other. Electrons largely control the chemical combination of atoms. Experiments in the early part of the 20th century made it clear that the negative electrons occupy the space outside the atomic nucleus, but their arrangement was completely unknown to Rutherford or other physicists of their time. This arrangement is now known and is the subject of Chapter 8.

EXERCISE 2.2 ***Describing Atoms***

If an atom were a macroscopic object with a radius of 100 m, it would approximately fill a football stadium. What would be the radius of the nucleus of such an atom? Can you think of an object that is about that size? ■

PORTRAIT OF A SCIENTIST

Ernest Rutherford (Oesper Collection in the History of Chemistry/University of Cincinnati)

Ernest Rutherford (1871–1937)

Lord Rutherford, one of the most interesting people in the history of science, was born in New Zealand in 1871 but went to Cambridge University in England to pursue his Ph.D. in physics in 1895. His original interest was in a phenomenon that we now call radio waves, and he apparently hoped to make his fortune in the field, largely so he could marry his fiancée back in New Zealand. His professor at Cambridge, J. J. Thomson, convinced him, however, to work on the newly discovered phenomenon of radioactivity. It was at Cambridge that he discovered α and β radiation, but he moved to McGill University in Canada in 1899. At McGill he did further experiments to prove that α radiation is actually composed of helium nuclei and that β radiation consists of electrons. (For this work he received the Nobel Prize in chemistry in 1908.)

At McGill, Rutherford was fortunate to have a talented student, Frederick Soddy, work with him. The two studied a radioactive gas coming from the radioactive element thorium. Their experiments showed that the gas was argon, which meant that they had made the first observation of the spontaneous disintegration of a radioactive element, one of the great discoveries of 20th century physics.

In 1903, Rutherford and his young wife visited with Pierre and Marie Curie in Paris, on the very day that Madame Curie received her doctorate in physics (see page 63). There was of course a celebration, and that evening while the party was in the garden of the Curies' home, Pierre Curie brought out a tube coated with a phosphor and containing a large quantity of radium in solution. The phosphor glowed brilliantly from the radiation given off by the radium. Rutherford later said the light was so bright he could clearly see Pierre Curie's hands were "in a very inflamed and painful state due to exposure to radium rays."

In 1907, Rutherford moved from Canada to Manchester University in England, and it was there that he performed the experiments that gave us the modern view of the atom. In 1919, he moved back to Cambridge and assumed the position formerly held by J. J. Thomson. Not only was Rutherford responsible for very important work in physics and chemistry, but he also guided the work of no fewer than ten future recipients of the Nobel Prize.

2.3 ATOMIC COMPOSITION

All atoms have neutrons in the nucleus (usually at least equal to the number of protons), except for the hydrogen atom, which has only a single proton in the nucleus.

The three primary constituents of atoms are **electrons, protons,** and **neutrons.** The **nucleus,** or core, of the atom is made up of protons and neutrons. Electrons are found in space about the nucleus (Figure 2.10). For an atom, which has no net electric charge, *the number of negatively charged electrons around the nucleus equals the number of positively charged protons in the nucleus.*

Atoms are extremely small; the radius of the typical atom is between 30 and 300 pm (3×10^{-10} m). To get a feeling for the incredible smallness of an atom, consider that one teaspoon of water (about 1 cm^3) contains about three times as many atoms as the Atlantic Ocean contains teaspoons of water.

All atoms of the same element have the same number of protons in the nucleus. This number is called the **atomic number,** and it is given the symbol **Z**. In the periodic table at the front of the book, the atomic number for each element is given above the element's symbol. A sodium atom, for example, has a nucleus containing 11 protons, so its atomic number is 11, and a uranium atom has 92 nuclear protons and $Z = 92$.

Just as our clocks and longitude are standardized relative to the time and longitude at Greenwich, England, so is a scale of atomic masses established relative to a standard. This standard is the mass of a carbon atom that has six protons

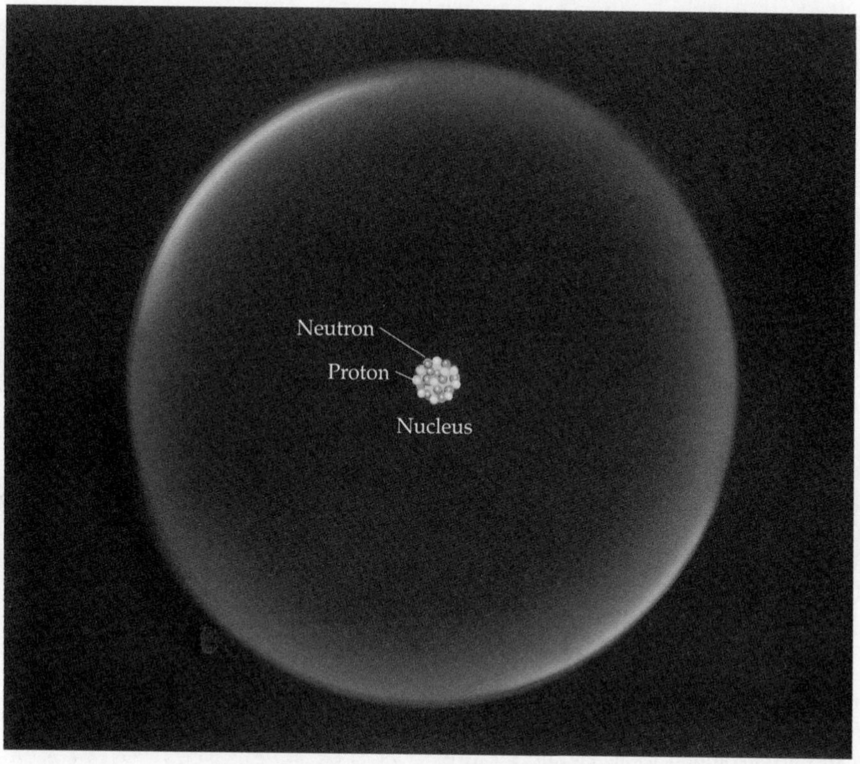

Figure 2.10 All atoms consist of one or more protons (positively charged) and usually at least as many neutrons (no charge) packed into an extremely small nucleus. Electrons (negatively charged) are arranged in space as a "cloud" about the nucleus. In an electrically neutral atom the number of electrons equals the number of protons.

and six neutrons in its nucleus. Such an atom is defined to have a mass of exactly 12 **atomic mass units** (or 12 amu), where 1 amu = $\frac{1}{12}$ the mass of a carbon atom having six protons and six neutrons in the nucleus. The mass of every other element is established relative to this mass. Thus, for example, experiment shows that an oxygen atom is, on the average, 1.33 times heavier than a carbon atom, so an oxygen atom has a mass of 1.33 × 12.0 amu, or 16.0 amu.

The amu is a unit of mass. 1 amu = $1.6605402 \times 10^{-27}$ kg

1amu = $\frac{1}{12}$ mass of Carbon atom

Masses of the basic atomic particles have been determined experimentally (Table 2.1). Notice that the proton and neutron have masses very close to 1 amu, whereas the electron is nearly 2000 times lighter.

Having established a relative scale of atomic masses, we can estimate the mass of any atom for which the nuclear composition is known. The proton and neutron have masses so close to 1 amu that the difference can often be ignored.

TABLE 2.1 Properties of Subatomic Particles[1]

Particle	Mass: *Grams*	Mass: *amu*	Charge	Symbol
Electron	9.109389×10^{-28}	0.0005485799	−1	${}^{0}_{-1}e$
Proton	1.672623×10^{-24}	1.007276	+1	${}^{1}_{1}p$
Neutron	1.674929×10^{-24}	1.008665	0	${}^{1}_{0}n$

[1]These constants and others in the book are taken from E. R. Cohen and B. N. Taylor: "Fundamental physical constants," *Physics Today,* Vol. 40, pp. BG11–BG15, 1987.

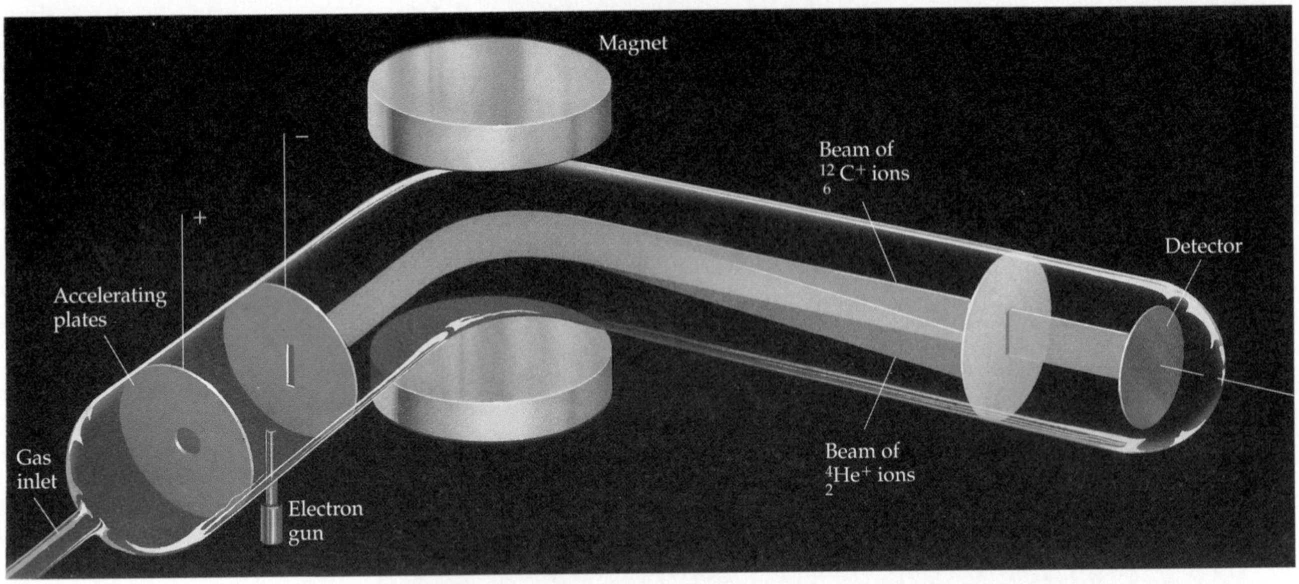

(a)

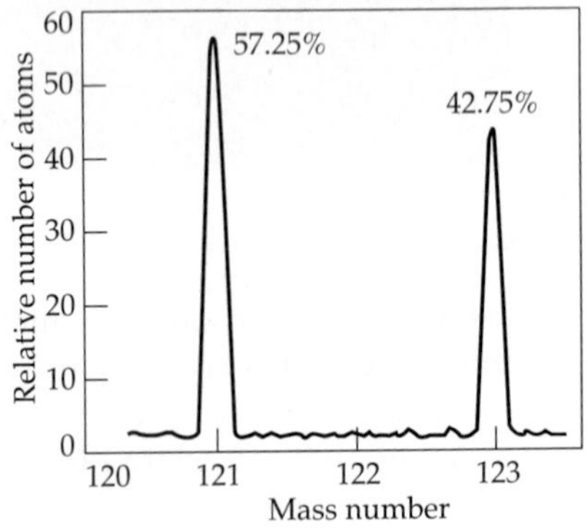

(b)

Figure 2.11 Mass spectrometer and spectrum. (a) The gas is injected into an evacuated tube. An electron beam ionizes a part of the sample by knocking electrons from the neutral atoms or molecules. Charged plates are arranged to accelerate positive ions toward the first slit and into the rest of the apparatus. Positive ions that pass the first slit move into a magnetic field perpendicular to their path, where they follow a curved path determined by the charge-to-mass ratio of the ion. A detector, behind the second slit, detects charged particles that pass through this slit. (Here the magnetic field was adjusted to allow carbon-12 ions to strike the second slit, whereas a beam of less massive helium-4 ions was bent too severely and missed the target slit.) (b) The result of separating the ions formed by the different isotopes of antimony in a mass spectrometer. The principal peak corresponds to the most abundant isotope, antimony-121. Percent relative abundance of the antimony isotopes is shown.

Electrons are so light that even a large number do not greatly affect the mass of the atom. Therefore, we only need to add up the number of *protons and neutrons* of an atom to estimate its mass. The result is called the **mass number** of that particular atom, a number given the symbol **A**. For example, a sodium atom has 11 protons and 12 neutrons in its nucleus; its mass number A is 23. The most common atom of uranium has 92 protons and 146 neutrons, so $A = 238$. With this information, we can symbolize an atom of known composition by the notation

$$\text{mass number} \longrightarrow \; {}^{A}_{Z}X \; \longleftarrow \text{element symbol}$$
$$\text{atomic number} \longrightarrow$$

(where the subscript Z is optional because the element symbol tells you what the atomic number must be). For example, the sodium atom described previously has the symbol $^{23}_{11}Na$ or just ^{23}Na. In words, we say "sodium-23."

The actual masses of atoms have been determined experimentally using mass spectrometers (Figure 2.11). It is always observed that, although the actual mass approximately equals the mass number, the actual mass is not an integral number (except for ^{12}C, which is exactly 12 by definition). For example, the mass of an iron atom with 32 neutrons, ^{58}Fe, is 57.9333 amu, slightly less than the mass number.

An obvious thing to do is to check the experimental masses by adding up all the masses of the protons, neutrons, and electrons of an atom to see if the experimental mass is obtained. This sum is slightly greater than the actual mass, however. This is not a mistake. Rather, the difference (sometimes called the *mass defect*) is related to the energy binding the particles of the nucleus together; we discuss this in more detail in Chapter 24.

EXAMPLE 2.1 *Atomic Composition*

How many neutrons are in an atom of platinum with a mass number of 195?

Solution Platinum has the symbol Pt, and its atomic number (shown in the periodic table inside the front cover) is 78. The atom therefore has 78 protons in the nucleus (and 78 electrons arranged outside the nucleus). The mass number of the atom is the sum of the number of protons and neutrons in the nucleus. Therefore,

$$\text{Mass number} = 195 = \text{number of protons} + \text{number of neutrons}$$
$$= 78 + \text{number of neutrons}$$
$$\text{Number of neutrons} = 195 - 78 = 117$$

EXERCISE 2.3 *Atomic Composition*

1. What is the mass number of a copper atom with 34 neutrons?
2. How many protons, neutrons, and electrons are there in a $^{59}_{28}Ni$ atom? ■

2.4 ISOTOPES

If we examine a natural sample of an element using a mass spectrometer (Figure 2.11) we find, in most cases, that all the atoms of that element do not have the same mass number. Take boron, for example. For many years boron-containing minerals such as borax have been mined in Death Valley, California. If the boron atoms of these minerals are examined, we find that, although all boron atoms have five nuclear protons, some atoms have five neutrons and others have six. That is, we find a collection of $^{10}_{5}B$ and $^{11}_{5}B$ atoms, which are called *isotopes.* **Isotopes** are *atoms having the same atomic number Z but a different mass number A.* In other words, isotopes are atoms of the same element, but they have different masses because they have different numbers of neutrons.

Most elements have at least two stable (nonradioactive) isotopes, but a few have only one isotope (aluminum, fluorine, and phosphorus, for example). Conversely, others have many isotopes (tin, for example, has ten stable isotopes). Generally, we refer to a particular isotope by giving its mass number (for example, uranium-238, ^{238}U), but some isotopes are so important that they have special names and symbols. Hydrogen atoms all have one proton. When that is the only nuclear particle, the element is called *protium,* or more usually "hydrogen." When one neutron is also present, the isotope $^{2}_{1}H$ is called *deuterium* or "heavy hydrogen" (symbol = D) (Table 2.2). The nucleus of radioactive hydrogen, $^{3}_{1}H$, or *tritium* (symbol = T), contains one proton and two neutrons.

If two atoms differ in the number of *protons* they contain, they are different *elements.* If they differ only in the number of *neutrons,* they are *isotopes.* It is the number of protons in the nucleus and the number of electrons outside the nucleus that largely determine the chemistry of the atom.

Frederick Soddy, Ernest Rutherford's assistant, coined the word "isotope" to describe the different forms of the same element.

TABLE 2.2 Masses of the Isotopes of Some Elements

Element	Symbol	Atomic Mass (amu)	Mass Number	Isotopic Mass (amu)	Natural Abundance (%)
Hydrogen	H	1.00794	1	1.0078	99.985
	D		2	2.0141	0.015
	T*		3	3.0161	0
Boron	B	10.811	10	10.0129	19.91
			11	11.0093	80.09
Oxygen	O	15.9994	16	15.9949	99.759
			17	16.9993	0.037
			18	17.9992	0.204
Nitrogen	N	14.00674	14	14.0031	99.63
			15	15.0001	0.37
Magnesium	Mg	24.305	24	23.9850	78.99
			25	24.9858	10.00
			26	25.9826	11.01

*Radioactive

Figure 2.12 Water containing ordinary hydrogen (^1_1H, protium) forms a solid that is less dense (d = 0.917 g/cm^3 at 0 °C) than liquid H_2O (d = 0.997 g/cm^3 at 25 °C) and so floats in the liquid. (Water is unique in this regard. The solid phase of virtually all other substances sinks in the liquid phase of that substance.) The same is true of D_2O, in which deuterium has been substituted for protium; "heavy ice" floats in "heavy water." "Heavy ice" is denser than ordinary water, however, so cubes made of D_2O sink in the liquid phase of H_2O. (The solid in the bottom of the glass is D_2O.) Substituting one isotope for another can have a profound effect. (C. D. Winters)

The substitution of one isotope of an element for another of that element in a compound can have interesting consequences (Figure 2.12). This is especially true when deuterium is substituted for hydrogen because the mass of deuterium is *double* that of hydrogen.

EXAMPLE 2.2 *Isotopes*

Silver has two isotopes, one with 60 neutrons and the other with 62 neutrons. What are the mass numbers and symbols of these isotopes?

Solution Silver has an atomic number of 47, so it has 47 protons in the nucleus. The two isotopes therefore have mass numbers of

Isotope 1: A = 47 protons + 60 neutrons = 107

Isotope 2: A = 47 protons + 62 neutrons = 109

The first isotope has a symbol $^{107}_{47}Ag$ and the second is $^{109}_{47}Ag$.

EXERCISE 2.4 *Isotopes*

Silicon has three isotopes with 14, 15, and 16 neutrons, respectively. What are the mass numbers and symbols of these three isotopes? ■

2.5 ATOMIC MASS

Boron atoms have two different isotopic masses: 10.0129 amu and 11.0093 amu. This means that the average mass of a collection of boron atoms is neither 10 nor 11 but somewhere in between, the actual value depending on the proportion of each kind of isotope.

The isotopic composition of an element is always expressed on a percentage basis in terms of the relative numbers of atoms of the various isotopes present.

$$\text{Percent abundance} = \frac{\text{number of atoms of a given isotope}}{\text{total number of atoms of all isotopes of that element}} \times 100\%$$

Isotope abundances can be determined using a mass spectrometer. Once the percent abundances and isotopic masses of each isotope are known, they can be used to calculate the average mass of atoms of that element. The boron isotopes, ^{10}B and ^{11}B, have percent abundances of 19.91% and 80.09%, respectively. This means that, if you could count out 10,000 boron atoms from an "average" natural sample, 1991 of them would have a mass of 10.0129 amu and 8009 of them would have a mass of 11.0093 amu. The average mass of a representative sample of atoms, expressed in atomic mass units, is called the **atomic mass.** For boron, the atomic mass is 10.81 amu, as is shown by the following calculation, in which the mass of each isotope is multiplied by its percent abundance:

Chemists usually use the term *atomic weight* of an element rather than "atomic mass." Although the quantity is more properly called a "mass" than a "weight," the term "atomic weight" is so commonly used that it has become accepted.

$$\text{Atomic mass} = \left(\frac{19.91}{100}\right) \cdot 10.0129 \text{ amu} + \left(\frac{80.09}{100}\right) \cdot 11.0093 \text{ amu} = 10.81 \text{ amu}$$

The atomic mass of each stable element has been determined by experiment, and it is these masses that appear in the periodic table in the front of the book. In the periodic table, each element's box contains the atomic number, the element symbol, and the atomic mass.

The periodic table entry for copper.

29 ⟵ *atomic number*
Cu ⟵ *symbol*
63.546 ⟵ *atomic mass*

EXAMPLE 2.3 ***Calculating Average Atomic Mass from Isotopic Abundances***

Bromine (used to make silver bromide, the important component of photographic film) has two naturally occurring isotopes, one with a mass of 78.918336 amu and a percent abundance of 50.69%. The other isotope, of mass 80.916289, has a percent abundance of 49.31%. Calculate the atomic mass of bromine.

Solution The atomic mass of any element is the average of the masses of all the isotopes in a representative sample. To calculate the atomic mass, you multiply the mass of each isotopes by its percent abundance divided by 100.

$$\text{Average atomic mass} = \left(\frac{\%\text{ abundance of isotope 1}}{100}\right)(\text{mass of isotope 1}) + \left(\frac{\%\text{ abundance of isotope 2}}{100}\right)(\text{mass of isotope 2}) + \cdots$$

For the bromine sample, the calculation is as follows:

$$\begin{aligned}\text{Average atomic mass of bromine} &= \text{atomic mass}\\ &= (0.5069)(78.918336 \text{ amu})\\ &\quad + (0.4931)(80.916289 \text{ amu})\\ &= 79.90 \text{ amu}\end{aligned}$$

EXERCISE 2.5 ***Calculating Atomic Mass***

Verify that the atomic mass of chlorine is 35.45 amu, given the following information:

^{35}Cl mass = 34.96885 amu, percent abundance = 75.77

^{37}Cl mass = 36.96590 amu, percent abundance = 24.23 ■

PROBLEM-SOLVING TIPS AND IDEAS

2.1 Calculating Your Grade or an Atomic Mass

In Example 2.3 you found the atomic mass of an element from the isotopic masses and their percent abundances. This calculation is done exactly the same way you calculate your average grade on quizzes in chemistry. The average value for a series of numbers can be found by adding the individual values and dividing by the number of values. For example, suppose you have two grades of 9 and three of 6 (out of 10) on five quizzes.

$$\text{Average} = \frac{9 + 9 + 6 + 6 + 6}{5} = 7.2$$

This is the same thing as

$$\text{Average} = (\tfrac{2}{5})(9) + (\tfrac{3}{5})(6) = (0.4)(9) + (0.6)(6) = 7.2$$

which means that $\frac{2}{5}$ of the time you had a 9, and $\frac{3}{5}$ of the time you had a 6. The fraction $\frac{2}{5}$ (or its decimal equivalent 0.4) is the "fractional abundance" of your grade. Expressed as a percentage, 40% of the time you received a 9, and 60% of the time you received a 6. Because more of your grades are 6s than 9s, the average grade is closer to 6. ■

2.6 THE PERIODIC TABLE

The periodic table of elements in Figure 2.13 and in the front of the book is one of the most useful tools in chemistry. Not only does it contain a wealth of information, but it can be used to organize many of the ideas of chemistry. Therefore, it is important that you become familiar with the main features and terminology of the periodic table. We shall refer to it often because much of this book is devoted to examining the chemical and physical properties of the elements and their interrelationships.

Figure 2.13 The periodic table of elements. Elements are listed in ascending order of atomic number. The following points are important: (1) Metals are shown in blue, the metalloids are green, and the nonmetals are yellow. (2) Periods are horizontal rows of elements and groups are vertical columns. (3) Groups are labeled by a number between 1 and 8 with a label of A (main group elements) or B (transition elements)—the system used most commonly at present in the United States. The new international system is to number the groups from 1 to 18. The periods are numbered 1 through 7. (4) Some groups have common names: Group 1A = alkali metals; Group 2A = alkaline earth metals; Group 7A = halogens; Group 8A = noble gases.

Metals

Metalloids

Nonmetals

1A (1)	2A (2)	3B (3)	4B (4)	5B (5)	6B (6)	7B (7)	8B (8)	(9)	(10)	1B (11)	2B (12)	3A (13)	4A (14)	5A (15)	6A (16)	7A (17)	8 (18)
Hydrogen 1 **H**																Hydrogen 1 **H**	Helium 2 **He**
Lithium 3 **Li**	Beryllium 4 **Be**											Boron 5 **B**	Carbon 6 **C**	Nitrogen 7 **N**	Oxygen 8 **O**	Fluorine 9 **F**	Neon 10 **Ne**
Sodium 11 **Na**	Magnesium 12 **Mg**											Aluminum 13 **Al**	Silicon 14 **Si**	Phosphorus 15 **P**	Sulfur 16 **S**	Chlorine 17 **Cl**	Argon 18 **Ar**
Potassium 19 **K**	Calcium 20 **Ca**	Scandium 21 **Sc**	Titanium 22 **Ti**	Vanadium 23 **V**	Chromium 24 **Cr**	Manganese 25 **Mn**	Iron 26 **Fe**	Cobalt 27 **Co**	Nickel 28 **Ni**	Copper 29 **Cu**	Zinc 30 **Zn**	Gallium 31 **Ga**	Germanium 32 **Ge**	Arsenic 33 **As**	Selenium 34 **Se**	Bromine 35 **Br**	Krypton 36 **Kr**
Rubidium 37 **Rb**	Strontium 38 **Sr**	Yttrium 39 **Y**	Zirconium 40 **Zr**	Niobium 41 **Nb**	Molybdenum 42 **Mo**	Technetium 43 **Tc**	Ruthenium 44 **Ru**	Rhodium 45 **Rh**	Palladium 46 **Pd**	Silver 47 **Ag**	Cadmium 48 **Cd**	Indium 49 **In**	Tin 50 **Sn**	Antimony 51 **Sb**	Tellurium 52 **Te**	Iodine 53 **I**	Xenon 54 **Xe**
Cesium 55 **Cs**	Barium 56 **Ba**	Lanthanum 57 ***La**	Hafnium 72 **Hf**	Tantalum 73 **Ta**	Tungsten 74 **W**	Rhenium 75 **Re**	Osmium 76 **Os**	Iridium 77 **Ir**	Platinum 78 **Pt**	Gold 79 **Au**	Mercury 80 **Hg**	Thallium 81 **Tl**	Lead 82 **Pb**	Bismuth 83 **Bi**	Polonium 84 **Po**	Astatine 85 **At**	Radon 86 **Rn**
Francium 87 **Fr**	Radium 88 **Ra**	Actinium 89 ****Ac**	Rutherfordium 104 **Rf**	Hahnium 105 **Ha**	Seaborgium 106 **Sg**	Nielsbohrium 107 **Ns**	Hassium 108 **Hs**	Meitnerium 109 **Mt**	110 **Discovered Nov. 1994**	111 **Discovered Dec. 1994**							

*Lanthanide Series

Cerium 58 **Ce**	Praseodymium 59 **Pr**	Neodymium 60 **Nd**	Promethium 61 **Pm**	Samarium 62 **Sm**	Europium 63 **Eu**	Gadolinium 64 **Gd**	Terbium 65 **Tb**	Dysprosium 66 **Dy**	Holmium 67 **Ho**	Erbium 68 **Er**	Thulium 69 **Tm**	Ytterbium 70 **Yb**	Lutetium 71 **Lu**

**Actinide Series

Thorium 90 **Th**	Protactinium 91 **Pa**	Uranium 92 **U**	Neptunium 93 **Np**	Plutonium 94 **Pu**	Americium 95 **Am**	Curium 96 **Cm**	Berkelium 97 **Bk**	Californium 98 **Cf**	Einsteinium 99 **Es**	Fermium 100 **Fm**	Mendelevium 101 **Md**	Nobelium 102 **No**	Lawrencium 103 **Lr**

Figure 2.14 Some common elements (C. D. Winters)

Group 1A: lithium (Li)

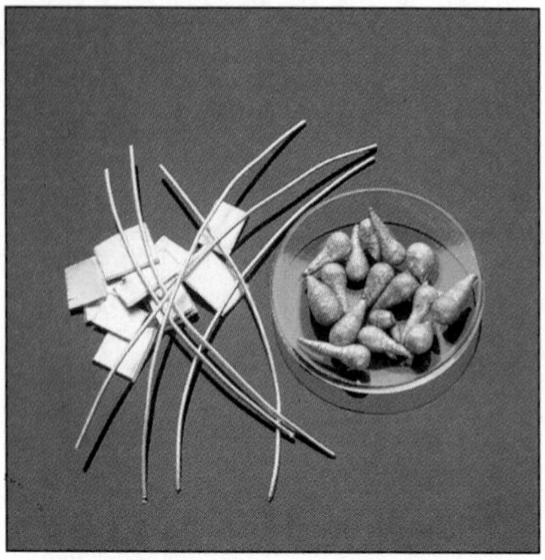

Group 3A: *left,* aluminum (Al); *right,* indium (In)

Group 5A: phosphorus (P)

Group 1A: sodium (Na)

Group 4A: silicon (Si)

Group 4A: *left,* carbon (C); *top,* lead (Pb); *right,* tin (Sn); *bottom,* silicon (Si)

Group 1A: potassium (K)

Group 6A: *left,* sulfur (S); *right,* selenium (Se)

Group 5A: nitrogen (N_2)

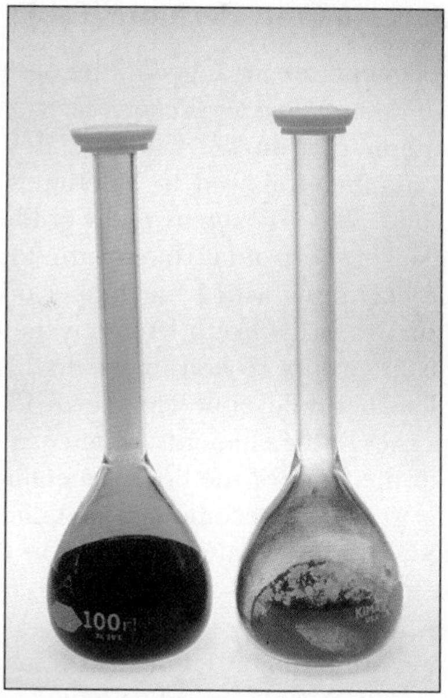

Group 7A: *left,* bromine (Br_2); *right,* iodine (I_2)

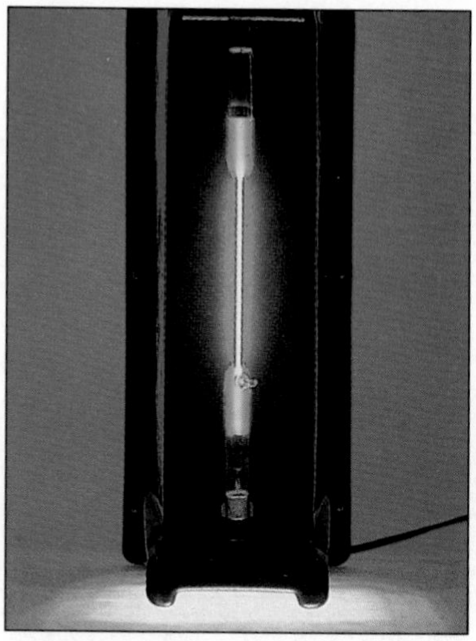

Group 8A: neon (Ne)

Fourth period transition metals:
left to right, Ti, V, Cr, Mn, Fe, Co, Ni, Cu

Group 1B, copper (Cu)

Group 1B, silver (Ag)

Group 1B, gold (Au)

Group 2B: *left,* zinc (Zn); *right,* mercury (Hg)

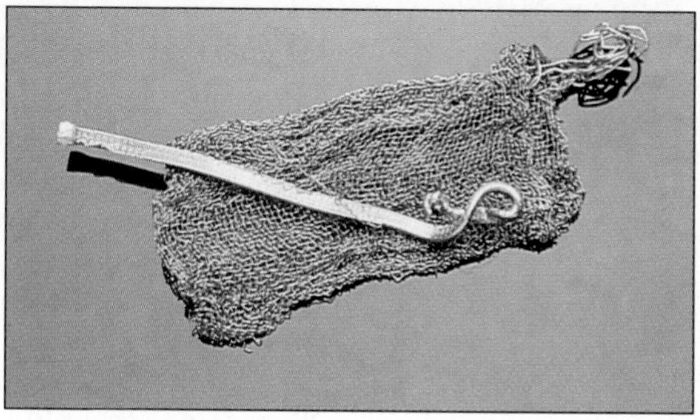

Group 8B: platinum (Pt)

Features of the Periodic Table

There is a movement to adopt a new set of group designations as an international standard; in this table the groups are numbered 1 through 18 from left to right. We shall use the "A/B table" in this text.

The elements are arranged in the periodic table in such a way that *elements having similar chemical and physical properties lie in vertical columns* called **groups.** The table commonly used in the United States has groups numbered 1 through 8, with each number followed by a letter A or B. Using this system, chemists often designate the *A groups* as **main group elements** and *B groups* as **transition elements.** The horizontal rows of the table are called **periods,** and they are numbered beginning with 1 for the period containing only H and He. For example, sodium, Na, is in Group 1A and is the first element in the third period. Mercury, Hg, is in Group 2B and in the sixth period.

The periodic table can be divided into several regions according to the properties of the elements, which can be classified as **metals** (blue in Figure 2.13 and in the front of the book), **metalloids** (green), and **nonmetals** (yellow). Elements gradually become less metallic as one moves from left to right across a period, and in the metalloid region their properties are intermediate between those of metals and nonmetals. Some elements are shown in Figure 2.14 on pages 78 and 79.

You are probably familiar with many properties of *metals* from your everyday experience. Metals are solids (except for mercury, page 78), conduct electricity, are ductile (they can be drawn into wires), are malleable (they can be rolled into sheets), and can form alloys (solutions of one or more metals in another metal). Iron, Fe, and aluminum, Al, are used in automobile parts because of their ductility and malleability and their low cost relative to other metals. Copper, Cu, is used in electric wiring because it conducts electricity better than most metals. Chromium is plated onto automobile parts because its metallic luster makes cars look better and also protects them from reacting with oxygen in the air.

Nonmetals have a wide variety of properties. Some are solids; bromine is a liquid; and some, like nitrogen and oxygen in the air, are gases at room temperature. With the exception of graphite, a form of carbon, nonmetals do not conduct electricity, which is one of the main factors that distinguishes them from metals. All nonmetals lie to the right of a zigzag line that passes between Al and Si, Ge and As, Sb and Te, and Po and At in Figure 2.13. Most of the elements next to this line have some properties that are typically metallic and other properties that are characteristic of nonmetals. These elements (B, Si, Ge, As, Sb, and Te) are called *metalloids* (or sometimes *semimetals*). Some of these elements are semiconductors of electricity and form the basis for the electronics revolution of the past several decades.

Elements intermediate in properties between metals and nonmetals have often been called *metalloids.* The Commission on Inorganic Nomenclature of the International Union of Pure and Applied Chemistry has recommended the term "semimetal," however.

Historical Development of the Periodic Table

Although the arrangement of the elements in the periodic table can now be understood on the basis of atomic structure, the table was originally developed from many, many experimental observations of elemental properties.

The historical development of the periodic table illustrates the way chemistry has developed: experimental observations lead to empirical correlations of properties and then to the prediction of results of further experiments. Once those predictions are tested, a theory can then be developed.

On the evening of February 17, 1869, at the University of St. Petersburg in Russia, a 35-year-old professor of general chemistry, Dmitri Ivanovitch Mendeleev (1834–1907) was writing a chapter of his soon-to-be-famous textbook on

PORTRAIT OF A SCIENTIST

Dmitri Ivanovitch Mendeleev (1834–1907)

Mendeleev was born in Tobolsk, Siberia, but was educated in St. Petersburg where he lived virtually all his life. He taught at St. Petersburg University and there wrote books and published his concept of chemical periodicity.

It is interesting that Mendeleev did little else with chemical periodicity after his initial articles. He went on to other interests, among them studying the natural resources of Russia and their commercial applications. In 1876, he visited the United States to study the fledgling oil industry and was much impressed with the industry but not with the country. He found Americans uninterested in science, and he felt the country carried on the worst features of European civilization.

By the end of the 19th century, political unrest was growing in Russia, and Mendeleev lost his position at the university. He was appointed chief of the Chamber of Weights and Measures for Russia, however, and established an inspection system for guaranteeing the honesty of weights and measures used in Russian commerce.

All pictures of Mendeleev show him with long hair. He made it a rule to cut his hair only once a year, in the spring, whether he had to appear at an important occasion or not.

Dmitri Ivanovitch Mendeleev (Oesper Collection in the History of Chemistry/University of Cincinnati)

chemistry. He had the properties of each element written on a separate card. While shuffling the cards trying to gather his thoughts before writing his manuscript, he realized that, if the elements were arranged in order of increasing atomic mass, certain properties were repeated several times! That is, he saw that a "periodicity" occurred in the properties of elements (Figure 2.15), and he summarized this in a table (Figure 2.16). He built the table by lining up the elements in a horizontal row in order of increasing atomic mass. Every time he came to an element with properties similar to one already in the row, he started a new row. The columns, then, contained elements with similar properties.

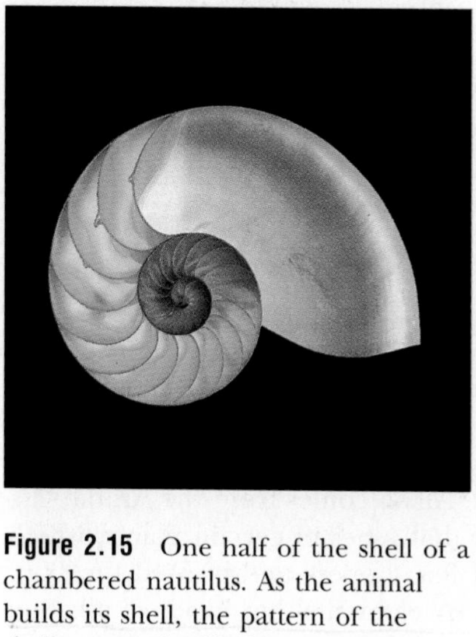

Figure 2.15 One half of the shell of a chambered nautilus. As the animal builds its shell, the pattern of the shell repeats itself with each revolution, an illustration of periodicity. (C. D. Winters)

TABELLE II

REIHEN	GRUPPE I. — R^2O	GRUPPE II. — RO	GRUPPE III. — R^2O^3	GRUPPE IV. RH^4 RO^2	GRUPPE V. RH^3 R^2O^5	GRUPPE VI. RH^2 RO^3	GRUPPE VII. RH R^2O^7	GRUPPE VIII. — RO^4
1	H = 1							
2	Li = 7	Be = 9,4	B = 11	C = 12	N = 14	O = 16	F = 19	
3	Na = 23	Mg = 24	Al = 27,3	Si = 28	P = 31	S = 32	Cl = 35,5	
4	K = 39	Ca = 40	— = 44	Ti = 48	V = 51	Cr = 52	Mn = 55	Fe = 56, Co = 59, Ni = 59, Cu = 63.
5	(Cu = 63)	Zn = 65	— = 68	— = 72	As = 75	Se = 78	Br = 80	
6	Rb = 85	Sr = 87	?Yt = 88	Zr = 90	Nb = 94	Mo = 96	— = 100	Ru = 104, Rh = 104, Pd = 106, Ag = 108.
7	(Ag = 108)	Cd = 112	In = 113	Sn = 118	Sb = 122	Te = 125	J = 127	
8	Cs = 133	Ba = 137	?Di = 138	?Ce = 140	—	—	—	— — — —
9	(—)	—	—	—	—	—	—	
10	—	—	?Er = 178	?La = 180	Ta = 182	W = 184	—	Os = 195, Ir = 197, Pt = 198, Au = 199.
11	(Au = 199)	Hg = 200	Tl = 204	Pb = 207	Bi = 208	—	—	
12	—	—	—	Th = 231	—	U = 240	—	— — — —

Figure 2.16 Dmitri Mendeleev's 1872 periodic table. The spaces marked with blank lines represent elements that Mendeleev deduced existed but were unknown at the time, so he left places for them in the table. The symbols at the top of the columns (e.g., R^2O and RH^4) are molecular formulas written in the style of the 19th century.

The most important feature of Mendeleev's table—and a mark of his genius and daring—was that he left empty spaces to retain the rationale of an ordered arrangement based on the periodic reoccurrence of similar properties; he deduced that these spaces would be filled by as-yet undiscovered elements. For example, in order of increasing atomic mass were copper (Cu), zinc (Zn), and then arsenic (As). If arsenic had been placed next to zinc, arsenic would have fallen under aluminum (Al). But arsenic forms compounds similar to those formed by phosphorus (P) and antimony (Sb), not aluminum. Mendeleev reasoned, therefore, that two as-yet undiscovered elements existed whose atomic masses were between zinc and arsenic. Arsenic therefore belonged in a position under phosphorus. The two missing elements were soon discovered: gallium (Ga) in 1875 and germanium (Ge) in 1886. In later years other gaps in Mendeleev's periodic table were filled as other predicted elements were discovered.

Not only did Mendeleev believe he could predict the existence of then unknown elements, but he also thought that he could detect inaccurate atomic masses. Given the chemical methods of that day, such inaccuracies were not unexpected. Therefore, when confronted with the elements tellurium (Te) and iodine (I), whose atomic masses were then taken to be 128 and 127, respectively, chemical similarities meant that he had to place Te in the same group with sulfur (Group 6A) and I in the same group as chlorine (Group 7A), even though this inverted their atomic mass order. He assumed the atomic mass of Te must be incorrect. Time has proved him wrong, however, since we know now that the atomic mass of tellurium is indeed slightly greater than that of iodine.

Tellurium and iodine are not the only pair of elements for which the element with a smaller atomic mass occurs later in the periodic table. Other reversed pairs in the modern periodic table include argon (Ar)–potassium (K) and cobalt (Co)–nickel (Ni).

The problem of the order of Te and I suggests that atomic mass is not the property that governs periodicity—another closely related property does this. This property was identified in 1913 by H. G. J. Moseley (1887–1915), a young English scientist working with Ernest Rutherford. Moseley bombarded many different metals with electrons in a cathode-ray tube and observed the x rays emitted by the metals. Most importantly, he found that the wavelengths of x rays emitted by a particular element are related in a precise way to the *atomic number* of that element. Based on his experiments, Moseley realized that other atomic properties may be similarly related to atomic number and not, as Mendeleev had believed, to atomic mass. Indeed, if the elements are arranged in order of increasing atomic number, the defects in the Mendeleev table are corrected. That is, the **law of chemical periodicity** should be restated as "the properties of the elements are periodic functions of *atomic number.*"

The wavelength of an x ray or of any other type of radiation is the distance between two crests or two troughs of the wave. See Chapter 7.

Chemistry and the Periodic Table

The vertical columns or groups of the periodic table contain elements having similar chemical and physical properties, and several groups of elements have distinctive names that are useful to know.

Elements in the leftmost column, **Group 1A,** are known as the **alkali metals** (except for hydrogen, which is not a metal). Alkali comes from the Arabic language; ancient Arabian chemists discovered that ashes of certain plants, which they called *al-qali,* gave water solutions that felt slippery and burned the skin. These ashes contain compounds of Group 1A elements that produce alkaline (basic) solutions. The elements themselves are very reactive, reacting with water to produce hydrogen and alkaline solutions (Figure 2.17). Because of their reactivity, these metals are found in nature only combined in compounds, never free.

The historical origins of the names of the chemical elements are quite interesting. See V. Ringnes, *Journal of Chemical Education,* Vol. 66, pp. 731–738, 1989.

A compound of sodium (Na), sodium chloride, has played an important role in history, for it is a fundamental part of the diet of humans and animals. Potassium (K) compounds are important plant nutrients. In addition to their similar reactivities, all these metallic elements form compounds with oxygen that have formulas like A_2O, where A represents the alkali metal: Li_2O, Na_2O, K_2O, Rb_2O, Cs_2O. Hydrogen also forms a compound having the same general formula: water, H_2O. This similarity of formulas was important to Mendeleev when he set up the periodic table (Figure 2.16).

The second group from the left, **Group 2A,** is also composed entirely of metals that occur naturally only in compounds. Except for beryllium (Be), these elements also react with water to produce alkaline solutions, and most of their oxides (such as CaO) form alkaline solutions; hence they are known as the **alkaline earth elements.** Magnesium (Mg) and calcium (Ca) are the sixth and fifth most abundant elements in the earth's crust, respectively. Calcium is especially well known, because it is one of the important elements in teeth and bones, and it occurs in vast limestone deposits. Calcium carbonate ($CaCO_3$) is the chief constituent of limestone and of corals, sea shells, marble, and chalk (Figure 2.18). Radium (Ra), the heaviest alkaline earth element, is radioactive and is used in radiation treatment of some cancers. When alkaline earth metals combine with oxygen, the general formula is EO, where E is one of the alkaline earth elements; for example, beryllium forms BeO and magnesium gives MgO.

Following Group 2A is a series of so-called **transition elements** that fill the fourth, fifth, and sixth periods in the center of the table. These are all metals. Some, like iron (Fe), are abundant in nature and important commercially. Others, like silver (Ag), gold (Au), and platinum (Pt), are much less abundant but also less reactive; they can be found in nature as the pure element and are coveted for their beauty and durability. Two rows at the very bottom of the table make room for the **lanthanides** (the series of elements following the element lanthanum) and the **actinides** (the series of elements following actinium). These are much less abundant and less important commercially, though some lanthanide compounds are used in color television picture tubes, uranium is the fuel for atomic power plants, and americium is used in smoke detectors.

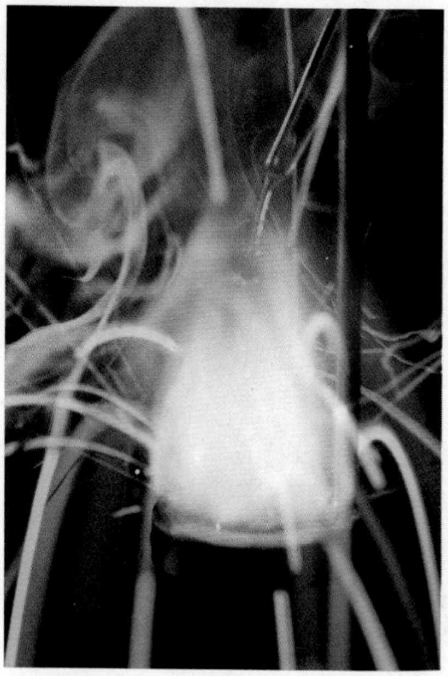

Figure 2.17 When water drops onto potassium, the reaction produces hydrogen, which burns in air.
(C. D. Winters)

Figure 2.18 Various forms of calcium carbonate, $CaCO_3$: a clear crystal of calcite, a seashell, and a piece of limestone.
(C. D. Winters)

Figure 2.19 Boron is found in borax, the chief ingredient of a washing powder called "20-Mule-Team Borax." The name comes from the fact that the borax was mined in Death Valley in California in the late 19th century, and teams of 20 mules each were used to haul out the heavy wagons loaded with borax. (C. D. Winters)

Gallium is becoming commercially more important. Gallium arsenide is an excellent semiconducting material of use in the electronics industry.

The next column to the right, **Group 3A,** has no special name. It contains a metalloid, boron (B), and four metals. Aluminum (Al) is the most abundant metal in the earth's crust at 8.3% by mass. It is exceeded in abundance only by the nonmetal oxygen (O; 45.5%) and the metalloid silicon (Si; 25.7%). These three elements are found combined in clays and other common minerals. Boron occurs in the mineral borax, which is mined in Death Valley, California. In the late 19th century the mineral was hauled out of the valley in wagons drawn by 20 mules, hence the name of a popular washing powder (Figure 2.19). Gallium, indium, and thallium are much less familiar because they have fewer important uses. All of these elements form oxygen compounds with the general formula X_2O_3, such as Al_2O_3.

Group 4A contains two nonmetals, carbon (C) and silicon (Si); a metalloid, germanium (Ge); and two metals, tin (Sn) and lead (Pb). Because of the change from nonmetallic to metallic behavior, more variation occurs in the properties of the elements from top to bottom of this group than in most; however, they all form oxygen compounds having the general formula XO_2. Carbon is the basis for the great variety of chemical compounds that make up living things. On earth it is found in carbonates like limestone (Figure 2.18), and in coal, petroleum, and natural gas—the fossil fuels. Silicon is the basis of many minerals, such as quartz and beautiful gem stones like amethyst (Figure 2.20). Tin and lead have been known for centuries, because they are easily smelted from their ores. Tin alloyed with copper makes bronze, which was used for centuries in utensils and weapons. Lead has been used in water pipes and paint, even though the element is quite toxic to humans. Indeed, one theory for the fall of the Roman empire contends that the Romans suffered from lead poisoning from the lead pipes they used to supply their water. In fact, the word "plumbing" comes from the Latin word for lead, *plumbum,* a name that is also the origin of the symbol (Pb) for lead.

Nitrogen, the first element in **Group 5A,** makes up about three fourths of earth's atmosphere, which accounts for nearly all the nitrogen at the earth's surface. The element is essential to life, so ways of fixing atmospheric nitrogen (forming compounds from the element) have been sought for at least a century. Nature accomplishes this easily in plants, but severe methods (high temperatures, for example) must be used in the laboratory to cause it to react with other elements. Phosphorus is also essential to life as an important constituent in

Figure 2.20 Quartz (*left*) is pure silicon dioxide, SiO_2. A trace of iron gives the solid a violet color (*right*), and the mineral is called amethyst. (C. D. Winters)

bones and teeth. The element glows in the dark, and so its name is based on Greek words meaning "light-bearing." Bismuth is the heaviest element that is not radioactive; all elements with atomic numbers greater than bismuth's (83) emit α-, β-, or γ-rays. As in Group 4A, elements at the top of this group are nonmetals, but a metal, Bi, is at the bottom. Nevertheless, all these elements form oxygen- or sulfur-containing compounds with the general formula E_2O_3 or E_2S_3. Examples include the brilliant yellow mineral orpiment (As_2S_3) and black stibnite (Sb_2S_3); the latter was used as a cosmetic by women in ancient societies.

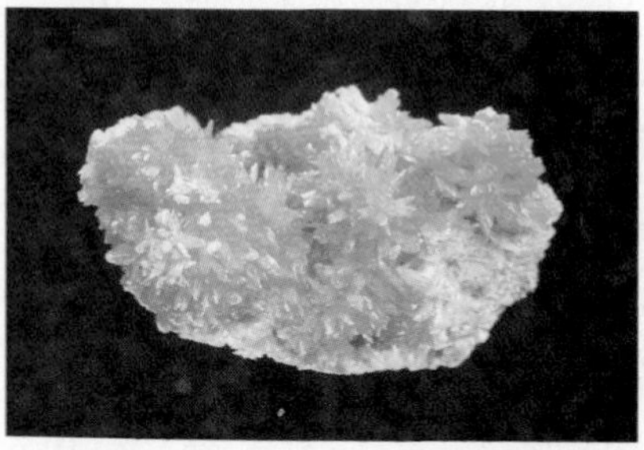

A sample of orpiment, As_2S_3, an example of the sulfur- (and oxygen-) containing compounds formed by Group 5A elements. (C. D. Winters)

Group 6A begins with oxygen, which constitutes about 20% of earth's atmosphere and combines readily with most other elements. Most of the energy that powers life on earth is derived from reactions in which oxygen combines with other substances. Sulfur has been known in elemental form since ancient times as brimstone or "burning stone." Sulfur, selenium, and tellurium are referred to collectively as *chalcogens* (from the Greek word, *khalkos,* for copper) because copper ores contain them. Their compounds are foul-smelling and poisonous; nevertheless sulfur and selenium are essential components of the human diet. Polonium was isolated in 1898 by Marie and Pierre Curie, who separated it from tons of a uranium-containing ore and named it for Madame Curie's native country, Poland. General formulas of oxygen compounds of these elements are EO_3 and EO_2.

At the far right of the periodic table are two groups composed entirely of nonmetals. Within each group the elements are quite similar, but they are completely different from elements in the other group. The **Group 7A** elements—fluorine, chlorine, bromine, and iodine—combine violently with alkali metals to form salts: table salt, NaCl, for example. The name for this group, **halogens,** comes from the Greek word *hals,* meaning salt, and *genes,* for forming. The halogens react with many other metals to form salts, and they also combine with most nonmetals. They are among the most reactive of all elements (Figure 2.21).

Be sure to notice the spelling of fluorine (u before o). A common misspelling is "flourine," which would be pronounced "flower-ene."

By contrast, the **Group 8** elements—helium, neon, argon, krypton, xenon, and radon—are the least reactive elements, are all gases, and are not very abundant on earth. Because of this, they were not discovered until the end of the 19th century. Helium, the second most abundant element in the universe after hydrogen, was discovered in the sun in 1868 by analysis of the solar spectrum. It was

(a) (b)

Figure 2.21 Like all the alkali metals, sodium reacts vigorously with halogens. (a) A flask containing yellow chlorine gas (Cl_2). (b) A piece of sodium was placed in the flask, and the sodium reacted with the chlorine to give table salt (sodium chloride, NaCl). The reaction evolves energy in the form of light and heat.
(C. D. Winters)

The Group 8 elements are also called the inert or rare gases.

not found on earth until 1895. (The name of the element comes from the Greek word for the sun, *helios.*) Until 1962, when a compound of xenon was first prepared, it was thought that none of these elements would combine chemically. This led to the name **noble gases** for this group, a term meant to denote their general lack of reactivity. Because its atoms are very light, helium is used in lighter-than-air craft such as blimps; neon and argon are used in advertising signs. Radon is radioactive and the cause of indoor air pollution problems when it seeps out of the ground into buildings.

EXERCISE 2.6 *The Periodic Table*

How many elements are in the third period of the periodic table? Give the name and symbol of each. Tell whether each element in the period is a metal, metalloid, or nonmetal. ■

Chemical Periodicity

Elements in a group have similar—but not identical—properties. Some properties of elements in a group differ by degree in a regular pattern. For example, the melting points of the Group 1A elements Li through Cs are (in degrees Celsius) 179, 98, 64, 39, and 28, respectively. Lithium reacts slowly with water, sodium reacts faster, and potassium still faster (Figure 2.17). For the elements at the bottom of Group 1A, just exposure to moist air produces a vigorous reaction or even explosion. All Group 1A elements are soft, silvery white, metallic solids, except cesium, which is a liquid on a warm day (see sodium in Figure 2.14).

Other properties differ to some degree but not in a regular pattern. For example, the densities (in grams per cubic centimeter) of the solids Li through Cs are 0.53, 0.97, 0.86, 1.53, and 1.87, respectively. Lithium, sodium, and potassium have densities less than 1.00 g/cm^3 (the approximate density of water at room temperature) and so float on water as they react vigorously.

Some properties are similar for every member of a group. For example, elements in a group generally react with other elements to form similar compounds. *This is the most useful and powerful inference that can be made from the periodic table.* Taking the alkali metals as an example, we find that they all can combine with halogens to give compounds with the formula MX, that is, one atom of metal (M) combines with one atom of the halogen (X) (Figure 2.21). Table salt, NaCl, and the compound in iodized salt, KI, are just two examples. This one observation from the periodic table gives us the ability to write formulas for compounds of cesium and rubidium, elements less familiar to us than the other members of the group. Hence, we can have confidence that the chlorides and bromides of cesium and rubidium must have the formulas CsCl, CsBr, RbCl, and RbBr, respectively.

Why do elements in the same group in the periodic table have similar chemical behavior? Why do metals and nonmetals have different properties? The answer is that all the elements in a group have atoms with similar electronic structures. (This is particularly true for the main group elements and the noble gases.) In fact, the chemical view of matter is primarily concerned with what happens to the electrons in atoms in the course of a chemical reaction. We shall therefore return to the matter of chemical periodicity once you have been introduced to the electronic structures of atoms in Chapter 8.

EXERCISE 2.7 ***Periodic Properties***

The element aluminum forms aluminum chloride, the formula for which is $AlCl_3$. Predict a formula for a compound formed from indium and fluorine. Compare the formulas of sodium oxide and sodium chloride. Can you see a relationship? A general formula for alkaline earth oxides was given during discussion of Group 2A elements. Use this and the relationship between chloride and oxide formulas to predict the formula for a compound formed from magnesium and chlorine. ■

2.7 THE MOLE: THE MACRO/MICRO CONNECTION

One of the most exciting parts of chemistry is the discovery of some new substance when one element reacts with another. But chemistry is also a quantitative science. When two chemicals react with each other, you want to know how many atoms of each are used so that a formula of the product can be established. This means we need some method of counting atoms, no matter how small they are. That is, there must be a way of connecting the macroscopic world, the world we can see, with the submicroscopic world of atoms. The solution to this problem is to define a convenient unit of matter that contains a known number of particles. The chemical counting unit that has come into use is the **mole.**

The word "mole" was apparently introduced in about 1896 by the German chemist Wilhelm Ostwald (1853–1932), who derived the term from the Latin

CURRENT ISSUES IN CHEMISTRY

Banning Chlorine

Elemental chlorine is produced in enormous quantities in the industrialized world. As depicted on the "chlorine family tree" shown here, it is the starting point for the production of at least 10,000 compounds that contain chlorine and that are important in our economy. In general the major uses of chlorine and chlorine-containing compounds are in controlling biological contaminants in public water supplies, as solvents for the preparation of pharmaceuticals, in making plastics, and in pulp and paper manufacturing.

Chlorine-containing compounds have also been used as bactericides and pesticides. Dichlorodiphenyltrichloroethane, commonly called DDT, was widely used to eradicate mosquitos that carried malaria. It was so effective that the World Health Organization once regarded shortages of the chemical as a threat to public health. Our experience with DDT, however, is one reason that environmental groups want to ban chlorine and chlorine-containing organic compounds.

Some chlorine-containing products. The white material is PVC (polyvinyl chloride) used to make the water and waste pipes in homes. (C. D. Winters)

The chemical properties of chlorine mean that many chlorine-containing compounds are very stable. Although this property makes them valuable in some applications, it may also mean that they persist in the environment. Chlorine-containing compounds accumulate in animals, reaching higher and higher concentrations the farther up the food chain you go. For example, small fish feed on waterplants or crustaceans that are contaminated; these fish are eaten by larger fish, which are eaten by still larger fish or birds. At each stage the concentration of DDT in the prey animal persists into the predator, which passes it on to the animal that preys on it. Biologists found that DDT accumulated in bald eagles and that this was a cause of the decline in the eagle population in the United States in the 1960s. The pesticide apparently interferes with the calcium carbonate structure of the eggs, creating thin shells; this caused the eagles' reproductive rate to drop severely; only 417 nesting pairs were counted in the lower 48 states in 1972. After DDT was banned for use in the United States in 1972, the eagle population began to recover, and 3747 nesting pairs were estimated in 1992.

Another source of environmental problems comes from the formation of toxic chlorine-containing compounds in the environment. For ex-

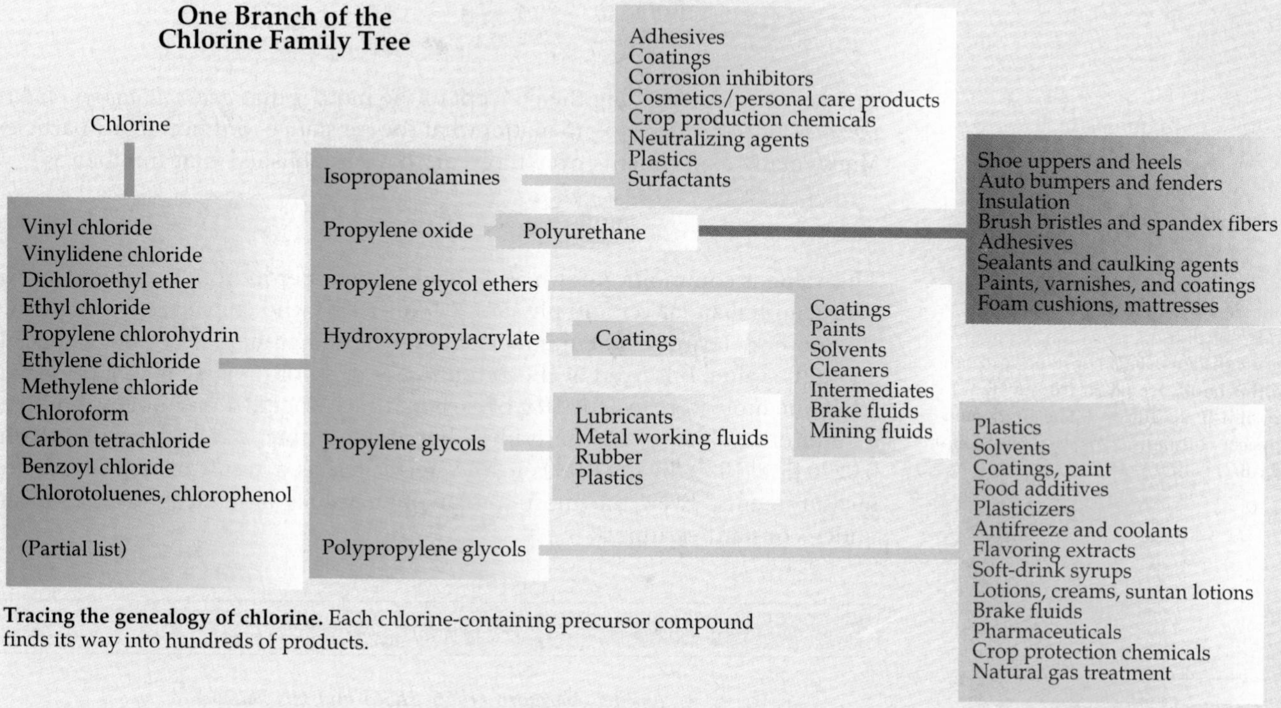

Tracing the genealogy of chlorine. Each chlorine-containing precursor compound finds its way into hundreds of products.

(I. Amato: Science, *Vol. 261, pp. 152–154, 1993.)*

ample, the largest use of chlorine gas is in bleaching paper. If the used chlorine is dumped into the effluent of a paper mill, it can combine with naturally occurring compounds to turn them into toxins. For this reason the paper industry has been working for some years to substitute processes that rely on oxygen, ozone, or peroxides to do the job that chlorine once did.

Organizations such as Greenpeace advocate a total ban on the use of chlorine in any form. They argue that, since alternatives exist for some chlorine-containing products, we should be safe rather than sorry. The organization has said that "There are no uses for chlorine that we regard as safe."

However, Mario Molina, who, with Sherwood Rowland, first recognized the effect of CFCs on the earth's ozone layer, believes that a total ban on chlorine is going too far. He is afraid that a total ban would deprive society of many useful materials while removing deleterious ones. As evidence of this, it has been noted that "of the nearly 400 new drugs approved for therapeutic use in humans since 1984, more than 60 are organochlorine compounds. Two of the ten most prescribed pharmaceuticals, Ceclor and Xanax, contain chlorine."* And then we must consider vancomycin, the only antibiotic effective against some *Staphylococcus* infections, or Lorstan, a new drug used to treat hypertension.

Mario Molina has said that "The underlying need is to make reasonable assessments of risk, and then tackle the problems that matter." Chlorine and its compounds are an example of the discussion of risks and benefits of chemistry in the Introduction. A spokesman for the chlorine industry said that "The stakes are high in this discussion. We must balance the risks of chlorine chemistry against not having access to chlorine and look carefully at the alternatives because they may have adverse impacts of their own."

*G.W. Gribble: *Chemical and Engineering News,* April 18, 1994, p. 18 and I. Amato: *Science,* Vol. 261, pp. 152–154, 1993.

word *moles,* meaning a "heap" or "pile." The mole, whose symbol is **mol,** is the SI base unit for measuring an *amount of substance* (Table 1.2) and is defined as follows:

> A **mole** is the amount of substance that contains as many elementary entities (atoms, molecules, or other particles) as there are atoms in *exactly* 12 g of the carbon-12 isotope.

The key to understanding the concept of the mole is that *one mole always contains the same number of particles,* no matter what the substance. But how many particles? Many, many experiments over the years have established that number as

$$1 \text{ mole} = 6.022136736 \times 10^{23} \text{ particles}$$

Avogadro's number is known to nine significant figures ($6.022136736 \times 10^{23}$), which is not unusual precision for many physical constants. This constant, and others in this book, are taken from E. R. Cohen and B. N. Taylor, "The fundamental physical constants," *Physics Today,* Vol. 40, pp. BG11–BG15, 1987.

This value is commonly known as **Avogadro's number** in honor of Amedeo Avogadro, an Italian lawyer and physicist (1776–1856) who conceived the basic idea (but never determined the number). Understand that there is nothing "special" about the value. It is fixed by the definition of the mole as exactly 12 g of carbon-12. If one mole were exactly 10 g of carbon, then Avogadro's number would have a different value. It is interesting that the number was just revised (from 6.022045×10^{23}) in the 1980s to the new value as a result of new and better measurements. Even the most fundamental values and ideas of science are under constant scrutiny.

PORTRAIT OF A SCIENTIST

Amedeo Avogadro, conte di Quaregna. (Chemical Heritage Foundation)

Amedeo Avogadro (1776–1856) and His Number

Lorenzo Romano Amedeo Carlo Avogadro, an Italian nobleman, was educated as a lawyer and practiced the profession for many years. About 1800, however, he turned to science and was the first professor in Italy in mathematical physics. In 1811, he first suggested the hypothesis, which we now call a law, that "equal volumes of gases under the same conditions have equal numbers of molecules." From this eventually came the concept of the mole. Avogadro did not see the acceptance of his ideas during his lifetime, and scientists were not convinced until the Italian chemist Stanislao Cannizzaro (1826–1910) described experiments proving them at the great chemical conference in Karlsruhe, Germany in 1860.

One of the great difficulties presented by Avogadro's number is comprehending its size. It may help to write it out in full as

$$6.022 \times 10^{23} = 602{,}200{,}000{,}000{,}000{,}000{,}000{,}000$$

or as

$$602{,}200 \times 1 \text{ million} \times 1 \text{ million} \times 1 \text{ million}$$

But, think of it this way: If you had Avogadro's number of unpopped popcorn kernels and poured them over the continental United States, the country would be covered in popcorn to a depth of 9 miles! Or, if you divided one mole of pennies equally among every man, woman, and child in the United States, each person could pay off the national debt (currently \$3.6 trillion or 3.6×10^{12}) and still have 20 trillion dollars left over for an ice cream cone or two.

How is Avogadro's number determined? It is clearly not possible to count all of the atoms in a mole. (If a computer counted 10 million atoms per second, it would take about 2 billion years to count all of the atoms in a mole.) At least four experimental methods can be used to do it, one of them being the measurement of the dimensions of the smallest repeating unit in a solid element (Chapter 13).

The mole is the chemist's six-pack or dozen; it is a counting unit. Many objects come in similar counting units. Shoes, socks, and gloves are sold by the pair, soft drinks by the six-pack, and eggs and donuts by the dozen. Atoms are counted by the mole.

As you saw in Section 2.3, the atomic mass scale is a relative scale, with the ^{12}C atom chosen as the standard, and the masses of all other atoms have been established by experiment and placed on this scale. For example, experiments show that a ^{16}O atom is 1.33 times heavier than a ^{12}C atom or that a ^{19}F atom is 1.58 times heavier than a ^{12}C atom. Because a mole of carbon-12 has a mass of exactly 12 g and contains 6.0221367×10^{23} atoms, and because *a mole of one kind of atom always contains the same number of particles as a mole of another kind of atom,* a mole of ^{16}O atoms has a mass in grams 1.33 times 12.0, or 16.0 g. Similarly, a mole of ^{19}F atoms is 1.58 times greater than 12.0 g, or 19.0 g.

Moles of Atoms, The Molar Mass

The mass in grams of 1 mol of atoms of any element (6.0221367×10^{23} atoms of that element) is the **molar mass** of that element. Molar mass is conventionally abbreviated with a capital italicized *M*, and is expressed in units of grams per mole (g/mol). It is *numerically equal to the atomic mass in atomic mass units.* Thus, based on experiments that take into account all isotopes of a particular element.

M = g/mol
amu of atom particle = g/mol of atom particle

$$\text{Molar mass of sodium (Na)} = \text{mass of exactly 1 mol of Na atoms}$$
$$= 22.9898 \text{ g/mol}$$
$$= \text{mass of } 6.02214 \times 10^{23} \text{ Na atoms}$$

$$\text{Molar mass of lead (Pb)} = \text{mass of exactly 1 mol of Pb atoms}$$
$$= 207.2 \text{ g/mol}$$
$$= \text{mass of } 6.022 \times 10^{23} \text{ Pb atoms}$$

The relative physical sizes of 1-mol quantities of some common elements are shown in Figure 2.22. Although each of these "piles of atoms" has a different volume and different mass, each contains 6.022×10^{23} atoms.

The mole concept is the cornerstone of quantitative chemistry. It is *essential* to be able to convert from moles to mass and from mass to moles. Dimensional analysis, which is described in Section 1.7 and in Appendix A, shows that this can be done in the following way:

MASS ⇄ MOLES CONVERSIONS

Moles to Mass	Mass to Moles
$\text{Moles} \cdot \frac{\text{grams}}{1 \text{ mol}} = \text{grams}$	$\text{Grams} \cdot \frac{1 \text{ mol}}{\text{grams}} = \text{moles}$
↑ molar mass	↑ 1/molar mass

For example, suppose you wish to use 0.35 mol of aluminum. What mass, in grams, of aluminum must you use? Using the molar mass of aluminum (27.0 g/mol),

$$0.35 \text{ mol Al} \cdot \frac{27.0 \text{ g}}{1 \text{ mol Al}} = 9.5 \text{ g of Al}$$

you find that 9.5 g of aluminum are required.

Figure 2.22 One mole quantities of common elements. Clockwise from the top left: copper beads (63.546 g); aluminum foil (26.982 g); lead shot (207.2 g); magnesium chips (24.305 g); chromium (51.996 g); and sulfur (32.066 g). Four of the samples are in 50-mL beakers. (C. D. Winters)

Look at the periodic table in the front of the book and notice that some atomic masses are known to more significant figures and decimal places than others. When using molar masses in a calculation, the convention followed in this book is to *use one more significant figure in the molar mass than in any of the other data.* For example, if you weigh out 16.5 g of carbon, you use 12.01 g/mol for the molar mass of C to find the moles of carbon present.

$$16.5\ \cancel{\text{g C}} \cdot \frac{1\ \text{mol C}}{12.01\ \cancel{\text{g C}}} = 1.37\ \text{mol C}$$

↑ Note that four significant figures are used in the molar mass of the element, but only three were in the sample mass.

A piece of elemental silicon on top of a wafer made of pure silicon. The tiny rectangles on the wafer are electronic circuits etched into the surface of the wafer. (C. D. Winters)

By using one more significant figure you guarantee that the precision of the molar mass is greater than the other numbers and does not limit the precision of the result.

EXAMPLE 2.4 *Mass to Moles*

How many moles are represented by 454 g of silicon, an element used in semiconductors? (454 g is equivalent to 1.00 lb)

Solution Use the periodic table in the front of the book to find the molar mass of silicon (28.09 g/mol). Convert the mass of silicon to its equivalent in moles.

$$454\ \cancel{\text{g}}\ \text{Si} \cdot \frac{1\ \text{mol Si}}{28.09\ \cancel{\text{g}}\ \text{Si}} = 16.2\ \text{mol Si}$$

This student is holding a beaker containing 518 g of lead in a 150-mL beaker. See Example 2.5.

EXAMPLE 2.5 *Moles to Mass*

What mass, in grams, is equivalent to 2.50 mol of lead (Pb)?

Solution For a conversion between mass and moles, you always need the molar mass, which is 207.2 g/mol in the case of lead. Thus, the number of grams of lead in 2.50 mol is

$$2.50\ \cancel{\text{mol}}\ \text{Pb} \cdot \frac{207.2\ \text{g Pb}}{1\ \cancel{\text{mol}}\ \text{Pb}} = 518\ \text{g Pb}$$

The photograph shows 518 g of lead shot in a 150-mL beaker.

EXAMPLE 2.6 *Mole Calculation*

The graduated cylinder in the photograph at the top of page 93 contains 25.4 cm^3 of mercury. If the density of mercury at 25 °C is 13.534 g/cm^3, how many moles of mercury are in the cylinder? How many atoms of mercury are there?

Solution As noted in **Problem Solving Tips and Ideas: 2.2,** volume and moles are not directly connected. You must first use the density to convert the volume to a mass, and then derive the quantity of mercury, in moles, from the mass. Finally, the number of atoms is obtained from the number of moles.

$$\boxed{\text{Volume, cm}^3} \xrightarrow[\text{use density}]{\times \frac{\text{g}}{\text{cm}^3}} \boxed{\text{Mass, g}} \xrightarrow[\text{use molar mass}]{\times \frac{\text{mol}}{\text{g}}} \boxed{\text{Moles}} \xrightarrow[\text{use Avogadro's number}]{\times \frac{\text{atoms}}{\text{mol}}} \boxed{\text{Atoms}}$$

Therefore, the volume of mercury is found to be equivalent to 344 g of mercury.

$$25.4 \text{ cm}^3 \text{ Hg} \cdot \frac{13.534 \text{ g Hg}}{1 \text{ cm}^3 \text{ Hg}} = 344 \text{ g Hg}$$

Knowing the mass, you can now find the quantity in moles.

$$344 \text{ g Hg} \cdot \frac{1 \text{ mol Hg}}{200.6 \text{ g Hg}} = 1.71 \text{ mol Hg}$$

Finally, because you know the relation between atoms and moles (Avogadro's number), you can now find the number of atoms in the sample.

$$1.71 \text{ mol Hg} \cdot \frac{6.022 \times 10^{23} \text{ atoms Hg}}{1 \text{ mol Hg}} = 1.03 \times 10^{24} \text{ atoms Hg}$$

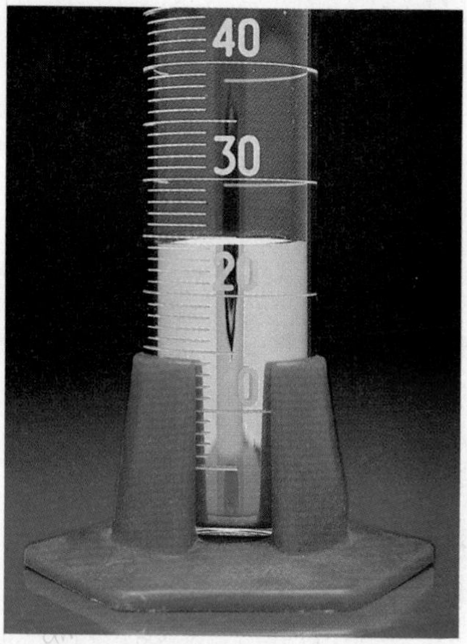

A graduated cylinder holding 25.4 cm^3 of mercury. See Example 2.6. (C. D. Winters)

EXERCISE 2.8 ***Mass/Mole Conversions***

1. What is the mass, in grams, of 2.5 mol of aluminum?
2. How many moles are represented by 454 g of sulfur? ■

EXERCISE 2.9 ***Atoms***

The density of platinum, Pt, is 21.45 g/cm^3. What is the volume (in cubic centimeters) of a piece of platinum that contains 1.0×10^{24} atoms? If the piece of metal is a square with a thickness of 0.10 cm, what is the length, in centimeters, of one side of the piece? ■

PROBLEM-SOLVING TIPS AND IDEAS

2.2 More on Unit Conversions

In Example 2.6 you wanted to find the quantity of mercury, in moles, in a given volume of the element, and then you wanted to know the number of atoms of mercury in that volume. You know that the mass of an element is directly related to the number of moles represented by that mass, and the number of atoms can be calculated once the number of moles is known.

$$\boxed{\text{Mass, g}} \xrightarrow[\text{use molar mass}]{\times \frac{\text{mol}}{\text{g}}} \boxed{\text{Moles}} \xrightarrow[\text{use Avogadro's number}]{\times \frac{\text{atoms}}{\text{mol}}} \boxed{\text{Atoms}}$$

The problem here was that you did not have the mass of mercury, only its volume. This is the reason you need the density of the substance, that is, the relation between volume and mass. Therefore, you first have to convert milliliters of mercury into grams of mercury using density as the conversion factor.

$$\boxed{\text{Volume, cm}^3} \xrightarrow[\text{use density}]{\times \frac{\text{g}}{\text{cm}^3}} \boxed{\text{Mass, g}}$$

As a final point, note that all three of the conversions in Example 2.6 involved a calculation of the form

$$\text{Given data} \cdot \frac{\text{Desired units}}{\text{Units of given data}} = \text{Answers in desired units}$$

(conversion factor) ■

CHAPTER HIGHLIGHTS

Having studied this chapter, you should be able to

- Explain the historical development of the atomic theory and identify some of the scientists who made important contributions (Sections 2.1–2.2).
- Describe electrons, protons, and neutrons, and the general structure of the atom (Section 2.3).
- Calculate the atomic mass of an element from isotopic abundances (Section 2.4).
- Define "isotope" and give the mass number and number of neutrons for a specific isotope (Section 2.4).
- Explain the difference between the atomic number and atomic mass of an element and find this information in the periodic table (Sections 2.3–2.5).
- Identify the periodic table location of groups, periods, metals, metalloids, nonmetals, alkali metals, alkaline earth metals, halogens, noble gases, and the transition elements (Section 2.6).
- Use the periodic table to predict properties of elements and formulas of simple compounds.
- Explain the concept of the mole and use this to find the number of moles of an element given the mass, or convert moles of an element to the mass of the element (Section 2.7).

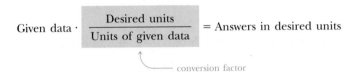

Mass → Moles $$\text{Mass, g}\left(\frac{1\ \text{mol}}{X\text{g}}\right) = \text{moles}$$

Moles → Mass $$\text{Moles}\left(\frac{X\text{g}}{1\ \text{mol}}\right) = \text{mass, g}$$

STUDY QUESTIONS

Review Questions

1. Which of John Dalton's ideas, which were first put forth early in the 19th century, are not true according to our present views of atomic theory?
2. What are the three fundamental particles from which atoms are built? What are their electric charges? Which of these particles constitute the nucleus of an atom? Which is the least massive particle of the three?
3. Define the atomic mass unit.

4. What is the difference between the mass number and the atomic number of an atom?
5. What did the discovery of radioactivity reveal about the structure of atoms?
6. What is the relationship between the work of J. J. Thomson and Robert Millikan? How was Ernest Rutherford's work related to J. J. Thomson's work?
7. If the nucleus of an atom were the size of a medium-sized orange (let us say with a diameter of about 6 cm), what would be the diameter of the atom?
8. Element 25 is found in the form of oxide "nodules" at the bottom of the sea. What is the name and symbol of this element?

A metal oxide "nodule" from the floor of the Pacific Ocean. (C. D. Winters)

9. The volcanic eruption of Mt. St. Helens in the state of Washington produced a considerable quantity of a radioactive element in the gaseous state. The element has atomic number 86. What are the symbol and name of this element?

The eruption of Mt. St. Helens in Washington state in May 1980. (Pat and Tom Leeson/Photo Researchers)

10. Lithium has two stable isotopes: ^{6}Li and ^{7}Li. One of them has an abundance of 92.5%, and the other has an abundance of 7.5%. Knowing that the atomic mass of lithium is 6.941, which is the more abundant isotope?
11. Primo Levi (1919–1987), an Italian writer and chemist, in his beautifully written autobiography, *The Periodic Table*, says of zinc that "it is not an element which says much to the imagination, it is gray and its salts are colorless, it is not toxic, nor does it produce striking chromatic reactions; in short, it is a boring metal. It has been known to humanity for two or three centuries, so it is not a veteran covered with glory like copper, nor even one of those newly minted elements which are still surrounded with the glamour of their discovery." From this, and from reading this chapter, make a list of the properties of zinc. For example, include in your list the position of the element in the periodic table, and tell how many electrons and protons an atom of zinc has. What is its atomic number and atomic mass? Zinc certainly is important in our economy. Can you think of any uses of the element? Check your dictionary (or a reference book such as *The Handbook of Chemistry and Physics*) and make a list of the uses of the element.
12. What was incorrect about Mendeleev's original concept of the periodic table? What is the modern "law of chemical periodicity," and how is this related to Mendeleev's ideas?
13. What is the difference between a group and a period in the periodic table?
14. Name and give symbols for (a) three elements that are metals, (b) four elements that are nonmetals, (c) and two elements that are metalloids. In each case locate the element in the periodic table by giving the group and period in which the element is found.
15. Name and give symbols for three transition metals in the fourth period. Look up each of your choices in a dictionary, and make a list of the properties that the dictionary entry gives. Also list the uses of the element if given by the dictionary.
16. Name three transition elements, a halogen, a noble gas, and an alkali metal.
17. Name two halogens. Look up each of your choices in a dictionary (or a reference book such as *The Handbook of Chemistry and Physics*), and make a list of the properties that the dictionary entry gives. Also list the uses of the element that might be given by the dictionary.
18. Name an element that was first discovered by Madame Curie. Give its name, symbol, and atomic number. Use a dictionary to find the origin of the name of this element.
19. If you divide Avogadro's number of pennies among the 250 million men, women, and children in the United States, and if each person could count one penny each second every day of the year for 8 h a day, how long would it take to count all of the pennies? How long would it take you to get bored doing this?
20. Why do you think it is more convenient to use some chemical counting unit such as the mole, when doing calculations rather than counting individual atoms?

NUMERICAL AND OTHER QUESTIONS

The Composition of Atoms

21. Give the mass number of each of the following atoms: (a) beryllium with 5 neutrons, (b) titanium with 26 neutrons, and (c) gallium with 39 neutrons.

22. Give the mass number of (a) an iron atom with 30 neutrons, (b) an americium atom with 148 neutrons, and (c) a tungsten atom with 110 neutrons.

23. Give the complete symbol ($^{A}_{Z}X$) for each of the following atoms: (a) sodium with 12 neutrons; (b) argon with 21 neutrons; and (c) gallium with 39 neutrons.

24. Give the complete symbol ($^{A}_{Z}X$) for each of the following atoms: (a) nitrogen with 8 neutrons; (b) zinc with 34 neutrons; and (c) xenon with 75 neutrons.

25. How many electrons, protons, and neutrons are there in an atom of (a) calcium-40, ^{40}Ca; (b) tin-119, ^{119}Sn; and (c) plutonium-244, ^{244}Pu?

26. How many electrons, protons, and neutrons are there in an atom of (a) carbon-13, ^{13}C; (b) chromium-50, ^{50}Cr; and (c) bismuth-205, ^{205}Bi?

27. Fill in the blanks in the table (one column per element).

Symbol	^{45}Sc	^{33}S	—	—
Number of protons	—	—	8	—
Number of neutrons	—	—	9	31
Number of electrons in the neutral atom	—	—	—	25

28. Fill in the blanks in the table (one column per element).

Symbol	^{65}Cu	^{37}Cl	—	—
Number of protons	—	—	34	—
Number of neutrons	—	—	46	46
Number of electrons in the neutral atom	—	—	—	36
Name of element	—	—	—	—

Isotopes

29. Radioactive americium-241 is used in household smoke detectors and in bone mineral analysis. Give the number of electrons, protons, and neutrons in an atom of americium-241.

30. The synthetic radioactive element, technetium, is used in many medical studies. Give the number of electrons, protons, and neutrons in an atom of technetium-99.

31. Which of the following are isotopes of element X, the atomic number for which is 9: $^{19}_{9}X$, $^{20}_{9}X$, $^{9}_{18}X$, and $^{21}_{9}X$?

32. Cobalt has three radioactive isotopes used in medical studies. Atoms of these isotopes have 30, 31, and 33 neutrons, respectively. Give the symbol for each of these isotopes.

Atomic Mass

33. Verify that the atomic mass of lithium is 6.94 amu, given the following information:
^{6}Li, exact mass = 6.015121 amu, percent abundance = 7.50%
^{7}Li, exact mass = 7.016003 amu, percent abundance = 92.50%

34. Verify that the atomic mass of magnesium is 24.31 amu, given the following information:
^{24}Mg, exact mass = 23.985042 amu, percent abundance = 78.99%
^{25}Mg, exact mass = 24.98537 amu, percent abundance = 10.00%
^{26}Mg, exact mass = 25.982593 amu, percent abundance = 11.01%

35. Gallium has two naturally occurring isotopes, ^{69}Ga and ^{71}Ga, with masses of 68.9257 amu and 70.9249 amu, respectively. Calculate the percent abundances of these isotopes of gallium.

36. Copper has two stable isotopes, ^{63}Cu and ^{65}Cu, with masses of 62.939598 amu and 64.927793 amu, respectively. Calculate the percent abundances of these isotopes of copper.

The Periodic Table

37. How many elements occur in Group 4A of the periodic table? Give the name and symbol of each of these elements. Tell whether each is a metal, nonmetal, or metalloid.

38. How many elements occur in the fourth period of the periodic table? Give the name and symbol of each of these elements. Tell if each is a metal, nonmetal, or metalloid.

39. Which period in the periodic table is as yet incomplete? What is the name given to the majority of these elements and what well-known property characterizes them?

40. How many periods of the periodic table have 8 elements, how many have 18 elements, and how many have 32 elements?

The Mole

41. Calculate the number of grams in
(a) 2.5 mol of boron
(b) 0.015 mol of oxygen
(c) 1.25×10^{-3} mol of iron
(d) 653 mol of helium

42. Calculate the number of grams in
(a) 6.03 mol of gold
(b) 0.045 mol of uranium
(c) 15.6 mol of Ne
(d) 3.63×10^{-4} mol of Pu

43. Calculate the number of moles represented by each of the following:
(a) 127.08 g of Cu
(b) 20.0 g of calcium
(c) 16.75 g of Al
(d) 0.012 g of potassium
(e) 5.0 mg of americium

44. Calculate the number of moles represented by each of the following:
(a) 16.0 g of Na
(b) 0.0034 g of platinum
(c) 1.54 g of P
(d) 0.876 g of arsenic
(e) 0.983 g of Xe

45. A chunk of sodium metal, Na, if thrown into a bucket of water, produces a dangerously violent explosion. If 50.4 g of sodium is used, how many moles of sodium does that represent? How many atoms? (*Caution:* Sodium is *very* reactive with water. The metal should be handled only by a knowledgeable chemist.)

46. Superman comes from the planet Krypton. If you have 0.00789 g of the gaseous element krypton how many moles does this represent? How many atoms?

47. Precious metals, such as gold and platinum, are sold in units of "troy ounces," where 1 troy ounce is 31.1 g. If you have a block of platinum with a mass of 15.0 troy ounces, how many moles of the metal do you have? What is the size of the block in cubic centimeters? (The density of platinum is 21.45 g/cm^3 at 20 °C.)

48. In an experiment, you need 0.125 mol of sodium metal. Sodium can be cut easily with a knife, so if you cut out a block of sodium, what should the volume of the block be in cubic centimeters? If you cut a perfect cube, what is the length of the edge of the cube? (The density of sodium is 0.968 g/cm^3.)

49. If you have a 35.67-g piece of chromium metal on your car, how many atoms of chromium are in the piece?

50. If you have a ring that contains 1.94 g of gold, how many atoms of gold are in the ring?

51. What is the average mass of one copper atom?

52. What is the average mass of one atom of titanium?

GENERAL QUESTIONS

53. Potassium has three naturally occurring isotopes, ^{39}K, ^{40}K, and ^{41}K, but ^{40}K has a very low natural abundance. Which of the other two is the more abundant? Briefly explain your answer.

54. Name three elements you encountered today. (Name only those that you have seen in their elemental form, not those combined into compounds.) Give the location of each of these elements in the periodic table by specifying the group and period in which it is found.

55. Figure 2.11 shows the mass spectrum of the isotopes of antimony. What are the symbols of the isotopes? Which is the more abundant isotope? How many protons, neutrons, and electrons does this isotope have? Without looking at a periodic table, give the approximate atomic mass of antimony.

56. The elements of Group 4A can combine to give compounds such as carbon tetrachloride, CCl_4. Give the formulas for the compounds formed between Cl and the other Group 4A elements, silicon and germanium.

57. When an athlete tears ligaments and tendons, they can be surgically attached to bone to keep them in place until they reattach themselves. A problem with current techniques, however, is that the screws and washers used are often too big to be positioned accurately or properly. A new titanium-containing device (see drawing) has therefore been invented to correct this problem.

(a) What are the symbol, atomic number, and atomic mass of titanium?

(b) In what group and period is it found? Name the other elements of its group.

(c) What chemical properties do you suppose make titanium an excellent choice for this and other surgical applications?

(d) Using a dictionary (or a reference book such as *The Handbook of Chemistry and Physics*), make a list of the properties of the element and its uses.

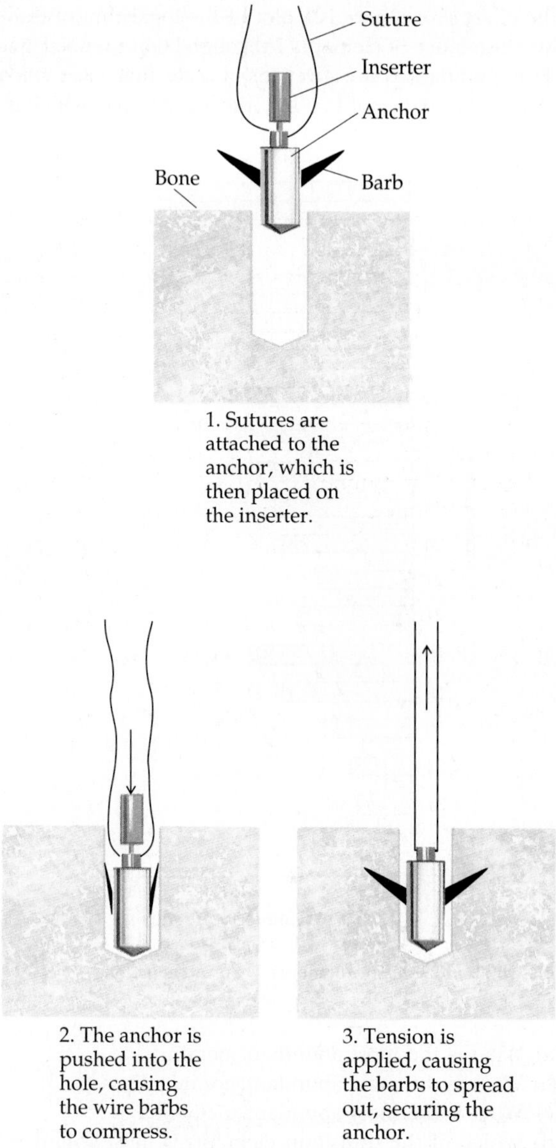
Suture
Inserter
Anchor
Bone
Barb
1. Sutures are attached to the anchor, which is then placed on the inserter.
2. The anchor is pushed into the hole, causing the wire barbs to compress.
3. Tension is applied, causing the barbs to spread out, securing the anchor.

A new method of anchoring ligaments and tendons to bone. (*New York Times,* July 1, 1992, p. D7.) (Redrawn with permission from the *New York Times.*)

58. The chart shown here is a plot of the logarithm of the relative abundance of elements 1 through 36 in the solar system. [The abundances are given on a scale that gives silicon a relative abundance of 1×10^6 (the logarithm of which is 6).]

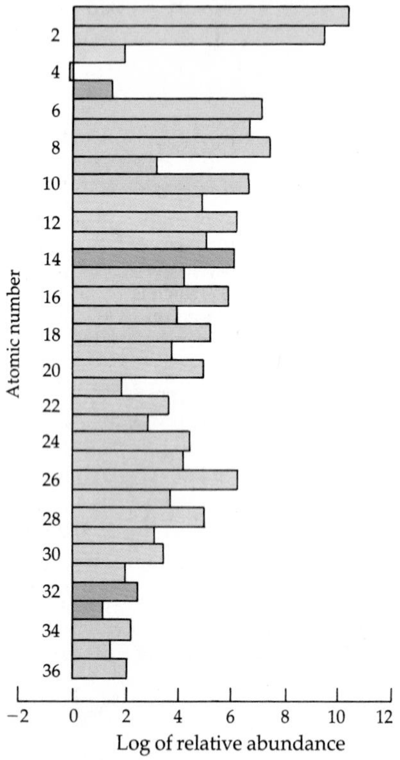

The relative abundances of elements 1–36 in the solar system.

(a) What is the most abundant metal?
(b) What is the most abundant nonmetal?
(c) What is the most abundant metalloid?
(d) Which of the transition elements is most abundant?
(e) How many halogens are considered on this plot and which is the most abundant?

59. The plot in the next column shows the variation in density with atomic number for the first 36 elements. Use this plot to answer the following questions:
(a) What three elements in this series have the highest values of density? What is their approximate density? Are these elements metals or nonmetals?
(b) Which element in the second period has the largest density? Which element in the third period has the largest density? What do these two elements have in common?
(c) Some elements have densities so low that they do not show up in the plot. What elements are these? What property do they have in common?

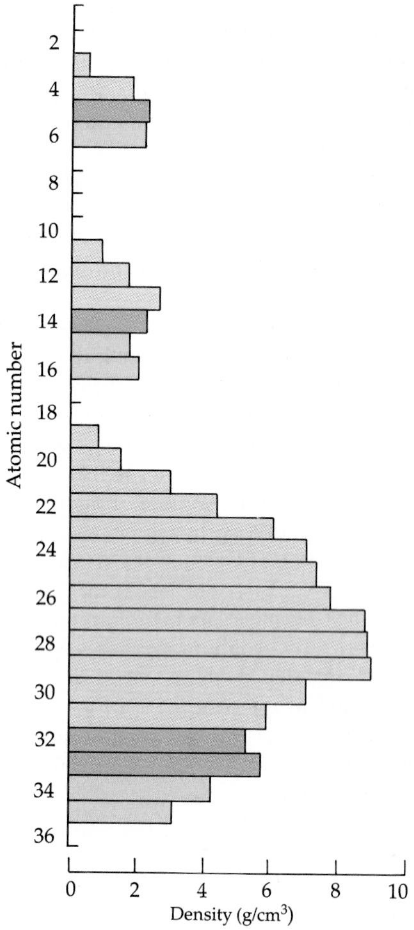

The densities of elements 1–36.

60. Gems and precious stones are measured in carats, a weight unit equivalent to 200. mg. If you have a 2.3-carat diamond in a ring, how many moles of carbon do you have? (Diamonds are pure carbon.)

61. The international markets in precious metals operate in the weight unit "troy ounce" (where 1 troy ounce is equivalent to 31.1 g). Assume platinum sells for \$420/troy ounce. (a) How many moles are in 1 troy ounce? (b) If you have \$5000 to spend, how many grams and how many moles of platinum can be purchased?

62. Gold prices fluctuate, depending on the international situation. If gold currently sells for \$389.80/troy ounce, how much must you spend to purchase 1.00 mol of gold? (1 troy ounce is equivalent to 31.1 g)

63. A piece of copper wire is 7.6 m long (about 25 ft) and has a diameter of 2.0 mm. If copper has a density of 8.92 g/cm^3, how many moles of copper and how many atoms of copper are in the piece of wire?

64. Dilithium is the fuel for the Starship *Enterprise*. Because its density is quite low, however, you need a large space to store a large mass. To estimate the volume required, we shall use the element lithium. If you need 256 mol for an interplanetary trip, what must the volume of a piece of lithium be? If the piece of lithium is a cube, what is the dimension of an edge of the cube? (The density for the element lithium is $0.534\ g/cm^3$ at 20 °C.)

65. Crossword Puzzle: In the 2 × 2 crossword shown here, each letter must be correct four ways: horizontally, vertically, diagonally, and by itself. Instead of words, use symbols of elements. When the puzzle is complete, the four spaces will contain the overlapping symbols of 10 elements. There is only one correct solution.

1	2
3	4

Horizontal

1–2: Two-letter symbol for a metal used in ancient times
3–4: Two-letter symbol for a metal that burns in air and is found in Group 5A

Vertical

1–3: Two-letter symbol for a metalloid
2–4: Two-letter symbol for a metal used in U.S. coins

Single squares: All one-letter symbols

1. A colorful nonmetal
2. Colorless gaseous nonmetal
3. An element that makes fireworks green
4. An element that has medicinal uses

Diagonal

1–4: Two-letter symbol for an element used in electronics
2–3: Two-letter symbol for a metal used with Zr to make wires for superconducting magnets

This puzzle first appeared in *Chemical and Engineering News,* Dec. 14, 1987 (p. 86) (submitted by S. J. Cyvin) and in *Chem Matters,* October, 1988.

66. Draw a picture showing the approximate positions of all protons, electrons, and neutrons in an atom of helium-4. Make certain that your diagram indicates both the number and position of each type of particle.

67. Use Avogadro's number and the molar mass of carbon to calculate the conversion factor between atomic mass units and grams.

CONCEPTUAL QUESTIONS

68. Calcium reacts with air to form a compound in which the ratio of calcium to oxygen atoms is 1 : 1; that is, the formula is CaO. Based on this fact, predict the formulas for compounds formed between oxygen and the Group 2A elements.

69. Compare the formulas of the compounds formed between Cl and Na, Mg, and Al. (See Exercise 2.7.) Can you see a pattern in these formulas? Describe this pattern. Based on this pattern, predict the formula of a compound formed between Si and Cl.

70. Again consider the plot of log abundance versus atomic number in Study Question 58. Can you uncover any relation between abundance and atomic number? Is there any difference between elements of even atomic number and those of odd atomic number?

71. Figure 2.17 shows how potassium reacts with water. The photos shown here depict what happens when magnesium and calcium are placed in water. Based on their relative reactivities, what might you expect to see when barium, another Group 2A element, is placed in water? Give the period in which each element (Mg, Ca, and Ba) is found. What correlation do you think you might find between the reactivity of these elements and their position in the periodic table?

Magnesium (*left*) and calcium (*right*) in water. (C. D. Winters)

72. Answer the following questions using the figures shown here. (Each question may have more than one answer.)

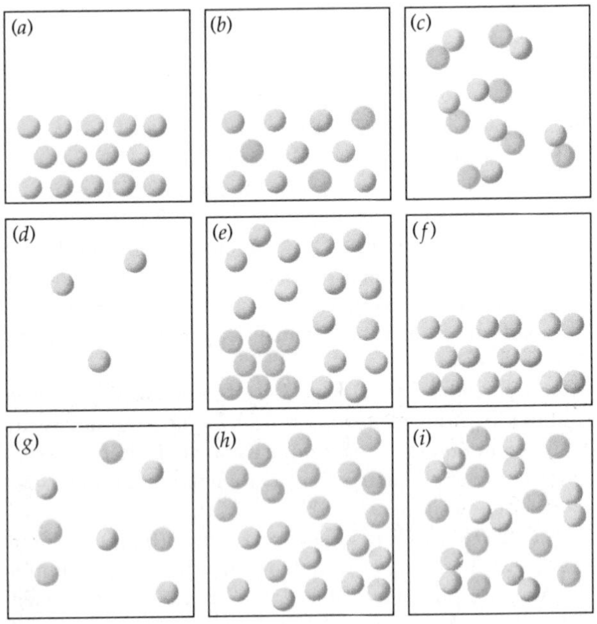

(a) Which represents submicroscopic particles in a sample of solid?
(b) Which represents submicroscopic particles in a sample of liquid?
(c) Which represents submicroscopic particles in a sample of gas?
(d) Which represents submicroscopic particles in a sample of an element?
(e) Which represents submicroscopic particles in a sample of a compound?
(f) Which represents submicroscopic particles of a solution?
(g) Which represents submicroscopic particles in a sample of a heterogeneous mixture?

73. Assume you have just discovered β-particles. What experiments would you carry out to show that β-particles are electrons?

74. Although carbon-12 is now used as the standard for atomic masses, this has not always been the case. Early attempts at classification used hydrogen as the standard, with the mass of hydrogen equal to 1.0000 amu. Later attempts defined atomic masses using oxygen (with a mass of 16.0000 amu). In each instance, the atomic masses of the other elements were defined relative to these masses.
(a) If H = 1.0000 amu was used as a standard for atomic masses, what would the atomic mass of oxygen be? What would be the value of Avogadro's number under these circumstances?
(b) Assuming the standard is O = 16.0000 amu, determine the value for the atomic mass of hydrogen and the value of Avogadro's number. (To answer this question, you need more precise data on current atomic masses: H, 1.00797 and O, 15.9994.)

75. Most standard analytical balances can measure accurately to the nearest 0.0001 g. Assume you weigh out a 2.0000-g sample of carbon. How many atoms are contained in this sample? Assuming the indicated accuracy of the measurement, what is the largest number of atoms that can be present in the sample?

76. When a sample of phosphorus burns in air, the compound P_4O_{10} forms. Assume one experiment showed that 0.744 g of phosphorus formed 1.704 g of P_4O_{10}. Use this information to determine the ratio of the atomic masses of phosphorus and oxygen (mass P/mass O). If the atomic mass of oxygen is assumed to be 16.000 amu, calculate the atomic mass of phosphorus.

77. An estimation of the radius of a lead atom:
(a) You are given a cube of lead that is 1.000 cm on each side. The density of lead is 11.35 g/cm^3. How many atoms of lead are contained in the sample?
(b) Atoms are spherical; therefore, the lead atoms in this sample cannot fill all the available space. As an approximation, assume that 60% of the space of the cube is filled with spherical lead atoms. Calculate the volume of one lead atom from this information. From the calculated volume (V), and the formula $V = \frac{4}{3}(\pi r^3)$, estimate the radius (r) of a lead atom.
(c) Assume the lead atoms along each side of the cube touch one another. How many atoms lie along one edge of the cube?

C H A P T E R 3

Molecules and Compounds

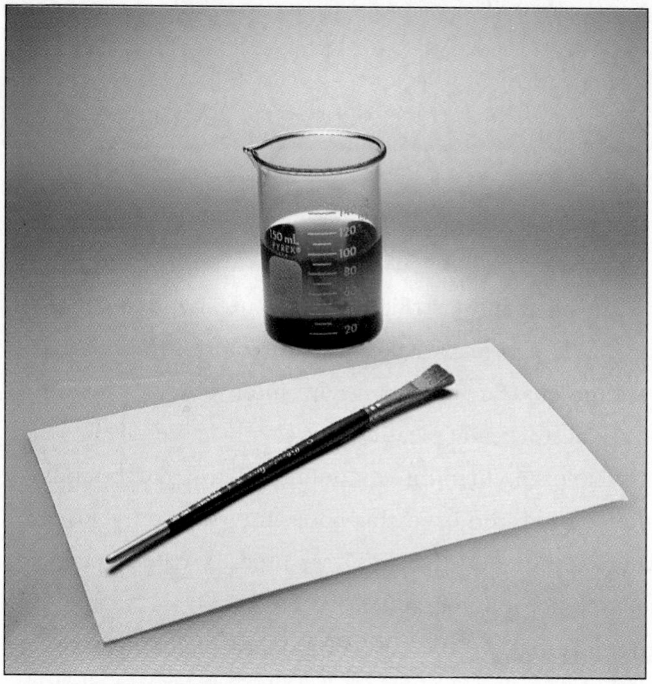

A Chemical Puzzler

Stir a red chemical compound with a formula of $CoCl_2 \cdot 6\ H_2O$ into a beaker of water to make a solution with a faint red color. Take some of this solution and write your message on a piece of paper. When the paper dries, the words are no longer visible. You have written a secret message! But warm the paper above a candle flame, and the words appear again. What is happening? What kind of chemical compounds are involved? Does this process, called a chemical reaction, happen with other compounds? Are the ingredients of our magic ink—hydrated cobalt chloride and water—essential to the process or are there others that would work?

Figure 3.1 "Circular Forms" by Robert Delauney. (The Bettmann Archives/ Newsphotos)

The French abstract painter Robert Delaunay (1885–1941) said that "color is form and subject," and you can see how well he used colors by looking at one of his paintings (Figure 3.1). But why is your eye drawn to the painting? Surely it is in part the brilliant colors—the reds, yellows, oranges, blues, and greens. Such paintings are interesting to chemists not only because they are aesthetically pleasing but also because all the colors are accounted for by colored chemical compounds. These compounds are also the basis of the colored photos in this book, the paint on your car, the color of dyes in your clothes, and the "ink" we used to write a secret message in the chapter opening photograph.

Many of the compounds that are the basis of modern pigments have been custom-made in the laboratory. Indeed, the synthesis of new compounds for a specific application is where much of the excitement of modern chemistry lies. This chapter introduces you to the major types of chemical compounds, and later chapters describe how many of these compounds can be made.

3.1 ELEMENTS THAT EXIST AS MOLECULES

One hundred and eleven elements are now known. Chemists have obtained pure samples of most of these elements; the only exceptions are the elements with the highest atomic numbers, elements that are radioactive and have very

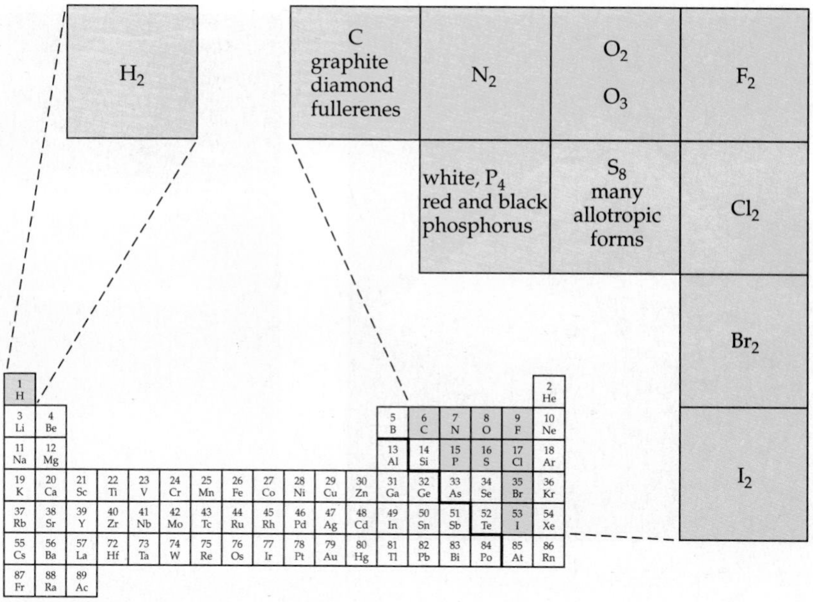

Figure 3.2 The nonmetallic elements that exist as polyatomic molecules. Notice that many of these elements have allotropic forms. Among them is sulfur, which has more allotropes than any other element.

short lifetimes. As you can see from Figure 2.14, most elements are metals and most are solids at room temperature. In the solid state, a metal consists of atoms packed closely together. This is the reason the density of metals is generally higher than the density of other elements.

Nonmetals may be solids, liquids, or gases at room temperature. Some, like the noble gases in Group 8, exist as atoms; others exist as molecules made up of two or more atoms. Some of these are two-atom, or diatomic, molecules and include hydrogen (H_2), nitrogen (N_2), oxygen (O_2), and the halogens of Group 7A (F_2, Cl_2, Br_2, and I_2). Others exist as molecules with more than two atoms (Figure 3.2); examples include sulfur and phosphorus. A sample of sulfur consists of S_8 molecules, and phosphorus is made up of P_4 molecules.

Oxygen, phosphorus, sulfur, and carbon are also interesting because each of these elements can exist in different forms called **allotropes.** The allotropes of oxygen are O_2, sometimes called dioxygen, and O_3, ozone. The latter is an unstable blue gas with a characteristic pungent odor. In fact, it was first detected by its odor, and its name comes from the Greek word, *ozein,* meaning "to smell." You may have noticed this peculiar odor during a thunderstorm because lightning bursts provide the energy to convert O_2 in the atmosphere to ozone.

The most common allotrope of elemental phosphorus consists of four-atom (tetratomic) molecules (P_4); it is known as white phosphorus (Figure 2.14). Two other allotropes, red phosphorus (Figure 3.3) and the less common black form, consist of more complex networks of phosphorus atoms.

The common lemon yellow form of sulfur consists of crown-shaped S_8 molecules (Figure 3.4). If this form is heated above 150 °C, however, the sulfur first melts and then the rings break open to give S_8 chains. These chains combine

Figure 3.3 White and red phosphorus are two allotropic forms of the element. White phosphorus consists of P_4 molecules; red phosphorus involves P_4 units bonded to one another in a chain. (C. D. Winters)

(a) (b)

Figure 3.4 Allotropes of sulfur. (a) At room temperature sulfur is a bright yellow solid. At the submicroscopic level it consists of eight-membered, puckered rings of sulfur atoms. (b) When sulfur is heated, the rings break open and eventually form chains of sulfur atoms called "plastic sulfur." (C. D. Winters)

with one another to give even longer chains, and the tangled mass of chains results in a viscous, or syrupy, liquid. When heated to an even higher temperature, the sulfur chains break into shorter lengths to give a liquid that, when poured slowly into water, results in a flexible orange-red thread, a form sometimes called "plastic sulfur."

Finally, pure carbon is found as extended networks of C atoms in two well-known forms: graphite and diamond (Figure 3.5). In graphite the atoms of carbon are arranged in flat sheets of interconnected, hexagonal rings in which each carbon atom is connected to three others in the same layer. These sheets of rings cling only weakly to one another, so one layer can slip easily over another. This explains why graphite is soft, is a good lubricant, and works in your pencil. (Pencil "lead" is not the element lead at all. Instead it is a composite of clay and graphite that leaves a trail of graphite sheets on the page as you write.) In addition, chemists and engineers have been able to form sheets of graphite into fibers and from them make extraordinarily strong composite materials now used to make tennis rackets, fishing rods, and the masts and booms of sail boats (Figure 3.6).

Figure 3.6 Some objects made of graphite fibers. (C. D. Winters)

The carbon atoms in diamonds are also arranged in six-sided rings, but each carbon atom is connected to *four* others, and these surround the central carbon atom at the corners of a tetrahedron (Figure 3.5). Two important consequences of this are that: (1) the atoms are again connected throughout the solid as in a graphite layer, but (2) the carbon atom rings cannot be flat. This structure causes diamonds to be extremely hard and denser than graphite (3.51 g/cm^3 for diamond and 2.22 g/cm^3 for graphite) and chemically even less reactive than graphite. Diamonds are also excellent conductors of heat. This property, combined with their hardness, means that they can be used on the tips of drills and other cutting tools. In fact, the demand for industrial diamonds is so great that over 60 tons of diamonds are made synthetically each year, in addition to those dug from the earth.

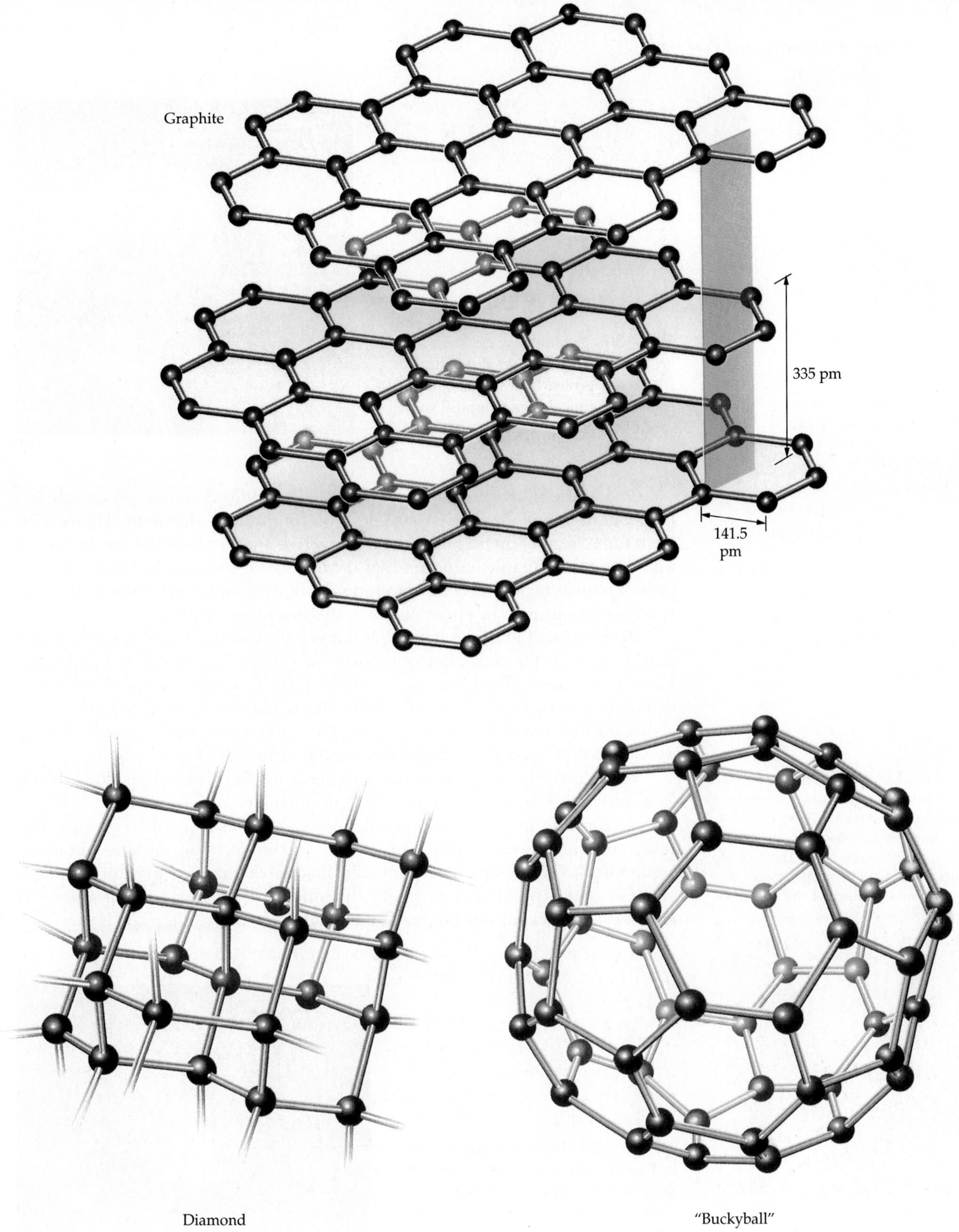

Figure 3.5 Three allotropes of carbon: graphite; diamond; and C_{60}, buckminsterfullerene, or "buckyball."

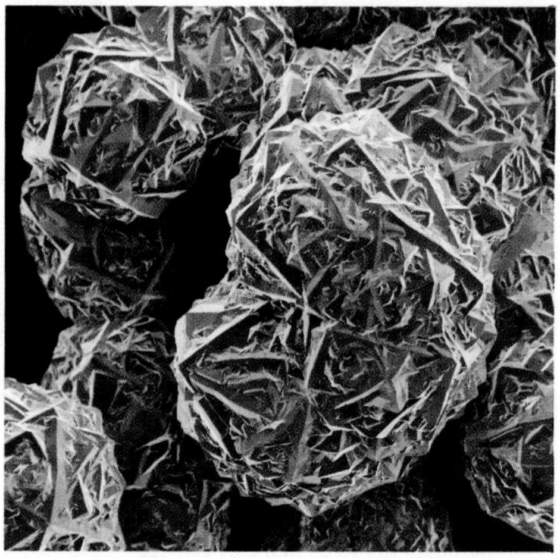

Figure 3.7 A scanning electron micrograph of a diamond thin film, which was grown by a technique called "chemical vapor deposition," or CVD. (Courtesy General Electric)

One of the most exciting developments in chemistry in the last decade has been the discovery that it is possible to grow thin films of diamond (Figure 3.7). Such films are very useful because the surface of another material can be coated with an extraordinarily hard film that conducts heat but not electricity. Indeed, some scientists believe that the development of diamond thin films is potentially the greatest advance in materials since the invention of plastic.

Diamond and graphite have been known for centuries, and it was believed that these were the only allotropic forms of carbon having well-defined structures. In the late 1980s, however, another form of carbon was identified as a component of black soot, the stuff that collects when carbon-containing materials are burned in a deficiency of oxygen. This substance is made up of molecules with 60 carbon atoms in a beautiful, regular structure (Figure 3.5).

The metalloid, boron, also forms spherical cages of boron atoms. This is the normal form of the solid element. These cages are linked to one another, however; they are not individual molecules, as in the case of C_{60} molecules. Cage-like structures are also common in the chemistry of silicon-oxygen compounds.

The structure of a C_{60} molecule is a spherical "cage" of carbon atoms. If you look carefully, you may recognize that it resembles a hollow soccer ball; the surface is made up of five-membered rings linked to six-membered rings (like those seen in the other forms of carbon and in so many carbon-containing compounds). The shape also reminded its discoverers of a structure called a geodesic dome invented years ago by the innovative American philosopher and engineer, R. Buckminster Fuller. The official name of the allotrope is therefore

The geodesic dome at Epcot Center in Florida. Such structures were designed originally by R. Buckminster Fuller, after whom the new class of carbon compounds is named. (Dana Hyde/Photo Researchers)

CURRENT ISSUES IN CHEMISTRY

Buckyballs, AIDS, and Modern Chemistry

A computer-generated model of the HIV-protease, with a buckyball (purple) binding to the active site. (*Science News,* August 7, 1993, p. 87.)

A recent discovery that buckyballs could turn out to be a weapon in the war against acquired immunodeficiency syndrome (AIDS) is interesting news, not only because it is a step in understanding how AIDS works chemically but also because it illustrates how modern chemistry is done.

The virus that causes AIDS, called the human immunodeficiency virus (HIV), reproduces by first growing a long protein chain. This chain is not used directly in reproduction. Rather, the chain is cut precisely into smaller segments that are used to reproduce the virus. The cutting is done by a large protein called an enzyme (named HIV-protease) that is made by the virus itself. The long protein chain is cut into smaller pieces within a hollow pocket or "active site" in HIV-protease.

One strategy for stopping AIDS is to block the reproduction of the HIV virus, and one way to do this is to plug up the active site in HIV-protease. There is a multigroup effort under way at the University of California at San Francisco based on drug design to come up with an inhibitor of the HIV-protease. Simon Friedman, a student working with Professor George L. Kenyon, was chatting late one evening with a friend, and she said "What are they going to try next, buckyballs?" Though this seemed an outlandish idea, Friedman said he "had stared at the structure of the protease on the computer enough" and knew that something the size and shape of a buckyball might work. He said "This led to my modeling of the potential interactions" of a buckyball with the molecules surrounding the cavity of the HIV-protease.

The problem was that C_{60} molecules are not soluble in the aqueous environment of the body. So, the San Francisco chemists turned to a group of chemists at the University of California at Santa Barbara, chemists already skilled in buckyball chemistry. Kenyon and Friedman asked them to chemically alter the C_{60} molecule to make it soluble in water. Professor Fred Wudl and his students at Santa Barbara accomplished that task and sent the compounds to San Francisco. In the meantime, another group, this one at Emory University in Georgia, had a similar idea, and they used Wudl's compounds as well.

Dr. Kenyon said, "We were astonished." The experiment worked the first time they tried it. Adding a special derivative of C_{60} paralyzed the HIV virus and rendered it noninfectious in human cells grown in the laboratory. But all of the people involved are cautious. One of them said that "This is not a drug for AIDS. [Rather], it appears to be the first practical biological application of buckyballs."

Finally, these results are a look into the way chemistry is now done. First, important problems are often quite complicated and cannot be solved by an individual working alone. Cooperation among many groups of people is essential. Second, molecular modeling by computer is increasingly important in gaining insight into molecules, their structures, and their interactions.

buckminsterfullerene, but chemists often call C_{60} molecules "buckyballs." We know now it is only one member of a larger family of carbon cages, generally called "fullerenes." Some fullerenes have fewer than 60 C atoms (called "buckybabies"), and some have an even larger number—C_{70} and the giant fullerenes, such as C_{240}, C_{540}, and C_{960}. The original buckyball, C_{60}, though, is the molecule found in greatest abundance in a pile of soot.

Professor Richard Smalley of Rice University, one of the discoverers of buckyballs, has said that "To a chemist [the discovery of buckyballs is] like Christmas." The fullerenes have extraordinary properties, and dozens of uses have been proposed: microscopic ball bearings, lightweight batteries, new lubricants, new plastics, antitumor therapy for cancer patients (by enclosing a radioactive atom within the cage), and many others. Indeed, one chemist described C_{60} as "a Swiss army knife of a molecule."*

EXERCISE 3.1 ***Allotropes***

Graphite feels slippery if rubbed between your fingers—a property that has led to its use as a lubricant. How is the microscopic structure of graphite related to its macroscopic properties? ■

3.2 MOLECULAR COMPOUNDS

The definition of a chemical formula is that "it is a collection of element symbols with right subscripts giving the relative numbers of atoms of different kinds of elements in the entity in question." P. Block, W. H. Powell, and W. C. Fernelius, *Inorganic Chemistry Nomenclature*, p. 16. Washington, DC, American Chemical Society, 1990.

The formula of sucrose can be written as $C_{12}(H_2O)_{11}$, making it seem that the compound is a "hydrate" of carbon. This is the origin of the name *carbohydrate.*

Compounds are pure substances that can be decomposed into two or more *different* pure substances (Section 1.3). You saw that sucrose, the sugar used in peanut brittle, is a compound with the **molecular formula** $C_{12}H_{22}O_{11}$. When sucrose is mixed with sulfuric acid, the sucrose decomposes in a chemical reaction to black carbon, an element, and colorless water, a compound (Figure 1.9).

$$1 \text{ molecule of } C_{12}H_{22}O_{11} \xrightarrow{\text{gives}} 12 \text{ atoms of C} + 11 \text{ molecules of } H_2O$$

Each molecule, the smallest unit of a molecular compound like sucrose, is changed into 12 *atoms* of carbon and 11 *molecules* of water. To describe this chemical change (or chemical reaction) on paper, the composition of each compound has been represented by its molecular formula, a shorthand way of expressing the number of atoms of each type in one molecule.

A striking feature of compounds is that the characteristics of the constituent elements are lost. Red phosphorus, shown in Figure 3.3, reacts violently with the element bromine, a foul-smelling, red-orange liquid (Figure 3.8) to produce a colorless liquid product. The formula of the product, PBr_3, shows that four atoms occur per molecule: one atom of phosphorus and three atoms of bromine. Similarly, the characteristic properties of elemental hydrogen, carbon, nitrogen, and oxygen are lost when the molecules shown in Table 3.1 are formed.

Ethanol and sucrose are examples of **organic compounds.** Such compounds invariably contain carbon and hydrogen and, like ethanol and sucrose, may also have one or more atoms of oxygen, nitrogen, sulfur, or phosphorus. Organic

*See R. F. Curl and R. E. Smalley, "Fullerenes," *Scientific American,* October, 1991, p. 54; and E. Edelson, "Buckyball, The magic molecule," *Popular Science,* August, 1991, p. 52. The latter article is a good discussion of the history of the discovery of buckyballs.

TABLE 3.1 Some Common Molecules and Their Formulas

Molecule	Formula	Representation
carbon dioxide	CO_2	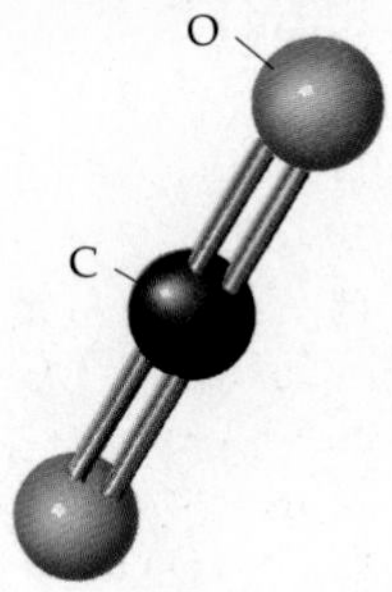
ammonia	NH_3	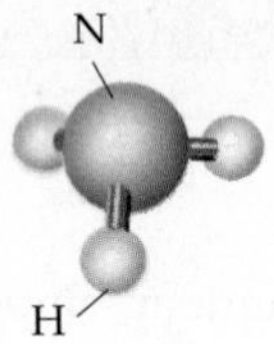
nitric acid	HNO_3	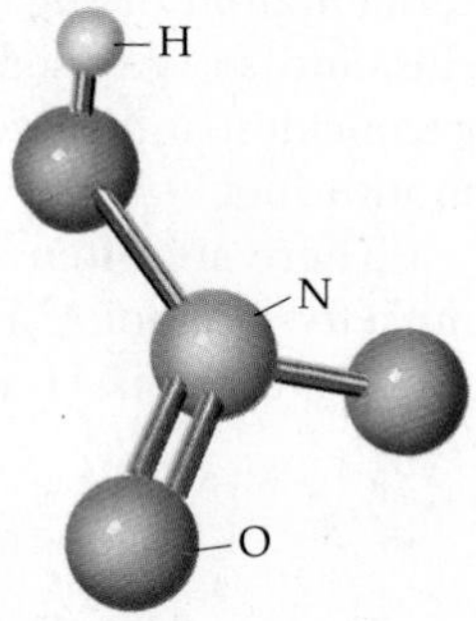
ethanol	C_2H_6O	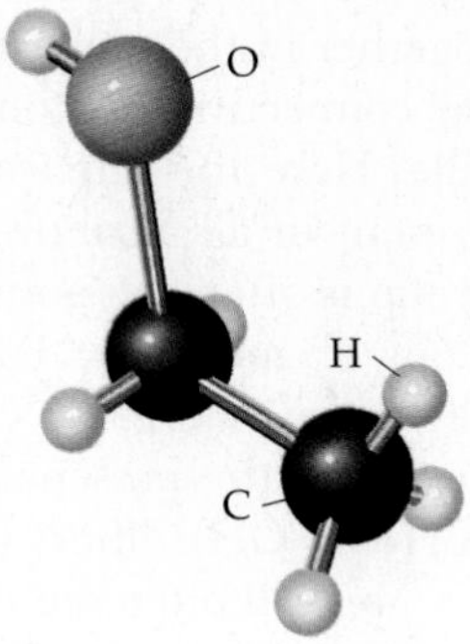

(a)

(b)

Figure 3.8 The reaction of phosphorus with bromine. (a) Red-orange liquid bromine, Br_2, in a graduated cylinder and red phosphorus in an evaporating dish. (b) When bromine is poured over the phosphorus, they react to give phosphorus tribromide, PBr_3, a colorless liquid. (C. D. Winters)

compounds are important because they are the basis of the clothes you wear, the food you eat, and the living organisms in your environment. They are also economically important. For example, over 700 million pounds of ethanol are synthesized every year in the United States, and additional quantities come from the fermentation of sugar, starch, or cellulose. Alcohols can dissolve other chemicals, and so are used in toiletries and pharmaceuticals. They are also used as germicides, antifreeze, and may soon serve more widely in the U.S. as an automotive fuel.

There are often several ways to write the formulas of compounds. The formula for ethanol, C_2H_6O, is a simple molecular formula denoting that there are two C atoms, six H atoms, and one O atom per molecule.

C_2H_6O	CH_3CH_2OH
molecular formula of ethanol	structural formula of ethanol

In the molecular formula of an organic compound the symbols of the elements are frequently written in alphabetical order, each with a subscript indicating the total number of atoms of that type in the molecule. The preceding formula on the right is a modified form showing how the atoms are grouped together in the molecule. Such formulas, called **structural formulas,** emphasize the connectivity of atoms and chemically important groups of atoms in the molecule. Here the important group is an OH attached to a C atom, a grouping present in all alcohols. This is useful information, as this so-called **functional group** is often the point of attack on the molecule where it reacts with another atom or molecule. Functional groups are responsible for the characteristic chemistry of organic compounds. For example, the chemistry of alcohols depends on the presence of an OH group attached to a C atom. Thus, ethanol (CH_3CH_2OH) differs little in its reactions from propanol ($CH_3CH_2CH_2OH$).

You will often see the formulas of molecules written to show more fully how the atoms of the molecule are connected to one another. You can determine the molecular formula of a compound from its structural formula by counting up the atoms.

STRUCTURAL FORMULAS

ethanol

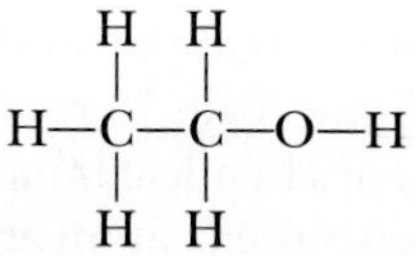

C_2H_6O
CH_3CH_2OH

cinnamaldehyde, the source of the aroma in cinnamon

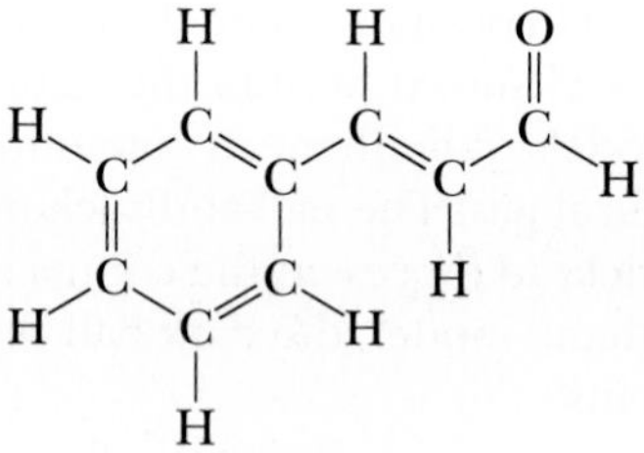

C_9H_8O

cisplatin, a cancer chemotherapy agent

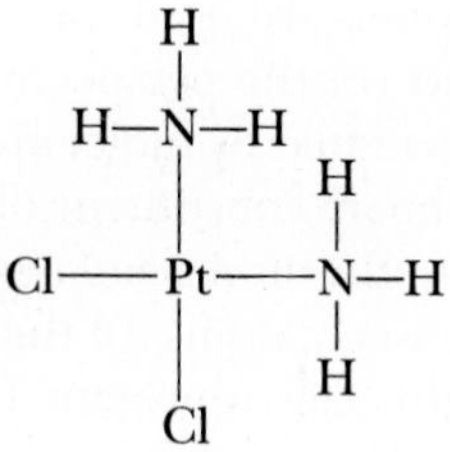

$Pt(NH_3)_2Cl_2$

The lines in these structural formulas represent **chemical bonds,** which are the forces that hold atoms together in molecules.

EXAMPLE 3.1 *Molecular Formulas*

The acrylonitrile molecule is the building block for acrylic plastics (such as Orlon and Acrilan). Its structural formula is shown here. What is the molecular formula for acrylonitrile?

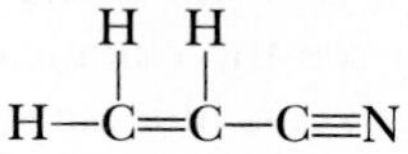

acrylonitrile

Solution Acrylonitrile has three C atoms, three H atoms, and one N atom. Therefore, its molecular formula is C_3H_3N.

EXERCISE 3.2 *Molecular Formulas*

The styrene molecule is the building block for polystyrene, the material familiar in drinking cups, building insulation, and packing material. Its structural formula is shown here. What is the molecular formula for styrene?

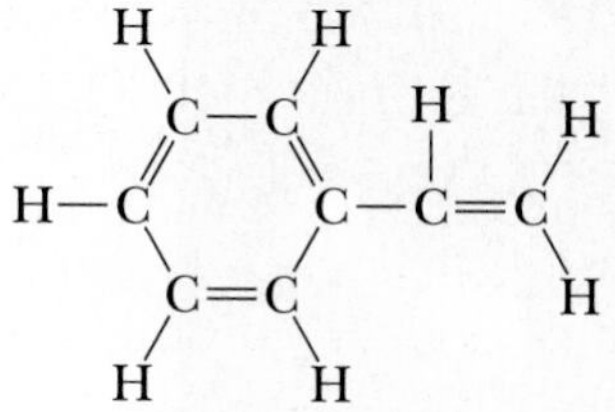

styrene

■

Molecular Models

The previous discussion illustrates the relation between symbols used in chemistry and the submicroscopic world they represent. In Table 3.1 and Figure 3.5, molecules are also represented by models that not only show the connections between atoms but also the positions of the atoms in space. Three types of models are used: (a) the "ball-and-stick" model, (b) the "space-filling" model, and (c) the perspective model shown here in the figure. These three types are illustrated by molecular models of methane, a common hydrocarbon that is the primary constituent of natural gas. The ball-and-stick model uses balls to represent the atoms and short sticks to represent the connections, or chemical bonds, between atoms. In the methane model, the dark ball represents the C atom and light balls represent H atoms.

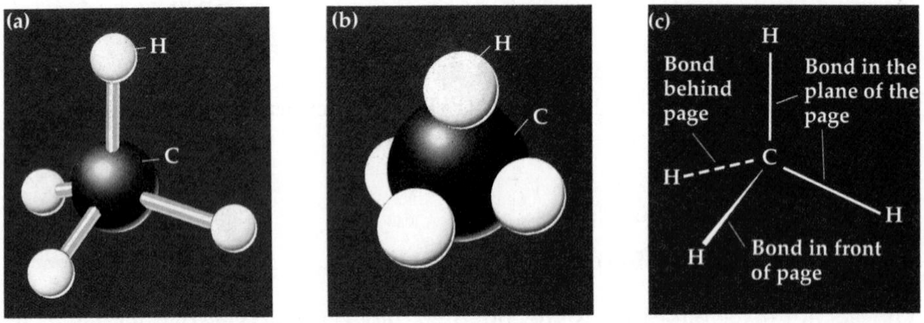

Three representations of the structure of methane. (a) A ball-and-stick model. (b) A space-filling model. (c) A perspective drawing.

The space-filling model is more realistic. It shows not only the spatial orientation of the atoms but also their relative sizes. The model shows better how the atoms fill space, but we do not see the chemical bonds holding the atoms together, and sometimes one atom can hide another from view.

Because it is not always possible or desirable to draw models such as those shown here in (a) and (b), we often use simple perspective drawings (c) to represent molecular structures on paper. Here we use a solid wedge (▬) to represent a bond extending out from the page and a dashed line (---) for a bond behind the page. A solid line represents bonds in the plane of the page.

Computers have made it increasingly simple to construct accurate models that can be displayed on a computer screen. A computer-generated, ball-and-stick model for ethanol is shown here, and you will see a number of such models in this book.

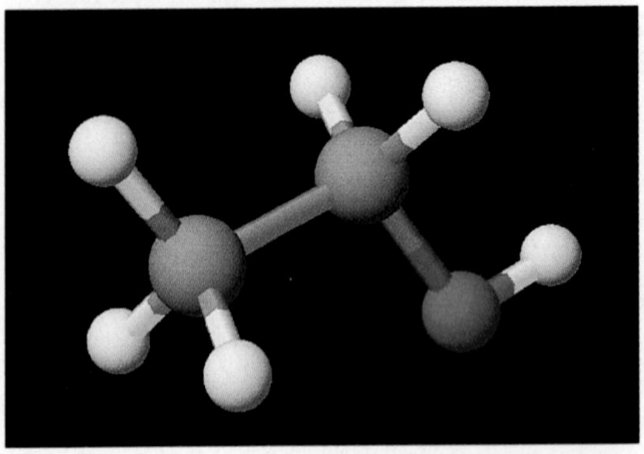

A computer-generated model of ethanol, CH_3CH_2OH. The C atoms are gray, the H atoms are white, and the O atom is red.

The reason we introduced the concept of molecular models at this point is that such models are extremely useful to chemists primarily because the structure of molecules is often crucial to their function (as outlined in the box *Current Issues in Chemistry: Buckyballs, AIDS, and Modern Chemistry*). We therefore often represent molecules with molecular models in this book. The methods you can use to predict the structures of molecules are outlined in detail beginning in Chapter 9.

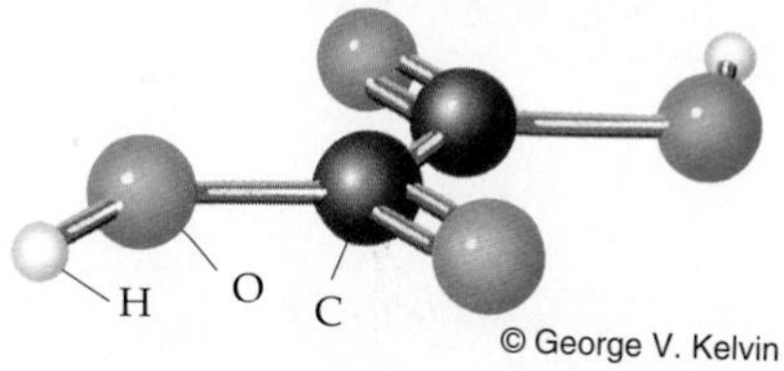

(a)

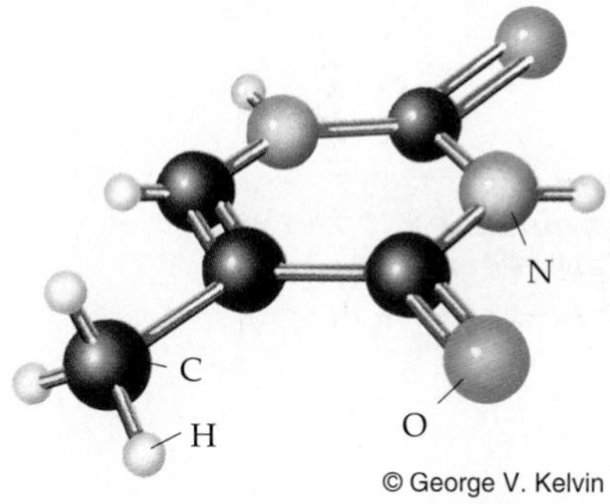

(b)

Two molecular formulas. (a) Oxalic acid, a compound found in many plants. (b) Thymine, one of the bases in DNA.

EXAMPLE 3.2 *Writing Molecular Formulas*

Write the molecular formula of each of the following compounds, and count the number of chemical bonds in each. The color codes for these models are: carbon = gray; hydrogen atoms = white; nitrogen atoms = blue; and oxygen atoms = red.

Solution

a. Oxalic acid has two C atoms, two H atoms, and four O atoms, giving a formula of $C_2H_2O_4$.

b. Thymine has five C atoms, six H atoms, two nitrogen atoms, and two oxygen atoms, giving a formula of $C_5H_6N_2O_2$.

EXERCISE 3.3 *Formulas of Molecules*

Glycine is an important amino acid, a constituent of many living things. Its structural formula is

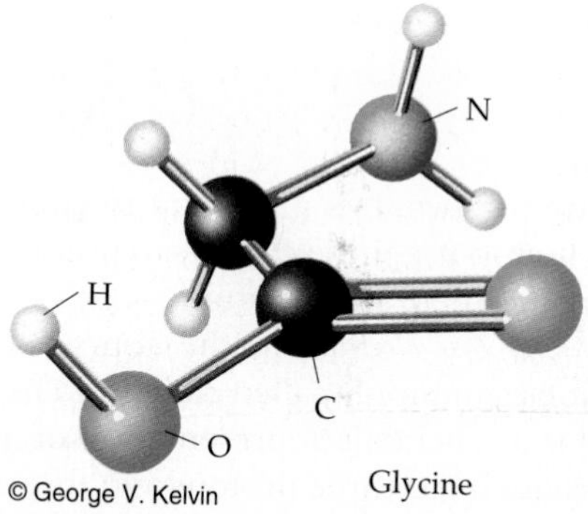

Glycine

What is its molecular formula? See Example 3.2 for the color coding of the model. ■

3.3 IONS

Atoms of almost all the elements can gain or lose electrons in chemical reactions to form **ions,** which are atoms or groups of atoms bearing a net charge. Many of the most important and familiar compounds are composed of ions and are thus known as **ionic compounds;** table salt (NaCl) and alum [$KAl(SO_4)_2 \cdot 12\ H_2O$],

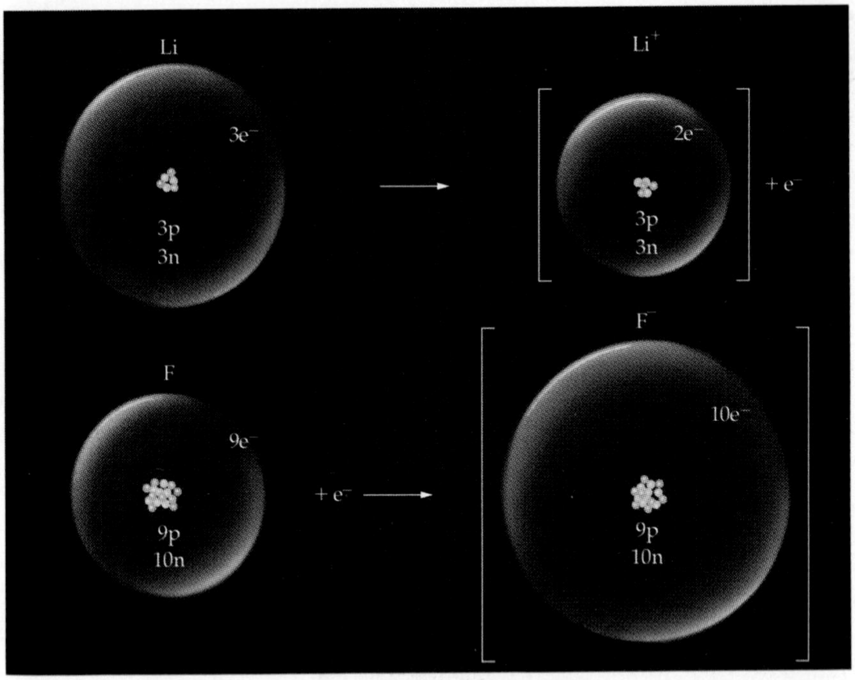

Figure 3.9 A lithium atom is electrically neutral because the number of positive charges (three protons) and negative charges (three electrons) are the same. When it loses one electron, it has one more positive charge than negative charge, so it has a net charge of 1+. We symbolize the resulting lithium cation as Li^+. A fluorine atom is similarly electrically neutral because it has nine protons and nine electrons. Because it is a nonmetal, it gains electrons. As an F^- anion, it has one more electron than it has protons, so it has a net charge of 1−.

which is used in making paper and pickles, are just two. It is important to know the charges on commonly encountered ions, how to use those charges to predict formulas of ionic compounds, and how to name those compounds.

A characteristic of *metals* is that their *atoms lose electrons* in the course of reactions to form ions with a *positive* charge, ions commonly called **cations.** (The name is pronounced "cat′-ion.") Figure 3.9 is a schematic representation of a lithium atom, which is electrically neutral because it has three protons and three electrons, losing one of its electrons to form a lithium ion. The ion now has one more positive charge than negative charge (three protons and two electrons); thus, it has a net charge of 1+. We symbolize the resulting lithium cation as Li^+. When more than one electron has been lost, the number is written with the charge sign; for example, the symbol for a calcium ion is Ca^{2+}.

In contrast with metals, *nonmetals frequently gain electrons to form ions with a negative charge* in the course of their reactions. Such ions are called **anions.** (The name is pronounced "ann′-ion.") Figure 3.9 depicts a fluorine atom with nine protons and nine electrons gaining an electron to form a fluoride ion (F^-), which has nine protons and ten electrons.

Charges on Monatomic Ions

Monatomic ions are single atoms that have lost or gained electrons. Typical charges on such ions are indicated in the periodic table in Figure 3.10. Be sure to notice that *metals of Groups 1A–3A form positive ions with a charge equal to the group number of the metal.*

It is *extremely* important that you know the ions commonly formed by the elements shown in Figure 3.10 as well as the common polyatomic ions given in Table 3.2.

Group	Metal Atom	Electrons Lost	Metal Ion
1A	Na (11 protons, 11 electrons)	1	Na^+ (11 protons, 10 electrons)
2A	Ca (20 protons, 20 electrons)	2	Ca^{2+} (20 protons, 18 electrons)
3A	Al (13 protons, 13 electrons)	3	Al^{3+} (13 protons, 10 electrons)

The transition metals also form cations. Unlike the elements in the main groups, though, no easily predictable pattern of behavior occurs for the transition metal cations. In addition, many of these metals form several different ions. An iron-containing compound, for example, may contain either Fe^{2+} or Fe^{3+} ions. Indeed, 2+ and 3+ ions are typical of many transition metals (Figure 3.10).

Group	Metal Atom	Electrons Lost	Metal Ion
7B	Mn (25 protons, 25 electrons)	2	Mn^{2+} (25 protons, 23 electrons)
8B	Fe (26 protons, 26 electrons)	2	Fe^{2+} (26 protons, 24 electrons)
8B	Fe (26 protons, 26 electrons)	3	Fe^{3+} (26 protons, 23 electrons)

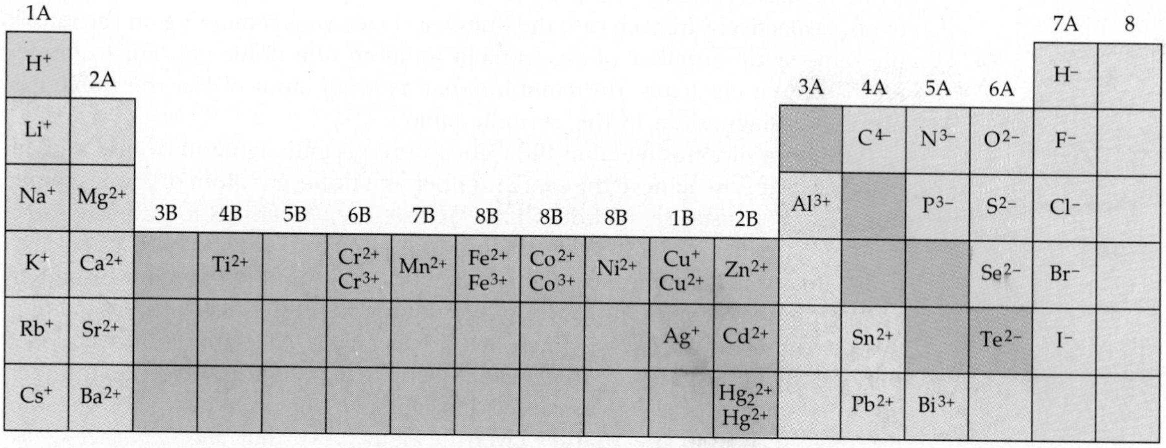

Figure 3.10 Charges on some common monatomic cations and anions. Note that metals usually form cations, in which the positive charge is given by the group number in the case of the main group metals (light blue). For transition metals (darker blue), the positive charge is variable, and other ions in addition to those illustrated are possible. Nonmetals generally form anions with a negative charge equal to 8 minus the group number.

Nonmetals often form ions with a negative charge equal to 8 minus the group number of the element because this represents the number of electrons gained by an atom of the element. For example, N is in Group 5A and so has a charge of 3− because an atom of N can gain three electrons.

Group	Nonmetal Atom	Electrons Gained	Nonmetal Ion
5A	N (7 protons, 7 electrons)	3 (= 8 − 5)	N^{3-} (7 protons, 10 electrons)
6A	S (16 protons, 16 electrons)	2 (= 8 − 6)	S^{2-} (16 protons, 18 electrons)
7A	Br (35 protons, 35 electrons)	1 (= 8 − 7)	Br^{-} (35 protons, 36 electrons)

H both + and −

Notice that hydrogen appears at two locations in Figure 3.10; the H atom can either lose or gain electrons, depending on the other atoms it encounters.

$$\text{H (1 proton, 1 electron)} \longrightarrow \text{H}^+ \text{ (1 proton, 0 electrons)} + e^-$$
$$\text{H (1 proton, 1 electron)} + e^- \longrightarrow \text{H}^- \text{ (1 proton, 2 electrons)}$$

Finally, the noble gases lose or gain electrons in only a few reactions. These elements are extremely stable chemically and have no common ions listed in Figure 3.10.

Valence Electrons

Another important idea is illustrated by the charges on many of the ions listed in Figure 3.10. Consider that the metals of Groups 1A, 2A, and 3A form ions having 1+, 2+, and 3+ charges; that is, their atoms have lost one, two, or three electrons, respectively. In each case the number of electrons remaining on the ions is the same as the number of electrons in an atom of a noble gas. For example, Mg^{2+} has ten electrons, the same number as in an atom of Ne, the noble gas preceding magnesium in the periodic table.

An atom of a nonmetal at the right in the periodic table must lose a great many electrons to achieve the same number as a noble gas atom of lower atomic number. (For instance, Cl would have to lose seven electrons to have the same number of electrons as Ne.) If an atom of a nonmetal were to *gain* a few electrons, however, it would have the same number as a noble gas atom of higher atomic number. For example, an oxygen atom has eight electrons. By gaining two per atom it forms O^{2-}, which has ten electrons, the same number as Ne. Because the noble gases are in periodic Group 8, eight minus the group number gives the number of electrons gained and hence the negative charge on the ion. It seems clear from the charges given in Figure 3.10 that *ions having the same number of electrons as a noble gas atom are especially favored when compounds form.*

While helping his introductory chemistry students learn the typical charges on ions and other facts of chemical combination, the American chemist G. N. Lewis (1875–1946) came up with the idea that electrons in atoms might be arranged in shells, with the nucleus at the center. Each shell could hold a characteristic number of electrons, and only those electrons in the outermost shell were involved when one atom combined with another. These outermost electrons came to be known as **valence electrons.** For example, a magnesium atom

The idea of valence electrons occupying the outermost shell in an atom and determining the numbers of different kinds of atoms that can combine is developed further in Chapter 8.

has two valence electrons (two more electrons than a neon atom), and, when these valence electrons are lost, an Mg^{2+} ion forms. An oxygen atom has six valence electrons (six more than a helium atom) and has room for two more (which would give it the same number as a neon atom); thus an O^{2-} ion is reasonable.

EXAMPLE 3.3 *Predicting Ion Charges*

Predict the charges on ions of aluminum and of sulfur.

Solution Aluminum is a metal in Group 3A of the periodic table, so it is predicted to lose three electrons to give the Al^{3+} cation.

$$Al \longrightarrow Al^{3+} + 3e^-$$

Sulfur is a nonmetal in Group 6A, so it is predicted to gain electrons to give an anion. The number of electrons gained is $8 - 6 = 2$. Therefore,

$$S + 2e^- \longrightarrow S^{2-}$$

EXERCISE 3.4 *Predicting Ion Charges*

Predict possible charges for ions formed from (1) K, (2) Se, (3) Be, (4) V, (5) Co, and (6) Cs. ■

Polyatomic Ions

A **polyatomic ion** contains two or more atoms bonded together, and the *aggregate* bears an electric charge. For example, carbonate ion, CO_3^{2-}, is a common polyatomic anion. It consists of one C atom and three O atoms with two units of negative charge. One of the most common polyatomic cations is NH_4^+, the ammonium ion. In this case, four H atoms surround an N atom, and the group bears a 1+ charge. *It is extremely important to know the names, formulas, and charges of the common polyatomic ions listed in Table 3.2.*

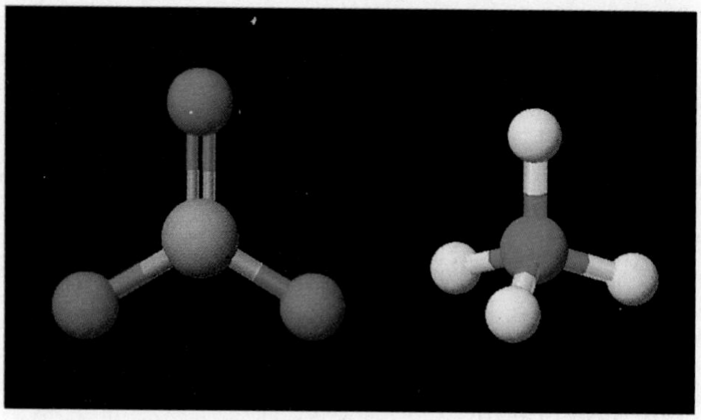

Computer-generated models of the carbonate ion, CO_3^{2-} (*left*) and the ammonium ion, NH_4^+ (*right*). The colors are C = gray, O = red, N = blue, and H = white.

TABLE 3.2 Names and Composition of Some Common Polyatomic Ions

Cation: Positive Ion	
NH_4^+	ammonium ion
Anions: Negative Ions	
Based on a Group 4A element	
CN^-	cyanide ion
$CH_3CO_2^-$	acetate ion
CO_3^{2-}	carbonate ion
HCO_3^-	hydrogen carbonate ion (or bicarbonate ion)
Based on a Group 5A element	
NO_2^-	nitrite ion
NO_3^-	nitrate ion
PO_4^{3-}	phosphate ion
HPO_4^{2-}	hydrogen phosphate ion
$H_2PO_4^-$	dihydrogen phosphate ion
Based on a Group 6A element	
OH^-	hydroxide ion
SO_3^{2-}	sulfite ion
SO_4^{2-}	sulfate ion
HSO_4^-	hydrogen sulfate ion (or bisulfate ion)
Based on a Group 7A element	
ClO^-	hypochlorite ion
ClO_2^-	chlorite ion
ClO_3^-	chlorate ion
ClO_4^-	perchlorate ion
Based on a transition metal	
CrO_4^{2-}	chromate ion
$Cr_2O_7^{2-}$	dichromate ion
MnO_4^-	permanganate ion

3.4 IONIC COMPOUNDS

Ions of opposite charge (Li^+ and F^- or Al^{3+} and O^{2-}, for example) are attracted to one another by *electrostatic forces* and form ionic compounds (see Figures 2.2 and 3.11), a major class of chemical compounds.

Let us first consider how to predict whether a compound will be ionic. As an extension of the guidelines on charges given in Figure 3.10, we can say the following:

1. *Metals* (the elements in *blue* in Figure 3.10) almost always form positive ions and form ionic compounds.
2. *Nonmetals* (the elements shown in *yellow* in Figure 3.10) give monatomic negative ions in ionic compounds *only* when combined with a metal.

A CLOSER LOOK *Ionic compounds and Coulomb's law*

The electrostatic force between oppositely charged ions is described by **Coulomb's law**

$$\text{Force of attraction} = k\frac{(n_+e)(n_-e)}{d^2}$$

where n_+ is the number of charges on the positive ion (e.g., 3 for Al^{3+}), n_- is the number of charges on the negative ion (e.g., 2 for O^{2-}), e is the charge on the electron (1.602×10^{-19} C), d is the distance between the centers of the ions (Figure 3.11), and k is a proportionality constant. The equation shows that *the force of attraction between oppositely charged ions increases as the charges on the ions become larger and as the distance between the ions becomes smaller.*

The strength of electrostatic forces directly influences the properties of compounds formed from ions. Because the force of attraction increases as the charges on the ions increases the attraction between ions with charges of 2+ and 2− is greater than ions with 1+ and 1− charges.

The closer the centers of the ions come to one another, the greater the force of attraction. This means that smaller ions of opposite charge will attract one another more strongly than larger ions.

These ideas will be very useful when we discuss the properties of ionic compounds later in this chapter and the solubility and other properties of solids later in the text.

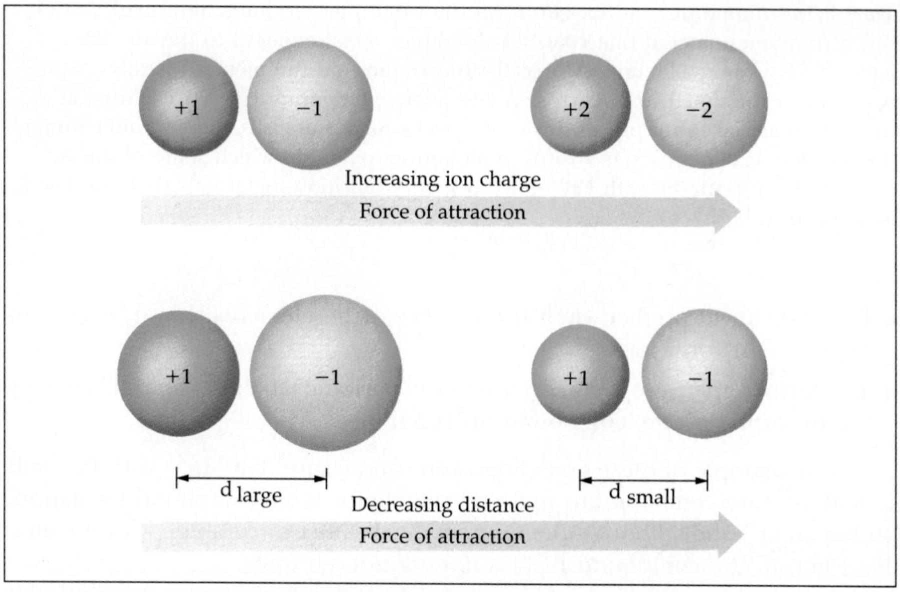

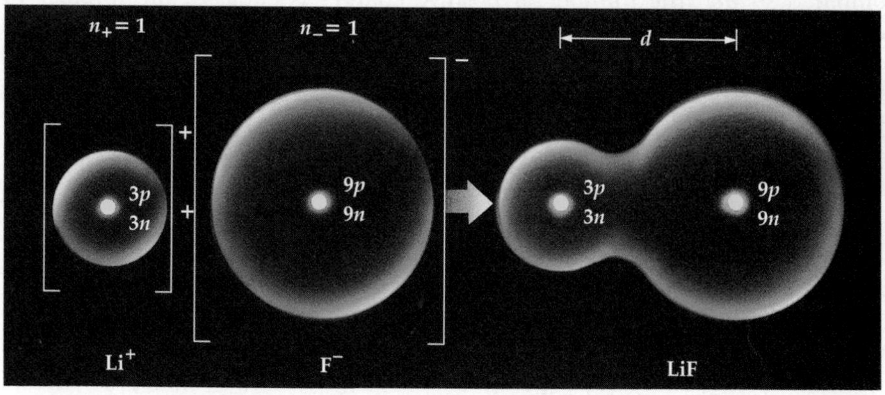

Figure 3.11 The formation of LiF from Li^+ and F^-. Here $n_+ = 1+$ for Li^+ and $n_- = 1-$ for F^-; d is the distance between the centers of the Li^+ and F^- ions.

Figure 3.12 Aluminum and oxygen form the white powder, aluminum oxide, Al_2O_3. This is the same material that coats all aluminum metal exposed to the air. The oxide forms an invisible, microscopically thin coating on the metal. It is this coating that protects the metal from corrosion and allows aluminum to be widely used as a structural material for airplanes, furniture, and cooking utensils, among other things. The red gem called a ruby is a form of aluminum oxide, in which some of the Al^{3+} ions have been replaced with Cr^{3+} ions. It is the transition metal ions that give rise to the color of rubies. (C. D. Winters)

3. It is difficult to predict when the *metalloids* (the elements shown in *green* in Figure 3.10) form ions.
4. The farther apart two elements are in the periodic table, the more likely they are to form an ionic compound on reaction.

As an example of these guidelines, you can be sure that Mg^{2+} with Br^- and K^+ with S^{2-} are combinations that are suitable for ionic compound formation. On the other hand, although the nonmetal chlorine can combine with the metalloid boron, the compound BCl_3 is not considered ionic.

When positive and negative ions combine to form an ionic compound, the outcome is an electrically neutral substance. This means you can predict the formulas of ionic compounds. If the ions and their charges are known, they are combined so that the number of positive charges is equal to the number of negative charges. Let us take the compound formed when aluminum interacts with oxygen as an example (Figure 3.12). Aluminum is a metal in Group 3A, so it loses three electrons to form the Al^{3+} cation. Oxygen is a nonmetal in Group 6A and so gains two electrons to form a 2− anion. To have a compound with the same number of positive and negative charges, two Al^{3+} ions [total charge is $6+ = 2 \times (3+)$] combine with 3 O^{2-} ions [total charge is $6- = 3 \times (2-)$]. This means 2 Al^{3+} + 3 O^{2-} give Al_2O_3.

Calcium ion has a 2+ charge because the metal is a member of Group 2A. It can combine with a number of anions to form ionic compounds such as those shown in the following table:

Ion Combination	Compound	Overall Charge on Compound
$Ca^{2+} + 2\ Cl^-$	$CaCl_2$	$(2+) + 2(1-) = 0$
$Ca^{2+} + CO_3^{2-}$	$CaCO_3$	$(2+) + (2-) = 0$
$3\ Ca^{2+} + 2\ PO_4^{3-}$	$Ca_3(PO_4)_2$	$3 \times (2+) + 2 \times (3-) = 0$

Notice that in writing all these formulas, *the symbol of the cation is always given first, followed by the symbol for the anion.* Also notice the use of parentheses when there is more than one polyatomic ion.

EXAMPLE 3.4 *Ionic Compounds*

For each of the following ionic compounds, write the symbols for the ions present and give the number of each: (1) $MgBr_2$, (2) Li_2CO_3, and (3) $Fe_2(SO_4)_3$.

Solution

1. $MgBr_2$ is composed of one Mg^{2+} ion and two Br^- ions. When a halogen such as bromine is combined only with a metal, you can assume the halogen is an anion with a charge of 1−. Magnesium is a metal in Group 2A and *always* has a charge of 2+ in its compounds.
2. Li_2CO_3 is composed of two lithium ions, Li^+, and one carbonate ion, CO_3^{2-}. To help you remember that carbonate has a 2− charge, you can see that Li is a Group 1A ion and so always has a 1+ charge in its compounds. Because the two 1+ charges neutralize the negative charge of the carbonate ion, the latter must be 2−.
3. $Fe_2(SO_4)_3$ comes from two Fe^{3+} ions and three sulfate ions, SO_4^{2-}. The way to rationalize this is to recall that sulfate is 2−. Because three sulfate ions occur (with a total charge of 6−), the two iron cations must have a total charge of 6+. This is possible only if each iron cation has a charge of 3+.

EXAMPLE 3.5 *Writing Formulas for Compounds Formed from Ions*

Write formulas for ionic compounds composed of an aluminum cation and each of the following anions: (1) fluoride ion, (2) sulfide ion, and (3) nitrate ion.

Solution First, the aluminum cation is predicted to have a charge of 3+ because Al is a metal in Group 3A.

1. Fluorine is a Group 7A element and so is a nonmetal. Its charge is predicted to be 1− (from 8 − 7 = 1). Therefore, we need 3 F^- ions (total charge for the three ions is 3−) to combine with one Al^{3+}. The formula of the compound is AlF_3.
2. Sulfur is a nonmetal in Group 6A and so forms a 2− anion. Thus, we need to combine 2 Al^{3+} ions [total charge is 6+ = 2(3+)] with 3 S^{2-} ions [total charge is 6− = 3(2−)]. The compound has the formula Al_2S_3.

3. The nitrate ion has the formula NO_3^- (Table 3.2). The question is therefore similar to the AlF_3 case, and the compound has the formula $Al(NO_3)_3$. Here we place parentheses around the NO_3 portion of the formula to show that three polyatomic NO_3^- ions are involved.

PROBLEM-SOLVING TIPS AND IDEAS

3.1 Formulas for Ions and Ionic Compounds

Writing formulas for ionic compounds takes practice, and it requires that you know the formulas and charges of the most common ions. The charges on monatomic ions are generally evident from the position of the element in the periodic table. As for polyatomic ions, at this point you simply have to learn their formulas and charges, especially for the most common ions, such as nitrate, sulfate, carbonate, phosphate, and acetate.

If you cannot remember the formula of a polyatomic ion, or you encounter an ion you have not seen before, you may be able to figure out its formula from the formula of one of its compounds. For example, suppose you are told that $NaCHO_2$ is sodium formate. You know that the sodium ion is Na^+, so the formate ion must be the remaining portion of the compound and it must have a charge of 1−, that is, the formate ion must be CHO_2^-. ■

EXERCISE 3.5 ***Ionic Compounds***

1. Give the number and identity of the constituent ions in each of the following ionic compounds: (a) NaF, (b) $Cu(NO_3)_2$, and (c) $NaCH_3CO_2$.
2. Iron is a transition metal and so can form ions with at least two different charges. Write the formulas of the compounds formed between iron and chlorine.
3. Write the formulas of all of the neutral ionic compounds that can be formed by combining the cations Na^+ and Ba^{2+} with the anions S^{2-} and PO_4^{3-}. ■

The Ionic Crystal Lattice

Ionic compounds are generally solids and have their ions arranged in extended three-dimensional networks such as that for sodium chloride (Figure 3.13). This regular array of positive and negative ions is called a **crystal lattice.**

Careful examination of the crystal lattice in Figure 3.13 reveals that each sodium ion is surrounded by six chloride ions, and each chloride ion is surrounded by six sodium ions. No discrete NaCl molecules exist in the sense that H_2 molecules exist in a sample of hydrogen or H_2O molecules exist in a sample of water. Although they do not contain molecules, ionic compounds do have well-defined ratios of one kind of ion to another, and these are represented by their formulas. For example, NaCl represents the simplest ratio of sodium ions to chloride ions in the lattice shown in Figure 3.13, namely 1:1. The combination of one Na^+ ion and one Cl^- ion is referred to as a **formula unit** of sodium chloride.

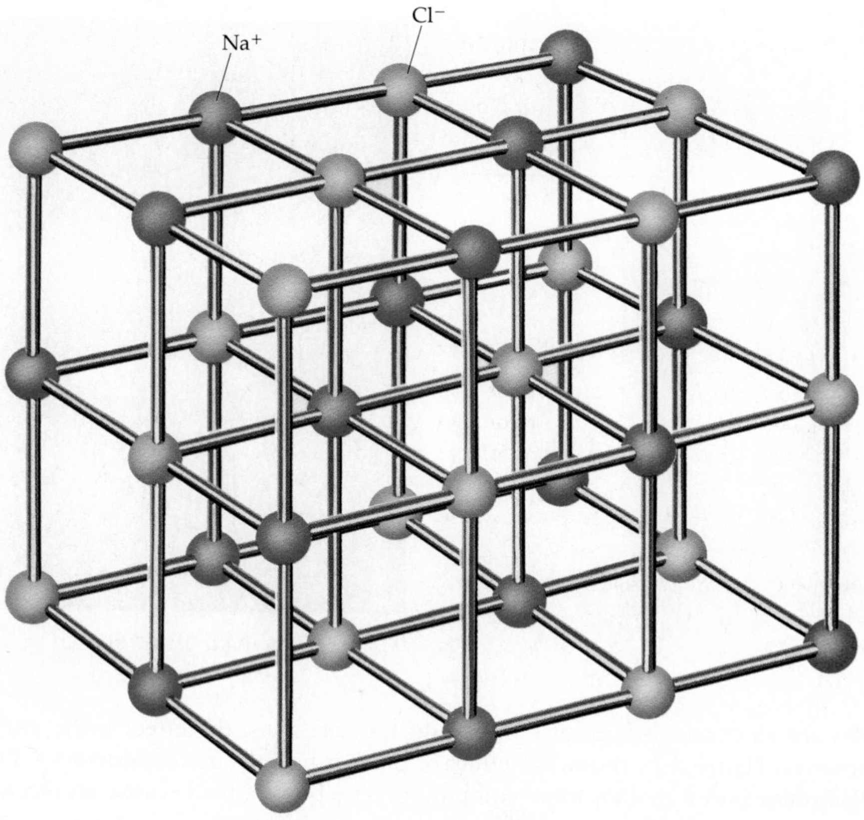

Figure 3.13 A model of a crystal of sodium chloride. The lines between the ions are not chemical bonds but simply reference lines to show the relationship of Na^+ (red) and Cl^- (green) in space.

Properties of Ionic Compounds

Compounds made up of ions have characteristic properties that can be understood in terms of the charges of the ions and their arrangement in a crystal lattice. Because each ion is surrounded by many oppositely charged nearest neighbors, each ion is held fairly tightly in its allotted location. At room temperature each ion can vibrate a little around its average location, but much more energy must be added before an ion can move fast and far enough to escape the confinement of its neighbors. A fairly high temperature is therefore required to melt an ionic compound, and ionic compounds are solids at room temperature. Moreover, the melting point correlates with the charges on the ions. For example, NaCl, which consists of 1+ and 1− ions, melts at 801 °C; Al_2O_3, with 3+ and 2− ions, melts at a much higher temperature, 2072 °C. (See the discussion in A Closer Look: Ionic Compounds and Coulomb's Law.) A high melting point is a useful property. For example, white Al_2O_3 is used in fire bricks, ceramics, and other materials that must withstand high temperatures. In combination with another ionic oxide, zirconium oxide (ZrO_2), aluminum oxide can be used to make very fine fibers that can withstand temperatures up to 1400 °C and can be incorporated into molten metals to strengthen the metal after it cools back to a solid.

In an ionic solid the ions are in fixed positions, but when the solid melts they can move out. Most ionic solids do not conduct electricity, but molten ionic compounds do. This happens because an electric current is a movement of charged particles from one place to another. In a metal wire the charged parti-

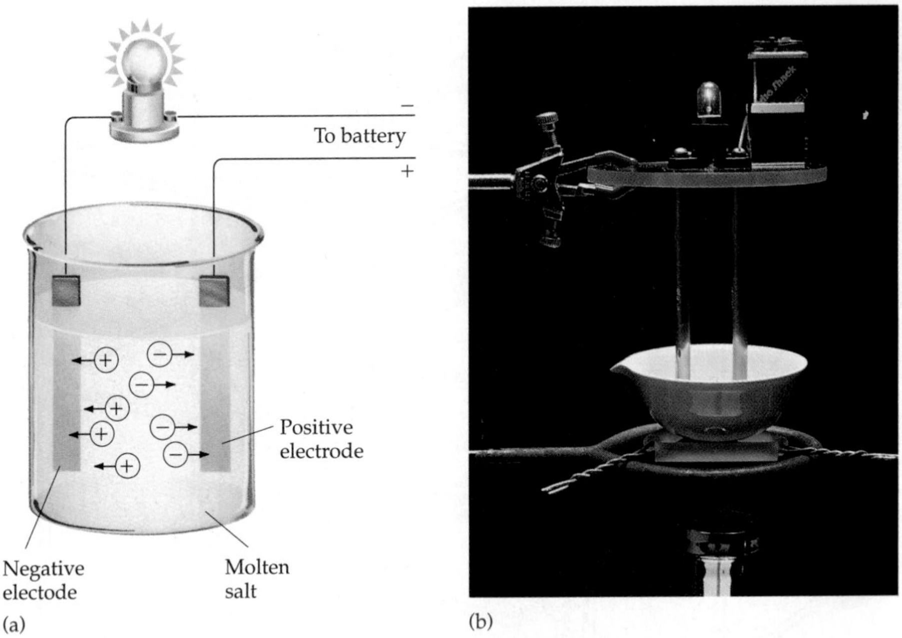

Figure 3.14 The conductivity of a molten salt. When an ionic compound is a solid, it does not conduct electricity. (a) When the salt is melted, however, ions are free to move and migrate to the electrodes dipping into the melt. (b) The lighted bulb shows that the electric circuit is complete. (b, C. D. Winters)

cles are electrons, but in an ionic liquid they are ions; the effect is the same, however. Figure 3.14 shows the effect of placing two electric conductors called **electrodes** into a molten ionic compound. The bulb lights because an electric circuit is completed by movement of ions through the molten substance.

Many ionic solids dissolve in water. The water that you drink every day and the oceans of the world contain small concentrations of many ionic substances that have been dissolved from solid materials in our environment (Table 3.3). Some of these, such as calcium ions, are necessary nutrients. The oceans contain much higher concentrations of substances such as NaCl that have dissolved in the water.

TABLE 3.3 Some Chemical Elements Dissolved in Seawater[1]

Element	Dissolved Species	Concentration[2] (mg/L)	Total Amount in the Oceans (tons)
Chlorine	Cl^-	1.95×10^4	2.57×10^{16}
Sodium	Na^+	1.08×10^4	1.42×10^{16}
Magnesium	Mg^{2+}	1.29×10^3	1.71×10^{15}
Sulfur	SO_4^{2-}	9.05×10^2	1.2×10^{15}
Calcium	Ca^{2+}	4.12×10^2	5.45×10^{14}
Potassium	K^+	3.80×10^2	5.02×10^{14}
Bromine	Br^-	67	8.86×10^{13}
Carbon	HCO_3^-, CO_3^{2-}, CO_2	28	3.7×10^{13}
Nitrogen	N_2, NO_3^-, NH_4^+	11.5	1.5×10^{13}
Strontium	Sr^{2+}	8	1.06×10^{13}

[1]Only the ten most abundant elements are listed. Data taken from *Seawater: Its Composition, Properties, and Behavior,* New York, Pergamon Press, 1992.

[2]Milligrams per liter of seawater is equivalent to parts per million.

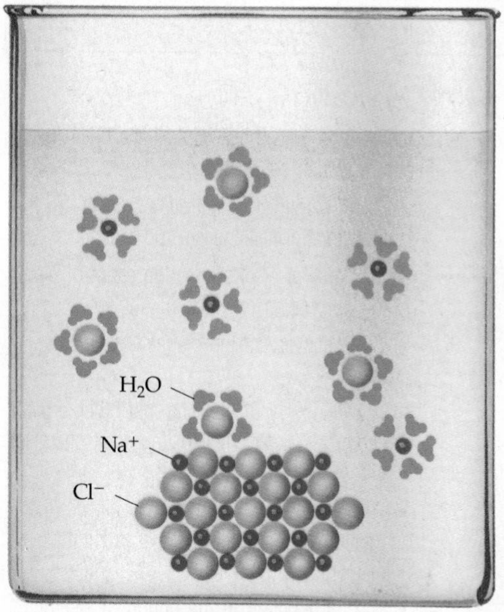

Figure 3.15 A model for the process of dissolving NaCl in water. The crystal dissociates to give Na^+ and Cl^- ions in aqueous (water) solution. These ions, which are cloaked in water molecules, are free to move about. Such solutions conduct electricity, so the substance dissolved is called an electrolyte.

Dissolving an ionic solid involves separating each ion from the oppositely charged ions that surround it in a crystal lattice, and this requires energy. Water is especially good at dissolving ionic compounds, however, because each water molecule has a positively charged end and a negatively charged end. Therefore, a water molecule can attract a positive ion to its negative end or it can attract a negative ion to its positive end. When an ionic compound dissolves, each negative ion becomes surrounded by water molecules with their positive ends pointing toward it, and each positive ion becomes surrounded by the negative ends of many water molecules (Figure 3.15).

The water-encased ions coming from a dissolved ionic compound are free to move about in the solution. Under normal conditions, the movement of ions is random, and the cations and anions from the dissolved ionic compound are dispersed uniformly throughout the solution. If two electrodes are placed in the solution and connected to a battery, however, the cations migrate through the solution to the negative electrode and anions move to the positive electrode. If a light bulb is inserted into the circuit as in Figure 3.14, the bulb lights, showing that the circuit is complete. Compounds whose aqueous solutions conduct electricity are called **electrolytes,** and all ionic compounds that are soluble in water are good electrolytes.

In ionic substances *each ion has its own characteristics, and these are different from the characteristics of the atom from which the ion was derived.* For example, a sodium ion, Na^+, has different properties from a sodium atom, Na, and a chloride ion, Cl^-, is very different from either a chlorine atom, Cl, or a molecule of chlorine, Cl_2. You eat sodium ions and chloride ions every day in table salt. However, you would never want to eat sodium, a metal that reacts vigorously with the saliva in your mouth, or chlorine (Cl_2), a yellow-green gas used as a disinfectant in municipal water supplies. This difference between ions and atoms or molecules is a corollary of the fact that compounds have properties that are very different from the properties of the elements that form them. We explore the chemical reactions of ions, particularly ions in solution, in the next chapter.

PORTRAIT OF A SCIENTIST

Michael Faraday (1791–1867)

Michael Faraday. (Oesper Collection in the History of Chemistry/University of Cincinnati)

The terms anion, cation, electrode, and electrolyte originated with Michael Faraday, one of the most influential men in the history of chemistry. Faraday was apprenticed to a bookbinder in London (England) when he was only 13. This suited him, however, as he enjoyed reading the books sent to the shop for binding. By chance, one of these was a small book on chemistry, and his appetite for science was whetted. He soon began performing experiments on electricity, and in 1812 a patron of the shop invited Faraday to accompany him to a lecture at the Royal Institution by one of the most famous chemists of his day, Sir Humphry Davy. Faraday was so intrigued by Davy's lecture that he wrote to ask Davy for a position as an assistant. Faraday was accepted and began work in 1813. His work was very fruitful, and Faraday was so talented that he was made the director of the laboratory of the Royal Institution about 12 years later.

It has been said that Faraday's contributions are so enormous that, had there been Nobel Prizes when he was alive, he would have received at least six. These could have been for discoveries such as

- Electromagnetic induction, which led to the first transformer and electric motor
- The laws of electrolysis (the effect of electric current on chemicals)
- The discovery of the magnetic properties of matter
- The discovery of benzene and other organic chemicals (which led to important chemical industries)
- The discovery of the "Faraday effect" (the rotation of the plane of polarized light by a magnetic field)
- The introduction of the notion of electric and magnetic fields

In addition to making discoveries that had profound effects on science, Faraday was an educator. He wrote and spoke about his work in memorable ways, especially in lectures to the general public that helped to popularize science.

EXERCISE 3.6 ***Chemical Formulas***

Explain some differences among the species Br_2, K, and KBr. Are these species elements or compounds? Are they ionic? ■

3.5 NAMES OF COMPOUNDS

Chemists try to use precise language to describe the world. Assigning clear, unambiguous names to compounds has always been a problem, however, and it is a problem that continues today as new, ever more complicated molecules are discovered. Nonetheless, rules do exist for naming the kinds of compounds you will read about in this text.

Naming Ionic Compounds

The name of an ionic compound is built from the names of the positive and negative ions in the compound.

Naming Positive Ions

With a few exceptions, the positive ions described in this text are metal ions. Positive ions are named by the following rules:

1. For a monatomic positive ion, that is, a metal cation, the name is that of the metal plus the word "ion." For example, we have already referred to Al^{3+} as the aluminum ion.
2. Some cases occur, especially in the transition series, in which a metal can form more than one type of positive ion. The most common practice is to indicate the charge of the ion by a Roman numeral in parentheses immediately following the ion's name (the Stock system). For example, Co^{2+} is the cobalt(II) ion, and Co^{3+} is the cobalt(III) ion.

Finally, you will encounter the nonmetal cation NH_4^+ or *ammonium* ion many times in this book, in the laboratory, and in your environment. Do not confuse the ammonium ion with the neutral ammonia molecule, NH_3.

Another cation that you will see on occasion is Hg_2^{2+}, the name of which is the mercury(I) ion. The reason for the Roman numeral (I) is that the ion is composed of two Hg^+ ions bonded together.

The Stock system was devised by Alfred Stock (1876–1946) a German chemist famous for his work on the hydrogen compounds of boron and silicon. Because he worked closely with mercury metal for many years, he contracted mercury poisoning, which leads to memory loss and severe physical disabilities.

Naming Negative Ions

Two types of negative ions must be considered: those having only one atom *(monatomic)* and those having several atoms *(polyatomic)*.

1. A *monatomic negative ion* is named by adding *-ide* to the stem of the name of the nonmetal element from which the ion is derived (see Figure 3.16). As a group, the anions of the Group 7A elements, the halogens, are called the *halide ions.*

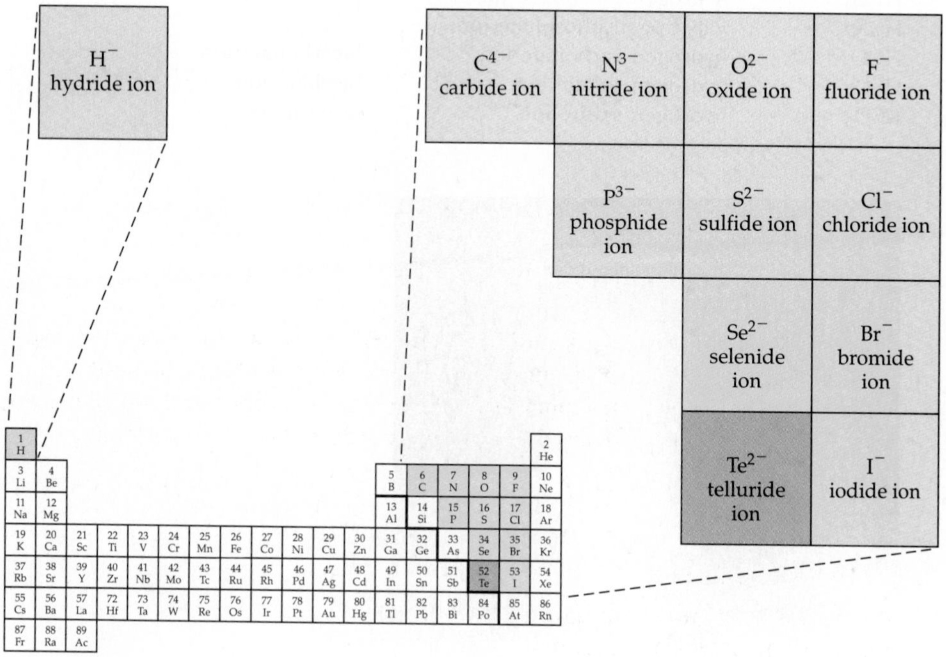

Figure 3.16 Names and charges of some monatomic ions of the nonmetals.

2. *Polyatomic negative ions* are quite common, especially those containing oxygen (called *oxoanions*). The names of some of the most common oxoanions are given in Table 3.2. Most of these names must simply be learned; however, some guidelines can help. For example, consider the pairs of ions below.

   NO_3^- is the nitr*ate* ion, whereas NO_2^- is the nitr*ite* ion
   SO_4^{2-} is the sulf*ate* ion, whereas SO_3^{2-} is the sulf*ite* ion

The oxoanion with the *greater number of oxygen atoms* is given the suffix *-ate*, and the oxoanion with the *smaller number of oxygen atoms* has the suffix *-ite*. For a series of oxoanions with more than two members, the ion with the largest number of oxygen atoms has the prefix *per-* and the suffix *-ate*. The ion with the smallest number of oxygen atoms has the prefix *hypo-* and the suffix *-ite*. The oxoanions containing chlorine are good examples.

ClO_4^-	*per*chlor*ate* ion
ClO_3^-	chlor*ate* ion
ClO_2^-	chlor*ite* ion
ClO^-	*hypo*chlor*ite* ion

Oxoanions that contain hydrogen are named by adding the word "hydrogen" before the name of the oxoanion. If two hydrogens are in the compound, we say *di*hydrogen. Many of these hydrogen-containing oxoanions have common names that are so often used that you should know them, too. For example, the hydrogen carbonate ion, HCO_3^-, is often named as the bicarbonate ion.

Ion	Systematic Name	Common Name
HPO_4^{2-}	hydrogen phosphate ion	
$H_2PO_4^-$	*di*hydrogen phosphate ion	
HCO_3^-	hydrogen carbonate ion	bicarbonate ion
HSO_4^-	hydrogen sulfate ion	bisulfate ion
HSO_3^-	hydrogen sulfite ion	bisulfite ion

Figure 3.17 Some common ionic compounds. Clockwise from the top are a box of salt (sodium chloride, NaCl), a clear crystal of calcite (calcium carbonate, $CaCO_3$), a pile of red $CoCl_2 \cdot 6\,H_2O$ [cobalt(II) chloride hexahydrate], and an octahedral crystal of fluorite (calcium fluoride, CaF_2). (C. D. Winters)

Naming Ionic Compounds

When naming ionic compounds, *the positive ion name is given first followed by the name of the negative ion.* Some examples are given in the table, and others are shown in Figure 3.17.

Ionic Compound	Ions Involved	Name
$CaBr_2$	Ca^{2+} and 2 Br^-	calcium bromide
$NaHSO_4$	Na^+ and HSO_4^-	sodium hydrogen sulfate
$(NH_4)_2CO_3$	2 NH_4^+ and CO_3^{2-}	ammonium carbonate
$Mg(OH)_2$	Mg^{2+} and 2 OH^-	magnesium hydroxide
$TiCl_2$	Ti^{2+} and 2 Cl^-	titanium(II) chloride
Co_2O_3	2 Co^{3+} and 3 O^{2-}	cobalt(III) oxide

EXERCISE 3.7 *Names and Formulas of Ionic Compounds*

1. Give the formula for each of the following ionic compounds. Use Table 3.2 and Figure 3.16.
   (a) Ammonium nitrate (d) Vanadium(III) oxide
   (b) Cobalt(II) sulfate (e) Barium acetate
   (c) Nickel(II) cyanide (f) Calcium hypochlorite
2. Name the following ionic compounds:
   (a) $MgBr_2$ (d) $KMnO_4$
   (b) Li_2CO_3 (e) $(NH_4)_2S$
   (c) $KHSO_3$ (f) CuCl and $CuCl_2$ ■

Naming Binary Compounds of the Nonmetals

Thus far we have described naming ions and ionic compounds. Another kind of compound comes from the combination of two nonmetals and is composed of molecules. These "two-element" or **binary compounds** of nonmetals can also be named in a systematic way.

Ionic structures are rare in binary, nonmetal compounds. The bonding in these compounds is generally said to be "covalent" and is described in Chapter 9.

Hydrogen forms binary compounds with all the nonmetals (except the noble gases). For compounds of oxygen, sulfur, and the halogens, the H atom is generally written first in the formula and is named first. The other nonmetal is named as if it were a negative ion.

Compound	Name
HF	hydrogen fluoride
HCl	hydrogen chloride
H_2S	dihydrogen sulfide

Hydrocarbons are binary compounds composed only of H and the nonmetal carbon. There are hundreds of such compounds, and one important class of hydrocarbons are the **alkanes,** all of which have the general formula C_xH_{2x+2} (Table 3.4). Notice that all are traditionally written with the C atom first followed by the H atom, and all have *-ane* as the second part of their name. When $x = 1$ to 4 the first part of the name is of historical origin; these are common names that just must be memorized (Figure 3.18).

Selected Hydrocarbons of the Alkane Family, C_xH_{2x+2}[1]

Name	Molecular Formula		State at Room Temperature
Methane	CH_4		
Ethane	C_2H_6		Gas
Propane	C_3H_8		
Butane	C_4H_{10}		
Pentane	C_5H_{12}	(pent- = 5)	
Hexane	C_6H_{14}	(hex- = 6)	
Heptane	C_7H_{16}	(hept- = 7)	Liquid
Octane	C_8H_{18}	(oct- = 8)	
Nonane	C_9H_{20}	(non- = 9)	
Decane	$C_{10}H_{22}$	(dec- = 10)	
Octadecane	$C_{18}H_{38}$	(octadec- = 18)	Solid
Eicosane	$C_{20}H_{42}$	(eicos- = 20)	

[1]This table lists only selected alkanes. Liquid compounds with 11 to 16 C atoms are also known, and other solid alkanes include $C_{17}H_{36}$ and $C_{19}H_{40}$.

When $x = 5$ or greater, the first part of the name tells how many carbon atoms are present. For example, the compound with six carbons is called *hex*ane.

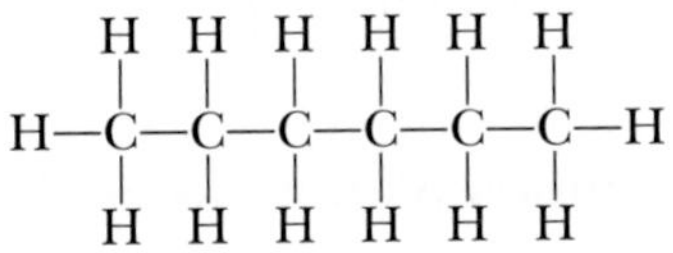

Some of the alkanes are economically important. Methane, the simplest alkane, is the major component of natural gas. It is also thought to be an important contributor to the "greenhouse effect." Propane, which can be a liquid

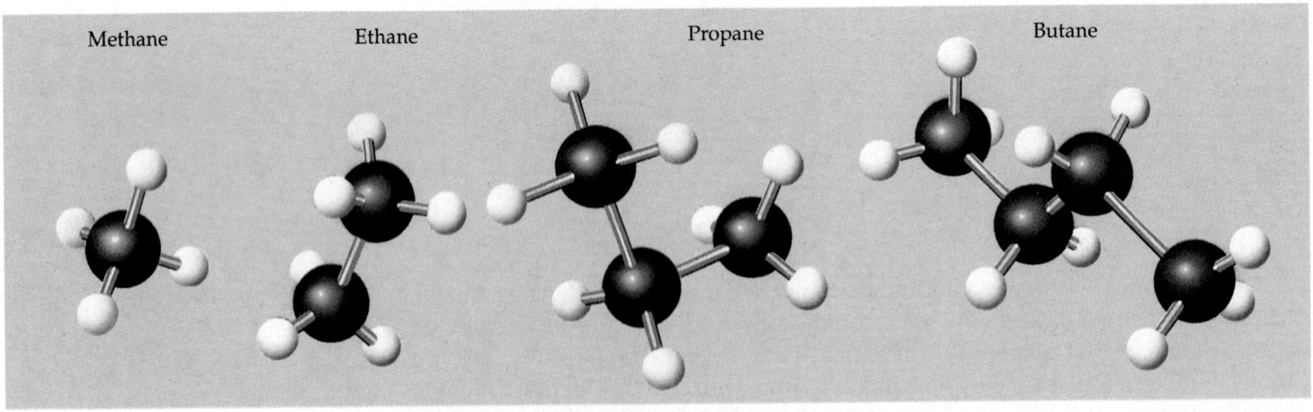

Figure 3.18 The structures of the four simplest hydrocarbons. Note that their names are of historical origin.

when it is under pressure, is the fuel used in backyard barbecues. You might also have used a camping stove fueled by butane, which is a liquid when under pressure.

Virtually all binary, nonmetal compounds are based on a combination of elements from Groups 4A–7A with one another or with hydrogen. The formula is generally written by putting the elements in order of increasing group number. When naming the compound, the number of atoms of a given type in the compound is designated with a prefix, such as "di-, tri-, tetra-, penta-," and so on.

Some camping stoves use the hydrocarbon butane as fuel. (C. D. Winters)

Compound	Systematic Name
NF_3	nitrogen trifluoride
NO	nitrogen monoxide
NO_2	nitrogen dioxide
N_2O	dinitrogen monoxide
N_2O_4	dinitrogen tetraoxide
PCl_3	phosphorus trichloride
PCl_5	phosphorus pentachloride
SF_6	sulfur hexafluoride
S_2F_{10}	disulfur decafluoride

Finally, many of the binary compounds of nonmetals were discovered years ago and have names so common they continue to be used. These names must simply be learned.

Compound	Common Name
H_2O	water
H_2S	hydrogen sulfide
NH_3	ammonia
N_2H_4	hydrazine
PH_3	phosphine
NO	nitric oxide
N_2O	nitrous oxide ("laughing gas")

Take note of the way formulas of binary nonmetal compounds are written. Alkane formulas are written with H following C, and the formulas of ammonia and hydrazine have H following N. Water and the hydrogen halides, however, have the H atom preceding O or the halogen atom. Tradition is the only explanation for such oddities in chemistry.

EXERCISE 3.8 ***Alkanes***

1. Use the general formula for alkanes to write the molecular formula for alkanes with 12 and 24 carbon atoms.
2. How many carbon and hydrogen atoms are in hexadecane? ■

EXERCISE 3.9 ***Naming Compounds***

1. Give the formula for each of the following binary, nonmetal compounds:
   (a) Carbon dioxide
   (b) Phosphorus triiodide
   (c) Sulfur dichloride
   (d) Xenon trioxide
   (e) Boron trifluoride
   (f) Dioxygen difluoride
   (g) Nonane
2. Name the following binary, nonmetal compounds:
   (a) N_2F_4 (b) HBr (c) SF_4 (d) ClF_3
   (e) BCl_3 (f) P_4O_{10} (g) C_7H_{16} ■

3.6 MOLECULES, COMPOUNDS, AND THE MOLE

The formula of a compound tells you the type of atoms or ions in the compound and the relative number of each. For example, in one molecule of phosphorus tribromide, PBr_3, one atom of P combines with three atoms of Br. But suppose you have Avogadro's number of P atoms (6.022×10^{23}) combined with the proper number of Br atoms. It is clear that three times as many Br atoms are required (18.07×10^{23} Br atoms) to give Avogadro's number of PBr_3 molecules. What masses of atoms are combined, and what is the mass of this many PBr_3 molecules?

P	combined with Br	in the compound PBr_3
6.022×10^{23} P atoms	$3 \times 6.022 \times 10^{23}$ Br atoms	6.022×10^{23} PBr_3 molecules
= 1.000 mol of P	= 3.000 mol of Br atoms	= 1.000 mol of PBr_3 molecules
= 30.97 g of P atoms	= 239.7 g of Br atoms	= 270.7 g of PBr_3 molecules

You may wish to review the mole concept (Section 2.7) at this point.

Because we know the number of moles of P and Br atoms, we know the masses of phosphorus and bromine that combine to form PBr_3. It follows from the law of the conservation of mass that the mass of PBr_3 is the sum of these masses (Section 2.1). That is, one mole of PBr_3 has a mass equivalent to the mass of one mole of P atoms plus 3 moles of Br atoms or 270.7 g. This is the **molar mass, *M*,** of PBr_3. Chemists often use the term "molecular weight" for molar mass; they mean the same thing.

Just as chemists often use atomic weight when they mean atomic mass, they often refer to the molecular weight of a compound. The latter term is very commonly used, but we shall generally use "molar mass" in this book.

Molecular Masses and Molar Masses

Element or Compound	Molar Mass (g/mol)	Average Mass of One Molecule[1] (grams)
O_2	32.00	5.314×10^{-23}
P_4	123.9	2.057×10^{-22}
NH_3	17.03	2.828×10^{-23}
H_2O	18.02	2.992×10^{-23}
CH_2Cl_2	84.93	1.410×10^{-22}

[1]See text for the calculation of the mass of one molecule.

Ionic compounds such as NaCl do not exist as individual molecules. Thus, no *molecular* formula can be given; rather, one can only write the simplest formula that shows the *relative* number of each kind of atom in a sample. Nonetheless, we talk about the molar mass of such compounds and calculate it from the simplest formula. To differentiate substances like NaCl that do not contain molecules, however, chemists sometimes refer to their *formula weight* instead of their molar mass or molecular weight.

Figure 3.19 is a photograph of one-mole quantities of several common compounds. To find the molar mass of any compound you need only add up the atomic masses for each element in one formula unit. As an example, let us find the molar mass of ammonium sulfate, $(NH_4)_2SO_4$, a compound used as a fertilizer, in flameproofing fabrics, in tanning leather, and in many other applica-

Figure 3.19 One-mole quantities of some ionic compounds. Clockwise from front right, they are: NaCl, white ($M = 58.44$ g/mol); $CuSO_4 \cdot 5\ H_2O$, blue ($M = 249.7$ g/mol); $NiCl_2 \cdot 6\ H_2O$, green ($M = 237.7$ g/mol); $K_2Cr_2O_7$, orange ($M = 294.2$ g/mol); and $CoCl_2 \cdot 6\ H_2O$, red ($M = 237.9$ g/mol). (C. D. Winters)

tions. It contains two ammonium ions, NH_4^+, and one sulfate ion, SO_4^{2-}. In one formula unit there are therefore two nitrogen atoms, eight hydrogen atoms, one sulfur atom, and four oxygen atoms, adding up to 132.15 g per mole of ammonium sulfate.

$$2 \text{ mol of N per mole of } (NH_4)_2SO_4 = 2 \text{ mol N} \cdot \frac{14.01 \text{ g N}}{1 \text{ mol N}} = 28.02 \text{ g N}$$

$$8 \text{ mol of H per mole of } (NH_4)_2SO_4 = 8 \text{ mol H} \cdot \frac{1.008 \text{ g H}}{1 \text{ mol H}} = 8.064 \text{ g H}$$

$$1 \text{ mol of S per mole of } (NH_4)_2SO_4 = 1 \text{ mol S} \cdot \frac{32.07 \text{ g S}}{1 \text{ mol S}} = 32.07 \text{ g S}$$

$$4 \text{ mol of O per mole of } (NH_4)_2SO_4 = 4 \text{ mol O} \cdot \frac{16.00 \text{ g O}}{1 \text{ mol O}} = 64.00 \text{ g O}$$

$$\text{Molar mass of } (NH_4)_2SO_4 = 132.15 \text{ g}$$

As was the case with elements, it is also very important to be able to convert the mass of the compound to the moles of the compound, or moles to mass. For example, if you have 454 g (one pound) of ammonium sulfate, how many moles of the compound do you have? The molar mass just calculated was 132.15 g, much less than one pound. In fact, it can be estimated before doing any detailed calculations that 454 g of ammonium sulfate should be between three or four moles. Now let us see what the numbers come out to be.

$$454 \cancel{\text{g}}\ (NH_4)_2SO_4 \cdot \frac{1 \text{ mol } (NH_4)_2SO_4}{132.2 \cancel{\text{g}}\ (NH_4)_2SO_4} = 3.43 \text{ mol } (NH_4)_2SO_4$$

The actual result is indeed between three and four moles.

The molar mass is the mass in grams of Avogadro's number of molecules or of formula units of an ionic compound (Figure 3.19). With this knowledge, it is possible to determine the number of molecules in any sample from its mass or even to determine the average mass of one molecule. For example, the average mass of one buckyball, C_{60} (Figure 3.5), is

When doing calculations we follow the convention of using one more significant figure in the molar mass than in any of the other data in a calculation (see page 92).

$$\frac{720.7 \text{ g } C_{60}}{1 \cancel{\text{mol}}\ C_{60}} \cdot \frac{1 \cancel{\text{mol}}\ C_{60}}{6.022 \times 10^{23} \text{ molecules}} = 1.197 \times 10^{-21} \text{ g/molecule}$$

The average masses of individual molecules in the table of molecular weights on page 132 were calculated in this way.

EXAMPLE 3.6 *Molar Mass and Moles*

You have 23.2 g of ethanol, C_2H_6O.

1. How many moles are represented by this mass of alcohol?
2. How many molecules of ethanol are in 23.2 g?
3. How many atoms of carbon are in 23.2 g of ethanol?
4. What is the average mass of one molecule of ethanol?

Solution The first step in any problem involving the conversion of mass and moles is to find the molar mass of the compound in question. Then we can find the number of molecules of ethanol and from that the atoms of carbon contained within that many molecules.

$$\text{Mass, g} \xrightarrow[\text{use molar mass}]{\times \frac{\text{mol}}{\text{g}}} \text{Moles} \xrightarrow[\text{use Avogadro's number}]{\times \frac{\text{molecules}}{\text{mol}}} \text{Molecules} \xrightarrow{\times \frac{\text{C atoms}}{\text{molecule}}} \text{Number of C atoms}$$

1. *Molar mass:*

$$2 \text{ mol of C per mole of alcohol} = 2 \text{ mol C} \cdot \frac{12.01 \text{ g C}}{1 \text{ mol C}} = 24.02 \text{ g C}$$

$$6 \text{ mol of H per mole of alcohol} = 6 \text{ mol H} \cdot \frac{1.008 \text{ g H}}{1 \text{ mol H}} = 6.048 \text{ g H}$$

$$1 \text{ mol O per mole of alcohol} = 1 \text{ mol O} \cdot \frac{16.00 \text{ g O}}{1 \text{ mol O}} = 16.00 \text{ g O}$$

$$\text{Molar mass of } C_2H_6O = 46.07 \text{ g/mol}$$

2. *Number of moles:* The molar mass expressed in units of grams per mole is the conversion factor in all mass-to-mole conversions.

$$23.2 \text{ g } C_2H_6O \cdot \frac{1 \text{ mol } C_2H_6O}{46.07 \text{ g } C_2H_6O} = 0.504 \text{ mol of } C_2H_6O$$

3. *Number of carbon atoms:* To find the number of C atoms in the alcohol sample, we first find the number of alcohol molecules. Here we convert 0.504 mol of C_2H_6O to the number of molecules of alcohol in that quantity of compound by using Avogadro's number.

$$0.504 \text{ mol } C_2H_6O \cdot \frac{6.022 \times 10^{23} \text{ molecules } C_2H_6O}{1 \text{ mol } C_2H_6O} = 3.04 \times 10^{23} \text{ molecules of } C_2H$$

You know that each molecule contains 2 carbon atoms, so you can find the number of carbon atoms in 23.2 g of the alcohol.

$$3.03 \times 10^{23} \text{ molecules } C_2H_6O \cdot \frac{2 \text{ C atoms}}{1 \text{ molecule } C_2H_6O} = 6.07 \times 10^{23} \text{ atoms of C}$$

4. *Mass of one molecule:* The units of the desired answer are grams per molecule, which indicates that you should multiply the starting unit of molar mass (grams per mole) by 1/Avogadro's number (mol/molecule), so that the unit "mol" cancels.

$$\frac{46.07 \text{ g } C_2H_6O}{1 \text{ mol}} \cdot \frac{1 \text{ mol}}{6.022 \times 10^{23} \text{ molecules}} = \frac{7.650 \times 10^{-23} \text{ g}}{1 \ C_2H_6O \text{ molecule}}$$

EXERCISE 3.10 ***Molar Mass and Moles to Mass Conversions***

1. Calculate the molar mass of (a) limestone, $CaCO_3$, and (b) caffeine, $C_8H_{10}N_4O_2$.
2. If you have 454 g of $CaCO_3$, how many moles does this represent?
3. To have 2.50×10^{-3} mol of caffeine, how many grams must you have? ■

3.7 DESCRIBING COMPOUND FORMULAS

It is often possible to predict the formula of a simple ionic compound, but thousands of compounds are of greater complexity. Given a sample of a compound, how can its formula be determined? The answer lies in chemical analysis, a major branch of chemistry that deals with the determination of molecular formulas and structures, among other things.

Percent Composition

According to the law of constant composition *any sample of a pure compound always consists of the same elements combined in the same proportions by mass.* Thus, it seems that there are at least two ways of expressing molecular composition: (a) in terms of the number of atoms of each type per molecule or (b) in terms of the mass of each element per mole of compound. Actually, there is at least one more way of expressing molecular composition, a method derived from (b). Composition can be given by the mass of each element in the compound relative to the total mass of the compound—that is, in terms of the mass percent of each element, or **percent composition** by mass. For ammonia, this is

Ways of Expressing Molecular Composition:

1. A formula giving the number of atoms of each element per molecule
2. Mass of each element per mole of compound
3. Mass of each element per 100 g of compound (percent composition)

$$\text{Mass percent N in } NH_3 = \frac{\text{mass of N in 1 mol of } NH_3}{\text{mass of 1 mol of } NH_3} = \frac{14.01 \text{ g N}}{17.030 \text{ g } NH_3} \times 100\%$$

$$= 82.27\% \text{ (or 82.27 g N per 100.0 g } NH_3)$$

$$\text{Mass percent of H in } NH_3 = \frac{3.024 \text{ g H}}{17.030 \text{ g } NH_3} \times 100\%$$

$$= 17.76\% \text{ H (or 17.76 g H per 100.0 g } NH_3)$$

EXERCISE 3.11 ***Percent Composition***

Express the composition of each compound below in terms of the mass of each element in 1.00 mol of compound and the mass percent of each element:

1. NaCl, sodium chloride
2. C_8H_{18}, octane
3. $(NH_4)_2SO_4$, ammonium sulfate ■

Empirical and Molecular Formulas

Now consider the *reverse* of the procedure just described: using relative mass or percent composition data to find a molecular formula. Let us say you know the identity of the elements in a sample and have determined the mass of each element in a given mass of compound (the percent composition) by chemical analysis. You can then calculate the relative number of moles of each element in one mole of compound and from this the relative number of atoms of each element in the compound. For example, for a compound composed of atoms of A and B, the steps from percent composition to a formula are

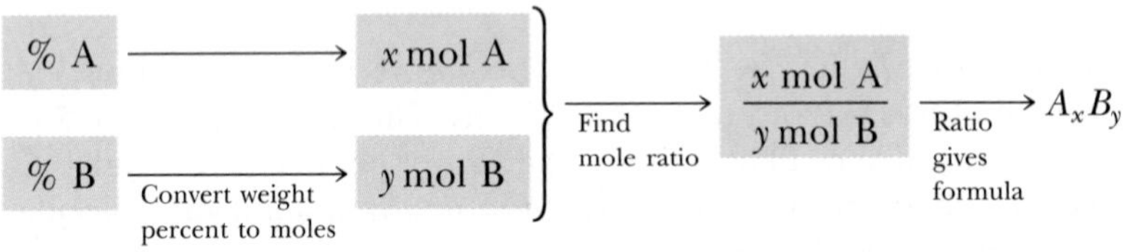

Consider hydrazine, a close relative of ammonia that is used to remove dissolved oxygen in hot water heating systems and to remove metal ions from polluted water. The mass percentages in a sample of hydrazine are 87.42% N and 12.58% H. Taking a 100.00-g sample of hydrazine, the percent composition data tell us that the sample contains 87.42 g of N and 12.58 g of H. The number of moles of each element in the 100.00-g sample is therefore

$$87.42\ \text{g N} \cdot \frac{1\ \text{mol N}}{14.007\ \text{g N}} = 6.241\ \text{mol of N}$$

$$12.58\ \text{g H} \cdot \frac{1\ \text{mol H}}{1.0079\ \text{g H}} = 12.48\ \text{mol of H}$$

Now we can use the number of *moles* of each element in 100.00 g of sample to find the number of moles of one element *relative* to the other. For hydrazine, this ratio is 2 mol of H to 1 of N,

$$\frac{12.48\ \text{mol H}}{6.241\ \text{mol N}} = \frac{2.000\ \text{mol H}}{1.000\ \text{mol N}} \longrightarrow NH_2$$

showing that 2 moles of H atoms exists for every mole of N atoms in hydrazine. Thus, in one molecule two atoms of H occur for every atom of N, that is, the simplest atom ratio is represented by the formula NH_2.

Percent composition data allow you to calculate the atom ratios in the compound. A *molecular* formula, however, must convey *two* pieces of information: (a) the *relative number* of atoms of each element in a molecule (the atom ratios) and (b) the *total number* of atoms in the molecule. For hydrazine you know now that twice as many H atoms occur as N atoms. This means the molecular formula could be NH_2. Because percent composition data give only the *simplest possible ratio* of atoms in a molecule, however, N_2H_4, N_3H_6, N_4H_8, and so forth also correspond to this percent composition. A formula such as NH_2, in which the atom ratio is the simplest possible, is called the **empirical formula.** In contrast, the *molecular formula* shows the *true number* of atoms of each kind in a molecule; it is derived by multiplying the empirical formula by a whole number.

To determine the molecular formula, the molar mass must be obtained from experiment. For example, experiments show that the molar mass of hydrazine is 32.0 g/mol, twice the formula mass of NH_2, which is 16.0 g/mol. This must mean that the molecular formula of hydrazine is two times the empirical formula of NH_2, that is, N_2H_4.

As another example of the usefulness of percent composition data, let us say that you collected the following information in the laboratory for the compound naphthalene, a substance commonly encountered in the form of "moth balls": % carbon = 93.71; % hydrogen = 6.29; molar mass = 128 g/mol. Using these data, you want to calculate the empirical and molecular formulas for the compound. The percent composition data tell you that 93.71 g of C and 6.29 g of H occur in a 100.00 g sample. You can therefore find the number of moles of each element in the sample.

$$93.71\ \cancel{\text{g C}} \cdot \frac{1\ \text{mol C}}{12.011\ \cancel{\text{g C}}} = 7.802\ \text{mol C}$$

$$6.29\ \cancel{\text{g H}} \cdot \frac{1\ \text{mol H}}{1.008\ \cancel{\text{g H}}} = 6.24\ \text{mol H}$$

This means that, in any sample of naphthalene, the ratio of moles of C to H is

$$\text{Mole ratio} = \frac{7.802\ \text{mol C}}{6.24\ \text{mol H}} = \frac{1.25\ \text{mol C}}{1.00\ \text{mol H}}$$

Now your task is to turn this decimal fraction into a whole-number ratio of C to H. To do this, recognize that 1.25 is the same as $1 + \frac{1}{4} = \frac{4}{4} + \frac{1}{4} = \frac{5}{4}$. Therefore, the ratio of C to H is

$$\text{Mole ratio} = \frac{\frac{5}{4}\ \text{mol C}}{1\ \text{mol H}} = \frac{5\ \text{mol C}}{4\ \text{mol H}}$$

and you know now that 5 C atoms occur for every 4 H atoms in naphthalene. Thus the simplest or *empirical formula* is C_5H_4. If C_5H_4 were the molecular formula, the molar mass would be 64 g/mol. However, your experiments gave the actual molar mass as 128 g/mol, twice the value for the empirical formula.

$$\frac{128\ \text{g/mol for naphthalene}}{64.0\ \text{g/mol of}\ C_5H_4} = 2.00\ \text{mol}\ C_5H_4\ \text{per mol of naphthalene}$$

The molecular formula is therefore $(C_5H_4)_2$ or $C_{10}H_8$.

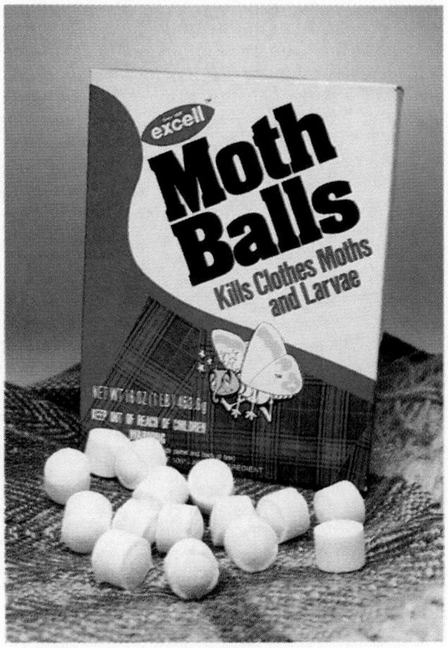

Naphthalene "moth balls." Naphthalene has an empirical formula of C_5H_4 and a molecular formula of $C_{10}H_8$. (C. D. Winters)

EXAMPLE 3.7 *Calculating a Formula from Percent Composition*

Vanillin is a common flavoring agent. It has a molar mass of 152 g/mol and is 63.15% C and 5.30% H; the remainder is oxygen. What are the empirical and molecular formulas for vanillin?

Solution This problem can be solved by following the "map" in the preceding text. First, find the number of moles of carbon and hydrogen in a 100.00-g sample of vanillin.

$$63.15\ \text{g C} \cdot \frac{1\ \text{mol C}}{12.011\ \text{g C}} = 5.258\ \text{mol C}$$

$$5.30\ \text{g H} \cdot \frac{1\ \text{mol H}}{1.008\ \text{g H}} = 5.26\ \text{mol H}$$

But what about oxygen? Because the total sample mass is 100.00 g, this means that the mass of oxygen is

$$100.00\ \text{g} = 63.15\ \text{g C} + 5.30\ \text{g H} + \text{mass of O}$$
$$\text{Mass of O} = 100.00\ \text{g} - (63.15\ \text{g C} + 5.30\ \text{g H})$$
$$\text{Mass of O} = 31.55\ \text{g O}$$

Now we can calculate moles of O as well.

$$31.55\ \text{g O} \cdot \frac{1\ \text{mol O}}{15.999\ \text{g O}} = 1.972\ \text{mol O}$$

To find the mole ratio, the best approach is to base the ratios on the element with the smallest number of moles present. In this case it is oxygen. Therefore,

$$\frac{\text{Mole C}}{\text{Mole O}} = \frac{5.258\ \text{mol C}}{1.972\ \text{mol O}} = \frac{2.666\ \text{mol C}}{1.000\ \text{mol O}} = \frac{2\frac{2}{3}\ \text{mol C}}{1\ \text{mol O}} = \frac{8\ \text{mol C}}{3\ \text{mol O}}$$

$$\frac{\text{Mole H}}{\text{Mole O}} = \frac{5.26\ \text{mol H}}{1.972\ \text{mol O}} = \frac{2.67\ \text{mol H}}{1.00\ \text{mol O}} = \frac{2\frac{2}{3}\ \text{mol H}}{1\ \text{mol O}} = \frac{8\ \text{mol H}}{3\ \text{mol O}}$$

Now we know that the empirical formula has the same number of atoms of C and H, and both are in an 8-to-3 ratio to O. Therefore, the empirical formula is $C_8H_8O_3$. This formula has a molar mass of 152.2 g/unit, which corresponds to the experimentally determined molar mass. Therefore, the molecular formula of vanillin, $C_8H_8O_3$, is the same as the empirical formula.

PROBLEM-SOLVING TIPS AND IDEAS

3.2 Finding Empirical and Molecular Formulas

- The experimental data available to find a formula may be in the form of percent composition or the masses of elements combined in some mass of compound. No matter what the starting point, the first step is always to convert masses of elements to moles.
- Be sure to use at least three significant figures when calculating empirical formulas. Fewer significant figures may give a misleading result.
- When finding atom ratios, it is easiest to divide the larger number of moles by the smaller one.
- Empirical and molecular formulas often differ for molecular compounds. In contrast, the formula of an ionic compound is generally the same as its empirical formula.

- To determine the molecular formula of a compound after calculating the empirical formula, the molar mass *must* be obtained by some experimental method. ■

EXERCISE 3.12 ***Empirical and Molecular Formulas***

Boron hydrides, compounds containing only boron and hydrogen, form a large class of compounds. One consists of 78.14% B and 21.86% H; its molar mass is 27.7 g/mol. What are the empirical and molecular formulas of this compound? ■

Determining and Using Formulas

The empirical formula of a compound can be calculated if the percent composition of the compound has been experimentally determined. Then, if the molar mass has also been determined by any of a number of experimental methods, the empirical formula can be converted to a molecular formula.

Tin metal and purple iodine will react to give an orange tin iodide (Figure 3.20).

$$\text{Sn metal} + I_2 \xrightarrow{\text{will give}} Sn_xI_y$$

The formula of Sn_xI_y can be determined by an experiment in which a mixture of weighed quantities of Sn and I_2 are heated in an organic solvent. The experiment is set up so that the quantity of Sn in the original mixture is far in excess of that needed to react with all the I_2 present. Therefore, after all the iodine has reacted and the orange tin-iodine compound has been formed and has dissolved in the organic solvent, unreacted solid tin metal can be separated from the solution by filtration (Figure 3.20). Let us say that the following data are collected in an experiment to find the values of x and y in Sn_xI_y:

A solvent is a component of a solution, usually the major component. Solutions having water as the solvent are perhaps most familiar. In Figure 3.20 the solvent, ethyl acetate, is a liquid organic compound.

(a)

(b)

Figure 3.20 The reaction of tin (Sn) and iodine (I_2) to give SnI_4.
(a) Weighed quantities of metallic tin (*right*) and the nonmetal iodine (*left*). The solvent for the reaction, ethyl acetate, is in the flask at the rear.
(b) The dark mixture of reactants in ethyl acetate is heated in the Erlenmeyer flask. After the reaction mixture is cooled, it is filtered to remove excess tin. The orange product, SnI_4 (on the filter paper at the front), precipitates and is recovered by filtration.
(C. D. Winters)

Mass of tin (Sn) in the original mixture	1.056 g
Mass of iodine (I_2) in the original mixture	1.947 g
Mass of tin (Sn) recovered after reaction	0.601 g

The first step is to find the mass of Sn that combined with 1.947 g of I_2. These masses are then converted to moles of Sn and I, and finally to the ratio of moles of Sn and I in the empirical formula.

Mass of Sn in the original mixture	1.056 g
Mass of Sn recovered after reaction	−0.601 g
Mass of Sn consumed in the reaction	0.455 g

$$0.455\ \text{g Sn} \cdot \frac{1\ \text{mol Sn}}{118.7\ \text{g Sn}} = 3.83 \times 10^{-3}\ \text{mol Sn used in reaction and appears in } Sn_xI_y$$

$$1.947\ \text{g } I_2 \cdot \frac{1\ \text{mol } I_2}{253.81\ \text{g } I_2} = 7.671 \times 10^{-3}\ \text{mol } I_2 \text{ used in reaction and appears in } Sn_xI_y$$

At this point we recognize that the iodine combines with tin as I atoms. Therefore, the moles of I combined with tin is

$$7.671 \times 10^{-3}\ \text{mol } I_2 \cdot \frac{2\ \text{mol I}}{1\ \text{mol } I_2} = 15.34 \times 10^{-3}\ \text{mol I}$$

Now we can find the ratio of Sn and I that combined.

$$\frac{15.34 \times 10^{-3}\ \text{mol I}}{3.83 \times 10^{-3}\ \text{mol Sn}} = \frac{4.00\ \text{mol I}}{1.00\ \text{mol Sn}}$$

The ratio of atoms of I to atoms of Sn is 4:1, which gives the empirical formula SnI_4. More experimental data are needed to find the molecular formula; such experiments show that the molecular formula is the same as the empirical formula in this case.

EXAMPLE 3.8 *Determining the Formula of a Binary Oxide*

Analysis shows that 0.586 g of potassium can combine with 0.480 g of O_2 gas to give a white solid with a formula of K_xO_y. What is the formula of the white solid?

Solution Our problem is to find the values of x and y in K_xO_y. To do this, we first need to find x, the number of moles of K, and y, the number of moles of O. We can then find the *ratio* of moles of K to moles of O to determine the chemical formula.

1. Calculate moles of K, x.

$$0.586\ \text{g K} \cdot \frac{1\ \text{mol K}}{39.10\ \text{g K}} = 0.0150\ \text{mol K}$$

2. Calculate moles of O, y.

$$0.480\ \text{g } O_2 \cdot \frac{1\ \text{mol } O_2}{32.00\ \text{g } O_2} \cdot \frac{2\ \text{mol O}}{1\ \text{mol } O_2} = 0.0300\ \text{mol O}$$

In this step we had to take into account that 1 mol of O_2 molecules contains 2 mol of O atoms, and it is the moles of O atoms that we need to know.

3. Determine the ratio of moles. This is usually done by dividing the larger number of moles by the smaller number of moles.

$$\frac{0.0300 \text{ mol O}}{0.0150 \text{ mol K}} = \frac{2 \text{ mol of O atoms}}{1 \text{ mol of K atoms}}$$

4. The empirical formula of the compound is KO_2.

The compound in this question, KO_2, is called potassium superoxide. When oxygen reacts with metals, three types of ionic compounds may be formed.

Compound Type	Example	Name	Oxygen-Containing Anion
Oxide	Li_2O	lithium oxide	O^{2-}
Peroxide	Na_2O_2	sodium peroxide	O_2^{2-}
Superoxide	KO_2	potassium superoxide	O_2^{-}

Peroxides and superoxides are used in self-contained breathing apparatuses and in space capsules because these oxides absorb CO_2 and evolve oxygen.

$$2\ Li_2O_2 + 2\ CO_2 \longrightarrow 2\ Li_2CO_3 + O_2$$

PROBLEM-SOLVING TIPS AND IDEAS

3.3 Finding Empirical and Molecular Formulas

The calculations in the text and in Examples 3.7 and 3.8 illustrate only a *few* of the ways of using chemical information to determine formulas. You should always focus on using the data to

1. Find the number of moles of atoms of the elements combined in a given mass of compound and then
2. Find the ratio of those moles

The data you are given may be mass percentages, masses of elements in a given amount of compound, or experimental information from a chemical reaction. ■

EXERCISE 3.13 ***Empirical Formula of a Binary Compound***

Gaseous chlorine was combined with 0.532 g of titanium, and 2.108 g of Ti_xCl_y was collected. What is the empirical formula of the titanium chloride? ■

Once the formula of a compound is known, you can begin to use the information in a variety of ways. One of those is to place the atoms in three-dimensional space and define the structure of the compound. This is an important part of chemistry because structural knowledge leads to insight into reactivity, a subject we shall come back to often in this text. Another use of molecular formulas is in quantitative chemistry. For example, because we knew the formula of the product of the reaction of phosphorus and bromine in Figure 3.8, we knew how much bromine was required to react completely with the phosphorus. Or, knowing the formula of a compound, we know how much product can be derived from it in a chemical reaction.

EXAMPLE 3.9 *Using Chemical Formulas*

What mass of copper(I) sulfide, Cu_2S, may be obtained from 2.00 kg of copper?

Solution Once a formula is known, the molar mass of the compound is known and is a vital piece of information. Here it is the link between the mass of copper and the mass of Cu_2S produced. Our plan for solving this problem therefore is first to find the moles of copper in 2.00 kg. Next, the formula Cu_2S tells us that two moles of Cu occur in each mole of Cu_2S. Moles of Cu_2S are therefore linked to the mass of copper.

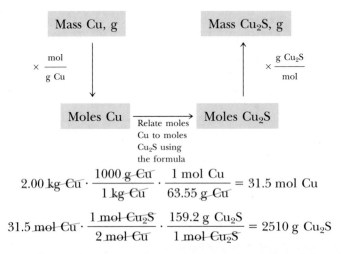

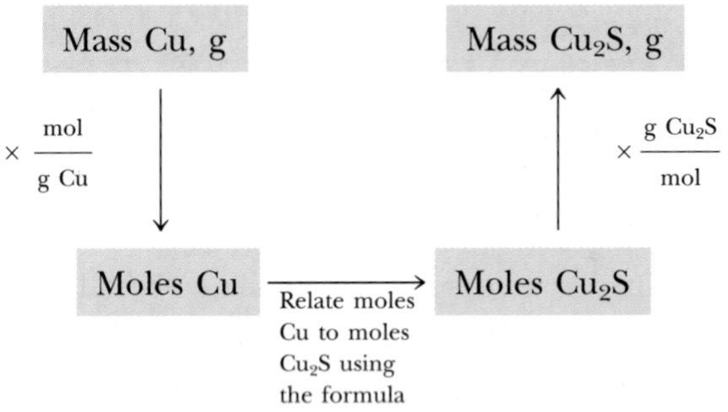

$$2.00\ \cancel{\text{kg Cu}} \cdot \frac{1000\ \cancel{\text{g Cu}}}{1\ \cancel{\text{kg Cu}}} \cdot \frac{1\ \text{mol Cu}}{63.55\ \cancel{\text{g Cu}}} = 31.5\ \text{mol Cu}$$

$$31.5\ \cancel{\text{mol Cu}} \cdot \frac{1\ \cancel{\text{mol Cu}_2\text{S}}}{2\ \cancel{\text{mol Cu}}} \cdot \frac{159.2\ \text{g Cu}_2\text{S}}{1\ \cancel{\text{mol Cu}_2\text{S}}} = 2510\ \text{g Cu}_2\text{S}$$

This is very similar to the approach we shall use to study weight relations in chemical reactions beginning in Chapter 5. There is an alternative in this case, however. Because the formula is known, we know the weight percent of copper in copper(I) sulfide. This tells us directly the number of grams of copper in 100 g of Cu_2S, so we can use this relationship to convert grams of copper to grams of copper(I) sulfide. The weight percent of Cu in Cu_2S is 79.83%. Therefore,

$$2.00 \times 10^3\ \cancel{\text{g Cu}} \cdot \frac{100.0\ \text{g Cu}_2\text{S}}{79.83\ \cancel{\text{g Cu}}} = 2510\ \text{g Cu}_2\text{S}$$

Either approach is effective and acceptable.

EXERCISE 3.14 *Using a Formula*

What mass of iron(III) oxide contains 50.0 g of iron? ■

3.8 HYDRATED COMPOUNDS

If ionic compounds are prepared in water solution and then isolated as solids, the crystals often have molecules of water trapped in the lattice. Compounds in which molecules of water are associated with the ions of the compound are called **hydrated compounds.** The beautiful green nickel(II) compound in Figure 3.19, for example, has a formula that is conventionally written as $NiCl_2 \cdot 6\ H_2O$.

TABLE 3.5 Some Common Hydrated Ionic Compounds

Compound	Systematic Name	Common Name	Uses
$Na_2SO_4 \cdot 10\ H_2O$	sodium sulfate decahydrate	Glauber's salt	cathartic
$Na_2CO_3 \cdot 10\ H_2O$	sodium carbonate decahydrate	washing soda	water softener
$Na_2S_2O_3 \cdot 5\ H_2O$	sodium thiosulfate pentahydrate	hypo	photography
$MgSO_4 \cdot 7\ H_2O$	magnesium sulfate heptahydrate	Epsom salt	cathartic, dyeing and tanning
$CaSO_4 \cdot 2\ H_2O$	calcium sulfate dihydrate	gypsum	wallboard
$CaSO_4 \cdot \frac{1}{2}\ H_2O$	calcium sulfate hemihydrate	plaster of Paris	casts, molds
$CuSO_4 \cdot 5\ H_2O$	copper(II) sulfate pentahydrate	blue vitriol	biocide

The dot between $NiCl_2$ and 6 H_2O indicates that six moles of water are associated with every mole of $NiCl_2$; it is equivalent to writing the formula as $NiCl_2(H_2O)_6$. The name of the compound, nickel(II) chloride *hexa*hydrate, reflects the presence of six moles of water. The molar mass of $NiCl_2 \cdot 6\ H_2O$ is 129.6 g/mol (for $NiCl_2$) plus 108.1 g/mol (for 6 H_2O) or 237.7 g/mol.

Hydrated compounds are much more common than you might realize (Table 3.5). The walls of your home may be covered with wallboard or "plaster board." These sheets contain hydrated calcium sulfate, or gypsum ($CaSO_4 \cdot 2\ H_2O$), as well as unhydrated $CaSO_4$, sandwiched between paper. Gypsum is a mineral that can be mined. Now however, it is commonly a byproduct in the manufacture of hydrofluoric acid and phosphoric acid, or it comes from cleaning sulfur dioxide from the exhaust gases of power plants.

If gypsum is heated to between 120 and 180 °C, the water is partly driven off to give $CaSO_4 \cdot \frac{1}{2}\ H_2O$, a compound commonly called "plaster of Paris" (Figure 3.21). If you have ever broken an arm or leg and had to have a cast, the cast may have been made of this partly hydrated calcium sulfate. It is an effective casting

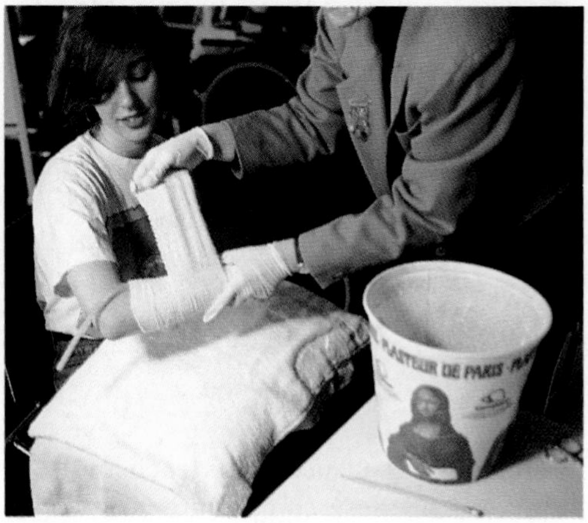

Figure 3.21 A plaster of Paris cast being applied to an arm. (C. D. Winters)

Figure 3.22 A crucible containing a hydrated compound ($CuSO_4 \cdot 5\ H_2O$) is heated to drive off all the water and leave anhydrous $CuSO_4$ (i.e., $CuSO_4$ without water of hydration). (C. D. Winters)

material because, when added to water, it forms a thick slurry that can be poured into a mold or spread out over a part of the body. As it takes on more water, the material increases in volume and forms a hard, inflexible solid. These properties also make plaster of Paris a useful material to artists, because the expanding compound fills a mold completely and makes a high quality reproduction.

There is no simple way to predict how much water will be involved in a hydrated compound; it must be determined experimentally. Such an experiment may involve heating the hydrated material so that all the water is released from the solid and evaporated (Figure 3.22). Only the **anhydrous** compound, a substance "without water," is left. The mass of the original, hydrated compound must equal the sum of the mass of the water driven away and the mass of the anhydrous compound left behind. For example, heating 100.00 g of copper sulfate pentahydrate gives the following results:

$$\begin{array}{llll}\text{Hydrated compound} & \longrightarrow \text{anhydrous compound} + & \text{water} \\ 100.00\text{ g } CuSO_4 \cdot 5\ H_2O & \longrightarrow \quad 63.92\text{ g } CuSO_4 & + 36.08\text{ g } H_2O\end{array}$$

The masses of the anhydrous compound and the water evolved can be determined by experiment. Converting these masses to moles and finding their mole ratio gives their ratio in the original, hydrated material.

Hydrated cobalt(II) chloride is the deep red solid in Figure 3.19. When heated it turns purple and then deep blue as it loses water to form anhydrous $CoCl_2$. On exposure to moist air, anhydrous $CoCl_2$ takes up water and is converted back into the red hydrated compound. It is this property that allows crystals of the blue compound to be used as a humidity indicator. (You may have seen them in a small bag packed with a piece of electronic equipment.) The compound also makes a good "invisible ink." A solution of cobalt(II) chloride in water is red, but if you write on paper with the solution it cannot be seen. When the paper is warmed, however, the cobalt compound dehydrates to give the deep blue anhydrous compound, and you can read the writing.

EXAMPLE 3.10 *Determining the Formula of a Hydrated Compound*

Suppose you want to know the value of x in red, hydrated cobalt(II) sulfate, $CoSO_4 \cdot x\ H_2O$, that is, the number of water molecules for each unit of $CoSO_4$. In the laboratory you weigh out 1.023 g of the solid and then heat it in a porcelain crucible (see Figure 3.22). After the water has been driven off completely, you are left with 0.603 g of blue, anhydrous cobalt(II) sulfate, that is, with pure $CoSO_4$.

$$1.023\text{ g } CoSO_4 \cdot x\ H_2O + \text{heat} \longrightarrow 0.603\text{ g } CoSO_4 + ?\text{ g } H_2O$$

Solution The information gathered in the laboratory gives the mass of water lost on heating.

Mass of hydrated compound	1.023 g
−Mass of anhydrous compound	−0.603 g
Mass of water	0.420 g

Because you want to know how many moles of H_2O occur for each mole of $CoSO_4$, the next step is to convert the masses of these compounds to moles.

$$0.420\ \cancel{\text{g H}_2\text{O}} \cdot \frac{1\ \text{mol H}_2\text{O}}{18.02\ \cancel{\text{g H}_2\text{O}}} = 0.0233\ \text{mol H}_2\text{O}$$

$$0.603\ \cancel{\text{g CoSO}_4} \cdot \frac{1\ \text{mol CoSO}_4}{155.0\ \cancel{\text{g CoSO}_4}} = 0.00389\ \text{mol CoSO}_4$$

The value of x is then determined from the mole ratio.

$$\frac{\text{moles H}_2\text{O}}{\text{moles CoSO}_4} = \frac{0.0233\ \text{mol H}_2\text{O}}{0.00389\ \text{mol CoSO}_4} = \frac{5.99\ \text{mol H}_2\text{O}}{1.00\ \text{mol CoSO}_4}$$

This tells us the water-to-$CoSO_4$ ratio is 6:1, so the formula of the hydrated compound is $CoSO_4 \cdot 6\ H_2O$, and its name is cobalt(II) sulfate hexahydrate.

EXERCISE 3.15 ***Determining the Formula of a Hydrated Compound***

Ruthenium is certainly one of the less familiar elements, but its chemistry is quite interesting. A good material to use to make other ruthenium compounds is $RuCl_3 \cdot x\,H_2O$. If you heat 1.056 g of this hydrated compound and find that only 0.838 g of $RuCl_3$ remains when all of the water has been driven off, what is the value of x? ■

CHAPTER HIGHLIGHTS

Having studied this chapter, you should be able to

- Interpret the meaning of **molecular formulas** and **structural formulas** (Section 3.1).
- Define **allotropes,** and give several examples (Section 3.2).
- List the elements that exist as diatomic molecules (Section 3.2).
- Recognize that metal atoms commonly lose one or more electrons to form positive ions **(cations),** while nonmetal atoms often gain electrons to form negative ions **(anions)** (see Figure 3.10).
- Recognize that the charge on a metal cation (other than the transition metals) is equal to the group number in which the element is found in the periodic table (M^{n+}, n = Group number) (Section 3.3). (Transition metal cations are typically 2+ or 3+, but other charges are observed.)
- Recognize that the negative charge on a single-atom or monatomic anion, X^{n-}, is given by $n = 8 -$ Group number (Section 3.3).
- Give the names or formulas of **polyatomic ions,** knowing their formulas or names, respectively (Table 3.2 and Section 3.3).
- Write the formulas for a number of ionic compounds (Section 3.4).
- Describe the properties of ionic compounds (Section 3.4).
- Name ionic compounds and simple binary compounds of the nonmetals (Section 3.5).
- Understand that the **molar mass** of a compound (often called the **molecular weight**) is the mass in grams of Avogadro's number of mole-

cules of the compound (Section 3.6). For ionic compounds, which do not consist of individual molecules, we often refer to the sum of atomic masses as the formula mass (or formula weight).

- Calculate the **molar mass** of a compound (Section 3.6).
- Calculate the number of moles of an element or compound that are represented by a given mass, and vice versa (Section 3.6).

$$\text{Mass of A, g} \underset{\times \frac{g}{mol}}{\overset{\times \frac{mol}{g}}{\rightleftarrows}} \text{Moles of A}$$

- Express molecular composition in terms of **percent composition** (Section 3.7).
- Use percent composition to determine the **empirical formula** of a compound (Section 3.7).
- Use experimental data to calculate the number of water molecules in a hydrated compound (Section 3.8).

STUDY QUESTIONS

Review Questions

1. A dictionary defines the word "compound" as a "combination of two or more parts." What are the "parts" of a chemical compound? Can you think of three pure (or nearly pure) compounds you have encountered today?
2. The Polish-born American chemist, Roald Hoffmann, has said that "Today chemistry is the science of molecules and their transformations." What does that statement mean in the context of your own experience?
3. Which of the following compounds might you properly call a hydrocarbon?
   (a) Ethanol, C_2H_6O
   (b) Ammonium sulfate, $(NH_4)_2SO_4$
   (c) Toluene, $C_6H_5CH_3$
4. Three elements composed of diatomic molecules are among the top ten chemicals produced in the United States (nitrogen, oxygen, and chlorine). For example, 10.1 billion kilograms of chlorine was made in 1992. By reading the newspapers or asking local citizens, can you find out how this element might be used in your community?
5. Ozone, O_3, is an allotrope of oxygen. It was reported in 1992 that "Scientific data obtained in the last six months have provided additional disturbing news about the decline in stratospheric ozone, and the consequent increase in surface exposure to increased ultraviolet radiation." Scan recent newspapers and magazines to find out what is happening in this area of scientific investigation. Is stratospheric ozone (i.e., ozone at very high altitudes) still on the decline? What are the United States and other countries doing to halt this decline in the ozone level?
6. The compound $(NH_4)_2CO_3$ consists of two polyatomic ions. What are the names and electric charges of these ions? What is the molar mass of this compound?
7. Octane is a member of the alkane family of hydrocarbons (see Table 3.2). For the sake of this problem, we assume that gasoline, a complex mixture of hydrocarbons, is represented by octane. If you fill the tank of your car with 68 L of gasoline (about 18 U.S. gal), how many grams of liquid have you put into the car? How many moles of octane are in the tank? How many molecules of octane? (The density of octane is 0.692 g/cm^3.)
8. What is the difference between an empirical and a molecular formula? Use the compound ethane, C_2H_6, to illustrate your answer.
9. Explain why 40 g of vanillin (whose structure is shown here) represents less mass than 3.0×10^{23} molecules of the compound.

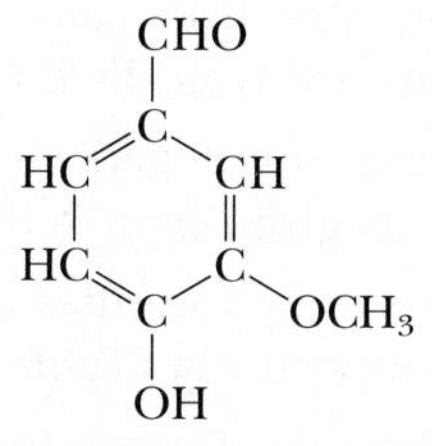

vanillin

10. Having read this chapter, can you explain how the "disappearing ink" in the Chemical Puzzler works (see the chapter opening photograph)? Use chemical changes in your explanation.

11. Potassium chloride, KCl, melts at 771 °C. In contrast, magnesium oxide, MgO, melts at 2826 °C. Using Coulomb's law, suggest a reason for the great difference in the melting points of KCl and MgO.
12. Two minerals, Cu_5FeS_4 and Cu_2S, can be considered as a source of copper. If you have a kilogram of each mineral, from which can you obtain the most copper?

Numerical and Other Questions

Molecular Formulas

13. Which of the following molecules contains more O atoms per molecule? More atoms of all kinds? (a) sucrose, $C_{12}H_{22}O_{11}$, or (b) glutathione, $C_{10}H_{17}N_3O_6S$ (the major, low-molecular-weight, sulfur-containing compound in plant or animal cells).

14. Write the molecular formula of each of the following compounds:
    (a) One of a series of compounds called the boron hydrides has four boron atoms and ten hydrogen atoms per molecule.
    (b) Vitamin C or ascorbic acid has six carbon atoms, eight hydrogen atoms, and six oxygen atoms per molecule.
    (c) A molecule of aspartame, an artificial sweetener, which has 14 carbon atoms, 18 hydrogen atoms, 2 nitrogen atoms, and 5 oxygen atoms.

15. Give the total number of atoms of each element in one formula unit of each of the following compounds:
    (a) CaC_2O_4 (d) $Pt(NH_3)_2Cl_2$
    (b) $C_6H_5CHCH_2$ (e) $K_4Fe(CN)_6$
    (c) $Cu_2CO_3(OH)_2$

16. Give the total number of atoms of each element in each of the molecules below.
    (a) $Co_2(CO)_8$ (d) $C_{10}H_9NH_2Fe$
    (b) $HOOCCH_2CH_2COOH$ (e) $C_6H_2CH_3(NO_2)_3$
    (c) $CH_3NH_2CHCOOH$

17. Write a molecular formula for the following two organic acids:

(a)

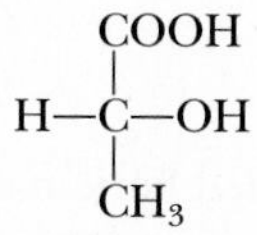

lactic acid

(b)

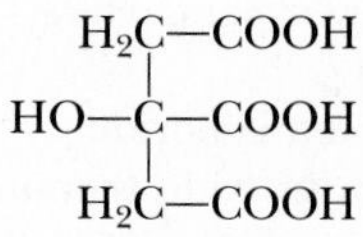

citric acid

18. Write a molecular formula for each of the following molecules:

(a)

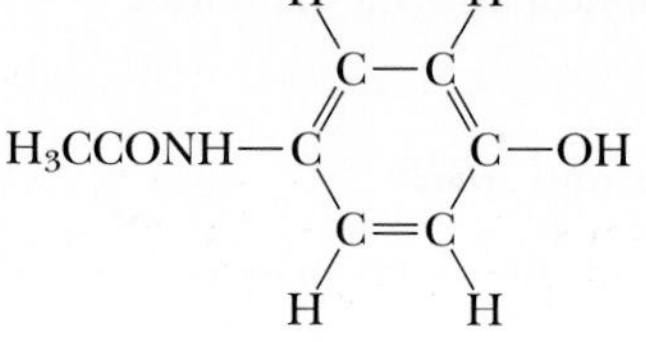

acetaminophen

(b)

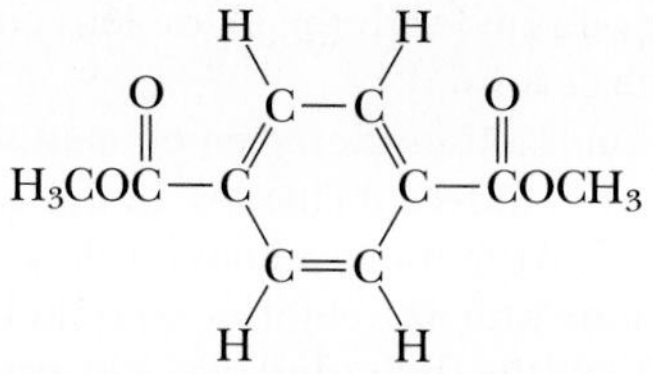

dimethyl terephthalate

Acetaminophen is an analgesic (found in such over-the-counter drugs as Tylenol). Dimethyl terephthalate is one of the two molecules used to make the polymers Dacron and Mylar.

Ions and Ion Charges

19. Predict the charges on the ions formed by aluminum and selenium.

20. Predict the charges of the ions in an ionic compound containing barium and bromine.

21. What charges are most commonly observed for ions of the following elements?
    (a) Magnesium (c) Iron
    (b) Zinc (d) Gallium

22. What charges are most commonly observed for ions of the following elements?
    (a) Selenium (c) Nickel
    (b) Fluorine (d) Nitrogen

23. Give the symbol, including the correct charge, for each of the following ions:
    (a) Strontium ion (e) Titanium(IV) ion
    (b) Aluminum ion (f) Hydrogen carbonate ion
    (c) Sulfide ion (g) Perchlorate ion
    (d) Cobalt(II) ion (h) Ammonium ion

24. Give the symbol, including the correct charge, for each of the following ions:
    (a) Permanganate ion (d) Nitrite ion
    (b) Dihydrogen phosphate ion (e) Sulfate ion
    (c) Phosphate ion (f) Sulfite ion

Ionic Compounds

25. For each of the following compounds, give the formula, charge, and number of each ion that makes up the compound:
    (a) K_2S (d) $Ca(ClO)_2$
    (b) $NiSO_4$ (e) $KMnO_4$
    (c) $(NH_4)_3PO_4$

26. For each of the following compounds, give the formula, charge, and number of each ion that makes up the compound:
    (a) $Ca(CH_3CO_2)_2$ (d) KH_2PO_4
    (b) $Co_2(SO_4)_3$ (e) $CuCO_3$
    (c) $Al(OH)_3$

27. Cobalt is a transition metal and so can form ions with at least two different charges. Write the formulas for the compounds formed with the oxide ions and each of two different cobalt ions.

28. Platinum is a transition element and so can form ions with at least two different charges. In this case, the ions are Pt^{2+} and Pt^{4+}. Write the formulas for the compounds of each of these ions with (a) chloride ions and (b) sulfide ions.

29. Which of the following are correct formulas for compounds? For those that are not, give the correct formula.
(a) AlCl (c) Ga_2O_3
(b) NaF_2 (d) MgS

30. Which of the following are correct formulas for compounds? For those that are not, give the correct formula.
(a) Ca_2O (c) Fe_2O_5
(b) $SrCl_2$ (d) K_2O

31. Solid magnesium oxide melts at 2800 °C. This property, combined with the fact that it is not an electric conductor, makes it an ideal heat insulator for electric wires in cooking ovens and toasters (see the photo). In contrast, solid NaCl melts at the relatively low temperature of 801 °C. What is the formula of magnesium oxide? Suggest a reason for its melting temperature being so much higher than that of NaCl.

Magnesium oxide, MgO, is used as an electrical insulator in heating elements. (C. D. Winters)

32. Assume you have an unlabeled bottle containing a white crystalline powder. The powder melts at 310 °C. You are told that it could be NH_3, NO_2, or $NaNO_3$. What do you think it is and why?

Naming Compounds

33. Name each of the following ionic compounds:
(a) K_2S (c) $(NH_4)_3PO_4$
(b) $NiSO_4$ (d) $Ca(ClO)_2$

34. Name each of the following ionic compounds:
(a) $Ca(CH_3CO_2)_2$ (c) $Al(OH)_3$
(b) $Co_2(SO_4)_3$ (d) KH_2PO_4

35. Give the formula for each of the following ionic compounds:
(a) Ammonium carbonate
(b) Calcium iodide
(c) Copper(II) bromide
(d) Aluminum phosphate
(e) Silver(I) acetate

36. Give the formula for each of the following ionic compounds:
(a) Calcium hydrogen carbonate
(b) Potassium permanganate
(c) Magnesium perchlorate
(d) Potassium hydrogen phosphate
(e) Sodium sulfite

37. Write the formulas for all the compounds that can be made by combining each of the cations with each of the anions listed here. Name each compound formed.

Cations	Anions
K^+	CO_3^{2-}
Ba^{2+}	Br^-
NH_4^+	NO_3^-

38. Write the formulas for all the compounds that can be made by combining each of the cations with each of the anions listed here. Name each compound formed.

Cations	Anions
Co^{2+}	SO_4^{2-}
Mg^{2+}	PO_4^{3-}
Li^+	S^{2-}

39. Give the name for each of the following binary, nonmetal compounds:
(a) NF_3 (c) BBr_3
(b) HI (d) C_6H_{14}

40. Give the name for each of the following binary, nonmetal compounds:
(a) C_8H_{18} (c) OF_2
(b) P_4S_3 (d) XeF_4

41. Give the formula for each of the following nonmetal compounds:
(a) Butane
(b) Dinitrogen pentaoxide
(c) Nonane
(d) Silicon tetrachloride
(e) Diboron trioxide (commonly called boric oxide)

42. Give the formula for each of the following nonmetal compounds:
(a) Bromine trifluoride (d) Pentadecane
(b) Xenon difluoride (e) Hydrazine
(c) Diphosphorus tetrafluoride

Molar Mass and Moles

43. Calculate the molar mass of each of the following compounds:
(a) Fe_2O_3, iron(III) oxide
(b) BF_3, boron trifluoride

(c) N_2O, dinitrogen monoxide (laughing gas)
(d) $MnCl_2 \cdot 4\ H_2O$, manganese(II) chloride tetrahydrate
(e) $C_6H_8O_6$, ascorbic acid or vitamin C

44. Calculate the molar mass of each of the following compounds:
(a) $B_{10}H_{14}$, a boron hydride once considered as a rocket fuel
(b) $C_6H_2(CH_3)(NO_2)_3$, trinitrotoluene (TNT), an explosive
(c) $PtCl_2(NH_3)_2$, a cancer chemotherapy agent called cisplatin
(d) $CH_3CH_2CH_2CH_2SH$, which has a skunk-like odor
(e) $C_{20}H_{24}N_2O_2$, quinine, used as an antimalarial drug

45. How many moles are represented by 1.00 g of each of the following compounds?
(a) CH_3OH, methanol
(b) Cl_2CO, phosgene, a poisonous gas
(c) NH_4NO_3, ammonium nitrate
(d) $MgSO_4 \cdot 7\ H_2O$, magnesium sulfate heptahydrate (epsom salt)

46. Assume you have 0.250 g of each of the following compounds. How many moles of each are present?
(a) $C_9H_8O_4$, aspirin
(b) $C_{14}H_{10}O_4$, benzoyl peroxide, used in acne medications
(c) $C_{20}H_{14}O_4$, phenolphthalein, a dye

47. Acrylonitrile, C_2H_3CN, is used to make acrylic plastics. If you have 2.50 kg of acrylonitrile, how many moles of the compound are present?

48. Acetone, $(CH_3)_2CO$, is an important industrial solvent. It was reported that 966 million kilograms of this organic compound was produced in 1991. How many moles does this represent?

49. An Alka-Seltzer tablet contains 324 mg of aspirin ($C_9H_8O_4$), 1904 mg of $NaHCO_3$, and 1000. mg of citric acid ($C_6H_8O_7$). (The last two compounds react with each other to provide the "fizz," bubbles of CO_2, when the tablet is put into water.)
(a) Calculate the number of moles of each substance in the tablet.
(b) If you take one tablet, how many molecules of aspirin are you consuming?

50. Some types of chlorofluorocarbons (CFCs) have been used as the propellant in spray cans of paint and other consumer products. The use of CFCs is being curtailed, however, because they are strongly suspected of causing environmental damage. If a spray can contains 250 g of CCl_2F_2, a CFC, how many molecules are you releasing into the air when you empty the can?

51. Sulfur trioxide, SO_3, is made in enormous quantities by combining oxygen and sulfur dioxide, SO_2. The trioxide is not usually isolated but is converted to sulfuric acid. If you have 1.00 kg of sulfur trioxide, how many moles does this represent? How many molecules? How many sulfur atoms? How many oxygen atoms?

52. Chlorofluorocarbons, or CFCs, are strongly suspected of causing environmental damage. A substitute may be CF_3CH_2F (which has no Cl atoms, the source of the environmental damage caused by CFCs). If you have 25.5 g of this new compound, how many moles does this represent? How many atoms of fluorine are contained in 25.5 g of the compound?

Percent Composition

53. Calculate the mass percent of each element in the following compounds:
(a) PbS, lead(II) sulfide, galena
(b) C_3H_8, propane, a hydrocarbon fuel
(c) $CoCl_2 \cdot 6\ H_2O$, a beautiful red compound. (See Figure 3.19)
(d) NH_4NO_3, ammonium nitrate, a fertilizer and an explosive.

54. Calculate the mass percent of each element in the following compounds:
(a) $MgCO_3$, magnesium carbonate
(b) C_6H_5OH, phenol, an organic compound used in some cleaners
(c) $CH_3CH_2CH_2CH_2CH_2CO_2H$, caproic acid, an organic acid with a characteristic goat-like odor.
(d) $C_{18}H_{27}NO_3$, capsaicin, the compound that gives the hot taste to chili peppers

55. Vinyl chloride, CH_2CHCl, is the basis of many important plastics (PVC) and fibers.
(a) Calculate the molar mass.
(b) Calculate the mass percent of each element in the compound.
(c) How many grams of carbon are in 454 g of vinyl chloride?

56. The copper-containing compound $Cu(NH_3)_4SO_4 \cdot H_2O$ is a beautiful blue solid. Calculate the molar mass of the compound and the mass percent of each element. How many grams of copper and how many of the water are in 10.5 g of the compound?

The blue copper compound $Cu(NH_3)_4SO_4 \cdot 5\ H_2O$. (C. D. Winters)

Empirical and Molecular Formulas

57. The empirical formula of succinic acid is $C_2H_3O_2$. Its molar mass is 118.1 g/mol. What is its molecular formula?

58. A well-known reagent in analytical chemistry, dimethylglyoxime, has the empirical formula C_2H_4NO. If its molar mass is 116.1 g/mol, what is the molecular formula of the compound?

59. Acetylene is a colorless gas that is used as a fuel in welding torches, among other things. It is 92.26% C and 7.74% H. Its molar mass is 26.02 g/mol. Calculate the empirical and molecular formulas.

60. There is a large family of boron-hydrogen compounds called boron hydrides. All have the formula B_xH_y, and all react with air and burn or explode. One member of this family contains 88.5% B; the remainder is hydrogen. Which of the following is its empirical formula: BH_3, B_2H_5, B_5H_7, B_5H_{11}, BH_2?

61. Nitrogen and oxygen form an extensive series of oxides with the general formula N_xO_y. One of them is a blue solid that comes apart, reversibly, in the gas phase. It contains 36.84% N. What is the empirical formula of this oxide?

62. Cumene is a hydrocarbon, a compound composed only of C and H. It is 89.94% carbon, and the molar mass is 120.2 g/mol. What are the empirical and molecular formulas of cumene?

63. Mandelic acid is an organic acid composed of carbon (63.15%), hydrogen (5.30%), and oxygen (31.55%). Its molar mass is 152.14 g/mol. Determine the empirical and molecular formulas of the acid.

64. An analysis of nicotine, a poisonous compound found in tobacco leaves, shows that it is 74.0% C, 8.65% H, and 17.35% N. Its molar mass is 162 g/mol. What are the empirical and molecular formulas of nicotine?

65. Cacodyl, a compound containing arsenic, was reported in 1842 by the German chemist Robert Wilhelm Bunsen. It has an almost intolerable garlic-like odor. Its molar mass is 210 g/mol, and it is 22.88% C, 5.76% H, and 71.36% As. Determine its empirical and molecular formulas.

66. The action of bacteria on meat and fish produces a poisonous compound called cadaverine. As its name and origin imply, it stinks! It is 58.77% C, 13.81% H, and 27.40% N. Its molar mass is 102.2 g/mol. Determine the molecular formula of cadaverine.

67. If "epsom salt," $MgSO_4 \cdot x\,H_2O$, is heated to 250 °C, all the water of hydration is lost. On heating a 1.687-g sample of the hydrate, 0.824 g of $MgSO_4$ remains. How many molecules of water occur per formula unit of $MgSO_4$?

68. The "alum" used in cooking is potassium aluminum sulfate hydrate, $KAl(SO_4)_2 \cdot x\,H_2O$ (see photo). To find the value of x, you can heat a sample of the compound to drive off all of the water and leave only $KAl(SO_4)_2$. Assume that you heat 4.74 g of the hydrated compound and that it loses 2.16 g of water. What is the value of x?

69. Elemental sulfur (1.256 g) is combined with fluorine, F_2, to give a compound with the formula SF_x, a very stable, colorless gas. If you have isolated 5.722 g of SF_x, what is the value of x?

70. A new compound containing xenon and fluorine was isolated by shining sunlight on a mixture of Xe (0.526 g) and F_2 gas. If all the xenon was consumed, and you isolated 0.678 g of the new compound, what is its empirical formula?

71. What mass of lead (II) sulfide (the mineral galena) is required to produce 2.00 kg of lead?

72. The mineral ilmenite has the formula $FeTiO_3$. What quantity of ilmenite is required if you wish to obtain 750 g of titanium? Or 750 g of iron?

73. Elemental phosphorus is made by heating calcium phosphate with carbon and sand in an electric furnace. How many kilograms of calcium phosphate must be used to produce 15.0 kg of phosphorus?

74. Chromium is obtained by heating chromium(III) oxide with carbon. If you want to produce 500. kg of chromium metal, what quantity of Cr_2O_3 (in kilograms) is required?

General Questions

75. Metallic sodium and chlorine gas are obtained by decomposing sodium chloride with an electric current. What masses of sodium and chlorine can be obtained from 2.00 metric tons of salt? (A metric ton is exactly 1000 kg.)

76. A drop of water is about 0.05 mL. How many molecules of water are in a drop of water? (Assume water has a density of 1.00 g/cm^3)

77. Write the molecular formula and calculate the molar mass for each of the molecules below. Which has the larger percentage of nitrogen? Of carbon?

(a)

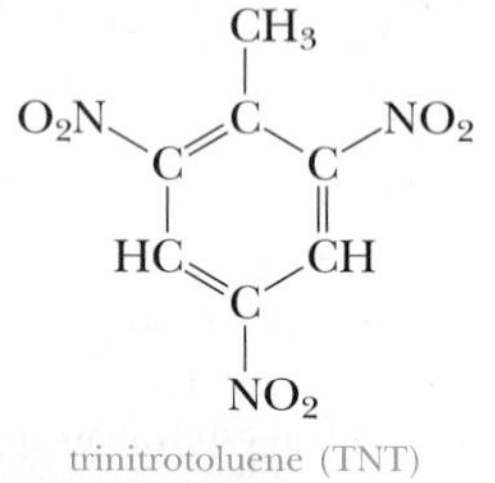

trinitrotoluene (TNT)

Alum crystals. (C. D. Winters)

(b)

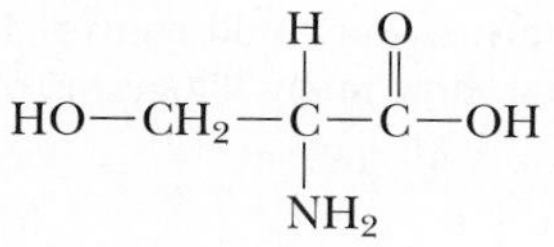

serine, an essential amino acid

78. Malic acid, an organic acid found in apples, contains C, H, and O in the following ratios: $C_1H_{1.50}O_{1.25}$. What is the empirical formula of malic acid?
79. What is the mass, in grams, of one molecule of the naturally occurring acid, oxalic acid, $C_2H_2O_4$?
80. What is the mass, in grams, of one molecule of the amino acid glycine, $NH_2CH_2CO_2H$?
81. Write the molecular formula and calculate the molar mass for the molecule shown here. What are the weight percentages of carbon, hydrogen, and sulfur in the compound? If you have 10.0 g of the compound, how many moles does this represent? How many grams of sulfur are in 10.0 g of the compound? Color code: carbon = gray; hydrogen = white; sulfur = yellow; and oxygen = red. (The molecule is called dimethyl sulfoxide, and it is commonly used as a solvent.)

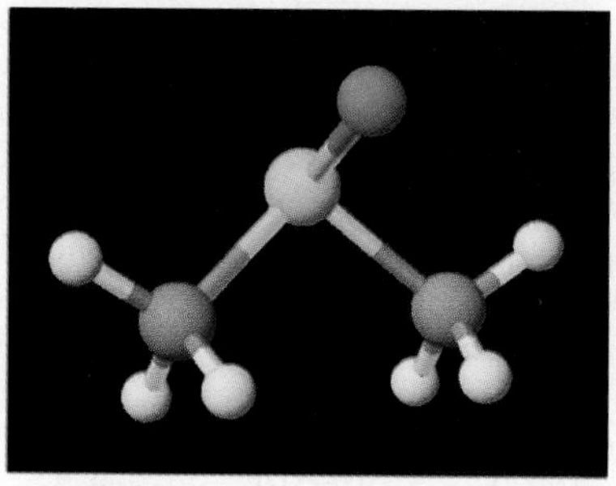

Computer-generated model of dimethyl sulfoxide.

82. Write the molecular formula and calculate the molar mass for trichloroacetic acid. How many grams of chlorine and of carbon are contained in 2.50 g of the compound? Color code: carbon = gray; hydrogen = white; chlorine = green; and oxygen = red. (The compound is called trichloroacetic acid. It is a relatively strong acid in water solution.)

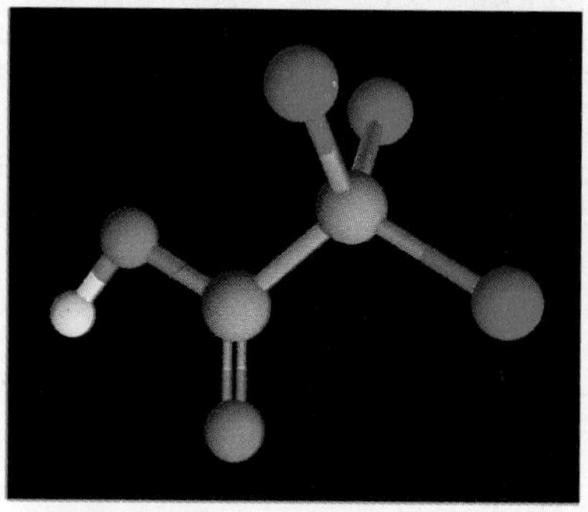

Computer-generated model of trichloroacetic acid.

83. The compound shown here contains carbon, hydrogen and iron. Color code: carbon = gray; hydrogen = white; iron = red. (The molecule is called ferrocene. It is representative of a large class of so-called organometallic compounds.)

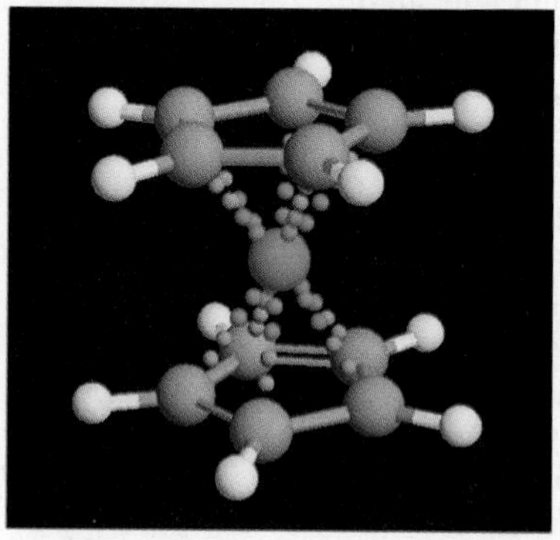

Computer-generated model of ferrocene.

(a) Write the molecular formula for the compound.
(b) How many grams of iron are in 0.150 g of the compound? How many iron atoms?
(c) Which element accounts for most of the mass of the molecule?

84. The molecule shown here is called β-D-ribose, and it is a member of the family of compounds called carbohydrates. Recall (page 108) that the reason these compounds are called carbohydrates is that their formulas may be written as $C_x(H_2O)_y$ or "hydrates of carbon." If 10.0 g of β-D-ribose are decomposed to carbon and water, how many grams of water are obtained? Color code: carbon = gray; hydrogen = white; oxygen = red.

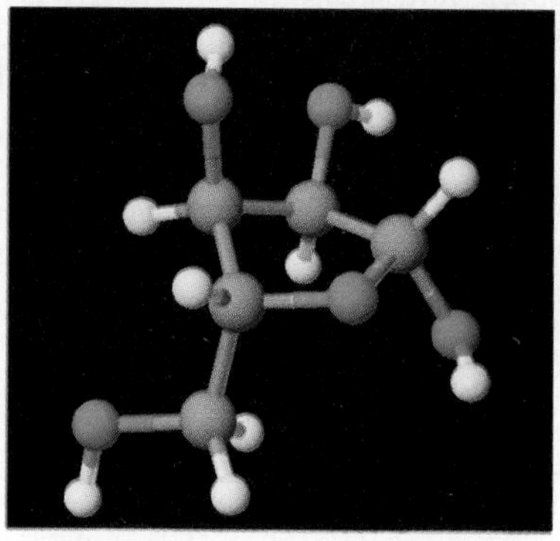

Computer-generated model of β-D-ribose.

85. Which of the following pairs of elements are likely to form ionic compounds when they react with one another? Write appropriate formulas for the ionic compounds you expect to form, and give the name of each.
(a) Chlorine and bromine
(b) Lithium and tellurium
(c) Sodium and argon
(d) Magnesium and fluorine
(e) Nitrogen and bromine
(f) Indium and sulfur
(g) Selenium and bromine

86. Name each of the following compounds, and tell which ones are best described as ionic:
(a) ClF_3 (f) OF_2
(b) NCl_3 (g) NaI
(c) $CaSO_4$ (h) Al_2S_3
(d) C_7H_{16} (i) PCl_5
(e) XeF_4 (j) K_3PO_4

87. Write the formula for each of the following compounds, and tell which ones are best described as ionic:
(a) Sodium hypochlorite
(b) Aluminum perchlorate
(c) Potassium permanganate
(d) Potassium dihydrogen phosphate
(e) Chlorine trifluoride
(f) Boron tribromide
(g) Calcium acetate
(h) Ammonium sulfite
(i) Disulfur dichloride
(j) Phosphorus trifluoride

88. Fluorocarbonyl hypofluorite is composed of 14.6% C, 39.0% O, and 46.3% F. If the molar mass of the compound is 82 g/mol, determine the empirical and molecular formulas of the compound.

89. Azulene, a beautiful blue hydrocarbon, has 93.71% C and a molar mass of 128.16 g/mol. What are the empirical and molecular formulas of azulene?

90. A major oil company has used a gasoline additive called MMT to boost the octane rating of its gasoline. What is the empirical formula of MMT if it is 49.5% C, 3.2% H, 22.0% O, and 25.2% Mn?

91. Direct reaction of iodine (I_2) and chlorine (Cl_2) produces an iodine chloride, I_xCl_y, a bright yellow solid. If you completely used up 0.678 g of iodine, and produced 1.246 g of I_xCl_y, what is the empirical formula of the compound? A later experiment showed the molar mass of I_xCl_y was 467 g/mol. What is the molecular formula of the compound?

92. Pepto-Bismol, which helps provide soothing relief for an upset stomach, contains 300. mg of bismuth subsalicylate, $C_{21}H_{15}Bi_3O_{12}$, per tablet. If you take two tablets for your stomach distress, how many moles of the "active ingredient" are you taking? How many grams of Bi are you consuming in two tablets?

93. Iron pyrite, often called "fool's gold," has the formula FeS_2 (see photo). If you could convert 15.8 kg of iron pyrite to iron metal, how many kilograms of the metal do you obtain?

Iron pyrite, or "fool's gold." (C. D. Winters)

94. Ilmenite is a mineral based on iron and titanium, $FeTiO_3$. If you have an ore containing ilmenite and the ore is 6.75% titanium, what is the mass (in grams) of ilmenite in 1.00 metric ton (exactly 1000 kg) of the ore?

95. Stibnite, Sb_2S_3, is a dark gray mineral from which antimony metal is obtained. If you have 1.00 kg of an ore that contains 10.6% antimony, what mass of Sb_2S_3 (in grams) is in the ore?

96. Which of the following is impossible?
(a) Silver foil that is 1.2×10^{-4} m thick.
(b) A sample of potassium that contains 1.784×10^{24} atoms
(c) Liquid water heated to a temperature of 345 K
(d) A gold coin of mass 1.23×10^{-3} kg
(e) 3.43×10^{-27} mol of S_8

97. Transition metals can combine with carbon monoxide (CO) to form compounds such as $Fe(CO)_5$ and $Co_2(CO)_8$. Assume that you combine 0.125 g of nickel with CO and isolate 0.364 g of $Ni(CO)_x$. What is the value of x?

98. A metal M forms a compound with the formula MCl_4. If the compound is 74.75% chlorine, what is the identity of M?

99. The mass of 2.50 mol of a compound with the formula ECl_4, in which E is a nonmetallic element, is 385 g. What is the molar mass of ECl_4? What is the identity of E?

100. In a reaction 2.04 g of vanadium combines with 1.93 g of sulfur to give a pure compound. What is the empirical formula of the product?

101. A mixture is made of two pure compounds: A (empirical formula = CH_4O) and B (empirical formula = C_2H_6O). If an analysis of the mixture indicates it is 49.9% carbon, how many grams of B are mixed with 1.00 g of A?

102. One experiment often done in general chemistry is heating a piece of copper in the presence of sulfur. The copper reacts with the sulfur to give a copper sulfide, and any excess sulfur vaporizes. Heating is usually continued until the mass of the sample no longer changes. Assume you have heated 1.2517 g of copper with an excess of sulfur. The copper sulfide thereby produced has a mass of 1.5723 g. What is the *precise* ratio of copper to sulfur atoms in the product? That is, how many copper atoms occur per sulfur atom? Is the ratio exactly a whole number? What should have been the final sample weight if the reaction had produced pure copper(I) sulfide (Cu_2S)?

103. The elements A and Z combine to produce two different compounds A_2Z_3 and AZ_2. If 0.15 mole of A_2Z_3 has a mass of 15.9 g, and 0.15 mole of AZ_2 has a mass of 9.3 g, what are the atomic masses of A and Z?

104. A mixture of nitrogen, oxygen, and carbon dioxide gases was analyzed and found to contain 22.5% N and 65.2% O by weight. The total moles of gas in the sample was 0.0108. What are the masses of N_2, O_2, and CO_2 in the sample?

105. The weight percent of oxygen in an oxide that has the formula MO_2 is 15.2%. What is the molar mass of this compound? What element or elements are possible for M?

106. The weight percent of hydrogen in a compound with the formula EH_3 is 8.9%. What is the atomic weight and probable identity of E?

Conceptual Questions

107. Draw diagrams in the boxes shown here to indicate the arrangement of submicroscopic particles of each substance. Consider each box to hold a very tiny portion of each substance. Each drawing should contain at least 16 particles, and it need not be three-dimensional.

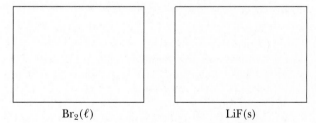

$Br_2(\ell)$ LiF(s)

108. Draw submicroscopic diagrams for each situation listed. Indicate atoms or monatomic ions as circles, indicate molecules or polyatomic ions by overlapping circles for the atoms that make up the molecule or ion, and distinguish among different kinds of atoms by labeling or shading the circles. In each case draw at least ten submicroscopic particles. Your diagrams can be two-dimensional.
(a) A crystal of solid sodium chloride
(b) The sodium chloride from (a) after it has been melted
(c) A sample of molten aluminum oxide, Al_2O_3

109. Assume that each of the ions listed is in aqueous (water) solution. Draw a submicroscopic picture of the ion and the water molecules in its immediate vicinity.
(a) Cl^- (c) Mg^{2+}
(b) K^+ (d) Al^{3+}

110. Of the ions listed in Question 109, which has the strongest attraction for water molecules in its vicinity? Why?

111. A sample of cobalt(II) chloride hexahydrate, a pink substance, was heated in a crucible and turned blue. Explain how you could use this observation to
(a) Keep track of changes in the weather.
(b) Test a series of compounds to see which ones were hydrates.
(c) Make something that indicates whether an electronic component had been exposed to moisture.
(d) Determine whether a substance that was burned in pure oxygen contained the element hydrogen.

112. When analyzed, an unknown compound gave these experimental results: C, 54.0%; H, 6.00%; and O, 40.0%. Four different students used these values to calculate the empirical formulas shown here. Which answer is correct? Why did some students not get the correct answer?
(a) $C_4H_5O_2$ (c) $C_7H_{10}O_4$
(b) $C_5H_7O_3$ (d) $C_9H_{12}O_5$

113. Two general chemistry students working together in the lab weigh out 0.832 g of $CaCl_2 \cdot 2\ H_2O$ into a crucible. After heating for a short time, and allowing the crucible to cool, these students determine that the sample has a mass of 0.739 g. They then do a quick calculation. On the basis of this calculation, what should they do next?
(a) Congratulate themselves on a job well done.
(b) Assume the bottle of $CaCl_2 \cdot 2\ H_2O$ was mislabeled; it actually contained something different.
(c) Heat the crucible again, and then reweigh it.

114. An empirical formula of $C_{5.0}H_{6.8}N_{2.0}$ is calculated from the following experimental data: C, 63.2 ± 0.6%; H, 7.2 ± 0.1%; and N, 29.6 ± 0.3%. Considering the accuracy of the experiment, can the formula be rounded off to $C_5H_7N_2$? Is another formula possible? Describe your reasoning.

Summary Questions

115. A piece of nickel foil, 0.550 mm thick and 1.25 cm square, is allowed to react with fluorine, F_2, to give a nickel fluoride. (a) How many moles of nickel foil were used? (The density of nickel is 8.908 g/cm^3.) (b) If you isolated 1.261 g of the nickel fluoride, what is its formula? (c) What is its name?

116. Uranium is used as a fuel, primarily in the form of uranium(IV) oxide, in nuclear power plants. This question considers some uranium chemistry.
   (a) A small sample of uranium metal (0.169 g) is heated to between 800 and 900 °C in air to give 0.199 g of a dark green oxide, U_xO_y. How many moles of uranium metal were used? What is the empirical formula of the oxide, U_xO_y? How many moles of U_xO_y must have been obtained?
   (b) The naturally occurring isotopes of uranium are ^{234}U, ^{235}U, and ^{238}U. Which is the most abundant?
   (c) If the hydrated compound $UO_2(NO_3)_2 \cdot ?\ H_2O$ is heated gently, the water of hydration is lost. If you have 0.865 g of the hydrated compound and obtain 0.679 g of $UO_2(NO_3)_2$ on heating, how many molecules of water of hydration were in each formula unit of the original compound?

James A. Cusumano

James A. Cusumano obtained a B.A. in 1964 and a Ph.D. in chemistry in 1967 from Rutgers University. He is the co-founder and chairman of Catalytica Inc. of Mountain View, California, a company that develops new catalysts, substances that speed up and control chemical reactions.

By 1990, it is estimated that the United States was producing about eight pounds per person per day of hazardous wastes and air pollutants. Cusumano believes that the most cost-effective, long term solution to pollution control is primary prevention—avoiding the formation of pollutants at the source. Therefore, Catalytica focuses on ways to replace hazardous raw materials with safer substances and to prevent pollution during the manufacturing process. The company's goals, and its method of operation, represent a new direction for the chemical industry.

A Basement Lab and Pop Music

James Cusumano grew up in Elizabeth, New Jersey, the first of 10 children of a postal worker and his homemaker wife. As a young child he had two loves: chemistry and music. His father wanted James to be a medical doctor. However, his father bought him a Gilbert Chemistry Set when he was 9. James set up a laboratory in his basement and decided he wanted to be a chemist. "On every birthday and Christmas all I ever wanted were chemistry books or chemicals. I started what I called O & O Research Laboratories when I was about 12, and I started to make ink. There was a grocery store around the corner. I told them I could make ink, and we could sell it and make a small profit. I guess they wanted to humor me so they let me do it. I made my own labels and packaging, and kids started to buy the stuff. Then I had my first problem. You had to put gum arabic in ink, and I put in too much. It clogged up all the pens, so I had to go back and clean the pens, and I realized I had the wrong formulation. That was my first experience with technical service."

"I got the bug to use chemistry to make other products such as cosmetics and spot removers. Then, in order to get chemicals, I wrote all over the United States to chemical companies asking for samples. I actually got visits from salesmen trying to sell 10,000 pounds of chemicals."

"I was also very interested in music, mainly because I wanted to start a band to play at proms. My dad had me take lessons from a band leader because I said I did not want to learn classical music. I just wanted to learn to play popular music to play at dances. The band leader taught me how to read chords and the lead notes and fake the rest. Then I started to write music. I went to New York City to sell it and started to make a hundred dollars at a time. Both chemistry and music were entrepreneurial, although I didn't think of it at that time. I learned to provide a service and got paid for it."

"When I was 16, one of the songs I sold was recorded ("Short Shorts Twist"), and I started making money. When I was 17 years old I was playing in Las Vegas, and I started making a significant amount of money and was helping my family. All the time my family said music is nice, but you have to continue school. So, I went to Rutgers to study chemistry."

"In my senior year I did an honors research project. That was my next exposure to chemistry and entrepreneurship. My professor said if you can solve this project [the direct combination of benzene (C_6H_6) and ammonia (NH_3) in the presence of oxygen to make the useful compound aniline ($C_6H_5NH_2$)] you may become a millionaire. I thought that was an exciting thing to do and so I worked with him. But I quickly found out the problem was too difficult to solve in the time I had. Nonetheless, that was my first exposure to catalysis."

Cusumano stayed at Rutgers to complete his Ph.D. in chemistry because he had completed some of the graduate work already and because he could continue to play music and could support himself rather luxuriously. Graduate fellowships in those days were only a few thousand dollars. However, he bought a brand new Cadillac from his earnings with music. "I was the only first year graduate student who had a Cadillac at Rutgers University," he said.

Starting a New Company

After his Ph.D. was completed, he went to work for Exxon. "I saw how industry solved problems from an economic point of view. I think industry is a good place to learn about catalysis because you have to worry about how to do science, how to do engineering, and how to make it work economically in the real world."

While at Exxon Cusumano met Ricardo Levy, who is now his partner at Catalytica, and Michel Boudart, a professor of chemical engineering at Stanford University. They came up with the idea of Catalytica Associates, a consulting company. Cusumano, Levy, and Boudart each put up $10,000, and started doing business in Levy's basement in 1974. "The first day we started work we got a contract with Merck, Sharpe, and Dohme in New Jersey. We agreed to help them with a problem on a compound that prevents worms in chickens."

Within a few weeks they had more contracts and decided to move to California. "In the period from 1974 to 1984 we built to about a 50-person business. Very comfortable and very profitable. We worked on about 200 projects with a hundred companies." The company continued to grow, and by the time they first offered stock for public sale in February 1993, they had raised many millions of dollars from venture capitalists. They also had begun to develop their own technology. They wanted to "use catalysis to create technology that is economically attractive and environmentally more friendly. We do that by designing a catalyst that takes the raw materials in a selective way directly to the product. That is the best situation. There are not many byproducts, so you don't have to get rid of them, and you don't have any waste."

Solving Environmental Problems

Catalytica has picked several areas where there are environmental problems and where the problems could be solved economically. Cusumano said they first turned to the problem of reducing nitrogen oxide (NO_x) emissions in gas-fired turbines used to generate electricity. When natural gas is burned, the temperature of the gas is about 1800 °C. The problem is that this gas cannot be put into the turbine because it will destroy the turbine blades. It must be at 1300 °C. In addition, at 1800 °C the N_2 and O_2 in the air combine to give NO_x, pollutants that must be removed. Catalytica's solution has been to develop a catalyst that allows natural gas to burn flamelessly at 1300 °C, where NO_x does not form. Furthermore, the cooler gas can be introduced into the turbine directly.

Catalytica's combustion system is being tested in conjunction with General Electric and is also being adapted to automobiles. In the near future, 2% of the cars in California and other states will have to be "zero emission vehicles." "Right now that means the electric car. The oil companies, Detroit, and others are not doing much with electric cars because battery technology is not such that it is possible to fuel a car that way. So, we have designed a completely different system that uses our technology. The consumer can use gasoline, and all that comes out the tail pipe is CO_2 and water. The electric car is zero pollution when it is on the highway, but not when you recharge it. It puts the pollution somewhere else."

As a result of recent environmental legislation, it is more difficult for petroleum refiners to produce high-octane gasoline economically without environmental risks. One way to make high-octane gasoline is to add "alkylates," alkanes [see Table 3.4] with special characteristics that make them burn efficiently. The problem now is that liquid hydrogen fluoride (HF) or concentrated sulfuric acid (H_2SO_4) is required to make alkylates. Indeed, some communities have passed laws that ban the use of HF in local chemical plants. Catalytica has developed a new process that replaces these highly acidic compounds with a solid catalyst that is safe enough to put on the kitchen counter.

Teamwork in Industry

The way technology is developed in companies in the United States is changing, and Catalytica demonstrates these new directions. "It used to be that chemists worked by themselves and came up with some interesting reactions. Then they would go to a chemical engineer, whom they had never met and who had no vested interest in the process of chemistry. The engineer would say 'that is not economically attractive' or 'it can't be done.' If they got past that group they would go to the marketing group, which may not have an interest in the product. Catalytica can't afford those kinds of mistakes. We have chemists work together from the beginning with chemical engineers and marketing people. They all have a vested interest in the technology and the business, and they work in parallel. If you put those things together you can increase the probability of success. That is extremely important for commercialized technology."

The Next Step

What does the future hold for the chemical industry? Cusumano believes that we will be using alkanes to make chemicals for the next 40 to 50 years. After that, he believes, the future is in biochemistry. "We will take things that grow and are renewable and convert them into chemicals and advanced materials and products. There is a little bit of that being done right now, but it is probably the next step. It will be a natural progression."

CHAPTER 4

Principles of Reactivity: Chemical Reactions

(C. D. Winters)

A Chemical Puzzler

Are you a tea drinker? Do you take it with one lump of sugar or two? With milk? What about a squeeze of lemon juice? If you use lemon juice, did you ever notice that the tea changes color when you add the lemon juice? Or have you ever had an upset stomach and used an Alka-Seltzer tablet to help settle it? You drop it into a glass of water, and the tablet fizzes. Bubbles of gas quickly rise in water. What is the connection between these observations? These are both chemical reactions, but what is the chemistry behind them?

Chemists have studied hundreds of thousands of reactions, and thousands more are waiting to be investigated. Hundreds of reactions are used by the chemical industry to manufacture the products we all use, and the mere act of reading this sentence involves an untold number of reactions in your body. Finally, thousands of reactions occur in our environment, both to our benefit and our harm. After looking at some basic principles of how to write chemical equations and the nature of compounds when they are dissolved in water, we classify some common reactions into a few types so that you can begin to predict what might happen in some reactions that you or even professional chemists have never seen before.

4.1 CHEMICAL EQUATIONS

To create the reactions shown in Figure 4.1, we cut ordinary kitchen aluminum foil into small pieces, dropped them into a beaker of liquid bromine in a laboratory fume hood, and moved back! The aluminum pieces soon began to burn with a brilliant light, and glowing globs of molten aluminum appeared in the beaker. The **chemical reaction** that occurred was the combination of aluminum atoms from the solid metal with Br_2 molecules from the liquid element to give a white solid consisting of Al_2Br_6 molecules. We can depict this using the following **balanced chemical equation** that shows the relative amounts of **reactants** (the substances combined in the reaction) and **products** (the substances obtained).

Al_2Br_6 has the empirical formula $AlBr_3$. The substance exists as a molecule, however, with the formula Al_2Br_6 in the gaseous or solid state.

$$\underset{\text{reactants}}{2\ \mathrm{Al(s)} + 3\ \mathrm{Br_2}(\ell)} \longrightarrow \underset{\text{product}}{\mathrm{Al_2Br_6(s)}}$$

(a) (b) (c)

Figure 4.1 Bromine Br_2, an orange-brown liquid, and aluminum metal (a) react so vigorously that the aluminum becomes molten and glows white hot (b). The vapor in (b) consists of vaporized Br_2 and some of the product, white Al_2Br_6. At the end of the reaction (c), the beaker is coated with aluminum bromide and the products of its reaction with atmospheric moisture. (*Note:* This reaction is dangerous! Under no circumstances should it be done except under properly supervised conditions.) (C. D. Winters)

PORTRAIT OF A SCIENTIST

Antoine Laurent Lavoisier (1743–1794)

On Monday, August 7, 1774, the Englishman Joseph Priestley (1733–1804) became the first person to isolate oxygen. He heated solid mercury(II) oxide, HgO, causing the oxide to decompose to mercury and oxygen.

$$2\ HgO(s) \longrightarrow 2\ Hg(\ell) + O_2(g)$$

Priestley did not immediately understand the significance of the discovery, but he mentioned it to the French chemist Antoine Lavoisier in October, 1774. One of Lavoisier's contributions to science was his recognition of the importance of exact scientific measurements and of carefully planned experiments, and he applied these methods to the study of oxygen. From this work he came to believe Priestley's gas was present in all acids and so he named it "oxygen," from the Greek words meaning "to form an acid." In addition, Lavoisier observed that the heat produced by a guinea pig when exhaling a given amount of carbon dioxide is similar to the quantity of heat produced by burning carbon to give the same amount of carbon dioxide. From this and other experiments he concluded that "Respiration is a combustion, slow it is true, but otherwise perfectly similar to that of charcoal." Although he did not understand the details of the process, this was an important step in the development of biochemistry.

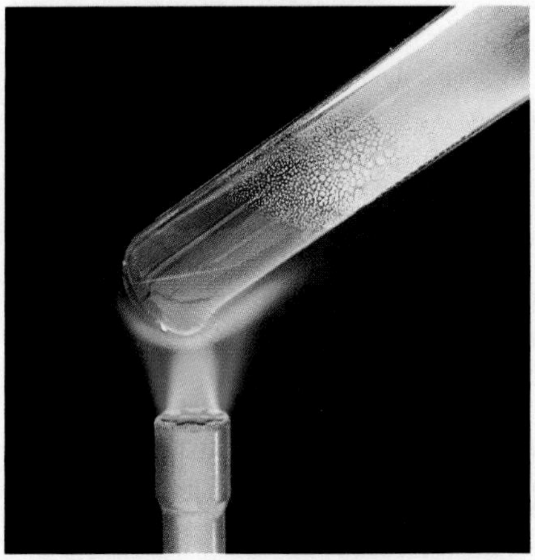

The decomposition of red mercury(II) oxide to give metallic mercury and oxygen. The mercury is seen as a film on the surface of the test tube. (C. D. Winters)

Lavoisier was a prodigious scientist, and the principles of naming chemical substances that he introduced are still in use today. Furthermore, he wrote a textbook in which he applied for the first time the principles of the conservation of matter to chemistry and used the idea to write early versions of chemical equations.

Because Lavoisier was an aristocrat, he came under suspicion during the Reign of Terror of the French Revolution. He was an investor in the Ferme Générale, the infamous tax-collecting organization in 18th century France. Tobacco was a monopoly product of the Ferme Générale, and it was common to cheat the purchaser by adding water to the tobacco, a practice that Lavoisier opposed. Nonetheless, because of his involvement with the Ferme, his career was cut short by the guillotine on May 8, 1794, on the charge of "adding water to the people's tobacco."*

*For an account of Lavoisier's life and his friendship with Benjamin Franklin, see Stephen Jay Gould: "The passion of Antoine Lavoisier." *Bully for Brontosaurus.* New York, Norton, 1991.

Lavoisier and his wife, as painted in 1788 by Jacques-Louis David. (The Metropolitan Museum of Art)

The physical states of the reactants and products are also indicated in an equation. The symbol (s) indicates a solid, (g) a gas, and (ℓ) a liquid. What the equation does *not* show are the conditions of the experiment or if any energy (in the form of heat or light) is involved. Lastly, a chemical equation does not tell you if the reaction happens very quickly or if it takes 100 years.

The energy involved in chemical reactions is the subject of Chapters 6 and 20. The study of the rate or speed of reactions—the science of kinetics—is the topic of Chapter 15.

In the 18th century, the great French scientist Antoine Lavoisier introduced the law of conservation of matter, which later became part of Dalton's atomic theory (Section 2.1). Lavoisier showed that *matter can neither be created nor destroyed.* This means that if you use 10 grams of reactants, then, if the reaction is complete, you must end up with 10 grams of products. Combined with Dalton's atomic theory, this also means that if 1000 atoms of a particular element react, then those 1000 atoms must appear in the products in some fashion. When applied to the reaction of aluminum and bromine, the conservation of matter means that 2 atoms of aluminum and 3 diatomic molecules of Br_2 (or 6 atoms of Br) are required to produce 1 molecule of Al_2Br_6 (in which two atoms of Al and six atoms of Br are combined).

$$2\ Al(s) + 3\ Br_2(\ell) \longrightarrow Al_2Br_6(s)$$

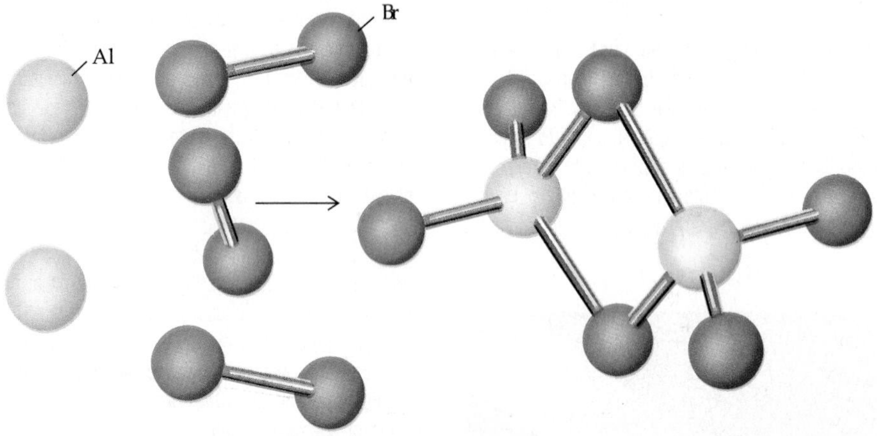

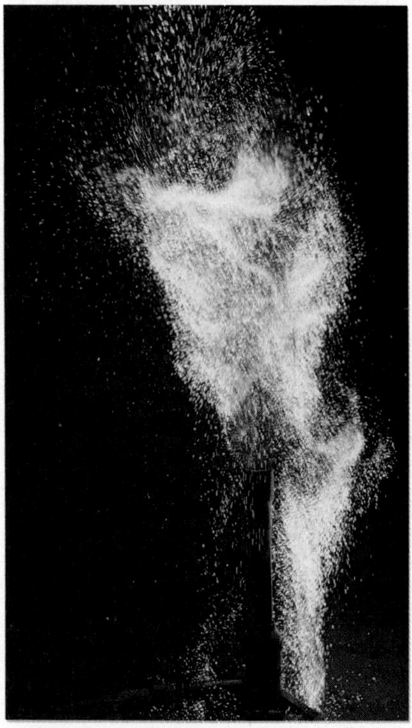

Figure 4.2 A stream of powdered iron is squirted into the flame of a Bunsen burner. The tiny particles of iron react with oxygen to form the iron oxides, FeO and Fe_2O_3. The flame and the energy released in the reaction heat the particles so they are white hot. (C. D. Winters)

The numbers in front of each substance in the balanced equation are put there to show how matter is conserved. They can be read as number of atoms (as in 2 atoms of Al), in terms of molecules (as in 3 molecules of Br_2), or in terms of moles (as in 1 mol of Al_2Br_6). Thus, the balanced equation for the reaction of aluminum and bromine tells us that 2 mol of aluminum react with 3 mol of bromine to produce 1 mol of Al_2Br_6. The relationship between the quantities of chemical reactants and products is called **stoichiometry** (pronounced ''stoy-key-AHM-uh-tree'') and the coefficients (or multiplying numbers) in a balanced equation are the **stoichiometric coefficients.**

Balanced chemical equations are fundamentally important in depicting the outcome of chemical reactions and to the quantitative understanding of chemistry. Thus, this chapter has two important goals.

- Learn to balance simple chemical equations (Sections 4.2 and 4.5).
- Learn a few general types of chemical reactions (precipitation, acid-base, gas-forming, oxidation-reduction) so that you are aware of the outcome of common reactions (Sections 4.6–4.10).

To reach the second goal, you need some tools. For this reason, you will also learn about the behavior of ionic compounds in aqueous solution (Section 4.3) and the identity and behavior of common acids and bases (Section 4.4).

EXERCISE 4.1 ***Chemical Reactions***

The reaction of iron with oxygen is shown in Figure 4.2. The equation for the reaction is

$$4\ Fe(s) + 3\ O_2(g) \longrightarrow 2\ Fe_2O_3(s)$$

1. What are the stoichiometric coefficients in this equation?
2. If you were to use 8000 atoms of Fe, how many molecules of O_2 are required to consume the iron completely? ■

Figure 4.3 A piece of magnesium ribbon burns in air to give the white solid, magnesium oxide, MgO. (C. D. Winters)

4.2 BALANCING CHEMICAL EQUATIONS

A chemical equation must be balanced before useful quantitative information can be obtained about a chemical reaction. Balancing equations ensures that the same number of atoms of each element appears on both sides of the equation. Many chemical equations can be balanced by trial and error, although some involve more trial than others.

One general class of chemical reactions is the reaction of metals or nonmetals with oxygen to give **oxides** of the general formula of M_xO_y. For example, iron reacts with oxygen to give iron(III) oxide (Figure 4.2),

$$4\ Fe(s) + 3\ O_2(g) \longrightarrow 2\ Fe_2O_3(s)$$

magnesium gives magnesium oxide (Figure 4.3),

$$2\ Mg(s) + O_2(s) \longrightarrow 2\ MgO(s)$$

and phosphorus, P_4, reacts vigorously with oxygen to give tetraphosphorus decaoxide, P_4O_{10} (Figure 4.4).

$$P_4(s) + O_2(g) \xrightarrow{\text{(unbalanced equation)}} P_4O_{10}(s)$$

The equations involving iron and magnesium are balanced as we have written them because the same number of metal and oxygen atoms occur on each side of the equation. Similarly, the phosphorus atoms are balanced in the $P_4 + O_2$ equation, but the oxygen atoms are not. To balance them, we must place a 5 in front of the O_2 on the left to give ten oxygen atoms on both sides of the equation.

$$P_4(s) + 5\ O_2(g) \xrightarrow{\text{(balanced equation)}} P_4O_{10}(s)$$

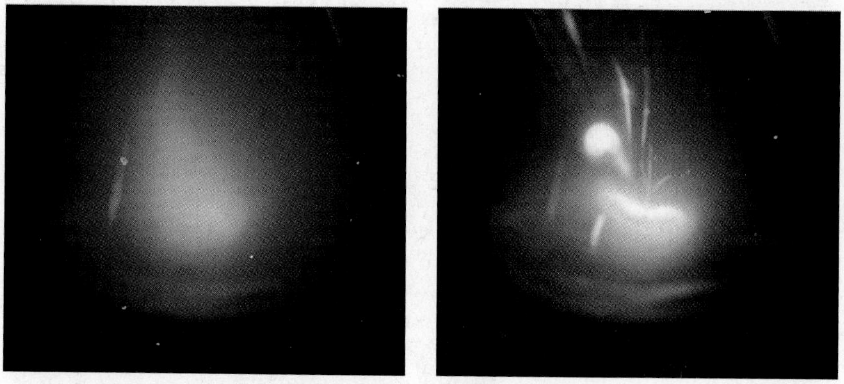

Figure 4.4 When white phosphorus is exposed to oxygen, it first glows *(left)*, a process called phosphorescence, and then bursts into flame *(right)*. To prevent this, white phosphorus is usually stored under water (Figure 3.3). (C. D. Winters)

The **combustion** or burning of a fuel in oxygen is accompanied by the evolution of heat and light. You are most familiar with combustion reactions such as the burning of octane, a component of gasoline.

$$2\ C_8H_{18}(g) + 25\ O_2(g) \longrightarrow 16\ CO_2(g) + 18\ H_2O(\ell)$$

In all combustion reactions involving oxygen, some or all the elements in the reactants end up as compounds containing oxygen, that is, as oxides. For hydrocarbons (Table 3.4) and for compounds containing only C, H, and O, the products of *complete* combustion are always carbon dioxide and water.

As an example of equation balancing, let us write the balanced equation for the complete combustion of propane, C_3H_8.

Step 1. *First, write down the correct formulas of the reactants and products.*

$$C_3H_8(g) + O_2(g) \xrightarrow{\text{(unbalanced equation)}} CO_2(g) + H_2O(\ell)$$

Here propane and oxygen are the reactants, and carbon dioxide and water are the products.

Step 2. *Balance the number of C atoms.* In combustion reactions it is usually best to balance the carbon atoms first and leave the oxygen atoms to the end (because the oxygen atoms are not all found in one compound). In this case three carbon atoms are in the reactants, so three must occur in the products. Three CO_2 molecules are therefore required on the right side.

$$C_3H_8 + O_2 \longrightarrow 3\ CO_2 + H_2O$$

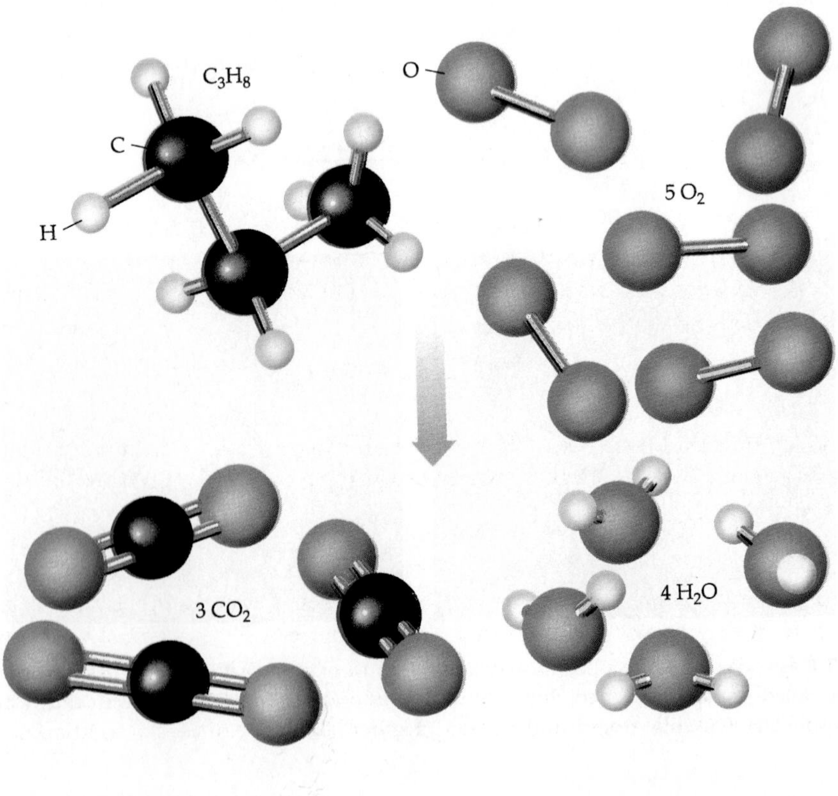

Step 3. *Balance the number of H atoms.* Eight H atoms are in the reactants. Each molecule of water has two hydrogen atoms, so four molecules of water provide the required eight hydrogen atoms on the right side.

$$C_3H_8 + O_2 \rightarrow 3\ CO_2 + 4\ H_2O$$

Step 4. *Balance the number of O atoms.* Ten oxygen atoms are on the right side ($3 \times 2 = 6$ in CO_2 plus $4 \times 1 = 4$ in water). Therefore, five O_2 molecules are needed to supply the required ten oxygen atoms.

$$C_3H_8 + 5\ O_2 \longrightarrow 3\ CO_2 + 4\ H_2O$$

Step 5. *Verify that the number of atoms of each element is balanced.* The equation shows 3 carbon atoms, 8 hydrogen atoms, and 10 oxygen atoms on each side.

The balancing of equations by trial and error is usually simple enough if you are organized in your approach. All equations that have the correct formulas for all reactants and products can be balanced. Finally, it is important to understand that *subscripts in the formulas of reactants and products cannot be changed to balance equations.* These subscripts identify these substances, and changing them changes the identity of the substance. For example, you cannot change CO_2 to CO to balance an equation because carbon monoxide, CO, is a very different compound from carbon dioxide, CO_2.

EXAMPLE 4.1 *Balancing the Equation for a Combustion Reaction*

Write the balanced equation for the combustion of butane, C_4H_{10}.

Solution

Step 1. *Write the correct formulas for reactants and products.* As is always the case for compounds containing only C, H, and O, the products of complete combustion are CO_2 and H_2O if the reaction goes to completion. Therefore, the unbalanced equation is

$$C_4H_{10}(g) + O_2(g) \longrightarrow CO_2(g) + H_2O(\ell)$$

Step 2. *Balance the C atoms.* Four carbon atoms in butane require the production of four CO_2 molecules.

$$C_4H_{10}(g) + O_2(g) \longrightarrow 4\ CO_2(g) + H_2O(\ell)$$

Step 3. *Balance the H atoms.* There are ten hydrogen atoms on the left, so five molecules of H_2O, each having two hydrogen atoms, are required on the right.

$$C_4H_{10}(g) + O_2(g) \longrightarrow 4\ CO_2(g) + 5\ H_2O(\ell)$$

Step 4. *Balance the O atoms.* As the reaction stands after step 3, there are 2 oxygen atoms on the left side and 13 on the right ($4 \times 2 = 8$ in CO_2 plus $5 \times 1 = 5$ in H_2O). That is, there is an even number of O atoms on the left and an odd number on the right. Because there cannot be an odd number of O atoms on the left (O atoms are found as O_2 molecules), multiply each coefficient on both sides of the equation by two so that there is now an even number of oxygen atoms (26) on the right side.

$$2\ C_4H_{10}(g) + \_\ O_2(g) \longrightarrow 8\ CO_2(g) + 10\ H_2O(\ell)$$

Now the O_2 on the left can be balanced by having 13 O_2 molecules on the left side of the equation.

$$2\ C_4H_{10}(g) + 13\ O_2(g) \longrightarrow 8\ CO_2(g) + 10\ H_2O(\ell)$$

Step 5. *Verify the result.* We see that 8 carbon atoms, 20 hydrogen atoms, and 26 oxygen atoms occur on each side of the equation.

EXERCISE 4.2 ***Balancing the Equation for a Combustion Reaction***

1. Pentane can burn completely in air to give carbon dioxide and water. Write a balanced equation for this combustion reaction. Refer to Table 3.4 for the formula of pentane.
2. Write a balanced chemical equation for the complete combustion of tetraethyllead, $Pb(C_2H_5)_4$ (which was used until recently as a gasoline additive). The products of combustion are $PbO(s)$, $H_2O(\ell)$, and $CO_2(g)$. ■

4.3 PROPERTIES OF COMPOUNDS IN AQUEOUS SOLUTION

Many reactions occur between solids and gases, or between two gases. The reactions in your body, however, occur in large part among substances dissolved in water, that is, in an **aqueous solution.** This is also true of many reactions that you will see in the laboratory and that occur all around you in nature. Therefore, to understand reactions in aqueous solutions, it is important first to understand something about the behavior of compounds in water. The focus here is on compounds that produce ions in aqueous solution.

If magnesium ribbon is put into an aqueous solution of the compound HCl, the mixture will bubble furiously as hydrogen gas is given off according to the following equation (Figure 4.5).

$$Mg(s) + 2\ HCl(aq) \longrightarrow MgCl_2(aq) + H_2(g)$$

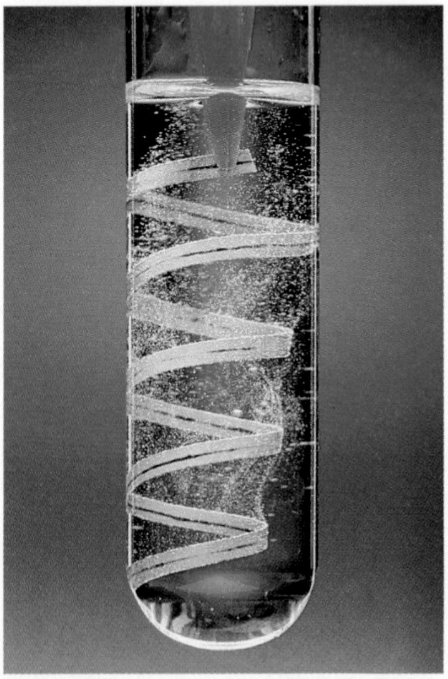

Figure 4.5 A ribbon of magnesium reacts with aqueous HCl to give H_2 gas and aqueous $MgCl_2$. (C. D. Winters)

In this balanced equation both HCl and $MgCl_2$ are followed by (*aq*) indicating that they are dissolved in water and are present as aqueous solutions. Also, these two substances exist as ions in the aqueous solution: $MgCl_2$ exists as $Mg^{2+}(aq)$ and $Cl^-(aq)$, and HCl exists as $H^+(aq)$ and $Cl^-(aq)$. But how do we know that ions are present in these aqueous solutions?

Ions in Aqueous Solution

A substance whose aqueous solution conducts electricity is called an **electrolyte** (Section 3.4). Ionic compounds, such as NaCl and $Ca(OH)_2$, can dissolve in water to varying degrees and produce aqueous solutions of ions. Because such solutions conduct electricity, ionic compounds are generally electrolytes.

Electrolytes can be classified as *strong* or *weak.* When sodium chloride and many other ionic compounds dissolve in water, the ions separate or dissociate

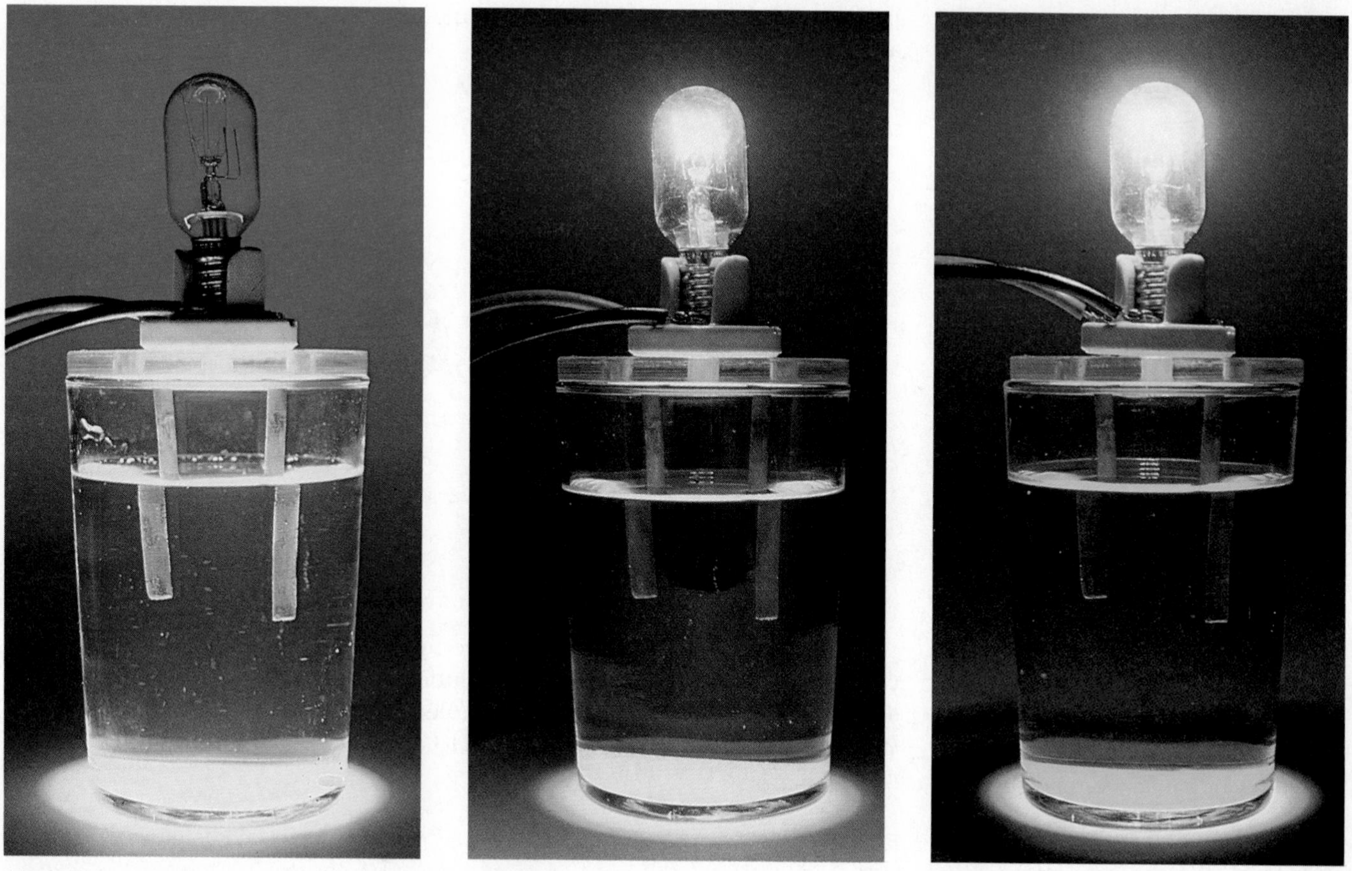

(a) (b) (c)

Figure 4.6 When an electrolyte is dissolved in water in the beaker and provides ions that are free to move about, the electric circuit is completed, and the light bulb included in the circuit glows. (See Figures 3.14 and 3.15.) (a) Pure water is a nonelectrolyte, and the bulb does not light. (b) In the case of a weak electrolyte, such as acetic acid, only a few of the dissolved molecules ionize. The bulb glows weakly, indicating that only a small current of electricity flows. (c) Dilute K_2CrO_4, potassium chromate, is a strong electrolyte. The bulb glows brightly, indicating that virtually every K_2CrO_4 unit has dissociated into its ions, K^+ and CrO_4^{2-}. (C. D. Winters)

completely. For every mole of NaCl that dissolves, one mole of Na^+ and one mole of Cl^- ions are found in solution.

$$\text{NaCl(aq)} \equiv \text{Na}^+\text{(aq)} + \text{Cl}^-\text{(aq)}$$
$$100\% \text{ dissociation} = \text{strong electrolyte}$$

The idea that salts such as NaCl dissociate *completely* to give only ions in solution is a simplification. In fact, there can be a measurable concentration of species such as NaCl(aq), species called "ion pairs."

Because ions are in high concentration in solution, the solution is a good conductor of electricity. Substances that dissociate completely in water and whose solutions conduct well are **strong electrolytes.**

Other substances produce only a small concentration of ions when they dissolve, and so are poor conductors of electricity; they are known as **weak electrolytes** (Figure 4.6). For example, when acetic acid—an important ingredient in vinegar—dissolves, only a few percent of the molecules are ionized to produce a cation and anion.

$$CH_3CO_2H(aq) \longrightarrow H^+(aq) + CH_3CO_2^-(aq)$$

acetic acid — hydrogen ion — acetate ion

< 5% ionization = weak electrolyte

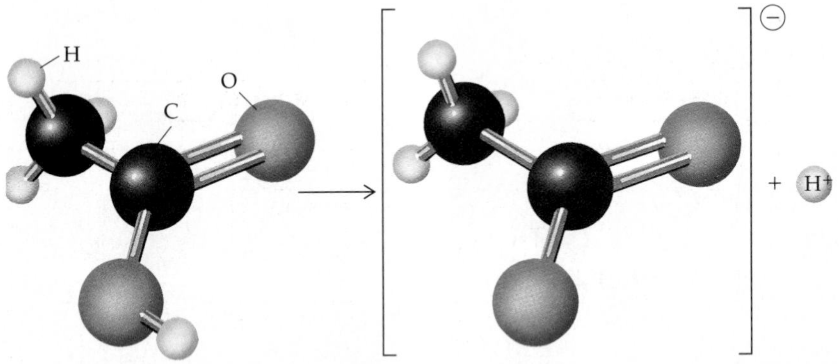

The term "dissociate" describes ionic compounds that separate into their constituent ions in aqueous solution. The term "ionize" describes nonionic or molecular compounds, like acetic acid, whose constituent molecules form ions in water.

Substances also occur that dissolve in water but do not ionize. These are called **nonelectrolytes** because their solutions do not conduct electricity (Figure 4.6). Nonelectrolytes are generally molecular compounds; examples include sugar ($C_{12}H_{22}O_{11}$), starch, ethanol (CH_3CH_2OH), and antifreeze (ethylene glycol, HOC_2H_4OH).

EXERCISE 4.3 ***Electrolytes***

Epsom salt, $MgSO_4 \cdot 7\ H_2O$, is sold in drugstores and used, as a solution in water, for various medical purposes. Methanol, CH_3OH, is dissolved in gasoline in the winter in colder climates to prevent the formation of ice in automobile fuel lines. Which of these compounds is an electrolyte and which is a nonelectrolyte? ■

Solubility of Ionic Compounds in Water

Table salt dissolves readily in water, but we cannot say from this that all ionic compounds dissolve to this extent in water. There are many that dissolve only to a small extent and still others that are essentially insoluble.* Fortunately, we can make some general statements about which types of ionic compounds are water-soluble.

Figure 4.7 lists a set of broad guidelines that can help predict whether a particular ionic compound will be soluble in water. For example, sodium nitrate, $NaNO_3$, contains both an alkali metal cation, Na^+, and the nitrate anion, NO_3^-. According to Figure 4.7, the presence of either of these ions ensures that the

*Chemists often refer to substances as soluble, partly soluble, or insoluble. These terms have no precise meaning, however. Usually what they mean is that the substance is soluble if a good deal of solid has visibly dissolved, when only a modest amount of solid has been added to the solvent. If only a small amount of solid has visibly dissolved, the solid is partly soluble. If no solid can be observed to dissolve, the substance is called insoluble or poorly soluble. Be aware, however, that even insoluble solids do dissolve to some extent (see Chapter 19).

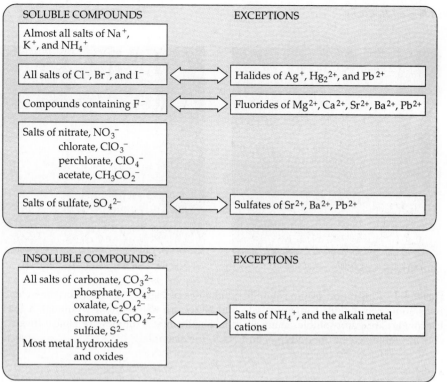

SOLUBLE COMPOUNDS	EXCEPTIONS
Almost all salts of Na^+, K^+, and NH_4^+	
All salts of Cl^-, Br^-, and I^-	Halides of Ag^+, Hg_2^{2+}, and Pb^{2+}
Compounds containing F^-	Fluorides of Mg^{2+}, Ca^{2+}, Sr^{2+}, Ba^{2+}, Pb^{2+}
Salts of nitrate, NO_3^- chlorate, ClO_3^- perchlorate, ClO_4^- acetate, $CH_3CO_2^-$	
Salts of sulfate, SO_4^{2-}	Sulfates of Sr^{2+}, Ba^{2+}, Pb^{2+}

INSOLUBLE COMPOUNDS	EXCEPTIONS
All salts of carbonate, CO_3^{2-} phosphate, PO_4^{3-} oxalate, $C_2O_4^{2-}$ chromate, CrO_4^{2-} sulfide, S^{2-} Most metal hydroxides and oxides	Salts of NH_4^+, and the alkali metal cations

Figure 4.7 Guidelines to predict the solubility of ionic compounds. If a compound contains *one of the ions* in the column on the left in the top chart, the compound is predicted to be at least moderately soluble in water. There are a few exceptions, and those are noted at the right. Poorly soluble ionic compounds are usually formed by the anions listed at the bottom of the chart, with the exceptions of compounds with NH_4^+ and the alkali metal cations.

compound is soluble in water. Furthermore, because ionic compounds that dissolve in water are electrolytes, $NaNO_3$ in aqueous solution really consists of the separated ions $Na^+(aq)$ and $NO_3^-(aq)$.

$NaNO_3$ in water consists of

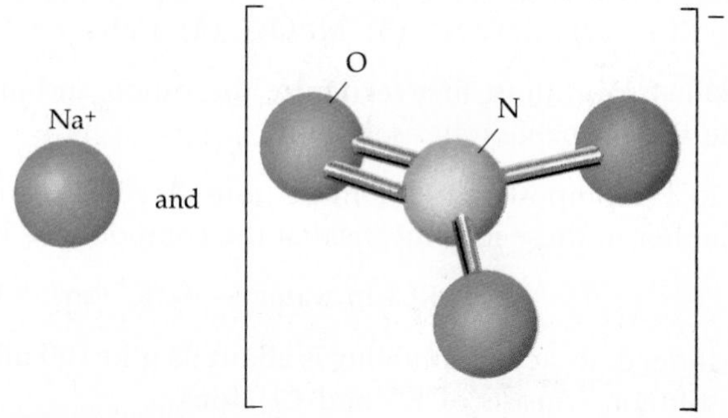

On the other hand, calcium hydroxide is poorly soluble in water. For example, if a spoonful of $Ca(OH)_2$ is placed in 100 mL of water, only 0.185 g or 0.00250 mol will dissolve at 0 °C.

$$0.00250 \text{ mol } Ca(OH)_2 \text{ dissolves in 100 mL water at 0 °C} \longrightarrow$$
$$0.00250 \text{ mol } Ca^{2+}(aq) + 2 \times 0.00250 \text{ mol } OH^-(aq)$$

When solid $Ca(OH)_2$ is placed in water, nearly all of it remains as the solid, and a heterogeneous mixture is the result (Figure 4.8). This is true of all metal hydroxides except those containing an alkali metal cation.

Other examples of common ionic compounds are given in Figure 4.8. Based on the guidelines, and the examples in Figure 4.8, you can assume an ionic compound will be moderately soluble if it contains *at least one* of the ions listed in the upper part of Figure 4.7.

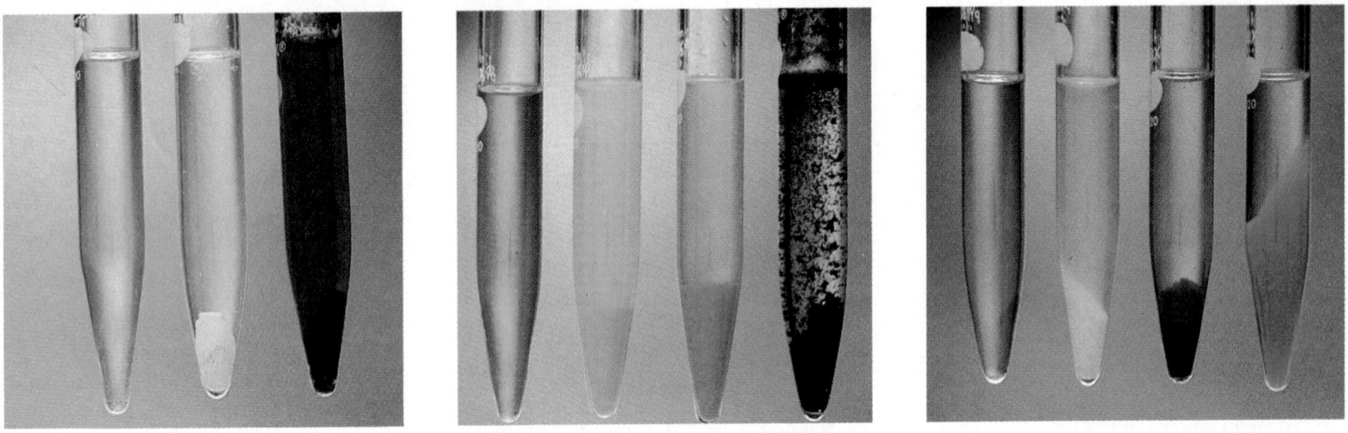

(a) $AgNO_3$, AgCl, AgOH (b) $(NH_4)_2S$, CdS, Sb_2S_3, PbS (c) NaOH, $Ca(OH)_2$, $Fe(OH)_3$, $Ni(OH)_2$

Figure 4.8 Illustration of the solubility guidelines of Figure 4.7. With certain exceptions, ionic compounds containing Cl^- and NO_3^- are water-soluble. Although a few hydroxides are water-soluble, many are not. Finally, metal sulfides are almost invariably insoluble, except for $(NH_4)_2S$ and Na_2S, for example.

EXAMPLE 4.2 *Solubility Guidelines*

Predict whether each of the following ionic compounds is likely to be water-soluble. If it is soluble, tell what ions exist in solution.
(1) KCl (2) $MgCO_3$ (3) MnO_2 (4) CaI_2

Solution You must first recognize the cation and anion involved and then decide the probable water-solubility.

1. KCl is composed of K^+ and Cl^- ions. According to Figure 4.7, the presence of either of these ions means that the compound is likely to be soluble in water.

$$\text{KCl in water} \longrightarrow K^+(aq) + Cl^-(aq)$$

Indeed, its actual solubility is about 35 g in 100 mL of water at 20 °C, and the solution consists of K^+ and Cl^- ions.

2. Magnesium carbonate is composed of Mg^{2+} and CO_3^{2-} ions. Mg^{2+} is in the alkaline earth group, a group that often does not form water-soluble compounds. The carbonate ion usually gives insoluble compounds (Figure 4.7), unless combined with an ion like Na^+ or NH_4^+. $MgCO_3$ is therefore predicted to be insoluble in water. (The actual solubility of $MgCO_3 \cdot 3\ H_2O$ is less than 0.2 g per 100 mL of water.)
3. Manganese(IV) oxide is composed of Mn^{4+} and O^{2-} ions. Again, Figure 4.7 suggests that oxides are soluble only when O^{2-} is combined with an alkali metal ion; Mn^{4+} is a transition metal ion, so MnO_2 is insoluble.
4. Calcium iodide is composed of Ca^{2+} and I^- ions. According to Figure 4.7 almost all iodides are soluble in water, so CaI_2 is water-soluble and produces calcium and iodide ions in water.

$$CaI_2 \text{ in water} \longrightarrow Ca^{2+}(aq) + 2\ I^-(aq)$$

A CLOSER LOOK *Soluble or Insoluble? Why Is It Important?*

Why do we care if compounds are soluble in water? As you will see later in this chapter, if a chemical reaction forms an insoluble compound from water-soluble materials, we can predict that the reaction will proceed from reactants to products. But there is a more important reason for our interest. The worldwide economy depends on metals and metal-containing compounds. An ordinary telephone, for example, contains more than 40 elements, the majority of them metals, such as titanium, chromium, iron, nickel, copper, zinc, and mercury. The minerals containing these metals must be mined from deposits in the earth, and the figure and table illustrate the point that most of them are compounds containing such anions as oxide or sulfide, whereas others exist as carbonates, phosphates, and silicates.

In terms of our current discussion, the important point is that these deposits would not be available in sufficient concentration to be mined if they were soluble in water. Over the centuries ground water would have dissolved the minerals and carried them away to the oceans. So, it is not surprising that minerals mined from the earth usually exist as compounds containing the anions in Figure 4.7 that lead to insoluble compounds.

Of course some elements, chiefly the alkali metals, exist as halides, such as NaCl and KCl. But these are obtained from the oceans or from brine wells. The latter are deposits in the earth that are concentrated aqueous solutions of a salt.

In some cases otherwise water-soluble compounds can be mined as the solid, and rock salt, or pure NaCl, is an example. These are generally found in very deep mines (averaging 1000 feet below the earth's surface) where water has not penetrated. Such deep mines are found in Louisiana, Texas, Ohio, New York, and Michigan. In fact, some salt mines that no longer contain minable deposits are now used to store equipment and archives. They are good storage places because they are so dry.

A deep salt mine. Here the salt is so deeply buried in the earth that water has not penetrated to that area for centuries. Consequently, the salt was not dissolved and carried away as a salt solution. (Morton Salt)

Some Ores of Important Metals

Metal	Ore
Iron	magnetite, Fe_3O_4
	hematite, Fe_2O_3
Titanium	ilmenite, $FeTiO_3$
	rutile, TiO_2
Copper	chalcocite, Cu_2S
	chalcopyrite, $CuFeS_2$
Molybdenum	molybdenite, MoS_2
Zinc	zinc blende, ZnS
Lead	galena, PbS
Mercury	cinnabar, HgS

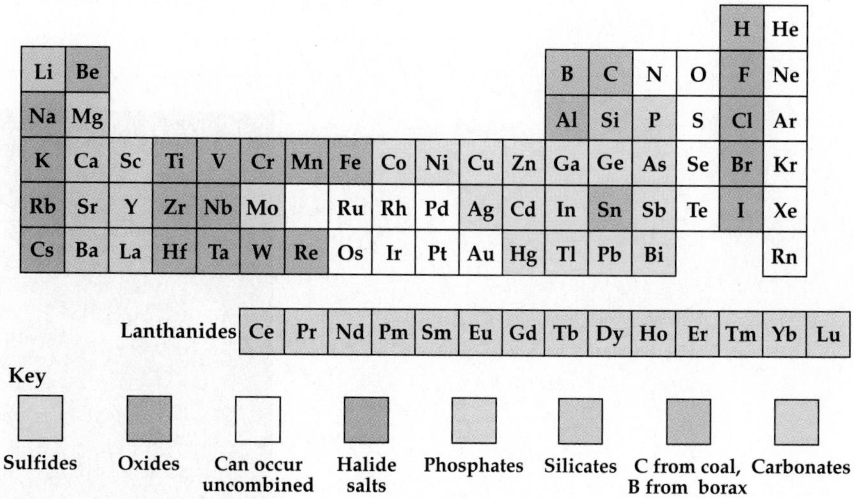

Sources of the elements. The transition metals are found uncombined or as oxides or sulfides. The lanthanides occur predominantly as phosphates. The blank spaces in the table are for elements that do not occur in nature (Tc, below Mn; Po, below Te; and At, below I).

Notice that the compound gives two I^- ions on dissolving in water. (A common misconception is that I_2 or some ion such as I_2^{2-} is found in solution. All halide-containing ionic compounds that dissolve in water produce F^-, Cl^-, Br^-, or I^- in aqueous solution.)

EXERCISE 4.4 ***Solubility of Ionic Compounds***

Predict whether each of the following ionic compounds is likely to be soluble in water. If soluble, write the formulas for the ions that will exist in aqueous solution.

(1) KNO_3 (2) $CaCl_2$ (3) CuO (4) $NaCH_3CO_2$ ■

4.4 ACIDS AND BASES

Acids and bases are two *very* important classes of compounds. Members of each class have a number of properties in common, some of which are related to properties of the other class. Solutions of acids change the colors of vegetable pigments in specific ways; for example, all acids change the color of litmus, a dye derived from certain lichens, from blue to red. And when you add lemon to tea, the acid in the lemon (citric acid) reacts with the compounds that give tea its color and change the color.

Similarly, bases affect pigments, but bases turn red litmus blue and make phenolphthalein pink. If an acid has made litmus red, adding a base eventually reverses the effect, making the litmus blue again. Thus, acids and bases seem to be opposites. A base can *neutralize* the effect of an acid, and an acid can neutralize the effect of a base.

Litmus and phenolphthalein are dyes that have different colors in acid and base. The latter is pink in base but colorless in acid.

Acids have other characteristic properties. They taste sour, produce bubbles of CO_2 gas when added to limestone, and dissolve many metals, producing a

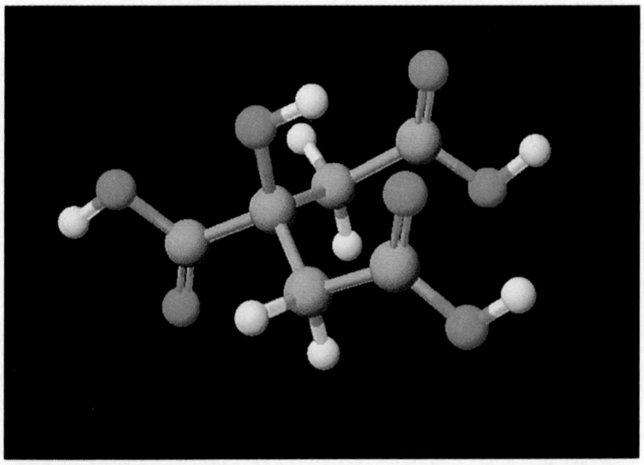

Citric acid, $C_6H_8O_7$. Notice the similarity of this structure to that of acetic acid in that the —CO_2H group is common to organic acids.

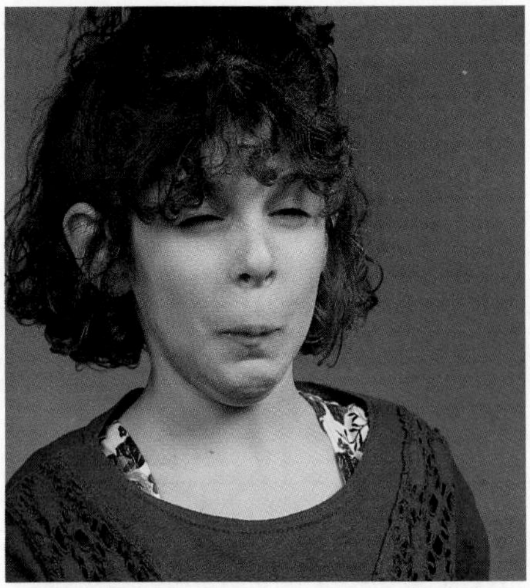

(a)

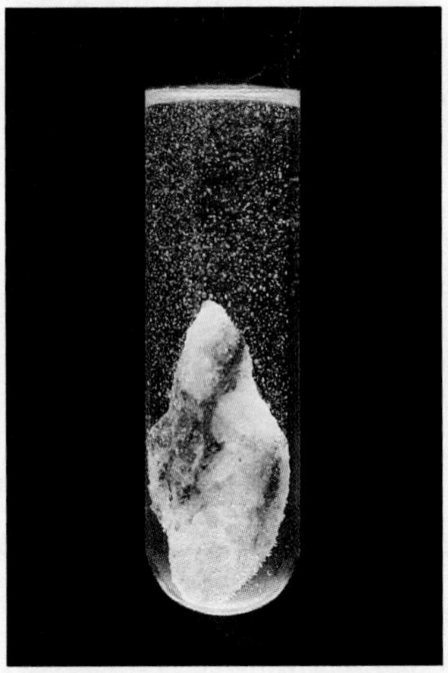

(b)

(c)

Figure 4.9 Illustration of the characteristic properties of acids. (a) A child chewing a piece of the extra sour gum. (b) A piece of limestone dissolving in acid. (c) A metal reacting with acid. (C. D. Winters)

flammable gas at the same time (Figure 4.9). Although tasting substances is *never* done in a chemistry laboratory, you have probably experienced the sour taste of citric acid, which is commonly found in fruits and is added to candies and soft drinks. Bases, in contrast, have a bitter taste. Bases do not often dissolve metals, but they do lead to insoluble compounds with many metal ions (Figure 4.10). Such precipitates can be made to dissolve by adding an acid, another case in which an acid counteracts a property of a base.

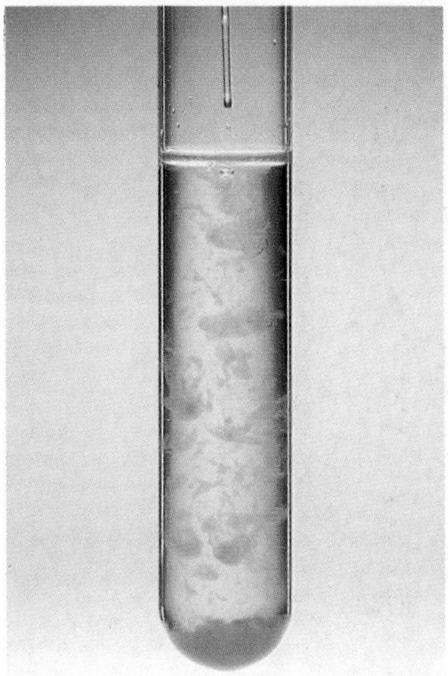

(a)

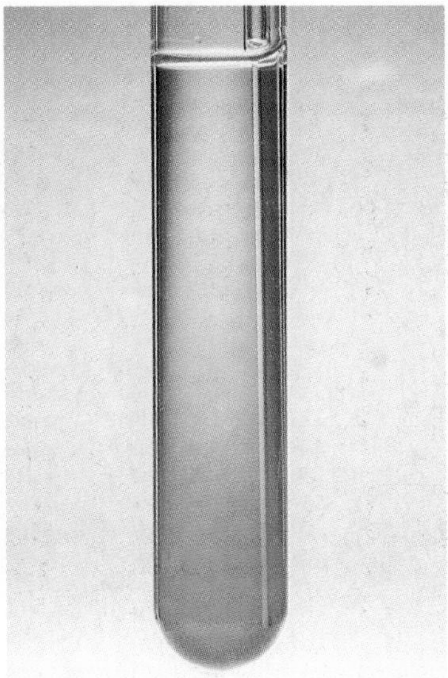

(b)

Figure 4.10 (a) Aqueous sodium hydroxide (NaOH) forms a precipitate of $Fe(OH)_3$ when added to an aqueous solution of iron(III) chloride.

$$FeCl_3(aq) + 3\ NaOH(aq) \longrightarrow Fe(OH)_3(s) + 3\ NaCl(aq)$$

(b) When aqueous HCl is added to the precipitate, the precipitate dissolves, forming water and soluble iron(III) chloride.

$$3\ HCl(aq) + Fe(OH)_3(s) \longrightarrow 3\ H_2O(\ell) + FeCl_3(aq)$$

TABLE 4.1 Common Acids and Bases

Strong Acids (strong electrolytes)		Strong Bases (strong electrolytes)	
HCl	hydrochloric acid	NaOH	sodium hydroxide
HNO_3	nitric acid	KOH	potassium hydroxide
$HClO_4$	perchloric acid	$Ca(OH)_2$	calcium hydroxide
H_2SO_4	sulfuric acid		
Weak acids (weak electrolytes)		**Weak base (weak electrolyte)**	
H_3PO_4	phosphoric acid	NH_3	ammonia
CH_3CO_2H	acetic acid		
H_2CO_3	carbonic acid		

The properties of acids can be interpreted in terms of some common feature of acid molecules, and a different feature can explain the properties of bases. *An* ***acid*** *is any substance that, when dissolved in pure water, increases the concentration of hydrogen ions, H^+, in the water.* One of the most common acids is hydrochloric acid, which ionizes in water to form a hydrogen ion, H^+, and a chloride ion (Cl^-).

$$HCl(aq) \longrightarrow H^+(aq) + Cl^-(aq)$$

hydrochloric acid
strong electrolyte
= 100% ionized

(See Table 4.1 for other common acids.) Because it is completely converted to ions in aqueous solution, HCl is a **strong acid** (and a strong electrolyte).

Sulfuric acid is the chemical produced in the largest quantity in the United States and in most industrialized countries. In 1993, 36 billion kg was produced, or about 160 kg per person. Most is now used to make phosphate fertilizers.

Some common acids, such as sulfuric acid, can provide more than one mole of H^+ per mole of acid.

$$H_2SO_4(aq) \longrightarrow H^+(aq) + HSO_4^-(aq)$$

sulfuric acid, 100% ionized → hydrogen ion + hydrogen sulfate ion

$$HSO_4^-(aq) \longrightarrow H^+(aq) + SO_4^{2-}(aq)$$

hydrogen sulfate ion, <100% ionized → hydrogen ion + sulfate ion

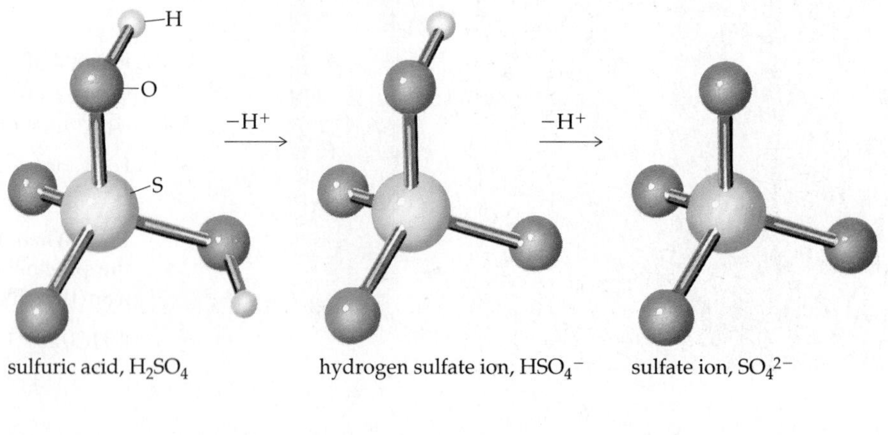

sulfuric acid, H_2SO_4 hydrogen sulfate ion, HSO_4^- sulfate ion, SO_4^{2-}

A CLOSER LOOK *H^+ Ions in Water*

The H^+ ion is a hydrogen atom that has lost its electron. Only the nucleus, a proton, remains. Because a proton is only about 1/10,000 as large as the average atom or ion, water molecules can approach very closely, and the proton and the water molecules are strongly attracted. At the very least the H^+ ion in water is better represented by the combination of H^+ and H_2O, **H_3O^+**, an ion called the **hydronium ion.** Experiments also show, however, other forms of the ion existing in water, one example being $[H_3O(H_2O)_3]^+$. Thus, the ionization of HCl actually provides a solution containing the hydrogen ion surrounded by some number of water molecules. We shall often use $H^+(aq)$ in this text to indicate the presence of hydronium and similar ions, but be aware that the real situation is much more complex.

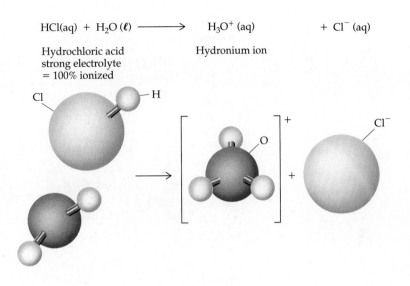

$$HCl(aq) + H_2O(\ell) \longrightarrow H_3O^+(aq) + Cl^-(aq)$$

Hydrochloric acid strong electrolyte = 100% ionized — Hydronium ion

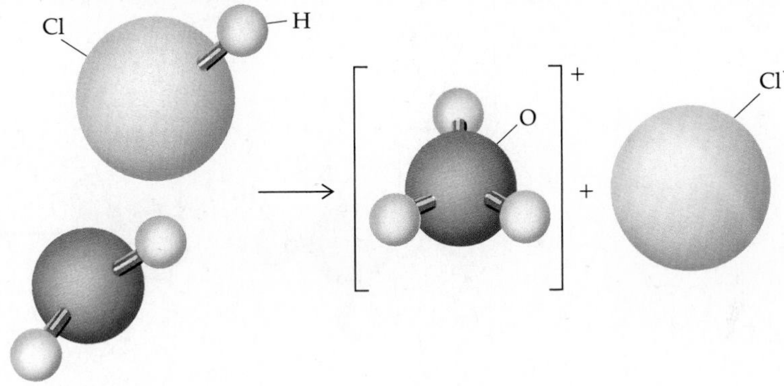

The first ionization reaction is essentially complete, so sulfuric acid is considered a strong acid (and, thus, a strong electrolyte as well). The hydrogen sulfate ion, like acetic acid, is, however, a weak electrolyte; that is, HSO_4^- is only partially ionized in aqueous solution. Both the hydrogen sulfate ion and acetic acid are therefore classified as **weak acids.**

*A **base** is a substance that increases the concentration of **hydroxide ion, OH^-,** when dissolved in pure water.* The properties that bases have in common are the properties of $OH^-(aq)$. Compounds that contain hydroxide ions, such as sodium or potassium hydroxide, are obvious bases. As ionic compounds they are strong electrolytes and **strong bases.**

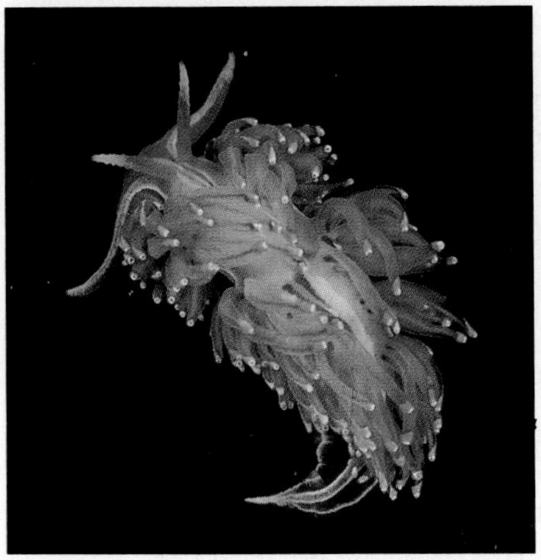

Sea slugs excrete sulfuric acid to defend themselves. (Peter J. Bryant/Biological Photo Service)

$$NaOH(s) \xrightarrow{H_2O} Na^+(aq) + OH^-(aq)$$

sodium hydroxide, base; strong electrolyte — H_2O = 100% dissociated — hydroxide ion

Ammonia, NH_3, is another very common base. Although the compound does not have an OH^- ion as part of its formula, it produces the OH^- ion on reaction with water.

$$NH_3(aq) + H_2O(\ell) \longrightarrow NH_4^+(aq) + OH^-(aq)$$

ammonia, base; weak electrolyte — ammonium ion — hydroxide ion

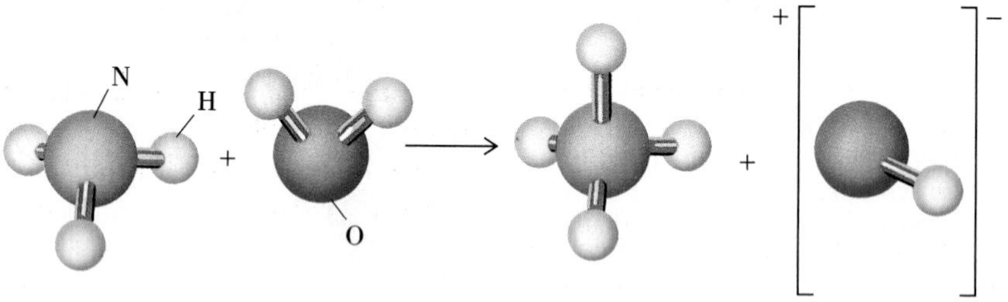

Only a small concentration of the ions is present, so ammonia is a **weak base** (and a weak electrolyte).

EXERCISE 4.5 *Acids and Bases*

1. What ions are produced when perchloric acid dissolves in water?
2. Calcium hydroxide is not very soluble in water. What little does dissolve, however, is dissociated. What ions are produced? ■

Oxides of Nonmetals and Metals

Each acid shown in Table 4.1 has one or more H atoms in the molecular formula that can be released to form H^+ ions in water. Less obvious examples of acids, though, are *oxides of nonmetals,* such as carbon dioxide and sulfur trioxide, which have no H atoms but which react with water to produce H^+ ions. Carbon dioxide, for example, dissolves in water to a small extent, and some of the dissolved molecules react with water to form the weak acid, carbonic acid. This acid then ionizes to a small extent to form the hydrogen ion, H^+, and the bicarbonate ion, HCO_3^-,

Dissolving CO_2: $$CO_2(g) + H_2O(\ell) \longrightarrow H_2CO_3(aq)$$ (carbonic acid)

H_2CO_3 ionization: $$H_2CO_3(aq) \longrightarrow H^+(aq) + HCO_3^-(aq)$$

Overall: $$CO_2(g) + H_2O(\ell) \longrightarrow H^+(aq) + HCO_3^-(aq)$$

Rainwater is always slightly acidic because of the carbonic acid produced from dissolved CO_2. The pH is about 5.7 (as compared with a neutral solution that has a pH of 7.0). See Chapter 17.

This reaction is very important in our environment. Carbon dioxide is normally found in small amounts in the atmosphere, so *rainwater is always slightly acidic.* Oxides like CO_2 that can react with water to produce H^+ ions are known as **acidic oxides.**

Oxides of sulfur and nitrogen are present in significant amounts in polluted air, and both of these nonmetal oxides can lead ultimately to acids and other pollutants. For example, SO_2 from human and natural sources can react with oxygen to give SO_3, which then forms sulfuric acid with water. Nitrogen dioxide reacts with water to give nitric and nitrous acids.

$$\underset{\text{from burning fossil fuels and from volcanoes}}{2\ SO_2(g)} + O_2(g) \longrightarrow 2\ SO_3(g)$$

$$SO_3(g) + H_2O(\ell) \longrightarrow \underset{\text{sulfuric acid}}{H_2SO_4(aq)}$$

$$2\ NO_2(g) + H_2O(\ell) \longrightarrow \underset{\text{nitric acid}}{HNO_3(aq)} + \underset{\text{nitrous acid}}{HNO_2(aq)}$$

Volcanoes are a major source of SO_2. Much of this finds its way to the oceans. It is estimated that 1.5×10^{14} g of SO_2 are dissolved in the oceans per year, most of it ending up as sulfate ions.

These reactions are the origin of acid rain in the United States, Canada, and other industrialized countries from the burning of fossil fuels, such as coal and gasoline. The gaseous oxides mix with other chemicals in the troposphere and with water, and the rain that falls is more acidic than if it contained only dissolved CO_2. When the rain falls on areas that cannot easily tolerate this greater than normal acidity, such as the northeastern parts of the United States and the eastern provinces of Canada, serious environmental problems can occur.

Oxides of metals can give basic solutions if they dissolve appreciably in water. Perhaps the best known example is calcium oxide, CaO, often called lime or quicklime. This metal oxide reacts with water to give calcium hydroxide, commonly called slaked lime. The latter compound, although not very soluble in water, dissolves sufficiently to give a strongly basic solution (Figure 4.11).

$$\underset{\text{lime}}{CaO(s)} + H_2O(\ell) \longrightarrow \underset{\text{slaked lime}}{Ca(OH)_2(s)}$$

Oxides like CaO that react with water to produce OH^- ions are known as **basic oxides.** Almost 15 billion kg of lime was produced in the United States in 1991 for use in the metals industry, in sewage and pollution control, in water treatment, in agriculture, and in the construction industry.

Figure 4.11 The white solid calcium oxide (lime) is a basic oxide because it reacts with water to produce the base $Ca(OH)_2$ (slaked lime). Water and a small quantity of CaO have been placed in both test tubes. The test tube at the right also contains the dye phenolphthalein, which turns red in the presence of a base. (C. D. Winters)

EXERCISE 4.6 ***Acidic and Base Oxides***

For each of the following, indicate whether you expect an acidic or basic solution when the compound dissolves in water.
(1) SeO_2 (2) MgO (3) P_4O_{10} ■

4.5 EQUATIONS FOR REACTIONS IN AQUEOUS SOLUTION: NET IONIC EQUATIONS

Let us again consider the reaction of magnesium with aqueous HCl (Figure 4.5).

$$Mg(s) + 2\ HCl(aq) \longrightarrow MgCl_2(aq) + H_2(g)$$

Hydrochloric acid and magnesium chloride both dissolve in water, and both are strong electrolytes. This means that the reactant solution really contains the ionization products of HCl,

$$HCl(aq) \xrightarrow{\text{consists of}} H^+(aq) + Cl^-(aq)$$

and the products include the ions of $MgCl_2(aq)$

$$MgCl_2(aq) \xrightarrow{\text{consists of}} Mg^{2+}(aq) + 2\ Cl^-(aq)$$

To be more informative, we could therefore rewrite the balanced equation as

$$Mg(s) + \underbrace{2\ H^+(aq) + 2\ Cl^-(aq)}_{\text{from 2 HCl(aq)}} \longrightarrow \underbrace{Mg^{2+}(aq) + 2\ Cl^-(aq)}_{\text{from } MgCl_2(aq)} + H_2(g)$$

Now two Cl^- ions appear both on the reactant and product sides of the equation. Such ions are often called **spectator ions** because they are not involved in the net reaction process; they only "look on" from the sidelines. Little chemical or stoichiometric information is lost therefore if the equation is written without them, and we can simplify the equation to

$$Mg(s) + 2\ H^+(aq) \longrightarrow Mg^{2+}(aq) + H_2(g)$$

The balanced equation that results from leaving out the spectator ions is the ***net ionic equation*** *for the reaction.* Here only the solid element (Mg), the molecule (H_2), and the ions [$H^+(aq)$, $Mg^{2+}(aq)$] that are involved in the *changes* that occur in the course of the reaction are included. That is, the *net ionic equation emphasizes the chemical changes* that occur.

Leaving out the spectator ions does not imply that Cl^- is totally unimportant in the Mg + HCl reaction. Indeed, H^+ or Mg^{2+} ions cannot exist alone in solution; a negative ion of some kind *must* be present to balance the positive ion charge. *Any* anion will do, however, as long as it forms water-soluble compounds with H^+ and Mg^{2+}. Thus, we could have used $HClO_4$ or HBr as the source of H^+, and then ClO_4^- or Br^- would be the spectator ion.

As a final point concerning net ionic equations, there must always be a *conservation of charge* as well as mass in a balanced chemical equation. Thus, in the Mg + HCl net ionic equation two positive electric charges occur on each side of the equation. On the left side two H^+ ions occur, each with a charge of 1+, for a total charge of 2+. This is balanced by a Mg^{2+} ion on the right side.

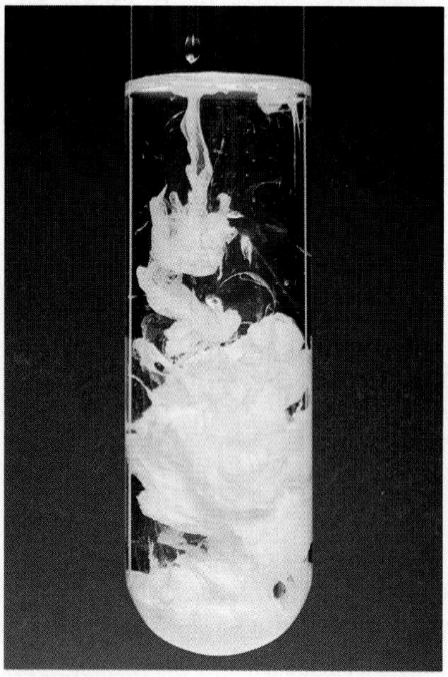

The reaction of silver nitrate and calcium chloride produces insoluble silver chloride and soluble calcium nitrate. See Example 4.3. (C. D. Winters)

EXAMPLE 4.3 *Writing and Balancing Net Ionic Equations*

Write a balanced, net ionic equation for the reaction of $AgNO_3$ and $CaCl_2$ to give AgCl and $Ca(NO_3)_2$.

Solution

Step 1. Write the complete, balanced equation using the correct formulas for reactants and products.

$$2\ AgNO_3 + CaCl_2 \longrightarrow 2\ AgCl + Ca(NO_3)_2$$

Step 2. Decide on the solubility of each compound using Figure 4.7. One general guideline is that nitrates are almost always soluble, so $AgNO_3$ and $Ca(NO_3)_2$ are water-soluble. Furthermore, with a few exceptions (e.g., AgCl), chlorides are water-soluble. We can therefore write

$$2\ AgNO_3(aq) + CaCl_2(aq) \longrightarrow 2\ AgCl(s) + Ca(NO_3)_2(aq)$$

Step 3. Recognizing that all soluble ionic compounds dissociate to form ions in aqueous solution, we have

$$AgNO_3(aq) \longrightarrow Ag^+(aq) + NO_3^-(aq)$$
$$CaCl_2(aq) \longrightarrow Ca^{2+}(aq) + 2\ Cl^-(aq)$$
$$Ca(NO_3)_2(aq) \longrightarrow Ca^{2+}(aq) + 2\ NO_3^-(aq)$$

This results in the following ionic equation:

$$2\ Ag^+(aq) + 2\ NO_3^-(aq) + Ca^{2+}(aq) + 2\ Cl^-(aq) \longrightarrow 2\ AgCl(s) + Ca^{2+}(aq) + 2\ NO_3(aq)$$

Step 4. Two spectator ions occur in the ionic equation (Ca^{2+} and NO_3^-), so these are eliminated to give the following net ionic equation:

$$2\ Ag^+(aq) + 2\ Cl^-(aq) \longrightarrow 2\ AgCl(s)$$

Notice that each species in the net equation is preceded by a coefficient of 2. Therefore, the equation can be simplified by dividing through by 2 to give

$$Ag^+(aq) + Cl^-(aq) \longrightarrow AgCl(s)$$

Step 5. Finally, notice that the sum of ion charges is the same on both sides of the equation. On the left, 1+ and 1− give zero; on the right the electric charge on AgCl is also zero.

EXERCISE 4.7 ***Net Ionic Equations***

Balance each of the following equations, and write net ionic equations:

1. $BaCl_2(aq) + Na_2SO_4(aq) \rightarrow BaSO_4(s) + NaCl(aq)$
2. Lead(II) nitrate reacts with potassium chloride to give lead(II) chloride and potassium nitrate. ■

4.6 TYPES OF REACTIONS IN AQUEOUS SOLUTION

Now that we have looked into the behavior of common compounds in water, we can turn to the second goal of this chapter—an exploration of the most common types of chemical reactions that occur in an aqueous environment. For example, we can decide what happens when an Alka-Seltzer is dropped into water (see the Chemical Puzzler at the beginning of the chapter).

Many of the reactions you will see in the chemistry laboratory occur in aqueous solution. Chemists are interested in such reactions not only because this is one way to make useful products, but also because these are the kinds of reactions that occur on the earth and in plants and animals. We therefore want to look into some common reaction patterns to see what their ''driving forces'' might be. In other words, how can you know when you mix two chemicals that they will combine to produce one or more new compounds?

Four important types of processes cause reactions to occur when reactants are mixed in aqueous solution. The first of these is **precipitation** (Figure 4.12) in which ions combine in solution to form an insoluble reaction product.

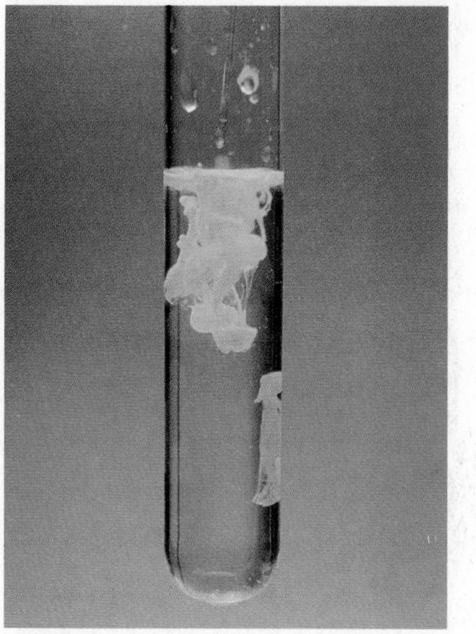

Figure 4.12 Adding a drop of aqueous potassium iodide to a solution of lead(II) nitrate leads to the formation of a yellow precipitate of lead(II) iodide and leaves water-soluble potassium nitrate in solution. (C. D. Winters)

Overall Equation

$$Pb(NO_3)_2(aq) + 2\ KI(aq) \longrightarrow PbI_2(s) + 2\ KNO_3(aq)$$

Net Ionic Equation

$$Pb^{2+}(aq) + 2\ I^-(aq) \longrightarrow PbI_2(s)$$

Acid-base reactions, in which H^+ ions and OH^- combine to form water, make up another important class of reactions.

Overall Equation

$$HNO_3(aq) + KOH(aq) \longrightarrow KNO_3(aq) + HOH(\ell)$$

Net Ionic Equation

$$H^+(aq) + OH^-(aq) \longrightarrow H_2O(aq)$$

This is the net ionic equation for all acid-strong base reactions. Weak acids are considered in Example 4.6.

Gas-forming reactions, chiefly between metal carbonates and acids (Figure 4.13), constitute a third type of reaction. One product from this type of reaction is carbonic acid, H_2CO_3, most of which decomposes to H_2O and CO_2. Carbon dioxide is the gas in the bubbles you see during this reaction.

Overall Equation

$$NiCO_3(s) + 2\ HNO_3(aq) \longrightarrow Ni(NO_3)_2(aq) + H_2CO_3(aq)$$

$$H_2CO_3(aq) \longrightarrow CO_2(g) + H_2O(\ell)$$

Net Ionic Equation

$$NiCO_3(s) + 2\ H^+(aq) \longrightarrow Ni^{2+}(aq) + CO_2(g) + H_2O(\ell)$$

Finally, **oxidation-reduction reactions** are a fourth class of chemical reactions. Here the important process is the transfer of electrons from one substance to another.

Overall Equation

$$Cu(s) + 2\ AgNO_3(aq) \longrightarrow Cu(NO_3)_2(aq) + 2\ Ag(s)$$

Net Ionic Equation

$$Cu(s) + 2\ Ag^+(aq) \longrightarrow Cu^{2+}(aq) + 2\ Ag(s)$$

In summary, four common "driving forces" are responsible for reactions in aqueous solution.

Reaction Type	Driving Force
Precipitation	Formation of an insoluble compound
Acid-base; neutralization	Formation of a salt and water
Gas-forming	Evolution of a water-insoluble gas such as CO_2
Oxidation-reduction	Transfer of electrons

Figure 4.13 The reaction of nickel carbonate with sulfuric acid. The reaction can produce large crystals of hydrated nickel(II) sulfate. The one in this photo were made by a student in a general chemistry laboratory.
(C. D. Winters)

These types of reactions are usually easy to recognize, but keep in mind that a reaction may have more than one driving force. For example, barium carbonate reacts readily with sulfuric acid to give barium sulfate, carbon dioxide, and water, a reaction that is both a precipitation and a gas-forming reaction.

$$BaCO_3(s) + H_2SO_4(aq) \longrightarrow BaSO_4(s) + H_2O(\ell) + CO_2(g)$$

Each of these reaction types is described in more detail in separate sections of this textbook.

4.7 PRECIPITATION REACTIONS

A precipitation reaction produces an insoluble product, a ***precipitate.*** Both the reactants and products are generally ionic compounds. Many precipitation reactions are possible because many positive ion-negative ion combinations give insoluble substances (see Figure 4.7). For example, because most chromates are insoluble, lead(II) chromate is easily precipitated by the reaction of a water-soluble lead(II) compound with a water-soluble chromate compound (Figure 4.14).

Figure 4.14 Adding a drop of aqueous potassium chromate to a solution of lead(II) nitrate leads to the formation of a yellow precipitate of lead(II) chromate and leaves water-soluble potassium nitrate in solution. (C. D. Winters)

Overall Equation

$$Pb(NO_3)_2(aq) + K_2CrO_4(aq) \longrightarrow PbCrO_4(s) + 2\ KNO_3(aq)$$

Net Ionic Equation

$$Pb^{2+}(aq) + CrO_4^{\ 2-}(aq) \longrightarrow PbCrO_4(s)$$

Almost all metal sulfides are insoluble in water. Therefore, if a soluble metal compound in nature comes in contact with a source of sulfide ions (say from a volcano, a natural gas pocket in the earth, or a "black smoker" in the ocean), the metal sulfide precipitates. In fact, this is how many metal sulfur-containing minerals such as iron pyrite (FeS_2, see Figure 1.7) are believed to have been formed.

Many reactions you see in nature and in the laboratory are precipitation reactions. Let us say you pour an aqueous solution of silver nitrate into one containing potassium bromide.

$$AgNO_3(aq) + KBr(aq) \longrightarrow ?$$

What do you see, and how do you express the result in a balanced chemical equation? From the information in Figure 4.7 you know that silver nitrate is water-soluble because it contains the NO_3^- ion. Similarly, potassium bromide is water-soluble because it contains K^+ and Br^- ions. This means these ions are released into solution when the compounds are dissolved, and they may recombine to form silver bromide and potassium nitrate.

$$AgNO_3(aq) + KBr(aq) \longrightarrow AgBr + KNO_3$$

Is either of the products insoluble? Does precipitation occur? Again referring to Figure 4.7, you see that salts containing the Br^- ion are generally soluble *except* for those involving Ag^+. Silver bromide, AgBr, is therefore *not* water-soluble. Potassium nitrate, KNO_3, however, is water-soluble because it contains both the K^+ and NO_3^- ions that generally lead to water-soluble compounds. Based on this information, it is possible now to write the overall and net ionic equations.

Overall Equation

$$AgNO_3(aq) + KBr(aq) \longrightarrow AgBr(s) + KNO_3(aq)$$

Net Ionic Equation

$$Ag^+(aq) + Br^-(aq) \longrightarrow AgBr(s)$$

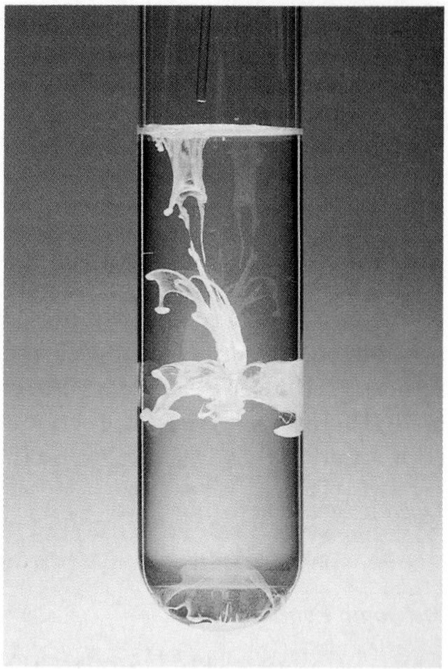

Figure 4.15 Adding aqueous potassium bromide, KBr, to aqueous silver nitrate, $AgNO_3$, leads to a precipitate of yellowish white silver bromide, AgBr, and an aqueous solution of potassium nitrate, KNO_3. Silver bromide is used in photographic film. When light strikes the film, the silver bromide is reduced to silver metal. (C. D. Winters)

This means that, on mixing the two water-soluble reactants, you should observe the immediate formation of a pale yellow cloud of silver bromide precipitate (Figure 4.15).

EXAMPLE 4.4 *Writing the Equation for a Precipitation Reaction*

Is an insoluble product formed when sodium carbonate is mixed with manganese(II) nitrate? If so, write the balanced equation and the net ionic equation.

Solution First, let us decide what ions are formed when sodium carbonate and manganese(II) nitrate are placed in water.

$$Na_2CO_3(aq) \longrightarrow 2\ Na^+(aq) + CO_3^{2-}(aq)$$
$$Mn(NO_3)_2(aq) \longrightarrow Mn^{2+}(aq) + 2\ NO_3^-(aq)$$

The only way that a reaction can occur is if the cations exchange partners, the sodium ion pairing with the nitrate ion (to give $NaNO_3$), and the manganese(II) ion pairing with the carbonate ion (to give $MnCO_3$). Are these compounds water-soluble? Is either one insoluble? $NaNO_3$ is soluble because most compounds containing the Na^+ and NO_3^- ions are water-soluble. In contrast, $MnCO_3$ is *not* soluble: most carbonates are insoluble (except when the cation is Na^+, K^+, or another alkali metal cation). Because an insoluble compound is produced, we can therefore predict that a reaction will occur, and the balanced equation for the reaction is

$$Na_2CO_3(aq) + Mn(NO_3)_2(aq) \longrightarrow 2\ NaNO_3(aq) + MnCO_3(s)$$

If the equation is rewritten showing the ions in solution,

$$2\ Na^+(aq) + CO_3^{2-}(aq) + Mn^{2+}(aq) + 2\ NO_3^-(aq) \longrightarrow 2\ Na^+(aq) + 2\ NO_3^-(aq) + MnCO_3(s)$$

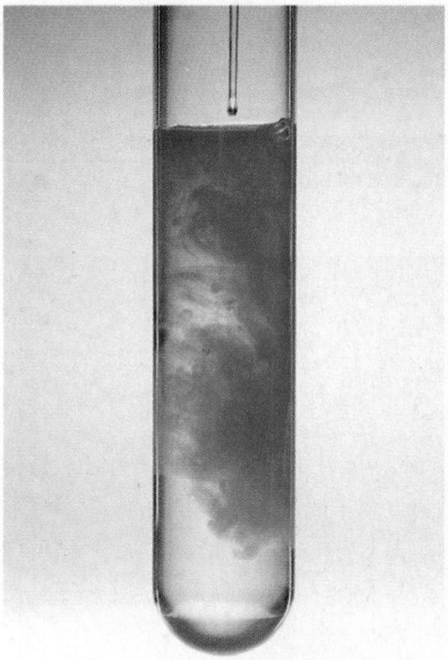

Aqueous silver nitrate reacts with aqueous potassium chromate. See Exercise 4.8.
(C. D. Winters)

it is clear that the sodium and nitrate ions are spectator ions. The net ionic equation must therefore be

$$Mn^{2+}(aq) + CO_3^{2-}(aq) \longrightarrow MnCO_3(s)$$

EXERCISE 4.8 ***Precipitation Reactions***

When aqueous silver nitrate is mixed with an aqueous solution of potassium chromate, K_2CrO_4, does a precipitate form? If so, write the balanced equation for the reaction, and then write the net ionic equation. ■

4.8 ACID-BASE REACTIONS

In general, acid-base reactions in aqueous solution produce a **salt** and water. For example,

$$\underset{\text{hydrochloric acid}}{HCl(aq)} + \underset{\text{sodium hydroxide}}{NaOH(aq)} \longrightarrow \underset{\text{sodium chloride}}{NaCl(aq)} + \underset{\text{water}}{H_2O(\ell)}$$

Here hydrochloric acid and sodium hydroxide give common table salt and water. Because of this, the word "salt" has come into our language to describe any *ionic compound whose cation comes from a base* (here Na^+ from NaOH) *and whose anion comes from an acid* (here Cl^- from HCl). Reaction of any of the acids listed in Table 4.1 with any of the bases listed there produces a salt and water (with the exception of ammonia; see Example 4.5).

Because hydrochloric acid is a strong electrolyte in water (Table 4.1), the complete ionic equation for the reaction of HCl(aq) and NaOH(aq) should be written as

CURRENT ISSUES IN CHEMISTRY

Black Smokers

The sun provides the energy to drive some of the chemical reactions fundamental to life on our planet, but some living things find other sources of energy. Abundant heat energy is available within the earth, and this fuels some very strange forms of life on the floor of the oceans.

In 1977, scientists were exploring the tectonic plates that form the earth's surface on the ocean floor thousands of feet down in the equatorial Pacific. Along the ridge that is the junction of two of these plates they found direct evidence for the great energy within the earth: thermal springs gushing a hot, black soup of minerals through "chimneys" in the ocean's floor. Water seeping into cracks in the thin surface along the ridge is superheated to between 300 and 400 °C by the magma of the earth's core that is close to the surface. This superhot water dissolves minerals in the crust and provides conditions for the conversion of sulfates in seawater to hydrogen sulfide, H_2S. When this hot water, now laden with dissolved minerals and rich in sulfides, gushes through the surface, it cools, and metal sulfides, such as those of copper, manganese, iron, zinc, and nickel, precipitate.

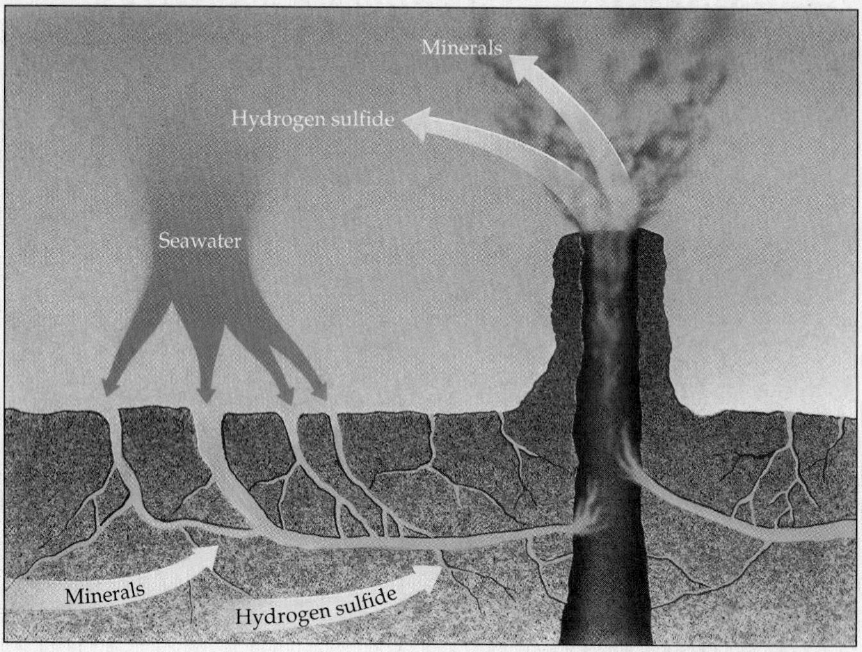

Hydrothermal vents form in the valleys created in the earth's surface where tectonic plates separate. The crust is thin in those regions, and water seeping through the crust becomes heated as hot as 400 °C. The hot water leaches minerals from the rocks. When the plume of hot water emerges from the sea floor, the ambient water temperature is only about 4 °C, so the minerals in the plume precipitate to form chimneys and smoke.

$$H_2S(aq) + Cu^{2+}(aq) \longrightarrow CuS(s) + 2\,H^+(aq)$$

Many metal sulfides are black, thus giving the appearance of "smoke" coming from the earth, and the vents have been called "black smokers." The smoke settles around the edges of the vent, and eventually forms a chimney of precipitated minerals.

The people studying the black smokers were amazed to discover that

$$\underbrace{H^+(aq) + Cl^-(aq)}_{\text{from HCl(aq)}} + \underbrace{Na^+(aq) + OH^-(aq)}_{\text{from NaOH(aq)}} \longrightarrow \underbrace{Na^+(aq) + Cl^-(aq)}_{\text{salt}} + \underbrace{H_2O(\ell)}_{\text{water}}$$

Because Na^+ and Cl^- ions appear on both sides of the equation, the net ionic equation is simply the combinations of the ions H^+ and OH^- to give water.

$$H^+(aq) + OH^-(aq) \longrightarrow H_2O(\ell)$$

This was taken with the sponson camera on DSV Alvin *on 19 June 1993 during dive #2618 at the Snake Pit Hydrothermal Vent Field on the Mid-Atlantic Ridge. The depth is more than 3,500 m; the temperature of the vent water is around 350 °C. The massive sulfides around the vent are covered with shrimp of the species* Rimicaris exoculata. *The coil shown at the lower right is part of a device to measure temperature.* (Photo by Steve Chamberlain, Syracuse University)

the chimneys were surrounded by dense fields of peculiar, primitive animals living in the hot, sulfide-rich environment. These smokers are under thousands of feet of water, and sunlight does not penetrate to these depths, so the animals have developed a way to live without the energy from sunlight. It is currently believed that they derive the energy needed to make the organic compounds on which they depend from the reaction of oxygen with hydrogen sulfide or the hydrogen sulfide ion, HS^-.

$$HS^-(aq) + 2\,O_2(g) \longrightarrow HSO_4^-(aq) + \text{energy}$$

See V. Tunnicliffe: *American Scientist,* Vol. 80, pp. 336–349, 1992, and R. A. Lutz and R. M. Haymon, *National Geographic,* Vol. 186, No. 5, pp 115–126, 1994, for more information.

Indeed, *this is always the net ionic equation for the reaction between any* ***strong*** *acid and any* ***strong*** *base.* Reactions between strong acids and strong bases are called **neutralization reactions** because, on completion of the reaction, the solution is neutral, neither acidic nor basic. The other ions (the cation of the base and the anion of the acid) remain unchanged. If the water is evaporated, however, the cation and anion form a solid salt. In the example above, NaCl can be obtained, whereas nitric acid, HNO_3, and NaOH give the salt sodium nitrate, $NaNO_3$.

(a)

(b)

Figure 4.16 Gypsum. (a) A sample of the white mineral gypsum is calcium sulfate dihydrate, $CaSO_4 \cdot 2\ H_2O$. (b) Coal is removed from the strip mine in the photograph, and burned in a nearby power plant (not shown). Sulfur oxides in the waste gas from the plant are converted to white calcium sulfate, $CaSO_4 \cdot 2\ H_2O$, which is then put back into a part of the mine from which the coal has been removed (at the center of the photograph). (C. D. Winters)

All acid-base reactions do not lead to a *neutral* solution. (See Chapter 18.) The term "neutralization" is generally used, however, when referring to acid-base reactions.

Overall Equation

$$HNO_3(aq) + NaOH(aq) \longrightarrow NaNO_3(aq) + H_2O(\ell)$$

Calcium oxide, lime, is inexpensive and is used in waste and pollution control. Indeed, one of its major uses is in "scrubbing" sulfur oxides from the exhaust gases of power plants fueled by coal and oil. The oxides of sulfur dissolve in water to produce acids (page 175), and these acids can react with a base. Lime produces the base calcium hydroxide when suspended in water. This suspension is sprayed into the exhaust stack of the plant where it reacts with the sulfur-containing acids. One such reaction is

$$Ca(OH)_2(s) + H_2SO_4(aq) \longrightarrow CaSO_4 \cdot 2\ H_2O(s)$$

The compound $CaSO_4 \cdot 2\ H_2O$, hydrated calcium sulfate, is found in the earth as the mineral gypsum (Table 3.5 and Figure 4.16a). Assuming the gypsum from a coal-burning power plant is not contaminated with compounds based on metals with higher atomic weights (e.g., Hg) or other pollutants, it is environmentally acceptable to put it into the earth (Figure 4.16b).

EXAMPLE 4.5 *Ammonia as a Base*

Write the balanced overall equation and the net ionic equation for the reaction of aqueous ammonia and nitric acid to produce the salt ammonium nitrate.

Solution The overall, balanced equation for the reaction is

$$\underset{\text{ammonia}}{NH_3(aq)} + \underset{\text{nitric acid}}{HNO_3(aq)} \longrightarrow \underset{\text{ammonium nitrate}}{NH_4NO_3(aq)}$$

Ammonia is a base because it reacts with water, to a small extent, to produce hydroxide ions.

$$NH_3(aq) + H_2O(\ell) \longrightarrow NH_4^+(aq) + OH^-(aq) \quad (1)$$

Nitric acid is a strong acid and produces hydrogen and nitrate ions. It is therefore written as $H^+(aq) + NO_3^-(aq)$. The OH^- ions produced by NH_3 react with the H^+ ions produced by HNO_3 to form water.

$$OH^-(aq) + H^+(aq) + NO_3^-(aq) \longrightarrow H_2O(\ell) + NO_3^-(aq) \quad (2)$$

Adding equations (1) and (2) gives the balanced, net ionic equation for the reaction.

$$NH_3(aq) + \cancel{H_2O(\ell)} \longrightarrow NH_4^+(aq) + \cancel{OH^-(aq)}$$
$$\underline{\cancel{OH^-(aq)} + H^+(aq) + \cancel{NO_3^-(aq)} \longrightarrow \cancel{H_2O(\ell)} + \cancel{NO_3^-(aq)}}$$
$$NH_3(aq) + H^+(aq) \longrightarrow NH_4^+(aq)$$

Notice that when the equations are summed, OH^-, NO_3^-, and H_2O are eliminated. The final result is that the net ionic equation for ammonia reacting with a strong acid is the transfer of H^+ from the acid to NH_3.

EXAMPLE 4.6 *Acid-Base Reactions*

Write the balanced, overall equation and the net ionic equation for the reaction of calcium hydroxide with acetic acid.

Solution Like many acid-base reactions, $Ca(OH)_2$ and CH_3CO_2H react to give a salt and water.

$$\underset{\text{calcium hydroxide}}{Ca(OH)_2(s)} + \underset{\text{acetic acid}}{2\ CH_3CO_2H(aq)} \longrightarrow \underset{\text{calcium acetate}}{Ca(CH_3CO_2)_2(aq)} + 2\ H_2O(\ell)$$

Here the salt is calcium acetate, which consists of one calcium ion, Ca^{2+}, and two acetate ions, $CH_3CO_2^-$. Notice that calcium hydroxide supplies two moles of OH^- ions per mole of $Ca(OH)_2$, and acetic acid supplies only one mole of H^+ ions per mole of the acid.

To write the net ionic equation we first recognize (a) that $Ca(OH)_2$ is not very soluble in water and (b) that acetic acid is a weak electrolyte. Neither of the reactants therefore consists of ions in solution. The product, calcium acetate, is a soluble ionic compound, consisting of Ca^{2+} and $CH_3CO_2^-$ ions in water. This means the net ionic equation is not very different from the overall equation.

$$Ca(OH)_2(s) + 2\ CH_3CO_2H(aq) \longrightarrow Ca^{2+}(aq) + 2\ CH_3CO_2^-(aq) + 2\ H_2O(\ell)$$

EXERCISE 4.9 *Acid-Base Reactions*

Write the balanced, overall equation and the net ionic equation for the reaction of magnesium hydroxide with hydrochloric acid. ■

4.9 GAS-FORMING REACTIONS

The driving force in some reactions is the formation of a gas. As mentioned when introducing the types of reactions (Section 4.6), an excellent example is the reaction of metal carbonates (or bicarbonates) with acids (Figure 4.13).

$$CaCO_3(s) + 2\ HCl(aq) \longrightarrow CaCl_2(aq) + H_2CO_3(aq)$$
$$\downarrow$$
$$H_2O(\ell) + CO_2(g)$$

A salt and H_2CO_3, carbonic acid, are *always* the products from an acid and a metal carbonate.

$$CO_3^{2-}(aq) + 2\ H^+(aq) \longrightarrow H_2CO_3(aq)$$

Carbonic acid is unstable, however, and much of it is rapidly converted to water and CO_2 gas. If the reaction is done in an open beaker, most of the gas bubbles out of the solution.

$$H_2CO_3(aq) \longrightarrow H_2O(\ell) + CO_2(g)$$

The Chemical Puzzler at the beginning of the chapter described a reaction between Alka-Seltzer and water. The bubbles that arise when the tablet is dropped into water come from the reaction of an acid (citric acid) with the hydrogen carbonate ion (or bicarbonate ion).

$$\underset{\text{citric acid}}{C_6H_8O_7(aq)} + \underset{\text{hydrogen carbonate ion}}{HCO_3^-(aq)} \longrightarrow C_6H_7O_7^-(aq) + H_2O(\ell) + CO_2(g)$$

The leavening agent in making biscuits is carbon dioxide, which comes from a gas-forming reaction. Baking powder consists of sodium hydrogen carbonate, $NaHCO_3$, and an acid, usually tartaric acid (a close relative of citric acid). In dry baking powder the ingredients are kept apart by using starch as a filler. When mixed into the moist batter, however, the acid and the hydrogen carbonate ion react to produce CO_2 (just like the reaction in Alka-Seltzer), and the biscuit dough rises.

$$\underset{\text{tartaric acid}}{C_4H_6O_6(aq)} + \underset{\text{hydrogen carbonate ion}}{HCO_3^-(aq)} \longrightarrow C_4H_5O_6^-(aq) + H_2O(\ell) + CO_2(g)$$

Acetic acid and other organic acids have CO_2H groups that can donate H^+ in water. Citric acid has three such groups (page 170) and tartaric acid has two.

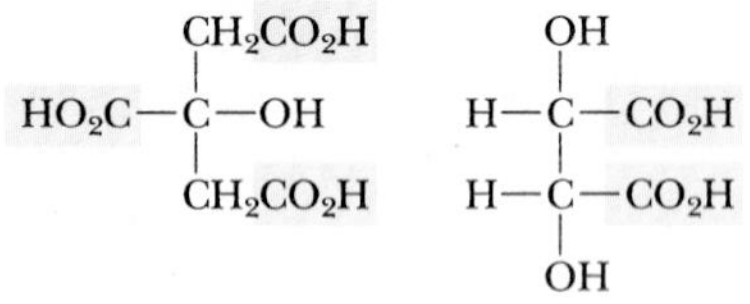

citric acid tartaric acid

EXAMPLE 4.7 *Gas-Forming Reactions*

Write a balanced equation for the reaction that occurs when nickel(II) carbonate is treated with sulfuric acid.

Solution As described in the text, when a metal carbonate (or metal hydrogen carbonate) is treated with an acid, the products are always water and CO_2, as well as a metal salt. The anion of the metal salt is the anion from the acid. Here the metal salt must be nickel(II) sulfate.

$$NiCO_3(s) + H_2SO_4(aq) \longrightarrow NiSO_4(aq) + H_2O(\ell) + CO_2(g)$$

EXERCISE 4.10 ***A Gas-Forming Reaction***

Cerussite, $PbCO_3$, is an important lead-containing mineral. Write a balanced equation that shows what happens when cerussite is treated with nitric acid. Give the name of each of the reaction products. ■

EXERCISE 4.11 ***Classifying Reactions***

Classify each of the reactions below as a precipitation, an acid-base reaction, or a gas-forming reaction. Predict the products of the reaction, and then balance the completed equation.

1. $CuCO_3(s) + H_2SO_4(aq) \rightarrow$
2. $Ba(OH)_2(s) + HNO_3(aq) \rightarrow$
3. $ZnCl_2(aq) + (NH_4)_2S(aq) \rightarrow$ ■

PROBLEM-SOLVING TIPS AND IDEAS

4.1 A Common Theme for Reactions in Aqueous Solution

All the reactions we have described to this point have been ones in which the ions of the reactants changed partners.

$$A^+B^- + C^+D^- \longrightarrow A^+D^- + C^+B^-$$

Thus, we might refer to them as **exchange reactions,** which gives us a good way of predicting the products of precipitation, acid-base, and gas forming reactions.

Precipitation

$$AgNO_3(aq) + KBr(aq) \longrightarrow AgBr(s) + KNO_3(aq)$$

Here the nitrate ions and bromide ions have exchanged places, and silver bromide precipitates.

Acid-Base

$$HNO_3(aq) + NaOH(aq) \longrightarrow NaNO_3(aq) + HOH(\ell)$$

In this reaction the nitrate and hydroxide ions have exchanged places, and water is one result. Acid-base reactions are also often referred to as **proton-transfer reactions.** That is, the H^+ ion transfers from the acid to the base, and water is the net result.

Gas-Forming

$$NiCO_3(s) + H_2SO_4(aq) \longrightarrow NiSO_4(aq) + H_2CO_3(aq)$$

$$H_2CO_3(aq) \longrightarrow H_2O(\ell) + CO_2(g)$$

In this gas-forming reaction the carbonate and sulfate ions have exchanged places, and H_2CO_3 is one result. This latter compound decomposes to water and the gas CO_2. ■

4.10 OXIDATION-REDUCTION REACTIONS

The terms "oxidation" and "reduction" come from reactions that have been known for centuries. Ancient civilizations learned how to change metal oxides and sulfides to the metal, that is, how to *reduce* ore to the metal. A modern example is the reduction of iron oxide with carbon monoxide to give iron metal.

$$Fe_2O_3(s) + 3\ CO(g) \longrightarrow 2\ Fe(s) + 3\ CO_2(g)$$

Here carbon monoxide removes oxygen from iron(III) oxide, so the iron(III) oxide is said to have been reduced.

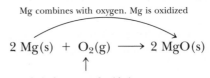

Fe_2O_3 loses oxygen and is reduced

$$Fe_2O_3(s) + 3\ CO(g) \longrightarrow 2\ Fe(s) + 3\ CO_2(g)$$

CO is the agent of reduction. It gains oxygen and is oxidized.

In this reaction carbon monoxide is the agent that brings about the reduction of iron ore to iron metal, so carbon monoxide is called the **reducing agent.**

When Fe_2O_3 is reduced by carbon monoxide, oxygen is removed from the iron ore and added to the carbon monoxide, which is "oxidized" by the addition of oxygen to give carbon dioxide. *Any process in which oxygen is added to another substance is an oxidation.* This too is a process known for centuries, and Figures 4.2 through 4.4 are excellent examples. In the reaction with magnesium, oxygen is the **oxidizing agent** because it is the agent responsible for the oxidation.

Mg combines with oxygen. Mg is oxidized

$$2\ Mg(s) + O_2(g) \longrightarrow 2\ MgO(s)$$

O_2 is the agent of oxidation

The reduction of iron ore to metallic iron with carbon or carbon monoxide is done on a massive scale worldwide. (Geneva Steel/Bruce Hansen)

The experimental observations outlined previously point to several important conclusions: (a) If one substance is oxidized, another substance in the same reaction must be reduced. For this reason, such reactions are often called oxidation-reduction reactions, or **redox reactions** for short. (b) The reducing agent is itself oxidized, and the oxidizing agent is reduced. (c) Oxidation is the opposite of reduction. For example, the reactions described previously show that the removal of oxygen is reduction and the addition of oxygen is oxidation.

Redox Reactions and Electron Transfer

Not all redox reactions involve oxygen. *All oxidation and reduction reactions, however, do involve transfer of electrons between substances.* When a substance *accepts electrons,* it is said to be **reduced** because there is a reduction in the electric charge on an atom of the substance. In the net ionic equation shown here, positively charged Ag^+ is reduced to uncharged Ag(s) on accepting electrons from copper metal. Because copper metal supplies the electrons and causes the Ag^+ ion to be reduced, Cu is called the **reducing agent** (Figure 4.17).

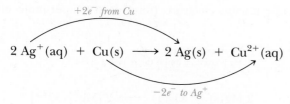

Ag^+ accepts electrons from Cu and is reduced to Ag;
Ag^+ is the oxidizing agent.

$+2e^-$ from Cu

$$2\,Ag^+(aq) + Cu(s) \longrightarrow 2\,Ag(s) + Cu^{2+}(aq)$$

$-2e^-$ to Ag^+

Cu donates electrons to Ag^+ and is oxidized to Cu^{2+};
Cu is the reducing agent.

Figure 4.17 The oxidation of copper metal by silver ion. A clean piece of copper wire is placed in a solution of silver nitrate, $AgNO_3$. With time, the copper reduces Ag^+ ions to silver metal crystals, and the copper metal is oxidized to Cu^{2+} ions. The blue color of the solution is due to the presence of aqueous copper(II) ion. (C. D. Winters)

When a substance *loses electrons,* the electric charge on an atom of the substance increases. The substance is said to have been **oxidized.** In our example, copper metal releases electrons on going to Cu^{2+}; because its electric charge has increased, it is said to have been oxidized. For this to happen, something must be available to take the electrons offered by the copper. In this case, Ag^+ is the electron acceptor, and its charge is reduced to zero in silver metal. Therefore, Ag^+ is the "agent" that causes Cu metal to be oxidized, so Ag^+ is the **oxidizing agent.** In every oxidation-reduction reaction, one reactant is reduced (and is therefore the oxidizing agent) and one reactant is oxidized (and is therefore the reducing agent). In summary,

If X loses one or more electrons, it is oxidized, and is the reducing agent.

$$X \longrightarrow X^{n+} + ne^-$$

If Y gains one or more electrons, it is reduced, and is the oxidizing agent.

$$Y + ne^- \longrightarrow Y^{n-}$$

In the reaction of magnesium and oxygen (Figure 4.3), oxygen is the oxidizing agent because it gains electrons (four electrons per molecule) on going to the oxide ion.

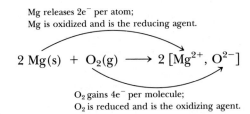

Mg releases $2e^-$ per atom;
Mg is oxidized and is the reducing agent.

$$2\ Mg(s) + O_2(g) \longrightarrow 2\ [Mg^{2+}, O^{2-}]$$

O_2 gains $4e^-$ per molecule;
O_2 is reduced and is the oxidizing agent.

In the same reaction, magnesium is the reducing agent because it releases two electrons per atom on forming the Mg^{2+} ion. All redox reactions can be analyzed in a similar manner.

Oxidation Numbers

How can you tell an oxidation-reduction reaction when you see one? How can you tell which substance has gained or lost electrons and so which one is the oxidizing or reducing agent, respectively? The answer is to look for a *change in the oxidation number of an element* in the course of the reaction. The **oxidation number** of an atom in a molecule or ion is defined as the electric charge an atom has or *appears* to have, as determined by some guidelines for assigning oxidation numbers (see "Guidelines for Determining Oxidation Numbers"). But first, two important points must be made:

The basis of the oxidation numbers can be understood after discussing chemical bonding. See Section 9.5.

- Except for monatomic ions, the oxidation number for an atom does *NOT* represent the actual electric charge on that atom. Nonetheless, oxidation numbers are useful because they are a way of identifying oxidation reduction reactions.

- It is often useful to know whether a reaction can be classified as an oxidation-reduction, an acid-base reaction, a precipitation reaction, or some other type of reaction. The reason for learning about oxidation numbers at this point is to be able to identify which reactions are oxidation-reduction processes.

Guidelines for Determining Oxidation Numbers

1. **Each atom in a pure element has an oxidation number of 0.** The oxidation number of Cu in metallic copper is 0 and is zero for each atom in I_2 or S_8.
2. **For ions consisting of a single atom, the oxidation number is equal to the charge on the ion.** Elements of periodic Groups 1A–3A form monatomic ions with a positive charge and oxidation number equal to the group number. Aluminum therefore forms Al^{3+}, and its oxidation number is +3. (See Section 3.3.)
3. **Fluorine is always −1 in compounds with other elements.**
4. **Cl, Br, and I are always −1 in compounds *except* when combined with O or F.** This means that Cl has an oxidation number of −1 in NaCl (in which Na is +1, as predicted by the fact that it is an element of Group 1A). In the ion ClO^-, however, the Cl atom has an oxidation number of +1 (and O has an oxidation number of −2; see Guideline 5).
5. **The oxidation number of H is +1 and of O is −2 in most compounds.** Although this statement applies to many, many compounds, a few important exceptions occur.
   - When H forms a binary compound with a metal, the metal forms a positive ion and H becomes a hydride ion, H^-. Thus, in CaH_2 the oxidation number of Ca is +2 (equal to the group number) and that of H is −1.
   - Oxygen can have an oxidation number of −1 in a class of compounds called *peroxides,* compounds based on the $O_2{}^{2-}$ ion. For example, in H_2O_2, hydrogen peroxide, H is assigned its usual oxidation number of +1, and so O is −1.
6. **The algebraic sum of the oxidation numbers in a neutral compound must be zero; in a polyatomic ion, the sum must be equal to the ion charge.** Examples of this rule are the previous compounds and others found in Example 4.8.

EXAMPLE 4.8 *Determining Oxidation Numbers*

Determine the oxidation number of the indicated element in each of the following compounds or ions:

1. Lithium in lithium oxide, Li_2O
2. Phosphorus in phosphoric acid, H_3PO_4

3. Manganese in the permanganate ion, MnO_4^-
4. Chromium in the dichromate ion, $Cr_2O_7^{2-}$
5. Carbon in toluene, C_7H_8

Solution

1. Li_2O is a neutral compound. The oxidation number of Li is +1, as predicted by its position in the periodic table. This means O has its "normal" oxidation number of −2.

$$Li_2O = (2Li^+)(O^{2-})$$

$$\begin{aligned}\text{Net charge} = 0 &= \text{oxidation number of Li} + \text{oxidation number of O}\\ &= 2(+1) + (-2)\end{aligned}$$

2. H_3PO_4 also has an overall charge of 0. If the oxygen atoms each have an oxidation number of −2 and that of the H atoms is +1, then the oxidation number of P must be +5.

$$H_3PO_4 = (3\ H^+)(P^{5+})(4\ O^{2-})$$

Net charge = 0

= sum of oxidation numbers of H atoms + oxidation number of P + sum of oxidation numbers for O atoms

= 3(+1) + (+5) + 4(−2)

3. The permanganate ion, MnO_4^-, has an overall charge of 1−. Because this compound is not a peroxide, O is assigned an oxidation number of −2, which means that Mn has an oxidation number of +7.

$$MnO_4^- = [(Mn^{7+})(4\ O^{2-})]^-$$

Net charge = 1−

= oxidation number of Mn + sum of oxidation numbers for O atoms

= (+7) + 4(−2)

4. Compounds containing dichromate ions, $Cr_2O_7^{2-}$, are widely used in the laboratory. Because the net charge on the ion is 2−, and O is assigned an oxidation number of −2, each Cr atom must have an oxidation number of +6.

$$Cr_2O_7^{2-} = [(2\ Cr^{6+})(7\ O^{2-})]^{2-}$$

Net charge = 2−

= sum of oxidation numbers for Cr atoms + sum of oxidation numbers for O atoms

= 2(+6) + 7(−2)

5. Toluene, a hydrocarbon, has the formula C_7H_8. Taking the oxidation number of H as +1, each carbon atom must have a fractional oxidation number of −8/7.

$$C_7H_8 = (7\ C^{8/7-})(8\ H^+)$$

Here we solve for the carbon oxidation number in the following way:

Net charge = 0

= sum of oxidation numbers for C atoms + sum of oxidation numbers for H atoms

= 7(oxidation number of C atom) + 8(+1)

C oxidation number = −8/7

Therefore, we have 7(−8/7) + 8(+1) = 0.

EXERCISE 4.12 ***Oxidation Numbers***

Assign an oxidation number to the underlined atom in each of the following molecules or ions.

1. $\underline{Fe}_2O_3$ 2. $H_2\underline{S}O_4$ 3. $\underline{C}O_3^{2-}$ 4. $\underline{C}_6H_6$ ■

Recognizing Oxidation-Reduction Reactions

Having learned some guidelines for determining oxidation numbers, you can tell which reactions can be classified as oxidation-reduction and which must be of some other type. In many cases, however, it will be obvious that the reaction is an oxidation-reduction because it involves a well-known oxidizing or reducing agent (Table 4.2).

Like oxygen, O_2, the halogens (F_2, Cl_2, Br_2, and I_2) are always oxidizing agents in their reactions with metals and most nonmetals. For example, Figure 2.20 illustrates the reaction of sodium metal with chlorine.

TABLE 4.2 Common Oxidizing and Reducing Agents

Oxidizing Agent	Reaction Product	Reducing Agent	Reaction Product
O_2, oxygen	O^{2-}, oxide ion	H_2, hydrogen	H^+, hydrogen ion or H combined in H_2O
Halogens F_2, Cl_2, Br_2, or I_2	Halide ion F^-, Cl^-, Br^-, or I^-	M, metals such as Na, K, Fe, and Al	M^{n+}, metal ions such as Na^+, K^+, Fe^{2+} or Fe^{3+}, and Al^{3+}
HNO_3, nitric acid	Nitrogen oxides, such as NO and NO_2	C, carbon (used to reduce metal oxides)	CO and CO_2
$Cr_2O_7^{2-}$, dichromate ion	Cr^{3+}, chromium(III) ion (in acid solution)		
MnO_4^-, permanganate ion	Mn^{2+}, manganese(II) ion (in acid solution)		

Na releases $1e^-$ per atom.
Oxidation number increases.
Na is oxidized and is the reducing agent.

$$2\ Na(s) + Cl_2(g) \longrightarrow 2\ [Na^+, Cl^-]$$

Cl_2 gains $2e^-$ per molecule.
Oxidation number decreases by 1 per Cl atom.
Cl_2 is reduced and is the oxidizing agent.

When halogens are combined with metals they always end up as halide ions, X^- (that is, as F^-, Cl^-, Br^-, or I^-) in the product. Like oxygen, the halogens are therefore always reduced (are oxidizing agents) in their reactions with metals.

Chlorine ends up as Cl^-, and, having acquired two electrons per Cl_2 molecule, it has been reduced and so is the oxidizing agent. (The oxidation number of Cl has decreased from 0 to -1.)

Be sure to notice that sodium begins as the element in this reaction, but it ends up as the Na^+ ion after combining with chlorine. (The Na oxidation number has increased from 0 to $+1$.) Thus, sodium is oxidized and is the reducing agent. Indeed, it is generally true that *metals are reducing agents.*

Chlorine is widely used as an oxidizing agent in water and sewage treatment. For example, it can remove hydrogen sulfide, H_2S, from drinking water by oxidizing the sulfide to insoluble, elemental sulfur. (Hydrogen sulfide has a characteristic "rotten egg" odor; it comes from the decay of organic matter or underground mineral deposits.)

$$8\ Cl_2(g) + 8\ H_2S(aq) \longrightarrow S_8(s) + 16\ HCl(aq)$$

Knowing easily recognized oxidizing and reducing agents allows you to predict whether a reaction is classified as an oxidation-reduction and, in some cases, to predict what the products might be. Tables 4.2 and 4.3, and the following list, are some guidelines.

- If an element or compound has combined with oxygen, the element or compound has been oxidized and O_2 is reduced. In the process the oxygen, O_2, is changed to the oxide ion, O^{2-} (as in a metal oxide) by adding electrons or is combined in a molecule, such as CO_2 or H_2O (as occurs in the combustion reaction of a hydrocarbon) in which its oxidation number is -2.
- If an element or compound has combined with a halogen, the element or compound has been oxidized. In the process the halogen, X_2, is changed to halide ions, X^-, by adding electrons, or it is combined in a molecule such as HCl where its oxidation number is -1. The halogen has therefore been reduced, and it is the oxidizing agent. Among the halogens, fluorine and chlorine are particularly strong oxidizing agents.

TABLE 4.3 Recognizing Oxidation-Reduction Reactions

	Oxidation	Reduction
In terms of oxidation number	Increase in oxidation number of an atom	Decrease in oxidation number of an atom
In terms of electrons	Loss of electrons by an atom	Gain of electrons by an atom
In terms of oxygen	Gain of oxygen	Loss of oxygen

Figure 4.18 Copper reacts vigorously with nitric acid to give brown NO_2 gas. The deep green solution contains copper(II) nitrate, $Cu(NO_3)_2$. (C. D. Winters)

- If a metal combines with another element or a compound, the metal has been oxidized. In the process, the metal has lost electrons to form a positive ion (as in metal oxides or halides, for example).

$$M \xrightarrow[\text{reducing agent}]{\text{oxidation reaction}} M^{n+} + ne^-$$

The metal, an electron donor, has therefore been oxidized and has functioned as a reducing agent. Most metals are reasonably good reducing agents, with metals from Groups 1A, 2A, and 3A (such as sodium, magnesium, and aluminum) being particularly good.

- Other common oxidizing and reducing agents are listed in Table 4.2, and some are described later on. When one of these agents takes part in a reaction, it is reasonably certain that it is a redox reaction.

Figure 4.18 illustrates the chemistry of one of the best oxidizing agents, nitric acid, HNO_3. Here the acid oxidizes copper metal to give copper(II) nitrate, and the nitrate ion is reduced to the brown gas NO_2. The net ionic equation for the reaction is

Notice that nitric acid, a common acid, can be both an oxidizing agent and an acid.

$$\underset{\text{reducing agent}}{Cu(s)} + \underset{\text{oxidizing agent}}{2\ NO_3^-(aq)} + 4\ H_3O^+(aq) \longrightarrow Cu^{2+}(aq) + 2\ NO_2(g) + 6\ H_2O(\ell)$$

Copper metal is clearly the reducing agent in the previous reaction; the metal has given up two electrons per metal atom to produce the Cu^{2+} ion. Nitrogen has been reduced from +5 (in the NO_3^- ion) to +4 (in NO_2); therefore, the nitrate ion in acid solution is an oxidizing agent.

The most common reducing agents in the laboratory are metals. The alkali and alkaline earth metals are quite strong reducing agents in most of their reactions. For example, potassium is capable of reducing water to H_2 gas (Figure 4.19).

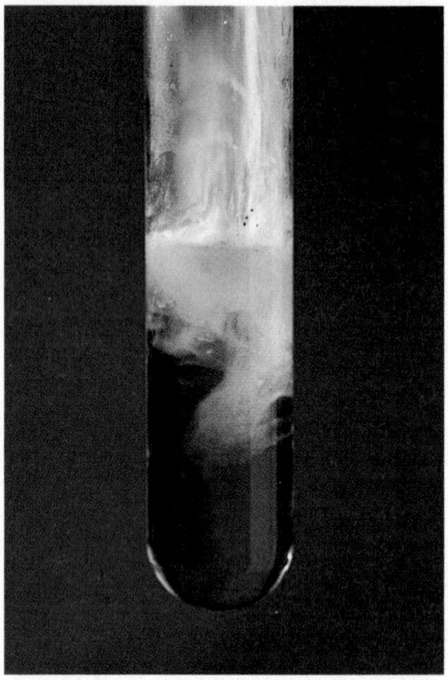

Figure 4.19 Potassium metal reacts very vigorously with water to give hydrogen gas and leave potassium hydroxide in solution. (C. D. Winters)

$$\underset{\text{reducing agent}}{2\ K(s)} + \underset{\text{oxidizing agent}}{2\ H_2O(\ell)} \longrightarrow 2\ KOH(aq) + H_2(g)$$

Aluminum metal is also a good reducing agent and is capable of reducing iron(III) oxide to iron metal in a reaction called the thermite reaction.

$$\underset{\text{oxidizing agent}}{Fe_2O_3(s)} + \underset{\text{reducing agent}}{2\ Al(s)} \longrightarrow 2\ Fe(s) + Al_2O_3(s)$$

Such a large quantity of heat is evolved in the reaction that the iron is produced in the molten state (Figure 4.20).

Hundreds of compounds are good oxidizing and reducing agents, and when mixed, they lead to a reaction. You should also be aware, however, that it is not a good idea to mix a strong oxidizing agent with a strong reducing agent; a violent reaction, even an explosion, may take place. This is the reason that chemicals are no longer stored on shelves in alphabetical order. This can be unsafe, because such an ordering may place a strong oxidizing agent next to a strong reducing agent.

As you inspect the equations for the oxidation-reduction reactions in this section notice that all are balanced. The same number of atoms of each element appears on both sides of the equation, and the net electric charge is the same on both sides. Balancing such equations, particularly when the reaction occurs in an acidic or basic solution, requires a special approach. This will be considered in Chapter 21 in the context of the discussion of electrochemistry.

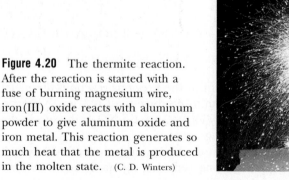

Figure 4.20 The thermite reaction. After the reaction is started with a fuse of burning magnesium wire, iron(III) oxide reacts with aluminum powder to give aluminum oxide and iron metal. This reaction generates so much heat that the metal is produced in the molten state. (C. D. Winters)

EXAMPLE 4.9 *Oxidation-Reduction Reactions*

In each reaction, decide which atom is undergoing a change in oxidation number, and then identify the oxidizing agent and the reducing agent.

1. $5\ Fe^{2+}(aq) + MnO_4^-(aq) + 8\ H^+(aq) \rightarrow 5\ Fe^{3+}(aq) + Mn^{2+}(aq) + 4\ H_2O(\ell)$

2. $H_2(g) + CuO(s) \rightarrow Cu(s) + H_2O(g)$

Solution

1. *The reaction of the iron(II) ion with permanganate ion* (Figure 4.21). The permanganate ion is described as an oxidizing agent in Table 4.2. This is because the Mn oxidation number in MnO_4^- is +7, and it decreases to +2 in the product, the Mn^{2+} ion. Because Mn has been reduced, the MnO_4^- ion is the oxidizing agent. Finally, the oxidation number of iron has increased from +2 to +3. That is, the Fe^{2+} ion is oxidized to Fe^{3+}. This means that the Fe^{2+} ion is the reducing agent (the supplier of electrons).

2. *The reaction of hydrogen gas with copper(II) oxide* (Figure 4.22). H_2 gas is a common reducing agent, widely used in the laboratory and in industry. The reason is that the oxidation number changes from 0 (in H_2) to +1 (in H_2O). Copper(II) oxide is the substance reduced here because the oxidation number of copper changes from +2 in CuO to 0 in copper metal.

Figure 4.21 The reaction of the purple permanganate ion (MnO_4^-, the oxidizing agent) with the iron(II) ion (Fe^{2+}, the reducing agent) in acidified aqueous solution gives the nearly colorless manganese(II) ion (Mn^{2+}) and iron(III) ion (Fe^{3+}). (C. D. Winters)

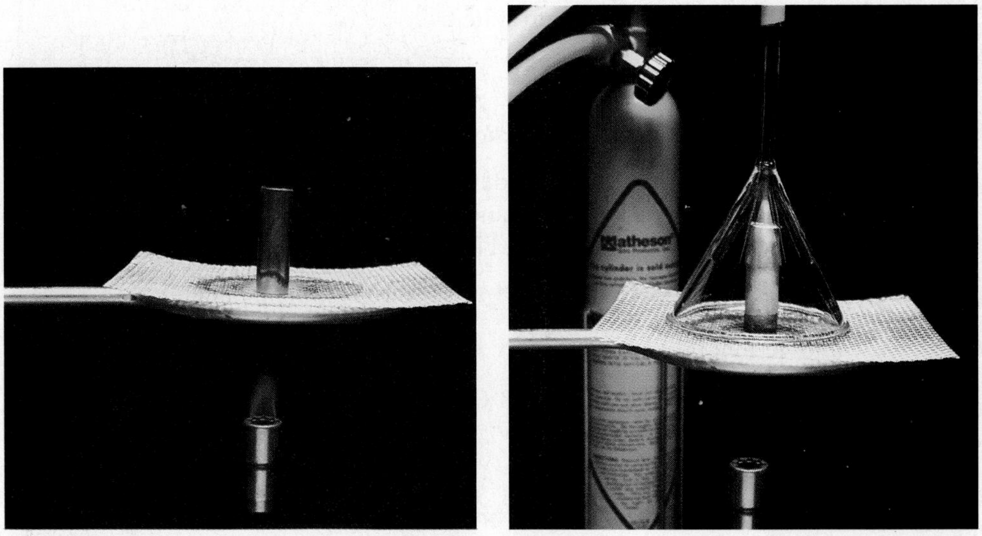

(a) (b)

Figure 4.22 Reduction of copper oxide with hydrogen. (a) A piece of copper has been heated in air to form a film of black copper(II) oxide on the surface. (b) When the hot copper metal, with its film of CuO, is placed in a stream of hydrogen gas (from the yellow tank at the rear), the oxide is reduced to copper metal, and water forms as the byproduct. (C. D. Winters)

EXAMPLE 4.10 *Types of Reactions*

Classify each of the following reactions as (a) precipitation, (b) acid-base, (c) gas-forming, or (d) oxidation-reduction.

1. $HNO_3(aq) + Ca(OH)_2(s) \rightarrow Ca(NO_3)_2(aq) + 2\ H_2O(\ell)$
2. $SO_4^{2-}(aq) + 2\ CH_2O(aq) + 2\ H^+(aq) \rightarrow H_2S(aq) + 2\ CO_2(g) + 2\ H_2O(\ell)$

Solution One of the best ways to differentiate reactions is to look first for a reaction that could be an oxidation-reduction, that is, a reaction in which the oxidation number of some substance changes. This is only the case in reaction (2). Writing this reaction with the oxidation number of each element indicated,

$$SO_4^{2-}(aq) + 2\ CH_2O(aq) + 2\ H^+(aq) \longrightarrow H_2S(aq) + 2\ CO_2(g) + 2\ H_2O(\ell)$$

Oxidation number	SO_4^{2-}	CH_2O	H^+	H_2S	CO_2	H_2O
	+6, −2	0, +1, −2	+1	+1, −2	+4, −2	+1, −2

you see that the oxidation number of S changes from +6 to −2, and that of C changes from 0 to +4. Therefore, sulfate, SO_4^{2-}, has been reduced (and is the oxidizing agent), and CH_2O has been oxidized (and is the reducing agent).

Notice that no such changes occur in oxidation number for the elements in reaction (1)

$$HNO_3(aq) + Ca(OH)_2(s) \longrightarrow Ca(NO_3)_2(aq) + 2\ H_2O(\ell)$$

Oxidation number	HNO_3	$Ca(OH)_2$	$Ca(NO_3)_2$	H_2O
	+1, +5, −2	+2, −2, +1	+2, +5, −2	+1, −2

Instead, reaction (1) is classified as an acid-base reaction. Here an acid (nitric acid, HNO_3) reacts with a base (calcium hydroxide, $Ca(OH)_2$) to give a salt (calcium nitrate, $Ca(NO_3)_2$) and water.

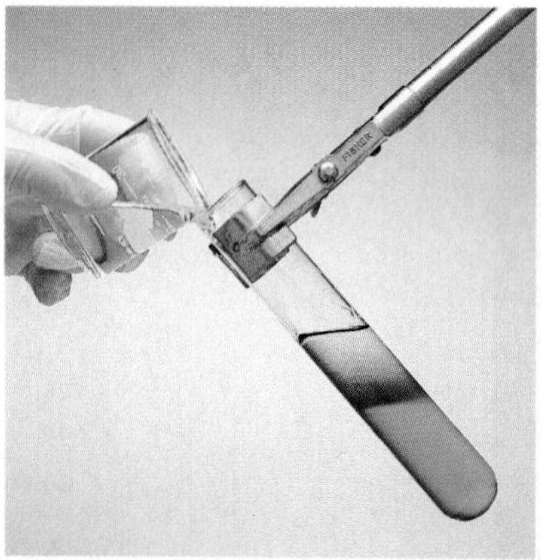

Figure 4.23 The reaction of ethanol with dichromate ion is the basis of the test used in a Breathalyzer. These devices are available in many stores and can be used to test for high levels of ethanol in the breath. If alcohol is present, the orange-red dichromate ion is changed to the green chromium(III) ion. Here ethanol is poured into a solution of potassium dichromate. Reaction has occurred in the top layer. (C. D. Winters)

EXERCISE 4.13 *Oxidation-Reduction Reactions*

The following reaction occurs in a device for testing the breath for the presence of alcohol. Identify the oxidizing and reducing agents and the substance oxidized and the substance reduced (Figure 4.23).

$$3\ C_2H_5OH(aq) + 2\ Cr_2O_7^{2-}(aq) + 16\ H^+(aq) \longrightarrow 3\ CH_3CO_2H(aq) + 4\ Cr^{3+}(aq) + 11\ H_2O(\ell)$$

ethanol; dichromate ion; orange-red; acetic acid; chromium(III) ion; green ■

EXERCISE 4.14 *Oxidation-Reduction Reactions*

Decide which of the following reactions are oxidation-reduction reactions. In each case explain your choice and identify the oxidizing and reducing agents.

1. $NaOH(aq) + HNO_3(aq) \rightarrow NaNO_3(aq) + H_2O(\ell)$
2. $Cu(s) + Cl_2(g) \rightarrow CuCl_2(s)$
3. $Na_2CO_3(aq) + 2\ HClO_4(aq) \rightarrow CO_2(g) + H_2O(\ell) + 2\ NaClO_4(aq)$
4. $2\ S_2O_3^{2-}(aq) + I_2(aq) \rightarrow S_4O_6^{2-}(aq) + 2\ I^-(aq)$ ■

CHAPTER HIGHLIGHTS

Now that you have finished studying this chapter, you should be able to

- Interpret the information conveyed by a **balanced chemical equation** (Section 4.1).
- Balance simple chemical equations (Section 4.2).
- Explain the differences between **electrolytes** and **nonelectrolytes** (Section 4.3).
- Predict the **solubility** of ionic compounds in water (Section 4.3).
- Recognize common **acids** and **bases** and understand their behavior in aqueous solution (Section 4.4).
- Recognize what ions are formed when an ionic compound dissolves in water (Sections 4.3–4.5).
- Write **net ionic equations** and show how to arrive at such an equation for a given reaction (Section 4.5).
- Use the ideas developed in Sections 4.1–4.5 as an aid in recognizing four of the common types of reactions that occur in aqueous solution and write balanced equations for such reactions (Section 4.6).

Reaction Type	**Driving Force**
Precipitation	Formation of an insoluble compound
Acid-base; neutralization	Formation of a salt and water
Gas-forming	Evolution of a water-insoluble gas such as CO_2
Oxidation-reduction	Transfer of electrons

- Predict the products of **precipitation reactions** (Section 4.7), the formation of an insoluble reaction product by the exchange of anions between the cations of the reactants.

$$Pb(NO_3)_2(aq) + 2\,KI(aq) \longrightarrow PbI_2(s) + 2\,KNO_3(aq)$$

- Predict the products of **acid-base reactions** involving common acids and bases (Section 4.8).

$$HNO_3(aq) + KOH(aq) \longrightarrow \underset{\text{potassium nitrate, a salt}}{KNO_3(aq)} + H_2O(\ell)$$

- Understand that the net ionic equation for the reaction of a strong acid with a strong base is $H^+(aq) + OH^-(aq) \rightarrow H_2O(\ell)$ (Section 4.8).
- Predict the products of **gas-forming reactions** (Section 4.9), the most common of which are those between a metal carbonate and an acid.

$$NiCO_3(s) + 2\,HNO_3(aq) \longrightarrow Ni(NO_3)_2(aq) + CO_2(g) + H_2O(\ell)$$

- Understand that the **oxidation number** of an element in a compound is the electric charge an atom has, or appears to have, when the electrons of the compound are counted according to a set of "Guidelines" (Section 4.10).

- Calculate oxidation numbers for elements in a compound (Section 4.10).
- Recognize **oxidation-reduction reactions** (often called **redox reactions**) (Section 4.10 and Table 4.3).

STUDY QUESTIONS

Review Questions

1. What information is provided by a balanced chemical equation?
2. In the reaction in Figure 4.1, how many molecules of Br_2 do you need for complete reaction if you have 2000 atoms of Al? How many molecules of Al_2Br_6 do you obtain from the reaction?
3. In the chapter find one example of each of the following reaction types: acid-base, precipitation, gas-forming, or oxidation-reduction reaction. Describe why each reaction exemplifies its class. Name the reactants and products of each reaction.
4. Find two examples in the chapter of an acid-base reaction. Write the balanced equation for each, and name the reactants and products.
5. What is an electrolyte? How can we differentiate between a weak and a strong electrolyte? Give an example of each.
6. Which of the following nickel(II) salts are soluble in water and which are insoluble: $Ni(NO_3)_2$, $NiCO_3$, $Ni_3(PO_4)_2$, and $NiCl_2$?
7. Name two acids that are strong electrolytes and one that is a weak electrolyte. Name two bases that are strong electrolytes and one that is a weak electrolyte.
8. Name the spectator ions in the reaction of nitric acid and magnesium hydroxide, and write the net ionic equation:

$$2\ H^+(aq) + 2\ NO_3^-(aq) + Mg(OH)_2(s) \longrightarrow 2\ H_2O(\ell) + Mg^{2+}(aq) + 2\ NO_3^-(aq)$$

   What type of reaction is this?
9. Name the water-insoluble product in each reaction:
   (a) $CuCl_2(aq) + H_2S(aq) \rightarrow CuS + 2\ HCl$
   (b) $CaCl_2(aq) + K_2CO_3(aq) \rightarrow 2\ KCl + CaCO_3$
   (c) $AgNO_3(aq) + NaI(aq) \rightarrow AgI + NaNO_3$
10. Find two examples of precipitation reactions in the chapter. Write balanced equations for these reactions, and name the reactants and products.
11. Find two examples of gas-forming reactions in the chapter. Write balanced equations for these reactions, and name the reactants and products.
12. Bromine is obtained from seawater by the following reaction:

$$Cl_2(g) + 2\ NaBr(aq) \longrightarrow 2\ NaCl(aq) + Br_2(\ell)$$

   What has been oxidized? What has been reduced? Name the oxidizing and reducing agents.
13. Oxidation-reduction reactions:
   (a) Explain the difference between oxidation and reduction. Give an example of each.
   (b) Explain the difference between an oxidizing agent and a reducing agent. Give an example of each.
14. What substances in the following list are common oxidizing or reducing agents: HNO_3, Na, NaCl, O_2, $CrCl_3$, and $KMnO_4$? Identify each such substance as an oxidizing or reducing agent. (Hint: See Table 4.2.)

Numerical and Other Questions

Balancing Equations

15. Balance the following equations:
   (a) $Cr(s) + O_2(g) \rightarrow Cr_2O_3(s)$
   (b) $Cu_2S(s) + O_2(g) \rightarrow Cu(s) + SO_2(g)$
   (c) $C_6H_5CH_3(\ell) + O_2(g) \rightarrow H_2O(\ell) + CO_2(g)$
16. Balance the following equations:
   (a) $Cr(s) + Cl_2(g) \rightarrow CrCl_3(s)$
   (b) $SiO_2(s) + C(s) \rightarrow Si(s) + CO(g)$
   (c) $Fe(s) + H_2O(g) \rightarrow Fe_3O_4(s) + H_2(g)$
17. Balance the following equations, and name the reaction products:
   (a) $MgO(s) + Fe(s) \rightarrow Fe_2O_3(s) + Mg(s)$
   (b) $AlCl_3(s) + H_2O(\ell) \rightarrow Al(OH)_3(s) + HCl(aq)$
   (c) $NaNO_3(s) + H_2SO_4(\ell) \rightarrow Na_2SO_4(s) + HNO_3(g)$
18. Balance the following equations, and name the reaction products:
   (a) $UO_2(s) + HF(\ell) \rightarrow UF_4(s) + H_2O(\ell)$
   (b) $NH_3(aq) + HNO_3(aq) \rightarrow N_2(g) + H_2O(\ell)$
   (c) $BF_3(g) + H_2O(\ell) \rightarrow HF(aq) + H_3BO_3(aq)$
19. Balance the following equations:
   (a) The synthesis of urea, a common fertilizer

$$CO_2(g) + NH_3(g) \longrightarrow CO(NH_2)_2(s) + H_2O(\ell)$$

   (b) Reactions used to make uranium(VI) fluoride for the enrichment of natural uranium

$$UO_2(s) + HF(aq) \longrightarrow UF_4(s) + H_2O(aq)$$
$$UF_4(s) + F_2(g) \longrightarrow UF_6(s)$$

(c) Reaction to make titanium(IV) chloride, which is then converted to titanium metal

$$TiO_2(s) + Cl_2(g) + C(s) \longrightarrow TiCl_4(\ell) + CO$$
$$TiCl_4(\ell) + Mg(s) \longrightarrow Ti(s) + MgCl_2(s)$$

20. Balance the following equations:
(a) Reaction to produce "superphosphate" fertilizer

$$Ca_3(PO_4)_2(s) + H_2SO_4(aq) \longrightarrow Ca(H_2PO_4)_2(aq) + CaSO_4(s)$$

(b) Reaction to produce diborane, B_2H_6

$$NaBH_4(s) + H_2SO_4(aq) \longrightarrow B_2H_6(g) + H_2(g) + Na_2SO_4(aq)$$

(c) Reaction to produce tungsten metal from tungsten(VI) oxide

$$WO_3(s) + H_2(g) \longrightarrow W(s) + H_2O(\ell)$$

Properties of Aqueous Solutions

(It may help to consult Figure 4.7 as you think about the questions in this section.)

21. Which compound or compounds in each of the following groups is (are) expected to be soluble in water?
(a) FeO, $FeCl_2$, and $FeCO_3$
(b) AgI, Ag_3PO_4, and $AgNO_3$
(c) NaCl, Li_2CO_3, and $KMnO_4$

22. Which compound or compounds in each of the following groups is (are) expected to be soluble in water?
(a) $PbSO_4$, $Pb(NO_3)_2$, and $PbCO_3$
(b) Na_2SO_4, $NaClO_4$, and $NaCH_3CO_2$
(c) AgBr, KBr, Al_2Br_6

23. Give the formula for
(a) A soluble compound containing the acetate ion
(b) An insoluble sulfide
(c) A soluble hydroxide
(d) An insoluble chloride

24. Give the formula for
(a) A soluble compound containing the chloride ion
(b) An insoluble hydroxide
(c) An insoluble carbonate
(d) A soluble nitrate-containing compound

25. Each compound below is water-soluble. What ions are produced in water?
(a) KI (c) $KHSO_4$
(b) K_2SO_4 (d) KCN

26. Each compound below is water-soluble. What ions are produced in water?
(a) KOH (c) $NaNO_3$
(b) $Mg(CH_3CO_2)_2$ (d) $(NH_4)_2SO_4$

27. Tell whether each of the following is water-soluble or not. If soluble, tell what ions are produced.
(a) $BaCl_2$ (c) $Pb(NO_3)_2$
(b) $Cr(NO_3)_2$ (d) $BaSO_4$

28. Tell whether each of the following is water-soluble or not. If soluble, tell what ions are produced.
(a) Na_2CO_3
(b) $CuSO_4$
(c) NiS
(d) $CaBr_2$

29. Name two water-soluble compounds containing Cu^{2+}. Name two water-insoluble Cu^{2+} compounds.

30. Name two water-soluble compounds containing Ba^{2+}. Name two water-insoluble Ba^{2+} compounds.

Acids and Bases

31. Write a balanced equation for the ionization of nitric acid in water.

32. Write a balanced equation for the ionization of perchloric acid in water.

33. Oxalic acid, which is found in certain plants, can provide two hydrogen ions in water. Write balanced equations (like those for sulfuric acid on page 172) to show how oxalic acid, $H_2C_2O_4$, can supply one and then a second H^+ ion.

34. Phosphoric acid can supply one, two, or three H^+ ions in aqueous solution. Write balanced equations (like those for sulfuric acid on page 172) to show this successive loss of hydrogen ions.

35. Write a balanced equation for reaction of the basic oxide, magnesium oxide, with water.

36. Write a balanced equation for the reaction of sulfur trioxide with water.

Writing Net Ionic Equations

37. Balance each of the following equations, and then write the net ionic equation:
(a) $Zn(s) + HCl(aq) \rightarrow H_2(g) + ZnCl_2(aq)$
(b) $Mg(OH)_2(s) + HCl(aq) \rightarrow MgCl_2(aq) + H_2O(\ell)$
(c) $HNO_3(aq) + CaCO_3(s) \rightarrow Ca(NO_3)_2(aq) + H_2O(\ell) + CO_2(g)$

38. Balance each of the following equations, and then write the net ionic equation:
(a) $(NH_4)_2CO_3(aq) + Cu(NO_3)_2(aq) \rightarrow CuCO_3(s) + NH_4NO_3(aq)$
(b) $Pb(OH)_2(s) + HCl(aq) \rightarrow PbCl_2(s) + H_2O(\ell)$
(c) $BaCO_3(s) + HCl(aq) \rightarrow BaCl_2(aq) + H_2O(\ell) + CO_2(g)$
(d) $HCl(aq) + MnO_2(s) \rightarrow MnCl_2(aq) + Cl_2(g) + H_2O(\ell)$

39. Balance each of the following equations, and then write the net ionic equation. Refer to Table 4.1 and Figure 4.7 for information on acids and bases and on solubility. Show states for all reactants and products (s, ℓ, g, aq).
(a) $Ba(OH)_2 + HNO_3 \rightarrow Ba(NO_3)_2 + H_2O$
(b) $BaCl_2 + Na_2CO_3 \rightarrow BaCO_3 + NaCl$
(c) $Na_3PO_4 + Ni(NO_3)_2 \rightarrow Ni_3(PO_4)_2 + NaNO_3$

40. Balance each of the following equations, and then write the net ionic equation. Refer to Table 4.1 and Figure 4.7 for

information on acids and bases and on solubility. Show states for all reactants and products (s, ℓ, g, aq).
(a) $ZnCl_2 + KOH \rightarrow KCl + Zn(OH)_2$
(b) $AgNO_3 + KI \rightarrow AgI + KNO_3$
(c) $NaOH + FeCl_2 \rightarrow Fe(OH)_2 + NaCl$

Types of Reactions in Aqueous Solution

41. Balance these equations and then classify each one as an acid-base reaction, a precipitation, or a gas-forming reaction:
(a) $K_2CO_3(aq) + Cu(NO_3)_2(aq) \rightarrow CuCO_3(s) + KNO_3(aq)$
(b) $Pb(NO_3)_2(aq) + HCl(aq) \rightarrow PbCl_2(s) + HNO_3(aq)$
(c) $MgCO_3(s) + HCl(aq) \rightarrow MgCl_2(aq) + H_2O(\ell) + CO_2(g)$

42. Balance each of these equations and then classify each one as an acid-base reaction, a precipitation, or a gas-forming reaction:
(a) $Ba(OH)_2(s) + HCl(aq) \rightarrow BaCl_2(aq) + H_2O(\ell)$
(b) $HNO_3(aq) + CoCO_3(s) \rightarrow Co(NO_3)_2(aq) + H_2O(\ell) + CO_2(g)$
(c) $Na_3PO_4(aq) + Cu(NO_3)_2(aq) \rightarrow Cu_3(PO_4)_2(s) + NaNO_3(aq)$

43. Balance each of these equations and then classify each one as an acid-base reaction, a precipitation, or a gas-forming reaction. Show states for the products (s, ℓ, g, aq), and then balance the completed equation. Write the net ionic equation.
(a) $MnCl_2(aq) + Na_2S(aq) \rightarrow MnS + NaCl$
(b) $K_2CO_3(aq) + ZnCl_2(aq) \rightarrow ZnCO_3 + KCl$
(c) $K_2CO_3(aq) + HClO_4(aq) \rightarrow KClO_4 + CO_2 + H_2O$

44. Balance each of these equations and then classify each one as an acid-base reaction, a precipitation, or a gas-forming reaction. Write the net ionic equation.
(a) $Fe(OH)_3(s) + HNO_3(aq) \rightarrow Fe(NO_3)_3 + H_2O$
(b) $FeCO_3(s) + HNO_3(aq) \rightarrow Fe(NO_3)_2 + CO_2 + H_2O$
(c) $FeCl_2(aq) + (NH_4)_2S(aq) \rightarrow FeS + NH_4Cl$
(d) $Fe(NO_3)_2(aq) + Na_2CO_3(aq) \rightarrow FeCO_3 + NaNO_3$

Precipitation Reactions

45. Balance the equation for the following precipitation reaction, and then write the net ionic equation. Indicate the state of each species (s, ℓ, aq, or g).

$$CdCl_2 + NaOH \longrightarrow Cd(OH)_2 + NaCl$$

46. Balance the equation for the following precipitation reaction, and then write the net ionic equation. Indicate the state of each species (s, ℓ, aq, or g).

$$Ni(NO_3)_2 + Na_2CO_3 \rightarrow NiCO_3 + NaNO_3$$

47. Predict the products of each precipitation reaction, and then balance the completed equation.
(a) $NiCl_2(aq) + (NH_4)_2S(aq) \rightarrow$
(b) $Mn(NO_3)_2(aq) + Na_3PO_4(aq) \rightarrow$

48. Predict the products of each precipitation reaction, and then balance the completed equation.
(a) $Pb(NO_3)_2(aq) + KBr(aq) \rightarrow$
(b) $Ca(NO_3)_2(aq) + KF(aq) \rightarrow$
(c) $Ca(NO_3)_2(aq) + Na_2C_2O_4(aq) \rightarrow$

49. Write an overall, balanced equation for the precipitation reaction that occurs when aqueous lead(II) nitrate is mixed with an aqueous solution of potassium hydroxide. Name each reactant and product.

50. Write an overall, balanced equation for the precipitation reaction that occurs when aqueous copper(II) nitrate is mixed with an aqueous solution of sodium carbonate. Name each reactant and product.

Acid-Base Reactions

51. Complete and balance the following acid-base reactions. Name the reactants and products.
(a) $CH_3CO_2H(aq) + Mg(OH)_2(s) \rightarrow$
(b) $HClO_4(aq) + NH_3(aq) \rightarrow$

52. Complete and balance the following acid-base reactions. Name the reactants and products.
(a) $H_3PO_4(aq) + KOH(aq) \rightarrow$
(b) $H_2C_2O_4(aq) + Ca(OH)_2(s) \rightarrow$
($H_2C_2O_4$ is oxalic acid, an acid capable of donating two H^+ ions.)

53. Write a balanced equation for the reaction of barium hydroxide with nitric acid to give barium nitrate, a compound used in pyrotechnics such as green flares.

54. Aluminum is obtained from bauxite, which is not a specific mineral but a name applied to a mixture of minerals. One of those minerals, which can dissolve in acids, is gibbsite, $Al(OH)_3$. Write a balanced equation for the reaction of gibbsite with sulfuric acid.

Rhodochrosite, a mineral that consists primarily of manganese(II) carbonate. See Study Question 55. (C. D. Winters)

Gas-Forming Reactions

55. The beautiful mineral rhodochrosite (previous page, bottom) is manganese(II) carbonate. Write an overall, balanced equation for the reaction of the mineral with hydrochloric acid. Name each reactant and product.

56. Many minerals are metal carbonates, and siderite is a mineral that consists largely of iron(II) carbonate. Write an overall, balanced equation for the reaction of the mineral with nitric acid, and name each reactant and product.

Oxidation Numbers

57. Determine the oxidation number of each element in the following ions or compounds:
(a) BrO_3^- (d) CaH_2
(b) $C_2O_4^{2-}$ (e) H_4SiO_4
(c) F_2 (f) SO_4^{2-}

58. Determine the oxidation number of each element in the following ions or compounds:
(a) SF_6 (d) N_2O_4
(b) $H_2AsO_4^-$ (e) $C_5H_8O_2$
(c) $C_2H_4O_2$ (f) XeO_4^{2-}

59. The following reaction can be used to prepare iodine in the laboratory (see photos below). Determine the oxidation number of each atom in the following equation:

$$2\ NaI(s) + 2\ H_2SO_4(aq) + MnO_2(s) \longrightarrow Na_2SO_4(aq) + MnSO_4(aq) + I_2(g) + 2\ H_2O(\ell)$$

60. Determine the oxidation number of each atom in the following reactions, which are used by the chemical industry to make acetic acid from the simple hydrocarbon C_2H_4:

$$\underset{\text{ethylene}}{C_2H_4(aq)} + \tfrac{1}{2}\ O_2(g) \longrightarrow \underset{\text{acetaldehyde}}{CH_3CHO(\ell)}$$

$$CH_3CHO(\ell) + \tfrac{1}{2}\ O_2(g) \longrightarrow \underset{\text{acetic acid}}{CH_3CO_2H(\ell)}$$

Oxidation-Reduction Reactions

61. Which of the following reactions are oxidation-reduction reactions? Explain your answer briefly. Classify the remaining reactions.
(a) $CdCl_2(aq) + Na_2S(aq) \rightarrow CdS(s) + 2\ NaCl(aq)$
(b) $2\ Ca(s) + O_2(g) \rightarrow 2\ CaO(s)$
(c) $Ca(OH)_2(s) + 2\ HCl(aq) \rightarrow CaCl_2(aq) + 2\ H_2O(\ell)$

62. Which of the following reactions are oxidation-reduction reactions? Explain your answer in each case. Classify the remaining reactions.
(a) $Zn(s) + 2\ NO_3^-(aq) + 4\ H^+(aq) \rightarrow Zn^{2+}(aq) + 2\ NO_2(g) + 2\ H_2O(\ell)$
(b) $Zn(OH)_2(s) + H_2SO_4(aq) \rightarrow ZnSO_4(aq) + 2\ H_2O(\ell)$
(c) $Ca(s) + 2\ H_2O(\ell) \rightarrow Ca(OH)_2(s) + H_2(g)$

63. In each of the following reactions, tell which reactant is oxidized and which is reduced. Designate the oxidizing agent and reducing agent.
(a) $2\ Mg(s) + O_2(g) \rightarrow 2\ MgO(s)$
(b) $C_2H_4(g) + 3\ O_2(g) \rightarrow 2\ CO_2(g) + 2\ H_2O(g)$
(c) $Si(s) + 2\ Cl_2(g) \rightarrow SiCl_4(\ell)$

Preparation of iodine. A mixture of sodium iodide and manganese(IV) oxide was placed in a flask in a hood *(left)*. On adding concentrated sulfuric acid *(right)*, brown gaseous I_2 was evolved. (See Study Question 59.) (C. D. Winters)

64. In each of the following reactions, tell which reactant is oxidized and which is reduced. Designate the oxidizing agent and reducing agent.
(a) $Ca(s) + 2\ HCl(aq) \rightarrow CaCl_2(aq) + H_2(g)$
(b) $Cr_2O_7^{2-}(aq) + 3\ Sn^{2+}(aq) + 14\ H^+(aq) \rightarrow 2\ Cr^{3+}(aq) + 3\ Sn^{4+}(aq) + 7\ H_2O(\ell)$
(c) $FeS(s) + 3\ NO_3^-(aq) + 4\ H^+(aq) \rightarrow 3\ NO(g) + SO_4^{2-}(aq) + Fe^{3+}(aq) + 2\ H_2O(\ell)$

GENERAL QUESTIONS

65. The mineral dolomite contains magnesium carbonate. Name the spectator ions in the reaction of magnesium carbonate and hydrochloric acid, and write the net ionic equation.

$$MgCO_3 + 2\ H^+(aq) + 2\ Cl^-(aq) \longrightarrow CO_2(g) + Mg^{2+}(aq) + 2\ Cl^-(aq) + H_2O(\ell)$$

What type of reaction is this?

66. Mg metal reacts readily with HNO_3, and the following compounds are all involved:

$$Mg(s) + HNO_3(aq) \longrightarrow Mg(NO_3)_2(aq) + NO_2(g) + H_2O(\ell)$$

(a) Balance the equation for the reaction.
(b) Name each compound.
(c) Write the net ionic equation for the reaction.
(d) What type of reaction is this?

67. The compound $(NH_4)_2S$ reacts with $Hg(NO_3)_2$ to give HgS and NH_4NO_3.
(a) Write the overall balanced equation for the reaction. Indicate the state (s or aq) for each compound.
(b) Name each compound.
(c) What type of reaction is this?

68. Complete and balance equations for the following combustion reactions:
(a) $C_{10}H_{10}Fe(s) + O_2(g) \rightarrow Fe_2O_3(s) + \cdots$
(b) $B_5H_9(\ell) + O_2(g) \rightarrow B_2O_3(s) + \cdots$
(c) $Si_2H_6(g) + O_2(g) \rightarrow SiO_2(s) + \cdots$

69. Azurite is a copper-containing mineral that often forms beautiful crystals (see photo). Its formula is $Cu_3(CO_3)_2(OH)_2$. Write a balanced equation for the reaction of this mineral with hydrochloric acid.

70. What species (atoms, molecules, or ions) are present in an aqueous solution of each of the following compounds?
(a) NH_3 (c) NaOH
(b) CH_3CO_2H (d) HBr

71. Classify each of the reactions as an acid-base reaction, a precipitation, or a gas-forming reaction. Show states for the products (s, ℓ, g, aq), and then balance the completed equation. Write the net ionic equation.
(a) $MnCl_2(aq) + Na_2S(aq) \rightarrow MnS + NaCl$
(b) $K_2CO_3(aq) + ZnCl_2(aq) \rightarrow ZnCO_3 + KCl$
(c) $K_2CO_3(aq) + HClO_4(aq) \rightarrow KClO_4 + CO_2 + H_2O$

72. Classify each of the reactions as an acid-base reaction, a precipitation, or a gas-forming reaction. Show states for the products (s, ℓ, g, aq), and then balance the completed equation. Write the net ionic equation.
(a) $Fe(OH)_3(s) + HNO_3(aq) \rightarrow Fe(NO_3)_3 + H_2O$
(b) $FeCO_3(s) + HNO_3(aq) \rightarrow Fe(NO_3)_2 + CO_2 + H_2O$
(c) $FeCl_2(aq) + (NH_4)_2S(aq) \rightarrow FeS + NH_4Cl$
(d) $Fe(NO_3)_2(aq) + Na_2CO_3(aq) \rightarrow FeCO_3 + NaNO_3$

73. Vitamin C is the simple compound $C_6H_8O_6$. One method for determining the amount of vitamin C in a sample is to react it with a solution of bromine, Br_2.

$$C_6H_8O_6(aq) + Br_2(aq) \longrightarrow 2\ HBr(aq) + C_6H_6O_6(aq)$$

What is oxidized and what is reduced in this reaction? Which substance is the oxidizing agent and which is the reducing agent?

74. Gold is dissolved from rock by treating the rock with sodium cyanide in the presence of oxygen.

$$4\ Au(s) + 8\ NaCN(aq) + O_2(g) + 2\ H_2O(\ell) \longrightarrow 4\ NaAu(CN)_2(aq) + 4\ NaOH(aq)$$

What is oxidized and what is reduced in this reaction? Which substance is the oxidizing agent and which is the reducing agent? (It may be useful to recall that the cyanide ion is CN^-.)

CONCEPTUAL QUESTIONS

75. The types of reactions described in this chapter can be used to prepare compounds. For example, insoluble barium chromate can be made by a precipitation reaction involving the soluble compounds $BaCl_2$ and K_2CrO_4.

$$BaCl_2(aq) + K_2CrO_4(aq) \longrightarrow BaCrO_4(s) + 2\ KCl(aq)$$

Azurite, a copper-containing mineral. (C. D. Winters)

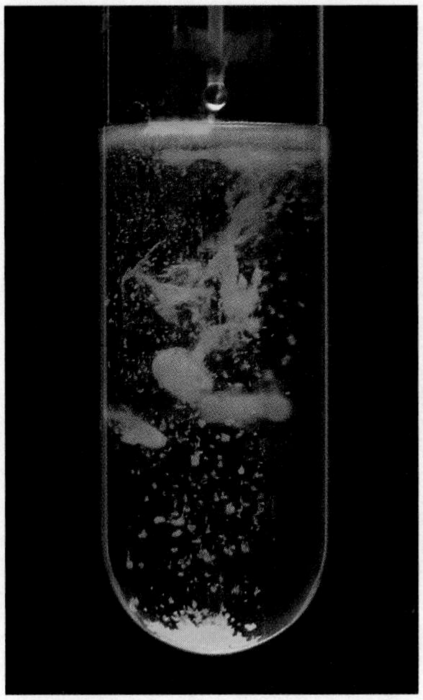

(a)

(b)

Preparation of barium chromate. (a) An aqueous solution of barium chloride is added from a dropper to a solution of potassium chromate, K_2CrO_4. (b) Barium chromate, $BaCrO_4$, precipitates and is separated from the solution by collecting it on a filter paper. (See Study Question 75.) (C. D. Winters)

The product, $BaCrO_4$, can be separated from water-soluble $BaCl_2$ by filtering the product mixture. The insoluble $BaCrO_4$ is trapped on the filter paper, and aqueous KCl passes through the paper. (See the photos above.) Suggest a precipitation reaction and a gas-forming reaction by which barium sulfate can be made.

76. Explain how to prepare zinc chloride by (a) an acid-base reaction or (b) a gas-forming reaction. The starting materials available are $ZnCO_3$, HCl, Cl_2, HNO_3, $Zn(OH)_2$, NaCl, $Zn(NO_3)_2$, and Zn. (See Question 75 for a discussion of compound preparation.)

77. The types of reactions described in this chapter occur in the world's oceans. For example, when water is anoxic (without dissolved oxygen), organic matter (abbreviated as CH_2O in the chemical equation) can be decomposed by reaction with sulfate ion, a major dissolved constituent in seawater.

$$2\ CH_2O(aq) + SO_4^{2-}(aq) \longrightarrow H_2S(aq) + 2\ HCO_3^-(aq)$$

What type reaction is this? Describe your reasoning.

The occurrence of such reactions affects other aspects of ocean chemistry. For example, the plot shown here depicts the concentration of Mn^{2+} ion at a location in the Black Sea. (Manganese reaches the oceans in rivers and in wind-blown dust. A certain amount also wells up from the sea floor.) The zero of the vertical axis is the point at which the O_2 concentration in the seawater has reached zero. No dissolved Mn^{2+} ion occurs above this point due to a reaction such as

$$Mn^{2+}(aq) + \tfrac{1}{2}\ O_2(g) + 2\ OH^-(aq) \longrightarrow MnO_2(s) + H_2O(\ell)$$

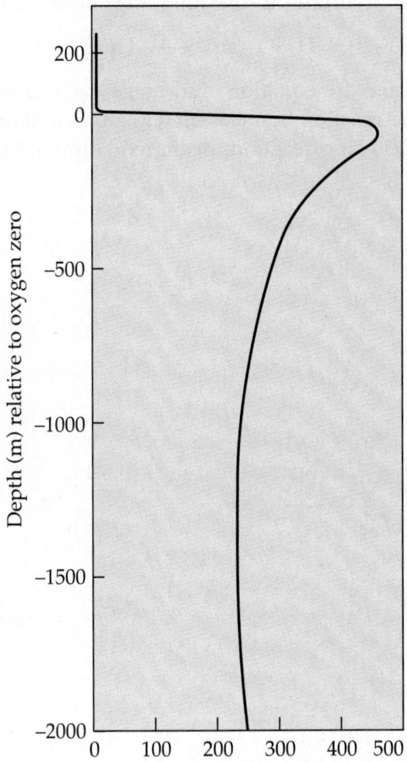

A concentration-depth profile for manganese(II) ion in the Black Sea. Redrawn with permission from *Ocean Chemistry and Deep-Sea Sediments,* New York, Pergamon Press, 1989, p. 56.

(Insoluble MnO_2 collects as manganese nodules [shown in the photo] on the floor of the ocean. The exact mechanism of their formation is still a mystery.) What type of reaction is this? Describe your reasoning.

Below the zero point on the vertical axis of the plot, the Mn^{2+} ion concentration declines because Mn^{2+} reacts with the hydrogen carbonate, HCO_3^-, to produce $MnCO_3$. Write a balanced equation for this reaction. Why does this deplete the concentration of Mn^{2+}?

A manganese nodule from the floor of the Pacific Ocean. These nodules consist primarily of manganese(IV) and iron(III) oxides. (C. D. Winters)

Summary Question

78. Much has been written recently about chlorofluorocarbons and their effect on our environment. Their manufacture begins with the preparation of HF from the mineral fluorspar according to the following *unbalanced* equation.

$$CaF_2(s) + H_2SO_4(aq) \longrightarrow HF(g) + CaSO_4(s)$$

(a) Balance the equation, and name each substance.
(b) Is the reaction best classified as a precipitation reaction, an acid-base reaction, or an oxidation-reduction?

The HF is combined with, for example, CCl_4 to make CCl_2F_2, called dichlorodifluoromethane, or CFC-12, and other chlorofluorocarbons.

$$2\ HF(g) + CCl_4(\ell) \longrightarrow CCl_2F_2(g) + 2\ HCl(g)$$

(c) Give the names of the compounds CCl_4 and HCl.
(d) Another chlorofluorocarbon produced in the reaction is composed of 8.74% C, 77.43% Cl, and 13.83% F. What is the empirical formula of the compound?

CHAPTER 5

Stoichiometry

A Chemical Puzzler

The reaction of vinegar with baking soda is an example of a gas-forming reaction. Vinegar is a solution of acetic acid, and baking soda is sodium hydrogen carbonate.

$$CH_3CO_2H(aq) + NaHCO_3(aq) \rightarrow NaCH_3CO_2(aq) + H_2O(\ell) + CO_2(g)$$

If you take a handful of baking soda, put it in a glass, and drop in vinegar, carbon dioxide gas bubbles out of the mixture. As you drop in more and more vinegar, more and more CO_2 gas is evolved. Eventually, though, the evolution of CO_2 stops. More vinegar does not lead to more CO_2 gas. Why?

The study of chemical reactions is the essence of chemistry, and one such reaction is that between an acidic oxide, SO_3, a base, $Ca(OH)_2$, and water.

$SO_3(g)$	$+\ Ca(OH)_2(s)$	$+\ H_2O(\ell)$	$\longrightarrow$	$CaSO_4 \cdot 2\ H_2O(s)$
acidic oxide	base			salt
sulfur trioxide	calcium hydroxide			calcium sulfate dihydrate
	(slaked lime)			(gypsum)

This is a useful reaction because it is one way of removing sulfur trioxide from the flue gas of coal- or oil-fired power-generating plants (Figure 4.16). This prevents the sulfur trioxide from being released into the atmosphere, where it would contribute to acid rain. To use this reaction effectively, we must consider it quantitatively. If you know how much sulfur trioxide is in the flue gas, how many tons of slaked lime are needed to remove the sulfur trioxide? How many tons of calcium sulfate are produced? And where will all of the calcium sulfate be put after it has been made? The principles used to answer at least the first two of these questions apply to all chemical reactions and are the subject of this chapter.

5.1 WEIGHT RELATIONS IN CHEMICAL REACTIONS: STOICHIOMETRY

A chemical equation shows the quantitative relationships between reactants and products (Section 4.1). These relationships, which are founded in the law of conservation of matter, are applied in the study of **stoichiometry.** The **stoichiometric coefficients** in a balanced equation give the relative numbers of atoms or molecules, or the numbers of moles, and therefore can also be used to calculate the masses of reactants and products.

A piece of phosphorus burns in chlorine to give phosphorus trichloride, PCl_3 (Figure 5.1), in an oxidation-reduction reaction (see Section 4.10). Sup-

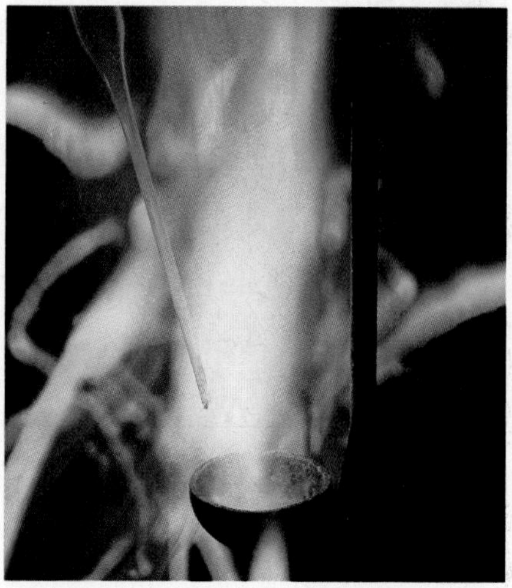

Figure 5.1 Phosphorus burns in chlorine gas with a bright flame to produce phosphorus trichloride, PCl_3. (C. D. Winters)

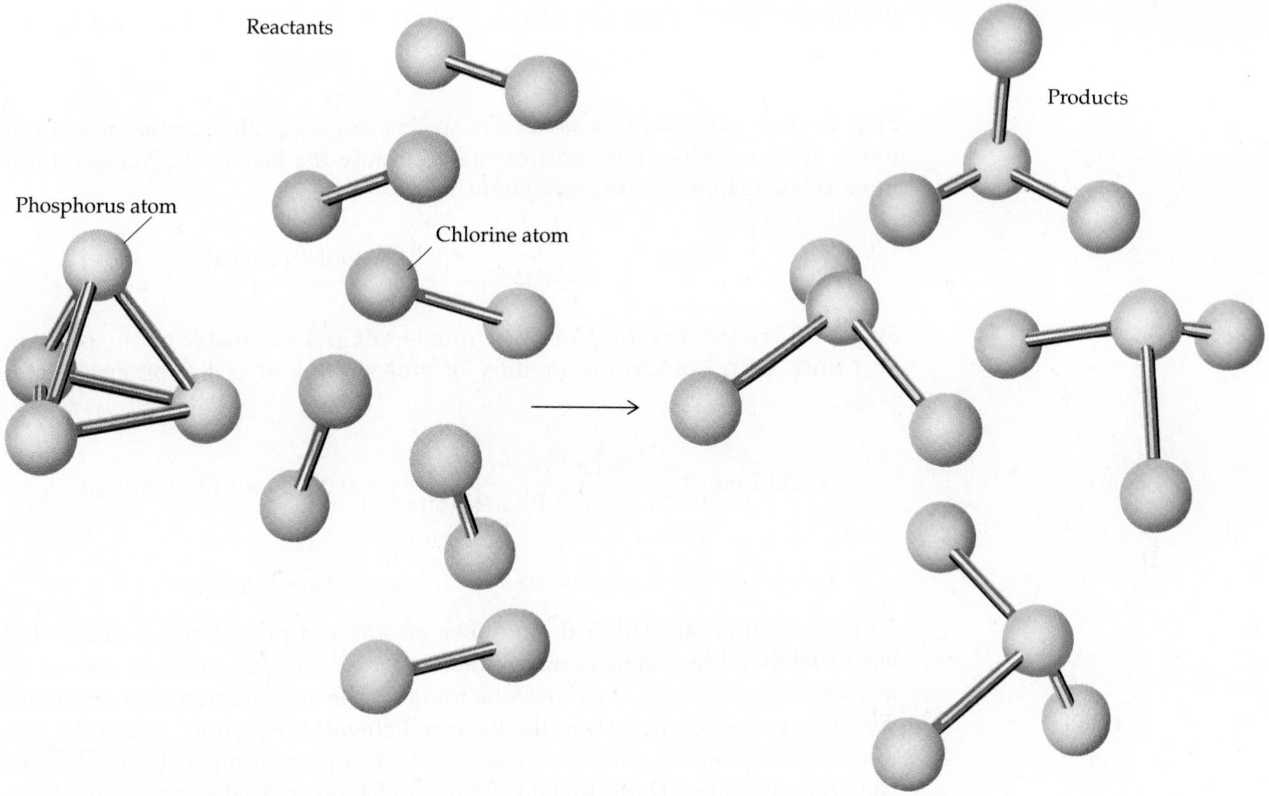

Reaction of phosphorus (P_4) with chlorine (Cl_2) to give phosphorus trichloride (PCl_3).

pose you use 1.00 mole of phosphorus (P_4, 124 g/mol) in this reaction. The following balanced equation shows that 6.00 moles or 425 g of Cl_2 must be used and that 4.00 moles or 549 g of PCl_3 can be produced.

Reactants			Product
$P_4(s)$	$+ 6\ Cl_2(g)$	$\longrightarrow$	$4\ PCl_3(s)$
1 molecule	6 molecules	$\longrightarrow$	4 molecules
1.00 mole	6.00 moles	$\longrightarrow$	4.00 moles
124 g	425 g	$\longrightarrow$	549 g (= 124 g + 425 g)

The balanced equation for the reaction of phosphorus and chlorine applies no matter how much P_4 is used. If 0.0100 mole of P_4 (1.24 g) is used, then 0.0600 mole of Cl_2 (4.25 g) is required, and 0.0400 mole of PCl_3 (5.49 g) can form. You can confirm by experiment that if 1.24 g of P_4 and 4.25 g of Cl_2 are used, then 5.49 g (= 1.24 g + 4.25 g) of PCl_3 can be produced.

Following this line of reasoning further, suppose you have a piece of phosphorus with a mass of 1.45 g. What mass of Cl_2 is required if all the phosphorus is to react? The following procedure leads to the solution.

Step 1. *Write the balanced equation* (using correct formulas for reactants and products). This is *always* the first step when dealing with chemical reactions.

$$P_4(s) + 6\ Cl_2(g) \longrightarrow 4\ PCl_3(\ell)$$

Step 2. *Calculate moles from masses.* From the mass of P_4 calculate the number of moles of P_4 available. This must be done because the balanced equation shows mole relationships, not mass relationships.

$$1.45\ \cancel{g\ P_4} \cdot \frac{1\ mol\ P_4}{123.9\ \cancel{g\ P_4}} = 0.0117\ mol\ P_4\ available$$

Step 3. *Use a stoichiometric factor.* The number of moles available of one reactant (P_4) must be related to the number of moles required of the other reactant (Cl_2).

$$0.0117\ mol\ P_4 \cdot \boxed{\frac{6\ mol\ Cl_2\ required}{1\ mol\ P_4\ available}} = 0.0702\ mol\ Cl_2\ required$$

↑

a stoichiometric factor (from the balanced equation)

To perform this calculation the number of moles of phosphorus available has been multiplied by a **stoichiometric factor,** a *mole ratio factor relating moles of the required reactant to moles of the available reactant.* The stoichiometric factor comes *directly* from the coefficients in the balanced chemical equation. This is the reason you must balance chemical equations before proceeding with calculations. Here the calculation shows that 0.0702 mol of Cl_2 is required to react with all the available phosphorus.

Step 4. *Calculate mass from moles.* From the number of moles of Cl_2 calculated in Step 3, you can now calculate the mass of Cl_2 required.

$$0.0702\ \cancel{mol\ Cl_2} \cdot \frac{70.91\ g\ Cl_2}{1\ \cancel{mol\ Cl_2}} = 4.98\ g\ Cl_2$$

Because the object of this example is to find the mass of Cl_2 required, the problem is solved.

You may also want to know the mass of PCl_3 that can be produced in the reaction of 1.45 g of phosphorus with the required mass of chlorine (4.98 g) when the reaction goes to completion, that is, when all of at least one of the reactants has been used completely. Because matter is conserved, this can be answered by adding the masses of P_4 and Cl_2 used (giving 1.45 g + 4.98 g = 6.43 g of PCl_3 produced). Alternatively, Steps 3 and 4 can be repeated, but with the appropriate stoichiometric factor and molar mass.

Step 3′. *Use a stoichiometric factor.* Relate the number of moles of available P_4 to the number of moles of PCl_3 that can be produced.

$$0.0117\ mol\ P_4 \cdot \boxed{\frac{4\ mol\ PCl_3\ produced}{1\ mol\ P_4\ available}} = 0.0468\ mol\ PCl_3\ produced$$

↑

a stoichiometric factor (from the balanced equation)

Step 4′. *Calculate mass from moles.* Convert moles of PCl_3 produced to a mass in grams.

$$0.0468 \text{ mol } PCl_3 \text{ produced} \cdot \frac{137.3 \text{ g } PCl_3}{1 \text{ mol } PCl_3} = 6.43 \text{ g } PCl_3$$

PROBLEM-SOLVING TIPS AND IDEAS

5.1 Stoichiometry Calculations

You are asked to determine what mass of product can be formed from a given mass of reactant. Keep in mind that it is not possible to calculate the mass of product in a single step. Instead, you must follow a route such as that illustrated here for the reaction of a reactant A to give the product B according to an equation such as $x\,A \rightarrow y\,B$.

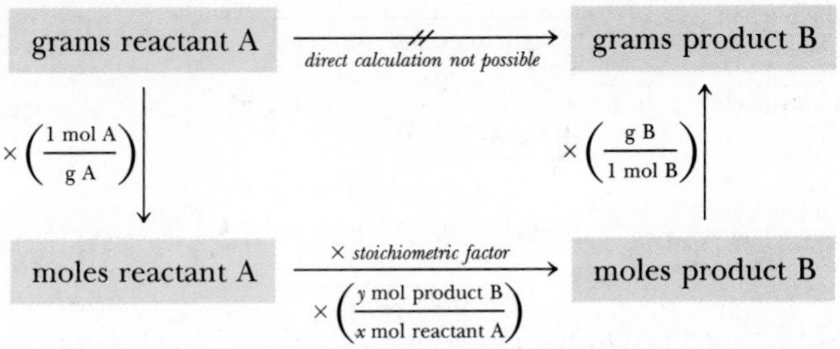

Here the mass of reactant A is converted to moles of A. Then, using the stoichiometric factor, you find moles of B. Finally, the mass of B is obtained by multiplying moles of B by its molar mass.

As you practice working with chemical stoichiometry, you will find that some minor differences arise from one situation to another, but do not let these differences obscure the fact that you must always follow the basic outline shown here. ■

EXAMPLE 5.1 *Weight Relations in Reactions*

Propane, C_3H_8, can be used as a fuel in your home, car, or barbecue grill because it is easily liquefied and transported. If 1.00 lb, or 454 g, of propane is

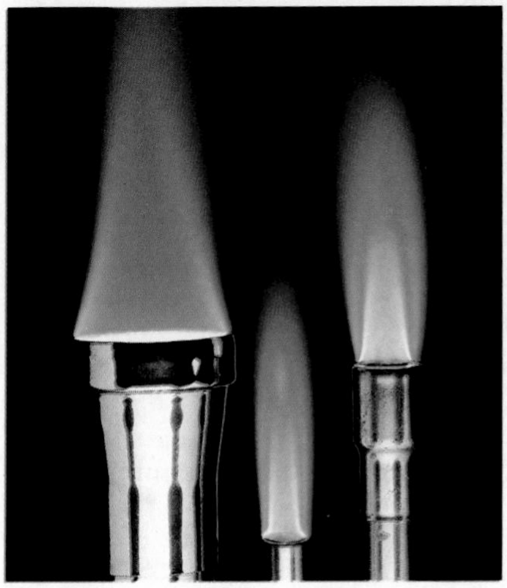

Here propane burns in laboratory Bunsen burners. (C. D. Winters)

burned, what mass of oxygen (in grams) is required for complete combustion, and what masses of carbon dioxide and water (in grams) are formed?

Solution

Step 1. Remember that the first step *must always* be to write a balanced equation.

$$C_3H_8(g) + 5\ O_2(g) \longrightarrow 3\ CO_2(g) + 4\ H_2O(g)$$

Having balanced the equation, you can perform the stoichiometric calculations. "Problem-Solving Tips and Ideas 5.1" suggests that you proceed in the following way:

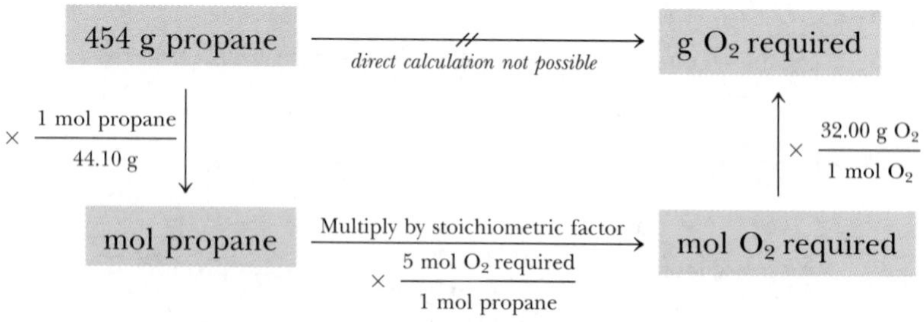

A direct calculation of the mass of O_2 required from the mass of propane is not possible. Instead, first find the moles of propane available, then relate this to moles of O_2 required using the stoichiometric factor. Finally, find the mass of O_2 required from the moles of O_2.

Step 2. Convert the mass of propane to moles.

$$454\ \cancel{g\ C_3H_8} \cdot \frac{1\ mol\ C_3H_8}{44.10\ \cancel{g\ C_3H_8}} = 10.3\ mol\ C_3H_8$$

Step 3. Use the stoichiometric factor to calculate moles of O_2 required.

$$10.3\ \cancel{mol\ C_3H_8} \cdot \frac{5\ mol\ O_2\ required}{1\ \cancel{mol\ C_3H_8}\ available} = 51.5\ mol\ O_2\ required$$

Step 4. Convert the number of moles of O_2 required to mass in grams.

$$51.5\ \cancel{mol\ O_2}\ required \cdot \frac{32.00\ g\ O_2}{1\ \cancel{mol\ O_2}} = \boxed{1650\ g\ O_2\ required}$$

Repeat Steps 3 and 4 to find the mass of CO_2 produced in the combustion. First, relate the number of moles of C_3H_8 available to the number of moles of CO_2 produced by using a stoichiometric factor.

$$10.3\ \cancel{mol\ C_3H_8\ available} \cdot \frac{3\ mol\ CO_2\ produced}{1\ \cancel{mol\ C_3H_8\ available}} = 30.9\ mol\ CO_2\ produced$$

Then convert the number of moles of CO_2 produced to the mass in grams.

$$30.9\ \cancel{mol\ CO_2} \cdot \frac{44.00\ g\ CO_2}{1\ \cancel{mol\ CO_2}} = \boxed{1360\ g\ CO_2}$$

Now, how can you find the mass of H_2O produced? You could go through Steps 3 and 4 again. It is easier, however, to recognize that the total mass of the reactants

$$454 \text{ g } C_3H_8 + 1650 \text{ g } O_2 = 2104 \text{ g of reactants}$$

must be the same as the total mass of products. The mass of water that can be produced is therefore

$$\text{Total mass of products} = 2104 \text{ g} = 1360 \text{ g } CO_2 \text{ produced} + ? \text{ g } H_2O$$

$$\text{Mass of } H_2O = 744 \text{ g}$$

EXERCISE 5.1 ***Weight Relations in Chemical Reactions***

What mass of carbon, in grams, can be consumed by 454 g of O_2 in a combustion to give carbon monoxide? What mass of CO can be produced?

$$2\ C(s) + O_2(g) \longrightarrow 2\ CO(g)$$ ■

5.2 REACTIONS IN WHICH ONE REACTANT IS PRESENT IN LIMITED SUPPLY

When carrying out a reaction, a chemist, or nature for that matter, rarely supplies the reactants in the exact stoichiometric ratio. Because the goal of a reaction is to produce the largest possible quantity of a useful compound from a given quantity of starting material, it is often the case that a large excess of one reactant is supplied to ensure that the more expensive reactant is completely converted. As an example, consider the preparation of cisplatin, $Pt(NH_3)_2Cl_2$, a compound used to treat cancer (see "Portrait of a Scientist: Barnett Rosenberg").

$$(NH_4)_2PtCl_4(s) + 2\ \underset{\text{ammonia}}{NH_3(aq)} \longrightarrow 2\ NH_4Cl(aq) + \underset{\text{cisplatin}}{Pt(NH_3)_2Cl_2(s)}$$

In this case, it makes sense to combine the more expensive chemical $(NH_4)_2PtCl_4$ (roughly \$40 per gram) with a much greater amount of the less expensive chemical, NH_3 (only pennies per gram), than is called for by the balanced equation. Thus, on completion of the reaction, all the $(NH_4)_2PtCl_4$ has been converted to product, and some NH_3 remains. How much $Pt(NH_3)_2Cl_2$ is formed? It depends on the amount of $(NH_4)_2PtCl_4$ present at the start, not on the amount of NH_3, because more NH_3 is present than was required by stoichiometry. A compound such as $(NH_4)_2PtCl_4$ in this example is called the **limiting reactant** because its amount determines or limits the amount of product formed.

PROBLEM-SOLVING TIPS AND IDEAS

5.2 A "Real-Life" Analogy to a Limiting Reactant Problem in Chemistry

As an analogy to a chemical "limiting reactant" situation, consider what happens if you want to make some cheeseburgers. Suppose you have enough meat to make 3 hamburger patties, and you also have $\frac{1}{2}$ dozen buns and 1 dozen slices of cheese. If each cheeseburger

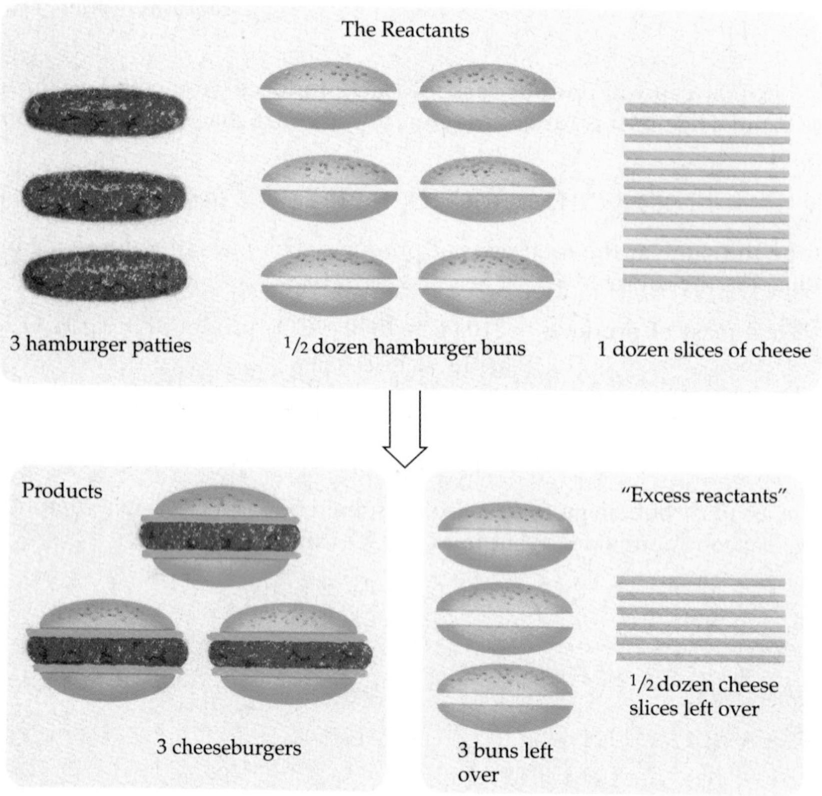

requires 1 bun, 2 slices of cheese, and 1 hamburger patty, you can make only 3 sandwiches. Three buns and 6 cheese slices are left over. The cheese and buns are "excess reactants," and the hamburger patties are the "limiting reactant" because the amount of hamburger meat has limited, or determined, the number of sandwiches that can be made. Furthermore, although meat is bought by the pound, buns by the dozen, and cheese by the slice, the combining units are the number of buns or cheese slices required by each hamburger patty, and your calculation must be done in those units, just as stoichiometry calculations are done in units of moles. ■

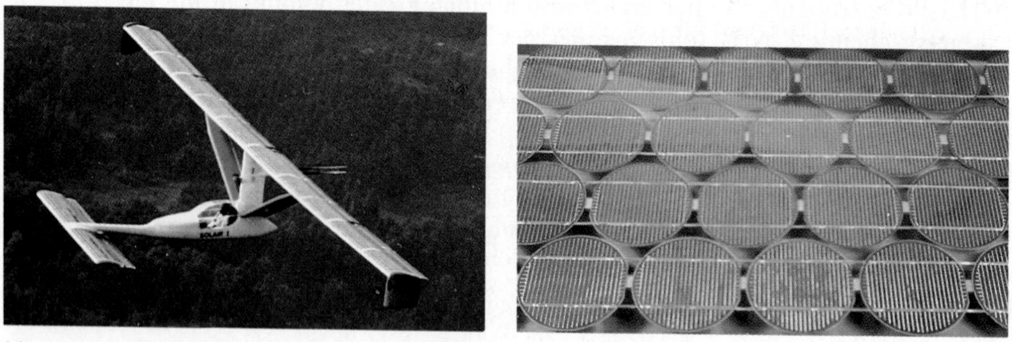

(a) (b)

Figure 5.2 (a) Pure silicon is used in solar cells on the wings of the solar-powered aircraft *Solair I.* (b) A closer view of the solar cells. (Courtesy of Wacker Silicones Corporation)

Another example of a reaction that involves a limiting reactant is the manufacture of the pure silicon that is used in computer chips or solar cells (Figure 5.2). The final step in the process is the reduction of purified, liquid silicon tetrachloride with very pure magnesium to give pure elemental silicon.

$$SiCl_4(\ell) + 2\ Mg(s) \longrightarrow Si(s) + 2\ MgCl_2(s)$$

The process of making pure Si starts with sand, SiO_2. This is reduced to impure Si, which is then made into $SiCl_4$. After purifying the $SiCl_4$, it is converted to very pure Si. Ultrapure silicon is used in silicon-based solar cells and computer chips are relatively expensive.

Suppose 225 g of $SiCl_4$ is mixed with 225 g of Mg. Are these reactants mixed in the correct stoichiometric ratio or is one of them in short supply? That is, will one of them limit the quantity of silicon that can be produced? If so, how much silicon can be formed if the reaction goes to completion? And how much of the excess reactant is left over when the maximum amount of silicon has been formed?

Because the quantities of both starting materials are given, the first step in answering these questions involves finding the number of moles of each.

$$225\ \cancel{g\ SiCl_4} \cdot \frac{1\ \text{mol}\ SiCl_4}{169.9\ \cancel{g\ SiCl_4}} = 1.32\ \text{mol}\ SiCl_4\ \text{available}$$

$$225\ \cancel{g\ Mg} \cdot \frac{1\ \text{mol Mg}}{24.31\ \cancel{g\ Mg}} = 9.26\ \text{mol Mg available}$$

Are these reactants present in the correct stoichiometric ratio as given by the balanced equation?

$$\text{Mole ratio of reactants as required by the balanced equation} = \frac{2\ \text{mol Mg}}{1\ \text{mol}\ SiCl_4}$$

$$\text{Mole ratio of reactants actually available} = \frac{9.26\ \text{mol Mg available}}{1.32\ \text{mol}\ SiCl_4\ \text{available}}$$

$$= \frac{7.02\ \text{mol Mg}}{1.00\ \text{mol}\ SiCl_4}$$

Dividing the moles of Mg available by moles of $SiCl_4$ available shows that the ratio is much larger than the (2 mol Mg/1 mol $SiCl_4$) ratio required by the balanced equation. This means more magnesium is available than is needed to react with all the available $SiCl_4$. Conversely, this means not enough $SiCl_4$ is present to react with all the available Mg. Therefore *$SiCl_4$ is the limiting reactant.* It is the equivalent of the hamburger patties in the cheeseburger analogy for limiting reactants.

It is generally best to divide the larger number by the smaller number.

Now that $SiCl_4$ has been shown to be the limiting reactant, we can calculate the mass of product expected based on the quantity of $SiCl_4$ available.

$$1.32\ \cancel{\text{mol}\ SiCl_4} \cdot \frac{1\ \cancel{\text{mol Si}}}{1\ \cancel{\text{mol}\ SiCl_4}} \cdot \frac{28.09\ \text{g Si}}{1\ \cancel{\text{mol Si}}} = 37.1\ \text{g Si}$$

The calculations showed that more than enough magnesium is available to react with the available silicon tetrachloride, $SiCl_4$. Magnesium is the "excess reactant," and it is possible to calculate the quantity of the metal that remains after all the $SiCl_4$ has been used. To do this, we first need to know how much magnesium is required to consume all the limiting reactant, $SiCl_4$.

$$1.32\ \cancel{\text{mol}\ SiCl_4}\ \text{available} \cdot \frac{2\ \text{mol Mg required}}{1\ \cancel{\text{mol}\ SiCl_4}\ \text{available}} = 2.64\ \text{mol Mg required}$$

PORTRAIT OF A SCIENTIST

Barnett Rosenberg and Platinum Compounds for Cancer Treatment

Barnett Rosenberg. (Doug Elbinger)

Dr. Barnett Rosenberg moved from New York City, where he had studied physics, to Michigan State University in East Lansing, Michigan, to establish a department of biophysics in 1961. It was there that he and his collaborators discovered that a compound that had been known for a century or more, $Pt(NH_3)_2Cl_2$, was effective in the treatment of certain types of cancers.

The discovery of the utility of this compound, now commonly known as *cisplatin,* in cancer therapy is a wonderful example of serendipity in science. The dictionary defines serendipity as "the faculty of making fortunate and unexpected discoveries by accident." In this case Rosenberg had set out to study a problem that had interested him for some time—the effect of electric fields on living cells—but the results of the experiment were quite different from what he expected. He and his students had passed an aqueous suspension of a live culture of *Escherichia coli* bacteria through an electric field between supposedly inert platinum plates. Much to their surprise they found that the growth of the cells was affected; cell division was no longer occurring. After careful experimentation, the effect on cell division was traced to tiny amounts of *cisplatin* produced by reaction of platinum with ammonium and chloride ions in the aqueous medium.

Much to the benefit of all of us, Rosenberg recognized that these laboratory results had implications in cancer chemotherapy, and subsequent experiments led to compounds now used to treat cancer patients. He recently said that the use of *cisplatin* has meant that "Testicular cancer went from a disease that normally killed about 80 percent of the patients, to one which is close to 95 percent curable. This is probably the most exciting development in the treatment of cancers that we have had in the past 20 years. It is now the treatment of first choice in ovarian, bladder, and osteogenic sarcoma [bone] cancers as well."

Because 9.26 moles of magnesium are available, the number of moles of excess magnesium can be calculated.

$$\text{Excess Mg} = 9.26 \text{ mol Mg available} - 2.64 \text{ mol Mg consumed}$$
$$= 6.62 \text{ mol Mg remain}$$

and then converted to a mass.

$$6.62 \text{ mol Mg} \cdot \frac{24.31 \text{ g Mg}}{1 \text{ mol Mg}} = 161 \text{ g Mg in excess of that required}$$

Finally, because 161 g of magnesium is left over, this means that 225 g − 161 g = 64 g of magnesium has been consumed.

PROBLEM-SOLVING TIPS AND IDEAS

5.3 Reactions Involving a Limited Reactant

The calculations involved in a "limiting reactant problem" are an extension of the model presented in "Problem-Solving Tips and Ideas 5.1." That is, when you mix quantities of two or more reactants and wish to know what quantity of product can be formed, you must first decide which reactant is the one that limits the outcome. Consider a reaction for which the balanced equation is

$$x\text{A} + y\text{B} \longrightarrow \text{product}$$

and the stoichiometric factor is x/y. You first find the number of moles of both A and B. Then, find the ratio of moles of A to moles of B.

$$\left.\begin{array}{l}\text{Mass of reactant A} \longrightarrow \text{moles of A} \\ \text{Mass of reactant B} \longrightarrow \text{moles of B}\end{array}\right\} \longrightarrow \frac{\text{moles of A available}}{\text{moles of B available}}$$

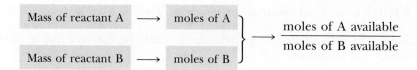

- If (mol A available/mol B available) is *greater than x/y,* then A is in excess and B is the limiting reactant.
- If (mol A available/mol B available) is *less than x/y,* then B is in excess and A is the limiting reactant.

Once the limiting reactant has been established, the remainder of the calculations follow the procedure outlined in "Problem-Solving Tips and Ideas 5.1." Remember that the quantity of product depends on the quantity of limiting reactant. ■

EXAMPLE 5.2 ***A Reaction Involving a Limiting Reactant***

- preventing further reaction because lack of limiting reactant.

Titanium tetrachloride, $TiCl_4$, is an important industrial chemical. For example, TiO_2, the material used as a white pigment in paper and paints, can be made from it. The tetrachloride can be made by combining titanium-containing ore (which is often impure TiO_2) with carbon and chlorine.

$$TiO_2(s) + 2\ Cl_2(g) + C(s) \longrightarrow TiCl_4(\ell) + CO_2(g)$$

If one begins with 125 g each of Cl_2 and C, but plenty of TiO_2-containing ore, which is the limiting reactant in this reaction? What quantity of $TiCl_4$, in grams, can be produced?

Solution As is always the case, when you have a mass of a pure element or compound, you should find the equivalent number of moles.

$$\text{Moles of } Cl_2 = 125\ \cancel{\text{g } Cl_2} \cdot \frac{1 \text{ mol } Cl_2}{70.91\ \cancel{\text{g } Cl_2}} = 1.76 \text{ mol } Cl_2$$

$$\text{Moles of C} = 125\ \cancel{\text{g C}} \cdot \frac{1 \text{ mol C}}{12.01\ \cancel{\text{g C}}} = 10.4 \text{ mol C}$$

Are the reactants present in a perfect stoichiometric ratio or is one of them present in a greater amount than necessary to react completely?

$$\frac{\text{Moles C available}}{\text{Moles } Cl_2 \text{ available}} = \frac{10.4 \text{ mol C}}{1.76 \text{ mol } Cl_2} = \frac{5.91 \text{ mol C}}{1.00 \text{ mol } Cl_2}$$

The stoichiometric factor required by the balanced equation is 1 mole of C to 2 moles of Cl_2, or 0.5 mol C/1 mol Cl_2. Because (5.91 mol C/1.00 mol Cl_2) is much larger than the stoichiometric factor, much more carbon is available than is required by the quantity of chlorine available. Conversely, not enough Cl_2 is present to use up all of the carbon, so *Cl_2 is the limiting reactant.*

Now that Cl_2 is known to be the limiting reactant, we can carry out the stoichiometry calculation to determine the quantity of $TiCl_4$ that can be produced using the approach in "Problem-Solving Tips and Ideas 5.1."

$$1.76 \text{ mol } Cl_2 \cdot \frac{1\ \cancel{\text{mol } TiCl_4}}{2 \text{ moi } Cl_2} \cdot \frac{189.7 \text{ g } TiCl_4}{1\ \cancel{\text{mol } TiCl_4}} = 167 \text{ g } TiCl_4$$

EXAMPLE 5.3 *A Reaction Involving a Limiting Reactant*

Methanol, CH_3OH, an excellent fuel, can be made by the reaction of carbon monoxide and hydrogen.

$$CO(g) + 2\ H_2(g) \longrightarrow \underset{\text{methanol}}{CH_3OH(\ell)}$$

Suppose 356 g of CO are mixed with 65.0 g of H_2. Which is the limiting reactant? What is the maximum mass of methanol that can be formed? What mass of the excess reactant remains after the limiting reactant has been consumed?

Solution As a first step, find the number of moles of each reactant.

$$\text{Moles of CO} = 356\ \text{g CO} \cdot \frac{1\ \text{mol CO}}{28.01\ \text{g CO}} = 12.7\ \text{mol CO}$$

$$\text{Moles of } H_2 = 65.0\ \text{g C} \cdot \frac{1\ \text{mol } H_2}{2.02\ \text{g } H_2} = 32.2\ \text{mol } H_2$$

Are these reactants, in the quantities available, present in a perfect stoichiometric ratio?

$$\frac{\text{Moles } H_2 \text{ available}}{\text{Moles CO available}} = \frac{32.2\ \text{mol } H_2}{12.7\ \text{mol CO}} = \frac{2.53\ \text{mol } H_2}{1\ \text{mol CO}}$$

The required mole ratio is 2 moles of H_2 to 1 mole of CO. Clearly more hydrogen is available than is required. Conversely, not enough CO is present to use up all of the hydrogen, so *CO is the limiting reactant.*

What is the maximum quantity of CH_3OH that can be formed? This calculation is based on the quantity of limiting reactant available.

$$12.7\ \text{mol CO} \cdot \frac{1\ \text{mol } CH_3OH \text{ formed}}{1\ \text{mol CO available}} \cdot \frac{32.04\ \text{g of } CH_3OH}{1\ \text{mol } CH_3OH}$$

$$= 407\ \text{g } CH_3OH \text{ formed}$$

What quantity of H_2 remains when all the CO has been converted to product? First, we must find the quantity of H_2 required to react with all the CO.

$$12.7\ \text{mol CO} \cdot \frac{2\ \text{mol } H_2}{1\ \text{mol CO}} = 25.4\ \text{mol } H_2 \text{ required}$$

Because 32.2 moles of H_2 is available, but only 25.4 moles is required by the limiting reactant, 6.8 moles of H_2 is in excess. This is equivalent to 14 g of H_2.

$$6.8\ \text{mol } H_2 \cdot \frac{2.02\ \text{g } H_2}{1\ \text{mol } H_2} = 14\ \text{g of } H_2 \text{ remain}$$

EXERCISE 5.2 *A Reaction Involving a Limiting Reactant*

You have 20.0 g of elemental sulfur, S_8, and 160 g of O_2. Which is the limiting reactant in the combustion of S_8 in oxygen to give SO_2 gas? What amount of which reactant (in moles) is left after complete reaction? What mass of SO_2, in grams, is formed in the complete reaction? ■

PROBLEM-SOLVING TIPS AND IDEAS

5.4 Reactions Involving a Limiting Reactant

In the preceding examples involving a limiting reactant, we first calculated moles of each reactant. Next, we compared the ratio of moles available to the stoichiometric factor from the balanced equation. This identified the limiting reactant.

There is another method that some of our students find works well for them. That is, calculate the mass (or moles) of product expected based on each reactant. The limiting reactant is that reactant that gives the smallest quantity of product. For example, refer to the $SiCl_4$ reaction with Mg on page 125. To confirm that $SiCl_4$ is the limiting reactant, calculate the quantity of elemental silicon that can be formed starting with (a) 1.32 mol of $SiCl_4$ and unlimited magnesium or (b) with 9.26 mol of Mg and unlimited $SiCl_4$.

1. Quantity of Si produced from 1.32 mol $SiCl_4$ and unlimited Mg

$$1.32\ \text{mol SiCl}_4\ \text{available} \cdot \frac{1\ \text{mol Si}}{1\ \text{mol SiCl}_4} \cdot \frac{28.09\ \text{g Si}}{1\ \text{mol Si}} = 37.1\ \text{g Si}$$

2. Quantity of Si produced from 9.26 g Mg and unlimited $SiCl_4$

$$9.26\ \text{mol Mg available} \cdot \frac{1\ \text{mol Si}}{2\ \text{mol Mg}} \cdot \frac{28.09\ \text{g Si}}{1\ \text{mol Si}} = 130.\ \text{g Si}$$

Comparing the quantities of silicon produced shows that the available $SiCl_4$ is capable of producing less silicon (37.1 g) than the available Mg (130 g). This confirms our conclusion that silicon tetrachloride, $SiCl_4$, is the limiting reactant.

As a final note, you will find this approach easier to use when there are more than two reactants, each present initially in some designated quantity. ■

5.3 PERCENT YIELD

The maximum quantity of product that can be obtained from a chemical reaction is the **theoretical yield.** There is invariably some waste in the isolation and purification of products, however; no matter how good a chemist you are, you will invariably "lose" small quantities of material along the way. For these reasons, the **actual yield** of a compound—the quantity of material you actually obtain in the laboratory or chemical plant—may be less than the theoretical yield. The efficiency of a chemical reaction and the techniques used to obtain the desired compound in pure form can be evaluated by calculating the ratio of the actual yield to the theoretical yield. We call the result the **percent yield** (Figure 5.3).

$$\text{Percent yield} = \frac{\text{actual yield}}{\text{theoretical yield}} \times 100\%$$

Suppose you made aspirin in the laboratory by the following reaction,

$$\underset{\text{salicylic acid}}{C_7H_6O_3(s)} + \underset{\text{acetic anhydride}}{C_4H_6O_3(\ell)} \longrightarrow \underset{\text{aspirin}}{C_9H_8O_4(s)} + \underset{\text{acetic acid}}{CH_3CO_2H(\ell)}$$

and that you began with 14.4 g of salicylic acid and an excess of acetic anhydride. Therefore, salicylic acid was the limiting reactant. If you obtain 6.26 g of aspirin,

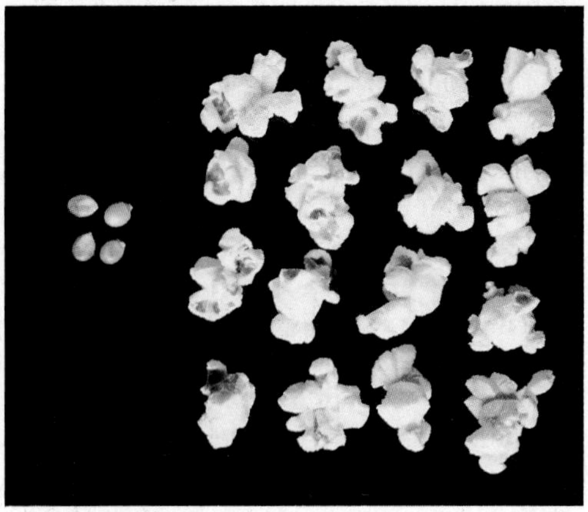

Figure 5.3 We began with 20 popcorn kernels and found that only 16 of them popped. The percent yield of popcorn from our "reaction" was (16/20) × 100%, or 80%.
(C. D. Winters)

what was the percent yield of this product? The first step is to find the number of moles of the limiting reactant, salicylic acid ($C_7H_6O_3$).

$$14.4\ \text{g}\ C_7H_6O_3 \cdot \frac{1\ \text{mol}\ C_7H_6O_3}{138.1\ \text{g}\ C_7H_6O_3} = 0.104\ \text{mol}\ C_7H_6O_3$$

Next, use the stoichiometric factor from the balanced equation to find the number of moles of aspirin expected based on the limiting reactant, $C_7H_6O_3$.

$$0.104\ \text{mol}\ C_7H_6O_3 \cdot \frac{1\ \text{mol aspirin}}{1\ \text{mol}\ C_7H_6O_3} = 0.104\ \text{mol aspirin expected}$$

The maximum quantity of aspirin that can be produced—the theoretical yield—is 0.104 mol. Because the quantity you measure in the laboratory is the mass of the product, it is customary to express the theoretical yield as a mass in grams.

$$0.104\ \text{mol aspirin} \cdot \frac{180.2\ \text{g aspirin}}{1\ \text{mol aspirin}} = 18.8\ \text{g aspirin}$$

Finally, with the actual yield known to be only 6.26 g, the percent yield of aspirin can be calculated.

$$\text{Percent yield} = \frac{6.26\ \text{g aspirin actually isolated}}{18.8\ \text{g aspirin expected}} \times 100\% = 33.3\%\ \text{yield}$$

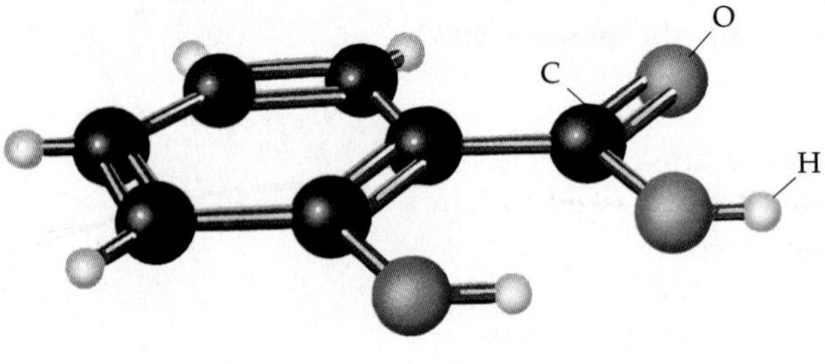

Salicylic acid

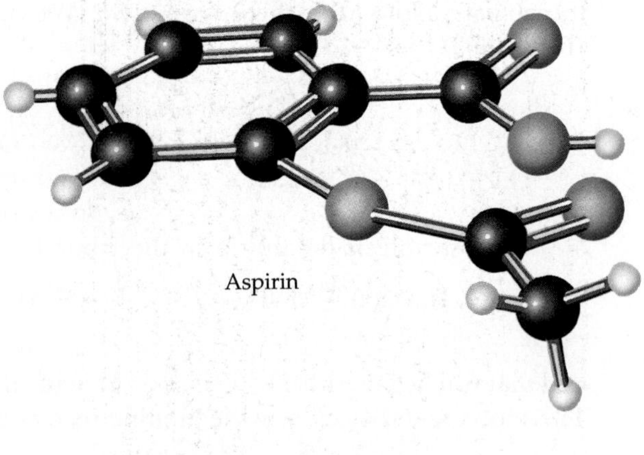

Aspirin

When a chemist makes a new compound or carries out a reaction, the percent yield is usually reported. It is useful information because it gives other chemists some idea of the quantity of product that can be reasonably expected from the reaction. But be aware that it is often difficult to obtain yields of 90% or better. It is impossible to keep track of every drop of liquid or crumb of solid, and reactions other than the desired one may occur. Moreover, many reactions simply do not go completely to products (as you may know from popping a cup of popcorn kernels).

EXERCISE 5.3 ***Percent Yield***

Professor Herbert C. Brown of Purdue University received the Nobel Prize in chemistry in 1979 for his work on the chemistry of diborane, B_2H_6, and its use in preparing new organic compounds. This gas can be prepared by the following reaction (which is carried out in a nonaqueous solvent, that is, a solvent other than water):

$$3\ NaBH_4 + 4\ BF_3 \longrightarrow 3\ NaBF_4 + 2\ B_2H_6$$

If you begin with 18.9 g of $NaBH_4$ (and excess BF_3), and you isolate 7.50 g of B_2H_6 gas, what is the percent yield of B_2H_6? ■

5.4 CHEMICAL EQUATIONS AND CHEMICAL ANALYSIS

With an increased awareness of environmental problems in recent years has come a need to know just what chemicals are in our environment and in what quantities. This need has made analytical chemistry even more important than before. Chemists in this field use their creativity to identify pure substances as well as to measure the quantities of components of mixtures. Analytical chemistry is often done now using instrumental methods (Figure 5.4), but classical chemical reactions and stoichiometry still play a central role.

Figure 5.4 A student using a modern analytical instrument, a mass spectrometer. (C. D. Winters)

Analysis of a Mixture

The analysis of mixtures is often very challenging. It can take a great deal of imagination to figure out how to use chemistry to determine what, and how much, is in the mixture. We can illustrate how analytical chemistry problems are solved, however, with a reasonably straightforward example.

Suppose you have a white powder that you know is a mixture of magnesium oxide (MgO) and magnesium carbonate ($MgCO_3$), and you are asked to find out what percent of the mixture is $MgCO_3$. Many metal carbonates decompose on heating to give metal oxides and carbon dioxide. For magnesium carbonate the reaction is

$$MgCO_3(s) \longrightarrow MgO(s) + CO_2(g)$$

If the powder is heated strongly, the magnesium carbonate in the mixture decomposes to MgO, and CO_2 is evolved as a gas. The solid left after heating the mixture consists only of MgO, and its mass is the sum of the mass of the MgO that was originally in the mixture *plus* the MgO remaining from the $MgCO_3$ decomposition. The difference in mass of the solid before and after heating gives the mass of CO_2 evolved and, by stoichiometry, the mass of $MgCO_3$ in the original mixture can be calculated.

Assume you have 1.599 g of a mixture of MgO and $MgCO_3$ and that heating evolves CO_2 and leaves 1.294 g of MgO. What was the weight percent of $MgCO_3$ in the original mixture? Knowing the mass of the mixture before and after heating, you can find the mass difference, which equals the mass of CO_2 evolved when the mixture was heated.

$$\begin{aligned} &\text{Mass of mixture } (\text{MgO} + \text{MgCO}_3) &&= 1.599 \text{ g} \\ - &\text{Mass after heating (pure MgO)} &&= \underline{-1.294 \text{ g}} \\ &\text{Mass of CO}_2 &&= 0.305 \text{ g} \end{aligned}$$

Now the mass of CO_2 can be converted to moles of CO_2.

$$0.305 \text{ g } \cancel{\text{CO}_2} \cdot \frac{1 \text{ mol CO}_2}{44.01 \text{ g } \cancel{\text{CO}_2}} = 0.00693 \text{ mol CO}_2$$

The balanced equation for the decomposition of $MgCO_3$ shows that for every mole of CO_2 given off on heating there was a mole of $MgCO_3$ in the mixture. Therefore, there must have been 0.00693 mole of $MgCO_3$ in the mixture, and the mass of $MgCO_3$ in the mixture was

$$0.00693 \cancel{\text{ mol CO}_2} \cdot \frac{1 \cancel{\text{ mol MgCO}_3}}{1 \cancel{\text{ mol CO}_2}} \cdot \frac{84.31 \text{ g MgCO}_3}{1 \cancel{\text{ mol MgCO}_3}} = 0.584 \text{ g MgCO}_3$$

This means the weight percent of magnesium carbonate in the mixture was

$$\frac{\text{Mass of MgCO}_3}{\text{Sample mass}} \cdot 100\% = \frac{0.584 \text{ g MgCO}_3}{1.599 \text{ g sample}} \cdot 100\% = 36.5\% \text{ MgCO}_3$$

EXAMPLE 5.4 *Analysis of a Mixture*

Butyllithium, LiC_4H_9, is a very reactive compound used by chemists to make new materials. One way to determine the quantity of LiC_4H_9 in a mixture is to add aqueous hydrochloric acid. The following reaction occurs:

$$\text{LiC}_4\text{H}_9 + \text{HCl(aq)} \longrightarrow \text{LiCl(aq)} + \text{C}_4\text{H}_{10}\text{(g)}$$

Suppose you have 5.606 g of a solution of LiC_4H_9 dissolved in benzene. On adding excess HCl(aq), 0.633 g of C_4H_{10} is evolved. What is the weight percent of LiC_4H_9 in the sample?

Solution Let us first calculate the moles of C_4H_{10} evolved.

$$0.633 \text{ g } \cancel{\text{C}_4\text{H}_{10}} \cdot \frac{1 \text{ mol C}_4\text{H}_{10}}{58.12 \text{ g } \cancel{\text{C}_4\text{H}_{10}}} = 0.0109 \text{ mol C}_4\text{H}_{10}$$

Next, calculate the moles and then the mass of LiC_4H_9 in the sample.

$$0.0109 \cancel{\text{ mol C}_4\text{H}_{10}} \cdot \frac{1 \cancel{\text{ mol LiC}_4\text{H}_9}}{1 \cancel{\text{ mol C}_4\text{H}_{10}}} \cdot \frac{64.06 \text{ g LiC}_4\text{H}_9}{1 \cancel{\text{ mol LiC}_4\text{H}_9}} = 0.698 \text{ g LiCl}_4\text{H}_9$$

Finally, calculate the weight percent of LiC_4H_9.

$$\frac{0.698 \text{ g } LiC_4H_9}{5.606 \text{ g sample}} \cdot 100\% = 12.4\%$$

EXERCISE 5.4 ***Chemical Analysis***

You have 2.357 g of a mixture of $BaCl_2$ and $BaCl_2 \cdot 2\ H_2O$. If experiment shows that the mixture has a mass of only 2.108 g after heating to drive off all the water of hydration in $BaCl_2 \cdot 2\ H_2O$, what is the weight percent of $BaCl_2 \cdot 2\ H_2O$ in the original mixture? ■

Determining the Empirical Formula of a Compound

The empirical formula of a compound can be determined if the percent composition of the compound is known (Section 3.7). But where do the percent composition data come from? Various methods are used, and many depend on reactions that transform the unknown but pure compound into known products. Assuming the reaction products can be isolated in pure form, the masses and the number of moles of each can be determined. Then, the moles of each product can be related to the number of moles of each element in the original compound. One method that works well for compounds that burn in oxygen is *analysis by combustion.* Each element (except oxygen) in the compound combines with oxygen to produce the appropriate oxide.

Consider an analysis of the hydrocarbon, methane, as an example of combustion analysis. A balanced equation for the combustion of methane shows that every mole of carbon in the original compound is converted to a mole of CO_2. Every mole of hydrogen in the original compound gives *half* a mole of H_2O. The carbon dioxide and water produced by the combustion are gases that can be

$CH_4(g) + 2\ O_2(g) \longrightarrow CO_2(g) + 2\ H_2O(g)$

1 mole C $\longrightarrow$ 1 mole CO_2

4 moles H $\longrightarrow$ 2 moles H_2O

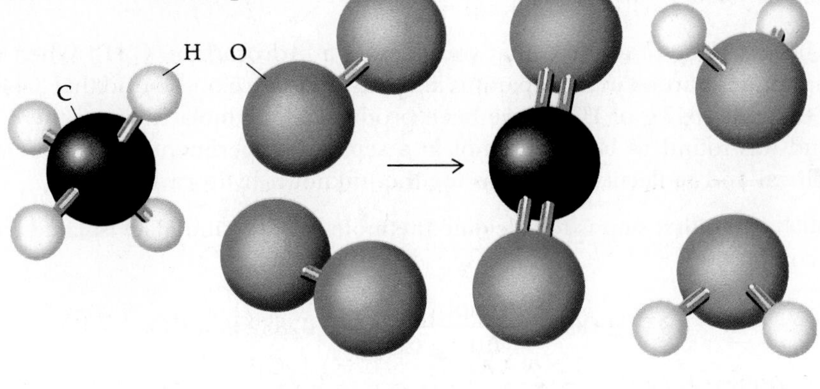

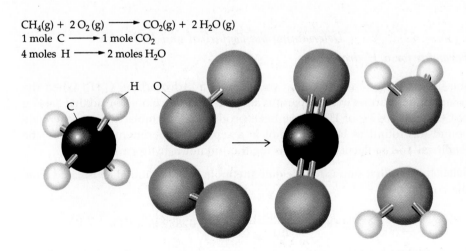

1 molecule CH_4 + 2 molecules $O_2 \longrightarrow$ 1 molecule CO_2 + 2 molecules H_2O

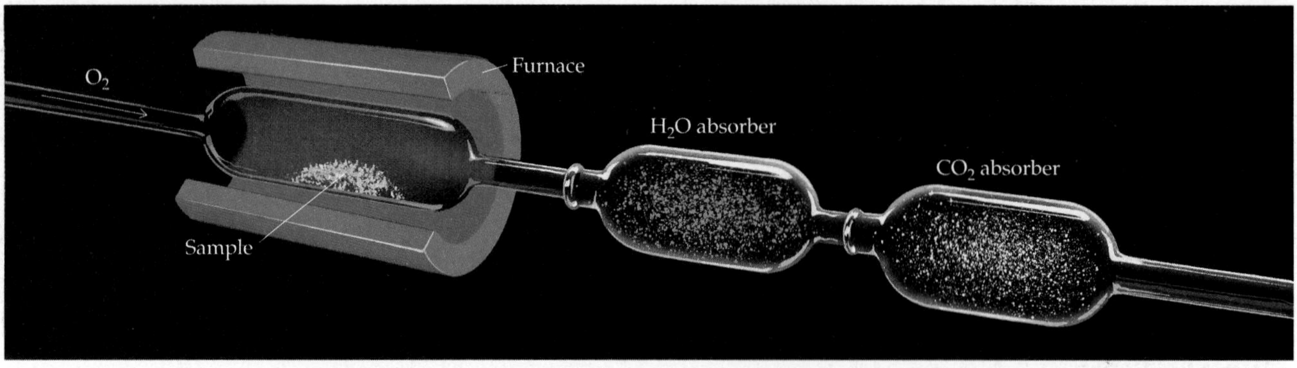

Figure 5.5 If a compound containing C and H is burned in oxygen, CO_2 and H_2O are formed, and the mass of each can be determined. The H_2O is absorbed by magnesium perchlorate, and the CO_2 is absorbed by finely divided NaOH supported on asbestos. The mass of each absorbent before and after combustion gives the masses of CO_2 and H_2O. Only a few milligrams of a combustible compound are needed for analysis by automated commercial equipment.

separated and their masses determined as illustrated in Figure 5.5. These masses can then be converted to the moles of C and H in CO_2 and H_2O, respectively, and the ratio of the moles of C and H in a sample of the original compound can be found. This ratio gives the empirical formula.

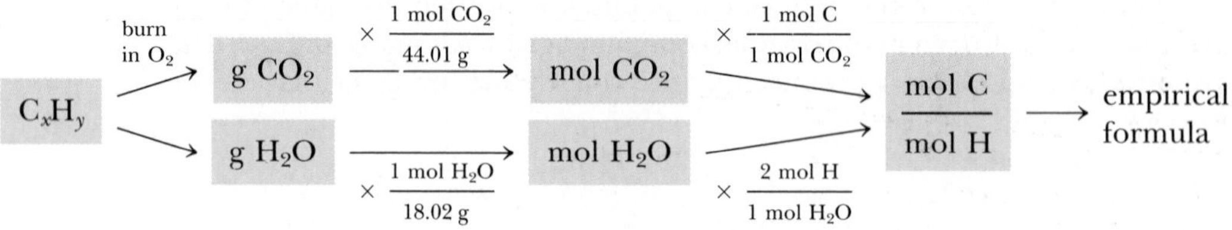

EXAMPLE 5.5 ***Determining the Empirical and Molecular Formulas for a Hydrocarbon***

You have 1.125 g of a liquid that you know is a hydrocarbon, C_xH_y. When the compound is burned in an apparatus like that in Figure 5.5, you find that 3.447 g of CO_2 and 1.647 g of H_2O have been produced. The molar mass of the compound was found to be 86.2 g/mol in a separate experiment. Determine the empirical and molecular formulas for the unknown hydrocarbon, C_xH_y.

Solution The first step is to calculate the moles of CO_2 and H_2O isolated from the combustion.

$$3.447 \text{ g } \cancel{CO_2} \cdot \frac{1 \text{ mol } CO_2}{44.010 \text{ g } \cancel{CO_2}} = 0.07832 \text{ mol } CO_2$$

$$1.647 \text{ g } \cancel{H_2O} \cdot \frac{1 \text{ mol } H_2O}{18.015 \text{ g } \cancel{H_2O}} = 0.09142 \text{ mol } H_2O$$

The previous discussion explains that for every mole of CO_2 isolated, one mole of C must have existed in the compound C_xH_y.

$$0.07832 \text{ mol } CO_2 \cdot \frac{1 \text{ mol C in } C_xH_y}{1 \text{ mol } CO_2} = 0.07832 \text{ mol C in } C_xH_y$$

and that 2 moles of H must have existed in the compound for every mole of H_2O isolated.

$$0.09142 \text{ mol } H_2O \cdot \frac{2 \text{ mol H in } C_xH_y}{1 \text{ mol } H_2O} = 0.1828 \text{ mol H in } C_xH_y$$

The original 1.125-g sample of compound therefore contained 0.07832 mol of C and 0.1828 mol of H. To determine the empirical formula of C_xH_y we find the ratio of moles of H to moles of C. (Here we use techniques outlined in Section 3.7.)

$$\frac{0.1828 \text{ mol H}}{0.07832 \text{ mol C}} = \frac{2.335 \text{ mol H}}{1 \text{ mol C}} = \frac{(2 + \frac{1}{3}) \text{ mol H}}{1 \text{ mol C}} = \frac{\frac{7}{3} \text{ mol H}}{1 \text{ mol C}} = \frac{7 \text{ mol H}}{3 \text{ mol C}}$$

The empirical formula of the hydrocarbon is therefore C_3H_7.

In a separate experiment, the molar mass was determined to be 86.2 g/mol. Comparing this with the molar mass calculated for the empirical formula,

$$\frac{\text{Molar mass from experiment}}{\text{Molar mass for } C_3H_7} = \frac{86.2 \text{ g/mol}}{43.1 \text{ g/mol}} = \frac{2}{1}$$

we find that the molecular formula is twice the empirical formula. That is, the molecular formula is $(C_3H_7)_2$, or C_6H_{14}.

As an aside, the determination of the molecular formula does not end the problem for a chemist. As is sometimes the case, the formula C_6H_{14} is appropriate for several distinctly different molecules, among the possibilities for which are

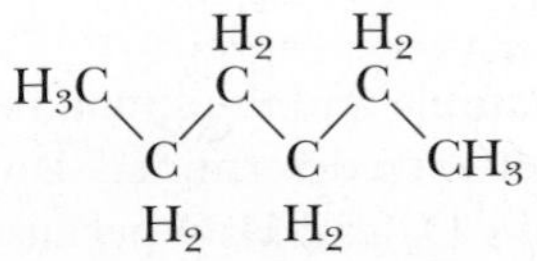

hexane
boiling point = 68.73 °C

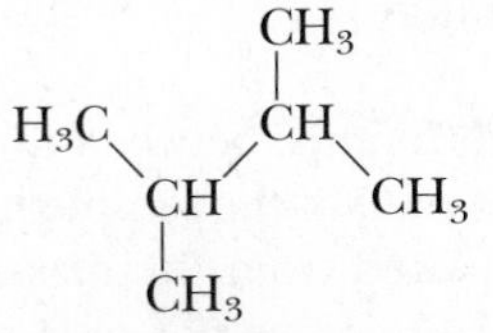

2,3-dimethylbutane
boiling point = 57.98 °C

To decide finally the identity of the unknown compound, more laboratory experiments are still necessary.

EXERCISE 5.5 ***Determining the Empirical and Molecular Formula for a Hydrocarbon***

A 0.523-g sample of C_xH_y was burned in air to give 1.612 g of CO_2 and 0.7425 g of H_2O. A separate experiment gave a molar mass of the unknown compound of 114 g/mol. Determine the empirical and molecular formulas for the hydrocarbon. ■

In Example 5.5 the moles of C and H found in the combustion products of the unknown compound were used directly to find the hydrogen-carbon ratio for the unknown compound. Alternatively, the moles of C and H could have been converted to the mass of each element and then to a weight percentage of the original 1.125-g sample. Using the data in Example 5.5, we have

$$0.1828\ \cancel{\text{mol H}} \text{ in } C_xH_y \cdot \frac{1.0079\ \text{g}}{1\ \cancel{\text{mol H}}} = 0.1842\ \text{g H}$$

$$0.07832\ \cancel{\text{mol C}} \text{ in } C_xH_y \cdot \frac{12.011\ \text{g}}{1\ \cancel{\text{mol C}}} = 0.9407\ \text{g C}$$

This leads to weight percentages of C and H in the 1.125-g sample of C_xH_y of 83.62% and 16.38%, respectively. These weight percentages can then be used to find the empirical formula by the methods of Section 3.7. Be sure to notice, though, that a calculation of the weight percentages is not necessary when information from an experiment is available that directly gives the moles of each element in a given sample mass. *Any method that gives the moles of each element in a given sample leads to an empirical formula.*

EXAMPLE 5.6 *Determining an Empirical Formula*

Suppose you isolate an acid from clover leaves and know that it contains only the elements C, H, and O. Burning 0.513 g of the acid in oxygen produces 0.501 g of CO_2 and 0.103 g of H_2O.

$$\underset{0.513\ \text{g}}{C_xH_yO_z(s)} + \text{some } O_2(g) \longrightarrow \underset{0.501\ \text{g}}{x\ CO_2(g)} + \underset{0.103\ \text{g}}{\frac{y}{2}\ H_2O(g)}$$

What is the empirical formula of the acid, $C_xH_yO_z$? Given that another experiment has shown that the molar mass of the acid is 90.04 g/mol, what is its molecular formula?

Solution The difference between this example and the situation in Example 5.5 is that the moles of O in the sample must also be determined. Unfortunately, this cannot be calculated from the masses of CO_2 and H_2O because the oxygen in CO_2 and the H_2O comes not only from $C_xH_yO_z$ but also from the O_2 used in combustion. Therefore, we shall pursue the following strategy:

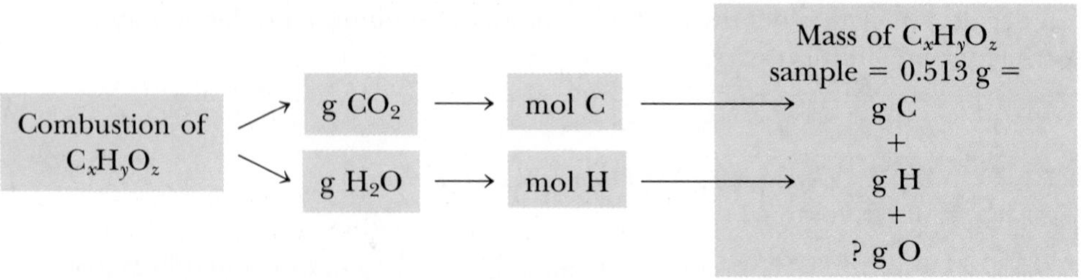

The masses of CO_2 and H_2O from combustion can be used to find the masses of C and H in the 0.513-g sample of $C_xH_yO_z$. Because the masses of C, H, and O must add up to 0.513 g, the mass of O can be found by subtraction. With the masses of all elements known, the number of moles of each can be calculated and the ratio of moles of each in $C_xH_yO_z$ determined.

The first step is to convert the masses of CO_2 and H_2O to moles.

$$0.501 \text{ g } CO_2 \cdot \frac{1 \text{ mol } CO_2}{44.01 \text{ g } CO_2} = 0.0114 \text{ mol } CO_2$$

and

$$0.103 \text{ g } H_2O \cdot \frac{1 \text{ mol } H_2O}{18.02 \text{ g } H_2O} = 0.00572 \text{ mol } H_2O$$

The moles of CO_2 and H_2O can now be converted to the masses of C and H that were in the original compound.

$$0.0114 \text{ mol } CO_2 \cdot \frac{1 \text{ mol C}}{1 \text{ mol } CO_2} \cdot \frac{12.01 \text{ g C}}{1 \text{ mol C}}$$
$$= 0.137 \text{ g C in } CO_2 \textit{ and} \text{ formerly in the acid sample}$$

$$0.00572 \text{ mol } H_2O \cdot \frac{2 \text{ mol H}}{1 \text{ mol } H_2O} \cdot \frac{1.008 \text{ g}}{1 \text{ mol H}} = 0.0115 \text{ g H in } H_2O \textit{ and} \text{ formerly in the acid sample}$$

↑ *Be sure to notice this stoichiometric factor*

These calculations reveal that the 0.513-g sample of acid contains 0.137 g of C and 0.0115 g of H; the remaining mass, 0.365 g, must be oxygen.

$$0.513\text{-g acid sample} - 0.137 \text{ g C} - 0.0115 \text{ g H} = 0.365 \text{ g O}$$

To find the empirical formula of the acid, you need only find the number of moles of each element in the acid sample.

$$0.137 \text{ g C} \cdot \frac{1 \text{ mol C}}{12.01 \text{ g C}} = 0.0114 \text{ mol C}$$

$$0.0115 \text{ g H} \cdot \frac{1 \text{ mol H}}{1.008 \text{ g H}} = 0.0114 \text{ mol H}$$

$$0.365 \text{ g O} \cdot \frac{1 \text{ mol O}}{16.00 \text{ g O}} = 0.0228 \text{ mol O}$$

Then, to find the mole ratio of elements, divide the number of moles of each element by the *smallest* number of moles.

$$\frac{0.0114 \text{ mol H}}{0.0114 \text{ mol C}} = \frac{1.00 \text{ mol H}}{1.00 \text{ mol C}} \quad \text{and} \quad \frac{0.0228 \text{ mol O}}{0.0114 \text{ mol C}} = \frac{2.00 \text{ mol O}}{1.00 \text{ mol C}}$$

The mole ratios show that, for every C atom in the molecule, one H atom and two O atoms occur. The *empirical formula* of the acid is therefore CHO_2, and the molar mass of the empirical formula unit is 45.02 g/mol.

Finally, to determine the molecular formula, the experimental molar mass of the compound and the molar mass of one empirical formula unit are compared.

$$\frac{90.04 \text{ g/mol unknown acid}}{45.02 \text{ g/mol } CHO_2} = \frac{2.000 \text{ mol } CHO_2}{1.000 \text{ mol unknown acid}}$$

Thus, the *molecular formula* of the acid is twice the empirical formula, that is, $C_2H_2O_4$. This compound is called oxalic acid and is widely distributed as the potassium and calcium salts in the leaves, roots, and rhizomes of various plants. It also occurs in human and animal urine, and calcium oxalate is a major constituent of kidney stones.

EXERCISE 5.6 ***Formula Determination from Combustion Analysis***

Vitamin C is composed of C, H, and O. Determine the empirical formula of vitamin C from the following data: burning 0.400 g of solid vitamin C in pure oxygen gives 0.600 g of CO_2 and 0.163 g of H_2O. ■

EXERCISE 5.7 ***Formula Determination from Combustion Analysis***

A molecule of a new compound is composed of only C, H, and Cr. When 0.178 g of the compound is burned in air, the products are CO_2 (0.452 g), H_2O (0.0924), and Cr_2O_3.

$$\underset{0.178\text{ g}}{C_xH_yCr_z(s)} + \text{some } O_2(g) \longrightarrow \underset{0.452\text{ g}}{x\, CO_2(g)} + \underset{0.0924\text{ g}}{\frac{y}{2}\, H_2O(g)} + \frac{z}{2}\, Cr_2O_3(s)$$

What is the empirical formula of the compound? ■

5.5 WORKING WITH SOLUTIONS

Many of the chemicals in your body or in a plant are dissolved in water, that is, they are in solution. Just as a living system does chemistry in solution, so too do chemists, and we need to do our work quantitatively. To accomplish this, we continue to use balanced equations and moles, but we measure volumes of solution rather than masses of solids, liquids, or gases. Solution concentration expressed as *molarity* relates the volume of solution in liters or milliliters to the amount of substance in moles.

Solution Concentration: Molarity

The concept of concentration is useful in many contexts. For example, about 4,900,000 people live in Wisconsin, and the state has a land area of roughly 56,000 square miles; therefore, the average concentration of people is about 88 per square mile. In chemistry the amount of solute dissolved in a given volume of solution can be found in the same way and is known as the **concentration** of the solution. Solution concentration is usually reported as *moles of solute per liter of solution;* this is called the **molarity** of the solution.

There are other ways of expressing solution concentration. In Table 3.3 the concentrations of ions in seawater are given in milligrams per liter of solution or parts per million.

$$\text{Molarity} = \frac{\text{moles of solute}}{\text{liters of solution}}$$

For example, if 58.4 g, or 1.00 mol, of NaCl is dissolved in enough water to give a total solution volume of 1.00 liter, the concentration, *c*, is 1.00 mole per liter, or 1.00 *molar.* This is often abbreviated as 1.00 M, where the capital M stands for "moles per liter."

$$c_{\text{molarity}} = 1.00 \text{ M} = [\text{NaCl}]$$

A formula in brackets means the concentration of the compound is given in moles per liter of solution.

Another common notation is to place the formula of the compound in square brackets; this implies that the concentration of the solute in moles of compound per liter of solution is being specified. Finally, note that chemists use the terms "moles per liter" and "molar" interchangeably.

It is important to notice that molarity refers to moles of solute per liter of *solution* (and not to liters of solvent). If one liter of water is added to one mole of a solid compound, the final volume probably will not be exactly one liter and the final concentration will not be exactly one molar (Figure 5.6). When making solutions of a given molarity, therefore, it is almost always the case that we dissolve the solute in a volume of solvent smaller than the desired volume of solution, then make up the final volume with more solvent.

The very effective oxidizing agent potassium permanganate, $KMnO_4$, which was used at one time as a germicide in the treatment of burns, is a common laboratory chemical. It is a shiny, purple-black solid that dissolves readily in water to give deep purple solutions. Suppose 0.435 g of $KMnO_4$ is dissolved in enough water to give 250. mL of solution (Figure 5.7). What is the molar concentration of $KMnO_4$? As is almost always the case, the first step is to convert the mass of material to moles.

$$0.435 \text{ g KMnO}_4 \cdot \frac{1 \text{ mol KMnO}_4}{158.0 \text{ g KMnO}_4} = 0.00275 \text{ mol KMnO}_4$$

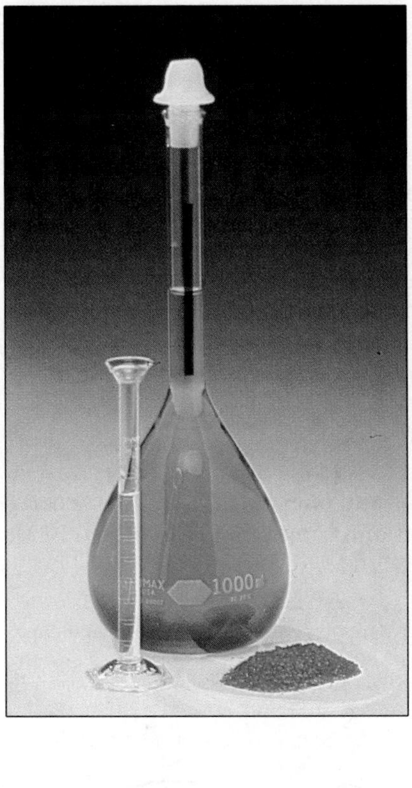

Figure 5.6 To make a 0.100-M solution of $CuSO_4$, 25.0 g, or 0.100 mol, of $CuSO_4 \cdot 5\,H_2O$ (the blue crystalline solid) was placed in a 1.00-L volumetric flask. For this photo, exactly 1.00 L of water was measured out and slowly added to the volumetric flask. When enough water had been added so that the *solution volume* was exactly 1.00 L, approximately 8 mL of water (the quantity in the small graduated cylinder) was left over. This emphasizes that molar concentrations are defined as moles per liter of solution and not per liter of water or other solvent. (C. D. Winters)

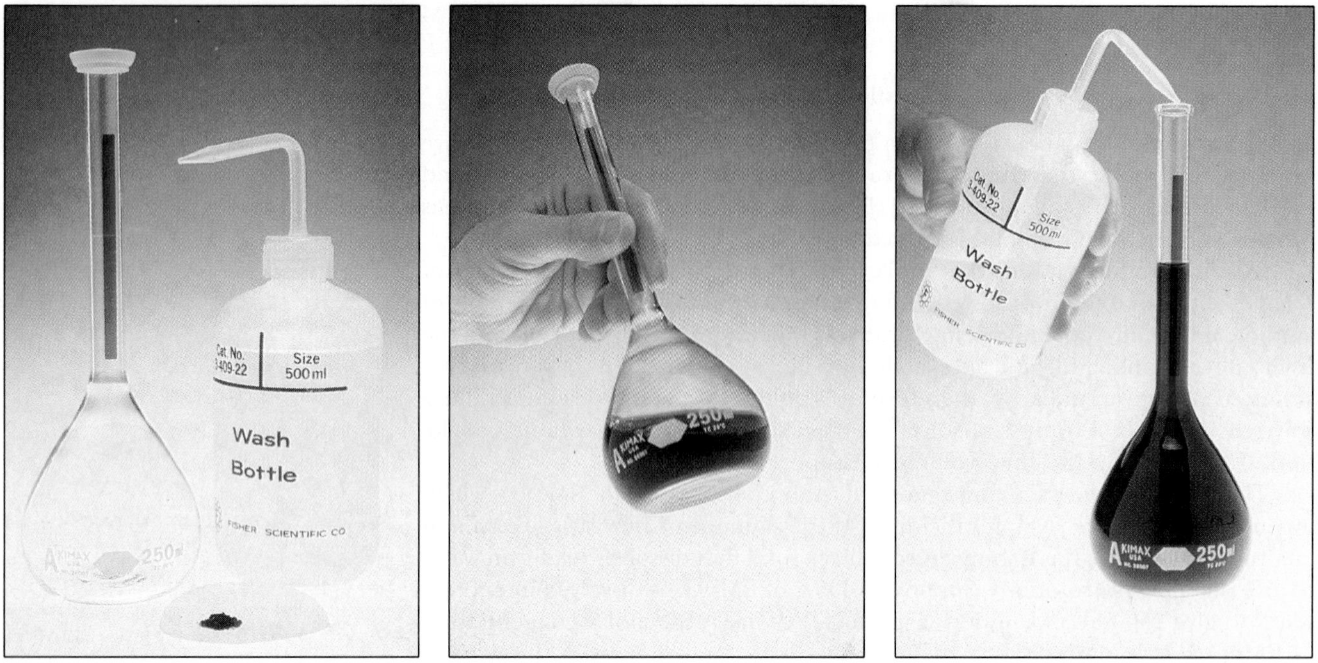

(a) (b) (c)

Figure 5.7 (a) A 0.0110-M solution of $KMnO_4$ is made by adding enough water to 0.435 g of $KMnO_4$ to make 0.250 L of solution. (b) To ensure the correct solution volume, the $KMnO_4$ is placed in a volumetric flask and dissolved in a small amount of water. After dissolving is complete, sufficient water is added to fill the flask to the mark on the neck. (c) The flask now contains 0.250 L of solution. (C. D. Winters)

Now that the number of moles of substance is known, this can be combined with the volume of solution—*which must be in liters*—to give the molarity. Because 250. mL is equivalent to 0.250 L,

$$\text{Molarity of KMnO}_4 = \frac{0.00275 \text{ mol KMnO}_4}{0.250 \text{ L solution}} = 0.0110 \text{ M}$$

The $KMnO_4$ concentration is 0.0110 molar, or 0.0110 M. This is useful information, but it is often equally useful to know the concentration of each type of ion in a solution. In Chapter 3 you learned that $KMnO_4$ dissociates completely into its ions, K^+ and MnO_4^-, when dissolved in water (Figure 5.8).

$$\underset{\text{100\% dissociation}}{\text{KMnO}_4(\text{aq}) \longrightarrow} \text{K}^+(\text{aq}) + \text{MnO}_4^-(\text{aq})$$

One mole of $KMnO_4$ provides one mole of K^+ and one mole of MnO_4^-. Accordingly, 0.0110 M $KMnO_4$ gives a concentration of K^+ in the solution of 0.0110 M; similarly, the concentration of MnO_4^- is 0.0110 M.*

*Chemists often use the expression "compound XY is __ M in solution." If XY is an ionic compound, however, it forms X and Y ions on dissolving; XY does not exist as such in solution. Thus, the expression "compound XY is __ M" is a way of saying that __ moles of XY units are dissolved in enough water to make one liter of solution.

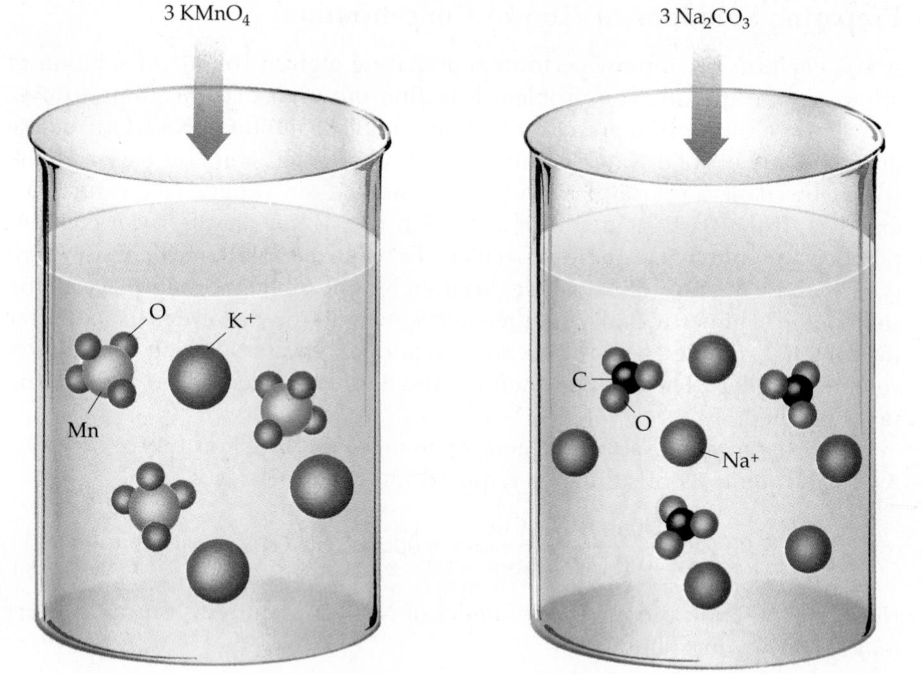

(a) $3\ KMnO_4 \longrightarrow 3\ K^+(aq) + 3\ MnO_4^-(aq)$ (b) $3\ Na_2CO_3 \longrightarrow 6\ Na^+(aq) + 3\ CO_3^{2-}(aq)$

Figure 5.8 Ion concentrations on dissolving ionic compounds. (a) When $KMnO_4$ dissolves in water, one mole of K^+ ions and one mole of MnO_4^- ions form for every mole of $KMnO_4$ dissolved. (b) Dissolving one mole of Na_2CO_3 produces two moles of Na^+ ions and one mole of CO_3^{2-} ions.

Another example of ion concentrations is provided by the dissociation of sodium carbonate, Na_2CO_3 (see Figure 5.8).

$$Na_2CO_3(aq) \longrightarrow 2\ Na^+(aq) + CO_3^{2-}(aq)$$

100% dissociation

If one mole of Na_2CO_3 is dissolved in enough water to make one liter of solution, the concentration of the sodium ion is $[Na^+] = 2$ M because the compound dissociates into two moles of Na^+ ions for each mole of Na_2CO_3 added. Because one mole of Na_2CO_3 yields one mole of CO_3^{2-}, the concentration of the carbonate ion is $[CO_3^{2-}] = 1$ M, and the total concentration of ions is 3 M.

EXERCISE 5.8 ***Solution Molarity***

Sodium bicarbonate, $NaHCO_3$, is used in baking powder formulations, in fire extinguishers, and in the manufacture of plastics and ceramics, among other things. If you have 26.3 g of the compound and dissolve it in enough water to make 200. mL of solution, what is the concentration of $NaHCO_3$? ■

- antidote for Acid burns.

EXERCISE 5.9 ***Ion Concentrations in Solution***

Both HCl and Na_2SO_4 are strong electrolytes when dissolved in water. Assume enough of each has been dissolved so that [HCl] = 1.0 M and $[Na_2SO_4]$ = 0.500 M. State the concentration of each ion and the total concentration of all ions in each solution. ■

Preparing Solutions of Known Concentration

A task chemists often must perform is preparing a given volume of solution of known concentration. The problem is to find out what mass of solute to use.

Suppose you wish to prepare 2.00 L of a 1.50-M solution of Na_2CO_3. You are given a bottle of solid Na_2CO_3 and some distilled water. You also have a 2.00-L volumetric flask, a special flask with a line marked on its neck (see Figures 5.6 and 5.7). If the flask is filled with a solution to this line (at 20 °C), it contains precisely the volume of solution specified. To make the solution, you must weigh the necessary quantity of Na_2CO_3 as accurately as possible, carefully place all the solid in the volumetric flask, and then add some water to dissolve the solid. After the solid has completely dissolved, you can add more water to bring the solution volume to 2.00 L. The solution then has the desired concentration and the volume specified.

But what mass of Na_2CO_3 is required to make the 2.00 L of 1.50-M Na_2CO_3? As usual, the moles of substance required must first be calculated.

$$2.00\,\cancel{\text{L}} \cdot \frac{1.50 \text{ mol } Na_2CO_3}{1.00\,\cancel{\text{L}} \text{ solution}} = 3.00 \text{ mol } Na_2CO_3 \text{ required}$$

Now that you know the number of moles of Na_2CO_3 required, you can convert moles to mass, in grams.

$$3.00\,\cancel{\text{mol } Na_2CO_3} \cdot \frac{106.0 \text{ g } Na_2CO_3}{1\,\cancel{\text{mol } Na_2CO_3}} = 318 \text{ g } Na_2CO_3$$

Thus, to prepare the desired solution, you should dissolve 318 g of Na_2CO_3 in enough water to make 2.00 L of solution.

EXERCISE 5.10 ***Preparing Solutions of Known Concentration***

An experiment in your laboratory requires 500. mL of a 0.0200-M solution of $KMnO_4$. You are given a bottle of solid $KMnO_4$, some distilled water, and a 500.-mL volumetric flask. Describe how to make the required solution. ■

Making a sodium carbonate solution as just described illustrates the most common way to create a solution of known concentration. Another method, however, is to *begin with a concentrated solution and add water to make it more dilute until the desired concentration is reached.** Many of the solutions prepared for your laboratory course are probably made by this *dilution* method. It is often more efficient to store a few liters of a concentrated solution and then add water to make it into many liters of a dilute solution.

A practical example of making a solution by dilution is mixing frozen concentrated orange juice with water to make orange juice of the right concentration to drink.

As an example of the dilution method, suppose you need 500 mL of 0.0010 M potassium dichromate, $K_2Cr_2O_7$, for use in chemical analysis. You have available a few liters of 0.100 M $K_2Cr_2O_7$ and some distilled water and glassware. How can you make the required 0.0010-M solution? The approach is to take some of the more concentrated $K_2Cr_2O_7$ solution, put it in a flask, and then add water until the $K_2Cr_2O_7$ is contained in a larger volume of water, that is, until it is less concentrated (or more dilute).

*This is true except for sulfuric acid solutions. When mixing water and sulfuric acid, the resulting solution becomes quite warm. If water is added to concentrated sulfuric acid in a flask, so much heat is evolved that the solution may boil over or splash and burn someone near the flask. To avoid this, chemists always add sulfuric acid to water when making a dilute acid solution.

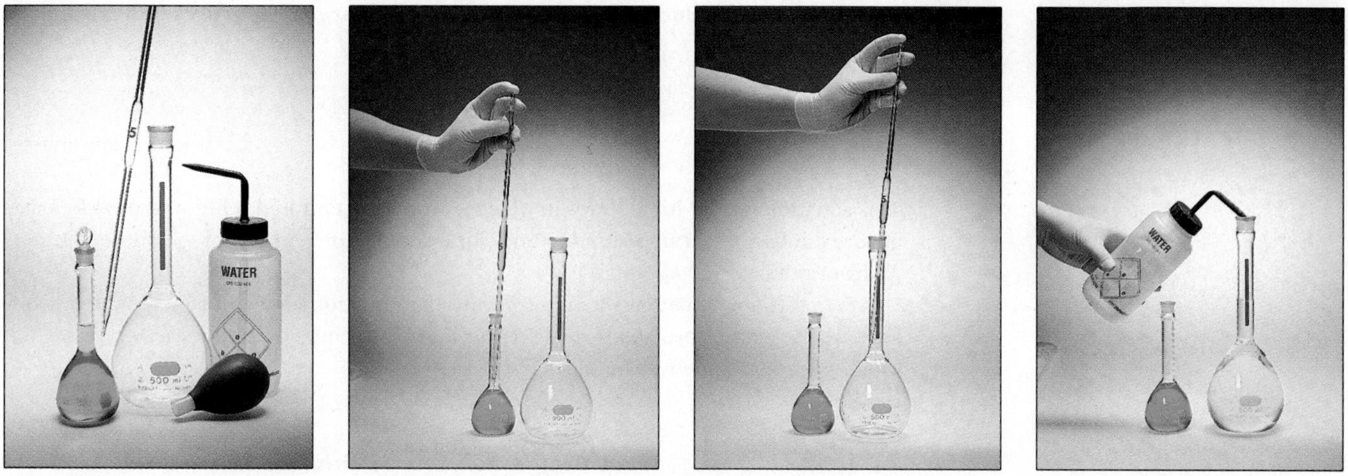

(a) (b) (c) (d)

Figure 5.9 (a) Equipment needed to make a solution by dilution. (b) A 5.0-mL sample of 0.100 M $K_2Cr_2O_7$ is withdrawn from a flask using a volumetric pipet. (c) The 5.0-mL sample is transferred to a 500.-mL volumetric flask. (d) The 500.-mL flask is filled with distilled water to the mark on the neck, and the concentration of the now diluted solution is 0.0010 M. (C. D. Winters)

The problem is to make a specified quantity of solution of known concentration. If the volume and concentration are known, then the number of moles of solute is also known. The number of moles of $K_2Cr_2O_7$ that must be in the final dilute solution is therefore

$$\text{Amount of } K_2Cr_2O_7 \text{ in final solution} = 0.500 \text{ L} \cdot 0.0010 \text{ mol/L} = 0.00050 \text{ mol } K_2Cr_2O_7$$

A more concentrated solution containing this number of moles of $K_2Cr_2O_7$ must be placed in the flask and then diluted to a total volume of 500. mL. The volume of 0.100 M $K_2Cr_2O_7$ that contains the required number of moles is 5.0 mL.

$$0.00050 \text{ mol } K_2Cr_2O_7 \cdot \frac{1.00 \text{ L}}{0.100 \text{ mol } K_2Cr_2O_7} = 0.0050 \text{ L, or } 5.0 \text{ mL}$$

Figure 5.9 illustrates how such a solution can be made.

PROBLEM-SOLVING TIPS AND IDEAS

5.5 Preparing a Solution by Dilution

A second look at the preparation of the $K_2Cr_2O_7$ solution suggests a simple way to remember how to do these calculations. The central idea is that the number of moles of $K_2Cr_2O_7$ in the final, dilute solution has to be equal to the number of moles of $K_2Cr_2O_7$ taken from the more concentrated solution. If c is the concentration (molarity) and V is the volume (and the subscripts d and c identify the dilute and concentrated solutions, respectively), the number of moles of solute in either solution can be calculated as follows:

$$\text{Amount of } K_2Cr_2O_7 \text{ in the final dilute solution} = c_d V_d = 0.00050 \text{ mol}$$

$$\text{Amount of } K_2Cr_2O_7 \text{ taken from the more concentrated solution} = c_c V_c = 0.00050 \text{ mol}$$

Because both *cV* products are equal to the same number of moles, we can use the following equation:

$$c_c V_c = c_d V_d$$

Moles of reagent in concentrated solution = moles of reagent in dilute solution

This equation is valid for all cases in which a more concentrated solution is used to make a more dilute one. It can be used to find, for example, the molarity of the dilute solution, c_d, from values of c_c, V_c, and V_d.

Students are sometimes tempted to use the equation above in stoichiometry problems when relating two solutions (Section 5.6). The equation applies *only* in cases for which the stoichiometric coefficient is 1/1. ■

EXAMPLE 5.7 *Preparing a Solution by Dilution*

Suppose you are doing an experiment to find out the quantity of iron in a vitamin pill. You must prepare a standard solution of iron(III) ion by placing 1.00 mL of 0.236 M iron(III) nitrate in a volumetric flask and diluting to exactly 100.0 mL. What is the concentration of the diluted iron(III) solution?

Solution From the volume and molarity of the original iron solution, you can find the number of moles of iron(III) in the first solution.

$$1.00 \times 10^{-3}\ \text{L} \cdot 0.236\ \text{mol/L} = 2.36 \times 10^{-4}\ \text{mol of iron(III)}$$

This number of moles is also found in the 100.0 mL of dilute solution. The new concentration is therefore found to be

$$\frac{2.36 \times 10^{-4}\ \text{mol of iron(III)}}{0.100\ \text{L}} = 2.36 \times 10^{-3}\ \text{M}$$

The solution was diluted by a factor of 100, so the new concentration is 100 times smaller than the original concentration.

EXERCISE 5.11 *Preparing a Solution by Dilution*

An experiment calls for you to use 250. mL of 1.00 M NaOH, but you are given a large bottle of 2.00 M NaOH. Describe how to make the 1.00 M NaOH in the desired volume. ■

EXERCISE 5.12 *Preparing a Solution by Dilution*

In one of your laboratory experiments you are given a solution of $CuSO_4$ that has a concentration of 0.15 M. If you mix 6.0 mL of this solution with enough water to have a total volume of 10.0 mL, what is the concentration of $CuSO_4$ in this new solution? ■

PROBLEM-SOLVING TIPS AND IDEAS

5.6 Moles, Volume, and Molarity

Another look at the discussion in this section suggests a useful form of the definition of molarity. That is,

$$\text{Moles solute} = c_{\text{molarity}} \text{ of solution (mol/L)} \cdot \text{volume of solution (L)}$$

$$\text{moles} = c_{\text{molarity}} \cdot V$$

As you practice working with solutions, you will find this form of the equation to be quite convenient. ■

5.6 STOICHIOMETRY OF REACTIONS IN AQUEOUS SOLUTION

General Solution Stoichiometry

A common type of reaction is that between a metal carbonate and an aqueous acid to give a salt and CO_2 gas, the type of reaction that occurs when you take a popular remedy for upset stomach (Figure 5.10) (Section 4.9).

$$\underset{\text{metal carbonate}}{CaCO_3(s)} + \underset{\text{acid}}{2\ HCl(aq)} \longrightarrow \underset{\text{salt}}{CaCl_2(aq)} + \underset{\text{water}}{H_2O(\ell)} + \underset{\text{carbon dioxide}}{CO_2(g)}$$

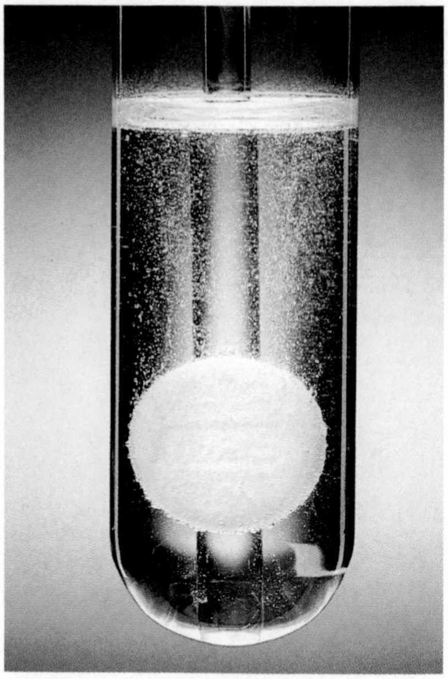

Figure 5.10 A commercial remedy for excess stomach acid, which contains a metal carbonate, reacts with aqueous hydrochloric acid, the acid present in the digestive system. The most obvious product is CO_2 gas. (C. D. Winters)

Suppose we want to know what mass of $CaCO_3$ is required to react completely with 25 mL of 0.750 M HCl. This can be solved in the same way as all the stoichiometry problems you have seen so far, except that the amount of one reactant is given in volume and concentration units. It is therefore necessary first to find the number of moles of HCl,

$$0.025\ \cancel{\text{L HCl}} \cdot \frac{0.750 \text{ mol HCl}}{1\ \cancel{\text{L HCl}}} = 0.019 \text{ mol HCl}$$

and then to relate this to the moles of $CaCO_3$ required.

$$0.019\ \cancel{\text{mol HCl}} \cdot \frac{1 \text{ mol } CaCO_3}{2\ \cancel{\text{mol HCl}}} = 0.0094 \text{ mol } CaCO_3$$

Finally, moles of $CaCO_3$ are converted to a mass in grams.

$$0.0094\ \cancel{\text{mol } CaCO_3} \cdot \frac{100.\text{ g } CaCO_3}{1\ \cancel{\text{mol } CaCO_3}} = 0.94 \text{ g } CaCO_3$$

Chemists do such calculations many times in the course of their work in research and product development. If you follow the general scheme outlined on page 236, and pay attention to the units on the numbers, you can successfully carry out any kind of stoichiometry calculations involving concentrations.

Zinc reacts readily with aqueous hydrochloric acid to give hydrogen gas and aqueous zinc chloride. (C. D. Winters)

EXAMPLE 5.8 ***Stoichiometry of a Reaction in Solution***

Metallic zinc reacts with aqueous solutions of acids such as HCl, as do many other metals.

$$Zn(s) + 2\ HCl(aq) \longrightarrow ZnCl_2(aq) + H_2(g)$$

Such reactions are often used to produce hydrogen gas for laboratory uses. If you have 10.0 g of zinc, what volume of 2.50 M HCl (in milliliters) do you need to convert completely the zinc to zinc chloride?

Solution The balanced equation for the reaction is given and shows that two moles of HCl are required for each mole of zinc. Once the number of moles of zinc available is known, the number of moles of HCl required can therefore be calculated. The volume of solution required is determined from that.

Begin by calculating moles of zinc available.

$$10.0 \text{ g Zn} \cdot \frac{1.00 \text{ mol Zn}}{65.39 \text{ g Zn}} = 0.153 \text{ mol Zn}$$

Next, use the stoichiometric factor "2 mol HCl/1 mol Zn" to find moles of HCl required.

$$0.153 \text{ mol Zn} \cdot \frac{2 \text{ mol HCl}}{1 \text{ mol Zn}} = 0.306 \text{ mol HCl required}$$

Finally, knowing the moles of HCl required and the concentration of the acid, you can calculate the volume of acid required.

$$0.306 \text{ mol HCl} \cdot \frac{1.00 \text{ L solution}}{2.50 \text{ mol HCl}} = 0.122 \text{ L of HCl solution or } 122 \text{ mL}$$

Because the answer is needed in milliliters, as a final step you convert liters to milliliters.

EXERCISE 5.13 ***Solution Stoichiometry***

As described previously, a common type of reaction is that between a metal carbonate and an aqueous acid to give a salt and CO_2 gas.

$$Na_2CO_3(aq) + 2\ HCl(aq) \longrightarrow 2\ NaCl(aq) + H_2O(\ell) + CO_2(g)$$

If you combine 50.0 mL of 0.450 M HCl and an excess of Na_2CO_3, what mass of NaCl (in grams) is produced? ■

PROBLEM-SOLVING TIPS AND IDEAS

5.7 Stoichiometry Calculations Involving Solutions

In "Problem-Solving Tips and Ideas 5.1" you learned about a common approach to stoichiometry problems. We can now modify that scheme for a reaction such as xA + yB → products.

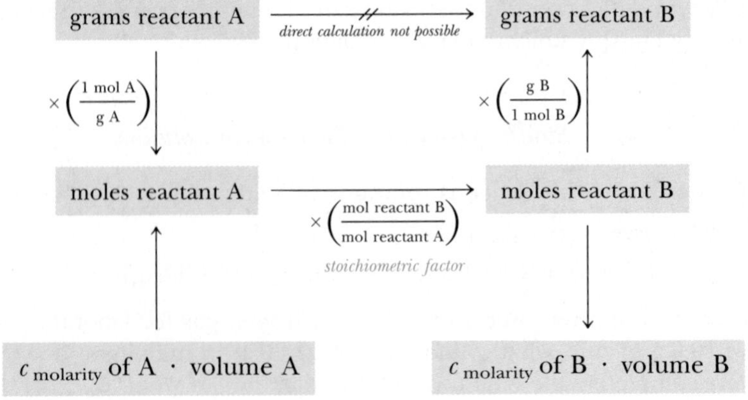

■

CURRENT ISSUES IN CHEMISTRY

Where Do Industrial Wastes Go?

Many of the reactions you saw in Chapter 4 and in this chapter involve acids. For instance, in Example 5.8 you saw how zinc reacts with hydrochloric acid. One reason for the emphasis on acid chemistry is that acids are some of the most important chemicals in our economy. Five acids are listed among the top 50 chemicals produced in the United States in 1993.

Acid	Billions of Kilograms	Rank
Sulfuric acid	36.43	1
Phosphoric acid	10.45	10
Nitric acid	7.74	14
Hydrochloric acid	2.93	26
Acetic acid	1.66	34

More than 15,000 companies use these acids to make other chemicals, to clean and refinish metals, to plate metals onto other metals or onto plastics, and in many other applications. A problem faced by all of these industries is what to do with acid-containing waste. For example, when acids are used to wash a metal surface, the washings contain unused acid along with ions of metals, such as copper(II), vanadium(III), silver(I), nickel(II), and lead(II). It is estimated that about 4 billion kg of acid-containing wastes are generated annually, and they cannot simply be flushed into the nearest lake or river. Not only would the acid damage aquatic life, but heavy metals are toxic to plants and animals.

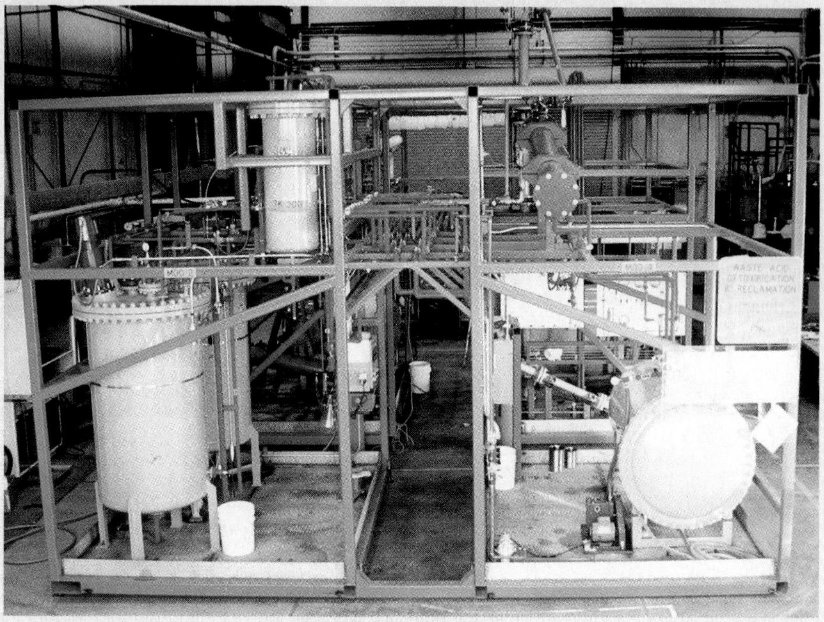

A transportable acid-recovery pilot plant for transforming metal-bearing spent acids into reusable acid and a reclaimable metal salt. (Courtesy of Viatec/Recovery Systems, Inc., Richmond, Washington)

Because of the enormous quantity of acid wastes generated every year, many laboratories have been involved in finding a way to recover unused acids and to remove heavy metals and other dissolved substances. A process recently developed at the Department of Energy's Pacific Northwest Laboratory holds promise for significantly reducing the volume and toxicity of acid waste.

Typically the waste from a chemical or metallurgical operation may contain sulfuric, phosphoric, and nitric acids, along with some heavy metal ions dissolved in the aqueous acid. When the mixture is heated, the acids vaporize, and the heavy metal ions remain in the liquid phase. The vapor is purified to yield a clean acid, in some cases cleaner than industrial-grade acids. The solution containing heavy metals is collected in a tank, and the metal ions are removed by adding salts of anions that form precipitates with the heavy metal ions. The metals can be reclaimed from these precipitates and can be sold or reused.

It is claimed that this new approach to recovering and detoxifying waste acids can lead to a significant cost saving for chemical industries.

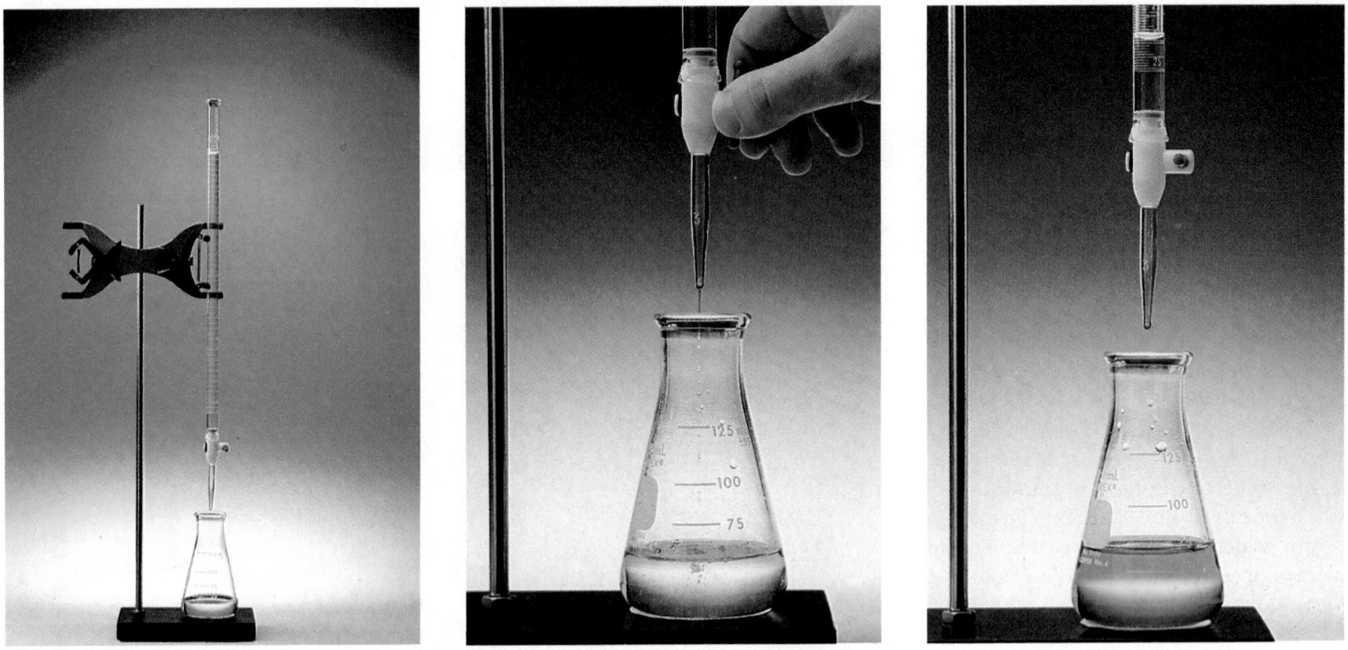

(a) (b) (c)

Figure 5.11 Titration of an acid in aqueous solution with a base. (a) A buret, a volumetric measuring device calibrated in divisions of 0.1 mL, is filled with an aqueous solution of a base of known concentration. (b) Base is added slowly from the buret to the solution. (c) A change in color of an indicator signals the equivalence point. (C. D. Winters)

Titrations

Your study of stoichiometry so far should have convinced you that if you know (1) the balanced equation for a reaction, and (2) the exact quantity of one of the reactants, then you can calculate the quantity of any other substance consumed or produced in the reaction. This is the essence of any technique of **quantitative chemical analysis,** the determination of the *quantity* of a given constituent in a mixture.*

Qualitative analysis is the determination of the identity of the constituents of a mixture.

An acid isolated from clover leaves was shown in Example 5.6 to have the molecular formula $H_2C_2O_4$. This compound, called oxalic acid, is commercially important in manufacturing paint and textiles, in metal treatment, and in photography.

The main use of oxalate ion is to remove calcium ions from aqueous solutions as insoluble calcium oxalate.

$$Ca^{2+}(aq) + H_2C_2O_4(aq) \longrightarrow CaC_2O_4(s) + 2\ H^+(aq)$$

Unfortunately, this reaction can also occur in your body and lead to kidney stones.

Suppose you are asked to analyze a sample of oxalic acid to ascertain its purity. Because the compound is an acid, it reacts with the base, sodium hydroxide, in aqueous solution according to the balanced equation (Section 4.8)

$$H_2C_2O_4(aq) + 2\ NaOH(aq) \longrightarrow Na_2C_2O_4(aq) + 2\ H_2O(\ell)$$

You can therefore tell how much oxalic acid is present in a given mass of sample if the following conditions are met in the reaction with NaOH:

1. You can tell when the amount of sodium hydroxide added is *just enough* to react with *all* the oxalic acid present in solution.
2. You know the volume of the sodium hydroxide solution added at the point of complete reaction.
3. You know the concentration of the sodium hydroxide solution.

These conditions are fulfilled in a **titration,** a procedure illustrated in the series of photographs in Figure 5.11. The solution containing oxalic acid is placed in a flask along with an acid-base indicator. An **indicator** is a dye that changes color when the reaction used for analysis is complete. In this case, the dye is colorless in acid solution but pink in basic solution. Aqueous sodium hydroxide of accurately known concentration is placed in a *buret,* a measuring cylinder that most commonly has a volume of 50.0 mL and is calibrated in 0.1-mL divisions. As the sodium hydroxide in the buret is added slowly to the acid solution in the flask, the acid reacts with the base according to the net ionic equation

$$H_2C_2O_4(aq) + 2\ OH^-(aq) \longrightarrow C_2O_4^{2-}(aq) + 2\ H_2O(\ell)$$

As long as some acid is present in solution, all the base supplied from the buret is consumed, and the indicator remains colorless. At some point, however, the number of moles of OH^- added exactly equals the number of moles of H^+ that can be supplied by the acid. This is called the **equivalence point.** To indicate when this point has been reached, an acid-base indicator, the dye previously mentioned, was added to the solution prior to titration. As soon as the slightest excess of base has been added, the solution becomes basic, and the dye changes color (Figure 5.12).

Figure 5.12 The juice of a red cabbage turns color when the acidity of a solution changes. When the solution is highly acidic, the juice gives the solution a red color. As the solution becomes less acid (more basic), the color changes from red to violet to yellow. Red cabbage juice is a natural acid-base indicator. (C. D. Winters)

When the equivalence point has been reached in a titration, the volume of base added since the beginning of the titration can be determined by reading the calibrated buret (see Figure 5.11). From this volume and the concentration of the base, the number of moles of base used can be found:

Moles of base added = concentration of base (mol/L) · volume of base (L)

Then, using the stoichiometric factor from the balanced equation, the number of moles of base added is related to the number of moles of acid present in the original sample.

EXAMPLE 5.9 *Acid-Base Titration*

Suppose you dissolve a 1.034-g sample of impure oxalic acid in some water, add an acid-base indicator, and titrate it with 0.485 M NaOH. The sample requires 34.47 mL of the NaOH solution to reach the equivalence point.

$$H_2C_2O_4(aq) + 2\ OH^-(aq) \longrightarrow C_2O_4{}^{2-}(aq) + 2\ H_2O(\ell)$$

What is the mass of oxalic acid, and what is its weight percent in the sample?

Solution The objective—to find the mass of oxalic acid from a knowledge of the volume and concentration of the NaOH solution—can be reached using the scheme in "Problem-Solving Tips and Ideas 5.7." First, calculate the number of moles of NaOH used in the reaction from the volume (in liters) and concentration (in moles per liter) of the NaOH solution.

$$0.03447\ \cancel{L} \cdot \frac{0.485\ \text{mol NaOH}}{1\ \cancel{L}} = 0.0167\ \text{mol NaOH}$$

The balanced, net ionic equation for the reaction shows that one mole of oxalic acid requires two moles of sodium hydroxide. This is the stoichiometric factor required for the calculation of moles of oxalic acid, and it gives the number of moles of oxalic acid present.

$$0.0167\ \cancel{\text{mol NaOH}} \cdot \frac{1\ \text{mol}\ H_2C_2O_4}{2\ \cancel{\text{mol NaOH}}} = 0.00836\ \text{mol}\ H_2C_2O_4$$

The mass of oxalic acid is found from the number of moles of the acid.

$$0.00836\ \cancel{\text{mol}\ H_2C_2O_4} \cdot \frac{90.04\ \text{g}\ H_2C_2O_4}{1\ \cancel{\text{mol}\ H_2C_2O_4}} = 0.753\ \text{g}\ H_2C_2O_4$$

This mass of oxalic acid represents 72.8% of the total sample mass.

$$\frac{0.753\ \text{g}\ H_2C_2O_4}{1.034\ \text{g sample}} \cdot 100\% = 72.8\%\ H_2C_2O_4$$

In Example 5.9 the concentration of the base used in the titration was given. In real life this usually has to be found by a prior measurement. The procedure by which the concentration of an analytical reagent is determined is called **standardization,** for which there are two general approaches.

One approach is to accurately weigh a sample of a pure, solid acid or base (known as a **primary standard**) and then titrate this sample with a solution of the base or acid to be standardized (Example 5.10). Another approach to standardizing a solution is to titrate it with another solution that is already standardized (Exercise 5.15). This is often done with standard solutions purchased from chemical supply companies.

EXAMPLE 5.10 *Standardizing an Acid by Titration*

Solutions of acids such as HCl can be standardized by using them to titrate a base such as Na_2CO_3, a solid that can be obtained in pure form, that can be weighed accurately, and that reacts completely with an acid.* Suppose that 0.263 g of Na_2CO_3 requires 28.35 mL of aqueous HCl for titration to the equivalence point. What is the molar concentration of the HCl?

Solution If we can find a way to determine the number of moles of HCl in 28.35 mL of the HCl solution, the molar concentration is known.

As usual, the balanced equation for the reaction is written first.

$$Na_2CO_3(aq) + 2\ HCl(aq) \longrightarrow 2\ NaCl(aq) + H_2O(\ell) + CO_2(g)$$

Next, the mass of Na_2CO_3 used as the standard is converted to moles.

$$0.263\ \cancel{g\ Na_2CO_3} \cdot \frac{1.00\ mol\ Na_2CO_3}{106.0\ \cancel{g\ Na_2CO_3}} = 0.00248\ mol\ Na_2CO_3$$

Knowing the number of moles of Na_2CO_3, the number of moles of HCl in the solution can be calculated by using the appropriate stoichiometric factor.

$$0.00248\ \cancel{mol\ Na_2CO_3} \cdot \frac{2\ mol\ HCl\ required}{1\ \cancel{mol\ Na_2CO_3}\ available} = 0.00496\ mol\ HCl$$

The 28.35-mL (or 0.02835 L) sample of aqueous HCl therefore contains 0.00496 mol of HCl, and so the concentration of the HCl solution is 0.175 M.

$$HCl\ concentration = \frac{0.00496\ mol\ HCl}{0.02835\ L} = 0.175\ M$$

EXERCISE 5.14 *Acid-Base Titration*

A 25.0-mL sample of vinegar requires 28.33 mL of a 0.953-M solution of NaOH for titration to the equivalence point. What mass (in grams) of acetic acid is in the vinegar sample, and what is the concentration of acetic acid in the vinegar?

$$\underset{\text{acetic acid}}{CH_3CO_2H(aq)} + NaOH(aq) \longrightarrow \underset{\text{sodium acetate}}{NaCH_3CO_2(aq)} + H_2O(\ell)$$

■

EXERCISE 5.15 *Standardization of a Base*

Hydrochloric acid, HCl, can be purchased from chemical supply houses in a solution with a concentration of 0.100 M, and such a solution can be used to

*You might think that solid NaOH would make a good primary standard. Solid NaOH is difficult to weigh accurately, however, because it is deliquescent, that is, it absorbs water from humid air. Furthermore, its solutions readily take up CO_2 from the air, thus changing the concentration of OH^-:

$$2\ NaOH(aq) + CO_2(g) \longrightarrow Na_2CO_3(aq) + H_2O(\ell)$$

standardize the solution of a base. If titrating 25.00 mL of a sodium hydroxide solution to the equivalence point requires 29.67 mL of 0.100 M HCl, what is the concentration of the base? ■

Acid-base titrations are extremely useful for quantitative chemical analysis, and some questions at the end of this chapter illustrate their scope. You should not get the impression, however, that a titration can only be done with an acid reacting with a base. Oxidation-reduction reactions, another class of reactions described in Chapter 4, lend themselves very well to chemical analysis by titration because many of these reactions go rapidly to completion in aqueous solution, and methods exist for finding their equivalence points.

EXAMPLE 5.11 *Using an Oxidation-Reduction Reaction in a Titration*

Suppose you wish to analyze an iron ore for its iron content. In this case, the iron in the ore can be converted quantitatively to the iron(II) ion, Fe^{2+}, in aqueous solution, and this solution can then be titrated with aqueous potassium permanganate, $KMnO_4$. The balanced net ionic equation for the reaction that occurs in the course of this titration is

$$\underset{\substack{\text{purple}\\\text{oxidizing agent}}}{MnO_4^-(aq)} + \underset{\substack{\text{colorless}\\\text{reducing agent}}}{5\ Fe^{2+}(aq)} + \underset{\text{colorless}}{8\ H^+(aq)} \longrightarrow \underset{\text{colorless}}{Mn^{2+}(aq)} + \underset{\text{pale yellow}}{5\ Fe^{3+}(aq)} + 4\ H_2O(\ell)$$

This is a useful analytical reaction, because it is easy to detect when all the iron(II) has reacted (Figure 5.13). The MnO_4^- ion is deep purple, but when it

(a) (b) (c)

Figure 5.13 Titration involving an oxidation-reduction reaction. Here a solution of Fe^{2+} ion (a reducing agent) is titrated with aqueous $KMnO_4$ (the oxidizing agent; in the buret) (a). As $KMnO_4$ is added to the solution, the iron(II) ion is oxidized, and the deep purple $KMnO_4$ solution is reduced (b). The products (Fe^{2+} and Mn^{2+}) are nearly colorless. Just past the equivalence point, when a slight excess of $KMnO_4$ has been added, the solution takes on a faint purple color (c). (C. D. Winters) reacts with Fe^{2+}, the color disappears because the reaction product, the Mn^{2+} ion, is colorless. Thus, as $KMnO_4$ is added from a buret, the purple color disappears as the solutions mix. When all the Fe^{2+} has been converted to Fe^{3+}, any additional $KMnO_4$ gives the solution a permanent purple color. Therefore, $KMnO_4$ solution is added from the buret until the initially colorless, Fe^{2+}-containing solution just turns a faint purple color, the signal that the equivalence point has been reached.

Let us assume that a 1.026-g sample of iron-containing ore requires 24.35 mL of 0.0195 M $KMnO_4$ to reach the equivalence point. What is the weight percent of iron in the ore?

Solution Because the volume and molar concentration of the $KMnO_4$ solution are known, the number of moles of $KMnO_4$ used in the titration can be calculated. Remembering first to change the volume to liters, we have

$$0.02435\,\cancel{L} \cdot \frac{0.0195 \text{ mol}}{1\,\cancel{L}} = 0.000475 \text{ mol } KMnO_4$$

Based on the balanced chemical equation for the reaction of $KMnO_4$ with Fe^{2+}, the number of moles of iron(II) that were present in the solution (and therefore the number of moles of iron in the ore sample) can be calculated.

$$0.000475\,\cancel{\text{mol } KMnO_4} \cdot \frac{5 \text{ mol } Fe^{2+}}{1\,\cancel{\text{mol } KMnO_4}} = 0.00237 \text{ mol } Fe^{2+}$$

The mass of iron is calculated from this,

$$0.00237\,\cancel{\text{mol Fe}} \cdot \frac{55.85 \text{ g Fe}}{1\,\cancel{\text{mol Fe}}} = 0.133 \text{ g iron (Fe)}$$

and, finally, the weight percent of iron in the ore can be calculated.

$$\frac{0.133 \text{ g iron}}{1.026 \text{ g sample}} \cdot 100\% = 12.9\% \text{ iron}$$

EXERCISE 5.16 ***Using an Oxidation-Reduction Reaction in a Titration***

Vitamin C, ascorbic acid, has the formula $C_6H_8O_6$. It is a reducing agent in addition to being an acid. One way to determine the ascorbic acid content of a vitamin pill is to react the ascorbic acid with *excess* iodine,

$$C_6H_8O_6(aq) + I_2(aq) \longrightarrow C_6H_6O_6(aq) + 2\ H^+(aq) + 2\ I^-(aq)$$

and then titrate the iodine that did *not* react with the ascorbic acid with sodium thiosulfate. The balanced, net ionic equation for the reaction occurring in the course of the titration is

$$I_2(aq) + 2\ S_2O_3^{2-}(aq) \longrightarrow 2\ I^-(aq) + S_4O_6^{2-}(aq)$$

50.00 mL of 0.0520 M I_2 was added to the sample containing ascorbic acid. After the reaction was complete, the I_2 unused in this reaction required 20.30 mL of 0.196 M $Na_2S_2O_3$ for titration to the equivalence point. Calculate the mass of ascorbic acid in the unknown sample. ■

Oxidation-reduction reactions, as well as acid-base reactions, are clearly useful in chemical analysis. No matter what type of reaction is used in analysis, however, you must remember that *before you use any reaction as a quantitative analytical method, you must know the balanced chemical equation* (or at least the stoichiometric relation between the key reactants and products) *to know the necessary stoichiometric factors.*

CHAPTER HIGHLIGHTS

Having studied this chapter, you should be able to

- Calculate the mass of one reactant or product from the mass of another reactant or product by using the balanced chemical equation (Section 5.1). The following general scheme is used for a reaction such as $xA \rightarrow yB$.

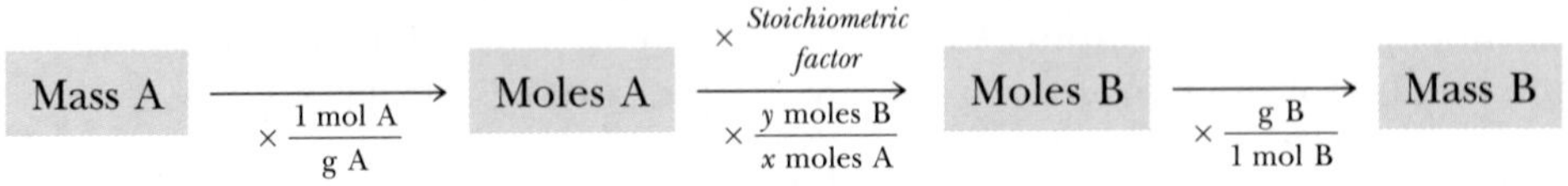

- Determine which of two reactants is the **limiting reactant** (Section 5.2).
- Explain the differences among **actual yield, theoretical yield,** and **percent yield,** and calculate percent yield (Section 5.3).

$$\text{Percent yield} = \frac{\text{actual yield (g)}}{\text{theoretical yield (g)}} \times 100\%$$

- Use stoichiometry principles to analyze a mixture or to find the empirical formula of an unknown compound (Section 5.4).
- Explain **molarity,** and calculate molarity values (Section 5.5). The concentration of a solute in a solution in units of molarity is

$$c_{\text{molarity}} = \frac{\text{moles of solute}}{\text{liters of solution}}$$

  A useful form of this equation is "moles solute = $c_{\text{molarity}} \cdot$ liters of solution."
- Describe how to prepare a solution of a given molarity from the solute and water or by dilution from a more concentrated solution (Section 5.5). When preparing a solution by diluting a more concentrated solution, the number of moles of solute remains the same. The product of the concentration and volume of the more concentrated solution (c) must therefore be the same as that for the diluted solution (d).

$$c_c \cdot V_c = c_d \cdot V_d$$

- Solve stoichiometry problems using solution concentrations (Section 5.6).
- Explain how a **titration** is carried out, explain **standardization,** and calculate concentrations or amounts of reactants from titration data (Section 5.6).

STUDY QUESTIONS

Numerical Questions

General Stoichiometry

1. If 10.0 g of carbon is combined with an exact, stoichiometric amount of oxygen (26.6 g) to get carbon dioxide, what mass, in grams, of CO_2 can be obtained? That is, what is the theoretical yield of CO_2?

2. Assume 16.04 g of methane, CH_4, is burned in oxygen.
   (a) What are the products of the reaction?
   (b) What is the balanced equation for this reaction?
   (c) What mass of O_2, in grams, is required for complete combustion?
   (d) What is the total mass of products expected?

3. If you want to synthesize 1.45 g of the semiconducting material GaAs, what masses of Ga and of As, in grams, are required?

4. Aluminum reacts with oxygen to give aluminum oxide.

$$4\ Al(s) + 3\ O_2(g) \longrightarrow 2\ Al_2O_3(s)$$

If you have 6.0 mol of Al, how many moles of O_2 are needed for complete reaction? What mass of Al_2O_3, in grams, can be produced?

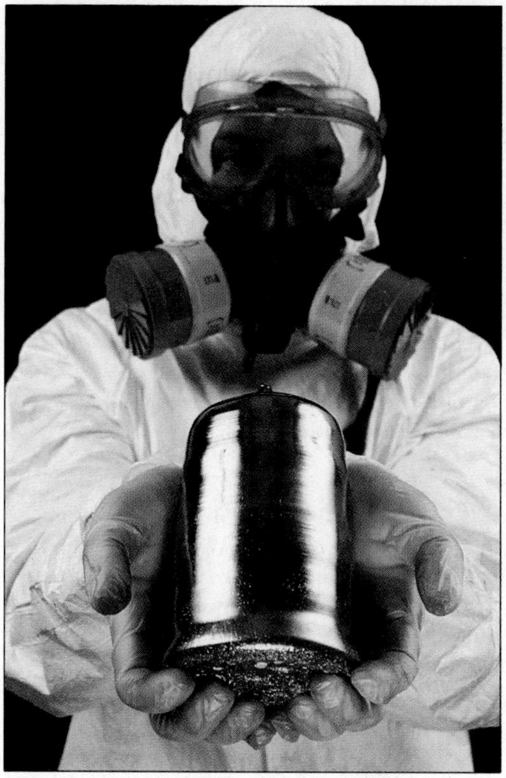

A freshly prepared ingot of gallium arsenide. (See Study Question 3.)

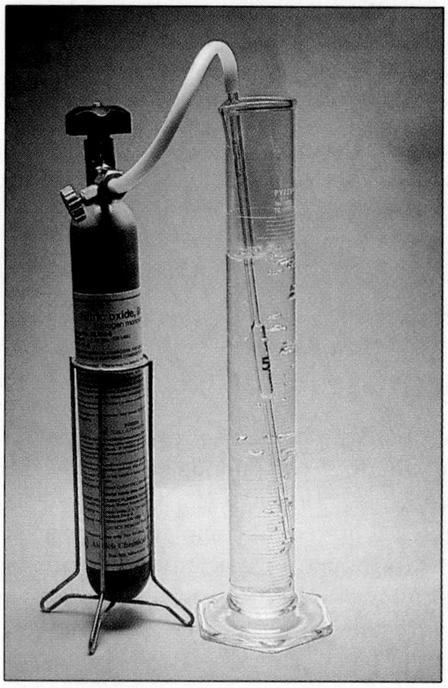

The gas NO is stored in a tank. It is bubbled through water, where it is evident the gas is colorless. As soon as the bubbles of NO enter the atmosphere, however, the NO is oxidized to brown NO_2. See Study Question 7. (C. D. Winters)

5. Cobalt metal reacts with hydrochloric acid according to the following *unbalanced* equation:

$$Co(s) + HCl(aq) \longrightarrow CoCl_2(aq) + H_2(g)$$

If you begin with 2.56 g of cobalt metal and excess hydrochloric acid, what mass of cobalt(II) chloride can be obtained? What mass of hydrogen gas can be obtained?

6. Like many metals, aluminum reacts with a halogen to give a metal halide (see Figure 4.1).

$$2\ Al(s) + 3\ Br_2(\ell) \longrightarrow Al_2Br_6(s)$$

If you begin with 2.56 g of Al, how many grams of Br_2 are required for complete reaction? How many grams of the white solid Al_2Br_6 are expected?

7. Nitrogen monoxide is oxidized in air to give brown nitrogen dioxide.

$$2\ NO(g) + O_2(g) \longrightarrow 2\ NO_2(g)$$

Starting with 2.2 mol of NO, how many moles and how many grams of O_2 are required for complete reaction? What mass of NO_2, in grams, is produced?

8. The final step in the manufacture of platinum metal (for use in automotive catalytic converters and other purposes) is the reaction

$$3\ (NH_4)_2PtCl_6(s) \longrightarrow 3\ Pt(s) + 2\ NH_4Cl(s) + 2\ N_2(g) + 16\ HCl(g)$$

If 12.35 g of $(NH_4)_2PtCl_6$ is heated, what mass of platinum metal, in grams, is expected? What mass of HCl is obtained as well?

9. The equation for one of the reactions in the process of reducing iron ore to the metal is

$$Fe_2O_3(s) + 3\ CO(g) \longrightarrow 2\ Fe(s) + 3\ CO_2(g)$$

(a) What is the maximum mass of iron, in grams, that can be obtained from 454 g (1.00 lb) of iron(III) oxide?
(b) What mass of CO is required to reduce the iron(III) oxide to iron metal?

10. Many metal halides react with water to produce the metal oxide (or hydroxide) and the appropriate hydrogen halide. For example,

$$TiCl_4(\ell) + 2\ H_2O(g) \longrightarrow TiO_2(s) + 4\ HCl(g)$$

(See the photograph below of this reaction.)
(a) Name the four compounds involved in this reaction.
(b) If you begin with 14.0 g of $TiCl_4$, what mass of water, in grams, is required for complete reaction, and what mass of each product is expected?

11. Gaseous sulfur dioxide, SO_2, can be removed from smokestacks by treatment with limestone and oxygen (see Figure 4.16).

$$2\ SO_2(g) + 2\ CaCO_3(s) + O_2(g) \longrightarrow 2\ CaSO_4(s) + 2\ CO_2(g)$$

(a) Name the compounds involved in this reaction.
(b) What mass of $CaCO_3$ is required to remove 150. g of SO_2?
(c) What mass of $CaSO_4$ is formed when 150. g of SO_2 is consumed completely?

Titanium tetrachloride, $TiCl_4$, is a liquid at standard temperature and pressure. When exposed to humid air it forms a dense fog of titanium(IV) oxide, TiO_2. (See Study Question 10.) (C. D. Winters)

12. Iron reacts with oxygen to give iron(III) oxide, Fe_2O_3.
(a) Write a balanced equation for this reaction.
(b) If an ordinary iron nail (assumed to be pure iron) has a mass of 5.58 g, what mass (in grams) of Fe_2O_3 does it produce if the nail is converted completely to this oxide?
(c) What mass of O_2 (in grams) is required for the reaction?

13. Careful decomposition of ammonium nitrate, NH_4NO_3, gives laughing gas (dinitrogen monoxide, N_2O) and water.
(a) Write a balanced equation for this reaction.
(b) What masses of N_2O and water can be obtained from 10.0 g of NH_4NO_3?

14. The cancer chemotherapy drug cisplatin, $Pt(NH_3)_2Cl_2$, can be made by reacting $(NH_4)_2PtCl_4$ with ammonia, NH_3. Besides cisplatin, the other product is NH_4Cl.
(a) Write a balanced equation for this reaction.
(b) To obtain 2.50 g of cisplatin, how many grams each of $(NH_4)_2PtCl_4$ and ammonia do you need?

Limiting Reactant

15. Aluminum chloride, $AlCl_3$, is an inexpensive reagent used in many industrial processes. It is made by treating scrap aluminum with chlorine according to the following balanced equation:

$$2\ Al(s) + 3\ Cl_2(g) \longrightarrow 2\ AlCl_3(s)$$

(a) Which reactant is limiting if 2.70 g of Al and 4.05 g of Cl_2 are mixed?
(b) What mass of $AlCl_3$ can be produced?
(c) What mass of the excess reactant remains when the reaction is completed?

16. The reaction of methane and water is one way of preparing hydrogen.

$$CH_4(g) + H_2O(g) \longrightarrow CO(g) + 3\ H_2(g)$$

If you begin with 995 g of CH_4 and 2510 g of water, what is the maximum possible yield of H_2?

17. Methanol, CH_3OH, is a clean-burning, easily handled fuel. It can be made by the direct reaction of CO and H_2 (obtained from heating coal with steam).

$$CO(g) + 2\ H_2(g) \longrightarrow CH_3OH(\ell)$$

Starting with a mixture of 12.0 g of H_2 and 74.5 g of CO, which is the limiting reactant? What mass of the excess reactant (in grams) remains after reaction is complete? What is the theoretical yield of methanol?

18. Disulfur dichloride, S_2Cl_2, is used to vulcanize rubber. It can be made by treating molten sulfur with gaseous chlorine.

$$S_8(\ell) + 4\ Cl_2(g) \longrightarrow 4\ S_2Cl_2(g)$$

Starting with a mixture of 32.0 g of sulfur and 71.0 g of Cl_2, which is the limiting reactant? What mass of S_2Cl_2 (in grams) can be produced? What mass of the excess reactant remains when the limiting reactant is consumed?

19. Ammonia gas can be prepared by the reaction of a basic oxide like calcium oxide with ammonium chloride, an acidic salt.

$$CaO(s) + 2\ NH_4Cl(s) \longrightarrow 2\ NH_3(g) + H_2O(g) + CaCl_2(s)$$

If 112 g of CaO and 224 g of NH_4Cl are mixed, what is the maximum possible yield of NH_3? What mass of the excess reactant remains after the maximum amount of ammonia has been formed?

20. Aspirin is produced by the reaction of salicylic acid and acetic anhydride (page 220).

$$\underset{\text{salicylic acid}}{C_7H_6O_3(s)} + \underset{\text{acetic anhydride}}{C_4H_6O_3(\ell)} \longrightarrow \underset{\text{aspirin}}{C_9H_8O_4(s)} + CH_3CO_2H(aq)$$

If you mix 100. g of each of the reactants, what is the maximum mass of aspirin that can be obtained?

Percent Yield

21. Ammonia gas can be prepared by the following reaction:

$$CaO(s) + 2\ NH_4Cl(s) \longrightarrow 2\ NH_3(g) + H_2O(g) + CaCl_2(s)$$

If 100. g of ammonia is obtained, but the theoretical yield is 136 g, what is the percent yield of this gas?

22. Diborane, B_2H_6, is a valuable compound in the synthesis of new organic compounds. One of the several ways this boron compound can be made is by the reaction

$$2\ NaBH_4(s) + I_2(s) \longrightarrow B_2H_6(g) + 2\ NaI(s) + H_2(g)$$

Suppose you use 1.203 g of $NaBH_4$ and excess iodine and obtain 0.295 g of B_2H_6. What is the percent yield of B_2H_6?

23. The reaction of zinc and chlorine has been used as the basis of a car battery.

$$Zn(s) + Cl_2(g) \longrightarrow ZnCl_2(s)$$

What is the theoretical yield of $ZnCl_2$ if 35.5 g of zinc is allowed to react with excess chlorine? If only 65.2 g of zinc chloride is obtained, what is the percent yield of the compound?

24. Disulfur dichloride, which has a revolting smell, can be prepared by directly combining S_8 and Cl_2, but it can also be made by the following reaction:

$$3\ SCl_2(\ell) + 4\ NaF(s) \longrightarrow SF_4(g) + S_2Cl_2(\ell) + 4\ NaCl(s)$$

Assume you begin with 5.23 g of SCl_2 and excess NaF. What is the theoretical yield of S_2Cl_2? If only 1.19 g of S_2Cl_2 is obtained, what is the percent yield of the compound?

Chemical Analysis

25. A mixture of $CuSO_4$ and $CuSO_4 \cdot 5\ H_2O$ has a mass of 1.245 g, but after heating to drive off all the water, the mass is only 0.832 g. What is the weight percent of $CuSO_4 \cdot 5\ H_2O$ in the mixture?

26. A sample of limestone and other soil materials is heated, and the limestone decomposes to give calcium oxide and carbon dioxide. A 1.506-g sample of limestone-containing material gives 0.711 g of CaO, in addition to gaseous CO_2, after being heated at a high temperature. What is the weight percent of $CaCO_3$ in the original sample?

27. A 1.25-g sample contains some of the very reactive com-

Dehydrating hydrated copper sulfate. (See Study Question 25.) (C. D. Winters)

pound $Al(C_6H_5)_3$. On treating the compound with aqueous HCl, 0.951 g of C_6H_6 is obtained.

$$Al(C_6H_5)_3(s) + 3\ HCl(aq) \longrightarrow AlCl_3(aq) + 3\ C_6H_6(\ell)$$

Assuming that $Al(C_6H_5)_3$ was converted completely to products, what is the weight percent of $Al(C_6H_5)_3$ in the original 1.25-g sample?

28. Bromine trifluoride reacts with metal oxides to evolve oxygen quantitatively. For example,

$$3\ TiO_2(s) + 4\ BrF_3(\ell) \longrightarrow 3\ TiF_4(s) + 2\ Br_2(\ell) + 3\ O_2(g)$$

Suppose you wish to use this reaction to determine the weight percent of TiO_2 in a sample of ore. To do this the O_2 gas from the reaction is collected. If 2.367 g of the TiO_2-containing ore evolves 0.143 g of O_2, what is the weight percent of TiO_2 in the sample?

Determination of Empirical Formulas

29. Styrene, the building block of polystyrene, is a hydrocarbon, a compound consisting only of C and H. If 0.438 g of styrene is burned in oxygen and produces 1.481 g of CO_2 and 0.303 g of H_2O, what is the empirical formula of styrene?

30. Mesitylene is a liquid hydrocarbon. If 0.115 g of the compound is burned in oxygen to give 0.379 g of CO_2 and 0.1035 g of H_2O, what is the empirical formula of mesitylene?

31. Propanoic acid, an organic acid, contains only C, H, and O. If 0.236 g of the acid burns completely in O_2, and gives 0.421 g of CO_2 and 0.172 g of H_2O, what is the empirical formula of the acid?

32. Quinone, a chemical used in the dye industry and in photography, is an organic compound containing only C, H, and O. What is the empirical formula of the compound if 0.105 g of the compound gives 0.257 g of CO_2 and 0.0350 g of H_2O when burned completely in oxygen?

33. Silicon and hydrogen form a series of compounds with the general formula Si_xH_y. To find the formula of one of them, a 6.22-g sample of the compound is burned in oxygen. On doing so, all of the Si is converted to 11.64 g of SiO_2 and all of the H to 6.980 g of H_2O. What is the empirical formula of the silicon compound?

34. To find the formula of a compound composed of iron and carbon monoxide, $Fe_x(CO)_y$, the compound is burned in pure oxygen, a reaction that proceeds according to the following, *unbalanced* equation:

$$Fe_x(CO)_y(s) + O_2(g) \longrightarrow Fe_2O_3(s) + CO_2(g)$$

If you burn 1.959 g of $Fe_x(CO)_y$ and obtain 0.799 g of Fe_2O_3 and 2.200 g of CO_2, what is the empirical formula of $Fe_x(CO)_y$?

Solution Concentration

35. Assume 6.73 g of Na_2CO_3 is dissolved in enough water to make 250. mL of solution. What is the molarity of the sodium carbonate? What are the molar concentrations of the Na^+ and CO_3^{2-} ions?

36. Some $K_2Cr_2O_7$, 2.335 g, is dissolved in enough water to make 500. mL of solution. What is the molarity of the potassium dichromate? What are the molar concentrations of the K^+ and $Cr_2O_7^{2-}$ ions?

37. What is the mass, in grams, of solute in 250. mL of a 0.0125 M solution of $KMnO_4$?

38. What is the mass, in grams, of solute in 100. mL of a 1.023×10^{-3} M solution of Na_3PO_4? What are the molar concentrations of the Na^+ and PO_4^{3-} ions?

39. What volume of 0.123 M NaOH, in milliliters, contains 25.0 g of NaOH?

40. What volume of 2.06 M $KMnO_4$, in liters, contains 322 g of solute?

41. If 4.00 mL of 0.0250 M $CuSO_4$ is diluted to 10.0 mL with pure water, what is the molarity of copper(II) sulfate in the diluted solution?

42. If you dilute 25.0 mL of 1.50 M hydrochloric acid to 500. mL, what is the molar concentration of the diluted acid?

43. If you need 1.00 L of 0.125 M H_2SO_4, which method below do you use to prepare this solution? Calculate the concentration of the sulfuric acid in each of the other cases as well.
(a) Dilute 36.0 mL of 1.25 M H_2SO_4 to a volume of 1.00 L.
(b) Dilute 20.8 mL of 6.00 M H_2SO_4 to a volume of 1.00 L.
(c) Add 950. mL of water to 50.0 mL of 3.00 M H_2SO_4.
(d) Add 500. mL of water to 500. mL of 0.500 M H_2SO_4.

44. If you need 300. mL of 0.500 M $K_2Cr_2O_7$, which method below do you use to prepare this solution? Calculate the concentration of the potassium dichromate in each of the other cases as well.
(a) Dilute 250. mL of 0.600 M $K_2Cr_2O_7$ to a volume of 300. mL.
(b) Add 50.0 mL of water to 250. mL of 0.250 M $K_2Cr_2O_7$.
(c) Dilute 125. mL of 1.00 M $K_2Cr_2O_7$ to a volume of 300. mL.
(d) Add 30.0 mL of 1.50 M $K_2Cr_2O_7$ to 270. mL of water.

45. For each solution tell what ions exist in aqueous solution, and give the concentration of each.
(a) 0.12 M $BaCl_2$ (c) 0.146 M $AlCl_3$
(b) 0.0125 M $CuSO_4$ (d) 0.500 M $K_2Cr_2O_7$

46. For each solution tell what ions exist in aqueous solution, and give the concentration of each.
(a) 0.25 M $(NH_4)_2SO_4$ (c) 0.056 M HNO_3
(b) 0.123 M Na_2CO_3 (d) 0.00124 M $KClO_4$

Stoichiometry of Reactions in Solution

47. How many grams of Na_2CO_3 are required for complete reaction with 25.0 mL of 0.155 M HNO_3?

$$Na_2CO_3(aq) + 2\ HNO_3(aq) \longrightarrow 2\ NaNO_3(aq) + CO_2(g) + H_2O(\ell)$$

48. How many milliliters of 0.125 M HNO_3 are required to react completely with 1.30 g of $Ba(OH)_2$?

$$2\ HNO_3(aq) + Ba(OH)_2(s) \longrightarrow Ba(NO_3)_2(aq) + 2\ H_2O(\ell)$$

49. One of the most important industrial processes in our economy is the electrolysis of brine solutions (aqueous solutions of NaCl). When an electric current is passed through an aqueous solution of salt, the NaCl and water produce $H_2(g)$, $Cl_2(g)$, and NaOH—all valuable industrial chemicals.

$$2\ NaCl(aq) + 2\ H_2O(\ell) \longrightarrow H_2(g) + Cl_2(g) + 2\ NaOH(aq)$$

What mass of NaOH can be formed from 10.0 L of 0.15 M NaCl? What mass of chlorine can be obtained?

50. One of the several ways diborane, B_2H_6, can be prepared is by the following reaction:

$$2\ NaBH_4(aq) + H_2SO_4(aq) \longrightarrow 2\ H_2(g) + Na_2SO_4(aq) + B_2H_6(g)$$

How many milliliters of 0.0875 M H_2SO_4 should be used to consume completely 1.35 g of $NaBH_4$? What mass of B_2H_6 can be obtained?

51. In the photographic developing process silver bromide is dissolved by adding sodium thiosulfate.

$$AgBr(s) + 2\ Na_2S_2O_3(aq) \longrightarrow Na_3Ag(S_2O_3)_2(aq) + NaBr(aq)$$

If you want to dissolve 0.250 g of AgBr, how many milliliters of 0.0138 M $Na_2S_2O_3$ should you add?

52. Hydrazine, N_2H_4, a base like ammonia, can react with an acid such as sulfuric acid.

$$2\ N_2H_4(aq) + H_2SO_4(aq) \longrightarrow 2\ N_2H_5^+(aq) + SO_4^{2-}(aq)$$

What mass of hydrazine reacts with 250. mL of 0.225 M H_2SO_4?

53. How many milliliters of 0.750 M $Pb(NO_3)_2$ solution are required to react completely with 1.00 L of 2.25 M NaCl solution? The balanced equation is

$$Pb(NO_3)_2(aq) + 2\ NaCl(aq) \longrightarrow PbCl_2(s) + 2\ NaNO_3(aq)$$

54. What volume, in milliliters, of 0.512 M NaOH is required to react completely with 25.0 mL of 0.234 M H_2SO_4?

55. What is the maximum mass, in grams, of AgCl that can be precipitated by mixing 50.0 mL of 0.025 M $AgNO_3$ solution with 100.0 mL of 0.025 M NaCl solution? Which reactant is in excess? What is the molar concentration of the excess reactant remaining in solution after the maximum mass of AgCl has been precipitated? (Assume the total volume after mixing is 150.0 mL.)

56. Suppose you mix 25.0 mL of 0.234 M $FeCl_3$ solution with 42.5 mL of 0.453 M NaOH. What is the maximum mass, in grams, of $Fe(OH)_3$ that precipitates? Which reactant is in excess? What is the molar concentration of the excess reactant remaining in solution after the maximum mass of $Fe(OH)_3$ has been precipitated?

Titrations

57. How many milliliters of 0.812 M HCl are required to titrate 1.33 g of NaOH to the equivalence point?

$$NaOH(aq) + HCl(aq) \longrightarrow NaCl(aq) + H_2O(\ell)$$

(a) (b)

(a) A precipitate of AgBr formed by adding $AgNO_3(aq)$ to $KBr(aq)$. (b) On adding $Na_2S_2O_3(aq)$, sodium thiosulfate, the solid AgBr dissolves. (See Study Question 51.) (C. D. Winters)

58. If 32.45 mL of HCl is used to titrate 2.050 g of Na_2CO_3 according to the following equation,

$$Na_2CO_3(aq) + 2\ HCl(aq) \longrightarrow 2\ NaCl(aq) + CO_2(g) + H_2O(\ell)$$

what is the molarity of the HCl?

59. How many milliliters of 0.955 M HCl are needed to titrate 2.152 g of Na_2CO_3 to the equivalence point?

$$Na_2CO_3(aq) + 2\ HCl(aq) \longrightarrow 2\ NaCl(aq) + CO_2(g) + H_2O(\ell)$$

60. Potassium acid phthalate, $KHC_8H_4O_4$ (molar mass = 204.22 g/mol), is used to standardize solutions of bases. The acidic anion reacts with strong bases (such as NaOH or KOH) according to the following net ionic equation:

$$HC_8H_4O_4^-(aq) + OH^-(aq) \longrightarrow C_8H_4O_4^{2-}(aq) + H_2O(\ell)$$

If a 0.902-g sample of potassium acid phthalate is dissolved in water and titrated to the equivalence point with 26.45 mL of NaOH, what is the molarity of the NaOH?

61. A noncarbonated soft drink contains an unknown amount of citric acid, $C_6H_8O_7$. If 100. mL of the soft drink requires 33.51 mL of 0.0102 M NaOH to neutralize the citric acid completely, how many grams of citric acid does the soft drink contain per 100. mL? The reaction of citric acid and NaOH is

$$H_3C_6H_5O_7(aq) + 3\ NaOH(aq) \longrightarrow Na_3C_6H_5O_7(aq) + 3\ H_2O(\ell)$$

62. Suppose you are given a 4.554-g sample that is a mixture of oxalic acid, $H_2C_2O_4$, and another solid that does not react with sodium hydroxide. If 29.58 mL of 0.550 M NaOH is required to titrate the oxalic acid in the 4.554-g sample to the equivalence point, what is the weight percent of oxalic acid in the mixture? Oxalic acid and NaOH react according to the equation

$$H_2C_2O_4(aq) + 2\ NaOH(aq) \longrightarrow Na_2C_2O_4(aq) + 2\ H_2O(\ell)$$

63. You are given a solid acid and told only that it could be citric acid or tartaric acid. To determine which acid you have you titrate a sample of the solid with NaOH. The appropriate reactions are:

Citric acid: $$H_3C_6H_5O_7(aq) + 3\ NaOH(aq) \longrightarrow Na_3C_6H_5O_7(aq) + 3\ H_2O(\ell)$$

Tartaric acid: $$H_2C_4H_4O_6(aq) + 2\ NaOH(aq) \longrightarrow Na_2C_4H_4O_6(aq) + 2\ H_2O(\ell)$$

You find that a 0.956-g sample requires 29.1 mL of 0.513 M NaOH for titration to the equivalence point. What is the unknown acid?

64. You have 0.954 g of an unknown acid, H_2A, which reacts with NaOH according to the balanced equation

$$H_2A(aq) + 2\ NaOH(aq) \longrightarrow Na_2A(aq) + 2\ H_2O(\ell)$$

If 36.04 mL of 0.509 M NaOH is required to titrate the acid to the equivalence point, what is the molar mass of the acid?

65. Vitamin C is the simple compound $C_6H_8O_6$. Besides being an acid, it is also a reducing agent. One method for determining the amount of vitamin C in a sample therefore is to titrate it with a solution of bromine, Br_2, a good oxidizing agent.

$$C_6H_8O_6(aq) + Br_2(aq) \longrightarrow 2\ HBr(aq) + C_6H_6O_6(aq)$$

Suppose a 1.00-g "chewable" vitamin C tablet requires 27.85 mL of 0.102 M Br_2 for titration to the equivalence point. How many grams of vitamin C are in the tablet?

66. To analyze an iron-containing compound, you convert all the iron to Fe^{2+} in aqueous solution and then titrate the solution with aqueous $KMnO_4$ according to the following balanced, net ionic equation:

$$MnO_4^-(aq) + 5\ Fe^{2+}(aq) + 8\ H^+(aq) \longrightarrow Mn^{2+}(aq) + 5\ Fe^{3+}(aq) + 4\ H_2O(\ell)$$

If a 0.598-g sample of the iron-containing compound requires 22.25 mL of 0.0123 M $KMnO_4$ for titration to the equivalence point, what is the weight percent of iron in the compound?

GENERAL QUESTIONS

67. Many metals react with halogens to give metal halides. For example, iron gives iron(III) chloride, $FeCl_3$, on reaction with chlorine gas.

The reaction of iron and chlorine. When hot steel wool is plunged into a flask filled with yellow chlorine gas, the steel wool glows brightly, and a cloud of brown iron(III) chloride forms. (C. D. Winters)

(a) Write the balanced chemical equation for the reaction.
(b) Beginning with 10.0 g of iron, what mass of Cl_2, in grams, is required for complete reaction? What quantity of $FeCl_3$, in moles and in grams, can be produced?
(c) If only 18.5 g of $FeCl_3$ is obtained, what is the percent yield?

68. Nitrogen gas can be prepared in the laboratory by the reaction of ammonia with copper(II) oxide according to the following *unbalanced* equation.

$$NH_3(g) + CuO(s) \longrightarrow N_2(g) + Cu(s) + H_2O(g)$$

If 26.3 g of gaseous NH_3 is passed over an excess of solid CuO, what mass of N_2, in grams, can be obtained?

69. In an experiment 1.056 g of a metal carbonate, containing an unknown metal M, is heated to give the metal oxide and 0.376 g CO_2.

$$MCO_3(s) + \text{heat} \longrightarrow MO(s) + CO_2(g)$$

What is the identity of the metal M?
(a) M = Ni (c) M = Zn
(b) M = Cu (d) M = Ba

70. An unknown metal reacts with oxygen to give the metal oxide, MO_2. Identify the metal based on the following information:

Mass of metal = 0.356 g

Mass of sample after converting metal completely to oxide = 0.452 g

71. The cancer chemotherapy agent, cisplatin, is made by the following reaction:

$$(NH_4)_2PtCl_4(s) + 2\ NH_3(aq) \longrightarrow 2\ NH_4Cl(aq) + Pt(NH_3)_2Cl_2(s)$$

Assume that 15.5 g of $(NH_4)_2PtCl_4$ is combined with 225 mL of 0.75 M NH_3 to make cisplatin.
(a) Which reactant is in excess and which is the limiting reactant?
(b) How many grams of cisplatin can be formed?
(c) After all the limiting reactant has been consumed and the maximum quantity of cisplatin has been formed, how many grams of the other reactant remain?

72. Uranium(VI) oxide reacts with bromine trifluoride to give uranium(IV) fluoride, an important step in the purification of uranium ore.

$$6\ UO_3(s) + 8\ BrF_3(\ell) \longrightarrow 6\ UF_4(s) + 4\ Br_2(\ell) + 9\ O_2(g)$$

If you begin with 365 g each of UO_3 and BrF_3, what is the maximum yield, in grams, of UF_4?

73. The important industrial chemical carbon tetrachloride, CCl_4, is made by the reaction of carbon disulfide with chlorine according to the following *unbalanced* equation:

$$CS_2(g) + Cl_2(g) \longrightarrow S_2Cl_2(\ell) + CCl_4(\ell)$$

What is the theoretical yield of CCl_4 if 125 g of CS_2 is mixed with 435 g of Cl_2? What mass of excess reactant remains when the reaction has gone to completion?

74. Aluminum bromide is a valuable laboratory chemical. What is the theoretical yield in grams of Al_2Br_6 if 25.0 mL of liquid bromine (density = 3.10 g/mL) and 12.5 g aluminum metal are used? Is any aluminum or bromine left over when the reaction has gone to completion? If so, what mass of which reactant remains?

$$2\ Al(s) + 3\ Br_2(\ell) \longrightarrow Al_2Br_6(s)$$

75. Boron forms an extensive series of compounds with hydrogen, all with the general formula B_xH_y. To analyze one of these compounds you burn it in air and isolate the boron in the form of B_2O_3 and the hydrogen in the form of water according to the following *unbalanced* equation:

$$B_xH_y(s) + \text{excess } O_2(g) \longrightarrow B_2O_3(s) + H_2O(g)$$

If 0.148 g of B_xH_y gives 0.422 g of B_2O_3 when burned in excess O_2, what is the empirical formula of B_xH_y?

76. Some compounds can be decomposed quantitatively with water or acid to give known compounds. Suppose you have a 0.643-g sample of a compound known to be composed of C, H, Al, and Cl. Furthermore, you know that it is composed of some number of CH_3 groups and chlorine atoms per aluminum atom. The formula could be written as $(CH_3)_xAlCl_y$. To find x and y you decompose the sample with acid in water. The CH_3 portion is evolved as methane gas, CH_4, and the aluminum and chloride ions remain in the water. The chloride ions are precipitated as AgCl by adding $AgNO_3$ to the solution. The data collected in the experiment are given here. What are the values of x and y?

$$\underset{0.643\text{ g}}{(CH_3)_xAlCl_y} \xrightarrow{\text{water}} x\ \underset{0.222\text{ g}}{CH_4(g)} + Al^{3+}(aq) + y\ Cl^-(aq)$$

$$Cl^-(aq) \xrightarrow{AgNO_3(aq)} \underset{0.996\text{ g}}{AgCl(s)}$$

77. What are the concentrations of ions in a solution made by diluting 10.0 mL of 2.56 M HCl with water to obtain 250. mL of solution?

78. One-half liter (500. mL) of 2.50 M HCl is mixed with 250. mL of 3.75 M HCl. Assuming the total solution volume after mixing is 750. mL, what is the concentration of hydrochloric acid in the resulting solution?

79. Diborane, B_2H_6, can be produced by the following reaction:

$$2\ NaBH_4(aq) + H_2SO_4(aq) \longrightarrow 2\ H_2(g) + Na_2SO_4(aq) + B_2H_6(g)$$

What is the maximum yield, in grams, of B_2H_6 that can be prepared starting with 250. mL of 0.0875 M H_2SO_4 and 1.55 g of $NaBH_4$?

80. Sodium thiosulfate, $Na_2S_2O_3$, is used as a "fixer" in black-and-white photography. Assume you have a bottle of sodium thiosulfate and want to determine its purity. The thiosulfate ion can be oxidized with I_2 according to the equation

$$I_2(aq) + 2\ S_2O_3^{2-}(aq) \longrightarrow 2\ I^-(aq) + S_4O_6^{2-}(aq)$$

This reaction occurs rapidly and quantitatively, and a simple method exists for observing when the reaction has reached the equivalence point. It can therefore be used as a method of analysis by titration. If you use 40.21 mL of 0.246 M I_2 in a titration, what is the weight percent of $Na_2S_2O_3$ in a 3.232-g sample of impure $Na_2S_2O_3$?

81. The lead content of a sample can be estimated by converting the lead to PbO_2

$$\text{Pb in sample} + \text{oxidizing agent} \longrightarrow PbO_2(s)$$

and then dissolving the PbO_2 in an acid solution of KI. This liberates I_2 according to the equation

$$PbO_2(s) + 4\ H^+(aq) + 2\ I^-(aq) \longrightarrow Pb^{2+}(aq) + I_2(aq) + 2\ H_2O(\ell)$$

The liberated I_2 is then titrated with $Na_2S_2O_3$ (see Study question 80).

$$I_2(aq) + 2\ S_2O_3^{2-}(aq) \longrightarrow 2\ I^-(aq) + S_4O_6^{2-}(aq)$$

The amount of titrated I_2 is related to the amount of lead in the sample. If 0.576 g of a lead-containing mineral requires 35.23 mL of 0.0500 M $Na_2S_2O_3$ for titration of the liberated I_2, what is the weight percent of lead in the mineral?

82. Gold can be dissolved from gold-bearing rock by treating the rock with sodium cyanide in the presence of oxygen.

$$4\ Au(s) + 8\ NaCN(aq) + O_2(g) + 2\ H_2O(\ell) \longrightarrow 4\ NaAu(CN)_2(aq) + 4\ NaOH(aq)$$

If you have exactly one metric ton (1 metric ton = 1000 kg) of gold-bearing rock, what volume of 0.075 M NaCN, in liters, do you need to extract the gold if the rock is 0.019% gold?

83. 25.0 mL of 1.50 M HNO_3 is added to 50.0 mL of 2.50 M NaOH. Which of the following answers best describes the composition of the mixture when the reaction between the HNO_3 and the NaOH is complete?
(a) $[HNO_3] = 0.500$ M; $[NaOH] = 1.17$ M; $[NaNO_3] = 0.500$ M
(b) $[HNO_3] = 0$ M; $[NaOH] = 1.17$ M; $[NaNO_3] = 0.500$ M
(c) $[HNO_3] = 1.25$ M; $[NaOH] = 0.750$ M; $[NaNO_3] = 4.00$ M
(d) $[HNO_3] = 0.015$ M; $[NaOH] = 0$ M; $[NaNO_3] = 1.67$ M

84. Cobalt(III) ion forms many compounds with ammonia. To find the formula of one of these compounds, you titrate the NH_3 in the compound with standardized acid.

$$Co(NH_3)_xCl_3(aq) + x\ HCl(aq) \longrightarrow x\ NH_4^+(aq) + Co^{3+}(aq) + (x+3)Cl^-(aq)$$

Assume that 23.63 mL of 1.500 M HCl is used to titrate 1.580 g of $Co(NH_3)_xCl_3$. What is the value of x?

85. You have done a laboratory experiment in which you obtained a new iron-containing compound. Its formula could be one of two possibilities: $K_3[Fe(C_2O_4)_3]$ or $K[Fe(C_2O_4)_2(H_2O)_2]$. (In each of these, the $C_2O_4^{2-}$ ion is the oxalate ion.) To find which is correct, you dissolve 1.356 g of the compound in acid, which converts the oxalate ion to oxalic acid,

$$2\ H^+(aq) + C_2O_4^{2-}(aq) \longrightarrow H_2C_2O_4(aq)$$

and then titrate the oxalic acid with potassium permanganate. The balanced, net ionic equation for the titration is

$$5\ H_2C_2O_4(aq) + 2\ MnO_4^-(aq) + 6\ H^+(aq) \longrightarrow 2\ Mn^{2+}(aq) + 10\ CO_2(g) + 8\ H_2O(\ell)$$

The titration requires 34.50 mL of 0.108 M $KMnO_4$. What is the correct formula of the iron-containing compound?

86. You wish to determine the weight percent of copper in a copper-containing alloy. After dissolving a sample of an alloy in acid, an excess of KI is added, and the Cu^{2+} and I^- ions undergo the reaction

$$2\ Cu^{2+}(aq) + 5\ I^-(aq) \longrightarrow 2\ CuI(s) + I_3^-(aq)$$

The liberated I_3^- is titrated with sodium thiosulfate according to the equation

$$I_3^-(aq) + 2\ S_2O_3^{2-}(aq) \longrightarrow S_4O_6^{2-}(aq) + 3\ I^-(aq)$$

If 26.32 mL of 0.101 M $Na_2S_2O_3$ is required for titration to the equivalence point, what is the weight percent of Cu in 0.251 g of the alloy?

87. Potassium perchlorate, $KClO_4$, is made in the following sequence of reactions:

$$Cl_2(g) + 2\ KOH(aq) \longrightarrow KCl(aq) + KClO(aq) + H_2O(\ell)$$
$$3\ KClO(aq) \longrightarrow 2\ KCl(aq) + KClO_3(aq)$$
$$4\ KClO_3(aq) \longrightarrow 3\ KClO_4(aq) + KCl(aq)$$

What mass of Cl_2 is needed to produce 1.0 kg of $KClO_4$?

88. The elements silver, molybdenum, and sulfur are combined to form Ag_2MoS_4. What is the maximum mass of Ag_2MoS_4 that can be obtained if 8.63 g of silver, 3.36 g of molybdenum, and 4.81 g of sulfur are combined?

89. Titanium(IV) oxide TiO_2, is heated in hydrogen gas to give water and a new titanium oxide, Ti_xO_y. If 1.598 g of TiO_2 produces 1.438 g of Ti_xO_y, what is the formula of the new oxide?

90. A mixture of the hydrocarbons C_8H_{18} and C_9H_{20} has a mass of 0.640 g. When the mixture is burned in O_2 and has been converted completely to CO_2 and H_2O, the mass of water collected is 0.904 g. What was the percent by weight of C_8H_{18} in the original mixture?

91. It is known that 1.2 g of element A reacts with exactly 3.2 g of oxygen to form an oxide, AO_x; 2.4 g of element A reacts with exactly 3.2 g of oxygen to form a second oxide AO_y.
(a) What is the ratio x/y?
(b) If $x = 2$, what is the identity of element A?

Conceptual Questions

92. A weighed sample of a metal is added to liquid bromine and allowed to react completely. The product substance is then separated from any leftover reactants and weighed. This experiment is repeated with several masses of the metal but with the same volume of bromine. The graph indicates the results. Explain why the graph has the shape that it does.

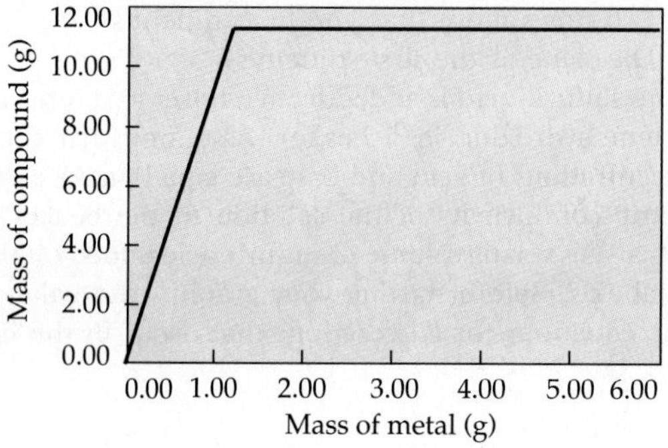

93. A series of experimental measurements like the ones described in Study Question 92 is carried out for the reaction of iron with bromine. The graph obtained is shown here. What is the empirical formula of the compound formed between iron and bromine? Write a balanced equation for the reaction between iron and bromine and name the product.

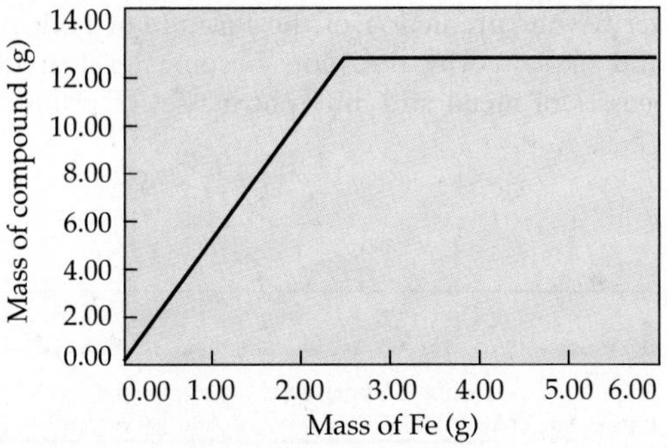

94. Ammonia is formed in a direct reaction of nitrogen and hydrogen.

$$N_2(g) + 3\ H_2(g) \longrightarrow 2\ NH_3(g)$$

The starting mixture is represented by the diagram, in which the blue circles represent N and the white circles represent H.

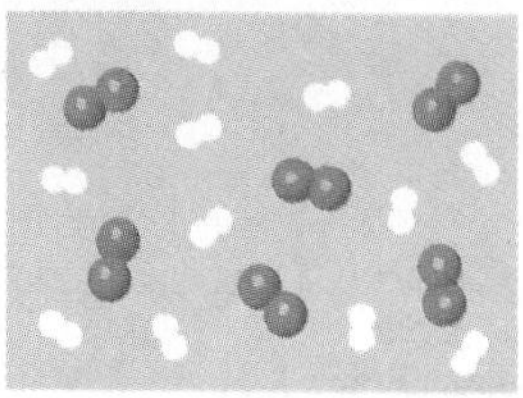

Which of the following represents the product mixture?

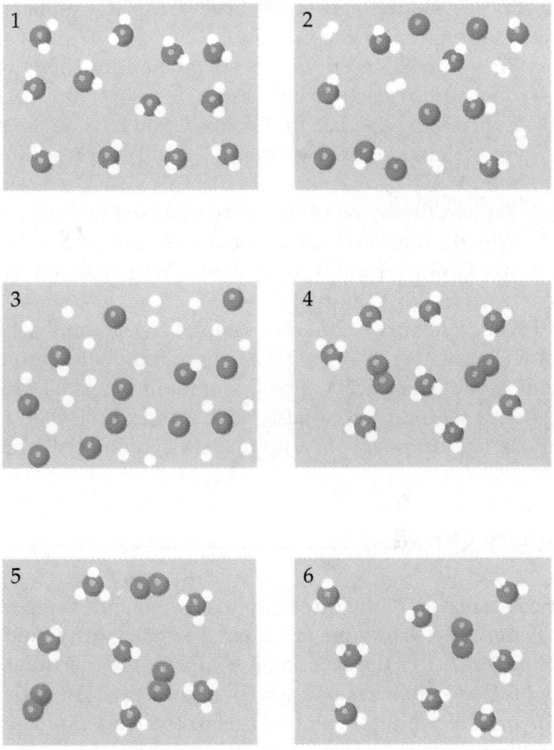

Which of the following is true?

(a) N_2 is the limiting reactant.
(b) H_2 is the limiting reactant.
(c) NH_3 is the limiting reactant.
(d) No reactant is limiting; they are present in the correct stoichiometric ratio.

95. Four groups of students from an introductory chemistry laboratory are studying the reactions of solutions of alkali metal halides with aqueous silver nitrate, $AgNO_3$.

Group A: NaCl   Group C: NaBr
Group B: KCl   Group D: KBr

All four groups dissolve 0.0040 mol of their particular salt in some water. Each then adds various masses of silver nitrate,

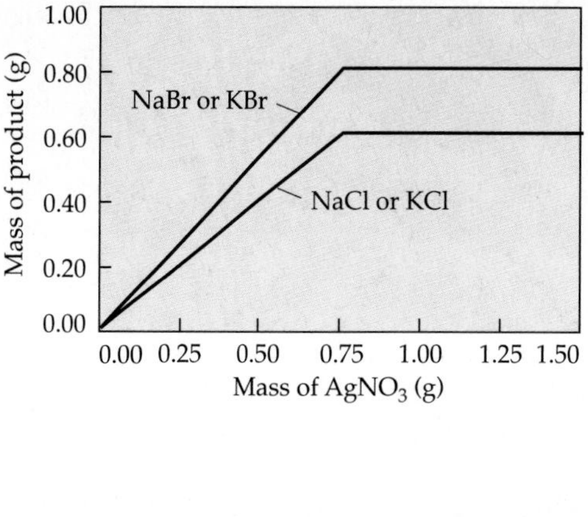

$AgNO_3$, to their solutions. After each group collects the silver halide precipitated in their reactions, the mass of this product is plotted versus the mass of $AgNO_3$ added. The results are given in the graph.

(a) Write the balanced, net ionic equation for the reaction observed by each group.

(b) Explain why the data for Groups A and B lie on the same line, whereas those for Groups C and D lie on a different line.

(c) Explain the shape of the curve observed by each group. Why do they level off at about 0.60 g of added $AgNO_3$ (for Groups A and B) or at about 0.80 g of added $AgNO_3$ (for Groups C and D)?

96. Two students titrate different samples of the same solution of HCl using 0.100 M NaOH solution and phenolphthalein indicator (Figure 5.11). The first student pipets 20.0 mL of the HCl solution into a flask, adds 20 mL of distilled water and a few drops of phenolphthalein solution, and titrates until a lasting pink color appears. The second student pipets 20.0 mL of the HCl solution into a flask, adds 60 mL of distilled water, and a few drops of phenolphthalein solution, and titrates to the first lasting pink color. Each student correctly calculates the molarity of a HCl solution. The second student's result is

(a) Four times less than the first student's

(b) Four times more than the first student's

(c) Two times less than the first student's

(d) Two times more than the first student's

(e) The same as the first student's

97. Dilute sulfuric acid is added from a buret to dilute, aqueous calcium hydroxide in a beaker. Assuming that the molar concentrations of acid and base are equal, make a graph of electric conductivity of the solution in the beaker on the vertical axis versus volume of sulfuric acid added on the horizontal axis. Before starting your graph, write balanced net ionic equations for all reactions that occur in the beaker.

Summary Question

98. Various masses of the three Group 2A elements, magnesium, calcium, and strontium, are allowed to react with liquid bromine, Br_2. After the reaction is complete the reaction product is freed of excess reactant(s) and weighed. In each case the mass of product is plotted against the mass of metal used in the reaction.

(a) Based on your knowledge of the reactions of metals with halogens, what product is predicted for each reaction? What is the name and formula for the reaction product in each case?

(b) Write a balanced equation for the reaction occurring each case.

(c) What kind of reaction, acid-base, oxidation-reduction, or some other type, occurs between the metals and bromine?

(d) Each plot shown in the graph illustrates that the mass of product increases with increasing mass of metal used, but the plot levels out at some point. Use these plots to verify your prediction of the formula of each product, and explain why the plots become level at different masses of metal and different masses of product.

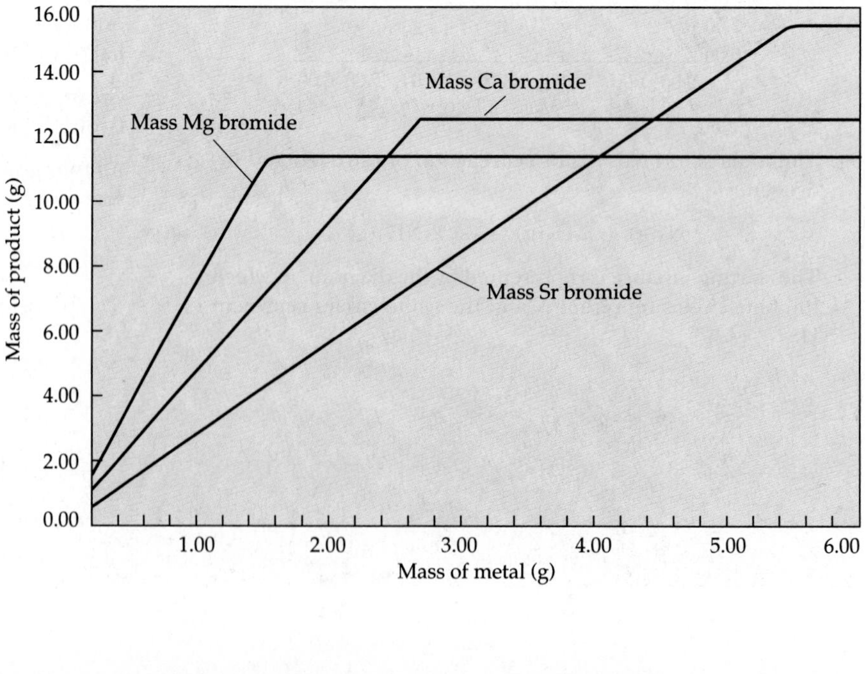

CHAPTER 6

Principles of Reactivity: Energy and Chemical Reactions

A Chemical Puzzler

Peanuts and peanut oil are organic material and so burn in air. How many burning peanuts would it take to boil a cup of water? How could you figure out how much thermal energy can be derived from a peanut or a handful of peanuts?

(C. D. Winters)

Some chemical reactions begin as soon as the reactants come into contact and continue until at least one reactant (the limiting reactant) is completely consumed. If a candy Gummi Bear, which is mostly sugar, is dropped into hot, molten potassium chlorate, the sugar reacts very vigorously with the potassium chlorate (Figure 6.1).

$$\underset{\text{sugar (reducing agent)}}{C_6H_{12}O_6(s)} + \underset{\text{potassium chlorate (oxidizing agent)}}{4\ KClO_3(\ell)} \longrightarrow 6\ CO_2(g) + 6\ H_2O(g) + 4\ KCl(s)$$

Or, if aluminum and bromine are combined, they react rapidly with evolution of a great deal of energy to give aluminum bromide (Figure 4.1). Other reactions happen much more slowly, but reactants are still converted almost completely to products. An example is the rusting of iron at room temperature. Given many year's time and enough flaking of iron(III) oxide from its surface, a piece of iron exposed to air will be converted completely to iron(III) oxide; in other words, it will rust away. Still other reactions are so slow at room temperature that even a lifetime is not enough to observe a

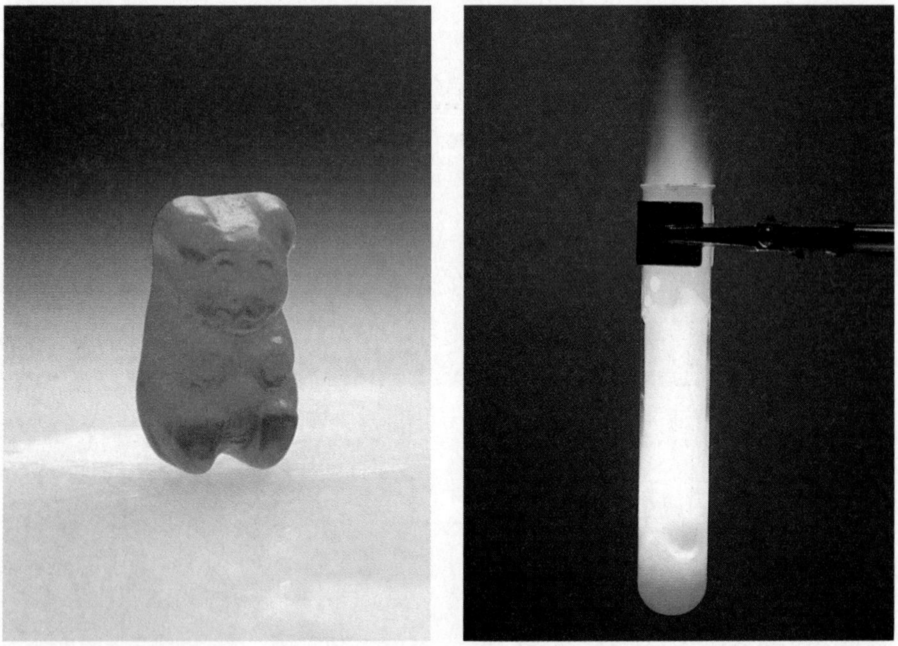

Figure 6.1 Some reactions begin as soon as the reactants come into contact and continue until one or more of the reactants have been consumed. Reactions such as these are said to be "product-favored." One example is the reaction of sugar (a reducing agent, here a Gummi Bear) with a good oxidizing agent (here molten potassium chlorate, $KClO_3$). (C. D. Winters)

measurable change. An example is combustion of gasoline, which burns rapidly at high temperatures but can be stored for long periods in direct contact with air—as long as the temperature is not raised by a spark or flame. Nevertheless, chemists are convinced that if one could wait long enough, gasoline and oxygen would be converted to CO_2 and H_2O.

Two questions about chemical reactions are important to chemists: *will a reaction occur* and, if it does, *how fast will it go.* In the present chapter we wish to investigate some aspects of the first question: how can we predict when a reaction will occur. We postpone discussing the factors that determine the speed of a reaction until Chapter 15.

We begin by categorizing reactions as **product-favored** or **reactant-favored systems.** If a system is product-favored, most of the reactants are eventually converted to products without outside intervention, although "eventually" may mean a very, very long time. Product-favored reactions are also those that usually, but not always, evolve energy. You already know something about product-favored reactions—they are the ones we described in Chapter 4. There you learned that the following kinds of reactions are predicted to favor the formation of products:

Product favored reactions.

- Reactions that lead to a precipitate
- Reactions in which an acid and a base combine to form water
- Reactions that produce a gas, such as the decomposition of a metal carbonate in acid
- Oxidation-reduction reactions, such as the combination of metals with oxygen or halogens to give metal oxides and halides, respectively, or the combustion of organic carbon-containing compounds

Some other reactions have virtually no tendency to occur by themselves. We label these reactions as **reactant-favored systems.** For example, the splitting of NaCl into its elements

Reactant-favored reaction: $2\ NaCl(s) \longrightarrow 2\ Na(s) + Cl_2(g)$

is exactly the opposite of the product-favored reaction of sodium and chlorine (see Figure 2.21).

Product-favored reaction: $2\ Na(s) + Cl_2(g) \longrightarrow 2\ NaCl(s)$

The splitting of NaCl is therefore reactant-favored. Evidence of this is that deposits of salt, NaCl, have existed on earth for millions of years without forming the elements Na and Cl_2. Without some continuous outside intervention, reactants in a reactant-favored system cannot be transformed into appreciable quantities of products.

What do we mean by outside intervention? Usually it is some flow of energy. For example, at very high temperatures, small but significant quantities of NO can be formed from air.

$$N_2(g) + O_2(g) \xrightarrow{+\text{energy}} 2\ NO(g)$$

Such high-temperature conditions are found in power plants and automobile engines, and a large number of such sources can produce enough NO and other nitrogen oxides to cause significant air pollution problems. Salt can be decomposed to its elements by providing heat to melt it and electric energy to separate the ions and form the elements.

$$2\ NaCl(\ell) \xrightarrow{\text{electric energy}} 2\ Na(\ell) + Cl_2(g)$$

In each case energy can cause a reactant-favored system to produce products.

Energy is a central idea when describing a reaction as product- or reactant-favored. For this reason it is useful to know something about energy and its interactions with matter. The most common of these interactions is the transfer of energy as heat or thermal energy when chemical reactions occur. Transfer of heat is a major theme of **thermodynamics,** the science of heat and work—the subject of this chapter.

Thermodynamics is a fundamental area of chemistry because understanding energy transfer in chemical reactions provides one means of predicting when reactions are product-favored, as described in Chapter 20.

Thermodynamics is also related to the problem of energy use in your home and to ways of conserving energy, to the question of the recycling of materials, and to the problem of current and future energy use in our economy. We shall spend considerable time describing energy transfer and the principles of thermodynamics as applied to simple chemical systems, but keep in mind the practical applications. We shall turn to a few of these at the end of the chapter.

6.1 ENERGY: ITS FORMS AND UNITS

Just what is energy, and where does the energy we use come from? **Energy** is defined as the capacity to do work, and work is something you experience all the time. If you climb a mountain, you do some work against the force of gravity as you carry yourself and your equipment up the mountain's side. You can do this work because you have the energy or capacity to do so, the energy having been provided by the food you have eaten. Food energy is chemical energy—energy stored in chemical compounds and released when the compounds undergo the chemical reactions of metabolism.

Figure 6.2 Water on the brink of a waterfall represents stored, or potential, energy, energy that can be used to generate electricity, for example.
(Image Enterprises/James Cowlin)

Energy can be assigned to one of two classes: kinetic or potential. An object has **kinetic energy** because it is moving. Examples are:

- *Thermal energy* of atoms, molecules, or ions in motion at the submicroscopic level. All matter has thermal energy because, according to the kinetic-molecular theory, the submicroscopic particles of matter are in constant motion (Section 1.1).
- *Mechanical energy* of a macroscopic object like a moving baseball or automobile
- *Electric energy* of electrons moving through a conductor
- *Sound,* which corresponds to compression and expansion of the spaces between molecules

Potential energy is energy that results from an object's position. Examples are:

- Chemical potential energy resulting from attractions among electrons and atomic nuclei in molecules
- Gravitational energy, such as that of a ball held well above the floor or that of water at the top of a waterfall (Figure 6.2)
- Electrostatic energy, such as that of positive and negative ions a small distance apart

Potential energy is stored energy—it can be converted into kinetic energy. For example as water droplets fall over a waterfall, the potential energy of the water is converted into kinetic energy, and the drops move faster and faster. Similarly, kinetic energy can be converted to potential energy. The kinetic energy of falling water can turn a turbine to produce electricity.

Temperature, Heat, and the Conservation of Energy

When you stand on a diving board, poised to dive into a swimming pool, you have considerable potential energy because of your position above the water. Once you jump off the board, some of that potential energy is converted progressively into kinetic energy (Figure 6.3), which depends on your mass (m) and velocity (v).

$$\text{Kinetic energy of a body in motion} = \frac{1}{2}mv^2 = \frac{1}{2}(\text{mass}) \cdot (\text{velocity})^2$$

During the dive, your mass is constant, whereas the force of gravity accelerates your body to move faster and faster as you fall, so your velocity and kinetic energy increase. This happens at the expense of potential energy. At the moment you hit the water, your velocity is abruptly reduced, and much of your kinetic energy is converted to mechanical energy of the water, which splashes as your body moves it aside by doing work on it. Eventually you float on the surface, and the water becomes still again. If you could see them, however, you would find that the water molecules were moving a little faster in the vicinity of your dive; that is, the temperature of the water would be a little higher.

This series of energy conversions, from potential to kinetic, and from kinetic to heat, illustrates the **law of energy conservation,** which states that energy can neither be created nor destroyed—the total energy of the universe is constant.

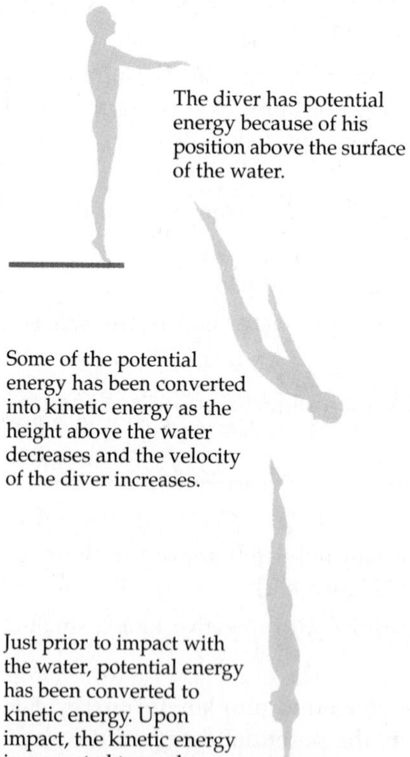
The diver has potential energy because of his position above the surface of the water.
Some of the potential energy has been converted into kinetic energy as the height above the water decreases and the velocity of the diver increases.
Just prior to impact with the water, potential energy has been converted to kinetic energy. Upon impact, the kinetic energy is converted to work.

Figure 6.3 Energy conservation and the interconversion of potential energy, kinetic energy, and work.

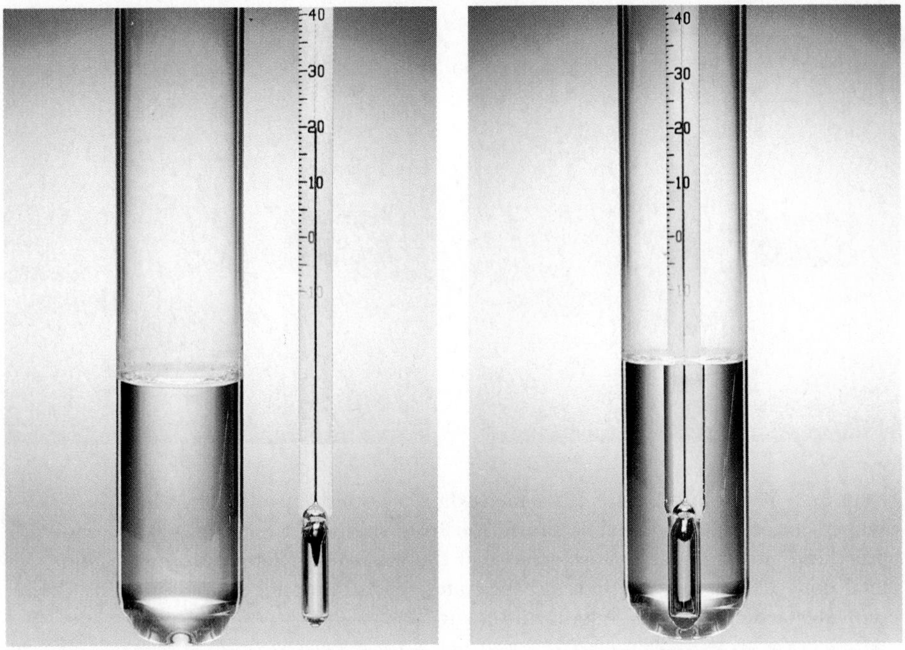

Figure 6.4 The mercury in a thermometer expands as heat is transferred to the liquid metal from warm water. (C. D. Winters)

This law summarizes the results of a great many experiments in which heat, work, and other forms of energy transfer have been measured and the total energy found to be the same before and after an event. The law of energy conservation is the reason we have been careful not to say that when a lump of coal is burned, its energy has been used up. What has been used up is an *energy resource:* the coal's capacity to transfer heat energy to its surroundings when it is burned. If the coal is burned in a power plant, its chemical energy can be changed to an equal quantity of energy in other forms. These are mainly electricity, which can be very useful, and thermal energy in the gases going up the smokestack. It is not the coal's energy, but rather its ability to store energy and release it in a form that people can use that has been used up.

Notice also that in our diving example the potential energy you originally had on the diving board ended up heating the water. Thermal energy, or heat, is associated with temperature, but heat is not the same as temperature. Transferring energy by heating an object increases the object's temperature, and the temperature increase can be measured with a thermometer. For example, Figure 6.4 shows a thermometer containing mercury. When the thermometer is placed into hot water, heat is transferred from the water to the thermometer. The increased energy of the mercury atoms means that they are moving about more rapidly, which slightly increases the volume of the spaces between the atoms. Consequently, the mercury expands (as most substances do when heated), and the upper end of the column of mercury rises higher in the thermometer tube.

Heat is not a substance that is contained in a body. Rather, it is observed as the interaction of a body with its surroundings.

Energy transfer, as heat, happens when two objects at different temperatures are brought into contact. Energy always transfers spontaneously from the hotter to the cooler object. For example, a piece of metal being heated in a Bunsen burner flame and a beaker containing cold water (Figure 6.5, *left*) are two objects with different temperatures. When the hot metal is plunged into the cold water (Figure 6.5, *right*), heat is transferred from the metal to the water until the two objects reach the same temperature. The quantity of heat may be sufficient to

Figure 6.5 Water in a beaker is warmed when a hotter object (a brass bar) is plunged into the water. Heat is transferred from the hotter metal bar to the cooler water (and enough heat is transferred that the water immediately around the hot metal boils). Eventually the bar and the water reach the same temperature (at which point they achieve thermal equilibrium). (C. D. Winters)

raise the temperature of the water to 100 °C immediately around the metal, and some water can boil.

Two important aspects of thermal energy should be understood:

- The more energy an atom or molecule has, the faster it moves.
- The total thermal energy in an object is the sum of the individual energies of all the atoms or molecules in that object.

The average speed of motion of atoms and molecules is related to the temperature. *The total thermal energy depends on temperature, the types of atoms or molecules, and the number of them in a sample.* For a given substance thermal energy depends on temperature and the amount of substance. Thus a cup of steaming coffee may contain less thermal energy than a bathtub full of warm water, even though the coffee is at a higher temperature.

Under most circumstances, objects in a given region, such as your room, are at about the same temperature. If an object such as a cup of coffee or tea is much hotter than this, it transfers energy by heating the rest of the objects in your room until the hot coffee cools off (and your room warms up a bit). If an object, say a cold glass of ice water, is much cooler than its surroundings in your room, heat is transferred to the water from everything else until it warms up (and your room cools off a little). Because the total amount of material in your room is much greater than in a cup of coffee or tea or a glass of ice water, the room temperature changes very little. Heat transfer occurs until everything is at the same temperature.

EXERCISE 6.1 *Energy*

You place a raw egg in a frying pan and fry it, over easy. As in jumping off a diving board, several kinds of energy are involved. Describe them and the changes they cause. ■

A CLOSER LOOK *Why Doesn't the Heat in a Room Cause Your Cup of Coffee to Boil?*

It is interesting (and useful) to think about *why* the heat in a room doesn't cause a cup of cold coffee to boil. According to the law of energy conservation, it could just as well happen that energy could be transferred from the rest of your room to a hot cup of coffee. The coffee would then get hotter and hotter, eventually boiling. But we know from experience that this never happens. There is a *directionality* in heat transfer: energy always transfers from hotter to colder, never the reverse. This directionality corresponds to a *spreading out of energy* over the greatest possible number of atoms or molecules. Whether a relatively small number of molecules in a hot cup of coffee cools by transferring energy to a large number of atoms and molecules surrounding the cup, or the large number of particles in the surrounding environment heats a glass of ice water by transferring some of their energy to relatively few molecules in the glass, the end result is that the thermal energy is spread evenly over the maximum number of molecules. Concentrating energy in only a few particles at the expense of many, or even concentrating energy over a large number of particles at the expense of a few, is highly unlikely and has never been observed on a macroscopic scale.

The idea that energy is spread over as many atoms and molecules as possible can be used to help us predict directionality in the conversion of chemical reactants to products, that is, to say whether a mixture of substances will be product-favored or reactant-favored. We shall return explicitly to this idea in Chapter 20.

Energy Units

Some of the units in which energy is measured were originally designed to measure heat. A **calorie** was originally defined as the quantity of energy (transferred by heating) that is required to raise the temperature of 1.00 g of pure liquid water by 1.00 degree Celsius, from 14.5 °C to 15.5 °C. The calorie represents a very small quantity of heat, and because we often work with larger quantities of matter, the **kilocalorie** is used instead. The kilocalorie (abbreviated kcal) is equivalent to 1000 calories.

1 kilocalorie = 1 kcal = 1000 calories

Most of us tend to think of heat as measured in calories, probably because we hear about dieting or read breakfast cereal boxes. The "calorie" used in these circumstances is the dietary Calorie, with a capital *C*, a unit equivalent to the kilocalorie. Thus, a breakfast cereal that gives you 100 Calories of nutritional energy per serving really provides 100 kcal, or 100×10^3 calories (calories with a small *c*).

Currently most chemists use the **joule** (J), the SI unit of energy. One calorie is now defined as equivalent to 4.184 J. The joule is preferred because it is derived directly from the units used in the calculation of mechanical energy (i.e., potential and kinetic energy). If a 2.0-kg object (about 4 pounds) is moving with a velocity of 1.0 meter per second (roughly 2 mph), the kinetic energy is

1 calorie = 4.184 joules

$$\text{Kinetic energy} = \frac{1}{2}mv^2 = \frac{1}{2}(2.0\text{ kg})(1.0\text{ m/s})^2 = 1.0\text{ kg}\cdot\text{m}^2/\text{s}^2 = 1.0\text{ J}$$

To give you some feeling for joules, suppose you are holding a six-pack of soft drink cans, each full of liquid. You drop the six-pack on your foot. Although you probably do not take time to calculate the kinetic energy at the moment of impact, it is about a calorie or two, that is, about 4 to 10 joules. In many countries that use standardized SI units, food energy is also measured in joules. For example, the packet of nonsugar sweetener shown in Figure 6.6 provides 16 kJ of nutritional energy, instead of the 140 kJ provided by two teaspoons of sugar.

Figure 6.6 A packet of artificial sweetener from Australia. The sweetener in the packet supplies 16 kJ of nutritional energy. It is equivalent to 2 level metric teaspoons of sugar, which supplies 140 kJ of nutritional energy. (C. D. Winters)

Like the calorie, the joule is often inconveniently small as a unit for many purposes in chemistry and other sciences, and so the **kilojoule (kJ),** which is equivalent to 1000 joules, is often used. For example, burning peanuts may produce 40,000 J or more per gram of peanut oil. Chemists find it more convenient to express large numbers such as these in kilojoules, that is, as 40 kJ/g in this case.

EXAMPLE 6.1 *Using Energy Units*

The average solar energy received by a horizontal surface in Madison, Wisconsin, in the summer is $2.3 \times 10^7\ \text{J/m}^2 \cdot \text{day}$. If a home in Madison has a horizontal roof that measures 10. m by 25 m, what quantity of energy, measured in kilojoules, strikes the roof?

Solution Because we know the quantity of energy received per square meter, we need only find the number of square meters of roof.

$$\text{Roof area} = (10.\ \text{m})(25\ \text{m}) = 250\ \text{m}^2$$

$$\text{Energy received} = 250\ \text{m}^2 \cdot \frac{2.3 \times 10^7\ \text{J}}{\text{m}^2 \cdot \text{day}} \cdot \frac{1\ \text{kJ}}{1000\ \text{J}} = 5.8 \times 10^6\ \text{kJ/day}$$

As an aside, it is also interesting to find the energy received in kilowatt-hours, the energy unit used by your local electric utility. Knowing that 1 kwh = 3.61×10^6 joules,

$$5.8 \times 10^9\ \text{J} \cdot \frac{1\ \text{kwh}}{3.61 \times 10^6\ \text{J}} = 1.6 \times 10^3\ \text{kwh}$$

we find that the roof receives about 1600 kwh of energy. This is equivalent to the energy from several 100-watt light bulbs burning for 24 hours per day for a year.

EXERCISE 6.2 *Energy Units*

1. A hot dog provides 160 Calories of energy. What is this energy in joules?
2. The energy used by a light bulb of *W* watts for *s* seconds is *Ws* joules. If you turn on a 75-watt bulb for 3.0 hours, how many joules of energy have been used?
3. The nonsugar sweetener in Figure 6.6 provides 16 kJ of nutritional energy. What is this energy in kilocalories? ■

6.2 SPECIFIC HEAT AND THERMAL ENERGY TRANSFER

In the 1770s Joseph Black of the University of Glasgow, Scotland, studied the nature of heat and was the first to distinguish clearly between the "hotness," or temperature, of an object and its heat capacity.

PORTRAIT OF A SCIENTIST

James P. Joule (1818–1889)

The joule is named for James Joule, the son of a wealthy brewer in Manchester, England, and a student of John Dalton (Chapter 1). The Joule family wealth, and a workshop in the brewery, gave James Joule the opportunity to pursue scientific studies. One of the "hot" topics of the day was the relation between heat and mechanical energy. Prior to Joule's studies on this subject, the French scientist Sadi Carnot (1796–1832) had studied steam engines for a very practical reason: Carnot thought France had been defeated by England in the Napoleonic wars because England was technologically superior, in part because steam engines were so central to England's powerful economy.

Scientists of the time held the view that heat was a massless fluid called caloric. Carnot and others believed the caloric fluid should be conserved in the operation of a steam engine, just as the water that turns a waterwheel to perform some work is conserved. But the caloric fluid wasn't apparently conserved. Instead, Carnot discovered a fundamental inefficiency in converting heat to work in a steam engine.

Joule began the process of disproving the concept of the caloric fluid. By doing very precise experiments, Joule showed that work could be converted *quantitatively* to heat. This was experimental evidence of the concept of the mechanical equivalence of heat, that heat and work can be interconverted, and that heat is not a massless fluid.

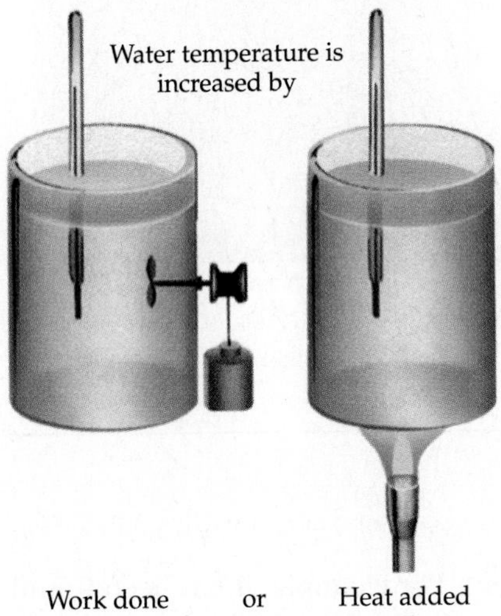

Experiments to show the relation between mechanical work and heat energy. (a) The temperature can be increased by doing work. (b) The temperature of water can be increased by adding heat from a chemical reaction in a gas burner.

James Prescott Joule (1818–1889). (Oesper Collection in the History of Chemistry/University of Cincinnati)

The **specific heat capacity,** *C*, of a substance is defined as the ratio of the heat supplied to some mass of the substance (say, 1.00 g) to the consequent rise in the substance's temperature. Mathematically, we express this as

$$\text{Specific heat capacity } (C) = \frac{\text{quantity of heat supplied}}{(\text{mass of object}) \cdot (\text{temperature change})} \quad \textbf{(6.1)}$$

$$\text{Specific heat capacity } (\text{J/g} \cdot \text{K}) = \frac{q(\text{J})}{[m(\text{g})] \cdot [\Delta T(\text{K})]}$$

where the *quantity of heat transferred by heating* is symbolized by ***q***. The symbol Δ (the capital Greek letter delta) means "change in." In chemistry we shall use heat in units of joules, masses in grams, and temperatures in kelvins. The specific heat of a substance, which is often just called its specific heat, is therefore expressed in units of joules per grams times kelvins (J/g · K).

The heat capacity of an object or quantity of a substance is the amount of heat required to raise its temperature 1 degree. The heat capacity *per gram* is the specific heat capacity or *specific heat.* The heat capacity per mole is the *molar heat capacity.*

PORTRAIT OF A SCIENTIST

Joseph Black 1728–1799

Joseph Black was a professor in Glasgow and Edinburgh in Scotland in the 18th century. Among his important studies was that of heat, which later led his student James Watt (1736–1819) to the improvement of the steam engine. Another of his students was Benjamin Rush (1745?–1813), who became the first professor of chemistry in the United States (at what is now the University of Pennsylvania).

Black was also one of the first scientists to be concerned about teaching chemistry to undergraduates, in part because the Industrial Revolution was under way, and students needed some knowledge of chemistry. In addition, he was interested in "scientific agriculture." His research on carbonates led to the use of lime to counteract excessive soil acidity.

Joseph Black (1728–1799). (Oesper Collection in the History of Chemistry/University of Cincinnati).

See Example 6.3 for one experimental approach to determining specific heat.

Specific heat is determined by experiment. For example, it has been found that 60.5 joules is required to change the temperature of 25.0 g of ethylene glycol (a compound used as antifreeze in automobile engines) by 1.00 K. This means the specific heat of the compound is

$$\text{Specific heat of ethylene glycol} = \frac{60.5 \text{ joules}}{(25.0 \text{ g})(1.00 \text{ K})} = 2.42 \text{ J/g} \cdot \text{K}$$

The specific heat of many substances have been determined experimentally, and a few values are listed in Table 6.1. The wide differences in the specific heat of substances can be used to distinguish them from each other, just as density can.

Notice in Table 6.1 that water has one of the highest known values of specific heat. For water, the specific heat is 4.184 J/g · K, whereas it is only about 0.45 J/g · K for iron or 0.84 J/g · K for common glass. That is, it takes about nine times as much heat to raise the temperature of a gram of water 1 K as it does for iron, or about five times as much heat is required to increase the temperature of a gram of water by the same amount as a gram of glass. The high specific heat of water also informs us that a considerable quantity of thermal energy must be transferred out of the substance before it cools down appreciably. For example, 1.00 g of water must give up 4.18 J of thermal energy to cool one kelvin, whereas 1.0 g of glass gives up only 0.84 J, or about one-fifth as much thermal energy as water for the same temperature change.

The very high specific heat of water is important to all of us, because many joules must be absorbed by a large body of water to raise its temperature just a degree or so. Conversely, a great deal of energy must be lost before the temperature of the water drops by more than a degree. Thus, a lake can store an enor-

TABLE 6.1 Specific Heat Values for Some Elements, Compounds, and Common Solids

Substance	Name	Specific Heat (J/g · K)
Elements		
Al	Aluminum	0.902
C	Graphite	0.720
Fe	Iron	0.451
Cu	Copper	0.385
Au	Gold	0.128
Compounds		
$NH_3(\ell)$	Ammonia	4.70
$H_2O(\ell)$	Water—liquid	4.184
$C_2H_5OH(\ell)$	Ethanol	2.46
$(CH_2OH)_2(\ell)$	Ethylene glycol (antifreeze)	2.42
$H_2O(s)$	Water—ice	2.06
$CCl_4(\ell)$	Carbon tetrachloride	0.861
$CCl_2F_2(g)$	Dichlorodifluoromethane (a chlorofluorocarbon)	0.598
Common Solids		
Wood		1.76
Cement		0.88
Glass		0.84
Granite		0.79

Figure 6.7 A practical example of specific heat capacity. The filling of an apple pie has a higher specific heat than the pie crust or the wrapper. (Notice the warning on the wrapper.) (C. D. Winters)

mous quantity of energy, and so bodies of water have a profound influence on our weather.

The equation defining specific heat informs us that the larger the specific heat, the more thermal energy a substance can store. Furthermore, substances with a high specific heat often cool down more slowly than those with a smaller specific heat. If you wrap some bread in aluminum foil and heat it in an oven, you know you can remove the foil after taking the bread from the oven without burning your fingers, even though the bread is very hot. This is also the reason that a chain of fast food restaurants warns you that the filling of an apple pie can be much warmer than the paper wrapper or the pie crust (Figure 6.7).

The rate at which an object cools is determined by its thermal conductivity. For most metals, low specific heat means a higher thermal conductivity.

Knowing the specific heat of a substance allows you to calculate the heat transferred to or from that substance if the mass of substance and its temperature change are also known. Conversely, you can calculate the temperature change that should occur when a given quantity of heat is transferred to or from a sample of known mass. The equation that defines specific heat allows you to do this, but it may be more convenient to rewrite it in the following form (Equation 6.2):

Heat transferred

$$= \text{(specific heat)(mass)(size of temperature change of the substance)} \quad \textbf{(6.2)}$$

$$q = (\text{specific heat in J/g} \cdot \text{K})(\text{mass in g})(\Delta T \text{ in kelvins})$$

For example, the thermal energy required to warm a 10.0-g piece of copper from 25 °C (298 K) to 325 °C (598 K) is

$$q = 10.0\,\cancel{g} \cdot \frac{0.385\,\text{J}}{\cancel{g} \cdot \text{K}} \cdot (598\,\text{K} - 298\,\text{K}) = 1160\,\text{J}$$

To this point, we have emphasized only the *quantity* of heat transferred. Equation 6.2, however, allows a demonstration not only of the quantity of heat but also the direction in which it is transferred. When ΔT is determined as in the previous calculation, that is, as

$$\Delta T = \text{final temperature} - \text{initial temperature}$$

it has an algebraic sign: positive (+) for T increase ($T_f > T_i$) and negative (−) for a decrease ($T_f < T_i$). If the temperature of the substance increases, ΔT has a positive (+) sign *and so does q*. This means heat was transferred *to* the substance (as in the example of heating a piece of copper). The opposite case, a decrease in the temperature of the substance, means ΔT has a negative (−) sign and so does q; heat was transferred *from* the substance.

Let us find the number of joules of heat energy transferred from a cup of coffee to your body and the surrounding air if the temperature of a cup of coffee held in your hand drops from 60.0 °C (333.2 K) to 37.0 °C (310.2 K) (normal body temperature). Assume the coffee has a mass of 250. g; also assume the specific heat of coffee is the same as water. The quantity of heat transferred can be obtained from Equation 6.2.

$$\text{Heat transferred} = q = \frac{4.184\,\text{J}}{\cancel{g} \cdot \text{K}}(250.\,\cancel{g})(\underset{\substack{\uparrow \\ \text{final temp.}}}{310.2\,\text{K}} - \underset{\substack{\uparrow \\ \text{initial temp.}}}{333.2\,\text{K}})$$

$$= -24.1 \times 10^3\,\text{J} = -24.1\,\text{kJ}$$

Notice that the final answer has a *negative* value. In this case heat energy is transferred from the coffee to the surroundings, and the temperature of the coffee decreases.

EXAMPLE 6.2 *Using Specific Heat*

A lake has a surface area of 2.6×10^6 square meters (about 1 square mile) and an average depth of 10. meters. What quantity of heat (in kilojoules) must be transferred to the lake water to raise the temperature by 1.0 °C? Assume the density of the water is 1.0 g/cm^3.

Solution To calculate the heat, we must know the mass of the lake water and therefore must first find the volume of water in cubic meters. After converting this volume to a volume in cubic centimeters, we can calculate the mass of water. With the mass known, the quantity of heat required can be found.

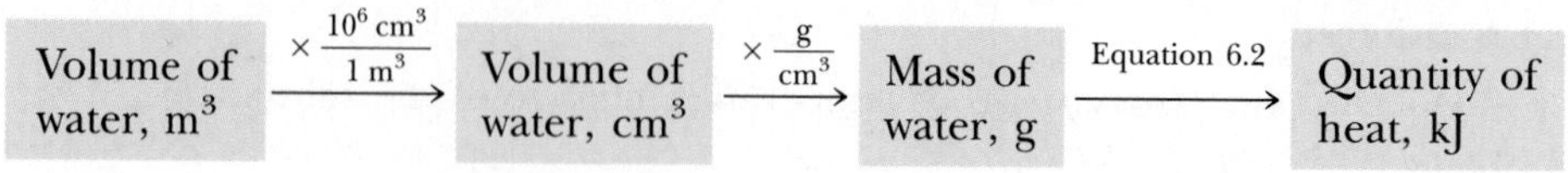

A CLOSER LOOK *Sign Conventions in Energy Calculations*

Whenever you take the difference between two quantities in chemistry, you should *always subtract the initial quantity from the final quantity*. A natural consequence of this convention is that the algebraic sign of the calculated result indicates an increase (+) or a decrease (−) in the quantity for the substance being studied. This is an important point, as you will see in other chapters of this book.

Thus far, we have described temperature changes and the direction of heat transfer. The table below summarizes the conventions used.

Be sure to understand that the sign of q is just a "signal" to tell the direction of heat transfer. Heat itself cannot be negative; it is simply a quantity of energy. As an example, consider your bank account. Assume you have \$26 in your account ($A_{initial}$), and after a withdrawal you have \$20 (A_{final}). The cash flow is thus

$$\text{Cash flow} = A_{final} - A_{initial} = \$20 - \$26 = -\$6$$

The negative sign on the \$6 indicates that a withdrawal has been made; the cash itself is not a negative quantity.

Thus, when we talk about heat we use an unsigned number. When we want to indicate the direction of transfer in a process, however, we shall attach a negative sign (heat transferred *from* the substance) or a positive sign (heat transferred *into* the substance) to the value of q.

Change in T of Object	Sign of ΔT	Sign of q	Direction of Heat Transfer
Increase	+	+	Heat transferred into object
Decrease	−	−	Heat transferred out of object

$$\text{Volume of water} = 2.6 \times 10^6\ \text{m}^2 \cdot 10.\ \text{m} \cdot \left(\frac{100\ \text{cm}}{1\ \text{m}}\right)^3 = 2.6 \times 10^{13}\ \text{cm}^3$$

$$\text{Mass} = 2.6 \times 10^{13}\ \text{cm}^3 \cdot \frac{1.0\ \text{g}}{\text{cm}^3} = 2.6 \times 10^{13}\ \text{g}$$

With the mass known, the quantity of heat required can be found using Equation 6.2, where ΔT is 1.0 °C or 1.0 K.

$$q = (2.6 \times 10^{13}\ \text{g})(4.184\ \text{J/g} \cdot \text{K})(1.0\ \text{K})$$
$$= 1.1 \times 10^{14}\ \text{J} \quad \text{or} \quad 1.1 \times 10^{11}\ \text{kJ}$$

EXAMPLE 6.3 *Determining a Specific Heat*

A 55.0-g piece of metal was heated in boiling water to 99.8 °C and then dropped into water in an insulated beaker (as in Figure 6.5). There is 225 mL of water (density = 1.00 g/mL) in the beaker, and its temperature before the metal was dropped in was 21.0 °C. The final temperature of the metal and water is 23.1 °C. What is the specific heat of the metal? Assume that no heat transfers through the walls of the beaker or to the atmosphere.*

*To simplify the situation, we also assume that the heat required to warm the walls of the beaker from 21.0 °C to 23.1 °C is negligible. This factor is taken into account in Section 6.9, however.

Solution The most important aspects of this problem to understand are the following:

- The water and the metal bar end up at the same temperature (T_{final} is the same for both).
- Because of the principle of energy conservation, the thermal energy transferred into the water on warming up and the thermal energy transferred out of the iron in cooling down are numerically equal.
- q_{metal} has a negative value because its temperature dropped as heat was transferred *out* of the metal.
- Conversely, q_{water} has a positive value because its temperature increased as heat was transferred *into* the water. Therefore, $q_{metal} = -q_{water}$.

Using the specific heat of water from Table 6.1 (and converting Celsius temperatures to kelvins), we have

$$q_{metal} = -q_{water}$$

$$(55.0 \text{ g})(C_{metal})(296.3 \text{ K} - 373.0 \text{ K}) = -(225 \text{ g})(4.184 \text{ J/g} \cdot \text{K})(296.3 \text{ K} - 294.2 \text{ K})$$

where $T_{initial}$ for the metal is 99.8 °C (373.0 K) and $T_{initial}$ for the water is 21.0 °C (294.2 K), and T_{final} for both metal and water is 23.1 °C (296.3 K). Solving this, we find

$$-4220 \text{ g} \cdot \text{K}(C_{metal}) = -1977 \text{ J}$$

$$C_{metal} = 0.469 \text{ J/g} \cdot \text{K}$$

EXERCISE 6.3 ***Using Specific Heat***

If 24.1 kJ is used to warm a piece of aluminum with a mass of 250. g, what is the final temperature of the aluminum if its initial temperature is 5.0 °C? The specific heat of aluminum is 0.902 J/g · K. ■

EXERCISE 6.4 ***Determining Specific Heat***

A 15.5-g piece of chromium, heated to 100.0 °C, is dropped into 55.5 g of water at 16.5 °C. The final temperature of the metal and the water is 18.9 °C. What is the specific heat of chromium? ■

EXERCISE 6.5 ***Heat Transfer Between Substances***

A piece of iron (400. g) is heated in a flame and then dropped into a beaker containing 1000. g of water. The original temperature of the water was 20.0 °C, but it is 32.8 °C after the iron bar was dropped in and both have come to the same temperature. What was the original temperature of the hot iron bar? ■

PROBLEM-SOLVING TIPS AND IDEAS

6.1 Calculating ΔT

Notice that specific heat values are given in units of J/g · K, where the temperature unit is the kelvin. *Virtually all calculations that involve temperature in chemistry are expressed in kelvins.*

In our calculations, however, we could have used Celsius temperatures. Why can we do this? Because the size of a kelvin degree and a Celsius degree are the same. The *difference* between two temperatures is therefore the same on both scales. For example, the difference between the boiling and freezing points of water is

$$\Delta T_{\text{Celsius}} = 100\ °\text{C} - 0\ °\text{C} = 100 \text{ Celsius degrees}$$

$$\Delta T_{\text{Kelvin}} = 373\ \text{K} - 273\ \text{K} = 100 \text{ kelvins}$$

So, in calculations that ask you to find the *difference* in temperatures, you can use either Celsius or kelvin temperatures. If, however, a calculation calls for you simply to use a temperature, you must use the temperature in kelvins.

6.2 Units of Specific Heat

In this textbook we have given specific heat values in units of J/g · K. This means that units of joules are divided by units of grams and kelvins. An alternative method of writing this is $\text{J g}^{-1}\ \text{K}^{-1}$. Recall that $\frac{1}{10}$ is the same as 10^{-1}, so 1/g is the same as g^{-1}, for example.

Specific heat values given in handbooks of chemistry (such as the *CRC Handbook of Chemistry and Physics*) are often *molar specific heat.* For example, liquid water has a specific heat of 4.184 J/g · K, or 75.4 J/mol · K.

$$4.184 \frac{\text{J}}{\cancel{\text{g}} \cdot \text{K}} \cdot \frac{18.02\ \cancel{\text{g}}}{1\ \text{mol}} = 75.4 \frac{\text{J}}{\text{mol} \cdot \text{K}}$$

6.3 Using the Concept of Energy Conservation

In Example 6.3 you learned that the thermal energy transferred from a hot piece of metal was equal to the thermal energy transferred to the water. We expressed this as $q_{\text{metal}} = -q_{\text{water}}$, where the negative sign reflects the fact that q_{metal} has a negative value ($q_{\text{metal}} = -9820$ J). Another way to look at this is to write the equation as $q_{\text{metal}} + q_{\text{water}} = 0$. This tells us that net value of the energy transferred must be zero; energy must be conserved. You may find this a useful point of view when working problems involving thermal energy transfer between objects. ■

6.3 ENERGY AND CHANGES OF STATE

So far we have described transfers of energy that occur between objects as a result of temperature differences. But energy transfers also occur when matter is transformed from one form to another in the course of a physical or chemical change.

When a solid melts, its atoms, molecules, or ions move about vigorously enough to break free of the constraints imposed by their neighbors in the solid. When a liquid boils, particles move much farther apart from one another. In both cases attractive forces among the particles must be overcome, which requires an input of energy. Joseph Black was the first to recognize that heat is associated with changes of state, which *always take place at constant temperature.*

H_2O 0°C heat of fusion 333 J/g

The quantity of heat required to melt ice at 0 °C is 333 J/g and is called the **heat of fusion** of the ice. You see in Figure 6.8 that the same quantity of heat (500. kJ) required to raise the temperature of a 1.00-kg block of iron from 0 °C

Fusion means melting. When a fuse in an electric circuit melts, it breaks the circuit and prevents current from flowing.

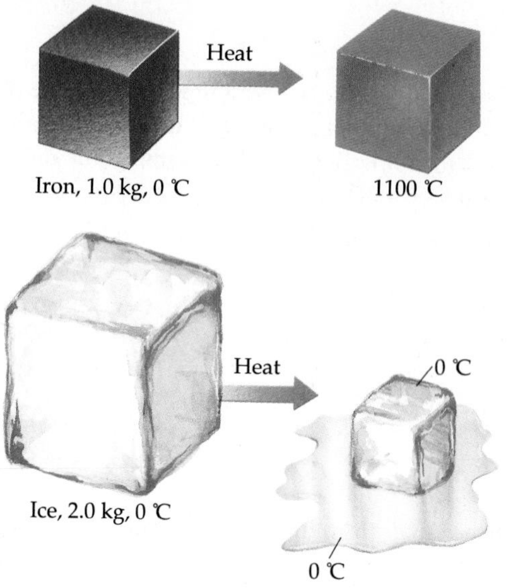

Figure 6.8 Heat transfer to an object can lead to a temperature or a phase change (or both). Here the same quantity of heat (500 kJ) has been transferred to a one-kilogram block of iron at 0 °C as has been transferred to a two-kilogram block of ice at 0 °C. The temperature of the iron block has increased by 1100 K and makes the block red hot. On the other hand, this quantity of heat has led only to the melting of 1.5 kg of ice, and the ice and melted water are still at 0 °C.

to 1100 °C, at which it glows red hot, leads only to the melting of 1.50 kg of ice, at which point the ice and the melted water both have a temperature of 0 °C.

$$500.\ \cancel{kJ} \cdot \frac{1000\ \cancel{J}}{\cancel{kJ}} \cdot \frac{1\text{ g ice melted at }0\ °\text{C}}{333\ \cancel{J}} = 1.50 \times 10^3\text{ g ice melted}$$

H_2O 100°C heat of vaporization 2260 J/g

The quantity of heat required to convert liquid water to vapor, called the **heat of vaporization,** is 2260 J/g at 100 °C. What quantity of water can be vaporized if 500. kJ of heat is transferred to water at 100 °C? Only 221 grams!

$$500.\ \cancel{kJ} \cdot \frac{1000\ \cancel{J}}{\cancel{kJ}} \cdot \frac{1\text{ g water vaporized}}{2260\ \cancel{J}} = 221\text{ g liquid water vaporized}$$

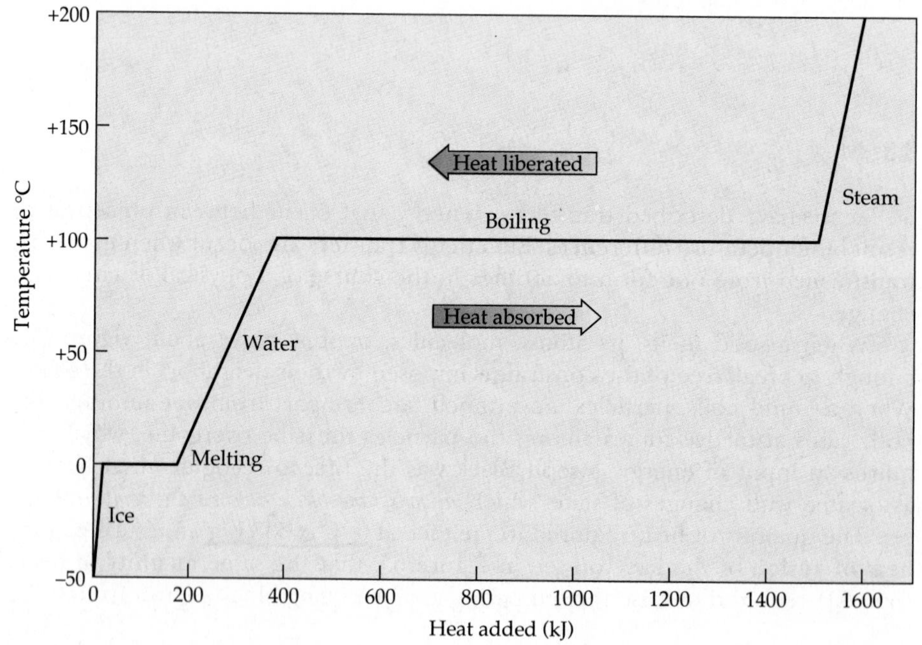

Figure 6.9 A graph illustrating the quantity of heat absorbed and the consequent temperature change as 500. g of water is warmed from −50 °C to 200 °C. The graph also illustrates the heat evolved when steam at 200 °C is cooled to −50 °C.

Figure 6.9 illustrates the quantity of heat absorbed and the consequent temperature change as 500. g of water is warmed from −50 °C to 200 °C. First, the temperature of the ice increases as heat is added. On reaching 0 °C, however, the temperature remains constant as sufficient heat is absorbed to melt the ice to liquid water. When all the ice has melted, the liquid absorbs heat and is warmed to 100 °C, the boiling point of water. The temperature is again constant as heat is absorbed to completely convert the liquid to vapor, that is, to steam. Any further heat absorbed heats the steam to 200 °C. The heat absorbed at each step is calculated in Example 6.4.

EXAMPLE 6.4 *Energy and Changes of State*

Calculate the quantity of heat involved in each step shown in Figure 6.9. That is, calculate the heat required to convert 500. g of ice at −50.0 °C to steam at 200 °C. The required specific heat is

State	Specific Heat (J/g · K)*
Ice	2.1
Water	4.2
Steam	2.0

Solution The quantity of heat required to warm ice at −50.0 °C to 0.0 °C is calculated by the methods of Section 6.2.

$$q \text{ (to warm ice to 0.0 °C)} = (2.1\ \text{J/g} \cdot \text{K})(500.\ \text{g})(273.2\ \text{K} - 223.2\ \text{K}) = 5.3 \times 10^4\ \text{J}$$

Next, the ice is melted at 0.0 °C to water at that temperature.

$$q \text{ (to melt ice)} = 500.\ \text{g} \cdot \frac{333\ \text{J}}{1\ \text{g ice}} = 1.67 \times 10^5\ \text{J}$$

After melting the ice at 0.0 °C, the water is heated from 0.0 °C to 100.0 °C.

$$q \text{ (to warm water to 100.0 °C)} = (4.2\ \text{J/g} \cdot \text{K})(500.\ \text{g})(373.2\ \text{K} - 273.2\ \text{K}) = 2.1 \times 10^5\ \text{J}$$

The water at 100.0 °C is evaporated to steam at 100.0 °C.

$$q \text{ (to evaporate water)} = 500.\ \text{g} \cdot \frac{2260\ \text{J}}{1\ \text{g water}} = 1.13 \times 10^6\ \text{J}$$

Steam is heated from 100.0 °C to 200.0 °C.

$$q \text{ (to heat steam to 200.0 °C)} = (2.0\ \text{J/g} \cdot \text{K})(500.\ \text{g})(473.2\ \text{K} - 373.2\ \text{K}) = 1.0 \times 10^5\ \text{J}$$

The total thermal energy required is the sum of the thermal energy required in each step.

$$q_{\text{total}} = 1.7 \times 10^6\ \text{J} \quad \text{or} \quad 1700\ \text{kJ}$$

*Specific heat changes with temperature. The values listed are average values for the temperature range.

Approximately 1700 kJ of thermal energy must be absorbed by the 500.-g sample to convert it from ice at −50.0 °C to steam at 200.0 °C. If the steam is to be converted back to ice, 1700 kJ of thermal energy must be liberated. The heat liberated, q, equals −1700 kJ.

EXERCISE 6.6 ***Changes of State***

What quantity of heat must be absorbed to warm 25.0 g of liquid methanol, CH_3OH, from 25.0 °C to its boiling point (64.6 °C) and then to evaporate the methanol completely at that temperature? The specific heat of liquid methanol is 2.53 J/g · K. The heat of vaporization of the compound is 2.00×10^3 J/g. ■

EXERCISE 6.7 ***Using a Change of State to Determine Specific Heat***

"ice calorimeter"

An "ice calorimeter" can be used to determine the specific heat of a metal. A piece of hot metal is dropped into a weighed quantity of ice. The quantity of heat given up by the metal can be determined from the amount of ice melted. Suppose a piece of metal with a mass of 9.85 g is heated to 100.0 °C and then dropped onto ice. When the metal's temperature has dropped to 0.0 °C, it is found that 1.32 g of ice has been melted to water at 0.0 °C. What is the specific heat of the metal? ■

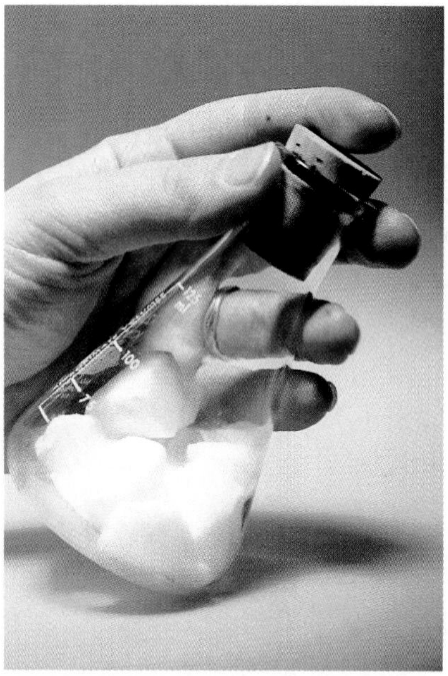

Figure 6.10 A system absorbing heat at constant volume. When heat is absorbed by solid CO_2 in a closed container, the pressure in the container increases as solid is converted to gas. Because no mechanical connection exists between the system and its surroundings, however, no work is done on the surroundings by the system. Therefore, q_v = heat absorbed at constant volume = ΔE. (C. D. Winters)

6.4 ENTHALPY

Carbon dioxide undergoes a change of state from solid to gas, a process called sublimation, at −78 °C.

$$CO_2(\text{solid, } -78\text{ °C}) \xrightarrow[\text{+ heat}]{} CO_2(\text{gas, } -78\text{ °C})$$

To describe the change from solid to gas in thermodynamic terms, we first need to extend the earlier discussion of heat as energy transferred between objects. In thermodynamics one of the "objects" is usually of primary concern. It may be a substance involved in a change of state or substances undergoing a reaction. It is called the **system.** The other "object" is called the **surroundings** and includes everything outside the system that can exchange energy with the system. In Figures 6.10 and 6.11 a CO_2 sample (both solid and vapor) is the *system* and interacts with its *surroundings*—the flask or plastic bag, the table on which they rest, and the air in the room surrounding the sample (and eventually the rest of the universe). A system may be contained within an actual physical boundary, such as a flask or a cell in your body. Alternatively, the boundary may be purely imaginary. For example, you could study the solar system within its surroundings, the rest of the galaxy. In any case, the system can always be defined precisely.

Suppose our system consists of solid and gaseous CO_2 at −78 °C, that the surroundings are also at −78 °C, and that one mole CO_2 (solid) is transformed into one mole CO_2 (gas). Just as the potential energy of a ball is greater when it is farther above the earth, which attracts it, the potential energy of a CO_2 molecule is greater when it is farther from another CO_2 molecule, again because an attractive force must be overcome when the two are separated. Thus, the energy

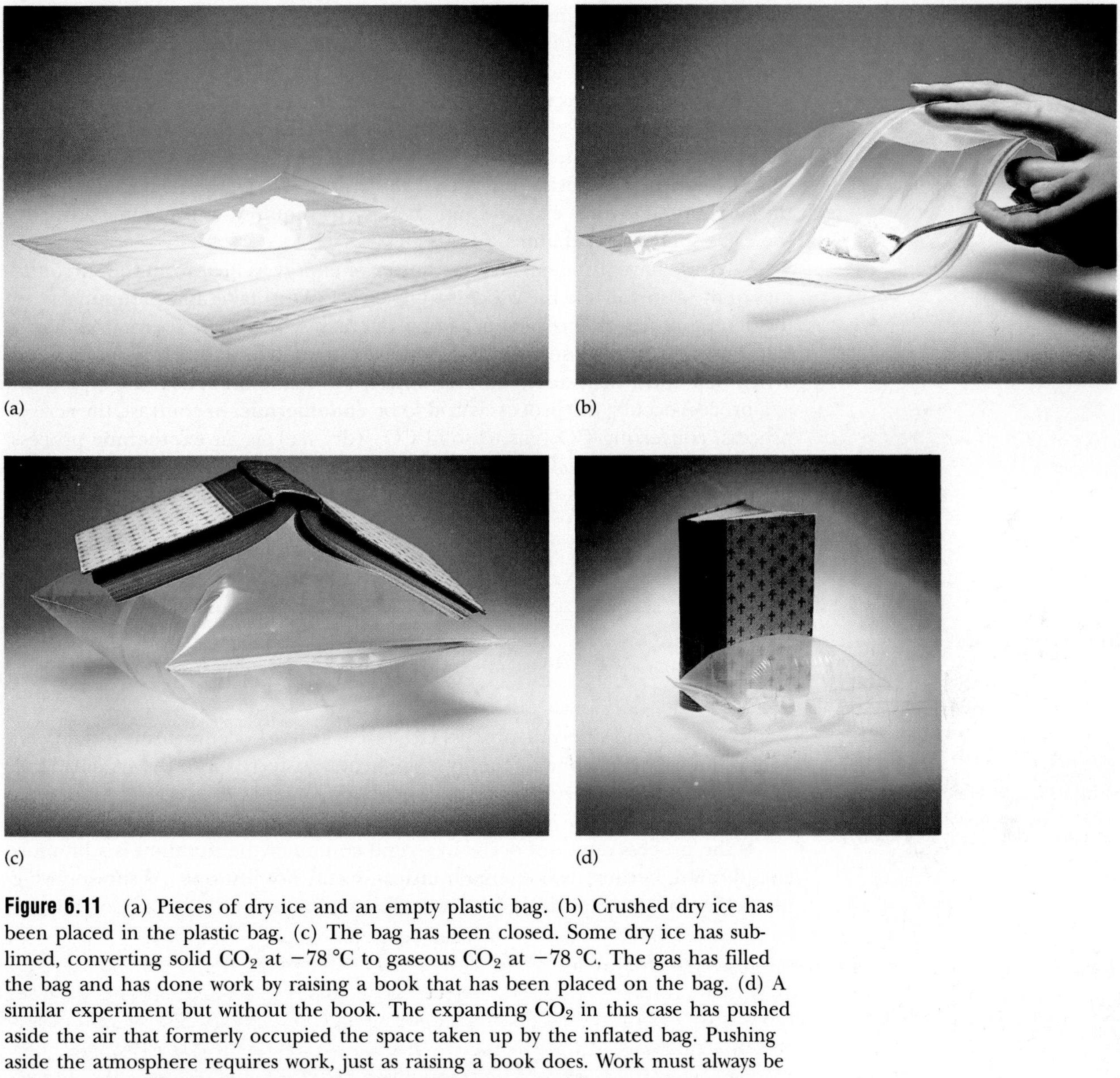

(a) (b) (c) (d)

Figure 6.11 (a) Pieces of dry ice and an empty plastic bag. (b) Crushed dry ice has been placed in the plastic bag. (c) The bag has been closed. Some dry ice has sublimed, converting solid CO_2 at −78 °C to gaseous CO_2 at −78 °C. The gas has filled the bag and has done work by raising a book that has been placed on the bag. (d) A similar experiment but without the book. The expanding CO_2 in this case has pushed aside the air that formerly occupied the space taken up by the inflated bag. Pushing aside the atmosphere requires work, just as raising a book does. Work must always be done to push aside the atmosphere by any reaction whose products occupy greater volume than the reactants. The heat absorbed by the system under these constant pressure conditions (q_p) is identified as the enthalpy change, ΔH. (C. D. Winters)

of a mole of CO_2(g) is greater than the energy of a mole of CO_2(s). Where does this energy come from? That depends on the circumstances.

Let us begin with the case in which the CO_2(s) and CO_2(g) are inside a *rigid container* so that the total volume cannot change (see Figure 6.10). The energy needed for one molecule to escape from solid to gas can be supplied by neighboring molecules, which consequently vibrate a little less about their position in the crystal lattice. As more and more molecules escape, the motion of remaining solid-state molecules diminishes more and more, which would lead to a lowering of the temperature. Atoms and molecules in the surroundings (the rigid container) are in contact with CO_2 molecules in the solid, however, and energy exchange can occur. As soon as the temperature of the solid CO_2 drops slightly below the temperature of the surroundings, more energy transfers into the solid CO_2 than transfers out, and the temperature of the solid CO_2 is restored to −78 °C. For the temperature not to drop below −78 °C, the heat transfer from the surroundings must exactly equal the energy required to overcome attractive forces as solid CO_2 sublimes.

What we have just described is transfer of energy as heat from the surroundings to the system while the process $CO_2(s) \rightarrow CO_2(g)$ takes place. In such a case the temperature of the system at first decreases, but then thermal energy is transferred from the surroundings and constant temperature is maintained. When heat must be transferred into a system to maintain constant temperature as a process occurs, the process is said to be **endothermic.** In contrast, the reverse process, converting CO_2 gas to solid CO_2 (dry ice), is an **exothermic** process; heat is transferred *out of* the sample of CO_2 to the surroundings when some of the $CO_2(g)$ condenses to a solid. In summary,

The prefix exo- means "out" and endo- means "in."

Phase Change	Direction of Heat Transfer	Sign of q_{system}	Type of Change
Solid CO_2 → CO_2 Gas	Surroundings → System	positive	*Endo*thermic
CO_2 Gas → Solid CO_2	System → Surrounding	negative	*Exo*thermic

When the endothermic process $CO_2(s) \rightarrow CO_2(g)$ occurs inside a rigid container, the only exchange of energy is the heat transferred into the system. Thus, by conservation of energy, the *heat absorbed at constant volume,* q_v, must equal the *change* in the energy of the system, ΔE.

The subscript v on q_v indicates that the quantity of heat has been determined at constant volume.

If the process does not occur in a rigid container the situation is a bit more complicated, because two energy transfers occur, not just one. As shown in Figure 6.11c, $CO_2(s)$ vaporizing into a flexible container can raise a book above a

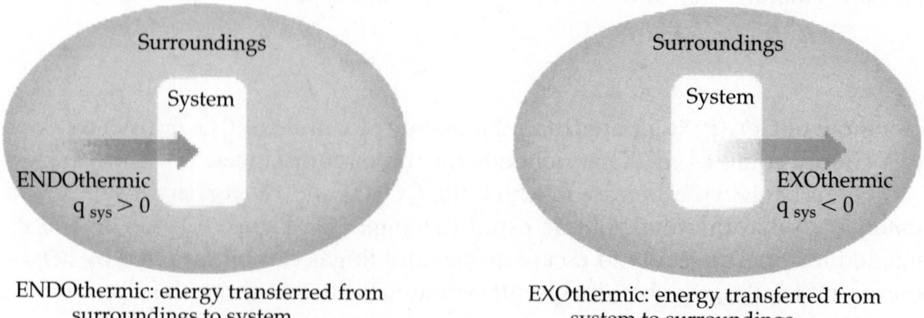

ENDOthermic: energy transferred from surroundings to system

EXOthermic: energy transferred from system to surroundings

A CLOSER LOOK *Internal Energy*

In chemistry, changes in internal energy, ΔE, that accompany a chemical or physical change are often the focus of attention. The internal energy in any given chemical system can be viewed as the sum of the potential and kinetic energy quantities associated with the system. Potential energy is the energy associated with the attractive and repulsive forces between all the nuclei and electrons in the system. This represents the energy associated with bonds in molecules, forces between ions, and also the forces between molecules in the liquid and solid state. Kinetic energy is the energy of motion of the atoms, ions, and molecules in the system; it is the sum of the energies associated with their translations, vibrations, and rotations.

Consider what happens when a chemical reaction takes place. A reaction rearranges the electrons and nuclei, so the potential energy of the products will be different than that of the reactants. Recall, however, that energy, like mass, is conserved in a reaction. If the potential energy associated with the products is lower than that of the reactants, then either energy is transferred out of the system into the surroundings, or the energy of the system is retained in the system as kinetic energy. In the latter case, the system becomes hotter. Conversely, if the internal energy of the products is higher than the energy of the reactants, then energy must either be added to the system or the energy needed is gained at the expense of the kinetic energy of the molecules in the system. In this case the temperature of the system decreases.

---

table top. Raising a book is an example of doing work. Not only is energy transferred into the solid CO_2 as heat, but energy is also transferred out of the CO_2 gas as work. Work is done whenever something is caused to move against an opposing force. In this case the gas has to work against the weight of the book to raise it.

Because energy is now transferred in two ways between system and surroundings, it is no longer true that $\Delta E = q$. Energy transferred in or out of the system by work, symbolized by w, must also be taken into account. The change in energy of the system is now

$$\Delta E = q + w \qquad (6.3)$$

Notice that w represents energy transferred, so it has a direction and a sign, just like q. The work w has a negative value if the system expends energy as work, and w has a positive value if the system receives energy in the form of work. For the process $CO_2(s) \rightarrow CO_2(g)$, w is negative because work is done on the surroundings when the atmosphere is pushed back.

That is, the energy of the system is changed by the quantity of energy transferred as heat and by the quantity of energy transferred by work. This is a statement of the **first law of thermodynamics,** also called the law of energy conservation. Another way of saying the same thing is that *the total amount of energy in the universe is constant.* Energy may be transferred as work or heat, but no energy can be lost, nor can heat or work be obtained from nothing. All the energy transferred between a system and its surroundings must be accounted for as heat and work. [This is true as long as no other energy transfer, such as the radiant energy of a glowing light stick (Figure 6.12) is involved.] The first law is important in all aspects of our lives; it plays a central role in our models of weather, in designing a power plant, or in understanding why diet and exercise together lead to weight loss.

Even if the book had not been on top of the plastic bag in Figure 6.11c, work would have been done by the expanding gas. This is because whenever a gas expands into the atmosphere it has to push back the atmosphere itself. Instead of raising a book, the expanding gas moves a part of the atmosphere. For the process shown in Figure 6.11d, the energy transferred from the surroundings has two effects: (1) it overcomes the forces holding the molecules together in the solid state at −78 °C, and (2) it does work on the atmosphere as the gas expands.

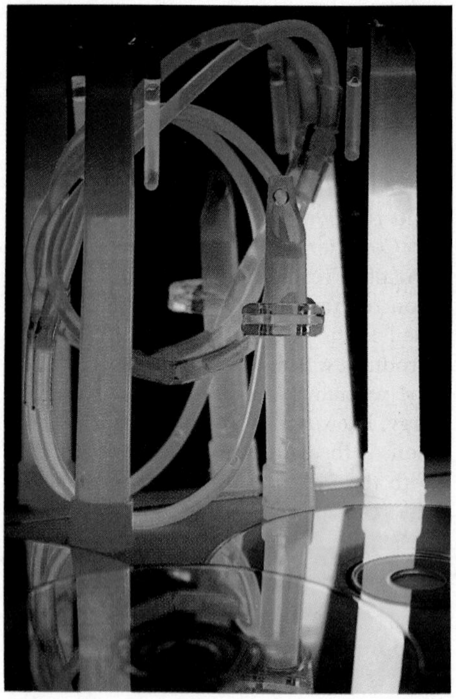

Figure 6.12 In "light sticks" a chemical reaction produces energy that is transferred to the surroundings in the form of light. (C. D. Winters)

Systems that convert heat into work are called heat engines: an example is the engine in an automobile, which converts heat resulting from the combustion of fuel into work to move the car forward.

The energy that allowed the system to perform work on its surroundings had to come from somewhere, and that "somewhere" was the energy, q, transferred as heat to the system from its surroundings. The system simply converted part of that heat into work. Therefore, the first law of thermodynamics tells us that the energy change, ΔE, for the CO_2 system must be less than q. The reason for this is that heat is transferred into the system from its surroundings (q has a positive sign), and work is done by the system on the surroundings (w has a negative sign). That is,

$$\Delta E = q + w = (q\text{, positive number}) + (w\text{, negative number})$$

and so

$$\Delta E < q$$

In plants and animals, as well as in the laboratory, reactions usually occur at constant pressure. *The heat transferred into (or out of) a system at constant pressure, q_p, equals a quantity called the* **enthalpy change,** symbolized by $\boldsymbol{\Delta H}$ ("delta H"). Thus,

The subscript p on q_p indicates that heat has been transferred at constant pressure.

$$\Delta H = q_p = \text{heat transferred to or from a system at constant pressure}$$

Because $\Delta E = q + w$, then at constant pressure $\Delta E = \Delta H + w$, showing that ΔH accounts for all the energy transferred except the amount that does the work of pushing back the atmosphere (which is usually small compared with ΔH). Even if pressure is not constant,

$$\begin{aligned}\Delta H &= \text{(enthalpy of the system at the end of a process)}\\ &\quad - \text{(enthalpy of the system at the start of the process)} \qquad \textbf{(6.4)}\\ &= H_{\text{final}} - H_{\text{initial}}\end{aligned}$$

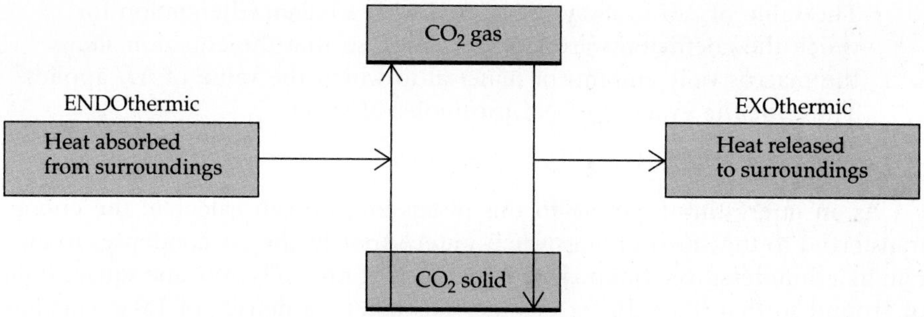

Figure 6.13 Enthalpy diagram for the interconversion of solid CO_2 and CO_2 gas at constant pressure.

If the enthalpy of the final system is greater than that of the initial system (as when solid CO_2 changes to CO_2 vapor at −78 °C), the enthalpy has increased (left side of Figure 6.13), and the process has a positive ΔH; q_p must also be positive and the process is **endo**thermic. Conversely, if the enthalpy of the final system is less than that of the initial system, heat has transferred out of the system, and ΔH is negative; the process is **exo**thermic (right side of Figure 6.13).

Negative enthalpy change: ΔH has a negative value; $\Delta H < 0$

$$\text{System} \xrightarrow{\text{Heat transfer}} \text{Surroundings}$$

Exothermic process with respect to system

Positive enthalpy change: ΔH has a positive value; $\Delta H > 0$

$$\text{Surroundings} \xrightarrow{\text{Heat transfer}} \text{System}$$

Endothermic process with respect to system

Another endothermic process is the evaporation of water to water vapor at 25 °C. The evaporation of one mole of water requires 44 kJ.

$$H_2O(\ell) \xrightarrow{+44.0\ \text{kJ}} H_2O(g) \qquad \Delta H_{\text{vaporization}} = +44.0\ \text{kJ/mol}$$

Change is endothermic
44.0 kJ of heat energy transferred from surroundings to the system (liquid H_2O)

But what about water vapor condensing to form liquid again? If 44.0 kJ of heat energy is required to break the attractions between H_2O molecules in one mole of the liquid so they can move into the gas phase, the same quantity of energy (44.0 kJ per mole) is regained when the molecules in the vapor condense to form the liquid. Condensation of water is exothermic; 44.0 kJ per mole is transferred from the water to the surroundings when a mole of H_2O molecules condenses to liquid.

The enthalpy of vaporization of any substance depends on the temperature at which it is measured. For water, $\Delta H_{\text{vaporization}}$ is 40.65 kJ/mol at the boiling point and 43.98 kJ/mol at 25 °C.

$$H_2O(g) \xrightarrow{-44.0\ \text{kJ}} H_2O(\ell) \qquad \Delta H_{\text{condensation}} = -44.0\ \text{kJ/mol}$$

Change is exothermic
44.0 kJ of heat energy transferred to surroundings from the system (H_2O vapor)

SUMMARY: *ΔH and Phase Changes*

Liquid → Solid + heat
Vapor → Liquid + heat

ΔH = negative number; change is **exo**thermic

Liquid + heat → Vapor
Solid + heat → Liquid

ΔH = positive number; change is **endo**thermic

Some key ideas that apply to all of thermodynamics may be summarized here.

- When heat transfer occurs (at constant pressure) from a system to its surroundings, as when a vapor condenses to a liquid or solid, the process is exothermic with respect to the system and $\Delta H(q_p)$ has a negative value. Conversely, when heat is absorbed from the surroundings, as when a solid or liquid changes to a vapor, the process is endothermic with respect to the system and $\Delta H(q_p)$ has a positive value.
- For changes that are the reverse of each other, the ΔH values are numerically the same, but their signs are opposite. Thus, for evaporation of water, $\Delta H = +44.0$ kJ/mol, whereas for the condensation of water $\Delta H = -44.0$ kJ/mol.
- The change in energy or enthalpy is directly proportional to the quantity of material undergoing a change. If two moles of water are evaporated, twice as much heat energy or 88.0 kJ is required.

One reason that water works to put out a fire is that the energy derived from combustion is used to vaporize water and is therefore not available to cause further combustion. (Photo Researchers, Inc.)

- The value of ΔH is always associated with a balanced equation for which the coefficients are read as moles, so that the equation shows the macroscopic amount of material to which the value of ΔH applies. Thus, for the evaporation of two moles of water

$$2\ H_2O(\ell) \longrightarrow 2\ H_2O(g) \qquad \Delta H = +88.0\ \text{kJ}$$

As an interesting footnote to this discussion, we can calculate the energy transferred to the surroundings when water vapor in the air condenses to give rain in a thunderstorm. Suppose that one inch of rain falls over one square mile of ground so that 6.6×10^{10} g of water has fallen (a density of 1.0 g/cm^3 has been assumed). The heat of vaporization of water at 25 °C is 44.0 kJ/mol. The quantity of heat transferred to the surroundings from water vapor condensation is therefore

$$6.6 \times 10^{10}\ \text{g water} \cdot \frac{1\ \text{mol}}{18.0\ \text{g}} \cdot \frac{44.0\ \text{kJ}}{1\ \text{mol}} = 1.6 \times 10^{11}\ \text{kJ}$$

This is about the same as the heat released when about 35 million kilograms (about 38,000 tons) of dynamite explodes! (The explosion of 1000 tons of dynamite is equivalent to 4.2×10^9 kJ.) This huge number tells you how much energy is "stored" in water vapor and why we think of storms as such great forces of energy in nature.

EXAMPLE 6.5 *Changes of State and* ΔH

The enthalpy change for the vaporization of methanol, CH_3OH, at 25 °C is 37.43 kJ/mol. What quantity of heat energy must be used to vaporize 25.0 g of methanol at 25 °C?

Solution To use the enthalpy of vaporization, we must first convert the mass of the compound to moles of compound.

$$25.0\ \text{g}\ CH_3OH \cdot \frac{1\ \text{mol}}{32.04\ \text{g}} = 0.780\ \text{mol}\ CH_3OH$$

Now the enthalpy of vaporization can be used to find the heat required.

$$0.780\ \text{mol}\ CH_3OH \cdot \frac{37.43\ \text{kJ}}{1\ \text{mol}\ CH_3OH} = 29.2\ \text{kJ}$$

EXERCISE 6.8 *Changes of State and* ΔH

The enthalpy change for the sublimation of solid iodine is 62.4 kJ/mol.

$$I_2(s) \longrightarrow I_2(g) \qquad \Delta H = 62.4\ \text{kJ}$$

1. What quantity of heat energy must be used to sublime 10.0 g of solid iodine?
2. If 3.45 g of iodine vapor condenses to solid iodine, what quantity of energy is involved? Is the process exo- or endothermic? ■

6.5 ENTHALPY CHANGES FOR CHEMICAL REACTIONS

The enthalpy change can be determined for any physical or chemical change. For a chemical reaction the products represent the "final system" and the reactants the "initial system." Therefore,

$$\Delta H = H_{products} - H_{reactants} \quad (6.5)$$

Like changes of state, chemical reactions can be exothermic or endothermic. Many of the reactions you have seen so far in photographs in this book [such as the Al + Br_2 reaction (Figure 4.1) and the reactions of elements with air (Figures 4.2 to 4.4)] are exothermic. To learn more about enthalpy changes that accompany chemical reactions, consider the decomposition of one mole of water vapor to its elements.

$$H_2O(g) \xrightarrow{+241.8\text{ kJ}} H_2(g) + \tfrac{1}{2}\,O_2(g) \qquad \Delta H = +241.8\text{ kJ}$$

Change is endothermic; ΔH is positive
Decomposition of 1 mol of water vapor requires 241.8 kJ of energy to be transferred in from the surroundings

(Note that to write an equation for the decomposition of one mole of H_2O, it is necessary to use a fractional coefficient for O_2. This is acceptable in thermochemistry because coefficients are always taken to mean moles and not molecules.) The left side of Figure 6.14 shows that the enthalpy of the products of this reaction is greater than that of the reactants. Water vapor would have to *absorb* 241.8 kJ of energy (at constant pressure) from the surroundings as it decomposes to its elements. That is, the enthalpy change for the *endo*thermic decomposition of water is $\Delta H = +241.8$ kJ per mole of water.

The decomposition of water can be reversed; hydrogen and oxygen can combine to form water. The quantity of heat energy involved in this oxidation-reduction reaction (Figure 6.14)

$$H_2(g) + \tfrac{1}{2}\,O_2(g) \xrightarrow{-241.8\text{ kJ}} H_2O(g) \qquad \Delta H = -241.8\text{ kJ}$$

Change is exothermic; ΔH is negative
Formation of 1 mol of water vapor transfers 241.8 kJ of energy to the surroundings

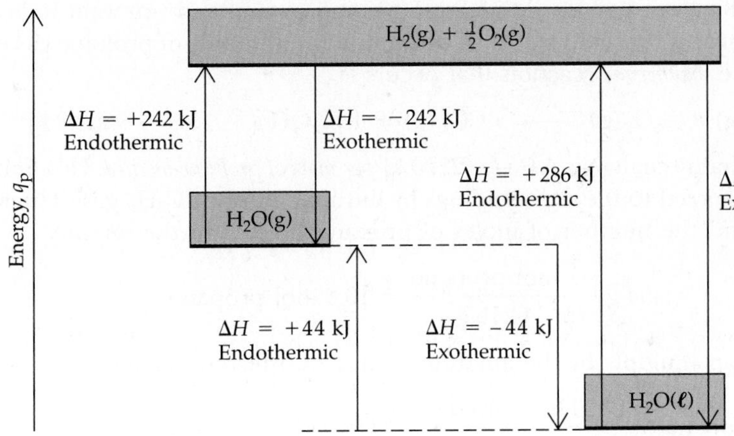

Figure 6.14 Enthalpy diagram for the interconversion of water vapor, $H_2O(g)$, liquid water, $H_2O(\ell)$, and the elements.

is the same as for the decomposition reaction except that the combination of the elements is *exo*thermic. That is, $\Delta H = -241.8$ kJ per mole of water vapor formed.

SUMMARY: Relationship of Reaction Heat and Enthalpy Change

Reactant → Product + heat
ΔH = negative number; reaction is **exo**thermic

Reactant + heat → Product
ΔH = positive number; reaction is **endo**thermic

As in the case of phase changes, some key ideas about enthalpy changes for reactions can be summarized:

- When heat is evolved or transferred (at constant pressure) to the surroundings by an exothermic reaction, ΔH has a negative value. Conversely, when heat is absorbed from the surroundings by an endothermic reaction, ΔH has a positive value.
- For chemical reactions that are the reverse of each other, the ΔH values are numerically the same, but their signs are opposite.
- The change in energy or enthalpy is directly proportional to the quantity of material undergoing a change. If 2 moles of water are decomposed, twice as much heat energy, or 483.6 kJ, is required.
- In thermodynamics, the value of ΔH is always associated with a balanced equation for which the coefficients are read as moles, so that the equation shows the macroscopic amount of material to which the value of ΔH applies. Thus, for the decomposition of 2 moles of water vapor

$$2\ H_2O(g) \longrightarrow 2\ H_2(g) + O_2(g) \qquad \Delta H = +483.6\ \text{kJ}$$

Because energy is transferred when a substance undergoes a change of state, the quantity of energy associated with a chemical reaction must depend on the physical state (solid, liquid, or gas) of the reactants and products. For example, the decomposition of one mole of *liquid* water to H_2 and O_2

$$H_2O(\ell) \longrightarrow H_2(g) + \tfrac{1}{2}\ O_2(g) \qquad \Delta H = +285.8\ \text{kJ}$$

requires more energy than the decomposition of one mole of water vapor, as shown in Figure 6.14. The heat of vaporization, 44 kJ/mol, must be added to the enthalpy change for the decomposition of $H_2O(g)$ to give the value for the decomposition of $H_2O(\ell)$.

Enthalpy changes for reactions have many practical applications. For instance, when enthalpies of combustion are known, the quantity of heat transferred by the combustion of a given mass of a fuel such as propane, C_3H_8, can be calculated. Suppose you are designing a heating system, and you want to know how much heat is provided by burning 454 grams (1 pound) of propane gas in a furnace. The *exothermic* reaction that occurs is

$$C_3H_8(g) + 5\ O_2(g) \longrightarrow 3\ CO_2(g) + 4\ H_2O(\ell) \qquad \Delta H = -2220\ \text{kJ}$$

and the enthalpy change is $\Delta H = -2220$ kJ *per mole of propane burned.* How much heat is transferred to the surroundings by burning 454 g of C_3H_8 gas? The first step is to find the number of moles of propane present in the sample.

$$454\ \cancel{\text{g}} \cdot \frac{1\ \text{mol propane}}{44.10\ \cancel{\text{g}}} = 10.3\ \text{mol propane}$$

Then you can multiply by the amount of heat transferred per mole of gas.

$$10.3\ \cancel{\text{mol propane}} \cdot \frac{2220\ \text{kJ evolved}}{1.00\ \cancel{\text{mol propane}}} = 22{,}900\ \text{kJ}$$

$$= \text{total heat evolved by 454 g of propane}$$

CURRENT ISSUES IN CHEMISTRY

MREs Are Heated in the FRH

If you have been in the Army or Marine Corps you know about MREs—meals ready to eat—and FRHs—Flameless Ration Heaters.

Soldiers need to eat when in the field, and so, until recently, they carried the canned C-rations made famous in World War II. Beginning with Operation Desert Storm, however, the 1990–1991 war to push Iraq out of Kuwait, and in the recent effort to feed starving people in Somalia, soldiers carried MREs with an FRH.

The main course of an MRE is a meal, such as chicken stew or spaghetti and meatballs, in a pouch made of plastic and aluminum foil. The food can be heated by dropping the pouch into a pot of boiling water or leaning it on the exhaust manifold of an engine. But there may not be time to boil water or wait for dinner to heat on an engine. The alternative is the FRH.

To heat an MRE a soldier drops the food pouch into a bag-like sleeve, slides in the FRH, which is enclosed in a thin plastic sleeve of its own, and then adds a small amount of water to the sleeve. The FRH contains magnesium metal, which combines with water to form magnesium hydroxide in a very exothermic, oxidation-reduction reaction.

$$Mg(s) + 2\ H_2O(\ell) \longrightarrow Mg(OH)_2(s) + H_2(g) + \text{heat}$$

The reaction generates enough heat to cook the meal without flame or smoke.

If you have never seen an MRE with its FRH, you might find a hand warmer in a camping store that works on the same principle. One type of hand warmer contains iron powder and other chemicals. It also works by an oxidation-reduction reaction, but in this case it is the reaction of iron with the O_2 of air.

$$4\ Fe(s) + 3\ O_2(g) \longrightarrow 2\ Fe_2O_3(s) + \text{heat}$$

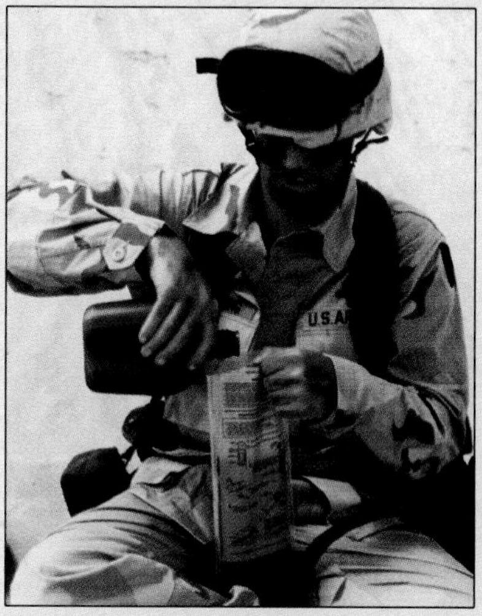

Soldier using an MRE. (U. S. Army Natick Research and Development Center)

This reaction is also very exothermic and can allow the hand warmer to maintain a temperature of 57 to 69 °C for several hours if the oxygen flow is somewhat restricted, as by keeping the warmer in one of your gloves.

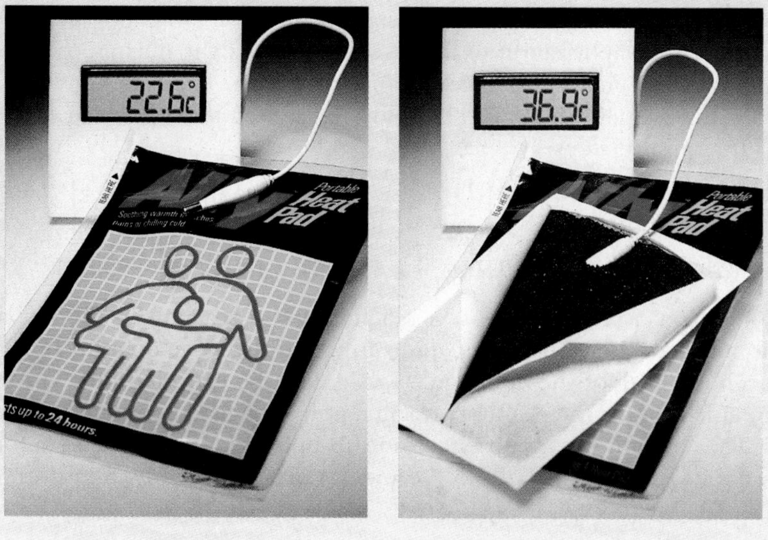

A hand warmer that uses the oxidation of iron as the source of heat. (C. D. Winters)

This is a substantial quantity of energy when you compare it with the fact that, when your body completely "burns" 454 g of milk, only about 1400 kJ is evolved, in part because milk is largely water.

EXERCISE 6.9 *Heat Energy Calculation*

How much heat energy is required to decompose 12.6 g of liquid water to the elements?

$$H_2O(\ell) \longrightarrow H_2(g) + \tfrac{1}{2}\,O_2(g) \qquad \Delta H = +285.8\text{ kJ}$$

■

6.6 HESS'S LAW

Knowing how much heat is transferred as a chemical process occurs is very important. First of all, the direction of heat transfer is an important clue that helps predict in which direction a chemical reaction will go because, *at room temperature, most exothermic reactions are product-favored.* Second, we can use ΔH to calculate the heat obtainable when a fuel is burned, as was done in Section 6.5. Third, when reactions are carried out on a larger scale, as in a chemical plant that produces sulfuric acid or ethylene as a raw material for plastics, the surroundings must have enough cooling capacity to prevent an exothermic reaction from overheating and possibly damaging the plant. We therefore want to know ΔH values for as many reactions as possible. For many reactions direct experimental measurements can be made by using a device called a calorimeter (Section 6.9), but for many other reactions this is not a simple task. Besides, it would be very time-consuming to measure values for every conceivable reaction, and it would take a great deal of space to tabulate so many values. Fortunately there is a better way. It is based on the fact that mass *and* energy are conserved in chemical reactions.

Energy conservation is the basis of **Hess's law,** which states that, *if a reaction is the sum of two or more other reactions, then ΔH for the overall process must be the sum of the ΔH values of the constituent reactions.* For example, as illustrated in the preceding section (see Figure 6.14) for the decomposition of *liquid* water into the elements $H_2(g)$ and $O_2(g)$ (with all substances at 25 °C), the two successive changes are (1) the vaporization of liquid water and (2) the decomposition of water vapor to the elements (with all substances at 25 °C). The equation and the ΔH value for the overall process can be found by adding the equations and the ΔH values for the two steps:

$$\begin{array}{lll} (1) & H_2O(\ell) \longrightarrow H_2O(g) & \Delta H_1 = +44.0\text{ kJ} \\ (2) & H_2O(g) \longrightarrow H_2(g) + \tfrac{1}{2}\,O_2(g) & \Delta H_2 = +241.8\text{ kJ} \\ \hline (1) + (2) & H_2O(\ell) \longrightarrow H_2(g) + \tfrac{1}{2}\,O_2(g) & \Delta H_{net} = +285.8\text{ kJ} \end{array}$$

Here, $H_2O(g)$ is a product of the first reaction and a reactant in the second. Thus, as in adding two algebraic equations in which the same quantity or term appears on both sides of the equation, $H_2O(g)$ can be canceled out. The net result is an equation for the overall reaction and its associated enthalpy change.

Hess's law is useful because it enables us to find the enthalpy change for a reaction or state change that cannot be measured conveniently. Suppose you want to know the enthalpy change for the formation of carbon monoxide, CO, from the elements, solid carbon (as graphite) and oxygen gas.

$$C(s) + \tfrac{1}{2}\,O_2(g) \longrightarrow CO(g) \qquad \Delta H = ?$$

Experimentally this is not easy to do; it is difficult to keep the carbon from burning completely to carbon dioxide. The way to solve this is to recognize that we can more easily determine by experiments the enthalpy change for the conversion of carbon to CO_2 and for the conversion of CO to CO_2.

$$C(s) + O_2(g) \longrightarrow CO_2(g) \qquad \Delta H = -393.5\text{ kJ}$$

$$CO(g) + \tfrac{1}{2}\,O_2(g) \longrightarrow CO_2(g) \qquad \Delta H = -283.0\text{ kJ}$$

As the diagram in Figure 6.15 makes clear, the formation of CO is the first of two steps in going from carbon to carbon dioxide. We know the enthalpy change for the second step (ΔH_2), and we know the enthalpy change for the net reaction (ΔH_{net}). Using Hess's law, therefore, we can solve for the enthalpy change for the first step (ΔH_1). This can be done readily, because ΔH for the net reaction is the sum of the ΔH values for the first and second steps.

Step 1	$C(s) + \tfrac{1}{2}\,O_2(g) \longrightarrow CO(g)$	$\Delta H_1 = ?$
Step 2	$CO(g) + \tfrac{1}{2}\,O_2(g) \longrightarrow CO_2(g)$	$\Delta H_2 = -283.0\text{ kJ}$
Net reaction	$C(s) + O_2(g) \longrightarrow CO_2(g)$	$\Delta H_{net} = -393.5\text{ kJ}$

$$\begin{aligned}\Delta H_{net} &= \Delta H_1 + \Delta H_2 \\ -393.5\text{ kJ} &= \Delta H_1 + (-283.0\text{ kJ}) \\ \Delta H_1 &= -110.5\text{ kJ}\end{aligned}$$

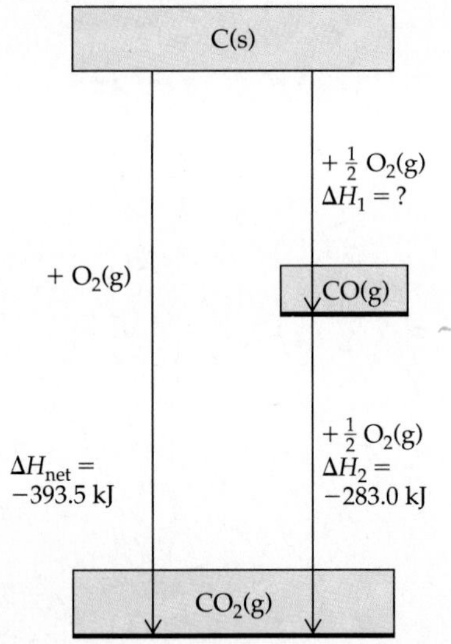

Figure 6.15 One mole of carbon dioxide can be formed from carbon and oxygen in a direct reaction ($\Delta H_{net} = -393.5$ kJ) or in two steps. Hess's law states that the sum of the enthalpy changes for the two steps must equal that for the direct reaction. That is, $\Delta H_1 + \Delta H_2 = \Delta H_{net}$. Knowing any two of these values, therefore, allows us to calculate the third.

EXAMPLE 6.6 *Using Hess's Law*

Suppose we want to know the enthalpy change for the formation of methane, CH_4, from solid carbon (as graphite) and hydrogen gas.

$$C(s) + 2\,H_2(g) \longrightarrow CH_4(g) \qquad \Delta H = ?$$

The enthalpy change for the direct combination of the elements would be extremely difficult to measure in the laboratory. We can measure ΔH, however, when the elements and methane burn in oxygen (Section 6.9).

	Reaction	ΔH (kJ)
(1)	$C(s) + O_2(g) \longrightarrow CO_2(g)$	−393.5
(2)	$H_2(g) + \tfrac{1}{2}\,O_2(g) \longrightarrow H_2O(\ell)$	−285.8
(3)	$CH_4(g) + 2\,O_2(g) \longrightarrow CO_2(g) + 2\,H_2O(\ell)$	−890.3

Solution The three reactions (1, 2, and 3), as they are now written, cannot be added together to obtain the equation for the formation of CH_4 from its elements. CH_4 should be a product, but it is a reactant in equation (3). The solution to this is to reverse equation (3). At the same time, the sign of ΔH for the reaction is reversed. If a reaction is exothermic in one direction (the combustion of methane evolves energy), its reverse must be endothermic.

(3)′ $CO_2(g) + 2\ H_2O(\ell) \longrightarrow CH_4(g) + 2\ O_2(g)$

$$\Delta H = -\Delta H_3 = +890.3\ \text{kJ}$$

Next, we see that two moles of $H_2O(\ell)$ is required as a product, whereas equation (2) is written for only one mole of water. We therefore multiply the stoichiometric coefficients in (2) by 2 and multiply the value of ΔH by 2.

2 × (2) $2\ H_2(g) + O_2(g) \longrightarrow 2\ H_2O(\ell)$

$$2\ \Delta H_2 = 2(-285.8\ \text{kJ}) = -571.6\ \text{kJ}$$

With these modifications, we can add the three equations to give the equation for the formation of methane from its elements.

	Reaction	**ΔH (kJ)**
(1)	$C(s) + O_2(g) \longrightarrow CO_2(g)$	−393.5 kJ
2 × (2)	$2\ H_2(g) + O_2(g) \longrightarrow 2\ H_2O(\ell)$	−571.6 kJ
(3)′	$CO_2(g) + 2\ H_2O(\ell) \longrightarrow CH_4(g) + 2\ O_2(g)$	+890.3 kJ
Net	$C(s) + 2\ H_2(g) \longrightarrow CH_4(g)$	−74.8 kJ

The solution to this problem is shown diagrammatically in Figure 6.16.

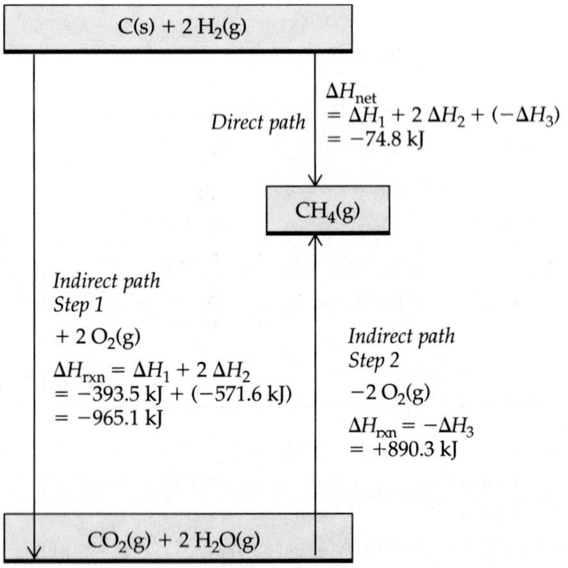

Figure 6.16 Methane may be formed directly from the elements. Alternatively, the carbon may be burned to $CO_2(g)$ and the hydrogen burned to water. Then CO_2 and H_2O are combined to form CH_4 (and the O_2 used in the combustion is returned). Hess's law states that the enthalpy change for the direct reaction (ΔH_{net}) is the sum of the enthalpy changes along the alternative path ($\Delta H_{net} = \Delta H_1 + 2\Delta H_2 + (-\Delta H_3)$). See Example 6.6 for a more complete designation of the ΔH values.

EXERCISE 6.10 *Using Hess's Law*

What is the enthalpy change for forming ethane, C_2H_6 from C(s) and $H_2(g)$?

$$2\ C(s) + 3\ H_2(g) \longrightarrow C_2H_6(g) \qquad \Delta H = ?$$

Use information from Example 6.6 together with the experimentally determined ΔH value for the combustion of ethane.

$$C_2H_6(g) + \tfrac{7}{2}\ O_2(g) \longrightarrow 2\ CO_2(g) + 3\ H_2O(\ell) \qquad \Delta H = -1559.7\ \text{kJ}$$ ■

EXERCISE 6.11 *Using Hess's Law*

Lead has been known and used for centuries. To obtain the metal, lead(II) sulfide (PbS; in the form of a common mineral called galena) is first roasted in air to form lead(II) oxide (PbO).

$$PbS(s) + \tfrac{3}{2}\ O_2(g) \longrightarrow PbO(s) + SO_2(g) \qquad \Delta H = -413.7\ \text{kJ}$$

and then lead(II) oxide is reduced with carbon to the metal.

$$PbO(s) + C(s) \longrightarrow Pb(s) + CO(g) \qquad \Delta H = +106.8\ \text{kJ}$$

What is the enthalpy change for the following reaction?

$$PbS(s) + C(s) + \tfrac{3}{2}\ O_2(g) \longrightarrow Pb(s) + CO(g) + SO_2(g)$$

Is the reaction exothermic or endothermic? How much energy, in kilojoules, is required (or evolved) when 454 g (1.00 pound) of PbS is converted to lead? ■

6.7 STATE FUNCTIONS

Hess's law works because the enthalpy change for a reaction is a **state function,** a quantity whose value is determined only by the state of the system. *The enthalpy change for a chemical or physical change does not depend on the path you choose to go from the initial conditions to the final conditions.* No matter how you go from reactants to products in a reaction the net heat evolved or required (at constant pressure) is always the same. This point is illustrated in Figure 6.15 for the formation of CO_2 from its elements and in Figure 6.16 for the formation of methane. For the reaction

$$C(s) + 2\ H_2(g) \longrightarrow CH_4(g) \qquad \Delta H = -74.8\ \text{kJ}$$

the enthalpy change for the direct reaction ($\Delta H_{direct} = -74.8$ kJ) is the same as the sum of the enthalpy changes along the indirect pathway that goes through carbon dioxide and water.

Many commonly measured quantities, such as the pressure of a gas, the volume of a gas or liquid, the temperature of a substance, and the size of your bank account, are state functions. You could have arrived at a current bank balance of \$25 by having deposited \$25, or you could have deposited \$100 and then withdrawn \$75. The volume of a balloon is also a state function. You can blow up a balloon to a large volume and then let some air out to arrive at the desired volume. Alternatively, you can blow up the balloon in stages, adding tiny amounts of air at each stage. The final volume does not depend on how you got there. For both bank accounts and balloons, an infinite number of ways exist for how to arrive at the final state, but the final value depends only on the size of the bank balance or the balloon, and not on the path taken from the initial to the final state.

Because enthalpy is a state function, in principle there is an absolute enthalpy for the reactants ($H_{\text{initial}} = H_{\text{reactants}}$) and for the products ($H_{\text{final}} = H_{\text{products}}$). The difference between these enthalpies is the change for the system.

$$\Delta H_{\text{reaction}} = H_{\text{final}} - H_{\text{initial}} = H_{\text{products}} - H_{\text{reactants}}$$

Because the reaction starts and finishes at the same place no matter which pathway is chosen, $\Delta H_{\text{reaction}}$ must always be independent of pathway. Unlike volume, temperature, pressure, energy, or a bank balance, however, the absolute enthalpy of a substance is not usually determined, only its change in a chemical or physical process. The thermal energy evolved or required in a chemical process, for example, is a reflection of the *difference* in enthalpy between the reactants and products. We determine only whether the enthalpy of the products is greater than or less than that of the reactants by some amount.

6.8 STANDARD ENTHALPIES OF FORMATION

Hess's law makes it possible for us to tabulate ΔH values for a few reactions and to derive a great many other ΔH values by adding together appropriate ones as described in Section 6.7. Because ΔH values depend on temperature and pressure, it is necessary to specify both of these when a table of data is made. Usually a pressure of 1 bar and a temperature of 25 °C are specified, although other conditions of temperature are used. The **standard state** of an element or a compound is the most stable form of the substance in the physical state in which it exists at 1 bar and the specified temperature. Thus, at 25 °C and a pressure of 1 bar, the standard state for hydrogen is the gaseous state, $H_2(g)$, and for sodium chloride, it is the solid state, NaCl(s). For carbon, which can exist in any one of three solid states at 1 bar and 25 °C, the most stable form, graphite, is selected as the standard.

In 1982, the International Union of Pure and Applied Chemistry chose a pressure of 1 bar for the standard state. This pressure is very close to the average atmospheric pressure observed near sea level (1 bar = 0.98692 standard atmosphere). See Chapter 12.

When a reaction occurs with all the reactants and products in their standard states under standard conditions, the observed, or calculated, enthalpy change is known as the **standard enthalpy change of reaction, $\Delta H°$,** where the superscript ° indicates standard conditions. All the reactions we have discussed to this point have followed this convention, so all the ΔH values should have ° attached.

It is common to use the term "heat of reaction" interchangeably with "enthalpy of reaction." Understand that it is only the heat of reaction at **constant pressure, q_p,** that is equivalent to the enthalpy change.

The standard enthalpy change of a reaction for the formation of one mole of a compound directly from its elements is called the **standard molar enthalpy of formation, $\Delta H_f°$,** where the subscript $_f$ indicates that one mole of the compound in question has been *formed* in its standard state from its elements, also in their standard states. Some of the reactions already discussed define standard molar enthalpies of formation.

$$H_2(g) + \tfrac{1}{2}\,O_2(g) \longrightarrow H_2O(\ell) \qquad \Delta H_f° = -285.8 \text{ kJ/mol}$$
$$C(s) + 2\,H_2(g) \longrightarrow CH_4(g) \qquad \Delta H_f° = -74.8 \text{ kJ/mol}$$

As another example, the following equation shows that 277.7 kJ is *evolved* if graphite, the standard state form for carbon, is combined with gaseous hydrogen and oxygen to form one mole of ethanol at 298 K.

$$2\ C(graphite) + 3\ H_2(g) + \tfrac{1}{2}\ O_2(g) \longrightarrow C_2H_5OH(\ell) \qquad \Delta H_f^\circ = -277.7\ \text{kJ/mol}$$

Finally, it is important to understand that a standard molar enthalpy of formation (ΔH_f°) is just a special case of an enthalpy change for a reaction. In contrast with the three previous reactions, the enthalpy change for the following reaction is *not* an enthalpy of formation; calcium carbonate has been formed from other compounds, not directly from its elements.

$$CaO(s) + CO_2(g) \longrightarrow CaCO_3(s) \qquad \Delta H_{rxn}^\circ = -178.3\ \text{kJ}$$

Therefore, the enthalpy change is given the more general symbol of ΔH_{rxn}°. The enthalpy change for the following reaction is also *not* a standard enthalpy of formation.

$$P_4(s) + 6\ Cl_2(g) \longrightarrow 4\ PCl_3(\ell) \qquad \Delta H_{rxn}^\circ = -1278.8\ \text{kJ}$$

Here a compound has been formed from its elements, but more than one mole of the compound has been formed, and so $\Delta H_{rxn}^\circ = 4\ \Delta H_f^\circ[PCl_3(\ell)]$.

Table 6.2 and Appendix K list values of ΔH_f°, obtained from the National Institute for Standards and Technology (NIST), for many other compounds. Be sure to notice that no values are listed in these tables for elements, such as C(graphite) or O_2(g). *Standard enthalpies of formation for the elements in their standard states are zero,* because forming an element in its standard state from the same element in its standard state involves no chemical or physical change.

To distinguish between enthalpies of formation and enthalpy changes for other kinds of reactions, we use the symbols ΔH_f° and ΔH_{rxn}°, respectively. The subscript $_{rxn}$ is short for "reaction."

Be sure to notice also that most ΔH_f° values are negative. That is, the process of forming most compounds from their elements is exothermic. Recall that we previously stated that *at room temperature most exothermic reactions are product-favored.* Thus, forming compounds from their elements (under standard conditions) is generally a product-favored process. Reactions of oxygen with metals are usually exothermic and product-favored

$$2\ Al(s) + \tfrac{3}{2}\ O_2(g) \longrightarrow Al_2O_3(s) \qquad \Delta H_f^\circ = -1675.7\ \text{kJ}$$

and carbon generally combines with other elements in reactions with negative enthalpy changes.

$$C(s) + O_2(g) \longrightarrow CO_2(g) \qquad \Delta H_f^\circ = -393.5\ \text{kJ}$$

$$2\ C(s) + 3\ H_2(g) + \tfrac{1}{2}\ O_2(g) \longrightarrow C_2H_5OH(\ell) \qquad \Delta H_f^\circ = -277.7\ \text{kJ}$$

One important exception to our general observation is ethyne (commonly called acetylene), a compound with a strongly endothermic standard enthalpy of formation.

$$2\ C(s) + H_2(g) \longrightarrow C_2H_2(g) \qquad \Delta H_f^\circ = +226.7\ \text{kJ}$$

It is interesting that acetylene is a good fuel and that it is the starting point for manufacturing many other carbon-containing compounds.

TABLE 6.2 Selected Standard Molar Enthalpies of Formation at 298 K

Substance	Name	Standard Molar Enthalpy of Formation (kJ/mol)
$Al_2O_3(s)$	Aluminum oxide	−1675.7
$BaCO_3(s)$	Barium carbonate	−1216.3
$CaCO_3(s)$	Calcium carbonate	−1206.9
$CaO(s)$	Calcium oxide	−635.1
$CCl_4(\ell)$	Carbon tetrachloride	−135.4
$CH_4(g)$	Methane	−74.8
$CH_3OH(\ell)$	Methanol	−238.7
$C_2H_5OH(\ell)$	Ethanol	−277.7
$CO(g)$	Carbon monoxide	−110.5
$CO_2(g)$	Carbon dioxide	−393.5
$C_2H_2(g)$	Ethyne (acetylene)	+226.7
$C_2H_4(g)$	Ethene (ethylene)	+52.3
$C_2H_6(g)$	Ethane	−84.7
$C_3H_8(g)$	Propane	−103.8
$C_4H_{10}(g)$	Butane	−888.0
$CuSO_4(s)$	Copper(II) sulfate	−771.4
$H_2O(g)$	Water vapor	−241.8
$H_2O(\ell)$	Liquid water	−285.8
$HF(g)$	Hydrogen fluoride	−271.1
$HCl(g)$	Hydrogen chloride	−92.3
$HBr(g)$	Hydrogen bromide	−36.4
$HI(g)$	Hydrogen iodide	+26.5
$KF(s)$	Potassium fluoride	−567.3
$KCl(s)$	Potassium chloride	−436.7
$KBr(s)$	Potassium bromide	−393.8
$MgO(s)$	Magnesium oxide	−601.7
$MgSO_4(s)$	Magnesium sulfate	−1284.9
$Mg(OH)_2(s)$	Magnesium hydroxide	−924.5
$NaF(s)$	Sodium fluoride	−573.6
$NaCl(s)$	Sodium chloride	−411.2
$NaBr(s)$	Sodium bromide	−361.1
$NaI(s)$	Sodium iodide	−287.8
$NH_3(g)$	Ammonia	−46.1
$NO(g)$	Nitrogen monoxide	+90.3
$NO_2(g)$	Nitrogen dioxide	+33.2
$PCl_3(\ell)$	Phosphorus trichloride	−319.7
$PCl_5(s)$	Phosphorus pentachloride	−443.5
$SiO_2(s)$	Silicon dioxide (quartz)	−910.9
$SnCl_2(s)$	Tin(II) chloride	−325.1
$SnCl_4(\ell)$	Tin(IV) chloride	−511.3
$SO_2(g)$	Sulfur dioxide	−296.8
$SO_3(g)$	Sulfur trioxide	−395.7

From "The NBS Tables of Chemical Thermodynamic Properties," 1982.

EXAMPLE 6.7 *Writing Equations to Define Enthalpies of Formation*

The standard enthalpy of formation of gaseous ammonia is −46.1 kJ/mol. Write the balanced equation for which the enthalpy of reaction is −46.1 kJ.

Solution The equation must show the formation of 1 mole of $NH_3(g)$ from the elements in their standard states; both N_2 and H_2 are gases at 25 °C and 1 bar. The correct equation is therefore

$$\tfrac{1}{2}\,N_2(g) + \tfrac{3}{2}\,H_2(g) \longrightarrow NH_3(g) \qquad \Delta H^\circ_{rxn} = \Delta H^\circ_f = -46.1\ \text{kJ}$$

EXERCISE 6.12 *Writing Equations to Define Enthalpies of Formation*

Write balanced equations to define the formation of ethanol and copper sulfate from their respective elements. Give the value of the standard molar enthalpy of formation for each. What is the value of ΔH°_{rxn} if 1.5 moles of ethanol are formed from the elements in their standard states? ■

Standard enthalpies of formation are very useful. For example, you can find the standard enthalpy change for any reaction *if the enthalpies of formation for all the reactants and products are known.* Suppose you are a chemical engineer and want to know how much heat is required to decompose calcium carbonate (limestone) to calcium oxide (lime) and carbon dioxide, with all substances at standard conditions.

$$CaCO_3(s) \longrightarrow CaO(s) + CO_2(g) \qquad \Delta H^\circ_{rxn} = ?$$

To do this, you find the following enthalpies of formation in a table such as Table 6.2 or Appendix K:

Compound	ΔH°_f (kJ/mol)
$CaCO_3(s)$	−1206.9
$CaO(s)$	−635.1
$CO_2(g)$	−393.5

and then use Equation 6.6 to find the standard enthalpy change for the reaction, ΔH°_{rxn}.

$$\text{Enthalpy change for a reaction} = \Delta H^\circ_{rxn} = \Sigma\,[\Delta H^\circ_f(\text{products})] - \Sigma[\Delta H^\circ_f(\text{reactants})] \qquad \textbf{(6.6)}$$

In Equation 6.6 the symbol Σ (the Greek letter *sigma*) means to "take the sum." Thus, to find ΔH°_{rxn} you add up the molar enthalpies of formation of the products and subtract from this sum the sum of the molar enthalpies of formation of the reactants. Applying this to the decomposition of limestone, we have

$$\begin{aligned}\Delta H^\circ_{rxn} &= \Delta H^\circ_f[CaO(s)] + \Delta H^\circ_f[CO_2(g)] - \Delta H^\circ_f[CaCO_3(s)]\\ &= 1\ \text{mol}(-635.1\ \text{kJ/mol}) + 1\ \text{mol}(-393.5\ \text{kJ/mol}) - 1\ \text{mol}(-1206.9\ \text{kJ/mol})\\ &= +178.3\ \text{kJ}\end{aligned}$$

A CLOSER LOOK Hess's Law and Equation 6.6

Equation 6.6 is a convenient way to apply Hess's law when the enthalpies of formation of all the reactants and products are known. Let us look again at the decomposition of calcium carbonate,

$$CaCO_3(s) \longrightarrow CaO(s) + CO_2(g) \qquad \Delta H^\circ_{rxn} = ?$$

and think about an alternative route from the reactant to the products. We can imagine the reaction as occurring by first breaking $CaCO_3$ up into its elements, and then recombining the elements in a different way to produce CO_2 and CaO.

The enthalpy change for each step in this process is known. Step 1 is the reverse of the equation for the formation of $CaCO_3$, so the enthalpy change is the negative of the enthalpy of formation of $CaCO_3$. For the other two steps, ΔH°_{rxn} is the same as the enthalpy of formation. Notice that the sum of reactions 1, 2, and 3 gives the equation for the net reaction. Most importantly, notice that ΔH°_{rxn} for the net reaction is the sum of the enthalpy changes for each step. Here

$$\Delta H^\circ_{rxn} = \Delta H^\circ_f[CaO(s)] + \Delta H^\circ_f[CO_2(g)] - \Delta H^\circ_f[CaCO_3(s)]$$

This is exactly the result given by applying Equation 6.6. The enthalpy change for the reaction is indeed the sum of the enthalpies of formation of the products minus that of the reactant.

	Reaction	ΔH°_{rxn}
Step 1.	$CaCO_3(s) \longrightarrow Ca(s) + C(s) + \frac{3}{2} O_2(g)$	$-\Delta H^\circ_f[CaCO_3(s)] = -(-1206.9\ kJ)$
Step 2.	$C(s) + O_2(g) \longrightarrow CO_2(g)$	$\Delta H^\circ_f[CO_2(g)] = -393.5\ kJ$
Step 3.	$Ca(s) + \frac{1}{2} O_2(g) \longrightarrow CaO(s)$	$\Delta H^\circ_f[CaO(s)] = -635.1\ kJ$
Net	$CaCO_3(s) \longrightarrow CaO(s) + CO_2(g)$	$\Delta H^\circ_{rxn} = +178.3\ kJ$

EXAMPLE 6.8 *Using Enthalpies of Formation*

Nitroglycerin is a powerful explosive, giving four different gases when detonated.

$$2\ C_3H_5(NO_3)_3(\ell) \longrightarrow 3\ N_2(g) + \tfrac{1}{2}\ O_2(g) + 6\ CO_2(g) + 5\ H_2O(g)$$

Given that the enthalpy of formation of nitroglycerin, ΔH°_f, is -364 kJ/mol, and consulting Table 6.2 for the enthalpies for the other compounds, calculate the energy (heat at constant pressure) liberated when 10.0 g of nitroglycerin is detonated.

Solution To solve this problem, we need the standard molar enthalpies of the products. Two are elements in their standard states (N_2 and O_2), so their values of ΔH°_f are zero. From Table 6.2, we have

$$\Delta H^\circ_f[CO_2(g)] = -393.5\ kJ/mol$$
$$\Delta H^\circ_f[H_2O(g)] = -241.8\ kJ/mol$$

The enthalpy change for the reaction can now be found using Equation 6.6:

$$\Delta H^\circ_{rxn} = 6\ mol \cdot \Delta H^\circ_f[CO_2(g)] + 5\ mol \cdot \Delta H^\circ_f[H_2O(g)] - 2\ mol \cdot \Delta H^\circ_f[C_3H_5(NO_3)_3(\ell)]$$
$$= 6\ mol(-393.5\ kJ/mol) + 5\ mol(-241.8\ kJ/mol) - 2\ mol(-364\ kJ/mol)$$
$$= -2842\ kJ$$

Based on the enthalpy change for the explosion of 2 moles of nitroglycerin, we can calculate the heat liberated by this exothermic reaction when only 10.0 g of nitroglycerin is used.

$$10.0 \text{ g nitroglycerin} \cdot \frac{1 \text{ mol nitroglycerin}}{227.1 \text{ g nitroglycerin}} = 0.0440 \text{ mol nitroglycerin}$$

$$0.0440 \text{ mol nitroglycerin} \cdot \frac{2842 \text{ kJ}}{2 \text{ mol nitroglycerin}} = 62.6 \text{ kJ}$$

In Example 6.8 we calculated the quantity of heat evolved by 10.0 g of reactant. Because we already knew the direction of heat transfer (ΔH was negative), it was not important or useful to include the algebraic sign in the final step. We only wanted to know the quantity of energy.

EXERCISE 6.13 ***Using Enthalpies of Formation***

Benzene, C_6H_6, is an important hydrocarbon. Calculate its enthalpy of combustion; that is, find the value of $\Delta H°$ for the following reaction using Equation 6.6.

$$C_6H_6(\ell) + \tfrac{15}{2}\ O_2(g) \longrightarrow 6\ CO_2(g) + 3\ H_2O(\ell)$$

The enthalpy of formation of benzene, $\Delta H_f°[C_6H_6(\ell)]$, is +49.0 kJ/mol. Use Table 6.2 for any other values you may need. ■

6.9 DETERMINING ENTHALPIES OF REACTION

Calorimetry

The heat evolved by a chemical reaction can be determined by a technique called **calorimetry.** When finding heats of combustion or the caloric value of foods, the measurement is often done in a *combustion calorimeter* (Figure 6.17). A weighed sample of a combustible solid or liquid is placed in a dish that is encased in a "bomb," a cylinder about the size of a large fruit juice can with heavy

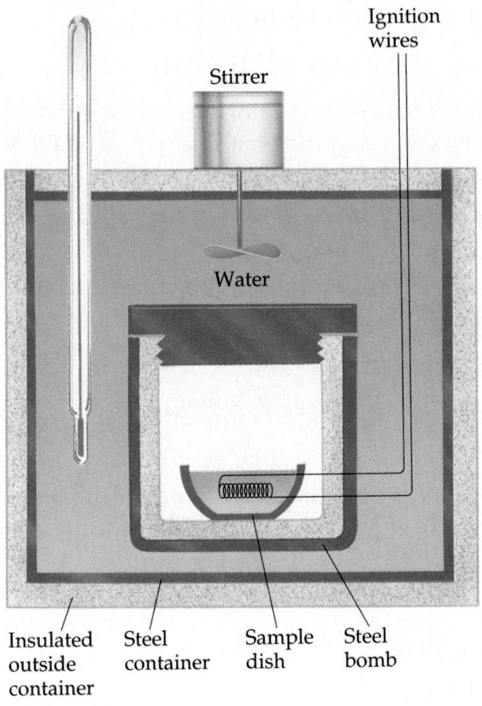

Figure 6.17 A combustion calorimeter. A combustible sample is burned in pure oxygen in a steel "bomb." The heat generated by the reaction warms the bomb and the water surrounding it. By measuring the temperature increase, the heat evolved by the reaction can be determined.

steel walls and ends. The bomb is then placed in a water-filled container with well-insulated walls. After filling the bomb with pure oxygen, the mixture of oxygen and sample is ignited, usually by an electric spark. The heat generated when the sample burns warms the bomb and the water around it, with both coming to the same temperature. In this configuration, the oxygen and the compound represent the *system,* and the bomb and water around it are the *surroundings.* From the law of energy conservation, we can say that

Heat transferred from the system = heat transferred into the surroundings

Heat evolved by the reaction = heat absorbed by water and bomb

$$q_{reaction} = -(q_{water} + q_{bomb})$$

where $q_{reaction}$ has a negative value because the combustion reaction is exothermic. The temperature change of the water, which is also equal to the change for the bomb, is measured. Using these experimental measurements, the total quantity of heat absorbed by the water and the bomb ($q_{water} + q_{bomb}$) can be calculated from the specific heat of the bomb and the water. According to the previous equation, this total gives the heat evolved by combustion of the compound.

Because the bomb is rigid, the heat transfer is measured at *constant volume* and is therefore equivalent to ΔE, the change in energy. As explained earlier (see Section 6.4), $\Delta E = q_v$, whereas the change in enthalpy is the heat evolved or required at constant pressure, that is, $\Delta H = q_p$. Because ΔE and ΔH are related in a relatively simple way, however, ΔH values can be calculated from ΔE values found in bomb calorimetry experiments.

EXAMPLE 6.9 *Determining the Enthalpy Change for a Reaction by Calorimetry*

Octane, C_8H_{18}, a primary constituent of gasoline, burns in air.

$$C_8H_{18}(\ell) + \tfrac{25}{2}\ O_2(g) \longrightarrow 8\ CO_2(g) + 9\ H_2O(\ell)$$

Suppose that a 1.00-g sample of octane is burned in a calorimeter that contains 1.20 kg of water. The temperature of the water and the bomb rises from 25.00 °C (298.15 K) to 33.20 °C (306.35 K). If the specific heat of the bomb, C_{bomb}, is known to be 837 J/K, calculate the heat transferred in the combustion of the 1.00-g sample of C_8H_{18}.

The specific heat of a bomb calorimeter, C_{bomb}, must be found in a separate experiment by measuring the change in T produced by burning a measured mass of a compound with a known ΔH of combustion.

Solution The heat that appears as a rise in temperature of the water surrounding the bomb is calculated as described in Section 6.2.

$$\begin{aligned} q_{water} &= (\text{specific heat of water})(m_{water})(\Delta T) \\ &= (4.184\ \text{J/g}\cdot\text{K})(1.20 \times 10^3\ \text{g})(306.35\ \text{K} - 298.15\ \text{K}) \\ &= +41.2 \times 10^3\ \text{J} \end{aligned}$$

The heat released by the reaction appears as a rise in temperature of the bomb and is calculated from the specific heat of the bomb (C_{bomb}, in units of joules per kelvin) and the temperature change, ΔT.

$$\begin{aligned} q_{bomb} &= C_{bomb} \cdot \Delta T \\ &= (837\ \text{J/K})(306.35\ \text{K} - 298.15\ \text{K}) \\ &= +6.86 \times 10^3\ \text{J} \end{aligned}$$

The total heat transferred by the reaction to its surroundings is equal to the negative of the sum of q_{water} and q_{bomb}. Thus,

$$\text{Total heat transferred by burning 1.00 g of octane} = q = -(41.2 \times 10^3\ \text{J} + 6.86 \times 10^3\ \text{J})$$

$$q = -48.1 \times 10^3\ \text{J, or } -48.1\ \text{kJ}$$

The experiment shows that 48.1 kJ of heat are evolved per gram of octane burned. Because the molar mass of octane is 114.2 g/mol, the heat transferred per mole is

$$\text{Heat transferred per mole} = (-48.1\ \text{kJ/g})(114.2\ \text{g/mol}) = -5.49 \times 10^3\ \text{kJ/mol}$$

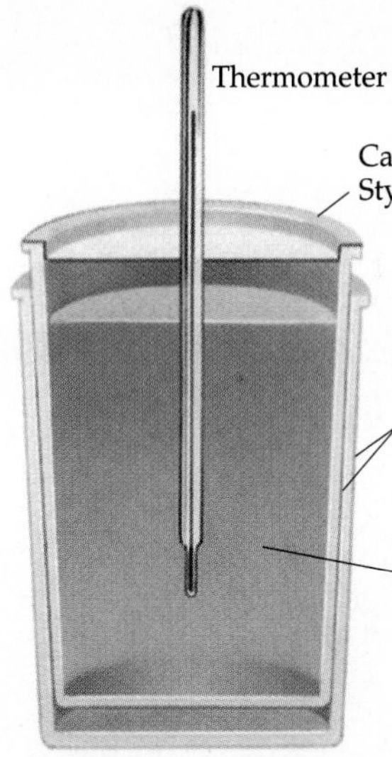

Figure 6.18 A coffee-cup calorimeter. Two Styrofoam coffee cups are placed one inside the other and covered with a lid. The exothermic reaction occurs in the aqueous solution in the container, and the cups provide enough insulation so that the heat is confined to the solution.

EXERCISE 6.14 ***Determining the Heat Released by a Reaction***

A 1.00-g sample of ordinary table sugar (sucrose, $C_{12}H_{22}O_{11}$) is burned in a combustion calorimeter. The temperature of 1.50×10^3 g of water in the calorimeter rises from 25.00 to 27.32 °C. If the specific heat of the bomb is 837 J/K and the specific heat of the water is 4.184 J/g · K, calculate (a) the heat evolved per gram of sucrose and (b) the heat evolved per mole of sucrose. ■

A bomb calorimeter is not always convenient to use, especially in the introductory chemistry laboratory. However, we can study reactions other than combustions in a "coffee-cup calorimeter" (Figure 6.18). Not only is it a simple, inexpensive device, but it operates at constant pressure. This means the heat evolved in a reaction in a coffee-cup calorimeter (q_p) is a measure of a reaction enthalpy (because $q_p = \Delta H$).

The specific heat of a metal can be determined by measuring the temperature change produced when a heated piece of metal is placed in a "coffee-cup calorimeter." See Example 6.3.

EXAMPLE 6.10 ***Using a Coffee-Cup Calorimeter***

Suppose you place 0.500 g of magnesium chips in a coffee-cup calorimeter and then add 100.0 mL of 1.00 M HCl. The reaction that occurs is

$$Mg(s) + 2\ HCl(aq) \longrightarrow H_2(g) + MgCl_2(aq)$$

The temperature of the solution increases from 22.2 to 44.8 °C. What is the enthalpy change for this reaction per mole of Mg? (Assume the specific heat of the solution is 4.20 J/g · K and that the density of the HCl solution is 1.00 g/mL.)

Solution Given the mass of solution, its specific heat, and ΔT, we can find the heat produced in the reaction. (The mass of the solution is the mass of the 100.0 mL of HCl plus the mass of magnesium: 100.5 g in this case.)

$$q = (100.5\ \text{g})(4.20\ \text{J/g}\cdot\text{K})(318.0\ \text{K} - 295.4\ \text{K}) = 9.54 \times 10^3\ \text{J}$$

This quantity of heat is produced by the reaction of 0.500 g of Mg. The amount produced by the reaction of 1.00 mol of Mg is

$$\text{Heat produced per mole} = \frac{9.54 \times 10^3\ \text{J}}{0.500\ \text{g}} \cdot \frac{24.31\ \text{g}}{1\ \text{mol}}$$

$$= 4.64 \times 10^5\ \text{J/mol, or } 464\ \text{kJ/mol}$$

This means the enthalpy change for the reaction of magnesium with aqueous HCl is $\Delta H = -464$ kJ/mol. Note that we have included a negative sign because the reaction is clearly exothermic.*

EXERCISE 6.15 ***Using a Coffee-Cup Calorimeter***

Assume you mix 200. mL of 0.400 M HCl with 200. mL of 0.400 M NaOH in a coffee-cup calorimeter. The temperature of the solutions before mixing was 25.10 °C; after mixing and allowing the reaction to occur, the temperature is 26.60 °C. What quantity of heat is produced on neutralizing one mole of acid? That is, what is the molar enthalpy of neutralization of the acid? Assume the densities of all solutions are 1.00 g/mL and that their specific heat is 4.20 J/g · K.) ■

Using Hess's Law

What experiments would you do to find the standard enthalpy of formation of solid calcium hydroxide?

$$Ca(s) + O_2(g) + H_2(g) \longrightarrow Ca(OH)_2(s) \qquad \Delta H_f^\circ = ?$$

We know from experience that the enthalpy change for the reaction cannot be measured directly in any convenient manner. It is therefore necessary to break down the formation of calcium hydroxide into a series of reactions that can be added together to give the equation for the formation of solid $Ca(OH)_2$ from the elements and whose enthalpy changes can be determined in a calorimeter (see Figure 6.18). We are guided in our choice of reactions by chemistry introduced in Chapter 4.

- Calcium is a metal and so reacts with oxygen to produce the oxide, CaO.

  $$Ca(s) + \tfrac{1}{2} O_2(g) \longrightarrow CaO(s) \qquad \Delta H^\circ_{rxn} = \Delta H_f^\circ[CaO(s)] = -635.1 \text{ kJ}$$

  The enthalpy change for this reaction can be determined in a calorimeter (or found in a table of ΔH_f° values).

- ΔH_f° for the formation of water from hydrogen and oxygen is well known.

  $$H_2(g) + \tfrac{1}{2} O_2(g) \longrightarrow H_2O(\ell) \qquad \Delta H^\circ_{rxn} = \Delta H_f^\circ[H_2O(\ell)] = -285.8 \text{ kJ}$$

- CaO is a basic oxide; it reacts with water to give calcium hydroxide, $Ca(OH)_2(s)$.

  $$CaO(s) + H_2O(\ell) \longrightarrow Ca(OH)_2(s) \qquad \Delta H^\circ_{rxn} = -65.2 \text{ kJ}$$

*When a ΔH value is stated for a chemical reaction, it is understood that the reactants and products are at the same temperature. In a calorimetry experiment, however, this condition is never met. In a coffee-cup calorimeter (unlike the bomb calorimeter) virtually all the energy produced remains within the system. Thus, the problem is to find how much heat would have to be transferred out of the system to keep its temperature unchanged. That amount of heat is precisely what was obtained when ΔH was calculated in this example. The negative sign is consistent with the transfer of heat *out* of the system.

Once CaO has been formed from the direct reaction of the metal with oxygen, the enthalpy change for its reaction with water can be determined in a calorimeter.

Adding the three reactions above together, we see that they give us the enthalpy of formation of solid $Ca(OH)_2$.

$$\begin{array}{ll} Ca(s) + \frac{1}{2} O_2(g) \longrightarrow CaO(s) & \Delta H_f^\circ[CaO(s)] = -635.1 \text{ kJ} \\ H_2(g) + \frac{1}{2} O_2(g) \longrightarrow H_2O(\ell) & \Delta H_f^\circ[H_2O(\ell)] = -285.8 \text{ kJ} \\ CaO(s) + H_2O(\ell) \longrightarrow Ca(OH)_2(s) & \Delta H^\circ_{rxn} = -65.2 \text{ kJ} \\ \hline Ca(s) + O_2(g) + H_2(g) \longrightarrow Ca(OH)_2(s) & \Delta H^\circ_{rxn} = -986.1 \text{ kJ} \end{array}$$

Many of the values of molar enthalpies of formation are found by very similar approaches.

EXERCISE 6.16 ***Finding an Enthalpy of Formation***

Using information in Table 6.2 and the following experimental enthalpy changes,

$$\begin{array}{ll} BaO(s) + H_2O(\ell) \longrightarrow Ba(OH)_2(s) & \Delta H^\circ_{rxn} = -105.4 \text{ kJ} \\ Ba(s) + \frac{1}{2} O_2(g) \longrightarrow BaO(s) & \Delta H_f^\circ[BaO(s)] = -553.5 \text{ kJ} \end{array}$$

calculate the enthalpy of formation for $Ba(OH)_2$ (s). ■

6.10 APPLICATIONS OF THERMODYNAMICS

Thermodynamics is the science of the transfer of energy as heat and work. As such it is related intimately to problems of energy use in our economy.

Available Energy Resources

What resources are currently available to supply energy for humans to use? Historically the first was *biomass,* material produced by living organisms that contains significant quantities of chemical potential energy. Biomass, mostly wood, still provides one third of the energy resources for some developing countries. *Fossil fuels*—petroleum, coal, and natural gas—made the Industrial Revolution possible and provide most of the energy resources used in the industrialized world today. For 200 years coal was the principal fuel for industrializing nations, only to be replaced by petroleum and natural gas around the middle of this century (Figure 6.19). Petroleum and natural gas are easier to handle and cleaner to use, but the price of petroleum shifts unpredictably with changing international conditions, and a significant amount of the world's petroleum is produced by nations that lack political stability.

Great quantities of coal are available, though, and attention is shifting back toward coal for energy and as a new source for many of the chemicals currently obtained from petroleum and natural gas.

Relatively small but still significant quantities of energy are now supplied by *hydroelectric* sources and *nuclear* energy, and still smaller quantities are provided in the form of direct *solar* energy, *geothermal* energy, and *wind* and ocean cur-

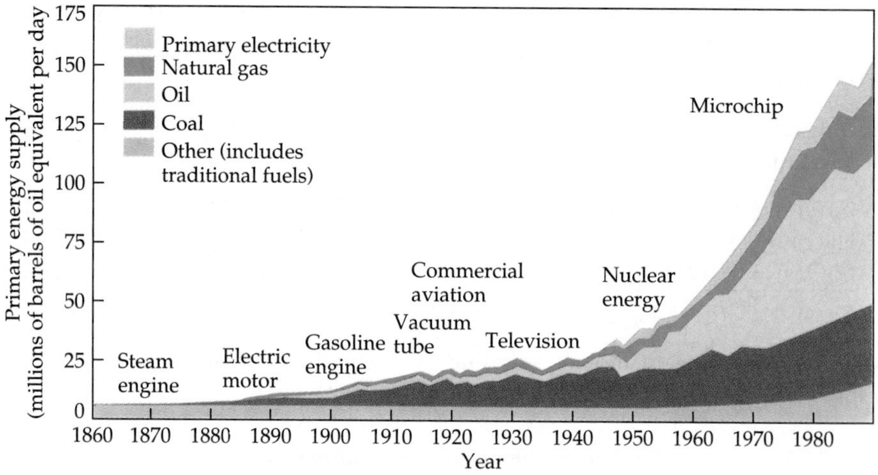

Figure 6.19 The rate of use of primary energy from 1860 to the present. (Primary energy exists in a crude form such as fossil fuels.) At the beginning of this period, wood was the predominant resource, but coal became more important with the advent of the steam engine and electric motor. With the invention of the automobile at the end of the 19th century, oil began to contribute. Coal continued its domination until the 1920s when it contributed more than 70% of the fuel used. Today coal meets about 26% of global energy needs, and oil is predominant. Natural gas use and electricity generated by hydroelectric and nuclear power plants have risen steadily since the 1940s. (For more information see G. R. Davis, *Scientific American,* September 1990, pp. 55–62.)

rents. These primary energy resources may be converted to electricity, a secondary resource, and a form of energy we find particularly useful.

Perhaps our most important resource is *solar energy,* energy available to us through biomass as a result of photosynthesis; fossil fuels that were formed from biomass as a result of changes that required millions of years; wind and water that store kinetic and potential energy; direct absorption of solar energy by our bodies and the materials around us; and, recently, photovoltaic cells capable of transforming solar energy into electric energy.

The United States, with only 5% of the world's population, accounts for 30% of the world's energy use. Only Canada uses more energy per capita. Since 1958, the United States has consumed more energy resources than it has produced.

Renewable energy resources, such as biomass or hydroelectric energy, can figure in long-range human energy planning, in contrast to fossil fuels, which, once used, are gone.

Energy in Industry

Thermodynamic principles can be used in a discussion of energy use in our industrial society and how it relates to chemistry. Much of the chemicals industry is involved in the production of basic materials from air, minerals, and petroleum, and in doing so it uses about 15% of the energy resources consumed by all industries (Figure 6.20). Any energy saving is therefore significant. Industries can save energy in three ways: (a) Lower the energy requirements of existing processes with energy-saving devices. (b) Introduce process refinements. (c) Develop new, lower energy methods of manufacture. The second and third of these are areas in which chemistry can most directly help.

For further reading on the use of energy in our society see the special issue of *Scientific American,* September 1990. See especially M. H. Ross and D. Steinmeyer, "Energy for Industry," page 89.

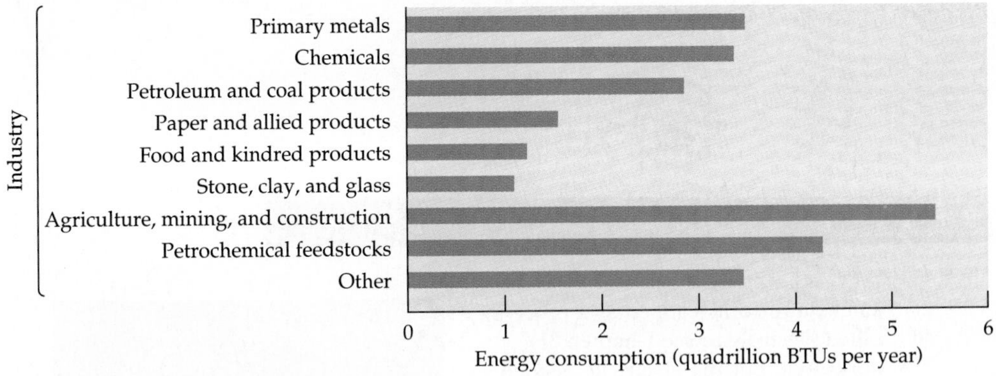

Figure 6.20 Industrial processes use about 37% of the energy resources in the United States, and the largest consumers are industries that convert raw materials. This figure shows the various uses made of the energy consumed by industry. (Units are in Btu, British thermal units, where $1 \text{ J} = 9.484 \times 10^{-4}$ Btu.)

About 16 billion kilograms of ammonia were produced in the United States in 1993. It is used directly as a fertilizer or converted to a variety of other products, such as nitric acid, urea (used as a fertilizer and in making plastics), and hydrazine (a widely used reducing agent). As described in Chapter 16, ammonia is made by the direct reaction of N_2 and H_2 (Figure 6.21).

$$N_2(g) + 3\ H_2(g) \longrightarrow 2\ NH_3(g) \qquad \Delta H^\circ_{rxn} = -92.2 \text{ kJ}$$

A recent refinement in the design of the reaction vessel raised the production rate of ammonia by 6% and reduced energy consumption by 5%. Of course an even bigger saving would come about if biotechnologists could make varieties of plants that could capture nitrogen directly from the air. Then we would not have to use ammonia as a fertilizer and would save the energy now used to make it. The manufacture of fertilizer consumes about 2% of all the energy resources used in industry, and this contributes to "greenhouse gases" that may cause global warming.

About 19 billion kilograms of ethylene, C_2H_4, were produced in the United States in 1993 from the dehydrogenation of alkanes such as ethane.

$$\underset{\text{ethane}}{C_2H_6(g)} \longrightarrow \underset{\text{ethylene}}{C_2H_4(g)} + H_2(g) \qquad \Delta H^\circ_{rxn} = +137.0 \text{ kJ}$$

Much of the product is used to make polyethylene, the plastic used in many consumer items. A significant amount, however, is used to manufacture other organic chemicals. Because ethylene production is the largest single consumer of energy resources in the chemical industry, great interest has arisen in improving the process to save energy and money. In this case, many small improvements in ethane-to-ethylene conversion have led to a 60% decline in the energy requirement per pound of ethylene produced since 1960. Even so, the energy resources used to make ethylene from ethane are four times the minimum required by the first law of thermodynamics (that is, four times 137.0 kJ/mol of ethylene), largely due to inefficiencies in energy transfer.

Modern buildings use acres of glass, so the energy cost of glass manufacture is an important consideration. Making glass requires large amounts of energy, first to melt the sand that makes up the glass and then to finish the glass. Here a development in manufacturing led to significant savings. Beginning in the 1960s, molten glass was poured onto baths of molten tin. Because tin melts at a temperature (232 °C) lower than glass, and because glass has a lower density

Figure 6.21 This world-class fertilizer complex in Nigeria consists of a 1000-metric-ton-a-day ammonia plant, a 1500-metric-ton-a-day urea plant, and a mixed fertilizer facility. As managing contractor on the project, M. W. Kellogg provided detailed engineering and is responsible for operations and maintenance under a partial ownership arrangement. (Courtesy of M. W. Kellogg, Inc.)

CURRENT ISSUES IN CHEMISTRY

The Hydrogen Economy

Considerable discussion has taken place about a potential "hydrogen economy" in which hydrogen gas provides energy for our homes, cars, and industries. The data in Table 6.3 show that hydrogen provides far more energy per gram than any other fuel listed. What are the advantages and disadvantages of using it as a fuel?

Currently the most visible use of hydrogen as a fuel is in the Space Shuttle. The cigar-shaped tank strapped to the shuttle contains 1.46×10^6 L (385,265 gallons) of hydrogen and 5.43×10^5 L (143,351 gallons) of oxygen. Just a few minutes into the flight, the fuel is exhausted, and the hydrogen and oxygen have been converted completely to nonpolluting water.

There are currently two major barriers to the use of hydrogen as an alternative to petroleum. First, an inexpensive way to make H_2 must be found that avoids the use of fossil fuels, which are now the major starting material for the production of hydrogen. Using electricity to break water down to its elements (a process called electrolysis; see Chapter 21) works well, but the electricity needed to do this is currently too expensive to produce hydrogen at a reasonable cost.

Solar cells may soon be cheap enough to provide the electric energy needed for water electrolysis, however. For example, the HYSOLAR project, carried out jointly by Saudi Arabia and Germany, uses solar energy to power an electrolysis facility that produces hydrogen. (See I. Dostrovsky, *Scientific American,* December, pp. 102–107, 1991.)

The second barrier to using hydrogen as a fuel is the need for a means of convenient storage. The space program has demonstrated that hydrogen can be stored relatively easily and safely as a liquid, even though cold temperatures and high pressures are required. The problem is that this may not be appropriate for a home or car. These uses will require alternative methods of storage, one of which is to store the H_2 as a simple compound with a metal. This compound—a metal hydride—would release the H_2 as a gas when heated. A prototype car run by hydrogen combustion in a slightly modified gasoline engine is shown in the photo. The hydrogen is stored as a metal hydride in a tank located where the gasoline tank would normally be.

H_2 is used as the fuel in the Space Shuttle. The cigar-shaped tank holds liquid H_2 and liquid O_2. (NASA)

A prototype car fueled by the combustion of hydrogen gas. The hydrogen is stored in a tank in the form of an iron-titanium hydride, and heat from the engine is used to release H_2 from the hydride. (Mercedes-Benz)

Figure 6.22 Plate glass is made by pouring molten glass onto a bath of molten tin. Because the surface of the liquid tin is smooth, as the glass cools the surface contacting the tin is completely smooth. The photo shows cooled plate glass emerging from the float-glass process. (Courtesy of PPG Industries)

than molten tin, the glass floats on top of the liquid tin and solidifies (Figure 6.22). This process makes the glass smooth on both sides and eliminates the energy cost of grinding and polishing the glass to make it smooth, which was the practice several decades ago.

Energy Resources for the Future

Figure 6.19 shows that petroleum currently provides about as much energy as coal and natural gas combined. This is largely because of the overwhelming importance of petroleum to fuel our cars and trucks. Of the fuels currently available in the United States for transportation, only those based on petroleum are relatively inexpensive, are transportable, and provide relatively high energy for a given amount of material.

We learned in the 1970s, however, that all that can change almost overnight. The present price of oil is about \$20 a barrel, but it went as high as \$35 a barrel in 1981.* At that time economists thought that it might reach as high as \$100 a barrel by the 1990s, and a rush started to discover new oil wells and to develop new sources of petroleum. However, there is currently a large surplus of oil-producing capacity in the world, and the price of oil has come down. This has had the effect of making it less economical for U.S. oil companies to drill for oil in this country and of providing less incentive for developing alternative fuels. All this makes yet another oil price increase likely in the future. In fact, some suspect that oil production in other countries will decline as reserves dwindle

*A barrel is equal to approximately 42 U.S. gallons.

TABLE 6.3 Energy Released by Combustion of Some Substances

Substances	Energy Released (kJ/g)
Hydrogen [to give $H_2O(\ell)$]	142
Gasoline	48
Crude petroleum	43
Typical animal fat	38
Coal	29.3
Charcoal	29
Paper	20
Dry biomass	16
Air-dried wood or dung	15

Data taken in part from J. Harte, *Consider a Spherical Cow,* University Science Books, Mill Valley, CA, 1988

and that prices will be sharply higher in the U.S. by the year 2000. So, once again we may be confronted with the necessity of searching for new sources of energy.

For the foreseeable future liquid fuels will be necessary for cars and trucks and, to some extent, for residential and commercial heating. But what about electricity generation? Presently this is done chiefly with coal and oil, but nuclear power plays a role, and may play a larger role in the future. And what about hydrogen as a fuel? The data in Table 6.3 show that the combustion of this element provides more energy per gram than any other fuel listed.

CHAPTER HIGHLIGHTS

When you have finished studying this chapter, you should be able to

- Describe the various forms of energy and the nature of heat and thermal energy transfer (Section 6.1).
- Know the difference between **kinetic** and **potential energy** (Section 6.1).
- Use the most common energy unit, the **joule** (Section 6.1)
- Use **specific heat** and the sign conventions for heat transfer (Section 6.2.).

  The amount of thermal energy transferred into or out of an object can be determined by using the equation

  $$q = \text{(mass of object in g)(specific heat in J/g}\cdot\text{K)(change in T in °C or K)}$$

  When q is *negative,* thermal energy has been transferred *out* of the object. When q is *positive,* thermal energy has been transferred *into* the object.
- Use **heat of fusion** and **heat of vaporization** to find the quantity of thermal energy involved in changes of state (Section 6.3).
- Recognize and use the language of thermodynamics: the **system** and its **surroundings; exothermic** and **endothermic** reactions; the **first law of thermodynamics** (the law of energy conservation); and **enthalpy** changes (Section 6.4).

- Understand the basis of the **first law of thermodynamics** (Section 6.4).

  When energy is transferred to or from a system, the change in energy for the system is given by $\Delta E = q + w$. Here q is the quantity of thermal energy transferred between the system and its surroundings; similarly, w is the work transferred. Both q and w are positive when heat or work, respectively, are transferred from the surroundings to the system.

- Recognize that when a process is carried out under constant pressure conditions, the heat transferred is equivalent to the enthalpy change, ΔH; that is, $q_p = \Delta H$ (where the subscript p means constant pressure) (Section 6.4).
- Understand that if the enthalpy change for a process is negative, heat has transferred from a system to its surroundings, and the process is **exothermic** with respect to the system (Section 6.4). Conversely, if ΔH is positive, heat has transferred from the surroundings to a system, and the process is **endothermic** with respect to the system.
- Use the fact that ΔH for a reaction is proportional to the quantity of material present (Sections 6.4–6.9).
- Apply **Hess's law** to find the enthalpy change for a reaction (Sections 6.6 and 6.9).
- Recognize various state functions, quantities whose value is determined only by the state of the system and not by the pathway by which that state was achieved (Section 6.7).
- Write a balanced chemical equation that defines the **standard molar enthalpy** of formation, ΔH_f°, for a compound (Section 6.8) and use equation 6.6 to calculate the enthalpy change for a reaction, ΔH_{rxn}°.

  Enthalpy change for a reaction $= \Delta H_{rxn}^\circ =$

  $$\Sigma\,[\Delta H_f^\circ(\text{products})] - \Sigma\,[\Delta H_f^\circ(\text{reactants})] \qquad \textbf{(6.6)}$$

  where Σ (the Greek letter *sigma*) means to "take the sum."

- Describe how to measure the quantity of heat energy transferred in a reaction by using **calorimetry** (Section 6.9).
- Discuss uses of energy in our economy and problems and opportunities for developing new energy resources (Section 6.10).

STUDY QUESTIONS

Review Questions

1. You pick up a 6-pack of soft drinks from the floor, but it slips from your hand and smashes into your foot. Comment on the work and energy involved in this sequence. What forms of energy are involved; at what stages of the process?
2. Based on your experience, when ice melts to liquid water is the process exothermic or endothermic? When liquid water freezes to ice at 0 °C is this exothermic or endothermic?
3. What is the first law of thermodynamics? Explain in words and use a mathematical expression.
4. For each of the following, define a system and its surroundings and give the direction of heat transfer:
   (a) Methane is burning in a gas furnace in your home.
   (b) Water drops, sitting on your skin after a dip in a swimming pool, evaporate.
   (c) Water, originally at 25 °C, is placed in the freezing compartment of a refrigerator.
   (d) Two chemicals are mixed in a flask sitting on a laboratory bench. A reaction occurs, and heat is evolved.

5. What is the value of the standard enthalpy of formation for any element under standard conditions?
6. Look at the enthalpies of formation of metal oxides in Table 6.2 or Appendix K and comment on your observations. Are oxidations of metals generally endothermic or exothermic?
7. A house is made of wood and glass. Assuming an equal amount of sunshine falls on a wooden wall and a piece of glass (of equal mass), which warms more? Explain briefly.
8. Criticize the following statements:
   (a) An enthalpy of formation refers to a reaction in which 1 mol of one or more reactants produces some quantity of product.
   (b) The standard enthalpy of formation of O_2 as a gas at 25 °C and a pressure of 1 bar is 15.0 kJ/mol.
   (c) The thermal energy transferred from 10.0 g of ice as it melts is $q = +6$ kJ.
9. Is the following reaction predicted to favor the reactants or products? Explain your answer briefly.

$$\text{Mg(s)} + \tfrac{1}{2}\,\text{O}_2\text{(g)} \longrightarrow \text{MgO(s)} \qquad \Delta H_f^\circ = -601.70 \text{ kJ}$$

10. Which of the following is not a state function?
   (a) The *volume* of a balloon.
   (b) The *time* it takes to drive from your home to your college or university.
   (c) The *temperature* of the water in a coffee cup.
   (d) The *potential energy* of a rock held in your hand.

Numerical Questions

Energy Units

11. A 2-in. piece of a two-layer chocolate cake with frosting provides 1670 kJ of energy. What is this in Calories?
12. If you are on a diet that calls for eating no more than 1200 Calories per day, how many joules would this be?
13. A parking lot in Los Angeles, California, receives an average of 2.6×10^7 J of solar energy per square meter per day in the summer. If the parking lot is 300. m long and 50.0 m wide, what is the total quantity of energy striking the area?
14. Your home loses heat in the winter through doors and windows and any poorly insulated walls. A sliding glass door (6 ft by 6½ feet with ½″ of insulating glass) allows 1.0×10^6 J/h to pass through the glass if the inside temperature is 22 °C (72 °F) and the outside temperature is 0 °C (32 °F). What quantity of heat, expressed in kilojoules, is lost per day? If 1 kwh of energy is equal to 3.61×10^6 J, how many kilowatt-hours of energy are lost per day through the door?

Specific Heat

15. 74.8 J of heat is required to raise the temperature of 18.69 g of silver from 10.0 to 27.0 °C. What is the specific heat of silver?
16. The specific heat of nickel is 0.445 J/g · K. How much heat is required to heat a 168-g piece of nickel from −15.2 to +23.6 °C?
17. Which requires more heat to warm from 22 °C to 85 °C, 50.0 g of water or 100. g of ethylene glycol (specific heat = 2.39 J/g · K)?
18. You hold a gram of copper in one hand and a gram of aluminum in the other. Each metal was originally at 0 °C. (Both metals are in the shape of a little ball that fits into your hand.) If they both take up heat at the same rate, which reaches your body temperature first?
19. How much heat energy in kilojoules is required to heat all the aluminum in a roll of aluminum foil (500. g) from room temperature (25 °C) to the temperature of a hot oven (255 °C)?
20. A 20.0-g piece of aluminum at 0.0 °C is dropped into a beaker of water. The temperature of the water drops from 90.0 to 75.0 °C. What quantity of heat energy did the piece of aluminum absorb?
21. Ethylene glycol, $(CH_2OH)_2$, is often used as an antifreeze in cars. Which requires more heat energy to warm from 25.0 to 100. °C, 250 g of water or 250 g of ethylene glycol?
22. Which requires more energy to warm from 0.0 to 29.5 °C, a 5.00-kg piece of granite or 5.00 kg of cement?
23. A 192-g piece of copper is heated to 100.0 °C in a boiling water bath and then dropped into a beaker containing 750. mL of water (density = 1.00 g/cm^3) at 4.0 °C. What is the final temperature of the copper and water after they come to thermal equilibrium? (The specific heat of copper is 0.385 J/g · K)
24. A 13.8-g piece of zinc was heated to 98.9 °C in boiling water and then dropped into 15.00 mL of water at 25.0 °C. When the water and metal come to thermal equilibrium, what is the temperature of the system? (Specific heat of zinc = 0.388 J/g · K)
25. When a mass of 182 g of gold at some temperature is added to 22.1 g of water at a temperature of 25.0 °C, the final temperature of the resulting mixture is 27.5 °C. If the specific heat of gold is 0.129 J/g · K, what was the initial temperature of the gold sample?
26. When 108 g of water at a temperature of 22.5 °C is mixed with 65.1 g of water at an unknown temperature, the final temperature of the resulting mixture is 47.9 °C. What was the temperature of the second sample of water?
27. A 150.0-g sample of a metal at 80.0 °C is placed in 150.0 g of water at 20.0 °C. The temperature of the final system (metal and water) is 23.3 °C. What is the specific heat of the metal?
28. A 237-g piece of molybdenum, initially at 100.0 °C, is dropped into 244 g of water at 10.0 °C. When the system comes to thermal equilibrium, the temperature is 15.3 °C. What is the specific heat of the metal?

Changes of State

29. The heat energy required to melt 1.00 g of ice at 0 °C is

333 J. If one ice cube has a mass of 62.0 g, and a tray contains 16 ice cubes, what quantity of energy is required to melt a tray of ice cubes at 0 °C?

30. Chloromethane, CH_3Cl, is used as a topical anesthetic. The temperature at which CH_3Cl liquid turns into a vapor (the boiling point) is −24.09 °C. What quantity of heat must be absorbed by the liquid to convert 150. g of liquid to a vapor at −24.09 °C? The heat of vaporization of CH_3Cl is 21.40 kJ/mol.

31. Calculate the quantity of heat required to convert the water in five ice cubes (60.1 g) from $H_2O(s)$ at 0.0 °C to $H_2O(g)$ at 100.0 °C. The heat of fusion of ice at 0 °C is 333 J/g; the heat of vaporization of liquid water at 100 °C is 2260 J/g.

32. Mercury, with a freezing point of −38.8 °C, is the only metal that is liquid at room temperature. How much heat energy (in joules) must be released by mercury if 1.00 mL of the metal is cooled from room temperature (23.0 °C) to −39 °C and then frozen to a solid? (The density of mercury is 13.6 g/cm^3. Its specific heat is 0.140 J/g · K, and its heat of fusion is 11.4 J/g.)

33. How much heat energy (in joules) is required to raise the temperature of 454 g of tin (1.00 lb) from room temperature (25.0 °C) to its melting point, 231.9 °C, and then melt the tin at that temperature? The specific heat of tin is 0.227 J/g · K, and the metal requires 59.2 J/g to convert the solid to a liquid.

34. Ethanol, C_2H_5OH, boils at 78.29 °C. How much heat energy (in joules) is required to heat 1.00 kg of this liquid from 20.0 °C to the boiling point and then change the liquid completely to a vapor at that temperature? (The specific heat of liquid ethanol is 2.44 J/g · K, and the enthalpy of vaporization is 38.56 kJ/mol.)

Enthalpy

35. Nitrogen monoxide has recently been found to be involved in a wide range of biological processes. The gas reacts with oxygen to give brown NO_2 gas.

$$2\ NO(g) + O_2(g) \longrightarrow 2\ NO_2(g) \qquad \Delta H^\circ_{rxn} = -114.1\ kJ$$

Is the reaction endothermic or exothermic? If 1.25 g of NO is converted completely to NO_2, what quantity of heat is absorbed or evolved?

36. Calcium carbide, CaC_2, is manufactured by reducing lime (CaO) with carbon at a high temperature. (The carbide is used to make acetylene, an industrially important organic chemical.)

$$CaO(s) + 3\ C(s) \longrightarrow CaC_2(s) + CO(g) \qquad \Delta H^\circ_{rxn} = +464.8\ kJ$$

Is the reaction endothermic or exothermic? If 10.0 g of CaO is allowed to react with an excess of carbon, what quantity of heat is absorbed or evolved by the reaction?

37. Isooctane (2,2,4-trimethylpentane) burns in air to give water and carbon dioxide.

$$2\ C_8H_{18}(\ell) + 25\ O_2(g) \longrightarrow 16\ CO_2(g) + 18\ H_2O(\ell) \qquad \Delta H^\circ_{rxn} = -10{,}922\ kJ$$

Is the combustion exothermic or endothermic? If you burn 1.00 L of the hydrocarbon (density = 0.6878 g/mL), what quantity of heat is involved?

38. Acetic acid is made by the combination of methanol and carbon monoxide.

$$CH_3OH(\ell) + CO(g) \longrightarrow CH_3CO_2H(\ell) \qquad \Delta H^\circ_{rxn} = -355.9\ kJ$$

Is the reaction exothermic or endothermic? If you produce 1.00 L of the acid (density = 1.044 g/mL), what quantity of heat is involved?

39. Methanol, CH_3OH, is a possible automobile fuel. The alcohol produces energy in a combustion reaction with O_2.

$$2\ CH_3OH(g) + 3\ O_2(g) \longrightarrow 2\ CO_2(g) + 4\ H_2O(\ell)$$

A 0.115-g sample of methanol evolves 1110 J when burned at constant pressure. What is the enthalpy change, ΔH°_{rxn}, for the reaction? What is the enthalpy change per mole of methanol (often called the molar heat of combustion)?

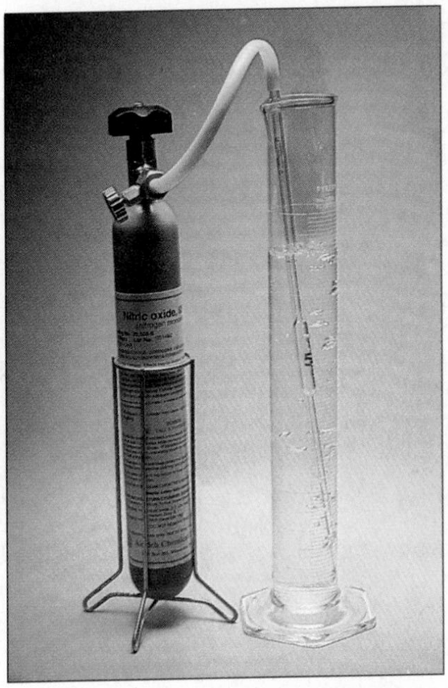

The gas NO is stored in a tank. When it is bubbled through water, the gas in the bubbles is colorless. As soon as the NO bubbles enter the atmosphere, however, colorless NO is oxidized to brown NO_2. See Study Question 35. (C. D. Winters)

40. White phosphorus, P_4, ignites in air to produce heat, light, and P_4O_{10} (Figure 4.4).

$$P_4(s) + 5\ O_2(g) \longrightarrow P_4O_{10}(s)$$

If 3.56 g of P_4 is burned, 37.4 kJ of heat is evolved at constant pressure. What is the enthalpy change for the combustion of 1 mol of P_4?

Hess's Law

41. Suppose you wish to know the enthalpy change for the formation of benzene, C_6H_6, and that the value is not available directly from any data tables.

$$6\ C(\text{graphite}) + 3\ H_2(g) \longrightarrow C_6H_6(\ell) \qquad \Delta H_f^\circ = ?$$

Use information from Example 6.6 and the enthalpy change for the combustion of benzene, which was determined experimentally, to calculate the standard molar enthalpy of formation of benzene.

$$2\ C_6H_6(\ell) + 15\ O_2(g) \longrightarrow 12\ CO_2(g) + 6\ H_2O(\ell)$$
$$\Delta H^\circ_{rxn} = -6534.8\text{ kJ}$$

42. Calculate the enthalpy change, ΔH°, for the formation of 1 mol of strontium carbonate (the material that gives the red color in fireworks) from its elements.

$$Sr(s) + C(\text{graphite}) + \tfrac{3}{2}\ O_2(g) \longrightarrow SrCO_3(s)$$

The experimental information available is:

$$Sr(s) + \tfrac{1}{2}\ O_2(g) \longrightarrow SrO(s) \qquad \Delta H_f^\circ = -592\text{ kJ}$$
$$SrO(s) + CO_2(g) \longrightarrow SrCO_3(s) \qquad \Delta H^\circ_{rxn} = -234\text{ kJ}$$
$$C(\text{graphite}) + O_2(g) \longrightarrow CO_2(g) \qquad \Delta H^\circ_{rxn} = -394\text{ kJ}$$

43. Using the reactions shown here, find the enthalpy change for the formation of PbO(s) from lead metal and oxygen gas.

$$Pb(s) + CO(g) \longrightarrow PbO(s) + C(s) \qquad \Delta H^\circ_{rxn} = -106.8\text{ kJ}$$
$$2\ C(s) + O_2(g) \longrightarrow 2\ CO(g) \qquad \Delta H^\circ_{rxn} = -221.0\text{ kJ}$$

If 250 g of lead reacts with oxygen to form lead(II) oxide, what quantity of heat (in kilojoules) is absorbed or evolved?

44. Suppose you wish to know the enthalpy change for the formation of liquid PCl_3 from the elements.

$$P_4(s) + 6\ Cl_2(g) \longrightarrow 4\ PCl_3(\ell) \qquad \Delta H_f^\circ = ?$$

This reaction cannot be carried out directly. Instead, the enthalpy change for the reaction of phosphorus and chlorine to give phosphorus pentachloride can be determined experimentally.

$$P_4(s) + 10\ Cl_2(g) \longrightarrow 4\ PCl_5(s) \qquad \Delta H^\circ_{rxn} = -1774.0\text{ kJ}$$

The enthalpy change for the reaction of phosphorus trichloride with more chlorine to give phosphorus pentachloride can also be measured.

$$PCl_3(\ell) + Cl_2(g) \longrightarrow PCl_5(s) \qquad \Delta H^\circ_{rxn} = -123.8\text{ kJ}$$

Use this information to calculate the enthalpy change for the formation of 1 mol of $PCl_3(\ell)$ from P_4 and Cl_2.

Standard Enthalpies of Formation

45. The standard molar enthalpy of formation of solid chromium(III) oxide is −1139.7 kJ/mol. Write the balanced equation for which the enthalpy of reaction is −1139.7 kJ.

46. The molar enthalpy of formation of methanol, $CH_3OH(\ell)$, is −238.7 kJ/mol. Write the balanced equation for which the enthalpy of reaction is −238.7 kJ.

47. The molar enthalpy of formation of glucose, $C_6H_{12}O_6(s)$, is −1260 kJ/mol.

(a) Is the formation of glucose from its elements exothermic or endothermic?

(b) Write a balanced equation depicting the formation of glucose from its elements and for which the enthalpy of reaction is −1260 kJ.

48. The molar enthalpy of formation of pyridine(ℓ), is +100.2 kJ/mol.

(a) Write the balanced equation for the formation of pyridine from its elements, that is, for the reaction for which the enthalpy of reaction is +100.2 kJ.

(b) Write the balanced equation for which the enthalpy change is +200.4 kJ.

49. Enthalpy changes have been determined experimentally for the following two reactions:

$$Pb(s) + Cl_2(g) \longrightarrow PbCl_2(s) \qquad \Delta H_f^\circ = -359.4\text{ kJ}$$
$$Pb(s) + 2\ Cl_2(g) \longrightarrow PbCl_4(\ell) \qquad \Delta H_f^\circ = -329.3\text{ kJ}$$

What is the enthalpy change for the reaction of lead(II) chloride with chlorine to give lead(IV) chloride?

$$PbCl_2(s) + Cl_2(g) \longrightarrow PbCl_4(\ell) \qquad \Delta H^\circ_{rxn} = ?$$

50. The enthalpy of formation of solid barium oxide, BaO, is −553.5 kJ/mol, and the enthalpy of formation of barium peroxide, BaO_2, is −634.3 kJ/mol. Calculate the enthalpy change for the reaction

$$BaO(s) + \tfrac{1}{2}\ O_2(g) \longrightarrow BaO_2(s)$$

51. An important step in the production of sulfuric acid is

$$SO_2(g) + \tfrac{1}{2}\ O_2(g) \longrightarrow SO_3(g)$$

It is also a key reaction in the formation of acid rain, beginning with the air pollutant SO_2. Using the data in Table 6.2, calculate the enthalpy change for the reaction. Is the reaction exothermic or endothermic?

52. In photosynthesis, the sun's energy brings about the combination of CO_2 and H_2O to form O_2 and a carbon-containing compound. In its simplest form, the reaction is

$$CO_2(g) + 2\ H_2O(\ell) \longrightarrow 2\ O_2(g) + CH_4(g)$$

Using the enthalpies of formation in Table 6.2, (a) calculate the enthalpy of reaction and (b) decide if the reaction is exothermic or endothermic.

53. The first step in the production of nitric acid from ammonia involves the oxidation of NH_3.

$$4\ NH_3(g) + 5\ O_2(g) \longrightarrow 4\ NO(g) + 6\ H_2O(g)$$

(a) Use the information in Table 6.2 or Appendix K to find the enthalpy change for this reaction. Is the reaction exothermic or endothermic?
(b) What quantity of heat is evolved or absorbed if 10.0 g of NH_3 are oxidized?

54. The Romans used calcium oxide (CaO) as mortar in stone structures. The CaO was mixed with water to give $Ca(OH)_2$, which slowly reacted with CO_2 in the air to give limestone.

$$Ca(OH)_2(s) + CO_2(g) \longrightarrow CaCO_3(s) + H_2O(g)$$

(a) Calculate the enthalpy change for the reaction above.
(b) What quantity of heat is evolved or absorbed if 1.00 kg of $Ca(OH)_2$ is allowed to react with a stoichiometric amount of CO_2?

55. Pure metals can often be prepared by reducing the metal oxide with hydrogen gas. For example,

$$WO_3(s) + 3\ H_2(g) \longrightarrow W(s) + 3\ H_2O(\ell)$$

(a) Calculate the enthalpy change for this reaction. (ΔH_f° for $WO_3(s)$ is -842.9 kJ/mol.)
(b) What quantity of heat is evolved or absorbed if 1.00 g of WO_3 is allowed to react with an excess of hydrogen gas?

56. Iron(III) oxide ("rust") can be produced from iron and oxygen in a sequence of reactions that can be written as

$$2\ Fe(s) + 6\ H_2O(\ell) \longrightarrow 2\ Fe(OH)_3(s) + 3\ H_2(g)$$
$$2\ Fe(OH)_3(s) \longrightarrow Fe_2O_3(s) + 3\ H_2O(\ell)$$
$$3\ H_2(g) + \tfrac{3}{2}\ O_2(g) \longrightarrow 3\ H_2O(\ell)$$

(a) What is the enthalpy change for each step?
(b) What is the enthalpy change for the overall process, the reaction of iron with oxygen to give iron(III) oxide?
(c) What quantity of heat is evolved or absorbed if 1.25 g of iron is converted to iron(III) oxide? (Values for ΔH_f° can be found in Table 6.2 and Appendix K. In addition, $\Delta H_f^\circ[Fe(OH)_3(s)] = -696.5$ kJ/mol.)

57. Naphthalene, $C_{10}H_8$, is burned in a calorimeter to find the enthalpy change for the combustion reaction.

$$C_{10}H_8(s) + 12\ O_2(g) \longrightarrow 10\ CO_2(g) + 4\ H_2O(\ell)$$

If $\Delta H^\circ_{rxn} = -5156.1$ kJ, what is the molar enthalpy of formation of naphthalene?

58. Styrene, C_8H_8, is burned in a calorimeter to find the enthalpy change for the combustion reaction.

$$C_8H_8(\ell) + 10\ O_2(g) \longrightarrow 8\ CO_2(g) + 4\ H_2O(\ell)$$

If $\Delta H^\circ_{rxn} = -4395.0$ kJ, what is the molar enthalpy of formation of styrene?

Calorimetry

59. How much heat energy (in kilojoules) is evolved by a reaction in a bomb calorimeter (Figure 6.17) in which the temperature of the bomb and water increases from 19.50 to 22.83 °C? The bomb has a heat capacity of 650. J/K; the calorimeter contains 320. g of water.

60. Sulfur (2.56 g) is burned in a bomb calorimeter with excess $O_2(g)$. The temperature increases from 21.25 to 26.72 °C. The bomb has a heat capacity of 923 J/K, and the calorimeter contains 815 g of water. Calculate the heat evolved, per mole of SO_2 formed, in the course of the reaction

$$S_8(s) + 8\ O_2(g) \longrightarrow 8\ SO_2(g)$$

61. You can find the amount of heat evolved in the combustion of carbon by carrying out the reaction in a combustion calorimeter. Suppose you burn 0.300 g of C(graphite) in an excess of $O_2(g)$ to give $CO_2(g)$.

$$C(graphite) + O_2(g) \longrightarrow CO_2(g)$$

The temperature of the calorimeter, which contains 775 g of water, increases from 25.00 to 27.38 °C. The heat capacity of the bomb is 893 J/K. What quantity of heat is evolved per mole of C?

62. Benzoic acid, $C_7H_6O_2$, occurs naturally in many berries. Suppose you burn 1.500 g of the compound in a combustion calorimeter and find that the temperature of the calorimeter increases from 22.50 to 31.69 °C. The calorimeter contains 775 g of water, and the bomb has a heat capacity of 893 J/K. How much heat is evolved per mole of benzoic acid?

63. Assume you mix 100.0 mL of 0.200 M CsOH with 50.0 mL of 0.400 M HCl in a coffee-cup calorimeter. The following reaction occurs:

$$CsOH(aq) + HCl(aq) \longrightarrow CsCl(aq) + H_2O(\ell)$$

The temperature of the solutions before mixing was 22.50 °C, and it rises to 24.28 °C after the acid-base reaction. What is the enthalpy of the reaction per mole of CsOH? (Assume the densities of the solutions are all 1.00 g/mL and the specific heat of the solutions are 4.2 J/g · K.)

64. Suppose you mix 125 mL of 0.250 M CsOH with 125 mL of 0.250 M HF in a coffee-cup calorimeter. The temperature of the original solution was 21.50 °C, and it rises to 24.40 °C after the reaction.

$$CsOH(aq) + HF(aq) \longrightarrow CsF(aq) + H_2O(\ell)$$

What is the enthalpy of reaction per mole of CsOH? Compare the result of this experiment with the one in Study Question 63. Does the reaction enthalpy depend on the fact that HCl is a strong acid (completely ionized in aqueous solution), whereas HF is a weak acid (incompletely ionized in aqueous solution)? (Assume the densities of the solutions are all 1.00 g/mL and the specific heats of the solutions are 4.2 J/g · K.)

65. An "ice calorimeter" can be used to determine the specific heat of a metal. A piece of hot metal is dropped into a weighed quantity of ice. The quantity of heat given up by the metal can be determined from the amount of ice melted. Suppose you heat a 50.0-g piece of metal to 99.8 °C and then drop it onto ice. When the metal's temperature has dropped to 0.0 °C, it is found that 7.33 g of ice had melted. What is the specific heat of the metal?

66. A 9.36-g piece of platinum is heated to 98.6 °C in a boiling water bath and then dropped onto ice. When the metal's temperature has dropped to 0.0 °C, it is found that 0.37 g of ice had melted. What is the specific heat of the metal?

General Questions

67. Which gives up more heat on cooling from 50 to 10 °C, 50.0 g of water or 100. g of ethanol (specific heat of ethanol = 2.46 J/g · K)?

68. The specific heat for copper metal is 0.385 J/g · K, and it is 0.128 J/g · K for gold. Assume you place 100. g of each metal, originally at 25 °C, in a boiling water bath at 100 °C. If each metal takes up heat at the same rate (the number of joules of heat absorbed per minute is the same), which piece of metal reaches 100 °C first?

69. The meals-ready-to-eat (MRE) in the military can be heated on a flameless heater (page 283). Assume the reaction in the heater is

$$Mg(s) + 2\ H_2O(\ell) \longrightarrow Mg(OH)_2(s) + H_2(g)$$

Calculate the enthalpy change under standard conditions (in joules) for this reaction. What quantity of magnesium is needed to supply the heat required to warm 25 mL of water ($d = 1.00$ g/mL) from 25 to 85 °C?

70. A commercial product called "Instant Car Kooler" contains 10% by weight ethanol, C_2H_5OH, and 90% by weight water. If the interior of your car is overheated, you spray the "Kooler" inside and the interior is cooled. It works because thermal energy is transferred to the alcohol and water from the warm air and evaporates the liquids. To drop the air temperature from 55 to 25 °C requires air to give up 3.6 kJ. How many grams of the ethanol-water mixture must be used to absorb this heat? (The enthalpy of vaporization for ethanol is 850 J/g and for water it is 2360 J/g.) (See "Car Cooler," *Chem Matters,* February, Vol. 11, p. 11, 1993.)

See Study Question 70. (C. D. Winters)

71. Calculate the *molar* specific heat (in J/mol · K) for the four metals in Table 6.1. What observation can you make about these values? Are they widely different or very similar? Using this information can you calculate the specific heat in J/g · K for silver? (The correct value for silver is 0.236 J/g · K.)

72. Suppose you add 100.0 g of water at 60.0 °C to 100.0 g of ice at 0.00 °C. Some of the ice melts and cools the warm water to 0.00 °C. When the ice-water mixture has come to a uniform temperature of 0 °C, how much ice has melted?

73. The combustion of diborane, B_2H_6, proceeds according to the equation

$$B_2H_6(g) + 3\ O_2(g) \longrightarrow B_2O_3(s) + 3\ H_2O(g)$$

and 1941 kJ of heat energy is liberated per mole of $B_2H_6(g)$ (at constant pressure). Calculate the molar enthalpy of formation of $B_2H_6(g)$ using this information, the data in Table 6.2, and the fact that ΔH_f° for $B_2O_3(s)$ is −1271.9 kJ/mol.

74. The following are two of the key reactions in the processing of uranium for use as fuel in nuclear power plants:

$$UO_2(s) + 4\ HF(g) \longrightarrow UF_4(s) + 2\ H_2O(g)$$
$$UF_4(s) + F_2(g) \longrightarrow UF_6(g)$$

(a) Calculate the enthalpy change, ΔH°, for each reaction using the data in Table 6.2, Appendix K, and the following:

Compound	**ΔH_f° (kJ/mol)**
$UO_2(s)$	−1085
$UF_4(s)$	−1914
$UF_6(g)$	−2147

(b) Calculate the enthalpy change for the overall conversion of UO_2 to UF_6.

$$UO_2(s) + 4\ HF(g) + F_2(g) \longrightarrow UF_6(g) + 2\ H_2O(g)$$

75. Given the following information, and the data in Table 6.2, calculate the molar enthalpy of formation for liquid hydrazine, N_2H_4.

$$N_2H_4(\ell) + O_2(g) \longrightarrow N_2(g) + 2\ H_2O(g) \qquad \Delta H^\circ_{rxn} = -534\ \text{kJ}$$

76. The combination of coke and steam produces a mixture called synthesis gas, which can be used as a fuel or as a starting material for other reactions. The equation for the production of coal gas is

$$C(s) + 2\ H_2O(g) \longrightarrow 2H_2(g) + CO_2(g)$$

Determine the standard enthalpy change for this reaction. What quantity of heat is involved if 1.0 metric ton (1000.0 kg) of carbon is converted to coal gas?

77. A catalytic converter in an automobile is used to convert unburned hydrocarbons, CO, and NO to CO_2, water, and nitrogen. Some reactions involving these exhaust gases are

(1) $2\ C_2H_2(g) + 5\ O_2(g) \longrightarrow 4\ CO_2(g) + 2\ H_2O(g)$

(2) $2\ CO(g) + O_2(g) \longrightarrow 2\ CO_2(g)$

(3) $2\ C_2H_2(g) + 10\ NO(g) \longrightarrow 4\ CO_2(g) + 2\ H_2O(g) + 5\ N_2(g)$

(4) $2\ NO(g) + 2\ CO(g) \longrightarrow N_2(g) + 2\ CO_2(g)$

(a) Calculate the enthalpy change for each step. (The required enthalpies of formation are found in Table 6.2.) Which steps are exothermic and which are endothermic?

(b) What quantity of heat is evolved or absorbed if 15.0 g of C_2H_2 is converted to products in the third step? If 15.0 g of $C_2H_2(g)$ is converted completely to CO_2 and water (Step 3), what quantity of heat is involved?

78. Ammonium nitrate decomposes exothermically to N_2O and water.

$$NH_4NO_3(s) \longrightarrow N_2O(g) + 2\ H_2O(g)$$

(a) If the enthalpy of formation of $N_2O(g)$ is 82.1 kJ/mol, how much heat is evolved (at constant pressure and under standard conditions)?

(b) If 8.00 kg of ammonium nitrate explodes, what quantity of heat is evolved (at constant pressure and under standard conditions)?

Decomposition of ammonium nitrate. This compound was involved in the explosive charge that damaged New York's World Trade Center in 1993. See Study Question 78. (C. D. Winters)

79. One method of producing H_2 on a large scale is the chemical cycle shown here.

Step 1: $SO_2(g) + 2\ H_2O(g) + Br_2(g) \longrightarrow H_2SO_4(\ell) + 2\ HBr(g)$

Step 2: $H_2SO_4(\ell) \longrightarrow H_2O(g) + SO_2(g) + \frac{1}{2}\ O_2(g)$

Step 3: $2\ HBr(g) \longrightarrow H_2(g) + Br_2(g)$

Using the table of standard enthalpies of formation in Appendix K, calculate $\Delta H°$ for each step. What is the equation for the *overall* process, and what is its enthalpy change? Is the overall process exothermic or endothermic?

80. Suppose you want to heat the air in your house with natural gas (CH_4). Assume your house has 275 m^2 (about 2800 ft^2) of floor area and that the ceilings are 2.50 m from the floors. The air in the house has a molar specific heat capacity of 29.1 J/mol · K. (The number of moles of air in the house can be found by assuming that the average molar mass of air is 28.9 g/mol and that the density of air at these temperatures is about 1.22 g/L.) How much methane do you have to burn to heat the air from 15.0 to 22.0 °C?

81. Methanol, CH_3OH, a compound that can be made relatively inexpensively from coal, is a promising substitute for gasoline. The alcohol has a smaller energy content than gasoline, but, with its higher octane rating, it burns more efficiently than gasoline in combustion engines. (It also has the added advantage of contributing to a lesser degree to some air pollutants.) Compare the heat of combustion *per gram* of CH_3OH and C_8H_{18} (isooctane), the latter being representative of the compounds in gasoline. ($\Delta H_f° = -259.2$ kJ/mol for isooctane.)

82. Camping stoves are fueled by propane (C_3H_8), butane ($C_4H_{10}(g)$, $\Delta H_f° = -125.6$ kJ/mol), gasoline, or ethanol (C_2H_5OH) (see photo). Calculate the heat of combustion per gram of each of these fuels. (Assume that gasoline is represented by isooctane, $C_8H_{18}(\ell)$, with $\Delta H_f° =$

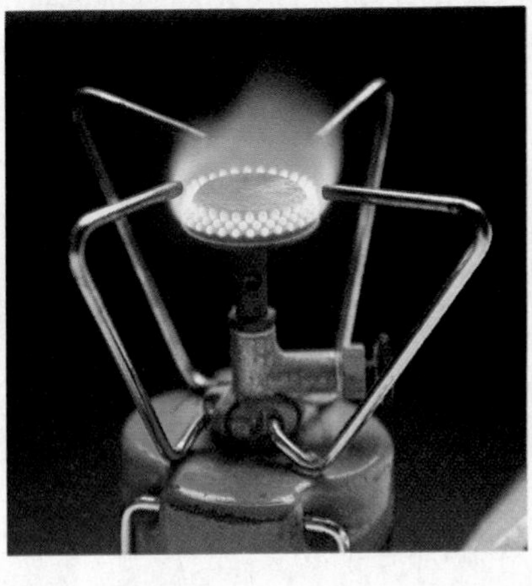

A camping stove that uses butane as a fuel. See Study Question 82. (C. D. Winters)

A control rocket on the Space Shuttle uses a hydrazine as the reducing agent and N_2O_4 as the oxidizer. See Study Question 83.

−259.2 kJ/mol.) Do you notice any great differences among these fuels? Are these differences related to their composition? (See T. Smith, "Camping Stoves," *Chem Matters,* Vol. 10, April, p. 7, 1992.)

83. Hydrazine and 1,1-dimethylhydrazine both react spontaneously with O_2 and can be used as rocket fuels.

$$\underset{\text{hydrazine}}{N_2H_4(\ell)} + O_2(g) \longrightarrow N_2(g) + 2\ H_2O(g)$$

$$\underset{\text{1,1-dimethylhydrazine}}{N_2H_2(CH_3)_2(\ell)} + 4\ O_2(g) \longrightarrow 2\ CO_2(g) + 4\ H_2O(g) + N_2(g)$$

The molar enthalpy of formation of liquid hydrazine is +50.6 kJ/mol, and that of liquid dimethylhydrazine is +48.9 kJ/mol. By doing appropriate calculations, decide whether the reaction of hydrazine or dimethylhydrazine with oxygen gives more heat *per gram* (at constant pressure). (Other enthalpy of formation data can be obtained from Table 6.2.)

84. A piece of metal with a mass of 27.3 g is heated to 98.90 °C and then dropped into 15.0 g of water at 25.00 °C. The final temperature of the system is 29.87 °C. What is the specific heat of the metal?

85. In the United States engineers express energy in units of Btu, which stands for British thermal unit. (1 Btu, which is equivalent to 1055 J, is the amount of heat required to raise the temperature of 1 lb of water from 59.5 to 60.5 °C.) If the average solar energy striking a horizontal surface in Washington, DC, in June is 2080 Btu per square foot, what is the energy in joules per square meter?

86. The heat of fusion of ice is 333 J/g. Suppose that three 100.-g ice cubes at 0 °C are dropped into 500. mL of tea initially at 20.0 °C. How much ice melts? Assume the tea is weak enough that its specific heat is the same as that of pure water.

87. Isomers are molecules with the same elemental composition but a different atomic arrangement. There are, for example, three ways of arranging the atoms in a compound with the formula C_4H_8. The double bond between carbon atoms occurs at a different location, or the arrangement of atoms around the double bond is different. The enthalpy of combustion of each isomer can be determined using a calorimeter.

Compound	Name	Enthalpy of Combustion (kJ/mol)
$H_2C{=}CH{-}CH_2{-}CH_3$	1-butene	−2696.7
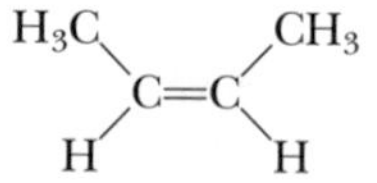	*cis*-2-butene	−2687.5
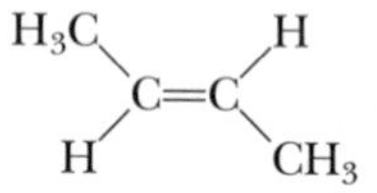	*trans*-2-butene	−2684.2

(a) Draw an enthalpy diagram such as that in Figure 6.16.

(b) What is the enthalpy change for the conversion of *cis*-2-butene to *trans*-2-butene?

(c) Knowing that the enthalpy of formation of 1-butene is −20.5 kJ/mol, calculate the enthalpy of formation values for both *cis*- and *trans*-2-butene.

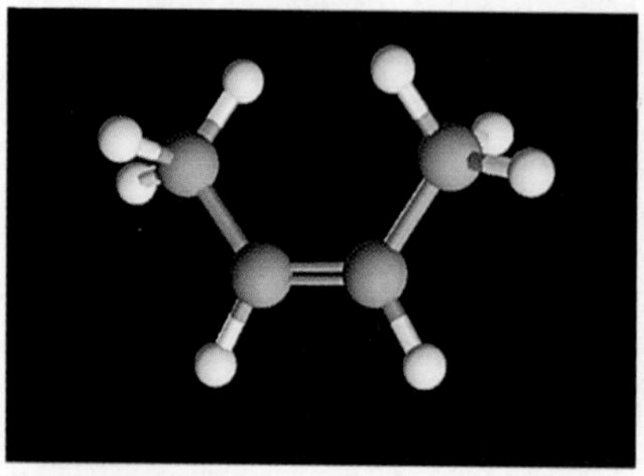
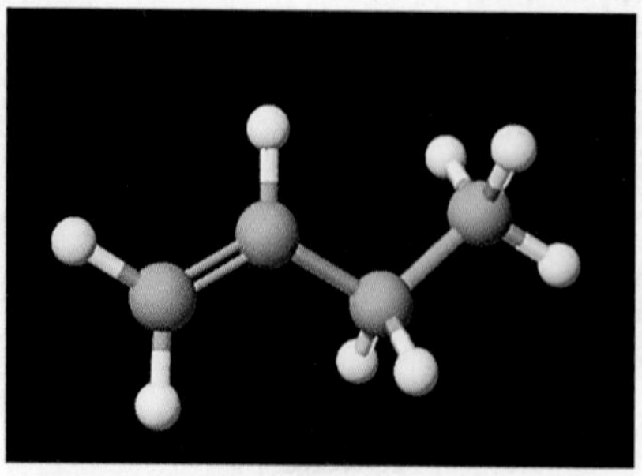
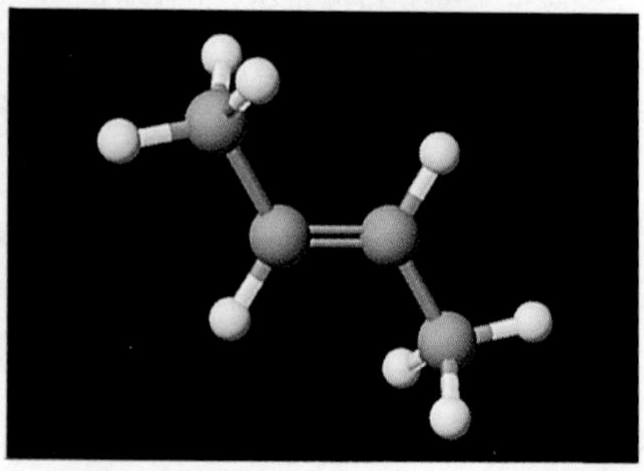

Models of three isomers of C_4H_8. See Study Question 87.

Conceptual Questions

88. Graphite is the standard state of carbon, so $\Delta H_f^\circ = 0$ kJ/mol. That of diamond, however, is +1.90 kJ/mol.

$$\text{C(graphite)} \longrightarrow \text{C(diamond)} \qquad \Delta H^\circ_{rxn} = +1.90 \text{ kJ}$$

Explain how to determine the enthalpy change for the conversion of graphite to diamond. Among the experimental information you have is the enthalpy change for the combustion of diamond.

$$\text{C(diamond)} + O_2(g) \longrightarrow CO_2(g) \qquad \Delta H^\circ_{rxn} = -395.4 \text{ kJ}$$

89. Listed in the table are the specific heat of a variety of metals. Suppose you want to work out a method of predicting the specific heat of a metal for which an experimental value is not available. Prepare a plot of the specific heat in the table versus the atomic weight of the metal. Does any relation exist between specific heat and atomic weight? Use this relation to predict the specific heat of platinum. (The specific heat for platinum is given in the literature as 0.133 J/g · K. What is the agreement between the predicted and actual values?

Metal	Specific Heat (J/g · K)
Chromium	0.450
Gold	0.128
Iron	0.451
Lead	0.127
Silver	0.236
Tin	0.227
Titanium	0.522

90. Suppose you are attending summer school and are living in a very old dormitory. The day is oppressively hot. There is no air conditioner, and you can't open the windows of your room because they are stuck shut from layers of paint. There is a refrigerator in the room, however. In a stroke of genius you open the door of the refrigerator, and cool air cascades out. The relief does not last long, though. Soon the refrigerator motor and condenser begin to run, and not long thereafter the room is hotter than it was before. Why did the room warm up?

91. You want to determine the value for the enthalpy of formation of $CaSO_4(s)$.

$$Ca(s) + \tfrac{1}{8} S_8(s) + 2\, O_2(g) \longrightarrow CaSO_4(s)$$

The reaction cannot be done directly. You know, however, that both calcium and sulfur react with oxygen to produce oxides in reactions that can be studied calorimetrically. You also know that the basic oxide CaO reacts with the acidic oxide SO_3 to produce $CaSO_4(s)$. Outline a method for determining ΔH_f° for $CaSO_4(s)$, and identify the information that must be collected by experiment. Using information in Table 6.2 confirm that ΔH_f° for $CaSO_4 = -1434.5$ kJ/mol.

Summary Question

92. Sulfur dioxide, SO_2, is a major pollutant in our industrial society, and it is often found in wine.

(a) In wine making, SO_2 is commonly added to kill microorganisms in the grape juice when it is put into vats before fermentation. Furthermore, it is used to neutralize byproducts of the fermentation process, enhance wine flavor, and prevent oxidation. Wine usually contains 80 to 150 ppm SO_2 (1 ppm = 1 part per million = 1 gram of SO_2 per 1 million grams of wine). The United States produced 440 million gal of wine in 1987. Assuming the density of wine is 1.00 g/cm^3, and that the wine contains 100. ppm of SO_2, how many grams and how many moles of SO_2 were contained in this wine?

(b) When SO_2 is given off by an oil- or coal-burning power plant, it can be trapped by reacting it with MgO in air to form $MgSO_4$.

$$MgO(s) + SO_2(g) + \tfrac{1}{2} O_2(g) \longrightarrow MgSO_4(s)$$

If 20. million tons of SO_2 are given off by coal-burning power plants each year, how much MgO must be supplied to remove all of this SO_2 (1 ton = 9.08×10^5 g)? How much $MgSO_4$ is produced?

(c) If ΔH_f° for $MgSO_4(s)$ is −2817.5 kJ/mol, how much heat (at constant pressure) is evolved or absorbed per mole of $MgSO_4$ by the reaction in part (b)?

(d) Sulfuric acid comes from the oxidation of sulfur, first to SO_2 and then to SO_3. The SO_3 is then absorbed by water to make H_2SO_4.

$$S(s) + O_2(g) \longrightarrow SO_2(g) \qquad \Delta H_f^\circ = -296.8 \text{ kJ}$$
$$SO_2(g) + \tfrac{1}{2} O_2(g) \longrightarrow SO_3(g) \qquad \Delta H^\circ_{rxn} = -98.9 \text{ kJ}$$
$$SO_3(g) + H_2O(\text{in 98\% } H_2SO_4) \longrightarrow H_2SO_4(\ell) \qquad \Delta H^\circ_{rxn} = -130.0 \text{ kJ}$$

The typical plant produces 750. tons of H_2SO_4 per day (1 ton = 9.08×10^5 g). Calculate the amount of heat produced by the plant per day.

CHAPTER 7

Atomic Structure

(C. D. Winters)

A Chemical Puzzler

The dishes in this photograph contain salts of various metal ions. Methanol is added and then set on fire. The heat of the burning methanol has vaporized some of the salts, as seen by the colors of the flames. The same effect is seen in fireworks. Sky rockets can explode to give red, green, yellow, and other colors. But why are the flames colored? Why do metal ions in the vapor phase emit light? Why does the color vary depending on the metal ion? How does this reflect the structure of the atom?

The colors of fireworks bursting in the night sky or of beautiful gemstones excite all of us. Have you ever wondered how these colors are produced? Part of the answer can be found in this chapter, and this will lead us to answers to still other questions that have intrigued chemists for much of this century. We shall see that these particular colors arise from ions of various salts used in producing fireworks or from the ions incorporated naturally in gemstones.

We know that chemical elements that exhibit similar properties are found in the same column of the periodic table. But why should this be so? The discovery of the electron, proton, and neutron (Section 2.2) prompted scientists to look for relationships between atomic structure and chemical behavior. As early as 1902 Gilbert N. Lewis (1875–1946) hit upon the idea that electrons in atoms might be arranged in shells, starting close to the nucleus and building outward. Lewis explained the similarity of chemical properties for elements in a given group by assuming that all the elements of that group have the same number of electrons in the outer shell. These are the **valence electrons,** first introduced in Section 3.3.

Lewis's model of the atom raises a number of questions. Where are the electrons located? Do they have different energies? Does any experimental evidence support this model? These questions were the reason for many of the experimental and theoretical studies that began around 1900 and continue to this day. This chapter and the next one outline the important results so far.

7.1 ELECTROMAGNETIC RADIATION

We are all familiar with water waves, and you may also know that some properties of radiation such as light can be described with the ideas of wave motion. These ideas came from the experiments of physicists in the 19th century, among them a Scot, James Clerk Maxwell (1831–1879). In 1864, he developed an elegant mathematical theory to describe all forms of radiation in terms of oscillating, or wave-like, electric and magnetic fields in space (Figure 7.1). Hence, radiation, such as light, microwaves, television and radio signals, and x-rays, is collectively called **electromagnetic radiation.** This view is important to our understanding of how electrons behave in atoms.

James Clerk Maxwell is regarded by many as the greatest physicist between Newton and Einstein.

Wave forms are illustrated with water waves in Figure 7.2. The **wavelength** of a wave is the distance between successive crests, or high points (or between successive troughs, or low points). This distance can be given in meters, nanometers, or whatever unit is convenient. The symbol for wavelength is the Greek letter λ (lambda).

Waves can also be characterized by their **frequency,** symbolized by the Greek letter ν (nu). For a wave traveling through some point in space, the frequency is

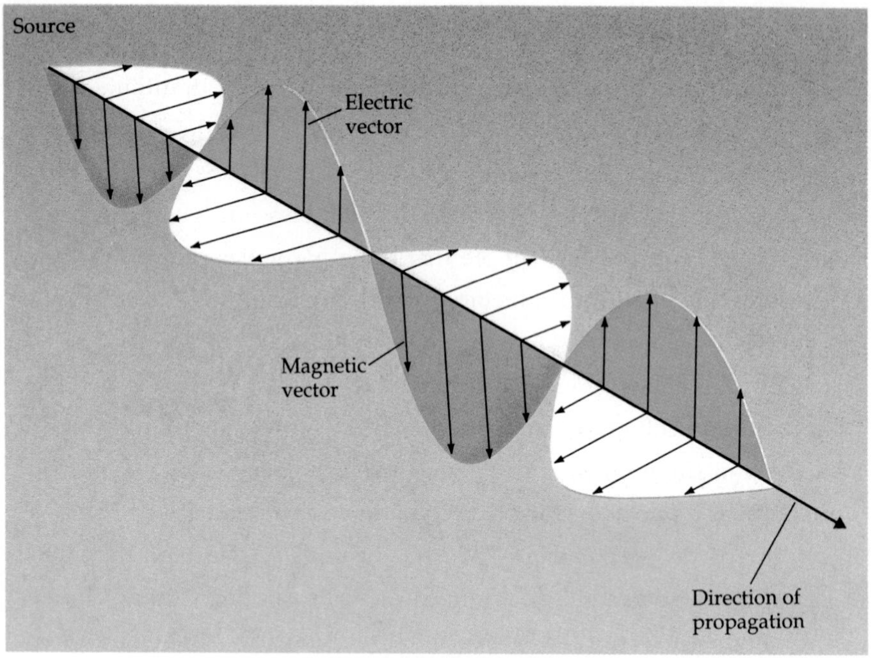

Figure 7.1 Electromagnetic radiation. In the 1860s James Clerk Maxwell developed the currently accepted theory that all forms of radiation are propagated through space as vibrating electric and magnetic fields, the fields being at right angles to one another. Each of the fields is described by a sine wave (because of the mathematical function describing the wave). Such oscillating fields emanate from vibrating charges in a source, such as a light bulb or radio antenna.

Maxwell developed the theory of electromagnetic radiation, but its existence was proved by the German experimentalist, Heinrich Hertz (1857–1894).

equal to the number of complete waves passing the point in that amount of time. Thus, we usually refer to the frequency as the number of cycles that pass per second (Figure 7.2). The unit for frequency is usually written as s^{-1} (standing for 1 per second, 1/s) and is now called the **hertz.**

If you have ever enjoyed water sports, you are familiar with the height of waves. In more scientific terms, the maximum height of a wave, as measured from the axis of propagation of the wave, is called the **amplitude.** In Figure 7.2, notice that the wave has zero amplitude at certain intervals along the wave. Points of zero amplitude, called **nodes,** always occur at intervals of $\lambda/2$ for standing waves such as those illustrated in Figure 7.2.

The concept of nodes is important in understanding atomic orbitals (Section 7.6).

Finally, the velocity of a moving wave is an important factor. As an analogy, consider cars in a traffic jam traveling bumper to bumper. If each car is 16 feet long, and if a car passes you every 4 seconds (that is, the frequency is 1 per 4 seconds or $\frac{1}{4}s^{-1}$), then the traffic is "moving" at the speed of (16 ft)($\frac{1}{4}s^{-1}$), or 4

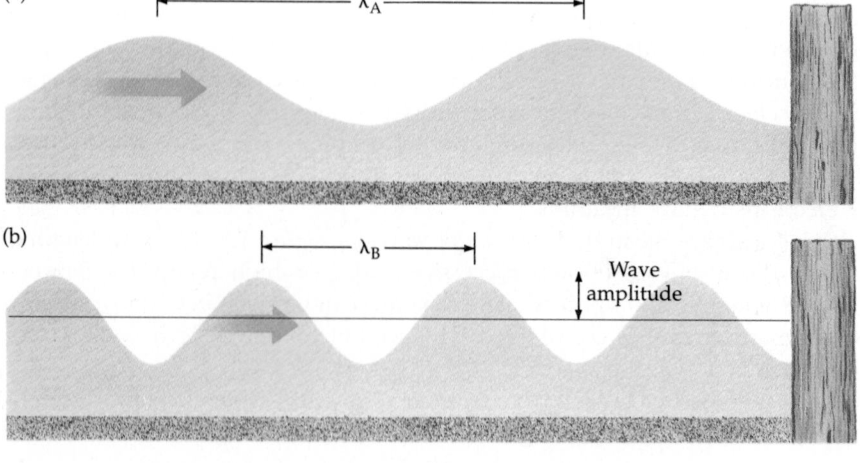

Figure 7.2 An illustration of wavelength and frequency with water waves. Both sets of waves are moving forward toward the post and are traveling with the same velocity. Wave (a) has a longer wavelength than wave (b) ($\lambda_A > \lambda_B$). Wave (a) has a lower frequency than (b) ($\nu_A < \nu_B$), because the number of times per second wave (a) hits the post is less than that of wave (b).

feet per second. This multiplication of length times frequency to give velocity is true for any periodic motion, including a wave.

$$\text{Wavelength (m)} \times \text{frequency (s}^{-1}) = \text{velocity (m/s)}$$
$$\lambda \quad \times \quad \nu \quad = \quad v$$

This same equation applies to electromagnetic radiation, where the product of the wavelength and frequency is equal to the velocity of light, *c*.

$$\lambda \cdot \nu = c \qquad \textbf{(7.1)}$$

The velocity of visible light and all other forms of electromagnetic radiation in a vacuum is a constant (symbolized by *c*) and is equal to 2.99792458×10^8 m/s (or 186,000 miles per second). This means that, if you know the wavelength of a light wave, you can readily calculate its frequency, and vice versa. For example, if you know that orange light has a wavelength of 625 nm, find the wavelength in meters and the frequency. Because the speed of light is expressed in meters per second, the wavelength in nanometers must be changed to meters before you can substitute into the equation $\lambda\nu = c = 2.998 \times 10^8$ m/s.

$$625 \text{ nm} \cdot \frac{1 \times 10^{-9} \text{ m}}{\text{nm}} = 6.25 \times 10^{-7} \text{ m}$$

$$\nu = \frac{c}{\lambda} = \frac{2.998 \times 10^8 \text{ m/s}}{6.25 \times 10^{-7} \text{ m}} = 4.80 \times 10^{14} \text{ s}^{-1}$$

We are bathed constantly in electromagnetic radiation, including the radiation you can see, visible light. As you know, visible light really consists of a spectrum of colors, ranging from red light at the long-wavelength end of the spectrum to violet light at the short-wavelength end (Figure 7.3). Visible light is,

The velocity of light passing through a substance (air, glass, water, etc.) depends on the chemical constitution of the substance and the wavelength of the light. This is the basis for using a glass prism to disperse light and is the explanation for rainbows. Sound velocity is also dependent on the material through which it passes.

You can remember the colors of visible light, *in order of decreasing wavelength,* by the famous phrase **ROY G BIV,** which stands for red, orange, yellow, green, blue, indigo, violet.

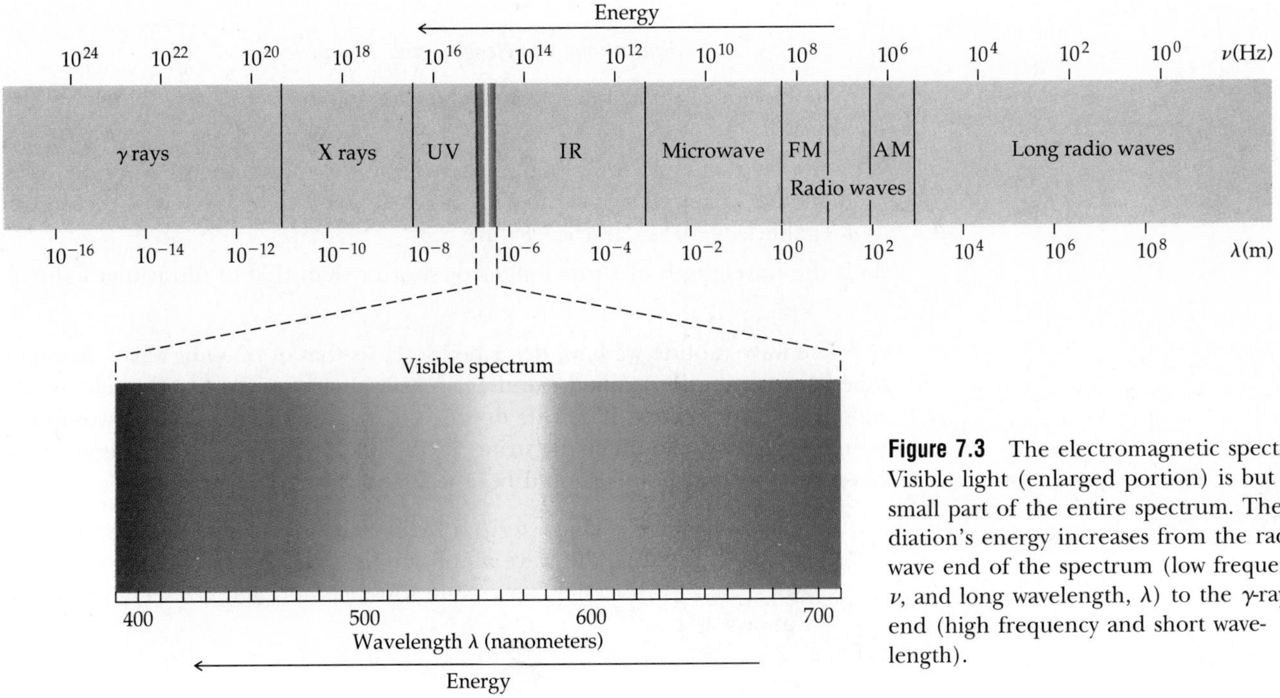

Figure 7.3 The electromagnetic spectrum. Visible light (enlarged portion) is but a small part of the entire spectrum. The radiation's energy increases from the radio-wave end of the spectrum (low frequency, ν, and long wavelength, λ) to the γ-ray end (high frequency and short wavelength).

however, only a small portion of the total electromagnetic spectrum. Ultraviolet (UV) radiation, the radiation that can lead to sunburn, has wavelengths shorter than those of visible light; x-rays and γ-rays, the latter emitted in the process of radioactive disintegration of some atoms, have even shorter wavelengths. At longer wavelengths than visible light, we first encounter infrared radiation (IR), the type that is sensed as heat. Longer still is the wavelength of the radiation in a microwave oven and that used in television and radio transmissions.

EXAMPLE 7.1 *Wavelength-Frequency Conversions*

The frequency of the radiation used in all microwave ovens sold in the United States is 2.45 GHz. (The unit GHz stands for "gigahertz"; 1 GHz is a billion cycles per second, or 10^9/s.) What is the wavelength (in meters) of this radiation? How much longer or shorter is this than the wavelength of orange light (625 nm)?

Solution The wavelength of microwave radiation in meters can be calculated directly from Equation 7.1.

$$\lambda = \frac{c}{\nu} = \frac{2.998 \times 10^8 \text{ m/s}}{2.45 \times 10^9 \text{ s}^{-1}} = 0.122 \text{ m}$$

$$\frac{\lambda(\text{microwaves})}{\lambda(\text{orange light})} = \frac{0.122 \text{ m}}{6.25 \times 10^{-7} \text{ m}} = 195{,}000$$

A comparison of the wavelengths of microwave radiation and orange light shows that the microwaves have wavelengths almost 200,000 times longer than those of orange light.

EXERCISE 7.1 *Radiation, Wavelength, and Frequency*

1. Which color in the visible spectrum has the highest frequency? Which has the lowest frequency?
2. Is the frequency of the radiation used in a microwave oven higher or lower than that from your favorite FM radio station (91.7 MHz) (where MHz = megahertz = 10^6 s^{-1})? (See Figure 7.3)
3. Is the wavelength of x-rays longer or shorter than that of ultraviolet light? ■

The wave motion we have described so far is that of *traveling* waves. Another type of wave motion, called **standing,** or **stationary,** waves, is also relevant to modern atomic theory. If you tie down a string at both ends, as you would the string of a guitar, and pluck the string, it vibrates as a standing wave (Figure 7.4). Several important points should be noted about standing waves:

- A standing wave is characterized by having two or more points of no movement; that is, the wave amplitude is zero at these points, called **nodes.** As with traveling waves, the distance between consecutive nodes is always $\lambda/2$.

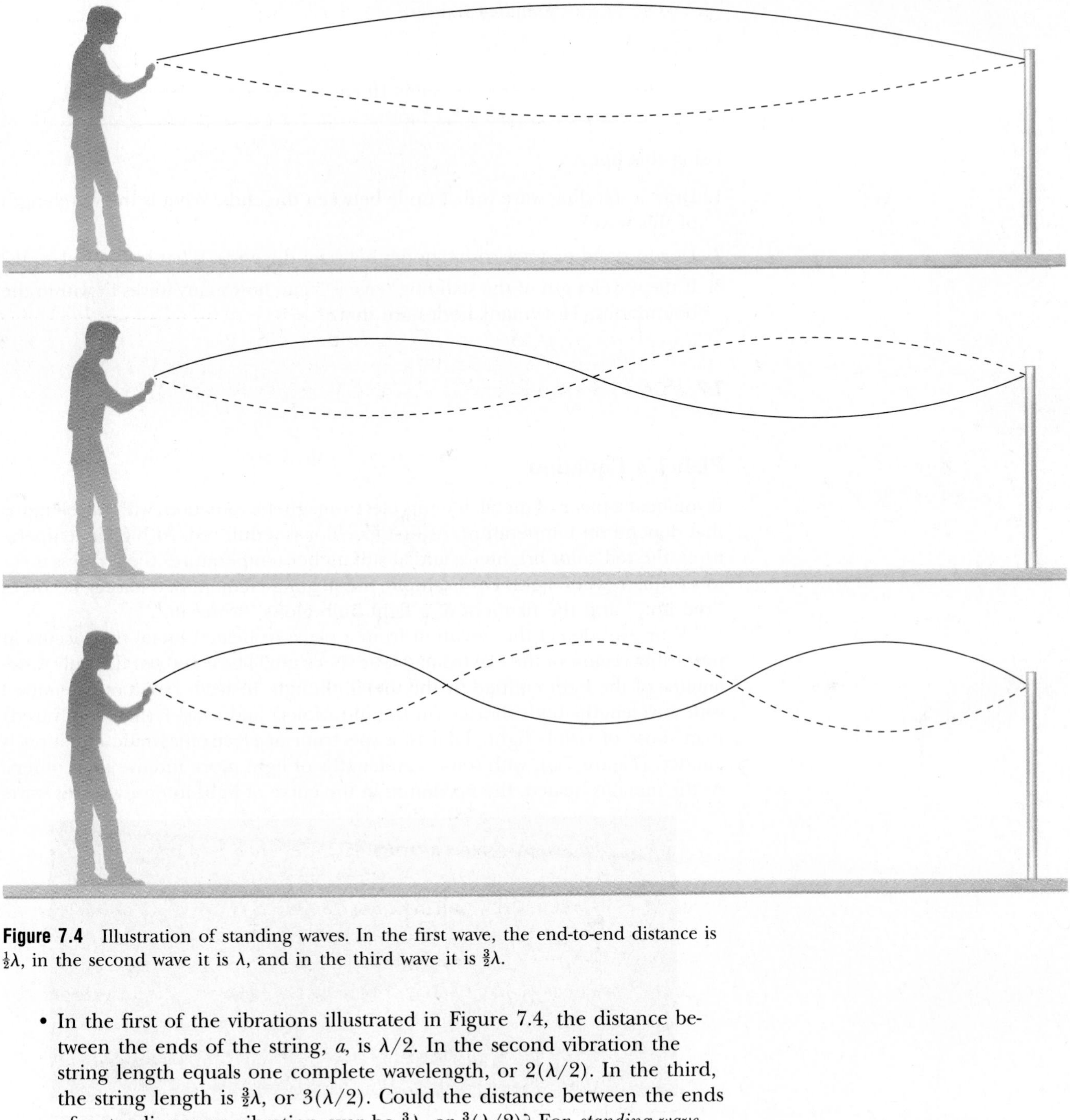

Figure 7.4 Illustration of standing waves. In the first wave, the end-to-end distance is $\frac{1}{2}\lambda$, in the second wave it is λ, and in the third wave it is $\frac{3}{2}\lambda$.

- In the first of the vibrations illustrated in Figure 7.4, the distance between the ends of the string, *a*, is $\lambda/2$. In the second vibration the string length equals one complete wavelength, or $2(\lambda/2)$. In the third, the string length is $\frac{3}{2}\lambda$, or $3(\lambda/2)$. Could the distance between the ends of a standing wave vibration ever be $\frac{3}{4}\lambda$, or $\frac{3}{2}(\lambda/2)$? For *standing wave vibrations,* both those you can see and those you cannot see, *only certain wavelengths are possible.* Because the standing wave ends must be nodes, the only allowed vibrations are those in which $a = n(\lambda/2)$ [where *a* is the distance from one end or "boundary" to the other, and *n* is an integer (1, 2, 3, . . .)]. This is an example of *quantization* in nature, a concept we turn to next.

nodes # are quantized:
$a = n(\lambda/2)$
dist int freq

EXERCISE 7.2 ***Standing Waves***

The line shown here is 10 cm long.

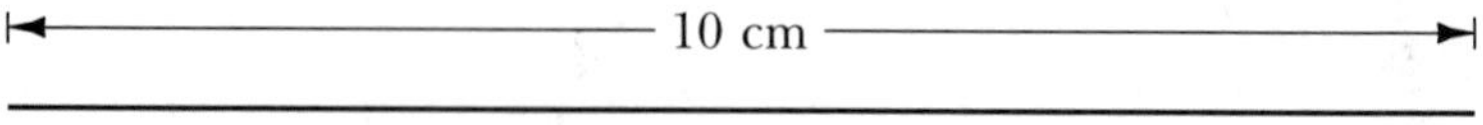

Using this line,

1. Draw a standing wave with 1 node between the ends. What is the wavelength of this wave?
2. Draw a standing wave with 2 nodes between the ends. What is its wavelength?
3. If the wavelength of the standing wave is 5 cm, how many waves fit within the boundaries? How many nodes are there? ■

7.2 PLANCK, EINSTEIN, ENERGY, AND PHOTONS

Planck's Equation

If you heat a piece of metal, it emits electromagnetic radiation, with wavelengths that depend on temperature. At first its color is a dull red. At higher temperatures, the red color brightens, and at still higher temperatures the redness turns to a brilliant white light. For example, the heating element of a toaster becomes "red hot," and the filament of a light bulb glows "white hot."

Your eyes detect the radiation from a piece of heated metal that occurs in the visible region of the electromagnetic spectrum. These are not the only wavelengths of the light emitted by the metal, though. Instead, radiation is emitted with wavelengths both shorter (in the ultraviolet) and longer (in the infrared) than those of visible light. That is, a spectrum of electromagnetic radiation is emitted (Figure 7.5), with some wavelengths of light more intense than others. As the metal is heated, the maximum in the curve of light intensity versus wave-

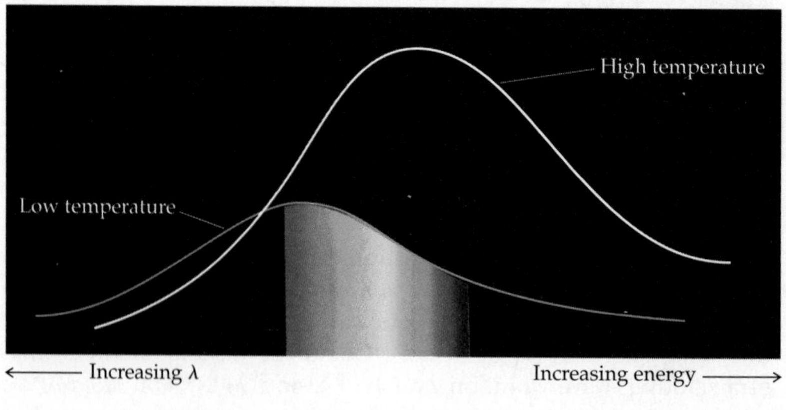

Figure 7.5 The spectrum of the radiation given off by a heated body. The red line is the spectrum for a "red hot" object, since the wavelength of maximum intensity occurs at the wavelength of red light. As the temperature of the object increases, the color of the object becomes more orange and then yellow, as the wavelength of the highest intensity radiation moves toward shorter wavelengths (toward the ultraviolet). At very high temperatures, the object is "white hot," and there is a comparable intensity of radiation at all wavelengths in the visible spectrum.

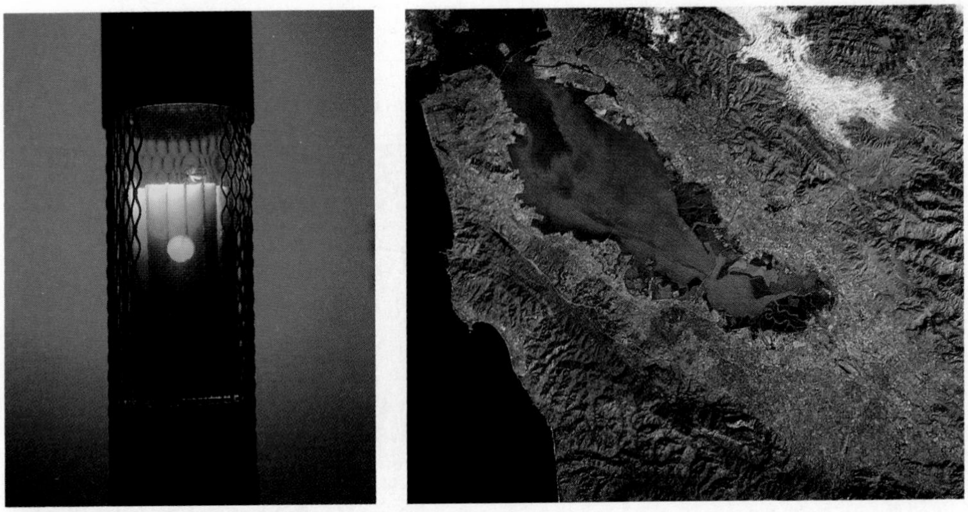

(a) (b)

Infrared radiation has wavelengths longer than light of the visible region of the spectrum. (a) A catalytic converter is used to make wood stoves burn more efficiently and evolve fewer pollutants. It glows red hot when heated. Infrared radiation is experienced as heat. (b) A photo of the San Francisco Bay area taken from a satellite. The film responds to wavelengths in the infrared region. (a: C. D. Winters; b: Earth Satellite Corp./Photo Researchers, Inc.)

length is shifted more and more to the ultraviolet region, so the color of the glowing object shifts from red to yellow and finally to white hot.

At the end of the 19th century, scientists were trying to explain the nature of the emissions from hot objects. They assumed that vibrating atoms in a heated object caused electromagnetic vibrations (light waves) to be emitted, and that those light waves could have any frequency along a continuously varying range. The classical wave theory predicted that as the object got hotter and acquired more energy, its color should shift to the blue and finally all the way to the violet and beyond, but no object was ever observed to do this. This became known as the "ultraviolet catastrophe."

In 1900, the German physicist, Max Planck (1858–1947), offered an explanation for the spectrum of a heated body, an explanation that contained the seeds of a revolution in scientific thought. He made what was at that time an incredible assumption: when an object emits radiation, there must be a minimum quantity of energy that can be emitted at any given time. That is, there is a small packet of energy such that no smaller quantity can be emitted, just as the atom is the smallest packet of an element. Planck called his packet of energy a **quantum.** He further stated that the energy of a quantum is related to the frequency of radiation by the equation

$$E_{\text{quantum}} = h \cdot \nu_{\text{radiation}} \quad \textbf{(7.2)}$$

Energy of a quantum of radiation = (Planck's constant) · (frequency of radiation)

The proportionality constant h is now called **Planck's constant,** in his honor. It has the value $6.6260755 \times 10^{-34}$ J · s.

Earlier in this section we calculated that orange light, with a wavelength of 625 nm, has a frequency of $4.80 \times 10^{14}\ \text{s}^{-1}$. Now we are able to calculate the energy of one quantum of orange light.

$$E = h\nu = (6.626 \times 10^{-34}\ \text{J} \cdot \text{s})(4.80 \times 10^{14}\ \text{s}^{-1}) = 3.18 \times 10^{-19}\ \text{J}$$

The theory based on Planck's work is called the **quantum theory.** Using this theory, Planck was able to calculate the number of quanta of each frequency that are emitted by a heated object. The number of quanta per second gives the intensity of the radiation, and because the frequency is related to the wave-

PORTRAIT OF A SCIENTIST

Max Planck (1858–1947)

Max Karl Ernst Ludwig Planck was raised in Munich, Germany, where his father was an important professor at the University. When still in his teens Planck decided to become a physicist in spite of the advice of the head of the physics department at Munich that "The important discoveries [in physics] have been made. It is hardly worth entering physics anymore." Fortunately, Planck did not take this advice, and he worked for a while at the University of Munich before going to Berlin to study thermodynamics. His interest in thermodynamics led him eventually to the ultraviolet catastrophe and to his revolutionary hypothesis. The discovery was announced two weeks before Christmas in 1900, and he was awarded the Nobel Prize in physics in 1918 for this work. Einstein later said it was a longing to find harmony and order in nature, a "hunger in his soul," that spurred Planck on.

Max Planck (Chemical Heritage Foundation)

An example of an automatic door opener operated by an electric eye device. (C. D. Winters)

length, Planck was able to calculate the spectrum of a hot object. His results agreed very well with experimentally measured spectra.

Einstein and the Photoelectric Effect

When a theory or model accurately predicts experimental results, the theory is usually regarded as valid. Surprisingly, Planck's theory was not well accepted at first, largely because of its radical assumption that energy is quantized. However, after Albert Einstein (1879–1955) used it to explain the photoelectric effect, Planck's quantum theory was firmly accepted.

The **photoelectric effect** occurs when light strikes the surface of a metal and electrons are ejected (Figure 7.6). The electrons ejected from the photocathode by the light move to the positively charged anode, and current flows in the cell. Because light causes current to flow, a photoelectric cell therefore acts as a light-activated switch. The automatic door openers in stores and elevators often work this way.

Experiments with photoelectric cells show that electrons are ejected from the surface only if light of some minimum frequency is used. If the frequency is lower than the minimum, no effect is observed, regardless of the light's intensity (brightness). If the frequency is above the minimum, however, increasing the light intensity causes more and more electrons to be ejected. Einstein decided all these observations could be explained by combining Planck's idea of energy quanta with the notion that light could be described not only as having wave-like properties but also as having particle-like properties. Einstein assumed these *massless* "particles," now called **photons,** carry the energy driven by Planck's law

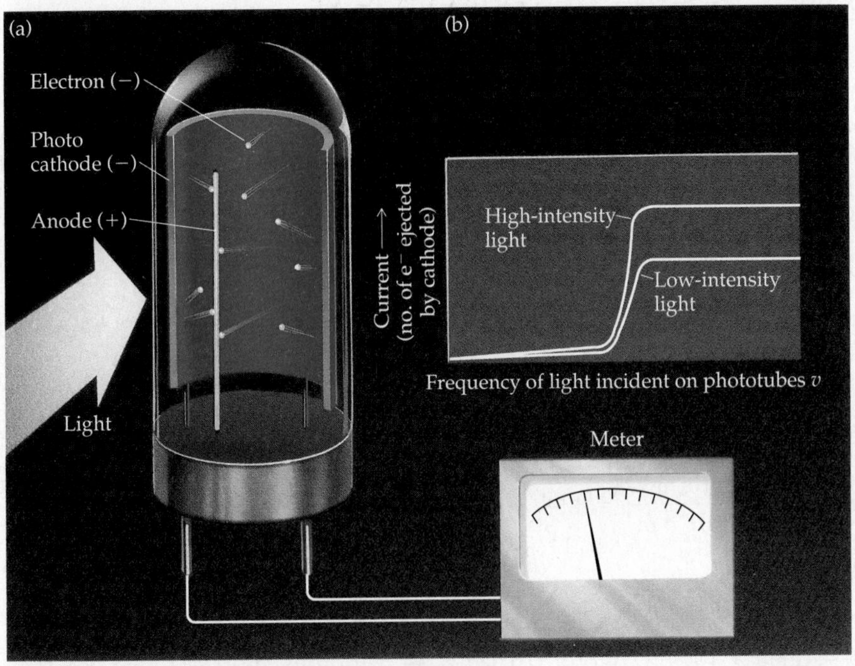

Figure 7.6 Photoelectric effect. (a) A photocell operates by the photoelectric effect. The main part of the cell is a photosensitive cathode. This is a material, usually a metal, that ejects electrons if struck by photons of light of sufficient energy. The ejected electrons move to the anode and a current flows in the cell. Such a device can be used as a switch in electric circuits. (b) As the frequency of the incident light increases, no current is observed until the critical frequency is reached. Light of this frequency and higher has enough energy to dislodge an electron from the surface of the photocathode. If higher intensity light is used, that is, light with a higher photon density, the only effect is to cause more electrons to be released from the surface; the onset of current is observed at the same frequency as with lower intensity light. When light of higher frequency than the minimum is used, the excess energy of the photon simply makes the electron escape the atom with greater velocity.

(Equation 7.2); that is, the energy of each photon in a stream of photons is proportional to the frequency of its wave.

The photoelectric effect can be explained readily using Einstein's proposal. It is easy to imagine that a high-energy particle would have to bump into an atom to cause the atom to lose an electron. It is also reasonable to accept the idea that an electron can be torn away from the atom only if some minimum amount of energy is used. If electromagnetic radiation can also be thought of as a stream of photons, as Einstein said, then the greater the intensity of light, the more photons there are. It then follows that the atoms of a metal surface do not lose electrons when the metal is bombarded by millions of photons if no individual photon has enough energy to remove an electron from an atom. Once the critical minimum energy (that is, minimum light frequency) is exceeded, the energy content of each photon is sufficient to displace an electron from a metal atom. Given this minimum energy, more photons having enough energy dislodge more electrons. Thus, the connection is made between light intensity and the number of electrons ejected after the minimum energy is exceeded.

Cesium is often used in "electric eye" devices such as automatic door openers because all visible wavelengths of light cause its atoms to emit electrons.

Einstein received the Nobel Prize in physics in 1921 for "services to theoretical physics, especially for the discovery of the law of the photoelectric effect."

Energy and Chemistry: Using Planck's Equation

Compact disc players use lasers that emit red light with a wavelength of 685 nm. What is the energy of one photon of this light? What is the energy of a mole of photons of the light? The first step is to convert the wavelength to the frequency of the radiation and then use this to calculate the energy per photon. Finally, the energy of a mole of photons is obtained by multiplying the energy per photon by Avogadro's number.

$$\lambda,\ \text{nm} \xrightarrow[\times \frac{10^{-9}\ \text{m}}{\text{nm}}]{} \lambda,\ \text{m} \xrightarrow[\nu = \frac{c}{\lambda}]{} \nu,\ \text{s}^{-1} \xrightarrow[E = h \cdot \nu]{} E,\ \text{J/photon} \xrightarrow[\times \text{Avogadro's number}]{} E,\ \text{J/mol}$$

$$685\ \text{nm}\ (10^{-9}\ \text{m/nm}) = 6.85 \times 10^{-7}\ \text{m}$$

$$\nu = \frac{2.998 \times 10^{8}\ \text{m/s}}{6.85 \times 10^{-7}\ \text{m}} = 4.38 \times 10^{14}\ \text{s}^{-1}$$

$$E = h\nu$$

$$= (6.626 \times 10^{-34}\ \text{J} \cdot \text{s/photon})(4.38 \times 10^{14}\ \text{s}^{-1}) = 2.90 \times 10^{-19}\ \text{J/photon}$$

$$E = (2.90 \times 10^{-19}\ \text{J/photon})(6.022 \times 10^{23}\ \text{photons/mol}) = 1.75 \times 10^{5}\ \text{J/mol}$$

The energy of a mole of photons of red light is equivalent to 175 kJ, and a mole of photons of blue light ($\lambda = 400$ nm) has an energy of about 300 kJ. These energies are in a range that can affect the bonds between atoms. It should not be surprising therefore that light can lead to chemical processes. For example, you may have seen cases in which sunlight causes paint or dye to fade or cloth to decompose.

The previous calculations show that, as the frequency of radiation increases, the energy of the radiation also increases (see Figure 7.3).

E increases ↑ as frequency increases ↑

Energy = Planck's constant · frequency

Similarly, as the wavelength of radiation decreases, the energy increases.

$$E = h \cdot \nu = \frac{h \cdot c}{\lambda}$$

E increases ↑

$$\text{Energy} = \frac{\text{Planck's constant} \cdot \text{speed of light}}{\text{wavelength}}$$

↓ as wavelength decreases

Therefore, photons of ultraviolet radiation—with wavelengths shorter than those of visible light—have higher energy than visible light. Because visible light has enough energy to affect bonds between atoms, it is obvious that ultraviolet light does as well. That is the reason that you should avoid ultraviolet radiation if you want to avoid a sunburn. In contrast, photons of infrared radiation—with wavelengths longer than those of visible light—have lower energy than visible light. You are aware of them as the heat given off by a glowing burner on an electric stove.

EXAMPLE 7.2 *Calculating Photon Energies*

Compare the energy of a mole of photons of red light from a laser (175 kJ/mol) with the energy of a mole of photons of x-radiation having a wavelength of 2.36 nm. Which has the greater energy? By what factor is one greater than the other?

Solution This calculation follows the scheme outlined in the text. First, we express the wavelength in meters.

$$2.36 \text{ nm} \cdot (1 \times 10^{-9} \text{ m/nm}) = 2.36 \times 10^{-9} \text{ m}$$

Next, calculate the frequency of the radiation.

$$\text{Frequency } (\nu) = \frac{c}{\lambda} = \frac{2.998 \times 10^{8} \text{ m/s}}{2.36 \times 10^{-9} \text{ m}} = 1.27 \times 10^{17} \text{ s}^{-1}$$

Now the energy can be calculated from Planck's equation,

$$E = h\nu$$
$$= (6.626 \times 10^{-34} \text{ J} \cdot \text{s/photon})(1.27 \times 10^{17} \text{ s}^{-1}) = 8.42 \times 10^{-17} \text{ J/photon}$$

(Alternatively, the wavelength in meters can be converted directly to energy using the equation $E = hc/\lambda$.) Finally, we can calculate the energy per mole of photons.

$$E = (8.42 \times 10^{-17} \text{ J/photon})(6.022 \times 10^{23} \text{ photons/mol}) = 5.07 \times 10^{7} \text{ J/mol}$$

The energy of a mole of x-ray photons, equivalent to 5.07×10^4 kJ/mol, is much larger—by a factor of 290—than the energy of a mole of photons of red light.

$$\frac{E \text{ of x-ray photons}}{E \text{ of photons of red light}} = \frac{5.07 \times 10^{4} \text{ kJ/mol}}{175 \text{ kJ/mol}} = 290.$$

EXERCISE 7.3 *Photon Energies*

Compare the energy of a mole of photons of blue light (4.00×10^2 nm) with the energy of a mole of photons of microwave radiation having a frequency of 2.45 Gigahertz (1 GHz = 10^9 s^{-1}). Which has the greater energy? By what factor is one greater than the other? ■

PROBLEM-SOLVING TIPS AND IDEAS

7.1 Energy, Frequency, and Wavelength

Be sure you remember the relationships between the energy, frequency, and wavelength of radiation.

$$E = h\nu, \qquad \text{and, because} \qquad \lambda\nu = c, \qquad E = \frac{hc}{\lambda}$$

It follows from these equations that

- High energy $\longrightarrow$ high frequency (large value of ν)
  $\longrightarrow$ short wavelength (small value of λ)
- low energy $\longrightarrow$ low frequency (small value of ν)
  $\longrightarrow$ long wavelength (large value of λ) ■

7.3 ATOMIC LINE SPECTRA AND NIELS BOHR

Atomic Line Spectra

The final piece of information that played a major role in developing the modern view of atomic structure is the observation of the properties of the light emitted by atoms after they absorb extra energy. The spectrum of white light, such as that from an incandescent light or from the sun, is the rainbow display of separated colors shown in Figure 7.7. Such a spectrum, containing light of all wavelengths, is called a **continuous spectrum.**

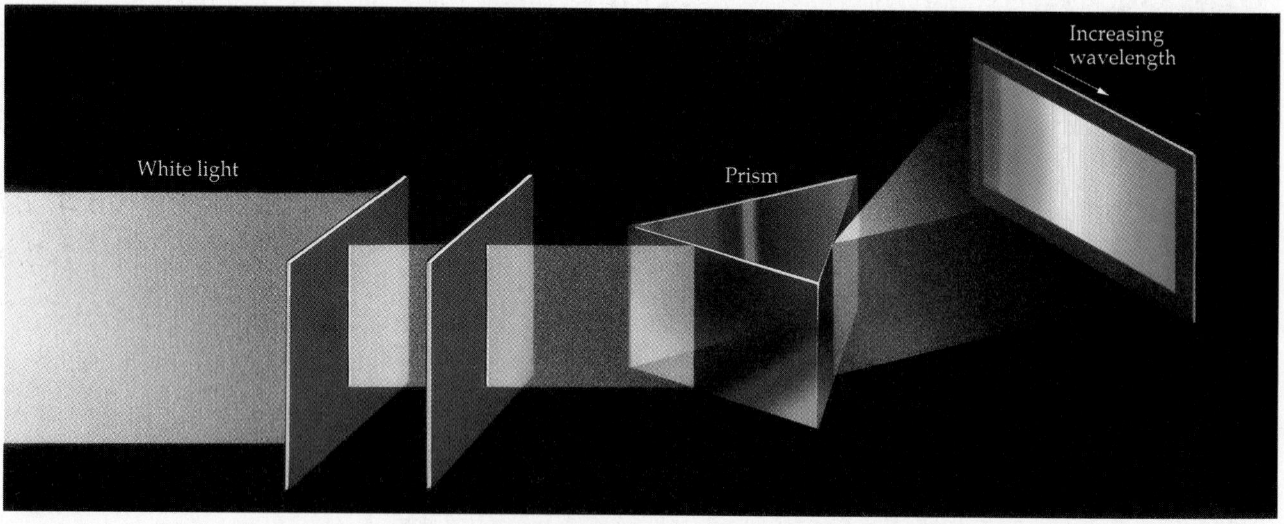

Figure 7.7 A spectrum of white light, produced by refraction in a prism. The light is observed with a spectroscope or spectrometer by passing the light through a narrow slit to isolate a thin beam, or line, of light. The beam is then passed through a device (a prism, or, in modern instruments, a diffraction grating) that separates the light into its component wavelengths. See also Figure 7.3.

Figure 7.8 Gas discharge signs of this type are often referred to as neon signs even though many do not contain that gas. Colors are emitted by excited atoms of noble gases: neon, reddish orange; argon, blue; helium, yellowish white. A helium-argon mixture emits an orange light, and a neon-argon mixture gives a dark lavender mixture. Mercury vapor, used in fluorescent lights, is also used in gas mixtures to obtain a wide range of colors. By using these various gas mixtures and colored glass tubes, most of the colors of the visible spectrum can be produced. (C. Purcell/Photo Researchers, Inc.)

If a high voltage is applied to an element in the gas phase at low pressure, the atoms absorb energy and are said to be "excited." The excited atoms emit light (Figure 7.8). An example of this phenomenon is a neon advertising sign, in which excited neon atoms emit orange-red light. When light from such a source passes through a prism onto a white surface, only a few colored lines are seen. This is called a **line emission spectrum** (Figure 7.9).

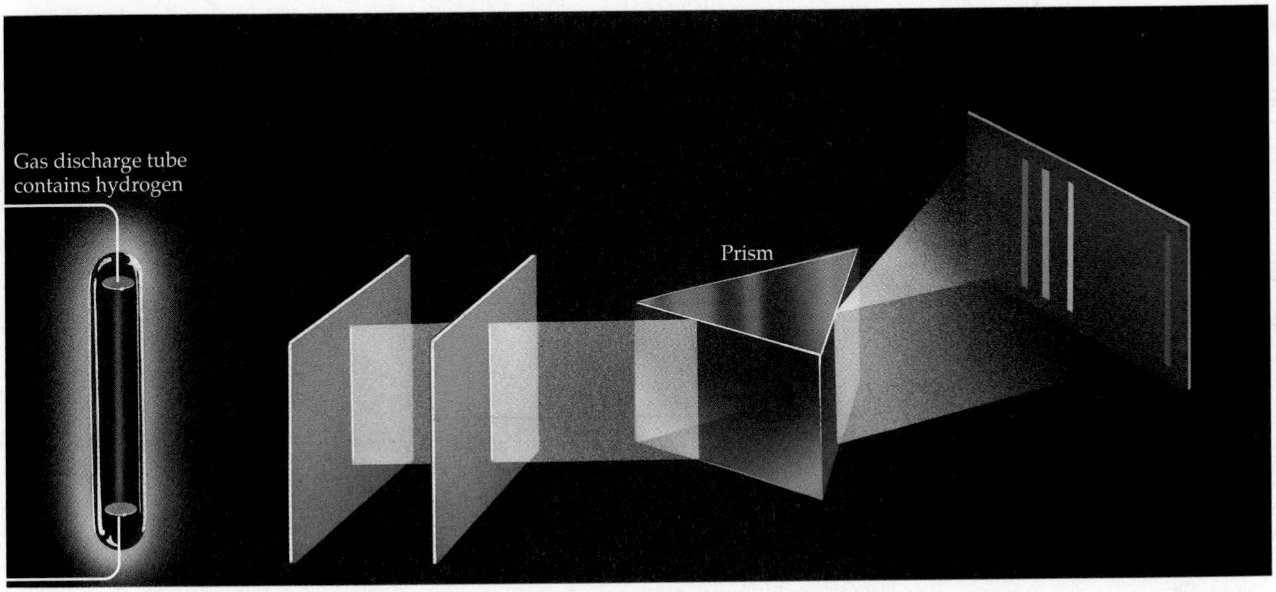

Figure 7.9 The line emission spectrum of excited hydrogen atoms. The emitted light is passed through a series of slits to isolate a narrow beam of light, which is then separated into its component wavelengths by a prism. A photographic plate or an instrument detects the separate wavelengths as individual lines. Hence, the name "line spectrum" for the light emitted by a glowing gas.

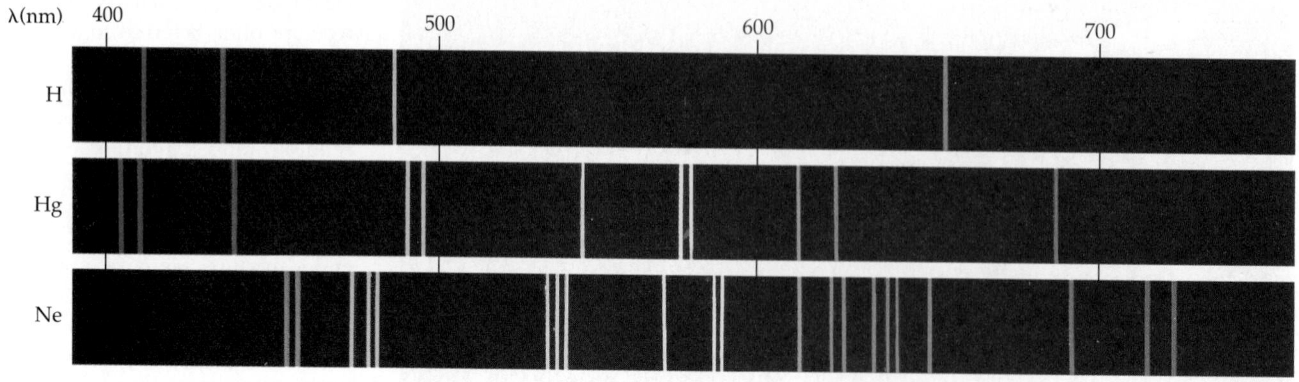

Figure 7.10 Line emission spectra of hydrogen, mercury, and neon. Excited gaseous elements produce characteristic spectra that can be used to identify the element as well as to determine how much element is present in a sample.

The line spectra of the visible light emitted by excited atoms of hydrogen, mercury, and neon are shown in Figure 7.10. Notice that every element has a unique line spectrum. The characteristic lines in the emission spectrum of an element can be used in chemical analysis, especially in metallurgy, both to identify the element and to determine how much of it is present.

One of the goals of scientists in the late 19th century was to explain why gaseous atoms emitted light of only certain frequencies. Attempts were made to find a mathematical relationship among the observed frequencies. It is always significant if experimental data can be related by a simple equation because a "regular" pattern of information implies a logical explanation for the observations. The first step in this direction came from Johann Balmer (1825–1898) and later Johannes Rydberg (1854–1919). They developed an equation from which it was possible to calculate the wavelength of the three lines of longest wavelength in the visible emission spectrum of hydrogen atoms, that is, the red, green, and blue lines in Figure 7.10. This equation, now called the Rydberg equation, was developed by trying to fit experimentally observed wavelengths to some mathematical relation.

special R constant for each element/compound

$$\frac{1}{\lambda} = R\left(\frac{1}{2^2} - \frac{1}{n^2}\right) \qquad n > 2 \qquad \textbf{(7.3)}$$

for H

The symbol m^{-1} means (1/meter), just the same as s^{-1} means 1/second.

Here n is an integer associated with each line, and R, now called the Rydberg constant, has the value $1.0974 \times 10^7\ m^{-1}$. If $n = 3$, the wavelength of the red line in the hydrogen spectrum is obtained (6.561×10^{-7} m, or 656.1 nm). If $n = 4$, the wavelength for the green line is obtained, and $n = 5$ gives the blue line. This group of visible lines (and others for which $n = 6, 7, 8$, and so on) is now called the **Balmer series** of lines.

Niels Bohr, a Danish physicist, provided the first connection between the spectra of excited atoms and the quantum ideas of Planck and Einstein. From Rutherford's work (Section 2.2), it was known that electrons are arranged in space outside of the atom's nucleus. For Bohr the simplest model of a hydrogen atom was to assume that the electron moved in a circular orbit around the nucleus. The problem with this model was that classical physics predicted that the atom could not exist this way. According to the theories at the time, the

PORTRAIT OF A SCIENTIST

Niels Bohr (1885–1962)

Niels Bohr was born in Copenhagen, Denmark. He earned a Ph.D. in physics in Copenhagen in 1911 and then went to work first with J. J. Thomson in Cambridge, England, and later with Ernest Rutherford in Manchester, England.* It was there that he began to develop the ideas that a few years later led to the publication of his theory of atomic structure and his explanation of atomic spectra. (For this work he received the Nobel Prize in 1922.) After working with Rutherford for a very short time, Bohr returned to Copenhagen, where he eventually became the director of the Institute for Theoretical Physics. Many young physicists carried on their work in this Institute, and seven of them later received Nobel Prizes for their studies in chemistry and physics. Among them were such well-known scientists as Werner Heisenberg, Wolfgang Pauli, and Linus Pauling. Element 107 was recently named nielsbohrium (Ns) in honor of Bohr and his work.

Niels Bohr (Chemical Heritage Foundation)

*Rutherford's contributions to chemistry and physics were described in Section 2.2. Bohr's life and his work are described by B. L. Haendler: *Journal of Chemical Education,* Vol. 59, p. 372, 1982. R. Moore: *Niels Bohr,* New York, Alfred Knopf, Inc., 1966; and A. Pais: *Niels Bohr's Times,* Oxford, Oxford University Press, 1991.

---

charged electron moving in the positive, electric field of the nucleus should lose energy. Eventually the electron would crash into the nucleus, much in the same way a satellite eventually crashes into the earth as the satellite loses energy by "rubbing up against" the earth's atmosphere. This is clearly not the case; if it were so, matter would eventually be destroyed.

To solve the "problem" of the stability of atoms, Bohr introduced the notion that the single electron of the hydrogen atom could occupy only certain orbits or energy levels. He identified the energy difference between the levels or orbits as a single quantum of energy. The energy of the electron in an atom is said to be "quantized." By combining this quantization postulate with the laws of motion from classical physics, Bohr showed that the energy possessed by the single electron in the nth orbit of the H atom is given by the simple equation

$$\text{Energy of the } n\text{th level} = E_n = -\frac{Rhc}{n^2} \qquad \textbf{(7.4)}$$

The *potential* energy of the electron is calculated by Equation 7.4.

R= 1.0974 * 10^7 m^-1

R is a proportionality constant, h is Planck's constant, and c is the velocity of light. Each allowed orbit was assigned a value of n, a unitless integer having values of 1, 2, 3, and so on (but not fractional values). This integer is now known as the **principal quantum number** for the electron.

In Bohr's model the radius of the circular orbits increases as n increases. As illustrated by Example 7.3, another consequence of the model and Equation 7.4

is that the energy of the electron becomes less negative as n increases. Thus, the orbit of lowest or most negative energy, with $n = 1$, is closest to the nucleus, and the electron of the hydrogen atom is normally in this energy level. An atom with its electrons in the lowest possible energy levels is said to be in the **ground state.**

Energy must be supplied to move the electron farther away from the nucleus because the positive nucleus and the negative electron attract each other. When the electron of a hydrogen atom occupies an orbit with n greater than 1, the atom has more energy than in its ground state (the energy is less negative) and is said to be in an **excited state.** The energies of the ground and several excited states are calculated in Example 7.3.

EXAMPLE 7.3 *Energies of the Ground and Excited States of the H atom*

Use Equation 7.4 to calculate the energies of the $n = 1$ and $n = 2$ states of the hydrogen atom in joules per atom and in kilojoules per mole. The values needed to use this equation are $R = 1.097 \times 10^7\ \text{m}^{-1}$, $h = 6.626 \times 10^{-34}\ \text{J} \cdot \text{s}$, and $c = 2.998 \times 10^8\ \text{m/s}$.

Solution When $n = 1$, the energy of an electron in a single H atom is

$$E_1 = -\frac{Rhc}{n^2} = -\frac{Rhc}{1^2} = -Rhc$$

$$= -Rhc = (1.097 \times 10^7\ \text{m}^{-1})(6.626 \times 10^{-34}\ \text{J} \cdot \text{s})(2.998 \times 10^8\ \text{m/s})$$

$$= -2.179 \times 10^{-18}\ \text{J/atom}$$

When $n = 2$, the energy is

$$E_2 = -\frac{Rhc}{2^2} = -\frac{E_1}{4} = -\frac{-2.179 \times 10^{-18}\ \text{J/atom}}{4} = -5.448 \times 10^{-19}\ \text{J/atom}$$

The conversion from joules per atom to kilojoules per mole is done using Avogadro's number and the relation 1 kJ = 1000 J.

$$E_1 = \left(\frac{-2.179 \times 10^{-18}\ \text{J}}{\text{atom}}\right) \cdot \left(\frac{6.022 \times 10^{23}\ \text{atoms}}{\text{mol}}\right) \cdot \left(\frac{1\ \text{kJ}}{1000\ \text{J}}\right) = -1312\ \text{kJ/mol}$$

Finally, because $E_2 = E_1/4$, we calculate E_2 to be -328.0 kJ/mol.

Notice that the calculated energies are both negative, with E_1 *more negative* than E_2. This arises because Bohr's equation reflects the fact that the energy of attraction between oppositely charged bodies (an electron and an atomic nucleus) depends on their charge and the distance between them. From Coulomb's law (Section 3.4), we know that the closer the electron is to the nucleus, the greater the energy of attraction; that is, the value of E is *more negative* as the distance becomes smaller, and chemists or physicists say that the energy is therefore lower (meaning *more negative*).

EXERCISE 7.4 *Electron Energies*

Calculate the energy of the $n = 3$ state of the H atom in (a) joules per atom and (b) kilojoules per mole. ■

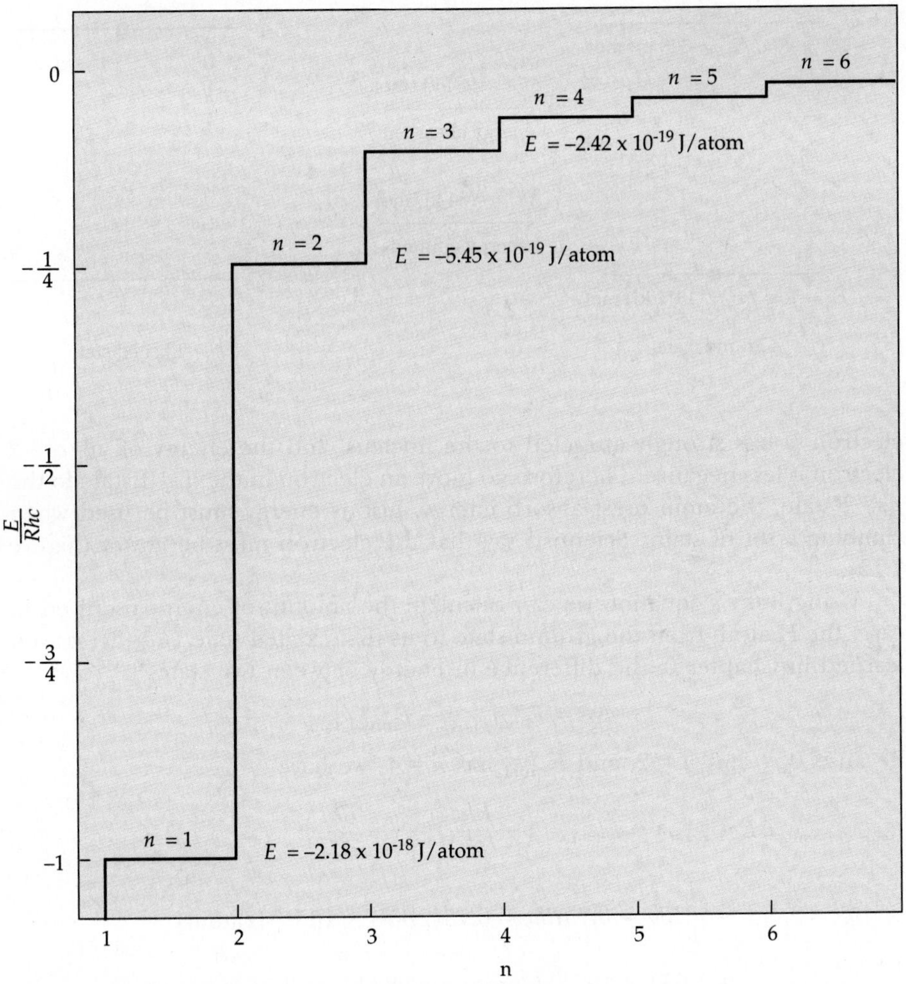

Figure 7.11 The energies of the electron in a hydrogen atom depend on the value of the principal quantum number n ($E_n = -Rhc/n^2$). Energies are given in joules per atom. Notice that the difference between successive energy states becomes smaller as n becomes larger.

You can think of the energy levels in the Bohr model as a set of stairs climbing out of the basement of an "atomic building" (where the energy of the H atom is -2.18×10^{-18} J/atom; see Example 7.3) to the ground level (where the energy is 0) (Figure 7.11). Each step represents a quantized energy level; as you climb the stairs, you can stop on any step but not between steps. Notice that the difference between this analogy and real stairs is that Bohr's stairs get closer and closer together as n increases.

A major assumption of Bohr's theory was that an electron in an atom would remain in its lowest energy level unless disturbed (an assumption shared with the quantum mechanical approach used today; Section 7.5). Energy is absorbed or evolved on changing from one energy level to another, and it is this idea that allowed Bohr to explain the spectra of excited gases.

When the H atom electron has $n = 1$, and so is in its ground state, the energy has a large negative value. As we "climb the stairs" to the $n = 2$ level, the

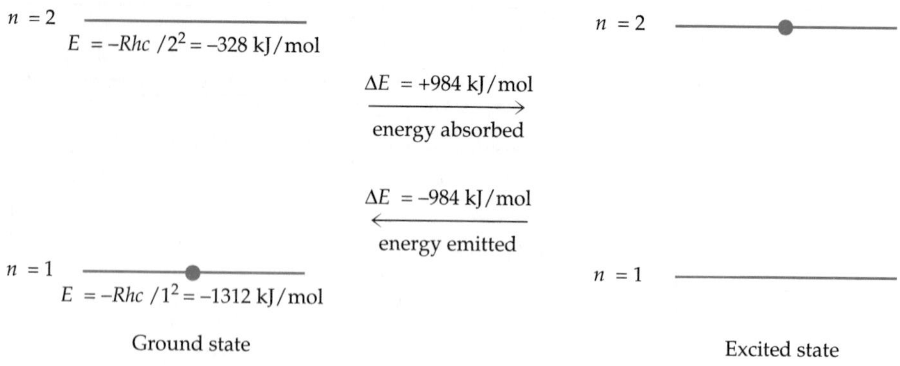

Figure 7.12 Absorption (ΔE positive) and emission (ΔE negative) of energy by the electron (·) in the H atom moving from the $n = 1$ state (the ground state) to the $n = 2$ state (the excited state) and back again.

electron is less strongly attracted to the nucleus, and the energy of an $n = 2$ electron is less negative. Therefore, to move an electron in the $n = 1$ state to the $n = 2$ state, the atom must absorb energy, just as energy must be used when climbing a set of stairs. Scientists say that the electron must be *excited* (Figure 7.12).

Using Bohr's equation we can calculate the amount of energy required to carry the H atom from the ground state to its first excited state ($n = 2$). As you learned in Chapter 6, the difference in energy between two states is

$$\Delta E = E_{\text{final state}} - E_{\text{initial state}}$$

Because E_{final} has $n = 2$, and E_{initial} has $n = 1$, we have

$$\Delta E = E_{n=2} - E_{n=1} = \left(-\frac{Rhc}{2^2}\right) - \left(-\frac{Rhc}{1^2}\right)$$

$$= \frac{3}{4}Rhc = 0.75Rhc = 0.75(2.179 \times 10^{-18}\ \text{J/atom})$$

$$= 1.634 \times 10^{-18}\ \text{J/atom, or 984 kJ/mol of H atoms}$$

where we used the energy of the $n = 1$ state, as calculated in Example 7.3, for the value of $-Rhc$. The amount of energy that must be *absorbed* by the atom so that an electron can move from the first to the second energy state is $0.75Rhc$, no more and no less. If $0.7Rhc$ is provided, no transition between states is possible. Energy levels in the H atom are quantized, with the consequence that only certain amounts of energy may be absorbed or emitted.

Moving an electron from a state of low n to one of higher n is an endothermic process; energy is absorbed, and the sign of the value of ΔE is positive. The opposite process, an electron "falling" from a level of higher n to one of lower n, therefore emits energy. For example, for a transition from $n = 2$ to $n = 1$,

$$\Delta E = E_{\text{final state}} - E_{\text{initial state}}$$

$$= E_{n=1} - E_{n=2} = \left(-\frac{Rhc}{1^2}\right) - \left(-\frac{Rhc}{2^2}\right)$$

$$= -\frac{3}{4}Rhc$$

The process is *exo*thermic; that is, 984 kJ must be *evolved* or *emitted* per mole of H atoms.

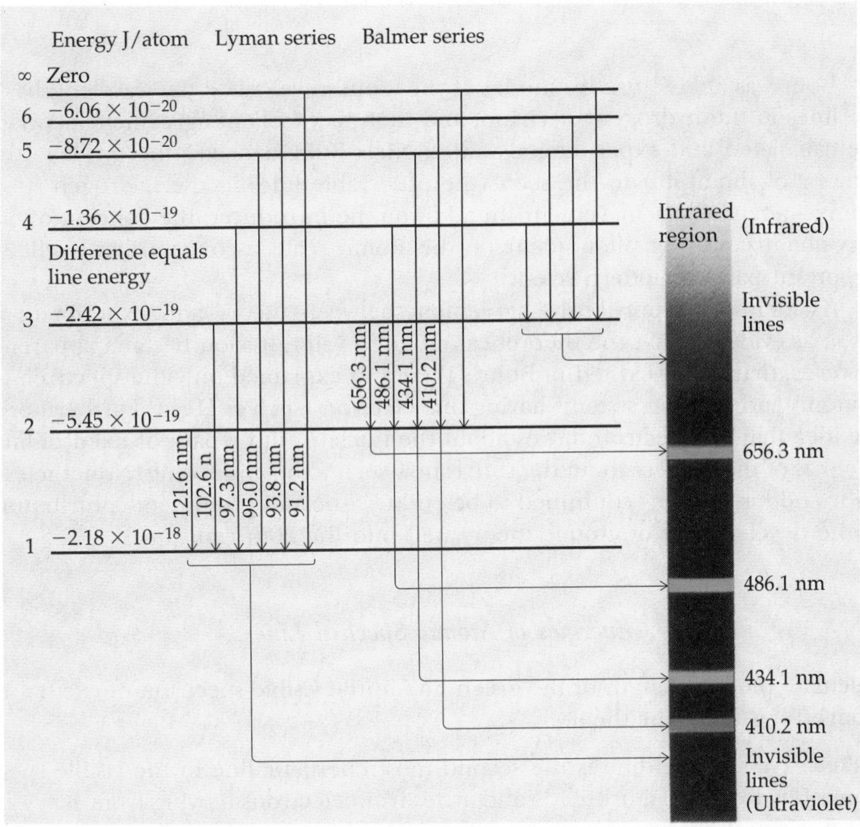

Figure 7.13 Some of the electronic transitions that can occur in an excited H atom. The lines in the ultraviolet region result from transitions to the $n = 1$ level. (This series of lines is called the Lyman series.) Transitions from levels with values of n greater than 2 to $n = 2$ occur in the visible region (Balmer series; see Figure 7.9). Lines in the infrared region result from transitions from levels with n greater than 3 or 4 to the $n = 3$ or 4 levels. (Only the series ending at $n = 3$ is illustrated.)

Depending on how much energy is added to a collection of H atoms, some atoms have their electrons excited from the $n = 1$ to the $n = 2$ or 3 or higher states. After absorbing energy, these electrons naturally move back down to lower levels (not necessarily directly to $n = 1$) and give back the energy the atom originally absorbed. That is, they *emit* energy in the process, and the energy is observed as light. This is the source of the lines observed in the emission spectrum of H atoms, and the same basic explanation holds for atoms of other elements.

For hydrogen, the series of emission lines having energies in the ultraviolet region (called the **Lyman series,** Figure 7.13) comes from electrons moving from states with n greater than 1 to that with $n = 1$. The series of lines that have energies in the visible region—the **Balmer series**—arises from electrons moving from states with $n = 3$ or greater to the state with $n = 2$.

In summary, we now recognize that the *origin of atomic spectra is the movement of electrons between quantized energy states.* If an electron is excited from a lower energy state to a higher one, then energy is absorbed, and an *absorption line* is seen. On the other hand, if an electron moves from a higher energy state to a lower one, energy is emitted and an *emission line* is observed. The energy of a given line in excited hydrogen atoms is

$$\Delta E = E_{\text{final}} - E_{\text{initial}}$$

$$= -Rhc\left(\frac{1}{n^2_{\text{final}}} - \frac{1}{n^2_{\text{initial}}}\right) \quad (7.5)$$

where Rhc is 1312 kJ/mol.

When Bohr's paper describing his ideas was first published in 1913, Einstein declared it to be "one of the great discoveries."

Bohr was able to use his model of the atom to calculate the wavelengths of the lines in the hydrogen spectrum, and there is excellent agreement between the calculated and experimental values. Niels Bohr had tied the unseen (the interior of the atom) to the seen (the observable lines in the hydrogen spectrum)—a fantastic achievement! In addition, he introduced the concept of energy quantization for phenomena on the atomic scale, a concept that is still an important part of modern science.

As mentioned previously, agreement between theory and experiment is taken as evidence that the theoretical model is valid. It soon became apparent, however, that a flaw existed in Bohr's theory. It explained only the spectrum of H atoms and of other systems having one electron (such as He^+). Furthermore, the idea that the electron moves about the nucleus with a path of fixed radius, like that of the planets about the sun, is now seen to be misleading. Nonetheless, Bohr and his students continued to be enthusiastic and prominent contributors to the development of atomic theory well into the 20th century.

EXAMPLE 7.4 *Energies of Atomic Spectral Lines*

Calculate the wavelength of the green line in the visible spectrum of excited H atoms using the Bohr theory.

Solution The green line is the second most energetic line in the visible spectrum of hydrogen (Figure 7.13) and arises from electrons moving from $n = 4$ to $n = 2$. Using Equation 7.5 where $n_{final} = 2$ and $n_{initial} = 4$, we have

$$\Delta E = -Rhc\left(\frac{1}{2^2} - \frac{1}{4^2}\right) = -Rhc(0.1875)$$

In the preceding text, we found that Rhc is 1312 kJ/mol, so the $n = 4$ to $n = 2$ transition involves an energy change of

$$\Delta E = -(1312 \text{ kJ/mol})(0.1875) = -246.0 \text{ kJ/mol}$$

The wavelength can now be calculated from the equation $E_{photon} = h\nu = hc/\lambda$.

$$\lambda = \frac{hc}{E_{photon}} =$$

$$\frac{(6.626 \times 10^{-34} \text{ J} \cdot \text{s/photon})(6.022 \times 10^{23} \text{ photons/mol})(2.998 \times 10^8 \text{ m})}{(246.0 \text{ kJ/mol})(10^3 \text{ J/kJ})}$$

$$= 4.863 \times 10^{-7} \text{ m}$$

$$= 4.863 \times 10^{-7} \text{ m } (10^9 \text{ nm/m}) = 486.3 \text{ nm}$$

The experimental value is 486.1 nm (see Figure 7.13). This represents excellent agreement between experiment and theory.

EXERCISE 7.5 *Energy of an Atomic Spectral Line*

The Lyman series of spectral lines for the H atom occurs in the ultraviolet region. They arise from transitions from higher levels down to $n = 1$. Calculate the frequency and wavelengths of the *least* energetic line in this series. ■

A CLOSER LOOK — *Experimental Evidence for Bohr's Theory*

Niels Bohr's model of the hydrogen atom was powerful because it could reproduce experimentally observed line spectra. But there was additional experimental confirmation.

If the electron in the hydrogen atom is moved from the ground state, where $n = 1$, to the energy level, where $n =$ infinity, the electron is considered to have been removed from the atom. That is, the atom has been ionized.

$$H(g) \longrightarrow H^+(g) + e^-$$

We can calculate the energy for this process from Equation 7.5 where $n_{final} = \infty$ and $n_{initial} = 1$.

$$\Delta E = -Rhc\left(\frac{1}{n^2_{final}} - \frac{1}{n^2_{initial}}\right)$$

$$= -Rhc\left(\frac{1}{\infty^2} - \frac{1}{1^2}\right) = +Rhc$$

Because $Rhc = 1312$ kJ/mol, the energy to move an electron from $n = 1$ to $n = \infty$ is 1312 kJ/mol of H atoms. We now call this the **ionization energy** of the atom (Section 8.6), and can measure it in the laboratory. The experimental value is known to be 1312 kJ/mol, in exact agreement with the result calculated from Bohr's theory!

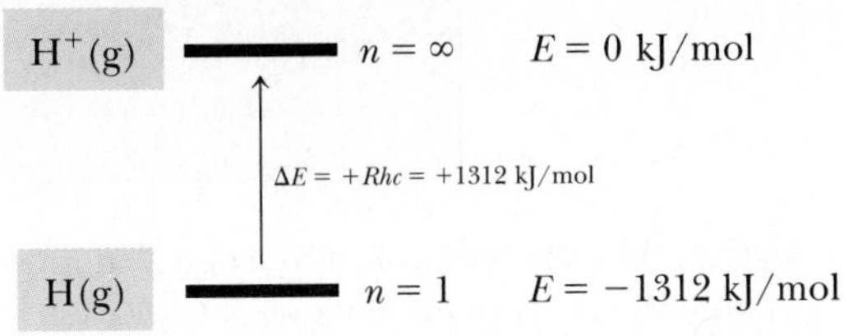

7.4 THE WAVE PROPERTIES OF THE ELECTRON

Einstein used the photoelectric effect to demonstrate that light, which is usually thought of as having wave properties, can also be thought about in terms of particles, or massless photons. This fact was pondered by Louis Victor de Broglie (1892–1987). If light can be considered as having both wave and particle properties, why doesn't matter behave similarly? That is, could a tiny object such as an electron, which we have considered a particle to this point, also exhibit wave properties in some circumstances? In 1925, de Broglie proposed that a free electron of mass m moving with a velocity v should have an associated wavelength given by the equation

$$\lambda = \frac{h}{mv} \qquad \textbf{(7.6)}$$

This idea was revolutionary because it linked the particle properties of the electron (m and v) with possible wave properties (λ). Experimental proof was soon produced. In 1927, C. J. Davisson and L. H. Germer, working at the Bell Telephone Laboratories in New Jersey, found that a beam of electrons was diffracted like light waves by the atoms of a thin sheet of metal foil and that de Broglie's relation was followed quantitatively (Figure 7.14). Because diffraction is an effect readily explained by the wave properties of light, it followed that electrons also can be described as waves under some circumstances.

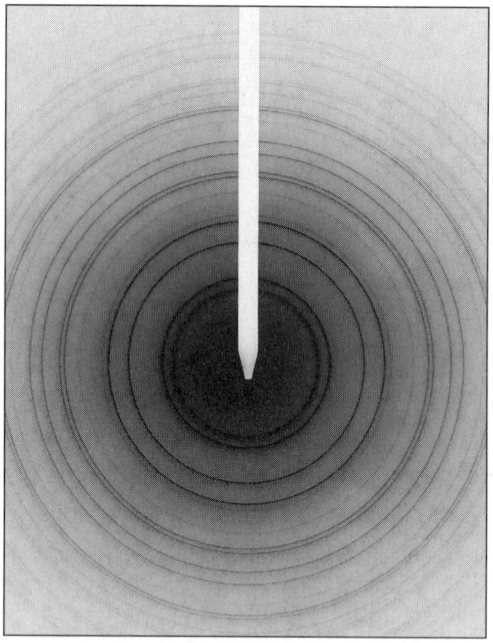

Figure 7.14 The electron diffraction pattern obtained for aluminum foil. (Donald Potter, Department of Metallurgy, University of Connecticut)

Recall that $h = 6.6260755 \times 10^{-34}$ J · s, a *very* small number.

De Broglie's equation suggests that any moving particle has an associated wavelength. Yet, if λ is to be large enough to measure, the product of m and v must be *very* small because h is so small. For example, a 114-g baseball traveling at 110 mph has a large mv product (5.6 kg · m/s) and therefore the incredibly small wavelength of 1.2×10^{-34} m! This tiny value is essentially meaningless because it cannot be measured with any instrument now available. It is only possible to observe wave-like properties for particles of extremely small mass, such as protons, electrons, and neutrons.

EXAMPLE 7.5 *Using De Broglie's Equation*

Calculate the wavelength associated with an electron of mass $m = 9.109 \times 10^{-28}$ g traveling at 40.0% of the velocity of light.

Solution First, consider the units involved. Wavelength is calculated from h/mv, where h is Planck's constant expressed as joules times seconds (J · s). As discussed in Chapter 6, 1 J = 1 kg · m²/s². Therefore, the mass must be used in kilograms and velocity in meters per second.

$$\text{Electron mass} = 9.109 \times 10^{-31}\ \text{kg}$$

$$\begin{aligned}\text{Electron velocity (40.0\% of light velocity)} &= (0.400)(2.998 \times 10^{8}\ \text{m/s}) \\ &= 1.20 \times 10^{8}\ \text{m/s}\end{aligned}$$

Substituting these values into de Broglie's equation, we have

$$\lambda = \frac{h}{mv} = \frac{6.626 \times 10^{-34}\ (\text{kg} \cdot \text{m}^2/\text{s}^2)(\text{s})}{(9.109 \times 10^{-31}\ \text{kg})(1.20 \times 10^{8}\ \text{m/s})} = 6.06 \times 10^{-12}\ \text{m}$$

A CLOSER LOOK *de Broglie and Electron Microscopes*

The suggestion made by Louis Victor de Broglie (1892–1987) that an electron could behave according to the physics of waves led to the modern technique of electron microscopy. From physics we know that the resolving power of a microscope depends on the wavelength of the light used to make the observation (for a given lens diameter). If visible light is used, then you can resolve things that are only fractions of a millimeter apart. What if a beam of electrons is used instead of a beam of photons of visible light? Let us say that you use a visible light microscope (with 550 nm light) and can resolve objects 0.005 mm apart. (The lower limit of resolution is about 500 nm for an optical microscope using visible light.) If you use the electron beam described in Example 7.5 (with $\lambda = 6.06 \times 10^{-3}$ nm), then objects only about 5×10^{-8} mm (0.05 nm) apart can be resolved. This is an improvement of 10^5 in resolving power over optical microscopes, and it means that we can observe objects almost on the molecular scale (where dimensions are in the range of 10^{-10} m). Electron diffraction microscopes have therefore become widely used tools to study matter.

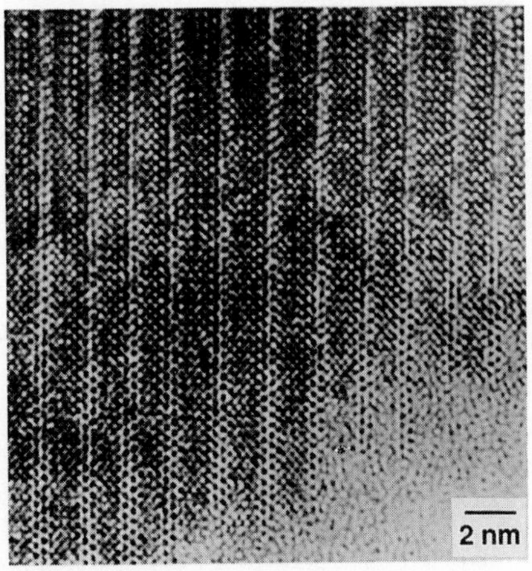

High-resolution transmission electron micrograph showing the atomic structure of a Bi-Sr-Ca-Cu-O high-temperature, ceramic oxide superconductor. (General Electric)

Louis Victor de Broglie (1892–1987). (Oesper Collection in the History of Chemistry/University of Cincinnati)

In nanometers, the wavelength is $(6.06 \times 10^{-12}\ \text{m})(1.00 \times 10^9\ \text{nm/m}) = 6.06 \times 10^{-3}\ \text{nm}$. This wavelength is only about 1/20 of the diameter of the H atom.

EXERCISE 7.6 ***De Broglie's Equation***

Calculate the wavelength associated with a neutron having a mass of 1.675×10^{-24} g and a kinetic energy of 6.21×10^{-21} J. (Recall that the kinetic energy of a moving particle is $E = mv^2/2$.) ■

7.5 THE WAVE MECHANICAL VIEW OF THE ATOM

In Copenhagen, Denmark, after World War I, Niels Bohr assembled a group of physicists who set out to derive a comprehensive theory for the behavior of electrons in atoms from the viewpoint of the electron as a particle. Erwin Schrödinger (1887–1961), an Austrian, independently worked toward the same goal, but he used de Broglie's hypothesis that an electron in an atom could be described by equations for wave motion. Although both Bohr and Schrödinger were successful in predicting some aspects of electron behavior, Schrödinger's approach gave correct results for some properties for which Bohr's failed. For this reason theoreticians today primarily use Schrödinger's concept. In any

Erwin Schrödinger (1887–1961) was born in Vienna, Austria, and studied at the University there. Following service in World War I as an artillery officer, he became a professor of physics at various universities; in 1928 he succeeded Max Planck as professor of theoretical physics at the University of Berlin. He shared the Nobel Prize in physics (with Paul Dirac) in 1933. (Oesper Collection in the History of Chemistry/University of Cincinnati)

event, the general theoretical approach to understanding atomic behavior, developed by Bohr, Schrödinger, and their associates, has come to be called **quantum mechanics,** or **wave mechanics.**

The Uncertainty Principle

Before you can appreciate Schrödinger's model for the behavior of electrons in atoms, you should know about a great debate that raged in physics early in the 20th century. De Broglie's suggestion that an electron can be described as having wave properties was confirmed by experiment (see Section 7.4). J. J. Thomson's experiment to measure the charge-to-mass ratio of an electron showed the particle-like nature of the electron (Figure 2.5). But how can an electron be described as *both* a particle and a wave? One can only conclude that *the electron has dual properties.* The result of a given experiment can be described *either* by the physics of waves *or* of particles; no single experiment can be done to show that the electron behaves *simultaneously* as a wave and a particle!

What does this *wave-particle duality* have to do with electrons in atoms? Werner Heisenberg (1901–1976) and Max Born (1882–1970) provided the answer. To explain the behavior of electrons in atoms, it seems most reasonable to assume that electrons have wave properties. If this is the case, Heisenberg concluded that it was impossible to fix *both* the position of an electron in an atom *and* its energy with any degree of certainty. If we attempt to determine either the location or the energy accurately, then the other is uncertain. In contrast, this is not the case in the world around you. For example, we can determine, with considerable accuracy, the energy of a moving car *and* its location.

Based on Heisenberg's idea, which we now call the **uncertainty principle,** Max Born proposed that the results of quantum mechanics should be interpreted as follows: *if we choose to know the energy of an electron in an atom with only a small uncertainty, then we must accept a correspondingly large uncertainty about its position in the space about the atom's nucleus.* In practical terms, this means the only thing we can do is calculate the likelihood, or **probability,** of finding an electron with a given energy within a given space. We turn to this viewpoint in the next section.

Schrödinger's Model of the Hydrogen Atom and Wave Functions

Schrödinger's model for the hydrogen atom was based on the premise that the electron can be described as a matter wave and not as a tiny particle orbiting the nucleus. Unlike Bohr's model, Schrödinger's approach resulted in an equation that is complex and mathematically difficult to solve except in simple cases. We need not be concerned with the mathematics, but the solutions to the equation—called **wave functions**—are chemically important. If we can understand the implications of these wave functions, then we will understand the modern view of the atom.

Wave functions, which are symbolized by the Greek letter ψ (psi), characterize the electron as a *matter wave.* The following important points can be made.

1. Only certain vibrations, called standing waves (see Figure 7.4), can be observed in a vibrating string. Similarly, the behavior of the electron in the atom is best described as a standing wave. Although electron motion is not as simple as that of a vibrating string, still *only certain wave functions are allowed.*

A CLOSER LOOK — *Heisenberg's Uncertainty Principle*

Heisenberg's uncertainty principle is expressed mathematically as

$$\Delta x \cdot \Delta(mv) > h$$

The uncertainty in the position of an electron in an atom (Δx) multiplied by the uncertainty in its momentum (Δmv) (which is related to its energy) must be larger than Planck's constant (h). This tells us that if we wish to know the momentum (or energy) of a very small object like an electron with great certainty ($\Delta mv \rightarrow 0$), then we must accept a great uncertainty in its position ($\Delta x \rightarrow \infty$).

Let us calculate the uncertainty in the position of an electron ($m = 9.11 \times 10^{-28}$ g) moving at 1.20×10^8 m/s (about 40% of the velocity of light). Assume the uncertainty in velocity is 0.100%. From Heisenberg's equation, we find $\Delta x > h/\Delta(mv)$.

$$\Delta x > \frac{(6.626 \times 10^{-34}\ \text{kg} \cdot \text{m}^2/\text{s}^2)(\text{s})}{(9.11 \times 10^{-31}\ \text{kg})(1.20 \times 10^8\ \text{m/s})(0.00100)}$$

$$> 6.06 \times 10^{-9}\ \text{m}$$

This uncertainty is about 6 nm, a very large distance considering that distances on the atomic and molecular scale are measured in nanometers.

Now let us compare the electron with an automobile. What is the position uncertainty for a car ($m = 1.00 \times 10^3$ kg) moving at 26.7 ± 0.0450 m/s (about 60 mph)?

$$\Delta x > \frac{(6.626 \times 10^{-34}\ \text{kg} \cdot \text{m}^2/\text{s}^2)(\text{s})}{(1.00 \times 10^3\ \text{kg})(0.0450\ \text{m/s})}$$

$$> 1.47 \times 10^{-35}\ \text{m}$$

This is such a small uncertainty that it cannot be measured by current instruments. This means we can know the position accurately of a large, moving object such as a car.

---

2. Each wave function ψ corresponds to an allowed energy for the electron. This is like Bohr's result for the hydrogen atom ($E_n = -Rhc/n^2$). For each integer n there is an atomic state characterized by its own wave function ψ and energy E_n.
3. Points 1 and 2 amount to saying that *the energy of the electron is quantized.* The concept of quantization enters Schrödinger's theory naturally with the basic assumption of an electron matter wave. This is in contrast with Bohr's theory in which quantization was imposed as a postulate at the start.
4. Each wave function ψ can only be interpreted in terms of the ideas of probability. The square of ψ gives the probability of finding the electron within a given region of space. Scientists refer to this as the **electron density** in a given region.

   Be sure to understand that the theory does not predict the exact position of the electron. Because Schrödinger's theory chooses to define the energy of the electron precisely, Heisenberg's uncertainty principle tells us this must result in a large uncertainty in electron position. This is why we can only describe the *probability* of the electron being at a certain point in space when in a given energy state.
5. The matter waves for the allowed energy states are also called **orbitals.** We shall say more about orbitals in a moment.
6. To solve Schrödinger's equation for an electron in three-dimensional space, three integer numbers—the quantum numbers n, ℓ, and m_ℓ—must be introduced. These quantum numbers may have only certain combinations of values, as outlined below. We shall use these combinations to define the energy states and orbitals available to the electron.

Werner Heisenberg (1901–1976) earned a Ph.D. in theoretical physics at the University of Munich in 1923 and then studied with Max Born and later Niels Bohr. He received the Nobel Prize in physics in 1932.
(Chemical Heritage Foundation)

Quantum Numbers

In three-dimensional space, three numbers are required to describe the location of an object in space (Figure 7.15). For the wave description of the electron in an atom, this requirement leads to the existence of **three quantum numbers n, ℓ, and m_ℓ.** Before looking into the meanings of the three quantum numbers, it is important to say that

- The quantum numbers n, ℓ, m_ℓ are all integers, but their values cannot be selected randomly.
- The three quantum numbers (and their values) are not parameters that scientists "dreamed up." Instead, when the behavior of the electron in the hydrogen atom is described as a matter wave, the quantum numbers are a natural consequence of that theory.

n, the principal quantum number = 1, 2, 3, . . . ∞

The principal quantum number n can have any integer value from 1 to infinity. As the name implies, it is the most important quantum number because the value of n is the primary factor in determining the energy of the electron. For the H atom the relation between the electron energy E and n is the same as given by Bohr's equation [$E_n = -Rhc/n^2$]. Because all the quantities in this equation are constants except n, the *energy of the electron in the H atom varies only with the value of n.*

The electron energy in an atom with more than one electron depends on n *and* other quantum numbers.

The value of n is also a measure of the most probable distance of the electron from the nucleus: the greater the value of n, the more probable it is that the electron is further from the nucleus.

Each electron is labeled according to its value of n. In atoms having more than one electron, two or more electrons may have the same n value. These electrons are then said to be in the same **electron shell.**

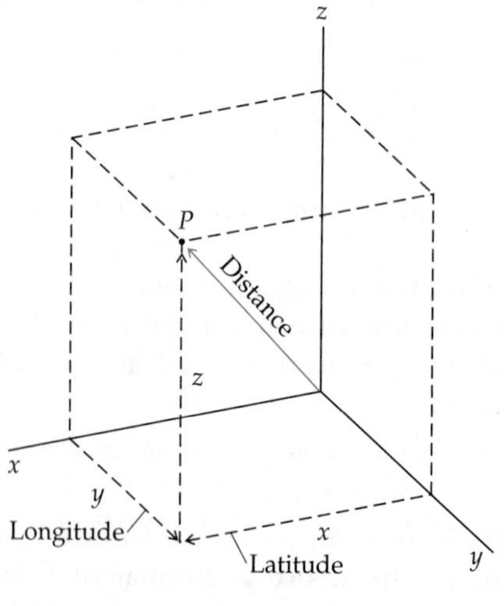

Figure 7.15 The cartesian coordinate system. The position of a point P in space can be specified by giving the x-, y-, and z-coordinates, or the distance of the point from the center of the coordinate system and its latitude and longitude.

ℓ, the angular momentum quantum number = 0, 1, 2, 3, . . . $n - 1$

The electrons of a given shell can be grouped into **subshells,** each subshell characterized by a different value of the quantum number ℓ and by a characteristic shape. For the nth shell, n different subshells are possible, each subshell corresponding to one of the n different values of ℓ. Each value of ℓ corresponds to a different *orbital shape,* or orbital type.

The value of n limits the number of subshells possible for the nth shell because ℓ can be no larger than $n - 1$. Thus, for $n = 1$, the rule tells us that ℓ must equal 0 and only 0. Because ℓ has only one value when $n = 1$, only one subshell is possible for an electron assigned to $n = 1$. When $n = 2$, ℓ can be either 0 or 1. Because two values of ℓ are now possible, there are two subshells in the $n = 2$ electron shell.

The values of ℓ are usually coded by letters according to the following scheme:

Value of ℓ	Corresponding Subshell Label
0	*s*
1	*p*
2	*d*
3	*f*

This means that a subshell with a label of $\ell = 1$ is called a "*p* subshell," and an orbital found in that subshell is called a "*p* orbital." Conversely, an electron assigned to a *p* subshell has an ℓ value of 1.

As an aside, it is interesting to take note of the origin of the letters used to designate subshells. Early studies of the spectra of elements other than hydrogen had more lines than could be explained by Bohr's theory. Scientists studying the spectrum of sodium atoms, for example, found four different types of lines, which they labeled *sharp, principal, diffuse,* and *fundamental.* To account for the additional lines, they believed a theory of the atom was needed that included energy subshells for electrons. When the notion of subshells arose from Schrödinger's model, the subshells were labeled as *s, p, d,* and *f.*

m_ℓ, the magnetic quantum number = 0, ±1, ±2, ±3, . . . ±ℓ

The magnetic quantum number, m_ℓ, specifies to which orbital within a subshell the electron is assigned. Orbitals in a given subshell differ only in their orientation in space, not in their shape.

The value of ℓ limits the integer values assigned to m_ℓ: m_ℓ can range from $+\ell$ to $-\ell$, with 0 included. For example, when $\ell = 2$, m_ℓ has five values: +2, +1, 0, −1, and −2. The number of values of m_ℓ for a given subshell ($= 2\ell + 1$) specifies the number of orientations that exist for the orbitals of that subshell.

Useful Information from Quantum Numbers

The three quantum numbers introduced thus far are a kind of "electronic zipcode." They tell us to which shell an electron is assigned (n), to which subshell within the shell (ℓ), and to which orbital within the subshell (m_ℓ). Their allowed values are summarized in Table 7.1.

Electrons in atoms are assigned to orbitals, which are grouped into subshells. Depending on the value of n, one or more subshells constitute an electron shell.

TABLE 7.1 Summary of the Quantum Numbers, Their Interrelationships, and the Orbital Information Conveyed

Principal Quantum Number Symbol = n Values = 1, 2, 3, . . . (Orbital Size, Energy)	Angular Momentum Quantum Number Symbol = ℓ Values = 0 . . . $n - 1$ (Orbital Shape)	Magnetic Quantum Number Symbol = m_ℓ Values = $-\ell$ . . . 0 . . . $+\ell$ (Orbital Orientations = Number of Orbitals in Subshell)	Number and Type of Orbitals in the Subshell (Number of Orbitals in Shell = number of values of $m_\ell = 2\ell + 1 = n^2$)
1	0	0	1 1*s* orbital (1 orbital of 1 type in the $n = 1$ shell)
2	0 1	0 +1,0,−1	1 2*s* orbital 3 2*p* orbitals (4 orbitals of 2 types in the $n = 2$ shell)
3	0 1 2	0 +1,0,−1 +2,+1,0,−1,−2	1 3*s* orbital 3 3*p* orbitals 5 3*d* orbitals (9 orbitals of 3 types in the $n = 3$ shell)
4	0 1 2 3	0 +1,0,−1 +2,+1,0,−1,−2 +3,+2,+1,0,−1,−2,−3	1 4*s* orbital 3 4*p* orbitals 5 4*d* orbitals 7 4*f* orbitals (16 orbitals of 4 types in the $n = 4$ shell)

When $n = 1$ the value of ℓ can only be 0, and so m_ℓ must also have a value of 0. This means that, in the electron shell closest to the nucleus, only one type of orbital or subshell exists, and that subshell consists of only a single orbital. This orbital is labeled "1*s*," the "1" conveying the value of *n* and "*s*" telling you that $\ell = 0$. *When $\ell = 0$, an s orbital is indicated, and only one s orbital can occur in a given electron shell.*

Electron subshells are labeled by first giving the value of *n* and then the value of ℓ in the form of its letter code. For $n = 1$ and $\ell = 0$, for example, the label is 1*s*.

When $n = 2$, ℓ can have two values (0 and 1), so two subshells or two types of orbitals occur in the second shell. One of these is the 2*s* subshell ($n = 2$ and $\ell = 0$), and the other is the 2*p* subshell ($n = 2$ and $\ell = 1$). Because the values of m_ℓ can be +1, 0, and −1 when $\ell = 1$, three *p*-type orbitals exist. Because all three have $\ell = 1$, they all have the same shape. They differ in their orientation in space, however, because one has $m_\ell = +1$, another has $m_\ell = 0$, and a third has $m_\ell = -1$. In summary, when $\ell = 1$, *p* orbitals are indicated, and three of them always occur. The converse is true as well: for *p* orbitals in any shell, ℓ is 1.

When $n = 3$, three subshells, or orbital types, are possible for an electron because ℓ has the values 0, 1, and 2. Because you see ℓ values of 0 and 1 again, you know that two of the subshells within the $n = 3$ shell are 3*s* (one orbital) and 3*p* (three orbitals). The third subshell is *d*, indicated by $\ell = 2$. Because m_ℓ has five values (+2, +1, 0, −1, and −2) when $\ell = 2$, five *d* orbitals (no more and no less) occur in the $\ell = 2$ subshell. Thus, whenever an electron shell has $n = 3$ or greater, one of the ℓ values is 2, indicating a set of five *nd* orbitals.

Besides *s*, *p*, and *d* orbitals, we occasionally need to refer to *f* electron orbitals, that is, orbitals for which $\ell = 3$. Seven such orbits exist because seven values of m_ℓ are possible when $\ell = 3$ (+3, +2, +1, 0, −1, −2, and −3).

No known elements have electrons assigned to subshells with ℓ greater than 3 in the ground state. If they existed, however, $\ell = 4$ would correspond to a *g* subshell, $\ell = 5$ to an *h* subshell, and so on. Such subshells might be important to consider, however, for an excited state atom.

EXERCISE 7.7 ***Using Quantum Numbers***

Complete the following statements:

1. When $n = 2$, the values of ℓ can be ______ and ______.
2. When $\ell = 1$, the values of m_ℓ can be ______, ______, and ______, and the subshell has the letter label ______.
3. When $\ell = 2$, the subshell is called a ______ subshell.
4. When a subshell is labeled *s*, the value of ℓ is ______, and m_ℓ has the value ______.
5. When a subshell is labeled *p*, ______ orbitals occur within the subshell.
6. When a subshell is labeled *f*, there are ______ values of m_ℓ, and ______ orbitals occur within the subshell. ■

7.6 THE SHAPES OF ATOMIC ORBITALS

The chemistry of an element and of its compounds is determined by the electrons in the element's atoms, particularly the electrons with the highest value of *n*, which are often called valence electrons (Section 3.3). The types of orbitals to which these electrons are assigned are also important, so we turn now to the question of orbital shape and orientation.

s Orbital

When an electron has $\ell = 0$, we often say the electron is assigned to, or "occupies," an *s* orbital. But what does this mean? What is an *s* orbital? What does it look like? To answer these questions, we begin with the wave function for an electron with $n = 1$ and $\ell = 0$, that is, with a 1*s* orbital. If we assume for the moment that the electron is a tiny particle and not a matter wave, and if we could photograph the 1*s* electron at one-second intervals for a few thousand seconds, the composite picture would resemble the drawing in Figure 7.16a. This resembles a cloud of dots, so chemists refer to such representations of electron orbitals as **electron cloud** pictures.

The fact that the density of dots is greater close to the nucleus (the electron cloud is denser close to the nucleus) indicates that the electron is most often found near the nucleus (or, conversely, it is less likely to be found farther away). Putting this in the language of quantum mechanics, we say that the greatest probability of finding the electron is in a tiny volume of space around the nucleus, whereas it is less probable that the electron is farther away. The "thinning" of the electron cloud at increasing distance, shown by the decreasing density of dots in Figure 7.16a, is illustrated in a different way in Figure 7.16b. Here we plotted the square of the wave function for the electron in a 1*s* orbital as

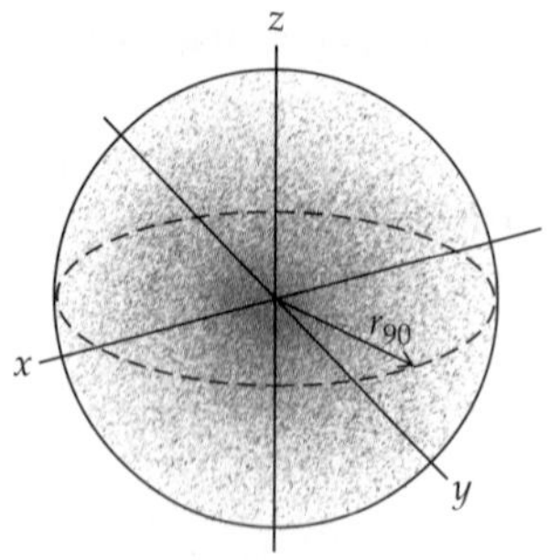

(a) Dot pattern of an electron with a 1s atomic orbital. Each dot represents the position of the electron at a different instant in time.

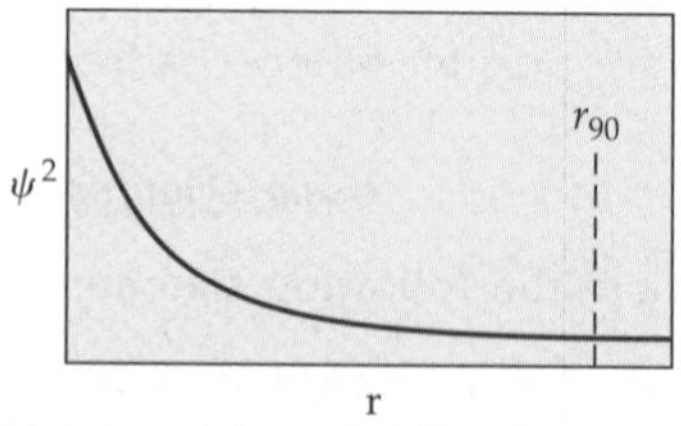

(b) A plot of the probability density as a function of distance for a one-electron atom with a 1s electron.

Figure 7.16 Different views of a 1*s* ($n = 1$ and $\ell = 0$) orbital. (a) Dot picture of an electron with a 1*s* orbital. Each dot represents the position of the electron at a different instant in time. Note that the dots cluster closest to the nucleus. r_{90} is the radius within which the electron is found 90% of the time. (b) A plot of the probability density as a function of distance for a one-electron atom with a 1*s* electron wave. (c) The surface of the sphere within which the electron is found 90% of the time for a 1*s* orbital. This surface is often called a "boundary surface." (A 90% surface was chosen arbitrarily. If the choice was the surface within which the electron is found 50% of the time, the sphere would be considerably smaller.)

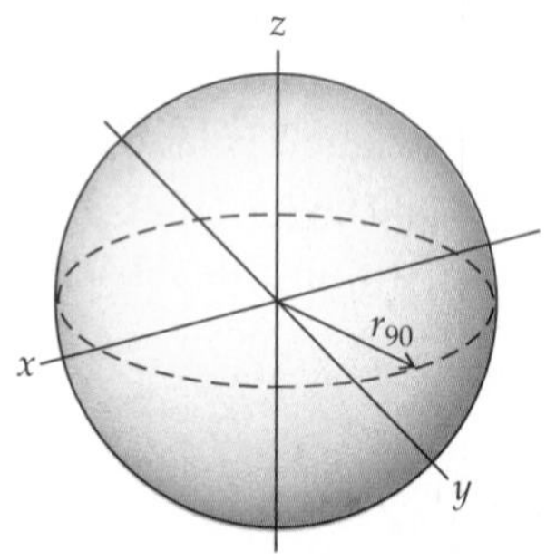

(c) The surface of the sphere within which the electron is found 90% of the time for a 1s orbital.

Another way to describe Figure 7.16b is to say that the electron behaves as a wave whose amplitude has the largest value very close to the nucleus.

a function of the distance of the electron from the nucleus. The units of ψ^2 at each point are 1/volume, so the numbers on the vertical axis of this plot represent the probability of finding the electron in each cubic nanometer, for example, at a given distance from the nucleus. For this reason, ψ^2 is called the **probability density.** For the 1*s* orbital, ψ^2 is very high for points immediately around the nucleus, but it drops off rapidly as the distance from the nucleus increases. Notice that the probability approaches but never reaches zero, even at very large distances.

For the 1*s* orbital, Figure 7.16a shows the electron is most likely found within a sphere with the nucleus at the center. No matter in which direction you proceed from the nucleus, the probability of finding an electron along a line in that direction drops off (Figure 7.16b). *The 1s orbital is spherical in shape.*

The visual image of Figure 7.16a is that of a cloud whose density is small at large distances from the center; there is no sharp boundary beyond which the electron is never found. However, *s* and other orbitals are often depicted as having a sharp boundary surface (Figure 7.16c), largely because it is easier to draw such pictures. To arrive at the diagram in Figure 7.16c we drew a sphere about the nucleus in such a way that the chance of finding the electron somewhere inside is 90%. Many misconceptions exist about pictures of this type. Understand that this surface is not real. The nucleus is surrounded by an "electron cloud" and not an impenetrable surface "containing" the electron. The electron is not distributed evenly throughout the volume enclosed by the surface; instead, it is most likely found nearer the nucleus.

From an analysis of ψ for *s* orbitals of different *n* values, we can arrive at the important conclusion that all orbitals labeled *s* are spherical in shape. In every case, *s* electrons can be found in a very small volume of space immediately around the nucleus. One important difference between *s* orbitals of different *n*, however, is that *the size of s orbitals increases as n increases* (Figure 7.17). Thus, the 1*s* orbital is more compact than the 2*s* orbital, which is more compact than the 3*s* orbital.

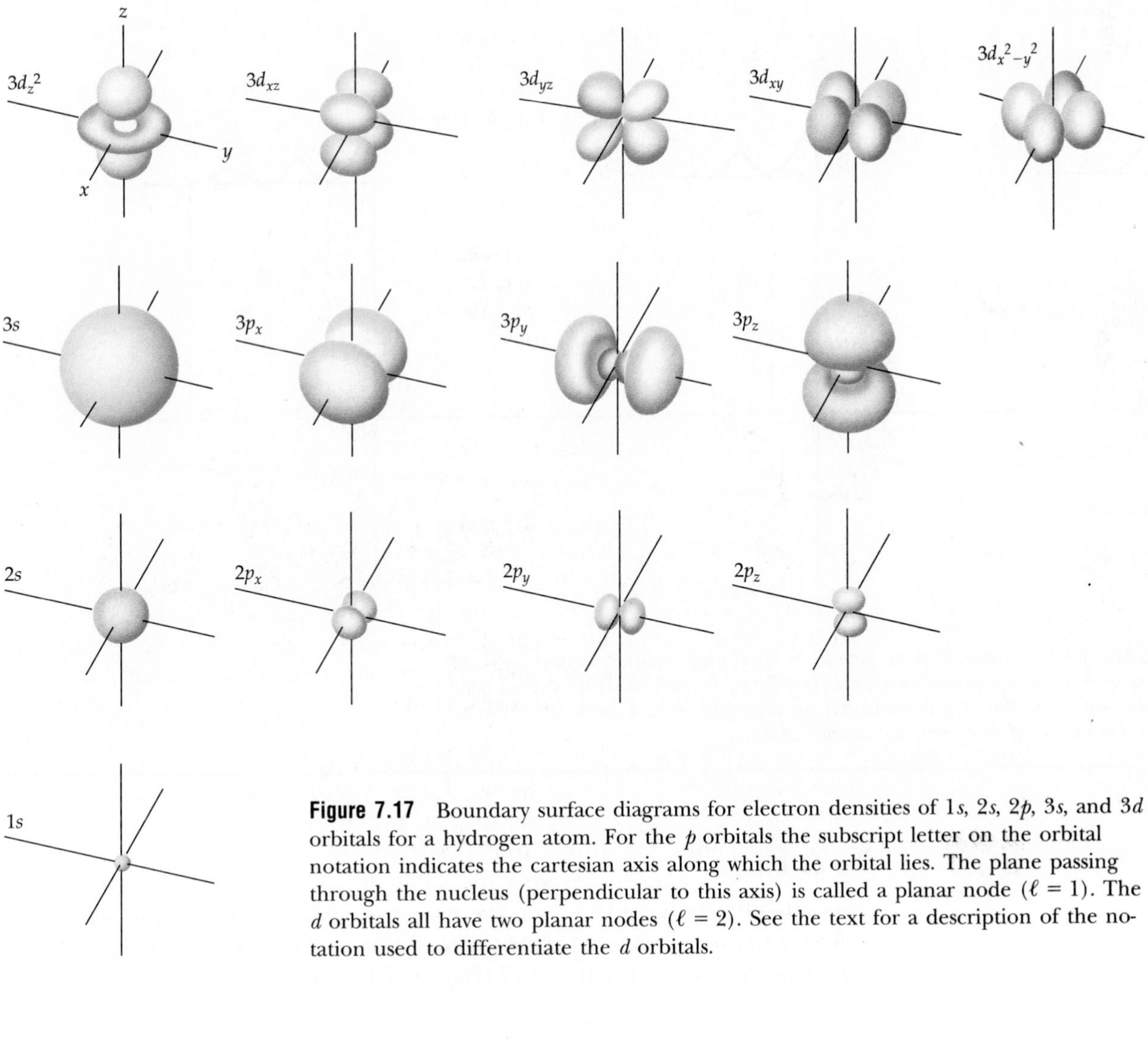

Figure 7.17 Boundary surface diagrams for electron densities of 1*s*, 2*s*, 2*p*, 3*s*, and 3*d* orbitals for a hydrogen atom. For the *p* orbitals the subscript letter on the orbital notation indicates the cartesian axis along which the orbital lies. The plane passing through the nucleus (perpendicular to this axis) is called a planar node ($\ell = 1$). The *d* orbitals all have two planar nodes ($\ell = 2$). See the text for a description of the notation used to differentiate the *d* orbitals.

A CLOSER LOOK Orbital Shapes and Spherical Nodes

Like the 1*s* orbital, a 2*s* orbital is also spherical in shape, but important differences exist between the two. First, the 2*s* electron cloud is larger than the 1*s* cloud; the point of maximum probability for the 2*s* electron is found slightly farther from the nucleus than that of the 1*s* electron (Figure 7.18). The second difference is seen in the figure shown here. When moving along a line away from the nucleus, the probability of finding the 2*s* electron drops off rapidly and actually reaches zero at 0.1058 nm. That is, a **node** occurs in the electron wave at this distance. Beyond this point, the probability rises again, passes through a maximum at 0.2116 nm, and then trails off at still larger distances. This happens no matter which direction you move away from the nucleus. If you could stand on the nucleus and look into space in any direction, you would always see a node at 0.1058 nm, and the points at which

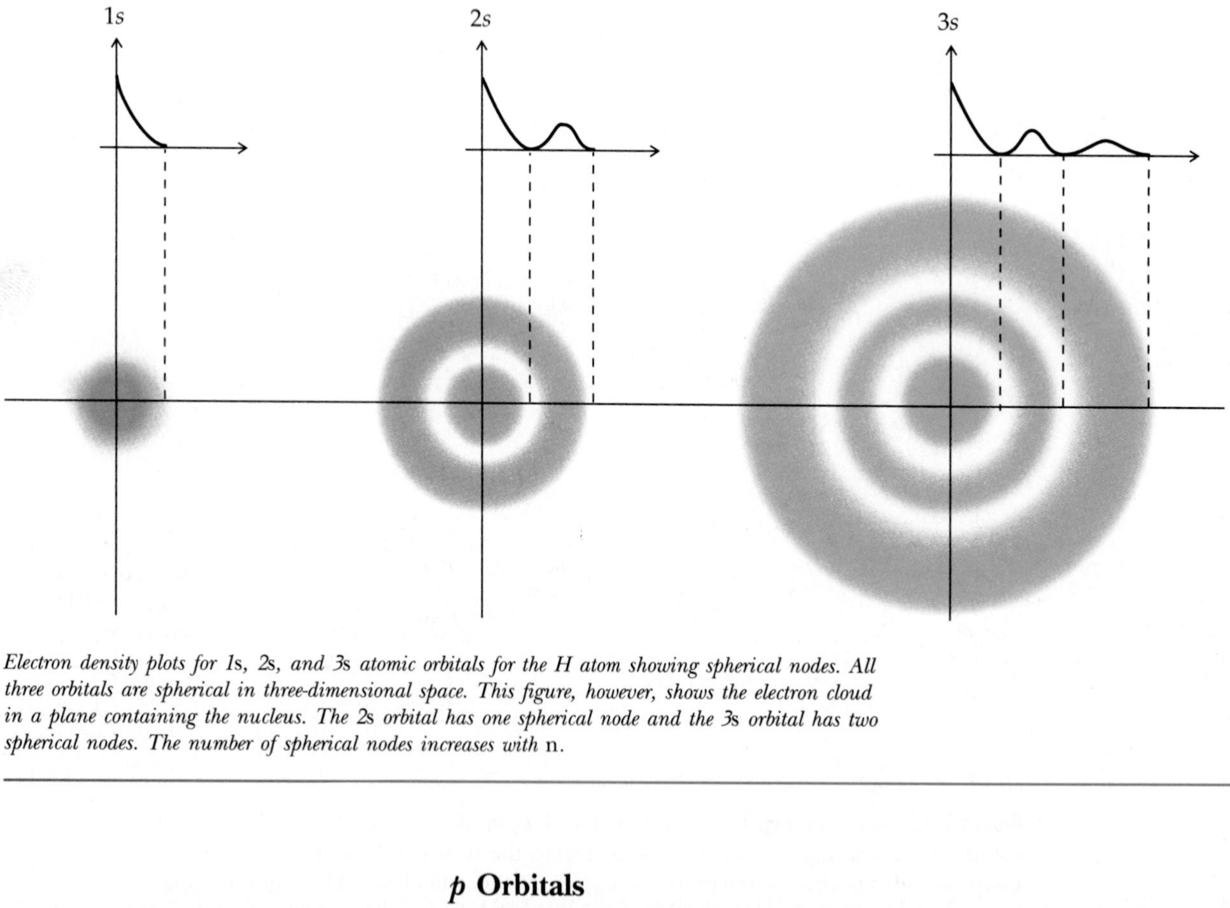

*Electron density plots for 1*s*, 2*s*, and 3*s* atomic orbitals for the H atom showing spherical nodes. All three orbitals are spherical in three-dimensional space. This figure, however, shows the electron cloud in a plane containing the nucleus. The 2*s* orbital has one spherical node and the 3*s* orbital has two spherical nodes. The number of spherical nodes increases with* n.

p Orbitals

Atomic orbitals for which $\ell = 1$ are called *p* orbitals and all have the same basic shape. *All p orbitals have one imaginary plane that slices through the nucleus and divides the region of electron density in half* (Figures 7.17 and 7.19).

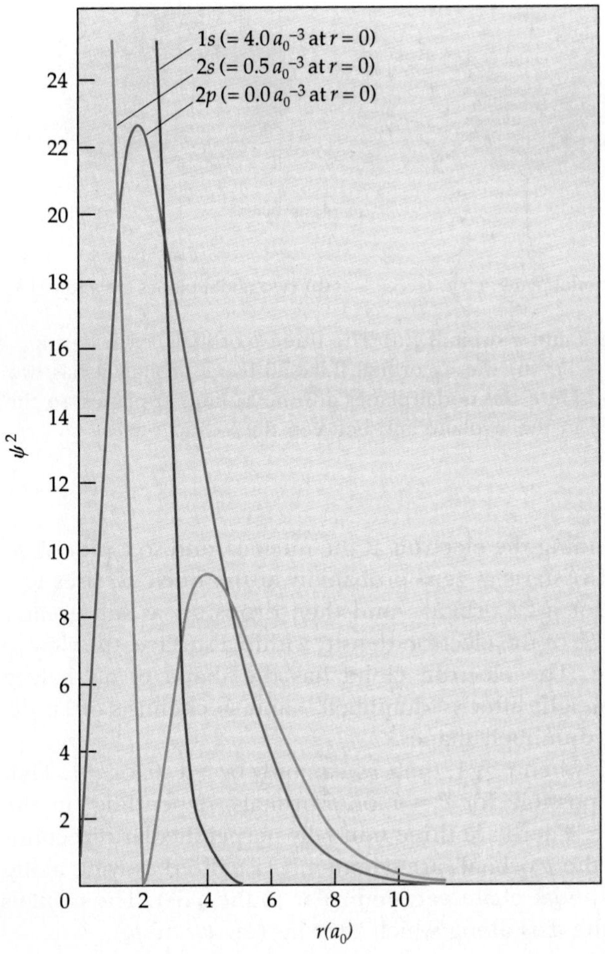

Figure 7.18 Accurate probability density plots for 1*s*, 2*s*, and 2*p*, and electron orbitals. The horizontal axis represents the distance from the nucleus in units of a_0 (atomic units), where a_0 = 0.0529 nm. Units of the vertical axis, which gives the probability or electron intensity, are $10^{-3}/a_0^3$. Note that the 1*s* electron wave has no node, but the 2*p* electron wave has a node at the nucleus. Also note that the 1*s* probability is so great very near the nucleus that the plot extends well beyond the top of the figure.

this happens describe the surface of a sphere. Because there is no probability of finding an electron on this surface, it is called a **spherical node.** The number of spherical nodes for any type of orbital is given by $(n - \ell - 1)$. For example, considering just *s* orbitals for which ℓ is always 0, you will find the following numbers of spherical nodes:

Orbital	Number of Spherical Nodes $(n - \ell - 1)$
1*s*	0
2*s*	1
3*s*	2
4*s*	3

Thus, as the value of *n* increases, the number of nodes in the electron wave increases. This means that all *s* orbitals are spherical, but the number of layers of electron density increases as *n* increases. The best analogy to this picture is an onion that has layer on layer of white or pink matter separated by a thin space.

The imaginary plane slicing through the nucleus is called a **nodal plane,** a planar surface on which there is zero probability of finding the electron. The electron can never be found in the nodal plane; the regions of electron density lie on either side of the nucleus. This means that, unlike *s* orbitals, for *p* orbitals

Nodal planes occur for all *p*, *d*, and *f* orbitals. In some cases, however, the "plane" is not flat and is better called a nodal *surface.*

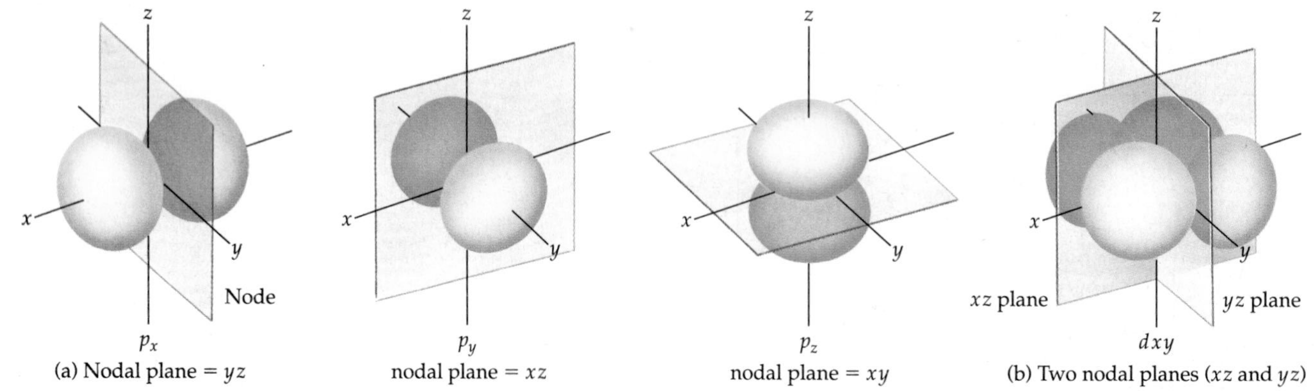

Figure 7.19 Nodal planes in p and d orbitals. (a) The three p orbitals, each having one nodal plane (because $\ell = 1$). (b) the d_{xy} orbital. Like all five d orbitals it has two nodal planes (because $\ell = 2$). Here the nodal planes are the xz- and yz-planes, so the regions of electron density lie in the xy-plane and between the x- and y-axes.

The amplitude of the electron wave is zero on a nodal plane.

there is no likelihood of finding the electron at the nucleus, and so a plot of ψ^2 versus distance (Figure 7.18) starts at zero probability at the nucleus, rises to a maximum (at 0.1058 nm for a $2p$ orbital), and then drops off at still greater distances. If you enclose 90% of the electron density within a surface, the view in Figure 7.17 is appropriate. The electron cloud has the shape of a "halved sphere" that resembles a weight lifter's "dumbbell," and so chemists often describe p orbitals as having dumbbell shapes.

According to Table 7.1, when $\ell = 1$, then m_ℓ can only be $+1$, 0, or -1. That is, three orientations are possible for $\ell = 1$ or p orbitals, depending on the location of the nodal plane. There are three mutually perpendicular directions in space (x, y, and z), and the p orbitals are commonly visualized as lying along those directions (with the nodal plane perpendicular to the axis). The orbitals are labeled according to the axis along which they lie (p_x, p_y, or p_z).

d Orbitals

The value of ℓ is equal to the number of nodal planes that slice through the nucleus. Thus, s orbitals, for which $\ell = 0$, have no nodal planes, and p orbitals, for which $\ell = 1$, have one planar nodal surface. It follows that the five d orbitals, for which $\ell = 2$, have two nodal surfaces, which results in four regions of electron density. The d_{xy} orbital, for example, lies in the xy-plane, and the two nodal planes are in the xz- and yz-planes. Two other orbitals, d_{xz} and d_{yz}, lie in planes defined by the xz- and yz-axes, respectively, and also have two, mutually perpendicular nodal planes.

Of the two remaining d orbitals, the $d_{x^2-y^2}$ orbital is easier to visualize. Like the d_{xy} orbital, the $d_{x^2-y^2}$ orbital results from two vertical planes slicing the electron density into quarters. Now, however, the planes bisect the x- and y-axes, so the regions of electron density lie along the x- and y-axes (Figure 7.17).

The final d orbital, d_{z^2}, has two main regions of electron density along the z-axis, but a "donut" of electron density also occurs in the xy-plane. This orbital has two nodal surfaces, but the surfaces are not flat. Think of an ice cream cone sitting with its tip at the nucleus. One of the electron clouds along the z-axis sits

Figure 7.20 One of seven possible f electron orbitals. Notice the presence of three nodal planes as required by an orbital with $\ell = 3$. (C. D. Winters)

inside the cone. If you have another cone pointing in the opposite direction from the first cone, again with its tip at the nucleus, another region of electron density fits inside this second cone. The region outside both cones defines the remaining, donut-shaped region of electron density.

f Orbitals

The seven f orbitals all have $\ell = 3$, meaning that three nodal surfaces slice through the nucleus, and so eight regions of electron density occur. This makes these orbitals less easily visualized, but one example is illustrated in Figure 7.20.

EXERCISE 7.8 ***Orbital Shapes***

1. What are the n and ℓ values for each of the following orbitals: 6*s*, 4*p*, 5*d*, and 4*f*?
2. How many nodal planes exist for a 4*p* orbital? For a 6*d* orbital? ■

Orbital Shape and Chemistry

We close with some problems to ponder. When an element is part of a molecule, are the orbitals the same? Do they have the same shapes? What does the shape of orbitals have to do with the chemistry of an element? These are questions we take up in the rest of the book. A few answers are in order, however.

Schrödinger's wave equation can be solved *exactly* for the hydrogen atom but not for heavier atoms or their ions. Nonetheless, chemists make the assumption that orbitals in other atoms are hydrogen-like, even when those atoms are part of a molecule. At least that is the best way chemists have found to interpret experimental information on how molecules react. For example, in the molecule shown on page 348, ethylene, C_2H_4, is attached to a platinum(II) ion that is also attached to three chloride ions. How is the ethylene attached to the platinum(II) ion? Why is it attached "sideways" to the metal ion? Chemists assume this is the case because the *shape* of a *d* orbital on the platinum(II) ion allows it to interact in just the right way with *p*-like orbitals on the carbon atoms.

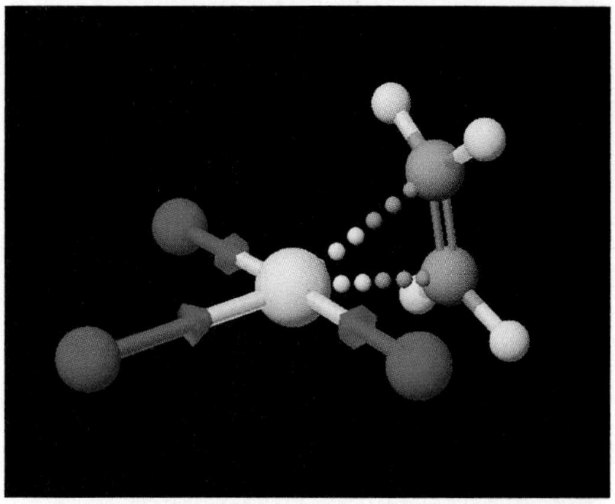

A model of the ion $(C_2H_4)PtCl_3^-$. Here the carbon atoms are gray, the hydrogen atoms are white, the platinum atom is silver, and the chlorine atoms are green. The dotted lines represent the chemical bond, which uses a Pt *d* orbital between Pt and C_2H_4. (The potassium salt of this ion, $K(C_2H_4)PtCl_3$, is widely known as "Zeise's salt" because it was discovered by Zeise in Copenhagen in about 1827.) (S. Young)

There is more to the story. Why do we assume a *d* orbital on the platinum(II) ion is key to the interaction with ethylene? Because the valence electrons of Pt^{2+} are located in those orbitals, a fact that will be clear after taking up the subject of the next chapter, the electron configurations of the elements and their ions.

CHAPTER HIGHLIGHTS

When you have finished studying this chapter, you should be able to

- Use the terms **wavelength, frequency, wave amplitude,** and **node** (Section 7.1).
- Use Equation 7.1 ($\lambda\nu = c$), which states that the product of the wavelength (λ) and frequency (ν) of electromagnetic radiation is equal to the velocity of light.
- Recall the relative wavelength (or frequency) of the various regions in the spectrum of electromagnetic radiation (Figure 7.3).
- Use the fact that the energy of a **photon,** a massless particle of radiation, is given by Planck's equation (Equation 7.2),

$$E = h\nu \tag{7.2}$$

  where h is Planck's constant (6.626×10^{-34} J · s). This is an extension of Planck's idea that energy at the atomic level is quantized (Section 7.2).
- Describe the Bohr model of the atom, how it can account for the emission line spectra of excited atoms, and the limitations of the model (Section 7.3).
- Understand that, in the Bohr model of the H atom, the electron can occupy only certain energy levels, each with an energy given by Equation 7.4

$$E_n = \frac{-Rhc}{n^2} \tag{7.4}$$

where R is the Rydberg constant, h is Planck's constant, c is the velocity of light, and n is an integer equal to or greater than 1 ($Rhc = 2.180 \times 10^{-18}$ J/atom or 1312 kJ/mol) (Section 7.3).

- Recall that Bohr postulated that if an electron moves to another state, the amount of energy absorbed or emitted in the process is equal to the difference in energy between the two states (Section 7.3).
- Understand that the modern view of the atom depends on the fact that particles on the atomic level obey the physics of waves (Section 7.4). The wavelength of an electron or any subatomic particle is given by de Broglie's equation (7.7),

$$\lambda = \frac{h}{mv} \qquad \textbf{(7.7)}$$

where m and v are the mass and velocity of the particle, respectively.

- Recognize the significance of wave or quantum mechanics in describing the modern view of atomic structure (Section 7.5).
- Understand that an **orbital** for an electron in an atom corresponds to an allowed energy of that electron. (The energy of the electron in the H atom can be calculated by the same equation derived by Bohr (Equation 7.4).)
- Understand that the position of the electron is not known with certainty; only the **probability** of the electron being within a given region of space can be calculated. This is the interpretation of the quantum mechanical model by Heisenberg and embodies his postulate called the **uncertainty principle.**
- Describe the allowed energy states of the electron in an atom using three quantum numbers n, ℓ, and m_ℓ (Section 7.5).

Quantum Number	**Name**	**Values**	**Orbital Property Described**
n	principal	1, 2, 3, . . . ∞	orbital size, energy, number of subshells, and number of orbital types in the shell
ℓ	angular momentum	0, 1, 2, . . . n − 1	orbital shape, number of orbitals in a subshell
m_ℓ	magnetic	$-\ell$ . . . 0 . . . $+\ell$	orbital orientation

- Describe the shapes of the orbitals (Section 7.6).

Value of ℓ	**Orbital Label**	**Nodal Planes**	**Orbital Shape**
0	*s*	0	spherical
1	*p*	1	"dumbbell" shape; 2 regions of electron density
2	*d*	2	4 regions of electron density
3	*f*	3	8 regions of electron density

STUDY QUESTIONS

Review Questions

1. Our modern view of atomic structure was developed through many experiments. Name at least three of these experiments, their outcomes, and the person most associated with that experiment. (You may have to review Chapter 2.)
2. Give the equation for each of the following important mathematical relations in this chapter:
   (a) The relationship between wavelength, frequency, and velocity of radiation
   (b) The relation between energy and frequency of radiation
   (c) The energy of an electron in a given energy state of the H atom
3. State Planck's equation in words and as a mathematical relation.
4. Name the colors of visible light beginning with that of highest energy.
5. Draw a picture of a standing wave, and use this to define the terms wavelength, amplitude, and node.
6. Which of the following best describes the importance of the photoelectric effect as explained by Einstein?
   (a) Light is electromagnetic radiation.
   (b) The intensity of a light beam is related to its frequency.
   (c) Light can be thought of as consisting of massless particles whose energy is given by Planck's equation, $E = h\nu$.
7. What is a photon? Explain how the photoelectric effect implies the existence of photons.
8. What are two major assumptions of Bohr's theory of atomic structure?
9. In what region of the electromagnetic spectrum is the Lyman series of lines found? The Balmer series?
10. Light is given off by a sodium- or mercury-containing streetlight when the atoms are excited in some way. The light you see arises for which of the following reasons:
    (a) Electrons moving from a given energy level to one of higher n
    (b) Electrons being removed from the atom, thereby creating a metal cation
    (c) Electrons moving from a given energy level to one of lower n
    (d) Electrons whizzing about the nucleus in an absolute frenzy
11. What is Heisenberg's uncertainty principle? Explain how it applies to our modern view of atomic structure.
12. How do we interpret the physical meaning of the square of the wave function? What are the dimensions of ψ^2?
13. What are the three quantum numbers used to describe an orbital? What property of an orbital is described by each quantum number? Specify the rules that govern the values of each quantum number.
14. On what does the amplitude of the electron wave in an atom depend?
15. Give the number of planar nodes for each orbital type: *s*, *p*, *d*, and *f*.
16. What is the maximum number of *s* orbitals found in a given electron shell? The maximum number of *p* orbitals? Of *d* orbitals? Of *f* orbitals?
17. Match the values of ℓ shown in the table with orbital type (*s*, *p*, *d*, or *f*).

ℓ Value	Orbital Type
3	______
0	______
1	______
2	______

18. Sketch a picture of the 90% boundary surface of an *s* orbital and the p_x orbital. Be sure the latter drawing shows why the *p* orbital is labeled p_x and not p_y, for example.

Numerical Questions

Electromagnetic Radiation

19. The colors of the visible spectrum, and the wavelengths corresponding to the colors, are given in Figure 7.3.
   (a) What colors of light involve less energy than green light?
   (b) Which color of light has photons of greater energy, yellow or blue?
   (c) Which color of light has the greater frequency, blue or green?

20. The regions of the electromagnetic spectrum are given in Figure 7.3. Answer the following questions based on this figure.
   (a) Which type of radiation involves less energy, x-rays or microwaves?
   (b) Which radiation has the higher frequency, radar or red light?
   (c) Which radiation has the longer wavelength, ultraviolet or infrared light?

21. The U.S. Navy has a system for communicating with submerged submarines. The system uses radio waves with a frequency of 76 s^{-1}. What is the wavelength of this radiation in meters? In miles?

22. An FM radio station has a frequency of 88.9 MHz (1 MHz = 10^6 Hz, or cycles per second). What is the wavelength of this radiation in meters?

23. Green light has a wavelength of approximately 5.0×10^2 nm. What is the frequency of this light? What is the energy in joules of one photon of green light? What is the energy in joules of 1.0 mol of photons of green light?

24. Violet light has a wavelength of about 380 nm. What is its frequency? Calculate the energy of one photon of violet light. What is the energy of 1.0 mol of violet photons? Compare the energy of photons of violet light with those of violet light. Which is more energetic and by what factor? (See Example 7.2.)

25. The most prominent line in the line spectrum of aluminum is found at 396.15 nm. What is the frequency of this line? What is the energy of one photon with this wavelength? Of 1 mol of these photons?

26. The most prominent line in the spectrum of magnesium is 285.2 nm. Others are found at 383.8 and 518.4 nm. In what region of the electromagnetic spectrum are these lines found? Which is the most energetic line? What is the energy of 1 mol of photons of the most energetic line? How much more energetic is a photon of this light compared with a photon associated with the least energetic line?

27. The most prominent line in the spectrum of mercury is found at 253.652 nm. Other lines are found at 365.015 nm, 404.656 nm, 435.833 nm, and 1013.975 nm.
(a) Which of these lines represents the most energetic light?
(b) What is the frequency of the most prominent line? What is the energy of one photon with this wavelength?
(c) Are any of the lines mentioned above found in the spectrum of mercury in Figure 7.10? What color or colors are these lines?

28. The most prominent line in the spectrum of neon is found at 865.438 nm. Other lines are found at 837.761 nm, 878.062 nm, 878.375 nm, and 1885.387 nm.
(a) Which of these lines represents the most energetic light?
(b) What is the frequency of the most prominent line? What is the energy of one photon with this wavelength?
(c) Are any of the lines mentioned above found in the spectrum of neon in Figure 7.10? What color or colors are these lines?

29. Place the following types of radiation in order of increasing energy per photon:
(a) Yellow light from a sodium lamp
(b) X-rays from an instrument in a dentist's office
(c) Microwaves in a microwave oven
(d) Your favorite FM music station at 91.7 MHz

30. Place the following types of radiation in order of increasing energy per photon:
(a) Radar signals
(b) Radiation within a microwave oven
(c) γ-Rays from a nuclear reaction
(d) Red light from a neon sign
(e) Ultraviolet radiation from a sun lamp

Photoelectric Effect

31. To cause a cesium atom on a metal surface to lose an electron, energy of 2.0×10^2 kJ/mol is required. Calculate the longest possible wavelength of light that can ionize a cesium atom. What is the region of the electromagnetic spectrum in which this radiation is found?

32. Assume you are an engineer designing a space probe to land on a distant planet. You wish to use a switch that works by the photoelectric effect. The metal you wish to use in your device requires 6.7×10^{-19} J/atom to remove an electron. You know that the atmosphere of the planet on which your device must work filters out all wavelengths of light less than 540 nm. Will your device work on the planet in question? Why or why not?

Atomic Spectra and The Bohr Atom

33. Consider only the following energy levels for the hydrogen atom.

____________ $n = 4$
____________ $n = 3$
____________ $n = 2$
____________ $n = 1$

The emission spectrum of an excited H atom consists of transitions between these levels.
(a) How many emission lines are possible, considering *only* the four quantum levels?
(b) Photons of the *highest energy* are emitted in a transition from the level with $n =$ __ to the level with $n =$ __.
(c) The emission line having the *longest wavelength* corresponds to a transition from the level with $n =$ __ to the level with $n =$ __.

34. Consider only the following energy levels for the H atom.

____________ $n = 5$
____________ $n = 4$
____________ $n = 3$
____________ $n = 2$
____________ $n = 1$

The emission spectrum of an excited H atom consists of transitions between these levels.
(a) How many emission lines are possible, considering *only* the five quantum levels?
(b) Photons of the *lowest frequency* are emitted in a transition from the level with $n =$ __ to the level with $n =$ __.
(c) The emission line having the *shortest wavelength* corresponds to a transition from the level with $n =$ __ to the level with $n =$ __.
(d) The emission line having the *highest energy* corresponds to a transition from the level with $n =$ __ to the level with $n =$ __.

35. If energy is absorbed by a hydrogen atom in its ground state, the atom is excited to a higher energy state. For example, the excitation of an electron from the level with $n = 1$ to the level with $n = 3$ requires radiation with a wavelength of 102.6 nm. Which of the following transitions requires radiation of *longer wavelength* than this?
(a) $n = 2$ to $n = 4$ (c) $n = 1$ to $n = 5$
(b) $n = 1$ to $n = 4$ (d) $n = 3$ to $n = 5$

36. The energy emitted when an electron moves from a higher energy state to one of lower energy in any atom can be observed as electromagnetic radiation.

(a) Which involves the emission of less energy in the H atom, an electron moving from $n = 4$ to $n = 2$ or an electron moving from $n = 3$ to $n = 2$?
(b) Which involves the emission of the greater energy in the H atom, an electron changing from $n = 4$ to $n = 1$ or an electron changing from $n = 5$ to $n = 2$? Explain fully.

37. Calculate the wavelength of light emitted when an electron changes from $n = 3$ to $n = 1$ in the H atom. In what region of the spectrum is this radiation found?

38. A line in the Balmer series of emission lines of excited H atoms has a wavelength of 410.2 nm. This series originates from electrons changing from high energy levels to the $n = 2$ level. To account for the 410.2 nm line, what is the value of n for the initial level?

39. An electron moves from the $n = 5$ to the $n = 1$ quantum level and emits a photon with an energy of 2.093×10^{-18} J. How much energy must the atom absorb to move an electron from $n = 1$ to $n = 5$?

40. Excited H atoms give off many emission lines. One series of lines, called the Pfund series, occurs in the infrared region. It results when an electron changes from higher levels to a level with $n = 5$. Calculate the wavelength and frequency of the *lowest energy* line of this series.

De Broglie and Matter Waves

41. An electron moves with a velocity of 2.5×10^8 cm/s. What is its wavelength?

42. A beam of electrons ($m = 9.11 \times 10^{-31}$ kg/electron) has an average speed of 1.3×10^8 m/s. What is the wavelength corresponding to the average speed?

43. Calculate the wavelength (in nanometers) associated with a 1.0×10^2-g golf ball moving at 30. m/s (about 67 mph). How fast must the ball travel to have a wavelength of 5.6×10^{-3} nm?

44. A rifle bullet (mass = 1.50 g) is moving with a velocity of 7.00×10^2 mph. What is the wavelength associated with this bullet?

Quantum Mechanics

45. Complete the following table:

Quantum Number	Atomic Property Determined by Quantum Number
______	Orbital size
______	Relative orbital orientation
______	Orbital shape

46. An orbital is designated 2*s*, and another orbital is designated 4*p*. Which is the larger orbital? Which has more planar nodes?

47. Answer the following questions:
(a) When $n = 4$, what are the possible values of ℓ?
(b) When ℓ is 2, what are the possible values of m_ℓ?
(c) For a 4*s* orbital, what are the possible values of n, ℓ, and m_ℓ?
(d) For a 4*f* orbital, what are the possible values of n, ℓ, and m_ℓ?

48. Answer the following questions:
(a) When $n = 4$, $\ell = 2$, and $m_\ell = -1$, to what orbital type does this refer? (Give the orbital label, such as 1*s*.)
(b) How many orbitals occur in the $n = 5$ electron shell? How many subshells? What are the letter labels of the subshells?
(c) If a subshell is labeled *g*, how many orbitals occur in the subshell? What are the values of m_ℓ?

49. A possible excited state of the H atom has the electron in a 4*p* orbital. List all possible sets of quantum numbers n, ℓ, and m_ℓ for this electron.

50. A possible excited state for the H atom has an electron in a 5*d* orbital. List all possible sets of quantum numbers n, ℓ, and m_ℓ for this electron.

51. How many subshells occur in the electron shell with the principal quantum number $n = 4$?

52. How many subshells occur in the electron shell with the principal quantum number $n = 5$?

53. Explain briefly why each of the following is not a possible set of quantum numbers for an electron in an atom.
(a) $n = 2$, $\ell = 2$, $m_\ell = 0$
(b) $n = 3$, $\ell = 0$, $m_\ell = -2$
(c) $n = 6$, $\ell = 0$, $m_\ell = 1$

54. Which of the following represent valid sets of quantum numbers? For a set that is invalid, explain briefly why it is not correct.
(a) $n = 3$, $\ell = 3$, $m_\ell = 0$
(b) $n = 2$, $\ell = 1$, $m_\ell = 0$
(c) $n = 6$, $\ell = 5$, $m_\ell = -1$
(d) $n = 4$, $\ell = 3$, $m_\ell = -4$

55. What is the maximum number of orbitals that can be identified by each of the following sets of quantum numbers? When "none" is the correct answer, explain your reasoning.
(a) $n = 4$, $\ell = 3$
(b) $n = 5$
(c) $n = 2$, $\ell = 2$
(d) $n = 3$, $\ell = 1$, $m_\ell = -1$

56. What is the maximum number of orbitals that can be identified by each of the following sets of quantum numbers? When "none" is the correct answer, explain your reasoning.
(a) $n = 3$, $\ell = 0$, $m_\ell = +1$
(b) $n = 5$, $\ell = 1$
(c) $n = 7$, $\ell = 5$
(d) $n = 4$, $\ell = 2$, $m_\ell = -2$

57. How many planar nodes are associated with each of the following orbitals? (a) 2*s*; (b) 5*d*; (c) 5*f*.

58. How many planar nodes are associated with each of the following atomic orbitals? (a) 4*f*; (b) 2*p*; and (c) 6*s*.

59. State which of the following orbitals cannot exist according to the quantum theory: 2*s*, 2*d*, 3*p*, 3*f*, 4*f*, and 5*s*. Briefly explain your answers.

60. State which of the following orbitals can exist and which

cannot according to the quantum theory: $3p$, $4s$, $2f$, $5g$, and $1p$. Briefly explain your answers.

61. Write the complete set of quantum numbers (n, ℓ, and m_ℓ) that quantum theory allows for each of the following orbitals: (a) $2p$, (b) $3d$, and (c) $4f$.

62. Write the complete set of quantum numbers (n, ℓ, and m_ℓ) that quantum theory allows for each of the following orbitals: (a) $5f$, (b) $4d$, and (c) $2s$.

63. A given orbital is labeled by the magnetic quantum number $m_\ell = -1$. This could not be a
(a) *g* orbital (d) *p* orbital
(b) *f* orbital (e) *s* orbital
(c) *d* orbital

64. A particular orbital has $n = 4$, $\ell = 2$, and $m_\ell = -2$. This orbital must be: (a) $3p$, (b) $4p$, (c) $5d$, or (d) $4d$.

General Questions

65. An AM radio station broadcasts at a frequency of $6.00 \times 10^5\ s^{-1}$ (KHz = 1 kilohertz = $1000\ s^{-1}$). What is the wavelength of this signal in meters? What is the energy of one photon of this frequency? Compare with the energy of a photon of red light with $\lambda = 685$ nm (page 322).

66. Radiation in the ultraviolet region of the electromagnetic spectrum is quite energetic. It is this radiation that causes dyes to fade and your skin to burn. If you are bombarded with 1 mol of photons with a wavelength of 300. nm, what amount of energy (in kilojoules per mole of photons) are you being subjected to?

67. Exposure to high doses of microwaves can cause damage. Estimate how many photons, with $\lambda = 12$ cm, must be absorbed to raise the temperature of your eye by 3.0 °C. Assume the mass of an eye is 10. g and its heat capacity is 4.0 J/g · K.

68. An advertising sign gives off red light and green light.
(a) Which light has the higher energy photons?
(b) One of the colors has a wavelength of 680 nm, and the other has a wavelength of 500 nm. Identify which color has which wavelength.
(c) Which light has the higher frequency?

69. When *Voyager 1* encountered the planet Neptune in 1989, the planet was approximately 2.7 billion miles (2.7×10^9 miles) from the earth. How long did it take for the television picture signal to reach earth from Neptune? (1 mile = 1.61 km.)

70. Assume your eyes receive a signal consisting of blue light, $\lambda = 470$ nm. The energy of the signal is 2.50×10^{-14} J. How many photons reach your eyes?

71. If sufficient energy is absorbed by an atom, an electron can be lost by the atom and a positive ion formed. The amount of energy required is called the ionization energy. In the H atom, the ionization energy is that required to change the electron from $n = 1$ to $n =$ infinity. (See A Closer Look: Experimental Evidence for Bohr's Theory, page 333.) Calculate the ionization energy for He^+ ions. Is the ionization energy of He^+ more or less than that of H? (Bohr's theory applies to He^+ because it, like the H atom, has a single electron. However, the electron energy is now given by $E = -Z^2Rhc/n^2$, where Z is the atomic number of helium.)

72. Hydrogen atoms absorb energy so that the electrons are excited to the $n = 7$ energy level. Electrons then undergo these transitions, among others: (a) $n = 7 \rightarrow n = 1$; (b) $n = 7 \rightarrow n = 6$; and (c) $n = 2 \rightarrow n = 1$. Which transition produces a photon with (i) the smallest energy; (ii) the highest frequency; (iii) the shortest wavelength?

73. What is the shortest wavelength photon an excited H atom can emit? Explain briefly.

74. A laser puts out 6.0×10^6 J/s. The heat capacity of solid and liquid iron averages 25 J/mol · K. Iron melts at 1538 °C and boils at 2861 °C. How many seconds are required for the laser to raise the temperature of 1.00×10^2 g of Fe to its boiling point from 25 °C? The heat of fusion of iron is 13.8 J/mol.

75. The square of the wave function for a $2p$ orbital is plotted against the distance from the nucleus in Figure 7.18. What is the approximate distance from the nucleus at which the maximum probability density is reached?

76. Rank the following orbitals in the H atom in order of increasing energy: $3s$, $2s$, $2p$, $4s$, $3p$, $1s$, and $3d$.

77. How many orbitals in an atom can have the following quantum number or designation?
(a) $3p$ (e) $6d$
(b) $4p$ (f) $5d$
(c) $4p_x$ (g) $5f$
(d) $n = 5$ (h) $7s$

78. Answer the following questions as a summary quiz on the chapter:
(a) The quantum number n describes the ____________ of an atomic orbital.
(b) The shape of an atomic orbital is given by the quantum number ____________.
(c) A photon of orange light has ____________ *(less* or *more)* energy than a photon of yellow light.
(d) The maximum number of orbitals that may be associated with the following set of quantum numbers $n = 4$ and $\ell = 3$ is ____________.
(e) The maximum number of orbitals that may be associated with the quantum number set $n = 3$, $\ell = 2$, and $m_\ell = -2$ is ____________.
(f) Label each of the orbital pictures with the appropriate letter:

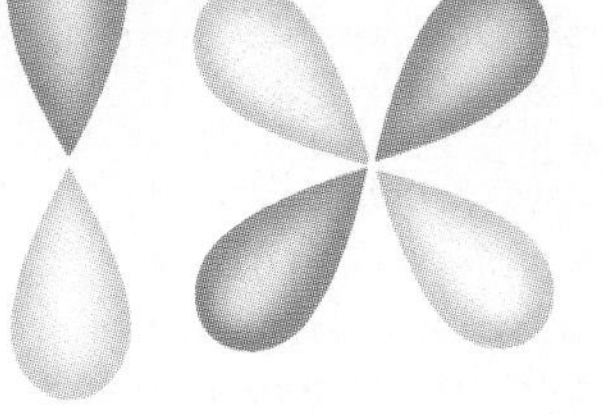

____________ ____________

(g) When $n = 5$, the possible values of ℓ are ___________.
(h) The maximum number of orbitals that can be assigned to the $n = 4$ shell is ___________.

79. Answer the following questions as a review of this chapter:
(a) The quantum number n describes the ___________ of an atomic orbital and the quantum number ℓ describes its ___________.
(b) When $n = 3$, the possible values of ℓ are ___________.
(c) What type of orbital corresponds to $\ell = 3$? ___________.
(d) For a $4d$ orbital, the value of n is ___________, the value of ℓ is ___________, and a possible value of m_ℓ is ___________.
(e) Each drawing represents a type of atomic orbital. Give the letter designation for the orbital, its value of ℓ, and specify the number of nodal planes.

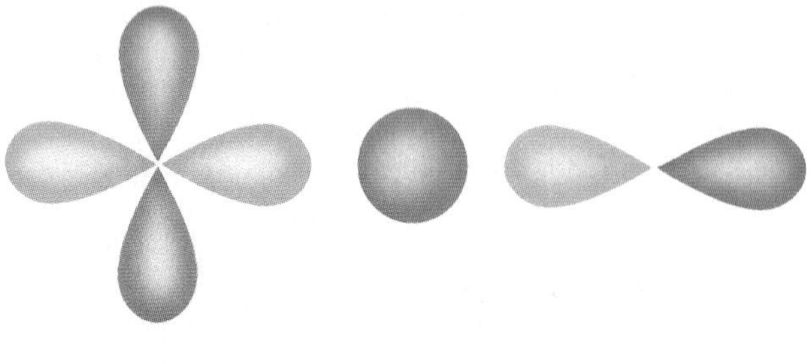

letter =	_____	_____	_____
ℓ value =	_____	_____	_____
nodal planes =	_____	_____	_____

(f) An atomic orbital that has three nodal planes is ___________.
(g) Which of the following orbitals cannot exist according to modern quantum theory: $2s$, $3p$, $2d$, $4g$, $5p$, $6p$?
(h) Which of the following is not a valid set of quantum numbers?

n	ℓ	m_ℓ
3	2	1
2	1	2
4	3	0

(i) What is the maximum number of orbitals that can be associated with each of the following sets of quantum numbers? (One possible answer is "none.")
(i) $n = 2$ and $\ell = 1$
(ii) $n = 3$
(iii) $n = 3$ and $\ell = 3$
(iv) $n = 2$, $\ell = 1$, and $m_\ell = 0$

80. Assume an electron is assigned to the $1s$ orbital in the H atom. Is the electron density zero at a distance of 0.40 nm from the nucleus? (Use Figure 7.18.)

81. Is the electron density of a $2p_z$ electron at a point on the x-axis 0.05 nm from the nucleus higher, lower, or the same when compared with the same distance on the z-axis?

82. Cobalt-60 is a radioactive isotope used in medicine for the treatment of certain cancers. It produces β-particles and γ-rays, the latter having energies of 1.173 and 1.332 MeV. (1 MeV = 1 million electronvolts and 1 eV = 9.6485×10^4 J/mol.) What are the wavelength and frequency of a γ-ray photon with an energy of 1.173 MeV?

Conceptual Questions

83. Bohr pictured the electrons of the atom as being located in definite orbits about the nucleus, just as the planets orbit the sun. Criticize this model in view of the quantum mechanical model.

84. What is the wave-particle duality? What are its implications in our modern view of atomic structure?

85. In what way does Bohr's model of the atom violate the uncertainty principle?

86. Suppose you live in a different universe where a different set of quantum numbers is required to describe the atoms of that universe. These quantum numbers have the following rules:

N	principal	1, 2, 3, . . . ∞
L	orbital	= N
M	magnetic	−1, 0, +1

How many orbitals are there altogether in the first three electron shells?

87. Which of these are observable?
(a) Position of electron in H atom
(b) Frequency of radiation emitted by H atoms
(c) Path of electron in H atom
(d) Wave motion of electrons
(e) Diffraction patterns produced by electrons
(f) Diffraction patterns produced by light
(g) Energy required to remove electrons from H atoms
(h) An atom
(i) A molecule
(j) A water wave

88. Derive the Rydberg equation, Equation 7.3, beginning with Bohr's energy equation (7.4).

89. In principle, is it possible to determine
(a) The energy of an electron in the H atom with high precision and accuracy?
(b) The position of a high-speed electron with high precision and accuracy?
(c) At the same time, both the position and energy of a high-speed electron with high precision and accuracy?

90. Imagine the nucleus of an H atom is located at the origin (the zero point) of an x, y, z graph.
(a) Assume that at the distance d, the probability of finding the $1s$ electron is 1.0×10^{-4} at $x = d$. What is the prob-

ability of finding the electron at $y = d$? Is the probability of finding the electron at $z = \frac{1}{2}d$ less than, greater than, or equal to 1.0×10^{-4}?

(b) The probability of finding a $2p_x$ electron at $x = d$ is 1×10^{-3}. What is the probability of finding the electron at $y = d$?

Summary Question

91. Technetium is not found naturally on earth; it must be synthesized in the laboratory. Nonetheless, because it is radioactive it has valuable medical uses. For example, the element in the form of sodium pertechnetate ($NaTcO_4$) is used in imaging studies of the brain, thyroid, and salivary glands and in renal blood flow studies, among other things.

(a) In what group and period of the periodic table is the element found?

(b) The valence electrons of technetium are found in the $5s$ and $4d$ subshells. What is a set of quantum numbers (n, ℓ, and m_ℓ) for one of the electrons of the $5s$ subshell?

(c) Technetium emits a γ-ray with an energy of 0.141 MeV. (1 MeV = 1 million electronvolts, where 1 eV = 9.6485×10^4 J/mol.) What are the wavelength and frequency of a γ-ray photon with an energy of 0.141 MeV?

(d) To make $NaTcO_4$, the metal is dissolved in nitric acid

$$7\ HNO_3(aq) + Tc(s) \longrightarrow HTcO_4(aq) + 7\ NO_2(g) + 3\ H_2O(\ell)$$

and the product, $HTcO_4$, is treated with NaOH to make $NaTcO_4$.

(i) Write a balanced equation for the reaction of $HTcO_4$ with NaOH.

(ii) If you begin with 4.5 mg of Tc metal, how much $NaTcO_4$ can be made? What mass of NaOH (in grams) is required to convert all of the $HTcO_4$ into $NaTcO_4$?

C H A P T E R 8

Atomic Electron Configurations and Chemical Periodicity

A Chemical Puzzler

Hot steel wool is plunged into a flask containing chlorine gas. Very quickly you see a brown cloud of iron(III) chloride form in the flask.

$$2\ Fe(s) + 3\ Cl_2(g) \longrightarrow 2\ FeCl_3(s)$$

This compound is ionic, being composed of iron(III) ions (Fe^{3+}) and chloride ions (Cl^-). The oxidation-reduction reaction that forms the compound requires that electrons be transferred from iron to chlorine. Why is iron such a good reducing agent? Why is chlorine such a good oxidizing agent? Why is the solid $FeCl_3$ product magnetic? Also, why do the iron(II) and iron(III) cations in the mineral magnetite make this mineral ferromagnetic?

We developed the modern view of the atom in the previous chapter. That model suggests that electrons can be arranged in shells in the space around the nucleus; each shell is distinguished by the quantum number n and consists of one or more subshells. Each subshell is described by the quantum number ℓ and contains one or more orbitals. The picture that emerges is a satisfying one that reflects the order of the natural world around us.

The model developed in Chapter 7 accurately describes atoms or ions, such as H and He^+, that have a single electron. To be useful, though, the model must be applicable to atoms with more than one electron, that is, to all the other known elements. Because the chemical properties of atoms depend on their electronic structure—the number and arrangement of electrons in the atom—one objective of this chapter is to develop a workable picture of the electronic structure of elements other than hydrogen.

Another objective of this chapter is to explore some of the physical properties of atoms: the ease with which atoms lose or gain electrons to form ions and the sizes of atoms and ions. These properties are directly related to the arrangement of electrons in atoms and thus to the chemistry of the elements and their compounds.

8.1 ELECTRON SPIN

Three quantum numbers (n, ℓ, and m_ℓ) allow us to define the orbital for an electron. To describe an electron in a multi-electron atom completely, however, one more quantum number, the **electron spin magnetic quantum number, m_s,** is required.

Around 1920, theoretical chemists realized that, because electrons interact with a magnetic field, there must be one more property to describe the electronic structure of atoms. It was soon verified experimentally that the electron behaves as though it has a spin, just as the earth has a spin. To understand this property and its relation to atomic structure, we should understand something of the general phenomenon of magnetism.

Magnetism

The needle of a compass at a given location on the earth always points in a given direction, no matter how the compass is moved. The needle is a magnet, such as a piece of iron. In 1600, William Gilbert concluded that the earth is also a large spherical magnet giving rise to a magnetic field that surrounds the planet (Figure 8.1). The compass needle is "drawn," or "attracted," into this magnetic field, one end of the needle pointing approximately to the earth's geographic north pole. Thus, we say the end of the compass needle pointing north is the

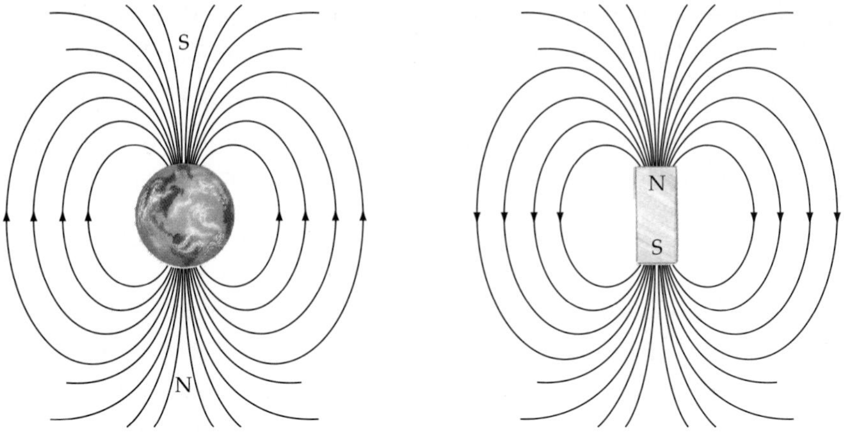

Figure 8.1 The magnetic fields of the earth and of a bar magnet. The magnetic field comes out of one end, arbitrarily called the "north magnetic pole," N, and loops toward the "south magnetic pole," S. The *geographic north pole* of the earth, named before the introduction of the term "magnetic pole," is the *magnetic south pole.*

magnet's "magnetic north pole," or simply its "north pole," designated N. The other end of the needle is its "south pole," designated S.

Identical magnetic poles (N-N or S-S) repel each other, and opposite poles (N-S) attract (Figure 8.2). Because the magnetic north pole of the compass needle points to the earth's geographic north pole, this must mean that this pole is actually the earth's magnetic south pole.

Paramagnetism and Unpaired Electrons

Most substances—chalk, sea salt, cloth—are slightly repelled by a strong magnet. They are said to be **diamagnetic.** In contrast, many metals and other compounds are attracted to a magnetic field. Such substances are generally called **paramagnetic,** and the magnitude of the effect can be determined with an apparatus such as that illustrated in Figure 8.3.

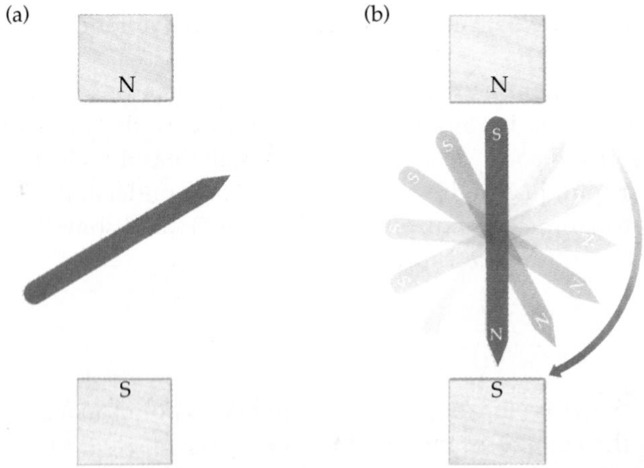

Figure 8.2 A magnetic needle in a strong magnetic field. The repulsion of similar poles (a) and attraction of opposite poles (b) cause the needle to line up as shown, the N pole of the needle facing the S pole of the magnet.

The magnetism of most paramagnetic materials is so weak that you can only observe it in the presence of a strong magnetic field. For example, the oxygen we breathe is paramagnetic; it sticks to the poles of a strong magnet (page 457). It does not cling to very weak magnets, however, such as those in the gadgets you stick on the door of your refrigerator. Other materials are so strongly magnetic, however, that you readily observe their effect. Examples include the mineral magnetite, Fe_3O_4, and the alloy called "Alnico" (for Al, Ni, and Co). These are often referred to as **ferromagnetic** materials, and it is these substances that are used to make the little magnetic devices you use in your home.

Paramagnetism and ferromagnetism arise from electron spins. An electron in an atom has the magnetic properties expected for a spinning, charged particle. What is important to us is the relation of spin to the arrangement of electrons in atoms. Experiments have shown that, if you place an atom with a single unpaired electron in a magnetic field, only two orientations are possible for the electron spin. That is, *electron spin is quantized.* One orientation is associated with a spin quantum number value of $m_s = +\frac{1}{2}$ and the other with an m_s value of $-\frac{1}{2}$ (Figure 8.4).

When one electron is assigned to an orbital in an atom, the electron's spin orientation can take either value of m_s. We observe experimentally that hydrogen atoms, each of which has a single electron, are paramagnetic; when an external magnetic field is applied, the electron magnets align with the field—like the needle of a compass—and experience an attractive force. In helium, two electrons are assigned to the same 1*s* orbital, and we can confirm by experiment that *helium is diamagnetic.* To account for this observation, we assume that the two

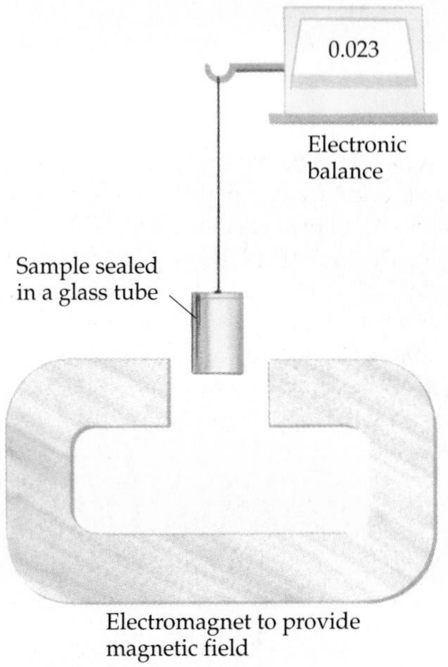

Figure 8.3 A magnetic balance used to measure the magnetic properties of a sample. The sample is first weighed with the electromagnet turned off. The magnet is then turned on and the sample reweighed. If the substance is paramagnetic, the sample is drawn into the magnetic field and the *apparent* weight increases.

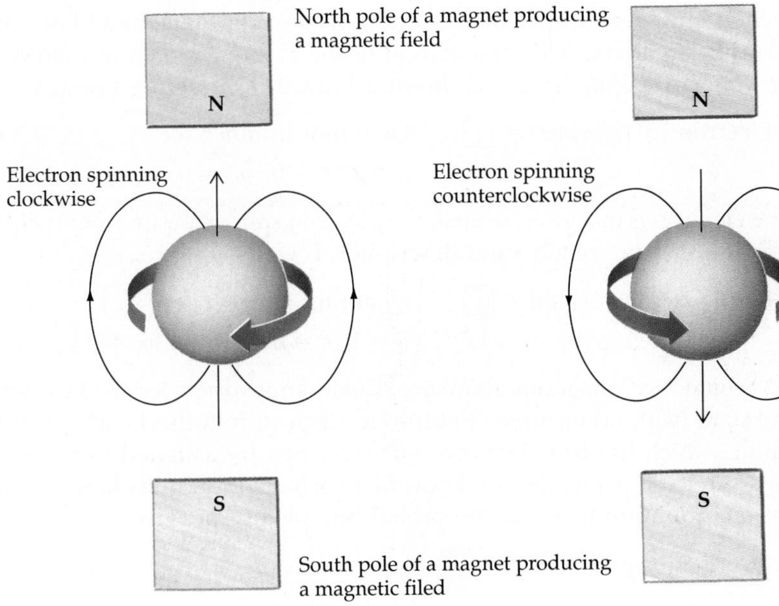

Figure 8.4 Quantization of electron spin. Because the electron acts as a "micromagnet" relative to a magnetic field, only two spins are possible. Any other position is forbidden. Therefore, we can say that the spin of an electron is quantized.

electrons assigned to the same orbital have opposite spin orientations; we say that their spins are **paired.** This means the magnetic field of one electron

$$\begin{matrix}\text{N}\\ \uparrow\\ \text{S}\end{matrix}$$

is "canceled out" by the magnetic field of the second

$$\begin{matrix}\text{S}\\ \downarrow\\ \text{N}\end{matrix}$$

of opposite spin.

In summary, *paramagnetism occurs in substances in which the constituent ions or atoms contain unpaired electrons.* Atoms in which all electrons are paired with partners of opposite spin are diamagnetic. This explanation opens the way to understanding the electron configurations of atoms with more than one electron.

8.2 THE PAULI EXCLUSION PRINCIPLE

To make the quantum theory consistent with experiment, the Austrian physicist Wolfgang Pauli (1900–1958) stated in 1925 his **exclusion principle:** *no two electrons in an atom can have the same set of four quantum numbers* (n, ℓ, m_ℓ, and m_s). This principle leads to yet another important conclusion, that *no atomic orbital can contain more than two electrons.*

Orbitals are not literally things or boxes in which electrons are placed. It is not conceptually correct to talk about electrons *in* or *occupying* orbitals, although it is commonly done for the sake of simplicity.

The $1s$ orbital of the H atom has the set of quantum numbers $n = 1$, $\ell = 0$, and $m_\ell = 0$. No other set is possible. If an electron is in this orbital, the electron spin direction must also be specified. Let us represent an orbital by a "box" and the electron by an arrow. A representation of the H atom is then as follows, in which the "electron spin arrow" is shown arbitrarily as pointing upward.

Electron in $1s$ orbital = [↑]	Quantum number set
$1s$	$n = 1,\ \ell = 0,\ m_\ell = 0,\ m_s = +\frac{1}{2}$

If only one electron is in a given orbital, the electron spin arrow may point either up or down. Thus, an equally valid description is

Electron in $1s$ orbital = [↓]	Quantum number set
$1s$	$n = 1,\ \ell = 0,\ m_\ell = 0,\ m_s = -\frac{1}{2}$

The "orbital box" diagrams above are equally appropriate for the H atom in its ground state (with no magnetic field): one electron is in the $1s$ orbital. For a helium atom, which has two electrons, *both* electrons are assigned to the $1s$ orbital. From the Pauli principle, you know that each electron must have a different set of quantum numbers, so the orbital box picture now is

Two electrons in the $1s$ orbital = [↑↓]

This electron has $n = 1,\ \ell = 0,\ m_\ell = 0,\ m_s = -\frac{1}{2}$

This electron has $n = 1,\ \ell = 0,\ m_\ell = 0,\ m_s = +\frac{1}{2}$

Each of the two electrons in the $1s$ orbital of an He atom has a *different set of the four quantum numbers.* The first three numbers of a set tell you this is a $1s$ orbital.

A CLOSER LOOK *Paramagnetism and Ferromagnetism*

Magnetic materials are relatively common, and many are important in our economy. They are found in stereo speakers and in telephone handsets, and magnetic oxides are used in recording tapes and computer disks.

The magnetic materials we use are ferromagnetic. The magnetic effect of ferromagnetic materials is much larger than for paramagnetic ones. Ferromagnetism occurs when the spins of unpaired electrons in a cluster of atoms (called a domain) in the solid align themselves in the same direction. Only the metals of the iron, cobalt, and nickel subgroups, as well as a few other metals such as neodymium, exhibit this property. They are also unique in that, once the domains are aligned in a magnetic field, the metal is permanently magnetized.

Many alloys exhibit greater ferromagnetism than do the pure metals themselves. One example is Alnico and another is an alloy of neodymium, iron, and boron.

Audio and video tapes are plastics coated with crystals of ferromagnetic Fe_2O_3, CrO_2, or another metal oxide. The recording head uses an electromagnetic field to create a varying magnetic field based on signals from a microphone. This magnetizes the tape as it passes through the head, the strength and direction of magnetization varying with the frequency of the sound to be recorded. When the tape is played back, the magnetic field of the moving tape induces a current, which is amplified and sent to the speakers.

Ferromagnetism is temperature-dependent. Substances gain stability by aligning electron spins within microscopic regions, or "domains," and domains can be aligned by placing the material in an external magnetic field. At higher temperatures, however, the atoms vibrate more and more within their places in the solid, and the spins can become randomly oriented. The temperature beyond which the energy is sufficient to overcome the aligning forces is called the Curie temperature, T_{Curie}. A substance can be ferromagnetic only below this temperature.

Material	T_{Curie}, **K**
Iron	1043
Cobalt	1388
Nickel	627
Gadolinium	293

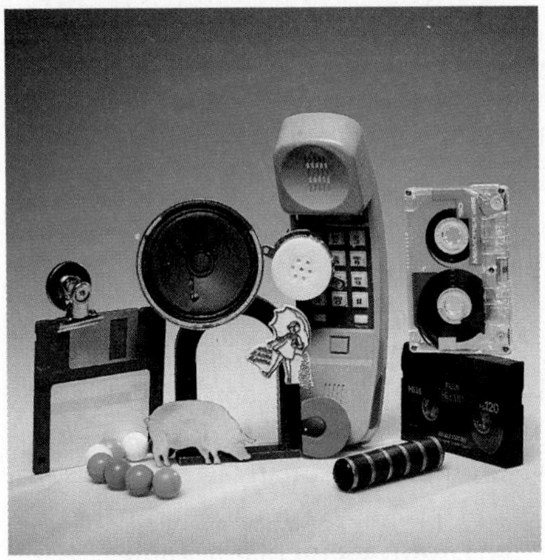

Photo of magnetic materials, such as cow magnet, refrigerator magnet, recording tape, computer disk, speaker magnet, telephone.

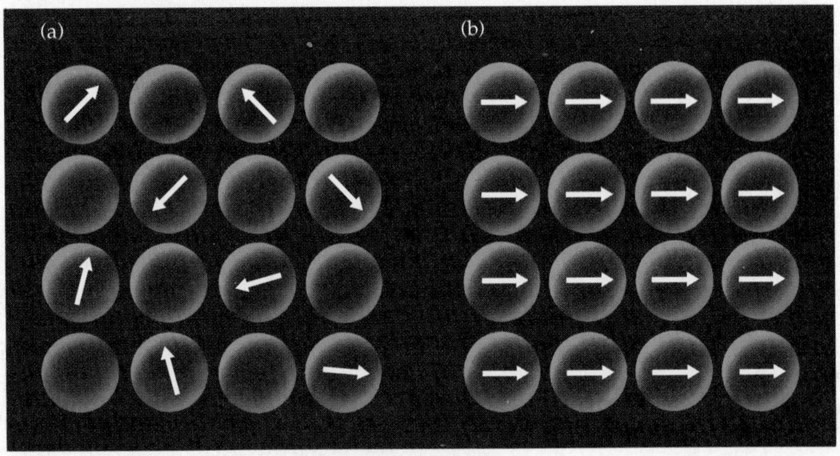

(a) Paramagnetism: the centers (atoms or ions) with magnetic moments are not aligned unless the substance is in a magnetic field. (b) Ferromagnetism: the spins of unpaired electrons in a cluster of atoms or ions align in the same direction.

There are only two choices for the fourth number, $m_s = +\frac{1}{2}$ or $-\frac{1}{2}$. Thus, *the 1s orbital, and any other atomic orbital, can be occupied by no more than two electrons, and these two electrons must have opposite spin directions.* The consequence is that the helium atom is diamagnetic, as experimentally observed.

The results expressed in this table were predicted by the Schrödinger theory and have been confirmed by experiment.

Note that n subshells occur in the nth shell.

TABLE 8.1 Number of Electrons Accommodated in Electron Shells and Subshells with $n = 1$ to 6

Electron Shell (n)	Subshells Available	Orbitals Available ($2\ell + 1$)	Number of Electrons Possible in Subshell [$2(2\ell + 1)$]	Maximum Electrons Possible for nth Shell ($2n^2$)
1	*s*	1	2	2
2	*s*	1	2	8
	p	3	6	
3	*s*	1	2	18
	p	3	6	
	d	5	10	
4	*s*	1	2	32
	p	3	6	
	d	5	10	
	f	7	14	
5	*s*	1	2	50
	p	3	6	
	d	5	10	
	f	7	14	
	*g**	9	18	
6	*s*	1	2	72
	p	3	6	
	d	5	10	
	*f**	7	14	
	*g**	9	18	
	*h**	11	22	

*These orbitals are not used in the ground state of any known element.

The number of orbitals in the nth electron shell is n^2, and the maximum number of electrons in the shell is $2n^2$.

The $n = 1$ electron shell in any atom can accommodate no more than two electrons. But what about the $n = 2$ shell? There are $n^2 = 4$ orbitals in the $n = 2$ shell: one *s* orbital and three *p* orbitals (Table 8.1). Because each orbital can be occupied by two electrons and no more, the 2*s* orbital is assigned to as many as two electrons, and the three 2*p* orbitals accommodate as many as six electrons, for a maximum of eight electrons with $n = 2$. This analysis is carried further for the electron shells normally observed in the known elements in Table 8.1.

8.3 ATOMIC SUBSHELL ENERGIES AND ELECTRON ASSIGNMENTS

Our goal is to understand the distribution of electrons in atoms with many electrons. These atoms can be ''built'' by assigning electrons to shells (defined by the quantum number n) of higher and higher energy. Within a given shell, electrons are assigned to orbitals within subshells (which are defined by the quantum number ℓ) of successively higher energy. In general, electrons are assigned in such a way that the total energy of the atom is as low as possible.

Experimental Evidence for Electron Configurations

Before learning the details of electron configurations, it is important to understand that our picture of the arrangement of electrons in atoms comes from experiment. The wave model of the atom is a reasonably successful attempt to rationalize these observations.

In Section 4.3 you learned that positive ions are formed by removing one or more electrons from an atom. The energy required in this process is called the **ionization energy** (see Section 8.6 for more details).

The ionization energy for the H atom was calculated from the Bohr theory on page 333.

1st ionization energy (899.4 kJ/mol) $\quad Be(g) \xrightarrow{+energy} Be^{+}(g) + e^{-}$

2nd ionization energy (1757.1 kJ/mol) $\quad Be^{+}(g) \xrightarrow{+energy} Be^{2+}(g) + e^{-}$

The second ionization energy is always larger than the first because the second electron is removed from a positively charged ion, whereas the first electron is removed from a neutral atom.

Ionization energies such as the ones for beryllium can be obtained experimentally. We can plot these data as the ratio of the second to the first ionization energy of an element versus atomic number (Figure 8.5). (Beryllium, for example, has a ratio of 1757.1/899.4 = 1.95.) This plot provides excellent evidence for the arrangement of electrons in shells and subshells of differing energy. It also shows the underlying periodicity of properties that characterizes the chemical elements.

- Lithium, sodium, and potassium each begin a new row of the periodic table, and you see that each has a much larger ratio of ionization energies than the other elements. We can conclude from this experimental observation that one electron (the one removed in the first ionization

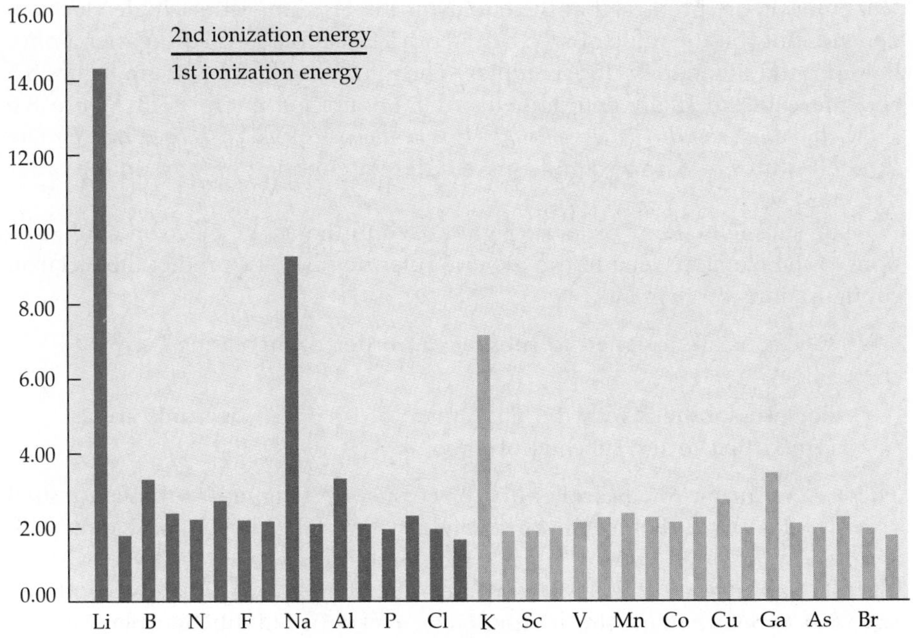

Figure 8.5 A plot of the ratio of the second and the first ionization energies versus atomic number for the first 36 elements.

step) occupies a new electron shell, n, whose energy is much higher than the shell used by the electrons of the elements in the previous row. The second ionization energy is much larger, mainly because the second electron must be removed from the $n - 1$ shell, and this shell lies much lower in energy than the nth shell.

1st ionization energy (513.3 kJ/mol) $Li(g) \longrightarrow Li^{+}(g) + e^{-}$ from $n = 2$ shell

2nd ionization energy (7298.0 kJ/mol) $Li^{+}(g) \longrightarrow Li^{2+}(g) + e^{-}$ from $n = 1$ shell

- Except for the alkali metals, most elements have about the same ionization energy ratio (values range from about 1.7 to 3). We can rationalize this by assuming that both electrons being removed from the elements beryllium to neon, for example, are in the same electron shell.
- The ratios for boron and aluminum are slightly larger than those of the other elements in their row in the periodic table (except for the relevant alkali metal). This indicates that a subshell has been completed at Be and Mg. Boron and aluminum are elements in which electrons are in a new subshell that has a slightly higher energy than the subshell used by the electrons of Be and Mg. Therefore, the first electron in boron, for example, is removed from one subshell in the $n = 2$ shell, and the second electron comes from a lower energy subshell within the $n = 2$ shell.

Let us now see how our experimental observations, and our explanations, agree with the currently accepted model of atomic structure.

Order of Subshell Energies and Assignments

Quantum theory predicts that the energy of the H atom, with a single electron, depends only on the value of n ($E = -Rhc/n^2$, Equation 7.4). For heavier atoms, however, the situation is more complex. The ionization energy plot in Figure 8.5 and the experimentally determined order of subshell energies in Figure 8.6 show that *the subshell energies of multielectron atoms depend on both n and ℓ.* The subshells with $n = 3$, for example, have different energies; they are in the order $3s < 3p < 3d$.

The subshell energy order in Figure 8.6 and the actual electron configurations of the elements lead to two general rules that help us predict the electron configurations of elements.

- Electrons are assigned to subshells in order of increasing "$n + \ell$" value.
- For two subshells with the same value of "$n + \ell$," electrons are assigned first to the subshell of lower n.

These rules mean, for example, that electrons are assigned to the $2s$ subshell ($n + \ell = 2 + 0 = 2$) before the $2p$ subshell ($n + \ell = 2 + 1 = 3$), or that they are assigned in the order $3s$ ($n + \ell = 3 + 0 = 3$) before $3p$ ($n + \ell = 3 + 1 = 4$) before $3d$ ($n + \ell = 3 + 2 = 5$). It also means that electrons fill the $4s$ subshell ($n + \ell = 4$) before filling the $3d$ subshell ($n + \ell = 5$). This filling order, which is summarized in Figure 8.7, has been amply verified by experiment.

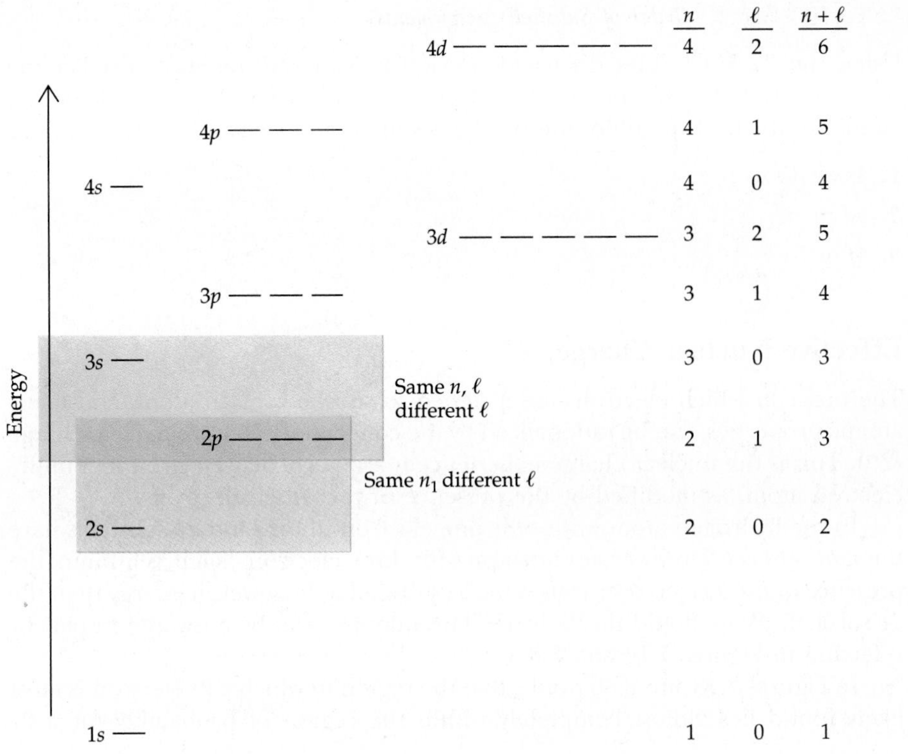

Figure 8.6 In a multielectron atom, energies of electron shells increase with increasing n, and subshell energies increase with increasing ℓ. (The energy axis is not to scale.) The subshells of a given shell lie in a given *band* of energies. The energy gaps between bands of subshell energies for a given shell become smaller as n increases. This means that the top of the band of energies for one quantum shell may eventually overlap the band of energies for the shell of next higher n.

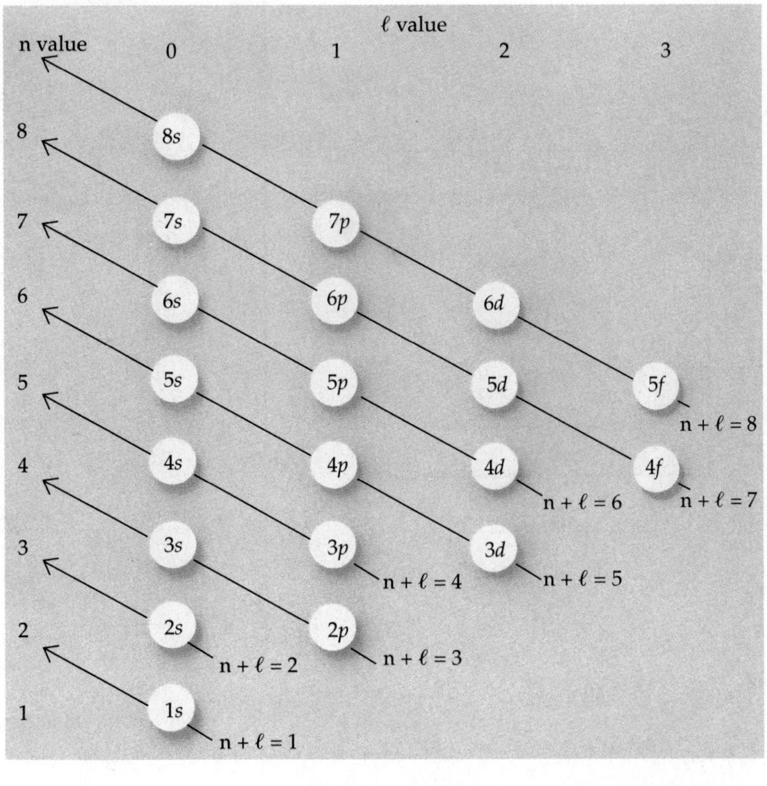

Figure 8.7 Subshells in atoms are filled in order of increasing $n + \ell$. When several subshells have the same value of $n + \ell$, the subshells are filled in order of increasing n. To use the diagram begin at $1s$ and follow the arrows of increasing $n + \ell$. (Thus, the order of filling is $1s \rightarrow 2s \rightarrow 2p \rightarrow 3s \rightarrow 3p \rightarrow 4s \rightarrow 3d$, and so on.)

EXERCISE 8.1 ***Order of Subshell Assignments***

Using the "$n + \ell$" rules discussed previously, you can generally predict the order of subshell assignments (the electron filling order) for a multielectron atom. To which of the following subshells should an electron be assigned first?

1. $4s$ or $4p$
2. $5d$ or $6s$
3. $4f$ or $5s$ ■

Effective Nuclear Charge, Z*

Effective nuclear charge, Z^*, is important because it can be used to rationalize electron configurations (Section 8.4) and the observed properties of atoms and their ions (Section 8.6).

The order in which electrons are assigned to subshells in an atom, and many atomic properties, can be rationalized by the concept of **effective nuclear charge (Z*)**. This is the nuclear charge experienced by a particular electron in a multielectron atom, as modified by the presence of the other electrons.

In the hydrogen atom, with only one electron, the $2s$ and $2p$ subshells have the same energy. However, for an atom with three electrons, such as lithium, the presence of the $1s$ electrons causes the $2s$ subshell to lie lower in energy than the $2p$ subshell. Why should this be true? This question can be answered in part by referring to Figures 7.18 and 8.8.

In Figures 7.18 and 8.8b notice that the region in which a $2s$ electron is most likely found lies almost completely within the region of probability for a $2p$

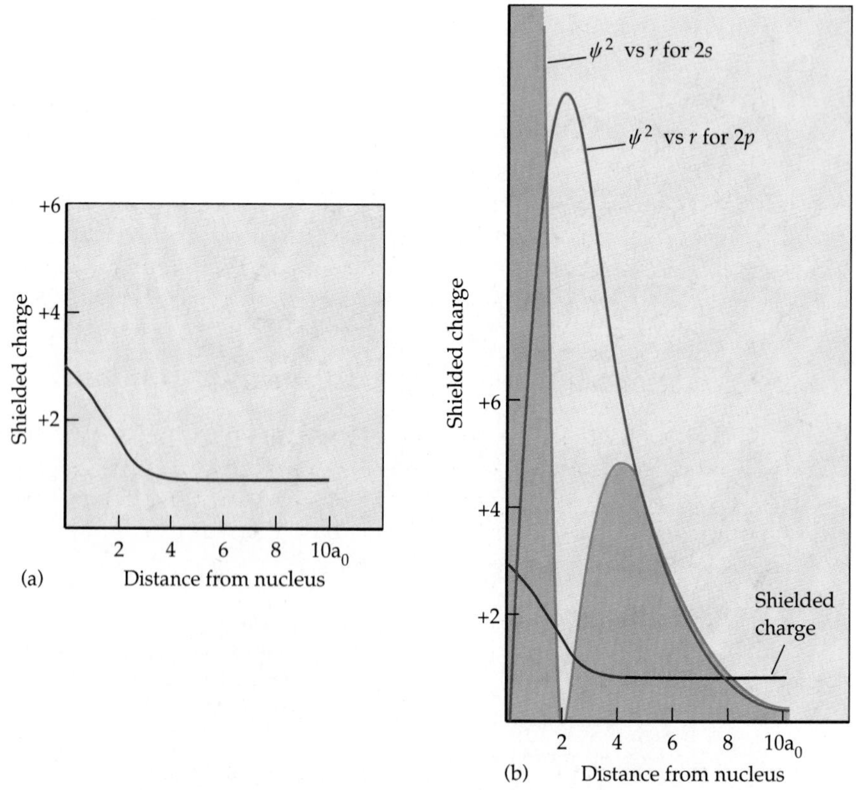

Figure 8.8 Shielded and effective nuclear charge. (a) The shielded nuclear charge is the positive charge an electron "feels" at a given distance from the nucleus. The shielded charge is shown for the third electron in the Li atom. (b) An electron in the $2s$ subshell has a higher probability of being in a region of high shielded charge than an electron in the $2p$ subshell.

electron. Just as importantly, these orbitals *penetrate* the region occupied by the 1*s* electrons, the 2*s* penetration being greater than for 2*p*. That is, an electron in a 2*s* subshell has a much higher probability of being very close to the nucleus than an electron in a 2*p* subshell. The third electron in lithium occupies the 2*s* subshell, rather than the 2*p* subshell, because the 2*s*-2*p* probability difference near the nucleus leads to a lower energy for a 2*s* than for a 2*p* electron.

Lithium, with a nucleus containing three protons, has two electrons in the 1*s* subshell. When the third electron is almost at the nucleus (regardless of its subshell type), Figure 8.8a shows that this electron experiences a charge of +3. The positive charge experienced by this third electron drops off rapidly, however, as it moves away from the nucleus; the charge finally reaches +1. The reason for this drop is that the negative charges of the two electrons in the 1*s* subshell shield, or screen, the effect of the positive nuclear charge, the screening effect becoming greater as the third electron moves outside the region occupied by the 1*s* electrons. The charge felt by any electron at any distance from the nucleus in an atom is called the *shielded nuclear charge.*

Figure 8.8b shows that an electron in a 2*s* subshell has a higher probability than an electron in the 2*p* subshell of being within the region of space where the shielded nuclear charge is larger than about +2. Thus, *on average,* an electron in a 2*s* subshell experiences a higher shielded nuclear charge than an electron in a 2*p* subshell. *The average shielded charge felt by the electron is called the* **effective nuclear charge, Z*.** For lithium, the 2*s* electron experiences a larger Z* than does a 2*p* electron. This means a 2*s* electron is more strongly attracted to the nucleus than a 2*p* electron, and the lithium atom has an electron configuration with a pair of electrons in the 1*s* subshell and a single 2*s* electron.

8.4 ATOMIC ELECTRON CONFIGURATIONS

The configurations of the elements through element 109 are given in Table 8.2. These are the *ground state electron configurations,* where electrons are found in the lowest energy shells, subshells, and orbitals available to them. In general, the guiding principle in assigning electrons to available orbitals is to do so in order of increasing "$n + \ell$" (Figure 8.7). Our emphasis, however, will be to connect the configurations of the elements with their position in the periodic table because this will ultimately allow us to organize a large number of chemical facts.

Electron Configurations of the Main Group Elements

Hydrogen, the first element in the periodic table, has one electron in a 1*s* orbital. One way to depict its electron configuration is with the **orbital box diagram** used earlier, but an alternative and more frequently used method is the **spectroscopic notation.** Using the latter method, the electron configuration of H is $1s^1$, or "one ess one."

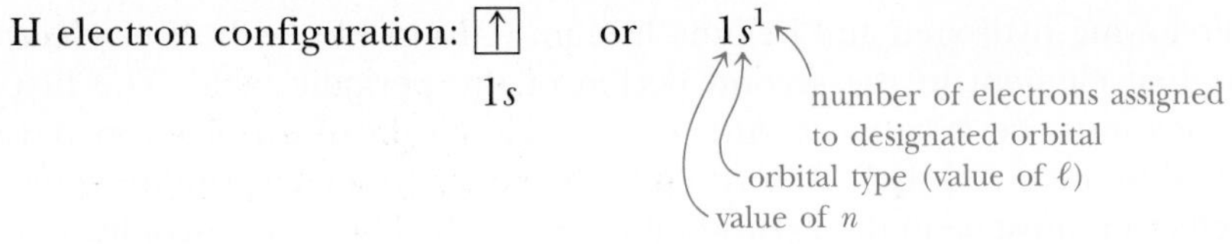

Orbital Box Diagram   Spectroscopic Notation

TABLE 8.2 Electron Configurations of Atoms in the Ground State

Z	Element	Configuration	Z	Element	Configuration	Z	Element	Configuration
1	H	$1s^1$	38	Sr	[Kr]$5s^2$	75	Re	[Xe]$4f^{14}5d^5 6s^2$
2	He	$1s^2$	39	Y	[Kr]$4d^1 5s^2$	76	Os	[Xe]$4f^{14}5d^6 6s^2$
3	Li	[He]$2s^1$	40	Zr	[Kr]$4d^2 5s^2$	77	Ir	[Xe]$4f^{14}5d^7 6s^2$
4	Be	[He]$2s^2$	41	Nb	[Kr]$4d^4 5s^1$	78	Pt	[Xe]$4f^{14}5d^9 6s^1$
5	B	[He]$2s^2 2p^1$	42	Mo	[Kr]$4d^5 5s^1$	79	Au	[Xe]$4f^{14}5d^{10}6s^1$
6	C	[He]$2s^2 2p^2$	43	Tc	[Kr]$4d^5 5s^2$	80	Hg	[Xe]$4f^{14}5d^{10}6s^2$
7	N	[He]$2s^2 2p^3$	44	Ru	[Kr]$4d^7 5s^1$	81	Tl	[Xe]$4f^{14}5d^{10}6s^2 6p^1$
8	O	[He]$2s^2 2p^4$	45	Rh	[Kr]$4d^8 5s^1$	82	Pb	[Xe]$4f^{14}5d^{10}6s^2 6p^2$
9	F	[He]$2s^2 2p^5$	46	Pd	[Kr]$4d^{10}$	83	Bi	[Xe]$4f^{14}5d^{10}6s^2 6p^3$
10	Ne	[He]$2s^2 2p^6$	47	Ag	[Kr]$4d^{10}5s^1$	84	Po	[Xe]$4f^{14}5d^{10}6s^2 6p^4$
11	Na	[Ne]$3s^1$	48	Cd	[Kr]$4d^{10}5s^2$	85	At	[Xe]$4f^{14}5d^{10}6s^2 6p^5$
12	Mg	[Ne]$3s^2$	49	In	[Kr]$4d^{10}5s^2 5p^1$	86	Rn	[Xe]$4f^{14}5d^{10}6s^2 6p^6$
13	Al	[Ne]$3s^2 3p^1$	50	Sn	[Kr]$4d^{10}5s^2 5p^2$	87	Fr	[Rn]$7s^1$
14	Si	[Ne]$3s^2 3p^2$	51	Sb	[Kr]$4d^{10}5s^2 5p^3$	88	Ra	[Rn]$7s^2$
15	P	[Ne]$3s^2 3p^3$	52	Te	[Kr]$4d^{10}5s^2 5p^4$	89	Ac	[Rn]$6d^1 7s^2$
16	S	[Ne]$3s^2 3p^4$	53	I	[Kr]$4d^{10}5s^2 5p^5$	90	Th	[Rn]$6d^2 7s^2$
17	Cl	[Ne]$3s^2 3p^5$	54	Xe	[Kr]$4d^{10}5s^2 5p^6$	91	Pa	[Rn]$5f^2 6d^1 7s^2$
18	Ar	[Ne]$3s^2 3p^6$	55	Cs	[Xe]$6s^1$	92	U	[Rn]$5f^3 6d^1 7s^2$
19	K	[Ar]$4s^1$	56	Ba	[Xe]$6s^2$	93	Np	[Rn]$5f^4 6d^1 7s^2$
20	Ca	[Ar]$4s^2$	57	La	[Xe]$5d^1 6s^2$	94	Pu	[Rn]$5f^6 7s^2$
21	Sc	[Ar]$3d^1 4s^2$	58	Ce	[Xe]$4f^1 5d^1 6s^2$	95	Am	[Rn]$5f^7 7s^2$
22	Ti	[Ar]$3d^2 4s^2$	59	Pr	[Xe]$4f^3 6s^2$	96	Cm	[Rn]$5f^7 6d^1 7s^2$
23	V	[Ar]$3d^3 4s^2$	60	Nd	[Xe]$4f^4 6s^2$	97	Bk	[Rn]$5f^9 7s^2$
24	Cr	[Ar]$3d^5 4s^1$	61	Pm	[Xe]$4f^5 6s^2$	98	Cf	[Rn]$5f^{10}7s^2$
25	Mn	[Ar]$3d^5 4s^2$	62	Sm	[Xe]$4f^6 6s^2$	99	Es	[Rn]$5f^{11}7s^2$
26	Fe	[Ar]$3d^6 4s^2$	63	Eu	[Xe]$4f^7 6s^2$	100	Fm	[Rn]$5f^{12}7s^2$
27	Co	[Ar]$3d^7 4s^2$	64	Gd	[Xe]$4f^7 5d^1 6s^2$	101	Md	[Rn]$5f^{13}7s^2$
28	Ni	[Ar]$3d^8 4s^2$	65	Tb	[Xe]$4f^9 6s^2$	102	No	[Rn]$5f^{14}7s^2$
29	Cu	[Ar]$3d^{10}4s^1$	66	Dy	[Xe]$4f^{10}6s^2$	103	Lr	[Rn]$5f^{14}6d^1 7s^2$
30	Zn	[Ar]$3d^{10}4s^2$	67	Ho	[Xe]$4f^{11}6s^2$	104	Rf	[Rn]$5f^{14}6d^2 7s^2$
31	Ga	[Ar]$3d^{10}4s^2 4p^1$	68	Er	[Xe]$4f^{12}6s^2$	105	Ha	[Rn]$5f^{14}6d^3 7s^2$
32	Ge	[Ar]$3d^{10}4s^2 4p^2$	69	Tm	[Xe]$4f^{13}6s^2$	106	Sg	[Rn]$5f^{14}6d^4 7s^2$
33	As	[Ar]$3d^{10}4s^2 4p^3$	70	Yb	[Xe]$4f^{14}6s^2$	107	Ns	[Rn]$5f^{14}6d^5 7s^2$
34	Se	[Ar]$3d^{10}4s^2 4p^4$	71	Lu	[Xe]$4f^{14}5d^1 6s^2$	108	Hs	[Rn]$5f^{14}6d^6 7s^2$
35	Br	[Ar]$3d^{10}4s^2 4p^5$	72	Hf	[Xe]$4f^{14}5d^2 6s^2$	109	Mt	[Rn]$5f^{14}6d^7 7s^2$
36	Kr	[Ar]$3d^{10}4s^2 4p^6$	73	Ta	[Xe]$4f^{14}5d^3 6s^2$			
37	Rb	[Kr]$5s^1$	74	W	[Xe]$4f^{14}5d^4 6s^2$			

We can use the same methods to describe the configurations of the other elements, and those for the first ten elements are illustrated in Table 8.3.

Following hydrogen and helium, lithium (Group 1A), with three electrons, is the first element in the second period of the periodic table. The first two electrons must be assigned to the $n = 1$ shell, so the third electron must be assigned to the $n = 2$ shell. According to the energy level diagram in Figure 8.6, that electron must be in the $2s$ subshell ($n + \ell = 2$). The spectroscopic notation, $1s^2 2s^1$, is read "one ess two, two ess one."

TABLE 8.3 Electron Configurations of Elements with Z = 1 to 10

	Electron Configuration		1s	2s	2p
		ℓ	0	0	1
		$n + \ell$	1	2	3
H	$1s^1$		[↑]	[ ]	[ ][ ][ ]
He	$1s^2$		[↑↓]	[ ]	[ ][ ][ ]
Li	$1s^2 2s^1$		[↑↓]	[↑]	[ ][ ][ ]
Be	$1s^2 2s^2$		[↑↓]	[↑↓]	[ ][ ][ ]
B	$1s^2 2s^2 2p^1$		[↑↓]	[↑↓]	[↑][ ][ ]
C	$1s^2 2s^2 2p^2$		[↑↓]	[↑↓]	[↑][↑][ ]
N	$1s^2 2s^2 2p^3$		[↑↓]	[↑↓]	[↑][↑][↑]
O	$1s^2 2s^2 2p^4$		[↑↓]	[↑↓]	[↑↓][↑][↑]
F	$1s^2 2s^2 2p^5$		[↑↓]	[↑↓]	[↑↓][↑↓][↑]
Ne	$1s^2 2s^2 2p^6$		[↑↓]	[↑↓]	[↑↓][↑↓][↑↓]

The position of lithium in the periodic table tells you its configuration immediately. All the elements of Group 1A (and 1B) have one electron assigned to an *s* orbital of the *n*th shell, for which *n* is the number of the period in which the element is found (Figure 8.9). For example, potassium is the first element in the $n = 4$ row (the fourth period), so potassium has the electron configuration of the element preceding it in the table (Ar) *plus* a final electron assigned to the 4*s* orbital.

Beryllium, in Group 2A, has two electrons in the 1*s* orbital plus two additional electrons. Figure 8.6 shows that the 2*s* orbital is appropriate, so the configuration of Be is $1s^2 2s^2$. *All elements of Group 2A have electron configurations of* [electrons of preceding noble gas] ns^2, where *n* is the period in which the element is found in the periodic table. Because all the elements of Group 1A have the configuration ns^1, and those in Group 2A have ns^2, these elements are called ***s*-block elements.**

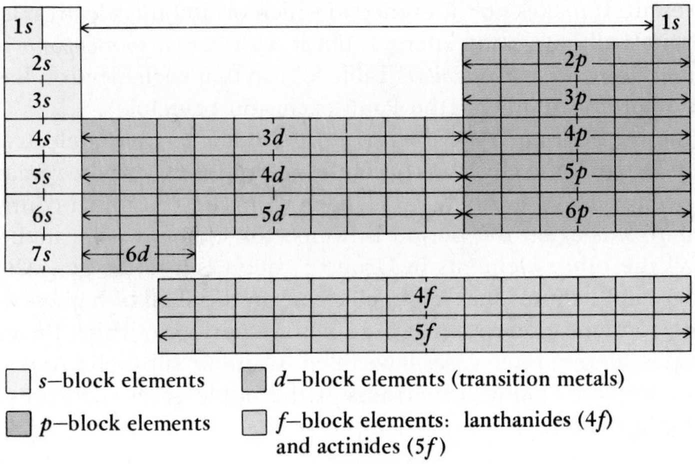

Figure 8.9 Electron configurations and the periodic table. The outermost electrons of an element are assigned to the indicated orbitals. See Table 8.2.

At boron (Group 3A) you first encounter an element in the block of elements on the right side of the periodic table. Because $1s$ and $2s$ orbitals are filled in a boron atom, the fifth electron must be assigned to a $2p$ orbital. In fact, *all the elements from Group 3A through Group 8 have electrons in p orbitals,* so these elements are sometimes called the ***p*-block elements.** All have the general configuration ns^2np^x, where x is the group number -2.

Carbon (Group 4A) is the second element in the p block, so a second electron is assigned to the $2p$ orbitals. For carbon in its lowest energy, or *ground state,* this electron *must* be assigned to either of the remaining p orbitals, and it *must* have the same spin direction as the first p electron.

In general, when electrons are assigned to p, d, or f orbitals, each successive electron is assigned to a different orbital of the subshell, and each electron has the same spin as the previous one; this pattern proceeds until the subshell is half full. Additional electrons must be assigned to half-filled orbitals. This procedure follows **Hund's rule,** which states that *the most stable arrangement of electrons is that with the maximum number of unpaired electrons,* all with the same spin direction. This arrangement makes the total energy of an atom as low as possible.

Hund's rule has been amply verified by experiments using equipment such as that depicted in Figure 8.3.

Notice that carbon is the second element in the p block of elements, so there must be two p electrons (besides the two $2s$ electrons already present in the $n = 2$ shell). Because C is in the second period of the table, the p orbitals involved are $2p$. Thus, you can immediately write the carbon electron configuration by referring to the periodic table: starting at H and moving from left to right across the successive periods, you write $1s^2$ to reach the end of period 1, and then $2s^2$ and finally $2p^2$ to bring the electron count to six. Carbon is in Group 4A of the periodic table because it has four electrons in the $n = 2$ shell.

Nitrogen is in the second period and in Group 5A; it is the third element in the p-block. Thus, it has a total of five electrons in the $n = 2$ shell; two of these are assigned to the $2s$ subshell, and three are assigned to the $2p$ subshell. Following Hund's rule, each electron in the $2p$ subshell occupies a different $2p$ orbital. All Group 5A atoms have a similar configuration: ns^2np^3, where n is the period in which the element is located in the periodic table.

Following nitrogen (Group 5A), oxygen (Group 6A) has a total of six electrons in its outer shell: two of the electrons are assigned to the $2s$ orbital, and, as it is the fourth element in the p block, the other four electrons are assigned to $2p$ orbitals. With nitrogen the $2p$ subshell was half full, containing one electron per orbital. Therefore, for oxygen the fourth $2p$ electron must pair up with one already present. It makes no difference to which orbital this electron is assigned (the $2p$ orbitals all have same energy), but it *must have a spin opposite to the other electron already assigned to that orbital* (Table 8.3) so that each electron has a different set of quantum numbers (the Pauli exclusion principle).

Fluorine is in Group 7A of the second period; it has seven electrons in the $n = 2$ shell. Two of these electrons occupy the $2s$ subshell, and the remaining five electrons occupy the $2p$ subshell. All halogen atoms have a similar configuration, ns^2np^5, where n is again the period in which the element is located.

Like all the other elements in Group 8, neon is a noble gas. All Group 8 elements (except helium) have eight electrons in the shell of highest n value, so all have the configuration ns^2np^6, where n is the period in which the element is found. That is, all the noble gases have filled ns and np subshells. As you will see, the nearly complete chemical inertness of the noble gases correlates with this electron configuration.

The next element after neon is sodium, and with it a new period is begun (recall Figure 8.5). Because sodium is the first element with $n = 3$, the added electron must be assigned to the $3s$ orbital. (Remember that all elements in Group 1A have the ns^1 configuration.) Thus, the complete electron configuration of sodium is that of neon (the preceding electron) plus one $3s$ electron.

$$\text{Na: } 1s^2 2s^2 2p^6 3s^1 \quad \text{or} \quad [\text{Ne}]3s^1$$

↑ noble gas notation represents core electrons

We have written the electron configuration for sodium in two ways, one in an abbreviated form that uses the **noble gas configuration.** The arrangement preceding the $3s$ electron is that of the noble gas neon, so, instead of writing out "$1s^2 2s^2 2p^6$," the completed electron shells are represented by placing the symbol of the corresponding noble gas in brackets.

The electrons included in the noble gas notation are often referred to as the **core electrons** of the atom. Not only is it a time-saving way to write electron configurations, but it also conveys the idea that the core electrons can generally be ignored when considering the chemistry of an element. The electrons beyond the core electrons, the $3s^1$ electron in the case of sodium, are the **valence electrons,** the electrons that determine the chemical properties of an element (Section 3.3).

EXAMPLE 8.1 *Electron Configurations*

Give the electron configuration of silicon, using the spectroscopic notation and the noble gas notation.

Solution Silicon, element 14, is the fourth element in the third period ($n = 3$),

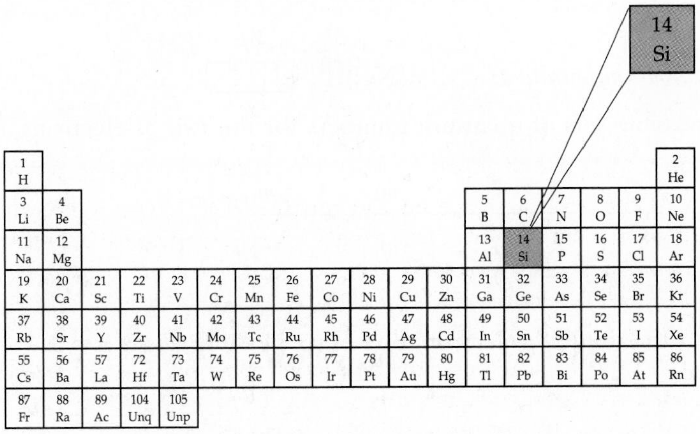

and it is in the p block. Therefore, the last four electrons in the atom have the configuration $3s^2 3p^2$. These are preceded by the completed shells $n = 1$ and $n = 2$, the electron arrangement for Ne. Therefore, the electron configuration of silicon is

$$1s^2 2s^2 2p^6 3s^2 3p^2 \quad \text{or} \quad [\text{Ne}]3s^2 3p^2$$

EXAMPLE 8.2 *Electron Configurations*

Give the electron configuration of sulfur, using the spectroscopic, noble gas, and orbital box notations.

Solution Sulfur, element 16, is the sixth element in the third period ($n = 3$), and it is in the *p* block. The last six electrons assigned to the atom therefore have the configuration $3s^2 3p^4$. These are preceded by the completed shells $n = 1$ and $n = 2$, the electron arrangement for Ne. Therefore, the electron configuration of sulfur is

Spectroscopic notation: $1s^2 2s^2 2p^6 3s^2 3p^4$

Noble gas notation: $[\mathrm{Ne}]3s^2 3p^4$

Orbital box notation: [Ne] 3s: [↑↓] 3p: [↑↓][↑][↑]

EXAMPLE 8.3 *Electron Configurations and Quantum Numbers*

Write the electron configuration for Al (using the noble gas notation) and give a set of quantum numbers for each of the electrons with $n = 3$ (the valence electrons).

Solution Aluminum is the third element in the third period. It therefore has three electrons with $n = 3$, and because Al is in the *p* block of elements, two of the electrons are assigned to 3*s* and the remaining electron is assigned to 3*p*. The element is preceded by the noble gas neon, so the electron configuration is $[\mathrm{Ne}]3s^2 3p^1$. Using a box notation, the configuration is

Aluminum configuration: [Ne] 3s: [↑↓] 3p: [↑][ ][ ]

The possible sets of quantum numbers for the two 3*s* electrons are

	n	ℓ	m_ℓ	m_s
For ↑	3	0	0	$+\frac{1}{2}$
For ↓	3	0	0	$-\frac{1}{2}$

and for the single 3*p* electron, one of six possible sets is $n = 3$, $\ell = 1$, $m_\ell = +1$, and $m_s = +\frac{1}{2}$.

EXERCISE 8.2 *Spectroscopic Notation, Orbital Box Diagrams, Quantum Numbers*

1. What element has the configuration $1s^2 2s^2 2p^6 3s^2 3p^5$?
2. Using the spectroscopic notation and a box diagram, show the electron configuration of chlorine.
3. Write one possible set of quantum numbers for the valence electrons of calcium. ■

Electron Configurations for the Transition Elements

The elements of the fourth through the sixth periods in the middle of the periodic table are those elements in which a *d* or *f* subshell is being occupied (Figure 8.9 and Table 8.2). Elements whose atoms are filling *d* subshells are often referred to as the **transition elements.** Those for which *f* subshells are filling are sometimes called the *inner transition elements* or, more usually, the **lanthanides** (filling 4*f* orbitals) and **actinides** (filling 5*f* orbitals).

In writing electron configurations, we follow the convention of writing the orbitals in order of increasing *n*. For a given *n*, the subshells are listed in order of increasing ℓ. Note that this is not always the filling order for the transition and inner transition elements.

According to Figure 8.9, which is based on experimentally determined electron configurations, the transition elements are always preceded by two *s*-block elements. Accordingly, scandium, the first transition element, has the configuration $[Ar]3d^14s^2$, and titanium follows with $[Ar]3d^24s^2$ (Table 8.4).

The expected configuration of the chromium atom is $[Ar]3d^44s^2$. However, the actual configuration has one electron assigned to *each* of the six available 3*d* and 4*s* orbitals: $[Ar]3d^54s^1$. This is explained by assuming that the 4*s* and the 3*d* orbitals have approximately the same energy at this point, thus giving rise to six orbitals of nearly the same energy. Each of the six valence electrons of chromium is assigned a separate orbital. This occurrence illustrates the fact that there are occasionally minor differences between the predicted and actual configurations. These have little or no effect on the chemistry of the element.

Following chromium, atoms of manganese, iron, and nickel have configurations that are expected on the basis of Figure 8.9. Copper, however, is slightly different than expected. This Group 1B element has a single electron in the 4*s* orbital, as expected from its group number; thus, the remaining ten electrons beyond the argon core are assigned to the 3*d* orbitals. Zinc ends the first transition series; the 4*s* and 3*d* orbitals are completely filled, as expected for Group 2B.

The fifth period ($n = 5$) follows the pattern of the fourth period with minor variations. The sixth period, however, includes the **lanthanide series** beginning with *lanthanum*, La. As the first element in the *d* block, lanthanum has the configuration $[Xe]5d^16s^2$. The next element, cerium (Ce), is set out in a separate row at the bottom of the periodic table, and it is with these elements that electrons

TABLE 8.4 Orbital Box Diagrams for the Elements Ca Through Zn

		3*d*	4*s*
Ca	$[Ar]4s^2$	[ ][ ][ ][ ][ ]	[↑↓]
Sc	$[Ar]3d^14s^2$	[↑][ ][ ][ ][ ]	[↑↓]
Ti	$[Ar]3d^24s^2$	[↑][↑][ ][ ][ ]	[↑↓]
V	$[Ar]3d^34s^2$	[↑][↑][↑][ ][ ]	[↑↓]
Cr*	$[Ar]3d^54s^1$	[↑][↑][↑][↑][↑]	[↑]
Mn	$[Ar]3d^54s^2$	[↑][↑][↑][↑][↑]	[↑↓]
Fe	$[Ar]3d^64s^2$	[↑↓][↑][↑][↑][↑]	[↑↓]
Co	$[Ar]3d^74s^2$	[↑↓][↑↓][↑][↑][↑]	[↑↓]
Ni	$[Ar]3d^84s^2$	[↑↓][↑↓][↑↓][↑][↑]	[↑↓]
Cu*	$[Ar]3d^{10}4s^1$	[↑↓][↑↓][↑↓][↑↓][↑↓]	[↑]
Zn	$[Ar]3d^{10}4s^2$	[↑↓][↑↓][↑↓][↑↓][↑↓]	[↑↓]

*These configurations do not follow the "$n + \ell$" rule.

are first assigned to *f* orbitals. This means the configuration of cerium is $[Xe]4f^1 5d^1 6s^2$. Moving across the lanthanide series, the pattern continues with some variation, with 14 electrons being assigned to the seven 5*f* orbitals by lutetium, Lu ($[Xe]4f^{14} 5d^1 6s^2$).

When you complete this section, you should be able to depict the electron configuration of any element in the *s* and *p* blocks, using the periodic table as a guide. Regarding the prediction of electron configurations for atoms of elements in the *d* and *f* blocks, you have seen that *minor* differences can occur between actual (Table 8.2) and predicted configurations. However, chemists are often more concerned with the chemistry of the ions formed by the elements, and ion configurations show no such "anomalies."

EXAMPLE 8.4 *Electron Configurations of the Transition Elements*

Using the spectroscopic and noble gas notations, give electron configurations for technetium (Tc) and osmium (Os). Base your answer on the positions of the elements in the periodic table. That is, for each element, find the preceding noble gas and then note the number of *s*, *d*, and *f* electrons that lead from the noble gas to the element.

Solution Proceeding along the periodic table, we come to the noble gas krypton, Kr, at the end of the $n = 4$ row before arriving at Tc (element 43) in the fifth period. Therefore, following the 36 electrons of Kr, 7 electrons remain to be assigned. According to the periodic table, two of these seven electrons are in the 5*s* orbital, and the remaining five are in 4*d* orbitals. Therefore, the technetium configuration is $[Kr]4d^5 5s^2$.

Osmium is a sixth period element, one that follows the lanthanide series of elements. After the last element in the $n = 5$ period, xenon, 22 elements occur, leading to Os. Of these, 2 are of the 6*s* type (Cs and Ba), 14 are of the 4*f* type (Ce through Lu), and the remaining 6 are of the 5*d* type (La through Os). Thus, the osmium configuration is $[Xe]4f^{14} 5d^6 6s^2$.

EXERCISE 8.3 *Electron Configurations*

Using a periodic table and without looking at Table 8.2, write electron configurations for the following elements: (a) P, (b) Zn, (c) Zr, (d) In, (e) Pb, and (f) U. Use the spectroscopic and noble gas notations. When you have finished, check your answers with Table 8.2. ■

8.5 ELECTRON CONFIGURATIONS OF IONS

A great deal of the chemistry of the elements is that of their ions. To form a cation from a neutral atom, the general rule is that one or more electrons are removed from the electron shell of highest *n*. If there is a choice of subshell within the *n*th shell, the electron or electrons of maximum ℓ are removed. These are the valence electrons of the atom. Thus, a sodium ion is formed by removing the $3s^1$ electron from the Na atom,

$$Na[1s^1 2s^2 2p^6 3s^1] \longrightarrow Na^+[1s^1 2s^2 2p^6] + e^-$$

and Al^{3+} is formed by removing three $n = 3$ electrons from an aluminum atom.

$$Al[1s^1 2s^2 2p^6 3s^2 3p^1] \longrightarrow Al^{3+}[1s^1 2s^2 2p^6] + 3e^-$$

The same general rule applies to transition metal atoms. This means that the titanium(II) cation has the configuration $[Ar]3d^2$, for example,

$$Ti[Ar]3d^2 4s^2 \longrightarrow Ti^{2+}[Ar]3d^2 + 2e^-$$

and the iron(II) and iron(III) cations have the configuration $[Ar]3d^6$ and $[Ar]3d^5$, respectively.

All common transition metal cations have electron configurations of the general type [noble gas core]$(n - 1)d^x$. It is very important to remember this as the chemical and physical properties of transition metal cations are determined by the presence of electrons in d orbitals.

Atoms and ions with unpaired electrons are paramagnetic, that is, they are capable of being attracted to a magnetic field (Section 8.1). Paramagnetism is important here because it provides experimental evidence that transition metal ions with charges of 2+ or greater have no ns electrons. For example, the Fe^{2+} ion is paramagnetic to the extent of 4 unpaired electrons, and the Fe^{3+} ion has five unpaired electrons. If three $3d$ electrons had been removed instead to form Fe^{3+}, the ion would still be paramagnetic but only to the extent of three unpaired electrons.

EXAMPLE 8.5 *Configurations of Transition Metal Ions*

Give the electron configuration for copper, Cu, and for its 1+ and 2+ ions. Are these ions paramagnetic? If so, how many unpaired electrons does each have?

Solution As illustrated in Table 8.4, copper has only one electron in the $4s$ orbital and ten electrons in $3d$ orbitals.

Cu: $[Ar]3d^{10}4s^1$

[↑↓]	[↑↓]	[↑↓]	[↑↓]	[↑↓]	[↑]
3*d*	3*d*	3*d*	3*d*	3*d*	4*s*

↓ −e (the 4*s* electron)

Cu^+: $[Ar]3d^{10}$

[↑↓]	[↑↓]	[↑↓]	[↑↓]	[↑↓]	[ ]
3*d*	3*d*	3*d*	3*d*	3*d*	4*s*

↓ $-e^-$ (one of a pair in a 3*d* orbital)

Cu^{2+}: $[Ar]3d^9$

[↑↓]	[↑↓]	[↑↓]	[↑↓]	[↑]	[ ]
3*d*	3*d*	3*d*	3*d*	3*d*	4*s*

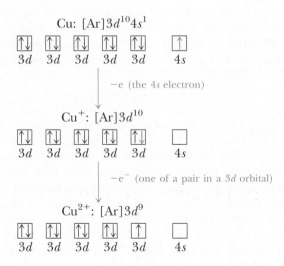

Copper(II) ions have one unpaired electron and so should be paramagnetic. In contrast, Cu^+ has no unpaired electrons, so the ion and its compounds are diamagnetic.

EXERCISE 8.4 ***Metal Ion Configurations***

Depict electron configurations for V^{2+}, V^{3+}, and Co^{3+}. Use orbital box diagrams and the noble gas notation. Are any of the ions paramagnetic? If so, give the number of unpaired electrons. ■

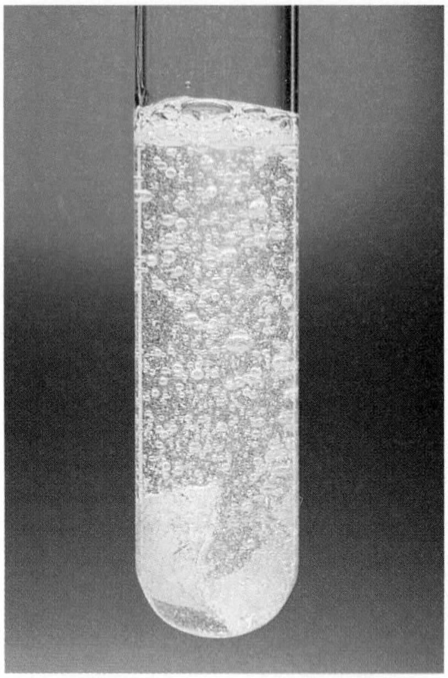

The reaction of the Group 2A metal calcium with water to produce the metal hydroxide ($Ca(OH)_2$) and H_2 gas. In the product, calcium is in the form of the Ca^{2+} ion. This reaction, and its products, are predicted by a knowledge of the positions of the elements in the periodic table.
(C. D. Winters)

8.6 ATOMIC PROPERTIES AND PERIODIC TRENDS

Around 1870, Dmitri Mendeleev (1834–1907) organized the first periodic table by studying the physical and chemical properties of the elements (Section 2.6). Although modified several times during the past 120 years, the periodic table still reflects Mendeleev's original understanding of chemical properties. We have used the table in this chapter as a guide to atomic and ionic electron configurations.

In Chapter 2 we described the division of the table into metallic elements, nonmetals, and a third category of elements called metalloids that share properties of both metals and nonmetals. Within these broad classes are groups of elements, each having a characteristic chemistry, for example, the alkali metals (Group 1A) and alkaline earth metals (Group 2A). In both cases, these metals react with water to produce H_2 and a metal hydroxide when exposed to water.

Group 1A $\quad Na(s) + H_2O(\ell) \longrightarrow NaOH(aq) + \frac{1}{2}\,H_2(g)$

Group 2A $\quad Ca(s) + 2\,H_2O(\ell) \longrightarrow Ca(OH)_2(aq) + H_2(g)$

Except for $Be(OH)_2$, the Group 1A and 2A metal hydroxides dissolve in water to give basic or alkaline solutions, a reaction so characteristic of these elements that this is the source of the group names. In all their chemical activity, these metals typically lose electrons to form 1+ (Group 1A) or 2+ (Group 2A) cations.

An objective of this section is to begin to show how atomic electron configurations are related to some physical properties of the elements and why those properties change in a reasonably predictable manner when moving down groups and across periods. Based on this knowledge, you should eventually be able to organize and predict new chemical and physical properties of elements and their compounds.

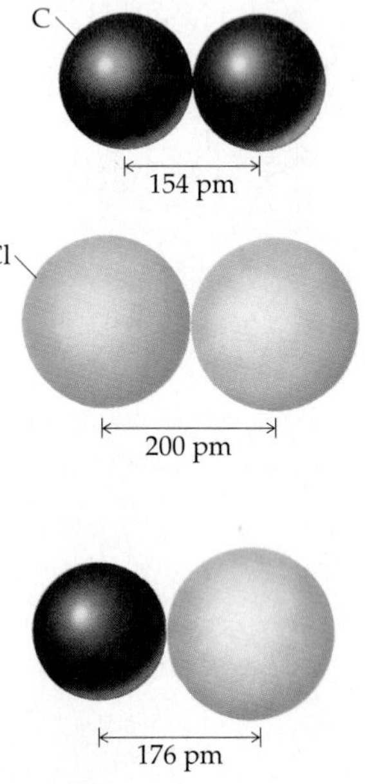

The sum of the atomic radii of C and Cl provides a good estimate of the C—Cl distance in a molecule having such a bond.

Atomic Size

Atoms are not like billiard balls. An electron orbital has no sharp boundary beyond which the electron never strays. How then can we define the size or radius of an atom? For atoms that form simple diatomic molecules, such as Cl_2, the atomic radius can be defined experimentally by dividing the distance between the centers of the two atoms by two. In the Cl_2 molecule, the distance from the center of one atom to the center of the other is 200 pm, which gives a Cl radius of 100 pm. Similarly, the C—C distance in diamond is 154 pm, so the radius of the carbon atom is 77 pm. To test these estimates, we can add them together to estimate the distance between Cl and C in CCl_4. The predicted distance of 177 pm agrees well with the experimentally measured C—Cl distance of 176 pm.

This approach can then be extended to other atomic radii. For example, the radii of O, C, and S can be estimated by measuring the O—H, C—Cl, and H—S

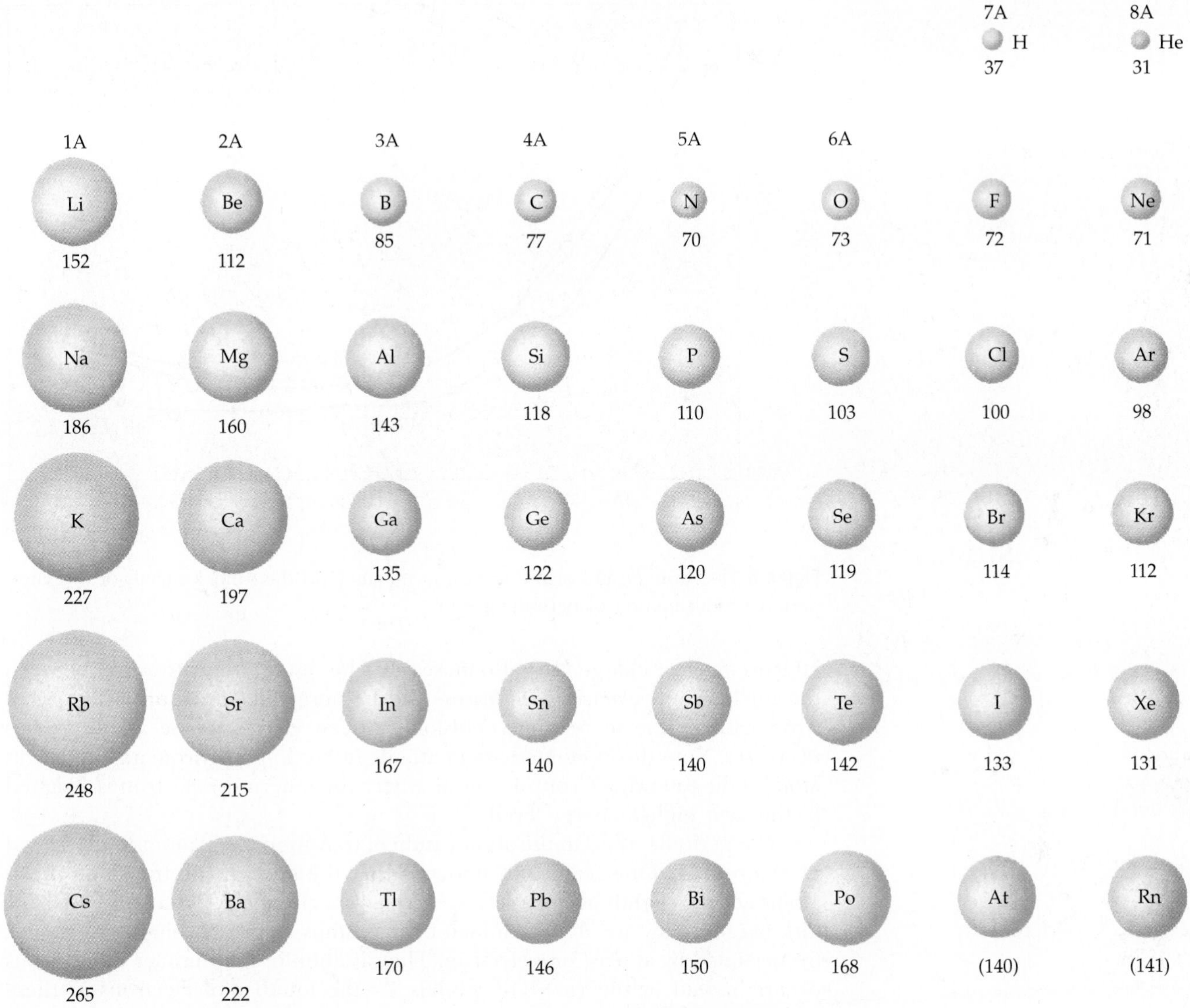

Figure 8.10 Atomic radii in picometers (1 pm = 1×10^{-12} m) for the main group elements.

distances in H_2O, CCl_4, and H_2S and then subtracting the H and Cl radii found from H_2 and Cl_2. Using this and other techniques, a reasonable set of atomic radii for main-group elements has been assembled (Figure 8.10).

For the main group elements, atomic radii increase going down a group in the periodic table and decrease going across a period. These trends can be rationalized as follows: (1) In going from the top to the bottom of a group in the periodic table, electrons are assigned to orbitals that are successively farther from the nucleus. As a result the atomic radii increase. (2) For a given period, the principal quantum number n of the valence orbitals stays the same. This means the radius of the orbitals to which the electrons are assigned is expected to remain approximately constant. As the atomic number increases, however, it is reasonable that the

When an atom is part of a molecule, the measured radius is the covalent radius. The radii in Figure 8.10 for some of the nonmetals are covalent radii.

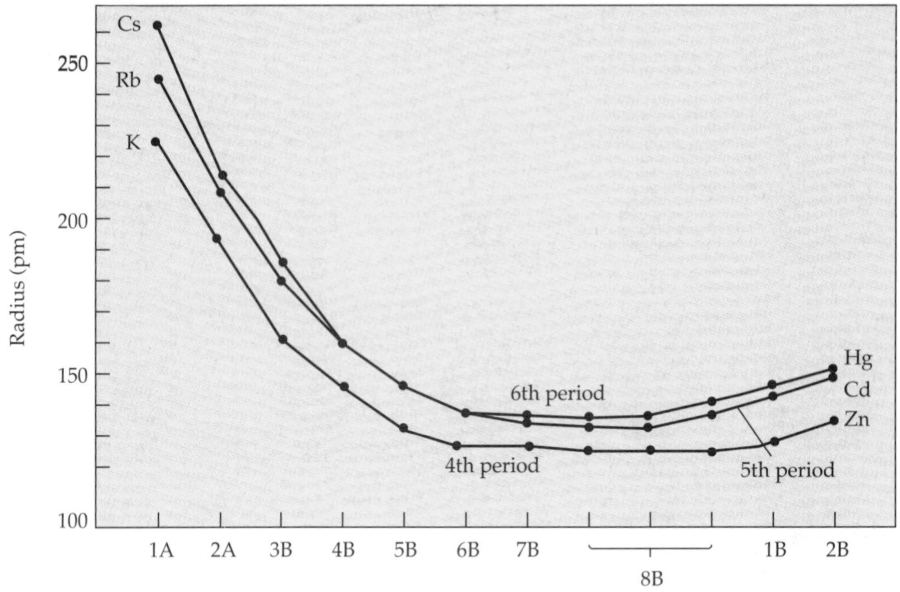

Figure 8.11 Atomic radii of the transition metals (and the *s*-block metals of the same periods) as a function of periodic group.

effective nuclear charge (Z^*) also increases. The result is that attraction between the nucleus and electrons increases, and because this attraction is somewhat stronger than the increasing repulsion between electrons, the atomic radius decreases. Note the large increase in atomic radius in going from any noble gas atom to the following Group 1A atom, where the outermost electron is assigned to the next higher energy level.

The periodic trend in the atomic radii of transition metal atoms is illustrated in Figure 8.11. One feature to notice is that the sizes of the transition metal atoms decline slightly across a series, especially beginning at Group 5B (V, Nb, or Ta), because they are all determined by the radius of an *ns* orbital ($n = 4$, 5, or 6), occupied by at least one electron. The variation in the number of electrons occurs instead in the $(n - 1)d$ orbitals. As the number of electrons in these $(n - 1)d$ orbitals increases, they increasingly repel the *ns* electrons, partly compensating for the increased nuclear charge across the periods. Consequently, the *ns* electrons experience only a slightly increasing nuclear attraction, and there is only a slight decline in radii until the small rise at Groups 1B and 2B due to the continually increasing electron-electron repulsions as the *d* subshell is filled.

Only a small decrease occurs in the size of the transition metals (and in the radii of their 2+ ions, for example). This has an important effect on their chemistry; the transition elements are more alike in their properties than are the main group elements. For example, the nearly identical radii of fifth- and sixth-period transition elements lead to difficult problems of metal recovery, which affects their price. The metals Ru, Os, Rh, Ir, Pd, and Pt are called the "platinum group metals" because they occur together in nature. Their radii and chemistry are so similar that their minerals are similar and are found in the same geologic zones.

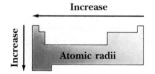

General trends in atomic radii of *s*- and *p*-block elements with position in periodic table.

EXERCISE 8.5 ***Periodic Trends in Atomic Radii***

Place the three elements Al, C, and Si in order of increasing atomic radius. ■

EXERCISE 8.6 *Estimating Atom-Atom Distances*

1. Using Figure 8.10, estimate the H—O and H—S distances in H_2O and H_2S, respectively.
2. If the interatomic distance in Br_2 is 228 pm, what is the radius of Br? Using this estimate, and that for Cl shown previously, estimate the distance between atoms in BrCl. ■

Ionization Energy

Ionization energy, the energy required to remove an electron from an atom in the gas phase, was introduced in Section 8.3 as one type of experimental evidence supporting the shell structure of the atom.

$$\text{Atom in ground state(g)} \longrightarrow \text{Atom}^+\text{(g)} + e^-$$
$$\Delta E \equiv \text{ionization energy, } IE$$

Recall from the discussion of the H atom in Section 7.3 that the process of ionization involves moving an electron from a given electron shell to a position outside the atom, that is, to n = infinity. Therefore, energy is *always* required to overcome the attraction of the nuclear charge. In accord with thermodynamic convention, the sign of the ionization energy is always positive.

Each atom (except H) can have a series of ionization energies, because more than one electron can always be removed. For example, the first three ionization energies of magnesium are

$$\underset{1s^22s^22p^63s^2}{\text{Mg(g)}} \longrightarrow \underset{1s^22s^22p^63s^1}{\text{Mg}^+\text{(g)}} + e^- \qquad IE(1) = 738 \text{ kJ/mol}$$

$$\underset{1s^22s^22p^63s^1}{\text{Mg}^+\text{(g)}} \longrightarrow \underset{1s^22s^22p^63s^0}{\text{Mg}^{2+}\text{(g)}} + e^- \qquad IE(2) = 1451 \text{ kJ/mol}$$

$$\underset{1s^22s^22p^6}{\text{Mg}^{2+}\text{(g)}} \longrightarrow \underset{1s^22s^22p^5}{\text{Mg}^{3+}\text{(g)}} + e^- \qquad IE(3) = 7733 \text{ kJ/mol}$$

Notice that removing each subsequent electron requires more and more energy, and the jump from the second [$IE(2)$] to the third [$IE(3)$] ionization energy is particularly large. Removing an electron from an atom increases the attractive force between the positively charged nucleus and the remaining electrons, and the ionization energy increases. In each ionization step, the remaining electrons of the product ion have a lower or more negative energy due to the increased attractive forces, and more energy is required to remove additional electrons.

The increase in energy required becomes especially great when removing an electron from a shell of lower n. In the preceding magnesium series, the outer electron in the second step has the same n and ℓ as that removed in the first step, so the second ionization energy [$IE(2)$] is larger simply because of increased attractive forces. In the third step, the outer electron is a $2p$ electron with a smaller n than the electron removed in the second step; we have dipped into a lower energy electron shell, and the ionization energy [$IE(3)$] increases greatly.

For main group (s- and p-block) elements, *first ionization energies generally increase across a period and decrease down a group* (Figures 8.12 and 8.13). The trend *across a period* is rationalized by assuming that effective nuclear charge, Z*, increases with increasing atomic number. Not only does this mean that the atomic radius decreases, but the energy required to remove an electron increases.

The very great difference in the second and third ionization energies for Mg is excellent experimental evidence for the existence of electronic shells in atoms. See Figure 8.5. Here the large ratio of (second IE/first IE) for Li, Na, and K indicates that the second electron is removed from the $n - 1$ shell.

The increase in size down a group overcomes the increase in Z* and leads to a decrease in IE.

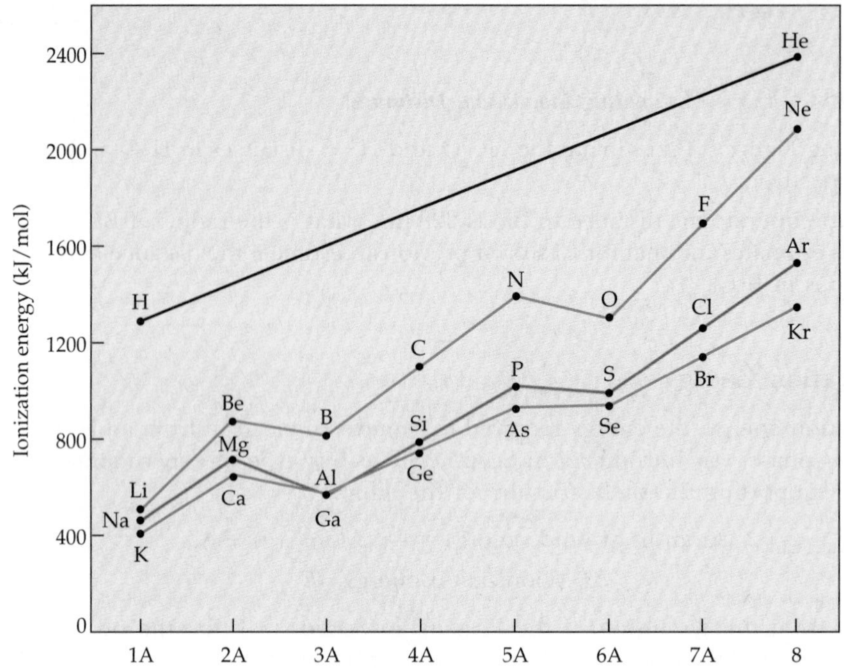

Figure 8.12 First ionization energies of the main group elements of the first four periods.

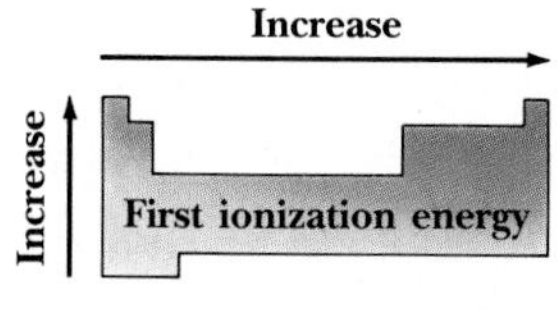

General trends in first ionization energies of A-group elements.

The trend in ionization energy values across a given period is not smooth, particularly in the second period. The reason for this is that the value of ℓ changes on going from *s*-block to *p*-block elements, from Be to B, for example. The $2p$ electrons are slightly higher in energy than the $2s$ electrons, and so the ionization energies decline.

On leaving B and moving on to C and then N, the effective nuclear charge increases, which again means an increase in ionization energy. The dip to lower ionization energy on passing from Group 5A to Group 6A is, however, especially noticeable for N and O. No change occurs in either n or ℓ, but electron-electron repulsions increase for the following reason. In Groups 3A–5A, electrons are assigned to separate p orbitals (p_x, p_y, and p_z). Beginning in Group 6A, however, two electrons are assigned to the same p orbital. Thus, beginning with this group

1A (1)	2A (2)	3B (3)	4B (4)	5B (5)	6B (6)	7B (7)	8B (8,9,10)			1B (11)	2B (12)	3A (13)	4A (14)	5A (15)	6A (16)	7A (17)	8 (18)
H 1312																	He 2371
Li 520	Be 899											B 801	C 1086	N 1402	O 1314	F 1681	Ne 2081
Na 496	Mg 738											Al 578	Si 786	P 1012	S 1000	Cl 1251	Ar 1521
K 419	Ca 599	Sc 631	Ti 658	V 650	Cr 652	Mn 717	Fe 759	Co 758	Ni 757	Cu 745	Zn 906	Ga 579	Ge 762	As 947	Se 941	Br 1140	Kr 1351
Rb 403	Sr 550	Y 617	Zr 661	Nb 664	Mo 685	Tc 702	Ru 711	Rh 720	Pd 804	Ag 731	Cd 868	In 558	Sn 709	Sb 834	Te 869	I 1008	Xe 1170
Cs 377	Ba 503	La 538	Hf 681	Ta 761	W 770	Re 760	Os 840	Ir 880	Pt 870	Au 890	Hg 1007	Tl 589	Pb 715	Bi 703	Po 812	At 890	Rn 1037

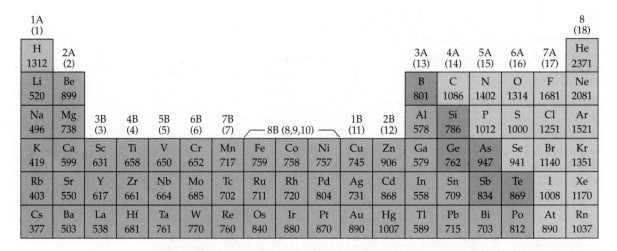

Figure 8.13 First ionization energies of the elements (kJ/mol).

TABLE 8.5 Electron Affinity Values for Some Elements (kJ/mol)*

H						
−72.77						
Li	Be	B	C	N	O	F
−59.63	0[†]	−26.7	−121.85	0	−140.98	−328.0
Na	Mg	Al	Si	P	S	Cl
−52.87	0	−42.6	−133.6	−72.07	−200.41	−349.0
K	Ca	Ga	Ge	As	Se	Br
−48.39	0	−30	−120	−78	−194.97	−324.7
Rb	Sr	In	Sn	Sb	Te	I
−46.89	0	−30	−120	−103	−190.16	−295.16
Cs	Ba	Tl	Pb	Bi	Po	At
−45.51	0	−20	−35.1	−91.3	−180	−270

*Data taken from H. Hotop and W. C. Lineberger: *Journal of Physical Chemistry, Reference Data,* Vol. 14, p. 731, 1985. (This paper also includes data for the transition metals.) Some values are known to more than two decimal places.

[†]Elements with an electron affinity of zero indicate that a stable anion A^- of the element does not exist in the gas phase.

the fourth *p* electron shares an orbital with another electron and thus experiences greater repulsion than it would if it had been assigned to an orbital of its own. Oxygen is an example.

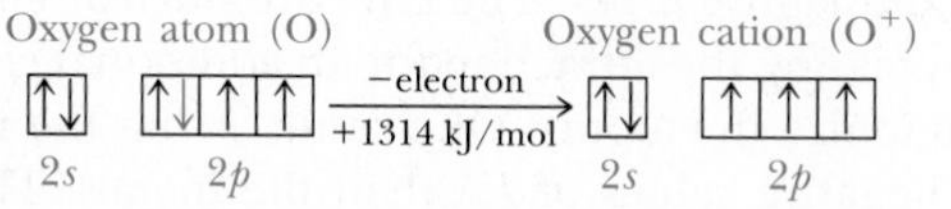

The greater repulsion experienced by the fourth 2*p* electron makes it easier to remove, thus giving each remaining *p* electron an orbital of its own.

Electron Affinity

Some atoms have an affinity, or "liking," for electrons and can acquire one or more electrons to form a negative ion. A measure of this **electron affinity,** *EA*, of an atom is the energy change occurring when an atom in the gas phase acquires an electron (Table 8.5).*

$$A(g) + e^-(g) \longrightarrow A^-(g) \qquad \Delta E \equiv \text{electron affinity, } EA$$

*The data in Table 8.5 are taken from H. Hotop and W. C. Lineberger: *Journal of Physical Chemistry, Reference Data,* Vol. 14, p. 731, 1985. *EA* is defined there as the difference between the energies of the ground state of $A(E_{gs})$ and its negative ion, $A^-(E_{ion})$: $EA(A) = E_{gs}(A) - E_{ion}(A^-)$. The quantity $EA(A)$ is positive if the stable ion A^- exists. The quantity we call electron affinity in this textbook is more properly called the *electron-gain enthalpy,* which is negative when a stable ion is formed. However, common usage equates the terms "electron affinity" and "electron-gain enthalpy."

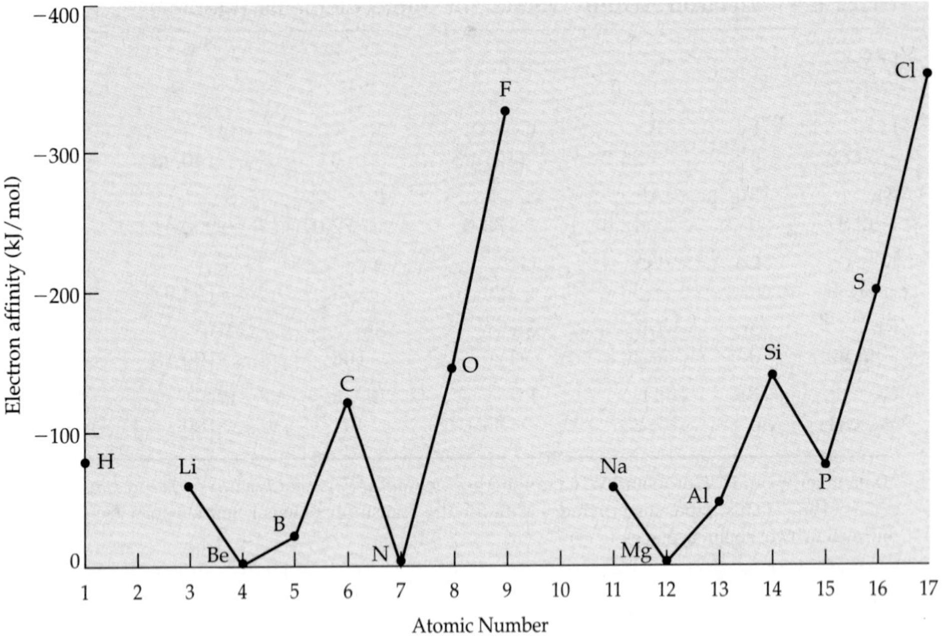

Figure 8.14 A plot of electron affinity (*EA*) against atomic number; the greater the affinity of an atom for an electron, the more negative the value of *EA*. The value of *EA* generally becomes more negative across a period.

For example, the electron affinity of fluorine is −328 kJ/mol because the element forms a stable anion, F^-. On the other hand, a stable beryllium anion does not exist, so a value of 0 is given in Table 8.5.

Periodic trends in electron affinity are closely related to those for ionization energy. An element with a high ionization energy generally has a high affinity for an electron. Thus, the values of *EA* become more negative on moving across a period (Table 8.5 and Figure 8.14). The effective nuclear charge of the atoms is increasing, thus increasing the attraction for an additional electron (and making it more difficult to ionize the atom). One result is that the nonmetals generally have much more negative values of *EA* than the metals. This of course agrees with our chemical experience, which tells us that metals generally do not form negative ions. On the other hand, elements at the right side of the periodic table often form anions (for example, O^{2-}, S^{2-}, F^-, and Cl^-).

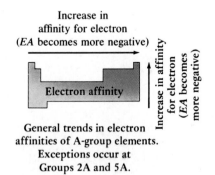

Figure 8.14 shows there are exceptions to the general trend in electron affinity values. For example, a beryllium atom has no affinity for an electron because the added electron in a Be^- ion must be assigned to a higher energy subshell (2*p*) than the valence electrons (2*s*) (see Figure 8.5). Nitrogen atoms also have no affinity for electrons. Here an electron pair must be formed when an N atom acquires an electron. Significant electron-electron repulsions occur in an N^- ion, making the ion much less stable. The increase in Z^* on going from carbon to nitrogen cannot overcome the effect of these electron-electron repulsions.

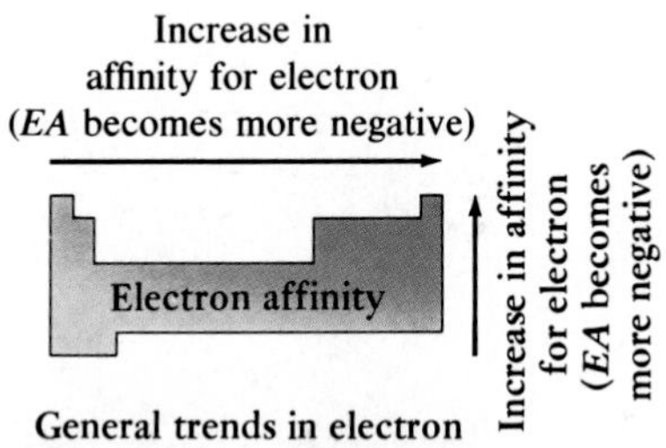

General trends in electron affinities of A-group elements. Exceptions occur at Groups 2A and 5A.

On descending a group of the periodic table, the affinity for an electron generally declines. Electrons are added farther and farther from the nucleus, so the attractive force between the nucleus and electrons decreases. Table 8.5 shows that this is the case for Cl, Br, and I or P, As, and Sb, for example. However, the affinity of the F atom for an electron is lower than that of chlorine (*EA* for F is less negative than *EA* for Cl); the same phenomenon is observed in Groups 3A through 6A as well. One explanation is that there are significant electron-electron repulsions in the F^- ion, which make the ion less stable. Adding an electron to the seven already present in the $n = 2$ shell of the F atom leads to considerable repulsion between electrons. Chlorine has a larger atomic volume than fluorine, so adding an electron does not result in such significant electron-electron repulsions in the Cl^- anion.

A severe drop in *EA* is also seen at P and other Group 5A elements on moving across a period. The *EA* of P is much lower than expected based on the increase observed for Si < S < Cl.

EXAMPLE 8.6 *Periodic Trends*

Compare the three elements C, O, and Si.

1. Place them in order of increasing atomic radius.
2. Which has the largest ionization energy?
3. Which has the most negative electron affinity?

Solution

1. *Atomic size.* Atomic radius declines on moving across a period, so oxygen must have a smaller radius than carbon. However, radius increases down a periodic group. Because C and Si are in the same group (Group 4A), Si must be larger than C. Therefore, in order of increasing size, the elements are O < C < Si.
2. *Ionization energy (IE).* Ionization energy generally increases across a period and decreases down a group; a large decrease in *IE* occurs from the second to the third period elements. Thus, the trend in ionization energies should be Si < C < O.
3. *Electron affinity (EA).* Electron affinity values generally become more negative across a period and less negative down a group. Therefore, the EA for O should be more negative than the EA for C. That is, O ($EA = -141.0$ kJ/mol) has a greater affinity for an electron than does C ($EA = -121.9$ kJ/mol).

It is very interesting to see that the *EA* for Si is more negative (−133.6 kJ/mol) than the *EA* of C. The reason for this is the effect of electron-electron repulsions; such repulsions are larger in the small C^- ion than in the larger Si^- ion.

EXERCISE 8.7 *Periodic Trends*

Compare the three elements Al, C, and Si.

1. Place the three elements in order of increasing atomic radius.
2. Rank the elements in order of increasing ionization energy. (Try to do this without looking at Table 8.5; then compare your estimates with the table.)
3. Which element, Al or Si, is expected to have the less negative electron affinity value? ■

Ion Sizes

Having considered the energies involved in forming positive and negative ions, let us now look at the periodic trends in their sizes.

Figure 8.15 shows clearly that the periodic trends in the sizes of a few common ions are the same as those for neutral atoms: positive or negative ions of the same group increase in size when descending the group. But pause for a mo-

Ion sizes are obtained from measurements of the solid-state structures of ionic compounds. See Chapter 13.

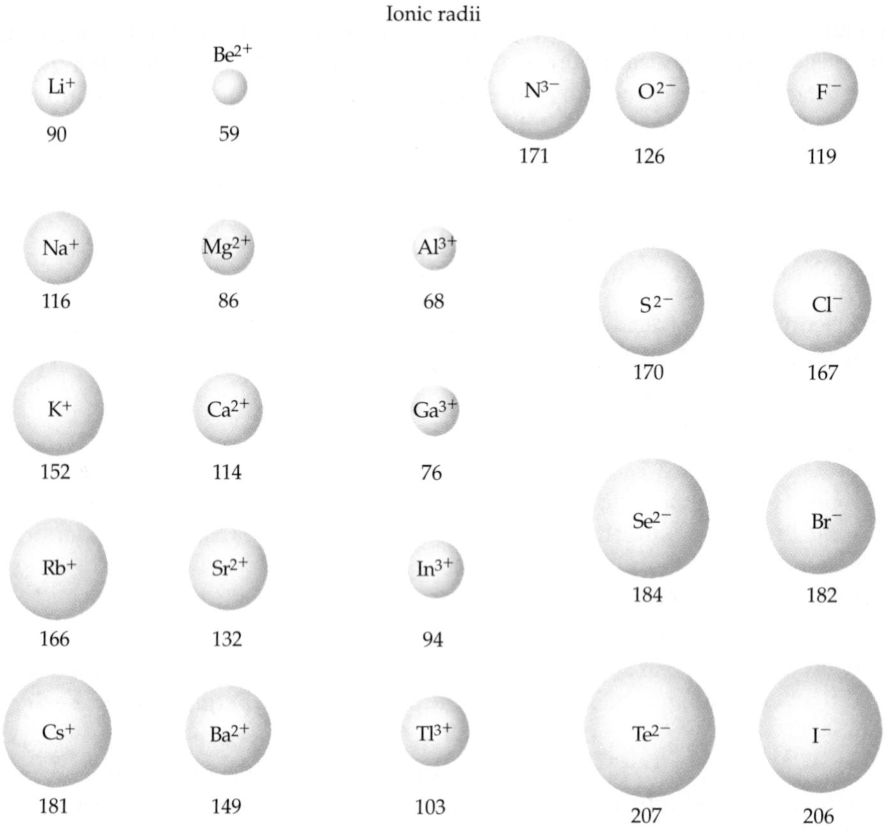

Figure 8.15 Relative sizes of some common ions. Radii are given in picometers (1 pm = 10^{-12} m).

ment and compare Figure 8.15 with Figure 8.10. When an electron is removed from an atom to form a cation, the size shrinks considerably; *the radius of a cation is always smaller than that of the atom from which it is derived.* For example, the radius of Li is 152 pm, whereas that of Li^+ is only 90 pm. This is understandable because when an electron is removed the attractive force of three protons is now exerted on only two electrons, and the remaining electrons contract toward the nucleus. The decrease in ion size is especially great when the last electron to be removed has a greater n (or smaller ℓ) than the new outer electron. This is the case for Li, for which the "old" outer electron was $2s$ and the "new" one is $1s$.

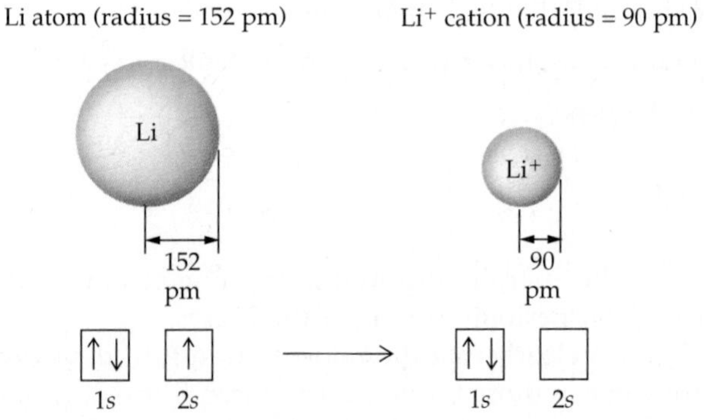

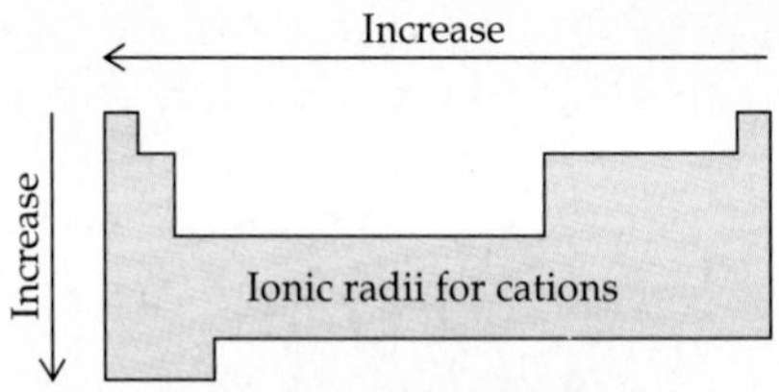

General trends in positive ion radii of A-group elements

The shrinkage will also be great when two or more electrons are removed, as for Al^{3+} where it is over 50%.

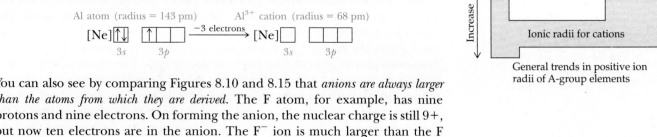

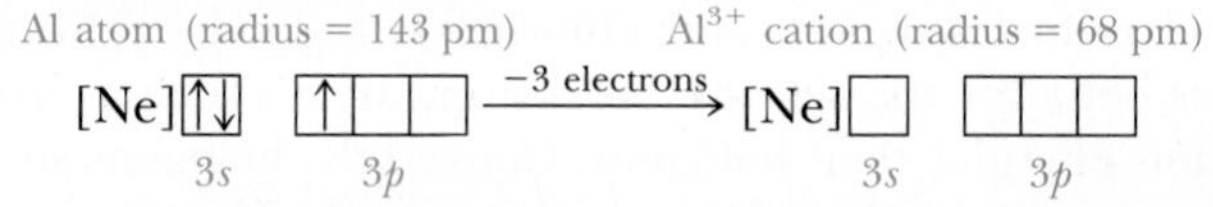

You can also see by comparing Figures 8.10 and 8.15 that *anions are always larger than the atoms from which they are derived.* The F atom, for example, has nine protons and nine electrons. On forming the anion, the nuclear charge is still 9+, but now ten electrons are in the anion. The F^- ion is much larger than the F atom because *increased electron-electron repulsions cause the atom to swell.*

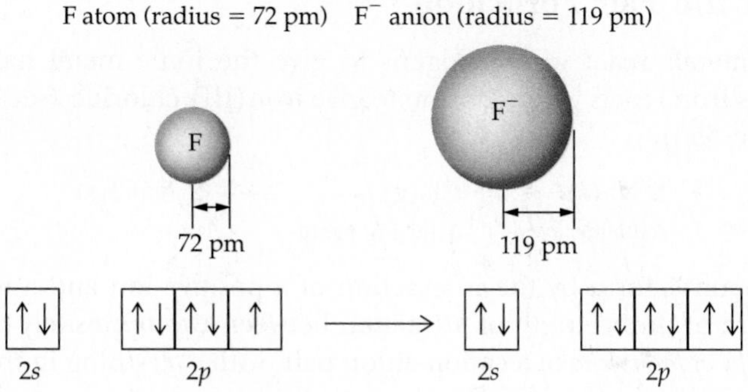

The oxide ion, O^{2-}, is **isoelectronic** with F^-, that is, they both have the same number of electrons (10). It is useful to compare the sizes of these and other isoelectronic ions across the periodic table. For example, consider O^{2-}, F^-, Na^+, and Mg^{2+}.

Ion	O^{2-}	F^-	Na^+	Mg^{2+}
Ionic radius (pm)	126	119	116	86
Number of nuclear protons	8	9	11	12
Number of electrons	10	10	10	10

All these ions have a total of ten electrons. The O^{2-} ion, however, has only 8 protons in its nucleus to attract these electrons, whereas F^- has 9, Na^+ has 11, and Mg^{2+} has 12. As the proton-electron ratio increases in a series of isoelectronic ions, the balance in electron-proton attraction and electron-electron repulsion shifts in favor of attraction, and the ion shrinks. As you can see in Figure 8.15, this is generally true for all isoelectronic series of ions.

EXERCISE 8.8 *Ion Sizes*

What is the trend in sizes of the ions N^{3-}, O^{2-}, and F^-? Briefly explain why this trend exists. ■

8.7 CHEMICAL REACTIONS AND PERIODIC PROPERTIES

In the *Chemical Puzzler* at the beginning of this chapter we wanted to know why iron is a good reducing agent, and chlorine is a good oxidizing agent. The reason for this behavior should now be clearer. Metals such as iron have much lower ionization energies than halogens. Conversely, halogens such as chlorine have much greater affinities for electrons than metals. Therefore, we expect iron to form a cation and chlorine an anion, the two combining to form an ionic compound. *Our objective now is to explore further the reactions of metals and halogens and their products in terms of the atomic properties and periodic trends described in this chapter. In many ways this is a continuation of the discussion of the formation of ions and ionic compounds that was begun in Chapter 3 (Sections 3.3 and 3.4).*

Energy of Ion Pair Formation

The alkali metals react with halogens to give the ionic metal halide (Figure 8.16), just as iron reacts with chlorine to give iron(III) chloride (see the *Chemical Puzzler,* page 356).

$$\underset{\text{reducing agent}}{2\ Na(s)} + \underset{\text{oxidizing agent}}{Cl_2(g)} \longrightarrow 2\ NaCl(s)$$

Ionic compounds form by the interaction of a positive ion and a negative ion. One measure of the strength of attraction between two oppositely charged ions is the *enthalpy of formation* of a cation-anion pair, with everything in the gas phase.

(a) (b)

Figure 8.16 The reaction of sodium and chlorine to give sodium chloride. (*left*) A flask of chlorine gas. (*right*) When sodium is added, the chlorine reacts with the metal to give NaCl. (C. D. Winters)

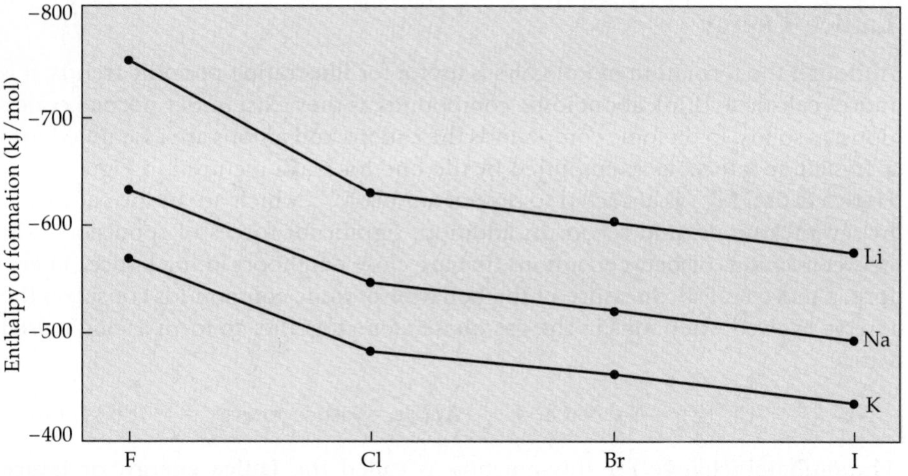

Figure 8.17 Enthalpies of formation of the alkali metal halides MX(g) from the ions M^+ and X^- in the gas phase.

The more negative the value of $\Delta H_{\text{formation}}$, the greater the energy of attraction between the ions of the ionic compound.

$$M^+(g) + X^-(g) \longrightarrow MX(g) \qquad \text{Energy} \equiv \Delta H_{\text{formation}}$$

$$Na^+(g) + Cl^-(g) \longrightarrow NaCl(g) \qquad \Delta H_{\text{formation}} = -552 \text{ kJ/mol}$$

The energy expressed as $\Delta H_{\text{formation}}$ can be estimated with a simple equation (8.1), an equation that is derived from Coulomb's law, the law that describes the energy of attraction between ions of opposite charge (page 119).

$$\Delta H_{\text{formation}} \propto -N\left(\frac{n_+ n_-}{d}\right) \qquad \textbf{(8.1)}$$

Here n_+ is the number of positive charges on the cation and n_- is the number of negative charges for the negative ion; for example, $n_+ = 2$ for Mg^{2+} and $n_- = 3$ for PO_4^{3-}. The variable d is the distance between the ion centers. Finally, we multiply by N, Avogadro's number, so that the energy is determined per mole. The important aspect of this equation is that the energy of attraction between ions of opposite charge depends

- Directly on the magnitude of the ion charges. The greater the ion charges, the greater the energy. That is, the energy is greater for Mg^{2+} and O^{2-} ions than for Na^+ and Cl^- ions.
- Inversely on the distance between the ions. As the distance between ions becomes greater, the energy declines. Because the distance depends on the sizes of the ions involved, the energy decreases as the ion size becomes greater.

We can see the effect of ion size by looking at the periodic trends in ion-ion interactions as given by a plot of $\Delta H_{\text{formation}}$ for the alkali metal halides (Figure 8.17). For example, values of $\Delta H_{\text{formation}}$ for the chlorides are in the order LiCl > NaCl > KCl. The reason is that the alkali metal ion sizes are in the order of $Li^+ < Na^+ < K^+$. The larger the positive ion, the less negative the value of $\Delta H_{\text{formation}}$, in accord with the effect of increasing the value of d in Equation 8.1. Also notice that as the halide ion increases in size ($F^- < Cl^- < Br^- < I^-$) the value of $\Delta H_{\text{formation}}$ becomes less negative.

TABLE 8.6 Lattice Energies of Some Ionic Compounds*

Compound	$\Delta H_{lattice}$ (kJ/mol)
LiF	−1037
LiCl	−852
LiBr	−815
LiI	−761
NaF	−926
NaCl	−786
NaBr	−752
NaI	−702
KF	−821
KCl	−717
KBr	−689
KI	−649
RbCl	−695

*D. Cubicciotti, *Journal of Chemical Physics,* Vol. 31, p. 1646, 1959.

Just as $\Delta H_{formation}$ depends on ion charge, so too does $\Delta H_{lattice}$. For example, the value of $\Delta H_{lattice}$ for $MgBr_2$ (−2390 kJ/mol) is much more negative than that of KBr (−689 kJ/mol) because of the 2+ charge on the magnesium ion (and because there are more ions per mole and therefore more interactions).

Lattice Energy

Although the formation of ion pairs is useful for illustrating periodic trends, it is more realistic to think about ionic compounds as they exist under normal conditions, as solids. In all ionic compounds the cations and anions are assembled into a crystalline lattice, as exemplified by the one for NaCl pictured in Figure 3.13. Here a cation M^{n+} is attracted to several anions X^{n-}, which are in turn attracted by several cations, and so on. In addition, significant forces of repulsion exist between cations or between anions that are close neighbors in the lattice. Therefore, a more realistic measure of the behavior of ionic compounds considers the energy evolved when ions in the gas phase come together to form a solid crystal lattice.

$$Na^+(g) + Cl^-(g) \longrightarrow NaCl(s) \qquad \Delta H_{rxn} = \text{lattice energy} = -786 \text{ kJ/mol}$$

The enthalpy change for this reaction is called the **lattice energy,** or **lattice enthalpy.** The lattice energy is controlled largely by the strength of attraction between the cations and anions, which we have measured with $\Delta H_{formation}$. The lattice energy of a compound cannot be measured directly. Instead, it is derived from other thermodynamic properties or is calculated from an equation derived from Equation 8.1. Our interest here is to examine periodic trends in lattice energies and to use them to explore the reasons some compounds are unknown.

The values in Table 8.6 show the periodic trends in lattice energies for alkali metal halides. Recall that ion sizes are in the order $Li^+ < Na^+ < K^+$ and $F^- < Cl^- < Br^- < I^-$ (Figure 8.15). Just as $\Delta H_{formation}$ depends on ion charge, so too does $\Delta H_{lattice}$. The lattice energy becomes less negative as the size of the halide ion becomes larger; for example, the lattice energy for LiI is less negative than that for LiF. The same trend is observed as the alkali metal cation becomes larger (e.g., the lattice energy for KF is less negative than that for LiF).

Sodium chloride has a standard molar enthalpy of formation of −411.2 kJ.

$$Na(s) + \tfrac{1}{2} Cl_2(g) \longrightarrow NaCl(s) \qquad \Delta H_f^\circ = -411.2 \text{ kJ/mol}$$

In Chapter 6 on thermodynamics, you learned that, if a reaction was considered to occur in a sequence of steps, and if the enthalpy change was known for each step, then the overall enthalpy change could be calculated (Hess's law, Section 6.6). To calculate the enthalpy of formation of solid sodium chloride, we break the overall process into a series of steps such as the following:

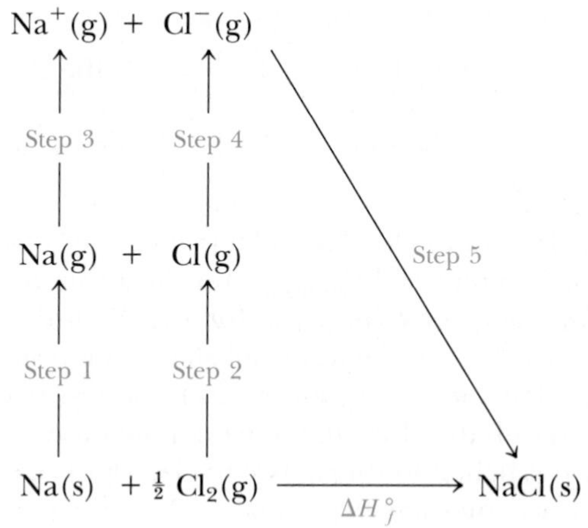

The sum of the enthalpy changes for Steps 1 through 5 is equal to the enthalpy of formation of NaCl(s). This approach to analyzing reaction energies is called a **Born-Haber cycle,** so named for Max Born and Fritz Haber, two German scientists prominent earlier in this century.

The enthalpy change for each step in the preceding reaction diagram is available from experiment or can be found in various tables.

Step 1 Enthalpy of formation of Na(g) = +107.32 kJ/mol (Appendix K)

Step 2 Enthalpy of formation of Cl(g) = +121.68 kJ/mol (Appendix K)

Step 3 Ionization energy for Na(g) = +496 kJ/mol (Figure 8.13)

Step 4 Electron affinity for Cl(g) = −349 kJ/mol (Table 8.5)

Step 5 Formation of NaCl(s) from the ions in the gas phase = −786 kJ/mol (Table 8.6)

$$\Delta H_f^\circ = \Delta H_{\text{Step 1}} + \Delta H_{\text{Step 2}} + \Delta H_{\text{Step 3}} + \Delta H_{\text{Step 4}} + \Delta H_{\text{Step 5}} = -410.\ \text{kJ/mol}$$

Notice in this calculation that the enthalpy changes for Steps 1, 2, and 3 are endothermic. Energy is required to vaporize solid sodium and then to ionize the vapor, as well as to separate two Cl atoms from a Cl_2 molecule. Conversely, the ΔH values for Steps 4 and 5 are both negative, because energy is evolved when a Cl atom acquires an electron and when Na^+ and Cl^- ions come together to form the solid crystalline lattice. Summing these values gives a calculated ΔH_f° (−410. kJ/mol) in very good agreement with the experimental value (−411.2 kJ/mol).

Lattice energies are useful for calculating thermodynamic values. For example, $\Delta H_{\text{lattice}}$ can be calculated from an equation much like 8.1. Thus, if ΔH_f° is known from experiment, and all of the values but one in the thermochemical cycle are known—an electron affinity, for example—this unknown value can be calculated.

EXERCISE 8.9 *Using Lattice Energies*

Calculate the molar enthalpy of formation, ΔH_f°, of solid sodium iodide using the approach outlined in the text. The required data are found in Appendix K and in Figure 8.13 and Tables 8.5 and 8.6. ■

Why Do Compounds Such As $NaCl_2$ and NaNe Not Occur?

Lattice energies are also useful in understanding chemistry, and you will see several more examples in this textbook. For the moment let us try to understand why a compound such as $NaCl_2$, in which sodium is present as the Na^{2+} ion, is quite unlikely to form. The formation of Na^{2+} (with the configuration $1s^22s^22p^5$) would require the loss of a second electron from sodium. (The first electron is lost from the $n = 3$ shell, but the second must be removed from the next lower shell, the $n = 2$ shell.) This requires a substantial amount of energy (4562 kJ/mol), so the total energy for Step 3 in the Born-Haber cycle on page 388 is (496 kJ + 4562 kJ). If we make the reasonable assumption that the lattice energy for $NaCl_2$ is at least double that of NaCl (the increase coming because the cation charge has doubled, and the size of Na^{2+} is less than that of Na^+), we would calculate a very positive value for the enthalpy of formation of $NaCl_2$ (about +3000 kJ/mol). A positive enthalpy change for a reaction means it is reactant-favored (Section 6.1), so it is highly unlikely that $NaCl_2$ would form.

Our analysis of $NaCl_2$ can give you some insight into many questions. One of the most important of these is why metals form cations with charges equal to their group numbers (Figure 3.10). That is, why is the group number of a main group metal related to the number of valence electrons for that element? The answer is that, when an ionic compound is formed, the lattice energy (ΔH for Step 5 in the Born-Haber cycle on page 388) must be large enough to offset the energy required to ionize the metal (ΔH for Step 3). To dip into an inner shell, to remove more than the valence electrons of the elements, would require such a large amount of energy (Step 3) that it could not be offset by the energy evolved in forming the lattice (Step 5). For this reason we can make the general statement that *main group metals form cations with an electron configuration equivalent to that of the nearest noble gas.*

Why are the Group 8 elements called the noble gases? The short answer is they are not very reactive—only xenon is known to form a number of well-defined compounds. But why aren't they reactive? If sodium is such a good reducing agent, why doesn't it reduce neon to Ne^- and form NaNe? Again, we can think about this question in terms of the Born-Haber cycle that was given for NaCl. Put Ne in place of Cl_2. Neon exists in the form of atoms, so $\Delta H_{\text{Step 2}}$ is not required. The electron affinity for Ne ($\Delta H_{\text{Step 4}}$) is expected to be extremely positive, however. The reason for this is that the electron must be placed in the next higher electron shell. Z* for Ne is relatively large (which explains its small size), but the $n = 3$ shell is so much higher in energy than the $n = 2$ shell that neon has no affinity for an electron. The lattice energy of NaNe is not large enough to overcome the large positive electron affinity of Ne, and NaNe does not exist.

The example of NaNe leads to the general statement that *when reacting with a reducing agent, nonmetals generally acquire enough electrons to achieve the electron configuration of the next higher noble gas.* To add additional electrons beyond that would require more energy than is returned by compound formation. This is the reason that chemists often speak of the tendency of atoms to achieve the **noble gas configuration** or of completing an **electron octet;** all noble gases have the configuration ns^2np^6 (plus any $(n-1)d^{10}$ and $(n-2)f^{14}$ electrons, as appropriate). The noble gas configuration, with a complete octet of electrons, represents a stable configuration for both positive and negative ions in their chemical compounds. This is a feature that we shall discuss further in the next chapter.

CHAPTER HIGHLIGHTS

When you have finished studying this chapter, you should be able to

- Classify substances as **paramagnetic** (attracted to a magnetic field; characterized by unpaired electron spins) or **diamagnetic** (not magnetic) (Section 8.1).
- Recognize that each electron in an atom has a different set of the four quantum numbers, n, ℓ, m_ℓ, and m_s, where m_s, the spin quantum number, has values of $+\frac{1}{2}$ or $-\frac{1}{2}$ (Section 8.2).
- Understand that the **Pauli exclusion principle** leads to the conclusion that no atomic orbital can be assigned to more than two electrons and that the two electrons in an orbital must have opposite spins (different values of m_s) (Section 8.2).
- Using the periodic table as a guide, depict electron configurations of the elements and monatomic ions by an **orbital box notation** or a **spectroscopic notation.** (In both cases, configurations can be abbreviated with the **noble gas notation**) (Sections 8.3 and 8.4).

- Recognize that electrons are generally assigned to the subshells of an atom in order of increasing subshell energy. In the H atom the subshell energies increase with increasing *n*, but, in a many-electron atom, the energies depend on both *n* and *ℓ* (Figure 8.6).
- When assigning electrons to atomic orbitals, apply the Pauli exclusion principle and **Hund's rule** (Section 8.3 and 8.4).
- Predict how properties of atoms—size, ionization energy (*IE*), and electron affinity (*EA*)—change on moving down a group or across a period of the periodic table (Section 8.6).

  The general periodic trends for these properties are
  (a) Atomic size: decreases across a period and increases down a group.
  (b) *IE*: increases across a period and decreases down a group.
  (c) *EA*: the affinity for an electron increases across a period and decreases down a group.
- Understand the concept of lattice energy and the role of atomic properties as well as lattice energy in determining the formulas of chemical compounds (Section 8.7).

STUDY QUESTIONS

Review Questions

1. Give the four quantum numbers, specify their allowed values, and tell what property of the electron they describe.
2. What is the Pauli exclusion principle?
3. Using lithium as an example, show the two methods of depicting electron configurations (orbital box diagram and spectroscopic notation).
4. What is Hund's rule? Give an example of its use.
5. What is the noble gas notation? Write an electron configuration using this notation.
6. Name an element of Group 3A. What does the group designation tell you about the electron configuration of the element?
7. Name an element of Group 6B. What does the group designation tell you about the electron configuration of the element?
8. What element is located in the fourth period of Group 4A? What does the element's location tell you about its electron configuration?
9. What element is located in the fifth period of Group 5B? What does the element's location tell you about its electron configuration?
10. What was Mendeleev's major contribution to the development of the concept of periodicity?
11. Tell what happens to atomic size, ionization energy, and electron affinity when proceeding across a period and down a group.
12. Consider the ionic compound MX. How does the enthalpy of formation change if (a) the size of M^+ increases, (b) the size of X^- increases, (c) the electron affinity of X decreases, or (d) the ionization energy of M decreases?

Numerical and Other Questions

Writing Electron Configurations

13. What are the electron configurations for Mg and Cl? Write these configurations using both the spectroscopic notation and orbital box diagrams.
14. What are the electron configurations for Al and S? Write these configurations using both the spectroscopic notation and orbital box diagrams.
15. Using the spectroscopic notation, give the electron configuration of vanadium, V. (The name of the element was derived from vanadis, a Scandinavian goddess.) Compare your answer with Table 8.2.
16. Using the spectroscopic notation, write the electron configurations for atoms of chromium and iron.
17. Depict the electron configuration of a germanium atom using the spectroscopic and noble gas notations.
18. Give the electron configuration for the noble gas element krypton using the spectroscopic notation.
19. Using the spectroscopic and noble gas notations, write electron configurations for atoms of the following elements and then check your answers with Table 8.2.
(a) Strontium, Sr, named for a town in Scotland
(b) Zirconium, Zr. This metal is exceptionally resistant to corrosion and so has important industrial applications. Moon rocks show a surprisingly high zirconium content compared with rocks on earth.
(c) Rhodium, Rh, used in jewelry and in catalysts in industry
(d) Tin, Sn. A metal used in the ancient world. Alloys of tin (solder, bronze, and pewter) are important.
20. Use the noble gas and spectroscopic notations to predict

electron configurations for the following metals of the third transition series.

(a) Tungsten, W. The element finds extensive use in the filaments of electric lamps and television tubes. It has the highest melting point of all the elements.

(b) Platinum, Pt, was used by pre-Columbian Indians in jewelry. It does not oxidize in air, no matter how high the temperature. Therefore, it is used to coat missile nose cones and in jet engine fuel nozzles.

21. The lanthanides, or rare earths, are now only "medium rare." All can be purchased for reasonable prices. Using the noble gas and spectroscopic notations, predict reasonable electron configurations for the following elements.

(a) Europium, Eu, is the most expensive of the rare earth elements.

(b) Ytterbium, Yb, was named for the village of Ytterby in Sweden where a mineral source of the element was found.

22. The actinide americium, Am, is a radioactive element that has found use in home smoke detectors. Depict its electron configuration using the noble gas and spectroscopic notations.

23. Predict reasonable electron configurations for the following elements of the actinide series of elements. Use the noble gas and spectroscopic notations.

(a) Plutonium, Pu, is best known as the fuel for nuclear weapons and a byproduct of nuclear power plant operation.

(b) Einsteinium, Es, was named for the famous physicist Albert Einstein.

24. Among the last elements of the periodic table are those with atomic numbers 104 through 111. The name rutherfordium, Rf, has been proposed for element 104 to honor the physicist Ernest Rutherford (page 70). Depict its electron configuration using the spectroscopic and noble gas notations.

25. Using orbital box diagrams, depict the electron configurations of the following ions: (a) Na^+, (b) Al^{3+}, and (c) Cl^-.

26. Using orbital box diagrams, depict the electron configurations of the following ions: (a) Mg^{2+}, (b) Si^{4+}, and (c) O^{2-}.

27. Using orbital box diagrams and the noble gas notation, depict the electron configurations of: (a) Ti, (b) Ti^{2+}, and (c) Ti^{4+}. Are either of the ions paramagnetic?

28. Using orbital box diagrams and the noble gas notation, depict the electron configurations of: (a) V, (b) V^{2+}, and (c) V^{5+}. Are either of the ions paramagnetic?

29. Element 25 can be found at the bottom of the sea in the form of oxide "nodules."

(a) Depict the electron configuration of this element using the noble gas notation and an orbital box diagram.

(b) Using an orbital box diagram, show the electrons beyond those of the preceding noble gas for the 2+ ion.

(c) Is the 2+ ion paramagnetic?

30. Cobalt commonly exists as 2+ and 3+ ions. Using orbital box diagrams and the noble gas notation, show electron configurations of these ions. Are these ions paramagnetic?

31. Ruthenium, whose compounds are used as catalysts in chemical reactions, has an electron configuration that does not fit the expected pattern (see Table 8.2).

(a) Based on its position in the periodic table, depict the electron configuration of Ru using the noble gas and spectroscopic notations. How does your predicted configuration differ from the actual configuration in Table 8.2?

(b) Using an orbital box notation (and the noble gas notation) depict the electron configuration of the ruthenium(III) ion. Can you arrive at the same ion configuration from either the actual or the predicted configuration of the element?

32. Platinum(II) ion is the central ion in cisplatin, $(NH_3)_2PtCl_2$, a cancer chemotherapy agent. Platinum has an electron configuration that does not fit the expected pattern (see Table 8.2).

(a) Based on its position in the periodic table, depict the electron configuration of Pt using the noble gas and spectroscopic notations. How does your predicted configuration differ from the actual configuration in Table 8.2?

(b) Using an orbital box notation (and the noble gas notation) depict the electron configuration of the platinum(II) ion. Can you arrive at the same ion configuration from either the actual or the predicted configuration of the element?

33. The rare earth elements, or lanthanides, commonly exist as 3+ ions. Using an orbital box diagram and the noble gas notation, show the electron configurations of the following:

(a) Sm and Sm^{3+} (samarium)

(b) Ho and Ho^{3+} (holmium)

34. Using an orbital box diagram (and the noble gas notation) show the electron configuration of uranium and of the uranium(IV) ion. Is either of these paramagnetic?

35. How many unpaired electrons do Co^{3+} and Ti^{2+} ions have? Is either of these paramagnetic?

36. Which of the following atoms or ions is paramagnetic and which are diamagnetic: Al, Al^{3+}, Mg, Co, Co^{3+}.

37. Are any of the 2+ ions of the elements Ti through Zn diamagnetic? Which 2+ ion has the greatest number of unpaired electrons?

38. Two elements in the first transition series (Sc through Zn) have four unpaired electrons in their 2+ ions. What elements fit this description?

Electron Configurations and Quantum Numbers

39. Depict the electron configuration for magnesium using the orbital box and noble gas notations. Give a complete set of four quantum numbers for each of the electrons beyond those of the preceding noble gas.

40. Depict the electron configuration for phosphorus using the orbital box and noble gas notations. Give one possible set of

four quantum numbers for each of the electrons beyond those of the preceding noble gas.

41. Using an orbital box diagram and the noble gas notation, show the electron configuration of titanium. Give one possible set of four quantum numbers for each of the electrons beyond those of the preceding noble gas.

42. Using an orbital box diagram and noble gas notation, show the electron configuration of gallium, Ga. Give a set of quantum numbers for the highest energy electron.

43. What is the maximum number of electrons that can be identified with each of the following sets of quantum numbers? In some cases, the answer may be "none." In such cases, explain why "none" is the correct answer.
(a) $n = 2$ and $\ell = 1$
(b) $n = 3$
(c) $n = 3$ and $\ell = 3$
(d) $n = 4$, $\ell = 1$, and $m_\ell = -1$, and $m_s = -\frac{1}{2}$
(e) $n = 5$, $\ell = 0$, $m_\ell = +1$

44. What is maximum number of electrons that can be associated with the following sets of quantum numbers? In one case, the answer is "none." Explain why this is true.
(a) $n = 4$, $\ell = 3$
(b) $n = 6$, $\ell = 1$, $m_\ell = -1$
(c) $n = 3$, $\ell = 3$, $m_\ell = -3$
(d) $n = 2$, $\ell = 1$, $m_\ell = 1$, $m_s = +\frac{1}{2}$

45. Explain briefly why each of the following is *not* a possible set of quantum numbers for an electron in an oxygen atom (in its ground state). In each case, change the incorrect value (or values) in some way to make the set valid.
(a) $n = 2$, $\ell = 2$, $m_\ell = 0$, $m_s = +\frac{1}{2}$
(b) $n = 2$, $\ell = 1$, $m_\ell = -1$, $m_s = 0$
(c) $n = 3$, $\ell = 1$, $m_\ell = +1$, $m_s = +\frac{1}{2}$

46. Explain briefly why each of the following is not a possible set of quantum numbers for an electron in an silicon atom (in its ground state). In each case, change the incorrect value (or values) in some way to make the set valid.
(a) $n = 4$, $\ell = 2$, $m_\ell = 0$, $m_s = 0$
(b) $n = 3$, $\ell = 1$, $m_\ell = -3$, $m_s = -\frac{1}{2}$
(c) $n = 3$, $\ell = 2$, $m_\ell = -1$, $m_s = +\frac{1}{2}$

Periodic Properties

47. Use the data in Figure 8.10 to estimate E—Cl bond distances when E is a Group 5A element.

48. Estimate the Xe—F bond distance in XeF_2 from the information in Figure 8.10. (Known Xe—F distances are in the range of 190 pm.)

49. Arrange the following elements in order of increasing size: Al, B, C, K, and Na. (Try doing it without looking at Figure 8.10, then check yourself by looking up the necessary atomic radii.)

50. Arrange the following elements in order of increasing size: Ca, Rb, P, Ge, and Sr. (Try doing it without looking at Figure 8.10, then check yourself by looking up the necessary atomic radii.)

51. Select the atom or ion in each pair that has the larger radius.
(a) Cl or Cl^-
(b) Al or N
(c) In or Sn

52. Select the atom or ion in each pair that has the larger radius.
(a) Cs or Rb
(b) O^{2-} or O
(c) Br or As

53. Which of the following groups of elements is arranged correctly in order of increasing ionization energy?
(a) C < Si < Li < Ne
(b) Ne < Si < C < Li
(c) Li < Si < C < Ne
(d) Ne < C < Si < Li

54. Arrange the following atoms in the order of increasing ionization energy: F, Al, P, and Mg.

55. Arrange the following atoms in the order of increasing ionization energy: Li, K, C, and N.

56. Arrange the following atoms in the order of increasing ionization energy: Si, K, As, and Ca.

57. Compare the elements Li, K, C, and N.
(a) Which has the largest atomic radius?
(b) Which has the most negative electron affinity?
(c) Place the elements in order of increasing ionization energy.

58. Compare the elements B, Al, C, and Si.
(a) Which has the most metallic character?
(b) Which has the largest atomic radius?
(c) Which has the most negative electron affinity?
(d) Place the three elements B, Al, and C in order of increasing first ionization energy.

59. Periodic trends. Explain each answer briefly.
(a) Place the following elements in order of increasing ionization energy: F, O, and S.
(b) Which has the largest ionization energy: O, S, or Se?
(c) Which has the most negative electron affinity: Se, Cl, or Br?
(d) Which has the largest radius: O^{2-}, F^- or F?

60. Periodic trends. Explain each answer briefly.
(a) Rank the following in order of increasing atomic radius: O, S, and F.
(b) Which has the largest ionization energy: P, Si, S, or Se?
(c) Place the following in order of increasing radius: Ne, O^{2-}, N^{3-}, F^-.
(d) Place the following in order of increasing ionization energy: Cs, Sr, Ba.

Lattice Energies

61. Using a Born-Haber cycle (page 388), calculate the molar enthalpy of formation of lithium chloride. How well does your calculation agree with the value of ΔH_f° [LiCl(s)] in Appendix K? (In addition to information from Appendix K and Tables 8.5, 8.6, and Figure 8.13, you need to know that the enthalpy of formation of Li(g) is +159.4 kJ/mol.)

62. Using a Born-Haber cycle (page 388), calculate the molar enthalpy of formation of lithium bromide. (In addition to

information from Appendix K and Tables 8.5, 8.6, and Figure 8.13, you need to know that the enthalpy of formation of Li(g) is +159.4 kJ/mol.) Compare your calculated value of ΔH_f° for LiBr(s) with that for LiCl(s) (as calculated in Study Question 61 and as given in Appendix K). Which is the larger of the two? Would you have predicted this difference? Base your answer on an analysis of the information in the Born-Haber cycles for LiCl and LiBr.

63. What is the trend in lattice energies when one uses heavier and heavier halide ions with a given alkali metal cation? Estimate the lattice energy of RbBr using the data in Table 8.6.

64. The molar enthalpies of formation for potassium chloride and potassium iodide are −436.7 kJ/mol and −327.9 kJ/mol, respectively. Based on a Born-Haber cycle, analyze why ΔH_f° for the iodide is less negative than ΔH_f° for the chloride.

General Questions

65. Element 109 (now named meitnerium) was produced in August, 1982, by a team at Germany's Institute for Heavy Ion Research. Depict its electron configuration using the spectroscopic and noble gas notations. Name another element found in the same group as 109.

66. A neutral atom has two electrons with $n = 1$, eight electrons with $n = 2$, eight electrons with $n = 3$, and two electrons with $n = 4$. Assuming this element is in its ground state, supply the following information: (a) atomic number; (b) total number of *s* electrons; (c) total number of *p* electrons; (d) total number of *d* electrons; and whether the element is (e) a metal, metalloid, or nonmetal.

67. Which of the following is *not* an allowable set of quantum numbers? Explain your answer briefly.

	n	ℓ	m_ℓ	m_s
(a)	2	0	0	$-\frac{1}{2}$
(b)	1	1	0	$+\frac{1}{2}$
(c)	2	1	−1	$-\frac{1}{2}$
(d)	4	3	+2	$-\frac{1}{2}$

68. What is the last orbital filled in the fictitious element with atomic number 120?

69. How many complete electron shells are there in element 71?

70. What element has a 2+ ion with the configuration $[Xe]4f^{14}5d^6$?

71. A possible excited state for the H atom has an electron in a 4*p* orbital. List all possible sets of quantum numbers (n, ℓ, m_ℓ, and m_s) for this electron.

72. Name the element corresponding to each characteristic below:
(a) The element with the electron configuration $1s^22s^22p^63s^23p^3$
(b) The element in the alkaline earth group that has the smallest atomic radius
(c) The element in Group 5A that has the largest ionization energy
(d) The element whose 2+ ion has the configuration $[Kr]4d^5$
(e) The element with the most negative electron affinity in Group 7A
(f) The element whose electron configuration is $[Ar]3d^{10}4s^2$

73. Two elements in the second transition series (Y through Cd) have four unpaired electrons in their 3+ ions. What elements fit this description?

74. Answer the questions below about the elements A and B, which have the electron configurations shown.

$$A = [Kr]5s^1 \qquad B = [Ar]3d^{10}4s^24p^4$$

(a) Is element A a metal, nonmetal, or metalloid?
(b) Which element has the greater ionization energy?
(c) Which element has the more negative electron affinity?
(d) Which element has larger atoms?

75. Answer the following questions about the elements with electron configurations below:

$$A = [Ar]4s^2 \qquad B = [Ar]3d^{10}4s^24p^5$$

(a) Is element A a metal, metalloid, or nonmetal?
(b) Is element B a metal, metalloid, or nonmetal?
(c) Which element is expected to have the larger ionization energy?
(d) Which element is the smaller of the two?

76. Which of the following ions are unlikely and why: Cs^+, In^{4+}, Fe^{6+}, Te^{2-}, Sn^{5+}, and I^-?

77. Place the following elements and ions in order of decreasing size: K^+, Cl^-, S^{2-}, and Ca^{2+}.

78. Rank the following in order of increasing ionization energy: Zn, Ca, Ca^{2+}, and Cl^-. Briefly explain your answer.

79. In general, as you move across a periodic table, the electron affinity of the elements becomes more negative. One exception to this trend, however, is the large decrease in the affinity for an electron when going from Group 4A elements to those in Group 5A. Explain why this decrease occurs.

80. Answer each of the following questions:
(a) Of the elements O, S, and F, which has the largest atomic radius?
(b) Which is larger, Cl or Cl^-?
(c) Which should have the largest difference between the first and second ionization energy: Si, Na, P, or Mg?
(d) Which has the largest ionization energy: O, S, or Se?
(e) Which of the following has the largest radius: Cl^-, O^{2-}, N^{3-}, or F^-?

81. Explain why the first ionization energy of Ca is greater than that of K, whereas the second ionization energy of Ca is lower than the second ionization energy of K.

82. The following are isoelectronic species: Cl^-, Ar, and K^+. Rank them in order of increasing (a) size, (b) ionization energy, and (c) affinity for an electron.

83. Compare the elements of Na, B, Al, and C with regard to the following properties:
(a) Which has the largest atomic radius?
(b) Which has the most negative electron affinity?

(c) Place the elements in order of increasing ionization energy.

84. The configuration for an element is given here.

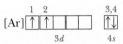

[Ar] [↑↓][↑][↑][↑][↑] (3*d*) [↑↓] (4*s*)

(a) What is the identity of the element with this configuration?
(b) Is a sample of the element paramagnetic or diamagnetic?
(c) How many unpaired electrons does the 3+ ion have?

85. The configuration of an element is given here.

[Ar] [↑ (1)][↑ (2)][ ][ ][ ] (3*d*) [↑↓ (3,4)] (4*s*)

(a) What is the identity of the element?
(b) In what group and period is the element found?
(c) Is the element a nonmetal, a main group element, a transition element, a lanthanide element, or an actinide element?
(d) Is the element diamagnetic or paramagnetic? If paramagnetic, how many unpaired electrons are there?
(e) Write a complete set of quantum numbers for electrons 1, 2, and 4.

Electron	n	ℓ	m_ℓ	m_s
1	______	______	______	______
2	______	______	______	______
4	______	______	______	______

(f) If two electrons are removed to form the 2+ ion, what two electrons are removed? Is the ion diamagnetic or paramagnetic?

86. Using a Born-Haber cycle (page 388), calculate the lattice energy of rubidium fluoride. (Use information in Appendix K, Tables 8.5 and 8.6, and Figure 8.13, and $\Delta H_f^\circ[\text{Rb(g)}] = +88.9$ kJ/mol and $\Delta H_f^\circ[\text{RbF(s)}] = -557.7$ kJ/mol.) Compare your calculated lattice energy for RbF with that for NaF. Which is the more negative of the two? Would you have predicted this difference?

Conceptual Questions

87. Explain why the sizes of atoms change when proceeding across a period of the periodic table.

88. Explain how the ionization energy of atoms changes and why the change occurs when proceeding down a group of the periodic table.

89. Explain why the sizes of transition metal atoms decrease only slightly across a period.

90. Write electron configurations to show the first two ionization processes for potassium. Explain why the second ionization energy is much greater than the first.

91. Predict which of the following elements has the greatest difference between the first and second ionization energy: Si, Na, P, and Mg. Explain your answer.

92. Why is the radius of Li^+ so much smaller than the radius of Li? Why is the radius of F^- so much larger than the radius of F?

93. Which ions in the following list are likely to be formed: K^{2+}, Cs^+, Al^{4+}, F^{2-}, and Se^{2-}? Do any of these ions have a noble gas configuration?

94. The ionization energies for the removal of the first electron in Si, P, S, and Cl are listed below. Briefly rationalize this trend.

Element	**First Ionization Energy (kJ/mol)**
Si	786
P	1012
S	1000
Cl	1251

95. Below is a plot of the density of the elements from Ca through Zn. Using your knowledge of the trends in element sizes on going across the periodic table, explain briefly why the density increases from Ca through Cu.

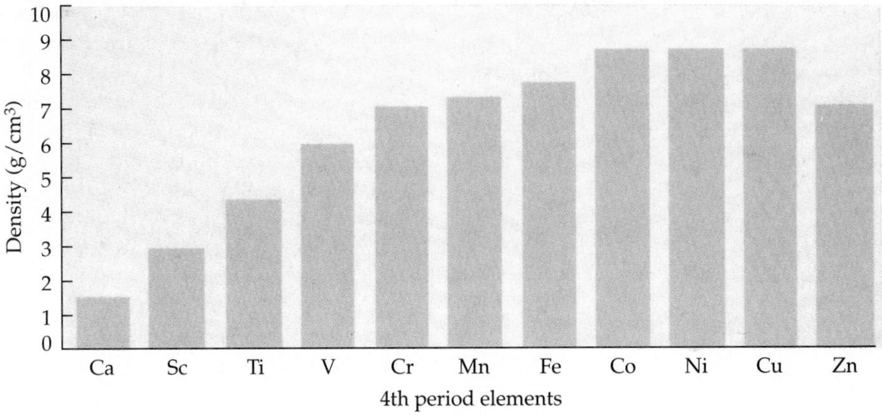

96. What arguments would you use to convince another student in general chemistry that MgO consists of the ions Mg^{2+} and O^{2-} and not the ions Mg^{+} and O^{-}? What experiments could be done to provide some evidence that the correct formulation of magnesium oxide is $Mg^{2+}O^{2-}$?

97. The reaction of cobalt metal with HCl gives $CoCl_2$, and the reaction with nitric acid gives $Co(NO_3)_3$. Using the magnetic behavior of these compounds, describe how to tell that $CoCl_3$ or $Co(NO_3)_2$ are not the reaction products.

98. Molar enthalpies of formation of three alkali metal fluorides are given here.

Metal Fluoride	**ΔH_f° (kJ/mol)**
LiF	−616.0
NaF	−573.6
KF	−567.3

(a) Analyze this trend in enthalpies of formation using a Born-Haber cycle. What feature of the properties of the alkali metals and their halides controls the trend?

(b) Explain why the enthalpies of formation of NaF and KF are very similar, but their lattice energies differ by 105 kJ/mol. $\Delta H_f^\circ[K(g)] = +89.2$ kJ/mol.

99. Give a plausible explanation for the observation that magnesium and chlorine give $MgCl_2$ and not $MgCl_3$.

Summary Question

100. When sulfur dioxide reacts with chlorine, the products are thionyl chloride, $SOCl_2$, and dichlorine oxide, Cl_2O.

$$SO_2(g) + 2\ Cl_2(g) \longrightarrow SOCl_2(g) + Cl_2O(g)$$

(a) Give the electron configuration for an atom of sulfur in the orbital box notation. Do *not* use the noble gas notation.

(b) Using the configuration given in part (a), write a set of quantum numbers for the highest energy electron in a sulfur atom.

(c) What element involved in this reaction (O, S, Cl) should have the smallest ionization energy? The smallest radius?

(d) Which should be smaller, the sulfide ion, S^{2-}, or the sulfur atom, S?

(e) If you want to make 675 g of $SOCl_2$, how many grams of Cl_2 are required?

(f) If you use 10.0 g of SO_2 and 20.0 g of Cl_2, what is the theoretical yield of $SOCl_2$?

(g) ΔH_{rxn}° for the reaction of SO_2 and Cl_2 is +164.6 kJ per mole of $SOCl_2$ produced. Using this and the table shown here, calculate the standard molar enthalpy of formation of $SOCl_2$.

Compound	**ΔH_f° (kJ/mol)**
Cl_2O	80.3
SO_2	−296.8

C H A P T E R 9

Bonding and Molecular Structure: Fundamental Concepts

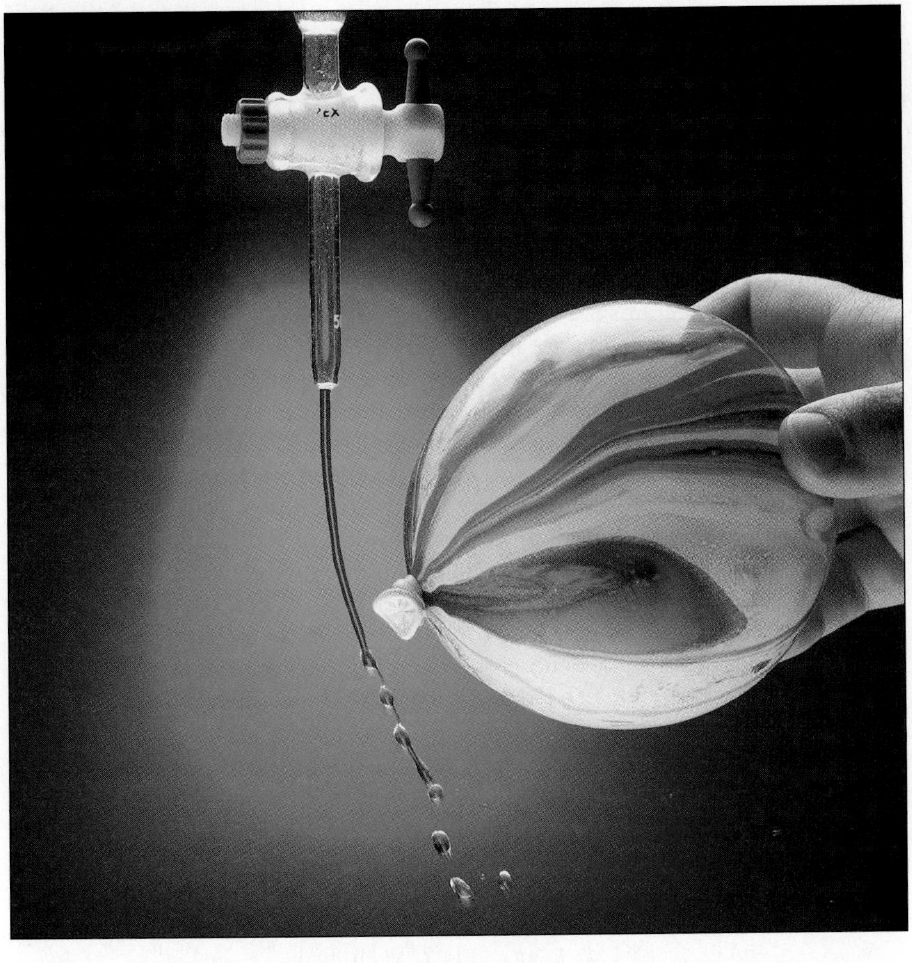

A stream of water is attracted to a balloon bearing a static charge. (C. D. Winters)

Chemical Puzzler

Water is affected by electric charges, and you can do an experiment to demonstrate the effect. You will need a nylon, rubber, or plastic comb, or a plastic knife, fork, or spoon from a fast-food restaurant; a water faucet and sink; and a piece of fur or cloth. Adjust the faucet to produce a thin stream of water (about 1–2 mm in diameter). Hold the comb horizontal to and a few centimeters lower than the end of the faucet. Bring the end of the comb close to (but not into) the stream of water. Does anything happen? Now rub the comb with the fur or cloth or run it through your hair to place a static electric charge on the comb. Immediately bring the comb close to the stream of water. What do you observe now? Finally, why is water affected by an electric charge?

Roald Hoffmann (1937–), the Polish-born American chemist, received the Nobel Prize in chemistry in 1981 for his work on chemical bonding, the force that binds atoms together to form molecules. Hoffmann said that "Chemistry is the science of molecules and their transformations. It is the science not so much of the . . . elements, but of the variety of molecules that can be built from them." This chapter is about two subjects, **molecular structure** and **molecular bonding.** Structure refers to the way atoms are arranged within molecules and polyatomic ions, and bonding defines the forces that hold adjacent atoms in a molecule together. As you will see, structure and bonding are closely interwoven in chemistry.

The answer to the "Chemical Puzzler" for this chapter will be obvious once you know the structure of water and how this confers special properties on water. Understanding the structures of larger molecules provides useful information about their chemical properties. Francis Crick and James D. Watson made one of the great discoveries of this century when they arrived at a satisfactory model for the structure of DNA (page 5). This and similar discoveries have focused the interest of modern chemists and biologists on the study of molecules and how their structures affect their function.

9.1 VALENCE ELECTRONS

The valence electrons of atoms were first introduced in Section 3.3 and described further in Sections 8.4 and 8.7. Because the valence electrons of atoms in a molecule or polyatomic ion are involved in bonding, we first want to examine the relationship between valence electrons and the periodic table.

It is possible to divide the electrons in an atom into two groups, the **valence electrons,** which participate in bonding to other atoms, and the **core electrons,** which do not. For main group elements, the valence electrons are the *s* and *p* electrons in the outermost shell. Core electrons include electrons from inner shells, which have a configuration identical to that of the noble gas preceding the element in the periodic table. In addition, the core electrons include electrons in the filled *d* orbitals for atoms in Groups 3A to 7A of the later periods. For transition elements, the valence electrons are those in the *ns* and $(n - 1)d$ orbitals, and the core electrons are those with a noble gas configuration in the underlying shells. The valence electrons for a few typical elements are shown in the table.

Element	Periodic Group	Total Configuration	Core Electrons	Valence Electrons
Na	1A	$[Ne]3s^1$	$1s^22s^22p^6 = [Ne]$	$3s^1$
Si	4A	$[Ne]3s^23p^2$	$1s^22s^22p^6 = [Ne]$	$3s^23p^2$
As	5A	$[Ar]3d^{10}4s^24p^3$	$1s^22s^22p^63s^23p^63d^{10} = [Ar]3d^{10}$	$4s^24p^3$
Ti	4B	$[Ar]3d^24s^2$	$1s^22s^22p^63s^23p^6 = [Ar]$	$3d^24s^2$

TABLE 9.1 Lewis Dot Symbols for Atoms

1A ns^1	2A ns^2	3A ns^2np^1	4A ns^2np^2	5A ns^2np^3	6A ns^2np^4	7A ns^2np^5	8 ns^2np^6
Li ·	· Be ·	· B ·	· C ·	· N ·	: O ·	: F ·	: Ne :
Na ·	· Mg ·	· Al ·	· Si ·	· P ·	: S ·	: Cl ·	: Ar :

These few examples show that *the number of valence electrons of each main group element is equal to the group number* (See Section 8.7). The fact that every element in a given group has the same number of valence electrons accounts for the similarity of chemical properties among members of the group.

The concept of valence electrons was first introduced by G. N. Lewis early in this century. He assumed that each noble gas atom had a completely filled outermost shell, which he regarded as a stable configuration because of the lack of reactivity of the noble gases. Because all noble gases (except He) have eight valence electrons, this observation is known as the **octet rule.** Lewis used the element's symbol to represent the atomic nucleus together with the core electrons. The valence electrons, represented by dots, are placed around the symbol one at a time until they are used up or until all four sides are occupied; any remaining electrons are paired with the ones already there. Chemists now refer to these pictures as **Lewis dot symbols,** or Lewis symbols, and the ones for the second and third period elements are shown in Table 9.1.

The Lewis symbol emphasizes the ns^2np^6 octet, the electron configuration of noble gases, as an especially stable arrangement. In fact, the bonding behavior of the main group elements can be considered the result of gaining, losing, or sharing valence electrons to achieve the same configuration as the nearest noble gas (Section 8.7). The octet rule is only a guideline, however, to which you will find exceptions. Nonetheless, it does provide a way of predicting the results of the most common reactions.

PORTRAIT OF A SCIENTIST

Gilbert Newton Lewis (1875–1946)

In a paper published in the *Journal of the American Chemical Society* in 1916, G. N. Lewis introduced the theory of the shared electron pair chemical bond. His idea revolutionized chemistry, and it is to honor his contribution that we often refer to "electron dot" structures as Lewis structures. However, he also made major contributions to other fields such as thermodynamics, isotope studies, and the interaction of light with substances. Of particular interest in this text is the extension of his theory of bonding to a generalized theory of acids and bases. This theory, often referred to as the Lewis acid-base theory, is described in Section 17.9.

G. N. Lewis was born in Massachusetts but raised in Nebraska. After earning his B.A. and Ph.D. at Harvard University, he began his academic career. In 1912, he was appointed chairman of the Chemistry Department at the University of California, Berkeley, and remained there the rest of his life. Lewis felt that a chemistry department should both teach and advance fundamental chemistry, and he was not only a productive researcher but also a teacher who profoundly affected his students. Among his ideas was the use of problem sets in teaching, an idea still in use today.

EXERCISE 9.1 ***Valence Electrons***

Give the number of valence electrons for Ca, As, and Br. Draw the Lewis dot symbol for Br. ■

9.2 CHEMICAL BOND FORMATION

Chemists use several theoretical models to describe the formation of a chemical bond. One approach takes into account the attraction between opposite electric charges and the repulsion between like charges (see Section 3.4). Let us see what happens as two hydrogen atoms come together to form a hydrogen molecule. When these atoms are widely separated, there is no interaction between them. If the two atoms approach each other closely enough for their electron clouds to interpenetrate or *overlap,* however, each atom's electrons are attracted to the other atom's nucleus (Figure 9.1). The potential energy of the system decreases, resulting in a net attraction between the two atoms (Figure 9.2). Calculations and experiment show that at 0.074 nm (74 pm) the potential energy reaches a minimum, and the H_2 molecule is most stable. This is the equilibrium *bond distance* in the H_2 molecule. At a smaller atom–atom distance, however, the repulsions between the nuclei of the two atoms, and between electrons of the two atoms, increase, and the potential energy curve rises steeply. The H_2 molecules is less stable when the distance between the atoms is very small.

What we have described is the formation of a **covalent bond** between two H atoms: the *sharing of electrons* between a pair of atoms. If, however, the electrons involved in bonding are strongly displaced toward one atom and away from the other, an **ionic bond** is formed. In the extreme, one or more valence electrons are transferred from one atom (the reducing agent) to another atom (the oxidizing agent), and a cation and anion are formed, respectively.

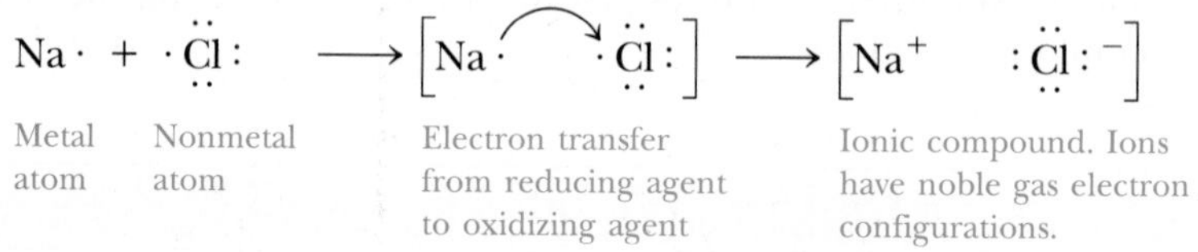

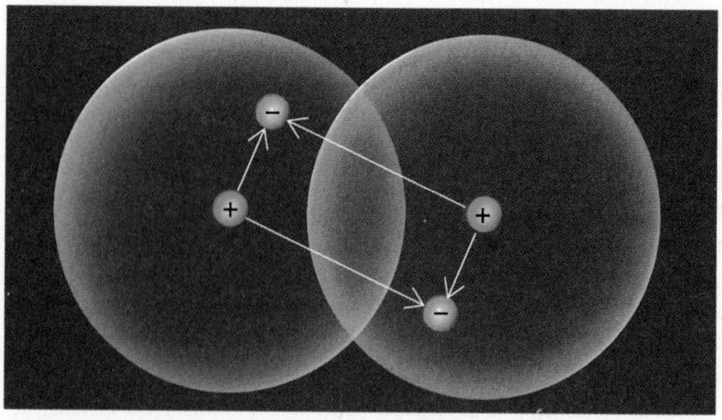

Figure 9.1 The formation of a covalent bond between two H atoms. One pair of electrons (one from each atom) moves into the internuclear region and is attracted to both H atom nuclei. It is this mutual attraction for 2 (or sometimes 4 or 6) electrons by two nuclei that leads to covalent bond formation.

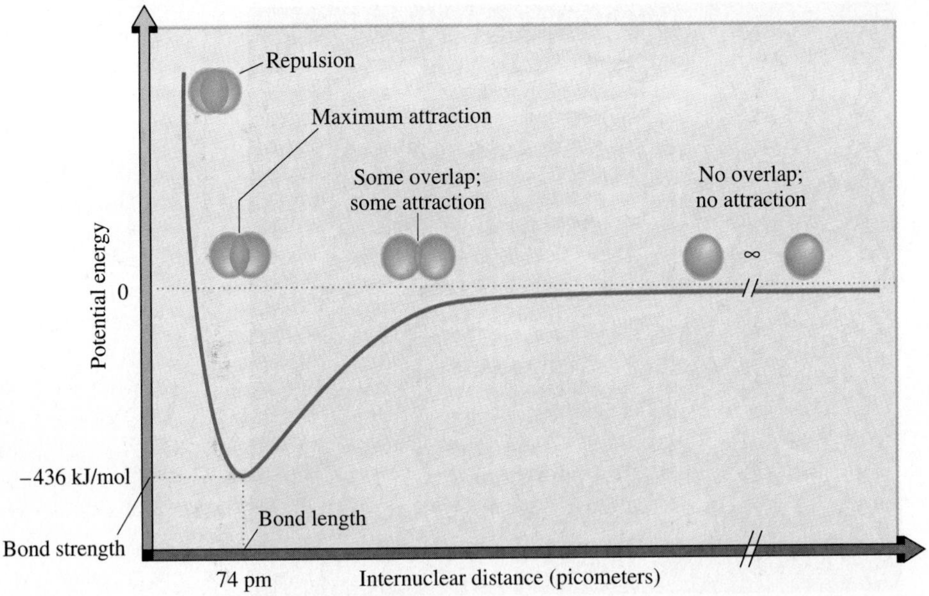

Figure 9.2 Energy change in the course of H—H bond formation from isolated H atoms. The most effective overlap of 1s orbitals occurs at an internuclear distance of 74 pm. If the overlap is less (the distance is more than 74 pm), the bond is weaker. If the overlap is greater (the distance is less than 74 pm), significant repulsions occur between the two nuclei, and the bond is weaker.

Ionic bonds generally involve metals from the left side of the periodic table interacting with nonmetals from the far right side. This follows from the atomic properties described in Chapter 8. Atoms near the noble gases in the periodic table form ions. That is, elements immediately following Group 8 (such as the alkali and alkaline earth metals) have low ionization energies and form cations with the configuration of the nearest noble gas. In contrast, elements immediately preceding Group 8 have high ionization energies and high electron affinities. They form anions, again with the configuration of the nearest noble gas (Section 8.6).

All ionic compounds are solids. Their structures consist of a three-dimensional array of positive and negative ions called a *crystal lattice* (Figure 9.3). In the lattice of NaCl, for example, one sodium cation, Na^+, is attracted by several chloride anions, Cl^-, which are in turn attracted by several cations, and so on. This means that attractive forces (as measured by the lattice energy, Section 8.7) extend throughout the lattice and lead to the observation that most ionic compounds melt only at temperatures much higher than room temperature.

If the electrons involved in the bond are more or less evenly distributed between the atoms, and the electrons are *shared* by two nuclei, the bond is called a **covalent bond.**

$$:\ddot{\underset{..}{I}}\cdot + \cdot\ddot{\underset{..}{Cl}}: \longrightarrow :\ddot{\underset{..}{I}}:\ddot{\underset{..}{Cl}}:$$

Covalent bonding generally occurs between nonmetals, elements that lie in the upper right corner of the periodic table. Compounds having covalent bonds range from very small molecules, such as H_2 and CH_4, to huge molecules, such as DNA and polyethylene, that are built of nonmetallic elements such as carbon, hydrogen,

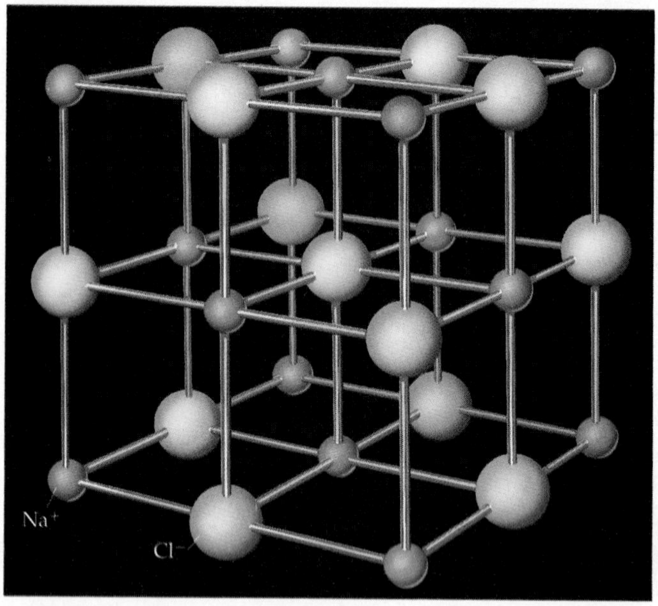

Figure 9.3 Sodium chloride crystal lattice. As described more fully in Chapter 13, the crystal lattice of NaCl, which is similar to that of many other ionic solids, consists of an extended and regular array of Na^+ and Cl^- ions, each ion surrounded by six ions of opposite charge.

nitrogen, and oxygen. Covalently bonded molecules can have a wide range of properties; some are gases, some are liquids, and many are solids.

This chapter describes covalent bonding. Some aspects of ionic bonding were discussed in Section 8.7, and more will be said in Chapter 13. As you read more about chemical bonds, keep in mind that ionic bonding (complete transfer of electrons) and covalent bonding (one or more electron pairs shared equally between two atoms) represent the extreme cases of bonding between atoms. In most compounds the bonding is intermediate between these two limiting cases.

9.3 COVALENT BONDING

Lewis Electron Dot Structures

Virtually all of the kinds of molecules of which we are built (proteins and nucleic acids), the foods we eat (carbohydrates, fats, and proteins), and the clothes we wear (cotton, wool, and synthetic fibers) consist of covalently bonded molecules.

A single covalent bond is formed when two atoms share a pair of electrons. The simplest examples are two-atom, or diatomic, molecules, such as H_2, F_2, or ICl. A crude but useful picture of the distribution of electrons in molecules—chemists call them **Lewis electron dot structures**—can be drawn by starting with Lewis dot symbols for atoms and arranging the valence electrons until each atom has a noble gas configuration (Table 9.2). For example, the Lewis structure for H_2 shows two electrons (two dots) shared between two hydrogen nuclei (H·) so that each H atom has the configuration of the noble gas helium. The two bonding electrons are often represented by a line instead of a pair of dots.

$$\mathrm{H:H} \quad \text{or} \quad \mathrm{H{-}H}$$

To obtain the Lewis structure for ICl, we begin with the Lewis dot symbols for I and Cl (both are in Group 7A, and each has 7 valence electrons). The dot symbol for each atom has a single unpaired electron, and those electrons are shared between the atoms to form a single covalent bond.

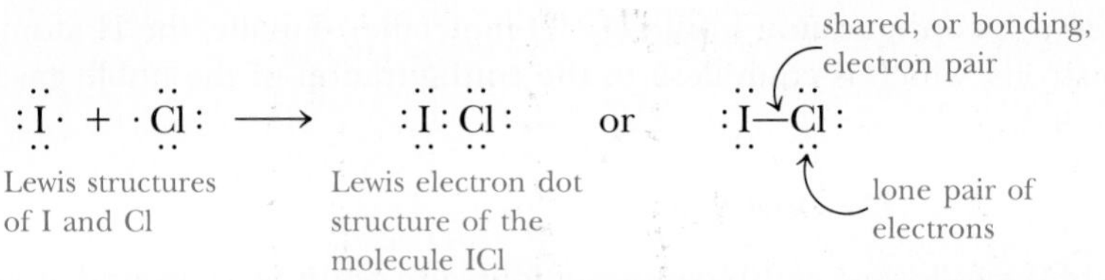

Each atom in the molecule now has a share in four pairs of electrons, so each has achieved a *noble gas configuration.* That is, *each atom is surrounded by an octet of electrons.*

The shared pair of electrons, represented by a pair of dots or a line between the two element symbols, are bonding electrons; these are called the bonding pair, or **bond pair.** The other six pairs in ICl are **nonbonding,** or **lone pairs,** of electrons. They are not involved in bonding, but instead occupy the same electron shell as the bonding electrons and are important in determining molecular shape (Section 9.5).

Other molecules in Table 9.2 have more than one covalent bond between atoms; that is, two or even three pairs of bonding electrons occur between atoms.

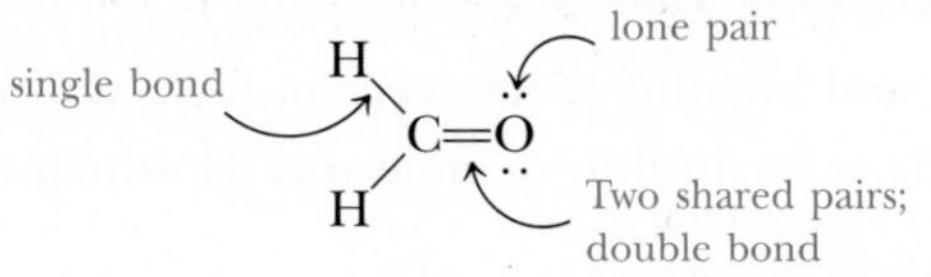

In this molecule, known as formaldehyde, a carbon-oxygen double bond occurs consisting of two electron pairs. Thus, the O atom is surrounded by four electron pairs (two bond pairs and two lone pairs) as is the carbon atom (four bond pairs). (The noble gas neon has four pairs of valence electrons, so C and O have

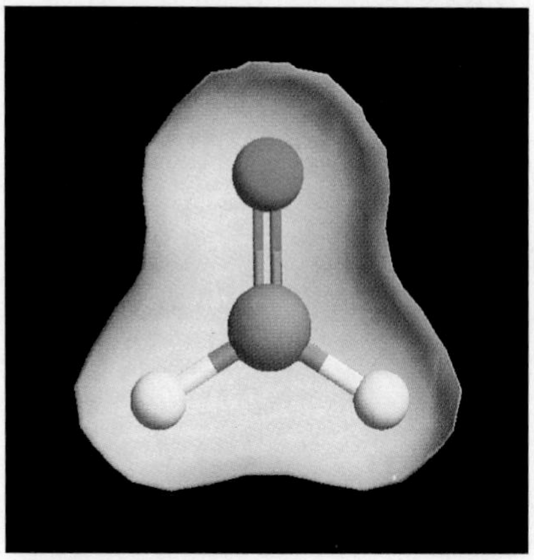

Figure 9.4 Computer-drawn model of the formaldehyde molecule, H_2CO (color code: C = gray, H = white, oxygen = red). The molecule is enclosed within a surface called an isovalue surface. (The front of the surface has been "cut away" so the molecule can be seen.) The electron density is the same at all points on this surface. (Susan Young)

TABLE 9.2 Examples of Lewis Structures by Periodic Group

	1A	4A	5A	6A	7A	8
Lewis structure	H—H	H—C—H (with H above and below C)	H—N̈—H (with H above N)	H—Ö: (with H above O)	H—F̈:	:N̈e:
Chemical formula	H_2	CH_4	NH_3	H_2O	HF	Ne
Bond pairs	1	4	3	2	1	0
Lone pairs	0	0	1	2	3	4
Total pairs of electrons at central atom	1	4	4	4	4	4
Lewis structure		C=C (with two H on each C)	:N≡N:	:C̈l—S̈: (with :C̈l: above S)	:C̈l—F̈:	
Chemical formula		C_2H_4	N_2	SCl_2	ClF	
Bond pairs		6	3	2	1	
Lone pairs		0	2	8	6	
Total pairs of electrons		6	5	10	7	

achieved this configuration in the H_2CO molecule.) Finally, the H atoms share two electrons, which is equivalent to the configuration of the noble gas helium (Figure 9.4).

EXERCISE 9.2 ***The Octet Rule***

Which of the following combinations of lone and bond pairs around atom A are consistent with the octet rule? That is, in which combinations is A surrounded by four pairs of electrons?

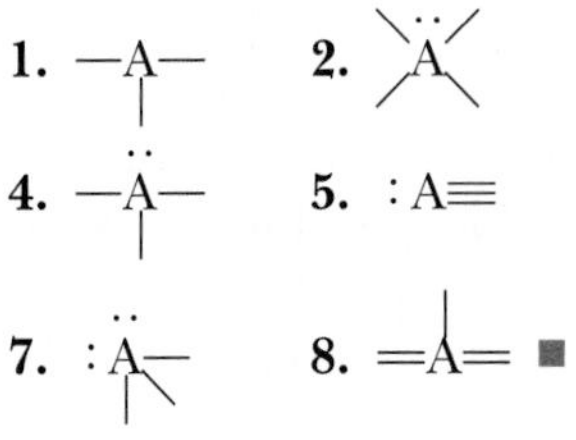

EXERCISE 9.3 ***Lewis Electron Dot Structures***

Use the Lewis structures in Table 9.2 to answer the following questions:

1. How many lone and bonding pairs occur in C_2H_4 and in SCl_2?

2. How many bonds occur in the N_2 molecule? How many lone pairs? ■

Drawing Lewis Structures

A knowledge of molecular structure is at the core of modern chemistry and biology. Therefore, it is important to learn to draw Lewis electron dot structure for molecules and polyatomic ions because, among other things, they can help to predict the structure of the molecule or ion. Figure 9.5 shows the guidelines used for drawing Lewis structures.

Ammonia, NH_3, is a valuable fertilizer and is fifth on the list of the most important chemicals produced in the United States. The H atoms must be terminal atoms, so the N atom is the central atom of the molecule. The total number of valence electrons is the sum of the group numbers: 5 (for N) + 3 (1 for each H) = 8 electrons, or four valence electron pairs. First, we form a single bond between each atom pair, using three of the four pairs of electrons available.

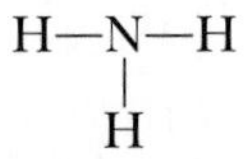

After formation of single covalent bonds, one pair of electrons remains, and this becomes a lone pair on the central atom.

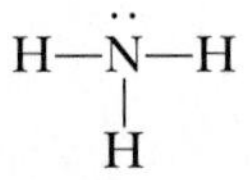

Each H atom now has a share in one pair of electrons as required, and the central N atom has a share in four electron pairs. In conclusion, three of the nitrogen pairs are shared bonding pairs, and the fourth is a lone electron pair.

The hypochlorite ion, ClO^-, is found in household bleach or in $Ca(ClO)_2$ (calcium hypochlorite), which is sold as a swimming pool disinfectant. The ion has 14 valence electrons: 7 (for Cl) + 6 (for O) + 1 (for the negative charge on the ion) = 14 electrons, or seven pairs. After forming the Cl—O bond, the six

Guidelines	*Example*	*Comments*
1. Predicting the arrangement of atoms within a molecule: • H is always an end or terminal atom. It is connected to only one other atom (Table 9.2). • The atom of lowest electron affinity in the molecule or ion is generally the central atom.	C is central in the molecule H_2CO.	Many simple ions or molecules contain O or a halogen. These atoms are often terminal atoms, except when combined with H. When O and F are combined, O is predicted to be central. However, Cl, Br, and I are central in species containing one of these halogens and O. Actually, a better criterion for the central atom is that it should be the atom of lowest electronegativity (see Section 9.4).
2. Valence Electrons. Find the total number of valence electrons in the molecule or ion by adding up the group numbers of the elements. For ions, *add* to the sum the ion charge for an anion and *subtract* from the sum the ion charge for a cation. The number of valence electron pairs is half the total number of valence electrons.	There are 12 valence electrons (or 6 pairs) because C is in Group 4A, O is in Group 6A, and the two H's are in Group 1A.	
3. Place one pair of electrons (to make a single bond) between each pair of bonded atoms.	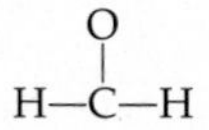	3 pairs have been used to make 3 bonds. (The way you draw the structure at this point is not important. Structures are considered in Section 9.5.)
4. Using the remaining pairs, place lone pairs about each terminal atom (except H) to satisfy the octet rule. If pairs are still left at this point, assign them to the central atom. If the central atom is from the third or higher period, it can accommodate more than four electron pairs.	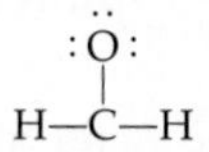	All 6 pairs have been placed in the molecule, but the C atom has a share in only 3 pairs.
5. If the central atom is not yet surrounded by four electron pairs, convert one or more terminal atom pairs to another bond.	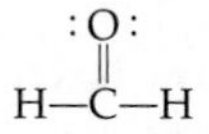	Use the general rule that when a double bond is formed, one or both of the atoms involved is usually C, N, O, S, and P. That is, bonds such as C=C, C=N, C=O, S=O, P=O, and so on are possible.

Figure 9.5 Guidelines for drawing Lewis electron dot structures for molecules and polyatomic ions.

remaining electron pairs are distributed around the "terminal" atoms. Both atoms now have a share in four electron pairs as required.

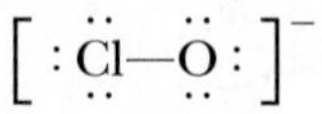

The NO_2^+ ion, called the nitronium ion, has 16 valence electrons:

	5 electrons from N	
+	12 electrons from O	
−	1 electron	(the positive charge of the ion comes from the fact that the combination of the N and two O atoms has lost 1 valence electron)

Total = 16 valence electrons (or 8 pairs)

The guidelines in Figure 9.5 state that N is likely to be the central atom because its electron affinity is lower than that of O. After forming two N—O bonds, the six remaining pairs of electrons are distributed on the terminal O atoms until each of these has a share in a total of four electron pairs.

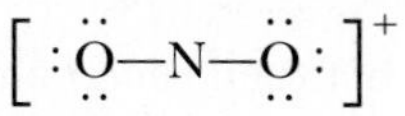

The central N atom now has a deficiency of two electron pairs. Thus, one lone pair of electrons on each O atom is converted to a bonding electron pair to give two NO double bonds.

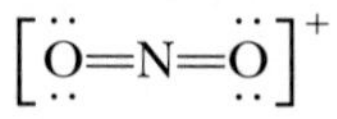

Each atom in the ion now has a share in four electron pairs. Nitrogen has a share in four bonding pairs, and each oxygen atom has two lone pairs and shares two bond pairs.

EXAMPLE 9.1 *Drawing Lewis Structures*

Draw Lewis structures for carbon monoxide, CO; carbon dioxide, CO_2; and phosgene, Cl_2CO.

Solution for CO

1. This is a diatomic molecule, so there is no "central" atom.
2. Total number of valence electrons = 4 (for C) + 6 (for O) = 10
3. Form a single covalent bond between C and O: C—O
4. Place the remaining pairs of electrons around the C and O:

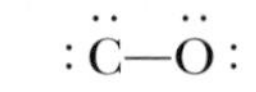

   Both C and O still require one more pair of electrons to have a completed octet.
5. Complete the octet around each atom by forming a triple bond between C and O.

   Each atom now has a share in three bonding pairs and one lone pair and so has a complete octet of electrons.

Solution for CO_2

1. The central atom is C, the atom with the lower affinity for an electron.
2. Total number of valence electrons = 16 = 4 for C + 2 × (6 for O)
3. Form single covalent bonds: O—C—O
4. Place the remaining electron pairs around the terminal atoms.

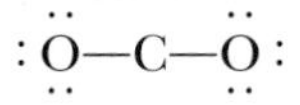

5. The C atom does not have a share in four electron pairs. We therefore use lone pairs on the O atoms to form carbon-oxygen double bonds.

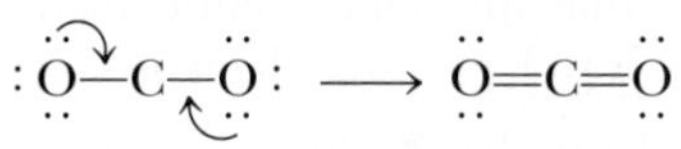

Solution for Cl_2CO

1. The central atom is C, the atom of lowest electron affinity.
2. The total number of valence electrons = 2 × 7 for Cl + 4 for C + 6 for O = 24
3. Form a single covalent bond between each pair of atoms. Three of 12 electron pairs are used.

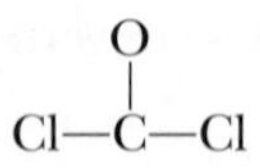

4. Place the remaining nine electron pairs around the terminal atoms (the two Cl's and the O atom), three pairs around each atom.

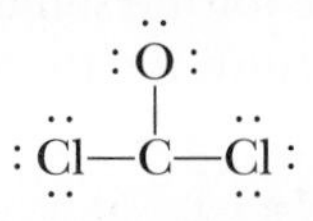

5. The central C atom does not have a completed octet, so a lone pair of electrons from the O atom is used to form another carbon-oxygen bond. All atoms now have a share in four pairs of electrons.

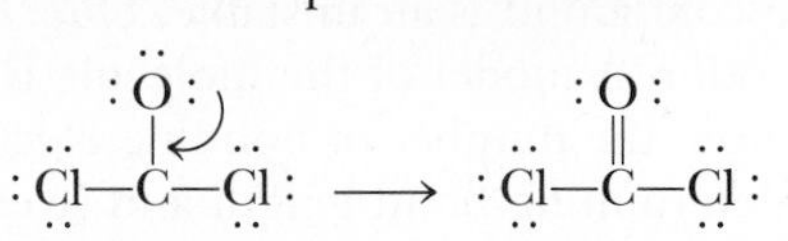

Notice that the octet rule was satisfied by forming a CO double bond and not a CCl double bond. With seven valence electrons, chlorine is inclined to form one bond, so as a general rule double bonds to this element are not formed. See Guideline 5 in Figure 9.5.

Example 9.1 raises some very useful points. First, notice the similarity between CO_2 (Example 9.1) and NO_2^+ (in the text preceding the example). Both have 16 valence electrons, and both have the same distribution of valence electrons. Such molecules are said to be **isoelectronic.** You will find it helpful to think in terms of isoelectronic molecules and ions. For example, if an O atom in CO_2 is replaced by N^- (both O and N^- have six valence electrons), the ion OCN^- is formed. Then, replacing O with S gives SCN^-, the thiocyanate ion. All of these species have 16 valence electrons.

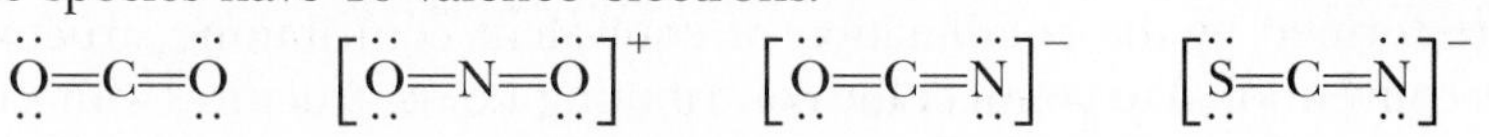

Isoelectronic 16-electron molecules and ions

Carbon monoxide, CO, is also isoelectronic with cyanide ion, CN^-. (Partly as a result, their chemistries are similar; both are very toxic, for example, because they bind to the iron of hemoglobin and block the uptake of oxygen.)

Also notice the similarity between H_2CO and Cl_2CO. Both have a C=O group at the center with two single covalent bonds to other atoms. This is a very important grouping in chemistry, and you will see many examples of such molecules. Just a few are

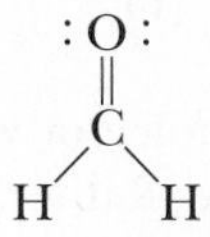

formaldehyde

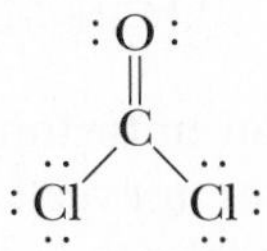

phosgene

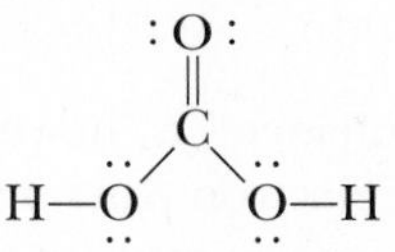

carbonic acid

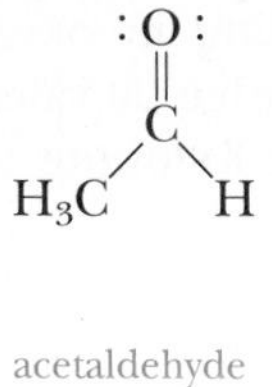

acetaldehyde

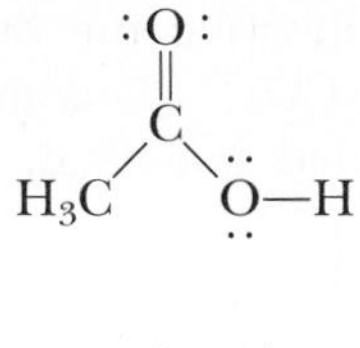

acetic acid

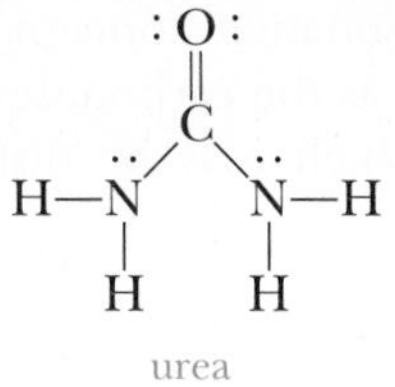

urea

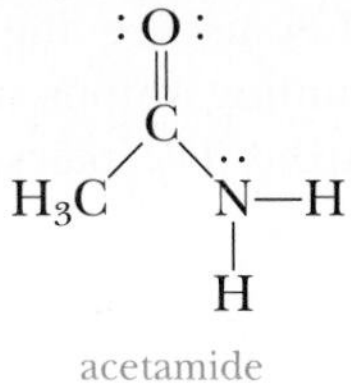

acetamide

EXERCISE 9.4 ***Drawing Lewis Structures***

Sketch a possible Lewis structure for the sulfate ion, SO_4^{2-}. Based on the discussion above, sketch a Lewis structure for the phosphate ion, PO_4^{3-}. ■

Resonance Structures

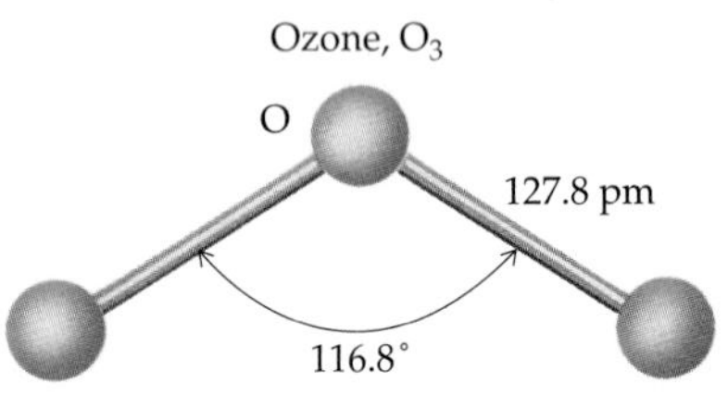

Ozone, O_3, protects the earth and its inhabitants from intense ultraviolet radiation from the sun. The compound is an unstable, blue, diamagnetic gas with a characteristic pungent odor. A model of the molecule is pictured here. As you will see in the next section, the number of bonding electron pairs between two atoms is important in determining bond length and strength. Ozone has equal O—O bond lengths, implying that an equal number of bond pairs occur on each side of the central O atom. Using the guidelines for drawing Lewis structures, however, you would come to a different conclusion. There are two possible ways of writing the Lewis structure for the molecule.

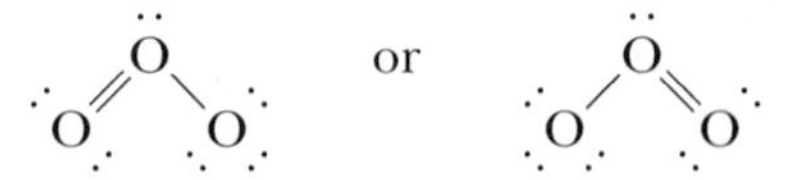

These structures are equivalent in that each structure has a double bond on one side of the central O atom and a single bond on the other side. If either were the actual structure of ozone, however, one bond would be shorter (O═O) than the other (O—O). The actual structure of ozone shows this is not the case, and Linus Pauling proposed the **theory of resonance** to reconcile the problem. The individual structures shown above for ozone are called **contributing structures.** They have identical patterns of covalent bonding and have equal energy. Pauling's theory combines the structures into a *composite* or **resonance hybrid,** a single structure formed by the combination of equivalent contributing structures.

It is conventional to connect the contributing Lewis structures with a double headed arrow, ↔, to indicate that the actual bonding is a composite of these structures. The contributing Lewis structures are often referred to as **resonance structures.** One structure can be formed from the other by moving a lone pair of electrons to form a bond, and a bond pair of electrons to become a lone pair.

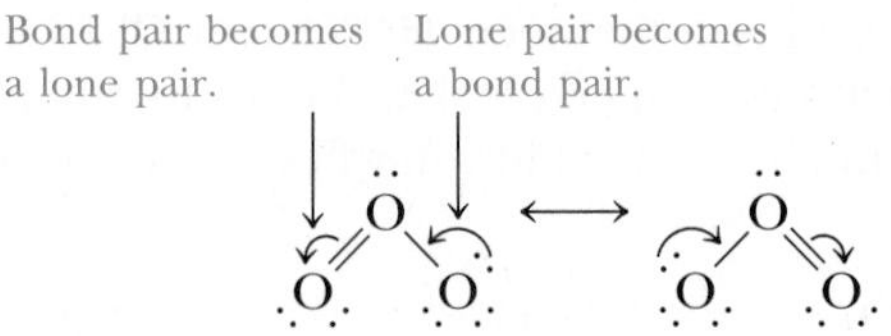

The term "resonance" is, however, an unfortunate choice of words; it implies a movement of electron pairs that does not really occur. Rather, this method of writing structures is an attempt to describe the actual electronic structure of the molecule or ion and yet retain Lewis structures with electron-pair bonds.

Let us use the resonance concept to describe the bonding in oxygen-containing anions such as the carbonate ion, CO_3^{2-}. This anion has 24 valence electrons (12 pairs), which may be distributed *initially* in the following way.

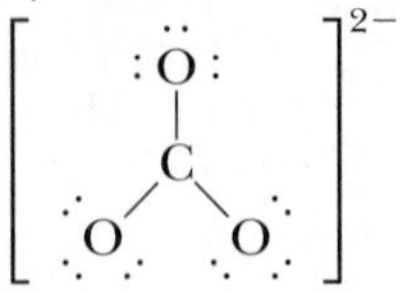

Because the carbon does not have a share in four pairs of electrons, a double bond is involved. Three equivalent representations show this,

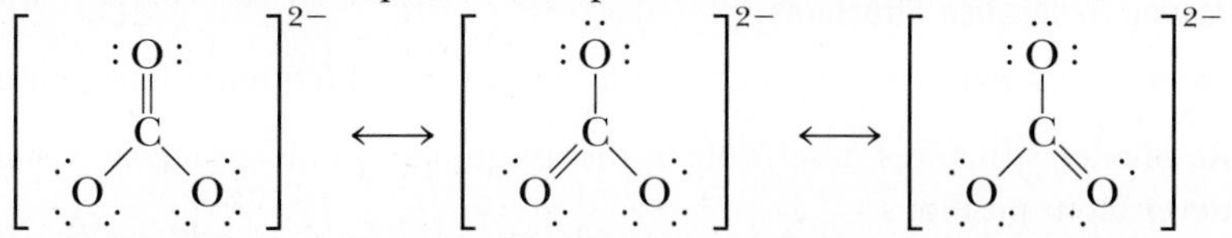

and these three Lewis structures contribute to the resonance hybrid structure. This representation is in good agreement with experimental results. All three carbon-oxygen bond distances are 129 pm, a distance intermediate between the C—O single bond (143 pm) and the C=O double bond (122 pm) distances.

EXAMPLE 9.2 *Drawing Resonance Structures*

Draw the resonance structures for the acetate ion, $CH_3CO_2^-$.

Solution The acetate ion has a total of 24 valence electrons, or 12 pairs:

$$2 \times 4 \text{ (for C atoms)} + 3 \times 1 \text{ (for 3 H atoms)} + 2 \times 6 \text{ (for 2 O atoms)} + 1 \text{ (for negative charge)} = 24$$

The ion has a carbon-carbon bond. One carbon atom is bonded to the two O atoms, and the other carbon atom is bonded to the three H atoms.

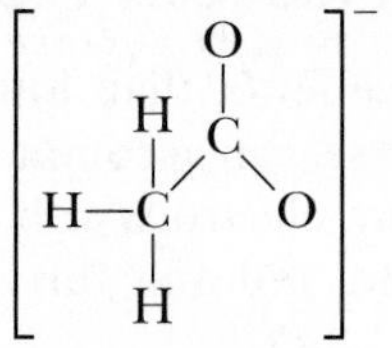

Six pairs of electrons are used to form single bonds between the pairs of atoms, and the remaining six pairs of electrons are placed as lone pairs around the O atoms.

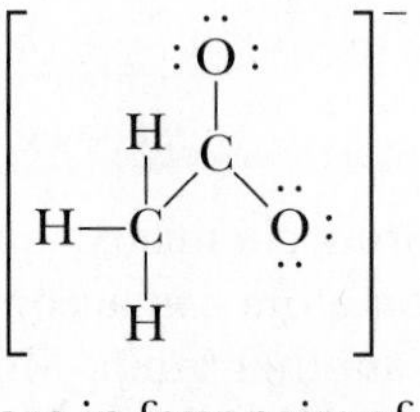

The C atom on the left has a share in four pairs of electrons. To allow the other C atom also to have a share in four pairs of electrons, two resonance structures can be written.

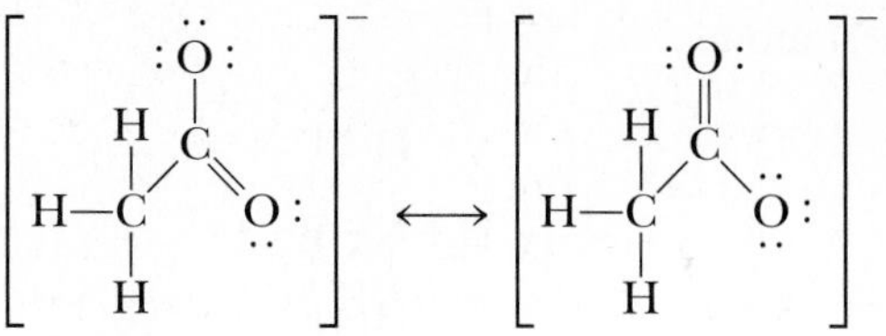

EXERCISE 9.5 *Drawing Resonance Structures*

Draw resonance structures for the nitrate ion, NO_3^-. ■

PROBLEM-SOLVING TIPS AND IDEAS

9.1 Drawing Resonance Structures

When drawing resonance structures, keep in mind these three important points:

- Resonance structures differ only in the assignment of electron-pair positions, never atom positions.
- Resonance structures differ in the number of bond pairs between a given pair of atoms.
- The actual molecular structure is a single hybrid structure formed by the combination of equivalent contributing structures. ■

Exceptions to the Octet Rule

Carbon, nitrogen, oxygen, and fluorine atoms are generally observed to share in four electron pairs when part of molecules and ions. Boron, however, is particularly noted for forming compounds in which it shares in only three electron pairs, and the main group elements beyond neon may share in more than four pairs.

Compounds in Which an Atom Has Fewer Than Eight Valence Electrons

There are compounds that have fewer than four pairs of electrons around the central atom. Hydrogen, of course, can accommodate at most two electrons in its valence shell, so it shares only two electrons with another atom. In BeH_2 only two pairs of electrons are around Be, and only three are around boron in BH_3 and BF_3.

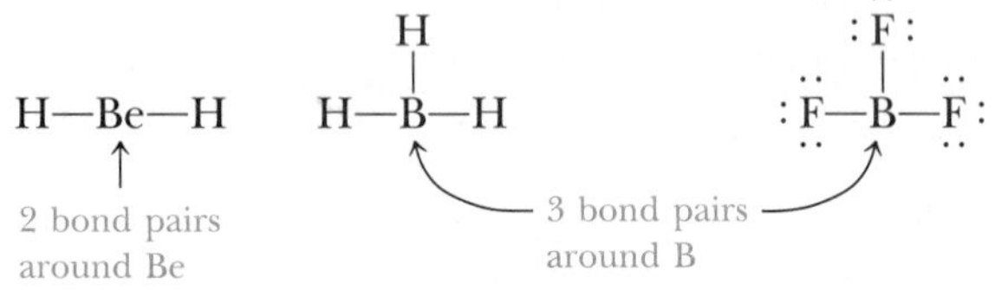

This situation is typical in boron chemistry, and it makes many boron compounds very reactive. The boron atom can accommodate a fourth pair, but only when that pair is provided by another atom. Molecules such as BF_3, therefore, interact readily with another molecule that has a nonbonding electron pair. Ammonia is such a molecule, and NH_3 and BF_3 react rapidly to form $H_3N—BF_3$. A bond is formed between the B and N atoms, where the bonding electron pair originated on the N atom.

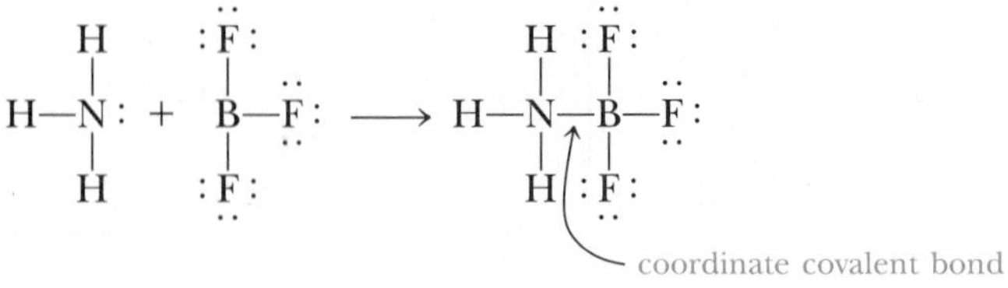

A coordinate covalent bond has traditionally been indicated by an arrow (N → B), which points away from the atom that is donating the electron pair.

A covalent bond in which the bonding pair originates on one of the bonded atoms is called a **coordinate covalent bond.**

Compounds in Which an Atom Has More Than Eight Valence Electrons

If an element of the third or higher periods is the central atom in a molecule or ion, it may be surrounded by more than four valence electron pairs (Table 9.3). Because such elements have *d* orbitals available, they can accommodate five, six, or even seven valence electrons pairs. For example, nitrogen only forms compounds such as NH_3, NH_4^+, and NF_3, whereas phosphorus, another Group 5A element, can accommodate five or even six valence electron pairs.

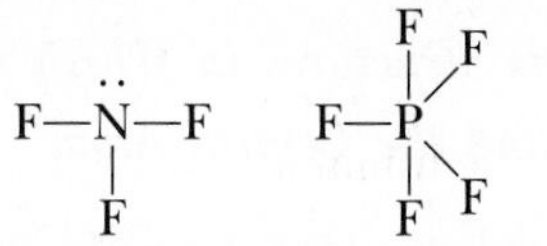

Nitrogen trifluoride and phosphorus pentafluoride.
The 3 lone pairs on each F atom are not shown.

When the authors of this book were undergraduates we were taught that the noble gases did not form chemical compounds and were presented with many theories about why they could not. Then some noble gas compounds were discovered in the early 1960s. One of the more intriguing examples of the early discoveries was XeF_2, which can be made by placing a flask containing xenon gas and fluorine gas in the sunlight and isolating the crystals of XeF_2 some weeks later. We can construct its electron dot structure by recognizing that Xe is the central atom and that the molecule has a total of 22 valence electrons: 8 (for Xe) + 14 (for 2 F atoms) = 22 electrons, or 11 pairs. We begin by forming two Xe—F covalent bonds, and then place three lone electron pairs around each of the two fluorine atoms.

No one knows why XeF_2 and other xenon compounds were not discovered years earlier. Some thought of the possibility, but they did not attempt their preparation or tried to do so in ways that would not work.

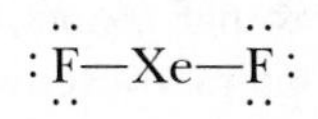

TABLE 9.3 Examples of Lewis Strucutures in Which the Central Atom Has More than Eight Electrons*

	Group 4A	Group 5A	Group 6A	Group 7A	Group 8
Atoms with five valence pairs		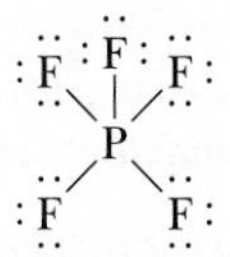	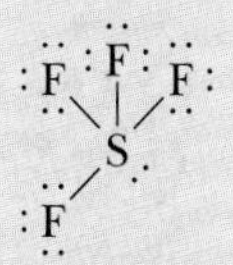		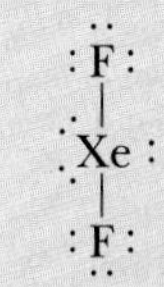
Bond pairs		5	4	3	2
Lone pairs		0	1	2	3
Atoms with six valence pairs	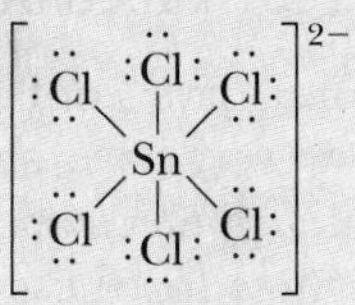	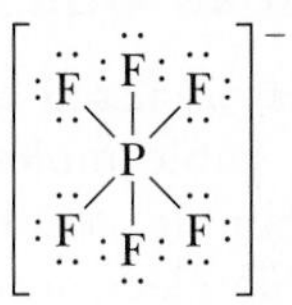	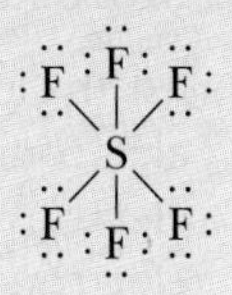	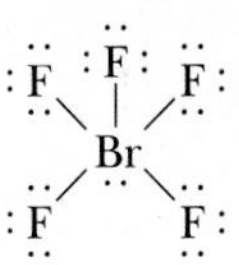	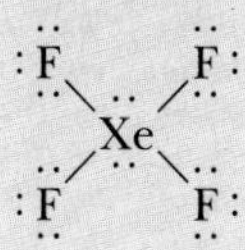
Bond pairs	6	6	6	5	4
Lone pairs	0	0	0	1	2

*In each case, the number of bond pairs and lone pairs about the central atom is given.

We have used eight electron pairs thus far, so 3 of the 11 original pairs remain. The three remaining pairs are placed on xenon, an atom that can accommodate more than a total of four electron pairs. Therefore, XeF_2 is a stable molecule with a total of five electron pairs (two bonding pairs and three lone pairs) around the central Xe atom.

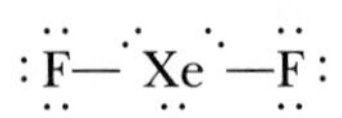

EXAMPLE 9.3 ***Lewis Structures in Which the Central Atom Has More Than Eight Electrons Around the Central Atom***

Sketch the Lewis structure of the $[ClF_4]^-$ ion.

Solution According to the guidelines in Figure 9.5, we proceed as follows:

1. The Cl atom is the central atom.
2. This ion has 36 valence electrons: (7 for Cl) + 4 × (7 for F) + 1 for ion charge, or 18 pairs.
3. Draw the ion with four single covalent Cl—F bonds.

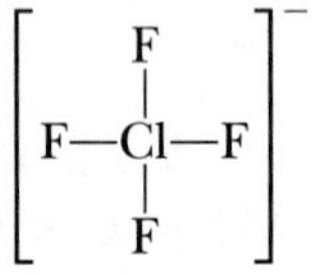

4. Place lone pairs on the terminal atoms. Because two electron pairs remain after placing lone pairs on the four F atoms, and because we know that Cl can accommodate more than four pairs, these two pairs are placed on the central Cl atom.

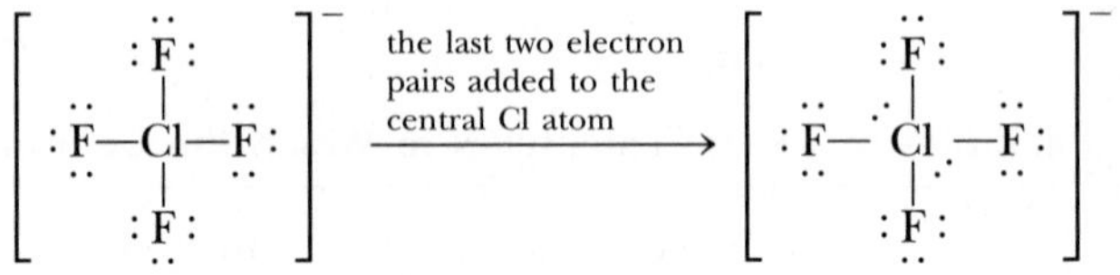

EXERCISE 9.6 ***Lewis Structures in Which the Central Atom Has More Than Eight Electrons Around the Central Atom***

Sketch the Lewis structure for $[ClF_2]^-$. How many lone pairs and bond pairs surround the Cl atom? ■

Molecules with an Odd Number of Electrons

All the molecules discussed to this point have contained only *pairs* of valence electrons. A few stable molecules, however, have an odd number of valence electrons. For example, NO has 11 valence electrons, and NO_2 has 17 valence electrons. Plausible electron dot structures for these molecules are

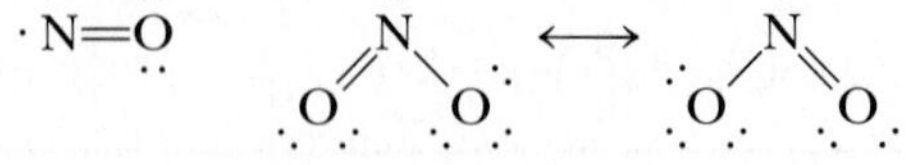

Both place the odd electron on the N atom.

Molecules such as NO and NO_2 are often called **free radicals** because of the presence of the unpaired electron. Atoms that contain unpaired electrons are also free radicals. How do these unpaired electrons affect reactivity? Simple free radicals such as H · and Cl · are very reactive and readily combine with other atoms to give molecules such as H_2, Cl_2, and HCl. Therefore, we expect free radical molecules to be more reactive than molecules with paired electrons, and most are. A free radical either combines with another free radical to form a molecule in which the electrons are paired, or it reacts with other molecules to produce new free radicals. These kinds of reactions are central to the formation of air pollutants. For example, when small amounts of NO and NO_2 are released from vehicle exhausts, the NO_2 decomposes in the presence of sunlight to give NO and O.

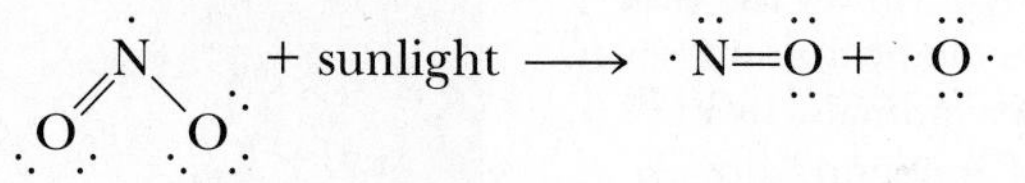

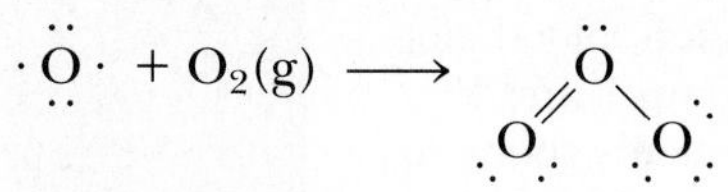

The free O atom then reacts with O_2 in the air to give ozone, O_3, an air pollutant that affects the respiratory system.

Free radicals also have the tendency to combine with themselves to form dimers. For example, when NO_2 gas is cooled it dimerizes to give colorless N_2O_4.

As expected, NO and NO_2 are paramagnetic because the odd electron is not paired with another electron. Experimental evidence indicates that the O_2 molecule is also paramagnetic (Section 8.1) with two unpaired electrons and a double bond. The predicted Lewis structure for O_2 shows a double bond, but all the electrons are paired. It is not possible to write a conventional Lewis structure for O_2 that agrees with experimental results.

Gaseous nitrogen dioxide, NO_2, is brown-orange. When the gas is cooled in ice, some NO_2 molecules join (or dimerize) to give N_2O_4, which is colorless, and the color of the gas mixture lightens. (C. D. Winters)

PROBLEM-SOLVING TIPS AND IDEAS

9.2 Drawing Lewis Electron Dot Structures

Remember these useful ideas when drawing Lewis structures:

- When multiple bonds are formed, one or both of the atoms involved is usually among the following: C, N, O, S, and P. Oxygen in particular has the ability to form multiple bonds with a variety of elements. Carbon forms many compounds having multiple bonds to another carbon or to N, O, or S.
- Nonmetals may form double and triple bonds but never quadruple bonds.
- Resonance structures differ only in the assignment of electron-pair positions, never atom positions.
- Resonance structures differ in the number of bond pairs between a given pair of atoms.
- Carbon, nitrogen, oxygen, and fluorine atoms are generally surrounded by four valence electron pairs.
- There are many exceptions to the octet rule, particularly with boron and elements of the third and higher periods. Elements of the third and higher periods may accommodate five, six, or even seven valence electron pairs. ■

CURRENT ISSUES IN CHEMISTRY

NO Is No Dud!

Nitrogen monoxide, NO, also known as nitric oxide, is the simplest, thermally stable odd-electron molecule known. As a consequence, it has been widely studied.

Some Physical Properties of NO

Melting point	−163.6 °C
Boiling point	−151.8 °C
ΔH_f°	90.2 kJ/mol
N—O bond distance	115 pm

NO is a colorless, paramagnetic gas that is moderately soluble in water. It can be synthesized by the reduction of a nitrite salt with iodide ion, for example,

$$KNO_2(aq) + KI(aq) + H_2SO_4(aq) \longrightarrow NO(g) + K_2SO_4(aq) + H_2O(\ell) + \tfrac{1}{2}\, I_2(aq)$$

and it is an intermediate in the industrial synthesis of nitric acid by the oxidation of ammonia.

The compound reacts rapidly with O_2 to form the reddish brown gas NO_2.

$$2\,NO(\text{colorless gas}) + O_2(g) \longrightarrow 2\,NO_2(\text{brown gas})$$

Because NO can be formed as a by-product of burning fossil fuels in air, it is assumed that the oxidation of NO is the source of NO_2 and nitrates in polluted air.

Imagine everyone's surprise when it was learned a few years ago that NO is synthesized in a biological process by animals as diverse as barnacles, fruit flies, horseshoe crabs, chickens, trout, and humans. As reported in *Chemical and Engineering News* in late 1993, "[NO] plays a role, often as a biological messenger, in an astonishing range of physiological processes in humans and other animals. Its expanding range of functions already include neurotransmission, blood clotting, blood pressure control, and a role in the immune system's ability to kill tumor cells and intracellular parasites."*

Nitroglycerin ($C_3H_5(NO_3)_3$)is an oily, colorless liquid. It is so unstable that, if it is dropped or struck in any way, it explodes, forming CO_2, H_2O, and NO_x gases. (Normally, it is dissolved in some absorbent material such as clay, and the result is dynamite.) In spite of this instability, nitroglycerin has been used for almost 90 years by physicians to regulate blood pressure, particularly in the treatment of chest pain (angina). It works by dilating blood vessels, which decreases the pressure and increases blood flow. The mechanism by which it led to muscle relaxation was only very recently discovered, and chemists have found that it involves NO.

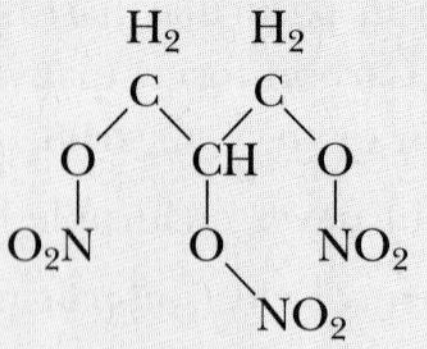

Nitroglycerin

*P. L. Feldman, O. W. Griffith, and D. J. Stuehr: *Chemical and Engineering News,* pp. 26–38, December 20, 1993.

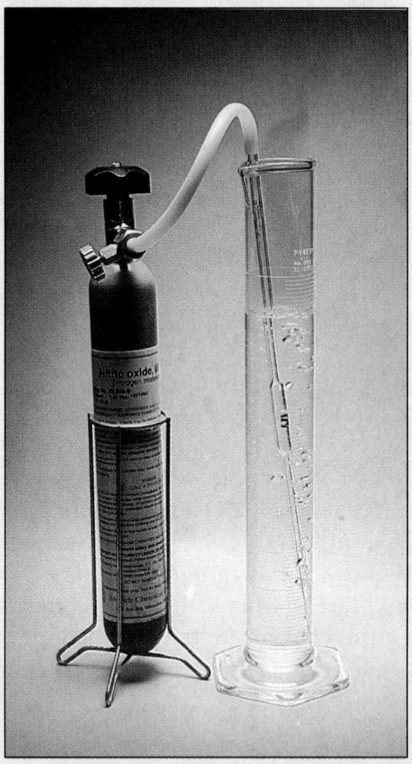

NO gas is bubbled through water. When the gas emerges into the air, the NO is rapidly oxidized to give brown-orange NO_2 gas. (C. D. Winters)

9.4 BOND PROPERTIES

Bond Order

The **order of a bond** is the number of bonding electron pairs shared by two atoms in a molecule. Various molecular properties can be understood by using this concept, including the distance between atoms *(bond length)* and the energy required to separate atoms from each other *(bond energy)*. In this book you will encounter bond orders between 1 and 3, as well as fractional bond orders.

The bond order is 1 when there is only a single covalent bond between a pair of atoms. Examples are the bonds in the following molecules:

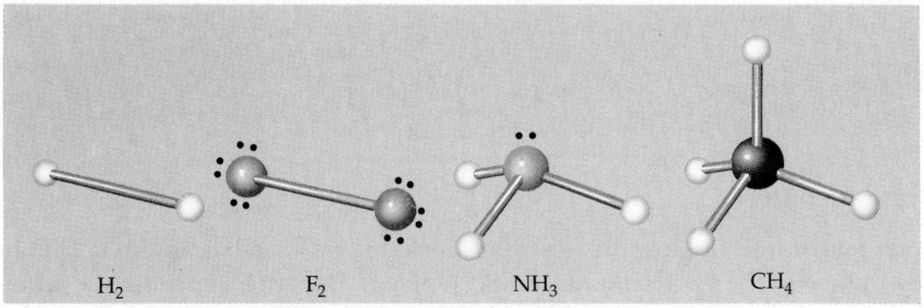

The bond order is 2 when two pairs are shared between two atoms. Examples are the C=O bonds in CO_2 and the C=C bond in ethylene, C_2H_4.

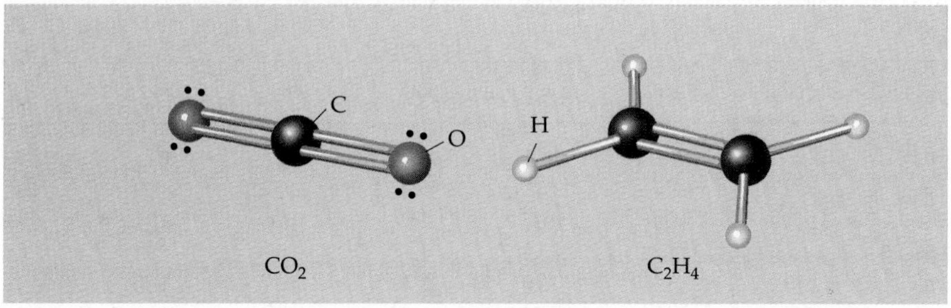

The bond order is 3 when two atoms are connected by three bonds. Examples are the carbon-carbon bond in acetylene, the carbon-oxygen bond in carbon monoxide, CO, and the carbon-nitrogen bond in the cyanide ion, CN^-.

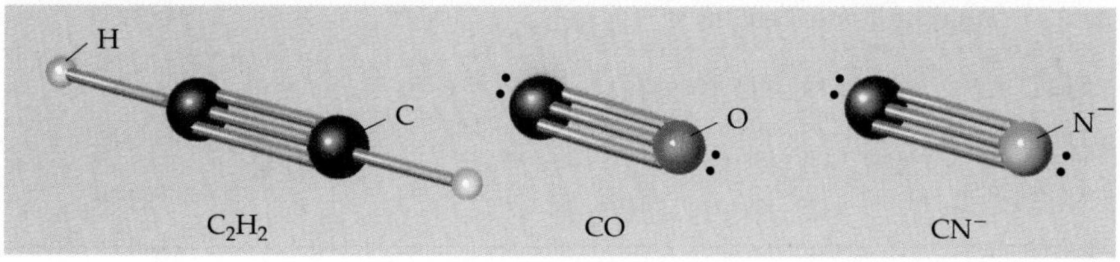

A *fractional bond order* is possible in molecules and ions having resonance structures. For example, what is the bond order for each oxygen-oxygen bond in O_3? Each resonance structure of O_3 has one O—O single bond and one O=O double bond, for a total of three shared bonding pairs accounting for two oxygen-oxygen links. If we define the bond order between any bonded pair of atoms as

$$\text{Bond order} = \frac{\text{number of shared pairs linking X and Y}}{\text{number of X-Y links in the molecule or ion}}$$

then the order is seen to be $\frac{3}{2}$, or one and one half for ozone.

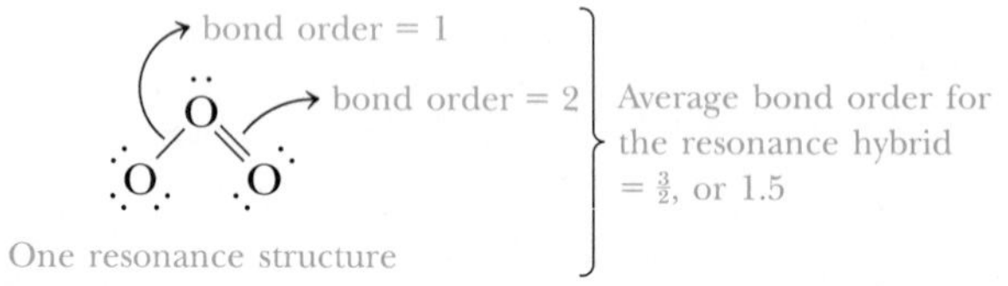

Bond Length

Bond length, the distance between the nuclei of two bonded atoms, is largely determined by the sizes of the atoms (Section 8.6). For given elements, the order of the bond then determines the final value of the distance.

TABLE 9.4 Some Approximate Single and Multiple Bond Lengths*

Single Bond Lengths

	Group										
	1A	*4A*	*5A*	*6A*	*7A*	*4A*	*5A*	*6A*	*7A*	*7A*	*7A*
	H	C	N	O	F	Si	P	S	Cl	Br	I
H	74	110	98	94	92	145	138	132	127	142	161
C		154	147	143	141	194	187	181	176	191	210
N			140	136	134	187	180	174	169	184	203
O				132	130	183	176	170	165	180	199
F					128	181	174	168	163	178	197
Si						234	227	221	216	231	250
P							220	214	209	224	243
S								208	203	218	237
Cl									200	213	232
Br										228	247
I											266

Multiple Bond Lengths

C=C	134	C≡C	121
C=N	127	C≡N	115
C=O	122	C≡O	113
N=O	115	N≡O	108

*In picometers (pm); 1 pm = 10^{-12} m.

Table 9.4 lists bond lengths for a number of common chemical bonds. It is important to recognize that these are *average* values. Variations in neighboring parts of a molecule can affect the length of a particular bond. For example, the C—H bond is 99.3 pm long in methane, CH_4, whereas it is only 95.9 pm long in acetylene, H—C≡C—H. Variation is possible by as much as 10% from the average values listed in Table 9.4.

Atom sizes vary in a fairly smooth way with the position of the element in the periodic table (Figure 8.10). When bonds of the same order are compared, the bond length is greater for the larger atoms. Thus, bonds involving carbon and another element increase along the series.

$$\xrightarrow[\text{Increase in bond distance}]{\text{C—N} < \text{C—C} < \text{C—P}}$$

Similarly, a C=O bond is shorter than a C=S bond, and a C≡N bond is shorter than a C≡C bond. Each of these trends can be predicted from the relative sizes shown here in the margin, and some common bond lengths are given in Table 9.4.

Relative sizes of atoms of Groups 4A, 5A, and 6A.

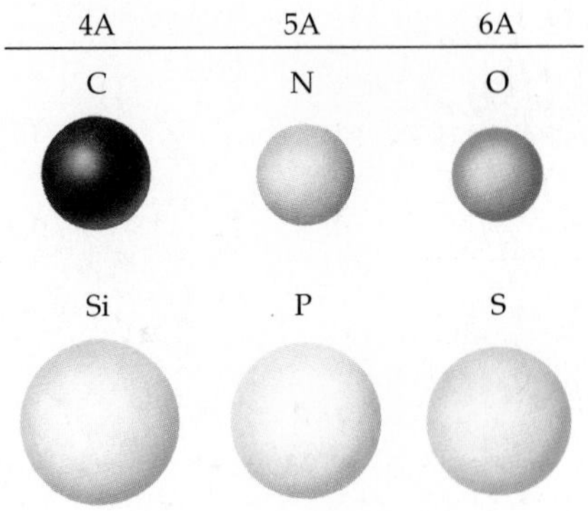

The effect of bond order is evident when bonds between the same two atoms are compared. For example, the bonds become shorter as the bond order increases in the series C—O, C=O, and C≡O.

Bond	C—O	C=O	C≡O
Bond Order	1	2	3
Bond Length (pm)	143	122	113

Adding a second bond to a C—O single bond to make a C=O shortens the bond by 21 pm. Adding yet another bond results in a further 9 pm reduction in bond length from C=O to C≡O. In general, double bonds between atoms are shorter than single bonds between the same set of atoms, and triple bonds between those same atoms are shorter still. As the electron density between the bonded atoms increases, the atoms are bonded more strongly, and the nuclei come closer together.

The carbonate ion, CO_3^{2-}, which has three equivalent resonance structures, has a CO bond order of $\frac{4}{3}$. (Four electron pairs link the central carbon with the three oxygen atoms.) Not surprisingly, this leads to a bond distance (129 pm) roughly intermediate between a C—O single bond (143 pm) and a C=O double bond (122 pm).

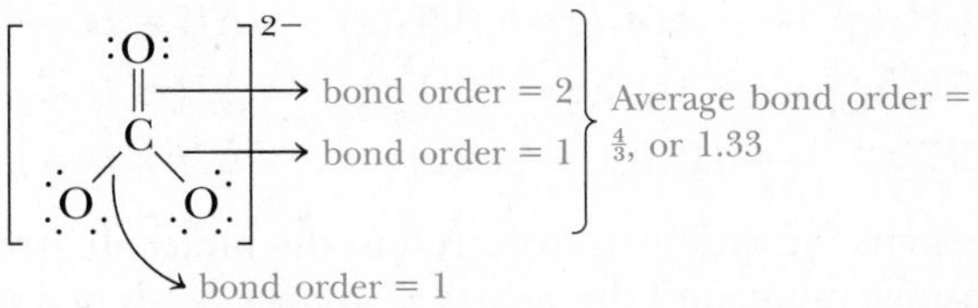

Benzene, C_6H_6, plays a significant role in chemistry. Not only is the compound an important solvent, but its H atoms can be replaced by other atoms or groups of atoms to give thousands of different compounds. Benzene is planar and symmetrical with all of the carbon-carbon bonds being the same length. Each carbon-carbon bond is 139 pm long, intermediate between the length of a

C—C single bond (154 pm) and a C=C double bond (134 pm). Chemists depict this with the resonance structures shown on the left in the figure or with the form on the right. The circle in the middle of the right-hand structure is meant to convey the idea that the six electrons forming the three double bonds are delocalized (spread over) the six carbon atoms of the ring.

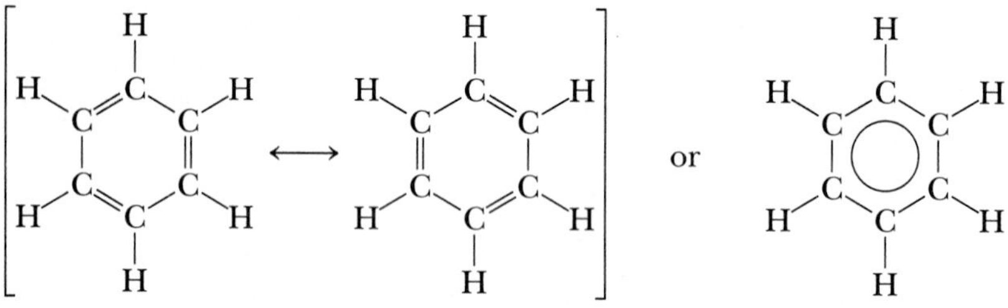

resonance structures

resonance hybrid

EXERCISE 9.7 ***Bond Distances and Bond Order***

1. Give the bond order of each of the following bonds and arrange them in order of decreasing bond distance: C=N, C≡N, and C—N.
2. Draw resonance structures for NO_2^-. What is the NO bond order in this ion? Consult Table 9.4 for N—O and N=O bond lengths. Compare these with the NO bond length in NO_2^- (124 pm). Account for any differences you observe. ■

Bond Energy

The **bond dissociation energy, *D*,** is the enthalpy change for breaking a bond in a molecule with the reactants and products in the gas phase under standard conditions. Suppose you wish to break the carbon-carbon bonds in ethane ($H_3C—CH_3$), ethylene ($H_2C=CH_2$), and acetylene (HC≡CH), for which the bond orders are 1, 2, and 3, respectively. For the same reason that the ethane C—C bond is the longest of the series, and the acetylene C≡C bond is the shortest, bond breaking requires the least energy for ethane and the most energy for acetylene.

$$\text{Molecule} \underset{\text{energy released}}{\overset{\text{energy supplied}}{\rightleftharpoons}} \text{Molecular fragments}$$

$$H_3C—CH_3(g) \longrightarrow H_3C(g) + CH_3(g) \qquad \Delta H = D = +347 \text{ kJ}$$

$$H_2C{=}CH_2(g) \longrightarrow H_2C(g) + CH_2(g) \qquad \Delta H = D = +611 \text{ kJ}$$

$$HC{\equiv}CH(g) \longrightarrow HC(g) + CH(g) \qquad \Delta H = D = +837 \text{ kJ}$$

Because D represents the energy transferred to the molecule from its surroundings, D has a positive value, and the *process of breaking bonds in a molecule is always endothermic.*

The amount of energy supplied to break carbon-carbon bonds in the molecules just discussed must be the same as the amount of energy released when the same bonds form. *The formation of bonds from atoms or radicals in the gas phase is always exothermic.* This means, for example, that ΔH for the formation of $H_3C—CH_3$ from two $CH_3(g)$ radicals is −347 kJ/mol.

$$H_3C(g) + CH_3(g) \longrightarrow H_3C—CH_3(g) \qquad \Delta H = -D = -347 \text{ kJ}$$

TABLE 9.5 Some Average Single- and Multiple-Bond Energies*

Single Bonds

	H	C	N	O	F	Si	P	S	Cl	Br	I
H	436	414	389	464	569	293	318	339	431	368	297
C		347	293	351	439	289	264	259	330	276	238
N			159	201	272		209		201	243?	
O				138	184	368	351		205		201
F					159	540	490	285	255	197?	
Si						176	213	226	360	289	
P							213	230	331	272	213
S								213	251	213	
Cl									243	218	209
Br										192	180
I											151

Multiple Bonds

N=N	418	C=C	611
N≡N	946	C≡C	837
C=N	615	C=O (in O=C=O)	803
C≡N	891	C=O (as in H_2C=O)	745
O=O (in O_2)	498	C≡O	1075

*In kilojoules per mole.

Some experimental bond energies are tabulated in Table 9.5. Notice the following important points.

- The energies listed are all *positive*. They are the energies required to break one mole of the bond in question. If you need the energy released when one mole of bonds is formed, the magnitude is the same, but the sign is *negative*.
- The energies are *average bond energies*. For example, a C—H bond has an average energy of 414 kJ/mol. This value may vary as much as 30 to 40 kJ/mol from molecule to molecule, however, just as bond lengths vary from one molecule to another.
- Bond energies are defined in terms of gaseous atoms or molecular fragments. If a reactant is in the solid or liquid state, you must include the energy required to convert it to a gas before using the bond energy values. For example, to know the energy required to convert liquid bromine to bromine atoms in the gas phase, we must first find the energy to convert liquid bromine to bromine vapor and then add on the energy required to break one mole of bromine-bromine bonds.

Liquid to vapor	$Br_2(\ell) \longrightarrow Br_2(g)$	$\Delta H = +30.9$ kJ
Bond breaking	$Br_2(g) \longrightarrow 2\ Br(g)$	$\Delta H = D = +192$ kJ
Total energy	$Br_2(\ell) \longrightarrow 2\ Br(g)$	$\Delta H_{total} = +223$ kJ

- Finally, be sure to notice that the bond energy increases with an increase in bond order.

In reactions between molecules, bonds in the reactants are broken and new bonds are formed in the products. If the total energy released when new bonds are formed exceeds the energy required to break the original bonds, the overall reaction is exothermic. If the opposite is true, then the overall reaction is endothermic. Let us see how this works in practice.

Natural oils can be converted to fats by a reaction called a *hydrogenation,* the addition of hydrogen. A simple example of this kind of reaction is the conversion of the hydrocarbon propene to propane.

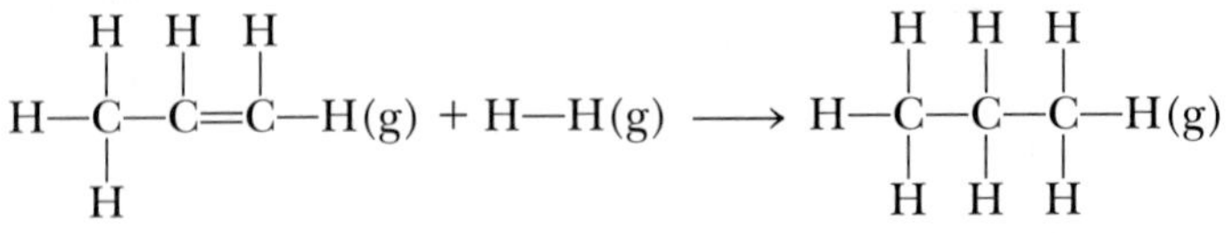

$$\underset{\text{propene}}{\text{H—C—C}=\text{C—H(g)}} + \text{H—H(g)} \longrightarrow \underset{\text{propane}}{\text{H—C—C—C—H(g)}}$$

If we knew the enthalpies of formation of propene and propane, we could use the technique described in Chapter 6 to find the enthalpy change for the reaction. Suppose, though, the values were not available. In such a case we can use bond energies to *estimate* the enthalpy change for the reaction. The first step is to examine the reactants and product to see what bonds are broken and what bonds are formed. For the reaction in question one C—C bond and six C—H bonds are *not* changed. Thus we need to focus only on the affected bonds.

Bonds broken: 1 mol of C=C bonds and 1 mol of H—H bonds

$$\text{H—C—C}=\text{C—H(g)} + \text{H—H(g)}$$

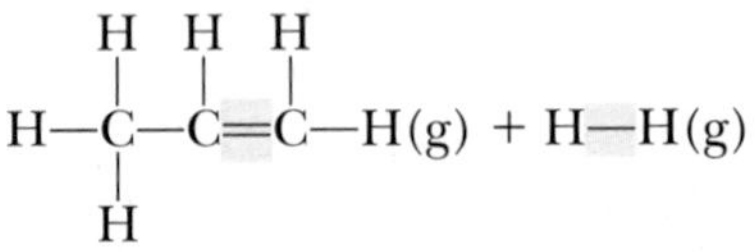

Energy required = 611 kJ for C=C bonds + 436 kJ for H—H bonds = 1047 kJ

Bonds formed: 1 mol of C—C bonds and 2 mol of C—H bonds

$$\text{H—C—C—C—H(g)}$$

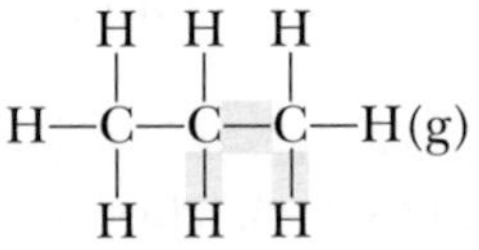

Energy evolved = 347 kJ for C—C bonds + 2 mol (414 kJ/mol for C—H bonds)
= 1175 kJ

When using bond energies to find the enthalpy change for a reaction, you should add up the energies of all the bonds broken and subtract from this the sum of the energies of the bonds formed.

$$\Delta H^{\circ}_{rxn} = \Sigma D(\text{bonds broken}) - \Sigma D(\text{bonds formed})$$

This equation tells you to multiply the bond energy for each bond broken by the number of bonds of that type, and add all of these up. Then, multiply the bond energy for each bond formed by the number of bonds of that type, and add all of these up. Because bond formation is exothermic, the quantity "ΣD(bonds formed)" is subtracted from the quantity "ΣD(bonds broken)." Therefore, for the hydrogenation reaction, we have

$$\Delta H^{\circ}_{rxn} = 1047 \text{ kJ} - 1175 \text{ kJ} = -128 \text{ kJ}$$

and the overall reaction is exothermic. (Using enthalpies of formation for propene and propane, we calculate $\Delta H^{\circ}_{rxn} = -123.8$ kJ, indicating that bond energy calculations can give acceptable results in many cases.)

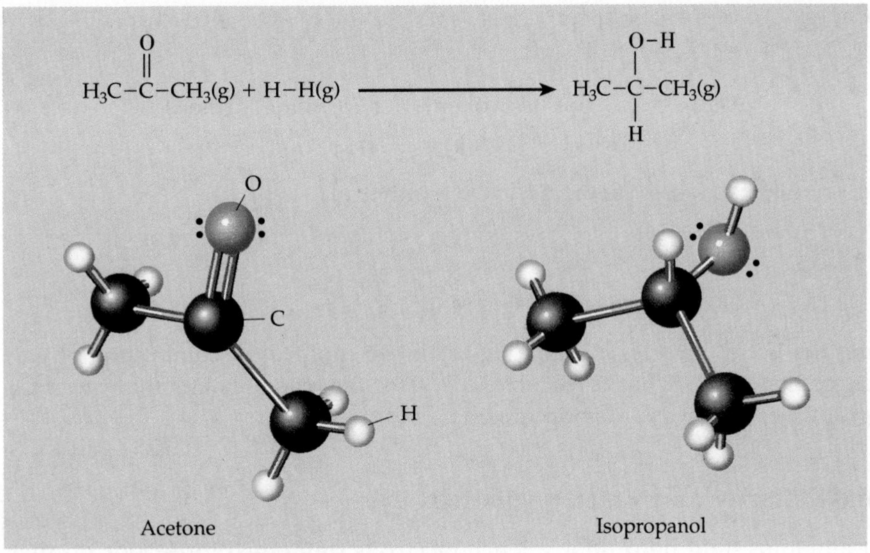

Acetone   Isopropanol

E X A M P L E 9.4 *Using Bond Energies*

Acetone, a common industrial solvent, can be converted to isopropanol, rubbing alcohol, by hydrogenation.

$$H_3C-\overset{\overset{\textstyle O}{\|}}{C}-CH_3(g) + H-H(g) \longrightarrow H_3C-\overset{\overset{\textstyle O-H}{|}}{\underset{\underset{\textstyle H}{|}}{C}}-CH_3(g)$$

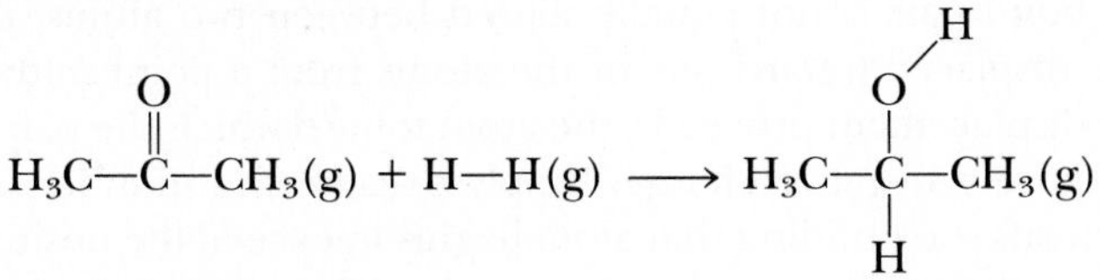

acetone   isopropanol

Predict the enthalpy change for the reaction using bond energies.

Solution   The first step is to examine the reactants and product to see what bonds are broken and what bonds are formed. For this reaction the two C—C bonds and six C—H bonds are not changed. We therefore need only focus on the bonds that have been broken in the reactants or formed in the products.

Bonds broken: 1 mol of C=O bonds and 1 mol of H—H bonds

$$H_3C-\overset{\overset{\textstyle O}{\|}}{C}-CH_3(g) + H-H(g)$$

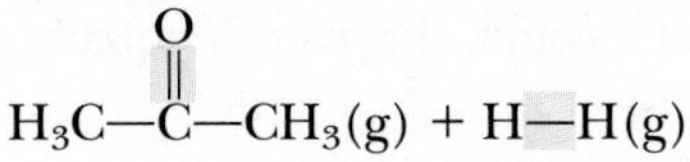

Energy required = 745 kJ for C=O bonds + 436 kJ for H—H bonds = 1181 kJ

Bonds formed: 1 mol of C—H bonds, 1 mol of C—O bonds, and 1 mol of O—H bonds

$$H_3C-\overset{\overset{\textstyle O-H}{|}}{\underset{\underset{\textstyle H}{|}}{C}}-CH_3(g)$$

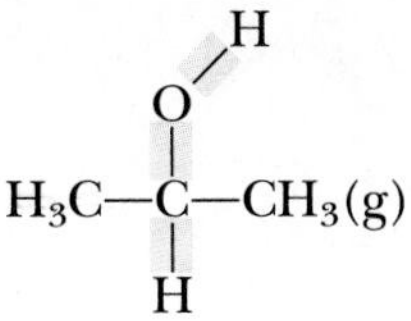

$$\text{Energy evolved} = 414 \text{ kJ for C—H} + 351 \text{ kJ for C—O} + 464 \text{ kJ for O—H}$$
$$= 1229 \text{ kJ/mol}$$

$$\Delta H^\circ_{rxn} = \Sigma\, D(\text{bonds broken}) - \Sigma\, D(\text{bonds formed})$$
$$= 1181 \text{ kJ} - 1229 \text{ kJ} = -48 \text{ kJ}$$

The overall reaction is predicted to be exothermic.

EXERCISE 9.8 ***Using Bond Energies***

Using the bond energies in Table 9.5, estimate the heat of combustion of gaseous methane, CH_4. That is, calculate ΔH°_{rxn} for the reaction of methane with O_2 to give water vapor and carbon dioxide gas. ■

Bond Polarity and Electronegativity

Not all atoms hold onto their valence electrons with equal force, nor do they take on electrons with equal ease (Section 8.6). That is, the elements have different values of ionization energy and electron affinity. This behavior carries over into molecules. It is generally true that, if two different kinds of atoms form a bond, one attracts the shared pair more strongly than the other. Only when two identical atoms form a bond can we presume that the bond pair is shared equally between the two atoms.

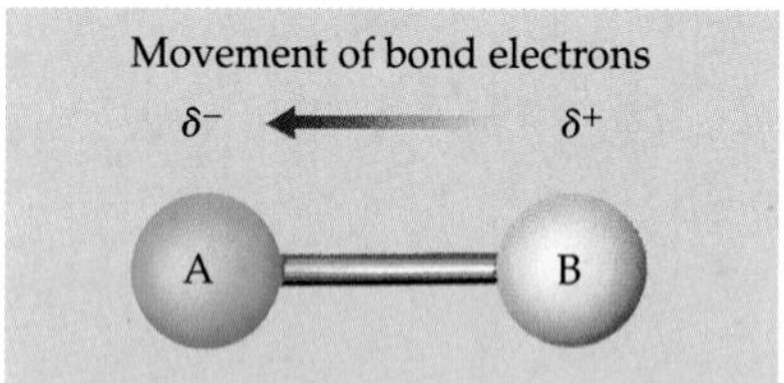

Polar bond between atoms A and B

When a bond pair is not equally shared between two atoms, the bonding electrons are displaced toward one of the atoms from a point midway between them. As the displacement proceeds, the atom toward which the pair is displaced begins to acquire a negative charge, simply because the number of negatively charged electrons surrounding that atom begins to exceed the positive charge of the nucleus. At the same time, the atom at the other end of the bond is being depleted in electron density and begins to acquire a positive charge. Thus, the bond acquires a positive end and a negative end; that is, it has *electric poles* and is called a **polar bond.** If the displacement of the bonding pair is essentially complete, the bond is ionic, and + and − symbols are written alongside the atom symbols in the Lewis drawings. When the displacement is less than complete, however, the bond is said to be a **polar covalent bond.** This polarity is indicated by writing $\delta+$ and $\delta-$ symbols alongside the atom symbols, where δ (the Greek letter "delta") stands for a partial charge. Three examples of common molecules that have polar bonds are shown in the figure. If no net displacement of the bonding electron pair occurs, the pair is equally shared, and the bond is **nonpolar covalent.**

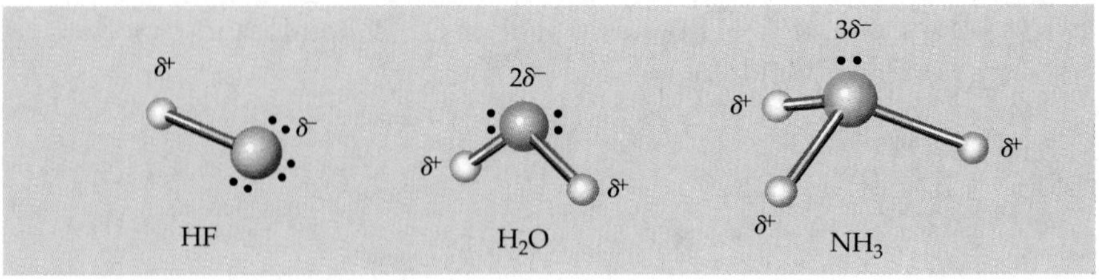

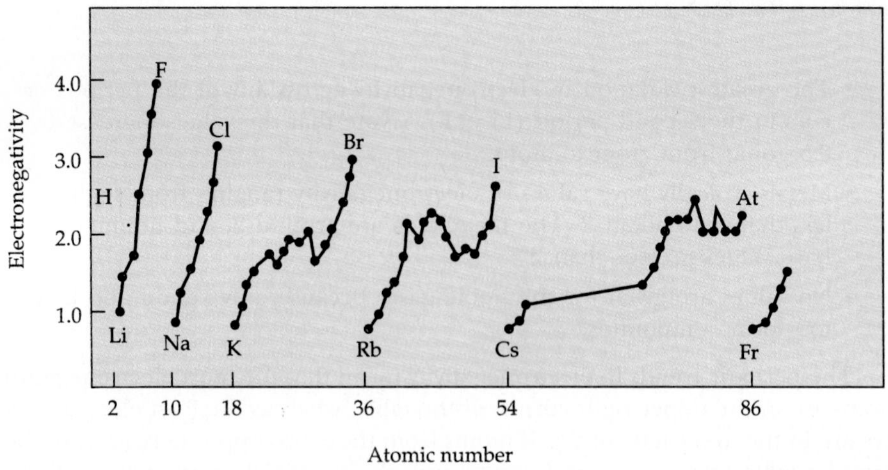

Figure 9.6 Values of the electronegativities of the elements plotted against atomic number.

In the 1930s, Linus Pauling analyzed the energies of bonds, and developed a scale of atom electronegativities (Figures 9.6 and 9.7). The **electronegativity (χ)** of an atom ***in a molecule*** is now taken as a *measure of the ability of the atom to attract electrons to itself.* Atom electronegativities are useful in deciding (a) if a bond is polar, (b) which atom of the bond is negative and which is positive, and (c) if one bond is more polar than another.

Although the trends in electronegativities are related to those for the ionization energies and electron affinities of the free atoms, *IE* and *EA* cannot be applied directly to bonding electrons.

As you examine Figures 9.6 and 9.7, notice the following special features and periodic trends:

- The element with the largest electronegativity (4.0) is fluorine in the upper right corner of the table. The element with the smallest value is francium, in the lower left corner. In general, electronegativity increases from left to right across a period and decreases down a group. These trends are clearly related to the trends in ionization energy and electron affinity.

<1.0	2.0 – 2.4
1.0 – 1.4	2.5 – 2.9
1.5 – 1.9	3.0 – 4.0

1 H 2.1																
3 Li 1.0	4 Be 1.5											5 B 2.0	6 C 2.5	7 N 3.0	8 O 3.5	9 F 4.0
11 Na 1.0	12 Mg 1.2											13 Al 1.5	14 Si 1.8	15 P 2.1	16 S 2.5	17 Cl 3.0
19 K 0.9	20 Ca 1.0	21 Sc 1.3	22 Ti 1.4	23 V 1.5	24 Cr 1.6	25 Mn 1.6	26 Fe 1.7	27 Co 1.7	28 Ni 1.8	29 Cu 1.8	30 Zn 1.6	31 Ga 1.7	32 Ge 1.9	33 As 2.1	34 Se 2.4	35 Br 2.8
37 Rb 0.9	38 Sr 1.0	39 Y 1.2	40 Zr 1.3	41 Nb 1.5	42 Mo 1.6	43 Tc 1.7	44 Ru 1.8	45 Rh 1.8	46 Pd 1.8	47 Ag 1.6	48 Cd 1.6	49 In 1.6	50 Sn 1.8	51 Sb 1.9	52 Te 2.1	53 I 2.5
55 Cs 0.8	56 Ba 1.0	57 La 1.1	72 Hf 1.3	73 Ta 1.4	74 W 1.5	75 Re 1.7	76 Os 1.9	77 Ir 1.9	78 Pt 1.8	79 Au 1.9	80 Hg 1.7	81 Tl 1.6	82 Pb 1.7	83 Bi 1.8	84 Po 1.9	85 At 2.1
87 Fr 0.8	88 Ra 1.0	89 Ac 1.1														

Figure 9.7 Electronegativity values in a periodic table arrangement.

- The greatest variation in electronegativity across any of the periods occurs in the second period (Li···F). (Note that the values increase by 0.5 going from atom to atom.)
- Metals typically have values of electronegativity ranging from slightly less than 1 to about 2. The metalloids are around 2, and nonmetals have values greater than 2.
- No values are given for the noble gases because only xenon and krypton form compounds.

The ionic character of a bond increases as the difference in atom electronegativities increases. A rough scale relating the difference in χ and % ionic character is as follows: 1.0, 20%; 1.5, 40%; 2.0, 60%; and 2.5, 80%.

The periodic trends in electronegativity mean that the most electronegative atoms are in the upper right corner of the table, whereas the least electronegative are in the lower left corner. If atoms from these two opposite regions of the periodic table form a chemical compound, their large electronegativity difference means that the more electronegative atom of the pair more strongly attracts the bonding electrons, and the bond is polar. If the electronegativity difference is large, the bond is nearly or completely ionic. For example, CsF is a highly polar (nearly ionic) combination because the elements are from opposite corners of the periodic table. The difference in χ values, $\Delta\chi$, is 3.2 [= 4.0 (for F) − 0.8 (for Cs)]. Based on electronegativity values, Cs is positive, F negative, and the bond is highly ionic.

EXAMPLE 9.5 *Estimating Bond Polarities*

For each of the following bond pairs, tell which is the more polar and indicate the negative and positive poles.

1. Li—F and Li—I

2. C—S and P—P

3. C═O and C═S

Solution

1. Li and F lie on opposite sides of the periodic table, with F in the extreme upper right corner (χ for Li = 1.0 and χ for F = 4.0). Similarly, Li and I are on opposite sides of the table, but I is at the bottom right corner of the table (χ for I = 2.5), indicating that it is much less electronegative than F. Therefore, although both bonds are expected to be strongly polar, with Li positive and the halide negative, the Li—F bond is much more polar than the Li—I bond. Indeed, LiF is almost as ionic as CsF.

$\delta+$ $\delta-$ Li—F $\Delta\chi = 3.0$, ionic	$\delta+$ $\delta-$ Li—I $\Delta\chi = 1.5$, moderately ionic

2. The P—P bond is nonpolar (or purely covalent) because the bond is between two atoms of the same kind. C is in the second period, and S is in the third period but closer to the right side of the table. Consequently, C and S both have the same electronegativity (2.5), and so their bond is also predicted to be nonpolar.

3. O lies above S in the periodic table, so O is more electronegative than S. This

PORTRAIT OF A SCIENTIST

Linus Pauling (1901–1994)

Linus Pauling was born in Portland, Oregon, in 1901, the son of a druggist. He earned a B.Sc. degree in chemical engineering from Oregon State College in 1922 and completed his Ph.D. in chemistry at the California Institute of Technology in 1925. Before joining Cal Tech as a faculty member, he traveled to Europe where he worked briefly with Erwin Schrödinger and Niels Bohr (Chapter 7). In chemistry he is best known for his work on chemical bonding. Indeed, his book on *The Nature of the Chemical Bond* has influenced several generations of scientists, and it was for this work that he was awarded the Nobel Prize in chemistry in 1954. Shortly after World War II, Pauling and his wife began a crusade to limit nuclear weapons, a crusade that came to fruition in the Limited Test Ban Treaty of 1963. For this effort, Pauling was awarded the 1963 Nobel Prize for peace. Never before had any person received two unshared Nobel Prizes.

(Oesper Collection in the History of Chemistry/ University of Cincinnati)

means the C═O bond is more polar than the C═S bond, which was predicted in part (2) to be nonpolar. For the C—O bond, O is the more negative atom of the polar bond, C—O. In this case, $\Delta\chi$ is 1.0, indicating a moderately polar bond.

EXERCISE 9.9 ***Bond Polarity***

For each of the following pairs of bonds, decide which is the more polar. For each polar bond, indicate the positive and negative poles. First make your prediction from the relative atom positions in the periodic table; then check your prediction by calculating $\Delta\chi$.

1. H—F and H—I
2. B—C and B—F
3. C—Si and C—S ■

Oxidation Numbers and Atom Formal Charges

There are a number of practical applications of the concept of electronegativity. For the moment, we illustrate two of them. First, electronegativity is useful in understanding the concept of oxidation numbers. Second, it can help us decide which of several possible molecular resonance structures contributes most significantly to the actual electronic structure of the molecule or ion.

Oxidation Numbers

We introduced oxidation numbers in Chapter 4 to keep track of electrons in oxidation-reduction reactions.

A free atom has a net electric charge of zero. If the atom is bound to another in a molecule, however, some of its valence electrons are shared with other atoms. *The oxidation number of an atom is the charge the atom would have if ALL of its bonds were considered to be completely ionic;* that is, all bonding pairs have been transferred to the most electronegative atom. For example, the rules introduced in Chapter 4 assumed the oxidation number of H in HF is +1, which leads to a value of −1 for the F atom. If the H—F bond were considered completely ionic, the bond pair would be transferred to the more electronegative atom, F, and the atom would become the F^- ion, leaving H as a simple H^+ ion.

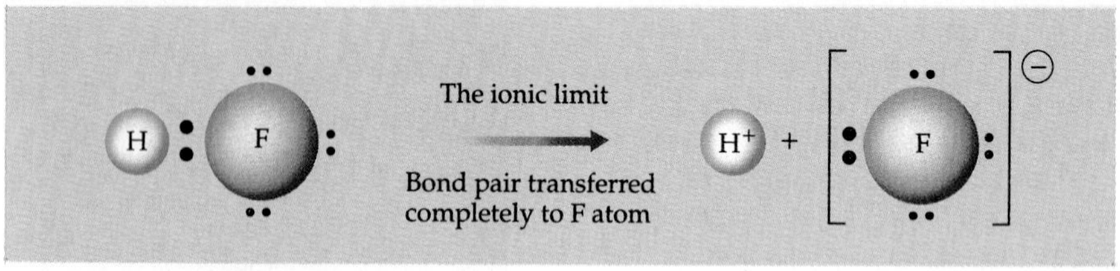

An atom in a compound or polyatomic ion that is bound to several oxygen or halogen atoms can have a very large oxidation number. As an example, consider the Lewis dot structure for the perchlorate ion, ClO_4^-. Because O atoms are more electronegative than a Cl atom, in the ionic limit the bonding electrons would cluster around the O atoms, leaving a Cl atom with a 7+ charge. Experiments show that highly charged ions such as Cl^{7+} voraciously attract electrons. It is inconceivable, therefore, that any species under normal conditions contains a Cl^{7+} ion in such a highly charged positive state. Instead of a 7+ charge, we say that Cl has an oxidation number of +7. Any charge in excess of ±3 on an atom chemical compound is *highly* improbable. This tells you that *oxidation numbers usually do not represent the real charge on an atom.* Nonetheless, oxidation numbers are useful to keep track of electrons in molecules and in chemical reactions (Chapter 4) and indicate the more positive and more negative atoms in a molecule. The latter helps to understand how molecules react to form new bonds.

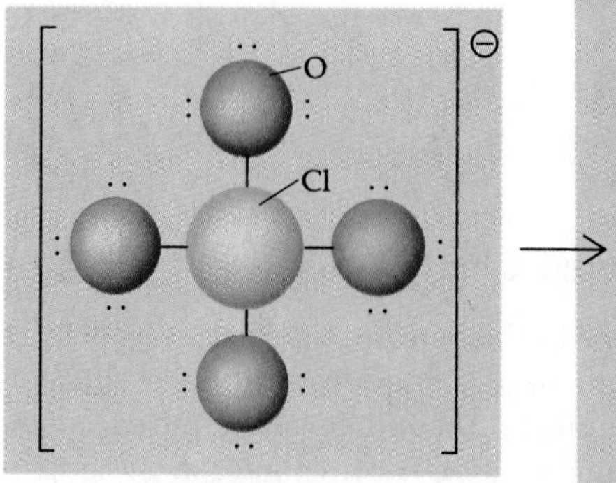

Lewis Structure

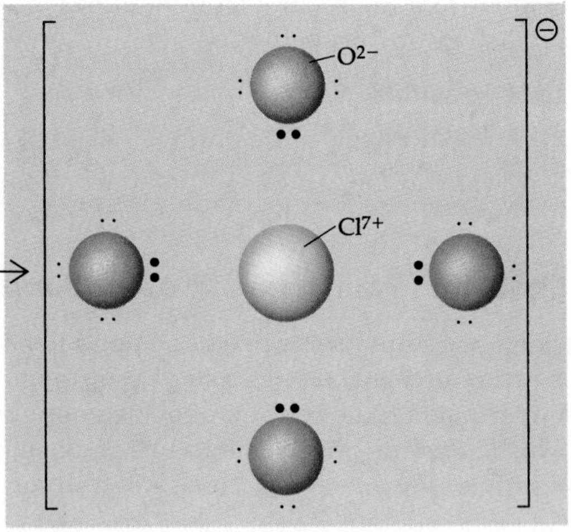

Ionic Limit for ClO_4^-

EXERCISE 9.10 ***Oxidation Numbers***

Using the approach just discussed, calculate the oxidation number for each atom in each of the following species. Do your results agree with the guidelines for deriving oxidation numbers given in Section 4.10?

1. SF_4

2. CO_3^{2-}

3. SO_3 ■

Formal Charges on Atoms

Oxidation numbers do not represent the real charge on an atom in a molecule. By *assuming that each bond pair is shared equally by two atoms,* however, we can make a more reasonable estimate of atom charges in molecules. These estimates are called **atom formal charges** and are given by the expression

$$\text{Atom formal charge} = \text{group number} - \text{number of lone pair electrons} - \frac{1}{2}(\text{number of bonding electrons})$$

Here we count the lone electrons around an atom and one half of the bonding electrons that it shares. To illustrate this, let us take one of the resonance structures for the nitrate ion, NO_3^-. Using the previous equation, we calculate a formal charge of 1− for the single bonded O atoms,

$$\text{Formal charge for single bonded O atom} = 6 - 6 - \frac{1}{2}(2) = 1-$$

a formal charge of 0 for the doubly bonded O atom,

$$\text{Formal charge for double bonded O atom} = 6 - 4 - \frac{1}{2}(4) = 0$$

and a formal charge 1+ for the central N atom.

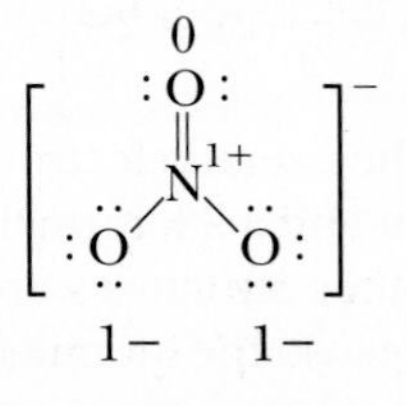

A resonance structure for the nitrate ion showing formal atom charges.

For comparison, the oxidation number of the O atoms is −2, and that of N is +5. Notice that the signs of the formal charges and oxidation numbers are the same (as is usually the case), but the magnitude of the formal charge is more realistic. Also notice that *the sum of the formal charges is equal to the ion charge.*

In the resonance structures for O_3 and CO_3^{2-} drawn in Section 9.3, all the possible resonance structures are equally likely; they are "equivalent" structures. Therefore, the molecule or ion has a symmetrical distribution of electrons over all the atoms involved—that is, its electronic structure consists of an equal "mixture," or "hybrid," of the resonance structures. In many other molecules, however, two or more resonance structures can be written that are not equivalent; this means that the electronic structure is more like one of the resonance structures than the others. Formal charges can be used together with the following two rules to decide which resonance structure is most stable and which is therefore the most important.

- Atoms in molecules (or ions) should have formal charges as small as possible. (This is the so-called *principle of electroneutrality*).
- A molecule (or ion) is most stable when any negative charge resides on the most electronegative atom.

Let us use formal charges on atoms to decide which of two resonance structures is more important for CO_2. For structure A, each atom has a formal charge of 0, a favorable situation. In B, however, one oxygen atom has a formal charge of 1+. This is an unfavorable condition for the very electronegative oxygen atom, so resonance structure B is thought to be of little importance.

Formal charges

Resonance structures

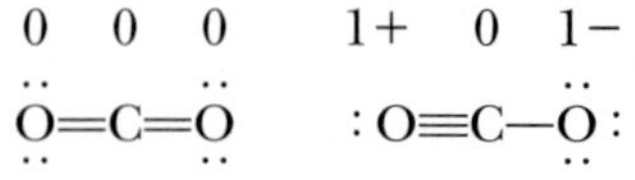

A B

Now use what you have learned with CO_2 to decide on the most reasonable resonance structure for the acid HOCN. The three resonance structures possible for HOCN are written with their atom formal charges as follows:

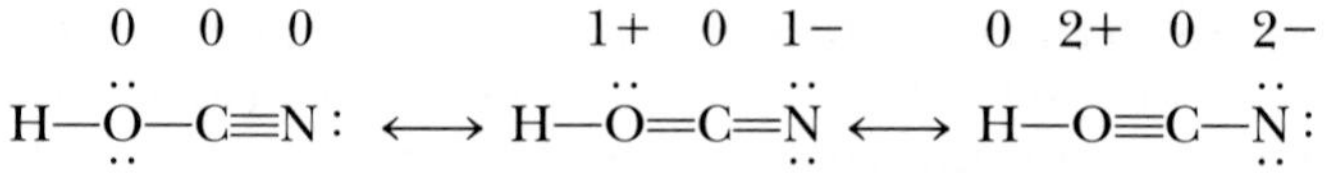

A B C

Examine Structure C as an example of the formal charge calculation for HOCN.

$$\text{Formal charge for O} = 6 - 0 - \frac{1}{2}(8) = 2+$$

$$\text{Formal charge for C} = 4 - 0 - \frac{1}{2}(8) = 0$$

$$\text{Formal charge for N} = 5 - 6 - \frac{1}{2}(2) = 2-$$

Structure C will not contribute significantly to the overall electronic structure of the ion. It has a 2− formal charge on the N atom and a 2+ formal charge on the O atom, whereas the formal charges in the other structures are 0 or 1±. Of structures A and B, A is the more significant because the atomic formal charges are all zero.

Sulfur dioxide, SO_2, is a common air pollutant near coal- and oil-burning power plants, and an enormous quantity is evolved by the world's active volcanoes. Let us consider the resonance structures for the molecule and calculate the atom formal charges.

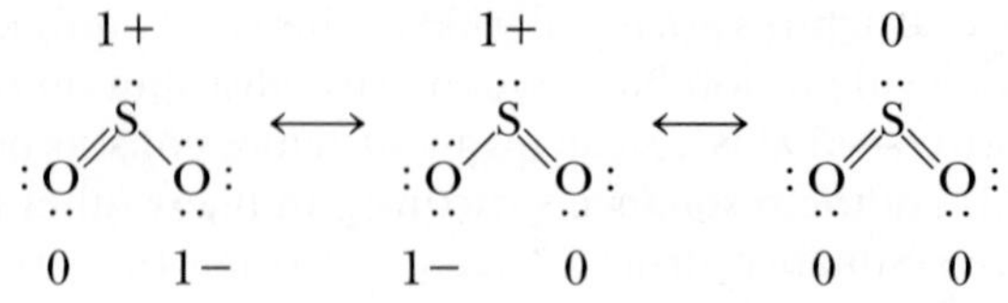

A B C

Both resonance structures A and B have formal charges of 1+ on the S atom and 1− on one of the O atoms. We can write a third resonance structure (C), however, by forming *two* SO double bonds, leading to zero formal charge on all three atoms. Structure C has more than four pairs of electrons around the S atom, but this is acceptable because elements from the third and higher periods may accommodate five or more pairs of electrons. Therefore, you often see resonance structures of molecules or ions with SO or PO double bonds because such structures minimize atom formal charges.

EXAMPLE 9.6 *Calculating Atom Formal Charges*

Calculate the atom formal charges for the S and O atoms for five of the ways of depicting the electronic structure of the sulfate ion.

Solution The sulfate ion, with a central S atom, has 32 valence electrons (six for each of the Group 6A atoms plus two for the negative charge). A number of structures can be written. (Note that B, C, and D have other resonance structures.)

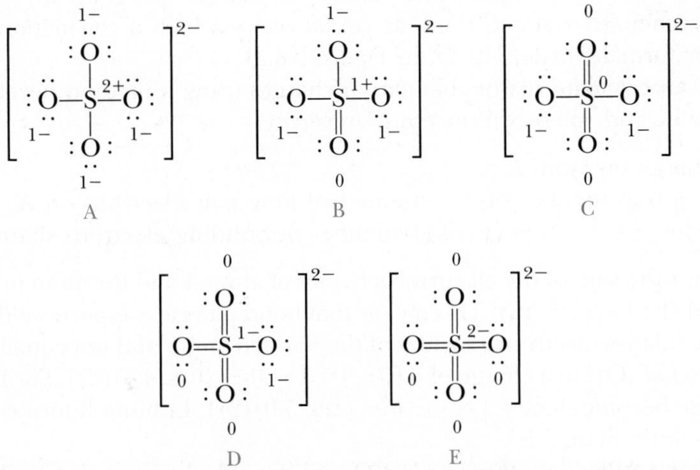

As an example, consider the calculation of formal charges in B.

$$\text{S atom formal charge} = 6 - 0 - \frac{1}{2}(10) = 1+$$

$$\text{Doubly bonded O atom formal charge} = 6 - 4 - \frac{1}{2}(4) = 0$$

$$\text{Single bonded O atom formal charge} = 6 - 6 - \frac{1}{2}(2) = 1-$$

Structures A and E contribute little to the overall electronic structure of the ion because both place a high charge on S. Furthermore, E has a 2− charge on S, a less electronegative atom than O. Structure C minimizes the charges on S and O, giving a formal charge of 0 for S and of 0 or 1− for the more electronegative O atoms. Therefore, structure C is the largest contributor to the overall electronic structure of the sulfate ion.

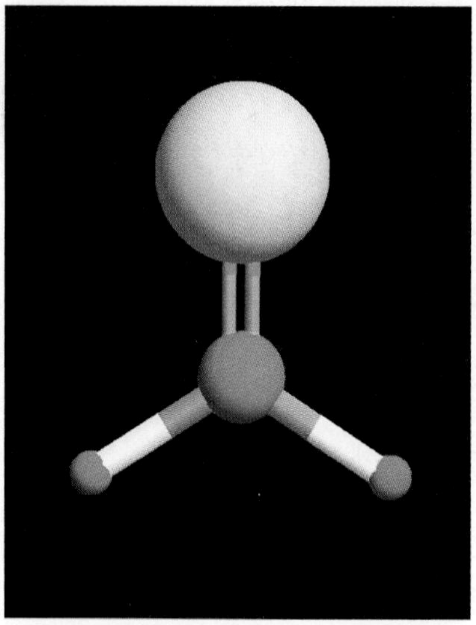

Figure 9.8 Computer generated representation of formaldehyde molecule showing the partial charges on the atoms. Red designates an atom with a positive charge, and yellow designates an atom with a negative charge. The relative sizes of the atoms represent their relative partial charges. Most of the positive charge resides on the C and H atoms, and most of the negative charge is on the O atom, as expected. Calculations by computer give H = +0.069, C = +0.14, and O = −0.28. (Susan Young)

EXERCISE 9.11 ***Atom Formal Charges***

The Lewis structures for the very reactive molecule H_2BF are depicted here. Boron often forms compounds having only three bond pairs and no lone pairs around the B atom. To have four electron pairs around boron, however, one might write Structure B. Calculate the formal charge on each atom in Structure B and comment on the relative importance of the two structures.

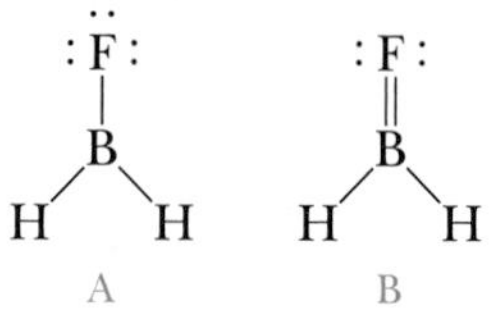

■

Atom Partial Charges

Because chemical reactions often occur by attack by some reagent at the more positive or more negative site in a molecule, chemists have long been interested in making reasonably accurate predictions of atom formal charges. Recently, computer programs that enable us to make such calculations have become more widely available, and, because the calculated charges are generally fractional numbers, chemists refer to them as *partial charges.* Such a calculation is illustrated for formaldehyde, H_2CO, in Figure 9.8.

You can calculate reasonable partial charges using a method developed by Professor Leland Allen of Princeton University.

Partial charge on atom A =

$$\text{group number of A} - \text{number of lone pair electrons on A} - (\chi_A/\Sigma\chi)(\text{number of bonding electrons shared by A})$$

where $\Sigma\chi$ is the sum of the electronegativities of atom A and the atom to which it is bonded (B) $(= \chi_A + \chi_B)$. This means that bond pairs are apportioned according to the relative electronegativities of the bonded atoms and not equally. When applied to CsF, Cs has a charge of 0.67+ $[= 1 - 0 - (0.8/4.8)(2)]$. For Li in LiF, the charge becomes 0.60+ $[= 1 - 0 - (1.0/5.0)(2)]$. Lithium fluoride is somewhat less ionic than CsF.

Applying Allen's method to H_2CO, you would obtain a partial charge of 0.33− on the oxygen atom $[= 6 - 4 - (3.5/6.0)(4)]$, and the partial charge on the carbon atom is 0.16+ $[= 4 - 0 - (2.5/6.0)(4) - (2.5/4.6)(2) - (2.5/4.6)(2)]$. Because the sum of the partial charges on O, C, and the 2 H's must equal zero, the partial charge on each H atom must be 0.09+. These are close to the charges calculated by a more sophisticated computer program. They are also more realistic than formal charges (which all = 0 in H_2CO) in that they better reflect the differing abilities of atoms in molecules to attract electrons.

9.5 MOLECULAR SHAPE

We can now turn to one of the most important aspects of chemistry, an understanding of the three-dimensional geometry of molecules and ions. The importance of this understanding cannot be overstated because many of the physical and chemical properties of compounds are tied to their structures. This is the reason we have used various structural models throughout the text.

A simple, reliable method for predicting the shapes of covalent molecules and polyatomic ions is the **valence shell electron-pair repulsion (VSEPR)** model,

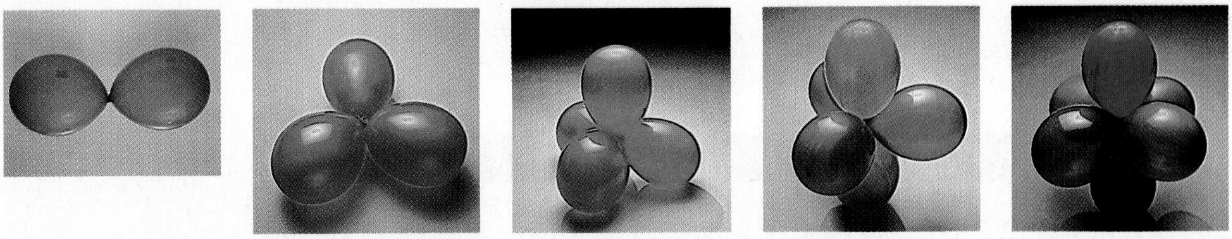

Figure 9.9 Balloon models of the electron pair geometries for two to six electron pairs, as predicted by VSEPR theory. (If two to six balloons of similar size and shape are tied together, they will naturally assume the geometries illustrated.) (C. D. Winters)

devised by Ronald J. Gillespie (1924–) and Ronald S. Nyholm (1917–1971). The VSEPR model is based on the idea that repulsions among the pairs of bonding or nonbonding electrons of an atom control the angles between bonds from that atom to other atoms surrounding it. The central atom and its core electrons are represented by the atom's symbol. This atomic core is surrounded by pairs of valence electrons, the number of pairs corresponding to the number of pairs of dots in the Lewis structure. The geometric arrangement of the electron pairs is predicted on the basis of their repulsions, and the geometry of the molecule or polyatomic ion depends on the numbers of lone pairs and bonding pairs.

It is important to note that the VSEPR model does not apply to molecules with a transition metal at the center.

How do repulsions among electron pairs result in different shapes? Imagine that a balloon represents each electron pair. Each balloon's volume represents a repulsive force that prevents other balloons from occupying the same space. When two, three, four, five, or six balloons are tied together at a central point (the central point represents the nucleus and core electrons of a central atom), the balloons form the shapes shown in Figure 9.9. These geometric arrangements minimize interactions among the balloons (electron-pair repulsions).

Central Atoms with Only Bond Pairs

The simplest application of VSEPR is to molecules in which all the electron pairs around the central atom are in single covalent bonds. Figure 9.10 illustrates the geometries predicted by the VSEPR model for molecules of the types AX_2 to AX_6 that contain only single covalent bonds, where A is the central atom.

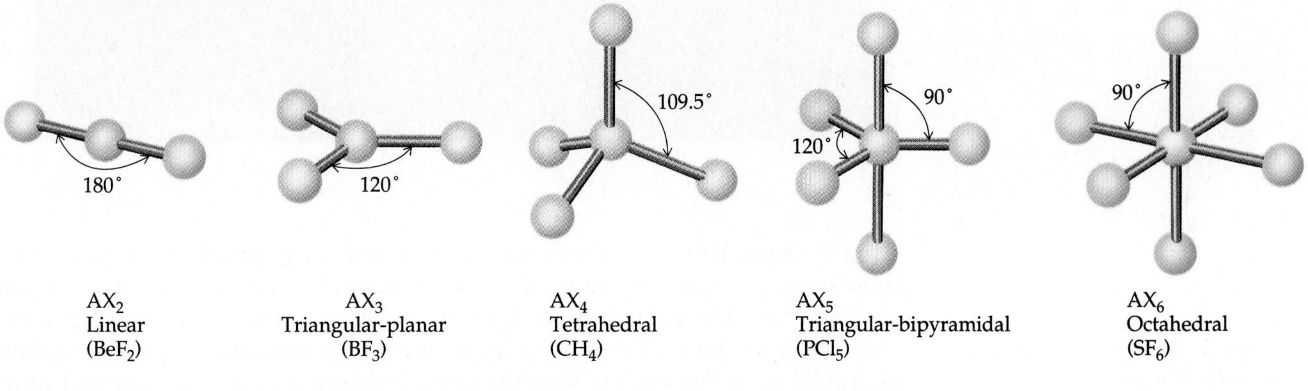

Figure 9.10 Geometries predicted by the VSEPR model for molecules of the types AX_2 to AX_6 that contain only single covalent bonds.

CURRENT ISSUES IN CHEMISTRY

Computer Molecular Modeling

As described in *The Introduction* (The Nature of Chemistry), modeling of molecules by computer is a widely used technique, and a practical example shows its usefulness.

A molecule that is needed by the chemical industry in the manufacture of high-strength polymers (see Chapter 11) is 2,6-diisopropylnaphthalene (2,6-DIPN). The problem is that the process used to make 2,6-DIPN also gives the closely related molecule 2,7-DIPN as a product. The latter molecule is not useful.

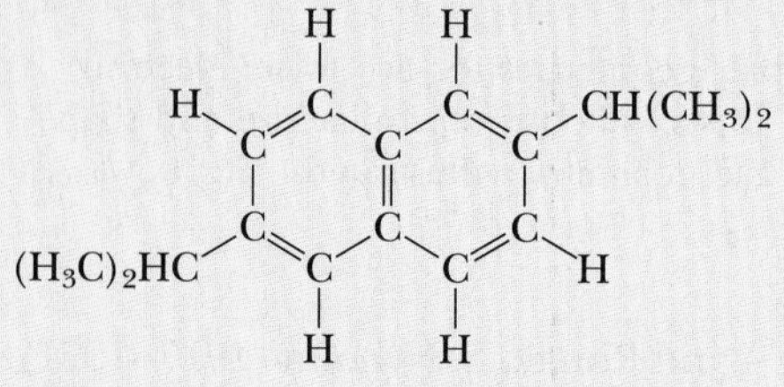

2,6-DIPN

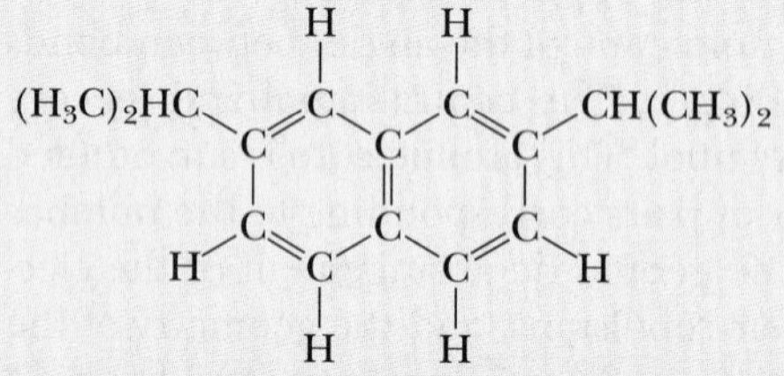

2,7-DIPN

This is wasteful of starting materials, and it means that chemical engineers must develop a way to separate the two compounds before the 2,6-DIPN can be used.

Chemists at Catalytica, Inc., in California have solved the problem by developing a method of raising the yield of 2,6-DIPN and of separating the 2,6 and 2,7 compounds. The separation method was worked out using computer molecular modeling. The figure below shows computer-generated models of the two compounds. If the two molecules are viewed from one end, it is clear that they fill space

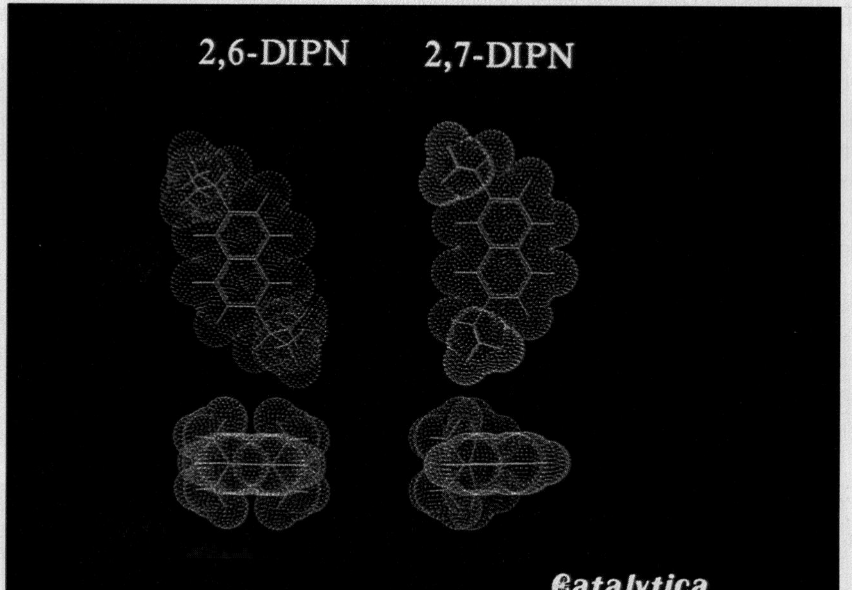

The word "triangular" is often used in place of trigonal. Both refer to objects at the corners of a triangle.

The **linear** geometry for two bond pairs and the **trigonal planar** geometry for three bond pairs involve a central atom that does not have an octet of electrons (Section 9.3). The central atom in a **tetrahedral** molecule obeys the octet rule with four bond pairs. The central atoms in **trigonal bipyramidal** and **octahedral** molecules have five and six bonding pairs, respectively, and are expected only

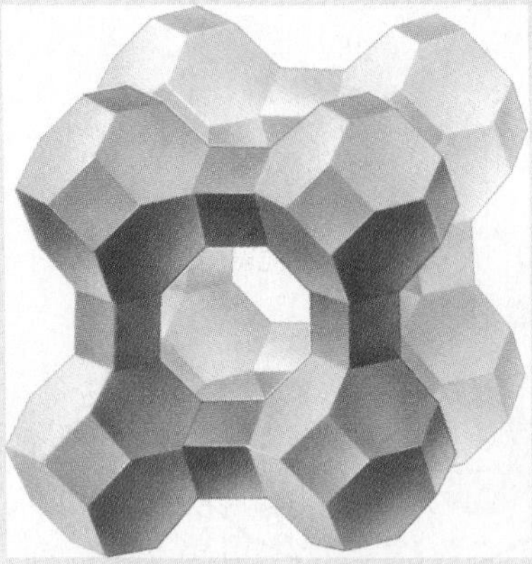

A naturally occurring zeolite is a network of aluminum, silicon, and oxygen atoms.

differently. Therefore, the Catalytica chemists used a "molecular sieve" that would recognize these different shapes. The sieve they chose belongs to a class of naturally occurring substances called zeolites. These are cages of atoms built from rings of silicon, aluminum, and oxygen atoms.

A computer model of a zeolite (called mordenite) shows that 2,6-DIPN—the desired molecule—fits comfortably into a channel in the zeolite, whereas 2,7-DIPN does not.

For more on the use of computer molecular modeling, see articles such as A. J. Olson and D. S. Goddsell, "Visualizing Biological Molecules," *Scientific American,* November, 1992, pages 76–81, and C. E. Bugg, W. M. Carson, and J. A. Montgomery, "Drugs by Design," Scientific American, December, 1993, pages 92–98.

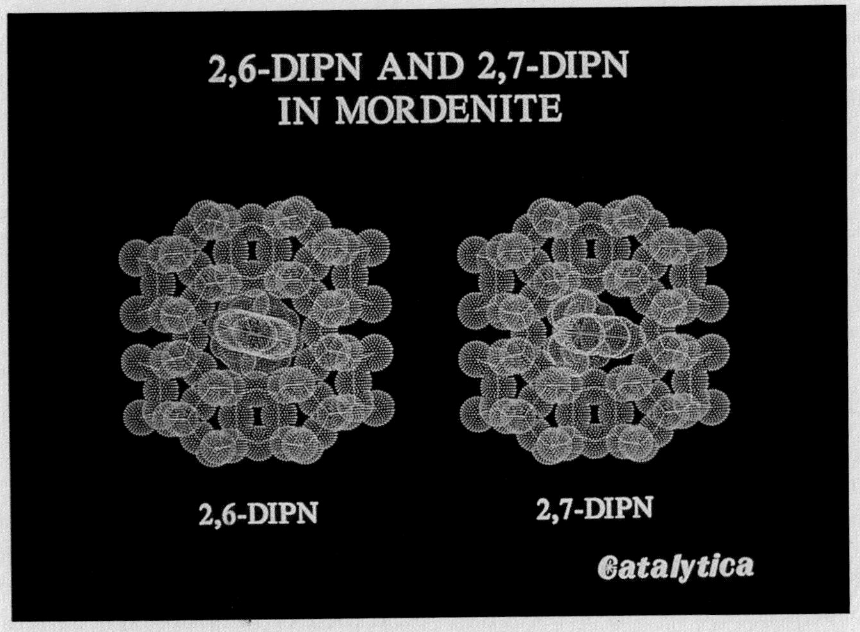

A computer model of 2,6-DIPN and 2,7-DIPN in a zeolite. Only the 2,6-DIPN fits comfortably into the channel in the zeolite.

when the central atom is an element in Period 3 or higher (Section 9.3). The geometries that are illustrated in Figure 9.10 are by far the most common in molecules and ions, and you should make yourself thoroughly familiar with them. The predicted bond angles agree with experimental values obtained from structural studies.

EXAMPLE 9.7 *Predicting Molecular Shapes*

Predict the shape of silicon tetrachloride, $SiCl_4$.

Solution The first step in predicting the shape of any molecule or ion is to draw the Lewis structure.

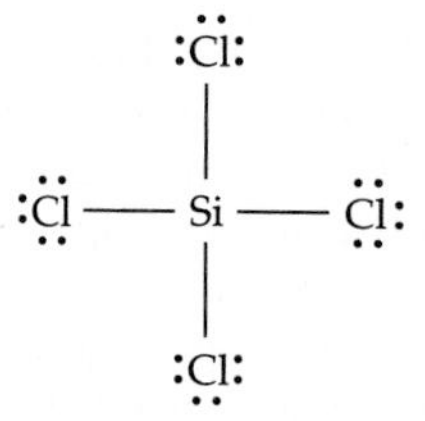

Lewis structure

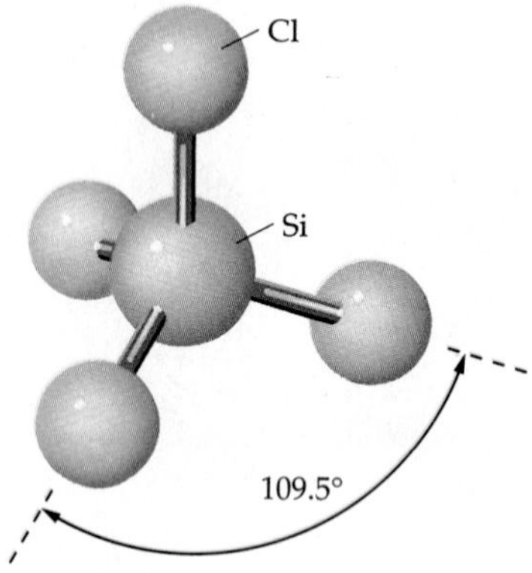

Molecular geometry

The Lewis structure does not need to be drawn in any particular way. We only need to convey the fact that it is surrounded by four single covalent Si—Cl bonds. However, this last part is the crucial information you need. Because four bond pairs form four single covalent bonds to Si, a tetrahedral structure is predicted for the $SiCl_4$ molecule, with Cl—Si—Cl bond angles of 109.5°. This prediction agrees with structural results for $SiCl_4$.

EXERCISE 9.12 *Predicting Molecular Shapes*

What is the shape of the dichloromethane molecule, CH_2Cl_2? Predict the Cl—C—Cl bond angle. ■

Central Atoms with Bond Pairs and Lone Pairs

How does the presence of lone pairs on the central atom affect the geometry of the molecule or polyatomic ion? The easiest way to visualize this situation is to return to the balloon model. Notice that the electron pairs do not have to be bonding pairs. We can predict the geometry of a molecule by applying the VSEPR model to the *total* number of electron pairs around the central atom. The shape predicted by this method, however, is the electron-pair geometry rather than the molecular geometry. We must then decide which positions are occupied by bond pairs and which by lone pairs. The **electron-pair geometry** around a central atom includes the spatial positions of all bond pairs and lone pairs of electrons, whereas the **molecular geometry** of a molecule or ion only involves the arrangement in space of its atoms. This distinction is necessary because only the atoms are located by structural techniques such as x-ray crystallography (Section

13.4), and positions occupied by lone pairs are not specified in the shapes of molecules. The success of the VSEPR model in predicting molecular shapes indicates that it is correct to account for the effects of lone pairs in this way. In other words, lone pairs of electrons around the central atom occupy spatial positions even though they are not included in the description of the shape of the molecule or ion.

Let's use the VSEPR model to predict the molecular geometry and bond angles in the NH_3 molecule, which has a lone pair on the central atom. First, draw the Lewis structure and count the total number of electron pairs around the central N atom.

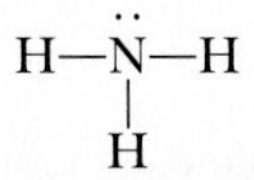

Because three bond pairs and one lone pair occur for a total of four pairs of electrons, we predict that the *electron-pair geometry* is tetrahedral. Draw a tetrahedron with N as the central atom and the three bond pairs represented by lines. The lone pair is drawn as follows to indicate its spatial position in the tetrahedron:

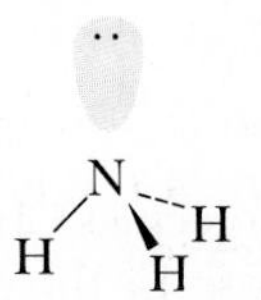

The *molecular geometry* is described as a trigonal pyramid because the nitrogen atom is at the apex of the pyramid, and the three hydrogen atoms form its trigonal base. (This can be seen by covering up the lone pair of electrons and looking at the molecular geometry.)

What is the predicted value for the H—N—H bond angle? Because the electron-pair geometry is tetrahedral, we would expect the H—N—H bond angle to be 109.5°. The experimentally determined bond angles in NH_3 are 107.5°, however. Does this different value indicate a flaw in the model, or could there be a difference between the spatial requirements of lone pairs and bond pairs? Actually, the latter is the case. Bond pairs are attracted into the bond region between atoms by the strong attractive forces of protons in two nuclei and are, therefore, relatively compact; in other words, we can think of them as "skinny." For a lone pair, however, only one nucleus attracts the electron pair, and this nuclear charge is not so effective in overcoming the normal repulsive forces between two negative electrons; as a result, lone pairs are "fat." Their increased volume spreads the lone pairs farther apart and squeezes the bond pairs closer together. Hence, the relative strength of repulsions is

Lone pair-lone pair > lone pair-bond pair > bond pair-bond pair

Using the balloon analogy, a lone pair is like a fatter balloon that takes up more room and squeezes the thinner balloons closer together.

Gillespie and Nyholm recognized the importance of the different spatial requirements of lone pairs and bond pairs and included this as part of their VSEPR model. For example, they used the VSEPR model to predict variations in

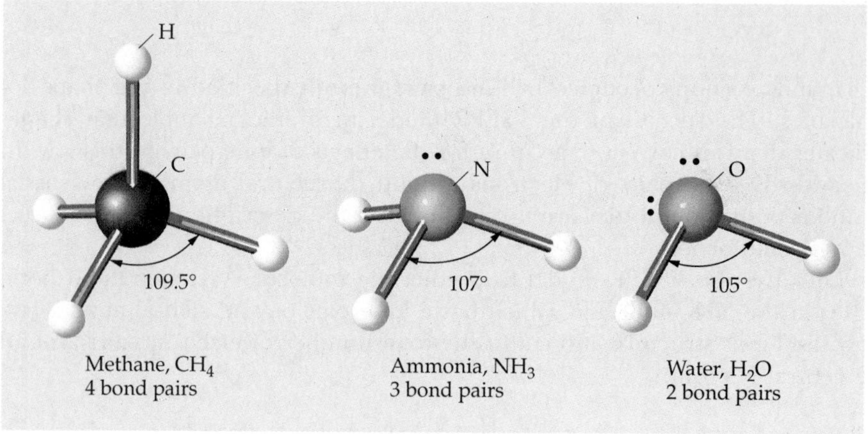

Figure 9.11 The geometries of methane, ammonia, and water. All have four electron pairs around the central atom, so all have a tetrahedral electron-pair geometry. (a) Methane has four bond pairs and so has a tetrahedral molecular shape. (b) Ammonia has three bond pairs and one lone pair, so it has a trigonal pyramidal molecular shape. (c) Water has two bond pairs and two lone pairs, so it has a bent or angular molecular shape. The decrease in bond angles in the series can be explained by the larger spatial requirements of the lone pairs, which squeeze the bond pairs closer together.

the bond angles in the series of molecules CH_4, NH_3, and H_2O. Notice in Figure 9.11 that the bond angles decrease in the series CH_4, NH_3, and H_2O as the number of lone pairs on the central atom increases. Figure 9.12 gives additional examples of electron-pair and molecular geometries for molecules and ions with three and four electron pairs around the central atom. Some experimentally determined bond angles are given for the examples.

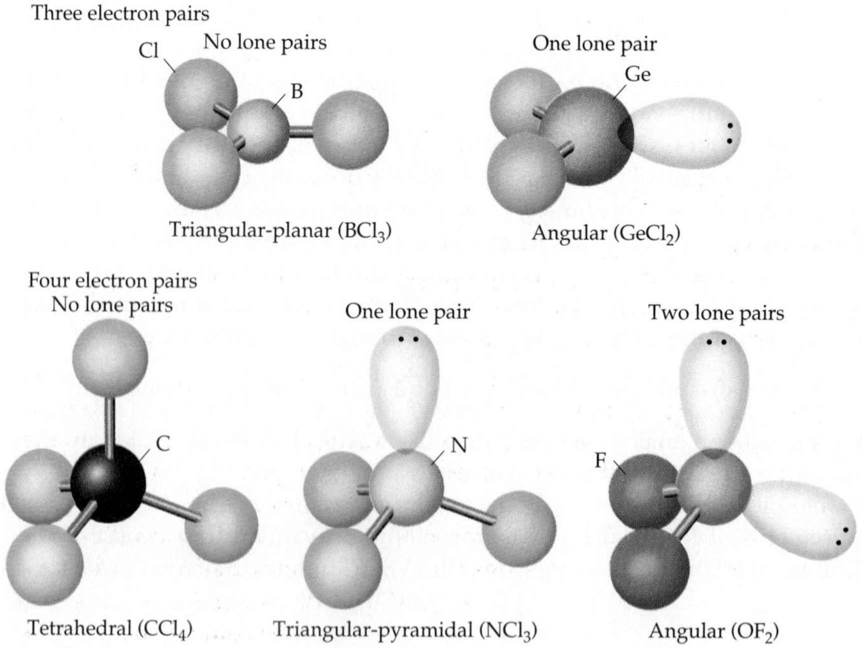

Figure 9.12 Some examples of electron-pair geometries and molecular shapes for molecules with three and four electron pairs around the central atom.

EXAMPLE 9.8 *Finding the Shapes of Molecules*

What are the molecular shapes of PH_3 and ClF_2^+?

Solution

1. The Lewis structure of phosphine, PH_3, shows that the central P atom is surrounded by four electron pairs, so the electron-pair geometry is tetrahedral. Because three of the four pairs are used to bond terminal atoms, the central P atom and the three H atoms form a trigonal *pyramidal* molecular shape like NH_3.

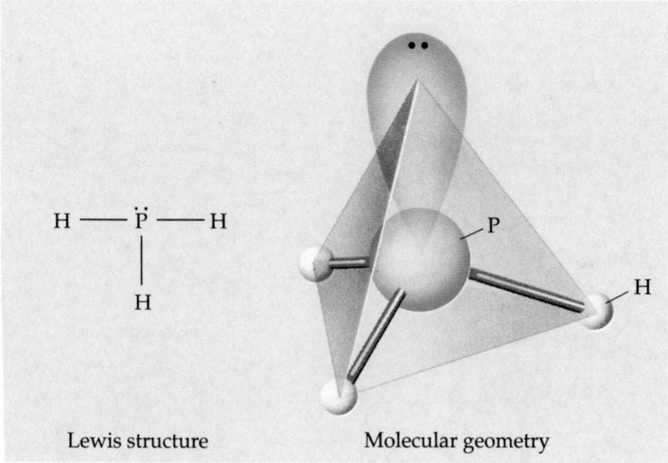

2. The ion ClF_2^+ has 20 valence electrons. The central Cl atom in ClF_2^+ is also surrounded by four electron pairs, so the electron-pair geometry is tetrahedral. Because only two of the four pairs are used to form bonds (as in H_2O), the shape of the ion is described as bent, or angular.

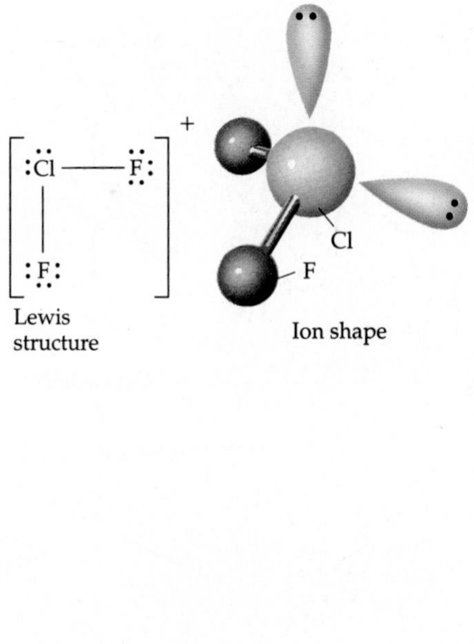

EXERCISE 9.13 *VSEPR and Molecular Shape*

Give the electron-pair geometry and molecular shape for BF_3 and BF_4^-. What is the effect on the molecular geometry of adding an F^- ion to BF_3 to give BF_4^-? ■

The situation becomes more complicated if the central atom has five or six electron pairs, some of which are lone pairs. Let us first consider the entries in Figure 9.13 for the case of five electron pairs. The three angles in the trigonal plane are all 120°. The angles between any of the pairs in this plane and an upper or lower pair are only 90°; thus, the trigonal bipyramidal structure has two sets of positions that are not equivalent. Because the positions in the trigonal plane lie in the equator of an imaginary sphere around the central atom, they are called the *equatorial* positions. The north and south poles are called the *axial* positions. Each equatorial position is closely flanked by only two other positions (the axial ones), and an axial position is closely flanked by three positions (the equatorial ones). This means that any lone pairs, which we assume to be fatter than bonding pairs, prefer to occupy equatorial positions rather than axial positions. For example, consider the ClF_3 molecule, which has three bond pairs and

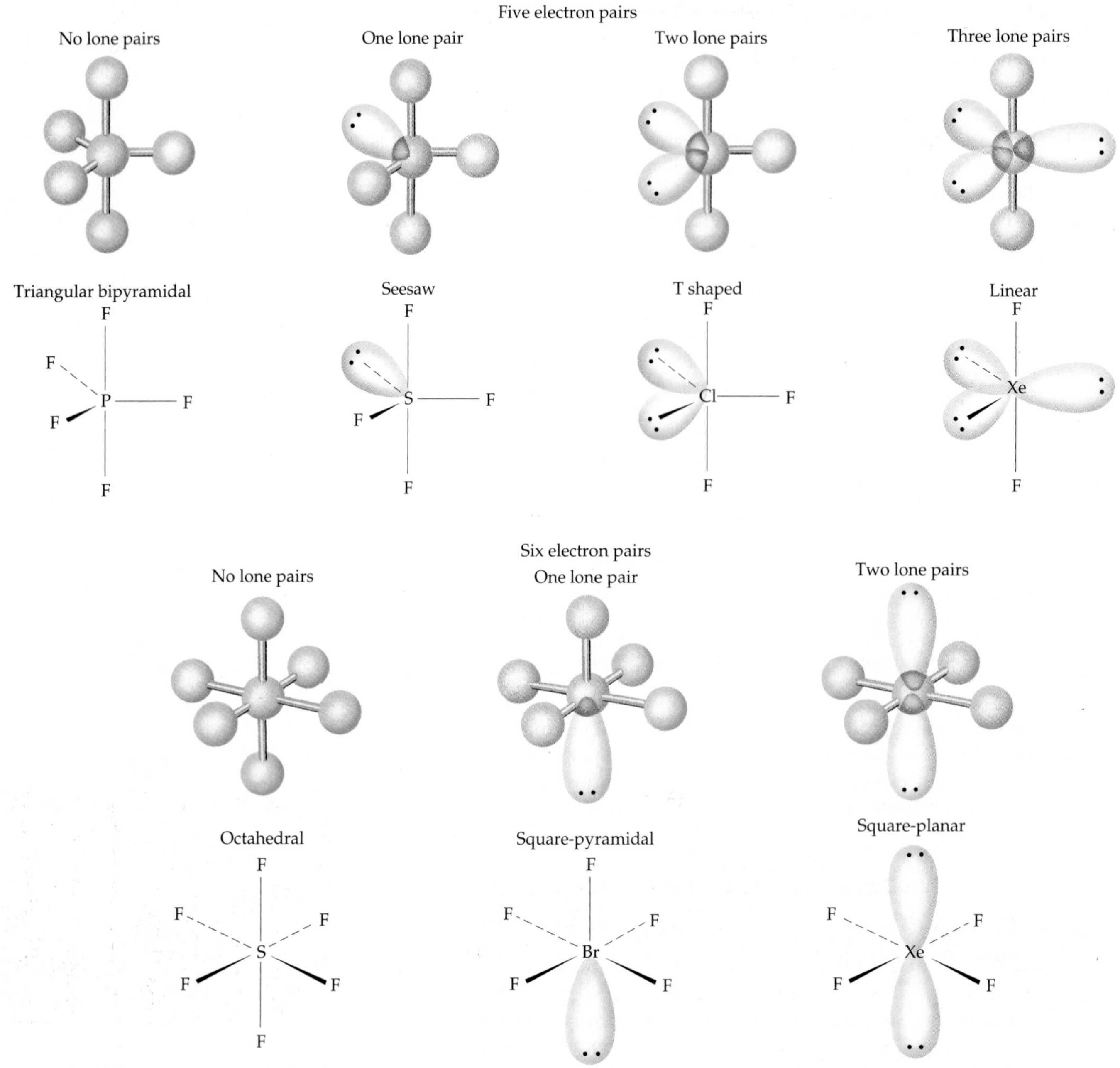

Figure 9.13 Electron-pair geometries and molecular shapes for molecules and ions with five or six electron pairs around the central atom.

two lone pairs. The two lone pairs in ClF_3 are equatorial; two bond pairs are axial and the third is allowed into the equatorial plane, so the molecular geometry is T-shaped (Figure 9.13).

All of the angles in the octahedron are 90°. Unlike the trigonal bipyramid, the octahedron has no distinct axial and equatorial positions; all positions are the same (Figure 9.10). Therefore, if the molecule has one lone pair, as in BrF_5, it makes no difference which apex it occupies. Usually the lone pair is drawn in the top or bottom position to make it easier to visualize the molecular geometry, which in this case is *square-pyramidal* (Figure 9.13).

An example of the power of the VSEPR model is the correct prediction of the shape of XeF_4. At one time the noble gases were not expected to form compounds because the gases have a stable octet of valence electrons. The existence of XeF_4 (Figure 9.14) challenged theorists because it could not be explained with existing bonding theories. Can the VSEPR model predict the correct geometry? The molecule has 36 valence electrons (8 from Xe and 28 from four F atoms). Eight electrons account for four bond pairs around Xe, and a total of 24 electrons occur in the lone pairs on the four F atoms. That leaves four electrons in two lone pairs on the Xe atom.

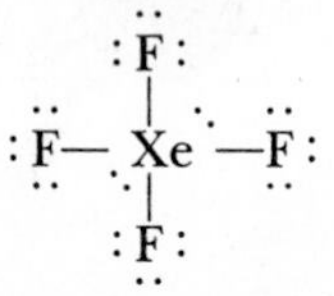

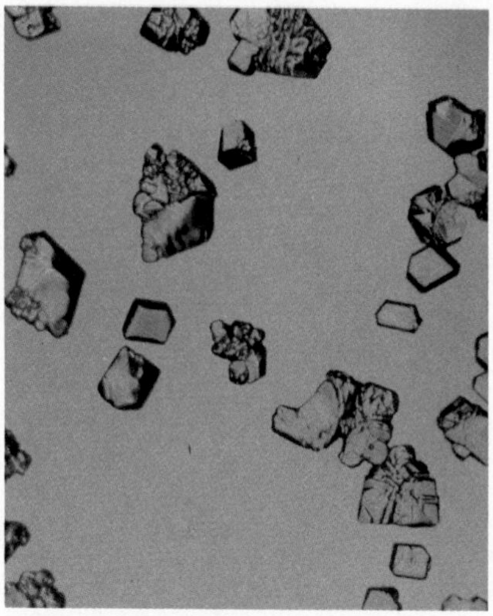

Figure 9.14 Crystals of xenon tetrafluoride, XeF_4. (Argonne National Laboratory)

Because Xe is in Period 5, it can accommodate more than an octet of electrons. The total of six electron pairs on Xe leads to a prediction of an octahedral electron-pair geometry. Where do you put the lone pairs? The lone pairs are best placed at opposite corners of the octahedron to provide them as much room as possible. The result is a *square-planar* molecular geometry for the XeF_4 molecule (cover the lone pairs in the XeF_4 drawing in Figure 9.13 to see this), and that shape agrees with experimental structure.

The VSEPR model has been useful in predicting the geometries of other xenon compounds. For example, the XeF_2 molecule is predicted to have a linear molecular geometry (Figure 9.13) because the three lone pairs occupy the equatorial positions of a trigonal bipyramid, and the two bond pairs are in the axial positions.

EXAMPLE 9.9 ***Predicting Molecular Shape***

What is the shape of the ICl_4^- ion?

Solution The ICl_4^- ion has 36 valence electrons. (Each atom is in Group 7A, giving a total of 35 electrons, and the negative charge adds 1 more.) This is isoelectronic with XeF_4, so the Lewis structure is similar. As in XeF_4, the two lone pairs require as much room as possible, so they are best placed above and below the molecular plane. The ICl_4^- ion therefore has a square-planar geometry.

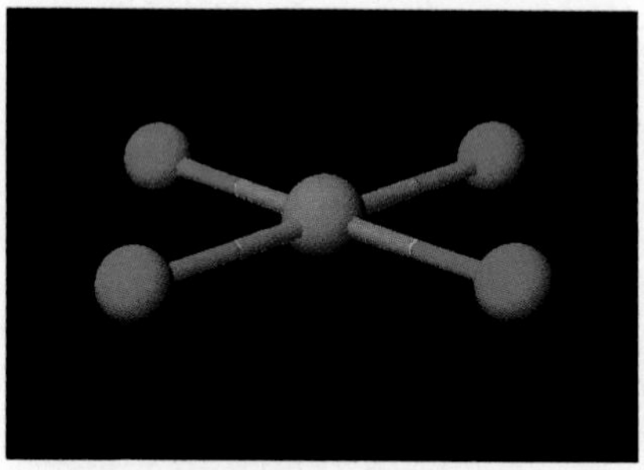

Shape of ICl_4^- ion. (Susan Young)

EXERCISE 9.14 ***Predicting Molecular Shape***

Draw the Lewis structure for ICl_2^-, and then decide on the geometry of the ion. ■

Multiple Bonds and Molecular Geometry

Although double bonds and triple bonds are shorter and stronger than single bonds (Section 9.4), they do not affect predictions of overall molecular shape. Why? Electron pairs involved in a multiple bond are all shared between the same two nuclei and therefore occupy the same region. Because they must remain in that region, two electron pairs in a double bond (or three in a triple bond) are like a single balloon, rather than two or three balloons. All electron pairs in a multiple bond count as one bond, and contribute to molecular geometry the same as a single bond does. For example, the C atom in CO_2 has no lone pairs and participates in two double bonds. Each double bond counts as one for the purpose of predicting geometry, so the structure of CO_2 is linear.

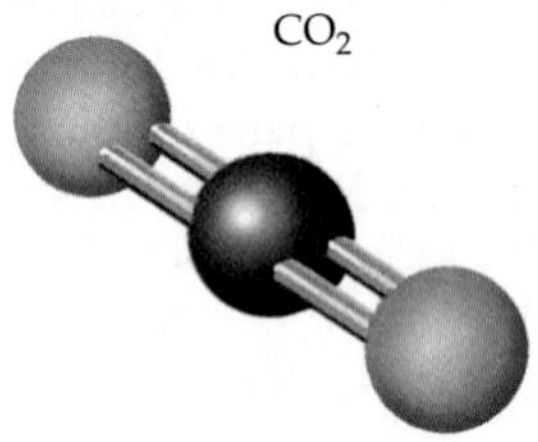

When resonance structures are possible, the geometry can be predicted from any of the Lewis resonance structures or from the resonance hybrid structure. For example, the geometry of the CO_3^{2-} ion is predicted to be trigonal planar because the carbon atom has three sets of bonds and no lone pairs.

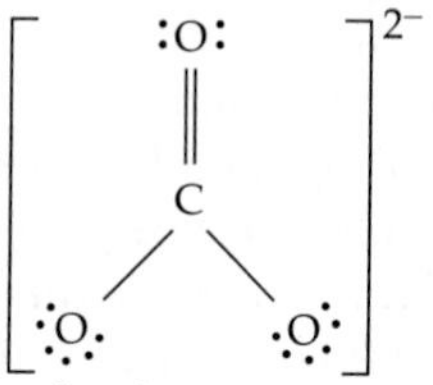

Lewis structure

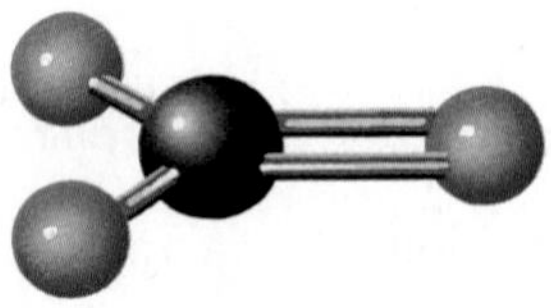

The NO_2^- ion is described as angular, or bent, because it has a lone pair on the central N atom and two bonds in the other two positions. A trigonal-planar electron-pair geometry is expected, and the geometry of the ion is angular, or bent.

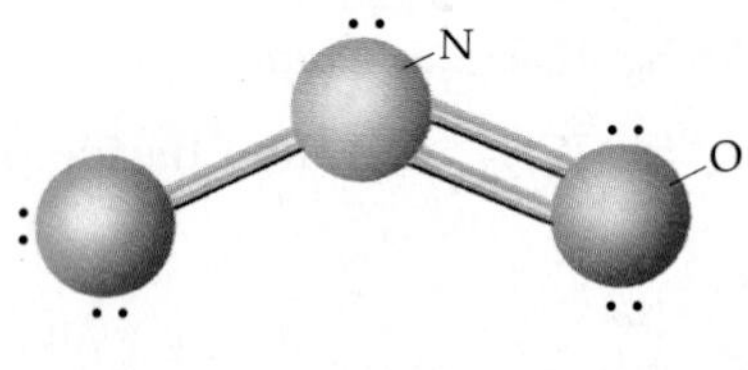

The techniques we have just outlined can be used to find the geometries of much more complicated molecules. Consider, for example, cysteine, one of the natural amino acids.

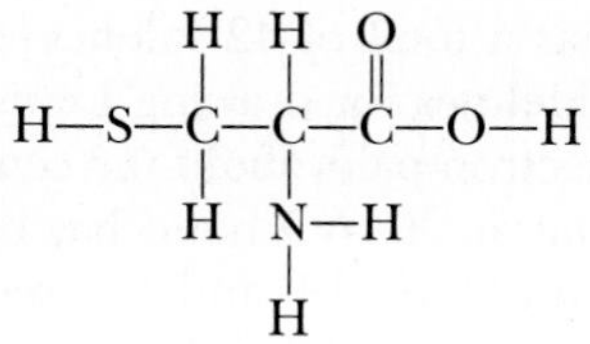

Cysteine, $HSCH_2CH(NH_2)CO_2H$

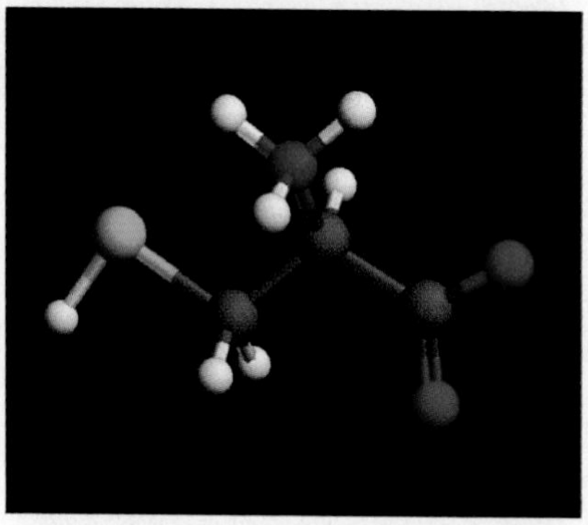

Figure 9.15 The structure of cysteine. This shows the molecule in its ionic form, where an H^+ ion has been transferred from the C—O—H group to the lone pair of the NH_2 group. The geometry around the N atom is tetrahedral because the atom is surrounded by four bonds.

We can get some idea of the overall structure of this molecule by first adding in any missing lone pair electrons and then considering the bond angles around each atom. First, a molecule with the formula $HSCH_2CH(NH_2)COOH$ must have 42 valence electrons, or 21 bonding and lone pair electrons. The bonds thus far located account for 14 of these pairs, so the remaining 7 pairs are distributed as lone pair electrons so that each atom has a share in four pairs.

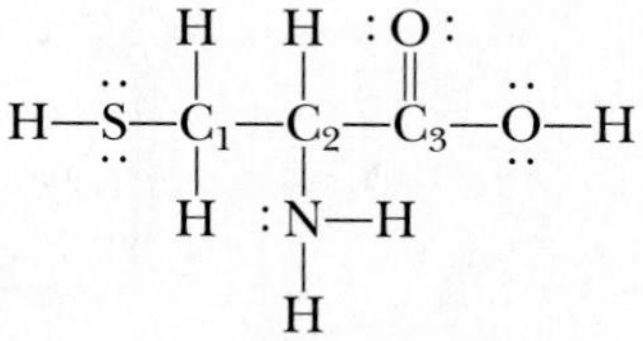

Now, we need to derive the bond angles. Four pairs of electrons occur around the S, N, C_1, and C_2 atoms, so the electron-pair geometry around each is tetrahedral. Thus, the H—S—C and H—N—H angles are predicted to be approximately 109°. The O atom in the grouping C—O—H also is surrounded by four pairs, and so this angle is likewise approximately 109°. Finally, the angle made by O—C_3—O is 120° because the electron-pair geometry around C_3 is planar and trigonal. The complete geometry is given by the computer-generated model shown in Figure 9.15.

EXAMPLE 9.10 *Finding the Shapes of Molecules and Ions*

What are the shapes of NO_3^- and $OXeF_4$?

Solution

1. The NO_3^- ion has the same number of valence electrons as the CO_3^{2-} ion; that is, nitrate and carbonate are isoelectronic. Thus, like the carbonate ion, the electron-pair geometry and molecular shape of the nitrate ion are trigonal-planar.

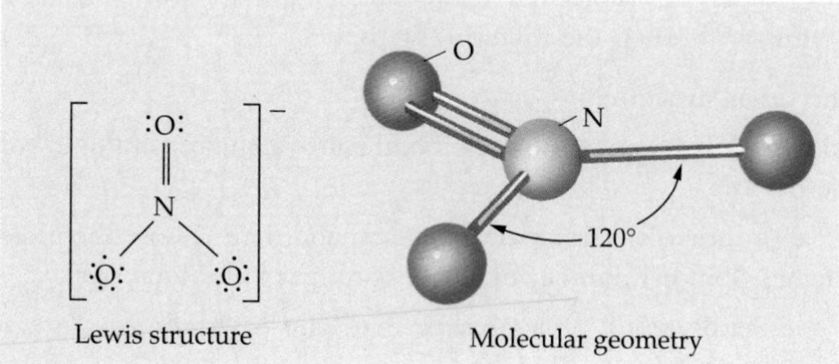

2. The $OXeF_4$ molecule has a total of 42 valence electrons: 8 for Xe + 6 for O + 4(7) for F. The guidelines for drawing Lewis structures lead to a structure with a total of six electron pairs about the central Xe atom, one of which is a lone pair. Notice that an Xe═O bond has been formed; otherwise the formal charges on Xe and O are 1+ and 1−, respectively. The six electron pairs about the Xe atom mean that the electron-pair geometry is octahedral. Only five of the structural pairs are bond pairs, so the molecular shape is square-pyramidal. The O and F atoms could in principle be placed so that either the O or the F is opposite the lone pair. The actual molecular geometry is that shown in the drawing.

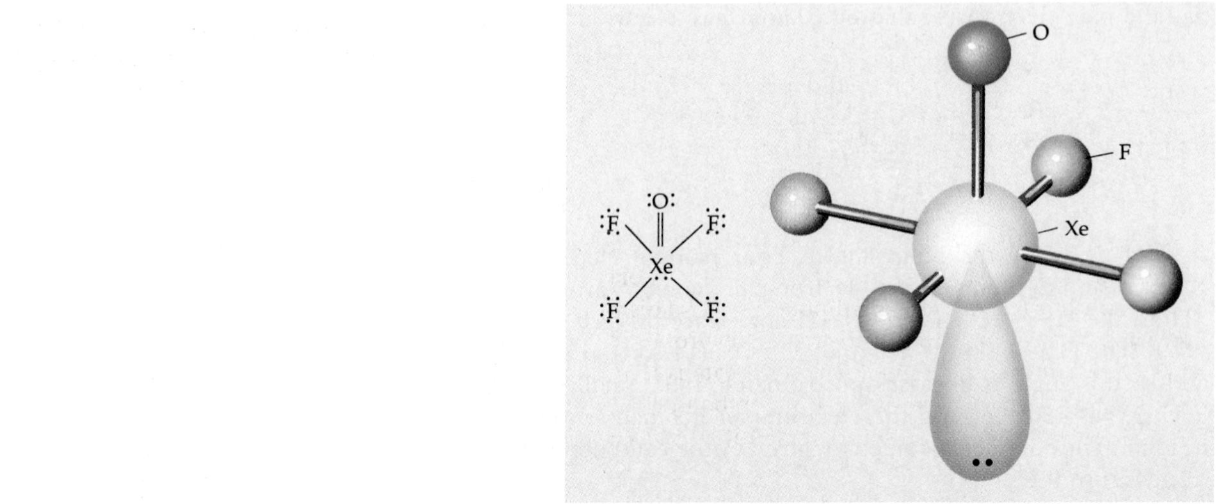

EXERCISE 9.15 ***Determining Molecular Shapes***

Use Lewis structures and the VSEPR model to determine the electron-pair and molecular geometries for (a) the phosphate ion, PO_4^{3-}; (b) the sulfite ion, SO_3^{2-}; and (c) IF_5. ■

PROBLEM-SOLVING TIPS AND IDEAS

9.3 Determining Molecular Structure

The shape of virtually any molecule or ion containing main group elements can be described by thinking through the following steps:

(1) Draw the Lewis structure.

(2) Determine the total number of single bond pairs (including multiple bonds counted as single pairs).

(3) Pick the appropriate electron-pair geometry, and then choose the molecular shape that matches the total number of single bond pairs and lone pairs.

(4) Predict the bond angles, remembering that lone pairs occupy more volume than bonding pairs do. ■

9.6 MOLECULAR POLARITY

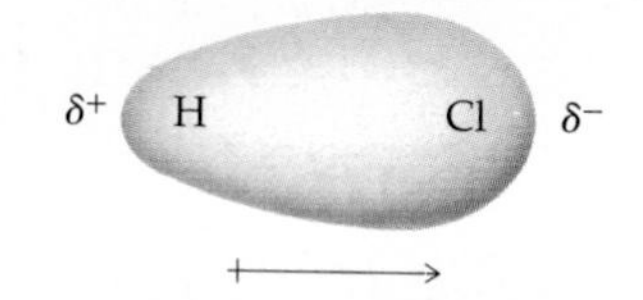

Figure 9.16 In a polar molecule the valence electron density has shifted slightly to one side of the molecule. To show the direction of charge transfer, that is, the direction of molecular polarity, we often use an arrow (+ → −), with the arrowhead pointing toward the negative end of the molecule and the plus sign at the positive end of the molecule.

The term "polar" was used in Section 9.4 to describe a bond in which one atom had a partial positive charge and the other a partial negative charge. Because most molecules have at least some polar bonds, however, molecules can also be polar. In a polar *molecule,* electron density accumulates toward one side of the molecule, giving that side a slight negative charge, $\delta-$, and leaving the other side with a slight positive charge of equal value, $\delta+$ (Figure 9.16).

Before describing the factors that determine whether a molecule is polar, let us look at the experimental measurement of the polarity of a molecule. Polar molecules experience a force in an electric field that tends to align them with the field (Figure 9.17). When the electric field is created by a pair of oppositely charged plates, the positive end of each molecule is attracted toward the negative plate, and the negative end is attracted toward the positive plate. The extent to which the molecules line up with the field depends on their **dipole moment,** μ, which is defined as the product of the magnitude of the partial charges ($\delta+$ and $\delta-$) and the distance by which they are separated. The SI unit of the dipole moment is the coulomb-meter, but they have traditionally been given in a derived unit called the debye (D) (where $1\ D = 3.34 \times 10^{-30}\ C \cdot m$). Dipole moments are determined experimentally, and typical values are listed in Table 9.6.

The force of attraction between the negative end of one polar molecule and the positive end of another (called a dipole-dipole force and discussed in Section 13.2) has an extraordinarily important effect on the properties of water and other polar substances. For example, molecular polarity is important in determining the temperatures at which liquid freezes or boils, if a liquid dissolves certain gases or solids or mixes with other liquids, if it adheres to glass or other solids, or how it may react with other substances.

To predict when a simple molecule will be polar, we need to consider if the molecule has polar bonds and how these bonds are positioned relative to one

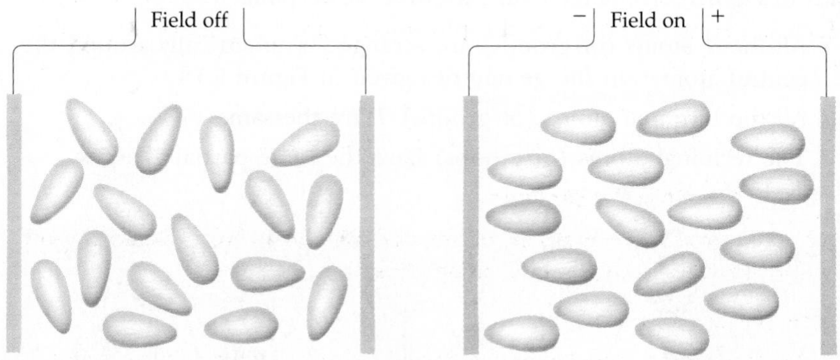

Figure 9.17 When placed between charged plates in an electric field polar molecules experience a force that tends to align them with the field. The negative end of the molecule is drawn to the positive plate, and vice versa. This property affects the capacitance of the plates (their ability to hold a charge) and provides a way to experimentally measure the magnitude of the dipole.

TABLE 9.6 Some Dipole Moments

Molecule (AB)	Moment, μ (D)	Geometry
HF	1.78	linear
HCl	1.07	linear
HBr	0.79	linear
HI	0.38	linear
H_2	0	linear
ClF	0.88	linear
BrF	1.29	linear
Molecule (AB_2)	**Moment, μ (D)**	**Geometry**
H_2O	1.85	bent
H_2S	0.95	bent
SO_2	1.62	bent
CO_2	0	linear
Molecule (AB_3)	**Moment, μ (D)**	**Geometry**
NH_3	1.47	trigonal-pyramidal
NF_3	0.23	trigonal-pyramidal
BF_3	0	trigonal-planar
Molecule (AB_4)	**Molecule, μ (D)**	**Geometry**
CH_4	0	tetrahedral
CH_3Cl	1.92	tetrahedral
CH_2Cl_2	1.60	tetrahedral
$CHCl_3$	1.04	tetrahedral
CCl_4	0	tetrahedral

another. To aid in this, consider molecules of the type CT_n with polar $C—T$ bonds in Figure 9.18. A molecule CT_n will *not* be polar if

- All the T atoms (or groups) are arranged symmetrically around the central atom C in the geometries given in Figure 9.18
- All the terminal atoms (or groups) T are the same
- The terminal atoms (or groups) have the same partial charges

Figure 9.18 Molecules of the type CT_n, where C is the central atom and T is the terminal atom. If all of the terminal atoms T are the same, molecules having these geometries are NOT polar.

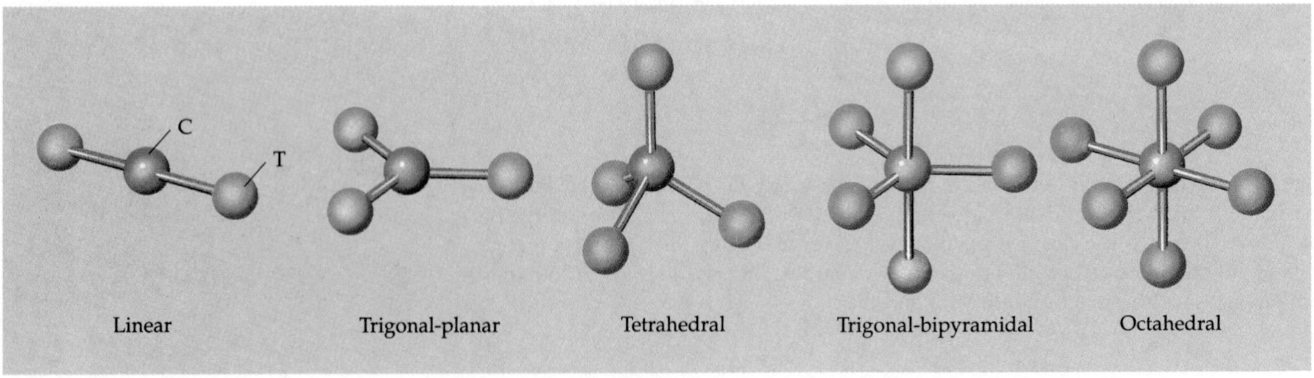

PORTRAIT OF A SCIENTIST

Peter Debye (1884–1966)

Peter Debye was born in the Netherlands, but he became professor of chemistry at Cornell University in 1940. He was noted for this work on x-ray diffraction, electrolyte solutions, and the properties of polar molecules. He received the Nobel Prize in chemistry in 1936 for his contributions to an understanding of molecular structure through his work on x-ray diffraction and dipole moments. Debye served as chairman of the Chemistry Department at Cornell from 1940 until 1950 and became a United States citizen in 1946. His research at Cornell on light scattering and other aspects of macromolecular chemistry moved him into both biochemistry and industrial polymer chemistry.

(Chemical Heritage Foundation)

On the other hand, if one of the terminal atoms T is different, and so has a different partial charge (δ), the molecule is polar. A molecule is also polar if the atoms are not symmetrically arranged.

Let us first consider a linear triatomic molecule such as carbon dioxide, CO_2. Here each C—O bond is polar, with the oxygen atom the negative end of the bond dipole. The terminal atoms are at the same distance from the C atom, they both have the same $\delta-$ charge, and they are symmetrically arranged around the central C atom (see CT_2 in Figure 9.18). Therefore, CO_2 has no *molecular* dipole, even though each bond is polar.

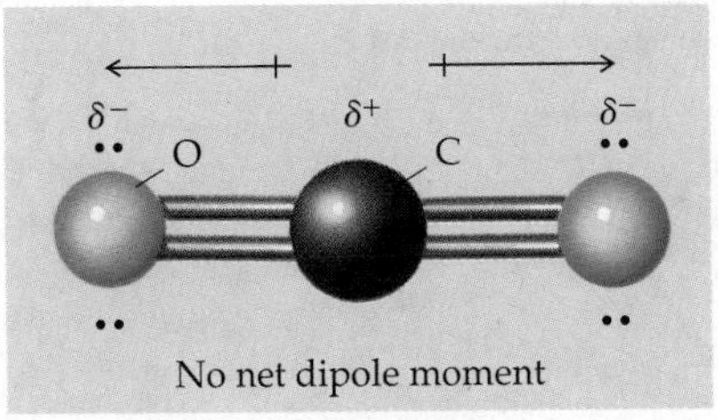

The three T atoms in a planar-trigonal molecule CT_3 are symmetrically arranged around the central atom because the T atoms are at the corners of an equilateral triangle; the angles between them are all the same (120°). In the planar-trigonal molecule BF_3, the B—F bonds are highly polar because F is much more electronegative than B (χ of B = 2.0 and χ of F = 4.0; see Figure 9.7). The molecule is nonpolar, however, because the three terminal atoms have the same $\delta-$ charge, are the same distance from the boron atom, and are arranged symmetrically around the central boron atom (see CT_3 in Figure 9.18).

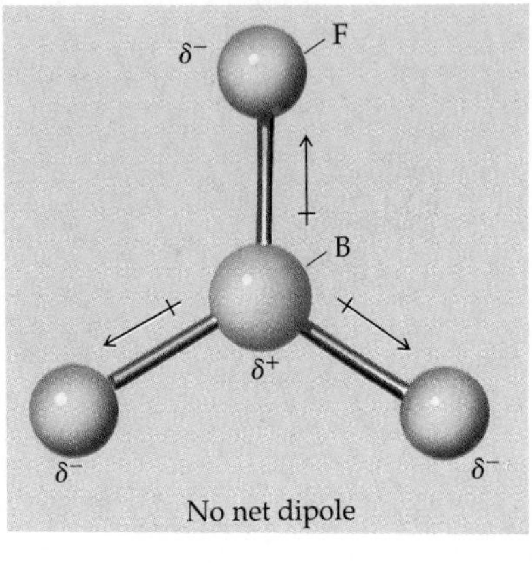

The planar-trigonal molecule phosgene, Cl_2CO, is polar ($\mu = 1.17$ D). Here the angles are all about 120°, so the O and Cl atoms are symmetrically arranged around the C atom. The electronegativities of the three atoms in the molecule differ, however: $\chi(O) > \chi(Cl) > \chi(C)$. Therefore, there is a net movement of electron density away from the center of the molecule, mostly toward the O atom.

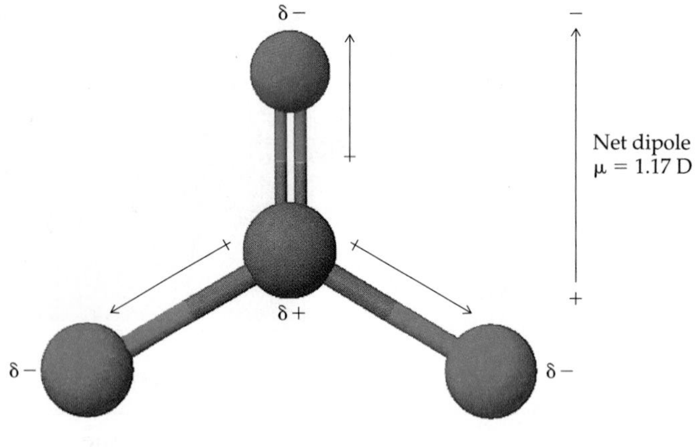

A simple tetrahedral molecule like carbon tetrachloride, CCl_4, is nonpolar because all the Cl atoms have the same $\delta-$ charge, have the same distance from the C atom, and are symmetrically distributed around the central carbon atom (see CT_4 in Figure 9.18). Changing one or more of the Cl atoms to H, as in $CHCl_3$ (chloroform), however, gives a polar molecule. In $CHCl_3$ the electronegativity for the H atom is less than that of the Cl atoms and the carbon-hydrogen distance is different from the carbon-chlorine distances.

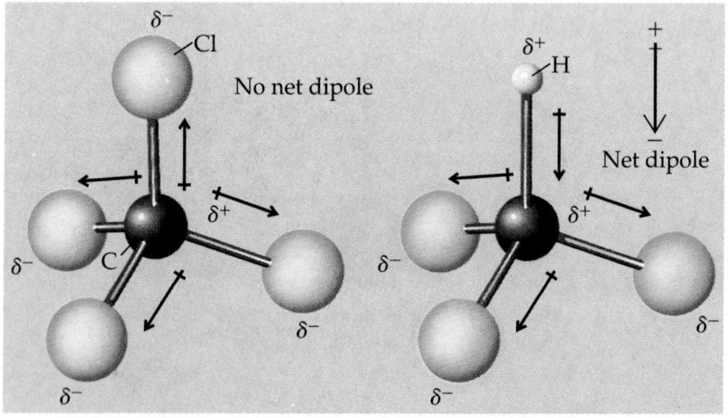

CCl_4 $CHCl_3$

Because the electronegativities of the atoms in $CHCl_3$ are Cl (3.0) > C (2.5) > H (2.1), the Cl atoms are on the more negative side of the molecule. This means the positive end of the molecular dipole is toward the H atom and the negative end toward the Cl atoms (Figure 9.19). Finally, we can conclude that only totally symmetrical tetrahedral molecules—where the atoms T in CT_4 (Figure 9.18) are all the same—are nonpolar (Table 9.6).

If one of the atoms of a tetrahedral molecule is replaced by a lone pair, as in NH_3, the electron pair geometry is still tetrahedral, but the molecular geometry is trigonal-pyramidal. Here the N—H bonds are polar because the electronegativity of H is less than that of N. The molecule is likewise polar because the N—H bonds are not arranged symmetrically.

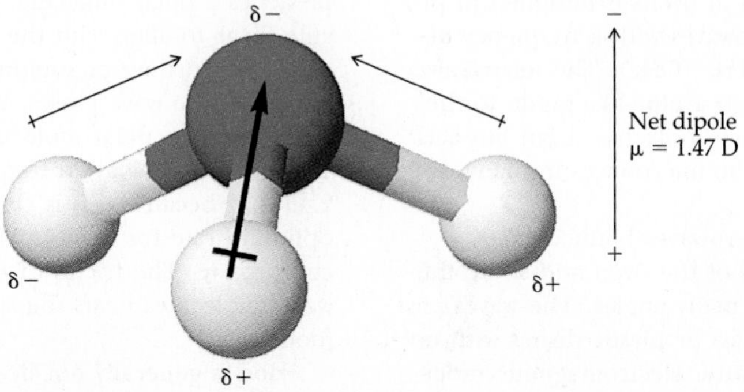

In general, all trigonal-pyramidal molecules must be polar because the lone pair of electrons leads to the placement of the *C—T* bonds on the opposite side of the molecule (Table 9.6).

Water has a tetrahedral electron-pair geometry and so is a bent, triatomic molecule. Because O has a larger electronegativity (3.5) than H (2.1), the O—H bonds are polar, with the H atoms having the same $\delta+$ charge. The H atoms are the same distance from the O atom at the center, but the H atoms are not symmetrically arranged. This means that electron density is drawn away from the H atoms, toward the O atom. Electron density accumulates on the O side of the molecule, making the molecule electrically "lopsided" and therefore polar (μ = 1.85 D).

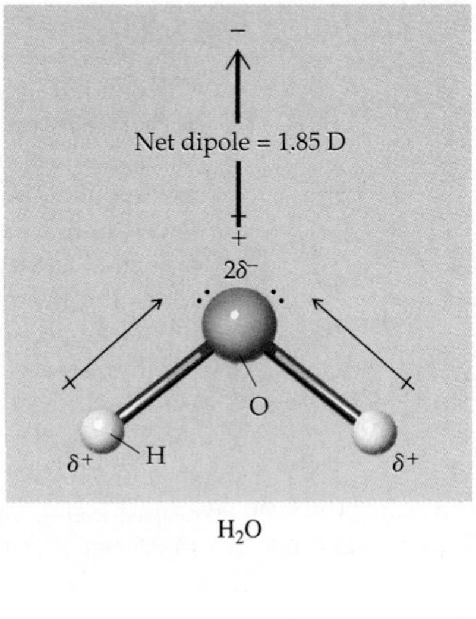

H_2O

In general, bent triatomic molecules are polar because the lone pairs lead to an arrangement where the terminal atoms are on one side of the molecule and the lone pairs are on the other side (Table 9.6).

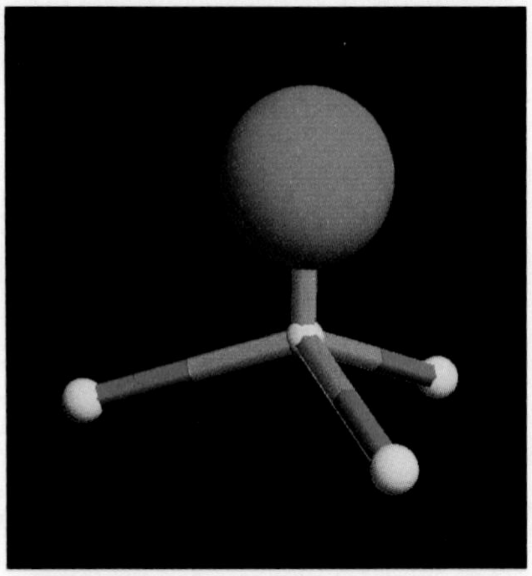

Figure 9.19 A computer-generated picture of chloroform, $CHCl_3$, where the atoms are colored according to their partial charge. Red designates an atom with a positive charge, and yellow designates an atom with a negative charge. The relative sizes of the atoms represent their relative partial charges. Most of the positive charge resides on the H atom, and most of the negative charge is on the Cl atoms, as expected. Calculations by computer give H = +0.16, C = −0.036, and Cl = −0.041 each. (Susan Young)

A CLOSER LOOK — *Cookin' with Polar Molecules*

Microwave ovens are common appliances in homes, dorm rooms, and offices. They work because water is polar.

Microwaves are generated in a device called a magnetron, a device invented during World War II as the heart of antiaircraft radar. The magnetron is a hollow cylinder with irregular walls, a rod-like cathode in the center, and a strong magnet positioned with north and south poles at opposite ends of the cylinder. An electric current flows from the cathode (which is electrically heated to help free electrons) across the air space to the cylinder wall that serves as the anode. As electrons begin to traverse this passage, the magnetic field forces them to move in circles around the cathode. The circular acceleration of the charged electrons creates electromagnetic waves. The magnetron in ovens is designed to produce microwaves with a frequency of 2.45 gigahertz (GHz). The microwaves flow through a pipe-like guide to the stirrer, which looks like a fan but acts to reflect the microwaves in many directions.

The microwaves bounce off the metal walls of the oven and strike the food from many angles. The waves pass through glass or plastic dishes with no effect. Because electromagnetic radiation consists of oscillating electric (and magnetic) fields (Figure 7.1), however, microwaves can affect mobile, charged particles such as dissolved ions or polar molecules. As each wave crest approaches a polar molecule, the molecule turns to align with the wave and continues turning or rotating as the trough of the wave passes. Water is the most common polar molecule in food. The microwaves have a frequency of 2.45 GHz because this is close to the optimum rate for making H_2O molecules rotate. The friction from rotating water molecules heats the surrounding food.

Food is generally not heated above 90 °C in a microwave oven because, as water boils away, the source of heat transfer is lost. Popcorn, however, must be heated in oil above 200 °C, though oil is not heated by microwaves as effectively as water. The food technologists who developed microwave popcorn solved this problem by adding a piece of metal foil (or metal-coated plastic film) to a paper bag. Microwaves are reflected by large metal surfaces but may be absorbed by small metal objects. The microwaves induce electric currents to flow back and forth through the metal, which quickly heats the foil above 200 °C.

Popcorn kernels contain starch, protein, and water sealed with a tight hull. Surrounded by oil at about 200 °C, the kernel heats up quickly, and the water and steam within rises far above the usual boiling point of water because the sealed hull keeps the contents under pressure. The pressurized steam transforms the starch grains into hot, gelatinized globules. When the hull finally ruptures—at 175 °C and a pressure of 9 atm—the expanding steam inflates the starch into the fluffy white form we love to eat with salt and butter.

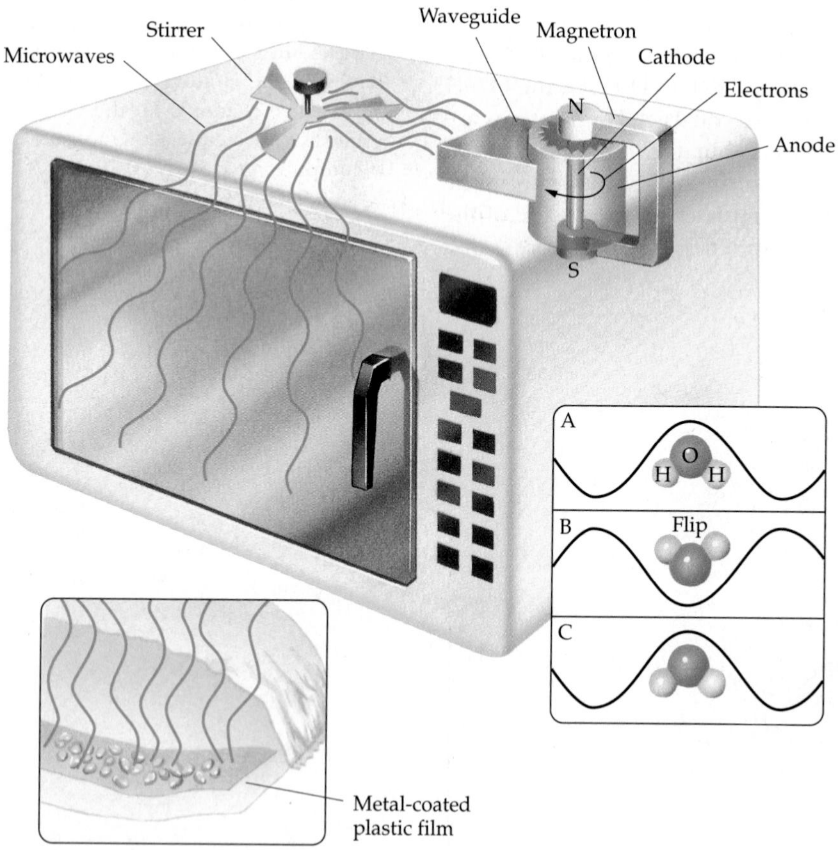

Adapted from J. Emsley: "Microwave Chemistry," *Chem Matters*, December 1993. Used with permission.

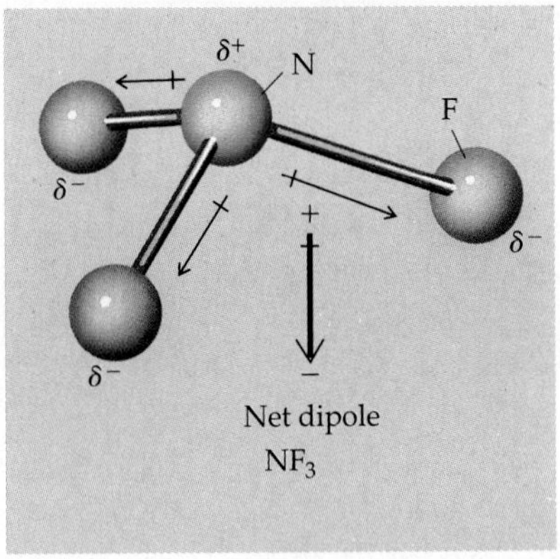

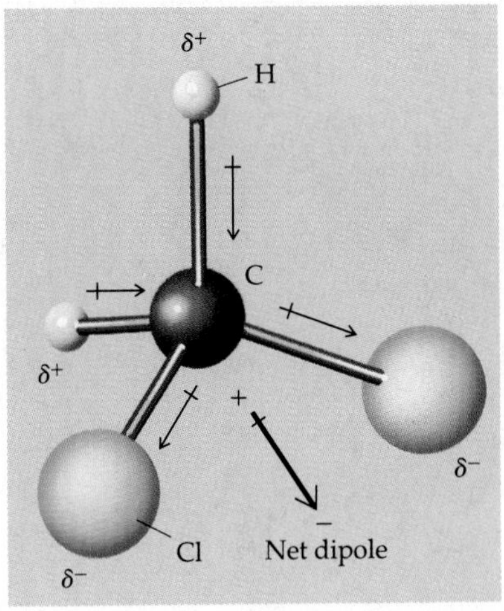

CH_2CL_2

EXAMPLE 9.11 *Molecular Polarity*

Are nitrogen trifluoride (NF_3), dichloromethane (CH_2Cl_2), and sulfur tetrafluoride (SF_4) polar or nonpolar? If polar, indicate the negative and positive sides of the molecule.

Solution NF_3 has the same pyramidal structure as NH_3.
Because F is more electronegative than N, each bond is polar, the more negative end being the F atom. This means that the NF_3 molecule as a whole is polar.*

In CH_2Cl_2 the electronegativities are in the order Cl (3.0) > C (2.5) > H (2.1). This means the bonds are polar,

$$\underset{\delta+\ \ \delta-}{\mathrm{H{-}C}} \qquad \underset{\delta+\ \ \delta-}{\mathrm{C{-}Cl}}$$

with a net movement of electron density away from the H atoms and toward the Cl atoms. Although the electron pair geometry around the C atom is tetrahedral, the polar bonds *cannot* be totally symmetric in their arrangement. This means the molecule must be polar, with the negative end toward the two Cl atoms and the positive end toward the two H atoms.

Sulfur tetrafluoride, SF_4, has an electron-pair geometry of a trigonal bipyramid (Figure 9.13). Because the lone pair occupies one of the positions, the S—F bonds are not arranged symmetrically. Furthermore, the S—F bonds are highly polar, the bond dipole having F as the negative end. (χ for S is 2.5 and χ for F is 4.0; see Figure 9.7.) Therefore, SF_4 is a polar molecule. The axial S—F bond dipoles cancel one another; they point in opposite directions. However, the equatorial S—F bond dipoles both point to one side of the molecule.

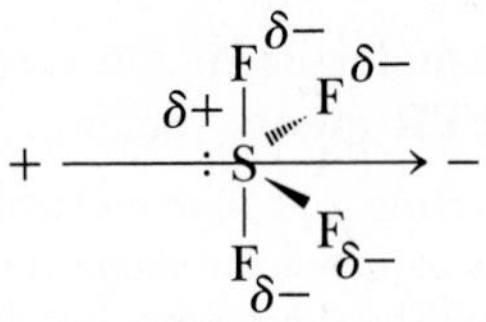

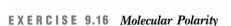

net dipole = 0.632 D

EXERCISE 9.16 *Molecular Polarity*

For each of the following molecules, decide whether the molecule is polar and which side is positive and which is negative: 1. $BFCl_2$, 2. NH_2Cl, and 3. SCl_2. ■

CHAPTER HIGHLIGHTS

When you have finished studying this chapter, you should be able to

- Decide on the number of **valence electrons** for a given element (Section 9.1).

*The dipole moment of NF_3 is much smaller (0.23 D) than that of NH_3 (1.47 D). This is because lone pairs have an effect. In NH_3 electron density moves away from H toward N and away from N toward the lone pair. These dipoles reinforce, and μ = 1.47 D. In NF_3, the N—F dipoles are directed oppositely from the N—lone pair dipole, so the net dipole is very small.

- Define the difference between **covalent** and **ionic bonding** (Section 9.2).
- Using the **octet rule,** draw **Lewis electron dot structures** for covalently bonded compounds (Section 9.3).
- Recognize that elements of Groups 2A and 3A do not always obey the octet rule, nor do elements of the third and higher periods of the periodic table (Section 9.3).
- Understand that when more than one acceptable Lewis structure can be written for a molecule or ion, these contributing structures are called **resonance structures.** These structures are combined into a composite **resonance hybrid,** which represents the actual electronic structure of the molecule (Section 9.3).
- Define and predict trends in **bond order, bond length,** and **bond dissociation energy** (Section 9.4).
- Use bond dissociation energies, D, in calculations (Section 9.4).

  Bond energies can be used to estimate reaction enthalpies by the expression $\Delta H^\circ_{rxn} = \Sigma D(\text{bonds broken}) - \Sigma D(\text{bonds made})$
- Use the concept of **electronegativity,** χ—the ability of an atom in a molecule to attract electrons—and recognize polar bonds in molecules (Section 9.4).
- Understand the difference between the oxidation number and the **formal charge** of an atom in a compound and use formal charge to decide which of several resonance structures is most correct (Section 9.4).
- Predict the shape or geometry of a molecule or ion using the **valence shell electron pair repulsion** or **VSEPR** theory (Section 9.5).

  According to the VSEPR theory the bonding and lone electron pairs surrounding an atom are oriented in space so as to make the angles between them as large as possible. The geometry described by these pairs is called the **electron-pair geometry.** The location of the bonding electron pairs defines the **molecular geometry.** The angle between bonded atoms depends on the presence and location of any lone pairs.
- Predict whether a molecule is polar (Section 9.6).

STUDY QUESTIONS

Review Questions

1. Give the number of valence electrons for Li, Sc, Zn, Si, and Cl.
2. Explain the difference between ionic and covalent bonding.
3. Refer to Table 9.2 and answer the following questions:
   (a) How many lone pairs and how many bond pairs are in ammonia (NH_3) and in hydrogen fluoride (HF)?
   (b) How many single bonds and how many double bonds are in N_2 and in C_2H_4?
4. In boron compounds the B atom often is not surrounded by four valence electron pairs. Illustrate this with BCl_3. Show how the molecule can obey the octet rule by forming a coordinate covalent bond with ammonia (NH_3).
5. Refer to Table 9.3 and answer the following questions:
   (a) Do any molecules have a second-period element as the central atom?
   (b) What is the maximum number of bond pairs and lone pairs that surround the central atom in any of these molecules?
6. Which of the following are odd-electron molecules: NO_2, SCl_2, NH_3, and NO_3?

7. Consider the following structures for the formate ion, HCO_2^-. What is the average C—O bond order in the ion?

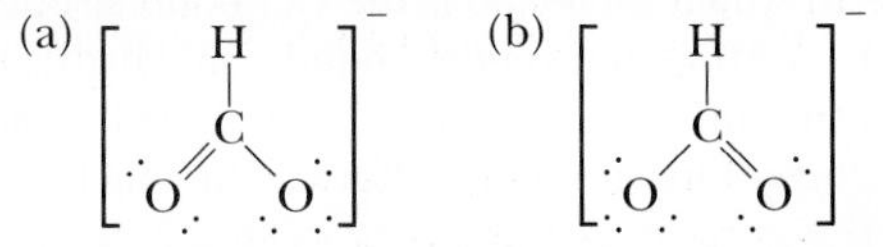

8. Give the bond order of the bonds in acetylene, H—C≡C—H.
9. Explain why the C—O bond order in the carbonate ion, CO_3^{2-}, is 1.33.
10. Consider a series of molecules in which the C atom is bonded by single bonds to atoms of second-period elements: C—O, C—F, C—N, C—C, and C—B. Place these bonds in order of increasing bond length.
11. Define bond dissociation energy. Does the enthalpy change for a bond-breaking reaction [such as H—Cl(g) → H(g) + Cl(g)] always have a positive sign, always have a negative sign, or may the sign vary? Explain briefly.
12. What is the relationship between bond order, bond length, and bond energy for a series of related bonds, say carbon-nitrogen bonds?
13. If you wished to calculate the enthalpy change for the reaction

$$\text{O=O(g)} + 2\ \text{H—H(g)} \longrightarrow 2\ \text{H—O—H(g)}$$

what bond energies do you need? Outline the calculation, being careful to show correct algebraic signs.
14. Define and give an example of a polar covalent bond. Give an example of a nonpolar bond.
15. Define electronegativity. Describe the difference between electronegativity and electron affinity.
16. Describe the trends in electronegativity in the periodic table.
17. Describe the difference between the oxidation number of an atom and its formal charge. Which is often the more realistic description of the charge on an atom in a molecule?
18. What is the principle of electroneutrality? How does this apply to the possible resonance structures of CO_2?
19. What is the VSEPR theory? What is the physical basis of the theory?
20. What is the difference between the electron-pair geometry and the molecular geometry of a molecule? Use the water molecule as an example in your discussion.
21. Designate the electron-pair geometry for each case of two to six electron pairs around a central atom.
22. What molecular geometries are possible for each of the following:

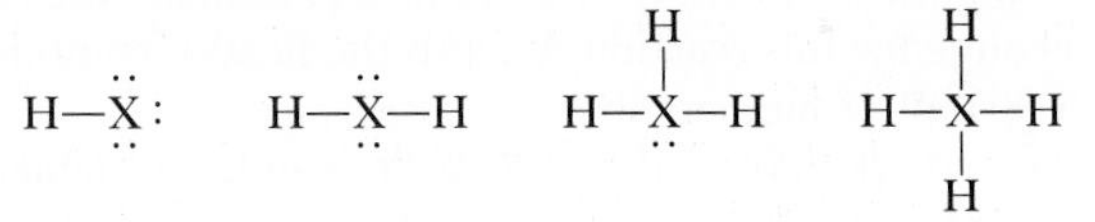

Give the H—X—H bond angle for each of the last three.
23. If you have three electron pairs around a central atom, how can you have a trigonal-planar molecule? A bent molecule? What bond angles are predicted in each case?
24. Draw a trigonal bipyramid of electron pairs. Designate the axial and equatorial pairs. Do axial and equatorial pairs also occur in an octahedron?
25. Does ammonia, NH_3, have a dipole moment? If so, what is the direction of the net dipole in NH_3?

Numerical and Other Questions

Valence Electrons

26. Give the periodic group number and number of valence electrons for each of the following atoms:
(a) N (c) Na (e) F
(b) B (d) Mg (f) S

27. Give the periodic group number and number of valence electrons for each of the following atoms:
(a) C (c) Ne (e) Se
(b) Cl (d) Si (f) Al

The Octet Rule

28. For each of the A groups of the periodic table, give the number of bonds an element is expected to form if it obeys the octet rule.

29. Which of the following elements can accommodate more than four valence electron pairs? That is, which can form compounds with five or six valence shell electron pairs?
(a) C (d) F (g) Se
(b) P (e) Cl (h) Sn
(c) O (f) B

Lewis Electron Dot Structures

30. Draw Lewis structures for the following molecules or ions:
(a) NF_3 (c) HOBr
(b) ClO_3^- (d) SO_3^{2-}

31. Draw Lewis structures for the following molecules or ions:
(a) Cl_2CO (c) NO_2^+
(b) BF_4^- (d) PO_4^{3-}

32. Draw Lewis structures for the following molecules:
(a) $CHClF_2$, one of the many chlorofluorocarbons (CFCs) that is no longer being used because of environmental problems.
(b) Formic acid, HCO_2H. The atomic arrangement is

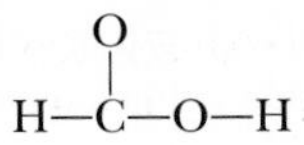

(c) Acetonitrile, H_3CCN
(d) Methanol, H_3COH

33. Draw Lewis structures for each of the following molecules:
(a) Tetrafluoroethylene (the molecule from which Teflon is built), F_2CCF_2.

(b) Vinyl chloride, H_2CCHCl, the molecule from which PVC plastics are made.

(c) Acrylonitrile, H_2CCHCN, the molecule from which materials such as Orlon are built.

(d) Methyltrichlorosilane, H_3CSiCl_3, a compound used in manufacturing "silicone" polymers.

34. Show all possible resonance structures for each molecule or ion.

(a) SO_2

(b) SO_3

(c) SCN^-

35. Show all possible resonance structures for each molecule or ion.

(a) Nitrate ion, NO_3^-

(b) Dinitrogen oxide (laughing gas), NNO

(c) Nitric acid,

$$H-O-N\begin{matrix}O\\\\O\end{matrix}$$

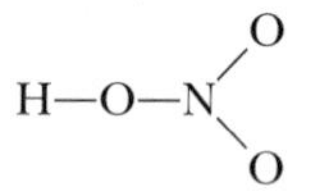

36. Draw Lewis structures for each of the following molecules or ions:

(a) BrF_3, (b) I_3^-, and (c) XeO_2F_2 (Xe is the central atom).

37. Draw Lewis structures for each of the following molecules or ions:

(a) BrF_5, (b) IF_3, (c) IBr_2^-

Bond Properties

38. Give the number of bonds for each of the following molecules. Tell the bond order for each bond. (Lewis structures for these molecules were drawn in previous Study Questions.)

(a) H_2CO (b) SO_3^{2-} (c) NO_2^+

39. Give the number of bonds for each of the following molecules. Tell the bond order for each bond. (Lewis structures for these molecules were drawn in previous Study Questions.)

(a) CN^- (b) Acetonitrile, H_3CCN (c) SO_3

40. In each pair of bonds below, predict which is the shorter. If possible, check your prediction by consulting Table 9.4.

(a) B—Cl or Ga—Cl

(b) C—O or Sn—O

(c) P—S or P—O

(d) The C=C or the C=O bond in acrolein, $H_2C=CH-C(H)=O$.

41. In each of the following pairs of bonds, predict which is the shorter. If possible, check your prediction by consulting Table 9.4.

(a) Si—N or P—O

(b) Si—O or C—O

(c) C—F or C—Br

(d) The C=C or the C≡N bond in acrylonitrile, $H_2C=CH-C≡N$.

42. Consider the carbon-oxygen bond in formaldehyde (H_2CO) and carbon monoxide (CO). In which molecule is the CO bond shorter? In which molecule is the CO bond stronger?

43. Compare the nitrogen-nitrogen bond in hydrazine, H_2NNH_2, with that in "laughing gas," N_2O. In which molecule is the nitrogen-nitrogen bond shorter? In which is the bond stronger?

44. Compare the nitrogen-oxygen bond lengths in NO_2^+ and in NO_3^-. In which ion is the bond longer? Explain briefly.

45. Compare the carbon-oxygen bond lengths in the formate ion, HCO_2^-, and in the carbonate ion, CO_3^{2-}. In which ion is the bond longer? Explain briefly.

Bond Energies and Reaction Enthalpies

46. Hydrogenation reactions, the addition of H_2 to a molecule, are widely used in industry to transform one compound into another (Section 11.2). For example, the molecule called butene (a member of a general class of compounds called "alkenes" because of the C=C double bond) is converted to butane (called an "alkane" because there are only C—H bonds and C—C single bonds) by addition of H_2.

$$H-\underset{\underset{H}{|}}{\overset{\overset{H}{|}}{C}}-\underset{\underset{H}{|}}{\overset{\overset{H}{|}}{C}}-\overset{\overset{H}{|}}{C}=\overset{\overset{H}{|}}{C}-H(g) + H_2(g) \longrightarrow$$

$$H-\underset{\underset{H}{|}}{\overset{\overset{H}{|}}{C}}-\underset{\underset{H}{|}}{\overset{\overset{H}{|}}{C}}-\underset{\underset{H}{|}}{\overset{\overset{H}{|}}{C}}-\underset{\underset{H}{|}}{\overset{\overset{H}{|}}{C}}-H\ (g)$$

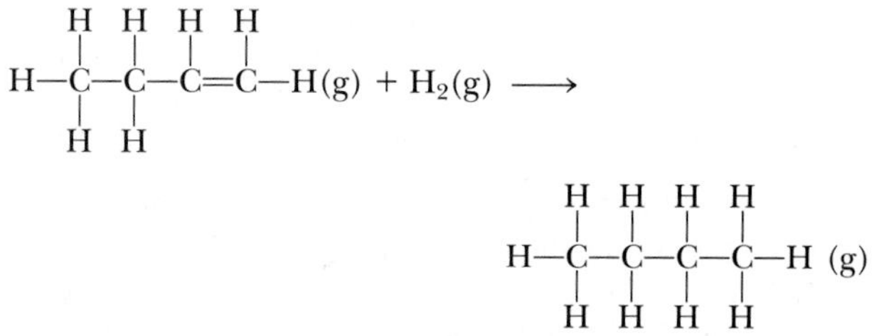

Use the bond energies of Table 9.5 to estimate the enthalpy change for this hydrogenation reaction.

47. In principle, dinitrogen monoxide, N_2O, can decompose to nitrogen and oxygen gas.

$$2\ N_2O(g) \longrightarrow 2\ N_2(g) + O_2(g)$$

Use bond energies to estimate the enthalpy change for this reaction.

48. Phosgene, Cl_2CO, is a highly toxic gas that was used as a weapon in World War I. Using the bond energies of Table 9.5, estimate the enthalpy change for the reaction of carbon monoxide and chlorine to produce phosgene.

$$CO(g) + Cl_2(g) \longrightarrow Cl_2CO(g)$$

49. The equation for the combustion of gaseous methanol is

$$2\ H_3COH(g) + 3\ O_2(g) \longrightarrow 2\ CO_2(g) + 4\ H_2O(g)$$

Using the bond energies in Table 9.5, estimate the enthalpy change for this reaction. What is the heat of combustion of 1.00 mol of methanol?

50. The compound oxygen difluoride is quite unstable, giving oxygen and HF on reaction with water.

$$OF_2(g) + H_2O(g) \longrightarrow O=O(g) + 2\ HF(g)$$

$$\Delta H°_{rxn} = -318\ kJ$$

Using bond energies, calculate the bond dissociation energy of the O—F bond in OF_2.

51. O atoms can combine with ozone to form oxygen.

$$O_3(g) + O(g) \longrightarrow 2\ O_2(g) \qquad \Delta H^\circ_{rxn} = -394\ kJ$$

Using ΔH°_{rxn} and the bond energy data in Table 9.5, *estimate* the bond energy for the oxygen-oxygen bond in ozone, O_3. How does your estimate compare with the energies of a O—O single bond and a O=O double bond? Does the oxygen-oxygen bond energy in ozone correlate with its bond order?

Electronegativity and Bond Polarity

52. In each pair of bonds, indicate the more polar bond and use an arrow to show the direction of polarity in each bond.
(a) C—O and C—N (c) P—H and P—N
(b) B—O and P—S (d) B—H and B—I

53. Given the bonds below, answer the questions that follow:
(a) C—N (b) C—H (c) C—Br (d) S—O
(i) Tell which atom is the more negatively charged in each bond.
(ii) Which is the most polar bond?

54. The molecule below is acrolein, the starting material for certain plastics.

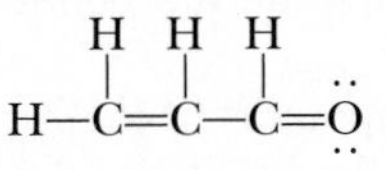

(a) Which bonds in the molecule are polar and which are nonpolar?
(b) Which is the most polar bond in the molecule? Which atom is the negative end of the bond dipole?

55. The molecule below is urea, a compound used in plastics and fertilizers.

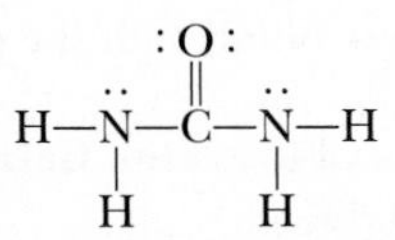

(a) Which bonds in the molecule are polar and which are nonpolar?
(b) Which is the most polar bond in the molecule? Which atom is the negative end of the bond dipole?

Oxidation Numbers and Atom Formal Charges

56. Using Lewis structures and atom electronegativities, give the oxidation number of each atom in the following molecules or ions.
(a) H_2O (d) N_2O
(b) H_2O_2 (e) ClO^-
(c) SO_2

57. Using Lewis structures and atom electronegativities, give the oxidation number of each atom in the following molecules.
(a) ClF_3 (c) OF_2
(b) XeF_2 (d) HCN

58. Calculate the formal charge on each atom in each of the following molecules or ions. In each case, compare the formal charge with the oxidation number.
(a) H_2O (c) NO_2^+
(b) CH_4 (d) HOF

59. Calculate the formal charge on each atom in each of the following molecules or ions. In each case, compare the formal charge with the oxidation number.
(a) ICl_2^- (c) ClO_2^-
(b) NH_3 (d) SF_4

60. Two resonance structures are possible for NO_2^-. Draw these structures, and then find the formal charge on each atom in each resonance structure.

61. Two resonance structures are possible for the formate ion, HCO_2^-. Draw these structures, and then find the formal charge on each atom in each resonance structure.

62. Three resonance structures are possible for dinitrogen oxide, N_2O (better known as "laughing gas").
(a) Draw the three resonance structures.
(b) Calculate the formal charge on each atom in each resonance structure.
(c) Based on formal charges, decide on the most reasonable resonance structure.

63. Nitric acid, HNO_3, has three resonance structures. One of them, however, contributes much less to the resonance hybrid than the other two. Sketch the three resonance structures, and assign a formal charge to each atom. Which one of your structures is the least important?

Molecular Geometry

64. Draw the Lewis structure for each of the following molecules or ions. Describe the electron-pair geometry and the molecular geometry.
(a) NH_2Cl (c) SCN^-
(b) Cl_2O (O is the central atom) (d) HOF

65. Draw the Lewis structure for each of the following molecules or ions. Describe the electron-pair geometry and the molecular geometry.
(a) ClF_2^+ (c) PO_4^{3-}
(b) $SnCl_3^-$ (d) CS_2

66. The following molecules or ions all have two oxygen atoms attached to a central atom. Draw the Lewis structure for each one, and then describe the electron-pair geometry and the molecular geometry. Comment on similarities and differences in the series.
(a) CO_2 (c) O_3 (e) SO_2
(b) NO_2^- (d) ClO_2^-

67. The following molecules or ions all have three oxygen atoms attached to a central atom. Draw the Lewis structure for each one, and then describe the electron-pair geometry and the molecular geometry. Comment on similarities and differences in the series.
(a) BO_3^{3-} (c) SO_3^{2-} (e) NO_3^-
(b) CO_3^{2-} (d) ClO_3^-

68. Draw the Lewis structure of each of the following molecules or ions. Describe the electron-pair geometry and the molecular geometry.
(a) ClF_2^- (c) ClF_4^-
(b) ClF_3 (d) ClF_5

69. Draw the Lewis structure of each of the following molecules or ions. Describe the electron-pair geometry and the molecular geometry.
(a) SiF_6^{2-} (c) SF_4
(b) PF_5 (d) XeF_4

70. Give approximate values for the indicated bond angles. (Be sure to account for all the bonding and lone pairs before making a decision on the bond angle.)
(a) O—S—O in SO_2
(b) F—B—F angle in BF_3
(c)

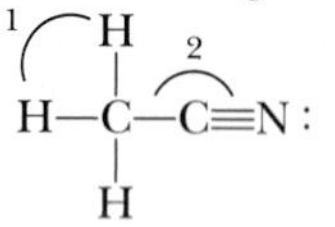

71. Give approximate values for the indicated bond angles. (Be sure to account for all the bonding and lone pairs before making a decision on the bond angle.)
(a) Cl—S—Cl in SCl_2
(b) N—N—O in N_2O
(c)

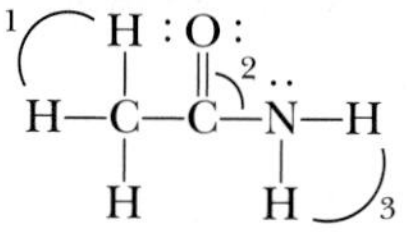

72. Acetylacetone has the structure below. Estimate the values of the indicated angles.

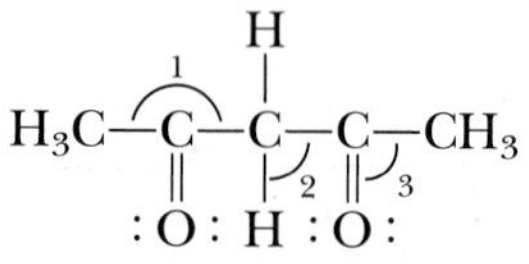

73. Phenylalanine is one of the natural amino acids. Estimate the values of the indicated angles.

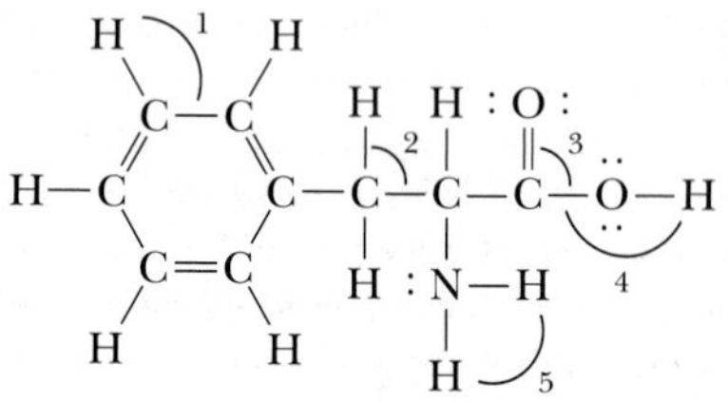

74. Give approximate values for the indicated bond angles.
(a) F—Se—F angles in SeF_4
(b) The O—S—F and the F—S—F bond angles (two are possible for each) in OSF_4 (the O atom is in an equatorial position)
(c) F—Br—F angles in BrF_5

75. Give approximate values for the indicated bond angles.
(a) F—S—F angles in SF_6
(b) F—Xe—F angle in XeF_2
(c) F—Cl—F in ClF_2^-

76. Which has the greater O—N—O bond angle, NO_2^- or NO_2^+? Explain your answer briefly.

77. Compare the F—Cl—F angles in ClF_2^+ and ClF_2^-. From Lewis structures determine the approximate bond angle in each ion. Explain which ion has the greater bond angle and why.

Molecular Polarity

78. Consider the following molecules:
(a) H_2O (b) NH_3 (c) CO_2 (d) ClF (e) CCl_4
(i) Which compound has the most polar bonds?
(ii) Which compounds in the list are *not* polar?
(iii) Which atom in ClF is more negatively charged?

79. Consider the following molecules:
(a) CH_4 (b) NCl_3 (c) BF_3 (d) CS_2
(i) Which compound has bonds with the greatest degree of polarity?
(ii) Which compounds in the list are *not* polar?

80. Which of the following molecules is (are) polar? For each polar molecule indicate the direction of polarity, that is, which is the negative and which is the positive side of the molecule?
(a) $BeCl_2$ (c) CH_3Cl
(b) HBF_2 (d) SO_3

81. Which of the following molecules is (are) *not* polar? In which molecule are there bonds with the greatest degree of polarity?
(a) CO (c) CF_4 (e) GeH_4
(b) BCl_3 (d) PCl_3

General Questions

82. Give the number of valence electrons for an atom in Groups 1A, 3A, and 4A.

83. Draw Lewis structures (and resonance structures where appropriate) for the following molecules and ions. What similarities and differences are there in this series? (a) CO_2, (b) N_3^-, and (c) OCN^-.

84. What are the orders of the N—O bonds in NO_2^- and NO_2^+? The nitrogen-oxygen bond length in one of these ions is 110 pm and in the other 124 pm. Which bond length corresponds to which ion? Explain briefly.

85. Urea is widely used as a fertilizer because of its high nitrogen content, so better methods for its production are always

being sought. Using Table 9.5, estimate the enthalpy change for a reaction to make urea, $(NH_2)_2CO$.

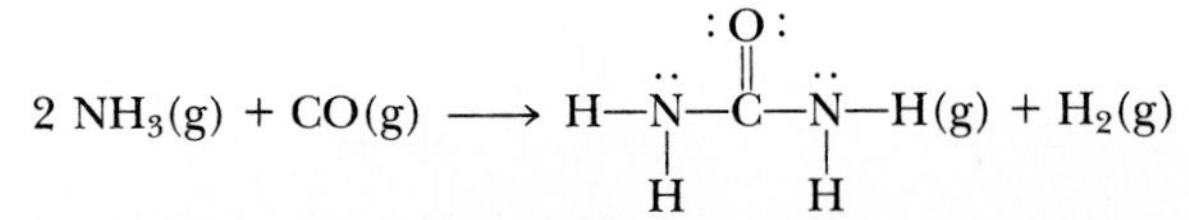

(Note: urea is a solid under normal conditions. The enthalpy change calculated here is for urea vapor.)

86. The molecule pictured here is acrylonitrile, the building block of the synthetic fiber Orlon.

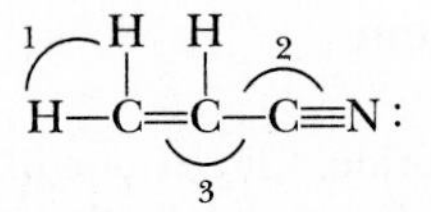

(a) Give the approximate values of angles 1, 2, and 3.
(b) Which is the shorter carbon-carbon bond?
(c) Which is the stronger carbon-carbon bond?
(d) Which is the most polar bond, and what is the negative end of the bond dipole?

87. Vanillin is the flavoring agent in vanilla extract and in vanilla ice cream. Its structure is drawn below.

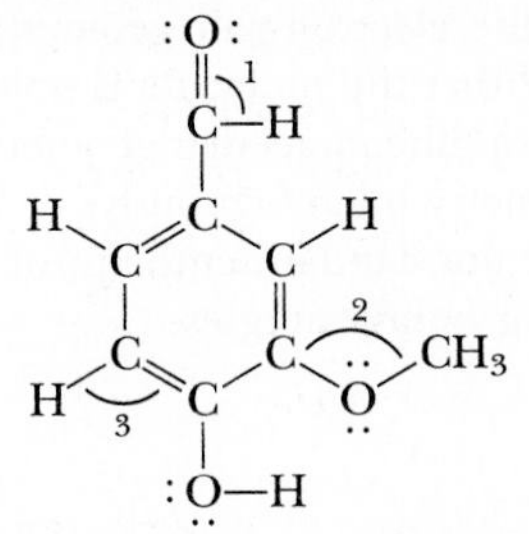

(a) Give values for the three bond angles indicated.
(b) Indicate the shortest carbon-oxygen bond in the molecule.
(c) Indicate the most polar bond in the molecule.

88. The following molecules or ions all have fluorine atoms attached to a central atom from Groups 3A through 7A. Draw the Lewis structure for each one, and then describe the electron-pair geometry and the molecular geometry. Comment on similarities and differences in the series.
(a) BF_3 (d) OF_2
(b) CF_4 (e) HF
(c) NF_3

89. The formula for nitryl chloride is $ClNO_2$. Draw the Lewis structure for the molecule, including all resonance structures. Describe the electron-pair and molecular geometries, and give values for all bond angles.

90. Given that the spatial requirements of lone pairs are much greater than those of bond pairs, explain why
(a) XeF_2 has a linear molecular structure and not a bent one.
(b) ClF_3 has a T-shaped structure and not a planar-triangular one.

91. The compound $C_2H_2Cl_2$ can exist in three forms. Do any of these have a net dipole moment? If yes, give the direction of the dipole moment.

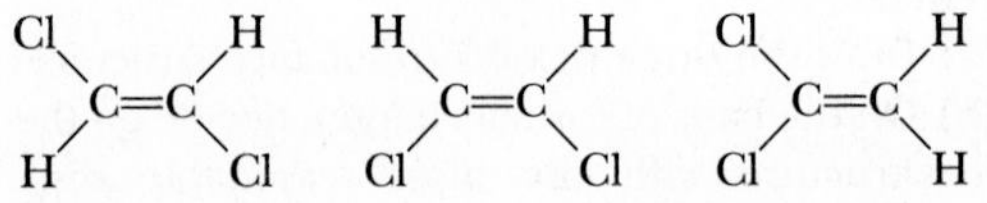

92. In 1962, Watson and Crick received the Nobel Prize for their simple but elegant model for the "heredity molecule" DNA (see page 2). The key to their structure (the "double helix") was an understanding of the geometry and bonding capabilities of nitrogen-containing bases such as the guanine molecule shown here.

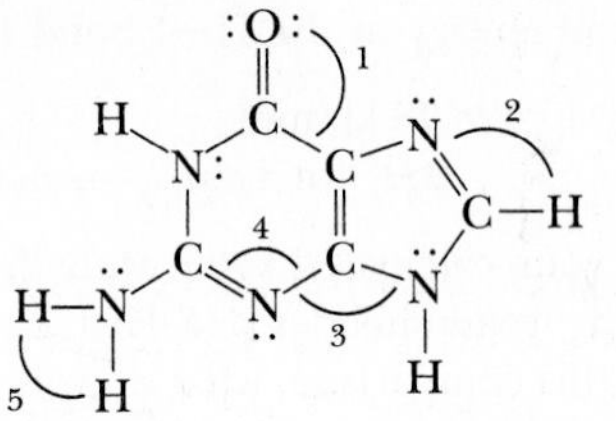

(a) Give approximate values for the indicated bond angles.
(b) Which are the most polar bonds in the molecule?

Conceptual Questions

93. The important molecule cyclohexane is made by adding H_2 to benzene.

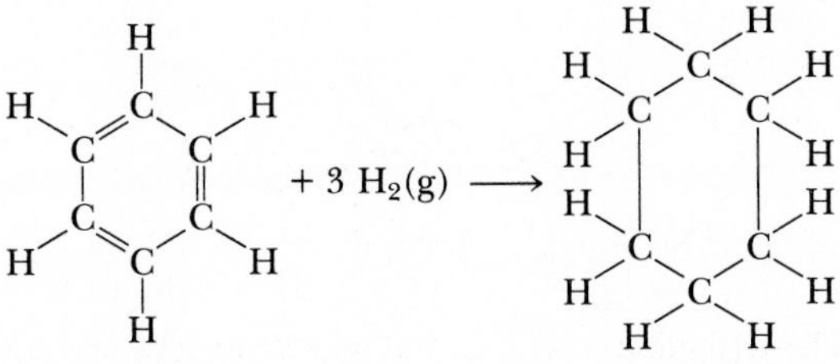

benzene cyclohexane

(a) Calculate the enthalpy of the reaction knowing that $\Delta H_f^\circ[C_6H_6(g)] = +82.8$ kJ/mol and $\Delta H_f^\circ[C_6H_{12}(g)] = -123.1$ kJ/mol.
(b) Benzene has two resonance structures, and this makes the C—C bond order 1.5. Assume, however, that the C_6 ring consists of three C=C bonds and three C—C bonds, and calculate the enthalpy of the hydrogenation reaction using bond energies.
(c) There should be a difference between the actual enthalpy of hydrogenation calculated in (a) and the one estimated from (b). This difference occurs because we assumed resonance in (a) but not in (b). What does this

tell you about the effect of resonance on the "stability" of a compound?

94. The cyanate ion, NCO^-, has the least electronegative atom, C, in the center. The very unstable fulminate ion, CNO^-, has the same elemental composition, but the N atom is in the center.

(a) Draw the three possible resonance structures of CNO^-.

(b) On the basis of formal charges, decide on the resonance structure with the most reasonable distribution of charge.

(c) Mercury fulminate is so unstable it is used in blasting caps. Can you offer an explanation for this instability? (*Hint:* Are the formal charges in any resonance structure reasonable in view of the relative electronegativities of the atoms?)

95. Use the following information, and data from Table 9.5, to estimate the energy of the H—I bond in HI(g):

$$\Delta H_f^\circ[\text{HI(g)}] = 26.48 \text{ kJ/mol}$$
$$\Delta H^\circ \text{ for } I_2(s) \longrightarrow I_2(g) \text{ is } 62.44 \text{ kJ/mol.}$$

Compare your calculated value with the bond energies of the other hydrogen halides (HF, HCl, and HBr) in Table 9.5. Based on this comparison, what do you conclude about the trend in bond energies in H—X as the atomic size of X increases?

96. Using Table 9.5, and the following thermodynamic information, estimate the bond dissociation energy for the B—F bond in BF_3.

$$B(s) \longrightarrow B(g) \qquad \Delta H_f^\circ[\text{B(g)}] = 563 \text{ kJ/mol}$$
$$B(s) + \tfrac{3}{2}F_2(g) \longrightarrow BF_3(g) \quad \Delta H_f^\circ[BF_3(g)] = -1137 \text{ kJ/mol}$$

97. Amides are an important class of organic molecules. They are usually drawn as sketched here, but another resonance structure is possible.

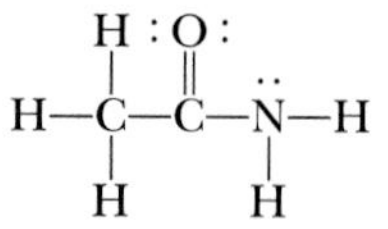

Draw that structure, and then suggest reasons why it is usually not pictured.

Summary Problem

98. Chlorine trifluoride, ClF_3, is one of the most reactive compounds known. It reacts violently with many substances generally thought to be inert and was used in incendiary bombs in World War II. It can be made by heating Cl_2 and F_2 in a closed container.

(a) Write a balanced equation to depict the reaction of Cl_2 and F_2 to give ClF_3.

(b) If you mix 0.71 g of Cl_2 with 1.00 g of F_2, what is the theoretical yield of ClF_3?

(c) Draw the electron dot structure of ClF_3.

(d) What is the electron-pair geometry for ClF_3?

(e) Knowing that the molecule is polar, what can you conclude about the molecular geometry? Does this establish the geometry unambiguously?

(f) Calculate the standard enthalpy of formation of gaseous ClF_3 using bond energies.

C H A P T E R **10**

Bonding and Molecular Structure: Orbital Hybridization, Molecular Orbitals, and Metallic Bonding

Chemical Puzzler

About 20% of the earth's atmosphere is made up of diatomic oxygen, O_2. As a gas it is colorless. When cooled to a very low temperature, however, the gas condenses to a pale blue liquid. For such a simple molecule, O_2 has an amazing property: it is paramagnetic. That is, it is attracted to a magnet (Section 8.1). This can be demonstrated easily by observing that liquid oxygen poured between the poles of a magnet remains suspended there. Paramagnetic behavior requires unpaired electrons, so we can conclude from this experiment that O_2 has one or more unpaired electrons. Shouldn't this fact be evident if we try to represent the bonding in O_2? A Lewis electron dot structure, however, does not contain unpaired electrons.

$$\ddot{\underset{..}{O}}=\ddot{\underset{..}{O}}$$

Lewis electron dot structure for the oxygen molecule.

This simple model fails to explain an undisputed experimental observation. The problem is to find a new approach to chemical bonding that more accurately describes the bonding in this molecule. (C. D. Winters)

The preceding chapters described ways chemists represent bonds between atoms and described some consequences of chemical bonding. This chapter outlines some ideas on how atoms use their orbitals in forming bonds.

Two common approaches are used in chemical bonding: **valence bond (VB) theory** and **molecular orbital (MO) theory.** The former was developed largely by Linus Pauling; the latter came from work by the American chemist Robert Sanderson Mulliken (1896–1986). Mulliken's approach was to combine pure atomic orbitals on each atom to derive molecular orbitals that are spread or delocalized over the molecule. The electrons of the molecule are assigned to these orbitals, and the molecular electron pairs are distributed over the molecule. In contrast, the valence bond approach is more closely tied to Lewis's idea of electron-pair bonds and of lone pairs of electrons localized on a particular atom.

Robert Mulliken was awarded the 1966 Nobel Prize in chemistry for the development of molecular orbital theory. Linus Pauling received the 1954 Prize for his work on the nature of the chemical bond.

Why are two theories used? Isn't one more correct than the other? Both give us good descriptions of the bonding in molecules and polyatomic ions, but they are used for different purposes. Valence bond theory is the easier to apply and leads to a good understanding of the bonding of molecules in their ground, or lowest, energy state. On the other hand, molecular orbital theory is most useful in understanding molecules in "excited states." For example, it helps us understand why compounds absorb light, and why some are red but others are blue or other colors of the rainbow. Chemists use the theory that best applies to the problem at hand.

10.1 VALENCE BOND THEORY

The description of covalent bonding begun in Chapter 9 can now be given in the terminology of valence bond theory. According to this theory, two atoms form a bond when both of the following conditions occur:

1. **Orbitals overlap** between two atoms. For example, if two H atoms approach each other closely enough, their 1*s* orbitals can partially occupy the same region of space (Figure 10.1).
2. A maximum of two electrons, of opposite spin, is present in the overlapping orbitals. Usually one electron is supplied by each of the two bonded atoms.

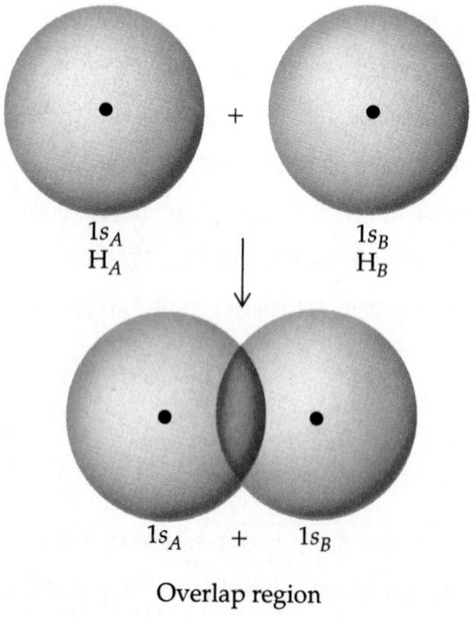

Figure 10.1 The overlap of 1*s* atomic orbitals to form the sigma (σ) bond of H_2.

Because of orbital overlap, the bonding pair of electrons is found within a region of space influenced by both nuclei. Both electrons are therefore simultaneously attracted to both atomic nuclei, leading to bonding.

As the extent of overlap between two orbitals increases, the strength of the bond increases. As shown in Figure 9.2 the potential energy drops as the two H atoms, originally far apart, come closer and closer. When the internuclear distance in the H_2 molecule is 74 pm, the energy of the system is reduced to a minimum. The maximum, *effective* overlap of 1*s* orbitals occurs at this distance. If the H atoms approach closer than 74 pm, however, repulsion increases between the two positively charged nuclei, and the energy of the system increases.

The overlap of two *s* orbitals, one from each of two atoms (Figure 10.1), leads to a **sigma (σ) bond:** the electron density of a sigma bond is greatest *along the axis* of the bond. Sigma bonds can also form by the overlap of an *s* orbital with a *p* orbital or by the head-to-head overlap of two *p* orbitals (Figure 10.2).

Because *s* orbitals are spherical, two H atoms can approach each other from any direction to form a strong sigma bond. Other types of orbital overlap ($s + p$ and $p + p$), however, are directional. For example, two *p* orbitals should overlap directly *along the axis* of the bond in order to form the strongest possible sigma bond (Figure 10.3).

Consider the HF molecule as an example of bond formation using the concepts of valence bond theory. We can account for the H—F sigma bond by overlap of a 1*s* orbital with a 2*p* orbital as in Figure 10.2a. The fluorine atom has seven valence electrons, one of them occupying a 2*p* orbital.

$$\text{H}\cdot \quad + \quad \cdot\ddot{\underset{..}{\text{F}}}\text{:} \quad \longrightarrow \quad \text{H—}\ddot{\underset{..}{\text{F}}}\text{:}$$

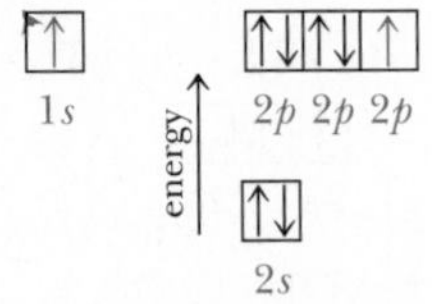

Covalent bond from the overlap of an H 1*s* orbital with an F 2*p* orbital. (See Figure 10.2.)

If this orbital overlaps the H 1*s* atomic orbital, a pair of electrons (one from H 1*s* and the other from F 2*p*) is now under the influence of both atomic nuclei, and bonding occurs.

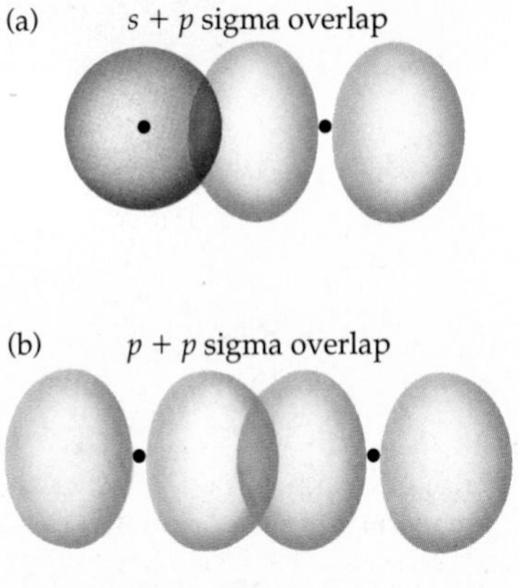

Figure 10.2 Sigma (σ) bond formation from *s* and *p* orbitals. (a) An *s* orbital overlaps a *p* orbital. (b) Two *p* orbitals overlap head to head.

Hybrid Orbitals

Although the simple picture of orbital overlap used to describe H_2 or HF works well, we often run into difficulty when molecules with more atoms are considered. Take methane, CH_4, as an example. A Lewis dot structure of methane shows that four C—H covalent sigma bonds must occur. Furthermore, VSEPR theory predicts, and experiments confirm, that the electron-pair geometry of the

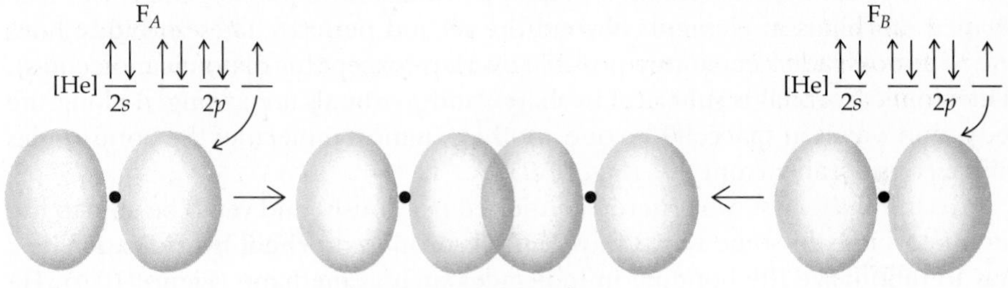

Figure 10.3 Sigma bond formation in F_2. Maximum overlap of atomic orbitals occurs when the orbitals overlap along the axis of the bond. If the orbitals do not overlap exactly head to head, the overlap is smaller. The bond is formed only when there is maximum, effective overlap.

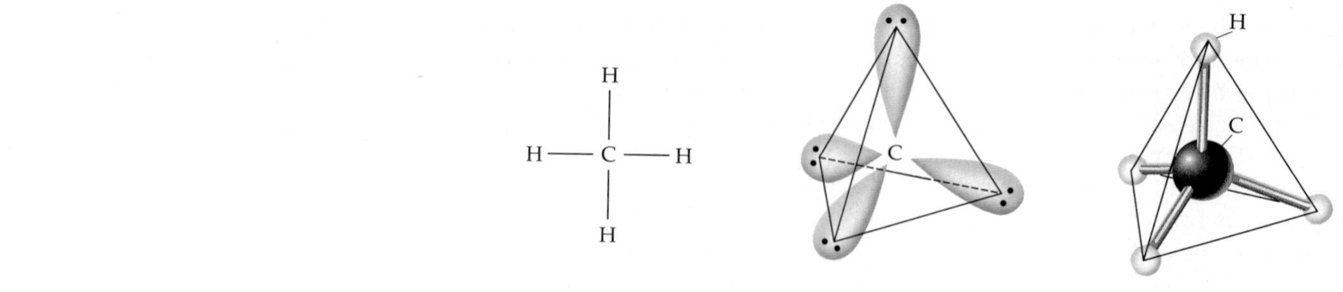

Lewis dot structure Electron-pair geometry Molecular geometry

C atom in CH_4 is tetrahedral. This means *four equivalent bonding electron pairs* must exist around the C atom, with an angle of 109° between the bond pairs. An isolated carbon atom, however, has only two unpaired electrons in its valence shell, and so might be expected to form only two bonds. In addition, these unpaired electrons are in atomic *p* orbitals, which are oriented 90° to one another.

The valence shell electron configuration of carbon

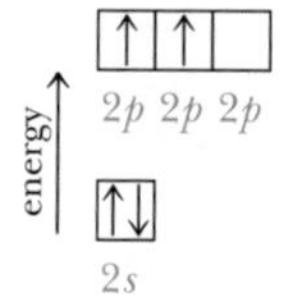

How can we account for four bonds oriented to the corners of a tetrahedron?

Explaining the bonding in any molecule or polyatomic ion requires the following concepts:

- Molecules and polyatomic ions adopt geometries that minimize the repulsions among valence shell electron pairs.
- Strong chemical bonds form by the maximum, effective overlap of atomic orbitals on a central atom with orbitals on the bonded atoms.

The orbitals on the central atom in a molecule or ion are the spherical *s* atomic orbital and the dumbbell-shaped *p* orbitals (and possibly the clover-leaf-shaped *d* orbitals in elements beyond the second period). These orbitals, however, cannot lead to maximum possible overlap (except for diatomic molecules). The atomic *s* orbital is spherical in shape, and *p* orbitals are arranged along the *x*-, *y*-, and *z*-axes in space, 90° to one another. In most molecules the bond angles differ considerably from 90° (Figure 9.10).

You may be familiar with hybridization in agriculture, where two varieties of a plant are mixed to provide a new variation that reflects some characteristics of each of its parents.

To fulfill the first two criteria mentioned previously, and yet to be able to use atomic orbitals in some way, Linus Pauling proposed **orbital hybridization** as a way to rationalize the bonding in molecules such as methane (Figure 10.4). He recognized that appropriate *s*, *p*, and *d* atomic orbitals on the central atom can be mixed to create a new set of orbitals—called **hybrid orbitals**—that are directed toward the atoms bonded to the central atom.

For the four C—H bonds of methane to have their maximum strength, there must be effective overlap of the four carbon orbitals with the H atom's *s* orbitals

Figure 10.4 Atomic orbitals can mix, or hybridize, to form hybrid orbitals. An analogy is mixing two different colors (left) to produce a third color, which is a "hybrid" of the original colors (middle). After mixing there are still two beakers (right), each containing the same volume of solution as before, but the color is a "hybrid" color. (C. D. Winters)

at the corners of a tetrahedron. Thus, Pauling suggested that the approach of the H atoms to the isolated C atom causes the four carbon *s* and *p* orbitals to combine or mix together. That is, chemists imagine that the C atom orbitals *hybridize* to provide four equivalent hybrid orbitals that point to the corners of a tetrahedron (Figure 10.5). We label *each* hybrid orbital as $\boldsymbol{sp^3}$ because the orbitals are the result of the combination of one *s* and three *p* orbitals on one atom. Each sp^3 hybrid orbital is identical in shape, and all are directed at tetrahedral angles.

Be sure to notice that *four* atomic orbitals produce *four* hybrid orbitals. The number of atomic orbitals used is *always* the same as the number of hybrid orbitals produced.

Orbital hybridization provides a theoretical explanation for the observation that CH_4, for example, has four equivalent bonds directed to the corners of a tetrahedron. Another outcome is that hybrid orbitals have greater magnitude in the region between nuclei than any of the atomic orbitals from which they are formed. This important observation means that greater overlap can be achieved between C and H in CH_4, for instance, and stronger bonds result with hybrid orbitals.

The following diagram illustrates the mixing of the carbon atom *s* and *p* orbitals to produce a new set of four orbitals, the hybrid orbital set. The four sp^3 hybrid orbitals have the same shape, and they all have the same energy, which is the weighted average of the parent *s* and *p* orbital energies. Because the orbitals have the same energy, electrons are assigned according to Hund's rule (Section 8.4).

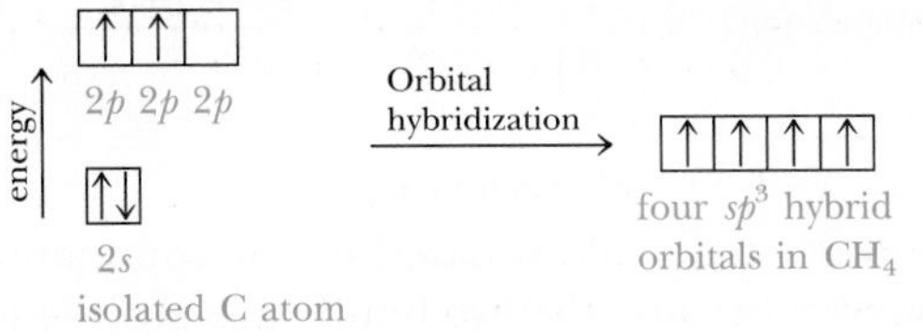

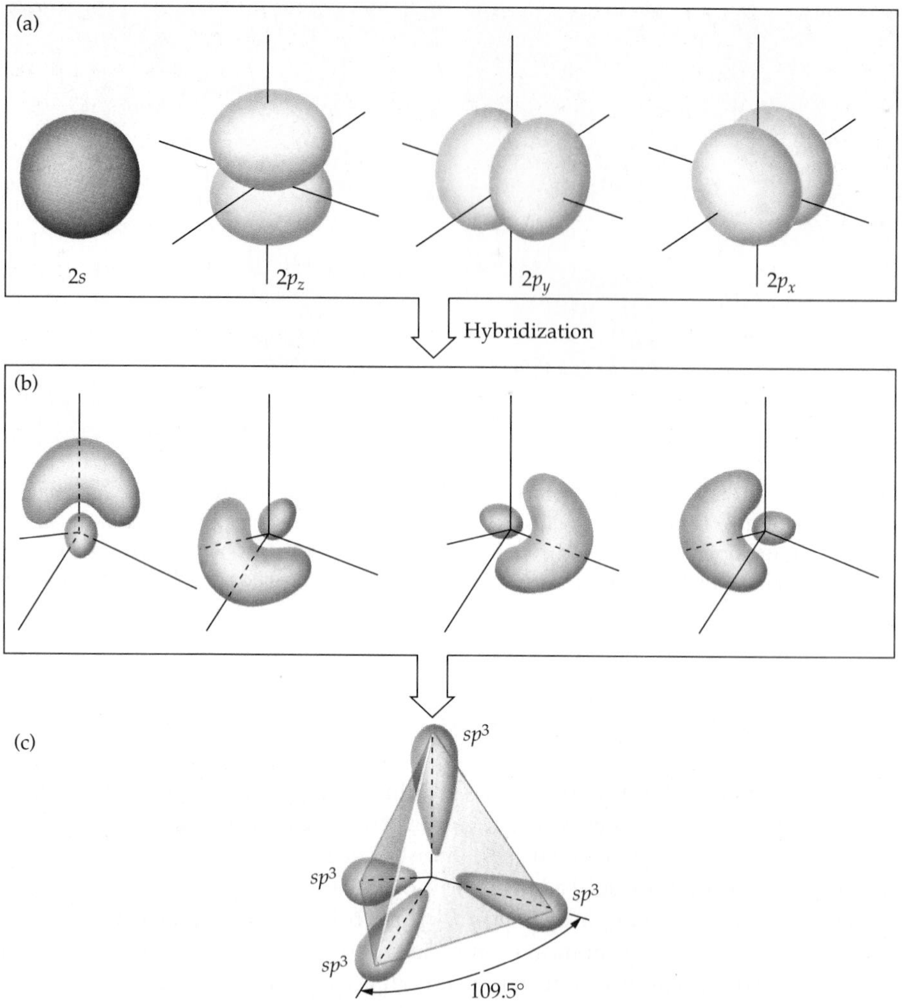

Figure 10.5 Hybrid orbital formation. (a) The four s and p atomic orbitals *on one atom.* (b) The s and p atomic orbitals are combined, or hybridized, to form four new sp^3 hybrid orbitals. (c) The four sp^3 hybrid orbitals are equivalent and are directed to the corners of a tetrahedron (angle 109.5°).

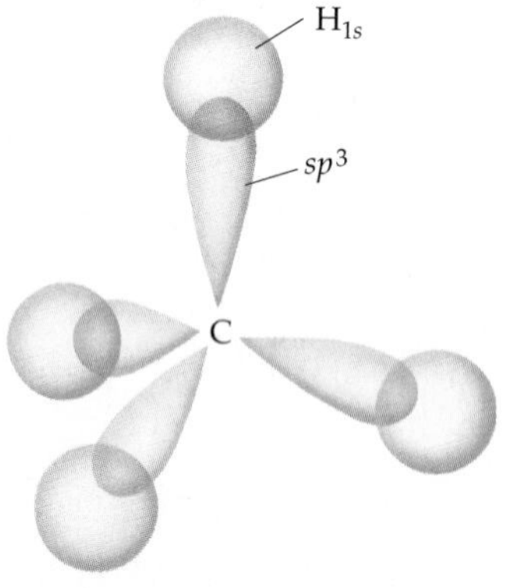

Figure 10.6 Bonding in methane, CH_4, according to the valence bond theory. A covalent C—H bond arises from the overlap of one of the four sp^3 hybrid orbitals on the C atom with the 1s orbital on a H atom.

The four sp^3 hybrid orbitals point in four different directions in space, 109.5° from one another. Overlap of each half-filled sp^3 hybrid orbital with a half-filled hydrogen 1s orbital gives *four equivalent C—H bonds arranged tetrahedrally,* as required by experimental evidence (Figure 10.6).

Hybrid orbitals can also be used to interpret the bonding and structure of such common molecules as H_2O and NH_3. An isolated O atom has two unpaired valence electrons as required for two bonds, but these are in 2p orbitals 90° apart.

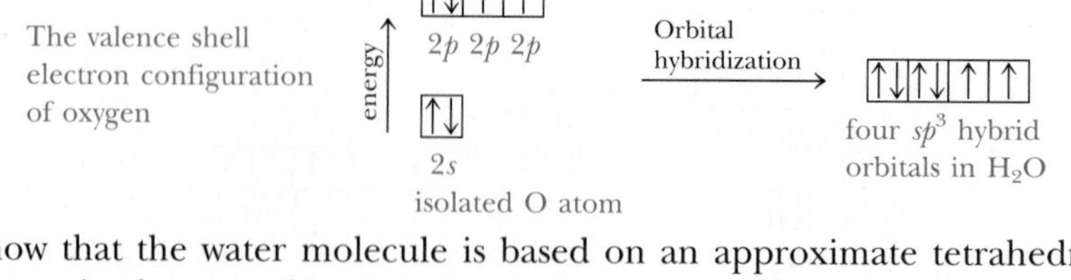

We know that the water molecule is based on an approximate tetrahedron of electron pairs, however, because the two bond pairs are 104.5° apart. If we allow the four $2s$ and $2p$ orbitals of oxygen to hybridize on approach of the H atoms, four sp^3 hybrid orbitals are created. Two of these orbitals are occupied by unpaired electrons and lead to the O—H sigma bonds. The other two orbitals contain the lone pairs of the water molecule (Figure 10.7).

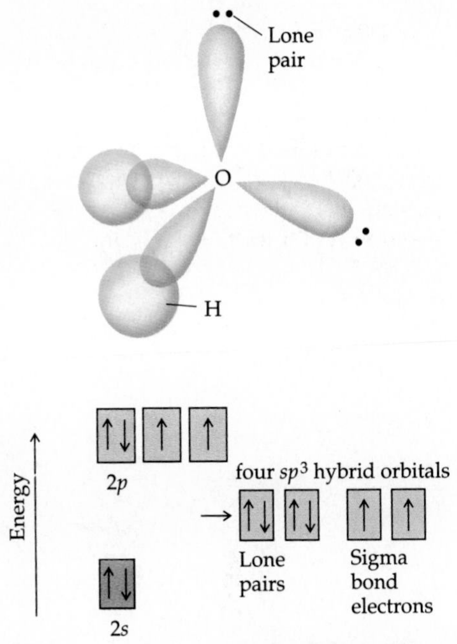

Figure 10.7 Bonding in the water molecule according to the valence bond theory. The O atom is sp^3 hybridized. The two O—H bonds arise from the overlap of an sp^3 hybrid orbital with an H atom $1s$ orbital. Nonbonding electron pairs are assigned to the two remaining sp^3 hybrid orbitals.

EXAMPLE 10.1 *Hybrid Orbitals in Bonding*

Describe the bonding in ammonia, NH_3, using orbital hybridization.

Solution The Lewis structure of ammonia shows that four valence shell electron pairs occur (three bond pairs and one lone pair), which must be at the corners of a tetrahedron. Because three of these pairs are bond pairs, the molecule has a trigonal-pyramidal molecular geometry.

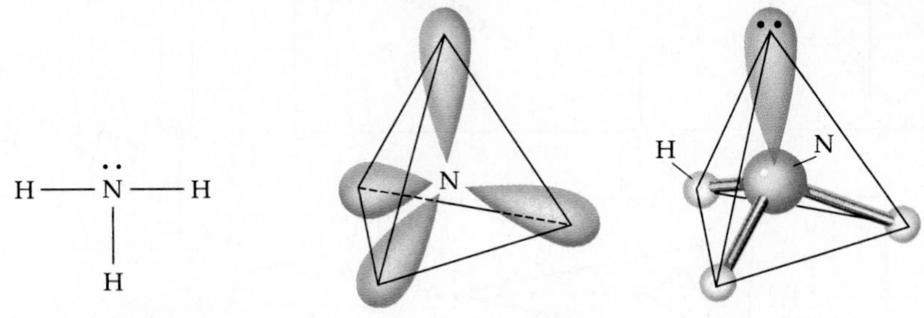

Lewis dot structure　　Electron-pair geometry　　Molecular geometry

To rationalize the bonding in NH_3, and an H—N—H angle of 107.5°, we invoke orbital hybridization. The *s* and *p* orbitals of nitrogen combine to give four sp^3 hybrid orbitals. One of these is assigned to the lone pair of electrons, and each of the other three is occupied by one unpaired electron. It is these singly occupied, hybrid orbitals that account for the N—H covalent bonds by overlapping with hydrogen $1s$ orbitals.

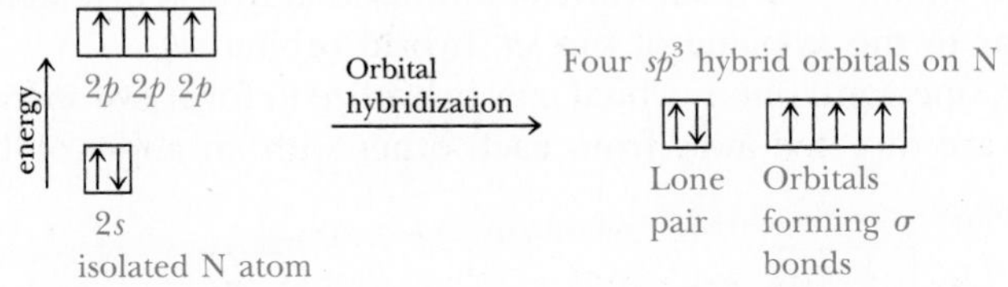

Linear and trigonal-planar geometries are also commonly observed. Orbital hybridization is again useful to describe the bonding in molecules and ions with those geometries. For example, an *s* and two of the *p* orbitals on the same atom can combine to form **three sp^2 hybrid atomic orbitals** that are directed to the corners of a planar triangle (Figure 10.8). For example, a boron atom can form three sp^2 orbitals in this manner.

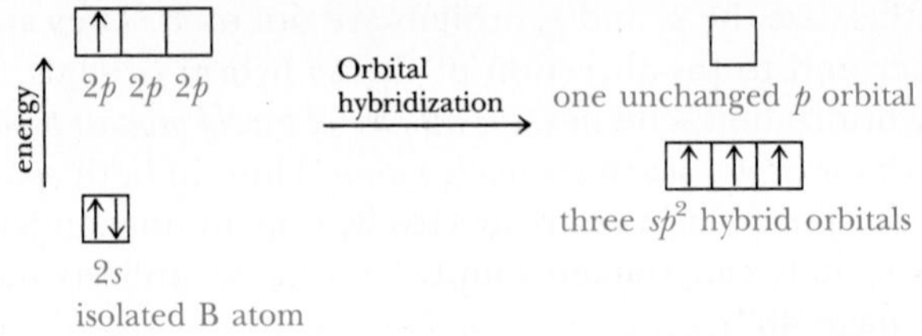

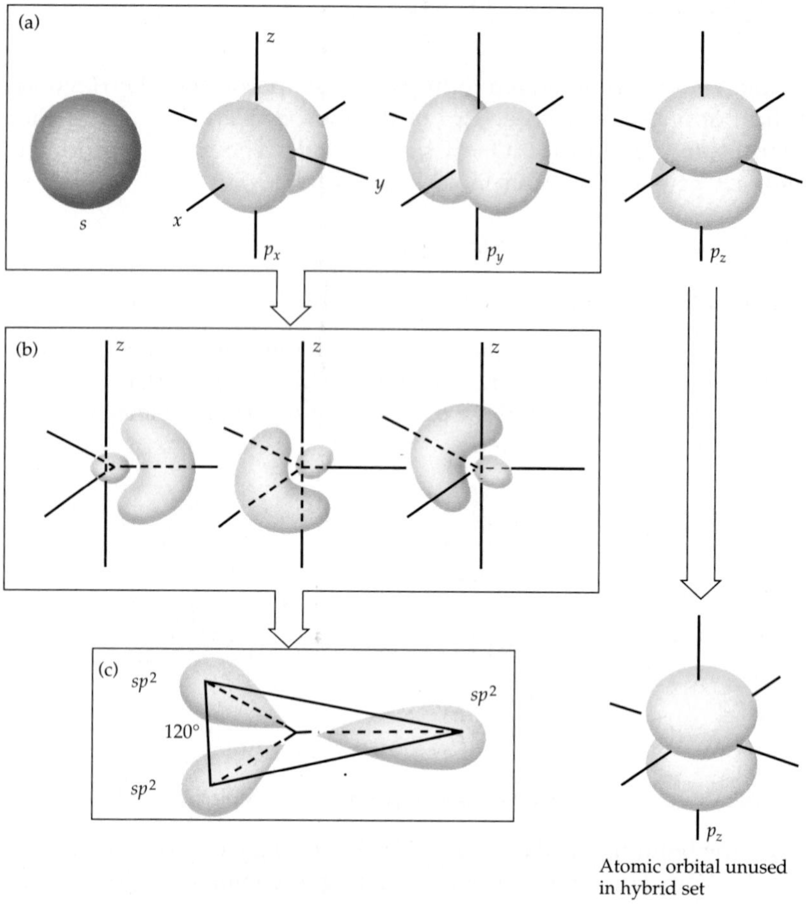

Figure 10.8 Formation of three sp^2 hybrid orbitals. An s orbital is combined with two p orbitals, say p_x and p_y, to form a set of three hybrid orbitals that lie in the imaginary xy plane. Each orbital is labeled sp^2, and the angle between them is 120°.

If the p_x and p_y orbitals are used in hybrid orbital formation, for example, the set of three hybrid orbitals lies in the imaginary xy-plane (Figure 10.8). Be sure to notice that the atomic p_z orbital, which was unused in hybrid orbital formation, is perpendicular to the xy-plane of the sp^2 hybrid orbital set.

Similarly, one s and one p orbital may hybridize to form **two *sp* hybrid atomic orbitals** that are directed away from each other with an angle of 180° (Figure 10.9).

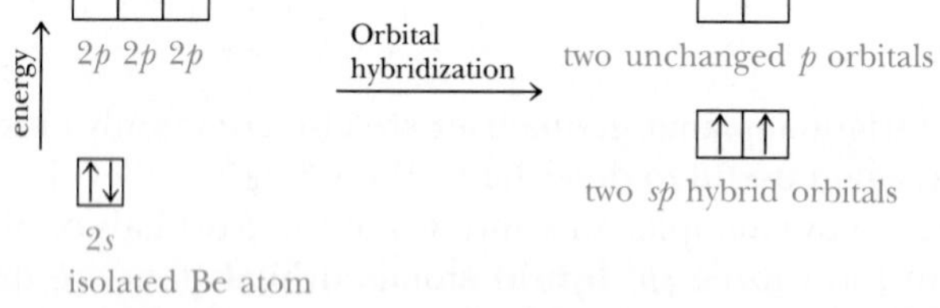

This scheme is appropriate for the beryllium atom in $BeCl_2$. Combining the s orbital with the p_x orbital gives the two sp hybrid orbitals that lie along the x-direction. In this case the p_y and p_z orbitals are not used. They are perpendicular to each other and to the direction of the sp hybrid orbitals.

In all the hybridization schemes, *the number of hybrid orbitals produced is equal to the number of atomic orbitals used in the combination.* Thus, in both sp^2 and sp hybridization, one or two pure p orbitals, respectively, remain unchanged after hybridization. These orbitals can remain empty or can be utilized in pi (π) bond formation (see page 469).

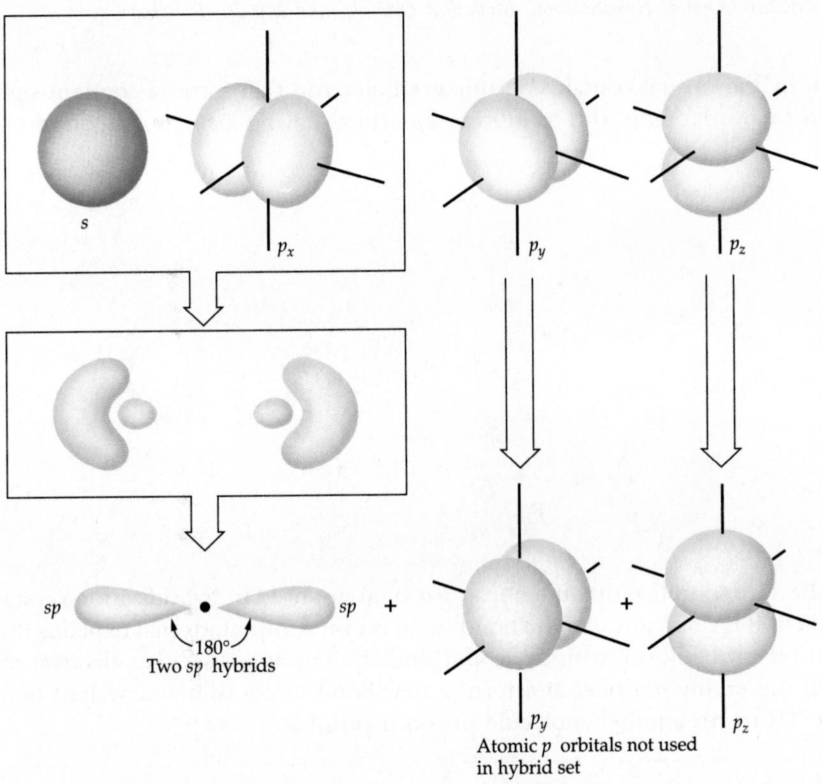

Figure 10.9 Formation of two *sp* hybrid orbitals. An *s* orbital is combined with a *p* orbital, say p_x, to form a set of two hybrid orbitals that lie along a straight line (the *x* axis). Each orbital is labeled *sp*, and the angle between them is 180°. Be sure to notice that the atomic p_y and p_z orbitals, which are perpendicular to the axis of the *sp* hybrid orbital set, are unused in this combination.

EXAMPLE 10.2 *Hybridization and Bonding*

Describe the bonding in BF_3 using orbital hybridization.

Solution The electron-pair and molecular geometry of BF_3 are both trigonal-planar. Three bonds, at the corners of a triangle, must be formed by the central atom. Thus, three boron atom orbitals (the *s* and the two *p* orbitals *in the molecular plane*) are hybridized to form three sp^2 orbitals.

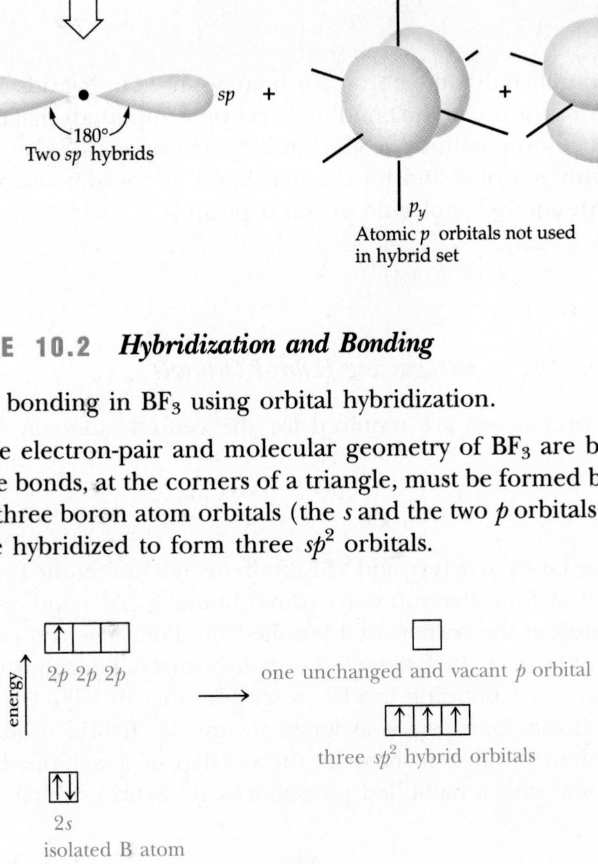

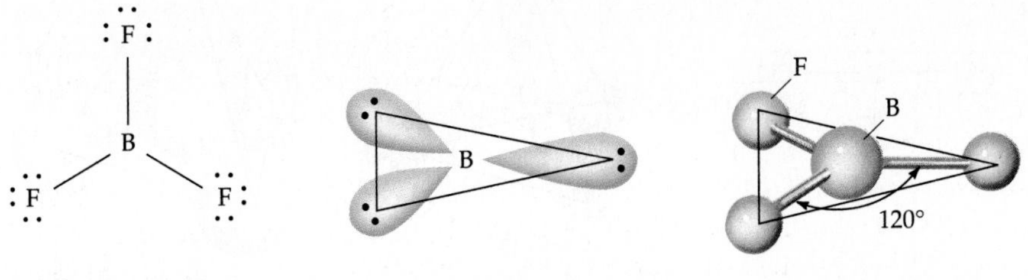

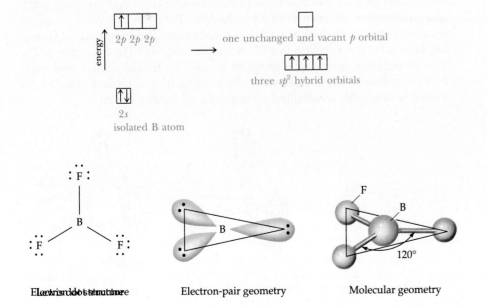

Electron dot structure   Electron-pair geometry   Molecular geometry

Each hybrid orbital contains an unpaired electron that forms a covalent sigma bond by overlapping with a fluorine $2p$ orbital having a single, unpaired electron.

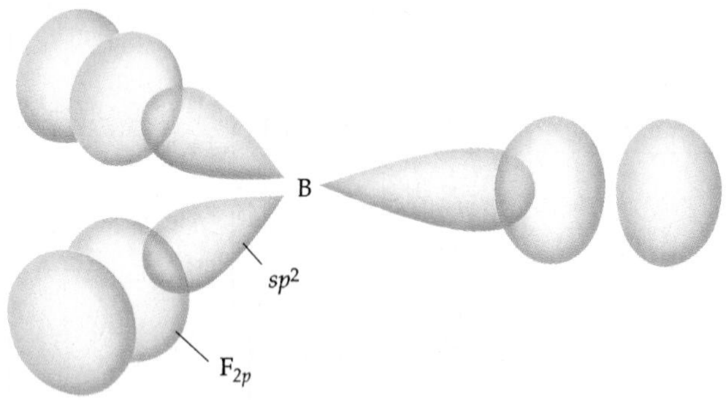

Be sure to notice that the boron p orbital not used in hybridization remains unfilled. It is this feature of the bonding in boron compounds that explains their great reactivity. Boron trifluoride and similar compounds seek an electron pair to fill the empty p orbital and form a new bond (a coordinate covalent bond, page 410) with another molecule or ion if possible.

EXAMPLE 10.3 *Recognizing Hybrid Orbitals*

What hybrid orbital sets are required for the central atoms in (1) PCl_3 and (2) BeH_2?

Solution

1. For PCl_3 the Lewis structure and VSEPR theory tell us that the P atom must be surrounded by four electron pairs (three bonding pairs and one lone pair) approximately at the corners of a tetrahedron. Four electron pairs arranged tetrahedrally require the P atom to be sp^3-hybridized. Phosphorus is in Group 5A; therefore, the bonding in PCl_3 resembles that in NH_3 (Example 10.1). The phosphorus lone pair is assigned to one sp^3 hybrid orbital, and each P—Cl covalent bond is formed by the overlap of a half-filled chlorine $3p$ atomic orbital with a half-filled phosphorus sp^3 hybrid orbital.

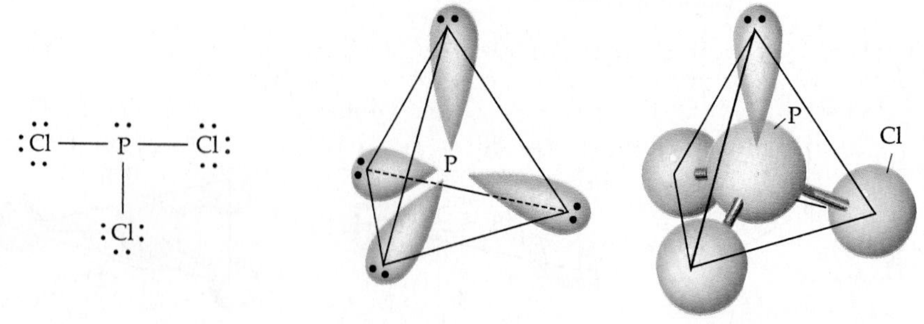

Lewis dot structure   Electron-pair geometry   Molecular geometry

2. Based on the Lewis structure of BeH_2, and VSEPR theory, the molecule is predicted to be linear. Linear geometry and two sigma bonds are achieved by forming two *sp* hybrid orbitals on Be (from the 2*s* and one of the 2*p* orbitals on Be, Figure 10.9) that lie along the H—Be—H axis.

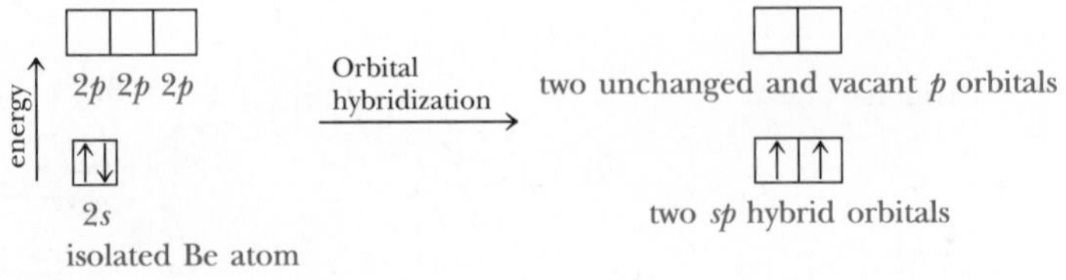

The unhybridized *p* orbitals of the Be atom are not filled in the BeH_2 molecule.

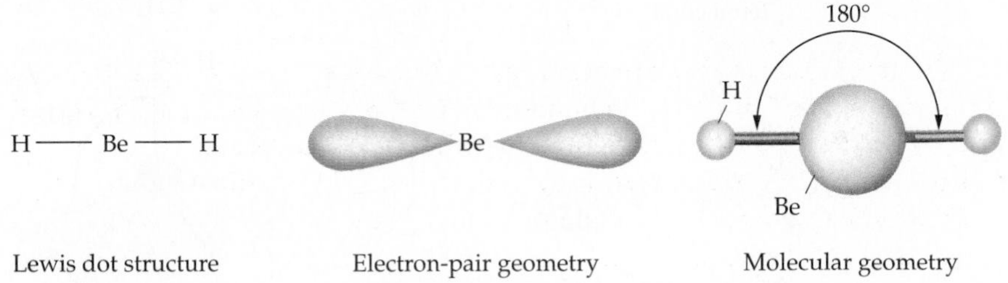

EXERCISE 10.1 ***Hybrid Orbitals and Bonding***

Describe the bonding in SCl_2 using hybrid orbitals. ■

Hybridization Involving *d* Orbitals

Elements of the third and higher periods can accommodate five or six valence shell electron pairs in molecules such as PF_5 and SF_6, respectively. To rationalize the bonding using valence bond theory, therefore, five or six atomic orbitals must be combined to form hybrid orbitals. Because only four atomic orbitals of the *s* and *p* type occur in a valence shell, the extra one or two orbitals required come from the *d* subshell of the same shell. Atomic *d* orbitals are considered to be valence shell orbitals for elements of the third and higher periods and so are available for hybrid formation. The hybrid orbital sets for five or six pairs are given in Table 10.1 and Figure 10.10, along with those sets already described.

TABLE 10.1 Hybrid Orbital Sets for Two to Six σ-Bonding or Lone Electron Pairs

Hybrid Orbital Set	Number of Hybrid Orbitals (Number of σ Bonds and Lone Pairs)	Geometry of Hybrid Orbitals	Example
sp	2	linear	Be in $BeCl_2$
sp^2	3	planar-trigonal	B in BF_3
sp^3	4	tetrahedral	C in CH_4; P in PF_3; O in H_2O
sp^3d	5	trigonal-bipyramidal	P in PF_5; S in SF_4
sp^3d^2	6	octahedral	S in SF_6; Br in BrF_5; Xe in XeF_4

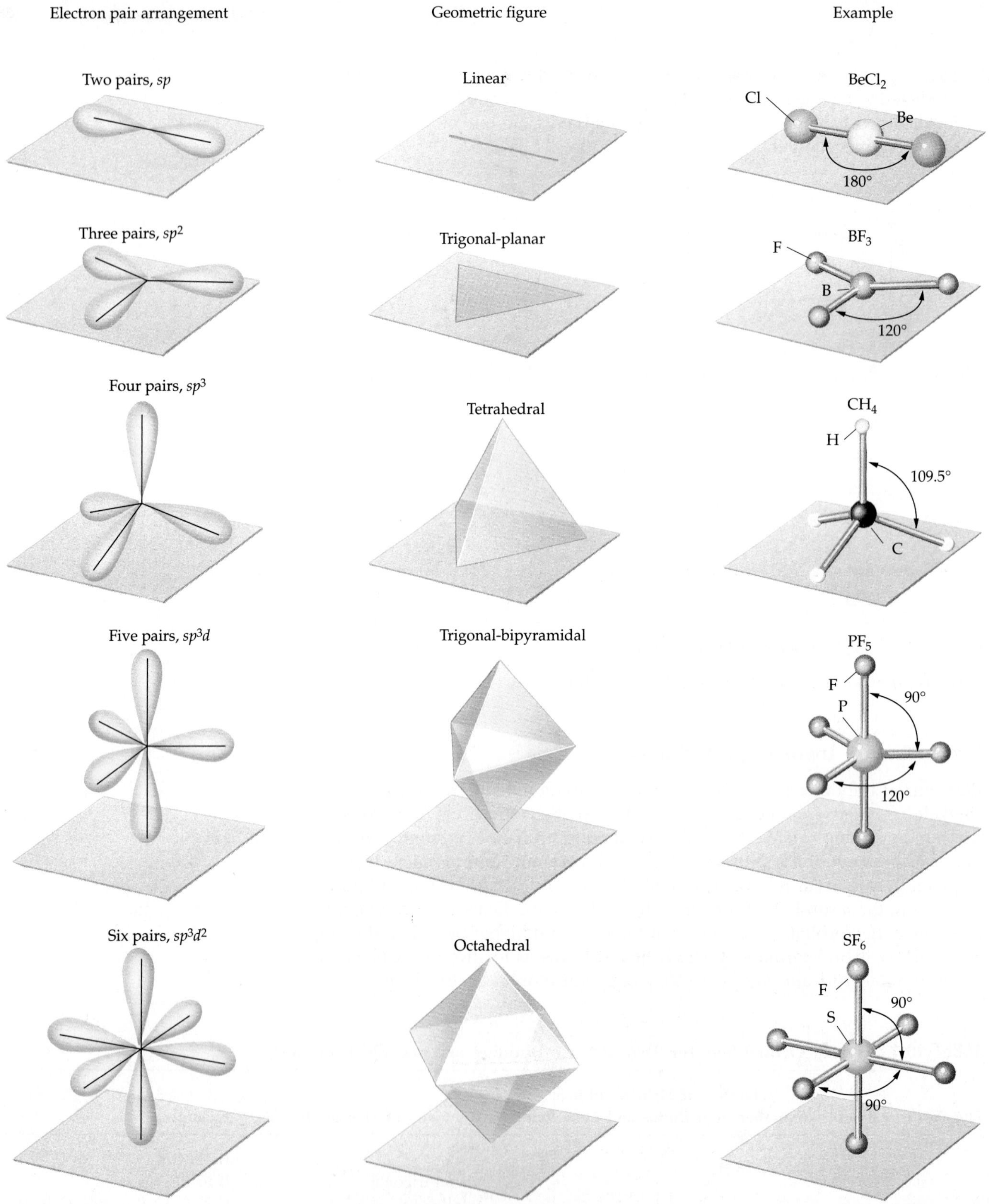

Figure 10.10 The geometry of the hybrid orbital sets for two to six structural electron pairs. In forming a hybrid orbital set, the *s* orbital is always used, plus as many *p* orbitals (and *d* orbitals) as required to give the necessary number of sigma bonding and lone pair orbitals.

EXAMPLE 10.4 ***Hybridization Involving* d *Orbitals***

Describe the bonding in PF_5 using hybrid orbitals.

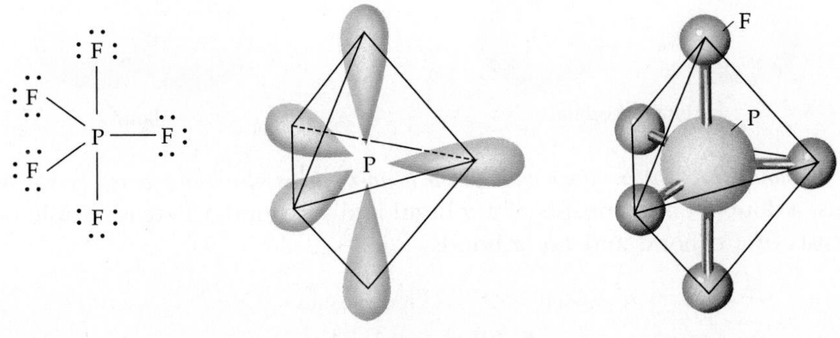

Solution The electron-pair and molecular geometry of PF_5 are trigonal-bipyramidal according to its Lewis structure and VSEPR theory. Five covalent bonds must point to the corners of a trigonal bipyramid. Therefore, the P atom must have a single electron in each of five hybrid orbitals. As illustrated in Figure 10.10, the hybrid scheme sp^3d is required.

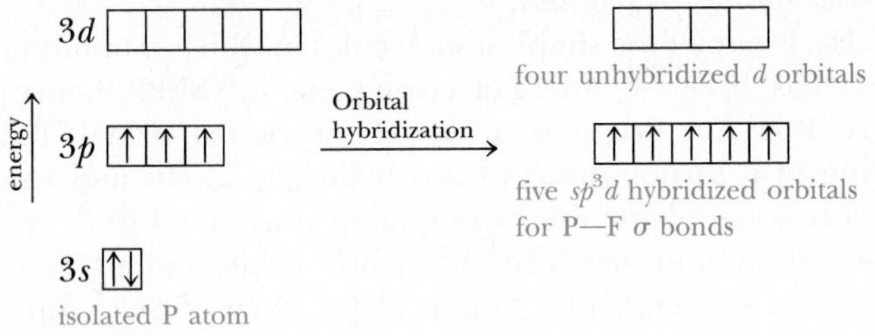

Each of the five P—F σ bonds involves overlap of one of the phosphorus sp^3d hybrid orbitals with a fluorine $2p$ orbital. (See Example 10.2 for an example of bonding to fluorine.)

EXERCISE 10.2 ***Hybridization Involving* d *Orbitals***

Describe the bonding in XeF_4 using hybrid orbitals. Remember to consider first the Lewis structure, then the electron-pair geometry (based on VSEPR theory), and then the molecular shape. ■

Multiple Bonding

There are two types of covalent bonds: **sigma (σ)** bonds and **pi (π)** bonds. According to valence bond theory, σ bonds arise from the overlap of atomic orbitals so that the bonding electrons lie along the bond axis. *Pi bonds come from the sideways overlap of* p *atomic orbitals,* meaning that the overlap region is above and below the internuclear axis, and the π-bonding electrons are above and below the bond axis.

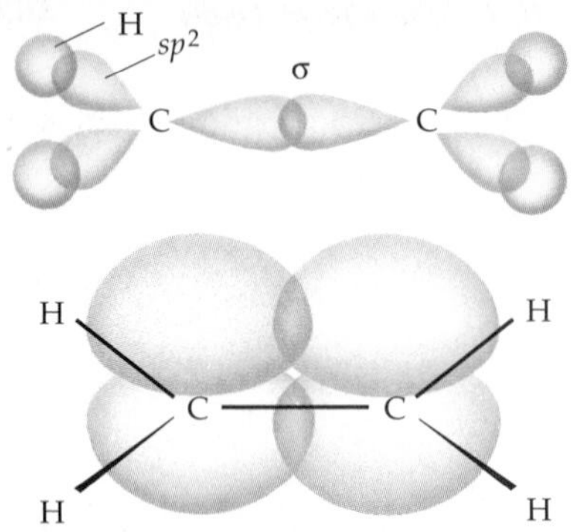

Figure 10.11 Bonding in ethylene, C_2H_4. (a) The C atom orbitals are hybridized to form three sp^2 hybrid orbitals, leaving one pure p orbital. Overlap of an sp^2 orbital from one C atom with an sp^2 hybrid orbital from the adjacent C atom gives the C—C sigma bond. The C—H bonds come from the overlap of a C atom sp^2 hybrid orbital with an H atom 1s orbital. (b) The pi (π) bond in C_2H_4 is formed from the sideways overlap of a p orbital on each atom as illustrated in the text.

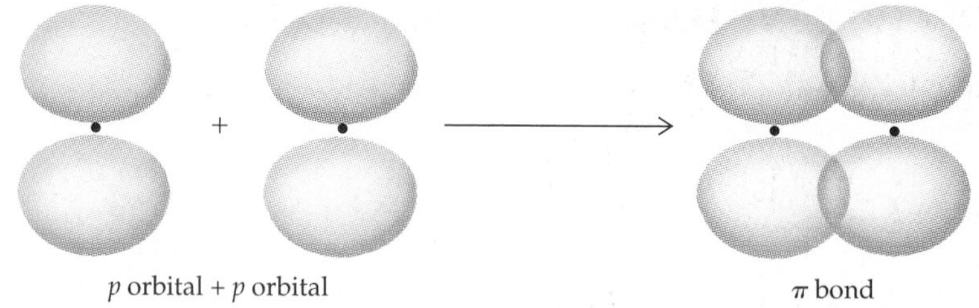

Pi bonds rarely occur alone without the bonded atoms also being joined by a σ bond. Thus, a double bond consists of a σ bond and a π bond, whereas a triple bond consists of a σ bond and *two* π bonds.

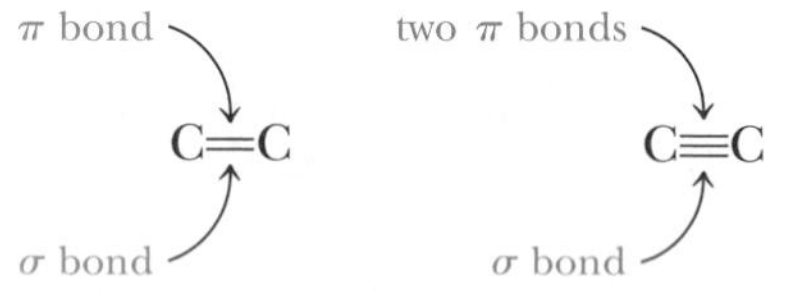

Because a π bond is formed from pure p atomic orbitals, one on each of two atoms, *a π bond may form only if unhybridized p orbitals remain on the bonded atoms,* in the correct orientation, after accounting for σ bond formation. Therefore, *when the Lewis structure shows multiple bonds, and when the atoms involved are from the second period, they are either sp^2- or sp-hybridized.*

Ethylene, C_2H_4, is one of the simplest molecules in which π bonding occurs. Each carbon atom has three σ-bonding electron pairs, so VSEPR theory predicts that the geometry about the carbon atoms in C_2H_4 is trigonal-planar. To account for the σ bonding of a carbon atom to two hydrogen atoms and to another carbon atom in a trigonal plane, the carbon atom is assumed to be sp^2-hybridized. Thus, each carbon atom has three sp^2 hybrid orbitals in a plane, and an unhybridized p orbital perpendicular to that plane. Each of these four orbitals contains one unpaired electron.

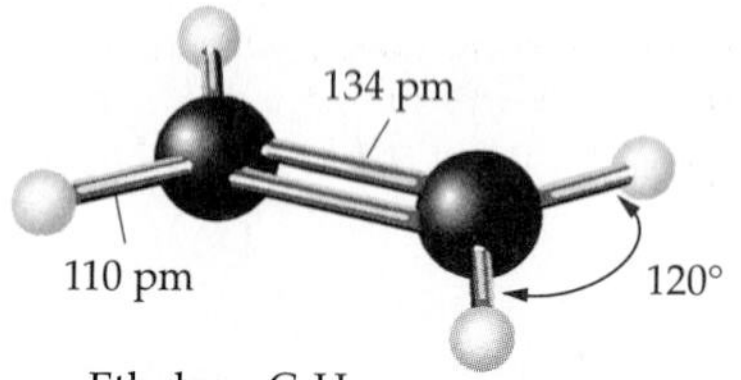

Ethylene, C_2H_4

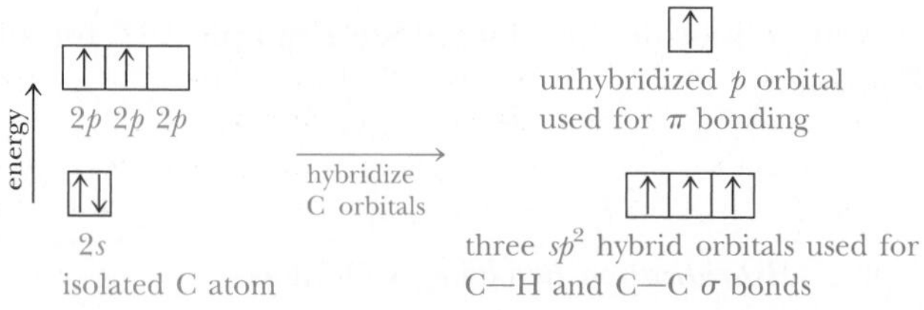

A π bond can form only if the σ orbitals of both C atoms are *all* in the same plane. If the plane containing the sp^2 hybrid orbitals of one C atom is twisted relative to the plane containing the sp^2 hybrid orbitals of the other C atom, the unhybridized p orbital on one C atom does not align with the unhybridized p orbital on the adjacent C atom, and a π bond does not form.

If the two C atoms approach each other, head-to-head overlap of an sp^2 hybrid orbital from each atom gives a σ bond, and sideways overlap of the half-filled, pure p orbitals forms a π bond (Figures 10.11 and 10.12). Thus, the C=C double bond in C_2H_4 consists of one σ bond and one π bond. To complete the bonding picture, each of the two remaining half-filled sp^2 hybrid orbitals on the C atoms overlaps with a half-filled hydrogen 1s orbital to form a C—H σ bond.

In Chapter 9 we pointed out that carbon can form multiple bonds with oxygen, sulfur, and nitrogen. To understand how such bonds are possible, consider formaldehyde, H_2CO, which has a carbon-oxygen π bond. This trigonal-planar molecule is based on an sp^2-hybridized carbon atom. The σ bonds to the O atom and the two H atoms form by overlap of these sp^2 hybrid orbitals with half-filled orbitals from the O and H atoms. In addition, an unhybridized p

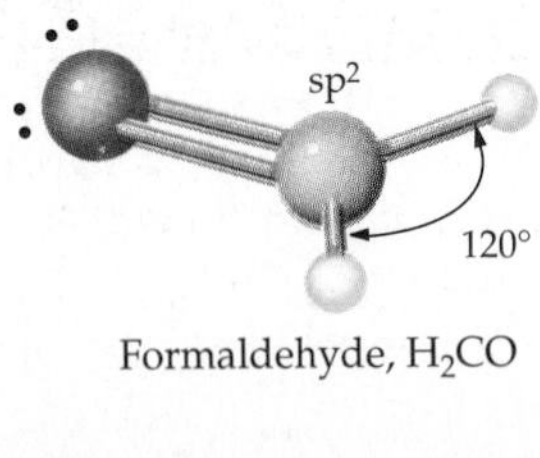

Formaldehyde, H_2CO

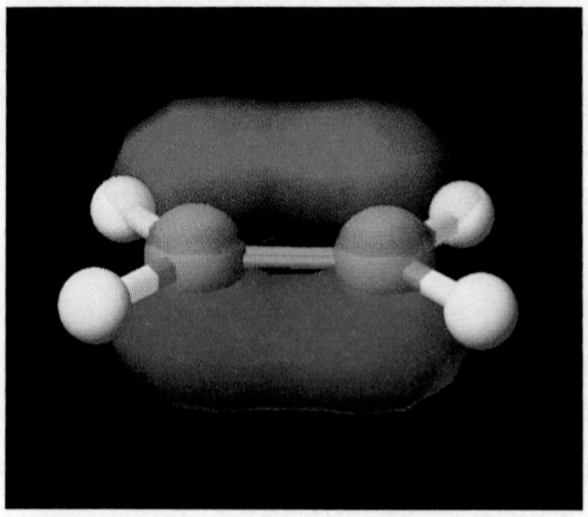

Figure 10.12 A computer-generated model of the π bond in ethylene. Note the lack of π electron density along the carbon-carbon axis. The electron density of a π bond lies above and below the carbon-carbon bond axis. (Susan Young)

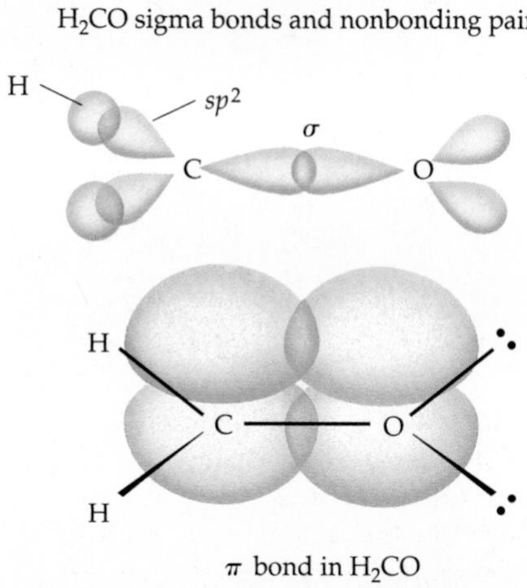

Figure 10.13 Bonding in formaldehyde, H_2CO. This model for the σ framework of the molecule is based on sp^2 hybridized C and O atoms. The C—O σ bond arises from the overlap of hybrid orbitals. Since a half-filled p atomic orbital remains on both C and O, these can overlap to form the carbon-oxygen π bond.

Another point of view on C—O bonding in H_2CO is to consider the O atom as unhybridized. The C—O sigma bond is formed by an unpaired electron in a p orbital, whereas the O atom lone pairs are electron pairs in a $2s$ and a $2p$ orbital. The π bond arises from an unpaired electron in the remaining p orbital.

orbital occurs perpendicular to the molecular plane (just as for the carbon atoms of C_2H_4). This p orbital is available for π bonding, this time with oxygen (Figure 10.13).

To complete the bonding picture in H_2CO, we might consider the O atom to be sp^2-hybridized as well. The two nonbonding pairs and a single electron occupy these hybrid orbitals. The latter is used to form the σ bond to carbon. The single electron in the unhybridized oxygen p orbital is available for C—O π bonding.

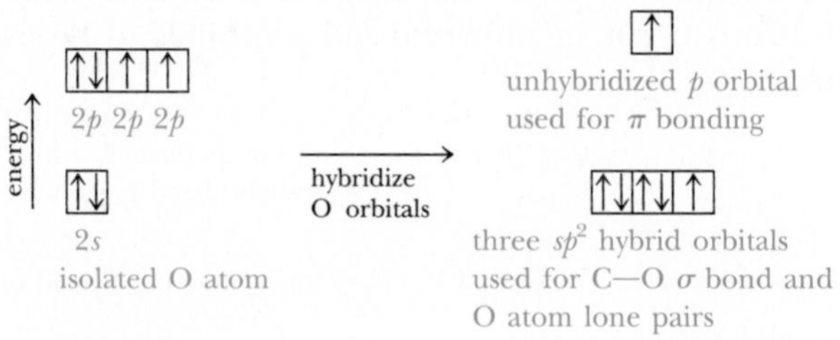

Acetylene, H—C≡C—H, is a simple molecule with a triple bond. Here one σ bond and two π bonds join the two carbon atoms. The electron-pair geometry around the carbon atoms is clearly linear, so each carbon atom is sp-hybridized. This means that two half-filled p orbitals remain on each carbon after hybridization and are available for π bond formation.

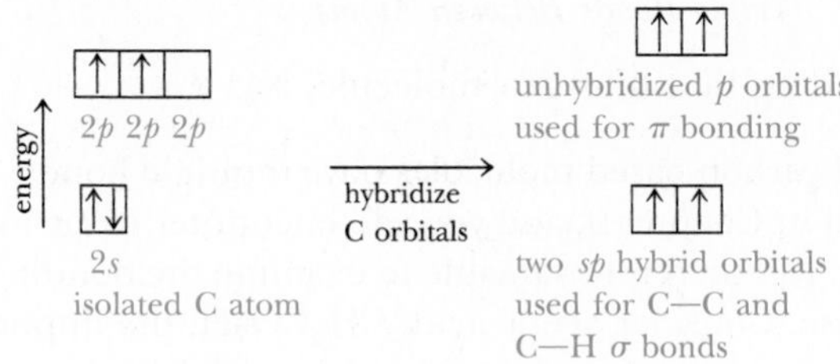

The orbitals about each C atom can be pictured as illustrated in Figure 10.9. The sp hybrid orbitals on each C atom are used for C—C and C—H sigma bond formation, and the pure p atomic orbitals overlap sideways to produce two carbon-carbon π bonds (Figure 10.14).

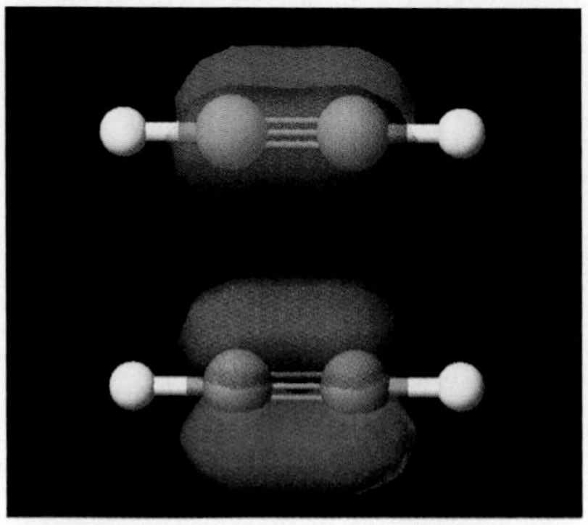

Figure 10.14 Computer-generated views of the π bonds in acetylene, C_2H_2. (Susan Young)

EXAMPLE 10.5 *The C≡O Triple Bond in Carbon Monoxide*

Describe the bonding in carbon monoxide, CO, using orbital hybridization.

Solution The Lewis structure

$$:C{\equiv}O:$$

informs us that it is useful to consider both the carbon and the oxygen as *sp*-hybridized. For electron "bookkeeping" purposes, one of these hybrid orbitals on each atom is initially assigned one electron for σ bond formation. The other *sp* hybrid orbital on each atom is then occupied by a nonbonding electron pair. Remaining electrons are then assigned to unhybridized orbitals.

	***sp*-hybridized C**	***sp*-hybridized O**
Unhybridized p *orbitals for pi bonding*	[↑][ ]	[↑↓][↑]
Two sp *hybrids*	[↑↓] lone pair [↑] sigma bond	[↑↓] lone pair [↑] sigma bond

Now a half-filled *sp* hybrid orbital occurs on each atom for sigma bond formation, as well as a lone pair on each atom assigned to an *sp* hybrid. Also a total of two pairs of electrons occur in unhybridized *p* orbitals to be used for two pi bonds as required.

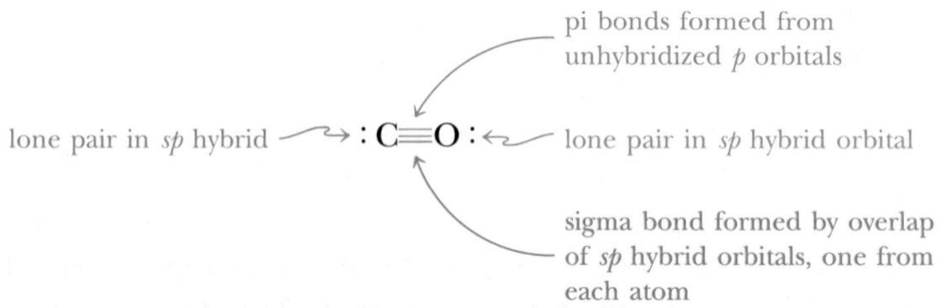

The orbital picture for CO is identical with that for HC≡CH in Figure 10.14 except that the C—H bond pairs become lone pairs.

EXERCISE 10.3 *Triple Bonds Between Atoms*

Describe the bonding in a nitrogen molecule, N_2. ■

Thousands of carbon-based molecules have multiple bonds. You already saw a number of them in Chapter 9, and you will encounter many more if you study chemistry further. It is therefore valuable to examine the bonding in a somewhat more complex case. Consider acetic acid, CH_3CO_2H, the important ingredient in vinegar. Its Lewis structure and a molecular model (with approximate bond angles) are shown here. The carbon atom of the CH_3 group must have a tetrahedral electron-pair geometry, so the C—H angle is approximately 109°. In valence bond terms, this means the carbon atom is sp^3-hybridized. The other carbon atom has a trigonal-planar electron-pair geometry, with substituents about 120° apart. This carbon must be sp^2-hybridized. Finally, we account for the C═O link

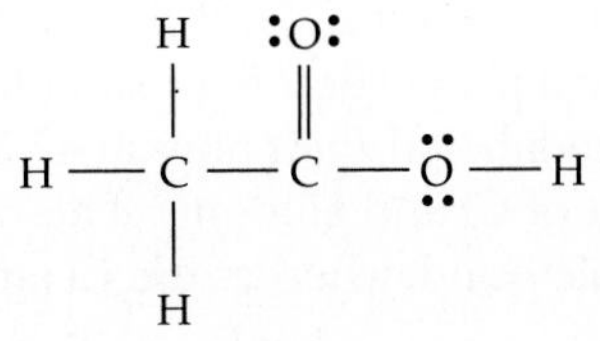

Lewis dot structure

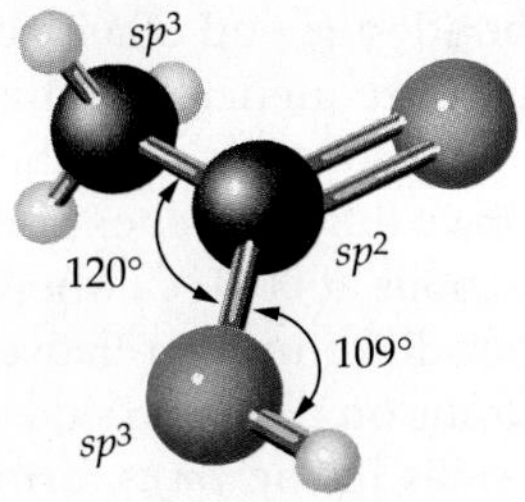

Molecular model

exactly as in the H_2CO molecule described earlier. Both the C atom and the O atom are taken as sp^2-hybridized, so an unhybridized p orbital remains on each atom to form the carbon-oxygen π bond.

EXERCISE 10.4 ***Bonding and Hybridization***

Analyze the bonding in acetonitrile, $H_3C{-}C{\equiv}N$. Estimate values for the H—C—H, H—C—C, and C—C—N angles, and indicate the hybridization of both carbon atoms and the nitrogen atom. ■

A Consequence of Multiple Bonding: Isomers

According to the kinetic theory of matter (Sections 1.1 and 12.6), molecules in the gas or liquid phase move rapidly and sometimes collide with one another. Atoms within molecules move as well; molecules constantly flex or vibrate along or around the bonds that hold the atoms together. If you have a flask of ethane gas, C_2H_6, the molecules are moving about in the flask, they are rotating in space, and parts of the molecule move or vibrate with respect to one another. One motion that is possible for the ethane molecule is the rotation of one end of the molecule with respect to the other. Indeed, essentially free rotation is possible around a σ bond in any molecule.

When two groups are bonded through a double bond, free rotation about the bond is not possible (under ordinary conditions). Because of the geometry

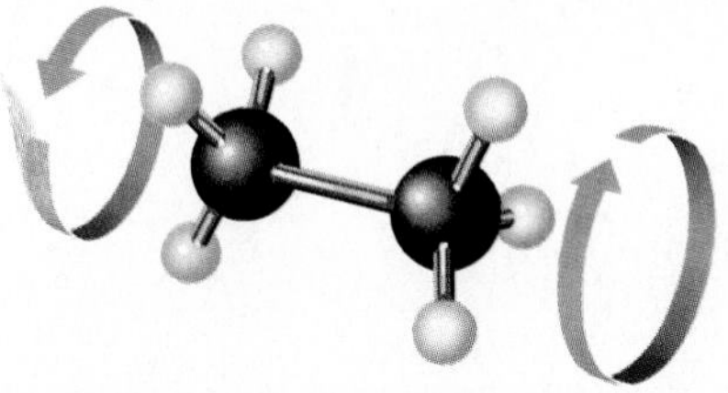

Free rotation occurs around the axis of a single (σ) bond.

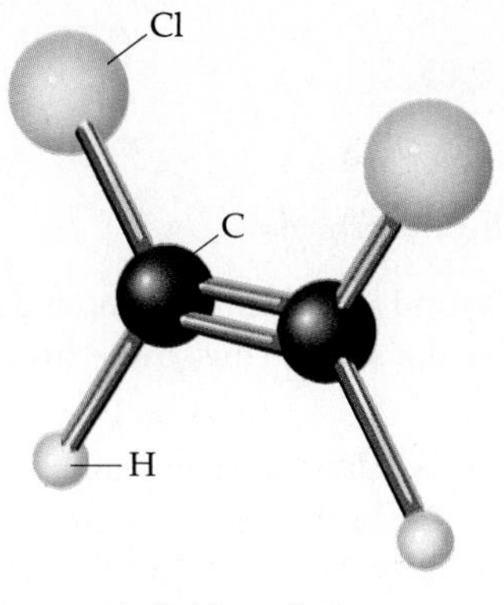

cis-dichloroethylene

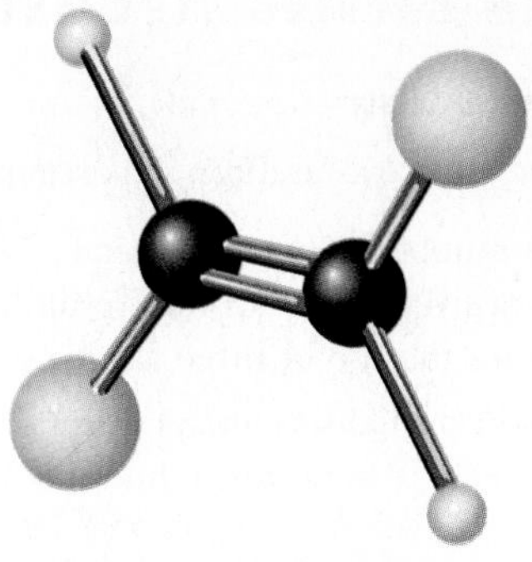

trans-dichloroethylene

of the π bond, one end of a carbon-carbon double bond is locked relative to the other. A consequence of this is that two compounds have the formula HClC=CHCl, for example. The compounds labeled *cis-* and *trans-*1,2,-dichloroethylene have different chemical and physical properties. For example, the form labeled *cis* boils at 60.3 °C, whereas the form labeled *trans* boils at 47.5 °C. Structurally, they differ in the relative placement of Cl and H atoms. The *cis* form has both Cl atoms on the same side of the double bond, whereas the Cl atoms are on opposite sides in the *trans* form.

Compounds like the *cis* and *trans* forms of HClC=CHCl are called **isomers.** In general, isomers are compounds that have the same formula but different structures. In some cases, two molecules with the same formula have the atoms connected differently. Thus, 1,1-dichloroethylene is an isomer of 1,2-dichloroethylene. These are called *structural* isomers. In other cases, such as *cis-* and *trans-*1,2-dichloroethylene, the atoms are connected in the same order but the atoms have different spatial arrangements. These are called *stereoisomers.*

Based on the structure of *cis-* and *trans-* HClC=CHCl, you would conclude that the *cis* form is polar, and the *trans* form is not (Section 9.6). It makes sense therefore that the polar *cis* form boils at the higher temperature.

Structural isomers

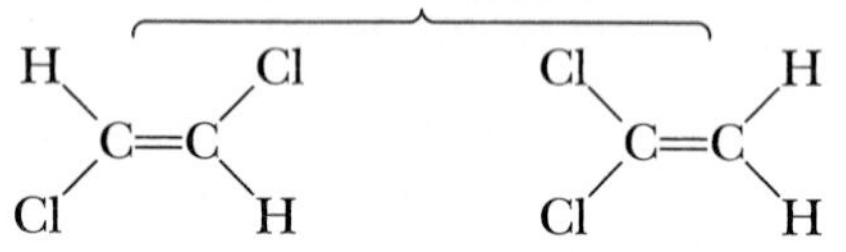

*trans-*1,2-dichloroethylene   1,1-dichloroethylene

The stereoisomers of HClC=CHCl are not easily interconverted. If one H—C—Cl were to rotate around the C=C double bond, the overlap of the p orbitals on each C atom must be broken, requiring 264 kJ/mol (= 611 kJ/mol − 347 kJ/mol).

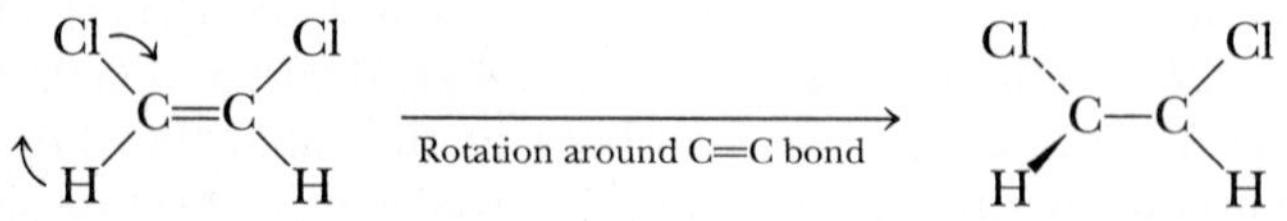

Energy of C=C bond = 611 kJ/mol   Energy of C—C bond = 347 kJ/mol

Energies of this magnitude are not available at ordinary temperatures. Rotation does occur under special conditions, however. Indeed, it is involved in the physiological process that allows us to see (see A Closer Look: Chemical Bonds and Vision).

PROBLEM-SOLVING TIPS AND IDEAS

10.1 Hybrid Orbitals

Remember these two useful points regarding hybrid orbitals:

- The number of atomic orbitals used in hybrid orbital formation, and the number of hybrid orbitals that result, is always the same. Thus, three hybrid orbitals require the use of three atomic orbitals.
- Hybrid orbital sets always include the s orbital plus as many p and d orbitals as are needed to produce the total number of hybrid orbitals required.

A CLOSER LOOK *Chemicals Bonds and Vision*

Rotation around a double bond occurs under special conditions and is not an esoteric problem of interest only to chemists. It occurs in the reactions that allow you to see. A yellow-orange compound, β-carotene, the natural coloring agent in carrots, breaks down in your body to produce vitamin A, and this compound is converted in the liver to a compound called 11-*cis*-retinal. In the retina of your eye 11-*cis*-retinal combines with the protein opsin to form a light-sensitive substance called rhodopsin, which is found in the rod cells of the eye. When light strikes the retina, enough energy is transferred to a rhodopsin molecule to allow rotation around a carbon-carbon double bond, transforming rhodopsin into metarhodopsin II, a molecule whose shape is quite different, as you can see from the structural formula shown here. This change in molecular shape causes a nerve impulse to be sent to your brain, and you see the light. Eventually metarhodopsin II reacts chemically to produce a different form of retinal, which is then converted back to vitamin A, and the cycle of chemical changes begins again. Decomposition of metarhodopsin II is not as rapid as its formation, however, and an image formed on the retina persists for a tenth of a second or so. This persistence of vision allows you to perceive movies and videos as continuously moving images, even though they actually consist of separate pictures, each captured on a piece of film for a thirtieth of a second.

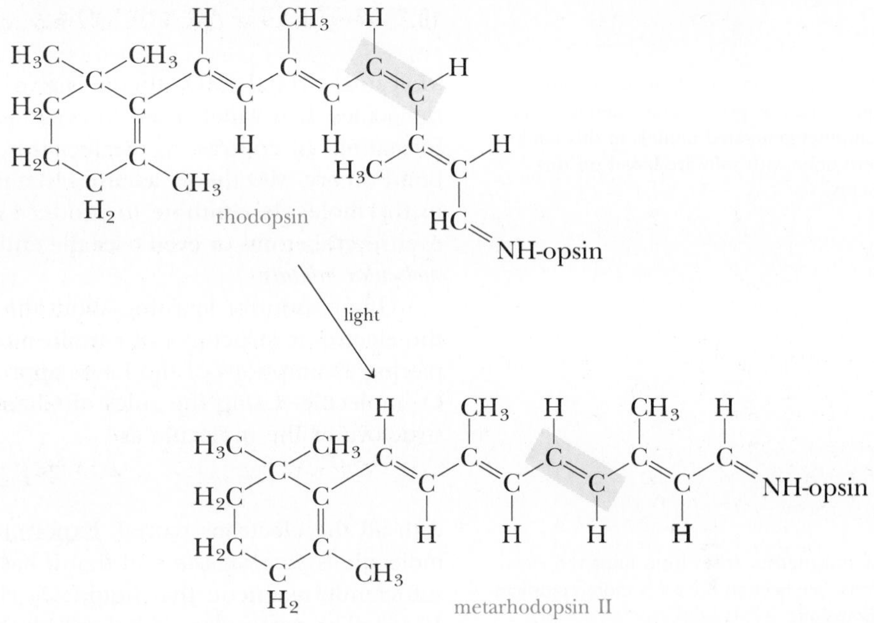

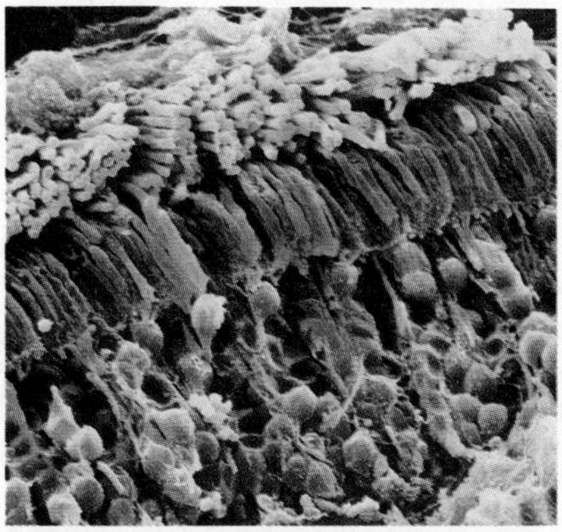

False-color scanning electron micrograph (SEM) of rod cells situated in the human retina. Two layers are visible here. The top layer (pinkish) has been severed, whereas in the bottom layer the rods are still connected to their nuclei (visible as soft purple spheres). About 130 million rod cells are situated in a retina, together with about 6.5 million cones (which are not visible here). Together they represent the photosensitive cells in our eyes. Rod cells are considered to be receptors for night vision, whereas cones are for daylight and color vision. Magnification is 1450 times at 6 × 7 cm size. (P. Motta, Photo Researchers, Inc.)

Hybrid Orbitals Required	Atomic Orbitals Combined	Geometry	*p* Orbitals Remaining for π Bonding
2	$s + p = sp$	linear	2
3	$s + p + p = sp^2$	planar-trigonal	1
4	$s + p + p + p = sp^3$	tetrahedral	0
5	$s + p + p + p + d = sp^3d$	trigonal-bipyramidal	0
6	$s + p + p + p + d + d = sp^3d^2$	octahedral	0

10.2 MOLECULAR ORBITAL THEORY

MO theory is the bonding theory most widely used by professional chemists. The computer-generated models in this book were done with software based on this theory.

Molecular orbital (MO) theory is an alternative way to view electron orbitals in molecules. It is widely used to account for molecular properties beyond bond formation. In contrast to the localized bond and lone pair orbitals of valence bond theory, MO theory assumes that pure *s* and *p* atomic orbitals of the atoms in the molecule combine to produce orbitals that are spread, or delocalized, over several atoms or even over the entire molecule. The new orbitals are called *molecular orbitals.*

One reason for learning about the MO concept is that it correctly predicts the electronic structures of certain molecules that do not follow the electron-pairing assumptions of the Lewis approach. The most common example is the O_2 molecule. Using the rules of Chapter 9, you would draw the electron dot structure of the molecule as

$$\ddot{\underset{..}{O}}=\ddot{\underset{..}{O}}$$

Paramagnetism arises from unpaired electrons. See Section 8.1 for a more complete discussion.

with all the electrons paired. Experiments clearly show, however, that the O_2 molecule is *paramagnetic* and that it has two *unpaired electrons* per molecule. It is sufficiently magnetic that liquid O_2 clings to the poles of a magnet (Figure 10.15). The molecular orbital approach can account for the paramagnetism of O_2. Valence bond theory cannot. To see how MO theory can apply to O_2 and other diatomic molecules, we shall first describe the four principles of the theory.

Principles of Molecular Orbital Theory

Valence bond theory assumes that one bond is formed by the combination of two atomic or hybrid orbitals, one from each of two adjacent atoms. In contrast, molecular orbital theory assumes that two molecular orbitals result from the combination of two atomic orbitals. In MO theory we begin with a given arrangement of atoms in the molecule at the known bond distances. We then determine the sets of molecular orbitals that can result from combining all the available orbitals on all the constituent atoms. These molecular orbitals more or less en-

Figure 10.15 Solid O_2 is paramagnetic and so clings to the poles of a magnet. The compound was cooled to a very low temperature to cause it to liquefy. Although not evident in this photo, liquid O_2 is light blue. (C. D. Winters)

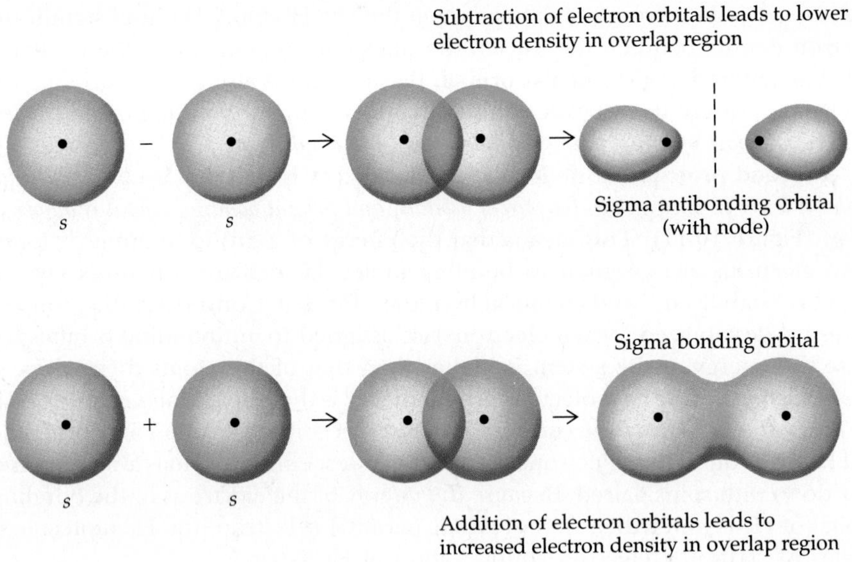

Figure 10.16 Formation of bonding and antibonding molecular orbitals from two *s* atomic orbitals on adjacent atoms. Notice the presence of a node in the antibonding orbital. (A node is a plane on which there is no probability of finding an electron.)

compass all the atoms of the molecule. A number of electrons equal to the total number of valence electrons for all the atoms in the molecule is assigned to these sets of molecular orbitals. Just as with orbitals in atoms, electrons are assigned to molecular orbitals according to the Pauli principle and Hund's rule.

With these ideas in mind, we can state the **first principle** of molecular orbital theory: *the number of molecular orbitals produced is always equal to the number of atomic orbitals brought by the atoms that have combined.* To see the consequence of this orbital conservation principle, consider first the H_2 molecule.

Bonding and Antibonding Molecular Orbitals in H_2

The principles of molecular orbital theory tell us that when the 1*s* orbitals of two hydrogen atoms overlap, *two* molecular orbitals result from the addition and subtraction of the overlapping atomic orbitals. In the molecular orbital resulting from addition of the overlapping atomic orbitals, the 1*s* regions of electron density *add* together to lead to an increased probability that electrons are found in the bond region (Figure 10.16).* In such an orbital electrons concentrate between the two hydrogen nuclei. This orbital is called a **bonding molecular orbital** and is the same as that described as a chemical bond by valence bond theory. Moreover, it is labeled a sigma (σ) orbital because the region of electron probability lies directly along the bond axis. We label this molecular orbital σ_{1s}. The subscript 1*s* designates the atomic orbitals that were the origin of the molecular orbital.

Because *two* combining atomic orbitals must produce *two* molecular orbitals, the other combination is constructed by *subtracting* one orbital from the other (Figure 10.16). When this happens the probability of finding an electron *between* the nuclei in the molecular orbital is *reduced,* and the electron density increases

*Orbitals are electron waves; therefore, a way to view molecular orbital formation is to assume that two electron waves, one from each atom of the bonded pair, interfere with one another. The interference can be constructive (to give a bonding MO) or destructive (to give an antibonding MO).

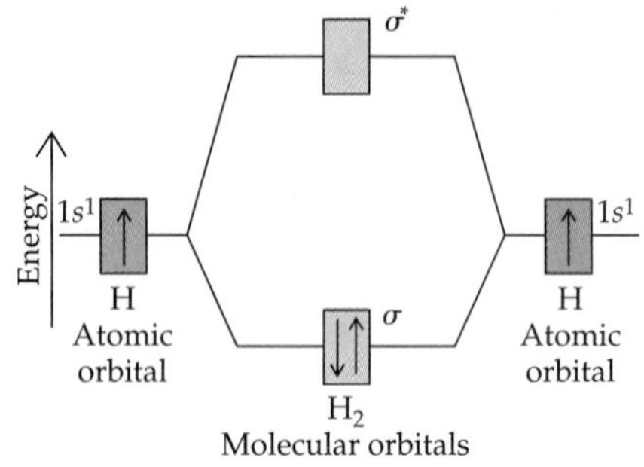

Figure 10.17 Energy level diagram for the molecular orbitals from two atomic 1*s* orbitals. The two electrons of H_2 are placed in the lowest energy σ_{1s} molecular orbital.

Antibonding orbitals have energies higher than the average energy of the atomic orbitals from which they were created. Their upward energy displacement is *slightly* greater than the downward displacement of the bonding orbitals.

in the regions away from the area between the two H atoms. Without significant electron density between the nuclei, the nuclei are repelled, and the orbital is called an **antibonding molecular orbital.** Because this is also a sigma orbital, it is labeled σ^*_{1s}, where the asterisk signifies that this is an antibonding orbital. *There is no counterpart of antibonding orbitals in valence bond theory.*

A **second principle** of molecular orbital theory is that the *bonding molecular orbital is lower in energy than the parent orbitals, and the antibonding orbital is higher in energy* (Figure 10.17). This means that the energy of a group of atoms is lower when electrons are assigned to bonding molecular orbitals. Chemists say the system is "stabilized," and chemical bonds are formed. Conversely, the group of atoms is "destabilized" when electrons are assigned to antibonding orbitals because the energy of the system is higher than that of the atoms themselves.

A **third principle** of molecular orbital theory is that the *electrons of the molecule are assigned to orbitals of successively higher energy* according to the Pauli principle and Hund's rule. Thus, electrons occupy the lowest energy orbitals available, and they do so with spins paired. Because the energy of the electrons in the bonding orbital of H_2 is lower than that of either parent 1*s* electron, the H_2 molecule is stable. We write the electron configuration of H_2 as $(\sigma_{1s})^2$.

Now let us combine two helium atoms to form dihelium, He_2. Because both He atoms have 1*s* valence orbitals, they combine to produce the same kind of molecular orbitals as in H_2. The four electrons of dihelium are assigned to these orbitals according to the scheme in Figure 10.18. The pair of electrons in σ_{1s} stabilizes He_2; however, the two electrons of σ^*_{1s} *de*stabilize the He_2 molecule. The energy decrease from the electrons in the σ_{1s} bonding molecular orbital is offset by the energy increase due to the electrons in the σ^*_{1s} antibonding molecular orbital. Thus, MO theory predicts that He_2 has no *net* stability, and laboratory experiments indeed show that two He atoms have no tendency to combine.

Bond Order

Bond order was defined in Chapter 9 as the *net number of bonding electron pairs linking a pair of atoms.* This same concept can be applied directly to molecular orbital theory, but now it is more convenient to calculate the bond order as

$$\text{Bond order} = \tfrac{1}{2}\,(\text{number of electrons in bonding molecular orbitals} - \text{number of electrons in antibonding molecular orbitals})$$

In the H_2 molecule, there are two electrons in a bonding orbital, so H_2 has a bond order of 1. In contrast, the effect of the σ_{1s} pair in He_2 is canceled by the effect of the σ^*_{1s} pair, so the bond order is 0.

Fractional bond orders are also possible. For example, even though He_2 does not exist, the He_2^+ ion has been detected. Its molecular orbital electron configuration is $(\sigma_{1s})^2(\sigma^*_{1s})^1$. Here there are two electrons in a bonding molecular orbital, but only one in an antibonding orbital. Molecular orbital theory therefore predicts that He_2^+ should have a net bond order of $\frac{1}{2}$ and explains the fact that the ion is stable enough to be observed.*

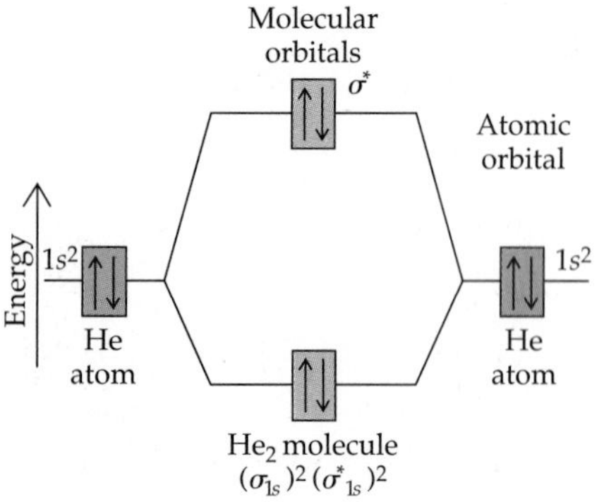

Figure 10.18 Molecular orbitals for the hypothetical dihelium molecule, He_2.

*Ions such as He_2^+ are only found in the gas phase using special instruments. They have never been observed in water or other solvents, nor does it seem they are likely to be. That is, there seems little chance that you would be able to isolate a salt such as dihelium fluoride, He_2F, although it is interesting to speculate on ways this might be done.

EXAMPLE 10.6 ***Molecular Orbitals and Bond Order***

Write the electron configuration of the H_2^- ion in molecular orbital terms. What is the bond order of the ion?

Solution This molecular ion has three electrons (one each from the H atoms plus one for the negative charge). Therefore, its configuration is $(\sigma_{1s})^2(\sigma^*_{1s})^1$, identical with the configuration for He_2^+. This means it also has a net bond order of $\frac{1}{2}$, and the ion is predicted to exist under special circumstances.

EXERCISE 10.5 ***Molecular Orbitals and Bond Order***

Write the configuration of the H_2^+ ion in molecular orbital terms. Compare the bond order of the ion with He_2^+ and H_2^-. Do you expect H_2^+ to exist? ■

Molecular Orbitals of Li_2 and Be_2

A **fourth principle** of molecular orbital theory is that *atomic orbitals combine to form molecular orbitals most effectively when the atomic orbitals are of similar energy.* This principle becomes important when we move past He_2 to Li_2, dilithium. A lithium atom has electrons in two orbitals of the *s* type (1*s* and 2*s*), so a 1*s*/2*s* combination is theoretically possible. Because the 1*s* and 2*s* orbitals are quite different in energy, however, the 1*s*/2*s* interaction cannot make an important contribution. Thus, the molecular orbitals can be considered to come only from 1*s*/1*s* and 2*s*/2*s* interactions (Figure 10.19). This means that the molecular orbital electron configuration of dilithium, Li_2, is $(\sigma_{1s})^2(\sigma^*_{1s})^2(\sigma_{2s})^2$. The bonding effect of the σ_{1s} electrons is canceled by the antibonding effect of σ^*_{1s}, so these pairs make no *net* contribution to bonding in Li_2. The result is that bonding in Li_2 is due to the electron pair assigned to the σ_{2s} orbital, and the *net* bond order is 1.

Is it really true that the Starship *Enterprise* is fueled with dilithium crystals?

The fact that the σ_{1s} and σ^*_{1s} electron pairs of Li_2 make no net contribution to bonding is exactly what you observed in drawing electron dot structures in Chapter 9: core electrons are ignored. In molecular orbital terms, core electrons are assigned to bonding and antibonding molecular orbitals that offset one another.

A diberyllium molecule, Be_2, is not expected to be stable. Neglecting the 1*s* core electrons, its electron configuration is

Be_2 MO Configuration: [core electrons]$(\sigma_{2s})^2(\sigma^*_{2s})^2$

$$\text{Bond order} = \tfrac{1}{2}(2 \text{ bonding electrons} - 2 \text{ antibonding electrons}) = 0$$

and you see that there are no *net* bonding electron pairs.

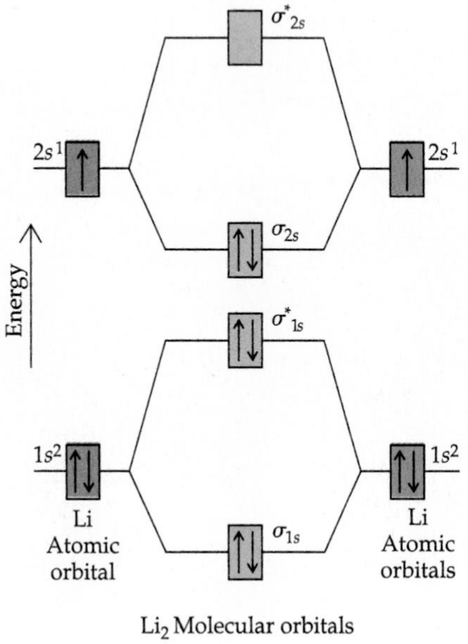

Figure 10.19 Energy level diagram for the combination of two atoms with 1*s* and 2*s* atomic orbitals. The electron configuration shown is for Li_2.

EXAMPLE 10.7 ***Molecular Orbitals in Diatomic Molecules***

Be_2 does not exist. But what about the Be_2^+ ion? Describe its electron configuration in molecular orbital terms and give the net bond order. Do you expect the ion to exist?

Solution The Be_2^+ molecular ion has only seven electrons (in contrast to eight for Be_2), of which four are core electrons. (The core electrons are assigned to σ_{1s}

and σ^*_{1s} molecular orbitals). The remaining three electrons are assigned to the σ_{2s} and σ^*_{2s} molecular orbitals (see Figure 10.19), so the MO electron configuration is

$$[\text{core electrons}](\sigma_{2s})^2(\sigma^*_{2s})^1$$

This means the net bond order is

$$\tfrac{1}{2}(2 \text{ bonding electrons} - 1 \text{ antibonding electron}) = \tfrac{1}{2}$$

and so Be_2^+ is predicted to exist.

EXERCISE 10.6 ***Molecular Orbitals in Diatomic Molecules***

In principle could you prepare a compound such as $NaLi_2$, in which the cation is Na^+ and the anion is Li_2^-? ■

Molecular Orbitals for Homonuclear Diatomic Molecules

A molecule formed from two identical atoms, such as H_2 or Li_2, is called a **homonuclear diatomic molecule.** With many of the principles of molecular orbital theory in place, we are ready to account for bonding in such important homonuclear diatomic molecules as N_2, O_2, and F_2. First, however, we need to see what types of molecular orbitals form when elements have both *s* and *p* valence orbitals.

Molecular Orbitals from Atomic *p* Orbitals

Six types of interactions are possible for two elements that both have *s* and *p* valence orbitals. Sigma-bonding and antibonding molecular orbitals are formed by *s* orbitals interacting as in Figure 10.16. Similarly, it is possible for a *p* orbital on one atom to interact with a *p* orbital on the other atom in head-to-head fashion to produce a pair of σ bonding and antibonding molecular orbitals (Figure 10.20). And finally, each atom has two *p* orbitals in planes perpendicular

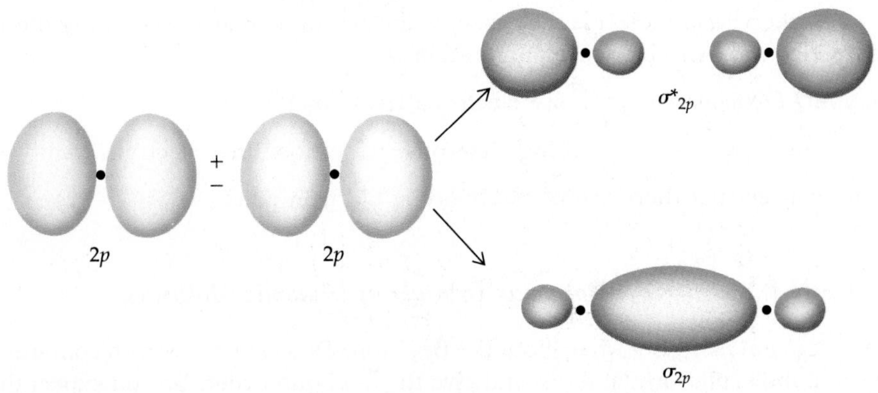

Figure 10.20 The sigma bonding (σ_{2p}) and antibonding (σ^*_{2p}) molecular orbitals that arise from 2*p* orbital overlap. Each orbital may accommodate two electrons. *p* orbitals in electron shells of higher *n* give molecular orbitals of the same basic shape.

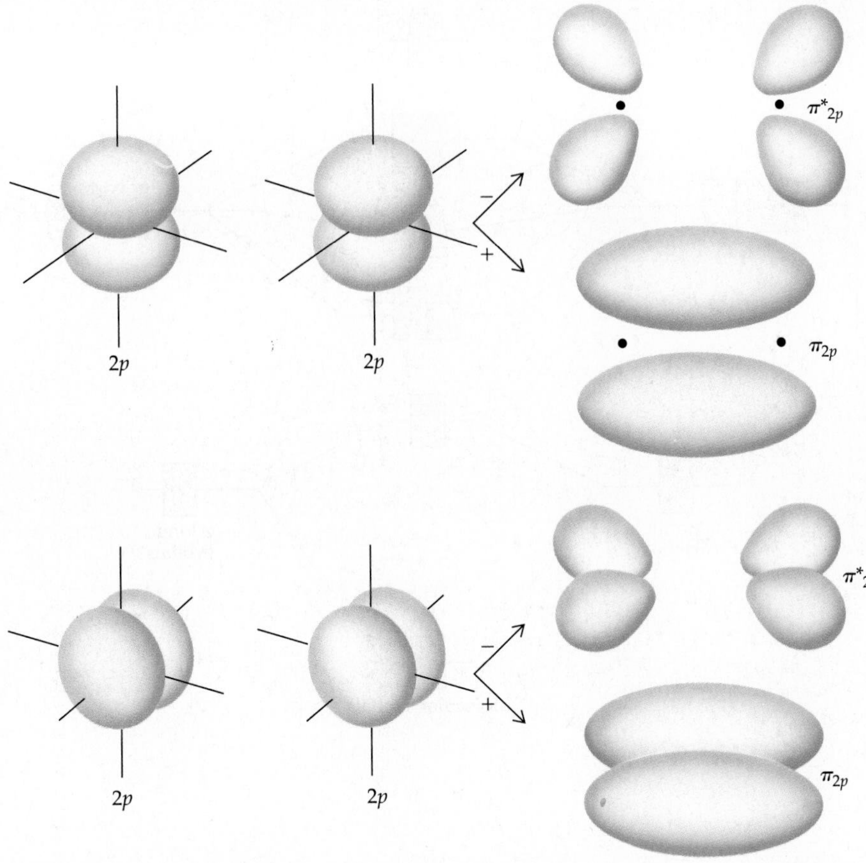

Figure 10.21 Pi (π) bonding and antibonding molecular orbitals. Sideways overlap of a pair of atomic *p* orbitals that lie along the same direction in space leads to a pi bonding (π_{2p}) and a pi antibonding (π^*_{2p}) molecular orbital. *p* orbitals in electron shells of higher *n* give molecular orbitals of the same basic shape.

to the σ bond connecting the two atoms. These *p* orbitals can interact sideways to give π bonding and antibonding molecular orbitals (Figure 10.21). Thus, two *p* orbitals on each atom produce *two* π bonding molecular orbitals (π_p) and *two* π antibonding molecular orbitals (π^*_p).

Electron Configurations for Some Homonuclear Diatomic Molecules

The orbital interactions just described lead to the energy level diagram in Figure 10.22. This allows us to explore the bonding in homonuclear diatomic molecules formed by combining two identical atoms having 1*s* core electrons and 2*s* and 2*p* valence orbitals. Electron assignments are given for the diatomic molecules B_2 through F_2 in Table 10.2, which has two important features.

First, notice the correlation between the electron configurations and the bond orders, bond lengths, and bond-dissociation energies at the bottom of Table 10.2. As the bond order between a pair of atoms increases, the atoms are more tightly bonded. This means that bond-dissociation energy should *increase*

The molecular orbitals formed by core electrons, σ_{1s} and σ^*_{1s}, do not lead to net bonding and are not included in Figure 10.22.

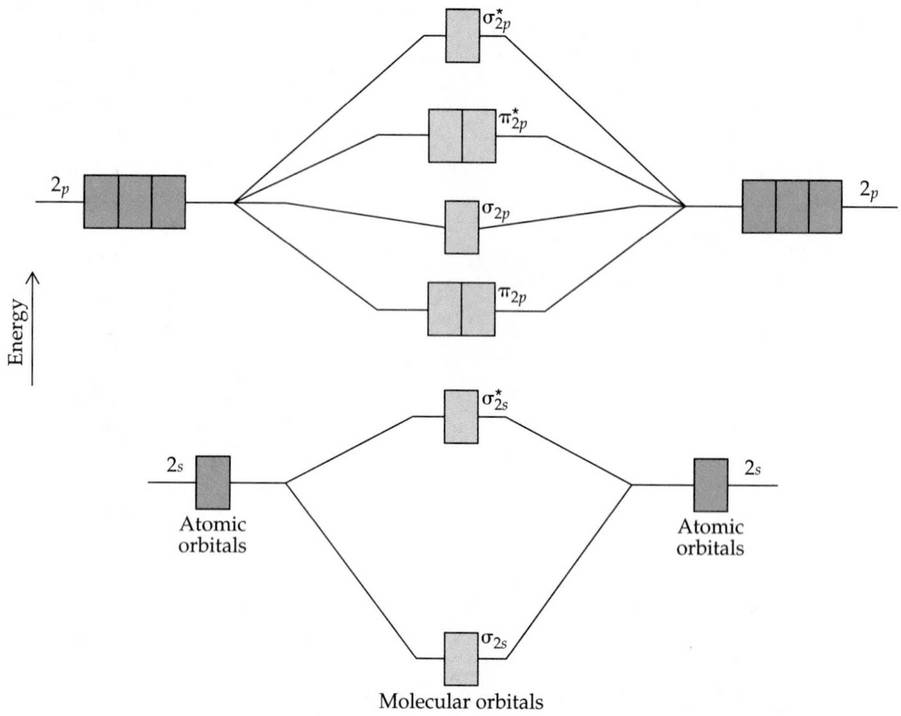

Figure 10.22 A molecular orbital energy level diagram for X_2, a homonuclear diatomic molecule of second period elements. Note that molecular orbitals formed by core electrons, σ_{1s} and σ^*_{1s}, do not lead to net bonding and are not included in the figure.

TABLE 10.2 Molecular Orbital Occupations and Physical Data for Homonuclear Diatomic Molecules of Second-Period Elements

	B_2	C_2	N_2	O_2	F_2
σ^*_{2p}	[ ]	[ ]	[ ]	[ ]	[ ]
π^*_{2p}	[ ][ ]	[ ][ ]	[ ][ ]	[↑][↑]	[↑↓][↑↓]
σ_{2p}	[ ]	[ ]	[↑↓]	[↑↓]	[↑↓]
π_{2p}	[↑][↑]	[↑↓][↑↓]	[↑↓][↑↓]	[↑↓][↑↓]	[↑↓][↑↓]
σ^*_{2s}	[↑↓]	[↑↓]	[↑↓]	[↑↓]	[↑↓]
σ_{2s}	[↑↓]	[↑↓]	[↑↓]	[↑↓]	[↑↓]
Bond order	One	Two	Three	Two	One
Bond-dissociation energy (kJ/mol)	290	620	946	498	159
Bond distance (pm)	159	131	110	121	143
Observed magnetic behavior (paramagnetic or diamagnetic)	Para	Dia	Dia	Para	Dia

A CLOSER LOOK *Molecular Orbitals for Compounds Having* p*-Block Elements*

Several features of the molecular orbital energy level diagram in Figure 10.22 might be described in more detail.

- The bonding and antibonding σ orbitals from 2*s* interactions are lower in energy than the σ and π MOs from 2*p* interactions. The reason is that 2*s* orbitals have a lower energy than 2*p* orbitals in the separated atoms.
- The separation of bonding and antibonding orbitals is greater for σ_{2p} than for π_{2p}. This happens because *p* orbitals overlap to a greater extent when they are oriented head to head (to give σ_{2p} MOs) than when they are side by side (to give π_{2p} MOs). The greater the orbital overlap, the greater the stabilization of the bonding MO and the greater the *destabilization* of the antibonding MO.

Figure 10.22 shows an energy ordering of molecular orbitals that you might not have expected, but there are reasons for this. A more sophisticated approach takes into account the "mixing" of *s* and *p* atomic orbitals, which have similar energies. This causes the σ_{2s} and σ_{2s}^* molecular orbitals to be lower in energy than otherwise expected, and the σ_{2p} and σ_{2p}^* orbitals to be pushed up in energy. For this reason the energy lowering and raising for the σ_{2s} and σ_{2s}^* orbitals (and for the σ_{2p} and σ_{2p}^* orbitals) in Figure 10.22 is not symmetrical with respect to the 2*s* and 2*p* atomic orbital energies.

Another refinement concerning Figure 10.22 is that, because *s* and *p* orbital mixing is important only for B_2, C_2, and N_2, the figure applies *strictly* only to these molecules. For O_2 and F_2, σ_{2p} is lower in energy than π_{2p}. *Nonetheless,* Figure 10.22 gives the correct bond order and magnetic behavior for these two molecules.

---

and bond distance should *decrease;* both are observed. Thus, N_2, with a bond order of 3, has the largest bond-dissociation energy and shortest bond distance.

Second, a major objective of this discussion of MO theory was an understanding of the bonding in dioxygen, O_2, because valence bond theory failed to predict the observed paramagnetism of the molecule. Dioxygen has twelve molecular valence electrons, so it has the molecular orbital configuration

$$[\text{core electrons}]\,(\sigma_{2s})^2(\sigma_{2s}^*)^2(\pi_{2p})^4(\sigma_{2p})^2(\pi_{2p}^*)^2$$

This configuration leads to a bond order of 2, just as predicted by valence bond theory. The MO configuration requires two unpaired electrons, however, *exactly* as determined by experiment. Thus, a simple molecular orbital picture leads to a view of the bonding in paramagnetic O_2 more in accord with experiment than is simple valence bond theory.

EXAMPLE 10.8 *Electron Configuration for a Homonuclear Diatomic Molecule-Ion*

When potassium reacts with O_2, potassium superoxide, KO_2, is produced. The superoxide ion is O_2^-. Write the molecular orbital electron configuration for the ion. Predict its bond order and magnetic behavior.

Solution Using the energy level diagram of Figure 10.22, the configuration of the ion is

$$O_2^- \quad [\text{core electrons}]\,(\sigma_{2s})^2(\sigma_{2s}^*)^2(\pi_{2p})^4(\sigma_{2p})^2(\pi_{2p}^*)^3$$

It is predicted to be paramagnetic to the extent of one unpaired electron, a prediction confirmed by experiment. The bond order is $1\frac{1}{2}$, because there are 8 bonding electrons and 5 antibonding electrons. This is a smaller bond order than in O_2, so we predict that the O—O bond length in O_2^- should be longer than in O_2. In fact, the superoxide ion has an O—O bond length of 134 pm, whereas that of O_2 is 121 pm.

EXAMPLE 10.9 ***Electron Configuration for a Heteronuclear Diatomic Molecule***

A number of simple molecules are formed from two elements of *different* kinds. Examples of **heteronuclear diatomic molecules** include NO, CO, and ClF. Although important differences do occur, it is nonetheless the case that the molecular orbital energy level diagram for *homo*nuclear diatomic molecules (Figure 10.22) can be used at least to judge the bond order and magnetic behavior of diatomic heteronuclear molecules. Let us do this for nitrogen monoxide, NO.

Solution Nitrogen monoxide has 11 molecular valence electrons. If these are assigned to the MOs for a homonuclear diatomic molecule, the molecular electron configuration is

$$[\text{core electrons}](\sigma_{2s})^2(\sigma^*_{2s})^2(\pi_{2p})^4(\sigma_{2p})^2(\pi^*_{2p})^1$$

The net bond order is

$$\tfrac{1}{2}(8 \text{ bonding electrons} - 3 \text{ antibonding electrons}) = 2\tfrac{1}{2}$$

The single, unpaired electron assigned to the π^*_{2p} molecular orbital means the molecule is *paramagnetic,* as predicted—and observed—for a molecule with an odd number of electrons.

EXERCISE 10.7 ***Molecular Electron Configurations***

The cations O_2^+ and N_2^+ are important components of the earth's upper atmosphere. Write the electron configuration of O_2^+. Predict its bond order and magnetic behavior. ■

Resonance and π Bonds

Ozone, O_3, is a simple triatomic molecule with equal oxygen-oxygen bond lengths. Equal X—O bond lengths are also observed in other similar molecules and ions such as SO_2, NO_2^-, and HCO_2^-.

Ozone, O_3

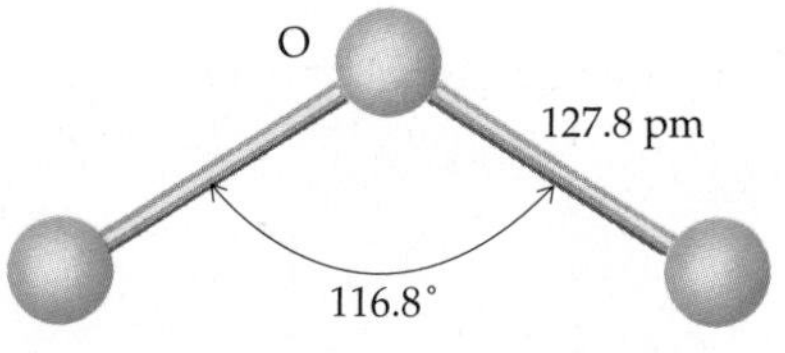

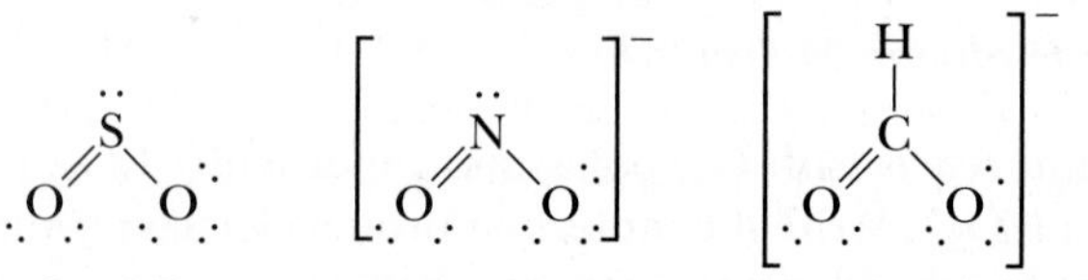

S—O = 143 pm, OSO = 119.5° | N—O = 124 pm, ONO = 115° | C—O = 127 pm, OCO = 124°

To rationalize these experimental observations, Linus Pauling proposed that the electronic structure of such molecules and ions is a hybrid of resonance structures (Section 9.3).

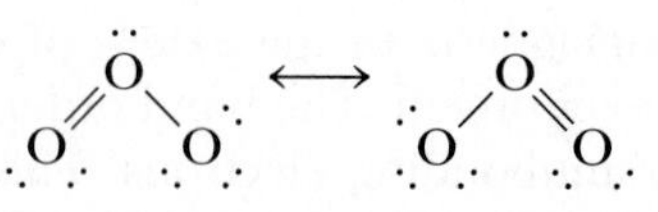

Molecular orbital theory provides another useful view of this problem.

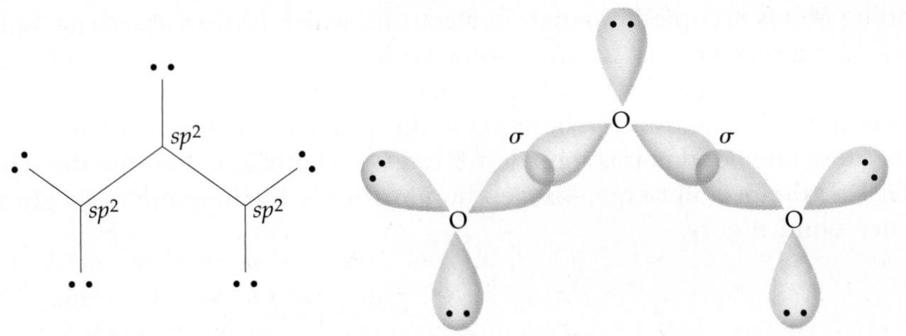

Top view of ozone showing σ bonds and nonbonding pairs in oxygen sp^2 hybrid orbitals

To visualize the bonding in ozone, let us begin by assuming that all three O atoms are sp^2-hybridized. The central atom uses sp^2 hybrid orbitals to form two σ bonds and to accommodate a lone pair. The terminal atoms use their sp^2 hybrids to form one σ bond and to accommodate two lone pairs.

The O_3 molecule has nine valence shell electron pairs to be accommodated, but the σ framework and lone pairs illustrated here account for only seven of these pairs. The π bonds in ozone arise from the two remaining pairs. Because we have assumed each O atom in O_3 is sp^2-hybridized, an unhybridized p orbital *perpendicular to the O_3 plane* remains on each of the three O atoms. The orbitals are in the correct orientation for the formation of π bonds. A principle of MO theory is that three atomic orbitals form three molecular orbitals; therefore, these three atomic p orbitals form three π molecular orbitals (Figure 10.23).

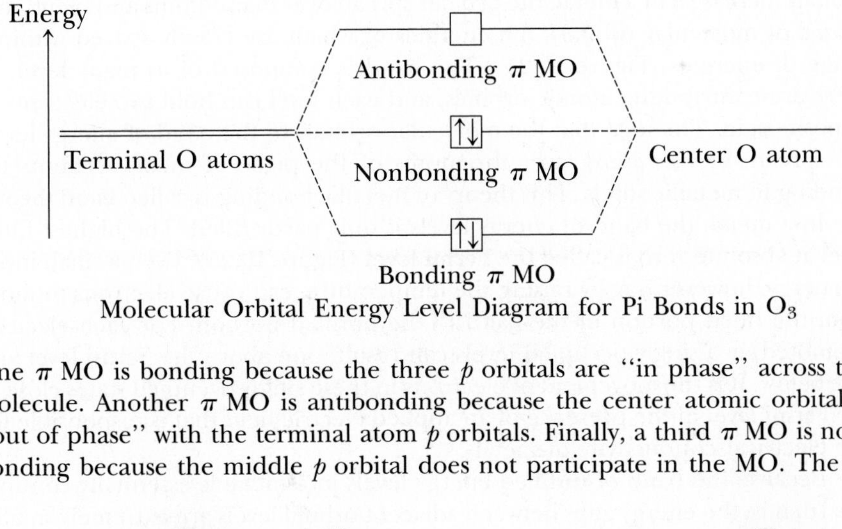

Molecular Orbital Energy Level Diagram for Pi Bonds in O_3

One π MO is bonding because the three p orbitals are "in phase" across the molecule. Another π MO is antibonding because the center atomic orbital is "out of phase" with the terminal atom p orbitals. Finally, a third π MO is non-bonding because the middle p orbital does not participate in the MO. The π-

Figure 10.23 π bonding in ozone. Each O atom in O_3 is sp^2 hybridized. As explained in the text, these hybrid orbitals are used to form the two O—O σ bonds and to accommodate five lone pairs. The two remaining electron pairs are accommodated in π molecular orbitals. Three atomic p orbitals, one on each O atom, form three π molecular orbitals, which accommodate two pairs of electrons. One of these pairs is π bonding. The other, in the nonbonding MO, is a lone pair.

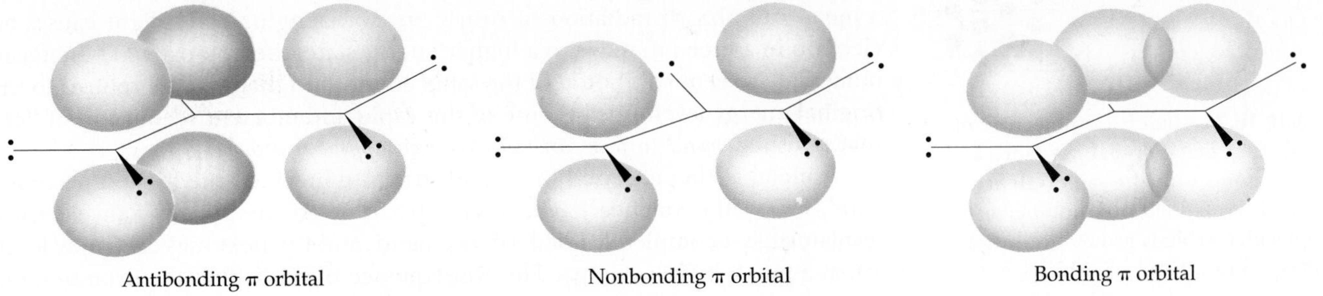

Antibonding π orbital   Nonbonding π orbital   Bonding π orbital

Molecular orbital theory generally provides a more useful picture of π bonding in molecules than valence bond theory. We shall return to this picture in Chapter 11 during the discussion of the chemistry of benzene.

bonding MO is occupied by a pair of electrons, which is delocalized, or "spread over," the molecule, just as the resonance hybrid implies. The π-nonbonding orbital is also occupied; however, as the name implies, occupation of this orbital neither helps nor hinders the bonding in the molecule. The π bond order of O_3 is $\frac{1}{2}$ because one bond pair is spread over two O—O linkages. Because the σ bond order is 1, the overall oxygen-oxygen bond order is $1\frac{1}{2}$, the same value given by valence bond theory.

10.3 METALS AND SEMICONDUCTORS

A 1-g crystal of lithium contains about 9×10^{22} atoms. This means that 36×10^{22} 2*s* and 2*p* valence atomic orbitals are available, which can form 36×10^{22} molecular orbitals.

The simple molecular orbital model used to describe diatomic molecules can be extended easily to describe the properties of metals and semiconductors.

Conductors, Insulators, and Band Theory

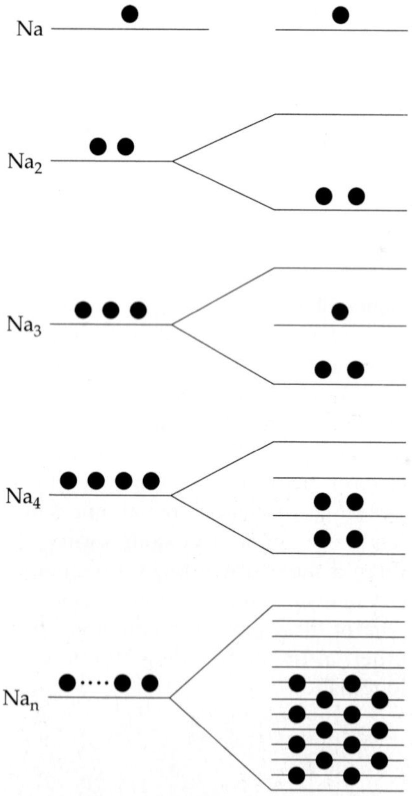

Figure 10.24 Bands of molecular orbitals in a metal crystal. As more and more atoms with the same valence orbitals are added, the number of molecular orbitals grows, until they merge into a band of orbitals.

Metal crystals can be viewed as "supermolecules" held together by *delocalized bonds* formed from the atomic orbitals of all the atoms in the crystal. Because even the tiniest piece of metal can contain a very large number of atoms, an even larger number of atomic orbitals is available to form molecular orbitals. For example, if four lithium atoms are in a group, and each Li atom contributes four orbitals (2*s* and 2*p*) to molecular orbital formation, 16 molecular orbitals are formed. If there are 400 lithium atoms, then 1600 molecular orbitals are formed, and so on. As the number of atoms increases, the number of possible molecular orbitals increases. In a metal, the orbitals spread over many atoms and blend into a *band* of molecular orbitals, the energies of which are closely spaced within a range of energies (Figure 10.24). The band is composed of as many levels as there are contributing atomic orbitals, and each level can hold two electrons of opposite spin. The idea that the molecular orbitals of the band of energy levels are spread, or *delocalized,* over the atoms of the piece of metal accounts for bonding in metallic solids. This theory of metallic bonding is called **band theory.**

In a metal, the band of energy levels is only partly filled. The highest filled level at absolute zero is called the **Fermi level** (Figure 10.25). Even a small input of energy, however, say by raising the temperature, can cause electrons to move from the filled portion of the band to the unfilled portion. For each electron promoted, two singly occupied levels can result, one above the Fermi level and one below. It is the movement of electrons in these singly occupied states close to the Fermi level in the presence of an applied electric field that is responsible for the electrical conductivity of metals.

Because the band of unfilled energy levels in a metal is essentially continuous (that is, the energy gaps between adjacent orbital levels are extremely small), a metal can absorb radiation of nearly any wavelength. When light causes an electron in a metal to move to a higher energy state, the now-excited system can immediately reemit a photon of the same energy and the electron returns to the original energy level. It is because of this rapid and efficient reemission of light that polished *metal surfaces are reflective and appear lustrous.*

A metal is characterized by a band structure in which the highest occupied band, called the **valence band,** is only partly filled. In contrast, an electrical **insulator** has a completely filled valence band, and the next higher, empty levels are at a much higher energy. The consequence of this is that the promotion of

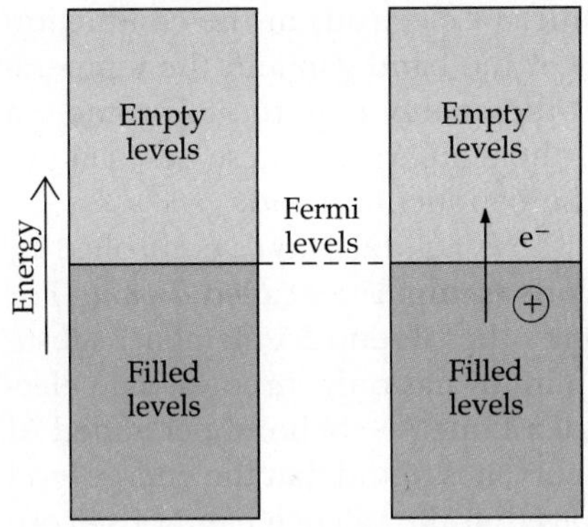

Figure 10.25 The partially filled band of "molecular orbitals" in a metal (left). The highest filled level is referred to as the Fermi level. The molecular orbitals are delocalized or spread over the metal, a fact that accounts for bonding in metals. Electrons can be promoted from the filled levels to empty levels by the input of modest amounts of energy (right). These electrons are freer to move in the now partially filled levels; this property accounts for the electrical and thermal conductivity of metals.

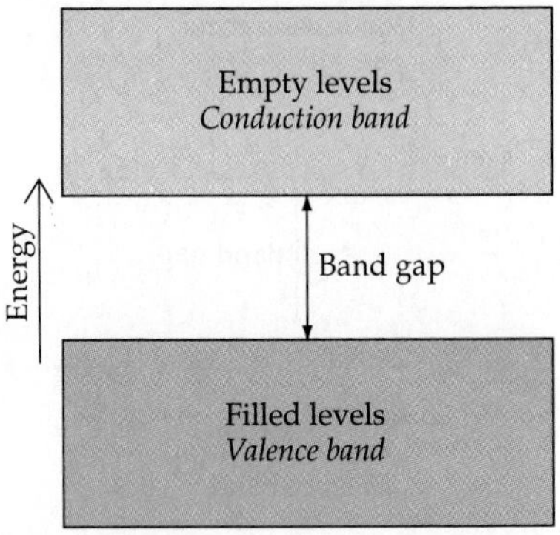

Figure 10.26 Band theory applied to semiconductors and insulators. In contrast to a metal, the band of filled levels (called the valence band) is separated from the band of empty levels (the conduction band) by the band gap. The band gap can range in energy from just a few kJ/mol to 500 kJ/mol or more.

an electron to a *slightly* higher energy is not possible, and the solid does not conduct electricity.

Diamonds, which are electrical insulators, are one form of pure carbon (Figure 3.5). Each C atom in a diamond is surrounded by four other C atoms at the corners of a tetrahedron. We normally think of the bonding in the solid as involving sp^3-hybridized C atoms forming four, localized covalent C—C bonds per carbon atom. Silicon and germanium, also elements of Group 4A, form diamond-like solids, and their bonding can be described similarly. An alternative view, however, is that the orbitals of each C, Si, or Ge atom form molecular orbitals that are delocalized over the solid. In this case, however, the continuous band of energy levels seen in a metal is split into two bands of levels: a lower energy, *filled* **valence band** of bonding molecular orbitals, and a higher energy, *unfilled* band of antibonding molecular orbitals. The latter is called the **conduction band,** and the two bands are separated in energy by an amount called the **band gap** (Figure 10.26).

Semiconductors

If an electron could be promoted from the valence band to the conduction band in a substance like diamond, metal-like properties could be observed. Diamonds are insulators, however, because the band gap between the valence and conduction bands is large, on the order of 500 kJ/mol, and it is difficult to promote electrons to the conduction band. **Semiconductors,** on the other hand, usually have band gaps in the range of 50 to 300 kJ/mol. At least a few electrons can be promoted to the conduction band by the input of modest amounts of energy, so electrical conduction can occur. The band gap narrows on descending through Group 4A; diamonds are insulators, whereas silicon and germanium are semiconductors. The consequence is that silicon is now widely used for the special properties that semiconductors bring to electrical devices.

Pure silicon belongs to a class of materials called **intrinsic** semiconductors. The usual view of such materials is that promoting an electron from the valence band to the conduction band creates a positive hole in the valence band. The semiconductor carries charge because the electrons in the conduction band migrate in one direction, and the positive holes in the valence band migrate in the opposite direction. (Positive holes "move" because an electron from an adjacent level can move into the hole, thus creating a "fresh" hole. In this respect, positive holes in the valence band move in the opposite direction from electrons in the conduction band.)

Various forms of elemental silicon, a semiconductor. (C. D. Winters)

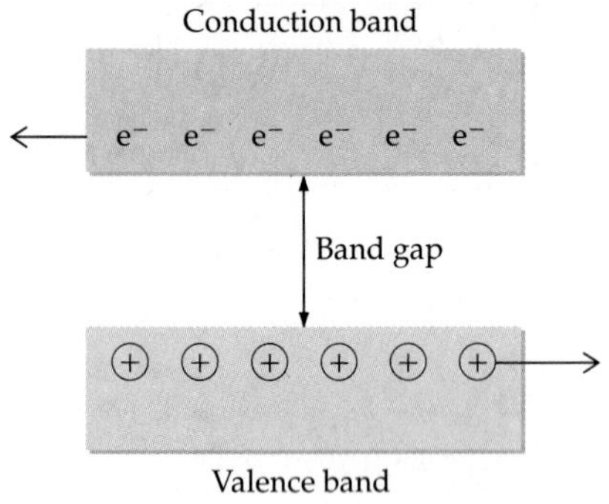

Positive and negative charge carriers in an intrinsic semiconductor

In an intrinsic semiconductor, the number of electrons in the conduction band is entirely governed by the magnitude of the band gap and the temperature. The smaller the band gap, the smaller the energy required to promote a significant number of electrons. Similarly, as the temperature increases, a larger number of electrons can be promoted to the conduction band.

Extrinsic semiconductors are materials whose conductivity is controlled by adding tiny numbers of atoms of different kinds as impurities called *dopants.* For example, suppose Al atoms (or atoms of some other Group 3A element) substitute for some Si atoms in solid silicon. Aluminum has only three valence electrons, whereas silicon has four; therefore, at least one Al—Si bond per added Al atom is deficient in electrons. Using band theory, it is found that the energy level associated with each Al—Si bond is not involved in the valence band of silicon; rather it forms a discrete level just above the valence band. This level is referred to as an *acceptor level* because it can accept electrons. The gap between the valence band and the acceptor level is usually quite small, so electrons are readily promoted from the valence band to the acceptor level. The positive holes left behind in the valence band are able to move under the influence of an electric potential, so an aluminum-doped semiconductor is called a *p*ositive hole, or *p-type semiconductor.*

Now suppose you add P atoms (or atoms of some other Group 5A element) to pure silicon. The material is still a semiconductor, but it is a *n*egative charge carrier, or *n-type semiconductor.* The reason for this is that phosphorus atoms have five valence electrons, one *more* than Si. The presence of these Group 5A atoms in the silicon therefore leads to a discrete, filled electron *donor level,* just below the conduction band. Electrons can be promoted readily from the donor level to the conduction band, and electrons in the conduction band carry the charge.

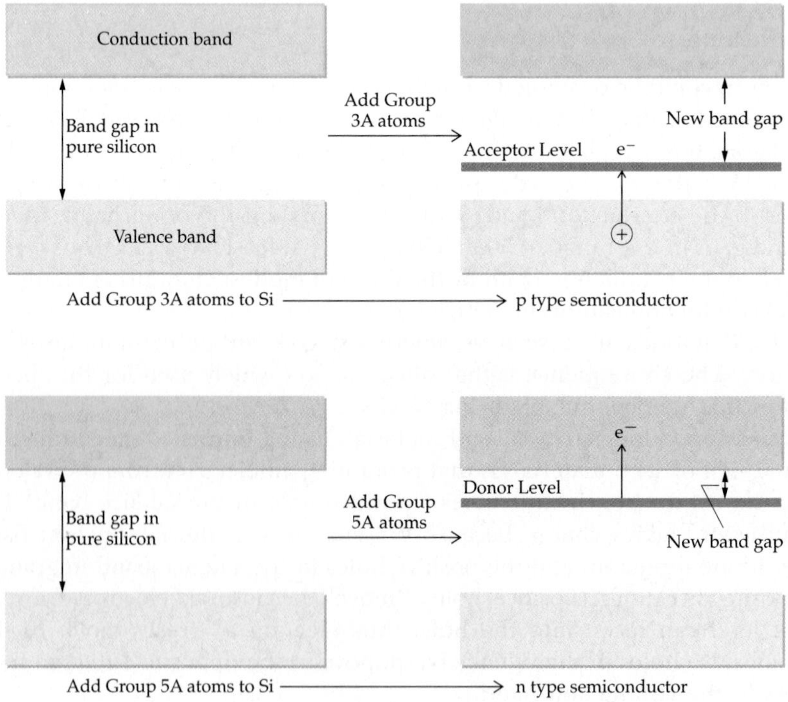

CHAPTER HIGHLIGHTS

When you have finished studying this chapter, you should be able to

- Describe the main features of the two most commonly used theories of chemical bonding, **valence bond theory** and **molecular orbital theory.**
- Recognize that valence bond theory views bonding as arising from the overlap of atomic orbitals on two atoms to give a bonding orbital with electrons localized between the bonded atoms (Section 10.1).

  In a sigma bond (σ), orbitals overlap head to head (*s* with *s*, *p* with *p*, or *s* with *p*), and the bonding electron density is concentrated along the bond axis. If *p* orbitals remain after sigma bond formation, one or two pi (π) bonds may form by the sideways overlap of *p* orbitals; π bonding electron density lies on both sides of the bond axis.
- Understand that molecular orbital theory assumes that *s* and *p* atomic orbitals of the atoms in the molecule combine to produce molecular orbitals that are generally delocalized over several atoms (Section 10.2).

  The number of molecular orbitals formed is always equal to the number of atomic orbitals brought by the combining atoms. Some of these molecular orbitals are **bonding** (and lie lower in energy than the energy of the parent atomic orbitals in the isolated atoms), and others are **antibonding** (and lie higher in energy). The electrons of the molecule are assigned to these orbitals, beginning with those of lowest energy, according to the Pauli principle and Hund's rule.
- Use the concept of **orbital hybridization** (Section 10.1).

Hybrid Orbital Set	Number of Hybrid Orbitals	Geometry
sp	Two	Linear
sp^2	Three	Planar-trigonal
sp^3	Four	Tetrahedral
sp^3d	Five	Trigonal-bipyramidal
sp^3d^2	Six	Octahedral

- Define **bond order** and describe other molecular properties using molecular orbital theory for diatomic molecules (Section 10.2).
- Appreciate **band theory** and how it is applied to solids, especially metals (Section 10.3).
- Understand the difference between a **conductor** of electricity, a **semiconductor,** and an **insulator** (Section 10.3).

STUDY QUESTIONS

Review Questions

1. What is the difference between a sigma (σ) and a pi (π) bond?
2. What is the maximum number of hybrid orbitals a carbon atom may form? What is the minimum number? Explain briefly.
3. What are the approximate angles between regions of electron density in sp, sp^2, and sp^3 hybrid orbital sets?
4. For each of the following electron-pair geometries, tell what hybrid orbital set is used: tetrahedral, linear, trigonal-planar, octahedral, trigonal-bipyramidal.

5. If an atom is *sp*-hybridized, how many pure *p* orbitals remain on the atom? How many pi bonds can the atom form?
6. What is the maximum number of hybrid orbitals nitrogen can form? Explain briefly.
7. What is the maximum number of hybrid orbitals a third-period element, say sulfur, can form? Explain briefly.
8. Give an example of a molecule with a central atom having more than four valence electron pairs. Tell what hybrid orbitals are used by the atom.
9. What is one important difference between molecular orbital theory and valence bond theory?
10. Describe four principles of molecular orbital theory.
11. Draw diagrams of the bonding and antibonding molecular orbitals of H_2, and tell how they differ.
12. What is meant by the order of a bond in terms of the molecular orbital theory? How are bond order, bond energy, and bond length related?
13. How is molecular orbital theory related to bonding in metals?
14. Explain briefly the theory of bonding in metals. What is the name usually applied to this theory?

Hybrid Orbitals

15. Draw the Lewis structure for OF_2. What are its electron-pair and molecular geometries? Describe the bonding in the molecule in terms of hybrid orbitals.

16. Draw the Lewis structure for the ClF_2^+ ion. What are its electron-pair and molecular geometries? Describe the bonding in the molecule in terms of hybrid orbitals.

17. Draw the Lewis structure for CH_2Cl_2. What are its electron-pair and molecular geometries? Describe the bonding in the molecule in terms of hybrid orbitals.

18. Draw the Lewis structure for BF_4^-. What are its electron-pair and molecular geometries? Describe the bonding in the ion in terms of hybrid orbitals.

19. Tell what hybrid orbital set is used by the underlined atom in each of the following molecules or ions:
(a) $\underline{B}Cl_3$ (b) $\underline{N}O_2^+$ (c) $H\underline{C}Cl_3$ (d) $H_2\underline{C}O$

20. Tell what hybrid orbital set is used by the underlined atom in each of the following molecules or ions:
(a) $\underline{C}S_2$ (b) $\underline{N}O_2^-$ (c) $\underline{N}O_3^-$ (d) $\underline{N}H_2^-$

21. Give the hybrid orbital set used by each of the indicated atoms in the molecules below.
(a) The carbon and oxygen in methanol, H_3COH.
(b) The carbon atoms in propene.

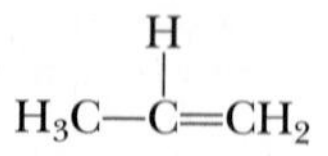

(c) The two carbon atoms and the nitrogen atom in the amino acid glycine.

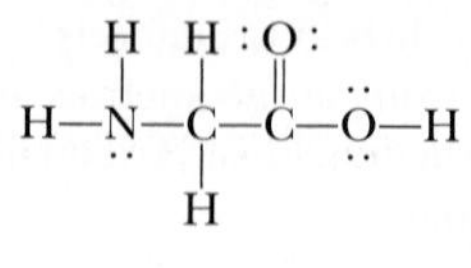

22. Give the hybrid orbital set used by each of the underlined atoms in the molecules below:

(a) 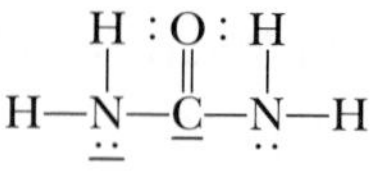

(b) 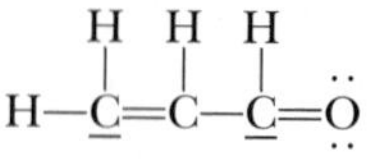

(c) 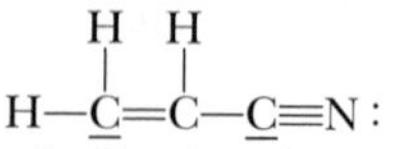

23. Give the Lewis structure and the electron-pair and molecular geometries for XeF_2. Describe the bonding in the molecule using hybrid orbitals for the central atom.

24. Give the Lewis structure, the electron-pair geometry, and the shape for the ion ICl_4^-. Describe the bonding in the ion using hybrid orbitals for the central atom.

25. Tell what hybrid orbital set is used by the underlined atom in each of the following molecules or ions:
(a) $\underline{S}F_6$ (b) $\underline{S}F_4$ (c) $\underline{I}Cl_2^-$

26. Tell what hybrid orbital set is used by the underlined atom in each of the following molecules or ions:
(a) $O\underline{Xe}F_4$ (b) $\underline{Cl}F_3$ (c) central Br in Br_3^-

27. What is the hybridization of the carbon atom in CO_2? Give a complete description of the σ and π bonding in this molecule.

28. What is the hybridization of the carbon atom in the NCO^- ion? Give a complete description of the σ and π bonding in this ion.

29. What is the hybridization of the carbon atom in phosgene, Cl_2CO? Give a complete description of the σ and π bonding in this molecule.

30. What is the hybridization of the sulfur atom in sulfur dioxide, SO_2? Give a complete description of the σ and π bonding in this molecule.

Molecular Orbital Theory

31. Hydrogen, H_2, can be ionized to give H_2^+. Write the electron configuration of the ion in molecular orbital terms. What is the bond order of the ion? Is the hydrogen-hydrogen bond stronger or weaker in H_2^+ than in H_2?

32. Give the electron configurations for the ions Li_2^+ and Li_2^- in molecular orbital terms. Compare the lithium-lithium bond order in the Li_2 molecule with its ions.

33. Calcium carbide, CaC_2, contains the acetylide ion, C_2^{2-}. Sketch the molecular orbital energy level diagram for the ion. How many net σ and π bonds does the ion have? What is the carbon-carbon bond order? How has the bond order changed on adding electrons to C_2 to obtain C_2^{2-}? Is the C_2^{2-} ion paramagnetic?

34. Oxygen, O_2, can acquire one or two electrons to give O_2^- (superoxide ion) or)O_2^{2-} (peroxide ion). Write the electron configuration for the ions in molecular orbital terms, and then compare them with the O_2 molecule on the follow-

ing basis: (a) magnetic character, (b) net number of σ and π bonds, (c) bond order, and (d) oxygen-oxygen bond length.

35. Assuming that we can apply the energy level diagram for homonuclear diatomic molecules (Figure 10.22) to heteronuclear diatomics, write the electron configuration for carbon monoxide, CO. Is the molecule diamagnetic or paramagnetic? What is the net number of σ and π bonds? What is the bond order?

36. The nitrosyl ion, NO^+, has an interesting chemistry. (a) Is NO^+ diamagnetic or paramagnetic? If paramagnetic, how many unpaired electrons does it have? (b) Assume that the molecular orbital diagram for a homonuclear diatomic molecule applies to NO^+. What is the highest energy molecular orbital occupied by an electron? (c) What is the N—O bond order? (d) If NO is ionized to form NO^+, does the N—O bond become stronger or weaker than in NO?

General Questions

37. Give the hybrid orbital set used by sulfur in each of the following molecules or ions:
(a) SO_2 (b) SO_3 (c) SO_3^{2-} (d) SO_4^{2-}
How does the hybrid orbital set change on adding O atoms to sulfur or on changing the charge on the species?

38. Give the Lewis structures of ClF_2^+ and ClF_2^-. What are the electron-pair and molecular geometries of each ion? What hybrid orbital set is used by Cl in each ion?

39. What is the hybridization of the nitrogen atom in the nitrate ion, NO_3^-? Describe the orbitals involved in the formation of an N═O bond.

40. The organic compound below is a member of a class known as oximes.

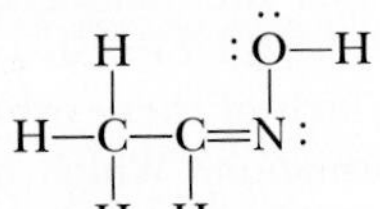

(a) What are the hybridizations of the two carbon atoms?
(b) What is the approximate C—N—O angle?

41. Acrolein, a component of photochemical smog, has a pungent odor and irritates eyes and mucous membranes.

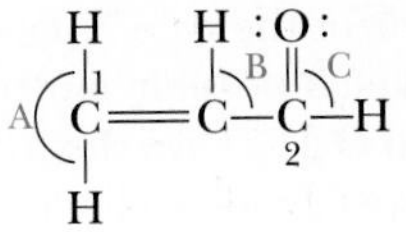

(a) What are the hybridizations of carbon atoms 1 and 2?
(b) What are the approximate values of angles A, B, and C?

42. The compound sketched here is acetylsalicylic acid, better known by its common name, aspirin. (See page 220.)

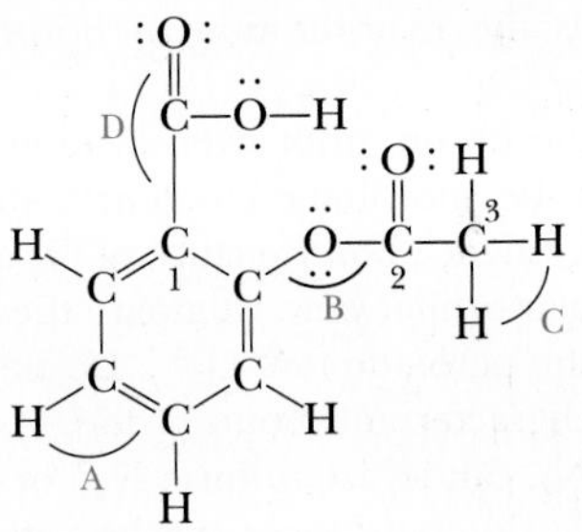

(a) How many π bonds are in aspirin? How many σ bonds?
(b) What are the approximate values of the angles marked A, B, C, and D?
(c) What hybrid orbitals are used by carbon atoms 1, 2, and 3?

43. Lactic acid is a natural compound found in sour milk.

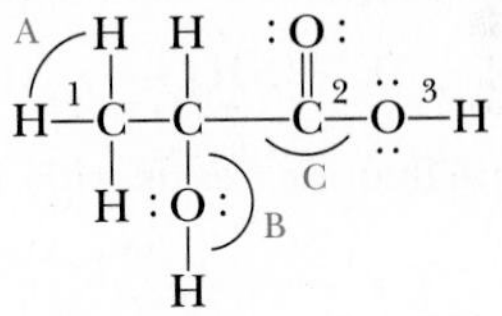

(a) How many π bonds are in lactic acid? How many σ bonds?
(b) Give the hybridization of each atom 1 through 3.
(c) Which OH bond is the shortest in the molecule?
(d) Give the approximate values of the bond angles A through C.

44. Histamine is found in normal body tissues and in blood and has the following structure:

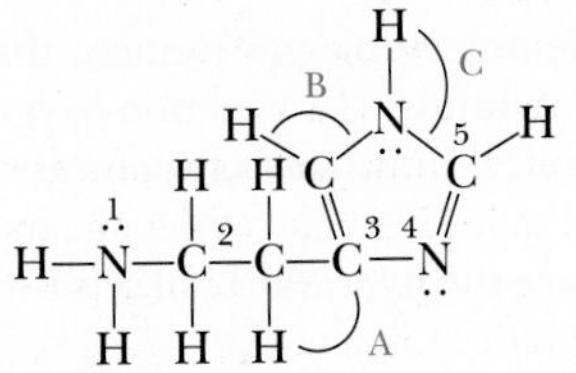

(a) How many σ and π bonds does the molecule have?
(b) Give the hybridizations of atoms 1 through 5.
(c) What are the approximate values of the bond angles A through C?

45. Boron trifluoride, BF_3, can accept a pair of electrons from another molecule to form a coordinate covalent bond (see Chapter 9), as in the following reaction with ammonia.

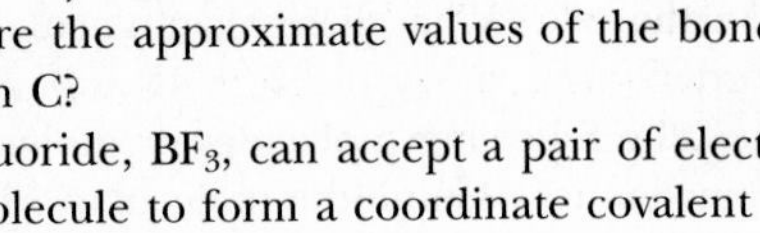

(a) What is the geometry about the boron atom in BF_3? In $H_3N—BF_3$?

(b) What is the hybridization of boron in the two compounds?
(c) Does the boron atom hybridization change on formation of the coordinate covalent bond?

46. The simple valence bond picture of O_2 does not agree with the molecular orbital view. Compare these two theories with regard to the peroxide ion, O_2^{2-}. Do they lead to the same magnetic character and bond order?

47. Nitrogen, N_2, can ionize to form N_2^+ or absorb an electron to give N_2^-. Compare these species with regard to
(a) Their magnetic character
(b) Net number of π bonds
(c) Bond order
(d) Bond length
(e) Bond strength

48. The ammonium ion is important in chemistry. Discuss any changes in hybridization and bond angles that may occur in the combination of ammonia and a proton.

$$H^+ + NH_3 \longrightarrow NH_4^+$$

49. Antimony pentafluoride reacts with HF according to the equation

$$2\ HF + SbF_5 \longrightarrow [H_2F]^+[SbF_6]^-$$

(a) What is the hybridization of Sb in the reactant and product?
(b) Draw a Lewis structure for H_2F^+. Is the ion linear or bent in structure? What is the hybridization of F in H_2F^+?

50. Iodine and oxygen form a complex series of ions, among them IO_4^- and IO_5^{3-}. Draw Lewis structures for these ions and specify their electron-pair geometries and the shapes of the ions. What is the hybridization of the I atom in these ions?

51. Xenon is the only noble gas element that forms well-characterized compounds. Two xenon-oxygen compounds are XeO_3 and XeO_4. Draw Lewis structures of each of these compounds and give their electron-pair and molecular geometries. What are the hybrid orbital sets used by xenon in these two oxides?

52. Which of the homonuclear diatomic molecules of the second-period elements (from Li_2 to Ne_2) are paramagnetic? Which ones have a bond order of 1? Which ones have a bond order of 2? What diatomic molecule has the highest bond order?

53. Which of the following molecules or molecule-ions should be paramagnetic? Assume the molecular orbital diagram in Figure 10.22 applies to all of them.
(a) NO (c) O_2^{2-} (e) CN
(b) OF^- (d) Ne_2^+

54. The sulfamate ion, $H_2N—SO_3^-$, can be thought of as having been formed from the amide ion, NH_2^-, and sulfur trioxide, SO_3.
(a) Sketch a structure for the sulfamate ion. Include estimates of bond angles.
(b) What changes in hybridization do you expect for N and S in the course of the reaction

$$NH_2^- + SO_3 \longrightarrow H_2N{-}SO_3^-$$

55. In many chemical reactions atom hybridization changes. In each of the following reactions, tell what change, if any, occurs to the underlined atom:
(a) $H_2C{=}\underline{C}H_2 + Cl_2 \rightarrow H_2ClC{-}CH_2Cl$
(b) $\underline{P}(CH_3)_3 + I_2 \rightarrow PI_2(CH_3)_3$
(c) $\underline{Xe}F_2 + F_2 \rightarrow XeF_4$
(d) $\underline{Sn}Cl_4 + 2\ Cl^- \rightarrow SnCl_6^{2-}$

56. The CN molecule has been found in interstellar space. Assuming the electronic structure of the molecule can be described using the molecular orbital energy level diagram in Figure 10.22, answer the following questions:
(a) What is the highest energy molecular orbital to which an electron or electrons may be assigned?
(b) What is the bond order of the molecule?
(c) How many net σ bonds are there? How many net π bonds?
(d) Is the molecule paramagnetic or diamagnetic?

57. The molecule CH_3NCS, the structure for which is shown here, is used as a pesticide.

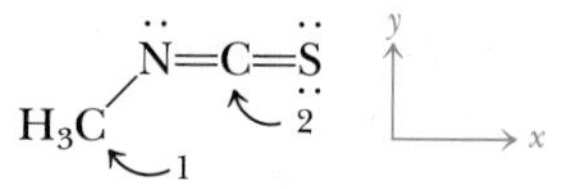

(a) Describe the hybrid orbital sets used by carbon 1, carbon 2, and the N atom.
(b) What is the C—N=C bond angle? The N=C=S bond angle?
(c) The C—N=C=S framework is planar. Assume these atoms lie in the plane of the paper and that this is identified as the *xy*-plane. Now consider the orbitals (s, p_x, p_y, and p_z) on carbon atom 2. Which of these orbitals are involved in hybrid orbital formation? Which of these orbitals are used in π bond formation with the neighboring N and S atoms?

Conceptual Questions

58. The elements of the second period from boron to oxygen form compounds of the type $X_2E—EX_2$, where X can be H or a halogen. Sketch possible molecular structures for B_2F_4, C_2H_4, N_2H_4, and O_2H_2. Give the hybridizations of E in each molecule and specify approximate X—E—E bond angles. What is the major difference between the carbon-containing compound and the other molecules?

59. When two amino acids react with each other, they form a linkage called an amide group, or a peptide link. (If more linkages are added, a protein, or polypeptide, is formed.)

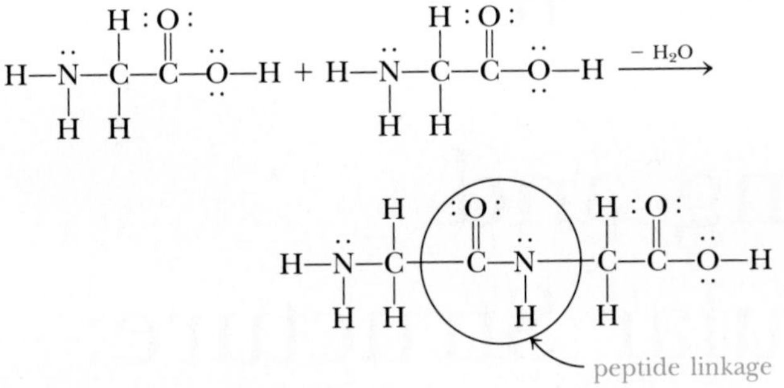

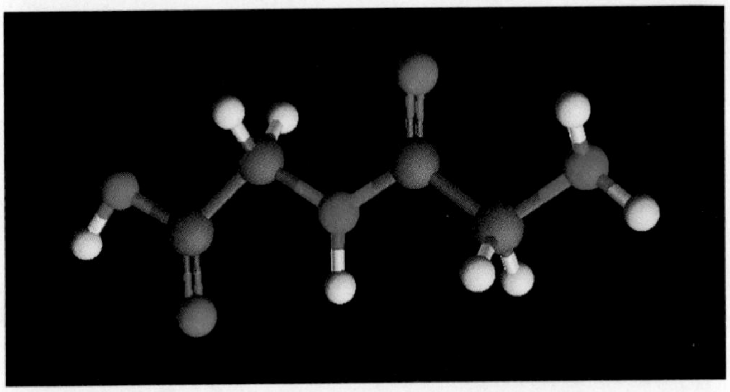

(a) What are the hybridizations of the C and N atoms in the peptide linkage?

(b) Is the structure illustrated the only resonance structure possible for the peptide linkage? If another resonance structure is possible, compare it with the one shown. Decide which is the more important structure.

(c) The computer-generated structure shown here, which contains a peptide linkage, shows that the linkage is flat. This is an important feature of proteins. Speculate on reasons that the CO—NH linkage is planar.

Summary Question

60. The compound whose structure is shown here is commonly called acetylacetone (and is often abbreviated *acac*). As shown it exists in two forms, one called the *enol* form and the other called the *keto* form.

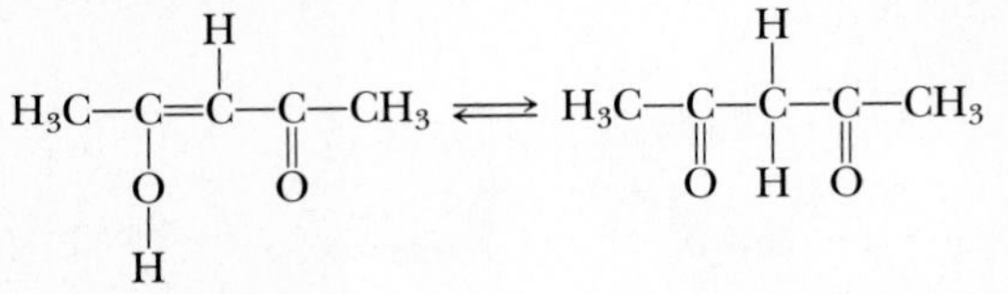

enol form *keto* form

While in the *enol* form, the molecule can lose H^+ from the —OH group to form the anion $[H_3C—C(O)═CH—C(O)—CH_3]^-$. One of the most interesting aspects of this anion is that one or more of them can react with a transition metal cation to give very stable, highly colored compounds.

(a) Using bond energies, calculate the approximate enthalpy change for the *enol* to *keto* change. Is the reaction predicted to be *exo-* or *endo*thermic?

(b) What is the hybridization of each atom (except H) in the enol form? What changes occur when this is transformed into the keto form?

(c) What is the electron-pair geometry and molecular geometry around each C atom in the keto and enol forms? What if any changes in hybridization occur when the keto form changes to the enol form?

(d) If you wanted to prepare 15.0 g of deep red $Cr(acac)_3$ using the following reaction:

$$CrCl_3 + 3\ CH_3C(OH){=}CH{-}C(O)CH_3{-}CH_3 + 3\ NaOH \longrightarrow Cr[CH_3C(O)CHC(O)CH_3]_3 + 3\ H_2O + 3\ NaCl$$

how many grams of the other reactants are needed?

C H A P T E R **11**

Bonding and Molecular Structure: Organic Chemistry

The Chemical Puzzler

Do you enjoy chewing bubble gum and blowing the largest possible bubble? Why is it possible to blow large bubbles from bubble gum but not regular gum? What is the chemical compound in bubble gum that allows you to do this? How it is made, and is this compound used in anything else?

Most of the compounds around you—and within you—involve covalent bonding. This very large and important class of compounds is based on carbon and is known as **organic compounds.** Saccharin is an example of an organic compound. Most of its framework is built of carbon atoms, but hydrogen, oxygen, nitrogen, and sulfur are present as well. The compound, discovered more than a hundred years ago, was the first artificial sweetener. It was discovered quite by accident at Johns Hopkins University by a student of Ira Remsen, the most famous American chemist of the 19th century. The student, a man named Fahlberg, noticed that the substance he had made and accidentally spilled on his hand, tasted sweet. He immediately recognized the significance of the discovery and patented it in 1885.

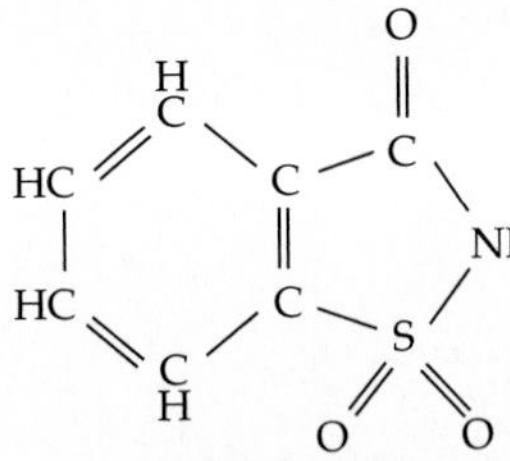

Saccharin

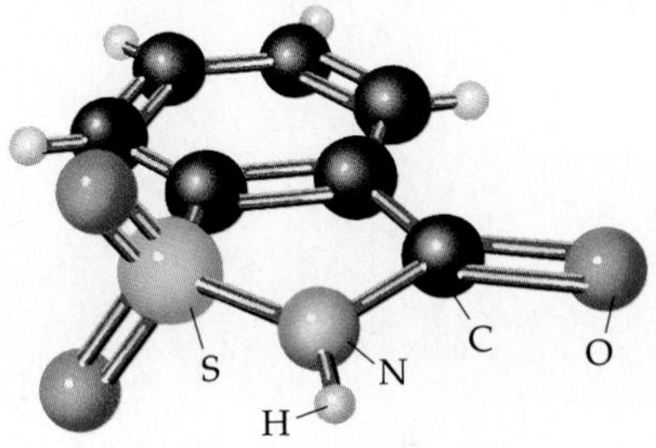

We begin the story of organic chemistry by focusing on carbon—carbon in the form of its allotropes: graphite, diamonds, and buckyballs (Figure 3.5). Graphite consists of many layers of carbon atoms. Each atom within a layer is bonded to three other carbon atoms at the corners of a planar triangle, thus forming an extended network of six-member carbon rings. In terms of what you learned about chemical bonding in Chapter 10, each carbon atom is assumed to be sp^2 hybridized. This means that there are carbon-carbon π bonds within a layer, but there is no covalent bonding between layers. The layers are attracted to one another, however, but only very weakly and so can slide readily over each other. This is the reason graphite is a lubricant and the most important ingredient in pencil "lead."

In contrast, each carbon atom in a diamond is surrounded by four carbon atoms, making a tetrahedron; each atom assumed to be sp^3-hybridized. This structure extends in three dimensions throughout the solid, again forming a rigid network of six-member carbon rings. This bonding arrangement makes diamonds extremely hard and chemically unreactive.

In one of the most spectacular discoveries of the past decade, chemists found new forms of carbon. The form with the formula C_{60} is now known affectionately as a buckyball. Here again each carbon atom is surrounded by three other carbon atoms, but they are arranged in five- or six-member rings, and the structure closes on itself, forming a perfect sphere of 60 carbon atoms. This arrangement is unique and was so unexpected that chemists are certain the molecule will have interesting applications (see Section 3.1).

Hydrocarbons, the simplest class of organic compounds, have a skeleton of carbon atoms bonded together in chains and rings. (The rings can be flat as in graphite or puckered as in diamond.) (Table 11.1). These compounds can be divided into at least four types, which in turn depend on the types of

TABLE 11.1 Some Types of Hydrocarbons*

Type of Hydrocarbon	Characteristic Features	Example
Alkanes	General formula C_nH_{2n+2} C—C single bonds All C atoms surrounded by 4 single bonds	CH_4, methane C_2H_6, ethane
Cyclic alkanes	General formula, C_nH_{2n}	C_6H_{12}, cyclohexane
Alkenes	General formula C_nH_{2n} C═C double bonds	$H_2C{=}CH_2$, ethylene
Alkynes	General formula C_nH_{2n-2}	HC≡CH, acetylene
Aromatics	Rings of C atoms with alternating single and double bonds leading to two or more stable resonance structures	C_6H_6, benzene 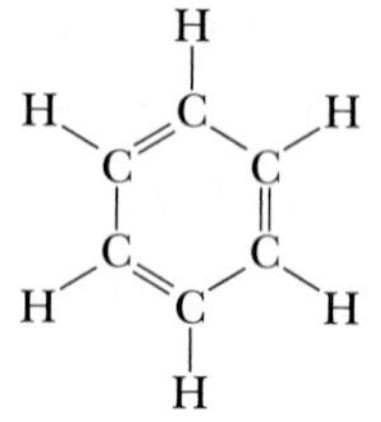

*See Table 3.4 for the names and formulas of alkanes.

chemical bonds found in them. The hydrocarbons range from gases to liquids to solids. Some, like the alkanes, are not very reactive chemically, whereas alkenes and alkynes in particular undergo a number of different reactions. Some have almost no odor, but others—most notably the aromatics—do have an odor, often pleasant, as you might gather from their name. Alkenes often have offensive odors. But two properties are shared by all hydrocarbons: none are water-soluble and all are flammable and so are potential fuels.

Organic chemistry is an important and fascinating subject that covers a vast intellectual territory. This chapter is therefore not an exhaustive study of the area. Certain types of compounds are not mentioned, and *very few* reactions are described. For a more in-depth analysis of this subject consult books such as W. H. Brown: *Organic Chemistry*. Philadelphia, Saunders College Publishing, 1995.

The early parts of this chapter describe the structural and reaction chemistry of the hydrocarbons. The later parts take up the bonding, structures, and a few important reactions of other major classes of organic compounds, such as alcohols, aldehydes, ketones, and carboxylic acids. This discussion should help you understand better the synthesis and properties of polymers and other modern materials, the subject of Section 11.8.

11.1 ALKANES

Alkanes, which are described in Tables 3.4 and 11.2, are typified by methane, the major component of natural gas. If a hydrogen atom of methane is replaced by a $—CH_3$ group, the new compound is ethane. If an H atom of ethane is replaced

TABLE 11.2 Some Alkanes with Unbranched Carbon Chains

Molecular Formula	Name	Boiling Point (°C)	Melting Point (°C)	State at Room Temp.
CH_4	Methane	−161	−184	
C_2H_6	Ethane	−88	−183	Gas
C_3H_8	Propane	−42	−188	
C_4H_{10}	Butane	−0.5	−138	
C_5H_{12}	Pentane	36	−130	
C_6H_{14}	Hexane	69	−94	
C_7H_{16}	Heptane	98	−91	
C_8H_{18}	Octane	126	−57	
C_9H_{20}	Nonane	150	−54	
$C_{10}H_{22}$	Decane	174	−30	
$C_{11}H_{24}$	Undecane	194.5	−25.6	Liquid
$C_{12}H_{26}$	Dodecane	214.5	−9.6	
$C_{13}H_{28}$	Tridecane	234	−6.2	
$C_{14}H_{30}$	Tetradecane	252.5	+5.5	
$C_{15}H_{32}$	Pentadecane	270.5	10	
$C_{16}H_{34}$	Hexadecane	287.5	18	
$C_{17}H_{36}$	Heptadecane	303	22.5	
$C_{18}H_{38}$	Octadecane	317	28	
$C_{19}H_{40}$	Nonadecane	330	32	
$C_{20}H_{42}$	Eicosane	205 (at 15 torr)	36.7	Solid

Alkanes are often called **saturated** compounds, meaning that each C atom is bound to four other atoms. The C atoms in alkanes are sp^3-hybridized.

by yet another $—CH_3$ group, propane results. Butane can be derived from propane by replacing an H atom of one of the chain-ending $—CH_3$ groups. A series of alkanes with unbranched carbon chains can be built up in this manner (see Figure 3.18).

The Physical and Chemical Properties of the Alkanes

Hydrocarbons are nonpolar, and four alkanes (methane, ethane, propane, and butane) are gases at room temperature. As the molecular weight of alkanes with unbranched carbon chains increases, however, the compounds are liquids or solids at standard temperature and pressure (see Table 11.2).

The fact that alkanes are nonpolar means that they do not significantly interact with polar water molecules. This explains why gasoline, which is composed mostly of alkanes, and water do not mix.

Alkanes burn readily in air to give CO_2 and water with very exothermic enthalpies of combustion. For this reason they are widely used as fuels.

$$CH_4(g) + 2\,O_2(g) \longrightarrow CO_2(g) + 2\,H_2O(\ell)$$

ΔH_f° (kJ/mol) −74.8 0 −393.5 −285.8

$\Delta H_{rxn}^\circ = -890.3$ kJ

Despite their ability to burn, however, alkanes are thought of as unreactive compounds. Their "unreactivity" is not due to thermodynamics; the enthalpy change for the combustion of methane shows that it is a product-favored reaction. Rather, it is a problem of chemical kinetics, a problem with the speed of the

A CLOSER LOOK *Writing Formulas and Drawing Structures*

You learned in Chapter 3 that there are various ways of presenting structures (page 112), and it is appropriate to return to that point again as we look at organic compounds. Consider butane, for example.

1. Molecular formula: C_4H_{10}. This type of formula gives information only on molecular composition.
2. Condensed formula: $CH_3CH_2CH_2CH_3$. This method of writing the formula gives some information on the way atoms are connected.
3. Structural formula: The following structure is an expanded version of the condensed formula (2):

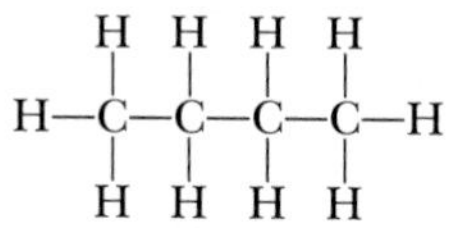

4. Structural formula: This method provides a more accurate representation of the three-dimensional structure of the molecule.

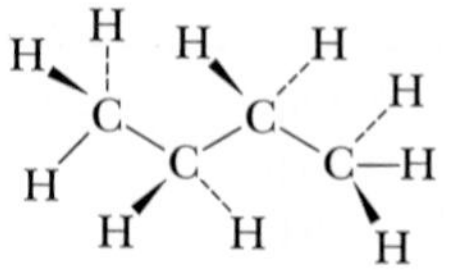

Structural formula

Shorthand representation

A shorthand way of drawing this structure is often used by chemists. It is understood that $—CH_3$ groups are at the ends of the chain and that the intermediate points are $—CH_2$ groups.

5. Ball-and-stick model: These models afford a better three-dimensional view than the formula in (4), which uses wedges and dashed lines for bonds (see Figures 3.5, 3.18 and others).
6. Space-filling model: This type of model, which we occasionally use in this book, gives perhaps the most accurate view of the structure of a molecule.

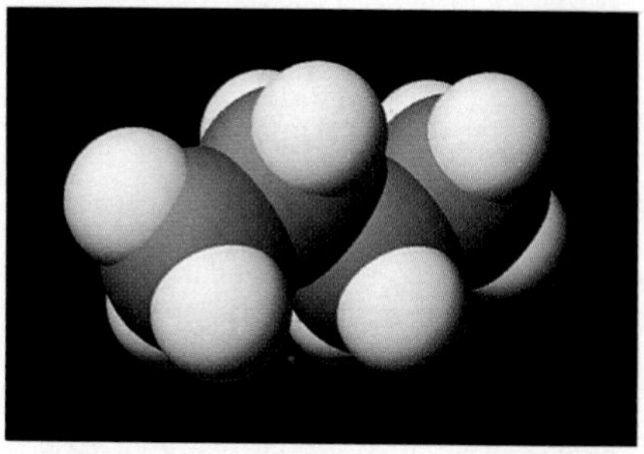

reaction (Chapter 15). Hydrocarbons and oxygen can coexist without reacting for centuries, but if the mixture is heated or a spark is struck, the reaction occurs rapidly. This apparent lack of reactivity is a challenging problem for chemists. Petroleum, which is a mixture of hydrocarbons that includes many alkanes, is an important raw material. The problem is to find ways of converting alkanes to other organic compounds selectively, easily, and at low cost.

Under the right conditions it is possible to replace one or more H atoms on an alkane by some other atom, say a halogen. For example, complete reaction of methane with chlorine in the presence of ultraviolet radiation or at high temperatures yields CCl_4, commonly known as carbon tetrachloride.

$$CH_4 \xrightarrow[UV]{Cl_2} CH_3Cl \xrightarrow[UV]{Cl_2} CH_2Cl_2 \xrightarrow[UV]{Cl_2} CHCl_3 \xrightarrow[UV]{Cl_2} CCl_4$$

	CH_3Cl	CH_2Cl_2	$CHCl_3$	CCl_4
Systematic name	chloro-methane	dichloro-methane	trichloro-methane	tetrachloro-methane
Common name	methyl chloride	methylene chloride	chloroform	carbon tetrachloride

Carbon tetrachloride was once widely used as a solvent for other organic compounds, as a dry cleaning fluid, and, because it does not burn, in fire extinguishers. Unfortunately, it is also known to cause liver damage and is a suspected carcinogen.

Another important use of CCl_4 has been as the starting material for the manufacture of chlorofluorocarbons (CFCs). In this process one or more of the

Cl atoms is replaced by F atoms from HF in the presence of an antimony compound ($SbFCl_4$).

$$CCl_4(\ell) + 2\ HF(g) \longrightarrow CCl_2F_2(g) + 2\ HCl(g)$$

Two of the most widely used CFCs have been CFC-11 (CCl_3F, trichlorofluoromethane) and CFC-12 (dichlorodifluoromethane, CCl_2F_2).

Chlorofluorocarbons have played an important role in our economy—until recently. In the 1970s M. T. Molina and F. S. Rowland discovered that CFCs could seriously deplete the layer of ozone, O_3, that protects the earth from the highly energetic ultraviolet radiation from the sun. Thus, the production of CFCs was banned as of January 1, 1996. Because CFCs and related compounds are so important economically, the chemical industry has worked feverishly to find a replacement. Fortunately, recent measurements suggest that hydrofluorocarbons such as CF_3CFH_2 (called HCFC-134a)—made by substituting some of the H atoms of ethane with F atoms—have none of the environmental effects of the CFCs but can substitute for them in refrigerators and air conditioners. Indeed it is estimated that more than 100 million pounds of HCFC-134a will have been made in the period 1992–1997 because the compound is used in automobile and home air conditioners and in commercial chillers.

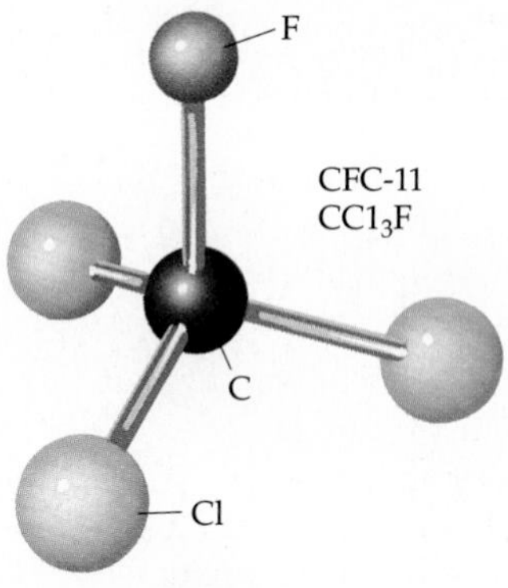

CFC-11
CCl_3F

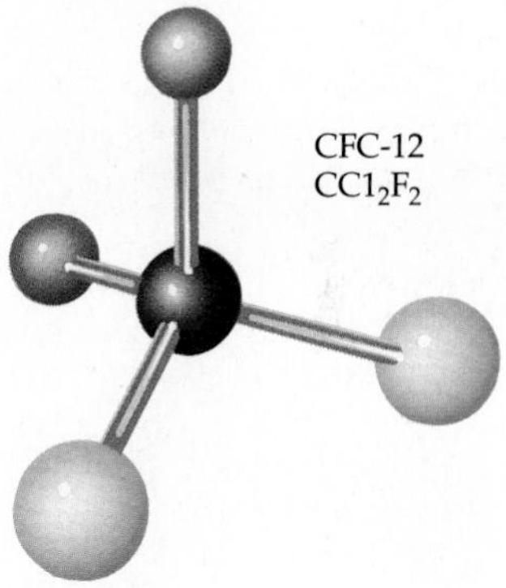
CFC-12
CCl_2F_2

Bonding and Structure in Alkanes

The most important structural feature of alkanes is that every C atom is surrounded by four other atoms at the corners of a tetahedron, the preferred geometry for an atom surrounded by four bonding electron pairs. (Chemists rationalize the bonding in the alkanes by assuming the C atoms are sp^3-hybridized.) It is important to recognize that this can only result in a zigzag, or nonlinear, chain of carbon atoms.

One interesting feature of alkane chemistry, which is also true for other types of organic compounds, is the existence of structural isomers. **Structural isomers** are compounds that have the same formula but that are structurally different because their atoms are connected differently (Section 10.1). For example, the atoms in a molecule with the formula C_4H_{10} can be connected in two

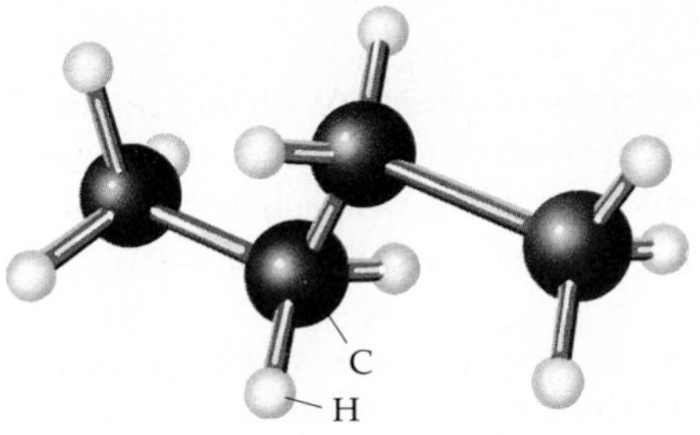

Butane, C_4H_{10}

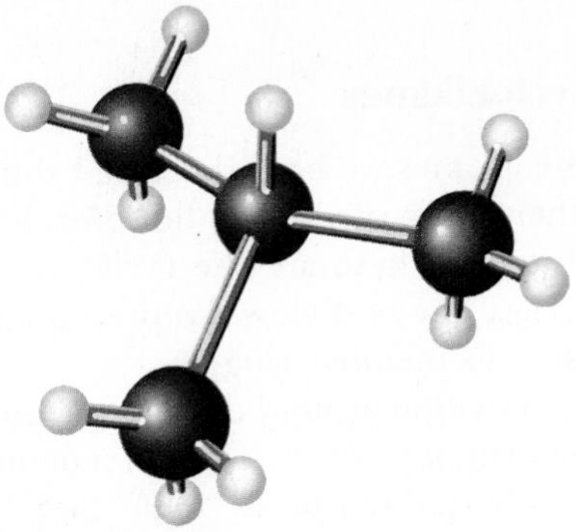
2-Methylpropane, C_4H_{10}

different ways. One alternative is called butane, and the other is named 2-methylpropane. (See the box "A Closer Look: Naming Alkanes.")

An alkane with the formula C_5H_{12} can have three isomers. In the middle molecule in the following illustration, the longest continuous chain of carbon atoms has four members, so the name is based on butane. Because a —CH_3, or methyl group, is attached to the second C atom in the chain, it is called 2-methylbutane. In the third case, the longest continuous chain of carbons has three members, and two —CH_3 groups are attached at the second carbon; therefore, this compound is named 2,2-dimethylpropane.

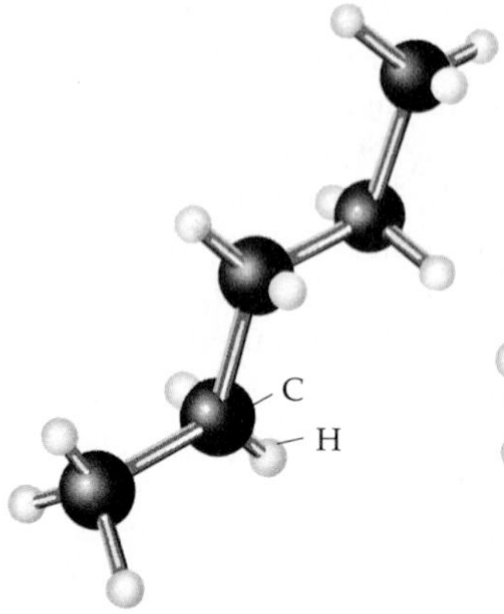

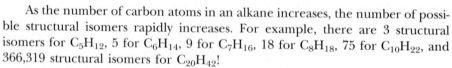

Pentane, C_5H_{12}

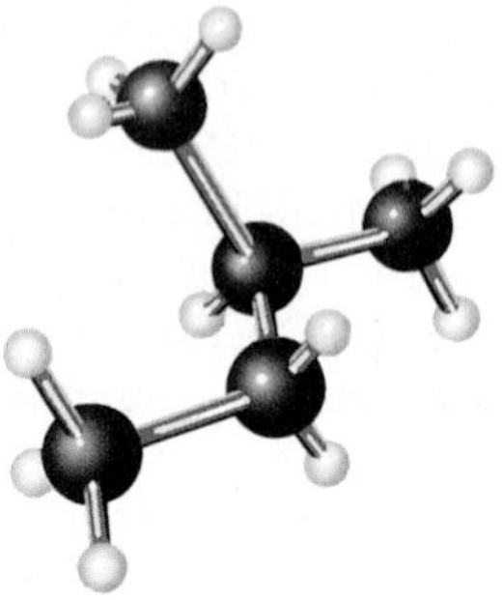
2-Methylbutane, C_5H_{12}

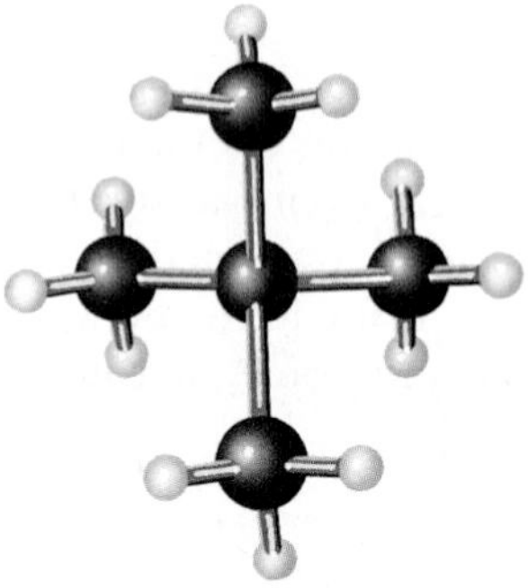
2,2-Dimethylpropane, C_5H_{12}

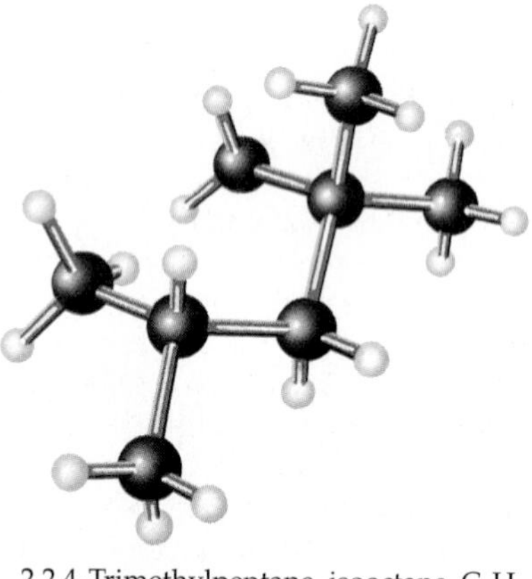
2,2,4-Trimethylpentane, isooctane, C_8H_{18}

As the number of carbon atoms in an alkane increases, the number of possible structural isomers rapidly increases. For example, there are 3 structural isomers for C_5H_{12}, 5 for C_6H_{14}, 9 for C_7H_{16}, 18 for C_8H_{18}, 75 for $C_{10}H_{22}$, and 366,319 structural isomers for $C_{20}H_{42}$!

One isomer of the alkane with the formula C_8H_{18} has the systematic name 2,2,4-trimethylpentane, but it is often called isooctane. It is the standard used in assigning octane ratings to gasoline. This compound is arbitrarily assigned an octane number of 100, and heptane, the straight-chain (unbranched) isomer of C_7H_{16}, is assigned a rating of 0. The antiknock performance of a particular gasoline is compared with that of various mixtures of 2,2,4-trimethylpentane and heptane, and an octane number is assigned on that basis. As the octane numbers of the two standards suggest, branched-chain hydrocarbons have better antiknock properties than straight-chain hydrocarbons.

Cycloalkanes

The alkanes we have described thus far consist of chains of carbon atoms. Another group of alkanes, however, is based on *rings* of tetrahedral, sp^3-hybridized C atoms. These are the cyclic, or *cyclo*alkanes, all of which have the formula C_nH_{2n}. The best known representative of this class is cyclohexane, C_6H_{12}, which has a six-member ring.

Rotation around carbon-carbon bonds is also possible in cycloalkanes. Here, however, it takes the form of a bending of the entire molecule. These forms are usually referred to as *"chair"* and *"boat" conformations* of the molecule because they resemble these shapes.

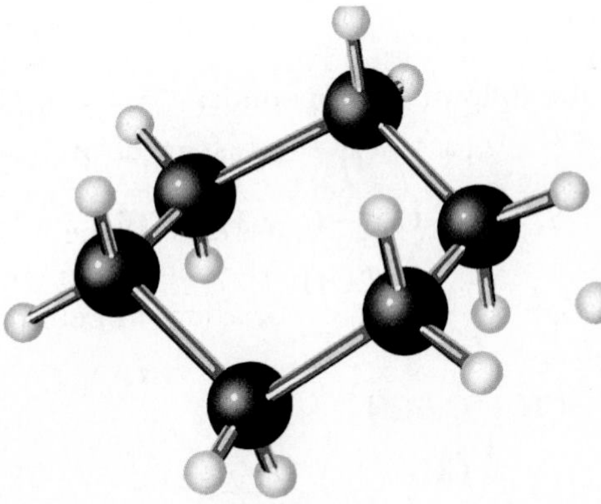

"Chair"
cyclohexane

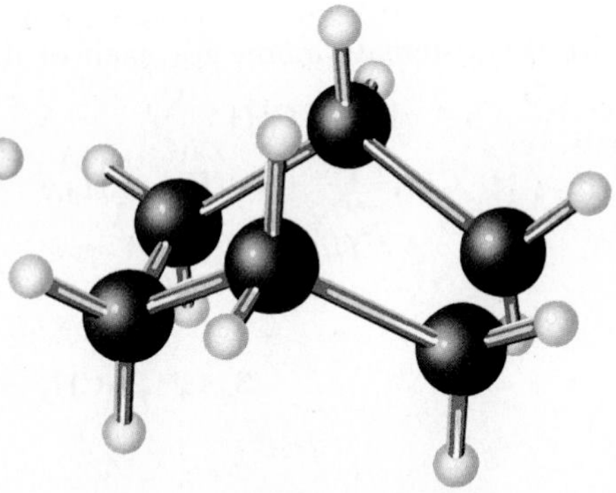

"Boat"
cyclohexane

As a final point, compare the structure of cyclohexane with that of diamond (page 105). In both cases the C atoms are arranged in a six-member ring. In diamond each C atom is attached to four C atoms. In cycloalkanes, each C atom is bound to two other C atoms—which leads to the carbon-based ring—and to two H atoms (or to other atoms in the case of a substituted cycloalkane).

A CLOSER LOOK *Naming Alkanes*

The *systematic* names of the alkanes—the names based on a widely accepted system of naming—were given in Table 11.2. These are specifically the names of the molecules in which the C atoms are attached to one another in a "straight" line. That is, the carbon chain does not branch. To name branching hydrocarbons, we need to examine the naming process more closely.

When an alkane loses an H atom thereby producing a piece of a molecule that can be attached elsewhere, the name of the piece thus produced is derived by dropping the suffix "-ane" and adding "-yl" to the alkane name. Thus, methane becomes the methyl group, $—CH_3$, and ethane becomes the ethyl group, $—C_2H_5$.

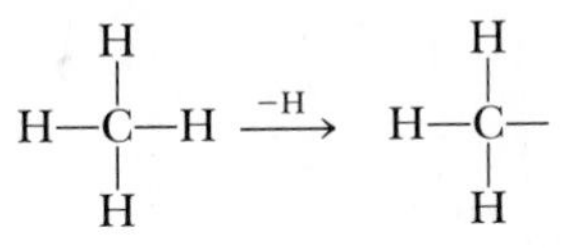

methane methyl group

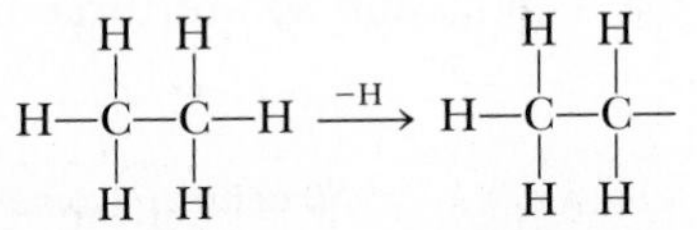

ethane ethyl group

The systematic names of alkanes are based on the number of carbon atoms in the longest *continuous* carbon chain. If that chain contains three carbons, the parent name is propane, if five carbons, it is pentane, and so on. If branches occur in the chain, the branch is treated as a substituent on that chain. For example, the compound shown here has a three-carbon chain with a $—CH_3$, or methyl, group attached at carbon 2. For this reason it is named 2-methylpropane.

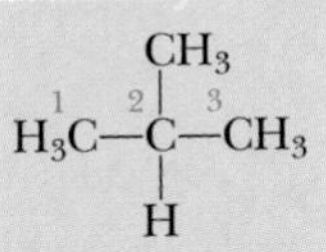

2-methylpropane

The next compound has a longer carbon chain of five carbons. The fact that they are not printed on the same straight line does not matter. Either methyl group at the right-hand end of the chain may be included in the longest continuous chain, the pentane chain. There are three methyl substituents, so the prefix is *tri*methyl. Two of the three methyl groups are attached to carbon atom 2 and one to carbon atom 4. By convention, the numbering begins at the end of the chain that provides the *lowest* possible set of numbers. This means the systematic name is 2,2,4-trimethylpentane (although its common name is isooctane).

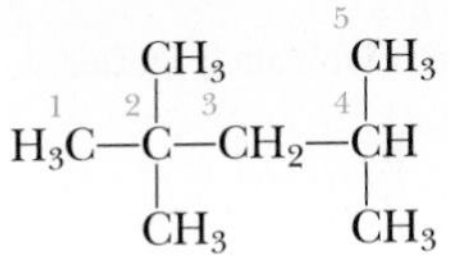

2,2,4-trimethylpentane, isooctane, C_8H_{18}

EXAMPLE 11.1 *Naming Alkanes*

Give the systematic name for each of the following compounds.

$$H_3C \quad CH_3$$
1. $CH_3-\underset{\underset{H}{|}}{\overset{\overset{}{|}}{C}}-\underset{\underset{H}{|}}{\overset{\overset{}{|}}{C}}-CH_2-CH_3$

2. $CH_3-CH_2-\underset{\underset{H}{|}}{\overset{\overset{H_2C-CH_3}{|}}{C}}-CH_2-CH_3$

3. $CH_3-CH_2-CH_2-\underset{\underset{CH_3}{|}}{\overset{\overset{CH_3}{|}}{C}}-CH_3$

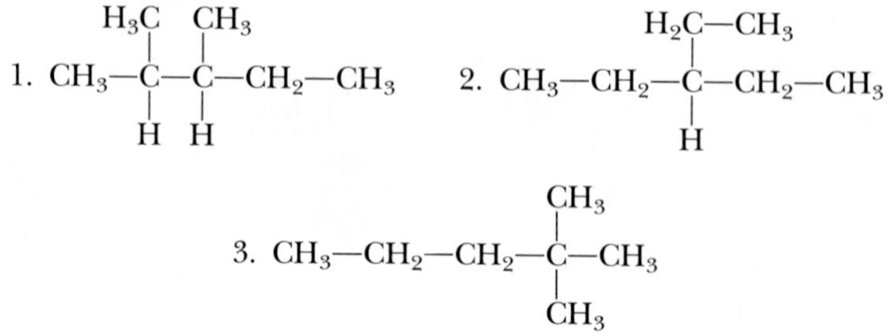

Solution

1. The longest continuous carbon chain has five carbon atoms, so the parent name is pentane. There are two methyl groups, one at carbon 2 and the other at carbon 3, so the systematic name is 2,3-dimethylpentane.

2. The longest continuous chain has five carbon atoms, so the parent name is again pentane. In this case, however, the substituent is an ethyl group ($-CH_2-CH_3$), so the systematic name is 3-ethylpentane.

3. The parent name is again pentane. The carbon atoms are numbered beginning at the end closest to the methyl substituents. The two methyl groups are therefore attached to carbon atom 2, and the systematic name is 2,2-dimethyl-pentane.

As a final point, notice that these three compounds have the same molecular formula, C_7H_{16}, and so are three of the nine possible structural isomers.

EXERCISE 11.1 *Structural Isomers*

Draw the structural isomers for C_6H_{14}. Give the systematic name for each. ■

EXERCISE 11.2 *Naming Alkanes*

The three compounds in Example 11.1 are all isomers with the molecular formula C_7H_{16}. What is the name of the unbranched alkane with this formula? ■

11.2 ALKENES AND ALKYNES

Compounds with double and triple bonds between the carbon atoms are often referred to as **unsaturated,** because one of the pairs of electrons in the multiple bond can be used to form bonds to newly attached atoms.

Alkenes are often called **olefins.**

$$\underset{H}{\overset{H}{>}}C=C\underset{H}{\overset{H}{<}} \xrightarrow{+X_2} H-\underset{\underset{H}{|}}{\overset{\overset{X}{|}}{C}}-\underset{\underset{H}{|}}{\overset{\overset{X}{|}}{C}}-H$$

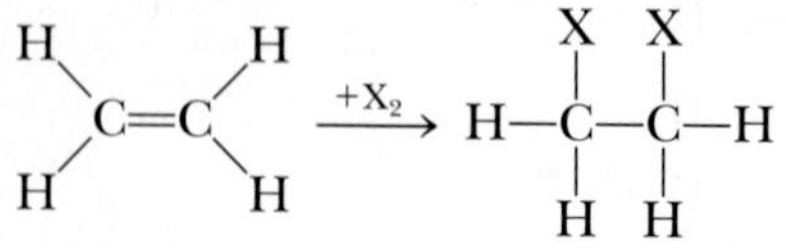

ethylene,
unsaturated

saturated

For this reason these compounds are often used by the chemical industry to synthesize new compounds. As such, they represent a class of organic compounds of considerable economic importance. One example is α-pinene, a naturally occurring alkene obtained by steam distilling the resin from various types of conifer trees. If α-pinene is treated with acid, it is converted to pine oil, which is used as a perfume and bactericide in many household products.

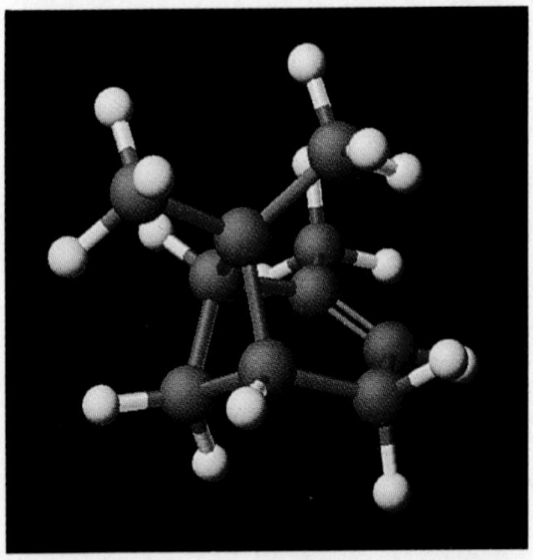

α-Pinene. It is interesting to note that an α-pinene molecule, with 10 C atoms and 16 H atoms, is built from two molecules of another alkene, isoprene, which has 5 C atoms and 8 H atoms. More about isoprene later in this section.

Naming Alkenes and Alkynes

The systematic names of alkenes and alkynes, such as those listed in Tables 11.3 and 11.4, are derived from the name of the corresponding alkane by dropping the "-ane" ending and adding "-ene" or "-yne" in its place. When necessary, the position of the double or triple bond is included in the prefix. To determine which position number to use, begin counting at the end of the carbon chain closest to the multiple bond.

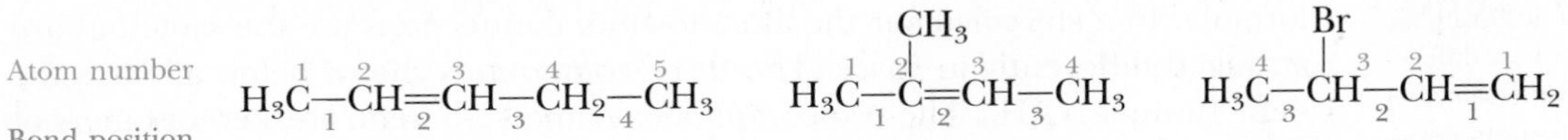

2-pentene 2-methyl-2-butene 3-bromo-1-butene

The simplest alkene, C_2H_4, and the simplest alkyne, C_2H_2, have the systematic names ethene and ethyne, respectively. However, they are almost universally known by their common names—ethylene and acetylene, respectively. We shall use the common names.

TABLE 11.3 Some Representative Alkenes

Structure	Systematic Name	Common Name	bp (°C)	mp (°C)
$CH_2{=}CH_2$	Ethene	Ethylene	−104	−169
$CH_3CH{=}CH_2$	Propene	Propylene	−48	−185
$CH_3CH_2CH{=}CH_2$	1-Butene	—	−6	−185
$CH_2{=}CHC(CH_3){=}CH_2$	2-Methyl-1,3-butadiene	Isoprene	34	−146

TABLE 11.4 Some Simple Alkynes

Structure	Systematic Name	Common Name	bp (°C)
$H-C\equiv C-H$	Ethyne	Acetylene	−75
$CH_3C\equiv CH$	Propyne	Methylacetylene	−23
$CH_3CH_2C\equiv CH$	1-Butyne	Ethylacetylene	9
$CH_3C\equiv CCH_3$	2-Butyne	Dimethylacetylene	27

Structure, Bonding, and Isomerism

The structures of alkenes and alkynes are predictable from the VSEPR theory, and are confirmed by experiment. In alkenes the substituents around a carbon atom that is doubly bonded to another carbon atom are arranged as a planar triangle. In alkynes the C≡C—R bond angle is 180°.

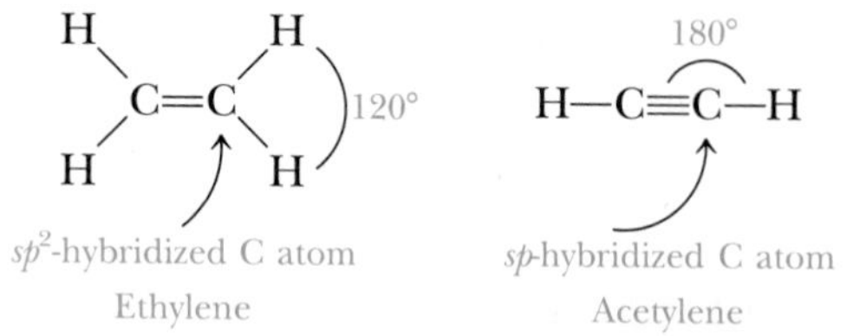

Valence bond theory would rationalize the structure of an alkene by assuming the C atom is sp^2-hybridized, with the pi bond having been formed from unhybridized p orbitals, one on each C atom (Section 10.1). Similarly, the C atoms involved in the triple bond in an alkyne are sp-hybridized; the two pi bonds formed from the two unhybridized p orbitals on each atom.

As outlined in Section 10.1, an important consequence of double bonding in alkenes is the possibility of stereoisomerism. Isomers always have the same formula. In a **stereoisomer** the atom-to-atom connections are the same but are arranged differently in space. The three compounds shown below all have the same formula, C_4H_8. The two compounds named 2-butene are stereoisomers of each other and are in turn structural isomers of 2-methylpropene.

Structural isomers have the same formulas, but their atoms are connected in a different order. Stereoisomers have the same formulas and the same atom-to-atom connections, but their atoms are arranged differently in space. See page 474.

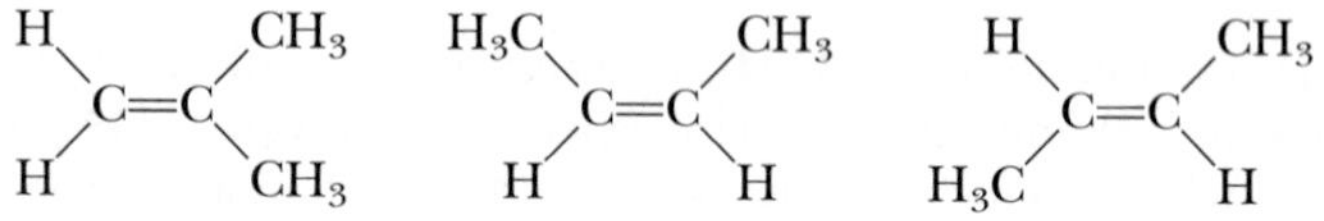

Name	2-Methylpropene (isobutene)	*cis*-2-Butene	*trans*-2-Butene
Boiling point	−6.95 °C	3.71 °C	0.88 °C
Melting point	−140.4 °C	−138.9 °C	−105.53 °C
Dipole moment	0.503 D	0.253 D	0 D
ΔH_f° (liq) (kJ/mol)	−16.9	−7.1	−11.4

Stereoisomers in which the two substituents are on the same side of the doubly bonded C atoms are called *cis;* those in which the substituents occur across the bond from each other are called *trans.* This form of stereoisomerism is therefore often called *cis-trans* isomerism.

The atoms attached at either end of a single or sigma bond can rotate freely relative to each other (see page 473). A consequence of pi bonding in alkenes, however, is that one end of the bond is "locked" with respect to the other end. Thus, *cis-* and *trans-*2-butene are distinctly different compounds as reflected by their physical properties. Be sure to notice that *trans-*2-butene is nonpolar, whereas both *cis-*2-butene and 2-methylpropene are polar.

The importance of *cis* and *trans* forms of alkenes is found in the chemistry of isoprene, the systematic name for which is 2-methyl-1,3-butadiene. When natural rubber is heated strongly in the absence of air it smells of isoprene, an observation that provided chemists with the first clue that rubber is composed of this

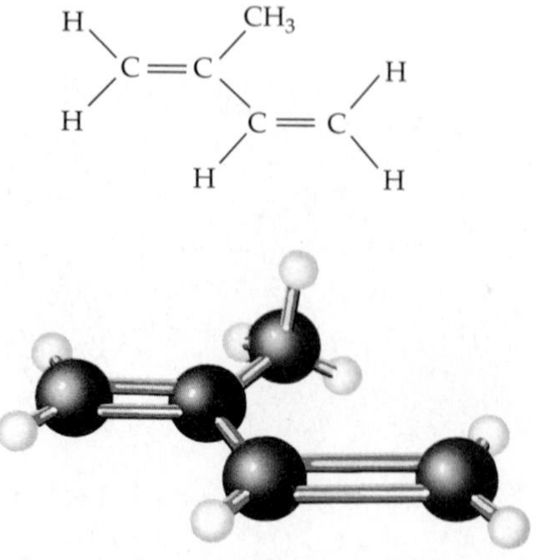

Isoprene,2-methyl-1,3-butadiene

building block. When chemists tried to link isoprene units together to make rubber, however, what they made was sticky and useless. It turned out that nature knew what it was doing. The rubber tree links isoprene units so that the links along the carbon chain are all *cis* to one another (Figure 11.1). Early laboratory attempts produced a material with a random mixture of *cis* and *trans* links, which had unsatisfactory properties.

But the story of rubber is even more interesting. If the sap from another species of rubber tree is heated, a type of rubber called *gutta-percha* is produced (Figure 11.2). Now all the links in the isoprene chain are *trans,* and the material is much harder. It is used for the covering on golf balls, for example.

Isoprene units are also part of the unsaturated carbon chain in carotene (Figure 11.3). Here eight isoprene units form a chain containing 11 double bonds; consequently, this structure has two important properties. First, the chain is not very flexible. Second, the pi electrons are loosely held by the molecule and can be excited by visible light. The light the molecule absorbs is in the blue-violet region of the visible spectrum, and so it appears orange-yellow to the observer. In fact, it is carotene that gives carrots their characteristic color.

Carotene or carotene-like molecules are partnered with chlorophyll in nature in the role of assisting in the harvesting of sunlight. Because typically about one carotene (or carotene-like) molecule occurs per chlorophyll molecule, very dark green leaves also have a high concentration of carotene. In the fall of the year, when chlorophyll molecules are destroyed, the green of the leaves fades away, and the yellows and reds of carotene and related molecules are seen.

Because carotene is a hydrocarbon, it is not soluble in water but is soluble in fat. Therefore, when a grazing animal eats grass or leaves, the animal also ingests carotene or related compounds. It is this that causes milk and butter to be slightly yellow in color.

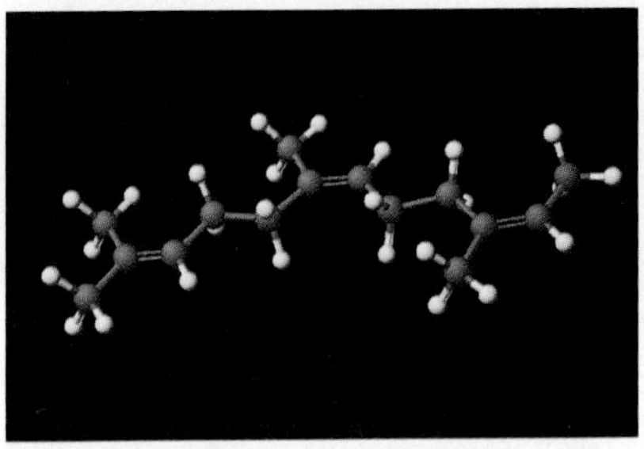

Figure 11.1 Natural rubber is a polymer (Section 11.8) of isoprene units. All the linkages in the carbon chain are *cis.*

The red color of tomatoes comes from a molecule very closely related to carotene. As the tomato ripens, the chlorophyll disintegrates and so the green color disappears to be replaced by the red of the carotene-like molecule.

(a) (b)

Figure 11.2 Rubber. (a) The sap that comes from this tree is a natural polymer of isoprene. (b) The covering on a golf ball consists of gutta-percha rubber. In this type of rubber all the links between isoprene units are *trans* (rather than *cis* as in softer rubber; Figure 11.1)

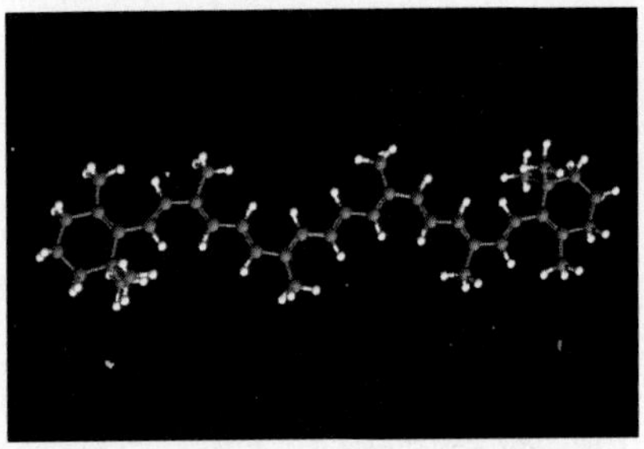

Figure 11.3 The carotene molecule is a hydrocarbon built from eight isoprene units.

Preparation of Alkenes and Alkynes

At one time most commercially useful organic compounds were derived from coal, and acetylene was used in the synthesis of other compounds. The following are common methods for preparing acetylene, C_2H_2.

From limestone

$$CaO(s) + 3\ C(s) \longrightarrow CaC_2(s) + CO(g)$$

$$CaC_2(s) + 2\ H_2O(\ell) \longrightarrow H{-}C{\equiv}C{-}H(g) + Ca(OH)_2(s)$$

From methane

$$2\ CH_4(g) \longrightarrow H{-}C{\equiv}C{-}H(g) + 3\ H_2(g)$$

Both involve high temperatures and are therefore costly. Thus, when relatively inexpensive ethylene became available from petroleum and natural gas about 50 years ago, acetylene was almost completely replaced by ethylene in industrial processes. Acetylene is still used, though, in oxyacetylene welding torches because the high heat of combustion of the hydrocarbon (Figure 11.4) allows for focused melting of metal.

$$C_2H_2(g) + \tfrac{5}{2}\ O_2(g) \longrightarrow 2\ CO_2(g) + H_2O(\ell)$$

	$C_2H_2(g)$	$O_2(g)$	$CO_2(g)$	$H_2O(\ell)$
ΔH_f° (kJ/mol)	226.7	0	−393.5	−285.8

$\Delta H^\circ_{rxn} = -1299.5$ kJ/mol of acetylene or -50.0 kJ/g

Ethylene and propene are prepared industrially on a large scale by using steam to "crack" the hydrocarbons found in natural gas and petroleum.

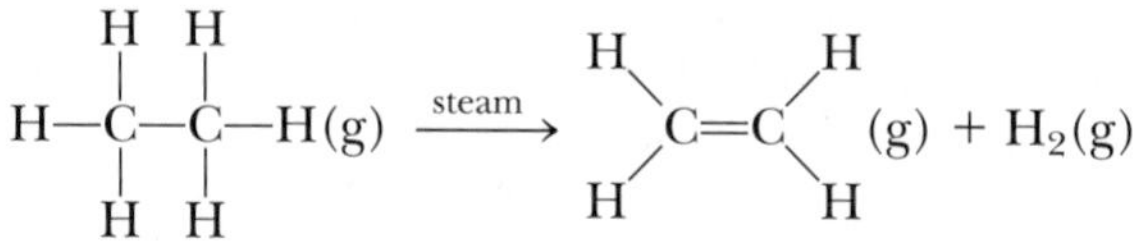

ethane ethylene

More than 18 billion kilograms of ethylene are produced annually in the United States, placing it in the top five chemicals produced. About half of this amount is used to make polyethylene (Section 11.8). The hydrogen produced by the reaction is also valuable because it is used to make ammonia, for example.

Figure 11.4 The reaction of acetylene with oxygen produces a very high temperature. A torch that burns this mixture is therefore used in welding metals. (C. D. Winters)

Addition Reactions

One of the most common reactions of alkenes (and alkynes) is the addition reaction in which reagents add to the carbon atoms of the C=C double bond.

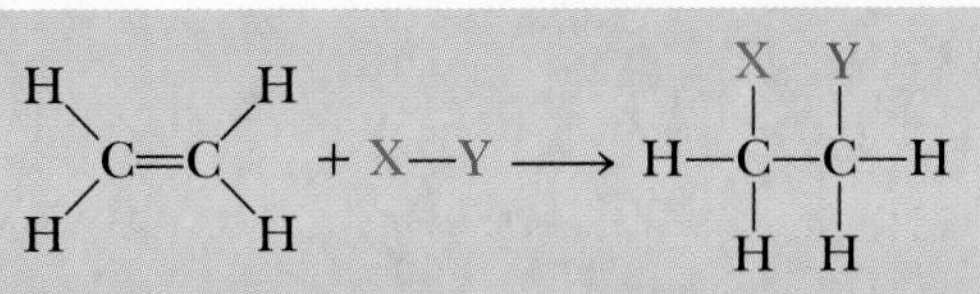

X—Y = H_2, H_2O, X_2, HX (where X is a halogen)

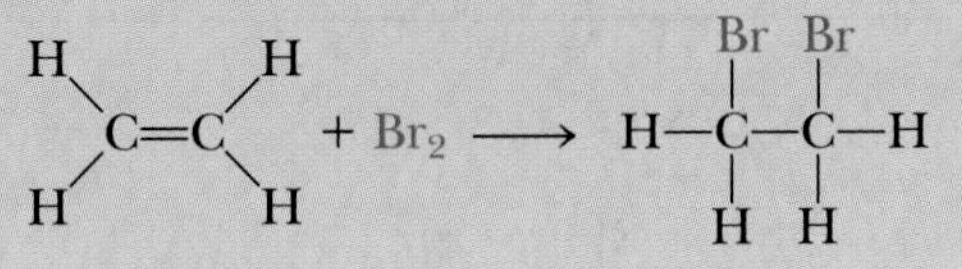

1,2-dibromoethane

If the reagent is hydrogen (X—Y = H_2), the reaction is called **hydrogenation,** and the product is an alkane. Specially prepared forms of metals, such as platinum, palladium, and rhodium, among others, are used to promote these reactions.

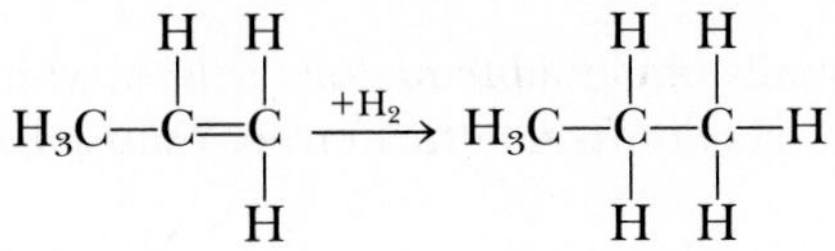

propene propane

You know about hydrogenation because the foods you eat are sometimes "partially hydrogenated." One brand of crackers has a label that says "Made with 100% pure vegetable shortening (partially hydrogenated soybean oil with hydrogenated cottonseed oil)." What does this mean? After describing more aspects of organic chemistry, we shall return to this question (Section 11.6).

EXERCISE 11.3 *Cis-Trans* Isomers

Draw the *cis* and *trans* isomers for 2-pentene. ■

EXERCISE 11.4 *Reactions of Alkenes*

1. Draw the structure of the compound derived from the reaction of HBr with ethylene.
2. Draw the structure for the compound that comes from the reaction of Br_2 with 2-butene. Name the compound. ■

11.3 AROMATIC COMPOUNDS

Benzene, C_6H_6, is a key molecule in chemistry. It is also the simplest member of a class of compounds called aromatics because they usually have a pleasant odor. Other members of this class include toluene and naphthalene (Table 11.5).

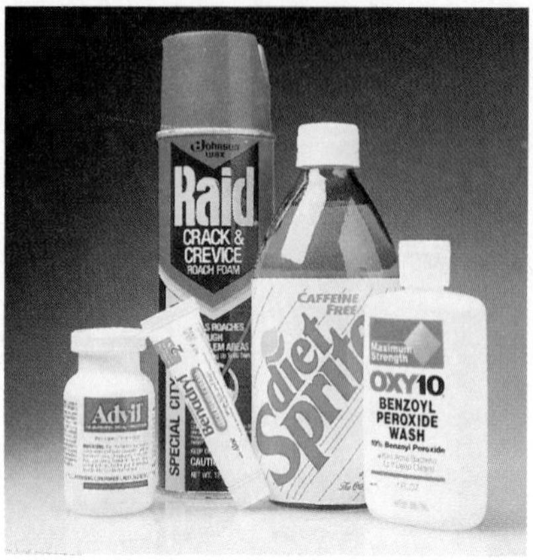

Some products containing compounds based on benzene. Examples are sodium benzoate in soft drinks, ibuprofen in Advil, and benzoyl peroxide in Oxy10. (C. D. Winters)

TABLE 11.5 Some Aromatic Compounds from Coal Tar

Name	Formula	Boiling Point (°C)	Melting Point (°C)	Solubility
Benzene	C_6H_6	80	+6	All insoluble in water
Toluene	$C_6H_5CH_3$	111	−95	
o-Xylene	$C_6H_4(CH_3)_2$	144	−27	
m-Xylene	$C_6H_4(CH_3)_2$	139	−54	
p-Xylene	$C_6H_4(CH_3)_2$	138	+13	
Naphthalene	$C_{10}H_8$	218	+80	

Benzene occupies a pivotal place in the history and practice of chemistry. Michael Faraday discovered the compound in 1825 as a byproduct of illuminating gas, itself produced by heating coal. Benzene is an important industrial chemical and is usually about 15th on the list of the top 50 chemicals produced in the United States every year. Not only is benzene an important solvent, but it is the starting point for making thousands of different compounds by replacing the H atoms of the ring.

Toluene was originally obtained from *Tolu balsam,* which is derived from the pleasant-smelling gum of a South American tree, *Toluifera balsamum.* This balsam

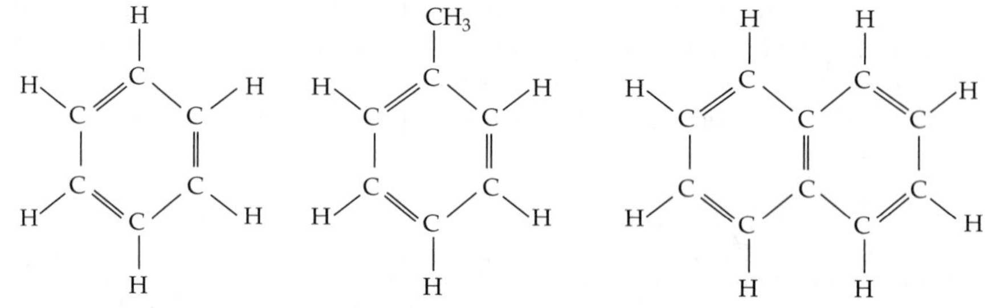

Benzene boiling point, 80°C — Toluene boiling point, 111°C — Naphthalene melting point, 80°C

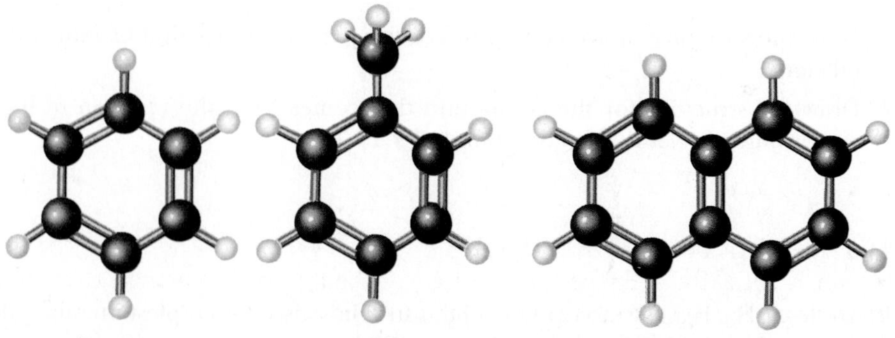

Benzene — Toluene — Naphthalene

has been used in cough syrups and perfumes. Naphthalene is the basic ingredient of "moth balls." You will also recognize now that saccharin (page 495) is a benzene derivative.

The formula of benzene suggested to 19th century chemists that it should be unsaturated, but, if viewed this way, its chemistry was perplexing. Whereas an alkene readily adds Br_2, for example, benzene does not react under these conditions. Benzene does react with Br_2 at higher temperatures in the presence of other chemicals, but the products are those of a substitution reaction and not an addition. The fact that only one monosubstitution product is ever obtained for a given reaction, along with other observations, strongly suggested to August Kekulé (1829–1896) that benzene has a planar and symmetrical ring structure in which all the carbon-carbon and carbon-hydrogen bonds are equivalent.

Substitution reaction

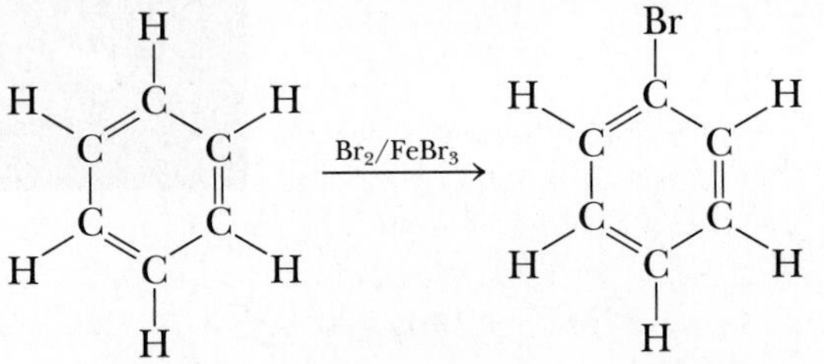

benzene bromobenzene

It was Kekulé who then suggested (in 1872) that the molecule could be represented by some combination of two structures, which we now call resonance structures.

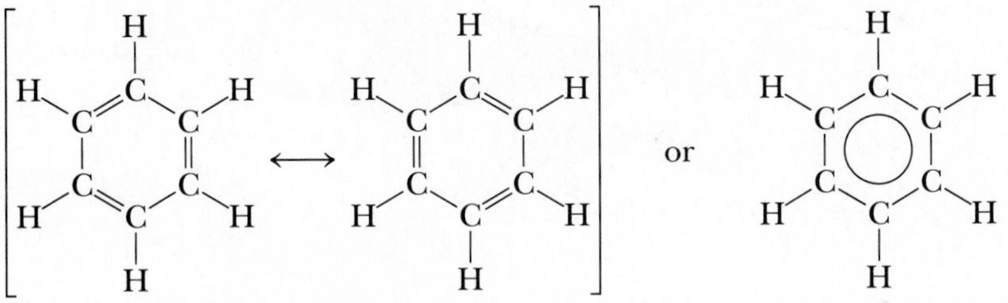

Resonance structures Resonance hybrid

Modern experimental methods have revealed that each carbon-carbon bond is 139 pm long, intermediate between the length of a C—C single bond (154 pm) and a C=C double bond (134 pm). Chemists often depict this with the resonance hybrid form on the right. The circle in the middle of the right-hand structure is meant to convey the idea that the pi-bonding electrons are delocalized, or spread, over the six carbon atoms of the ring. Finally, benzene is often drawn using the following shorthand notation

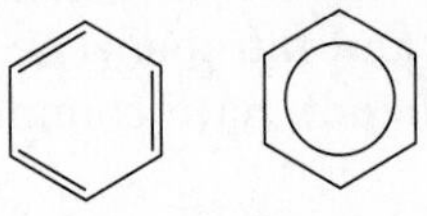

Representations of benzene, C_6H_6

in which neither the C atoms of the ring or the attached H atoms are shown. However, whenever another group has been substituted for one or more hydrogen atoms, the substituents will be shown.

Bonding in Benzene

The structure of benzene was deduced from its chemistry, but an image from a scanning tunneling microscope confirms it is a six-member ring. In this image each ring appears as a raised bump with a slight depression in the middle (a).

To understand the bonding in benzene let us begin by assuming that each carbon atom is sp^2-hybridized. Two of the hybrid orbitals are used to form sigma bonds with neighboring carbon atoms, and the third sp^2 hybrid orbital is used to form a sigma bond to an H atom.

After accounting for carbon-carbon and carbon-hydrogen sigma bonding, six *p* orbitals remain, one on each C atom (b). Each orbital is occupied by a single electron. These six orbitals form three pi-bonding molecular orbitals and three pi-antibonding molecular orbitals. Six electrons are to be placed in these orbitals, so the three pi-bonding molecular orbitals are fully occupied by electrons. A more detailed analysis of bonding, using a computer, shows that the electrons are distributed evenly over the molecule (c, d).

(b)

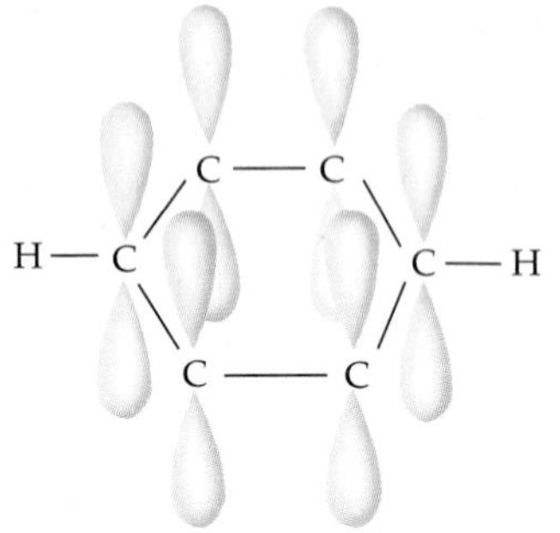

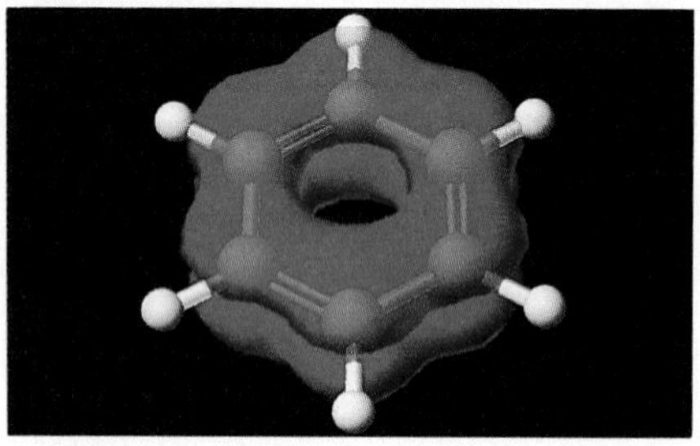

(d)

(c)

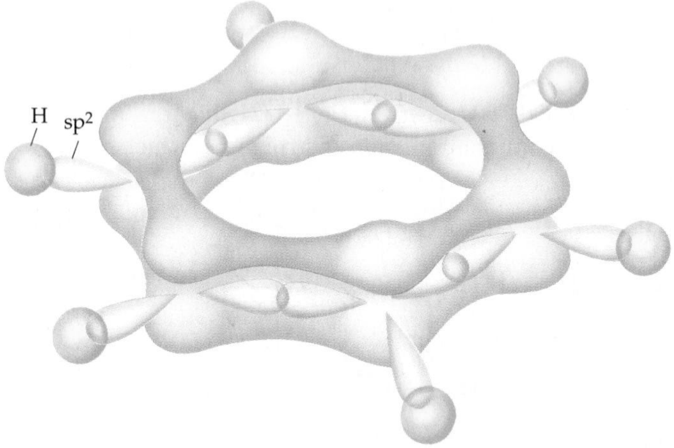

(a) Scanning tunneling microscope image of benzene molecules. (IBM Almaden)

Naming Aromatic Compounds

Most aromatic compounds are named as derivatives of benzene, examples of which include chlorobenzene and nitrobenzene. A few, however, such as toluene, *m*-xylene, styrene, and phenol, have common names.

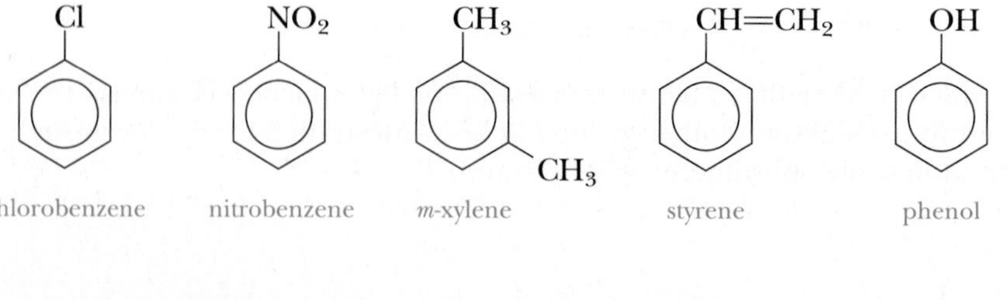

chlorobenzene nitrobenzene *m*-xylene styrene phenol

When a benzene ring bears two substituents, as in *m*-xylene, the relative positions are often indicated by the terms *ortho, meta,* and *para.* When writing the name of the compound, usually only the first letter of one of these terms is given.

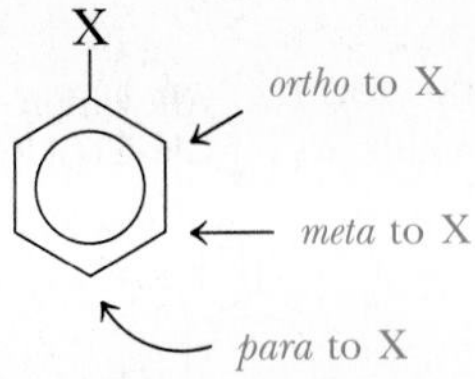

Some examples of this naming system are as follows:

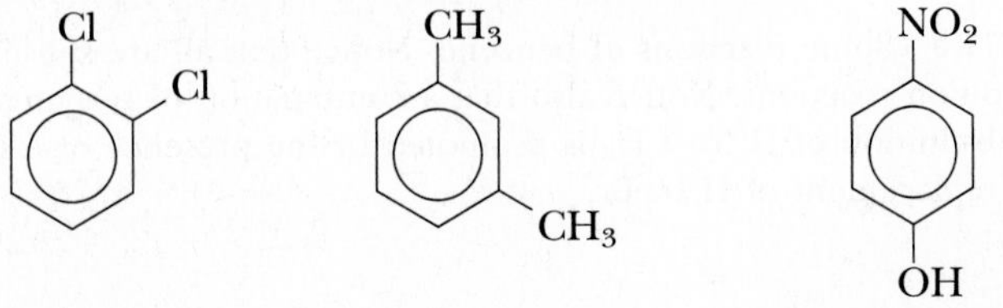

Systematic name:	1,2-dichlorobenzene	1,3-dimethylbenzene	1,4-hydroxynitrobenzene
Common name:	*o*-dichlorobenzene	*m*-xylene	*p*-nitrophenol

Notice that when more than one different substituent occurs, they are named in alphabetical order.

Finally, if more than two substituents occur, the *ortho, meta, para* system does not work, and the substituent positions are numbered. The lowest possible set of numbers is used.

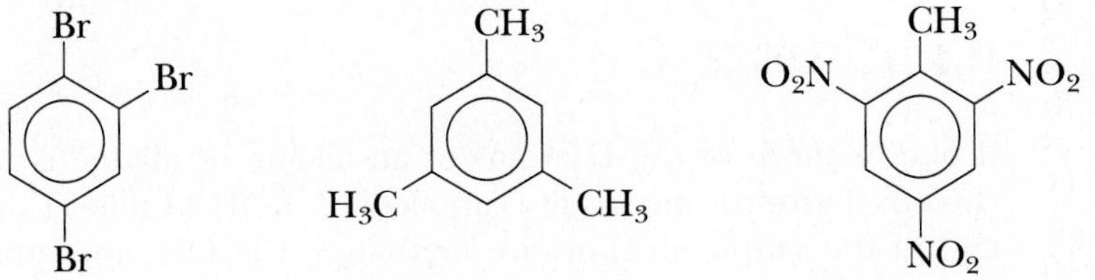

Systematic name:	1,2,4-tribromobenzene	1,3,5-trimethylbenzene	1-methyl-2,4,6-trinitrobenzene
Common name:	none	mesitylene	2,4,6-trinitrotoluene (TNT)

EXERCISE 11.5 ***Naming Aromatic Compounds***

Naming derivatives of benzene can quickly become complicated, so only a few examples are given here.

1. Give systematic names for

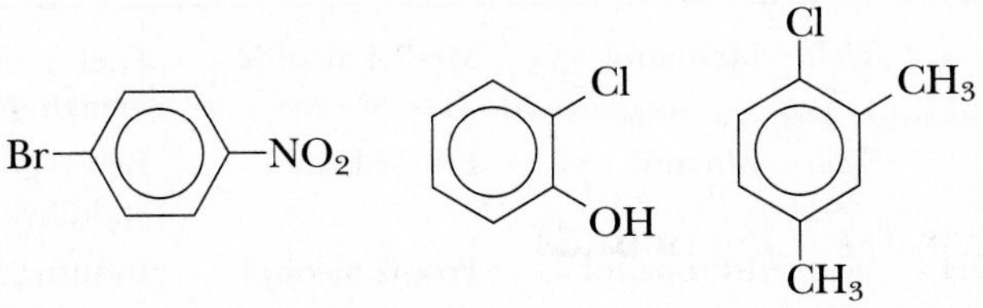

2. Draw a structure for *o*-chloronitrobenzene. ■

The Chemistry of Benzene

In the box "A Closer Look: Bonding in Benzene" evidence was presented for the delocalization of the pi-bonding electrons over all six C atoms, a conclusion that profoundly affects the chemistry of benzene. Instead of addition reactions typical of compounds with carbon-carbon double bonds such as alkenes, ben-

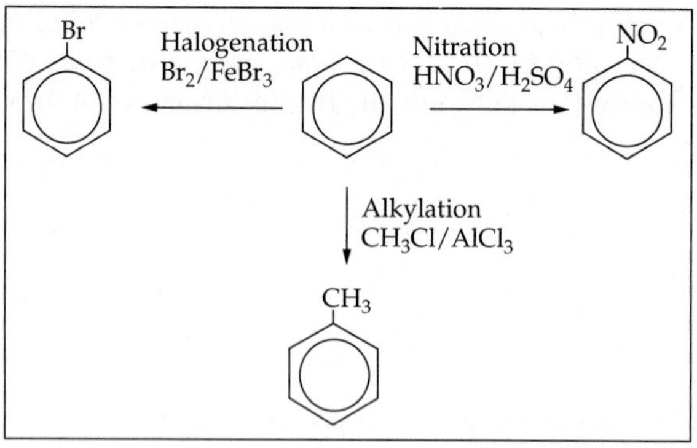

Figure 11.5 Some reactions of benzene. Notice that all are substitution reactions and not addition reactions. Notice also that a combination of reagents is often needed. The substitution of H for CH_3 is promoted by the presence of $AlCl_3$, and $FeBr_3$ assists in the replacement of H by Br.

zene and other aromatic compounds undergo substitution reactions. For example, it is possible to substitute a halogen atom, a nitro group, or an alkyl or other hydrocarbon grouping for one or more of the H atoms of benzene (Figure 11.5).

11.4 ALCOHOLS

If one or more of the H atoms of an alkane or alkene is replaced by an OH (hydroxyl group), the result is an alcohol, ROH (Table 11.6). The most important of the simple alcohols are methanol, CH_3OH, and ethanol, CH_3CH_2OH.

TABLE 11.6 Some Important Alcohols

Condensed Formula	bp (°C)	Systematic Name	Common Name	Use
CH_3OH	65.0	Methanol	Methyl alcohol	Fuel, gasoline additive, making formaldehyde
CH_3CH_2OH	78.5	Ethanol	Ethyl alcohol	Beverages, gasoline additive, solvent
$CH_3CH_2CH_2OH$	97.4	1-Propanol	Propyl alcohol	Industrial solvent
$CH_3CH(OH)CH_3$	82.4	2-Propanol	Isopropyl alcohol	Rubbing alcohol
$HOCH_2CH_2OH$	198	1,2-Ethanediol	Ethylene glycol	Antifreeze
$CH_2(OH){-}CH(OH){-}CH_2(OH)$	290	1,2,3-Propanetriol	Glycerol (glycerin)	Moisturizer in foods

CURRENT ISSUES IN CHEMISTRY

New Life for Natural Gas

Your home may be heated with natural gas, which consists largely of methane, CH_4. Although widely used, it is also widely wasted! The world's known and projected reserves of this, the most simple hydrocarbon, are 250,000 trillion liters, the energy equivalent of 1.5 trillion barrels of oil. Unfortunately, much of it is in areas such as Southeast Asia or in the northern reaches of Canada that are far removed from the centers of fuel consumption. Methane can be carried as a gas in pipelines, but this method is expensive if the distances are great. Alternatively, it can be liquefied and carried by ship, but this procedure risks a fiery catastrophe. The drawbacks to transportation mean that methane is often simply burned or "flared off" when it comes out of the ground with oil.

One solution to making methane useful is to convert it, where it is found, to a more readily transportable liquid such as methanol, CH_3OH. The methanol can then be used directly as a fuel, or used to make other chemicals.

It has been known for some time that methane can be converted to carbon monoxide and hydrogen,

$$CH_4(g) + H_2O(g) \longrightarrow CO(g) + 3\ H_2(g)$$

and this mixture of gases can readily be turned into methanol in another step.

$$CO(g) + 2\ H_2(g) \longrightarrow CH_3OH(\ell)$$

Unfortunately, the first step in this process is a high-temperature (>900 °C), energy-intensive process. In 1993, however, Catalytica, Inc. of California announced that methane can be converted to methanol with a 43% yield by using a mercury(II) salt as a catalyst (a substance that accelerates a reaction without being consumed in the reaction) in the presence of sulfuric acid and water.

$$CH_4(g) + \tfrac{1}{2}\ O_2(g) \xrightarrow[180\ °C]{Hg(II)\ salt,\ H_2SO_4,\ H_2O} CH_3OH(\ell)$$

Although problems potentially exist in a process that uses toxic mercury compounds and highly corrosive sulfuric acid, the discovery is important to consumers of energy and chemical worldwide.

Just as exciting is another discovery regarding methane. Chemical engineers at the University of Minnesota have found that methane can in fact be converted to CO and H_2 under very mild conditions. They simply found the right catalyst! The photograph shown here illustrates what happens when the engineers flowed a room-temperature mixture of methane and oxygen through a heated, sponge-like ceramic disk coated with platinum or rhodium. Rather than oxidizing the methane all the way to water and carbon dioxide, the process produces a hot mixture of CO and H_2, which can be converted in good yield to methanol.

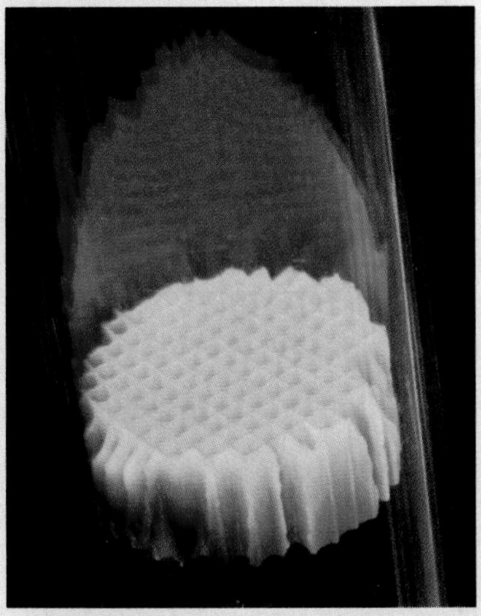

Methane flowing through a catalyst. (Schmidt/University of Minnesota)

For more information on these discoveries see *Science News*, January 16, 1993, page 36; and *Science*, January 15, 1993, pages 340–346.

More than 450 million kilograms of methanol is produced in the United States annually, and most of it is used to make formaldehyde (CH_2O) and acetic acid (CH_3CO_2H), both of which are components of polymers and are important chemicals in their own right. In addition, methanol is used as a solvent and as a deicer in gasoline The latter property depends on the fact that methanol is much like water, and so can interact strongly with water (Section 13.2). Metha-

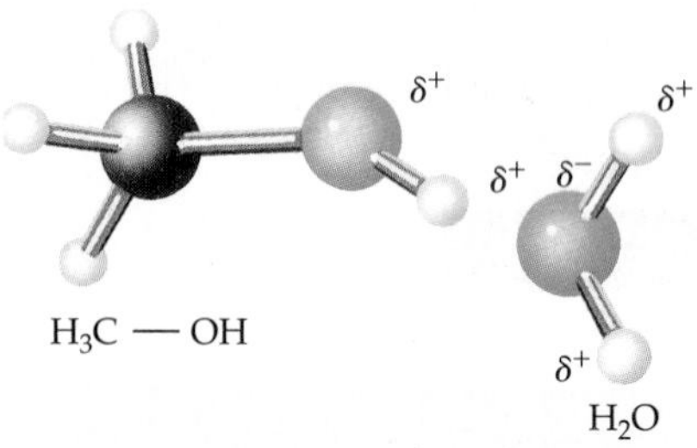

Methanol forms hydrogen bonds with water.

nol is often called "wood alcohol" because it was originally produced by heating wood in the absence of air. It is also found in new wine, where it contributes to the odor or bouquet of the wine. Like ethanol, methanol causes intoxication, but it is poisonous largely because the human body converts it to formic acid and formaldehyde. These compounds attack the cells of the retina, which can lead to permanent blindness.

Methanol is also used as a fuel in high-powered racing cars and would be used more widely if it could be made more cheaply. A method for doing this has recently been developed, however, so you may see methanol available for automobile fuel in the future.

Methanol, CH_3OH, is used as the fuel in cars of the type that race in Indianapolis. (Bernard Asset/Photo Researchers, Inc.)

Ethanol is one of the most important chemicals produced by nature and by the chemical industry. It is the "alcohol" of alcoholic beverages and is prepared for this purpose by the fermentation of sugar from a variety of plant sources. For many years, industrial alcohol, which is used as a solvent and as a starting material for the synthesis of other compounds, was also made by fermentation. In the last several decades, however, it became cheaper to make ethanol from petroleum byproducts, specifically by the *addition* of water to ethylene in the presence of a catalyst.

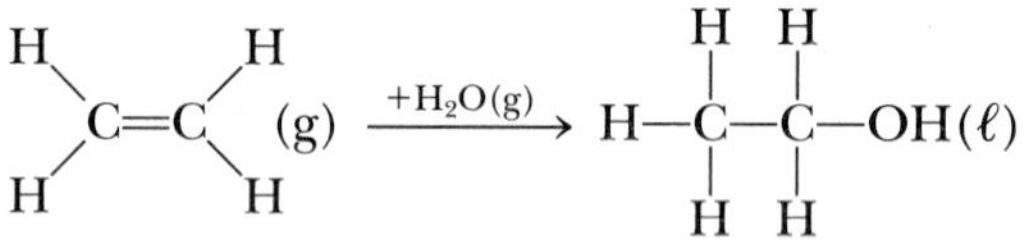

ethylene ethanol

Naming Alcohols

Systematic names of alcohols are derived from the names of the corresponding alkanes by dropping the "-e" ending and adding "-ol." Thus, the systematic name for CH_3CH_2OH is ethanol, although you will often hear chemists call it ethyl alcohol. When necessary, a numerical prefix is added to designate the position of the hydroxyl group (see Table 11.6).

Alcohols can be classified into three categories based on the number of carbons bonded to the C atom bearing the —OH group. *Primary alcohols* have one carbon and two hydrogen atoms attached, whereas *secondary alcohols* have two carbons and one hydrogen atom attached. *Tertiary alcohols* have three carbon atoms attached to the C atom bearing the —OH group.

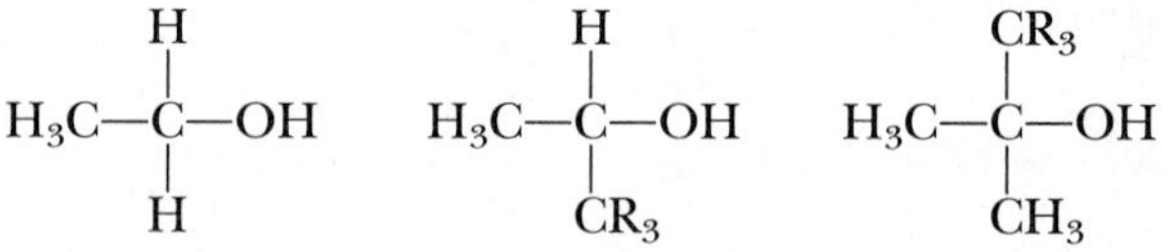

Primary alcohol Secondary alcohol Tertiary alcohol

Alcohols having two or more —OH groups are also known. Two of the most important polyhydric alcohols are ethylene glycol and glycerol. Ethylene glycol is familiar as automobile antifreeze. Glycerol is about 60% as sweet as cane sugar and so is often used in confectioneries; its greatest use, however, is as a softener in soaps and lotions. When glycerol reacts with nitric acid it produces nitroglyc-

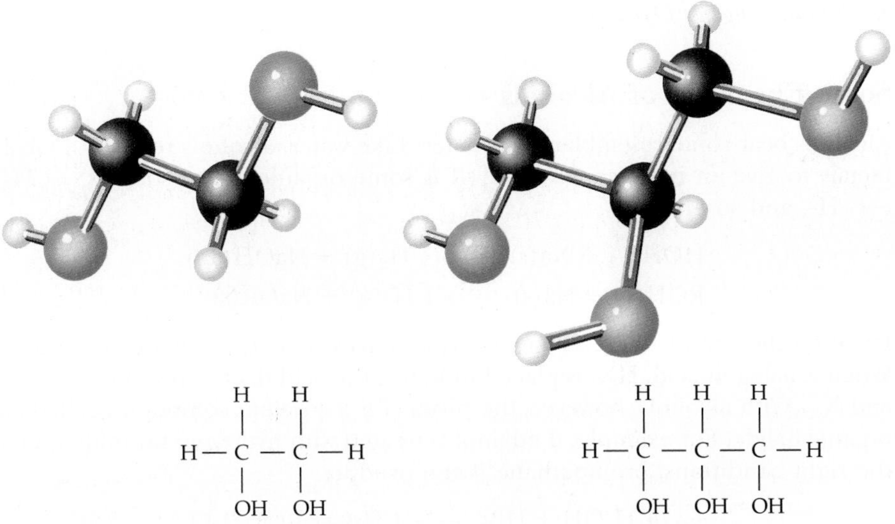

Systematic name	1,2-Ethanediol	1,2,3-Propanetriol
Common name	Ethylene glycol	Glycerol or glycerine

erine, the explosive component of dynamite (Figure 11.6). Nitroglycerine is also used, however, to treat a common heart condition called angina.

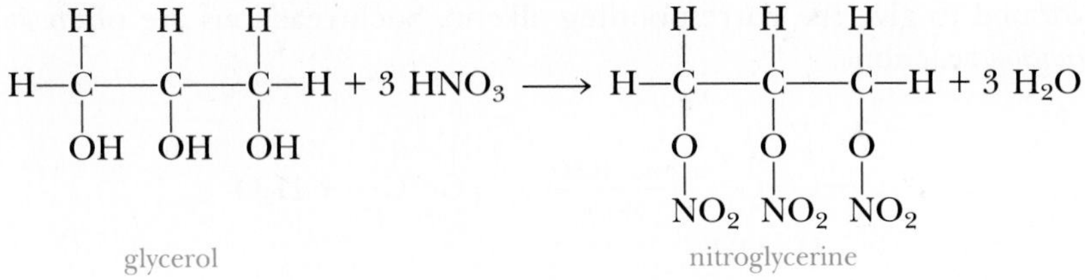

EXERCISE 11.6 ***Structures and Names of Alcohols***

1. What are the systematic names for $CH_3CH_2CH_2OH$ and $CH_3CH(OH)CH_3$?
2. Draw structural formulas for 3-pentanol and 3-methyl-2-butanol.
3. Tell whether each alcohol in part 2 is primary, secondary, or tertiary. ■

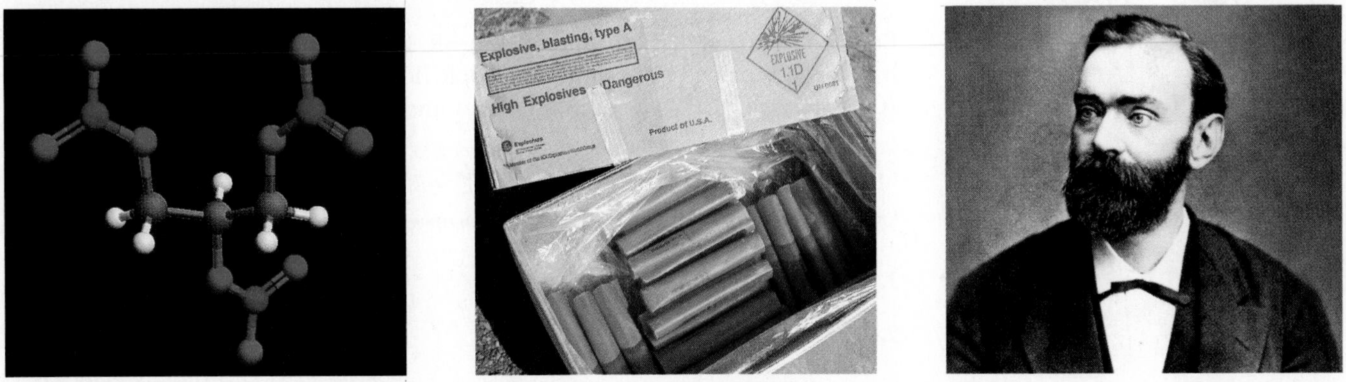

(a) (b) (c)

Figure 11.6 Nitroglycerine and dynamite. (a) The structure of nitroglycerine, $C_3H_5O_9N_3$. A mechanical shock causes the compound to explode to give H_2O, CO_2, and N_2. (b) Dynamite contains nitroglycerine along with a mixture of other substances such as ammonium nitrate, sulfur, and inert solids. (c) The fortune of Alfred Nobel (1833–1896) was built on the manufacture of dynamite and is now used to fund the annual Nobel Prizes. (a, Susan Young; b, C. D. Winters; c, The Bettmann Archive)

Some Chemistry of Alcohols

Alcohols bear some resemblance to water. Like water, alcohols react with alkali metals to give an ionic compound. (R is some organic group such as $—CH_3$, $—C_2H_5$, and so on.)

$$HOH(\ell) + Na(s) \longrightarrow \tfrac{1}{2}\,H_2(g) + NaOH(aq)$$
$$ROH(\ell) + Na(s) \longrightarrow \tfrac{1}{2}\,H_2(g) + NaOR(s)$$

On the other hand, there are major differences between alcohols and water. When a halogen acid, HX, is placed in water, the acid dissociates to give H_3O^+ and X^-. With alcohols, however, the product is a covalent compound called an organic halide. For example, if ethanol is treated with hydrogen bromide under the right conditions, bromoethane is the product.

$$CH_3CH_2OH + HBr \longrightarrow CH_3CH_2Br + H_2O$$

Here the —OH group of the alcohol is substituted by a bromine atom; hence, such reactions are often referred to as *substitution* reactions.

Concentrated sulfuric acid has a great affinity for water. When alcohols (with H and OH on adjacent carbon atoms) are treated with this acid, the alcohol is dehydrated to give the corresponding alkene. Such reactions are often called *elimination* reactions.

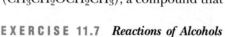

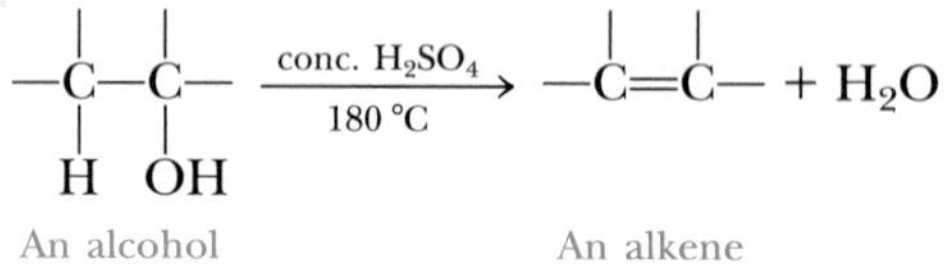

(Notice that this is just the reverse of the reaction used to make ethanol from ethylene.) By modifying the conditions slightly, a type of organic compound called an *ether* can be produced instead of an alkene.

$$2\ ROH \xrightarrow[\text{heat}]{\text{conc. } H_2SO_4} R—O—R + H_2O$$

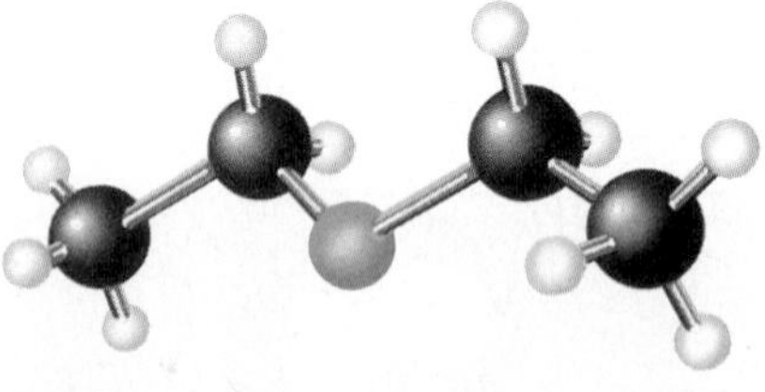

Diethyl ether, $CH_3CH_2OCH_2CH_3$

Ethers have the general formula R—O—R. It is as though both H atoms of water have been replaced by organic groups, R. The best known ether is diethyl ether ($CH_3CH_2OCH_2CH_3$), a compound that until recently was used as an anesthetic.

EXERCISE 11.7 ***Reactions of Alcohols***

Complete each of the following reactions:

1. $CH_3CH_2CH_2OH \xrightarrow{H_2SO_4/180\ °C}$
2. $CH_3CH_2CH_2OH \xrightarrow{HI}$ ■

11.5 COMPOUNDS WITH A CARBONYL GROUP

The object in Figure 11.7 is a breath tester and is used to determine if the concentration of alcohol in exhaled breath is high. The tester relies on the oxidation of alcohol (Figure 4.23). Indeed, a general property of primary and secondary alcohols is that they are reducing agents.

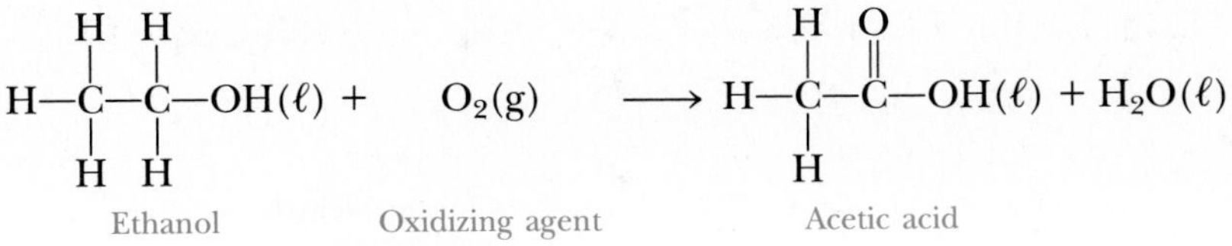

The oxidation of a primary alcohol such as ethanol occurs in two steps: first it is oxidized to a type of compound called an aldehyde and then finally to a carboxylic acid. The air-oxidation of alcohol in wine produces wine vinegar. Because acids have a sour taste, the word vinegar means "sour wine" (from the French *vin aigre*)

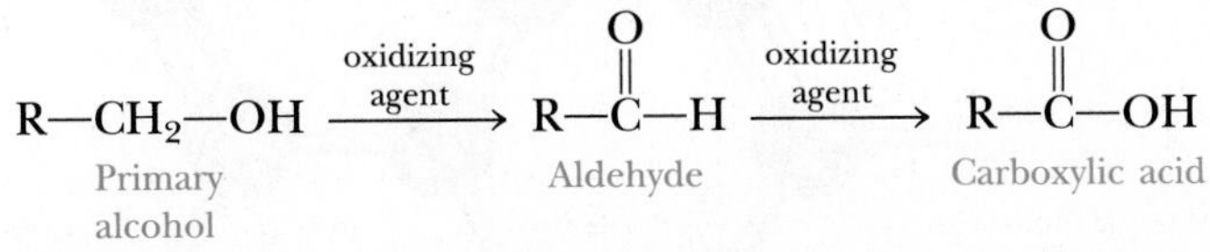

In contrast, oxidation of a secondary alcohol results only in a ketone, a member of another general class of compounds.

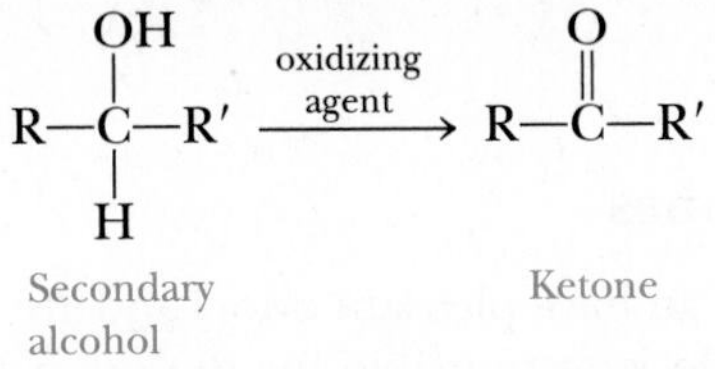

(R and R′ are organic groups. They may be the same or different.)

The oxidizing agents required for these reactions are common reagents, such as $KMnO_4$ and $K_2Cr_2O_7$. Indeed, potassium dichromate is the oxidizing agent used in the breath tester in Figure 11.7.

A look at the preceding reactions suggests a general conclusion: aldehydes, ketones, and carboxylic acids all have a double-bonded carbon-oxygen grouping called a carbonyl group. Aldehydes have one organic group and one H attached to a carbonyl group. Ketones have two organic groups attached. Carboxylic acids incorporate a carbonyl group into a carboxyl group in which the C of C=O is attached to an —OH group and an organic group.

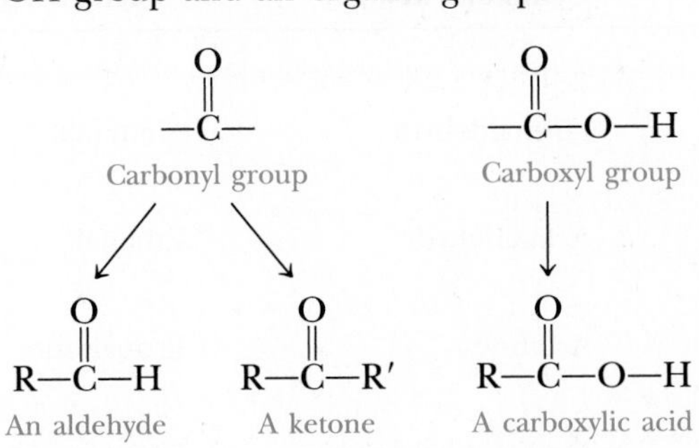

In each case, the carbonyl group is the center of a trigonal-planar arrangement of atoms, so the central carbon atom is sp^2-hybridized. Carbon-oxygen pi bonding is described in Section 10.1.

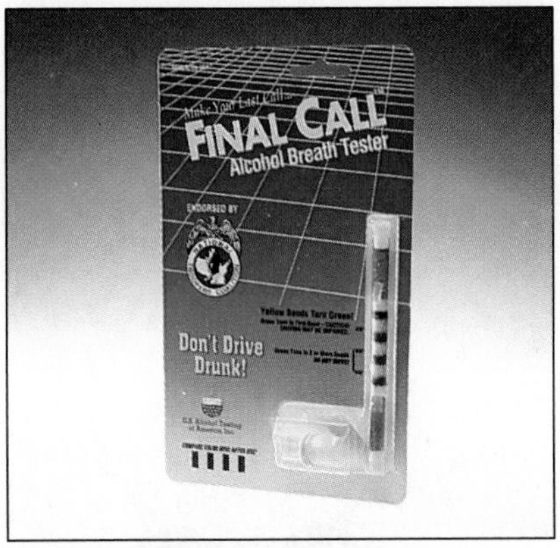

Figure 11.7 A device for testing the breath for the presence of ethanol. The breath tester works because ethanol can be oxidized to acetaldehyde and then to acetic acid by $K_2Cr_2O_7$, potassium dichromate. The yellow-orange dichromate turns green-blue as it is reduced to the chromium(III) ion by ethanol in acid solution. See Figure 4.23. (C. D. Winters)

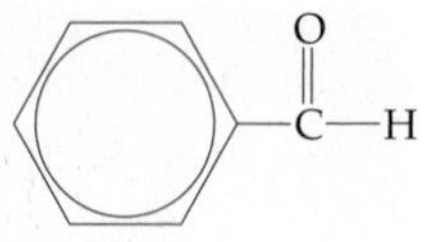

Benzaldehyde,
an aldehyde

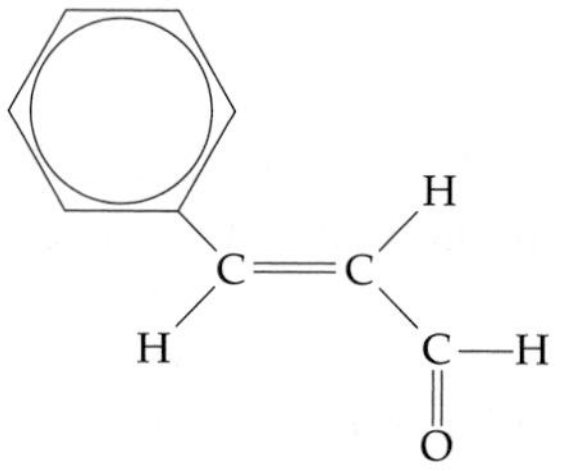

Cinnamaldehyde,
an aldehyde

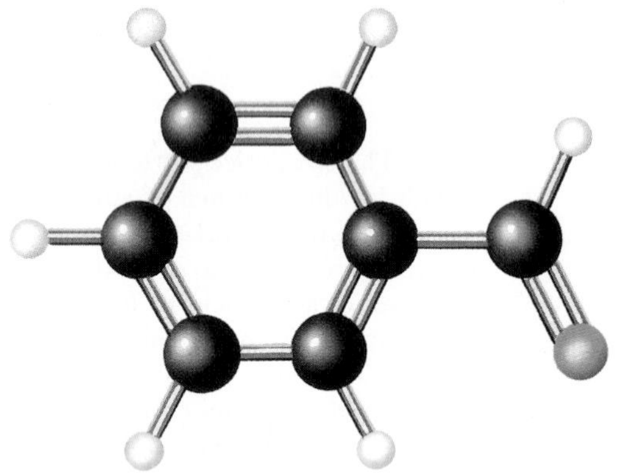

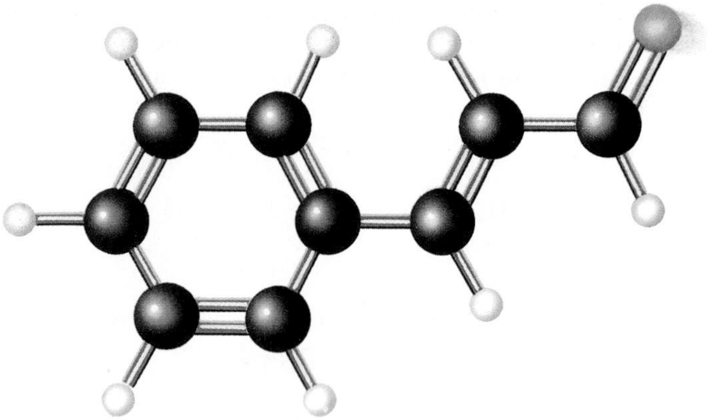

Aldehydes and Ketones

Aldehydes and ketones can have pleasant odors and are often used as the basis of fragrances. Benzaldehyde is responsible for the odor of almonds and cherries, and cinnamaldehyde is found in the bark of the cinnamon tree. The ketone *p*-hydroxyphenyl-2-butanone is responsible for the odor of ripe raspberries (a favorite of the authors of this book).

As with other classes of organic compounds, simple aldehydes generally have common names (Table 11.7). Systematic names, however, are formed by drop-

The odors of almonds and cinnamon are due to aldehydes, whereas the odor of fresh raspberries comes from a ketone. (C. D. Winters)

TABLE 11.7 Simple Aldehydes and Ketones

Structure	Common Name	Systematic Name	bp (°C)
$\mathrm{H\overset{O}{\overset{\|}{C}}H}$	Formaldehyde	Methanal	−21
$\mathrm{CH_3\overset{O}{\overset{\|}{C}}H}$	Acetaldehyde	Ethanal	21
$\mathrm{CH_3\overset{O}{\overset{\|}{C}}CH_3}$	Acetone	Propanone	56
$\mathrm{CH_3\overset{O}{\overset{\|}{C}}CH_2CH_3}$	Methyl ethyl ketone	Butanone	80
$\mathrm{CH_3CH_2\overset{O}{\overset{\|}{C}}CH_2CH_3}$	Diethyl ketone	3-Pentanone	102

ping the final "-e" from the name of the parent alkane and adding "-al." Thus, the systematic name of the compound H_2CO, commonly called formaldehyde, is methanal.

The simplest ketone, CH_3COCH_3, is commonly called acetone. The systematic name, however, is derived by beginning with the name of the alkane that has the same number of carbon atoms. The final "-e" is dropped and "-one" added. Thus, acetone should be named propanone.

Aldehydes and ketones are the oxidation products of primary and secondary alcohols, respectively. It is therefore reasonable that primary and secondary alcohols can be formed by *reducing* the aldehyde or ketone using common reducing agents, such as $NaBH_4$ or $LiAlH_4$.

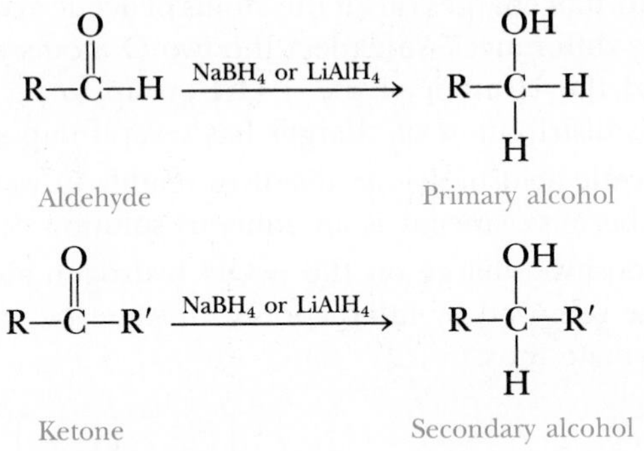

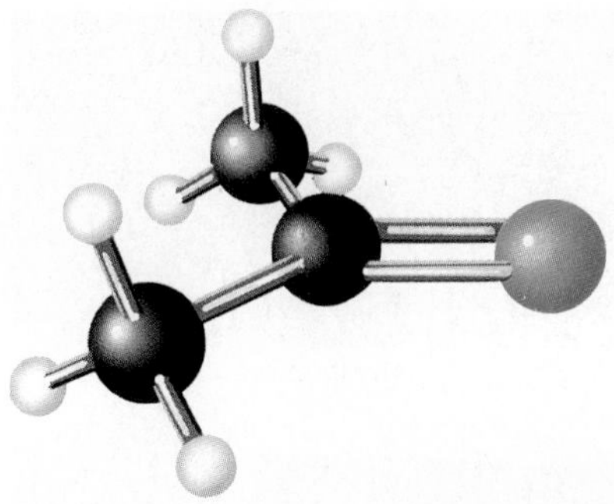

Acetone, CH_3COCH_3

EXERCISE 11.8 ***Structures of Aldehydes and Ketones***

Draw structural formulas for (1) 2-pentanone and (2) butanal. ■

EXERCISE 11.9 ***Names of Aldehydes and Ketones***

Name the following compounds:

1. $CH_3CH_2CH_2\overset{O}{\overset{\|}{C}}CH_2CH_3$ 2. $CH_3CH_2CH_2CH_2\overset{O}{\overset{\|}{C}}H$ ■

EXERCISE 11.10 ***Reactions of Aldehydes and Ketones***

Complete the following reactions:

1. $CH_3CH_2\overset{O}{\overset{\|}{C}}H \xrightarrow{NaBH_4 \text{ or } LiAlH_4} ?$

2. $(CH_3)_2CH\overset{O}{\overset{\|}{C}}H \xrightarrow{Na_2Cr_2O_7} ?$ ■

Carboxylic Acids

Sulfuric acid is the most widely produced inorganic acid, and acetic acid, CH_3CO_2H, is the most important organic acid. About 1.5 billion kilograms of acetic acid is produced annually in the United States. It is widely used to make plastics and synthetic fibers, as a fungicide, and as the starting material for preparing dietary supplements.

The taste of sourdough bread is due to the presence of acetic acid (CH_3CO_2H) and lactic acid ($CH_3CH(OH)CO_2H$). (C. D. Winters)

Acetic acid is produced in bread when leavened by the yeast *Saccharomyces exigus.* Another group of bacteria, *Lactobacillus sanfrancisco,* metabolizes maltose, and it excretes acetic acid and lactic acid (see below), acids that give sourdough bread its peculiar, sour taste.

Carboxylic acids are the end product of alcohol oxidation. For years acetic acid was therefore made by oxidizing the alcohol produced by fermentation. Now, however, acetic acid is generally made by combining carbon monoxide and methanol in the presence of a catalyst.

$$\underset{\text{methanol}}{CH_3OH(\ell)} + CO(g) \longrightarrow \underset{\text{acetic acid}}{CH_3CO_2H(\ell)}$$

Although the formal charges on all the atoms of acetic acid are zero, because of electronegativity differences we expect the two O atoms to be slightly negatively charged, and the H atom of the —OH group to be positively charged (Figure 11.8). This distribution of charges has several important implications:

- The polar acetic acid molecule dissolves readily in water, which you already know because vinegar is an aqueous solution of acetic acid.
- The strong positive charge on the —OH hydrogen atom means that the H can be removed by interaction with water to produce the hydronium and acetate ions.

The most important interaction between acetic acid and water is hydrogen bonding, which is described in Section 13.2.

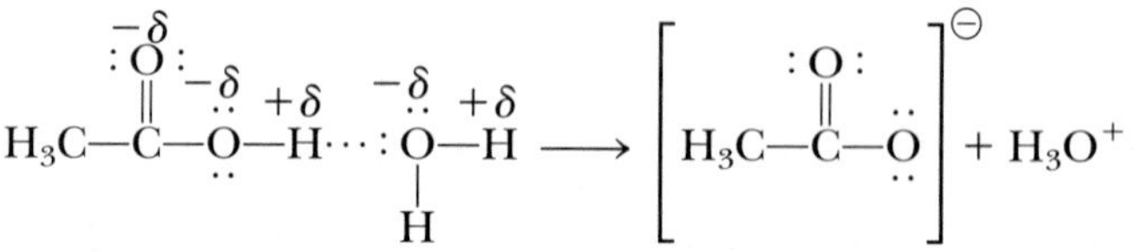

Other common organic acids include formic acid, lactic acid, and oxalic acid (Tables 11.8 and 11.9). All are found naturally. Formic acid is one component of the venom of stinging ants, and oxalic acid occurs in significant concentrations in leafy green plants, such as rhubarb and spinach. Lactic acid is particularly widespread in nature, being found in sour milk, pickles, and sauerkraut, among others. Benzoic acid is found in most berries.

In general the longer chain carboxylic acids have unpleasant odors.

TABLE 11.8 Some Simple Carboxylic Acids

Structure	Common Name	Systematic Name	bp (°C)
$HC(=O)OH$	Formic acid	Methanoic acid	101
$CH_3C(=O)OH$	Acetic acid	Ethanoic acid	118
$CH_3CH_2C(=O)OH$	Propionic acid	Propanoic acid	141
$CH_3(CH_2)_2C(=O)OH$	Butyric acid	Butanoic acid	163
$CH_3(CH_2)_3C(=O)OH$	Valeric acid	Pentanoic acid	187

TABLE 11.9 Some Other Naturally Occurring Carboxylic Acids

Name	Structure	Natural Source
Benzoic acid	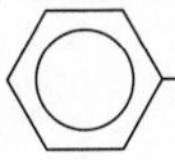$-CO_2H$	Berries
Citric acid	$HO_2C-CH_2-C(OH)(CO_2H)-CH_2-CO_2H$	Citrus fruits
Lactic acid	$CH_3-CH(OH)-CO_2H$	Sour milk
Malic acid	$HO_2C-CH_2-CH(OH)-CO_2H$	Apples
Oleic acid	$CH_3(CH_2)_7-CH{=}CH-(CH_2)_7-CO_2H$	Vegetable oils
Oxalic acid	HO_2C-CO_2H	Rhubarb, spinach, cabbage, tomatoes
Stearic acid	$CH_3(CH_2)_{16}-CO_2H$	Animal fats
Tartaric acid	$HO_2C-CH(OH)-CH(OH)-CO_2H$	Grape juice, wine

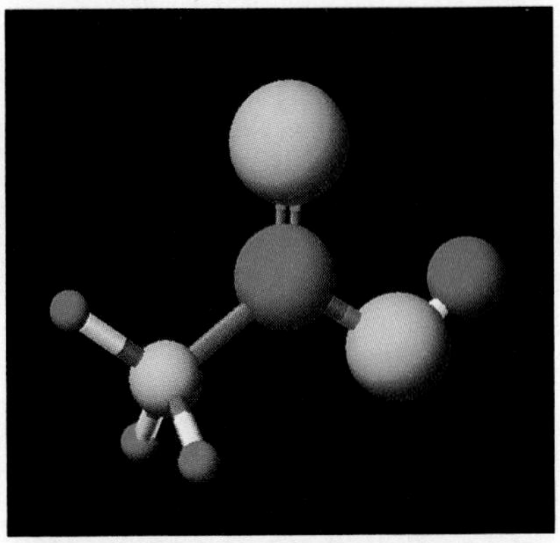

Figure 11.8 Acetic acid, CH_3CO_2H. This computer-generated molecular model of acetic acid shows the calculated partial charges on the atoms of the molecule. A yellow sphere indicates a negative charge, and a red sphere indicates a positive charge. The relative sizes of the spheres indicate the relative size of the charge. Notice that the H atom of the —OH group is positive. This hydrogen atom can interact with water to form a hydronium ion. (Susan Young)

Carboxylic acids often have common names derived from the Latin or other name for the source of the acid (see Table 11.8). Because formic acid is found in ants, its name comes from the Latin word for ant *(formica)*. Butyric acid gives rancid butter its unpleasant odor, and the name is related to the Latin word for butter *(butyrum)*. As seen in Table 11.8, the systematic names of acids are formed by dropping the "e" on the name of the corresponding alkane and adding "oic" or "ic."

A CLOSER LOOK *Muscle Cramps and Hangovers*

The formula for lactic acid, $C_3H_6O_3$, is that of a carbohydrate, a class of compounds that includes the sugars. Fresh milk can become contaminated easily with bacteria that metabolize the milk sugar, lactose, use it for energy, and excrete lactic acid. The acid causes the droplets of fat in the milk to coalesce, and the milk curdles. The manufacture of yogurt and cheese is a controlled version of this process.

When you exercise, your body uses glucose (from glycogen) as an energy source. Glucose metabolism produces pyruvic acid, CH_3COCO_2H, which is burned aerobically (in the presence of O_2) to give carbon dioxide and water. If you need a sudden burst of energy, as in a sprint race, your muscles may be short of oxygen, so the pyruvic acid is degraded anaerobically (without oxygen) to give energy and lactic acid. It is the buildup of this acid in your muscles that makes you feel tired.

Lactic acid is produced in your body as part of your normal metabolism, and it is in turn removed from your body by your liver. If you drink an alcoholic beverage, your liver also metabolizes the alcohol. If you have had too much to drink, however, lactic acid metabolism may not be efficient, allowing the lactic acid to build up in your body and creating a feeling of fatigue.

(See Trevor Smith: *Chem Matters*, February 1989, p. 4.)

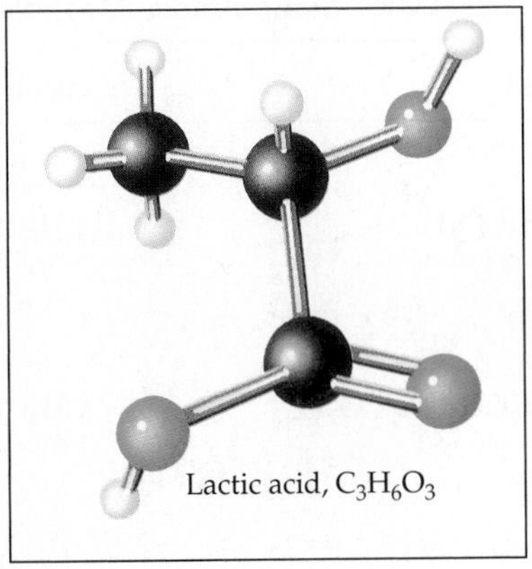

Esters

The reaction of a carboxylic acid and an alcohol is often called esterification.

One of the most important reactions of carboxylic acids occurs with alcohols in the presence of a strong acid and produces an ester (Table 11.10).

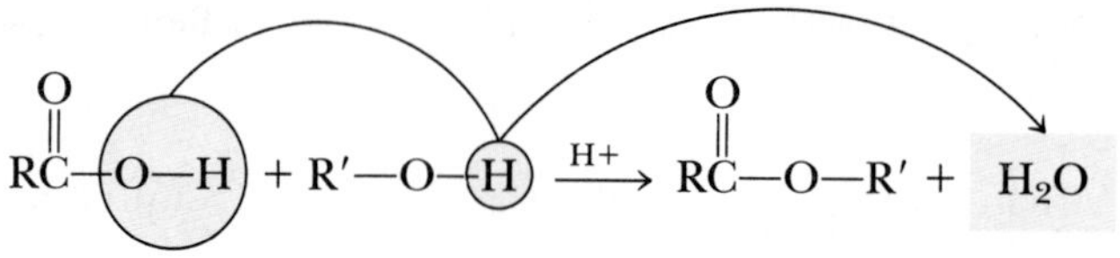

Carboxylic acid   Alcohol   Ester

$$\underset{\text{acetic acid}}{CH_3C(=O)-O-H} + \underset{\text{ethanol}}{CH_3CH_2OH} \xrightarrow{H^+} \underset{\text{ethyl acetate}}{CH_3C(=O)-O-CH_2CH_3} + H_2O$$

The two-part name of an ester is given by (1) the name of the alkyl group from the alcohol and (2) the name of the carboxylate group derived from the acid. In general, ester names are of the form "alkyl carboxylate." In the previous reaction acetic acid has combined with ethanol (commonly called ethyl alcohol) to give ethyl acetate.

Perhaps the most important reaction of esters is their *hydrolysis* (literally, a reaction with water), a reaction that is the reverse of the formation of the ester. This reaction is generally done in the presence of a base such as NaOH and produces the alcohol and a *salt* of the carboxylic acid.

$$\underset{\text{Ester}}{RC(=O)-O-R'} + NaOH \xrightarrow[\text{in water}]{\text{heat}} \underset{\text{Carboxylate salt}}{RC(=O)-O^-Na^+} + \underset{\text{Alcohol}}{R'OH}$$

$$\underset{\text{ethyl acetate}}{CH_3C(=O)-O-CH_2CH_3} + NaOH \xrightarrow[\text{in water}]{\text{heat}} \underset{\text{sodium acetate}}{CH_3C(=O)-O^-Na^+} + \underset{\text{ethanol}}{CH_3CH_2OH}$$

TABLE 11.10 Some Acids, Alcohols, and Their Esters

Acid	Alcohol	Ester	Odor of Ester
CH_3CO_2H acetic acid	$CH_3CH(CH_3)CH_2CH_2OH$ 3-methyl-1-butanol	$CH_3C(=O)OCH_2CH_2CH(CH_3)CH_3$ 3-methylbutyl acetate	Banana
$CH_3CH_2CH_2CO_2H$ butanoic acid	$CH_3CH_2CH_2CH_2OH$ 1-butanol	$CH_3CH_2CH_2C(=O)OCH_2CH_2CH_2CH_3$ butyl butanoate	Pineapple
$CH_3CH_2CH_2CO_2H$ butanoic acid	 benzyl alcohol	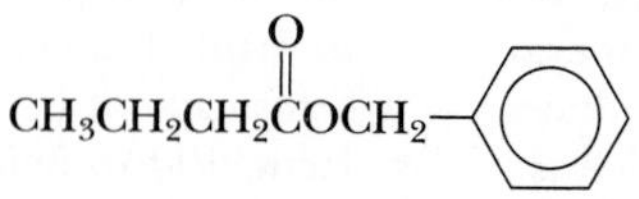 benzyl butanoate	Rose

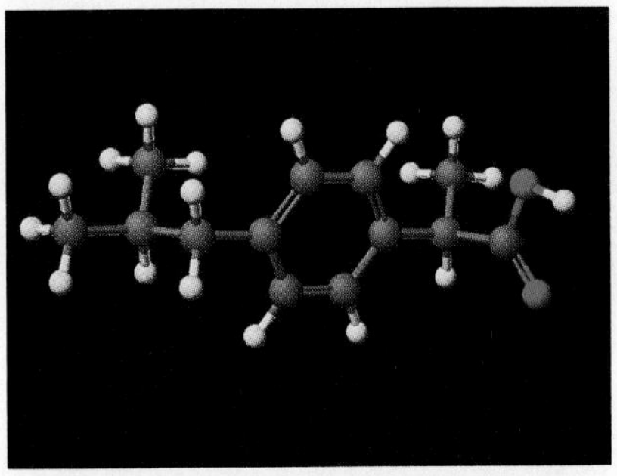

Ibuprofen (above) and aspirin (page 220) are common analgesics. Both are carboxylic acids, and aspirin is an ester of salicylic acid.

The carboxylic acid can be recovered from this process if the sodium salt, for example, is treated with a strong acid such as HCl.

$$CH_3\overset{\overset{O}{\|}}{C}{-}O^-Na^+(aq) + HCl(aq) \longrightarrow CH_3\overset{\overset{O}{\|}}{C}{-}OH(aq) + NaCl(aq)$$

sodium acetate — acetic acid

The hydrolysis of esters is sometimes called a *saponification reaction* (from the Latin word *sapo,* meaning soap) because the hydrolysis of a special type of ester, a fat, produces a soap (see Section 11.6).

Unlike the acids from which they are derived, esters often have pleasant odors (Table 11.10). Typical examples are methyl salicylate or "oil of wintergreen" and benzyl acetate. Methyl salicylate is derived from salicylic acid, the parent compound of aspirin.

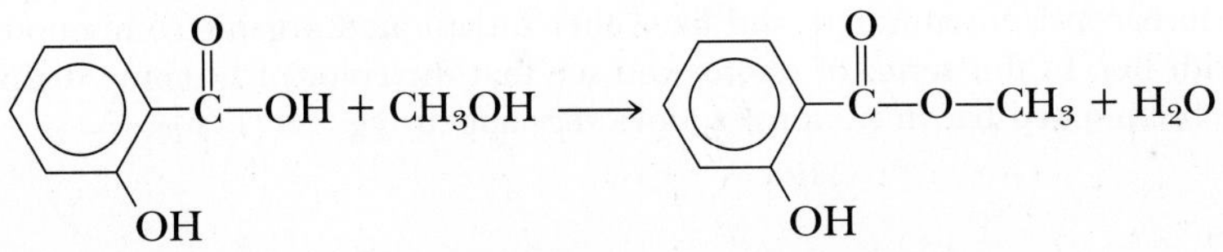

salicylic acid — methanol — methyl salicylate (oil of wintergreen)

EXERCISE 11.11 ***Naming Acids and Esters***

Give a name for each of the following:

1. $CH_3(CH_2)_4\overset{\overset{O}{\|}}{C}OH$ 2. $CH_3\overset{\overset{O}{\|}}{C}OCH_2CH_2CH_3$ ■

EXERCISE 11.12 ***Reactions of Acids***

Complete the following reactions:

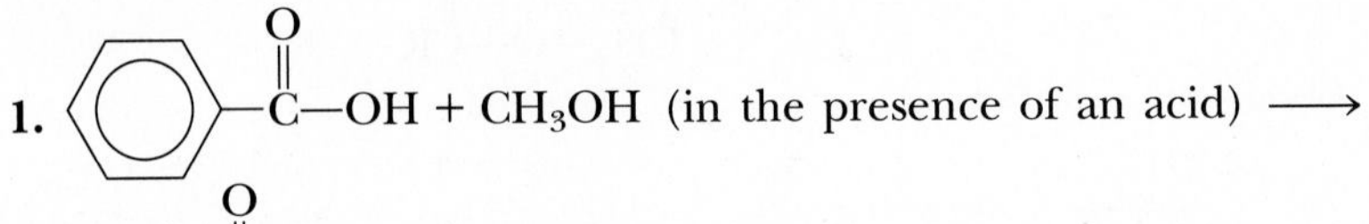

2. $CH_3\underset{\underset{CH_3}{|}}{C}H\overset{\overset{O}{\|}}{C}OH + KOH \longrightarrow$ ■

11.6 FATS AND OILS

Fats and oils are structurally and chemically similar; they are triesters of long-chain "fatty acids" with 1,2,3-propanetriol (glycerol). The —R groups, which can be the same or different within the same molecule, can be saturated or

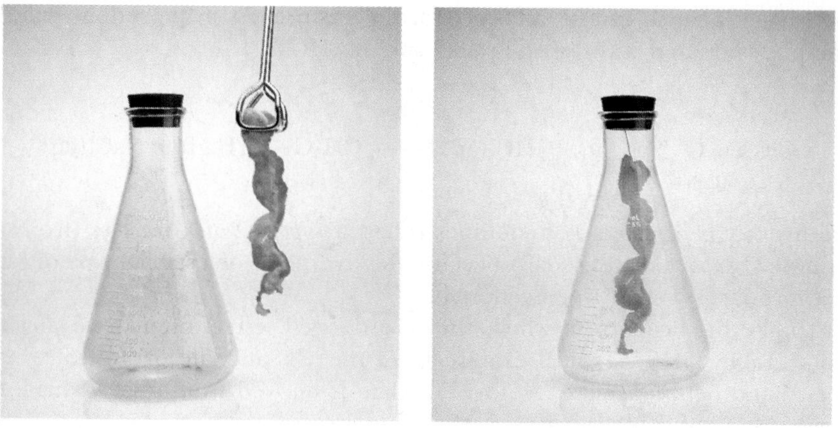

The fat in bacon is unsaturated, and like other unsaturated organic compounds, it reacts with Br_2. In this series of photos you see that the color of bromine disappears, and the brominated bacon no longer looks very appetizing. (C. D. Winters)

The word "fat" is used generically, like "alcohol" or "acid." All fats are triesters of glycerol, but they are often subdivided into two types depending on whether they are solids (fats) or liquids (oils) at room temperature.

unsaturated organic groups, that is, they may contain one or more carbon-carbon double bonds. If the group is unsaturated it may be monounsaturated or polyunsaturated, depending on whether one or more double bonds occur. Some common fatty acids are given in Table 11.11.

The reaction of glycerol with nitric acid to give nitroglycerin is similar to the glycerol/fatty acid reaction.

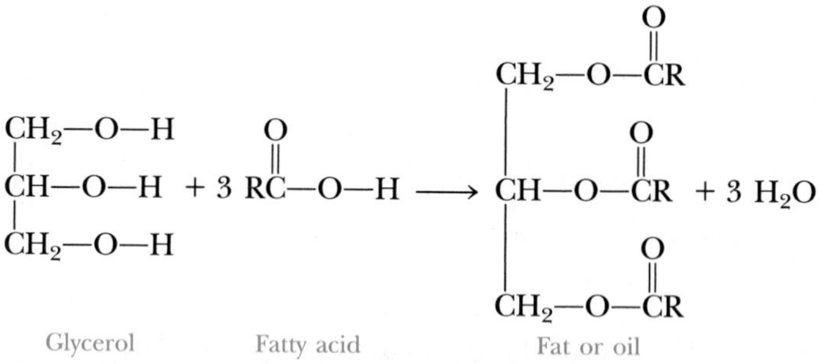

TABLE 11.11 Common Fatty Acids

Name	*Number of C Atoms*	*Formula*
Saturated Acids		
Butyric	C_4	$CH_3CH_2CH_2CO_2H$
Lauric	C_{12}	$CH_3(CH_2)_{10}CO_2H$
Myristic	C_{14}	$CH_3(CH_2)_{12}CO_2H$
Palmitic	C_{16}	$CH_3(CH_2)_{14}CO_2H$
Stearic	C_{18}	$CH_3(CH_2)_{16}CO_2H$
Unsaturated Acids		
Oleic	C_{18}	$CH_3(CH_2)_7CH{=}CH(CH_2)_7COOH$
Linolenic	C_{18}	$CH_3CH_2CH{=}CHCH_2CH{=}CHCH_2CH{=}CH(CH_2)_7COOH$

Percentage of Calories from Fat

French fries	45
Ice cream	45
Chocolate chips	56
Peanut butter	77
Hot dogs	82
Cream cheese	90
Olives	98
Butter and margarine	100

Recall (Chapter 6) that 1 Calorie = 1000 calories = 4184 J.

Whether the fatty acid is saturated or unsaturated has a great effect on the physical properties of the triester. In general, saturated fatty acids lead to fats (solids), and unsaturated fatty acids are found in oils (liquids). The reason for this is that the saturated stearic acid chain is flexible because all the carbon atoms are singly bonded to their neighbors (Figure 11.9). As a consequence the C_{18} chains can roll up into little balls, making the triester molecules a compact package. These can then pack close to one another, resulting in a solid. In contrast, the presence of a double bond in a fatty acid such as oleic acid (Figure 11.10) means the chain is less flexible. Oleic acid cannot bend around the double bond in the middle of the chain and so forms triesters that are less compact than those of saturated fatty acids. The result is an oil.

Plant seeds often contain oils that serve as a food source for the developing embryo of the plant. Olive oil is about 75% monounsaturated fatty acid, principally oleic acid. Linolenic acid, which has three carbon-carbon double bonds, is found in corn, cottonseeds, and soybeans, and is especially abundant in rapeseeds. Rapeseed oil is used in cooking oils and margarine.

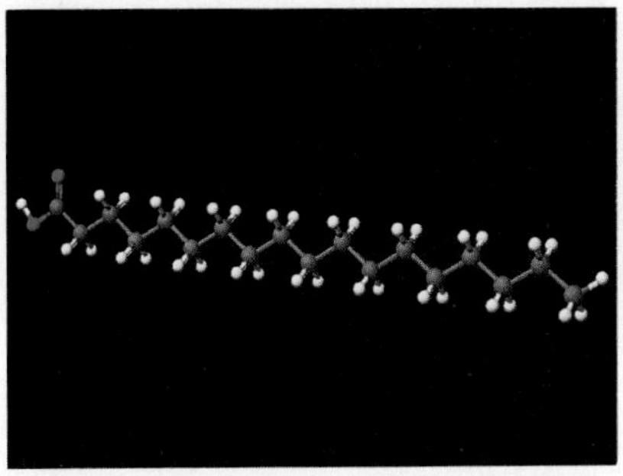

Figure 11.9 The structure of stearic acid, a saturated fatty acid.

Fat in Food

The amount of fat in a food is important because the fat supplies about 38 kJ (or 9 Cal) of energy per gram. In general, Americans are overweight and consume too much fatty food. In fact, fat makes up about 40% of the energy intake in the average American diet, although nutritionists think it should be no higher than about 30%. Unfortunately, a large percentage of the calorie content of many of our favorite foods (see table in the margin) is from fat.

It is also important from a dietary viewpoint whether the fatty acid groups in foods are saturated or unsaturated. Not only is it recommended that you consume no more than 30% of your calories in the form of fat, but nutritionists also believe that no more than a third of this should be from saturated fat. So, if it is better to consume unsaturated fats than saturated ones, why then do food companies hydrogenate oil to reduce their unsaturation (Section 11.2)?

There are several reasons for this. First, the double bonds in unsaturated fatty acids are reactive, and oxygen can attack the molecule at this point. When an oil is oxidized, unpleasant odors and flavors develop. Therefore, reducing the number of double bonds reduces the likelihood of spoiling. Second, hydrogenating an oil makes it less liquid. Food processors often need a solid fat to improve a food's quality. (For example, if liquid vegetable oil were used in a cake icing, the icing would slide off the cake.) Rather than use animal fat, which also contains cholesterol, the manufacturer turns to a hydrogenated or partially hydrogenated oil.

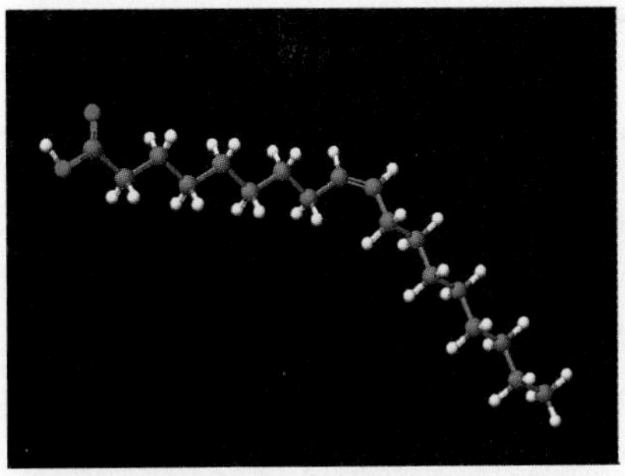

Figure 11.10 The structure of oleic acid, an unsaturated fatty acid. Notice the presence of a carbon-carbon double bond in the middle of the chain, which means that the chain cannot bend around these carbon atoms. The result is that glycerol esters of this acid are generally oils. (Susan Young)

Soap

Hydrolysis of an ester in the presence of a base breaks the ester down into its component parts—an alcohol and a salt of the acid. This reaction has often been called a **saponification reaction,** because the treatment of fat or oil gives glycerol and a salt of a long-chain carboxylic acid such as sodium stearate. Compounds such as sodium stearate that have an ionic end and a long-chain hydrocarbon end are soaps. As you will see in Section 14.4, the ionic end allows them to interact with water, while the hydrocarbon end enables them to mix with oily or greasy substances.

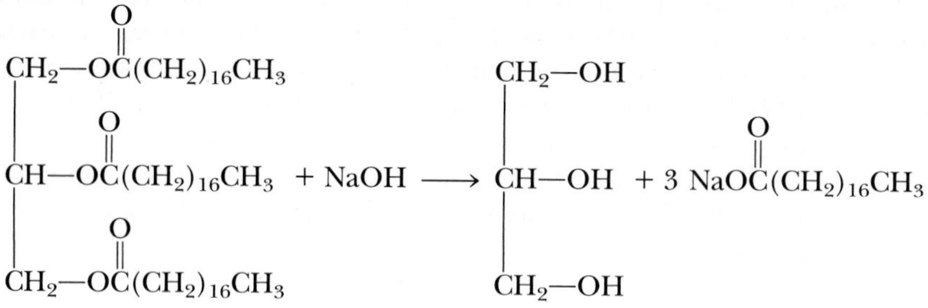

Glyceryl stearate, a fat — Glycerol — Sodium stearate, a soap

EXERCISE 11.13 ***Fats and Oils***

Draw the structure for glyceryl tripalmitate. When this triester is saponified, what are the products? ■

11.7 AMINES AND AMIDES

Amines are derivatives of ammonia in which one or more of the H atoms of NH_3 has been replaced with an organic group.

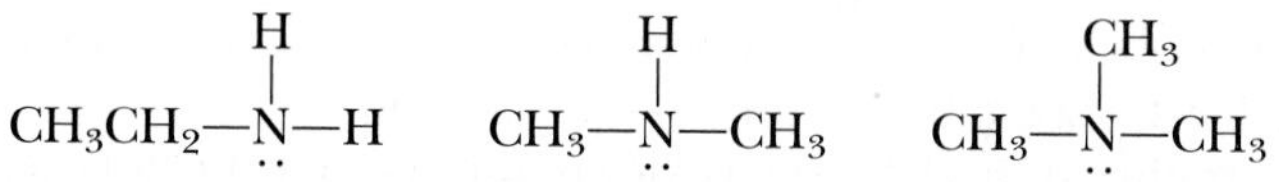

Primary amine ethylamine — Secondary amine dimethylamine — Tertiary amine trimethylamine

Structurally, they are like ammonia in that all have a trigonal-pyramidal molecular geometry around the N atom with approximately tetrahedral bond angles, and all have a lone pair of electrons on the N atom. A consequence of this lone pair is that amines, like ammonia, are bases.

$$RNH_2(aq) + H_2O(\ell) \longrightarrow RNH_3^+(aq) + OH^-(aq)$$
$$(CH_3)_3N(aq) + HCl(aq) \longrightarrow [(CH_3)_3NH^+]Cl^-(aq)$$

Amines usually have very offensive odors. You know what the odor is if you have ever smelled decaying fish. Two appropriately named amines, putrescine and cadaverine, add to the odor of urine and are present in bad breath.

$H_2N—CH_2CH_2CH_2CH_2—NH_2$
Putrescine
(1,4-butanediamine)

$H_2N—CH_2CH_2CH_2CH_2CH_2—NH_2$
Cadaverine
1,5-pentanediamine

Methylamine, CH_3NH_2

A few amines are water-soluble, but most are not. All amines are bases, however, and react with acids to give salts, many of which are water-soluble. For example, the amine, procaine, is very poorly soluble in water.

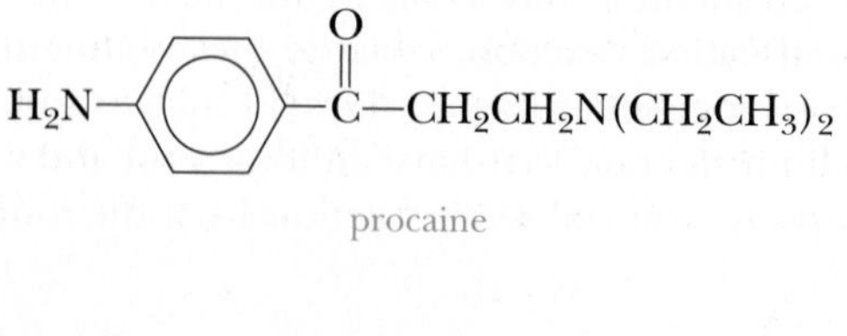

procaine

Procaine reacts with hydrochloric acid, however, to give a salt of the type $[RNH_3^+]Cl^-$, and the resulting salt, procaine hydrochloride, is very water-soluble (1 g dissolves in 1 mL of water). Procaine hydrochloride is sold as a local anesthetic and is commonly known by its widely used brand name, Novocain.

Just as an ester can be viewed as being derived from a carboxylic acid and an alcohol, an **amide** can be said to be derived from a carboxylic acid and an amine.

$$\underset{\text{Carboxylic acid}}{R\overset{\overset{\displaystyle O}{\|}}{C}\text{—}OH} + \underset{\text{Amine}}{H\text{—}NR_2} \longrightarrow HOH + \underset{\text{Amide}}{R\overset{\overset{\displaystyle O}{\|}}{C}\text{—}NR_2}$$

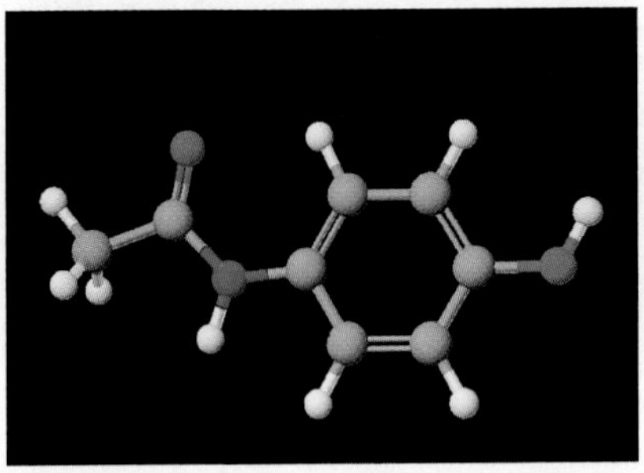

Tylenol, N-acetyl-p-aminophenol

The amide grouping is important in some synthetic polymers (Section 11.8) and in many naturally occurring compounds, especially in proteins. You have already seen it in saccharin (page 495), and it is an important part of N-acetyl-*p*-aminophenol, an analgesic known by the generic name acetaminophen and sold under the names Datril, Momentum, and Tylenol, among others. This compound was apparently discovered by accident when a common organic compound called acetanilide (like acetaminophen but without the —OH group) was mistakenly put into a prescription for a patient. Acetanilide acts as an analgesic, but it can be toxic. Placing an —OH group *para* to the amide group makes the compound nontoxic and an effective headache remedy.

Accidental discoveries have always played a significant role in chemistry. Artificial sweeteners (saccharin, cyclamate, and aspartame), Teflon, and polyethylene were discovered by accident as was the use of lithium carbonate for the treatment of manic depression. See R. M. Roberts: *Serendipity*. John Wiley and Sons, Inc., 1989.

11.8 SYNTHETIC ORGANIC POLYMERS

It is impossible for us to get through a day without using a dozen or more synthetic organic **polymers.** The word "polymer" means "many parts" (from the Greek, *poly* and *meros*). Polymers are made by chemically joining together many small molecules into one giant molecule or macromolecule with molecular weights ranging from thousands to millions. The small molecules used to synthesize polymers are called **monomers.**

An excellent discussion of the polymer industry is given in P. J. Chenier: *Survey of Industrial Chemistry*, 2nd ed., New York, VCH Publishers, 1992.

In spite of their familiarity, the extensive use of synthetic polymers is a recent development. Synthetic resins such as Bakelite, fibers such as rayon, and plastics such as celluloid were made early in this century. Most of the products with which you are familiar, however, became available only in the last half of the 20th century. In 1976, synthetic polymers outstripped steel as the most widely used material in the United States.

The polymer industry uses many different ways to classify polymers. For example, they can be classified by their response to heating. **Thermoplastics** undergo reversible changes when heated and cooled. When heated repeatedly, they soften and flow; when cooled, they harden again. Polyethylene and polystyrene are thermoplastics. **Thermosetting plastics** are plastic when first heated, but they form a highly cross-linked structure when heated further, and when reheated, they cannot be softened and reformed without extensive degradation. The Formica used for kitchen counter tops is an example of a thermosetting plastic.

Another classification of polymers depends on their end uses as **plastics, fibers, elastomers, coatings,** and **adhesives.** This classification is the basis of the production data published regularly by *Chemical and Engineering News* (Table 11.12). To give you an idea of the great size of the industry, the average produc-

TABLE 11.12 Production of Synthetic Polymers

Category	1993 Production (millions of kilograms)
Plastics	
Thermosetting resins	3,117
(e.g., polyester: 574)	
Thermoplastic resins	23,846
(e.g., high-density polyethylene: 9912)	
Synthetic rubber (elastomers)	1,993
(e.g., styrene butadiene: 817)	
Synthetic fibers	4,221
(e.g., nylon: 2664)	
TOTAL	33,177

tion of synthetic polymers in the United States is approaching 150 kg per person annually. Approximately 80% of the organic chemical industry is devoted to the production of synthetic polymers, and about half of the top 50 chemicals produced in the United States are used in making plastics, fibers, and synthetic rubber.

Finally, synthetic polymers can be also classified according to their method of synthesis. **Addition polymers** are made by directly joining monomer units. **Condensation polymers** are made by combining monomer units in such a way that a small molecule, usually water, is split out between them. We adopt this approach in this textbook.

Addition Polymers

Addition polymerization, also called **chain growth polymerization,** is characterized by the fact that intermediates in the process are often free radicals or similar highly reactive species that cannot be isolated. Once polymer growth has started, monomer units add to the growing chains very rapidly; the molecular weight of the chain builds up quickly.

There are several major types of thermoplastic polymers—high- and low-density polyethylene, polypropylene, polystyrene, and polyvinyl chloride, among others—all of which are addition polymers (Table 11.13). Over 23 million kilograms of these polymers were produced in 1993.

Polyethylene

The monomer for addition polymers normally contains one or more double bonds. The simplest monomer of this group is ethylene, C_2H_4. When ethylene is heated to between 100 and 250 °C at a pressure of 1000 to 3000 atm in the presence of a catalyst, polymers with molecular weights up to several million may be formed. The polymerization of ethylene begins with breaking the pi bond of the carbon-carbon double bond to leave an unpaired electron that is a reactive

TABLE 11.13 Ethylene Derivatives That Undergo Addition Polymerization

Formula	Monomer Common Name	Polymer Name (Trade Names)	Uses	U.S. Polymer Production (Metric tons/Yr)*
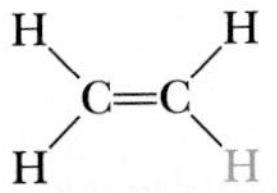	Ethylene	Polyethylene (Polythene)	Squeeze bottles, bags, films, toys and molded objects, electric insulation	7 million
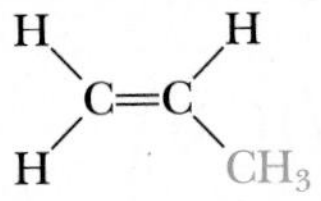	Propylene	Polypropylene (Vectra, Herculon)	Bottles, films, indoor–outdoor carpets	1.2 million
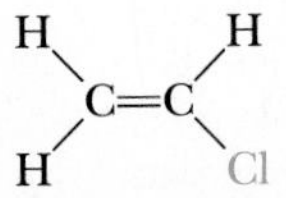	Vinyl chloride	Poly(vinyl chloride) (PVC)	Floor tile, raincoats, pipe	1.6 million
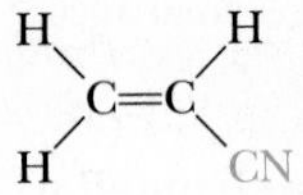	Acrylonitrile	Polyacrylonitrile (Orlon, Acrilan)	Rugs, fabrics	0.5 million
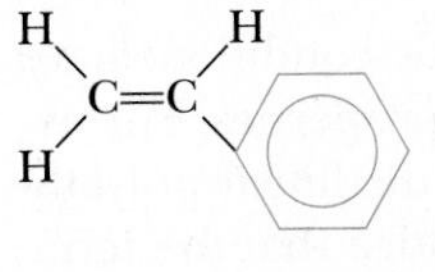	Styrene	Polystyrene (Styrene, Styrofoam, Styron)	Food and drink coolers, building material insulation	0.9 million
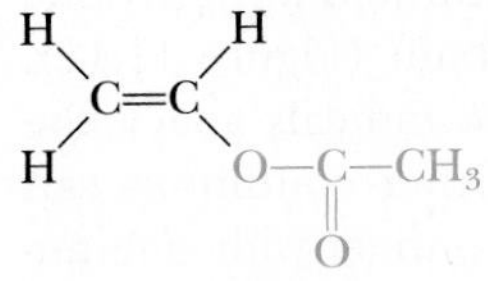	Vinyl acetate	Poly(vinyl acetate) (PVA)	Latex paint, adhesives, textile coatings	200,000
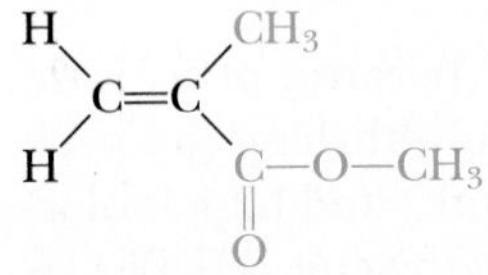	Methyl methacrylate	Poly(methyl methacrylate) (Plexiglas, Lucite)	High-quality transparent objects, latex paints, contact lenses	200,000
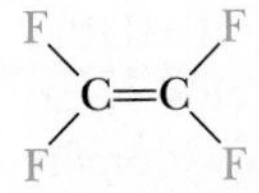	Tetrafluoroethylene	Polytetrafluoroethylene (Teflon)	Gaskets, insulation, bearings, pan coatings	6,000

*One metric ton = 1000 kg

site at each end of the molecule. This step, called the **initiation** step of the polymerization, can be accomplished with chemicals such as organic peroxides that are unstable and break apart to give free radicals (· OR). These react with ethylene to produce new free radicals.

An organic peroxide, RO—OR, produces free radicals, RO · , each with an unpaired electron. See page 412.

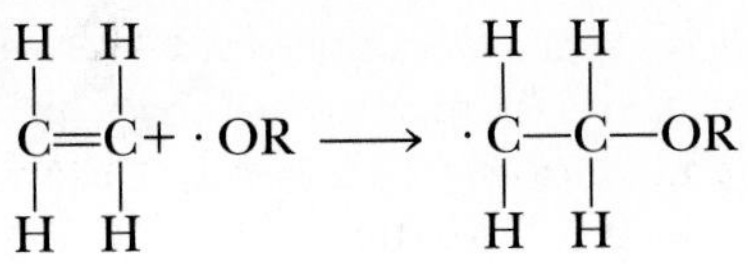

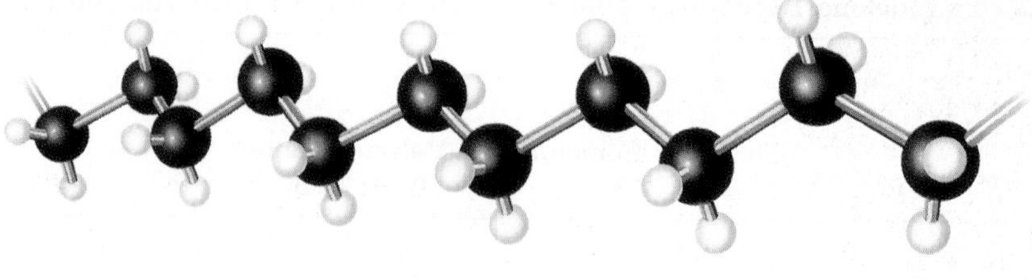

Figure 11.11   A model of linear poly-ethylene.

The growth of the polyethylene chain begins as the initial free radical forms a bond to another ethylene molecule, leaving another unpaired electron to bond with yet another ethylene molecule. For example,

$$\underset{\substack{H \ H \\ | \ | \\ H \ H}}{C{=}C} + \cdot \underset{\substack{H \ H \\ | \ | \\ H \ H}}{C{-}C}{-}OR \longrightarrow \cdot \underset{\substack{H \ H \ H \ H \\ | \ | \ | \ | \\ H \ H \ H \ H}}{C{-}C{-}C{-}C}{-}OR \xrightarrow{n\ CH_2{=}CH_2} \left(\underset{\substack{H \ H \\ | \ | \\ H \ H}}{C{-}C}\right)_n$$

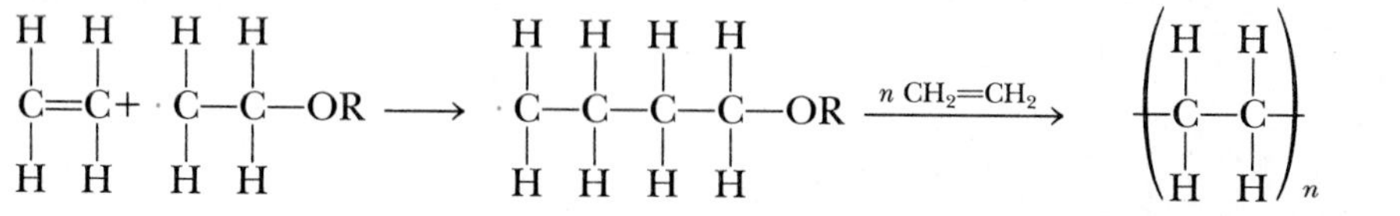

polyethylene
n ranges from 1000
to 50,000

In the process, the unsaturated hydrocarbon monomer, ethylene, is changed to a saturated hydrocarbon polymer, polyethylene.

Polyethylenes formed under various pressures and catalytic conditions have different molecular structures and hence different physical properties. For example, chromium oxide as a catalyst yields almost exclusively the linear polyethylene—a polymer with no branches on the carbon chain. Notice that the tetrahedral arrangement of bonds around each carbon in the saturated polyethylene chain creates a zigzag arrangement of bonds along the chain (Figure 11.11).

If ethylene is heated to 230 °C at a high pressure, free radicals attack the chain at random positions, causing irregular branching. Other conditions can lead to cross-linked polyethylene, in which the branches connect with one another (Figure 11.12).

Polyethylene is the world's most widely used polymer because of its wide range of properties (Figure 11.13). Long, linear chains of polyethylene can pack closely together and give a material of high density (0.97 g/mL) and high molecular weight. This material, referred to as high-density polyethylene (HDPE), is hard, tough, and rigid. The plastic milk bottle is a typical application of HDPE.

If the chains of polyethylene branch, they cannot pack together closely, so the resulting material has a lower density (0.92 g/mL) and is called low-density

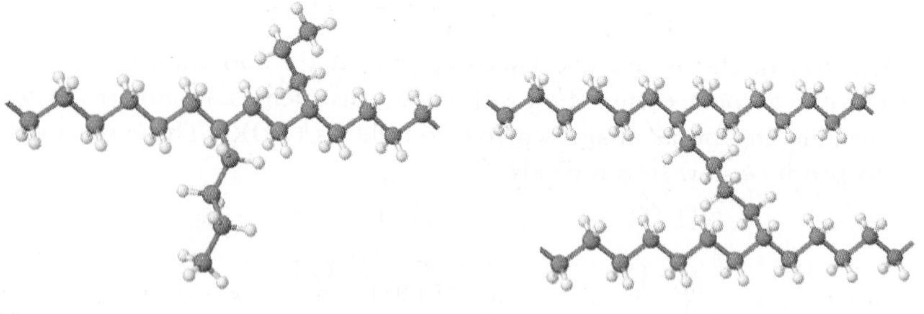

Figure 11.12   Models of (a) branched low-density polyethylene (LDPE) and (b) cross-linked polyethylene (CLPE).

(a)   (b)

(a)

(b)

Figure 11.13 Polyethylene. (a) Poly-ethylene film is used in wrappings. It is produced by extruding the molten plastic through a ring-like gap and inflating the film like a balloon.
(b) Some common products packaged in high density polyethylene. The symbol for recyclable high density poly-ethylene materials has a 2 in the center and the letters HDPE under the symbol.   (a, The Stock Market; b, C. D. Winters)

polyethylene (LDPE). This material is soft and flexible, so materials such as sandwich bags are made of LDPE. Finally, if the linear chains of polyethylene are treated so as to cause cross links to form between them a cross-linked polyethyl-ene (CLPE) results. Cross-linked polymers are very tough materials; the plastic caps on milk bottles are made from CLPE.

In 1993, the United States chemical indus-try produced 5470 million kg of LDPE and 4500 million kg of HDPE. This represents about 30% of the total synthetic polymer production.

Polystyrene, Polyvinyl Chloride, and Other Addition Polymers

If one of the H atoms of ethylene is changed to another group, addition poly-mers can still be made, but they have different properties depending on the nature of the group labeled X (see Table 11.13).

$$n \begin{array}{c} H \\ \backslash \\ C \\ / \\ H \end{array}\!\!=\!\!\begin{array}{c} H \\ / \\ C \\ \backslash \\ X \end{array} \longrightarrow \left(\begin{array}{cc} H & H \\ | & | \\ C\!-\!C \\ | & | \\ H & X \end{array}\right)_n$$

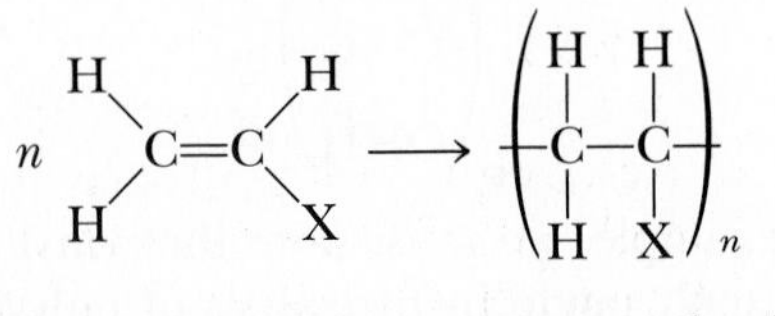

Polystyrene is made by polymerizing styrene, and n is about 5700.

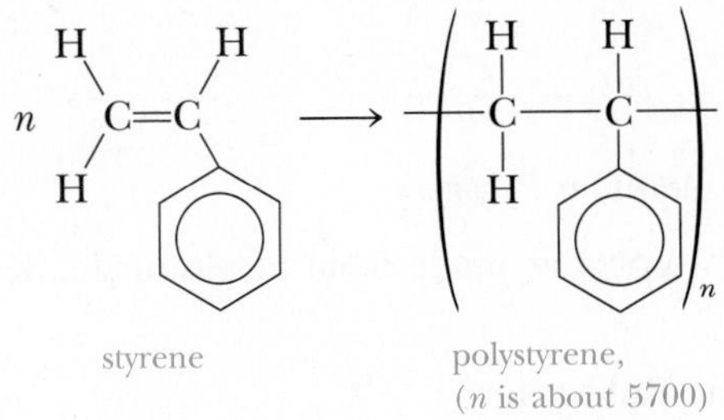

styrene

polystyrene,
(n is about 5700)

The resulting polymer is a clear, hard, colorless solid at room temperature, but it can be molded easily at 250 °C. More than 2 billion kilograms of polystyrene are produced in the United States each year to make food containers, toys, electric parts, insulating panels, appliance and furniture components, and many other items. A major use of polystyrene is in the production of Styrofoam by a process known as "expansion molding." In this process, polystyrene beads are placed in a mold and heated with steam or hot air. The beads, which are 0.25 to 1.5 mm in diameter, contain 4% to 7% by weight of a low-boiling liquid such as pentane. The steam causes the low-boiling liquid to vaporize and expand the beads; as the foamed particles expand, they are molded in the shape of the mold cavity. Styrofoam is used for egg cartons, meat trays, coffee cups, and packing material (Figure 11.14).

Figure 11.14 Some common consumer articles made of polystyrene. (C. D. Winters)

Poly(vinyl chloride) (PVC), used in floor tile, garden hoses, plumbing pipes, and trash bags, is made from vinyl chloride (see Table 11.13), which has a chlorine atom substituted for one of the hydrogen atoms in ethylene (Figure 11.15).

The numerous variations in substituents, length, branching, and cross linking make it possible to produce a variety of properties for each type of addition polymer. Chemists and chemical engineers can fine-tune the properties of the polymer to match the desired properties by appropriate selection of monomer and reaction conditions, thus accounting for the widespread and growing use of polymers.

EXAMPLE 11.2 *Addition Polymers*

Draw the structural formula of the repeating unit for polyvinyl alcohol (Figure 11.16).

Solution The monomer for this polymer is vinyl alcohol, the structure of which can be drawn by analogy with those of other vinyl derivatives in Table 11.13.

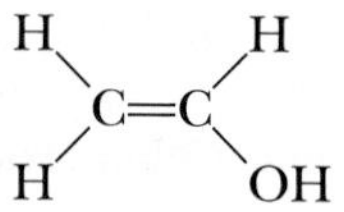

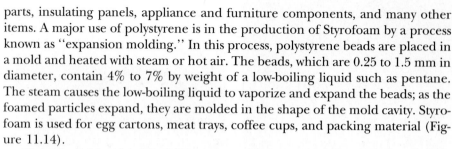

The repeating unit in the polymer has the structure drawn below.

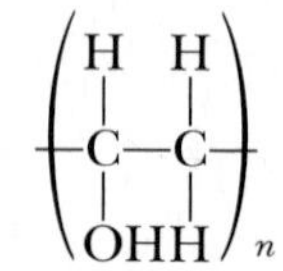

As a footnote to this example, we must note that vinyl alcohol does not exist. Polyvinyl alcohol is actually made by hydrolysis of polyvinyl acetate (a polymer used in paints and adhesives), which comes from the polymerization of vinyl acetate.

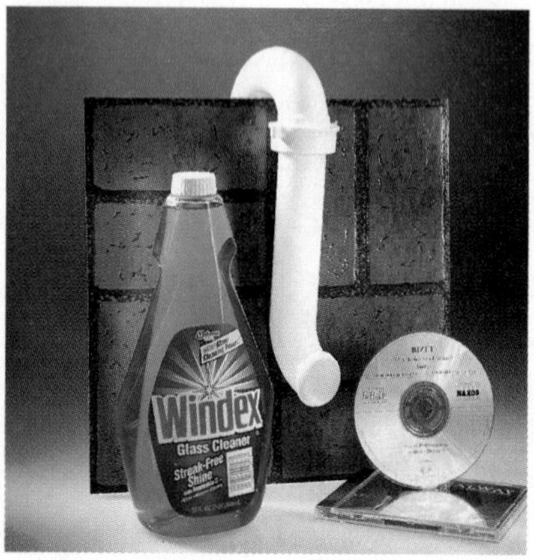

Figure 11.15 Some products made of polyvinyl chloride. The symbol for recyclable PVC materials has a 3 in the center of the triangle of curved arrows with a V under the symbol. (C. D. Winters)

EXERCISE 11.14 *Addition Polymers*

Draw three units of a polymer made from acrylonitrile. See Table 11.13. ■

Natural and Synthetic Rubbers

Natural rubber was first introduced into Europe in 1740, but it remained a curiosity until 1823 when Charles Mackintosh invented a method of waterproofing cotton cloth with an aqueous suspension of rubber. The mackintosh, as rain coats are even now called, became extremely popular despite a major problem: natural rubber is notably weak, and it is thermoplastic, soft and tacky when warm but brittle at low temperature. In 1839, after five years of work on natural rubber, the American inventor, Charles Goodyear (1800–1860), discovered that heating gum rubber with sulfur produces a material that is no longer sticky but is still elastic, water-repellent, and resilient.

Vulcanized rubber, as the type of rubber Goodyear discovered is now known, contains short chains of sulfur atoms that bond together the polymer chains of the natural rubber.

$$-CH_2-\underset{\underset{CH_3}{|}}{C}=CH-CH_2-CH_2-\underset{\underset{CH_3}{|}}{C}=CH-CH_2-$$

$$-CH_2-\underset{\overset{CH_3}{|}}{C}=CH-CH_2-CH_2-\underset{\overset{CH_3}{|}}{C}=CH-CH_2- \quad \xrightarrow{\text{heat with sulfur}}$$

$$-CH_2-\underset{\overset{CH_3}{|}}{C}=CH-\underset{\underset{S}{|}}{CH}-CH_2-\underset{\overset{CH_3}{|}}{C}=CH-CH_2-$$

$$-CH_2-\underset{\underset{CH_3}{|}}{C}=CH-\underset{\overset{S}{|}}{CH}-CH_2-\underset{\underset{CH_3}{|}}{C}=CH-CH_2-$$

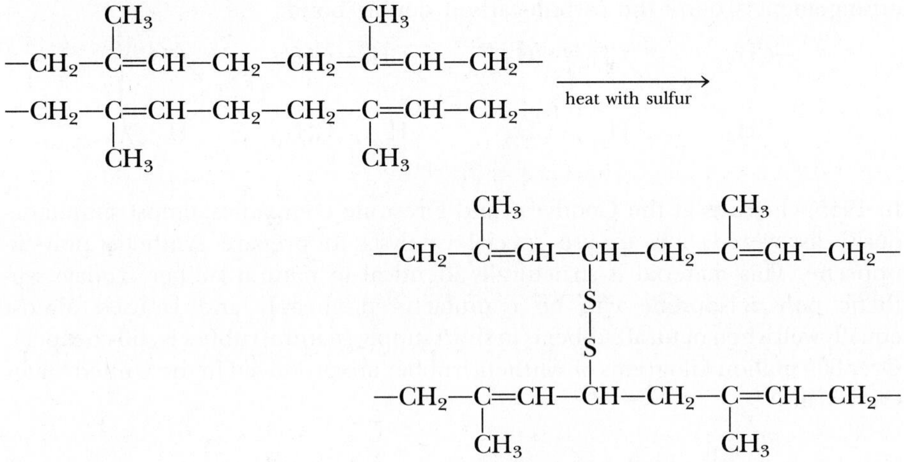

The sulfur chains help to align the polymer chains, so the material does not undergo a permanent change when stretched but springs back to its original shape and size when the stress is removed (Figure 11.17). As described earlier, substances that behave this way are called **elastomers.**

In later years chemists searched for ways to make a synthetic rubber so we would not be completely dependent on imported natural rubber during emergencies, such as during the first years of World War II. In the mid-1920s, German chemists polymerized butadiene (obtained from petroleum and structurally similar to isoprene, but without the methyl group side chain).

$$n \quad \underset{H}{\overset{H}{>}}C=C\underset{\underset{H}{}}{\overset{H}{<}}C=C\underset{H}{\overset{H}{<}} \quad \xrightarrow[\text{polymerization}]{\text{addition}}$$

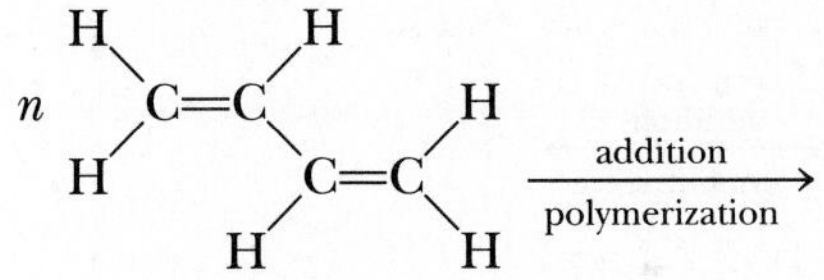

1,3-butadiene

$$-CH_2\underset{\underset{H}{}}{\overset{}{\diagdown}}C=C\underset{H}{\overset{CH_2}{\diagup}}\left(CH_2\underset{\underset{H}{}}{\overset{}{\diagdown}}C=C\underset{H}{\overset{CH_2}{\diagup}}\right)_n CH_2\underset{\underset{H}{}}{\overset{}{\diagdown}}C=C\underset{H}{\overset{CH_2}{\diagup}}-$$

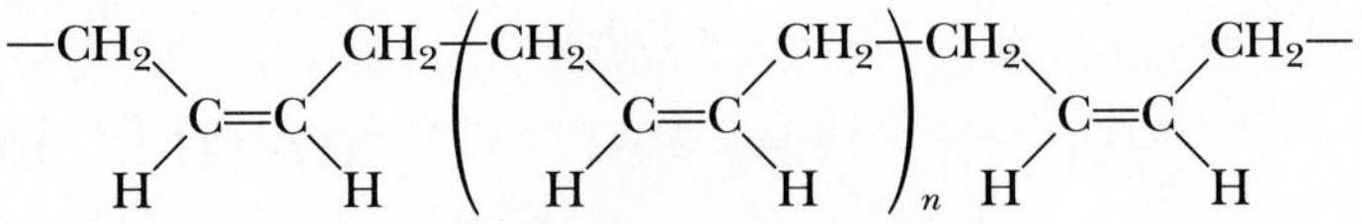

polybutadiene

Polybutadiene is used in the production of tires, hoses, and belts. Almost 500 million kilograms of the polymer were made in the United States in 1993.

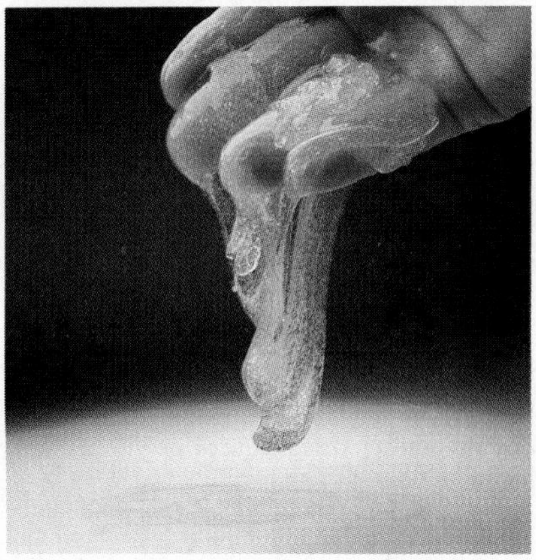

Figure 11.16 Polyvinyl alcohol, derived from vinyl alcohol ($CH_2{=}CHOH$) is used for, among other things, a coating on grease-proof paper and a thickener for foods. When an aqueous suspension of polyvinyl alcohol is mixed with boric acid, $B(OH)_3$ (or its sodium salt), an ester forms and leads to "cross-linking" of the polymer. The resulting viscous substance is known as "slime."

$$\left(\begin{matrix} H & H \\ | & | \\ -C & -C- \\ | & | \\ H & O \end{matrix}\right)_n$$

$$B-OH$$

$$\left(\begin{matrix} H & O \\ | & | \\ -C & -C- \\ | & | \\ H & H \end{matrix}\right)_n$$

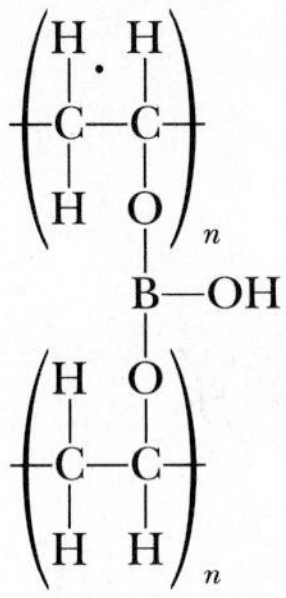

Polyvinyl alcohol cross-linked
with boric acid, $B(OH)_3$.

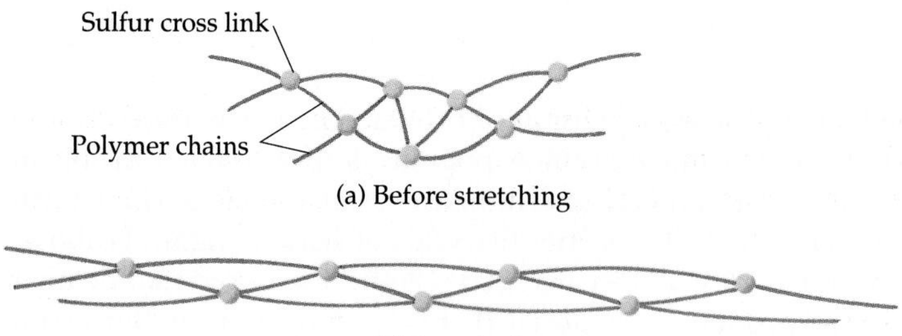

(a) Before stretching

(b) Stretched

Figure 11.17 Stretched and unstretched vulcanized rubber.

See the earlier discussion of polyisoprene on page 504.

As described in Section 11.2, the behavior of natural rubber is due to the specific molecular geometry within the polymer chain. Natural rubber has a *cis* arrangement around the carbon-carbon double bond.

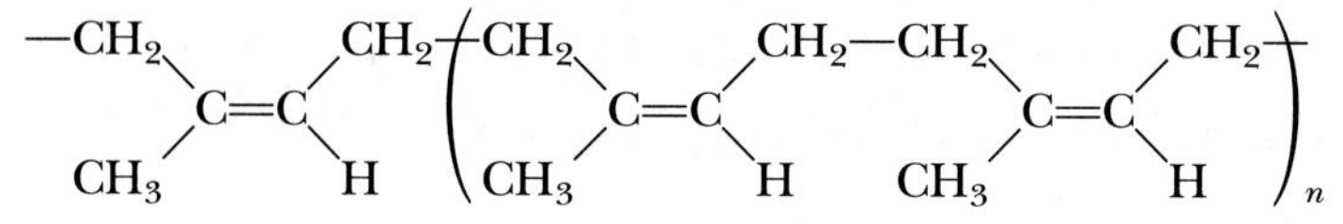

poly-*cis*-isoprene

In 1955, chemists at the Goodyear and Firestone companies almost simultaneously discovered how to use special catalysts to prepare synthetic poly-*cis*-isoprene. This material is structurally identical to natural rubber. Today, synthetic poly-*cis*-isoprene can be manufactured cheaply and is used almost equally well when natural rubber is in short supply (natural rubber is still cheaper). Over 800 million kilograms of synthetic rubber are produced in the United States every year.

Copolymers

Many commercially important addition polymers are **copolymers,** polymers obtained by polymerizing a mixture of two or more monomers. A copolymer of styrene with butadiene is the most important synthetic rubber produced in the United States. A 3-to-1 mole ratio of butadiene to styrene is used to make styrene-butadiene rubber (SBR). More than 1.4 million tons of SBR is produced each year in the United States for making tires. And a little is left over to make *bubble gum.* The stretchiness of gum once came from natural rubber, but SBR is now used to help you blow bubbles.

The use of SBR in bubble gum is described in *Chem Matters,* October, 1994.

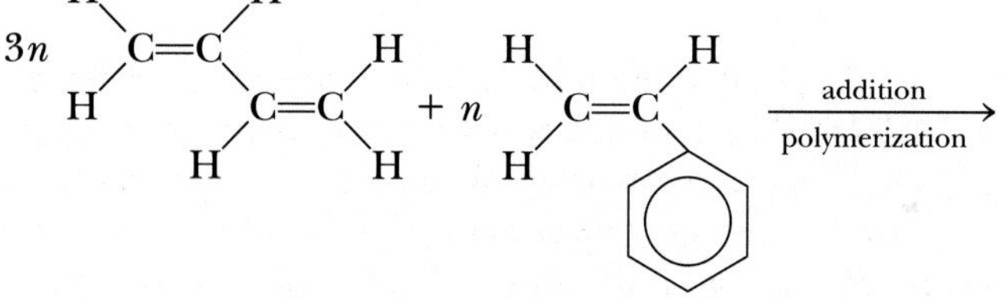

1,3-butadiene styrene

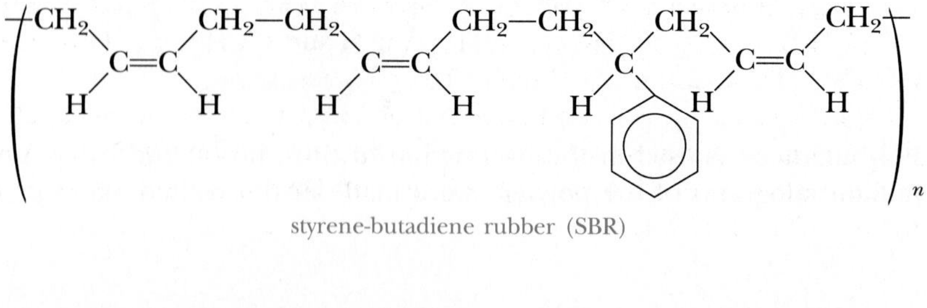

styrene-butadiene rubber (SBR)

Other important copolymers are made by polymerizing mixtures of ethylene and propylene or acrylonitrile, butadiene, and styrene. Saran Wrap is an example of a copolymer of vinyl chloride with 1,1-dichloroethylene.

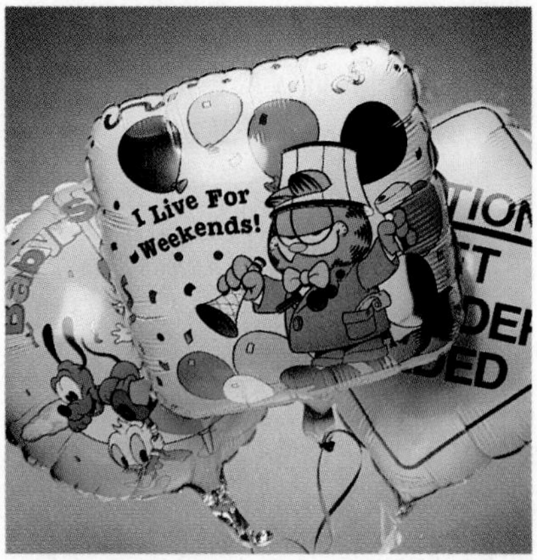

Condensation Polymers

A chemical reaction in which two molecules react by splitting out or eliminating a small molecule is called a **condensation reaction.** The reactions of alcohols with carboxylic acids to give esters (Section 11.5) are examples of condensation. This important type of chemical reaction does not depend on the presence of a double bond in the reacting molecules. Rather, it requires the presence of two different kinds of reactive, or **functional,** groups on two different molecules. If each reacting molecule has two functional groups, both of which can react, it is possible for condensation reactions to produce long-chain polymers.

Polyesters

A molecule with two carboxylic acid groups, such as terephthalic acid, and another molecule with two alcohol groups, such as ethylene glycol, can react with each other at both ends.

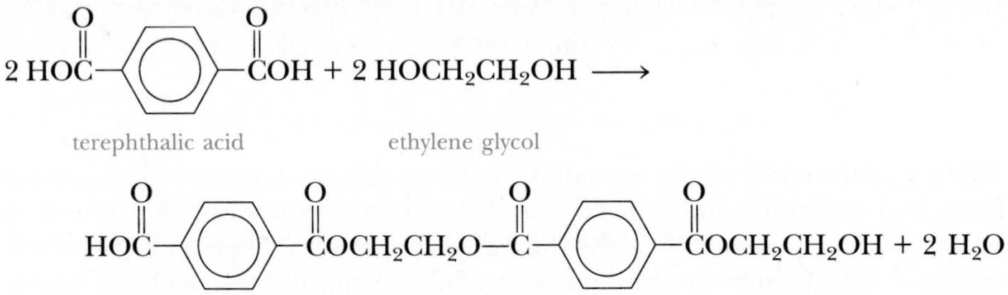

Figure 11.18 Mylar, a polyester, can be made into extremely thin films. Because the film has very tiny pores, it is used to make balloons that can be inflated and filled with helium; the helium atoms diffuse through the pores in the film only very slowly.
(C. D. Winters)

$$2\ HO\overset{O}{\overset{\|}{C}} \!\!-\!\!\bigcirc\!\!-\!\!\overset{O}{\overset{\|}{C}}OH + 2\ HOCH_2CH_2OH \longrightarrow$$

terephthalic acid   ethylene glycol

$$HO\overset{O}{\overset{\|}{C}}\!\!-\!\!\bigcirc\!\!-\!\!\overset{O}{\overset{\|}{C}}OCH_2CH_2O\!\!-\!\!\overset{O}{\overset{\|}{C}}\!\!-\!\!\bigcirc\!\!-\!\!\overset{O}{\overset{\|}{C}}OCH_2CH_2OH + 2\ H_2O$$

If *n* molecules of acid and alcohol react in this manner, the process will continue until a large polymer molecule, known as a **polyester,** is produced.

$$-O\!\!\left(\!\!\overset{O}{\overset{\|}{C}}\!\!-\!\!\bigcirc\!\!-\!\!\overset{O}{\overset{\|}{C}}OCH_2CH_2O\!\!\right)_{\!\!n}$$

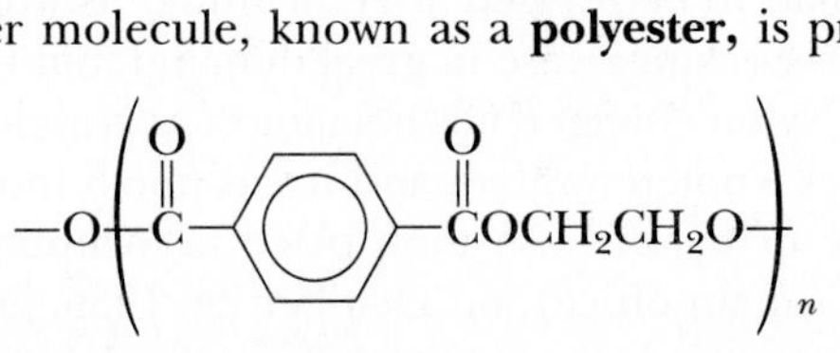

More than 2 million tons of polyethylene terephthalate, commonly referred to as PET, is produced in the United States each year for use in making beverage bottles, apparel, tire cord, film for photography and magnetic recording, food packaging, coatings for microwave and conventional ovens, and home furnishing. A variety of trade names are associated with the various applications. Polyester textile fibers are marketed under such names as Dacron and Terylene. Films of the same polyester, when magnetically coated, are used to make audio and TV tapes. This film, Mylar, has unusual strength and can be rolled into sheets one-thirtieth the thickness of a human hair (Figure 11.18).

The inert, nontoxic, noninflammatory, and non-blood-clotting characteristics of Dacron polymers make Dacron tubing an excellent substitute for human blood vessels in heart bypass operations, and Dacron sheets are used as temporary skin for burn victims.

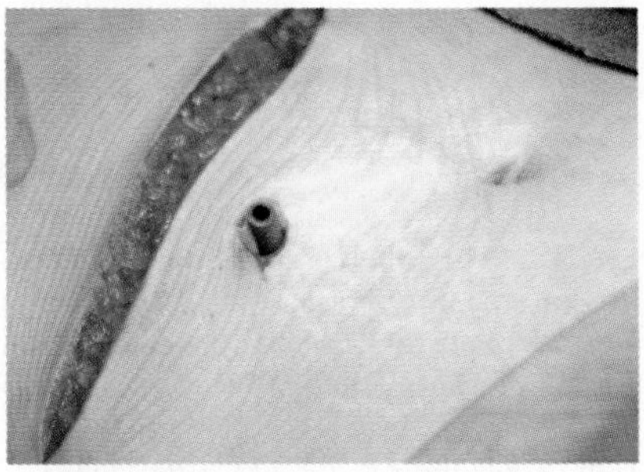

The use of Dacron in medicine. During gall bladder surgery, a Dacron sheet is placed on the skin.
(Comstock, Inc)

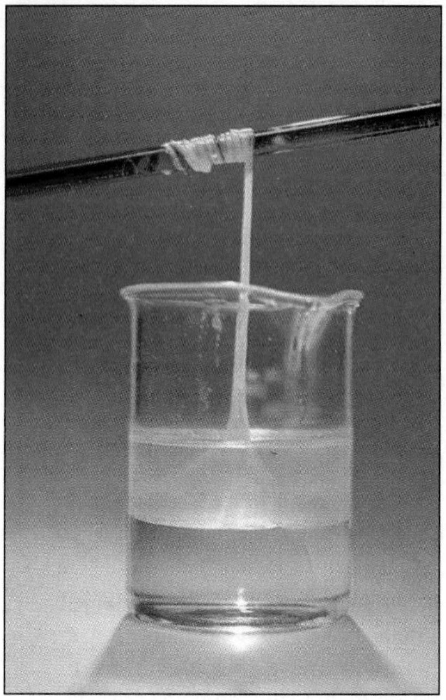

Figure 11.19 Nylon-6,6. Hexamethylenediamine is dissolved in water *(bottom layer),* and a derivative of adipic acid (adipoyl chloride) is dissolved in hexane *(top layer).* The two compounds react at the interface between the layers to form nylon, which is being wound onto a stirring rod.
(C. D. Winters)

Polyamides

Another useful and important type of condensation reaction is that between a carboxylic acid and a primary amine to form an amide (Section 11.7). Polymers are produced when diamines (compounds containing two —NH_2 groups) react with dicarboxylic acids (compounds containing two —CO_2H groups). Reactions of this type yield a group of polymers that may have affected society to a greater extent than any other type. These are the **polyamides,** or nylons.

In 1928, the Du Pont Company embarked on a program of basic research headed by Dr. Wallace Carothers (1896–1937), who came to Du Pont from the Harvard University faculty. His research interests were high-molecular-weight compounds, such as rubber, proteins, and resins. In February, 1935, his research yielded a product known as Nylon-6,6 (Figure 11.19), prepared from adipic acid (a diacid) and hexamethylenediamine (a diamine).

$$\underset{\text{adipic acid}}{n\,\mathrm{HO{-}\overset{\overset{\displaystyle O}{\|}}{C}{-}(CH_2)_4{-}\overset{\overset{\displaystyle O}{\|}}{C}{-}OH}} + \underset{\text{hexamethylenediamine}}{n\,\mathrm{H_2N{-}(CH_2)_6{-}NH_2}} \longrightarrow$$

$$\mathrm{{-}\overset{\overset{\displaystyle O}{\|}}{C}{-}(CH_2)_4{-}\overset{\overset{\displaystyle O}{\|}}{C}}\left(\mathrm{\underset{\underset{\displaystyle H}{|}}{N}{-}(CH_2)_6{-}\underset{\underset{\displaystyle H}{|}}{N}{-}\overset{\overset{\displaystyle O}{\|}}{C}{-}(CH_2)_4{-}\overset{\overset{\displaystyle O}{\|}}{C}}\right)_n\mathrm{\underset{\underset{\displaystyle H}{|}}{N}{-}(CH_2)_6{-}\overset{\overset{\displaystyle H}{|}}{N}{-}} + n\,\mathrm{H_2O}$$

nylon-6,6
(The amide groups are outlined for emphasis.)

This material could easily be extruded into fibers that were stronger than natural fibers and chemically more inert. The discovery of nylon jolted the American textile industry at almost precisely the right time. Natural fibers were not meeting the needs of 20th-century Americans. Silk was not durable and was expensive, wool was scratchy, linen crushed easily, and cotton did not have a high-fashion image. All four had to be pressed after cleaning. As women's hemlines rose in the mid-1930s, silk stockings were in great demand, but they were very expensive and short-lived. Nylon changed all that almost overnight. It could be knitted into the sheer hosiery women wanted, and it was much more durable than silk. The first public sale of nylon hose took place in Wilmington, Delaware (the location of Du Pont's main office), on October 24, 1939, just before the start of World War II. The war caused all commercial use of nylon to be abandoned until 1945, however, because the industry turned to making parachutes and other materials for the military. It was not until about 1952 that nylon could again meet the demands of the textile industry.

See D. A. Hounshell and J. K. Smith, Jr.: "Nylon drama," *Invention and Technology,* Fall, 1988, pp. 40–55 for a very good accounting of the history of nylon.

Hair, wool, and silk are all natural polymers that closely resemble nylon.

Figure 11.20 illustrates why nylons make such good fibers. To have good tensile strength (an ability to resist tearing), the chains of atoms in a polymer should be able to attract one another, but not so strongly that the plastic cannot be initially extended to form the fibers. Ordinary covalent chemical bonds linking the chains together would be too strong. There is an attraction, however, between the N-bonded H atoms of one chain (which have a partial positive charge) and the C-bonded O atoms of another chain (which have a partial negative charge). Although these bonds have a strength that is only about a tenth that of an ordinary covalent bond, they link the chains in the desired manner.

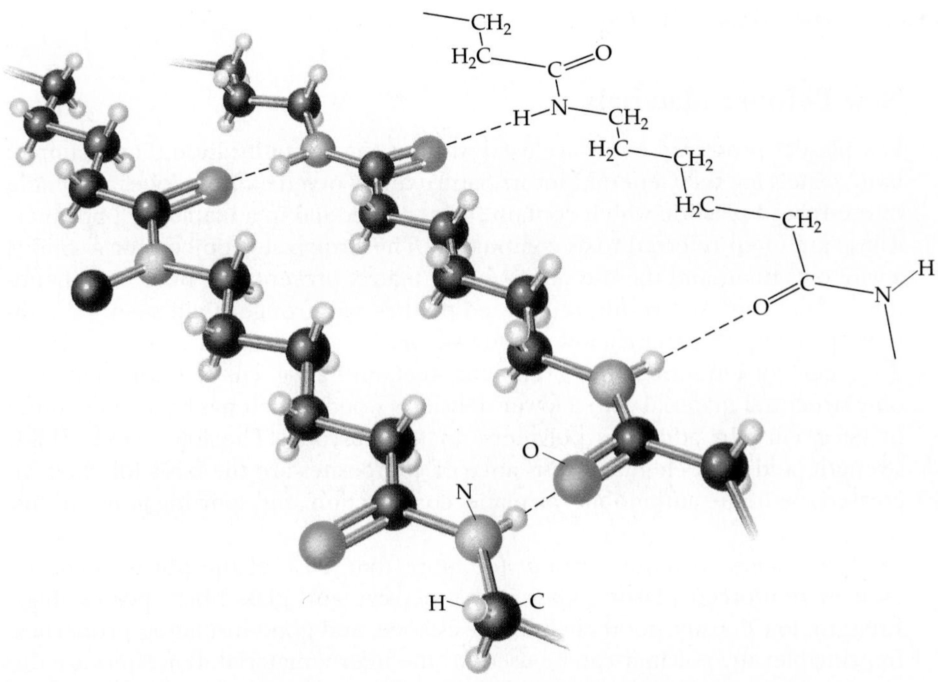

Figure 11.20 Bonding occurs between chains of nylon. A carbonyl oxygen (with a slight negative charge) on one chain interacts with an amine hydrogen (with a slight positive charge) on a neighboring chain.

Kevlar, another polyamide, is used to make bulletproof vests (Figure 11.21) and fireproof garments. Kevlar is made from *p*-phenylenediamine and terephthalic acid.

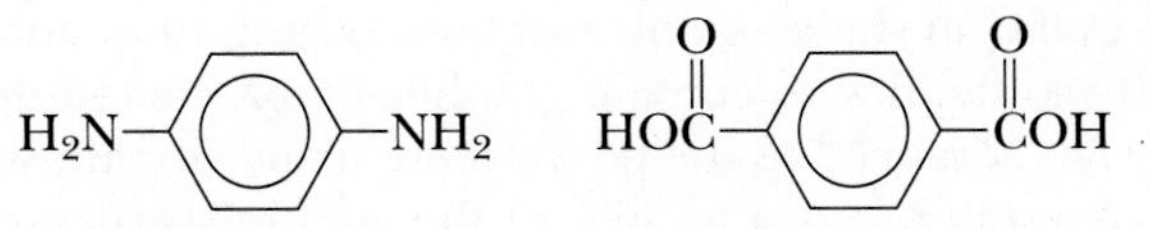

p-phenylenediamine　　terephthalic acid

EXAMPLE 11.3 ***Condensation Polymers***

Write the repeating unit of the condensation polymer obtained by combining $HO_2CCH_2CH_2CO_2H$ (succinic acid) and $H_2NCH_2CH_2NH_2$ (1,2-ethylenediamine).

Solution A condensation polymer composed of a diacid and a diamine forms by loss of water between monomers to give an amide bond. The repeating unit is therefore

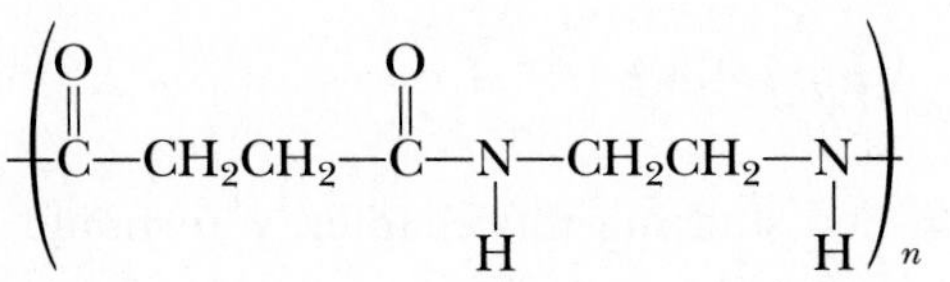

EXERCISE 11.15 ***Condensation Polymers***

Draw the structure of the repeating unit in the condensation polymer obtained from the reaction of terephthalic acid with ethylene glycol. Is this a polyamide or a polyester? ■

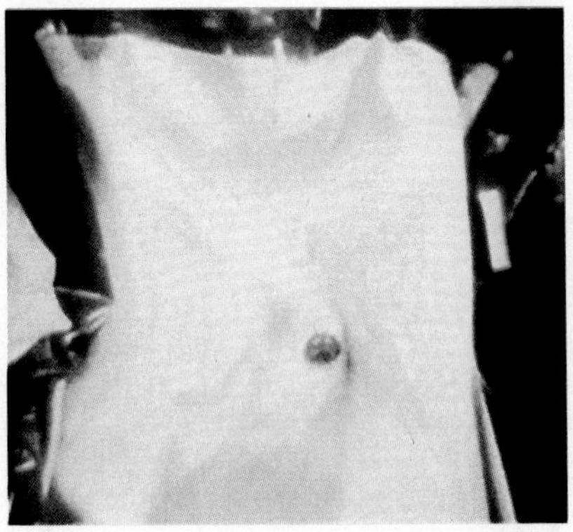

Figure 11.21 This vest, made of Kevlar fiber, has deflected a bullet.
(E. I. DuPont de Nemours)

Polymer composite materials are increasingly used. One example is the graphite-epoxy composite used for tennis rackets, golf clubs, sail boat masts and booms, and aircraft parts. (Jerry Watcher/Photo Researchers, Inc.)

New Polymer Materials

Few plastics produced today are used without some modification. For example, body panels for the General Motors Saturn and Corvette automobiles are made of **reinforced plastics,** which contain fibers embedded in a matrix of a polymer. These are often referred to as **composites.** The strongest geometry for a solid is a wire or a fiber, and the use of a polymer matrix prevents the fiber from bending or buckling. As a result, reinforced plastics are stronger than steel. In addition, the composites have a low density—from 1.5 to 2.25 g/cm^3, compared with 2.7 g/cm^3 for aluminum, 7.9 g/cm^3 for steel, and 2.5 g/cm^3 for concrete. The only structural material with a lower density is wood, which has an average value of 0.5 g/cm^3. In addition, polymers do not corrode. The low density, high strength, and high chemical resistance of composites are the basis for their increased use in the automobile, airplane, construction, and sporting goods industries.

Glass fibers currently account for more than 90% of the fibrous material used in reinforced plastics; glass is inexpensive and glass fibers possess high strength, low density, good chemical resistance, and good insulating properties. In principle, any polymer can be used for the matrix material. Polyesters are the number one polymer matrix at the present time, so glass-reinforced polyester composites are used in structural applications, such as boat hulls, airplanes, missile casings, and automobile body panels.

Epoxy resin is a polymeric ether.

Other fibers and polymers have been used, and the trend is toward increased utilization of composites in automobiles and aircraft. For example, a composite of fibers made of graphite in an epoxy matrix is used in the construction of the Lear jet. Graphite-epoxy composites are used in a number of sporting goods, such as golf-club shafts, tennis racquets, fishing rods, and skis. The F-16 military aircraft was the first to contain graphite-epoxy composite material, and the technology has advanced to the point where many aircraft, such as the F-18, use graphite composites for up to 26% of the aircraft's structural weight. This percentage is expected to increase to between 40 and 50% in future aircraft.

Although few automobiles have exterior body panels made of plastics, most contain a number of components that are plastic. Examples include bumpers, trim, light lenses, grilles, dashboards, seat covers, steering wheels—enough plastics to account for an average of 250 lb per car. The increased emphasis on improving fuel efficiency will lead to greater amounts of plastics in the construction of future automobiles, in both interior components and exterior body panels. General Motors predicts that during the 1990s it will manufacture one million plastic-body automobiles per year.

CHAPTER HIGHLIGHTS

When you have finished studying this chapter, you should be able to

- Identify an alkane, a hydrocarbon with the general formula C_nH_{2n+2} (Section 11.1).
- Draw structural formulas for unbranched- and branched-chain alkanes (Section 11.1).
- Give systematic names for unbranched- and branched-chain alkanes (Section 11.1).

- Draw the structure of a cycloalkane (Section 11.1).
- Identify an alkene (C_nH_{2n}) or an alkyne (C_nH_n) (Section 11.2).
- Give systematic names for alkenes or alkynes (Section 11.2).
- Describe the bonding in alkenes and alkynes (Section 11.2).
- Draw structural formulas for *cis* and *trans* stereoisomers of alkenes (Section 11.2).
- Draw the structural formula for the product of an addition reaction of an alkene or alkyne with hydrogen, water, halogens, or hydrogen halides (Section 11.2).
- Draw the structural formula for a derivative of benzene and give its systematic name (Section 11.3).
- Describe the bonding in benzene and related aromatic compounds (Section 11.3).
- Recognize alcohols and give their systematic names (Section 11.4).
- Write chemical equations for common reactions of alcohols (reactions with alkali metals or sulfuric acid) (Section 11.4).
- Recognize the difference between compounds with a carbonyl group (aldehydes, ketones, carboxylic acids, and esters), give systematic names for these compounds, and draw their structural formulas (Section 11.5).
- Write chemical equations for oxidation and reduction reactions of aldehydes and ketones (Section 11.5).
- Describe the bonding and structure of carboxylic acids (Section 11.5).
- Write a chemical equation for the preparation of an ester (Section 11.5).
- Write an equation for the reaction of an ester with sodium hydroxide (Section 11.5).
- Draw the structural formula of a fat or oil (Section 11.6).
- Draw the structural formula of an amine or an amide (Section 11.7).
- Define important terms in polymer chemistry (Section 11.8). (For example, you should be able to define monomer and polymer and know the difference between a thermosetting plastic and a thermoplastic.)
- Describe the different methods of making polymers (addition and condensation polymers) (Section 11.8).
- Draw the repeating unit of common addition and condensation polymers (Section 11.8).

STUDY QUESTIONS

Review Questions

Alkanes

1. What is the name of the straight- or unbranched-chain alkane with the formula C_8H_{18}?
2. What is the molecular formula for an alkane with 13 carbon atoms?

3. Which of the following is an alkane? Which could be a cycloalkane?
(a) C_2H_4, (b) C_5H_{10}, (c) $C_{14}H_{30}$, (d) C_7H_8

4. Three of the nine possible structural isomers for the alkane C_7H_{16} are given in Example 11.1. Draw three more of these isomers and give the systematic name of each.

5. Give the systematic name for the following alkane. Draw a structural isomer of the compound and give its name.

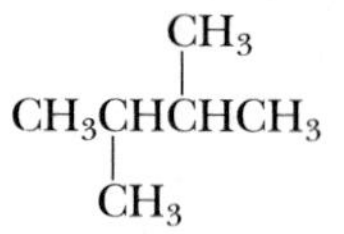

6. Give the systematic name for the following alkane. Draw a structural isomer of the compound and give its name.

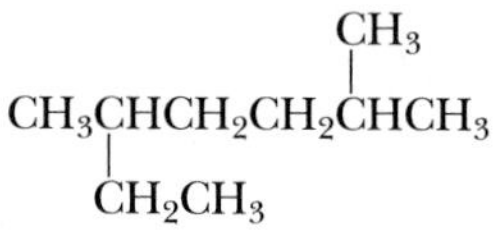

7. Draw the structure of each of the following compounds:
(a) 2,3-Dimethylpentane
(b) 2,4-Dimethyloctane
(c) 3-Ethylhexane
(d) 2-Methyl-3-ethylhexane

8. Draw the structure of each of the following compounds:
(a) 2,2-Dimethylhexane
(b) 3,3-Diethylpentane
(c) 2-Methyl-3-ethylheptane
(d) 3-Propylhexane

Alkenes and Alkynes

9. Draw structures for the *cis* and *trans* isomers of (a) 1,2-dichloroethene and (b) 3-methyl-3-hexene.

10. Draw structural formulas for
(a) 3-Hexyne
(b) 1-Pentyne
(c) 3-Pentene
(d) 3-Methyl-1-butene

11. A hydrocarbon with the formula C_5H_{10} can be either an alkene or a cycloalkane.
(a) Draw a structure for each of the isomers possible for C_5H_{10}, assuming it is an alkene. Give the systematic name of each isomer you have drawn.
(b) Draw a structure for the cycloalkane having the formula C_5H_{10}.

12. What structural requirement is necessary for an alkene to have *cis* and *trans* isomers? Can *cis* and *trans* isomers exist for an alkyne?

13. Draw the structure and give the systematic name for the products of the following reactions:
(a) $CH_3CH_2CH{=}CH_2 + Br_2 \rightarrow$
(b) $CH_3CH{=}CHCH_3 + H_2 \rightarrow$

14. Draw the structure and give the systematic name for the products of the following reactions:
(a) 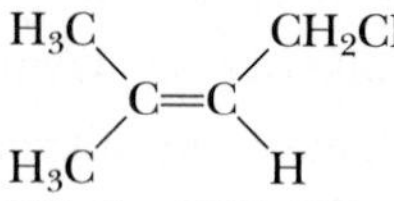 $+ H_2 \rightarrow$
(b) $CH_3C{\equiv}CCH_2CH_3 + Br_2 \rightarrow$

Benzene and Aromatic Compounds

15. Draw structural formulas for the following compounds:
(a) *p*-Dichlorobenzene (alternatively called 1,4-dichlorobenzene)
(b) *m*-Bromotoluene (alternatively called 1-bromo-3-methylbenzene)
(c) *p*-Diethylbenzene

16. Give the systematic name for the following compounds:

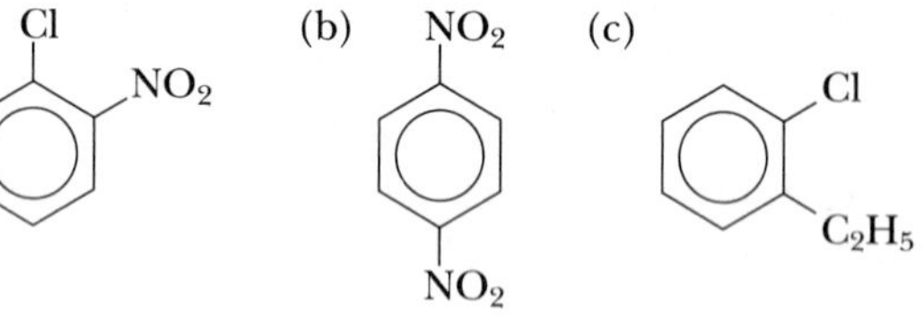

17. Show how to prepare propylbenzene from benzene and an appropriate propyl derivative.

18. Draw the resonance structures for naphthalene, $C_{10}H_8$.

Alcohols

19. Give the systematic name for each of the following alcohols and tell if each is primary, secondary, or tertiary:
(a) $CH_3CH_2CH_2OH$ (b) $CH_3CH_2CH_2CH_2OH$
(c) 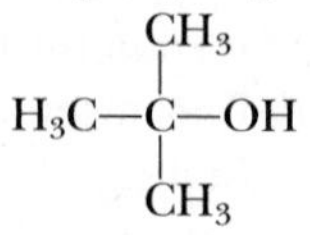 (d) 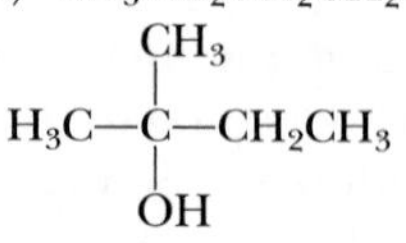

20. Draw structural formulas for the following alcohols:
(a) 1-Pentanol and 2-pentanol
(b) 2-Methyl-1-hexanol
(c) 3,3-Dimethyl-2-butanol

21. Draw structural formulas for all the alcohols with the formula $C_4H_{10}O$. Give the systematic name of each.

22. Draw structural formulas for all the alcohols with the formula $C_5H_{12}O$. Give the systematic name of each.

23. Complete the following reactions:
(a) $CH_3CH_2CH_2CH_2OH \xrightarrow{H_2SO_4/180\ °C}$
(b) $CH_3CH_2CH_2OH + Na \rightarrow$
(c) $CH_3CH_2CH_2CH_2OH + HBr \rightarrow$

24. Tell how to prepare each of the following compounds from an alcohol and other appropriate reagents.
(a) 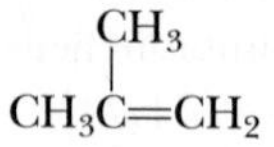

(b) 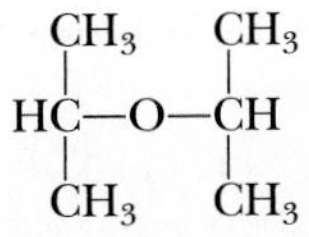

Compounds with a Carbonyl Group

25. Draw structural formulas for (a) 2-hexanone, (b) pentanal, and (c) hexanoic acid.

26. Give systematic names for

(a) 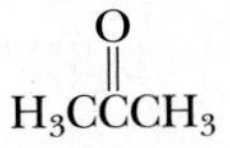 (b) 

(c) 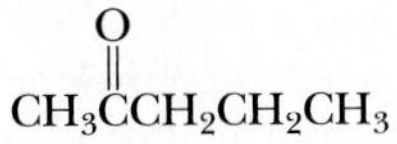

27. Give systematic names for

(a) 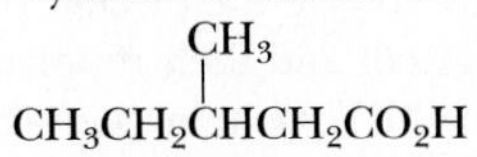

(b) 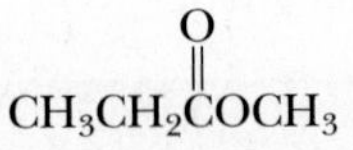

(c) 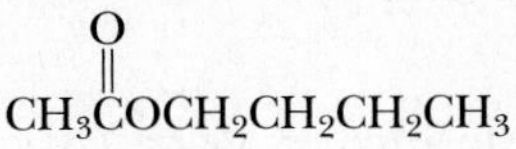

(d) 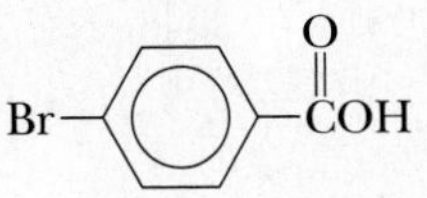

28. Draw structural formulas for the following acids and esters:
   (a) 2-Methylhexanoic acid
   (b) Pentyl butanoate (which has the odor of apricots)
   (c) Octyl acetate (which has the odor of oranges)

29. Give the structural formula and systematic name for the product from each of the following reactions:
   (a) Propanal and $KMnO_4$
   (b) Propanal and $LiAlH_4$
   (c) 2-Butanone and $LiAlH_4$
   (d) 2-Butanone and $KMnO_4$

30. Describe how to prepare 2-pentanol beginning with the appropriate ketone.

31. Describe how to prepare propyl propanoate beginning with appropriate materials.

32. Give the name and structure of the product of the reaction of benzoic acid and 2-propanol.

33. Draw structural formulas and give the names for the products of the following reaction:

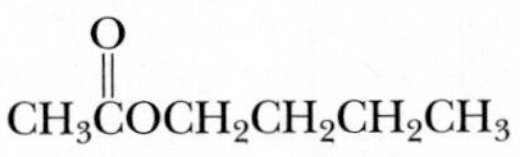 + NaOH (followed by an acid)

34. Draw structural formulas and give the names for the products of the following reaction:

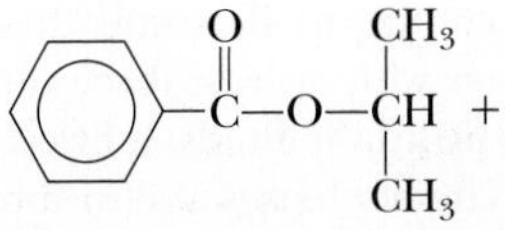 + NaOH (followed by acid)

Polymers

35. Polyvinyl acetate is the binder in water-based paints.
   (a) Show a portion of this polymer with three monomer units.
   (b) Describe how to make polyvinyl alcohol from polyvinyl acetate.

36. Polychloroprene is made from a chlorinated butadiene pictured here.

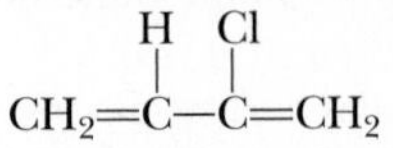

The monomer for polychloroprene

One of the names under which it is sold is neoprene. Draw the structure of polychloroprene with at least three monomer units.

37. Saran is a copolymer of 1,1-dichloroethene and chloroethene (vinyl chloride). Draw a possible structure for this polymer.

38. The structure of methyl methacrylate is given in Table 11.13. Draw a polymethyl methacrylate (PMMA) polymer that has four monomer units. Is the polymer chain flat or does it twist in some manner? (PMMA has excellent optical properties and is used to make hard contact lenses.)

General Questions

39. In addition to the structural isomerism of alkanes and the *cis-trans* isomerism of some alkenes, other types of isomers are found in organic chemistry. For example, dimethyl ether (CH_3OCH_3) and ethanol (CH_3CH_2OH) are isomers; they have the same molecular formula but their chemical functionality is quite different.
   (a) Draw all of the isomers possible for C_3H_8O. Give the systematic name of each and tell into what class of compound it fits.
   (b) Draw the structural formula for an aldehyde and a ketone with the molecular formula C_4H_8O. Give the systematic name of each.

40. Draw the structural formula for each of the nine possible isomers of the chlorinated alkanes $C_4H_8Cl_2$. Name each compound.

41. Draw the structure of glyceryl trilaurate. When this triester is saponified, what are the products?

42. Silicone polymers have a silicon-oxygen backbone. They can be made by beginning with dimethyldichlorosilane, $(CH_3)_2SiCl_2$. This is allowed to react with water (hydrolyzed) to replace Si—Cl bonds with Si—OH bonds. The resulting $(CH_3)_2Si(OH)_2$ molecules form the polymer by eliminating a molecule of water between two units to form a Si—O—Si

linkage. Draw structural formulas to illustrate these reactions. Draw a silicone polymer with at least three monomer units. (Silicones are used as hydraulic fluids, lubricants, and sealants. They are found in contact lenses and in tubes used in medical applications.)

43. Dimethyl ether, H_3COCH_3, can bind to BF_3 by forming a coordinate covalent bond involving an O atom lone pair of electrons and the B atom of BF_3.
(a) Draw the Lewis structure for the compound that results from forming a coordinate covalent bond between dimethyl ether and BF_3.
(b) What is the hybridization of the O atom before and after bond formation? For the B atom before and after bond formation?
(c) What is the F—B—F bond angle before and after reaction?

44. Give structural formulas and systematic names for the three structural isomers of trimethylbenzene, $C_6H_3(CH_3)_3$.

45. On page 495 the carbon atoms in graphite were described as sp^2 hybridized and it was stated that there are carbon-carbon pi bonds within a layer. Draw a portion of a graphite layer, say six rings, and describe the bonding in this fragment.

46. Voodoo lilies depend on carrion beetles for pollination. Because carrion beetles are attracted to dead animals, and because dead and putrefying animals give off the horrible smelling amine cadaverine, the lily likewise releases cadaverine (and the closely related compound putrescine). A decarboxylase enzyme converts the naturally occurring amino acid lysine to cadaverine.

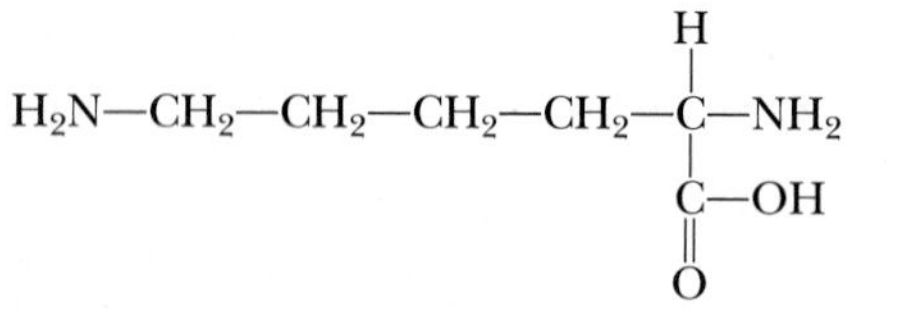

lysine

What group of atoms must be replaced in lysine to make cadaverine? (Lysine is essential to human nutrition but is not synthesized in the human body.)

47. Complete the following reactions. Name the organic reactant and product in each reaction.
(a) $CH_3CH_2CH{=}CH_2 + Br_2 \rightarrow$
(b) $HC{\equiv}CCH_3 + H_2$ (in the presence of a catalyst) $\rightarrow$
(c) $HC{\equiv}CCH_3 + 2\ H_2$ (in the presence of a catalyst) $\rightarrow$
(d) $CH_3CH{=}CHCH_3 + H_2O \rightarrow$

48. When 1-propanol is treated with concentrated sulfuric acid, two products are possible. Draw structural formulas for the products and name each one.

49. Kevlar is a polyamide made from *p*-phenylenediamine and terephthalic acid. Draw the repeating unit of the Kevlar polymer.

50. Identify the eight isoprene units in the structure of carotene.

51. In the text it is noted that benzoic acid occurs in many berries. When humans eat berries, however, benzoic acid is converted to hippuric acid in the body by reaction with the amino acid glycine.

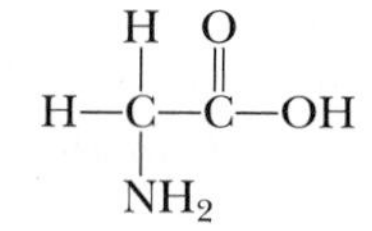

glycine

Draw the structure of hippuric acid knowing that it is an amide formed by reaction of the carboxylic acid group of benzoic acid and the amino group (—NH_2) of glycine. Why is hippuric acid referred to as an acid?

52. A well-known company selling outdoor clothing has recently introduced jackets made of recycled polyethylene terephthalate (PET), the principal material in many soft drink bottles. Another company makes new PET fibers by treating recycled bottles with methanol to give the diester, dimethylterephthalate, and ethylene glycol and then repolymerizing these compounds to give new PET. Write a chemical equation to show how the reaction of PET with methanol can give dimethylterephthalate and ethylene glycol.

The students are wearing jackets made from recycled PET soda bottles. (C. D. Winters)

Conceptual Questions

53. Write balanced equations for the combustion of ethane and ethanol. Which compound has the more negative enthalpy change for combustion? If ethanol is assumed to be partially oxidized ethane, what effect does this have on the heat of combustion?

54. One of the resonance structures for pyridine is illustrated here. Draw another resonance structure for the molecule. Comment on the relationship between this compound and benzene.

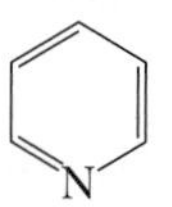

pyridine

55. Describe a simple chemical test to tell the difference between $CH_3CH_2CH_2CH{=}CH_2$ and its isomer cyclopentane.

56. Describe a simple chemical test to tell the difference between 2-propanol and its isomer methyl ethyl ether.

57. Consider *cis-* and *trans*-2-butene and compare these compounds with butane.

(a) *Trans*-2-butene can be converted to *cis*-2-butene by twisting the molecule around the C=C bond. A considerable expenditure of energy is required, however (approximately 264 kJ/mol).

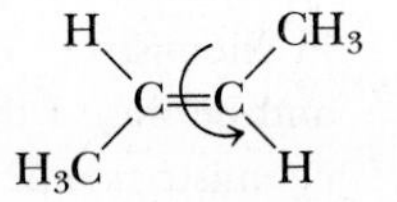

Explain why energy is required for this bond rotation.

(b) It was stated in the text that free rotation occurs around carbon-carbon single bonds. Thus, one end of the butane molecule can rotate relative to the other end around the bond joining the middle two C atoms.

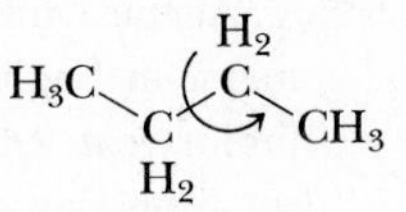

An expenditure of energy is in fact required, however (about 3 kJ/mol), although much less than in the case of *trans*-2-butene. Inspect the space-filling model of butane on page 498 and suggest a reason for the fact that rotation around the middle C—C bond requires some energy.

58. Plastics make up about 20% of the volume of landfills. There is therefore considerable interest in reusing or recycling these materials. To identify common plastics, a set of universal symbols is now used, five of which are illustrated here.

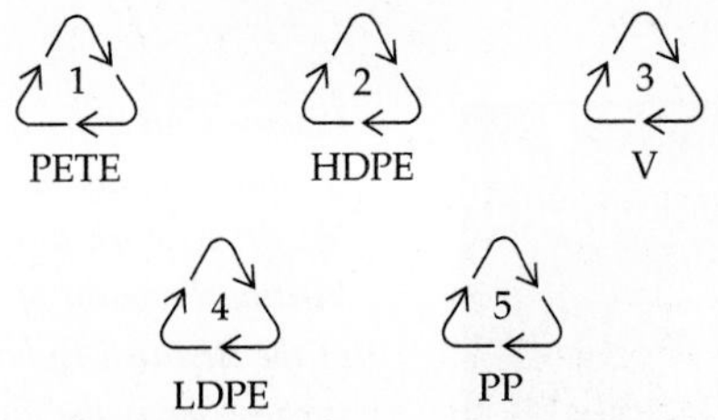

They symbolize low- and high-density polyethylene, polyvinyl chloride, polypropylene, and polyethylene terephthalate.

(a) Tell which symbol belongs to which type of plastic.

(b) Find an item in the grocery or drug store made from each of these plastics.

(c) Some properties of several plastics are listed here. Based on this information, describe how to separate samples of these plastics from one another.

Plastic	**Density (g/cm^3)**	**Melting Point (°C)**
Polypropylene	0.91	170
High-density polyethylene	0.96	135
Poly(ethylene terephthalate)	1.34–1.39	245

Summary Question

59. Maleic acid is prepared by the catalytic oxidation of benzene. It is an unsaturated *di*carboxylic acid, that is, it has *two* carboxylic acid groups.

(a) Combustion of 0.125 g of the acid gives 0.190 g of CO_2 and 0.0388 g of H_2O. Calculate the empirical formula of the acid.

(b) A 0.261-g sample of the acid requires 34.60 mL of 0.130 M NaOH for complete titration (so that the H^+ ions from both carboxylic acid groups are used). What is the molecular formula of the acid?

(c) Draw a Lewis structure for the acid.

(d) Describe the hybridization used by the C atoms.

(e) What are the bond angles around each C atom?

i n t e r v i e w

Jeanette Grasselli Brown

Jeanette Grasselli Brown was born in Cleveland, Ohio, the daughter of Hungarian immigrants who only received an eighth-grade education. As a child, she didn't learn to speak English until she entered the first grade at the age of six.

Today, Grasselli Brown is one of the world's most respected authorities on infrared and Raman spectroscopy—nondestructive techniques for "fingerprinting" molecular structures by analyzing energy changes characterized by spectra. During her 38-year career with BP America (previously The Standard Oil Company), she developed a Raman cell that led to a better understanding of the structure and subsequent improvements in catalysts used to convert propene to acrylonitrile, the monomer used to produce the acrylates found in acrylic carpets and tires.

After graduating summa cum laude from Ohio University in 1950 with a Bachelor's degree in chemistry, Grasselli Brown joined The Standard Oil Co., and ultimately was named Director of Corporate Research, Analytical and Environmental Sciences. She earned a Master's degree from Case Western Reserve University in 1958, and also holds six honorary doctoral degrees. In 1988, she left BP America to become a Distinguished Visiting Professor of Chemistry and Director of the Research Enhancement at Ohio University. She is the author of more than 70 publications, six books, and one patent. She has earned numerous awards, including the 1986 Garvan Medal, presented by the American Chemical Society (ACS) to the nation's outstanding woman chemist. Most recently, she received the 1993 Award in Analytical Chemistry from the ACS.

Grasselli Brown is also noted for her contributions to professional societies and public policy-making committees. She served on the National Research Council's Board on Chemical Science and Technology, the group that prepared the famous Opportunities in Chemistry *report in 1985. She is immediate past chair of the U.S. National Committee for IUPAC, the International Union of Pure and Applied Chemistry. Currently, she is one of only seven U.S. scientists serving on a White House-appointed committee examining science and technology agreements between the United States and Japan.*

An Early Appreciation for Learning

My father truly was a self-made industrial scientist. He had come to America, worked in a foundry, learned the business, and then decided to strike out on his own. During World War II, he launched a small aluminum foundry and began supplying sand-cast aluminum parts to the auto industry. Because I grew up around a foundry, I had access to technology as a child.

My brother and I were both encouraged to pursue college educations. We were told that a college education would open new doors of opportunity and fulfillment. In those days, an accepted career for a girl was teaching or nursing. I was very fond of reading, so I planned to study English and teach. Then I took a chemistry course in high school. One day, my teacher asked me, "Why aren't you thinking of majoring in chemistry?" He was very persuasive, and because I thoroughly enjoyed my chemistry studies I took his advice. With his help, I received a scholarship at Ohio University, and I was launched on my college career and what has become a life-long love affair with chemistry and with learning.

At the University, I was a double major in English and chemistry. In my junior year, I began to work for Professor Jesse Day as a laboratory assistant doing metallography work. This was a perfect fit for me because we were using microscopes to learn about the physical structure of aluminum alloys, which would in turn relate to their properties. My studies at the University were beginning to gel with what I had learned from my father about his foundry business!

After I graduated in 1950, I received an excellent offer from Alcoa, in Pittsburgh. But my younger brother was suffering from Hodgkin's disease, and I wanted to be in Cleveland, close to him. So I accepted an offer from The Standard Oil Co. in my home town.

IR and Raman Sectroscopy

The position with Standard Oil led to an entirely new and incredibly exciting direction in my life and career. The company had just acquired an infrared spectrometer—a new technology back in 1950.

As early as the 19th century, people had studied the interaction of matter with light in the infrared region. Yet, no commercial applications for this knowledge were developed until World War II, when German planes were bombing England. The British were intrigued by how the German planes could fly such long distances toward their targets. They thought the German aircraft fuel might contain some unique constituents. Using infrared spectroscopy, the British identified important molecular components of fuels and worked to synthesize them.

It was that research, plus a parallel effort to make synthetic rubber, that led to the tremendous popularity of infrared spectroscopy. The natural source of rubber, the *Hevea brasiliensis* tree of South America, was expensive to harvest and transport. There was an industrial need to provide a synthetic rubber for automobile and aircraft tires. Infrared spectroscopy was used to follow the course of these early syntheses and to verify that the proper structure had been produced.

In the infrared region, selected frequencies of light correspond to various vibrational frequencies (and therefore to different energy levels) of a molecule. When a molecule vibrates by stretching, rocking, twisting, or by other motions of the chemical bonds, infrared radiation corresponding to the vibrational frequency is absorbed and the energy level of the molecule is raised. By measuring the specific frequencies of light that are absorbed, one can obtain a spectrum. This spectrum is characteristic of the masses of the atoms involved in the vibration and the strength of the bonds between them. Thus, an infrared spectrum provides a unique "fingerprint" of a molecule. Infrared spectra can be obtained (nondestructively) on solids, liquids, and gases, and the qualitative and quantitative composition of materials can be determined quickly and positively. Because of this, infrared spectroscopy has been widely applied in industry, academia, medicine, and even in forensic science.

Infrared spectroscopy is often coupled with a technique known as Raman spectroscopy. Based on the effect discovered by Chandrasekara V. Raman (1880–1970), this method involves the scattering of light by molecules. Scattered light results from collisions of photons with molecules and results in a change in the energy of the molecule. This gives rise to the Raman effect. The energy exchange is equivalent to the amount of energy absorbed by the molecule in the infrared spectroscopy effect, making the two techniques highly complementary.

A Breakthrough in Catalytic Science

During my industrial career I found many interesting applications for infrared and Raman spectroscopy, many of which were of real value to the company. Among these, and perhaps of greatest importance, was our work in catalysis research. A key objective of industrial researchers in petroleum companies of the 1950s and 1960 was to "upgrade" or convert low-valued chemicals found in crude oil to more valuable chemicals. This was of course the foundation of the "petrochemical" industry. This industry was already proficient in the use of catalysts to upgrade hydrocarbons in crude oil to higher octane fuels and quality lubricants. So, catalytic processes for petrochemical research were a natural extension. The catalysts were composed of inorganic compounds on an inert support (such as silica or alumina). The supports strongly absorbed in the infrared region, making it difficult to study the structure of the active catalyst. However, the supports did not absorb in the Raman! My colleagues and I therefore developed a special, heated cell that allowed us to use Raman spectroscopy to examine the supported catalyst while it was at the temperature of the reaction. We could stimulate the conditions in an actual chemical plant and follow reactions as they occurred. This technology helped our company gain a real competitive advantage in designing better catalysts. This is an excellent example of how exciting, challenging, and rewarding industrial research can be.

Contributions to Society

While I've spent many years working in the laboratory, I'm equally proud of my work within the larger scientific community. In 1986, I was invited by the Academy of Sciences to give a talk in Hungary, where my parents were born. It was very emotional to go back and be reminded of the sights and smells and sounds of my childhood. I immediately formed strong friendships with chemist peers in Hungary. We work together now, to organize international conferences, and to publish Hungarian research. Many scientists in the former Eastern bloc would like to connect better with industry, but they don't know how. I'd like to transfer my knowledge of industry/academic interactions to them.

I have also worked for many organizations based in Washington, D.C., such as the National Institute of Standards and Technology and the National Research Council. Most recently, I was appointed to serve on a joint high-level advisory panel established by the Trade and Technology Act of 1988. Our objective is to evaluate current science and technology agreements between the U.S. and Japan, and to suggest new agreements that would enhance trade.

Issues of women in the workplace are another big concern for me. I was fortunate to work for a far-sighted company. BP America routinely recruited outstanding women scientists. I helped establish part-time policies and flex-time schedules so that women could continue to work and have a family.

The percentage of women in the workplace today is still not representative of the general population, but I think that's changing. As long as there's opportunity, women will keep rising through the ranks of academia and industry. But we have to keep working on it!

C H A P T E R **12**

Gases

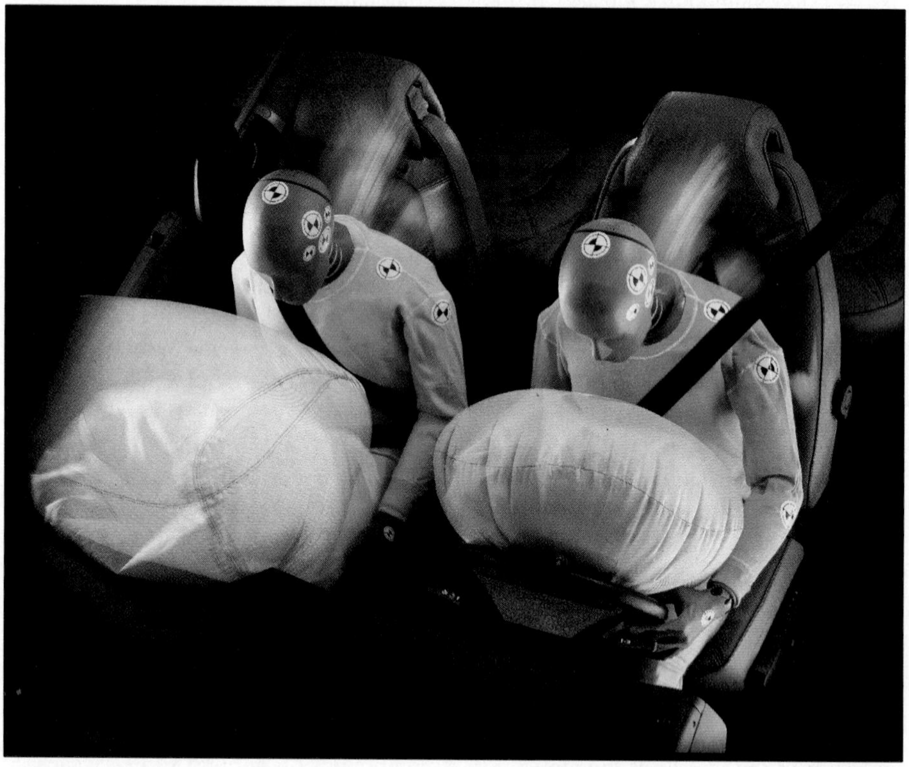

(Saab Cars USA, Inc.)

A Chemical Puzzler

Americans are involved in an alarming number of automobile accidents, and more than 50,000 of us are killed every year. Many would be saved from death or injury, though, if all cars were equipped with air bags. Automobile air bags, or the "supplemental restraint system" (SRS) as they are often called, depend on some basic chemistry. When a device in your car senses that the car has decelerated very rapidly, your seat belts are tightened, a chemical explodes inside a bag hidden in the steering wheel or dashboard of the car (or both), and the bag is rapidly inflated. Your life is saved by the expansion of a gas. What is the chemistry involved in operating an air bag? What gas is involved? Does the operation of the air bag depend on temperature? If the air in the car is very cold, as on a winter day in Minnesota, is the bag as effective? What if the accident happens on a high mountain road where the pressure of the atmosphere is much lower than at sea level? Is the bag more or less effective?

The Chemical Puzzler poses a number of questions about the operation of the air bag, a safety device now common in many makes of automobiles. The air bag in your automobile is rapidly inflated in the event of an accident by the explosion of sodium azide (NaN_3), which releases nitrogen gas,*

$$2\ NaN_3(s) \longrightarrow 2\ Na(s) + 3\ N_2(g)$$

and your forward momentum is slowed by your seat belt and cushioned by the air bag. If an air bag has a volume of 35 L, how much sodium azide is required to generate the required volume of gas? Does the amount of sodium azide required to produce 35 L of gas depend on the temperature of the gas? There are many questions, and a number can be answered by a study of the properties of gases.

Aside from understanding automobile air bags, at least three other reasons exist for studying gases. First, gas behavior is well understood and can be expressed in terms of simple mathematical models. One objective of scientists is to develop models of natural phenomena, and a study of gas behavior introduces you to this approach. Second, some common elements and compounds exist in the gaseous state under normal conditions of pressure and temperature (Table 12.1). Furthermore, many common liquids can be vaporized, and the properties of these vapors are important. Finally, our gaseous atmosphere provides one means of transferring energy and material throughout the globe, and it is the source of life-giving chemicals.

TABLE 12.1 Some Common Gaseous Elements and Compounds

(at 1 atm pressure and 25 °C)

He	CO
Ne	CO_2
Ar	CH_4
Kr	C_2H_4
Xe	C_3H_8
H_2	HF
O_2	HCl
O_3 (ozone)	HBr
N_2	HI
F_2	NO
Cl_2	NO_2
	H_2S

A gas is a substance that is normally in the gaseous state at ordinary pressures and temperatures. A vapor is the gaseous form of a substance that is normally a liquid or solid at ordinary pressures and temperatures. Thus, we often speak of helium gas and water vapor.

12.1 THE PROPERTIES OF GASES

To describe the gaseous state, chemists have learned that only four quantities are needed: (1) the *quantity* of gas, n (in moles); (2) the *temperature* of the gas, T (in kelvins); (3) the *volume* of gas, V (in liters); and (4) the *pressure* of the gas, P (in atmospheres). Let us examine the last of these, the concept of pressure and its units.

Gas Pressure

If you blow up a balloon, the rubber skin stretches and becomes taut. But put too much air into it, and the rubber skin breaks and the air escapes with explosive force. The tightness of the balloon's skin is caused by the force of the gas molecules striking the surface inside the balloon. The force per unit area is the pressure of the gas.

Pressure units

$$\begin{aligned} 1\text{ atm} &= 760\text{ mm Hg (exactly)} \\ &= 760\text{ torr} \\ &= 101.3\text{ kilopascals (kPa)} \\ &= 1.013\text{ bar} \\ &= 14.7\text{ lb/in.}^2\text{ (psi)} \end{aligned}$$

and

$$1\text{ bar} = 10^5\text{ Pa (exactly)}$$

*The equation for the explosion of sodium azide shows that sodium metal is one product. This metal surely reacts with atmospheric oxygen or water to produce Na_2O_2 or NaOH, respectively. For more information see W. L. Bell: "The chemistry of air bags." *Journal of Chemical Education,* Vol. 67, p. 61, 1990.

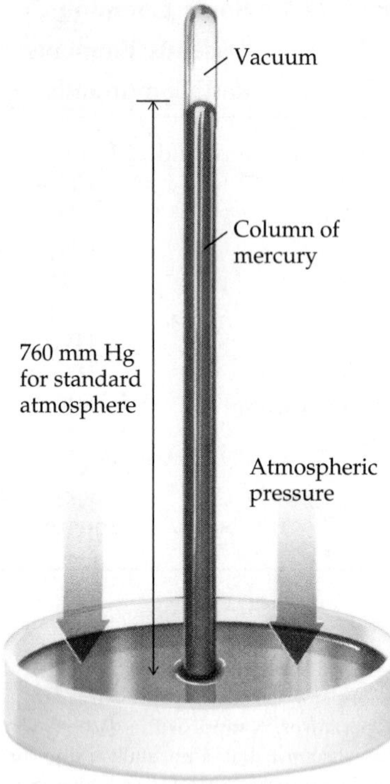

Figure 12.1 A mercury barometer. The pressure of the atmosphere on the surface of the mercury in the dish is balanced by the downward pressure exerted by the mercury in the column.

An old-fashioned pump that depends on atmospheric pressure to lift water from a well.

A gas, such as our atmosphere, exerts a pressure on every surface it contacts, no matter what the direction of contact. Atmospheric pressure can be measured with a barometer, which can be made by filling a tube with a liquid and inverting the tube in a dish containing the same liquid. Figure 12.1 shows a mercury-filled barometer. At sea level the height of the mercury column is about 760 mm above the surface of the mercury in the dish. The pressure exerted by the mercury column is balanced by the pressure at the bottom of the column of air above the dish—a column of gas that extends to the top of the atmosphere. Pressure exerted by a mercury barometer is usually reported in units of millimeters of mercury **(mm Hg),** a unit sometimes called the **torr** in honor of Evangelista Torricelli (1608–1647), who invented the mercury barometer in 1643. The **standard atmosphere (atm)** is defined as

$$1 \text{ standard atmosphere} = 1 \text{ atm} = 760 \text{ mm Hg (exactly)}$$

The SI unit of pressure is the **pascal (Pa),** named for the French mathematician and philosopher Blaise Pascal (1623–1662). It is the only pressure unit that is defined directly in terms of force per unit area.

$$1 \text{ pascal (Pa)} = 1 \text{ newton/meter}^2$$

This is a very small unit compared with ordinary pressures, so the **kilopascal (kPa)** is more often used. The relationship among the units listed above is

$$1 \text{ atm} = 760 \text{ mm Hg} = 101.325 \times 10^3 \text{ Pa} = 101.325 \text{ kPa}$$

Finally, atmospheric pressures are sometimes reported in the unit called the **bar,** where 1 bar = 100,000 Pa. Therefore,

$$1 \text{ atm} = 1.013 \text{ bar}$$

The thermodynamic data in Chapter 6 and in Appendix K are given for gas pressures of 1 bar.

Any liquid can be used in a barometer, but, as described in *A Closer Look: Measuring Gas Pressures,* the height of the column depends on the density of the liquid. A comparison of mercury, with a density of 13.6 g/cm^3, and water (density = 1 g/cm^3) indicates that if a barometer were filled with water, the column would be almost 34 feet high! Obviously we don't use water-filled barometers, but this principle is important to keep in mind when digging a well. A simple hand pump can create a reduced pressure at the top of a well casing that allows atmospheric pressure on the water at the bottom of a well to cause the water to rise upward. When the water reaches the surface, it flows out. Submersible pumps are now used in very deep water wells, but a well with a simple surface pump cannot be deeper than about 33 feet.

EXAMPLE 12.1 *Pressure Unit Conversions*

Convert a pressure of 635 mm Hg into its corresponding value in units of atmospheres (atm), bars, and kilopascals (kPa).

Solution The relationship between the units mm Hg and atm is: 1 atm = 760. mm Hg. Notice that the given pressure is less than 760. mm Hg. so the equivalent in atmospheres is less than 1 atm.

$$635 \cancel{\text{mm Hg}}\left(\frac{1.00 \text{ atm}}{760. \cancel{\text{mm Hg}}}\right) = 0.836 \text{ atm}$$

The pressure of a gas is a measure of the force that it exerts on its container. **Force** is the physical quantity that causes a change in the motion of a mass if it is free to move. For instance, gravity is the force the earth exerts on all objects near it, causing the same acceleration (change in velocity with time) for all objects. The gravitational force is also known as the **weight** of the object.

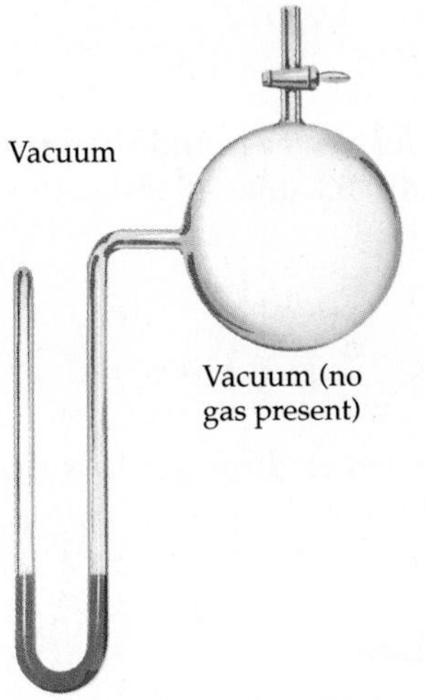

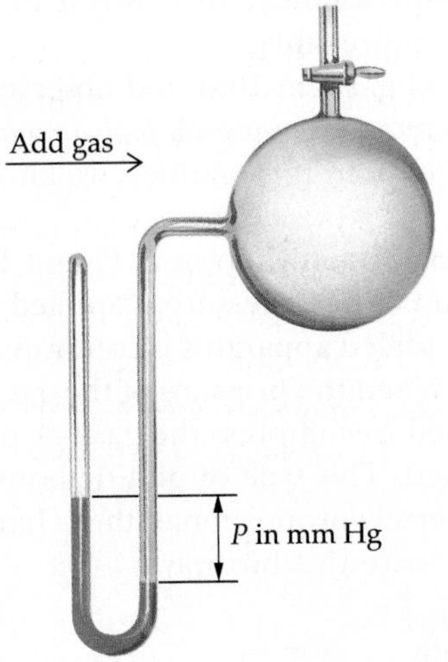

Newton's first law states that force = mass × acceleration. The SI units of mass and acceleration are the kilogram (kg) and the meter per second squared (m/s^2), respectively, so the units of force are kg · m/s^2. This derived unit is given the name **newton (N).**

Pressure is defined as the force exerted on an object divided by the area over which the force is exerted.

$$\text{Pressure} = \frac{\text{force}}{\text{area}}$$

This book, for example, weighs a little more than 1.8 kg and so exerts a force of about 20 N. It has an area of approximately 530 cm^2 when it lies flat on a surface. Therefore, the pressure exerted by the book on a surface is about 400 N/m^2.

To see how we can use a barometer (see Figure 12.1) to measure pressure, consider the pressure that the mercury in the column exerts on the mercury in the dish. This pressure is, by definition, the weight of the mercury column divided by the cross-sectional area of the tube (that is, as weight per square centimeter, for example). Because weight is proportional to mass, we can write the pressure as shown below in Equation 1.

The mass of an object is equal to its volume times its density (see Chapter 1), so the pressure *P* exerted by the mercury in the vertical column is calculated as in Equation 2. This means that the pressure exerted by the column of mercury in a barometer (see Figure 12.1) depends only on the height of the mercury column and not on its area. Because this pressure is equal to the atmospheric pressure outside the beaker, it is natural to measure atmospheric pressure (or the pressure of any other gas) by measuring the height of the column of mercury it can support. A common unit of pressure is the **millimeter of mercury (mm Hg).**

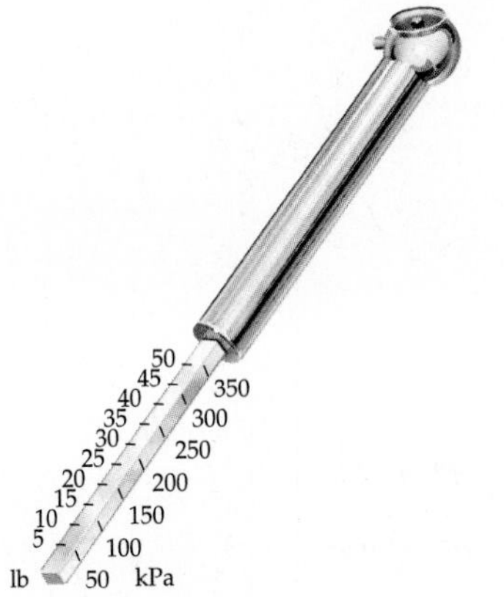

A tire gauge. The "lb" scale measures air pressure in pounds per square inch (psi).

Gas pressures are also often measured in the laboratory with a U-tube manometer. The closed side of a mercury-filled, U-shaped glass tube is evacuated so that no gas remains to press on the mercury surface on that side. The other side is open to the gas whose pressure is to be measured. The gas pressure (in mm Hg) is read directly as the difference in mercury levels in the closed and open sides.

You may have used a tire gauge to check the pressure in your car or bike tires. Such gauges indicate the pressure in pounds per square inch (psi), and sometimes in kilopascals. Both of these scales refer to the pressure *in excess* of atmospheric pressure. (A flat tire is not a vacuum; it contains air with a pressure of 1 atm.) A tire containing air at a pressure of 3.4 atm has a gauge pressure of 2.4 atm; the tire gauge shows 2.4 atm × (14.7 lb/in.^{2})/atm, or 35 psi. On the kilopascal scale, the pressure is 2.4 atm × 101 kPa/atm = 242 kPa.

Equation 1

$$\text{Pressure} = \frac{\text{weight of mercury column}}{\text{cross-sectional area of column}} \propto \frac{\text{mass of mercury column}}{\text{area of column}}$$

The symbol ∝ means "proportional to."

Equation 2

$$P \propto \frac{\text{volume of mercury column} \times \text{density of Hg}}{\text{area of column}}$$

$$\propto \frac{(\text{height of column} \times \text{area of column}) \times \text{density of Hg}}{\text{area of column}}$$

$$\propto \text{height of column} \times \text{density of mercury}$$

The relationship between the unit atm and bars is: 1 atm = 1.013 bar. Therefore, we have

$$0.836 \text{ atm}\left(\frac{1.013 \text{ bar}}{\text{atm}}\right) = 0.847 \text{ bar}$$

Finally, the relationship between the unit mm Hg and kPa is: 101.325 kPa = 760. mm Hg. Therefore,

$$635 \text{ mm Hg}\left(\frac{101.325 \text{ kPa}}{760. \text{ mm Hg}}\right) = 84.7 \text{ kPa}$$

EXERCISE 12.1 ***Pressure Unit Conversions***

Rank the following pressures in decreasing order of magnitude (largest first, smallest last): 75 kPa, 250 mm Hg, 0.83 bar, and 0.63 atm. ■

12.2 THE GAS LAWS: THE EXPERIMENTAL BASIS

Experimentation in the 17th and 18th centuries led to three gas laws that provide the basis for our understanding of gas behavior.

A bicycle pump works by compressing a sample of atmospheric air into a smaller volume. As a consequence, the pressure of the sample is larger. You experience Boyle's law because you can feel the increasing pressure of the gas as you press down on the plunger.

The Compressibility of Gases: Boyle's Law

When you pump up the tires of your bicycle, the pump squeezes the air into a smaller volume. This property is called **compressibility.** In contrast to gases, liquids and solids are almost completely incompressible.

Robert Boyle studied the compressibility of gases in 1661 and observed that *the volume of a fixed amount of gas at a given temperature is inversely proportional to the pressure exerted on the gas.* Because all gases behave in this manner, we know this principle as **Boyle's law.**

Boyle's law can be demonstrated by the experiment shown in Figure 12.2. A hypodermic syringe is filled with air and sealed. When pressure is applied to the movable plunger of the syringe, the air in the sealed apparatus is compressed; its volume decreases, and its pressure increases. When the pressure of the gas in the syringe—as measured by the mass of lead used to compress the gas—is plotted as a function of $1/V$, a straight line is observed. This type of plot demonstrates that the volume and pressure of a gas are inversely proportional; they change in opposite directions. Mathematically, we can write this two ways.

$$P \propto \frac{1}{V} \quad \text{or equivalently} \quad V \propto \frac{1}{P}$$

Both show that, for a given quantity of gas at a fixed temperature, the gas volume decreases if the pressure increases. Conversely, if the pressure is lowered, then the gas volume increases.

Boyle's experimentally determined relation can be put into a useful mathematical form as follows. When two quantities are proportional to each other, they can be equated if a proportionality constant, here called C_B, is introduced.

PORTRAIT OF A SCIENTIST

Robert Boyle (1627–1691)

Robert Boyle was born in Ireland, in a home that still stands, as the 14th and last child of the first Earl of Cork. He first published his studies of gases in 1660, and a book, *The Sceptical Chymist,* was published in 1680. Although Boyle was the first to define elements in modern terms, he retained medieval views about what the elements were. For example, Boyle thought that gold was not an element, but rather a metal that could be formed from other metals. Boyle was also a physiologist—he was the first to show that the healthy human body has a constant temperature. Not everyone applauded all aspects of Boyle's work. Isaac Newton, a young man when Boyle was at the peak of his career, questioned the correctness of Boyle's ideas.

Robert Boyle. (Oesper Collection in the History of Chemistry/University of Cincinnati)

Thus,

$$P = C_B\left(\frac{1}{V}\right) \qquad \text{or} \qquad PV = C_B \tag{12.1}$$

The last relation, $PV = C_B$, is known as Boyle's law: *For a given quantity of gas at a given temperature, the product of pressure and volume is a constant,* where the constant C_B is determined by the quantity of gas (in moles) and the temperature (in kelvins). This means that if the pressure-volume product is known for one set of conditions (P_1 and V_1), it is known for all other conditions of pressure and volume (P_2 and V_2), (P_3 and V_3), and so on for that gas sample.

$$P_1V_1 = P_2V_2 = P_3V_3 \cdots$$

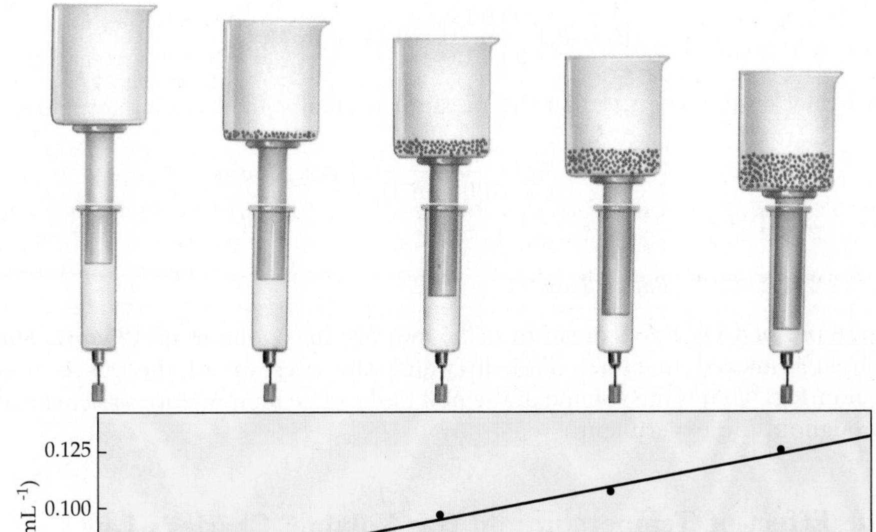

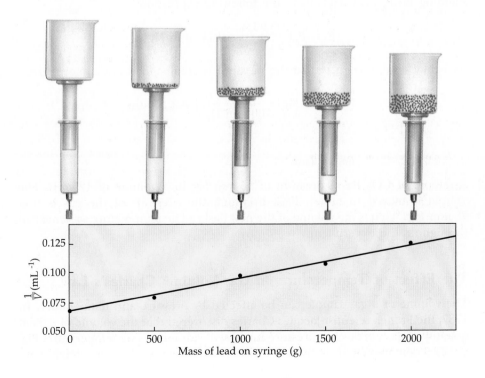

Figure 12.2 A simple way to demonstrate Boyle's law. A syringe, containing some air, is sealed and then lead shot is added to the beaker on top of the barrel. As the mass of lead increases, the air in the syringe compresses, and the pressure of the air in the syringe increases. A plot of (1/volume of air) versus the mass of lead shows that a linear relationship exists between $1/V$ and P (as measured by the mass of lead). (See D. Davenport; *Journal of Chemical Education,* Vol. 39, p. 252, 1962.)

This form of Boyle's law is useful when we want to know, for example, what happens to the volume of a given quantity of gas when the pressure changes (at a constant temperature).

EXAMPLE 12.2 *Boyle's Law*

A sample of gaseous nitrogen in an automobile air bag has a pressure of 745 mm Hg in a 35.0-L bag. If this sample is transferred to a 25.0-L bag with the same temperature as before, what is the pressure of the gas in the new bag?

Solution You will find it is quite useful to make a table of the information provided.

Original Conditions	Final Conditions
$P_1 = 745$ mm Hg	$P_2 = ?$
$V_1 = 35.0$ L	$V_2 = 25.0$ L

You know that $P_1V_1 = P_2V_2$; Therefore,

$$P_2 = \frac{P_1V_1}{V_2} = \frac{(745 \text{ mm Hg})(35.0 \text{ L})}{(25.0 \text{ L})} = 1040 \text{ mm Hg}$$

Remember that the *essence of Boyle's law is that P and V change in opposite directions.* Indeed, this suggests a way to solve the problem without going through the algebraic manipulation of $P_1V_1 = P_2V_2$. That is, because you know that the volume has decreased, you know the new pressure (P_2) must be greater than the old pressure (P_1); thus, P_1 must be multiplied by a volume factor that has a value *greater than 1* to reflect the fact that P_2 must be greater than P_1. Simply understanding Boyle's law leads us to the following expression:

$$P_2 = P_1\left(\frac{35.0 \text{ L}}{25.0 \text{ L}}\right) = 1040 \text{ mm Hg}$$

As a final note, recognize that the pressure is greater than one atmosphere.

$$1040 \text{ mm Hg}\left(\frac{1 \text{ atm}}{760 \text{ mm Hg}}\right) = 1.37 \text{ atm}$$

EXERCISE 12.2 *Boyle's Law*

A sample of CO_2 has a pressure of 55 mm Hg in a volume of 125 mL. The sample is moved to a new flask in which the pressure of the gas is now 78 mm Hg. What is the volume of the new flask? (The temperature was constant throughout the experiment.) ■

The Effect of Temperature on Gas Volume: Charles's Law

The volume of a gas sample can be affected by pressure and temperature. In 1787, the French scientist Jacques Charles discovered that the volume of a fixed quantity of gas at constant pressure increased with increasing temperature. Figure 12.3 demonstrates how the volumes of two different gas samples change with

PORTRAIT OF A SCIENTIST

Jacques Alexandre César Charles (1746–1823)

The French chemist Charles was most famous in his lifetime for his experiments in ballooning. The first such flights were made by the Montgolfier brothers in June 1783, using a large spherical balloon made of linen and paper and filled with hot air. In August 1783, however, a different group, supervised by Jacques Charles, tried a different approach. Exploiting his recent discoveries in the study of gases, Charles decided to inflate the balloon with hydrogen gas. Because hydrogen would escape easily from a paper bag (see Section 12.7), Charles made a bag of silk coated with a rubber solution. Inflating the bag to its final diameter took several days and required nearly 500 pounds of acid and 1000 pounds of iron to generate the hydrogen gas. A huge crowd watched the ascent on August 27, 1783. The balloon stayed aloft for almost 45 minutes and traveled about 15 miles. When it landed in a village, however, the people were so terrified they tore it to shreds.

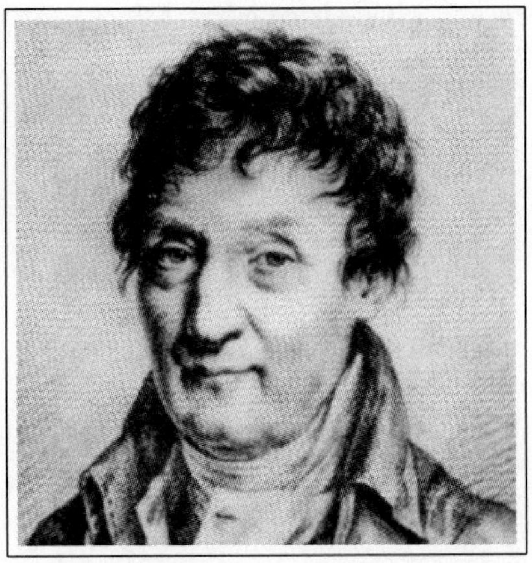

Jacques Charles. (The Bettmann Archive)

temperature (at a constant pressure). When the plots of volume versus temperature are extended toward lower temperatures, they all reach zero volume at a common temperature, −273.15 °C. Of course gases do not actually reach zero volume at this temperature; they liquefy well above that temperature (as indicated by the dotted line in Figure 12.3).

In 1848, William Thomson, also known as Lord Kelvin, proposed that it would be convenient to have a temperature scale in which the zero point was −273.15 °C. This temperature scale has been named for Lord Kelvin, and the units of the scale are known as kelvins. The Kelvin degree, which has been adopted as the SI unit for temperature measurement, is equivalent in size to the Celsius degree. When the Kelvin temperature scale is used, the volume-temperature relationship, now known as **Charles's law,** is as follows: *If a given quantity of gas is held at a constant pressure, its volume is directly proportional to the absolute temperature.* This can be expressed by the relation

$$V \propto T$$

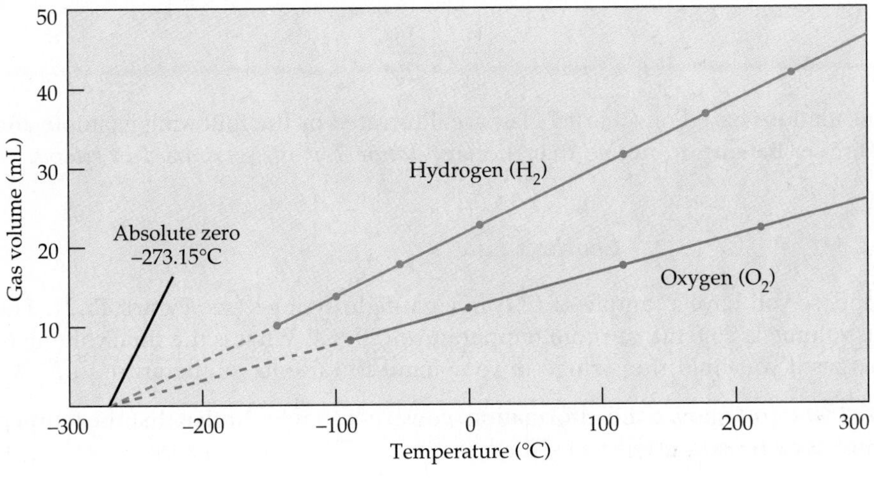

Figure 12.3 Charles's law. The volumes of two different samples of gases decrease with decreasing temperature (at constant pressure). These graphs (as well as those of all other gases) intersect the temperature axis at about −273 °C. (The dotted lines indicate that the substance is actually liquid in this temperature range. If the substance were a gas, its volume-temperature relationship would be described by the dotted line.)

(a) (b) (c)

Figure 12.4 A dramatic illustration of Charles's law. Some air-filled balloons are placed in liquid nitrogen (at 77 K) (a). The volume of the gas is dramatically reduced at this temperature. After all the balloons have been placed in the liquid nitrogen (b), they are poured out again and reinflate to their original volume when warmed back to room temperature (c). (C. D. Winters)

or, using a proportionality constant, C_C, we can write the equation

$$V = C_C \times T$$

or

$$C_C = \frac{V}{T} \qquad \textbf{(12.2)}$$

Charles's law is illustrated in an extreme manner in Figure 12.4.

The quotient V/T is always equal to the same constant (for a given sample at a specified pressure). This leads to the conclusion that the volume of a sample of gas doubles when T, the Kelvin temperature, is doubled (at a constant pressure), for example. That is, if we know the volume and temperature of a gas sample (V_1 and T_1), we can find the new volume V_2 at a new temperature T_2 from the equation

$$\frac{V_1}{T_1} = \frac{V_2}{T_2}$$

Calculations based on Charles's law are illustrated by the following example and exercise. Be sure to notice that the *temperature T must be expressed in kelvins.*

EXAMPLE 12.3 *Charles's Law*

Suppose you have a sample of CO_2 in a gas-tight syringe (see Figure 12.2). The gas volume is 25.0 mL at room temperature (20 °C). What is the final volume of the gas if you hold the syringe in your hand to raise its temperature to 37 °C?

Solution To organize the information, construct a table. Notice that the temperature *must* be converted to kelvins.

Original Conditions	Final Conditions
$V_1 = 25.0$ mL	$V_2 = ?$
$T_1 = 20 + 273 = 293$ K	$T_2 = 37 + 273 = 310.$ K

You know that $V_1/T_1 = V_2/T_2$, so you can rearrange the equation to solve for V_2 and substitute the information given.

$$V_2 = T_2\left(\frac{V_1}{T_1}\right) = (310.\text{ K})\left(\frac{25.0\text{ mL}}{293\text{ K}}\right) = 26.5\text{ mL}$$

As expected, the volume of gas increases with an increase in temperature. Once again, this suggests another logical way to solve the problem. The new volume (V_2) must be equal to the old volume (V_1) multiplied by a temperature fraction that is *greater than 1* to reflect the effect of the increase in temperature. That is, V_1 is increased to V_2 by the temperature increase.

$$V_2 = V_1\left(\frac{310.\text{ K}}{293\text{ K}}\right)$$

EXERCISE 12.3 ***Charles's Law***

A balloon is inflated with helium to a volume of 45 L at room temperature (25 °C). If the balloon is inflated with the same quantity of helium on a very cold day (−10 °C), what is the new volume of the balloon? ■

As a final note on Boyle's and Charles's laws, it is important to point out that neither law depends on the identity of the gas being studied. These laws reflect the properties of all gases and describe the behavior of any gaseous substance, regardless of its identity.

Equal Volumes of Gases Contain Equal Numbers of Molecules (at Constant *T* and *P*): Gay-Lussac and Avogadro

The beginning of the 19th century was an exciting time for chemistry, with work being done by John Dalton, Jacques Charles, Amedeo Avogadro, and the Frenchman Joseph Gay-Lussac. Gay-Lussac experimented with the reactions of gases and found that volumes of gases always combine with one another in the ratio of small whole numbers, as long as the volumes are measured at the same temperature and pressure. This statement is now referred to as Gay-Lussac's *law of combining volumes.* As an example, this means that 100 mL of H_2 gas combines exactly with 50 mL of O_2 gas to give exactly 100 mL of H_2O vapor if all the gases are measured at the same T and P (Figure 12.5).

Gay-Lussac's law remained only a summary of experimental observations until it was explained by the work of the Italian physicist and lawyer, Amedeo Avogadro. In 1811 Avogadro published his ideas, now known as *Avogadro's hypothesis,* that *equal volumes of gases under the same conditions of temperature and pressure have equal numbers of molecules.* Applied to Gay-Lussac's experiment, this means that 100 mL of H_2 molecules in the gas phase must have twice the number of

PROPERTY OF SHINKO CHENG
(415) 878-4492 '96

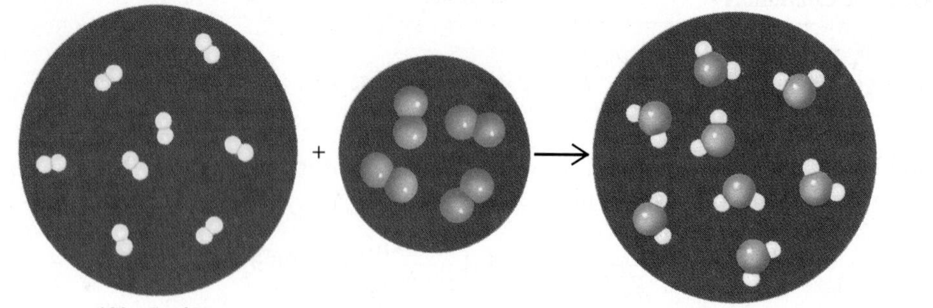

Figure 12.5 An illustration of Gay-Lussac's law: volumes of gases, at the same temperature and pressure, combine with one another in the ratio of small whole numbers. Here, one volume of O_2, say 50 mL at 100 °C and 1 atm pressure, combines with two volumes of H_2 gas (100 mL) to give two volumes (100 mL) of H_2O vapor, both of the latter gases also measured at 100 °C and 1 atm pressure.

molecules as 50 mL of O_2, and when combined by reaction they should produce 100 mL of H_2O molecules.

Avogadro's law follows from Avogadro's hypothesis: *The volume of a gas, at a given temperature and pressure, is directly proportional to the quantity of gas.* Thus, V is proportional to n, the number of moles of gas, so

$$V \propto n \qquad \text{or} \qquad V = C_A n \tag{12.3}$$

where C_A is the proportionality constant. The essence of Avogadro's law is that the volume and number of moles of a gas change in the same direction. As the quantity of gas increases, the volume of the container must increase (at constant T and P). This is the reason that a balloon inflates as you puff more air into it.

EXAMPLE 12.4 *Avogadro's Law*

Ammonia can be made directly from the elements.

$$N_2(g) + 3\,H_2(g) \longrightarrow 2\,NH_3(g)$$

If one begins with 15.0 L of $H_2(g)$ at a given T and P, what volume of $N_2(g)$ is required for complete reaction (at the same T and P)? What is the theoretical yield of NH_3, in liters?

Solution Because gas volume is proportional to the number of moles of a gas, we can use volumes instead of moles and treat this just as we did stoichiometry problems in Chapter 5.

1. Calculate the volume of N_2 required (in liters).

$$15.0\text{ L } H_2(g) \text{ available}\left(\frac{1\text{ L } N_2(g)\text{ required}}{3\text{ L } H_2(g)\text{ available}}\right) = 5.00\text{ L } N_2(g)\text{ required}$$

2. Calculate the volume of NH_3 produced (in liters).

$$15.0\text{ L } H_2(g) \text{ available}\left(\frac{2\text{ L } NH_3(g)\text{ produced}}{3\text{ L } H_2(g)\text{ available}}\right) = 10.0\text{ L } NH_3\text{ produced}$$

EXERCISE 12.4 *Avogadro's Law*

Methane burns in oxygen to give the usual products, CO_2 and H_2O.

$$CH_4(g) + 2\,O_2(g) \longrightarrow 2\,H_2O(g) + CO_2(g)$$

If 22.4 L of gaseous CH_4 is burned, what volume of O_2 is required for complete combustion? What volumes of H_2O and CO_2 are produced? Assume all gases are measured at the same temperature and pressure. ■

12.3 THE IDEAL GAS LAW

A balloon inflates because, as you blow into it, you are introducing a greater and greater mass of air. This increases the volume of the balloon. (The pressure also increases because, as the balloon inflates, the skin of the balloon is stretched.)

Four interrelated quantities can be used to describe the state of a gas: pressure, volume, temperature, and quantity (moles). The three gas laws described in Section 12.2 tell how one gas property is affected as another is changed, assuming the other properties remain fixed. In summary, these laws are

Boyle's Law	Charles's Law	Avogadro's Law
$V \propto \frac{1}{P}$	$V \propto T$	$V \propto n$
(constant T, n)	(constant P, n)	(constant T, P)

If all three laws are combined, the result is

$$V \propto \frac{nT}{P} \quad \text{or} \quad V = R\left(\frac{nT}{P}\right)$$

The proportionality constant, labeled ***R***, is called the **gas constant.** This constant, a number that connects P, V, T, and n, is a "universal constant," a number that you can always use to interrelate the properties of a gas. Rearranging the combined equation gives

$$PV = nRT \tag{12.4}$$

This last equation is called the **ideal gas law** because it describes the state of a so-called "ideal" gas. As you will learn in Section 12.9, there is no such thing as an "ideal" gas. However, real gases at pressures around an atmosphere or less and temperatures around room temperature usually behave close enough to ideality that $PV = nRT$ is quite an adequate description of their behavior.

To use the equation $PV = nRT$, we need a value for R. Many experiments show that under conditions of **standard temperature and pressure (STP)** (a gas temperature of 0 °C, or 273.15 K, and a pressure of exactly 1 atm) exactly 1 mol of gas occupies **22.414 L,** a quantity called the **standard molar volume.** Substituting these values into the ideal gas law, we can solve for R.

$$R = \frac{PV}{\text{nT}} = \frac{(1.0000 \text{ atm})(22.414 \text{ L})}{(1.0000 \text{ mol})(273.15 \text{ K})} = 0.082057 \frac{\text{L} \cdot \text{atm}}{\text{K} \cdot \text{mol}}$$

With a value for R, we can now use the ideal gas law to perform some useful calculations.

EXAMPLE 12.5 *The Ideal Gas Law*

The nitrogen gas in an air bag, with a volume of 35 L, exerts a pressure of 850 mm Hg at 25 °C. How many moles of N_2 are in the air bag?

Solution First, let us write down the information provided.

$$V = 35\ \text{L} \qquad P = 850\ \text{mm Hg} \qquad T = 25\ °\text{C} \qquad n = ?$$

To use the ideal gas law with R having units of (L · atm/K · mol), the pressure must be expressed in atmospheres and the temperature in kelvins. Therefore,

$$P = 850\ \text{mm Hg}\left(\frac{1\ \text{atm}}{760.\ \text{mm Hg}}\right) = 1.1\ \text{atm}$$

$$T = 25 + 273 = 298\ \text{K}$$

Now rearrange the ideal gas law to solve for the number of moles, n,

$$n = \frac{PV}{RT}$$

and substitute into this equation.

$$n = \frac{(1.1\ \cancel{\text{atm}})(35\ \cancel{\text{L}})}{(0.082057\ \cancel{\text{L}} \cdot \cancel{\text{atm}}/\cancel{\text{K}} \cdot \text{mol})(298\ \cancel{\text{K}})} = 1.6\ \text{mol}\ N_2$$

Notice that units of atmosphere, liters, and kelvins cancel to leave the answer in units of moles.

EXERCISE 12.5 ***The Ideal Gas Law***

The balloon used by Charles in his historic flight in 1783 was filled with about 1300 mol of H_2. If the temperature of the gas was 20. °C, and its pressure was 750 mm Hg, what was the volume of the balloon? ■

The ideal gas law is useful for calculating one of the four properties of a gas when the other three are known. Many times, however, we need to know what happens to a gas when a change occurs, in one, two, or even three of the parameters P, V, and T. Because R is a universal constant for gases, a quantity of a gas, n_1, exerting a pressure P_1 in a volume V_1 at a temperature T_1 must obey the equation

$$R = \frac{P_1 V_1}{n_1 T_1}$$

If the quantity of the gas remains the same, but the other conditions change to new values (to P_2, V_2, and T_2), we write the same expression for R

$$R = \frac{P_2 V_2}{n_2 T_2}$$

(where $n_2 = n_1$ in this case). Under either set of conditions, the quotient PV/nT is equal to universal constant R. Therefore,

$$\frac{P_1 V_1}{n_1 T_1} = \frac{P_2 V_2}{n_2 T_2} \qquad \textbf{(12.5)}$$

This is sometimes called the **general gas law,** or **combined gas law.**

EXAMPLE 12.6 *The General Gas Law*

Helium-filled balloons are used to carry scientific instruments high into the atmosphere. Suppose that a balloon is launched when the temperature is 22.5 °C and the barometric pressure is 754 mm Hg. If the balloon's volume is 4.19×10^3 L, what will it be at a height of 20 miles, where the pressure is 76.0 mm Hg and the temperature is −33.0 °C?

Solution Assume that no gas escapes from the balloon. Then only T, P, and V change. With this in mind, we begin by setting out the information given in a table. Notice that *temperatures must be used in kelvins; this is always true when using any of the gas laws.*

Initial Conditions	Final Conditions
$V_1 = 4.19 \times 10^3$ L	$V_2 = ?$ L
$P_1 = 754$ mm Hg	$P_2 = 76.0$ mm Hg
$T_1 = 22.5$ °C (295.7 K)	$T_2 = -33.0$ °C (240.2 K)
$n_1 = n_2$	

We rearrange the general gas law (Equation 12.5) to calculate the new volume V_2. (Notice that n_1 and n_2 cancel each other.)

$$V_2 = \left(\frac{T_2}{P_2}\right) \times \left(\frac{P_1V_1}{T_1}\right) = V_1 \times \frac{P_1}{P_2} \times \frac{T_2}{T_1}$$

$$= 4.19 \times 10^3 \text{ L} \times \frac{754 \text{ mm Hg}}{76.0 \text{ mm Hg}} \times \frac{240.2 \text{ K}}{295.7 \text{ K}} = 3.38 \times 10^4 \text{ L}$$

Notice that the ratio of pressures is the same whether they are given in units of mm Hg or atm.

The pressure decreases by almost a factor of 10, which should lead to a volume increase. This increase is partly offset by the fact that the temperature drops, which should lead to a volume decrease. On balance, the volume increases because the pressure drops so substantially.

The reasoning we used in this example could have been used to solve the problem without setting up the general gas law equation and doing an algebraic manipulation. The effect of a pressure decrease should be a volume increase. Therefore, we should have multiplied V_1 by a pressure factor greater than 1. Similarly, a drop in temperature leads to a volume decrease. For this reason we should multiply V_1 by a temperature factor less than 1.

PROBLEM-SOLVING TIPS AND IDEAS

12.1 Units and the General Gas Law

When using R = 0.082057 L · atm/K · mol, the volume must be in units of liters, pressure in units of atmospheres, and temperature in kelvins. When using the general gas law (Equation 12.5), however, it is not absolutely necessary to make unit changes *except* for expressing T in kelvins. The reason for this is that the general gas law considers *ratios* of P and V. Nonetheless, as you work gas law problems it is probably best to get into the habit of always using atmospheres and liters (as well as kelvins) so that you do not make a mistake. ■

Figure 12.6 Gas density. The two balloons are filled with nearly equal quantities of gas at the same temperature and pressure. The blue balloon contains low-density hydrogen ($d =$ 0.08 g/L), and the red balloon contains argon, a higher density gas ($d =$ 1.8 g/L). In comparison, the density of dry air at 1 atm pressure and 25 °C is about 1.2 g/L. (C. D. Winters)

EXERCISE 12.6 ***The General Gas Law***

You have a 20.-L cylinder of helium at a pressure of 150 atm and at 30 °C. How many balloons can you fill, each with a volume of 5.0 L, on a day when the atmospheric pressure is 755 mm Hg and the temperature is 22 °C? ■

As a final comment on the general gas law, notice that, for a given quantity of gas in a constant volume, the gas pressure is directly proportional to the absolute temperature.

$$P_2 = P_1\left(\frac{T_2}{T_1}\right) \text{ when } V_1 = V_2$$

When you drive an automobile for some distance, you know the tire pressure goes up. The reason for this is that friction warms the tires, and because the tire volume is nearly constant, the pressure increases.

The Density of Gases

The density of a gas at a given temperature and pressure (Figure 12.6) is a useful quantity. Let us see how this is related to the ideal gas law. By rearranging the expression $PV = nRT$, we have

$$\frac{n}{V} = \frac{P}{RT}$$

Because the number of moles (n) of any compound is given by its mass (m) divided by its molar mass (M), we can rewrite the equation as

$$\frac{m}{M \cdot V} = \frac{P}{RT}$$

Because density (d) is defined as mass/volume (m/V), we can rearrange the preceding equation to relate density of gas, pressure, molar mass, and temperature.

$$d = \frac{m}{V} = \frac{PM}{RT} \qquad \textbf{(12.6)}$$

Now we have an equation relating the density of a gas to its molar mass at a given temperature and pressure.

EXAMPLE 12.7 ***Gas Density and Molar Mass***

The density of an unknown gas is 1.23 g/L at STP. Calculate its molar mass.

Solution List the information given in a short table, recalling that "STP" is shorthand for a pressure of 1 atm and a temperature of 0 °C.

$d = 1.23$ g/L

$T =$ standard temperature $= 273.15$ K

P = standard pressure = 1.000 atm

R = 0.082057 L · atm/K · mol

M = ?

The equation for gas density (Equation 12.6) is rearranged to solve for molar mass (M).

$$M = \frac{dRT}{P}$$

$$= \frac{(1.23\ \text{g/L})(0.082057\ \text{L} \cdot \text{atm/K} \cdot \text{mol})(273.15\ \text{K})}{1.000\ \text{atm}} = 27.6\ \text{g/mol}$$

EXERCISE 12.7 ***Gas Density Calculation***

Calculate the density of dry air at 15.0 °C and 1.00 atm if its molar mass (average) under those conditions is 28.96 g/mol. ■

This discussion of gas density has some important practical implications. From the equation $d = M(P/RT)$, we know that gas density is directly proportional to the molar mass (for gases at the same T and P). Dry air, with an average molar mass of about 29 g/mol, has a density of about 1.2 g/L (at 1 atm and 25 °C). This means that gases or vapors with molar masses greater than 29 g/mol have densities larger than 1.2 g/L (at 1 atm and 25 °C). Therefore, gases such as CO_2, SO_2, and gasoline vapor settle along the ground if released into the atmosphere. Conversely, gases such as H_2, He, CO, CH_4 (methane), and NH_3 rise if released into the atmosphere. The release of methyl isocyanate (H_3CNCO) into the air on December 3, 1984, in Bhopal, India, killed several hundred people and injured thousands more because the toxic vapors were heavier than air and settled close to the ground where they could be inhaled.

The carbon dioxide gas from a fire extinguisher is more dense than air. When the CO_2 emerges from the tank, it cools significantly. Water vapor in the air is condensed in the plume of CO_2 gas.

Calculating the Molar Mass of a Gas from *P*, *V*, and *T* Data

When a new compound is isolated in the laboratory, one of the first things to be done is to determine its molar mass. Of the number of ways of doing this, one uses a mass spectrometer such as the one described in Figure 2.11. If the compound is in the gas phase, a classical method of determining the molar mass is to measure the pressure exerted by a given mass of the gas at a given temperature.

EXAMPLE 12.8 ***Calculating the Molar Mass of a Gas from* P, V, *and* T *Data***

Let us assume you have done an experiment to determine the empirical formula of a compound now used to replace CFCs in air conditioners. Your results give a formula of CHF_2. Now you need the molar mass of the compound to be able to know the molecular formula. Therefore, you do another experiment in which you find that a 0.100-g sample of the unknown compound has a pressure of 70.5 mm Hg in a 256-mL container at 22.3 °C. What is the molar mass of the compound? What is the formula of the molecule?

Solution Let us begin by organizing the data.

$$V = 256 \text{ mL or } 0.256 \text{ L}$$

$$T = 22.3\ °\text{C, or } 295.5 \text{ K}$$

$$P = 70.5 \text{ mm Hg}\left(\frac{1 \text{ atm}}{760.}\right) = 0.0928 \text{ atm}$$

We can find the molar mass of the gas by first using the ideal gas law to calculate n, the number of moles of gas equivalent to 0.100 g of the gas.

$$n = \frac{PV}{RT} = \frac{(0.0928 \text{ atm})(0.256 \text{ L})}{(0.082057 \text{ L} \cdot \text{atm/K} \cdot \text{mol})(295.5 \text{ K})} = 9.80 \times 10^{-4} \text{ mol}$$

Now you know that 0.100 g of gas is equivalent to 9.80×10^{-4} mol. Therefore,

$$\text{Molar mass} = \frac{0.100 \text{ g}}{9.80 \times 10^{-4} \text{ mol}} = 102 \text{ g/mol}$$

With this result, we can compare the experimentally determined molar mass with the mass of a mole of gas with the empirical formula CHF_2.

$$\frac{\text{Experimental molar mass}}{\text{Mass of 1 mol of } CHF_2} = \frac{102 \text{ g/mol}}{51.0 \text{ g/formula unit}} = 2 \text{ formula units of } CHF_2$$

Therefore, the formula of the compound is $C_2H_2F_4$, a compound known as HCFC-134, a replacement for chlorofluorocarbons in home and automobile air conditioners.

You may wish to use an alternative approach to the calculation of this formula. Recall that the density of a gas gives the molar mass if the temperature and pressure are known (Equation 12.6). In this case, the density is

$$d = \frac{0.100 \text{ g}}{0.256 \text{ L}} = 0.391 \text{ g/L}$$

If the equation $d = PM/RT$ is solved for the molar mass (M), we have

$$M = \frac{dRT}{P} = \frac{(0.391\ \text{g/L})(0.082057\ \text{L}\cdot\text{atm/K}\cdot\text{mol})(295.5\ \text{K})}{0.0928\ \text{atm}} = 102\ \text{g/mol}$$

EXERCISE 12.8 ***Molar Mass From* P, V, *and* T *Data***

A 0.105-g sample of a gaseous compound has a pressure of 560. mm Hg in a volume of 125 mL at 23.0 °C. What is its molar mass? ■

12.4 THE GAS LAWS AND CHEMICAL REACTIONS

Many important chemical reactions, such as the industrial production of ammonia or chlorine, involve gases.

$$N_2(g) + 3\ H_2(g) \longrightarrow 2\ NH_3(g)$$
$$2\ NaCl(aq) + 2\ H_2O(\ell) \longrightarrow 2\ NaOH(aq) + H_2(g) + Cl_2(g)$$

It is important to know how to deal with such reactions quantitatively, so the examples that follow explore chemical stoichiometry calculations involving gases. The scheme in Figure 12.7 connects these calculations for gas reactions with those done in Chapter 4.

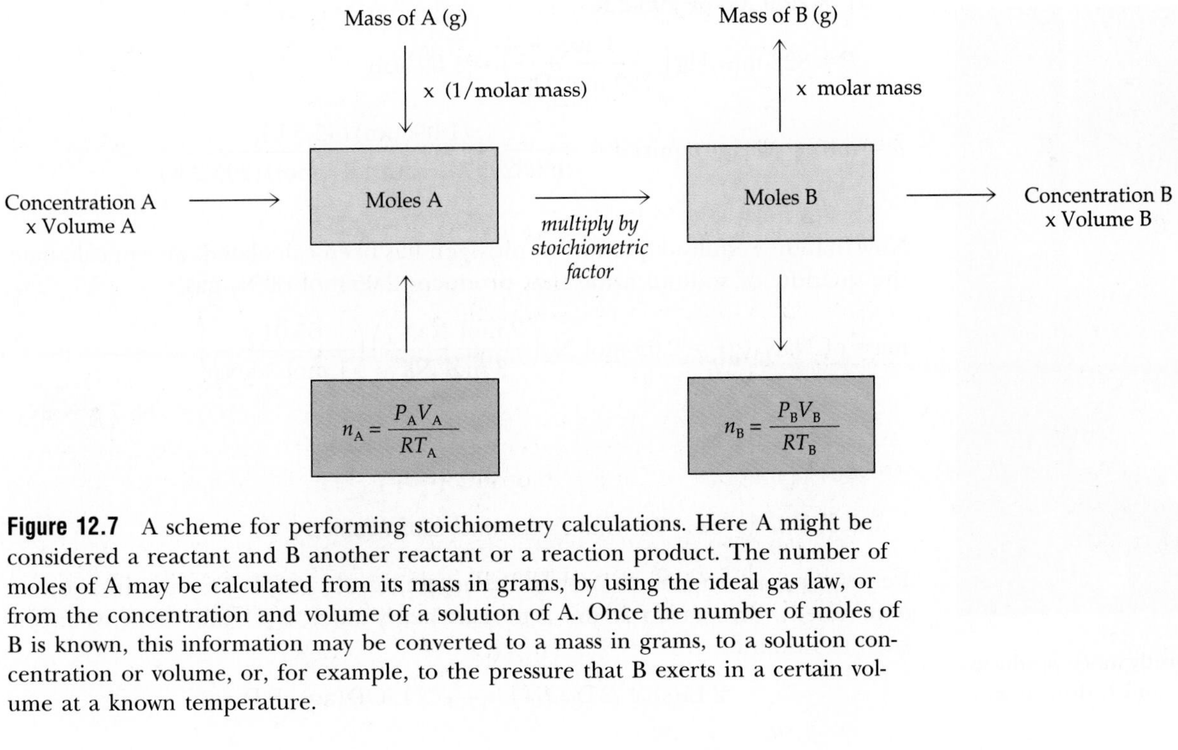

Figure 12.7 A scheme for performing stoichiometry calculations. Here A might be considered a reactant and B another reactant or a reaction product. The number of moles of A may be calculated from its mass in grams, by using the ideal gas law, or from the concentration and volume of a solution of A. Once the number of moles of B is known, this information may be converted to a mass in grams, to a solution concentration or volume, or, for example, to the pressure that B exerts in a certain volume at a known temperature.

EXAMPLE 12.9 *Gas Laws and Stoichiometry*

Let us say you are asked to design an air bag for a car. You know that the bag should be filled with gas with a pressure higher than atmospheric pressure, say 828 mm Hg, at a temperature of 22.0 °C. The bag has a volume of 45.5 L. What quantity of sodium azide, NaN_3, should be used to generate the required quantity of gas? The gas-producing reaction is

$$2\ NaN_3(s) \longrightarrow 2\ Na(s) + 3\ N_2(g)$$

Solution The general logic to be followed here can be outlined as follows:

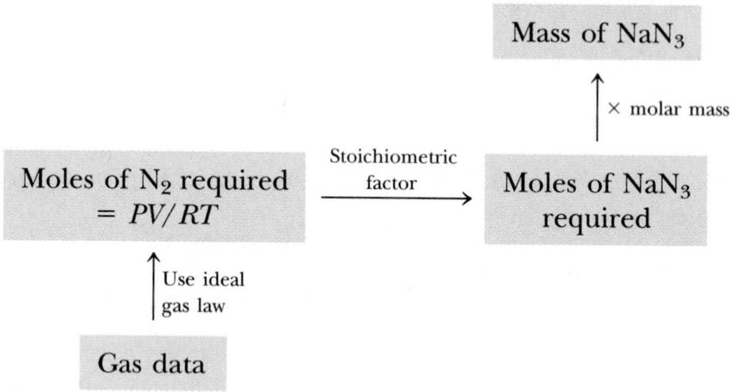

The first step must be to find the quantity of gas required, in moles, so that this can be related to the quantity of sodium azide required.

$$V = 45.5\ \text{L}$$

$$T = 22.0\ °\text{C, or } 295.2\ \text{K}$$

$$P = 828\ \text{mm Hg}\left(\frac{1\ \text{atm}}{760\ \text{mm Hg}}\right) = 1.09\ \text{atm}$$

$$n = \text{moles of } N_2 \text{ required} = \frac{(1.09\ \text{atm})(45.5\ \text{L})}{(0.082057\ \text{L} \cdot \text{atm/K} \cdot \text{mol})(295.2\ \text{K})}$$

$$= 2.05\ \text{mol } N_2$$

Now that the required quantity of nitrogen has been calculated, we can calculate the quantity of sodium azide that produces 2.05 mol of N_2 gas.

$$\text{mass of } NaN_3(g) = 2.05\ \text{mol } N_2\left(\frac{2\ \text{mol } NaN_3}{3\ \text{mol } N_2}\right)\left(\frac{65.01\ \text{g}}{1\ \text{mol } NaN_3}\right)$$

$$= 88.7\ \text{g } NaN_3$$

Lithium reacting with water produces lithium hydroxide and hydrogen gas.

EXAMPLE 12.10 *Gas Laws and Stoichiometry*

Let us suppose we wish to prepare some deuterium gas, D_2, for use in another experiment. One way to do this is to react heavy water, D_2O, with an active metal such as lithium.

$$2\ Li(s) + 2\ D_2O(\ell) \longrightarrow 2\ LiOD(aq) + D_2(g)$$

If we combine 0.125 g of Li metal with 15.0 mL of D_2O (d = 1.11 g/mL), what quantity of D_2 (in moles) can be prepared? If dry D_2 gas is captured in a 1450-mL flask at 22.0 °C, what is the pressure of the gas? (Deuterium has an atomic weight of 2.0147 g/mol.)

Solution We are combining two reactants with no guarantee that they are in the correct stoichiometric ratio. Therefore, this is a *limiting reactant problem,* and we need to find the number of moles of each substance and then see if one of them is present in limited amount. Once the limiting reactant is known, we can find the quantity of D_2 produced and then calculate its pressure under the conditions given.

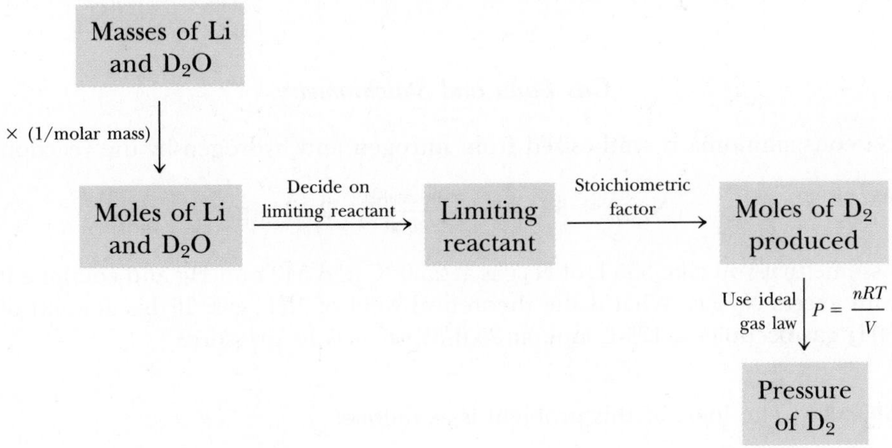

1. Calculate moles of Li.

$$0.125 \text{ g Li}\left(\frac{1 \text{ mol}}{6.941 \text{ g}}\right) = 0.0180 \text{ mol Li}$$

2. Calculate moles of D_2O.

$$15.0 \text{ mL } D_2O\left(\frac{1.11 \text{ g}}{\text{mL}}\right)\left(\frac{1 \text{ mol } D_2O}{20.03 \text{ g}}\right) = 0.831 \text{ mol } D_2O$$

3. Decide which reactant is the limiting reactant.

$$\text{Ratio moles of reactants available} = \frac{0.831 \text{ mol } D_2O}{0.0180 \text{ mol Li}} = \frac{46.2 \text{ mol } D_2O}{1 \text{ mol Li}}$$

The balanced equation shows that the ratio should be 1 mol of D_2O for 1 mol of Li. Therefore, Li is the limiting reactant, and the remainder of our calculations are based on the moles of Li available.

4. Use the limiting reactant to calculate the quantity of D_2 produced.

$$0.0180 \text{ mol Li}\left(\frac{1 \text{ mol } D_2 \text{ produced}}{2 \text{ mol Li}}\right) = 0.00900 \text{ mol } D_2 \text{ produced}$$

5. Calculate the pressure of D_2.

$P = ?$

$V = 1450$ mL, or 1.45 L

$T = 22.0$ °C, or 295.2 K

$n = 0.00900$ mol D_2

$$P = \frac{nRT}{V} = \frac{(0.00900 \text{ mol})(0.082057 \text{ L} \cdot \text{atm/K} \cdot \text{mol})(295.2 \text{ K})}{1.45 \text{ L}} = 0.15 \text{ atm}$$

$$0.150 \text{ atm}\left(\frac{760 \text{ mm Hg}}{1 \text{ atm}}\right) = 114 \text{ mm Hg}$$

EXAMPLE 12.11 *Gas Laws and Stoichiometry*

Gaseous ammonia is synthesized from nitrogen and hydrogen by the reaction

$$N_2(g) + 3\,H_2(g) \xrightarrow[500\ °C]{\text{iron catalyst}} 2\,NH_3(g)$$

Assume that you take 355 L of H_2 gas at 25.0 °C and 542 mm Hg and combine it with excess N_2 gas. What is the theoretical yield of NH_3 gas? If this amount of NH_3 gas occupies a 125-L tank at 25.0 °C, what is its pressure?

Solution The logic of this problem is as follows:

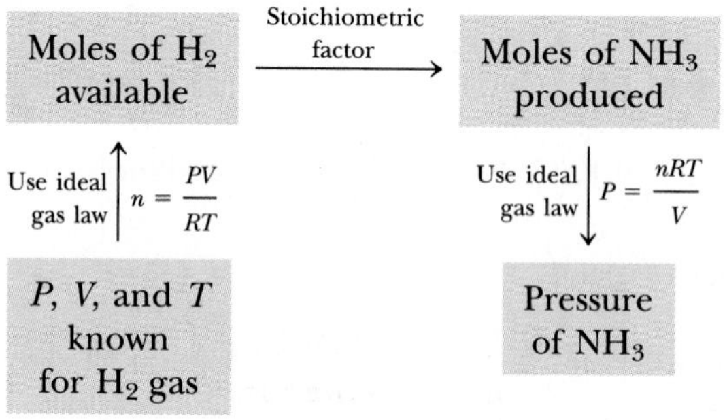

1. Calculate moles of H_2 gas available.

$P = (542 \text{ mm Hg})/(760. \text{ mm Hg/atm}) = 0.713$ atm

$V = 355$ L

$T = 25.0$ °C $+ 273.15 = 298.2$ K

$$\text{Moles of } H_2 = \frac{(0.713 \text{ atm})(355 \text{ L})}{(0.082057 \text{ L} \cdot \text{atm/K} \cdot \text{mol})(298.2 \text{ K})} = 10.3 \text{ mol } H_2$$

2. Calculate the number of moles of NH_3 that can be produced.

$$10.3 \text{ mol } H_2\left(\frac{2 \text{ mol } NH_3 \text{ produced}}{3 \text{ mol } H_2 \text{ available}}\right) = 6.87 \text{ mol } NH_3 \text{ produced}$$

3. Calculate the pressure of 6.87 mol of ammonia gas under the specified conditions.

$n = 6.87$ mol

$P = ?$

$V = 125$ L

$T = 25.0\ °C + 273.15\ K = 298.2\ K$

$$\text{Pressure of } NH_3 = \frac{nRT}{V} = \frac{(6.87 \text{ mol } NH_3)(0.082057 \text{ L} \cdot \text{atm/K} \cdot \text{mol})(298.2 \text{ K})}{125 \text{ L}}$$

$$= 1.35 \text{ atm}$$

EXERCISE 12.9 ***Gas Laws and Stoichiometry***

Gaseous oxygen reacts with aqueous hydrazine to produce water and gaseous nitrogen according to the balanced equation

$$N_2H_4(aq) + O_2(g) \longrightarrow 2\ H_2O(\ell) + N_2(g)$$

If a solution contains 180 g of N_2H_4, what is the maximum volume of O_2 that reacts with the hydrazine if the O_2 is measured at a barometric pressure of 750 mm Hg and room temperature (21 °C)? ■

12.5 GAS MIXTURES AND PARTIAL PRESSURES

The air you breathe is a blend of nitrogen, oxygen, carbon dioxide, water vapor, and small amounts of other gases (Table 12.2). Atmospheric pressure is the sum of the pressures exerted by the combination of these individual gases. The same is true of every gas mixture.

John Dalton was the first to observe that *the pressure of a mixture of gases is the sum of the pressures of the different components of the mixture.* Because the pressure of each individual gas in the mixture is called its **partial pressure,** Dalton's observation is known as **Dalton's law of partial pressures.** This is a consequence of the fact that the ideal gas law applies to all gases regardless of composition. There is,

TABLE 12.2 Components of Atmospheric Dry Air

Constituent	Molar Mass	Mole Percent	Partial Pressure at STP (atm)
N_2	28.01	78.084	0.78084
O_2	32.00	20.946	0.20946
CO_2	44.01	0.033	0.00033
Ar	39.95	0.934	0.00934

Average molar mass of dry air = 28.960 g/mol

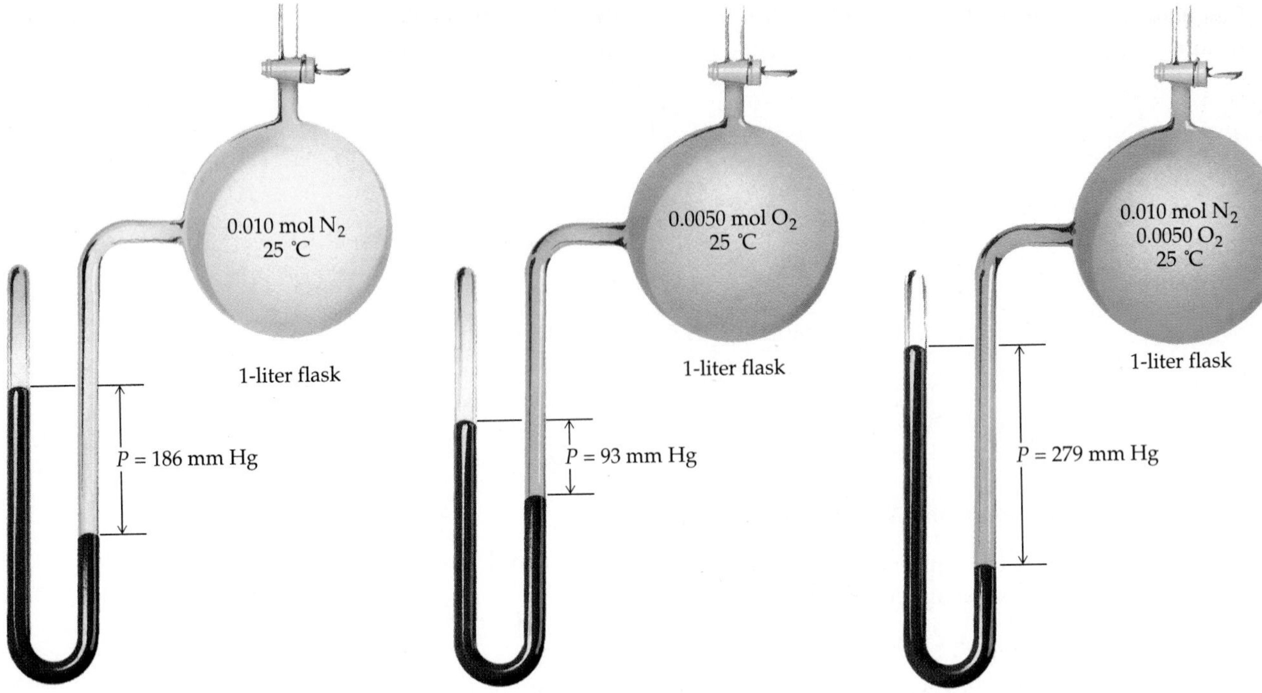

Figure 12.8 Dalton's law. *Left and middle:* 0.010 mol of N_2 in a 1.0-L flask at 25 °C exerts a pressure of 186 mm Hg, and 0.0050 mol of O_2 in a 1.0-L flask at 25 °C exerts a pressure of 93 mm Hg. *Right:* the N_2 and O_2 samples are mixed in the same 1.0-L flask at 25 °C. The total pressure, 279 mm Hg, is the sum of the pressures each gas would have if it were alone in the flask.

after all, no quantity in the ideal gas law that depends in any way on the chemical constitution of the gas molecules. Only by doing experiments dealing with the chemical properties of a gas can you tell a mixture of gases from a pure gas.

Suppose you fill a flask with a mixture of nitrogen and oxygen, in which the total pressure is P_{total} at some temperature, say 25 °C (Figure 12.8). The ideal gas law applied to this mixture is

$$P_{total}V = n_{total}RT$$

where n_{total} is the total number of moles of gas in the flask.

$$n_{total} = n_{N_2} + n_{O_2}$$

Substituting this into the equation for an ideal gas, we have

$$P_{total}V = (n_{N_2} + n_{O_2})RT$$

If we rearrange this equation slightly, and expand, we have a useful result.

$$P_{total} = \frac{n_{total}RT}{V} = \frac{(n_{N_2} + n_{O_2})RT}{V} = n_{N_2}\left(\frac{RT}{V}\right) + n_{O_2}\left(\frac{RT}{V}\right)$$

Remembering that $P = nRT/V$, we finally see that the total pressure of a mixture of gases is the sum of the pressure of each individual gas.

$$P_{total} = P_{N_2} + P_{O_2}$$

In a mixture of gases, all of the constituents have the same volume and temperature. A consequence of this is that *the pressure of one constituent of a mixture is directly proportional to the number of moles of that constituent.* Let us use this idea to extend the usefulness of Dalton's law.

You have a mixture of several gases, A, B, C, and so on. From the ideal gas law we know that the pressure of any individual component is $P = nRT/V$. Therefore, we can write an equation for the ratio of the partial pressure of component A relative to the total pressure of the mixture.

$$\frac{P_A}{P_{total}} = \frac{n_A(RT/V)}{n_{total}(RT/V)} = \frac{n_A}{n_{total}}$$

Notice that the ratio of pressures is the same as the ratio of moles of gas A to the total number of moles of gas. This ratio, n_A/n_{total}, is called the **mole fraction** of A, and it is usually given the symbol X_A.

$$\frac{P_A}{P_{total}} = \frac{n_A}{n_{total}} = X_A \qquad (12.7)$$

In general, the mole fraction of any component A of a mixture is

$$X_A = \text{mole fraction of A} = \frac{n_A}{n_{total}} = \frac{\text{moles of A}}{\text{moles of A} + \text{moles of B} + \text{moles of C} + \cdots} \qquad (12.8)$$

In the gas mixture illustrated in Figure 12.8, the mole fractions of nitrogen and oxygen are

$$X_{N_2} = \frac{0.010 \text{ mol } N_2}{0.010 \text{ mol } N_2 + 0.0050 \text{ mol } O_2} = 0.67$$

$$X_{O_2} = \frac{0.0050 \text{ mol } O_2}{0.010 \text{ mol } N_2 + 0.0050 \text{ mol } O_2} = 0.33$$

Two-thirds of the mixture is nitrogen and one-third is oxygen. Because these gases are the two components of a mixture, the sum of the mole fractions must equal 1.

$$X_{N_2} + X_{O_2} = 0.67 + 0.33 = 1.00$$

In all cases, the sum of the mole fractions of the components of a mixture must be 1.

$$X_A + X_B + X_C + \cdots = 1$$

It is useful to apply the idea of mole fractions to the composition of air. Equation 12.9 informs us that the partial pressure of a gas in a mixture is equal to the product of its mole fraction and the total pressure.

$$P_A = X_A P_{total} \qquad (12.9)$$

Partial pressure of gas A = (mole fraction of A) (total pressure)

If this is applied to a sample of dry air, with a total barometric pressure of 745 mm Hg, the partial pressure of each gas can be calculated.

$$\text{Partial pressure } N_2 = (0.78084)(745 \text{ mm Hg}) = 582 \text{ mm Hg}$$

$$\text{Partial pressure of } O_2 = (0.20946)(745 \text{ mm Hg}) = 156 \text{ mm Hg}$$

Here we multiplied the total pressure by the mole fraction of each gas as given in Table 12.2.

Figure 12.9 A gas-mixing manifold used by an anesthesiologist to prepare a gas mixture to keep a patient unconscious during an operation. By proper mixing, the anesthetic gas can be slowly added to the breathing mixture. Near the end of the operation, the anesthetic gas can be replaced by air of normal composition or by pure oxygen.

EXAMPLE 12.12 *Partial Pressures of Gases*

Halothane has the formula $C_2HBrClF_3$. It is a nonflammable, nonexplosive, and nonirritating gas that is a commonly used inhalation anesthetic (Figure 12.9). Suppose you mix 15.0 g of halothane vapor with 23.5 g of oxygen gas. If the total pressure of the mixture is 855 mm Hg, what is the partial pressure of each gas?

Solution One way to solve this is to recognize that the partial pressure of a gas is given by the total pressure of the mixture multiplied by the mole fraction of the gas. Therefore, we first calculate the mole fractions of halothane and of O_2.

1. Calculate mole fractions

$$\text{Moles } C_2HBrClF_3 = 15.0 \text{ g}\left(\frac{1 \text{ mol}}{197.4 \text{ g}}\right) = 0.0760 \text{ mol}$$

$$\text{Moles } O_2 = 23.5 \text{ g}\left(\frac{1 \text{ mol}}{32.00 \text{ g}}\right) = 0.734 \text{ mol } O_2$$

$$\text{Mole fraction of } C_2HBrClF_3 = \frac{0.0760 \text{ mol } C_2HBrClF_3}{0.810 \text{ total moles}} = 0.0938$$

Because the sum of the mole fraction of halothane and of O_2 must equal 1.000, the mole fraction of oxygen is 0.906.

$$X_{\text{halothane}} + X_{\text{oxygen}} = 1.000$$
$$0.0938 + X_{\text{oxygen}} = 1.000$$
$$X_{\text{oxygen}} = 0.906$$

2. Calculate partial pressures.

$$\text{Partial pressure of halothane} = 0.0938 \cdot P_{\text{total}} = 0.0938 \text{ (855 mm Hg)} = 80.2 \text{ mm Hg}$$

Because

$$P_{\text{halothane}} + P_{\text{oxygen}} = 855 \text{ mm Hg}$$
$$P_{\text{oxygen}} = 855 \text{ mm Hg} - 80.2 \text{ mm Hg} = 775 \text{ mm Hg}$$

EXERCISE 12.10 *Partial Pressures*

The halothane-oxygen mixture described in Example 12.12 is placed in a 5.00-L tank at 25.0 °C. What is the total pressure (in millimeters of mercury) of the gas mixture in the tank? What are the partial pressures (in millimeters of mercury) of the gases? ■

An application of Dalton's law that you may see in your laboratory course occurs in experiments in which you generate a gas such as N_2 or H_2 and collect it by displacing water from a container (Figure 12.10). Here the total pressure in the collecting flask is the sum of the pressures of the N_2 or H_2 *plus* the pressure of the water vapor. Many liquids such as water evaporate to some extent, and the vapor exerts a pressure that depends on the temperature of the system.

The concept of vapor pressure is described in Section 13.3. The vapor pressure of water is given at different temperatures in Appendix E.

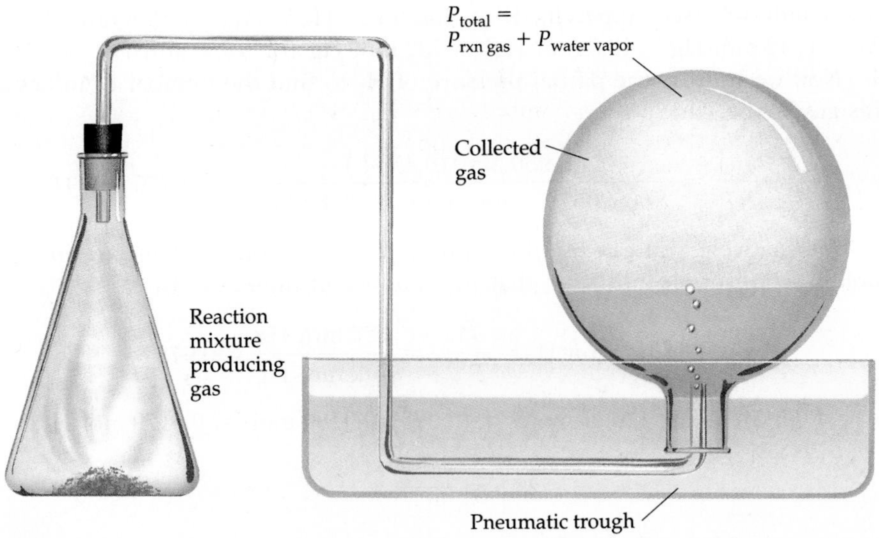

Figure 12.10 Collecting a gas over water. The pressure of the generated gas forces some of the water from the collection flask. Thus, the pressure of the collected gas (P_{total}) represents the sum of the pressure of the gas from the reaction ($P_{rxn\ gas}$) plus the pressure generated by the water vapor from the evaporation of water in the collection flask ($P_{water\ vapor}$). See Example 12.13.

EXAMPLE 12.13 *Using Partial Pressures: Collecting a Gas over Water*

Small quantities of H_2 gas can be prepared in the laboratory by the following reaction.

$$Mg(s) + 2\,HCl(aq) \longrightarrow MgCl_2(aq) + H_2(g)$$

Assume you carried out this experiment and collected 456 mL of H_2 gas as illustrated in Figure 12.10. The temperature of the gas mixture was 22.0 °C, and the total pressure of gas in the flask was 742 mm Hg. How many total moles of gas (hydrogen + water vapor) were in the flask? How many moles of H_2 did you prepare?

Solution

Step 1. Calculate the total number of moles of gas.

P = total pressure = 742 mm Hg = 0.976 atm

V = 0.456 L

T = 22.0 °C = 295.2 K

n = total moles of gas = moles H_2 + moles H_2O vapor = ?

$$n = \frac{PV}{RT} = \frac{(0.976\text{ atm})(0.456\text{ L})}{(0.082057\text{ L}\cdot\text{atm/K}\cdot\text{mol})(295.2\text{ K})} = 0.0184\text{ mol gas}$$

Step 2. Calculate the number of moles of H_2 gas. The total pressure in the collecting flask, 742 mm Hg, is the sum of the partial pressures of the two gases in the flask.

$$P_{total} = 742\text{ mm Hg} = P_{H_2} + P_{H_2O}$$

To find the number of moles of H_2, you must know the partial pressure of H_2. The partial pressure of water vapor over liquid water can be obtained from tables

Hydrogen gas is being generated by the reaction of magnesium metal with aqueous HCl. The gas is collected by the displacement of water. (C. D. Winters)

in a handbook (see Appendix E in this text). Here P_{H_2O} is 19.8 mm Hg, so P_{H_2} = (742 mm Hg − 19.8 mm Hg) = 722 mm Hg (or 0.950 atm).

Now we can use the partial pressure of H_2 to find the number of moles of this gas in the flask.

$$n_{H_2} = \frac{P_{H_2}V}{RT} = \frac{(0.950\ \text{atm})(0.456\ \text{L})}{(0.082057\ \text{L}\cdot\text{atm/K}\cdot\text{mol})(295.2\ \text{K})} = 0.0179\ \text{mol H}_2$$

Alternatively, you can use the ratio of P_{H_2} to the total pressure to find the mole fraction of H_2 and from that the number of moles of H_2.

$$\text{Mole fraction H}_2 = \frac{P_{H_2}}{P_{total}} = \frac{722\ \text{mm Hg}}{742\ \text{mm Hg}} = 0.973$$

$$\text{Moles H}_2 = X_{H_2}\cdot\text{total moles} = (0.973)(0.0184\ \text{mol}) = 0.0179\ \text{mol H}_2$$

EXERCISE 12.11 ***Partial Pressures of Gases***

In an experiment similar to that in Figure 12.10, 352 mL of gaseous nitrogen is collected in a flask over water at a temperature of 24.0 °C. The total pressure of the gases in the flask is 742 mm Hg. What mass of N_2 is collected? (See Appendix E for water vapor pressure.) ■

12.6 THE KINETIC-MOLECULAR THEORY OF GASES

Figure 12.11 Illustration of the movement of gas molecules in the gas phase. Open dishes of aqueous ammonia and HCl are placed side by side. When molecules of NH_3 and HCl escape from the dishes and encounter one another, you observe the formation of ammonium chloride, NH_4Cl.

So far we have discussed the macroscopic properties of gases. Now we turn to the *kinetic-molecular theory,* a description of the behavior of gases at the molecular or atomic level. A qualitative introduction to the theory was given in Section 1.1, where it was mentioned that the theory applies to liquids and solids as well as to gases. In this section we discuss its application to gases and how the gas laws can be interpreted in terms of molecular motions.

If your friend is wearing a perfume or a pizza smells good, how do you know it? In scientific terms, we know that molecules of perfume or the aroma-causing molecules of food enter the gas phase and drift through space until they reach the cells of your body that react to odors. The same thing happens in the laboratory when bottles of aqueous ammonia and hydrochloric acid sit side by side (Figure 12.11). Molecules of the two compounds enter the gas phase and drift along until they encounter one another, combine, and form a cloud of tiny particles of ammonium chloride.

If you changed the temperature of the environment of the containers in Figure 12.11 and compared the times needed for the cloud of ammonium chloride to form, you find that the time is longer at lower temperatures. The speed at which molecules move depends on the temperature. In fact, it can be shown that the average kinetic energy, $\overline{KE}$, of a collection of gas molecules depends **only** on the temperature, a concept expressed by the following relation

$$\overline{KE} \propto T \qquad (12.10)$$

where the horizontal bar over KE indicates an *average value.* The kinetic energy of only one molecule is given by

$$KE = \frac{1}{2}(\text{mass})(\text{speed})^2 = \frac{1}{2}mu^2 \tag{12.11}$$

where u is the speed of the molecule. If you could measure the speeds of the trillions of molecules in a gas sample, however, you would find that some have a speed of u_1, some have a different speed u_2, and so on. The *average* speed, $\overline{u}$, of the molecules is therefore

$$\overline{u} = \frac{n_1 u_1 + n_2 u_2 + \cdots}{N} \tag{12.12}$$

where n_x is the number of molecules with the speed u_x and N is the total number of molecules. This means the average kinetic energy of many molecules, $\overline{KE}$, is related to the *average of the* ***squares*** *of their speeds,* $\overline{u^2}$ (called the "mean square speed"), by the equation

$$\overline{KE} = \frac{1}{2}m\overline{u^2} \tag{12.13}$$

Now, because $\overline{KE}$ is proportional to T and to $m\overline{u^2}$, T must be proportional to $m\overline{u^2}$ as well, and so we can say that

$$\frac{1}{2}m\overline{u^2} = CT \tag{12.14}$$

where C is a proportionality constant. We shall come back to this useful relationship in a moment.

Another observation you have made is that *gases are compressible.* For example, you can "pump up" a tire by forcing more gas into it. By contrast, it is difficult to squeeze solids or liquids to force them to occupy a smaller volume. This must mean that the *distance between gas particles (atoms or molecules) is very large relative to the actual size of the particles,* whereas the opposite is true in solids and liquids.

Finally, we noted in Chapter 1 that *gases occupy completely the volume available to them,* in contrast with liquids and solids. As described in Section 13.2, molecules in the vapor phase can condense to form liquids and solids if the forces of attraction *between* molecules—called *intermolecular forces*—are sufficiently large. The fact that gases exist and that they completely occupy the volume available to them is evidence that intermolecular forces in the gas phase must be vastly weaker than in liquids or solids.

The experimental observations we have just described are the principal tenets of the **kinetic-molecular theory.**

1. Gases consist of molecules whose separation is much greater than the size of the molecules themselves.

2. The molecules of a gas are in continual, random, and rapid motion.

3. The average kinetic energy of gas molecules is determined by the gas temperature. *All molecules, regardless of their mass, have the same average kinetic energy at the same temperature.*

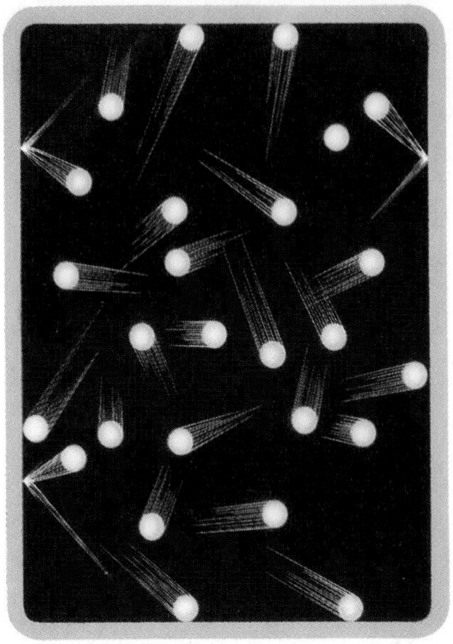

Figure 12.12 Gas pressure is caused by the bombardment of the container walls by gas molecules.

4. Gas molecules collide with one another and with the walls of their container, but they do so without loss of energy.*

These four points describe all gases. In addition, in an **ideal gas** there are no intermolecular forces and the gas particles occupy no volume. Assuming a gas is ideal, the ideal gas law can be derived from the mathematical expression of the ideas above. As you will see below, however, there is no such thing as an ideal gas, even though some do come close to this behavior.

Kinetic-Molecular Theory and the Gas Laws

The gas laws, which come from experiment, can be explained by the kinetic-molecular theory. The starting place is to write equations describing how pressure arises from collisions of gas molecules with the walls of the container holding the gas (Figure 12.12). The force developed by these collisions depends on the number of collisions and the average force per collision. When the temperature of a gas is increased, the average force of a collision on the walls of its container increases because the average kinetic energy of the molecules increases. Also, because the speed increases with temperature, more collisions occur per second. Thus, the collective force per square centimeter is greater, and the pressure increases. Mathematically, this means that P is proportional to T when n and V are constant.

Increasing the number of molecules of a gas at a fixed temperature and volume does not change the average collision force, but it does increase the number of collisions occurring per second. Thus, the pressure increases, and so we can say that P is proportional to n when V and T are constant.

If the pressure is not allowed to increase when either the number of molecules of gas or the temperature is increased, the volume of the container (the area over which the collisions can take place) must increase. This is expressed by stating that V is proportional to nT when P is constant, a statement that is a combination of *Avogadro's law* and *Charles's law.*

Finally, if the temperature is constant, the average impact force of molecules of a given mass with the container walls must be constant. If n is kept constant while the volume of the container is made smaller, however, the *number of collisions* with the container walls per second must increase. This must mean that the pressure increases, and so P is proportional to $1/V$ when n and T are constant, as stated by *Boyle's law.*

Distribution of Molecular Speeds

The relative number of molecules that have a given speed can be measured experimentally. Figure 12.13 graphs the number of molecules versus their speed; the higher a point on the curve, the greater the number of molecules having that speed.

Two important conclusions should be drawn from Figure 12.13. First, you see that some molecules are fast (have a high kinetic energy), and others are

*Scientists call such collisions "perfectly elastic"; colliding billiard balls come close to having elastic collisions, but basketballs have inelastic collisions. In a perfectly elastic collision of a fast and a slow molecule, the fast one slows down and the slow one speeds up, but the sum of the kinetic energies of the two molecules is the same after the collisions as before.

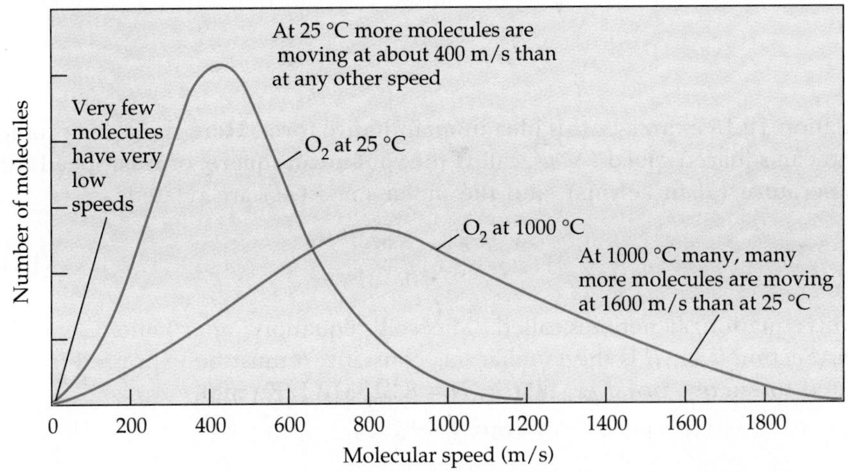

Figure 12.13 A plot of the relative number of molecules with a given speed versus that speed. The curve is called a Boltzmann distribution of molecular speeds. The red curve shows the effect of a temperature increase on the distribution of speeds.

slow (have a low kinetic energy). There is, however, a most common speed, which corresponds to the maximum in the distribution curve. For oxygen gas at 25 °C, for example, the maximum in the curve comes at a speed of 400 m/s, and most of the molecules have speeds within the range from 200 m/s to 700 m/s. (Also notice that the curves are not symmetrical around their maxima. This means that the average speed is a little faster than the most common speed.)

The second conclusion to draw from Figure 12.13 is that, as the temperature increases, the most common speed goes up, and the number of molecules traveling very fast increases a great deal. The increase in average speed as a function of increasing temperature is expressed by Equation 12.14. Be sure to notice, though, that even though the curve for the higher temperature is "flatter" and broader than the one at a lower temperature, the *areas under the curves are the same because the number of molecules in the sample is fixed.**

Finally, a relationship exists among molecular mass, average speed, and temperature. The average kinetic energy, $\overline{KE}$, is fixed by the temperature, so the product $\frac{1}{2}m\overline{u^2}$ (Equation 12.13) is also fixed. Because two gases with two different molecular masses must have the same average kinetic energy at the same temperature, the heavier gas molecules must have a lower average speed (Figure 12.14).

Curves of speed (or energy) versus number of molecules, such as those in Figure 12.13, are often called **Boltzmann distribution curves.** They are named after Ludwig Boltzmann (1844–1906), an Austrian mathematician who did much work on the kinetic theory of gases. See also Chapter 20.

The distribution of speeds, energies, or other molecular properties, such as that illustrated by Figure 12.13, is an important concept in science. You will see the idea come up again in later chapters, particularly in Chapter 13 on the behavior of liquids.

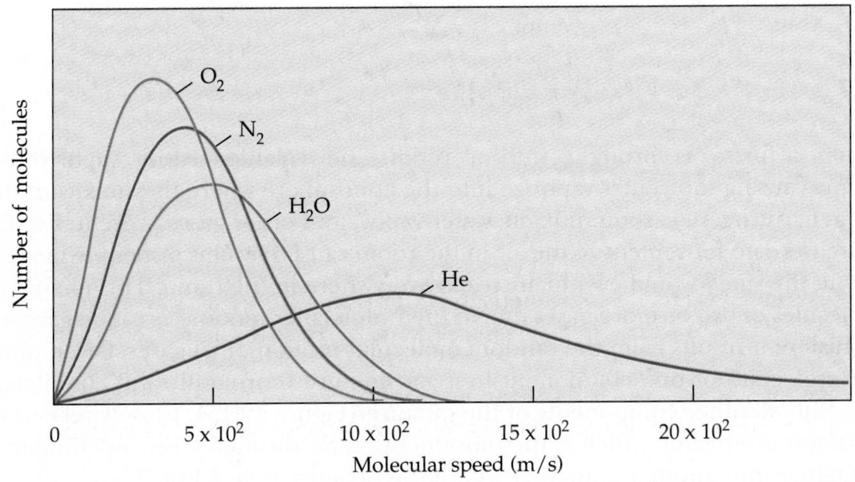

Figure 12.14 Effect of molecular mass on the Boltzmann distribution curve at a given temperature. Notice the similarity of the effect of increasing mass in this figure to the effect of decreasing temperature in Figure 12.13.

*If you have taken some calculus, you know that the area under a distribution curve reflects the total number of objects on which the curve is based. For a given number of objects, the shape of the distribution curve may change, but not the area under the curve.

Equation 12.15 expresses this idea in quantitative form. Here the square root of the mean squared speed ($\sqrt{\overline{u^2}}$, called the **root-mean-square, or rms, speed**), the temperature (T, in kelvins), and the molar mass (M) are related.

$$\sqrt{\overline{u^2}} = \sqrt{\frac{3RT}{M}} \tag{12.15}$$

In this equation, sometimes called "Maxwell's equation" after James Clerk Maxwell (Section 7.1), R is the familiar gas constant. R must be expressed in units related to energy, however; that is, $\mathbf{R = 8.314510\ J/K \cdot mol.}$

The gas constant R can be expressed in different units. If P is measured in SI units of pascals ($kg/m \cdot s^2$), and V is measured in m^3, then R has the value

$$R = \frac{(1.01325 \times 10^5\ kg/m \cdot s^2)(22.414 \times 10^{-3}\ m^3)}{(1\ mol)(273.15\ K)}$$

$$= 8.3145\ kg \cdot m^2/s^2 \cdot mol \cdot K$$

Because 1 $kg \cdot m^2/s^2$ is 1 joule, the gas constant is also

$$R = 8.3145\ J/mol \cdot K$$

EXAMPLE 12.14 *Molecular Speed*

Calculate the rms speed of oxygen molecules at 25 °C.

Solution We must use Equation 12.15 with M expressed in units of kilograms per mole because R is in units of $J/K \cdot mol$, and $1\ J = 1\ kg \cdot m^2/s^2$. Thus, the molar mass of O_2 is 32.0×10^{-3} kg/mol.

$$\sqrt{\overline{u^2}} = \sqrt{\frac{3RT}{M}} = \sqrt{\frac{3(8.3145\ J/K \cdot mol)(298\ K)}{32.0 \times 10^{-3}\ kg/mol}} = \sqrt{2.32 \times 10^5\ J/kg}$$

To obtain the answer in meters per second, we use the relation $1\ J = 1\ kg \cdot m^2/s^2$, which means we have

$$\sqrt{\overline{u^2}} = \sqrt{2.32 \times 10^5\ kg \cdot m^2/kg \cdot s^2} = \sqrt{2.32 \times 10^5\ m^2/s^2}$$
$$= 482\ m/s$$

This speed is equivalent to about 1100 miles per hour!

EXERCISE 12.12 *Molecular Speeds*

Calculate the rms speeds of helium atoms and N_2 molecules at 25 °C. ■

12.7 DIFFUSION AND EFFUSION

When a pizza is brought into a room, the volatile—easily vaporized—aroma-causing molecules vaporize into the atmosphere where they mix with the oxygen, nitrogen, carbon dioxide, water vapor, and other gases present. Even if there were no movement of the air in the room caused by fans or people moving about, the smell would eventually reach everywhere in the room. This mixing of molecules of two or more gases due to their molecular motions is called **gaseous diffusion;** it results from the random molecular motion of all gases. Given time, the molecules of one component in a gas mixture thoroughly and completely mix with all other components of the mixture (Figure 12.15). Closely related to diffusion is effusion, which is the movement of gas through a tiny opening in a container into another container where the pressure is very low (Figure 12.16).

Thomas Graham (1805–1869), a Scottish chemist, studied the effusion of gases and found experimentally that the rates of effusion of two gases—the quantity of material moving from one place to another in a given amount of

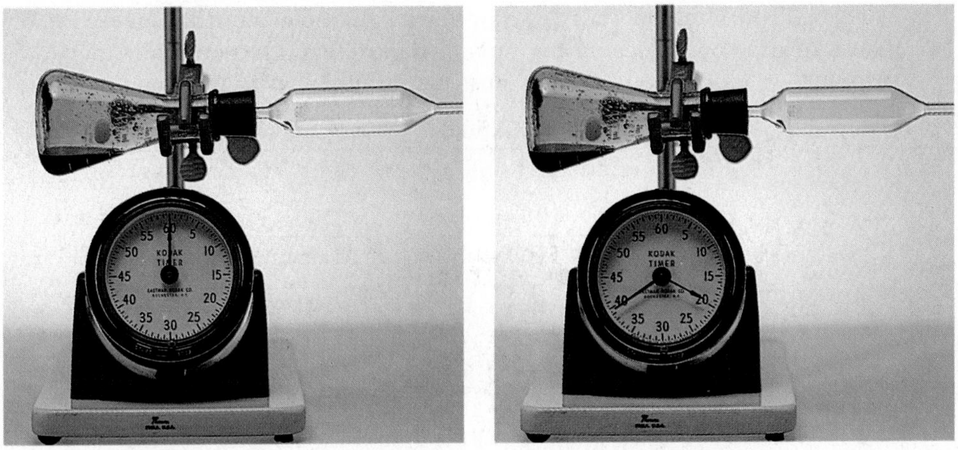

Figure 12.15 Brown NO_2 gas diffuses out of the flask in which it was generated and into the attached tube in a matter of minutes. (The NO_2 was made by reacting copper with nitric acid in an oxidation-reduction reaction. You may notice blue-green crystals of another reaction product, copper(II) nitrate, on the inside of the flask.)

time—were inversely proportional to the square roots of their molar masses at the same temperature and pressure.

$$\frac{\text{Rate of effusion of gas 1}}{\text{Rate of effusion of gas 2}} = \sqrt{\frac{\text{molar mass of gas 2}}{\text{molar mass of gas 1}}} \qquad \textbf{(12.16)}$$

This is now known as **Graham's law.** It is readily derived from Maxwell's equation by recognizing that the rate of effusion depends on the speed of the molecules. Therefore, the ratio of the rms speeds is the same as the ratio of the effusion rates.

$$\frac{\text{Rate of effusion of gas 1}}{\text{Rate of effusion of gas 2}} = \frac{\sqrt{\overline{u^2}} \text{ of gas 1}}{\sqrt{\overline{u^2}} \text{ of gas 2}} = \sqrt{\frac{3RT/(M \text{ of gas 1})}{3RT/(M \text{ of gas 2})}}$$

Canceling out like terms, we have the simple expression developed by Graham.

In Figure 12.16 you see that a greater relative number of H_2 molecules effuse in a given time than N_2 molecules. Graham's law now allows a quantitative comparison of their relative rates.

$$\frac{\text{Rate of effusion of } H_2}{\text{Rate of effusion of } N_2} = \sqrt{\frac{M \text{ of } N_2}{M \text{ of } H_2}} = \sqrt{\frac{28.0 \text{ g/mol}}{2.02 \text{ g/mol}}} = \frac{3.72}{1}$$

This calculation tells you that H_2 molecules effuse through the barrier 3.72 times faster than N_2 molecules.

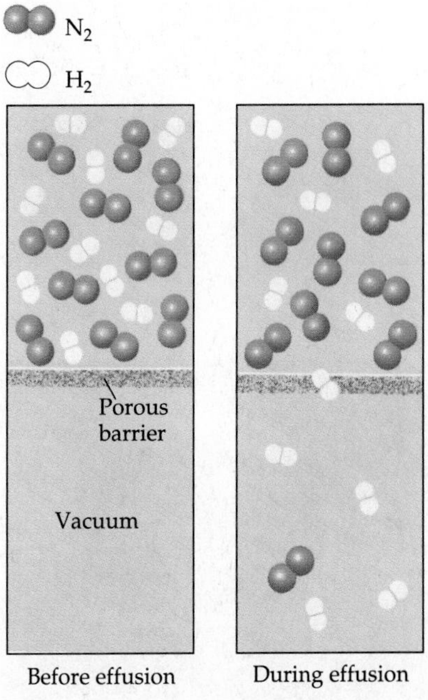

Figure 12.16 An illustration of the effusion of gas molecules through the pores of a membrane or other porous barrier. Lighter molecules with greater average speeds strike the membrane more often and pass more rapidly through it than the heavier, slower molecules at the same temperature.

EXAMPLE 12.15 *Graham's Law of Effusion*

Tetrafluoroethylene, C_2F_4, effuses through a barrier at the rate of 4.6×10^{-6} mol/h. An unknown gas, consisting only of boron and hydrogen, effuses at the rate of 5.8×10^{-6} mol/h under the same conditions. What is the molar mass of the unknown gas?

Solution From Graham's law we know that a light molecule effuses more rapidly than a heavier one. Because the unknown gas effuses more rapidly than C_2F_4 (M = 100. g/mol), the unknown must have a molar mass less than 100.

$$\frac{\text{Rate of effusion of unknown}}{\text{Rate of effusion of } C_2F_4} = \sqrt{\frac{M \text{ of } C_2F_4}{M \text{ of unknown gas}}}$$

$$\frac{5.8 \times 10^{-6} \text{ mol/h}}{4.6 \times 10^{-6} \text{ mol/h}} = \sqrt{\frac{100.\text{ g/mol}}{M \text{ of unknown}}}$$

$$1.3 = \sqrt{\frac{100.\text{ g/mol}}{M \text{ of unknown}}}$$

To solve for the unknown molar mass, square both sides of the equation.

$$1.6 = \frac{100.\text{ g/mol}}{M \text{ of unknown}}$$

and rearrange to find M for the unknown.

$$M = 63 \text{ g/mol}$$

The boron hydrogen compound corresponding to this molar mass is B_5H_9, which is called pentaborane.

EXERCISE 12.13 ***Graham's Law***

A sample of pure methane, CH_4, is found to effuse through a porous barrier in 1.50 min. Under the same conditions, an equal number of molecules of an unknown gas effuse through the barrier in 4.73 min. What is the molar mass of the unknown gas? ■

12.8 SOME APPLICATIONS OF THE GAS LAWS AND KINETIC-MOLECULAR THEORY

Rubber Balloons and Why They Leak

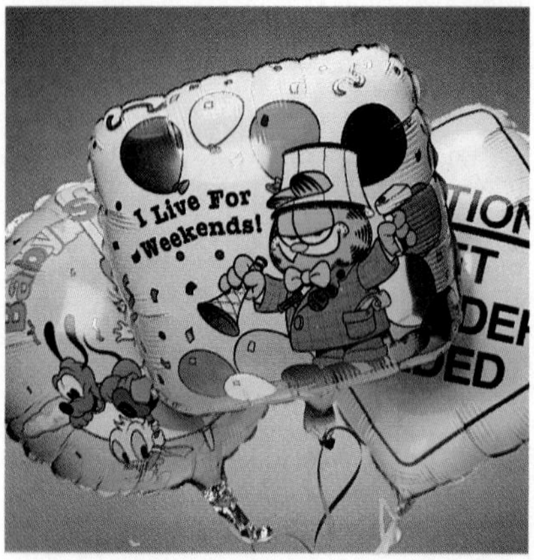

Gas will effuse much less rapidly out of a balloon made of Mylar (a polyethylene terephthalate) than out of a rubber balloon.

Graham's law allows us to explain the earlier observation concerning the balloon used by Charles in 1783 (see page 553). In the brief biography of Charles, we mentioned that he filled the balloon with H_2 gas, but he had to go to extraordinary lengths to keep the gas in the balloon. Unlike the first hot air balloons that were made of paper, Charles had to use silk and coat it with rubber. The reason for this is simple: at any temperature at which He, N_2, and O_2 are gases, the lighter weight H_2 molecules have a higher average speed than the N_2 or O_2 molecules of air, so H_2 molecules effuse rapidly through paper, a more porous material than silk coated with rubber.

For much the same reason that Charles used special materials in his hydrogen-filled balloon, you should be cautious when buying a helium-filled, rubber balloon at a carnival. Lightweight He atoms have a higher rms speed than the heavier N_2 or O_2 molecules of air, so He atoms can effuse rapidly through the balloon wall, and the balloon deflates. The newer balloons made of Mylar film (a polyethylene terephthalate), however, keep He gas enclosed much longer because the film has much smaller pores than rubber and the He atoms cannot escape as easily.

Deep Sea Diving

Another application of the gas laws arises in deep sea diving (Figure 12.17). To explore the floor of the oceans, scientists have built diving ships to operate at great depths. These ships are filled with an atmosphere of oxygen and helium at high pressures, but the oxygen in these ships has a lower mole fraction than in normal sea-level air. It has been found that our bodies function best when the partial pressure of oxygen is about 0.21 atm. If the pressure of gas you breathe is 2 atm, for example, and the composition of the gas is that of normal air, the partial pressure of oxygen is about 0.4 atm. Such high oxygen partial pressures are toxic, so the gas that divers breathe must have a lower mole fraction of O_2.

Not only is the fraction of O_2 reduced in diving gases, but the nitrogen of normal air is replaced by helium. Nitrogen is more soluble in blood and body fluids at high pressures and leads to *nitrogen narcosis,* a condition similar to alcohol intoxication. Helium is less soluble in blood and is thus more suitable to dilute the O_2. Although helium solves one problem, however, it creates another. Divers in such an atmosphere feel decidedly chilly, even though the temperature may be a normally comfortable 70 °C. This can be explained by the kinetic-molecular theory. Helium atoms move with an average velocity about 2.6 times greater than that of N_2 molecules at the same temperature (see Exercise 12.12). When gas molecules strike the human body, they carry away some of the heat generated by metabolism. Because helium atoms move so much more rapidly than N_2 molecules, the helium atoms collide more frequently with the body and are thus more efficient in carrying away heat energy.

Another side effect of using helium in the diver's atmosphere is that the divers sound like Alvin, the Chipmunk! This happens because the vocal cords vibrate faster in an atmosphere less dense than air, and the pitch of the voice is raised.

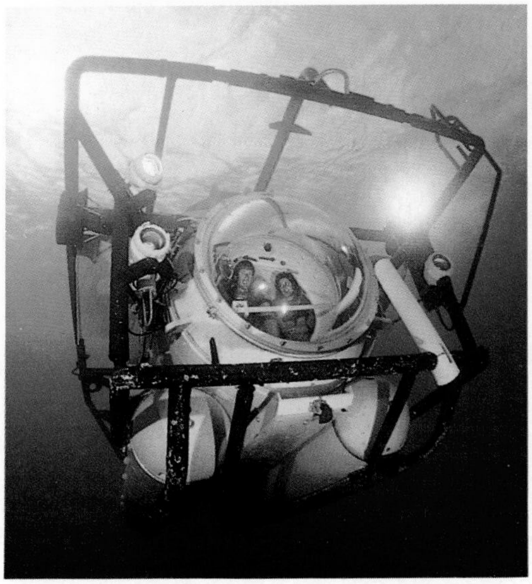

Figure 12.17 Some deep-sea diving vehicles use an atmosphere of oxygen and helium. (Photo Researchers, Inc.)

Separation of Uranium Isotopes

In the race to construct atomic explosives during World War II, considerable quantities of uranium enriched in the ^{235}U isotope were needed. One technique that was developed used gas diffusion (Figure 12.18). To separate the ^{235}U isotope from the more abundant ^{238}U isotope, all the uranium was converted to UF_6, one of the few uranium compounds that is volatile at a reasonable temperature. (UF_6 boils at about 56 °C.) When UF_6 vapor is allowed to diffuse through a series of porous membranes, the lighter molecules of $^{235}UF_6$ diffuse more rapidly than the heavier $^{235}UF_6$ molecules. Graham's law allows us to calculate the difference in rates.

$$\frac{\text{Rate of diffusion of } ^{235}UF_6}{\text{Rate of diffusion of } ^{238}UF_6} = \sqrt{\frac{352.041 \text{ g/mol}}{349.034 \text{ g/mol}}} = 1.00430$$

This calculation shows that the lighter $^{235}UF_6$ molecules diffuse about 0.4% faster than the heavier $^{238}UF_6$ molecules. When this process is repeated many thousands of times through successive porous membranes, the uranium is enriched in ^{235}U. The first bomb-grade uranium was prepared this way at the Oak Ridge, Tennessee, gaseous diffusion plant during World War II.

Figure 12.18 The gaseous diffusion plant for separation of uranium isotopes at the Department of Energy's K-25 site in Oak Ridge, Tennessee. Interior view of processing equipment shows the arrangement of piping, compressors, motors, and large tank-shaped "diffusers." (Notice the worker wearing a hardhat and standing in front of the tank marked STG-2.) Compressors pump volatile UF_6 into the diffuser vessels where, at each stage, the uranium is slightly enriched in the fissionable uranium-235 isotope. For use as fuel in nuclear power plants, the uranium is enriched to approximately 3% ^{235}U (compared with 0.7% for uranium found in the natural state).

12.9 NONIDEAL BEHAVIOR: REAL GASES

If you are working with a gas at approximately room temperature and at pressures of 1 atmosphere or less, the ideal gas law is remarkably successful in relating the quantity of gas and its pressure, volume, and temperature. At higher pressures or lower temperatures, however, serious deviations from the ideal gas law are often observed. The origin of these deviations is easily understood in terms of the breakdown of some assumptions of the kinetic theory of gases.

At standard temperature and pressure (STP), the volume occupied by a single molecule is very small relative to its share of the total gas volume. Recall that 6.02×10^{23} molecules occur in a mole and that a mole of gas occupies 22.4 L (22.4×10^{-3} m^3) at STP. The volume, V, that each molecule has to move around is given by

$$V = \frac{22.4 \times 10^{-3}\ \text{m}^3}{6.023 \times 10^{23}\ \text{molecules}} = 3.72 \times 10^{-26}\ \text{m}^3\ \text{molecule}$$

If this volume is assumed to be a sphere, then the radius, r, of the sphere is about 2000 pm. The radius of the smallest gaseous substance, the helium atom, is 31 pm (see Figure 8.10), so the helium atom has about the same amount of space to move around in as a pea has inside a basketball. Now suppose the pressure has increased significantly, to 1000 atm. The volume available to each molecule is a sphere with a radius of only about 200 pm, which means the situation is now like that of a pea inside a sphere a bit larger than a Ping-Pong ball. The important point is that the volume occupied by the gas molecules themselves is no longer negligible at higher pressures, which violates the first tenet of the kinetic-molecular theory. The kinetic-molecular theory and the ideal gas law are concerned with the volume available to the molecules to move about, not the volume of the molecules themselves. The experimentally determined volume, however, must include both. At high pressures, therefore, the measured volume is larger than predicted by $PV = nRT$.

Figure 12.19 All gases can be liquefied at some temperature, even the nitrogen in the air around us. The boiling point of liquid nitrogen is −196 °C. (The fact that nitrogen gas can be changed to *liquid* nitrogen also shows that gas molecules cannot have zero volume.) (C. D. Winters)

Another assumption of the kinetic-molecular theory was that collisions between molecules are elastic, an assumption that implies that the atoms or molecules of the gas never stick to one another by some type of force. This is clearly nonsense as well. All gases can be liquefied (Figure 12.19) at some temperature—although some require a very low temperature—and the only way this can happen is if the atoms or molecules cling together because of the forces between the molecules, that is, by **intermolecular forces** (Section 13.2). Even at temperatures well above the liquefaction temperature of the gas these forces can be appreciable, and they can lead to measurable deviations from ideal gas behavior. This means that when a molecule is about to hit the wall of its container, most other molecules are farther away from the wall and therefore pull it away from the wall (Figure 12.20). This attraction causes the molecule to hit the wall with less impact—the collision is softer than if no attraction existed among the molecules. Because all collisions are softer, the observed gas pressure (P_{obs}) is less than that predicted by the ideal gas law (P_{ideal}). Although this effect can be observed in gases at pressures around 1 atmosphere, it becomes particularly pronounced when the pressure is high.

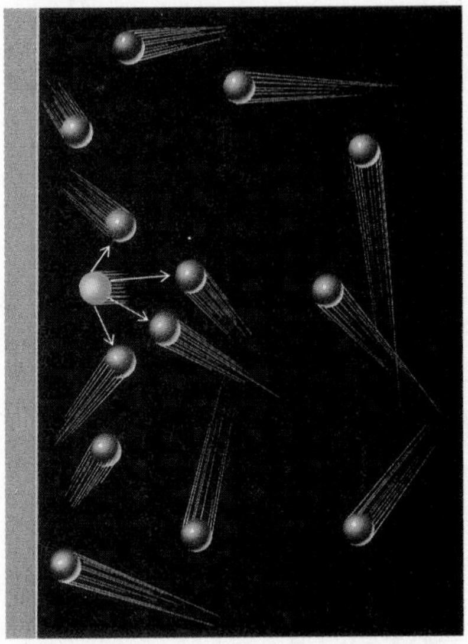

Figure 12.20 A gas molecule strikes the walls of a container with less force due to the attractive forces between it and its neighbors.

The Dutch physicist Johannes van der Waals (1837–1923) studied the breakdown of the ideal gas law equation and developed another equation to describe gases at high pressure or gases in which effects of intermolecular forces can be appreciable. Under such circumstances, the so-called van der Waals equation allows better predictions than the ideal gas law.

The van der Waals equation

$$\left[P + a\left(\frac{n}{V}\right)^2\right]\left[V - bn\right] = nRT$$

correction for intermolecular forces (the $a(n/V)^2$ term)

correction for molecular volume (the bn term)

a and b = van der Waals constants **(12.17)**

Although Equation 12.17 might seem complicated at first glance, the terms in square brackets are simply those of the ideal gas law, each corrected for the effects discussed previously.

The pressure correction term, $a(n/V)^2$, allows for intermolecular forces. The term n/V is the concentration of the gas in the absence of the association of atoms or molecules due to intermolecular forces. Because the actual gas pressure is lower than the ideal pressure (as calculated from $PV = nRT$) owing to intermolecular forces, $a(n/V)^2$ is added to the observed pressure P_{obs} to make this portion of the equation equal to P in $PV = nRT$. The constant a, which is determined experimentally, typically has values in the range 0.01 to 10 atm · (L/mol)^2; however, the values have no simple relation to other molecular properties.

The bn term in van der Waals's equation corrects the container volume to a smaller value, the free volume available to the gas molecules. Here n is the number of moles of gas, and b is an experimental quantity that gives the correction (per mole) for the molecular volume. Typical values of b range from 0.01 to 0.1 L/mol, roughly increasing with increasing molecular size (Table 12.3).

As an example of the importance of these corrections, consider a sample of 8.00 mol of chlorine gas, Cl_2, in a 4.00-L tank at 27.0 °C. The ideal gas law leads you to expect a pressure of 49.2 atm. A more realistic estimate of the pressure, obtained from the van der Waals equation, is only 29.5 atm, 20 atm less than the ideal pressure!

TABLE 12.3 Some van der Waals Constants

Substance	a(atm · L^2/mol^2)	b(L/mol)
He	0.034	0.0237
Ar	1.34	0.0322
H_2	0.244	0.0266
N_2	1.39	0.0391
O_2	1.36	0.0318
Cl_2	6.49	0.0562
CO_2	3.59	0.0427
H_2O	2.25	0.0428

EXERCISE 12.14 ***van der Waals's Equation***

Using both the ideal gas law and van der Waals's equation calculate the pressure expected for 10.0 mol of helium gas in a 1.00-L container at 25 °C ■

CHAPTER HIGHLIGHTS

When you have finished studying this chapter, you should be able to

- Describe how pressure measurements are made (Section 12.1).
- Use the units of pressure, especially millimeters of mercury (mm Hg), atmospheres, and pascals (Section 121.).

  The standard atmosphere is the pressure capable of supporting a column of mercury 760 mm high. Therefore, 1 atm = 760 mm Hg. Also 1 atm = 101.3 kPa, or 1.013 bar. Conditions of **STP, standard temperature and pressure,** are 273 K and 1 atm.

- Understand the basis of the gas laws and how to use these laws (Section 12.2).

  Boyle's law specifies that the pressure of a gas is inversely proportional to its volume ($P \propto 1/V$) at constant T and n. Charles's law tells us that gas volume is directly proportional to temperature ($V \propto T$) at constant n and P. Finally, Avogadro's law specifies that gas volume is directly proportional to its quantity ($V \propto n$) at constant T and P.

- Understand the ideal gas law and how to use this equation (Section 12.3).

  If the three gas laws are combined into one statement, the **ideal gas law** (**$PV = nRT$**) results. Because PV/nT is always equal to R, the **general gas law** (or **combined gas law**) can be derived from the ideal gas law.

  $$\frac{P_1 V_1}{n_1 T_1} = \frac{P_2 V_2}{n_2 T_2} \tag{12.5}$$

  It is usually applied to find out what happens when a given quantity of gas (so $n_1 = n_2$) undergoes a change in P, V, and T conditions.

- Calculate the molar mass of a compound from a knowledge of the pressure of a known quantity of a gas in a given volume at a known temperature (Section 12.3).
- Apply the gas laws to a study of the stoichiometry of reactions (Section 12.4).
- Use Dalton's law (Section 12.5).

  Dalton's law of partial pressures specifies that the total pressure of a mixture of gases is the sum of the **partial pressures** of the individual gases (A, B, C, . . .) in the mixture.

  $$P_{total} = P_A + P_B + P_C + \cdots$$

  The partial pressure of a gas in a mixture (P_A) is given by its mole fraction (X_A) times the total pressure of the mixture (P_{total}).

  $$P_A = X_A P_{total}$$

  The mole fraction of a component (A) of a mixture is defined as the number of moles of A divided by the total moles of all components.

- Apply the **kinetic-molecular theory,** a theory of gas behavior at the molecular level (Section 12.6).

  One important aspect of the theory is that the average kinetic energy of gas molecules ($\overline{KE}$), which is proportional to the mass of the gas and to $\overline{u^2}$ (the average of the squares of the molecular speeds)

  $$\overline{KE} = \frac{1}{2}m\overline{u^2}$$

  is proportional to the temperature of the gas.

  $$\overline{KE} \propto T$$

  Because $\overline{KE}$ is determined by temperature, heavier molecules (large m) must move with a slower average speed (u) than lighter molecules (small m) at a given temperature.

- Understand the phenomena of diffusion and effusion and how to use Graham's law (Section 12.6).

  Graham's law of effusion states that the rates of effusion of two gases are inversely proportional to the square roots of their molar masses at the same temperature and pressure.

  $$\frac{\text{Rate of effusion of gas 1}}{\text{Rate of effusion of gas 2}} = \sqrt{\frac{\text{molar mass of gas 2}}{\text{molar mass of gas 1}}} \qquad \textbf{(12.16)}$$

- Appreciate the fact that gases usually do not behave as ideal gases (Section 12.9).

  The ideal gas law assumes, among other things, that gas molecules have no volume and that they do not interact with one another by **intermolecular forces.** Neither assumption is completely correct, however, because a *real gas* behaves in a more complex way than an ideal gas. One description of the behavior of real gases is found in the **van der Waals equation.**

STUDY QUESTIONS

Review Questions

1. Name the three gas laws that interrelate P, V, and T. Explain the relationships in words and in equations.
2. What conditions are represented by STP? What is the volume of a mole of ideal gas under these conditions?
3. Show how to calculate the molar mass of a gas from P, V, and T measurements and other information.
4. Explain how to determine the density of a gas from P, V, and T data and other information.
5. State Avogadro's law. Relate your discussion to the formation of water vapor from its elements.

   $$2\,H_2(g) + O_2(g) \longrightarrow 2\,H_2O(g)$$

   For example, if 2 mol of H_2 is used at STP, how many liters of O_2 is required at STP and how many liters of H_2O is produced at STP?
6. State Dalton's law. If the air you breathe is 78% N_2 and 22% O_2 (on a mole basis), what is the mole fraction of O_2? What is the partial pressure of O_2 when the atmospheric pressure is 748 mm Hg?
7. What are the basic assumptions of the kinetic-molecular theory? Which of these assumptions are most nearly correct and which are violated to some extent by a real gas?
8. Explain Boyle's law on the basis of kinetic-molecular theory.
9. In van der Waals's equation, what properties of a real gas are accounted for by the constants a and b?
10. State Graham's law in words and in equation form. If gas A effuses four times more rapidly than gas B at 25 °C, does this ratio of rates of effusion increase, decrease, or remain the same as the temperature of the gas increases?

Numerical Questions

Measuring Pressure

11. Gas pressure can be expressed in units of millimeters of mercury, torrs, atmospheres, bars, and pascals (although millimeters of mercury and atmospheres are used exclusively in this book). Do the following unit conversions.
(a) 725 mm Hg to atmospheres
(b) 0.67 atm to millimeters of mercury
(c) 740 mm Hg to kilopascals
(d) 0.75 atm to kilopascals
(e) 745 mm Hg to bars
(f) 125 mm Hg to torr
(g) 12 mm Hg to kilopascals

12. The average barometric pressure at an altitude of 10 km is 210 mm Hg. Express this pressure in atmospheres, bars, and kilopascals.

13. If the mercury levels in a U-tube manometer are as illustrated here, express the gas pressure in millimeters of mercury, atmospheres, torrs, and kilopascals.

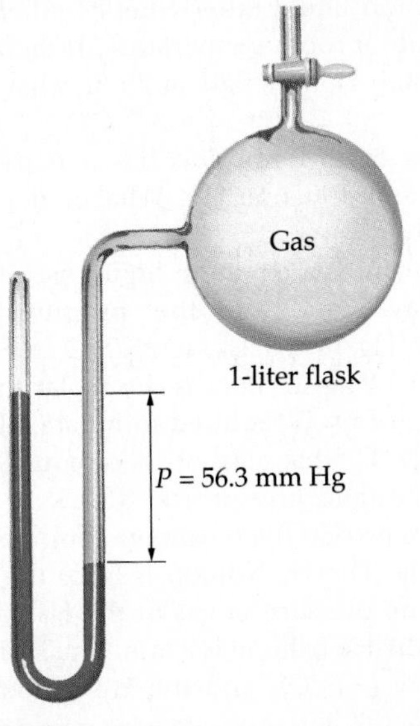

14. If the pressure of a gas is 95.0 kPa, what is the difference (in millimeters of mercury) in the mercury levels in the U-tube manometer? (See the figure accompanying Study Question 13.)

The Gas Laws

15. A sample of CO_2 gas is placed in a 125-mL flask where it exerts a pressure of 67.5 mm Hg. What is the pressure of this gas sample when it is transferred to a 500.-mL flask at the same temperature?

16. A sample of nitrogen gas has a pressure of 56.5 mm Hg in a 250.-mL flask. The sample is transferred to a new flask where it exerts a pressure of 23.6 mm Hg at the same temperature. What is the volume of this new flask?

17. You have 3.5 L of helium at a temperature of 22.0 °C. What volume would the helium occupy at 37 °C? (The pressure of the helium sample is constant.)

18. A 25.0-mL sample of gas is enclosed in a gas-tight syringe (Figure 12.2) at 22 °C. If the syringe is immersed in an ice bath (0 °C), what is the new gas volume, assuming that the pressure is held constant?

19. Water can be made by combining gaseous O_2 and H_2. If you begin with 1.0 L of $H_2(g)$ at 380 mm Hg and 25 °C, how many liters of $O_2(g)$ do you need for complete reaction if the O_2 gas is also measured at 380 mm Hg and 25 °C?

20. Methane, CH_4, burns in air according to the equation

$$CH_4(g) + 2\,O_2(g) \longrightarrow CO_2(g) + 2\,H_2O(g)$$

How many liters of O_2 are required for complete reaction with 5.2 L of CH_4? How many liters of H_2O vapor are produced? Assume all gases are measured at the same temperature and pressure.

21. You have a sample of gas in a flask with a volume of 350. mL. At 25.5 °C the pressure of the gas is 135 mm Hg. If you decrease the temperature to 0.0 °C, what is the gas pressure at the lower temperature?

22. Imagine that you live in a small cabin with an interior volume of 150 m^3. On a cold morning the indoor temperature is 10 °C, but by afternoon the sun has warmed the cabin air to 18 °C. Because air expands (to maintain a constant pressure) as the temperature increases, and because the cabin is not sealed, some air has leaked out of the cabin. How many cubic meters of air have been forced out of the cabin by the sun's warming? How many liters?

23. A sample of gas occupies 135 mL at 22.5 °C; the pressure is 165 mm Hg. What is the pressure of the gas sample when it is placed in a 250.-mL flask at a temperature of 0.0 °C?

24. Assume that you place some octane in the cylinder of an automobile engine. The cylinder has a volume of 250. cm^3, and the pressure of gaseous octane is 3.50 atm in the hot engine (250 °C). What would be the pressure in the cylinder if you lower the temperature of the automobile engine

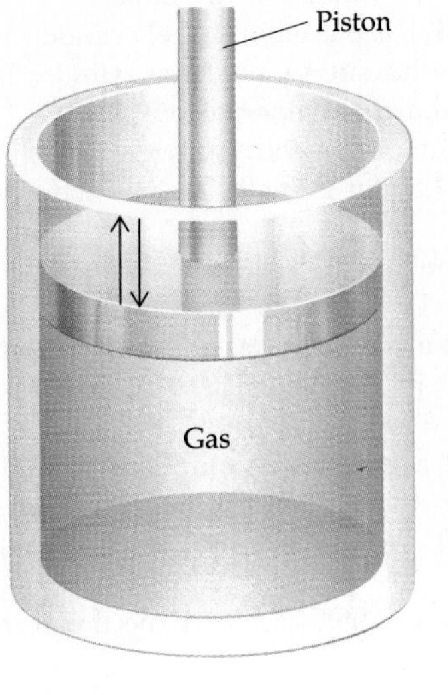

A representation of a cylinder in an automobile engine. As the gas expands, the piston moves up.

to room temperature (25 °C) and change the volume of the cylinder to 500. cm^3?

25. You have a sample of CO_2 in a flask (A) with a volume of 154 mL. At 22.5 °C, the pressure of the gas is 136.5 mm Hg. To find the volume of another flask (B), you move the CO_2 to that flask and find that its pressure is now 94.3 mm Hg at 245 °C. What is the volume of flask B?

26. A sample of gas was contained in a 245-mL flask at a temperature of 23.5 °C; the gas pressure is found to be 48.5 mm Hg. The gas is moved to a new flask, which is immersed in ice and water, and which has a volume of 68 mL. What is the pressure of the gas in the smaller flask at the new temperature?

27. One of the cylinders of an automobile engine has a volume of 400. cm^3. The engine takes in air at a pressure of 1.00 atm and a temperature of 15 °C and compresses it to a volume of 50.0 cm^3 at 77 °C. What is the final pressure of the gas in the cylinder? (The ratio of before and after volumes, in this case 400:50, or 8:1, is called the compression ratio.)

28. Assume that a balloon needs to displace at least 1.00×10^5 L of air. You fill the balloon with helium to a volume of 1.05×10^5 L on the ground where the pressure is 737 mm Hg and the temperature is 16.0 °C. When the balloon ascends to a height of 2 miles where the pressure is only 600. mm Hg and the temperature is −33 °C, does the balloon still displace the required 1.00×10^5 L of air?

The Ideal Gas Law

29. A 1.25-g sample of CO_2 is contained in a 850.-mL flask at 22.5 °C. What is the pressure of the gas?

30. 30.0 kg of helium is placed in a balloon. What is the volume of the balloon if the final pressure is 1.20 atm and the temperature is 22 °C?

31. To find the volume of a flask, it is first evacuated so that it contains no gas at all. Next, 4.4 g of CO_2 is introduced into the flask. On warming to 22 °C, the gas exerts a pressure of 635 mm Hg. What is the volume of the flask??

32. 1.50 g of hexane, C_6H_{14}, is placed in a steel cylinder. What is the pressure of the hexane vapor if the cylinder has a volume of 250. cm^3 and the temperature is 250 °C?

33. If you have a 150.-L tank of gaseous nitrogen, and the gas exerts a pressure of 41.8 mm Hg at 25 °C, how many moles of nitrogen are in the tank?

34. How many grams of helium are required to fill a 5.0-L balloon to a pressure of 1.1 atm at 25 °C?

35. A 0.982-g sample of an unknown gas exerts a pressure of 700. mm Hg in a 450.-mL container at 23 °C. What is the molar mass of the gas?

36. A 0.0125-g sample of a gas with an empirical formula of CHF_2 is placed in a 165-ml flask. It has a pressure of 13.7 mm Hg at 22.5 °C. What is the molecular formula of the compound?

37. A hydrocarbon, C_xH_y, is 82.66% carbon. Experiment shows that 0.218 g of the hydrocarbon has a pressure of 374 mm Hg in a 185-mL flask at 23 °C. What are the empirical and molecular formulas of the compound?

38. A boron hydride, with the general formula B_xH_y, is 14.4% hydrogen. If 0.0818 g of the compound exerts a pressure of 191 mm Hg at 22.5 °C in a 125-mL flask, what are the empirical and molecular formulas of the boron hydride?

39. Analysis of a chlorofluorocarbon, CCl_xF_y, shows that it is 11.79% C and 69.57% Cl. In another experiment you find that 0.107 g of the compound fills a 458-mL flask at 25 °C with a pressure of 21.3 mm Hg. What is the molecular formula of the compound?

40. Five compounds are in the family of sulfur-fluorine compounds with the general formula S_xF_y. One of these compounds is 25.23% S. If you place 0.0955 g of the compound in a 89-mL flask at 45 °C, the pressure of the gas is 83.8 mm Hg. What is the molecular formula of S_xF_y?

41. Forty miles above the earth's surface the temperature is 250 K and the pressure is only 0.20 mm Hg. What is the density of air in grams per liter at this altitude? (Assume the molar mass of air is 29 g/mol.)

42. A common liquid ether [diethyl ether, $(C_2H_5)_2O$] vaporizes easily at room temperature. If the vapor has a pressure of 233 mm Hg in a flask at 25 °C, what is the density of the vapor?

43. Gaseous methyl fluoride has a density of 0.259 g/L at 400. K and 190. mm Hg. What is the molar mass of the compound?

44. Chloroform is a common liquid used in the laboratory. It vaporizes readily. If the pressure of chloroform is 195 mm Hg in a flask at 25.0 °C and the density of the vapor is 1.25 g/L, what is the molar mass of chloroform?

45. If 12.0 g of O_2 is required to inflate a balloon to a certain size at 27 °C, what mass of O_2 is required to inflate it to the same size (and pressure) at 81 °C?

46. You have two gas-filled balloons, one containing He and the other H_2. The H_2 balloon is twice the size of the He balloon. The pressure of gas in the H_2 balloon is 1 atm, and that in the He balloon is 2 atm. The H_2 balloon is outside in the snow (−5 °C), and the He balloon is inside a warm building (23 °C).
   (a) Which balloon contains the greater number of molecules?
   (b) Which balloon contains the greater mass of gas?

47. Assume that a bicycle tire containing 0.406 mol of air will burst if its internal pressure reaches 7.25 atm, at which time the internal volume is 1.52 L. To what temperature, in degrees Celsius, does the air in the tire need to be heated to cause a blowout?

48. The temperature of the atmosphere on Mars can be as high as 27 °C at the equator at noon, and the atmospheric presure is about 8 mm Hg. If a spacecraft could collect 10. m^3 of this atmosphere, compress it to a small volume, and send it back to earth, how many moles would the sample contain?

Gas Laws and Stoichiometry

49. Iron reacts with acid to produce iron(II) chloride and hydrogen gas.

$$Fe(s) + 2\,HCl(aq) \longrightarrow FeCl_2(aq) + H_2(g)$$

The H_2 gas from the reaction of 1.0 g of iron with excess hydrochloric acid is collected in a 15.0-L flask at 25 °C. What is the pressure of the H_2 gas in this flask?

50. Silane, SiH_4, reacts with O_2 to give silicon dioxide and water.

$$SiH_4(g) + 2\,O_2(g) \longrightarrow SiO_2(s) + 2\,H_2O(\ell)$$

A 5.20-L sample of SiH_4 gas at 500 mm Hg pressure and 25 °C is allowed to react with O_2 gas. What volume of O_2 gas, in liters, is required for complete reaction, if the oxygen has a pressure of 500 mm Hg at 25 °C?

51. Sodium azide, the explosive compound in automobile air bags, decomposes according to the equation

$$2\,NaN_3(s) \longrightarrow 2\,Na(s) + 3\,N_2(g)$$

What mass of sodium azide is required to provide the nitrogen to inflate a 25.0-L bag to a pressure of 1.3 atm at 25 °C?

52. The hydrocarbon benzene burns to give CO_2 and water vapor.

$$2\,C_6H_6(g) + 15\,O_2(g) \longrightarrow 12\,CO_2(g) + 6\,H_2O(g)$$

If a 0.095-g sample of benzene burns completely in O_2, what is the pressure of water vapor in a 4.75-L flask at 30.0 °C? If the O_2 gas needed for complete combustion is contained in a 4.75-L flask at 22 °C, what is its pressure?

53. Hydrazine reacts with O_2 according to the equation

$$N_2H_4(g) + O_2(g) \longrightarrow N_2(g) + 2\,H_2O(\ell)$$

Assume that the O_2 needed for the reaction is in a 450-L tank at 23 °C. What must the oxygen pressure be in the tank to have enough oxygen to consume 1.00 kg of hydrazine completely?

54. Oxygen masks use canisters containing potassium superoxide. The superoxide consumes the CO_2 exhaled by a person and replaces it with oxygen.

$$4\,KO_2(s) + 2\,CO_2(g) \longrightarrow 2\,K_2CO_3(s) + 3\,O_2(g)$$

What mass of KO_2, in grams, is required to use up 8.90 L of CO_2 at 22.0 °C and 767 mm Hg?

55. Assume that you wish to convert 1.0×10^3 g of uranium metal to gaseous UF_6 by reaction with gaseous fluorine. The fluorine gas required in the reaction is contained in a 50.0-L tank with a pressure of 8.0 atm at 25 °C. When all of the uranium metal has been converted to UF_6, what is the pressure of the fluorine remaining in the tank at 25 °C?

56. $Ni(CO)_4$ can be made by reacting finely divided nickel with gaseous CO. If you have CO in a 1.50-L flask at a pressure of 418 mm Hg at 25.0 °C, what is the maximum number of grams of $Ni(CO)_4$ that can be made?

57. Iron forms a series of compounds of the type $Fe_x(CO)_y$. If you heat the compounds in air, they decompose to Fe_2O_3 and CO_2 gas. After heating a 0.142-g sample of $Fe_x(CO)_y$, you isolate the CO_2 in a 1.50-L flask at 25 °C. The pressure of the gas is 44.9 mm Hg. What is the formula of $Fe_x(CO)_y$?

58. Group 2A metal carbonates decompose to the metal oxide and CO_2 on heating.

$$MCO_3(s) \longrightarrow MO(s) + CO_2(g)$$

You heat 0.158 g of a white, solid carbonate of a Group 2A metal and find that the evolved CO_2 has a pressure of 69.8 mm Hg in a 285-mL flask at 25 °C. What metal is *M*?

59. Potassium superoxide reacts with CO_2 to give oxygen gas.

$$4\,KO_2(s) + 2\,CO_2(g) \longrightarrow 2\,K_2CO_3(s) + 3\,O_2(g)$$

If you combine 16.0 g of KO_2 with the CO_2 in a 4.00-L tank, in which the gas pressure is 1.24 atm at 23 °C, which reactant is consumed completely? If the O_2 gas is captured from the reaction, what is its pressure in a 2.50-L flask at 25 °C?

60. If you place 2.25 g of solid silicon in a 6.56-L flask that contains CH_3Cl with a pressure of 585 mm Hg at 25 °C, how many grams of $(CH_3)_2SiCl_2(g)$, dimethyldichlorosilane, can be formed?

$$Si(s) + 2\,CH_3Cl(g) \longrightarrow (CH_3)_2SiCl_2(g)$$

What pressure of $(CH_3)_2SiCl_2(g)$ do you expect in this same flask at 95 °C on completion of the reaction? (Dimethyldichlorosilane is one starting material used to make silicones—polymeric substances used as lubricants, antistick agents, and in water-proofing caulk.)

Gas Mixtures

61. Helium (0.56 g) and hydrogen gas are mixed in a flask at room temperature. The partial pressure of He is 150 mm Hg and that of H_2 is 25 mm Hg. How many grams of H_2 are present?

62. What is the total pressure in atmospheres of a gas mixture that contains 1.0 g of H_2 and 8.0 g of Ar in a 3.0-L container at 27 °C? What are the partial pressures of the two gases?

63. A cylinder of compressed gas is labeled "Composition: 4.5% H_2S, 3.0% CO_2, balance N_2." The pressure gauge attached to the cylinder reads 46 atm. Calculate the partial pressure of each gas, in atmospheres, in the cylinder.

64. A halothane-oxygen mixture ($C_2HBrClF_3 + O_2$) can be used as an anesthetic. Assume that a tank containing such a mixture has the following partial pressures: *P*(halothane) = 170 mm Hg and $P(O_2)$ = 570 mm Hg.
   (a) What is the ratio of the number of the moles of halothane to the number of moles of O_2?
   (b) If the tank contains 160 g of O_2, how many grams of $C_2HBrClF_3$ are present?

65. A collapsed balloon is filled with He to a volume of 12 L at a pressure of 1.0 atm. Oxygen (O_2) is then added so that the final volume of the balloon is 26 L with a total pressure of 1.0 atm. The temperature, constant throughout, is equal to 20 °C.

(a) How many grams of He does the balloon contain?
(b) What is the final partial pressure of He in the balloon?
(c) What is the partial pressure of O_2 in the balloon?
(d) What is the mole fraction of each gas?

66. A miniature volcano can be made in the laboratory with ammonium dichromate. When ignited it decomposes in a fiery display.

$$(NH_4)_2Cr_2O_7(s) \longrightarrow N_2(g) + 4\,H_2O(g) + Cr_2O_3(s)$$

If 5.0 g of ammonium dichromate is used, and if the gases from this reaction are trapped in a 3.0-L flask at 23 °C, what is the total pressure of the gas in the flask? What are the partial pressures of N_2 and H_2O?

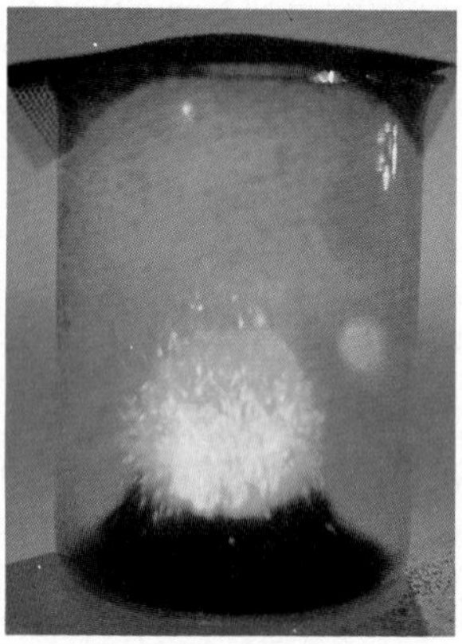

Ammonium dichromate decomposes exothermically to give nitrogen gas, water vapor, and green chromium(III) oxide.

67. Dichlorine oxide is a powerful oxidizing agent that is used to bleach wood pulp and to treat municipal water supplies. It is made by the reaction

$$SO_2(g) + 2\,Cl_2(g) \longrightarrow SOCl_2(g) + Cl_2O(g)$$

If you put SO_2 in a flask so that its pressure is 125 mm Hg at 22 °C, and if you add Cl_2 gas to this same flask, what should the Cl_2 partial pressure be in order to have the correct stoichiometric ratio of SO_2 to Cl_2?

68. As shown in the reaction in Study Question 67, dichlorine oxide is prepared by the reaction of sulfur oxide and chlorine. If you react 0.235 g of SO_2 with a stoichiometric amount of Cl_2, what is the total pressure of the reaction products in a 250.-mL flask at a temperature of 85 °C after the SO_2 and Cl_2 have been used up?

69. A sample of nitrogen gas is collected over water at 18 °C (Figure 12.10). If the barometric pressure in the laboratory is 747 mm Hg, what is the partial pressure of the dry nitrogen gas in the sample? (See Appendix E for water vapor pressure data.)

70. You are given 1.56 g of a mixture of $KClO_3$ and KCl. When heated, the $KClO_3$ decomposes to KCl and O_2,

$$2\,KClO_3(s) \longrightarrow 2\,KCl(s) + 3\,O_2(g)$$

and 327 mL of O_2 is collected over water at 19 °C. The total pressure of the gas in the collection flask is 735 mm Hg. What is the weight percentage of $KClO_3$ in the sample? (See Appendix E for water vapor pressure data.)

71. If equal masses of O_2 and N_2 are placed in separate containers of equal volume at the same temperature, which of the following statements is true? If false, tell why it is false.
(a) The pressure in the N_2-containing flask is greater than that in the flask containing O_2.
(b) More molecules are in the flask containing O_2 than in the one containing N_2.

72. Suppose you have two pressure-proof steel cylinders of equal volume, one containing CO and the other acetylene, C_2H_2.
(a) If you have 1 kg of each compound, in which cylinder is the pressure greater at 25 °C?
(b) Now suppose the CO cylinder has twice the pressure of the acetylene cylinder at 25 °C. Which cylinder contains the greater number of molecules?

Kinetic-Molecular Theory

73. You are given two flasks of equal volume. Flask A contains H_2 at 0 °C and 1 atm pressure. Flask B contains CO_2 gas at 0 °C and 2 atm pressure. Compare these two gases with respect to each of the following:
(a) Average kinetic energy per molecule
(b) Average molecular velocity
(c) Number of molecules
(d) Mass of gas

74. Equal masses of gaseous N_2 and Ar are placed in separate flasks of equal volume, both gases being at the same temperature. Tell whether each of the following statements is true or false. Briefly explain your answer in each case.
(a) More molecules of N_2 are present than atoms of Ar.
(b) The pressure is greater in the Ar flask.
(c) Ar atoms have a greater velocity than the N_2 molecules.
(d) The molecules of N_2 collide more frequently with the walls of the flask than do the atoms of Ar.

75. If the average speed of an oxygen molecule is 4.28×10^4 cm/s at 25 °C, what is the average speed of a CO_2 molecule at the same temperature?

76. Calculate the rms speed for CO molecules at 25 °C. What is the ratio of its speed to that of Ar atoms at the same temperature?

77. Place the following gases in order of increasing average molecular speed at 25 °C: (a) Ar, (b) CH_4, (c) N_2, and (d) CH_2F_2.

78. The reaction of SO_2 with Cl_2 to give dichlorine oxide was described previously.

$$SO_2(g) + 2\,Cl_2(g) \longrightarrow SOCl_2(g) + Cl_2O(g)$$

All of the compounds involved in the reaction are gases. Place them in order of increasing average speed.

Diffusion and Effusion

79. Argon gas is ten times as dense as helium gas at the same temperature and pressure. Which gas effuses faster? How much faster?

80. In each pair of gases below, tell which effuses faster:
(a) CO_2 or F_2
(b) O_2 or N_2
(c) C_2H_4 or B_2H_6
(d) two CFCs: $CFCl_3$ or $C_2Cl_2F_4$

81. A gas whose molar mass you wish to know effuses through an opening at a rate only one third as great as the effusion rate of helium. What is the molar mass of the unknown gas?

82. A sample of uranium fluoride is found to effuse at the rate of 17.7 mg/h. Under comparable conditions, gaseous I_2 effuses at the rate of 15.0 mg/h. What is the molar mass of the uranium fluoride? (*Caution:* rates must be used in units of moles per time.)

Nonideal Gases

83. In the text it is stated that the pressure of 8.00 mol of Cl_2 in a 4.00-L tank at 27.0 °C is 29.5 atm. Using the van der Waals equation, verify this result and compare it with the pressure expected from the ideal gas law.

84. You want to store 165 g of CO_2 gas in a 12.5-L tank at room temperature (25 °C). Calculate the pressure the gas would have using (a) the ideal gas law and (b) van der Waals's equation.

General Questions

85. Complete the table.

	atm	mm Hg	kPa	bar
Standard atmosphere	1			
Partial pressure of N_2 in the atmosphere		593		
Tank of compressed H_2				133
Atmospheric pressure at the top of Mt. Everest			33.7	

86. You want to fill a tank with CO_2 gas at 865 mm Hg and 25 °C. The tank is 20.0 m long with a 10.0-cm radius. What mass of CO_2 (in grams) is required?

87. Acetaldehyde is a common liquid that vaporizes readily. A pressure of 331 mm Hg is observed in a 125-mL flask at 0.0 °C, and the density of the vapor is 0.855 g/L. What is the molar mass of acetaldehyde?

88. On combustion, 1.0 L of a gaseous compound of hydrogen, carbon, and nitrogen produces 2.0 L of CO_2, 3.5 L of H_2O vapor, and 0.50 L of N_2 at STP. What is the empirical formula of the compound?

89. To what temperature, in degrees Celsius, must a 25.5-mL sample of oxygen at 90 °C be cooled for its volume to shrink to 21.5 mL? Assume that the pressure and mass of the gas are constant.

90. Methane is burned in a laboratory Bunsen burner to give CO_2 and water vapor. Methane gas is supplied to the burner at the rate of 5.0 L/min (at a temperature of 28 °C and a pressure of 773 mm Hg). At what rate must oxygen be supplied to the burner (at a pressure of 742 mm Hg and a temperature of 26 °C)?

91. You have a sample of helium gas at −33 °C, and you want to increase the average speed of helium atoms by 10.0%. To what temperature should the gas be heated to accomplish this?

92. You have a 550-mL tank of gas with a pressure of 1.56 atm at 24 °C. You thought the gas was pure carbon monoxide gas, CO, but you later found it was contaminated by small quantities of gaseous CO_2 and O_2. Analysis shows that the tank pressure is only 1.34 atm (at 24 °C) if the CO_2 is removed and that 0.0870 g of O_2 can be removed chemically. What are the masses of CO and CO_2 in the tank, and what were the partial pressures of each of the three gases in the 550-mL tank at 24 °C?

93. A 3.0-L bulb containing He at 145 mm Hg is connected by a valve to a 2.0-L bulb containing Ar at 355 mm Hg. (See the figure at the bottom of this page.) Calculate the partial pressure of each gas and the total pressure after the valve between the flasks is opened.

94. A 1.0-L flask contains 10.0 g each of O_2 and CO_2 at 25 °C.
(a) Which gas has the greater partial pressure, O_2 or CO_2, or are they the same?
(b) Which molecules have the greater average speed, or are they the same?
(c) Which molecules have the greater average kinetic energy, or are they the same?

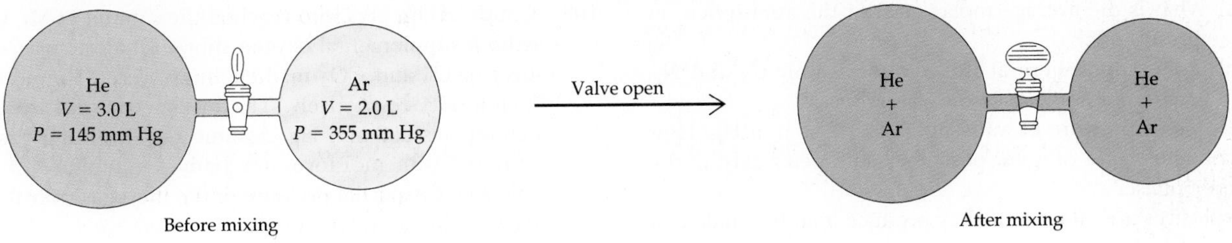

95. Which of the following gas samples contains the largest number of gas molecules? Which contains the smallest number? Which represents the largest mass of gas?
(a) 1.0 L of H_2 at STP
(b) 1.0 L of Ar at STP
(c) 1.0 L of H_2 at 27 °C and 760 mm Hg
(d) 1.0 L of He at 0 °C and 900 mm Hg

96. Two flasks, each with a volume of 1.00 L, contain O_2 gas with a pressure of 380 mm Hg. Flask A is at 25 °C, and flask B is at 0 °C. Which flask contains the greater number of O_2 molecules?

97. You are given a solid mixture of $NaNO_2$ and NaCl and are asked to analyze it for the amount of $NaNO_2$ present. To do so you allow it to react with sulfamic acid, HSO_3NH_2, in water according to the equation

$$NaNO_2(aq) + HSO_3NH_2(aq) \longrightarrow NaHSO_4(aq) + H_2O(\ell) + N_2(g)$$

What is the weight percentage of $NaNO_2$ in 1.232 g of the solid mixture if reaction with sulfamic acid produces 295 mL of N_2 gas? The gas was collected over water (Figure 12.10) at a temperature of 21.0 °C and with the barometric pressure equal to 736.0 mm Hg.

98. Silane, SiH_4, reacts with O_2 to give silicon dioxide and water

$$SiH_4(g) + 2\,O_2(g) \longrightarrow SiO_2(s) + 2\,H_2O(g)$$

If you mix SiH_4 with O_2 in the correct stoichiometric ratio, and if the total pressure of the mixture is 120 mm Hg, what are the partial pressures of SiH_4 and O_2? When the reactants have been completely consumed, what is the total pressure in the flask?

99. Chlorine trifluoride, ClF_3, is a valuable reagent because it can be used to convert metal oxides to metal fluorides.

$$6\,NiO(s) + 4\,ClF_3(g) \longrightarrow 6\,NiF_2(s) + 2\,Cl_2(g) + 3\,O_2(g)$$

(a) How many grams of NiO can react with ClF_3 gas if the gas has a pressure of 250 mm Hg at 20 °C in a 2.5 L-flask?
(b) If the ClF_3 described in part (a) is completely consumed, what are the partial pressures of Cl_2 and of O_2 in the 2.5-L flask at 20 °C (in millimeters of mercury)? What is the total pressure in the flask?

100. The density of air 20 km above the earth's surface is 92 g/m^3. The pressure of the atmosphere is 42 mm Hg and the temperature is −63 °C.
(a) What is the average molar mass of the atmosphere at this altitude?
(b) If the atmosphere at this altitude is only O_2 and N_2, what is the mole fraction of each gas?

101. The vapor pressure of water at 25 °C is 23.8 mm Hg. How many molecules of water per cubic centimeter exist in the vapor phase?

102. Explosives are effective if they produce a large number of gaseous molecules as products. Nitroglycerin, for example, detonates according to the equation

$$2\,C_3H_5N_3O_9(s) \longrightarrow 6\,CO_2(g) + 3\,N_2(g) + 5\,H_2O(g) + \tfrac{1}{2}O_2(g)$$

If 1.0 g of nitroglycerin explodes, calculate the volume the product gases would occupy if their total pressure is 1.0 atm at 5.0×10^2 °C. Compare with the volume the product gases would occupy at 25 °C and 1.0 atm pressure.

103. Chlorine dioxide, ClO_2, reacts with fluorine to give a new gas that contains the elements Cl, O, and F. In an experiment you find that 0.150 g of the gas has a pressure of 17.2 mm Hg in a 1850-mL flask at 21 °C. What is the identity of this unknown gas?

104. Phosphine gas, PH_3, is toxic when it reaches a concentration of 7×10^{-5} mg/L. To what pressure does this correspond at 25 °C?

105. The Starship *Enterprise* is not really fueled by dilithium crystals! Rather, it uses a mixture of diborane, B_2H_6, and oxygen. The two react according to the equation

$$B_2H_6(g) + 3\,O_2(g) \longrightarrow B_2O_3(s) + 3\,H_2O(g)$$

(a) If all of the gases involved in the reaction above are at the same temperature, what are their relative rms speeds?
(b) Diborane and O_2 are contained in separate fuel tanks of equal volume and at the same temperature. The pressure in the O_2 tank is 45 atm. If the B_2H_6 tank is to contain an amount of B_2H_6 that reacts exactly with the O_2, what must the pressure be in the B_2H_6 tank (in atmospheres)?

106. A xenon fluoride can be prepared by heating a mixture of Xe and F_2 gases to a high temperature in a pressure-proof container. Assume that xenon gas is added to a 0.25-L container until its pressure is 0.12 atm at 0.0 °C. Fluorine gas is then added until the total pressure is 0.72 atm at 0.0 °C. After the reaction is complete, the xenon is consumed completely and the pressure of the F_2 remaining in the container is 0.36 atm at 0 °C. What is the empirical formula of the compound prepared from Xe and F_2?

107. A balloon at the circus is filled with helium gas to a gauge pressure of 22 mm Hg at 25 °C. The volume of the gas is 300. mL, and the barometric pressure is 755 mm Hg. How many moles of helium are in the balloon? (Remember that gauge pressure = total pressure − barometric pressure. See page 549.)

108. A study of climbers who reached the summit of Mt. Everest without supplemental oxygen showed that the partial pressures of O_2 and CO_2 in their lungs were 35 mm Hg and 7.5 mm Hg, respectively. The barometric pressure at the summit was 253 mm Hg. Assume that the lung gases are saturated with moisture at a body temperature of 37 °C. Calculate the partial pressure of the inert gas (mostly nitrogen) in the climbers' lungs.

The barometric pressure is very low on the summit of a very high mountain (about $\frac{1}{3}$ atm), and so the partial pressure of oxygen is very low. (K. Gunnar/FPG International)

109. Assume that you have a glass tube 50. cm long. You allow some $NH_3(g)$ to diffuse along the tube from one end and some $HCl(g)$ to diffuse along the tube from the opposite end. The gases meet at some point and form solid NH_4Cl. (See the photos accompanying this question.) Estimate the distance along the tube where the gases meet and react to form solid NH_4Cl. (Note that Graham's law applies less strictly to diffusion. It gives only an estimate of the distance here.)

Ammonia and HCl are injected into the bent tube through the rubber caps at opposite ends of the tube. The gases diffuse along the tube and eventually meet, where they form a white ring of solid NH_4Cl (just below the bend on the right hand tube.) (C. D. Winters)

110. Acetylene can be made by allowing calcium carbide to react with water

$$CaC_2(s) + 2\,H_2O(\ell) \longrightarrow C_2H_2(g) + Ca(OH)_2(s)$$

You react 2.65 g of CaC_2 with excess water. If you collect the acetylene over water (Figure 12.10) and find that the gas (acetylene and water vapor) has a volume of 795 mL at 25.2 °C at a barometric pressure of 735.2 mm Hg, what is the percent yield of acetylene?

111. Assume that you have synthesized a new compound of boron and hydrogen, B_xH_y. Its empirical and molecular formulas can be determined from the following information.
(a) To determine its empirical formula, you burn it in pure O_2 to give B_2O_3 and H_2O. The water from the combustion of 0.492 g of B_xH_y has a mass of 0.540 g.
(b) A 0.0631-g sample of boron hydride quickly evaporates into a flask with a volume of 120. mL; the gas exerts a pressure of 98.6 mm Hg at 23 °C.

112. Chlorine trifluoride, one of the most reactive compounds known, is made by reacting chlorine and fluorine.

$$Cl_2(g) + 3\,F_2(g) \longrightarrow 2\,ClF_3(g)$$

Assume that you mix 0.71 g of Cl_2 with 1.00 g of F_2 in a 258-mL flask at 23 °C. What are the partial pressures of each of the two reactants before reaction? What is the partial pressure of the product and any leftover reactant after reaction in the 258-mL flask at 23 °C? What is the total pressure in the flask at the end of the reaction? Is the final total pressure more or less than the initial total pressure?

Conceptual Questions

113. State whether each of the following properties is characteristic of all gases, some gases, or no gas:
(a) Transparent to light
(b) Colorless
(c) Unable to pass through a filter paper
(d) More difficult to compress than water
(e) Odorless
(f) Settles on standing

114. State whether each of the following samples of matter is a gas. If not enough information is provided for you to decide, write "insufficient information."
(a) A material is in a steel tank at 100 atm pressure. When the tank is opened to the atmosphere, the material suddenly expands, increasing its volume by 10%.
(b) A 1.0-mL sample of the material weighs 8.2 g.
(c) When a material is released from a point 30 ft below the surface of a lake at sea level (equivalent in pressure to about 760 mm Hg), it rises rapidly to the surface, at the same time doubling its volume.
(d) A material is transparent and pale green in color.
(e) One cubic meter of material contains as many molecules as 1 m^3 of air at the same temperature and pressure.

115. The formula of water in early chemistry books was given as HO. The volume relationships in the reaction of hydrogen and oxygen were known at the time, however, and they can be represented, using the old formula for water, as

$$2\ H_? + O_? \longrightarrow 2\ HO$$

Balance the equation by substituting numbers for the question marks. Do not change anything else. If the formula of water was really HO, what is the simplest formula for oxygen and for hydrogen?

116. Does the effect of intermolecular attraction on the properties of a gas become more significant or less significant if
(1) The gas is compressed to a smaller volume at constant temperature?
(2) More gas is forced into the same volume at constant temperature?
(3) The temperature of the gas is raised at constant pressure?

117. Each of the four tires of a car is filled with a different gas. Each tire has the same volume and each is filled to the same pressure, 3.0 atm, at 25 °C. One tire contains 116 g of air, another tire has 80.7 g of neon, another tire has 16.0 g of helium, and the fourth tire has 160. g of an unknown gas.
(a) Do all four tires contain the same number of gas molecules? If not, which one has the greatest number of molecules?
(b) How many times heavier is a molecule of the unknown gas than an atom of helium?
(c) In which tire do the molecules have the largest kinetic energy? The greatest average speed?

118. In the Chemical Puzzler at the beginning of this chapter we asked if the operation of the air bag depended on temperature or the altitude of the car. Can you comment on these questions now?

Summary Questions

119. Chlorine gas (Cl_2) is used as a disinfectant in municipal water supplies, although chlorine dioxide (ClO_2) and ozone are becoming more widely used. ClO_2 is better than Cl_2 because it leads to fewer chlorinated byproducts, which are themselves pollutants.
(a) How many valence electrons are in ClO_2?
(b) The chlorite ion, ClO_2^-, is obtained by reducing ClO_2. Draw a possible electron dot structure for the ion ClO_2^-. (Cl is the central atom.)
(c) What is the hybridization of the central Cl atom in ClO_2^-? What is the shape of the ion?
(d) Which species do you suppose has the larger bond angle, O_3 or ClO_2^-? Explain briefly.
(e) Chlorine dioxide, ClO_2, a yellow-green gas, can be made by the reaction of chlorine with sodium chlorite.

$$2\ NaClO_2(s) + Cl_2(g) \longrightarrow 2\ NaCl(s) + 2\ ClO_2(g)$$

Assume that you react 15.6 g of $NaClO_2$ with chlorine gas, which has a pressure of 1050 mm Hg in a 1.45-L flask at 22 °C. How many grams of ClO_2 can be produced? What is the pressure of the ClO_2 gas in a 1.25-L flask at 25 °C?

120. The sodium azide required for automobile air bags is made by the reaction of sodium metal with dinitrogen oxide in liquid ammonia as a solvent.

$$3\ N_2O(g) + 4\ Na(s) + NH_3(\ell) \longrightarrow NaN_3(s) + 3\ NaOH(s) + 2\ N_2(g)$$

(a) You have 65.0 g of sodium and a 35.0-L flask of N_2O gas (with a pressure of 2.12 atm at 23 °C). What is the maximum possible yield (in grams) of NaN_3?
(b) Draw a Lewis structure for the azide ion. Include all possible resonance structures. Which resonance structure is most likely?
(c) What is the shape of the azide ion?

CHAPTER 13

Bonding and Molecular Structure: Intermolecular Forces, Liquids, and Solids

A Chemical Puzzler

In the hot days of summer, you often want a tall, cool glass of water with ice. You go to the refrigerator and take some ice cubes from a tray. In passing, you notice that the frozen cubes bulge out over the top of the dividers. There seems to be a greater volume of ice in the tray than the amount of water you poured in. You drop the ice cubes into the water, and enjoy the cool drink. But another attribute of ice is important and is so common that you may overlook it: ice cubes float in liquid water. This property of water is shared by *very* few other substances in the universe. In fact, it is so strange that we want to know why it happens. And while we are at it, let us try to understand why water expands when it is frozen. The photos show a glass jar filled with water and sealed. When frozen, the water expanded and broke the jar.

We are surrounded by gases, liquids, and solids: the gases of the atmosphere; liquid water in the oceans, lakes, and growing plants; and solid earth. Of the 111 known elements, the vast majority are solids. Only 11 are gases under normal conditions (H_2, N_2, O_2, F_2, Cl_2, and the noble gases), and fewer still are liquids (Hg and Br_2). The behavior of gases was explored in Chapter 12, so we turn now to the other phases of matter. You will find this a useful chapter because it explains, among other things, why your body cools when you sweat, how bodies of water can influence local climate, why diamonds are hard but graphite is slippery, and something about the internal structure of beautifully crystalline solids.

Our objective in this chapter is to describe the liquid and solid states of matter in more detail and, in particular, the forces responsible for their formation. Figure 13.1 illustrates how this chapter is organized.

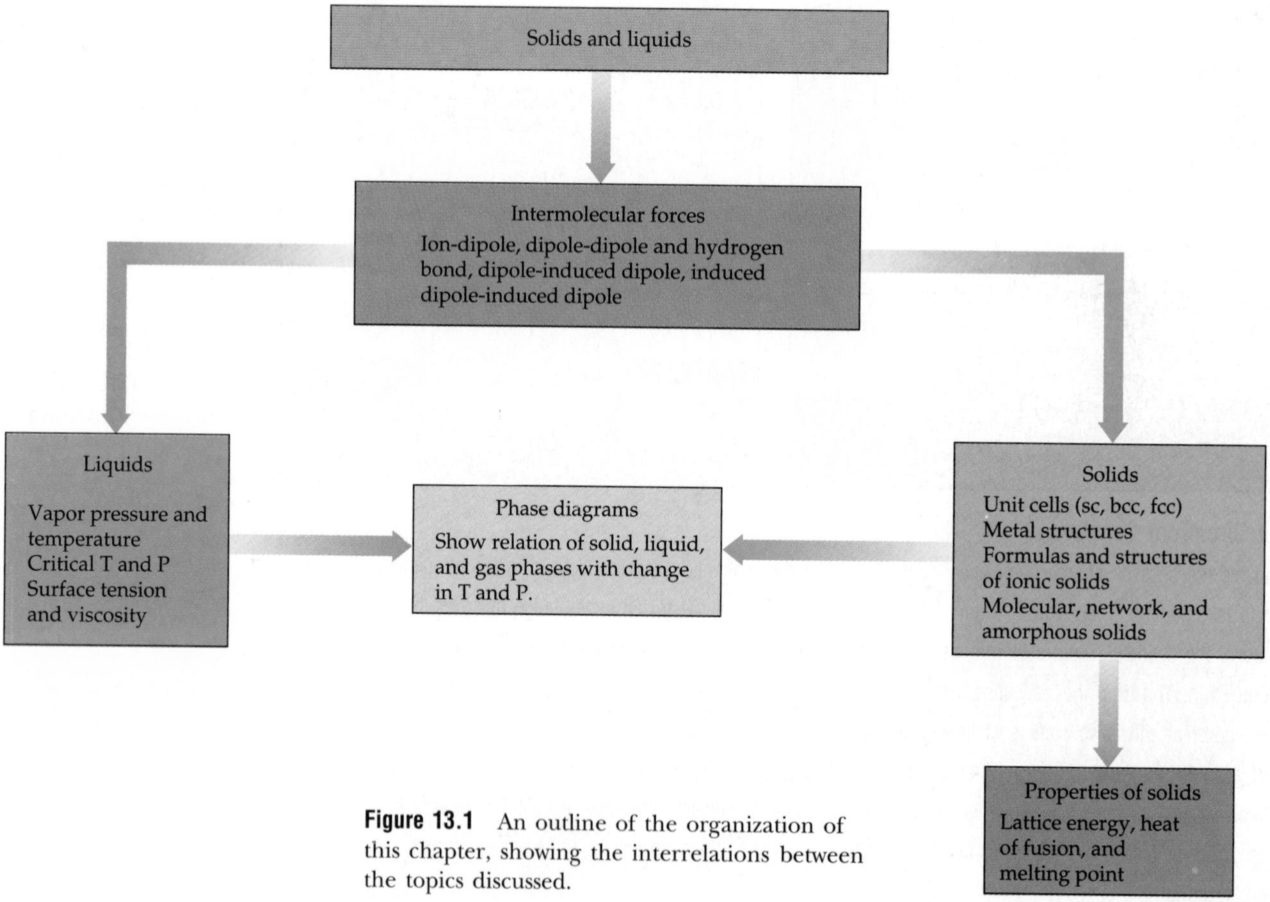

Figure 13.1 An outline of the organization of this chapter, showing the interrelations between the topics discussed.

13.1 PHASES OF MATTER AND THE KINETIC-MOLECULAR THEORY

The kinetic-molecular theory of gases (see Section 12.6) assumes (1) that gas molecules or atoms are widely separated; (2) that no forces of attraction exist between them; (3) that the molecules are in continual, random, and rapid motion; and (4) that their kinetic energy is determined only by the gas temperature. It is *only* on the first two points that gases differ significantly from liquids and solids.

Gases, unlike liquids and solids, can be compressed because so much space exists between gas molecules. In fact, a compressed gas could ultimately become a liquid, occupying much less space than the gas. Figure 13.2 shows a flask containing about 300 mL of liquid nitrogen. If all this liquid were allowed to evaporate to a gas, it would fill a very large balloon to a pressure of 1 atm at room temperature. This illustrates that *much* more space exists between molecules in the gas phase than in the liquid phase.

In contrast with the change in volume when converting from a gas to a liquid, no dramatic change occurs in volume when a liquid is converted to a solid. Figure 13.3 shows liquid and solid benzene side by side, and you see they are not appreciably different in volume. This means that the atoms in the liquid are packed together about as tightly as the atoms in the solid phase.

Each molecule or atom in the liquid or solid phase is clearly closer to a neighbor than in the gas phase. In fact, they cannot come much closer together. That is, the **compressibility** of liquids and solids, their change in volume with change in pressure, is small compared with that of gases. The air-fuel mixture in your car's engine is routinely compressed by a factor of about 10 before it is ignited. In contrast, the volume of liquid water changes only by 0.005% per atmosphere of pressure applied to it.

Figure 13.2 Liquid N_2. If this sample of liquid nitrogen were evaporated to N_2 gas at 25 °C, the gas would fill a large balloon (>200 L) to a pressure of 1 atm, which illustrates the fact that much more space exists between molecules in the gas phase than in the liquid phase. The fact that nonpolar N_2 molecules can be condensed to form a liquid also illustrates the importance of intermolecular forces. (C. D. Winters)

Figure 13.3 The same volume of liquid benzene is placed in two test tubes, and one tube is frozen. The solid and liquid phases have almost the same volume, showing that the molecules are packed together almost as tightly in the liquid phase as they are in the solid phase. (C. D. Winters)

The kinetic-molecular theory of gases assumes there are no forces attracting molecules to one another, so gases can expand infinitely to fill their containers uniformly and completely. In contrast, stronger **intermolecular forces** exist in liquids and solids (Section 13.2), and these attractive forces in a liquid or solid prevent a significant expansion of a liquid or solid. In addition, these forces account for the fact that liquids (and solids if they are powdered) can be poured from one container to another, they account for the strange properties of water, and for many properties of biological molecules. We therefore begin with an exploration of intermolecular forces and their importance and then move to a description of the liquid and solid phases.

13.2 INTERMOLECULAR FORCES

Intermolecular forces are the attractive forces between molecules or between ions and molecules. Without such interactions, all substances would be ideal gases.

Intermolecular forces, particularly those *not* involving ions, are known collectively as **van der Waals's forces,** named for the physicist who developed the equation to describe the behavior of real gases (Section 12.9).

The various types of intermolecular forces, which are listed in Table 13.1, involve interactions between ions and polar molecules, between polar molecules, and between molecules in which a dipole can be induced or created. All these types of forces are important in nature.

TABLE 13.1 Summary of Intermolecular Forces

Type of Interaction	Principal Factors Responsible for Interaction Energy	Approximate Magnitude (kJ/mol)
Ion — Dipole 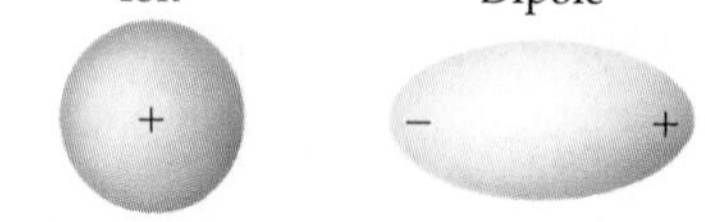	Ion charge; dipole moment	40-600
Dipole — Dipole (including H-bonding) 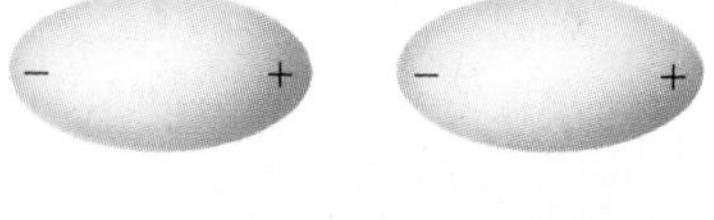	Dipole moment	5-25
Dipole — Induced dipole 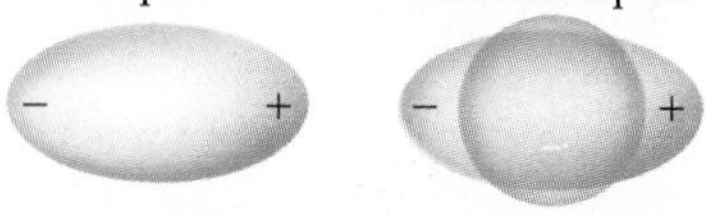	Dipole moment; polarizability	2-10
Induced dipole — Induced dipole 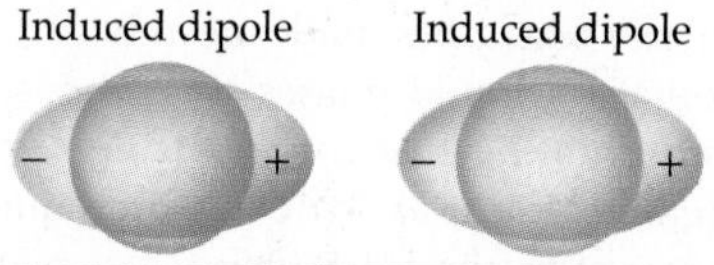	Polarizability	0.05-40

(Arrow pointing up from the bottom row to the top row: Increasing strength of interaction)

To put the discussion of intermolecular forces in perspective, recall that energies of attraction between two oppositely charged ions can amount to many hundreds of kilojoules per mole (Section 8.7). Covalent bonds have energies that are generally in the range of 100 to 400 kJ/mol. These energies usually far exceed those for the intermolecular forces to be described later.

Interactions Between Ions and Molecules with a Permanent Dipole

A dipole consists of separated positive and negative charges (Section 9.6), so a positive cation or negative anion can be attracted to one of these charges and repelled by the other (see Table 13.1). The force involved in the attraction between a positive or negative ion and a polar molecule is less than that for ion-ion attractions (Section 8.7) but greater than for any other intermolecular force we shall describe later. As described for ion-ion attractions in Chapters 3 and 8, the strength of ion-dipole attractions depends on

- *The distance between the ion and the dipole:* the closer the ion and dipole, the stronger the attraction.
- *The charge on the ion:* the higher the ion charge, the stronger the attraction.
- *The magnitude of the dipole:* the greater the magnitude of the dipole, the stronger the attraction.

Water is a familiar example of a polar molecule, a molecule with positive and negative electric poles (Figure 9.17). Thus, if a water molecule encounters an ion, a force of attraction exists between them. When positive and negative charges are attracted to one another, energy is released as the particles come together to form a bond (Section 8.7). For example, if gaseous sodium ions were to plunge into water, a strong interaction would occur between the ions and the water dipoles; the reaction would be strongly exothermic.

$$Na^+(g) + x\,H_2O(\ell) \longrightarrow [Na(H_2O)_x]^+(aq) \;(x \text{ probably} = 6) \qquad \Delta H_{rxn} = -405 \text{ kJ}$$

(This experiment cannot be done directly. Instead, the energy change for the reaction must be determined from indirect measurements.)

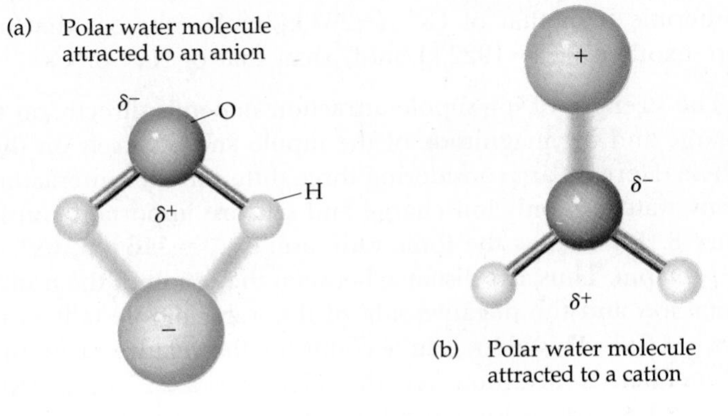

A metal ion bound by water molecules is said to be **hydrated.** (It is experimentally difficult to measure the number of water molecules of hydration; however, six is a good estimate for most cations.) The energy change for this process is called the **heat,** or **enthalpy, of hydration.** If some polar solvent molecule other than water is involved, the ion is more generally said to be **solvated,** and we use the term **enthalpy of solvation.**

In the case of ion-dipole interactions, the attractive force dependson $1/d^2$, where d is the distance between the center of the ion and the oppositely charged "pole" of the dipole. As the ion radius becomes larger, the enthalpy of hydration therefore becomes less exothermic, as illustrated by the series of alkali metal cations.

Cation	Ion Radius (pm)	Enthalpy of Hydration (kJ/mol)
Li^+	90	−515
Na^+	116	−405
K^+	152	−321
Rb^+	166	−296
Cs^+	181	−263

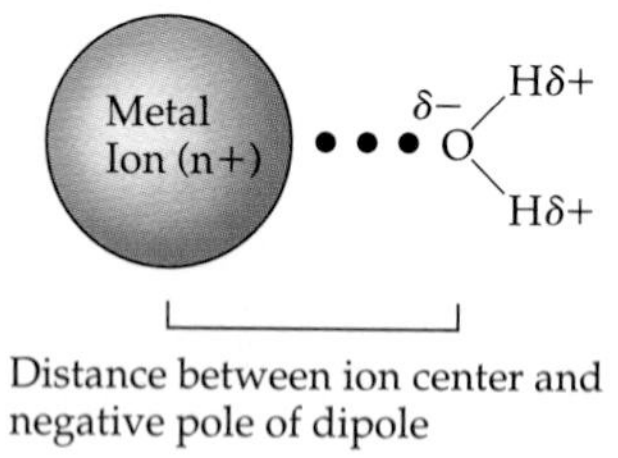

It is interesting to compare these values with the enthalpy of hydration of the proton, the H^+ ion, which is estimated to be −1090 kJ/mol. This extraordinarily large value is due to the tiny size of the hydrogen ion, which means the positive charge is spread over a very small volume. A result is that the H^+ ion is highly hydrated in water, and we commonly represent it as the hydronium ion, H_3O^+, even though the structure is probably more complicated than this.

As described in Section 3.8, many of the solids you use in the laboratory are hydrated salts that have formulas such as $BaCl_2 \cdot 2\ H_2O$ or $CoCl_2 \cdot 6\ H_2O$. Generally it is the metal ion that interacts with water. Therefore, the formula of $CoCl_2 \cdot 6\ H_2O$, for example, is better written as $[Co(H_2O)_4Cl_2] \cdot 2\ H_2O$, because four of the six water molecules are associated with the Co^{2+} ion by ion-dipole attractive forces (Figure 13.4).

(a)

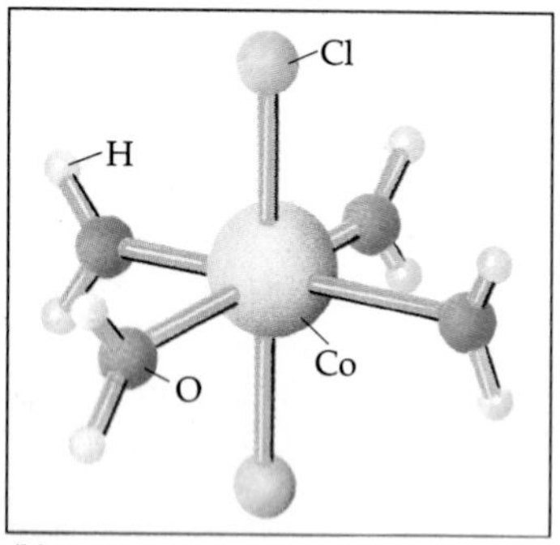

(b)

Figure 13.4 Hydrated cobalt(II) chloride, $CoCl_2 \cdot 6\ H_2O$. (a) The solid salt. (b) The structure of $[Co(H_2O)_4Cl_2]$. The four water molecules are bound to the Co^{2+} ion by ion-dipole forces. (*a*, C. D. Winters)

EXAMPLE 13.1 *Hydration Energy*

Explain why the enthalpy of hydration of Na^+ (−405 kJ/mol) is only somewhat more exothermic than that of Cs^+ (−263 kJ/mol), whereas that of Mg^{2+} is much more exothermic (−1922 kJ/mol) than that of Na^+ or Cs^+.

Solution The strength of ion-dipole attraction depends directly on the size of the ion charge and the magnitude of the dipole and inversely on the distance between them. Here we are considering three different ions interacting with the same solvent, water, so only ion charge and size are important considerations. From Figure 8.15 we know the ionic radii are: $Na^+ = 116$ pm, $Cs^+ = 181$ pm, and $Mg^{2+} = 86$ pm. Thus, the distance between the center of the positive charge on the metal ion and the negative side of the water dipole falls in this order: $Mg^{2+} < Na^+ < Cs^+$. This alone can account for the relative magnitudes of the hydration energies. Notice, however, that Mg^{2+} has a 2+ charge, whereas the other ions are 1+. Just as ion-ion attractions increase with increasing charge

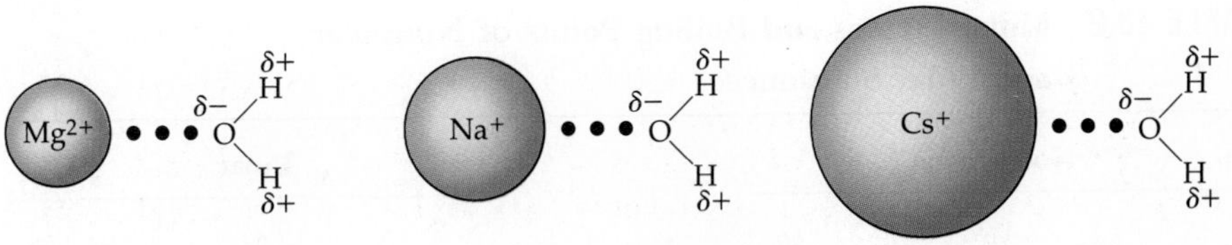

(page 119), the greater charge on Mg^{2+} also leads to a greater force of ion-dipole attraction than for a 1+ ion. This means the hydration energy for Mg^{2+} is *much* more negative than for the other two ions.

EXERCISE 13.1 ***Hydration Energy***

Which should have the more negative hydration energy, Ca^{2+} or Ba^{2+}? Explain briefly. ■

Interactions Between Molecules with Permanent Dipoles

Molecules containing permanent dipoles are formed when atoms of different electronegativity bond together unsymmetrically. When a polar molecule encounters another polar molecule, of the same or different kind, they can interact; the positive end of one molecule is attracted to the negative end of the other polar molecule (Figure 13.5).

In general, energy is released when molecules interact with one another,

$$\text{Gas} \underset{\substack{\Delta H_{\text{vaporization}} \\ \text{energy absorbed by sample}}}{\overset{\substack{\text{energy released by sample} \\ \Delta H_{\text{condensation}}}}{\rightleftharpoons}} \text{Liquid}$$

and this is one reason you must cool a gas to convert it to a liquid: the heat evolved on interaction has to be removed (Section 6.3).* (For example, HCl gas must be cooled to −85 °C to make liquid HCl.) Conversely, energy is required to separate interacting dipoles. For this reason liquid water or any other substance

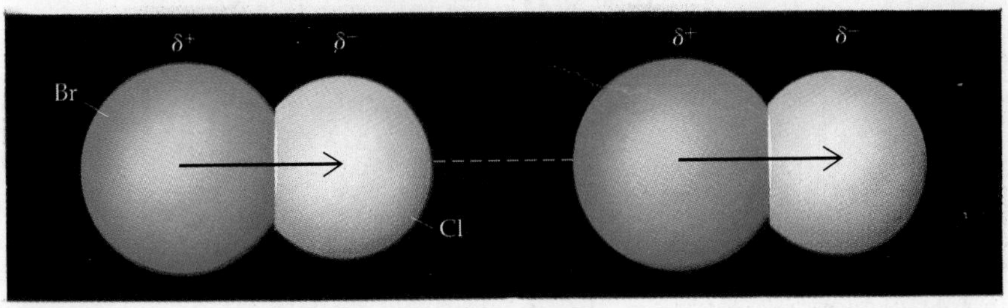

Figure 13.5 Dipole-dipole attractions between two molecules of BrCl. The Cl atom is more electronegative than Br and attracts electrons away from the Br atom, which becomes slightly positive. The negative end of one molecule is attracted to the positive end of the other. The dipole-dipole force is indicated by a dotted line.

*Another reason to cool a gas to liquefy it is to reduce the kinetic energy of the molecules.

TABLE 13.2 Molar Masses and Boiling Points of Nonpolar and Polar Substances

Nonpolar			Polar		
	M (*g/mol*)	*bp* (°*C*)		*M* (*g/mol*)	*bp* (°*C*)
N_2	28	−196	CO	28	−192
SiH_4	32	−112	PH_3	34	−88
GeH_4	77	−90	AsH_3	78	−62
Br_2	160	59	ICl	162	97

in the liquid state must be heated to convert it to the vapor phase, wherein the molecules interact only weakly or not at all. The logical conclusion, as reflected by the data in Table 13.2, is that the boiling points of liquids should reflect whether the molecules of the liquid are polar. In general, the boiling point of a compound with a permanent dipole is higher than that of a nonpolar compound (both having about the same molar mass).

The ability of one substance to dissolve another is considered in more detail in Chapter 14.

The presence of a permanent dipole also influences solubility. Water, for example, dissolves readily in ethanol (C_2H_5OH) because there is a strong interaction between these two polar molecules. In contrast, water does not dissolve in gasoline to any appreciable extent, because the hydrocarbon molecules (such as octane, $CH_3CH_2CH_2CH_2CH_2CH_2CH_2CH_3$) that constitute gasoline are not polar. Water-hydrocarbon attractions are so weak that they cannot disrupt the stronger water-water attractions.

Observations of solubility are the basis of the old saying that "like dissolves like." This means that polar molecules dissolve in a polar solvent, and nonpolar molecules dissolve in a nonpolar solvent. Polar ethylene glycol dissolves in water and the mixture is used as antifreeze in automobiles, and nonpolar cooking oil (Section 11.6) dissolves in a nonpolar solvent, such as gasoline or CCl_4 (see Figure 13.6).

(a)

(b)

Figure 13.6 Intermolecular forces and solubility or "like dissolves like." (a) Ethylene glycol, $C_2H_4(OH)_2$, is a polar molecule and dissolves readily in the polar solvent water. (b) Oil and water do not mix owing to the vastly different intermolecular forces within each liquid, and because the forces of attraction between oil and water are not sufficient to disrupt the forces within liquid water. (C. D. Winters)

Hydrogen Bonding

When a hydrogen atom is attached to a very small electronegative atom X, the interaction between the H—X bond dipole and polar molecules is greater than expected for ordinary dipole-dipole attractions. This strong interaction is called **hydrogen bonding** because it occurs *only* when H is part of the interacting dipoles.

Hydrogen bonding can be considered a special form of dipole-dipole force.

Bond dipoles arise between atoms as a result of a difference in electronegativity (Section 9.4). The electronegativities of N (3.0), O (3.5), and F (4.0) are among the highest of all the elements, whereas that of H (2.1) is much less. The covalent bonds N—H, O—H, and F—H are therefore extremely polar. Electron density shifts toward N, O, or F, causing this atom to take on a *partial negative charge.* As a result, the H atom bonded to nitrogen, oxygen, or fluorine acquires a *partial positive charge.* If an atom of an adjacent molecule (or even one in the same molecule) is the negative end of a bond dipole and has one or more lone pairs of electrons, an unusually strong attraction can occur. The H atom is a bridge between two electronegative atoms: X and Y.

$$\overset{\delta-}{\text{X}}—\overset{\delta+}{\text{H}}\cdots\overset{\delta-}{\text{Y}}$$

hydrogen bond (the dotted line between H and Y)

Hydrogen bonds are strongest if the neighboring atom is a small, highly electronegative atom. This means the strongest hydrogen-bonding interactions are those shown in the table.

An important reason for the unusual strength of the hydrogen bond compared with dipole-dipole interactions is the small size of the H atom. This means the positive charge of the X—H dipole is concentrated in a very small volume.

Types of Hydrogen Bonds

N—H···N—	O—H···N—	F—H···N—
N—H···O—	O—H···O—	F—H···O—
N—H···F—	O—H···F—	F—H···F—

Here the H atom is attached to an electronegative atom X (N, O, or F), and a hydrogen bond (designated by the dotted line) exists with another electronegative atom (again, N, O, or F). Hydrogen bonding can also occur if the neighboring atom Y is Cl or S, but interaction with these larger atoms is usually much weaker. Hydrogen bonds may occur between molecules (*inter*molecular bonds) or, if, the geometry allows, within molecules (*intra*molecular bonds).

One of the best examples of hydrogen bonding is found in hydrogen fluoride, which, in the solid state, is a zigzag chain of hydrogen-bonded HF molecules (Figure 13.7).

Further illustrating of the effects of hydrogen bonding are the different properties of an alcohol, ethanol (which contains an O—H bond), and an ether,

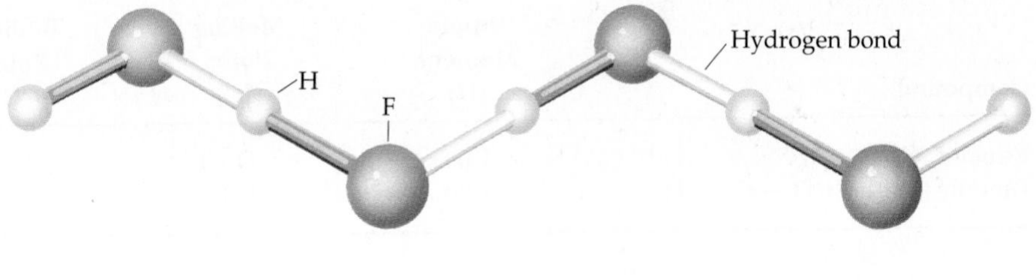

Figure 13.7 Hydrogen bonding between HF molecules. A lone pair of electrons on an F atom in one molecule attracts an H atom in an adjacent molecule. This repeats over and over, forming a chain of HF molecules.

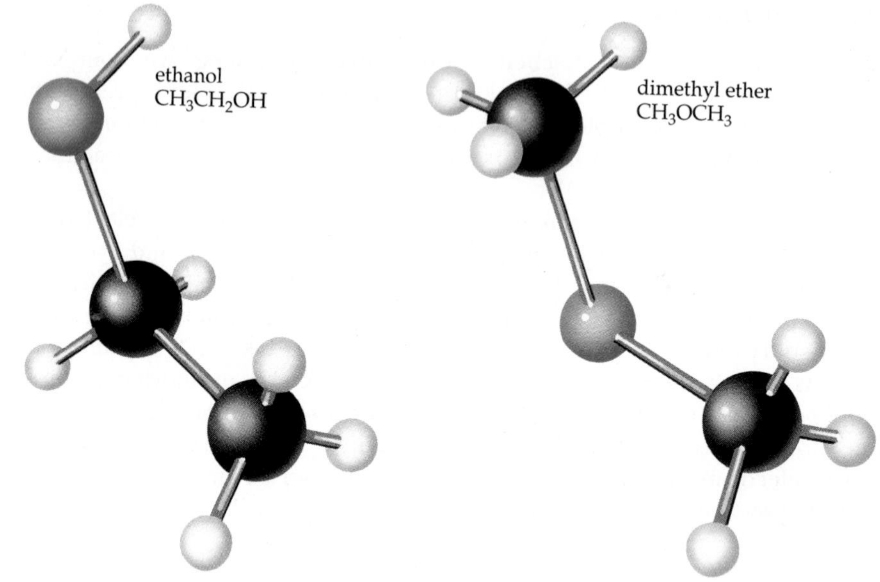

dimethyl ether (which contains only C—H and C—O bonds). These molecules have the same elemental composition and molar mass, and their dipole moments are similar. Their melting and boiling points, however, are vastly different (Table 13.3). When a solid melts or a liquid boils, molecules are being separated from one another; intermolecular forces are being disrupted. Therefore, the fact that the temperature required to melt or boil ethanol is higher than for dimethyl ether means that stronger forces exist between the molecules of ethanol in the liquid and solid than between dimethyl ether molecules. Ethanol molecules have O—H bonds, so intermolecular hydrogen bonding is possible. No such bonds exist in the ether, so hydrogen bonding cannot occur.

The boiling point data for several series of structurally analogous molecules in Figure 13.8 are another illustration of the consequences of hydrogen bonding. Consider, for example, the hydrogen compounds of Groups 4A through 7A. The principles described in Chapter 9 allow us to predict that the compounds of Group 4A (CH_4, SiH_4, GeH_4, SnH_4) are all tetrahedral, nonpolar molecules. The plot in Figure 13.8 shows that their boiling points increase with increasing molecular mass (due to increased intermolecular dispersion forces as explained on page 609). This same effect is also operating for the heavier molecules of the hydrogen compounds of elements of Groups 5A, 6A, and 7A. The boiling points of NH_3, H_2O, and HF, however, certainly do *not* follow these trends because

TABLE 13.3 A Comparison of the Physical Properties of Ethanol and Dimethyl Ether

Compound	Dipole Moment (D)	Melting Point (°C)	Boiling Point (°C)
Ethanol, C_2H_5—OH	1.69	−114.1	78.29
Dimethyl ether, CH_3—O—CH_3	1.30	−141.5	−24.8

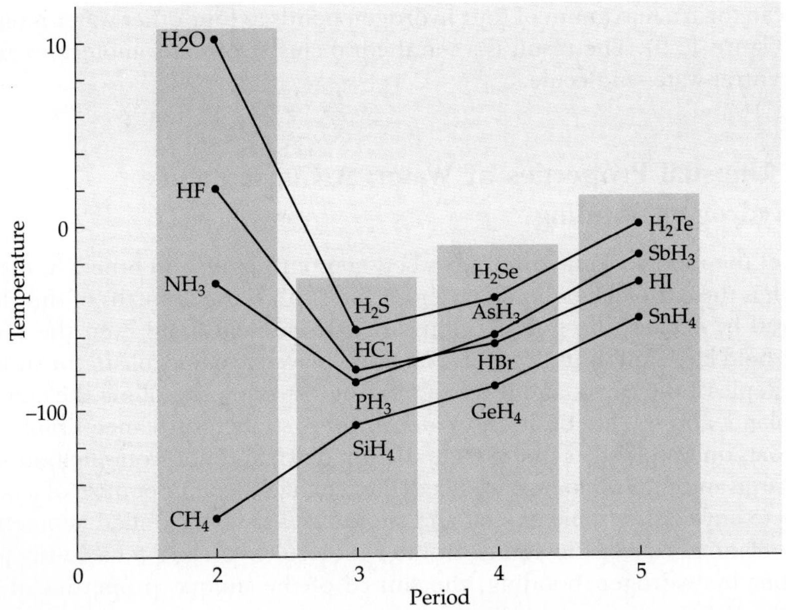

Figure 13.8 The boiling points of some simple hydrogen-containing compounds. Lines connect molecules containing atoms from the same periodic group. The effect of hydrogen bonding is apparent in the high boiling points of H_2O, HF, and NH_3.

hydrogen bonding is so strong within the liquid phase of these substances. If you extrapolate the line established by H_2S, H_2Se, and H_2Te, for example, water might be expected to boil at around −90 °C. The boiling point of water, however, is almost 200 °C more than the expected temperature, clearly indicating a strong degree of intermolecular attraction.

In liquid and solid water, in which the molecules are close enough to interact, the hydrogen atom of one water molecule is attracted to the lone pair of electrons on the oxygen atom of an adjacent water molecule. Because each hydrogen atom can form a hydrogen bond to an oxygen atom in another water molecule, and because each oxygen atom has two lone pairs, each water mole-

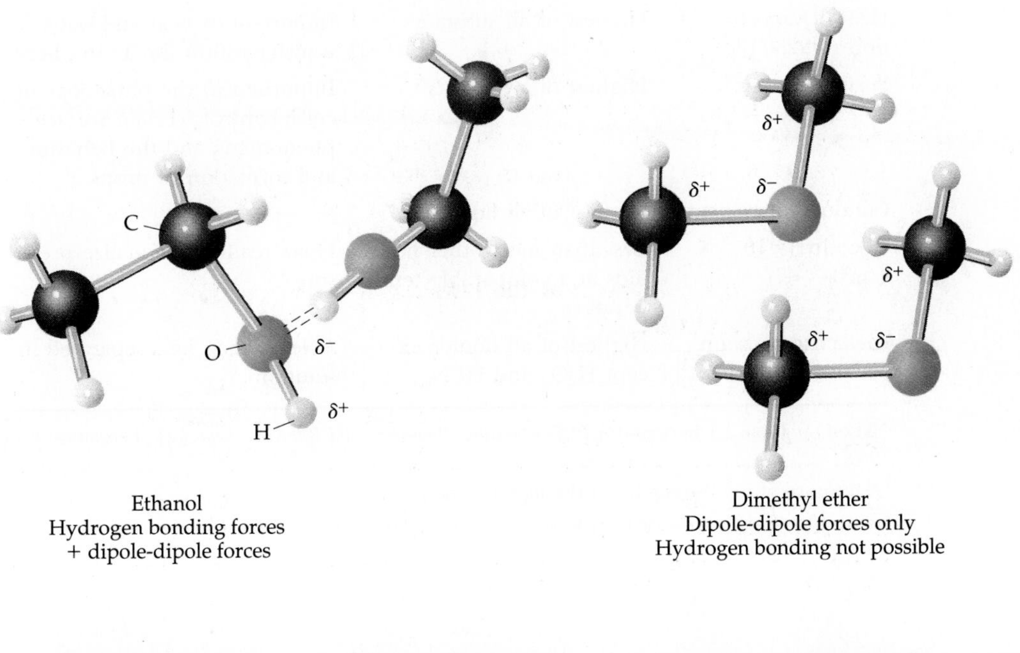

cule can form a maximum of four hydrogen bonds to four other water molecules (see Figure 13.9). The result is a tetrahedral cluster of water molecules around the central water molecule.

The Unusual Properties of Water: A Consequence of Hydrogen Bonding

One of the most striking differences between our planet and others in the solar system is the existence of liquid water on the earth. Three fourths of the globe is covered by oceans, the polar regions are vast ice fields, and even the soil and rocks hold large amounts of water. Water has played a major role in the history of the people of the earth, and it is a significant factor in controlling the climate of our planet and the life on it. But water is a very strange substance (Table 13.4). Ice floats on water, lakes freeze from the top down and not from the bottom up, very large amounts of energy are needed to change the temperature of a body of water to any considerable extent, and snow flakes have a six-sided geometry. All these features reflect the ability of water molecules to cling tenaciously to one another by hydrogen bonding, the source of the unique properties of water (Figure 13.9).

TABLE 13.4 Anomalous Properties of Water*

Property	Comparison with Other Substances	Importance in Physical and Biological Environment
Specific heat ($=4.18\ J/g \cdot K$)	Highest of all liquids and solids except NH_3	Prevents rapid temperature changes; heat transfer by water movements is very large; tends to maintain body temperature
Heat of fusion ($=333\ J/g$)	Highest except for NH_3	Thermostatic effect at freezing point due to absorption or release of heat
Heat of vaporization ($=2250\ J/g$)	Highest of all substances	Important in heat and water transfer within the atmosphere
Surface tension† ($=7.2 \times 10^9\ N/m$)	Highest of all liquids	Important in the physiology of cells; controls certain surface phenomena and the behavior and formation of drops
Conduction of heat	Highest of all liquids	
Viscosity‡ ($=10^{-3}\ N \cdot s/m^2$)	Less than most other liquids at a comparable temperature	Flows readily to equalize pressure
Dielectric constant ($=80$ at 20 °C)	Highest of all liquids except H_2O_2 and HCN	Able to keep ions separated in solution

*Based on Table 1.1 in *Seawater: Its Composition, Properties, and Behavior,* New York, Pergamon Press, 1989.

†A measure of the "strength" of the liquid surface.

‡A measure of the resistance to distortion or flow of a liquid. The greater the viscosity the less readily the liquid flows.

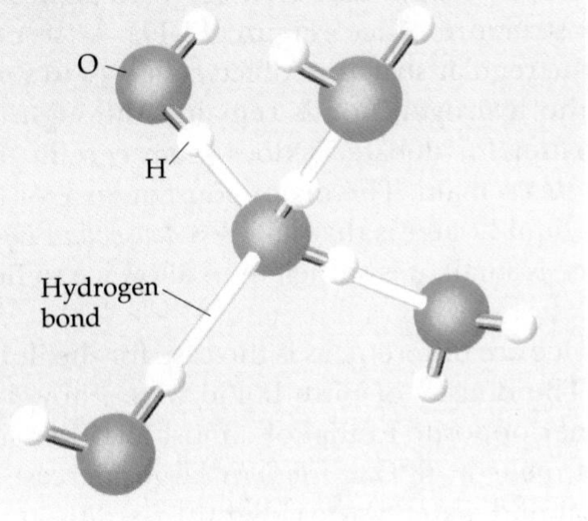

Figure 13.9 Hydrogen bonding between water molecules. Each water molecule can participate in four hydrogen bonds. Two involve its two H atoms, and the other two involve the two lone pairs on the O atom.

Each of the water molecules hydrogen bonded to a central water molecule can in turn hydrogen bond to yet more water molecules. This could lead to a complicated, random structure; however, nature often exhibits a tendency to achieve great regularity, and ice is beautifully regular (Figure 13.10). It is an open-cage structure in which the oxygen atoms of hydrogen-bonded water molecules are arranged at the corners of puckered, six-sided rings, or hexagons. Each edge of a six-sided ring consists of a normal O—H covalent bond and a hydrogen

Many examples of six-sided rings exist in nature. For example, the C atoms in diamond and in graphite are arranged in six-sided rings (Figure 3.5).

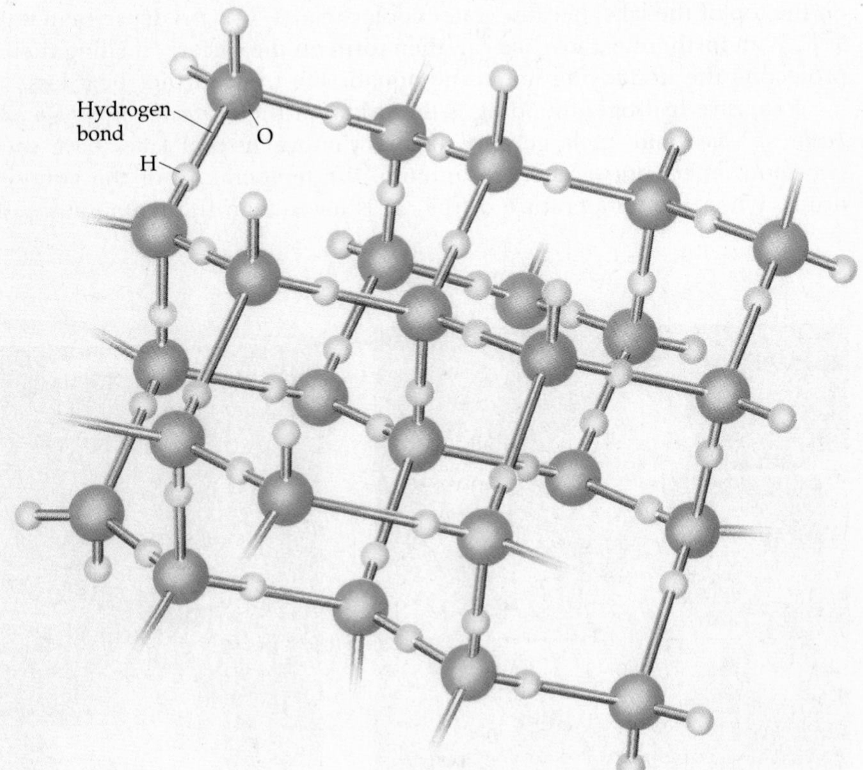

Figure 13.10 The structure of ice. Each oxygen atom is covalently bound to two hydrogen atoms and hydrogen bonded to two other hydrogen atoms. The hydrogen bonds are longer than the covalent O—H bonds. Due to hydrogen bonding, ice has a more open structure than liquid water. Because the water molecules are farther apart, on average, in ice than in liquid water, ice is less dense than liquid water.

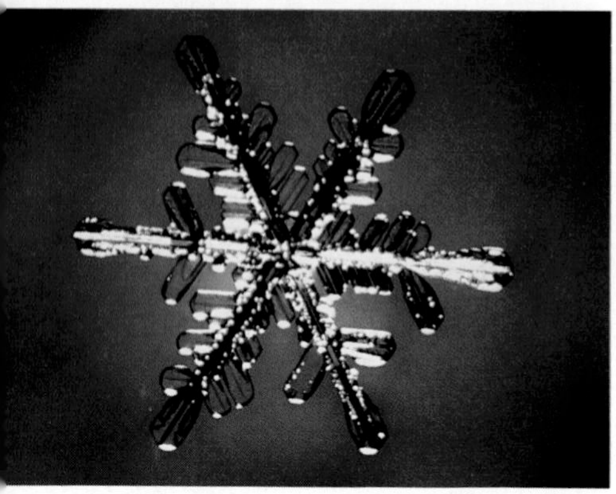

Figure 13.11 The six-sided, or hexagonal, geometry of a snowflake is a reflection of the six-sided rings formed when water molecules hydrogen bond to form ice. (NCAR/Tom Stack and Associates)

bond. It is interesting that snowflakes are always based on six-sided figures, a reflection of the internal molecular structure of ice (Figure 13.11).

When ice melts to liquid water, the regular structure of ice collapses to some extent, but approximately 85% of the hydrogen bonds remain. Although the structure of liquid water is still a matter for debate, it does seem certain that some ice-like clumps of water molecules remain. The major consequence of the open structure of ice compared with liquid water is that *ice is less dense than liquid water.* Although this density difference is small, it is sufficient to allow ice to float on top of water (Figure 13.12).

The densities of liquid water and ice are different, as is the case for the liquid and solid phases of any substance. The density of pure liquid water, however, changes with temperature in a manner opposite to that of almost all other substances (Figure 13.13). When water melts at 0 °C a *relatively* large increase in density occurs, and when the now liquid water warms slightly, its density increases even more. This unique process occurs because, as the ice-like structure of very cold water begins to break down, the liquid becomes more compact. The density reaches a maximum at 4 °C, but it begins to decline again at higher temperatures as more and more hydrogen bonds are broken and the structure loosens as the kinetic energy of the water molecules increases.

Because of the way that water density changes with temperature, *lakes do not freeze solidly from the bottom up in the winter.* When lake water cools with the approach of winter, the density increases, the cooler water sinks, and the warmer water rises. This means the water and dissolved material turn over until all the water reaches 4 °C, the maximum density. (This process carries oxygen-rich water to the lake bottom to restore the oxygen used during the summer, and it brings nutrients to the top layers of the lake.) As the water cools further, it stays on the top of the lake, because water cooler than 4 °C is *less* dense than water at 4 °C. With further heat loss, ice can then form on the surface, floating there and protecting the underlying water and aquatic life from further heat loss.

Extensive hydrogen bonding is the origin of the *extraordinarily high heat capacity of water,* and, in large part, this is why oceans and lakes have such an enormous effect on weather. In autumn, the temperature of the atmosphere drops. When the temperature of the air is lower than the temperature of the

Figure 13.12 Ice made from water floats in liquid water because solid water is less dense than liquid water, a very unusual occurrence (*left*). The photo at the right shows liquid benzene (C_6H_6) with a chunk of solid benzene lying on the bottom. Solid benzene is *more* dense than the liquid and so sinks in the liquid. This is the behavior exhibited by virtually all known substances *except* water. (C. D. Winters)

ocean or lake, the ocean or lake gives up heat to the atmosphere, and the air temperature drop is moderated. Furthermore, so much heat must be given off for each degree drop in temperature (because so many hydrogen bonds form) that the decline in water temperature is gradual. For this reason the temperature of the ocean or a large lake is generally higher than the average air temperature until late in the fall.

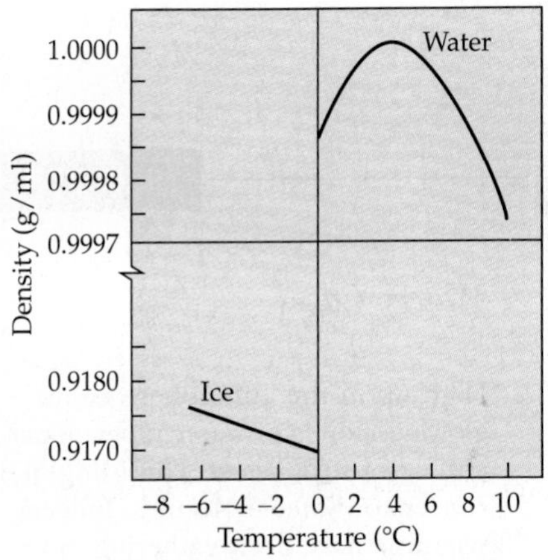

Figure 13.13 The temperature dependence of the densities of ice and liquid water. Notice that the scale is exaggerated. The change in density is quite small, only about 0.0001 g/mL per degree. Pure liquid water has a density of 1.00 g/cm^3 at 4 °C, whereas ice at 0 °C has a density of 0.917 g/cm^3. The lower density of ice than water means that ice floats in fresh water. In contrast, the density of seawater increases with falling temperature right down to the freezing point. This is a crucial distinction between freshwater and sea water, and it has a profound effect on oceanic circulation processes and on the formation of sea ice.

Dispersion Forces: Interactions Involving Induced Dipoles

Dispersion forces are the most common of all intermolecular forces, and are found in all molecular substances. Such forces are electrostatic in nature and arise from attractions involving *induced dipoles.* They explain why nonpolar I_2 molecules form a solid under normal conditions and are able to dissolve in water.

Interactions Between Polar and Nonpolar Molecules

Oxygen dissolves in water to a very small extent; a typical concentration is about 10 ppm (or about 0.001% by weight). This is important because microorganisms use dissolved oxygen to convert organic substances dissolved in water to simpler compounds. The quantity of oxygen required to oxidize a given quantity of organic material is called the **biochemical oxygen demand (BOD).** Highly polluted water often has a high concentration of organic matter and so has a high BOD.

How can nonpolar O_2 molecules dissolve in a polar substance like water? The answer is that polar molecules such as water can *induce* a dipole in molecules that do not have a permanent dipole. To see how this can occur, picture a polar water molecule approaching a nonpolar molecule such as O_2 (Figure 13.14). The electron cloud of an isolated O_2 molecule is distributed, on average, sym-

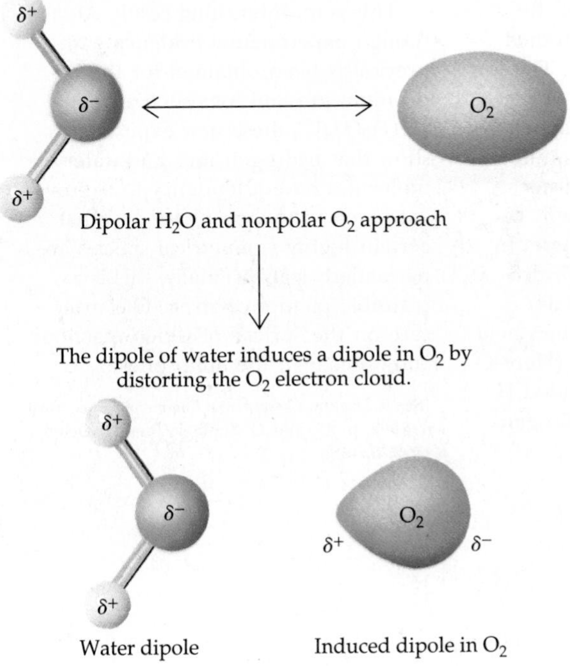

Figure 13.14 A polar water molecule can induce a dipole in nonpolar O_2.

CURRENT ISSUES IN CHEMISTRY

Watery Cages

High up in the atmosphere eerie, glowing clouds of water vapor apparently grow from seeds consisting of clusters of water molecules. Indeed, scientists have been gathering more and more evidence that liquid water consists, not of single isolated molecules, but rather of clusters of molecules. At a given instant almost every molecule in a glass of water is bonded to four other molecules through hydrogen bonds. But these are not stable clusters. Rather, other molecules continually join a given cluster, forcing one of the original molecules out, which then goes on to bind to another cluster.

Evidence for clusters of H_2O molecules is even stronger in the gas phase. A. W. Castleman, Jr., and his students at the Pennsylvania State University forced highly pressurized water through a nozzle into a vacuum. They fired a beam of electrons into the vapor, ionizing it, and then studied the composition of the vapor. The ion $[H(H_2O)_{21}]^+$ "just leaps out of the spectrum [of species in the vapor]," Castleman said. Based on computer modeling, they believe the cluster must be highly regular, probably resembling the picture shown here. In the center of the cluster is a hydronium ion, H_3O^+, and clustered around it are 20 water molecules held together by hydrogen bonds. (Here O atoms are blue, hydrogen-bonded H atoms are green, and non-hydrogen-bonded H atoms are red.)

Noctilucent clouds over an Antarctic research station. (Doug Allan/Photo Researchers, Inc.)

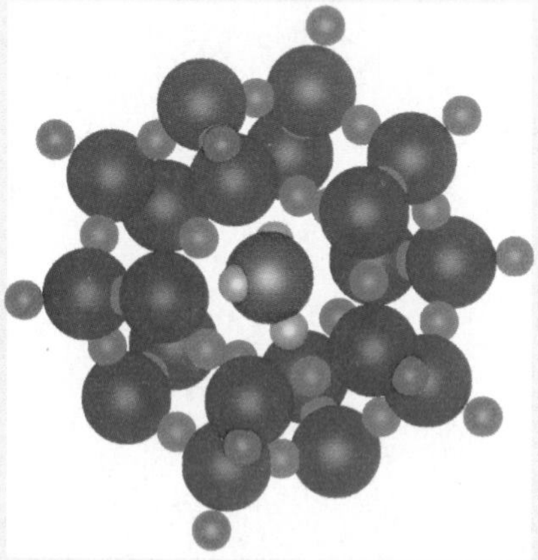

(A. Welford Castleman, Jr.)

This is an interesting result. Although experimental evidence had previously been obtained for the hydronium ion and ions such as $[H(H_2O)_8]^+$, these new experiments show that hydrogen ions and water molecules have a tendency to aggregate even in the gas phase and that certain highly symmetrical species are particularly stable. Finally, this is a plausible picture of an ice-like structure on the surface of which reactions could occur in the atmosphere.

(See R. Dagani: *Chemical and Engineering News,* April 8, 1992, p. 47; and C. Zimmer: *Discover,* October 1992, p. 103.)

TABLE 13.5 The Solubility of Some Gases in Sea Water*

Gas	Molar Mass (g/mol)	Solubility at 20 °C (g gas/100 g water)†
H_2	2.01	0.000160
N_2	28.0	0.000190
O_2	32.0	0.000434
Cl_2	70.9	0.729

*Data taken from J. A. Dean, *Lange's Handbook of Chemistry,* 14th ed. New York: McGraw-Hill, 1992.
†Pressure of gas + pressure of water vapor = 760 mm Hg.

metrically between the bonded O atoms. As the negative end of the polar H_2O molecule approaches, however, the O_2 electron cloud moves away to reduce repulsion between the O_2 cloud and the negative end of the H_2O dipole, and the O_2 molecule itself becomes polar. That is, a dipole has been *induced* in the otherwise nonpolar O_2 molecule, and H_2O and O_2 molecules are now attracted to one another, although only weakly. Oxygen dissolves in water because a force of attraction exists between a permanent dipole and an induced dipole (*dipole-induced dipole forces*).

The process of inducing a dipole is called **polarization,** and the degree to which the electron cloud of an atom (such as Ne or Ar) or a molecule (such as O_2, N_2, or I_2) can be distorted and a dipole induced depends on the **polarizability** of the atom or molecule. Although this property is difficult to measure experimentally, it makes sense that the valence electrons of atoms or molecules with large, extended electron clouds, such as I_2, can be polarized or distorted more readily than gases, such as He or H_2, in which the valence electrons are close to the nucleus and more tightly held. In general, the more electrons in a molecule or atom, the more easily the electrons can be polarized because they are far from the restraining forces of atomic nuclei.

The table of the solubilities of common gases in sea water (Table 13.5) illustrates the effect of interactions between a dipole and an induced dipole. Here you see a trend to higher solubility with increasing mass. As the molecular mass of the gas increases, either the number of valence electrons increases *or* the valence electrons are less tightly held. Therefore, the ease of polarization of the electron cloud generally increases with mass, and, because a dipole is more readily induced as the polarizability increases, the strength of dipole-induced dipole interactions generally increases with mass. And finally, because the solubility of substances such as O_2 depends on the strength of the dipole-induced dipole interaction, the solubility of nonpolar substances in polar solvents generally increases with mass.

Interactions Between Nonpolar Molecules

The last of the intermolecular forces to be considered is that between two nonpolar molecules (see Table 13.1). Because a dipole can be induced in a nonpolar molecule, two nonpolar molecules can bind to each other by *induced dipole/induced dipole forces.* Such forces can range from very weak to relatively strong.

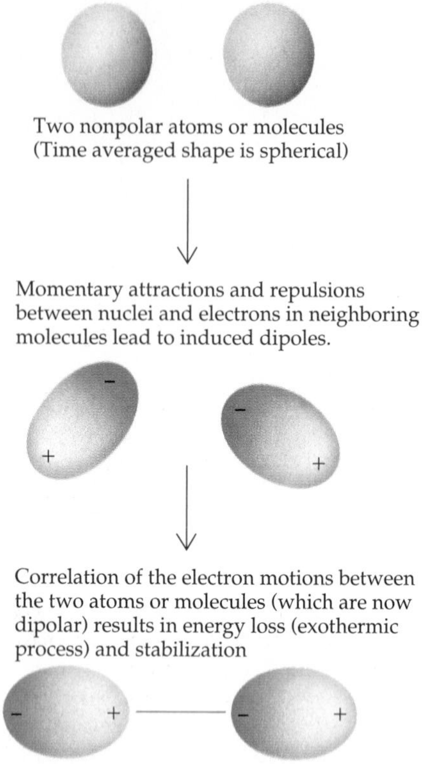
Two nonpolar atoms or molecules
(Time averaged shape is spherical)
Momentary attractions and repulsions between nuclei and electrons in neighboring molecules lead to induced dipoles.
Correlation of the electron motions between the two atoms or molecules (which are now dipolar) results in energy loss (exothermic process) and stabilization

Figure 13.15 Interaction between two induced dipoles.

Nonpolar molecules such as I_2 must attract one another because I_2 is a solid at room temperature; without intermolecular forces of some kind it would be a gas. Similarly, N_2, O_2, and the noble gases can all be liquefied at low temperatures (see Figure 13.2).

In Table 13.6 you see a clear trend to higher boiling points of nonpolar gases with increasing molar mass, largely because *dispersion forces generally become stronger with increasing size and therefore mass* (and stronger intermolecular forces mean that a higher temperature is required to break down these forces and allow the molecules to leave the liquid and go into the vapor phase).

To understand how two nonpolar molecules can attract each other, remember that the electrons in atoms or molecules are in a state of constant motion. On average, the electron cloud around an atom is spherical. When two nonpolar atoms or molecules approach each other, however, attractions or repulsions between their electrons and nuclei can lead to distortions in their electron clouds. That is, dipoles can be induced momentarily in neighboring atoms or molecules, and it is these induced dipoles that can lead to intermolecular attraction (Figure 13.15).

TABLE 13.6 Molar Enthalpy of Vaporization and Vapor Pressures for Common Compounds*

Compound	Molar Mass (g/mol)	ΔH_{vap} (kJ/mol)†	Boiling Point (°C) (Vapor pressure = 760 mm Hg)
Polar Compounds			
HF	20.0	25.2	19.7
HCl	36.5	16.2	−84.8
HBr	80.9	19.3	−66.4
HI	127.9	19.8	−35.6
NH_3	17.0	23.3	−33.3
H_2O	18.0	40.7	100.0
SO_2	64.1	24.9	−10.0
Nonpolar Compounds			
CH_4 (methane)	16.0	8.2	−161.5
C_2H_6 (ethane)	30.1	14.7	−88.6
C_3H_8 (propane)	44.1	19.0	−42.1
C_4H_{10} (butane)	58.1	22.4	−0.5
He	4.0	0.08	−268.9
Ne	20.2	1.7	−246.1
Ar	39.9	6.4	−185.9
Xe	131.3	12.6	−108.0
H_2	2.0	0.90	−252.9
N_2	28.0	5.6	−195.8
O_2	32.0	6.8	−183.0
F_2	38.0	6.6	−188.1
Cl_2	70.9	20.4	−34.0
Br_2	159.8	30.0	58.8

*Data taken from D. R. Lide, *Basic Laboratory and Industrial Chemicals,* Boca Raton, FL, CRC Press, 1993.

†ΔH_{vap} is given at the normal boiling point of the liquid.

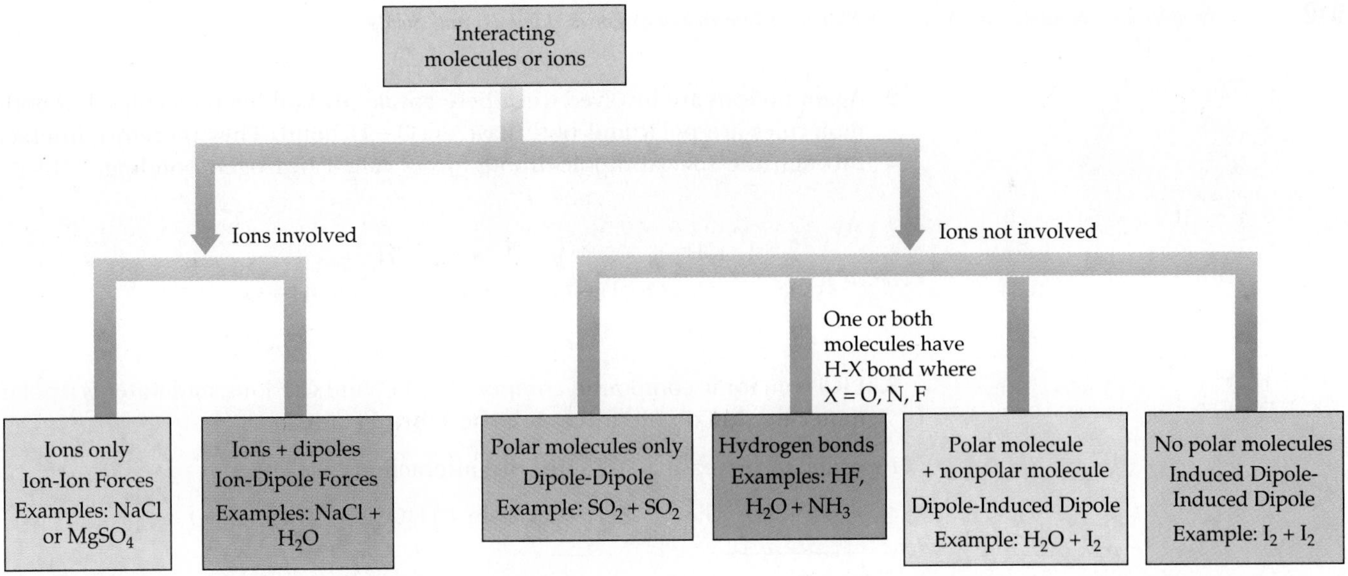

Figure 13.16 Deciphering intermolecular forces. A scheme to help decide what types of intermolecular forces are important in a given chemical system. Recall that dispersion forces, which arise from induced dipoles and can range from weak to relatively strong, can be found even in polar molecules.

Finally, it is important to state again that dispersion forces are found in all molecules, both polar and nonpolar. For example, polar HCl molecules are bound to one another through a combination of dipole-dipole forces *and* dispersion forces. Dispersion forces, however, are the only ones among nonpolar molecules. This point is made in Figure 13.16, a diagram that outlines the common intermolecular forces. The purpose of the diagram is to help you decide what types of forces are appropriate for a given set of molecules.

EXAMPLE 13.2 *Intermolecular Forces*

Decide what type of intermolecular force is involved in each case, and place them in order of increasing strength of interaction.

1. In liquid methane, CH_4
2. In a mixture of water and methanol, H_2O and CH_3OH
3. In a solution of LiCl in water

Solution To answer this question it is helpful to follow the outline in Figure 13.16.

1. No ions are involved with CH_4, a simple molecule with covalent bonds. To decide if dipole-dipole forces or an induced dipole is involved, you must first decide if the molecule is polar. In this case a Lewis structure and the VSEPR theory (Chapter 9) show the molecule has a tetrahedral shape.

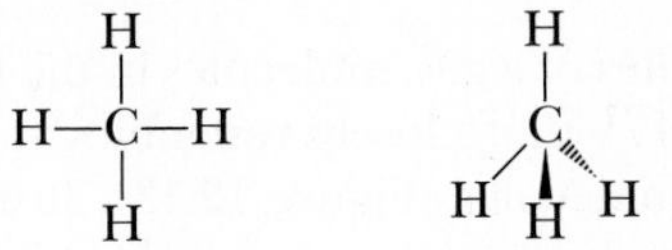

Lewis structure — Molecular shape = tetrahedral, Nonpolar molecule

The molecule cannot therefore be polar. The only way methane molecules can interact is through induced dipole forces.

2. Again no ions are involved with these covalently bonded molecules, but both molecules are polar and both have an O—H bond. They therefore interact through the special dipole-dipole forces called hydrogen bonding.

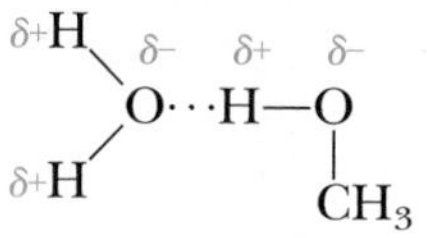

and

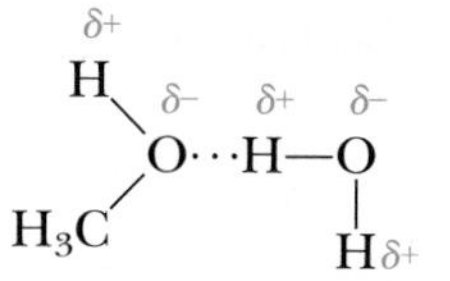

3. LiCl is an ionic compound composed of Li^+ and Cl^- ions, and water is a polar molecule. Ion-dipole forces are therefore involved.

In order of increasing strength, the interactions are

$$\text{liquid } CH_4 < H_2O \text{ in } CH_3OH < LiCl \text{ in } H_2O$$

EXERCISE 13.2 ***Intermolecular Forces***

Decide what type of intermolecular force is involved with (1) liquid O_2; (2) hydrated $MgSO_4$; (3) O_2 in H_2O. Place the interactions in order of increasing strength. ■

13.3 PROPERTIES OF LIQUIDS

Gases condense to liquids when the molecules no longer have sufficient kinetic energy to overcome the intermolecular forces. Substances that we commonly observe as liquids—water, alcohol, and gasoline—all condense at temperatures above room temperature. Other substances such as the nitrogen in the air, however (see Figure 13.2), require much lower temperatures.

About 70% of our planet is covered by liquid water, and our lives are shaped by the unique properties of this compound, especially in the liquid state. Other liquids, such as the oils and fluids in cars and airplanes, are important in our industrial economy.

Enthalpy of Vaporization

Molecules in the liquid state are relatively close together, but they have a regular structure only in very small regions (Figure 1.3). Furthermore, because the molecules are close together and yet are moving in all directions, confined liquids can transmit applied pressure equally in all directions. This property is the basis of the hydraulic fluids that operate automotive brakes and airplane control surfaces.

Furthermore, like the molecules of a gas, molecules in the liquid phase have a range of energies (Figure 13.17) that closely resembles the distribution of energies you saw earlier for gas molecules (Figure 12.13). Just as for gases, the average kinetic energy depends only on temperature: the higher the temperature, the higher the average kinetic energy and the greater the relative number of molecules with high kinetic energy.

Some of the molecules in the liquid phase have more kinetic energy than the potential energy of the intermolecular attractive forces holding the liquid

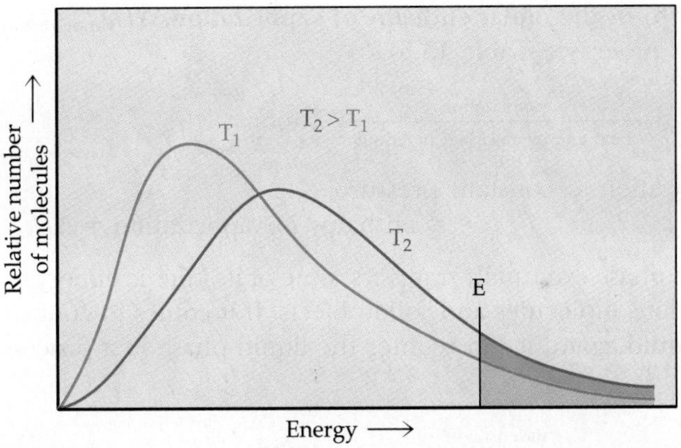

Figure 13.17 The distribution of molecular energies in the liquid phase. T_2 is a higher temperature than T_1, so there is a greater proportion of molecules having the energy marked E at the higher temperature.

 Relative number of molecules having enough energy to evaporate at lower temperature, T_1

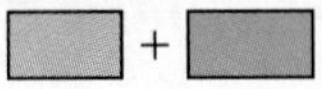 Relative number of molecules having enough energy to evaporate at higher temperature, T_2

molecules to one another. If such a molecule is at the surface of the liquid, and is moving in the right direction, it will break free of its neighbors and enter the gaseous phase (Figure 13.18). This process is called **vaporization,** or **evaporation.** Because high-energy molecules leave the liquid and take some of their energy with them, the vaporization process can continue only if additional energy is supplied to the liquid to produce more molecules with the minimum kinetic energy (say *E* in Figure 13.17) to vaporize. For this reason, vaporization is always an endothermic process, and the heat energy required to vaporize a sample at a given temperature is the **heat of vaporization** (expressed in energy per

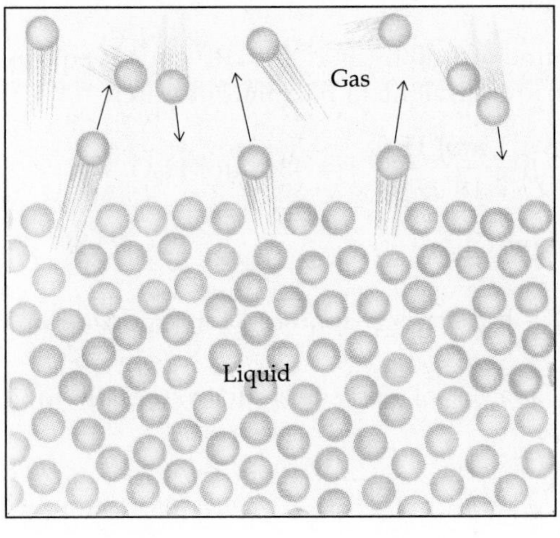

Figure 13.18 Molecules in the liquid and gas phases. All the molecules in both phases are moving, although the distances traveled in the liquid before collision with another molecule are small. Some of the liquid-phase molecules are moving with a kinetic energy large enough to overcome the intermolecular forces in the liquid and escape to the gas phase. At the same time, some molecules of the gas re-enter the liquid surface.

unit of mass; see Chapter 6) or the **molar enthalpy of vaporization,** $\Delta H_{vaporization}$ (expressed in energy per mole; see Table 13.6).

$$\text{Liquid} \xrightarrow[\text{heat energy absorbed by liquid}]{\text{vaporization}} \text{Vapor}$$

Heat required for vaporization at constant pressure
= enthalpy of vaporization = ΔH_{vap}

A molecule in the gas phase eventually transfers some of its kinetic energy by colliding with slower gaseous molecules and solid objects. If it comes in contact with the surface of the liquid again, it can reenter the liquid phase in a process called **condensation.**

$$\text{Vapor} \xrightarrow[\text{heat energy released by vapor}]{\text{condensation}} \text{Liquid}$$

Heat released at constant pressure = ΔH_{cond}

Because new intermolecular bonds are made, condensation is exothermic; for condensation to occur, the heat energy must be removed. The enthalpy change for condensation is equal but opposite in sign to the enthalpy of vaporization. For example, the enthalpy change for the vaporization of 1 mol of liquid water is +40.7 kJ. On condensing water vapor to liquid water at 100 °C, $\Delta H_{cond} = -\Delta H_{vap} = -40.7$ kJ/mol.

EXAMPLE 13.3 *Enthalpy of Vaporization*

You put 1.00 L of water (about 4 cupsful) in a pan at 100 °C, and it slowly evaporates. How much heat must have been absorbed (at constant pressure) by the water as it vaporized?

Solution Three pieces of information are necessary to solve this problem:

1. ΔH_{vap} for water = +40.7 kJ/mol at 100 °C
2. The density of water at 100 °C = 0.958 g/cm^3. (This is needed because ΔH_{vap} is expressed in units of kilojoules per mole, so you first must find the mass of water then the number of moles.)
3. Molar mass of water = 18.02 g/mol

Given the density of water, a volume of 1.00 L (or 1.00 × 10^3 cm^3) is equivalent to 958 g, and this mass is in turn equivalent to 53.2 mol of water.

$$1.00\ \cancel{\text{L}}\left(\frac{1000\ \cancel{\text{mL}}}{1\ \cancel{\text{L}}}\right)\left(\frac{0.958\ \cancel{\text{g}}}{1\ \cancel{\text{mL}}}\right)\left(\frac{1\ \text{mol H}_2\text{O}}{18.02\ \cancel{\text{g}}}\right) = 53.2\ \text{mol H}_2\text{O}$$

Therefore, the amount of energy required is

$$53.2\ \cancel{\text{mol}}\left(\frac{40.7\ \text{kJ}}{\cancel{\text{mol}}}\right) = 2.16 \times 10^3\ \text{kJ} = \text{heat energy required for vaporization}$$

2160 kJ is equivalent to about one quarter of the energy in your daily food intake.

EXERCISE 13.3 ***Enthalpy of Vaporization***

The molar enthalpy of vaporization of wood alcohol, CH_3OH (methanol), is 35.21 kJ/mol at 64.6 °C. How much energy is required to evaporate 1.00 kg of this alcohol? ■

Figure 13.19 Your body sweats to rid itself of excess heat. The higher energy water molecules evaporate, leaving lower energy molecules. The molecules that remain correspond to a lower average temperature. (C. D. Winters)

An enormous amount of heat is required to convert liquid water to water vapor (see Example 13.3 and Table 13.4), a fact that is important to your environment and your physical well-being. When you exercise vigorously, your body responds by sweating to rid itself of the excess heat. The sweat, which is mostly water, can be evaporated by the input of heat energy. The source of this energy is the heat generated by your muscles, so the evaporation of sweat removes the excess heat, and your body is cooled.

The heat of vaporization and condensation also play an important role in our weather. For example, if enough water condenses from the air to fall as an inch of rain on an acre of ground, the heat released is over 200 million kilojoules! This is equivalent to about 50 tons of exploded dynamite, that is, to the energy released by a small bomb.

Finally, it is interesting to look at trends in ΔH_{vap} values for related molecules in Table 13.6. The boiling points of nonpolar liquids (such as the hydrocarbons, atmospheric gases, and the halogens) tend to increase with increasing atomic or molecular mass, a reflection of increasing intermolecular dispersion forces (see Tables 13.2 and 13.6 and Figure 13.8). Similarly, the boiling points and enthalpies of vaporization of the heavier hydrogen halides (HX, where X = Cl, Br, and I) increase with increasing molecular mass. For these molecules, hydrogen bonding is not as important as in HF, so dipole-dipole and dispersion forces take over, the latter becoming increasingly important with increasing mass. Lastly, notice the very high heat of vaporization of water and hydrogen fluoride that comes from *extensive* hydrogen bonding.

Vapor Pressure

If you put some water in an open beaker, eventually the water evaporates completely. If you put some water in a sealed flask (Figure 13.20), however, the liquid evaporates only until the rate of vaporization equals the rate of condensation. When this occurs, a state of **dynamic equilibrium** is established.

The state of equilibrium is a concept used throughout chemistry and one we shall return to often. We signal this situation by connecting the two states or the reactants and products by a set of double arrows, $\rightleftarrows$.

$$\text{Liquid} \rightleftarrows \text{Vapor}$$

Molecules continue to move from the liquid to the vapor phase, and some in the vapor phase move back to the liquid phase; the mass of material in each phase is constant over time. (In contrast to the closed flask, the water in an open beaker can never come to equilibrium with gas phase water molecules; air movement and gas diffusion quickly remove the water vapor from the vicinity of the liquid surface.)

When liquid/vapor equilibrium has been established, the pressure exerted by the water vapor is called the **equilibrium vapor pressure** (often, though not correctly, called just the vapor pressure). The equilibrium vapor pressure of any

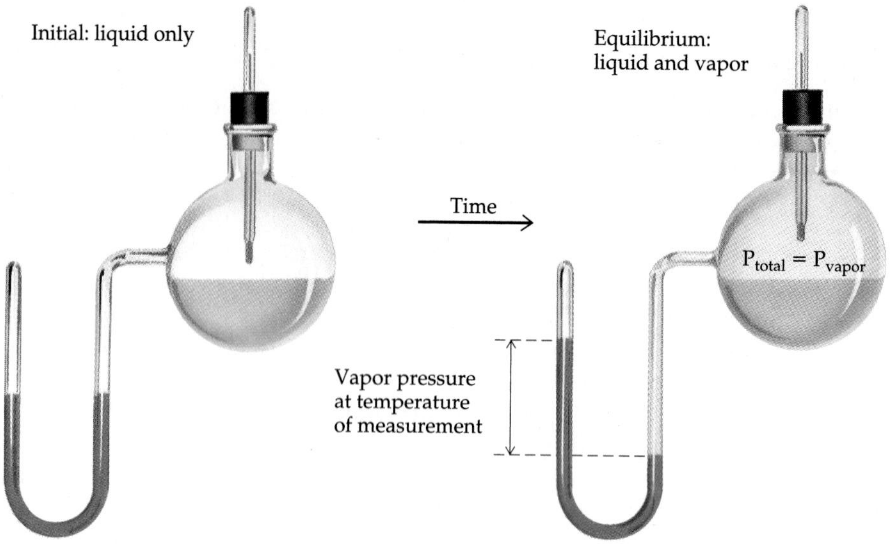

Figure 13.20 A volatile liquid is placed in an evacuated flask (*left*). In the beginning, no molecules of liquid are in the vapor phase. In a short time, however, some of the molecules in the vapor phase exert a pressure (*right*). If the pressure is measured when the liquid and vapor are in equilibrium, the pressure is called the equilibrium vapor pressure.

substance is a measure of the tendency of its molecules to escape from the liquid phase and enter the vapor phase *at a given temperature.* We often refer to this tendency as the **volatility** of the compound. The higher the equilibrium vapor pressure at a given temperature, the more **volatile** the compound.

Because the average energy of a molecule in the liquid phase is a function of temperature (see Figure 13.17), molecules move more rapidly and the rate of vaporization increases as the temperature rises. The equilibrium vapor pressure must therefore also increase with temperature (Figure 13.21). Any point along a vapor pressure versus temperature curve for a compound (see Figure 13.21) represents a pressure and temperature at which liquid and vapor are in equilibrium. For example, at 25 °C the equilibrium vapor pressure of water is 24 mm Hg, whereas it is 149 mm Hg at 60 °C. This means that if water is placed

At the conditions of *T* and *P* given by any point on the curves in Figure 13.21, pure liquid and its vapor are in dynamic equilibrium. If the *T* and *P* define a point *not* on the curve, the system is *not* at equilibrium.

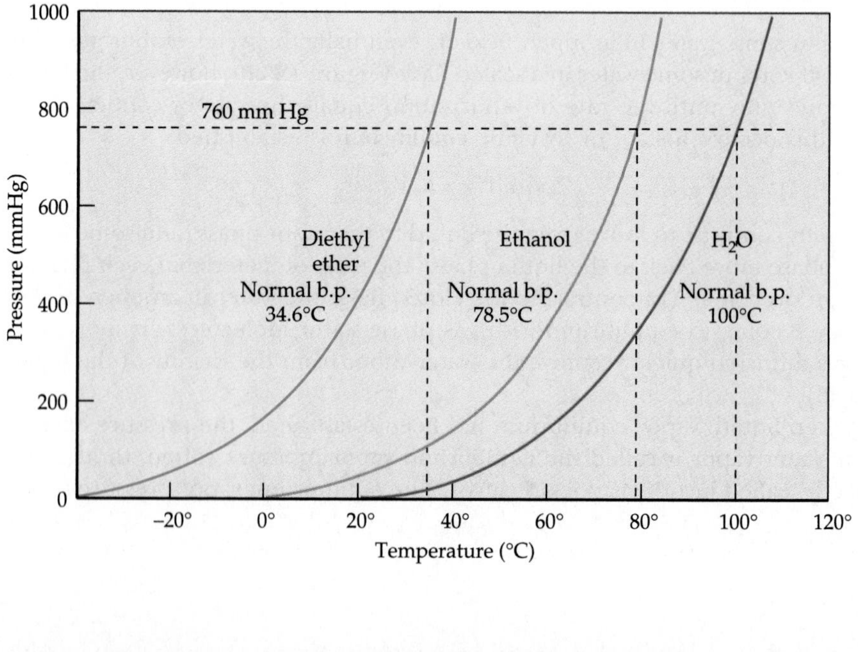

Figure 13.21 Vapor pressure curves for diethyl ether ($C_2H_5OC_2H_5$), ethanol (CH_3CH_2OH), and water. Each curve represents the conditions of T and P at which the two phases (pure liquid and its vapor) are in equilibrium. (P is the vapor pressure of the liquid at a temperature T.) The compounds exist as liquids under conditions of T and P defined by the points to the left of a curve, and they exist as vapor for all temperatures and pressures to the right of a curve.

in an evacuated flask maintained at 60 °C, liquid will evaporate until the pressure exerted by the vapor is 149 mm Hg. (If not enough water is put in the flask, it is possible that all the liquid will evaporate before this equilibrium pressure is reached.)

It is obvious from Figure 13.21 that the relation between temperature and the vapor pressure of a pure liquid is not a straight line. The more complicated curved relation was studied in the 19th century by the German physicist R. Clausius and the Frenchman B. P. E. Clapeyron, who showed that the temperature was related to the vapor pressure by the equation

$$\ln P = -\Delta H_{vap}/RT + C$$

The term $\ln P$ is the natural logarithm of the vapor pressure, T is the Kelvin temperature at which P is measured, ΔH_{vap} is the enthalpy of vaporization of the liquid, R is the ideal gas constant, and C is a constant characteristic of the compound in question. This equation means that enthalpies of vaporization can be obtained from the measurement of vapor pressures at various temperatures.

EXAMPLE 13.4 *Vapor Pressure*

You place 2.00 L of water in your dormitory room, which has a volume of 4.33×10^4 L. You seal the room and wait for the water to evaporate. Will it all evaporate at 25 °C? (At 25 °C the density of water is 0.997 g/ml, and its vapor pressure is 23.8 mm Hg.)

Vapor pressures of water at various temperatures are given in Appendix E.

Solution One approach to solving this problem is to calculate the quantity of water that must evaporate in order to exert a pressure of 23.8 mm Hg in a volume of 4.33×10^4 L at 25 °C. We shall therefore use the ideal gas law to find the quantity of water vapor in moles.

$$n = \frac{PV}{RT} = \frac{\left(\frac{23.8\ \text{mm Hg}}{760\ \text{mm Hg/atm}}\right)(4.33 \times 10^4\ \text{L})}{(0.0821\ \text{L} \cdot \text{atm/K} \cdot \text{mol})(298\ \text{K})} = 55.4\ \text{mol}\ H_2O$$

$$55.4\ \text{mol}\ H_2O\left(\frac{18.02\ \text{g}}{\text{mol}}\right) = 999\ \text{g}\ H_2O$$

$$999\ \text{g}\ H_2O\left(\frac{1\ \text{mL}}{0.997\ \text{g}}\right) = 1.00 \times 10^3\ \text{mL}\ H_2O$$

This calculation shows that only half of the available water needs to evaporate to achieve an equilibrium water vapor pressure of 23.8 mm Hg at 25 °C in the dorm room.

EXERCISE 13.4 *Vapor Pressure Curves*

Look at the vapor pressure curve for ethanol in Figure 13.21.

1. What is the approximate vapor pressure of the alcohol at 40 °C?
2. Does a liquid/vapor equilibrium occur at 60 °C when the pressure is 600 mm Hg? If not, what must the temperature be to achieve this pressure? ■

EXERCISE 13.5 *Vapor Pressure*

If you seal 0.50 g of pure water in an evacuated 5.0-L flask and heat the whole assembly to 60 °C, is the pressure equal to or less than the equilibrium vapor pressure of water at this temperature? What if you use 2.0 g of water? Under either set of conditions is any liquid water left in the flask, or does all of the water evaporate? ■

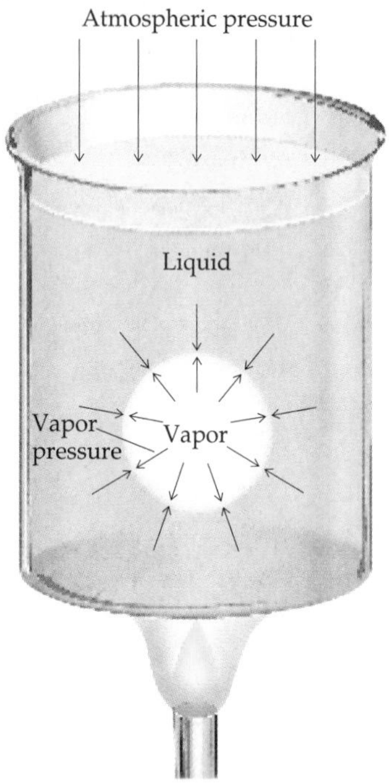

Figure 13.22 A bubble of vapor forms in a liquid. When the vapor pressure of the liquid is equal to the external pressure (atmospheric pressure when a liquid is in a container open to the atmosphere), the liquid boils.

Boiling Point

If you have a beaker of water open to the atmosphere, the mass of the atmosphere is pressing down in the surface. As heat is added, more and more water can evaporate, pushing the molecules of the atmosphere aside. If enough heat is added, a temperature is eventually reached at which the vapor pressure of the liquid equals the atmospheric pressure; bubbles of vapor begin to form in the liquid, and the liquid *boils* (Figure 13.22).

The boiling point of a liquid is the temperature at which the vapor pressure is equal to the external pressure, and, if the external pressure is 1 atm, the temperature is designated the **normal boiling point.** Normal boiling points of some liquids are listed in Table 13.6. Be sure to notice the direct relationship between normal boiling point, enthalpy of vaporization, and intermolecular forces. We have already pointed out that stronger intermolecular forces lead to larger values of ΔH_{vap} and, therefore, to higher boiling points (see Figure 13.8).

The normal boiling point of water at sea level is 100 °C. But, if you live at higher altitudes, such as in Salt Lake City, Utah, where the barometric pressure is less than 1 atm (about 650 mm Hg), water boils at lower temperatures. (The curve for the equilibrium vapor pressure of water in Figure 13.21 shows that the temperature is about 95 °C when $P = 650$ mm Hg). Cooks know that food has to be cooked a bit longer in Salt Lake City or Denver to achieve the same effect as in New York City at sea level.

To shorten cooking time in Denver, or anywhere else, use a pressure cooker. This is a sealed pot that allows water vapor to build up to pressures slightly greater than the external, or atmospheric, pressure. The boiling point of the water increases, and foods cook faster.

Critical Temperature and Pressure

The vapor pressure of a liquid continues to increase with temperature up to the **critical point,** where the curve of vapor pressure versus temperature comes to an abrupt halt (Figure 13.23). The temperature at which this occurs is the **critical**

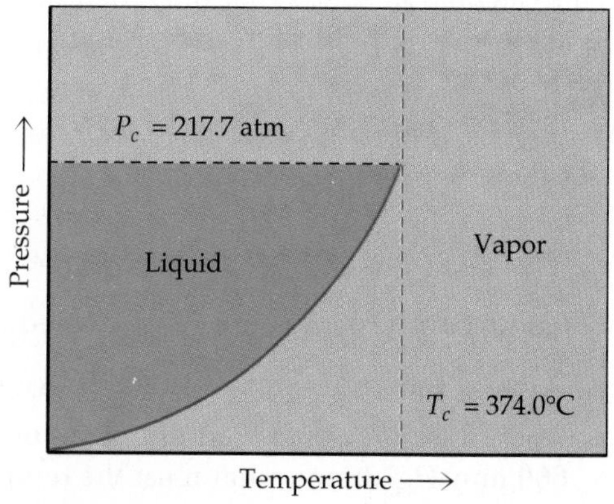

Figure 13.23 The vapor pressure curve for water near the critical temperature, T_c. This temperature is "critical" because, beyond this point, vapor cannot be converted to liquid no matter how high the pressure exerted on the vapor.

TABLE 13.7 Critical Temperatures and Pressures for Common Compounds*

Compound	T_c (°C)	P_c (atm)	P_c (MPa)†
CH_4 (methane)	−82.6	45.4	4.60
C_2H_6 (ethane)	32.3	49.1	4.98
C_3H_8 (propane)	96.7	41.9	4.25
C_4H_{10} (butane)	152.0	37.3	3.78
CCl_2F_2 (CFC-12)	111.8	40.9	4.14
NH_3	132.4	112.0	11.35
H_2O	374.0	217.7	22.06
SO_2	157.7	77.8	7.88

*Data taken from D. R. Lide, *BasicLaboratory and Industrial Chemicals,* Boca Raton, FL, CRC Press, 1993.

†1 MPa = 1000 kPa = 10 bar = 9.86923 atm

temperature, T_c, and the corresponding vapor pressure is the **critical pressure, P_c.** The significance of this point is that, if the temperature exceeds T_c, all the liquid molecules have sufficient kinetic energy to separate from one another, regardless of the pressure, and a conventional liquid no longer exists. Instead the substance is often called a supercritical gas. A **supercritical gas** is one at a pressure so high that its density resembles a liquid's, while its viscosity (ability to flow) remains close to that of a gas.

A knowledge of critical temperatures and pressures (see Table 13.7) is important in designing air conditioners, for example. These devices cool a room by using the room heat to vaporize a liquid. To make the process continuous, however, the vapor is condensed back to the liquid state by compressing the vapor to a pressure higher than their vapor pressure (as long as the temperature of the vapor is less than T_c). *Chlorofluorocarbons* (CFCs) have been widely used as the fluid in air conditioners and refrigerators because some of them can be liquefied at reasonable pressures. For example, CCl_3F has a T_c of 198 °C and a P_c of 43.5 atm. This means CCl_3F vapor can be converted to a liquid by applying modest pressures at any temperature below 198 °C, a temperature far above room temperature. You will also notice that propane and butane, which are widely used as fuels in camping stoves, backyard grills, and home heating and cooking, can be readily liquefied.

Chlorofluorocarbons (CFCs) have been widely used in air conditioners and refrigerators because they often have useful values of T_c and P_c and were once thought to be chemically inert. Recently, however, it was found that they are reactive under certain conditions and are responsible for damage to the "ozone layer" in the upper stratosphere. They are therefore banned from use and are being replaced by compounds such as CF_3CH_2F, which has the desirable properties of CFCs without containing Cl atoms.

In the supercritical state, certain fluids, such as water and carbon dioxide, take on unexpected properties, such as the ability to dissolve normally insoluble materials. Supercritical CO_2 is especially useful. This fluid does not dissolve water or polar compounds such as sugar, but it does dissolve nonpolar oils, which constitute many of the flavoring or odor-causing compounds in foods. For example, food companies use supercritical CO_2 to extract caffeine from coffee.

Surface Tension, Capillary Action, and Viscosity

Molecules at the surface of a liquid behave differently from those in the interior because molecules in the interior interact with molecules all around them (Figure 13.24). In contrast, surface molecules are affected only by those below the

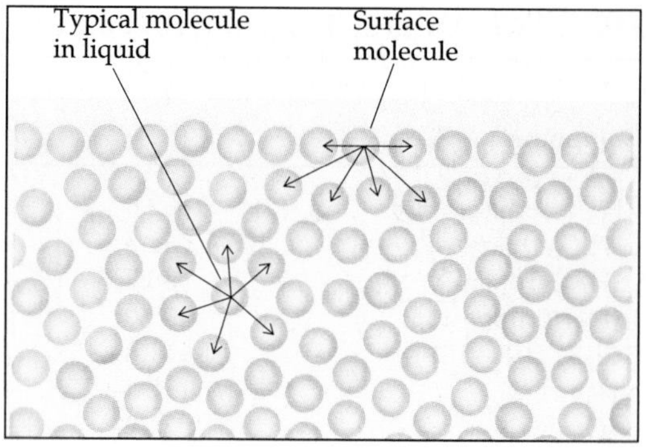

Figure 13.24 There is a difference between the forces acting on a molecule within the liquid phase and those acting on a molecule at the surface of the liquid.

surface layer. This phenomenon leads to a *net* inward force of attraction on the surface molecules, contracting the surface and making it behave as though it had a "skin." The "toughness" of the skin of a liquid is measured by its **surface tension,** the energy required to break through the surface or to disrupt a liquid drop and spread the material out as a film. It is surface tension that causes water drops to be spheres and not little cubes, for example (Figure 13.25), because the sphere has a smaller surface area than any other shape of the same volume.

Capillary action is closely related to surface tension. When a small-diameter glass tube is placed in water, the water rises in the tube, just as water rises in a piece of paper in water (Figure 13.26). Because polar Si—O bonds occur on the surface of glass, polar water molecules are attracted by **adhesive** forces between the two different substances. These forces are strong enough that they can compete with the **cohesive** forces between the water molecules themselves. Thus, some water molecules can *adhere* to the walls, and other water molecules are attracted to these and build a "bridge" back into the liquid. In addition, the surface tension of the water (from cohesive forces) is great enough to pull the liquid up the tube, so the water level rises in the tube. The rise continues until the various forces (adhesion between water and glass, cohesion between water molecules, and the force of gravity on the water column) are in equilibrium. It is these forces that lead to the characteristic concave, or downward-curving, **meniscus** that you see for water in a drinking glass or in a laboratory test tube (Figure 13.27).

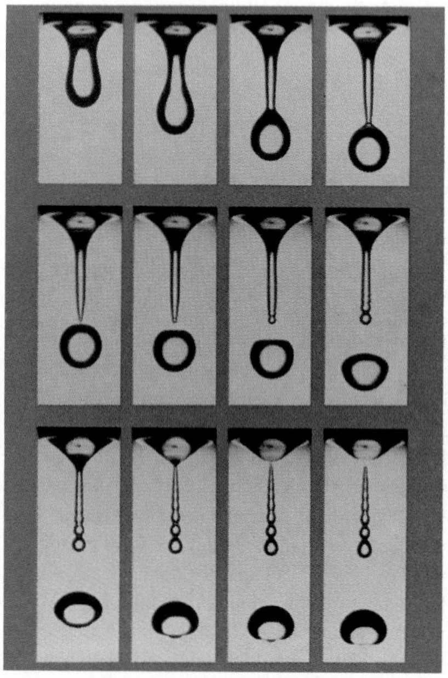

Figure 13.25 This series of photographs shows the different stages of fission when a water drop falls from a circular plate 12.5 cm in diameter. The drop was illuminated by a strobe light of 5-μs duration. The total time for this sequence was less than 0.05 second. (S. R. Nagel/James Frank Institute/University of Chicago)

Of course some liquids exist, such as mercury, for which cohesive forces (high surface tension) are much greater than adhesive forces between mercury and glass, for example. Mercury does not climb the walls of a glass capillary, and when it is placed in a tube it has a convex, or upward-curving, meniscus (see Figure 13.27).

Finally, one additional, but important property of liquids is their **viscosity,** their resistance to flow. When you turn over a glassful of water, it empties quickly, but it takes much more time to empty a glassful of honey or rubber cement. Although intermolecular forces play a significant role in determining viscosity, other factors are clearly also at work. For example, olive oil consists of molecules with very long chains of CH_2 units and O atoms (see Chapter 11), and it is about 70 times more viscous than ethyl alcohol (ethanol), a molecule with only a three-atom chain ($H_3C—CH_2—OH$). The long-chain molecules of natural oils are floppy and become entangled with one another; the longer the chain

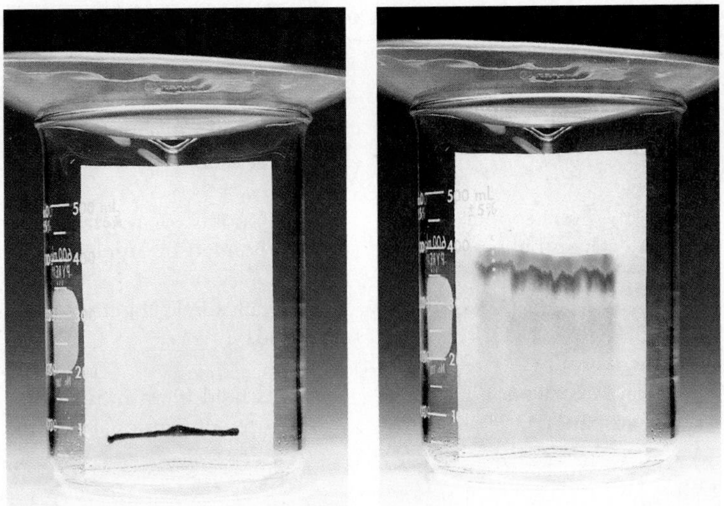

Figure 13.26 Capillary action. Polar water molecules are attracted to the O—H bonds in paper fibers, and water rises in the paper. If a line of ink is placed in the path of the rising water, the different components of the ink are attracted differently to the water and paper and are separated in a process called *chromatography*. (C. D. Winters)

Figure 13.27 The meniscus formed by two different liquids. For water, the cohesive forces between molecules are partly overcome by the adhesive forces between water and glass, producing a concave meniscus for water. In contrast, the cohesive forces within mercury are significantly greater than the adhesive forces between mercury and glass, producing a convex meniscus for mercury. (C. D. Winters)

the greater the tangling and the greater the viscosity. Also, longer chains have greater intermolecular forces because there are more atoms to attract one another, each atom contributing to the total force.

EXERCISE 13.6 ***Viscosity***

Glycerol is used in cosmetics. Do you expect its viscosity to be larger or smaller than the viscosity of ethanol, CH_3CH_2OH? Why or why not?

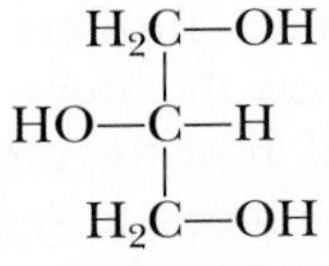

glycerol

■

Figure 13.28 Types of solids. The clear crystal and white crystal are ionic solids, a form of calcium carbonate called calcite. The irregular copper-colored solid is in fact copper as it occurs naturally in the earth. The large, round bar is silicon, a network solid, and the purple crystals are amethyst, another network solid. Finally, the other solids in the picture are glass and polyethylene, both amorphous solids. (C. D. Winters)

13.4 METALLIC AND IONIC SOLIDS

Many kinds of solids exist in the world around us (Figure 13.28). Solid-state chemistry is one of the booming areas of science, especially because of exciting developments in new materials. In this section we can only give you a *very* brief introduction to some of the principles of this exciting area.

To organize solid-state chemistry, we might classify different types of solids as in Table 13.8. This section describes the solid-state structures of some common metallic and ionic solids, with some information on a few other kinds of solids as well.

TABLE 13.8 Structures and Properties of Various Types of Solid Substances

Type	Examples	Structural Units
Ionic	NaCl, K_2SO_4, $CaCl_2$, $(NH_4)_3PO_4$	Positive and negative ions; no discrete molecules
Metallic	Iron, silver, copper, other metals and alloys	Metal atoms (or positive metal ions surrounded by an electron sea)
Molecular	H_2, O_2, I_2, H_2O, CO_2, CH_4, CH_3OH, CH_3CO_2H	Molecules held together by covalent bonds
Network	Graphite, diamond, quartz, feldspars, mica	Atoms held in an infinite one-, two-, or three-dimensional network
Amorphous (glassy)	Glass, polyethylene, nylon	Covalently bonded networks with no long-range regularity

continued

Crystal Lattices and Unit Cells of Metal Atoms

In both gases and liquids, molecules move continually and randomly, and rotate and vibrate as well. Because of this movement, no extensive, orderly arrangement of molecules in the gas or liquid state is possible, resulting in the major difference between liquids and solids. In solids, the molecules, atoms, or ions cannot move (although they vibrate and occasionally rotate). Long-range order, a characteristic of the solid state, is therefore possible. It is to these orderly arrangements that we first turn our attention.

The beautiful, external regularity of a crystal of salt or a metal suggests that it has an *internal* symmetry as well. Indeed, all crystal lattices (whether of metals or ionic solids) are built of **unit cells,** the smallest, repeating unit that has all of the symmetry characteristics of the way the atoms are arranged. The external appearance of the solid is a reflection of the unit cell.

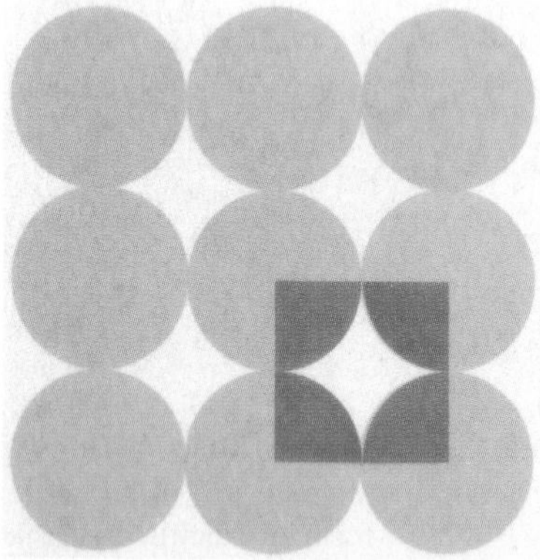

Figure 13.29 Unit cells for a "two-dimensional solid" made from flat, circular objects. Here the unit cell is a square. Each circle at a corner of a square contributes one fourth of its area to the area inside the square. Thus, there is a *net* of one circle per unit cell.

To understand the idea of a unit cell, look at the repeating pattern of circles in Figure 13.29. A "unit" cell could be drawn around each atom, but this unit cell would not reveal how the circles are arranged with respect to one another; that is, it would not show the symmetry of the arrangement of circles. A way to draw a cell that shows this is to draw lines from the center of one circle to the centers of neighboring circles; a square unit cell results. Furthermore, because each of the four circles contributes a quarter of itself to the square contents, a *net* of one circle is located *within* each square. This unit cell reveals the pattern of circles, and the entire pattern can be built by joining identical square unit cells edge to edge, each cell containing the equivalent of one circle.

Solids can be built by piling up three-dimensional unit cells like building blocks, as the artist M. C. Escher illustrated in a well-known drawing (Figure 13.30). The corners defining each unit cell in simple solids represent identical environments for an ion in an ionic solid, a metal atom in a metallic solid, or a molecule in a molecular solid. Such points are equivalent to each other, and collectively they define the **crystal lattice.**

TABLE 13.8 *continued*

Forces Holding Units Together	Typical Properties
Ionic; attractions among charges on positive and negative ions	Hard; brittle; high melting point; poor electric conductivity as solid, good as liquid; often water-soluble
Metallic; electrostatic attraction among metal ions and electrons	Malleable; ductile; good electric conductivity in solid and liquid; good heat conductivity; wide range of hardness and melting points
Dispersion forces, dipole-dipole forces, hydrogen bonds	Low to moderate melting points and boiling points; soft; poor electric conductivity in solid and liquid
Covalent; directional electron-pair bonds	Wide range of hardnesses and melting points (3-dimensional bonding > 2-dimensional bonding > 1-dimensional bonding); poor electric conductivity, with some exceptions
Covalent; directional electron-pair bonds	Noncrystalline; wide temperature range for melting; poor electric conductivity, with some exceptions

To construct crystal lattices, nature uses seven, three-dimensional unit cells that differ from one another in the different relative lengths of their sides and the different angles at which their edges meet (Figure 13.31). We shall be concerned here, however, only with **cubic unit cells,** cells with equal edges that meet at 90° angles (as in Escher's drawing in Figure 13.30). Cubic unit cells are easily visualized and are very common in nature. Within the cubic class, three important cell symmetries occur: **simple cubic (sc), body-centered cubic (bcc),** and **face-centered cubic (fcc)** (Figure 13.32). All three have *eight* identical atoms or ions at the corners of a cube. The body-centered (bcc) and face-centered (fcc) arrangements, however, differ from the simple cube (sc) in having additional atoms, of the same type as those at the corners, at other locations. The **bcc** structure is called "body-centered" because it has one additional atom at the center of the cube or "body." The **fcc** arrangement is called "face-centered" because it has, in the center of each of the six cube faces, an ion or atom of the same type as the corner ions or atoms.

All three cubic unit cells have identical atoms or ions at the corners of the cube. When the cubes pack together to make a macroscopic, three-dimensional crystal, an atom or ion at a corner is therefore shared among eight cubes (Figure 13.33a). Because of this, only one eighth of each corner atom or ion is actually *within* a given unit cell. Because a cube has eight corners, and because one eighth of the atom or ion at each corner "belongs to" a particular unit cell, the result is that there is one net atom or ion within that unit cell.

$$(8 \text{ corners of a cube}) \times (\tfrac{1}{8} \text{ of each corner atom or ion is within a unit cell}) = 1 \text{ net atom or ion within a unit cell}$$

A simple cubic arrangement has one net atom or ion within the unit cell.

A body-centered cube has an additional atom, and it is wholly within the unit cell at the cube's center. Because the center atom is present in addition to those at the cube corners, *the body-centered cubic arrangement has a net of two atoms within the unit cell.*

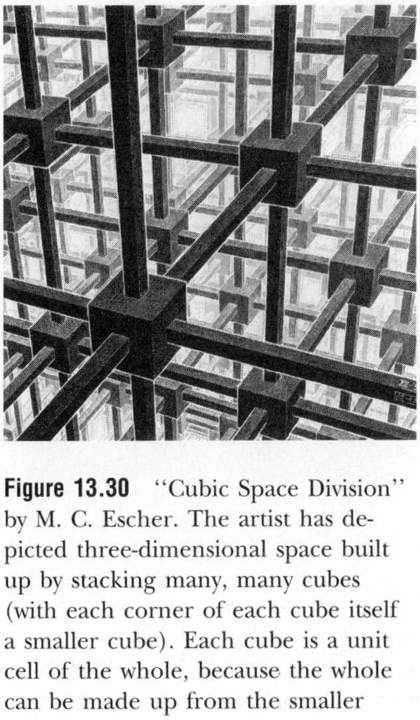

Figure 13.30 "Cubic Space Division" by M. C. Escher. The artist has depicted three-dimensional space built up by stacking many, many cubes (with each corner of each cube itself a smaller cube). Each cube is a unit cell of the whole, because the whole can be made up from the smaller cubes. (M. C. Escher Foundation, Baarn, The Netherlands)

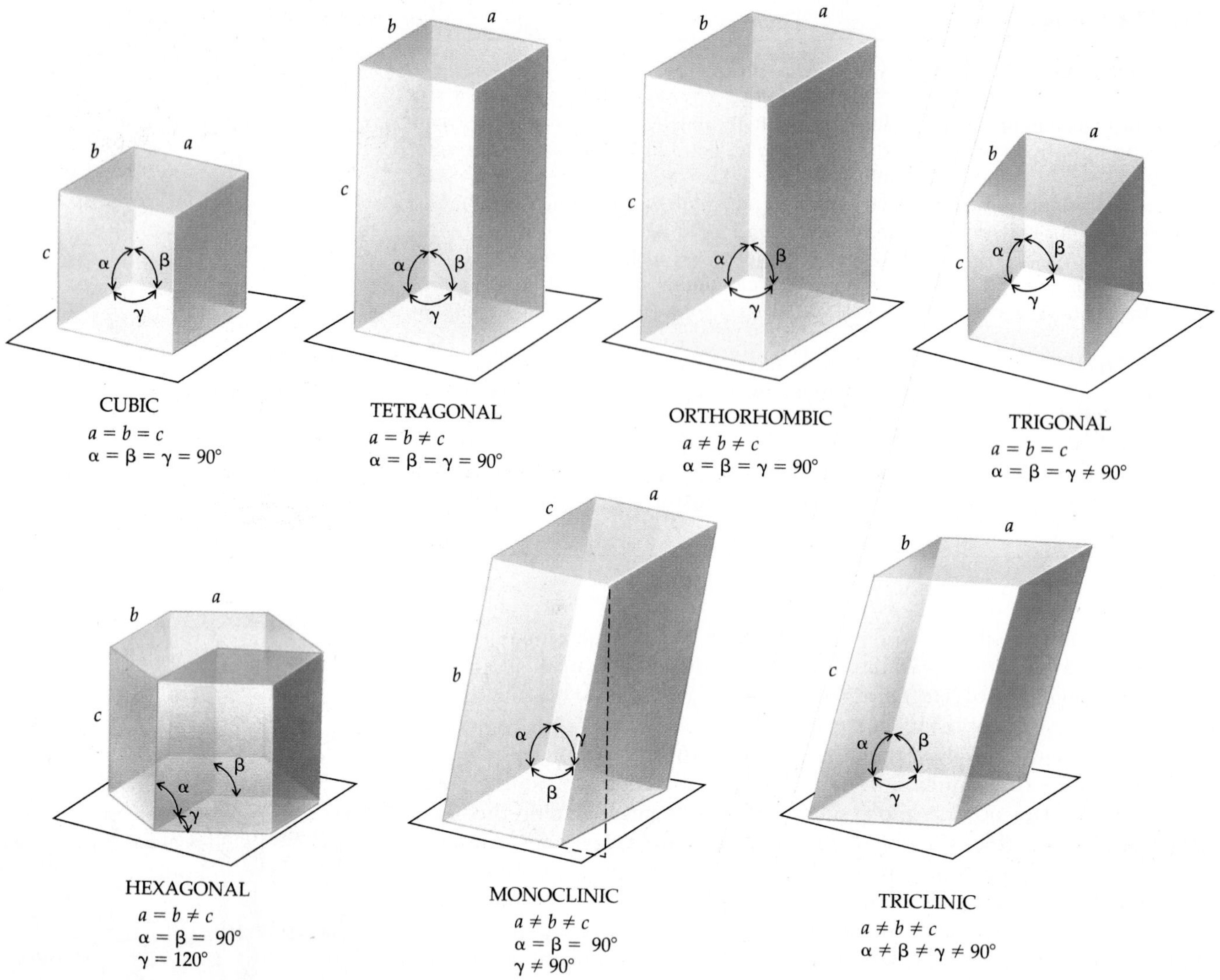

Figure 13.31 Shapes of the unit cells of the seven crystal systems.

The fcc arrangement is one of the most important. Here there is an atom or ion in each of the six faces of the cube. One half of each of these belongs to the unit cell (Figure 13.33b). Therefore,

(6 faces of a cube) × ($\frac{1}{2}$ of an atom or ion is within a unit cell)
= 3 net face-centered atoms or ions within a unit cell

This means *the face-centered cubic arrangement has a net of four atoms or ions within the unit cell,* a net of one contributed by the corner atoms or ions and another three contributed by the ones centered in the six faces.

A laboratory technique, x-ray crystallography, can be used to determine the structure of a crystalline substance. Once the structure is known, the information can be combined with other experimental information to calculate such useful parameters as the radius of an atom. This approach is outlined in Example 13.5.

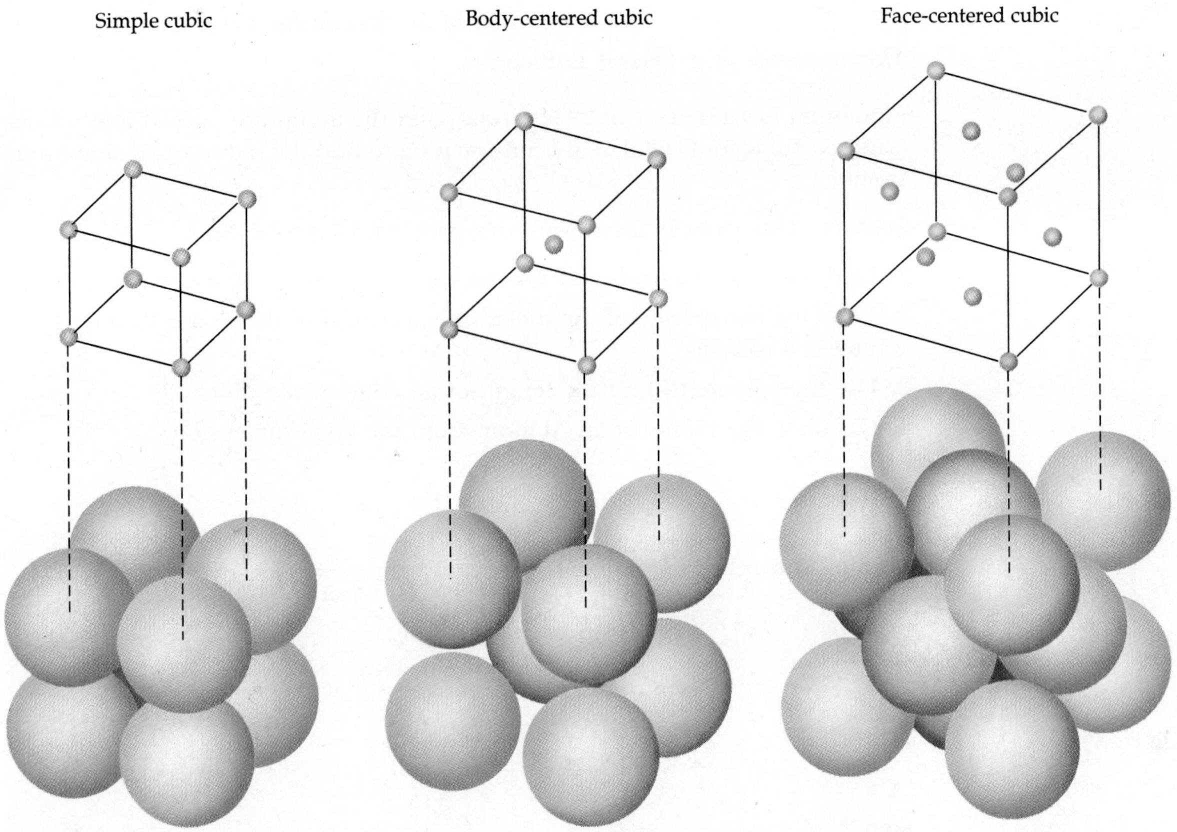

Figure 13.32 The three different types of cubic unit cells. The *top row* shows the lattice points of the three cells. In the *bottom row* the points are replaced with space-filling spheres representing the atoms or ions of the lattice. All spheres—no matter what their color—represent identical atoms or ions centered on the lattice points. Notice that the spheres at the corners of the body-centered and face-centered cubes do not touch one another. Rather, each corner atom in the body-centered cell touches the center atom, and each corner atom in the face-centered cell touches spheres in the three adjoining faces.

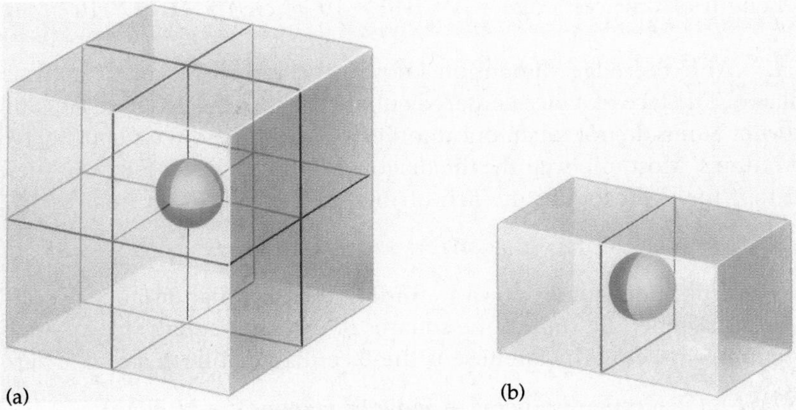

Figure 13.33 Atom sharing at cube corners and faces. (a) In any cubic lattice each corner atom (or ion) is shared equally among eight cubes; one eighth of each atom (or ion) is within a particular cube. (b) In a face-centered lattice, each atom (or ion) in a cube face is shared equally between two cubes. Each atom (or ion) of this type contributes one half of itself to a given cube.

EXAMPLE 13.5 *Determination of an Atomic Radius from Measurements of a Crystal Lattice*

Aluminum has a density of 2.699 g/cm^3, and the atoms are packed into a face-centered cubic unit cell. Use this information to find the radius of an aluminum atom.

Solution Our strategy for solving this problem is as follows:

1. Find the mass of a unit cell from the knowledge that it is face-centered cubic.
2. Combine the density of the metal with the mass of the unit cell to find the unit cell volume.
3. Use the volume to find the length of an edge of the unit cell.
4. Calculate the radius of an Al atom from the edge dimension.

Step 1.

$$\text{Mass of unit cell} = \left(\frac{26.98\ \text{g}}{\text{mol}}\right)\cdot\left(\frac{1\ \text{mol}}{6.022 \times 10^{23}\ \text{atoms}}\right)\cdot\left(\frac{4\ \text{atoms}}{\text{unit cell}}\right)$$

molar mass; Avogadro's number; Number of atoms in fcc unit cell

$$= 1.792 \times 10^{-22}\ \text{g/unit cell}$$

Step 2.

$$\text{Volume of unit cell} = \left(\frac{1.792 \times 10^{-22}\ \text{g}}{\text{unit cell}}\right)\cdot\left(\frac{1\ \text{cm}^3}{2.699\ \text{g}}\right)$$

Mass of unit cell; Density of Al

$$= 6.640 \times 10^{-23}\ \text{cm}^3/\text{unit cell}$$

Step 3. Now, recall that the volume of a cube = (length of edge)^3. To find the length of an edge, therefore, we need to find the cube root of the cube volume.

$$\text{Length of unit cell edge} = \sqrt[3]{6.640 \times 10^{-23}\ \text{cm}^3} = 4.049 \times 10^{-8}\ \text{cm}$$

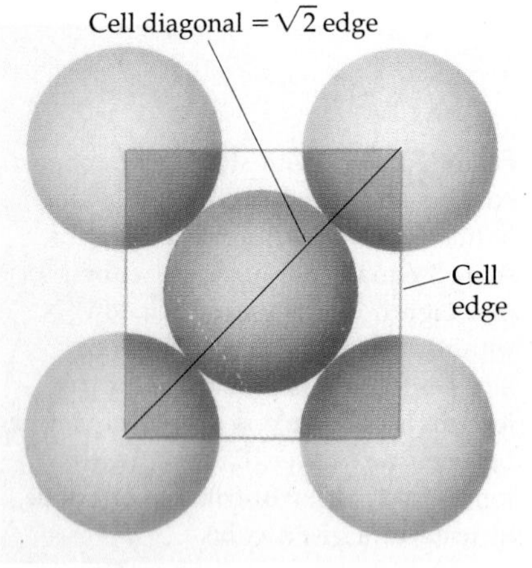

Step 4. With the edge dimension known, the radius of an Al atom can be calculated. The face of a face-centered cube is drawn here. Notice that, although the corner atoms do not touch one another, the center atom touches each of the corner atoms. Most importantly, the diagonal distance across a face of the cell is equal to 4 times the radius of each of the circles. That is, in this case

$$\text{Cell face diagonal} = 4 \times (\text{Al atom radius})$$

So, our problem comes down to finding the diagonal distance. Recall that, for a right triangle, the sum of the squares of the sides equals the square of the hypotenuse, where the hypotenuse is the face diagonal in this case. Therefore,

$$(\text{Diagonal distance})^2 = (\text{edge})^2 + (\text{edge})^2 = 2(\text{edge})^2$$

Taking the square root of both sides, we have

$$\text{Diagonal distance} = \sqrt{2} \times (\text{cell edge})$$
$$= \sqrt{2} \times (4.049 \times 10^{-8}\text{ cm}) = 5.727 \times 10^{-8}\text{ cm}$$

If this distance is divided by 4, we have the Al atom radius.

$$\text{Al atom radius} = (5.727 \times 10^{-8}\text{ cm})/4 = 1.432 \times 10^{-8}\text{ cm}$$

Atomic dimensions are usually expressed in picometers. The radius in this unit is therefore 143.2 pm.

$$1.432 \times 10^{-8}\text{ cm} \times \left(\frac{1\text{ m}}{100\text{ cm}}\right)\left(\frac{1\text{ pm}}{1 \times 10^{-12}\text{ m}}\right) = 143.2\text{ pm}$$

This is in excellent agreement with the radius given in Table 8.10.

EXAMPLE 13.6 ***The Structure of Solid Iron***

Iron has a density of 7.8740 g/cm^3, and the radius of an iron atom is 126 pm. Verify that the structure of the solid is a body-centered cube.

Solution As outlined in the text, a body-centered cubic unit cell contains a net of two atoms. The objective is to use the density of iron and the atom radius to calculate the net number of atoms in the unit cell. If there are two atoms, the unit cell is bcc. Our strategy can therefore be laid out as follows:

1. Use the atomic radius to find the length of an edge of the cube.
2. Calculate the unit cell volume from the edge dimension.
3. Combine the unit cell volume with the experimental density to calculate the mass of the unit cell.
4. Compare the mass of one iron atom with the mass of the unit cell. The unit cell mass should equal twice the mass of one Fe atom.

Step 1. In a body-centered cubic unit cell the atoms of the lattice touch only along the diagonal line running from a corner at the "top" of the cube to the opposite corner at the "bottom" of the cube. (As in the fcc structure in Example 13.5 the corner atoms do not touch one another.) The distance along this diagonal is equal to 4 times the radius of an atom and is also equal to $\sqrt{3}$ times the length of the edge of the cube.
The reason for this is as follows:

$$\text{Diagonal distance across cube face} = \sqrt{2} \times (\text{cell edge})$$
$$(\text{Diagonal distance across the cube})^2 = (\text{cell edge})^2 + (\text{face diagonal})^2$$
$$= (\text{cell edge})^2 + [\sqrt{2} \times (\text{cell edge})]^2$$
$$= 3(\text{edge})^2$$

Taking the square root of both sides, we have

$$\text{Diagonal distance across the cube} = \sqrt{3} \times (\text{cell edge})$$

The cell diagonal distance is equal to 4 times the radius of an atom. Therefore,

$$4 \times \text{atom radius} = \sqrt{3} \times (\text{cell edge})$$

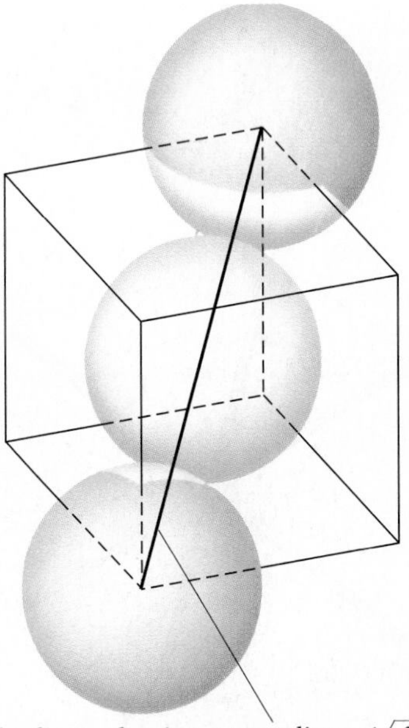

Cube diagonal = 4 × atom radius = $\sqrt{3}$ edge

Body-centered cube with two opposite corner atoms and center atom.

and so the length of an edge of the unit cell in this case is

$$\text{Cell edge} = \frac{4 \times \text{atom radius}}{\sqrt{3}} = \frac{4 \times 126 \text{ pm}}{\sqrt{3}} = 291 \text{ pm}$$

Expressing the length in centimeters, we have

$$291 \text{ pm} \cdot \left(\frac{1 \text{ m}}{1 \times 10^{12} \text{ pm}}\right) \cdot \left(\frac{100 \text{ cm}}{1 \text{ m}}\right) = 2.91 \times 10^{-8} \text{ cm}$$

Step 2. Calculate the unit cell volume

$$\text{Unit cell volume} = (2.91 \times 10^{-8} \text{ cm})^3 = 2.46 \times 10^{-23} \text{ cm}^3$$

Step 3. Calculate the mass of the unit cell from the cell volume and density.

$$\text{Mass of the unit cell} = (2.46 \times 10^{-23} \text{ cm}^3)\left(\frac{7.8740 \text{ g}}{\text{cm}^3}\right) = 1.94 \times 10^{-22} \text{ g}$$

Step 4. Calculate the mass of an iron atom and compare it with the mass of a unit cell.

$$\text{Mass of one Fe atom} = \left(\frac{55.85 \text{ g}}{1 \text{ mol}}\right) \cdot \left(\frac{1 \text{ mol}}{6.022 \times 10^{23} \text{ atoms}}\right)$$

$$= 9.274 \times 10^{-23} \text{ g/atom}$$

$$\frac{1.94 \times 10^{-22} \text{ g/unit cell}}{9.274 \times 10^{-23} \text{ g/atom}} = 2.09 \text{ atoms/unit cell}$$

Our calculations show that effectively two atoms occur per unit cell, as required by a body-centered cubic cell. The iron crystal is therefore body-centered cubic.

EXERCISE 13.7 ***Determination of an Atomic Radius from Measurements of a Crystal Lattice***

Gold is a face-centered cubic unit cell. The density of the solid is 19.32 g/cm³. Calculate the radius of a gold atom. ■

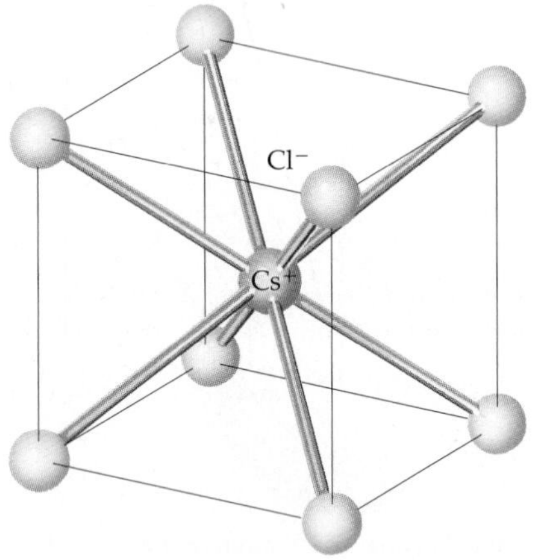

Cesium chloride (CsCl) crystal structure.

Structures and Formulas of Ionic Solids

The lattices of many ionic compounds are built by taking a simple cubic or face-centered cubic lattice of spherical ions of one type and placing ions of opposite charge in the holes left in the lattice. The number and location of the holes that are filled are the keys to understanding the relation between the lattice structure and the formula of a salt.

The hole in a simple cube is in the center of the cell, and this is the only type of space available. The ionic compound cesium chloride, CsCl, adopts just this structure: simple cubes of Cl^- ions with a Cs^+ ion in the center of each cube.

The structure of NaCl, sodium chloride, represents one of the most common ways of constructing an ionic solid (Figure 13.34). You can view the salt as having been built from the larger Cl^- ions packed as a face-centered cube with

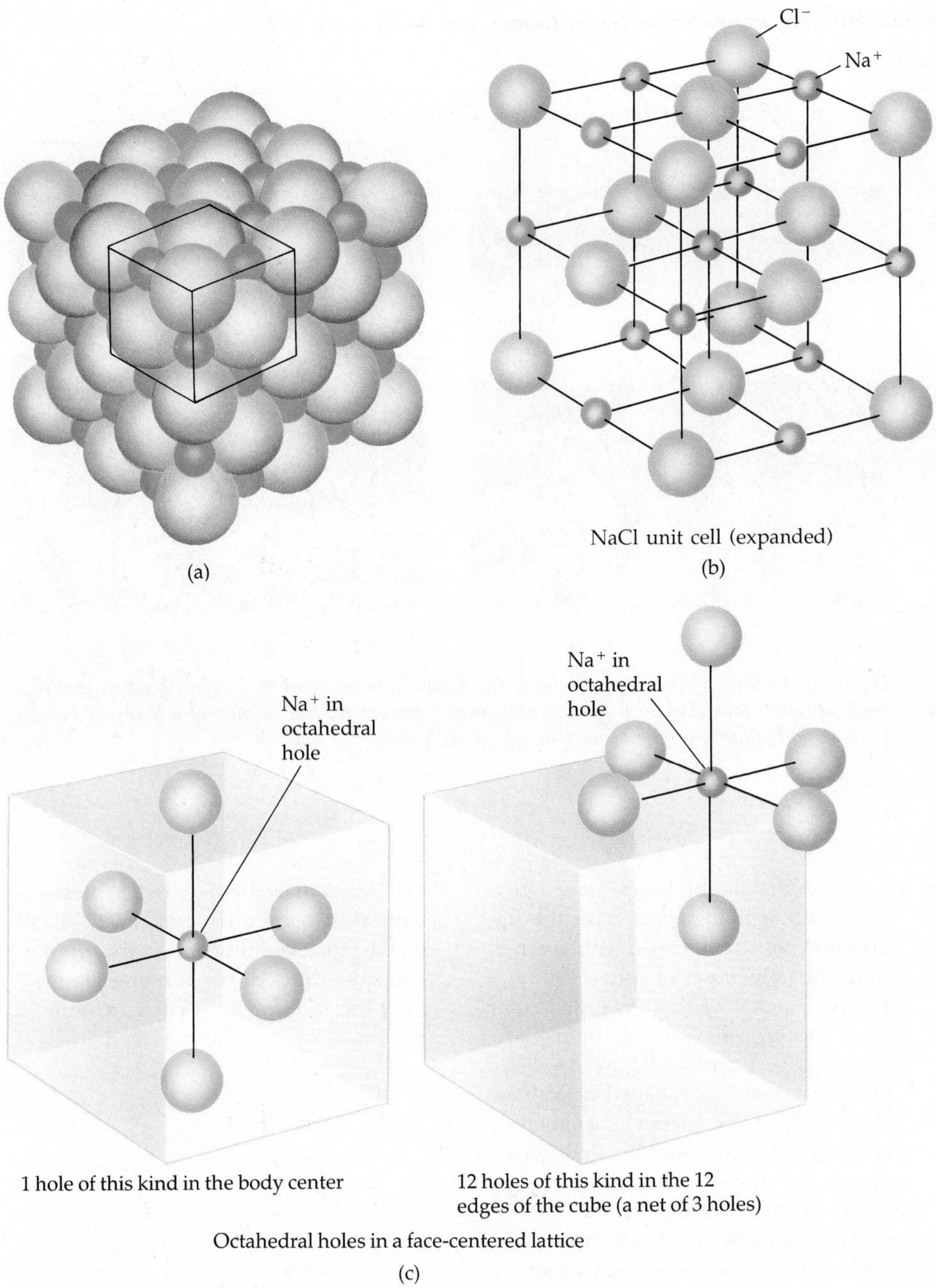

Figure 13.34 Sodium chloride. (a) An extended view of a sodium chloride lattice. The smaller Na^+ ions (red) are packed into a face-centered cubic lattice of larger Cl^- ions (green). One unit cell is outlined. (b) An expanded view of the NaCl face-centered cubic unit cell. The lines only represent the connections between lattice points. (c) A close-up view of the octahedral holes in the lattice.

the smaller Na^+ ions filling all the so-called *octahedral* lattice holes.* The holes are said to be octahedral because each Na^+ is surrounded by six Cl^- ions at the corners of an octahedron.

To understand the relation between the structure of the unit cell of a salt and the formula of the salt we need to consider the positions of the ions and the fraction belonging to the unit cell. Let us do that for NaCl, one unit cell of which

In general, we think of building ionic lattices out of the larger anions, say Cl^-, and then placing the smaller cations, Na^+, for example, in some or all of the holes that remain.

*The octahedral holes do not represent all of the available holes. So-called *tetrahedral* holes also occur. We shall not pursue this further here; however, you should be aware that the Na^+ ions occupy *only* octahedral holes and not a random mixture of octahedral and tetrahedral holes.

A CLOSER LOOK Packing Atoms and Ions into Unit Cells

It is a ''rule'' of nature that things are done as efficiently as possible. Imagine that atoms or ions are tiny, hard spheres like marbles. If these atomic ''marbles'' are packed as efficiently as possible, we find that the fcc lattice uses 74% of the available space, whereas spheres packed as a simple cube use only 52% of the available space, clearly a less efficient arrangement. You can appreciate this fact to some extent by looking at Figure A, a two-dimensional example. Here you see the two, basic ways of packing hard, spherical ions or atoms together in one layer. In one arrangement (Figure Aa), the spheres are at the corners of a square and each touches *four* other spheres. In the other (Figure Ab), each sphere touches *six* others at the corners of a hexagon. In either arrangement, little holes are left between atoms in the layer (and this space is greater in Figure Aa than in Ab). *The way the are arranged in the layers and the way the layers are stacked, one on top of another, is the key to the structures of metallic and ionic solids.*

To fill three-dimensional space, layers of atoms or ions are stacked one on top of the other. If you start with the square arrangement (Figure Aa) for the first layer and simply stack the next layer of spherical ions or atoms *directly* on top of the first, the arrangement resembles Escher's work in Figure 13.30, an extended array of cubic unit cells.

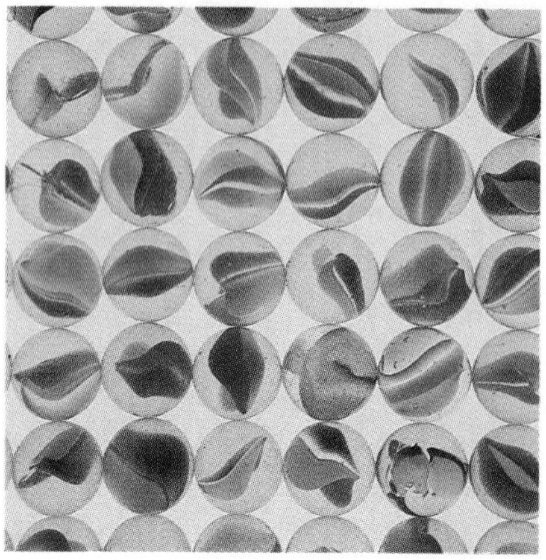

(a)

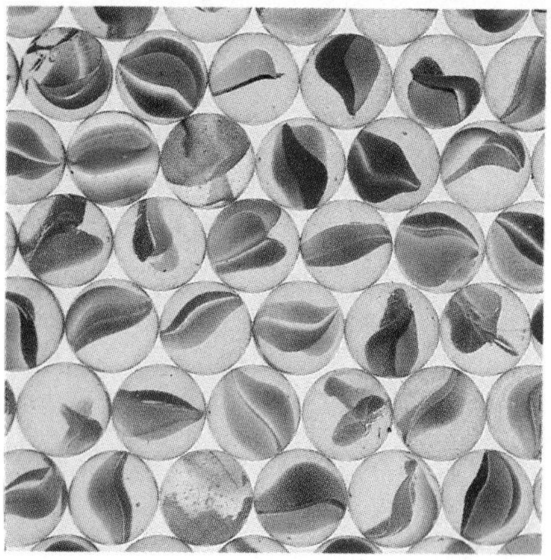

(b)

Figure A *Packing of spheres in one layer. (a) Marbles are arranged in a square pattern, and each marble contacts four others. (b) Each marble contacts six others at the corners of a hexagon. Pattern (b) is a more efficient arrangement than (a).* (C. D. Winters)

Two other methods are more efficient ways of filling space than the simple cube, and both begin with the hexagonal arrangement of spheres in Figure Ab. One of these is called **cubic close packing** and leads to the fcc lattice described earlier (Figure B). The other arrangement is called **hexagonal close packing.** This latter arrangement is known in nature, but we shall only discuss the simple cubic and face-centered cubic arrangements.

The atoms of metals are commonly packed as: (1) body-centered cubes, (2) face-centered cubes; or (3) hexagonally close-packed structures. The alkali metals, for example, are body-centered cubic, and nickel, copper, and aluminum are face-centered cubic; magnesium is hexagonal close-packed.

is illustrated in Figure 13.34b. Formula and structure are related by counting the net number of ions that define the unit cell and then finding the number of ions of opposite charge that are located in the lattice holes. Here we consider NaCl as a face-centered cubic lattice of Cl^- ions with smaller Na^+ ions in the octahedral lattice holes. A face-centered cubic lattice of atoms or ions of type X has a *net* of four X atoms or ions within the unit cell. Therefore, the NaCl lattice has four net Cl^- ions within the unit cell.

The Na^+ ions are in octahedral holes in the Cl^- lattice. Figure 13.34c shows that one octahedral hole occurs at the cube's center. Furthermore, three net

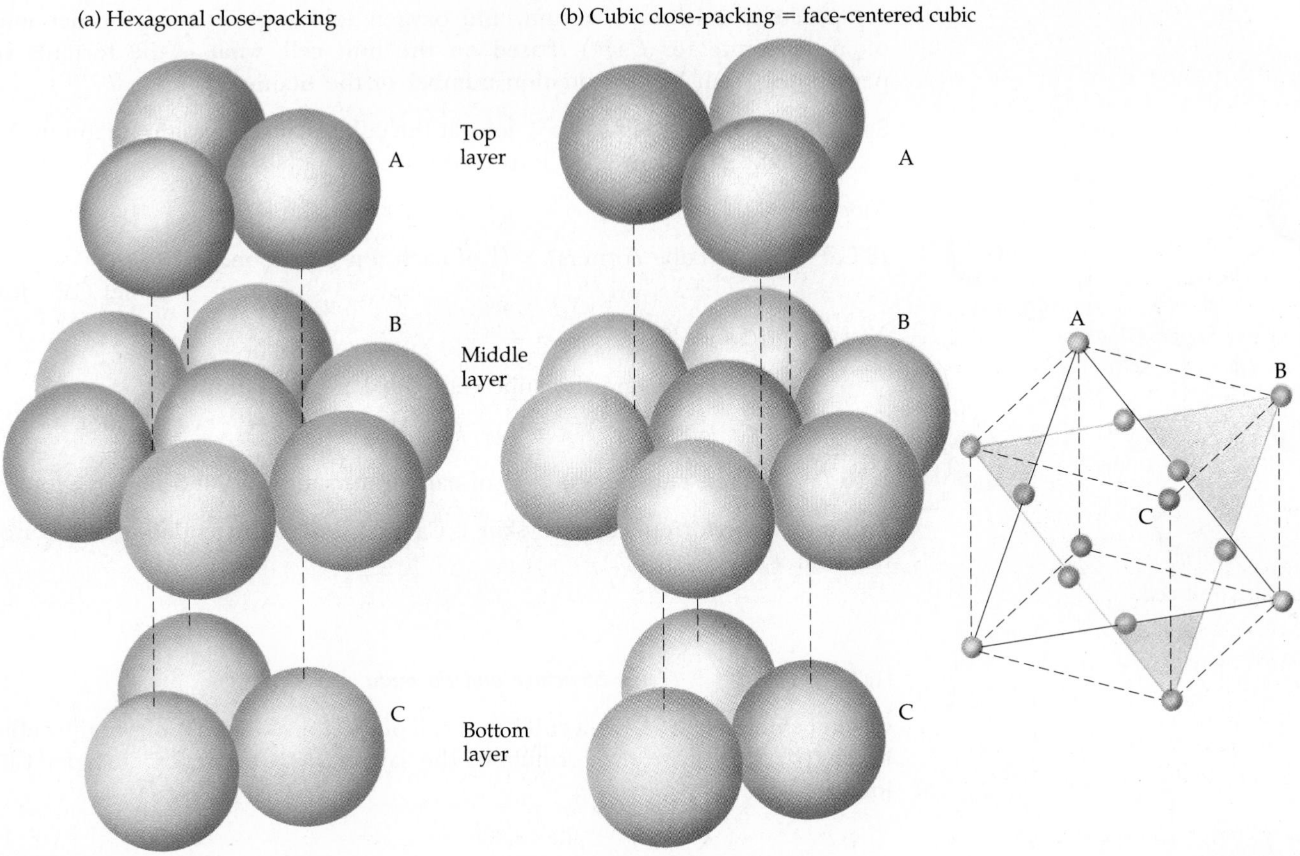

Figure B *The two most efficient ways to pack atoms or ions in crystalline materials. In both arrangements the packing within each layer is the hexagonal pattern in Figure A2. (a) In the* **hexagonal close-packed (hcp)** *arrangement, the layers placed above and below a given layer fit into the same depressions on either side of the middle layer. In a three-dimensional crystal, the layers repeat in their pattern in the manner ABABAB. . . . Atoms in each A layer are directly above the ones in another A layer; the same holds for the B layers. (b) In* **cubic close-packing (ccp),** *the atoms of "top" layer (A) rest in one set of depressions in the "middle" layer (B), and those of the "bottom" layer (C) rest in the remaining depressions. In a crystal, the pattern is repeated ABCABCABC. . . . By turning the whole crystal, you can see that the ccp arrangement is just the face-centered cubic structure.*

octahedral holes are accounted for by Na^+ ions in the cube's edges. We arrive at this number because there are 12 cube edges, and one fourth of each ion sharing an edge is *within* the unit cell being considered. Thus, the total number of octahedral holes, each occupied by a Na^+ ion in NaCl, is

$$(1 \text{ octahedral hole at cube center}) + (12 \text{ edges})(\tfrac{1}{4} \text{ octahedral hole per edge}) = 4 \text{ octahedral holes}$$

The number of Cl^- lattice ions and the number of Na^+ ions in octahedral holes is both 4, so a unit cell of NaCl has a 1:1 ratio of Na^+ and Cl^- ions as required.

EXAMPLE 13.7 *Ionic Structure and Formula*

One unit cell of the mineral perovskite is illustrated here. The compound is composed of calcium, titanium, and oxygen (although in nature other ions often substitute for Ca^{2+}). Based on the unit cell, what is the formula of perovskite? What is the oxidation number of the titanium ion?

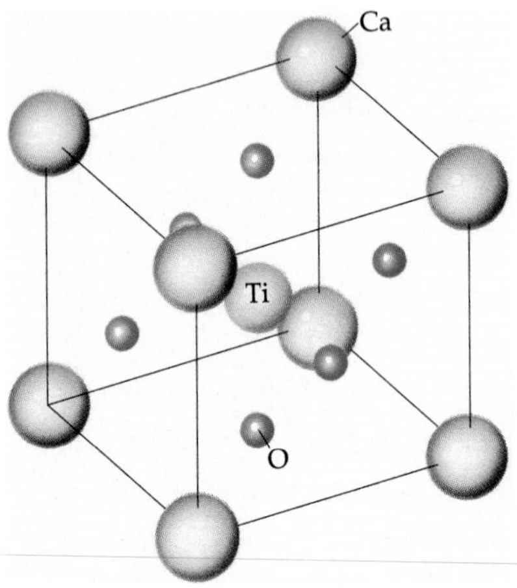

Perovskite unit cell.

Solution The unit cell has Ca^{2+} ions at the cube corners, a titanium ion in the center of the cell, and oxide ions in the face centers.

Number of Ca^{2+} ions:

(8 Ca^{2+} ions at cube corners) × ($\frac{1}{8}$ of each ion inside unit cell)
= 1 net Ca^{2+} ion

Number of titanium ions:

1 ion is in the cube center = 1 net titanium ion

Number of O^{2-} ions:

(6 O^{2-} ions in cube faces) × ($\frac{1}{2}$ of each ion inside cell) = 3 net O^{2-} ions

This means the formula of perovskite is $CaTiO_3$, and so the oxidation number of titanium is +4.

EXERCISE 13.8 *Ionic Structure and Formula*

Cesium chloride, CsCl, has a cubic unit cell of Cl^- ions with Cs^+ ions in the cubic holes (p. 628). Prove the formula of the salt must have one Cs^+ ion per Cl^- ion. ■

In summary, compounds with the formula MX often occur in either of two common crystal structures: (1) M^{n+} ions occupying *all* of the cubic holes of a simple cubic X^{n-} lattice, or (2) M^{n+} ions in all of the octahedral holes in a face-centered cubic X^{n-} lattice.* CsCl is a good example of (1) (Exercise 13.8), and NaCl is the usual example of (2). Indeed, chemists and geologists in particular have observed that the sodium chloride, or "rock salt," structure is adopted by many classes of compounds, most especially by all the alkali metal halides (except CsCl, CsBr, and CsI), all the oxides and sulfides of the alkaline earth metals, and all the oxides of formula MO of the transition metals of the fourth period. Finally, the formulas of compounds in general must be reflected in the structure of their unit cells (Example 13.7); therefore, the formula can always be derived from the unit cell structure.

Different atoms or ions have different radii (see Chapter 8). Because metallic or ionic solids form by packing atoms or ions as closely as possible, the size of the unit cell and the density of the solid depend on the radii of the atoms or ions involved.

*Although a metal may have a body-centered cubic structure, ionic solids do *not* take this form. Some may say, incorrectly, that CsCl is body-centered cubic, but it is more properly a simple cubic lattice of Cl^- ions with Cs^+ ions in the cubic holes. The adjectives "simple," "body-centered," and "face-centered" tell only the arrangement of one ion type, *not both.*

EXAMPLE 13.8 *Calculating the Density of an Ionic Compound from Unit Cell Dimensions*

The radius of the Na^+ ion is 116 pm, and the radius of Cl^- is 167 pm. From these radii, and the knowledge of the structure of NaCl, calculate the density of salt in grams per cubic centimeter.

Solution The density of a solid is the mass of a unit cell divided by the volume of the cell. Let us first find the volume of the unit cell and then its mass.

Step 1. Calculate the volume of the NaCl unit cell.

One face of the face-centered NaCl unit cell appears as shown in the margin. The Cl^- ions define the lattice, and the Na^+ and Cl^- ions along each edge just touch one another. This means that one edge of the unit cell is equal to one Cl^- ion radius plus twice the radius of Na^+ plus another Cl^- ion radius, or

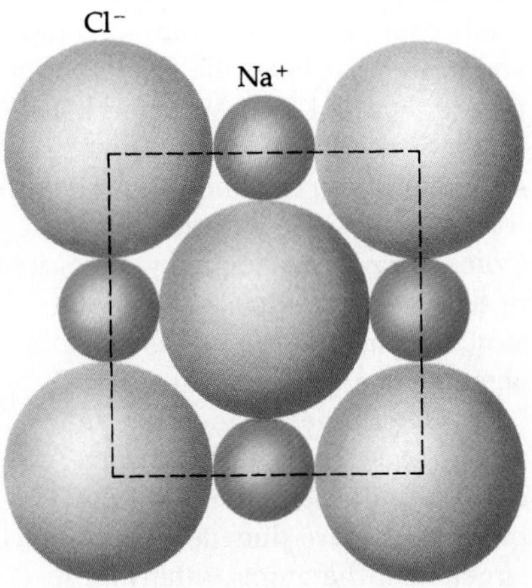

One face of the unit cell of NaCl.

$$\text{NaCl unit cell edge} = 167 \text{ pm} + 2(116 \text{ pm}) + 167 \text{ pm} = 566 \text{ pm}$$

Because the crystal is cubic, the volume of the unit cell is the cube of the edge.

$$\text{Volume of unit cell} = (\text{edge})^3 = (566 \text{ pm})^3 = 1.81 \times 10^8 \text{ pm}^3$$

Converting this to cubic centimeters, we have

$$\text{Volume in cubic centimeters} = 1.81 \times 10^8 \text{ pm } (10^{-10} \text{ cm/pm})^3 = 1.81 \times 10^{-22} \text{ cm}^3$$

Step 2. Calculate the mass of the NaCl unit cell.

As outlined previously in the text, the unit cell contains four Na^+ and four Cl^- ions. We can obtain the mass of one formula unit of NaCl from the molar mass of NaCl and Avogadro's number.

$$\left(\frac{58.44 \text{ g}}{1 \text{ mol NaCl}}\right) \times \left(\frac{1 \text{ mol NaCl}}{6.022 \times 10^{23} \text{ formula units}}\right) = 9.704 \times 10^{-23} \text{ g/formula unit}$$

Because there are four NaCl "formula units" per unit cell, the cell has a mass of

$$\left(\frac{9.704 \times 10^{-23} \text{ g}}{\text{formula unit}}\right) \times \left(\frac{4 \text{ NaCl formula units}}{1 \text{ unit cell}}\right) = 3.882 \times 10^{-22} \text{ g/unit cell}$$

Step 3. Calculate the density of NaCl from the mass and volume of a unit cell.

$$\text{Density} = \left(\frac{3.882 \times 10^{-22} \text{ g}}{1 \text{ unit cell}}\right) \times \left(\frac{1 \text{ unit cell}}{1.81 \times 10^{-22} \text{ cm}^3}\right) = 2.14 \text{ g/cm}^3$$

The experimental density is 2.164 g/cm^3.

EXERCISE 13.9 *Density from Unit Cell Dimensions*

KCl crystallizes in the same crystal structure as NaCl. Using the ion sizes in Figure 8.15, calculate the density of KCl. ■

A CLOSER LOOK — *Using X-Rays to Determine Crystal Structure*

How do chemists determine the distance between the atoms in a metal, the ions in a salt, or the atoms of a molecule? In 1912, Max von Laue, a German physicist, found that crystalline solids could be used to diffract x-rays, and, somewhat later, the English scientists (father and son) William H. and W. Lawrence Bragg showed that x-ray diffraction by crystalline solids could be used to determine distances between atoms. **X-ray crystallography,** the science of determining atomic-scale crystal structures, is now used extensively by chemists.

To determine distances on an atomic scale we have to use some type of probe to locate atoms or ions. The probe must have dimensions not much larger than the atoms, otherwise it would pass over them unperturbed. The probe must therefore have a size of only a few picometers; x-rays, high-energy photons with a wavelength in this range, meet the requirement. Crystalline solids are used because the atoms or molecules must be immobilized so they can be located with x-rays and because the ions, atoms, or molecules of the crystal must be arranged in an orderly manner.

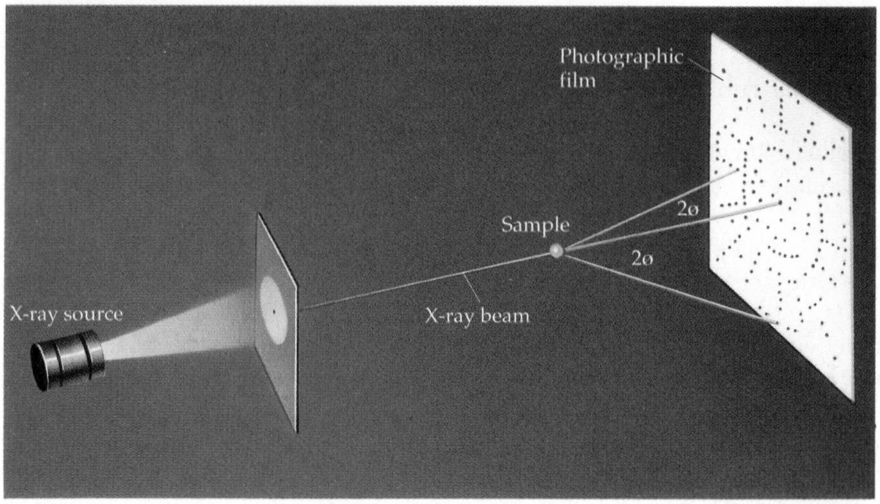

Figure A *In the x-ray diffraction experiment a beam of x-rays is directed at a crystalline solid. The photons of the x-ray beam are scattered by the atoms of the solid. The angle of scattering depends on the locations of the atoms in the crystal. The scattered x-rays are detected by a photographic film or an electronic detector.*

To "see" radiation interact with atoms we must rely on a change in the radiation as its passes through the crystal. The change is observed as a scattering, or "diffraction," of the photons of radiation from their original path (Figure A). (The x-ray experiment is similar in principle to the rainbow effect you see if you hold a compact disc at the correct angle in the light. The disc has a series of closely spaced lines on its surface, and light is diffracted as it strikes the grooves.) To determine the location of the scattered x-ray photons,

PROBLEM-SOLVING TIPS AND IDEAS

13.1 A Summary of Cubic Unit Cells

It is important to know the number of atoms, ions, or formula units in each type of unit cell.

- A *simple cube* of atoms or ions always contains 1 net atom or ion.
- A *body-centered cubic unit cell* of atoms always contains 2 net atoms within the cell.
- A *face-centered cubic unit cell* of X atoms (or ions) always contains 4 *net* X atoms (or ions) within the cell.

Finally, you are reminded that the body-centered cube is not adopted by ionic compounds. The CsCl lattice is a simple cube of Cl^- ions (or Cs^+ ions) with a Cs^+ ion (or Cl^- ion) in the center of the cube. ■

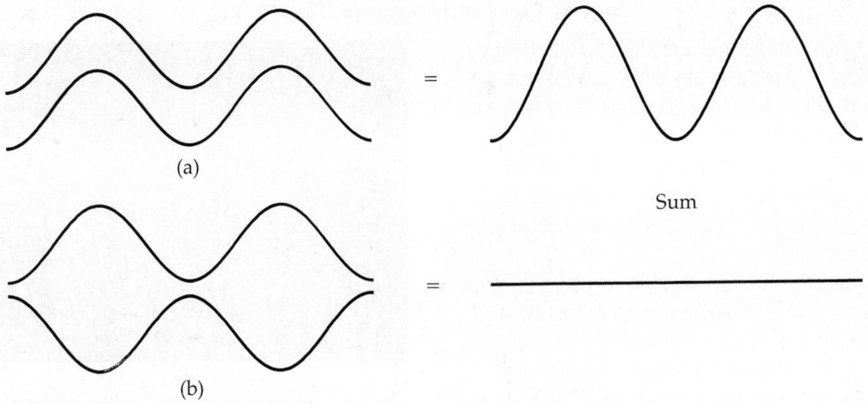

Figure B *(a) Constructive interference occurs when two in-phase waves combine to produce a wave of greater amplitude. (b) Destructive interference results from the combination of two waves of equal magnitude that are exactly out of phase. The result is zero amplitude.*

a photographic film or electronic detector is placed around the crystal. When an x-ray photon strikes an atom, the energy of the photon causes the atom's electrons to oscillate so the atom becomes like a radio station antenna, rebroadcasting the photon to the detector as the electron oscillations die out.

X-ray crystallography structural information depends on the fact that crystals scatter x-rays depending on the locations of the atoms within the crystal. To understand this property consider the wave properties of photons. You know that waves have peaks and valleys at different points in space. If two waves meet at some point, and the peak of one wave meets the valley of another, the waves cancel each other at that point. If, however, the waves meet peak to peak, they reinforce each other, and, if this occurs at the surface of the detector, the radiation produces a detectable signal or points on a photographic film. In contrast, at points on the detector where scattered waves meet and cancel, no signal is detectable (Figure B).

For the photon waves to meet peak to peak at some point on the detector, they must leave the atoms of the crystal with synchronized oscillations, that is, with their peaks and valleys in unison. This condition is met when the x-rays are scattered at special values of the angle θ (Figure A), an angle related to the distances between atoms in the solid.

The experiment as it is really done is of course more complicated than we have described. Nonetheless, with modern instruments and computers, chemists and physicists can usually determine quite readily the location of atoms in a crystal and the distances between them. Indeed, the technique has provided so much structural information during the past 20 years or so that the science of chemistry has itself been revolutionized. Many of the structural models you have seen in this book are based on the results of x-ray crystallography and related techniques such as electron diffraction.

PROBLEM-SOLVING TIPS AND IDEAS

13.2 Calculations with Unit Cells

In Example 13.5 we used the density of a solid, and a knowledge of the cell type, to find the atomic radius.

Mass of 1 atom or 1 formula unit —(× Number of formula units per unit cell)→ Mass of unit cell —(Divide by density)→ Volume of unit cell → Dimensions of cell or atom or ion radius

In Example 13.6 this procedure was reversed. We verified the cell type from the density of the solid and the atomic radius. The point is that density is often the key piece of information.

$$\text{Density} = \frac{\text{mass}}{\text{volume}}$$

PORTRAIT OF A SCIENTIST

Dorothy Crowfoot Hodgkin.

Dorothy Crowfoot Hodgkin (1910–1994)

A recent biography of Dorothy Crowfoot Hodgkin said that "more than any other scientist, she personified the transformation of crystallography from a black art into an indispensable scientific tool. . . . She made not one brilliant breakthrough but a series of them, deciphering the structure of one medically important substance after another."

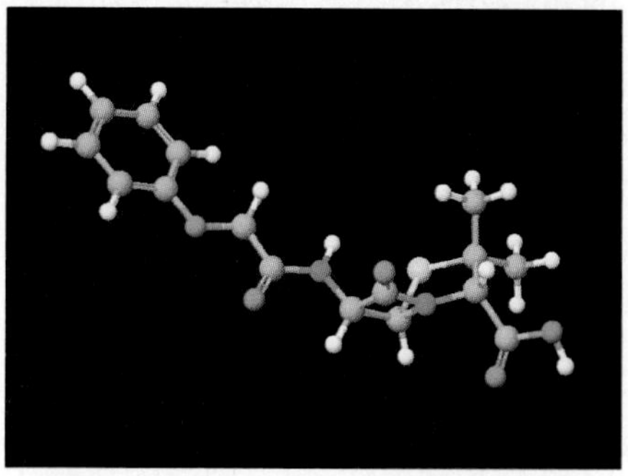

The structure of penicillin V. Carbon = gray; hydrogen = white; oxygen = red; nitrogen = blue; and sulfur = yellow.) (Susan Young)

Dorothy Crowfoot Hodgkin was born in Egypt, then a British colony, where her father was a supervisor of schools and ancient monuments. She went to school in England, however, and graduated from Oxford University in 1931. (At the time Oxford University had only one woman student for every five men.) She was fascinated by the structures of crystals, and so went to Cambridge University where she worked with John D. Bernal of the Mineralogical Institute. After only a year, Oxford invited her back as a chemistry instructor, and she stayed there until her retirement in 1977.

Her first major achievement was the determination of the structure of penicillin, which was completed in 1945. After World War II she began the work that earned her the 1964 Nobel Prize in chemistry, the determination of the structure of vitamin B_{12}, the factor used to treat pernicious anemia. She also attempted some of the first studies of viruses. Another of her major accomplishments, and one which occupied much of her life, was the determination of the structure of insulin.

Having this information we therefore can identify several typical calculations in solid-state chemistry.

- Knowledge of the unit cell type and composition (which gives unit cell mass), in addition to the density, provides the unit cell volume, from which atomic or ionic radii can be found (Example 13.5).
- Knowledge of the density, cell type, and atomic or ionic radii (which gives the unit cell volume) provides the mass of the unit cell. This can be used to verify the cell type (as in Example 13.6).
- Knowledge of the unit cell type and composition (which gives unit cell mass) as well as atomic or ionic radii (which gives cell volume) provides the density of the solid (as in Example 13.8). ■

13.5 MOLECULAR AND NETWORK SOLIDS

Thus far we have described the structures of metals and of simple ionic solids. But many other types of solids exist, a number of which are important in the revolution in new materials that has taken place during the past decade. We describe two of them: molecular and network solids.

Molecular Solids

Covalently bonded molecules, such as H_2O, I_2, CO_2, and many others condense to form solids at temperatures for which the intermolecular forces are sufficiently strong to keep the molecules in place. The structure of ice is particularly interesting and was described in Section 13.2 (Figure 13.10).

Recently, there has been a great deal of interest in the solid-state structure of C_{60}, the compound known as the buckyball (Section 3.1), in part because the spherical C_{60} molecules can only be bound to one another in the solid state by simple dispersion forces. Figure 13.35 is a scanning tunneling microscope (STM) image of C_{60} packed onto the surface of GaAs at 470 K. The flat area at the left of the picture clearly shows close-packed C_{60} molecules with a face-centered cubic unit cell. (The length of the unit cell side is about 142 pm.)

The fact that buckyballs pack into a regular cubic unit cell has caught the imagination of some researchers at Sandia National Laboratories. It was found that gaseous molecules such as H_2 and O_2 fit neatly into the holes in the close-packed C_{60} lattice. Under pressure, as many as half of the available holes are filled. When the pressure is released, the gas oozes out slowly. This has led to the notion that solid C_{60} could be used as a gas-storage device or as a selective gas filter. The smaller atmospheric gases such as O_2 and N_2 could be separated from larger, gaseous molecules such as CH_4 because only small gases are absorbed by C_{60}.

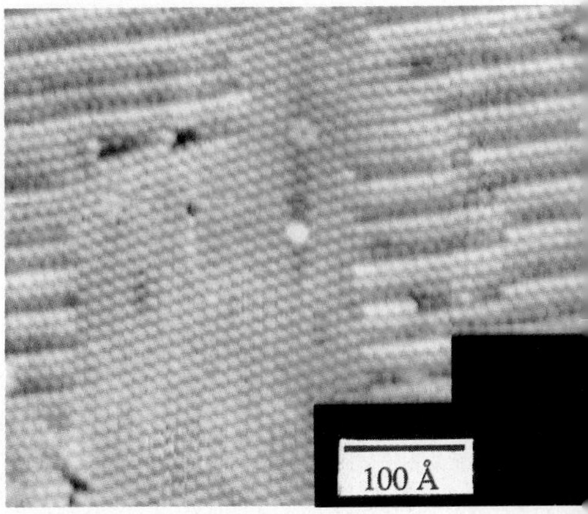

Figure 13.35 An STM image of close-packed C_{60} molecules. The flat region at the left is the close-packed surface of the face-centered lattice of C_{60} molecules. The C_{60} molecules are arranged in a face-centered unit cell with an edge dimension of 142 pm. (Courtesy John H. Weaver, University of Minnesota, *Science,* Vol. 253, pp 429–433, July 26, 1991)

Network Solids

A number of solids are composed of networks of covalently bonded atoms; two excellent examples, **graphite** and **diamond,** were described in Section 3.1 (Figures 3.5 and 13.36).

Diamonds have a low density ($d = 3.51\ g/cm^3$), but they are also the hardest material and the best conductor of heat known. They are transparent to visible light and infrared and ultraviolet radiation. They are electrically insulating but behave as semiconductors with some advantages over silicon. What more could a scientist want in a material—except a cheap, practical way to make it!

In the 1950s, scientists at General Electric in Schenectady, New York, achieved something alchemists had attempted for centuries, the synthesis of diamonds from carbon-containing materials, including wood or peanut butter. Their technique is to heat graphite to a temperature of 1500 °C in the presence of a metal, such as nickel or iron, and under a pressure of 50,000 to 65,000 atm. Under these conditions, the carbon dissolves in the metal and slowly forms diamonds. Over $500 million worth of diamonds are made this way annually, much of them used for abrasives and diamond-coated cutting tools.

The high-pressure synthesis of diamonds is expensive, and the diamonds are not entirely pure nor crystalline enough for making semiconductor devices. Even from the beginning, therefore, the search was on for better ways to make diamonds. The most promising method so far seems to be a low-pressure process called "chemical vapor deposition" (CVD). A mixture of hydrogen and a carbon-containing gas such as methane (CH_4) is decomposed by heating to about 2200 °C (with microwaves or a hot wire). Carbon atoms deposit on a silicon plate or other material in the chamber and slowly build up a film of tiny diamonds (Figure 3.7).

Figure 13.36 Diamonds are the hardest material and the best conductor of heat known, are transparent to visible light and infrared and ultraviolet radiation, and can behave as semiconductors. (C. D. Winters)

Thus far the CVD process is much more expensive than the high-pressure process. The promise of CVD diamond thin films is so enormous, however, that development is rapid in the U.S., Japan, and the Soviet Union. Cutting tools with diamond thin films are being test marketed, and high-quality loudspeakers with diamond-coated tweeters are already on the market. A polycrystalline diamond heat sink for electronic devices is now on the market, and a company in Ohio has even reported that they can apply a diamond coating to sunglasses to make them scratch-resistant.

A recent National Research Council report said that "success in deposition of diamond coatings on a variety of substrates at practical growth rates is one of the most important technological developments in the past decade. The ultimate economic impact of this technology may well outstrip that of high-temperature superconductors." Some say that the development of synthetic diamond films is potentially the greatest advance in materials since the invention of plastics.

Silicates, compounds of silicon and oxygen, are another important class of network solids. You know them in the form of materials such as sand, quartz, talc, and mica, or as a major constituent of rocks such as granite (Figure 13.37). Another type of silicate, a zeolite, was mentioned as a "molecular filter" in the separation of organic compounds (page 432). The details of silicate structures are outlined in Section 22.5.

Amorphous and imperfect solids are other classes of solids. Amorphous solids include glass, also a silicate, and organic polymers (Section 11.9). Imperfect solids include semiconductors (Section 10.3).

13.6 THE PHYSICAL PROPERTIES OF SOLIDS

We know now that the outward shape of a crystalline solid is a reflection of its internal structure. But what about the temperatures at which solids melt, their hardness, or their solubility in water? All of these and many other physical properties of solids are also of interest to chemists, geologists, engineers, and others. A few such properties are explored here.

The **melting point** of a solid is the temperature at which the crystal lattice collapses and the solid is converted to liquid. Just as in the case of the liquid/vapor transformation, melting requires energy, an energy that in Chapter 6 we referred to as the **heat of fusion** (expressed in joules per gram) or that we can call the **enthalpy of fusion** (expressed in kilojoules per mole).

$$\text{Solid} \underset{\text{heat energy evolved by sample}}{\overset{\text{heat energy absorbed by sample}}{\rightleftharpoons}} \text{Liquid}$$

$$\text{Heat energy absorbed on melting} = \text{Enthalpy of fusion} = \Delta H_{\text{fusion}}\ (\text{kJ/mol})$$

$$\text{Heat energy evolved on freezing} = \text{Enthalpy of crystallization} = -\Delta H_{\text{fusion}}\ (\text{kJ/mol})$$

Enthalpies of fusion can range from just a few thousand joules per mole to many thousands per mole (Table 13.9). A low enthalpy of fusion certainly means that the solid melts at a low temperature, whereas high melting points reflect high enthalpies of fusion. Figure 13.38 shows the enthalpies of fusion of most of the metals of the periodic table relative to one another (see also Table 13.9). Just a few of the interesting features of the figure are

Figure 13.37 Naturally occurring silicates: clear quartz, sand, sheets of mica, light green talc, and sandstone. (C. D. Winters)

TABLE 13.9 Melting Points and Enthalpies of Fusion of Some Solids

Compound	Melting Point (°C)	Enthalpy of Fusion (kJ/mol)	Type of Intermolecular Forces
		Metals	
Hg	−39	2.29	Metal bonding (see Section 10.3)
Na	98	2.60	
Al	660	10.7	
Ti	1668	20.9	
W	3422	35.2	
		Molecular solids: Nonpolar molecules	
O_2	−219	0.440	Dispersion forces only
F_2	−220	0.510	
Cl_2	−102	6.41	
Br_2	−7.2	10.8	
		Molecular solids: Polar molecules	
HCl	−114	1.99	All three HX molecules have dipole-dipole forces enhanced by dispersion forces that increase with size and molar mass.
HBr	−87	2.41	
HI	−51	2.87	
H_2O	0	6.02	Hydrogen bonding
		Ionic solids	
NaF	996	33.4	All ionic solids have extended ion-ion interactions. Note the general trend is the same as that for lattice energies (see Section 8.7 and Figure 13.39)
NaCl	801	28.2	
NaBr	747	26.1	
NaI	660	23.6	

1. Metals that have notably low melting points, such as the alkali metals, also have very low enthalpies of fusion.
2. Mercury, a liquid at room temperature, has an enthalpy of fusion of only 2.29 kJ/mol
3. Metals of the third transition series, such as tungsten, have extraordinarily high enthalpies of fusion and very high melting points. In fact, tungsten has the highest melting point of all the known elements except for carbon, and the uses of tungsten reflect this. For example, pure tungsten is used as the filament—the glowing element—in light bulbs; no other material has been found to work better since the invention of light bulbs in 1908.

The melting temperature of a solid can convey a great deal of information. Table 13.9 gives you some data for several basic types of compounds: (1) metals, (2) polar and nonpolar, low-molecular-weight molecules, and (3) ionic solids. In general, low-molecular-weight, nonpolar substances that form molecular solids have low melting points. The melting point increases, however, as the size and molar mass increase because dispersion forces become stronger with increasing molar mass. Thus, increasing amounts of energy are required to break down the intermolecular forces in the solid, a principle reflected in an increasing enthalpy of fusion.

The melting point of an impure solid (one consisting of an intimate mixture of two or more components) is lower than that of a pure solid (Chapter 14). Therefore, if you measure the melting point of a solid, you can assess its relative purity. For this reason, melting point determinations are a common laboratory operation.

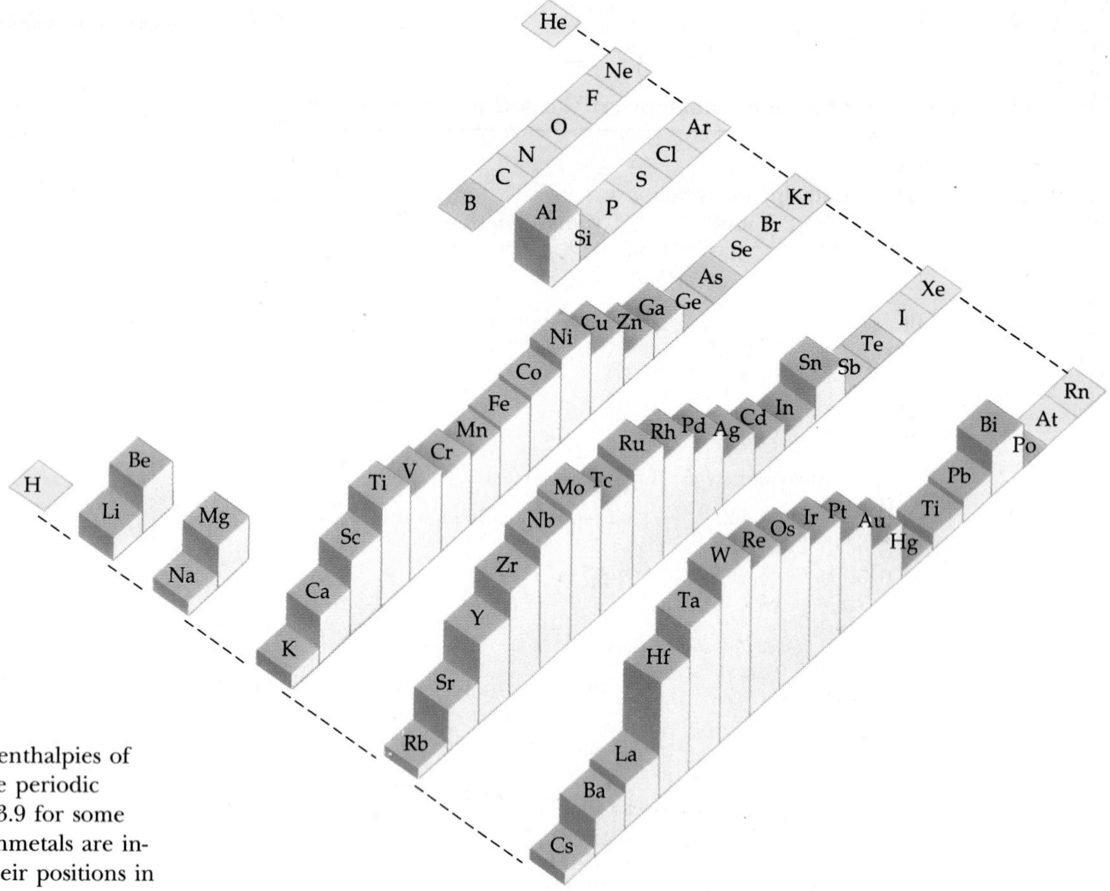

Figure 13.38 Relative enthalpies of fusion for metals in the periodic table. Refer to Table 13.9 for some values. (Symbols of nonmetals are included only to show their positions in the table.)

The lattice energy of an ionic solid is the energy evolved when separate ions in the gas phase come together to form a crystalline solid. See Section 8.7 and Table 8.6.

Ionic compounds always have higher melting points and heats of fusion than molecular solids (Table 13.9). This phenomenon is due to very strong ion-ion forces in ionic solids, forces that are reflected in high lattice energies (Section 8.7). Because ion-ion forces depend on ion size (as well as ion charge), there is a relationship between lattice energy and the position of the metal or halogen in the periodic table (Figure 13.39). That is, as the cation size increases from Li^+ to Cs^+, the lattice energy declines for compounds of a given halide ion. A similar decline occurs as the halide ion size increases from F^- to I^- when salts of a given cation are considered. As illustrated by the data in Table 13.9, the decline in lattice energy with increasing ion size is reflected in a decrease in melting point and enthalpy of fusion.

Molecules can escape directly from the solid to the gas phase by **sublimation** (Figure 13.40),

$$\text{Solid} \longrightarrow \text{Gas} \qquad \text{Heat energy required at constant pressure} = \Delta H_{\text{sublimation}}$$

an endothermic process. The heat energy required in this process is called the enthalpy of sublimation. Water, which has a molar enthalpy of sublimation of 51 kJ/mol, can be converted from solid ice to water vapor quite readily. One of the best examples of the use of this property is the frost-free refrigerator. During certain times, the freezer compartment is warmed slightly. Molecules of water freed from the surface of the ice (the vapor pressure of ice is 4.60 mm Hg at 0 °C) are removed in a current of air blown through the freezer. Other common substances that sublime are I_2 and Dry Ice, solid CO_2.

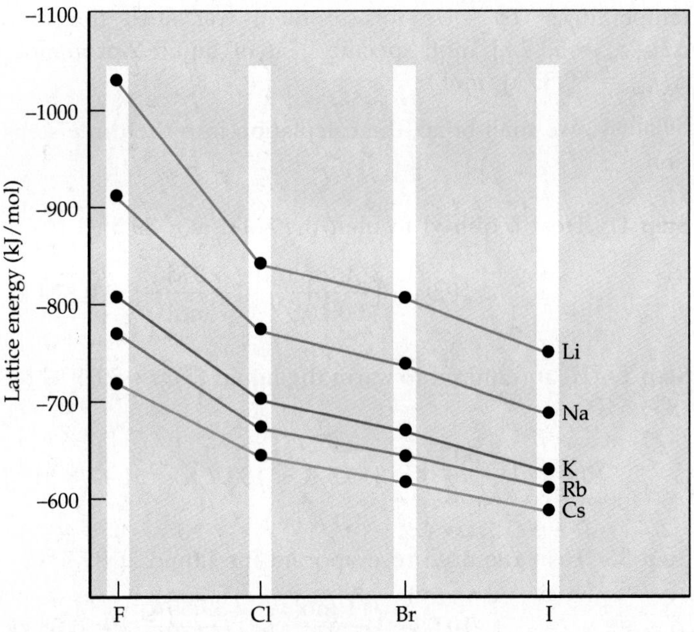

Figure 13.39 Trends in the lattice energies of the alkali metal halides.

Figure 13.40 Sublimation is the conversion of a solid directly to its vapor. Here naphthalene ($C_{10}H_8$) sublimes when heated with warm water. (C. D. Winters)

EXAMPLE 13.9 *Heat of Vaporization and Fusion*

Rubbing alcohol, or 2-propanol, is an organic compound that freezes at −89.5 °C and boils at 82.3 °C.

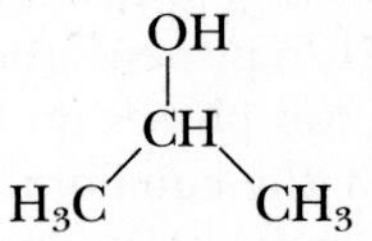

2-propanol, an alcohol

What quantity of heat is required to melt 10.0 g of 2-propanol at −89.5 °C, heat the resulting liquid to the boiling point, and then evaporate the liquid at this temperature? To solve this problem you need the following information: ΔH_{fusion} = 5.37 kJ/mol; specific heat of liquid 2-propanol = 2.60 J/g · K; and ΔH_{vap} = 39.85 kJ/mol.

Solution We shall break the calculation into the three steps in the transformation.

Step 1. Heat required to melt the solid at −89.5 °C.

$$10.0\ \text{g} \cdot \left(\frac{1\ \text{mol}}{60.10\ \text{g}}\right) \cdot \left(\frac{5.37\ \text{kJ}}{\text{mol}}\right) = 0.894\ \text{kJ}$$

Step 2. Heat required to warm the liquid from −89.5 °C (183.7 K) to +82.3 °C (355.5 K).

$$10.0\ \text{g} \cdot \left(\frac{2.60\ \text{J}}{\text{g} \cdot \text{K}}\right) \cdot (355.5\ \text{K} - 183.7\ \text{K}) = 4.47 \times 10^3\ \text{J, or } 4.47\ \text{kJ}$$

Step 3. Heat required to evaporate the liquid at 82.3 °C.

$$10.0\ \text{g} \cdot \left(\frac{1\ \text{mol}}{60.10\ \text{g}}\right) \cdot \left(\frac{39.85\ \text{kJ}}{\text{mol}}\right) = 6.63\ \text{kJ}$$

Step 4. Total heat required = 11.99 kJ

EXERCISE 13.10 ***Heat of Fusion***

Calculate the heat required to melt 100.0 g of water at its melting point. Compare this with that required to melt 100.0 g of the hydrocarbon octane (C_8H_{18}, melting point = 56.8 °C; ΔH_{fusion} = 20.65 kJ/mol). Is there a relation between the calculated heats and the types of intermolecular forces involved? ■

13.7 CHANGES IN STRUCTURE AND PHASE

When a solid is heated, several things can happen. The obvious possibility is that the solid can eventually melt to a liquid, and another is that it might sublime to give the vapor.

Phase Diagrams

Phase diagrams are used to illustrate the relation between phases of matter as pressure and temperature are changed. Each line in a phase diagram, such as the one for water in Figure 13.41, represents the conditions of T and P at which equilibrium exists between the two phases on either side of the line.

As described in Section 13.3, the equilibrium vapor pressure for water rises with temperature along the curve AD, and the vapor pressure reaches 760 mm Hg at 100 °C, the normal boiling point. As the temperature of a liquid decreases, so does its vapor pressure until point A, the **triple point,** is reached (at

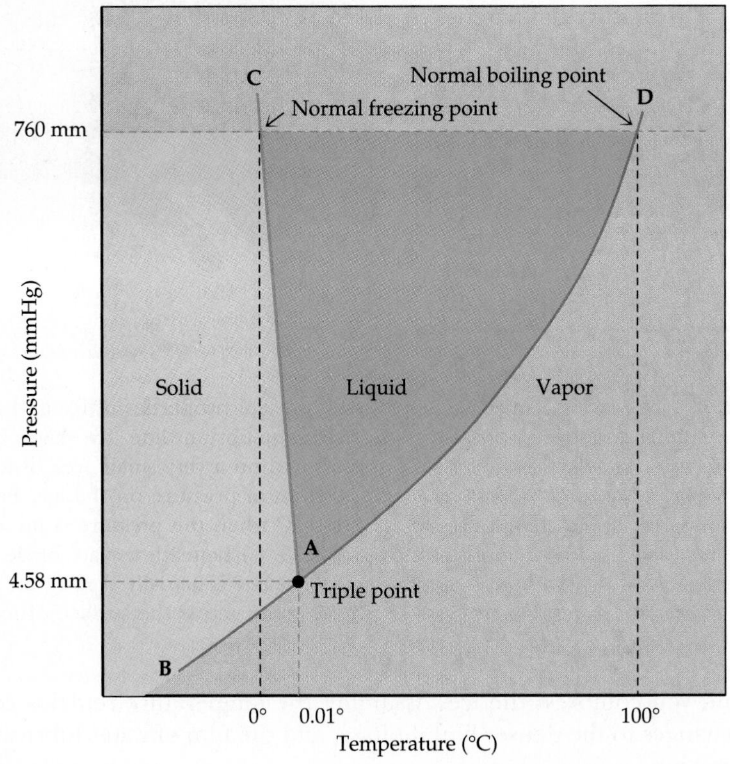

Figure 13.41 Phase diagram for water. Note that the scale is greatly exaggerated.

$P = 4.58$ mm Hg and $T = 0.01$ °C for water). Here, all three phases (solid, liquid, and vapor) are in equilibrium.

At temperatures and pressures below the triple point, solid water (ice) is in equilibrium with water vapor. The equilibrium vapor pressure of ice subliming to vapor at a given temperature is given by the line AB.

The curve AC shows the conditions of pressure and temperature at which solid/liquid equilibrium exists. (Because no vapor pressure is involved here, the pressure referred to is the external pressure on the liquid.) Water is highly unusual, because this line has a *negative slope.* That is, the higher the external pressure, the lower the melting point. The change for water is approximately 0.01 °C for each increase in pressure of 1 atm.

Of all the thousands of substances for which phase diagrams are known, only water, bismuth, and antimony have solid/liquid equilibrium lines with a negative slope.

The negative slope of the water solid/liquid equilibrium line can be explained from our knowledge of the structure of water and ice. When pressure increases on an object, common sense tells you the object will be forced into a smaller volume and thus become denser. Because liquid water is denser than ice (due to the open-lattice structure of ice), *ice and water in equilibrium respond to increased pressure (at constant T) by melting ice to form more water;* the same mass of water requires less volume. This property is illustrated in Figure 13.42 using a solid/liquid equilibrium line with a greatly exaggerated slope.

Figure 13.42 also helps to explain why you can skate on ice. When skating, your body weight is concentrated on the very small area below the skate blades. Therefore, your weight per unit area, the pressure you exert on the ice, is very

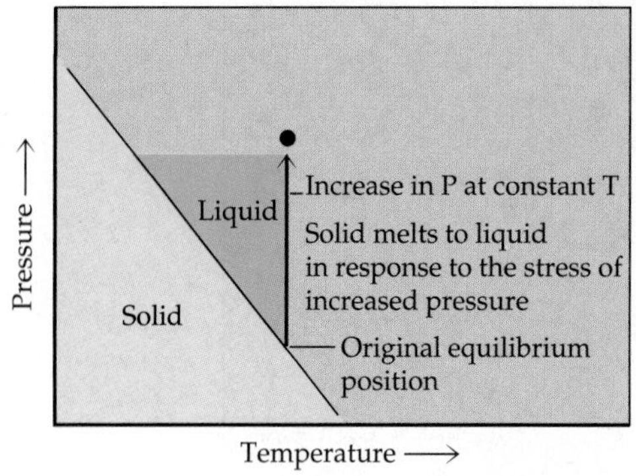

Figure 13.42 Ice skating is made easier by the unusual properties of ice and water, and the resulting negative slope of the ice/water equilibrium line. Ice skates have very thin blades, so a skater's weight is concentrated on a very small area of ice. Thus, a person of average weight can exert 500 atm of pressure on the ice. Because the melting point of ice changes by about −0.01 °C when the pressure is increased by 1 atm, the ice has a lower melting point (about −5 °C) beneath a skate blade. This effect, combined with frictional heating, means a skater is actually skating on a film of water under the skates, which helps the blade glide across the surface of ice.

large, and you compress the ice. Assuming the temperature remains constant, the ice changes to the denser liquid phase, and the film of water lubricates your sliding motion.

The basic features of the phase diagram for CO_2 (Figure 13.43) are the same as those of water, except the CO_2 solid/liquid equilibrium line has a *positive* slope. Solid CO_2 is denser than the liquid, so the solid sinks to the bottom in a container of liquid CO_2. Also notice that solid CO_2 sublimes directly to CO_2 gas

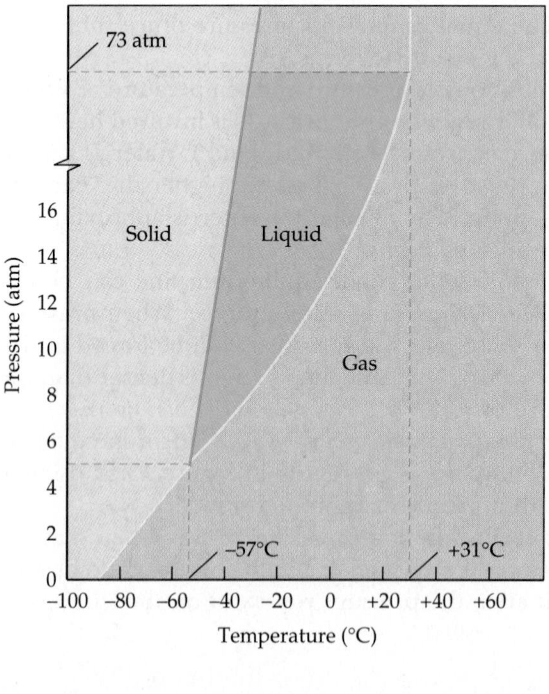

Figure 13.43 A phase diagram for carbon dioxide, showing the critical pressure and temperature.

at 1 atm pressure. For this reason CO_2 is called *Dry Ice;* it looks like water-ice but does not melt.

CHAPTER HIGHLIGHTS

Having studied this chapter, you should be able to

- Use the kinetic-molecular theory to define the difference in solids, liquids, and gases (Section 13.1).
- Describe the different **intermolecular interactions** in liquids and solids (Section 13.2).
- Identify when an **ion-dipole** interaction is likely (Section 13.2). Such forces depend on the distance between the ion and the dipole, the magnitude of the dipole, and the charge on the ion (Example 13.1).
- Tell when an ion is likely to be hydrated (that is, bound to polar water molecules) (Section 13.2).
- Tell when two molecules can interact through a **dipole-dipole** force of attraction or when **hydrogen bonding** may occur. The latter occurs most strongly when H is attached to O, N, or F (Section 13.2 and Example 13.2).
- Explain how **hydrogen bonding** affects the properties of water (Section 13.2).
- Know when the interaction between molecules can only involve **induced dipoles** (dispersion forces) (Section 13.2).
- Explain the processes of evaporation and condensation of a liquid or its vapor and use the **enthalpy,** or **heat, of vaporization** in calculations (Section 13.3).
- Define and use the concept of the **equilibrium vapor pressure** of a liquid and its relation to the boiling point of a liquid (Section 13.3).

  Evaporation of a liquid continues at a given temperature until a **dynamic equilibrium** is established at which point the rate of evaporation equals the rate at which vapor molecules condense, or reenter the liquid phase. At equilibrium, the pressure exerted by the vapor phase molecules is called the **equilibrium vapor pressure.** More **volatile** compounds have higher equilibrium vapor pressures at a given temperature. If the vapor pressure equals the atmospheric, or external, pressure, the liquid boils. When the external pressure is 1 atm, the temperature is the **normal boiling point.** The equilibrium vapor pressure of a liquid increases with temperature.
- Describe the phenomena of **critical temperature,** T_c, and **critical pressure,** P_c, of a substance (Section 13.3).
- Describe how intermolecular interactions affect the **cohesive forces** between identical liquid molecules, the energy necessary to break through the surface of a liquid **(surface tension),** and the resistance to flow, or **viscosity,** of liquids (Section 13.3).
- Characterize different types of solids: **metallic** (e.g., copper), **ionic** (e.g., NaCl and CaF_2), **molecular** (e.g., water and I_2), **network** (e.g., diamonds), and **amorphous** (e.g., glass and many synthetic polymers) (Sections 13.4 and 13.5 and Table 13.8).

- Describe the three types of cubic unit cells: **simple cubic** (sc), **body-centered cubic** (bcc), and **face-centered cubic** (fcc) (Section 13.4).

  All three types are utilized by metals, whereas only the sc and fcc arrangements occur for ionic compounds.
- Perform calculations that relate the characteristics of solids and their unit cells (density, cell dimensions, ion or atom radius, and unit cell type) (Section 13.4 and Examples 13.5, 13.6, and 13.8).
- Derive the formula of an ionic compound from its unit cell (Section 13.4 and Example 13.7).
- Define the **heat,** or **enthalpy, of fusion** and use this in a calculation (Section 13.6).
- Identify the different points (triple point, normal boiling point, freezing point) and regions (solids, liquid, vapor) of a **phase diagram,** and use the diagram to evaluate the vapor pressure of a liquid or the relative densities of liquid and solid (Section 13.6).

STUDY QUESTIONS

Review Questions

1. Name the types of forces that can be involved between two molecules and between ions and molecules.
2. Explain how a water molecule can interact with a molecule such as CO_2. What intermolecular force is involved?
3. Explain why intermolecular interactions can occur in an aqueous solution of copper(II) chloride.
4. Hydrogen bonding is most important when H is attached to certain very electronegative atoms. Which are those atoms?
5. Explain how hydrogen bonding leads to the decline in density of water from 4 °C to solid ice at 0 °C.
6. Explain why the specific heat of water is so large compared with many other liquids. How does this affect weather?
7. Explain why the special properties of ice allow you to skate on ice.
8. Why is the vapor pressure of water at 25 °C (23.8 mm Hg) so much lower than the vapor pressure of methanol, CH_3OH, at 25 °C (127 mm Hg)?
9. Explain, in terms of intermolecular forces, the trends in boiling points of the molecules EH_3, where E is a Group 5A element (see Figure 13.8).
10. Explain why the viscosity of an oil is so much greater than the viscosity of liquid benzene, C_6H_6.
11. What are the three types of cubic unit cells? Explain their similarities and differences.
12. Sketch the body-centered cubic unit cell of potassium metal. What is the net number of potassium atoms within the unit cell?
13. Define the lattice energy of an ionic solid. How is it related to the size of the ions and their charge? How is it related to the melting point of an ionic solid?
14. Sketch the phase diagram for water. Label the normal boiling point, melting point, and triple point, and show what regions of temperature and pressure are appropriate to solid, liquid, and vapor.
15. Explain why the solid/liquid equilibrium line in the water phase diagram has a negative slope.

Intermolecular Forces

16. When KCl dissolves in water, what type of attractive forces must be overcome in the liquid water? What type of forces must be overcome in the solid KCl? What type of attractive forces cause KCl to be able to dissolve in liquid water?

17. What intermolecular force(s) must be *overcome* to
    (a) Melt ice
    (b) Melt solid I_2
    (c) Remove the water of hydration from $MnCl_2 \cdot 4\,H_2O$.
    (d) Convert liquid NH_3 to NH_3 vapor

18. One example of a hydrated salt is $NiSO_4 \cdot 6\,H_2O$. What kind of attractive force is responsible for binding the water to the nickel ions of nickel sulfate?

19. When I_2 dissolves in methanol, CH_3OH, what type of forces must be overcome within the solid I_2 to allow it to dissolve? What type of forces must be disrupted between CH_3OH molecules when I_2 dissolves? What type of forces exist between I_2 and CH_3OH molecules in solution?

20. Describe the intermolecular force that must be overcome in converting each of the following from a liquid to a gas:
(a) Liquid O_2
(b) Mercury
(c) CH_3I (methyl iodide)
(d) CH_3CH_2OH (ethanol)

21. Describe the intermolecular force that must be overcome in converting each of the following from a liquid to a gas:
(a) CO_2
(b) NH_3
(c) $CHCl_3$
(d) CCl_4

22. Rank the following in order of increasing strength of intermolecular forces in the pure substances. Which do you think might exist as gases at 25 °C and 1 atm: Ne, CH_4, CO, and CCl_4?

23. Rank the following in order of increasing strength of intermolecular forces in the pure substances. Which do you think might exist as gases at 25 °C and 1 atm: $CH_3CH_2CH_2CH_3$ (butane), CH_3OH (methanol), and He?

24. Explain why the boiling point of H_2S is lower than that of water.

25. Tell which member of each of the following pairs of compounds has the higher boiling point:
(a) O_2 or N_2
(b) SO_2 or CO_2
(c) HF or HI
(d) SiH_4 or GeH_4

26. The normal boiling point of CH_3Cl (chloromethane) is −24 °C, whereas that for CH_3I is 42.4 °C. Which compound has the stronger intermolecular forces in the liquid phase? What type of intermolecular forces are involved?

27. Consider the following four compounds: (a) SCl_2; (b) NH_3; (c) CH_4; (d) CO. Place the four compounds in order of increasing boiling point.

28. Which of the following compounds would be expected to form intermolecular hydrogen bonds in the liquid state?
(a) CH_3OCH_3 (dimethyl ether)
(b) CH_4
(c) HF
(d) CH_3CO_2H (acetic acid)
(e) Br_2
(f) CH_3OH (methanol)

29. Which of the following compounds would be expected to form intermolecular hydrogen bonds in the liquid state?
(a) H_2Te
(b) HCO_2H (formic acid)
(c) HI
(d)

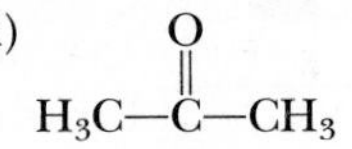

acetone

30. Which compound in each of the following pairs of salts is more likely found with a hydrated cation? Briefly explain your reasoning in each case.
(a) LiCl or CsCl
(b) $NaNO_3$ or $Mg(NO_3)_2$
(c) RbCl or $NiCl_2$

31. When salts of Mg^{2+}, Na^+, and Cs^+ are placed in water, the positive ion is hydrated (as is the negative ion). Which ion is most strongly hydrated? Which ion is least strongly hydrated?

Liquids

32. The enthalpy of vaporization of liquid mercury is 59.11 kJ/mol. What quantity of heat is required to vaporize 0.500 mL of mercury at 357 °C, its normal boiling point? (The density of mercury is 13.6 g/mL.)

33. Ethanol, CH_3CH_2OH, has a vapor pressure of 59 mm Hg at 25 °C. What quantity of heat energy is required to evaporate 125 mL of the alcohol at 25 °C? The enthalpy of vaporization of the alcohol at 25 °C is 42.32 kJ/mol. The density of the liquid is 0.7849 g/mL.

34. Answer the following questions using Figure 13.21:
(a) What is the equilibrium vapor pressure of diethyl ether at room temperature (approximately 20 °C)?
(b) Place the three compounds in Figure 13.21 in order of increasing intermolecular forces.
(c) If the pressure in a flask is 400 mm Hg, and if the temperature is 40 °C, which of these three compounds are liquids and which are gases?

35. Answer the following questions using Figure 13.21:
(a) What is the approximate equilibrium vapor pressure of water at 60 °C? Compare your answer with the data in Appendix E.
(b) At what temperature does water have an equilibrium vapor pressure of 600 mm Hg?
(c) Compare the equilibrium vapor pressures of water and ethanol at 70 °C.

36. Assume you seal 0.1 g of diethyl ether (Figure 13.21) in an evacuated 100.-mL flask. (There are no molecules of any other gas in the flask.) If the flask is held at 30 °C, what is the approximate gas pressure in the flask? If the flask is placed in an ice bath, does additional liquid ether evaporate or does some ether condense to a liquid?

37. Refer to Figure 13.21 as an aid in answering these questions:
(a) You put some water at 60 °C in a plastic milk carton and seal the top very tightly so gas cannot enter or leave the carton. What happens when the water cools?
(b) If you put a few drops of liquid diethyl ether on your hand, does it evaporate completely or remain a liquid?

38. Vapor pressure curves for CS_2 (carbon disulfide) and CH_3NO_2 (nitromethane) are drawn here.

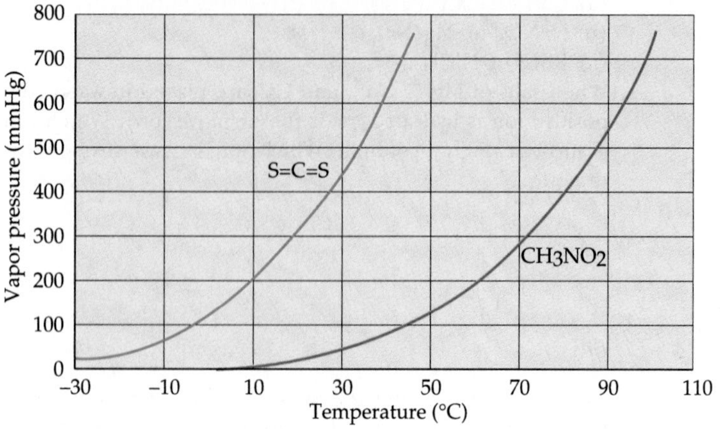

(a) What are the vapor pressures of CS_2 and CH_3NO_2 at 40 °C?
(b) What type of intermolecular forces exist in the liquid phase of each compound?
(c) What is the normal boiling point of CS_2? Of CH_3NO_2?
(d) At what temperature does CS_2 have a vapor pressure of 600 mm Hg?
(e) At what temperature does CH_3NO_2 have a vapor pressure of 60 mm Hg?

39. Answer each of the following questions with *increase, decrease,* or *not change.*
(a) If the intermolecular forces decrease in going from one liquid to another, the normal boiling point will ______.
(b) If the intermolecular forces increase in going from one liquid to another, the vapor pressure will ______.
(c) If the surface area of a liquid increases, the vapor pressure will ______.
(d) If the temperature of a liquid is decreased, the equilibrium vapor pressure will ______.

40. The simple hydrocarbons methane (CH_4) and ethane (C_2H_6) cannot be liquefied at room temperature, no matter how high the pressure. Propane (C_3H_8), the next compound in the series of simple alkanes, has a critical pressure of 42 atm and a critical temperature of 96.7 °C. Can this compound be liquefied at room temperature? Can you think of a place where you would find liquefied propane?

41. Carbon dioxide can be converted to a supercritical fluid. Its critical temperature and pressure are 304.2 K and 72.8 atm, respectively. Can carbon monoxide (T_c = 132.9 K; P_c = 34.5 atm) be liquefied at or above room temperature? Explain briefly.

Metallic and Ionic Solids

42. Outline a two-dimensional unit cell for the pattern shown here. If the black squares are labeled A and the white squares are B, what is the simplest formula for a "compound" based on this pattern? Assume this pattern extends infinitely.

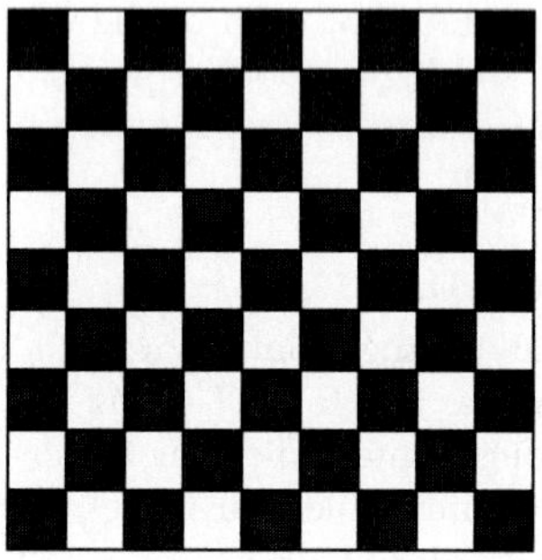

43. Outline a two-dimensional unit cell for the pattern shown here. If the black squares are labeled A and the white squares are B, what is the simplest formula for a "compound" based on this pattern? Assume this pattern extends infinitely.

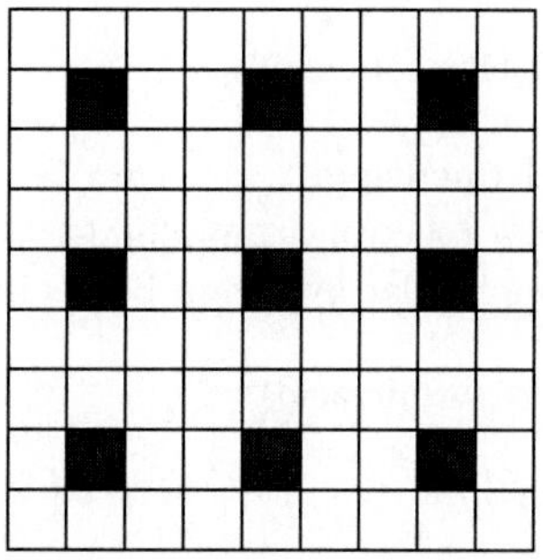

44. Calcium metal crystallizes in a face-centered cubic unit cell. The density of the solid is 1.54 g/cm^3. What is the radius of a calcium atom?

45. Solid zinc has a simple cubic unit cell. The density of zinc is 7.14 g/cm^3. Use this information to calculate the radius of a zinc atom.

46. Vanadium metal has a density of 6.11 g/cm^3. Assuming the vanadium atomic radius is 132 pm, is the vanadium unit cell simple cubic, body-centered cubic, or face-centered cubic?

47. The density of copper metal is 8.95 g/cm^3. If the radius of a copper atom is 127.8 pm, is the copper unit cell simple cubic, body-centered cubic, or face-centered cubic?

48. The metal hydride LiH has a density of 0.77 g/cm^3. The edge of the unit cell is 408.6 pm. If it is assumed that the H^- ions define the lattice points, does the compound have a face-centered cubic or a simple cubic unit cell?

49. Thallium(I) chloride, TlCl, crystallizes in either a simple cubic or a face-centered cubic unit cell of Cl^- ions with Tl^+ ions in the lattice holes. The density of the solid is 7.00 g/cm^3, and the edge of the unit cell is 385 pm. What is the unit cell geometry?

50. One way of viewing the unit cell of perovskite was illustrated in Example 13.7. Another way is shown here. Prove that this view also leads to a formula of $CaTiO_3$.

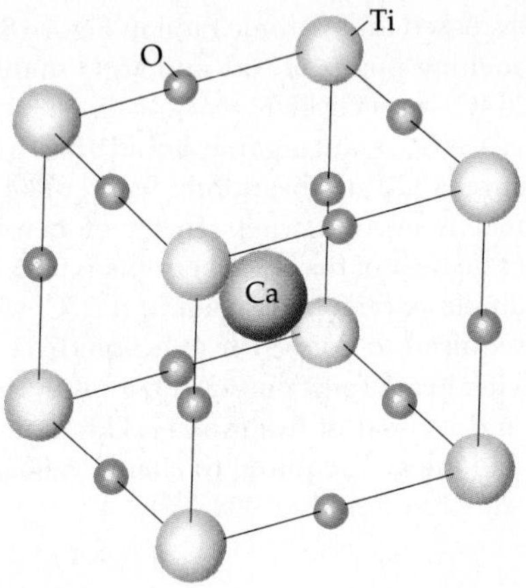

51. Rutile, TiO_2, crystallizes in a structure characteristic of many other ionic compounds. How many formula units of TiO_2 are in the unit cell illustrated here? (The oxide ions marked by an x are wholly within the cell; the others are in the cell faces.)

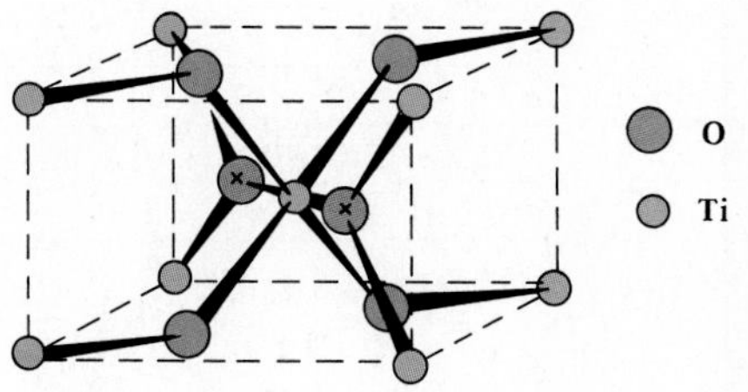

52. Based on the unit cell pictured here, what is the formula of zinc blende? Zinc blende is the main source of zinc. Many other compounds crystallize in this structure, including many important semiconductors such as GaAs.

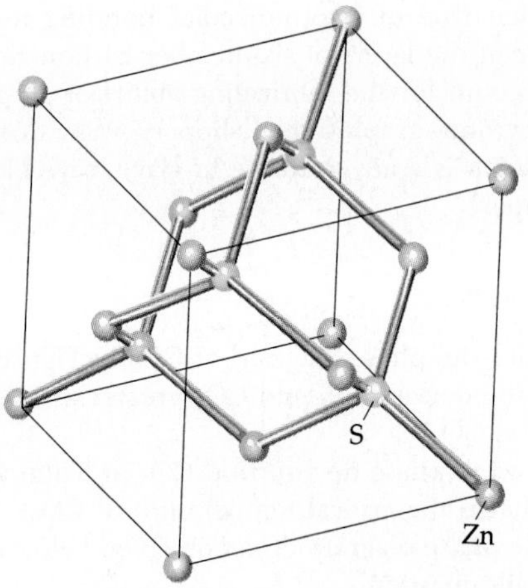

53. Cuprite is a semiconductor. Oxide ions are at the cube corners and in the cube center. Copper ions are wholly within the unit cell. What is the formula of cuprite? What is the oxidation number of copper?

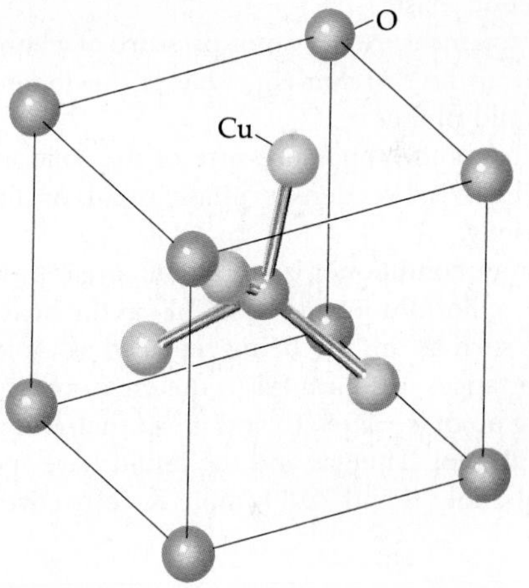

Molecular and Network Solids

54. A picture of the structure of the diamond unit cell is shown here. How many carbon atoms are in one unit cell? If the density of diamond is 3.51 g/cm^3, what is (a) the volume of the unit cell and (b) the length of the edge?

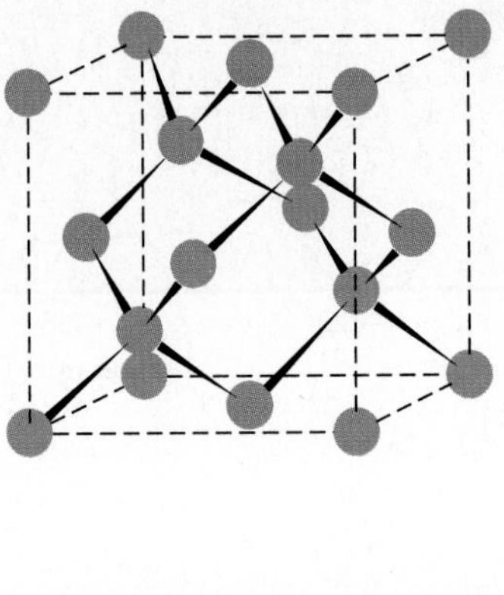

55. Figure 3.5 is a drawing of graphite's structure.
   (a) What type of intermolecular bonding forces exist between the layers of six-member carbon rings?
   (b) Account for the lubricating ability of graphite. That is, why does graphite feel slippery? Why does pencil lead (which is really graphite in clay) leave black marks on paper?

Phase Changes

56. Consider the phase diagram of CO_2 in Figure 13.43.
   (a) Is the density of liquid CO_2 greater or less than that of solid CO_2?
   (b) In what phase do you find CO_2 at 5 atm and 0 °C?
   (c) What is the critical temperature of CO_2?

57. Use the phase diagram of xenon given below to answer the following questions:
   (a) In what phase is xenon found at room temperature and 1.0 atm pressure?
   (b) If the pressure exerted on a xenon sample is 0.75 atm, and the temperature is −114 °C, in what phase does the xenon exist?
   (c) If you measure the vapor pressure of a liquid xenon sample to be 380 mm Hg, what is the temperature of the liquid phase?
   (d) What is the vapor pressure of the solid at −122 °C?
   (e) Which is the denser phase, solid or liquid? Explain briefly.

58. If your air conditioner is more than a year or two old, it may use the chlorofluorocarbon CCl_2F_2 as the heat transfer fluid. (CFCs such as this are being replaced as rapidly as possible by substances less harmful to the environment.) Its normal boiling point is −29.8 °C, and the enthalpy of vaporization is 20.11 kJ/mol. The gas and the liquid have specific heats of 117.2 J/mol · K and 72.3 J/mol · K, respectively. How much heat must be evolved when 20.0 g of CCl_2F_2 is cooled from +40 °C to −40 °C?

59. Liquid ammonia, $NH_3(\ell)$, was once used in home refrigerators as the heat transfer fluid. The specific heat of the liquid is 4.7 J/g · K and that of the vapor is 2.2 J/g · K. The enthalpy of vaporization is 23.33 kJ/mol at the boiling point. If you heat 10. kg of liquid ammonia from −50.0 °C to its boiling point of −33.3 °C, and then on to 0.0 °C, how much heat energy must you supply?

Physical Properties of Solids

60. Which compound should have the higher lattice energy, $MgCl_2$ or $BaCl_2$? Briefly explain your reasoning.

61. Which compound should have the higher lattice energy, BeO or BaO? Briefly explain your reasoning.

62. For the pair of compounds CsF and CsI, tell which compound is expected to have the higher melting point, and briefly explain, based on the lattice energies of the two compounds.

63. The ions of NaF and MgO are isoelectronic, and the internuclear distances are about the same (235 pm and 212 pm, respectively, based on the ionic radii in Figure 8.15). Why then are the melting points of NaF and MgO so different (996 °C and 2826 °C, respectively)?

64. Benzene, C_6H_6, is an organic liquid that freezes at 5.5 °C (see Figure 13.3) to beautiful, feather-like crystals. How much heat is evolved when 15.5 g of benzene freezes at 5.5 °C? (The heat of fusion of benzene is 9.95 kJ/mol.) If the 15.5-g sample is remelted, again at 5.5 °C, what quantity of heat is required to convert it to a liquid?

65. The specific heat of silver is 0.235 J/g · K. Its melting point is 962 °C, and its heat of fusion is 11.3 kJ/mol. What quantity of heat, in joules, is required to change 5.00 g of silver from solid at 25 °C to liquid at 962 °C?

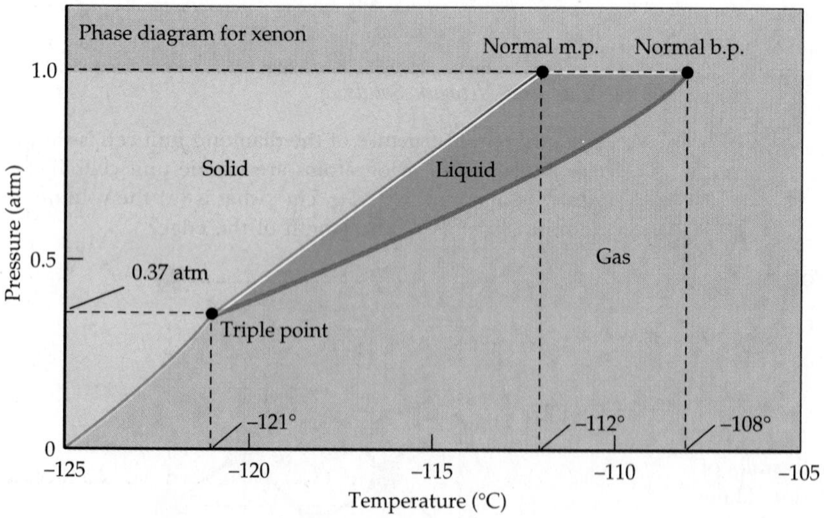

General Questions

66. Rank the following substances in order of increasing strength of intermolecular forces: (a) Ar, (b) CH_3OH, (c) CO_2, and (d) CaO.

67. What type of intermolecular forces are important in the liquid phase of (a) C_2H_6 and (b) $(CH_3)_2CHOH$?

68. Construct an approximate phase diagram for O_2 from the following information: normal boiling point, 90.18 K; normal melting point, 54.8 K; and triple point 54.34 K (at a pressure of 2 mm Hg). Very roughly estimate the vapor pressure of liquid O_2 at −196 °C, the lowest temperature easily reached in the laboratory. Is the density of liquid O_2 greater or less than that of solid O_2?

69. Cooking oil floats on water. From this observation, what conclusions can you draw regarding the polarity, or hydrogen bonding ability, of molecules found in cooking oil?

70. Acetone, $(CH_3)_2C{=}O$, is a common laboratory solvent. It is usually contaminated with water, however. Why does acetone absorb water so readily? Draw molecular structures showing how water and acetone can interact. What intermolecular force(s) is (are) involved in the interaction?

71. Rationalize the observation that $CH_3CH_2CH_2OH$ (1-propanol) has a boiling point of 97.2 °C, whereas a compound with the same empirical formula, methyl ethyl ether ($CH_3CH_2{-}O{-}CH_3$), boils at 7.4 °C.

72. Many common salts are hydrated. Examples include $CoCl_2 \cdot 6\ H_2O$ and $CuSO_4 \cdot 5\ H_2O$. Which of the following two salts is more likely to be hydrated, $BeSO_4$ or $BaSO_4$?

73. A unit cell of cesium chloride is shown on page 628. The density of the solid is 3.99 g/cm^3, and the radius of the Cl^- ion is 167 pm. What is the radius of the Cs^+ ion in the center of the cell? (Although it is not quite true, for the sake of this calculation we shall assume that the Cs^+ ion touches all of the corner Cl^- ions and all ions in the face of the cell touch one another.)

74. If you place 1.0 L of ethanol (C_2H_5OH) in a room that is 3.0 m long, 2.5 m wide, and 2.5 m high, does all the alcohol evaporate? If some liquid remains, how much remains? The vapor pressure of ethanol at 25 °C is 59 mm Hg, and the density of the liquid at this temperature is 0.785 g/cm^3.

75. The polar liquid methanol, CH_3OH, is placed in a glass tube. Is the meniscus of the liquid concave or convex?

76. Liquid ethylene glycol, $HOCH_2CH_2OH$, is one of the main ingredients in commercial antifreeze. Do you predict its viscosity to be greater or less than that of ethanol, C_2H_5OH?

77. Calcium fluoride is the well-known mineral fluorite (page 49). It is known that each unit cell contains four Ca^{2+} ions and eight F^- ions (because the F^- ions fill all the so-called "tetrahedral" holes in a face-centered cubic lattice of Ca^{2+} ions). The edge of the CaF_2 unit cell is 5.46295×10^{-8} cm in length. The density of the solid is 3.1805 g/cm^3. Use this information to calculate Avogadro's number.

78. Select the substance in each of the following pairs that should have the higher boiling point:
(a) Br_2 or ICl
(b) Neon or krypton
(c) C_2H_5OH (ethanol) or 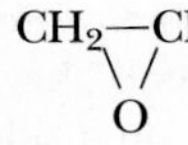
ethylene oxide
(d) 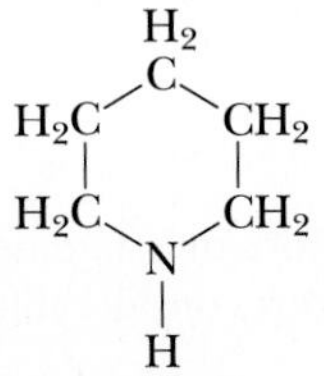
piperidine
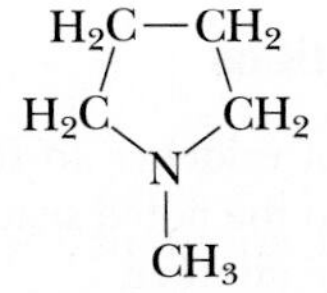
N-methylpyrrolidine

79. You can get some idea of how efficiently spherical atoms or ions are packed in a three-dimensional solid by seeing how well circular atoms pack in two dimensions. Using the drawings shown here, prove that B is a more efficient way to pack circular atoms than A. A unit cell of A contains portions of four circles and one hole. In B, packing coverage can be calculated by looking at a triangle that contains portions of three circles and one hole. Show that A fills about 80% of the available space, whereas B fills closer to 90% of the available space.

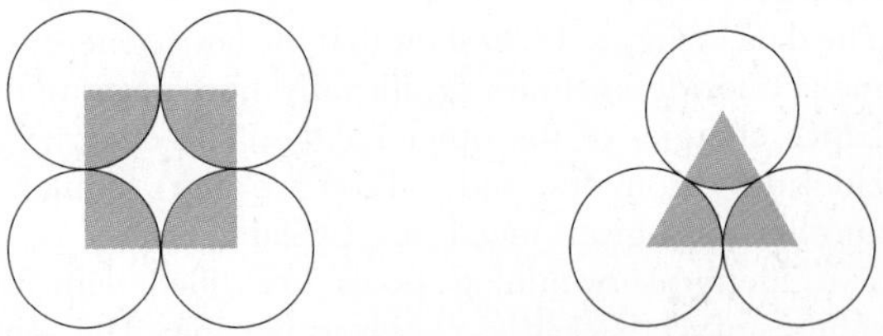
A B

80. If a simple cubic unit cell is formed so that the spherical atoms or ions just touch one another along the edge, calculate the percentage of empty space within the unit cell. (Recall that the volume of a sphere is $\frac{4}{3}(\pi r^3)$, where r is the radius of the sphere.)

81. If the lakes and oceans of another planet are filled with ammonia, what quantity of energy is evolved (in joules) when 1.00 mol of liquid ammonia cools from −33.3 °C (its boiling point) to −43.3 °C? (The specific heat of liquid NH_3 is 4.70 J/g · K.) Compare this with the quantity of heat evolved by 1.00 mol of liquid water cooling by exactly 10 °C.

82. Account for these facts:
(a) Although ethanol (C_2H_5OH) (bp, 80 °C) has a higher molar mass than water (bp, 100 °C), the alcohol has a lower boiling point.
(b) Salts of the ion HF_2^- are known.
(c) Mixing 50 mL of ethanol with 50 mL of water produces a solution with a volume less than 100 mL.

83. Mercury and many of its compounds are dangerous poisons if breathed, swallowed, or even absorbed through the skin. The liquid metal has a vapor pressure of 0.00169 mm Hg at 24 °C. If the air in a small room is saturated with mercury vapor, how many atoms of mercury vapor occur per cubic meter? Assume the room is 4 m square and 2.5 m high (about 13 ft on a side with a ceiling at about 10 ft).

Conceptual Questions

84. Cite two pieces of evidence to support the statement that water molecules in the liquid state exert considerable attractive forces on one another.

85. Can $CaCl_2$ have a unit cell like that of sodium chloride? Explain.

86. Why is it not possible for a salt with the formula M_3X (Na_3PO_4, for example) to have a face-centered cubic lattice of X anions and M cations?

87. Suggest a reason that nonpolar CO_2 is so much more soluble in sea water than O_2 or N_2.

88. During thunderstorms in the Midwest, very large hailstones can fall from the sky. (Some are the size of golf balls!) To preserve some of these stones, we put them in the freezer compartment of a frost-free refrigerator. Our friend, who is a chemistry student, tells us to use an older model that is not frost-free. Why?

89. The data in Figure 13.39 show that the lattice energy of alkali metal fluorides is always significantly more negative than the lattice energies of the other halides. The energies for the chlorides, bromides, and iodides are more similar to one another for a given metal ion. Explain.

90. Two identical swimming pools are filled with uniform spheres of ice packed as closely as possible. The spheres in the first pool are the size of grains of sand; those in the second pool are the size of oranges. The ice in both pools melts. In which pool, if either, is the water level higher? (Neglect any differences in filling space at the planes next to the walls and bottom.)

91. The figure shown in part c is a plot of vapor pressure versus temperature for dichlorodifluoromethane, CCl_2F_2. The heat of vaporization of the liquid is 165 J/g, and the specific heat of the liquid is about 1.0 J/g · K.

(a) What is the normal boiling point of CCl_2F_2?

(b) A steel cylinder containing 25 kg of CCl_2F_2 in the form of liquid and vapor is set outdoors on a warm day (25 °C). What is the approximate pressure of the vapor in the cylinder?

(c) The cylinder valve is opened and CCl_2F_2 vapor gushes out of the cylinder. Soon, however, the flow becomes much slower, and the outside of the cylinder is coated with ice frost. When the valve is closed, and the cylinder is reweighed, it is found that 20 kg of CCl_2F_2 is still in the cylinder. Why is the flow fast at first? Why does it slow down long before the cylinder is empty? Why does the outside become icy?

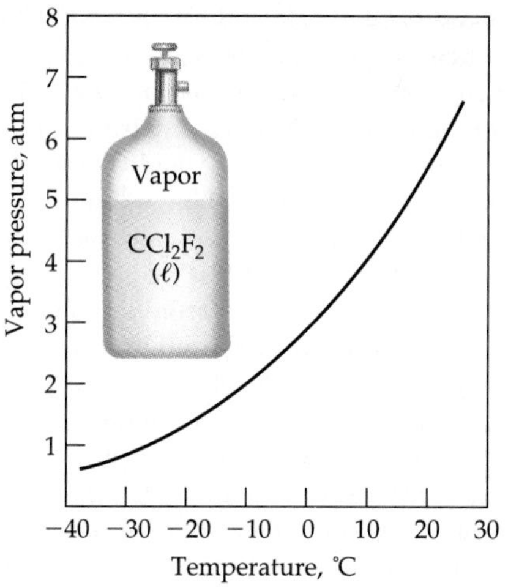

(d) Which of the following procedures is effective in emptying the cylinder rapidly (and safely)? (1) Turn the cylinder upside down and open the valve. (2) Cool the cylinder to −78 °C in Dry Ice and open the valve. (3) Knock off the top of the cylinder, valve and all, with a sledge hammer.

92. The pattern shown here is a two-dimensional model for a crystal. The small spheres represent cations (A), and the large spheres represent anions (B). Sketch at least two alternative unit cells. How many net ions of each type are within your unit cells? What is the formula of the salt?

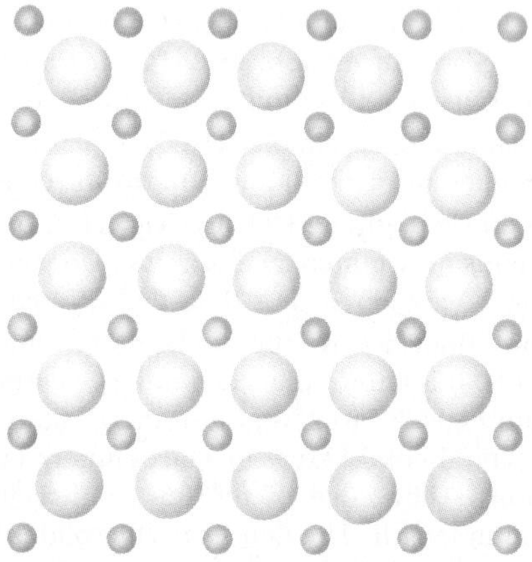

Summary Question

93. Sulfur dioxide, SO_2, is found in polluted air. It comes from the combustion of fossil fuels and from factories that convert certain metal-containing ores to metals or metal oxides.

(a) Draw the Lewis structure for SO_2. Describe the O—S—O angle and molecular geometry of the compound.

(b) What types of forces are responsible for binding SO_2 molecules to one another in the solid or liquid phase?

(c) Using the information in the table, place the compounds listed in order of increasing intermolecular forces.

Compound	Normal Boiling Point (°C)
SO_2	−10.05
NH_3	−33.33
CH_4	−161.48
H_2O	100.0

(d) Airborne sulfur dioxide is one of the compounds responsible for "acid rain." It is thought that the $SO_2(g)$ first oxidizes to $SO_3(g)$ in air. What is the standard enthalpy change for this process? What is the standard enthalpy change for the combination of $SO_3(g)$ and water to give $H_2SO_4(aq)$?

CHAPTER 14

Solutions and Their Behavior

(a) (b) (c)

A Chemical Puzzler

If an egg is placed in dilute acetic acid (vinegar), the acid reacts with the calcium carbonate of the shell (Figure a), but the membrane around the egg remains intact.

$$CaCO_3(s) + 2\ CH_3CO_2H(aq) \longrightarrow Ca^{2+}(aq) + 2\ CH_3CO_2^{-}(aq) + H_2O(\ell) + CO_2(g)$$

If the egg, without its shell, is placed in pure water, the egg swells (Figure b). If the egg is placed in a solution with a high solute concentration, however (a mixture of equal volumes of water and corn syrup, for example), it shrivels dramatically (Figure c). This same effect is observed when vegetables or meat are cured in brine, a concentrated solution of NaCl. If you put a fresh cucumber into brine, water flows out of its cells and into the brine, leaving behind a shriveled vegetable. With the proper spices, however, a shriveled cucumber is a delicious pickle!

A **solution** is a homogeneous mixture of two or more substances in a single phase. It is usual to think of the component present in largest amount as the **solvent** and the other component as the **solute.** Most of the solutions that first come to mind probably involve water as the solvent: apple cider, soft drinks, and beer. Some consumer products (lubricating oils, gasoline, household cleaners), however, involve a liquid other than water as the solvent. Still other solutions do not involve a liquid solvent at all. The air you breathe is a solution of nitrogen, oxygen, carbon dioxide, water vapor, and other gases. Glass objects are all around you. Although glass is variously referred to as an amorphous solid or a supercooled liquid, it is a solution of metal oxides (Na_2O and CaO, among others) in silicon dioxide. Finally, the solder that is used to make connections in your calculator or computer is a solid solution of tin, lead, and other metals. Although many types of solutions exist, our objective in this chapter is to develop an understanding of gases, liquids, and solids dissolved in *liquid* solvents.

Common sense tells you that adding a solute to a pure liquid must change the properties of the liquid, because the intermolecular forces are changed or disrupted. Indeed, this is the reason some solutions are made. For instance, adding antifreeze to your car's coolant water prevents the coolant from boiling over in the summer and freezing up in the winter. As we will describe shortly, the changes in freezing and boiling points from pure solvent to solution are called **colligative properties.** The *magnitude of colligative properties*—changes in freezing point, boiling point, and vapor pressure from solvent to solution, as well as solution osmotic pressure—***ideally** depend only on the number of solute particles per solvent molecule and not on the nature of the solute or solvent.*

Three major topics are covered in this chapter. First, because colligative properties depend on the relative number of solvent and solute particles in solution, convenient ways of describing solution concentration in these terms are required. Second, we consider how and why the solutions form on the molecular level. This gives us some insight into the third topic, the colligative properties themselves.

14.1 UNITS OF CONCENTRATION

To define the colligative properties of a solution, we need ways of measuring solute concentrations that reflect the number of molecules or ions of solute per molecule of solvent. Molarity, the concentration unit useful in stoichiometry calculations, does not work with colligative properties. Here's why.

Molarity (M) is defined as the number of moles of solute *per liter of solution.*

$$\text{Molar concentration of solute A (M)} = \frac{\text{moles of A}}{\text{liters of solution}}$$

For example, the flask on the right side in Figure 14.1 contains a 0.10 *molar* aqueous solution of potassium chromate that was made by adding enough water to 0.10 mol of K_2CrO_4 (19.4 g) to make 1.0 L of solution. No attention was given to how much solvent (water) was actually added. If 1.00 L of water had been added to 19.4 g of K_2CrO_4, however, the volume of solution would be 1.00 L *only* if the solute and solvent together take up exactly 1.00 L of volume. As illustrated by the flask on the left side in Figure 14.1, this is certainly not the case here, and it is almost never true. Making up a solution to have a particular volume does not ensure that the number of molecules of solvent is known.

Several concentration units do reflect the number of molecules or ions of solute per solvent molecule: molality, mole fraction, weight percent, and parts per million.

The **molality** of a solution is defined as the number of moles of solute *per kilogram* of solvent.

Notice that a concentration in molality is indicated by a small italicized "*m*," and molarity is signaled with a regular capital M.

$$\text{Molality of A } (m) = \frac{\text{moles of A}}{\text{kilograms of solvent}} \quad \textbf{(14.1)}$$

The solution in the flask on the left side of Figure 14.1, for example, has a molality of

$$\text{Molality of } K_2CrO_4 = \frac{19.4\text{ g}(1\text{ mol}/194\text{ g})}{1.00\text{ kg water}} = 0.100\ m$$

Notice that different quantities of water were used to make the 0.10 M and 0.10 *m* solutions of K_2CrO_4 in Figure 14.1. This means that the *molarity and molality of a given solution cannot be the same* (although the difference is negligibly small when the solution is quite dilute, say less than 0.01 M).

Mole fraction was first used in connection with gas mixtures (Section 12.5).

The *mole fraction* of a component A, X_A, of a solution is defined as the number of moles of A divided by the total number of moles of all of the components of the solution. For component A in a solution containing the components A, B, and C, the mole fraction of A is

$$\text{Mole fraction of A } (X_A) = \frac{\text{moles of A}}{\text{moles of A} + \text{moles of B} + \text{moles of C}} \quad \textbf{(14.2)}$$

As an example, consider a solution that contains 1.00 mol (46.1 g) of ethanol, C_2H_5OH, in 9.00 mol (162 g) of water. Here the mole fraction of alcohol is 0.100 and that of water is 0.900.

$$X_{\text{alcohol}} = \frac{1.00\text{ mol alcohol}}{1.00\text{ mol alcohol} + 9.00\text{ mol water}} = 0.100$$

$$X_{\text{water}} = \frac{9.00\text{ mol water}}{1.00\text{ mol alcohol} + 9.00\text{ mol water}} = 0.900$$

Notice that the sum of the mole fractions of all components in the solution equals exactly 1, a fact that is true for all solutions.

$$X_{\text{water}} + X_{\text{alcohol}} = 1.000$$

Expressing a concentration as a **weight percentage*** is straightforward because it is simply the mass of solute per 100 g of solution.

Wt% of A =

$$\frac{\text{grams of A}}{\text{grams of A + grams of other solute + grams of solvent}} \times 100\% \qquad \textbf{(14.3)}$$

Our alcohol/water mixture contains 46.1 g of alcohol in 162 g of water, so the total solution mass is 208 g, and the weight % of alcohol is

$$\text{Wt\% alcohol} = \frac{46.1 \text{ g alcohol}}{46.1\text{g alcohol} + 162 \text{ g water}} \times 100\%$$

$$= 22.2\% \text{ alcohol by weight}$$

Natural solutions are often very dilute. Environmental chemists, biologists, geologists, oceanographers, and others therefore use units such as milligrams per liter of solution (mg/L) or parts per million (ppm). Because water at 25 °C has a density of 1.0 g/mL, a concentration of 1.0 mg/L is equivalent to 1.0 mg of solute in 1000 g of water or to 1.0 g of solute in 1,000,000 g of water.

$$\frac{1 \text{ mg solute}}{1 \text{ L solution}} \approx \frac{1.0 \text{ mg solute}}{1000 \text{ g water}} = \frac{1.0 \times 10^{-3} \text{ g solute}}{1000 \text{ g water}} = \frac{1.0 \text{ g solute}}{1.0 \times 10^{6} \text{ g water}}$$

That is, 1.0 mg/L is equivalent to 1 ppm.

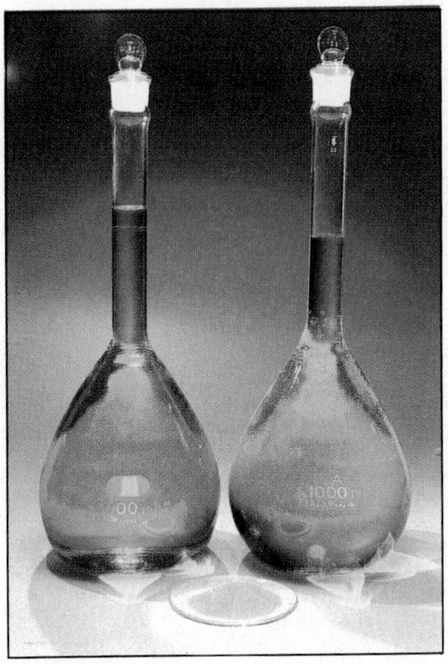

Figure 14.1 Molarity and molality. The photo shows a 0.100 *molal* solution (0.100 *m*) of potassium chromate (flask on the left) and a 0.100 *molar* solution (0.100 M) (flask on the right). In the flask on the right, 0.100 mol (19.4 g) of K_2CrO_4 was mixed with enough water to make 1.0 L of solution. (The volumetric flask was filled to the mark on its neck, an indication the volume is exactly 1.0 L.) Exactly 1000 g (1.00 kg) of water was added to 0.100 mol of K_2CrO_4 in the flask on the left. Adding 1.00 kg of water leads to a solution clearly having a volume greater than 1.00 L. (The small pile of yellow solid in front of the flasks is 0.100 mol of K_2CrO_4.) (C. D. Winters)

EXAMPLE 14.1 *Calculating Mole Fraction, Molality, and Weight Percent*

Assume you add 1.2 kg of $C_2H_4(OH)_2$, ethylene glycol, as an antifreeze to 4.0 kg of water in the radiator of your car. What are the mole fraction, molality, and weight percent of the ethylene glycol?

Solution 1.2 kg of ethylene glycol (molar mass = 62.1 g/mol) is equivalent to 19 mol, and 4.0 kg of water represents 220 mol.

Mole fraction:

$$X_{\text{glycol}} = \frac{19 \text{ mol glycol}}{19 \text{ mol glycol} + 220 \text{ mol water}} = 0.080$$

Molality:

$$\text{Molality} = \frac{19 \text{ mol glycol}}{4.0 \text{ kg water}} = 4.8\ m$$

Weight percentage:

$$\text{Wt\%} = \frac{1.2 \times 10^{3} \text{ g glycol}}{1.2 \times 10^{3} \text{ g glycol} + 4.0 \times 10^{3} \text{ g water}} \times 100\% = 23\%$$

---

*Strictly speaking, this is a *mass* percentage; however, long usage has made "weight percentage" the standard term.

EXAMPLE 14.2 ***Parts per Million***

You dissolve 560 g of $NaHSO_4$ in a swimming pool that contains 4.5×10^5 L of water at 25 °C. What is the concentration of sodium ion in parts per million? (Sodium hydrogen sulfate is used to adjust the pH of the pool water.)

Solution As a first step, we calculate the quantity of sodium ions (grams) in 560 g of $NaHSO_4$.

$$560 \text{ g NaHSO}_4\left(\frac{1 \text{ mol NaHSO}_4}{120. \text{ g NaHSO}_4}\right)\left(\frac{1 \text{ mol Na}^+}{1 \text{ mol NaHSO}_4}\right)\left(\frac{23.0 \text{ g}}{1 \text{ mol Na}^+}\right) = 110 \text{ g Na}^+$$

Then, we can use the mass of sodium ions added to the pool to find the number of milligrams per liter, which is equivalent to parts per million.

$$\frac{110 \text{ g Na}^+ (1000 \text{ mg/1 g})}{4.5 \times 10^5 \text{ L}} = 0.24 \text{ mg/L} = 0.24 \text{ ppm}$$

EXERCISE 14.1 ***Mole Fraction, Molality, and Weight Percent***

If you dissolve 10.0 g, or about 1 heaping teaspoonful, of sugar, $C_{12}H_{22}O_{11}$, in a cup of water (250. g), what are the mole fraction, molality, and weight percentage of sugar? ■

EXERCISE 14.2 ***Parts Per Million***

Sea water has a sodium ion concentration of 1.08×10^4 ppm. If the sodium is dissolved as sodium chloride, how many grams of NaCl are in each liter of sea water? ■

14.2 THE SOLUTION PROCESS

Both ammonium chloride and sodium hydroxide dissolve readily in water, but the water cools down when NH_4Cl dissolves, and it warms up when NaOH dissolves. On the other hand, $CaCO_3$ dissolves in water only to a small extent. Why do these things happen? What controls the solution process?

When a solute is dissolved in a solvent, the attractive forces between solute and solvent particles must be great enough to overcome the attractive forces (intermolecular forces) within the pure solvent and within the pure solute. As described in Section 13.2, when solutes are dissolved, they become *solvated* if solvent molecules are sufficiently attracted to solute molecules or ions (usually by dipole-dipole or ion-dipole forces). If water is the solvent, *solvation* is called more specifically *hydration.*

Solubility is the maximum amount of material that can dissolve in a given amount of solvent at a given temperature to produce a stable solution. In Section 4.3 you learned some guidelines for the solubility of common ionic compounds, but we can now be somewhat more quantitative. For example, it was suggested in Figure 4.7 that common nitrates and chlorides are usually soluble, except for the chlorides of Ag^+, Pb^{2+}, and $Hg_2{}^{2+}$. About 950 g of silver nitrate, $AgNO_3$, will dissolve in 100 mL of water at 100 °C, and so this is classified as a *soluble* salt. In contrast, only 0.00217 g of silver chloride, AgCl, dissolves in

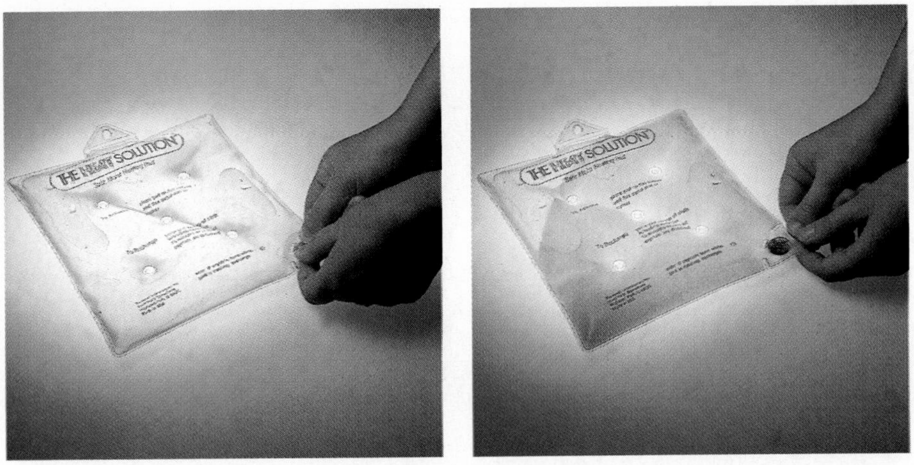

Figure 14.2 An example of a supersaturated solution is a "heat pack" called "The Heat Solution." The bag contains a supersaturated solution of sodium acetate ($NaCH_3CO_2$). If the solution is disturbed (by adding a small crystal of the salt or by some mechanical means [hitting the pack with your hand]), hydrated sodium acetate begins to precipitate (photo on the right). As described below, solutions of most salts release heat when the salt precipitates, and the temperature of the mixture rises to essentially the temperature of the "freezing point" of the salt (48 °C in this case). (You can make your own "heat pack" by warming a mixture of sodium thiosulfate, $Na_2S_2O_3 \cdot 5\ H_2O$ (440 g; available in photo stores) and 2 tbsp of water. Once a solution has formed and has cooled slowly to room temperature, dropping in a small crystal of the salt causes the salt to precipitate from solution, and the temperature rises to almost 50 °C.) (See G. Marsella: "Hot and cold packs," *Chem Matters,* February, p. 12, 1987.) (C. D. Winters)

100 mL of water at 100 °C, and so this solid is said to be insoluble, even though it does have some degree of solubility.*

When the maximum amount of solute has been dissolved in a solvent and equilibrium is attained, the solution is said to be **saturated.** *The concentration of dissolved solute in a saturated solution is a measure of the solubility of the solute.* If the solute concentration is less than the saturation amount, the solution is **unsaturated.** On the other hand, in some instances solutions can temporarily contain more solute than the saturation amount, and these are called **supersaturated.** Figure 14.2 illustrates what happens when a supersaturated solution is disturbed.

Liquids Dissolving in Liquids

If two liquids mix to an appreciable extent to form a solution they are said to be **miscible.** In contrast, **immiscible** liquids do *not* mix to form a solution; they exist as separate layers in contact with one another (Figure 13.6).

*When can you properly say that one substance is soluble and another is insoluble? There are no good rules about this, because there are degrees of solubility; nothing is totally without some solubility at some temperature. Chemists frequently say something is insoluble, partly soluble, or soluble. Usually what they mean is that something is soluble if they can see that a good deal of the solid has gone into solution. If only some solid has visibly dissolved, the solid is partly soluble. If no dissolution of any solid is observed, it is said to be insoluble or sparingly soluble.

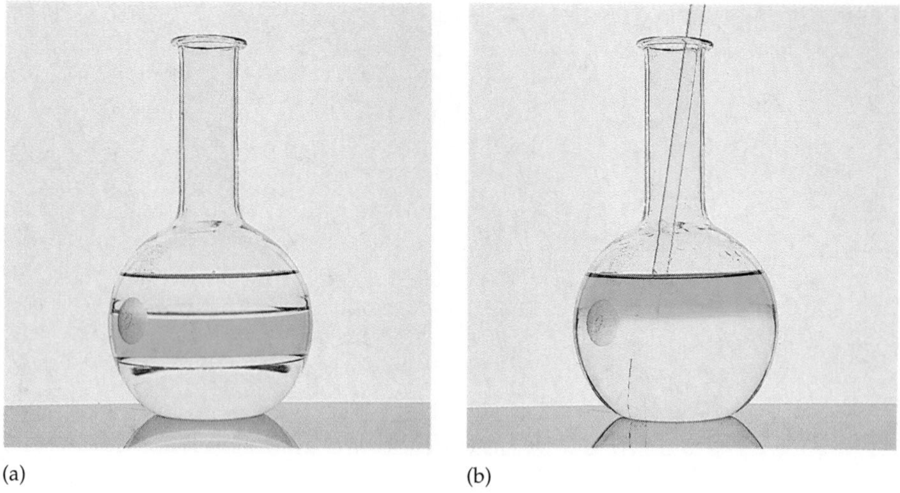

Figure 14.3 Miscibility. (a) The colorless, denser bottom layer is nonpolar carbon tetrachloride, CCl_4. The middle layer is a solution of $CuSO_4$ in water, and the colorless, less dense top layer is nonpolar octane ($CH_3CH_2CH_2CH_2CH_2CH_2CH_2CH_3$). The weak intermolecular interactions between the two nonpolar molecules and water cannot overcome the very strong forces between water molecules and allow them to be miscible with water. (b) After stirring the mixture, the two nonpolar liquids form a homogeneous mixture, because they are miscible with each other. This layer of mixed liquids is now on the bottom because its density is greater than that of the water layer. (C. D. Winters)

The nonpolar liquids octane (C_8H_{18}) and carbon tetrachloride (CCl_4) are miscible; they mix in all proportions to form a homogeneous solution (Figure 14.3). On the other hand, polar water and nonpolar octane are immiscible. You would observe that the less dense liquid, octane, simply floats as a layer on top of the denser water layer (Figure 14.3b). Finally, polar ethanol molecules (C_2H_5OH) dissolve in all proportions in water; beer, wine, and other alcoholic beverages contain amounts of alcohol ranging from just a few percent to more than 50%. It is these observations that have led to the familiar rule: *like dissolves like.* That is, two or more nonpolar liquids frequently are miscible with each other, just as are two or more polar liquids. Liquids of different types do not mix to any appreciable degree.

What is the molecular basis for the "like dissolves like" guideline? Molecules of pure octane or pure benzene, both of which are nonpolar, are held together in the liquid phase by weak dispersion forces (Section 13.2). Similar weak forces can exist between an octane molecule and a CCl_4 molecule, and these molecules are attracted to one another.

In contrast to the CCl_4/octane solution, water and hydrocarbons do *not* mix (Figure 14.3). This observation can be analyzed with the following scheme, one that may be applied to any solvent/solute interaction.

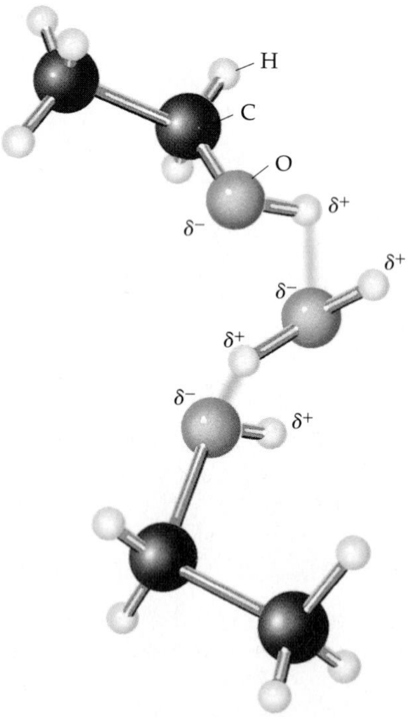

Water dissolves readily in ethanol due to hydrogen bonding between the molecules.

$H_2O(g)$	+	octane(g)	Forming intermolecular bonds. Enthalpy change is negative. →	$[H_2O/\text{octane}](g)$
↑ Enthalpy of vaporization of water (positive)		↑ Enthalpy of vaporization of octane (positive)		Enthalpy change on condensing to a liquid (negative) ↓
$H_2O(\ell)$	+	octane(ℓ)	Enthalpy of solution, $\Delta H_{solution}$ → May be positive or negative	$[H_2O/\text{octane}](\ell)$

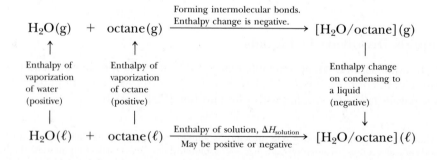

The enthalpy of solution, $\Delta H_{solution}$, is the net heat energy involved in the process of solution formation (at constant pressure). According to Hess's law (Section 6.6), the energy involved in a complex process can be analyzed by breaking the process into simpler steps, whose energies are known. In this case we can find $\Delta H_{solution}$, in principle, by adding up the energies required to vaporize both the solute and solvent, the enthalpy change when the two are attracted by intermolecular forces in the gas phase, and then the enthalpy change when the mixture condenses to form a liquid solution. The important point here is that the enthalpy changes for vaporizing octane and, in particular, water are much more positive than the enthalpy changes for the other steps. Therefore, the value of the overall enthalpy change, $\Delta H_{solution}$, is likely to be positive, reflecting an endothermic process. In Chapter 6 it was noted that an endothermic process is often not product-favored. In large part, this is the reason that polar and nonpolar liquids do not mix well.

Although the enthalpy of solution is important in determining solubility, another quantity, the *entropy* of solution, is also a contributing factor. This is considered in Chapter 20.

Solids Dissolving in Liquids

Solid I_2 is an example of a nonpolar molecular solid, the molecules being held together by weak intermolecular dispersion forces. To dissolve this nonpolar solid in a nonpolar liquid, say CCl_4, the attractions between I_2 molecules and between CCl_4 must be disrupted. The enthalpy of solution is close to zero, however, because the energy required to separate solid I_2 molecules and liquid CCl_4 molecules is approximately returned by the formation of new intermolecular attractions between I_2 and CCl_4. Thus, nonpolar I_2 is soluble in nonpolar CCl_4 (Figure 14.4).

Sucrose, a sugar, is also a molecular solid. In this case, the molecules are polar and interact in the solid phase through hydrogen bonds, the same kind of attraction that occurs between sugar and water molecules. Hydrogen bonding between sugar and water is strong enough that the energy evolved here can be

(a) (b)

Figure 14.4 Water, carbon tetrachloride (CCl_4), and iodine. (a) Water (a polar molecule) and CCl_4 (a nonpolar molecule) are immiscible, and the less dense water layer is found on top of the denser CCl_4 layer. A small amount of iodine dissolves in water to give a brown solution (*top*). (b) The mixture in (a) has been stirred. The nonpolar molecule I_2 is more soluble in nonpolar CCl_4, as indicated by the fact that I_2 dissolves preferentially in CCl_4 to give a purple solution. (C. D. Winters)

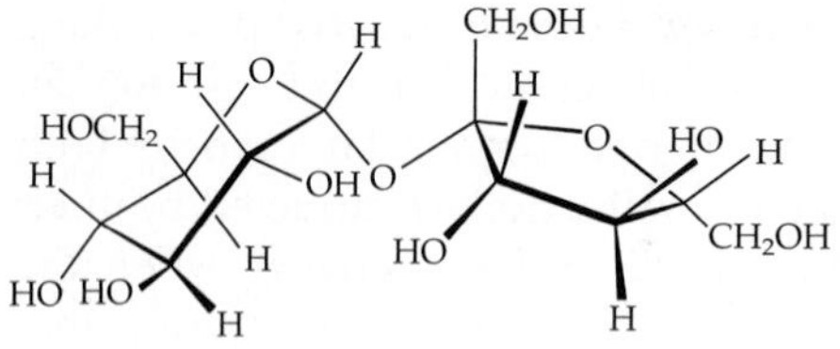

Sucrose, or cane sugar.

supplied to disrupt the sugar/sugar and water/water interactions. On the other hand, sugar is not soluble at all in CCl_4 or other nonpolar liquids. A nonpolar liquid does not interact sufficiently well with sugar to cause it to dissolve.

Network solids include graphite, diamond, and quartz sand (SiO_2), and your intuition tells you they do not dissolve readily in water. After all, where would all the beaches be if sand dissolved readily in water? The normal covalent chemical bonding in network solids is simply too strong to be replaced by weaker hydrogen-bonding attraction to water dipoles.

Dissolving Ionic Solids

Common ionic solids are not soluble to any appreciable extent in nonpolar solvents such as CCl_4, and they vary greatly in their solubility in water (Figure 4.8). At 20 °C, 100 g of water dissolves 74.5 g of $CaCl_2$, but only 0.0014 g of limestone, $CaCO_3$, can be dissolved no matter how hard you try. It is useful to explore the reasons for the widely different solubilities.

Alkali metal halides dissolve in water because strong ion-dipole forces lead to strong ion hydration and help to break down the cation-anion attraction in the crystal lattice (Figure 3.15). There are two ways to analyze the overall energy involved here. First, the enthalpy of solution ($\Delta H_{solution}$) can be estimated from the lattice energy of the metal halide (Section 8.7) and the enthalpy of hydration of the gaseous ions to form hydrated ions ($\Delta H_{hydration}$).

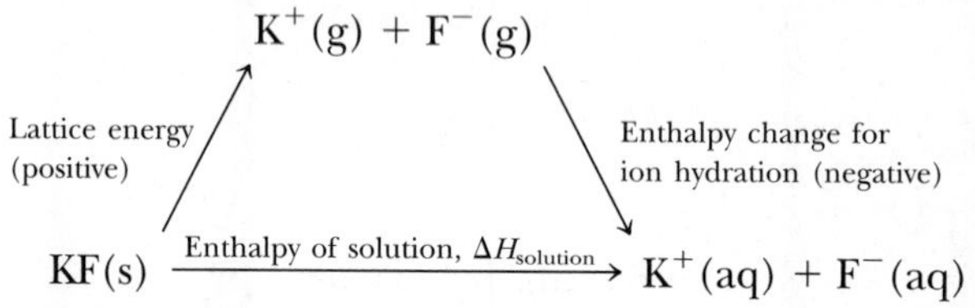

From Hess's law (Chapter 6) you know the enthalpy of solution is the sum of the enthalpies of the processes along the path from reactants [KF(s)] to products [K^+(aq) + F^-(aq)].

$$\Delta H_{solution} = -\text{lattice energy} + \text{enthalpy of hydration}$$

$$\Delta H_{solution} \text{ for KF(s)} = 821 \text{ kJ/mol} + (-819 \text{ kJ/mol}) = +2 \text{ kJ/mol}$$

Using this approach we can estimate that the enthalpy change for dissolving KF is about zero. What is important here is to put this fact in the context of the data in Table 14.1, which lists the water solubility of a series of simple salts with their calculated enthalpy of solution. Here you see that there is a rough correlation between the value of $\Delta H_{solution}$ (as estimated previously) and solubility in water: as $\Delta H_{solution}$ becomes more and more negative, the solubility increases. Rubid-

TABLE 14.1 Estimated Enthalpy of Solution and Water Solubility of Some Ionic Compounds

Compound	Lattice Energy (kJ/mol)	$\Delta H_{hydration}$ (kJ/mol)	$\Delta H_{solution}$ (kJ/mol)	Solubility in H_2O (g/100 mL)
AgCl	−912	−851	+61	0.000089 (10 °C)
NaCl	−786	−760.	+26	35.7 (0 °C)
LiF	−1037	−1005	+32	0.3 (18 °C)
KF	−821	−819	+2	92.3 (18 °C)
RbF	−789	−792	−3	130.6 (18 °C)

ium fluoride is very soluble in water and has an exothermic enthalpy of solution. Silver chloride is quite insoluble and has a very positive value of $\Delta H_{solution}$. Solubility is favored when the energy required to break down the lattice of the solid (the lattice energy) is smaller or roughly equal to the energy given off when the ions are hydrated ($\Delta H_{hydration}$).

The approach outlined here for predicting water solubility from lattice energy and enthalpy of hydration is useful, because the process can be visualized. One can "see" a crystal lattice flying apart into ions in the gas phase and then these ions plunging into water. Unfortunately, reasonably good values for lattice energy and $\Delta H_{hydration}$ are available for relatively few salts, so a more general process for obtaining the enthalpy of solution is needed. Consider again the example of dissolving KF. Hess's law suggests that $\Delta H^\circ_{solution}$ can be estimated from enthalpies of formation.

$$\text{KF(s)} \longrightarrow \text{KF(aq)}$$
$$\Delta H^\circ_{solution} = \Delta H^\circ_f[\text{KF(aq)}] - \Delta H^\circ_f[\text{KF(s)}]$$

These values are available from tables published by the National Institute of Standards and Technology (NIST).

$$\Delta H^\circ_f[\text{KF(s)}] = -526.3 \text{ kJ/mol} \qquad \Delta H^\circ_f[\text{KF(aq)}] = -585.0 \text{ kJ/mol}$$

and they give a value for $\Delta H^\circ_{solution}$ of −58.7 kJ, suggesting that the compound should be water-soluble.* Data useful for calculating solution enthalpies for other compounds are given in Table 14.2.

TABLE 14.2 Data for Calculating Enthalpy of Solution

Compound	ΔH°_f(s) (kJ/mol)	ΔH°_f(aq, 1 *m*) (kJ/mol)
LiF	−616.0	−611.1
NaF	−573.6	−572.8
KF	−526.3	−585.0
RbF	−557.7	−583.8
LiCl	−408.6	−445.6
NaCl	−411.2	−407.3
KCl	−436.7	−419.5
RbCl	−435.4	−418.3
NaOH	−425.6	−470.1
NH_4NO_3	−365.6	−339.9

EXERCISE 14.3 ***Calculating Enthalpy of Solution***

1. Use the data in Table 14.1 to compare the enthalpies of solution for AgCl and RbF. Comment on the relation between the enthalpy of solution and solubility of these two salts.
2. Use the data in Table 14.2 to calculate the enthalpy of solution of ammonium nitrate. Is it expected to be water-soluble? ■

In Exercise 14.3(2) you calculated the enthalpy of solution of NH_4NO_3, and you should have found it to be quite positive. The energy required to dissolve

*It is not unusual to see small differences in thermochemical data calculated by different methods. The difference in this case comes from complicating factors that are not important to our overall conclusion.

Figure 14.5 A commercially available "cold pack." A typical cold pack consists of a sealed bag of water inside a bag of ammonium nitrate. When the bags are squeezed firmly, the inner pouch breaks and the water and NH_4NO_3 mix. The endothermic solution process causes the water temperature to drop to about 5 °C. (C. D. Winters)

this salt comes from its surroundings, so the water cools dramatically when the solid solute is added. This is the basis of the "chemical cold packs" that you may have seen used to reduce swelling after an injury (Figure 14.5).

Factors Affecting Solubility: Pressure and Temperature

Biochemists and physicians, among others, are interested in the solubility of gases such as CO_2 and O_2 in water or body fluids, and all scientists—especially geologists—need to know about the solubility of solids in various solvents. Pressure and temperature are two external factors that may control such processes. Both affect the solubility of gases in liquids, whereas normally only temperature is important in determining the solubility of solids in liquids.

Henry's law holds quantitatively only for gases that do not interact chemically with the solvent. It does not work perfectly for NH_3, for example, which gives small concentrations of NH_4^+ and OH^- in water or for CO_2, which reacts with water to form carbonic acid.

Dissolving Gases in Liquids: Henry's Law

The solubility of a gas in a liquid is directly proportional to the gas pressure. As the gas pressure increases, so does its solubility. This is a statement of **Henry's law,**

$$S_g = k_H P_g \tag{14.4}$$

where S_g is the gas solubility, P_g is the partial pressure of the gaseous solute, and k_H is Henry's law constant, a constant *characteristic of the solute and solvent.* Notice that Henry's law applies only to gases dissolving in liquids; pressure has little if any effect on the solubility of solids in liquids or liquids in liquids.

Carbonated soft drinks are good examples of Henry's law. They are packed under pressure in a chamber filled with carbon dioxide gas, some of which dissolves in the drink. When the can or bottle is opened, the partial pressure of CO_2 above the solution drops, which causes the concentration of CO_2 in solution to drop, and gas bubbles out of the solution (Figure 14.6). The same process can happen with gases dissolved in your blood if you are an underwater diver (Figure 14.7).

TABLE 14.3 Henry's Law Constants (25 °C)*

Gas	k_H (M/mm Hg)
N_2	8.42×10^{-7}
O_2	1.66×10^{-6}
CO_2	4.48×10^{-5}

*For solubility in H_2O. From W. Stumm and J. J. Morgan: *Aquatic Chemistry,* p. 109. New York, Wiley, 1981.

EXAMPLE 14.3 ***Using Henry's law***

What is the concentration of O_2 in a fresh water stream in equilibrium with air at 25 °C? Atmospheric pressure is 1.0 atm. Express the answer in grams of O_2 per liter of water.

Solution Henry's law can be used to calculate the solubility of oxygen. First, however, the partial pressure of O_2 in air must be calculated. From Chapter 12 you know that air is 21% O_2, which means the mole fraction of O_2 is 0.21. If the total pressure is 1.0 atm, the partial pressure of O_2 is 160 mm Hg.

$$P(O_2) = 1.0 \text{ atm}\left(\frac{760 \text{ mm Hg}}{1 \text{ atm}}\right)(0.21) = 160 \text{ mm Hg}$$

From Henry's law, we have

$$\text{Solubility of } O_2 = \left(\frac{1.66 \times 10^{-6} \text{ M}}{\text{mm Hg}}\right)(160 \text{ mm Hg}) = 2.65 \times 10^{-4} \text{ M}$$

This concentration can be expressed in grams per liter using the molar mass of O_2.

$$\text{Solubility of } O_2 = \left(\frac{2.65 \times 10^{-4} \text{ mol}}{\text{L}}\right)\left(\frac{32.0 \text{ g}}{\text{mol}}\right) = 0.00848 \text{ g/L}$$

This concentration of O_2 (8.48 mg/L) is quite low, but it is sufficient to provide the oxygen required by aquatic life.

Figure 14.6 Illustration of Henry's law. The greater the partial pressure of CO_2 over the soft drink in a bottle of soda, the greater the amount of CO_2 dissolved. More CO_2 is dissolved in the closed bottle than in the bottle open to the atmosphere. (C. D. Winters)

Figure 14.7 If you dive to any appreciable depth using scuba gear, you must be concerned with the solubility of gases in your blood. Nitrogen is soluble in blood, so if you breathe a high-pressure mixture of O_2 and N_2 deep under water, the concentration of N_2 in your blood can be appreciable. If you ascend too rapidly, the nitrogen is released as the pressure decreases and forms bubbles in the blood. This affliction, sometimes called the "bends," is painful and can be fatal if blood, forced out of capillaries and the brain, for example, is deprived of oxygen. To partly circumvent this, deep-sea divers sometimes use a helium-oxygen mixture instead of nitrogen-oxygen, because helium is not nearly as soluble in blood as nitrogen. (Brian Parker/Tom Stack and Associates)

Carbon Dioxide, Seashells, and Killer Lakes

Carbon dioxide is such a simple molecule, but it plays important roles in your body, in our environment, and in our economy.

Exhale—and CO_2 is among the gases coming from your lungs because CO_2 is produced in the "burning" of carbohydrates. In the following equation CH_2O is used to represent a carbohydrate.

$$O_2(g) + CH_2O \longrightarrow CO_2(g) + H_2O(\ell)$$

Some of the CO_2 from this "combustion" is not exhaled but is dissolved in body fluids. As described previously, CO_2 can react with water to produce the weak acid H_2CO_3. This in turn loses H^+ to give the hydrogen carbonate ion, HCO_3^-, in solution. These substances act together as a "buffer" (see Chapter 18) to control the pH of your blood and body tissues.

Carbon dioxide is also the end product of fermentation, the slow burning of carbohydrates. When some beverages, such as beer and champagne, are made, the gas is allowed to dissolve in the liquid. When a bottle of the beverage is opened, some of the dissolved CO_2 comes out of solution, and you see it as a foamy head on beer or bubbles in champagne. The presence of dissolved CO_2 in many beverages is also part of the reason they are acidic.

More rapid burning of carbohydrates or of hydrocarbons—as in the burning of forests or the combustion of fossil fuels—is the source of an ever-increasing amount of CO_2 in earth's atmosphere. The concentration of CO_2 has steadily increased since the beginning of the Industrial Revolution, and the increase is accelerating. Since just 1957 the amount of CO_2 in the atmosphere has increased about 6%.

It has been estimated that the mean global surface temperature has increased about 0.5 °C since the late 19th century and that the mean sea level has risen 10 to 15 cm in that period, in part due to melting polar ice and in part due to the thermal expansion of surface water. This warming of the planet comes about because CO_2 is a "greenhouse" gas, that is, it traps heat from the earth that would normally be dissipated into space. These observations have led to speculation that the planet may become overly warm, with harmful environmental effects.

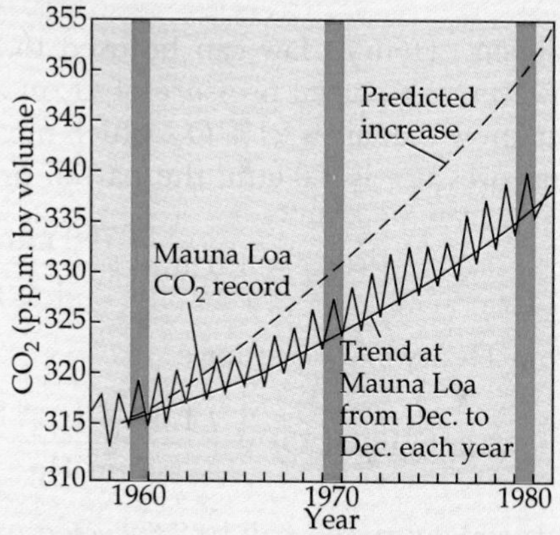

The concentration of atmospheric CO_2 since 1958. This plot is based on data collected at the Mauna Loa Observatory in Hawaii, which is located far from urban areas where carbon dioxide levels are high because of factories, power plants, and motor vehicles. There is a seasonal fluctuation because photosynthesis consumes CO_2 in the summer. The dotted line is the CO_2 concentration predicted on the basis of fossil fuel consumption. (The concentration of CO_2 is given in parts per million, that is, in units of milligrams per kilogram of air.)

The increase in atmospheric CO_2 concentration has not been as great as predicted, however, because of

EXERCISE 14.4 ***Using Henry's Law***

Henry's law constant for CO_2 in water at 25 °C is 4.48×10^{-5} M/mm Hg. What is the concentration of CO_2 in water when the partial pressure is $\frac{1}{3}$ atm? (Note that CO_2 reacts with water to give traces of H_3O^+ and HCO_3^-. The reaction occurs to such a small extent, however, that Henry's law is obeyed at low CO_2 partial pressures.) ■

Temperature Effects on Solubility: Le Chatelier's Principle

You know from experience that much more gas is released from a can of soda that is opened when it is warm than when it is cold. This illustrates the fact that *temperature affects solubility.*

Lake Nyos in Cameroon, the site of a natural disaster. In 1986 a huge bubble of CO_2 escaped from the lake and asphyxiated more than 1700 people. (Courtesy of George Kling)

"sinks," mechanisms in the environment that absorb the gas. Some is used up in increased plant respiration. In fact, plant ecologists have found that a doubling of the CO_2 level can cause silver maple trees, for example, to grow 61% faster.

Much of the CO_2 on the earth is tied up in the form of carbonates such as calcium carbonate, $CaCO_3$, the primary constituent of seashells and corals. Indeed, about 85% of the carbon on earth is in the form of "carbonate sediments." As CO_2 enters the atmosphere much of it finds its way to this sink, and the increase in atmospheric concentration is not as large as expected.

Large amounts of carbon dioxide can dissolve in underground water, and people have enjoyed sparkling mineral water from natural springs for centuries. But dissolved CO_2 can be deadly. In the African nation of Cameroon in 1986 a huge bubble of CO_2 gas escaped from Lake Nyos and moved down a river valley at 20 m/s (about 45 mph). Because CO_2 is denser than air, it hugged the ground and displaced the air in its path. More than 1700 people suffocated. The CO_2 came from springs of carbonated groundwater at the bottom of the lake. Because the lake is so deep, the CO_2 mixed little with the upper layers of water, and the bottom layer became supersaturated with CO_2. When this delicate situation was changed, perhaps because of an earthquake or landslide, the CO_2 came out of the lake water just like it does when a can of soda is opened.

Carbon dioxide is in the top 20 chemicals made in the United States, where it is used largely for refrigeration in the form of solid CO_2, "Dry Ice," and to carbonate beverages. You will see it as a product or reactant in many other reactions in this book.

Gases that dissolve to an appreciable extent in solvents usually do so in an exothermic process. That is, the enthalpy of solution is negative.

$$\text{Gas} + \text{liquid solvent} \longrightarrow \text{saturated solution} + \text{heat energy}$$
$$\Delta H_{rxn} = \Delta H_{solution} \text{ (where } \Delta H_{solution} \text{ has a negative value)}$$

The dissolving process continues until a saturated solution is formed. Gases continue to dissolve in a liquid until the dissolving process is *exactly* counterbalanced by gas molecules coming out of solution with the *consumption* of heat.

$$\text{Saturated solution} + \text{heat energy} \longrightarrow \text{gas} + \text{liquid solvent}$$
$$\Delta H_{rxn} = -\Delta H_{solution} \text{ (where } \Delta H_{solution} \text{ has a negative value)}$$

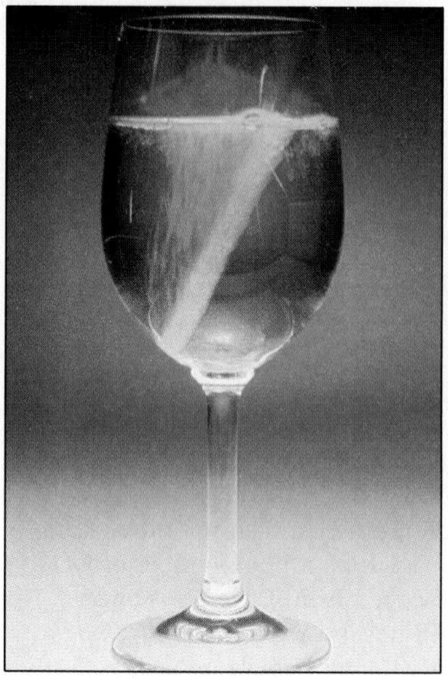

Figure 14.8 A warm glass rod is placed in a glass of ginger ale. The heat energy of the rod is absorbed by a cold solution of CO_2 in water and causes the CO_2 to be less soluble. The Henry's law constant for CO_2 in water is 4.48×10^{-5} M/mm Hg at 25 °C, whereas it drops 20% to 3.6×10^{-5} M/mm Hg at 100 °C. (C. D. Winters)

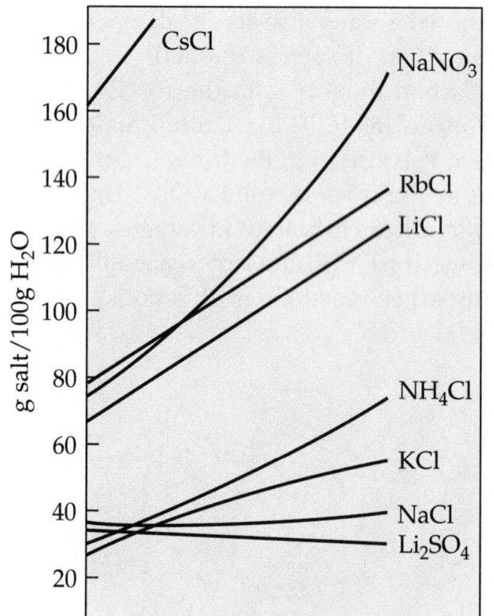

Figure 14.9 The temperature dependence of the solubility of some ionic compounds in water.

At this point the solution is saturated with gas, and *equilibrium* is established. At equilibrium, as many molecules come out of solution as dissolve in a given period. We depict this situation by showing the original process with arrows running in both directions, $\rightleftarrows$.

$$\text{Gas} + \text{liquid solvent} \rightleftarrows \text{saturated solution} + \text{heat energy}$$

To understand how temperature affects solubility, we turn to **Le Chatelier's principle,** which states that *a change in any of the factors determining an equilibrium causes the system to adjust to reduce or counteract the effect of the change.* For example, if a solution of a gas in a liquid is heated, the equilibrium shifts in an attempt to absorb some of the added heat energy. That is, the reaction

$$\text{Gas} + \text{liquid solvent} \underset{}{\overset{\text{Exothermic process, } \Delta H_{rxn} \text{ is negative}}{\rightleftarrows}} \text{saturated solution}$$

Added heat energy
Equilibrium shifts left

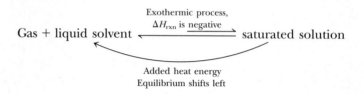

shifts back to the left because heat energy can be consumed in the process that gives free gas molecules and pure solvent (Figure 14.8). For this reason a dissolved gas always becomes less soluble with increasing temperature.

When solids dissolve in liquids, the process can be endothermic. Here, when equilibrium is reached, as many pairs of Na^+ and Cl^- ions leave the surface of a dissolving salt crystal, for example, and return to the surface of the solid in a given period. If the process is altered by adding heat energy, the reaction is displaced to absorb the added energy. Le Chatelier's principle predicts that the solubility of the salt *increases with increasing temperature.*

$$\text{Solid} + \text{liquid solvent} \underset{}{\overset{\text{Endothermic process, } \Delta H_{rxn} \text{ is positive}}{\rightleftarrows}} \text{saturated solution}$$

Added heat energy
Equilibrium shifts right

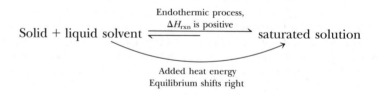

This phenomenon is true for *many* ionic compounds, as illustrated in Figure 14.9.

Some salts also exist that have a negative enthalpy of solution (the process is *exo*thermic), and they become *less soluble with increasing temperature.* The solubility curve for Li_2SO_4 in Figure 14.9 is just one example of this behavior.

In view of the discussion in Chapter 13 of trends in lattice energies (see Figure 13.39), it is interesting to notice the trend in solubilities of the alkali metal chlorides in Figure 14.9. In general, the smaller the lattice energy of a salt, the greater its solubility. Thus, the solubilities of alkali metal chlorides increase in the order NaCl < KCl < RbCl < CsCl. Lithium chloride is clearly an exception. In fact, lithium salts often have properties that do not reflect the general trends in Group 1A.

14.3 COLLIGATIVE PROPERTIES

When water contains dissolved sodium chloride, the vapor pressure of water over the solution is different from that of pure water, as is the freezing point of the solution, its boiling point, and its osmotic pressure. These changes between pure solvent and solution are called the **colligative properties** of the solution.

Changes in Vapor Pressure: Raoult's Law

Molecules of water as well as ions or molecules from the solute are at the surface of an aqueous solution (Figure 14.10). Water molecules can leave the liquid and enter the gas phase, exerting a vapor pressure. Not as many water molecules are at the surface as in pure water, however, because some of them have been displaced by dissolved ions or molecules. Not as many water molecules are available to leave the liquid's surface, therefore, and the vapor pressure is *lower* than that of pure water at a given temperature. From this we can conclude that the vapor pressure of the solvent at the surface, $P_{solvent}$, must be proportional to the relative number of solvent molecules in a solution, that is, to the *mole fraction* of solvent. Thus, because $P_{solvent} \propto X_{solvent}$, we can write

$$P_{solvent} = X_{solvent}K \quad \textbf{(14.5)}$$

(where K is a constant). This equation tells you, for example, that if only half as many solvent molecules are present at the surface of a solution as at the surface of the pure liquid, then the vapor pressure of the solvent in the solution is only half as great as that of the pure solvent at the same temperature.

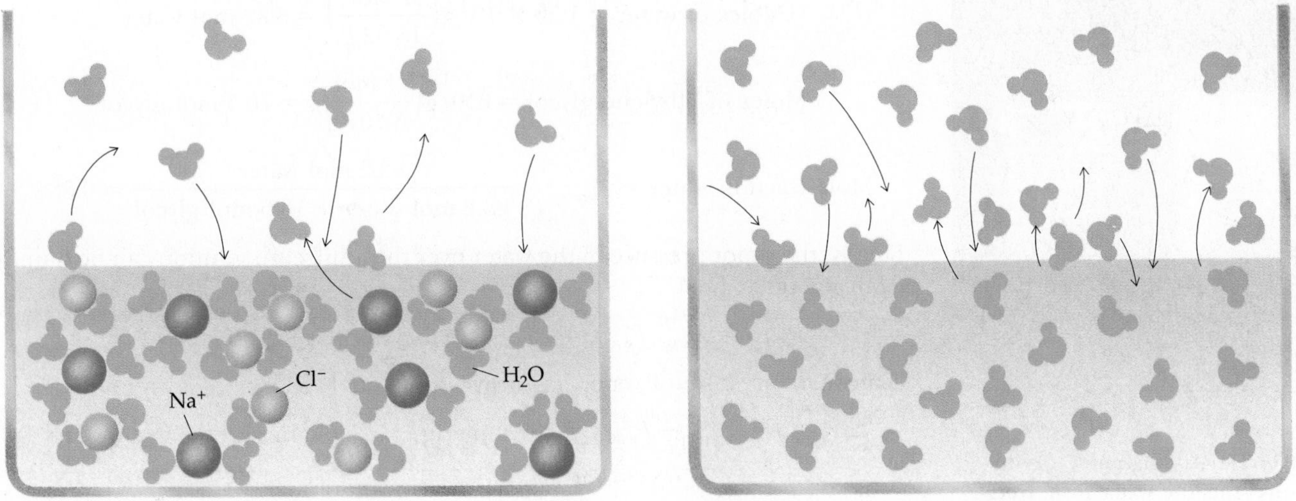

Figure 14.10 Sea water (a) is an aqueous solution of sodium chloride and many other salts. The vapor pressure of water over an aqueous solution is not as large as the vapor pressure of water over pure water (b) at the same temperature.

If we are dealing *only* with pure solvent, Equation 14.5 becomes

$$P^\circ_{\text{solvent}} = X_{\text{solvent}}K$$

where P°_{solvent} is the vapor pressure of the pure solvent and X_{solvent} is equal to 1. This means that $P^\circ_{\text{solvent}} = K$; that is, the constant K is just the vapor pressure of the pure solvent. Substituting K into Equation 14.5, we arrive at an equation called **Raoult's law.**

François M. Raoult (1830–1901) was a professor of chemistry at the University of Grenoble in France. Like the ideal gas law, Raoult's law is a description of a simplified model of a solution. *An ideal solution is one that obeys Raoult's law.* Although most solutions are not ideal, just as most gases are not ideal, Raoult's law is used as a good approximation to solution behavior.

$$P_{\text{solvent}} = X_{\text{solvent}} \cdot P^\circ_{\text{solvent}} \qquad \textbf{(14.6)}$$

This equation can be used, for example, to calculate the vapor pressure of the solvent over a solution containing a nonvolatile solute. Or by measuring the vapor pressures of the pure solvent and of the solvent over the solution, the mole fraction of solvent can be calculated.

EXAMPLE 14.4 *Using Raoult's Law*

Ethylene glycol is used in hydraulic brake fluids, as a solvent in the paint industry, in ball-point pen ink, and in synthetic fibers, among other things. The dialcohol dissolves easily in water due to extensive hydrogen bonding. See Figure 13.6.

The compound 1,2-ethanediol, HOC_2H_4OH, is commonly called ethylene glycol. Among other things, it is a common ingredient of automobile antifreeze (Figure 14.11). If 650. g of ethylene glycol is dissolved in 1.50 kg of water (a 30.2% solution, a commonly used antifreeze solution), what is the vapor pressure of the water over the solution at 90 °C? (The vapor pressure of pure water at 90 °C is 525.8 mm Hg, as given in Appendix E.)

Solution After calculating the number of moles of water and glycol, we calculate the mole fraction of the water and then use this to find the vapor pressure of water over the solution.

$$\text{Moles of water} = 1.50 \times 10^3\,\text{g}\left(\frac{1\ \text{mol}}{18.02\ \text{g}}\right) = 83.2\ \text{mol water}$$

$$\text{Moles of ethylene glycol} = 650.\,\text{g}\left(\frac{1\ \text{mol}}{62.07\ \text{g}}\right) = 10.5\ \text{mol glycol}$$

$$\text{Mole fraction water} = X_{\text{water}} = \frac{83.2\ \text{mol water}}{83.2\ \text{mol water} + 10.5\ \text{mol glycol}} = 0.888$$

Finally, the vapor pressure of the water over the antifreeze solution can be found from Raoult's law.

$$P_{\text{water}} = X_{\text{water}}P^\circ_{\text{water}} = (0.888)(525.8\ \text{mm Hg}) = 467\ \text{mm Hg}$$

The vapor pressure of water has dropped about 11%.

$$\Delta P_{\text{water}} = P_{\text{water}} - P^\circ_{\text{water}} = 467\ \text{mm Hg} - 525.8\ \text{mm Hg} = -59\ \text{mm Hg}$$

Figure 14.11 Adding antifreeze to water prevents the water from freezing. Here a jar of pure water (*left*) and a jar of water to which automobile antifreeze had been added (*right*) were kept overnight in the freezing compartment of a home refrigerator. (C. D. Winters)

EXERCISE 14.5 *Using Raoult's Law*

Assume you dissolve 10.0 g of sugar ($C_{12}H_{22}O_{11}$) into 225 mL (225 g) of water and warm the water to 60 °C. What is the vapor pressure of the water over this solution? ■

EXAMPLE 14.5 ***Calculating a Molar Mass Using Raoult's Law***

Eugenol is the active component of oil of cloves. It has an empirical formula of C_5H_6O. You dissolve 0.144 g of the compound in 10.00 g of benzene, C_6H_6. The vapor pressure of the benzene solution is 94.35 mm Hg at 25 °C, down from 95.00 mm Hg for pure benzene at this temperature. Use this information to calculate the molar mass of eugenol and its molecular formula.

Solution Knowing the vapor pressure of the pure solvent (benzene) and that of the solvent over the solution, you can calculate the mole fraction of benzene from Raoult's law, Equation 14.6.

$$P^{\circ}_{\text{benzene}} = 95.00 \text{ mm Hg} \qquad P_{\text{benzene}} = 94.35 \text{ mm Hg}$$

$$X_{\text{benzene}} = \frac{P_{\text{benzene}}}{P^{\circ}_{\text{benzene}}} = 0.9932$$

The mole fraction of benzene is defined as moles of benzene per total moles. Because the moles of benzene are

$$10.00 \text{ g } C_6H_6\left(\frac{1 \text{ mol}}{78.115 \text{ g}}\right) = 0.1280 \text{ mol } C_6H_6$$

we have the following expression:

$$X_{\text{benzene}} = 0.9932 = \frac{0.1280 \text{ mol benzene}}{0.1280 \text{ mol benzene} + ?\text{ mol eugenol}}$$

Solving this expression for moles of eugenol produces 8.764×10^{-4} mol. This means the molar mass of eugenol is

$$\frac{0.144 \text{ g eugenol}}{8.764 \times 10^{-4} \text{ mol eugenol}} = 164 \text{ g/mol}$$

Finally, because C_5H_6O has a mass of 82 g per formula unit, the molecular formula of eugenol is $(C_5H_6O)_2$, or $C_{10}H_{12}O_2$.

EXERCISE 14.6 ***Using Raoult's Law to Determine Molar Mass***

If you add 0.454 g of nitroglycerin to 100. g of benzene at 25 °C, the vapor pressure of the solution is 94.85 mm Hg. What is the molar mass of nitroglycerin? (The vapor pressure of pure benzene at 25 °C is 95.00 mm Hg.) ■

Boiling Point Elevation

According to Raoult's law (Equation 14.6) *the vapor pressure of the solvent over a solution must be lower than that of the pure solvent.* This leads to a prediction that the boiling point of a solvent must rise when a nonvolatile solute is added.

Suppose you have a solution of a nonvolatile solute in the volatile solvent benzene; the solute concentration is 0.200 mol in 100. g of benzene (C_6H_6), or $X_{\text{solute}} = 0.135$. Using Raoult's law, we find that the vapor pressure of benzene at 60 °C drops from 400. mm Hg for the pure solvent to 346 mm Hg for the solution, as follows:

The mole fraction of benzene is

$$X_{\text{benzene}} = 1 - 0.135 = 0.865$$

and its vapor pressure is given by Raoult's law.

$$P_{\text{benzene}} = X_{\text{benzene}}P^{\circ}_{\text{benzene}} = (0.865)(400.\text{ mm Hg}) = 346\text{ mm Hg}$$
$$\Delta P_{\text{benzene}} = 346\text{ mm Hg} - 400.\text{ mm Hg} = 54\text{ mm Hg}$$

This point is marked on the vapor pressure curve in Figure 14.12. Now, what is the vapor pressure when the temperature of the solution is raised another 10 °C? $P^{\circ}_{\text{benzene}}$ becomes larger with increasing temperature, so P_{benzene} must also become larger. This point, and ones calculated in the same way for other temperatures, defines a new vapor pressure curve for the solution (the lower curve in Figure 14.12).

What are the consequences of lowering the solvent vapor pressure due to the presence of a nonvolatile solute? The normal boiling point of a volatile liquid is the temperature at which its vapor pressure is equal to 1 atm, or 760 mm Hg (Section 13.3). Because the vapor pressure is lowered when a solute is added, the vapor pressure curve for the solution in Figure 14.12 reaches a pressure of 760 mm Hg at a temperature *higher* than the normal boiling point of the solvent. This is the reason that *the boiling point of a solution is* ***raised*** *relative to that of the pure solvent.*

How large is the elevation in boiling point? This depends on the mole fraction of the solvent and, ultimately, on the mole fraction or concentration of the solute. Fortunately, there is a simple relationship that connects the increase in boiling point with the *molality* of the solute (Equation 14.7).

$$\text{Elevation in boiling point, } \Delta t_{\text{bp}} = K_{\text{bp}} \cdot m_{\text{solute}} \qquad (14.7)$$

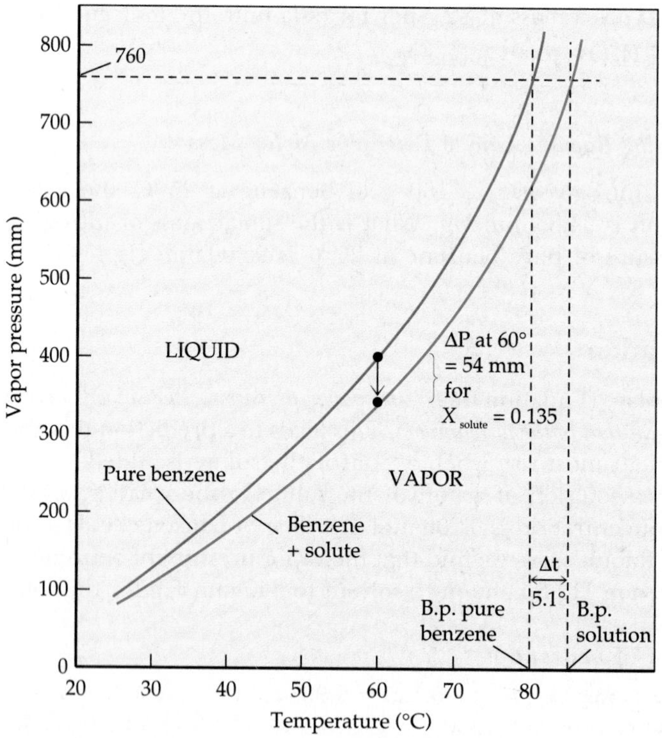

Figure 14.12 Lowering of the vapor pressure of a volatile solvent, benzene, by addition of a nonvolatile solute. (There is 0.200 mol of solute dissolved in 0.100 kg of solvent.) At 60 °C the vapor pressure of the benzene was lowered by 54 mm Hg from 400. mm Hg to 346 mm Hg. At the normal boiling point of benzene, 80.1 °C, the vapor pressure is lowered from 760 mm Hg to 657 mm Hg. To boil, the vapor pressure of the benzene over the solution must be 760 mm Hg, so the solution boils 5.1 °C higher than the pure solvent, that is, at 85.2 °C.

TABLE 14.4 Some Boiling Point Elevation and Freezing Point Depression Constants

Solvent	Normal bp (°C) pure solvent	K_{bp} (°C/m)	Normal fp (°C) pure solvent	K_{fp} (°C/m)
Water	100.00	+0.5121	0.0	−1.86
Benzene	80.10	+2.53	5.50	−5.12
Camphor	207	+5.611	179.75	−39.7
Chloroform ($CHCl_3$)	61.70	+3.63	—	—

The proportionality constant, K_{bp}, is called the **boiling point elevation constant.** It corresponds to the elevation caused by a solute concentration of one molal and has a different value for each solvent (see Table 14.4).

EXAMPLE 14.6 *Boiling Point Elevation*

In Example 14.5 you learned that eugenol (oil of cloves) has a formula of $C_{10}H_{12}O_2$. What is the boiling point of a solution in which 0.144 g of this compound is dissolved in 10.0 g of benzene?

Solution You cannot calculate the boiling point directly, only its change. Because you know K_{bp} (Table 14.4), you only need to calculate the molality, m, to use Equation 14.7. In this case you have 8.77×10^{-4} mol of eugenol

$$0.144 \text{ g eugenol}\left(\frac{1 \text{ mol eugenol}}{164.2 \text{ g eugenol}}\right) = 8.77 \times 10^{-4} \text{ mol eugenol}$$

and 0.0100 kg of solvent, so the solute has a concentration of 0.0877 m,

$$\frac{8.77 \times 10^{-4} \text{ mol eugenol}}{0.01000 \text{ kg benzene}} = 8.77 \times 10^{-2}\ m$$

Thus, the boiling point of the benzene rises 0.222 °C.

$$\Delta t_{bp} = (2.53\ °\text{C}/m)(0.0877\ m) = 0.222\ °\text{C}$$

Because the boiling point *rises* relative to that of the pure solvent, the boiling point of the solution is 80.10 °C + 0.222 °C = 80.32 °C.

EXERCISE 14.7 *Boiling Point Elevation*

What quantity of ethylene glycol, $C_2H_4(OH)_2$, must be added to 100. g of water to raise the boiling point by 1.0 °C? Express the answer in grams. ■

PROBLEM-SOLVING TIPS AND IDEAS

14.1 Using Colligative Properties to Determine Molar Masses

Early in this book you learned you could calculate a molecular formula from an empirical formula if you knew the molar mass. But how do you know the molar mass of an unknown compound? An experiment must be done to find this crucial piece of information, and one way to do this is to use a colligative property of a solution of the compound. If the compound is soluble in a solvent of appreciable vapor pressure, or with a reasonable K_{bp} or K_{fp}, the molar mass can then be determined. All approaches use the same basic logic.

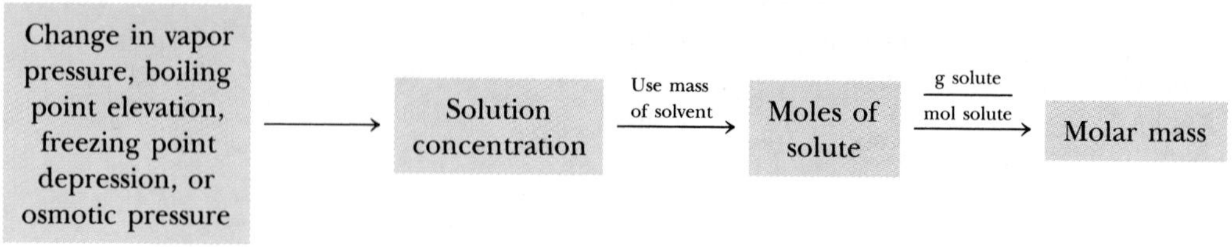

There are disadvantages to using colligative properties. An ideal solution requires that the solute concentration be low (say under 0.10 m). Because $\Delta P_{solvent}$ can sometimes be small for a dilute solution, this means that measuring the change is difficult. It is easier to measure small changes in boiling point or freezing point, but determining molar mass in a boiling solvent requires that the compound be nonvolatile and stable at the boiling point. To use the depression in solvent freezing point means the unknown compound must also be soluble in a cold solvent. Osmotic pressure (page 680) is often used for compounds having a high molar mass because a readily measurable osmotic pressure is observed even for very dilute solutions. ■

EXAMPLE 14.7 *Determining Molar Mass by Boiling Point Elevation*

Determine the molar mass of oil of wintergreen (methyl salicylate) from the following data: 1.25 g of the oil is dissolved in 100. g of benzene. The boiling point of the solution is 80.31 °C.

Solution The calculations done when using colligative properties to determine a molar mass always follow the pattern outlined in Problem-Solving Tips and Ideas, 14.1. Here we first use the boiling point elevation to calculate the solution concentration.

$$\text{Boiling point elevation} = 80.31\ °\text{C} - 80.10\ °\text{C} = 0.21\ °\text{C}$$

$$\text{Molality of solution} = \frac{\Delta t_{bp}}{K_{bp}} = \frac{0.21\ °\text{C}}{2.53\ °\text{C}/m} = 0.083\ m$$

The quantity of solute in the solution is calculated from the solution concentration.

$$\text{Moles of solute} = \left(\frac{0.083\ \text{mol}}{1.00\ \text{kg solvent}}\right)(0.100\ \text{kg solvent}) = 0.0083\ \text{mol solute}$$

Now we can combine the moles of solute with its mass.

$$\frac{1.25\ \text{g}}{0.0083\ \text{mol}} = 150\ \text{g/mol}$$

Methyl salicylate has the formula $C_8H_8O_3$ (page 523) and a molar mass of 152.14 g/mol.

EXERCISE 14.8 ***Determining Molar Mass by Boiling Point Elevation***

Crystals of the beautiful blue hydrocarbon, azulene (0.640 g), which has an empirical formula of C_5H_4, are dissolved in 100. g of benzene. The boiling point of the solution is 80.23 °C. What is the molecular formula of azulene? ■

The elevation of the boiling point of a solvent on adding a solute has many practical consequences. One of them is the summer protection your car's engine receives from "all-season" antifreeze (see Figure 14.11). The main ingredient of commercial antifreeze is 1,2-ethanediol (HOC_2H_4OH), commonly called ethylene glycol. The car's radiator and cooling system are sealed to keep the coolant under pressure, so that it will not vaporize at normal engine temperatures. When the air temperature is high in the summer, however, the radiator could still "boil over" if it were not protected with "antifreeze." By adding this nonvolatile liquid, the solution in the radiator has a higher boiling point than that of pure water.

Freezing Point Depression

Another consequence of dissolving a solute in a solvent is that the freezing point of the solution is lower than that of the pure solvent. For an ideal solution, the depression of the freezing point is given by an equation similar to that for the elevation of the boiling point:

$$\text{Freezing point depression, } \Delta t_{fp} = K_{fp} \cdot m_{solute} \qquad \textbf{(14.8)}$$

where K_{fp} is the freezing point depression constant. A few values of the constant for various solvents are given in Table 14.4.

The practical aspects of freezing point changes from pure solvent to solution are similar to those for boiling point elevation. The very name of the liquid you add to the radiator in your car, antifreeze, indicates its purpose (see Figure 14.11). The label on the container of antifreeze tells you, for example, to add 6 qt of antifreeze to a 12-qt cooling system in order to lower the freezing point to −34 °C and raise the boiling point to +226 °C.

EXAMPLE 14.8 ***Freezing Point Depression***

What volume of ethylene glycol, HOC_2H_4OH, must be added to 5.50 kg of water to lower the freezing point of the water from 0.0 °C to −10.0 °C? (This is approximately the situation in your car.)

Solution The solution concentration and freezing point depression are related by Equation 14.8. The solute concentration (molality) in a solution with a freezing point depression of −10.0 °C is therefore

$$\text{Solute concentration } (m) = \frac{\Delta t_{fp}}{K_{fp}} = \frac{-10.0\text{ °C}}{-1.86\text{ °C}/m} = 5.38\ m$$

Because we wish to use 5.50 kg of water, we need 29.6 mol of glycol.

$$\left(\frac{5.38 \text{ mol glycol}}{1.00 \text{ kg water}}\right)(5.50 \text{ kg water}) = 29.6 \text{ mol glycol}$$

The molar mass of glycol is 62.07 g/mol, so the mass required is

$$29.6 \text{ mol glycol}\left(\frac{62.07 \text{ g}}{1 \text{ mol}}\right) = 1840 \text{ g glycol}$$

The density of ethylene glycol is 1.11 kg/L, so the volume of antifreeze to be added is 1.84 kg · (1 L/1.11 kg) = 1.66 L.

EXERCISE 14.9 ***Freezing Point Depression***

Some people have summer homes on a lake or in the woods. In the northern United States, these summer homes are usually closed up for the winter. When doing so the owners "winterize" the plumbing by putting antifreeze in the toilet tanks, for example. Is adding 500. g of HOC_2H_4OH to a tank containing 3.00 kg of water sufficient to keep the water from freezing at −25 °C? ■

Why Is a Solution Freezing Point Depressed?

Imagine the freezing process this way. When the temperature is held at the freezing point of a pure solvent, freezing begins with a few molecules clustering together to form a tiny amount of solid. More molecules of the liquid move to the surface of the solid, and the solid grows. Recall that a quantity of heat, the heat of fusion, is evolved; as long as this heat energy is removed, solidification continues. If the heat is not removed, however, the opposing processes of freezing and melting can come into equilibrium; at this point, the number of molecules moving from solid to liquid is the same as the number moving from liquid to solid in a given time.

But what happens in a freezing *solution?* Again, a few molecules of solvent cluster together to form some solid. More and more solvent molecules join them, and the solid phase, which is pure solid solvent, continues to grow as long as the heat of fusion is removed. At the same time some solvent molecules are returning to the liquid from the solid. The freezing and melting processes can come into equilibrium when the numbers of molecules moving in the two directions are the same in a given time. There is a problem, however. The liquid layer next to the solid contains solute molecules or ions, whereas the solid is pure solvent. (This is analogous to the situation at the solution/vapor interface in Figure 14.10.) If the temperature is held at the normal freezing point of the pure solvent, the number of molecules of the solid (pure solvent) entering the liquid phase must therefore be greater than the number of solvent molecules leaving the solution and depositing on the solid in a given time. Why? For the same reason the vapor pressure of a solution is lower than that of the pure solvent: solute molecules have replaced some solvent molecules in the liquid at the liquid/vapor or liquid/solid interface. Thus, to have the same number of solvent molecules moving in each direction (solid → liquid and liquid → solid) in a given time, the temperature must be lowered to slow down movement from solid

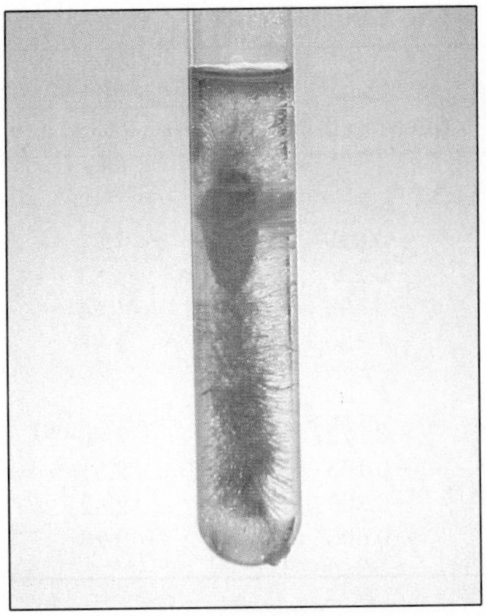

Figure 14.13 A purple dye was dissolved in water, and the solution was frozen slowly. When a solution freezes, the solvent solidifies as the pure substance. Thus, pure ice formed along the walls of the tube, and the dye stayed in solution. The solution became more concentrated as more and more solvent was frozen out, and the resulting solution had a lower and lower freezing point. When equilibrium was reached, pure, colorless ice had formed along the walls of the tube with concentrated solution in the center of the tube. (C. D. Winters)

to liquid. That is, the freezing temperature of a solution must be less than that of the pure liquid solvent.

As a solution freezes, solvent molecules are removed from the liquid phase and are deposited on the solid. This means that the concentration of the solute in the liquid solution increases, and the solution freezing point further declines as a result. A solution therefore does not have a sharply defined freezing point. However, we usually take the freezing point of a solution as the point at which solid solvent crystals first begin to appear.

When aqueous solutions freeze, the fact that the solid is pure ice and the solution is more concentrated than before (Figure 14.13) can be put to some practical use. One way to purify a liquid solvent in the laboratory is to freeze it slowly, saving the solid (pure solvent) and throwing away any remaining solution (now more concentrated with impurities). Early Americans (and some contemporary ones as well) knew that a drink called ''apple jack'' can be made by cooling apple cider. Fermenting cider produces a small amount of alcohol. If the cider is cooled, some of the water freezes to pure ice, leaving a solution higher in alcohol content.

Colligative Properties of Solutions Containing Ions

The lowering of the freezing point of a solution is important if you live in the northern United States. It is a common practice to scatter salt on snowy or icy roads or sidewalks. When the sun shines on the snow or patch of ice, a small amount is melted and the water dissolves some of the salt. The freezing point of the solution is thereby lowered. The solution ''eats'' its way through the ice, breaking it up, and the icy patch is no longer dangerous for drivers or for people walking.

Salt is used on roads because it is inexpensive and dissolves readily in water. Salt is also important because it is an electrolyte; it dissolves to give ions in solution.

$$NaCl(s) \longrightarrow Na^+(aq) + Cl^-(aq)$$

TABLE 14.5 Freezing Point Depression of Some Ionic Solutions

Mass %	m (mol/kg)	Δt_{fp} (calculated, °C)	Δt_{fp} (measured, °C)	$\frac{\Delta t_{fp}\text{, measured}}{\Delta t_{fp}\text{, calculated}}$
NaCl				
0.0700	0.0120	−0.0223	−0.0433	1.94
0.500	0.0860	−0.160	−0.299	1.87
1.00	0.173	−0.322	−0.593	1.84
2.00	0.349	−0.649	−1.186	1.83
Na_2SO_4				
0.0700	0.00493	−0.00917	−0.0257	2.80
0.500	0.0354	−0.0658	−0.165	2.51
1.00	0.0711	−0.132	−0.320	2.42
2.00	0.144	−0.268	−0.606	2.26

Remember that colligative properties depend, not on *what* is dissolved, but only on the *number of particles of solute* per solvent particle. Let us look at some experimental data (Table 14.5) to see the effect of the dissociation of salt and of another electrolyte, Na_2SO_4, on the solution freezing point.

The first thing to notice in Table 14.5 is that the measured freezing point depression is much larger than that calculated from Equation 14.9. The last column of the table shows that, for NaCl, Δt_{fp} is at least 1.8 times larger than expected based on Equation 14.9, and Δt_{fp} for Na_2SO_4 is even larger, approaching 3 times larger than expected for dilute solutions. This peculiarity was discovered by Raoult in 1884 and studied in detail by Jacobus Hendricus van't Hoff in 1887. Later in that same year Svante Arrhenius provided the explanation: electrolytes exist in solution as ions. Thus, a 0.100 m solution of NaCl really contains two solutes, 0.100 m Na^+ and 0.100 m Cl^-. What we should use to calculate the freezing point depression is the *total* molality.

$$m_{total} = m_{Na^+} + m_{Cl^-} = (0.100 + 0.100)\,\text{mol/kg} = 0.200\ \text{mol/kg}$$
$$\Delta t_{fp} = (-1.86\ °\text{C}/m)(0.200\ m) = -0.372\ °\text{C}$$

Similarly, for Na_2SO_4,

$$\underset{0.100\ \text{mol}}{Na_2SO_4(aq)} \longrightarrow \underset{0.200\ \text{mol}}{2\ Na^+(aq)} + \underset{0.100\ \text{mol}}{SO_4^{2-}(aq)}$$

$$m_{total} = m_{Na^+} + m_{SO_4^{2-}} = (0.200 + 0.100)\,\text{mol/kg} = 0.300\ \text{mol/kg}$$
$$\Delta t_{fp} = (-1.86\ °\text{C}/m)(0.300\ m) = -0.558\ °\text{C}$$

To calculate the freezing point depression, we should first find the total molality of the solute by multiplying the *apparent* molality—the number you get from the mass and molar mass of the compound—by the number of ions in the formula: 2 for NaCl, 3 for Na_2SO_4, 4 for $LaCl_3$, 5 for $Al_2(SO_4)_3$, and so on.

If solute ionization is ignored, Δt_{fp} can be calculated from Equation 14.9 using the apparent molality of the solution. The ratio of the experimentally observed value of Δt_{fp} to the value calculated from the apparent molality is called the **van't Hoff factor** and is represented by *i*.

$$i = \frac{\Delta t_{fp}\text{, measured}}{\Delta t_{fp}\text{, calculated}} = \frac{\Delta t_{fp}\text{, measured}}{K_{fp} \cdot m}$$

or

$$\Delta t_{fp}\text{, measured} = K_{fp} \cdot m \cdot i$$

The numbers in the last column of Table 14.5 are van't Hoff factors. The factor is approximately the same for all colligative properties; vapor pressure lowering, boiling point elevation, freezing point depression, and osmotic pressure are all larger for electrolytes than for nonelectrolytes of the same molality.

The van't Hoff factor approaches a whole number (2, 3, and so on) only in very dilute solutions. In more concentrated solutions, judging by the experimental freezing point depressions in Table 14.5, there are many fewer ions than expected. This behavior, which is typical of all ionic compounds, results from strong attractions and repulsions between ions. On average, a positive ion has more negative ions than positive ions in its vicinity, and a negative ion has more positive ions than negative ions in its vicinity. The net result is the same as if some of the positive and negative ions were paired, and the colligative properties are lower than would be calculated from the total molality of all the ions. Indeed, in more concentrated solutions, and especially in solvents less polar than water, many of the ions can be associated into *ion pairs* and even larger clusters.

EXAMPLE 14.9 *Freezing Point and Ionic Solutions*

A 0.00200 *m* aqueous solution of an ionic compound $Co(NH_3)_4(NO_2)_2Cl$ freezes at 0.00732 °C. How many moles of ions does 1 mol of this salt give on being dissolved?

Solution First, let us calculate the freezing point depression expected for a 0.00200 *m* solution, assuming the salt does not dissociate into ions.

$$\Delta t_{fp}\text{, calculated} = K_{fp} \cdot m = (-1.86\ °C/m)(0.00200\ m) = -3.72 \times 10^{-3}\ °C$$

Now compare the calculated freezing point depression with the measured depression. This gives us the van't Hoff factor.

$$i = \frac{\Delta t_{fp}\text{, measured}}{\Delta t_{fp}\text{, calculated}} = \frac{-7.32 \times 10^{-3}\ °C}{-3.72 \times 10^{-3}\ °C} = 1.97 \approx 2$$

It appears that 1 mol of salt gives 2 mol of ions. In this case, the ions are $[Co(NH_3)_4(NO_2)_2]^+$ and Cl^-.

EXERCISE 14.10 *Freezing Point and Ionic Compounds*

Calculate the freezing point of 500. g of water that contains 25.0 g of NaCl. Assume the van't Hoff factor *i* is 1.85 for NaCl. ■

Osmosis

See the Chemical Puzzler. An egg contains a high concentration of protein, mostly albumin. The membrane around the egg is semipermeable to water. Water moves into the egg when it is placed in pure water but out of the egg if it is placed in a concentrated sugar solution.

Osmosis is the movement of solvent molecules through a semipermeable membrane from a region of lower solute concentration to a region of higher solute concentration. This movement can be demonstrated with a simple experiment. The beaker in Figure 14.14 contains pure water, and a concentrated sugar solution is in the bag and tube. The liquids are separated by a **semipermeable membrane,** a thin sheet of material (such as a vegetable tissue or cellophane) through which only certain types of molecules can pass; here, water molecules can pass but larger sugar molecules (or hydrated ions) cannot (Figure 14.15). When the experiment begins, the liquid levels in the beaker and the tube are the same. Over time, however, the level of the sugar solution inside the tube rises, the level of pure water in the beaker falls, and the sugar solution becomes steadily more dilute. After a while, no further net change occurs; equilibrium is reached.

From a molecular point of view, the semipermeable membrane does not present a barrier to the movement of water molecules, so they move through the membrane in both directions. When a solution contains large sugar molecules or hydrated ions, however, not as many water molecules strike the membrane in a given time on the solution side as on the pure-water side. Thus, over a given time, more water molecules pass through the membrane from the pure water side to the solution side than in the opposite direction. In effect, water molecules tend to move from regions of high water concentration (low solute concentration) to regions of low water concentration (high solute concentration). The same is true for any solvent, as long as the membrane passes solvent molecules but not solute molecules.

Why does the system eventually reach equilibrium? It is evident that the solution in the tube in Figure 14.14 can never reach zero sugar concentration,

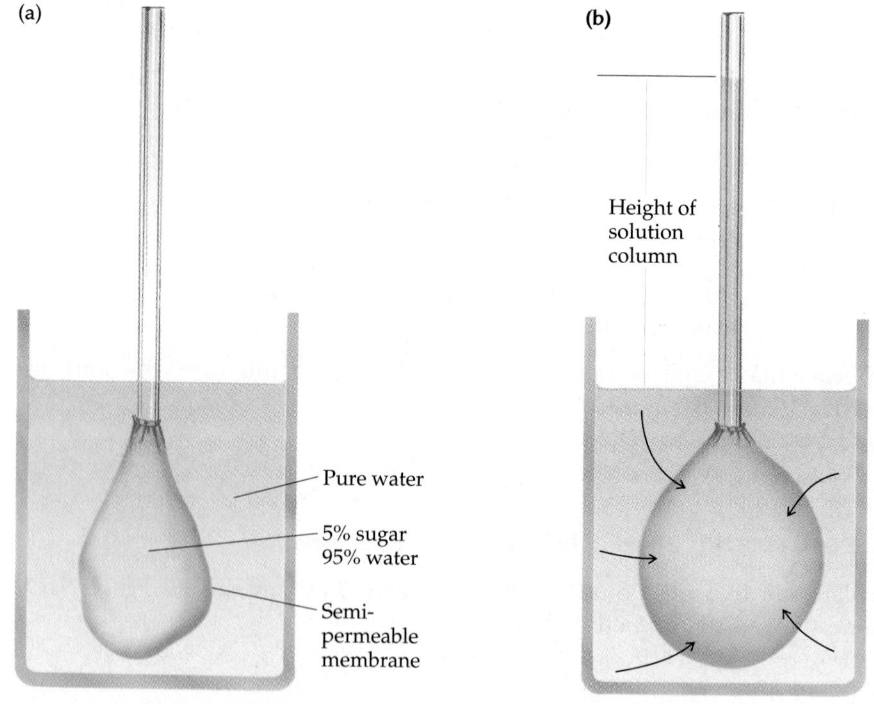

Figure 14.14 The process of osmosis. (a) The bag attached to the tube contains a solution that is 5% sugar and 95% water. The beaker contains pure water. The bag is made of a material that is selectively permeable (semipermeable); it allows water but not sugar molecules to pass through. (b) Over time, water flows from a region of low solute concentration (pure water) to one of higher solute concentration (the sugar solution). Flow continues until the pressure exerted by the column of solution in the tube above the water level in the beaker is great enough to result in equal rates of passage of water molecules in both directions. The height of the column of solution (b) is a measure of the osmotic pressure, Π.

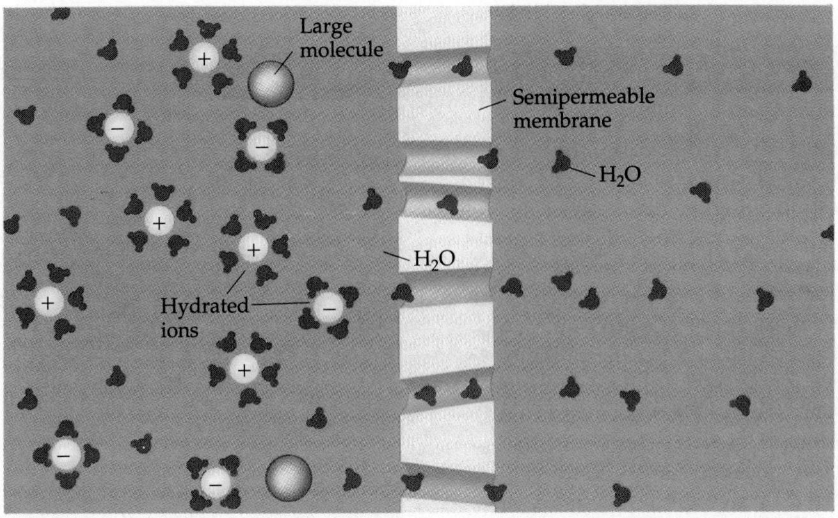

Figure 14.15 Osmotic flow through a membrane that is selectively permeable (semipermeable) to water. Dissolved substances such as hydrated ions or large sugar molecules cannot diffuse through the membrane. (This membrane is shown acting as a molecular sieve. Other types of membranes operate in a different manner, but the net effect is the same.)

which would be required to equalize the number of water molecules moving through the membrane in each direction in a given time. The answer lies in the fact that the solution moves higher and higher in the tube as osmosis continues and water moves into the sugar solution. Eventually the pressure exerted by this height of solution counterbalances the pressure of the water moving through the membrane from the pure-water side, and no further net movement of water occurs. An equilibrium of forces is achieved. The pressure created by the column of solution when the system is at equilibrium is called the **osmotic pressure, Π,** and a measure of this pressure is the difference in height between the solution in the tube and the level of pure water in the beaker (see Figure 14.14).

The osmotic pressure of sea water, which contains about 35 g of dissolved salts per kilogram of sea water, is about 27 atm.

From experimental measurements on *dilute* solutions, it is known that osmotic pressure (Π, in atmospheres) and concentration (c, in moles per liter) are related by the equation

$$\Pi = cRT \tag{14.9}$$

where R is the gas constant (0.082057 L · atm/K · mol) and T is the absolute temperature. This equation is analogous to the ideal gas law ($PV = nRT$), with Π taking the place of P and c being equivalent to n/V.

According to the osmotic pressure equation, the pressure exerted by a 0.10 M solution of particles at 25 °C is

$$\Pi = (0.10\ \text{mol/L})(0.0821\ \text{L} \cdot \text{atm/K} \cdot \text{mol})(298\ \text{K}) = 2.4\ \text{atm}$$

Because pressures on the order of 10^{-3} atm are easily measured, concentrations of about 10^{-4} M can be determined. This makes osmosis an ideal method for measuring the molar masses of very large molecules, many of which are of biological importance.

CURRENT ISSUES IN CHEMISTRY

The Chemistry of Survival

You have always dreamed of sailing in the South Pacific so you take a year out of college and hitch a ride on a 35-foot sailboat bound from San Francisco to Fiji. Somewhere west of Hawaii, however, your small boat is rammed by a whale. As the boat is sinking, you have just enough time to climb into the life raft.

Now your thoughts turn to survival, and the main problem is water. Your body loses as much as 2.5 L a day of water in exhaled breath, in sweat, and in urine and feces. This water has to be replenished. You can survive for only 12 days without water. If you drink a half liter per day, you can survive for 20 to 24 days, and 1 L of water a day will enable you to survive indefinitely, provided you can find some food.

In his epic poem, *The Ancient Mariner,* Samuel Coleridge laments that, on the ocean, there is "Water, water every where, Nor any drop to drink." The human body can use only fresh water. You cannot drink the salt water of the ocean because the salt concentration is higher than that of the cells in your body. Thus, salt water causes water to flow out of the cells into the bloodstream. For this reason one of the most important pieces of equipment in a life raft is the portable device for removing the salt from ocean water (Figure A), which works on the principle of *reverse osmosis.* If the osmotic pressure can be counterbalanced by an external pressure, then the osmotic flow of water can be reversed. The problem is that the osmotic pressure of sea water is about 27 atm. To remove any useful amount of fresh water from sea water requires the application of at least twice that pressure to the semipermeable membrane.

The reverse osmosis device in the figure was designed so that an individual can apply enough pressure to generate 4.5 L of fresh water in an hour. In fact, in 1989 William and Simonne Butler survived in a life raft in the Pacific using such a hand-pumped reverse osmosis water purifier. They were adrift with only a fishing hook, a piece of line, and a knife, in addition to the reverse-osmosis device. They were able to make about 2.5 L of water a day and caught fish for food. Although they lost some weight from their ordeal, they were in fairly good health when the Costa Rican Coast Guard rescued them after 66 days.

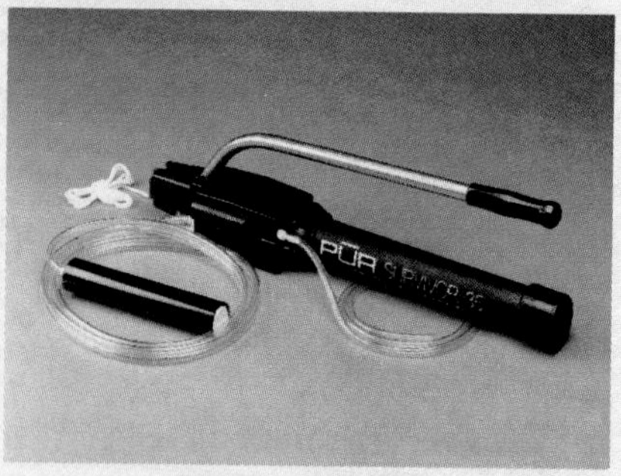

This hand-operated water desalinator works by reverse osmosis. It weighs just 7 lb and can produce 4.5 L of pure water from sea water in an hour. (Courtesy of Recovery Engineering, Inc.)

(See J. Alper: "Survival at sea," *Chem Matters,* p. 4, October 1992.)

EXAMPLE 14.10 *Osmotic Pressure and Molar Mass*

β-Carotene is the most important of the A vitamins. Its molar mass can be determined by measuring the osmotic pressure generated by a given mass of carotene dissolved in the solvent chloroform. Calculate the molar mass of carotene if 7.68 mg, dissolved in 10.0 mL of chloroform, gives an osmotic pressure of 26.57 mm Hg at 25.00 °C.

Solution The concentration of β-carotene in chloroform can be calculated from the osmotic pressure (Equation 14.9).

$$\text{Concentration (M)} = \frac{\Pi}{RT} = \frac{(26.57 \text{ mm Hg})(1 \text{ atm}/760 \text{ mm Hg})}{(0.082057 \text{ L}\cdot\text{atm/K}\cdot\text{mol})(298.15 \text{ K})}$$

$$= 1.429 \times 10^{-3} \text{ mol/L}$$

Now the quantity of β-carotene dissolved in 10.0 mL of solvent can be calculated.

$$(1.429 \times 10^{-3} \text{ mol/L})(0.0100 \text{ L}) = 1.429 \times 10^{-5} \text{ mol}$$

This quantity of β-carotene (1.429×10^{-5} mol) is equivalent to 7.68 mg (7.68×10^{-3} g). This gives us the molar mass.

$$\frac{7.68 \times 10^{-3}\ \text{g}}{1.429 \times 10^{-5}\ \text{mol}} = 537\ \text{g/mol}$$

β-Carotene is a hydrocarbon (which consists of eight isoprene units; see page 505) with the formula $C_{40}H_{56}$.

EXERCISE 14.11 *Osmotic Pressure and Molar Mass*

144 mg of aspartame, the artificial sweetener, is dissolved in exactly 25 mL of water. The osmotic pressure observed at 25 °C is 364 mm Hg. What is the molar mass of aspartame? ■

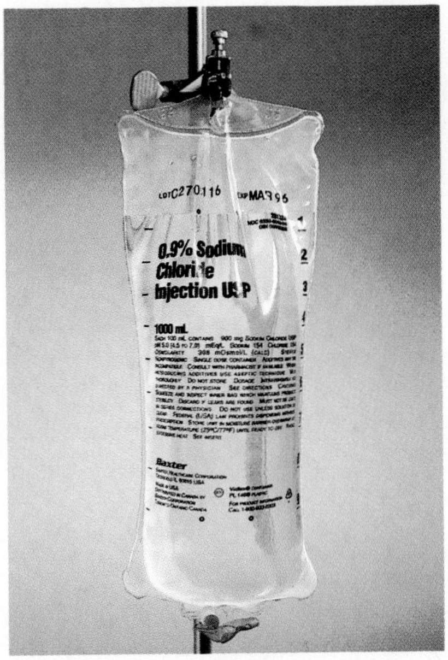

An isotonic saline solution. (C. D. Winters)

Osmosis is of great practical significance, especially for people in the health professions. Patients who become dehydrated through illness often need to be given water and nutrients intravenously. But water cannot simply be dripped into a patient's vein. Rather, the intravenous solution must have the same overall solute concentration as the patient's blood: the solution must be isoosmotic, or *isotonic.* If pure water was used, the inside of a blood cell would have a higher solute concentration (lower water concentration), and water would flow into the cell. This *hypotonic* situation would cause the red blood cells to burst (Figure 14.16). The opposite situation, *hypertonicity,* occurs if the intravenous solution is more concentrated than the contents of the blood cell. In this case the cell would lose water and shrivel up. To combat this, a dehydrated patient is rehydrated in the hospital with a sterile saline solution that is 0.16 M NaCl, a solution that is isotonic with body fluids.

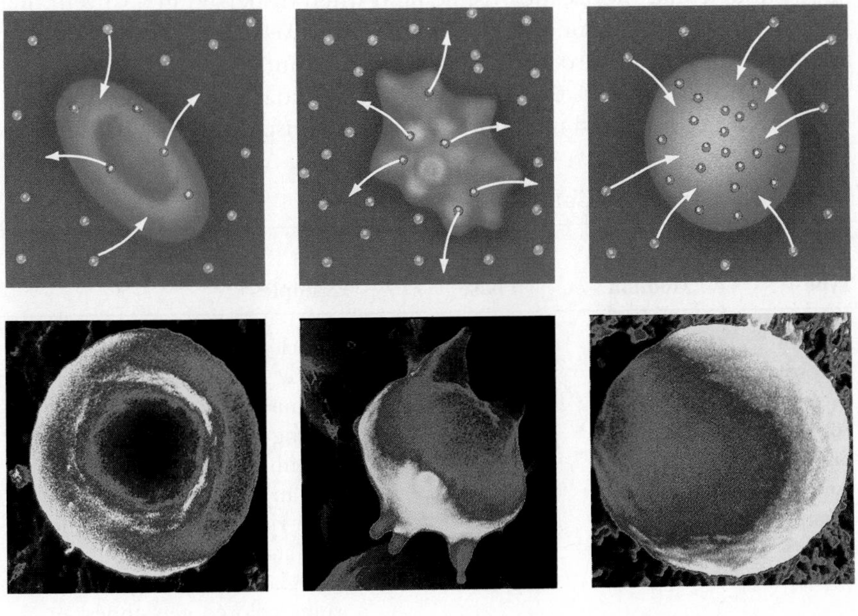

(a) Isotonic solution (b) Hypertonic solution (c) Hypotonic solution

Figure 14.16 Osmosis and living cells. (a) A cell placed in an *isotonic* solution. The net movement of water in and out of the cell is zero because the concentration of solutes inside and outside the cell is the same. (b) In a *hypertonic* solution, the concentration of solutes outside the cell is greater than that inside. There is a net flow of water out of the cell, causing the cell to dehydrate, shrink, and perhaps die. (c) In a *hypotonic* solution, the concentration of solutes outside of the cell is less than that inside. There is a net flow of water into the cell, causing the cell to swell and perhaps to burst. (See the Chemical Puzzler, page 654.) (Photos by David Phillips/Science Source/Photo Researchers, Inc.)

14.4 COLLOIDS

The solutions discussed so far represent only one type of a broad spectrum of mixtures. We defined a solution broadly as a homogeneous mixture of two or more substances in a single phase. To this we should add that, *in a true solution, no settling of the solute should be observed, and the solute particles should be in the form of ions or relatively small molecules.* Thus, common salt and sugar form true solutions in water.

You are also familiar with *suspensions,* which result, for example, if a handful of fine sand is added to water and shaken vigorously. Sand particles are still visible and gradually settle to the bottom of the beaker or bottle. **Colloidal dispersions** represent a state intermediate between a solution and a suspension. Colloids include many of the foods you eat and the materials around you; among them are Jello, milk, fog, and porcelain (see Table 14.6).

Around 1860, the British chemist Thomas Graham found that substances such as starch, gelatin, glue, and albumin from eggs diffused only very slowly when placed in water, compared with sugar or salt. In addition, the former substances differ significantly in their ability to diffuse through a thin membrane: sugar molecules diffuse through many membranes, but the very large molecules that make up starch, gelatin, glue and albumin do not. Moreover, Graham found that he could not crystallize these latter substances, whereas he could crystallize sugar, salt, and other materials that form true solutions. Therefore, Graham coined the word *colloid* (from the Greek, meaning "glue") to describe a class of substances distinctly different from sugar and salt and similar materials. We now know we can crystallize some colloidal substances, albeit with difficulty, so there really is no sharp dividing line between these classes. Colloids do, however, have the following distinguishing characteristics. (1) It is generally true that colloidal materials have very high molecular masses; this is certainly true of human cells and of proteins such as hemoglobin that have molar masses in the thousands. (2) The particles of a colloid are relatively large (say 1000 nm), large enough that they scatter visible light when dispersed in a solvent, making the mixture appear cloudy (Figure 14.17). (3) Even though colloidal particles are large, they are not so large that they settle out.

Graham also gave us the words *sol* for a colloidal solution (a dispersion of a solid substance in a fluid medium) and *gel* for a dispersion that has developed a

TABLE 14.6 Types of Colloids

Type	Dispersing Medium	Dispersed Phase	Examples
Aerosol	Gas	Liquid	Fog, clouds, aerosol sprays
Aerosol	Gas	Solid	Smoke, airborne viruses, automobile exhaust
Foam	Liquid	Gas	Shaving cream, whipped cream
Emulsion	Liquid	Liquid	Mayonnaise, milk, face cream
Sol	Liquid	Solid	Gold in water, milk of magnesia, mud
Foam	Solid	Gas	Foam rubber, sponge, pumice
Gel	Solid	Liquid	Jelly, cheese, butter
Solid sol	Solid	Solid	Milk glass, many alloys such as steel, some colored gemstones

(a)

(b)

Figure 14.17 Colloidal suspensions or dispersions scatter light, a phenomenon known as the Tyndall effect. (a) Dust in the air scatters the light coming through the trees in a forest along the Oregon coast. (b) A narrow beam of light from a laser is passed through an NaCl solution (*left*) and then a colloidal mixture of gelatin and water (*right*). [(a) Bob Pool/Tom Stack and Associates. (b) C. D. Winters]

structure that prevents it from being mobile. For example, Jello is a sol when the solid is first mixed with boiling water, but it becomes a gel when cooled. Other examples of gels are the gelatinous precipitates of $Al(OH)_3$, $Fe(OH)_3$, and $Cu(OH)_2$ (Figure 14.18).

Colloidal dispersions consist of finely divided particles that, as a result, have a very high surface area. For example, if you have one-millionth of a mole of colloidal particles, each assumed to be a sphere with a diameter of 200 nm, the total surface area of the particles would be on the order of 100 million square centimeters or the size of several football fields. Therefore, many of the properties of colloids depend on the properties of surfaces.

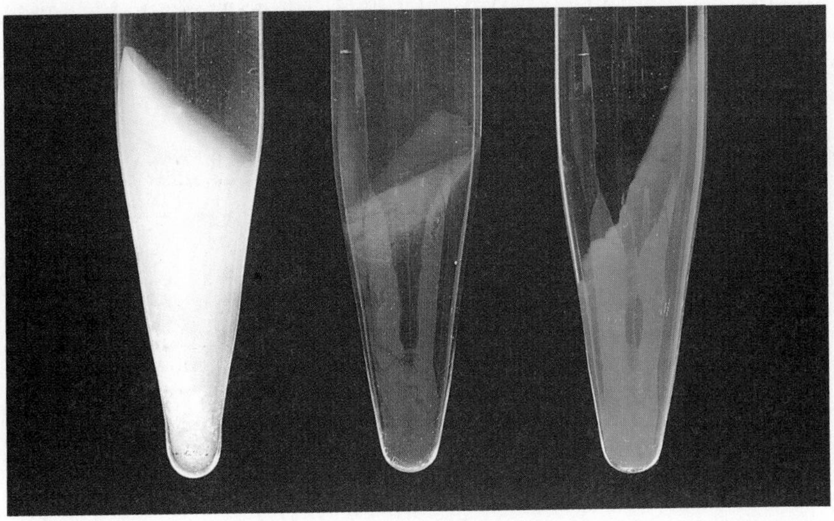

Figure 14.18 Three gelatinous precipitates: (*left*) $Al(OH)_3$; (*middle*) $Fe(OH)_3$, and (*right*) $Cu(OH)_2$. (C. D. Winters)

Types of Colloids

Colloids are classified according to the state of the dispersed phase and the dispersing medium (or continuous phase). Table 14.6 lists several types of colloids and gives examples of each.

Colloids with water as the dispersing medium can be classified as **hydrophobic** (from the Greek, meaning "water-fearing") or **hydrophilic** ("water-loving"). A hydrophobic colloid is one in which only weak attractive forces exist between the water and the surface of the colloidal particles. Examples are dispersions of metals and of nearly insoluble salts in water. When salts like AgCl precipitate (for example, from the reaction of $AgNO_3$ and NaCl), the result is often a colloidal dispersion. Because the reaction occurs too rapidly for ions to gather from long distances and make one large crystal, the ions can only form small particles.

Why do hydrophobic colloids exist? Why don't the particles come together (coagulate) and form larger particles? The answer seems to be that the colloid particles carry electric charges of like sign. An AgCl particle, for example, absorbs Ag^+ ions if they are present in substantial concentration; there is an attraction between Ag^+ ions in solution and Cl^- ions on the surface of the particle. The particle thus becomes positively charged. (In the same manner, a particle can accumulate a negative charge in a solution having a high concentration of negative ions, such as Cl^-.) Each colloid particle surrounds itself with a layer of ions, so the particles repel one another and are prevented from coming together to form a precipitate (Figure 14.19).

A stable hydrophobic colloid can be made to coagulate by introducing ions into the dispersing medium. Milk is a colloidal suspension of hydrophobic particles. When milk ferments, lactose (milk sugar) is converted to lactic acid, which forms hydrogen ions and lactate ions. The protective charge on the surfaces of the colloidal particles is overcome, and the milk coagulates; the milk solids come together in clumps called "curds."

Soil particles carried by water in rivers and steams are hydrophobic colloids. When river water carrying large amounts of colloidally suspended particles meets sea water with its high concentration of salts, the particles coagulate to form the silt seen at the mouth of the river (Figure 14.20). This is also the reason that municipal water treatment plants add aluminum salts such as $Al_2(SO_4)_3$ to water. In aqueous solution, aluminum ions exist as $[Al(H_2O)_6]^{3+}$ and even more complicated ions, which neutralize the charge on the hydrophobic colloidal soil particles, precipitating the soil particles out from the water.

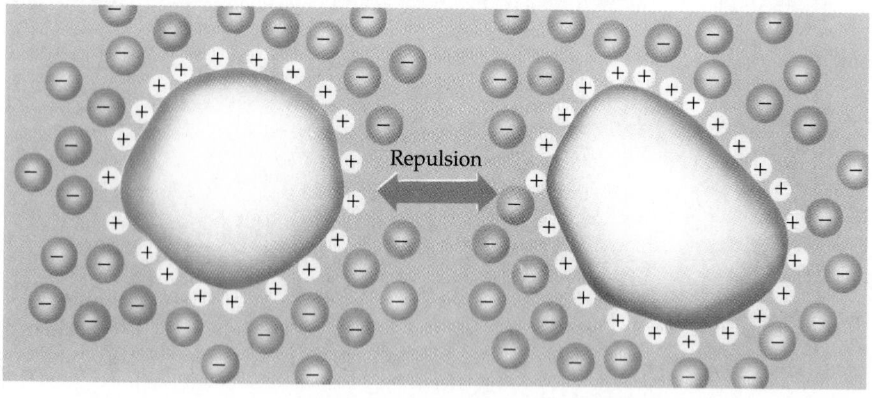

Figure 14.19 A hydrophobic colloid is stabilized by positive ions absorbed onto each particle and an atmosphere of negative ions surrounding the particle.

Figure 14.20 Silt forms at a river delta as colloidal soil particles come in contact with salt water in the ocean. (Here the Ashley and Cooper Rivers empty into the Atlantic Ocean at Charleston, SC.) The high concentration of ions in salt water causes the colloidal soil particles to coagulate. (NASA/Peter Arnold, Inc.)

Hydrophilic colloids are strongly attracted to water molecules. They have groups, such as —OH and —NH_2, on their surfaces (Figure 14.21). These groups form strong hydrogen bonds to water, thus stabilizing the colloid. Proteins and starch are important examples, and homogenized milk is the most familiar example.

Emulsions are colloidal dispersions of one liquid in another, such as oil or fat in water. You find emulsions in familiar forms: salad dressing, mayonnaise, and milk, among others. If vegetable oil and vinegar are mixed to make a salad dressing, the mixture quickly separates into two layers because the nonpolar oil molecules do not interact with the polar molecules of water and acetic acid. So why are milk and mayonnaise apparently homogeneous mixtures that do not separate into layers? The answer is that they contain an **emulsifying agent,** such as soap or a protein. Lecithin is a protein found in egg yolks, so mixing egg yolks with oil and vinegar stabilizes the colloidal dispersion known as mayonnaise. To understand how this works, we look into the functioning of soaps and detergents.

One of the best ways to find out about the properties of food in terms of chemistry is in *The Cookbook Decoder,* a cookbook written by a chemist, A. Grosser (New York, Beaufort Books, Inc., 1981).

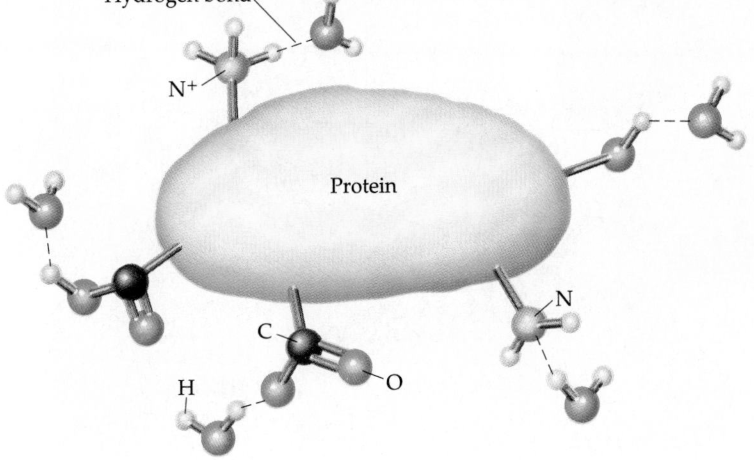

Figure 14.21 A hydrophilic colloidal particle is stabilized by hydrogen bonding to water.

Surfactants

A sodium soap is a solid at room temperature, whereas potassium soaps are usually liquid.

Soaps and detergents are emulsifying agents. Soap is made by heating a fat with sodium or potassium hydroxide (Section 11.6).

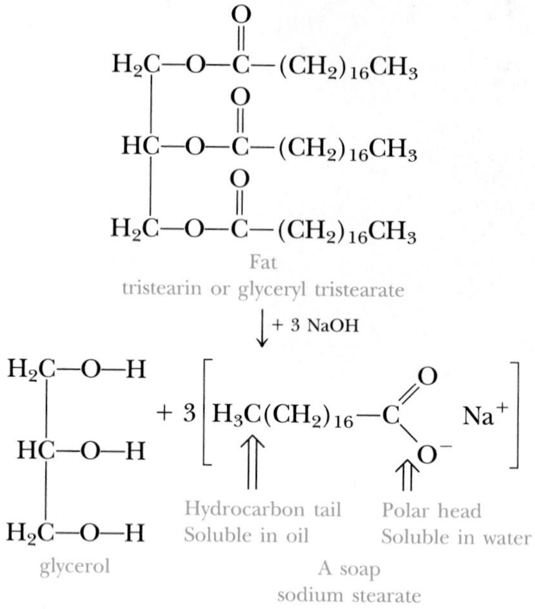

The resulting fatty acid anion has a split personality: it has a nonpolar, hydrophobic hydrocarbon tail that is soluble in other similar hydrocarbons and a polar, hydrophilic head that is soluble in water.

Oil cannot be simply washed away from dishes or clothing with water because oil is a nonpolar liquid that is insoluble in polar water. Instead, we must add soap to the water to clean away the oil. The nonpolar molecules of the oil

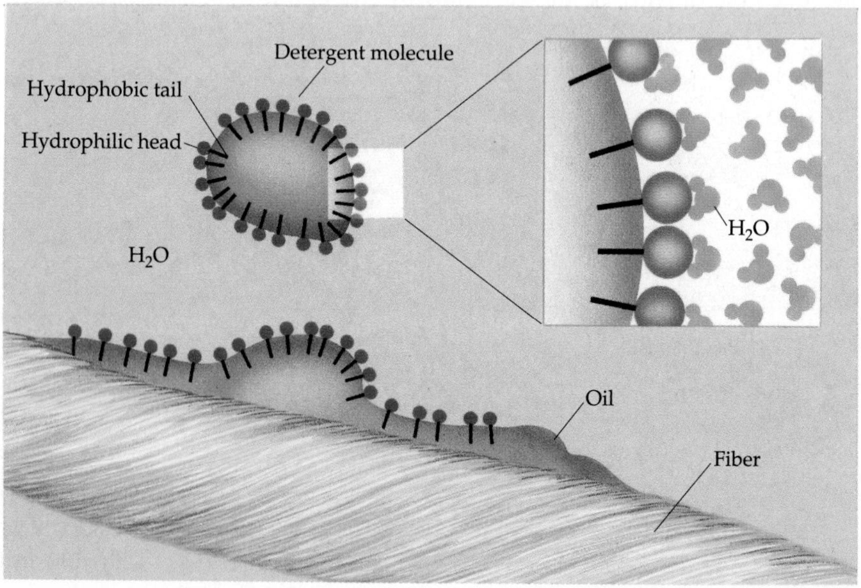

Figure 14.22 The cleaning action of soap depends on the structure of the soap. Soap molecules interact strongly with water through the charged, hydrophilic end of the molecule. The long, hydrocarbon end of the molecule is hydrophobic but can bind through dispersion forces with other hydrocarbons such as oil.

Figure 14.23 The surface tension of water is affected by adding a detergent. At the left, sulfur (density = 2.1 g/cm^3) is carefully placed on the surface of water (density 1.0 g/cm^3). The surface tension of the water keeps the denser sulfur afloat. Several drops of detergent are then placed on the surface of the water (*right*). The surface tension of the water is reduced, and the sulfur sinks to the bottom of the beaker. (C. D. Winters)

interact with the hydrocarbon tails of the soap molecules, leaving the polar heads of the soap to interact with surrounding water molecules. The oil and water then mix (Figure 14.22). If the oily material on a piece of clothing or a dish also contain some dirt particles, the dirt can now be washed away.

Substances such as soaps that affect the properties of surfaces, and so affect the interaction between two phases, are called surface-active agents, or **surfactants,** for short (Figure 14.23). A surfactant that is used for cleaning has come to be called a **detergent.** One function of a surfactant is to lower the surface tension of water, which enhances the cleansing action of the detergent.

The soap and detergent industry is enormous. Roughly 25 to 30 million tons of household and toilet soaps, and synthetic and soap-based laundry detergents are produced annually worldwide.

Many detergents used in the home and industry are synthetic. One example is sodium lauryl benzenesulfonate, a biodegradable compound.

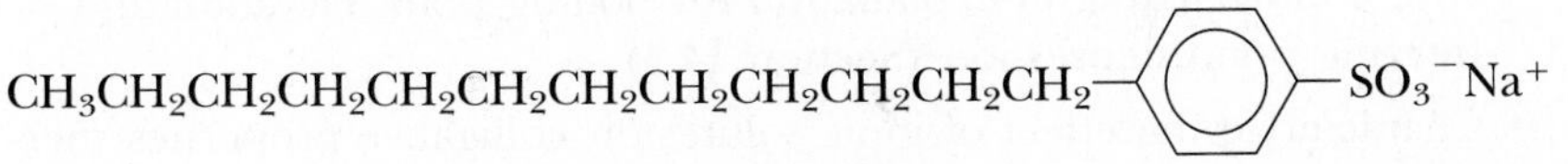

sodium lauryl benzenesulfonate

In general, synthetic detergents use the sulfonate, $-SO_3^-$ group as the polar head instead of carboxylate, $-CO_2^-$ because the $-CO_2^-$ head allows the soap to form an insoluble precipitate with any Ca^{2+} or Mg^{2+} ions that are present in water. ''Hard'' water is characterized by high concentrations of these ions, which combined with carboxylates are the source of bathtub ring and ''tattle-tale'' gray. The synthetic sulfonate detergents have the advantage of not forming such precipitates because their calcium salts are more soluble in water.

CHAPTER HIGHLIGHTS

When you have finished studying this chapter, you should be able to

- Define the terms **solution, solvent, solute,** and **colligative properties,** among others (Section 14.1).
- Use the following concentration units: **molality, molarity, mole fraction, weight percent,** and **parts per million** (Section 14.1).
- Describe the process of dissolving a solute in a solvent, including the energy changes that may occur (Section 14.2).
- Understand the difference between a **saturated, unsaturated,** and **supersaturated** solution (Section 14.2).
- Explain the difference between a **miscible** and an **immiscible** liquid in a liquid/liquid solution (Section 14.2).
- Understand the relation of **lattice energy** and **enthalpy of hydration** to the **enthalpy of solution** for an ionic solute (Section 14.2).
- Describe the effect of pressure and temperature on the solubility of a solute (Section 14.2).
- Use **Henry's law** to calculate the solubility of a gas in a solvent (Section 14.2).
- Apply **Le Chatelier's principle** to the change in solute solubility with temperature changes (Section 14.2).
- Calculate the mole fraction of a solute or solvent ($X_{solvent}$) or the effect of a solute on solvent vapor pressure ($P_{solvent}$) using **Raoult's law** (Section 14.3).

$$\text{Raoult's law: } P_{solvent} = X_{solvent} \cdot P^{\circ}_{solvent} \quad \textbf{(14.6)}$$

(where $P^{\circ}_{solvent}$ is the vapor pressure of the pure solvent at a specified temperature.)

- Calculate the boiling point elevation or freezing point depression caused by a solute in a solvent (Section 14.3).

$$\text{Elevation in boiling point, } \Delta t_{bp} = K_{bp} \cdot m_{solute} \quad \textbf{(14.7)}$$

$$\text{Depression of freezing point, } \Delta t_{fp} = K_{fp} \cdot m_{solute} \quad \textbf{(14.8)}$$

- Use colligative properties to determine the molar mass of a solute (Section 14.3 and Problem-Solving Tips and Ideas, 14.1).
- Give a molecular-level explanation for boiling point elevation and freezing point depression (Section 14.3).
- Characterize the effect of ionic solutes on colligative properties (Section 14.3).
- Use the **van't Hoff *i* factor** in calculations involving colligative properties (Section 14.3).
- Give a molecular-level explanation for osmosis (Section 14.3).
- Use osmotic pressure (Π) in calculations of solute concentration or the determination of the molar mass of a solute (Section 14.3).

$$\Pi = cRT \quad \textbf{(14.9)}$$

(where c = concentration in moles per liter, R is the gas constant, and T is the temperature in kelvins).

- Recognize the difference between a homogeneous solution, a suspension, and a **colloid** or colloidal dispersion (Section 14.4).
- Characterize **hydrophobic** and **hydrophilic** colloids (Section 14.4).
- Describe the action of a **surfactant** (Section 14.4).

STUDY QUESTIONS

Review Questions

1. Is the following statement true or false? If false, change it to make it true. "Colligative properties depend on the nature of the solvent and solute and on the concentration of the solute."
2. Name the four colligative properties described in this chapter. Write the mathematical expression that describes each of these.
3. Define molality and tell how it differs from molarity.
4. How is the solubility of a gas or solid related to its enthalpy of solution?
5. Explain the relation between enthalpy of solution of a gas or solid and the temperature dependence of solubility.
6. How are the aqueous solubilities of NaCl, KCl, RbCl, and CsCl in Figure 14.9 related to their lattice energies?
7. Name three effects that govern the solubility of a gas in water.
8. If you dissolve equal molar amounts of NaCl and $CaCl_2$ in water, the calcium salt lowers the freezing point of the water almost 1.5 times as much as the NaCl. Why?
9. Explain why a cucumber shrivels up when it is placed in a concentrated solution of salt.
10. Explain the differences between a colloidal dispersion, a suspension, and a true solution.
11. Explain the difference between a sol and a gel. Give an example of each.
12. Explain how a surfactant, such as a soap or detergent, functions.

Numerical Questions

Concentration Units

13. Assume you dissolve 2.56 g of malic acid, $C_4H_6O_5$, in half a liter of water (500.0 g). Calculate the molarity, molality, mole fraction, and weight percent of acid in the solution.

14. Camphor, a white solid with a pleasant odor, is extracted from the roots, branches, and trunk of the camphor tree. Assume you dissolve 45.0 g of camphor ($C_{10}H_{16}O$) in 500. mL of ethanol, C_2H_5OH. Calculate the molarity, molality, mole fraction, and weight percent of camphor in this solution. (The density of ethanol is 0.785 g/mL.)

15. Fill in the blanks in the table.

Compound	Molality	Weight Percent	Mole Fraction
KI	0.15	____	____
C_2H_5OH	____	3.0%	____
$C_{12}H_{22}O_{11}$	0.10	____	____

16. Fill in the blanks in the table.

Compound	Molality	Weight Percent	Mole Fraction
NaCl	____	10.0	____
CH_3CO_2H	0.0083	____	____
$C_2H_4(OH)_2$	____	15.0	____

17. You want to prepare a solution that is 0.200 m in $NaNO_3$. How many grams of the salt must you add to 250. g of water? What is the mole fraction of $NaNO_3$ in the resulting solution?

18. You want to prepare a solution that is 0.0512 m in K_2CO_3. How many grams of the salt must you add to 500. g of water? What is the mole fraction of K_2CO_3 in the solution?

19. You want to prepare a solution of ethylene glycol, $C_2H_4(OH)_2$, that has a glycol mole fraction of 0.125. How many grams of the glycol should you combine with 950. g of water? What is the molality of the solution?

20. You wish to prepare an aqueous solution of glycerol, $C_3H_5(OH)_3$, that has a mole fraction of the organic compound of 0.093. How many grams of glycerol must you combine with 500. g of water to make this solution? What is the molality of the solution?

21. Fill in the blanks in the following table.

Compound	Grams Compound	Grams Water	Molality	Mole Fraction of Compound
K_2CO_3	____	250.	0.0125	____
C_2H_5OH	13.5	150.	____	____
$NaNO_3$	____	555	____	0.0934

22. Fill in the blanks in the following table.

Compound	Grams Compound	Grams Water	Molality	Mole Fraction of Compound
$AgNO_3$	_____	800.	0.0245	_____
$C_2H_4(OH)_2$	_____	250.	_____	0.0545
$Pt(NH_3)_2Cl_2$	0.0075	200.	_____	_____

23. Hydrochloric acid is sold as a concentrated aqueous solution. If the molarity of commercial HCl is 12.0 and its density is 1.18 g/cm^3, calculate
(a) The molality of the solution
(b) The weight percent of HCl in the solution

24. Concentrated sulfuric acid has a density of 1.84 g/cm^3 and is 95.0% by weight H_2SO_4. What is the molality of this acid? What is its molarity?

25. A 10.7 *m* solution of NaOH has a density of 1.33 g/cm^3 at 20 °C. Calculate
(a) The mole fraction of NaOH
(b) The weight percentage of NaOH
(c) The molarity of the solution

26. Concentrated aqueous ammonia has a molarity of 14.8 and a density of 0.90 g/cm^3. What is the molality of the solution? Calculate the mole fraction and weight percentage of NH_3.

27. You dissolve 2.00 g of $Ca(NO_3)_2$ in 750 g of water. What is the molality of $Ca(NO_3)_2$? What is the total molality of the ions? (That is, what is the total concentration of Ca^{2+} and NO_3^- ions?)

28. You want a solution that is 0.100 *m* in ions. How many grams of Na_2SO_4 must you dissolve in 150. g of water? (Assume total dissociation of the ionic solid.)

29. The average lithium ion concentration in sea water is 0.18 ppm. What is the molality of Li^+ in sea water?

30. Silver ion has an average concentration of 28 parts per billion (ppb) in U.S. water supplies.
(a) What is the molality of the silver ion?
(b) If you want 1.0×10^2 g of silver and could recover it chemically from water supplies, how many liters of water must you treat? (Assume the density of water is 1.0 g/cm^3.)

The Solution Process

31. You make a saturated solution of NaCl at 25 °C. No solid is present in the beaker holding the solution. What can be done to increase the amount of dissolved NaCl in this solution? You know that NaCl has an endothermic enthalpy of solution. That is,

$$NaCl(s) + heat \longrightarrow NaCl(aq)$$
$$\Delta H_{solution}[NaCl] = +3.9 \text{ kJ/mol}$$

(a) Add more solid NaCl.
(b) Raise the temperature of the solution.
(c) Raise the temperature of the solution and add some NaCl.
(d) Lower the temperature of the solution and add some NaCl.

32. Potassium nitrate dissolves in water in the following amounts per 100. g of water: (a) 31.6 g at 20 °C, (b) 85.5 g at 50 °C, and (c) 202 g at 90 °C. Is the enthalpy of solution of KNO_3 expected to be positive or negative?

33. The ionic compound ammonium chloride can be used as a "chemical cold pack." When the solid is mixed with water, the solution becomes very cold. Calculate the enthalpy of formation of $NH_4Cl(aq)$ [ΔH_f° (NH_4Cl, 1 *m*)] knowing that $NH_4Cl(s)$ has an enthalpy of solution of +14.8 kJ/mol. Should the compound become more or less soluble as the temperature increases? Explain briefly.

34. Some ionic compounds evolve heat when mixed with water. In fact, dissolving $CaCl_2$ is so exothermic that it is used in a commercial "hot pack." Calculate the enthalpy of the reaction $CaCl_2(s) \rightarrow Ca^{2+}(aq) + 2\ Cl^-(aq)$ from $\Delta H_f^\circ[CaCl_2(s)] = -795$ kJ/mol and $\Delta H_f^\circ[CaCl_2(aq)] = -877.89$ kJ/mol. Should the compound become more or less soluble as the temperature increases? Explain briefly.

35. In which of the following solvents should Na^+ have the most negative enthalpy of solvation? Explain briefly.
(a) $H_2O(\ell)$ (c) $CS_2(\ell)$
(b) $CCl_4(\ell)$ (d) $CH_3CH_2CH_2CH_2CH_2CH_2CH_3(\ell)$

36. Which anion should have the most negative enthalpy of hydration? Explain briefly.
(a) I^- (c) Cl^-
(b) Br^- (d) F^-

Henry's Law

37. The partial pressure of O_2 in your lungs varies from 25 mm Hg to 40 mm Hg. How much O_2 can dissolve in water at 25 °C if the partial pressure of O_2 is 40. mm Hg?

38. Henry's law constant for O_2 in water at 25 °C is 1.66×10^{-6} M/mm Hg. Which of the following is a reasonable constant when the temperature is 50 °C? Explain the reason for your choice.
(a) 8.80×10^{-7} M/mm Hg
(b) 1.66×10^{-6} M/mm Hg
(c) 3.40×10^{-6} M/mm Hg
(d) 8.40×10^{-5} M/mm Hg

39. A soft drink has an aqueous CO_2 concentration of 0.0506 M at 25 °C. What is the pressure of CO_2 gas in the drink?

40. Hydrogen gas has a Henry's law constant of 1.07×10^{-6} M/mm Hg at 25 °C when dissolving in water. If the total pressure of gas (H_2 gas plus water vapor) over water is 1.0 atm, what is the concentration of H_2 in the water in grams per milliliter? (See Appendix E for the vapor pressure of water.)

Vapor Pressure Changes

41. Some ethylene glycol ($C_2H_4(OH)_2$, 35.0 g) is dissolved in half a liter of water (500.0 g). The vapor pressure of water at 32 °C is 35.7 mm Hg. What is the vapor pressure of the water/glycol solution at 32 °C? (The glycol is assumed to be nonvolatile.)

42. Urea, $OC(NH_2)_2$, is widely used in fertilizers and plastics. The compound is quite soluble in water; 1.00 g can be dissolved in 1.00 mL of water. Assuming the density of water is 1.00 g/mL and that you dissolve 9.00 g of urea in 10.0 mL of water, what is the vapor pressure of the solution at 24 °C?

43. Pure ethylene glycol, $C_2H_4(OH)_2$, is added to 2.00 kg of water in the cooling system of a car. The vapor pressure of the water in the system when the temperature is 90 °C is 457 mm Hg. How many grams of glycol are added? (Assume that ethylene glycol is not volatile at this temperature. See Appendix E for the vapor pressure of water.)

44. Pure iodine (100. g) is dissolved in 300. g of CCl_4 at 65 °C. Given that the vapor pressure of CCl_4 at this temperature is 531 mm Hg, what is the vapor pressure of the CCl_4/I_2 solution at 65 °C? (Assume that I_2 does not contribute to the vapor pressure.)

45. A quantity (10.0 g) of a nonvolatile solute is dissolved in 100. g of benzene (C_6H_6). The vapor pressure of pure benzene at 30 °C is 121.8 mm Hg, and that of the solution is 113.0 mm Hg at the same temperature. What is the molar mass of the solute?

46. A solution prepared from 20.0 g of a nonvolatile solute in 154 g of the solvent carbon tetrachloride (CCl_4) has a vapor pressure of 504 mm Hg at 65 °C. (The vapor pressure of pure CCl_4 is 531 mm Hg at 65 °C.) What is the molar mass of the solute?

Boiling Point Elevation

47. What is the boiling point of a solution composed of 15.0 g of urea, $OC(NH_2)_2$, in 0.500 kg of water?

48. Verify the result in Figure 14.12 that 0.200 mol of a nonvolatile solute in 100. g of benzene (C_6H_6) produces a solution whose boiling point is 85.2 °C.

49. What is the boiling point of a solution composed of 15.0 g of $CHCl_3$ and 0.515 g of the nonvolatile solute acenaphthalene, $C_{12}H_{10}$, a component of coal tar?

50. What is the boiling point of a solution composed of 0.755 g of caffeine ($C_8H_{10}O_2N_4$) in 100. g of benzene?

51. A solution of glycerol, $C_3H_5(OH)_3$, in 750. g of water has a boiling point of 104.3 °C at a pressure of 760 mm Hg. How many grams of glycerol are in the solution? What is the mole fraction of the solute?

52. Phenanthrene, $C_{14}H_{10}$, is an aromatic hydrocarbon. If you dissolve some phenanthrene in 50.0 g of benzene, the boiling point of the solution is 80.51 °C. How many grams of the hydrocarbon must have been dissolved?

53. Arrange the following solutions in order of increasing boiling point:
(a) 0.10 *m* KCl (c) 0.080 *m* $MgCl_2$
(b) 0.10 *m* sugar

54. Arrange the following aqueous solutions in order of increasing boiling point:
(a) 0.20 *m* ethylene glycol (nonvolatile, nonelectrolyte)
(b) 0.12 *m* $(NH_4)_2SO_4$
(c) 0.10 *m* $CaCl_2$
(d) 0.12 *m* KNO_3

55. Butylated hydroxyanisole (BHA) is used as an antioxidant in margarine and other fats and oils; it prevents oxidation and improves the shelf life of the food. What is the molar mass of BHA if 0.640 g of the compound, dissolved in 25.0 g of chloroform, produces a solution whose boiling point is 62.22 °C?

56. You add 0.255 g of an orange, crystalline compound whose *empirical* formula is $C_{10}H_8Fe$ to 11.12 g of benzene. The boiling point of the benzene rises from 80.10 °C to 80.26 °C. What is the molar mass and molecular formula of the compound?

57. Anthracene is a hydrocarbon obtained from coal. The empirical formula of anthracene is C_7H_5. To find its molecular formula you dissolve 0.500 g in 30.0 g of benzene. The boiling point of the pure benzene is 80.10 °C, whereas the solution has a boiling point of 80.34 °C. What is the molecular formula of anthracene?

58. Benzyl acetate is one of the active components of oil of jasmine. If 0.125 g of the compound is added to 25.0 g of chloroform ($CHCl_3$), the boiling point of the solution is 61.82 °C. What is the molar mass of benzyl acetate?

Freezing Point Depression

59. A mixture of ethanol (C_2H_5OH) and water has a freezing point of −16.0 °C.
(a) What is the molality of the alcohol?
(b) What is the weight percent of alcohol in the solution?

60. Some ethylene glycol, $C_2H_4(OH)_2$, is added to your car's cooling system along with 5.0 kg of water.
(a) If the freezing point of the water/glycol solution is −15.0 °C, how many grams of $C_2H_4(OH)_2$ must have been added?
(b) What is the boiling point of the solution?

61. If 52.5 g of LiF is dissolved in 300. g of water, what is the expected freezing point of the solution? (Assume the van't Hoff *i* factor for LiF is 2.)

62. If you have ever made homemade ice cream, you know that you cool the milk and cream by immersing the container in ice and a concentrated solution of rock salt (NaCl) in water. If you want to have a water/salt solution that freezes at −10. °C, how many grams of NaCl must you add to 3.0 kg of water? (Assume the van't Hoff *i* factor for NaCl is 1.85.)

63. An aqueous solution contains 0.180 g of an unknown, nonionic solute and 50.0 g of water. The solution freezes at −0.040 °C. What is the molar mass of the solute?

64. The organic compound called aluminon is used as a reagent to test for the presence of the aluminum ion in aqueous solution. A solution of 2.50 g of aluminon in 50.0 g of

water freezes at −0.197 °C. What is the molar mass of aluminon?

65. You have isolated a new compound whose empirical formula is $(C_2H_5)_2AlF$. You know, however, that the molecular form of the compound consists of two of these units combined to give $[(C_2H_5)_2AlF]_2$ or four of them to give $[(C_2H_5)_2AlF]_4$. To answer the question, you dissolve 0.125 g of the compound in 15.65 g of benzene. The freezing point of the solution is 5.41 °C (whereas the freezing point of pure benzene is 5.50 °C). Which molecular formula is correct?

66. The freezing point of pure biphenyl ($C_{12}H_{10}$) is found to be 70.03 °C. If 0.100 g of naphthalene is added to 10.0 g of biphenyl, the freezing point of the mixture is found to be 69.40 °C. If K_{fp} for biphenyl is −8.00 °C/*m*, what is the molar mass of naphthalene?

67. List the following aqueous solutions in order of increasing melting point: (a) 0.1 *m* sugar; (b) 0.1 *m* NaCl; (c) 0.08 *m* $CaCl_2$; and (d) 0.04 *m* Na_2SO_4. (The last three are all assumed to dissociate completely into ions in water.)

68. Arrange the following aqueous solutions in order of decreasing freezing point: (a) 0.20 *m* ethylene glycol (nonvolatile, nonelectrolyte), (b) 0.12 *m* K_2SO_4, (c) 0.10 *m* $MgCl_2$, (d) 0.12 *m* KBr. (The last three are all assumed to dissociate completely into ions in water.)

Osmosis

69. A solution contains 3.00% phenylalanine ($C_9H_{11}NO_2$) and 97.00% water by mass. Assume the phenylalanine is nonionic and nonvolatile. Find (a) the freezing point of the solution, (b) the boiling point of the solution, and (c) the osmotic pressure of the solution at 25 °C. In your view, which of these is most easily measurable in the laboratory?

70. Estimate the osmotic pressure of human blood at 37 °C. It is isotonic with a 0.16 M NaCl solution. Assume the van't Hoff *i* factor is 1.9 for NaCl.

71. An aqueous solution containing 1.00 g of bovine insulin (a protein, not ionized) per liter has an osmotic pressure of 3.1 mm Hg at 25 °C. Calculate the molar mass of bovine insulin.

72. Calculate the osmotic pressure of a 0.0120 M solution of NaCl in water at 0 °C. Assume the van't Hoff *i* factor is 1.94 for this solution.

Colloids

73. When solutions of $BaCl_2$ and Na_2SO_4 are mixed, a cloudy liquid is produced. After a few days, a white solid is observed on the bottom of the beaker with a clear liquid above it.
(a) Write a balanced equation for the reaction that occurs.
(b) Why is the solution cloudy at first?
(c) What happens during the few days of waiting?

74. The dispersed phase of a certain colloidal dispersion consists of spheres of diameter 1.0×10^2 nm.
(a) What is the volume $[(V = (\frac{4}{3}\pi r^3)]$ and surface area $(A = 4\pi r^2)$ of each sphere?
(b) How many spheres have a total volume of 1.0 cm^3?
(c) What is the total surface area of these spheres in square meters?

General Questions

75. Solution Properties
(a) Which solution is expected to have the higher boiling point, 0.10 *m* Na_2SO_4 or 0.15 *m* sugar?
(b) The enthalpy of solution of sodium hydroxide is strongly *exo*thermic. Is solid NaOH more or less soluble as the temperature is increased?
(c) For which aqueous solution is the vapor pressure of water higher, 0.30 *m* NH_4NO_3 or 0.15 *m* Na_2SO_4?

76. Arrange the following aqueous solutions in order of (i) increasing vapor pressure of water and (ii) increasing boiling point.
(a) 0.35 *m* $C_2H_4(OH)_2$ (nonvolatile solute)
(b) 0.50 *m* sugar
(c) 0.20 *m* KBr (a strong electrolyte)
(d) 0.20 *m* Na_2SO_4 (a strong electrolyte)

77. Dimethylglyoxime [DMG, $(CH_3CNOH)_2$] is used as a reagent to precipitate nickel ion. Assume that 53.0 g of DMG is dissolved in 500. g of ethanol (C_2H_5OH).
(a) What is the mole fraction of DMG?
(b) What is the molality of the solution?
(c) The vapor pressure of pure alcohol is 760. mm Hg at 78.4 °C (the normal boiling point). What is the vapor pressure of the alcohol over the solution at 78.4 °C?
(d) What is the boiling point of the solution? (DMG does not produce ions in solution.) (K_{bp} for ethanol = +1.22 °C/*m*)

78. One of life's great pleasures is making homemade ice cream in the summer. Fresh milk and cream, sugar, and

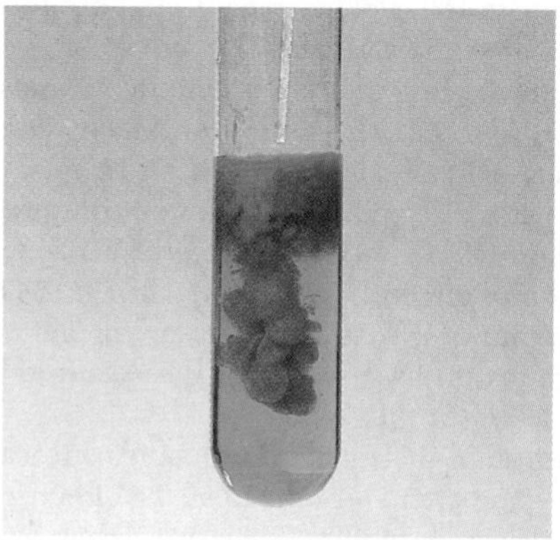

The red, insoluble compound formed between nickel(II) ion and dimethylglyoxime (DMG) is precipitated when the DMG is added to a basic solution of Ni^{2+}(aq). See Study Question 77. (C. D. Winters)

fruit are churned in a bucket suspended in an ice/water mixture, the freezing point of which has been lowered by adding rock salt. One manufacturer of home ice cream freezers recommends that you add 2.50 lb (1130 g) of rock salt (NaCl) to 16.0 lb of ice (7250 g) in a 4-qt freezer. For the solution when this mixture melts calculate
(a) The weight of percentage of NaCl
(b) The mole fraction of NaCl
(c) The molality of the solution

79. Water at 25 °C has a density of 0.997 g/cm^3. Calculate the molality and molarity of pure water at this temperature.

80. How much N_2 can dissolve in water at 25 °C if the N_2 partial pressure is 585. mm Hg?

81. The organic compound aluminon is used as a reagent to test for the presence of the aluminum ion in your unknowns in the laboratory. A solution of 2.50 g of aluminon in 50.0 g of water freezes at −0.197 °C. (K_{fp} for water is −1.86 °C/*m*.) What is the molar mass of the compound?

82. Consider the following aqueous solutions: (i) 0.20 *m* $C_2H_4(OH)_2$ (nonvolatile, nonelectrolyte); (ii) 0.10 *m* $CaCl_2$; (iii) 0.12 *m* KBr; and (iv) 0.12 *m* Na_2SO_4.
(a) Which solution has the highest boiling point?
(b) Which solution has the lowest freezing point?
(c) Which solution has the highest water vapor pressure?

83. Solution Properties
(a) Potassium nitrate dissolves in water in the following amounts per 100. g of water: 31.6 g at 20 °C and 202 g at 90 °C. Is the enthalpy of solution of KNO_3 expected to be positive or negative?
(b) Which solution is expected to have the *higher* boiling point, 0.20 *m* KBr or 0.30 *m* sugar?
(c) Which aqueous solution has the *lower* freezing point, 0.12 *m* NH_4NO_3 or 0.10 *m* Na_2CO_3?

84. Instead of using NaCl to melt the ice on your sidewalk, you decide to use $CaCl_2$. (Like NaCl, $CaCl_2$ is a strong electrolyte.) If you add 35.0 g of $CaCl_2$ to 150. g of water, what is the freezing point of the solution? (Assume $i = 3$ for $CaCl_2$, even though it is less than 3.)

85. Hexachlorophene is used in germicidal soap. What is its molar mass if 0.640 g of the compound, dissolved in 25.0 g of chloroform, produces a solution whose boiling point is 61.93 °C?

86. The smell of ripe raspberries is due to *p*-hydroxylphenyl-2-butanone, which has the empirical formula C_5H_6O. To find its molecular formula, you dissolve 0.135 g in 25.0 g of chloroform, $CHCl_3$. The boiling point of the solution is 61.82 °C. What is the molecular formula of the solute?

87. The solubility of NaCl in water at 100 °C is 39.1g/100. g of water. Calculate the boiling point of this solution. (Assume $i = 1.85$ for NaCl.)

88. The organic salt $(C_4H_9)_4N^+ClO_4^-$ dissolves in chloroform. How many grams of the salt must be dissolved if the boiling point of a solution of the salt in 25.0 g of chloroform is 63.20 °C? (Assume the salt completely dissociates into its ions in solution; that is, $i = 2$.)

89. The Henry's law constant for Cl_2 in water is 8.2×10^{-5} M/mm Hg at 25 °C. What mass of Cl_2 can be dissolved in 100. g of water when the total pressure (pressure of water vapor plus the pressure of chlorine gas) is 800. mm Hg?

90. If a volatile solute is added to a volatile solvent, both substances contribute to the vapor pressure over the solution. Assuming an ideal solution, the vapor pressure of each is given by Raoult's law, and the total vapor pressure is simply the sum of the vapor pressure of each component. A solution, assumed to be ideal, is made from 1.0 mol of toluene ($C_6H_5CH_3$) and 2.0 mol of benzene (C_6H_6). The vapor pressures of the pure solvents are 22 mm Hg and 75 mm Hg, respectively, at 20 °C. What is the total vapor pressure of the mixture?

91. A solution is made by adding 50.0 mL of ethanol (C_2H_5OH) to 50.0 mL of water.
(a) What are the mole fractions of each component? (The densities of ethanol and water are 0.785 g/mL and 1.00 g/mL, respectively.)
(b) What is the total vapor pressure over the solution? (See Study Question 90.) The vapor pressure of water at 20 °C is 17.5 mm Hg and that of ethanol is 43.6 mm Hg.

92. A 2.0% (by mass) aqueous solution of novocainium chloride ($C_{13}H_{21}ClN_2O_2$) freezes at −0.237 °C. Calculate the van't Hoff *i* factor. How many moles of ions are in solution per mole of compound?

93. A solution is 4.00% (by mass) maltose and 96.00% water. It freezes at −0.229 °C.
(a) Calculate the molar mass of maltose (which is not an ionic compound).
(b) The density of the solution is 1.014 g/mL. Calculate the osmotic pressure of the solution.

94. The enthalpy of solution of $KClO_3$ in water is +41 kJ/mol. Does the solubility of $KClO_3$ increase or decrease as the temperature increases?

95. The solubility of ammonium formate, NH_4CHO_2, in water is 102 g/100 g of water at 0 °C and 546 g/100 g of water at 80 °C. A solution is prepared by dissolving NH_4CHO_2 in 200 g of water until no more dissolves at 80 °C. The solution is then cooled to 0 °C. How many grams of NH_4CHO_2 precipitate? (Assume no water evaporates and that the solution is not supersaturated.)

96. The lattice energy of $CaCl_2$ is −2258 kJ/mol. Its enthalpy of solution is −83 kJ/mol. Calculate the enthalpy of hydration for $Ca^{2+}(g) + 2\ Cl^-(g) \rightarrow Ca^{2+}(aq) + 2\ Cl^-(aq)$.

97. Cigars are best stored in a "humidor" at 18 °C and 55% relative humidity. This means the pressure of water vapor should be 55% of the vapor pressure of pure water at the same temperature. The proper humidity can be maintained by placing a solution of glycerol [$C_3H_5(OH)_3$] and water in the humidor. Calculate the percentage by mass of glycerol that lowers the vapor pressure of water to the desired value. (Assume that the vapor pressure of glycerol is zero.)

98. A 2.00% solution of H_2SO_4 in water freezes at -0.796 °C.
   (a) Calculate the van't Hoff i factor.
   (b) Which of the following best represents sulfuric acid in a dilute aqueous solution? H_2SO_4, $H^+ + HSO_4^-$, or $2\ H^+ + SO_4^{2-}$.

99. A compound is known to be a potassium salt, KX. If 4.00 g of the salt is dissolved in 100. g of water, the solution freezes at -1.28 °C. Which of the elements of Group 7A is X?

100. The following table lists the concentrations of the principal ions in sea water.

Ion	Concentration (ppm)
Cl^-	1.95×10^4
Na^+	1.08×10^4
Mg^{2+}	1.29×10^3
SO_4^{2-}	9.05×10^2
Ca^{2+}	4.12×10^2
K^+	3.80×10^2
Br^-	67

   (a) Calculate the freezing point of water.
   (b) Calculate the osmotic pressure of sea water at 25 °C. What is the minimum pressure needed to purify sea water by reverse osmosis?

101. A tree is 10. m tall.
   (a) What must be the total molarity of solutes if the sap rises to the top of the tree by osmotic pressure at 20 °C? Assume the groundwater outside the tree is pure water and that the density of the sap is 1.0 g/mL. (1 mm Hg = 13.6 mm H_2O)
   (b) If the only solute in the sap is sucrose, $C_{12}H_{22}O_{11}$, what is its percentage by mass?

102. An aqueous solution containing 10.0 g of starch per liter gives an osmotic pressure of 3.8 mm Hg at 25 °C.
   (a) What is the molar mass of starch? (Because not all starch molecules are identical, the result is an average.)
   (b) What is the freezing point of the solution? Would it be easy to determine the molecular weight of starch by measuring the freezing-point depression? (Assume that the molarity and molality are the same for this solution.)

Conceptual Questions

103. Examine the photographs in the Chemical Puzzler that opens this chapter. Provide an explanation for your observations.

104. Solutions of salts have boiling points higher than those calculated using the equation $\Delta t_{bp} = K_{bp} \cdot m$ (where m is the molality of the salt). Briefly explain this observation.

105. A solution of 5.00 g of acetic acid in 100. g of benzene freezes at 3.37 °C. A solution of 5.00 g of acetic acid in 100. g of water freezes at -1.49 °C. Find the molecular weight of acetic acid from each of these experiments. What can you conclude about the state of the acetic acid molecules dissolved in each of these solvents? Recall the discussion of hydrogen bonding in Section 13.2 and propose a structure for the species in benzene solution.

106. Explain how each of the following affects the solubility of an ionic compound.
   (a) Lattice energy
   (b) Enthalpy of hydration of the anion
   (c) Enthalpy of hydration of the cation

107. Account for the fact that alcohols, such as methanol (CH_3OH) and ethanol (C_2H_5OH), are quite miscible with water, whereas an alcohol with a long carbon chain such as octanol ($C_8H_{17}OH$) is poorly soluble in water.

108. A protozoan (single-celled animal) that normally lives in the ocean is placed in fresh water. Will it shrivel or burst? Explain briefly.

109. Starch contains C—C, C—H, C—O, and O—H bonds. Hydrocarbons have only C—C and C—H bonds. Both starch and hydrocarbons can form colloidal dispersions in water. Which dispersion is classified as hydrophobic? Which is hydrophilic? Explain briefly.

Summary Questions

110. A newly synthesized compound containing boron and fluorine is 22.1% boron. Dissolving 0.146 g of the compound in 10.0 g of benzene gives a solution with a vapor pressure of 94.16 mm Hg at 25 °C. (The vapor pressure of pure benzene at this temperature is 95.26 mm Hg). In a separate experiment, it is found that the compound does not have a dipole moment.
   (a) What is the molecular formula for the compound?
   (b) Draw a Lewis structure for the molecule and suggest a possible molecular structure. Give the bond angles in the molecule and the hybridization of the boron atom.

111. In chemical research we often send newly synthesized compounds to commercial laboratories for analysis. These laboratories determine the weight percent of C and H by burning the compound and collecting the evolved CO_2 and H_2O. They determine the molar mass by measuring the osmotic pressure of a solution of the compound. Calculate the empirical and molecular formulas of a compound, C_xH_yCr, given the following information:
   (a) The compound contains 73.94% C and 8.27% H; the remainder is chromium.
   (b) At 25 °C, the osmotic pressure of a solution containing 5.00 mg of the unknown dissolved in 100. mL of chloroform is 3.17 mm Hg.

CHAPTER 15

Principles of Reactivity: Chemical Kinetics

A Chemical Puzzler

Many cars are now equipped with catalytic converters in their exhaust systems. Internal combustion engines produce gases, such as NO and CO, in addition to the usual products of combustion (CO_2 and H_2O). Because NO and CO contribute to air pollution, the exhaust gases are sent through a catalytic converter to change them to nonpolluting gases, such as N_2, O_2, CO_2, and others. How does a catalytic converter work? In spite of the fact that the converter is the site of continuous chemical reactions when the engine is running, the converter does not need to be refilled. The chemicals in the converter do not run out. Why?

The goal of the chapters titled *Principles of Reactivity* is to answer some important questions: whether a reaction can occur when two or more substances are mixed, how rapid the reaction is, and how far it has gone toward products when all change has apparently ceased. The principles of thermodynamics, which were first described in Chapter 6, provide a partial answer to the first question. That is, exothermic reactions usually favor formation of products. Unfortunately, this criterion is insufficient in a practical sense: a reaction may be product-favored but still require such a long time to occur that it would take several human lifetimes to produce appreciable quantities of products. Such a reaction is of little use if you are trying to make and sell a product. This problem is the concern of chemical kinetics.

The principles of thermodynamics are expanded on in Chapter 20. Predicting how far a reaction can proceed is the subject of Chapter 16.

Chemical kinetics, the subject of this chapter, can be divided into two parts. The first, at the macroscopic level, is the study of rates of reactions: what the rate of reaction means; how to determine a rate by experiment; and how factors, such as the concentrations of reactants and temperature, influence rates. The second part of this subject looks at reactions at the submicroscopic level. Here, the concern is with **reaction mechanisms,** the detailed pathways taken by atoms and molecules as a reaction proceeds.

15.1 RATES OF CHEMICAL REACTIONS

The concept of rate is encountered in many nonchemical circumstances. We refer to the speed of an automobile in terms of the distance traveled per hour (kilometers per hour), to the rate of flow of water from a faucet in volume per minute (liters per minute), or the rate of growth of the population in terms of the number of births per day. In each case a change is measured over an interval of time. The rate of a chemical reaction is similarly defined. The **rate of a reaction** refers to the change in concentration of a substance per unit of time. During a chemical reaction, the concentration of the reactant or reactants decreases with time, and the concentration of the products increases.

An easy way to gauge the speed of an automobile is to measure how far it travels during a time interval. Two measurements are made: distance traveled and time. The speed is the distance traveled divided by the time elapsed, or Δdistance/Δtime. If an automobile travels 2.4 miles in 4.5 min (or 0.075 h), its speed is (2.4 miles/0.075 h), or 32 miles per hour.

Chemical reaction rates are measured the same way. To determine the rate of a chemical reaction, two quantities—concentration and time—are measured. The rate of the reaction can then be described as the decrease in concentration of a reactant, or the increase in concentration of a product, per unit time, Δconcentration/Δtime.

The concentration of a substance undergoing reaction can be determined by a variety of methods. It can be related to such measurable quantities as pressure (for a reaction accompanied by a change in the number of moles of gas),

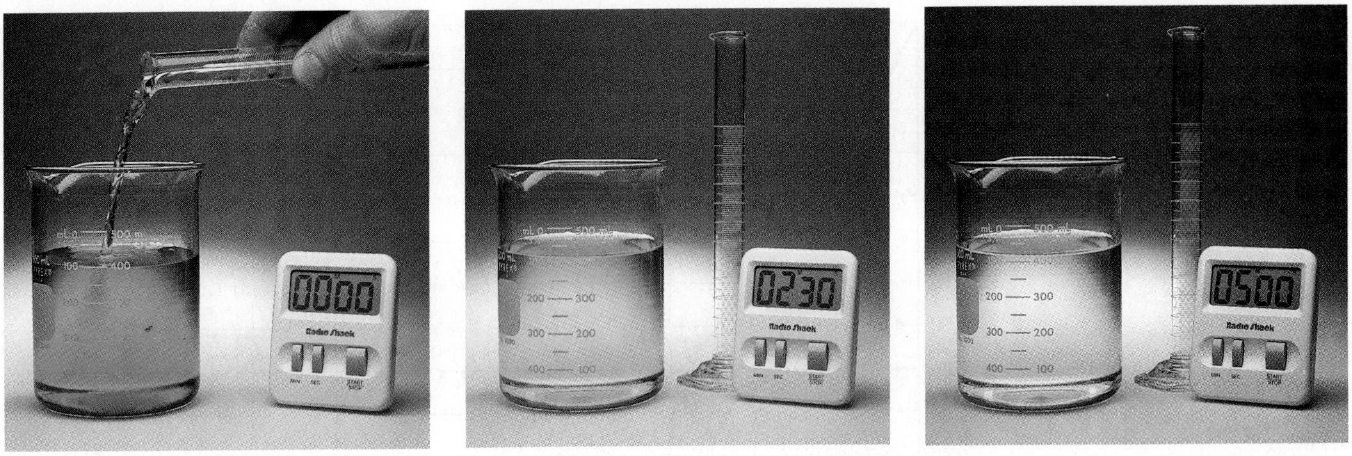

(a) (b) (c)

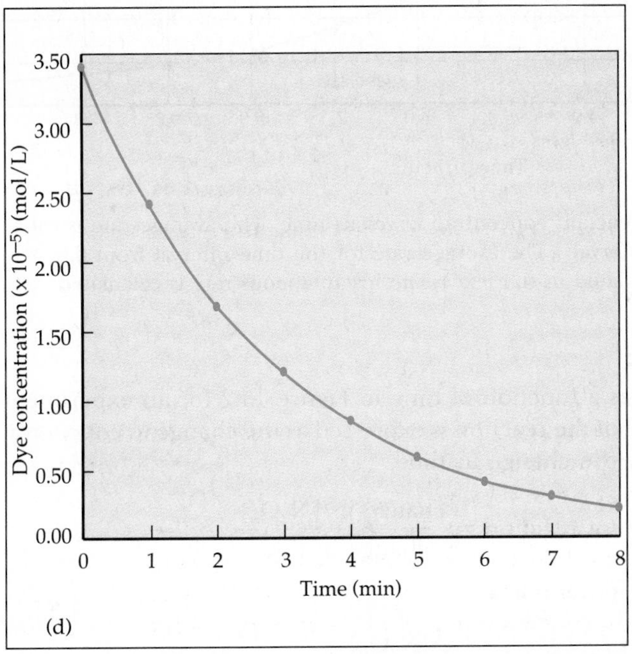

(d)

Figure 15.1 A few drops of blue food dye were added to water followed by a solution of bleach. (Initially, the concentration of dye was about 3.4×10^{-5} M, and the bleach [NaOCl] concentration was about 0.034 M.) The dye faded as it reacted with the bleach (as seen in the three photos, a, b, and c). The color change was followed by a spectrophotometer (d). (a–c, C. D. Winters)

color, pH, and electric conductivity (Figure 15.1). Changes in the concentration of a substance usually lead to a corresponding change in one or more of these quantities. If measurements are made at definite time intervals, the rate of the reaction at any concentration can be obtained.

An example of rate determination is the decomposition of dinitrogen pentaoxide dissolved in liquid carbon tetrachloride.

$$N_2O_5(\text{solvent}) \longrightarrow 2\ NO_2(\text{solvent}) + \tfrac{1}{2}\ O_2(g)$$

Here the pressure of O_2 gas can be used as an indicator of reaction rate because the increase in O_2 pressure is related to the decrease in the concentration of N_2O_5. For every $\frac{1}{2}$ mol of O_2 formed, 1 mol of N_2O_5 disappears. The concentra-

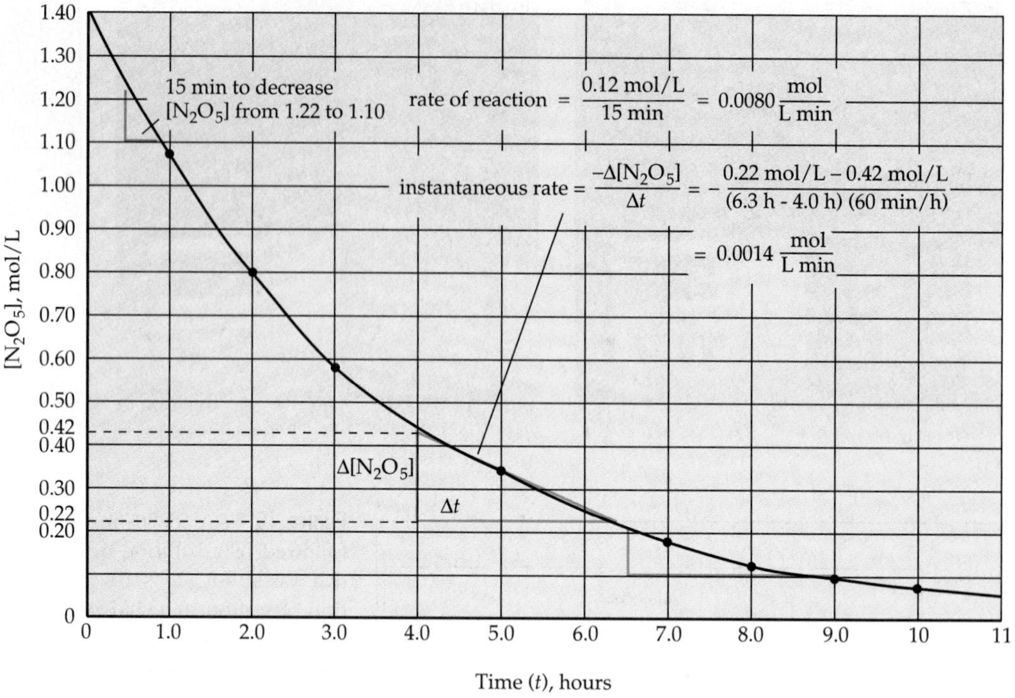

Figure 15.2 The plot of reactant concentration versus time. The average rate is calculated for a 15-minute interval. (The average rate for the time interval from 6.5 hours to 9.0 hours is calculated in the text.) The instantaneous rate is calculated when $[N_2O_5] = 0.34$ M.

tion of N_2O_5 is plotted as a function of time in Figure 15.2 for an experiment done at 30.0 °C. The rate of the reaction is expressed as the change in concentration of N_2O_5 divided by the change in time,

$$\text{Rate of reaction} = \frac{\text{change in } [N_2O_5]}{\text{change in time}}$$

or more briefly,

The Greek letter Δ (delta) means that a *change* in some quantity has been measured. As usual Δ = final − initial.

$$\text{Rate} = -\frac{\Delta[N_2O_5]}{\Delta t}$$

The minus sign is needed because the concentration of N_2O_5 decreases with time. For example, the rate between 40. min and 55 min (see Figure 15.2) is given by

$$-\frac{\Delta[N_2O_5]}{\Delta t} = -\frac{(1.10 \text{ mol/L}) - (1.22 \text{ mol/L})}{55 \text{ min} - 40. \text{ min}} = +\frac{0.12 \text{ mol/L}}{15 \text{ min}} = 0.0080 \frac{\text{mol}}{\text{L} \cdot \text{min}}$$

If the rate of the N_2O_5 decomposition reaction were expressed in terms of the rate of appearance of NO_2, it would be *twice* the rate of disappearance of N_2O_5. The balanced chemical equation tells us that 2 mol of NO_2 is formed from 1 mol of N_2O_5.

$$\text{Rate} = \frac{\Delta[NO_2]}{\Delta t} = \frac{0.0080 \text{ mol } N_2O_5 \text{ consumed}}{\text{L} \cdot \text{min}} \cdot \frac{2 \text{ mol } NO_2 \text{ formed}}{1 \text{ mol } N_2O_5 \text{ consumed}}$$

$$= 0.016 \frac{\text{mol } NO_2 \text{ formed}}{\text{L} \cdot \text{min}}$$

In terms of O_2, the rate is

$$\text{Rate} = \frac{\Delta[O_2]}{\Delta t} = \frac{0.0080 \text{ mol } N_2O_5 \text{ consumed}}{\text{L} \cdot \text{min}} \cdot \frac{\frac{1}{2} \text{ mol } O_2 \text{ formed}}{1 \text{ mol } N_2O_5 \text{ consumed}}$$

$$= 0.0040 \frac{\text{mol } O_2 \text{ formed}}{\text{L} \cdot \text{min}}$$

Figure 15.2 shows that the rate of the reaction decreases as the reactant concentration decreases. The concentration of N_2O_5 drops sharply at the beginning of the reaction, but it decreases more slowly near the end of the reaction. We can verify this by comparing the rate of disappearance of N_2O_5 calculated previously for a 15-min interval during the first hour (where the concentration dropped by 0.12 mol/L) with the rate of reaction calculated for the time interval from 6.5 h to 9.0 h (when the concentration also dropped by 0.12 mol/L; see Figure 15.2).

$$-\frac{\Delta[N_2O_5]}{\Delta t} = -\frac{(0.10 \text{ mol/L}) - (0.22 \text{ mol/L})}{540 \text{ min} - 390 \text{ min}} = +\frac{0.12 \text{ mol/L}}{150 \text{ min}} = 0.00080 \frac{\text{mol}}{\text{L} \cdot \text{min}}$$

The rate has dropped to 1/10 of its previous value.

The rate calculated by this procedure is the **average rate** over the chosen time interval. We might also ask what the **instantaneous rate** is at a single point in time. The instantaneous rate is determined by drawing a line tangent to the concentration-time curve at a particular time (see Figure 15.2), and obtaining the rate from the slope of this line. For example, when $[N_2O_5] = 0.34$ mol/L and $t = 5.0$ h, the rate is

$$\text{Rate at } 0.34 \text{ M} = -\frac{\Delta[N_2O_5]}{\Delta t} = -\frac{(0.22 \text{ mol/L}) - (0.42 \text{ mol/L})}{(6.3 \text{ h} - 4.0 \text{ h}) \cdot (60. \text{ min/h})}$$

$$= +\frac{0.20 \text{ mol/L}}{140 \text{ min}} = 0.0014 \frac{\text{mol}}{\text{L} \cdot \text{min}}$$

This shows that 0.0014 mol/L of N_2O_5 is consumed per minute.

The difference between an average rate and an instantaneous rate has an analogy in the speed of an automobile. In the previous example, the car traveled 2.4 miles in 4.5 min for an average speed of 32 miles per hour. At any instant in time, however, the car may have moved much slower or much faster, as indicated by the car's speedometer.

EXAMPLE 15.1 *Reaction Rates and Stoichiometry*

Give the relative rates for disappearance of reactants and formation of products for the following reaction:

$$4\, PH_3(g) \longrightarrow P_4(g) + 6\, H_2(g)$$

Solution In this reaction 4 mol of PH_3 disappears when 1 mol of P_4 and 6 mol of H_2 are formed. To equate rates, we must divide $\Delta[\text{reagent}]/\Delta t$ by the stoichiometric coefficient in the balanced equation.

$$\text{Reaction rate} = -\frac{1}{4}\left(\frac{\Delta[PH_3]}{\Delta t}\right) = +\frac{\Delta[P_4]}{\Delta t} = +\frac{1}{6}\left(\frac{\Delta[H_2]}{\Delta t}\right)$$

$$\text{Reaction rate} = -\frac{1}{4}(\text{rate of change of } [PH_3])$$

$$= \text{rate of change of } [P_4]$$

$$= \frac{1}{6}(\text{rate of change of } [H_2])$$

Because 4 mol of PH_3 disappears for every mole of P_4 formed, the rate of formation of P_4 can only be one-fourth of the rate of disappearance of PH_3. Similarly, P_4 must appear at only one-sixth of the rate that H_2 appears.

EXERCISE 15.1 ***Reaction Rates and Stoichiometry***

What are the relative rates of appearance or disappearance of each product and reactant, respectively, in the decomposition of nitrosyl chloride, NOCl?

$$2\ NOCl(g) \longrightarrow 2\ NO(g) + Cl_2(g)$$

EXERCISE 15.2 ***Rate of Reaction***

Sucrose decomposes to fructose and glucose in acid solution. A plot of the concentration of sucrose as a function of time is given here. What is the rate of change of the sucrose concentration over the first 2 h? What is the rate of change over the last 2 h? Estimate the instantaneous rate at 4 h.

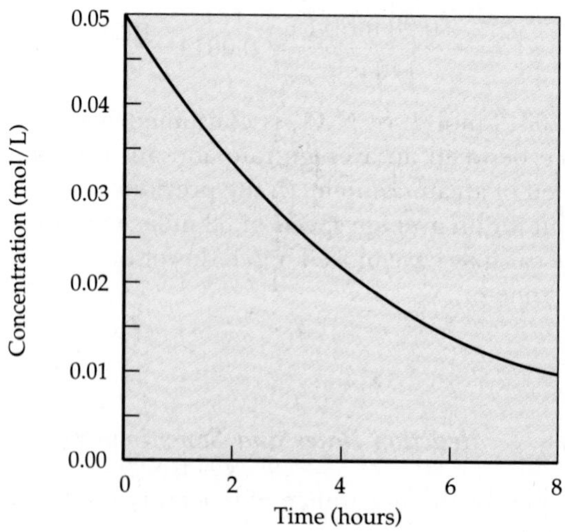

15.2 REACTION CONDITIONS AND RATE

For a chemical reaction to occur, reactant molecules must come together so that atoms can be exchanged or rearranged. Atoms and molecules are mobile in the gas phase or in solution, and so reactions are often carried out in a mixture of gases or between solutes in a solution. Under these circumstances, three factors affect the speed of a reaction. Each is described in detail in the sections that follow.

- *Concentrations of reactants* (Section 15.3). An Alka-Seltzer tablet contains sodium bicarbonate and citric acid, which react to give CO_2 gas and other products when the tablet is placed in water (Figure 15.3). The reaction is faster when the tablet is in pure water than when it is in ethanol containing only a small amount of water.
- *Temperature* (Section 15.5). In your own experience you know that heating things can bring about more rapid reaction. That is why, for example, we cook food, and in the laboratory we often heat the reactants to make reactions occur faster.
- *Catalysts* (Section 15.7). Catalysts are substances that accelerate chemical reactions but are not themselves transformed. For example, hydrogen peroxide, H_2O_2, can decompose to water and oxygen.

$$2\ H_2O_2(\ell) \longrightarrow O_2(g) + 2\ H_2O(\ell)$$

If the peroxide is stored in a cool place in a clean plastic container, it is reasonably stable for many months; the rate of the decomposition reaction is extremely slow. But, in the presence of a manganese salt, an iodide-containing salt, or a biological substance called an enzyme, an energetic reaction occurs as shown by vigorous bubbling and the rapid escape of steam (Figure 15.4). In fact, an insect called a bombardier

Figure 15.3 The speed of reaction can depend on the concentrations of the reactants. Here an Alka-Seltzer is placed in pure water (*right*) or in ethanol to which a trace of water has been added (*left*). The reaction rate is much greater when the concentration of water is high. (Alka-Seltzer requires water for reaction. The ingredients do not react in alcohol.) (C. D. Winters)

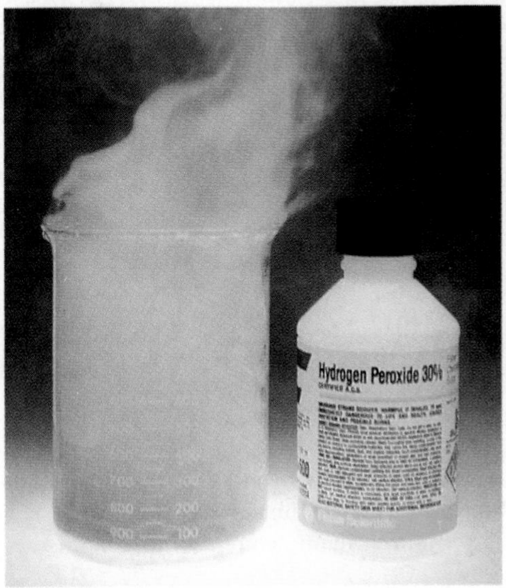

Figure 15.4 Decomposition of hydrogen peroxide can be accelerated by the catalyst MnO_2. Here a 30% solution of H_2O_2 is dropped onto the black solid MnO_2. The peroxide rapidly decomposes to O_2 and H_2O. The water is given off as steam because of the high heat of reaction. (C. D. Winters)

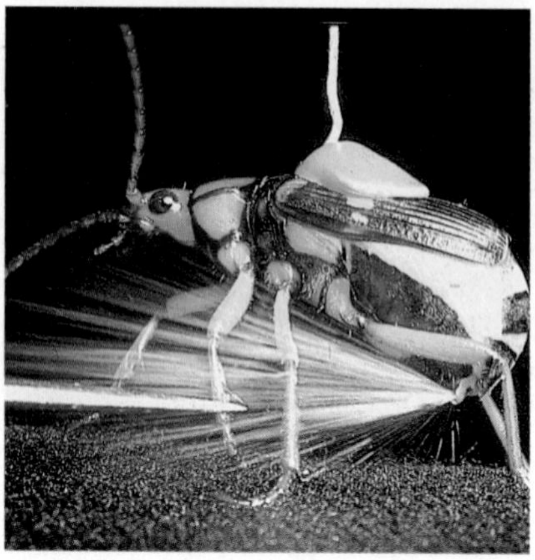

Figure 15.5 A bombardier beetle uses the catalyzed decomposition of hydrogen peroxide as a defense mechanism. The heat of the reaction lets the insect eject steam and other irritating chemicals with explosive force.
(Thomas Eisner with Daniel Ansehansley)

beetle uses the reaction as its defense mechanism (Figure 15.5). By combining the organic compound hydroquinone with a peroxide in the presence of an enzyme, it produces a stream of superheated steam and an irritating chemical to spray on its enemies.

If a reaction involves a solid, the *surface area of the solid* can affect the rate (Figure 15.6). Because molecules must come in contact to react, only molecules on the surface of a solid can react with another chemical reagent. It is reasonable, then, that the rate depends on the number of molecules at the surface of a solid, which depends on the surface area. The smaller the particles of the solid, the more of the material that is exposed on the surface and the faster the reaction.

15.3 EFFECT OF CONCENTRATION ON REACTION RATE

It is often possible to change the rate of a reaction by changing the concentrations of reactants. One goal in studying kinetics is to determine the effect of concentration on the rate. The effect can be determined by measuring the rate of a reaction in experiments in which there are different concentrations of each reactant (and temperature is held constant). For example, when such experiments were carried out for the decomposition of N_2O_5 to NO_2 and O_2, it was discovered that, if the concentration of N_2O_5 is doubled, the reaction rate is likewise doubled. If the concentration of N_2O_5 is halved, then the reaction rate is halved. This relation between concentration and time gives a plot of $[N_2O_5]$

(a) (b)

Figure 15.6 The combustion of lycopodium powder. (a) The spores of this common moss burn only with difficulty when piled in a dish. (b) If the surface area is increased, and a finely divided powder is sprayed into a flame, combustion is rapid.
(C. D. Winters)

versus time with the shape shown in Figure 15.2, which shows that the reaction rate is directly proportional to the reactant concentration.

$$\text{Rate of reaction} \propto [N_2O_5]$$

(where the symbol $\propto$ means "proportional to.") We previously calculated that the rate of disappearance of N_2O_5 is 0.0014 mol/L · min when $[N_2O_5]$ = 0.34 mol/L. You can use Figure 15.2, however, to show that the rate is three times faster (0.0042 mol/L · min) when $[N_2O_5]$ is three times larger (1.0 mol/L).

Other relationships between reaction rate and reactant concentration are also encountered. Some reactions are known in which the rate increases much faster than the increase in concentration, whereas in others the rate decreases as the concentration increases.

The Rate Equation

The relationship between reactant concentration and reaction rate is expressed by an equation called a **rate equation,** or **rate law.** For the N_2O_5 decomposition reaction the rate equation is

$$\text{Rate} = k[N_2O_5]$$

where the proportionality constant k is called the **rate constant.** This rate equation tells us that the reaction rate is proportional to the concentration of the reactant.

In general, for a homogeneous reaction such as

$$aA + bB \xrightarrow{C} xX$$

where C is a catalyst, the rate equation has the form

$$\text{Rate} = k[A]^m[B]^n[C]^p$$

The rate equation expresses the fact that the rate of reaction is proportional to the reactant concentrations (and perhaps the catalyst concentration), each concentration being raised to some power. *It is important to recognize that the exponents, m, n, and p in this case, are not necessarily the stoichiometric coefficients for the balanced chemical equation.* The exponents *must be determined by experiment.* (The exponents m, n, and p are often positive, whole numbers, but they can be negative numbers, fractions, or zero.)

As an example of a rate equation, consider the equation for the decomposition of hydrogen peroxide in the presence of a catalyst such as iodide ion.

$$2\ H_2O_2(aq) \xrightarrow{I^-(aq)} 2\ H_2O(\ell) + O_2(g)$$

Experiments show that the reaction has the following rate equation:

$$\text{Rate} = k[H_2O_2][I^-]$$

Here the exponent on each concentration term is 1, even though the stoichiometric coefficient of H_2O_2 is 2 and I^- does not even appear in the balanced equation.

A catalyst does not appear in the balanced, overall equation for the reaction. It is not consumed by the reaction. It may appear in the rate expression, however. See Section 15.7.

The Rate Constant *k*

The rate constant *k* is a proportionality constant that relates rate and concentration *at a given temperature; it must be evaluated by experiment.* It is an important quantity because, once it is known, it enables you to find the reaction rate for a new set of concentrations. To see how to use *k*, consider the substitution of Cl^- ion by water in the cancer chemotherapy agent cisplatin, $Pt(NH_3)_2Cl_2$.

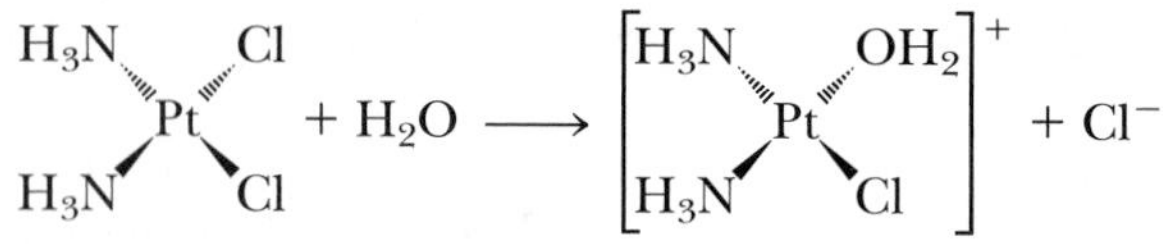

The rate expression for this reaction is

$$\text{Rate} = k[Pt(NH_3)_2Cl_2]$$

and the rate constant *k* is 0.090/h. A knowledge of *k* allows us to calculate the rate at a particular reactant concentration, for example when $[Pt(NH_3)_2Cl_2] =$ 0.018 mol/L.

$$\text{Rate} = (0.090/\text{h})(0.018\ \text{mol/L}) = 0.0016\ \text{mol/L} \cdot \text{h}$$

The Order of a Reaction

The **order** with respect to a particular reactant is the exponent of its concentration term in the rate expression, and the **total reaction order** is the sum of the exponents on all concentration terms. The rate equation for the decomposition of N_2O_5

$$2\ N_2O_5 \longrightarrow 4\ NO_2 + O_2$$
$$\text{Rate} = k[N_2O_5]$$

has an exponent of 1 on $[N_2O_5]$, which means the reaction is first order with respect to N_2O_5. If the concentration of N_2O_5 is doubled, the rate of reaction doubles. If the concentration of N_2O_5 is one fourth of its original value, then the rate is one fourth of the initial rate. Using Figure 15.2 we see that the initial rate is about 0.0058 (mol/L · min) when $[N_2O_5] = 1.40$ mol/L. When the concentration drops to 0.35 mol/L, however, one fourth of the original value, the reaction rate is about 0.0014 (mol/L · min), one fourth of the original rate.

The reaction of NO and chlorine has been studied at 50 °C, and the rate equation is

$$2\ NO(g) + Cl_2(g) \longrightarrow 2\ NOCl(g)$$
$$\text{Rate} = k[NO]^2[Cl_2]$$

The reaction is second order in NO, first order in Cl_2, and third order overall. Experimental data show that the reaction rate is 1.43×10^{-6} mol/L · s when $[NO] = [Cl_2] = 0.250$ mol/L. If $[Cl_2]$ is held constant, however, and [NO] is doubled to 0.500 mol/L, then the reaction rate increases by a factor of four to 5.72×10^{-6} mol/L · s.

Finally, the decomposition of ammonia on a platinum surface at 856 °C is interesting because it is zero order.

$$2\ NH_3(g) \longrightarrow N_2(g) + 3\ H_2(g)$$

This means the reaction rate is independent of NH_3 concentration.

$$\text{Rate} = k[NH_3]^0 = k$$

Notice that the orders with respect to the reactants can be equal to the stoichiometric coefficients of the reactants. However, it is often the case that they are not. For example, the reaction between bromate and bromide ions in acid solution

$$BrO_3^-(aq) + 5\ Br^-(aq) + 6\ H^+(aq) \longrightarrow 3\ Br_2(aq) + 3\ H_2O(\ell)$$

is first-order in bromate and bromide ions and second-order in hydrogen ion.

$$\text{Rate} = k[BrO_3^-][Br^-][H^+]^2$$

The reaction order for a reaction is important because it gives some insight into the most interesting question of all—how the reaction occurs. This is described further in Section 15.6.

Determination of the Rate Equation

The relation between rate and concentration must be determined experimentally. One way to do this is the method of initial rates. The **initial rate** is the reaction rate during the first few percent of reaction. It can be measured by mixing the reactants and determining $\Delta[\text{product}]/\Delta t$ or $-\Delta[\text{reactant}]/\Delta t$ after 1% to 2% of the limiting reactant has been consumed. Measuring the rate only during the initial stage of a reaction is convenient because initial concentrations are known, and it avoids possible complications arising from interference by reaction products or the occurrence of other reactions.

As an example of the determination of a reaction rate by this method, consider the reaction of a base, say sodium hydroxide, with methyl acetate, an organic compound and a widely used industrial solvent. The reaction produces acetate ion and methanol.

$$\underset{\text{methyl acetate}}{CH_3\overset{\overset{\displaystyle O}{\|}}{C}{-}O{-}CH_3} + OH^- \longrightarrow \underset{\text{acetate ion}}{CH_3\overset{\overset{\displaystyle O}{\|}}{C}{-}O^-} + \underset{\text{methanol}}{CH_3OH}$$

Because reaction rates change with temperature, several experiments were done at the same temperature, and the data in the table were collected.

Experiment	Initial Concentrations *[CH₃CO₂CH₃]*	*[OH⁻]*	Initial Reaction Rate (mol/L · s)
1	0.050 M	0.050 M	0.00034
	↓ no change	↓ × 2	↓ × 2
2	0.050 M	0.10 M	0.00069
	↓ × 2	↓ no change	↓ × 2
3	0.10 M	0.10 M	0.00137

When the initial concentration of either $CH_3CO_2CH_3$ or OH^- is doubled, and the concentration of the other reactant is held constant, the reaction rate doubles. This rate doubling shows that the rate for the reaction is directly proportional to the concentration of *both* $CH_3CO_2CH_3$ and OH^-. The reaction is first order in each of these reactants, and the rate equation that reflects these experimental observations is

$$\text{Rate} = k[CH_3CO_2CH_3][OH^-]$$

From this equation we can also conclude that doubling *both* concentrations at the same time causes the rate to go up by a factor of 4. What happens, however, if one concentration is doubled and the other halved? The rate equation tells us the rate does not change, and that is what is observed!

For the methyl acetate rate equation, or any other, the value for k, the rate constant, can be found by substituting values of rate and concentration for one experiment into the rate equation. To find k for the methyl acetate/hydroxide ion reaction, data from the first experiment can be substituted, for example,

$$\text{Rate} = 0.00034 \text{ mol/L} \cdot \text{s} = k(0.050 \text{ mol/L})(0.050 \text{ mol/L})$$

$$k = \frac{0.00034 \text{ mol/L} \cdot \text{s}}{(0.050 \text{ mol/L})(0.050 \text{ mol/L})} = 0.136 \text{ L/mol} \cdot \text{s}$$

and we calculate that k is 0.136 L/mol · s at 25 °C.

EXAMPLE 15.2 *Determining a Rate Equation for the NO + O_2 Reaction*

The rate of the reaction of nitrogen monoxide and oxygen

$$2\ NO(g) + O_2(g) \longrightarrow 2\ NO_2(g)$$

was measured at 25 °C starting with various concentrations of NO and O_2, and the data in the table were collected.

	Initial Concentrations (mol/L)		
Experiment	*[NO]*	*[O_2]*	**Initial Rate (mol/L · s)**
1	0.020	0.010	0.028
2	0.020	0.020	0.057
3	0.020	0.040	0.114
4	0.040	0.020	0.227
5	0.010	0.020	0.014

Based on these data, what is the rate equation? What is the value of the rate constant k?

Solution In the first three experiments the concentration of NO is constant, whereas the O_2 concentration increases from 0.010 to 0.020 to 0.040 mol/L. Each time the O_2 concentration is doubled, the initial rate also doubles. For example, when $[O_2]$ is doubled from 0.020 to 0.040 mol/L, the initial rate increases by a factor of 2 from 0.057 to 0.114 mol/L · s. This means that the initial rate is directly proportional to $[O_2]$.

In Experiments 2, 4, and 5 the O_2 concentration is constant, whereas [NO] varies. From Experiment 2 to 4, [NO] is doubled, whereas the initial rate increases by a factor of 4, or 2^2.

$$\frac{\text{Experiment 4 rate}}{\text{Experiment 2 rate}} = \frac{0.227 \text{ mol/L} \cdot \text{s}}{0.057 \text{ mol/L} \cdot \text{s}} = \frac{4}{1}$$

This same result is found on comparing Experiments 4 and 5, and it means that the initial rate is proportional to the *square* of [NO]. The rate equation is therefore

$$\text{Rate} = k[O_2][NO]^2$$

The rate constant k can be found by inserting data for one of the experiments into the rate equation. For Experiment 1, for example,

$$\text{Rate} = 0.028 \text{ mol/L} \cdot \text{s} = k(0.010 \text{ mol/L})(0.020 \text{ mol/L})^2$$

$$k = \frac{0.028 \text{ mol/L} \cdot \text{s}}{(0.010 \text{ mol/L})(0.020 \text{ mol/L})^2} = 7.0 \times 10^3 \text{ L}^2/\text{mol}^2 \cdot \text{s}$$

A better value of k is obtained by calculating the value for k for each experiment and then averaging the values. Once this value is available, it can be used to calculate the initial rate for any set of NO and O_2 concentrations.

EXERCISE 15.3 ***Interpreting a Rate Law***

The rate equation for the reduction of NO to N_2 with hydrogen is

$$2\ NO(g) + 2\ H_2(g) \longrightarrow N_2(g) + 2\ H_2O(g)$$
$$\text{Rate} = k[NO]^2[H_2]$$

1. What is the order of the reaction with respect to the NO? With respect to H_2?
2. If the concentration of NO is doubled, what happens to the reaction rate?
3. If the concentration of H_2 is halved, what happens to the reaction rate? ■

EXERCISE 15.4 ***Using Rate Laws***

The rate constant k is 0.090/h for the reaction

$$Pt(NH_3)_2Cl_2 + H_2O \longrightarrow [Pt(NH_3)_2(H_2O)Cl]^+ + Cl^-$$

and the rate equation is Rate = $k[Pt(NH_3)_2Cl_2]$. Calculate the rate of reaction when the concentration of $Pt(NH_3)_2Cl_2$ is 0.020 M. What is the rate of change in the concentration of Cl^- under these conditions? ■

EXERCISE 15.5 ***Reaction Order***

In the following reaction, a Co—Cl bond is replaced by a Co—OH_2 bond.

$$[Co(NH_3)_5Cl]^{2+}(aq) + H_2O(\ell) \longrightarrow [Co(NH_3)_5H_2O]^{3+}(aq) + Cl^-(aq)$$
$$\text{Rate} = k\{[Co(NH_3)_5Cl]^{2+}\}^m$$

Using the data in the table, find the value of m in the rate equation and calculate k.

Experiment	Initial Concentration of $[Co(NH_3)_5Cl]^{2+}$	Initial Rate (mol/L · min)
1	1.0×10^{-3}	1.3×10^{-7}
2	2.0×10^{-3}	2.6×10^{-7}
3	3.0×10^{-3}	3.9×10^{-7}
4	1.0×10^{-2}	1.3×10^{-6} ■

15.4 RELATIONSHIPS BETWEEN CONCENTRATION AND TIME

It is sometimes useful or important to know how long a reaction must proceed to reach a predetermined concentration of some reagent, or what the reactant and product concentrations will be after some time has elapsed. One way to do this is to collect experimental data and construct a curve such as that shown in Figure 15.2. This can be inconvenient and time-consuming, however. A simpler approach is to determine the reaction order for each reagent by experiment and then derive an equation that relates concentration and time. This equation can then be used to calculate a concentration at any time, or vice versa.

First-Order Reactions

Suppose the reaction R → products is first order. This means the reaction rate is directly proportional to the concentration of R raised to the first power, or, mathematically,

$$\text{Rate} = -\frac{\Delta[\text{R}]}{\Delta t} = k[\text{R}]$$

Using the methods of calculus, this equation can be transformed into a very useful equation that relates the reactant concentration and time. (This equation is often called the **integrated rate equation** because integral calculus is used in its derivation.)

$$\ln \frac{[\text{R}]_t}{[\text{R}]_0} = -kt \qquad \textbf{(15.1)}$$

Here $[\text{R}]_0$ is the concentration of the reactant at time $t = 0$. ($t = 0$ does not need to correspond to the actual beginning of the experiment; it can be the time when instrument readings were started, for example.) $[\text{R}]_t$ is the concentration at a later time. In words this equation says

$[\text{R}]_t/[\text{R}]_0$ is the fraction of material *remaining* after the specified time period.

$$\text{Natural logarithm}\left(\frac{\text{concentration of R after some time}}{\text{concentration of R at start of experiment}}\right) = -(\text{rate constant})(\text{elapsed time})$$

See Appendix A for a discussion of logarithms and their use.

Notice the negative sign in the equation. The ratio $[\text{R}]_t/[\text{R}]_0$ is less than 1 because $[\text{R}]_t$ is always less than $[\text{R}]_0$. This means the logarithm of $[\text{R}]_t/[\text{R}]_0$ is negative, and so the other side of the equation must also have a negative sign. Equation 15.1 is useful in three ways:

- If $[R]_t/[R]_0$ is measured in the laboratory after some amount of time has elapsed, then k can be calculated.
- If $[R]_0$ and k are known, then the concentration of material expected to remain after a given amount of time can be calculated.
- If k is known, Equation 15.1 can be used to calculate the time elapsed until R reaches some predetermined concentration.

Finally, notice that the ratio $[R]_t/[R]_0$ is dimensionless. This means that k for first-order reactions is independent of the units chosen for concentration and so [R] can be expressed in any convenient quantity unit—moles per liter, moles, grams, number of atoms, number of molecules, or pressure.

EXAMPLE 15.3 *The First-Order Rate Equation*

Cyclopropane, C_3H_6, has been used in a mixture with oxygen as an anesthetic. (This practice has diminished greatly, however, because the compound is very flammable.) The compound is known to rearrange to propene, a different molecule of the same formula.

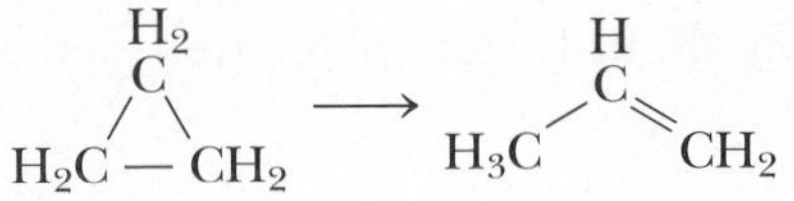

cyclopropane → propene

$$\text{Rate} = k[\text{cyclo}] \qquad k = 5.4 \times 10^{-2}\ \text{h}^{-1}$$

If the initial concentration of cyclopropane is 0.050 mol/L, how many hours must elapse for the concentration of the compound to drop to 0.010 mol/L?

Solution The first-order rate equation applied to this reaction is

$$\ln \frac{[\text{cyclo}]_t}{[\text{cyclo}]_0} = -kt$$

where $[\text{cyclo}]_t$ is 0.010 mol/L, $[\text{cyclo}]_0$ is 0.050 mol/L, and k is given above.

$$\ln \frac{[0.010]}{[0.050]} = -(5.4 \times 10^{-2}\text{h}^{-1})t$$

$$\frac{-\ln(0.20)}{5.4 \times 10^{-2}\text{h}^{-1}} = t$$

$$\frac{-(-1.61)}{5.4 \times 10^{-2}\text{h}^{-1}} = t$$

$$t = 30.\ \text{h}$$

EXAMPLE 15.4 *Using the First-Rate Equation*

Hydrogen peroxide decomposes in dilute sodium hydroxide at 20 °C in a first-order reaction.

$$2\ H_2O_2(aq) \longrightarrow 2\ H_2O(\ell) + O_2(g)$$

$$\text{Rate} = k[H_2O_2] \qquad k = 1.06 \times 10^{-3}\ \text{min}^{-1}$$

If the initial concentration of H_2O_2 is 0.020 mol/L, what is the concentration of the peroxide after exactly 100 min?

Solution Here Equation 15.1 is written as

$$\ln \frac{[H_2O_2]_t}{[H_2O_2]_0} = -kt$$

where $[H_2O_2]_0$, k, and t are known. To calculate $[H_2O_2]_t$, the peroxide concentration after 100. min, it is convenient to rearrange the equation to

$$\ln [H_2O_2]_t - \ln [H_2O_2]_0 = -kt$$

Substituting into this equation, we have

$$\begin{aligned} \ln [H_2O_2]_t - \ln (0.020) &= -(1.06 \times 10^{-3}\ \text{min}^{-1})(100.\ \text{min}) \\ \ln [H_2O_2]_t - (-3.91) &= -(1.06 \times 10^{-1}) \\ \ln [H_2O_2]_t &= -3.91 - 0.106 \\ &= -4.02 \end{aligned}$$

Taking the antilogarithm of -4.02 (that is, the inverse ln of -4.02 or $e^{-4.02}$), we find the concentration of hydrogen peroxide after 100. min is $[H_2O_2]_t =$ 0.018 mol/L.

EXERCISE 15.6 *Using the First-Order Rate Equation*

Sucrose, a sugar, decomposes in acid solution to the simpler sugars glucose and fructose. The rate expression is Rate = k[sucrose], where $k = 0.21\ \text{h}^{-1}$. If the original concentration of sucrose is 0.010 mol/L, what is its concentration after 5.0 h? ■

PROBLEM-SOLVING TIPS AND IDEAS

15.1 Using Logarithms and Your Calculator

This text uses natural logarithms (ln x) because this is the form in which most mathematical equations are derived.

1. To find the logarithm of a number, enter the number in your calculator and press the key marked "ln x" or just "ln." For example,

$$\ln 2.00 = 0.693$$

2. To find the number corresponding to a given logarithm, enter the number into your calculator and press the e^x key. What number x corresponds to the logarithm -0.245?

$$\begin{aligned} \ln x &= -0.245 \\ x &= e^{-0.245} \quad (\text{or } x = \text{inverse ln of } -0.245) \\ &= 0.783 \end{aligned}$$

See Appendix A for further discussion of logarithms, their use, and their relation to the proper number of significant figures. ■

Second-Order Reactions

Suppose the reaction R → products is second order. The rate equation is

$$\text{Rate} = -\frac{\Delta[\text{R}]}{\Delta t} = k[\text{R}]^2$$

Again using the methods of calculus, this equation can be transformed into an equation that relates reactant concentration and time.

$$\frac{1}{[\text{R}]_t} - \frac{1}{[\text{R}]_0} = kt \qquad \textbf{(15.2)}$$

As before, $[\text{R}]_0$ is the concentration of the reactant at the time $t = 0$, and $[\text{R}]_t$ is the concentration at a later time. This same equation applies to a reaction such as $\text{R}_1 + \text{R}_2 \rightarrow$ products, where the rate equation is

$$\text{Rate} = k[\text{R}_1][\text{R}_2]$$

and the special condition is met that the initial concentrations of R_1 and R_2 are the same.

EXAMPLE 15.5 ***Using the Second-Order Concentration/Time Equation***

The gas phase decomposition of HI

$$2\ \text{HI(g)} \longrightarrow \text{H}_2\text{(g)} + \text{I}_2\text{(g)}$$

has the rate equation

$$\text{Rate} = k[\text{HI}]^2$$

where $k = 30.$ L/mol · min at 443 °C. How much time does it take for the concentration of HI to drop from 0.010 mol/L to 0.0050 mol/L at 443 °C?

Solution Here $[\text{R}]_0 = 0.010$ mol/L and $[\text{R}]_t = 0.0050$ mol/L. Substituting into Equation 15.2, we have

$$\frac{1}{0.0050\ \text{mol/L}} - \frac{1}{0.010\ \text{mol/L}} = (30.\ \text{L/mol} \cdot \text{min})t$$

$$(2.0 \times 10^2\ \text{L/mol}) - (1.0 \times 10^2\ \text{L/mol}) = (30.\ \text{L/mol} \cdot \text{min})t$$

$$t = 3.3\ \text{min}$$

EXERCISE 15.7 ***Using the Second-Order Concentration/Time Equation***

Using the rate constant for HI decomposition given in Example 15.5, calculate the concentration of HI after 10. min if $[\text{HI}]_0 = 0.010$ mol/L. ■

Zero-Order Reactions

For a zero-order reaction of the kind R → products, the rate equation is

$$\text{Rate} = -\frac{\Delta[\text{R}]}{\Delta t} = k[\text{R}]^0 = k$$

This equation leads to the integrated rate equation

$$[\text{R}]_0 - [\text{R}]_t = kt \qquad \textbf{(15.3)}$$

where k is expressed in units of moles per (liters × seconds).

Graphical Methods for Determining Reaction Order and the Rate Constant

The equations relating concentration and time for zero-, first-, and second-order reactions are all very different. Nonetheless, they suggest a convenient way to determine the order of a reaction and its rate constant. If each of these equations is rearranged slightly, it is apparent that each is the equation for a straight line, $y = a + bx$, where b is the slope of the line and a is the intercept (the value of y when x is zero). ($x = t$ in each case.)

Plot for the equation $y = a + bx$
Slope = b and Intercept = a

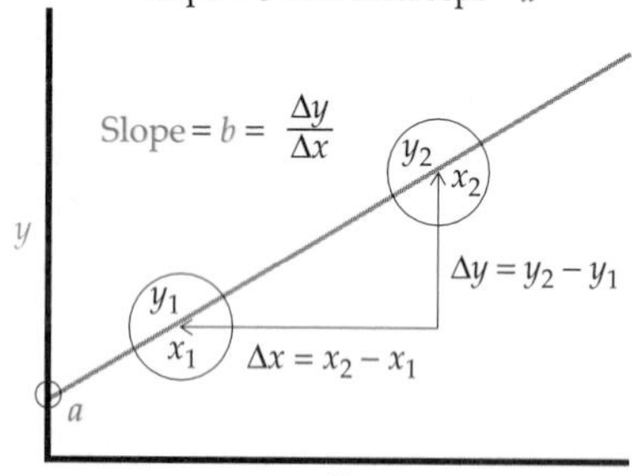

Zero Order	First Order	Second Order
$[\text{R}]_t = [\text{R}]_0 - kt$	$\ln [\text{R}]_t = \ln [\text{R}]_0 - kt$	$\frac{1}{[\text{R}]_t} = \frac{1}{[\text{R}]_0} + kt$
$\downarrow y \quad \downarrow a \quad \downarrow bx$	$\downarrow y \quad \downarrow a \quad \downarrow bx$	$\downarrow y \quad \downarrow a \quad \downarrow bx$

The decomposition of ammonia on a platinum surface was previously mentioned as a zero-order reaction,

$$2\ NH_3(g) \longrightarrow N_2(g) + 3\ H_2(g) \qquad \text{rate} = k[NH_3]^0 = k$$

which means the reaction is independent of NH_3 concentration. The rate constant k is given by the slope of the line when the concentration of R, $[\text{R}]_t$, is plotted (on the y, or vertical, axis) against time t(on the x, or horizontal, axis) (Figure 15.7). The straight line obtained proves that this reaction can only be zero order. The slope of the line is found by selecting any two points on the line and reading off the coordinates as indicated by the arrows in Figure 15.7. The slope = $-k$, so in this case

A line slanting downward from the left has a negative slope.

$$-k = -1.5 \times 10^{-6}\ \text{mol/L} \cdot \text{s}$$
$$k = 1.5 \times 10^{-6}\ \text{mol/L} \cdot \text{s}$$

The intercept of the line at $t = 0$ is equal to $[\text{R}]_0$.

The plot of concentration versus time for a first-order reaction is always a curved line (see Figure 15.2). Plotting ln [reactant] versus time, however, always produces a straight line with a negative slope when the reaction is first order in that reactant. Consider the decomposition of hydrogen peroxide (see Example 15.4).

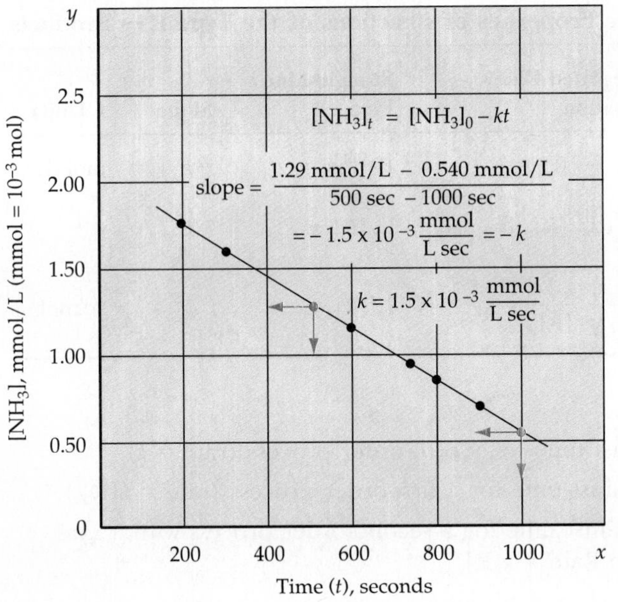

Figure 15.7 Plot of the concentration of ammonia, $[NH_3]_t$, against time for the decomposition of NH_3 [$2\ NH_3(g) \rightarrow N_2(g) + 3\ H_2(g)$] on a metal surface at 856 °C, a zero-order reaction.

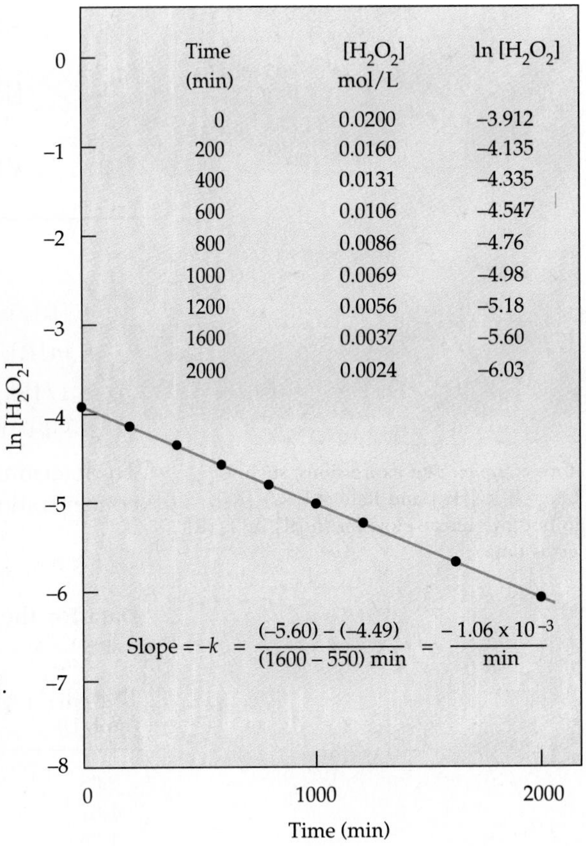

Figure 15.8 Kinetic plot for the decomposition of hydrogen peroxide [$2\ H_2O_2(aq) \rightarrow 2\ H_2O(\ell) + O_2(g)$]. A plot of ln [$H_2O_2$] against time is a straight line with a negative slope, indicating a first-order reaction. The rate constant $k = -\text{slope}$.

$$2\ H_2O_2(aq) \longrightarrow 2\ H_2O(\ell) + O_2(g)$$

$$\text{Rate} = k[H_2O_2] \qquad k = 1.06 \times 10^{-3}\ \text{min}^{-1}$$

The natural logarithm of [H_2O_2], ln [H_2O_2], is plotted against time in Figure 15.8. We observe a straight line, which unquestionably shows that the reaction is first order in H_2O_2. Notice that the slope is negative because the $-kt$ term has a negative sign and k is positive.

The decomposition of NO_2 is a second-order process,

$$2\ NO_2(g) \longrightarrow 2\ NO(g) + O_2(g)$$

$$\text{Rate} = k[NO_2]^2$$

and the concentration-time data for the reaction have been plotted as $1/[NO_2]$ versus time (Figure 15.9). This gives a straight line as predicted for a second-order process. Here the slope of the line is equal to k.

These examples of commonly observed rate expressions illustrate a useful way to determine reaction order (see Table 15.1; next page). A straight line is observed *only* when

TABLE 15.1 Characteristic Properties of Reactions of the Type R → Products

Order	Rate Equation	Integrated Rate Equation	Straight-Line Plot	Slope	*k* Units
0	$k[R]^0$	$[R]_0 - [R]_t = kt$	$[R]_t$ vs. t	$-k$	mol/L · s
1	$k[R]^1$	$\ln([R]_t/[R]_0) = -kt$	$\ln [R]_t$ vs. t	$-k$	s^{-1}
2	$k[R]^2$	$\frac{1}{[R]_t} - \frac{1}{[R]_0} = kt$	$1/[R]_t$ vs. t	k	L/mol · s

- $[R]_t$ is plotted against time for a zero-order process (rate = k).
- $\ln[R]_t$ is plotted against time for a first-order process (rate = $k[R]_t$).
- $1/[R]_t$ is plotted against time for a second-order process with a rate equation of the type Rate = $k[R]_t^2$.

More complex rate expressions, such as Rate = $k[R_1][R_2]$ and Rate = $k[R_1]^2[R_2]$, do not give linear plots for ln [R] or 1/[R] versus time.

To determine the reaction order, therefore, a chemist plots the experimental concentration-time data in different ways until a straight-line plot is achieved.

EXERCISE 15.8 ***Using Graphical Methods***

Data for the decomposition of N_2O_5 in a particular solvent at 45 °C are as follows:

$[N_2O_5]_t$ (mol/L)	t (min)
2.08	3.07
1.67	8.77
1.36	14.45
0.72	31.28

Plot $[N_2O_5]_t$, $\ln [N_2O_5]_t$, and $1/[N_2O_5]_t$ against time t. What is the order of the reaction? What is the rate constant for the reaction? ■

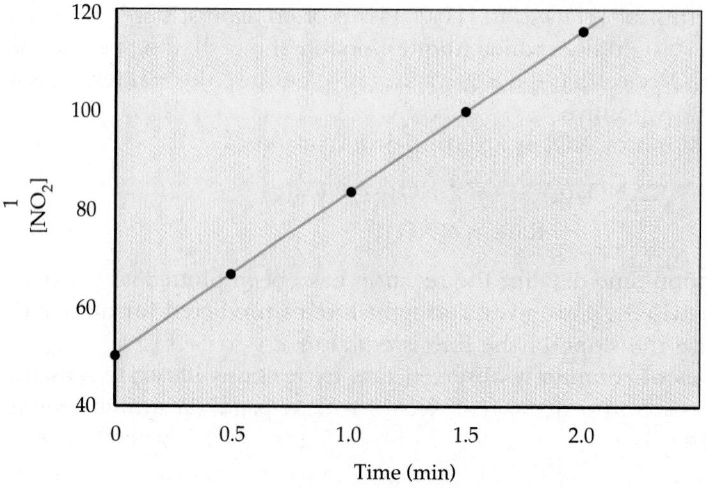

Figure 15.9 Concentration versus time curve for the decomposition of NO_2 to NO and O_2. The data are plotted as $1/[NO_2]$ versus time. For a rate expression of the form "rate = $k[A]^2$," a straight line is observed when $1/[A]$ is plotted against time.

Half-Life and First-Order Reactions

The **half-life,** $t_{1/2}$, of a reaction is the time required for the concentration of a reactant to decrease to one half its initial value. It indicates the stability of a reactant; the longer the half-life, the greater the stability of the reactant or reactants. The form of the relationship between half-life and initial concentration is different for different reaction orders.

For a reactant R in a reaction that is first order in R, $t_{1/2}$ is the time when

$$[R]_t = \frac{1}{2}[R]_0 \qquad \text{or} \qquad \frac{[R]_t}{[R]_0} = \frac{1}{2}$$

where $[R]_0$ is the initial concentration and $[R]_t$ is the concentration after half of the reactant has been consumed. To find $t_{1/2}$ we use the concentration-time equation (Equation 15.1) and proceed as follows. Taking the first-order rate equation,

$$\ln \frac{[R]_t}{[R]_0} = -kt$$

and substituting the fact that $[R]_t/[R]_0 = \frac{1}{2}$ when $t = t_{1/2}$, we have

$$\ln\left(\frac{1}{2}\right) = -kt_{1/2}$$

or

$$\ln 2 = 0.693 = kt_{1/2}$$

Rearranging this equation, we come to the very useful equation that relates half-life and the first-order rate constant (where ln 2 = 0.693).

$$t_{1/2} = \frac{0.693}{k} \qquad \textbf{(15.4)}$$

Notice that both k and $t_{1/2}$ are independent of concentration for first-order reactions.

To illustrate the concept of half-life for first-order reactions, the concentration of H_2O_2 has been plotted as a function of time for the decomposition of hydrogen peroxide (Figure 15.10).

$$2\ H_2O_2(aq) \longrightarrow H_2O(\ell) + O_2(g)$$
$$\text{Rate} = k[H_2O_2]$$

Because the rate constant k for this reaction is 1.06×10^{-3}/min, the half-life is

$$t_{1/2} = \frac{0.693}{k} = \frac{0.693}{1.06 \times 10^{-3}\ \text{min}^{-1}} = 654\ \text{min}$$

If the initial concentration of H_2O_2 is 0.020 M, the concentration has dropped to one half of 0.020 M, or 0.010 M, after 654 min. Notice that the concentration drops again by half after another 654 min. That is, after two half-lives (1308 min) the concentration is only $(\frac{1}{2}) \cdot (\frac{1}{2}) = (\frac{1}{2})^2 = \frac{1}{4}$, or 25% of the initial concentration, or 0.0050 M. After three half-lives (1962 min), the concentration has now dropped to $(\frac{1}{2}) \cdot (\frac{1}{2}) \cdot (\frac{1}{2}) = (\frac{1}{2})^3 = \frac{1}{8}$, or 12.5% of the initial value; here $[H_2O_2] = 0.0025$ M.

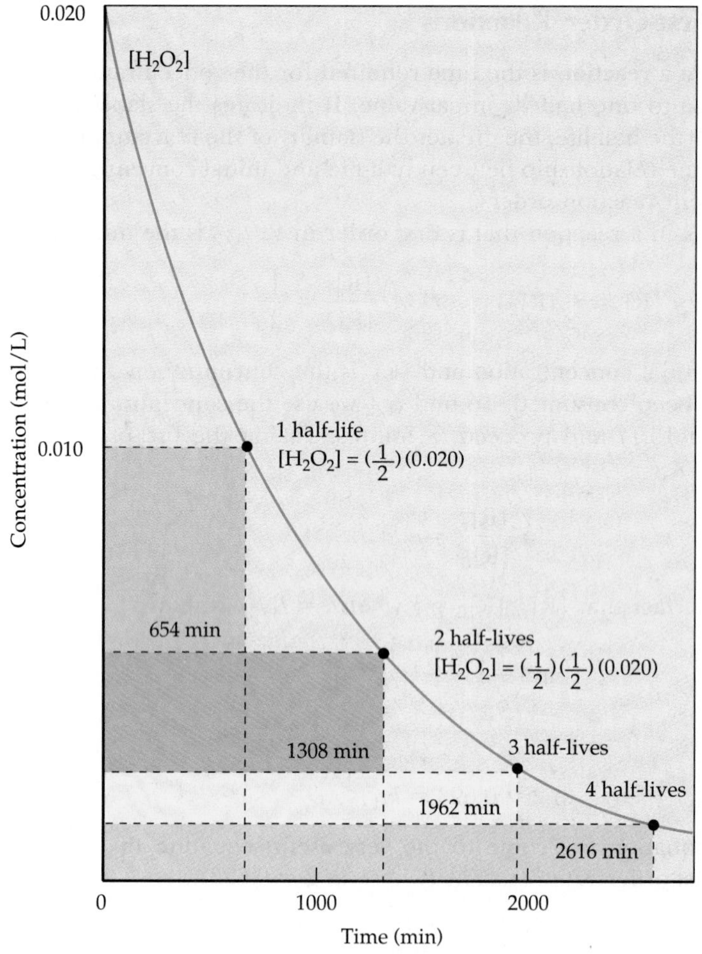

Figure 15.10 The concentration versus time curve for the disappearance of H_2O_2 ($k = 1.06 \times 10^{-3}\ \text{min}^{-1}$) (See Figure 15.8). At 654 minutes ($t_{1/2}$) the initial concentration of H_2O_2 has halved. During each successive interval of 654 minutes the concentration again halves.

One reason for introducing the concept of half-life is that it makes it easier to visualize the magnitude of a reaction rate. Radioactive elements, for example, all decay by a first-order process (see Chapter 24). Americium is used in smoke detectors and in medicine for the treatment of certain malignancies. The isotope of americium with a mass number of 241, ^{241}Am, has a rate constant k for decay of 0.0016 year^{-1}. In contrast, radioactive iodine-125, which is used for studies of thyroid functioning, has a rate constant for decay of 0.011 day^{-1}. Which element decays faster? After calculating half-lives, it is easier to see that ^{125}I decays much more rapidly ($t_{1/2}$ = 63. day) than ^{241}Am ($t_{1/2}$ = 430 year).

EXAMPLE 15.6 *Half-Life of a First-Order Process*

Sucrose, $C_{12}H_{22}O_{11}$, decomposes to fructose and glucose in acid solution with the rate law

$$\text{Rate} = k[\text{sucrose}] \qquad k = 0.208\ \text{h}^{-1} \text{ at } 25\ °\text{C}$$

Find the half-life of sucrose under these conditions. Calculate the time required for 87.5% of the initial concentration of sucrose to disappear.

A CLOSER LOOK *Half-Life and Reaction Orders*

Many processes of practical interest are first order, especially the decay of radioactive elements. Because k and $t_{1/2}$ are directly related, $t_{1/2}$ is often used as a way of describing the stability of a compound or a radioactive isotope. It also provides a simple way of (1) determining the quantity of reactant remaining at any time after the reaction has started (q_t) or (2) the time required to reduce some quantity (q_0) of reactant to a smaller quantity (q_t). Consider again the example of H_2O_2 decomposition in the text. After two half-lives, the concentration remaining was

$$[H_2O_2] = q_t = [0.020 \text{ M}] \cdot \left(\frac{1}{2}\right) \cdot \left(\frac{1}{2}\right)$$
$$= 0.0050 \text{ M}$$

This equation can be generalized to a very useful equation, where the exponent $t/t_{1/2}$ is the number of half-life periods through which the reaction has progressed.

$$q_t = q_0 \left(\frac{1}{2}\right)^{t/t_{1/2}}$$

If we take the logarithm of both sides, the same equation can be written as

$$\ln q_t = \ln q_0 + \frac{t}{t_{1/2}} \ln \left(\frac{1}{2}\right)$$

Although less widely used, half-lives of zero- and second-order processes are also easily defined. Substituting $t_{1/2}$ for t and $\frac{1}{2}[R]_0$ for $[R]_t$ in Equation 15.3 for a zero-order reaction,

$$[R]_0 - \frac{1}{2}[R]_0 = kt_{1/2}$$

and this gives the equation

$$t_{1/2} \text{ for a zero-order reaction} = \frac{[R]_0}{2k}$$

Making the same substitution in Equation 15.2 for a second-order reaction

$$\frac{1}{\frac{1}{2}[R]_0} - \frac{1}{[R]_0} = kt_{1/2}$$

leads to the following equation:

$$\frac{1}{[R]_0} = kt_{1/2}$$

Notice that the half-life expressions for zero- and second-order processes include the initial concentration, whereas that for a first-order process does not. This means that half-lives for zero- and second-order processes vary with initial concentration; each successive half-life requires a different amount of time than the one before it.

Solution The half-life for the change is

$$t_{1/2} = \frac{0.693}{k} = \frac{0.693}{0.208 \text{ h}^{-1}} = 3.33 \text{ h}$$

As illustrated for a different reaction in Figure 15.10, the concentration of a reactant in a first-order process has dropped to 12.5% of its original value after three half-lives. This is the same as saying that 87.5% of the reactant has been consumed. The time required for sucrose to reach this concentration is 3 × 3.33 h = 9.99 h.

EXERCISE 15.9 ***Half-Life of a First-Order Process***

The rate constant for the transformation of cyclopropane to propene (see Example 15.3) is 5.40×10^{-2} h^{-1}. What is the half-life of the reaction? What fraction of the cyclopropane remains after 51.2 h? What fraction remains after 18.0 h? ■

15.5 A MICROSCOPIC VIEW OF REACTIONS

"Reactant concentrations can affect reaction rates" is a macroscopic observation. We can also describe reaction rates in terms of what happens in the microscopic world of atoms and molecules by using the kinetic theory of matter (Section 12.6). Our goal is to find explanations for the factors that govern reaction rates. We shall base our explanations on the **collision theory** of reaction rates, which assumes that molecules must collide with one another in order to react.

Effect of the Nature of the Reactants on Reaction Rate—Activation Energy

According to kinetic theory, molecules in a gas or liquid move rapidly and frequently bump into one another. Atoms within molecules move as well; molecules constantly flex or vibrate along or around the bonds that hold the atoms together. If a molecule has enough energy, the arrangement of atoms can be changed, resulting in a different molecule. An example is the conversion of *cis*-2-butene to *trans*-2-butene. The experimentally determined rate equation is

$$\text{Reaction rate} = k[\textit{cis}\text{-2-butene}]$$

which shows that the reaction is first order in the reactant, *cis*-2-butene. As seen in the models, the *cis* form can become the *trans* form if one half of the *cis* molecule is twisted relative to the other. Thus, it is a reasonable hypothesis that the molecular pathway or reaction mechanism that converts *cis*-2-butene to *trans*-2-butene involves twisting the molecule (see Sections 10.1 and 11.2).

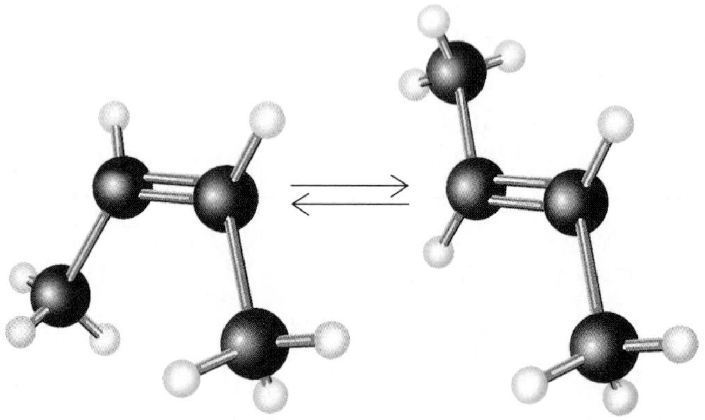

We know from experiment, though, that carbon-carbon double bonds are like springs. Springs can be stretched, twisted, and bent, although it takes the input of energy to do so. Consequently, some kinetic energy must be converted to potential energy when one end of the *cis*-2-butene molecule twists relative to the other, just as it would if a spring were stretched or bent. At room temperature most of the molecules do not have enough kinetic energy to twist far enough so that *cis*-2-butene can be changed to *trans*-2-butene, and so *cis*-2-butene can be kept in a sealed flask at room temperature for a long time without any appreciable quantity of *trans*-2-butene being formed.

Figure 15.11 shows a plot of potential energy versus the angle of twist in *cis*- and *trans*-2-butene. The potential energy of a *cis*-2-butene molecule is 435×10^{-21} J higher when one end is twisted by 90° from the initial, flat molecule. This is similar to the increased potential energy that an object like a car has at the top of a hill compared with its energy at the bottom. Just as a car cannot reach the top of a hill unless it has enough energy, a molecule cannot reach the top of the "hill" for a reaction unless it has enough energy. Notice that the top of the hill can be approached from either side, and from the top a twisted molecule can go downhill energetically to either the *cis* or the *trans* form.

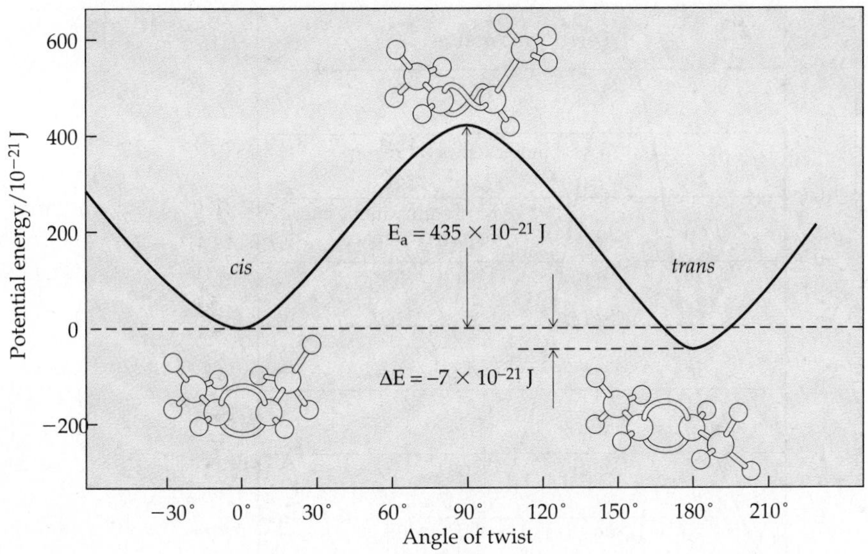

Figure 15.11 An energy profile for the conversion of *cis*-2-butene to *trans*-2-butene. The double bond between the two central atoms resists twisting. When enough energy is added, however, the molecule can twist around the central carbon-carbon bond. When the potential energy has risen to 435×10^{-21} J/molecule, the angle between the ends of the molecule is 90 °. A molecule of *cis*-2-butene must have at least this much energy before it can convert to *trans*-2-butene.

An interesting relationship shown in Figure 15.11 connects kinetics and thermodynamics. The energy of the product, a molecule of *trans*-2-butene, is 7×10^{-21} J lower than that of the reactant, a molecule of *cis*-2-butene. This means that the *cis* → *trans* reaction is exothermic by 7×10^{-21} J/molecule, which translates to 4 kJ/mol. Conversely, *cis*-2-butene is higher in energy than *trans*-2-butene by 7×10^{-21} J/molecule, and so the reverse reaction requires that 4 kJ/mol be absorbed from the surroundings; it is *endothermic*. The energy hill that has to be climbed when the reverse reaction occurs is $(435+7) \times 10^{-21}$ J, or 442×10^{-21} J high (=266 kJ/mol).

Every chemical reaction has an energy barrier that must be surmounted if molecules are to react. The heights of such barriers vary greatly—from almost zero to hundreds of kilojoules per mole. For similar reactions at a given temperature, the higher the energy barrier the slower the reaction. The minimum energy required to surmount the barrier is called the **activation energy, E_a,** for the reaction. For the *cis*-2-butene → *trans*-2-butene reaction the activation energy is 435×10^{-21} J/molecule or 262 kJ/mol. In general, *the fact that reaction rates vary from extremely fast to extremely slow is due to differences in activation energies.*

The energy available by heat transfer from the surroundings to a reaction at room temperature is only about 2–3 kJ/mol. The reactions described have activation energies much higher than this.

Another factor affecting the rate of a reaction is that not all molecules at a given temperature have the same kinetic energy (see Section 12.6). Some have a very low energy, some a very high energy, but most molecules have some intermediate energy. That is, there is a distribution of molecular energies, as illustrated in Figure 15.12. If this diagram is applied to the transformation of *cis*- to *trans*-2-butene, the shaded area represents the number of molecules having a kinetic energy of at least 266 kJ/mol. If the minimum kinetic energy required for

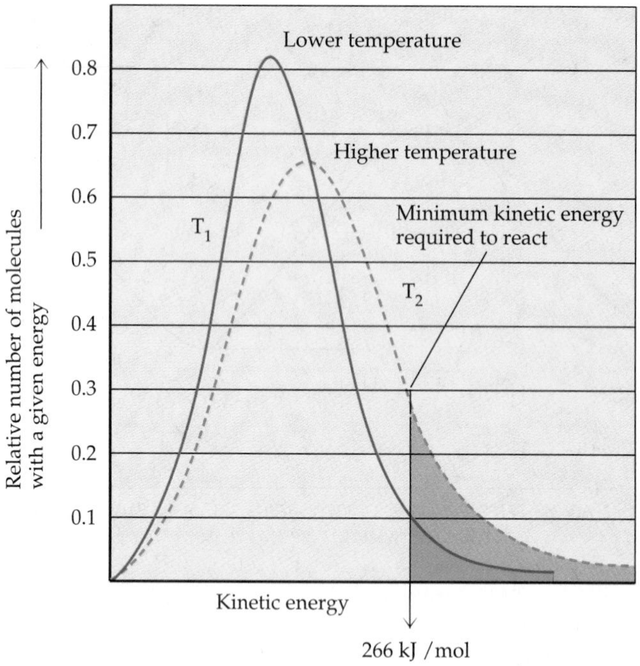

Figure 15.12 Kinetic-energy distribution curve. The vertical axis gives the relative number of molecules possessing the energy indicated on the horizontal axis. The shaded area represents the fraction of molecules having the minimum energy required for reaction. In the case of the *cis*- to *trans*-2-butene reaction, comparatively few molecules have sufficient energy at room temperature (*red area*). At a higher temperature, however, a larger fraction of molecules has the minimum energy required (*red area* + *blue area*).

reaction were only 200 kJ/mol, let us say, then many more molecules would have the required energy, and the reaction rate would be much higher.

Effect of Reactant Concentrations on Reaction Rate

Now let us consider the effect of concentration. Suppose a flask contains 0.005 mol/L of *cis*-2-butene vapor at room temperature. The molecules have a wide range of energies, but only a few of them have enough energy at this temperature to get over the activation energy barrier. Now suppose that another flask contains 0.010 mol/L of *cis*-2-butene, that is, the concentration of molecules is twice that of the first flask. If both flasks are at the same temperature, both have the same fraction of molecules with enough energy to cross over the barrier. In the flask containing twice as many molecules, however, there must be twice as many molecules crossing the barrier at any given time. The rate of the *cis* → *trans* reaction is twice as great, therefore. That is, reaction rate is proportional to the concentration of *cis*-2-butene, which is exactly a statement of the rate equation Rate = k[*cis*-2-butene].

There are many reactions in which two molecules of different kinds must collide with one another. One example is the reaction of nitrogen monoxide and ozone

$$NO(g) + O_3(g) \longrightarrow NO_2(g) + O_2(g)$$

for which the experimental rate equation is

$$\text{Reaction rate} = k[NO][O_3]$$

Here the reaction involves the collision of an NO molecule and an O_3 molecule, but there is still only a single step. Because the molecules must collide to exchange atoms, however, the rate depends on the number of collisions per unit time. Figure 15.13a represents a flask containing one NO molecule (the red ball) and several O_3 molecules (the blue balls). Within a given time period, the NO molecule collides with, let us say, five O_3 molecules. If the concentration of

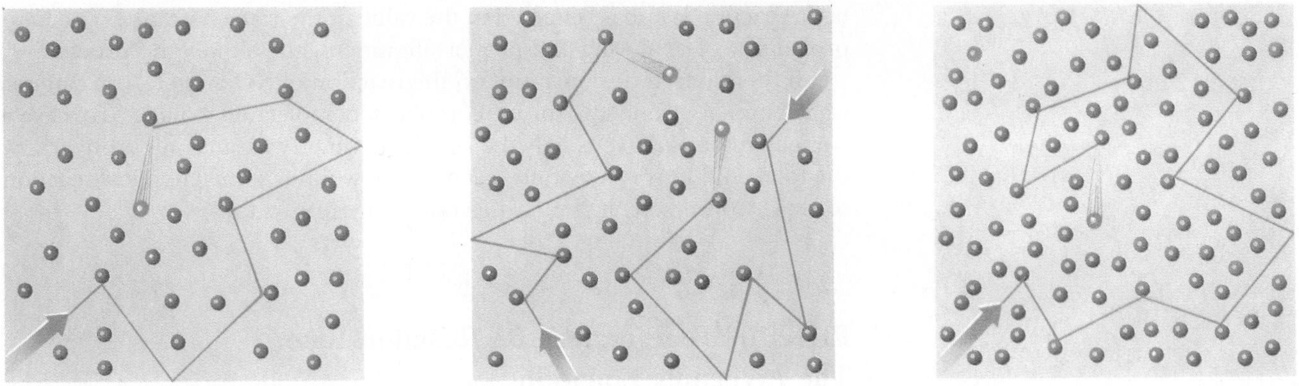

Figure 15.13 The effect of concentration on the frequency of molecular collisions. (a) A single red molecule moves among 50 blue molecules and collides with five of them per second. (b) Two red molecules now move among 50 blue molecules, and there are 10 red-blue collisions per second. (c) If the number of blue molecules is doubled to 100, the frequency of the red-blue collisions is also doubled, to 10 per second. The number of collisions is thus proportional to the concentration of *both* red *and* blue molecules.

NO molecules is doubled to two in the flask (Figure 15.13b), each NO collides with five different O_3 molecules, and so the total number of collisions with O_3 molecules in the same period of time is now ten. Doubling the concentration of NO has doubled the rate. The same thing would happen if the O_3 concentration were doubled (Figure 15.13c). This description of the NO + O_3 reaction applies to all processes involving the collision of two molecules A and B. Such reactions have the general rate equation Reaction rate = k[A][B], where the reaction is first order in each of the two reactants.

Effect of Molecular Orientation on Reaction Rate

Another interesting aspect of the NO + O_3 reaction is that the molecules must collide in the proper orientation for the reaction to be effective. Having a sufficiently high energy is necessary, but this is not sufficient to ensure that reactants will form products. In the case of the reaction between NO and O_3, the N of NO must come together with one of the O atoms at the end of the O_3 molecule (Figure 15.14). This so-called ''steric factor'' is important in determining how

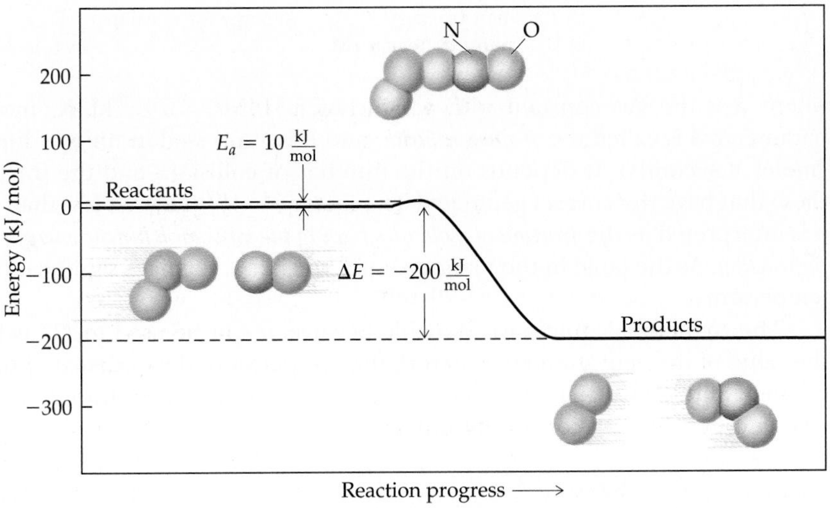

Figure 15.14 The exothermic reaction of NO + O_3. As NO and O_3 approach one another, energy is needed to squeeze them together until the N atom of NO and one of the O atoms from O_3 are close enough to form a bond. Also notice that the NO and O_3 must come together in the proper orientation; one of the terminal O atoms of O_3 must approach the N of NO. Reaction is not possible if O_3 approaches the O atom of NO, or if the central O atom of O_3 is the one that makes contact. This ''steric factor'' is important in determining how large the reaction rate can be. The more difficult it is to achieve the proper alignment, the slower the reaction is.

fast a reaction is and is reflected in the value of the rate constant *k*. The lower the probability of achieving the proper alignment, the slower the process.

If there is a steric constraint on the reaction of NO and O_3, you can imagine what happens when two more complicated molecules collide. To have them come together in just exactly the correct geometry means only a tiny fraction of the total collisions can lead to reaction. No wonder some chemical reactions are so slow. Conversely, it is amazing that some are so fast!

Effect of Temperature on Reaction Rate: The Arrhenius Equation

The most common way to speed up a reaction is to increase the temperature. A mixture of natural gas (methane) and air is stable for centuries at room temperature; the reaction to give CO_2 and H_2O is extraordinarily slow. If a lighted match is brought up to the mixture, however, the temperature is raised, the reaction rate increases, and the gas ignites. Thereafter the heat evolved by the combustion reaction maintains a high temperature, and the reaction continues at a rapid rate.

Reaction rates increase with temperature because higher temperature means a greater fraction of reactant molecules have enough energy to cross the activation energy barrier. Consider again the conversion of *cis*- to *trans*-2-butene (see Figure 15.11). At room temperature relatively few *cis*-2-butene molecules possess the minimum kinetic energy required for reaction. As the temperature increases, however, more and more molecules have enough energy (see Figure 15.12), and the reaction rate increases.

Svante Arrhenius (1859–1927) was a Swedish chemist who, among other things, derived the relation between the rate constant and temperature from experiment.

Reaction rates depend on the energy and number of collisions between reacting molecules, on whether the collisions have the correct geometry, and on the temperature. These requirements are summarized by the **Arrhenius equation**

Temperature (K)	Value of $e^{-E_a/RT}$ for E_a = 40 kJ
298	9.7×10^{-8}
400	5.9×10^{-6}
600	3.3×10^{-4}
800	2.4×10^{-3}

$$k = \text{reaction rate constant} = Ae^{-E_a/RT} \quad (15.5)$$

A: frequency of collisions with correct geometry when reactant concentrations = 1M

$e^{-E_a/RT}$: fraction of molecules with minimum energy for reaction

where R is the gas constant with a value of 8.314510×10^{-3} kJ/K · mol. The parameter A is called the *frequency factor,* and it is expressed in units of liters per (moles × seconds). It depends on the number of collisions and the fraction of these that have the correct geometry. The factor $e^{-E_a/RT}$ is always less than 1, and it is interpreted as the *fraction of molecules having the minimum kinetic energy required for reaction.* As the table in the margin shows, the factor changes significantly with temperature.

An interpretation of the significance of A goes beyond the level of this text. One aspect of A that you should remember, however, is that it becomes smaller as the reactants become larger.

The Arrhenius equation is valuable because it can be used to (1) calculate the value of the activation energy from the temperature dependence of the rate constant and (2) calculate the rate constant for a given temperature if the activation energy and A factor are known.

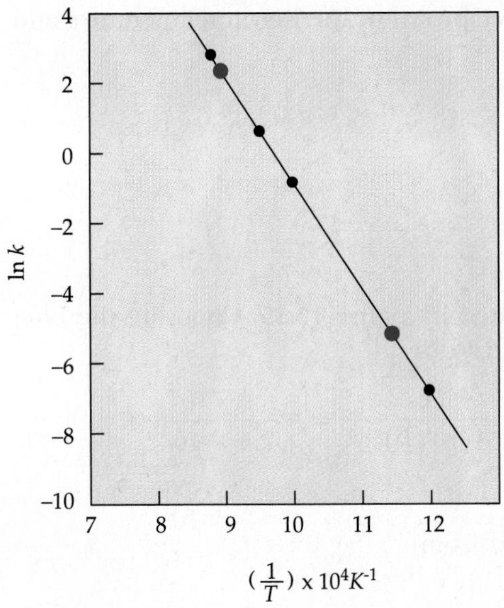

Figure 15.15 A plot of ln k versus $1/T$ for the reaction

$$2\ N_2O(g) \longrightarrow 2\ N_2(g) + O_2(g)$$

The slope of the line gives E_a as outlined in Example 15.7.

If we take the natural logarithm of each side of Equation 15.5, we have

$$\ln k = \ln A - \left(\frac{E_a}{RT}\right)$$

and, if we rearrange this slightly, it becomes the equation for a straight line relating ln k to $(1/T)$.

$$\ln k = \ln A + \left[-\frac{E_a}{R}\left(\frac{1}{T}\right)\right] \longleftarrow \text{Arrhenius equation}$$

$$y \quad = \quad a \quad + \quad bx \quad \longleftarrow \text{equation for straight line} \qquad \textbf{(15.6)}$$

This means that, if the natural logarithm of k (ln k) is plotted versus $1/T$, the result is a downward-sloping line with a slope of $(-E_a/R)$ (Figure 15.15). So, now we have a means to calculate E_a from experimental values of k at several temperatures, a calculation illustrated in Example 15.7.

EXAMPLE 15.7 *Determination of E_a from the Arrhenius Equation*

Using the experimental data shown here, calculate the activation energy E_a for the reaction

$$2\ N_2O(g) \longrightarrow 2\ N_2(g) + O_2(g)$$

Temperature (K)	k [L/mol · s]
1125	11.59
1053	1.67
1001	0.380
838	0.0011

Solution The first step is to find the reciprocal of the Kelvin temperature and the natural logarithm of k.

$(1/T)\ K^{-1}$	$\ln k$
8.889×10^{-4}	2.4501
9.497×10^{-4}	0.513
9.990×10^{-4}	−0.968
11.9×10^{-4}	−6.81

These data are then plotted as illustrated in Figure 15.15. Choosing the blue points on the graph, the slope is found to be

$$\text{Slope} = \frac{\Delta \ln k}{\Delta(1/T)} = \frac{2.0 - (-5.6)}{(9.0 - 11.5)(10^{-4})\text{K}^{-1}} = -\frac{7.6}{2.5 \times 10^{-4}}\text{K}$$

$$= -3.0 \times 10^4\ \text{K}$$

The activation energy is then evaluated from

$$\text{Slope} = -\frac{E_a}{R}$$

$$-3.0 \times 10^4\ \text{K} = -\frac{E_a}{8.31 \times 10^{-3}\ \text{kJ/K} \cdot \text{mol}}$$

$$E_a = 250\ \text{kJ/mol}$$

In addition to the graphical method for evaluating k illustrated in Example 15.7, we can obtain E_a algebraically. Knowing k at two different temperatures, we can write the equation for each of these conditions.

$$\ln k_2 = \ln A - \left(\frac{E_a}{RT_2}\right) \quad \text{and} \quad \ln k_1 = \ln A - \left(\frac{E_a}{RT_1}\right)$$

If one of these equations is subtracted from the other, we have

$$\ln k_2 - \ln k_1 = \ln\frac{k_2}{k_1} = -\frac{E_a}{R}\left[\frac{1}{T_2} - \frac{1}{T_1}\right] \qquad (15.7)$$

an equation from which E_a can be obtained. This equation now suggests an alternate use, however. The rate constant k_2 can be calculated for another temperature T_2, if E_a, k_1, and T_1 are known. The first of these situations is modeled in Example 15.8.

A good rule of thumb is that reaction rates double for every 10 °C rise in temperature in the vicinity of room temperature (for reactions with E_a around 50 kJ/mol).

EXAMPLE 15.8 *Calculating E_a from the Temperature Dependence of k*

Using values of k determined at two different temperatures, calculate the value of E_a for the decomposition of HI.

$$2\ HI(g) \longrightarrow H_2(g) + I_2(g)$$
$$k = 2.15 \times 10^{-8}\ \text{L/mol} \cdot \text{s at 650. K}$$
$$k = 2.39 \times 10^{-7}\ \text{L/mol} \cdot \text{s at 700. K}$$

Solution Here we have k_1 at T_1 and k_2 at T_2, and so we use Equation 15.7.

$$\ln \frac{2.39 \times 10^{-7}\ \text{L/mol} \cdot \text{s}}{2.15 \times 10^{-8}\ \text{L/mol} \cdot \text{s}} = -\frac{E_a}{8.315 \times 10^{-3}\ \text{kJ/K} \cdot \text{mol}}\left[\frac{1}{700.\ \text{K}} - \frac{1}{650.\ \text{K}}\right]$$

$$\ln (11.1) = -\frac{E_a}{8.315 \times 10^{-3}\ \text{kJ/K} \cdot \text{mol}}(-0.000110/\text{K})$$

$$E_a = 182\ \text{kJ/mol}$$

EXERCISE 15.10 ***Calculating E_a from the Temperature Dependence of k***

The colorless gas N_2O_4 decomposes to the brown gas NO_2 (nitrogen dioxide) in a first-order reaction.

$$N_2O_4(g) \longrightarrow 2\ NO_2(g)$$

The rate constant $k = 4.5 \times 10^3\ s^{-1}$ at 274 K and $1.00 \times 10^4\ s^{-1}$ at 283 K. What is the energy of activation, E_a? ■

15.6 REACTION MECHANISMS

We come now to one of the more important reasons to study the rates of reactions: to understand the **reaction mechanism,** the sequence of bond-making and bond-breaking steps that occurs during the conversion of reactants to products. You have seen that the rate equation for a reaction can be determined by experiment. Based on the rate equation, chemists can often make an educated guess about the mechanism.

In some reactions the conversion of reactants to products occurs in a single step. Nitrogen monoxide and ozone react in this manner (see Figure 15.14).

$$NO(g) + O_3(g) \longrightarrow NO_2(g) + O_2(g)$$

Most chemical reactions, however, involve a sequence of steps. An example is the reaction of bromine and nitrogen monoxide.

$$Br_2(g) + 2\ NO(g) \longrightarrow 2\ BrNO(g)$$

It is unlikely that three molecules of just the correct type will collide in just the right orientation for reaction. Thus, it is more reasonable that this reaction occurs in a sequence of steps (Figure 15.16). For example, Br_2 and NO are believed to combine to produce a transient species, Br_2NO.

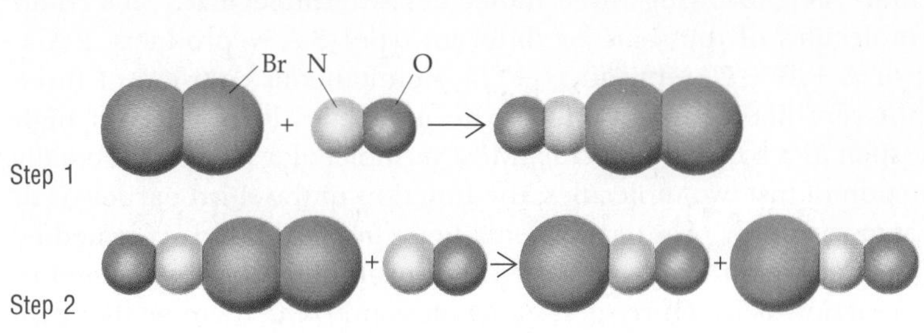

Figure 15.16 A representation of the proposed two-step mechanism by which NO and Br_2 are converted to BrNO.

Step 1 $$Br_2(g) + NO(g) \longrightarrow Br_2NO(g)$$

This then reacts with another NO to produce the final products.

Step 2 $$Br_2NO(g) + NO(g) \longrightarrow 2\ BrNO(g)$$

Adding Steps 1 and 2 yields the net reaction

$$Br_2(g) + 2\ NO(g) \longrightarrow 2\ BrNO(g)$$

Each of the steps in the reaction sequence is called an **elementary step,** which is defined as a chemical equation that describes an assumed single molecular event, such as the formation or rupture of a chemical bond or the displacement of atoms as a result of a molecular collision. Each step has its own activation energy barrier E_a and rate constant k, and the steps *must* add up to give the balanced equation for the overall reaction. The set of steps that satisfactorily explain the kinetic properties of a chemical reaction constitutes a reaction mechanism.

The mechanism of a reaction is postulated from experimental data. To see how this is done, we first describe three types of elementary steps.

Molecularity of Elementary Steps

Reactions occur by collisions of molecules, and elementary steps are classified by the number of reactant molecules (or ions, atoms, or free radicals) that come together in a step. This number is called its **molecularity.** The order of a reaction can be a fractional number, whereas the molecularity of a step is always a whole, positive number. When one molecule is the reactant in a step, the reaction is a **unimolecular** process. A **bimolecular** elementary process involves two molecules. These may be identical molecules (A + A → products) or different ones (A + B → products). For example, a two-step mechanism has been proposed for the decomposition of ozone

$$2\ O_3(g) \longrightarrow 3\ O_2(g)$$

Step 1 (unimolecular) $$O_3(g) \longrightarrow O_2(g) + O(g)$$

Step 2 (bimolecular) $$O_3(g) + O(g) \longrightarrow 2\ O_2(g)$$

It involves a unimolecular step followed by a bimolecular step.

An elementary step involving three molecules is **termolecular.** This could involve three molecules of the same or different type (3 A → products; 2 A + B → products; or A + B + C → products). The simultaneous collision of three molecules is not very likely, unless one of the molecules involved is in high concentration, such as a solvent molecule. Most termolecular processes actually involve the reaction of just two molecules; the function of the third particle is to carry away the excess energy produced when a new chemical bond is formed by the first two molecules in an exothermic step. For example, N_2 is unchanged in a termolecular reaction between oxygen molecules and oxygen atoms that produces ozone in the upper atmosphere.

$$O(g) + O_2(g) + N_2(g) \longrightarrow O_3(g) + \text{energetic } N_2(g)$$

The probability that four or more molecules may simultaneously combine with sufficient kinetic energy and proper orientation is so small that reaction molecularities greater than 3 are never proposed.

Rate Equations for Elementary Steps

The rate equation for a reaction *cannot* be predicted from its overall stoichiometry. In contrast, *the rate equation of an elementary step is given by the product of the rate constant and the concentrations of the reactants in that step.* This means we can write the rate equation for any elementary step, as in the examples in the table.

Elementary Step	Molecularity	Rate Equation
$A \rightarrow$ product	Unimolecular	Rate $= k[A]$
$A + B \rightarrow$ product	Bimolecular	Rate $= k[A][B]$
$A + A \rightarrow$ product	Bimolecular	Rate $= k[A]^2$
$2\ A + B \rightarrow$ product	Termolecular	Rate $= k[A]^2[B]$

For example, this means the rate laws for each of the two steps in the decomposition of ozone are

$$\text{Rate for (unimolecular) step 1} = k[O_3]$$
$$\text{Rate for (bimolecular) step 2} = k'[O_3][O]$$

where the two rate constants k and k' do not have the same value.

Molecularity and Reaction Order

The molecularity of an elementary step and its order are the same. A unimolecular elementary step must be first order, a bimolecular elementary step must be second order, and a termolecular elementary step must be third order; the converse is also true for *elementary* reactions. The relation between molecularity and order, however, is emphatically *not* true for the *overall* reaction. If you discover experimentally that a reaction is first order, you cannot conclude that it occurs in a single, unimolecular elementary step. An example of this is the decomposition of N_2O_5.

$$2\ N_2O_5(g) \longrightarrow 4\ NO_2(g) + O_2(g)$$

Here the rate equation is Rate $= k[N_2O_5]$, but chemists are fairly certain the mechanism involves a series of both unimolecular and bimolecular steps. Similarly, a second-order rate equation does not imply the reaction occurs in a single, bimolecular elementary step.

To see how the experimentally observed rate equation *for the overall reaction* is connected with a possible mechanism or sequence of elementary steps requires some chemical intuition. This is the subject of the next section.

EXAMPLE 15.9 *Elementary Steps*

The hypochlorite ion undergoes self oxidation-reduction to give the chlorate, ClO_3^-, and chloride ions.

$$3\ ClO^-(aq) \longrightarrow ClO_3^-(aq) + 2\ Cl^-(aq)$$

It is thought that the reaction occurs in two steps.

Step 1 $ClO^-(aq) + ClO^-(aq) \longrightarrow ClO_2^-(aq) + Cl^-(aq)$

Step 2 $ClO_2^-(aq) + ClO^-(aq) \longrightarrow ClO_3^-(aq) + Cl^-(aq)$

What is the molecularity of each step? Write the rate equation for each reaction step. Show that the sum of these reactions gives the equation for the net reaction.

Solution Because two ions are involved in each step, each step is bimolecular. The rate equation for any elementary step is the product of the concentrations of the reactants. In this case, the rate equations are

Step 1 $\text{Rate} = k[ClO^-]^2$

Step 2 $\text{Rate} = k'[ClO^-][ClO_2^-]$

Finally, on adding the equations for the two elementary steps we see that the ClO_2^- ion is a product of the first step and a reactant in the second step. It therefore cancels out, and we are left with the stoichiometric equation for the overall reaction.

Step 1 $ClO^-(aq) + ClO^-(aq) \longrightarrow \cancel{ClO_2^-}(aq) + Cl^-(aq)$

Step 2 $\cancel{ClO_2^-}(aq) + ClO^-(aq) \longrightarrow ClO_3^-(aq) + Cl^-(aq)$

Sum of steps $3\ ClO^-(aq) \longrightarrow ClO_3^-(aq) + 2\ Cl^-(aq)$

EXERCISE 15.11 *Elementary Steps*

Nitrogen monoxide is reduced by hydrogen to give water and nitrogen,

$$2\ NO(g) + 2\ H_2(g) \longrightarrow N_2(g) + 2\ H_2O(g)$$

and one possible mechanism to account for this reaction is

$$2\ NO(g) \rightleftharpoons N_2O_2(g)$$
$$N_2O_2(g) + H_2(g) \longrightarrow N_2O(g) + H_2O(g)$$
$$N_2O(g) + H_2(g) \longrightarrow N_2(g) + H_2O(g)$$

What is the molecularity of each of the three steps? What is the rate equation for the third step? Show that the sum of these elementary steps is the net reaction. ■

Reaction Mechanisms and Rate Equations

The rate equation is determined by experiment. The mechanism is a good guess (a hypothesis) about the way the reaction occurs. Several mechanisms can correspond to the same experimental rate equation, so a postulated mechanism can be quite wrong and can provoke disputes between scientists.

One of the most interesting areas of chemistry is the study of the mechanisms of chemical reactions, especially those in biological systems. Now the ideas are in place to begin to see how chemists make the connection between the *experimental* rate equation and the reaction mechanism.

Imagine that a reaction takes place whose mechanism involves two sequential steps and that we know the rates of both steps.

$$\textbf{Elementary Step 1} \qquad A + B \xrightarrow[\substack{\text{Slow, } E_a \text{ large} \\ 0.001 \text{ reaction/s}}]{k_1} X + M$$

$$\textbf{Elementary Step 2} \qquad M + A \xrightarrow[\substack{\text{Fast, } E_a \text{ small} \\ 100 \text{ reactions/s}}]{k_2} Y$$

$$\textbf{Overall Reaction} \qquad 2\,A + B \longrightarrow X + Y$$

In the first reaction A and B come together and slowly produce one of the products (X) plus another reaction species, M. Almost as soon as M is formed, however, it is rapidly consumed by reaction with an additional molecule of A. Thus, the products X and Y are the result of two elementary steps.

One of the most important aspects of kinetics to understand is that products can never be produced at a rate faster than the rate of the slowest step. *The rate of the overall reaction is limited by and is exactly equal to the combined rates of all elementary steps up through the slowest step in the mechanism.* The slowest elementary step of a sequence is therefore called the **rate-determining step,** or rate-limiting step. You are already familiar with rate-determining steps. No matter how fast you shop in the supermarket, it always seems that the time it takes to finish is determined by the wait in the checkout line.

In the example above, the rate-determining elementary step—the first step—is bimolecular and so has the rate equation

$$\text{Rate} = k_1[\text{A}][\text{B}]$$

where k_1 is the rate constant for that step. The overall reaction follows this same second-order rate equation because the slower step in the mechanism is the first step and is bimolecular.

Now let us apply these ideas to the mechanism of a real reaction. Experiment shows that the reaction of nitrogen dioxide with fluorine

$$\textbf{Overall Reaction} \qquad 2\,NO_2(g) + F_2(g) \longrightarrow 2\,FNO_2(g)$$

has a second-order rate equation.

$$\text{Reaction rate} = k[\text{NO}_2][\text{F}_2]$$

The experimental rate equation immediately rules out the possibility that the reaction occurs in a single step because if the overall stoichiometric equation were an elementary step, the rate equation would be

$$\text{Reaction rate} = k[\text{NO}_2]^2[\text{F}_2]$$

and this rate equation does *not* agree with experiment. There must therefore be at least two steps in the mechanism, and the rate-determining elementary step must involve NO_2 and F_2 in a 1:1 ratio. The simplest possibility is identical with our previous example, the hypothetical reaction 2 A + B → products.

$$\textbf{Elementary Step 1} \quad \text{Slow} \qquad NO_2(g) + F_2(g) \xrightarrow{k_1} FNO_2(g) + F(g)$$

$$\textbf{Elementary Step 2} \quad \text{Fast} \qquad NO_2(g) + F(g) \xrightarrow{k_2} FNO_2(g)$$

$$\textbf{Overall Reaction} \qquad 2\,NO_2(g) + F_2(g) \longrightarrow 2\,FNO_2(g)$$

At this introductory level you cannot be expected to derive reaction mechanisms. Given a mechanism, however, you can decide if it is in agreement with experiment.

That is, NO_2 and F_2 first produce one molecule of the product (FNO_2) plus an F atom, and the F atom then reacts with additional NO_2 to give one more molecule of product. If we assume that the first, bimolecular step is rate-determining, its rate equation would be Rate = $k_1[NO_2][F_2]$, exactly the rate equation observed experimentally. The experimental rate constant is therefore the same as k_1.

The F atom in the first step of the NO_2/F_2 reaction is called an intermediate. A **reaction intermediate** is produced in one step of a reaction sequence, but it is consumed in a subsequent step. As a result it does not appear in the net stoichiometric equation. Reaction intermediates usually have a very fleeting existence, but they occasionally have long enough lifetimes to be observed.

EXAMPLE 15.10 *Elementary Steps and Reaction Mechanisms*

Oxygen atom transfer from nitrogen dioxide to carbon monoxide produces nitrogen monoxide and carbon dioxide.

$$NO_2(g) + CO(g) \longrightarrow NO(g) + CO_2(g)$$

It has the following rate equation at temperatures less than 500 K.

$$\text{Rate} = k[NO_2]^2$$

Can this reaction occur in one bimolecular step whose stoichiometry is the same as the overall reaction?

Solution If the reaction occurred simply by the collision of one NO_2 molecule with one CO molecule (that is, the equation for the overall reaction and the equation for the single elementary step are the same), the rate equation would be

$$\text{Rate} = k[NO_2][CO]$$

This does not agree with experiment, however, so the mechanism must involve more than a single step. In fact, the reaction is thought to occur in two, bimolecular steps involving an intermediate, NO_3.

Elementary Step 1:	Slow, rate-determining	$2\ NO_2(g) \longrightarrow NO_3(g) + NO(g)$
Elementary Step 2:	Fast	$NO_3(g) + CO(g) \longrightarrow NO_2(g) + CO_2(g)$
Overall Reaction		$NO_2(g) + CO(g) \longrightarrow NO(g) + CO_2(g)$

The first, or rate-determining, step indeed has a rate equation that agrees with experiment.

EXERCISE 15.12 *Elementary Steps and Reaction Mechanisms*

The Raschig reaction produces the industrially important reducing agent hydrazine, N_2H_4, from NH_3 and OCl^- in basic, aqueous solution. A proposed mechanism is

Step 1 Fast $NH_3(aq) + OCl^-(aq) \longrightarrow NH_2Cl(aq) + OH^-(aq)$

Step 2 Slow $NH_2Cl(aq) + NH_3(aq) \longrightarrow N_2H_5^+(aq) + Cl^-(aq)$

Step 3 Fast $N_2H_5^+(aq) + OH^-(aq) \longrightarrow N_2H_4(aq) + H_2O(\ell)$

1. What is the overall stoichiometric equation?
2. Which step of the three is rate-determining?
3. Write the rate equation for the rate-determining elementary step.
4. What reaction intermediates are involved? ■

We have described only one type of the many possible reaction mechanisms. Many reactions have quite complex mechanisms. As an example, you might wish to consider some of the reactions that are thought to lead to the much-discussed hole in the earth's ozone layer. Although many interrelated processes combine in this phenomenon, scientists now understand some of them reasonably well, and a few are described in the box on page 738.

Reaction mechanisms are described further in Section 16.7.

PROBLEM-SOLVING TIPS AND IDEAS

15.2 A Summary of the Principles of Rate Equations and Reaction Mechanisms

The connection between an experimental rate equation and the hypothesis of a reaction mechanism is important in chemistry.

1. Experiments must first be performed that define the effect of reactant concentrations on the rate of the reaction. This gives the experimental rate equation.
2. A mechanism for the reaction is proposed on the basis of the experimental rate equation; the principles of stoichiometry, molecular structure, and bonding; general chemical experience, and intuition.
3. The *proposed* reaction mechanism is used to *derive* a rate equation. This rate equation must contain only those species present in the overall chemical reaction and not the reaction intermediates. If the derived and experimental rate equations are the same, the postulated mechanism *may* be a reasonable hypothesis of the reaction sequence.
4. If more than one mechanism can be proposed, and they all predict derived rate equations in agreement with experiment, then more experiments must be done. ■

15.7 CATALYSTS AND REACTION RATE

Let us again consider the conversion of *cis*- to *trans*-2-butene, this time as an example of the operation of a catalyst.

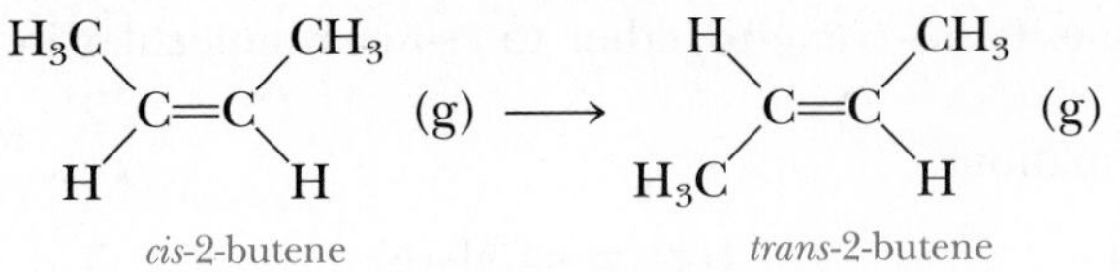

cis-2-butene *trans*-2-butene

As described earlier, the rate equation for the reaction is

$$\text{Reaction rate} = k[cis\text{-2-butene}]$$

(see Figure 15.11).

If a trace of gaseous molecular iodine, I_2, is added to *cis*-2-butene, the iodine acts as a catalyst, accelerating the change to *trans*-2-butene. The iodine is neither consumed nor produced in the overall reaction and so does not appear in the overall balanced equation. Because the reaction rate depends on the concentration of I_2, however, this is a term in the rate equation:

$$\text{Rate} = k[cis\text{-2-butene}][I_2]^{1/2}$$

The rate of the conversion of *cis*- to *trans*-2-butene changes because the presence of I_2 somehow changes the reaction mechanism. The best hypothesis is that iodine molecules first dissociate to form iodine atoms.

Step 1 I_2 dissociation

$$\tfrac{1}{2}I_2(g) \rightleftharpoons I(g)$$

(This equation has a coefficient of $\frac{1}{2}$ for I_2 to emphasize that only one of the two I atoms from the I_2 molecule is needed in subsequent steps of the mechanism.) An iodine atom then attaches to the *cis*-2-butene molecule, breaking one of the bonds between the carbon atoms and allowing the ends of the molecule to twist freely relative to each other.

Step 2 Attachment of I atom to *cis*-2-butene

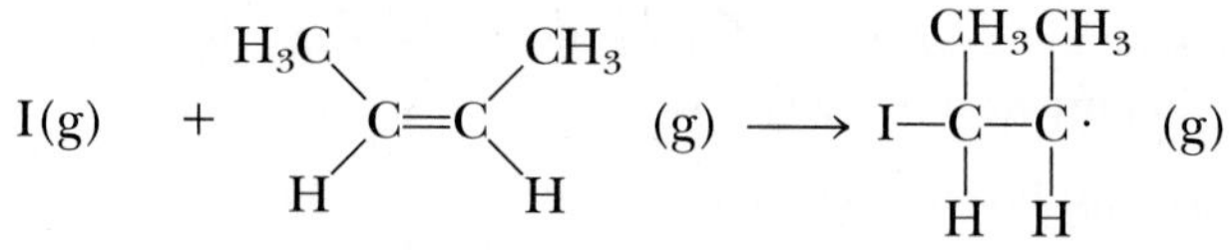

cis-2-butene

Step 3 Rotation around the C—C bond

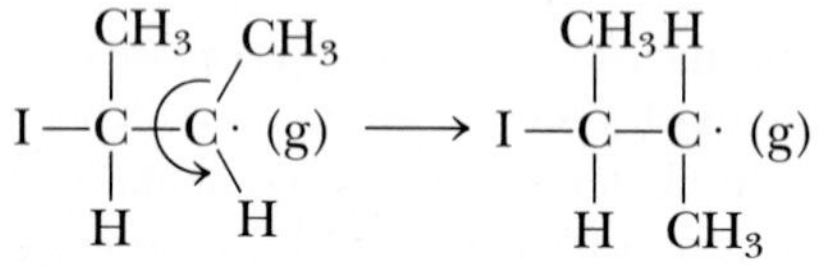

Step 4 Loss of an I atom and re-formation of the carbon-carbon double bond.

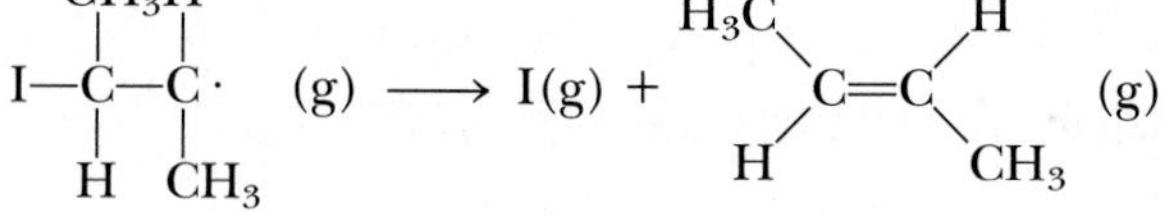

trans-2-butene

After the double bond re-forms to give *trans*-2-butene and the iodine atom falls away, two iodine atoms come together to re-form molecular iodine.

Step 5 I_2 formation

$$I(g) \rightleftharpoons \tfrac{1}{2}I_2(g)$$

Four important points are associated with this mechanism.

- Notice that I_2 dissociates to atoms and then re-forms. To an "outside" observer the concentration of I_2 is unchanged; it is not involved in the

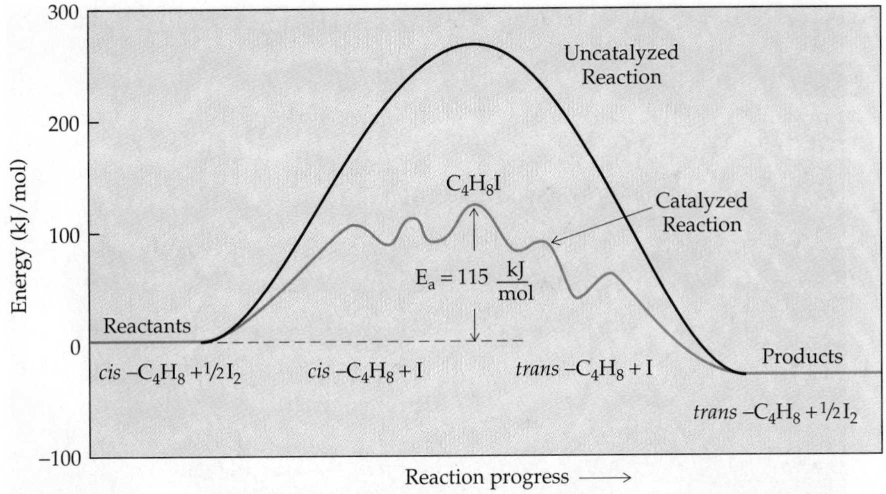

Figure 15.17 A catalyst accelerates a reaction by altering the mechanism so that the activation energy is lowered. With a smaller barrier to overcome, there are more reacting molecules with sufficient energy to surmount the barrier, and the reaction occurs more readily. The energy profile for the uncatalyzed conversion of *cis* to *trans*-2-butene is shown by the black curve, and that for the iodine-catalyzed reaction is represented by the red curve. Notice that the shape of the barrier has changed because the mechanism has changed. See the text for a description of the steps involved.

balanced, stoichiometric equation even though it has appeared in the rate equation. This is generally true of catalysts.

- Figure 15.17 shows that the activation energy barrier to reaction is changed (because the mechanism changed), and it is significantly lower. Thus, the reaction rate has gone up. In fact, dropping the activation energy from 262 kJ/mol for the uncatalyzed reaction to 115 kJ/mol for the catalyzed process makes the catalyzed reaction 10^{15} times faster!
- The catalyzed mechanism has five reaction steps, and the diagram of energy versus reaction progress (see Figure 15.17) has five energy barriers (five humps appear in the curve).
- The catalyst I_2 and the reactant *cis*-2-butene are both in the gas phase during the reaction.

When a catalyst is present in the same phase as the reacting substance, as in the iodine-catalyzed transformation of *cis*-2-butene, it is called a **homogeneous catalyst.**

Catalysis in Industry and the Environment

An expert in the field of industrial chemistry has said that ''Every year more than a trillion dollars' worth of goods is manufactured with the aid of manmade catalysts. Without them, fertilizers, pharmaceuticals, fuels, synthetic fibers, solvents, and surfactants would be in short supply. Indeed, 90 percent of all manufactured items use catalysts at some stage of production.'' The major areas of catalyst use are in petroleum refining, industrial production of chemicals, and environmental controls. We shall look at only a few examples here.

Virtually every industrial reaction uses a **heterogeneous catalyst,** one that is present in a different phase from the reactants being catalyzed. An example is the manganese compound in Figure 15.4. Heterogeneous catalysts are used in industry because they are more easily separated from the products and leftover reactants than are homogeneous catalysts. (Enzymes, the substances that catalyze reactions in plants and animals, are often homogeneous catalysts, however.)

See the interview with James Cusumano (page 155) where the importance of catalysis in solving environmental problems is described.

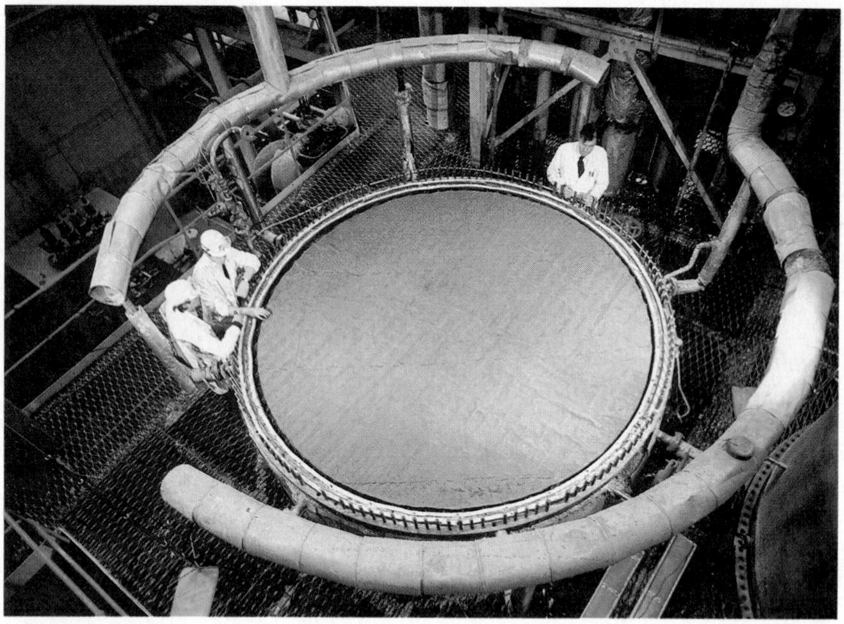

Figure 15.18 The platinum-rhodium gauze catalyst used for the oxidation of ammonia in the manufacture of nitric acid. (Johnson Matthey)

About 10 billion kilograms of polyethylene (see Section 11.8) is produced annually in the United States. The catalysts used to make these and similar polymers are often called Ziegler-Natta catalysts and were named for Karl Ziegler, a German chemist, and Guilio Natta, an Italian chemist. They shared the Nobel Prize in chemistry in 1963 for this work.

Catalysts for chemical processing are generally metal-based and often contain precious metals, such as platinum and palladium. In the United States more than $600 million worth of such catalysts are employed annually by the chemical-processing industry, almost half of them in the preparation of polymers such as polyethylene (Section 11.8).

About 7 billion kilograms of nitric acid is made annually in the United States using the Ostwald process, the first step of which involves the controlled oxidation of ammonia over a Pt-containing catalyst (Figure 15.18)

$$4\ NH_3(g) + 5\ O_2(g) \xrightarrow{\text{Pt-containing catalyst}} 4\ NO(g) + 6\ H_2O(g) \qquad \Delta H^\circ_{rxn} = -905.5\ kJ$$

followed by further oxidation of NO to NO_2.

$$2\ NO(g) + O_2(g) \longrightarrow 2\ NO_2(g) \qquad \Delta H^\circ_{rxn} = -114.1\ kJ$$

In a typical plant, a mixture of air with 10% NH_3 is passed very rapidly over the catalyst at high pressure and at about 850 °C. Roughly 96% of the ammonia is converted to NO_2, making this one of the most efficient industrial catalytic reactions. The final step is to absorb the NO_2 into water to give the acid and NO, the latter being recycled into the process.

$$3\ NO_2(g) + H_2O(\ell) \longrightarrow 2\ HNO_3(aq) + NO(g) \qquad \Delta H^\circ_{rxn} = -138.2\ kJ$$

Acetic acid, CH_3CO_2H, has a place in the organic chemicals industry comparable to that of sulfuric acid in the inorganic chemicals industry; about 1.6 billion kilograms of acetic acid was made in the United States in 1993. Acetic acid is used widely in industry to make plastics and synthetic fibers, as a fungicide, and as the starting material for preparing many dietary supplements. One way of synthesizing the acid is an excellent example of homogeneous catalysis. A rhodium-based compound catalyzes the combination of carbon monoxide and methanol, both inexpensive chemicals, to form acetic acid.

$$\underset{\text{methanol}}{CH_3OH} + CO \xrightarrow{\text{Rh-containing catalyst}} \underset{\text{acetic acid}}{CH_3\overset{\overset{\displaystyle O}{\|}}{C}-OH}$$

The role of the rhodium-containing catalyst in this reaction is to bring the reactants together and allow them to rearrange to products. The first step in the process is the reaction of the alcohol with hydrogen iodide to give CH_3I

$$CH_3OH + HI \longrightarrow CH_3I + H_2O$$

which then reacts with the catalyst, a molecule containing a rhodium(I) ion and CO. This gives a new molecule with CH_3, I, and CO attached to the metal center. Acetic acid is the product after these fragments rearrange and the intermediate reacts with water.

$$\{Rh(CH_3)(CO)I\} + H_2O \longrightarrow \text{Rh catalyst} + HI + CH_3CO_2H$$

Besides producing acetic acid, this final step regenerates the Rh-containing catalyst and produces HI, which is then available to react with more CH_3OH to begin a new catalytic cycle.

The largest growth in catalyst use is predicted to be in emissions control for both automobiles and power plants. This market consumes very large quantities of platinum group metals: platinum, palladium, rhodium, and iridium. Some 9200 kg of platinum, and 1320 kg each of palladium and rhodium, were sold in the United States in the first half of 1991 for automotive uses. In contrast, chemical processing used only 1370 kg of all three metals, and the petroleum industry used 1780 kg of platinum and rhodium.

The purpose of the catalysts in the exhaust system of an automobile is to ensure that the combustion of carbon monoxide and hydrocarbons is complete (Figure 15.19)

$$2\ CO(g) + O_2(g) \xrightarrow{\text{Pt-NiO catalyst}} 2\ CO_2(g)$$

$$\underset{\substack{\text{isooctane}\\ \text{a component of gasoline}}}{2\ C_8H_{18}(g)} + 25\ O_2(g) \xrightarrow{\text{Pt-NiO catalyst}} 16\ CO_2(g) + 18\ H_2O(g)$$

and to convert nitrogen oxides to molecules less harmful to the environment. At the high temperature of combustion, some N_2 from the air reacts with O_2 to give

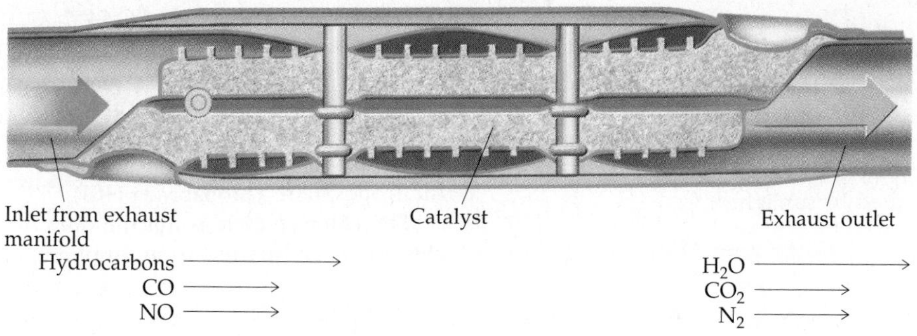

Figure 15.19 Cross-sectional view showing the flow of gases through an automobile catalytic converter.

Depletion of Stratospheric Ozone

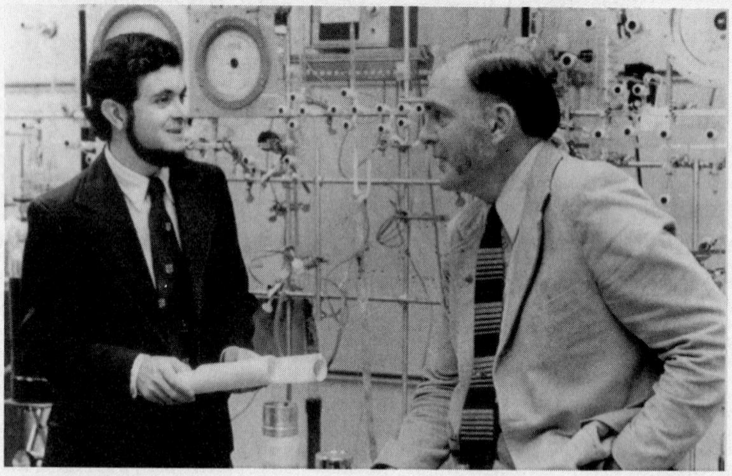

F. Sherwood Rowland (left) and Mario J. Molina (right). These scientists first recognized the potential for the depletion of the earth's atmosphere by CFCs. Rowland is Professor of Chemistry at the University of California, Irvine, and Molina is Professor of Environmental Sciences at the Massachusetts Institute of Technology. (The Bettmann Archive & Newsphotos)

Much of our life in the United States today depends on refrigeration. We cool our homes, cars, offices, and shopping centers with air conditioners. We preserve our food and medicines with refrigerators. Until very recently all of these refrigeration units used chlorofluorocarbons (CFCs) as the heat-exchanging fluid. That situation has now changed dramatically, in part because of laboratory studies of the kinetics of the reactions that CFCs might undergo in the stratosphere.

Nonflammable, nontoxic CFCs such as CCl_2F_2 (called CFC-12; see Section 11.1) were discovered by scientists at the Frigidaire Division of General Motors in 1928. By 1988, the total, worldwide consumption of CFCs was over a billion kilograms annually. In the U. S. almost 5000 businesses produced CFC-related goods and services worth more than \$28 billion a year, and there are more than 700,000 CFC-related jobs. CFCs were used in the U. S. mostly as refrigerants, foam-blowing agents for polystyrene and polyurethane, aerosol propellants, and industrial solvents.

It is ironic that the very properties that led to the first use of CFCs are now causing worldwide concern. Once gaseous CFCs are released into the troposphere, that part of the earth's atmosphere ranging from the surface to an altitude of about 10 km, no mechanism is known for their destruction. They therefore rise to the stratosphere where they are eventually destroyed by solar radiation. This destruction has significant consequences for our environment, as first recognized by M. J. Molina and F. S. Rowland in 1974.*,† From laboratory experiments, they predicted that continued use of CFCs would lead eventually to a significant depletion of the ozone layer around the earth. The reason this is worrisome is that, for every 1% loss of ozone from the stratosphere, an additional 2% of the sun's high-energy ultraviolet radiation can reach the earth's surface. This could result in increases in skin cancer, damage to plants, and other effects that we do not suspect now.

Ozone is produced in the stratosphere when high-energy ultraviolet radiation causes the photodissociation of oxygen to give O atoms, which react with O_2 molecules to produce ozone, O_3.

$$O_2(g) + \text{radiation } (\lambda < 280 \text{ nm}) \longrightarrow 2\,O(g)$$

$$O(g) + O_2(g) \longrightarrow O_3(g)$$

The ozone produced by this mechanism in the stratosphere is quite abundant (10 ppm), which is fortunate because O_3 is also photodissociated by sunlight,

$$O_3(g) \rightleftharpoons O(g) + O_2(g)$$

and the O atoms produced react with more O_2 to regenerate O_3. The process keeps 95% to 99% of the sun's ultraviolet radiation from reaching the earth's surface.

The problem with CFCs is that they disrupt the protective ozone layer by a "chlorine catalytic cycle." The CFCs rise to the stratosphere where a C—Cl bond is broken by a high-energy photon. The Cl atom attacks ozone to give the 13-electron molecule ClO, chlorine oxide. This would not necessarily be a problem, but the ClO can react with an O atom to give O_2 and regenerate a Cl atom. The Cl atom can then destroy still another O_3, and so on and on in a "catalytic cycle." The net reaction is the destruction of a significant quantity of ozone. In fact, it is estimated that each Cl atom can destroy as many as 100,000 ozone molecules before the Cl atom is inactivated or returned to the troposphere (probably as HCl).

The chlorine cycle is not the only chlorine chemistry in the stratosphere.

In fact, at least two other major kinds of reactions are believed to *interfere* with ozone loss. In one case ClO reacts with nitrogen monoxide, NO, to release a Cl atom and form nitrogen dioxide. The NO_2 goes on to regenerate a molecule of ozone, however. In another reaction, ClO forms chlorine nitrate ($ClONO_2$), a compound that at least temporarily acts as a "chlorine reservoir." Eventually, though, this compound also breaks apart and frees Cl atoms to resume their ozone destruction.

If these interference reactions are important, CFCs might have only a minimal effect on earth's ozone layer. In the early spring in the southern hemisphere, however, the ozone layer over the Antarctic is significantly depleted, a fact clearly illustrated in satellite images of ozone concentration done in the late 1980s.‡ One theory used to explain this observation involves the high-altitude clouds common over the Antarctic continent in the winter. Chlorine nitrate could condense in these extremely cold clouds, and chlorine atoms could also be trapped as HCl. Rowland and Molina estimate that one out of every three or four collisions of $ClONO_2$ with HCl-containing ice crystals leads to a reaction such as

$$ClONO_2 + HCl \longrightarrow HNO_3 + Cl_2$$

The first sunlight of spring that warms the clouds can trigger the release of atomic chlorine by photodissociating chlorine molecules. Because nitrogen oxides are trapped in the clouds as nitric acid, the "chlorine catalytic cycle" can run unchecked for five or six weeks in the spring.

Whatever the theories for the springtime Antarctic ozone loss, the problem is real, and people around the world have taken steps to halt any further deterioration. Chemical companies in the United States have halted CFC production and are actively searching for substitutes.§ Some states now require recycling of CFCs used in automobile air-conditioners or have banned them outright. In January, 1989, 24 nations signed the *Montreal Protocol on Substances That Deplete the Ozone Layer,* which calls for reductions in production and use of certain CFCs. A meeting in Copenhagen, Denmark, in 1992 led to a complete ban on CFC production and use by 1996.

But there will be trade-offs. For example, CFC substitutes now available are less efficient as refrigerants, so it is estimated that appliances will use 3% more electricity in the United States, which will increase consumer costs. Furthermore, because electricity is mostly generated by burning fossil fuels, the amount of CO_2 evolved will increase, which in turn will contribute to the "greenhouse" effect.

CFCs and their relation to ozone depletion is just one more example of the risks and benefits problem first described in the Introduction to this book (page 12). In this case scientists and citizens have concluded that the risks of CFCs outweigh the benefits.

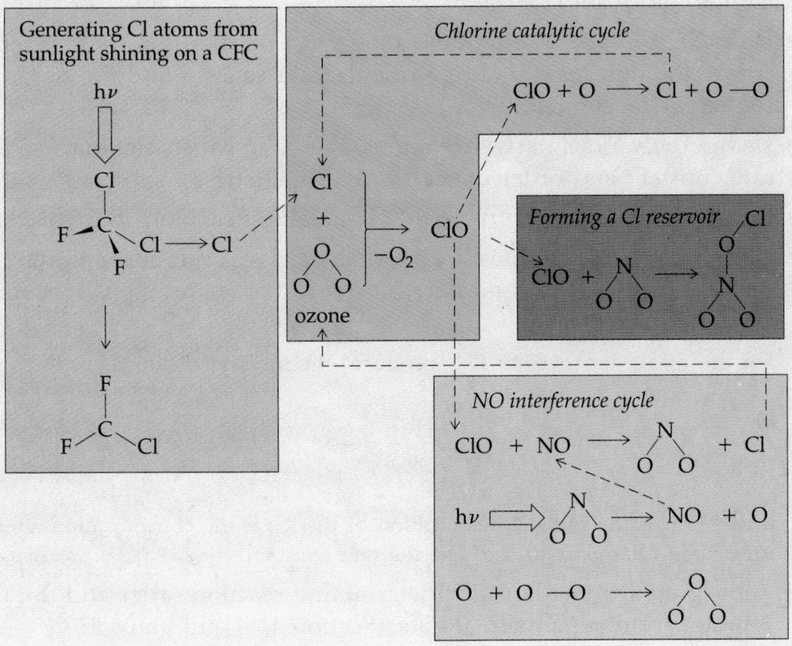

*See also M. J. Molina and F. S. Rowland: *Nature,* Vol. 249, p. 810, 1974; F. S. Rowland: *American Scientist,* Vol. 77, p. 219, 1989.

†F. S. Sherwood and M. J. Molina: "Ozone depletion: 20 years after the alarm," *Chemical and Engineering News,* pp. 8–13, August 15, 1994.

‡R. S. Stolarski, *Scientific American,* Vol. 258, p. 30, 1988.

§L. E. Manzer, *Science.* Vol. 249, p. 31, 1990.

NO, a serious air pollutant. Nitrogen oxide is unstable and should revert to N_2 and O_2. Its rate of reversion is slow, however. Fortunately, catalysts have been developed that greatly speed this reaction.

$$2\ NO(g) \xrightarrow{\text{catalyst}} N_2(g) + O_2(g)$$

The role of the heterogeneous catalyst in the preceding reactions is probably to weaken the bonds of the reactants and to assist in product formation.

EXERCISE 15.13 ***Catalysis***

Which of the following statements is (are) true? If any are false, change the wording to make them true.

1. The concentration of a homogeneous catalyst may appear in the rate equation.
2. A catalyst is always consumed in the overall reaction.
3. A catalyst must always be in the same phase as the reactants.

■

CHAPTER HIGHLIGHTS

Having studied this chapter, you should be able to

- Explain the concept of reaction rate (Section 15.1).
- Derive the average and instantaneous rate of a reaction from experimental information (Section 15.1).
- Describe the various conditions that affect reaction rate (i.e., reactant concentrations, temperature, presence of a catalyst, and the state of the reactants) (Section 15.2).
- Define the various parts of a rate equation and their significance (the rate constant and order of reaction) (Section 15.3).
- Derive a rate equation from experimental information (Section 15.3).
- Describe and use the relationships between reactant concentration and time for zero-order, first-order, and second-order reactions (Section 15.4 and Table 15.2).

  For the first-order reaction R → products, the integrated rate law is

$$\ln \frac{[R]_t}{[R]_0} = -kt \qquad \textbf{(15.1)}$$

  where $[R]_0$ is the initial concentration of R, $[R]_t$ is the concentration after some time t has elapsed, and k is the rate constant.

- Apply graphical methods for determining reaction order and the rate constant from experimental data (Section 15.4 and Table 15.2).
- Use the concept of half-life ($t_{1/2}$), especially for first-order reactions (Section 15.4 and Table 15.22).

For a first-order reaction, the half-life is inversely proportional to *k*.

$$t_{1/2} = \frac{0.693}{k} \quad (15.3)$$

- Describe the collision theory of reaction rates (Section 15.5).
- Appreciate the relation of activation energy (E_a) to the rate and thermodynamics of a reaction (Section 15.5).
- Use collision theory to describe the effect of reactant concentration on reaction rate (Section 15.5).
- Understand the effect of molecular orientation on reaction rate (Section 15.5).
- Describe the effect of temperature on reaction rate using the theories of reaction rates and the Arrhenius equation (Equation 15.5 and Section 15.5).
- Use the Arrhenius equation (Equation 15.5) to calculate the activation energy from experimental data (Section 15.5).
- Understand the concept of reaction mechanism (the sequence of bond-making and bond-breaking steps that occurs during the conversion of reactants to products) and the relation of the mechanism to the overall, stoichiometric equation for a reaction (Section 15.6).
- Describe the elementary steps of a mechanism and give their molecularity (Section 15.6).
- Define the rate-determining step in a mechanism and describe any reaction intermediates (Section 15.6).
- Describe the functioning of a catalyst and its effect on the activation energy and mechanism of a reaction (Section 15.7).
- Define homogeneous and heterogeneous catalysts (Section 15.7).

STUDY QUESTIONS

Review Questions

1. Which of the following can be used to determine the rate equation for a chemical reaction?
   (a) Theoretical calculations
   (b) Measuring the rate of the reaction as a function of the concentration of the reacting species
   (c) Measuring the rate of the reaction as a function of temperature
2. Describe four conditions that determine the rate of a chemical reaction.
3. Refer to Figure 15.2. After 2.0 h, what is the concentration of NO_2? Of O_2?
4. Using the rate equation Rate = $k[A]^2[B]$, define the order of the reaction with respect to A and B and the total order of the reaction.
5. If a reaction has the experimental rate equation Rate = $k[A]^2$, explain what happens to the rate when the concentration of A is tripled. When the concentration of A is halved?
6. A reaction has the experimental rate equation Rate = $k[A]^2[B]$. If the concentration of A is doubled, and the concentration of B is halved, what happens to the reaction rate?
7. Write the equation that allows us to find the concentration of reactant as a function of time for a first-order reaction. Define each term in the equation.
8. After five half-life periods for a reaction, what fraction of reactant remains?
9. If you plot 1/[reactant] versus time and observe a straight line, what is the order of the reaction? If ln [reactant] is plotted versus time, and a straight line of negative slope is observed, what is the order of the reaction?
10. Draw a reaction energy diagram for a one-step exothermic

process. Mark the activation energies of the forward and reverse processes and describe how you can calculate the net energy change for the reaction.

11. Explain how collision theory accounts for the temperature dependence of reaction rates.
12. Write the Arrhenius equation and define each part of the equation. What experimental information is required if you wish to use this equation to calculate the activation energy of a reaction?
13. What is the "mechanism" of a chemical reaction? In your discussion, define the terms "elementary step" and "rate-determining step."
14. Define the term "molecularity" in terms of the collision theory.
15. Define the term "reaction intermediate." Give an example using one of the reaction mechanisms described in the text.
16. What is the effect of a catalyst on the mechanism of a reaction?
17. Explain the difference between a homogeneous and a heterogeneous catalyst. Give an example of each.

Numerical Questions and Problems

Reaction Rates

18. Give the relative rates of disappearance of reactants and formation of products for each of the following reactions:
(a) $2\ O_3(g) \rightarrow 3\ O_2(g)$
(b) $2\ HOF(g) \rightarrow 2\ HF(g) + O_2(g)$
(c) $2\ NO(g) + Br_2(g) \rightarrow 2\ BrNO(g)$

19. Consider the reaction

$$N_2(g) + 3\ H_2(g) \longrightarrow 2\ NH_3(g)$$

At the instant N_2 is reacting at a rate of 0.25 mol/L · min, at what rates are H_2 disappearing and NH_3 forming?

20. Experimental data are listed in the table here for the hypothetical reaction $A \rightarrow 2\ B$.

Time (s)	[A] (mol/L)
0.00	1.000
10.0	0.833
20.0	0.714
30.0	0.625
40.0	0.555

(a) Plot these data, connect the points with a smooth line, and calculate the rate of change of [A] for each 10-s interval from 0 to 40 s. Why does the rate of change decrease from one time interval to the next?
(b) How is the rate of change of [B] related to the rate of change of [A] in the same time interval? Calculate the rate of change of [B] for the time interval from 10 to 20 s.
(c) What is the instantaneous rate when [A] = 0.75 mol/L?

21. A compound called phenyl acetate reacts with water according to the equation

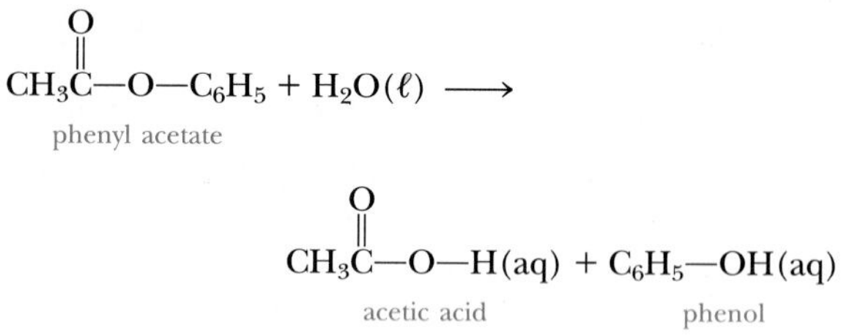

and the data in the table are collected at 5 °C.

Time (min)	[Phenyl acetate] (mol/L)
0	0.55
0.25	0.42
0.50	0.31
0.75	0.23
1.00	0.17
1.25	0.12
1.50	0.085

(a) Plot these data, and describe the shape of the curve observed. Compare this with Figure 15.2.
(b) Calculate the rate of change of [phenyl acetate] during the period of 0.20 min to 0.40 min and then during the time period 1.2 min to 1.4 min. Compare the values and tell why the one is smaller than the other.
(c) What is the rate of change of [phenol] during the time period 1.00 min to 1.25 min?
(d) What is the instantaneous rate at 15 s? At 35 s?

Concentration and Rate Equations

22. The reaction between ozone and nitrogen dioxide has been studied at 231 K.

$$2\ NO_2(g) + O_3(g) \longrightarrow N_2O_5(s) + O_2(g)$$

Experiment shows the reaction is first order in both NO_2 and O_3.
(a) Write the rate equation for the reaction.
(b) How does tripling the concentration of NO_2 affect the reaction rate?
(c) How does halving the concentration of O_3 affect the reaction rate?

23. Nitrosyl bromide, NOBr, is formed from NO and Br_2.

$$2\ NO(g) + Br_2(g) \longrightarrow 2\ NOBr(g)$$

Experiment shows that the reaction is second order in NO and first order in Br_2.
(a) Write the rate equation for the reaction.

(b) How does the reaction rate change if the concentration of Br_2 is tripled?
(c) What happens to the reaction rate when the concentration of NO is doubled?

24. Data are given in the table at 660 K for the reaction

$$2\,NO(g) + O_2(g) \longrightarrow 2\,NO_2(g)$$

Reactant Concentration (mol/L) [NO]	[O₂]	Rate of Disappearance of NO (mol/L · s)
0.020	0.010	1.0×10^{-4}
0.040	0.010	4.0×10^{-4}
0.020	0.040	4.0×10^{-4}

(a) Write the rate equation for the reaction.
(b) Calculate the rate constant.
(c) Calculate the rate (expressed in moles per liter × seconds) at the instant when [NO] = 0.045 M and $[O_2]$ = 0.025 M.
(d) At the instant when O_2 is reacting at the rate 5.0×10^{-4} mol/L · s, at what rate is NO reacting and NO_2 forming?

25. The reaction

$$2\,NO(g) + 2\,H_2(g) \longrightarrow N_2(g) + 2\,H_2O(g)$$

has been studied at 904 °C.

Reactant Concentration (mol/L) [NO]	[H₂]	Rate of Appearance of N₂ (mol/L · s)
0.420	0.122	0.136
0.210	0.122	0.0339
0.210	0.244	0.0678
0.105	0.488	0.0339

(a) Write the rate equation for the reaction.
(b) Calculate the rate constant at 904 °C.
(c) Find the rate of appearance of N_2 at the instant when [NO] = 0.550 M and $[H_2]$ = 0.199 M.

Concentration-Time Equations

26. The decomposition of N_2O_5 in nitric acid is a first-order reaction. It takes 4.26 min at 55 °C to decrease 2.56 mg of N_2O_5 to 2.50 mg. Find k in minutes^{-1} and seconds^{-1}.

27. The rate equation for the reaction of sucrose in water

$$C_{12}H_{22}O_{11}(aq) + H_2O(\ell) \longrightarrow 2\,C_6H_{12}O_6$$

is rate = $k[C_{12}H_{22}O_{11}]$. After 2.57 h at 27 °C, 5.00 g/L of sucrose has decreased to 4.50 g/L. Find k.

28. The transformation of cyclopropane to propene was described in Example 15.3. If the initial concentration of cyclopropane is 0.050 M, how many hours must elapse for the concentration to drop to 0.025 M? (The first-order rate constant is $5.4 \times 10^{-2}\ h^{-1}$.)

29. The decomposition of SO_2Cl_2 is first order in SO_2Cl_2,

$$SO_2Cl_2(g) \longrightarrow SO_2(g) + Cl_2(g)$$

and it has a rate constant of 0.17 h^{-1}. If the initial concentration of SO_2Cl_2 is 1.25×10^{-3} M, how long does it take for the concentration to drop to 0.31×10^{-3} M?

30. The rate constant for the decomposition of nitrogen dioxide

$$NO_2(g) \longrightarrow NO(g) + \tfrac{1}{2}\,O_2(g)$$

with a laser beam is 3.40 L/mol · min. Find the time in seconds needed to decrease the concentration of NO_2 from 2.00 mol/L to 1.50 mol/L.

31. Ammonium cyanate, NH_4NCO, rearranges in water to give urea, $(NH_2)_2CO$.

$$NH_4NCO(aq) \longrightarrow (NH_2)_2CO(aq)$$

The rate equation is Rate = $k[NH_4NCO]^2$ where k = 0.0113 L/mol · min. If the original concentration of NH_4NCO is 0.458 mol/L, how much time elapses before the concentration is reduced to 0.300 mol/L?

Half-Life

32. For a reaction with the rate equation Rate = k[A], and $k = 3.33 \times 10^{-6}\ h^{-1}$, what is the half-life of the reaction? How long does it take for the concentration of A to drop from 1.0 M to 0.20 M?

33. The decomposition of N_2O_5 (to give NO_2 and O_2) follows the rate equation

$$\text{Rate} = k[N_2O_5]$$

where k is $5.0 \times 10^{-4}\ s^{-1}$ at a particular temperature.
(a) What is the half-life of the reaction?
(b) How long does it take for the N_2O_5 concentration to drop to one tenth of its original value?

34. The decomposition of phosphine, PH_3, proceeds according to the equation

$$4\,PH_3(g) \longrightarrow P_4(g) + 6\,H_2(g)$$

It is found that the reaction has the rate equation Rate = $k[PH_3]$. If the half-life is 37.9 s, how much time is required for three fourths of the PH_3 to decompose?

35. The decomposition of SO_2Cl_2 is first order in SO_2Cl_2, and it has a half-life of 4.1 h.

$$SO_2Cl_2(g) \longrightarrow SO_2(g) + Cl_2(g)$$

If you begin with 1.6×10^{-3} mol of SO_2Cl_2 in a flask, how many hours elapse before the quantity of SO_2Cl_2 has decreased to 2.00×10^{-4} mol?

36. The rate constant for the decomposition of gaseous azomethane

$$CH_3N{=}NCH_3(g) \longrightarrow N_2(g) + C_2H_6(g)$$

is 40.8 min^{-1} at 425 °C. Find the number of moles of $CH_3N{=}NCH_3$ and N_2 in a flask 0.0500 min after 2.00 g of $CH_3N{=}NCH_3$ has been added.

37. The compound $Xe(CF_3)_2$ is unstable; its half-life is 30. min. 7.50 mg of $Xe(CF_3)_2$ is placed in a flask. Later, 0.25 mg of $Xe(CF_3)_2$ is found in the flask. How long was the sample in the flask?

38. Hypofluorous acid, HOF, is very unstable, decomposing in a first-order reaction to give HF and O_2 with a half-life of only 30. min at room temperature.

$$HOF(g) \longrightarrow HF(g) + \tfrac{1}{2}\,O_2(g)$$

If the partial pressure of HOF in a 1.00-L flask is initially 100. mm Hg at 25 °C, what is the total pressure in the flask and the partial pressure of HOF after 30. min? After 45 min?

39. We know that the decomposition of SO_2Cl_2 is first order in SO_2Cl_2,

$$SO_2Cl_2(g) \longrightarrow SO_2(g) + Cl_2(g)$$

with a half-life of 245 min at 600 K. If you begin with a partial pressure of SO_2Cl_2 of 25 mm Hg in a 1.0-L flask, what is the partial pressure of each reactant and product after 245 min? What is the partial pressure of each reactant after 12 h?

40. The radioactive isotope copper-64 (^{64}Cu) is used in the form of copper(II) acetate to study Wilson's disease. The isotope has a half-life of 12.7 h. What quantity of copper(II) acetate remains after two days and 16 h? After five days?

41. Radioactive gold-198 is used as the metal in the diagnosis of liver problems. The half-life of this isotope is 2.7 days. If you begin with 5.6 mg of the isotope, what quantity remains after 1.0 day?

Graphical Analysis of Rate Equations and k

42. Common sugar, sucrose, reacts in dilute acid solution to give the simpler sugars glucose and fructose. Both of the simple sugars have the same formula, $C_6H_{12}O_6$.

$$C_{12}H_{22}O_{11}(aq) + H_2O(\ell) \longrightarrow 2\,C_6H_{12}O_6(aq)$$

The rate of this reaction has been studied in acid solution, and the data in the table were obtained.

Time (min)	**$[C_{12}H_{22}O_{11}]$ (mol/L)**
0	0.316
39	0.274
80	0.238
140	0.190
210	0.146

(a) Plot the data above as ln [sucrose] versus time and 1/[sucrose] versus time. What is the order of the reaction?

(b) Write the rate equation for the reaction and calculate the rate constant *k*.

(c) Estimate the concentration of sucrose after 175 min.

43. Data for the reaction of phenyl acetate with water are given in Study Question 21. Plot these data as ln [phenyl acetate] and 1/[phenyl acetate] versus time. What is the order of the reaction with respect to phenyl acetate? Determine the rate constant from your plot and calculate the half-life for the reaction.

44. Data for the first-order decomposition of dinitrogen oxide

$$2\,N_2O(g) \longrightarrow 2\,N_2(g) + O_2(g)$$

on a gold surface at 900 °C are given. Find the rate constant by graphing these data in an appropriate manner (Table 15.2). Write the rate equation and find the decomposition rate at 900 °C when $[N_2O] = 0.035$ mol/L.

Time (min)	**$[N_2O]$ (mol/L)**
15.0	0.0835
30.0	0.0680
80.0	0.0350
120.0	0.0220

45. The thermal decomposition of ammonia has been studied at high temperatures.

$$NH_3(g) \longrightarrow NH_2(g) + H(g)$$

The data in the table were collected at 2000 K.

Time (h)	**$[NH_3]$ (mol/L)**
0	8.000×10^{-7}
25	6.75×10^{-7}
50	5.84×10^{-7}
75	5.15×10^{-7}

Plot the appropriate concentration expression (Table 15.2) against time to find the order of the reaction. Find the rate constant for the reaction from the slope of the line. Use the given data and the appropriate integrated rate equation to check your answer.

Kinetics and Energy

46. For the hypothetical reaction A + B → C + D, the activation energy is 32 kJ/mol. For the reverse reaction (C + D → A + B), the activation energy is 58 kJ/mol. Is the reaction A + B → C + D exothermic or endothermic?

47. The reaction of H_2 molecules with F atoms has an activation energy of 8 kJ/mol and an enthalpy change of −133 kJ.

$$H_2(g) + F(g) \longrightarrow HF(g) + H(g)$$

Draw a diagram like Figure 15.14 for this process.

48. Calculate the activation energy (E_a) for the reaction

$$N_2O_5(g) \longrightarrow 2\ NO_2(g) + \tfrac{1}{2}\ O_2(g)$$

from the observed rate constants: k at 25 °C = $3.46 \times 10^{-5}\ s^{-1}$ and k at 55 °C = $1.5 \times 10^{-3}\ s^{-1}$.

49. If the rate constant for a reaction triples in value when the temperature rises from 300. K to 310. K, what is the activation energy of the reaction?

50. Data for the following reaction

$$Mn(CO)_5(CH_3CN)^+ + NC_5H_5 \longrightarrow Mn(CO)_5(NC_5H_5)^+ + CH_3CN$$

have been collected.

***k*, min^{-1}**	***T*(K)**
0.0409	298
0.0818	308
0.157	318

(a) Calculate E_a from a plot of ln k versus $1/T$.
(b) Calculate A and then find k at 311 K.

51. The energy of activation for the reaction

$$C_4H_8(g) \longrightarrow 2\ C_2H_4(g)$$

is 260 kJ/mol. At 800 K, $k = 0.0315\ sec^{-1}$. Find k at 850 K.

Mechanisms

52. Each of the following equations represents an elementary step. Write the rate equation for each elementary step and give the molecularity of the step.
(a) $NO(g) + NO_3(g) \rightarrow 2\ NO_2(g)$
(b) $Cl(g) + H_2(g) \rightarrow HCl(g) + H(g)$
(c) $(CH_3)_3CBr(aq) \rightarrow (CH_3)_3C^+(aq) + Br^-(aq)$

53. The reaction between chloroform ($CHCl_3$) and chlorine gas proceeds in a series of three elementary steps.

Step 1 Fast, reversible $\quad Cl_2(g) \rightleftharpoons 2\ Cl(g)$

Step 2 Slow $\quad CHCl_3(g) + Cl(g) \longrightarrow CCl_3(g) + HCl(g)$

Step 3 Fast $\quad CCl_3(g) + Cl(g) \longrightarrow CCl_4(g)$

Overall $\quad CHCl_3(g) + Cl_2(g) \longrightarrow CCl_4(g) + HCl(g)$

(a) Which of the steps is rate-determining?
(b) Write the equation for the rate-determining step.
(c) What is the molecularity of each step?

54. The ozone, O_3, in the earth's upper atmosphere decomposes according to the equation $2\ O_3(g) \rightarrow 3\ O_2(g)$. The mechanism of the reaction is thought to proceed through an initial fast, reversible step and then a slow second step.

Step 1 Fast, reversible $\quad O_3(g) \rightleftharpoons O_2(g) + O(g)$

Step 2 Slow $\quad O_3(g) + O(g) \longrightarrow 2\ O_2(g)$

(a) Which of the steps is the rate-determining step?
(b) Write the rate equation for the rate-determining step.
(c) What is the molecularity of each step?

55. Iodide ion is oxidized in acid solution by hydrogen peroxide.

$$H_2O_2(aq) + 2\ H^+(aq) + 2\ I^-(aq) \longrightarrow I_2(aq) + 2\ H_2O(\ell)$$

A proposed mechanism is

Step 1 $\quad H_2O_2(aq) + I^-(aq) \xrightarrow{\text{Slow}} H_2O(\ell) + OI^-(aq)$

Step 2 $\quad H^+(aq) + OI^-(aq) \xrightarrow{\text{Fast}} HOI(aq)$

Step 3 $\quad HOI(aq) + H^+(aq) + I^-(aq) \xrightarrow{\text{Fast}} I_2(aq) + H_2O(\ell)$

(a) Show that the three elementary steps add up to give the overall, stoichiometric equation.
(b) What is the molecularity of each step?
(c) For this mechanism to be consistent with kinetic data, what must be the experimental rate equation?
(d) Identify any intermediates in the elementary steps in this reaction.

Catalysis

56. Which of the following statements is (are) true?
(a) The concentration of a homogeneous catalyst may appear in the rate equation.
(b) A catalyst is always consumed in the reaction.
(c) A catalyst must always be in the same phase as the reactants.
(d) A catalyst can change the course of a reaction and allow different products to be produced.
(e) A catalyst can cause a change in which elementary step is rate-determining for a particular reaction.

57. What is the effect of a catalyst on
(a) The activation energy
(b) The enthalpy of the reactants and products?

58. Carbonic anhydrase is a biological catalyst, an enzyme. It catalyzes the hydration of CO_2.

$$CO_2(g) + H_2O(\ell) \longrightarrow H_2CO_3(aq)$$

This is a critical reaction involved in the transfer of CO_2 from tissues to the lung via the bloodstream. One enzyme molecule hydrates 10^6 molecules of CO_2 per second. How many kilograms of CO_2 are hydrated in 1 h in one liter of solution by 5×10^{-6} M enzyme?

59. Biological reactions nearly always occur in the presence of enzymes, which are very powerful catalysts. For example, the enzyme catalase that acts on peroxides reduces the activation energy from 72 kJ/mol (in the uncatalyzed reaction) to 28 kJ/mol at 298 K. What is the increase in *k*? Assume A remains constant.

General Questions

60. The transfer of an O atom from NO_2 to CO has been studied at 540 K

$$CO(g) + NO_2(g) \longrightarrow CO_2(g) + NO(g)$$

and the following data were collected. Use them to
(a) Write the rate equation.
(b) Determine the reaction order with respect to each reactant.
(c) Calculate the rate constant, making certain to give the correct units for *k*.

Initial Concentration (mol/L)		Initial Rate
[CO]	*[NO₂]*	(mol/L · h)
5.1×10^{-4}	0.35×10^{-4}	3.4×10^{-8}
5.1×10^{-4}	0.70×10^{-4}	6.8×10^{-8}
5.1×10^{-4}	0.18×10^{-4}	1.7×10^{-8}
1.0×10^{-3}	0.35×10^{-4}	6.8×10^{-8}
1.5×10^{-3}	0.35×10^{-4}	10.2×10^{-8}

61. Ammonium cyanate, NH_4NCO, rearranges in water to give urea, $(NH_2)_2CO$.

$$NH_4NCO(aq) \longrightarrow (NH_2)_2CO(aq)$$

Using the data in the table,

Time (min)	$[NH_4NCO]$ (mol/L)
0	0.458
45.0	0.370
107	0.292
230.	0.212
600.	0.114

(a) Decide if the reaction is first or second order.
(b) Calculate *k*.
(c) Calculate the half-life of ammonium cyanate under these conditions.
(d) Calculate the concentration of NH_4NCO after 12.0 h.

62. Nitrogen oxides, NO_x (a mixture of NO and NO_2 collectively designated as NO_x), play an essential role in the production of pollutants found in photochemical smog. The average half-life for the removal of NO_x in the smokestack emissions in a large city during daylight is 3.9 h.
(a) Starting with 1.50 mg in an experiment, what quantity of NO_x remains after 5.25 h? (Assume the reaction is first order.)
(b) How many hours of daylight must elapse to decrease 1.50 mg of NO_x to 2.50×10^{-6} mg?

63. At temperatures less than 500 K the reaction between carbon monoxide and nitrogen dioxide

$$NO_2(g) + CO(g) \longrightarrow CO_2(g) + NO(g)$$

has the rate equation Rate = $k[NO_2]^2$. Which of the three mechanisms suggested here best agrees with the experimentally observed rate equation?

Mechanism 1 Single, elementary step

$$NO_2 + CO \longrightarrow CO_2 + NO$$

Mechanism 2 Two steps

Slow $\quad NO_2 + NO_2 \longrightarrow NO_3 + NO$

Fast $\quad NO_3 + CO \longrightarrow NO_2 + CO_2$

Mechanism 3 Two steps

Slow $\quad NO_2 \longrightarrow NO + O$

Fast $\quad CO + O \longrightarrow CO_2$

64. Chlorine atoms are thought to lead to the destruction of the earth's ozone layer by the following sequences of reactions:

$$Cl + O_3 \longrightarrow ClO + O_2$$
$$ClO + O \longrightarrow Cl + O_2$$

where the O atoms in the second step come from the decomposition of ozone by sunlight.

$$O_3(g) \rightleftharpoons O(g) + O_2(g)$$

What is the net equation on summing these three equations? Why does this lead to ozone loss in the stratosphere? What is the role played by Cl in this sequence of reactions? What name is given to species such as ClO?

65. Nitryl fluoride, an explosive compound, can be made by treating nitrogen dioxide with fluorine.

$$2\ NO_2(g) + F_2(g) \longrightarrow 2\ NO_2F(g)$$

Use the rate data in the table to answer the following questions:
(a) Write the rate equation for the reaction.

(b) What is the order of reaction with respect to each component of the reaction?
(c) What is the numerical value of the rate constant k?

Experiment	Initial Concentrations (mol/L) $[NO_2]$	$[F_2]$	$[NO_2F]$	Initial Rate (mol/L · s)
1	0.001	0.005	0.001	2×10^{-4}
2	0.002	0.005	0.001	4×10^{-4}
3	0.006	0.002	0.001	4.8×10^{-4}
4	0.006	0.004	0.001	9.6×10^{-4}
5	0.001	0.001	0.001	4×10^{-5}
6	0.001	0.001	0.002	4×10^{-5}

66. A catalyst can cause a reaction to proceed along a different pathway with a lower activation energy. If the catalyst lowers the activation energy by 5 kJ/mol, how much faster is the reaction, if both occur at 298 K (and the frequency factor A does not change)?

67. Tell whether the following statements are true or false. If false, rewrite the sentence to make it correct..
(a) The rate-controlling elementary step in a reaction is the slowest step in a mechanism.
(b) It is possible to change the rate constant by changing the temperature.
(c) As a reaction proceeds at constant temperature, the rate remains constant.
(d) A reaction that is third order overall must involve more than one step.

68. The decomposition of dinitrogen pentaoxide

$$2\ N_2O_5(g) \longrightarrow 4\ NO_2(g) + O_2(g)$$

has the rate equation Rate = $k[N_2O_5]$. It has been found experimentally that the decomposition is 20% complete in 6.0 h at 300. K. Calculate the rate constant and the half-life at 300. K.

69. The data in the table give the temperature dependence of the rate constant for the reaction

$$2\ N_2O_5(g) \longrightarrow 4\ NO_2(g) + O_2(g)$$

Plot these data in the appropriate way to derive the activation energy for the reaction.

T (K)	k (s^{-1})
338	4.87×10^{-3}
328	1.50×10^{-3}
318	4.98×10^{-4}
308	1.35×10^{-4}
298	3.46×10^{-5}
273	7.87×10^{-7}

70. The gas phase reaction

$$2\ N_2O_5(g) \longrightarrow 2\ N_2O_4(g) + O_2(g)$$

has an activation energy of 103 kJ, and the rate constant is 0.0900 min^{-1} at 328.0 K. Find the rate constant at 338.0 K.

71. Egg protein from the albumin is precipitated when an egg is cooked in boiling water. The E_a for this first-order reaction is 520 kJ/mol. Estimate the time to prepare a 3-min egg at an altitude at which water boils at 90 °C.

72. The decomposition of gaseous dimethyl ether at ordinary pressures is first order. Its half-life is 25.0 min at 500 °C.

$$CH_3OCH_3(g) \longrightarrow CH_4(g) + CO(g) + H_2(g)$$

(a) Starting with 8.00 g, what mass remains (in grams) after 125 min and after 145 min?
(b) Calculate the time in minutes required to decrease 7.60 ng (nanograms) to 2.25 ng.
(c) What fraction of the original dimethyl ether remains after 150 min?

73. Three mechanisms are proposed for the gas-phase reaction of NO with Br_2 to give BrNO:

Mechanism 1

$$NO(g) + NO(g) + Br_2(g) \longrightarrow 2\ BrNO(g)$$

Mechanism 2

Step 1 $\quad NO(g) + Br_2(g) \rightleftharpoons Br_2NO(g)$

Step 2 $\quad Br_2NO(g) + NO(g) \longrightarrow 2\ BrNO(g)$

Mechanism 3

Step 1 $\quad NO(g) + NO(g) \rightleftharpoons N_2O_2(g)$

Step 2 $\quad N_2O_2(g) + Br_2(g) \longrightarrow 2\ BrNO(g)$

(a) Write the balanced equation for the net reaction.
(b) What is the molecularity for each step in each mechanism.
(c) What are the intermediates formed in Mechanisms 2 and 3?

74. An enthalpy diagram is given here for the absorption and dissociation of O_2 on a platinum surface.

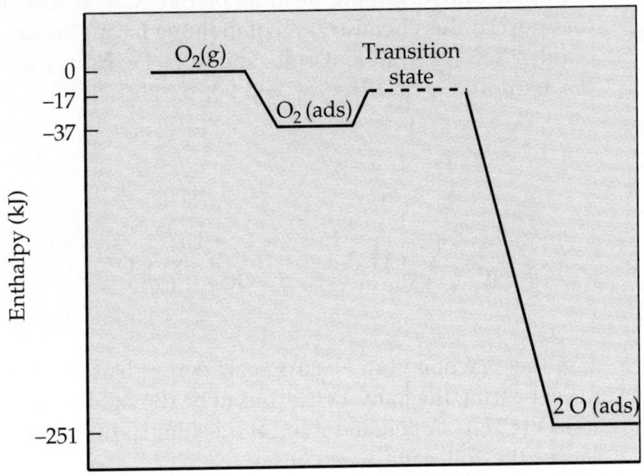

(a) What is ΔH for
   (1) $O_2(g) \rightarrow O_2(\text{adsorbed})$
   (2) $O_2(\text{adsorbed}) \rightarrow 2\ O(\text{adsorbed})$
   (3) $O_2(g) \rightarrow 2\ O(\text{adsorbed})$?
(b) What is E_a for $O_2(\text{adsorbed}) \rightarrow 2\ O(\text{adsorbed})$?

75. Radioactive iodine-131, which has a half-life of 8.04 days, is used in the form of sodium iodide to treat cancer of the thyroid. If you begin with 25.0 mg of $Na^{131}I$, what quantity of the material remains after 1 month (31 days)?

Conceptual Questions

76. A common danger confronts workers in grain elevators, coal mines, and flour mills but not in rock quarries. Identify and explain the danger.

77. Radioactive isotopes are often used as "tracers" to follow an atom through a chemical reaction, and the following is an example. Acetic acid reacts with methanol by eliminating a molecule of water and forming methyl acetate (see Chapter 11). Explain how to use the radioactive isotope ^{18}O to show whether the oxygen atom in the water comes from the —OH of the acid or the —OH of the alcohol.

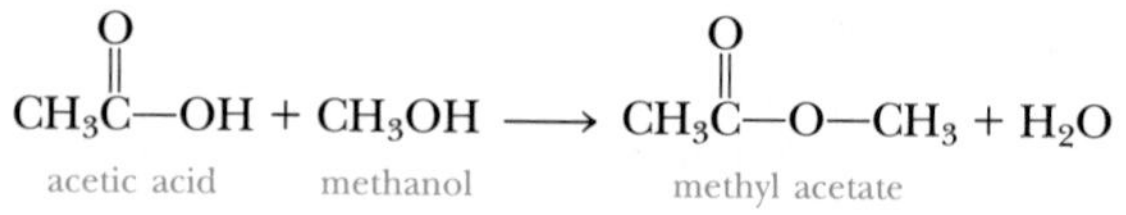

78. Hydrogenation reactions, processes wherein H is added to a molecule, are usually catalyzed. An excellent catalyst is a very finely divided metal suspended in the reaction solvent. Tell why finely divided rhodium, for example, is a much more efficient catalyst than a small block of the metal.

79. It is instructive to use a mathematical model in connection with Question 78. Suppose you have 1000 blocks, each of which is 1.0 cm on a side. If all 1000 of these blocks are stacked to give a cube that is 10. cm on a side, what fraction of the 1000 blocks have at least one surface on the outside surface of the cube? Now divide the 1000 blocks into eight equal piles of blocks and form them into eight cubes, 5.0 cm on a side. Now what fraction of the blocks have at least one surface on the outside of the cubes? How does this mathematical model pertain to Question 78?

Summary Questions

80. The radioactive noble gas radon has been the focus of much attention lately because it can be found in homes (see Chapter 24). Radon-222 (^{222}Rn) has a half-life of 3.82 days. Assume radon gas is in the basement of a home (with the dimensions 12 m by 7.0 m by 3.0 m) and that the gas has a partial pressure of 1.0×10^{-6} mm Hg.
(a) How many atoms of ^{222}Rn are in a liter of air in the basement?
(b) If the radon gas is not replenished in the basement, how many atoms of ^{222}Rn remain per liter of air after 30. days?

81. The substitution of CO in $Ni(CO)_4$ was studied some years ago (in the nonaqueous solvents toluene and hexane) and led to an understanding of some of the general principles that govern the chemistry of compounds having metal–CO bonds. (See J. P. Day, F. Basolo, and R. G. Pearson: *Journal of the American Chemical Society,* Vol. 90, p. 6927, 1968.)

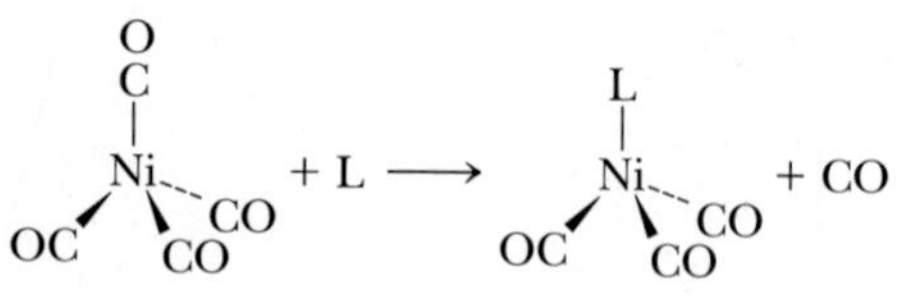

L in this reaction is an electron-pair donor, like CO, but the atom bearing the pair of electrons to be donated is usually P, as in $P(CH_3)_3$. A detailed study of the kinetics of the reaction led to the following mechanism:

Slow $\quad Ni(CO)_4 \longrightarrow Ni(CO)_3 + CO$

Fast $\quad Ni(CO)_3 + L \longrightarrow Ni(CO)_3L$

(a) What is the molecularity of each of the elementary reactions?
(b) It was found that doubling the concentration of $Ni(CO)_4$ increased the reaction rate by a factor of 2. Doubling the concentration of L had no effect on the reaction rate. Based on this information, write the rate equation for the reaction. Does this agree with the mechanism above?
(c) The experimental rate constant for the reaction, when $L = P(C_6H_5)_3$, is $9.3 \times 10^{-3}\ s^{-1}$ at 20 °C. If the initial concentration of $Ni(CO)_4$ is 0.025 M, what is the concentration of the product after 5.0 min?
(d) $Ni(CO)_4$ is formed by reacting nickel metal with carbon monoxide. If you have 750 mL of CO at a pressure of 1.50 atm at 22 °C, and the CO is combined with 0.125 g of nickel metal, how many grams of $Ni(CO)_4$ can be formed? If CO remains after reaction, what is its pressure in the 750 mL flask at 29 °C?
(e) An excellent way to make pure nickel metal is to decompose $Ni(CO)_4$ in a vacuum at a temperature slightly higher than room temperature. What is the enthalpy change for the decomposition reaction

$$Ni(CO)_4 \longrightarrow Ni(s) + 4\ CO(g)$$

if the molar enthalpy of formation of $Ni(CO)_4$ gas is −602.91 kJ/mol?

CHAPTER 16

Principles of Reactivity: Chemical Equilibria

A Chemical Puzzler

In the Chemical Puzzler for Chapter 4 you saw that heating pink $CoCl_2 \cdot 6\ H_2O$ caused the compound to lose some water to give a blue compound. Here $CoCl_2 \cdot 6\ H_2O$ and HCl were dissolved in water to give a pink solution. When the solution was heated the color turned blue (*left*), whereas the pink color was restored when the solution was cooled (*right*). The change is reversible. The color changes to pink when the heated solution is cooled, and the color changes to blue when the cold solution is warmed. The *endothermic* change occurring is

$$\underset{\text{pink}}{[Co(H_2O)_6]^{2+}(aq)} + 4\ Cl^-(aq) \rightleftharpoons \underset{\text{blue}}{CoCl_4^{2-}(aq)} + 6\ H_2O(\ell)$$

Why is this change reversible? Could we have predicted the effect of heat on this reaction?

(C. D. Winters)

Because the concept of equilibrium is fundamental in chemistry, this and the next three chapters explore this concept as applied to chemical reactions. Equilibrium is not, however, peculiar to chemistry. You participate in social situations and live in an economy that represent an equilibrium of competing forces. You and your family or your roommate have arrived at some arrangements that keep your personal relationships as smooth as possible. That is, you have achieved an equilibrium. Of course this equilibrium can be easily upset when a stress is applied to the arrangement, as when another child is added to the family or another roommate moves in. In international affairs, countries achieve an equilibrium of competing interests, an equilibrium that can be upset when, for example, the currency of one country changes in value or if a third country intervenes in the relation-concentrations of reactants and products at equilibrium are a measure of the relative intrinsic tendency of chemical reactions to proceed from reactants to products.

Lake Nyos' (a lake in Africa) explosive loss of CO_2 is an interesting example of the effect of outside influences on an equilibrium. See page 666.

Our goal in this chapter is to explore the consequences of the reversibility of chemical reactions, that in a closed system a state of equilibrium is eventually achieved between reactants and products and that outside forces can affect that equilibrium. A major result of this exploration will be an ability to describe "chemical reactivity" in quantitative terms. As you will see, the concentrations of reactants and products at equilibrium are a measure of the relative intrinsic tendency of chemical reactions to proceed from reactants to products.

16.1 THE NATURE OF THE EQUILIBRIUM STATE

If you have ever visited a limestone cave, you must surely have been impressed with the beautiful limestone stalactites and stalagmites, which are made chiefly of calcium carbonate (Figure 16.1). How did these evolve?

The key aspect of the formation of stalactites and stalagmites is the reversibility of chemical reactions. Calcium carbonate is found in underground deposits in the form of limestone, a leftover of ancient oceans. If water seeping through the limestone contains dissolved CO_2, a reaction occurs in which the mineral dissolves, giving an aqueous solution of Ca^{2+} and HCO_3^- ions.

$$CaCO_3(s) + CO_2(aq) + H_2O(\ell) \longrightarrow Ca^{2+}(aq) + 2\ HCO_3^-(aq)$$

When the mineral-laden water reaches a cave, the reverse reaction occurs, with CO_2 being evolved into the cave and solid $CaCO_3$ being deposited.

$$Ca^{2+}(aq) + 2\ HCO_3^-(aq) \longrightarrow CaCO_3(s) + CO_2(g) + H_2O(\ell)$$

The concept of reversibility was first introduced in Chapter 13 when we discussed phase equilibria.

Dissolving and reprecipitating limestone can be illustrated by a laboratory experiment with some soluble salts containing the Ca^{2+} and HCO_3^- ions (say

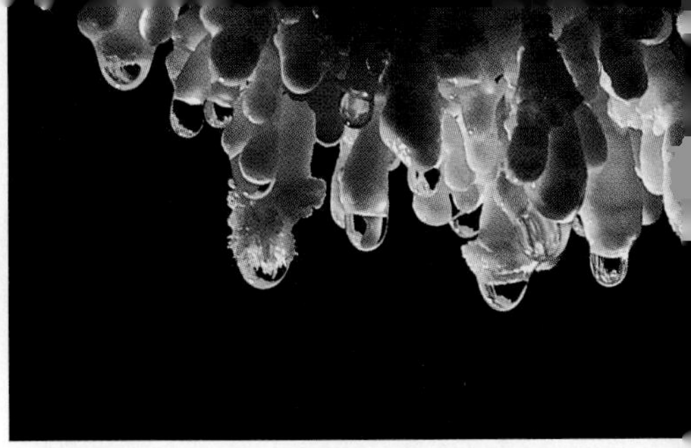

$CaCl_2$ and $NaHCO_3$). If you put the salts in an open beaker of water in the laboratory, you soon see bubbles of CO_2 gas and solid $CaCO_3$. As CO_2 gas escapes from the solution into the air, solid $CaCO_3$ precipitates. If you drop a piece of dry ice (solid CO_2) into the solution, however, the solid $CaCO_3$ redissolves (Figure 16.2). This experiment illustrates an important feature of chemical reactions: In principle, *all chemical reactions are reversible.*

Figure 16.1 Calcium carbonate stalactites cling from the ceiling of a cave, and stalagmites grow up from the floor. The chemistry producing these formations is an excellent illustration of the reversibility of chemical reactions. (Arthur N. Palmer)

Now let us carry out the reaction of calcium carbonate, water, and carbon dioxide in a different way. Suppose a solution of Ca^{2+} and HCO_3^- is placed in a closed container, from which the CO_2 gas cannot escape. The reaction producing $CaCO_3$ and CO_2 continues for a while, but eventually no further change occurs. If we were to examine the reaction system, we would find that Ca^{2+}, HCO_3^-, CO_2, and H_2O coexist in the system. It looks as though the reaction has stopped, but this is not the case. Instead, the reaction has reached equilibrium.

$$Ca^{2+}(aq) + 2\ HCO_3^-(aq) \rightleftharpoons CaCO_3(s) + CO_2(g) + H_2O(\ell)$$

The set of double arrows, $\rightleftharpoons$, in an equation indicates that the reaction is reversible and, in general chemistry courses, is often a signal to you that the reaction is to be studied using the concepts of chemical equilibria.

How might the equilibrium condition be described at the molecular level? As we begin the reaction in a closed container, only Ca^{2+} and HCO_3^- are present, and they react to give the products ($CaCO_3$, CO_2, and H_2O) at some rate. As

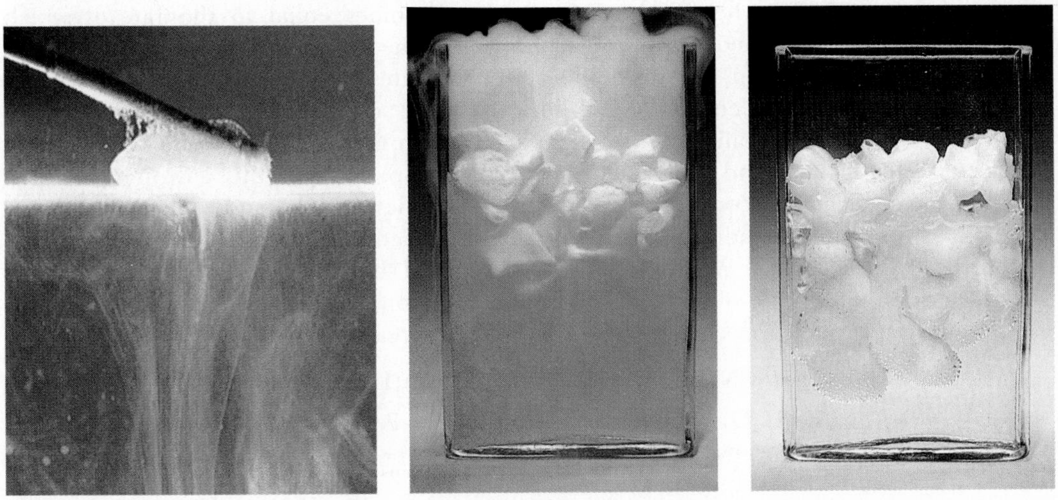

(a) (b) (c)

Figure 16.2 Carbonate equilibria in the $CO_2/Ca(OH)_2$ system. (a) Solid CO_2 is held at the top of a solution of $Ca(OH)_2$, and $CaCO_3$ is precipitated.

$$CO_2(aq) + H_2O(\ell) \rightleftharpoons H_2CO_3(aq)$$
$$H_2CO_3(aq) + Ca(OH)_2(aq) \rightleftharpoons CaCO_3(s) + 2\ H_2O(\ell)$$

(b) When more dry ice (solid CO_2) is added, the concentration of dissolved CO_2 builds up and more $CaCO_3$ precipitates. (An indicator dye has been added to the solution that turns blue because the solution is basic from the $Ca(OH)_2$.) (c) When the concentration of dissolved CO_2 is great enough, $CaCO_3$ redissolves according to the equation

$$CaCO_3(s) + CO_2(aq) + H_2O(\ell) \rightleftharpoons Ca^{2+}(aq) + 2\ HCO_3^-(aq)$$

(The indicator is now very pale yellow, denoting an acidic solution.) (C. D. Winters) the reactants are used up, the rate of reaction slows. The reaction products, $CaCO_3$, CO_2, and H_2O begin to combine, however, with a rate that increases as their concentration increases. Eventually the rate of the forward reaction, the formation of $CaCO_3$, and the rate of the reverse reaction, the redissolving of $CaCO_3$, become equal. With $CaCO_3$ being formed and redissolving at the same rate, no further macroscopic change is observed. The system is at equilibrium.

This description of events at the submicroscopic level defines a basic principle of chemical equilibria. Chemical equilibria are *dynamic.* When the system is at equilibrium, the forward and reverse reactions continue, but they take place at *equal rates.*

Figure 16.3 The reaction of aqueous iron(III) ion and aqueous thiocyanate ion, SCN^-. The colorless solutions are mixed to give the red-orange ion $[Fe(SCN)]^{2+}$, which is in equilibrium with the reactants. (C. D. Winters)

A common introductory chemistry experiment is the study of the equilibrium that exists in the reaction of aqueous iron(III) ion, which we designate as $Fe(H_2O)^{3+}$ for simplicity, and aqueous thiocyanate ion, SCN^-.

$$\underset{\text{Nearly colorless}}{Fe(H_2O)^{3+}(aq)} + \underset{\text{Colorless}}{SCN^-(aq)} \rightleftharpoons \underset{\text{Red-Orange}}{Fe(SCN)^{2+}(aq)} + H_2O(\ell)$$

When colorless solutions of these two ions are mixed (Figure 16.3), the SCN^- ion rapidly replaces water on the Fe^{3+} ion to give a red-orange compound in which the SCN^- ion is bonded to Fe^{3+}. As the concentration of $Fe(SCN)^{2+}$ begins to build up, this ion reacts with water to release SCN^- and reverts to aqueous Fe^{3+}. Eventually, the rate at which SCN^- replaces H_2O on Fe^{3+} to form $Fe(SCN)^{2+}$ (the "forward" reaction) becomes equal to the rate at which $Fe(SCN)^{2+}$ sheds SCN^- ion to go back to the simple ions in solution (the "reverse" reaction). At this point—when the rates of the forward and reverse reactions become equal—equilibrium has been achieved.

When equilibrium has been achieved in the $Fe(H_2O)^{3+} + SCN^-$ reaction, the forward and reverse reactions do not stop. To prove this, a drop of an aqueous solution containing radioactive SCN^- ion can be added to the solution. (The ion is made radioactive by using radioactive ^{14}C in place of normal, nonradioactive ^{12}C.) When the solution is sampled shortly after adding the radioactive SCN^- ion, we observe that the radioactive ion is incorporated into $Fe(SCN)^{2+}$, an observation explained by the following reactions:

$$Fe(SCN)^{2+}(aq) + H_2O(\ell) \rightleftharpoons Fe(H_2O)^{3+}(aq) + SCN^-(aq)$$
$$Fe(H_2O)^{3+}(aq) + S^*CN^-(aq) \rightleftharpoons Fe(S^*CN)^{2+}(aq) + H_2O(\ell)$$

*The symbol S^*CN^- means the thiocyanate ion contains radioactive ^{14}C.

The only way for the radioactive SCN^- ion to be incorporated into the red-orange $Fe(SCN)^{2+}$ ion is if the reaction with water is dynamic and reversible, and continues even at equilibrium.

Not only are equilibrium processes dynamic and reversible, but, for a specific reaction, the nature of the equilibrium state is the same, no matter what the direction of approach. Let us say you measure the concentrations at equilibrium of acetic acid and the ions that come from its ionization in aqueous solution.

$$\underset{\text{acetic acid}}{CH_3CO_2H(aq)} + H_2O(\ell) \rightleftharpoons \underset{\text{acetate ion}}{CH_3CO_2^-(aq)} + \underset{\text{hydronium ion}}{H_3O^+(aq)}$$

The nature of the equilibrium state is the same whatever the direction of approach only if the number of atoms of each element used is the same in the forward and reverse directions.

Because acetic acid is a weak acid, the concentrations of the acetate and hydronium ion are small. Now mix sodium acetate and hydrochloric acid

$$NaCH_3CO_2(aq) + HCl(aq) \longrightarrow CH_3CO_2H(aq) + NaCl(aq)$$

a reaction that has the net ionic equation

$$\underset{\text{acetate ion}}{CH_3CO_2^-(aq)} + \underset{\text{hydronium ion}}{H_3O^+(aq)} \rightleftharpoons \underset{\text{acetic acid}}{CH_3CO_2H(aq)} + H_2O(\ell)$$

and measure the equilibrium concentrations. Assuming that you began with, say, 1 mol of acetic acid in the first experiment and 1 mol each of sodium acetate and HCl in the second experiment (all in the same volume), the concentrations of acetic acid, acetate ion, and hydronium ion at equilibrium will be identical.

16.2 THE EQUILIBRIUM CONSTANT

One way to describe the equilibrium position of a chemical reaction is to give equilibrium concentrations of the reactants and products. The **equilibrium constant expression** relates concentrations of reactants and products at equilibrium at a given temperature to a numerical constant. This expression is an indication of the amounts of reactants and products that are present when equilibrium is achieved.

The notion that equilibrium concentrations of reactants and products are related in a simple manner is easy to prove by experiments such as that for the reaction of hydrogen and iodine to produce hydrogen iodide.

$$H_2(g) + I_2(g) \rightleftharpoons 2\ HI(g)$$

A very large number of experiments have shown that at equilibrium the ratio

$$\frac{[HI]^2}{[H_2][I_2]} = 55.64$$

is always the same within experimental error for all experiments done at 425 °C. For example, assume enough H_2 and I_2 has been placed in a flask so that the concentration of each is 0.0175 mol/L at 425 °C. Over time, the concentrations of H_2 and I_2 decline and that of HI increases; a state of equilibrium is attained eventually. If the gases in the flask are then analyzed, the result is $[H_2] = [I_2] =$ 0.0037 mol/L, whereas [HI] = 0.0276 mol/L.

Square brackets around a chemical formula indicate molar concentrations. [HI] indicates the HI concentration in moles per liter.

Initial and Equilibrium Concentrations (mol/L)

Equation	**$H_2(g)$**	**+ $I_2(g)$**	**$\rightleftharpoons$ 2 HI(g)**
Initial concentration	0.0175	0.0175	0
Change in concentration as reaction proceeds to equilibrium	−0.0138	−0.0138	+0.0276
Equilibrium concentration	0.0037	0.0037	0.0276

Putting these equilibrium concentration values into the preceding expression

$$\frac{[HI]^2}{[H_2][I_2]} = \frac{(0.0276)^2}{(0.0037)(0.0037)} = 56$$

gives a ratio of about 56 (or 55.64 if the experimental information contains more significant figures). This ratio is always the same for all experiments at 425 °C, no matter from which direction the reaction is approached (mixing H_2 and I_2 or allowing HI to decompose) and no matter what the initial concentrations. If we began with 2.0 mol of H_2 and 2.0 mol of I_2, equilibrium is reached with concentrations of reactants and products that satisfy the equilibrium constant expression ($[H_2] = [I_2] = 0.42$ mol/L; $[HI] = 3.16$ mol/L).

Hundreds of experiments on many different chemical systems have proved that equilibrium constants can be calculated from the following general expression. For the general reaction

$$a\text{A} + b\text{B} \rightleftharpoons c\text{C} + d\text{D}$$

the equilibrium concentrations of reactants and products are always related by the equilibrium constant expression

The equilibrium constant K does not have units.

$$\text{Equilibrium constant} = K = \frac{\overset{\text{product concentrations}}{[\text{C}]^c[\text{D}]^d}}{\underset{\text{reactant concentrations}}{[\text{A}]^a[\text{B}]^b}} \quad \textbf{(16.1)}$$

Product concentrations appear in the numerator and reactant concentrations in the denominator of Equation 16.1. Each concentration is raised to the power of its stoichiometric coefficient in the balanced equation. The value of the constant K depends on the particular reaction and on the temperature.

The value of an equilibrium constant for a given reaction provides information about the extent of reaction when equilibrium has been achieved. After learning a few characteristics of the equilibrium expression, we shall turn to that topic.

Writing Equilibrium Constant Expressions

Reactions Involving Solids and Water

You should know a few rules about writing equilibrium constant expressions for chemical reactions. For example, the oxidation of solid, yellow sulfur produces colorless sulfur dioxide gas (Figure 16.4).

$$\tfrac{1}{8}\,S_8(s) + O_2(g) \rightleftharpoons SO_2(g)$$

Figure 16.4 Sulfur burns in oxygen to give sulfur dioxide gas, SO_2. (C. D. Winters)

Following the general principle that products appear in the numerator and reactants in the denominator, you write

$$K' = \frac{[SO_2]}{[S_8]^{1/8}[O_2]}$$

Because sulfur is a molecular solid and because the concentration of molecules within any solid is fixed, the sulfur concentration is not changed either by reaction or by addition or removal of some solid. Furthermore, it is an experimental fact that the equilibrium concentrations of O_2 and SO_2 are not changed by the amount of sulfur, as long as some solid sulfur is present at equilibrium. By convention, therefore, chemists do not include the concentrations of any *solid* reactants or products in the equilibrium constant expression. You should therefore write the equilibrium expression for the sulfur and oxygen reaction as

$$K = \frac{[SO_2]}{[O_2]}$$

where $K = 4.2 \times 10^{52}$ (at 25 °C).

There are special considerations for reactions occurring in aqueous solution. Consider ammonia, which is a weak base owing to its interaction with water.

$$NH_3(aq) + H_2O(\ell) \rightleftharpoons NH_4^+(aq) + OH^-(aq)$$

Water is the solvent in this reaction. Its concentration is large for a dilute solution of solute, and it is unchanged by the reaction. For this reason the molar concentration of water, like that of solids, is not included in the equilibrium constant expression. Thus, we write

$$K = \frac{[NH_4^+][OH^-]}{[NH_3]}$$

where $K = 1.8 \times 10^{-5}$ (at 25 °C). We see a similar situation for weak acids in aqueous solution. Formic acid, HCO_2H, is a weak acid because it can only partially transfer an H^+ ion to water to form the hydronium ion.

$$HCO_2H(aq) + H_2O(\ell) \rightleftharpoons HCO_2^-(aq) + H_3O^+(aq)$$

In this case, we write

$$K = \frac{[HCO_2^-][H_3O^+]}{[HCO_2H]}$$

where $K = 1.8 \times 10^{-4}$ (at 25 °C).

Expressing Concentrations: K_c and K_p

The concentrations in the equilibrium constant expression are usually given in moles per liter (M), so the equilibrium constant symbol K usually has a subscript c for concentration, as in K_c. Equilibrium constant expressions for reactions involving gases, however, can also be written in terms of the partial pressures of the gaseous reactants and products. The rearranged form of the ideal gas law $P = (n/V)RT$ tells us that the partial pressures of gaseous substances are directly proportional to their molar concentrations (n/V). When an equilibrium constant expression is written in terms of partial pressures, K is given a subscript p, as in K_p. In some cases the numerical values of K_c and K_p are identical, but they are usually different.

In this book we shall generally use equilibrium constants when concentrations are given in moles per liter, K_c, unless we are discussing a gas-phase reaction in which it is more appropriate to use partial pressures and K_p. The box "A Closer Look: Equilibrium Constant Expressions for Gases" shows you how to express K in terms of gas pressures for reactions involving gases and how K_c and K_p are related.

A CLOSER LOOK — *Equilibrium Constant Expressions for Gases—K_c and K_p*

Many metal carbonates such as limestone decompose on heating to give the metal oxide and CO_2 gas.

$$CaCO_3(s) \rightleftharpoons CaO(s) + CO_2(g)$$

The equilibrium condition for this reaction can be expressed either in terms of the number of moles per liter of CO_2, $K_c = [CO_2]$, or in terms of the pressure of CO_2, $K_p = P_{CO_2}$. From the ideal gas law, you know that

$$P = \left(\frac{n}{V}\right)RT =$$
(concentration in moles per liter) $\cdot$ RT

For this reaction, we can therefore say that $K_p = [CO_2]RT$. Because $K_c = [CO_2]$, the interesting conclusion is that $K_p = K_c(RT)$. That is, the *values* of K_p and K_c are *not* the same; K_p for the decomposition of calcium carbonate is the product of K_c and the factor RT.

The equilibrium constant in terms of partial pressures, K_p, is known for the reaction of N_2 and H_2 to produce ammonia.

$$N_2(g) + 3\ H_2(g) \rightleftharpoons 2\ NH_3(g)$$

$$K_p = \frac{P^2_{NH_3}}{P_{N_2}P^3_{H_2}} = 5.8 \times 10^5 \text{ at } 25\ °C$$

Does K_c, the equilibrium constant in terms of concentrations, have the same value as or a different value from K_p? We can answer this by substituting for each pressure in K_p the equivalent expression [moles per liter] $\cdot$ (RT). That is,

$$K_p = \frac{\{[NH_3](RT)\}^2}{\{[N_2](RT)\}\{[H_2](RT)\}^3} = \frac{[NH_3]^2}{[N_2][H_2]^3} \cdot \frac{1}{(RT)^2} = \frac{K_c}{(RT)^2}$$

Solving for K_c, we find

$$K_p = 5.8 \times 10^5 = \frac{K_c}{[(0.08206)(298)]^2}$$

$$K_c = 3.5 \times 10^8$$

Once again you see that K_p and K_c are not the same but are related by some function of RT.

Looking carefully at these and other examples, we find that, in general,

$$K_p = K_c(RT)^{\Delta n} \quad \textbf{(16.2)}$$

where Δn is the change in the number of moles of gas evolved as the reaction goes from reactants to products.

Δn = total moles of gaseous products − total moles of gaseous reactants

For the decomposition of $CaCO_3$,

$$\Delta n = 1 - 0 = 1$$

whereas the value of Δn for the ammonia synthesis is

$$\Delta n = 2 - 4 = -2$$

EXERCISE 16.1 ***Writing Equilibrium Constant Expressions***

Write equilibrium constant expressions for each of the following reactions:

1. $PCl_5(g) \rightleftharpoons PCl_3(g) + Cl_2(g)$
2. $Cu(OH)_2(s) \rightleftharpoons Cu^{2+}(aq) + 2\ OH^-(aq)$
3. $Cu(NH_3)_4{}^{2+}(aq) \rightleftharpoons Cu^{2+}(aq) + 4\ NH_3(aq)$
4. $CH_3CO_2H(aq) + H_2O(\ell) \rightleftharpoons CH_3CO_2{}^-(aq) + H_3O^+(aq)$ ■

Manipulating Equilibrium Expressions

Chemical equations can be balanced using different sets of stoichiometric coefficients. For example, the oxidation of carbon can give carbon monoxide,

$$C(s) + \tfrac{1}{2}\ O_2(g) \rightleftharpoons CO(g)$$

and the equilibrium constant expression for this reaction as written is

$$K_1 = \frac{[CO]}{[O_2]^{1/2}} = 4.6 \times 10^{23} \text{ at } 25\ °C$$

You can write the chemical equation equally well, however, as

$$2\ C(s) + O_2(g) \rightleftharpoons 2\ CO(g)$$

and the equilibrium constant is now

$$K_2 = \frac{[CO]^2}{[O_2]} = 2.1 \times 10^{47}$$

When you compare the two equilibrium expressions you find that $K_2 = K_1^2$; that is,

$$K_2 = \frac{[CO]^2}{[O_2]} = \left\{\frac{[CO]}{[O_2]^{1/2}}\right\}^2 = K_1^2$$

In general, when the stoichiometric coefficients of a balanced equation are multiplied by some factor, the equilibrium constant for the new equation (K_{new}) is the old equilibrium constant (K_{old}) *raised to the power of the multiplication factor.* In the case of the oxidation of carbon, the second equation was obtained by multiplying the first equation by 2. Therefore, K_2 is the *square* of K_1.

Closely related to the effect of using a new set of stoichiometric coefficients is what happens to K when a chemical equation is reversed. Compare the values of K for formic acid transferring an H^+ ion to water

$$HCO_2H(aq) + H_2O(\ell) \rightleftharpoons HCO_2^-(aq) + H_3O^+(aq)$$

$$K_1 = \frac{[HCO_2^-][H_3O^+]}{[HCO_2H]}$$

where $K = 1.8 \times 10^{-4}$ (at 25 °C), with the opposite reaction, the gain of an H^+ ion by the formate ion, HCO_2^-.

$$HCO_2^-(aq) + H_3O^+(aq) \rightleftharpoons HCO_2H(aq) + H_2O(\ell)$$

$$K_2 = \frac{[HCO_2H]}{[HCO_2^-][H_3O^+]}$$

where $K = 5.6 \times 10^3$ (at 25 °C). Here $K_2 = 1/K_1$ (or $K_2 = K_1^{-1}$ because reversing an equation has the same effect as multiplying the coefficients in a balanced equation by −1). The equilibrium constants for a reaction and its reverse are always the *reciprocals* of one another.

It is often necessary to add two equations together to obtain the equation for a net process. As an example, consider the reaction that takes place when silver chloride dissolves in water (to a very small extent). Ammonia is then added to the solution, and the ammonia reacts with the silver ion to form a water-soluble polyatomic ion, $Ag(NH_3)_2^+$ (Figure 16.5). Adding the equation for the process of dissolving solid AgCl to the equation for the reaction of Ag^+ ion with ammonia gives the equation for the net reaction—the process of dissolving solid AgCl in the presence of aqueous ammonia. (All equilibrium constants are given at 25 °C.)

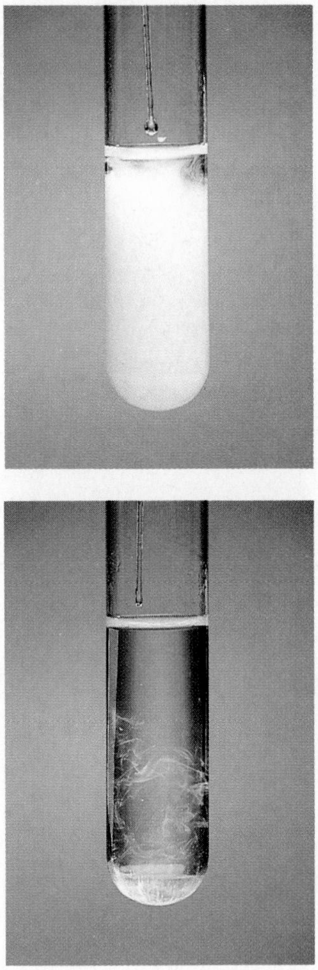

Figure 16.5 A precipitate of AgCl(s) is suspended in water (top). When aqueous ammonia is added, the ammonia reacts with the trace of silver ion in the solution, and the silver chloride dissolves (bottom). (C. D. Winters)

$$AgCl(s) \rightleftharpoons Ag^+(aq) + Cl^-(aq) \qquad K_1 = [Ag^+][Cl^-] = 1.8 \times 10^{-10}$$

$$Ag^+(aq) + 2\ NH_3(aq) \rightleftharpoons Ag(NH_3)_2^+(aq) \qquad K_2 = \frac{[Ag(NH_3)_2^+]}{[Ag^+][NH_3]^2} = 1.6 \times 10^7$$

Net reaction:

$$AgCl(s) + 2\ NH_3(aq) \rightleftharpoons Ag(NH_3)_2^+(aq) + Cl^-(aq)$$

Multiplying the equilibrium constants for the two reactions, K_1 by K_2, gives the equilibrium constant for the net reaction, K_{net}.

$$K_{net} = K_1 \cdot K_2 = [Ag^+][Cl^-] \cdot \frac{[Ag(NH_3)_2{}^+]}{[Ag^+][NH_3]^2} = \frac{[Ag(NH_3)_2{}^+][Cl^-]}{[NH_3]^2}$$

$$= K_1 K_2 = 2.9 \times 10^{-3}$$

In general, when two or more equations are added to produce a net equation, the equilibrium constant for the net equation is the *product* of the equilibrium constants for the added equations.

EXAMPLE 16.1 *Manipulating Equilibrium Constant Expressions*

A mixture of nitrogen, hydrogen, and ammonia can be brought to equilibrium. When the equation is written using whole-number coefficients, as follows, the value of K_c is 3.5×10^8 at 25 °C.

Equation 1

$$N_2(g) + 3\ H_2(g) \rightleftharpoons 2\ NH_3(g) \qquad K_1 = 3.5 \times 10^8$$

However, the equation can also be written as

Equation 2

$$\tfrac{1}{2}\ N_2(g) + \tfrac{3}{2}\ H_2(g) \rightleftharpoons NH_3(g) \qquad K_2 = ?$$

What is the value of K_2, the equilibrium constant for Equation 2? What is the value of K_3, the equilibrium constant for the reverse of Equation 1, that is, the decomposition of ammonia to the elements?

Equation 3

$$2\ NH_3(g) \rightleftharpoons N_2(g) + 3\ H_2(g) \qquad K_3 = ?$$

Solution To see the relation between K_1 and K_2, let us first write the equilibrium expressions for these two balanced equations.

$$K_1 = \frac{[NH_3]^2}{[N_2][H_2]^3} = 3.5 \times 10^8$$

and

$$K_2 = \frac{[NH_3]}{[N_2]^{1/2}[H_2]^{3/2}}$$

Writing the expressions makes it clear that K_1 is the square of K_2; that is, $K_1 = K_2{}^2$. The answer to our question is therefore

$$K_2 = \sqrt{K_1} = \sqrt{3.5 \times 10^8} = 1.9 \times 10^4$$

Equation 3 is the reverse of Equation 1, and its equilibrium expression is

$$K_3 = \frac{[N_2][H_2]^3}{[NH_3]^2}$$

In this case, K_3 is the reciprocal of K_1; that is, $K_3 = 1/K_1$.

$$K_3 = \frac{[NH_3]^2}{[N_2][H_2]^3} = \frac{1}{K_1} = \frac{1}{3.5 \times 10^8} = 2.9 \times 10^{-9}$$

As a final comment, notice that the production of ammonia from the elements has a large equilibrium constant. As expected, the reverse reaction, the decomposition of ammonia to its elements, has a small equilibrium constant.

EXERCISE 16.2 ***Manipulating Equilibrium Constant Expressions***

The conversion of oxygen to ozone has a very small equilibrium constant.

$$\tfrac{3}{2}\, O_2(g) \rightleftharpoons O_3(g) \qquad K_c = 2.5 \times 10^{-29}$$

1. What is the value of K_c when the equation is written using whole-number coefficients?

$$3\, O_2(g) \rightleftharpoons 2\, O_3(g)$$

2. What is the value of K_c for the conversion of ozone to oxygen?

$$2\, O_3(g) \rightleftharpoons 3\, O_2(g)$$ ■

EXERCISE 16.3 ***Manipulating Equilibrium Constant Expressions***

The following equilibrium constants are given at 500 K:

$$H_2(g) + Br_2(g) \rightleftharpoons 2\, HBr(g) \qquad K_p = 7.9 \times 10^{11}$$
$$H_2(g) \rightleftharpoons 2\, H(g) \qquad K_p = 4.8 \times 10^{-41}$$
$$Br_2(g) \rightleftharpoons 2\, Br(g) \qquad K_p = 2.2 \times 10^{-15}$$

Calculate K_p for the reaction of H and Br atoms to give HBr.

$$H(g) + Br(g) \rightleftharpoons HBr(g) \qquad K_p = ?$$ ■

PROBLEM-SOLVING TIPS AND IDEAS

16.1 Writing and Manipulating Equilibrium Constant Expressions

This section is an important one. You should now know

1. How to write an equilibrium constant expression from the balanced equation, recognizing that the concentrations of pure solids and of liquids used as solvents do not appear in the expression
2. That when the stoichiometric coefficients in a balanced equation are changed by a factor of n, $K_{new} = (K_{old})^n$
3. That when a balanced equation is reversed, $K_{new} = 1/K_{old}$
4. That when several balanced equations are added to obtain a net, balanced equation, $K_{net} = K_1 \cdot K_2 \cdot K_3 \cdot \ldots$
5. That the value of K depends on the way the equilibrium concentrations are expressed (in moles per liter or P). ■

The Meaning of the Equilibrium Constant

The value of the equilibrium constant indicates whether a reaction is product- or reactant-favored. In addition, it can be used to calculate how much product will be present at equilibrium, which is valuable information to chemists and chemical engineers.

A large value of K means that reactants are converted largely to products when equilibrium has been achieved. That is, the products are strongly favored over the reactants at equilibrium. An example is the NO + O_3 reaction described in Section 15.5.

$K \gg 1$: Reaction is *product-favored;* equilibrium concentrations of products are greater than equilibrium concentrations of reactants.

$$NO(g) + O_3(g) \rightleftharpoons NO_2(g) + O_2(g) \qquad K_c = 6 \times 10^{34} \text{ at } 25\text{ °C}$$

$$K_c = \frac{[NO_2][O_2]}{[NO][O_3]}$$

Because $K_c \gg 1$, $[NO_2][O_2] \gg [NO][O_3]$. The very large value of K indicates that, if stoichiometric amounts of NO and O_3 are mixed in a flask and allowed to come to equilibrium, virtually none of the reactants will be found.

TABLE 16.1 Selected Equilibrium Constants

Reaction	Equilibrium Constant K_c (at 25 °C)
Nonmetal Reactions	
$\frac{1}{8} S_8(s) + O_2(g) \rightleftharpoons SO_2(g)$	4.2×10^{52}
$2 H_2(g) + O_2(g) \rightleftharpoons 2 H_2O(g)$	3.2×10^{81}
$N_2(g) + 3 H_2(g) \rightleftharpoons 2 NH_3(g)$	3.5×10^{8}
$N_2(g) + O_2(g) \rightleftharpoons 2 NO(g)$	1.7×10^{-3} (at 2300 K)
Weak Acids and Bases	
$HCO_2H(aq) + H_2O(\ell) \rightleftharpoons HCO_2^-(aq) + H_3O^+(aq)$ formic acid	1.8×10^{-4}
$CH_3CO_2H(aq) + H_2O(\ell) \rightleftharpoons CH_3CO_2^-(aq) + H_3O^+(aq)$ acetic acid	1.8×10^{-5}
$H_2CO_3(aq) + H_2O(\ell) \rightleftharpoons HCO_3^-(aq) + H_3O^+(aq)$ carbonic acid	4.2×10^{-7}
$NH_3(aq) + H_2O(\ell) \rightleftharpoons NH_4^+(aq) + OH^-(aq)$ ammonia (weak base)	1.8×10^{-5}
"Insoluble" Solids	
$CaCO_3(s) \rightleftharpoons Ca^{2+}(aq) + CO_3^{2-}(aq)$	3.8×10^{-9}
$AgCl(s) \rightleftharpoons Ag^+(aq) + Cl^-(aq)$	1.8×10^{-10}

Essentially all will have been converted to NO_2 and O_2. A chemist would say "the reaction has gone to completion."

Conversely, a small K (as in the formation of ozone from oxygen) means that very little of the reactants have formed products when equilibrium has been achieved. In other words, the reactants are favored over the products at equilibrium.

$K \ll 1$: Reaction is *reactant-favored;* equilibrium concentrations of reactants are greater than equilibrium concentrations of products.

$$\tfrac{3}{2}\,O_2(g) \rightleftharpoons O_3(g) \qquad K_c = 2.5 \times 10^{-29} \text{ at } 25\ °C$$

$$K_c = \frac{[O_3]}{[O_2]^{3/2}}$$

Because $K_c \ll 1$, $[O_3] \ll [O_2]^{3/2}$. The very small value of K indicates that, if O_2 is placed in a flask, *very* little O_2 will have been converted to O_3 when equilibrium is achieved.

Equilibrium constants for a few reactions are given in Table 16.1. These reactions occur to widely varying extents, as shown by the wide range of values of K.

EXERCISE 16.4 ***The Equilibrium Constant and Extent of Reaction***

Solid AgCl and AgBr were each placed in 1.0 L of water in separate beakers. Are the following reactions product- or reactant-favored? When equilibrium is achieved, in which beaker will the concentration of silver ion be larger?

$$AgCl(s) \rightleftharpoons Ag^+(aq) + Cl^-(aq) \qquad K_c = 1.8 \times 10^{-10}$$

$$AgBr(s) \rightleftharpoons Ag^+(aq) + Br^-(aq) \qquad K_c = 3.3 \times 10^{-13}$$

■

16.3 THE REACTION QUOTIENT

In Chapter 13 you first encountered phase diagrams, graphic representations of the equilibria involved in transformations between the phases of matter. For any liquid, the plot of temperature versus equilibrium vapor pressure is a curved line (Figure 13.21). This line defines all the conditions of T and P at which equilibrium can exist. There can be no equilibrium in a system whose temperature and vapor pressure do not represent a set of points on the line.

The equilibrium situation for a chemical reaction can also be represented graphically for some simple cases. For example, consider a reaction such as the transformation of butane to isobutane (2-methylpropane).

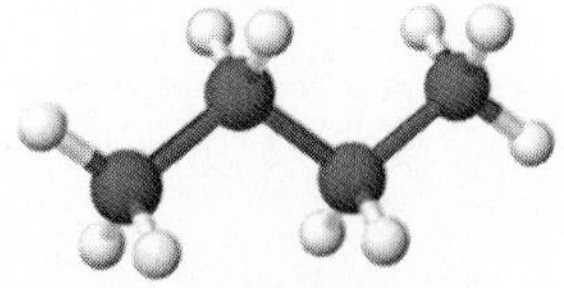
Butane

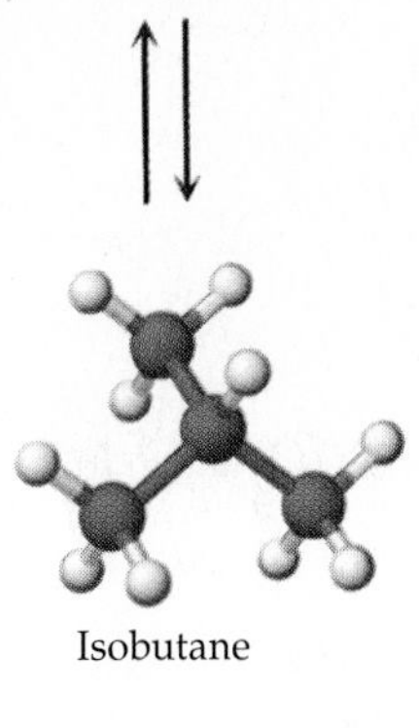
Isobutane

$$K_c = \frac{[\text{isobutane}]}{[\text{butane}]} = 2.50 \text{ at } 298\text{ K}$$

If the concentration of one of the compounds is known, then only one value of the other concentration will satisfy the equilibrium constant expression. For example, if [butane] is 1.0 mol/L, the equilibrium concentration of isobutane

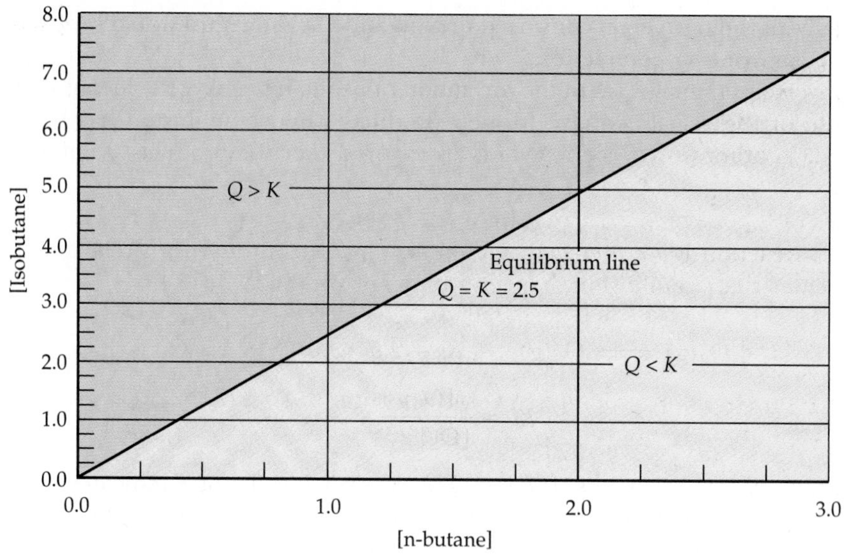

Figure 16.6 A plot of the concentrations of butane and isobutane that satisfy the expression $Q =$ [isobutane]/[butane] for the equilibrium reaction butane $\rightleftarrows$ isobutane. When $Q = 2.5$, $Q = K$, the equilibrium constant. If $Q < K$, the tan portion of the diagram, then the system is not at equilibrium, and reactants are further converted to products. If $Q > K$, the system is again not at equilibrium, but products must revert to reactants to establish equilibrium.

must be 2.5 mol/L. If [butane] were changed to 0.80 M, then [isobutane] at equilibrium must be

$$[\text{Isobutane}] = K_c\,[\text{butane}] = 2.50\ (0.80\ \text{M}) = 2.0\ \text{M}$$

Similar plots can be made for other types of equilibrium expressions, but they are more complicated.

Choosing a number of possible values for [butane] and solving for the allowed value of [isobutane] eventually leads to the points plotted in Figure 16.6. That is, the equilibrium expression $K =$ [isobutane]/[butane] is just the equation of a straight line (with a slope equal to K).

The graphical treatment of the simple equilibrium expression $K =$ [isobutane]/[butane] is useful for two reasons. First, Figure 16.6 illustrates the fact that there is an infinite number of possible *sets* of equilibrium concentrations of reactant and product (all the points along the equilibrium line). Second, sets of concentrations for butane and isobutane that do *not* lie along the line in Figure 16.6 do *not* satisfy the equilibrium condition (just as in the case of phase equilibria; Figure 13.21). If this is the case, the system attempts to shift to a new set of concentrations that will satisfy the equilibrium condition.

Any point in Figure 16.6, whether on or off the line, can be defined by the ratio [isobutane]/[butane]. This ratio is given the general name of the *reaction quotient* Q, and it is equal to the equilibrium constant K when the reaction is at equilibrium. Suppose you have a system composed of 1.5 mol/L of butane and 1.5 mol/L of isobutane (at 298 K). This means that the ratio of concentrations, Q, is

$$Q = \frac{[\text{isobutane}]}{[\text{butane}]} = \frac{1.5}{1.5} = 1.0$$

a ratio given by a point in the tan portion of Figure 16.6. This set of concentrations clearly does *not* represent an equilibrium system because $Q < K$. It is certain, however, that some of the butane will change eventually into isobutane,

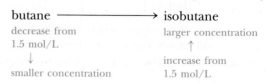

butane $\longrightarrow$	isobutane
decrease from 1.5 mol/L	larger concentration
$\downarrow$	$\uparrow$
smaller concentration	increase from 1.5 mol/L

thereby lowering [butane] and raising [isobutane]. Indeed, this transformation continues until the ratio [isobutane]/[butane] = 2.5; that is, until $Q = K$, and the ratio of concentrations in the system is represented by a point on the line in Figure 16.6. This process is somewhat analogous to the changes that must occur when a liquid and its vapor are not in equilibrium.

What happens when there is too much isobutane in the system relative to the amount of butane? Suppose [isobutane] = 5.00 M but [butane] is only 1.25 M. Now the reaction quotient Q is greater than $K (Q > K)$.

$$Q = \frac{[\text{isobutane}]}{[\text{butane}]} = \frac{5.00}{1.25} = 4.00$$

This is represented by a point in the blue portion of Figure 16.6. The system is again not at equilibrium, but it can proceed to an equilibrium state by converting isobutane to butane

butane ⟵	isobutane
larger concentration	decrease from
↑	5.00 mol/L
increase from	↓
1.25 M	smaller concentration

Only when Q has a value of 2.5 ($Q = K$) is the system finally at equilibrium.

For any reaction

$$a\text{A} + b\text{B} \rightleftharpoons c\text{C} + d\text{D}$$

the **reaction quotient Q** is defined by the equation

$$Q_c = \frac{[\text{C}]^c[\text{D}]^d}{[\text{A}]^a[\text{B}]^b}$$

The expression for Q_c has the *appearance* of the equilibrium expression, but Q_c differs from K_c in that the concentrations in the expression are *not necessarily* equilibrium concentrations. The following general statements can be made regarding the relationship between K and Q:

The Relationship Between the Reaction Quotient Q and the Equilibrium Constant K

1. **If $Q < K$, the system is not at equilibrium and Reactants → Products.**
   The ratio of product concentrations to reactant concentrations is too small. More reactants must be converted to products (thus increasing Q) to achieve equilibrium (when $Q = K$).
2. **If $Q = K$, the system is at equilibrium.**
3. **If $Q > K$, the system is not at equilibrium and Products → Reactants.**
   The ratio of product concentrations to reactant concentrations is too large. To reach equilibrium, products must be converted to reactants (thus decreasing the value of Q until $Q = K$).

We will often make use of these ideas in these chapters on chemical equilibria (see Section 16.5, for example).

EXAMPLE 16.2 ***The Reaction Quotient***

Molecules of the brown gas nitrogen dioxide, NO_2, combine to form the colorless gas N_2O_4.

$$2\ NO_2(g) \rightleftharpoons N_2O_4(g) \qquad K_c = 170 \text{ at } 298\text{ K}$$

Suppose the concentration of NO_2 is 0.015 M and the concentration of N_2O_4 is 0.025 M. Is Q_c larger than, smaller than, or equal to K_c? If the system is not at equilibrium, in which direction will the reaction proceed to achieve equilibrium?

Solution The equilibrium constant expression for the reaction is

$$K_c = \frac{[N_2O_4]}{[NO_2]^2} = 170$$

If the reactant and product concentrations are substituted into the reaction quotient expression, we have

$$Q_c = \frac{(0.025)}{(0.015)^2} = 110$$

The value of Q_c is less than the value of K_c ($Q_c < K_c$), so the reaction is not at equilibrium. It must proceed to equilibrium by converting more NO_2 to N_2O_4, thus increasing $[N_2O_4]$ and decreasing $[NO_2]$ until $Q_c = K_c$.

EXERCISE 16.5 ***The Reaction Quotient***

Answer the following questions regarding the butane/isobutane equilibrium ($K_c = 2.5$ at 298 K) and Figure 16.6.

1. Is the system at equilibrium when [butane] = 0.97 M and [isobutane] = 2.18 M? If it is not at equilibrium, in which direction will the reaction proceed to achieve equilibrium?
2. Is the system at equilibrium when [butane] = 0.75 M and [isobutane] = 2.60 M? If it is not at equilibrium, in which direction will the reaction proceed to achieve equilibrium? ■

EXERCISE 16.6 ***The Reaction Quotient***

At 2300 K the equilibrium constant, K_c, for the formation of NO(g)

$$N_2(g) + O_2(g) \rightleftharpoons 2\ NO(g)$$

is 1.7×10^{-3}. You have a vessel in which the concentration of N_2 at 2300 K is 0.50 mol/L, that of O_2 is 0.25 mol/L, and that of NO is 4.2×10^{-3} mol/L. Is the system at equilibrium? If not, predict which way the reaction will proceed to achieve equilibrium. ■

16.4 CALCULATING AN EQUILIBRIUM CONSTANT

When the values of all the concentrations are known *at equilibrium,* calculating an equilibrium constant simply involves substituting the data into the equilibrium expression, which is illustrated by the following example.

EXAMPLE 16.3 *Calculating an Equilibrium Constant*

A mixture of SO_2, O_2, and SO_3 is allowed to reach equilibrium at 852 K. The equilibrium concentrations are $[SO_2] = 3.61 \times 10^{-3}$ mol/L, $[O_2] = 6.11 \times 10^{-4}$ mol/L, and $[SO_3] = 1.01 \times 10^{-2}$ mol/L. Calculate K_c, at 852 K, for the reaction

$$2\ SO_2(g) + O_2(g) \rightleftharpoons 2\ SO_3(g)$$

Solution First, write the equilibrium constant expression in terms of concentrations.

$$K_c = \frac{[SO_3]^2}{[SO_2]^2[O_2]}$$

Next, substitute the experimental information into the expression and calculate K_c.

$$K_c = \frac{(1.01 \times 10^{-2})^2}{(3.61 \times 10^{-3})^2(6.11 \times 10^{-4})} = 1.28 \times 10^4$$

More commonly, an experiment provides information only on the initial quantities of reactants and the concentration at equilibrium of only one of the reactants or of one of the products. Equilibrium concentrations of the rest of the reactants and products must then be inferred from the balanced chemical equation. The remainder of this section describes calculations of this type.

Consider the oxidation of sulfur dioxide to sulfur trioxide.

$$2\ SO_2(g) + O_2(g) \rightleftharpoons 2\ SO_3(g)$$

The expression for K_c was given in Example 16.3. Now, let us suppose that, in an experiment to determine K_c for this reaction, you place 1.00 mol of SO_2 and 1.00 mole O_2, but no SO_3, into a 1.00-L flask. You cannot tell from this information how much of the SO_2 and O_2 will react on reaching equilibrium or how much SO_3 will then form. You only know that the system will contain a mixture of SO_2, O_2, and SO_3 at equilibrium.

The balanced chemical equation for the oxidation of SO_2 helps to define the situation at equilibrium. Let x be the number of moles of O_2 consumed as the reaction proceeds to equilibrium. The balanced equation tells us that 2 mol of SO_2 is consumed for each mole of O_2 consumed. Therefore, $2x$ mol of SO_2 must be consumed at the same time. Because 2 mol of SO_3 forms for each mole of O_2 consumed, $2x$ moles of SO_3 form when x mol of O_2 is consumed. The key idea is that all the changes can be expressed in terms of a single unknown (here we have used x) and the known coefficients in the balanced equation. These numbers can be displayed in the form of an *equilibrium table.*

Equation	**$2\ SO_2$**	**+**	**O_2**	**$\rightleftharpoons 2\ SO_3$**
Initial moles	1.00		1.00	0
Change on proceeding to equilibrium	$-2x$		$-x$	$+2x$
Moles at equilibrium	$1.00 - 2x$		$1.00 - x$	$2x$

With the equilibrium table constructed, we only need to measure experimentally the equilibrium quantity of any one of the reactants or products. This can be used to find the value of *x* and therefore the equilibrium concentrations of all the substances in the reaction.

EXAMPLE 16.4 *Calculating an Equilibrium Constant*

In a 1.00-L flask at 1000 K we place 1.00 mol of SO_2 and 1.00 mol of O_2. When equilibrium is achieved, 0.925 mol of SO_3 is formed. Calculate K_c, at 1000 K, for the reaction

$$2\ SO_2(g) + O_2(g) \rightleftharpoons 2\ SO_3(g)$$

Solution First, write the equilibrium constant expression in terms of concentrations.

$$K_c = \frac{[SO_3]^2}{[SO_2]^2[O_2]}$$

Next, the equilibrium table described previously tells us that the quantity of SO_3 formed as the reaction proceeds to equilibrium is equal to the SO_2 consumed in the reaction. Furthermore, from the balanced equation we know that the quantity of O_2 consumed is equal to half of the quantity of SO_3 formed. We can therefore construct the equilibrium table for this case as shown.

Equation	$2\ SO_2$ +	O_2	$\rightleftharpoons 2\ SO_3$
Initial moles	1.00	1.00	0
Change on proceeding to equilibrium	−0.925	−0.925/2	+0.925
Moles at equilibrium	0.075	0.537	0.925
Concentration at equilibrium (M)	0.075	0.537	0.925

Remember that concentrations are needed in the equilibrium constant expression. The final line of the table therefore gives the concentration of each substance at equilibrium. With these now known, it is possible to calculate K_c.

$$K_c = \frac{[SO_3]^2}{[SO_2]^2[O_2]} = \frac{(0.925)^2}{(0.075)^2(0.537)} = 2.8 \times 10^2$$

EXAMPLE 16.5 *Calculating an Equilibrium Constant*

One mol of ethanol and 1.00 mol of acetic acid are dissolved in water and kept at 100 °C. The volume of the solution is 250 mL. At equilibrium, 0.25 mol of acetic acid is consumed in producing ethyl acetate. Calculate K_c at 100 °C for the reaction

$$\underset{\text{ethanol}}{C_2H_5OH(aq)} + \underset{\text{acetic acid}}{CH_3CO_2H(aq)} \rightleftharpoons \underset{\text{ethyl acetate}}{CH_3CO_2C_2H_5(aq)} + H_2O(\ell)$$

Solution At equilibrium, 0.25 mol of acid is consumed, so 0.25 mol of ethanol must also be consumed and 0.25 mol of ethyl acetate forms. The equilibrium table is therefore as shown.

Equation	C_2H_5OH	+ CH_3CO_2H ⇌	$CH_3CO_2C_2H_5$ + H_2O
Initial moles	1.00	1.00	0
Change	−0.25	−0.25	+0.25
Moles at equilibrium	0.75	0.75	0.25
Concentration at equilibrium	$\frac{0.75 \text{ mol}}{0.250 \text{ L}}$ = 3.0 M	$\frac{0.75 \text{ mol}}{0.250 \text{ L}}$ = 3.0 M	$\frac{0.25 \text{ mol}}{0.250 \text{ L}}$ = 1.0 M

The concentration of each substance at equilibrium is now known, and K_c can be calculated.

$$K_c = \frac{[CH_3CO_2C_2H_5]}{[C_2H_5OH][CH_3CO_2H]} = \frac{1.0}{(3.0)(3.0)} = 0.11$$

EXAMPLE 16.6 *Calculating an Equilibrium Constant*

Suppose a tank initially contains H_2S with a pressure of 10.00 atm at 800 K. When the reaction

$$2\ H_2S(g) \rightleftharpoons 2\ H_2(g) + S_2(g)$$

has come to equilibrium, the partial pressure of S_2 vapor is 2.0×10^{-2} atm. Calculate K_p.

Solution The concentration of a gas at equilibrium can be expressed in moles per liter or as a partial pressure. Recall from our previous discussion (page 756) that partial pressures of gases are proportional to concentration ($P = (n/V)RT$). The equilibrium expression that we want to evaluate is, therefore,

$$K_p = \frac{P^2_{H_2} \cdot P_{S_2}}{P^2_{H_2S}}$$

Let us set up an equilibrium table that expresses the equilibrium partial pressures of each gas in terms of a single unknown quantity, x.

Equation	2 H_2S ⇌	2 H_2 +	S_2
Initial pressure	10.0	0	0
Change in pressure	$-2x$	$+2x$	$+x$
Equilibrium partial pressure (in terms of x)	$10.00 - 2x$	$2x$	x
Equilibrium partial pressures (actual values)	10.00 − 2(0.020)	2(0.020)	0.020

In the table we designated the quantity of S_2 formed at equilibrium as x. The balanced equation informs us that for every mole (or atmosphere pressure) of S_2 formed, 2 mol (or 2 atm) of H_2 also forms, and 2 mol (or 2 atm) of H_2S is consumed. Because x is known from experiment to be 0.020 atm in this case, the partial pressure of each gas is known at equilibrium, and K_p can be calculated.

$$K_p = \frac{P^2_{H_2} \cdot P_{S_2}}{P^2_{H_2S}} = \frac{[2(0.020)]^2 \cdot (0.020)}{[10.00 - 2(0.020)]^2} = 3.2 \times 10^{-7}$$

EXERCISE 16.7 ***Calculating an Equilibrium Constant***

A mixture of H_2 (9.838×10^{-4} mol) and I_2 (1.377×10^{-3} mol) is sealed in a quartz tube and kept at 350 °C for a week. During this time the reaction

$$H_2(g) + I_2(g) \rightleftharpoons 2\ HI(g)$$

comes to equilibrium. The tube is broken open, and 4.725×10^{-4} mol of I_2 is found.

1. Calculate the number of moles of H_2 and HI present at equilibrium.
2. Assume the volume of the tube is 10.0 mL. Calculate K_c for the reaction.
3. Is the value of K_c different if the volume of the tube is 20.0 mL? ■

EXERCISE 16.8 ***Calculating an Equilibrium Constant***

A solution is prepared by dissolving 0.050 mol of diiodocyclohexane, $C_6H_{10}I_2$, in the solvent CCl_4. The total solution volume is 1.00 L. When the reaction

$$C_6H_{10}I_2 \rightleftharpoons C_6H_{10} + I_2$$

comes to equilibrium at 35 °C, the concentration of I_2 is 0.035 mol/L.

1. What are the concentrations of $C_6H_{10}I_2$ and C_6H_{10} at equilibrium?
2. Calculate K_c, the equilibrium constant. ■

16.5 USING EQUILIBRIUM CONSTANTS IN CALCULATIONS

Another type of situation is more common than those described in Section 16.4. That is, you may know the value of K_c as well as the initial number of moles, concentrations, or partial pressures of reactants or of reactants and products and need to find the quantities present at equilibrium. Once again we solve these problems by constructing a table that expresses the unknown chemical quantities in terms of one unknown, say x.

EXAMPLE 16.7 ***Calculating a Concentration from an Equilibrium Constant***

The equilibrium constant K_c for

$$H_2(g) + I_2(g) \rightleftharpoons 2\ HI(g) \qquad K_c = 55.64$$

has been determined at 425 °C. If 1.00 mol each of H_2 and I_2 is placed in a 0.500-L flask at 425 °C, what are the equilibrium concentrations of H_2, I_2, and HI?

Solution Having written the balanced chemical equation, the next step is to write the equilibrium constant expression.

$$K_c = 55.64 = \frac{[HI]^2}{[H_2][I_2]}$$

Next, we set up an equilibrium table to find a way to express the equilibrium concentrations of H_2, I_2, and HI in terms of a single unknown, x. Notice that the quantities in the last three lines of the table are concentrations. Here we define x as the quantity of H_2 or of I_2 that is consumed in the reaction. For this reason, $2x$ is the quantity of HI produced because the stoichiometric factor is (2 mol HI per 1 mol H_2).

Equation	H_2	+	I_2	$\rightleftharpoons$ 2 HI
Initial moles	1.00		1.00	0
Initial concentrations	$\frac{1.00\text{ mol}}{0.500\text{ L}}$ = 2.00 M		$\frac{1.00\text{ mol}}{0.500\text{ L}}$ = 2.00 M	0
Change in concentrations	$-x$		$-x$	$+2x$
Equilibrium concentrations	$2.00 - x$		$2.00 - x$	$2x$

Now the equilibrium concentrations can be substituted into the equilibrium expression.

$$K_c = 55.64 = \frac{[HI]^2}{[H_2][I_2]} = \frac{(2x)^2}{(2.00 - x)(2.00 - x)} = \frac{(2x)^2}{(2.00 - x)^2}$$

In this case, the unknown quantity x can be found by taking the square root of both sides of the equation,

$$\sqrt{K_c} = 7.459 = \frac{2x}{2.00 - x}$$

and then solving for x.

$$7.459(2.00 - x) = 2x$$
$$14.9 - 7.459x = 2x$$
$$14.9 = 9.459x$$
$$x = 1.58$$

With x known, we can now solve for the equilibrium concentrations of the reactants and products.

$$[H_2] = [I_2] = 2.00 - x = 0.42\text{ M}$$

$$[HI] = 2x = 3.16\text{ M}$$

It is always wise to verify these values by substituting them back into the equilibrium expression to see if your calculated K_c agrees with the one given in the problem. In this case $(3.16)^2/(0.42)^2 = 57$. The slight discrepancy with the given value, $K_c = 55.64$, is due to the fact that we know $[H_2]$ and $[I_2]$ to only two significant figures.

EXAMPLE 16.8 *Calculating a Concentration from an Equilibrium Constant*

The reaction

$$N_2(g) + O_2(g) \rightleftharpoons 2\text{ NO}(g)$$

contributes to air pollution whenever a fuel is burned in air at a high temperature, as in a gasoline engine. At 1500 K, $K_c = 1.0 \times 10^{-5}$. A sample of air is heated to 1500 K. Before any reaction occurs, $[N_2] = 0.80$ mol/L and $[O_2] = 0.20$ mol/L. Calculate the equilibrium concentration of NO.

Solution As in previous problems, we write the equilibrium expression and then set up a table of equilibrium concentrations.

$$K_c = \frac{[NO]^2}{[N_2][O_2]} = 1.0 \times 10^{-5}$$

Equation	N_2	+	O_2	$\rightleftharpoons$ 2 NO
Initial concentrations	0.80		0.20	0
Change in concentration	$-x$		$-x$	$+2x$
Equilibrium concentrations	$0.80 - x$		$0.20 - x$	$2x$

Next, the equilibrium concentrations are substituted into the equilibrium constant expression.

$$1.0 \times 10^{-5} = \frac{(2x)^2}{(0.80 - x)(0.20 - x)}$$

This quadratic equation can be rearranged into the standard form ($ax^2 + bx + c = 0$) (see Appendix A).

$$(1.0 \times 10^{-5})(0.80 - x)(0.20 - x) = 4x^2$$

$$(1.0 \times 10^{-5})(0.16 - 1.00x + x^2) = 4x^2$$

$$(4 - 1.0 \times 10^{-5})x^2 + 1.0 \times 10^{-5}\,x - 0.16 \times 10^{-5} = 0$$

$$ax^2 \quad + \quad bx \quad + \quad c \quad = 0$$

We see the rearranged equation is of the general form $ax^2 + bx + c = 0$, where

$$a = (4 - 1.0 \times 10^{-5}) \qquad b = 1.0 \times 10^{-5} \qquad c = -0.16 \times 10^{-5}$$

Such equations can be solved by the quadratic formula given in Appendix A. Using this, you will find two roots to this equation:

$$x = 6.3 \times 10^{-4} \quad \text{or} \quad x = -6.3 \times 10^{-4}$$

Because x stands for the quantity of N_2 lost on proceeding to equilibrium, and $2x$ is the quantity of NO formed, the negative root is physically meaningless. Thus, we choose the positive root and use this to calculate the equilibrium concentrations of the reactants and products.

$$[N_2] = 0.80 - 6.3 \times 10^{-4} \approx 0.80 \text{ M}$$

$$[O_2] = 0.20 - 6.3 \times 10^{-4} \approx 0.20 \text{ M}$$

$$[NO] = 2x = 1.3 \times 10^{-3} \text{ M}$$

This result brings us to an important point. Notice in this example that the quantity of NO formed is very small, as are the quantities of N_2 and O_2 consumed. Indeed, the concentrations of N_2 and O_2 have not changed (to two significant figures) on proceeding to equilibrium. In situations in which the change in the reactant concentration is very small, equilibrium calculations can be simplified greatly by making an assumption. Here we assume that x is very small relative to 0.80 or 0.20, which means that $0.80 - x \approx 0.80$ and $0.20 - x \approx 0.20$. The equilibrium expression therefore becomes

$$1.0 \times 10^{-5} = \frac{(2x)^2}{(0.80)(0.20)}$$

$$1.6 \times 10^{-6} = 4x^2$$

$$x = 6.3 \times 10^{-4}$$

The value of x obtained using the assumption is the same as from the quadratic equation.

How do you know when you can assume that the unknown quantity x is so small that you can write a simplified equation that does not require the quadratic formula? In most equilibrium calculations, a quantity (x) may be ignored if it is less than 5% of the smallest quantity initially present (here 0.20). In this example, x from the approximate equation was 6.3×10^{-4}, and it is $[(6.3 \times 10^{-4})/0.20]100\% = 0.32\%$ of the smallest quantity initially present. In general, when solving equilibrium problems of this type you (a) assume that the unknown (x) is small and solve the simplified equation. Then (b) compare it with the smallest quantity available. If the result is less than 5% of the smallest quantity, then you need not solve the full equation using the quadratic formula.

For a further discussion of such calculations, see Section 17.7 and Appendix A.

The 5% figure chosen here is somewhat arbitrary. It is chosen because it is close to the accuracy of the data used in the problems in this book. See Appendix A.

EXERCISE 16.9 *Calculating a Concentration from an Equilibrium Constant*

At 1000 K, $K_c = 33$ for the reaction

$$H_2(g) + I_2(g) \rightleftharpoons 2\,HI(g)$$

H_2 and I_2 are initially present at equal concentrations, 6.00×10^{-3} mol/L for each. Find the concentration of each reactant and product at equilibrium. ■

EXERCISE 16.10 *Calculating a Concentration from an Equilibrium Constant*

Graphite and carbon dioxide are kept at constant volume at 1000 K until the reaction

$$C(\text{graphite}) + CO_2(g) \rightleftharpoons 2\,CO(g)$$

comes to equilibrium. At this temperature, $K_c = 0.021$. The initial concentration of CO_2 is 0.012 mol/L. Calculate the equilibrium concentration of CO. ■

PROBLEM-SOLVING TIPS AND IDEAS

16.2 How to Assign the Unknown *x*

In Example 16.7 we set up the equilibrium table shown here.

$$H_2(g) + I_2(g) \rightleftharpoons 2\,HI(g)$$

Equation	H_2	+	I_2	$\rightleftharpoons$ 2 HI
Initial concentrations	2.00 M		2.00 M	0
Change in concentrations	$-x$		$-x$	$+2x$
Equilibrium concentrations	$2.00 - x$		$2.00 - x$	$2x$

There we assigned the unknown x to the quantity of H_2 and I_2 that is consumed on proceeding to equilibrium. We could also have made the unknown quantity of HI formed at equilibrium and given this the symbol y. The table would then have been as shown.

Equation	H_2	+	I_2	$\rightleftarrows$ 2 HI
Initial concentrations	2.00 M		2.00 M	0
Change in concentrations	$\frac{-y}{2}$		$\frac{-y}{2}$	$+y$
Equilibrium concentrations	$2.00 - \frac{y}{2}$		$2.00 - \frac{y}{2}$	y

These new values can be substituted into the usual equilibrium expression.

$$K_c = 55.64 = \frac{[HI]^2}{[H_2][I_2]} = \frac{(y)^2}{(2.00 - y/2)(2.00 - y/2)}$$

Solving for y, you find [HI] = y = 3.16, exactly the same result achieved in Example 16.7 (where [HI] = $2x$). We conclude that the unknown may be assigned in any of several ways when solving a problem. The important point is to define clearly the meaning of the unknown. ■

16.6 DISTURBING A CHEMICAL EQUILIBRIUM: Le Chatelier's Principle

There are three common ways in which a chemical reaction at equilibrium may be disturbed: (1) a change in temperature, (2) a change in the concentration of a reactant or product, and (3) a change in volume (Table 16.2). If you try to raise the temperature of the system (by adding heat energy), a chemical reaction occurs that acts to cool the system (by using heat energy). If additional reactant or product is added to a reaction at equilibrium, the system responds by using up some of what has been added. In a system in which one or more of the reactants or products is a gas, decreasing the available volume leads to a pressure increase; the system reacts so as to decrease the pressure. These outcomes can be predicted by **Le Chatelier's principle:** A change in any of the factors that determine the equilibrium conditions of a system causes the system to change in such a manner as to *reduce* or *counteract* the effect of the change.

Le Chatelier's principle was applied to the solubility of substances in Section 14.2.

Effect of Temperature Change on Equilibria

Temperature effects on phase equilibria and on the solubility of solids in water and other solvents were described in Chapters 13 and 14. For example, you found that an increase in temperature leads to an increase in the vapor pressure of a liquid (Section 13.3) and to an increase in the solubility of most salts (Section 14.2). As in the case of the solubility of salts, you can make a qualitative prediction about the effect of a temperature change on the equilibrium position in a chemical reaction if you know whether the reaction is exothermic or endothermic. As an example, consider the endothermic reaction of N_2 and O_2 to give NO.

$$N_2(g) + O_2(g) \rightleftharpoons 2\,NO(g) \qquad \Delta H^\circ_{rxn} = +180.5\text{ kJ}$$

$$K_c = \frac{[NO]^2}{[N_2][O_2]} \qquad K = 4.5 \times 10^{-31} \text{ at } 298\text{ K}$$

$$K = 6.7 \times 10^{-10} \text{ at } 900\text{ K}$$

$$K = 1.7 \times 10^{-3} \text{ at } 2300\text{ K}$$

We are surrounded by N_2 and O_2, but you know that they do not react appreciably at room temperature. The position of equilibrium lies almost completely to the left. However, if a mixture of N_2 and O_2 is heated above 700 °C, as in an automobile engine, the equilibrium shifts toward NO, and a greater concentration of NO can exist in equilibrium with reactants. The preceding experimental equilibrium constants show that [NO] increases and $[N_2]$ and $[O_2]$ decrease as the temperature increases at equilibrium.

To see why the value of *K* can change with temperature, consider the N_2/O_2 reaction further. The enthalpy change for the reaction is +180.5 kJ, so we might imagine heat as a "reactant." Le Chatelier's principle informs us that input of energy (as heat) causes the equilibrium to shift in a direction that counteracts this input. The way to counteract the energy input here is to use up some of the added heat by consuming N_2 and O_2 and producing more NO. An increase in temperature must therefore be accompanied by increased production of NO and consumption of N_2 and O_2. Because this raises the value of the numerator in the *K* expression, and lowers the denominator, *K* must also increase in value.

A more quantitative analysis of temperature effects on equilibria can be made using thermodynamics. See Chapter 20.

K does not change unless the temperature changes.

TABLE 16.2 Effects of Disturbances on Equilibrium and *K*

Disturbance	Change as Mixture Returns to Equilibrium	Effect on Equilibrium	Effect on *K*
Addition of reactant	Some of added reactant is consumed	Shift to right	No change
Addition of product	Some of added product is consumed	Shift to left	No change
Decrease in volume, increase in pressure	Pressure decreases	Shift toward fewer gas molecules	No change
Increase in volume, decrease in pressure	Pressure increases	Shift toward more gas molecules	No change
Rise in temperature	Heat energy is consumed	Shift in the endothermic direction	Change
Drop in temperature	Heat energy is generated	Shift in exothermic direction	Change

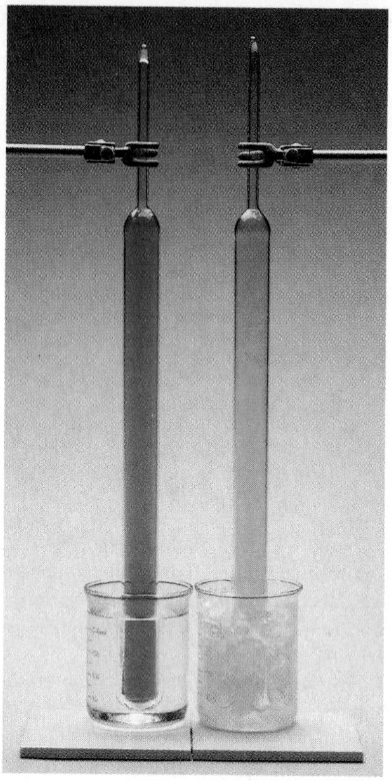

Figure 16.7 Effect of temperature on the N_2O_4/NO_2 equilibrium. The tubes in the photograph both contain a mixture of NO_2 and N_2O_4. As predicted by Le Chatelier's principle, the equilibrium favors colorless N_2O_4 at lower temperatures. This is clearly seen in the tube at the right, where the gas in the ice bath at 0 °C is only slightly brown because there is only a small partial pressure of the brown gas NO_2. At 50 °C (the tube at the left), the equilibrium is shifted toward NO_2 shown by the dark brown color. (Marna G. Clarke)

Two molecules of the brown gas NO_2 combine to form colorless N_2O_4, and an equilibrium between these compounds is readily achieved in a closed system (Figure 16.7).

$$2\ NO_2(g) \rightleftharpoons N_2O_4(g) \qquad \Delta H° = -57.2\ kJ$$

$$K_c = \frac{[N_2O_4]}{[NO_2]^2} \qquad K_c = 170 \text{ at } 298\ K$$

$$K_c = 1300 \text{ at } 273\ K$$

Here the reaction is *exo*thermic, so we might imagine heat as a reaction "product." By lowering the temperature of the reaction, as in Figure 16.7, some heat is removed. According to Le Chatelier's principle, this consumption of heat can be counteracted if the reaction produces more heat by combination of NO_2 to give more N_2O_4. Thus, the equilibrium concentration of NO_2 declines, that of N_2O_4 increases, and the values of K become larger as the temperature declines.

Based on the examples in this section, it can be stated that *increasing* the temperature of a system at equilibrium causes a reaction in the direction that results in *absorption* of heat energy; *decreasing* the temperature causes a reaction in the direction that results in *evolution* of heat energy.

EXERCISE 16.11 ***Le Chatelier's Principle***

Consider the effect of temperature changes on the following equilibria.

1. A mixture of three gases is in equilibrium.

$$2\ NOCl(g) \rightleftharpoons 2\ NO(g) + Cl_2(g) \qquad \Delta H°_{rxn} = +77.1\ kJ$$

Does the concentration of NOCl increase or decrease at equilibrium as the temperature of the system is increased?

2. Does the concentration of SO_3 increase or decrease when the temperature increases?

$$2\ SO_2(g) + O_2(g) \rightleftharpoons 2\ SO_3(g) \qquad \Delta H°_{rxn} = -198\ kJ$$

■

Effect of the Addition or Removal of a Reagent

If the concentration of a reactant or product is changed from its equilibrium value at a *given temperature,* the reaction *must* shift to a new equilibrium position for which the reaction quotient is still equal to K. To illustrate this phenomenon, we return to the case of the butane/isobutane equilibrium and Figure 16.8.

$$\text{butane} \rightleftharpoons \text{isobutane}$$

$$K_c = \frac{[\text{isobutane}]}{[\text{butane}]} = 2.50 \text{ at } 25\ °C$$

Suppose the equilibrium concentration of isobutane in a 1.0-L flask is 1.25 M and that of butane is 0.50 M. Now add 1.50 mol of butane to the flask. What happens to the concentrations of butane and isobutane as the system moves to reestablish equilibrium? Le Chatelier's principle allows us to make a qualitative prediction. In this case, after adding butane, more of this compound is now present than can be in equilibrium with 1.25 M isobutane. That is, the reaction quotient Q is now smaller than K.

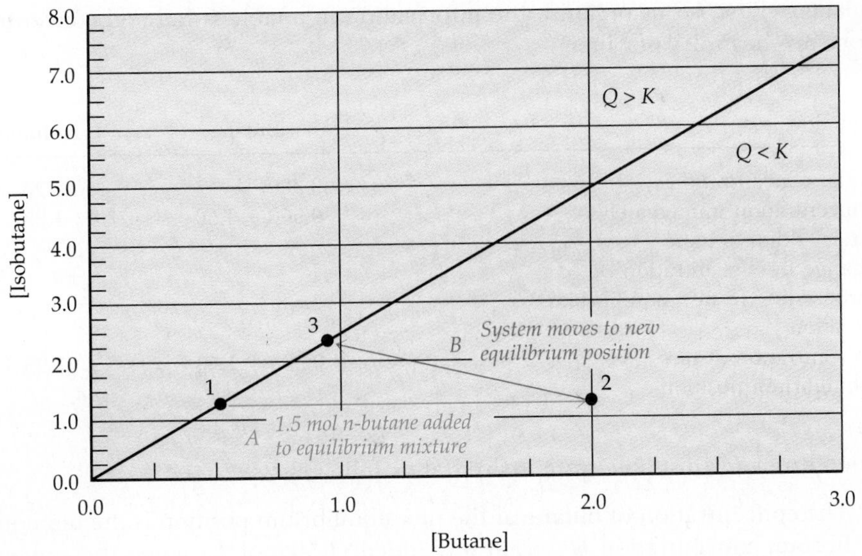

Figure 16.8 The effect of adding a reagent to a system at equilibrium. The butane/isobutane system was at equilibrium when [butane] = 0.50 M and [isobutane] = 1.25 M (point 1). If 1.50 mol/L of butane is added, the system moves to point 2 ($Q < K$). Equilibrium is restored by moving along line B to a new equilibrium position (point 3) where [butane] = 0.93 M and [isobutane] = 2.32 M. Line B has a slope of 1 because, in restoring equilibrium, each mole of butane consumed produces a mole of isobutane.

After adding excess butane to an equilibrium mixture

$$Q = \frac{[\text{isobutane}]}{[\text{butane}]} = \frac{1.25 \text{ mol/L present}}{0.50 \text{ mol/L present} + 1.50 \text{ mol/L added}} < K_c(=2.50)$$

As you learned in Section 16.3, when $Q < K$, the system is not at equilibrium. The equilibrium can be reestablished, however, if the concentration of the product increases and the concentration of the reactant decreases. In Figure 16.8, we moved from the equilibrium line, point 1, to the region where $Q < K$. To return to equilibrium, more isobutane is formed by consuming butane, 1 mol/L of butane being consumed for every mole per liter of isobutane being formed. As explained in the caption to Figure 16.8, and in Example 16.9 (to follow), equilibrium is reestablished when Q again equals 2.5 (at point 3), but now the concentrations of butane and isobutane are different from those before adding excess butane.

EXAMPLE 16.9 *The Effect of Concentration Change on an Equilibrium*

Here we wish to work out an algebraic solution to the problem of disturbing the butane/isobutane equilibrium. Assume equilibrium has been established in a 1.00-L flask with [butane] = 0.500 mol/L and [isobutane] = 1.25 mol/L. Then 1.50 mol/L of butane is added. What are the new equilibrium concentrations of butane and isobutane?

Solution First, let us organize the information in a table (where all concentrations are in moles per liter).

Equation	Butane	$\rightleftharpoons$ Isobutane
Initial concentration	0.500	1.25
Concentration immediately after adding butane	0.500 + 1.50	1.25
Change in concentration on proceeding to new equilibrium position	$-x$	$+x$
Concentration at new equilibrium position	$0.500 + 1.50 - x$	$1.25 + x$

The entries in this table were arrived at as follows:

1. The concentration of butane at the new equilibrium position is the old equilibrium concentration *plus* what was added (1.50 mol/L) *minus* the concentration of butane that must be converted to isobutane to reestablish equilibrium. The quantity of butane that is converted is unknown as yet and so is designated as x.
2. The quantity of isobutane at the new equilibrium position is the quantity that was already present (1.25 mol/L) plus the quantity formed (x mol/L) on proceeding to the new equilibrium position.

Having defined [butane] and [isobutane] at the new equilibrium position, and remembering that K is a constant (=2.50) throughout, we can write

$$K_c = \frac{[\text{isobutane}]_{\text{new}}}{[\text{butane}]_{\text{new}}} = 2.50 = \frac{1.25 + x}{0.500 + 1.50 - x}$$

$$2.50\ (0.500 + 1.50 - x) = 1.25 + x$$

$$3.75 = 3.50x$$

$$x = 1.07 \text{ mol/L}$$

We now know that the new equilibrium position is established at

$$[\text{butane}] = 0.500 + 1.50 - x = 0.93 \text{ mol/L}$$

$$[\text{isobutane}] = 1.25 + x = 2.32 \text{ mol/L}$$

as confirmed by the graphical analysis in Figure 16.8. Be sure to verify that [isobutane]/[butane] = 2.32/0.93 = 2.5.

EXERCISE 16.12 ***The Effect of Concentration Change on an Equilibrium***

Equilibrium exists between butane and isobutane when [butane] = 0.20 M and [isobutane] = 0.50 M. What are the equilibrium concentrations of butane and isobutane if 2.00 mol/L of isobutane is added to the original mixture? First try to estimate the new equilibrium concentrations from Figure 16.8 and then calculate the exact values. ■

The Effect of Volume Change on Gas-Phase Equilibria

For a reaction that involves gases, what happens to equilibrium concentrations or pressures if the size of the container is changed? (This occurs, for example, when fuel and air are compressed in an automobile engine.) To answer this question, recall that concentrations are expressed in moles per liter. If the volume of a gas changes, the concentration therefore must also change, and the equilibrium concentrations can change. As an example, once again consider the equilibrium

$$\underset{\text{brown gas}}{2\ NO_2(g)} \rightleftharpoons \underset{\text{colorless gas}}{N_2O_4(g)} \qquad K_c = \frac{[N_2O_4]}{[NO_2]^2} = 170 \text{ at } 298 \text{ K}$$

What happens to this equilibrium if the volume of the flask holding the gases is suddenly halved? Because the concentration of a gas increases as the volume available to the gas decreases, the immediate result is that the concentrations of both gases double. This means, however, that the system is no longer at equilibrium because the quotient $[N_2O_4]/[NO_2]^2$ cannot be equal to 170. For example, assume equilibrium is established when $[N_2O_4]$ is 0.0280 mol/L and $[NO_2]$ is 0.0128 mol/L. If both concentrations double when the volume is halved, $[N_2O_4]$ is now 0.0560 mol/L and $[NO_2]$ is 0.0256 mol/L. The reaction quotient Q under these circumstances is $(0.0560)/(0.0256)^2 = 85.5$, a value clearly less than K. Because Q is less than K, the quantity of product must therefore increase at the expense of the reactants, and the equilibrium shifts in favor of N_2O_4.

$$2\ NO_2(g) \rightleftharpoons N_2O_4(g)$$

$$\xrightarrow[\text{Equilibrium shifts right}]{\text{Decrease volume of container}}$$

This means that one molecule of N_2O_4 is formed by consuming two molecules of NO_2. The concentration of NO_2 decreases twice as fast as that of N_2O_4 increases until the reaction quotient, $[N_2O_4]/[NO_2]^2$, is once again equal to K.

The conclusions for the NO_2/N_2O_4 equilibrium can be generalized. For any reaction involving gases, the stress of a volume *decrease* (a pressure increase) is counterbalanced by a shift in the equilibrium to the side of the reaction with the *fewer* number of molecules of gases. For a volume increase (a pressure decrease), the opposite situation results: The equilibrium shifts to the side of the reaction with the greater number of molecules of gases.

EXERCISE 16.13 ***Effect of Concentration and Volume Changes on Equilibria***

The formation of ammonia from its elements is an important industrial process.

$$N_2(g) + 3\ H_2(g) \rightleftharpoons 2\ NH_3(g)$$

1. Does the equilibrium shift to the left or the right when extra H_2 is added? When extra NH_3 is added?
2. What is the effect on the position of the equilibrium when the volume of the system is increased? Does the equilibrium shift to the left or to the right, or is the system unchanged? ■

CURRENT ISSUES IN CHEMISTRY

Ammonia Production

In 1898 William Ramsay, the discoverer of the noble gases, pointed out a potential catastrophe for the world: a looming shortage of nitrogen-containing fertilizers. Nitrogen was usually obtained from naturally occurring salts such as $NaNO_3$ or from nitrogen-containing compounds in bird droppings. Ramsay predicted, however, that these sources would be depleted and, because of the importance of fertilizer in growing food, there would be a worldwide disaster by the middle of the 20th century. That this has not occurred is due to the work

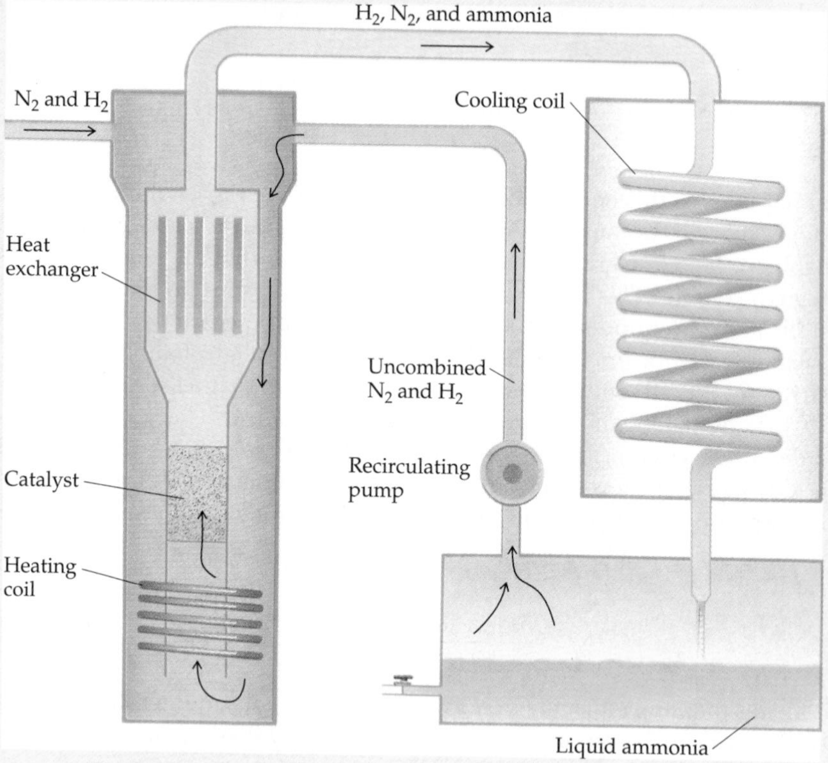

The Haber process for the synthesis of ammonia. A mixture of N_2 and H_2 gas is pumped over the catalytic surface to produce ammonia. The NH_3 is collected as a liquid (bp, −33 °C), and the uncombined N_2 and H_2 are recycled in the catalytic chamber.

16.7 EQUILIBRIUM, KINETICS, AND REACTION MECHANISMS

When describing the dynamic nature of the equilibrium involving aqueous Fe^{3+} and SCN^- (Section 16.1), we noted that equilibrium is established when the rate at which the product forms is equal to the rate at which the product reverts to reactants. By exploring this idea further we shall have a better insight into chemical equilibrium and will also be able to apply these ideas to the mechanisms of chemical reactions.

Reaction Rates and Equilibrium

Figure 16.9 shows the results of experiments done more than 100 years ago (!) on the reaction

$$H_2(g) + I_2(g) \rightleftharpoons 2\ HI(g)$$

of the German chemist Fritz Haber (1868–1934). Haber received the Nobel Prize in 1918 for working out a method of synthesizing ammonia directly from its elements.

$$N_2(g) + 3\ H_2(g) \rightleftharpoons 2\ NH_3(g)$$

At 25 °C:

$$K_c \text{ (calculated)} = 3.5 \times 10^8, \quad \Delta H^\circ_{rxn} = -92.2 \text{ kJ}$$

At 450 °C:

$$K_c \text{ (experimental)} = 0.16, \quad \Delta H^\circ_{rxn} = -111.3 \text{ kJ}$$

Ammonia is now one of the most important compounds in commerce. It is not only a widely used fertilizer but also a base, and so is often in the "top five" chemicals produced in the United States. In 1994, about 16 billion kilograms were produced at a cost of about $150 per ton.

The manufacture of ammonia is a good illustration of the principles of chemical equilibria and kinetics.

- The reaction is strongly exothermic and so is predicted to be product-favored ($K > 1$ at 25 °C). The reaction is slow at this temperature, however, so it is carried out at a higher temperature to increase the reaction rate.
- Although the reaction rate increases with temperature, the equilibrium constant declines, as predicted by Le Chatelier's principle for an exothermic reaction. Thus, for a given concentration of starting material, the equilibrium concentration of NH_3 is smaller at higher temperatures.
- To increase the equilibrium concentration of NH_3, the reaction is carried out at a higher pressure. This does not change the value of K, but an increase in pressure can be compensated by converting N_2 and H_2 to NH_3; two moles of NH_3 gas exert less pressure than a total of four moles of gaseous reactants ($N_2 + 3\ H_2$) in the same size container.
- A catalyst is used to increase the reaction rate further. An effective catalyst for the Haber process is Fe_3O_4 mixed with KOH, SiO_2, and Al_2O_3 (all inexpensive chemicals). Because the catalyst is not effective below 400 °C, the optimum temperature is about 450 °C.

The horizontal axis represents the time elapsed since the beginning of the experiment. The vertical axis represents the composition of the reaction mixture, from a mixture of H_2 and I_2 at the bottom to pure HI at the top. One series of experiments (the blue curve) begins with equal numbers of moles of H_2 and I_2 but no HI. As time passes the "forward" reaction, $H_2 + I_2 \rightarrow 2\ HI$, occurs; the concentrations of H_2 and I_2 decrease, and the concentration of HI increases. Because the rate of a reaction depends on the concentrations of the reactants, the rate of the forward reaction decreases.

The HI formed in the forward reaction can also decompose to H_2 and I_2 in the "reverse" reaction, $2\ HI \rightarrow H_2 + I_2$. The rate of the reverse reaction is initially zero, because no HI is present. As the concentration of HI increases, however, the rate of the reverse reaction increases. As the forward rate decreases, and the reverse rate increases, the net rate of production of HI decreases. This decrease is apparent in Figure 16.9 in the leveling off of the blue curve. When the rates of the forward and reverse reactions are equal, there is no longer any net change in concentrations. The reactions have not stopped; the molecules

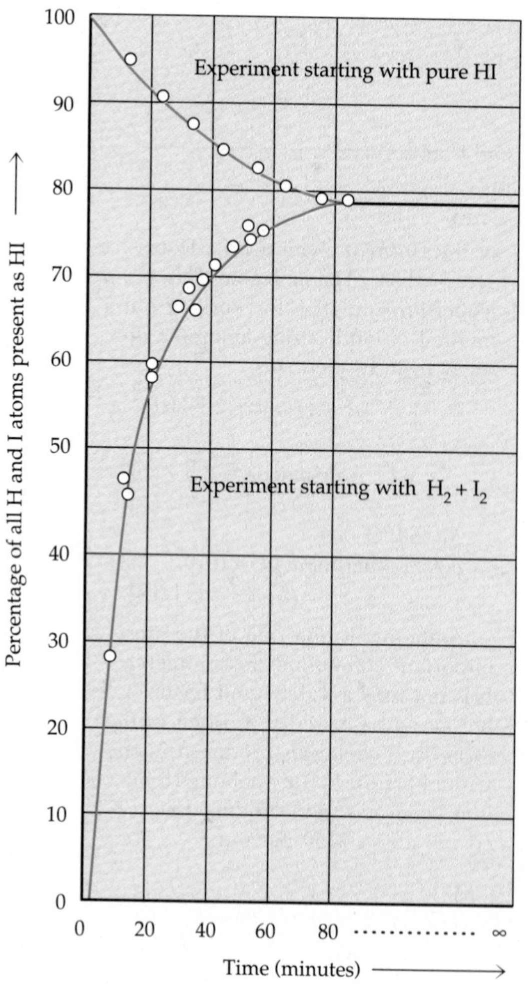

Figure 16.9 Experimental data for the reaction

$$H_2(g) + I_2(g) \rightleftharpoons 2\ HI(g)$$

at 445 °C. (Taken from M. Bodenstein: *Zeitschrift für Physikalische Chemie,* Vol. 13, p. 111, 1894.)

have not been told they no longer need to react. The forward and reverse reactions are still going on, but they occur at equal rates. The mixture is at equilibrium.

Another series of experiments (the red curve in Figure 16.9) starts with pure HI. Now the rate of decomposition of HI (2 HI → H_2 + I_2) decreases with time, whereas the rate of formation of HI from H_2 and I_2 increases with time. The curve levels off at the same equilibrium composition as in the experiments starting with H_2 and I_2.

In summary, the condition of equilibrium is recognized experimentally when concentrations of reactants and products cease to change. As long as concentrations are still changing with time, a mixture has not reached equilibrium.

Reaction Mechanisms and Chemical Equilibria

All the reaction mechanisms described in Section 15.6 had a rate-determining elementary step followed by one or more fast steps. Many other reactions occur, however, in which the reactants rapidly produce some intermediates that subsequently react in a slow step. One example of such a reaction is the oxidation of nitrogen monoxide by oxygen

$$2\ NO(g) + O_2(g) \longrightarrow 2\ NO_2(g) \qquad \text{rate} = k[NO]^2[O_2]$$

The reaction has an experimental rate law that is third order overall. This could imply a mechanism in which all three reacting molecules collide in a single, termolecular step. Experimental evidence, however, supports the existence of a reaction intermediate, and so the process is thought to occur in two elementary steps. Unlike the examples of mechanisms in Section 15.6, the second step here is rate-determining.

Elementary Step 1 *Fast, equilibrium:*

$$NO(g) + O_2(g) \underset{k_{-1}}{\overset{k_1}{\rightleftharpoons}} \underset{\text{intermediate}}{OONO(g)}$$

Elementary Step 2 *Slow, rate-determining:*

$$NO(g) + OONO(g) \xrightarrow{k_2} 2\ NO_2(g)$$

Overall reaction

$$2\ NO(g) + O_2(g) \longrightarrow 2\ NO_2(g)$$

This means the rate law predicted for the reaction should be

$$\text{Rate} = k_2[NO][OONO]$$

This rate law cannot be compared directly with the experimental rate law because it contains the concentration of a transient intermediate, OONO. The experimental rate law cannot include reaction intermediates whose concentrations cannot be measured. We therefore want to express the postulated rate law in a way that eliminates the intermediate. To do this, we look carefully at the equilibrium in the first, very rapid step.

At the beginning of the reaction NO and O_2 react rapidly and produce the intermediate OONO with a rate constant k_1.

$$\text{Rate of production of OONO} = k_1[NO][O_2]$$

Because the intermediate is consumed only very slowly in the second step, much of the OONO reverts to NO and O_2 before it can be consumed in the second step.

$$\text{Rate of reversion of OONO to NO and } O_2 = k_{-1}[OONO]$$

As NO and O_2 form OONO, their concentration drops, so the rate of the forward reaction decreases. At the same time, the concentration of OONO builds up, so the rate of the reverse reaction increases. Eventually, the rates of the forward and reverse reactions become the same, and the first elementary step reaches a state of equilibrium. Because the forward and reverse reactions in the first step are so much faster than the second elementary step, equilibrium is established before any significant amount of OONO is consumed by NO to give NO_2. The state of equilibrium for the first step remains throughout the lifetime of the overall reaction; it is broken only when the overall reaction is nearly complete.

Because equilibrium is established when the rates of the forward and reverse reactions are the same, this means

$$\text{Rate of forward reaction} = \text{rate of reverse reaction}$$
$$k_1[\text{NO}][\text{O}_2] = k_{-1}[\text{OONO}]$$

Rearranging this equation, we find

Notice that the equilibrium constant is the ratio of the rate constants for the forward (k_1) and reverse reactions (k_{-1}).

$$\frac{k_1}{k_{-1}} = \frac{[\text{OONO}]}{[\text{NO}][\text{O}_2]} = K_c$$

Of course the quotient $[\text{OONO}]/[\text{NO}][\text{O}_2]$ is just the equilibrium constant K_c for the first elementary step; this conclusion allows us to fit the mechanism for the NO/O_2 reaction to the experimentally observed rate law. From the equilibrium constant expression we know that

$$[\text{OONO}] = K_c[\text{NO}][\text{O}_2]$$

If $K_c[\text{NO}][\text{O}_2]$ is substituted for [ONOO] in the rate law for the rate-determining elementary step, we have

$$\text{Rate} = k_2[\text{NO}][\text{OONO}] = k_2[\text{NO}]\{K_c[\text{NO}][\text{O}_2]\}$$

which becomes

$$\text{Rate} = k_2K_c[\text{NO}]^2[\text{O}_2]$$

Because both k_2 and K_c are constants, their product is another constant k', and we have

$$\text{Rate} = k'[\text{NO}]^2[\text{O}_2]$$

This is exactly the rate law that was derived from experiment, so we believe we have postulated a reasonable mechanism. It is not the *only* possible mechanism, however. As described Example 16.10, at least one other mechanism is consistent with the same experimental rate law.

EXAMPLE 16.10 *Reaction Mechanisms Involving an Equilibrium Step*

The NO + O_2 reaction described in the text could also occur by the following mechanism.

Elementary Step 1 *Fast, equilibrium:*

$$\text{NO(g)} + \text{NO(g)} \underset{k_{-1}}{\overset{k_1}{\rightleftharpoons}} \underset{\text{intermediate}}{\text{N}_2\text{O}_2\text{(g)}}$$

Elementary Step 2 *Slow, rate-determining:*

$$\text{N}_2\text{O}_2\text{(g)} + \text{O}_2\text{(g)} \xrightarrow{k_2} 2\ \text{NO}_2\text{(g)}$$

Overall reaction $$2\ \text{NO(g)} + \text{O}_2\text{(g)} \longrightarrow 2\ \text{NO}_2\text{(g)}$$

Show that this mechanism also leads to the experimental rate law, Rate = $k[\text{NO}]^2[\text{O}_2]$.

A CLOSER LOOK — *Eating, Equilibrium, and Kinetics*

The rates of almost all biological reactions are influenced by catalysts—called enzymes—that often accelerate reactions by a factor of at least a million. Many experiments have established that the following mechanism can account for this.

Step 1:

$$\underset{\text{Enzyme}}{E} + \underset{\text{Substrate}}{S} \rightleftharpoons \underset{\text{Enzyme-substrate intermediate}}{ES}$$

Step 2:

$$ES \xrightarrow{k} E + \underset{\text{Product}}{P}$$

In the first step the enzyme E combines reversibly with a reactant (called a *substrate,* S) to form an intermediate, ES, which forms the product P in the second step. Step 2 is rate-determining for many enzymatic reactions. The rate law is therefore

$$\text{Rate} = k[\text{ES}]$$

Because an intermediate does not appear in an experimental rate law, we can apply the ideas in Section 16.7 to find the relation of the intermediate to other species in the reaction.

$$K_c \text{ for Step } 1 = \frac{[\text{ES}]}{[\text{E}][\text{S}]}$$

$$[\text{ES}] = K_c[\text{E}][\text{S}]$$

This leads to the following rate law:

$$\text{Rate} = kK_c[\text{E}][\text{S}]$$

If the substrate concentration is very high relative to the enzyme concentration, the equilibrium in Step 1 is then shifted almost completely to the right. Very little free enzyme is present; virtually all is tied up as the enzyme-substrate intermediate. This means that [ES] is essentially equal to the initial concentration of the enzyme $[\text{E}]_0$. This fact is important because the rate of reaction reaches its maximum value under these conditions and becomes independent of the substrate concentration. The order of the reaction is therefore zero in S, and the rate law becomes

$$\text{Rate}_{\text{max}} = kK_c[\text{E}]_0[\text{S}]^0 = k'[\text{E}]_0$$

That is, the maximum rate is proportional to the initial enzyme concentration.

This rate law applies to many biological reactions, such as the degradation of glucose. This reaction is an important energy source. At its normal concentrations in the blood, the enzyme that catalyzes this reaction, hexokinase, is saturated with the substrate, glucose, and the rate becomes independent of the glucose concentration as shown in the figure. Initially the rate rises rapidly, but at normal blood glucose concentrations (about 3×10^{-3} M), the rate becomes independent of the glucose concentration. Addition of glucose at this point does not increase the rate. As a consequence, our food consumption rates have little effect on the rates of most of our metabolic reactions.

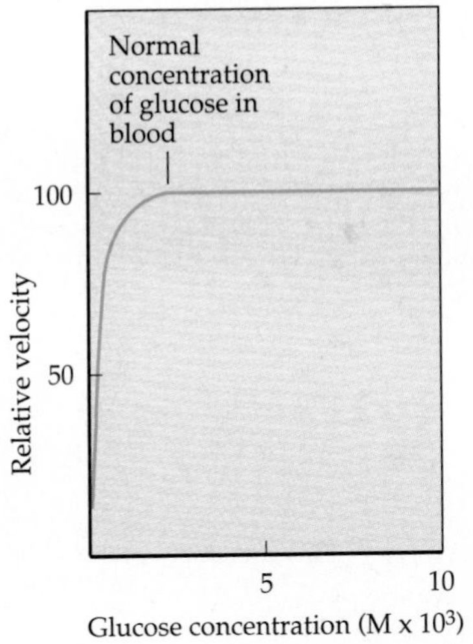

A plot of the glucose concentration versus the relative reaction rate in the enzyme-catalyzed degradation of glucose. The plot shows the reaction is zero-order with respect to glucose.

(Adapted from F. Brescia, J. Arents, H. Meislich, et al.: *General Chemistry,* p. 759. New York, Harcourt Brace Jovanovich, 1988.)

Solution The rate law for the rate-determining elementary step is

$$\text{Rate} = k_2[\text{N}_2\text{O}_2][\text{O}_2]$$

The compound N_2O_2 is an intermediate, which is consumed in the slow step and cannot appear in the final derived rate law. Therefore, we again need to express the postulated rate law in a way that eliminates the intermediate. As in the previous text, we recognize that $[\text{N}_2\text{O}_2]$ is related to [NO] by the equilibrium constant.

$$K_c = \frac{k_1}{k_{-1}} = \frac{[\text{N}_2\text{O}_2]}{[\text{NO}]^2}$$

If this is solved for $[N_2O_2]$, we have $[N_2O_2] = K_c[NO]^2$. When this is substituted into the derived rate law

$$\text{Rate} = k_2\{K_c[NO]^2\}[O_2]$$

the resulting equation is identical to the experimental rate law where $k_2K_c = k$.

The $NO + O_2$ reaction has an experimental rate law that suggests two possible mechanisms. The challenge is to decide which is correct. In this case further experimentation showed the presence of the species OONO as a short-lived intermediate, thus providing more evidence for the mechanism described previously in the text.

EXERCISE 16.14 ***Reaction Mechanisms Involving an Equilibrium Step***

The experimental rate equation for the gas-phase reaction $H_2(g) + I_2(g) \longrightarrow 2\,HI(g)$ is

$$\text{Rate} = k[H_2][I_2]$$

The formation of HI was considered for a long time to be typical of many reactions in which the mechanism is simply the combination of two molecules. It is now known, however, that the mechanism proceeds through a process involving atoms. One possible mechanism is

Elementary Step 1 *Fast, equilibrium:*

$$I_2(g) \rightleftharpoons 2\,I(g)$$

Elementary Step 2 *Slow:*

$$2\,I(g) + H_2(g) \longrightarrow 2\,HI(g)$$

Do these two elementary steps add up to the overall, stoichiometric equation? Does the mechanism predict the experimental rate law? Based on your knowledge of collision theory, is this a reasonable mechanism? ■

16.8 IS THERE LIFE AFTER EQUILIBRIUM?

The chemical reactions in a living plant have not achieved equilibrium. When equilibrium is attained, the plant literally dies. (C. D. Winters)

When the battery "runs down" in your calculator or car, you say that the battery is "dead." But it hasn't "died." As will be explained in Chapter 21, the reactions in the battery have just come to a state of equilibrium. It is for this reason that many of the chemical reactions that are interesting and important to us all occur under *nonequilibrium conditions:* the biochemical reactions in plants and animals, the production of elements in the stars, and the synthesis of useful materials. It is important for us to know how far away from equilibrium a system may be, how strong the tendency or drive is toward equilibrium and the rate of progress in that direction, how the system may be prevented from reaching equilibrium, and, perhaps, how to derive useful work from the system as it moves toward equilibrium. Although it is useful to know about the state of a system when it achieves equilibrium, an understanding of systems *not* at equilibrium is also necessary.

CHAPTER HIGHLIGHTS

Having studied this chapter, you should be able to

- Understand the nature and characteristics of the state of equilibrium: Chemical reactions are reversible; equilibria are dynamic; and the nature of the equilibrium state is the same, no matter what the direction of approach (Section 16.1).
- Write an equilibrium constant expression for any chemical reaction (Section 16.2). For the general reaction

$$a\text{A} + b\text{B} \rightleftharpoons c\text{C} + d\text{D}$$

the equilibrium concentrations of reactants and products are always related by Equation 16.1.

$$\text{Equilibrium constant} = K = \frac{\overbrace{[\text{C}]^c[\text{D}]^d}^{\text{product concentrations}}}{\underbrace{[\text{A}]^a[\text{B}]^b}_{\text{reactant concentrations}}} \quad (16.1)$$

- Recognize that the concentrations of solids and solvents (e.g., water) are not included in equilibrium constant expressions (Section 16.2).
- Appreciate the fact that equilibrium concentrations may be expressed in terms of reactant and product concentrations (expressed in moles per liter), and K is then designated as K_c. Alternatively, concentrations of gases may be represented by partial pressures, and K for such cases is designated K_p (Section 16.2).
- Know how K changes as different stoichiometric coefficients are used in a balanced equation or if the equation is reversed (Section 16.2).
- Know that, when two chemical equations are added to give a net chemical equation, the value of K for the net equation is the product of the values of K for the summed equations (Section 16.2).
- Recognize that a large value of K ($K \gg 1$) means the reaction is product-favored, and the product concentrations are greater than the reactant concentrations at equilibrium. A small value of K ($K \ll 1$) indicates a reactant-favored reaction in which the product concentrations are smaller than the reactant concentrations at equilibrium (Section 16.2).
- Apply the idea of the reaction quotient (Q) to decide if a reaction is at equilibrium ($Q = K$), or if there will be a net conversion of reactants to products ($Q < K$) or products to reactants ($Q > K$) to attain the state of equilibrium (Section 16.3).
- Calculate an equilibrium constant given the reactant and product concentrations at equilibrium (Section 16.4).
- Use equilibrium constants to calculate the concentration of a reactant or product at equilibrium (Section 16.5).
- Apply Le Chatelier's principle to predict the effect of a disturbance on a chemical equilibrium: a change in temperature, a change in concen-

tration, or a change in volume or pressure for a reaction involving gases (Section 16.6 and Table 16.2).

- Understand the effect on a reaction mechanism and on the kinetics of a reaction if one step in the mechanism involves a chemical equilibrium (Section 16.7).

STUDY QUESTIONS

Review Questions

1. Name three important features of the equilibrium condition.
2. Decide if each of the following statements is true or false. If false, change the wording to make it true.
   (a) The magnitude of the equilibrium constant is always independent of temperature.
   (b) When two chemical equations are added to give a net equation, the equilibrium constant for the net equation is the product of the equilibrium constants of the summed equations.
   (c) The equilibrium constant for a reaction has the same value as *K* for its reverse.
   (d) Only the concentration of CO_2 appears in the equilibrium constant expression for the reaction

$$CaCO_3(s) \rightleftharpoons CaO(s) + CO_2(g)$$

   (e) For the reaction

$$CaCO_3(s) \rightleftharpoons CaO(s) + CO_2(g)$$

   the value of *K* is the same whether the amount of CO_2 is expressed in moles per liter or as gas pressure.
3. Neither $PbCl_2$ nor PbF_2 is appreciably soluble in water. If solid $PbCl_2$ and solid PbF_2 are placed in equal amounts of water in separate beakers, in which beaker is the concentration of Pb^{2+} greater? Equilibrium constants for these solids dissolving in water are

$$PbCl_2(s) \rightleftharpoons Pb^{2+}(aq) + 2\ Cl^-(aq) \qquad K = 1.7 \times 10^{-5}$$

$$PbF_2(s) \rightleftharpoons Pb^{2+}(aq) + 2\ F^-(aq) \qquad K = 3.7 \times 10^{-8}$$

4. How does the reaction quotient for a reaction differ from the equilibrium constant for that reaction?
5. If the reaction quotient is smaller than the equilibrium constant for a reaction such as A $\rightleftharpoons$ B, does this mean that reactant A continues to disappear to form B, or does B form A, as the system moves to equilibrium?
6. The decomposition of calcium carbonate

$$CaCO_3(s) \rightleftharpoons CaO(s) + CO_2(g)$$

   is an endothermic process. Using Le Chatelier's principle, explain how increasing the temperature affects the equilibrium. If more $CaCO_3$ is added to the flask in which this equilibrium exists, how is the equilibrium affected? What if some additional CO_2 is placed in the flask?

Numerical and Other Questions

Writing and Manipulating Equilibrium Constant Expressions

7. Write equilibrium constant expressions for the following reactions. For gases, use either pressures or concentrations.
   (a) $2\ H_2O_2(g) \rightleftharpoons 2\ H_2O(g) + O_2(g)$
   (b) $PCl_3(g) + Cl_2(g) \rightleftharpoons PCl_5(g)$
   (c) $CO(g) + \frac{1}{2}\ O_2(g) \rightleftharpoons CO_2(g)$
   (d) $C(s) + CO_2(g) \rightleftharpoons 2\ CO(g)$
   (e) $FeO(s) + CO(g) \rightleftharpoons Fe(s) + CO_2(g)$
8. Write equilibrium constant expressions for the following reactions. For gases, use either pressures or concentrations.
   (a) $3\ O_2(g) \rightleftharpoons 2\ O_3(g)$
   (b) $SiH_4(g) + 2\ O_2(g) \rightleftharpoons SiO_2(g) + 2\ H_2O(g)$
   (c) $Ni(s) + 4\ CO(g) \rightleftharpoons Ni(CO)_4(g)$
   (d) $(NH_4)_2CO_3(s) \rightleftharpoons 2\ NH_3(g) + CO_2(g) + H_2O(g)$
   (e) $BaSO_4(s) \rightleftharpoons Ba^{2+}(aq) + SO_4^{2-}(aq)$
9. Consider the following equilibria involving $SO_2(g)$ and their corresponding equilibrium constants.

$$SO_2(g) + \tfrac{1}{2}\ O_2(g) \rightleftharpoons SO_3(g) \qquad K_1$$

$$2\ SO_3(g) \rightleftharpoons 2\ SO_2(g) + O_2(g) \qquad K_2$$

   Which of the following expressions relates K_1 to K_2?
   (a) $K_2 = K_1^2$
   (b) $K_2^2 = K_1$
   (c) $K_2 = K_1$
   (d) $K_2 = \frac{1}{K_1}$
   (e) $K_2 = \frac{1}{K_1^2}$
10. How is the equilibrium constant K_c for the reaction of hydrazine and chlorine trifluoride

$$N_2H_4(g) + \tfrac{4}{3}\ ClF_3(g) \rightleftharpoons 4\ HF(g) + N_2(g) + \tfrac{2}{3}\ Cl_2(g)$$

   related to K'_c for the reaction written in the following way?

$$3\ N_2H_4(g) + 4\ ClF_3(g) \rightleftharpoons 12\ HF(g) + 3\ N_2(g) + 2\ Cl_2(g)$$

   (a) $K_c = K'_c$
   (b) $K_c = \frac{1}{K'_c}$
   (c) $3K_c = K'_c$
   (d) $(K_c)^3 = K'_c$
   (e) $K_c = (K'_c)^3$

11. Calculate K_c for the reaction

$$Fe(s) + H_2O(g) \rightleftharpoons FeO(s) + H_2(g)$$

given the following information:

$$H_2O(g) + CO(g) \rightleftharpoons H_2(g) + CO_2(g) \qquad K_c = 1.6$$
$$Fe(s) + CO_2(g) \rightleftharpoons FeO(s) + CO(g) \qquad K_c = 1.5$$

12. Calculate K_c for the reaction

$$SnO_2(s) + 2\ CO(g) \rightleftharpoons Sn(s) + 2\ CO_2(g)$$

given the following information:

$$SnO_2(s) + 2\ H_2(g) \rightleftharpoons Sn(s) + 2\ H_2O(g) \qquad K_c = 8.12$$
$$H_2(g) + CO_2(g) \rightleftharpoons CO(g) + H_2O(g) \qquad K_c = 0.771$$

13. The equilibrium constant K_c for the reaction

$$CO_2(g) \rightleftharpoons CO(g) + \tfrac{1}{2}\ O_2(g)$$

is 6.66×10^{-12} at 1000 K. Calculate K_c for the reaction

$$2\ CO(g) + O_2(g) \rightleftharpoons 2\ CO_2(g)$$

14. The equilibrium constant K_c for the reaction

$$H_2(g) + Cl_2(g) \rightleftharpoons 2\ HCl(g)$$

at 500 K is 4.8×10^{10}. Calculate K_c for
(a) $\tfrac{1}{2}\ H_2(g) + \tfrac{1}{2}\ Cl_2(g) \rightleftharpoons HCl(g)$
(b) $HCl(g) \rightleftharpoons \tfrac{1}{2}\ H_2(g) + \tfrac{1}{2}\ Cl_2(g)$

The Reaction Quotient

15. A mixture of SO_2, O_2, and SO_3 at 1000 K contains the gases at the following concentrations: $[SO_2] = 5.0 \times 10^{-3}$ mol/L, $[O_2] = 1.9 \times 10^{-3}$ mol/L, and $[SO_3] = 6.9 \times 10^{-3}$ mol/L. Which way will the reaction

$$2\ SO_2(g) + O_2(g) \rightleftharpoons 2\ SO_3(g)$$

go to reach equilibrium? K_c for the reaction is 279.

16. $K_c = 5.6 \times 10^{-12}$ at 500 K for the reaction

$$I_2(g) \rightleftharpoons 2\ I(g)$$

A mixture kept at 500 K contains I_2 at a concentration of 0.020 mol/L and I at a concentration of 2.0×10^{-8} mol/L. Which way must the reaction go to reach equilibrium?

17. The equilibrium constant K_c for the reaction

$$2\ NOCl(g) \rightleftharpoons 2\ NO(g) + Cl_2(g)$$

is 3.9×10^{-3} at 300 °C. A mixture contains the gases at the following concentrations: $[NOCl] = 5.0 \times 10^{-3}$ mol/L, $[NO] = 2.5 \times 10^{-3}$ mol/L, and $[Cl_2] = 2.0 \times 10^{-3}$ mol/L. Is the reaction at equilibrium? If not, in which direction does the reaction move to come to equilibrium?

18. The reaction

$$2\ NO_2(g) \rightleftharpoons N_2O_4(g)$$

has an equilibrium constant, K_c, of 170 at 25 °C. If analysis of the system shows that 2.0×10^{-3} mol of NO_2 is present in a 10.-L flask along with 1.5×10^{-3} mol of N_2O_4, is the system at equilibrium? If it is not at equilibrium, does the concentration of NO_2 increase or decrease as the system proceeds to equilibrium?

Calculating Equilibrium Constants

19. A mixture of SO_2, O_2, and SO_3 at 1000 K contains the gases at the following concentrations: $[SO_2] = 3.77 \times 10^{-3}$ mol/L, $[O_2] = 4.30 \times 10^{-3}$ mol/L, and $[SO_3] = 4.13 \times 10^{-3}$ mol/L. Calculate the equilibrium constant K_c for the reaction

$$2\ SO_2(g) + O_2(g) \rightleftharpoons 2\ SO_3(g)$$

20. The reaction

$$PCl_5(g) \rightleftharpoons PCl_3(g) + Cl_2(g)$$

was examined at 250 °C. At equilibrium $[PCl_5] = 4.2 \times 10^{-5}$ mol/L, $[PCl_3] = 1.3 \times 10^{-2}$ mol/L, and $[Cl_2] = 3.9 \times 10^{-3}$ mol/L. Calculate K_c for the reaction.

21. Hydrogen and carbon dioxide react at a high temperature to give water and carbon monoxide.

$$H_2(g) + CO_2(g) \rightleftharpoons H_2O(g) + CO(g)$$

(a) Laboratory measurements at 986 °C show that there is 0.11 mol each of CO and water vapor and 0.087 mol each of H_2 and CO_2 at equilibrium in a 1.0-L container. Calculate the equilibrium constant for the reaction at 986 °C.
(b) If there is 0.050 mol each of H_2 and CO_2 in a 2.0-L container at equilibrium at 986 °C, what amounts of CO(g) and $H_2O(g)$, in moles, are present?

22. The reaction

$$C(s) + CO_2(g) \rightleftharpoons 2\ CO(g)$$

occurs at high temperatures. At 700 °C, a 2.0-L flask contains 0.10 mol of CO, 0.20 mol of CO_2, and 0.40 mol of C at equilibrium.
(a) Calculate K_c for the reaction at 700 °C.
(b) Calculate K_c for the reaction, also at 700 °C, if the amounts at equilibrium in the 2.0-L flask are 0.10 mol of CO, 0.20 mol of CO_2, and 0.80 mol of C.

23. At a very high temperature, water vapor is 10.% dissociated into $H_2(g)$ and $O_2(g)$. (That is, 10.% of the original water has been transformed into products, and 90.% remains.)

$$H_2O(g) \rightleftharpoons H_2(g) + \tfrac{1}{2}\ O_2(g)$$

Assuming a water concentration of 2.0 mol/L before dissociation, calculate the equilibrium constant K_c.

24. Two moles of hydrogen iodide is placed in a 1.0-L container at a certain temperature. The hydrogen iodide partially dissociates according to the equation

$$2\ HI(g) \rightleftharpoons H_2(g) + I_2(g)$$

If 20.% of the HI has dissociated at equilibrium, calculate K_c.

25. Three moles of pure SO_3 are placed in an 8.00-L flask at 1150 K. At equilibrium, 0.58 mol of O_2 has formed. Calculate K_c for the reaction at 1150 K.

$$2\ SO_3(g) \rightleftharpoons 2\ SO_2(g) + O_2(g)$$

26. At 600 K a mixture of CO and Cl_2 has the following concentrations: $[CO] = 0.0102$ mol/L and $[Cl_2] = 0.00609$ mol/L. When the reaction

$$CO(g) + Cl_2(g) \rightleftharpoons COCl_2(g)$$

comes to equilibrium, $[Cl_2] = 0.00301$ mol/L.

(a) Calculate the concentrations of CO and $COCl_2$ at equilibrium.

(b) Calculate K_c.

27. When solid ammonium carbamate sublimes, it dissociates completely into ammonia and carbon dioxide according to the equation

$$N_2H_6CO_2(s) \rightleftharpoons 2\ NH_3(g) + CO_2(g)$$

At 25 °C, experiment shows that the total pressure of the gases in equilibrium with the solid is 0.116 atm. What is the equilibrium constant K_p?

28. Ammonium iodide dissociates reversibly to ammonia and hydrogen iodide if the salt is heated to a sufficiently high temperature.

$$NH_4I(s) \rightleftharpoons NH_3(g) + HI(g)$$

Some ammonium iodide is placed in a flask, which is then heated to 400 °C. If the total pressure in the flask when equilibrium has been achieved is 705 mm Hg, what is the value of K_p (when partial pressures are in atmospheres)?

Using Equilibrium Constants in Calculations

29. The hydrocarbon C_4H_{10} can exist in two forms: butane and isobutane. The value of K_c for the interconversion of the two forms is 2.5 at 25 °C.

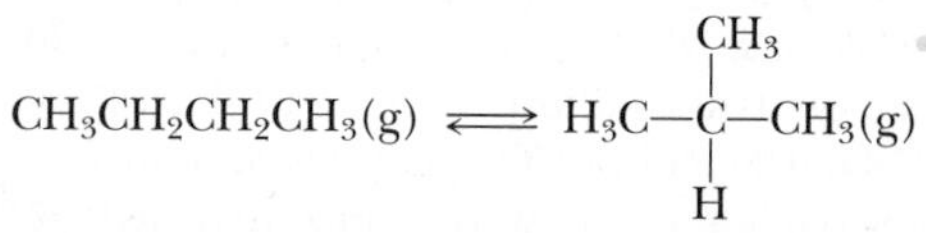

butane   isobutane

If you place 0.017 mol of butane in a 0.50-L flask at 25 °C and allow equilibrium to be established, what are the equilibrium concentrations of the two forms of butane?

30. Cyclohexane, C_6H_{12} (a hydrocarbon), can isomerize, or change, into methylcyclopentane, a compound of the same formula but with a different molecular structure

$$\underset{\text{cyclohexane}}{C_6H_{12}(g)} \rightleftharpoons \underset{\text{methylcyclopentane}}{C_5H_9CH_3(g)}$$

The equilibrium constant has been estimated to be 0.12 at 25 °C. If you had originally placed 0.045 mol of cyclohexane in a 2.8-L flask, what are the concentrations of cyclohexane and methylcyclopentane when equilibrium is established?

31. Carbonyl bromide, $COBr_2$, decomposes to CO and Br_2 with an equilibrium constant K_c of 0.190 at 73 °C.

$$COBr_2(g) \rightleftharpoons CO(g) + Br_2(g)$$

If 0.015 mol of $COBr_2$ is in a 2.5-L flask at equilibrium, what are the concentrations of CO and Br_2 at equilibrium?

32. The equilibrium constant for the reaction

$$Pb(\ell) + H_2O(g) \rightleftharpoons PbO(s) + H_2(g)$$

is 1.3×10^{-4} at 1000 K. A mixture of gases initially contains H_2O at 1.0 atm and H_2 at 1.0×10^{-3} atm at 1000 K. The mixture is kept in contact with molten lead and solid PbO at 1000 K until the reaction has come to equilibrium. What are the partial pressures of H_2O and H_2 at equilibrium? Which way did the reaction move to come to equilibrium?

33. The equilibrium constant K_c for the reaction

$$N_2O_4(g) \rightleftharpoons 2\ NO_2(g)$$

at 25 °C is 5.88×10^{-3}. Suppose 20.0 g of N_2O_4 is placed in a 5.00-L flask at 25 °C. Calculate (a) the number of moles of NO_2 present at equilibrium and (b) the percentage of the original N_2O_4 that is dissociated.

34. The equilibrium constant K_p for $N_2O_4 \rightleftharpoons 2\ NO_2(g)$ is 0.15 at 25 °C. If the pressure of N_2O_4 at equilibrium is 0.85 atm, what is the total pressure of the gas mixture ($N_2O_4 + NO_2$) at equilibrium?

35. The equilibrium constant K_c for the decomposition of $COBr_2$ is 0.190 at 73 °C.

$$COBr_2(g) \rightleftharpoons CO(g) + Br_2(g)$$

1.06 g of $COBr_2(g)$ is placed in a 8.0-L flask and heated to 73 °C.

(a) What are the concentrations of reactants and products when equilibrium is established?

(b) What is the total pressure in the flask at equilibrium at 73 °C?

36. At 450 °C, 3.60 mol of ammonia is placed in a 2.00-L vessel and allowed to decompose to the elements.

$$2\ NH_3(g) \rightleftharpoons N_2(g) + 3\ H_2(g)$$

If the experimental value of K_c is 6.3 for this reaction at this temperature, calculate the equilibrium concentration of each reagent.

37. The equilibrium constant K_c for the dissociation of iodine

$$I_2(g) \rightleftharpoons 2\ I(g)$$

is 3.76×10^{-3} at 1000 K. Suppose 1.00 mol of I_2 is placed in a 2.00-L flask at 1000 K. What are the concentrations of I_2 and I when the system comes to equilibrium?

38. The value of K_c for the dissociation of PCl_5

$$PCl_5(g) \rightleftharpoons PCl_3(g) + Cl_2(g)$$

is 33.3 at 760 K. What are the equilibrium concentrations of all compounds at 760 K if 1.50 g of PCl_5 and 15.0 g of PCl_3 are placed in a flask with a volume of 36.3 mL?

Disturbing a Chemical Equilibrium: Le Chatelier's Principle

39. The value of K_p for the following reaction is 0.16 at 25 °C. The enthalpy change for the reaction at standard conditions is +16.1 kJ.

$$2\ NOBr(g) \rightleftharpoons 2\ NO(g) + Br_2(g)$$

Predict the effect of the following changes on the position of the equilibrium; that is, state which way the equilibrium will shift (left, right, or no change) when each of the following changes is made:
(a) Adding more $Br_2(g)$
(b) Removing some NOBr(g)
(c) Decreasing the temperature
(d) Increasing the container volume

40. The oxidation of NO to NO_2

$$2\ NO(g) + O_2(g) \rightleftharpoons 2\ NO_2(g)$$

is exothermic. Predict the effect of the following changes on the position of the equilibrium; that is, state which way the equilibrium will shift (left, right, or no change) when each of the following changes is made:
(a) Adding more $O_2(g)$
(b) Adding more $NO_2(g)$
(c) Increasing the volume of the reaction flask
(d) Lowering the temperature

41. Consider the isomerization of butane with an equilibrium constant of $K_c = 2.5$.

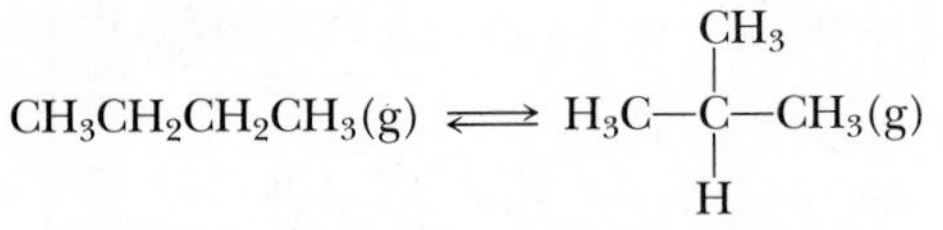

butane isobutane

The system is originally at equilibrium with [butane] = 1.0 M and [isobutane] = 2.5 M.
(a) If 0.50 mol/L of isobutane is suddenly added and then the system shifts to a new equilibrium position, what is the equilibrium concentration of each gas?
(b) If 0.50 mol/L of butane is added, and the system shifts to a new equilibrium position, what is the equilibrium concentration of each gas?

42. The value of K_c for the decomposition of ammonium hydrogen sulfide is 1.8×10^{-4} at 25 °C.

$$NH_4HS(s) \rightleftharpoons NH_3(g) + H_2S(g)$$

(a) When the pure salt decomposes in a flask, what are the equilibrium concentrations of NH_3 and H_2S?
(b) If NH_4HS is placed in a flask already containing 0.020 mol/L of NH_3 and then the system is allowed to come to equilibrium, what are the equilibrium concentrations of NH_3 and H_2S?

Equilibrium, Kinetics, and Reaction Mechanisms

43. Hydrogen and carbon monoxide react to give formaldehyde under certain conditions.

$$H_2(g) + CO(g) \longrightarrow HCHO(g)$$

The mechanism proposed for this reaction is

Step 1 *Fast, reversible:*

$$H_2 \rightleftharpoons 2\ H$$

Step 2 *Slow:*

$$H + CO \longrightarrow HCO$$

Step 3 *Fast:*

$$H + HCO \longrightarrow HCHO$$

What rate law is derived from this mechanism?

44. The experimental rate equation for the reaction

$$H_2(g) + I_2(g) \rightleftharpoons 2\ HI(g)$$

is Rate = $k[H_2][I_2]$. Does the following mechanism satisfy the experimental rate equation?

Step 1 *Fast, reversible:*

$$I_2 \rightleftharpoons 2\ I$$

Step 2 *Fast, reversible:*

$$I + H_2 \rightleftharpoons IH_2$$

Step 3 *Slow:*

$$IH_2 + I \longrightarrow 2\ HI$$

45. Several mechanisms have been proposed for the reduction of NO.

$$2\ H_2(g) + 2\ NO(g) \longrightarrow N_2(g) + 2\ H_2O(g)$$

The experimental rate law is Rate = $k[NO]^2[H_2]$. Indicate which of the following mechanisms is consistent with this rate law.

Mechanism 1

Step 1 *Slow:*

$$H_2 + NO \longrightarrow H_2O + N$$

Step 2 *Fast:*

$$N + NO \longrightarrow N_2 + O$$

Step 3 *Fast:*

$$O + H_2 \longrightarrow H_2O$$

Mechanism 2

Step 1 *Fast, reversible:*

$$2\ NO \rightleftharpoons N_2O_2$$

Step 2 *Slow:*

$$N_2O_2 + H_2 \longrightarrow H_2O + N_2O$$

Step 3 *Fast:*

$$N_2O + H_2 \longrightarrow N_2 + H_2O$$

46. A possible mechanism for the exothermic reaction

$$2\ NO(g) + O_2(g) \longrightarrow 2\ NO_2(g) \qquad \Delta H^\circ_{rxn} = -114.1\ kJ$$

is

Step 1 *Fast, reversible:*

$$NO + NO \rightleftharpoons N_2O_2$$

Step 2 *Slow (rate constant = k_2):*

$$N_2O_2 + O_2 \longrightarrow 2\ NO_2$$

(a) Derive the experimental rate law.
(b) Step 1 is an exothermic reaction. In contrast with the usual behavior, the rate of the reaction decreases with temperature. Offer an explanation for this observation based on this mechanism.

General Questions

47. The equilibrium constant K_c for the reaction

$$N_2(g) + O_2(g) \rightleftharpoons 2\ NO(g)$$

is 1.7×10^{-3} at 2300 K.
(a) What is K_p for this reaction?
(b) What is K_c for the reaction when written as

$$\tfrac{1}{2}\ N_2(g) + \tfrac{1}{2}\ O_2(g) \rightleftharpoons NO(g)$$

(c) What is K_c for the reaction

$$2\ NO(g) \rightleftharpoons N_2(g) + O_2(g)$$

48. If K_p for the formation of phosgene, Cl_2CO, is 6.5×10^{11} at 25 °C

$$CO(g) + Cl_2(g) \rightleftharpoons Cl_2CO(g)$$

what is the value of K_p for the dissociation of phosgene?

$$Cl_2CO(g) \rightleftharpoons CO(g) + Cl_2(g)$$

49. The equilibrium constant K_c for the reaction

$$N_2(g) + O_2(g) \rightleftharpoons 2\ NO(g)$$

is 1.7×10^{-3} at 2300 K. The gases in this reaction have the following partial pressures in a reaction vessel at 2300 K: $P(N_2) = 0.50$ atm, $P(O_2) = 0.25$ atm, and $P(NO) = 4.2 \times 10^{-3}$ atm. Is the system at equilibrium? If not, which way does the reaction proceed to equilibrium.

50. One mole of Br_2 is placed in a 1.00-L flask and heated to 1756 K, a temperature at which the halogen dissociates to atoms.

$$Br_2(g) \rightleftharpoons 2\ Br(g)$$

If Br_2 is 1.0% dissociated at this temperature, calculate K_c.

51. Calculate the equilibrium constant K_c at 25 °C for the reaction

$$2\ NOCl(g) \rightleftharpoons 2\ NO(g) + Cl_2(g)$$

using the following information. In one experiment 2.00 mol of NOCl is placed in a 1.00-L flask, and the concentration of NO after equilibrium is achieved is 0.66 mol/L.

52. Equal numbers of moles of H_2 gas and I_2 vapor are mixed in a flask and heated to 700 °C. The initial concentration of each gas is 0.0088 mol/L, and 78.6% of the I_2 is consumed when equilibrium is achieved according to the reaction

$$H_2(g) + I_2(g) \rightleftharpoons 2\ HI(g)$$

Calculate K_c for this reaction.

53. The total pressure for a mixture of N_2O_4 and NO_2 is 1.5 atm. Calculate the partial pressure of each gas in the mixture.

$$2\ NO_2(g) \rightleftharpoons N_2O_4(g) \qquad K_p = 6.75\ (\text{at } 25\ °C)$$

54. Ammonium hydrogen sulfide decomposes on heating.

$$NH_4HS(s) \rightleftharpoons NH_3(g) + H_2S(g)$$

"This is a lovely old song that tells of a young woman who leaves her cottage, and goes off to work. She arrives at her destination, and places some solid NH_4HS in a flask containing 0.50 atm of ammonia, and attempts to determine the pressures of ammonia and hydrogen sulfide when equilibrium is reached." (Sidney Harris)

If K_p is 0.11 at 25 °C (when the partial pressures are measured in atmospheres), what is the total pressure in the flask at equilibrium?

55. Two molecules of gaseous acetic acid can form a dimer through hydrogen bonds.

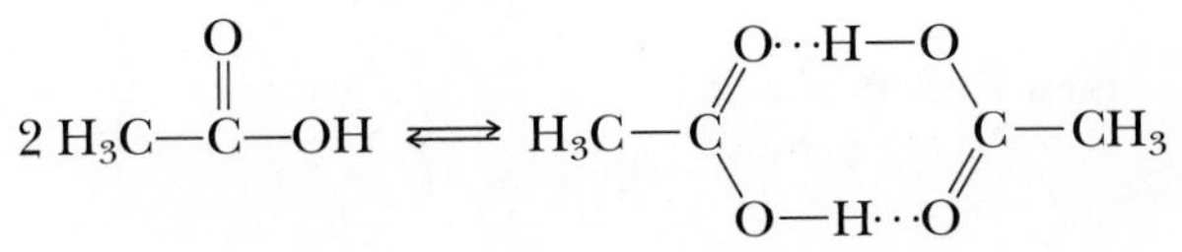

The equilibrium constant K_c at 25 °C has been determined to be 3.2×10^4. Assume that acetic acid is present initially at a concentration of 5.4×10^{-4} mol/L at 25 °C and that no dimer is present initially.

(a) What percentage of acetic acid is converted to the dimer?

(b) The energy of the hydrogen bonds is 132 kJ/mol. As the temperature goes up, in which direction does the equilibrium shift?

56. The equilibrium constant for the butane/isobutane isomerization reaction is 2.5 at 25 °C. (See Questions 29 and 41.) If 1.75 mol of butane and 1.25 mol of isobutane are mixed, is the system at equilibrium? If not, when it proceeds to equilibrium which reagent increases in concentration? Calculate the concentrations of the two compounds when the system reaches equilibrium.

57. The equilibrium constant K_c for the reaction

$$2\ SO_2(g) + O_2(g) \rightleftharpoons 2\ SO_3(g)$$

is 279 at 1000 K. Calculate K_p for the reaction.

58. At 250 °C $K_p = 0.039$ for the reaction

$$2\ NOCl(g) \rightleftharpoons 2\ NO(g) + Cl_2(g)$$

Calculate K_c for the reaction.

59. Zinc carbonate dissolves very poorly in water ($K_c = 1.5 \times 10^{-11}$).

$$ZnCO_3(s) \rightleftharpoons Zn^{2+}(aq) + CO_3^{2-}(aq)$$

If some solid $ZnCO_3$ is placed in water, what are the molar concentrations of Zn^{2+} and CO_3^{2-} when equilibrium is achieved?

60. At 2300 K the equilibrium constant for the formation of NO(g) is $K_c = 1.7 \times 10^{-3}$.

$$N_2(g) + O_2(g) \rightleftharpoons 2\ NO(g)$$

(a) If analysis shows that the concentrations of N_2 and O_2 are both 0.25 M, and that of NO is 0.0042 M, is the system at equilibrium?

(b) If the system is not at equilibrium, in which direction does the reaction proceed?

(c) When the system is at equilibrium, what are the equilibrium concentrations?

61. At 1800 K, oxygen dissociates very slightly into its atoms.

$$O_2(g) \rightleftharpoons 2\ O(g) \qquad K_p = 1.2 \times 10^{-10}$$

If you place 1.0 mol of O_2 in a 10.-L vessel and heat it to 1800 K, how many O atoms are present in the flask?

62. A reaction important in smog formation is

$$O_3(g) + NO(g) \rightleftharpoons O_2(g) + NO_2(g) \qquad K_c = 6.0 \times 10^{34}$$

(a) If the initial concentrations are $[O_3] = 1.0 \times 10^{-6}$ M, $[NO] = 1.0 \times 10^{-5}$ M, $[NO_2] = 2.5 \times 10^{-4}$ M, and $[O_2] = 8.2 \times 10^{-3}$ M, is the system at equilibrium? If not, in which direction does the reaction proceed?

(b) If the temperature is increased, as on a very warm day, do the concentrations of the products increase or decrease? (*Hint:* You may have to calculate the enthalpy change for the reaction to discover if it is exothermic or endothermic.)

63. The equilibrium reaction $N_2O_4(g) \rightleftharpoons 2\ NO_2(g)$ has been thoroughly studied (see Figure 16.7). If the total pressure in a flask containing NO_2 and N_2O_4 gas at 25 °C is 1.50 atm, and the value of K_p at this temperature is 0.148, what fraction of the N_2O_4 has dissociated to NO_2? What happens to the fraction dissociated if the volume of the container is increased so that the total equilibrium pressure falls to 1.00 atm?

64. The equilibrium constants for the dissociation of three complexes of trimethylborane, $L:B(CH_3)_3$, have been determined.

$$L:B(CH_3)_3(g) \rightleftharpoons L(g) + B(CH_3)_3(g)$$

L	**K_p (atm) at 100 °C**
$(CH_3)_3P$	0.128
$(CH_3)_3N$	0.472
H_3N	4.62

(a) If you begin an experiment by placing 0.010 mol of each complex in a flask, which has the largest partial pressure of $B(CH_3)_3$ at 100 °C?

(b) If 0.73 g (0.010 mol) of $H_3N:B(CH_3)_3$ is placed in a 100.-mL flask and heated to 100 °C, what is the partial pressure of each gas in the equilibrium mixture and what is the total pressure? What is the percent dissociation of $H_3N:B(CH_3)_3$?

65. Sulfuryl chloride, SO_2Cl_2, is a compound with very irritating vapors; it is used as a reagent in the synthesis of organic compounds. When heated to a sufficiently high temperature it decomposes to SO_2 and Cl_2.

$$SO_2Cl_2(g) \rightleftharpoons SO_2(g) + Cl_2(g) \qquad K_c = 0.045 \text{ at } 375\ °C$$

(a) In a 1.00-L flask you place 6.70 g of SO_2Cl_2 and heat it to 375 °C. What is the concentration of each of the compounds in the system when equilibrium is achieved? What fraction of SO_2Cl_2 has dissociated?

(b) What are the concentrations of SO_2Cl_2, SO_2, and Cl_2 at equilibrium in the 1.00-L flask at 375 °C if you begin with a mixture of SO_2Cl_2 (6.70 g) and Cl_2 (1.00 atm)? What fraction of SO_2Cl_2 has dissociated?

(c) Do the fractions of SO_2Cl_2 in parts (a) and (b) agree with your expectation based on Le Chatelier's principle?

66. The reaction between chloroform and chlorine gas is thought to proceed as follows:

Step 1 *Fast, equilibrium:*

$$Cl_2(g) \rightleftharpoons 2\ Cl(g)$$

Step 2 *Slow:*

$$CHCl_3(g) + Cl(g) \rightleftharpoons CCl_3(g) + HCl(g)$$

Step 3 *Fast:*

$$CCl_3(g) + Cl(g) \rightleftharpoons CCl_4(g)$$

Overall reaction

$$CHCl_3(g) + Cl_2(g) \rightleftharpoons CCl_4(g) + HCl(g)$$

(a) Write the rate law for the rate-determining step.
(b) Show that the mechanism shown here agrees with the experimental rate law Rate = $k[CHCl_3][Cl_2]^{1/2}$.

67. The ozone layer in the earth's upper atmosphere is important in shielding the earth from very harmful ultraviolet radiation. The ozone, O_3, decomposes according to the equation

$$2\ O_3(g) \longrightarrow 3\ O_2(g)$$

The mechanism of the reaction is thought to proceed through an initial fast equilibrium and a slow step.

Step 1 *Fast, reversible:*

$$O_3(g) \rightleftharpoons O_2(g) + O(g)$$

Step 2 *Slow:*

$$O_3(g) + O(g) \longrightarrow 2\ O_2(g)$$

(a) Which of the steps is rate-determining? Write the rate law for this step.
(b) Show that the mechanism shown here agrees with the experimental rate law Rate = $k[O_3]^2/[O_2]$.

68. Many biochemical reactions are catalyzed by acids. A typical mechanism consistent with the experimental rate equation, Rate = $k[X][HA]^{1/2}$, in which HA is the acid and X is the reactant, is

Step 1 *Fast, reversible:*

$$HA \rightleftharpoons H^+ + A^-$$

Step 2 *Fast, reversible:*

$$X + H^+ \rightleftharpoons XH^+$$

Step 3 *Slow:*

$$XH^+ \longrightarrow \text{products}$$

HA is the only source of H^+ and A^-. Is the measured rate independent of the acid strength? Explain.

69. Hemoglobin (Hb) can form a complex with either O_2 or CO. For the reaction

$$O_2Hb(aq) + CO(g) \rightleftharpoons COHb(aq) + O_2(g)$$

at body temperature, K is about 200. If the ratio [COHb]/[O_2Hb] comes close to 1, death is probable. What partial pressure of CO in the air is likely to be fatal? Assume the partial pressure of O_2 is 0.2 atm.

Conceptual Questions

70. 1-Butene can change, or isomerize, to *trans*-2-butene in the gas phase.

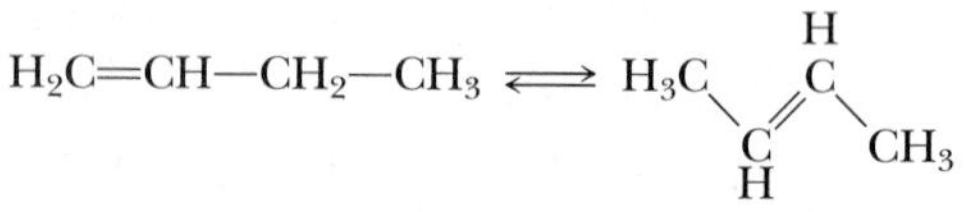

1-butene *trans*-2-butene

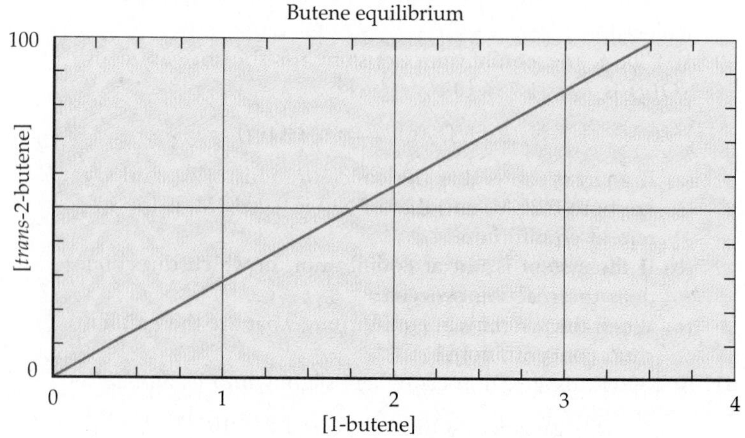

An equilibrium plot for this reaction is given in the figure.

(a) What is the value of K for the equilibrium?

(b) If $\Delta H°$ for the reaction is -11.5 kJ, does the value of K increase or decrease as the temperature increases?

(c) Use the equilibrium plot to find the equilibrium concentrations of reactant and product when 1.20 mol/L of 1-butene is added to an equilibrium mixture, whose concentration of 1-butene is 0.60 mol/L.

(d) Consider the transformation of 1-butene to *cis*-2-butene.

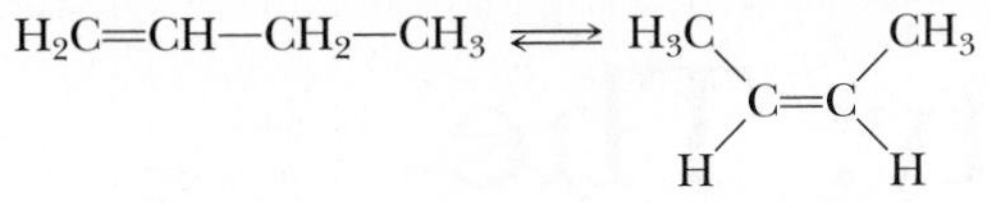

1-butene cis-2-butene

The value of K_c for this reaction is 8.84. Construct an equilibrium plot similar to the one for 1-butene and *trans*-2-butene. Compare the slope of the line in this plot with the slope of the line for the 1-butene/*trans*-2-butene equilibrium.

(e) What is the equilibrium constant for the transformation of *trans*-2-butene to *cis*-2-butene?

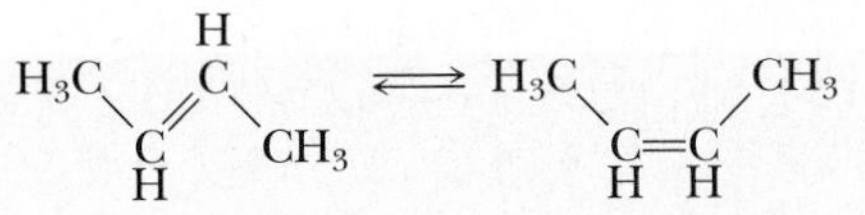

trans-2-butene *cis*-2-butene

Suppose an equilibrium mixture of the *cis*- and *trans*-2-butene has 0.010 mol/L *cis*-2-butene and an appropriate concentration of *trans*-2-butene. If an additional 0.010 mol/L of *cis*-2-butene is added, what are the new equilibrium concentrations of the hydrocarbons?

71. A sample of liquid water is sealed in a container. Over time some of the liquid evaporates, but equilibrium is reached eventually. At this point you can measure the equilibrium vapor pressure of the water. Is the process $H_2O(g) \rightleftarrows H_2O(\ell)$ a dynamic equilibrium? Explain the changes that take place in reaching equilibrium in terms of the rates of the competing processes of evaporation and condensation.

72. An ice cube is placed in a beaker of water at 20 °C. The ice cube partially melts, and the temperature of the water is lowered to 0 °C. At this point, both ice and water are at 0 °C, and no further change is apparent. Is the system at equilibrium? Is this a dynamic equilibrium? That is, are events still occurring at the molecular level? Suggest an experiment to test whether this is so. (*Hint:* Consider using D_2O.)

Summary Question

73. Nitrosyl bromide, NOBr, is prepared by the direct reaction of NO and Br_2.

$$2\ NO(g) + Br_2(g) \longrightarrow 2\ NOBr(g)$$

but the compound dissociates readily at room temperature.

$$NOBr(g) \rightleftharpoons NO(g) + \tfrac{1}{2}\ Br_2(g)$$

(a) If you mix 3.50 g of NO and 9.67 g of Br_2, how many grams of NOBr can be prepared?

(b) If N is the central atom of nitrosyl bromide, draw the electron dot structure for the molecule.

(c) What is the electron pair geometry of NOBr? What is its molecular geometry? Is the molecule polar?

(d) Some NOBr is placed in a flask at 25 °C and allowed to dissociate. The total pressure at equilibrium is 190 mm Hg, and the compound is found to be 34% dissociated. What is the value of K_p?

CHAPTER 17

Principles of Reactivity: The Chemistry of Acids and Bases

A Chemical Puzzler

Some foods and household products are decidedly acidic and others are basic. From your experience can you tell which ones belong to which category? How can you test a substance to tell if it is acidic or basic? What kinds of molecules make a substance acidic or basic?

(C. D. Winters)

Acids and bases are everywhere in our environment. Volcanoes and hot springs can be strongly acidic due to the presence of SO_2 and HCl. Basic carbonate ions are present in most natural waters along with such other substances as borate, phosphate, arsenate, silicate, and ammonia. Foods contain natural acids, and your body's metabolism is regulated by reactions between acids and bases. Thus, it is no wonder that acids and bases have been studied for hundreds of years.

Volcanoes spew tons of hydrochloric acid, sulfur dioxide, and other acidic compounds into the earth's atmosphere. (Pat and Tom Leeson/Photo Researchers, Inc.)

17.1 ACIDS, BASES, AND ARRHENIUS

The word ''acid'' comes from the Latin word *acidus,* meaning sour. It was probably applied originally to vinegar, but it was later the name given to other substances that had a sour taste. The term ''alkali'' is derived from an Arabic word for the ashes that come from burning certain plants. Potash, or potassium carbonate, is one product of this process, and because water solutions of potash feel soapy and taste bitter, ''alkali'' was later applied more generally to substances having those properties. Finally, the word ''salt,'' which finds its roots in many languages, probably originally meant sea salt, or sodium chloride, although it has a broader meaning in chemistry.

See the previous discussion of acids and bases on page 170.

With time the terms ''acid'' and ''alkali'' were applied more broadly. Robert Boyle, who is most famous for his work with gases, wrote in 1684 that alkalis give soapy solutions, restore vegetable colors reddened by acids, and react with acids to give what he called ''indifferent salts.'' Acids, on the other hand, he characterized by their sour taste, their ability to be corrosive, to redden blue vegetable colors, and to lose all these properties when brought into contact with alkalis.

Robert Boyle's remark is an early, explicit mention of the use of acid-base indicators. The juice of red cabbage is an excellent indicator of acidity (see Figure 5.12).

By the 18th century, salts were recognized as the product of the reaction of an acid and an alkali. About 1750, the Frenchman, Rouelle, contended that a natural salt was formed by an acid reacting with any substance capable of serving as a ''base for [the salt]—a water-soluble alkali, an 'earth,' a metal, and so on.'' Thus, the word *base* entered chemistry's vocabulary, and we now recognize reactions such as the following as salt producers (Figure 17.1).

$$2\ HCl(aq) + CaCO_3(s)\ \text{(limestone, an ``earth'')} \longrightarrow CaCl_2(aq) + H_2O(\ell) + CO_2(g)$$

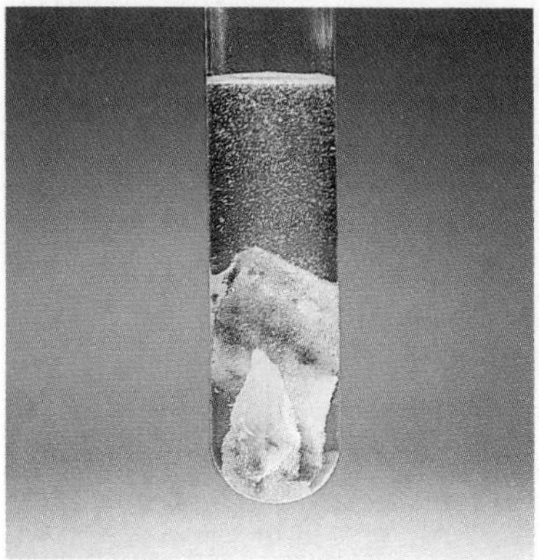

Figure 17.1 A salt-producing reaction, the reaction of aqueous HCl with limestone, $CaCO_3$. (C. D. Winters)

Because of their universal importance, more general ways of classifying and explaining acid-base behavior and reactions have been developed over the past 100 years. One of the earlier concepts still in use was proposed by the Swedish chemist Svante Arrhenius (1859–1927): an acid is a substance that contains hydrogen and releases a hydrogen ion (H^+) as one of the products of ionic dissociation in water.

An Arrhenius base is a compound that produces hydroxide ions in water. Thus, sodium hydroxide is clearly a base, but so is ammonia. Even though ammonia does not contain a hydroxide ion in its formula, it produces hydroxide ion when dissolved in water.

Arrhenius's Definition of Acids and Bases

	Acid	Base
General Reaction	$H_xB \rightarrow x\,H^+ + B^{x-}$	$M(OH)_y \rightarrow M^{y+} + y\,OH^-$
Examples	$HCl(aq) \rightarrow H^+(aq) + Cl^-(aq)$	$NaOH(aq) \rightarrow Na^+(aq) + OH^-(aq)$
		$NH_3(aq) + H_2O(\ell) \rightarrow NH_4^+(aq) + OH^-(aq)$

An Arrhenius acid and an Arrhenius base react in a neutralization reaction to produce a salt and water.

$$HCl(aq) + NaOH(aq) \longrightarrow NaCl(aq) + H_2O(\ell)$$
$$2\ HNO_3(aq) + Fe(OH)_2(s) \longrightarrow Fe(NO_3)_2(aq) + 2\ H_2O(\ell)$$

Arrhenius's concept is limited to aqueous solutions because it refers to ions (H^+ and OH^-) derived from water. A truly general concept of acids and bases should be appropriate to other solvents, however. More general models of acid-base behavior have been developed, and we shall examine them in this chapter.

17.2 THE HYDRONIUM ION AND WATER AUTOIONIZATION

The unusual properties of water are a recurring theme of this book (see Sections 4.4 and 13.2). Because the acid and base reactions we want to discuss take place in water, and the acid-base reactions in your body happen in your aqueous interior, we come again to the behavior of water.

Ions are hydrated in water, owing to the attraction between an ion and polar water molecules (Section 13.1). Thus, we typically use formulas, such as $Na^+(aq)$ and $Cl^-(aq)$, when dealing with ions in aqueous solution. The hydrogen ion, H^+, is a special case of this, however. Although we often write the formula for the species as $H^+(aq)$, the formula $H_3O^+(aq)$ better represents the actual structure of the hydrated hydrogen ion in aqueous solution. The ion is called the **hydronium ion.** Its name and structure are analogous with the ammonium ion, NH_4^+, which forms when a hydrogen ion adds to a molecule of ammonia.

See *Current Issues in Chemistry: Watery Cages* (Chapter 13) for the structure of a hydronium ion enclosed in a cage of water molecules.

An acid such as HCl does not need to be present for the hydronium ion to exist in water. In fact, two water molecules can interact with each other to produce a hydronium ion and a hydroxide ion by proton transfer from one water molecule to the other.

$$2\,H_2O(aq) \rightleftharpoons H_3O^+(aq) + OH^-(aq)$$

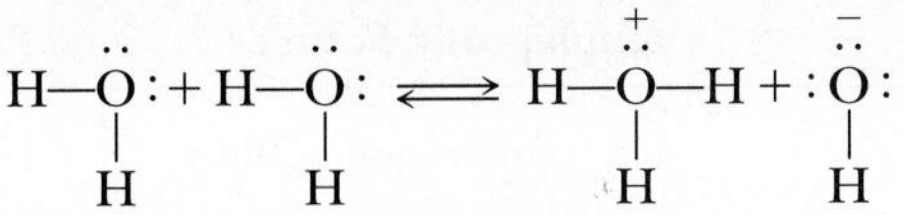

The existence of the so-called **autoionization** reaction of water was demonstrated many years ago by Friedrich Kohlrausch (1840–1910). He found that, even after water is painstakingly purified, it still conducts electricity to a very small extent, because autoionization produces very low concentrations of H_3O^+ and OH^- ions even in the purest water. Water autoionization is the cornerstone of our concepts of aqueous acid-base behavior.

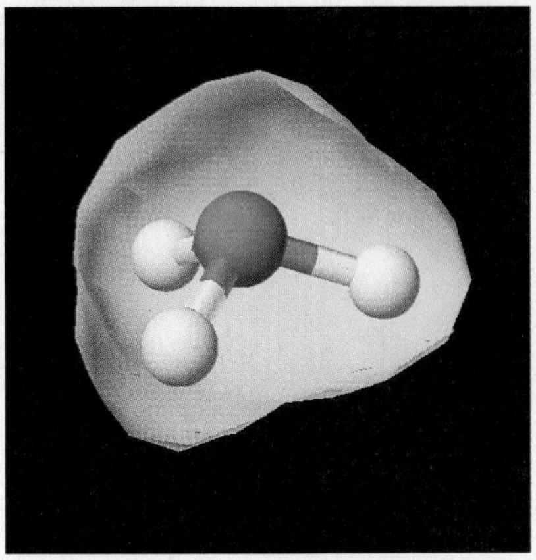

The hydronium ion, H_3O^+. Here the ion is enclosed within a surface that shows the distribution of electrons in the ion. (Susan Young)

17.3 THE BRØNSTED CONCEPT OF ACIDS AND BASES

In 1923, Johannes N. Brønsted (1879–1947) in Copenhagen (Denmark) and Thomas M. Lowry (1874–1936) in Cambridge (England) independently suggested a new concept of acid and base behavior. They proposed that an **acid** is any substance that can *donate a proton* to any other substance. Thus, acids can be neutral compounds such as nitric acid,

The Brønsted theory of acid-base behavior is often—and perhaps more fairly—called the Brønsted-Lowry theory.

$$\underset{\text{Acid}}{HNO_3(aq)} + H_2O(\ell) \longrightarrow H_3O^+(aq) + NO_3^-(aq)$$

or they can be cations or anions.

$$\underset{\text{Acid}}{NH_4^+(aq)} + H_2O(\ell) \longrightarrow H_3O^+(aq) + NH_3(aq)$$

$$\underset{\text{Acid}}{H_2PO_4^-(aq)} + H_2O(\ell) \longrightarrow H_3O^+(aq) + HPO_4^{2-}(aq)$$

According to Brønsted and Lowry, a **base** is a substance that can *accept a proton* from another substance. These can be neutral compounds,

$$\underset{\text{Base}}{NH_3(aq)} + H_2O(\ell) \longrightarrow NH_4^+(aq) + OH^-(aq)$$

or anions.

Few if any cations are Brønsted bases.

$$\underset{\text{Base}}{CO_3^{2-}(aq)} + H_2O(\ell) \longrightarrow HCO_3^-(aq) + OH^-(aq)$$

$$\underset{\text{Base}}{PO_4^{3-}(aq)} + H_2O(\ell) \longrightarrow HPO_4^{2-}(aq) + OH^-(aq)$$

A wide variety of Brønsted acids are known, and you are familiar with many of them (Table 4.1). Acids, such as HF, HCl, HNO_3, and CH_3CO_2H (acetic acid), are all capable of donating one proton and so are called **monoprotic acids.** Other acids, however, are capable of donating two or more protons, and these are called **polyprotic acids** (Table 17.1). An example is sulfuric acid.

$$H_2SO_4(aq) + H_2O(\ell) \longrightarrow H_3O^+(aq) + HSO_4^-(aq)$$
$$HSO_4^-(aq) + H_2O(\ell) \longrightarrow H_3O^+(aq) + SO_4^{2-}(aq)$$

TABLE 17.1 Polyprotic Acids and Bases

Acid Form	Amphiprotic Form	Base Form
H_2S (hydrosulfuric acid or hydrogen sulfide)	HS^- (hydrogen sulfide ion)	S^{2-} (sulfide ion)
H_3PO_4 (phosphoric acid)	$H_2PO_4^-$ (dihydrogen phosphate ion)	HPO_4^{2-} (hydrogen phosphate ion)
$H_2PO_4^-$ (dihydrogen phosphate ion)	HPO_4^{2-} (hydrogen phosphate ion)	PO_4^{3-} (phosphate ion)
H_2CO_3 (carbonic acid)	HCO_3^- (hydrogen carbonate or bicarbonate ion)	CO_3^{2-} (carbonate ion)
$H_2C_2O_4$ (oxalic acid)	$HC_2O_4^-$ (hydrogen oxalate ion)	$C_2O_4^{2-}$ (oxalate ion)

Just as there are acids that can donate more than one proton, there are **polyprotic bases** that can accept more than one proton. The anions of polyprotic acids are polyprotic bases, examples of which include SO_4^{2-}, PO_4^{3-}, CO_3^{2-}, and $C_2O_4^{2-}$. This behavior is illustrated by the carbonate ion.

$$\underset{\text{Base}}{CO_3^{2-}(aq)} + H_2O(\ell) \longrightarrow HCO_3^-(aq) + OH^-(aq)$$

$$\underset{\text{Base}}{HCO_3^-(aq)} + H_2O(\ell) \longrightarrow H_2CO_3(aq)$$

Water is amphiprotic in its autoionization reaction. One molecule of H_2O acts as a base and the other acts as an acid.

Amphiprotic substances also exist. These are molecules or ions that can behave either as a Brønsted acid or base, and one of the best examples is water. Water acts as a base in the presence of HCl, in that H_2O accepts a proton from the acid,

$$\underset{\text{Acid}}{HCl(aq)} + \underset{\text{Base}}{H_2O(\ell)} \longrightarrow H_3O^+(aq) + Cl^-(aq)$$

and it is an acid when donating a proton to ammonia.

$$\underset{\text{Acid}}{H_2O(\ell)} + \underset{\text{Base}}{NH_3(aq)} \longrightarrow NH_4^+(aq) + OH^-(aq)$$

The amphiprotic ions HPO_4^{2-} and HCO_3^- are especially important in biochemistry.

Other amphiprotic substances you have encountered thus far include such important anions as HCO_3^- (bicarbonate) and $H_2PO_4^-$ (dihydrogen phosphate).

$$\underset{\text{Acid}}{H_2PO_4^-(aq)} + \underset{\text{Base}}{H_2O(\ell)} \longrightarrow H_3O^+(aq) + HPO_4^{2-}(aq)$$

$$\underset{\text{Base}}{H_2PO_4^-(aq)} + \underset{\text{Acid}}{H_2O(\ell)} \longrightarrow H_3PO_4(aq) + OH^-(aq)$$

EXERCISE 17.1 ***Brønsted Acids and Bases***

1. Write a balanced equation for the reaction that occurs when H_3PO_4, phosphoric acid, donates a proton to water to form the dihydrogen phosphate ion.
2. Write a balanced equation for the reaction that occurs when the cyanide ion, CN^-, accepts a proton from water to form HCN. Is CN^- a Brønsted acid or base?

3. Write balanced equations for the two stepwise reactions that occur when $H_2C_2O_4$, oxalic acid, acts as a polyprotic acid in aqueous solution. ■

Conjugate Acid-Base Pairs

Let us expand on the Brønsted notion of acids and bases. In each of the chemical equations written so far a *proton has been transferred* to or from water. For example, the hydrogen carbonate ion (bicarbonate ion) can act as an acid and can transfer a proton to water, which accepts the proton and acts as a Brønsted base, producing the hydronium ion and carbonate ion.

$$\underset{\text{Acid}}{HCO_3^-(aq)} + \underset{\text{Base}}{H_2O(\ell)} \rightleftharpoons \underset{\text{Acid}}{H_3O^+(aq)} + \underset{\text{Base}}{CO_3^{2-}(aq)}$$

The hydronium ion, however, is a proton donor, an acid, and the carbonate ion was described previously as a proton acceptor, a base. Thus, the H_3O^+ ion can transfer a proton to the CO_3^{2-} ion to re-form the HCO_3^- ion and H_2O. The reaction is reversible and is written as an equilibrium reaction.

There is another important observation to make in the reaction of hydrogen carbonate with water. The hydrogen carbonate and carbonate ions are related to one another by the loss or gain of H^+, as are H_2O and H_3O^+. *A pair of compounds or ions that differ by the presence of one H^+ unit* is called a **conjugate acid-base pair.** We say that CO_3^{2-} is the conjugate base of the acid HCO_3^-, and HCO_3^- is the conjugate acid of the base CO_3^{2-}.

The notion of an equilibrium (denoted by $\rightleftharpoons$) involving conjugate acids and bases is a fundamental principle of the Brønsted theory.

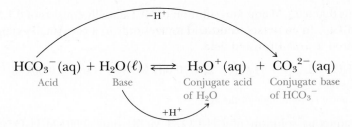

$-H^+$

$$\underset{\text{Acid}}{HCO_3^-(aq)} + \underset{\text{Base}}{H_2O(\ell)} \rightleftharpoons \underset{\text{Conjugate acid of } H_2O}{H_3O^+(aq)} + \underset{\text{Conjugate base of } HCO_3^-}{CO_3^{2-}(aq)}$$

$+H^+$

Every acid-base reaction involving H^+ transfer has two conjugate acid-base pairs. To convince yourself of this, look at the reactions above and those in Table 17.2.

TABLE 17.2 Conjugate Acid-Base Pairs

Name	Acid 1	Base 2		Base 1	Acid 2*
Hydrogen chloride	HCl	$+ H_2O$	$\longrightarrow$	Cl^-	$+ H_3O^+$
Nitric acid	HNO_3	$+ H_2O$	$\longrightarrow$	NO_3^-	$+ H_3O^+$
Hydrogen carbonate	HCO_3^-	$+ H_2O$	$\rightleftharpoons$	CO_3^{2-}	$+ H_3O^+$
Acetic acid	CH_3CO_2H	$+ H_2O$	$\rightleftharpoons$	$CH_3CO_2^-$	$+ H_3O^+$
Hydrogen cyanide	HCN	$+ H_2O$	$\rightleftharpoons$	CN^-	$+ H_3O^+$
Hydrogen sulfide	H_2S	$+ H_2O$	$\rightleftharpoons$	HS^-	$+ H_3O^+$
Ammonia	H_2O	$+ NH_3$	$\rightleftharpoons$	OH^-	$+ NH_4^+$
Carbonate ion	H_2O	$+ CO_3^{2-}$	$\rightleftharpoons$	OH^-	$+ HCO_3^-$
Water	H_2O	$+ H_2O$	$\rightleftharpoons$	OH^-	$+ H_3O^+$

*Acid 1 and base 1 are a conjugate pair, as are base 2 and acid 2.

With this background, let us now turn to the question of the strengths of Brønsted acids and bases and to the problem of deciding whether a given acid-base reaction is product- or reactant-favored.

EXERCISE 17.2 ***Conjugate Acids and Bases***

1. In the following reaction, identify the acid on the left and its conjugate base on the right. Similarly, identify the base on the left and its conjugate acid on the right.

$$HBr(aq) + NH_3(aq) \rightleftharpoons NH_4^+(aq) + Br^-(aq)$$

2. What is the conjugate base of H_2S? Of NH_4^+?
3. What is the conjugate acid of NO_3^-? Of HPO_4^{2-}? ■

Relative Strengths of Acids and Bases

In water some acids are better proton donors than others, and some bases are better proton acceptors than others. For example, a dilute solution of hydrochloric acid consists largely of $H_3O^+(aq)$ and $Cl^-(aq)$ ions; the acid is nearly 100% ionized, and so it is considered a *strong* Brønsted acid.

$$HCl(aq) + H_2O(\ell) \longrightarrow H_3O^+(aq) + Cl^-(aq)$$

Strong acid (≈100% ionized)
$[H_3O^+] \approx$ initial concentration of the acid

This means that a 0.1 M aqueous solution of HCl actually consists of 0.1 M H_3O^+ and 0.1 M Cl^-. In contrast, acetic acid ionizes only to a very small extent and so is considered a *weak* Brønsted acid.

$$0.1\ M\ CH_3CO_2H(aq) + H_2O(\ell) \rightleftharpoons 0.001\ M\ H_3O^+(aq) + 0.001\ M\ CH_3CO_2^-(aq)$$

Weak acid (<100% ionized)
$[H_3O^+] \ll$ initial concentration of the acid

A 0.1 M aqueous solution of CH_3CO_2H yields only 0.001 M $H_3O^+(aq)$ and 0.001 M $CH_3CO_2^-(aq)$. About 99% of the acetic acid is *not* ionized.

The oxide ion is a *very* strong Brønsted base in aqueous solution. Indeed, it is so strong that it does not exist in water. The ion reacts completely with water to produce hydroxide ion,

$$O^{2-}(aq) + H_2O(\ell) \longrightarrow 2\ OH^-(aq)$$

Strong base
$[OH^-] = 2 \times$ (initial concentration of O^{2-})

and we write the following equation for the dissolving of lithium oxide, for example.

$$Li_2O(s) + H_2O(\ell) \longrightarrow 2\ Li^+(aq) + 2\ OH^-(aq)$$

In contrast, aqueous ammonia and the carbonate ion produce only a very small concentration of OH^- ion and are classed as weak Brønsted bases.

$$NH_3(aq) + H_2O(\ell) \rightleftharpoons NH_4^+(aq) + OH^-(aq)$$

$$CO_3^{2-}(aq) + H_2O(\ell) \rightleftharpoons HCO_3^-(aq) + OH^-(aq)$$

Weak bases
$[OH^-] \ll$ initial concentration of base

In the Brønsted model, an acid donates a proton and produces a conjugate base. This model also informs us that, in general, *the stronger the acid, the weaker its conjugate base.* Aqueous HCl, for example, is a strong acid because it has a strong tendency to donate a proton to water and produce its conjugate base Cl^-. In this reaction water acts as a base and accepts the proton from HCl to produce H_3O^+, the conjugate acid of water. Because HCl is a strong acid, the reaction proceeds almost completely to the right; essentially no HCl is left intact in the solution at equilibrium.

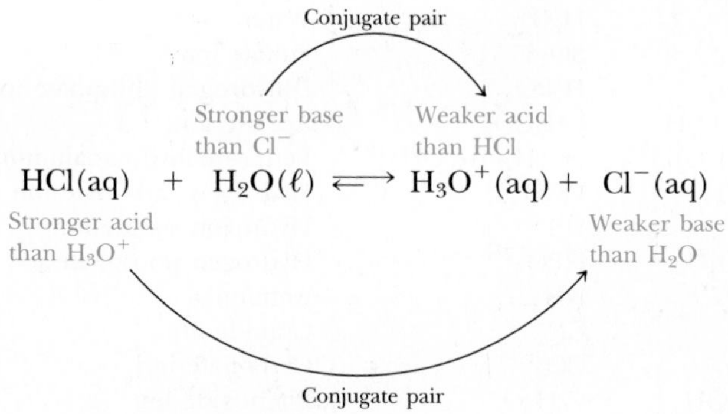

In the case of a strong acid, the acid and base on the left side of the balanced equation are stronger than the conjugate acid and base on the right. The stronger acid and base react to give predominantly the weaker acid and base. Of the two acids here, HCl and H_3O^+, HCl is better able to donate a proton. Of the two bases, H_2O and Cl^-, water must be the stronger base, and it wins out in the competition for the proton. The equilibrium lies far to the right, and we have denoted this by using arrows of unequal length.

Acetic acid is the classic example of a weak acid. Acetic acid ionizes to a very small extent in water. Thus, of the two acids present in aqueous acetic acid (CH_3CO_2H and H_3O^+), the hydronium ion is the stronger. Of the two bases (H_2O and acetate ion, $CH_3CO_2^-$), the acetate ion must be the stronger. At equilibrium, the solution consists mostly of acetic acid with only a small concentration of acetate ion and hydronium ion. Again, the equilibrium lies toward the side of the reaction having the weaker acid and base.

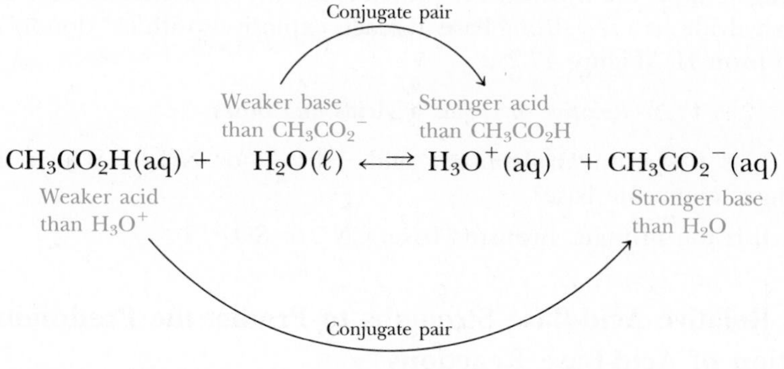

These two examples of the relative extent of acid-base reactions illustrate an important principle in Brønsted acid-base theory: all proton transfer reactions proceed *from the stronger* acid-base pair *to the weaker* acid-base pair.

TABLE 17.3 Relative Strengths of Acids and Bases

Conjugate Acid		Conjugate Base	
Name	*Formula*	*Formula*	*Name*
Perchloric acid	$HClO_4$	ClO_4^-	Perchlorate ion
Sulfuric acid	H_2SO_4	HSO_4^-	Hydrogen sulfate ion
Hydrochloric acid	HCl	Cl^-	Chloride ion
Nitric acid	HNO_3	NO_3^-	Nitrate ion
Hydronium ion	H_3O^+	H_2O	Water
Hydrogen sulfate ion	HSO_4^-	SO_4^{2-}	Sulfate ion
Phosphoric acid	H_3PO_4	$H_2PO_4^-$	Dihydrogen phosphate ion
Acetic acid	CH_3CO_2H	$CH_3CO_2^-$	Acetate ion
Hexaaquaaluminum ion	$[Al(H_2O)_6]^{3+}$	$[Al(H_2O)_5(OH)]^{2+}$	Pentaaquahydroxoaluminium ion
Carbonic acid	H_2CO_3	HCO_3^-	Hydrogen carbonate ion
Hydrogen sulfide	H_2S	HS^-	Hydrogen sulfide ion
Dihydrogen phosphate ion	$H_2PO_4^-$	HPO_4^{2-}	Hydrogen phosphate ion
Ammonium ion	NH_4^+	NH_3	Ammonia
Hydrogen cyanide	HCN	CN^-	Cyanide ion
Hydrogen carbonate ion	HCO_3^-	CO_3^{2-}	Carbonate ion
Phenol	C_6H_5OH	$C_6H_5O^-$	Phenoxide ion
Water	H_2O	OH^-	Hydroxide ion
Ethanol	C_2H_5OH	$C_2H_5O^-$	Ethoxide ion
Ammonia	NH_3	NH_2^-	Amide ion
Methylamine	CH_3NH_2	CH_3NH^-	Methylamide ion
Hydrogen	H_2	H^-	Hydride ion
Methane	CH_4	CH_3^-	Methide ion

← Increasing Acid Strength (upward) · Increasing Base Strength (downward) →

Figure 17.2 The basic properties of the hydride ion, H^-. Calcium hydride, CaH_2, is the source of H^- ion. This ion is such a powerful proton acceptor that it reacts vigorously with water, a proton donor, to give H_2. (C. D. Winters)

$$H^-(aq) + H_2O(\ell) \longrightarrow H_2(g) + OH^-(aq)$$

Acids and bases can be ordered on the basis of their relative abilities to donate or accept protons in aqueous solution, and we have done so for a few Brønsted acids and bases in Table 17.3. At the top on the left are the stronger acids; those substances strongly donate protons, and their conjugate bases are extremely weak. The opposite is true for acids at the bottom left of the table. For example, the H_2 molecule can be considered an acid in the sense that it can conceivably donate a proton (H^+) and form the conjugate base H^-, the hydride ion. Hydrogen H_2, however, is an exceedingly weak acid. We know this because its conjugate base, the hydride ion (which is known in substances such as NaH, sodium hydride) is a *very* strong base. It reacts explosively with H^+ donors such as H_2O to form H_2 (Figure 17.2).

EXERCISE 17.3 ***Relative Strengths of Acids and Bases***

1. Which is the stronger Brønsted acid, HCO_3^- or NH_4^+? Which has the stronger conjugate base?
2. Which is the stronger Brønsted base, CN^- or SO_4^{2-}? ■

Using Relative Acid-Base Strengths to Predict the Predominant Direction of Acid-Base Reactions

The chart of acids and bases in Table 17.3 can be used to predict whether the equilibrium lies predominantly to the left or the right in an acid-base reaction. The examples that follow show you how to do this.

EXAMPLE 17.1 *Predicting the Direction of Acid-Base Reactions*

Write a balanced equation for the reaction that occurs between each acid-base pair in water. Decide whether the equilibrium lies predominantly to the left or right.

1. Acetic acid, CH_3CO_2H, and sodium cyanide, NaCN. Is HCN, a poisonous acid, formed to a significant extent?
2. Ammonium chloride, NH_4Cl, and sodium carbonate, Na_2CO_3.

Solution

1. **Reaction of CH_3CO_2H and NaCN**
The conjugate base of acetic acid is the acetate ion, $CH_3CO_2^-$. The other reactant, NaCN, is a water-soluble salt that forms Na^+ and CN^- ions in water. Sodium ion is not listed in Table 17.3 because it does not react appreciably with water. The CN^- ion, however, is a base with HCN as its conjugate acid. The net ionic equation for the reaction between CH_3CO_2H and CN^- can therefore be written as

$$CH_3CO_2H(aq) + CN^-(aq) \rightleftharpoons HCN(aq) + CH_3CO_2^-(aq)$$

To decide to which side the equilibrium lies, left or right, we can compare the two acids (or the two bases) in the reaction. According to Table 17.3, HCN is a weaker acid than CH_3CO_2H, and $CH_3CO_2^-$ is a weaker base than CN^-. In a sense, we can view the situation as a competition between two bases for the available H^+ ion. The stronger base is expected to win out.

Because all Brønsted acid-base reactions move predominantly toward the weaker acid and base, the reaction favors the $HCN/CH_3CO_2^-$ pair at equilibrium.

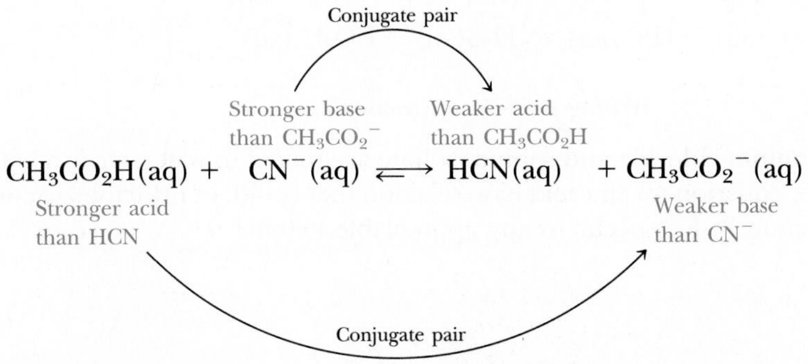

This result tells you that you certainly would *not* want to mix acetic acid and a cyanide ion salt because HCN is the product. The average fatal dose of HCN is 50 to 60 mg, or about 0.002 mol.

2. **Reaction of NH_4Cl and Na_2CO_3**
Ammonium chloride dissociates to form NH_4^+ and Cl^- ions in water. The ammonium ion is a weak acid

$$NH_4^+(aq) + H_2O(\ell) \rightleftharpoons H_3O^+(aq) + NH_3(aq)$$

and the chloride ion is a weak base.

$$Cl^-(aq) + H_2O(\ell) \rightleftharpoons HCl(aq) + OH^-(aq)$$

As you will see in Section 17.8, we can disregard the base properties of ClO_4^-, Cl^-, and NO_3^- in aqueous solution. These ions do not contribute to the acidity or basicity of a solution.

As shown in Table 17.3, the OH^- ion is a *much* stronger base than Cl^-, and HCl is a much stronger acid than H_2O. The equilibrium lies so far to the left that we can assume this reaction does not occur. In general, any anion X^- that is a weaker base than OH^- *does not react appreciably with water* to give HX and OH^-. Thus, for NH_4Cl, we only need to be concerned with the NH_4^+ ion.

Sodium carbonate, Na_2CO_3, dissociates in water to give the ions Na^+ and CO_3^{2-}. As in the reaction of CH_3CO_2H and NaCN, we can ignore any contribution from the Na^+ ion. The carbonate ion, however, is a base (Table 17.3).

$$CO_3^{2-}(aq) + H_2O(\ell) \rightleftharpoons HCO_3^-(aq) + OH^-(aq)$$

Having decided the ammonium ion is an acid and the carbonate ion is a base, we can write the equation for their reaction.

$$\underset{\text{Acid}}{NH_4^+(aq)} + \underset{\text{Base}}{CO_3^{2-}(aq)} \rightleftharpoons \underset{\text{Conjugate acid}}{HCO_3^-(aq)} + \underset{\text{Conjugate base}}{NH_3(aq)}$$

Ammonium ion is a stronger acid than HCO_3^-, and CO_3^{2-} is a stronger base than NH_3. Some reaction does therefore occur. You know this because you can detect the characteristic odor of ammonia over the solution. (If you open a bottle of *solid* ammonium carbonate, $(NH_4)_2CO_3$, you can also smell ammonia.)

EXERCISE 17.4 ***Predicting the Direction of Acid-Base Reactions***

For each of the following reactions, predict whether the equilibrium lies predominantly to the left or to the right:

1. $HSO_4^-(aq) + NH_3(aq) \rightleftharpoons NH_4^+(aq) + SO_4^{2-}(aq)$
2. $HCO_3^-(aq) + HS^-(aq) \rightleftharpoons H_2S(aq) + CO_3^{2-}(aq)$ ■

EXERCISE 17.5 ***Writing Acid-Base Reactions***

If ammonium chloride and sodium sulfate are mixed in water, write a balanced, net ionic equation for the acid-base reaction that could, in principle, occur. Does the reaction in fact occur to any appreciable extent? ■

17.4 STRONG ACIDS AND BASES

The acids in Table 17.3 are listed in descending order of their ability to donate a proton. *The hydronium ion is actually the strongest acid that can exist in water.* Notice, however, that several acids are listed even higher in the table than H_3O^+. How can this be? When they are placed in water, strong acids such as HCl, HBr, HI, HNO_3, H_2SO_4, and $HClO_4$ (see Table 4.1) ionize *completely* to form H_3O^+ and their conjugate base by reacting with water.

$$\underset{\text{Strong acid, 100\% ionization for loss of first H atom as } H_3O^+}{H_2SO_4(aq)} + H_2O(\ell) \longrightarrow H_3O^+(aq) + \underset{\text{Weak conjugate base of } H_2SO_4}{HSO_4^-(aq)}$$

CURRENT ISSUES IN CHEMISTRY

Sulfuric Acid—A Gauge of Business Conditions

Sulfuric acid is a colorless, syrupy liquid with a density of 1.84 g/mL and a boiling point of 270 °C. It has several desirable properties that have led to its widespread use: It is generally less expensive than other acids, is a strong acid, can be handled in steel containers, reacts readily with many organic compounds to produce useful products, and reacts readily with lime (CaO), the least expensive and most readily available base.

The acid is made by the *contact process.* The first step is combustion of sulfur in air to give sulfur dioxide (Figure 16.4). (Sulfur dioxide is also obtained from the smelting of copper and lead ores.)

$$\tfrac{1}{8}S_8(s) + O_2(g) \longrightarrow SO_2(g) \qquad \Delta H^\circ_{rxn} = -296.8 \text{ kJ}$$

Sulfur is found in very pure form in underground deposits along the coast of the United States in the Gulf of Mexico. It is recovered by pumping superheated steam into the sulfur beds to melt the sulfur. The molten sulfur is brought to the surface by means of compressed air. (F. Grehan/Photo Researchers, Inc.)

This gas is then combined with more oxygen, in the presence of a catalyst, to give sulfur trioxide,

$$SO_2(g) + \tfrac{1}{2}O_2(g) \longrightarrow SO_3(g) \qquad \Delta H^\circ_{rxn} = -98.9 \text{ kJ}$$

which can give sulfuric acid when absorbed in water.

$$SO_3(g) + H_2O(\ell) \longrightarrow H_2SO_4(aq)$$

For some years sulfuric acid has been the chemical produced in the largest amount in the United States (and in many other industrialized countries). About 40 billion kg (40 million metric tons) was made in the United States in 1994. Some economists have said this production is a measure of a nation's industrial strength or general business conditions. The reason is that the acid is used in so many ways. Currently, over two thirds of the production is used in the phosphate fertilizer industry, which makes "superphosphate" fertilizer by treating phosphate rock with sulfuric acid.

$$Ca_{10}F_2(PO_4)_6(s) + 7\,H_2SO_4(aq) + 3\,H_2O(\ell) \longrightarrow 3\,Ca(H_2PO_4)_2 \cdot H_2O(s) + 7\,CaSO_4(s) + 2\,HF(g)$$

The remainder is used to make pigments, explosives, alcohol, pulp and paper, detergents, and as a component in storage batteries.

Thus, H_2SO_4 does not exist in water; rather, only H_3O^+ and HSO_4^- are present because H_3O^+ is the strongest acid species that can exist in aqueous solution. Acids that ionize 100% in aqueous solution are called **strong acids.** Because strong acids are converted completely to H_3O^+ and the appropriate anion in water, they appear to be the same strength. We say that water "levels" their acidic character to the level of H_3O^+.

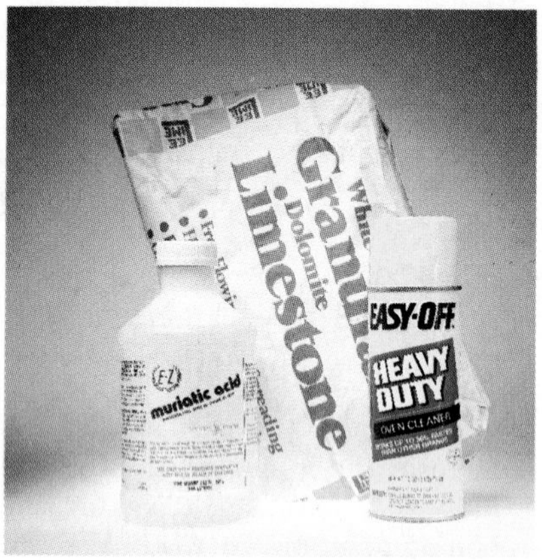

Some household chemicals are strong acids or bases. Oven cleaners often contain sodium hydroxide, and muriatic acid is aqueous HCl. Garden lime forms $Ca(OH)_2$ on adding water. (C. D. Winters)

Just as no acids can be stronger in water than H_3O^+, no base can be stronger than OH^- in aqueous solution. The strong bases in Table 17.3 are species lower than OH^- in the right column: $C_2H_5O^-$, NH_2^-, H^-, and CH_3^- all react completely with water to produce OH^- (see Figure 17.2).

$$NH_2^-(aq) + H_2O(\ell) \longrightarrow NH_3(aq) + OH^-(aq)$$

amide ion
Strong base

Weak conjugate acid of NH_2^-

About 17 billion kg of lime was made in the United States in 1993. $Ca(OH)_2$ is not very soluble in water. Only 0.17 g dissolves in 100 g of water at 10 °C.

Commonly encountered strong bases, however, are hydroxides or oxides of Group 1A or 2A metals, with NaOH and KOH perhaps most familiar. Although not as soluble as NaOH in water, $Ca(OH)_2$ is among the most inexpensive bases known. Billions of kilograms of CaO, or lime, are made annually by decomposing limestone, $CaCO_3$, at temperatures between 800 and 1000 °C.

$$CaCO_3(s) \longrightarrow CaO(s) + CO_2(g)$$

About 12 billion kg of NaOH, or caustic soda, is made every year in the United States, and even very pure laboratory grade material costs less than \$10/kg. On the other hand, $CsOH \cdot H_2O$ costs about \$800/kg!

In a reaction characteristic of many metal oxides, CaO forms slaked lime, or $Ca(OH)_2$, when mixed with water.

$$CaO(s) + H_2O(\ell) \longrightarrow Ca(OH)_2(s)$$

Clear, aqueous solutions of slaked lime are known as *limewater.*

17.5 WEAK ACIDS AND BASES

Very few acids and bases strongly donate or accept protons, respectively. *The vast majority of acids and bases are weak.*

The relative strength of an acid or base can be expressed quantitatively with an **equilibrium constant.** For the general *weak acid* HA, for example, we can write

$$HA(aq) + H_2O(\ell) \rightleftharpoons H_3O^+(aq) + A^-(aq)$$

$$K_a = \frac{[H_3O^+][A^-]}{[HA]} \quad (17.1)$$

where K has a subscript a to indicate that it is an equilibrium constant for a weak acid in water. The value of K is less than 1 for a weak acid, indicating that the product of the equilibrium concentrations of the hydronium ion and the conjugate base of the weak acid is smaller than the equilibrium concentration of the weak acid.

Similarly, we can write the equilibrium expression for a *weak base* B in water. Here we label K with a subscript b. Its value is also less than 1.

$$B(aq) + H_2O(\ell) \rightleftharpoons BH^+(aq) + OH^-(aq)$$

$$K_b = \frac{[BH^+][OH^-]}{[B]} \quad (17.2)$$

Some acids and bases are ordered on the basis of their relative abilities to donate or accept protons in aqueous solution in Table 17.3. This has been done for a larger number of substances in Table 17.4, where each acid and base is listed with its value of K_a or K_b, respectively. The following are important ideas concerning Table 17.4.

- As in Table 17.3, acids are listed at the left and their conjugate bases are on the right.
- A large value of K indicates products are strongly favored, whereas a small value of K indicates the reactants are favored.
- The strongest acids are at the upper left (see also Table 17.3). They have the largest K_a values. K_a values become smaller on descending the chart as the acid strength declines.
- The strongest bases are at the lower right. They have the largest K_b values. K_b values become larger on descending the chart as base strength increases.
- The weaker the acid, the stronger its conjugate base. That is, the smaller the value of K_a, the larger the value of K_b.
- The K_a values suggest a way to classify acids and their conjugate bases.

Acid Strength	K_a	Conjugate Base Strength	K_b
Strong	> 1	very weak	$< 10^{-16}$
Weak	1 to 10^{-16}	weak	1 to 10^{-16}
Very weak	$< 10^{-16}$	strong	> 1

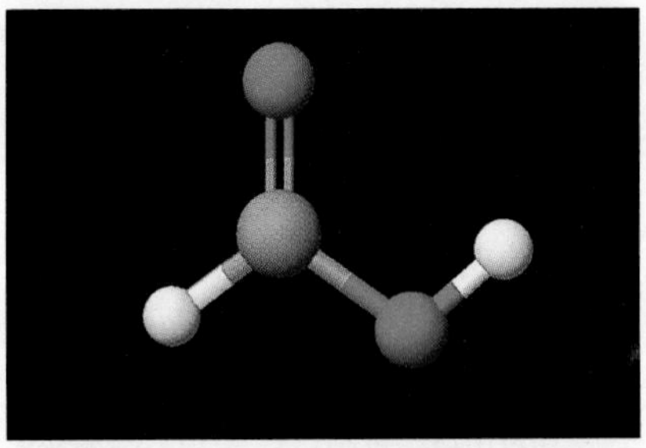
(a)

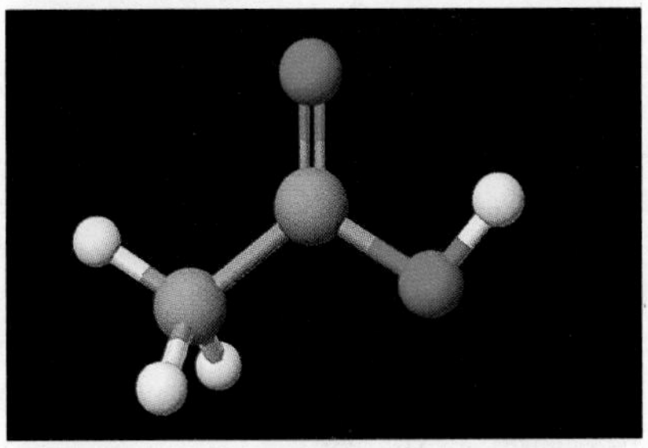
(b)

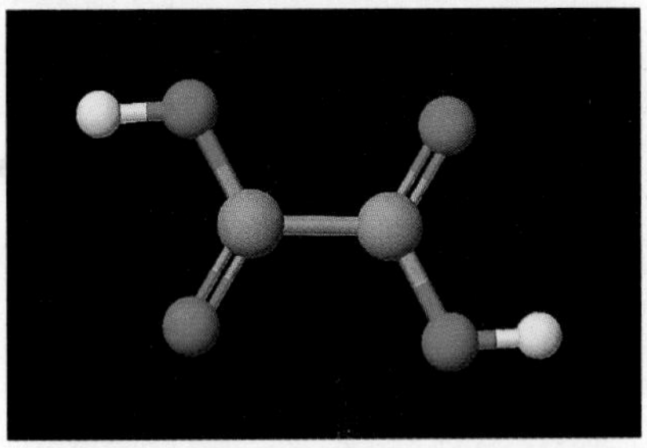
(c)

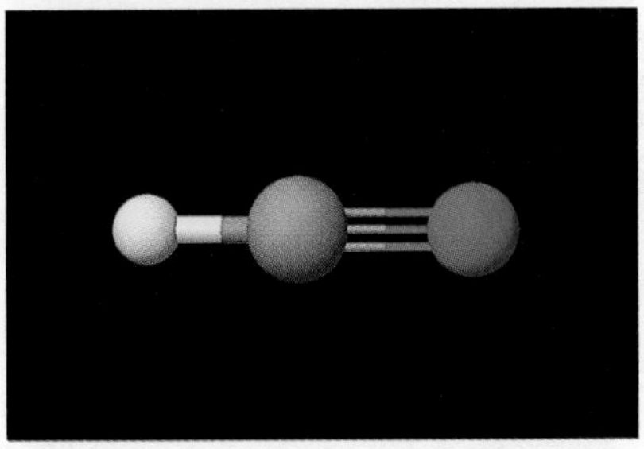
(d)

Figure 17.3 Some weak acids. (a) Formic acid, HCO_2H. (b) Acetic acid, CH_3CO_2H. (c) Oxalic acid, $H_2C_2O_4$. (d) Hydrocyanic acid, HCN. Note the similarity in structure of the organic acids. All have the carboxylic acid group, $—CO_2H$. (Susan Young)

Some Weak Acids

Neutral Molecules as Acids

A few neutral molecules with an ionizable hydrogen atom are strong Brønsted acids, but the vast majority are weak. Several weak acids are pictured in Figure 17.3, and a number are listed in Table 17.4.

TABLE 17.4 Ionization Constants for Some Acids and Their Conjugate Bases

Acid Name	Acid	K_a	Base	K_b	Base Name
Perchloric acid	$HClO_4$	large	ClO_4^-	very small	perchlorate ion
Sulfuric acid	H_2SO_4	large	HSO_4^-	very small	hydrogen sulfate ion
Hydrochloric acid	HCl	large	Cl^-	very small	chloride ion
Nitric acid	HNO_3	large	NO_3^-	very small	nitrate ion
Hydronium ion	H_3O^+	55.5	H_2O	1.8×10^{-16}	water
Sulfurous acid	H_2SO_3	1.2×10^{-2}	HSO_3^-	8.3×10^{-13}	hydrogen sulfite ion
Hydrogen sulfate ion	HSO_4^-	1.2×10^{-2}	SO_4^{2-}	8.3×10^{-13}	sulfate ion
Phosphoric acid	H_3PO_4	7.5×10^{-3}	$H_2PO_4^-$	1.3×10^{-12}	dihydrogen phosphate ion
Hexaaquairon(III) ion	$Fe(H_2O)_6^{3+}$	6.3×10^{-3}	$Fe(H_2O)_5OH^{2+}$	1.6×10^{-12}	pentaaquahydroxoiron(III) ion
Hydrofluoric acid	HF	7.2×10^{-4}	F^-	1.4×10^{-11}	fluoride ion
Nitrous acid	HNO_2	4.5×10^{-4}	NO_2^-	2.2×10^{-11}	nitrite ion
Formic acid	HCO_2H	1.8×10^{-4}	HCO_2^-	5.6×10^{-11}	formate ion
Benzoic acid	$C_6H_5CO_2H$	6.3×10^{-5}	$C_6H_5CO_2^-$	1.6×10^{-10}	benzoate ion
Acetic acid	CH_3CO_2H	1.8×10^{-5}	$CH_3CO_2^-$	5.6×10^{-10}	acetate ion
Propanoic acid	$CH_3CH_2CO_2H$	1.3×10^{-5}	$CH_3CH_2CO_2^-$	7.7×10^{-10}	propanoate ion
Hexaaquaaluminum ion	$Al(H_2O)_6^{3+}$	7.9×10^{-6}	$Al(H_2O)_5OH^{2+}$	1.3×10^{-9}	pentaaquahydroxoaluminum ion
Carbonic acid	H_2CO_3	4.2×10^{-7}	HCO_3^-	2.4×10^{-8}	hydrogen carbonate ion
Hexaaquacopper(II) ion	$Cu(H_2O)_6^{2+}$	1.6×10^{-7}	$Cu(H_2O)_5OH^+$	6.25×10^{-8}	pentaaquahydroxocopper(II) ion
Hydrogen sulfide	H_2S	1×10^{-7}	HS^-	1×10^{-7}	hydrogen sulfide ion
Dihydrogen phosphate ion	$H_2PO_4^-$	6.2×10^{-8}	HPO_4^{2-}	1.6×10^{-7}	hydrogen phosphate ion
Hydrogen sulfite ion	HSO_3^-	6.2×10^{-8}	SO_3^{2-}	1.6×10^{-7}	sulfite ion
Hypochlorous acid	$HClO$	3.5×10^{-8}	ClO^-	2.9×10^{-7}	hypochlorite ion
Hexaaqualead(II) ion	$Pb(H_2O)_6^{2+}$	1.5×10^{-8}	$Pb(H_2O)_5OH^+$	6.7×10^{-7}	pentaaquahydroxolead(II) ion
Hexaaquacobalt(II) ion	$Co(H_2O)_6^{2+}$	1.3×10^{-9}	$Co(H_2O)_5OH^+$	7.7×10^{-6}	pentaaquahydroxocobalt(II) ion
Boric acid	$B(OH)_3(H_2O)$	7.3×10^{-10}	$B(OH)_4^-$	1.4×10^{-5}	tetrahydroxoborate ion
Ammonium ion	NH_4^+	5.6×10^{-10}	NH_3	1.8×10^{-5}	ammonia
Hydrocyanic acid	HCN	4.0×10^{-10}	CN^-	2.5×10^{-5}	cyanide ion
Hexaaquairon(II) ion	$Fe(H_2O)_6^{2+}$	3.2×10^{-10}	$Fe(H_2O)_5OH^+$	3.1×10^{-5}	pentaaquahydroxoiron(II) ion
Hydrogen carbonate ion	HCO_3^-	4.8×10^{-11}	CO_3^{2-}	2.1×10^{-4}	carbonate ion
Hexaaquanickel(II) ion	$Ni(H_2O)_6^{2+}$	2.5×10^{-11}	$Ni(H_2O)_5OH^+$	4.0×10^{-4}	pentaaquahydroxonickel(II) ion
Hydrogen phosphate ion	HPO_4^{2-}	3.6×10^{-13}	PO_4^{3-}	2.8×10^{-2}	phosphate ion
Water	H_2O	1.8×10^{-16}	OH^-	55.5	hydroxide ion
Hydrogen sulfide ion*	HS^-	1×10^{-19}	S^{2-}	1×10^5	sulfide ion
Ethanol	C_2H_5OH	very small	$C_2H_5O^-$	large	ethoxide ion
Ammonia	NH_3	very small	NH_2^-	large	amide ion
Hydrogen	H_2	very small	H^-	large	hydride ion
Methane	CH_4	very small	CH_3^-	large	methide ion

Increasing Acid Strength ↑ (acid column); Increasing Base Strength ↓ (base column)

*The values of K_a for HS^- and K_b for S^{2-} are estimates.

Cations as Weak Acids

The ammonium ion is an example of a cation acting as a weak Brønsted acid.

$$NH_4^+(aq) + H_2O(\ell) \rightleftharpoons NH_3(aq) + H_3O^+(aq) \qquad K_a = 5.6 \times 10^{-10}$$

When the salt of a metal cation is placed in water, the metal ion becomes hydrated, as explained in Section 13.2. In fact, the interaction is usually sufficiently strong that the ion is surrounded by as many as six water molecules, $[M(H_2O)_6]^{n+}$, where M represents a metal ion with a charge of $n+$. Metal ions with a charge of 2+ or 3+, such as Al^{3+} and many transition metal ions, produce acidic, aqueous solutions.

$$[Cu(H_2O)_6]^{2+}(aq) + H_2O(\ell) \rightleftharpoons [Cu(H_2O)_5(OH)]^+(aq) + H_3O^+(aq) \qquad K_a = 1.6 \times 10^{-7}$$

With a K_a value of 1.6×10^{-7}, the aqueous Cu^{2+} ion is less acidic than acetic acid, but more acidic than $H_2PO_4^-$ (see Table 17.4).

Figure 17.4 Baking powder contains the weak acid $Ca(H_2PO_4)_2$, calcium dihydrogen phosphate. This can react with the hydrogen carbonate ion in baking soda ($NaHCO_3$) to form HPO_4^{2-} ion as well as carbon dioxide and water. Here solid KH_2PO_4 is added to a solution of $NaHCO_3$.
(C. D. Winters)

Anions as Weak Acids

Six anionic Brønsted acids are listed in Table 17.4. The dihydrogen phosphate anion, $H_2PO_4^-$, for example, is the acid in baking powder,

$$H_2PO_4^-(aq) + H_2O(\ell) \rightleftharpoons HPO_4^{2-}(aq) + H_3O^+(aq) \qquad K_a = 6.2 \times 10^{-8}$$

where its function is to provide a hydrogen ion to produce CO_2 from baking soda ($NaHCO_3$) (Figure 17.4).

$$\underset{\text{Weak acid}}{H_2PO_4^-(aq)} + \underset{\text{Weak base}}{HCO_3^-(aq)} \rightleftharpoons \underset{\text{Conjugate acid of bicarbonate ion}}{H_2CO_3(aq)} + \underset{\text{Conjugate base of } H_2PO_4^-}{HPO_4^{2-}(aq)}$$

$$H_2CO_3(aq) \rightarrow CO_2(g) + H_2O(\ell)$$

Some Weak Bases

A variety of weak bases are important in chemistry, and some are listed in Table 17.4.

Neutral Molecules as Bases

Ammonia, NH_3, which is perhaps the best known weak base, produces a very small amount of hydroxide ion when NH_3 accepts a proton from water.

$$NH_3(aq) + H_2O(\ell) \rightleftharpoons NH_4^+(aq) + OH^-(aq) \qquad K_b = 1.8 \times 10^{-5}$$

The compound is also the parent of a large series of compounds called amines, in which the H atoms of NH_3 are replaced by some other substituent. For example, if the substituent is the methyl group, CH_3, we have

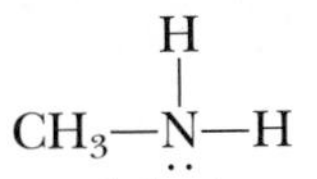

methylamine
$K_b = 5.0 \times 10^{-4}$

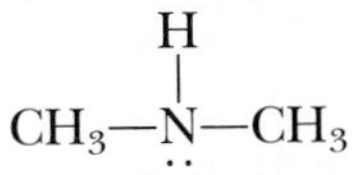

dimethylamine
$K_b = 7.4 \times 10^{-4}$

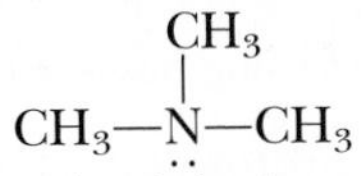

trimethylamine
$K_b = 7.4 \times 10^{-5}$

Many foods and household products contain weak acids such as acetic acid (in vinegar), citric acid (in fruit juices), and the hydrogen sulfate ion (in sodium hydrogen sulfate used as a pH adjuster for swimming pools).
(C. D. Winters)

Common cations from Groups 1A and 2A, such as Na^+, K^+, Ca^{2+}, and Mg^{2+}, do not contribute to the acidity of solutions. See Example 17.1 and Section 17.8.

Other examples of weak bases that are neutral molecules include nicotine, caffeine, and the antimalarial compound quinine (see Section 11.7).

Anions as Weak Bases

You have already seen examples of anions acting as Brønsted bases in aqueous solution. For example, the cyanide ion, CN^-, the conjugate base of the weak acid HCN, produces a measurable concentration of hydroxide ion in water,

$$\underset{\text{Base}}{CN^-(aq)} + \underset{\text{Acid}}{H_2O(\ell)} \rightleftharpoons \underset{\text{Conjugate acid of } CN^-}{HCN(aq)} + \underset{\text{Conjugate base of water}}{OH^-(aq)}$$

$$K_b = \frac{[HCN][OH^-]}{[CN^-]} = 2.5 \times 10^{-5}$$

Cyanide ion is a weaker base than hydroxide ion, so the equilibrium is predicted to lie to the left. Nonetheless, according to Table 17.4, CN^- is intermediate in base strength between two well-known bases, NH_3 and CO_3^{2-}, so the CN^- ion clearly reacts with water to a small extent to produce a basic solution. In general, *the conjugate base of a weak acid produces a basic solution in water.*

EXERCISE 17.6 *Weak Acids and Bases*

1. Lactic acid, $CH_3CHOHCO_2H$, has a value of K_a of 1.4×10^{-4}. Where does this fit in Table 17.4?

$$CH_3CHOHCO_2H(aq) + H_2O(\ell) \rightleftharpoons H_3O^+(aq) + CH_3CHOHCO_2^-(aq)$$

2. Write a balanced equation for the lactate ion, $CH_3CHOHCO_2^-$, functioning as a Brønsted base in water. ■

Many common products contain weak bases. Examples include aqueous ammonia, detergents, phosphates in fertilizers, and citrates and benzoates in soft drinks. (C. D. Winters)

PROBLEM-SOLVING TIPS AND IDEAS

17.1 Strong or Weak?

How can you tell whether an acid or base is weak? The easiest way is to remember those few that are strong (see Table 4.1 and the following short table), and all others are probably weak. See also Sections 17.3 and 17.4. Some common **strong acids** are

Hydrohalic acids: HCl, HBr, and HI

Nitric acid: HNO_3

Sulfuric acid: H_2SO_4 (for loss of first H^+ only)

Perchloric acid: $HClO_4$

Some common **strong bases** are

Group 1A hydroxides: LiOH, NaOH, KOH

Group 2A hydroxides: $Ca(OH)_2$, $Sr(OH)_2$, and $Ba(OH)_2$ ■

17.6 WATER AND THE pH SCALE

The stronger a Brønsted acid or base, the larger the concentration of $H_3O^+(aq)$ or $OH^-(aq)$ for a given concentration of acid or base, respectively. If these concentrations could be measured quantitatively, we would have a way to compare acid and base strengths.

The Water Ionization Constant, K_w

Water autoionizes, transferring a proton from one water molecule to another and producing a hydronium ion and a hydroxide ion.

$$2\,H_2O(\ell) \rightleftharpoons H_3O^+(aq) + OH^-(aq)$$

Because the hydroxide ion is a much stronger base than water, and the hydronium ion is a much stronger acid than water (Table 17.4), the equilibrium lies far to the left side. In fact, in pure water at 25 °C only about 2 out of a billion water molecules are ionized at any instant. To express this idea more quantitatively, we can write the equilibrium constant expression

$$K = \frac{[H_3O^+][OH^-]}{[H_2O]^2}$$

Recall from Section 16.2 that in pure water or in dilute aqueous solutions (say 0.1 M solute or less), the concentration of water can be considered to be a constant (55.5 M). For this reason $[H_2O]^2$ is included in the constant K, and the equilibrium constant expression becomes

$$K[H_2O]^2 = [H_3O^+][OH^-]$$

$$K_w = [H_3O^+][OH^-] \qquad (17.3)$$

This equilibrium constant is given a special symbol $\boldsymbol{K_w}$, and is known as the **ionization constant for water.** In pure water, the transfer of a proton between two water molecules leads to one H_3O^+ and one OH^-. Because this is the only source of these ions, we know that $[H_3O^+]$ must equal $[OH^-]$ in pure water. Electrical conductivity measurements of pure water show that $[H_3O^+] = [OH^-] = 1.0 \times 10^{-7}$ M at 25 °C, and so

$$K_w = [H_3O^+][OH^-] = (1.0 \times 10^{-7})(1.0 \times 10^{-7}) = 1.0 \times 10^{-14}$$

The hydronium ion and hydroxide ion concentrations in pure water are *both* 1.0×10^{-7} at 25 °C, and the water is said to be **neutral.** If some acid or base is added to pure water, however, the equilibrium

$$2\,H_2O(\ell) \rightleftharpoons H_3O^+(aq) + OH^-(aq)$$

is disturbed. Adding acid raises the concentration of the H_3O^+ ions. To oppose this increase, Le Chatelier's principle (Section 16.6) predicts that a small fraction of the H_3O^+ ions reacts with OH^- ions to form water. This lowers $[OH^-]$ until the product of $[H_3O^+]$ and $[OH^-]$ is again equal to 1.0×10^{-14} at 25 °C. Similarly, adding a base to pure water raises the OH^- ion concentration. Le Chatelier's principle predicts that some of the added OH^- ions react with H_3O^+ ions present in the solution, thereby lowering $[H_3O^+]$ until the value of the product $[H_3O^+][OH^-]$ equals 1.0×10^{-14} at 25 °C.

The equation $K_w = [H_3O^+][OH^-]$ is valid in pure water and in any aqueous solution. K_w is temperature-dependent; the autoionization reaction is endothermic, so K_w increases with temperature.

°C	K_w
10	0.29×10^{-14}
15	0.45×10^{-14}
20	0.68×10^{-14}
25	1.01×10^{-14}
30	1.47×10^{-14}
50	5.48×10^{-14}

Thus, for aqueous solutions at 25 °C, we can say that

- In a neutral solution $[H_3O^+] = [OH^-]$

  Both are equal to 1.0×10^{-7} M
- In an acidic solution $[H_3O^+] > [OH^-]$.

  $[H_3O^+] > 1.0 \times 10^{-7}$ M and $[OH^-] < 1.0 \times 10^{-7}$ M
- In a basic solution $[H_3O^+] < [OH^-]$.

  $[H_3O^+] < 1.0 \times 10^{-7}$ M and $[OH^-] > 1.0 \times 10^{-7}$ M

EXAMPLE 17.2 *Ion Concentrations in a Solution of a Strong Base*

If you have a 0.0010 M aqueous solution of NaOH, what are the hydroxide and hydronium ion concentrations?

Solution NaOH is a strong base and so is 100% dissociated into ions in water to give an initial concentration of OH^- of 0.0010 M. Before adding NaOH, the water initially contained a small concentration of OH^- (10^{-7} M) from the autoionization of water.

To solve this problem, we take the point of view that the water ionization reaction occurs only *after* excess OH^- has been added (in the form of NaOH). Water ionization gives equal concentrations of H_3O^+ and OH^-, and both are equal to the unknown quantity x. Thus, the *net* OH^- equilibrium concentration must be (0.0010 M + OH^- from water) = $(0.0010 + x)$.

Equation	$2\ H_2O(\ell) \rightleftharpoons$	$H_3O^+(aq)$	$+\ OH^-(aq)$
Before ionization (M)		0	0.0010
Change in concentrations on proceeding to equilibrium		$+x$	$+x$
After equilibrium is achieved (M)		x	$0.0010 + x$

We can solve for x from the expression for K_w.

$$K_w = 1.0 \times 10^{-14} = [H_3O^+][OH^-] = (x)(0.0010 + x)$$

On expanding this equation to solve for x, it is found to be a quadratic equation that can be solved by the usual methods (see Appendix A and Example 16.8). A useful approximation, however, can be made that simplifies the calculation. According to Le Chatelier's principle, the water autoionization equilibrium is suppressed by the presence of the OH^- ion from NaOH. Thus, the value of x, the concentration of H_3O^+ and OH^- coming from water, must be *smaller* than 10^{-7}. This means that x in the term $(0.0010 + x)$ can be ignored. (Following the usual rules for significant figures, the sum of 0.0010 and 1.0×10^{-7} is 0.0010.) For this reason, we can write the following approximate, but nearly correct expression:

$$K_w = 1.0 \times 10^{-14} = [H_3O^+][OH^-] \approx (x)(0.0010)$$

and so $x = [H_3O^+]$ in the presence of 0.0010 M NaOH $\approx 1.0 \times 10^{-11}$ M.

As a final step, it is useful to check our approximation:

$$[H_3O^+][OH^-] = (1.0 \times 10^{-11})(0.0010 + 1.0 \times 10^{-11}) \approx 1.0 \times 10^{-14}$$

The calculated product of the H_3O^+ and OH^- ion concentrations is effectively K_w, so the approximation is valid.

EXERCISE 17.7 ***Hydronium Ion Concentration in a Solution of a Strong Acid***

Gaseous HCl (0.0020 mol) is bubbled into 5.0×10^2 mL of water to make an aqueous HCl solution. What are the concentrations of H_3O^+ and OH^- in this solution? ■

The Connection Between the Ionization Constants for an Acid and Its Conjugate Base

Table 17.4 informs us that, for a series of acids, the strength of their conjugate bases increases as the acid strength decreases. Now that you have been introduced to K_w, the autoionization constant for water, this relationship can be expressed mathematically,

$$K_a \cdot K_b = K_w \quad (17.4)$$

where K_a is the ionization constant for a weak acid, and K_b is the ionization constant for its conjugate base. As the value of K_a decreases, the value of K_b must increase because their product is a constant (at a given temperature).

To see the origin of Equation 17.4, suppose you add the equation for the ionization of a weak acid, say HCN, and the equation for the hydrolysis of its conjugate base, CN^-.

$$\begin{array}{lll} HCN(aq) + H_2O(\ell) \rightleftharpoons H_3O^+(aq) + CN^-(aq) & & K_a = 4.0 \times 10^{-10} \\ CN^-(aq) + H_2O(\ell) \rightleftharpoons HCN(aq) + OH^-(aq) & & K_b = 2.5 \times 10^{-5} \\ \hline 2\ H_2O(\ell) \rightleftharpoons H_3O^+(aq) + OH^-(aq) & & K_w = 1.0 \times 10^{-14} \end{array}$$

The result is the equation for the autoionization of water. Recall from Section 16.2 that the equilibrium constant for a reaction that is the sum of two others is the product of the equilibrium constants for the summed reactions. Therefore,

$$K_a \cdot K_b = \left(\frac{[H_3O^+][CN^-]}{[HCN]}\right)\left(\frac{[HCN][OH^-]}{[CN^-]}\right) = [H_3O^+][OH^-] = K_w$$

Equation 17.4 is useful because K_b can be calculated from a knowledge of K_a. For example, the value of K_b for the cyanide ion in Table 17.4 was calculated from the value of K_a for its conjugate acid, HCN.

$$K_b = \frac{K_w}{K_a \text{ for HCN}} = \frac{1.0 \times 10^{-14}}{4.0 \times 10^{-10}} = 2.5 \times 10^{-5}$$

EXERCISE 17.8 ***Using the Equation*** $K_a \cdot K_b = K_w$

K_a for lactic acid, $CH_3CHOHCO_2H$, is 1.4×10^{-4} (see Exercise 17.6). What is K_b for the conjugate base of this acid, $CH_3CHOHCO_2^-$? Where does this base fit in Table 17.4? ■

The pH Scale

A second way of expressing hydronium ion concentration avoids using very small numbers or exponential notation. This is the pH scale, a widely used method of expressing acidity. The **pH** of a solution is defined as *the negative of the base-10 logarithm (log) of the hydronium ion concentration.*

In general, pX = −log X, where X can be any measureable quantity.

$$pH = -\log[H_3O^+] \tag{17.5}$$

In a similar way, the pOH of a solution is defined as the negative of the base-10 logarithm of the hydroxide ion concentration.

$$pOH = -\log[OH^-]$$

In pure water, the hydronium and hydroxide ion concentrations are both 1.0×10^{-7} M. Therefore, for pure water at 25 °C

$$\begin{aligned} pH &= -\log(1.0 \times 10^{-7}) = -[\log(1.0) + \log(10^{-7})] \\ &= -[(0.00) + (-7)] \\ &= 7.00 \end{aligned}$$

In the same way, you can show that the pOH of pure water is also 7.00 at 25 °C.

If we take the negative logarithms of both sides of the expression $K_w = [H_3O^+][OH^-]$, we obtain another useful equation.

$$\begin{aligned} K_w = [H_3O^+][OH^-] &= 1.0 \times 10^{-14} \\ -\log([H_3O^+][OH^-]) &= -\log(1.0 \times 10^{-14}) \\ -\log([H_3O^+]) + (-\log[OH^-]) &= 14.00 \end{aligned}$$

$$pH + pOH = 14.00 \tag{17.6}$$

The sum of the pH and pOH of a solution must be equal to 14.00 at 25 °C.

EXAMPLE 17.3 *Calculating pH*

An aqueous solution contains 0.700 g of NaOH and has a volume of 500. mL. What is its pH?

Solution As described in Example 17.2, sodium hydroxide is totally dissociated in aqueous solution to Na^+ and OH^- ions. The first step is to find the concentration of the OH^- ions.

$$0.700 \text{ g NaOH}\left(\frac{1 \text{ mol NaOH}}{40.00 \text{ g NaOH}}\right) = 0.0175 \text{ mol NaOH}$$

$$[OH^-] = \frac{0.0175 \text{ mol}}{0.500 \text{ L}} = 0.0350 \text{ M}$$

Here the OH^- ion concentration is 0.0350 M from the NaOH *plus* an additional amount from the autoionization of water. As we concluded in Example 17.2, however, the contribution to the OH^- ion concentration from water is *exceedingly* small, so we assume the solution has $[OH^-] = 0.0350$ M. Substituting this into the K_w expression, we have

$$K_w = [H_3O^+][OH^-] = [H_3O^+](0.0350) = 1.0 \times 10^{-14}$$
$$[H_3O^+] = 2.9 \times 10^{-13}\ \text{M}$$
$$pH = -\log [H_3O^+] = -\log (2.9 \times 10^{-13}) = -(-12.54) = 12.54$$

Notice that the same result is obtained if the pOH is calculated from the OH^- ion concentration.

$$pOH = -\log [OH^-] = -\log (0.0350) = -(-1.456) = 1.456$$

With the pOH known, pH can be found from Equation 17.6.

$$pH + pOH = 14.00$$
$$pH + 1.456 = 14.00$$
$$pH = 12.54$$

EXAMPLE 17.4 ***pH and Hydronium Ion Conversions***

(a) Suppose the hydronium ion concentration in vinegar is 1.6×10^{-3} M. Calculate its pH.

(b) The pH of seawater is 8.30. Calculate $[H_3O^+]$ and $[OH^-]$.

Solution (a) The pH of vinegar is found from Equation 17.5.

$$pH = -\log [H_3O^+] = -\log (1.6 \times 10^{-3}) = -(-2.80) = 2.80$$

(b) The pH of seawater is 8.30. Therefore,

$$pH = 8.30 = -\log [H_3O^+]$$
$$\log [H_3O^+] = -pH = -8.30$$

We can find $[H_3O^+]$ by finding the antilog of the negative of the pH.

$$[H_3O^+] = 10^{-pH} = 10^{-8.30} = 5.0 \times 10^{-9}\ \text{M}$$

There are two ways to calculate the concentration of the OH^- ion. You can use Equation 17.4

$$K_w = [H_3O^+][OH^-] = 1.0 \times 10^{-14} \text{ at } 25\ °C$$

and calculate $[OH^-]$ from it.

$$(5.0 \times 10^{-9})[OH^-] = 1.0 \times 10^{-14}$$
$$[OH^-] = 2.0 \times 10^{-6}\ \text{M}$$

Alternatively, you can calculate the pOH from pH + pOH = 14.00, and then convert pOH to $[OH^-]$. Try this to see if you do indeed obtain 2.0×10^{-6} M.

PROBLEM-SOLVING TIPS AND IDEAS

17.2 Calculating and Using pH

- An alternative and useful form of the definitions of pH and pOH is

$$[H_3O^+] = 10^{-pH} \qquad [OH^-] = 10^{-pOH}$$

This is useful because you can key −pH (or −pOH) into your calculator, for example, and then press the 10^x key (or its equivalent on your calculator) to find the hydronium ion (or hydroxide ion) concentration.

- Significant figures and logarithms:
The digits to the left of the decimal point in a pH represent a power of 10. Only the digits to the right of the decimal are significant. In Example 17.3,

$$pH = -\log(2.9 \times 10^{-13}) = -\log(2.9) + (-\log(10^{-13})) = -0.46 + 13.00 = 12.54$$

there are two digits to the right of the decimal in 12.54 because the H_3O^+ concentration had two significant figures. ■

A 0.0040 M solution of HCl, a strong acid, has a pH of 2.40 (Exercise 17.7), whereas a 0.0350 M solution of NaOH, a strong base, has a pH of 12.54 (Example 17.3). Because the pH of pure water is 7.00 at 25 °C,

Solutions with pH less than 7.00 (at 25 °C) are acidic; solutions with pH greater than 7.00 are basic. Solutions with pH = 7.00 at 25 °C are considered neutral.

The relation between acidity, basicity, and pH or pOH is illustrated graphically in the chart below. To give you a feeling for the pH scale, the approximate pH values for some common aqueous solutions are given in Figure 17.5.

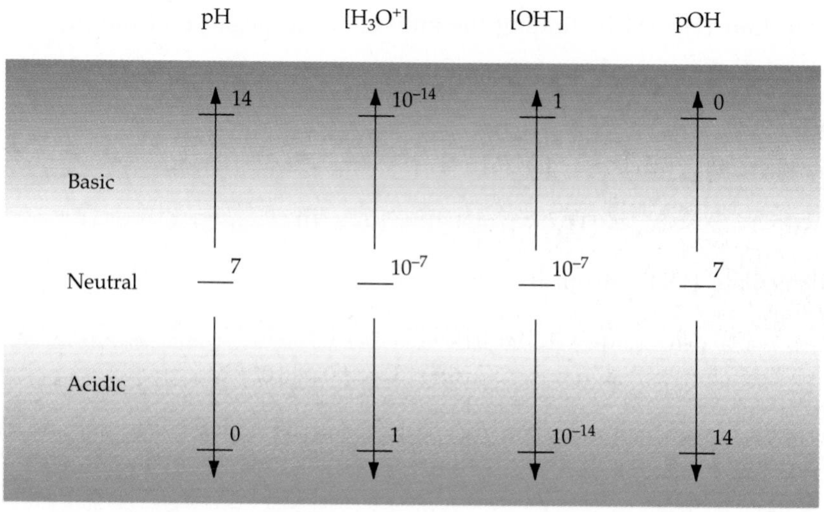

The relation between hydronium and hydroxide ion concentrations and pH and pOH.

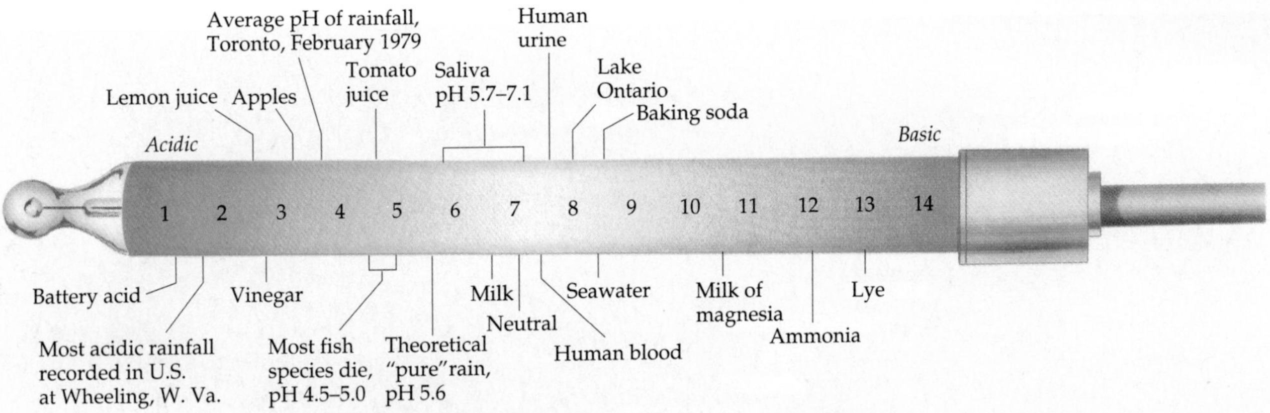

Figure 17.5 The pH of some common aqueous solutions. The scale is superimposed on the drawing of a pH electrode used in the measurement of pH by an instrumental method.

EXERCISE 17.9 ***pH and Hydronium Ion Conversions***

The pH of a diet soda is 4.32 at 25 °C. What are the hydronium and hydroxide ion concentrations in the soda? ■

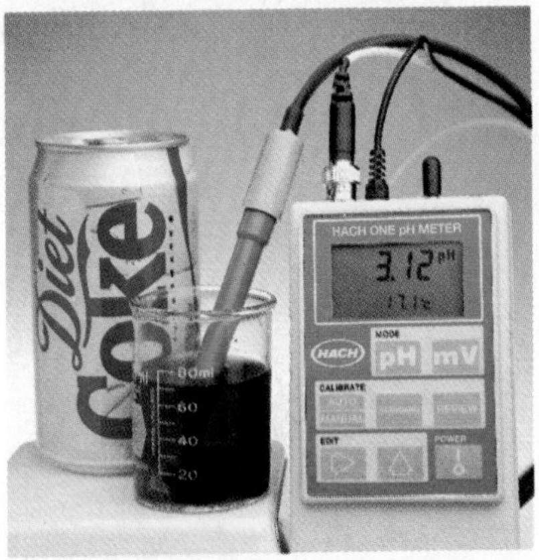

Soft drinks are often quite acidic, as indicated by this diet drink, which has a pH of about 3.1. (C. D. Winters)

Determining pH

The pH of a solution can be determined approximately using an **indicator,** *a substance that changes color in some known pH range* (Figures 17.6 and 17.7). Recall that acids were originally defined by the fact that they made certain vegetable dyes turn red. Indicators, such as phenolphthalein, are often large molecules derived from plants.

Figure 17.6 A collection of household products. Each solution contains a few drops of a chemical called a pH indicator (in this case a "universal indicator"). A color of yellow or red indicates a pH less than 7. A color of green to purple indicates a pH greater than 7.

Some acid-base indicators

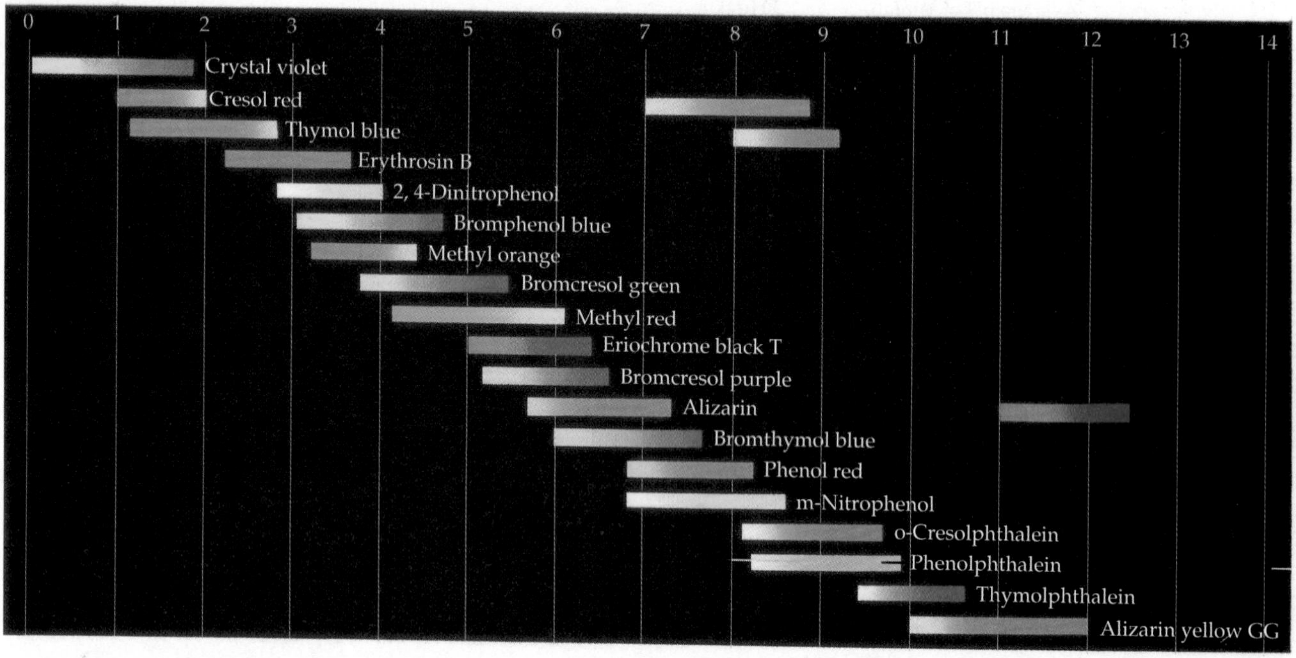

Figure 17.7 Some common acid-base indicators. The color changes occur over a range of pH values. Notice that a few indicators have two color changes over two different pH ranges. (Hach Company)

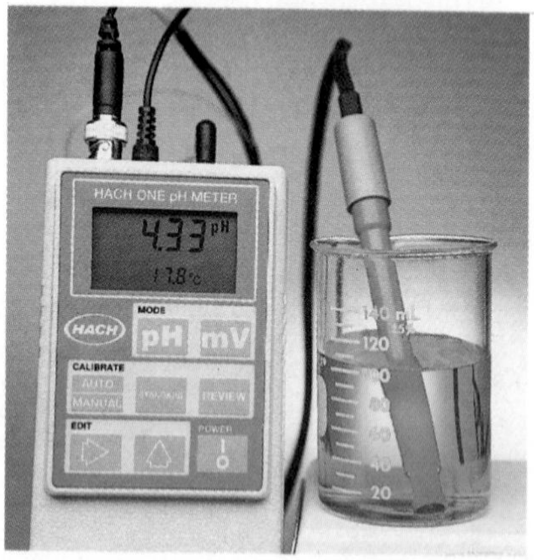

Figure 17.8 An electronic instrument for the measurement of pH. The blue solution is 0.10 M $CuSO_4$, and the pH of 4.5 illustrates the fact that many metal ions give weakly acidic solutions (see Section 17.8). (C. D. Winters)

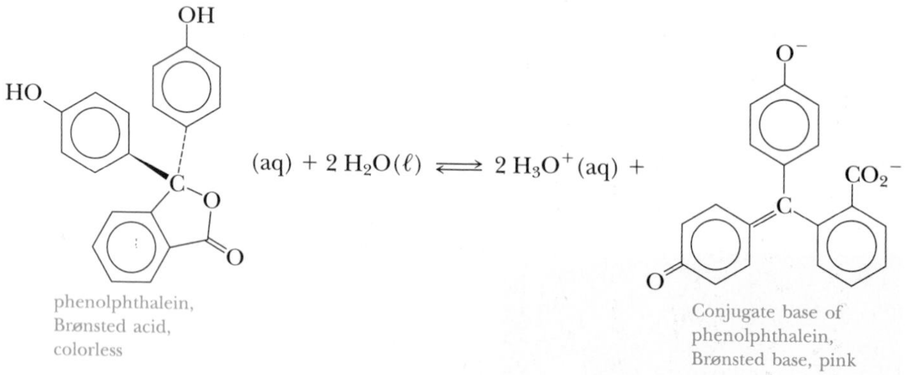

These dyes, or indicators, can exist in conjugate acid and base forms, and they have different colors depending on the form in which they are found. Common litmus paper is impregnated with a natural plant juice that is red in solutions more acidic than about pH = 5 but blue when the pH exceeds about 8.2 in basic solution.

Although a pH may be approximated by using an appropriate chemical indicator, a modern pH meter is far preferable for accurately determining pH (Figure 17.8).

17.7 EQUILIBRIA INVOLVING WEAK ACIDS AND BASES

Calculating K_a or K_b from Initial Concentrations and Measured pH

The K_a and K_b values found in Table 17.4 and in the more extensive tables in Appendices F and G were all determined by experiment. There are several ways to approach the problem, but one method is to determine the pH of the solution. The following example illustrates the method.

EXAMPLE 17.5 *Calculating a K_a from a Measured pH*

Lactic acid is a monoprotic acid that occurs naturally in sour milk and arises from metabolism in the human body. (See Section 11.5 and A Closer Look: Muscle Cramps and Hangovers, page 521.) A 0.10 M aqueous solution of lactic acid, $CH_3CHOHCO_2H$, has a pH of 2.43. What is the value of K_a for lactic acid?

Solution The equation for the equilibrium interaction of lactic acid with water is

$$\underset{\text{lactic acid}}{CH_3CHOHCO_2H(aq)} + H_2O(\ell) \rightleftharpoons H_3O^+(aq) + \underset{\text{lactate ion}}{CH_3CHOHCO_2^-(aq)}$$

$$K_a \text{ (lactic acid)} = \frac{[H_3O^+][CH_3CHOHCO_2^-]}{[CH_3CHOHCO_2H]}$$

To calculate K_a, we need the equilibrium concentration of each species. The pH of the solution directly tells us the equilibrium concentration of $[H_3O^+]$, and we can derive the others ($[CH_3CHOHCO_2H]$ and $[CH_3CHOHCO_2^-]$) from this. We therefore begin by converting the pH to $[H_3O^+]$.

$$[H_3O^+] = 10^{-pH} = 10^{-2.43} = 3.7 \times 10^{-3}\text{ M}$$

The equilibrium table shown here gives the initial concentrations of the important species in solution, describes what happens as the reaction proceeds to equilibrium, and then shows the equilibrium concentrations.

Equation	$CH_3CHOHCO_2H$ + H_2O $\rightleftharpoons$	H_3O^+ +	$CH_3CHOHCO_2^-$
Before ionization (M)	0.10	0	0
Change on proceeding to equilibrium	$-x$	$+x$	$+x$
After equilibrium is achieved (M)	$0.10 - x$	x	x

The following points can be made concerning the equilibrium table:

1. Hydronium ion, H_3O^+, is present in solution from lactic acid ionization *and* from water autoionization. Le Chatelier's principle informs us that the H_3O^+ added to the water by lactic acid suppresses the H_3O^+ coming from the water autoionization. Because measurement of the pH tells us the *total* $[H_3O^+]$ is 3.7×10^{-3} M, and because $[H_3O^+]$ from water must be less than 10^{-7} M, the pH is almost completely a reflection of H_3O^+ from lactic acid. (For weak acids

and bases, this approximation can *almost* always be made. The exception is when x is near 10^{-7}; that is, when the equilibrium pH is in the 6 to 8 range. Think through each case to be certain.)

2. The quantity x represents the equilibrium concentrations of hydronium ion and lactate ion. That is, at equilibrium $[H_3O^+] \approx [C_3H_5O_3^-] = x = 3.7 \times 10^{-3}$ M.
3. By stoichiometry, x is also the quantity of acid that ionized on proceeding to equilibrium.

Bearing these points in mind, we can calculate K_a for lactic acid.

$$K_a = \frac{[H_3O^+][CH_3CHOHCO_2^-]}{[CH_3CHOHCO_2H]} = \frac{(3.7 \times 10^{-3})(3.7 \times 10^{-3})}{0.10 - 0.0037} = 1.4 \times 10^{-4}$$

Comparing this value of K_a with others in Table 17.4, you see that lactic acid can be classed as a moderately weak acid.

EXERCISE 17.10 ***Calculating a*** $\mathbf{K_a}$ ***Value from a Measured pH***

A solution prepared from 0.10 mol of propanoic acid dissolved in sufficient water to give 1.0 L of solution has a pH of 2.94. Determine K_a for propanoic acid. The acid ionizes according to the balanced equation

$$CH_3CH_2CO_2H(aq) + H_2O(\ell) \rightleftharpoons H_3O^+(aq) + CH_3CH_2CO_2^-(aq)$$ ■

There is an important point to notice in Example 17.5. The lactic acid concentration at equilibrium was given by "original acid concentration − quantity of acid ionized." In this example the expression was (0.10 − 0.0037). By the usual rules governing significant figures, (0.10 − 0.0037) is equal to 0.10. The acid is weak, so very little of it ionizes (approximately 4%), and the value of $[CH_3CHOHCO_2H]$ at equilibrium is essentially equal to its initial value. The approximation that the denominator in the K_a expression for a weak acid, HA,

$$HA(aq) + H_2O(\ell) \rightleftharpoons H_3O^+(aq) + A^-(aq)$$

$$K_a = \frac{[H_3O^+][A^-]}{[HA]_0 - [H_3O^+]} \approx \frac{[H_3O^+][A^-]}{[HA]_0}$$

is just $[HA]_0$, the initial acid concentration, is useful in calculations involving weak acids and bases. Error analysis shows that

> The approximation that $[\text{acid}]_{\text{equilibrium}}$ is effectively equal to $[\text{acid}]_{\text{initial}}$ ($= [HA]_0$) is valid whenever $[HA]_0$ is greater than or equal to $100 \cdot K_a$.

The use of this approximation is illustrated in the next several examples.

Calculating Equilibrium Concentrations and pH from Initial Concentrations and K_a or K_b

Knowing values of the equilibrium constants for weak acids and bases enables us to calculate the pH of a solution of a weak acid or base.

EXAMPLE 17.6 *Calculating Equilibrium Concentrations and pH from* K_a

Benzoic acid occurs free and combined in nature. For example, most berries contain up to 0.05% of this acid by weight, and sodium benzoate is often used as a preservative in soft drinks.

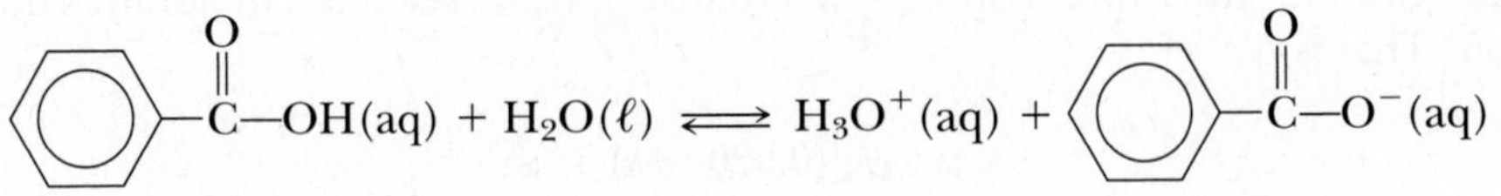

benzoic acid ($C_6H_5CO_2H$) — benzoate ion ($C_6H_5CO_2^-$)

Calculate the pH of a 0.020 M solution of benzoic acid if $K_a = 6.3 \times 10^{-5}$.

Solution As usual, we organize the information in a table.

Equation	$C_6H_5CO_2H$ + H_2O	⇌ H_3O^+	+ $C_6H_5CO_2^-$
Before ionization (M)	0.020	0	0
Change on proceeding to equilibrium	$-x$	$+x$	$+x$
After equilibrium is achieved (M)	$(0.020 - x)$	x	x

Every mole of benzoic acid that ionizes gives 1 mol of hydronium ion and 1 mol of benzoate ion, so $[H_3O^+] = [C_6H_5CO_2^-] = x$ at equilibrium. (Here we ignored any H_3O^+ that arises from water ionization; see Example 17.5.) Furthermore, stoichiometry tells us that the quantity of acid ionized is also x. Thus, the benzoic acid concentration at equilibrium is

$$[C_6H_5CO_2H] = \text{initial acid concentration} - \text{quantity of acid that ionized}$$
$$= [C_6H_5CO_2H]_0 - x$$
$$= 0.020 - x$$

Substituting these equilibrium concentrations into the K_a expression, we have

$$K_a = \frac{[H_3O^+][C_6H_5CO_2^-]}{[C_6H_5CO_2H]} = \frac{(x)(x)}{(0.020 - x)} = 6.3 \times 10^{-5}$$

As described in the text just previous to this Example, we know that x is small compared with 0.020 (because $[HA]_0 \geq 100 \cdot K_a$), so that

$$K_a = 6.3 \times 10^{-5} \approx \frac{x^2}{0.020}$$

This means that

$$x = \sqrt{K_a(0.020)} = 0.0011 \text{ M}$$

and we find that

$$[H_3O^+] = [C_6H_5CO_2^-] = 0.0011 \text{ M}$$

and

$$[C_6H_5CO_2H] = (0.020 - x) = 0.019 \text{ M}$$

Finally, the pH of the solution is found to be

$$pH = -\log (1.1 \times 10^{-3}) = 2.96$$

Now, let's think about the result. Because benzoic acid is weak, we made the approximation that $(0.020 - x) \approx 0.020$. If we do *not* make the approximation and work with the expression $K_a = x^2/(0.020 - x)$, we see it is a quadratic equation. That is,

$$K_a(0.020 - x) = x^2$$
$$0.020\ K_a - K_a x = x^2$$
$$x^2 + K_a x - 0.020\ K_a = 0$$

is a quadratic equation of the type $ax^2 + bx + c = 0$, where $a = 1$, $b = K_a$, and $c = -0.20K_a$. Such an expression can be solved by the quadratic formula or by the method of successive approximations (see Appendix A). Either method gives $x = [H_3O^+] = 1.1 \times 10^{-3}$, the same answer to two significant figures that we obtained from the "approximate" expression $x = \sqrt{K_a(0.020)}$.

EXERCISE 17.11 ***Calculating Equilibrium Concentrations and pH from K$_a$***

What are the equilibrium concentrations of acetic acid, the acetate ion, and H_3O^+ for a 0.10 M solution of acetic acid ($K_a = 1.8 \times 10^{-5}$)? What is the pH of the solution? ■

In the previous example, we calculated the pH of an aqueous solution of a weak acid. You know it is weak because K_a was so small that $[H_3O^+] \ll [HA]_0$. Another way to say this is that only a small percentage of the acid ionized to produce hydronium ion in water. Indeed, in the case of a 0.020 M solution of benzoic acid, the percentage ionization is less than 10%.

$$\text{Percentage ionized} = \frac{\text{quantity of acid ionized}}{\text{initial acid concentration}} \times 100\% = \frac{0.0011\ \text{M}}{0.020\ \text{M}} \times 100\% = 5.5\%$$

This is the reason the hydronium ion concentration for a weak acid often can be found to a good approximation from

$$K_a \approx \frac{[H_3O^+ \text{ from weak acid}][\text{conjugate base}]}{[\text{acid}]_{\text{initial}}}$$

In general, the hydronium ion concentration obtained from this approximate expression is the same (to 2 significant figures) as that from a more exact solution to the problem whenever the weak acid is no more than about 10% ionized (and the approximation given on page 820 can be used).

EXAMPLE 17.7 ***Calculating Equilibrium Concentrations and pH from K_a***

Formic acid, HCO_2H, was first obtained in 1670 as a product of the destructive distillation of ants, whose Latin genus name is *Formica.* The acid is moderately weak, with $K_a = 1.8 \times 10^{-4}$.

$$HCO_2H(aq) + H_2O(\ell) \rightleftharpoons H_3O^+(aq) + HCO_2^-(aq)$$

If you have a 0.0010 M solution of the acid, what is the pH of the solution? What is the concentration of formic acid at equilibrium?

Solution The usual equilibrium table is given here. Notice that we have again made the reasonable approximation that H_3O^+ from water ionization can be ignored.

Equation	HCO_2H	$+ H_2O \rightleftharpoons$	H_3O^+	$+ HCO_2^-$
Before ionization (M)	0.0010		0	0
Change on proceeding to equilibrium	$-x$		$+x$	$+x$
After equilibrium is achieved (M)	$0.0010 - x$		x	x

Substituting the values in the table into the K_a expression we have

$$K_a = \frac{[H_3O^+][HCO_2^-]}{[HCO_2H]} = \frac{(x)(x)}{(0.0010 - x)} = 1.8 \times 10^{-4}$$

Formic acid is a weak acid because it has a value of K_a much less than 1. In this situation, however, $[HA]_0$ is *not* greater than $100 \cdot K_a$. In fact, 0.0010 is *less than* $100 \cdot (1.8 \times 10^{-4})$, so the usual approximation is not reasonable. If we do solve for the $[H_3O^+]$ using the approximate expression $x \approx \sqrt{K_a(0.0010)}$, we obtain a value of x $(= [H_3O^+] = [HCO_2^-])$ of 4.2×10^{-4} M. But a check of the percentage ionization of the acid

$$\text{Percentage ionized} = \frac{4.2 \times 10^{-4}}{0.0010} \times 100\% = 42\%$$

shows clearly that 0.0010 M formic acid is *much* more than 10% ionized, so the answers from the approximate and exact solution to the problem differ significantly. This means we have to find the equilibrium concentrations by solving the "exact" expression.

$$K_a = \frac{[H_3O^+][HCO_2^-]}{[HCO_2H]} = \frac{(x)(x)}{(0.0010 - x)} = 1.8 \times 10^{-4}$$

$$x^2 + (1.8 \times 10^{-4})x + (-1.8 \times 10^{-4})(0.0010) = 0$$

Here $a = 1$, $b = 1.8 \times 10^{-4}$, and $c = -1.8 \times 10^{-7}$ in the quadratic expression $ax^2 + bx + c = 0$. Solving for x, we find

$$x = \frac{-(1.8 \times 10^{-4}) \pm \sqrt{(1.8 \times 10^{-4})^2 - 4(1)(-1.8 \times 10^{-7})}}{2(1)}$$

$$= 3.4 \times 10^{-4}\text{ M} \quad \text{and} \quad -5.2 \times 10^{-4}\text{ M}$$

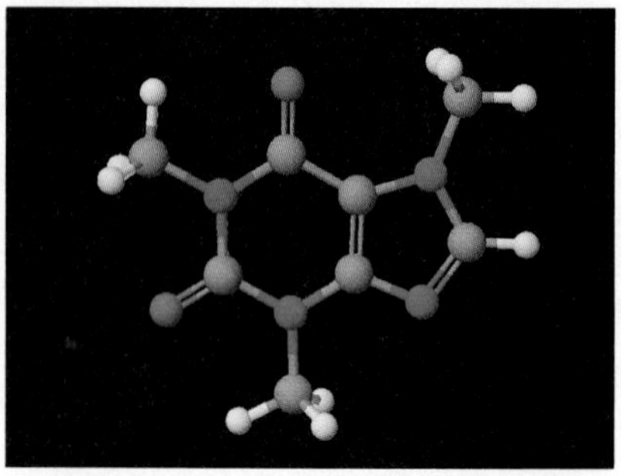

Caffeine, a naturally occurring base, $C_8H_{10}N_4O_2$. (C atoms are gray, H atoms are white, N atoms are blue, and O atoms are red.) (Susan Young)

Because the negative root of the equation is chemically meaningless, we use the positive root and find that

$$[H_3O^+] = [HCO_2^-] = 3.4 \times 10^{-4}\ M$$

and so

$$[HCO_2H] = 0.0010 - x = 0.0007\ M$$

The pH of the formic acid solution is therefore

$$pH = -\log(3.4 \times 10^{-4}) = 3.47$$

a pH indicating an acidic solution.

It is important to notice in this problem that the approximate solution failed because (1) the acid concentration is small and (2) the acid is not all that weak. These made invalid the approximation that $[HA]_{equilibrium} \approx [HA]_{initial}$.

EXERCISE 17.12 ***Calculating Equilibrium Concentrations and pH from K_a***

What are the equilibrium concentrations of HF, the fluoride ion, and H_3O^+ when a 0.015 M solution of HF is allowed to come to equilibrium? ■

The examples solved thus far have involved weak acids. Weak bases, however, are widely found in nature. Examples include caffeine and nicotine. Caffeine, which is found in coffee, tea, and cola nuts, is a well-known stimulant. Pure nicotine, which is extracted from tobacco, is highly toxic. It is used as an insecticide in the form of its salt with sulfuric acid.

The next example describes the calculation of the pH of an aqueous solution of the weak base pyridine. Notice that nicotine is a derivative of pyridine.

EXAMPLE 17.8 ***Calculating the pH of an Aqueous Solution of a Weak Base***

Pyridine, a weak base, was discovered in coal tar in 1846.

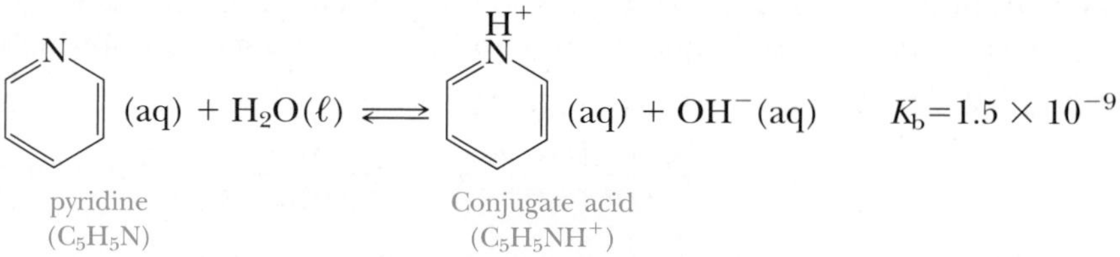

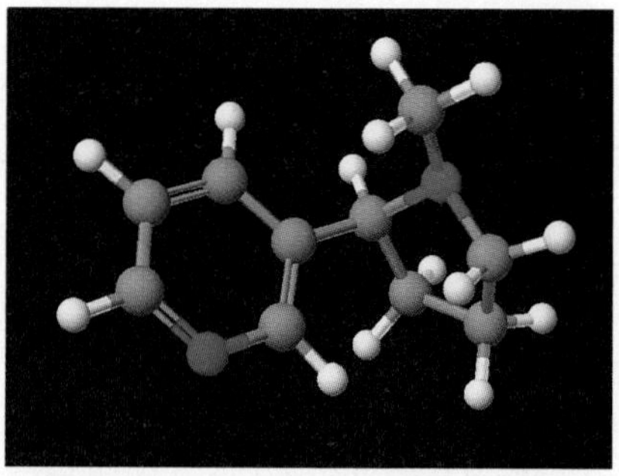

Nicotine, a naturally occurring base, $C_{10}H_{14}N_2$. (C atoms are gray, H atoms are white, and N atoms are blue.) (Susan Young)

What is the pH of a 0.010 M aqueous solution of pyridine?

Solution The equilibrium expression for this weak base is

$$K_b = 1.5 \times 10^{-9} = \frac{[C_5H_5NH^+][OH^-]}{[C_5H_5N]}$$

and the concentrations of the base and ionic products are

Equation	C_5H_5N	$+ H_2O \rightleftharpoons$	$C_5H_5NH^+$	$+ OH^-$
Before ionization (M)	0.010		0	0
Change on proceeding to equilibrium	$-x$		$+x$	$+x$
After equilibrium is achieved (M)	$0.010 - x$		x	x

If the expressions for the equilibrium concentrations are substituted into the K_b expression, we have

$$K_b = 1.5 \times 10^{-9} = \frac{(x)(x)}{0.010 - x}$$

an expression similar to the one used to find the hydronium ion concentration in a weak acid solution. Two points should be noticed here.

1. We ignored any OH^- ion that can arise from the autoionization of water. The reasons are similar to those for ignoring H_3O^+ from water in solutions of weak acids.

2. Pyridine is a very weak base, and it is probably a good assumption that it is less than 10% ionized in solution. We therefore solve the approximate expression for x,

$$K_b = 1.5 \times 10^{-9} = \frac{(x)(x)}{0.010}$$

and this gives a value of

$$x = [OH^-] = [C_5H_5NH^+] = 3.9 \times 10^{-6}\ M$$

The hydroxide ion concentration is indeed *much* less than the original concentration of pyridine; that is, $(0.010\ M - 3.9 \times 10^{-6})$ is nearly equal to 0.010 M. Less than 1% of the base has reacted with water to give ions, and the approximate solution is valid.

3. This example shows that the "rule of thumb" that was used with weak acids can be applied to weak bases. That is, if $[Base]_0 \geq 100 \cdot K_b$, then the approximation that $[OH^-] = \sqrt{K_b[Base]_0}$ can be used.

The objective of the problem was to find the pH of the solution.

$$pOH = -\log [OH^-] = -\log (3.9 \times 10^{-6}) = 5.41$$
$$pH = 14.00 - pOH = 14.00 - 5.41 = 8.59$$

The solution is weakly basic.

EXERCISE 17.13 ***Calculating the pH of an Aqueous solution of a Weak Base***

What is the pH of a 0.025 M solution of ammonia, NH_3? ■

A CLOSER LOOK *Stomach Acidity, or "I Ate Too Much!"*

As soon as food reaches your stomach, acidic gastric juices are released by glands in the mucous lining of the stomach. The high acidity (the pH is about 1!) is due to dissolved hydrochloric acid and is needed for the enzyme pepsin to catalyze the digestion of proteins in foods. Hydrogen ions are produced in blood plasma by the ionization of dissolved CO_2. These ions are transported through the stomach lining along with Cl^-, which provides charge balance. Because the stomach wall contains protein, it is reasonable to wonder why the stomach does not digest itself. In fact, it sometimes does, producing a hole—an ulcer. Most often, however, the stomach wall resists the attack of H^+, thanks to a protective layer of mucous-producing cells. These cells prevent the H^+ and Cl^- ions from diffusing back into the blood plasma. Ordinarily, these cells are being continuously sloughed off and replaced at the rate of about half a million cells per minute.

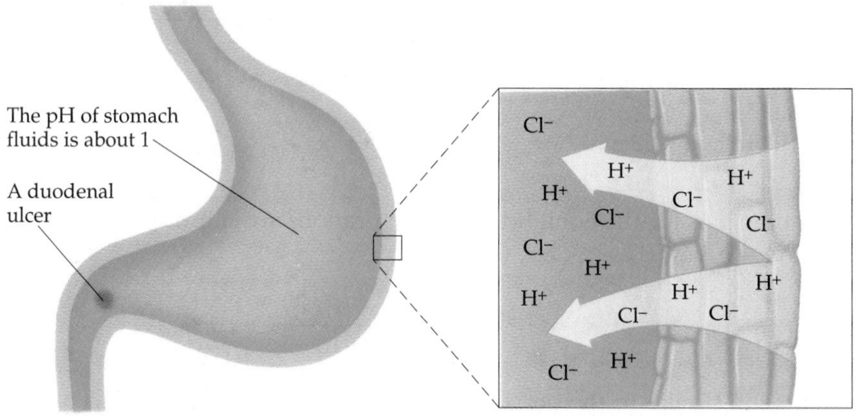

The lining of the stomach contains cells that secrete a solution of hydrochloric acid. The pH of the solution is about 1. Some people's stomachs produce far more acid than is needed for the primary digestion of food. When this happens, an ulcer can form.

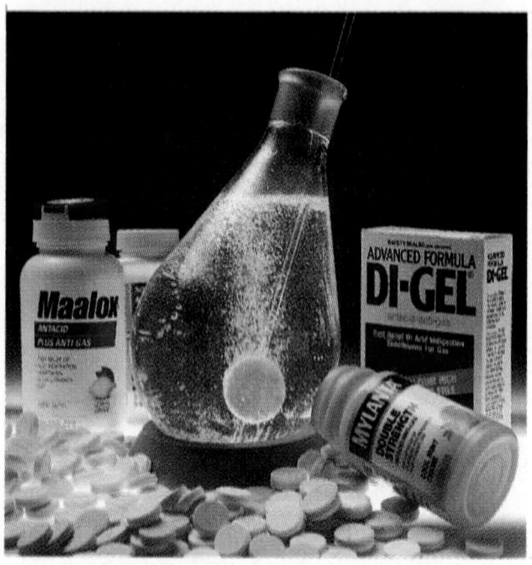

A variety of commercial antacids, all of which contain bases such as sodium bicarbonate, magnesium hydroxide, aluminum hydroxide, and calcium carbonate. The photo also shows the reaction of an Alka-Seltzer tablet. Notice that one product is CO_2, which comes from the reaction of the sodium bicarbonate and the citric acid in the tablet. (C. D. Winters)

When you eat too much food, or when your stomach is irritated by very spicy food, it responds with an outpouring of acid, and the pH is lowered to the point where discomfort is felt. Heartburn is a frequent symptom of excess acidity, and it can be alleviated by an antacid. The reaction of milk of magnesia is typical,

$$Mg(OH)_2(s) + 2\ H_3O^+(aq) \longrightarrow Mg^{2+}(aq) + 4\ H_2O(\ell)$$

but there are many other antacids, such as Tums and Di-Gel ($CaCO_3$), Alka-Seltzer ($NaHCO_3$), Amphogel ($Al(OH)_3$), Rolaids ($NaAl(OH)_2CO_3$), and Maalox (a combination of $Mg(OH)_2$ and $Al(OH)_3$).

The stomach wall can also be damaged by the action of aspirin. Aspirin, acetylsalicylic acid, is a weak carboxylic acid (Section 11.5) with $K_a = 3.27 \times 10^{-4}$.

At the pH of the stomach, most of the conjugate base, the acetylsalicylate ion, reacts with H^+, leaving aspirin largely un-ionized. The equilibrium lies to the left. In the neutral form aspirin molecules are able to penetrate the stomach wall, and, once inside, are in a region of lower acidity. They are therefore able to ionize and produce H^+. The accumulation of H^+ in the wall can cause bleeding, but the amount of blood lost per aspirin tablet is not generally harmful.

(Adapted from F. Brescia, J. Arents, H. Meislich, et al.: *General Chemistry,* 5th ed., San Diego, Harcourt Brace Jovanovich, 1988.)

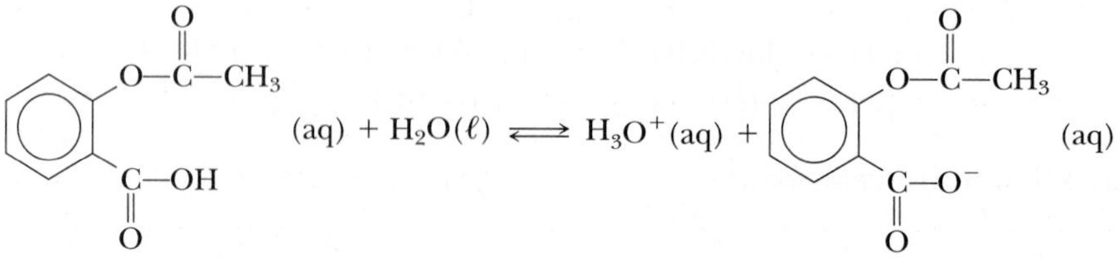

acetylsalicylic acid, aspirin

acetylsalicylate ion Conjugate base

PROBLEM-SOLVING TIPS AND IDEAS

17.3 Using K_a and K_b

The mathematical approach used to this point can be summarized using the hypothetical weak acid HA. (The same ideas apply to weak bases.)

$$HA(aq) + H_2O(\ell) \rightleftharpoons H_3O^+(aq) + A^-(aq)$$

The equilibrium constant expression, *in terms of equilibrium concentrations,* is

$$K_a = \frac{[H_3O^+][A^-]}{[HA]}$$

where $[H_3O^+]$ is the hydronium ion concentration from the reaction of the weak acid with water and from the autoionization of water. When the pH falls outside the range between 6 and 8, we can assume that H_3O^+ ion from water autoionization does not contribute significantly. This also means that the equilibrium concentration of HA must be $[HA] = [HA]_0 - [H_3O^+]$, where $[HA]_0$ is the original concentration of HA. Therefore, we can write K_a in a form that allows us to solve for $[H_3O^+]$ if we know $[HA]_0$.

$$K_a = \frac{[H_3O^+]^2}{[HA]_0 - [H_3O^+]}$$

Solving for $[H_3O^+]$ generally requires the use of the quadratic formula or the method of successive approximations (Appendix A). This is not necessary, however, if we make the approximation that $[HA]_0 - [H_3O^+] \approx [HA]_0$, an approximation that is valid when $[HA]_0 \geq 100 \cdot K_a$. ■

17.8 ACID-BASE PROPERTIES OF SALTS: HYDROLYSIS

Many compounds in nature and in consumer products are salts. A salt is an ionic compound that could have been formed by the reaction of an acid with a base; a salt's positive ions come from the base, and its negative ions come from the acid. Common table salt is one example, but others include potassium benzoate ($NaC_6H_5CO_2$) and potassium citrate ($K_3C_6H_5O_7$) in soft drinks, sodium bicarbonate in stomach antacids, and various phosphates in fertilizers (Figure 17.9).

What is important to us here is that many salts produce acidic or basic aqueous solutions by hydrolysis. A **hydrolysis** reaction is said to have occurred when a salt dissolves in water and leads to changes in the H_3O^+ and OH^- concentrations of the water. These are of course just some of the reactions we have been talking about. For example, you have already seen that the ammonium ion and some metal ions—most notably ions such as Al^{3+}, Pb^{2+}, and transition metal ions of 2+ and 3+ charge—increase the hydronium ion concentration in aqueous solution (see Section 17.5 and Figure 17.8). Furthermore, we described the fact that the anionic portion of a weak acid—the conjugate base of the weak acid—reacts with water to give a measurable concentration of hydroxide ion. For example, the benzoate ion, the conjugate base of the weak acid benzoic acid (Table 17.4), hydrolyzes to give a basic solution and re-form the weak acid.

The word "hydrolysis" literally means "breaking a substance apart" (-lysis) by water (hydro-).

$$C_6H_5CO_2^-(aq) + H_2O(\ell) \rightleftharpoons C_6H_5CO_2H(aq) + OH^-(aq) \qquad K_b = 1.6 \times 10^{-10}$$

benzoate ion — Weak base; benzoic acid — Conjugate acid

Figure 17.9 Many common household products contain salts. Some of these salts are neutral (such as NaCl), some are acidic (such as ammonium salts), but many are basic (such as carbonates, bicarbonates, and benzoates). (Benzoates are salts of the organic acid benzoic acid and are often found in soft drinks.) (C. D. Winters)

Example 17.9 illustrates the calculation of the pH of an aqueous solution of a salt where the anion is the conjugate base of a weak acid.

EXAMPLE 17.9 *Calculating the pH of an Aqueous Solution of the Salt of a Weak Acid*

Sodium hypochlorite, NaClO, is used as a source of chlorine in some laundry bleaches, swimming pool disinfectants, and water treatment plants. Estimate the pH of a 0.015 M solution of NaClO. (Use Table 17.4 to obtain K_b for ClO^-.)

Solution Sodium hypochlorite consists of a sodium ion and a hypochlorite ion. The Na^+ ion does not hydrolyze in water (page 803), but the ClO^- ion is the conjugate base of a weak acid and so is expected to produce a basic solution.

$$\underset{\text{hypochlorite ion, Weak base}}{ClO^-(aq)} + H_2O(\ell) \rightleftharpoons \underset{\text{hypochlorous acid, Conjugate acid of } ClO^-}{HClO(aq)} + OH^-(aq)$$

$$K_b = 2.9 \times 10^{-7} = \frac{[HClO][OH^-]}{[ClO^-]}$$

The concentrations of the hypochlorite ion, hypochlorous acid, and hydroxide ion are given in the equilibrium table.

Equation	ClO^-	$+ H_2O \rightleftharpoons$	HClO	$+ OH^-$
Concentration before ionization (M)	0.015		0	0
Change in concentration on proceeding to equilibrium	$-x$		$+x$	$+x$
Concentration after equilibrium is achieved (M)	$0.015 - x$		x	x

The hypochlorite ion is a very weak base, as indicated by the very small value of K_b. It is therefore reasonable to assume x is small compared with 0.015 in $(0.015 - x)$, and so we can write

$$K_b = 2.9 \times 10^{-7} \approx \frac{x^2}{0.015}$$

Solving for x gives $x = 6.6 \times 10^{-5}$. Because $(0.015 - 6.6 \times 10^{-5}) \approx 0.015$, our assumption that x is negligible is justified. At equilibrium, therefore, the concentrations are

$$[HClO] = [OH^-] = 6.6 \times 10^{-5} \quad \text{and} \quad [ClO^-] = 0.015 \text{ M}$$

Finally, the pH of the solution is

$$K_w = 1.0 \times 10^{-14} = [H_3O^+][OH^-] = [H_3O^+]\,(6.6 \times 10^{-5})$$

$$[H_3O^+] = 1.5 \times 10^{-10} \text{ M}$$

$$pH = 9.82$$

The pH indicates a basic solution, as expected for the conjugate base of a weak acid.

Products that contain an aqueous solution of sodium hypochlorite, NaClO. The hypochlorite ion, ClO^-, is a weak base. See Example 17.9. (C. D. Winters)

EXERCISE 17.14 ***Calculating the pH of an Aqueous Salt Solution***

The ammonium ion is the conjugate acid of the weak base ammonia.

$$NH_4^+(aq) + H_2O(\ell) \rightleftharpoons H_3O^+(aq) + NH_3(aq)$$

What is the pH of a 0.50 M solution of ammonium chloride? ■

With the knowledge you have gained in this chapter of the types of Brønsted acids and bases, you can predict with some assurance whether a given salt will be acidic or basic. To summarize, the bases listed above water on the right side of Table 17.4 are extremely weak and do not affect the pH of an aqueous solution. Acids listed below water on the left side of Table 17.4 are extremely weak; they do not make a solution acidic. This can be used to make the following predictions:

(a) A salt such as $NaNO_3$ gives a neutral, aqueous solution because Na^+ does not hydrolyze to an appreciable extent and because NO_3^- is the conjugate base of a strong acid and also does not hydrolyze appreciably.

(b) An aqueous solution of K_3PO_4 should be basic because PO_4^{3-} is the conjugate base of the weak acid HPO_4^{2-}, whereas K^+ does not hydrolyze appreciably.

(c) An aqueous solution of $FeCl_2$ should be weakly acidic, because Fe^{2+} hydrolyzes to give an acidic solution (Section 17.5), whereas Cl^- is the conjugate base of the strong acid HCl and so does not hydrolyze to an appreciable extent.

Some additional explanation is needed concerning salts of amphiprotic anions, such as HCO_3^- and $H_2PO_4^-$. Because they have an ionizable hydrogen, they can act as acids,

$$\underset{\text{Acid}}{HCO_3^-(aq)} + H_2O(\ell) \rightleftharpoons CO_3^{2-}(aq) + H_3O^+(aq) \qquad K_a = 4.8 \times 10^{-11}$$

but they are also the conjugate bases of weak acids.

$$\underset{\text{Base}}{HCO_3^-(aq)} + H_2O(\ell) \rightleftharpoons H_2CO_3(aq) + OH^-(aq) \qquad K_b = 2.4 \times 10^{-8}$$

Whether the solution is acidic or basic depends on the relative size of K_a and K_b. In the case of an aqueous solution of the hydrogen carbonate anion, K_b is larger than K_a, so $[OH^-]$ is larger than $[H_3O^+]$, and a solution of a salt such as $NaHCO_3$ is basic.

Is a solution of sodium dihydrogen phosphate, NaH_2PO_4, weakly acidic or weakly basic?

Finally, what happens if you have a salt based on an acidic cation and a basic anion? One example is ammonium fluoride. Here the ammonium ion decreases the pH, and the fluoride ion increases the pH.

$$NH_4^+(aq) + H_2O(\ell) \rightleftharpoons H_3O^+(aq) + NH_3(aq) \quad K_a(NH_4^+) = 5.6 \times 10^{-10}$$

$$F^-(aq) + H_2O(\ell) \rightleftharpoons HF(aq) + OH^-(aq) \qquad K_b(F^-) = 1.4 \times 10^{-11}$$

Because $K_a(NH_4^+) > K_b(F^-)$, the ammonium ion is a stronger acid than fluoride ion is a base. The resulting solution should be slightly acidic.

In general, for a salt that has an acidic cation and a basic anion, the pH of the solution will be determined by the ion that is the stronger acid or base of the two.

TABLE 17.5 Acid-Base Properties of Typical Ions in Aqueous Solution

	Neutral		Basic			Acidic
Anions	Cl^- Br^- I^-	NO_3^- ClO_4^-	$CH_3CO_2^-$ HCO_2^- CO_3^{2-} S^{2-} F^-	CN^- PO_4^{3-} HCO_3^- HS^- NO_2^-	SO_4^{2-} HPO_4^{2-} SO_3^{2-} ClO^-	HSO_4^- $H_2PO_4^-$ HSO_3^-
Cations	Li^+ Na^+ K^+	Mg^{2+} Ca^{2+} Ba^{2+}	None			Al^{3+} NH_4^+ Transition metal ions

All the necessary information to determine the pH of a salt is in Table 17.4. We have also summarized it, however, in a concise manner in Table 17.5 and in the following statements:

Cation	Anion	pH of the Solution
From strong base (Na^+)	From strong acid (Cl^-)	= 7 (neutral)
From strong base (K^+)	From weak acid ($CH_3CO_2^-$)	>7 (basic)
From weak base (NH_4^+)	From strong acid (Cl^-)	<7 (acidic)
From any weak base (BH^+)	From any weak acid (A^-)	Depends on relative strengths of acid and base

Will a solution of nickel(II) acetate, $Ni(CH_3CO_2)_2$, be slightly acidic or slightly basic?

EXERCISE 17.15 ***Predicting the pH of Salt Solutions***

For each salt below, predict whether the pH will be greater than, less than, or equal to 7.

(a) NaCl

(b) $FeCl_3$

(c) NH_4NO_3

(d) Na_2HPO_4 ■

17.9 POLYPROTIC ACIDS AND BASES

As described in Section 17.3 and Table 17.1, some important acids are capable of donating more than one proton and are therefore called polyprotic. Indeed, many occur in nature. Examples include citric acid and malic acid, the latter of which is found in apples. Phosphoric acid is commonly used in the food industry. It ionizes in three steps.

First ionization step

$$H_3PO_4(aq) + H_2O(\ell) \rightleftharpoons H_3O^+(aq) + H_2PO_4^-(aq) \qquad K_{a1} = 7.5 \times 10^{-3}$$

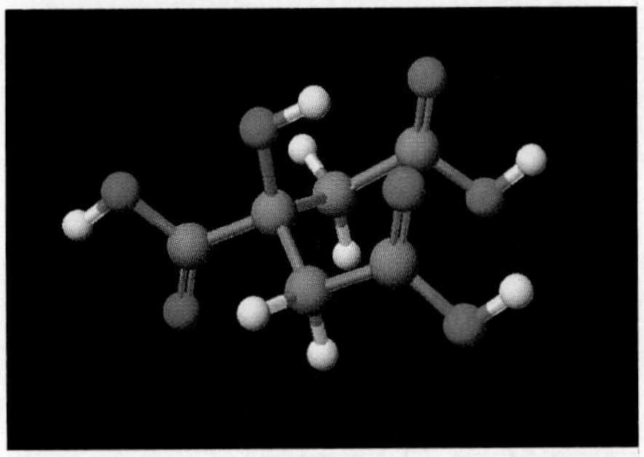

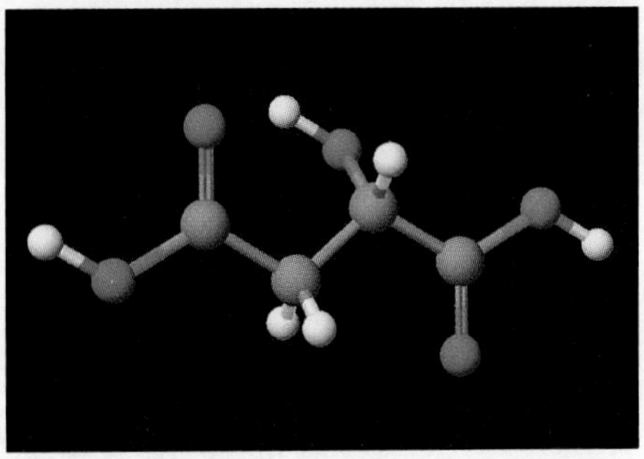

Naturally occurring polyprotic acids include (left) citric acid, $C_6H_8O_7$, and (right) malic acid, $C_4H_6O_5$. (C atoms are gray, H atoms are white, and O atoms are red.)

Second ionization step

$$H_2PO_4^-(aq) + H_2O(\ell) \rightleftharpoons H_3O^+(aq) + HPO_4^{2-}(aq) \qquad K_{a2} = 6.2 \times 10^{-8}$$

Third ionization step

$$HPO_4^{2-}(aq) + H_2O(\ell) \rightleftharpoons H_3O^+(aq) + PO_4^{3-}(aq) \qquad K_{a3} = 3.6 \times 10^{-13}$$

Notice, however, that the K_a values for each successive step become smaller and smaller because it is more difficult to remove H^+ from a negatively charged ion, such as $H_2PO_4^-$, than from a neutral molecule, such as H_3PO_4. Furthermore, the larger the negative charge of the anionic acid, the more difficult it is to remove H^+.

For many inorganic acids, such as phosphoric acid, carbonic acid, and hydrogen sulfide, each successive loss of a proton is about 10^4 to 10^6 times more difficult than the previous ionization step. This means that the first ionization step of a polyprotic acid produces up to about a million times more H_3O^+ than the second step. A consequence is that, for many inorganic polyprotic acids, the pH of the solution depends primarily on the hydronium ion generated *in the first ionization step;* the hydronium ion produced in the second step can be ignored.

The principles used to describe the behavior of polyprotic acids apply to their conjugate bases. This is illustrated by the calculation of the pH of a solution containing the carbonate ion (Example 17.10).

EXAMPLE 17.10 *Calculating the pH of the Solution of a Polyprotic Base*

The carbonate ion, CO_3^{2-}, is important in the environment. It is a base in water, forming the hydrogen carbonate ion, which in turn can form carbonic acid.

$$CO_3^{2-}(aq) + H_2O(\ell) \rightleftharpoons HCO_3^-(aq) + OH^-(aq) \qquad K_{b1} = 2.1 \times 10^{-4}$$

$$HCO_3^-(aq) + H_2O(\ell) \rightleftharpoons H_2CO_3(aq) + OH^-(aq) \qquad K_{b2} = 2.4 \times 10^{-8}$$

What is the pH of a 0.10 M solution of Na_2CO_3?

Solution The fact that the second ionization constant K_{b2} is so much smaller than the first ionization constant K_{b1} means that hydroxide ion concentration in the solution results almost entirely from the first step. So, let us assume that *only* the first step occurs and calculate the resulting concentrations.

Equation	CO_3^{2-}	$+ H_2O \rightleftharpoons$	HCO_3^-	$+ OH^-$
Before ionization (M)	0.10		0	0
Change on proceeding to equilibrium	$-x$		$+x$	$+x$
After equilibrium is achieved (M)	$0.10 - x$		x	x

The concentration of OH^- ($=x$) can be derived from the expression

$$K_{b1} = 2.1 \times 10^{-4} = \frac{[HCO_3^-][OH^-]}{[CO_3^{2-}]} = \frac{x^2}{0.10 - x}$$

Because K_{b1} is relatively small, it is reasonable to make the approximation that $(0.10 - x) \approx 0.10$, and so

$$x = [HCO_3^-] = [OH^-] \approx \sqrt{K_{b1}\,(0.10)} = 4.6 \times 10^{-3}\text{ M}$$

Using this value of $[OH^-]$ to calculate the pH, we find

$$pOH = -\log\,(4.6 \times 10^{-3}) = 2.34$$
$$pH = 11.66$$

and the concentration of the hydrogen carbonate ion is

$$[HCO_3^-] = 0.10 - 0.0046 \approx 0.10\text{ M}$$

This last result shows that the approximation was indeed justified.

Next, let us turn to the problem of obtaining the concentration of carbonic acid, H_2CO_3, a product of the second ionization step.

Equation	HCO_3^-	$+ H_2O \rightleftharpoons$	$H_2CO_3 +$	OH^-
Before ionization (M)	4.6×10^{-3}		0	4.6×10^{-3}
Change on proceeding to equilibrium	$-y$		$+y$	$+y$
After equilibrium is achieved (M)	$(4.6 \times 10^{-3} - y)$		y	$(4.6 \times 10^{-3} + y)$

Because K_{b2} is so small, the second step occurs to a *much* smaller extent than the first step. This means the amount of H_2CO_3 and OH^- produced in the second step ($=y$) is *much* smaller than 10^{-3} M. Therefore, it is reasonable that both $[HCO_3^-]$ and $[OH^-]$ are very close to 4.6×10^{-3} M.

$$K_{b2} = 2.4 \times 10^{-8} = \frac{[H_2CO_3][OH^-]}{[HCO_3^-]} \approx \frac{(y)(4.6 \times 10^{-3})}{4.6 \times 10^{-3}}$$

Because $[HCO_3^-]$ and $[OH^-]$ have nearly identical values, they cancel from the expression, and we find that $[H_2CO_3]$ is simply equal to K_{b2}.

$$y = [H_2CO_3] = K_{b2} = 2.4 \times 10^{-8}$$

EXERCISE 17.16 ***Calculating the pH of the Solution of a Polyprotic Acid***

Suppose you have a 0.10 M solution of oxalic acid, $H_2C_2O_4$. What is the pH of the solution? What is the concentration of the oxalate ion, $C_2O_4^{2-}$? ■

PROBLEM-SOLVING TIPS AND IDEAS

17.4 Polyprotic Acids or Bases

For polyprotic acids, H_2A, in which K_{a1} and K_{a2} are different by 10^3 or more,

$$HA(aq) + H_2O(\ell) \rightleftharpoons H_3O^+(aq) + HA^-(aq) \qquad K_{a1}$$

$$HA^-(aq) + H_2O(\ell) \rightleftharpoons H_3O^+(aq) + A^{2-}(aq) \qquad K_{a2}$$

we can say that

- The hydronium ion comes largely from the first ionization step and is given by $[H_3O^+] = \sqrt{K_{a1} \cdot [H_2A]_0}$ (where $[H_2A]_0$ is the original concentration of the diprotic acid).
- The hydronium ion produced in the second ionization step is extremely small. The hydronium ion concentration of the solution is therefore effectively equal to $[HA^-]$. From this it follows that $[A^{2-}] = K_{a2}$.
- These same conclusions apply to aqueous equilibria of polyprotic bases (anions of polyprotic acids). ■

17.10 MOLECULAR STRUCTURE, BONDING, AND ACID-BASE BEHAVIOR

One of the most interesting aspects of chemistry is the correlation between molecular structure, bonding, and observed properties. Here it is useful to analyze the connection between the structure and bonding of some acids and their relative strengths.

When an acid HA dissociates in water, we can think of the process as the sum of a series of steps. (For simplicity, we ignore the solvation of the individual species because this can have *about* the same effect in analogous cases.)

1. H—A bond breaking. (One bonding electron is retained by each partner.)

$$H—A \longrightarrow H\cdot + \cdot A$$

2. Loss of an electron by H to form H^+

$$H\cdot \longrightarrow H^+ + e^-$$

3. Gain of an electron by A to form A^-

$$A\cdot + e^- \longrightarrow A^-$$

Because we are only interested in relative acidities, we can further simplify our analysis by ignoring the second step, one common to all acid dissociations. Thus, to get some insight into the relative strengths of several acids we can compare

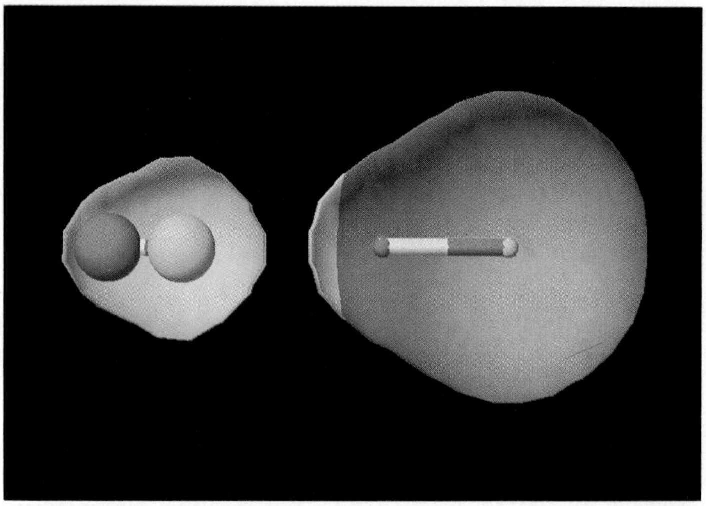

Computer-generated models of HF (left) and HBr (right) showing the relative partial electric charges on the atoms. (Red spheres represent positive atoms and yellow spheres represent negative atoms. The size of the sphere reflects the relative charge.) In HF the H atom on the left is positive (+0.29) and the F atom on the right is negative (−0.29). There is a substantial charge on each atom, and so the molecule is quite polar. In HBr, the electric charges are much smaller than in HF (as reflected by the much smaller spheres). Both molecules are surrounded by a surface that represents the electron density contour. Notice the much larger electron cloud around the Br atom than around the F atom (the right-most atom in each case). (Susan Young)

the H—A bond strengths in the acids (Step 1) and the relative electron affinities of the fragment A (Step 3). In general, the more easily the H—A bond is broken and the greater the electron affinity of A, the greater the relative strength of the acid. These two effects can work together (a weak H—A bond and a high electron affinity for A) to lead to a strong acid. It is often observed, however, that they work in opposite directions, so the balance between the effects controls acidity.

Let us use this approach with the binary hydrogen halides, HA. Their relative acid strengths are known to be in the order shown in the table.

	—Increasing acid strength→			
	HF	*HCl*	*HBr*	*HI*
H—A bond strength (kJ/mol)	569	431	368	297
Electron affinity of A (kJ/mol)	−328	−349	−325	−295

Here it is evident that H—A bond strengths largely control the relative acidities of these compounds. The strongest acid, HI, has the weakest H—A bond. If the electron affinity of A were the controlling factor, then HCl would have been the strongest acid.

Next, let us examine acids in which the proton is always bonded to the same element, but other changes are made in the molecule. Oxoacids fit this description, and two series are of interest.

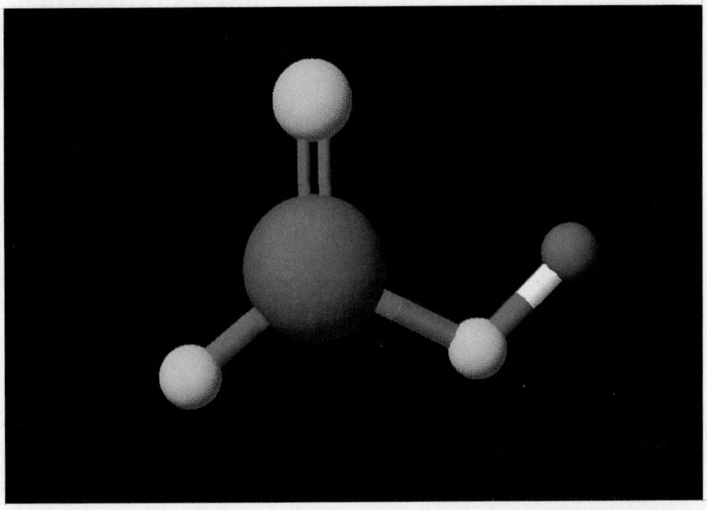

Figure 17.10 This computer-generated model of nitric acid, HNO_3, shows the relative partial electric charges on the atoms. (Red spheres represent positive atoms and yellow spheres represent negative atoms. The size of the spheres reflects the relative charge.) The central N atom is positively charged as is the acidic H atom. (Susan Young)

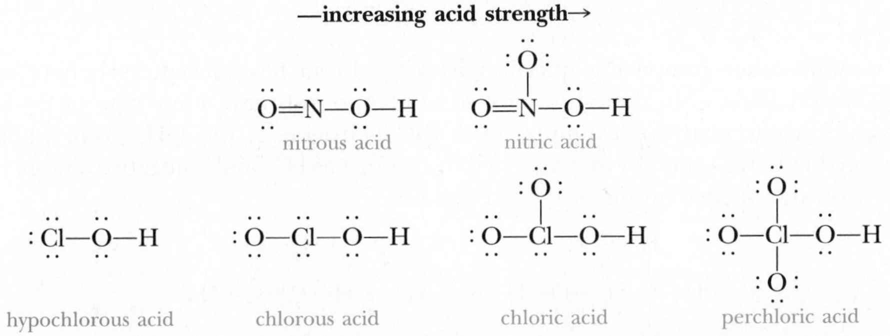

From the trends seen in the nitrogen- and chlorine-based oxoacids, it is logical to conclude that the greater the number of O atoms attached to the central atom in the acid, the stronger the acid. This increase in acidity is due to the **inductive effect,** the attraction of electrons from adjacent bonds by a more electronegative atom. As more and more electronegative O atoms are attached, the inductive effect is stronger, the O—H bond is more strongly polarized ($^{\delta-}$O—H$^{\delta+}$), and the bond is more readily broken (Figure 17.10).

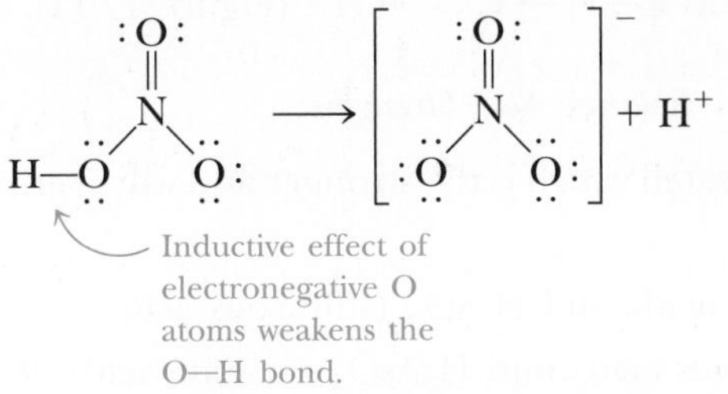

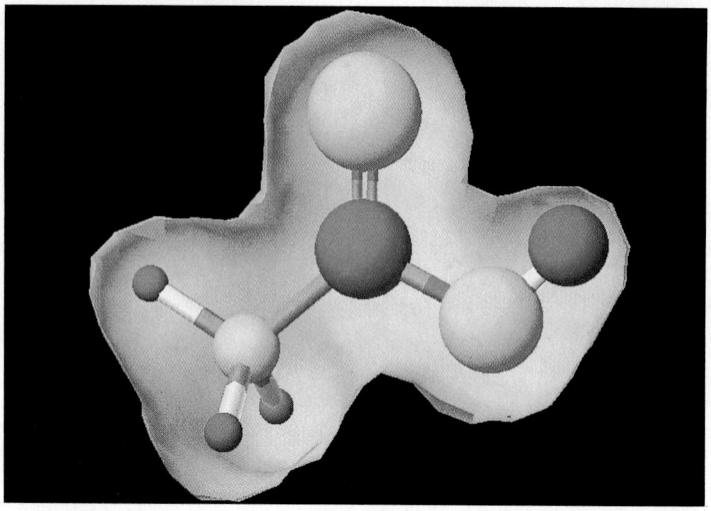

Figure 17.11 Computer molecular modeling methods support the notion that it is the OH hydrogen atom that is donated when acetic acid, CH_3CO_2H, functions as an acid. (Here the molecule is enclosed within a surface that shows the distribution of electrons in the molecule. Red spheres represent positive atoms and yellow spheres represent negative atoms. The size of the spheres reflects the relative charge.) The OH hydrogen bears a positive partial charge (+0.24), due to inductive effect of O atoms, whereas the CH hydrogen atoms are essentially uncharged. (Susan Young)

One other interesting question is: Why do so few substances behave as Brønsted acids, even though hundreds of compounds have some type of E—H bond? Acetic acid, for example, loses the hydrogen of the OH group when placed in water (and not that attached to carbon) as H^+, and a negative charge is left on the oxygen atoms of the acetate ion.

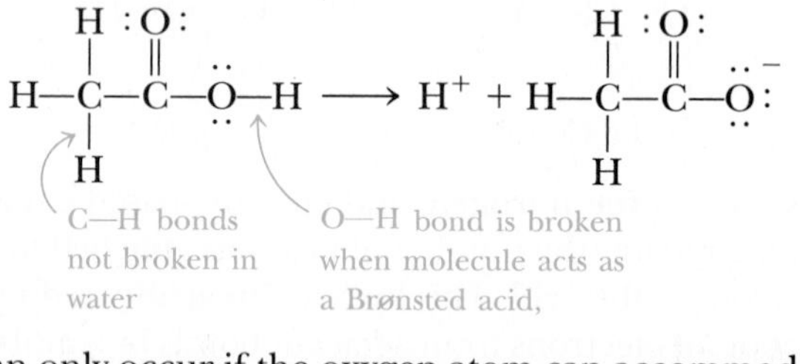

Such cleavage can only occur if the oxygen atom can accommodate the negative charge; because oxygen has a high electronegativity, it can accept this charge relatively easily. In contrast, the C—H hydrogen of acetic acid (and many other molecules) is not dissociated as H^+ in the presence of water because the carbon atom is not sufficiently electronegative to accommodate the negative charge left if the bond breaks as $C—H \rightarrow C:^- + H^+$ (Figure 17.11).

EXERCISE 17.17 ***Relative Acid Strengths***

In each pair of acids, tell which is the stronger and why. Verify your answers using Appendix F.

1. H_2SO_4 (sulfuric acid) and H_2SO_3 (sulfurous acid)
2. H_3AsO_3 (arsenious acid) and H_3AsO_4 (arsenic acid) ■

17.11 THE LEWIS CONCEPT OF ACIDS AND BASES

The theory of acid-base behavior advanced by Brønsted and Lowry in the 1920s works well for aqueous solutions. A more general theory, however, was developed by Gilbert N. Lewis in the 1930s. His theory is based on the sharing of an electron pair between an acid and a base rather than the proton transfer idea of Brønsted and Lowry. A **Lewis acid** is a substance that can accept a pair of electrons from another atom to form a new bond, and a **Lewis base** is a substance that can donate a pair of electrons to another atom to form a new bond. This means that an acid-base reaction in the Lewis sense can occur if there is a molecule (or ion) with a pair of electrons that can be donated and a molecule (or ion) that can accept an electron pair.

G. N. Lewis is the same scientist who developed the first concepts of the electron pair bond. See page 399 for a short biography of this important American scientist.

$$\underset{\text{Acid}}{A} + \underset{\text{Base}}{B:} \longrightarrow \underset{\text{Adduct, or complex}}{B:A}$$

The result is often called an acid-base **adduct,** or **complex.** In Section 9.3 this type of chemical bond was called a **coordinate covalent bond.**

A simple example of a Lewis acid-base reaction is the formation of the hydronium ion from H^+ and water. The H^+ ion has no electrons in its valence, or 1*s*, orbital, but the water molecule has two unshared pairs of electrons (located in sp^3 hybrid orbitals). One of the pairs can be shared between H^+ and water, thus forming an O—H bond. A similar interaction occurs between H^+ and the base ammonia to form the ammonium ion. Such reactions are very common. In general, they involve Lewis acids that are cations or neutral molecules with an available, empty valence orbital and bases that are anions or neutral molecules with a lone electron pair.

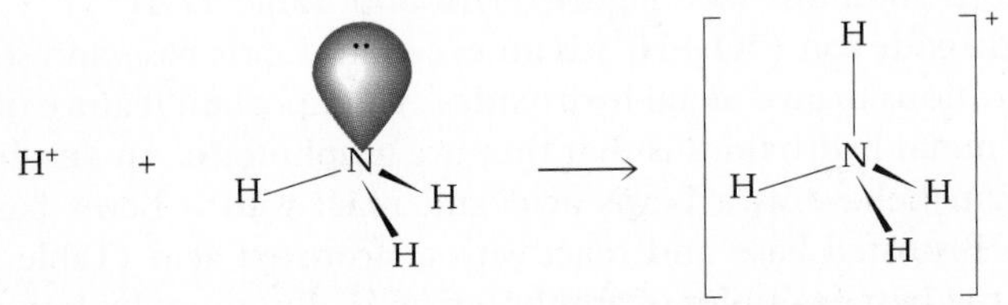

Acid Base

Cationic Lewis Acids

All metal cations are potential Lewis acids. Not only are electron pairs attached to their positive charge, but all have at least one empty orbital. This empty orbital can overlap the orbital bearing the electron pair of the base and can thereby form a two-electron chemical bond. Consider the beryllium ion, Be^{2+}, which can form four donor-acceptor bonds.

$$Be^{2+}(aq) + 4\ H_2O(\ell) \longrightarrow [Be(H_2O)_4]^{2+}(aq)$$

Water molecules are electron-pair donors or Lewis bases, so Be^{2+} and H_2O form an acid-base adduct, and the four empty orbitals on beryllium mean that up to four such bonds can form. Once formed, the 2+ charge of the beryllium ion means that the electrons of the $H_2O—Be^{2+}$ bond are very strongly attracted to the cation. As a result, the O—H bonds of the bound water molecules are polar-

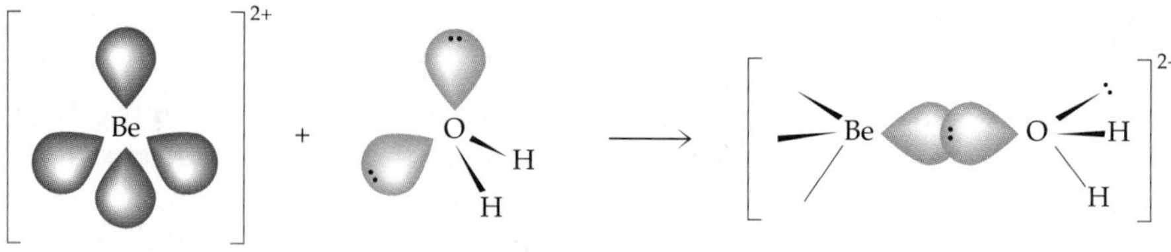

Be^{2+}, sp^3 hybridized, tetrahedral

H_2O, sp^3 hybridized, tetrahedral

The orbital overlap that leads to one of the $H_2O \rightarrow Be$ donor-acceptor bonds.

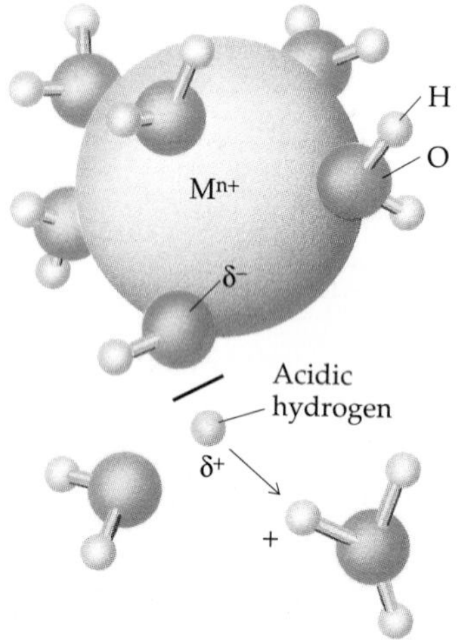

Figure 17.12 The attraction between water molecules and a positively charged metal ion is strong enough that a bound water molecule may lose a proton to the surrounding water and form an acidic solution.

ized, because the oxygen of water, in relinquishing electrons to Be^{2+}, demands more from its O—H bonds.

$$(H_2O)_3Be^{+2}\text{—}:\ddot{O}^{\delta-}\text{—}H^{\delta+} \quad (\text{with } H^{\delta+} \text{ also bonded to O})$$

The net effect is that the O—H bond in a coordinated water molecule is weakened and the H atom is lost as a proton more readily than in an uncoordinated water molecule (Figure 17.12). Thus, the $[Be(H_2O)_4]^{2+}$ complex ion functions as a Brønsted acid or proton donor.

$$[Be(H_2O)_4]^{2+}(aq) + H_2O(\ell) \rightleftharpoons [Be(H_2O)_3(OH)]^{+}(aq) + H_3O^{+}(aq)$$

This is the inductive effect described on page 835, and it explains why metal cations such as Al^{3+} and 2+ and 3+ transition metal cations generally form acidic aqueous solutions (see Figure 17.12 and Table 17.4).

The hydroxide ion ($:\ddot{O}\text{—}H^-$) is an excellent Lewis base and so binds readily to metal cations to give metal hydroxides. An important feature of the chemistry of some metal hydroxides is that they are **amphoteric.** An amphoteric metal hydroxide can behave as a Lewis acid and react with a Lewis base, or it can behave as a Brønsted base and react with a Brønsted acid (Table 17.6).

One of the best examples of amphoterism is aluminum hydroxide, $Al(OH)_3$ (Figure 17.13). Adding OH^- to a precipitate of $Al(OH)_3$ produces the water-soluble $[Al(OH)_4]^-$ ion. Here $Al(OH)_3$ acts as a Lewis acid, and the hydroxide ion is a Lewis base.

$$\underset{\text{Lewis acid}}{Al(OH)_3(s)} + \underset{\text{Lewis base}}{OH^-(aq)} \longrightarrow [Al(OH)_4]^-(aq)$$

TABLE 17.6 Some Common Amphoteric Metal Hydroxides

Hydroxide	Reaction as a Base	Reaction as an Acid
$Al(OH)_3$	$Al(OH)_3(s) + 3\ H_3O^+(aq) \longrightarrow Al^{3+}(aq) + 6\ H_2O(\ell)$	$Al(OH)_3(s) + OH^-(aq) \longrightarrow [Al(OH)_4]^-(aq)$
$Zn(OH)_2$	$Zn(OH)_2(s) + 2\ H_3O^+(aq) \longrightarrow Zn^{2+}(aq) + 4\ H_2O(\ell)$	$Zn(OH)_2(s) + 2\ OH^-(aq) \longrightarrow [Zn(OH)_4]^{2-}(aq)$
$Sn(OH)_4$	$Sn(OH)_4(s) + 4\ H_3O^+(aq) \longrightarrow Sn^{4+}(aq) + 8\ H_2O(\ell)$	$Sn(OH)_4(s) + 2\ OH^-(aq) \longrightarrow [Sn(OH)_6]^{2-}(aq)$
$Cr(OH)_3$	$Cr(OH)_3(s) + 3\ H_3O^+(aq) \longrightarrow Cr^{3+}(aq) + 6\ H_2O(\ell)$	$Cr(OH)_3(s) + OH^-(aq) \longrightarrow [Cr(OH)_4]^-(aq)$

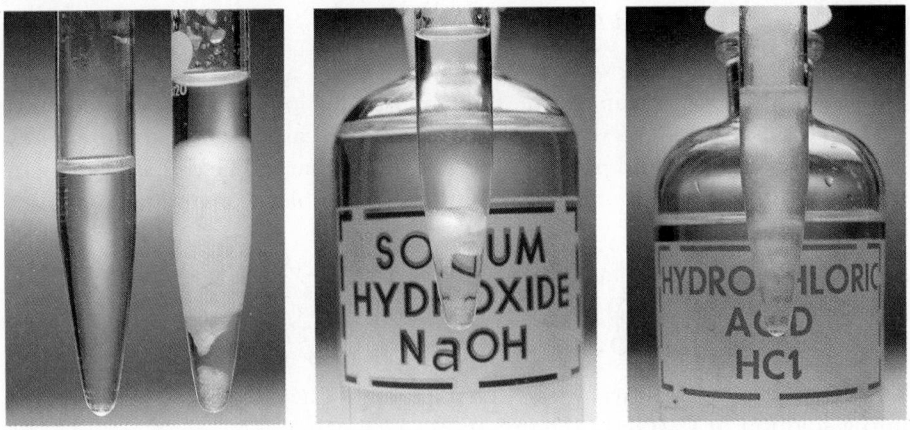

(a) (b) (c)

Figure 17.13 The amphoteric nature of $Al(OH)_3$. (a) Adding aqueous ammonia to a soluble salt of Al^{3+} leads to a precipitate of $Al(OH)_3$. (b) Adding a strong base (NaOH) to $Al(OH)_3$ dissolves the precipitate. Here aluminum hydroxide acts as a Lewis acid toward the Lewis base OH^- and forms the soluble sodium salt of the complex ion $Al(OH)_4^-$. (c) Freshly precipitated $Al(OH)_3$ is dissolved as strong acid (HCl) is added. In this case $Al(OH)_3$ acts as a Brønsted base and forms a soluble aluminum salt and water. (C. D. Winters)

If acid is added to the $Al(OH)_3$ precipitate, it again dissolves. This time, however, the aluminum hydroxide is a Brønsted base.

$$Al(OH)_3(s) + 3\,H_3O^+(aq) \longrightarrow Al^{3+}(aq) + 6\,H_2O(\ell)$$

Many metal ions also form complexes with the Lewis base ammonia. For example, silver ion readily forms a water-soluble, colorless complex ion in aqueous ammonia. Indeed, this complex ion is so stable that the insoluble compound AgCl can be dissolved in aqueous ammonia (Figure 16.5).

$$AgCl(s) + 2\,NH_3(aq) \longrightarrow [H_3N\text{—}Ag\text{—}NH_3]^+(aq) + Cl^-(aq)$$

Light blue aqueous copper(II) ions also react with ammonia to produce a beautiful, deep blue complex ion with four ammonia molecules surrounding each metal ion (Figure 17.14).

$$Cu^{2+}(aq) + 4\,NH_3(aq) \rightleftharpoons \left[\begin{array}{c} \ddot{N}H_3 \\ \downarrow \\ H_3N{:}\longrightarrow Cu \longleftarrow {:}NH_3 \\ \uparrow \\ \ddot{N}H_3 \end{array}\right]^{2+}(aq)$$

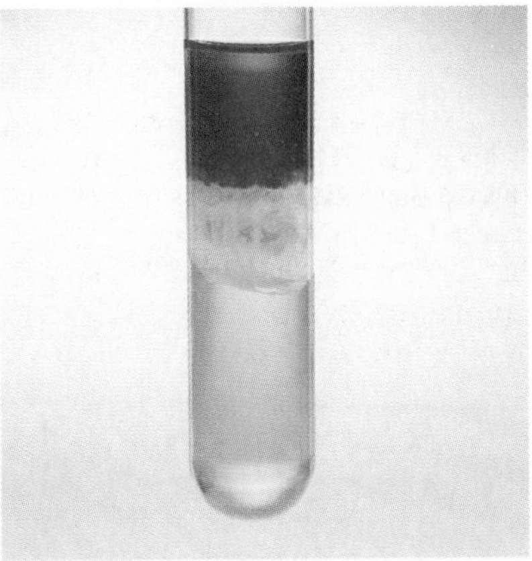

Figure 17.14 The Lewis acid-base complex ion $[Cu(NH_3)_4]^{2+}$. Here aqueous ammonia was added to aqueous $CuSO_4$ (the light blue solution at the bottom of the tube). The small concentration of OH^- in $NH_3(aq)$ first formed insoluble blue-white $Cu(OH)_2$ (the solid in the middle of the tube). With additional NH_3, however, the deep blue, soluble complex ion formed (the solution at the top of the tube). (C. D. Winters)

Molecular Lewis Acids

Lewis's acid-base concept accounts nicely for the fact that oxides of nonmetals behave as acids (Section 4.4). Two important examples are carbon dioxide and sulfur dioxide.

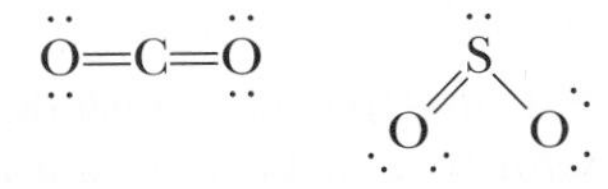

Because oxygen is electronegative, the C=O bonds in CO_2 are polarized away from carbon and toward oxygen. This causes the carbon atom to be slightly deficient in electrons and positive. The negatively charged Lewis base OH^- can therefore attack the carbon to give, ultimately, the bicarbonate ion.

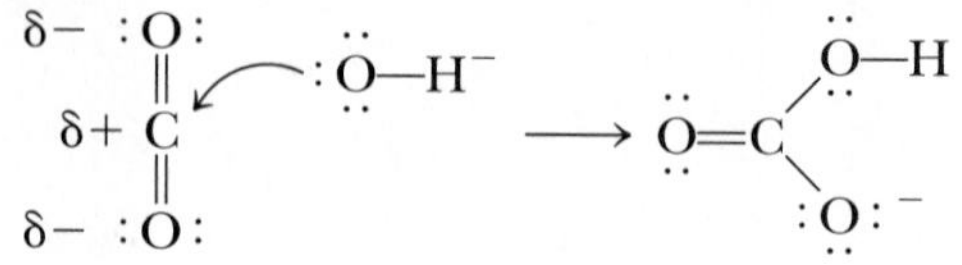

This reaction also accounts for the precipitation of $CaCO_3$ when CO_2 is bubbled into a solution of $Ca(OH)_2$ (Figure 16.2),

$$\underset{\text{Lewis base}}{Ca(OH)_2(s)} + \underset{\text{Lewis acid}}{CO_2(aq)} \longrightarrow CaCO_3(s) + H_2O(\ell)$$

a reaction that we can properly call a Lewis acid-base reaction.

EXERCISE 17.18 ***Lewis Acids and Bases***

Describe each of the following as a Lewis acid or a Lewis base:

1. PH_3

2. BCl_3

3. H_2S

4. HS^- ■

CHAPTER HIGHLIGHTS

Having studied this chapter, you should be able to

- Define and use the three main acid-base theories (Sections 17.1, 17.3, and 17.11).

Concept	Acid	Base
Arrhenius	source of H^+	source of OH^-
Brønsted	H^+ donor	H^+ acceptor
Lewis	electron-pair acceptor	electron-pair donor

- Recognize common **monoprotic** and **polyprotic** acids and bases and write balanced equations for their ionization in water (Section 17.3).
- Appreciate when a substance can be **amphiprotic** (Section 17.3).
- Recognize the **Brønsted acid** and base in a reaction and identify the conjugate partner of each (Section 17.3).
- Use Table 17.3 to decide on the relative strengths of acids and bases.
- Write acid-base reactions and decide whether they are product- or reactant-favored (Section 17.3 and Table 17.3).
- Understand the concept of **water autoionization** and its role in Brønsted acid-base chemistry (Sections 17.2 and 17.3).

- Identify common strong acids and bases (Section 17.4).
- Recognize some common weak acids and understand that they can be neutral molecules (such as acetic acid), cations (such as NH_4^+), hydrated metal ions (such as $Fe(H_2O)_6^{2+}$), or anions (such as HCO_3^-) (Section 17.5 and Table 17.4).
- Recognize some common weak bases and understand that they can be neutral molecules (such as NH_3) or anions (such as CO_3^{2-}) (Section 17.5 and Table 17.4).
- Calculate the pH of a solution from a knowledge of the hydronium ion or hydroxide ion concentration (Section 17.6).
- Use the pH of a solution to calculate the hydronium ion or hydroxide ion concentration (Section 17.6).
- Calculate the equilibrium constant for a weak acid (K_a) or weak base (K_b) from experimental information (such as pH, $[H_3O^+]$, or $[OH^-]$) (Section 17.7).
- Use the equilibrium constant and other information to calculate the pH of a solution of a weak acid or weak base (Section 17.7).
- Describe the acid-base properties of salts and calculate the pH of a solution of a salt of a weak acid or of a weak base (Section 17.8).
- Calculate the pH of a solution of a polyprotic acid (Section 17.9).
- Appreciate the connection between the structure of a compound and its acidity or basicity (Section 17.10).
- Characterize a compound as a Lewis base (an electron-pair donor) or Lewis acid (an electron-pair acceptor) (Section 17.11).

STUDY QUESTIONS

Review Questions

1. Outline the main ideas of the Arrhenius, Brønsted, and Lewis theories of acids and bases. How does the Lewis theory relate to the Brønsted theory?
2. Write a balanced equation depicting the autoionization of water. What is the experimental evidence for this reaction?
3. Write balanced chemical equations showing that phosphoric acid is a polyprotic acid.
4. Write balanced chemical equations showing that the hydrogen sulfate ion, HSO_4^-, is an amphiprotic substance.
5. Designate the Brønsted acid and the base on the left side of each of the following equations, and designate the conjugate partner of each on the right side:

$$HNO_2(aq) + H_2O(\ell) \rightleftharpoons H_3O^+(aq) + NO_2^-(aq)$$
$$NH_4^+(aq) + CN^-(aq) \rightleftharpoons NH_3(aq) + HCN(aq)$$

6. Show that water can be a Brønsted base and a Lewis base.
7. If you have 0.1 M solutions of each of the following Brønsted acids, in which solution is the hydronium ion concentration larger? In which solution is the pH higher?

$$NH_4^+(aq) + H_2O(\ell) \rightleftharpoons H_3O^+(aq) + NH_3(aq) \quad K_a = 5.6 \times 10^{-10}$$
$$HCO_2H(aq) + H_2O(\ell) \rightleftharpoons H_3O^+(aq) + HCO_2^-(aq) \quad K_a = 1.8 \times 10^{-4}$$

8. If you have 0.1 M solutions of each of the following Brønsted bases, in which solution is the hydroxide ion concentration larger? In which solution is the pH higher?

$$CN^-(aq) + H_2O(\ell) \rightleftharpoons HCN(aq) + OH^-(aq) \quad K_b = 2.5 \times 10^{-5}$$
$$CH_3NH_2(aq) + H_2O(\ell) \rightleftharpoons CH_3NH_3^+(aq) + OH^-(aq) \quad K_b = 5.0 \times 10^{-4}$$

9. If Ni^{2+} exists as $[Ni(H_2O)_6]^{2+}$ in aqueous solution, write a balanced equation to show how hydrolysis of the ion leads to an acidic solution.
10. Define the term "amphoteric." Give an example.
11. Which should be the stronger acid and why: H_2SeO_4 or H_2SeO_3?

Numerical and Other Questions

The Brønsted Concept of Acids and Bases

12. Write the formula and give the name of the conjugate base of each of the following acids:
(a) HCN
(b) HSO_4^-
(c) HF
(d) HNO_2
(e) HCO_3^-

13. Write the formula and give the name of the conjugate acid of each of the following bases:
(a) NH_3
(b) HCO_3^-
(c) HS^-
(d) Br^-
(e) HSO_4^-

14. What are the products for each of the following acid-base reactions? Indicate the acid and its conjugate base and the base and its conjugate acid.
(a) $HNO_3 + H_2O \rightarrow$
(b) $HSO_4^- + H_2O \rightarrow$
(c) $H_3O^+ + F^- \rightarrow$

15. What are the products for each of the following acid-base reactions? Indicate the acid and its conjugate base and the base and its conjugate acid.
(a) $HClO_4 + H_2O \rightarrow$
(b) $NH_4^+ + H_2O \rightarrow$
(c) $NH_2^- + H_2O \rightarrow$
(d) $HCO_3^- + OH^- \rightarrow$

16. Dissolving K_2CO_3 in water gives a basic solution. Write a balanced equation showing how the carbonate ion is responsible for this effect.

17. Dissolving ammonium bromide in water gives an acidic solution. Write a balanced equation showing how this can occur.

18. Write balanced equations showing how the HPO_4^{2-} ion of sodium monohydrogen phosphate, Na_2HPO_4, can be a Brønsted acid or a Brønsted base.

19. Write balanced equations showing how the hydrogen oxalate ion, $HC_2O_4^-$, can be both a Brønsted acid and a Brønsted base.

20. In each of the following acid-base reactions, identify the Brønsted acid and base on the left, and their conjugate partners on the right.
(a) $HCO_2H(aq) + H_2O(\ell) \rightleftharpoons HCO_2^-(aq) + H_3O^+(aq)$
(b) $H_2S(aq) + NH_3(aq) \rightleftharpoons HS^-(aq) + NH_4^+(aq)$
(c) $HSO_4^-(aq) + OH^-(aq) \rightleftharpoons SO_4^{2-}(aq) + H_2O(\ell)$

21. In each of the following acid-base reactions, identify the Brønsted acid and base on the left, and their conjugate partners on the right.
(a) $CH_3CO_2H(aq) + C_5H_5N(aq) \rightleftharpoons CH_3CO_2^-(aq) + C_5H_5NH^+(aq)$
(b) $N_2H_4(aq) + HSO_4^-(aq) \rightleftharpoons N_2H_5^+(aq) + SO_4^{2-}(aq)$
(c) $[Al(H_2O)_6]^{3+}(aq) + OH^-(aq) \rightleftharpoons [Al(H_2O)_5OH]^{2+}(aq) + H_2O(\ell)$

22. Several acids are listed here with their respective equilibrium constants:

$HF(aq) + H_2O(\ell) \rightleftharpoons H_3O^+(aq) + F^-(aq)$ $\quad K_a = 7.2 \times 10^{-4}$

$NH_4^+(aq) + H_2O(\ell) \rightleftharpoons H_3O^+(aq) + NH_3(aq)$ $\quad K_a = 5.6 \times 10^{-10}$

$CH_3CO_2H(aq) + H_2O(\ell) \rightleftharpoons H_3O^+(aq) + CH_3CO_2^-(aq)$ $\quad K_a = 1.8 \times 10^{-5}$

(a) Which is the strongest acid? Which is the weakest?
(b) What is the conjugate base of the acid HF?
(c) Which acid has the weakest conjugate base?
(d) Which acid has the strongest conjugate base?

23. Several acids are listed here with their respective equilibrium constants:

$C_6H_5OH(aq) + H_2O(\ell) \rightleftharpoons H_3O^+(aq) + C_6H_5O^-(aq)$ $\quad K_a = 1.3 \times 10^{-10}$

$HCO_2H(aq) + H_2O(\ell) \rightleftharpoons H_3O^+(aq) + HCO_2^-(aq)$ $\quad K_a = 1.8 \times 10^{-4}$

$HC_2O_4^-(aq) + H_2O(\ell) \rightleftharpoons H_3O^+(aq) + C_2O_4^{2-}(aq)$ $\quad K_a = 6.4 \times 10^{-5}$

(a) Which is the strongest acid? Which is the weakest?
(b) Which acid has the weakest conjugate base?
(d) Which acid has the strongest conjugate base?

24. Several bases are listed here with their respective K_b values:

$NH_3(aq) + H_2O(\ell) \rightleftharpoons NH_4^+(aq) + OH^-(aq)$ $\quad K_b = 1.8 \times 10^{-5}$

$C_5H_5N(aq) + H_2O(\ell) \rightleftharpoons C_5H_5NH^+(aq) + OH^-(aq)$ $\quad K_b = 1.5 \times 10^{-9}$

$N_2H_4(aq) + H_2O(\ell) \rightleftharpoons N_2H_5^+(aq) + OH^-(aq)$ $\quad K_b = 8.5 \times 10^{-7}$

(a) Which is the strongest base? Which is the weakest?
(b) What is the conjugate acid of C_5H_5N?
(c) Which base has the strongest conjugate acid? Which has the weakest?

25. Several bases are listed here with their respective K_b values:

$HS^-(aq) + H_2O(\ell) \rightleftharpoons H_2S(aq) + OH^-(aq)$ $\quad K_b = 1 \times 10^{-7}$

$CN^-(aq) + H_2O(\ell) \rightleftharpoons HCN(aq) + OH^-(aq)$ $\quad K_b = 2.5 \times 10^{-5}$

$NO_2^-(aq) + H_2O(\ell) \rightleftharpoons HNO_2(aq) + OH^-(aq)$ $\quad K_b = 2.2 \times 10^{-11}$

(a) Which is the strongest base? Which is the weakest?
(b) What is the conjugate acid of HS^-?
(c) Which base has the strongest conjugate acid? Which has the weakest?

26. State which of the following ions or compounds has the strongest conjugate base and briefly explain your choice:
(a) HSO_4^-
(b) CH_3CO_2H
(c) $HClO$

27. Which of the following compounds has the strongest conjugate acid? Briefly explain your choice.
(a) CN^-
(b) NH_3
(c) SO_4^{2-}

Writing Acid-Base Reactions

28. Ammonium chloride and sodium dihydrogen phosphate, NaH_2PO_4, are mixed in water. Using Table 17.3, write a balanced equation for the acid-base reaction that could, in principle, occur. Does the reaction occur to any significant extent?

29. Acetic acid, CH_3CO_2H, and sodium hydrogen carbonate, $NaHCO_3$, are mixed in water. Using Table 17.3, write a balanced equation for the acid-base reaction that could, in principle, occur. Does the reaction occur to any significant extent?

30. For each reaction shown here, predict whether the equilibrium lies predominantly to the left or to the right. Explain your prediction briefly.
(a) $H_2S(aq) + CO_3^{2-}(aq) \rightleftharpoons HS^-(aq) + HCO_3^-(aq)$
(b) $HCN(aq) + SO_4^{2-}(aq) \rightleftharpoons CN^-(aq) + HSO_4^-(aq)$
(c) $CN^-(aq) + NH_3(aq) \rightleftharpoons HCN(aq) + NH_2^-(aq)$
(d) $SO_4^{2-}(aq) + CH_3CO_2H(aq) \rightleftharpoons HSO_4^-(aq) + CH_3CO_2^-(aq)$

31. For each reaction shown here, predict whether the equilibrium lies predominantly to the left or to the right. Explain your prediction briefly.
(a) $NH_4^+(aq) + Br^-(aq) \rightleftharpoons NH_3(aq) + HBr(aq)$
(b) $HPO_4^{2-}(aq) + CH_3CO_2^-(aq) \rightleftharpoons PO_4^{3-}(aq) + CH_3CO_2H(aq)$
(c) $NH_2^-(aq) + H_2O(\ell) \rightleftharpoons NH_3(aq) + OH^-(aq)$
(d) $Fe(H_2O)_6^{3+}(aq) + HCO_3^-(aq) \rightleftharpoons Fe(H_2O)_5(OH)^{2+}(aq) + H_2CO_3(aq)$

pH Calculations

32. A certain table wine has a pH of 3.40. What is the hydronium ion concentration of the wine? Is it acidic or basic?

33. Milk of magnesia has a pH of 10.5. What is the hydronium ion concentration of the solution? What is the hydroxide ion concentration? Is the solution acidic or basic?

34. What is the pH of a 0.0013 M solution of HNO_3? What is the hydroxide ion concentration of the solution?

35. What is the pH of a 1.2×10^{-4} M solution of KOH? What is the hydronium ion concentration of the solution?

36. What is the pH of a 0.0015 M solution of $Ca(OH)_2$?

37. The pH of a solution of $Ba(OH)_2$ is 10.66 at 25 °C. What is the hydroxide ion concentration in the solution? If the solution volume is 250. mL, how many grams of $Ba(OH)_2$ must have been dissolved?

38. Make the following interconversions. In each case, tell whether the solution is acidic or basic.

	pH	$[H_3O^+]$	$[OH^-]$
(a)	1.00	______	______
(b)	10.50	______	______
(c)	______	1.3×10^{-5}	______
(d)	______	______	2.3×10^{-4}

39. Make the following interconversions. In each case, tell whether the solution is acidic or basic.

	pH	$[H_3O^+]$	$[OH^-]$
(a)	______	6.7×10^{-8}	______
(b)	______	______	2.2×10^{-7}
(c)	5.25	______	______
(d)	______	2.5×10^{-2}	______

Using pH to Calculate Ionization Constants

40. A 2.5×10^{-3} M solution of an unknown acid has a pH of 3.80 at 25 °C.
(a) What is the hydronium ion concentration of the solution?
(b) Is the acid a strong acid, a moderately weak acid (K_a of about 10^{-5}), or a very weak acid (K_a of about 10^{-10})?

41. A 0.015 M solution of an unknown base has a pH of 10.09.
(a) What are the hydroxide and hydronium ion concentrations of this solution?
(b) Is the base a strong base, a moderately weak base (K_b of about 10^{-5}), or a very weak base (K_b of about 10^{-10})?

42. A 0.015 M solution of hydrogen cyanate, HOCN, has a pH of 2.67.
(a) What is the hydronium ion concentration in the solution?
(b) What is the ionization constant K_a for the acid?

43. A 0.10 M solution of chloroacetic acid, $ClCH_2CO_2H$, has a pH of 1.95. Calculate the K_a for the acid.

44. A 0.025 M solution of hydroxylamine has a pH of 9.11. What is the value of K_b for this weak base?

$$H_2NOH(aq) + H_2O(\ell) \rightleftharpoons H_3NOH^+(aq) + OH^-(aq)$$

45. Methylamine, CH_3NH_2, is a weak base.

$$CH_3NH_2(aq) + H_2O(\ell) \rightleftharpoons CH_3NH_3^+(aq) + OH^-(aq)$$

If the pH of a 0.065 M solution of the amine is 11.70, what is K_b?

Using Ionization Constants to Calculate pH

46. The ionization constant of a very weak acid HA is 4.0×10^{-9}. Calculate the equilibrium concentrations of H_3O^+, A^-, and HA in a 0.040 M solution of the acid.

47. What are the equilibrium concentrations of H_3O^+, acetate ion, and acetic acid in a 0.20 M aqueous solution of acetic acid (CH_3CO_2H)?

48. If you have a 0.025 M solution of HCN, what are the equilibrium concentrations of H_3O^+, CN^-, and HCN? What is the pH of the solution?

49. Phenol, C_6H_5OH, is a weak organic acid.

$$C_6H_5OH(aq) + H_2O(\ell) \rightleftharpoons C_6H_5O^-(aq) + H_3O^+(aq) \quad K_a = 1.3 \times 10^{-10}$$

Although somewhat toxic to humans, it is used as a disinfectant and in the manufacture of plastics. If you dissolve 0.780 g of the acid in enough water to make 500. mL of solution, what is the equilibrium hydronium ion concentration? What is the pH of the solution?

50. Propanoic acid, $CH_3CH_2CO_2H$, ionizes in water according to the equation

$$CH_3CH_2CO_2H(aq) + H_2O(\ell) \rightleftharpoons CH_3CH_2CO_2^-(aq) + H_3O^+(aq)$$

If you dissolve 0.588 g of the acid in enough water to make 250. mL of solution, what is the equilibrium hydronium ion concentration? What is the pH of the solution?

51. The hydrogen phthalate ion, $C_8H_5O_4^-$, is a weak acid with $K_a = 2.0 \times 10^{-7}$.

$$C_8H_5O_4^-(aq) + H_2O(\ell) \rightleftharpoons C_8H_4O_4^{2-}(aq) + H_3O^+(aq)$$

If you dissolve 2.55 g of potassium hydrogen phthalate, $KC_8H_5O_4$, in enough water to make 250. mL of solution, what is the pH of the solution?

52. Benzoic acid, $C_6H_5CO_2H$, has a K_a of 6.3×10^{-5}, whereas that of a derivative of this acid, 4-chlorobenzoic acid ($ClC_6H_4CO_2H$), is 1.0×10^{-4}.
(a) Which is the stronger acid?
(b) For a 0.010 M solution of each of these monoprotic acids, which has the higher pH?

53. Place the following acids (1) in order of increasing strength and (2) in order of increasing pH assuming you have a 0.10 M solution of each acid:
(a) 4-Chlorobenzoic acid, $ClC_6H_4CO_2H$ ($K_a = 1.0 \times 10^{-4}$)
(b) Bromoacetic acid, $BrCH_2CO_2H$ ($K_a = 1.3 \times 10^{-3}$)
(c) Trimethylammonium ion, $(CH_3)_3NH^+$ ($K_a = 1.6 \times 10^{-10}$)

54. A hypothetical weak base MOH has $K_b = 5.0 \times 10^{-4}$ for the reaction

$$MOH(aq) \rightleftharpoons M^+(aq) + OH^-(aq)$$

Calculate the equilibrium concentrations of MOH, M^+, and OH^- in a 0.15 M solution of MOH.

55. What are the equilibrium concentrations of NH_3, NH_4^+, and OH^- in a 0.15 M solution of aqueous ammonia? What is the pH of the solution?

56. The weak base methylamine, CH_3NH_2, has $K_b = 5.0 \times 10^{-4}$. It reacts with water according to the equation

$$CH_3NH_2(aq) + H_2O(\ell) \rightleftharpoons CH_3NH_3^+(aq) + OH^-(aq)$$

Calculate the equilibrium hydroxide ion concentration in a 0.25 M solution of the base. What are the pH and pOH of the solution?

57. Calculate the pH of a 0.12 M aqueous solution of the base aniline, $C_6H_5NH_2$ ($K_b = 4.0 \times 10^{-10}$).

58. Calculate the pH of a 0.0010 M aqueous solution of HF.

59. A solution of hydrofluoric acid, HF, has a pH of 2.30. Calculate the equilibrium concentrations of HF, F^-, and H_3O^+, and calculate the amount of HF originally dissolved per liter.

Acid-Base Properties of Salts

60. If each of the salts listed here is dissolved in water to give a 0.10 M solution, which solution has the highest pH? Which has the lowest pH?
(a) Na_2S
(b) Na_3PO_4
(c) NaH_2PO_4
(d) NaF
(e) $NaCH_3CO_2$ (sodium acetate)
(f) $AlCl_3$

61. Which of the following common food additives gives a basic solution when dissolved in water?
(a) $NaNO_3$ (used as a meat preservative)
(b) $NaC_6H_5CO_2$ (sodium benzoate; used as a soft drink preservative)
(c) Na_2HPO_4 (used as an emulsifier in the manufacture of pasteurized cheese)

62. Calculate the hydronium ion concentration and pH in a 0.20 M solution of the salt ammonium chloride, NH_4Cl.

63. Calculate the hydronium ion concentration and pH in a 0.015 M solution of the salt sodium acetate, $NaCH_3CO_2$.

64. Sodium cyanide is the salt of the weak acid HCN. Calculate the concentration of H_3O^+, OH^-, HCN, and Na^+ in a solution prepared by dissolving 10.8 g of NaCN in 500. mL of pure water at 25 °C.

65. The sodium salt of propanoic acid, $NaCH_3CH_2CO_2$, is used as an antifungal agent by veterinarians. Calculate the equilibrium concentration of H_3O^+ and OH^-, and the pH, for a solution of 0.10 M $NaCH_3CH_2CO_2$. (See Table 17.4 for the properties of $CH_3CH_2CO_2^-$ ion.)

Conjugate Acid-Base Pairs

66. The organic base aniline forms its conjugate acid, the anilinium ion, when treated with HCl.

$$\underset{\text{aniline}}{C_6H_5NH_2(aq)} + HCl(aq) \longrightarrow \underset{\text{anilinium ion}}{C_6H_5NH_3^+(aq)} + Cl^-(aq)$$

(a) If K_b for aniline is 4.0×10^{-10}, what is K_a for the anilinium ion?
(b) What is the pH of a 0.080 M solution of anilinium hydrochloride, $C_6H_5NH_3Cl$?

67. The local anesthetic Novocain is the hydrogen chloride salt of an organic base, procaine.

$$\underset{\text{procaine}}{C_{13}H_{20}N_2O_2(aq)} + HCl(aq) \longrightarrow \underset{\text{Novocain}}{[HC_{13}H_{20}N_2O_2]^+Cl^-(aq)}$$

$$K_a \text{ (Novocain)} = 1.4 \times 10^{-9}$$

(a) What is K_b for procaine?
(b) What is the pH of a 0.0015 M solution of Novocain?

68. Saccharin ($HC_7H_4NO_3S$) is a weak acid with $K_a = 2.1 \times 10^{-12}$ at 25 °C. It is used in the form of sodium saccharide, $NaC_7H_4NO_3S$. What is the pH of a 0.10 M solution of sodium saccharide at 25 °C?

69. Pyridine is a weak organic base and readily forms a salt with hydrochloric acid.

$$\underset{\text{pyridine}}{C_5H_5N(aq)} + HCl(aq) \longrightarrow \underset{\text{pyridinium ion}}{C_5H_5NH^+(aq)} + Cl^-(aq)$$

What is the pH of a 0.025 M solution of pyridinium hydrochloride, $[C_5H_5NH]^+Cl^-$?

Polyprotic Acids and Bases

70. Sulfurous acid, H_2SO_3, is a weak acid capable of providing two H^+ ions.
(a) What is the pH of a 0.45 M solution of H_2SO_3?
(b) What is the equilibrium concentration of the sulfite ion, SO_3^{2-}, in the 0.45 M solution of H_2SO_3?

71. Ascorbic acid (vitamin C, $C_6H_8O_6$) is a diprotic acid ($K_{a1} = 6.8 \times 10^{-5}$ and $K_{a2} = 2.7 \times 10^{-12}$). What is the pH of a solution that contains 5.0 mg of acid per milliliter of solution?

72. Hydrazine, N_2H_4, can interact with water in two stages.

$$N_2H_4(aq) + H_2O(\ell) \rightleftharpoons N_2H_5^+(aq) + OH^-(aq) \qquad K_{b1} = 8.5 \times 10^{-7}$$

$$N_2H_5^+(aq) + H_2O(\ell) \rightleftharpoons N_2H_6^{2+}(aq) + OH^-(aq) \qquad K_{b2} = 8.9 \times 10^{-16}$$

(a) What is the concentration of OH^-, $N_2H_5^+$, and $N_2H_6^{2+}$ in a 0.010 M aqueous solution of hydrazine?
(b) What is the pH of the 0.010 M solution of hydrazine?

73. Ethylenediamine, $H_2N—C_2H_4—NH_2$, can interact with water in two steps, forming OH^- in each step (see Appendix G.) If you have a 0.15 M aqueous solution of the amine, calculate the concentration of $[H_3N—C_2H_4—NH_3]^{2+}$ and OH^-.

Lewis Acids and Bases

74. Decide if each of the following substances should be classified as a Lewis acid or base:
(a) Mn^{2+}
(b) CH_3NH_2
(c) H_2NOH in the reaction

$$H_2NOH(aq) + HCl(aq) \longrightarrow [H_3NOH]Cl(aq)$$

(d) SO_2 in the reaction

$$SO_2(g) + BF_3(g) \rightleftharpoons O_2S—BF_3(s)$$

(e) $Zn(OH)_2$ in the reaction

$$Zn(OH)_2(s) + 2\ OH^-(aq) \rightleftharpoons Zn(OH)_4^{2-}(aq)$$

75. Decide if each of the following substances should be classified as a Lewis acid or a Lewis base:
(a) BCl_3
(b) $H_2N—NH_2$, hydrazine
(c) CN^- in the reaction

$$Au^+(aq) + 2\ CN^-(aq) \rightleftharpoons [Au(CN)_2]^-(aq)$$

(d) Fe^{3+}

76. Trimethylamine, $(CH_3)_3N$, is a common reagent. It interacts readily with diborane, B_2H_6. The latter dissociates to BH_3, and this forms a complex with the amine, $(CH_3)_3N—BH_3$. Is the BH_3 fragment a Lewis acid or a Lewis base?

77. Carbon monoxide forms complexes with low-valence metals. For example, $Ni(CO)_4$ and $Fe(CO)_5$ are well known. Carbon monoxide also forms complexes with the iron(II) ion in hemoglobin, thus preventing the hemoglobin from acting normally to take up oxygen. Is CO a Lewis acid or a Lewis base?

78. Draw a Lewis dot structure of ICl_3. Does it function as a Lewis acid or base in reacting with Cl^- to form ICl_4^-? What are the likely structures of ICl_3 and ICl_4^-? What are the likely hybridizations used by I in ICl_3 and ICl_4^-?

79. Carbon dioxide reacts with water to form the bicarbonate ion (page 840). It can also react with the oxide ion, O^{2-}, to form the carbonate ion. Show that the reaction

$$CO_2 + O^{2-} \longrightarrow CO_3^{2-}$$

is a Lewis acid-base reaction.

General Questions

80. If a salt of the hydride ion (H^-), NaH, is put in water, it reacts almost explosively with water to form $H_2(g)$ (see Figure 17.2).

$$H^- + H_2O \longrightarrow H_2 + OH^-$$

Specify the Lewis acid and Lewis base in this reaction. Is the resulting aqueous solution acidic or basic?

81. Liquid ammonia autoionizes just as water does.
(a) Write a balanced equation for the autoionization process of liquid ammonia.
(b) What is the conjugate acid of NH_3? The conjugate base?
(c) If $NaNH_2$ dissolves in liquid ammonia, is the solution acidic or basic?

82. About this time, you may be wishing you had an aspirin. Aspirin is an organic acid with a K_a of 3.27×10^{-4} for the reaction

$$HC_9H_7O_4(aq) + H_2O(\ell) \rightleftharpoons C_9H_7O_4^-(aq) + H_3O^+(aq)$$

You have two tablets, each having 0.325 g of aspirin (mixed with a neutral "binder" to hold the tablet together), which you dissolve in a glass of water (200. mL). What is the pH of the solution?

83. Consider the following ions: CO_3^{2-}, Br^-, S^{2-}, and ClO_4^-
(a) Which of these anions will give a basic solution in water?
(b) Which one is the strongest base?
(c) Write a chemical equation for the reaction of each basic anion with water.

84. Hydrogen sulfide, H_2S, and sodium acetate, $NaCH_3CO_2$, are mixed in water. Using Table 17.3, write a balanced equation for the acid-base reaction that could, in principle, occur. Does the reaction occur to any significant extent?

85. For each of the following reactions, predict whether the equilibrium lies predominantly to the left or to the right. Explain your prediction briefly.
(a) $HCN(aq) + CO_3^{2-}(aq) \rightleftharpoons CN^-(aq) + HCO_3^-(aq)$
(b) $HCO_3^-(aq) + SO_4^{2-}(aq) \rightleftharpoons CO_3^{2-}(aq) + HSO_4^-(aq)$
(c) $HSO_4^-(aq) + CH_3CO_2^-(aq) \rightleftharpoons SO_4^{2-}(aq) + CH_3CO_2H(aq)$
(d) $Co(H_2O)_6^{2+}(aq) + HCO_3^-(aq) \rightleftharpoons Co(H_2O)_5(OH)^+(aq) + H_2CO_3(aq)$

86. The base $Ca(OH)_2$ is almost insoluble in water; only 0.50 g can be dissolved in 1.0 L of water at 25 °C. If the dissolved substance is completely dissociated into its constituent ions, what is the pH of the solution?

87. A monoprotic acid HX has $K_a = 1.3 \times 10^{-3}$. Calculate the equilibrium concentration of HX and H_3O^+ and the pH for a 0.010 M solution of the acid.

88. *m*-Nitrophenol, a weak acid, can be used as a pH indicator, because it is yellow at a pH above 8.6 and colorless at a pH below 6.9. If the pH of a 0.010 M solution of the compound is 3.44, calculate the K_a of the compound.

89. The ionization constant K_w for water at body temperature (37 °C) is 2.5×10^{-14}. What are the concentrations of H_3O^+ and OH^- at this temperature? How do these differ from the concentrations at 25 °C?

90. The butylammonium ion, $C_4H_9NH_3^+$, has a K_a of 2.3×10^{-11}.

$$C_4H_9NH_3^+(aq) + H_2O(\ell) \rightleftharpoons H_3O^+(aq) + C_4H_9NH_2(aq)$$

(a) Calculate K_b for the conjugate base, $C_4H_9NH_2$ (butylamine).
(b) Place the butylammonium ion and its conjugate base in Table 17.4. Name an acid weaker than $C_4H_9NH_3^+$ and a base stronger than $C_4H_9NH_2$.
(c) What is the pH of a 0.015 M solution of the butylammonium ion?

91. The base ethylamine ($CH_3CH_2NH_2$) has a K_b of 4.3×10^{-4}. A related base, ethanolamine ($HOCH_2CH_2NH_2$), has a $K_b = 3.2 \times 10^{-5}$.
(a) If you have a 0.10 M solution of each, which has the higher pH?
(b) Which of the two bases is stronger? Calculate the pH of its 0.10 M solution.

92. For each of the following salts, predict whether an aqueous solution has a pH less than, equal to, or greater than 7:
(a) $NaHSO_4$
(b) NH_4Br
(c) $KClO_4$
(d) Na_2CO_3
(e) $(NH_4)_2S$
(f) $NaNO_3$
(g) Na_2HPO_4
(h) LiBr
(i) $FeCl_3$

93. Chloroacetic acid, $ClCH_2CO_2H$, is a moderately weak acid ($K_a = 1.40 \times 10^{-3}$). Suppose you dissolve 94.5 mg of the acid in 100. mL of water. What is the pH of the solution?

94. Hydroxylamine, NH_2OH, has a K_b of 6.6×10^{-9}. What are the pH and pOH of a 0.051 M solution of the base?

95. To what volume should 100. mL of any weak acid HA with a concentration 0.20 M be diluted in order to double the percentage dissociation?

96. Given the following solutions:

0.1 M NH_3	0.1 M NH_4Cl
0.1 M Na_2CO_3	0.1 M $NaCH_3CO_2$ (sodium acetate)
0.1 M NaCl	0.1 M $NH_4CH_3CO_2$ (ammonium acetate)
0.1 M CH_3CO_2H	

(a) Which of the solutions are acidic?
(b) Which of the solutions are basic?
(c) Which of the solutions is most acidic?

97. Arrange the following 0.1 M solutions in order of increasing pH:
(a) NaCl
(b) NH_4Cl
(c) HCl
(d) $NaCH_3CO_2$ (sodium acetate)
(e) KOH

98. Arrange the following 1.0 M solutions in order of increasing pH:
(a) NaCl
(b) NH_3
(c) NaCN
(d) HCl
(e) NaOH
(f) CH_3CO_2H

99. Nicotinic acid, $C_6H_5NO_2$, is found in minute amounts in all living cells, but appreciable amounts occur in liver, yeast, milk, adrenal glands, white meat, and corn. Whole wheat flour contains about 60. μg/g of flour. One gram of the acid dissolves in 60. mL of water and gives a pH of 2.70. What is the approximate value of K_a for the acid?

100. Oxalic acid, $H_2C_2O_4$, is a moderately weak acid capable of losing two protons. What is the pH of a 0.25-M solution of the acid? What are the concentrations of the $HC_2O_4^-$ and $C_2O_4^{2-}$ ions?

$$H_2C_2O_4(aq) + 2\ H_2O(\ell) \rightleftharpoons 2\ H_3O^+(aq) + C_2O_4^{2-}(aq)$$

101. Nicotine, $C_{10}H_{14}N_2$, has two basic nitrogen atoms, and both can react with water to give a basic solution.

$$Nic(aq) + H_2O(\ell) \rightleftharpoons NicH^+(aq) + OH^-(aq)$$
$$NicH^+(aq) + H_2O(\ell) \rightleftharpoons NicH^{2+}(aq) + OH^-$$

K_{b1} is 7.0×10^{-7} and K_{b2} is 1.1×10^{-10}. Calculate the approximate pH of the 0.020 M solution.

102. Iodine, I_2, is much more soluble in a water solution of potassium iodide, KI, than it is in pure water. The anion found in solution is I_3^-. Write an equation for this reaction, indicating the Lewis acid and Lewis base.

103. Sulfur dioxide reacts with O^{2-} to form the sulfite ion and SO_3 reacts with S^{2-} to form the thiosulfate ion, $S_2O_3^{2-}$. Each can be considered the reaction of a Lewis acid and Lewis base.
(a) Write balanced equations to depict the formation of sulfite and thiosulfate ions.
(b) Write electron dot structures for each species and show which act as Lewis acids and which as Lewis bases.

104. A hydrogen atom in the organic base pyridine, C_5H_5N, can be substituted by various atoms or groups to give XC_5H_4N, where X is an atom such as Cl or a group such as CH_3. The following table gives K_a values for the conjugate acids of a variety of substituted pyridines.

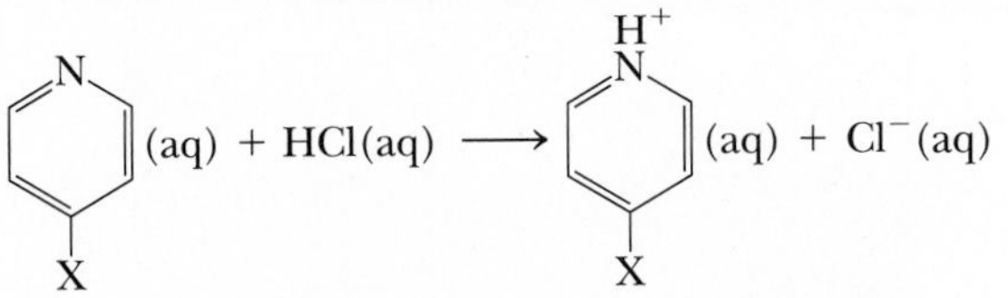

Substituted pyridine — Conjugate acid

Atom or Group X	K_a of Conjugate Acid
NO_2	5.9×10^{-2}
Cl	1.5×10^{-4}
H	6.8×10^{-6}
CH_3	1.0×10^{-6}

(a) Suppose each conjugate acid is dissolved in sufficient water to give a 0.050 M solution. Which solution has the highest pH? The lowest pH?
(b) Which of the substituted pyridines is the strongest Brønsted base? Which one is the weakest Brønsted base?

105. Let us consider a salt of a weak base and a weak acid such as ammonium cyanide. Both NH_4^+ and CN^- ion hydrolyze in aqueous solution, but the net reaction can be considered as a proton transfer from NH_4^+ to CN^-.

$$NH_4^+(aq) + CN^-(aq) \rightleftharpoons NH_3(aq) + HCN(aq)$$

(a) Show that the equilibrium constant for this reaction, K_{total}, is

$$K_{total} = \frac{K_w}{K_a K_b}$$

where K_a is the dissociation constant for the weak acid HCN and K_b is the constant for the weak base NH_3.
(b) Prove that the hydronium ion concentration in this solution must be given by

$$[H_3O^+] = \sqrt{\frac{K_w K_a}{K_b}}$$

(c) What is the pH of a 0.15 M solution of ammonium cyanide?

Conceptual Questions

106. Acetic acid is a weak Brønsted acid. It is interesting to compare the strength of this acid with a related series of acids where the H atoms of the CH_3 group in acetic acid are replaced by Cl.

Acid	K_a
CH_3CO_2H	1.8×10^{-5}
$ClCH_2CO_2H$	1.4×10^{-3}
Cl_2CHCO_2H	3.3×10^{-2}
Cl_3CCO_2H	2.0×10^{-1}

(a) What trend in acid strength do you observe as H is successively replaced by Cl? Can you suggest a reason for this trend?
(b) Suppose each of the acids above was present as a 0.10 M aqueous solution. Which has the highest pH? The lowest pH?

107. The behavior of the acid-base indicator phenolphthalein is described in Section 17.6. Using Le Chatelier's principle, explain why the dye exists as a colorless molecule in acidic solution, but it is deep red in basic solution.

108. Even when dissolved in 100% sulfuric acid, perchloric acid behaves as an acid.
(a) Write a balanced equation showing how perchloric acid can transfer a proton to sulfuric acid.
(b) Draw a Lewis electron dot structure for sulfuric acid.
(c) How can sulfuric acid function as a base?

109. Discuss the validity of this statement: Strong acids leveled by water have conjugate bases weaker than water.

Summary Questions

110. Equilibrium constants can be measured for the dissociation of Lewis acid-base complexes such as the dimethyl ether complex of BF_3, $(CH_3)_2O—BF_3$. The value of K (here K_p) for the reaction

$$(CH_3)_2O—BF_3(g) \rightleftharpoons BF_3(g) + (CH_3)_2O(g)$$

is 0.17.

(a) Tell which product is the Lewis acid and which is the Lewis base.
(b) What is the F—B—F angle in BF_3? In $(CH_3)_2O—BF_3$?
(c) What is the hybridization of O in $(CH_3)_2O—BF_3$? Of the boron atom?
(d) If you place 1.00 g of the complex in a 565-mL flask at 25 °C, what is the total pressure in the flask? What are the partial pressures of the Lewis acid, the Lewis base, and the complex?

111. Sulfanilic acid, which is used in making dyes, is made by reacting aniline with sulfuric acid.

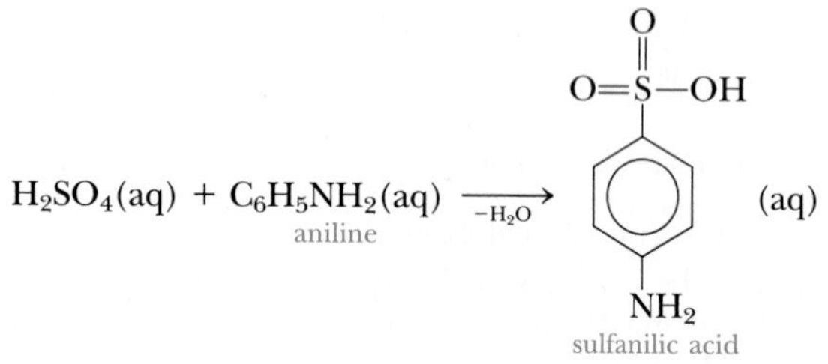

(a) If you want to prepare 150.0 g of sulfanilic acid, and you expect only an 85% yield, how many milliliters of aniline should you take as a starting material (density of aniline = 1.02 g/mL)?
(b) Give approximate values for the following bond angles in sulfanilic acid: C—N—H; H—N—H; O—S—C; and O—S—O.
(c) Sulfanilic acid has a K_a value of 5.9×10^{-4}. The sodium salt of the acid, $Na(H_2NC_6H_4SO_3)$, is quite soluble in water. If you dissolve 1.25 g of the salt in water to give 100. mL of solution, what is the pH of the solution?

CHAPTER 18

Principles of Reactivity: Reactions Between Acids and Bases

A Chemical Puzzler

Because of severe anxiety, you can suffer from alkalosis. That is, the pH of your blood is too high. This can lead to serious medical problems such as muscle spasms and convulsions. One cure is to breathe into a paper bag. Knowing that the pH of your blood is maintained by the weak acid carbonic acid and its conjugate base, the hydrogen carbonate ion, why does this cure for alkalosis work?

Reactions of acids with bases are all around you and within you. The pH of the oceans and your blood is controlled by the chemistry of carbonic acid. Well over 50 million tons of fertilizers such as ammonium nitrate, made by reaction of the base ammonia with nitric acid, are manufactured every year. The acid in "acid rain" comes from the interaction of gaseous, nonmetal oxides, such as CO_2, SO_2, and nitrogen oxides, with water, and this acid rain falls to the earth where it is neutralized by reaction with minerals.

This chapter continues the exploration of acid-base chemistry with particular emphasis on the results of acid-base reactions, the control of such reactions, and some reactions of practical concern.

18.1 ACID-BASE REACTIONS

Using tables such as Table 17.3, the direction of an acid-base reaction can be predicted. For example, acetic acid (CH_3CO_2H) should react to a significant extent with ammonia in water.

$$CH_3CO_2H(aq) + NH_3(aq) \rightleftharpoons NH_4^+(aq) + CH_3CO_2^-(aq)$$

Remember that reactions always proceed in the direction of the weaker acid-base pair, and you find here that NH_4^+ is a weaker acid than CH_3CO_2H and that $CH_3CO_2^-$ is a weaker base than NH_3. We can verify this because the overall reaction is the sum of three other reactions whose equilibrium constants are known.

Ionization of the acid:

$$CH_3CO_2H(aq) + H_2O(\ell) \rightleftharpoons H_3O^+(aq) + CH_3CO_2^-(aq) \qquad K_a = 1.8 \times 10^{-5}$$

Ionization of the base:

$$NH_3(aq) + H_2O(\ell) \rightleftharpoons NH_4^+(aq) + OH^-(aq) \qquad K_b = 1.8 \times 10^{-5}$$

Union of hydronium ion and hydroxide ion:

$$H_3O^+(aq) + OH^-(aq) \rightleftharpoons 2\ H_2O(\ell) \qquad K = \frac{1}{K_w} = 1.0 \times 10^{14}$$

Net reaction:

$$CH_3CO_2H(aq) + NH_3(aq) \rightleftharpoons NH_4^+(aq) + CH_3CO_2^-(aq) \qquad K_{net} = \frac{K_a K_b}{K_w} = 3.2 \times 10^4$$

The formation of NH_4^+ and $CH_3CO_2^-$ is actually driven by coupling the $CH_3CO_2H + NH_3$ reaction to a second, strongly driven reaction. This happens because some of the products of one reaction are reactants in a second reaction—in this case the formation of H_2O from H_3O^+ and OH^-—thereby removing these products as they are formed. This kind of "coupling reaction" is very common in living systems and is one of the very important roles of adenosine triphosphate (ATP) in our bodies.

In Section 16.1 you learned that, when equations are added to find a net equation, K_{net} is the product of the K values for each of the summed equations. In this case K_{net} is fairly large: 32,000. It is large because, although the acid and base are both weak, the H_3O^+ and OH^- ions that they produce are very strong and are "swept up" by the water formation reaction with its extraordinarily high equilibrium constant of 10^{14}. Thus, as suggested by Le Chatelier's principle, the acid and base ionization reactions shift much further to the right than either would go if it occurred alone, and the overall process has a very large equilibrium constant. It is a product-favored reaction.

The reaction of acetic acid and ammonia is a good example of a weak acid reacting with a weak base, and so is the reaction of citric acid and bicarbonate ion in Figure 18.1. These reactions, however, are examples of only one of four possible types (Type 4) of acid-base reaction in aqueous solution.

Figure 18.1 A commercial remedy for excess stomach acid. The bubbles are carbon dioxide, CO_2, from the reaction between a Brønsted acid (citric acid, $C_6H_8O_7$) and a Brønsted base (HCO_3^-, from sodium hydrogen carbonate). (C. D. Winters)

Type of Acid-Base Reaction	Strength of Acid	Strength of Base
1	Strong	Strong
2	Strong	Weak
3	Weak	Strong
4	Weak	Weak

The outcome of each of these combinations is considered in turn.

The Reaction of a Strong Acid with a Strong Base

Strong acids and bases are effectively 100% ionized in solution (Section 17.4); therefore, if we mix HCl and NaOH, we can write the equation

$$H_3O^+(aq) + Cl^-(aq) + Na^+(aq) + OH^-(aq) \longrightarrow 2\ H_2O(\ell) + Na^+(aq) + Cl^-(aq)$$

which leads to the *net ionic equation*

$$H_3O^+(aq) + OH^-(aq) \longrightarrow 2\ H_2O(\ell) \qquad K = \frac{1}{K_w} = 1.0 \times 10^{14}$$

As you learned in Chapter 4, the net ionic equation for the reaction of any strong base with any strong acid is always simply the union of hydronium ion and hydroxide ion to give water. The enormous value of K for the reaction shows that it is, for all practical purposes, quantitatively complete. Thus, if equal numbers of moles of NaOH and HCl are mixed, the result is just a solution of NaCl in water. Because the constituents of NaCl, Na^+ and Cl^- ions, arise from a strong base and a strong acid, respectively, they produce a *neutral* aqueous solution (Table 17.5). For this reason reactions of strong acids and bases have come to be called "neutralizations."

The Reaction of a Strong Acid with a Weak Base

The reaction between HCl and ammonia is a good example of a strong acid/weak base reaction. Here we assume that HCl is 100% ionized in solution and that the H_3O^+ ion produced by the acid reacts with the OH^- ion from the weak base.

$$NH_3(aq) + H_2O(\ell) \rightleftharpoons NH_4^+(aq) + OH^-(aq) \qquad K_b = 1.8 \times 10^{-5}$$

$$H_3O^+(aq) + OH^-(aq) \rightleftharpoons 2\ H_2O(\ell) \qquad K = \frac{1}{K_w} = 1.0 \times 10^{14}$$

$$H_3O^+(aq) + NH_3(aq) \rightleftharpoons H_2O(\ell) + NH_4^+(aq) \qquad K_{net} = \frac{K_b}{K_w} = 1.8 \times 10^9$$

The overall equilibrium constant is quite large, and the reaction proceeds essentially to completion. After reaction of equal molar quantities of HCl and NH_3, the solution contains the salt ammonium chloride, NH_4Cl. The Cl^- ion is the conjugate base of the strong acid HCl and so has no effect on the solution pH (Table 17.5). The NH_4^+ ion, however, is the conjugate acid of the weak base NH_3, so it produces an acidic solution.

$$NH_4^+(aq) + H_2O(\ell) \rightleftharpoons H_3O^+(aq) + NH_3(aq)$$
$$K_a \text{ for } NH_4^+ \text{ (from Table 17.4)} = 5.6 \times 10^{-10}$$

Two important conclusions can be drawn from this example.

- Mixing equal molar quantities of a strong acid and a weak base gives an acidic solution.
- The overall reaction is the reverse of the reaction that defines K_a for the conjugate acid of the weak base, so its K is just $1/K_a$ for the conjugate acid of the weak base in the reaction. Here $K_a = 5.6 \times 10^{-10}$, so $K_{net} = 1/(5.6 \times 10^{-10}) = 1.8 \times 10^9$.

In Chapter 5 you learned that the amount of acid in solution, for example, can be determined by titration with a base (Figure 5.11). The point at which the moles of OH^- supplied by the base is equal to the moles of H_3O^+ that can be supplied by the acid is called the **equivalence point.**

EXAMPLE 18.1 *pH at the Equivalence Point of a Strong Acid/Weak Base Reaction*

Suppose you mix exactly 100 mL of 0.10 M HCl with exactly 50 mL of 0.20 M NH_3. What is the pH of the resulting solution?

Solution The balanced, net ionic equation for the HCl/NH_3 reaction is given in the text preceding this example.

$$H_3O^+(aq) + NH_3(aq) \rightleftharpoons H_2O(\ell) + NH_4^+(aq)$$

Ammonium ion is a weak Brønsted acid, which means that the pH of the solution depends on K_a for this ion and on its concentration.

Here we are mixing 0.010 mol of HCl (0.100 L × 0.10 M) with 0.010 mol of NH_3 (0.050 L × 0.20 M NH_3); 0.010 mol of NH_4Cl is produced. Because the solution volume after reaction is the total of the volumes of the reacting solutions, exactly 150 mL, the product ion concentrations are

$$[NH_4^+] = [Cl^-] = 0.010 \text{ mol}/0.150 \text{ L} = 0.067 \text{ M}$$

Equation	$H_3O^+(aq)$ +	$NH_3(aq)$ ⇌	$H_2O(\ell)$ + $NH_4^+(aq)$
Concentrations before (M)	0.10	0.20	0
Quantity before	0.010 mol	0.010 mol	0
Quantity after	0	0	0.010 mol
Concentrations after (M)	0	0	0.067

Now we can solve for the hydronium ion concentration of the solution using the value of K_a for ionization of NH_4^+.

Equation	$NH_4^+(aq)$	$+ H_2O(\ell) \rightleftharpoons$	$H_3O^+(aq)$	$+ NH_3(aq)$
Concentrations before NH_4^+ hydrolysis (M)	0.067		0	0
Change in concentrations on proceeding to equilibrium	$-x$		$+x$	$+x$
Concentrations at equilibrium (M)	$(0.067 - x)$		x	x

Substituting the equilibrium concentrations into the equation for K_a, we have

$$K_a = 5.6 \times 10^{-10} = \frac{[NH_3][H_3O^+]}{[NH_4^+]} = \frac{(x)(x)}{0.067 - x} \approx \frac{x^2}{0.067}$$

Let us make the approximation that $(0.067 - x) \approx 0.067$, and so the hydronium ion concentration is

$$[H_3O^+] = 6.1 \times 10^{-6}\ M$$

and the pH is 5.21. The solution is clearly acidic, as anticipated for a solution containing the conjugate acid of a weak base.

Note that the approximation $(0.067 - x) \approx 0.067$ is valid. Subtracting 6.1×10^{-6} from 0.067 indeed makes no difference in the final answer.

EXERCISE 18.1 ***pH at the Equivalence Point of a Strong Acid/Weak Base Reaction***

Aniline, $C_6H_5NH_2$, is a weak organic base first discovered in 1826. If you mix exactly 50 mL of 0.20 M HCl with 0.93 g of aniline, are the acid and base completely consumed? What is the pH of the resulting solution? (The K_a for $C_6H_5NH_3^+$ is 2.4×10^{-5}.)

$$HCl(aq) + C_6H_5NH_2(aq) \rightleftharpoons C_6H_5NH_3^+(aq) + Cl^-(aq)$$ ■

The Reaction of a Weak Acid with a Strong Base

Consider the weak acid formic acid, HCO_2H, and its reaction with the strong base NaOH.

$$HCO_2H(aq) + H_2O(\ell) \rightleftharpoons H_3O^+(aq) + HCO_2^-(aq) \qquad K_a = 1.8 \times 10^{-4}$$

$$H_3O^+(aq) + OH^-(aq) \rightleftharpoons 2\ H_2O(\ell) \qquad K = \frac{1}{K_w} = 1.0 \times 10^{14}$$

$$HCO_2H(aq) + OH^-(aq) \rightleftharpoons H_2O(\ell) + HCO_2^-(aq) \qquad K_{net} = \frac{K_a}{K_w} = 1.8 \times 10^{10}$$

Once again the equilibrium constant for the overall process is large, so the net reaction goes essentially to completion. Assuming that equal molar quantities of acid and base were mixed, the final solution contains only sodium formate ($NaHCO_2$), a salt that is 100% dissociated in water. However, although the Na^+ ion is the cation of a strong base and so gives a neutral solution, the formate ion is the anion of a weak acid and is therefore a base (Table 17.4). The solution is basic because of the hydrolysis of the formate ion.

$$H_2O(\ell) + HCO_2^-(aq) \rightleftharpoons HCO_2H(aq) + OH^-(aq)$$

$$K_b \text{ (from Table 17.4)} = 5.6 \times 10^{-11}$$

Again there are two important conclusions.

- Mixing equal molar quantities of a strong base with a weak acid produces a salt whose anion is the conjugate base of the weak acid. The solution is *basic,* with the pH depending on K_b for the anion.
- The net reaction between a strong base and a weak acid is the reverse of the reaction that defines K_b for the conjugate base of the weak acid. Therefore, K for the net reaction is $1/K_b$ for the conjugate base of the weak acid (here $K_{net} = 1/K_b = 1/5.6 \times 10^{-11} = 1.8 \times 10^{10}$).

EXAMPLE 18.2 *pH at the Equivalence Point of a Strong Base/Weak Acid Reaction*

Suppose you mix 50. mL of 0.10 M NaOH with 50. mL of 0.10 M formic acid, HCO_2H. What is the pH of the resulting solution?

Solution The net ionic equation for the acid-base reaction that occurs is given in the text.

$$HCO_2H(aq) + OH^-(aq) \rightleftharpoons H_2O(\ell) + HCO_2^-(aq)$$

Here we mix 0.0050 mol of HCO_2H (0.050 L × 0.10 M) with 0.0050 mol of NaOH; 0.0050 mol of $NaHCO_2$ is produced. Because 50. mL of each reagent was mixed, $NaHCO_2$ is dissolved in 1.0×10^2 mL of water. The concentration of this salt is therefore $[NaHCO_2] = 0.050$ M.

With the salt concentration known, we can solve for the pH of the solution. From the discussion preceding this example we know the final solution is basic due to the hydrolysis of the conjugate base of formic acid.

Equation	$H_2O(\ell) + HCO_2^-(aq)$	$\rightleftharpoons HCO_2H(aq)$	$+ OH^-(aq)$
Concentrations before HCO_2^- hydrolysis (M)	0.050	0	0
Change in concentrations on proceeding to equilibrium	$-x$	$+x$	$+x$
Concentrations at equilibrium (M)	$(0.050 - x)$	x	x

In Example 18.2 $[OH^-]$ is given as 0 before equilibrium is established. It is actually 10^{-7} M, but as described in Chapter 17, hydroxide and hydronium ion from water autoionization can generally be ignored in calculations of this type. This same simplification is made in the other examples in this chapter.

Therefore, we can write

$$K_b = 5.6 \times 10^{-11} = \frac{[HCO_2H][OH^-]}{[HCO_2^-]} = \frac{(x)(x)}{0.050 - x} \approx \frac{x^2}{0.050}$$

As described in Example 17.9, we make the approximation that $(0.050 - x) \approx 0.050$, and so the hydroxide ion concentration is

$$[OH^-] = 1.7 \times 10^{-6} \text{ M}$$

This gives a H_3O^+ concentration of 6.0×10^{-9} M and a pH of 8.23. The solution is basic, as predicted.

As a final comment, note that the approximation that $(0.050 - x) \approx 0.050$ is valid. Subtracting 1.7×10^{-6} from 0.050 indeed makes no difference in the final answer.

EXERCISE 18.2 ***pH at the Equivalence Point of a Strong Base/Weak Acid Reaction***

What volume of 0.100 M NaOH, in milliliters, is required to react completely with 0.976 g of the weak, monoprotic acid benzoic acid ($C_6H_5CO_2H$)? What is the pH of the solution after reaction? (See Table 17.4 for the K_b for the benzoate ion, $C_6H_5CO_2^-$.) ■

PROBLEM-SOLVING TIPS AND IDEAS

18.1 The pH of the Solution at the Equivalence Point of an Acid-Base Reaction

Finding the pH at the equivalence point for an acid-base reaction always involves three calculation steps. There are no shortcuts. Consider the reaction of a weak acid, HA, with a strong base as an example. (The same principles apply to other acid-base reactions.)

$$HA(aq) + OH^-(aq) \longrightarrow A^-(aq) + H_2O(\ell)$$

Step 1 *Stoichiometry problem*

At the equivalence point, the acid and base have been consumed completely to leave a solution of A^-, the conjugate base of the weak acid. Use the principles of stoichiometry to calculate (a) the quantity of base added, (b) the quantity of acid consumed, and (c) the quantity of conjugate base formed.

Step 2 *Calculate the concentration of conjugate base, [A^-]*

Recognize that the volume of the solution is the sum of the original volume of acid solution plus the volume of base solution added.

Step 3 *pH at the equivalence point*

Use the concentration of conjugate base from Step 2 and the value of K_b for the base (A^-) to calculate the concentration of OH^- in the solution.

$$H_2O(\ell) + A^-(aq) \rightleftharpoons HA(aq) + OH^-(aq) \qquad K = K_b \text{ for } A^-$$ ■

The Reaction of a Weak Acid with a Weak Base

If acetic acid, a weak acid, is mixed with ammonia, a weak base, the following reaction occurs (see page 850).

$$CH_3CO_2H(aq) + NH_3(aq) \rightleftharpoons NH_4^+(aq) + CH_3CO_2^-(aq)$$

$$K_{net} = 3.2 \times 10^4$$

If equal molar quantities of acid and base are mixed, the resulting solution contains only ammonium acetate, $NH_4CH_3CO_2$. Is this solution acidic or basic?

TABLE 18.1 Characteristics of Acid-Base Reactions

Type	Example	Net Ionic Equation	K	Species Present after Equal Molar Amounts are Mixed; pH
Strong acid + strong base	$HCl + NaOH$	$H_3O^+(aq) + OH^-(aq) \rightleftharpoons 2\ H_2O(\ell)$	1.0×10^{14}	Cl^-, Na^+, pH = 7, neutral
Strong acid + weak base	$HCl + NH_3$	$H_3O^+(aq) + NH_3(aq) \rightleftharpoons NH_4^+(aq) + H_2O(\ell)$	1.8×10^9	Cl^-, NH_4^+, pH < 7, acidic
Weak acid + strong base	$HCO_2H + NaOH$	$HCO_2H(aq) + OH^-(aq) \rightleftharpoons HCO_2^-(aq) + H_2O(\ell)$	1.8×10^{10}	HCO_2^-, Na^+, pH > 7, basic
Weak acid + weak base	$HCO_2H + NH_3$	$HCO_2H(aq) + NH_3(aq) \rightleftharpoons HCO_2^-(aq) + NH_4^+(aq)$	3.2×10^4	HCO_2^-, NH_4^+, pH depends on K_b and K_a of conjugate base and acid.

Ammonium ion is the conjugate acid of a weak base, and so should contribute to the solution acidity.

$$NH_4^+(aq) + H_2O(\ell) \rightleftharpoons H_3O^+(aq) + NH_3(aq) \qquad K_a = 5.6 \times 10^{-10}$$

On the other hand, acetate ion ($CH_3CO_2^-$) is the conjugate base of a weak acid; it should make the solution basic.

$$CH_3CO_2^-(aq) + H_2O(\ell) \rightleftharpoons CH_3CO_2H(aq) + OH^-(aq) \qquad K_b = 5.6 \times 10^{-10}$$

In Section 17.8 you learned that the pH of a solution of the conjugate acid and base of a weak base and acid depends on the relative values of K_a and K_b. Here $K_a = K_b$. This tells us that equal quantities of H_3O^+ and OH^- are formed, so the solution is expected to be neutral. When a weak acid and a weak base react in equal molar amounts in solution, the pH of the solution depends on the *relative K values of the conjugate base and acid.*

A Summary of Acid-Base Reactions

When writing an acid-base reaction, we first pay attention to whether the reactants are strong or weak. When one reactant is strong and the other one weak, the pH of the solution after mixing equal molar amounts of acid and base is controlled by the conjugate partner of the weak acid or weak base. This means the equilibrium constant for the acid-base reaction is the reciprocal of the K for ionization of the conjugate base or acid. These relationships are summarized in Table 18.1.

EXERCISE 18.3 ***pH of a Solution of a Salt of a Weak Base and Weak Acid***

Suppose you mix exactly 50 mL of 0.10 M acetic acid and 50 mL of 0.10 M pyridine (C_5H_5N), a weak base. (K_a for the pyridinium ion, $C_5H_5NH^+$, is 6.7×10^{-6}, and K_b for acetate ion is 5.6×10^{-10}.) Is the solution acidic or basic? ■

18.2 THE COMMON ION EFFECT

One objective of this chapter is to follow the course of events during an acid-base titration. Before doing this, though, we have to explore the nature of solutions of acids and bases that contain another solute, in particular an ion that is "common" to the acid or base equilibrium.

Lactic acid is a weak acid found in sour milk, apples and other fruit, beer and wine, and several plants.

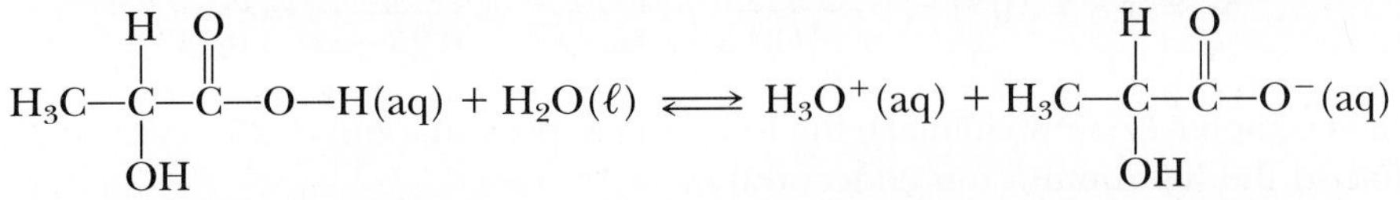

lactic acid ($C_3H_6O_3$) lactate ion ($C_3H_5O_3^-$)

$K_a = 1.3 \times 10^{-4}$

Suppose a sample containing lactic acid is analyzed by titrating the acid with the strong base NaOH.

$$C_3H_6O_3(aq) + OH^-(aq) \rightleftharpoons C_3H_5O_3^-(aq) + H_2O(\ell)$$

lactic acid lactate ion

At the equivalence point in the titration, the acid has been consumed and converted completely to its conjugate base, the lactate ion. *Before* the equivalence point, however, some lactate ion is present in solution along with as yet unreacted lactic acid. Thus, before the equivalence point, the *weak acid is present* along with *some amount of its conjugate base.* If we halt the titration at some intermediate stage before the equivalence point, we observe that the ionization of the remaining lactic acid is affected by the lactate ion present. This phenomenon is called the **common ion effect** because an ion (here the lactate ion) "common" to the ionization of the acid is present in an amount greater than that produced by simple acid ionization. In an acid ionization, the presence of some conjugate base in solution does, according to Le Chatelier's principle, limit the extent to which the acid ionizes and thus affects the pH of the solution (Figure 18.2).

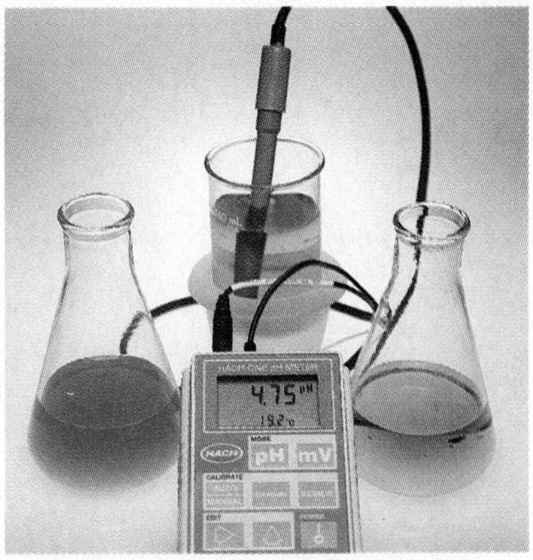

Figure 18.2 The common ion effect. An acid (0.25 M acetic acid, left) is mixed with a base (0.10 M sodium acetate, right). The pH meter shows the resulting solution (center) has a lower hydronium ion concentration (a pH of about 5) than the acetic acid solution (pH is about 2.7). (Universal indicator is red in low pH, yellow in slightly acidic media, and green in neutral to weakly basic media.) (C. D. Winters)

EXAMPLE 18.3 *The Common Ion Effect*

What is the pH of 0.25 M aqueous acetic acid? What is the pH of the solution after adding sodium acetate to make the solution 0.10 M in the salt? (Ignore any change in volume on adding the sodium acetate.)

Solution Let us first determine the pH of the 0.25 M acid solution using the approach outlined in Chapter 17. Then we can turn to the effect of adding a "common ion."

Step 1 *Determine the pH of 0.25 M acetic acid.*

Equation	$CH_3CO_2H(aq) + H_2O(\ell) \rightleftharpoons$	$H_3O^+(aq)$ +	$CH_3CO_2^-(aq)$
Concentration before CH_3CO_2H ionizes (M)	0.25	0	0
Change in concentration on proceeding to equilibrium	$-x$	$+x$	$+x$
Concentration at equilibrium (M)	$(0.25 - x)$	x	x

The appropriate equilibrium expression is

$$K_a = 1.8 \times 10^{-5} = \frac{[CH_3CO_2^-][H_3O^+]}{[CH_3CO_2H]} = \frac{(x)(x)}{0.25 - x} \approx \frac{x^2}{0.25}$$

As in Chapter 17, we shall make the following approximation: $(0.25 - x) \approx 0.25$, and so the hydronium ion concentration is

$$[H_3O^+] = 2.1 \times 10^{-3}\ M$$

(The approximation is valid because subtracting 2.1×10^{-3} from 0.25 indeed makes no difference in the final answer.) The hydronium ion concentration leads to a pH of 2.67.

Step 2 *Effect of added "common ion"*

Sodium acetate, $NaCH_3CO_2$, is very soluble in water and is 100% dissociated into its ions, Na^+ and $CH_3CO_2^-$ (acetate ion). As you learned in Chapter 17, Na^+ ion has no effect on the pH of a solution (see Table 17.5). On the other hand, $CH_3CO_2^-$ is the conjugate base of the weak acid CH_3CO_2H, so it should contribute some excess OH^- ions to the solution,

$$\underset{\text{acetate ion}}{CH_3CO_2^-(aq)} + H_2O(\ell) \rightleftharpoons \underset{\text{acetic acid}}{CH_3CO_2H(aq)} + OH^-(aq)$$

and thus reduce the hydronium ion concentration below 2.1×10^{-3} M, the concentration found if the solution contains only acetic acid.

We solve for the net hydronium ion concentration in the following way. First, imagine the solution contains only the weak acid (CH_3CO_2H, 0.25 M) and its conjugate base ($CH_3CO_2^-$, 0.10 M) and that neither has yet reacted with water to generate H_3O^+ or OH^-, respectively.

Step 1 *Concentrations in the acetic acid/sodium acetate mixture **before** ionization of the acid*

$$\underset{0.25\ M}{CH_3CO_2H(aq)} + H_2O(\ell) \rightleftharpoons \underset{0}{H_3O^+(aq)} + \underset{0.10\ M}{CH_3CO_2^-(aq)}$$

Now imagine that the acid ionizes to give H_3O^+ and $CH_3CO_2^-$, both in the amount y, in the presence of the added $NaCH_3CO_2$.

Step 2 *Concentrations in the acetic acid/sodium acetate mixture **after** ionization of the acid*

$$\underset{(0.25 - y)\ M}{CH_3CO_2H(aq)} + H_2O(\ell) \rightleftharpoons \underset{y\ M}{H_3O^+(aq)} + \underset{(0.10\ M + y)\ M}{CH_3CO_2^-(aq)}$$

Without the added salt $NaCH_3CO_2$, which provides the "common ion" $CH_3CO_2^-$, ionization of the acid would have produced H_3O^+ and $CH_3CO_2^-$, both in the amount x. We know from Le Chatelier's principle, however, that y must be less than x ($y < x$). Therefore, the concentration table for the ionization of acetic acid in the presence of the "common ion" acetate ion is

	$[CH_3CO_2H]$	$[H_3O^+]$	$[CH_3CO_2^-]$
Concentration before CH_3CO_2H ionizes (M)	0.25	0	0.10
Change in concentration on proceeding to equilibrium	$-y$	$+y$	$+y$
Concentration at equilibrium (M)	$0.25 - y$	y	$0.10 + y$

and the appropriate equilibrium expression is

$$K_a = 1.8 \times 10^{-5} = \frac{[H_3O^+][CH_3CO_2^-]}{[CH_3CO_2H]} = \frac{(y)(0.10 + y)}{0.25 - y} \approx \frac{(y)(0.10)}{0.25}$$

Let us begin by asking what approximation converts the "exact" expression above into the more easily solved "approximate" expression. We know that the amount of $CH_3CO_2^-$ formed by ionization in a 0.25 M solution of acetic acid is 0.0021 M, and Le Chatelier's principle predicts that the acid produces even less acetate ion in the presence of the "common ion." It is therefore reasonable to assume that $(0.10 + y)$ M $\approx$ 0.10 M and that $(0.25 - y)$ M $\approx$ 0.25 M. Solving the "approximate" expression, we find that

$$y = [H_3O^+] = [CH_3CO_2^-] \text{ added to solution by } CH_3CO_2H \text{ ionization}$$
$$\approx 4.5 \times 10^{-5} \text{ M}$$

pH = 4.35

This is indeed a very small value for $[H_3O^+]$ and for the concentration of $CH_3CO_2^-$ that is added to the solution by acetic acid ionization. Our approximations are reasonable. (Furthermore, we obtain the same answer from the "exact expression" using the method of successive approximations.)

Finally, compare the situations before and after adding acetate ion, the common ion. The pH before adding sodium acetate was 2.67. However, the acidity decreased (the pH increased to 4.35) after adding the weak base sodium acetate.

EXERCISE 18.4 ***The Common Ion Effect***

Assume you have a 0.30 M solution of formic acid (HCO_2H) and add enough sodium formate ($NaHCO_2$) to make the solution 0.10 M in the salt. Calculate the pH of the formic acid solution before and after adding sodium formate. ■

PROBLEM-SOLVING TIPS AND IDEAS

18.2 Solving Common Ion Problems

The key idea of common ion problems is that the weak acid HA, for example, ionizes to produce *less* H_3O^+ in the presence of the common ion A^- than it would have in the absence of A^-. Thus, adding A^- to a solution of the weak acid HA causes the pH to be higher than in the absence of A^-. This is easy to remember because adding a base to any solution raises the pH of the solution. ■

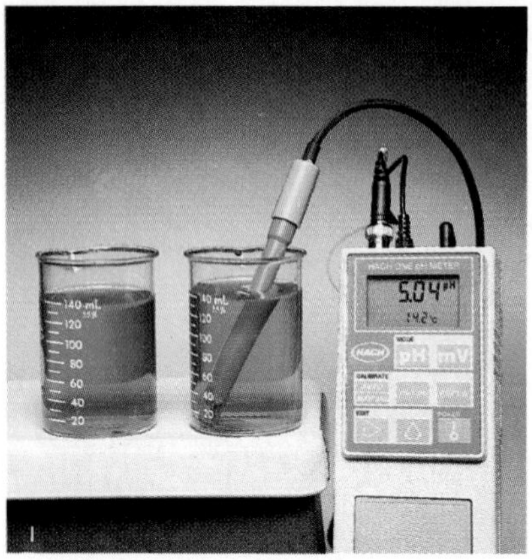

(a)

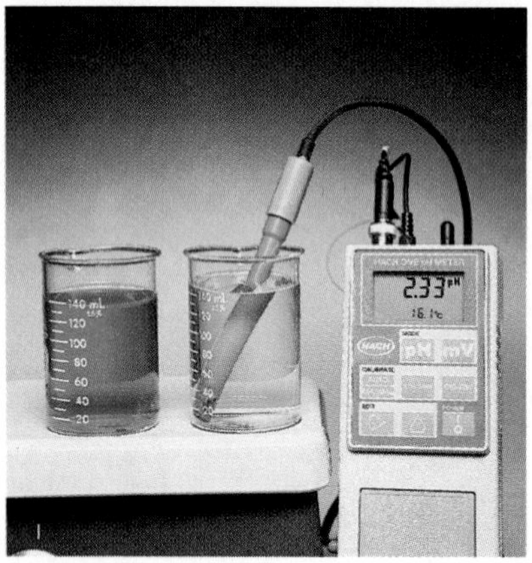

(b)

Figure 18.3 Buffer solutions. (a) The pH electrode indicates the pH of water that contains a trace of acid (and bromphenol blue indicators; see Figure 17.7). The solution at the left is a buffer having a pH of about 7. (b) When 5 mL of 0.10 M HCl is added to each solution, the pH of the water drops several units, whereas the pH of the buffer stays constant (as implied by the fact that the indicator color has not changed). (C. D. Winters)

18.3 BUFFER SOLUTIONS

The normal pH of human blood is 7.4. Experiment clearly shows that the addition of a small quantity of strong acid or base to blood, say 0.01 mol to a liter of blood, leads to a change in pH of only about 0.1 pH units (Figure 18.3). In comparison, if you add 0.01 mol of HCl to 1.0 L of pure water, the pH drops from 7 to 2, whereas addition of 0.01 mL of NaOH increases the pH from 7 to 12. Blood, and many other body fluids, are said to be *buffered;* that is, their pH is resistant to change on addition of a strong acid or base.

In general, two species are required in a **buffer solution,** one (an acid) capable of reacting with added OH^- ions and another (a base) that can consume added H_3O^+ ions. An additional requirement is that the acid and base do not react with each another. This means that a buffer is usually prepared from roughly equal quantities of a *conjugate* acid-base pair: (a) a weak acid and its conjugate base (acetic acid and acetate ion, for example), or (b) a weak base and its conjugate acid (ammonia and ammonium ion, for example). Some systems commonly used in the laboratory are given in Table 18.2.

To see how a buffer works, let us consider an acetic acid/acetate ion buffer. Acetic acid, the weak acid, is needed to consume any added hydroxide ion.

$$CH_3CO_2H(aq) + OH^-(aq) \longrightarrow CH_3CO_2^-(aq) + H_2O(\ell) \qquad K = 1.8 \times 10^9$$

The equilibrium constant for the reaction is very large because OH^- is a much stronger base than acetate, $CH_3CO_2^-$, as we know from Table 17.4. This means that any OH^- entering the solution from an outside source is consumed completely. In a similar way, any hydronium ion added to the solution reacts with the acetate ion present in the buffer.

$$H_3O^+(aq) + CH_3CO_2^-(aq) \longrightarrow H_2O(\ell) + CH_3CO_2H(aq) \qquad K = 5.6 \times 10^4$$

The equilibrium constant for this reaction is also quite large because H_3O^+ is a much stronger acid than CH_3CO_2H, again as suggested by the ionization constants given in Table 17.4.

Now that we have established that a buffer solution should effectively remove small amounts of added acid or base, work through the following example, which shows more quantitatively how the pH of a solution can be maintained.

TABLE 18.2 Some Commonly Used Buffer Systems

Weak Acid	Conjugate Base	Acid K_a	Useful pH Range
Phthalic acid $C_6H_4(CO_2H)_2$	Hydrogen phthalate ion $C_6H_4(CO_2H)(CO_2)^-$	1.3×10^{-3}	1.9–3.9
Acetic acid CH_3CO_2H	Acetate ion $CH_3CO_2^-$	1.8×10^{-5}	3.7–5.8
Dihydrogen phosphate ion $H_2PO_4^-$	Hydrogen phosphate ion HPO_4^{2-}	6.2×10^{-8}	6.2–8.2
Hydrogen phosphate ion HPO_4^{2-}	Phosphate ion PO_4^{3-}	3.6×10^{-13}	11.3–13.3

EXAMPLE 18.4 *A Buffer Solution*

Calculate the pH change that occurs when 1.0 mL of 1.0 M HCl is added to (1) 1.0 L of pure water and (2) 1.0 L of acetic acid/sodium acetate buffer with $[CH_3CO_2H] = 0.70$ M and $[CH_3CO_2^-] = 0.60$ M.

Solution

Step 1 *Adding Acid to Pure Water.* 1.0 mL of 1.0 M HCl represents 0.0010 mol of acid. If this is added to 1.0 L of pure water, the hydrogen ion concentration is 1.0×10^{-3} M. This means the H_3O^+ concentration is raised from 10^{-7} to 10^{-3}, so the pH falls from 7 to 3.

Step 2 *pH of Acetic Acid/Acetate Ion Buffer Solution.* We first need to determine the pH of the buffer solution before any HCl is added. The balanced equation connecting the species in solution is just the ionization of acetic acid.

Equation	$CH_3CO_2H(aq)$ + $H_2O(\ell)$ ⇌	$H_3O^+(aq)$ +	$CH_3CO_2^-(aq)$
Concentrations before CH_3CO_2H ionizes (M)	0.70	0	0.60
Change in concentrations on proceeding to equilibrium	$-x$	$+x$	$+x$
Concentrations at equilibrium (M)	$0.70 - x$	x	$0.60 + x$

and the appropriate equilibrium expression is

$$K_a = 1.8 \times 10^{-5} = \frac{[H_3O^+][CH_3CO_2^-]}{[CH_3CO_2H]} = \frac{(x)(0.60 + x)}{0.70 - x} \approx \frac{(x)(0.60)}{0.70}$$

From Example 18.3, we know that the value of x is very small with respect to 0.70 or 0.60, so we can use the "approximate expression" to find x, the hydronium ion concentration. Therefore, on rearranging the "approximate" expression, we have

$$[H_3O^+] = x = \frac{[CH_3CO_2H]}{[CH_3CO_2^-]} K_a \approx \frac{(0.70)}{(0.60)}(1.8 \times 10^{-5}) = 2.1 \times 10^{-5}\text{ M}$$

$$\text{pH} = 4.68$$

Step 3 *Add HCl to the Acetic Acid/Acetate Ion Buffer Solution.* HCl is a strong acid that is 100% ionized in water. Therefore, we are only concerned with the fact that it supplies H_3O^+ and reacts *completely* with the base in the solution according to the following equation:

$$H_3O^+(aq) + CH_3CO_2^-(aq) \rightleftharpoons H_2O(\ell) + CH_3CO_2H(aq)$$

	H_3O^+ from added HCl	$CH_3CO_2^-$ from buffer	CH_3CO_2H from buffer
Moles before reaction	0.0010	0.600	0.700
Change in moles from complete reaction	−0.0010	−0.001	+0.001
Moles after reaction	0	0.599	0.701
Concentrations after reaction (M)	0	0.598	0.700

Because the added HCl reacts *completely* with acetate ion to produce acetic acid, the solution is once again a buffer containing only the weak acid and its salt. Therefore, we only need to consider the equilibrium for the ionization of the weak acid in the presence of its common ion, $CH_3CO_2^-$, as in Step 2 above.

Equation	$CH_3CO_2H(aq)$ + $H_2O(\ell)$ ⇌	$H_3O^+(aq)$ +	$CH_3CO_2^-(aq)$
Concentrations before CH_3CO_2H ionizes (M)	0.700	0	0.598
Change in concentrations on proceeding to equilibrium	$-y$	$+y$	$+y$
Concentrations at equilibrium (M)	$(0.700 - y)$	y	$0.598 + y$

As usual, we make the approximation that y, the amount of H_3O^+ formed by ionizing acetic acid in the presence of acetate ion, is small compared with 0.700 M or 0.598 M. Therefore, we can write

$$[H_3O^+] = y = \frac{[CH_3CO_2H]}{[CH_3CO_2^-]}K_a \approx \frac{(0.700)}{(0.598)}(1.8 \times 10^{-5}) = 2.1 \times 10^{-5}\ M$$

$$pH = 4.68$$

Within the number of significant figures allowed, the pH *does not change* in the buffer solution after adding HCl, even though it changed by 4 units when 1 mL of 1.0 M HCl was added to 1.0 L of pure water. Buffer solutions do indeed "buffer." In this case the acetate ion consumed the added hydronium ion, and the solution thus resisted a change in pH.

EXERCISE 18.5 ***Buffer Solutions***

Calculate the pH of 0.500 L of a buffer solution composed of 0.50 M formic acid (HCO_2H) and 0.70 M sodium formate ($NaHCO_2$) before and after adding 10.0 mL of 1.0 M HCl. ■

General Expressions for Buffer Solutions

In Example 18.4 we solved for the hydronium ion concentration of the acetic acid/acetate ion buffer solution by rearranging the K_a expression to give

$$[H_3O^+] = \frac{[CH_3CO_2H]}{[CH_3CO_2^-]}K_a$$

This result can be generalized for any buffer. For a buffer solution based on a *weak acid and its conjugate base,* the hydronium ion concentration is given by

$$[H_3O^+] = \frac{[\text{acid}]}{[\text{conjugate base}]} K_a \qquad (18.1)$$

Because the equilibrium concentrations of acid and conjugate base are always nearly the same as their initial concentrations in a buffer solution, these are the concentrations used in this expression to calculate the hydronium ion concentration.

When the usual expression for K_b is rearranged, the hydroxide ion concentration in a buffer composed of a *weak base and its conjugate acid* (e.g., NH_3 and NH_4^+) is

$$[OH^-] = \frac{[\text{base}]}{[\text{conjugate acid}]} K_b \qquad (18.2)$$

The Henderson-Hasselbalch Equation

The general forms for calculating the hydronium ion or hydroxide ion concentrations of buffer solutions are useful in finding the pH of such solutions and in deciding how to prepare buffers. They are often written in a different form, however. If we take the negative logarithm of each side of the equation for $[H_3O^+]$, for example, we have

$$-\log [H_3O^+] = \left\{-\log \frac{[\text{acid}]}{[\text{conjugate base}]}\right\} + (-\log K_a)$$

You know that $-\log [H_3O^+]$ is defined as pH, and from Chapter 17 you recall that the definition can be extended to other quantities. Thus, $-\log K_a$ is equivalent to pK_a. Furthermore, because

$$-\log \frac{[\text{acid}]}{[\text{conjugate base}]} = +\log \frac{[\text{conjugate base}]}{[\text{acid}]}$$

The preceding equation can be rewritten as

$$\text{pH} = pK_a + \log \frac{[\text{conjugate base}]}{[\text{acid}]} \qquad (18.3)$$

This equation, known as the **Henderson-Hasselbalch equation,** shows clearly that the pH of the solution of a weak acid and its conjugate base is controlled primarily by the strength of the acid (as expressed by pK_a). The "fine" control of the pH is then given by the relative amounts of conjugate base and acid. When the concentrations of conjugate base and acid are the same in a solution, the ratio of [conjugate base]/[acid] is 1. The log of 1 is zero, so $\text{pH} = pK_a$ under these circumstances. If there is more of the conjugate base in the solution than acid, for example, then $\text{pH} > pK_a$.

The Henderson-Hasselbalch equation is valid when the ratio of [conjugate base]/[acid] is no larger than 10 and no smaller than 0.1. Many handbooks of chemistry list acid ionization constants in terms of pK_a values, so the approximate pH values of possible buffer solutions are readily apparent.

EXAMPLE 18.5 ***Using the Henderson-Hasselbalch Equation***

You dissolve 2.00 g of benzoic acid ($C_6H_5CO_2H$) and 2.00 g of sodium benzoate ($NaC_6H_5CO_2$) in enough water to make 1.00 L of solution. Calculate the pH of the solution using the Henderson-Hasselbalch equation.

Solution Our first objective is to calculate the pK_a of the acid. Appendix F (and Table 17.4) gives K_a for benzoic acid as 6.3×10^{-5}. Therefore,

$$-\log (6.3 \times 10^{-5}) = pK_a = 4.20$$

Next, we need the concentrations of the acid (benzoic acid) and its conjugate base (benzoate ion).

$$2.00 \text{ g benzoic acid} \left(\frac{1 \text{ mol}}{122.1 \text{ g}}\right) = 0.0164 \text{ mol benzoic acid}$$

$$2.00 \text{ g sodium benzoate} \left(\frac{1 \text{ mol}}{144.1 \text{ g}}\right) = 0.0139 \text{ mol sodium benzoate}$$

Because the solution volume is 1.00 L, the concentrations are [benzoic acid] = 0.0164 M and [sodium benzoate] = 0.0139 M. Therefore,

$$pH = 4.20 + \log \frac{0.0139}{0.0164}$$

$$= 4.20 + \log (0.848) = 4.13$$

Notice that the pH is less than the pK_a because the ratio of conjugate base to acid concentration was less than 1.

EXERCISE 18.6 ***Using the Henderson-Hasselbalch Equation***

Suppose you dissolve 15.0 g of $NaHCO_3$ and 18.0 g of Na_2CO_3 in enough water to make 1.00 L of solution. Use the Henderson-Hasselbalch equation to calculate the pH of the solution. (Consider this as a solution of the weak acid HCO_3^- and its conjugate base, the CO_3^{2-} ion.) ■

Preparing Buffer Solutions

There are two obvious requirements for a buffer solution. First, it should have the capacity to control the pH after the addition of reasonable amounts of acid and base. That is, there must be a large enough concentration of acetic acid in an acetic acid/acetate ion buffer, for example, to consume all of the hydroxide ion that may be added and still control the pH (see Example 18.4). Buffers are usually prepared from 0.10 M to 1.0 M solutions of reagents. Any buffer, however, loses its capacity if too much strong acid or base is added.

The second requirement for a buffer solution is that it should control the pH at the desired value. The equation for hydronium ion concentration of an acid buffer

$$[H_3O^+] = \frac{[\text{acid}]}{[\text{conjugate base}]} K_a$$

shows us how to prepare a buffer solution of given pH. First, we want [acid] ≈ [conjugate base] so that the solution can buffer equal amounts of added acid or base. Given this, we want to choose an acid whose K_a is near the $[H_3O^+]$ we want. The exact value of $[H_3O^+]$ is then achieved by adjusting the acid/conjugate base ratio. The following example illustrates this approach.

EXAMPLE 18.6 *Preparing a Buffer Solution*

Suppose you wish to prepare a buffer solution to maintain the pH at 4.30. A list of possible acids (and their conjugate bases) is shown in the table.

Acid	Conjugate Base	K_a
HSO_4^-	SO_4^{2-}	1.2×10^{-2}
CH_3CO_2H	$CH_3CO_2^-$	1.8×10^{-5}
HCN	CN^-	4.0×10^{-10}

Which combination should be selected and what should the ratio of acid to conjugate base be?

Solution The desired hydronium ion concentration is 5.0×10^{-5} M,

$$\text{pH} = 4.30, \qquad \text{so } [H_3O^+] = 10^{-\text{pH}} = 10^{-4.30} = 5.0 \times 10^{-5}\text{ M}$$

Of the acids given, only acetic acid (CH_3CO_2H) has a K_a value close to that of the desired $[H_3O^+]$. Therefore, you need only adjust the ratio $[CH_3CO_2H]/[CH_3CO_2^-]$ to achieve the desired hydronium ion concentration.

$$[H_3O^+] = 5.0 \times 10^{-5} = \frac{[CH_3CO_2H]}{[CH_3CO_2^-]}(1.8 \times 10^{-5})$$

To satisfy the previous expression, the ratio $[CH_3CO_2H]/[CH_3CO_2^-]$ must be 2.8 to 1.

$$\frac{[CH_3CO_2H]}{[CH_3CO_2^-]} = \frac{5.0 \times 10^{-5}}{1.8 \times 10^{-5}} = \frac{2.8}{1}$$

Therefore, if you add 0.28 mol of acetic acid and 0.10 mol of sodium acetate (or any other pair of molar quantities in the ratio 2.8/1) to enough water to make 1 L of solution, the solution constitutes a buffer with a pH of 4.30.

EXERCISE 18.7 *Preparing a Buffer Solution*

Using an acetic acid/sodium acetate buffer solution, what ratio of acid to conjugate base will you need to maintain the pH at 5.00? Explain how you would make up such a solution. ■

Example 18.6 raises one more important point concerning buffer solutions. The hydronium ion concentration depends on the ratio of [acid]/[conjugate base] (or the hydroxide ion concentration depends on [base]/[conjugate acid]). Although we write these ratios in terms of reagent concentrations, it is not concentrations that are important. Rather, the *relative number of moles* of acid and conjugate base is important. Because both reagents are dissolved in the same solution, their concentrations depend on the same solution volume. In

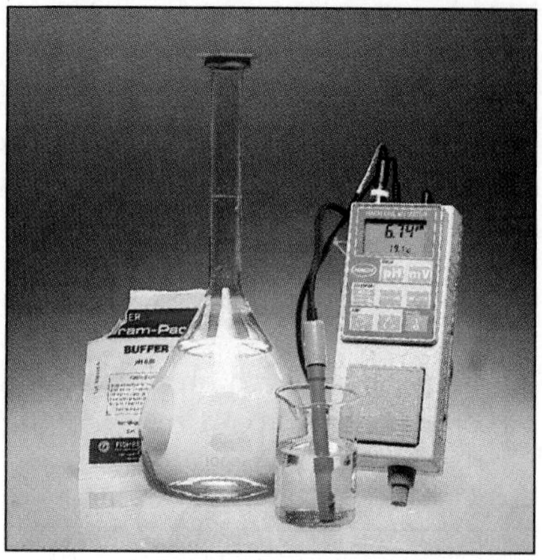

Figure 18.4 A commercial buffer solution. The solid acid and conjugate base in the packet are mixed with water to give a solution with the indicated pH. (C. D. Winters)

Example 18.6, the ratio 2.8/1 for acetic acid and sodium acetate implied that 2.8 mol of the acid and 1.0 mol of sodium acetate were dissolved per liter.

$$\frac{[CH_3CO_2H]}{[CH_3CO_2^-]} = \frac{2.8 \text{ mol/L}}{1.0 \text{ mol/L}} = \frac{2.8 \text{ mol}}{1.0 \text{ mol}}$$

Alternatively, 1.4 mol of acetic acid and 0.50 mol of sodium acetate, or any reasonably large amounts of these reagents that give a ratio of 2.8/1, could have been used. Furthermore, because the actual concentration is not important, the acid and its conjugate base could have been dissolved in any reasonable amount of water. The mole ratio of acid to its conjugate base is maintained, no matter what the solution volume may be. This is the reason that commercially available buffer solutions are sold as the premixed dry ingredients. To use them you only need to mix the ingredients in some volume of pure water (Figure 18.4).

PROBLEM-SOLVING TIPS AND IDEAS

18.3 The pH of a Buffer Solution

Because the pH of a buffer depends on the pK_a of the acid and on the ratio of moles of acid and conjugate base, the pH of a buffer solution does not change on dilution.

Suppose you wish to know the pH of a buffer solution composed of 0.10 M NH_3 and 0.050 M NH_4Cl. The same result is achieved if you consider this a solution of a weak base (NH_3) and its conjugate acid (NH_4^+)

$$NH_3(aq) + H_2O(\ell) \rightleftharpoons NH_4^+(aq) + OH^-(aq)$$

$$[OH^-] = \frac{[NH_3]}{[NH_4^+]} K_b = \frac{0.10}{0.050}(1.8 \times 10^{-5}) = 3.6 \times 10^{-5}$$

$$pH = 9.55$$

or as a solution of a weak acid (NH_4^+) and its conjugate base (NH_3).

$$NH_4^+(aq) + H_2O(\ell) \rightleftharpoons NH_3(aq) + H_3O^+(aq)$$

$$[H_3O^+] = \frac{[NH_4^+]}{[NH_3]} K_a = \frac{0.050}{0.10}(5.6 \times 10^{-10}) = 2.8 \times 10^{-10}$$

$$pH = 9.55$$

All buffer solutions may be treated in a similar manner. ■

18.4 ACID-BASE TITRATION CURVES

In Section 5.6 we described acid-base titrations, a method for the accurate analysis of the amount of acid or base in a sample. In the current chapter you learned something more about acid-base titrations. For example, you know now that the pH at the equivalence point is 7, and the solution is truly "neutral," only when a strong acid is titrated with a strong base and vice versa. If one of the substances being titrated is weak, then the pH at the equivalence point is not 7. We found that (see Table 18.1)

1. Weak acid + strong base → pH > 7 at equivalence point due to hydrolysis of the conjugate base of the weak acid
2. Strong acid + weak base → pH < 7 at equivalence point due to hydrolysis of the conjugate acid of the weak base.

Titrations combining a weak acid and weak base are generally not done because the equivalence point cannot be accurately judged.

To give us more insight into the properties of weak acids and bases, we want to describe how the pH of a solution of an acid or base changes as it is being titrated with a base or acid. To illustrate this, the following examples explore two of the most common situations: (1) the titration of a strong base with a strong acid and (2) the titration of a weak acid with a strong base.

EXAMPLE 18.7 *The Titration of a Strong Base with a Strong Acid*

You begin with 50.0 mL of a 0.100 M solution of HCl. Calculate the pH of the solution as 0.100 M NaOH is slowly added. Plot the results with the pH of the resulting solution on the vertical axis and milliliters of base added on the horizontal axis.

Solution

Step 1 *pH before adding base.* HCl is a strong acid, so $[H_3O^+] = 0.100$ M, which means the pH in the beginning is 1.000.

Step 2 *pH after adding 10.0 mL of 0.100 M NaOH.* To find this answer, we must know the number of moles of H_3O^+ in the solution *after* the base is added. We therefore calculate the quantity of acid in the beginning and subtract from it the quantity of acid that reacts with the base according to the net ionic equation

$$H_3O^+(aq) + OH^-(aq) \longrightarrow 2\ H_2O(\ell)$$

Initially, the number of moles of $H_3O^+ = (0.0500\ \text{L})(0.100\ \text{M}) = 0.00500$ mol
The number of moles of OH^- added $= (0.0100\ \text{L})(0.100\ \text{M}) = 0.00100$ mol

Equation	Moles H_3O^+	Moles OH^-
Before reaction	0.00500 in 50.0 mL solution	0.00100 mol in 10.0 mL solution
Change on reaction	−0.00100 mol	−0.00100 mol
After reaction	0.00400 mol in 60.0 mL solution	0 mol
Concentrations after reaction	0.0667 M	0 M

Here you see that 0.00100 mol of NaOH consumed 0.00100 mol of HCl to leave 0.00400 mol of HCl in the solution. Because the solutions were combined, the total volume after reaction is the sum of the combining volumes, 60.0 mL in this case. Therefore, the hydronium ion concentration after reaction is

$$[\mathrm{H_3O^+}] = \frac{0.00400\ \mathrm{mol}}{0.0600\ \mathrm{L}} = 0.0667\ \mathrm{M}$$

and the pH of the solution is 1.18.

Step 3 *pH after any volume of base is added, up to the equivalence point.* In general, the hydronium ion concentration up to the equivalence point can be calculated from the equation

$$[\mathrm{H_3O^+}] = \frac{\text{original moles acid} - \text{total moles base added}}{\text{volume acid} + \text{volume base added}}$$

For instance, after 45.0 mL of NaOH has been added, the acid concentration is

$$[\mathrm{H_3O^+}] = \frac{0.00500\ \mathrm{mol\ H_3O^+} - (0.0450\ \mathrm{L})(0.100\ \mathrm{M})}{0.0500\ \mathrm{L} + 0.0450\ \mathrm{L}} = 0.00526\ \mathrm{M}$$

and the pH is 2.279.

When the volume of added base has reached 49.5 mL, only 0.50 mL away from the equivalence point, the pH indicates the solution is still quite acidic.

$$[\mathrm{H_3O^+}] = \frac{0.00500\ \mathrm{mol\ H_3O^+} - (0.0495\ \mathrm{L})(0.100\ \mathrm{M})}{0.0500\ \mathrm{L} + 0.0495\ \mathrm{L}} = 5.03 \times 10^{-4}\ \mathrm{M}$$

$$\mathrm{pH} = 3.299$$

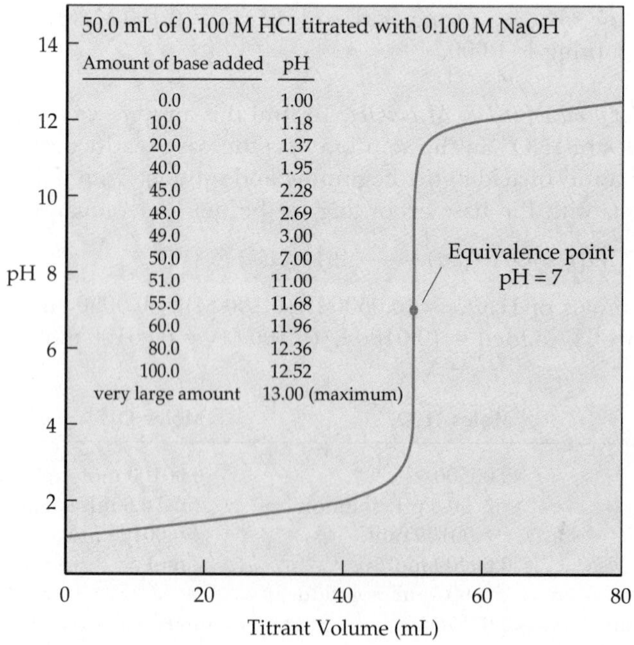
50.0 mL of 0.100 M HCl titrated with 0.100 M NaOH
Amount of base added
pH
0.0 1.00
10.0 1.18
20.0 1.37
40.0 1.95
45.0 2.28
48.0 2.69
49.0 3.00
50.0 7.00
51.0 11.00
55.0 11.68
60.0 11.96
80.0 12.36
100.0 12.52
very large amount 13.00 (maximum)
Equivalence point
pH = 7
pH
14
12
10
8
6
4
2
0
20
40
60
80
Titrant Volume (mL)

Figure 18.5 The change in pH as a strong acid is titrated with a strong base. In this case 50.0 mL of 0.100 M HCl is titrated with 0.100 M NaOH. The pH at the equivalence point is 7.0 (at 25 °C), as is always the case for the reaction of a strong acid with a strong base.

Step 4 *pH at the equivalence point.* This is the reaction of a strong acid with a strong base. Therefore, the pH is 7.00 at the equivalence point (at 25 °C).

Step 5. *pH beyond the equivalence point.* No acid remains in the solution after the equivalence point. NaOH is added to a solution that contains only NaCl, a neutral salt. Therefore, the pH depends on the concentration of OH^- from added NaOH. For example, after only 1.0 mL of base has been added beyond the equivalence point (total volume of added base = 51.0 mL), the pH is 11.00.

$$[OH^-] = \frac{\text{moles excess base}}{\text{total volume}} = \frac{(0.0010\ \text{L})(0.100\ \text{M})}{0.0500\ \text{L acid} + 0.0510\ \text{L base}} = 9.9 \times 10^{-4}\ \text{M}$$

$$\text{pOH} = 3.00 \qquad \text{and so the} \qquad \text{pH} = 11.00$$

The results of these calculations are summarized in the table and titration curve in Figure 18.5.

EXERCISE 18.8 ***Titration of a Strong Acid with a Strong Base***

For the titration outlined in Example 18.7, verify the pH calculated for the addition of (1) 40.0 mL of NaOH and (2) 60.0 mL of NaOH. What is the pH after 49.9 mL of NaOH is added? ■

The pH at each point in the titration of the strong acid HCl with the strong base NaOH is illustrated in Figure 18.5. As calculated in Example 18.7, the pH rises slowly until very close to the equivalence point. Then the pH increases very rapidly, rising 7 units (the H_3O^+ concentration decreases by a factor of 10 million!) when only a very small amount of base (perhaps a drop or two) is added. After the equivalence point is passed, only a small further rise in pH is seen.

The titration of a weak acid with a strong base is somewhat different from the strong acid/strong base titration just described. To illustrate, let us look carefully at the titration curve for the titration of 100.0 mL of 0.100 M acetic acid with 0.100 M NaOH (Figure 18.6).

$$CH_3CO_2H(aq) + NaOH(aq) \longrightarrow NaCH_3CO_2(aq) + H_2O(\ell)$$

Four points, or regions, on this curve are especially interesting:

1. The pH before titration begins
2. The pH at the midpoint of the titration
3. The pH at the equivalence point
4. The pH when base is added beyond the equivalence point

The pH before any base is added (2.87) is found in the usual way (Example 17.6) for the ionization of a weak acid. At the equivalence point of the titration, the solution consists simply of sodium acetate, and so the pH (8.72) is controlled by the acetate ion, the conjugate base of acetic acid (see Example 18.2).

Now let us determine the pH at the midpoint of the titration, that is, at the point at which half of the acid has been consumed. As NaOH is added to the acetic acid, the base is consumed and sodium acetate is produced. This means that at every point between the beginning of the titration (when only pure acetic

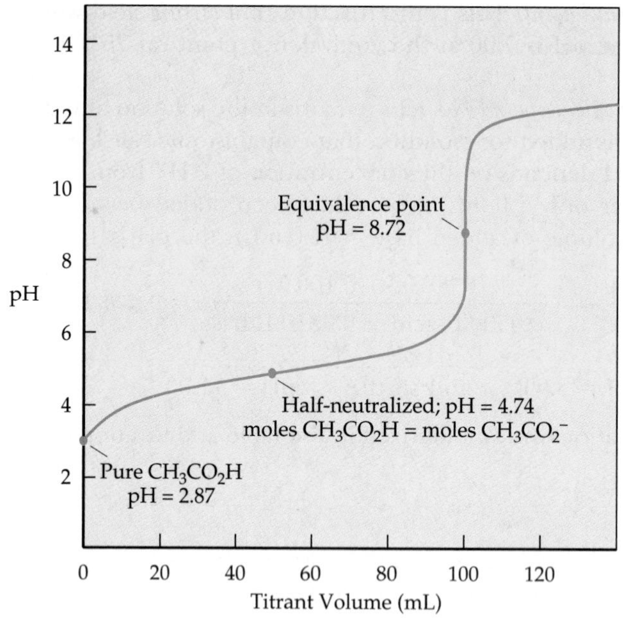

Figure 18.6 The change in pH during the titration of a weak acid with a strong base: 100. mL of 0.100 M acetic acid (CH_3CO_2H) is titrated with 0.100 M NaOH. Note especially that (1) Acetic acid is a weak acid so the pH of the original solution is 2.87. (2) The pH at the point at which half the acid has reacted with base is equal to the pK_a for the acid (pH = pK_a = 4.74), a general observation in the titration of a weak acid with a strong base. (3) At the equivalence point the solution consists of acetate ion, a weak base; therefore, the pH at this point is 8.72.

acid is present) and the equivalence point (where only sodium acetate is present) the solution contains *both* acetic acid *and* its salt, sodium acetate. These are the components of a buffer solution, and the hydronium ion concentration can be found from the following expression:

$$[H_3O^+] = \frac{[CH_3CO_2H]}{[CH_3CO_2^-]} K_a$$

Thus, at any point in the titration (before the equivalence point), the hydronium ion concentration can be calculated from

$$[H_3O^+] = \frac{[CH_3CO_2H \text{ remaining}]}{[CH_3CO_2^- \text{ formed}]} K_a$$

Now, what happens when exactly half of the acid has been consumed by base? At this point half of the acid has been converted to the conjugate base, $CH_3CO_2^-$. Half of the acid remains. Therefore, $[CH_3CO_2H] = [CH_3CO_2^-]$, and

$$[H_3O^+] \text{ at the midpoint of the titration} = K_a$$

In the particular case of the titration of acetic acid with strong base, $[H_3O^+] = 1.8 \times 10^{-5}$ M at the midpoint, and so the pH is 4.74. This result is of general importance, however, because we can say to a good approximation that

- At the midpoint in any weak acid/strong base titration, $[H_3O^+] = K_a$ (or pH = pK_a) of the weak acid.
- At the midpoint in any weak base/strong acid titration, $[OH^-] = K_b$ (or pOH = pK_b) of the weak base.

You can see the value of these conclusions. If you perform an acid-base titration in which one of the components is weak, and you record the pH values at each step, you can readily determine K for the weak acid or base.

As a final point concerning the titration of a weak acid with a strong base, we want to know the pH *after* the equivalence point. In the case of the CH_3CO_2H/NaOH titration, the solution consists of sodium acetate *plus* any NaOH that was added in excess after passing the equivalence point. Both acetate ion and hydroxide ion are bases, so the total concentration of OH^- is the sum of that from excess NaOH *plus* that produced by the hydrolysis of the acetate ion.

$$CH_3CO_2^-(aq) + H_2O(\ell) \rightleftharpoons CH_3CO_2H(aq) + OH^-(aq)$$

The OH^- concentration from the hydrolysis reaction is very small, however, compared with that from excess NaOH, so, after the equivalence point

$$[OH^-] = \frac{\text{moles excess OH}^- \text{ from NaOH}}{\text{total volume in liters}}$$

EXAMPLE 18.8 ***Titration of Acetic Acid with Sodium Hydroxide***

What is the pH of the solution when 90.0 mL of 0.100 M NaOH has been added to 100.0 mL of 0.100 M acetic acid (see Figure 18.6)?

Solution The pH is desired at a point after the midpoint of the titration (pH = 4.74) but before the equivalence point (pH = 8.72). The solution is a buffer solution, containing some unreacted acetic acid plus some sodium acetate formed in the reaction of the acid with sodium hydroxide. Let us first calculate the quantities of reactants before reaction (= concentration × volume) and then use the principles of stoichiometry to calculate the quantities of reactants and products after reaction.

Equation	$CH_3CO_2H(aq)$ +	$OH^-(aq)$ ⟶	$CH_3CO_2^-(aq) + H_2O(\ell)$
Quantity before reaction (mol)	0.0100	0.00900	0
Change occurring on reaction	−0.00900	−0.00900	+0.00900
Quantity after reaction (mol)	0.0010	0	0.00900

This is a buffer solution, whose hydronium ion concentration can be found from the equation

$$[H_3O^+] = \frac{[CH_3CO_2H \text{ remaining}]}{[CH_3CO_2^- \text{ formed}]} K_a$$

Recall that the ratio of moles of acid and conjugate base is the same as the ratio of their concentrations. Therefore,

$$[H_3O^+] = \frac{0.0010 \text{ mol}}{0.00900 \text{ mol}} \cdot 1.8 \times 10^{-5} = 2.0 \times 10^{-6} \text{ M}$$

and this gives a pH of 5.70, in agreement with Figure 18.6.

EXERCISE 18.9 ***Titration of a Weak Acid with a Strong Base***

The titration of 0.100 acetic acid with 0.100 M NaOH is described in the text. What is the pH of the solution when 35.0 mL of the base has been added? ■

PROBLEM-SOLVING TIPS AND IDEAS

18.4 Calculating the pH at Various Stages of an Acid-Base Reaction

Finding the pH at some point in an acid-base reaction always involves three or four calculation steps. Consider the titration of a weak base with a strong acid as an example. (The same principles apply to other acid-base titrations.)

$$H_3O^+(aq) + B(aq) \longrightarrow BH^+(aq) + H_2O(\ell)$$

Step 1 *Stoichiometry problem*
Up to the equivalence point, acid is consumed completely to leave a solution containing some base (B) and its conjugate acid (BH^+). Use the principles of stoichiometry to calculate (a) the quantity of acid added, (b) the quantity of base consumed, (c) the quantity of base remaining, and (d) the quantity of conjugate acid (BH^+) formed.

Step 2 *Calculate the concentrations of base, [B], and conjugate acid, [BH⁺]*
Recognize that the volume of the solution at any point is the sum of the original volume of the base solution plus the volume of acid solution added.

Step 3 *Calculate pH before the equivalence point*
At any point before the equivalence point, the solution is a buffer solution because the base and its conjugate acid are present. Calculate $[OH^-]$ using the concentrations from Step 2 and the value of K_b for the weak base (see Equation 18.2).

Step 4 *pH at the equivalence point*
If the reaction has gone to the equivalence point, calculate the concentration of the conjugate acid using the procedure of Steps 1 and 2 and use this value with the value of K_a for this acid to calculate the concentration of H_3O^+ in the solution. ■

The titrations illustrated thus far involve a monoprotic acid (HA) reacting with a base, such as NaOH or NH_3. There are, however, many polyprotic acids and bases. One example is oxalic acid, $H_2C_2O_4$.

$$H_2C_2O_4(aq) + H_2O(\ell) \rightleftharpoons HC_2O_4^-(aq) + H_3O^+(aq) \qquad K_{a1} = 5.9 \times 10^{-2}$$
$$HC_2O_4^-(aq) + H_2O(\ell) \rightleftharpoons C_2O_4^-(aq) + H_3O^+(aq) \qquad K_{a2} = 6.4 \times 10^{-5}$$

As illustrated in Figure 18.7, complications arise when an aqueous solution of the acid is titrated with sodium hydroxide, for example. Here 100. mL of 0.100 M oxalic acid is titrated with 0.100 M NaOH. The first significant rise in pH is experienced after 100 mL of base has been added, indicating that the first proton of the acid has been titrated.

$$H_2C_2O_4(aq) + OH^-(aq) \rightleftharpoons HC_2O_4^-(aq) + H_2O(\ell)$$

The rise in pH from 50 to 150 mL of added base is not well defined, however. The reason for this is that the solution contains not only $HC_2O_4^-$ but also the dianion $C_2O_4^{2-}$. These ions are, respectively, a weak acid and its conjugate base,

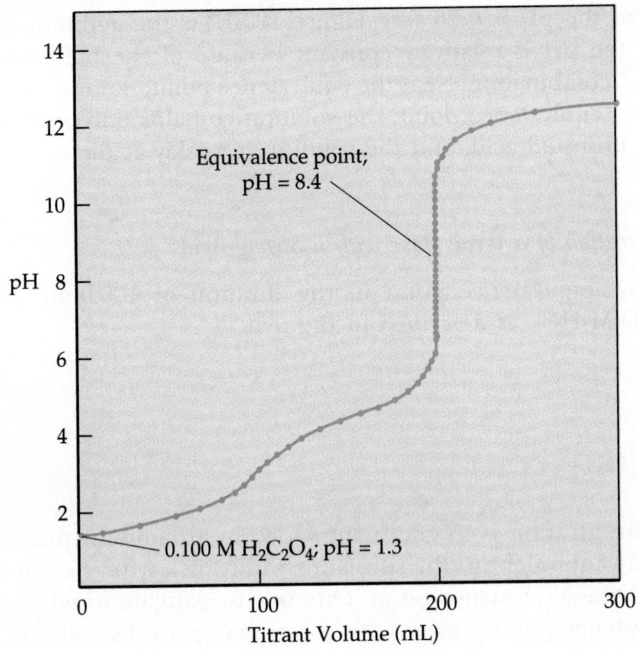

Figure 18.7 The curve for the titration of 100. mL of 0.100 M oxalic acid ($H_2C_2O_4$, a weak diprotic acid) with 0.100 M NaOH. The first rise in pH (at 100 mL) occurs when the first H atom of $H_2C_2O_4$ is titrated, and the second rise (at 200 mL) occurs at the completion of the reaction

$$HC_2O_4^-(aq) + OH^-(aq) \longrightarrow C_3O_4^{2-}(aq) + H_2O(\ell)$$

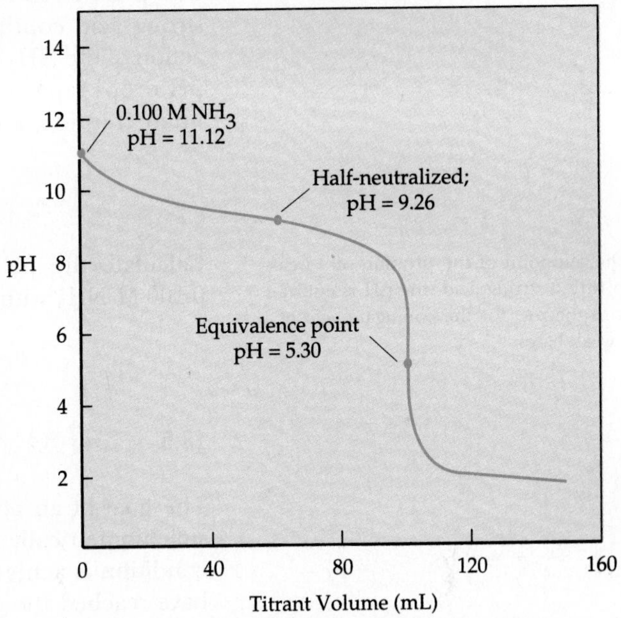

Figure 18.8 The change in pH during the titration of a weak base (100.0 mL of 0.100 M NH_3) with a strong acid (0.100 M HCl). The pH at the half-neutralization point is equal to the pK_a for the conjugate acid (NH_4^+) of the weak base (NH_3) (pH = pK_a = 9.26). At the equivalence point the solution contains the NH_4^+ ion, a weak acid, so the pH is about 5.

in other words, the components of a buffer. Thus, the species in the solution resist a rapid rise in pH. Only when the second proton of oxalic acid is titrated,

$$HC_2O_4^-(aq) + OH^-(aq) \rightleftharpoons C_2O_4^{2-}(aq) + H_2O(\ell)$$

does the pH increase significantly. The pH at this second equivalence point is controlled by the K_b value for $C_2O_4^{2-}$, the conjugate base of the hydrogen oxalate ion.

$$C_2O_4^{2-}(aq) + H_2O(\ell) \rightleftharpoons HC_2O_4^-(aq) + OH^-(aq)$$

$$K_b = K_w/K_{a2} = 1.6 \times 10^{-10}$$

Calculation of the pH at the equivalence point indicates that it should be about 8.5, as observed.

Finally, it is useful to consider the titration of a weak base with a strong acid. Figure 18.8 illustrates the case for the titration of 100.00 mL of 0.100 M NH_3 with 0.100 M HCl.

$$NH_3(aq) + HCl(aq) \longrightarrow NH_4^+(aq) + Cl^-(aq)$$

Notice that the pH of the original 0.100 M NH_3 solution is 11.12, as expected for a weak base. At the midpoint of the titration, half of the original ammonia has been converted to ammonium chloride, and half of the ammonia remains. The hydroxide ion concentration at the midpoint is therefore

$$[OH^-] \text{ at midpoint} = \frac{[NH_3]}{[NH_4^+]} K_b$$

$$[OH^-] = K_b = 1.8 \times 10^{-5} \text{ M}$$

The pOH is 4.74 and so the pH is 9.26 (see Figure 18.8). As the addition of strong acid continues, the pH is relatively constant because of the buffering action of the NH_3/NH_4^+ combination. Near the equivalence point, however, the pH drops rapidly. At the equivalence point, the solution contains only ammonium chloride, a weak Brønsted acid, and the solution is weakly acidic.

EXERCISE 18.10 ***Titration of a Weak Base with a Strong Acid***

At the midpoint of the titration of a weak base with a strong acid, the pH is equivalent to the pK_a for the conjugate acid of the weak base.

Calculate the pH at the equivalence point in the titration of 100.0 mL of 0.100 M NH_3 with 0.100 M HCl, as described in the text. ■

18.5 ACID-BASE INDICATORS

The goal of an acid-base titration is to add a substance in an amount that is stoichiometrically exactly equivalent to the substance with which it reacts. This condition is achieved at the equivalence point. One way to estimate when you have reached the equivalence point is to observe some change in the solution. Using a pH meter you can see when the pH of the solution changes by a number of units (see Figures 18.5 and 18.6), or you can add a few drops of an indicator, a dye that has different colors in solutions of different pH (see Figure 17.7). The dye's color change is said to occur at the *end point* of the titration, and we try to make the difference between the end and equivalence points negligible.

An acid-base indicator (HInd) is usually an organic dye that is itself a weak acid.

$$\text{HInd(aq)} + H_2O(\ell) \rightleftharpoons H_3O^+\text{(aq)} + \text{Ind}^-\text{(aq)}$$

$$K_a \text{ for the indicator} = \frac{[H_3O^+][\text{Ind}^-]}{[\text{HInd}]}$$

The acid form of the compound (HInd) has one color, and the conjugate base (Ind^-) has another (Figure 17.7). According to Le Chatelier's principle, addition of H_3O^+ or OH^- to an indicator causes one color or the other to appear, depending on whether HInd or Ind^- is the predominant species. The idea is to try to choose an indicator with a K_a near that of the acid being titrated and one whose color changes strongly at the equivalence point. If the color change is strong on going from an acidic to a basic medium, we need add only a tiny bit of the indicator to the solution and then titrate both acids at the same time.

If the K_a expression for the indicator is rearranged

$$[H_3O^+] = \frac{[\text{HInd}]}{[\text{Ind}^-]}K_a \quad \text{or} \quad \frac{[H_3O^+]}{K_a} = \frac{[\text{HInd}]}{[\text{Ind}^-]}$$

it is apparent that a tiny amount of an indicator can in fact reveal the pH of a solution by changing the $[\text{HInd}]/[\text{Ind}^-]$ ratio. If a drop or two of some dilute indicator is added in an acid-base titration, the ratio of $[\text{HInd}]/[\text{Ind}^-]$ is controlled by the hydronium ion concentration of the solution. As $[H_3O^+]$ changes in the course of a titration, the ratio of $[\text{HInd}]/[\text{Ind}^-]$ must also change in order to maintain the equality in the equation above. When the H_3O^+ concentration is high, [HInd] must therefore be large and [Ind^-] small. The indicator color is that of HInd. In the case of the common indicator phenolphthalein, the dye is colorless at high concentrations of H_3O^+ or low pH, whereas the indicator thymol blue is red under these conditions (Figure 17.7). When the pH increases, and [H_3O^+] declines, the ratio [HInd]/[Ind^-] must also decrease. Thus, the concentration of Ind^- increases, and the color of Ind^- is seen. Phenolphthalein, for example, is red above pH 10, whereas thymol blue is yellow between pH 3 and 8 and purple at pH of about 9.

In principle, the color of the indicator changes when [H_3O^+] = K_a of the indicator (because [HInd] = [Ind^-] at this point), and so you might think that you can accurately determine [H_3O^+] by carefully observing the color change. In practice, your eyes are not quite that good. Usually, you see the color of HInd when [HInd]/[Ind^-] is about 10/1, and you see the color of Ind^- when [HInd]/[Ind^-] is about 1/10. This means the color change is observed over a hydronium ion concentration interval of about 100, which corresponds to 2 pH units. This is not really a problem, however, as you can see in Figures 18.6 and 18.7; on passing through the equivalence point of these titrations the pH changes by as many as 7 units.

As Figure 17.7 shows, a variety of indicators is available, each changing color in a different pH range. If you are analyzing a weak acid or base by titration, you must choose an indicator that changes color in a range that includes the pH to be observed at the equivalence point. This means that an indicator that changes color in the pH range 7 ± 2 should be used for a strong acid/strong base titration. On the other hand, the pH at the equivalence point in the titration of a weak acid with a strong base is greater than 7; therefore, you should use an indicator that changes color at a pH of about 8.

EXERCISE 18.11 ***Indicators***

Use Figure 17.7 to decide which indicator is best to use in the titration of NH_3 with HCl in Example 18.1. ■

CHAPTER HIGHLIGHTS

When you have finished reading this chapter, you should be able to

- Predict the pH of an acid-base reaction at its equivalence point (Section 18.1).

Acid	Base	pH
Strong	Strong	= 7 (neutral)
Strong	Weak	<7 (acidic)
Weak	Strong	>7 (basic)
Weak	Weak	Depends on *K* values of conjugate base and acid

- Calculate the pH at the equivalence point in the reaction of a strong acid with a strong base or weak base, or in the reaction of a strong base with a weak acid (see Examples 18.1 and 18.2).
- Predict the effect of the presence of a "common ion" on the pH of the solution of a weak acid or base (Section 18.2).

- Describe the functioning of buffer solutions (Section 18.3).
- Calculate the pH of a buffer solution before and after adding excess acid or base (Section 18.3 and Example 18.4).
- Use the Henderson-Hasselbalch equation (Equation 18.3) to calculate the pH of a buffer solution of given composition (Section 18.3).
- Use the Henderson-Hasselbalch equation to predict the change in pH when the K_a of the acid in a buffer changes or when the composition of the buffer changes.
- Describe how a buffer solution of a given pH can be prepared (Section 18.3).
- Calculate the pH at any point in an acid-base titration (Section 18.4).
- Understand the differences between the titration curves for a strong acid/strong base titration versus cases in which one of the substances is weak (Section 18.4).
- Describe how an indicator functions in an acid-base titration (Section 18.5).

STUDY QUESTIONS

Review Questions

1. What is the equivalence point of an acid-base reaction?
2. Decide if the pH is equal to 7, less than 7, or greater than 7 when the following reactions occur using equal molar amounts of acid and base:
   (a) A weak base with a strong acid
   (b) A strong base with a strong acid
   (c) A strong base with a weak acid
3. Sketch the general shape of the titration curve when (a) a strong base is titrated with a strong acid and (b) a weak base is titrated with a strong acid. In each case indicate if the pH at the equivalence point is less than 7, equal to 7, or greater than 7.
4. Briefly describe how a buffer solution can control the pH of a solution when strong acid is added. Use the NH_3/NH_4Cl buffer as an example.
5. Briefly describe how a buffer solution can control the pH of a solution when strong base is added. Use a solution of acetic acid and sodium acetate as an example.
6. Briefly explain the difference between the equivalence point of a titration and its end point.
7. Use the Henderson-Hasselbalch equation to decide if the pH of a buffer solution based on a weak acid and its conjugate increases or decreases when
   (a) The ionization constant of the weak acid increases
   (b) The acid concentration is decreased relative to the concentration of its conjugate base

Acid-Base Reactions

8. Calculate the value of the equilibrium constant for the reaction of benzoic acid ($C_6H_5CO_2H$) with NaOH. Does the equilibrium lie predominantly to the left or to the right side of the reaction?

$$C_6H_5CO_2H(aq) + OH^-(aq) \rightleftharpoons H_2O(\ell) + C_6H_5CO_2^-(aq)$$

9. What is the value of the equilibrium constant for the reaction of the weak base aniline ($C_6H_5NH_2$) with HCl? Does equilibrium lie predominantly to the left or to the right side of the reaction?

$$C_6H_5NH_2(aq) + H_3O^+(aq) \rightleftharpoons C_6H_5NH_3^+(aq) + H_2O(\ell)$$

10. Calculate the hydronium ion concentration and pH of the solution that results when 22.0 mL of 0.10 M acetic acid, CH_3CO_2H, is mixed with 22.0 mL of 0.10 M NaOH.

11. Calculate the hydronium ion concentration and the pH when 50.0 mL of 0.40 M NH_3 is mixed with 50.0 mL of 0.40 M HCl.

12. For each of the following cases, decide whether the pH is less than 7, equal to 7, or greater than 7:
   (a) Equal volumes of 0.10 M acetic acid, CH_3CO_2H, and 0.10 M KOH are mixed.

(b) 25 mL of 0.015 M NH_3 is mixed with 25 mL of 0.015 M HCl.
(c) 100. mL of 0.0020 M HNO_3 is mixed with 50. mL of 0.0040 M NaOH.

13. For each of the following cases, decide whether the pH is less than 7, equal to 7, or greater than 7:
(a) 25 mL of 0.45 M H_2SO_4 is mixed with 25 mL of 0.90 M NaOH.
(b) 15 mL of 0.050 M formic acid, HCO_2H, is mixed with 15 mL of 0.050 M NaOH.
(c) 25 mL of 0.15 M $H_2C_2O_4$ (oxalic acid) is mixed with 25 mL of 0.30 M NaOH. (Both H^+ ions of oxalic acid are titrated.)

14. Phenol, C_6H_5OH, is a weak organic acid that has been used as a disinfectant. Suppose 0.515 g of the compound is dissolved in exactly 100 mL of water. The resulting solution is titrated with 0.123 M NaOH.

$$C_6H_5OH(aq) + OH^-(aq) \longrightarrow C_6H_5O^-(aq) + H_2O(\ell)$$

What are the concentrations of all of the following ions at the equivalence point: Na^+, H_3O^+, OH^-, and $C_6H_5O^-$? What is the pH of the solution?

15. Assume you dissolve 0.235 g of the weak acid benzoic acid ($C_6H_5CO_2H$) in exactly 100 mL of water and then titrate the solution with 0.108 M NaOH.

$$C_6H_5CO_2H(aq) + OH^-(aq) \longrightarrow C_6H_5CO_2^-(aq) + H_2O(\ell)$$

What are the concentrations of all of the following ions at the equivalence point: Na^+, H_3O^+, OH^-, and $C_6H_5CO_2^-$? What is the pH of the solution at the equivalence point?

16. You require 36.78 mL of 0.0105 M HCl to reach the equivalence point in the titration of 25.0 mL of aqueous ammonia. What are the concentrations of H_3O^+, OH^-, and NH_4^+ at the equivalence point? What is the pH of the solution at the equivalence point? What was the concentration of NH_3 in the original ammonia solution?

17. A solution of the weak base aniline, $C_6H_5NH_2$, in 25.0 mL of water requires 25.67 mL of 0.175 M HCl to reach the equivalence point. What are the concentrations of H_3O^+, OH^-, and $C_6H_5NH_3^+$ at the equivalence point? What is the pH of the solution at the equivalence point? What was the concentration of aniline in the original solution?

The Common Ion Effect and Buffer Solutions

18. Does the pH of the solution increase, decrease, or stay the same when you
(a) Add solid ammonium chloride to a dilute aqueous solution of NH_3?
(b) Add solid sodium acetate to a dilute aqueous solution of acetic acid?
(c) Add solid KCl to a dilute aqueous solution of KOH?

19. Does the pH of the solution increase, decrease, or stay the same when you
(a) Add solid sodium oxalate, $Na_2C_2O_4$, to 50.0 mL of 0.015 M oxalic acid, $H_2C_2O_4$?
(b) Add solid ammonium chloride to 100 mL of 0.016 M HCl?
(c) Add 20.0 g of NaCl to 1.0 L of 0.10 M sodium acetate, $NaCH_3CO_2$?

20. Does the pH of the solution increase, decrease, or stay the same when you
(a) Add 10.0 mL of 0.10 M HCl to 25.0 mL of 0.10 M NH_3?
(b) Add 25.0 mL of 0.050 M NaOH to 50.0 mL of 0.050 M acetic acid?

21. Does the pH of the solution increase, decrease, or stay the same when you
(a) Add 25.0 mL of 0.050 M NaOH to 25.0 mL of 0.075 M oxalic acid, $H_2C_2O_4$?
(b) Add 1.0 mL of 0.10 M HCl to 25.0 mL of 0.10 M pyridine, a weak base?

22. What is the pH of a buffer solution that is 0.20 M with respect to ammonia, NH_3, and 0.20 M with respect to ammonium chloride, NH_4Cl?

23. What is the pH of a buffer solution if 100.0 mL of the solution is 0.15 M with respect to acetic acid and contains 1.56 g of sodium acetate, $NaCH_3CO_2$?

24. What is the pH of the buffer solution that results when 2.2 g of NH_4Cl, ammonium chloride, is added to 250 mL of 0.12 M NH_3? Is the final pH lower or higher than the pH of the original ammonia solution?

25. Lactic acid ($CH_3CHOHCO_2H$) is found in sour milk, in sauerkraut, and in muscles after activity (see page 521). If 2.75 g of sodium lactate, $NaCH_3CHOHCO_2$, is added to 500. mL of 0.100 M lactic acid, what is the pH of the resulting buffer solution? Is the final pH lower or higher than the pH of the lactic acid solution? (K_a for lactic acid $= 1.4 \times 10^{-4}$)

26. How many grams of ammonium chloride, NH_4Cl, must be added to exactly 500 mL of 0.10 M NH_3 solution to give a solution with a pH of 9.0?

27. How many grams of sodium acetate, $NaCH_3CO_2$, must be added to 1.00 L of 0.10 M acetic acid to give a solution with a pH of 4.5?

28. If a buffer solution is made of 12.2 g of benzoic acid ($C_6H_5CO_2H$) and 7.20 g sodium benzoate ($NaC_6H_5CO_2$) in exactly 250 mL of solution, what is the pH of the buffer? If the solution is diluted to exactly 500 mL with pure water, what is the pH of the solution?

29. If a buffer solution is prepared from 5.15 g of NH_4NO_3 and 0.10 L of 0.15 M NH_3, what is the pH of the solution? What is the new pH if the solution is diluted with pure water to a volume of exactly 500 mL?

30. Which of the following combinations is the best to buffer the pH at approximately 9?
(a) HCl/NaCl
(b) ammonia/ammonium nitrate [NH_3/NH_4Cl]
(c) acetic acid/sodium acetate [CH_3CO_2H/$NaCH_3CO_2$]

31. Many natural processes can be studied in the laboratory but

only in an environment of controlled pH. Which of the following combinations is the best choice to buffer the pH at approximately 7?
(a) H_3PO_4/NaH_2PO_4
(b) NaH_2PO_4/Na_2HPO_4
(c) Na_2HPO_4/Na_3PO_4

32. A buffer solution was prepared by adding 4.95 g of sodium acetate, $NaCH_3CO_2$, to 250. mL of 0.150 M acetic acid, CH_3CO_2H.
(a) What is the pH of the buffer?
(b) What is the pH of 100. mL of the buffer solution if you add 80. mg of NaOH to the solution?

33. You dissolve 0.425 g of NaOH in 2.00 L of a solution that has $[H_2PO_4^-] = [HPO_4^{2-}] = 0.132$ M. What was the pH of the solution before adding NaOH? After adding NaOH?

34. A buffer solution is prepared by adding 0.125 mol of ammonium chloride to 500. mL of 0.500 M solution of ammonia.
(a) What is the pH of the buffer?
(b) If 0.0100 mol of HCl gas is bubbled into 500. mL of the buffer, what is the new pH of the solution?

35. What is the pH change when 20.0 mL of 0.100 M NaOH is added to 80.0 mL of a buffer solution consisting of 0.169 M NH_3 and 0.183 M NH_4Cl?

Using the Henderson-Hasselbalch Equation

36. What is the value of pK_a for acetic acid? Use the Henderson-Hasselbalch equation to calculate the pH of a solution that has an acetic acid concentration of 0.050 M and a sodium acetate concentration of 0.075 M.

37. What is the value of pK_a for the ammonium ion? Use the Henderson-Hasselbalch equation to calculate the pH of a solution that has an ammonium chloride concentration of 0.050 M and an ammonia concentration of 0.045 M.

38. The pK_a for formic acid is 3.74.
(a) What is the pH of a solution that has a formic acid concentration of 0.050 M and a sodium formate concentration of 0.035 M?
(b) What must the ratio of acid to conjugate base be in order to increase the pH by 0.5 relative to the solution in part a?

39. The pK_a for the ammonium ion is 9.25.
(a) What is the pH of a solution that has an ammonia concentration of 0.133 M and an ammonium chloride concentration of 0.106 M?
(b) What must the ratio of acid to conjugate base be in order to decrease the pH by 0.5 relative to the solution in part a?

Titration Curves and Indicators

40. Without doing detailed calculations, sketch the curve for the titration of 30.0 mL of 0.10 M NaOH with 0.10 M HCl. Indicate the approximate pH at the beginning of the titration and at the equivalence point. What is the total solution volume at the equivalence point?

41. Without doing detailed calculations, sketch the curve for the titration of 50 mL of 0.050 M pyridine, C_5H_5N (a weak base) with 0.10 M HCl. Indicate the approximate pH at the beginning of the titration and at the equivalence point. What is the total solution volume at the equivalence point?

42. You titrate 25.0 mL of 0.11 M NH_3 with 0.10 M HCl.
(a) What is the pH of the NH_3 solution before the titration begins?
(b) What is the pH at the equivalence point?
(c) What is the pH at the midpoint of the titration?
(d) What indicator in Figure 17.7 is best for detecting the equivalence point?
(e) Calculate the pH of the solution after adding 5.00, 15.0, 20.0, 22.0, and 30.0 mL of the acid. Combine this information with the information from the preceding parts and plot the titration curve.

43. Aniline, $C_6H_5NH_2$, is a weak base used extensively in the dye industry. Its conjugate acid, anilinium hydrochloride, $[C_6H_5NH_3]Cl$, can be titrated with a strong base such as NaOH. Assume 50.0 mL of 0.100 M anilinium hydrochloride is titrated with 0.185 M NaOH. (K_a for anilinium hydrochloride is 2.4×10^{-5}.)
(a) What is the pH of the $[C_6H_5NH_3]Cl$ solution before the titration begins?
(b) What is the pH at the equivalence point?
(c) What is the pH at the midpoint of the titration?
(d) What indicator in Figure 17.7 is best for detecting the equivalence point?
(e) Calculate the pH of the solution after adding 5.00, 10.0, 20.0, and 30.0 mL of the base. Combine this information with that from the preceding parts and plot the titration curve.

44. Using Figure 17.7, suggest an indicator to use in each of the following titrations:
(a) The weak base pyridine is titrated with HCl.
(b) Formic acid is titrated with NaOH.
(c) Hydrazine is titrated with HCl.

45. Using Figure 17.7, suggest an indicator to use in each of the following titrations:
(a) Na_2CO_3 is titrated to HCO_3^- with HCl.
(b) Hypochlorous acid is titrated with NaOH.
(c) Trimethylamine is titrated with HCl.

General Questions

46. You dissolve 1.00 mol of propanoic acid and 0.40 mol of NaOH in enough water to make 1.00 L of solution.
(a) Write a balanced equation to depict the reaction that can occur.
(b) How many moles of acid and of its conjugate base are present after the reaction?
(c) Calculate the pH of the solution.
(d) Does the pH increase, decrease, or remain the same if 0.40 g of NaOH is added to the solution?

47. You add 12.5 mL of 4.15 M acetic acid to 25.0 mL of 1.00 M NaOH. Calculate the hydronium ion concentration and pH of the resulting solution.

48. Calculate the concentrations of NH_4^+, OH^-, NH_3, and Na^+ in a solution that was originally 0.040 M NaOH and 0.20 M in NH_3.

49. The quantity of 25.0 mL of an aqueous solution of formic acid, HCO_2H, requires 25.67 mL of 0.275 M NaOH to reach the equivalence point. What are the concentrations of H_3O^+, OH^-, and HCO_2^- at the equivalence point? What is the pH of the solution at the equivalence point? What was the concentration of formic acid in the original solution?

50. What is the pH of a 0.160 M acetic acid solution? If 56.8 g of sodium acetate is added to 1.50 L of the 0.160 M acetic acid solution, what is the new pH of the solution?

51. You dissolve 0.515 g of phenol, C_6H_5OH, a weak organic acid, in exactly 100 mL of water. The resulting solution is titrated with 0.123 M NaOH.

$$C_6H_5OH(aq) + OH^-(aq) \longrightarrow C_6H_5O^-(aq) + H_2O(\ell)$$

What are the concentrations of all of the following ions at the equivalence point: Na^+, H_3O^+, OH^-, and $C_6H_5O^-$? What is the pH at the equivalence point?

52. Calculate the pH at the equivalence point in a titration of 25.0 mL of 0.120 M formic acid, HCO_2H, with 0.105 M NaOH.

53. A 500.-mL solution contains 0.150 mol of $NaNO_2$ and 0.200 mol of HNO_2. How many more grams of $NaNO_2$ must be added to this solution to have a pH of 4.00?

54. A buffer solution has an acetic acid concentration of 0.50 M and a sodium acetate concentration of 0.88 M.
(a) What is the pH of the buffer solution?
(b) If 0.10 mol of NaOH is added to 1.00 L of the buffer solution, what is the pH after addition?

55. How many moles of HCl must be added to 1.00 L of a buffer made from 0.150 M NH_3 and 10.0 g of NH_4Cl to decrease the pH by 1 unit?

56. *Buffer capacity* is defined as the number of moles of a strong acid or strong base that are required to change the pH of 1 L of the buffer solution by 1 unit. What is the buffer capacity of a solution that is 0.10 M in acetic acid and 0.10 M in sodium acetate?

57. Arrange the following solutions in order of increasing pH (all reagents are 0.10 M):
(a) NaCl
(b) NH_3
(c) $CH_3CO_2H/NaCH_3CO_2$
(d) HCl
(e) NH_3/NH_4Cl
(f) CH_3CO_2H

58. You titrate 25.0 mL of a 0.0256 M aqueous solution of aniline, $C_6H_5NH_2$ (a weak base) with 0.0195 M HCl.
(a) What is the pH of the aniline solution before the titration begins?
(b) What is the pH at the equivalence point of the titration?
(c) What is the pH at the midpoint of the titration?
(d) What indicator in Figure 17.7 is best for detecting the equivalence point?
(e) Calculate the pH of the solution after adding 5.00, 10.0, 15.0, 20.0, 24.0, and 30.0 mL of the acid. Combine this information with that from above and plot the titration curve.

59. A 0.30 M solution of a weak acid HA has a pH of 2.25. What is the pH of an equimolar solution of HA and the Na^+ salt of the conjugate base, NaA?

60. The pH of human blood is controlled by several buffer systems, among them the reaction

$$H_2PO_4^-(aq) + H_2O(\ell) \rightleftharpoons H_3O^+(aq) + HPO_4^{2-}(aq)$$

Calculate the ratio of $[H_2PO_4^-]/[HPO_4^{2-}]$ in normal blood having a pH of 7.40.

61. Prove that the pH at the midpoint in the titration of a weak base with a strong acid is pH = 14.00 + log K_b (at 25 °C).

62. When 16.0 mL of 0.20 M benzoic acid is mixed with 32.0 mL of 0.10 M NH_3, what is the pH of the resulting solution? See Study Question 17–105 for a method of calculating the hydronium ion concentration.

63. You dissolve 0.221 g of trimethylamine, $(CH_3)_3N$, in 50.0 mL of water, and the resulting solution is titrated with 0.100 M HCl.
(a) What is the pH of the original amine solution?
(b) What is the pH of the solution at the midpoint of the titration?
(c) What is the pH at the equivalence point?
(d) Sketch a rough titration curve for the titration and decide on a suitable indicator (Figure 17.7).

64. Hydroxylamine is a weak base that readily forms salts such as $[NH_3OH]Cl$, hydroxylamine hydrochloride. This compound is used as a reducing agent in photography and as an antioxidant in soaps. Assume you have 25.0 mL of a 0.155 M solution of $[NH_3OH]Cl$ and titrate it with 0.108 M NaOH.
(a) What is the pH of the $[NH_3OH]Cl$ solution before the titration begins?
(b) What is the pH at the equivalence point?
(c) What is the pH at the midpoint of the titration?

65. The weak base ethanolamine, $HOCH_2CH_2NH_2$, can be titrated with HCl.

$$HOCH_2CH_2NH_2(aq) + H_3O^+(aq) \longrightarrow HOCH_2CH_2NH_3^+(aq) + H_2O(\ell)$$

Assume you have 25.0 mL of a 0.010 M solution of ethanolamine and titrate it with 0.0095 M HCl. If the pH at the midpoint of the titration is 9.50, what is the pH at the equivalence point?

66. The pH at the second equivalence point in the titration of 0.100 M oxalic acid with 0.100 M NaOH was given as approximately 8.4 on page 873. Perform suitable calculations to verify this result.

67. Three titration curves are pictured here, each of which represents the titration of 100. mL of 0.050 M acid (or base) with 0.050 M base (or acid). Four types of titration are possible: (a) a strong acid titrated with a strong base; (b) a strong base titrated with a strong acid; (c) a weak acid titrated with a strong base; or (d) a weak base titrated with a strong acid. Indicate for each diagram the type of titration and briefly describe your reasoning.

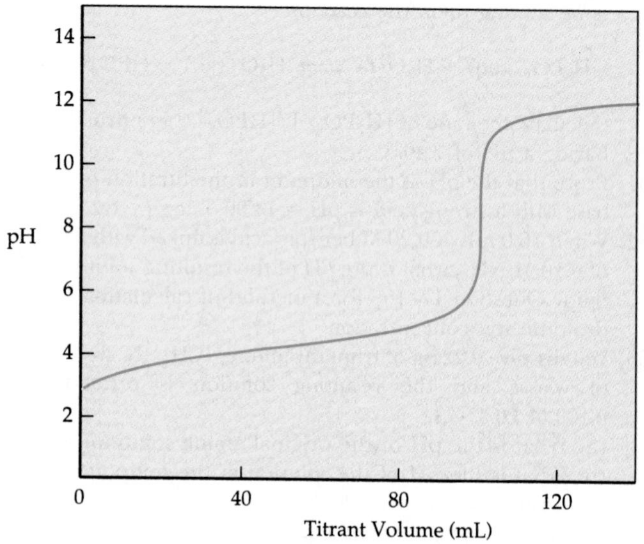

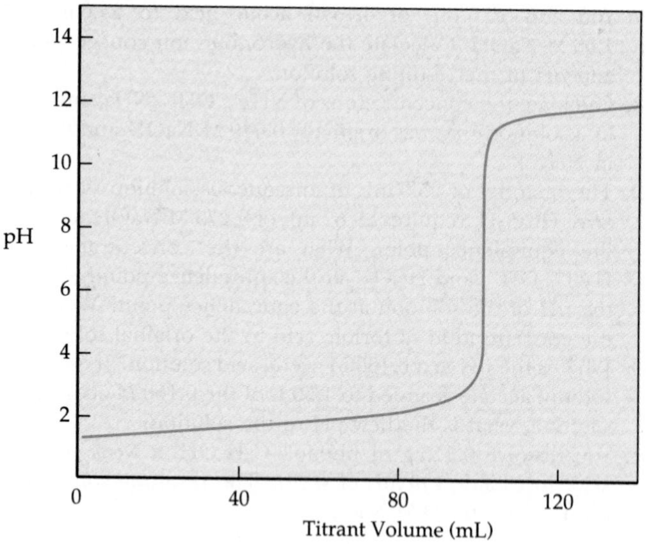

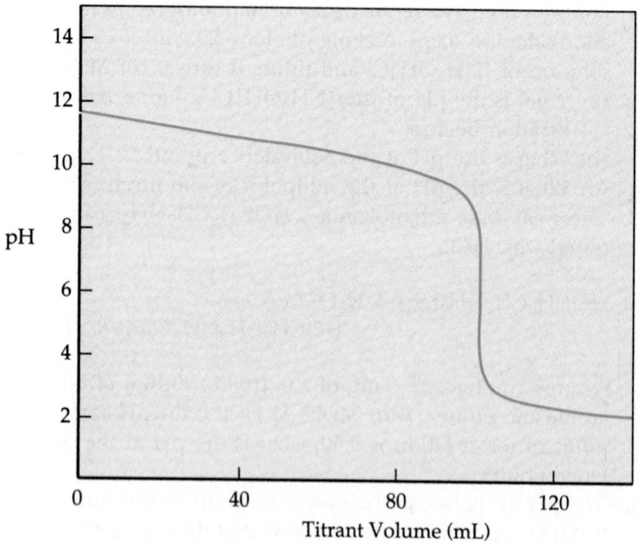

Conceptual Questions

68. Two acids, each approximately 10^{-2} M in concentration, are titrated separately with a strong base. The acids show the following pH values at the equivalence point: HA, pH = 9.5 and HB, pH = 8.5.
(a) Which is the stronger acid, HA or HB?
(b) Which of the conjugate bases, A^- or B^-, is the stronger base?

69. During active exercise, lactic acid ($K_a = 1.4 \times 10^{-4}$) is produced in the muscle tissues (see page 521). At the pH of the body (pH = 7.4), which form is primarily present: nonionized lactic acid [$CH_3CH(OH)CO_2H$] or the lactate ion [$CH_3CH(OH)CO_2^-$]?

70. Composition diagrams, commonly known as "alpha plots," are often used to visualize the species in a solution of an acid or base as the pH is varied. The diagram for 0.100 M acetic acid is shown here.

The plot shows how the fraction [= alpha (α)] of acetic acid and its conjugate base changes as the pH increases. It is another way of viewing the relative concentrations of acetic acid and acetate ion as a strong base is added to a solution of acetic acid in the course of a titration.
(a) Explain why the fraction of acetic acid declines and that of acetate ion increases as the pH increases.
(b) Which species predominates at a pH of 4, acetic acid or acetate ion? What is the situation at a pH of 6?
(c) Consider the point where the two lines cross. The fraction of acetic acid in the solution is 0.5, and so is that of acetate ion. That is, the solution is half acid and half conjugate base; their concentrations are equal. At this point the graph shows the pH is 4.75. Explain why the pH at this point is 4.75.

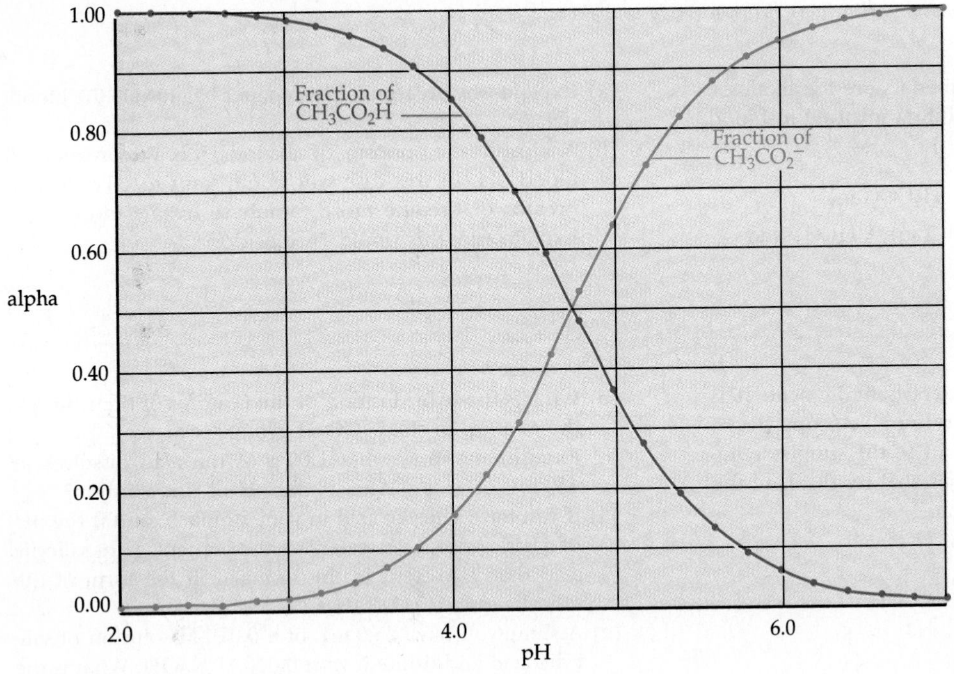

Alpha plot for acetic acid.

71. The composition diagram, or alpha plot, for the important acid-base system of carbonic acid, H_2CO_3, is illustrated here. (See Study Question 70 for more information on such diagrams.)
   (a) Explain why the fraction of bicarbonate ion, HCO_3^-, rises and then falls as the pH increases.
   (b) What is the composition of the solution when the pH is 6.0? When the pH is 10.0?
   (c) If you want to buffer a solution at a pH of 11.0, what should be the ratio of HCO_3^- to CO_3^{2-}?

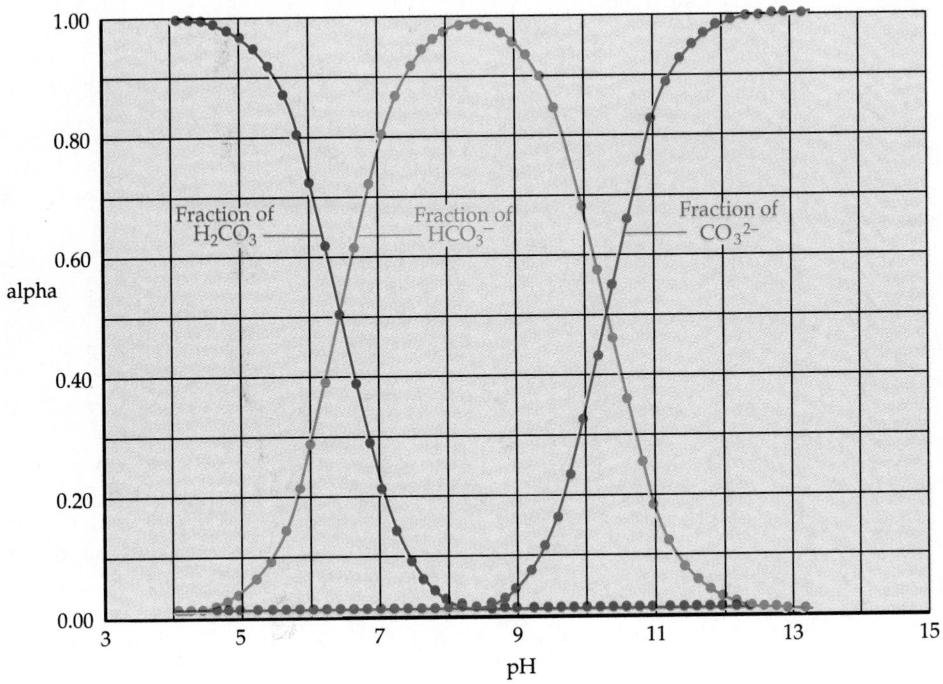

Alpha plot for carbonic acid.

72. The chapter-opening Puzzler described a cure for alkalosis, an elevation of blood pH. The equilibria involved in blood pH control are

$$CO_2(g) + H_2O(\ell) \rightleftharpoons H_2CO_3(aq)$$

$$H_2CO_3(aq) + H_2O(\ell) \rightleftharpoons H_3O^+(aq) + HCO_3^-(aq)$$

(a) Explain why breathing into a paper bag lowers the blood pH.

(b) Acidosis is the opposite of alkalosis; it is a depression of blood pH. In this case you would want to take deeper breaths or breathe more rapidly to exhale more CO_2. Explain why this would cure acidosis.

Summary Question

73. The chemical name for aspirin is acetylsalicylic acid. It is believed that the analgesic and other desirable properties of aspirin are due not to the aspirin but to the simpler compound salicylic acid, $C_6H_4(OH)CO_2H$, that results from the breakdown of aspirin in the stomach.

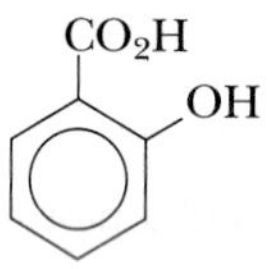

salicylic acid, $K_a = 1.1 \times 10^{-3}$

(a) Give approximate values for the following bond angles in the acid: (i) C—C—C in the ring; (ii) C—C═O; (iii) either of the C—O—H angles; and (iv) C—C—H.

(b) What is the hybridization of the C atoms of the ring? Of the C atom in the $—CO_2H$ group?

(c) Experiment shows that 1.00 g of the acid dissolves in 460 mL of water. What is the pH of this solution?

(d) If you have salicylic acid in your stomach, and if the pH of gastric juice is 2.0, calculate the percentage of salicylic acid that is present in the stomach in the form of the salicylate ion, $C_6H_4(OH)CO_2^-$.

(e) Assume you have 25.0 mL of a 0.014 M solution of salicylic acid and titrate it with 0.010 M NaOH. What is the pH at the midpoint of the titration? What is the pH at the equivalence point?

CHAPTER 19

Principles of Reactivity: Precipitation Reactions

Chemical Puzzler

Some caves are formed by dissolving limestone, $CaCO_3$, by organic acids and carbon dioxide in water that trickles through the ground. However, the stalactites and stalagmites that you see in caves are formed by the reprecipitation of the limestone. How are these processes connected? (Arthur Palmer)

Beginning in Chapter 4 you were introduced to several types of reactions, among them precipitation reactions. A precipitation reaction is an exchange reaction in which one of the products is an insoluble compound,

$$MA(aq) + BX(aq) \longrightarrow MX(s) + BA(aq)$$

$$CaCl_2(aq) + Na_2CO_3(aq) \longrightarrow CaCO_3(s) + 2\ NaCl(aq)$$

A moderately soluble compound has a solubility somewhat greater than about 0.01 mol/L, whereas soluble compounds dissolve to the extent of at least 0.1 mol/L.

that is, a compound having a solubility of less than about 0.01 mol of dissolved material per liter of solution. If you stir calcium carbonate (as the mineral calcite) into pure water, only about 6 mg dissolve per liter at 25 °C. No wonder that sea shells or corals, which are mostly calcium carbonate, do not dissolve appreciably in the sea (Figure 19.1). On the other hand, you know that "sea salt," NaCl, is very soluble in water.

It will be helpful if you review Figure 4.7, the guidelines for predicting which compounds are soluble in water and which are not.

How do you know when to predict an insoluble compound as the product of a reaction—and why should you care? In Chapter 4 we described some guidelines for making predictions (Figure 4.7), and we also discussed some important minerals that are insoluble compounds—compounds formed by precipitation reactions. In this chapter we want to make our estimates of solubility more quantitative, and we want to explore the conditions under which some compounds precipitate and others do not. Finally, we want to discuss some more practical aspects of chemical equilibria, this time applied to insoluble substances.

Figure 19.1 Sea shells do not dissolve appreciably in sea water because the shells are composed mostly of insoluble calcium carbonate. (C. D. Winters)

19.1 THE SOLUBILITY PRODUCT CONSTANT, K_{sp}

Silver bromide, AgBr, is used in photographic film. This water-insoluble salt can be made by adding a water-soluble silver salt ($AgNO_3$) to an aqueous solution of a bromide-containing salt (KBr). The *net ionic equation* for the reaction that occurs is

$$Ag^+(aq) + Br^-(aq) \longrightarrow AgBr(s)$$

If some of the precipitated AgBr is placed in pure water, some salt eventually dissolves, and an equilibrium is established.

$$AgBr(s) \rightleftharpoons Ag^+(aq, 5.7 \times 10^{-7}\ M) + Br^-(aq, 5.7 \times 10^{-7}\ M)$$

When the AgBr has dissolved to the greatest extent possible, the solution is said to be *saturated* (see Section 14.2), and experiment shows that the concentrations of the silver and bromide ions in the solution are each about 5.7×10^{-7} M at

25 °C. The extent to which an insoluble salt dissolves can be expressed in terms of the equilibrium constant for the dissolving process. In this case,

$$K = [Ag^+(aq)][Br^-(aq)]$$

Recall from Chapter 16 that the "concentration" of solids does not appear in an equilibrium constant expression.

When silver bromide dissolves in water, one mole of Ag^+ ions and one mole of Br^- ions are produced for every mole of AgBr dissolved.* The solubility of silver bromide can therefore be determined if the concentration of *either* the silver ion *or* the bromide ion is measured. This means that the previous equilibrium expression tells us that the *product* of the two concentrations measures the *solubility* of the solid compound and is a *constant.* Hence, this constant has come to be called the **solubility product constant,** and it is often designated by K_{sp}.

$$K_{sp} = [Ag^+(aq)][Br^-(aq)]$$

Because the concentrations of both Ag^+ and Br^- are 5.7×10^{-7} M when silver bromide is in equilibrium with its ions, K_{sp} is

$$K_{sp} = [5.7 \times 10^{-7}][5.7 \times 10^{-7}] = 3.3 \times 10^{-13} \text{ (at 25 °C)}$$

As with any equilibrium constant, K_{sp} values change with temperature, the direction of change depending on the enthalpy of solution of the solid. For example, calcite is about twice as soluble at 2 °C as at 25 °C because it has an exothermic enthalpy of solution.

The solubility product constant K_{sp} always has the form

$$A_xB_y(s) \rightleftharpoons xA^{y+}(aq) + yB^{x-}(aq) \qquad K_{sp} = [A^{y+}]^x[B^{x-}]^y$$

This means that you write K_{sp} expressions for the dissolving of other salts as follows

$$CaF_2(s) \rightleftharpoons Ca^{2+}(aq) + 2\,F^-(aq) \qquad K_{sp} = [Ca^{2+}][F^-]^2 = 3.9 \times 10^{-11}$$

$$Ag_2SO_4(s) \rightleftharpoons 2\,Ag^+(aq) + SO_4{}^{2-}(aq) \qquad K_{sp} = [Ag^+]^2[SO_4{}^{2-}] = 1.7 \times 10^{-5}$$

The numerical values of K_{sp} for a few salts are given in Table 19.1, but many more values are collected in Appendix H. Notice that all are given for a temperature of 25 °C.

Finally, be sure not to confuse *solubility* with the *solubility product.* The solubility of a salt is the quantity present in a unit amount of a saturated solution, expressed in moles per liter, grams per 100 mL, or other units. The *solubility product* is an equilibrium constant. Nonetheless, there is a connection between them: if either is known, the other can be calculated.

TABLE 19.1 K_{sp} Values for Some Insoluble Salts

Compound	K_{sp} at 25 °C
$CaCO_3$	3.8×10^{-9}
$SrCO_3$	9.4×10^{-10}
$BaCO_3$	8.1×10^{-9}
$BaSO_4$	1.1×10^{-10}
CaF_2	3.9×10^{-11}
CdS	3.6×10^{-29}
PbS	8.4×10^{-28}
CuS	8.7×10^{-36}
HgS	3.0×10^{-53}
AgCl	1.8×10^{-10}
AgBr	3.3×10^{-13}
AgI	1.5×10^{-16}
Ag_2CrO_4	9.0×10^{-12}
$PbCl_2$	1.7×10^{-5}
$PbCrO_4$	1.8×10^{-14}
Hg_2Cl_2	1.1×10^{-18}

EXERCISE 19.1 *Writing K_{sp} Expressions*

Write K_{sp} expressions for the following insoluble salts and look up numerical values for the constant in Appendix H.

1. $BaSO_4$
2. BiI_3
3. Ag_2CO_3 ■

*Dissolving ionic solids is actually a complex process. It almost always involves more than just the simple equilibria shown here. As a result, you should realize that simple K_{sp} calculations can be inaccurate. In particular, the calculated solubility can be incorrect, especially for salts having ions with charges larger than 1+ or 1−. It has been noted that "the solubility product [is more properly] a description, not of solubility, but simply of the ionic concentrations in the saturated solution." (Meites, L., Pode, J. S. F., and Thomas, H. C. *Journal of Chemical Education,* p. 667, 1966.)

19.2 DETERMINING K_{sp} FROM EXPERIMENTAL MEASUREMENTS

In practice, solubility product constants are determined by careful laboratory measurements using various chemical and spectroscopic methods. We shall not go into these methods but shall assume that the given ion concentrations have been measured experimentally.

EXAMPLE 19.1 K_{sp} *from Solubility Measurements*

The solubility of silver iodide, AgI, is 1.22×10^{-8} mol/L at 25 °C. Calculate K_{sp} for AgI.

Solution When silver iodide dissolves in solution, the following reaction and K_{sp} expression are appropriate.

$$AgI(s) \rightleftharpoons Ag^+(aq) + I^-(aq) \qquad K_{sp} = [Ag^+][I^-]$$

We know from experiment that the solubility of AgI is 1.22×10^{-8} mol/L at 25 °C. Because each mole of AgI that dissolves provides a mole of Ag^+ ions and a mole of I^- ions, the concentration of each ion is 1.22×10^{-8} M. Therefore,

$$K_{sp} = [Ag^+][I^-] = (1.22 \times 10^{-8})(1.22 \times 10^{-8}) = 1.49 \times 10^{-16}$$

EXAMPLE 19.2 K_{sp} *from Solubility Measurements*

Lead(II) chloride dissolves to a slight extent in water.

$$PbCl_2(s) \rightleftharpoons Pb^{2+}(aq) + 2\,Cl^-(aq)$$

Calculate K_{sp} if the lead ion concentration is found to be 1.62×10^{-2} mol/L.

Solution When $PbCl_2$ dissolves to a small extent in water, the balanced equation shows that the concentration of Cl^- ion must be twice the Pb^{2+} ion concentration. If

$$[Pb^{2+}] = 1.62 \times 10^{-2}\text{ M} \qquad \text{then} \qquad [Cl^-] = 2 \times [Pb^{2+}] = 3.24 \times 10^{-2}\text{ M}$$

This means the solubility product constant is

$$K_{sp} = [Pb^{2+}][Cl^-]^2 = (1.62 \times 10^{-2})(3.24 \times 10^{-2})^2 = 1.70 \times 10^{-5}$$

EXERCISE 19.2 K_{sp} *from Solubility Measurements*

The barium ion concentration, $[Ba^{2+}]$, in a saturated solution of barium fluoride is 7.5×10^{-3} M. Calculate the K_{sp} for BaF_2.

$$BaF_2(s) \rightleftharpoons Ba^{2+}(aq) + 2\,F^-(aq)$$

■

19.3 ESTIMATING SALT SOLUBILITY FROM K_{sp}

The K_{sp} values for many insoluble salts have been determined through a variety of laboratory measurements (Appendix H). These are an invaluable aid because they can be used to estimate the solubility of a solid salt or to determine if a solid will precipitate if solutions of its anion and cation are mixed. We turn now to methods that can be used to estimate the solubility of a salt from its K_{sp} value. Later we shall see how to use these predictions to plan the separation of ions that are mixed in solution (Section 19.6).

EXAMPLE 19.3 *Solubility from K_{sp}*

The K_{sp} for $CaCO_3$ (as the mineral calcite) is 3.8×10^{-9} at 25 °C. Calculate the solubility of calcium carbonate in pure water in (1) moles per liter and (2) grams per liter.

Equilibria involving the carbonate ion are actually more complex than can be presented here. The solubility of $CaCO_3$ is best defined only in terms of $[Ca^{2+}]$ because the hydrolysis of $CO_3{}^{2-}$ increases the solubility of $CaCO_3$. See Section 19.8.

Solution The equation for the solubility of $CaCO_3$ is

$$CaCO_3(s) \rightleftharpoons Ca^{2+}(aq) + CO_3{}^{2-}(aq)$$

and the equilibrium expression is $K_{sp} = [Ca^{2+}][CO_3{}^{2-}]$. When 1 mol of calcium carbonate dissolves, 1 mol of Ca^{2+} and 1 mol of $CO_3{}^{2-}$ ions are produced. Thus, the solubility of $CaCO_3$ can be measured by determining the concentration of *either* Ca^{2+} *or* $CO_3{}^{2-}$. Let us denote the solubility of $CaCO_3$ (in moles per liter) by x; that is, x moles of $CaCO_3$ dissolve per liter. Therefore, both $[Ca^{2+}]$ and $[CO_3{}^{2-}]$ must also equal x at equilibrium.

Equation	**$CaCO_3(s) \rightleftharpoons$ $Ca^{2+}(aq)$**	**+ $CO_3{}^{2-}(aq)$**
Initial concentration (M)	0	0
Change on proceeding to equilibrium	$+x$	$+x$
Equilibrium concentration (M)	x	x

Because K_{sp} is the product of the calcium and carbonate ion concentrations, K_{sp} is the square of the solubility x,

$$K_{sp} = [Ca^{2+}][CO_3{}^{2-}] = 3.8 \times 10^{-9} = (x)(x) = x^2$$

and so the value of x is

$$x = \sqrt{3.8 \times 10^{-9}} = 6.2 \times 10^{-5}\ \text{M}$$

The solubility of $CaCO_3$ in pure water is 6.2×10^{-5} mol/L. To find its solubility in grams per liter, we need only multiply by the molar mass of $CaCO_3$.

$$\text{Solubility in grams per liter} = (6.2 \times 10^{-5}\ \text{mol/L})(100.\ \text{g/mol}) = 0.0062\ \text{g/L}$$

EXAMPLE 19.4 *Solubility from K_{sp}*

Knowing that the K_{sp} of MgF_2 is 6.4×10^{-9}, calculate the solubility of the salt in (1) moles per liter and (2) grams per liter.

Solution As usual, we begin by writing the equilibrium equation and the K_{sp} expression.

$$MgF_2(s) \rightleftharpoons Mg^{2+}(aq) + 2\,F^-(aq) \qquad K_{sp} = [Mg^{2+}][F^-]^2$$

The next problem is to define the salt solubility in terms that allow us to solve the K_{sp} expression for this value. From the balanced equation we know that, if 1 mol of MgF_2 dissolves, 1 mol of Mg^{2+} and 2 mol of F^- appear in the solution. This means that the MgF_2 solubility is equivalent to the concentration of Mg^{2+} in the solution. If the solubility of MgF_2 is given by the unknown quantity x, then $[Mg^{2+}] = x$ and $[F^-] = 2x$.

Equation	**$MgF_2(s) \rightleftharpoons$ $Mg^{2+}(aq)$**	**+ 2 $F^-(aq)$**
Initial concentration (M)	0	0
Change on proceeding to equilibrium	$+x$	$+2x$
Equilibrium concentration (M)	x	$2x$

Substituting these values into the K_{sp} expression, we find

$$K_{sp} = [Mg^{2+}][F^-]^2 = (x)(2x)^2 = 4x^3$$

Solving the equation for x,

$$x = \sqrt[3]{\frac{K_{sp}}{4}} = \sqrt[3]{\frac{6.4 \times 10^{-9}}{4}} = 1.2 \times 10^{-3}\text{ M}$$

we find that 1.2×10^{-3} mol of MgF_2 dissolves per liter to give

$$[Mg^{2+}] = x = 1.2 \times 10^{-3}\text{ M}$$
$$[F^-] = 2x = 2 \times (1.2 \times 10^{-3}) = 2.4 \times 10^{-3}\text{ M.}$$

Finally, the solubility of MgF_2 in grams per liter is

$$(1.2 \times 10^{-3}\text{ mol/L})(62.3\text{ g/mol}) = 0.075\text{ g } MgF_2\text{/L}$$

EXERCISE 19.3 ***Salt Solubility from K_{sp}***

Using the values of K_{sp} in Appendix H, calculate the solubility of CuI and of $Mg(OH)_2$ in moles per liter. ■

PROBLEMS-SOLVING TIPS AND IDEAS

19.1 Seeing Double?

Problems like Example 19.4 (and Example 19.2) often provoke such questions as "Aren't you counting things twice when you multiply x by 2 and then square it as well?" in the expression $K_{sp} = (x)(2x)^2$. The answer is no. The quantities that belong in the equation are the *concentration* of Mg^{2+} and the *concentration* of F^-.

$$K_{sp} = (\text{concentration of } Mg^{2+})(\text{concentration of } F^-)^2$$

From the formula of the compound we know that the concentration of F^- must be *twice* that of magnesium: if $[Mg^{2+}] = x$ then $[F^-]$ must equal $2x$. Having defined the ion concentration, we then square the concentration of F^- as the equilibrium expression demands. ■

The relative solubilities of salts can often be deduced by comparing values of solubility product constants, but you must be careful! For example, the K_{sp} for silver chloride is

$$AgCl(s) \rightleftharpoons Ag^+(aq) + Cl^-(aq) \qquad K_{sp} = 1.8 \times 10^{-10}$$

whereas that for silver chromate is

$$Ag_2CrO_4(s) \rightleftharpoons 2\,Ag^+(aq) + CrO_4^{2-}(aq) \qquad K_{sp} = 9.0 \times 10^{-12}$$

In spite of the fact that silver chromate has a numerically smaller K_{sp} value, it is about 10 times more soluble than silver chloride. If you solve for their solubilities as in the previous examples, you find Solubility (AgCl) = 1.34×10^{-5} mol/L, whereas Solubility (Ag_2CrO_4) = 1.3×10^{-4} mol/L. Direct comparisons of solubility on the basis of K_{sp} values can *only* be made for salts having the same ion ratio. This means, for example, that you can directly compare solubilities of 1:1 salts such as the silver halides by comparing their K_{sp} values. The K_{sp} values for the compounds of Ag^+ with Cl^-, Br^-, and I^- are in the order

$$AgI\ (K_{sp} = 1.5 \times 10^{-16}) < AgBr\ (K_{sp} = 3.3 \times 10^{-13}) < AgCl\ (K_{sp} = 1.8 \times 10^{-10})$$

and so their solubilities are: S(AgI) < S(AgBr) < S(AgCl). Similarly, you can compare 1:2 salts such as the lead halides.

$$PbI_2\ (K_{sp} = 8.7 \times 10^{-9}) < PbBr_2\ (K_{sp} = 6.3 \times 10^{-6}) < PbCl_2\ (K_{sp} = 1.7 \times 10^{-5})$$

$$S\ (PbI_2) < S\ (PbBr_2) < S(PbCl_2)$$

increase in K_{sp} and increase in solubility →

EXERCISE 19.4 ***Comparing Salt Solubilities***

Using K_{sp} values, tell which salt in each pair is more soluble in water.

1. AgCl or AgCN
2. $Mg(OH)_2$ or $Ca(OH)_2$
3. $MgCO_3$ or $CaCO_3$ ■

An interesting example of the practical consequences of relative solubilities of insoluble salts occurs in the sea. One group of marine animals, the pteropods, forms the mineral aragonite, one type of $CaCO_3$. Most marine organisms, however, form calcite, another type of calcium carbonate. Measurements at 2 °C and at depths of several thousand meters show that aragonite is about 1.5 times more soluble than calcite. This suggests why pteropod shells are not found at all beyond depths of a few hundred meters in the Pacific Ocean, whereas substantial calcite deposits are found to depths of 3500 m. The solubility of $CaCO_3$ in either mineral form is given by the reaction

$$CaCO_3(s) + H_2O(\ell) + CO_2(g) \rightleftharpoons Ca^{2+}(aq) + 2\,HCO_3^-(aq)$$

Because this is a dynamic equilibrium, aragonite (the more soluble form) dissolves, and the less soluble form (calcite) precipitates.

A cluster of aragonite needles formed by precipitation in Jewel Cave, Wind Cave National Park (South Dakota). Aragonite is a form of calcium carbonate. (Arthur N. Palmer)

Figure 19.2 Some common minerals that are insoluble salts: *(left)* fluorite, CaF_2; *(right)* calcite, $CaCO_3$; and *(center)* iron pyrite, or "fool's gold," FeS_2. (C. D. Winters)

19.4 PRECIPITATION OF INSOLUBLE SALTS

Metal-bearing ores often contain the metal in the form of an insoluble salt (Figure 19.2), and, to complicate matters, the ores often contain several such metal salts. Virtually all industrial methods for separating metals from their ores involve dissolving the metal salts to obtain the metal ion or ions in solution. The solution is then usually concentrated in some manner, and a precipitating agent is added to precipitate selectively only one type of metal ion as an insoluble salt. In the case of nickel, the ion can be precipitated as insoluble nickel(II) sulfide or nickel(II) carbonate.

$$NiS(s) \rightleftharpoons Ni^{2+}(aq) + S^{2-}(aq) \qquad K_{sp} = 3.0 \times 10^{-21}$$
$$NiCO_3(s) \rightleftharpoons Ni^{2+}(aq) + CO_3^{2-}(aq) \qquad K_{sp} = 6.6 \times 10^{-9}$$

The final step in obtaining the metal itself is to reduce the ion to the metal either chemically or electrochemically (Chapter 21).

The goal of this section is to work out methods for determining if a precipitate forms under a given set of conditions. For example, if Ag^+ and Cl^- are present at some given concentrations, does AgCl precipitate from the solution?

K_{sp} and the Reaction Quotient Q

Silver chloride, like silver bromide, is used in photographic films. It dissolves to a very small extent in water and has a correspondingly small value of K_{sp}.

$$AgCl(s) \rightleftharpoons Ag^+(aq) + Cl^-(aq) \qquad K_{sp} = [Ag^+][Cl^-] = 1.8 \times 10^{-10}$$

But let us look at the problem from the other direction: If a solution contains Ag^+ and Cl^- ions at some concentrations, does AgCl precipitate from solution? This is the same question we asked in Section 16.3 when we wanted to know if a given mixture of reactants and products was an equilibrium mixture or if the reactants continued to form products or if products reverted to reactants. The solution there was to calculate the **reaction quotient *Q*.**

See Section 16.3 for a discussion of the reaction quotient. In the case of equilibria involving insoluble solids, some chemists call the reaction quotient the "ion product."

Recall that the reaction quotient expression is the same as that for the equilibrium constant. For silver chloride, Q is given by

$$Q = [Ag^+][Cl^-]$$

The difference between Q and K is that the concentrations required in the reaction quotient expression may or may not be those at equilibrium.

Based on the discussion in Section 16.3, we can reach the following important conclusions for a slightly soluble salt such as AgCl:

1. **If $Q = K_{sp}$, the system is at equilibrium.**
   When the product of the ion concentrations is equal to K_{sp}, the solution is **saturated.** No more solid AgCl will dissolve.
2. **If $Q < K_{sp}$, the system is not at equilibrium; the solution is *not* saturated.**
   This means one of two things: (a) If solid AgCl is present, more will dissolve until equilibrium is achieved (when $Q = K_{sp}$). (b) If solid AgCl is not already present, more $Ag^+(aq)$ or more $Cl^-(aq)$ (or more of both) can be added to the solution until precipitation of solid AgCl begins (when $Q = K_{sp}$).
3. **If $Q > K_{sp}$, the system is not at equilibrium; the solution is *super*saturated.**

The concentrations of Ag^+ and Cl^- in solution are too high, and AgCl will precipitate (until $Q = K_{sp}$).

EXAMPLE 19.5 ***Solubility and the Reaction Quotient***

Some solid AgCl has been placed in a beaker of water. After some time, experiment shows that the concentrations of Ag^+ and Cl^- are each 1.2×10^{-5} mol/L. Has the system reached equilibrium? That is, is the solution saturated? If not, will more AgCl dissolve?

Solution The solution to the problem is to compare Q and K_{sp}.

$$Q = [Ag^+][Cl^-] = (1.2 \times 10^{-5})(1.2 \times 10^{-5}) = 1.4 \times 10^{-10}$$
$$K_{sp} = 1.8 \times 10^{-10}$$

Here Q is less than K_{sp}. The solution is therefore not yet saturated, and AgCl will continue to dissolve until $Q = K_{sp}$, at which point $[Ag^+] = [Cl^-] = 1.3 \times 10^{-5}$ M. That is, an additional 0.1×10^{-5} mol of AgCl will dissolve per liter (for a total of about 1.9 mg).

EXERCISE 19.5 ***Solubility and the Reaction Quotient***

Solid PbI_2 ($K_{sp} = 8.7 \times 10^{-9}$) is placed in a beaker of water. After some time, the lead(II) concentration is measured and found to be 1.1×10^{-3} M. Has the system reached equilibrium? That is, is the solution saturated? If not, will more PbI_2 dissolve? ■

K_{sp} and Precipitation Reactions

With some knowledge of the reaction quotient, we can decide (1) if a precipitate forms when the ion concentrations are known or (2) what concentrations of ions are required to begin the precipitation of an insoluble salt.

EXAMPLE 19.6 ***Deciding Whether a Precipitate Forms***

Suppose the concentration of aqueous nickel(II) ion in a solution is 1.5×10^{-6} M. If enough Na_2CO_3 is added to make the solution 6.0×10^{-4} M in carbonate ion, CO_3^{2-}, does precipitation of $NiCO_3$ occur ($K_{sp} = 6.6 \times 10^{-9}$)? If not, does it occur if the concentration of CO_3^{2-} is raised by a factor of 100?

Solution The insoluble salt $NiCO_3$ dissolves according to the balanced equation

$$NiCO_3(s) \rightleftharpoons Ni^{2+}(aq) + CO_3^{2-}(aq)$$

and the solubility product expression is $K_{sp} = [Ni^{2+}][CO_3^{2-}]$. When the concentrations of nickel(II) and carbonate ion are those stated earlier,

$$Q = [Ni^{2+}][CO_3^{2-}] = (1.5 \times 10^{-6})(6.0 \times 10^{-4}) = 9.0 \times 10^{-10}$$
$$Q\ (9.0 \times 10^{-10}) < K_{sp}\ (6.6 \times 10^{-9})$$

It is evident that the reaction quotient is less than the value of K_{sp}. The solution is therefore not yet saturated, and precipitation does *not* occur.

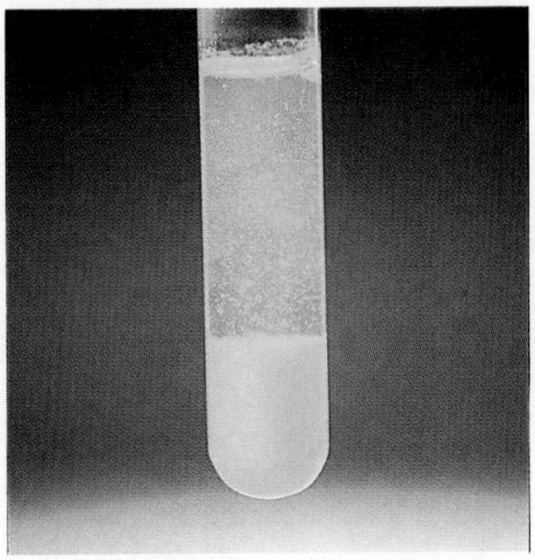

Lead(II) iodide is precipitated by mixing solutions of lead(II) nitrate and potassium iodide. See Exercise 19.5. (C. D. Winters)

If $[CO_3^{2-}]$ is increased by a factor of 100, it becomes 6.0×10^{-2}. Under these circumstances, Q is 9.0×10^{-8}.

$$Q = [Ni^{2+}][CO_3^{2-}] = (1.5 \times 10^{-6})(6.0 \times 10^{-2}) = 9.0 \times 10^{-8}$$

a value larger than K_{sp}. Precipitation of $NiCO_3$ occurs and continues until the Ni^{2+} and CO_3^{2-} concentrations have declined to the point at which their product is equal to K_{sp}.

EXERCISE 19.6 ***Deciding Whether a Precipitate Forms***

If the concentration of strontium ion is 2.5×10^{-4} M, does precipitation of $SrSO_4$ occur when enough of the soluble salt Na_2SO_4 is added to make the solution 2.5×10^{-4} M in SO_4^{2-}? K_{sp} for $SrSO_4$ is 2.8×10^{-7}. ■

Now that we know how to decide if a precipitate forms when the concentration of each ion is known, let us turn to the problem of deciding how much of the precipitating agent is required to begin the precipitation of an ion at a given concentration level.

EXAMPLE 19.7 ***Ion Concentrations Required to Begin Precipitation***

The concentration of barium ion, Ba^{2+}, in a solution is 0.010 M.

1. What concentration of sulfate ion, SO_4^{2-}, is required to just begin precipitating $BaSO_4$ (Figure 19.3)?
2. When the concentration of sulfate ion in the solution reaches 0.015 M, what concentration of barium ion remains in solution?

Solution As usual, let us begin by writing the balanced equation for the equilibrium that exists when $BaSO_4$ precipitates and the K_{sp} expression.

$$BaSO_4(s) \rightleftharpoons Ba^{2+}(aq) + SO_4^{2-}(aq) \qquad K_{sp} = [Ba^{2+}][SO_4^{2-}] = 1.1 \times 10^{-10}$$

1. From the K_{sp} expression, we know that, when the product of the ion concentrations exceeds 1.1×10^{-10}, that is, when $Q > K_{sp}$, precipitation occurs. The Ba^{2+} ion concentration is known (0.010 M), so the SO_4^{2-} ion concentration that leads to precipitation can be calculated.

$$K_{sp} = 1.1 \times 10^{-10} = [0.010][SO_4^{2-}]$$

$$[SO_4^{2-}] = \frac{K_{sp}}{[Ba^{2+}]} = \frac{1.1 \times 10^{-10}}{0.010} = 1.1 \times 10^{-8}\ \text{M}$$

The result tells us that if the sulfate ion is just *slightly* greater than 1.1×10^{-8} M, $BaSO_4$ begins to precipitate; $Q = [Ba^{2+}][SO_4^{2-}]$ would then be greater than K_{sp}.

2. If the sulfate ion concentration is increased to 0.015 M, the maximum concentration of Ba^{2+} ion that can exist in solution (in equilibrium with $BaSO_4$) is

$$[Ba^{2+}] = \frac{K_{sp}}{[SO_4^{2-}]} = \frac{1.1 \times 10^{-10}}{0.015} = 7.3 \times 10^{-9}\ \text{M}$$

(a) (b) (c)

Figure 19.3 Barium sulfate is quite insoluble in water ($K_{sp} = 1.1 \times 10^{-10}$) (see Example 19.7). (a) Mixing aqueous solutions of barium chloride and sulfuric acid gives white, insoluble $BaSO_4$. (b) A sample of the mineral barite, which is mostly barium sulfate. (C. D. Winters) (c) Barium sulfate is opaque to x-rays; therefore, if you drink a "cocktail" containing $BaSO_4$ it does not dissolve in your stomach or intestines. Instead, it progresses through your digestive organs and can be followed by x-ray analysis. This photo is an x-ray film of a gastrointestinal tract after a person ingested barium sulfate. It is fortunate that $BaSO_4$ is so insoluble, because water- and acid-soluble barium salts are toxic. (Susan Leavines/Science Source/Photo Researchers, Inc.)

The fact that the barium ion concentration is so small under these circumstances means that the Ba^{2+} ion has been essentially completely removed from solution. (It began at 0.010 M and has dropped by a factor of about 10 million.) For all practical purposes, the Ba^{2+} ion precipitation is complete.

EXERCISE 19.7 ***Ion Concentrations Required to Begin Precipitation***

What is the minimum concentration of I^- that can cause precipitation of PbI_2 from a 0.050 M solution of $Pb(NO_3)_2$? K_{sp} for PbI_2 is 8.7×10^{-9}. What concentration of Pb^{2+} ions remains in solution when the concentration of I^- is 0.0015 M? ■

EXAMPLE 19.8 ***K_{sp} and Precipitations***

Suppose you mix 100. mL of 0.0200 M $BaCl_2$ with 50.0 mL of 0.0300 M Na_2SO_4. Does $BaSO_4$ precipitate ($K_{sp} = 1.1 \times 10^{-10}$) (see Figure 19.3)?

Solution To answer this question, we need to know the concentrations of the Ba^{2+} and SO_4^{2-} ions. We know these concentrations in the original solutions but not in the solution after they have been mixed. Therefore, we first calculate these new concentrations in the total solution volume of 150. mL.

$$[Ba^{2+}] = \frac{(0.0200\ \text{mol/L})(0.100\ \text{L})}{0.150\ \text{L}} = 0.0133\ \text{M after mixing}$$

$$[SO_4^{2-}] = \frac{(0.0300\ \text{mol/L})(0.0500\ \text{L})}{0.150\ \text{L}} = 0.0100\ \text{M after mixing}$$

Now the reaction quotient can be calculated,

$$Q = [Ba^{2+}][SO_4^{2-}] = (0.0133)(0.0100) = 1.33 \times 10^{-4}$$

and we see it is greater than K_{sp}. Therefore, precipitation of $BaSO_4$ occurs.

EXERCISE 19.8 ***K_{sp} and Precipitations***

You have 100.0 mL of 0.0010 M silver nitrate. Does AgCl precipitate if you add 5.0 mL of 0.025 M HCl? ■

PROBLEM-SOLVING TIPS AND IDEAS

19.2 Does a Precipitate Form?

One difficulty students have when solving a problem such as "If the concentrations of the ions Ag^+ and Cl^- are 1.0×10^{-5} M, does a precipitate of AgCl form?" is the temptation to write the equation

$$Ag^+(aq) + Cl^-(aq) \rightleftharpoons AgCl(s)$$

Instead, approach *all* problems involving "insoluble" salts from the viewpoint of the salt *dissolving* in water,

$$AgCl(s) \rightleftharpoons Ag^+(aq) + Cl^-(aq)$$

and use the K_{sp} expression with the appropriate ion concentrations. ■

19.5 SOLUBILITY AND THE COMMON ION EFFECT

The test tube on the left in Figure 19.4 contains a precipitate of silver acetate, $AgCH_3CO_2$, in water. The solution is saturated, so there are silver ions and acetate ions in solution in equilibrium with solid silver acetate.

$$AgCH_3CO_2(s) \rightleftharpoons Ag^+(aq) + CH_3CO_2^-(aq)$$

But what happens if some excess silver ions are added, say by adding silver nitrate? Le Chatelier's principle (Section 16.6) suggests that more precipitate should form because a product ion has been added, thus causing the equilibrium position to shift to the left. This is indeed observed, as illustrated by the test tube on the right in Figure 19.4.

The ionization of weak acids and bases is affected by the presence of an ion common to the equilibrium process (Section 18.2), and the effect of adding silver ion to the saturated silver acetate solution is another example of the *common ion effect.* In the case of slightly soluble salts, the effect is always to lower the salt solubility.

EXAMPLE 19.9 ***The Common Ion Effect and Salt Solubility***

The solubility of AgCl is 1.3×10^{-5} M (0.0019 g/L). If some AgCl is placed in 1.00 L of a 0.55-M solution of NaCl, how many grams of AgCl dissolve?

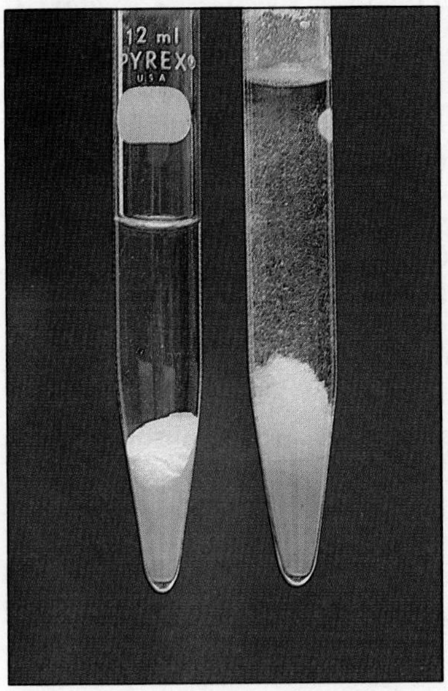

Figure 19.4 The common ion effect. The test tube at the left contains a saturated solution of silver acetate, $AgCH_3CO_2$. When 1 M $AgNO_3$ is added to the test tube, the equilibrium

$$AgCH_3CO_2(s) \rightleftharpoons Ag^+(aq) + CH_3CO_2^-(aq)$$

shifts to the left, as evidenced by the test tube at the right, in which more solid silver acetate has formed. (C. D. Winters)

Solution In *pure* water, the solubility of AgCl is equal to either $[Ag^+]$ or $[Cl^-]$.

$$\text{Solubility of AgCl in pure water} \equiv [Ag^+] \text{ or } [Cl^-] = \sqrt{K_{sp}} = 1.3 \times 10^{-5}\text{ M}$$

In water already containing the Cl^- ion, however, Le Chatelier's principle predicts that the solubility is *less than* 1.3×10^{-5}. In addition, the solubility can now be expressed only in terms of the concentration of the silver ion.

$$\text{Solubility of AgCl in presence of added } Cl^- \equiv [Ag^+] = x < 1.3 \times 10^{-5}\text{ M}$$

$$AgCl(s) \rightleftharpoons Ag^+(aq) + Cl^-(aq)$$

←excess Cl^- shifts equilibrium to the left—

To solve for the solubility of AgCl in the presence of added Cl^- ion, set up the usual table, which shows the concentrations of Ag^+ and Cl^- when equilibrium is attained.

Equation	**AgCl(s)** $\rightleftharpoons$ **Ag⁺(aq) +**	**Cl⁻(aq)**
Initial concentration (M)	0	0.55
Change on proceeding to equilibrium	$+x$	$+x$
Equilibrium concentration (M)	x	$(x + 0.55)$

Some AgCl dissolves in the presence of chloride ion and produces Ag^+ and Cl^- ion concentrations of x mol/L. Some chloride ion was already present, however, so the *total chloride ion concentration* is the amount coming from AgCl (equals x) *plus* what was already there (0.55 M).

Now the K_{sp} expression can be written using the equilibrium concentrations from the table.

$$K_{sp} = 1.8 \times 10^{-10} = [Ag^+][Cl^-] = (x)(x + 0.55)$$

and rearranged to

$$x^2 + 0.55x - K_{sp} = 0$$

This is a quadratic equation and can be solved by the methods in Appendix A. The easiest approach, however, is to make the approximation that x is *very* small with respect to 0.55; that is, we assume it makes only a negligible difference to the answer if we assume that $(x + 0.55) \approx 0.55$. This is a very reasonable assumption because we know that the solubility equals 1.3×10^{-5} M *without* the common ion Cl^-, and it is even smaller in the presence of added Cl^-. Therefore,

$$K_{sp} = 1.8 \times 10^{-10} \approx (x)(0.55)$$

$$x = [Ag^+] \approx 3.3 \times 10^{-10}\ \text{M} \qquad \text{or} \qquad 4.7 \times 10^{-8}\ \text{g/L}$$

As predicted by Le Chatelier's principle, the solubility of AgCl in the presence of added Cl^- is less (3.3×10^{-10} M) than in pure water (1.3×10^{-5} M).

As a final step, let us check the approximation we made by substituting the approximate value of x into the exact expression $K_{sp} = (x)(x + 0.55)$. Then, if the product $(x)(x + 0.55)$ is the same as the given value of K_{sp}, the approximation is valid.

$$K_{sp} = (x)(x + 0.55) = (3.3 \times 10^{-10})(3.3 \times 10^{-10} + 0.55) \approx 1.8 \times 10^{-10}$$

EXERCISE 19.9 ***The Common Ion Effect and Salt Solubility***

Calculate the solubility of $BaSO_4$ (1) in pure water and (2) in the presence of 0.010 M $Ba(NO_3)_2$. K_{sp} for $BaSO_4$ is 1.1×10^{-10}. ■

PROBLEM-SOLVING TIPS AND IDEAS

19.3 Solubility in the Presence of a Common Ion

To calculate the solubility of a slightly soluble salt in the presence of an ion common to the equilibrium, let us say $BaSO_4$ in the presence of added Ba^{2+} ion, proceed in the following way:

1. The concentration of the common ion, here Ba^{2+}, is known.
2. Assume some of the slightly soluble salt dissolves, producing Ba^{2+} and $SO_4{}^{2-}$ in this case. The equilibrium constant (K_{sp}) for the dissolving process is known.
3. To solve the K_{sp} expression, the concentration of the common ion is assumed to be equal to that of the added, soluble salt. The concentration of the ion other than the common ion is a measure of the solubility of the salt. For the $BaSO_4$ case

$$K_{sp} = [Ba^{2+}][SO_4{}^{2-}] = [\text{common ion}][\text{ion that reflects salt solubility}]$$ ■

There are two important ideas to be learned from Example 19.9. First, the solubility of AgCl in the presence of a common ion was reduced by a factor of about 10^5, in accordance with Le Chatelier's principle. Second, we made the approximation that the amount of common ion added to the solution was very large compared with the amount of that ion coming from the insoluble salt, which allowed us to simplify our calculations. This is almost always the case, but you should *always* check the approximation.

EXAMPLE 19.10 *The Common Ion Effect and Salt Solubility*

Calculate the solubility of silver chromate, Ag_2CrO_4, at 25 °C in (1) pure water and (2) in the presence of 0.0050 M K_2CrO_4 solution.

Solution The balanced equation defining silver chromate's solubility is

$$Ag_2CrO_4(s) \rightleftharpoons 2\,Ag^+(aq) + CrO_4^{2-}(aq)$$
$$K_{sp} = [Ag^+]^2[CrO_4^{2-}] = 9.0 \times 10^{-12}$$

Part 1 *Solubility in pure water.* Because there is a 1:1 stoichiometric ratio for moles of CrO_4^{2-} in solution to moles of Ag_2CrO_4 dissolved, we define the solubility of this salt in terms of $[CrO_4^{2-}] = x$. Thus, x is not only the concentration of CrO_4^{2-} in solution but is also the number of moles per liter of Ag_2CrO_4 dissolved.

Equation	$Ag_2CrO_4(s) \rightleftharpoons$	$2\,Ag^+(aq)$ +	$CrO_4^{2-}(aq)$
Before Ag_2CrO_4 begins to dissolve (M)		0	0
Change on proceeding to equilibrium		$+2x$	$+x$
After equilibrium is achieved (M)		$2x$	x

This must mean that $[Ag^+] = 2x$, and so the K_{sp} expression can be written as

$$K_{sp} = 9.0 \times 10^{-12} = [Ag^+]^2[CrO_4^{2-}] = (2x)^2(x) = 4x^3$$
$$x^3 = 2.3 \times 10^{-12}\ M$$
$$x = [CrO_4^{2-}] = \sqrt[3]{2.3 \times 10^{-12}} = 1.3 \times 10^{-4}\ M$$

This means the ion concentrations at equilibrium in pure water are

$$[Ag^+] = 2x = 2.6 \times 10^{-4}\ M \quad \text{and} \quad [CrO_4^{2-}] = x = 1.3 \times 10^{-4}\ M$$

We know now that 1.3×10^{-4} mol of Ag_2CrO_4 dissolves per liter.

Do not be confused about the use of $(2x)^2$, where the factor 2 appears both as the multiplier of x and the exponent of $(2x)$. This occurs because $[CrO_4^{2-}]$ is defined as x, and the stoichiometry demands that $[Ag^+]$ be defined as $2x$. This is substituted into the K_{sp} expression in place of $[Ag^+]$. The K_{sp} expression then demands that $[Ag^+]$ or the equivalent expression $2x$ be squared. See Problem-Solving Tips and Ideas, 19.1.

Part 2 *Solubility in a solution containing chromate ion.* In the presence of excess chromate ion from dissolved K_2CrO_4, the concentration of Ag^+ is less than in pure water. To calculate this concentration, let us again assume the solubility of Ag_2CrO_4 is equivalent to the amount of CrO_4^{2-} ion that appears in solution from the dissolving process. Now, however, this quantity is assigned a different unknown value, y. Thus the concentration of Ag^+ ion at equilibrium is $2y$ and that of CrO_4^{2-} ion is y *plus* the amount of CrO_4^{2-} already in the solution.

Equation	$Ag_2CrO_4(s) \rightleftharpoons$	$2\,Ag^+(aq)$ +	$CrO_4^{2-}(aq)$
Before Ag_2CrO_4 begins to dissolve (M)		0	0.0050
Change on proceeding to equilibrium		$+2y$	$+y$
After equilibrium is achieved (M)		$2y$	$(y + 0.0050)$

Substituting the equilibrium amounts into the K_{sp} expression, we have

$$K_{sp} = 9.0 \times 10^{-12} = [Ag^+]^2[CrO_4^{2-}] = (2y)^2(y + 0.0050)$$

Again make the approximation that y is very small with respect to 0.0050, and so $(y + 0.0050) \approx 0.0050$. (This is probably reasonable because $[CrO_4^{2-}]$ is 0.00013 M *without* added chromate ion, and it is certain that y is even smaller in the presence of extra chromate ion.) Therefore, the approximate expression is

$$K_{sp} = 9.0 \times 10^{-12} = [Ag^+]^2[CrO_4^{2-}] = (2y)^2(0.0050)$$

and so y is

$$y = [CrO_4^{2-}] \text{ from } Ag_2CrO_4 = 2.1 \times 10^{-5}\text{ M}$$

This means the silver ion concentration in the presence of the common ion is

$$[Ag^+] = 2y = 4.2 \times 10^{-5}\text{ M}$$

This silver ion concentration is less than its value in pure water (2.6×10^{-4}), and you again see the result of adding an ion "common" to the equilibrium.

As a final step, check the approximation. Substitute y back into the expression $K_{sp} = (2y)^2(0.0050 + y)$ to see if K_{sp} calculated from the expression agrees with the given K_{sp}.

$$K_{sp} = [2(2.1 \times 10^{-5})]^2(0.0050 + 0.000021) = 8.9 \times 10^{-12}$$

Here you see that the calculated K_{sp} is only about 1% different from the given K_{sp}. The approximation is valid.

EXERCISE 19.10 ***The Common Ion Effect and Salt Solubility***

Calculate the solubility of $Zn(CN)_2$ at 25 °C (1) in pure water and (2) in the presence of 0.10 M KCN. K_{sp} for $Zn(CN)_2$ is 8.0×10^{-12}. ■

19.6 SOLUBILITY, ION SEPARATIONS, AND QUALITATIVE ANALYSIS

In many courses in introductory chemistry a portion of the laboratory work is devoted to the qualitative analysis of aqueous solutions, the identification of anions and metal cations. The purpose of such laboratory work is (1) to introduce you to some basic chemistry of various ions and (2) to illustrate how the principles of chemical equilibria can be applied.

Assume you have an aqueous solution that contains some or all of the following metal ions: Ag^+, Pb^{2+}, Cd^{2+}, and Cu^{2+}. Your objective is to separate the ions from one another so that each type of ion ends up in a separate test tube; the presence or absence of the ion can then be established. As a first step in this process, you want to find one reagent that forms a precipitate with one or more of the cations and leaves the others in solution. This is done by comparing K_{sp} values for salts of cations with various anions (say S^{2-}, OH^-, Cl^-, or SO_4^{2-}), looking for an anion that gives insoluble salts for some cations but not others.

Looking over the list of solubility products in Appendix H, you notice that all of the ions in our example solution form very insoluble sulfides (Ag_2S, PbS, CdS, and CuS). However, only two of them form insoluble chlorides: AgCl and $PbCl_2$. Thus, your "magic reagent" for partial cation separation could be aqueous HCl, which forms precipitates with two of the ions while the other two

Some Insoluble Sulfides and Chlorides

Compound	K_{sp} at 25 °C
Ag_2S	7.9×10^{-51}
PbS	3.2×10^{-28}
CdS	1.0×10^{-27}
CuS	7.9×10^{-37}
AgCl	1.8×10^{-10}
$PbCl_2$	1.7×10^{-5}

(a) (b) (c)

Figure 19.5 Ion separations by solubility difference. (a) The solution contains nitrate salts of Ag^+, Pb^{2+}, Cd^{2+}, and Cu^{2+}. (The Cu^{2+} ion in water is light blue; the others are colorless.) (b) Aqueous HCl is added in an amount sufficient to precipitate completely AgCl and $PbCl_2$ (both white solids). (c) The blue solution, now containing only Cu^{2+} and Cd^{2+}, is poured into another test tube, leaving white, solid AgCl and $PbCl_2$ in the first test tube. (C. D. Winters)

remain in solution (Figure 19.5). Now you are left with the task of separating AgCl and $PbCl_2$ from each other and aqueous Cd^{2+} and Cu^{2+} from each other (Figure 19.5). The separation of AgCl and $PbCl_2$ is not difficult because $PbCl_2$ dissolves in hot water, but AgCl remains insoluble.

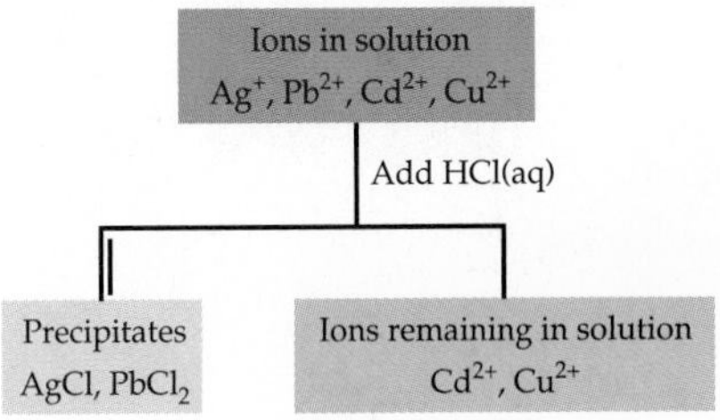

EXAMPLE 19.11 *Separation of Two Ions by Difference in Solubility*

A solution contains Cl^- (as NaCl, 0.010 M) and CrO_4^{2-} (as K_2CrO_4, 0.0010 M). A solution of $AgNO_3$ is added slowly (without changing the volume of the original solution appreciably).

1. Which precipitates first, AgCl or Ag_2CrO_4?
2. We find that AgCl precipitates before Ag_2CrO_4. What is the Cl^- ion concentration when Ag_2CrO_4 begins to precipitate?

Solution As usual, we first write the equations for dissolving the insoluble salts and look up their K_{sp} values.

$$AgCl(s) \rightleftharpoons Ag^+(aq) + Cl^-(aq) \qquad K_{sp} = 1.8 \times 10^{-10}$$

$$Ag_2CrO_4(s) \rightleftharpoons 2\,Ag^+(aq) + CrO_4^{2-}(aq) \qquad K_{sp} = 9.0 \times 10^{-12}$$

1. To discover which anion precipitates first, calculate the concentration of Ag^+ required to just begin the precipitation of each salt.
   (a) To *begin* to precipitate AgCl when $[Cl^-] = 0.010$ M

$$K_{sp} = [Ag^+][Cl^-] = 1.8 \times 10^{-10}$$

$$[Ag^+] = \frac{1.8 \times 10^{-10}}{0.010} = 1.8 \times 10^{-8}\text{ M}$$

(b) To *begin* to precipitate Ag_2CrO_4 when $[CrO_4^{2-}] = 0.0010$ M

$$K_{sp} = [Ag^+]^2[CrO_4^{2-}] = 9.0 \times 10^{-12}$$

$$[Ag^+] = \sqrt{\frac{9.0 \times 10^{-12}}{0.0010}} = 9.5 \times 10^{-5}\ \text{M}$$

These calculations show that much less silver ion is needed to begin the precipitation of AgCl than to begin the precipitation of Ag_2CrO_4. Therefore, AgCl precipitates before Ag_2CrO_4 as $AgNO_3$ is added slowly to the solution.

2. Ag_2CrO_4 begins to precipitate when $[Ag^+] = 9.5 \times 10^{-5}$. The concentration of Cl^- when the Ag^+ ion concentration has this value can be calculated using the K_{sp} expression for AgCl.

$$[Cl^-] = \frac{K_{sp}}{[Ag^+]} = \frac{1.8 \times 10^{-10}}{9.5 \times 10^{-5}} = 1.9 \times 10^{-6}\ \text{M}$$

This means the percentage of Cl^- ion still in solution when Ag_2CrO_4 just begins to precipitate is

$$\frac{1.9 \times 10^{-6}\ \text{M}}{0.010\ \text{M}} \times 100\% = 0.019\%$$

Thus, if the AgCl precipitate is removed from the reaction mixture (by filtration) just before the Ag_2CrO_4 begins to precipitate, 99.98% of the Cl^- ion is by then separated from the CrO_4^{2-} ion.

EXERCISE 19.11 ***Separation of Two Ions by Difference in Solubility***

Under the right circumstances, aqueous Ag^+ can be separated from aqueous Pb^{2+} by the difference in the solubilities of their chloride salts, AgCl and $PbCl_2$.

1. If you begin with both metal ions having a concentration of 0.0010 M, which ion precipitates first as an insoluble chloride on adding HCl?
2. What is the concentration of the metal ion that precipitates first, just before the second metal chloride begins to precipitate? ■

EXERCISE 19.12 ***Schemes for Ion Separation***

The cations of each of the following pairs appear together in one solution:

1. Ag^+ and Bi^{3+} 2. Fe^{2+} and K^+

You may add only one reagent to precipitate one cation and not the other. Consult the solubility product table in Appendix H to decide whether to use Cl^-, S^{2-}, or OH^- as the precipitating ion in each case. (The precipitating ions are introduced in the form of HCl, $(NH_4)_2S$, or NaOH, for example.) ■

19.7 SIMULTANEOUS EQUILIBRIA

In many instances two or more reactions occur at the same time in a solution, all of them being described as equilibrium processes. Chemists characterize such situations as examples of **simultaneous equilibria.**

One example of simultaneous equilibria is the case in which a reagent is added to a saturated solution of an insoluble salt, converting the salt to another, even less soluble salt. Consider two common lead compounds, lead(II) chloride and lead(II) chromate.

$$PbCl_2,\quad K_{sp} = 1.7 \times 10^{-5} \qquad \text{and} \qquad PbCrO_4,\quad K_{sp} = 1.8 \times 10^{-14}$$

If you add a few drops of K_2CrO_4 to a small amount of a white precipitate of $PbCl_2$ and shake the mixture, the solid changes to yellow $PbCrO_4$. That this is possible is evident from the equilibrium constant, K_{net}, for the process

$$PbCl_2(s) + CrO_4^{2-}(aq) \rightleftharpoons PbCrO_4(s) + 2\ Cl^-(aq) \qquad K_{net} = 9.4 \times 10^8$$

This reaction is the sum of two reactions whose equilibrium constants are known.

$$PbCl_2(s) \rightleftharpoons Pb^{2+}(s) + 2\ Cl^-(aq) \qquad K_1 = K_{sp} = 1.7 \times 10^{-5}$$

$$Pb^{2+}(aq) + CrO_4^{2-}(aq) \rightleftharpoons PbCrO_4(s) \qquad K_2 = \frac{1}{K_{sp}} = \frac{1}{1.8 \times 10^{-14}}$$

The equilibrium constant for the overall reaction is the product of the equilibrium constants for the summed reactions. That is, $K_{net} = K_1 \cdot K_2 = 9.4 \times 10^8$. This very large value indicates that the reaction should proceed from left to right.

Simultaneous equilibria are very important in many operations in the laboratory and in nature. The next sections explore more examples.

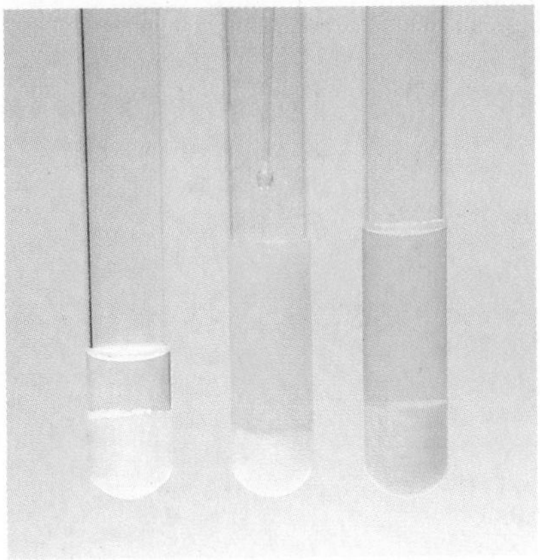

The test tube at the left contains a precipitate of white $PbCl_2$. In the middle test tube, a solution of yellow K_2CrO_4 has been added to insoluble white $PbCl_2$. The test tube at the right shows what happens when the $PbCl_2$ precipitate is stirred in the presence of K_2CrO_4. The white solid $PbCl_2$ has been transformed into less soluble yellow $PbCrO_4$. (C. D. Winters)

EXERCISE 19.13 ***Simultaneous Equilibria***

Silver forms many insoluble salts. Which is more soluble, AgCl or AgBr? If you add sufficient bromide ion to an aqueous suspension of AgCl(s), can AgCl be converted to AgBr? To answer this, derive the value of the equilibrium constant for

$$AgCl(s) + Br^-(aq) \rightleftharpoons AgBr(s) + Cl^-(aq)$$

■

19.8 SOLUBILITY AND pH

The next time you are tempted to wash a supposedly insoluble salt down the kitchen or laboratory drain, stop and consider the consequences. Many metal ions, such as lead, chromium, and mercury, are *toxic* in the environment. Even if the so-called insoluble salt does not appear to dissolve, its solubility in water may be greater than you think, in part owing to the possibility of hydrolysis of the anion of the salt (Figure 19.6).

Lead sulfide, PbS, is found in nature as the mineral galena. Let us consider what happens if a trace of lead(II) sulfide dissolves in water. First, the insoluble salt dissolves to an extremely small extent.

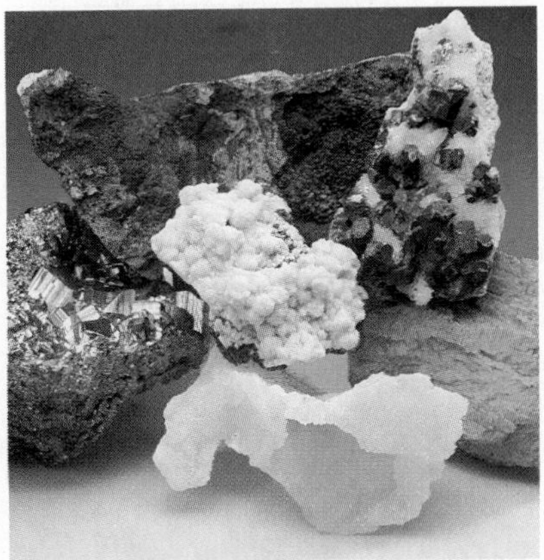

Figure 19.6 Many common minerals have basic anions and so are more soluble in acid than in pure water. Examples include black galena (PbS), white calcite ($CaCO_3$), fool's gold (FeS_2), blue azurite ($2\ CuCO_3 \cdot Cu(OH)_2$), yellow orpiment (As_2S_3), and pink rhodochrosite ($MnCO_3$). (C. D. Winters)

Step 1 *The solubility of PbS in water*

$$PbS(s) \rightleftharpoons Pb^{2+}(aq) + S^{2-}(aq)$$

One product of the reaction is a sulfide ion, an anion even more basic than the OH^- ion (Table 17.4). Therefore, the sulfide ion is extensively hydrolyzed.

Step 2 *The hydrolysis of sulfide ion*

$$S^{2-}(aq) + H_2O(\ell) \rightleftharpoons HS^-(aq) + OH^-(aq) \qquad K_b = 1 \times 10^5$$

The overall process is the sum of two simultaneous equilibria,

$$PbS(s) + H_2O(\ell) \rightleftharpoons Pb^{2+}(aq) + HS^-(aq) + OH^-(aq)$$

which has an equilibrium constant of 3.2×10^{-28}. Thus, because of the sulfide ion hydrolysis, metal sulfides are more soluble in water than expected from the simple ionization of the salt.

The lead sulfide example leads to the general observation that any salt containing an anion that is the conjugate base of a weak acid dissolves in water to a greater extent than given by K_{sp}. This means that salts of phosphate, acetate, carbonate, and cyanide, as well as sulfide, can be affected, because all of these ions undergo the general hydrolysis reaction

$$X^-(aq) + H_2O(\ell) \rightleftharpoons HX(aq) + OH^-(aq)$$

in aqueous solution.

The possibility that the anion of an insoluble salt undergoes hydrolysis leads to yet another useful, general conclusion: If a strong acid is added to a water-insoluble salt such as $CaCO_3$, OH^- ion from X^- ion hydrolysis is removed (by formation of water). This shifts the X^- ion hydrolysis reaction further to the right; the weak acid HX is formed, and the salt dissolves (Figure 19.7). For example, calcium carbonate dissolves readily when hydrochloric acid is added, and some of the reactions occurring are

$$CaCO_3(s) \rightleftharpoons Ca^{2+}(aq) + CO_3^{2-}(aq) \qquad K = K_{sp} = 3.8 \times 10^{-9}$$

$$CO_3^{2-}(aq) + H_2O(\ell) \rightleftharpoons HCO_3^-(aq) + OH^-(aq) \qquad K = K_b = 2.1 \times 10^{-4}$$

$$OH^-(aq) + H_3O^+(aq) \rightleftharpoons 2\ H_2O(\ell) \qquad K = 1/K_w = 1/1 \times 10^{-14}$$

$$CaCO_3(s) + H_3O^+(aq) \rightleftharpoons Ca^{2+}(aq) + HCO_3^-(aq) + H_2O(\ell) \qquad K_{net} = K_{sp}(K_b)(1/K_w) = 79.8$$

In the presence of strong acid, the hydrogen carbonate ion reacts further with acid to give H_2CO_3, which then reverts to CO_2 gas and water.

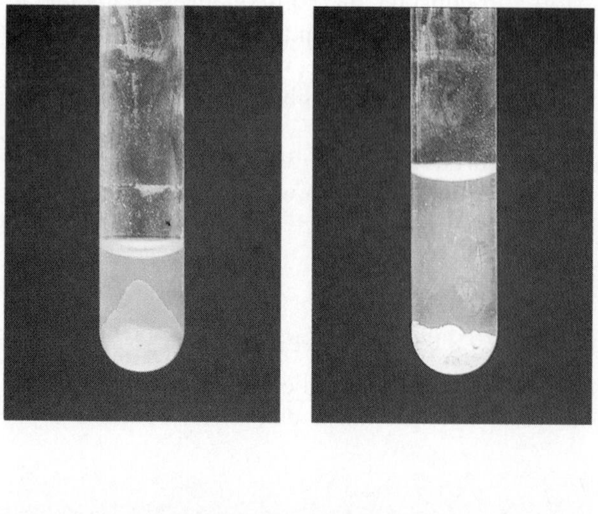

Figure 19.7 Dissolving precipitates in acid. *(Left)* A precipitate of AgCl (white) and Ag_3PO_4 (yellow). *(Right)* Adding a strong acid dissolves Ag_3PO_4 and leaves insoluble AgCl. (C. D. Winters)

A CLOSER LOOK

Equilibria Involving Sulfide Ions

Equilibria involving sulfide ions have long presented a difficult problem, especially in view of the hydrolysis of the strongly basic sulfide ion, and the uncertainty in the equilibrium constant for this reaction. It has been argued, however, that the solubility of metal sulfides should be treated as in the PbS example in the text. That is, a metal sulfide such as zinc sulfide dissolves according to the equation

$$ZnS(s) + H_2O(\ell) \rightleftharpoons Zn^{2+}(aq) + OH^-(aq) + HS^-(aq)$$

and that the equilibrium constant for this reaction, its solubility product constant, K_{sp}, is actually

$$K_{sp} = [Zn^{2+}][OH^-][HS^-] = 2 \times 10^{-25}$$

In practice, we are interested in dissolving metal sulfides in acid solution. In the presence of a strong acid both OH^- and HS^- are protonated to form H_2O and H_2S, respectively. Therefore, in strong acid zinc sulfide dissolves to some extent to give zinc ions, H_2O, and H_2S,

$$ZnS(s) + 2\,H_3O^+(aq) \rightleftharpoons Zn^{2+}(aq) + H_2S(aq) + 2\,H_2O(\ell)$$

and the equilibrium constant for the reaction is called K_{spa}, for K_{sp} in acid.

$$K_{spa} = \frac{[Zn^{2+}][H_2S]}{[H_3O^+]^2} = 2 \times 10^{-4}$$

The value of K_{spa} is considerably larger than K_{sp} because the equilibrium is shifted strongly to the right due to the protonation of the two basic anions, OH^- and HS^-. Indeed, it is generally true that K_{spa} values are about 10^{21} larger than K_{sp} values for metal sulfides, and sulfides for which K_{spa} is greater than about 10^{-2} are readily soluble in strong acids.

Metal Sulfide	K_{sp}	K_{spa}
HgS (black)	2.0×10^{-53}	2×10^{-32}
CuS	7.9×10^{-37}	6×10^{-16}
PbS	3.2×10^{-28}	3×10^{-7}
ZnS	2.0×10^{-25}	2×10^{-4}
FeS	7.9×10^{-19}	6×10^{2}
MnS	3.2×10^{-14}	3×10^{7}

For more details see R. J. Myers, *J. Chem. Educ.*, Vol. 63, p. 687, 1986.

$$H_2CO_3(aq) \rightleftharpoons CO_2(g) + H_2O(\ell) \qquad K \approx 10^5$$

The CO_2 bubbles out of solution, and the equilibrium is moved even further to the right (see Figure 4.9)

Carbonates are generally soluble in strong acids, and so are many metal sulfides

$$FeS(s) + 2\,H_3O^+(aq) \rightleftharpoons Fe^{2+}(aq) + H_2S(aq) + 2\,H_2O(\ell)$$

and metal hydroxides.

$$Mg(OH)_2(s) + 2\,H_3O^+(aq) \rightleftharpoons Mg^{2+}(aq) + 4\,H_2O(\ell)$$

In general, the solubility of a salt containing the conjugate base of a weak acid is increased by *addition of a stronger acid* to the solution. In contrast, the salts are not soluble in strong acid if the anion is the conjugate base of a strong acid. For example, AgCl is not soluble in strong acid

$$AgCl(s) \rightleftharpoons Ag^+(aq) + Cl^-(aq) \qquad K_{sp} = 1.8 \times 10^{-10}$$

$$H_3O^+(aq) + Cl^-(aq) \rightleftharpoons HCl(aq) + H_2O(\ell) \qquad K < 1$$

because Cl^- is a *very* weak base (Table 17.3) and so is not removed by the strong acid H_3O^+. You can see that this same conclusion applies to insoluble salts of Br^- and I^-.

19.9 SOLUBILITY AND COMPLEX IONS

Complex ions are described in more detail in Chapter 23. The bonding in complex ions can be thought of as the interaction of a Lewis acid (the metal ion) with a Lewis base, the bound water or ammonia molecules. See also Section 17.11.

The metal ions we have been discussing exist in aqueous solution as complex ions. That is, they consist of the metal ion and water molecules, bound into a single entity. The negative end of the polar water molecule, the oxygen atom, is attracted to positive metal ion (Section 17.11). Indeed, any negative ion or Lewis base, such as ammonia, can be attracted to a metal ion.

As you shall see in Chapter 23, complex ions are important in chemistry. They are the basis of such biologically important substances as hemoglobin and vitamin B_{12}. For our present purposes, they are important in that a water-soluble complex ion of a metal ion can often be formed in preference to an insoluble salt. For example, adding sufficient ammonia to a precipitate of AgCl causes the insoluble salt to dissolve because the water-soluble complex ion $Ag(NH_3)_2^+$ is preferentially formed (Figure 19.8)

$$AgCl(s) + 2\,NH_3(aq) \rightleftharpoons Ag(NH_3)_2^+(aq) + Cl^-(aq)$$

We can view dissolving AgCl(s) in this way as a two-step process. First, AgCl dissolves minimally in water giving $Ag^+(aq)$ and $Cl^-(aq)$ ions. Then, the $Ag^+(aq)$ ion combines with NH_3 to give the ammonia complex. Lowering the $Ag^+(aq)$ concentration through complexation with NH_3 shifts the solubility equilibrium to the right, and more solid AgCl dissolves.

$$AgCl(s) \rightleftharpoons Ag^+(aq) + Cl^-(aq) \qquad K_{sp} = 1.8 \times 10^{-10}$$
$$Ag^+(aq) + 2\,NH_3(aq) \rightleftharpoons Ag(NH_3)_2^+(aq) \qquad K = 1.6 \times 10^7$$

This is another example of simultaneous equilibria. This is similar to the case in which an even more insoluble precipitate is formed from a more soluble one

Figure 19.8 Forming and dissolving precipitates. (a) AgCl is precipitated by adding NaCl(aq) to $AgNO_3(aq)$. (b) The precipitate of AgCl dissolves on adding aqueous NH_3 to give water-soluble $Ag(NH_3)_2^+$. (c) The silver-ammonia complex ion is changed to insoluble AgBr on adding NaBr(aq). (d) The precipitate of AgBr is dissolved on adding $Na_2S_2O_3(aq)$. The product is the water-soluble complex ion $Ag(S_2O_3)_2^{3-}$.

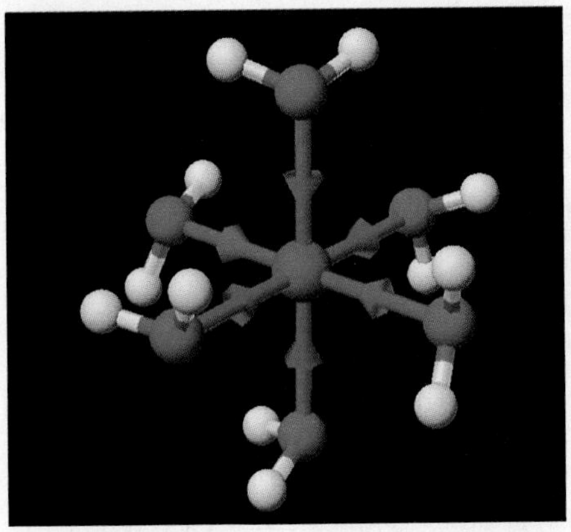

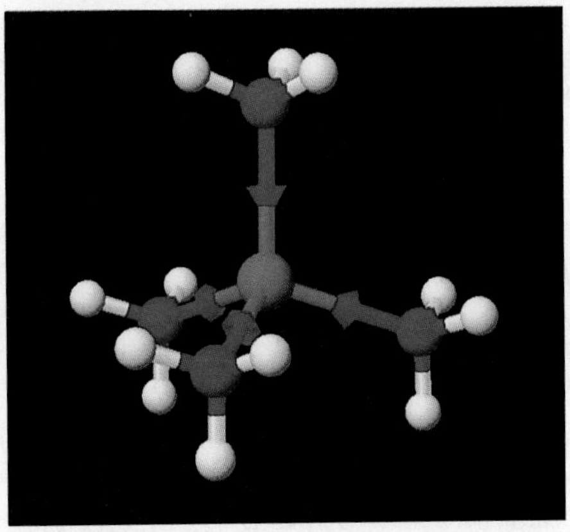

Two complex ions, $Fe(H_2O)_6^{2+}$ (left) and $Cu(NH_3)_4^{2+}$ (right). (Susan Young)

or hydrolysis of an anion of an insoluble salt leads to greater than expected solubility (page 901).

The equilibrium constant for the formation of a complex ion is called a formation constant, $K_{formation}$. For the silver-ammonia complex, the value of the constant is

$$Ag^+(aq) + 2\,NH_3(aq) \rightleftharpoons Ag(NH_3)_2^+(aq) \qquad K_{formation} = 1.6 \times 10^7$$

The large value of the equilibrium constant means that the equilibrium lies well to the right and provides the driving force for dissolving AgCl. If we combine $K_{formation}$ with K_{sp} to give the net equilibrium constant for dissolving AgCl, we obtain

$$AgCl(s) + 2\,NH_3(aq) \rightleftharpoons Ag(NH_3)_2^+(aq) + Cl^-(aq)$$

$$K_{net} = K_{sp} \cdot K_{formation} = (1.8 \times 10^{-10})(1.6 \times 10^7) = 2.9 \times 10^{-3}$$

AgCl is much more soluble in the presence of ammonia than in pure water.

Formation constants have been measured for the formation of many other complex ions, making it possible to compare the stabilities of various complex ions by comparing values of their formation constants. For silver(I) ion, a few other values are

Formation Equilibrium	$K_{formation}$
$Ag^+(aq) + 2\,Cl^-(aq) \rightleftharpoons AgCl_2^-(aq)$	2.5×10^5
$Ag^+(aq) + 2\,S_2O_3^{2-}(aq) \rightleftharpoons Ag(S_2O_3)_2^{3-}(aq)$	2.0×10^{13}
$Ag^+(aq) + 2\,CN^-(aq) \rightleftharpoons Ag(CN)_2^-(aq)$	5.6×10^{18}

Values of $K_{formation}$ for other complex ions are given in Appendix I.

Figure 19.8 shows what happens as complex ions are formed. Beginning with a precipitate of AgCl, adding aqueous ammonia dissolves the precipitate to give the soluble complex ion $Ag(NH_3)_2^+$. Silver bromide is even more stable than $Ag(NH_3)_2^+$, so AgBr forms in preference to the complex ion on adding bromide ion. If thiosulfate ion, $S_2O_3^{2-}$, is then added, however, $K_{formation}$ for $Ag(S_2O_3)_2^{3-}$ shows that it is very stable, and this complex ion forms in preference to AgBr.

Another conclusion that comes from the table of formation constants is that K_{sp} calculations are not very accurate. One reason is that such calculations ignore complex ion formation. For example, in the presence of excess Cl^- ion, the solubility of AgCl is larger than expected because the complex ion $AgCl_2^-$ can be formed.

EXAMPLE 19.12 *Dissolving Precipitates Using Complex Ion Formation*

How many moles of ammonia must be added to dissolve 0.050 mol of AgCl suspended in 1.0 L of water?

Solution As described in the preceding text, the reaction that occurs is

$$AgCl(s) + 2\,NH_3(aq) \rightleftharpoons Ag(NH_3)_2^+(aq) + Cl^-(aq) \quad K_{net} = 2.9 \times 10^{-3}$$

If 0.050 mol/L of AgCl is to be dissolved, 0.050 mol/L of the complex ion $Ag(NH_3)_2^+$ and 0.050 mol/L of Cl^- ion are formed. If we wish to have these concentrations at equilibrium, what must the concentration of ammonia be?

$$K_{net} = 2.9 \times 10^{-3} = \frac{[Ag(NH_3)_2^+][Cl^-]}{[NH_3]^2}$$

Solving for $[NH_3]$, we find it is 0.93 mol/L. Therefore, to dissolve AgCl, enough ammonia must be added to form the complex ion with Ag^+ ion (2×0.050 mol = 0.10 mol), and additional ammonia must then be added to bring the concentration to 0.93 mol/L. Therefore, a total of (0.10 mol + 0.93 mol) or 1.03 mol of NH_3 must be added.

EXERCISE 19.14 *Dissolving Precipitates Using Complex Ion Formation*

Does 100. mL of 4.0 M aqueous ammonia completely dissolve 0.010 mol of AgCl suspended in 1.0 L of pure water? ■

19.10 EQUILIBRIA IN THE ENVIRONMENT: CARBON DIOXIDE AND CARBONATES

Corals are composed mostly of calcium carbonate. (C. D. Winters)

Our biosphere is a complicated mixture of carbon-containing compounds, some being created, some being transformed, and others being decomposed at any moment. The compound that links these processes and their materials or products is carbon dioxide, which is produced in the biological process of respiration and consumed by photosynthetic organisms. Although CO_2 constitutes only about 0.0325% of the atmosphere, a recent estimate is that the earth's atmosphere contains 700 billion tons of carbon in the form of the gas.

Our atmosphere normally contains about 0.0003 atm of CO_2, and when this is in equilibrium with CO_2 dissolved in water, the concentration of aqueous H_2CO_3 is about 10^{-5} M. Although this solubility is small, the oceans are thought to hold roughly 60 times as much CO_2 as the atmosphere. In the oceans, however, the CO_2 is in the form of (1) dissolved gas or carbonic acid; (2) one of the ionization products of H_2CO_3; (3) solid metal carbonates, such as $CaCO_3$ or $MgCO_3$; and (4) organic matter.

$H_2CO_3(aq) + H_2O(\ell) \rightleftharpoons H_3O^+(aq) + HCO_3^-(aq)$ $K_{a1} = 4.2 \times 10^{-7}$

$HCO_3^-(aq) + H_2O(\ell) \rightleftharpoons H_3O^+(aq) + CO_3^{2-}(aq)$ $K_{a2} = 4.8 \times 10^{-11}$

Because H_2CO_3 is a weak acid, solutions of CO_2 in pure water are slightly acidic, and, because K_{a2} is much smaller than K_{a1}, the pH can be estimated from the first equilibrium reaction (Section 17.9). Using $[H_2CO_3] \approx 10^{-5}$ M, the hydronium ion concentration is about 2.2×10^{-6} M, and the pH is therefore about 5.6. This means that the rain that falls even in a nonpolluted environment should be slightly acidic.

The oceans of the earth contain enormous quantities of calcium carbonate produced by various sea creatures. There is, therefore, the additional equilibrium of calcium carbonate to consider.

$$CaCO_3(s) \rightleftharpoons Ca^{2+}(aq) + CO_3^{2-}(aq) \qquad K_{sp} = 3.8 \times 10^{-9}$$

When this is included, sea water saturated with carbon dioxide is found to have a pH of 8.2 ± 0.3. The slightly alkaline character of sea water arises from the fact that the carbonate ion from the mineral calcite is the conjugate base of a weak acid, the bicarbonate ion, and so carbonate hydrolysis produces OH^-.

$$CO_3^{2-}(aq) + H_2O(\ell) \rightleftharpoons HCO_3^-(aq) + OH^-(aq) \qquad K_b = 2.1 \times 10^{-4}$$

The most important aspect of the carbonate equilibrium system is that it buffers the pH of sea water. The addition of acids by undersea volcanic activity or other natural processes is countered primarily by carbonic acid formation (and subsequent loss of carbon dioxide to the atmosphere).

$$HCO_3^-(aq) + H_3O^+(aq) \rightleftharpoons H_2CO_3(aq) + H_2O(\ell)$$

and secondarily by bicarbonate formation.

$$H_3O^+(aq) + CO_3^{2-}(aq) \rightleftharpoons HCO_3^-(aq) + H_2O(\ell)$$

On the other hand, an increase in hydroxide ion concentration is counteracted by the reaction

$$OH^-(aq) + HCO_3^-(aq) \rightleftharpoons H_2O(\ell) + CO_3^{2-}(aq)$$

This reaction of course leads to an increase in the carbonate concentration. If the sea water is above the saturation limit of calcium carbonate, limestone is precipitated (Figures 19.9 and 19.10).

Figure 19.9 The system $CO_2/Ca(OH)_2$ in water. Dry Ice (solid CO_2) is added to saturated $Ca(OH)_2(aq)$. (The bromthymol blue indicator shows the solution is basic.) The cloudiness in the solution indicates $CaCO_3(s)$ is just beginning to precipitate.

$$CO_2(g) + H_2O(\ell) \rightleftharpoons H_2CO_3(aq)$$

$$H_2CO_3(aq) + Ca(OH)_2(aq) \rightleftharpoons CaCO_3(s) + 2\,H_2O(\ell)$$

As the concentration of dissolved CO_2 builds up, more $CaCO_3$ precipitates. When the concentration of dissolved CO_2 is large enough, $CaCO_3$ dissolves according to the equation

$$CaCO_3(s) + CO_2(g) + H_2O(\ell) \rightleftharpoons Ca^{2+}(aq) + 2\,HCO_3^-(aq)$$

It is just this reaction that leads to the dissolution of limestone in CO_2-saturated ground water. Reversal then leads to the reprecipitation of limestone in caves (See Figure 16.2 for a complete set of photographs of this reaction).

Figure 19.10 Redwall limestone formation in the Grand Canyon. (John Kotz)

Carbon dioxide is being generated in ever-increasing amounts, due in part to the increase in the population of the earth, in part to the clearing of forests (and thus to less use of CO_2 in photosynthesis), and in part to increased combustion of fossil fuels. Indeed, there is a fear that this could lead to a global warming trend caused by the "greenhouse effect."

Fortunately, the amount of CO_2 in the atmosphere is not increasing as rapidly as might be expected, largely because the ocean is a great CO_2 sink. As the partial pressure of CO_2 increases, CO_2 solubility increases, and it is estimated that the sea has absorbed roughly half of the increase in CO_2. Although this should lead in turn to an increase in hydronium ion concentrations, it can be controlled through reaction with carbonate ion. Furthermore, as the upper layers of sea water containing dissolved CO_2 are mixed with the lower layers in contact with carbonate-containing sediments, hydronium ion can be removed by a reaction such as

$$H_3O^+(aq) + CaCO_3(s) \rightleftharpoons HCO_3^-(aq) + Ca^{2+}(aq) + H_2O(\ell)$$

The problem is that ocean mixing is relatively slow, requiring times on the order of 1000 years.

19.11 CHEMICAL EQUILIBRIA: AN EPILOGUE

The chemistry of CO_2 and carbonates in the environment, which is similar to their chemistry in the human body, illustrates the vastness of the subject of chemical equilibria and their importance. We are reminded again, however, that many of the chemical reactions that are interesting and important to us all occur under *nonequilibrium conditions:* the biochemical reactions in plants and animals, the production of elements in the stars, and the synthesis of useful materials. Although many processes in living entities can be modeled using the concepts of equilibrium theory, it is the movement toward a state of equilibrium that is the driving force of life.

CHAPTER HIGHLIGHTS

Now that you have finished studying this chapter, you should be able to

- Write the equilibrium constant expression—the solubility product constant, K_{sp}—for any insoluble salt (Section 19.1).
- Calculate K_{sp} values from experimental data (Section 19.2).
- Estimate the solubility of a salt from the value of K_{sp} (Section 19.3).
- Use K_{sp} values to decide the order of precipitation of two or more insoluble salts (Section 19.4).
- Decide if a precipitate forms when the ion concentrations are known (Section 19.4).
- Calculate the ion concentrations that are required to begin the precipitation of an insoluble salt (Section 19.4).
- Calculate the solubility of a salt in the presence of a common ion (Section 19.5).

- Use K_{sp} values to devise a method of separating ions in solution from one another (Section 19.6).
- Calculate the equilibrium constant for the net reaction for a situation in which two or more equilibrium processes are occurring in solution (Section 19.7).
- Understand that hydrolysis increases the solubility of a salt when the anion is the conjugate base of a weak acid (Section 19.8).
- Recognize that the solubility of insoluble salts containing basic anions is affected by the pH of the solution (Section 19.8).
- Understand that the formation of a complex ion can increase the solubility of an insoluble salt (Section 19.9).

STUDY QUESTIONS

Review Questions

1. Explain why $[CaCO_3]$ does not appear in the K_{sp} expression for

$$CaCO_3(s) \rightleftharpoons Ca^{2+}(aq) + CO_3^{2-}(aq)$$

2. What is a "reaction quotient" and how does it differ from an equilibrium constant? Use the following equilibrium in your discussion:

$$AgCl(s) \rightleftharpoons Ag^+(aq) + Cl^-(aq)$$

3. In Appendix H, find two salts that have K_{sp} values less than 1×10^{-40}.
   (a) Write balanced equations to show the equilibria existing when the compounds dissolve in water.
   (b) Write the K_{sp} expression for each compound.
4. Explain the terms "saturated," "not saturated" (or "undersaturated"), and "supersaturated." Use an equilibrium such as

$$CaF_2(s) \rightleftharpoons Ca^{2+}(aq) + 2\ F^-(aq)$$

to illustrate your answer.

5. What is the common ion effect? Use the equilibrium

$$Fe(OH)_2(s) \rightleftharpoons Fe^{2+}(aq) + 2\ OH^-(aq)$$

in your discussion.

6. Explain why the solubility of Ag_3PO_4 can be greater in water than is calculated from the K_{sp} of the salt.
7. Silver chloride is insoluble in water. If aqueous ammonia is added, however, the precipitate of AgCl dissolves. Explain.

Numerical and Other Questions

Solubility Guidelines

The following questions are a review of the solubility guidelines discussed in Section 4.3.

8. Name two insoluble salts of each of the following ions:
   (a) Cl^-
   (b) Zn^{2+}
   (c) Fe^{2+}
9. Name two insoluble salts of the following ions:
   (a) SO_4^{2-}
   (b) Ni^{2+}
   (c) Br^-
10. Using the table of solubility guidelines (Figure 4.7), predict whether each of the following is insoluble or soluble in water:
   (a) $(NH_4)_2S$
   (b) $ZnCO_3$
   (c) FeS
   (d) $BaSO_4$
11. Using the table of solubility guidelines (Figure 4.7), predict whether each of the following is insoluble or soluble in water:
   (a) $Zn(NO_3)_2$
   (b) $Al(OH)_3$
   (c) $PbCl_2$
   (d) CuS
12. Which of the following mixtures of salts should lead to a precipitate when 0.10 M solutions are mixed? Give the formula of the precipitate.
   (a) $NaI(aq) + AgNO_3(aq)$
   (b) $KCl(aq) + Pb(NO_3)_2(aq)$
   (c) $CuCl_2(aq) + Mg(NO_3)_2(aq)$
13. Each of the following pairs of salts is soluble in water. State whether a precipitate can form when you add one salt to the other in aqueous solution, and give the formula of the precipitate.
   (a) $NaCl + AgNO_3$
   (b) $KOH + MgCl_2$
   (c) $KOH + NaCl$
   (d) $KCl + Pb(NO_3)_2$

Writing Solubility Product Constant Expressions

14. For each of the following salts, (i) write a balanced equation showing the equilibrium occurring when the salt is added to water and (ii) write the K_{sp} expression. Give the value for the K_{sp} of each salt.
(a) AgCN
(b) $PbCO_3$
(c) AuI_3

15. For each of the following salts, (i) write a balanced equation showing the equilibrium occurring when the salt is added to water and (ii) write the K_{sp} expression. Give the value for the K_{sp} of each salt.
(a) $PbSO_4$
(b) BaF_2
(c) Ag_2SO_4

Calculating K_{sp}

16. When 1.55 g of solid thallium(I) bromide is added to 1.00 L of water, the salt dissolves to a small extent.

$$TlBr(s) \rightleftharpoons Tl^+(aq) + Br^-(aq)$$

The thallium(I) and bromide ions in equilibrium with TlBr each have a concentration of 1.8×10^{-3} M. What is the value of K_{sp} for TlBr?

17. When 250 mg of CaF_2, calcium fluoride, is added to 1.00 L of water, the salt dissolves to a very small extent.

$$CaF_2(s) \rightleftharpoons Ca^{2+}(aq) + 2\,F^-(aq)$$

At equilibrium, the concentration of Ca^{2+} is found to be 2.1×10^{-4} M. What is the value of K_{sp} for CaF_2?

18. Calcium hydroxide, $Ca(OH)_2$, dissolves in water to the extent of 0.93 g/L. What is the K_{sp} of $Ca(OH)_2$?

$$Ca(OH)_2(s) \rightleftharpoons Ca^{2+}(aq) + 2\,OH^-(aq)$$

19. At 20 °C, a saturated aqueous solution of silver acetate, $Ag(CH_3CO_2)$, contains 1.0 g dissolved in 100.0 mL of solution. Calculate K_{sp} for silver acetate.

$$AgCH_3CO_2(s) \rightleftharpoons Ag^+(aq) + CH_3CO_2^-(aq)$$

20. At 25 °C, 34.9 mg of Ag_2CO_3 dissolves in 1.0 L of pure water.

$$Ag_2CO_3(s) \rightleftharpoons 2\,Ag^+(aq) + CO_3^{2-}(aq)$$

What is the solubility product constant for this salt?

21. You place 2.75 g of barium fluoride in 1.00 L of pure water at 25 °C. After equilibrium has been established,

$$BaF_2(s) \rightleftharpoons Ba^{2+}(aq) + 2\,F^-(aq)$$

the fluoride ion concentration is 0.0150 M. What is the K_{sp} of BaF_2?

22. Solid $Ca(OH)_2$ (1.234 g) is placed in 1.00 L of pure water at 25 °C. The pH of the solution is found to be 12.40. Estimate the K_{sp} for $Ca(OH)_2$.

23. You add 0.979 g of $Pb(OH)_2$ to 1.00 L of pure water at 25 °C. The pH is 8.92. Estimate the K_{sp} for $Pb(OH)_2$.

Estimating Salt Solubility from K_{sp}

The K_{sp} values required by the problems that follow are found in Table 19.1 or in Appendix H.

24. Estimate the solubility of silver cyanide in (a) moles per liter and (b) grams per liter in pure water at 25 °C.

$$AgCN(s) \rightleftharpoons Ag^+(aq) + CN^-(aq)$$

25. What is the molar concentration of $Au^+(aq)$ in a saturated solution of AuCl in pure water at 25 °C?

$$AuCl(s) \rightleftharpoons Au^+(aq) + Cl^-(aq)$$

26. The K_{sp} of radium sulfate, $RaSO_4$, is 4.2×10^{-11}. If 25 mg of radium sulfate is placed in 100. mL of water, how many milligrams of the salt dissolve?

27. If 50. mg of lead(II) sulfate is placed in 250. mL of pure water, what percentage of the solid dissolves?

28. Estimate the solubility of magnesium fluoride, MgF_2, in (a) moles per liter and (b) grams per liter of pure water.

$$MgF_2(s) \rightleftharpoons Mg^{2+}(aq) + 2\,F^-(aq)$$

29. Estimate the solubility of lead(II) bromide in (a) moles per liter and (b) grams per liter of pure water.

30. Use K_{sp} values to decide which compound in each of the following pairs is the more soluble:
(a) AgBr or AgSCN
(b) $SrCO_3$ or $SrSO_4$
(c) MgF_2 or CaF_2
(d) AgI or HgI_2

31. Use K_{sp} values to decide which compound in each of the following pairs is the more soluble:
(a) $PbCl_2$ or $PbBr_2$
(b) BiI_3 or $Bi(OH)_3$
(c) HgS or FeS
(d) $Fe(OH)_2$ or $Zn(OH)_2$

32. Rank the following compounds in order of increasing solubility in water: Na_2CO_3, BaF_2, $BaCO_3$, and Ag_2CO_3.

33. Rank the following compounds in order of increasing solubility in water: AgI, HgS, PbI_2, $PbSO_4$, NH_4NO_3.

34. If you place 5.0 mg of $NiCO_3$ in 1.0 L of pure water does all the salt dissolve before equilibrium can be established, or does some salt remain undissolved?

35. If you place the following amounts in pure water, does all the salt dissolve before equilibrium can be established, or does some salt remain undissolved?
(a) 5.0 mg of MgF_2 in 125 mL of pure water
(b) 0.50 g of CaF_2 in 100. mL of pure water

Precipitations

36. Sodium carbonate is added to a solution in which the concentration of Ni^{2+} ion is 0.0024 M. (a) Does precipitation of $NiCO_3$ occur when the concentration of the carbonate ion is 1.0×10^{-6} M or (b) when it is 100 times greater (or 1.0×10^{-4} M)? The equilibrium process involved is

$$NiCO_3(s) \rightleftharpoons Ni^{2+}(aq) + CO_3^{2-}(aq)$$

37. You have a solution that has a lead(II) concentration of 0.0012 M. If enough soluble chloride-containing salt is added so that the Cl^- concentration is 0.010 M, does $PbCl_2$ precipitate? The equilibrium process involved is

$$PbCl_2(s) \rightleftharpoons Pb^{2+}(aq) + 2\,Cl^-(aq)$$

38. If the concentration of Zn^{2+} in 10. mL of water is 1.6×10^{-4} M, does zinc hydroxide, $Zn(OH)_2$, precipitate when 4.0 mg of NaOH is added?

39. You have 100. mL of a solution that has a lead(II) concentration of 0.0012 M. Does $PbCl_2$ precipitate when 1.20 g of solid NaCl is added?

40. If the concentration of Mg^{2+} in sea water is 1350 mg/L, what OH^- concentration is required to precipitate $Mg(OH)_2$?

41. A sample of hard water contains about 2.0×10^{-3} M Ca^{2+}. A soluble fluoride-containing salt such as NaF is added to "fluoridate" the water (to aid in the prevention of dental caries). What is the maximum concentration of F^- that can be present *without* precipitating CaF_2?

42. If you mix 50. mL of 0.0012 M $BaCl_2$ with 25 mL of 1.0×10^{-6} M H_2SO_4 does a precipitate of $BaSO_4$ form? The equilibrium process involved is

$$BaSO_4(s) \rightleftharpoons Ba^{2+}(aq) + SO_4^{2-}(aq)$$

43. Does a precipitate of $Mg(OH)_2$ form when 25.0 mL of 0.010 M NaOH is combined with 75.0 mL of a 0.10 M solution of magnesium chloride?

44. If you mix 10. mL of 0.0010 M $Pb(NO_3)_2$ with 5.0 mL of 0.015 M HCl, does $PbCl_2$ precipitate?

45. If 100. mL of 0.10 M H_2SO_4 is added to 50.0 mL of 0.00013 M $BaCl_2$, does $BaSO_4$ precipitate?

46. The cations Ba^{2+}, Sr^{2+}, and Pb^{2+} can all be precipitated as very insoluble sulfates. If you add a soluble sulfate-containing salt (say Na_2SO_4) to a solution containing all of these metal ions, each with a concentration of 0.10 M, in what order are the sulfates precipitated?

47. Hydrogen iodide, HI, is added slowly to a solution 0.10 M in each of the following ions: Pb^{2+}, Ag^+, and Hg_2^{2+}. The insoluble salts PbI_2, AgI, and Hg_2I_2 eventually form. In what order do these salts precipitate?

48. You often work with salts of Fe^{3+}, Pb^{2+}, and Al^{3+} in the laboratory. All are found in nature, and all are important economically. If you have a solution containing these three ions, each at a concentration of 0.1 M, what is the order in which their hydroxides precipitate as aqueous NaOH is slowly added?

49. Alkaline earth metal ions can be precipitated as insoluble carbonates. If you have a solution of Mg^{2+}, Ca^{2+}, Sr^{2+}, and Ba^{2+} ions, all with the same concentration, what is the order in which their carbonates are precipitated as sodium carbonate is added slowly?

Common Ion Effect

50. Calculate the molar solubility of silver thiocyanate, AgSCN, in pure water and in water containing 0.010 M NaSCN.

51. Calculate the solubility of silver carbonate, Ag_2CO_3, in moles per liter, in pure water. Compare this with the molar solubility of Ag_2CO_3 in 225 mL of water to which 0.15 g of Na_2CO_3 has been added.

52. Calculate the solubility, in milligrams per milliliter, of silver phosphate, Ag_3PO_4, in (a) pure water and (b) in water that is 0.020 M in $AgNO_3$.

53. What is the solubility, in grams per milliliter, of BaF_2 (a) in pure water and (b) in water containing 5.0 mg/mL of KF?

Separations

54. To separate Ca^{2+} and Mg^{2+} ions from one another, ammonium oxalate, $(NH_4)_2C_2O_4$, is added to a solution that is 0.020 M in both metal ions. If the concentration of the oxalate ion is adjusted properly, the metal oxalates can be precipitated separately. In this case CaC_2O_4 precipitates before MgC_2O_4. The chemical equilibria involved are:

$$MgC_2O_4(s) \rightleftharpoons Mg^{2+}(aq) + C_2O_4^{2-}(aq)$$
$$CaC_2O_4(s) \rightleftharpoons Ca^{2+}(aq) + C_2O_4^{2-}(aq)$$

(a) What concentration of oxalate ion, $C_2O_4^{2-}$, precipitates the maximum amount of Ca^{2+} ion without precipitating Mg^{2+}?

(b) What concentration of Ca^{2+} remains in solution when Mg^{2+} just begins to precipitate?

55. In principle, the ions Ba^{2+} and Ca^{2+} can be separated by the difference in the solubility of their fluorides, BaF_2 and CaF_2. If you have a solution that is 0.10 M in both Ba^{2+} and Ca^{2+}, CaF_2 begins to precipitate first as fluoride ion is slowly added to the solution.

(a) What concentration of fluoride ion will precipitate the maximum amount of Ca^{2+} ion without precipitating BaF_2?

(b) What concentration of Ca^{2+} remains in solution when BaF_2 just begins to precipitate?

56. A solution contains 0.10 M iodide ion, I^-, and 0.10 M carbonate ion, CO_3^{2-}.

(a) If solid $Pb(NO_3)_2$ is slowly added to the solution, which salt precipitates first, PbI_2 or $PbCO_3$?

(b) What is the concentration of the first ion that precipitates (CO_3^{2-} or I^-) when the second, or more soluble, salt begins to precipitate?

57. A solution contains Ca^{2+} and Pb^{2+} ions, both at a concentration of 0.010 M. You wish to separate the two ions from each other as completely as possible by precipitating one but not the other using aqueous Na_2SO_4 as the precipitating agent.

(a) Which precipitates first as sodium sulfate is added, $CaSO_4$ or $PbSO_4$?

(b) What is the concentration of the first ion that precipitates (Ca^{2+} or Pb^{2+}) when the second, or more soluble, salt begins to precipitate?

58. Each of the following pairs of ions is found together in aqueous solution. Using the table of solubility product constants in Appendix H, devise a way to separate these ions by precip-

itating one of them as an insoluble salt and leaving the other in solution.
(a) Ba^{2+} and Na^+
(b) Bi^{3+} and Cd^{2+}

59. Each of the following pairs of ions is found together in aqueous solution. Using the table of solubility product constants in Appendix H, devise a way to separate these ions by adding one reagent to precipitate one of them as an insoluble salt and leave the other in solution.
(a) Cu^{2+} and Ag^+
(b) Al^{3+} and Fe^{3+}

Simultaneous Equilibria and Complex Ions

60. Solid silver bromide, AgBr, can be dissolved by adding concentrated aqueous ammonia to give the water-soluble silver-ammonia complex ion.

$$AgBr(s) + 2\ NH_3(aq) \rightleftharpoons Ag(NH_3)_2^+(aq) + Br^-(aq)$$

Show that this equation is the sum of two other equations, one for dissolving AgBr to give its ions and the other for the formation of the $Ag(NH_3)_2^+$ ion from Ag^+ and NH_3. Calculate K_{net} for the overall reaction.

61. Solid gold(I) chloride, AuCl, is dissolved when excess cyanide ion, CN^-, is added to give a water-soluble complex ion.

$$AuCl(s) + 2\ CN^-(aq) \rightleftharpoons Au(CN)_2^-(aq) + Cl^-(aq)$$

Show that this equation is the sum of two other equations, one for dissolving AuCl to gives its ions and the other for the formation of the $Au(CN)_2^-$ ion from Au^+ and CN^-. Calculate K_{net} for the overall reaction. (See Appendices H and I for the required equilibrium constants.)

62. What is the equilibrium constant for the following reaction?

$$AgCl(s) + I^-(aq) \rightleftharpoons AgI(s) + Cl^-(aq)$$

Does the equilibrium lie predominantly to the left or right? Does AgI form if iodide ion, I^-, is added to a saturated solution of AgCl?

63. Calculate the equilibrium constant for the following reaction:

$$Zn(OH)_2(s) + 2\ CN^-(aq) \rightleftharpoons Zn(CN)_2(s) + 2\ OH^-(aq)$$

Does the equilibrium lie predominantly to the left or right? Can zinc hydroxide be transformed into zinc cyanide by adding a soluble salt of the cyanide ion? (See Appendices H and I for the required equilibrium constants.)

64. Does 5.0 mL of 2.5 M NH_3 dissolve 1.0×10^{-4} mol of AgBr suspended in 1.0 L of pure water?

65. Can you completely dissolve 15.0 mg of AuCl in 100.0 mL of water if you add 15.0 mL of 6.00 M NaCN? (See Study Question 61.)

Solubility and pH

66. Which of the following barium salts should be soluble in a strong acid such as HCl: $Ba(OH)_2$, $BaSO_4$, or $BaCO_3$?

67. Which of the following silver salts should be soluble in a strong acid: AgI, Ag_2CO_3, or Ag_3PO_4?

68. Which of the following bismuth salts should be soluble in a strong acid: BiI_3, $Bi_2(CO_3)_3$, or $BiPO_4$?

69. Suggest a method for separating a precipitate consisting of a mixture of CuS and $Cu(OH)_2$.

General Questions

70. Limestone, $CaCO_3$, can exist in two mineral forms: calcite and aragonite. In pure water, calcite has a K_{sp} of 3.8×10^{-9}, whereas the K_{sp} of aragonite is 6.0×10^{-9}. Which is the more soluble in pure water?

71. If you place 1.00×10^2 mg of $BaSO_4$ in 1.0 L of water, approximately how many milligrams of $BaSO_4$ dissolve at 25 °C? (See Figure 19.3.)

72. To make up an unknown solution for the students in your laboratory, you dissolve 1.0×10^{-5} mol of $AgNO_3$ in 1.0 L of water. Unfortunately, you used a liter of tap water instead of distilled water. If the chloride ion concentration in tap water is 2.0×10^{-4} M, is your error revealed immediately by the formation of a white precipitate of AgCl?

73. The alkaline earth metal ions can be precipitated from aqueous solution by addition of fluoride ion.
(a) If you have a 0.015 M solution of Ba^{2+}, what is the minimum concentration of F^- ion necessary to begin precipitation of BaF_2?
(b) If you started with 100.0 mL of the 0.015 M Ba^{2+}-containing solution, how many milligrams of NaF are required to just begin precipitation?
(c) After adding 0.50 g of NaF to 100.0 mL of 0.015 M Ba^{2+}-containing solution, what is the concentration of Ba^{2+} remaining in solution?

74. The alkaline earth cations Mg^{2+}, Ca^{2+}, and Ba^{2+} can be precipitated as their fluoride carbonate or sulfate salts. If you have a solution of these three ions, each with a concentration of 0.001 M, in what order do these ions precipitate as their insoluble sulfates as you slowly add sulfuric acid to the solution?

75. The solubilities of lead and zinc compounds are often similar. For example, the K_{sp} of $Pb(OH)_2$ is 2.8×10^{-16} and that of $Zn(OH)_2$ is 4.5×10^{-17}. If Pb^{2+} and Zn^{2+} are present in a solution, each with a concentration of 1.0×10^{-6} M, and if the solution has a pH of 8.0, does either $Pb(OH)_2$ or $Zn(OH)_2$ precipitate?

76. Suppose you mix 15.0 mL of 0.010 M $CaCl_2$ and 25.0 mL of 0.0010 M NaOH. Does $Ca(OH)_2$ precipitate? If so, how many grams of $Ca(OH)_2$ form?

77. Predict the order in which the following salts will be precipitated from solution as I^- ion is slowly added (in the form of NaI) to a solution containing Ag^+, Bi^{3+}, and Pb^{2+}. The metal ions are initially 0.10 M.

78. Zinc hydroxide, $Zn(OH)_2$, is a relatively insoluble base. A saturated solution has a pH of 8.65. Calculate the K_{sp} for $Zn(OH)_2$.

79. Although it is a violent poison if swallowed, mercury(II) cya-

nide, $Hg(CN)_2$, has been used as a topical (skin) antiseptic.
(a) What is the molar solubility of this salt in pure water?
(b) How many milligrams of $Hg(CN)_2$ dissolve per liter of pure water?
(c) How many milliliters of water are required to dissolve 1.0 g of the salt?

80. A saturated solution of $Mg(OH)_2$ has a pH of 10.49. Use this information to calculate the K_{sp} of $Mg(OH)_2$.

81. If 0.581 g of solid $Mg(OH)_2$ is added to 1.00 L of pure water, what quantity dissolves? If the solution is buffered at a pH of 5.00, does more of the hydroxide dissolve? If so, how much?

82. Suppose you mix 2.00 g of $AgNO_3$ and 3.00 g of K_2CrO_4 in enough water to make 50.0 mL of solution. What is the concentration of the ions—Ag^+, NO_3^-, K^+, and CrO_4^{2-}—in the final solution? The K_{sp} for Ag_2CrO_4 is 9.0×10^{-12}.

83. The Ca^{2+} ion in hard water is often precipitated as $CaCO_3$ by adding soda ash, Na_2CO_3. If the calcium ion concentration in hard water is 0.010 M, and if the Na_2CO_3 is added until the carbonate ion concentration is 0.050 M, what percentage of the calcium ion has been removed from the water? (You may ignore hydrolysis of the carbonate ion.)

84. Photographic film is coated with crystals of AgBr suspended in gelatin. Light exposure leads to the reduction of some of the silver ions to metallic silver. Unexposed AgBr is dissolved with sodium thiosulfate in the fixing step.

$$AgBr(s) + 2\ S_2O_3^{2-}(aq) \rightleftharpoons Ag(S_2O_3)_2^{3-}(aq) + Br^-(aq)$$

(a) Using the appropriate K_{sp} and $K_{formation}$ values in Appendices H and I, calculate the equilibrium constant for the dissolving process.
(b) If you want to dissolve 1.0 g of AgBr in 1.0 L of solution, how many grams of $Na_2S_2O_3$ must be added?

85. Does 5.0 mL of 2.5 M NH_3 dissolve 25 mg of AgI suspended in 1.0 L of pure water?

86. A test tube contains insoluble ZnS and 0.10 M H_2S. A strong acid such as HCl(aq) is added. What is the concentration of the Zn^{2+} ion when the pH of the solution is adjusted to 1.50? (See *A Closer Look: Equilibria Involving Sulfide Ions* for the equilibrium constant expression for ZnS dissolving in acid.)

87. A test tube contains 5.0 mg of FeS in 10. mL of water. Assume the solution also contains H_2S with a concentration of 0.10 M. What should the pH be if all of the FeS is to dissolve?

$$FeS(s) + 2\ H_3O^+(aq) \rightleftharpoons Fe^{2+}(aq) + H_2S(aq) + 2\ H_2O(\ell) \qquad K_{spa} = 6 \times 10^2$$

(See *A Closer Look: Equilibria Involving Sulfide Ions;* page 903)

Conceptual Questions

88. Explain how changing pH affects the solubility of $Ni(OH)_2$.

89. Explain how changing pH affects the solubility of Ag_3PO_4.

90. Suppose some cyanide ion, CN^-, is added to AgCN in equilibrium with its ions. Two reactions are possible:

$$AgCN(s) \rightleftharpoons Ag^+(aq) + CN^-(aq)$$
$$AgCN(s) + CN^-(aq) \rightleftharpoons Ag(CN)_2^-(aq)$$

Do the concentrations of Ag^+ and CN^- increase, decrease, or not change?

91. Explain why $PbCO_3$ can dissolve in strong acid but $PbCl_2$ cannot.

92. Barium carbonate dissolves to some extent in the presence of carbon dioxide.

$$BaCO_3(s) + CO_2(g) + H_2O(\ell) \rightleftharpoons Ba^{2+}(aq) + 2\ HCO_3^-(aq)$$
$$K = 4.5 \times 10^{-5}$$

(a) How is the solubility of barium carbonate affected by the pressure of CO_2?
(b) How is the solubility of barium carbonate affected by a decrease in the pH?

93. Two common constituents of kidney stones are calcium phosphate, $Ca_3(PO_4)_2$, and hydrated calcium oxalate, $Ca(C_2O_4)_2 \cdot H_2O$. Are kidney stones containing these salts more likely to form when the urine is acidic or when it is basic?

Summary Question

94. Aluminum hydroxide reacts with phosphoric acid to give aluminum phosphate, $AlPO_4$. The solid exists in many of the same crystal forms as SiO_2 and is used industrially as the basis of adhesives, binders, and cements.
(a) Write a balanced equation for the reaction of aluminum chloride and phosphoric acid.
(b) If you begin with 152 g of aluminum chloride, and 3.0 L of 0.750 M phosphoric acid, how many grams of $AlPO_4$ can be isolated?
(c) If you place 25.0 g of $AlPO_4$ in enough pure water to have a volume of exactly one liter, what are the concentrations of Al^{3+} and PO_4^{3-} at equilibrium?
(d) Does the solubility of $AlPO_4$ increase or decrease on adding HCl? Explain briefly.
(e) If you mix 1.50 L of 0.0025 M Al^{3+} (in the form of $AlCl_3$) with 2.50 L of 0.035 M Na_3PO_4, does a precipitate of $AlPO_4$ form? If so, how many grams of $AlPO_4$ precipitate?

C H A P T E R 20

Principles of Reactivity: Entropy and Free Energy

A Chemical Puzzler

Some crystals of potassium permanganate are placed in pure water. Over time, and without stirring, the solute disperses through the water. Why does this mixing occur? What does this have to do with chemical reactions? (Charles D. Winters)

Some chemical and physical changes take place by themselves, given enough time. If you stretch a rubber band, it snaps back spontaneously and quickly. If you put a spoonful of sugar in your coffee or tea, it dissolves, and the molecules distribute themselves evenly throughout the liquid. When water is added to sodium, the metal reacts, usually very rapidly (Figure 20.1). These changes are all said to be spontaneous, or *product-favored,* although they do proceed at different speeds.

To control chemical reactions in order to produce new pharmaceuticals or new polymers or to prevent unwanted reactions in our environment, we must understand why some chemical reactions are product-favored and others are not. We must understand how to predict when a reaction will be product-favored and how to control that tendency if necessary. The objective of this chapter is to describe a method of predicting when reactions will be product- or reactant-favored.

Figure 20.1 A product-favored process, the reaction of sodium metal with water to give sodium hydroxide and hydrogen

$$2\,Na(s) + 2\,H_2O(\ell) \longrightarrow 2\,NaOH(aq) + H_2(g)$$

20.1 SPONTANEOUS REACTIONS AND SPEED: THERMODYNAMICS VERSUS KINETICS

As described in Chapter 6, a *product-favored* chemical reaction is one in which most of the reactants can eventually be converted to products, given sufficient time. The reactions of most alkali and alkaline earth metals are examples of product-favored reactions, and the reaction of sodium with water is no exception (see Figure 20.1; see also Figure 2.17).

$$\xrightarrow{\text{product favored}}$$
$$2\,Na(s) + 2\,H_2O(\ell) \longrightarrow 2\,NaOH(aq) + H_2(g)$$

The contrasting term, *reactant-favored* reaction, may be misleading. One of the most widely discussed reactions in the past few years is the decomposition of ozone, O_3, to O atoms and O_2 molecules, a reaction that can occur by the action of photons of sunlight in the 280- to 310-nm range of wavelengths.

$$\xleftarrow{\text{reactant favored}}$$
$$O_3(g) \longrightarrow O(g) + O_2(g)$$

This reaction is reactant-favored. This does not mean it does *not* occur at all; rather, it means that when equilibrium is achieved, not many molecules of O_3 have broken down into products. The equilibrium constant is *much* less than 1.

$$K = \frac{[O][O_2]}{[O_3]} = 0.0063 \text{ at } 25\ ^\circ C$$

The oxidation of H_2 by O_2 to form water is another product-favored reaction, but it occurs *only* if you ignite the mixture (Figure 20.2). The mere presence of oxygen is not enough; H_2 gas can stay in contact with air a long time if you are careful not to set off the reaction. This means there is a difference between the speed of a reaction and whether it is product-favored. **Thermody-**

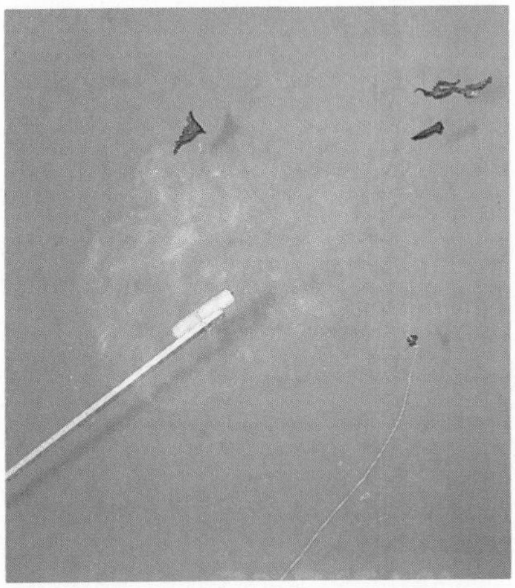

Figure 20.2 A lighted candle was held to a balloon containing H_2 gas. When the balloon burst, the H_2 gas mixed with O_2 in the air, and the mixture burned when ignited by the candle. The product-favored reaction that occurred was

$$2\,H_2(g) + O_2(g) \longrightarrow 2\,H_2O(g)$$

A CLOSER LOOK *A Review of Concepts of Thermodynamics*

System: The part of the universe (the specific atoms, molecules, or ions) under study

Surroundings: The rest of the universe exclusive of the system

Exothermic: Thermal energy transfer proceeds from the system to the surroundings.

Endothermic: Thermal energy transfer proceeds from the surroundings to the system.

First law of thermodynamics: The law of the conservation of energy, $\Delta E = q + w$. The change in the energy of a system is equal to the heat transferred to the system and the work done on the system.

Enthalpy change: The thermal energy transferred at constant pressure

State function: A quantity whose changed value is determined only by the initial and final values

Standard conditions: $P = 1$ bar and T is usually 298 K.* Material in solution has a concentration of 1 molal. (Note that in a dilute solution $m \approx$ M.)

Enthalpy of formation: The enthalpy change occurring when a compound is formed from its elements, ΔH_f°. Elements in their standard states have $\Delta H_f^\circ = 0$.

*Note that standard pressure for thermodynamic quantities is 1 bar, where 1 bar = 0.98692 atm. See Chapters 6 and 12.

namics, the subject of this chapter, is the science of energy transfer, and it helps us predict whether a reaction can occur *given enough time.* Thermodynamics can tell us nothing about the *speed* of the reaction, however. The study of the rates of reactions, and why some are fast and others are slow, is called **kinetics;** it was the topic of Chapter 15.

Beginning with Chapter 6, we have often used ideas of energy differences and energy transfer. In Chapter 6 we described the energy involved in chemical reactions, in Chapters 7 through 11 the concern was with energy on the atomic and molecular level, and in Chapters 13 and 14 we discussed the energy involved in changes in physical state. Now, we want to define more completely the energy changes that cause some chemical and physical changes to be product-favored.

We end this section with a note on nomenclature. Chemists often use the term *spontaneous* to refer to a product-favored reaction. A *nonspontaneous* process is reactant-favored. However, the word "spontaneous" often brings with it the idea that something not only happens but happens quickly, thus mixing up the concepts of thermodynamics and kinetics. For this reason the terms product- and reactant-favored are generally used in this book to separate clearly the concept of reaction probability from that of reaction speed.

20.2 DIRECTIONALITY OF REACTIONS: ENTROPY

On page 289 we stated that at room temperature most exothermic reactions are product-favored and therefore are useful for carrying out chemical transformations. The explanation for this is very similar to our explanation of the one-way transfer of energy from a hotter to a colder object (page 263). When an exothermic reaction takes place, chemical potential energy that had been stored in relatively few atoms and molecules (the reactants) spreads over many more atoms and molecules (the products, and especially the surroundings). Because a great many more atoms and molecules are in the surroundings than in the reactants and products, it is always true that after an exothermic reaction, energy is distributed more randomly—dispersed over a much larger number of atoms and molecules—than it was before.

But why is energy dispersal favored? The answer lies in probability. It is much more probable that energy will be dispersed than that it will be concentrated. When energy is added to an atom or molecule, it must be added in small increments or units. These units of energy are easily transferred from one atom or molecule to another. To explain energy dispersal and probability, consider a very small sample of matter consisting of two atoms, A and B, and suppose that this sample contains two units of energy. The energy can be distributed in three ways over the two atoms: (1) atom A could have both units of energy; (2) atom A and atom B could each have one; or (3) atom B could have both. Label these three situations as A^2, AB, and B^2.

Now suppose that atoms A and B come into contact with two other atoms, C and D, which initially have no energy. We want to distribute the two units of energy from A and B over all four atoms. Now there are 10 equally probable arrangements: A^2, AB, AC, AD, B^2, BC, BD, C^2, CD, and D^2. Only three of these—A^2, AB, and B^2—have all the energy concentrated in atoms A and B, which was the original situation.

The situation just described corresponds to sample A + B being initially at a higher temperature than sample C + D (because the atoms A and B have more kinetic energy). (Recall that for a given substance thermal energy depends on temperature and the amount of substance.) When all four atoms are in contact, there are only 3 chances out of 10 that all the energy will remain in atoms A and B; that is, that the temperature of A + B will remain high and the temperature of C + D will remain low. There are 7 chances out of 10 (AC, AD, BC, BD, C^2, CD, and D^2) that at least some energy will be dispersed from sample A + B to sample C + D. When this dispersal of energy occurs, we say that the samples A + B and C + D have gone from a situation where the energy was more highly ordered to one where it is more disordered.

The probability that energy will become disordered becomes overwhelming when large numbers of atoms or molecules are involved. For example, suppose that atoms A and B had been brought into contact with a mole of other atoms. There would still be only 3 arrangements in which all the energy is associated with atoms A and B, but there would be many, many more arrangements (more than 10^{47}) in which all the energy had been transferred to other atoms. In such a case it is almost certain that energy will be transferred. If energy can be dispersed over a much larger number of atoms or molecules, it will be.

A situation in which energy is concentrated in only a few atoms or molecules is unlikely. In fact, this is something we value highly because a substance in which a large quantity of chemical potential energy is concentrated among relatively few atoms or molecules is an energy resource. Examples are coal, oil, wood, and natural gas.

Just as there is a tendency for highly concentrated energy to disperse, highly concentrated matter also tends to disperse. For example, suppose a gas is confined within a single flask connected through a stopcock to a second flask of equal size from which all gas molecules have been removed (Figure 20.3). What happens if the stopcock is opened? Experience teaches us that the gas in the original flask will expand to fill the entire volume available to it.

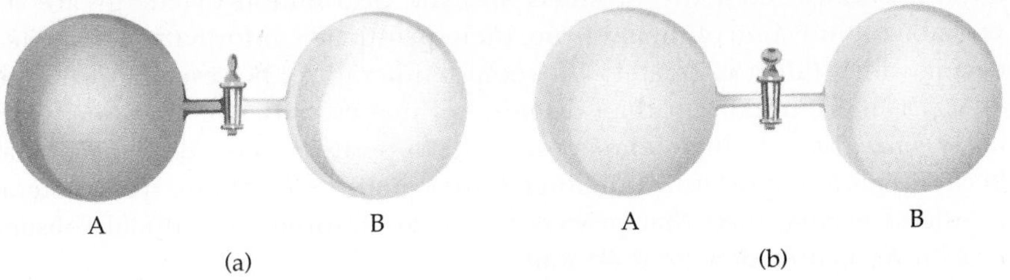

Figure 20.3 The expansion of a gas is a highly probable process. (a) A gas is confined in one flask, A. There are no atoms or molecules of gas in B. (b) When the valve between the flasks is opened, the gas atoms or molecules rush into flask B and eventually distribute themselves evenly between A and B.

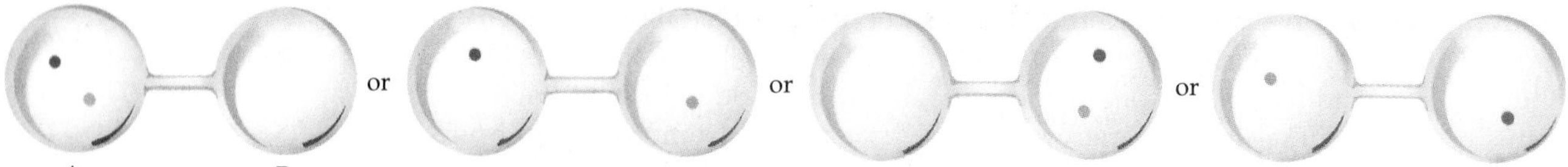

Figure 20.4 When two molecules are placed in two flasks of equal volume, four different arrangements are possible.

Dispersal of the gas can be analyzed in the same way as dispersal of energy. Suppose a single gas molecule occupies the two-flask system with the stopcock open. The molecule moves around within one flask until it hits the opening connecting it to the other flask; then it occupies the second flask for a while before returning to the first. Because the volumes are equal and the molecule's motion is random, it spends half its time in each flask, on average. The probability is $\frac{1}{2}$ that it can be found in flask A and $\frac{1}{2}$ that it is in flask B. Now consider two molecules within the same system. There are four equally probable arrangements, as shown in Figure 20.4, but only one of them has both molecules in flask A. The probability that both molecules are in flask A is thus $\frac{1}{4}$ or $(\frac{1}{2})^2$. By making a similar diagram you can verify that for three molecules there are eight arrangements; only one of these corresponds to all three molecules in flask A, giving a probability of $\frac{1}{8}$, or $(\frac{1}{2})^3$. In general (with the stopcock open), the probability that all molecules will be in the same flask is $(\frac{1}{2})^n$, where n is the number of molecules.

As the number of molecules increases, the probability of them all being in the same flask, a more highly ordered state, becomes very, very small. For example, suppose that 1.00 mol of gas is in flask A. The number of molecules in 1.00 mol is 6.02×10^{23}, and so the probability is $(\frac{1}{2})^{6.02\times10^{23}}$. This is an incredibly small number! It is so highly improbable that all molecules stay in flask A that it is absolutely certain that many molecules will move from flask A to flask B.

The two-flask system we have just described is a simple example of a product-favored system—the final arrangement with gas in both flasks is *much* more probable than the initial one, and so the process occurs of its own accord. On the other hand, if we want to reverse the process by concentrating all the molecules into flask A, an outside influence such as a pump is required—the pump could do work on the gas to force it into a highly improbable arrangement. The work done by the pump is stored in the gas and can be used later for some other purpose.

To summarize, there are two ways that the final state of a system can be more probable than the initial one: (1) having energy dispersed over a greater number of atoms and molecules and (2) having the atoms and molecules themselves more disordered. If both of these happen, then a reaction is definitely product-favored because both the products and the distribution of energy are more probable. If only one of them is true, then quantitative information is needed to decide which effect is greater. (At room temperature, however, disordering of energy is more important than dispersal of matter, and *most exothermic reactions are product-favored.* At high temperatures the opposite is true—dispersal of matter becomes more important.) If neither matter nor energy is more spread out after a process occurs, then that process is reactant-favored—the initial substances remain no matter how long we wait.

Entropy: A Measure of Matter Dispersal or Disorder

The dispersal or disorder in a sample of matter can be measured with a calorimeter (Figure 6.17), the same instrument needed to measure the enthalpy change when a reaction occurs. The result is a thermodynamic function called **entropy** and symbolized by *S*. Measurement of entropy depends on the assumption that in a perfect crystal at the absolute zero of temperature (0 K, or −273.15 °C) all translational motion ceases and there is no disorder; this sets the zero of the entropy scale.

The third law of thermodynamics: If the entropy of each element in some crystalline state is taken as zero at 0 K, every substance has a finite entropy.

When energy is transferred to matter in very small increments, so that the temperature change is very small, the entropy change can be calculated as $\Delta S = q/T$, the heat absorbed divided by the absolute temperature at which the change occurs. By starting as close as possible to absolute zero and repeatedly introducing small quantities of energy, an entropy change can be determined for each small increase of temperature. These entropy changes can then be added to give the total (or absolute) entropy of a substance at any desired temperature.

Though it is impossible to cool anything all the way to absolute zero, it is possible to get very close, and there are ways of estimating the disorder that is already in a substance near 0 K; thus, accurate entropy values can be obtained for many substances.

The results of such measurements for several substances at 298 K are given in Table 20.1. These are standard molar entropy values, and so they apply to one mole of each substance at the standard pressure of 1 bar (= 0.98692 atm) and are expressed in units of joules per kelvin per mole (J/K · mol).

TABLE 20.1 Some Standard Molar Entropy Values at 298 K

Compound or Element	Entropy, $S°$ (J/K · mol)
C(graphite)	5.7
C(g)	158.1
CH_4(g)	186.3
C_2H_6(g)	229.6
C_3H_8(g)	269.9
CH_3OH(ℓ)	126.8
CO_2(g)	213.7
Ca(s)	41.4
Ar(g)	154.7
H_2(g)	130.7
O_2(g)	205.1
N_2(g)	191.6
H_2O(g)	188.8
H_2O(ℓ)	69.9
HCl(g)	186.9
F_2(g)	202.8
Cl_2(g)	223.1
Br_2(ℓ)	152.2
I_2(s)	116.1
NaF(s)	51.5
MgO(s)	26.9
$CaCO_3$(s)	92.9

From *The NBS Tables of Chemical Thermodynamic Properties*, American Institute of Physics, New York, 1982.

PORTRAIT OF A SCIENTIST

Ludwig Boltzmann (1844–1906)

Ludwig Boltzmann was an Austrian mathematician and physicist who gave us a useful interpretation of entropy (and who also did much of the work on the kinetic theory of gases). Engraved on his tombstone in Vienna is his equation relating entropy and "chaos," $S = k \log W$ (where k is a fundamental constant of nature, now called Boltzmann's constant). Boltzmann said the symbol W was related to the number of ways that atoms or molecules can be arranged in a given state, always keeping their total energy fixed. His equation tells us, therefore, that if there are only a few ways to arrange the atoms of a substance—that is, if there are only a few places in which we can put our atoms or molecules—then the entropy is low. On the other hand, the entropy is high if there are many possible arrangements ($W \gg 1$), that is, if the level of chaos is high.

The tombstone of Boltzmann with the equation S = k *log* W *engraved on it.* (Oesper Collection in the History of Chemistry/University of Cincinnati)

Additional entropy values are given in Appendix K.

Some interesting and useful generalizations can be drawn from the data given in Table 20.1.

- *When comparing the same or very similar substances, entropies of gases are much larger than those of liquids, which are larger than for solids.*
  In a solid the particles can only vibrate around lattice positions. When a solid melts, its particles are freer to move around, and molar entropy increases. When a liquid vaporizes, restrictions due to forces between the particles nearly disappear, and another large entropy increase occurs. For example, the entropies (in J/K · mol) of I_2(solid), Br_2(liquid), and Cl_2(gas) are 116.1, 152.2, and 223.0, respectively, and the entropies of C(s, graphite) and C(g) are 5.7 and 158.1, respectively.
- *Entropies of more complex molecules are larger than those of simpler molecules, especially in a series of closely related compounds.*
  In a more complicated molecule there are more ways for the atoms to be arranged in three-dimensional space and hence greater entropy. Entropies (in J/K · mol) for methane (CH_4), ethane (CH_3CH_3), and propane ($CH_3CH_2CH_3$) are 186.3, 229.6, and 269.9, respectively. For atoms or molecules of similar molar mass, we have Ar, CO_2, and $CH_3CH_2CH_3$ with entropies of 154.7, 213.7, and 269.9 J/K · mol, respectively.

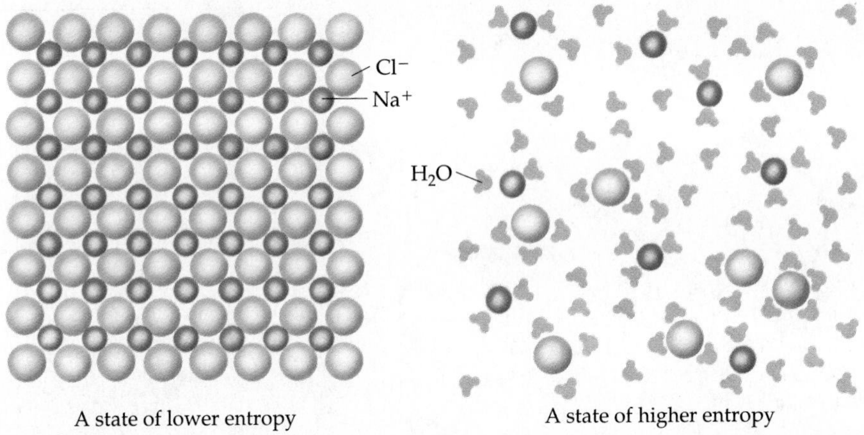

Figure 20.5 There is a large increase in entropy when a highly ordered solid dissolves in a solvent.

- *Entropies of ionic solids become larger as the attractions among the ions become weaker.*
  The weaker the forces between ions, the easier it is for the ions to vibrate about their lattice positions. Examples are MgO(s) and NaF(s) with entropies of 26.8 and 51.5 J/K mol, respectively; the 2+ and 2− charges on the magnesium ions and oxide ions result in greater attractive forces and hence lower entropy.

See the discussion of coulombic forces of attraction between oppositely charged ions on page 119.

- *Entropy usually increases when a pure liquid or solid dissolves in a solvent.*
  Because Table 20.1 refers only to pure substances, no example values are available; however, matter usually becomes more dispersed or disordered when a substance dissolves and different kinds of molecules mix together (Figure 20.5).
- *Entropy increases when a dissolved gas escapes from a solution.*
  Although gas molecules are dispersed among solvent molecules in solution, the very large entropy increase that occurs on changing from the liquid to the gas phase results in a higher entropy for separated gas and liquid than for the mixture.

The previous generalizations can be used to predict whether there is an increase in disorder of the substances involved when reactants are converted to products. (Such predictions are much easier to make for entropy changes than for enthalpy changes.) For the processes $H_2O(s) \rightarrow H_2O(\ell)$ and $H_2O(\ell) \rightarrow H_2O(g)$, we expect an entropy increase in each case because water molecules in the solid are more ordered than in the liquid and much more ordered than in the gas (Figure 20.6). This is confirmed by entropy measurements. At 273.15 K, for example, ice can be converted to water very slowly, and there is no temperature change. The quantity of energy transferred as heat is the heat of fusion (6.02 kJ/mol). Moreover, the transfer of energy is into the water, so $q = +6020\,\text{J/mol}$. This gives an entropy change of $+22\,\text{J/K} \cdot \text{mol}$.

$$\Delta S \text{ for } H_2O(s) \longrightarrow H_2O(\ell) = \frac{q}{T} = \frac{+6020\,\text{J/mol}}{273.15\,\text{K}} = +22.0\,\text{J/K} \cdot \text{mol}$$

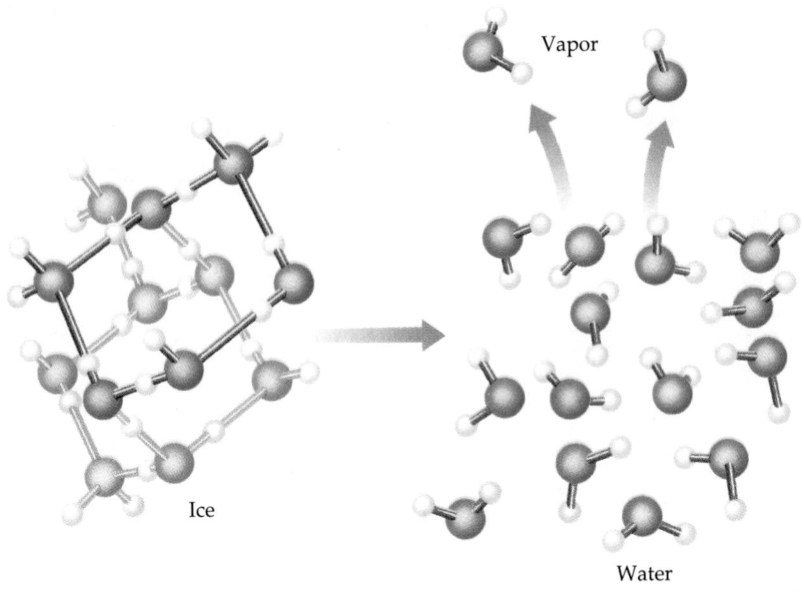

Figure 20.6 Increasing disorder in the melting of ice to liquid water and the evaporation of the liquid to vapor. The entropy change is positive.

Similarly, boiling water requires that energy be transferred into the water, so q must again be positive, and so must ΔS. Here $q = +40.7$ kJ/mol (the heat of vaporization of water at the boiling point), and ΔS is $+109$ J/K · mol.

$$\Delta S \text{ for } H_2O(\ell) \longrightarrow H_2O(g) = \frac{q}{T} = \frac{+40{,}700 \text{ J/mol}}{373.15 \text{ K}} = +109 \text{ J/K} \cdot \text{mol}$$

For the decomposition of iron(III) oxide to its elements,

$$2\ Fe_2O_3(s) \longrightarrow 4\ Fe(s) + 3\ O_2(g)$$

we would also predict an increase in entropy, because 3 mol of gaseous oxygen is present in the products, whereas the reactant is a solid. Because gases usually have much higher entropies than solids or liquids, gaseous substances are most important in determining entropy changes. Indeed, $\Delta S° = +551.7$ J/K in this case.

An example in which a decrease in entropy would be predicted is

$$2\ CO(g) + O_2(g) \longrightarrow 2\ CO_2(g)$$

Here there are 3 mol of gaseous substances (2 of CO and 1 of O_2) at the beginning of the reaction but only 2 mol of gaseous substances at the end. Two moles of gas (of whatever kind) contain less entropy than 3 mol of gas, and so ΔS is negative. The actual value is $\Delta S° = -173.0$ J/K.

Another example of decreased entropy is the process

$$Ag^+(aq) + Cl^-(aq) \longrightarrow AgCl(s)$$

Here the reactant ions are free to move about among water molecules in aqueous solution, but those same ions are held in a crystal lattice in the solid, a situation with much greater constraint. As a result, $\Delta S° = -33.1$ J/K.

EXERCISE 20.1 ***Calculating ΔS Values for Phase Changes***

The enthalpy of vaporization of benzene (C_6H_6) is 30.9 kJ/mol at the boiling point of 80.1 °C. Calculate the entropy change for benzene going from liquid to vapor and from vapor to liquid at 80.1 °C. ■

EXERCISE 20.2 *Entropy*

For each of the following processes, predict whether you would expect entropy to be greater for the products than for the reactants. Explain how you arrived at your prediction.

1. $CO_2(g) \rightarrow CO_2(s, \text{Dry Ice})$
2. $KCl(s) \rightarrow KCl(aq)$
3. $MgCO_3(s) + \text{heat} \rightarrow MgO(s) + CO_2(g)$ ■

Entropy and the Second Law of Thermodynamics

Experience with many, many chemical reactions and other processes in which energy is transferred has led to the **second law of thermodynamics,** which states that *the total entropy of the universe is continually increasing.* Whenever anything happens, matter, energy, or both become more dispersed or disordered. This means that a *product-favored reaction* is accompanied by an *increase in the entropy* of the universe; $\Delta S_{\text{universe}}$ is positive. Evaluating whether this will happen during a proposed chemical reaction allows us to predict whether or not reactants form appreciable quantities of products.

Such an evaluation can be made relatively easily if we consider only a carefully specified situation—a set of standard conditions as defined on page 288. Once a prediction has been made, corrections can be applied to account for differences from the standard conditions. Predicting whether a reaction is product-favored involves two steps: (1) calculating how much entropy is created by dispersal of matter; and (2) calculating how much entropy is created by dispersal of energy. Both calculations are done assuming that reactants at standard conditions are converted completely to products at standard conditions.

As an example of predicting the product favorability of a reaction, let us consider a process that might be used to manufacture liquid methanol for use as automobile fuel.

$$CO(g) + 2\ H_2(g) \longrightarrow \underset{\text{methanol}}{CH_3OH(\ell)}$$

We base our prediction on having 1 mol $CO(g)$ and 2 mol $H_2(g)$ as reactants, each at a pressure of 1 bar. We further assume that the product is 1 mol of liquid methanol, also at a pressure of 1 bar. If the total entropy is predicted to be higher after the product has been produced, then the reaction is product-favored under these conditions and might be useful. If not, perhaps some other conditions could be used, or perhaps we should consider some other methanol-producing reaction altogether.

Calculating the Entropy Change for a System

To calculate the entropy change due to dispersal of matter in the course of a reaction, ΔS_{system}, we assume that each reactant and each product is present in the amount required by its stoichiometric coefficient. All substances are assumed to be at standard pressure and at the temperature specified, so that values from Table 20.1 (or Appendix K) apply. Then we can add up all the entropies of the products and subtract the entropies of the reactants to see whether there is an increase or decrease in entropy.

Notice that the equation for calculating ΔS°_{system} has the same form as that for calculating ΔH° for a reaction (Equation 6.6).

$$\Delta S^\circ_{system} = \Sigma\, S^\circ(\text{products}) - \Sigma\, S^\circ(\text{reactants}) \qquad (20.1)$$

The entropy change for the reacting system, ΔS°_{system}, is the change occurring when reactants in their standard states are converted completely to products in their standard states. Thus, for the conversion of 1 mol of CO(g) and 2 mol of H_2(g) to 1 mol of $CH_3OH(\ell)$, we have

$$CO(g) + 2\,H_2(g) \longrightarrow CH_3OH(\ell)$$

$$\begin{aligned}\Delta S^\circ_{system} &= S^\circ[CH_3OH(\ell)] - \{S^\circ[CO(g)] + 2\,S^\circ[H_2(g)]\}\\ &= 1\text{ mol }(126.8\text{ J/K}\cdot\text{mol})\\ &\quad - \{1\text{ mol }(197.6\text{ J/K}\cdot\text{mol}) + 2\text{ mol }(130.7\text{ J/K}\cdot\text{mol})\}\\ &= -332.2\text{ J/K}\end{aligned}$$

Notice that this calculation gives the entropy change for the reaction *system*. In this case the decrease in entropy is large because 3 mol of gaseous material is converted to 1 mol of a more complicated, but liquid-phase product.

EXAMPLE 20.1 *Calculating an Entropy Change for a Chemical Reaction*

Nitrogen dioxide is formed from nitrogen monoxide and oxygen in a product-favored reaction at 25 °C (page 414). Determine the standard entropy change, ΔS°, for the reaction, ΔS°_{rxn} (= ΔS°_{system}).

Solution As in any problem involving a chemical reaction, first write the balanced equation.

$$2\,NO(g) + O_2(g) \longrightarrow 2\,NO_2(g)$$

Next, subtract the entropies of the reactants from the entropies of the products (Appendix K), paying careful attention to scale each entropy by the number of moles of reactant or product involved.

$$\begin{aligned}\Delta S^\circ_{rxn} &= (2\text{ mol }NO_2)(240.1\text{ J/K}\cdot\text{mol})\\ &\quad - [(2\text{ mol NO})(210.8\text{ J/K}\cdot\text{mol}) + (1\text{ mol }O_2)(205.1\text{ J/K}\cdot\text{mol})]\\ &= -146.5\text{ J/K}\end{aligned}$$

or -73.25 J/K for the formation of 1 mol of NO_2.

Notice that the sign of the entropy change is negative. This is largely due to the fact that the chemical reaction began with 3 mol of gaseous reactants and ended with 2 mol of gaseous product.

EXERCISE 20.3 *Calculating an Entropy Change for a Chemical Reaction*

The active ingredient in a popular antacid remedy is $CaCO_3$ (more familiar as the main component of chalk or limestone). Using the entropy values in Table 20.1, calculate the entropy change for the formation of $CaCO_3$(s) from the elements. (Use graphite as the standard state of carbon.) Is the sign of the entropy change for the formation of $CaCO_3$(s) positive or negative? Account for the increase or decrease in entropy in the formation of this compound in terms of your notion of disorder in chemical systems. ■

Calculating the Entropy Change in the Surroundings

Predicting the product favorability of a reaction involves calculating the entropy created by the dispersal of matter *and* of energy (page 918). We have done the first of these calculations for the formation of $CH_3OH(\ell)$ from CO and H_2O.

$$CO(g) + 2\ H_2(g) \longrightarrow CH_3OH(\ell) \qquad \Delta S^\circ_{system} = -332.2\ J/K$$

Next, the entropy created by the dispersal of energy by this chemical reaction—the entropy change for the surroundings—can be evaluated by calculating ΔH° for the reaction (from tables such as Table 6.2 or Appendix K) and by assuming that this quantity of energy is transferred to or from the surroundings. If the energy transfer is slow and occurs at a constant temperature, the entropy change for the surroundings can be calculated as

$$\Delta S^\circ_{surroundings} = \frac{q_{surroundings}}{T} = \frac{-\Delta H_{system}}{T}$$

The minus sign in this equation comes from the fact that $-\Delta H_{system}$ is the energy transferred *out of* the system and hence equals the energy transferred *into* the surroundings, $q_{surroundings}$.

The equation states that for an exothermic reaction (ΔH_{system} has a negative value) there will be an increase in entropy of the surroundings. For the proposed methanol-producing reaction $\Delta H^\circ_{system} = -128.14$ kJ (calculated from Table 6.2), and so the entropy change is

$$\Delta S^\circ_{surroundings} = -\frac{(-128.14\ kJ)(1000\ J/kJ)}{298\ K} = +430.\ J/K$$

Calculating the Total Entropy Change for System and Surroundings

The entropy changes for the system and its surroundings have been calculated for the reaction

$$CO(g) + 2\ H_2(g) \longrightarrow CH_3OH(\ell)$$
$$\Delta S^\circ_{system} = -332.2\ J/K \qquad \Delta S^\circ_{surroundings} = +430.\ J/K$$

The *total* entropy change for a process, referred to as $\Delta S_{universe}$ (the entropy change for the universe), is the sum of the entropy change for the system (ΔS_{system}) and the entropy change for the surroundings ($\Delta S_{surroundings}$). (We assume that nothing else but our reaction happens, and so there are no other entropy changes.) For the formation of methanol from CO(g) and $H_2O(g)$, this total entropy change is

$$\Delta S^\circ_{universe} = \Delta S^\circ_{system} + \Delta S^\circ_{surroundings} = (-332.2 + 430.)\ J/K = 98\ J/K$$

The reaction is accompanied by an increase in the entropy of the universe. It follows from the second law of thermodynamics that, if we had CO(g) and $H_2(g)$, each at 1 bar pressure and all in contact with one another, they would react to form $CH_3OH(\ell)$. The process is product-favored and might be useful for manufacturing methanol.

Predictions of the sort we have just made by calculating $\Delta S_{universe}$ can also be made qualitatively, without calculations, if we know whether or not a reaction is

TABLE 20.2 Predicting Whether a Reaction Is Product-Favored

ΔH_{system}	ΔS_{system}	Product-Favored
−, exothermic	+, less order	Yes
−, exothermic	−, more order	Depends on T and relative magnitudes of ΔH and ΔS for the system but generally product-favored at lower T.
+, endothermic	+, less order	Depends on T and relative magnitudes of ΔH and ΔS for the system but generally product-favored at higher T.
+, endothermic	−, more order	No

exothermic and if we can predict whether matter is dispersed when the reaction takes place. A reaction is sure to be product-favored if it is *exothermic* and proceeds *from* a state of *order to* one of *disorder.* Also, a reaction is certainly ***not*** product-favored if it is *endothermic* and there is a *decrease in entropy* for the system. Two other possible cases are indicated in Table 20.2, but they are more difficult to predict without quantitative information.

As examples of such predictions, consider gas-producing reactions of solids. They are product-favored because they are exothermic and produce highly disordered gases.

$$CaCO_3(s) + 2\ HCl(aq) \longrightarrow CaCl_2(aq) + H_2O(\ell) + CO_2(g) + \text{heat}$$

$$\Delta H^\circ_{rxn} = -15.2\ \text{kJ} \qquad \Delta S^\circ_{rxn} = 137.6\ \text{J/K}$$

Similarly, combustion reactions are product-favored because they are exothermic and produce a larger number of product molecules from a few reactant molecules.

$$\underset{\text{butane}}{2\ C_4H_{10}(g)} + 13\ O_2(g) \longrightarrow 8\ CO_2(g) + 10\ H_2O(g) + \text{heat}$$

$$\Delta H^\circ_{rxn} = -5315.1\ \text{kJ} \qquad \Delta S^\circ_{rxn} = 2362.8\ \text{J/K}$$

But what about a reaction such as the production of ethylene, C_2H_4, from ethane, C_2H_6? The reaction is very endothermic, although the entropy change is predicted to be positive.

$$C_2H_6(g) \longrightarrow H_2(g) + C_2H_4(g)$$
$$\Delta H^\circ_{rxn} = +136.94\ \text{kJ} \qquad \Delta S^\circ_{rxn} = +120.6\ \text{J/K}$$

So, the enthalpy change suggests the reaction should not be product-favored, whereas the entropy change suggests the opposite. Which is the more important? Calculating $\Delta S_{surroundings}$ involves dividing the enthalpy change by temperature, and ΔH° does not change much as temperature increases. Therefore, the higher the temperature the less important the ΔH° term is. At room temperature ΔH° is usually the deciding factor, and so the ethylene-producing reaction is not expected to be product-favored. This is indeed the case at 25 °C. To make this reaction work in industry, chemical engineers have designed a special process at about 1000 °C. At this higher temperature the $\Delta S°$ term is more important, and more products can be produced.

One goal of chemists is often to take small molecules and assemble them into larger molecules that can be sold for much more than the cost of the reactants. An example is the system described earlier, assembling CO and H_2 into methanol, CH_3OH,

$$CO(g) + 2\ H_2(g) \longrightarrow CH_3OH(\ell)$$
$$\Delta H°_{rxn} = -128.14\ kJ \qquad \Delta S°_{system} = -332.2\ J/K \qquad \Delta S°_{universe} = 98\ J/K$$

and then turning the methanol into gasoline (that is, into molecules such as octane, C_8H_{18}). The problem with this is that we are fighting a losing battle with entropy. The way to get this to work, however, is to increase the entropy somewhere else in the universe. Indeed, Roald Hoffmann, who shared the 1981 Nobel Prize in chemistry, has said that "One amusing way to describe synthetic chemistry, the making of molecules that is at the intellectual and economic center of chemistry, is that it is the local defeat of entropy."*

EXERCISE 20.4 ***Is a Reaction Product- or Reactant-Favored?***

Classify the following reactions as one of the four types of reactions summarized in Table 20.2.

Reaction	$\Delta H°_{rxn}$ (298 K) kJ	$\Delta S°_{system}$ (298 K) J/K
(a) $CH_4(g) + 2\ O_2(g) \rightarrow 2\ H_2O(\ell) + CO_2(g)$	−890.3	−243.0
(b) $2\ Fe_2O_3(s) + 3\ C(graphite) \rightarrow 4\ Fe(s) + 3\ CO_2(g)$	+467.9	+560.3
(c) $C(graphite) + O_2(g) \rightarrow CO_2(g)$	−393.5	+2.9
(d) $N_2(g) + 3\ Cl_2(g) \rightarrow 2\ NCl_3(g)$	−460	−275 ■

EXERCISE 20.5 ***Is a Reaction Product- or Reactant-Favored?***

Is the direct reaction of hydrogen and chlorine to give hydrogen chloride gas predicted to be product-favored or reactant-favored?

$$H_2(g) + Cl_2(g) \longrightarrow 2\ HCl(g)$$

Answer the question by calculating the values for $\Delta S°_{system}$ and $\Delta S°_{surroundings}$ (at 298 K) and then summing them to determine $\Delta S°_{universe}$. ■

20.3 GIBBS FREE ENERGY

Calculations of the sort done in the previous section would be simpler if we did not have to evaluate separately the entropy change of the surroundings from a table of $\Delta H_f°$ values and the entropy change of the system from a table of S° values. A new thermodynamic function, defined by J. Willard Gibbs (1839–1903), a professor at Yale University, solved this dilemma. In Gibbs's honor this function is now called the **Gibbs free energy** and given the symbol *G*.

J. Willard Gibbs (1839–1903). (Burndy Library/Courtesy AIP Emilio Segre Visual Archives)

**American Scientist,* Nov–Dec. 1987, pp. 619–621.

In the previous section we showed that the total entropy change accompanying a chemical reaction carried out slowly at constant temperature and pressure is

$$\Delta S_{\text{universe}} = \Delta S_{\text{surroundings}} + \Delta S_{\text{system}}$$
$$= \frac{-\Delta H_{\text{system}}}{T} + \Delta S_{\text{system}}$$

Multiplying through this equation by $-T$, the result is

$$-T\Delta S_{\text{universe}} = \Delta H_{\text{system}} - T\Delta S_{\text{system}}$$

Gibbs defined his free energy function so that $-T\Delta S_{\text{universe}}$ is equal to the change in the free energy of the system, ΔG_{system}. That is,

$$\Delta G_{\text{system}} = -T\Delta S_{\text{universe}} = \Delta H_{\text{system}} - T\Delta S_{\text{system}}$$

Under standard conditions, the equation becomes

$$\Delta G^\circ_{\text{system}} = \Delta H^\circ_{\text{system}} - T\Delta S^\circ_{\text{system}} \tag{20.2}$$

This equation, called the Gibbs free energy equation, is one of the most important equations in all of science. Because Gibbs stated that $\Delta G_{\text{system}} = -T\Delta S_{\text{universe}}$, the free energy of the system must decrease if the entropy of the universe increases. This provides a way of predicting whether a reaction will be product-favored, a way that depends only on the system and not on a calculation of the entropy change for the universe. It also allows us to answer the question unanswered by Table 20.2: what happens when both $\Delta H^\circ_{\text{rxn}}$ and $\Delta S^\circ_{\text{rxn}}$ have the same sign?

- If the reaction is exothermic (negative $\Delta H^\circ_{\text{system}} = \Delta H^\circ_{\text{rxn}}$) and if the entropy of the system increases (positive $\Delta S^\circ_{\text{system}} = \Delta S^\circ_{\text{rxn}}$), then $\Delta G^\circ_{\text{system}}$ ($= \Delta G^\circ_{\text{rxn}}$) must be negative and the reaction is product-favored.
- If ΔH° and ΔS° have the same sign, then the magnitude of T and the relative magnitudes of the enthalpy and entropy changes determine whether ΔG° is negative and the reaction is product-favored.
- If ΔH° is positive and ΔS° is negative, then ΔG° must be positive, and the reaction cannot be product-favored under standard conditions and at the temperature for which the data were tabulated.

These conclusions are the same ones previously tabulated in Table 20.2. The Gibbs function is useful because it allows us to make a decision about the favorability of a reaction, especially in cases where both the enthalpy change and entropy change have the same algebraic sign. Let us turn to the various ways this valuable tool can be used.

Calculating $\Delta G^\circ_{\text{rxn}}$, the Free Energy Change for a Reaction

Enthalpy and entropy changes can be calculated for chemical reactions using values of ΔH°_f and S° for substances in the reaction. Then, $\Delta G^\circ_{\text{rxn}}$ ($= \Delta G^\circ_{\text{system}}$)

can be found from the resulting values of ΔH°_{rxn} and ΔS°_{rxn} using Equation 20.2, as illustrated in the following example and exercise.

EXAMPLE 20.2 *Calculating ΔG°_{rxn} from ΔH°_{rxn} and ΔS°_{rxn}*

Calculate the standard free energy change for the formation of methane at 298 K.

$$\text{C(graphite)} + 2\ \text{H}_2\text{(g)} \longrightarrow \text{CH}_4\text{(g)}$$

Solution The following values for ΔH°_f and S° are provided in Appendix K.

	C(graphite)	H_2(g)	CH_4(g)
ΔH°_f (kJ/mol)	0	0	−74.8
S° (J/K · mol)	5.7	130.7	186.3

From these values, we can find both ΔH° and ΔS° for the reaction.

$$\begin{aligned}\Delta H^\circ_{rxn} &= \Delta H^\circ_f\,[\text{CH}_4\text{(g)}] - \{\Delta H^\circ_f\,[\text{C(graphite)}] + 2\ \Delta H^\circ_f\,[\text{H}_2\text{(g)}]\}\\ &= -74.8\ \text{kJ} - (0 + 0)\\ &= -74.8\ \text{kJ}\end{aligned}$$

$$\begin{aligned}\Delta S^\circ_{rxn} &= S^\circ\,[\text{CH}_4\text{(g)}] - \{S^\circ\,[\text{C(graphite)}] + 2\ S^\circ\,[\text{H}_2\text{(g)}]\}\\ &= 186.3\ \text{J/K}\cdot\text{mol} - [1\ \text{mol}(5.7\ \text{J/K}\cdot\text{mol}) + 2\ \text{mol}\ (130.7\ \text{J/K}\cdot\text{mol})]\\ &= -80.8\ \text{J/K}\end{aligned}$$

Both the enthalpy change and the entropy change for this reaction are negative. In Table 20.2 this is a case when the reaction is predicted to be product-favored at "low temperature." These values alone do not tell us if the temperature is low enough, however. By combining them in the Gibbs equation, and calculating ΔG°_{rxn} for a temperature of 25 °C, we can predict with certainty the outcome of the reaction.

$$\begin{aligned}\Delta G^\circ_{rxn} &= \Delta H^\circ_{rxn} - T\Delta S^\circ_{rxn}\\ &= -74.8\ \text{kJ} - (298\ \text{K})(-80.8\ \text{J/K})(1\ \text{kJ}/1000\ \text{J})\\ &= -74.8\ \text{kJ} - (-24.1\ \text{kJ})\\ &= -50.7\ \text{kJ}\end{aligned}$$

ΔG°_{rxn} is negative at 298 K, so the reaction is predicted to be product-favored.

In this case the product $T\Delta S^\circ$ is negative and smaller than ΔH°_{rxn} because the entropy change is relatively small. Chemists call this an "enthalpy-controlled reaction" because the exothermic nature of the reaction overcomes the decline in entropy of the system.

An enthalpy-driven reaction has a relatively large and negative ΔH and a small, positive ΔS.

EXERCISE 20.6 *Calculating ΔG°_{rxn} from ΔH°_{rxn} and ΔS°_{rxn}*

Using values of ΔH°_f and S° to find ΔH°_{rxn} and ΔS°_{rxn}, respectively, calculate the free energy change for the formation of 1 mol of NH_3(g) from the elements at standard conditions (and 25 °C): $\frac{1}{2}\ N_2(g) + \frac{3}{2}\ H_2(g) \rightarrow NH_3(g)$. ■

TABLE 20.3 Standard Molar Free Energies of Formation for Some Substances at 298 K

Element or Compound	ΔG_f° (kJ/mol)
$H_2(g)$	0
$O_2(g)$	0
$N_2(g)$	0
C(graphite)	0
C(diamond)	2.9
CO(g)	−137.2
$CO_2(g)$	−394.4
$CH_4(g)$	−50.7
$H_2O(g)$	−228.6
$H_2O(\ell)$	−237.1
$NH_3(g)$	−16.5

Standard Free Energy of Formation, ΔG_f°

In Example 20.2 and Exercise 20.6, the standard free energy change was calculated for the formation of 1 mol of a compound from its elements, with all reactants and products in the standard state. Therefore, the calculated ΔG_{rxn}° can be identified as the **standard molar free energy of formation** for methane, ΔG_f° [CH_4(g)], or ammonia, ΔG_f° [NH_3(g)]. These and a few other values of ΔG_f° are listed in Table 20.3, and many more values are given in Appendix K. Be sure to notice that $\Delta G_f^\circ = 0$ for graphite and for other elements. Elements in their standard states have ΔG_f° values of 0, for the same reason that they have ΔH_f° values of 0.

Just as ΔH_{rxn}° can be calculated from standard enthalpies of formation, the free energy change for a reaction can also be found from values of ΔG_f° by the general equation

$$\Delta G_{reaction}^\circ = \Sigma\, \Delta G_f^\circ \text{ (products)} - \Sigma\, \Delta G_f^\circ \text{ (reactants)} \qquad \textbf{(20.3)}$$

The example and exercises that follow illustrate how this is done.

EXAMPLE 20.3 *Calculating ΔG_{rxn}° from ΔG_f°*

Calculate the free energy change for the combustion of methane from the standard free energies of formation of the products and reactants.

Solution We first write the balanced equation for the reaction and then find the value of $\Delta G_{f}^\circ$for each reactant and product.

$$CH_4(g) + 2\ O_2(g) \longrightarrow 2\ H_2O(g) + CO_2(g)$$

ΔG_f° (kJ/mol) −50.7 0 −228.6 −394.4

Be sure to notice that the free energy of formation values are given for 1 mol. Each value must be multiplied by the number of moles involved.

$$\begin{aligned}\Delta G_{rxn}^\circ &= 2\ \Delta G_f^\circ [H_2O(g)] + \Delta G_f^\circ [CO_2(g)] - \{\Delta G_f^\circ [CH_4(g)] + 2\ \Delta G_f^\circ [O_2(g)]\} \\ &= 2 \text{ mol } (-228.6 \text{ kJ/mol}) + 1 \text{ mol } (-394.4 \text{ kJ/mol}) - \\ &\qquad [1 \text{ mol } (-50.7 \text{ kJ/mol}) + 2 \text{ mol } (0 \text{ kJ/mol})] \\ &= -800.9 \text{ kJ}\end{aligned}$$

ΔG_{rxn}° has a large, negative value, indicating that the reaction is product-favored under standard conditions.

EXERCISE 20.7 *Calculating ΔG_{rxn}° from ΔG_f°*

1. Write a balanced chemical equation depicting the formation of gaseous carbon dioxide (CO_2) from its elements.
2. What is the standard free energy of formation of 1.00 mol of CO_2 gas?
3. What is the standard free energy change for the reaction when 2.5 mol of CO_2 gas is formed from the elements? ■

EXERCISE 20.8 ***Calculating ΔG°_{rxn} from ΔG°_f***

Calculate the standard free energy change for the combustion of 1.00 mol of benzene, $C_6H_6(\ell)$, to give $CO_2(g)$ and $H_2O(g)$. ■

PROBLEM-SOLVING TIPS AND IDEAS

20.1 Using ΔG°_f

Be aware that Equation 20.3 is simply a shortcut that can be used to calculate the free energy change of a reaction when the free energies of formation (ΔG°_f) of the reactants and products are known. Do not, however, let this obscure the fact that it is the balance of the enthalpy and entropy changes, as well as the temperature, that determines the value of ΔG°_{rxn} as expressed by the Gibbs equation (Equation 20.2). ■

As a final thought in this section, we might ask what is meant by the term "free" energy? Does it mean that we can get something for nothing? No, the free energy change of a reaction is a measure of the *maximum magnitude of the net useful work* that can be obtained from a reaction. Consider the formation of methane again.

$$C(\text{graphite}) + 2\ H_2(g) \longrightarrow CH_4(g)$$

$$\Delta H^\circ_{rxn} = -74.8\ \text{kJ} \quad \text{and} \quad \Delta S^\circ_{rxn} = -80.8\ \text{J/K}$$

$$\Delta G^\circ_{rxn} = -74.8\ \text{kJ} - (298\ \text{K})(-80.8\ \text{J/K})(1\ \text{kJ}/1000\ \text{J})$$

$$= -74.8\ \text{kJ} + 24.1\ \text{kJ} = -50.7\ \text{kJ}$$

The reaction is exothermic; it has an enthalpy change of −74.8 kJ. Part of this thermal energy, though, is used to bring order to the system (and so the entropy declines). The amount of thermal energy diverted to this is $T\Delta S^\circ = 24.1$ kJ. Therefore, only 50.7 kJ of energy is "free" or available for useful work.

Figure 20.7 The product-favored reaction of potassium with water

$$K(s) + H_2O(\ell) \longrightarrow KOH(aq) + \tfrac{1}{2}H_2(g)$$

is driven by both enthalpy and entropy changes, and so the free energy change for the reaction is less than zero. ($\Delta H^\circ_{rxn} = -196.5$ kJ; $\Delta S^\circ_{rxn} = 22.9$ J/K; and $\Delta G^\circ_{rxn} = -203.4$ kJ.)

Product-Favored or Reactant-Favored?

The enthalpy and entropy changes for a chemical reaction depend on the reactants and products and may be positive or negative. These in turn influence the sign of the free energy change and whether the reaction is product- or reactant-favored (Table 20.2 and Examples 20.2 and 20.3). Let us look at some further predictions that are based on the calculation of the free energy change for a reaction.

If the free energy change of the system decreases in a process (ΔG_{system} is negative), then the process is product-favored. Because $\Delta G^\circ = \Delta H^\circ - T\Delta S^\circ$, a process is certainly product-favored if the enthalpy of the system declines *and* its entropy increases. This is the case for the reaction of potassium with water illustrated by Figure 20.7 or for the combustion of carbon.

	C(graphite) +	**O_2(g)** ⟶	**CO_2(g)**	**Overall**
ΔH°_f (kJ/mol)	0	0	−393.5	−393.5
ΔS° (J/K · mol)	5.7	205.1	213.7	2.9
ΔG°_f (kJ/mol)	0	0	−394.4	−394.4

The free energy change for this reaction can be calculated in either of two ways. You can add up all of the ΔG_f° values for the products and subtract the sum of those for the reactants.

$$\Delta G^\circ_{rxn} = \Sigma\, \Delta G_f^\circ\,(\text{products}) - \Sigma\, \Delta G_f^\circ\,(\text{reactants}) = -394.4\text{ kJ} - (0 + 0)$$
$$= -394.4\text{ kJ}$$

Alternatively, you can calculate the enthalpy and entropy changes for the reaction, ΔH°_{rxn} and ΔS°_{rxn}, and combine them using the Gibbs equation.

$$\begin{aligned}\Delta G^\circ_{rxn} &= \Delta H^\circ_{rxn} - T\Delta S^\circ_{rxn}\\ &= -393.5\text{ kJ} - (298)(2.9\text{ J/K})(1\text{ kJ}/1000\text{ J})\\ &= -393.5\text{ kJ} - 0.86\text{ kJ}\\ &= -394.4\text{ kJ}\end{aligned}$$

Both the enthalpy and the entropy changes contribute to making this reaction product-favored. The reaction liberates 393.5 kJ of heat energy (which makes the surroundings more disordered), and the small, positive entropy change also contributes slightly to the disordering of the universe.

Another possibility is that a reaction is endothermic (ΔH°_{rxn} is positive), and the entropy of the system decreases (ΔS°_{rxn} is negative). This always leads to a positive value of ΔG°_{rxn}, and the prediction that the reaction is reactant-favored. This is certainly true for the conversion of graphite to diamond.

The conversion of graphite to diamond is thermodynamically unfavorable at all temperatures. The process can, however, be carried out under high pressure in the presence of a catalyst.

	C(graphite) $\longrightarrow$	C(diamond)	Overall for Reaction
ΔH_f° (kJ/mol)	0	1.9	+1.9
S° (J/K · mol)	5.7	2.4	−3.3

$$\Delta G^\circ_{rxn} = +1.9\text{ kJ} - (298\text{ K})(-3.3\text{ J/K})(1\text{ kJ}/1000\text{ J}) = 2.9\text{ kJ}$$

If the entropy decreases (ΔS° is negative), $-T\Delta S^\circ$ is a positive quantity and the reaction can be spontaneous only if ΔH° is large and negative and outweighs the positive ($-T\Delta S^\circ$) term. Such cases are called *enthalpy-driven* reactions, an example of which is the formation of salt.

	2 Na(s) +	Cl_2(g) $\longrightarrow$	2 NaCl(s)	Overall for Reaction
ΔH_f° (kJ/mol)	0	0	2(−411.2)	−822.4
S° (J/K · mol)	2(51.2)	223.1	2(72.1)	−181.3

$$\Delta G^\circ_{rxn} = -822.4\text{ kJ} - (298\text{ K})(-181.3\text{ J/K})(1\text{ kJ}/1000\text{ J}) = -768.4\text{ kJ}$$

The free energy change for the reaction is a large negative number, clearly indicating the reaction is product-favored under standard conditions, something you already knew from Figure 2.21. However, although the reaction generates 822.4 kJ of heat energy (which is used to disorder the surroundings), only 768.4 kJ is available to disorder the universe. That is, 54 kJ of energy at 25 °C (= $T\Delta S^\circ$) is not "free," having been used to create some order in the system. Thus, the formation of salt is *enthalpy-driven,* as are many such reactions.

The formation of a highly-ordered, crystalline lattice maximizes attractive forces between ions. Such substances generally have a large negative value of ΔH_f°. See Section 8.7.

Now let us consider a case in which both the enthalpy and entropy changes are positive. The only way that the reaction can be spontaneous is for $-T\Delta S°$ to be large enough that $\Delta G°$ is negative. This is true for a number of salts, particularly those with low ion charges, when dissolving in water. Because the salt begins as a highly ordered solid with low entropy, dissolving it in water gives a jumble of ions with a high entropy.

	$NH_4NO_3(s)$ $\longrightarrow$	$NH_4NO_3(aq, 1\ m)$	**Overall for Process**
$\Delta H_f°$ (kJ/mol)	−365.6	−339.9	+25.7
$S°$ (J/K · mol)	151.1	259.8	+108.7

$$\Delta G°_{rxn} = +25.7\text{ kJ} - (298\text{ K})(+108.7\text{ J/K})(1\text{ kJ}/1000\text{ J}) = -6.7\text{ kJ}$$

The reaction is product-favored, and the enthalpy and entropy changes for the reaction clearly show that the reaction is *entropy-driven*. That is, although the process is endothermic (and you would feel an obvious cooling effect if you held your hand on a beaker or cold pack containing dissolving ammonium nitrate), this is outweighed by the large increase in entropy of the system, and the reaction is product-favored under standard conditions. The disordering of the system when the solid dissolves is a very potent driving force in this case.

Entropy is often the *"force"* that drives the mixing of any two substances, liquid or gas, with each other. Chapter 14 described the formation of ideal solutions, ones in which the forces between solute molecules and between solvent molecules are the same as between solute and solvent molecules. Because the energies of the two kinds of molecules *cannot* change on forming an ideal solution, it is the increase in entropy experienced by the solute and solvent molecules as they mix that provides the driving force.

See the Chemical Puzzler for this chapter.

EXERCISE 20.9 ***Predicting the Outcome of a Reaction***

Using free energies of formation in Appendix K, calculate $\Delta G°_{rxn}$ for (1) the formation of $CaCO_3(s)$ from $CO_2(g)$ and $CaO(s)$ and (2) the decomposition of $CaCO_3(s)$ to give $CO_2(g)$ and $CaO(s)$. Which reaction is product-favored under standard conditions? ■

Free Energy and Temperature

If a reaction has a positive enthalpy change *and* a positive entropy change, the only way it can be product-favored under standard conditions is if $-T\Delta S°$ is large enough to outweigh $\Delta H°$. This can happen in two ways: the entropy change can be positive and large (as is the case in dissolving NH_4NO_3 and many other ionic solids in water), or the entropy change can be positive and the temperature high. The latter is in fact one of the reasons reactions are often carried out at high temperatures. Let us consider an example of this case.

Our economy is based in large measure on the production of iron, and we can think of at least three different ways to reduce iron(III) oxide to metallic iron. One way is to heat iron(III) oxide and hope it decomposes to iron and oxygen.

$$Fe_2O_3(s) \longrightarrow 2\ Fe(s) + \tfrac{3}{2}O_2(g) \qquad \Delta G°_{rxn} = +742\text{ kJ}$$

We can obtain some idea of the feasibility of this by calculating ΔG°_{rxn}. Using the data in Appendix K, we find that ΔH°_{rxn} is +824.2 kJ and ΔS°_{rxn} is +275 J/K. The enthalpy change means the reaction could be reactant-favored, whereas the entropy change means it could be product-favored. Unfortunately, ΔH°_{rxn} is so positive that it cannot be outweighed by $-T\Delta S^\circ$ (= −82 kJ at 25 °C) at any reasonable temperature, and ΔG°_{rxn} at room temperature is +742 kJ at 298 K!

Another way to reduce iron(III) oxide is the so-called **thermite reaction.** Here the reactant-favored decomposition of $Fe_2O_3(s)$ to Fe(s)

$$Fe_2O_3(s) \longrightarrow 2\ Fe(s) + \tfrac{3}{2}O_2(g) \qquad \Delta G^\circ_{rxn} = +742\ kJ$$

is coupled with the highly product-favored oxidation of aluminum to Al_2O_3.

$$2\ Al(s) + \tfrac{3}{2}O_2(g) \longrightarrow Al_2O_3(s) \qquad \Delta G^\circ_{rxn} = -1582\ kJ$$

The sum of these reactions is

$$Fe_2O_3(s) + 2\ Al(s) \longrightarrow 2\ Fe(s) + Al_2O_3(s) \qquad \Delta G^\circ_{rxn} = -840.\ kJ$$
$$\Delta H^\circ_{rxn} = -851.5\ kJ \qquad \Delta S^\circ_{rxn} = -37.5\ J/K$$

This is certainly a product-favored process (Figure 20.8). The unfavorable ΔS°_{rxn} at 298 K ($-T\Delta S^\circ$ = +11.2 kJ) is swamped by a very large and negative enthalpy change. That means that a large amount of heat is produced, so much in fact that the products are raised to the melting point of iron (1530 °C), and white hot, molten metal streams out of the reaction. The reaction has been applied to welding procedures, but it is unfortunately not practical for the production of iron on a large scale. The cost of producing the aluminum to use as a reducing agent is much larger than the value of the iron produced.

(a) (b) (c)

Figure 20.8 The thermite reaction. (a) After starting the reaction with a fuse of burning magnesium wire, iron(III) oxide reacts with aluminum powder to give aluminum oxide and iron. (b) The Fe_2O_3/Al reaction generates so much heat that the iron is produced in the molten state. (c) It has dropped out of the clay pot, which originally contained the reactants, and burned through a sheet of iron placed under the pot. (C. D. Winters)

The usual method of reducing iron(III) oxide to iron metal is to use carbon or carbon monoxide, both inexpensive reducing agents. The decomposition of Fe_2O_3 is coupled with the highly product-favored combustion of carbon or carbon monoxide.

$$C(\text{graphite}) + O_2(g) \longrightarrow CO_2(g)$$

$$\Delta H^\circ_{rxn} = -393.5 \text{ kJ} \qquad \Delta S^\circ_{rxn} = -2.9 \text{ J/K} \qquad \Delta G^\circ_{rxn} = -394.4 \text{ kJ}$$

That is, the overall process for reduction by carbon can be thought of as the sum of two reactions.

$$\begin{array}{lll} 2\ Fe_2O_3(s) \longrightarrow 4\ Fe(s) + 3\ O_2(g) & 2(\Delta G^\circ_{rxn} = +742 \text{ kJ}) \\ 3\ C(\text{graphite}) + 3\ O_2(\text{graphite}) \longrightarrow 3\ CO_2(g) & 3(\Delta G^\circ_{rxn} = -394.4 \text{ kJ}) \\ \hline 2\ Fe_2O_3(s) + 3\ C(\text{graphite}) \longrightarrow 4\ Fe(s) + 3\ CO_2(s) & \Delta G^\circ_{rxn} = +300.8 \text{ kJ} \end{array}$$

$$\Delta H^\circ_{rxn} = +467.9 \text{ kJ} \qquad \Delta S^\circ_{rxn} = +560.3 \text{ J/K}$$

In practice, Fe_2O_3 is not reduced directly with C. Instead, the carbon is burned to give CO, which then acts as the reducing agent. See Chapter 23.

Even though the entropy is large, the $-T\Delta S^\circ$ term ($= -167$ kJ at 25 °C) is not large enough to offset the very unfavorable (positive) enthalpy change at 25 °C. Why then is this process used in industry? Precisely because the large, positive entropy change allows the reaction to become spontaneous at higher temperature. Let us calculate the *minimum* temperature T at which ΔG°_{rxn} is no longer positive, that is, the temperature at which it is zero.

$$\Delta G^\circ_{rxn} = \Delta H^\circ_{rxn} - T\Delta S^\circ_{rxn}$$

$$0 = +467.9 \text{ kJ} - T(0.5603 \text{ kJ/K})$$

$$T = 835 \text{ K, or } 562\ ^\circ\text{C}$$

The free energy change for the reaction becomes 0 at 562 °C, and at higher temperatures it is negative. Thus, the reaction can become product-favored by raising the temperature to a point easily reached in an industrial furnace.*

EXERCISE 20.10 ***Temperature and Free Energy Change***

Is the reduction of magnesia, MgO, with carbon a product-favored process at 25 °C? If not, at what temperature does it become so?

$$MgO(s) + C(\text{graphite}) \longrightarrow Mg(s) + CO(g)$$

■

20.4 THERMODYNAMICS AND THE EQUILIBRIUM CONSTANT

The free energy change for a reaction, ΔG°, is the increase or decrease in free energy as the reactants in their standard states are converted *completely* to the products in their standard states. But complete conversion is not often observed in practice. A product-favored reaction proceeds largely to products, but some reactants may remain at equilibrium. A reactant-favored reaction proceeds only

*To calculate ΔG at a much higher temperature, we used the enthalpy and entropy values appropriate for 25 °C. This is not entirely correct, because ΔH and ΔS are somewhat temperature-dependent. Their dependency on T, however, is *much* smaller than that of ΔG (as long as the temperature change does not include a phase change). Thus, our calculation provides an estimate of the point at which the reaction becomes spontaneous.

partially to products before achieving equilibrium. The question now is how $\Delta G°$ is related to the conditions at equilibrium.

To answer this, let us consider what happens as a reaction proceeds from reactants to products at constant temperature and pressure. When the reactants are mixed, the system becomes more disordered, and the entropy of the system increases. Then, as the reaction proceeds, heat energy might be evolved, further increasing the entropy of the system and its surroundings. When the system reaches equilibrium, products or reactants may predominate. Rarely, however, would reactants be converted *completely* to products. Under these conditions the free energy change for the reaction is not equal to $\Delta G°$ but to ΔG (without the superscript ° that signifies standard conditions). The relationship between $\Delta G°$ and ΔG is

$$\Delta G = \Delta G° + RT \ln Q$$

where R is the universal gas constant, T is the temperature in kelvins, and Q is the reaction quotient (Section 16.3). That is, for the general reaction of A and B giving products C and D

$$a\text{A} + b\text{B} \rightleftharpoons c\text{C} + d\text{D}$$

the reaction quotient, Q, is

$$Q = \frac{[\text{C}]^c[\text{D}]^d}{[\text{A}]^a[\text{B}]^b}$$

You learned in Chapter 16 that if $Q = K$ (or $Q/K = 1$), then the system is at equilibrium. The reaction is neither product-favored nor reactant-favored at *equilibrium,* and $\Delta G = 0$. Equilibrium is characterized by the inability to do work.

Because $\Delta G = 0$ and $Q = K$ at equilibrium, the equation $\Delta G = \Delta G° + RT \ln Q$ leads to

$$0 = \Delta G° + RT \ln K \qquad \text{(at equilibrium)}$$

Rearranging the last equation leads to a useful relationship between the standard free energy change for a reaction and **thermodynamic equilibrium constant, K.**

$$\Delta G° = -RT \ln K \qquad \textbf{(20.4)}$$

See Section 16.2 for a discussion of K_p and K_c.

For equilibria involving only gases, the thermodynamic equilibrium constant is K_p. For those that involve compounds in solution, it is equal to K_c.

The relationship between free energy and equilibrium is illustrated by Figure 20.9. The left side of each diagram gives the total free energy of the reactants, and the right side gives the total free energy of the products. The difference between the free energy of the pure reactants and the pure products is $\Delta G°$, which, like K, depends only on temperature and is a constant for a given reaction.

The relationship between $\Delta G°$ and K in Equation 20.4 informs us that, when $\Delta G°$ is negative, K must be greater than 1. Furthermore, the more negative the value of $\Delta G°$, the larger the equilibrium constant. We say that products are favored over reactants or that products are more stable than reactants. This situation is illustrated by Figure 20.9a. The opposite case is illustrated by Figure 20.9b. Here $\Delta G°$ is positive, the reaction is reactant-favored, and K must be less

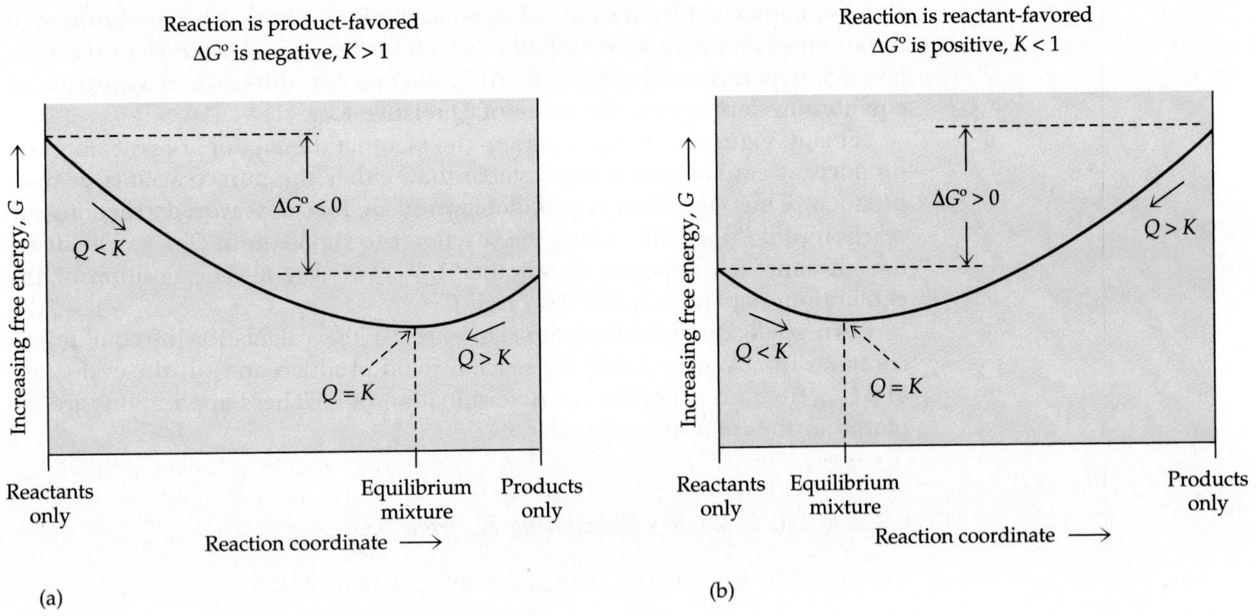

Figure 20.9 The variation in free energy for a reversible reaction carried out at constant temperature. The standard free energy change for a reaction, ΔG°_{rxn}, is the change in free energy for the *complete* conversion of reactants in their standard states to products in their standard states. In (a) ΔG°_{rxn} is negative, and the products are more stable than reactants. The reaction is product-favored and $K > 1$. In (b) ΔG°_{rxn} is positive, and the reactants are more stable than products. The reaction is reactant-favored, and $K < 1$. In both cases, the mixture of reactants and products at equilibrium is more stable than pure products or pure reactants. The position of the equilibrium depends on the ΔG°_{rxn} and T. Comparing Q and K at any point gives the direction of approach to equilibrium. If Q is not equal to K, the reaction runs "downhill" until $Q = K$.

than 1 (when $K < 1$, ln K is negative, and the term $-RT \ln K$ is positive). Now the reactants are more stable than the products. Finally, it is possible that ΔG° is 0, so K would be equal to 1. This extremely rare situation would mean that $[C]^c[D]^d = [A]^a[B]^b$ (for the reaction $aA + bB \rightleftarrows cC + dD$) at equilibrium. These relationships can be summarized as follows:

ΔG°	K	**Product Formation**
$\Delta G^\circ < 0$	$K > 1$	Products favored over reactants at equilibrium
$\Delta G^\circ = 0$	$K = 1$	At equilibrium when $[C]^c[D]^d = [A]^a[B]^b$; very rare
$\Delta G^\circ > 0$	$K < 1$	Reactants favored over products at equilibrium

In Section 16.3 we described what happens when a reaction is not at equilibrium.

- When $Q < K$ the reaction continues to proceed from reactants to products until equilibrium is achieved.
- When $Q > K$ the reaction proceeds from products to reactants until equilibrium is achieved.

These situations are also illustrated by Figure 20.9. If $Q < K$, equilibrium is approached from left to right on each of the curves in the figure. If $Q > K$, equilibrium is approached from right to left on each of the curves. The composition of the equilibrium mixture, and whether or not the reaction is product- or reactant-favored, depends on the value of $\Delta G°$, whereas the direction of approach to equilibrium depends on the value of Q relative to K.

Finally, Figure 20.9 also illustrates the fact that the mixture of reactants and products at equilibrium is more stable than either the pure reactants or pure products. Thus, in either a product-favored or reactant-favored reaction, the reaction runs "downhill" along the reaction coordinate until $Q = K$. The direction of approach depends on whether $Q > K$ or $Q < K$. The position of the equilibrium depends on the $\Delta G°$ and T.

Two applications of Equation 20.4 are (i) the calculation of equilibrium constants from values of ΔG_f° for reactants and products and (ii) the evaluation of ΔG°_{rxn} from an experimental determination of K. These applications are explored in the following examples.

EXAMPLE 20.4 *Calculating K_p from ΔG°_{rxn}*

The standard free energy change for the reaction, ΔG°_{rxn},

$$N_2(g) + 3\,H_2(g) \rightleftharpoons 2\,NH_3(g)$$

is −32.9 kJ. Calculate the equilibrium constant for this reaction at 25 °C.

Solution In this case, we need only substitute the appropriate values into Equation 20.4, taking care that the units of ΔG°_{rxn} are the same as those of RT.

$$\begin{aligned} \Delta G^\circ_{rxn} &= -RT \ln K_p \\ (-32.9\text{ kJ})(1000\text{ J/kJ}) &= -(8.3145\text{ J/K}\cdot\text{mol})(298\text{ K}) \ln K_p \\ \ln K_p &= 13.3 \\ K_p &= e^{13.3} \\ &= 6 \times 10^5 \end{aligned}$$

The equilibrium constant has a very large value, which means the equilibrium position lies very far to the product side at 25 °C. (To find K_p using a calculator, enter 13.3 and then strike the key labeled "e^x" or "inv(erse) ln x." See Appendix A for more information.)

EXAMPLE 20.5 *Calculating ΔG°_{rxn} from K_p*

Calculate ΔG°_{rxn} for the decomposition of ammonium chloride at 25 °C from $K_p = 1.1 \times 10^{-16}$.

$$NH_4Cl(s) \rightleftharpoons NH_3(g) + HCl(g)$$

Solution Here we substitute the value of K_p into Equation 20.4 at a temperature of 25 °C (=298 K).

$$\begin{aligned} \Delta G^\circ_{rxn} &= -RT \ln K_p = -(8.3145\text{ J/K}\cdot\text{mol})(298\text{ K}) \ln(1.1 \times 10^{-16}) \\ &= 9.1 \times 10^4\text{ J} \\ &= 91\text{ kJ} \end{aligned}$$

Using values of ΔG_f° from Appendix K you can verify that the free energy change is indeed +91 kJ.

EXAMPLE 20.6 *Uses of ΔH_f°, S°, ΔG_f° and the Calculation of K*

In Chapter 4 you studied in a qualitative way the driving forces of chemical reactions. One of these is the formation of a gaseous product that can escape when the reaction is open to the surroundings. Let us look at such a reaction from the quantitative, thermodynamic point of view.

$$MgCO_3(s) \longrightarrow MgO(s) + CO_2(g)$$

1. Is the reaction product-favored at room temperature?
2. Is the reaction enthalpy-driven or entropy-driven?
3. What is the value of K_p at 25 °C?
4. At what temperature does $K_p = 1$?
5. Does high temperature make the reaction more or less product-driven?

Solution To answer the first two questions we need to know ΔG°_{rxn} and its sign, as well as ΔH°_{rxn}. Using data from Appendix K, we have

$$\begin{aligned}\Delta H^\circ_{rxn} &= \Delta H_f^\circ[MgO(s)] + \Delta H_f^\circ[CO_2(g)] - \Delta H_f^\circ[MgCO_3(s)]\\ &= 1\text{ mol }(-601.7\text{ kJ/mol}) + 1\text{ mol }(-393.5\text{ kJ/mol})\\ &\quad - 1\text{ mol }(-1095.8\text{ kJ/mol})\\ &= +100.6\text{ kJ}\end{aligned}$$

$$\begin{aligned}\Delta S^\circ_{rxn} &= S^\circ[MgO(s)] + S^\circ[CO_2(g)] - S^\circ[MgCO_3(s)]\\ &= 1\text{ mol }(26.9\text{ J/K}\cdot\text{mol}) + 1\text{ mol }(213.7\text{ J/K}\cdot\text{mol}) - 1\text{ mol }(65.7\text{ J/K}\cdot\text{mol})\\ &= +174.9\text{ J/K}\end{aligned}$$

With the enthalpy and entropy changes for reaction known, we can combine them to calculate the free energy change for the reaction.

$$\begin{aligned}\Delta G^\circ_{rxn} &= \Delta H^\circ_{rxn} - T\Delta S^\circ_{rxn} = 100.6\text{ kJ} - (298\text{ K})(174.9\text{ J/K})(1\text{ kJ/1000 J})\\ &= 48.5\text{ kJ}\end{aligned}$$

The decomposition of magnesium carbonate to give carbon dioxide is reactant-favored at 298 K. What little reaction that does occur at 298 K is entropy-driven because the entropy change is positive. The enthalpy change is positive and too large to be offset by the increase in entropy to make the reaction product-favored at this temperature.

Having found ΔG°_{rxn}, K_p can now be calculated at 298 K from

$$\Delta G^\circ_{rxn} = -RT\ln K_p$$

or

$$\ln K_p = -\frac{\Delta G^\circ_{rxn}}{RT} = -\frac{48{,}500\text{ J}}{(8.3145\text{ J/K}\cdot\text{mol})(298\text{ K})} = -19.6$$

$$K_p = e^{-19.6} = 3 \times 10^{-9}$$

This means that $K_p = P_{CO_2} = 3 \times 10^{-9}$, or that the partial pressure of CO_2 is 3×10^{-9} atm at equilibrium at 25 °C. The partial pressure is extremely small because ΔG°_{rxn} has such a large, positive value.

At what temperature does $K_p = 1$? From the preceding analysis you know that ΔS°_{rxn} drives the reaction. This means that, at a sufficiently high temperature, the negative value of $-T\Delta S^\circ$ could become large enough to outweigh the inhibiting effect of a positive ΔH°_{rxn}, and ΔG°_{rxn} would become negative. A balance between ΔH° and $-T\Delta S^\circ$ is reached when $\Delta G^\circ = 0$, that is, when $K = 1$. The breakeven temperature is

$$T = \frac{\Delta H^\circ}{\Delta S^\circ} = \frac{100.6 \text{ kJ}}{0.1749 \text{ kJ/K}} = 575.2 \text{ K (or 302.0 °C)}$$

At approximately 300 °C the equilibrium constant is 1; that is, $P_{CO_2} = 1$ atm. At still higher temperatures, the $-T\Delta S^\circ$ term increasingly dominates the ΔH° term, and the yield increases because K increases. This means the answer to question 5 is "more product-favored," and this is true for any reaction with $\Delta S^\circ_{rxn} > 0$.

EXERCISE 20.11 ***Calculating K from the Free Energy Change***

Calculate K_p at 298 K from the value of ΔG°_{rxn} for the reaction

(a) $S(s) + O_2(g) \rightarrow SO_2(g)$

(b) $CaCO_3(s) \rightarrow CaO(s) + CO_2(g)$ ■

EXERCISE 20.12 ***Coupling Chemical Reactions***

Tin(IV) oxide can be reduced to tin metal using carbon as the reducing agent.

$$SnO_2(s) + C(s) \longrightarrow Sn(s) + CO_2(g)$$

Show that this process is the sum of two reactions, the oxidation of carbon to give CO_2, and the decomposition of SnO_2 to give Sn and O_2. What are the values of ΔG°_{rxn} for the separate reactions? What is ΔG°_{rxn} for the overall process? Is the loss of oxygen by tin more or less product-favored when carbon is used as the reducing agent? ■

20.5 THERMODYNAMICS AND TIME

With this chapter, we brought together the **three laws of thermodynamics.**

First law: The total energy of the universe is a constant.

Second law: The total entropy of the universe is always increasing.

Third law: The entropy of a pure, perfectly formed crystalline substance at absolute zero is zero.

Some cynic long ago paraphrased the first two laws into simpler statements. The first law is a statement that "You can't win!", and the second law tells you that "You can't break even either!" Yet another interpretation of the second law is Murphy's Law that "Things always tend to go wrong."

The second law tells us the entropy of the universe increases in a product-favored process. A snowflake spontaneously melts in a warm room, but you never see a glassful of water molecules spontaneously reassembling into snowflakes in a cold place. Molecules of your perfume or cologne spontaneously diffuse throughout a room, but they don't spontaneously collect again on your body. With time, all natural processes result in chaos. This is what scientists mean when they say that the second law is an expression in physical—as opposed to psychological—form of what we call *time.* In fact, entropy has been called "time's arrow."

Neither of the first two laws of thermodynamics has ever been or can be proven. It is just that there never has been a single, concrete example showing otherwise. No less a scientist than Albert Einstein once remarked that thermodynamic theory ". . . is the only physical theory of the universe content [which], within the framework of applicability of its basic concepts, will never be overthrown."

Einstein's statement does not mean that people have not tried (and are continuing to try) to disprove the laws of thermodynamics. Someone is always claiming to have invented a machine that performs useful work without expending energy—a perpetual motion machine. Although such a machine was actually granted a patent recently by the U.S. Patent Office (presumably the patent examiner had not had a course in thermodynamics), no workable perpetual motion machine has ever been demonstrated; the laws of thermodynamics are safe.

The second law demands that disorder increases with time. Because all natural processes take place as time progresses and result in increased disorder, it is evident that entropy and time "point" in the same direction.

If you are interested in the theories of the origin of the universe, and in "time's arrow," read *A Brief History of Time, From the Big Bang to Black Holes,* by Stephen W. Hawking, Bantam Books, New York, 1988.

CHAPTER HIGHLIGHTS

For many practical reasons chemists would like to be able to predict if a reaction favors the products or the reactants. That is the main objective of this chapter. To achieve that we have introduced the concept of entropy and have described the interplay of enthalpy and entropy changes for chemical reactions. A summary of the concepts in this chapter is given in the following table:

ΔH°_{rxn}	ΔS°_{rxn}	Sign of ΔG°_{rxn}	K	Reaction Outcome
−, exothermic	+, less order	−	>1	Product-favored at all T
−, exothermic	−, more order	− or +	Depends on T	Depends on T and on relative magnitude of ΔH°_{rxn} and ΔS°_{rxn}. Generally product-favored at *lower* T.
+, endothermic	+, less order	+ or −	Depends on T	Depends on T and on relative magnitude of ΔH°_{rxn} and ΔS°_{rxn}. Generally product-favored at *higher* T.
+, endothermic	−, more order	+	<1	Reactant-favored at all T.

When you have finished studying this chapter, you should be able to

- Describe the difference between the information provided by kinetics and thermodynamics (Section 20.1).
- Understand that entropy is a measure of matter dispersal or disorder (Section 20.2).
- Predict the sign of the entropy change for a reaction or change in state (Section 20.2).

- Calculate the entropy change for a change of state or for a chemical reaction (Section 20.2 and Equation 20.1).
- Use entropy and enthalpy changes to predict whether a reaction is product- or reactant-favored (Section 20.2 and Table 20.2).
- Understand the connection between enthalpy and entropy changes for a reaction and the Gibbs free energy change (Section 20.3, Table 20.2, and the summary table on page 941).
- Calculate the change in free energy for a reaction from the enthalpy and entropy changes (Equation 20.2) or from the standard free energy of formation of reactants and products (ΔG_f°, Equation 20.3) (Section 20.3).
- Describe the relationship between the free energy change for a reaction and its equilibrium constant (Section 20.4 and Equation 20.4).
- Show that a reactant-favored reaction can become product-favored if coupled with another reaction that is strongly product-favored (Section 20.4).

STUDY QUESTIONS

Many of these questions require thermodynamic data. If the required data are not given in the question, consult the tables in this chapter or Appendix K.

Review Questions

1. State the three laws of thermodynamics.
2. What is meant by a "product-favored chemical reaction?"
3. Criticize the following statements:
   (a) The entropy increases in all product-favored reactions.
   (b) A reaction with a negative free energy change ($\Delta G_{rxn}^\circ < 0$) is predicted to be product-favored with rapid transformation of reactants to products.
   (c) All product-favored processes are exothermic.
   (d) Endothermic reactions are never product-favored.
4. Decide if each of the following statements is true or false. If false, rewrite to make it true.
   (a) The entropy of a substance increases on going from the liquid to the vapor state at any temperature.
   (b) An exothermic reaction is always product-favored.
   (c) Reactions with a positive ΔH_{rxn}° and a positive ΔS_{rxn}° can never be product-favored.
   (d) Reactions with $\Delta G_{rxn}^\circ < 0$ have an equilibrium constant greater than 1.
   (e) When the equilibrium constant of a reaction is less than 1, then ΔG_{rxn}° is less than 0.
5. Explain why the entropy of the system increases on dissolving solid NaCl in water $\{S^\circ[NaCl(s)] = 72.1\ J/K \cdot mol$ and $S^\circ[NaCl(aq)] = 115.5\ J/K \cdot mol\}$.

Entropy

6. Which substance has the higher entropy in each of the following pairs?
   (a) A sample of Dry Ice (solid CO_2) at −78 °C or CO_2 vapor at 0 °C
   (b) Sugar, as a solid or dissolved in a cup of tea
   (c) Two 100-mL beakers, one containing 1 mol of pure water and the other containing 1 mol of pure alcohol, or a beaker containing a mixture of the water and alcohol
7. Which substance has the higher entropy in each of the following pairs?
   (a) A sample of pure silicon (to be used in a computer chip) or a piece of silicon containing a trace of some other atoms, such as B or P
   (b) An ice cube or liquid water, both at 0 °C
   (c) A sample of pure solid I_2 or iodine vapor, both at room temperature
8. By comparing the formulas or states for each pair of compounds, decide which is expected to have the higher entropy at the same temperature.
   (a) $KCl(s)$ or $AlCl_3(s)$
   (b) $CH_3I(\ell)$ or $CH_3CH_2I(\ell)$
   (c) $NH_4Cl(s)$ or $NH_4Cl(aq)$
9. By comparing the formulas or states for each pair of compounds, decide which is expected to have the higher entropy at the same temperature.
   (a) $NaCl(s)$ or $MgCl_2(s)$
   (b) $CH_3NH_2(g)$ or $(CH_3)_2NH(g)$
   (c) $Au(s)$ or $Hg(\ell)$

10. Calculate the entropy change, $\Delta S°$, for each of the following changes and comment on the sign of the change:
(a) C(diamond) $\rightarrow$ C(graphite)
(b) Na(g) $\rightarrow$ Na(s)
(c) $Br_2(\ell) \rightarrow Br_2(g)$

11. Calculate the entropy change, $\Delta S°$, for each of the following changes and comment on the sign of the change:
(a) $NH_4Cl(s) \rightarrow NH_4Cl(aq)$
(b) $C_2H_5OH(\ell) \rightarrow C_2H_5OH(g)$
(c) $CCl_4(g) \rightarrow CCl_4(\ell)$

12. The enthalpy of vaporization of liquid diethyl ether, $(C_2H_5)_2O$, is 26.0 kJ/mol at the boiling point of 35.0 °C. Calculate ΔS for (a) liquid to vapor and (b) vapor to liquid at 35.0 °C.

13. Calculate the entropy change, ΔS, for the vaporization of ethanol, C_2H_5OH, at the normal boiling point of the pure alcohol, 78.0 °C. The enthalpy of vaporization of the alcohol is 39.3 kJ/mol.

Reactions and Entropy Change

14. Calculate the standard molar entropy change for the formation of gaseous propane (C_3H_8) at 25°C.

$$3\ C(\text{graphite}) + 4\ H_2(g) \longrightarrow C_3H_8(g)$$

15. Calculate the standard molar entropy change for the formation of silicon hydride (silane) at 25 °C.

$$Si(s) + 2\ H_2(g) \longrightarrow SiH_4(g)$$

16. Calculate the standard molar entropy change for the formation of each of the following compounds from the elements at 25 °C:
(a) $H_2O(\ell)$
(b) $Mg(OH)_2(s)$
(c) $PbCl_2(s)$

17. Calculate the standard molar entropy change for the formation of each of the following compounds from the elements at 25 °C:
(a) ICl(g)
(b) $COCl_2(g)$
(c) $CaCO_3(s)$

18. Calculate the standard entropy change for each of the following reactions at 25 °C:
(a) $2\ Al(s) + 3\ Cl_2(g) \rightarrow 2\ AlCl_3(s)$
(b) $C_2H_5OH(\ell) + 3\ O_2(g) \rightarrow 2\ CO_2(g) + 3\ H_2O(g)$

19. Calculate the standard molar entropy change for each of the following reactions at 25 °C:
(a) $Ca(s) + 2\ H_2O(\ell) \rightarrow Ca(OH)_2(aq) + H_2(g)$
(b) $Na_2CO_3(s) + 2\ HCl(aq)$
$\rightarrow 2\ NaCl(aq) + H_2O(\ell) + CO_2(g)$

20. What are the signs of the enthalpy and entropy changes for the splitting of water to give gaseous hydrogen and oxygen, a process that requires considerable energy? Is this reaction likely to be product-favored or not? Explain your answer briefly.

21. Cyclohexane is produced by adding hydrogen gas to benzene.

$$\underset{\text{benzene}}{C_6H_6(\ell)} + 3\ H_2(g) \longrightarrow \underset{\text{cyclohexane}}{C_6H_{12}(\ell)}$$

The enthalpies of formation are

$$\Delta H_f°[C_6H_6(\ell)] = +49.0\ \text{kJ/mol}$$
$$\Delta H_f°[C_6H_{12}(\ell)] = -156.4\ \text{kJ/mol}$$

Is this reaction likely to be product-favored or reactant-favored? Explain your reasoning briefly.

22. Classify each of the reactions according to one of the four reaction types summarized in Table 20.2.
(a) $Fe_2O_3(s) + 2\ Al(s) \rightarrow 2\ Fe(s) + Al_2O_3(s)$
$\Delta H° = -851.5$ kJ $\Delta S° = -37.5$ J/K
(b) $N_2(g) + 2\ O_2(g) \rightarrow 2\ NO_2(g)$
$\Delta H° = 66.4$ kJ $\Delta S° = -122$ J/K

23. Classify each of the reactions according to one of the four reaction types summarized in Table 20.2.
(a) $C_6H_{12}O_6(s) + 6\ O_2(g) \rightarrow 6\ CO_2(g) + 6\ H_2O(\ell)$
$\Delta H° = -673$ kJ $\Delta S° = 60.4$ J/K
(b) $MgO(s) + C(\text{graphite}) \rightarrow Mg(s) + CO(g)$
$\Delta H° = 491.18$ kJ $\Delta S° = 197.67$ J/K

24. Is the combustion of ethane, C_2H_6, likely to be a product-favored reaction?

$$C_2H_6(g) + \tfrac{7}{2}\ O_2(g) \longrightarrow 2\ CO_2(g) + 3\ H_2O(g)$$

Answer the question by calculating the value of $\Delta S°_{\text{universe}}$. Required values of $\Delta H_f°$ and $S°$ are in Appendix K. Does your calculated answer agree with your preconceived idea of this reaction?

25. In the discussion of "meals ready to eat" (MRE's) on page 283, it was noted that the reaction of magnesium with water provides the heat.

$$Mg(s) + 2\ H_2O(\ell) \longrightarrow Mg(OH)_2(s) + H_2(g)$$

Is this reaction in fact predicted to be product-favored? Answer the question by calculating the value of $\Delta S°_{\text{universe}}$. Required values of $\Delta H_f°$ and $S°$ are in Appendix K. Does your calculated answer agree with your preconceived idea of this reaction?

Free Energy

26. Using values of $\Delta H_f°$ and $S°$, calculate $\Delta G°_{\text{rxn}}$ for each of the following reactions:
(a) $Sn(s) + 2\ Cl_2(g) \rightarrow SnCl_4(\ell)$
(b) $NH_3(g) + HCl(g) \rightarrow NH_4Cl(s)$
Which of the values of $\Delta G°_{\text{rxn}}$ that you have just calculated corresponds to a standard free energy of formation, $\Delta G_f°$? In those cases, compare your calculated values with the values of $\Delta G_f°$ tabulated in Appendix K. Which of these reactions is (are) predicted to be product-favored? Are the reactions enthalpy- or entropy-driven?

27. Using values of ΔH_f° and S°, calculate ΔG°_{rxn} for each of the following reactions:
(a) $Ca(s) + 2\,H_2O(\ell) \rightarrow Ca(OH)_2(aq) + H_2(g)$
(b) $6\,C(graphite) + 3\,H_2(g) \rightarrow C_6H_6(\ell)$
Which of the values of ΔG°_{rxn} that you have just calculated corresponds to a standard free energy of formation, ΔG_f°? In those cases, compare your calculated values with the values of ΔG_f° tabulated in Appendix K. Which of these reactions is (are) predicted to be product-favored? Are the reactions enthalpy- or entropy-driven?

28. Using values of ΔH_f° and S°, calculate the standard molar free energy of formation, ΔG_f°, for each of the following compounds:
(a) $CS_2(g)$
(b) $N_2H_4(\ell)$
(c) $COCl_2(g)$
Compare your calculated values of ΔG_f° with those listed in Appendix K. Which reactions are predicted to be product-favored?

29. Using values of ΔH_f° and S°, calculate the standard molar free energy of formation, ΔG_f°, for each of the following compounds:
(a) $Mg(OH)_2(s)$
(b) $NOCl(g)$
(c) $Na_2CO_3(s)$
Compare your calculated values of ΔG_f° with those listed in Appendix K. Which reactions are predicted to be product-favored?

30. Write a balanced equation that depicts the formation of 1 mol of $Fe_2O_3(s)$ from its elements. What is the standard free energy of formation of 1.00 mol of $Fe_2O_3(s)$? What is the value of ΔG°_{rxn} when 454 g (1 lb) of $Fe_2O_3(s)$ is formed from the elements?

31. Hydrazine is used to remove dissolved oxygen from the water in hot-water heating systems.

$$N_2H_4(\ell) + O_2(g) \longrightarrow 2\,H_2O(\ell) + N_2(g)$$

What is the value of ΔG°_{rxn} when 1.00 mol of N_2H_4 is oxidized? What is the value of ΔG°_{rxn} for the oxidation of 1.00 kg of hydrazine?

32. Using values of ΔG_f°, calculate ΔG°_{rxn} for each of the following reactions. Which are predicted to be product-favored?
(a) $Ca(s) + Cl_2(g) \rightarrow CaCl_2(s)$
(b) $2\,HgO(s) \rightarrow 2\,Hg(\ell) + O_2(g)$
(c) $NH_3(g) + 2\,O_2(g) \rightarrow HNO_3(\ell) + H_2O(\ell)$

33. Using values of ΔG_f°, calculate ΔG°_{rxn} for each of the following reactions. Which are predicted to be product-favored?
(a) $HgS(s) + O_2(g) \rightarrow Hg(\ell) + SO_2(g)$
(b) $2\,H_2S(g) + 3\,O_2(g) \rightarrow 2\,H_2O(g) + 2\,SO_2(g)$
(c) $SiCl_4(g) + 2\,Mg(s) \rightarrow 2\,MgCl_2(s) + Si(s)$

34. What is the value of ΔG_f° for $BaCO_3(s)$? You know that $\Delta G^\circ_{rxn} = +218.1$ kJ for the reaction

$$BaCO_3(s) \longrightarrow BaO(s) + CO_2(g)$$

and other data are available in Appendix K.

35. What is the value of ΔG_f° for $TiCl_2(s)$? You know that $\Delta G^\circ_{rxn} = -272.8$ kJ for

$$TiCl_2(s) + Cl_2(g) \longrightarrow TiCl_4(\ell)$$

and other data are available in Appendix K.

36. Hydrogenation, the addition of hydrogen to an organic compound, is a reaction of considerable industrial importance. Calculate ΔH°, ΔS°, and ΔG° at 25 °C for the hydrogenation of octene, C_8H_{16}, to give octane, C_8H_{18}. Is the reaction product- or reactant-favored under standard conditions?

$$C_8H_{16}(g) + H_2(g) \longrightarrow C_8H_{18}(g)$$

The information in the table is required, in addition to data in Appendix K.

Compound	**ΔH_f°(kJ/mol)**	**S°(J/K · mol)**
Octene, C_8H_{16}	−82.93	462.8
Octane, C_8H_{18}	−208.45	463.6

37. Synthesis gas, a mixture of H_2 and CO, can be converted to methane, CH_4.

$$3\,H_2(g) + CO(g) \longrightarrow CH_4(g) + H_2O(g)$$

Calculate ΔH°, ΔS°, and ΔG° at 25 °C for the reaction. Is it predicted to be product- or reactant-favored under standard conditions?

Thermodynamics and Equilibrium Constants

38. The formation of NO(g) from its elements

$$\tfrac{1}{2}\,N_2(g) + \tfrac{1}{2}\,O_2(g) \longrightarrow NO(g)$$

has a standard free energy change, ΔG_f°, of +86.57 kJ/mol at 25 °C. Calculate K_p at this temperature. Comment on the connection between the sign of ΔG° and the magnitude of K_p.

39. Methanol, CH_3OH, is now widely used as a fuel in race cars such as those that compete in the Indianapolis 500 (see Chapter 6). The liquid fuel can be formed using the reaction

$$C(graphite) + \tfrac{1}{2}\,O_2(g) + 2\,H_2(g) \longrightarrow CH_3OH(\ell)$$

Calculate K_p for the formation of methanol at 25 °C. Comment on the connection between the sign of ΔG° and the magnitude of K_p.

40. Ethylene reacts with hydrogen to produce ethane.

$$H_2C{=}CH_2(g) + H_2(g) \longrightarrow H_3C{-}CH_3(g)$$

(a) Using the data in Appendix K, calculate ΔG° for the reaction at 25 °C. Is the reaction predicted to be product-favored under standard conditions?
(b) Calculate K_p from ΔG°_{rxn}. Comment on the connection between the sign of ΔG° and the magnitude of K_p.

41. Use the data in Appendix K to calculate $\Delta G°$ and K_p at 25 °C for the reaction

$$2\ HBr(g) + Cl_2(g) \rightleftharpoons 2\ HCl(g) + Br_2(\ell)$$

Comment on the connection between the sign of $\Delta G°$ and the magnitude of K_p.

42. Titanium(IV) chloride, $TiCl_4(\ell)$, is produced by the reaction of carbon and chlorine with TiO_2.

$$TiO_2(s) + C(s) + 2\ Cl_2(g) \longrightarrow TiCl_4(\ell) + CO_2(g)$$

The reaction can be thought of as occurring in two steps: the reduction of TiO_2 with carbon

$$TiO_2(s) + C(s) \longrightarrow Ti(s) + CO_2(g)$$

and the oxidation of titanium metal to the product.

$$Ti(s) + 2\ Cl_2(g) \longrightarrow TiCl_4(\ell)$$

(a) Calculate $\Delta G°_{rxn}$ and K_p for each of the two steps.
(b) Calculate $\Delta G°_{rxn}$ and K_p for the overall process. How are these related to the values of the free energy change and K_p for the two steps in the process?
(c) Is the overall reaction enthalpy- or entropy-driven?

43. Insoluble silver chloride can be dissolved in the presence of excess chloride ion. The equation for the overall reaction is

$$AgCl(s) + Cl^-(aq) \rightleftharpoons AgCl_2^-(aq)$$

(a) Show that the overall reaction is the sum of two others: the ionization of AgCl(s) to give silver(I) and chloride ions, and the formation of $AgCl_2^-(aq)$ from $Ag^+(aq)$ and $Cl^-(aq)$ ions.
(b) Calculate the equilibrium constant for the overall process from the equilibrium constants for the two steps. (Constants for the two steps are found in Appendices H and I.)
(c) Calculate the free energy change for each step and for the overall reaction. Required values are in Appendix K or in the following list:

Species	**$\Delta G_f°$ (kJ/mol)**
$Ag^+(aq)$	+77.1
$Cl^-(aq)$	−131.2
$AgCl_2^-(aq)$	−215.4

How is $K_{overall}$ related to $\Delta G°_{overall}$? Is the overall reaction product-favored?

General Questions

44. Sulfur burns in air according to the equation

$$S(s) + O_2(g) \longrightarrow SO_2(g)$$

Without doing calculations, predict the signs of $\Delta H°$ and $\Delta S°$ for the reaction. Next, verify your prediction with a calculation.

45. Calculate the entropy change involved in the formation of 1.0 mol of each of the following gaseous hydrocarbons under standard conditions. (Use graphite as the standard state of carbon.)

(a) H—C≡C—H
acetylene

(b) 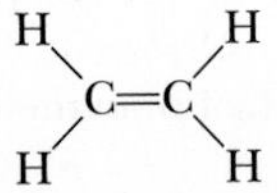
ethylene

(c) 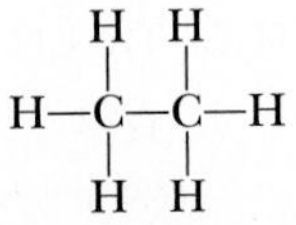
ethane

What trend do you see in these values? Does $\Delta S°$ increase or decrease on adding H atoms?

46. For each of the following processes, give the algebraic sign of $\Delta H°$, $\Delta S°$, and $\Delta G°$. No calculations are necessary; use your common sense.
(a) The splitting of liquid water to give gaseous oxygen and hydrogen, a process that requires a considerable amount of energy.
(b) The explosion of dynamite, a mixture of nitroglycerin, $C_3H_5N_3O_9$, and diatomaceous earth, gives gaseous products, such as water, CO_2, and others; much heat is evolved.
(c) The combustion of gasoline in the engine of your car, as exemplified by the combustion of octane.

$$2\ C_8H_{18}(g) + 25\ O_2(g) \longrightarrow 16\ CO_2(g) + 18\ H_2O(g)$$

47. Yeast can produce ethanol by the fermentation of glucose, the basis for the production of most alcoholic beverages.

$$C_6H_{12}O_6(aq) \longrightarrow 2\ C_2H_5OH(\ell) + 2\ CO_2(g)$$

Calculate $\Delta H°$, $\Delta S°$, and $\Delta G°$ for the reaction. Is the reaction product- or reactant-favored? (In addition to the thermodynamic values in Appendix K, you need the following for $C_6H_{12}O_6(aq)$: $\Delta H_f° = -1260.0$ kJ/mol; $S° = 289$ J/K · mol; and $\Delta G_f° = -918.8$ kJ/mol.)

48. Elemental boron, in the form of thin fibers, can be made by reducing a boron halide with H_2.

$$BCl_3(g) + \tfrac{3}{2}\ H_2(g) \longrightarrow B(s) + 3\ HCl(g)$$

The standard enthalpy of formation of $BCl_3(g)$ is −403.8 kJ/mol, and its entropy, $S°$, is 290 J/K · mol. The entropy, $S°$, for B(s) is 5.86 J/K · mol. Calculate $\Delta H°$, $\Delta S°$, and $\Delta G°$ at 25 °C for this reaction. Is it predicted to be product-favored under standard conditions? If product-favored, is it enthalpy-driven or entropy-driven?

49. The equilibrium constant, K_p, for $N_2O_4(g) \rightleftharpoons 2\ NO_2(g)$ is 0.14 at 25 °C. Calculate $\Delta G°$ from this constant, and compare your calculated value with that determined from the $\Delta G_f°$ values in Appendix K.

50. Most metal oxides can be reduced with hydrogen to the pure metal. (Although such reactions work well, it is an expensive method and not used often for large-scale preparations.) The reduction of iron(II) oxide

$$FeO(s) + H_2(g) \longrightarrow Fe(s) + H_2O(g)$$

has an equilibrium constant of 0.422 at 700 °C. Estimate ΔG°_{rxn}.

51. The equilibrium constant for the butane-isobutane equilibrium at 25 °C is 2.5.

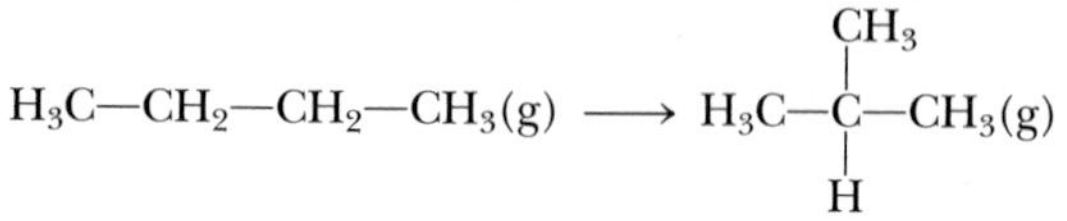

butane isobutane

Calculate ΔG°_{rxn} at this temperature in kilojoules per mole.

52. Almost 5 billion kg of benzene, C_6H_6, is made each year. It is used as a starting material for many other compounds and as a solvent (although it is also a carcinogen, and its use is therefore restricted). One compound that can be made from benzene is cyclohexane, C_6H_{12}.

$$C_6H_6(g) + 3\,H_2(g) \longrightarrow C_6H_{12}(g)$$

$\Delta H^\circ_{rxn} = -206.1$ kJ $\Delta S^\circ_{rxn} = -363.12$ J/K

Is this reaction predicted to be product-favored under standard conditions at 25 °C? Is the reaction enthalpy- or entropy-driven?

53. Iodine, I_2, dissolves readily in carbon tetrachloride with an enthalpy change that is effectively zero.

$$I_2(s) \longrightarrow I_2 \text{ (in } CCl_4 \text{ solution)}$$

What is the sign of ΔG°_{rxn}? Is the dissolving process entropy-driven or enthalpy-driven? Explain briefly.

54. A crucial reaction for the production of synthetic fuels is the conversion of coal and steam to H_2.

$$C(s) + H_2O(g) \longrightarrow CO(g) + H_2(g)$$

(a) Calculate ΔG°_{rxn} for this reaction at 25 °C assuming C(s) is graphite.
(b) Calculate K_p for the reaction at 25 °C.
(c) Is the reaction predicted to be product-favored under standard conditions? If not, at what temperature does it become so?

55. Calculate ΔG°_{rxn} for the decomposition of sulfur trioxide to sulfur dioxide and oxygen.

$$2\,SO_3(g) \longrightarrow 2\,SO_2(g) + O_2(g)$$

(a) Is the reaction product-favored under standard conditions at 25 °C?
(b) If the reaction is not product-favored at 25 °C, is there a temperature at which it becomes so?
(c) What is the equilibrium constant for the reaction at 1500 °C?

56. Methanol is relatively inexpensive to produce. Much consideration has been given to using it as a precursor to other fuels such as methane, which could be obtained by the decomposition of the alcohol.

$$CH_3OH(\ell) \longrightarrow CH_4(g) + \tfrac{1}{2}\,O_2(g)$$

(a) What are the sign and magnitude of the entropy change for the reaction? Does the sign of ΔS° agree with your expectation? Explain briefly.
(b) Is the reaction product-favored under standard conditions at 25 °C? Use thermodynamic values to prove your answer.
(c) If not product-favored at 25 °C, at what temperature does the reaction become product-favored?

57. Photosynthetic bacteria carry out the synthesis of high free energy compounds such as glucose from CO_2 and H_2O using light as the source of energy.

$$6\,CO_2(g) + 6\,H_2O(\ell) \longrightarrow C_6H_{12}O_6(aq) + 6\,O_2(g)$$

The free energy change is +2870 kJ/mol of glucose. In the deep ocean where there is no light, however, this same synthesis can apparently be done by bacteria using hydrogen sulfide as the energy source. Show that, by adding the following reaction

$$H_2S(g) + \tfrac{1}{2}\,O_2(g) \longrightarrow H_2O(\ell) + S(s)$$

to the glucose synthesis reaction above, sufficient free energy is produced so that the overall process

$$24\,H_2S(g) + 6\,CO_2(g) + 6\,O_2(g) \longrightarrow C_6H_{12}O_6(aq) + 18\,H_2O(\ell) + 24\,S(s)$$

is product-favored. (ΔG°_f for glucose is −918.8 kJ/mol). (For further information see S. Krishnamurthy, *Journal of Chemical Education,* Vol. 58, p. 981, 1981).

58. Some metal oxides can be decomposed to the metal and oxygen under reasonable conditions. Is the decomposition of silver(I) oxide product-favored at 25 °C?

$$2\,Ag_2O(s) \longrightarrow 4\,Ag(s) + O_2(g)$$

If not, can it become so if the temperature is raised? At what temperature is the reaction product-favored?

59. Mercury vapor is dangerous because it can be ingested into the lungs. Use the following data to calculate the entropy change for the process $Hg(\ell) \rightarrow Hg(g)$ at 298 K.

	ΔH°_f, kJ/mol	ΔG°_f, kJ/mol
$Hg(\ell)$	0	0
$Hg(g)$	61.317	31.85

Then, estimate the temperature at which K_p for the process is equal to (a) 1.00 and (b) 1/760. What is the vapor pressure at each of these temperatures? (Experimental vapor pressures are 1 mm Hg at 126.2 °C and 1 atm at 356.6 °C.)

Conceptual Questions

60. Calculate the entropy change for dissolving HCl gas in water. Is the sign of $\Delta S°$ what you expected? Why or why not?

61. Why is the standard state entropy of Br_2 greater than that of I_2?

62. For the reaction

$$CaCO_3(s) \rightleftharpoons CaO(s) + CO_2(g)$$

is the reaction product-favored at a low temperature or at a high temperature?

63. Sulfur undergoes a phase transition between 80 and 100 °C.

$$S_8(\text{rhombic}) \longrightarrow S_8(\text{monoclinic})$$

$$\Delta H°_{rxn} = 3.213 \text{ kJ/mol and } \Delta S°_{rxn} = 8.7 \text{ J/K}$$

(a) Estimate $\Delta G°$ for the transition at 80 and 100 °C. What do these results tell you about the stability of the two forms of sulfur at each of these temperatures?
(b) Calculate the temperature at which $\Delta G° = 0$. What is the significance of this temperature?

Summary Questions

64. Consider the formation of NO(g) from its elements.

$$N_2(g) + O_2(g) \longrightarrow 2\ NO(g)$$

(a) Use the free energy data in Appendix K to calculate K_p at 25 °C. Is the reaction product-favored at this temperature?
(b) Assume that $\Delta H°_{rxn}$ and $\Delta S°_{rxn}$ are nearly constant with temperature and calculate $\Delta G°_{rxn}$ at 700 °C. Estimate K_p from the new value of $\Delta G°_{rxn}$ at 700. °C. Is the reaction product-favored at 700. °C?
(c) Using K_p at 700. °C, calculate the equilibrium partial pressures of the three gases if you mix 1.00 atm each of N_2 and O_2.

65. Calculate $\Delta G_f°$ for HI(g) at 350 °C, given the following equilibrium partial pressures: $P(H_2) = 0.132$ atm, $P(I_2) = 0.295$ atm, and $P(HI) = 1.61$ atm. At 350 °C and 1 atm, the stable form of I_2 is a gas.

$$\tfrac{1}{2}\,H_2(g) + \tfrac{1}{2}\,I_2(g) \rightleftharpoons HI(g)$$

66. Silver(I) oxide can be formed by the reaction of silver metal and oxygen.

$$4\ Ag(s) + O_2(g) \longrightarrow 2\ Ag_2O(s)$$

(a) Calculate $\Delta H°_{rxn}$, $\Delta S°_{rxn}$, and $\Delta G°_{rxn}$ for the reaction.
(b) What is the pressure of O_2 in equilibrium with Ag and Ag_2O at 25 °C?
(c) At what temperature does the pressure of O_2 in equilibrium with Ag and Ag_2O equal 1.00 atm?

CHAPTER 21

Principles of Reactivity: Electron Transfer Reactions

A Chemical Puzzler

Batteries are used in a wider and wider variety of appliances. Some batteries "run down" and are discarded. Others we can regenerate by attaching them to a source of current. How do batteries work? Why can some be regenerated and others not?

In 1994, the California Air Resources Board determined that by 1998, 2% of new cars and light trucks in California are to be zero-emissions vehicles (ZEVs); that is, they should give off absolutely no volatile organic compounds, nitrogen oxides, or carbon monoxide. By 2001, 5% of the vehicles should be ZEVs, and 10% must meet that requirement by 2003. About a dozen states—including such populous states as Maryland, Massachusetts, New Jersey, and New York—have indicated that they will follow California's lead. How will this be accomplished? It is likely that electric vehicles (EVs) will be the only type of vehicle that is capable of meeting the California standards in the near future. As a result, the California standards have given impetus to auto manufacturers and technology companies around the world to develop new methods of supplying electric power. This can only be done by batteries or fuel cells, which in turn depend on electron transfer reactions and electrochemistry, the subject of this chapter.

Hydrogen-powered ZEVs were mentioned by James A. Cusumano in his interview (page 155). (See also page 300.)

21.1 OXIDATION-REDUCTION REACTIONS

Oxidation-reduction reactions—also called *redox reactions*—occur by electron transfer and constitute a major class of chemical reactions (Section 4.10). Because examples of redox reactions occur everywhere, you experience their consequences daily. Corrosion, for example, occurs by redox reactions. The iron and steel in cars, bridges, and buildings can oxidize to rust, and aluminum structures can corrode (Figure 21.1 and Section 21.7). Many biological processes depend on electron transfer reactions. For example, the oxygen you take in as you breathe is converted ultimately to water and carbon dioxide. The oxidation number of the oxygen in the product molecules (H_2O and CO_2) is -2, so electrons must have been transferred to O_2 molecules to cause their reduction. Where did the electrons come from? At least in the final step they are transferred to O_2 from hemoglobin, a large iron-containing molecule. Other biological electron transfer processes include the conversion of water to O_2 in green plants by

A battery-powered test vehicle. (General Motors)

Figure 21.1 The formation of rust on objects made of iron and steel can eventually destroy their structural integrity. (C. D. Winters)

photosynthesis, and the conversion of atmospheric N_2 by bacteria to a usable form of nitrogen, such as NH_4^+.

$$N_2(g) + 8\ H_3O^+(aq) + 6\ e^- \longrightarrow 2\ NH_4^+(aq) + 8\ H_2O(\ell)$$

A **battery** is an **electrochemical cell,** or a collection of such cells, that produces a current or flow of electrons at a constant voltage as a result of an electron transfer reaction. The power to run your calculator or computer or to start your car, or even to propel your ZEV, comes from a battery.

Redox reactions are important in manufacturing chemicals and producing metals. For example, many metals are commercially prepared or purified by the direct application of electricity in a process called **electrolysis,** the use of electric energy to produce a chemical change such as the reduction of copper ions to copper metal.

$$Cu^{2+}(aq) + 2\ e^- \longrightarrow Cu(s)$$

Other metals, such as iron, are prepared using chemical reducing agents (see page 188).

$$Fe_2O_3(s) + 3\ CO(g) \longrightarrow 2\ Fe(s) + 3\ CO_2(g)$$

Chemists have only recently begun to understand the way in which electrons are transferred from one site to another in redox reactions. Moreover, the prevention of corrosion, the construction of more powerful batteries, and the plating of metals using electricity are now better understood. The general subject is fascinating because it applies to so many problems of practical interest.

To understand the general subject of electron transfer reactions, and electrochemistry in particular, we organize the subject as follows: Equations for electron transfer reactions can appear complicated at first glance, so we first describe some special techniques for balancing such equations. This is followed by a description of electrochemical cells in which an electric current is produced by an electron transfer reaction—cells that are sometimes called **voltaic cells,** or **galvanic cells.** This then leads to a discussion of batteries and fuel cells and then to the process of corrosion and its prevention. Finally, we describe **electrolysis** and some important industrial processes.

Voltaic and galvanic cells are named for Count Alessandro Volta (1745–1827), who studied current-producing reactions, and Dr. Luigi Galvani (1737–1798), who made early studies of animal electricity and produced electricity chemically.

You may wish to review Section 4.10 and recall the definitions of the terms "oxidation," "reduction," "reducing agent," "oxidizing agent," and "oxidation number."

Some photos of redox reactions are found earlier in the book: Na/H_2O, p. 7; K/H_2O, pp. 30, 83, 196; electrolysis of $SnCl_2$, p. 34; Na/Cl_2, pp. 86, 386; Ca/H_2O, pp. 99, 376; Al/Br_2, p. 159; sugar/$KClO_3$, p. 256.

Balancing Equations for Oxidation-Reduction Reactions

Balancing equations for redox reactions may appear to be a formidable task in some cases. Fortunately, though, there are systematic ways of doing so. One approach is illustrated by the equation for the reaction of aqueous copper(II) ions and metallic zinc (Figure 21.2). Here a piece of zinc is immersed in an aqueous solution of copper sulfate. After a time, the blue color of the aqueous Cu^{2+} ion fades, and copper metal "plates out," or forms a coating on the zinc strip. In addition, the zinc strip slowly disappears. What happened?

The fate of the Cu^{2+} ion in the test tube in Figure 21.2 is probably obvious; the blue color of aqueous copper(II) ion fades as it is reduced to copper metal. We depict this by the equation

$$Cu^{2+}(aq) + 2\ e^- \longrightarrow Cu(s)$$

Cu^{2+} gains electrons, is reduced, and is the oxidizing agent.

Because we observe that the zinc metal disappears, it must be the source of the electrons that cause reduction of Cu^{2+}, and we depict this by the equation

$$Zn(s) \longrightarrow Zn^{2+}(aq) + 2\,e^-$$

Zn loses electrons, is oxidized, and is the reducing agent.

Zinc is the reducing agent because it donates electrons and forms aqueous Zn^{2+} in the process.

The equation for the net chemical reaction occurring in the test tube is the sum of the equations for the two **half-reactions:** one for the oxidation of zinc and one for the reduction of copper(II) ion.

Oxidation:	$Zn(s) \longrightarrow Zn^{2+}(aq) + 2\,e^-$
Reduction:	$Cu^{2+}(aq) + 2\,e^- \longrightarrow Cu(s)$
Net reaction:	$Zn(s) + Cu^{2+}(aq) \longrightarrow Cu(s) + Zn^{2+}(aq)$

Notice that the equation for each half-reaction is balanced for mass; one atom of each kind appears on each side of the equation. There is also a *charge balance,* that is, the algebraic sum of charges on one side of the equation equals the algebraic sum of the charges on the other side. (Here both sides have a net charge of 2+.) Because the equation for the net reaction is the sum of the balanced equations for the half-reactions, the net equation is likewise balanced for mass and charge.

The preceding redox reaction illustrates the general approach to balancing equations for oxidation-reduction reactions, which is explored further in the following examples.

Figure 21.2 An oxidation-reduction reaction. A strip of zinc metal was placed in a solution of copper(II) sulfate (*left*), and the zinc reacts with the copper(II) ions to give copper metal and zinc ions in solution.

$$Zn(s) + Cu^{2+}(aq) \longrightarrow Zn^{2+}(aq) + Cu(s)$$

Copper metal accumulates on the zinc strip, and the blue color of aqueous copper(II) ions fades as copper(II) ions disappear from solution (*middle* and *right*). (C. D. Winters)

EXAMPLE 21.1 *Balancing an Equation for an Oxidation-Reduction Reaction*

Balance the equation for the reaction of silver(I), Ag^+, with copper (Figure 4.17).

$$Cu(s) + Ag^+(aq) \longrightarrow Ag(s) + Cu^{2+}(aq)$$

Solution

Step 1. *Recognize the reaction as an oxidation-reduction process.* Here the oxidation number for silver changes from +1 to 0, and that for copper changes from 0 to +2.

Step 2. *Separate the process into half-reactions.*

Reduction (decrease in Ag oxidation number): $Ag^+(aq) \longrightarrow Ag(s)$

Oxidation (increase in Cu oxidation number): $Cu(s) \longrightarrow Cu^{2+}(aq)$

Step 3. *Balance each half-reaction for mass.*

Both half-reactions are already balanced for mass; the same number of atoms of each element appear on each side.

Step 4. *Balance each half-reaction for charge.* The equation is balanced for charge by adding electrons to the more positive side of the half-reaction.

$Ag^+(aq) + e^- \longrightarrow Ag(s)$ Each Ag^+ ion acquires an electron. Silver(I) is the oxidizing agent and is reduced.

$Cu(s) \longrightarrow Cu^{2+}(aq) + 2\ e^-$ Each Cu atom loses two electrons. Copper is the reducing agent and is oxidized.

In each case the net electric charge on each side of the equation is 0.

Step 5. *Multiply each half-reaction by an appropriate factor.* The reducing agent must donate as many electrons as the oxidizing agent acquires. Here one atom of copper produces two electrons, whereas one Ag^+ ion acquires one electron. Because two Ag^+ ions are required to consume the two electrons produced by a Cu atom, the Ag^+/Ag half-reaction is multiplied by 2.

$$2\ Ag^+(aq) + 2\ e^- \longrightarrow 2\ Ag(s)$$

Step 6. *Add the half-reactions to produce the overall balanced equation.*

$$\begin{array}{rcl} Cu(s) & \longrightarrow & Cu^{2+}(aq) + 2\ e^- \\ 2\ Ag^+(aq) + 2\ e^- & \longrightarrow & 2\ Ag(s) \\ \hline Cu(s) + 2\ Ag^+(aq) & \longrightarrow & 2\ Ag(s) + Cu^{2+}(aq) \end{array}$$

Step 7. *Check the overall equation to ensure that both mass and charge balance.* Here two silver atoms or ions and one copper atom or ion appear on each side, and the net charge on each side is 2+. The equation is balanced.

EXERCISE 21.1 ***Balancing an Equation for an Oxidation-Reduction Reaction***

Balance the equation

$$Cr^{2+}(aq) + I_2(aq) \longrightarrow Cr^{3+} + I^-(aq)$$

Write the balanced half-reactions and the balanced net ionic equation. Identify the oxidizing agent, the reducing agent, the substance oxidized, and the substance reduced. ■

A problem often encountered when balancing equations for reactions in aqueous solution is that water and either hydrogen ion or hydroxide ion can enter into the reaction. In acidic conditions, H^+ may be a reactant or a product, and in basic conditions it is possible for OH^- to participate. It may also be necessary to use the *pair* of species H^+ and H_2O to balance equations for reactions in acid solution. Similarly, the *pair* OH^- and H_2O may be needed to balance equations for reactions in basic solution. Examples 21.2 through 21.4 show how to determine when these species are needed and how to place them in the equation.

We balance equations using H^+ rather than H_3O^+ because it is much simpler. If desired, the equation can be adjusted later to include hydronium ions by adding the same number of H_2O molecules to each side of the equation.

EXAMPLE 21.2 *Balancing Equations for Oxidation-Reduction Reactions in Acid Solution*

Balance the net ionic equation for the reaction of permanganate ion with oxalic acid in acid solution.

$$H_2C_2O_4(aq) + MnO_4^-(aq) \longrightarrow Mn^{2+}(aq) + CO_2(g)$$

Solution

Step 1. *Recognize the reaction as an oxidation-reduction.* The oxidation number of Mn changes from +7 in MnO_4^- to +2 in Mn^{2+}, and C changes from +3 in $H_2C_2O_4$ to +4 in CO_2.

Step 2. *Separate the overall process into half-reactions.*

Oxidation: $H_2C_2O_4(aq) \longrightarrow CO_2(g)$ — $H_2C_2O_4$ is the reducing agent.

Reduction: $MnO_4^-(aq) \longrightarrow Mn^{2+}(aq)$ — MnO_4^- is the oxidizing agent.

Step 3. *Balance each half-reaction for mass.* Begin by balancing all atoms except O and H; these are always the last to be balanced because they often appear in more than one reactant or product.

Oxalic acid half-reaction: First, balance the C atoms in this half-reaction.

$$H_2C_2O_4(aq) \longrightarrow 2\ CO_2(g)$$

The number of O atoms is now also balanced, but the number of H atoms is not. Because the reaction occurs in acidified solution, however, H^+ ions are present. In *acid solution,* a mass balance for H may be achieved by *adding H^+* to the side of the equation *deficient in H* atoms. Here two H^+ ions are added to the right side.

$$H_2C_2O_4(aq) \longrightarrow 2\ CO_2(aq) + 2\ H^+(aq)$$

Permanganate half-reaction: The Mn atoms are already balanced, but an oxygen-containing species must be added to the right side to achieve an O-atom balance.

$$MnO_4^-(aq) \longrightarrow Mn^{2+}(aq) + \text{(need 4 O atoms)}$$

In *acid solution, add H_2O to the side requiring O atoms.* One H_2O molecule is added for each O atom required. In this case, four H_2O molecules must be added to the right side.

$$MnO_4^-(aq) \longrightarrow Mn^{2+}(aq) + 4\ H_2O(\ell)$$

This means that eight unbalanced H atoms are now on the right. Again, H atom balance is achieved by adding H^+ ions to the side deficient in H atoms. Here eight H^+ ions are added to the left side of the equation.

$$8\ H^+(aq) + MnO_4^-(aq) \longrightarrow Mn^{2+}(aq) + 4\ H_2O(\ell)$$

To balance O in acid solution, add one H_2O to the oxygen-deficient side for each O required and then add two H^+ ions to the other side of the equation for each H_2O added.

Step 4. *Balance the half-reactions for charge.* The mass-balanced $H_2C_2O_4$ equation has a net charge of 0 on the left side and 2+ on the right. Therefore, 2 e^- are added to the more positive right side.

$$H_2C_2O_4(aq) \longrightarrow 2\ CO_2(g) + 2\ H^+(aq) + 2\ e^-$$

The mass-balanced MnO_4^- half-reaction has a charge of 7+ on the left and 2+ on the right. Therefore 5 e^- are added to the more positive left side.

$$5\ e^- + 8\ H^+(aq) + MnO_4^-(aq) \longrightarrow Mn^{2+}(aq) + 4\ H_2O(\ell)$$

Step 5. *Multiply the half-reactions by appropriate factors.* The reducing agent must donate as many electrons as the oxidizing agent consumes. The $H_2C_2O_4$ half-reaction should therefore be multiplied by 5, and the MnO_4^- half-reaction by 2. Now 5 mol of the reducing agent ($H_2C_2O_4$) provides 10 mol of electrons, which are consumed by 2 mol of the oxidizing agent (MnO_4^-).

$$5\ [H_2C_2O_4(aq) \longrightarrow 2\ CO_2(g) + 2\ H^+(aq) + 2\ e^-] \quad \text{Reducing agent}$$

$$2[5\ e^- + 8\ H^+(aq) + MnO_4^-(aq) \longrightarrow Mn^{2+}(aq) + 4\ H_2O(\ell)] \quad \text{Oxidizing agent}$$

Step 6. *Add the half-reactions to give the balanced, overall equation.*

$$5\ H_2C_2O_4(aq) \longrightarrow 10\ CO_2(g) + 10\ H^+(aq) + 10\ e^-$$

$$10\ e^- + 16\ H^+(aq) + 2\ MnO_4^-(aq) \longrightarrow 2\ Mn^{2+}(aq) + 8\ H_2O(\ell)$$

$$5\ H_2C_2O_4(aq) + 16\ H^+(aq) + 2\ MnO_4^-(aq) \longrightarrow 10\ CO_2(g) + 10\ H^+(aq) + 2\ Mn^{2+}(aq) + 8\ H_2O(\ell)$$

Step 7. *Cancel common reactants and products.* Here 16 H^+ ions are on the left and 10 H^+ on the right. This means a net of 6 H^+ ions are consumed in the reaction.

$$5\ H_2C_2O_4(aq) + 6\ H^+(aq) + 2\ MnO_4^-(aq) \longrightarrow 10\ CO_2(g) + 2\ Mn^{2+}(aq) + 8\ H_2O(\ell)$$

Step 8. *Check the final result to ensure mass and charge balance.*

Mass balance: 2 Mn, 28 O, 10 C, and 16 H

Charge balance: 4+ on both sides

EXAMPLE 21.3 *Balancing Equations for Oxidation-Reduction Reactions in Acid Solution*

Balance the net ionic equation for the reaction of the organic compound ethanol, C_2H_5OH, with the dichromate ion in acid solution (see Figure 4.23).

$$\underset{\text{ethanol}}{C_2H_5OH(aq)} + \underset{\text{dichromate ion; orange-red}}{Cr_2O_7^{2-}(aq)} \longrightarrow \underset{\text{acetic acid}}{CH_3CO_2H(aq)} + \underset{\text{chromium(III) ion; green}}{Cr^{3+}(aq)}$$

Solution

Step 1. *Recognize the reaction as an oxidation-reduction process.* Here Cr changes from +6 to +3 (so $Cr_2O_7^{2-}$ is reduced), and C changes from −2 to 0 (and so C_2H_5OH is oxidized).

Step 2. *Break the overall equation into half-reactions.*

Oxidation: $C_2H_5OH(aq) \longrightarrow CH_3CO_2H(aq)$ C_2H_5OH is the reducing agent.

Reduction: $Cr_2O_7^{2-}(aq) \longrightarrow Cr^{3+}(aq)$ $Cr_2O_7^{2-}$ is the oxidizing agent.

Step 3. *Balance each half-reaction for mass.* Here we start with the C_2H_5OH half-reaction. The C atoms are balanced, but O and H are not. The first step is therefore to add H_2O to the O-deficient left side,

$$C_2H_5OH(aq) + H_2O(\ell) \longrightarrow CH_3CO_2H(aq)$$

and then to balance H by adding H^+ to the right side of the equation (two H^+ to balance the H_2O added to the left and two H^+ to balance the H^+ "lost" by C_2H_5OH).

$$C_2H_5OH(aq) + H_2O(\ell) \longrightarrow CH_3CO_2H(aq) + 4\ H^+(aq)$$

Next, turn to the $Cr_2O_7^{2-}$ half-reaction. The Cr atoms should be balanced first,

$$Cr_2O_7^{2-}(aq) \longrightarrow 2\ Cr^{3+}(aq)$$

and then H_2O is added to the O-deficient right side.

$$Cr_2O_7^{2-}(aq) \longrightarrow 2\ Cr^{3+}(aq) + 7\ H_2O(\ell)$$

Finally, the H atoms are balanced by placing H^+ on the H-deficient left side.

$$14\ H^+(aq) + Cr_2O_7^{2-}(aq) \longrightarrow 2\ Cr^{3+}(aq) + 7\ H_2O(\ell)$$

Step 4. *Balance the half-reactions for charge.*

$$C_2H_5OH(aq) + H_2O(\ell) \longrightarrow CH_3CO_2H(aq) + 4\ H^+(aq) + 4\ e^-$$
$$6\ e^- + 14\ H^+(aq) + Cr_2O_7^{2-}(aq) \longrightarrow 2\ Cr^{3+}(aq) + 7\ H_2O(\ell)$$

This confirms that C_2H_5OH is the reducing agent (electron donor) and $Cr_2O_7^{2-}$ is the oxidizing agent (electron acceptor).

Step 5. *Multiply the balanced half-reactions by appropriate factors.*

$$3[C_2H_5OH(aq) + H_2O(\ell) \longrightarrow CH_3CO_2H(aq) + 4\ H^+(aq) + 4\ e^-]$$
$$2[6\ e^- + 14\ H^+(aq) + Cr_2O_7^{2-}(aq) \longrightarrow 2\ Cr^{3+}(aq) + 7\ H_2O(\ell)]$$

Three moles of ethanol produce 12 mol of electrons, which are then consumed by 2 mol of dichromate ion.

Step 6. *Add the balanced half-reactions.*

$$3\ C_2H_5OH(aq) + 3\ H_2O(\ell) \longrightarrow 3\ CH_3CO_2H(aq) + 12\ H^+(aq) + 12\ e^-$$
$$12\ e^- + 28\ H^+(aq) + 2\ Cr_2O_7^{2-}(aq) \longrightarrow 4\ Cr^{3+}(aq) + 14\ H_2O(\ell)$$

$$3\ C_2H_5OH(aq) + 3\ H_2O(\ell) + 28\ H^+(aq) + 2\ Cr_2O_7^{2-}(aq) \longrightarrow 3\ CH_3CO_2H(aq) + 12\ H^+(aq) + 4\ Cr^{3+}(aq) + 14\ H_2O(\ell)$$

Step 7. *Eliminate common reactants and products.* Water and H^+ ions appear on both sides of the overall equation in Step 6. The equation can therefore be

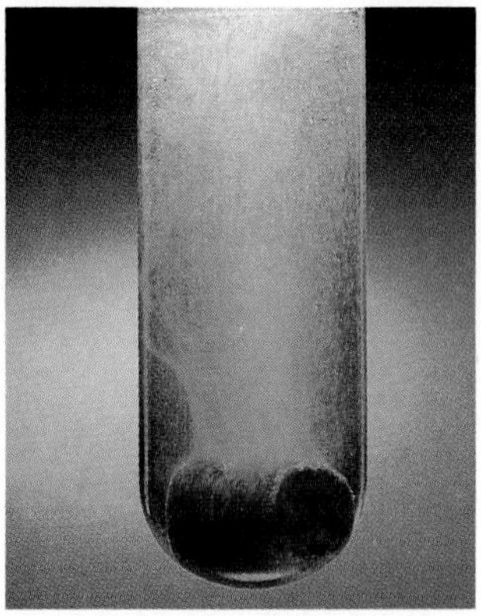

Metallic cobalt reacts with nitric acid to give pink $Co(NO_3)_3$, H_2, and nitrogen oxides, among them NO_2. (C. D. Winters)

simplified by recognizing that a *net* of 11 H_2O appears on the right, and a *net* of 16 H^+ is on the left. The final, balanced net ionic equation is

$$3\ C_2H_5OH(aq) + 16\ H^+(aq) + 2\ Cr_2O_7^{2-}(aq) \longrightarrow 3\ CH_3CO_2H(aq) + 4\ Cr^{3+}(aq) + 11\ H_2O(\ell)$$

Step 8. *Check the final result for mass and charge balance.*

Mass balance: 6 C, 17 O, 34 H, and 4 Cr

Charge balance: 12+ on both sides

EXERCISE 21.2 ***Balancing Equations for Oxidation-Reduction Reactions in Acid Solution***

Cobalt metal reacts with nitric acid to give a cobalt(III) salt and NO_2 gas. The unbalanced net ionic equation is

$$Co(s) + NO_3^-(aq) \longrightarrow Co^{3+}(aq) + NO_2(g)$$

1. Balance the equation for the reaction in acid solution.
2. Identify the oxidizing and reducing agents and the substance oxidized and the substance reduced. ■

Examples 21.2 and 21.3 illustrate the technique of balancing equations for redox reactions that occur in acid solution. Under those conditions H^+ ions or the H^+/H_2O pair can be used to achieve balanced equations. Conversely, in basic solution, only OH^- ions or the OH^-/H_2O pair can be used.

EXAMPLE 21.4 ***Balancing Equations for Oxidation-Reduction Reactions in Basic Solution***

Bismuth(III) ion, Bi^{3+}, is reduced to bismuth metal by tin(II) in basic solution.

$$SnO_2^{2-}(aq) + Bi(OH)_3(s) \longrightarrow Bi(s) + SnO_3^{2-}(aq)$$

Balance the equation for this reaction.

Solution

Step 1. *Verify that the reaction is an oxidation-reduction process.* The oxidation number of Bi changes from +3 to 0 and that for Sn changes from +2 to +4.

Step 2. *Separate the overall reaction into half-reactions.*

Reduction: $$Bi(OH)_3(s) \longrightarrow Bi(s)$$

Oxidation: $$SnO_2^{2-}(aq) \longrightarrow SnO_3^{2-}(aq)$$

Step 3. *Balance each half-reaction for mass.* In basic solution, OH^- can be a reactant or a product. Hydroxide ion is a product of the reduction of $Bi(OH)_3$ to bismuth metal.

$$Bi(OH)_3 \longrightarrow Bi(s) + 3\ OH^-(aq)$$

In the half-reaction involving tin, the Sn atoms are balanced, but the left side is deficent in oxygen.

$$(1\ \text{O atom required}) + SnO_2^{2-}(aq) \longrightarrow SnO_3^{2-}(aq)$$

In basic solution, two H-containing species exist: OH^- and H_2O. However, OH^- is oxygen-rich compared with H_2O. (Half of the atoms are O in OH^-, whereas only one third of them are O in H_2O). Therefore, OH^- can be thought of as a supplier of oxide ion (and also of water), as suggested by the following hypothetical, balanced equation:

Another way of balancing equations in basic solution is to treat the half-reaction as if it occurs in acid solution and then add OH^- ions to both sides of the equation to "neutralize" the OH^-. See Problem-Solving Tips and Ideas 21.1.

$$2\ OH^- \longrightarrow [O^{2-}] + H_2O$$

Two OH^- ions are therefore added to the O-deficient side to supply the one O atom needed, and H_2O is added to the other side to balance the H atoms.

$$2\ OH^-(aq) + SnO_2^{2-}(aq) \longrightarrow SnO_3^{2-}(aq) + H_2O(\ell)$$

Notice that *two* OH^- ions are required to supply *one* O atom and that *one* H_2O molecule must be added to the other side for *every O atom supplied by* OH^-.

Step 4. *Balance the half-reactions for charge.*

To balance O in a redox process in basic solution, add 2 OH^- to the O-deficient side and 1 H_2O to the other side, according to the balanced equation $2\ OH^- \rightarrow [O^{2-}] + H_2O$.

$$3\ e^- + Bi(OH)_3(s) \longrightarrow Bi(s) + 3\ OH^-(aq)$$
$$2\ OH^-(aq) + SnO_2^{2-}(aq) \longrightarrow SnO_3^{2-}(aq) + H_2O(\ell) + 2\ e^-$$

This confirms that $Bi(OH)_3$ is the oxidizing agent (electron acceptor) and SnO_2^{2-} is the reducing agent (electron donor).

Step 5. *Multiply the half-reactions by appropriate factors to balance the number of electrons donated and accepted.*

$$2[3\ e^- + Bi(OH)_3(s) \longrightarrow Bi(s) + 3\ OH^-(aq)]$$
$$3[2\ OH^-(aq) + SnO_2^{2-}(aq) \longrightarrow SnO_3^{2-}(aq) + H_2O(\ell) + 2\ e^-]$$

Step 6. *Add the half-reactions.*

$$6\ e^- + 2\ Bi(OH)_3(s) \longrightarrow 2\ Bi(s) + 6\ OH^-(aq)$$
$$6\ OH^-(aq) + 3\ SnO_2^{2-}(aq) \longrightarrow 3\ SnO_3^{2-}(aq) + 3\ H_2O(\ell) + 6\ e^-$$
$$\overline{3\ SnO_2^{2-}(aq) + 2\ Bi(OH)_3(s) \longrightarrow 2\ Bi(s) + 3\ SnO_3^{2-}(aq) + 3\ H_2O(\ell)}$$

Step 7. *Simplify by eliminating reactants and products common to both sides.* This was already done in Step 6.

Step 8. *Check the final result.*

Mass balance: 2 Bi, 3 Sn, 12 O, and 6 H

Charge balance: 6− on both sides

EXERCISE 21.3 ***Balancing Equations for Oxidation-Reduction Reactions in Basic Solution***

Batteries based on the reduction of sulfur appear very promising.* One being studied currently involves the reaction of sulfur with aluminum in a basic environment.

$$Al(s) + S(s) \longrightarrow Al(OH)_3(s) + HS^-(aq)$$

1. Balance the equation, showing each balanced half-reaction.
2. Identify the oxidizing and reducing agents, the substance oxidized, and the substance reduced. ■

*(See D. Peramunage and S. Licht: *Science,* Vol. 261, p. 1029, 1993.)

The preceding examples illustrate one method for balancing equations for redox reactions. The steps you go through are always the same, but *many variations* in detail exist, especially in the mass-balance step, Step 3. Only a few of these variations can be illustrated here, so the best way to learn to balance redox equations is to practice.

PROBLEM-SOLVING TIPS AND IDEAS

21.1 Balancing Equations for Oxidation-Reduction Reactions

- Never just add O^{2-} ions, O atoms, or O_2 molecules to balance oxygen in an equation. (Molecular O_2 should only appear if it is known to be a reactant or product.)
- Oxide ion cannot exist in water. Nonetheless, it often appears as though O^{2-} is the product of a reduction half-reaction in an acid solution or is required in an oxidation reaction in a basic solution. To balance such half-reactions, you can treat them as though (a) the product O^{2-} ion is converted to H_2O in acid solution by reaction with H^+ ion ($O^{2-} + 2\ H^+ \rightarrow H_2O$) or (b) the required O^{2-} ion is produced by OH^- in basic solution, with water as another product ($2\ OH^- \rightarrow O^{2-} + H_2O$).
- You may only use the pairs H^+/H_2O in acid or OH^-/H_2O in base (or H^+ or OH^- alone if appropriate).
- Another method of handling reactions in basic solution is to treat the half-reaction as if it had occurred in acid and then add OH^- to both sides of the equation to "neutralize" the H^+. For example, for the half-reaction $SnO_2^{2-}(aq) \rightarrow SnO_3^{2-}(aq)$, first balance the equation in acid solution,

$$H_2O(\ell) + SnO_2^{2-}(aq) \longrightarrow SnO_3^{2-}(aq) + 2\ H^+(aq) + 2\ e^-$$

and then add enough OH^- ions to both sides of the equation so that the H^+ ions are converted to water. Here two OH^- ions would be added and the equation is balanced after eliminating excess water molecules.

$$H_2O(\ell) + 2\ OH^-(aq) + SnO_2^{2-}(aq) \longrightarrow SnO_3^{2-}(aq) + 2\ H^+(aq) + 2\ OH^-(aq) + 2\ e^-$$

$$2\ OH^-(aq) + SnO_2^{2-}(aq) \longrightarrow SnO_3^{2-}(aq) + H_2O(\ell) + 2\ e^-$$

- Never add H atoms or H_2 molecules to balance hydrogen. (Molecular H_2 should only appear if it is known to be a reactant or product.) Hydrogen balance is achieved with H^+ and H_2O in acid or with H_2O and OH^- in basic solution.

- Be sure to write all the charges on any ions involved. Failing to include the correct charge is the most common error seen on student papers. ■

21.2 CHEMICAL CHANGE LEADING TO ELECTRIC CURRENT

Metallic zinc reacts readily with aqueous copper(II) ion to produce aqueous zinc(II) ion and copper metal (see Figure 21.2).

$$Zn(s) + Cu^{2+}(aq) \longrightarrow Zn^{2+}(aq) + Cu(s)$$

This may be interesting, but it would also be useful if the electrons transferred from zinc to copper(II) ion could be employed to power a computer or car. The problem with just dropping a piece of zinc into a solution of copper(II) ions is that the electrons provided by the zinc move directly to the aqueous Cu^{2+} ions on contact. In order to use the reaction as the basis of a battery, zinc metal and Cu^{2+} ions must be placed in separate containers (Figure 21.3). Electrons can then pass from the zinc **electrode,** a conductor of electrons, out of the solution into the external wire, through the device to be powered, and on to an electrode dipping into the solution of Cu^{2+} ions. Copper metal is then "plated out" onto the electrode in the beaker containing $Cu^{2+}(aq)$.

An electrode conducts electrons into and out of a solution. It is most often a metal plate or wire or a piece of graphite.

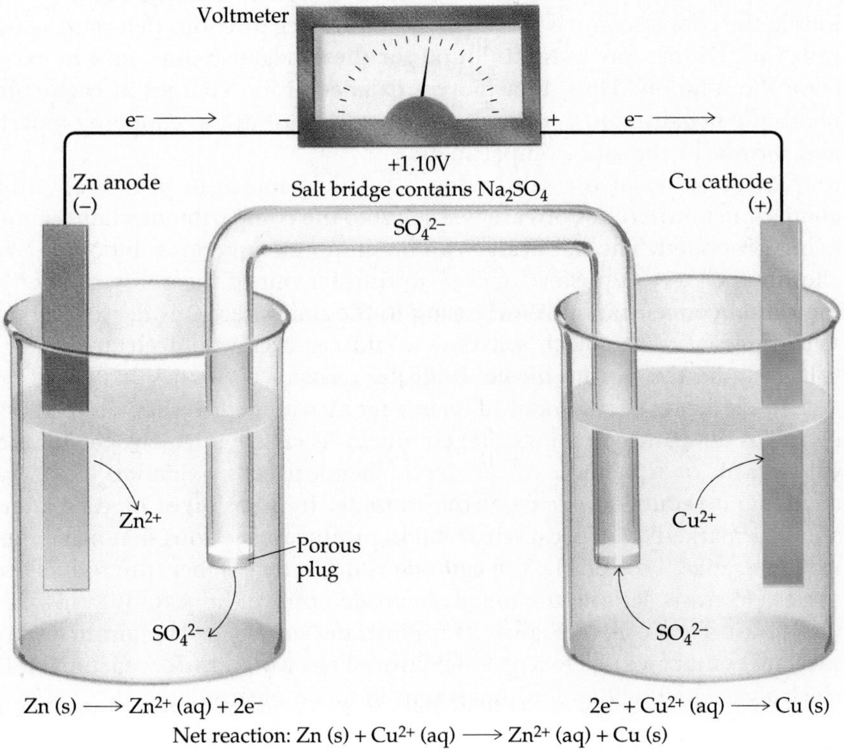

Figure 21.3 A voltaic cell using $Cu^{2+}(aq)/Cu(s)$ and $Zn^{2+}(aq)/Zn(s)$ half-cells. A voltage of 1.10 V is generated if the cell is set up under the conditions shown. Electrons flow through the external wire from the Zn electrode (anode) to the Cu electrode (cathode). A salt bridge provides a connection between the half-cells for ion flow, allowing SO_4^{2-} ions to flow from the copper to the zinc compartment.

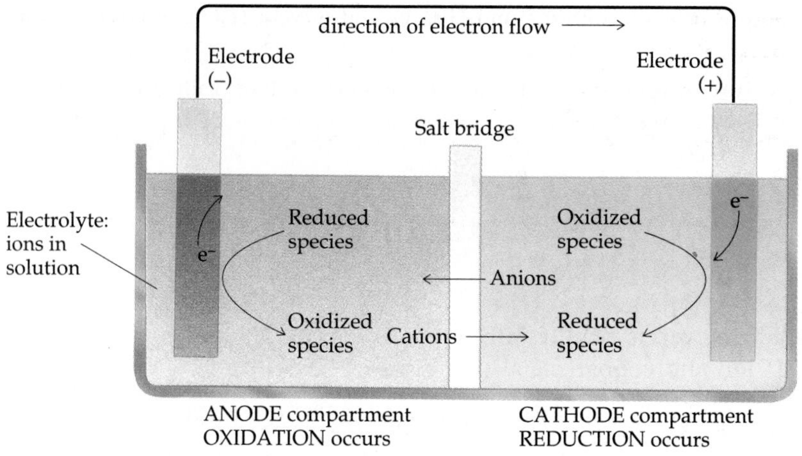

Figure 21.4 A summary of the terminology used in voltaic cells. Notice that negative charge moves in a circle through the cell and external wire. Electrons move from the negative electrode (anode, site of oxidation) to the positive electrode (cathode, site of reduction) in the external wire, and anions move from the cathode compartment to the anode compartment in the cell. Ions of the electrolyte carry charge from one electrode to the other.

The arrangement we just described works *only* if a **salt bridge,** a device for maintaining a balance of ion charges in the cell compartments, is also included. When the Zn electrode provides electrons to the wire, $Zn^{2+}(aq)$ enters the solution in the Zn compartment (see Figure 21.3), and negative ions must be found to balance these newly generated positive charges. Similarly, the loss of Cu^{2+} ions in the copper compartment leaves behind negative ions that were associated with Cu^{2+}. Some way must be found for these negative ions, now in excess, to leave the solution. Thus, to achieve a balance of ion charges in each compartment, the negative ion concentration must *decrease* in the copper compartment and *increase* in the zinc compartment.

The salt bridge also allows cation flow. Cations (Na^+ from the salt bridge) move into the copper compartment in Figure 21.3 (to replenish the Cu^{2+} consumed in the electrode reaction) and out of the zinc compartment (because Zn^{2+} ions are being added to the solution by the electrode reaction).

The function of the salt bridge is to allow anions to pass freely from the compartment where cations are being lost to the compartment where cations are being generated. The salt bridge can be simply an aqueous solution of Na_2SO_4, allowing $SO_4{}^{2-}$ to transfer (or Na^+ to transfer out of the compartment where the cation concentration is increasing to the one where it is decreasing). If the salt bridge were removed, ion flow would cease as would electron flow. The voltage indicated on the meter would be zero.

Strictly speaking, a battery is a collection of cells that produces a current or flow of electrons at a constant voltage as a result of an oxidation-reduction reaction.

An oxidizing agent and a reducing agent arranged so they can react only if electrons flow through an outside conductor is called an **electrochemical cell,** a **voltaic cell,** or a **battery.** In *all* electrochemical cells **oxidation** occurs at the **anode,** and **reduction** occurs at the **cathode.** In a flashlight or car battery the anode is marked "–" because oxidation produces electrons that make the electrode negative. Conversely, the cathode is marked "+" because reduction consumes electrons, leaving the metal electrode positive. Important terms are summarized in Figure 21.4. Figure 21.5 illustrates an electrochemical cell built of substances that give rise to a product-favored electron transfer reaction and of an **electrolyte** which allows ion movement between electrodes.

To remember that **A**node and **O**xidation are paired as are **C**athode and **R**eduction, note the alphabetic orders:

Anode	⟷	Oxidation
↑		↑
(A before C)		(O before R)
↓		↓
Cathode	⟷	Reduction

Finally, it is useful to point out that we often talk about electric circuits for the simple reason that electric charges—electrons or ions—always flow in a "circle." As shown in Figures 21.3 and 21.4 electrons move through the external circuit from reducing agent (Zn) to oxidizing agent (Cu^{2+}); the negative ions complete the circle through the salt bridge (from Cu^{2+} to Zn^{2+}).

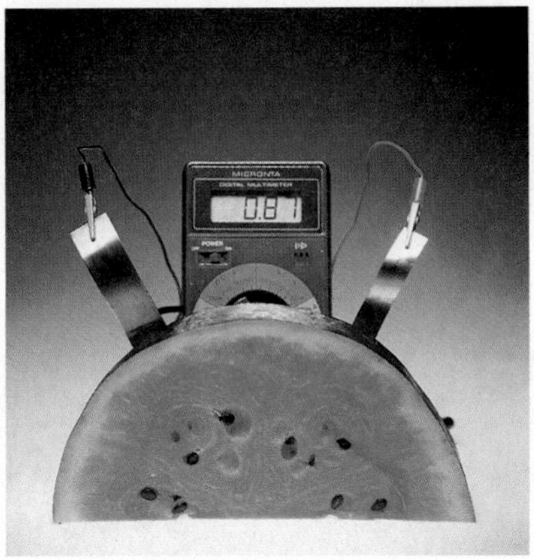

Figure 21.5 An electrochemical cell can be made by inserting copper and zinc electrodes into almost any conductive material. (C. D. Winters)

EXAMPLE 21.5 *Electrochemical Cells*

A simple voltaic cell has been assembled with Ni(s) and $Ni(NO_3)_2$(aq) in one compartment and Cd(s) and $Cd(NO_3)_2$(aq) in the other. An external wire connects the two electrodes, and a salt bridge containing $NaNO_3$ connects the two solutions. The net reaction is

$$Cd(s) + Ni^{2+}(aq) \longrightarrow Ni(s) + Cd^{2+}(aq)$$

What half-reaction occurs at each electrode? Which is the anode, and which is the cathode? What is the direction of electron flow in the external wire and of anion flow in the salt bridge?

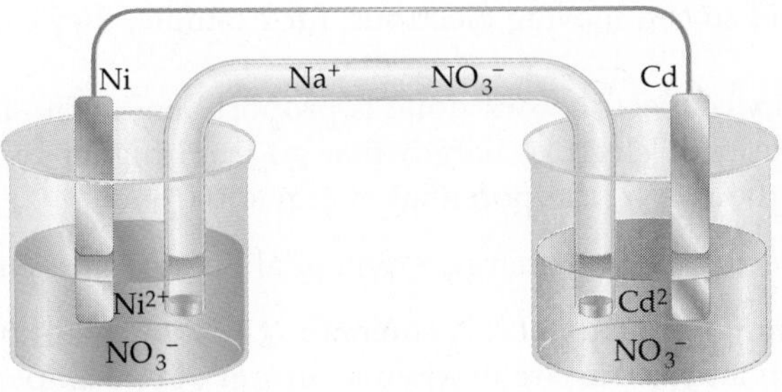

Solution Based on the net ionic equation we know Cd(s) is the reducing agent, and Ni^{2+}(aq) is the oxidizing agent. Thus, the half-reactions are

Anode, oxidation: $Cd(s) \longrightarrow Cd^{2+}(aq) + 2\ e^-$

Cathode, reduction: $Ni^{2+}(aq) + 2\ e^- \longrightarrow Ni(s)$

Electrons flow from their source (the oxidation of Cd) through the wire to the electrode where they are used to reduce Ni^{2+}(aq).

Because Cd^{2+} ions are formed in the anode compartment, anions must move into that compartment from the salt bridge. The Ni^{2+} concentration in the cathode compartment is decreasing, so anions move out of that compartment into the salt bridge. The "circle" of flow of negative charge is complete: Electrons flow from Cd to Ni, and negative ions move from Ni to Cd.

Although we have described the salt bridge as a device that allows anion flow to maintain electric neutrality, it also allows cation flow. Cations (here Na^+ from the salt bridge) move into the nickel compartment (to replenish the Ni^{2+} consumed in the electrode reaction) and out of the cadmium compartment (because Cd^{2+} ions are being added to the solution by the electrode reaction).

EXERCISE 21.4 ***Electrochemical Cells***

A voltaic cell has been assembled with the net reaction

$$Ni(s) + 2\ Ag^+(aq) \longrightarrow Ni^{2+}(aq) + 2\ Ag(s)$$

Give the half-reactions for this electron transfer process, indicating whether each is an oxidation or reduction and deciding which happens at the anode and which at the cathode. What is the direction of electron flow in an external wire connecting the two electrodes? If a salt bridge connecting the cell compartments contains KNO_3, what is the direction of flow of the nitrate ions? ■

21.3 ELECTROCHEMICAL CELLS AND POTENTIALS

Electrons generated at the site of oxidation (the anode) of a cell are thought to be "driven," or "pushed," toward the cathode by an **electromotive force,** or **emf.** This force is due to the difference in electric potential energy of an electron at the two electrodes. Just as a ball rolls downhill in response to a difference in gravitational potential energy, an electron moves from an electrode of higher electric potential energy to one of lower electric potential energy. A moving ball can do work and so can moving electrons; for example, they can power a computer.

The quantity of electric work done is proportional to the number of electrons (the quantity of electric charge) that go from higher to lower potential energy and to the size of the potential energy difference.

$$\text{Electric work} = \text{charge} \times \text{potential energy difference}$$

Charge is measured in coulombs. A **coulomb** (C) is the quantity of charge that passes a point in an electric circuit when a current of one ampere flows for one second. The charge on a single electron is very small (1.6022×10^{-19} C), so it takes 6.24×10^{18} electrons to produce just one coulomb of charge. Electric potential energy difference is measured in volts. The volt is defined so that one joule of work is performed when one coulomb of charge passes through a potential difference of one volt.

$$\text{Volt} = \frac{1\ \text{joule}}{1\ \text{coulomb}} \quad \text{or} \quad \text{Joule} = 1\ \text{volt} \times 1\ \text{coulomb}$$

The maximum work that can be accomplished by an electrochemical cell is equal to the product of the cell potential and the charge passing through the circuit. The cell potential depends in turn on the substances that make up the cell, whether they are gases or solutes in solution, and on their concentration. The quantity of charge depends on the quantity of reactants consumed. You know that if two batteries produce the same voltage, the one with the greater quantity of reactants is capable of more work (Figure 21.6).

Because the potential of a cell depends on the concentrations of reactants and products, **standard conditions** have been defined for electrochemical mea-

Figure 21.6 These dry cell batteries all produce 1.5 V. Some are larger than others, however, because they contain more oxidizing and reducing agents and so can produce more electrical work. (C. D. Winters)

surements. These are the same as for standard enthalpies and free energies of formation (Sections 6.8 and 20.3). When the reactants and products are present as pure solids or in solution at a concentration of 1.0 M, or as gases at 1.0 bar, the measured cell potential is the **standard potential, $E°$.** Unless specified otherwise, all values of $E°$ are given at 25 °C (298 K). By definition, cell potentials for product-favored electrochemical reactions are *positive.* For example, the cell illustrated in Figure 21.3 has a potential of +1.10 V at 25 °C.

$E°$ and $\Delta G°$

The standard potential $E°$ is a quantitative measure of the tendency of the reactants in their standard states to proceed to products in their standard states and so is the standard free energy change for a reaction, $\Delta G°_{rxn}$. The exact relation between them is

$$\Delta G°_{rxn} = -nFE° \qquad \textbf{(21.1)}$$

where n is the number of moles of electrons transferred between oxidizing and reducing agents in a balanced redox reaction, and F is the **Faraday constant,** 9.6485309×10^4 J/V · mol.

The Faraday constant is named in honor of Michael Faraday, the first to investigate quantitatively the relation between chemistry and electricity. See page 126 and Section 21.9.

The reaction of Zn(s) and Cu^{2+}(aq) produces a current, so you readily conclude that it is a product-favored reaction as written

$$Zn(s) + Cu^{2+}(aq) \longrightarrow Zn^{2+}(aq) + Cu(s) \qquad E° = +1.10\ V$$

Because spontaneous reactions have a negative free energy change, $\Delta G°_{rxn}$, the negative sign in Equation 21.1 confirms that *all product-favored* electron transfer reactions have a *positive* $E°$.

If the direction of the reaction is reversed, the sign of $\Delta G°_{rxn}$ and so the sign of $E°$ are reversed. Thus, if we write the equation for the reduction of Zn^{2+} by Cu,

$$Cu(s) + Zn^{2+}(aq) \longrightarrow Cu^{2+}(aq) + Zn(s) \qquad E° = -1.10\ V$$

this equation depicts a reaction that is *not* product-favored. It must have a positive $\Delta G°_{rxn}$ and a negative $E°$. *Reactant-favored* reactions have a *negative* $E°$.

When a reaction is reversed, the magnitudes of $\Delta G°$ and $E°$ remain the same, but their signs are reversed [(+ → −) or (− → +)].

EXAMPLE 21.6 ***The Relation Between* E° *and* ΔG°*rxn***

The reaction of zinc metal with copper(II) ions (see Figure 21.3) has a standard cell potential $E°$ of +1.10 V at 25 °C. Calculate $\Delta G°_{rxn}$ for the reaction.

$$Zn(s) + Cu^{2+}(aq) \longrightarrow Zn^{2+}(aq) + Cu(s)$$

Solution To obtain $\Delta G°_{rxn}$ we substitute the given $E°$ value into Equation 21.1.

$$\Delta G°_{rxn} = -(2.00 \text{ mol electrons transferred})\left(\frac{9.65 \times 10^4 \text{ J}}{\text{V} \cdot \text{mol}}\right)(1.10 \text{ V})\left(\frac{1 \text{ kJ}}{1000 \text{ J}}\right)$$

$$= -212 \text{ kJ}$$

EXERCISE 21.5 ***The Relation Between* E° *and* ΔG°*rxn***

The following reaction has an $E°$ value of −0.76 V. Calculate $\Delta G°_{rxn}$, and tell whether the reaction is product-favored or reactant-favored.

$$H_2(g) + 2\ H_2O(\ell) + Zn^{2+}(aq) \longrightarrow Zn(s) + 2\ H_3O^+(aq)$$ ■

Calculating the Potential $E°$ of an Electrochemical Cell

An oxidation-reduction reaction is the sum of two half-reactions, one for oxidation and the other for reduction (see Section 21.1). For the $Zn(s)/Cu^{2+}(aq)$ reaction in Figure 21.3, for example,

Anode, oxidation:	$Zn(s)$	$\longrightarrow$	$Zn^{2+}(aq) + 2\ e^-$
Cathode, reduction:	$Cu^{2+}(aq) + 2\ e^-$	$\longrightarrow$	$Cu(s)$
Net process:	$Zn(s) + Cu^{2+}(aq)$	$\longrightarrow$	$Cu(s) + Zn^{2+}(aq)$

It would be very helpful if we could *predict* the standard potential for this reaction, and it makes sense that it might be the sum of the potentials of the half-reactions. The problem is that potentials for isolated half-reactions cannot be obtained directly because the potential measures the potential energy difference for electrons in two different chemical environments. We can always measure the standard potential for any half-reaction in combination with some other, standard half-reaction, however. Indeed, chemists decided that the standard half-cell reaction against which all others are measured is the **standard hydrogen electrode** (SHE),

$$2\ H_3O^+(aq) + 2\ e^- \longrightarrow H_2(g, 1 \text{ atm}) + 2\ H_2O(\ell) \qquad E° = 0.00 \text{ V}$$

and a *potential of 0.00 V* has been assigned to this half-reaction. (This value has no physical meaning in itself, just as the half-reaction alone has no meaning.) To measure the potential for any half-reaction, we make that reaction one side of an electrochemical cell and the H_2/H_3O^+ half-reaction the other side. Molecular hydrogen, H_2, is a reducing agent, and $H_3O^+(aq)$ is an oxidizing agent. Thus, when the standard hydrogen electrode is paired with another half-cell in an electrochemical cell, the H_2/H_3O^+ half-reaction can be either an oxidation or a reduction. If the other half-cell contains a better reducing agent than H_2, then H_3O^+ is reduced to H_2.

H_3O^+ reduced:

$$2\ H_3O^+(aq, 1\ M) + 2\ e^- \longrightarrow H_2(g, 1\ atm) + 2\ H_2O(\ell) \qquad E° = 0.00\ V$$

If the other half-cell contains a better oxidizing agent than H_3O^+, then H_2 is oxidized.

H_2 oxidized:

$$H_2(g, 1\ atm) + 2\ H_2O(\ell) \longrightarrow 2\ H_3O^+(aq, 1\ M) + 2\ e^- \qquad E° = 0.00\ V$$

In either direction, the H_2/H_3O^+ half cell has a potential of 0.00 V. The measured potential of the electrochemical cell is then *assigned* as the potential of the half-cell being studied. To demonstrate the strategy for determining half-cell potentials, consider the following examples.

Figure 21.7 illustrates a cell in which one compartment contains the H_2/H_3O^+ reaction mixture, and the other compartment has a zinc electrode dipping into a solution of 1 M Zn^{2+}. The compartments are connected by an external wire (for electron flow) and a salt bridge (for ion flow).

The *measured potential* of an electrochemical cell is *always positive.* The device used to measure potentials, a voltmeter, is designed to give a positive potential only when the positive terminal (+) of the voltmeter is connected to the positive electrode and the negative terminal (−) to the negative electrode. In this way we not only measure the potential, but we find the sign of each of the electrodes. Thus, when a voltmeter is attached to the cell shown in Figure 21.7, it is found that the H_2/H_3O^+ electrode is positive, the zinc electrode is negative, and the

When a redox reaction involves an ion or compound that cannot be made into solid electrodes, a chemically inert conductor of electricity can be used. Platinum and gold are frequently used as inert electrodes.

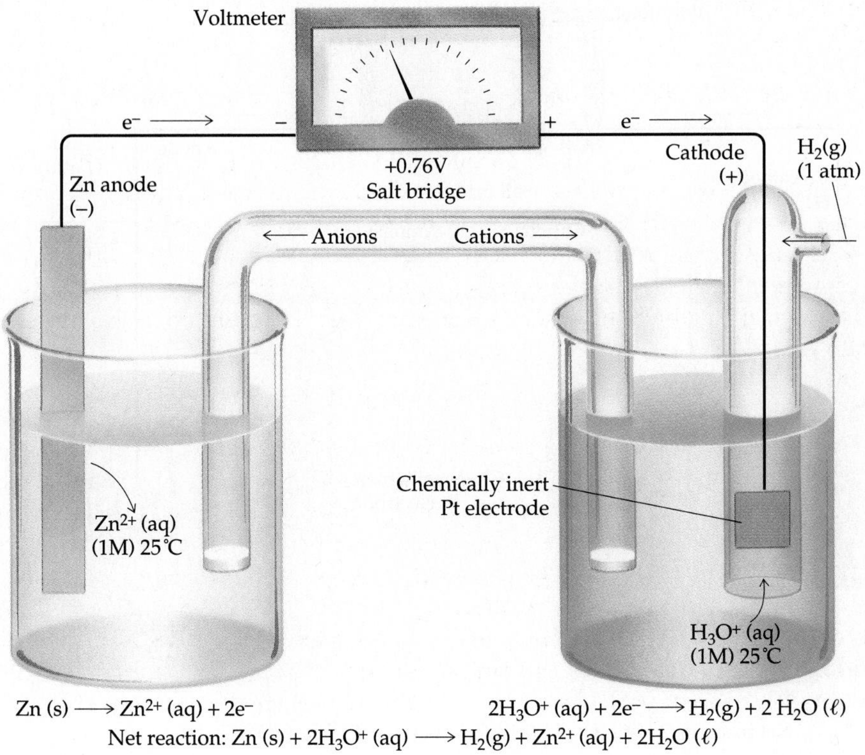

Figure 21.7 An electrochemical cell using $Zn^{2+}(aq)/Zn(s)$ and $H_3O^+(aq)/H_2(g)$ half-cells. A voltage of +0.76 V is generated when the cell is set up under the conditions shown at 25 °C. Electrons flow from the Zn electrode (anode) to the $H_3O^+(aq)/H_2(g)$ electrode (cathode) to produce $Zn^{2+}(aq)$ and $H_2(g)$. Zinc is the reducing agent, and $H_3O^+(aq)$ is the oxidizing agent. (Because the species in the $H_3O^+(aq)/H_2(g)$ half-reaction cannot be fashioned into a solid electrode, electrons are transferred using a piece of platinum foil.)

See the photo of the reaction of Zn with acid on page 235.

measured potential is $E° = +0.76$ V. Although both Zn and H_2 can potentially function as reducing agents here, the observation that the Zn electrode is negative means that Zn is the source of electrons, and so we know that Zn is a *better reducing agent* than H_2 gas. We therefore observe that the zinc electrode dissolves as zinc metal is oxidized to Zn^{2+} ions, and H_2 gas is formed from the reduction of H_3O^+.

Anode, oxidation:	$Zn(s) \longrightarrow Zn^{2+}(aq, 1\ M) + 2\ e^-$	$E° = ?$ V
Cathode, reduction:	$2\ H_3O^+(aq, 1\ M) + 2\ e^- \longrightarrow H_2(g, 1\ atm) + 2\ H_2O(\ell)$	$E° = 0.00$ V
Net reaction:	$Zn(s) + 2\ H_3O^+(aq, 1\ M) \longrightarrow Zn^{2+}(aq) + H_2(g, 1\ atm) + 2\ H_2O(\ell)$	$E°_{net} = +0.76$ V

The + sign of $E°_{net}$ correctly reflects the fact that the overall reaction is product-favored as written (and we already knew this from the fact that zinc reacts readily with acid). Because the potential for the H_2/H_3O^+ half-cell is 0.00 V, the $E°$ for the Zn/Zn^{2+} half-cell must be +0.76 V.

The potential produced by an electrochemical cell is the sum of the potentials of the oxidizing half-reaction and the reducing half-reaction.

$$Zn(s) \longrightarrow Zn^{2+}(aq, 1\ M) + 2\ e^- \qquad E° = +0.76\ V$$

What is $E°$ for the Cu/Cu^{2+} half-cell in our original electrochemical cell in Figure 21.3? In Figure 21.8 this half-cell is shown coupled with the standard H_2/H_3O^+ half-cell. The measured cell potential is +0.34 V, the H_2/H_3O^+ half-cell is negative, and the Cu/Cu^{2+} electrode is positive. Furthermore, the concentration of Cu^{2+} ions declines, and metallic copper forms. All these experimental observations confirm that Cu^{2+} is being reduced and that H_2 must be the reducing agent, the source of electrons. Because the reducing agent, H_2, is oxidized to H_3O^+, the H_2 electrode must be the anode (and is negatively charged). Further-

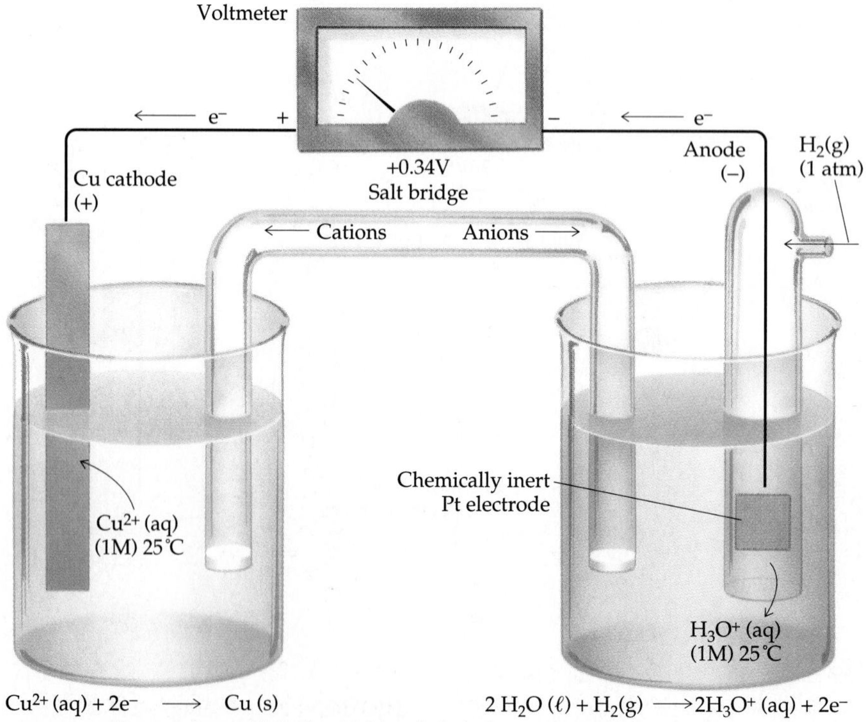

Figure 21.8 An electrochemical cell using $Cu^{2+}(aq)/Cu(s)$ and $H_3O^+(aq)/H_2(g)$ half-cells. A voltage of +0.34 V is generated when the cell is set up as pictured at 25 °C. Electrons flow from the $H_3O^+(aq)/H_2(g)$ electrode (the anode) to the copper electrode (cathode) to produce copper and hydronium ions. Therefore, $H_2(g)$ is the reducing agent and $Cu^{2+}(aq)$ is the oxidizing agent.

more, because Cu^{2+} ions are the acceptors of electrons (the oxidizing agent), this electrode is the cathode (and is positively charged). Most importantly, it is apparent that $H_2(g)$ is a better reducing agent than Cu(s), and the appropriate half-reactions and net reaction are

Anode, oxidation:	$H_2(g, 1\ atm) + 2\ H_2O(\ell) \longrightarrow 2\ H_3O^+(aq, 1\ M) + 2\ e^-$	$E° = 0.00\ V$
Cathode, reduction:	$Cu^{2+}(aq, 1\ M) + 2\ e^- \longrightarrow Cu(s)$	$E° = ?V$
Net reaction:	$Cu^{2+}(aq, 1\ M) + H_2(g, 1\ atm) + 2\ H_2O(\ell) \longrightarrow Cu(s) + 2\ H_3O^+(aq, 1\ M)$	$E°_{net} = +0.34\ V$

The half-cell potential for Cu^{2+} (aq, 1 M) + 2 $e^- \rightarrow$ Cu(s) must be +0.34 V at 25 °C.

The potential of a cell in which Zn(s) reduces Cu^{2+}(aq) to Cu(s) can now be calculated, because we have $E°$ values for the half-reactions involved.

Anode, oxidation:	$Zn(s) \longrightarrow Zn^{2+}(aq, 1\ M) + 2\ e^-$	$E° = +0.76\ V$
Cathode, reduction:	$Cu^{2+}(aq, 1\ M) + 2\ e^- \longrightarrow Cu(s)$	$E° = +0.34\ V$
Net reaction:	$Cu^{2+}(aq, 1\ M) + Zn(s) \longrightarrow Cu(s) + Zn^{2+}(aq, 1\ M)$	$E°_{net} = +1.10\ V$

This is an important result. Two half-cell potentials were measured independently against the same standard H_2/H_3O^+ half-cell, and the sum of the two potentials is equal to the experimentally measured $E°$ (see Figure 21.3.) Now we can use a similar technique with any of the preceding half-cells (Cu/Cu^{2+}, Zn/Zn^{2+}, or H_2/H_3O^+) as reference cells and determine $E°$ values for hundreds of other possible half-cells.

EXAMPLE 21.7 *Determining a Half-Reaction Potential*

The cell illustrated here has a potential of $E° = +0.51$ V at 25 °C. The net ionic equation for the cell reaction is

$$Zn(s) + Ni^{2+}(aq, 1\ M) \longrightarrow Zn^{2+}(aq, 1\ M) + Ni(s)$$

Which electrode is the anode, and which is the cathode? What are the polarities of the electrodes? What is the value of $E°$ for the half-cell $Ni^{2+}(aq) + 2\ e^- \rightarrow Ni(s)$?

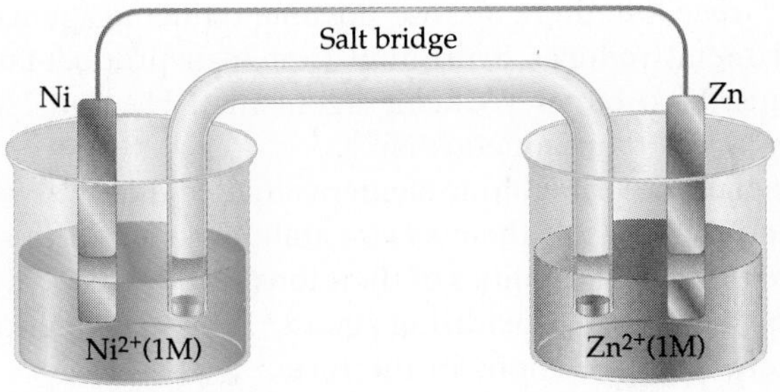

Solution Zn(s) is oxidized to Zn^{2+}(aq), so the Zn electrode is the anode and is negatively charged because it is the source of electrons. Nickel(II) ions are reduced to Ni metal at the Ni electrode, so this is the cathode, and it is positive.

Because the overall cell potential is known, and the potential for the $Zn(s)/Zn^{2+}(aq)$ half-cell is known, $E°$ for the nickel half-cell can be calculated.

Anode, oxidation:	$Zn(s) \longrightarrow Zn^{2+}(aq) + 2\,e^-$	$E° = +0.76$ V
Cathode, reduction:	$Ni^{2+}(aq) + 2\,e^- \longrightarrow Ni(s)$	$E° = ?$ V
Net reaction:	$Zn(s) + Ni^{2+}(aq) \longrightarrow Zn^{2+}(aq) + Ni(s)$	$E°_{net} = +0.51$ V

At 25 °C the value of $E°$ for the reaction $Ni^{2+}(aq) + 2\,e^- \rightarrow Ni(s)$ is -0.25 V.

EXERCISE 21.6 ***Determining a Half-Reaction Potential***

Given that the reduction of aqueous copper(II) with iron metal has an $E°_{net}$ value of $+0.78$ V, what is $E°_{rxn}$ for the half-cell $Fe(s) \rightarrow Fe^{2+}(aq, 1\,M) + 2\,e^-$?

$$Fe(s) + Cu^{2+}(aq, 1\,M) \longrightarrow Fe^{2+}(aq, 1\,M) + Cu(s)$$

■

21.4 USING STANDARD POTENTIALS

The standard potential $E°$ for the reduction of Ni^{2+} to metallic nickel is -0.25 V (see Example 21.7). What does this mean? This tells us that if Ni^{2+} is used as an oxidizing agent, coupled with H_2 as the reducing agent,

Cathode, reduction:	$Ni^{2+}(aq, 1\,M) + 2\,e^- \longrightarrow Ni(s)$	$E° = -0.25$ V
Anode, oxidation:	$H_2(g, 1\,atm) + 2\,H_2O(\ell) \longrightarrow 2\,H_3O^+(aq, 1\,M) + 2\,e^-$	$E° = 0.00$ V
Net reaction:	$Ni^{2+}(aq, 1\,M) + H_2(g, 1\,atm) + 2\,H_2O(\ell) \longrightarrow Ni(s) + 2\,H_3O^+(aq, 1\,M)$	$E°_{net} = -0.25$ V

the reaction is *reactant-favored* under standard conditions; $E°_{net}$ is negative so $\Delta G°$ is positive. From thermodynamics, you know that if a reaction is not product-favored in the direction written ($\Delta G°$ is positive), the reverse reaction is product-favored ($\Delta G°$ is negative). Therefore, the reaction

$$Ni(s) + 2\,H_3O^+(aq, 1\,M) \longrightarrow Ni^{2+}(aq, 1\,M) + H_2(g, 1\,atm) + 2\,H_2O(\ell) \qquad E°_{net} = +0.25\text{ V}$$

is product-favored, and $E°_{net}$ for the reaction is positive. Just as acid-base reactions always move in the direction of the weaker acid-base pair (see Section 17.3), redox reactions move toward the weaker oxidizing agent/reducing agent pair. In the Ni/H_3O^+ reaction, there are two possible reducing agents: Ni(s) and H_2. Because nickel metal reduces hydronium ion in a product-favored reaction, nickel metal must be a better reducing agent than H_2, and H_3O^+ must be a better oxidizing agent than aqueous Ni^{2+}.

Thus far we have described four elements that could act as reducing agents: Zn, Cu, H_2, and Ni. What are their *relative* abilities to act in this manner? Conversely, what are the relative abilities of their ions (Zn^{2+}, Cu^{2+}, H_3O^+, and Ni^{2+}) to act as electron acceptors or oxidizing agents? If we write half-reactions involving these elements and their ions in the form

$$\text{Oxidized form} + n\,e^- \longrightarrow \text{Reduced form}$$

and list them in order of descending $E°$ values, we will have placed the oxidizing agents in descending order of their ability to attract electrons.

Increasing strength as oxidizing agent ↑	Half-reaction	Increasing strength as reducing agent ↓	
	$Cu^{2+}(aq, 1\ M) + 2\ e^- \longrightarrow Cu(s)$		$E° = +0.34$ V
	$2\ H_3O^+(aq, 1\ M) + 2\ e^- \longrightarrow H_2(g, 1\ atm) + 2\ H_2O(\ell)$		$E° = 0.00$ V
	$Ni^{2+}(aq, 1\ M) + 2\ e^- \longrightarrow Ni(s)$		$E° = -0.25$ V
	$Zn^{2+}(aq, 1\ M) + 2\ e^- \longrightarrow Zn(s)$		$E° = -0.76$ V

The value of $E°$ becomes more negative down the series. This means that $Cu^{2+}(aq)$ is the best oxidizing agent of the substances on the left; that is, Cu^{2+} shows the greatest tendency to be reduced. Conversely, Zn^{2+} is the worst oxidizing agent; it has the least tendency to be reduced. Of the substances on the right, Zn(s) is the best reducing agent (best electron donor), because $E°$ for the half-reaction

$$Zn(s) \longrightarrow Zn^{2+}(aq, 1\ M) + 2\ e^- \qquad E° = +0.76$$

has the most positive value. By the same reasoning, Cu is the worst reducing agent.

The preceding table of half-reaction potentials tells us that, at standard conditions, the following reactions are product-favored:

Increasing oxidizing ability ↑		Increasing reducing ability ↓	
	Cu^{2+} can oxidize H_2, Ni, and Zn		H_2 can reduce Cu^{2+}
	H_3O^+ can oxidize Ni and Zn		Ni can reduce H_3O^+ and Cu^{2+}
	Ni^{2+} can oxidize Zn		Zn can reduce Ni^{2+}, H_3O^+, and Cu^{2+}

Each of these reactions has a positive $E°_{net}$ (and a negative $\Delta G°$). The reduction of $Cu^{2+}(aq)$ with Ni further illustrates the point.

Oxidation:	$Ni(s) \longrightarrow Ni^{2+}(aq) + 2\ e^-$	$E° = +0.25$ V
Reduction:	$Cu^{2+}(aq) + 2\ e^- \longrightarrow Cu(s)$	$E° = +0.34$ V
Net reaction:	$Ni(s) + Cu^{2+}(aq) \longrightarrow Ni^{2+}(aq) + Cu(s)$	$E°_{net} = +0.59$ V

The positive $E°$ (and negative $\Delta G°$) for the reaction confirms that it is product-favored.

Many more values of $E°$ are given in Appendix J.

We have just created a small portion of a **table of standard reduction potentials** (Table 21.1). Like the table of conjugate acids and bases in Table 17.4, Table 21.1 is *very* useful. Some important points concerning this table are

- The $E°$ values are for reactions written in the form "oxidized form + electrons → reduced form." The species on the left side of the reaction is an oxidizing agent, and the species on the right is a reducing agent. All potentials are therefore for reduction reactions.
- When writing the reaction "reduced form → oxidized form + electrons," the sign of $E°$ is reversed, but the value of $E°$ is unaffected. Thus,

$$Li(s) \longrightarrow Li^+(aq) + e^- \qquad E° = +3.045\ V$$

TABLE 21.1 Standard Reduction Potentials in Aqueous Solution at 25 °C*

Reduction Half-Reaction		$E°$ (V)
$F_2(g) + 2\ e^-$	$\longrightarrow 2\ F^-(aq)$	+2.87
$H_2O_2(aq) + 2\ H_3O^+(aq) + 2\ e^-$	$\longrightarrow 4\ H_2O(\ell)$	+1.77
$PbO_2(s) + SO_4^{2-}(aq) + 4\ H_3O^+(aq) + 2\ e^-$	$\longrightarrow PbSO_4(s) + 6\ H_2O(\ell)$	+1.685
$MnO_4^-(aq) + 8\ H_3O^+(aq) + 5\ e^-$	$\longrightarrow Mn^{2+}(aq) + 12\ H_2O(\ell)$	+1.52
$Au^{3+}(aq) + 3\ e^-$	$\longrightarrow Au(s)$	+1.50
$Cl_2(g) + 2\ e^-$	$\longrightarrow 2\ Cl^-(aq)$	+1.360
$Cr_2O_7^{2-}(aq) + 14\ H_3O^+(aq) + 6\ e^-$	$\longrightarrow 2\ Cr^{3+}(aq) + 21\ H_2O(\ell)$	+1.33
$O_2(g) + 4\ H_3O^+(aq) + 4\ e^-$	$\longrightarrow 6\ H_2O(\ell)$	+1.229
$Br_2(\ell) + 2\ e^-$	$\longrightarrow 2\ Br^-(aq)$	+1.08
$NO_3^-(aq) + 4\ H_3O^+(aq) + 3\ e^-$	$\longrightarrow NO(g) + 6\ H_2O(\ell)$	+0.96
$OCl^-(aq) + H_2O(\ell) + 2\ e^-$	$\longrightarrow Cl^-(aq) + 2\ OH^-(aq)$	+0.89
$Hg^{2+}(aq) + 2\ e^-$	$\longrightarrow Hg(\ell)$	+0.855
$Ag^+(aq) + e^-$	$\longrightarrow Ag(s)$	+0.80
$Hg_2^{2+}(aq) + 2\ e^-$	$\longrightarrow 2\ Hg(\ell)$	+0.789
$Fe^{3+}(aq) + e^-$	$\longrightarrow Fe^{2+}(aq)$	+0.771
$I_2(s) + 2\ e^-$	$\longrightarrow 2\ I^-(aq)$	+0.535
$O_2(g) + 2\ H_2O(\ell) + 4\ e^-$	$\longrightarrow 4\ OH^-(aq)$	+0.40
$Cu^{2+}(aq) + 2\ e^-$	$\longrightarrow Cu(s)$	+0.337
$Sn^{4+}(aq) + 2\ e^-$	$\longrightarrow Sn^{2+}(aq)$	+0.15
$2\ H_3O^+(aq) + 2\ e^-$	$\longrightarrow H_2(g) + 2\ H_2O(\ell)$	0.00
$Sn^{2+}(aq) + 2\ e^-$	$\longrightarrow Sn(s)$	−0.14
$Ni^{2+}(aq) + 2\ e^-$	$\longrightarrow Ni(s)$	−0.25
$V^{3+}(aq) + e^-$	$\longrightarrow V^{2+}(aq)$	−0.255
$PbSO_4(s) + 2\ e^-$	$\longrightarrow Pb(s) + SO_4^{2-}(aq)$	−0.356
$Cd^{2+}(aq) + 2\ e^-$	$\longrightarrow Cd(s)$	−0.40
$Fe^{2+}(aq) + 2\ e^-$	$\longrightarrow Fe(s)$	−0.44
$Zn^{2+}(aq) + 2\ e^-$	$\longrightarrow Zn(s)$	−0.763
$2\ H_2O(\ell) + 2\ e^-$	$\longrightarrow H_2(g) + 2\ OH^-(aq)$	−0.8277
$Al^{3+}(aq) + 3\ e^-$	$\longrightarrow Al(s)$	−1.66
$Mg^{2+}(aq) + 2\ e^-$	$\longrightarrow Mg(s)$	−2.37
$Na^+(aq) + e^-$	$\longrightarrow Na(s)$	−2.714
$K^+(aq) + e^-$	$\longrightarrow K(s)$	−2.925
$Li^+(aq) + e^-$	$\longrightarrow Li(s)$	−3.045

Increasing strength of oxidizing agents (↑, left column); Increasing strength of reducing agents (↓, right column)

*In volts (V) versus the standard hydrogen electrode.

- All the half-reactions are reversible. For example, aqueous H_3O^+ is reduced to $H_2(g)$ in Figure 21.7, whereas $H_2(g)$ is oxidized to H_3O^+ in Figure 21.8.
- The more positive the value of $E°$ for the reactions in Table 21.1, the better the oxidizing ability of the ion or compound on the left side of the reaction. This means $F_2(g)$ *is the best oxidizing agent in the table.*

$$F_2(g, 1\ atm) + 2\ e^- \longrightarrow 2\ F^-(aq, 1\ M) \qquad E° = +2.87\ V$$

The ion at the bottom left corner of the table, $Li^+(aq)$, is the poorest oxidizing agent because its $E°$ is the most negative. The oxidizing agents in the table (ions, elements, and compounds at the left) *increase* in strength *from the bottom to the top of the table.*

- The more negative the value of the reduction potential $E°$ the less likely the reaction occurs as a reduction, and the more likely the reverse reaction occurs (as an oxidation). Thus, Li(s) is the strongest reducing agent in the table, and F^- is the weakest reducing agent. The reducing agents in the table (the ions, elements, or compounds at the right) *increase* in strength *from the top to the bottom.*
- The reaction between any substance on the left in this table (an oxidizing agent) with any substance lower than it on the right (a reducing agent) is product-favored under standard conditions.
- The algebraic sign of the half-reaction potential is the sign of the electrode when it is attached to the H_2/H_3O^+ standard cell. (See Figures 21.7 and 21.8.)
- Electrochemical potentials depend on the nature of the reactants and products and their concentrations, not on the quantities of material used. Changing the stoichiometric coefficients for a half-reaction therefore does not change the value of $E°$. For example, the reduction of Fe^{3+} has an $E°$ of +0.771 V, whether the reaction is written as

$$Fe^{3+}(aq, 1\ M) + e^- \longrightarrow Fe^{2+}(aq, 1\ M) \qquad E° = +0.771\ V$$

or as

$$2\ Fe^{3+}(aq, 1\ M) + 2\ e^- \longrightarrow 2\ Fe^{2+}(aq, 1\ M) \qquad E° = +0.771\ V$$

The volt is defined as "energy/charge." Multiplying a reaction by some number causes *both* the energy and the charge to be multiplied by that number. Thus, the ratio "energy/charge = volt" does not change.

EXAMPLE 21.8 *Predicting* E° *and the Direction of a Redox Reaction*

Decide if each of the following reactions is product-favored in the direction written, and calculate $E°_{net}$.

1. $2\ Al(s) + 3\ Sn^{4+}(aq) \longrightarrow 2\ Al^{3+}(aq) + 3\ Sn^{2+}(aq)$

2. $2\ Cl^-(aq) + Br_2(\ell) \longrightarrow 2\ Br^-(aq) + Cl_2(aq)$

Solution Reaction (1) is predicted to be product-favored as written (and thus to have a positive $E°$). As indicated in Table 21.1, Al(s) is a stronger reducing agent than Sn^{2+}, and Sn^{4+} is a stronger oxidizing agent than Al^{3+}. This is verified by the positive value of $E°$.

$$\begin{array}{lr} 2\ [Al(s) \longrightarrow Al^{3+}(aq) + 3\ e^-] & E° = +1.66\ V \\ 3\ [Sn^{4+}(aq) + 2\ e^- \longrightarrow Sn^{2+}(aq)] & E° = +0.15\ V \\ \hline 2\ Al(s) + 3\ Sn^{4+}(aq) \longrightarrow 3\ Sn^{2+}(aq) + 2\ Al^{3+}(aq) & E°_{net} = +1.81\ V \end{array}$$

Reaction (2) is predicted to be reactant-favored. Based on the reduction potentials in Table 21.1, we know that Br_2 is a weaker oxidizing agent than Cl_2 and Cl^- is a weaker reducing agent than Br^-. The negative sign of $E°_{net}$ for the reaction as written confirms this.

$$\begin{array}{lr} 2\ Cl^-(aq) \longrightarrow Cl_2(aq) + 2\ e^- & E° = -1.360\ V \\ Br_2(\ell) + 2\ e^- \longrightarrow 2\ Br^-(aq) & E° = +1.08\ V \\ \hline 2\ Cl^-(aq) + Br_2(aq) \longrightarrow 2\ Br^-(aq) + Cl_2(aq) & E°_{net} = -0.28\ V \end{array}$$

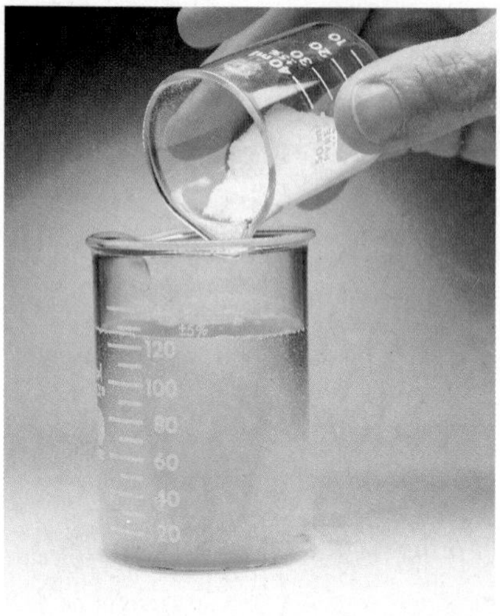

The reaction of a peroxide (such as H_2O_2) with aqueous bromide ion produces Br_2, as evidenced by the brown color of the solution. The reaction is used to generate bromine to disinfect water. (See Exercise 21.7.) (C. D. Winters)

If the reaction is written in the opposite direction, however,

$$2\ Br^-(aq) + Cl_2(aq) \longrightarrow 2\ Cl^-(aq) + Br_2(\ell) \qquad E^\circ_{net} = +0.28\ V$$

then the process is product-favored. Indeed, it can be observed that chlorine dissolved in water oxidizes bromide ion to bromine.

EXERCISE 21.7 ***Predicting E° and the Direction of a Redox Reaction***

Is the following reaction product-favored under standard conditions? What is the value of E°_{net}?

$$H_2O_2(aq) + 2\ H_3O^+(aq) + 2\ Br^-(aq) \longrightarrow Br_2(\ell) + 4\ H_2O(\ell)$$ ■

EXAMPLE 21.9 ***Constructing an Electrochemical Cell***

Using the half-reactions $Fe(s)/Fe^{2+}(aq)$ and $Cu(s)/Cu^{2+}(aq)$, construct an electrochemical cell, and predict its standard potential E°.

Solution First, let us decide which is the reducing agent and which is the oxidizing agent. From Table 21.1 we conclude that Fe(s) is a better reducing agent than Cu(s), and $Cu^{2+}(aq)$ is a better oxidizing agent than $Fe^{2+}(aq)$. Therefore, Fe(s) reduces $Cu^{2+}(aq)$ to copper metal.

Oxidation:	$Fe(s) \longrightarrow Fe^{2+}(aq) + 2\ e^-$	$E^\circ = +0.44\ V$
Reduction:	$Cu^{2+}(aq) + 2\ e^- \longrightarrow Cu(s)$	$E^\circ = +0.34\ V$
Net reaction:	$Fe(s) + Cu^{2+}(aq) \longrightarrow Fe^{2+}(aq) + Cu(s)$	$E^\circ_{net} = +0.78\ V$

For a cell that operates under standard conditions, we need to have a solid iron electrode dipping into a 1.0 M solution of $Fe^{2+}(aq)$. In another compartment we can have a solid copper electrode (or an electrode of some other electric conductor, like platinum or graphite) dipping into a solution of 1.0 M $Cu^{2+}(aq)$. An external wire connects the two electrodes, and a salt bridge allows for ion flow. (Because virtually all metal ions form soluble nitrate salts, we can use $Cu(NO_3)_2$ and $Fe(NO_3)_2$ in the cell compartments and $NaNO_3$ in the salt bridge.) When the cell is assembled, electrons flow from the Fe(s) anode to the Cu(s) cathode, and anions (NO_3^- in this case) flow from the $Cu^{2+}(aq)$ compartment to the $Fe^{2+}(aq)$ compartment.

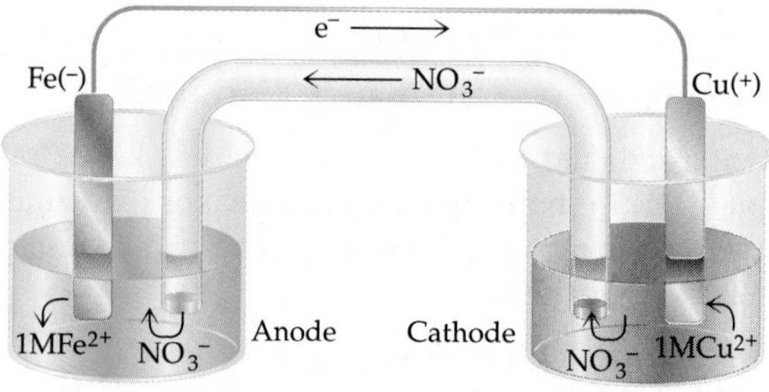

EXERCISE 21.8 ***Constructing an Electrochemical Cell***

Draw a diagram of an electrochemical cell using the half-cells $Zn(s)/Zn^{2+}(aq)$ and $Al(s)/Al^{3+}(aq)$. Decide first on the net reaction, and predict its $E°$ value. Show the direction of electron flow in the external wire and the directions of ion flow in the salt bridge. Tell which compartment is the anode and which is the cathode. ■

Walter Nernst (1864–1941) was a German physicist and chemist known for his work relating to the third law of thermodynamics. (Francis Simon, AIP Niels Bohr Library)

21.5 ELECTROCHEMICAL CELLS AT NONSTANDARD CONDITIONS

Oxidation-reduction reactions in the real world rarely occur under standard conditions. Even if the cell started out with all dissolved species at 1 M concentration, these would change as the reaction progressed; reactant concentrations decrease and those of the products increase. How can we define the potential of cells under *non*standard conditions?

The Nernst Equation

The standard cell potential, $E°$, is the potential measured under standard conditions, that is, with all dissolved substances having a concentration of 1.0 mol per liter. These are almost never the conditions in a real electrochemical cell, however, so how can we predict the potential under nonstandard conditions, E? The answer is that the standard potential, $E°$, can be corrected by a factor that includes the temperature of the reaction, the number of moles of electrons transferred between oxidizing and reducing agents in a balanced redox equation (n), and the concentrations of reactants and products. This relationship is called the **Nernst equation,**

$$E = E° - (RT/nF) \ln Q$$

where Q is the reaction quotient (page 763), F is the Faraday constant (9.6485309×10^4 J/V · mol), and R is the gas constant (8.314510 J/K · mol). When T is 298 K, we can write a modified form of the Nernst equation that we find useful in practical chemical applications.

$$E = E° - \frac{0.0257\text{ V}}{n} \ln Q \quad \text{at } 25\ °\text{C} \qquad (21.2)$$

The Nernst equation is often written using the base-10 logarithm of Q.

$$E = E° - \frac{0.0592\ V}{n} \log Q$$

This equation allows us to find the potential produced by a cell under nonstandard conditions or to find the concentration of a reactant or product by measuring the potential produced by a cell.

Before using the Nernst equation to derive a numerical answer, let us explore some of its consequences. Taking the reaction

$$Zn(s) + Ni^{2+}(1.0\text{ M, aq}) \longrightarrow Zn^{2+}(1.0\text{ M, aq}) + Ni(s) \qquad E°_{net} = +0.51\text{ V}$$

what happens to the cell potential if, for example, $[Ni^{2+}]$ is 1.0 M whereas $[Zn^{2+}]$ is only 0.0010 M? In this case Q is much smaller than 1.

$$Q = \frac{[Zn^{2+}]}{[Ni^{2+}]} = \frac{0.0010}{1.0} = 0.0010$$

When you take the logarithm of a number less than 1, the result is a negative number ($\ln 0.001 = -6.91$). Because the "correction factor" in the Nernst equation is subtracted from $E°$, the combination of negative signs means that E is more positive than $E°$ in this case. When the concentrations of products are low relative to the reactant concentrations in a product-favored reaction, the cell potential is more positive than $E°$; the reaction becomes even more product-favored.

EXAMPLE 21.10 *Using the Nernst Equation*

Determine the cell potential at 25 °C for

$$Fe(s) + Cd^{2+}(aq) \longrightarrow Fe^{2+}(aq) + Cd(s)$$

when (1) $[Fe^{2+}] = 0.010$ M and $[Cd^{2+}] = 1.0$ M and (2) $[Fe^{2+}] = 1.0$ M and $[Cd^{2+}] = 0.010$ M.

Solution To calculate a nonstandard potential, we first need the standard potential for the cell reaction, $E°_{net}$.

Oxidation:	$Fe(s) \longrightarrow Fe^{2+}(aq) + 2\,e^-$	$E° = +0.44$ V
Reduction:	$Cd^{2+}(aq) + 2\,e^- \longrightarrow Cd(s)$	$E° = -0.40$ V
Net reaction:	$Fe(s) + Cd^{2+}(aq) \longrightarrow Fe^{2+}(aq) + Cd(s)$	$E°_{net} = +0.04$ V

Next, let us substitute $E°$ and the conditions for solution (1) into the Nernst equation.

$$E = E° - \frac{0.0257\ \text{V}}{n} \ln \frac{[Fe^{2+}]}{[Cd^{2+}]}$$

As in chemical equilibrium calculations (see Chapter 16), the concentration of a solid does not enter into the expression. Now, using $n = 2$ (the number of moles of electrons transferred), and the ion concentrations given earlier,

$$E = +0.04\ \text{V} - \frac{0.0257\ \text{V}}{2} \ln \frac{0.010}{1.0} = +0.10\ \text{V}$$

The cell potential E is larger than $E°_{net}$, so the tendency to transfer electrons from Fe(s) to $Cd^{2+}(aq)$ is greater than under standard conditions.

For the condition described by (2), the Nernst equation is

$$E = +0.04\ \text{V} - \frac{0.0257\ \text{V}}{2} \ln \frac{1.0}{0.010} = -0.02\ \text{V}$$

The cell potential E is now negative. This means the reaction, as originally written, is reactant-favored. Therefore, the reaction is product-favored in the other direction. Under these conditions, cadmium metal will reduce iron(II) ion.

$$Fe^{2+}(aq, 1.0\ \text{M}) + Cd(s) \longrightarrow Fe(s) + Cd^{2+}(aq, 0.010\ \text{M}) \qquad E_{net} = +0.02\ \text{V}$$

EXERCISE 21.9 *Using the Nernst Equation*

Calculate E_{net} for the following reaction:

$$2\ Ag^+(aq, 0.80\ M) + Hg(\ell) \longrightarrow 2\ Ag(s) + Hg^{2+}(0.0010\ M, aq)$$

Is the reaction product-favored or reactant-favored under these conditions? How does this compare with the reaction under standard conditions? ■

$E°$ and the Equilibrium Constant

The cell potential, and even the reaction direction, can change when the concentrations of products and reactants change (see Example 21.10). Thus, as reactants are converted to products in any product-favored reaction, the value of E_{net} must decline from its initial positive value to eventually reach zero. A *potential of zero* means that no *net* reaction is occurring; it is an indication that the cell has reached *equilibrium.* Thus, when $E_{net} = 0$, the Q term in the Nernst equation is equivalent to the equilibrium constant K for the reaction. So, when equilibrium has been attained, Equation 21.2 can be rewritten as

$$E = 0 = E° - \frac{0.0257\ V}{n} \ln K$$

which rearranges to

$$\ln K = \frac{nE°}{0.0257\ V} \quad \text{at } 25\ °C \qquad (21.3)$$

This is an extremely useful equation, because it tells us that the equilibrium constant for a reaction can be obtained from a calculation or measurement of $E°_{net}$.

EXAMPLE 21.11 **E° *and Equilibrium Constants***

Calculate the equilibrium constant for the reaction

$$Fe(s) + Cd^{2+}(aq) \longrightarrow Fe^{2+}(aq) + Cd(s) \qquad E°_{net} = +0.04\ V$$

What are the equilibrium concentrations of the Fe^{2+} and Cd^{2+} ions if each began with a concentration of 1.0 M?

Solution The reaction was found to have a value of $E°_{net}$ of +0.04 V in Example 21.10. Therefore, substituting into Equation 21.3, we have

$$\ln K = \frac{(2.00)(0.04\ V)}{0.0257\ V}$$

$$= 3.1$$

$$K = 20$$

The concentrations of Fe^{2+} and Cd^{2+} when the cell has reached equilibrium are given by the equilibrium expression.

$$K = 20 = [Fe^{2+}]/[Cd^{2+}]$$

Because the cell began at standard conditions, the original concentrations of both ions were 1.0 M. As the reaction proceeded to equilibrium, x mol/L of Cd^{2+} was consumed and x mol/L of Fe^{2+} was produced. Therefore,

$$K = 20 = \frac{1.0 + x}{1.0 - x}$$

Solving this we find $x = 0.9$ M. Thus, the equilibrium concentrations are

$$[Fe^{2+}] = 1.0 + x = 1.9 \text{ M} \quad \text{and} \quad [Cd^{2+}] = 1.0 - x = 0.10 \text{ M}$$

EXERCISE 21.10 E° *and Equilibrium Constants*

In Exercise 21.9 you calculated E_{net} for

$$2\ Ag^+(aq, 0.80\ M) + Hg(\ell) \longrightarrow 2\ Ag(s) + Hg^{2+}(0.0010\ M, aq)$$

What is the equilibrium constant for this reaction? ■

21.6 BATTERIES AND FUEL CELLS

The voltaic cells we described to this point can produce a useful potential, but the potential declines rapidly as the reactant concentrations decline. For this reason, there has been great interest over the years in the design of usable batteries, voltaic cells that deliver current at a constant potential. As we depend more and more on portable computers, cellular phones, personal information devices, and pagers—and if there is any hope of producing a useable electric car—it is ever more important that lightweight, long-lived batteries be developed.

Batteries can be classified as primary and secondary. **Primary batteries** use oxidation-reduction reactions that cannot be reversed easily, so when the reactants are used up, the battery is "dead" and is discarded. **Secondary batteries** are often called **storage batteries,** or **rechargeable batteries.** The reactions in these batteries can be reversed; the battery can be "recharged."

Fuel cells are another type of electrochemical device. In a battery the oxidizing and reducing agents are held within a closed container. In contrast, the reactants in a fuel cell are supplied from an outside source.

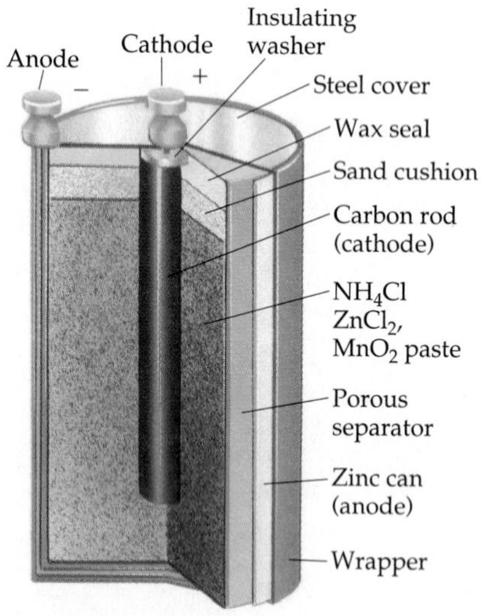

Figure 21.9 A diagrammatic representation of the Leclanché dry cell. It consists of a zinc anode (the battery container), a graphite cathode, and an electrolyte consisting of a moist paste of MnO_2, NH_4Cl, and $ZnCl_2$.

Primary Batteries

The common **dry cell battery,** invented by Georges Leclanché in 1866, is the energy source in toys, flashlights, and remote controllers for TVs, among other things. This type of battery contains a carbon rod electrode inserted into a moist paste of NH_4Cl, $ZnCl_2$, and MnO_2 in a zinc can that serves as the anode (Figure 21.9).

Anode, oxidation: $Zn(s) \longrightarrow Zn^{2+}(aq) + 2\ e^-$

The electrons produced reduce the ammonium ion to ammonia and hydrogen at the carbon cathode.

Cathode, reduction: $2\ NH_4^+(aq) + 2\ e^- \longrightarrow 2\ NH_3(g) + H_2(g)$

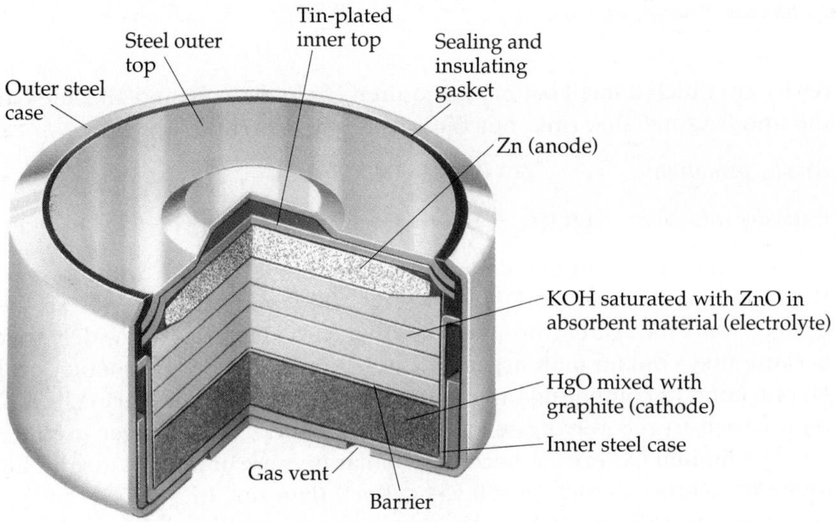

Figure 21.10 The mercury battery. The reducing agent is zinc, and the oxidizing agent is mercury(II) oxide.

The products of the cathode reaction are gases and would cause the sealed dry cell to explode if they were not removed. This is the reason for the manganese(IV) oxide, an oxidizing agent that consumes the hydrogen,

$$2\ MnO_2(s) + H_2(g) \longrightarrow Mn_2O_3(s) + H_2O(\ell)$$

and the ammonia is taken up by zinc(II) ion.

$$Zn^{2+}(aq) + 2\ NH_3(g) + 2\ Cl^-(aq) \longrightarrow Zn(NH_3)_2Cl_2(s)$$

All these reactions lead to the following net process and produce a potential of 1.5 V:

Net reaction: $2\ MnO_2(s) + 2\ NH_4Cl(s) + Zn(s) \longrightarrow Mn_2O_3(s) + H_2O(\ell) + Zn(NH_3)_2Cl_2(s)$

Unfortunately, there are at least two disadvantages to this battery. If current is drawn from the battery rapidly, the gaseous products cannot be consumed rapidly enough, and the potential drops. Furthermore, a slow reaction between the zinc electrode and ammonium ion leads to further deterioration, and the battery has a poor "shelf life."*

The somewhat more expensive "alkaline" battery is now commonly used because it avoids some of the problems of dry cell batteries. An **alkaline battery** produces 1.54 V, and the key reaction is again the oxidation of zinc, this time under alkaline or basic conditions. The oxidation or anode reaction is

Anode, oxidation: $Zn(s) + 2\ OH^-(aq) \longrightarrow ZnO(s) + H_2O(\ell) + 2\ e^-$

and the electrons produced are consumed by reduction of manganese(IV) oxide at the cathode.

Cathode, reduction: $2\ MnO_2(s) + H_2O(\ell) + 2\ e^- \longrightarrow Mn_2O_3(s) + 2\ OH^-(aq)$

In contrast to the Leclanché battery, no gases are formed in the alkaline battery, and there is no decline in potential under high current loads.

Mercury batteries are a close relative of alkaline batteries (Figure 21.10) and are typically used in calculators, cameras, watches, heart pacemakers, and other

*The shelf life of a dry cell is also affected by temperature. You can double or triple the shelf life of a battery if you store it in a refrigerator at about 4 °C.

Figure 21.11 Lithium batteries are useful where light weight and high current densities are useful. (C. D. Winters)

devices in which a small battery is required. As in dry cells and alkaline batteries, the anode is metallic zinc, but the cathode is mercury(II) oxide.

Anode, oxidation: $Zn(s) + 2\ OH^-(aq) \longrightarrow ZnO(s) + H_2O(\ell) + 2\ e^-$

Cathode, reduction: $HgO(s) + H_2O(\ell) + 2\ e^- \longrightarrow Hg(\ell) + 2\ OH^-(aq)$

These materials are tightly compacted powders separated by a moist paste of HgO containing some NaOH or KOH. Moistened paper serves as the "salt bridge," and the battery produces 1.35 V. These batteries are widely used, but, because they contain mercury, they can lead to some environmental problems. Mercury and its compounds are poisonous, so if possible mercury cells should be reprocessed to recover the metal when the battery is no longer useful.

The **lithium battery** has become popular because of its light weight. Lithium, which has a lower density ($d = 0.534\ g/cm^3$) than zinc ($d = 7.14\ g/cm^3$), is used as the anode (Figure 21.11). Another advantage is that lithium is a stronger reducing agent than zinc (Table 21.1), so lithium batteries produce about 3 V.

Secondary Batteries

The Leclanché cell and alkaline and mercury batteries no longer produce a current when the chemicals inside have reached equilibrium conditions. At that point they must be discarded. In contrast, secondary, or **storage,** batteries can be recharged, some of them hundreds of times. The original reactant concentrations can be restored by reversing the net cell reaction using an external source of electric energy.

A 12-V commercial lead storage battery contains six cells.

An automobile battery—the **lead storage battery**—is perhaps the best example. Such batteries are used to supply the energy to the engine starter of a car, but once the engine is running, the battery is recharged by current from the car's alternator. There are two types of electrodes in a lead storage battery (Figure 21.12): one made of porous lead (the reducing agent) and the other of compressed, insoluble lead(IV) oxide (the oxidizing agent). The electrodes, arranged alternately in a stack and immersed in aqueous sulfuric acid, are sepa-

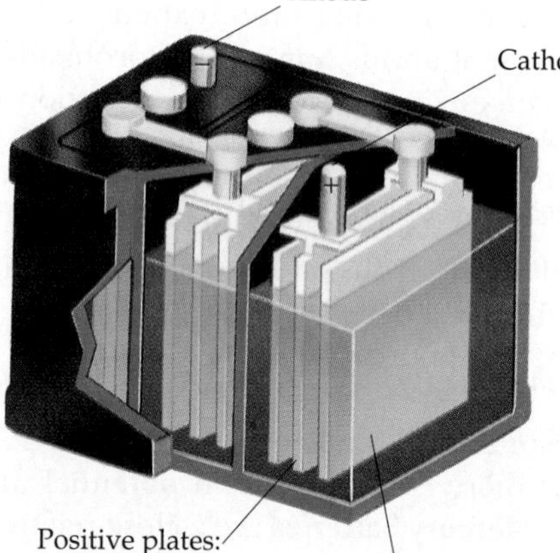

Figure 21.12 The lead storage battery.

rated by thin fiberglass sheets. When the cell acts as a supplier of electric energy, the lead electrode is oxidized to insoluble lead(II) sulfate.

Anode, oxidation: $Pb(s) + SO_4^{2-}(aq) \longrightarrow PbSO_4(s) + 2\ e^-$ $\quad E° = +0.356\ V$

The electrons move through the external circuit to the lead(IV) oxide electrode where they cause reduction of PbO_2.

Cathode, reduction:
$$PbO_2(s) + 4\ H_3O^+(aq) + SO_4^{2-}(aq) + 2\ e^- \longrightarrow PbSO_4(s) + 6\ H_2O(\ell) \quad E° = +1.685\ V$$

The net reaction supplies electric energy but leaves both electrodes coated with an adhering film of white lead(II) sulfate and consumes sulfuric acid.

Net process: $Pb(s) + PbO_2(s) + 2\ H_2SO_4(aq) \longrightarrow 2\ PbSO_4(s) + 2\ H_2O(\ell) \quad E° = +2.041\ V$

Figure 21.13 The nickel-cadmium or "ni-cad" battery. (C. D. Winters)

A lead storage battery can usually be recharged by supplying electric energy to reverse the net process. The film of $PbSO_4$ is converted back to metallic lead and PbO_2, and sulfuric acid is regenerated.

Lead storage batteries are large and heavy and produce a relatively low power for their mass, so many attempts have been and are being made to improve on them. Nonetheless, they do produce a relatively constant 2 V and a large initial current. This combination has been hard to beat, and they remain in widespread use.

Rechargeable, lightweight **nickel-cadmium** ("ni-cad") batteries (Figure 21.13) are used in a variety of cordless appliances, such as telephones and video camcorders. These have the advantage that the oxidizing and reducing agent can be regenerated easily when recharged, and they produce a nearly constant potential.

Anode, oxidation: $Cd(s) + 2\ OH^-(aq) \longrightarrow Cd(OH)_2(s) + 2\ e^-$

Cathode, reduction: $NiO(OH)(s) + H_2O(\ell) + e^- \longrightarrow Ni(OH)_2(s) + OH^-(aq)$

Like mercury batteries, ni-cad batteries should be disposed of with care because cadmium and its compounds are toxic.

Fuel Cells

Storage batteries contain the chemicals necessary for a reversible electrochemical reaction. A current is produced by a product-favored reaction until the chemicals reach equilibrium. Reversal of the reaction by an external current source restores the original reactants. A **fuel cell** is also an electrochemical device for converting chemical energy, provided by a fuel and an oxidant, into electricity. In contrast to a storage battery, though, a fuel cell does not need to involve a reversible reaction; the reactants are supplied to the cell as needed from an external source.

The best known fuel cell is the hydrogen-oxygen cell (Figure 21.14) used in the Gemini, Apollo, and Space Shuttle programs. The net cell reaction is the oxidation of hydrogen with oxygen to give water. Rather than allowing these gases to react directly and produce energy in the form of heat, they are made to react in such a way that the energy produced can be tapped by an electric device. A stream of H_2 gas is pumped onto the anode of the cell, and pure O_2 gas is

On April 13, 1970, Apollo 13 was on its way to a landing on the moon when an O_2 tank in the fuel cell exploded, wrecking part of the craft. Not only did this reduce the electricity supply, but it also limited the supply of water. With considerable skill, the pilots brought the craft back to earth several days later. (See D. Scott, "Lost in Space," *Chem Matters*, February, 1994, pages 4–8.)

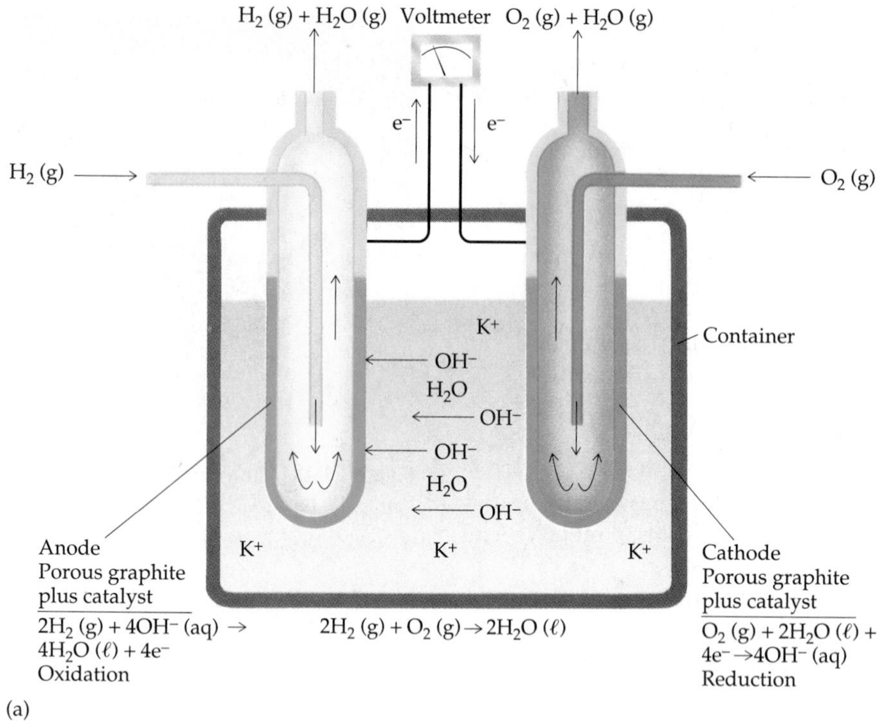

Figure 21.14 Fuel cells. (a) Schematic of an H_2/O_2 fuel cell. Oxidation of H_2 occurs in the anode chamber, and reduction of O_2 occurs in the cathode chamber. The water produced is often used for drinking purposes. (b) A hydrogen-oxygen fuel cell. Three of these units provide the power for the Space Shuttle. (Courtesy of International Fuel Cells Corporation)

directed to the cathode. Because the cell contains concentrated KOH, the following reactions occur under basic conditions:

Anode, oxidation: $2\ H_2(g) + 4\ OH^-(aq) \longrightarrow 4\ H_2O(\ell) + 4\ e^-$

Cathode, reduction: $O_2(g) + 2\ H_2O(\ell) + 4\ e^- \longrightarrow 4\ OH^-(aq)$

Net reaction: $2\ H_2(g) + O_2(g) \longrightarrow 2\ H_2O(\ell)$ $\quad E = 0.9$ V at 70–140 °C

The product, water, is swept out of the cell as a vapor in the hydrogen and oxygen stream and can be purified for drinking purposes.

Fuel cells have been valuable in the space program because they are lightweight and highly efficient (see Figure 21.14). The fuel cells on board the Space Shuttle deliver the same power as batteries weighing 10 times more. On a typical seven-day mission, the Shuttle fuel cells consume 1500 lb of hydrogen and generate 190 gal of drinking water.

21.7 CORROSION: REDOX REACTIONS IN THE ENVIRONMENT

Corrosion is the deterioration of metals, usually with loss of metal to a solution in some form, by a product-favored oxidation-reduction reaction. The corrosion of iron, for example, is the conversion of the metal to red-brown rust, hydrated iron(III) oxide [$Fe_2O_3 \cdot H_2O$], and other products (see Figure 21.1). This is significant because 25% of the annual steel production in the United States is estimated to be for the replacement of material lost to corrosion.

Zero-Emissions Vehicles

California has mandated that 10% of the vehicles registered in the state must be zero-emissions vehicles (ZEVs) by 2003. How can this be accomplished? One way is to make gasoline or diesel engines emissions-free, a very difficult engineering problem. Another is to use electric vehicles powered by batteries or fuel cells.

The large car manufacturers have already designed prototype electric cars (page 300), and some use the tried-and-true lead-acid, or lead storage, battery. Years of development have perfected that battery, but other types, such as the sodium-sulfur battery, are being developed. The sodium-sulfur battery being used by Ford has the advantage that it allows about three times the driving range of a lead storage battery (for the same battery weight). Two electrons are transferred between sodium and sulfur in the Na-S battery,

Anode, oxidation:

$$2\ Na \longrightarrow 2\ Na^+ + 2\ e^-$$

Cathode, reduction:

$$3\ S + 2\ e^- \longrightarrow S_3^{2-}$$

Net reaction:

$$2\ Na + 3\ S \longrightarrow Na_2S_3$$

so the mass of reactants consumed per mole of electrons is 71 g/mol e^-. Contrast that with a lead storage battery

$$Pb(s) + PbO_2(s) + 2\ H_2SO_4(aq) \longrightarrow 2\ PbSO_4(s) + 2\ H_2O(\ell)$$

that requires 321 g of reactants per mole of electrons. Unfortunately, although sodium-sulfur batteries have a weight advantage, they are currently expensive and must be heated to 300 °C. The battery can be kept this hot if the car is driven each day, but if it cools, the battery is dead.

As the table shows, lead storage batteries deliver the least power per kg of battery weight. But the table and photograph show another problem—the power available from any type of battery is much less than is available from an equivalent mass of gasoline.

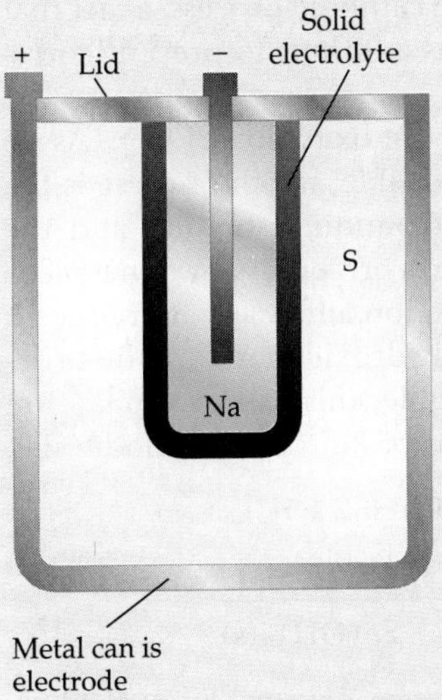

The sodium-sulfur battery. Sodium is oxidized at the anode to Na^+, and the electrons flow to the cathode, where sulfur is reduced to the S_3^{2-} ion. Ion flow within the battery is through a solid electrolyte salt bridge, which is made of beta-alumina ($Na_2O \cdot 11\ Al_2O_3$). The battery operates at 300–350 °C. This high temperature is required to make the beta-alumina conductive and to melt the reactants.

Amount of Energy Stored per Kilogram of Battery Mass

Chemical System	**Watt-hour/kg (1 Wh = 3600 J)**
lead-acid	18–56
nickel-cadmium	33–70
sodium-sulfur	80–140
lithium polymer	150
gasoline-air	12,200

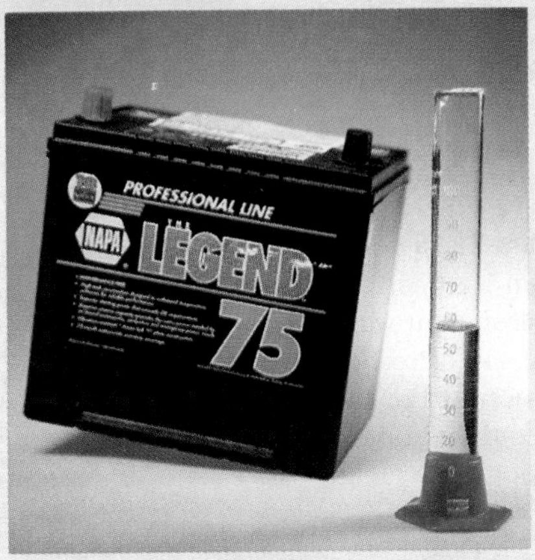

Gram for gram, gasoline packs more energy than a fully charged lead storage battery. A 15-kg lead battery has the same amount of stored energy as 59 mL of gasoline. This illustrates the problem for designers of ZEVs that use batteries for their motive power. (C. D. Winters)

Fuel cells are another possible solution. The hydrogen-oxygen fuel cell is well developed and has a maximum efficiency of 83%. In practice, though, it is likely that the efficiency may only be about 50% to 65%. This type of fuel cell is employed in the Space Shuttle (see Figure 21.14) and uses KOH as the electrolyte. It develops high current densities and can start cold. One problem, though, is that the fuel (H_2) and oxidant (O_2) must be free of CO_2, which could be a problem in large-scale automotive use. Hydrogen fuel cells with a phosphoric acid electrolyte are currently used to provide electricity and heat in buildings, but they operate at high temperatures, and are probably not practical for cars because of their weight.

The gasoline engines used in cars today are the product of years of engineering development. If we are to develop a useful, efficient, zero-emissions propulsion system within about 10 years, a significant effort will be required by many chemists and engineers in universities and industry.

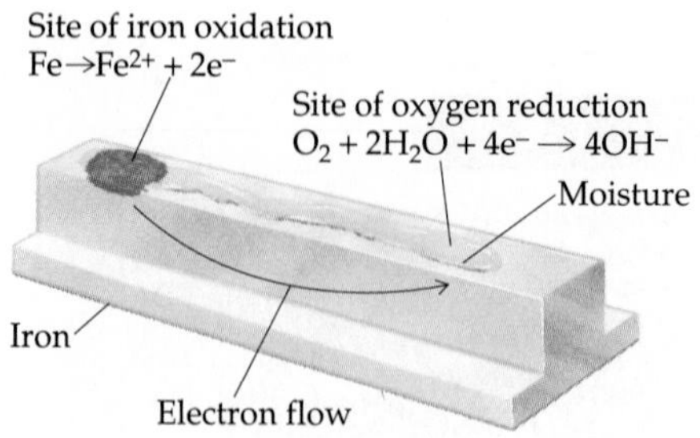

Figure 21.15 The reactions that occur when iron corrodes in an aqueous environment with oxygen present. The site of oxidation of iron may be different from the site of oxygen reduction because electrons can flow through the metal.

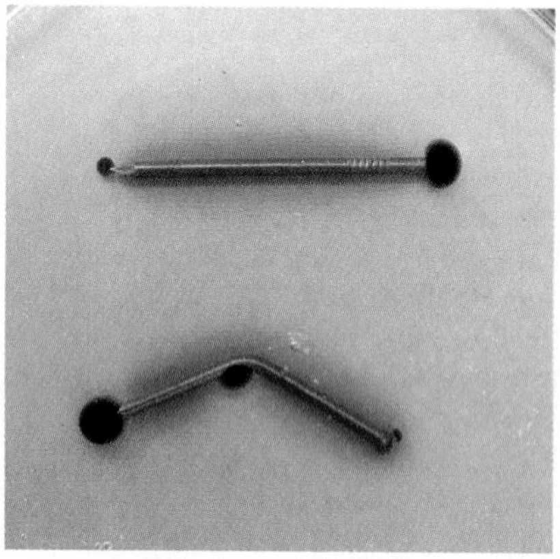

Figure 21.16 Corroding iron nails. Two nails were placed in an agar gel, which also contained the indicator phenolphthalein and $[Fe(CN)_6]^{3-}$. The nails began to corrode and gave Fe^{2+} ions at the tip or where the nail is bent. That these points are the anode is indicated by the blue-green color of Prussian blue, a complex of $[Fe(CN)_6]^{3-}$ and Fe^{2+}. The remainder of the nail is the cathode, where oxygen is reduced in water to give OH^- (see Figure 21.15). The presence of the OH^- ion is indicated by the red color of the indicator. (C. D. Winters)

For corrosion to occur at the surface of a metal, there must be anodic areas where the metal can be oxidized to metal ions as electrons are produced,

Anode, oxidation: $M(s) \longrightarrow M^{n+}(aq) + n\ e^-$

and cathodic areas where the electrons are consumed by any or all of several possible half-reactions, such as

Cathode, reduction:

$$2\ H_3O^+(aq) + 2\ e^- \longrightarrow H_2(g) + 2\ H_2O(\ell)$$
$$2\ H_2O(\ell) + 2\ e^- \longrightarrow H_2(g) + 2\ OH^-(aq)$$
$$O_2(g) + 2\ H_2O(\ell) + 4\ e^- \longrightarrow 4\ OH^-(aq)$$

Anodic areas occur at cracks in the oxide coating, at boundaries between phases, or around impurities. The cathodic areas occur at the metal oxide coating, at less reactive metallic impurity sites, or around other metal compounds such as sulfides.

The other requirements for corrosion are (1) an electric connection between the anode and cathode and (2) an electrolyte with which both anode and cathode are in contact. Both requirements are easily fulfilled, as seen in Figures 21.15 and 21.16.

If the relative rates of the anodic and cathodic corrosion reactions could be measured independently, the anodic reaction would be found to be the faster. When the two reactions are coupled to each other as in a corroding metal, however, the overall rate can only be that of the slower process. This generally means that corrosion is controlled by the rate of the cathodic process, a fact that helps to explain the chemistry of corroding systems and how to prevent corrosion.

In the corrosion of iron, the anodic reaction is the oxidation of iron. As we just noted, however, three cathodic reactions are possible; which of these is the fastest depends, in part, on the acidity of the surrounding solution and the amount of oxygen present. When little or no oxygen is present—as when a piece of iron is buried in soil such as moist clay—hydrogen ion and water are reduced. As indicated in the preceding reactions, $H_2(g)$ and hydroxide ions are the products. Because iron(II) hydroxide is relatively insoluble, this can precipitate on the metal surface and inhibit the further formation of Fe^{2+} at the anodic site.

Anode, oxidation:	$Fe(s) \longrightarrow Fe^{2+}(aq) + 2\ e^-$
Cathode, reduction:	$2\ H_2O(\ell) + 2\ e^- \longrightarrow H_2(g) + 2\ OH^-(aq)$
Net reaction:	$Fe(s) + 2\ H_2O(\ell) \longrightarrow H_2(g) + Fe(OH)_2(s)$

Corrosion under oxygen-free conditions is slow for two reasons. First, H_2O reduction is slow. Second, a coating of insoluble $Fe(OH)_2$, which forms from the products of the electron transfer reaction, inhibits further reaction.

If both water and O_2 are present, the chemistry of iron corrosion is somewhat different, and the corrosion reaction is about 100 times faster than without oxygen.

Anode, oxidation:	$2\ [Fe(s) \longrightarrow Fe^{2+}(aq) + 2\ e^-]$
Cathode, reduction:	$O_2(g) + 2\ H_2O(\ell) + 4\ e^- \longrightarrow 4\ OH^-(aq)$
Net reaction:	$2\ Fe(s) + 2\ H_2O(\ell) + O_2(g) \longrightarrow 2\ Fe(OH)_2(s)$

If oxygen is not freely available, further oxidation of the iron(II) hydroxide is limited to the formation of magnetic iron oxide (which can be thought of as a mixed oxide of Fe_2O_3 and FeO).

$$6\ Fe(OH)_2(s) + O_2(g) \longrightarrow \underset{\text{green hydrated magnetite}}{2\ Fe_3O_4 \cdot H_2O(s)} + 4\ H_2O(\ell)$$

$$Fe_3O_4 \cdot H_2O(s) \longrightarrow H_2O(\ell) + \underset{\text{black magnetite}}{Fe_3O_4(s)}$$

It is the black magnetite that you find coating an iron object that has corroded by resting in moist soil. On the other hand, if the iron object has free access to oxygen and water, as in the open or in flowing water, red-brown iron(III) oxide forms (Figure 21.1).

$$4\ Fe(OH)_2(s) + O_2(g) \longrightarrow \underset{\text{red-brown}}{2\ Fe_2O_3 \cdot H_2O(s)} + 2\ H_2O(\ell)$$

This is the familiar rust you see on cars and buildings and the substance that colors the water red in some mountain streams or in your home.

Other substances in air and water can assist in corrosion (Figure 21.17). Chlorides, from sea air or from salt spread on the roads in winter, are notorious. Because the chloride ion is relatively small, it can diffuse into and through a protective metal oxide coating. Metal chlorides, which are more soluble than metal oxides or hydroxides, can then form. These chloride salts leach back through the oxide coating, and a path is now open for oxygen and water to further attack the underlying metal. This is the reason you often see small pits on the surface of a corroded metal.

There are many methods for stopping a metal object from corroding, some more effective than others, but none totally successful. The general approaches are (1) to inhibit the anodic process, (2) to inhibit the cathodic process, or (3) to do both. The usual method is **anodic inhibition,** attempting to directly prevent the oxidation reaction by painting the metal surface or by allowing a thin oxide film to form. More recently developed methods are illustrated by the following reaction:

$$2\ Fe(s) + 2\ Na_2CrO_4(aq) + 2\ H_2O(\ell) \longrightarrow Fe_2O_3(s) + Cr_2O_3(s) + 4\ NaOH(aq)$$

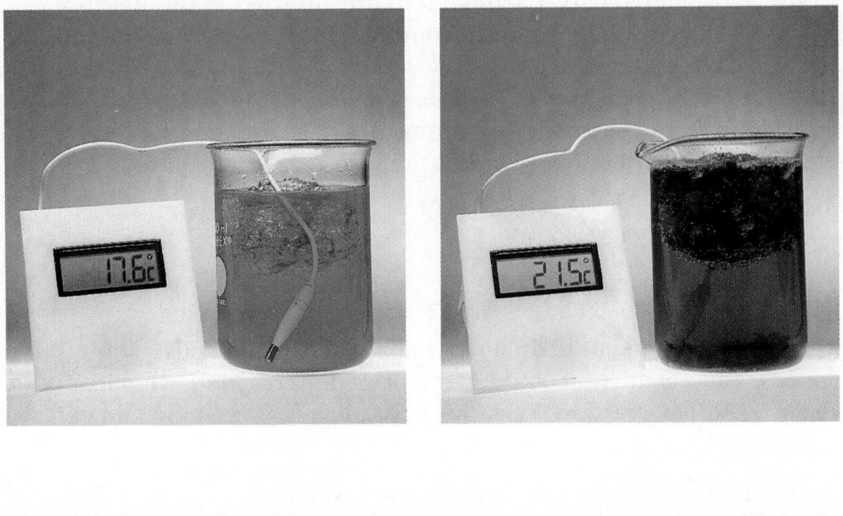

Figure 21.17 A ball of aluminum foil is added to a solution of copper(II) nitrate and sodium chloride. Normally, the coating of chemically inert Al_2O_3 on the surface of aluminum protects the metal from further oxidation. In the presence of the Cl^- ion, however, the coating of Al_2O_3 is breached, and aluminum reduces copper(II) ions to copper metal. The reaction is rapid and is so exothermic that the water can boil on the surface of the foil. (Notice that the blue color of copper(II) ions has faded as these ions are consumed in the reaction.) (C. D. Winters)

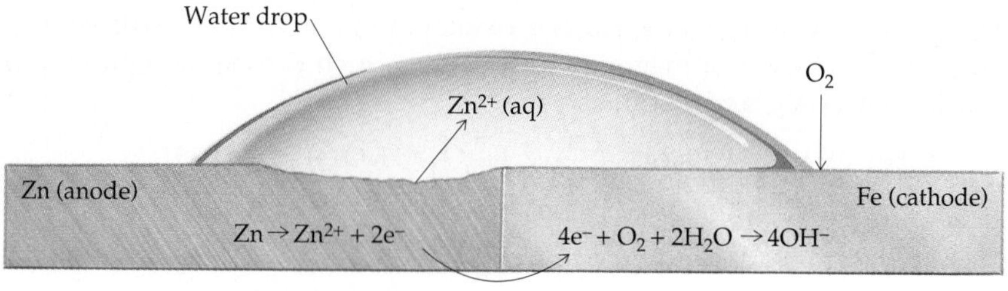

Figure 21.18 Cathodic protection of an iron-containing object. The iron is coated with a film of zinc, a metal more easily oxidized than iron. Therefore, the zinc acts as an anode and forces iron to become the cathode, thereby preventing the corrosion of the iron.

An iron surface is oxidized by a chromium(VI) salt to give iron(III) and chromium(III) oxides. These form a coating impervious to O_2 and water, and further atmospheric oxidation is inhibited.

There are several other ways to inhibit metal oxidation, one of which is to force the metal to become the cathode, instead of the anode, in an electrochemical cell. Hence, this is called **cathodic protection.** This is usually done by attaching another, more readily oxidized metal. The best example of this is **galvanized iron,** iron that has been coated with a thin film of zinc (Figure 21.18). The $E°$ for zinc oxidation is considerably more positive than $E°$ for iron oxidation (see Table 21.1), so the zinc metal film is oxidized before any of the iron. Thus, the zinc coating forms what is called a **sacrificial anode.** Another reason for using a zinc coating on iron is that, when the zinc is corroded, $Zn(OH)_2$ forms on the surface. This hydroxide is even less soluble than $Fe(OH)_2$, so the insoluble hydroxide film further slows corrosion.

21.8 ELECTROLYSIS: CHEMICAL CHANGE FROM ELECTRIC ENERGY

An electric current can be produced by a chemical change. Equally important, however, is the opposite process, **electrolysis,** the use of an electric current to bring about chemical change.

What happens if a pair of inert electrodes, which are connected to a battery, are inserted into a bath of molten NaCl (Figure 21.19)? (Because the NaCl is molten, the Na^+ and Cl^- ions are free to move in the melt; see Figure 3.14.) The external battery, or other source of electric potential, acts as an "electron pump," and electrons flow from this source into one of the electrodes, thereby giving it a negative charge. Sodium ions are attracted to this negative electrode and are reduced when electrons from the electrode are accepted, making the electrode the cathode. The battery simultaneously draws electrons from the other electrode, giving it a positive electric charge. Chloride ions are attracted to this electrode and surrender electrons. Because oxidation has occurred, this is the anode. Thus, the following reactions have occurred in molten NaCl.

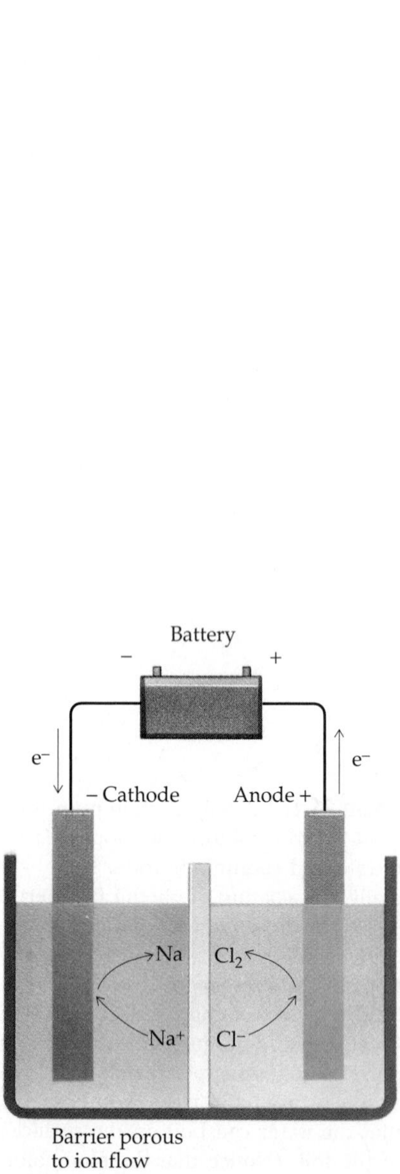

Figure 21.19 Electrolysis of molten sodium chloride. See also Figure 3.14.

Anode, oxidation: $2\ Cl^- \longrightarrow Cl_2(g) + 2\ e^-$

Cathode, reduction: $2\ Na^+ + 2\ e^- \longrightarrow 2\ Na(\ell)$

Net reaction: $2\ Cl^- + 2\ Na^+ \longrightarrow 2\ Na(\ell) + Cl_2(g) \qquad E_{net} \approx -4\ V$

The half-cell potentials in Table 21.1 apply only to aqueous solutions. If we use these to *estimate* the potential for the preceding reaction, we obtain a value of about −4 V. The reaction is clearly not product-favored in the direction written, and this is the reason that an external battery has been attached. The battery, with a potential greater than 4 V, forces the reactant-favored reaction to occur by "pumping" electrons in the proper direction.

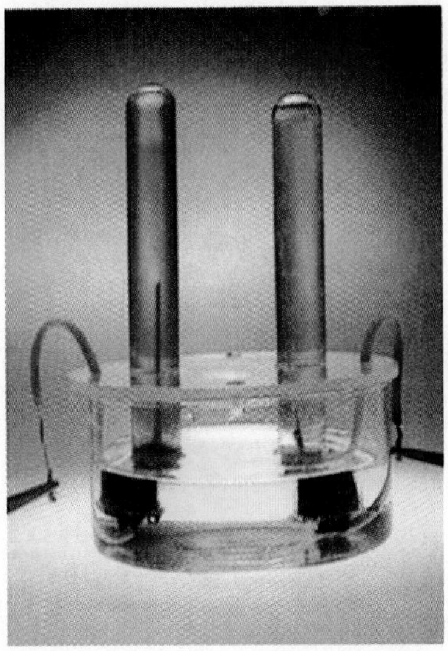

Figure 21.20 When a water solution of NaI is electrolyzed, the characteristic brownish color of I_2 appears at the anode (*left*). At the cathode (*right*) H_2O molecules are reduced to $H_2(g)$ and OH^- ions, which turn phenolphthalein pink. (C. D. Winters)

PROBLEM-SOLVING TIPS AND IDEAS

21.2 A Summary of Electrochemical Terminology

Whether you are describing a voltaic cell or an electrolysis cell, the terms *anode* and *cathode* always refer to the electrodes at which oxidation and reduction occur, respectively. The polarity of the electrodes is reversed, however, in a battery or electrolysis cell.

Type of Cell	Electrode	Function	Polarity
Battery	Anode	Oxidation	—
	Cathode	Reduction	+
Electrolysis	Anode	Oxidation	+
	Cathode	Reduction	—

In a voltaic cell, the negative electrode is the one at which electrons are produced. In an electrolysis cell, the negative electrode is the one onto which the external source is "pumping" the electrons. ■

What if an aqueous solution of a salt, say sodium iodide, is used instead of a molten salt (without water) (Figure 21.20)? With water now present, are Na^+ and I^- ions reduced and oxidized, respectively, or is water involved? Possible *reduction reactions* are

$$Na^+(aq) + e^- \longrightarrow Na(s) \qquad E° = -2.71 \text{ V}$$
$$2\ H_2O(\ell) + 2\ e^- \longrightarrow H_2(g) + 2\ OH^-(aq) \qquad E° = -0.83 \text{ V}$$
$$2\ H_3O^+(aq) + 2\ e^- \longrightarrow 2\ H_2O(\ell) + 2\ H_2(g) \qquad E° = 0.00 \text{ V}$$

When several reactions at an electrode are possible, the cathode in this case, we must consider not only which is the most easily reduced (best oxidizing agent) but also which is reduced most *rapidly*. Complications usually occur when currents are large—as in a commercial electrolysis cell—and when reactant concentrations are small. In this case H_2 and OH^- are clearly observed as products. This is reasonable because water is certainly reduced more readily than sodium. Furthermore, because the hydronium ion concentration is only about 10^{-7} M in an NaCl solution, the middle reaction is the best description of the net change occurring at the cathode.*

For the oxidation processes possible in aqueous sodium iodide, we need to compare the two reactions

$$2\ I^-(aq) \longrightarrow I_2(aq) + 2\ e^- \qquad E° = -0.535 \text{ V}$$
$$6\ H_2O(\ell) \longrightarrow O_2(g) + 4\ H_3O^+(aq) + 4\ e^- \qquad E° = -1.23 \text{ V}$$

*All of the electrode potentials for the reactions discussed here should be corrected to nonstandard conditions. Further, one should take into account the kinetic phenomenon of *overpotential*. Without adding these complications, however, $E°$ values do allow us to make some *estimates* of the reactions that occur under electrolysis conditions.

(A third possibility, the oxidation of OH^- ion, is not considered due to the very low concentration of the ion.) The iodide ion is the more readily reduced species. The best description of the chemistry that occurs on electrolysis of aqueous NaI is therefore the reactions

Anode, oxidation:	$2\ I^-(aq) \longrightarrow I_2(aq) + 2\ e^-$	$E° = -0.535\ V$
Cathode, reduction:	$2\ H_2O(\ell) + 2\ e^- \longrightarrow H_2(g) + 2\ OH^-(aq)$	$E° = -0.83\ V$
Net reaction:	$2\ I^-(aq) + 2\ H_2O(\ell) \longrightarrow H_2(g) + 2\ OH^-(aq) + I_2(aq)$	$E°_{net} = -1.37\ V$

The products are hydrogen, hydroxide ion, and iodine, all of which are easily identified in the experiment in Figure 21.20.

What happens if an aqueous solution of some other metal halide such as $CuCl_2$ is electrolyzed? As before, consult Table 21.1, and, considering all possible reactions, find the oxidation and reduction reactions that require the smallest potential. In this case aqueous Cu^{2+} ion is *much* more easily reduced ($E° = +0.34\ V$) than water ($E° = -0.83\ V$) at the cathode, so copper metal is produced. At the anode, two oxidations are possible: $Cl^-(aq)$ to $Cl_2(g)$ and H_2O to $O_2(g)$. Experiments show that chloride ion is generally oxidized more rapidly than water, so the reactions occurring on electrolysis of aqueous copper(II) chloride are (Figure 21.21)

Anode, oxidation:	$2\ Cl^-(aq) \longrightarrow Cl_2(g) + 2\ e^-$	$E° = -1.36\ V$
Cathode, reduction:	$Cu^{2+}(aq) + 2\ e^- \longrightarrow Cu(s)$	$E° = +0.34\ V$
Net reaction:	$Cu^{2+}(aq) + 2\ Cl^-(aq) \longrightarrow Cu(s) + Cl_2(g)$	$E°_{net} = -1.02\ V$

Again, this process is of obvious commercial importance. The copper used in wiring, in coins, and for other purposes is purified by electrolysis (Chapter 23).

A useful general principle can be derived from the preceding discussion. If an electric current is passed through a solution the electrode reactions are most likely those requiring the least potential (and those that occur most rapidly). In water, this means that a substance will be reduced if it has a reduction potential *less negative* than about $-0.8\ V$, the potential for the reduction of pure water. A check of Table 21.1 and Appendix J shows that this includes commercially useful

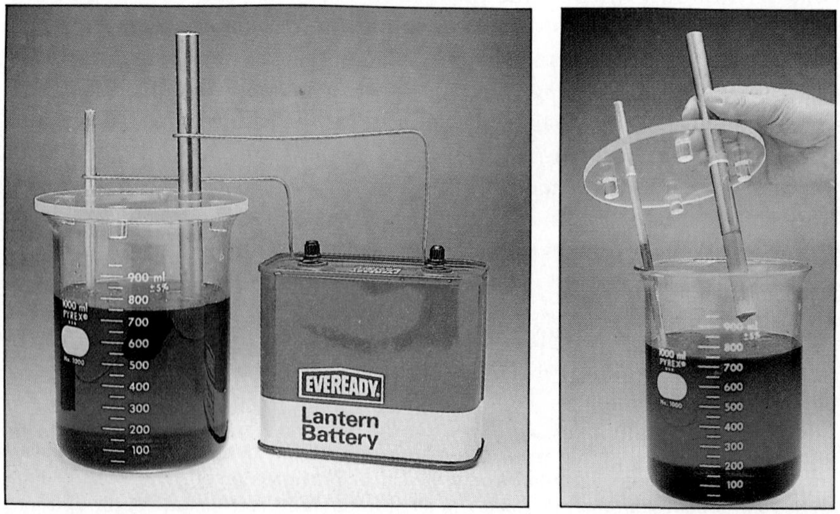

Figure 21.21 The electrolysis of aqueous copper(II) chloride. Copper metal is "plated out" at the cathode (negative electrode, site of reduction), and Cl_2 gas is evolved at the anode (positive electrode, site of oxidation).

metals, such as Pt, Cu, Ag, Au, and Cd. Indeed, the electrolysis method is used to coat or "plate" other materials with these metals. If a substance has a reduction potential *more negative* than about −0.8 V, then only water is reduced. Substances falling into this category include Na, K, Mg, and Al. To produce these metals requires methods other than the reduction of their ions in aqueous solution.

EXAMPLE 21.12 *Electrolysis of Aqueous NaOH*

Predict what happens when an electric current is passed through aqueous sodium hydroxide.

Solution First, list all the species in solution. In this case they are Na^+, OH^-, and H_2O. Next, use Table 21.1 to decide which of the species can be oxidized and which can be reduced, and note the potential of each possible reaction.

Reductions:

$$Na^+(aq) + e^- \longrightarrow Na(s) \qquad E° = -2.71\ V$$

$$2\ H_2O(\ell) + 2\ e^- \longrightarrow H_2(g) + 2\ OH^-(aq) \qquad E° = -0.83\ V$$

Oxidation:

$$4\ OH^-(aq) \longrightarrow O_2(g) + 2\ H_2O(\ell) + 4\ e^- \qquad E° = -0.40\ V$$

It is evident that water is reduced to H_2 at the cathode and OH^- is oxidized to O_2 at the anode. The cell reaction is $2\ H_2O(\ell) \rightarrow 2\ H_2(g) + O_2(g)$, and the potential under standard conditions is −1.23 V.

EXERCISE 21.11 *Electrolysis of Salts*

Predict the results of passing an electric current through each of the following solutions:

1. Molten NaBr
2. Aqueous NaBr
3. Aqueous $SnCl_2$ (See page 34 for a photo of the electrolysis of $SnCl_2$.) ■

21.9 COUNTING ELECTRONS

Metallic silver is produced at the cathode in the electrolysis of aqueous $AgNO_3$, the reaction being $Ag^+(aq) + e^- \rightarrow Ag(s)$. One mol of electrons is required to produce 1 mol of silver from 1 mol of silver ions. In contrast, 2 mol of electrons is required to produce 1 mol of copper.

$$Cu^{2+}(aq) + 2e^- \longrightarrow Cu(s)$$

It follows that, if the number of moles of electrons flowing through the electrolysis cell could be measured, the number of moles of silver or copper produced could be calculated. Conversely, if the amount of silver or copper produced is known, then the number of moles of electrons used could be calculated.

The number of moles of electrons consumed or produced in an electron transfer reaction is usually obtained by measuring the current flowing in the external electric circuit in a given time. The **current** flowing in an electric circuit is the amount of charge (in units of coulombs) passed per unit time, and the usual unit for current is the **ampere.**

$$\text{Current, } I \text{ (amps)} = \frac{\text{electric charge (coulombs, C)}}{\text{time (seconds, s)}} \qquad (21.4)$$

Michael Faraday (1791–1867) first explored the quantitative aspects of electricity (page 126). In his honor scientists have defined the Faraday constant, 96,485.31 C/mol, as the *charge carried by one mole of electrons* (see page 973). The current passing through an electrochemical cell, and the time the current flowed, are easily measured with modern instruments. The charge that passed through the cell can therefore be obtained by multiplying the current (in amperes) by the time (in seconds). Knowing the charge, and using the Faraday constant as a conversion factor, the number of moles of electrons that passed through an electrochemical cell can be calculated.

EXAMPLE 21.13 *Using the Faraday Constant*

A current of 1.50 amp is passed through a solution containing silver ions for 15.0 min. The voltage is such that silver deposited at the cathode. What mass of silver, in grams, is deposited?

$$Ag^+(aq) + e^- \longrightarrow Ag(s)$$

Solution From the half-reaction we know that if 1 mol of electrons passed through the cell, then 1 mol of silver was deposited. To find the number of moles of electrons, we need to know the total electric charge passed through the cell. The charge can be calculated from experimental measurements of the current (in amps) and the time the current flowed. Thus, the logic of the calculation is

Time (s) × current (amps) → Charge, C → Mol e⁻ → Mol Ag → Mass Ag

1. Calculate the charge (number of coulombs) passed in 15.0 min.

$$\begin{aligned}\text{Charge (coulombs, C)} &= \text{current (amps)} \times \text{time (seconds)}\\ &= 1.50 \text{ amps } (15.0 \text{ min})(60.0 \text{ s/min})\\ &= 1.35 \times 10^3 \text{ C}\end{aligned}$$

2. Calculate the number of moles of electrons.

$$1.35 \times 10^3 \text{ C}\left(\frac{1 \text{ mol e}^-}{9.65 \times 10^4 \text{ C}}\right) = 1.40 \times 10^{-2} \text{ mol e}^-$$

3. Calculate the number of moles of silver that passed through the cell and then the mass of silver deposited.

$$1.40 \times 10^{-2} \text{ mol e}^-\left(\frac{1 \text{ mol Ag}}{1 \text{ mol e}^-}\right)\left(\frac{107.9 \text{ g Ag}}{1 \text{ mol Ag}}\right) = 1.51 \text{ g Ag}$$

EXAMPLE 21.14 *Using the Faraday Constant*

One of the half-reactions occurring in the lead storage battery is

$$Pb(s) + SO_4^{2-}(aq) \longrightarrow PbSO_4(s) + 2\,e^-$$

If a battery delivers 1.50 amp, and if its lead electrode contains 500. g of lead, how long can current flow before the lead in the electrode is consumed?

Solution Here we begin with current and the mass of material used in the reaction, and we want to calculate a time. This is the reverse of Example 21.13.

Mass Pb → Mol Pb → Mol e^- → Charge, C → Time (s)

1. Calculate the quantity of lead available (mol).

$$500.\ \text{g}\left(\frac{1\ \text{mol}}{207.2\ \text{g}}\right) = 2.41\ \text{mol Pb}$$

2. Calculate the number of moles of electrons passed through the cell, assuming all the lead was consumed.

$$2.41\ \text{mol Pb}\left(\frac{2\ \text{mol e}^-}{1\ \text{mol Pb}}\right) = 4.83\ \text{mol e}^-$$

3. Calculate the charge carried by the electrons that passed through the cell.

$$4.83\ \text{mol e}^-\left(\frac{9.65 \times 10^4\ \text{C}}{1\ \text{mol e}^-}\right) = 4.66 \times 10^5\ \text{C}$$

4. Use the charge passed through the cell and the measured current to calculate the time.

$$4.66 \times 10^5\ \text{C} = 1.50\ \text{amp} \times \text{time (s)}$$
$$\text{Time} = 3.10 \times 10^5\ \text{s (or 86.2 h)}$$

EXERCISE 21.12 ***Using the Faraday Constant***

In the commercial production of sodium by electrolysis, the cell operates at 7.0 V and a current of 25×10^3 amp. How many grams of sodium can be produced in one hour? ■

21.10 THE COMMERCIAL PRODUCTION OF CHEMICALS BY ELECTROCHEMICAL METHODS

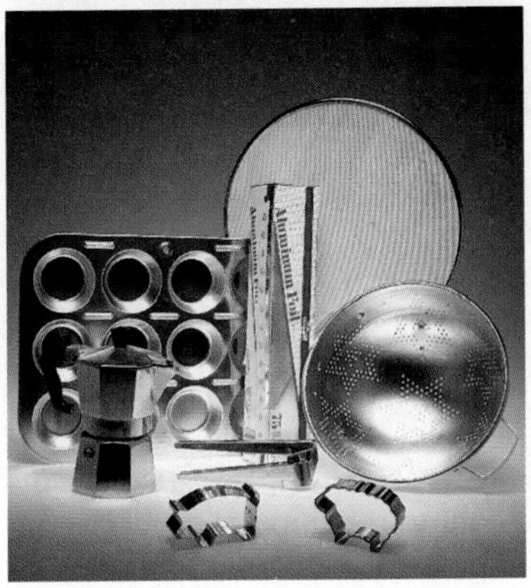

Aluminum

Aluminum is the third most abundant element in the earth's crust and has many important uses in our economy. You probably know it best in its use in the kitchen as a food wrapper, a use demonstrating its excellent formability. Just as importantly, aluminum has a low density and excellent corrosion resistance; the latter property comes from the fact that a transparent, chemically inert film of aluminum oxide, Al_2O_3, clings tightly to the metal's surface. It is these properties that have led to the many other uses of aluminum in aircraft parts, ladders, and automobile parts, for example.

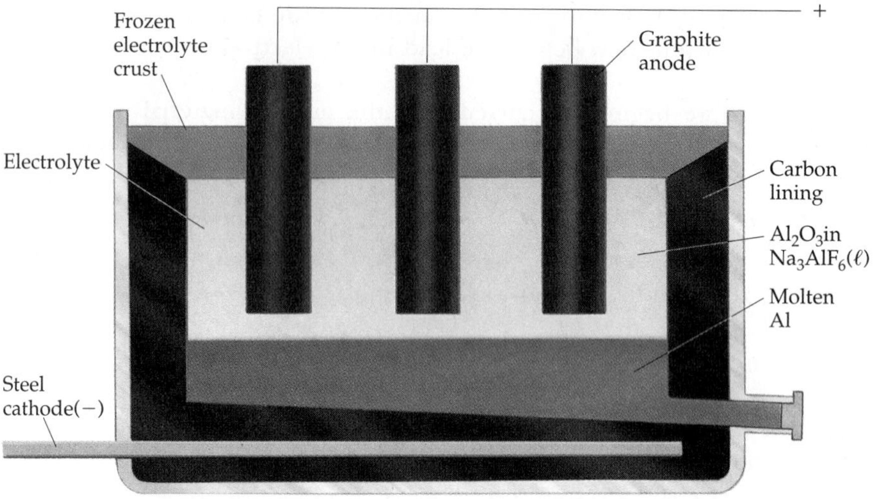

Figure 21.22 Industrial production of aluminum by electrolysis. Purified aluminum-containing ore (bauxite), essentially Al_2O_3, is mixed with cryolite (Na_3AlF_6) to give a mixture that melts at a lower temperature than Al_2O_3 alone. The aluminum-containing materials are reduced at the graphite cathode to give molten aluminum. Oxygen is produced at the carbon anode, and the gas reacts slowly with the carbon to give CO_2, leading to eventual destruction of the electrode.

As with some other commercially important metals, the history of the development of a practical method for aluminum production is interesting. Aluminum was originally made in the 19th century by reducing $AlCl_3$ with sodium,

$$3\ Na(s) + AlCl_3(s) \longrightarrow Al(s) + 3\ NaCl(s)$$

but only at a very high cost. It was therefore considered a precious metal, chiefly used in jewelry. In fact, in the 1855 Paris Exposition, some of the first aluminum metal produced was exhibited along with the crown jewels of France. Napoleon III saw its possibilities for military use, however, and commissioned studies on improving its production. The French had a ready source of aluminum-containing ore, bauxite, and in 1886 a 23-year-old Frenchman, Paul Heroult, conceived the electrochemical method in use today. In an interesting coincidence, an American, Charles Hall, who was only 22 at the time, announced his invention of the identical process in the same year. Hence, the commercial process is now known as the Hall-Heroult process.

The essential features of the Hall-Heroult process are illustrated in Figure 21.22. The aluminum-containing ore, chiefly in the form of Al_2O_3, is mixed with cryolite, Na_3AlF_6. The mixture is melted at about 980 °C and electrolyzed using graphite electrodes. Aluminum is produced at the cathode and oxygen at the anode. The cells operate at the very low potential of 4.0 to 5.5 V, but at a current of 50,000 to 150,000 amp. Each kilogram of aluminum requires 13 to 16 kilowatt-hours (kwh) of energy, exclusive of that required to heat the furnace. This is the reason there is so much interest in recycling soft drink cans and other aluminum objects. The recycled metal can be purified and made into new materials at a fraction of the cost of making aluminum from the ore.

Charles Martin Hall (1863–1914) was only 22 years old in 1886 when he discovered the electrolytic process for extracting aluminum from Al_2O_3 in a woodshed behind his family's home in Oberlin, Ohio. Following his discovery, Hall went on to found the company that eventually became the Aluminum Corporation of America. He died a millionaire in 1914.

Chlorine and Sodium Hydroxide

Chlorine is used to treat water and sewage and in the production of organic chemicals, such as pesticides and vinyl chloride, the building block of plastics called PVCs (polyvinyl chloride). In 1993, chlorine was eighth on the list of chemicals produced in the largest amounts in the United States, with about 11

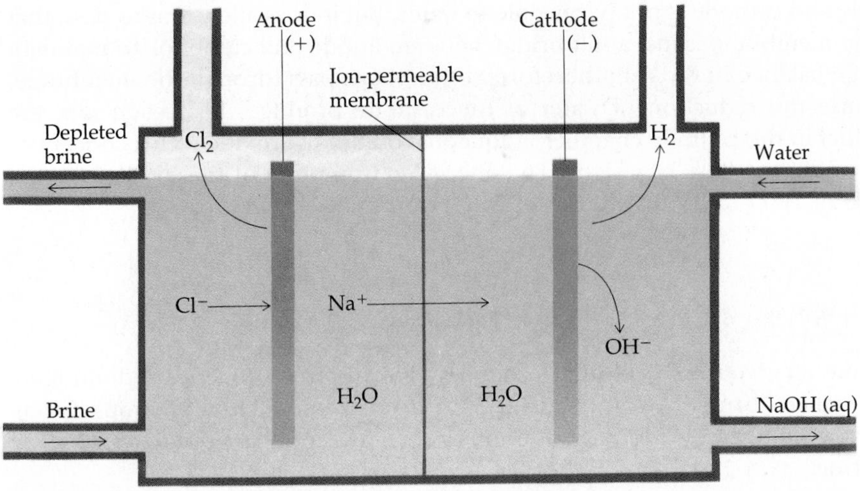

Figure 21.23 A simplified drawing of a membrane cell for the production of NaOH and Cl_2 gas from a saturated, aqueous solution of NaCl (brine). Here the anode and cathode compartments are separated by a water-impermeable but ion-conducting membrane. A widely used membrane is made of Nafion, a fluorine-containing polymer that is a relative of polytetrafluoroethylene (Teflon). Pure brine is fed into the anode compartment and dilute sodium hydroxide or water into the cathode compartment. Overflow pipes carry the evolved gases and NaOH away from the chambers of the electrolysis cell.

billion kg having been made. Almost all Cl_2 is made by electrolysis, with 95% coming from the electrolysis of brine, a saturated aqueous solution of NaCl. The other product coming from these cells, NaOH, is equally valuable, and almost 12 billion kg was produced in the United States in 1993.

See the essay on page 88 regarding the current controversy over chlorine use.

Three different electrolysis methods are used to convert aqueous NaCl to Cl_2 and NaOH: the mercury process, the diaphragm process, and the membrane process. As the name implies, the mercury process uses liquid mercury as the cathode. Although this process is still used, largely because it produces high-purity sodium hydroxide and Cl_2 gas, the problems of mercury pollution from these cells are enormous. As a consequence, mercury electrolysis cells are being phased out in favor of other methods, chiefly the membrane process.

A very simple diagram of a membrane electrolysis cell is shown in Figure 21.23. The anode reaction is the oxidation of chloride ion to Cl_2 gas

$$2\ Cl^-(aq) \longrightarrow Cl_2(g) + 2\ e^-$$

and the cathode process is the reduction of water.

$$2\ H_2O(\ell) + 2\ e^- \longrightarrow H_2(g) + 2\ OH^-(aq)$$

Activated titanium is used for the anode, and stainless steel or nickel are preferred for cathodes. In the membrane cell, brine is introduced into the anode chamber, and chloride ion oxidation occurs. The membrane separating the

A chlor-alkali plant. Each membrane cell shown contains many anodes and cathodes arranged in series. (Oxychem)

anode and cathode is not permeable to water, but it does allow ions to pass; that is, the membrane is just a salt bridge between anode and cathode. To maintain charge balance in the cell, therefore, sodium ions pass through the membrane. Because the reduction of water at the cathode produces hydroxide ion, the product in the cathode chamber is aqueous sodium hydroxide with a concentration of 20% to 35% by weight. The energy consumption of these cells is in the range of 2000 to 2500 kwh/ton of NaOH produced.

One of the greatest difficulties in creating economical membrane cells has been in finding membrane material that has a reasonable cost. It must be stable at high salt concentrations, stable to a high difference in pH from one side to the other, and stable to such strong oxidizing agents as Cl_2.

EXAMPLE 21.15 *Chlorine Production*

Assume an electrolysis cell that produces chlorine from aqueous sodium chloride (called "brine") operates at 4.6 V (with a current of 3.0×10^5 amp). Calculate the number of kilowatt-hours of energy required to produce 1.00 kg of chlorine.

Solution The *watt* is a unit of electric power. It describes the *rate* of energy consumption or production; that is, the watt has units of energy/time and is defined as 1 watt = 1 joule/second. From page 962 you know that 1 joule = 1 volt · coulomb. Therefore,

$$1 \text{ joule} = 1 \text{ watt} \cdot \text{second} = 1 \text{ volt} \cdot \text{coulomb}$$

A kilowatt-hour is the expenditure of 1000 W for 1 h, so

$$(1000 \text{ W})(1 \text{ h})\left(\frac{3600 \text{ s}}{\text{h}}\right)\left(\frac{1 \text{ J/s}}{\text{W}}\right) = 3.60 \times 10^6 \text{ J per kilowatt-hour}$$

To calculate the number of kilowatt-hours, therefore, we need to calculate the energy involved (in joules), and the chain of calculations leading to this is

$$\text{Mass of Cl}_2 \longrightarrow \text{Moles of Cl}_2 \longrightarrow \text{Charge required} \longrightarrow \text{Energy}$$

The reaction producing chlorine gas is

$$2\ Cl^-(aq) \longrightarrow Cl_2(g) + 2\ e^-$$

Therefore, to produce 1.00 kg of Cl_2, 2.72×10^6 C is required.

$$1.00 \times 10^3 \text{ g}\left(\frac{1 \text{ mol Cl}_2}{70.91 \text{ g}}\right)\left(\frac{2 \text{ mol e}^-}{1 \text{ mol Cl}_2}\right)\left(\frac{9.65 \times 10^4 \text{ C}}{1 \text{ mol e}^-}\right) = 2.72 \times 10^6 \text{ C}$$

The energy required is

$$\text{Energy (J)} = (4.6 \text{ V})(2.72 \times 10^6 \text{ C}) = 1.3 \times 10^7 \text{ J}$$

Finally, the power required is

$$1.3 \times 10^7 \text{ J}\left(\frac{1 \text{ kwh}}{3.60 \times 10^6 \text{ J}}\right) = 3.5 \text{ kwh}$$

EXERCISE 21.13 *Producing Sodium*

Sodium metal is produced by electrolysis from molten sodium chloride. The cell operates at 7.0 V with a current of 25×10^3 amp. How many kilowatt-hours of electricity is used to produce 1.00 kg of sodium metal? ■

CHAPTER HIGHLIGHTS

Having studied this chapter, you should be able to

- Define and use the terms **battery,** or **electrochemical cell; fuel cell; electrolysis; electrode; electrolyte; salt bridge; anode;** and **cathode.**
- Balance equations for oxidation-reduction reactions in acidic or basic solutions using the half-reaction approach (Section 21.1).
- Explain the workings of an electrochemical cell (which half-reaction occurs at the anode, which at the cathode, the polarity of the electrodes, the direction of electron flow in the external connection, and the direction of ion flow in the salt bridge) (Section 21.2).
- Appreciate the meaning of the **standard electrode potential** $E°$, and its connection to the free energy change $\Delta G°$, for a cell reaction (Section 21.3).
- Recognize that product-favored reactions have a positive $E°$, whereas reactant-favored reactions have negative $E°$ (Section 21.3).
- Know that the **standard hydrogen electrode** ($E° = 0.00$ V) is the standard against which all half-reaction potentials are measured (Section 21.3).
- Appreciate the method by which standard potentials of half-reactions can be determined (Section 21.3).
- Use Table 21.1, the **table of standard reduction potentials** (Section 21.4).
- Understand that when a half-reaction or net electrochemical reaction is reversed, the sign of $E°$ is reversed but its value does not change (Section 21.4).
- Recognize that, as the value of $E°$ for a reduction half-reaction becomes more negative, the ion or molecule becomes a better oxidizing agent (the substance on the left is more readily reduced) (Section 21.4).
- Apply the idea that the reaction between any substance on the left in Table 21.1 (an oxidizing agent) and any substance lower than it on the right (a reducing agent) is product-favored under standard conditions (Section 21.4).
- Recognize that electrochemical potentials depend on the nature of the reactants and products and their concentrations, not on the quantities of material used (Sections 21.4 and 21.5).
- Predict the sign and value of $E°_{net}$ for a redox reaction (Section 21.4).
- Use the **Nernst equation** (Equation 21.2) to calculate the cell potential under nonstandard conditions (Section 21.5).
- Calculate the equilibrium constant for a reaction from the value of $E°$ (Equation 21.3 and Section 21.5).
- Recognize the difference between **primary** and **secondary batteries** (Section 21.6).
- Appreciate the chemistry and advantages and disadvantages of dry

cells, alkaline batteries, mercury batteries, lithium batteries, lead storage batteries, and ni-cad batteries (Section 21.6).

- Understand the difference between batteries and fuel cells (Section 21.6).
- Understand reactions involved in **corrosion** and how **anodic** and **cathodic protection** can inhibit corrosion (Section 21.7).
- Describe the difference between electrolysis of an electrolyte and the operation of a galvanic or voltaic cell (Section 21.8).
- Identify the reactions that occur in the electrolysis of a molten salt (Section 21.8).
- Characterize the reactions occurring on electrolysis of an electrolyte in water (Section 21.8).
- Use the relationship between current (I, amp), electric charge (coulombs, C), and time (seconds, s) (Equation 21.4) and use the Faraday constant (96.5×10^4 C/mol e^-) (Section 21.9).
- Describe electrochemical methods for the production of aluminum, chlorine, and sodium hydroxide (Section 21.10).

STUDY QUESTIONS

Review Questions

1. In each of the following reactions, identify the substance oxidized and the substance reduced. Identify the oxidizing agent and the reducing agent.
   (a) $2\ Al(s) + 3\ Cl_2(g) \rightarrow 2\ AlCl_3(s)$
   (b) $FeS(s) + 3\ NO_3^-(aq) + 4\ H_3O^+(aq) \rightarrow$
   $3\ NO(g) + SO_4^{2-}(aq) + Fe^{3+}(aq) + 6\ H_2O(\ell)$
2. Explain the function of a salt bridge in an electrochemical cell.
3. Decide if each of the following statements is true or false. If false, rewrite it to make it a correct statement.
   (a) Oxidation always occurs at the anode of an electrochemical cell.
   (b) The anode of a battery is the site of reduction and is negative.
   (c) The potential of a cell does not change with temperature.
   (d) All product-favored oxidation-reduction reactions have negative E°_{net}.
4. Decide which phrase (a through d) best completes the sentence. A product-favored oxidation-reduction reaction has
   (a) A positive ΔG° and a positive E°
   (b) A negative ΔG° and a positive E°
   (c) A positive ΔG° and a negative E°
   (d) A negative ΔG° and a negative E°
5. Decide which phrase (a through c) best completes the sentence. In Table 21.1, Zn(s) can
   (a) Oxidize Fe(s) and Cd(s)
   (b) Reduce Al^{3+} and Mg^{2+}
   (c) Reduce Cd^{2+} and Ag^+
6. Decide if each of the following statements is true or false. If false, rewrite it to make it a correct statement.
   (a) The value of an electrode potential doubles when the half-reaction is multiplied by a factor of 2. That is, E° for ($2\ Li^+ + 2\ e^- \rightarrow 2\ Li$) is twice that for ($Li^+ + e^- \rightarrow Li$).
   (b) Al is the strongest reducing agent listed in Table 21.1.
   (c) The equilibrium constant for an oxidation-reduction reaction can be calculated using the Nernst equation.
   (d) Changing the concentrations of dissolved substances does not change the potential observed for an electrochemical cell.
7. What are the advantages and disadvantages of lead storage batteries?
8. How does a fuel cell differ from a battery?
9. Explain why the products of electrolysis of molten NaCl differ from those obtained from aqueous NaCl.
10. Describe the electrochemical method for the manufacture of Cl_2 and NaOH.
11. What is the difference between anodic and cathodic protection against corrosion? Explain how each works.

Numerical and Other Questions

Balancing Equations for Redox Reactions

Balance equations for the *half-reactions* in Questions 12 through 17. Tell whether the reactant is an oxidizing or reducing agent and if the overall process is an oxidation or reduction. Unless noted otherwise, all are carried out in acid solution, meaning that H^+ or H^+ and H_2O may be used to balance the equation.

12. (a) $Cr(s) \rightarrow Cr^{3+}(aq)$
(b) $AsH_3(g) \rightarrow As(s)$
(c) $VO_3^-(aq) \rightarrow V^{2+}(aq)$

13. (a) $Br_2(\ell) \rightarrow Br^-(aq)$
(b) $VO^{2+}(aq) \rightarrow V^{3+}(aq)$
(c) $U^{4+}(aq) \rightarrow UO_2^+(aq)$

14. (a) $Cr_2O_7^{2-}(aq) \rightarrow Cr^{3+}(aq)$
(b) $CH_3CHO(aq) \rightarrow CH_3CO_2H(aq)$
(c) $Bi^{3+}(aq) \rightarrow HBiO_3(aq)$

15. (a) $HOI(aq) \rightarrow I^-(aq)$
(b) $NO(g) \rightarrow HNO_2(aq)$
(c) $C_6H_5CH_3(aq) \rightarrow C_6H_5CO_2H(aq)$

16. The half-reactions here are in basic solution. You may need to use OH^- or the OH^-/H_2O pair to balance the equation.
(a) $Sn(s) \rightarrow Sn(OH)_4^{2-}(aq)$
(b) $MnO_4^-(aq) \rightarrow MnO_2(s)$
(c) $ClO^-(aq) \rightarrow Cl^-(aq)$

17. The half-reactions here are in basic solution. You may need to use OH^- or the OH^-/H_2O pair to balance the equation.
(a) $CrO_2^-(aq) \rightarrow CrO_4^{2-}(aq)$
(b) $Br_2(\ell) \rightarrow BrO_3^-(aq)$
(c) $Ni(OH)_2(s) \rightarrow NiO_2(s)$

Use the half-reaction method to balance the equations in Questions 18 through 23. For reactions in acid solution you may need to use H^+ or H^+ and H_2O to balance the equation.

18. Reaction (a) is in neutral solution. Reactions (b) and (c) are in acid solution.
(a) $Cl_2(aq) + Br^-(aq) \rightarrow Br_2(aq) + Cl^-(\ell)$
(b) $Sn(s) + H^+(aq) \rightarrow Sn^{2+}(aq) + H_2(g)$
(c) $Zn(s) + VO^{2+}(aq) \rightarrow Zn^{2+}(aq) + V^{3+}(aq)$

19. Reaction (a) is in neutral solution. Reactions (b) and (c) are in acid solution.
(a) $Hg^{2+}(aq) + Cu(s) \rightarrow Cu^{2+}(aq) + Hg(\ell)$
(b) $MnO_2(s) + Cl^-(aq) \rightarrow Mn^{2+}(aq) + Cl_2(g)$
(c) $Zn(s) + NO_3^-(aq) \rightarrow Zn^{2+}(aq) + N_2O(g)$

20. The reactions here are in acid solution.
(a) $Ag^+(aq) + HCHO(aq) \rightarrow Ag(s) + HCO_2H(aq)$
(b) $H_2S(aq) + Cr_2O_7^{2-}(aq) \rightarrow S(s) + Cr^{3+}(aq)$
(c) $Zn(s) + VO_3^-(aq) \rightarrow V^{2+}(aq) + Zn^{2+}(aq)$

21. The reactions here are in acid solution.
(a) $MnO_4^-(aq) + HSO_3^-(aq) \rightarrow Mn^{2+}(aq) + SO_4^{2-}(aq)$
(b) $Cr_2O_7^{2-}(aq) + Fe^{2+}(aq) \rightarrow Cr^{3+}(aq) + Fe^{3+}(aq)$
(c) $Ag(s) + NO_3^-(aq) \rightarrow NO_2(g) + Ag^+(aq)$

22. The reactions here are in basic solution. You may need to add OH^- or OH^- and H_2O to balance the equation.
(a) $Zn(s) + ClO^-(aq) \rightarrow Zn(OH)_2(s) + Cl^-(aq)$
(b) $ClO^-(aq) + CrO_2^-(aq) \rightarrow Cl^-(aq) + CrO_4^{2-}(aq)$
(c) $Br_2(\ell) \rightarrow Br^-(aq) + BrO_3^-(aq)$*

*This is a disproportionation reaction: one substance, Br_2, functions both as the reducing and the oxidizing agent.

23. The reactions here are in basic solution. You may need to add OH^- or OH^- and H_2O to balance the equation.
(a) $Fe(OH)_2(s) + CrO_4^{2-}(aq) \rightarrow Fe_2O_3(s) + Cr(OH)_4^-(aq)$
(b) $PbO_2(s) + Cl^-(aq) \rightarrow ClO^-(aq) + Pb(OH)_3^-(aq)$
(c) $Al(s) + OH^-(aq) \rightarrow Al(OH)_4^-(aq) + H_2(g)$

Electrochemical Cells and Cell Potentials

24. In principle, the reaction of chromium and iron(II) ion can be used to build an electrochemical cell.

$$2\ Cr(s) + 3\ Fe^{2+}(aq) \longrightarrow 2\ Cr^{3+}(aq) + 3\ Fe(s)$$

(a) Write the half-reactions involved.
(b) Which half-reaction is an oxidation, and which is a reduction?
(c) Which half-reaction occurs in the anode compartment and which in the cathode compartment?

25. Chlorine gas, Cl_2, can oxidize zinc metal in a reaction that has been suggested as the basis of a battery.
(a) Write the half-reactions involved.
(b) Which half-reaction is an oxidation, and which is a reduction? Which half-reaction occurs in the anode compartment and which in the cathode compartment?

26. Suppose the half-cells $Cu^{2+}(aq)/Cu(s)$ and $Sn^{2+}(aq)/Sn(s)$ are to be used as the basis of a battery. If the polarity of the copper electrode is found to be positive and that of the tin electrode is negative, write the half-reactions that occur in each half-cell. Decide which is the oxidation and which is the reduction. Identify which half-reaction occurs at the anode and which at the cathode.

27. In principle, the half cells $Fe^{2+}(aq)/Fe(s)$ and $O_2(g)/H_2O(\ell)$ (in acid solution) can be used as the basis of a battery. If the polarity of the iron electrode is found to be negative, write the half-reactions that occur in the cell. Identify which is the oxidation, and which is the reduction. Decide which half-reaction occurs at the anode and which at the cathode.

28. The standard potential for the reaction of Mg(s) with $I_2(s)$ is +2.91 V. What is the standard free energy change $\Delta G°$ for the reaction?

29. The standard voltage $E°$ for the reaction of Zn(s) and $Cl_2(g)$ is +2.12 V (see Study Question 25). What is the standard free energy change $\Delta G°$ for the reaction?

30. Calculate the value of $E°$ for each of the following reactions. Decide if each is product-favored in the direction written.
(a) $2\ I^-(aq) + Zn^{2+}(aq) \rightarrow I_2(s) + Zn(s)$
(b) $Zn^{2+}(aq) + Ni(s) \rightarrow Zn(s) + Ni^{2+}(aq)$
(c) $2\ Cl^-(aq) + Cu^{2+}(aq) \rightarrow Cu(s) + Cl_2(g)$

31. Calculate the value of $E°$ for each of the following reactions. Decide if each is product-favored in the direction written.
(a) $Br_2(\ell) + Mg(s) \rightarrow Mg^{2+}(aq) + 2\ Br^-(aq)$
(b) $Sn^{2+}(aq) + 2\ Ag^+(aq) \rightarrow Sn^{4+}(aq) + 2\ Ag(s)$
(c) $2\ Zn(s) + O_2(g) + 2\ H_2O(\ell) \rightarrow 2\ Zn^{2+}(aq) + 4\ OH^-(aq)$

32. Balance each of the following unbalanced equations, then calculate the standard potential $E°$, and decide whether

each is product-favored as written. (Half-reaction potentials are found in Appendix J.)
(a) $Sn^{2+}(aq) + Ag(s) \rightarrow Sn(s) + Ag^+(aq)$
(b) $Zn(s) + Sn^{4+}(aq) \rightarrow Sn^{2+}(aq) + Zn^{2+}(aq)$
(c) $I_2(s) + Br^-(aq) \rightarrow I^-(aq) + Br_2(\ell)$

33. Balance each of the following unbalanced equations, then calculate the standard potential $E°$, and decide whether each is product-favored as written. (Half-reaction potentials are found in Appendix J.)
(a) $Ce^{4+}(aq) + Cl^-(aq) \rightarrow Ce^{3+}(aq) + Cl_2(g)$
(b) $Cu(s) + NO_3^-(aq) + H_3O^+(aq) \rightarrow Cu^{2+}(aq) + NO(g) + H_2O(\ell)$
(c) $Fe^{2+}(aq) + Cr_2O_7^{2-}(aq) + H_3O^+(aq) \rightarrow Fe^{3+}(aq) + Cr^{3+}(aq)$

34. Consider the following half-reactions:

Half-Reaction	***E°* (V)**
$Cl_2(g) + 2\,e^- \rightarrow 2\,Cl^-(aq)$	+1.36
$I_2(s) + 2\,e^- \rightarrow 2\,I^-(aq)$	+0.535
$Pb^{2+}(aq) + 2\,e^- \rightarrow Pb(s)$	−0.126
$V^{2+}(aq) + 2\,e^- \rightarrow V(s)$	−1.18

(a) Which is the weakest oxidizing agent in the list?
(b) Which is the strongest oxidizing agent?
(c) Which is the strongest reducing agent?
(d) Which is the weakest reducing agent?
(e) Does Pb(s) reduce $V^{2+}(aq)$ to V(s)?
(f) Does I_2 oxidize $Cl^-(aq)$ to $Cl_2(g)$?
(g) Name the elements or ions that can be reduced by Pb(s).

35. Consider the following half-reactions:

Half-Reaction	***E°* (V)**
$Ce^{4+}(aq) + e^- \rightarrow Ce^{3+}(aq)$	+1.61
$Ag^+(aq) + e^- \rightarrow Ag(s)$	+0.80
$Hg_2^{2+}(aq) + 2\,e^- \rightarrow 2\,Hg(\ell)$	+0.79
$Sn^{2+}(aq) + 2\,e^- \rightarrow Sn(s)$	−0.14
$Ni^{2+}(aq) + 2\,e^- \rightarrow Ni(s)$	−0.25
$Al^{3+}(aq) + 3\,e^- \rightarrow Al(s)$	−1.66

(a) Which is the weakest oxidizing agent in the list?
(b) Which is the strongest oxidizing agent?
(c) Which is the strongest reducing agent?
(d) Which is the weakest reducing agent?
(e) Does Sn(s) reduce $Ag^+(aq)$ to Ag(s)?
(f) Does $Hg(\ell)$ reduce $Sn^{2+}(aq)$ to Sn(s)?
(g) Name the ions that can be reduced by Sn(s).
(h) What metals can be oxidized by $Ag^+(aq)$?

36. Use the table of half-reaction potentials in Study Question 34 to answer the following questions:
(a) Which reaction leads to the maximum positive standard potential?
(b) If the $I_2(s)/I^-(aq)$ half-cell is combined with the $V^{2+}(aq)/V(s)$ half-cell, write an equation for the product-favored reaction that occurs. What is the value of $E°_{net}$ for the reaction?

37. Use the table of half-reaction potentials in Study Question 35 to answer the following questions:
(a) Which reaction leads to the maximum positive standard potential?
(b) If the $Ni^{2+}(aq)/Ni(s)$ half-cell is combined with the $Hg_2^{2+}(aq)/Hg(\ell)$ half-cell, write the equation for the product-favored reaction that occurs. What is its standard potential?
(c) Write the equation for the product-favored reaction that occurs when the half-reactions $Ag^+(aq)/Ag(s)$ and $Ce^{4+}(aq)/Ce^{3+}(aq)$ are combined. What is the value of $E°_{net}$ for this reaction?

38. Assume that you assemble an electrochemical cell based on the half-reactions $Zn^{2+}(aq)/Zn(s)$ and $Ag^+(aq)/Ag(s)$.
(a) Write the equation for the product-favored reaction that occurs in the cell and calculate $E°_{net}$.
(b) Which electrode is the anode and which is the cathode?
(c) Diagram the components of the cell.
(d) If you use a silver wire as an electrode, is it the anode or cathode?
(e) Do electrons flow from the Zn electrode to the Ag electrode, or vice versa?
(f) If a salt bridge containing $NaNO_3$ connects the two half-cells, in which direction do the nitrate ions move, from the zinc to the silver compartment, or vice versa?

39. An electrochemical cell uses Al(s) and $Al^{3+}(aq)$ in one compartment and Ag(s) and $Ag^+(aq)$ in the other.
(a) Write a balanced equation for the product-favored reaction that occurs in this cell, and calculate $E°_{net}$.
(b) Which is the better reducing agent, Ag or Al?
(c) Which is the anode, and which is the cathode? Indicate the polarity of each electrode.
(d) Sodium nitrate is in the salt bridge connecting the two half-cells. In which direction do the nitrate ions flow, from Al to Ag or Ag to Al?

Cells Under Nonstandard Conditions, E° and K

40. Calculate the voltage delivered by an electrochemical cell, using the following reaction, if all dissolved species are 0.10 M:

$$2\,Fe^{3+}(aq) + 2\,I^-(aq) \longrightarrow 2\,Fe^{2+}(aq) + I_2(s)$$

Compare the voltage under non-standard conditions with $E°_{net}$, the standard voltage for the reaction.

41. Using the following reaction, calculate the voltage delivered by an electrochemical cell if all dissolved species are 0.015 M:

$$2\,Fe^{2+}(aq) + H_2O_2(aq) + 2\,H_3O^+(aq) \longrightarrow 2\,Fe^{3+}(aq) + 4\,H_2O(\ell)$$

Compare the voltage under non-standard conditions with $E°_{net}$, the standard voltage for the reaction.

42. An electrochemical cell is constructed of one half-cell in which a silver wire dips into an aqueous solution of $AgNO_3$. The other half-cell consists of an inert platinum wire in an aqueous solution of Fe^{2+} and Fe^{3+}.
(a) What reaction occurs in this cell?
(b) What is E°_{net}?
(c) If $[Ag^+] = 0.10$ M, but $[Fe^{2+}]$ and $[Fe^{3+}]$ are both 1.0 M, what is the value of E_{net}? Is the net cell reaction still that in part (a)? If not, what is the net reaction under the new conditions?

43. An electrochemical cell is constructed of one half-cell in which a silver wire dips into a 1.0 M aqueous solution of $AgNO_3$. The other half-cell consists of a zinc electrode in a solution of 1.0 M $Zn(NO_3)_2$.
(a) What reaction occurs in the cell under standard conditions?
(b) What is E°_{net}?
(c) What is the polarity of the Zn electrode?
(d) Do electrons flow from Zn to Ag or from Ag to Zn?
(e) If Ag^+ has a concentration of 0.50 M, but Zn^{2+} is held at 1.0 M, what is the voltage of the cell? Compare this with E°_{net} and comment on the effect of the change in silver concentration.

44. Calculate equilibrium constants for the following reactions:
(a) $2\ Fe^{3+}(aq) + 2\ I^-(aq) \rightarrow 2\ Fe^{2+}(aq) + I_2(s)$
(b) $I_2(s) + 2\ Br^-(aq) \rightarrow 2\ I^-(aq) + Br_2(aq)$

45. Calculate equilibrium constants for the following reactions:
(a) $Zn^{2+}(aq) + Ni(s) \rightarrow Zn(s) + Ni^{2+}(aq)$
(b) $Cu(s) + 2\ Ag^+(aq) \rightarrow Cu^{2+}(aq) + 2\ Ag(s)$

Electrolysis, Electrical Energy, and Power

46. If you wish to convert 1.00 g of $Au^{3+}(aq)$ ion to Au(s) in a "gold-plating" process, how long must you electrolyze a solution if the current passing through the circuit is 2.00 amp?

47. If you electrolyze a solution of $Ni^{2+}(aq)$ to form Ni(s), and use a current of 0.15 amp for 10 min, how many grams of Ni(s) is produced?

48. A current of 2.50 amp is passed through a solution of $Ni(NO_3)_2$ for 2.00 h. What mass of nickel is deposited at the cathode?

49. A current of 0.0125 amp is passed through a solution of $CuCl_2$ for 2.00 h. What mass of copper is deposited at the cathode, and what mass of chlorine gas is produced at the anode?

50. An old method of measuring the current flowing in a circuit was to use a "silver coulometer." The current passed first through a solution of $Ag^+(aq)$ and then into another solution containing an electroactive species. The amount of silver metal deposited at the cathode was weighed, and, if the time was noted, the current could be calculated. If 0.052 g of Ag was deposited during a 450-s experiment, what was the current flowing in the circuit?

51. As noted in the Study Question 50, a "silver coulometer" was used in the past to measure the current flowing in an electrochemical cell. Suppose you found that the current flowing through an electrolysis cell deposited 0.089 g of Ag metal at the cathode after exactly 10 min. If this same current then passed through a cell containing gold(III) ion in the form of $[AuCl_4]^-$, how much gold was deposited at the cathode in that electrolysis cell?

52. The basic reaction occurring in the cell in which Al_2O_3 and aluminum salts are electrolyzed is $Al^{3+} + 3\ e^- \rightarrow Al(s)$. If the cell operates at 5.0 V and 1.0×10^5 amp, how many grams of aluminum metal can be produced in an 8.0-h day?

53. The vanadium(II) ion can be produced by electrolysis of a vanadium(III) salt in solution. How long must you carry out an electrolysis if you wish to convert completely 0.125 L of 0.015 M $V^{3+}(aq)$ to $V^{2+}(aq)$ if the current is 0.268 amp?

54. The reactions occurring in a lead storage battery are given in Section 21.6. A typical battery might be rated at "50 ampere-hours." This means it has the capacity to deliver 50. amps for 1.0 h (or 1.0 amp for 50. h). If it does deliver 1.0 amp for 50. h, how many grams of lead are consumed to accomplish this?

55. A battery can be built using the reaction between Al metal and O_2 from the air. If the Al anode of this battery consists of 84 g of aluminum, how many hours can the battery produce 1.0 amp of electricity (assuming an unlimited supply of O_2)?

56. A lead storage battery (Section 21.6) operates at 12.0 V and is rated at "100 ampere-hours." This term means it can deliver 1.0 amp of current for 100. h. What is the power rating of the battery in watts?

57. An electrolysis cell for aluminum production operates at 5.0 V and a current of 1.0×10^5 amp. Calculate the number of kilowatt-hours of energy required to produce 1 metric ton (1.0×10^3 kg) of aluminum.

58. Electrolysis of brine leads to chlorine gas. In 1993, 10.9 billion kg of Cl_2 was manufactured. How many kilowatt-hours of energy must have been used to produce this amount of chlorine from NaCl? Assume the electrolysis cells operate at 4.6 V and 3.0×10^5 amp.

59. Electrolysis of molten NaCl is done in cells operating at 7.0 V and 4.0×10^4 amp. How much Na(s) and Cl_2(g) can be produced in one day in such a cell? What is the energy consumption in kilowatt-hours? (Assume 100% efficiency.)

60. An aqueous solution of KI is placed in a beaker with two inert platinum electrodes. When the cell is attached to an external battery, electrolysis occurs.
(a) Write the half-reaction occurring at the anode. What is the polarity of the electrode?
(b) Write the half-reaction occurring at the cathode. What is the polarity of the electrode?
(c) If 0.050 amp passes through the cell for 5.0 h, how many grams of each product (H_2, I_2, and KOH) are expected?

61. An aqueous solution of $NiCl_2$ is placed in a beaker with two inert platinum electrodes. When the cell is attached to an external battery, electrolysis occurs.
(a) Write the half-reaction occurring at the anode. What is the polarity of the electrode?

(b) Write the half-reaction occurring at the cathode? What is the polarity of the electrode?
(c) If 0.0250 amp flows through the cell for 1.25 h, how many grams of each product is expected?

62. Describe what happens at the anode and at the cathode when electrolyzing each of the following solutions:
(a) KBr(aq)
(b) NaF(molten)
(c) NaF(aq)

63. Describe what happens at the anode and at the cathode when electrolyzing each of the following solutions:
(a) $NiBr_2(aq)$
(b) KI(aq)
(c) $CdCl_2(aq)$

General Questions

64. Balance the following equations for redox reactions that occur in acid solution:
(a) $I^-(aq) + Br_2(\ell) \rightarrow IO_3^-(aq) + Br^-(aq)$
(b) $U^{4+}(aq) + MnO_4^-(aq) \rightarrow Mn^{2+}(aq) + UO_2^+(aq)$
(c) $I^-(aq) + MnO_2(s) \rightarrow Mn^{2+}(aq) + I_2(s)$

65. Balance the following equations for redox reactions that occur in basic solution:
(a) $CN^-(aq) + CrO_4^{2-}(aq) \rightarrow OCN^-(aq) + Cr(OH)_4^-(aq)$
(b) $Co^{2+}(aq) + OCl^-(aq) \rightarrow Co(OH)_3(s) + Cl^-(aq)$

66. The half-cells $Ni(s)/Ni^{2+}(aq)$ and $Cd(s)/Cd^{2+}(aq)$ are assembled into a battery.
(a) Write a balanced equation for the reaction occurring in the cell.
(b) What is oxidized, and what is reduced? What is the reducing agent, and what is the oxidizing agent?
(c) Which is the anode, and which is the cathode? What is the polarity of the Cd electrode?
(d) What is E°_{net} for the cell?
(e) What is the direction of electron flow in the external wire?
(f) If the salt bridge contains KNO_3, toward which compartment do the NO_3^- ions migrate?
(g) Calculate the equilibrium constant for the net reaction.
(h) If the concentration of Cd^{2+} is reduced to 0.010 M, and $[Ni^{2+}] = 1.0$ M, what is the voltage produced by the cell? Is the net reaction still the reaction given in part (a)?
(i) If 0.050 amp is drawn from the battery, how long can it last if you begin with 1.0 L of each of the solutions, and each was initially 1.0 M in dissolved species? The electrodes each weigh 50.0 g in the beginning.

67. You are told to assemble an electrochemical cell, with one half-cell being $Cl_2(g)/Cl^-(aq)$. The other half-cell could be $Al^{3+}(aq)/Al(s)$, $Mg^{2+}(aq)/Mg(s)$, or $Zn^{2+}(aq)/Zn(s)$. Which of the metal ion/metal combinations would you choose so as to produce the largest possible positive E°_{net}? Write a balanced equation for the reaction you have chosen.

68. Mendelevium, Md, was the first actinide element found to exist as a stable 2+ ion in aqueous solution. The equilibrium constant for the reaction

$$V^{2+}(aq) + Md^{3+}(aq) \longrightarrow Md^{2+}(aq) + V^{3+}(aq)$$

is 15 at 25 °C. Calculate E° for the half-reaction $Md^{3+}(aq) + e^- \rightarrow Md^{2+}(aq)$.

69. The standard potential for the lead storage cell is 2.04 V. What is the equilibrium constant for the reaction?

70. The standard potential E° for the standard lead storage cell is 2.041 V. Calculate the potential of the battery when the sulfuric acid concentration is 6.00 M at 25 °C.

71. A current of 0.0100 amp is passed through a solution of rhodium sulfate. The only reaction at the cathode is the deposition of rhodium metal. After 3.00 h, 0.038 g of Rh has been deposited. What is the charge on the rhodium ion, Rh^{x+}?

72. Suppose you use $Cu^{2+}(aq)/Cu(s)$ and $Sn^{2+}(aq)/Sn(s)$ half-cells as the basis of an electrochemical cell. If the cell starts at standard conditions, and 0.400 amp of current flows for 48.0 h, what are the concentrations of the dissolved species at this point? What is the potential of the cell after 48.0 h? (Assume 1.00 L of solution.)

73. The total charge that can be delivered by a large dry cell before its voltage drops too low is usually about 35 amp-h. (One amp-hour is the charge that passes through a circuit when 1 amp flows for 1 h.) What mass of Zn is consumed when 35 amp-h of charge is drawn from the cell?

74. A magnesium bar with a mass of 5.0 kg is attached to a buried iron pipe to protect the pipe from corrosion.
(a) Explain how the magnesium protects the pipe.
(b) If a current of 0.030 amp flows between the bar and the pipe, how many years elapse before the magnesium is entirely consumed?

75. A proposed automobile battery involves the reaction of Zn(s) and $Cl_2(g)$ to give $ZnCl_2$. If you want such a battery to operate 10. h and deliver 1.5 amp of current, what is the minimum mass of zinc that the anode must contain?

76. In principle, a battery can be made from aluminum metal and chlorine gas.
(a) Write a balanced equation for the reaction that occurs in a battery using $Al^{3+}(aq)/Al(s)$ and $Cl_2(g)/Cl^-(aq)$ half-reactions.
(b) Tell which half-reaction occurs at the anode and which at the cathode. What are the polarities of these electrodes?
(c) Calculate the standard potential E°_{net} for the battery.
(d) If the pressure of $Cl_2(g)$ is only 0.50 atm, how is the voltage of the battery affected? Is it more positive or more negative than E°_{net}? Calculate E_{net}.
(e) If you want the battery to deliver a current of 0.75 amp, how long can it operate if the aluminum electrode contains 30.0 g of Al? (Assume an unlimited supply of chlorine.)

77. Batteries are listed by their "cranking power." This is the amount of current the battery can produce for 30 s; a typical value is 450 amp. How many coulombs flow through the bat-

tery in 30. s? If this is a lead storage battery, how much lead (Pb) is consumed in 30. s?

78. An expensive but lighter alternative to the lead storage battery is the silver-zinc battery. Zinc is the reducing agent and silver oxide the oxidizing agent.

$$Ag_2O(s) + Zn(s) + H_2O(\ell) \longrightarrow Zn(OH)_2(s) + 2\ Ag(s)$$

The electrolyte is 40% KOH, and silver/silver oxide electrodes are separated from zinc/zinc hydroxide electrodes by a plastic sheet permeable to hydroxide ion. Under normal operating conditions, the battery has a potential of 1.59 V. (R. C. Plumb, *J Chem Ed* Vol. 50, p. 857, 1973.)

(a) How much energy can be produced per gram of reactants in the silver/zinc battery? Assume the battery produces a current of 0.10 amp.

(b) How much energy can be produced per gram of reactants in the standard lead storage battery? Assume the battery produces a current of 0.10 amp at 2.0 V.

(c) Which battery produces the greater energy per gram of reactants?

79. Fluorinated organic compounds are important commercially, because they are used as herbicides, flame retardants, and fire-extinguishing agents, among other things. A reaction such as

$$CH_3SO_2F + 3\ HF \longrightarrow CF_3SO_2F + 3\ H_2$$

is carried out electrochemically in liquid HF as the solvent.

(a) If you electrolyze 150 g of CH_3SO_2F, how many grams of HF is required and how many grams of each product can be isolated?

(b) Is H_2 produced at the anode or cathode of the electrolysis cell?

(c) A typical electrolysis cell operates at 8.0 V and a low current such as 250 amp. How many kilowatt-hours of energy does one such cell consume in 24 h?

Conceptual Questions

80. Four metals, A, B, C, and D, exhibit the following properties:

(a) Only A and C react with 1.0 M hydrochloric acid to give $H_2(g)$.

(b) When C is added to solutions of the ions of the other metals, metallic B, D, and A are formed.

(c) Metal D reduces B^{n+} to give metallic B and D^{n+}.

Based on the information above, arrange the four metals in order of increasing ability to act as reducing agents.

81. Which of the following is the best way to store partly used steel wool? Which is the least conducive to rusting?

(a) Keep it immersed in plain tap water?

(b) Keep it immersed in soapy water (a basic solution)?

(c) Just leave it in the kitchen sink?

82. The free energy change for a reaction, ΔG°_{rxn}, is the maximum energy that can be extracted from the process, whereas ΔH°_{rxn} is the total chemical potential energy change. The efficiency of a fuel cell is the ratio of these two quantities.

$$\text{Efficiency} = \frac{\Delta G^\circ_{rxn}}{\Delta H^\circ_{rxn}} \cdot 100\%$$

Consider the hydrogen-oxygen fuel cell where the net reaction is

$$H_2(g) + \tfrac{1}{2}\ O_2(g) \longrightarrow H_2O(\ell)$$

(a) Calculate the efficiency of the fuel cell under standard conditions.

(b) Calculate the efficiency of the fuel cell if the product is water vapor instead of liquid water.

(c) Does the efficiency depend on the reaction product? Explain why or why not.

83. Why does iron corrode well in solutions with a high concentration of dissolved CO_2?

Summary Questions

84. A hydrogen/oxygen fuel cell operates on simple reaction, $H_2(g) + \tfrac{1}{2}\ O_2(g) \longrightarrow H_2O(\ell)$.
If the cell is designed to produce 1.5 amp of current, and if the hydrogen is contained in a 1.0-L tank at 200. atm pressure at 25 °C, how long can the fuel cell operate before the hydrogen runs out? (Assume the supply of O_2 is unlimited.)

85. Living organisms derive energy from the oxidation of food, typified by glucose.

$$C_6H_{12}O_6(aq) + 6\ O_2(g) \longrightarrow 6\ CO_2(g) + 6\ H_2O(\ell)$$

Electrons in this redox process are transferred from glucose to oxygen in a series of at least 25 steps. It is interesting to calculate the total daily current flow in a typical organism and the rate of energy expenditure (power). (See T. P. Chirpich, *Journal of Chemical Education,* Vol. 52, p. 99, 1975.)

(a) The molar enthalpy of combustion of glucose is −2800 kJ. If you are on a typical daily diet of 2400 Calories, how many moles of glucose must be consumed in a day if glucose is assumed to be the only source of energy? How many moles of O_2 must be consumed in the oxidation process?

(b) How many moles of electrons must be supplied to reduce the amount of O_2 calculated in part (a)?

(c) Based on the answer in part (b), calculate the current flowing, per second, in your body from the combustion of glucose.

(d) If the average standard potential in the electron transport chain is 1.0 V, what is the rate of energy expenditure in watts?

CHAPTER 22

The Chemistry of the Main Group Elements

A Chemical Puzzler

Aluminum (*right*) does not react with nitric acid whereas copper (*left*) reacts vigorously to give NO_2 gas and $Cu(NO_3)_2$. Why does aluminum metal not react with strongly oxidizing nitric acid?

(C. D. Winters)

The abundances of the elements in the solar system are shown plotted against atomic number in Figure 22.1. What do you see? Hydrogen and helium are clearly the most abundant—they make up about 97% of the mass of the universe. Most of the mass of our solar system is in the sun, and its primary components are hydrogen and helium. After that, you notice that lithium, beryllium, and boron are very low in abundance, but carbon is very high. From that point on, the abundances generally decline as the atomic number increases. The elements of the A groups of the periodic table—often referred to as the main group, or representative, elements—are more abundant, on the whole, than the transition elements, which begin with scandium. Except for the high abundance of iron and nickel, the trend continues downward.

On the earth, analysis to determine composition is limited to the crust, the outer shell of the planet, and its atmosphere. Here, oxygen, silicon, and aluminum together make up over 80% of the mass of the earth's crust. Altogether eight main group elements are among the 10 most abundant elements in earth's crust. Oxygen and nitrogen are the primary components of the atmosphere, and water is highly abundant on the surface, underground, and as vapor in the atmosphere. Most rocks and minerals are compounds of the main group elements. The chemistry of limestone ($CaCO_3$), which contains calcium, carbon, and oxygen, was described in Section 19.10. Sand and quartz (Figure 2.20) are composed of silicon and oxygen, and many other

Abundances in earth's crust:

O	49.5%	Na	2.6%
Si	25.7%	K	2.4%
Al	7.4%	Mg	1.9%
Fe	4.7%	H	0.9%
Ca	3.4%	Ti	0.6%

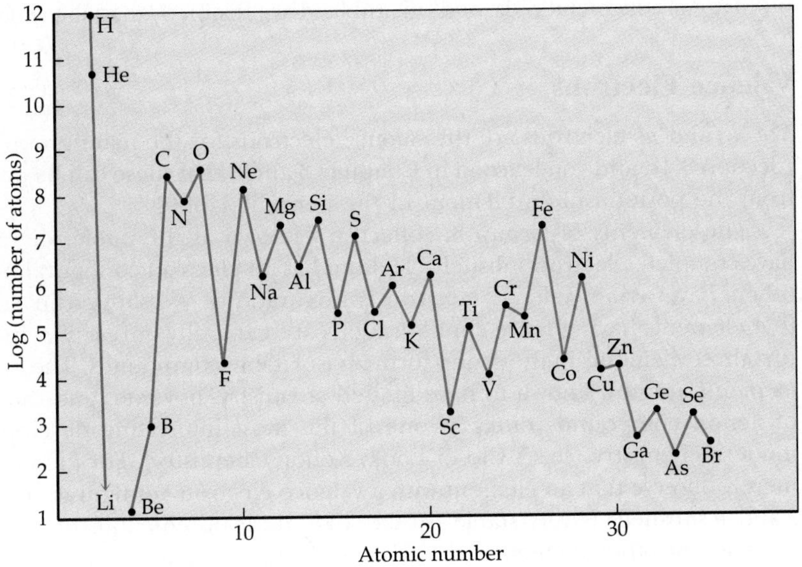

Figure 22.1 The abundances of the elements in the solar system. The scale is relative to the number of H atoms.

common minerals, such as $CaSO_4 \cdot 2\ H_2O$ and fluorite, CaF_2 (Figure 4.16 and page 49), are composed of main group elements.

As a further indication of the importance of the main group elements, we need only look at the top 10 chemicals produced by the U.S. chemical industry in 1993 (see inside back cover). All are main group elements or their compounds.

Because main group elements and their compounds are of such great economic importance—and have an interesting chemistry—this chapter is a brief tour of these elements. Chapter 23 is devoted to the transition elements.

Figure 22.2 Group 4A elements illustrate the trend to increasing metallic character on descending a periodic group: nonmetallic carbon; silicon, a semimetal; and the metals tin and lead. (Carbon is shown here in the form of a graphite crucible, and silicon is a round bar. Lead is present as bullets, a toy, and a sphere, and tin is present as chips of the metal.) (C. D. Winters)

22.1 THE PERIODIC TABLE—A GUIDE TO THE ELEMENTS

You have already seen that there are remarkable similarities among the properties of the elements in a given group. Indeed, it was this type of information that guided Mendeleev in creating the periodic table. As illustrated in Figure 2.16, Mendeleev placed elements in groups based on the stoichiometries of their common oxygen and hydrogen compounds.

We see further evidence of similarities of the elements in a given group with reference to the highest oxidation states of the elements and by some of their common oxides, oxoanions, and hydrides (Table 22.1).

A knowledge of trends of metallic character is particularly useful as we survey these elements. You saw in Chapters 2 and 8 that the Group 1A elements, the alkali metals, are the most metallic elements in the periodic table. In contrast, elements on the far right are nonmetals, and in between are the metalloids (semimetals) (Figure 2.14). You also saw that metallic character generally increases from the top of a group to the bottom. This is particularly well illustrated in Group 4A; carbon at the top of the group is a nonmetal, silicon and germanium are classic metalloids, and tin and lead are distinctly metallic (Figure 22.2).

Valence Electrons

The *ns* and *np* electrons are the valence electrons for the main group elements (Section 8.4), and you learned in Chapters 3 and 8 that these can be determined from the position of an element in the periodic table.

As mentioned on page 80, there is a controversy in the chemical community about the numbering of groups in the periodic table. One argument in favor of retaining the use of A and B group designations is the ease it provides in identifying the number of valence electrons.

The elements of Group 8, collectively known as the **noble,** or **rare gases,** have complete electron subshells. Helium has an electron configuration $1s^2$; the others have ns^2np^6 valence electron configurations. We showed in Section 8.7 that elements having these configurations are generally unreactive. Indeed, the first three elements in the group form no isolatable compounds. The other three elements are now known to have limited chemistry, however, and the discovery of xenon compounds ranks as one of the most interesting developments in modern chemistry (see A Closer Look: Xenon Chemistry). For the moment, we merely observe that an element with a valence electron configuration with filled *s* and *p* subshells is very stable—a fact that guides predictions of the chemical behavior of other elements that often react in a way to achieve a "noble gas configuration."

TABLE 22.1 Similarities within Periodic Groups*

Group	1A	2A	3A	4A	5A	6A	7A
Common oxide	M_2O	MO	M_2O_3	AO_2	A_2O_5	AO_3	A_2O_7
Common hydride	MH	MH_2	MH_3	AH_4	AH_3	AH_2	AH
Highest oxidation state	+1	+2	+3	+4	+5	+6	+7
Common oxoanion				CO_3^{2-} SiO_4^{4-}	NO_3^- PO_4^{3-}	SO_4^{2-}	ClO_4^-

*M denotes a metal and A denotes a nonmetal.

Ionic Compounds of Main Group Elements

The elements in Groups 1A and 2A invariably form positively charged ions whose electron configurations are the same as that for the previous noble gas. Thus, the commonly encountered compounds of these elements (for example, NaCl, $CaCO_3$) can immediately be identified as ionic. We may expect all compounds of these elements to have typical ionic properties: they will be crystalline solids with high melting points and conduct electricity in the molten state (see Section 3.4).

A CLOSER LOOK *Xenon Chemistry*

While studying the chemistry of PtF_6 in 1962, Neil Bartlett noticed, quite accidentally, that exposing this compound to air gave $[O_2^+][PtF_6^-]$. Just as importantly, he recognized that the ionization energy of O_2 and Xe were comparable. So, he proceeded to react Xe and PtF_6 and found evidence of compound formation. By analogy to the oxygen compound, Bartlett first characterized the solid he obtained as $[Xe^+][PtF_6^-]$, although he later found that the material was in fact a mixture of several xenon compounds. Soon thereafter other scientists reported that xenon forms several stable xenon fluorides, XeF_2, XeF_4, and XeF_6 (Figure 9.14), and an extensive array of compounds was eventually prepared.

Attempts to extend this chemistry to other noble gases have had limited success. The chemistry of radon is probably similar to that of xenon but cannot be studied easily due to the radioactivity of this element. Krypton fluorides are unstable and decompose rapidly.

Most chemists never considered the possibility that a noble gas compound might not only exist but be quite stable. One worker commented, retrospectively, that this could be characterized as the "closed shell—closed mind" syndrome.

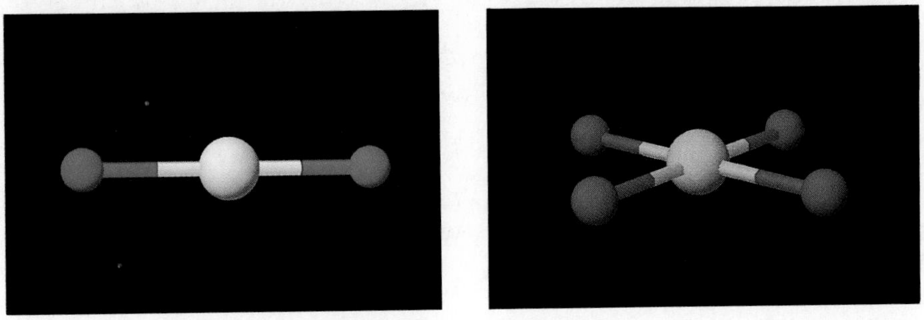

The structures of two xenon compounds, linear XeF_2 and square planar XeF_4. (Susan Young)

TABLE 22.2 Some Reactions of Group 1A and 2A Metals with Halogens or Oxygen

Metal	Nonmetal	Product
K(s), Group 1A	$Br_2(\ell)$, Group 7A	KBr(s), ionic
Ba(s), Group 2A	$Cl_2(g)$, Group 7A	$BaCl_2(s)$, ionic
Al(s), Group 3A	$F_2(g)$, Group 7A	$AlF_3(s)$, ionic
Na(s), Group 1A	$S_8(s)$, Group 6A	$Na_2S(s)$, ionic
Mg(s), Group 2A	$O_2(g)$, Group 6A	MgO(s), ionic

The elements in Groups 6A and 7A can reach a noble gas configuration by adding electrons. Thus, in many reactions, the Group 7A elements (halogens) form anions with a 1− charge (halide ions, F^-, Cl^-, Br^-, I^-), and the Group 6A elements form anions with a 2− charge (O^{2-}, S^{2-}, Se^{2-}, Te^{2-}). Unlike the Group 1A and 2A metals, however, other avenues are open to the nonmetallic elements. Covalent compounds formed in reactions between two or more nonmetals are well known. Thus, a description of the chemistry of main group elements is in two parts: first, the formation of ionic compounds in reactions with metals; second, the covalent compounds from reactions with other nonmetals.

The chemistry of Group 3A elements parallels the behavior of elements in Groups 1A and 2A. A majority of compounds of Group 3A elements contain 3+ ions. For example, many compounds of aluminum contain the Al^{3+} ion. In Group 5A chemistry, ions are also encountered; for example, the nitride ion, N^{3-}, has an argon electron configuration. It has been noted, however, that the energy required to form highly charged cations and anions is large (Section 8.7), which means that their formation is often unfavorable relative to other possible modes of behavior.

Some covalent chemistry also occurs in Group 3A, especially with boron, and nitrogen chemistry is dominated by covalent compounds. Consider ammonia, NH_3; the ammonium ion, NH_4^+; the various nitrogen oxides; nitric acid, HNO_3; and the nitrate ion, NO_3^-, all of which involve covalent bonding from nitrogen to the second element.

With this brief overview, we can now predict the results of reactions between elements of Groups 1A, 2A, and 3A and elements of Group 6A or 7A (Table 22.2). In general, we expect the metal to be oxidized and the nonmetal reduced to form ionic compounds. The structures of such compounds will be an infinite lattice of positive and negative ions.

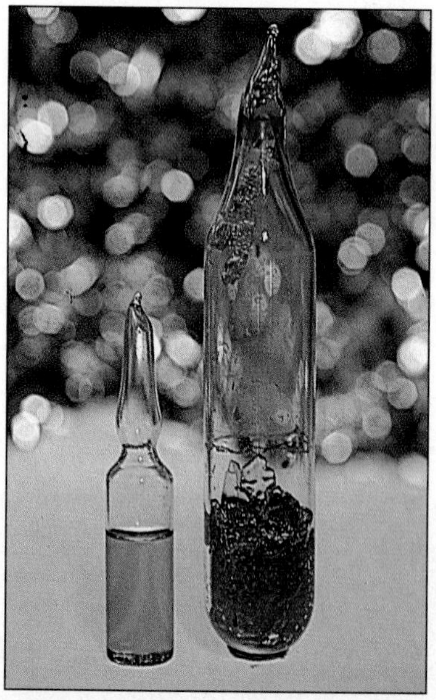

Liquid BBr_3 (*left*) and solid BI_3 (*right*). Both have covalent bonds between boron and the halogen atom. (C. D. Winters)

EXAMPLE 22.1 *Reactions of Group 1A and 2A elements*

Give the formula and name for the product in each of the following reactions. Write a balanced chemical equation for the reaction.

1. $Ca(s) + S_8(s)$
2. $Rb(s) + I_2(s)$

3. Lithium and chlorine

4. Aluminum and oxygen

Solution Group 1A elements are converted to 1+ ions, Group 2A elements are converted to 2+ ions, and Group 3A elements are converted to 3+ ions. In their reactions with metals, halogen atoms add a single electron to give anions with a 1− charge, whereas Group 6A elements are predicted to add two electrons to form anions with a 2− charge (although sometimes they do not react in this simple fashion). These predictions are based on the assumption that ions are formed with electron configurations of the nearest noble gas.

Balanced Equation	Product Name
1. $8\ Ca(s) + S_8(s) \rightarrow 8\ CaS(s)$	Calcium sulfide
2. $2\ Rb(s) + I_2(s) \rightarrow 2\ RbI(s)$	Rubidium iodide
3. $2\ Li(s) + Cl_2(g) \rightarrow 2\ LiCl(s)$	Lithium chloride
4. $4\ Al(s) + 3\ O_2(g) \rightarrow 2\ Al_2O_3(s)$	Aluminum oxide

EXERCISE 22.1 ***Main Group Element Chemistry***

Write a balanced chemical equation for the formation of the following compounds from the elements:

1. NaBr

2. CaSe

3. $AlCl_3$

4. K_2O ■

Covalent Compounds and Electron Configurations

You have already seen many reactions between different nonmetals such as that between carbon and excess oxygen to form CO_2. The products of these reactions are molecular compounds in which the atoms share electron pairs.

The valence electron configuration of an element controls the stoichiometry of its compounds as well as the charge on ions. Involving all the valence electrons in the formation of compounds is a frequent and reasonable occurrence in main group element chemistry. This being so, we should not be surprised to discover halogen compounds whose stoichiometries reflect the fact that the core element has the highest possible oxidation number. For example, fluorine forms ionic compounds with elements of Groups 1A through 3A, whereas the compounds are covalent with highly polar bonds for Groups 4A through 7A. In every case the central element has its maximum oxidation number.

Group	Compound
1A	NaF
2A	MgF_2
3A	AlF_3
4A	SiF_4
5A	PF_5
6A	SF_6
7A	IF_7

EXAMPLE 22.2 ***Predicting Formulas for Compounds of Main Group Elements***

What formula is predicted for each of the following:

1. The product of a reaction between germanium and excess oxygen.

2. The product of the reaction of arsenic and fluorine.
3. A neutral compound composed of phosphorus and chlorine.
4. The anion of selenic acid.

Solution

1. Germanium is in Group 4A. We therefore predict that this element will form an oxide of formula GeO_2, in which germanium has an oxidation number of +4.
2. Arsenic, in Group 5A, reacts vigorously with fluorine to form AsF_5, in which arsenic has an oxidation number of +5.
3. PCl_5 is the product formed when phosphorus reacts with excess chlorine.
4. Selenium is below sulfur in the periodic table, and its chemistry is likely to be similar. Thus, Se is oxidized to SeO_3 with excess oxygen. This species is an acid anhydride (like SO_3) and forms selenic acid, H_2SeO_4, with water. The anion of this acid, the selenate ion, has the formula SeO_4^{2-}. Selenium's reaction with oxygen, the acid properties of the oxide, and the formula for selenate ion are all analogous to sulfur chemistry.

EXERCISE 22.2 ***Predicting Formulas for Main Group Compounds***

Write formulas for

1. Hydrogen telluride
2. Sodium arsenate
3. Selenium hexachloride
4. Perbromic acid ■

We should expect many similarities among elements in the same periodic group. Such analogies allow us to extrapolate from simple compounds that are easily recognized to analogous compounds of less common elements. You already know examples of this. Water, H_2O, is the simplest hydrogen compound of Group 6A, and you expect to find hydrogen compounds with similar stoichiometries with other elements in this group. Indeed, H_2S, H_2Se, and H_2Te are well known.

EXAMPLE 22.3 ***Predicting Formulas***

Predict formulas for

1. A compound of hydrogen and phosphorus.
2. The hypobromite ion.
3. Germane (the simplest hydrogen compound of germanium).

Solution

1. Phosphine, PH_3, has a stoichiometry similar to ammonia, NH_3.
2. Hypobromite ion, OBr^-, is similar to hypochlorite ion, OCl^-, the anion derived by deprotonation of hypochlorous acid.

3. GeH_4 is analogous to CH_4 and SiH_4.

EXERCISE 22.3 ***Predicting Formulas***

Identify a second-period compound that has a formula and structure analogous to each of the following:

1. PH_4^+
2. S_2^{2-}
3. P_2H_4
4. PF_3 ■

EXAMPLE 22.4 ***Recognizing Incorrect Formulas***

One formula is incorrect in each of the following groups. Pick out the incorrect formula and indicate why it is unlikely.

1. $CsSO_4$, KCl, $NaNO_3$, Li_2O
2. MgO, CaI_2, Ba_2SO_4, $CaCO_3$
3. CO, CO_2, CO_3
4. PF_3, PF_4^+, PF_2, PF_6^-

Solution

1. $CsSO_4$. Sulfate ion has a 2− charge, so this formula would require a Cs^{2+} ion. Cesium, in Group 1A, forms only 1+ ions. In addition, if you add up the valence electrons for all of the atoms you find it to be an odd number.
2. Ba_2SO_4. This formula requires a Ba^+ ion. The ion charge does not equal the group number.
3. CO_3. Given that O has an oxidation number of −2, CO_3 would lead to a carbon oxidation number of +6. Carbon is in Group 4A, however, and can have a maximum oxidation state of +4.
4. PF_2. This species has an odd number of electrons.

EXERCISE 22.4 ***Recognizing Incorrect Formulas***

Explain why compounds with the following formulas do not exist: ClO, Na_2Cl, $CaCH_3CO_2$, C_3H_7. ■

PROBLEM-SOLVING TIPS AND IDEAS

22.1 Nonexistent Compounds?

It may seem that the array of known chemical compounds is without limit. Not all combinations of elements are acceptable, however. It may help to be able to recognize when a formula is incorrect or when a combination of elements would not be found together in a compound. Let us list some useful hints about what compounds do *not* exist as stable entities. All are based on the ideas outlined in Chapter 8.

- The three lightest noble gases do not form any compounds; only Xe has an extensive chemistry.
- Group 1A elements never form 2+ or 3+ ions; Group 2A metals never form 1+ ions.
- Very few covalent compounds contain an odd number of electrons. The same holds true of polyatomic ions. (You only encounter three exceptions: NO and NO_2, and later in this chapter, ClO_2). If you count the electrons in a formula, and the number is odd, then the formula is probably wrong.
- The maximum oxidation state of an element in a compound is equal to the number of the group. ■

22.2 HYDROGEN

The discovery of hydrogen was described on page 57.

Hydrogen is ninth in abundance in the earth's crust, 0.9% by mass, where it occurs primarily in water and in fossil fuels. In its earliest history, H_2 was mainly used as a fuel. In the middle of the 19th century it was found that heating soft coal (in the absence of air) gave a gas that could be used for cooking and lighting. This gas, called coal gas, contains about 20% H_2 along with several lightweight hydrocarbons.

Because coal gas was useful, new methods were sought for its production. It was found that injecting water into a bed of red hot coke produces a mixture of H_2 and CO. This mixture is known as water gas or synthesis gas (syngas).

$$\underset{\text{coke}}{C(s)} + H_2O(g) \longrightarrow \underbrace{H_2(g) + CO(g)}_{\text{water gas}}$$

A modern plant for the production of syngas, located in South Africa. Coal, oxygen, and steam are heated under pressure to produce a mixture of hydrogen, carbon monoxide, methane, and carbon dioxide. Hydrogen sulfide and nitrogen are usually formed in small quantities, because coal usually contains some sulfur and nitrogen. After purification of syngas, the mixture of H_2, CO, and CH_4 is used to manufacture various hydrocarbons. (Fluor Engineers, Inc.)

Water gas burns cleanly and can be handled readily. The amount of heat produced, however, is only about half that from the combustion of coal gas, and its flame is nearly invisible. Furthermore, carbon monoxide is highly toxic and has no odor. Despite its hazards, water gas was used as a cooking gas to some extent until about 1950, but only after adding another material to make the flame luminous and a malodorous compound so that leaks would be detected. Although no longer used as a fuel, there is renewed interest in syngas because recent chemical research has shown that it can be used to manufacture hydrocarbons.

Synthesis of Hydrogen Gas

About 300 billion L (STP) of hydrogen gas is produced worldwide in a year, and virtually all is used immediately in other processes. The largest quantity of hydrogen is produced by the catalytic steam re-formation of hydrocarbons. This process uses methane, CH_4, as the primary starting material. Methane reacts with steam at high temperature to give CO and H_2.

$$CH_4(g) + H_2O(g) \longrightarrow 3\ H_2(g) + CO(g) \qquad \Delta H^\circ_{rxn} = +206\ \text{kJ}$$

The reaction is rapid in the 900 °C to 1000 °C range and goes nearly to completion. More hydrogen can be obtained in a second step in which the CO that is formed reacts with more water. This so-called water gas shift reaction is run at 400 °C to 500 °C and is slightly exothermic.

$$H_2O(g) + CO(g) \longrightarrow H_2(g) + CO_2(g) \qquad \Delta H° = -41 \text{ kJ}$$

The CO_2 formed in the process is removed by reaction with CaO (to give $CaCO_3$), thus leaving fairly pure hydrogen.

Electrolysis of water is the cleanest method of H_2 production (Figure 22.3), and it provides a valuable byproduct, high-purity O_2. Electric energy is quite expensive, however, so this method is not generally used.

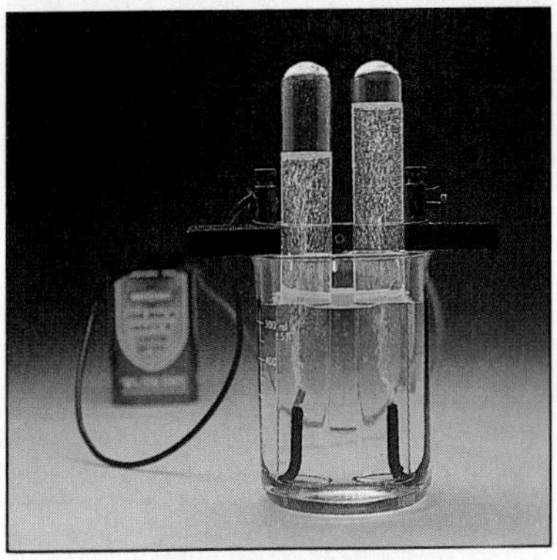

Figure 22.3 Electrolysis of a dilute aqueous solution of H_2SO_4, to give H_2 (left test tube) and O_2 (right test tube). (C. D. Winters)

Because hydrogen is such a valuable commodity, there is considerable interest in finding a way to split water into hydrogen and oxygen using thermal energy. At temperatures over 1000 °C, H_2O, H_2, and O_2 exist at equilibrium. Achieving such high temperatures is quite costly, however, so many approaches have been suggested for achieving this result at lower temperature using a sequence of reactions. One example is the following:

Step 1, at 750 °C:	$CaBr_2 + H_2O \longrightarrow 2\ HBr + CaO$
Step 2, at 100 °C:	$Hg + 2\ HBr \longrightarrow HgBr_2 + H_2$
Step 3, at 25 °C:	$HgBr_2 + CaO \longrightarrow HgO + CaBr_2$
Step 4, at 500 °C:	$HgO \longrightarrow Hg + \frac{1}{2}O_2$
Net reaction:	$H_2O(\ell) \longrightarrow H_2(g) + \frac{1}{2}\ O_2(g)$

Even though water is essentially free, thermal processes to produce hydrogen are not yet economical relative to those using natural gas or coal as a starting material.

A number of reactions can be used in the laboratory to form hydrogen (Table 22.3), one of the simplest of which is the reaction of a metal with an acid (see Figure 4.9). In 1783, Charles (of Charles's law) used the reaction of sulfuric acid with iron to produce the hydrogen for a lighter-than-air balloon.

The reaction of aluminum with NaOH (Figure 22.4) also generates hydrogen as one product. During World War II, this method was used to obtain hydrogen to inflate small balloons for weather observation and to raise radio antennas. Metallic aluminum was plentiful because it came from damaged aircraft. Finally, reaction (3) in Table 22.3 is perhaps the most efficient way to synthesize H_2 in the laboratory (Figure 17.2). It is also useful for removing traces of water from liquid compounds that do not have a reactive $-OH$ group.

Figure 22.4 The reaction of aluminum with aqueous NaOH produces hydrogen and $NaAl(OH)_4$. (C. D. Winters)

TABLE 22.3 Methods for Preparing H_2 in the Laboratory

1. Metal + Acid → metal salt + H_2

   $Mg(s) + 2\ HCl(aq) \rightarrow MgCl_2(aq) + H_2(g)$

2. Metal + H_2O or base → metal hydroxide or oxide + H_2

   $2\ Na(s) + 2\ H_2O(\ell) \rightarrow 2\ NaOH(aq) + H_2(g)$

   $2\ Fe(s) + 3\ H_2O(\ell) \rightarrow Fe_2O_3(s) + 3\ H_2(g)$

   $2\ Al(s) + 2\ KOH(aq) + 6\ H_2O(\ell) \rightarrow 2\ KAl(OH)_4(aq) + 3\ H_2(g)$

3. Metal hydride + H_2O → metal hydroxide + H_2

   $CaH_2(s) + 2\ H_2O(\ell) \rightarrow Ca(OH)_2(s) + 2\ H_2(g)$

Hydrogen burns in an atmosphere of bromine to give hydrogen bromide, HBr, a covalent hydride in which H has an oxidation number of +1. (C. D. Winters)

Properties of Hydrogen

Under standard conditions, hydrogen is a colorless gas. Its very low boiling point, 20.7 K, reflects its nonpolar character and low molecular mass. It is, of course, the least dense gas known.

Hydrogen combines chemically with virtually every other element except the noble gases. Three different types of hydrogen-containing binary compounds are known.

Ionic Metal Hydrides

Ionic metal hydrides form when H_2 reacts with Group 1A and 2A metals.

$$2\ Na(s) + H_2(g) \longrightarrow 2\ NaH(s)$$
$$Ca(s) + H_2(g) \longrightarrow CaH_2(s)$$

These ionic compounds contain the hydride ion, H^-, in which hydrogen is in the -1 oxidation state.

Covalent Hydrides

Covalent hydrides are formed with electronegative elements, such as the nonmetals carbon, nitrogen, oxygen, and fluorine. Here the formal oxidation number of the hydrogen atom is +1.

$$N_2(g) + 3\ H_2(g) \longrightarrow 2\ NH_3(g)$$
$$F_2(g) + H_2(g) \longrightarrow 2\ HF(g)$$

Interstitial Hydrides

Hydrogen is absorbed by many metals forming interstitial hydrides, in which hydrogen atoms reside in the spaces between metal atoms (called interstices) in the crystal lattice. For example, when a piece of palladium metal is used as an electrode for the electrolysis of water, the metal can soak up a thousand times its volume of hydrogen (at STP). Most interstitial hydrides are nonstoichiometric, that is, the ratio of metal and hydrogen does not involve whole numbers. When interstitial hydrides are heated, H_2 can be driven out. Thus, these materials can be used to store H_2, the same as a sponge can store water. This is one way to store hydrogen for use as a fuel in automobiles (page 300).

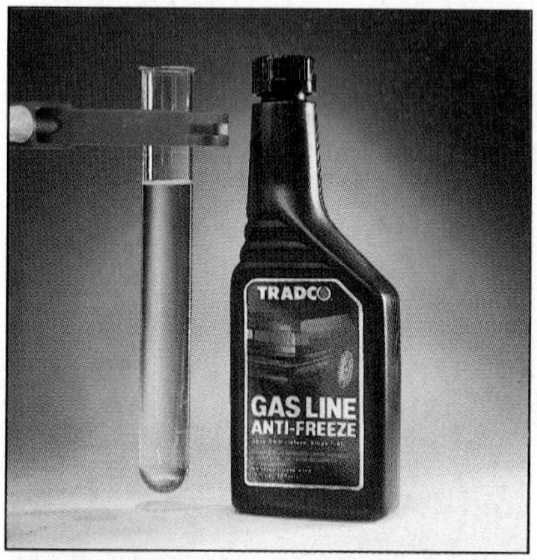

Methanol is often added to gasoline to prevent "fuel line freeze" in automobiles in the winter. The alcohol forms hydrogen bonds with water, thus preventing the water from freezing and clogging the gas line. (C. D. Winters)

Some Uses of Hydrogen

By far the largest use of H_2 gas is in the production of ammonia, NH_3, by the Haber process (Section 16.6).

$$N_2(g) + 3\ H_2(g) \longrightarrow 2\ NH_3(g)$$

A large amount is also used to make methanol, CH_3OH (page 513).

$$2\ H_2(g) + CO(g) \longrightarrow CH_3OH(\ell) \qquad \Delta H^\circ_{rxn} = -128.2\ kJ$$

Almost 4.8 billion kg of this colorless liquid was produced in 1993. Methanol is used as an additive in gasoline because oxygen-containing compounds cause gasoline to burn more cleanly. In addition, methanol is often added to gasoline in cold weather to prevent "fuel line freeze"; in this use, its function is to dissolve traces of water that often contaminate gasoline.

22.3 SODIUM AND POTASSIUM

Sodium and potassium are the sixth and seventh most abundant elements in earth's crust, 2.6% and 2.4%, respectively, by mass. Both metals, as well as the other Group 1A elements, are highly reactive with oxygen, water, and other oxidizing agents (Figure 2.17). In all cases, compounds of the Group 1A metals contain the element as a 1+ ion.

The name "sodium" comes from caustic soda, NaOH. The Latin name for caustic soda (NaOH), *natrium,* furnished the symbol Na. The name "potassium" is derived from the word "potash" (literally, the ashes from a fire pot). Potash (K_2CO_3) can be extracted from ashes with water. The Latin word for potash is *kalium,* and hence we get the symbol K for this element.

The solubility guidelines indicate that most sodium and potassium compounds are water-soluble (Figure 4.7), so it is not surprising that sodium and potassium salts are found on earth either in solution, in the oceans, or in underground deposits that are the residue of ancient seas (page 169). To a much smaller extent, these elements also are found in minerals, like Chilean saltpeter ($NaNO_3$) and borax ($Na_2B_4O_7 \cdot 10\ H_2O$).

Within the earth's crust, sodium and potassium are about equally abundant; however, sea water contains about 2.8% NaCl but only about 0.8% KCl. Why this great difference, given that compounds of these elements have similar solubilities? The answer lies in the fact that potassium is an important factor in plant growth. Much of the potassium in ground water is taken up by plants. Most plants contain four to six times as much potassium as sodium. You may be aware that potassium is one of the three components in most fertilizers.

Some NaCl is essential in the diet of humans and animals because many biological functions are controlled by the concentrations of Na^+ and Cl^- ions. Animals travel great distances to reach a salt lick, and farmers often place large blocks of salt in fields for cattle. The fact that salt has been important for a long time is evident in surprising ways. We are paid a "salary" for work done; this word is derived from the Latin word *salarium,* which meant "salt money" because Roman soldiers were paid in salt. We still talk about "salting away" money for a rainy day, a term related to the practice of preserving meat by salting it.

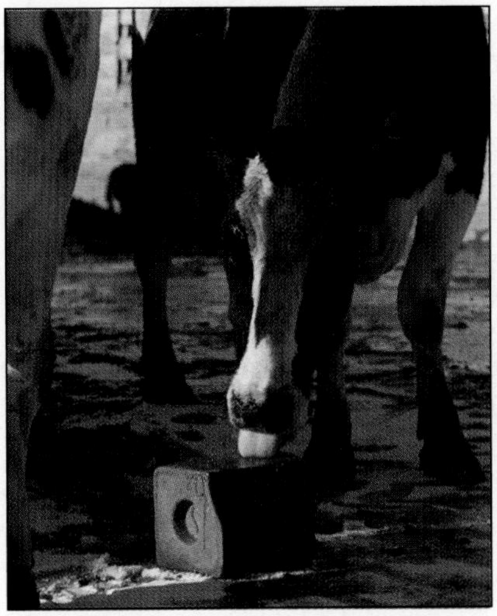

All animals, including humans, need a certain amount of salt in their diets. Sodium ion is important in maintaining electrolyte balance and in regulating osmotic pressure. (C. D. Winters)

Preparation and Properties of Sodium and Potassium

The pure metals were prepared first by the English scientist Sir Humphry Davy (1778–1829) in 1807 by electrolyzing the molten carbonates Na_2CO_3 and K_2CO_3. Sodium is still produced by electrolysis, although molten NaCl is now used (Chapter 21). Common chemical species are not strong enough reducing agents to convert Na^+ to the metal, so electrolysis is the only viable method of preparation. Potassium can be made by electrolysis also, but there are problems with this method, not the least of which is that molten potassium is soluble in the molten salt, making separation difficult. The preferred preparation of potassium uses the reaction of sodium vapor with molten KCl.

$$Na(g) + KCl(\ell) \longrightarrow K(g) + NaCl(\ell)$$

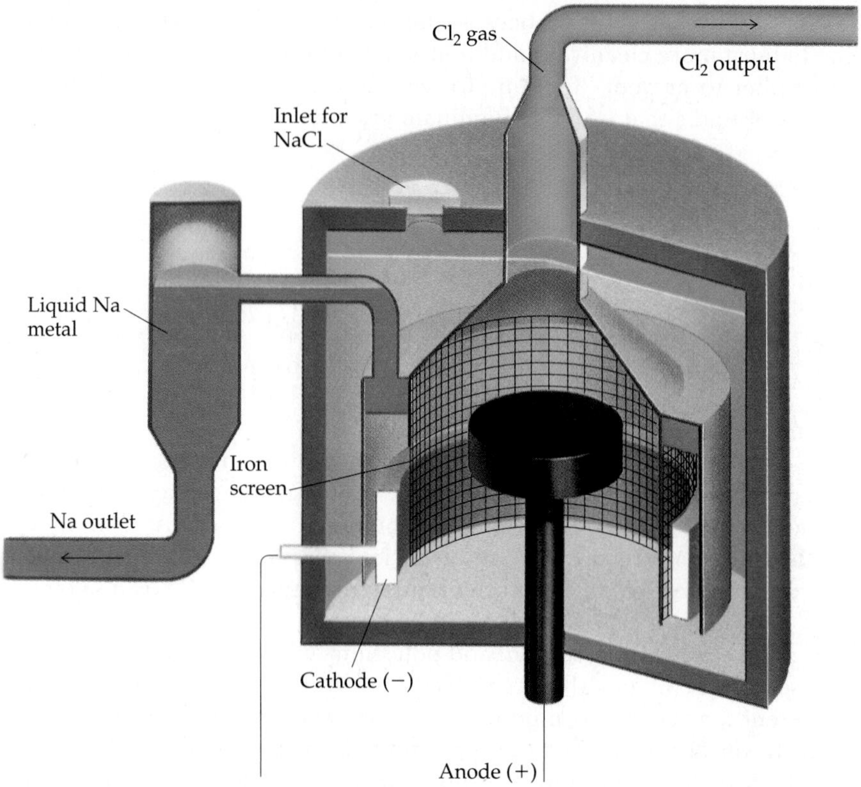

A Downs cell for producing sodium metal by electrolysis. The circular iron cathode is separated from the graphite anode by an iron screen. Because the cell operates at about 600 °C, the sodium is produced at the cathode in the molten state. The liquid metal has a low density, so it floats on the top of the liquid salt. The chlorine produced at the anode bubbles out of the cell and is collected.

This reaction is a good example of the importance of understanding chemical equilibria. The equilibrium constant is less than 1, indicating that reactants are favored. The potassium, however, is removed continuously from the reaction, shifting the equilibrium to the right.

Both sodium and potassium are silvery metals that are soft and easily cut with a knife (Figure 2.14). Their densities are just a bit less than water. Melting points for both elements are quite low, 93.5 °C for sodium and 65.65 °C for potassium. As with all alkali metals, these elements are highly reactive. When exposed to air, the metal surface is quickly coated with an oxide film. Consequently, the metals must be stored in a way to avoid contact with air; this is often done by placing them in kerosene or mineral oil.

The melting points of the alkali metals decline on moving down the group. Cesium melts at 28.5 °C. Thus, Cs, Ga (m.p. 29.8 °C), and Hg are the only metals that are liquids at or near room temperature.

The Group 1A metals are highly reactive. The reaction with water generates an aqueous solution of the metal hydroxide and hydrogen (page 7 and Figure 2.17),

$$2\ Na(s) + H_2O(\ell) \longrightarrow 2\ Na^+(aq) + 2\ OH^-(aq) + H_2(g)$$

and reaction with any of the halogens yields a metal halide (Figure 2.21).

$$2\ Na(s) + Cl_2(g) \longrightarrow 2\ NaCl(s)$$
$$2\ K(s) + Br_2(\ell) \longrightarrow 2\ KBr(s)$$

Chemistry also produces some surprises. Group 1A metal oxides, M_2O, are known, but they are not the principal products of reactions between the Group 1A elements and oxygen. The primary product of the reaction of sodium and oxygen is sodium peroxide, Na_2O_2, and not sodium oxide, Na_2O. The principal product from the reaction of potassium and oxygen is KO_2, potassium superoxide.

$$2\ Na(s) + O_2(g) \longrightarrow Na_2O_2(s)$$
$$K(s) + O_2(g) \longrightarrow KO_2(s)$$

Recall the oxygen species $[O_2^+]$, $[O_2^-]$, and $[O_2^{2-}]$ from the earlier discussion on molecular orbital theory (Chapter 10). Oxygen-oxygen bond orders of 2.5, 1.5, and 1.0 were calculated for these species based on MO theory.

Both Na_2O_2 and KO_2 are ionic, with Group 1A cations paired with either the peroxide ion (O_2^{2-}) or the superoxide ion O_2^-. These are not just laboratory curiosities. They are used in oxygen generation devices in places where people are confined, such as submarines, aircraft, and space craft. When a person breathes, for every liter of O_2 inhaled, 0.82 L of CO_2 is exhaled. A requirement of an O_2 generation system, therefore, is that it should produce a larger volume of O_2 than the volume of CO_2 taken in. This requirement is met with peroxides and superoxides.

$$4\ KO_2(s) + 2\ CO_2(g) \longrightarrow 2\ K_2CO_3(s) + 3\ O_2(g)$$

Sodium Compounds of Commercial Importance

Electrolysis of aqueous sodium chloride (Chapter 21) is the basis of the chlor-alkali industry, one of the largest chemical industries in the United States. The major commercial products from this process are chlorine and sodium hydroxide.

$$2\ NaCl(aq) + 2\ H_2O(\ell) \longrightarrow 2\ NaOH(aq) + H_2(g) + Cl_2(g)$$

In 1993, 11.7 billion kg of NaOH and 11.1 billion kg of Cl_2 were produced in the United States.

Sodium carbonate is another commercially important compound. In 1993, 9.0 billion kg of Na_2CO_3 was produced in the United States, making it the 12th ranked industrial chemical. This compound has the common name *soda ash,* or *washing soda,* and it has been obtained, since prehistoric times, from naturally occurring deposits of $Na_2CO_3 \cdot 10\ H_2O$. *Trona,* $Na_2CO_3 \cdot NaHCO_3 \cdot 2\ H_2O$, however, estimated at 6×10^{10} tons, was recently discovered in Wyoming, and virtually all Na_2CO_3 in the United States now comes from that source. About 40% of the soda ash is used in the manufacture of glass, but large amounts are also used in water treatment, in pulp and paper manufacture, and in cleaning materials.

Soda ash mined from Searles Lake, California saline deposits. (Jack Dermid/Photo Researchers, Inc.)

As described on page 88 the use of chlorine has come under attack from environmental groups; therefore, considerable interest has arisen in manufacturing sodium hydroxide without the co-product chlorine, as in the case of salt electrolysis. This has led to a revival of the old "lime-soda process," which produces NaOH from inexpensive lime (CaO) and soda (Na_2CO_3).

$$Na_2CO_3(aq) + CaO(s) + H_2O(\ell) \longrightarrow 2\ NaOH(aq) + CaCO_3(s)$$

The calcium carbonate byproduct is filtered off and is recycled into the process by heating it (calcining) to recover lime.

$$CaCO_3(s) \longrightarrow CaO(s) + CO_2(g)$$

Sodium bicarbonate, $NaHCO_3$, is produced in small amounts from soda ash. You are probably more aware of the bicarbonate under the common name **baking soda.** Not only is $NaHCO_3$ used in cooking, but it is also added in small amounts to table salt. This is because NaCl is often contaminated with small amounts of $MgCl_2$. The magnesium salt is hygroscopic; that is, it picks water up from the air and, in doing so, causes the NaCl to clump on damp days. Adding $NaHCO_3$ converts $MgCl_2$ to magnesium carbonate, a nonhygroscopic salt.

$$MgCl_2(s) + 2\ NaHCO_3(s) \longrightarrow MgCO_3(s) + 2\ NaCl(s) + H_2O(\ell) + CO_2(g)$$

22.4 CALCIUM AND MAGNESIUM

The Group 2A elements are called *alkaline earths.* The "earth" part of the group name is left over from the days of medieval alchemy. To alchemists, any solid substance that did not melt and was not changed by fire into another substance was called an "earth." Various compounds of Group 1A and 2A elements that were known in those times, such as NaOH and CaO, were alkaline according to the experimental tests of the alchemists: they had a bitter taste and could be shown to neutralize acids. Group 1A compounds, however, melted in a fire or combined with the clay containers in which they were heated. Melting points of most Group 2A compounds are very high, due to the strong attractive forces between the M^{2+} cation with the anions present (page 119). For example, CaO melts at 2572 °C, a temperature well beyond the range of an ordinary fire.

Calcium compounds such as lime (CaO) were known and used in ancient times. Calcium metal, however, was first prepared in 1808 by Sir Humphry Davy, who also prepared magnesium, strontium, and barium in the same year. As with sodium and potassium, Davy's preparation of these elements was accomplished by electrolysis of a molten salt.

The rocks in one section of the Grand Canyon are largely dolomite (the darker band of rocks), a mixture of $MgCO_3$ and $CaCO_3$. (James Cowlin/Image Enterprises)

The great abundance of calcium and magnesium on earth leads to their occurrence in plants and animals, and both elements form many commercially important compounds. It is on this chemistry that we want to focus our attention.

Similar to the Group 1A elements, the Group 2A elements are very reactive, so they are only found in nature combined with other elements. Unlike Group 1A metals, however, many compounds of the Group 2A elements have low water-solubility. This explains their common occurrence in various minerals. Limestone ($CaCO_3$), gypsum ($CaSO_4 \cdot 2\ H_2O$), and fluorspar (CaF_2) are examples of common calcium-containing minerals. Magnesite ($MgCO_3$), talc or soapstone ($3\ MgO \cdot 4\ SiO_2 \cdot H_2O$), and asbestos ($3\ MgO \cdot 4\ SiO_2 \cdot 2\ H_2O$) are magnesium-containing minerals, whereas dolomite is a compound with both elements ($MgCO_3 \cdot CaCO_3$).

Limestone, a sedimentary rock, is found widely on the surface of the earth (Figure 19.10). These deposits are the fossilized remains of marine life and are the most common of the *calcite* forms of the compound. You also know other forms of calcite, however. Marble is fairly pure calcite formed by crystallization of $CaCO_3$ under high pressure. High-quality deposits of marble are found in Italy and in the United States in Vermont, Georgia, and Colorado. Another form of calcite is *Icelandic spar,* which forms large, clear crystals (Figure 2.18).

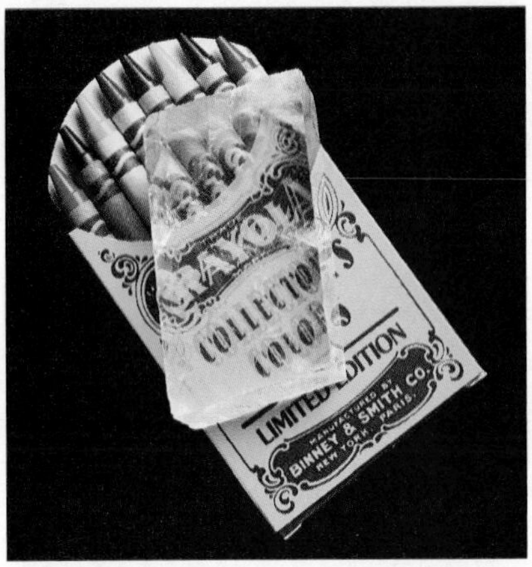

A crystal of Iceland spar ($CaCO_3$) displays birefringence. That is, an image is doubled on passing the crystal. (C. D. Winters)

Properties of Calcium and Magnesium

Calcium and magnesium are fairly high-melting, silvery metals. The chemical properties of Group 2A elements present few surprises. All are oxidized by a wide range of oxidizing agents to form ionic compounds that contain the M^{2+} ion. For example, these elements combine with halogens to form MX_2, and with oxygen or sulfur to form MO or MS (see Figure 4.3 for reaction of Mg and O_2). All except beryllium react with water to form hydrogen and the metal hydroxide, $M(OH)_2$. With acids, hydrogen is evolved and a salt of the metal and the anion of the acid results.

Metallurgy of Magnesium

Several hundred thousand tons of magnesium are produced annually because the low density of the metal (1.74 g/cm^3) makes it useful in lightweight alloys. For example, most aluminum used today contains about 5% magnesium to improve its mechanical properties and to make it more resistant to corrosion. Other alloys having more magnesium than aluminum are used when a high strength-to-weight ratio is needed and when corrosion resistance is important, such as in aircraft and automotive parts and in lightweight tools.

Calcium metal and warm water react to form hydrogen gas and calcium hydroxide. (C. D. Winters)

Although there are many magnesium-containing minerals, most magnesium is obtained from sea water, in which it is present in a concentration of about 0.05 M. To obtain the element, magnesium is first precipitated from sea water as the relatively insoluble hydroxide (K_{sp} for $Mg(OH)_2 = 1.5 \times 10^{-11}$). The base used in this reaction, CaO, is prepared from sea shells, $CaCO_3$. Heating $CaCO_3$ gives CO_2 and CaO, and addition of water to CaO gives calcium hydroxide. When $Ca(OH)_2$ is added to sea water, $Mg(OH)_2$ precipitates.

$$Mg^{2+}(aq) + Ca(OH)_2(s) \longrightarrow Mg(OH)_2(s) + Ca^{2+}(aq)$$

Magnesium hydroxide is isolated by filtration and then neutralized with hydrochloric acid.

$$Mg(OH)_2(s) + 2\ HCl(aq) \longrightarrow MgCl_2(aq) + 2\ H_2O(\ell)$$

After evaporating the water, solid magnesium chloride is left. The anhydrous $MgCl_2$ melts at 708 °C, and the molten salt is electrolyzed to give the metal and chlorine.

$$MgCl_2(\ell) \longrightarrow Mg(s) + Cl_2(g)$$

Apatite, a mineral with the general formula 3 $Ca_3(PO_4)_2 \cdot CaX_2$ (X = F, Cl, OH). (Bill Tronca/Tom Stack and Associates)

Calcium Compounds of Commercial Importance

The most important fluoride of the alkaline earth metals is fluorspar, CaF_2, but fluorapatite ($CaF_2 \cdot 3\ Ca_3(PO_4)_2$) is becoming increasingly important as a commercial source of fluorine.

Almost half of the CaF_2 mined is used in the steel industry where it is added to the mixture of materials that are melted to make crude iron. The CaF_2 serves to remove some impurities and improves the separation of molten metal from slag, the layer of silicate impurities and byproducts that comes from reducing iron ore to the metal (Chapter 23).

The other use of fluorspar is in the manufacture of hydrofluoric acid by reaction of the mineral with concentrated sulfuric acid.

$$CaF_2(s) + H_2SO_4(\ell) \longrightarrow 2\ HF(g) + CaSO_4(s)$$

Hydrogen fluoride is very reactive and is an extremely important chemical. It is used to make cryolite, Na_3AlF_6, a material needed in aluminum production, and fluorocarbons such as polytetrafluoroethylene (Teflon).

Apatites are collectively referred to as phosphate rock. Over 100 million tons are mined annually; Florida alone accounts for about one-third of the world's output. Much of this rock is converted to phosphoric acid by reaction with sulfuric acid. Phosphoric acid is used to manufacture a multitude of products (fertilizers and detergents), and its reaction products are found in baking powder, in frozen fish, and in many other food products.

$$\underset{\text{fluorapatite}}{CaF_2 \cdot 3\ Ca_3(PO_4)_2(s)} + 10\ H_2SO_4(aq) \longrightarrow 10\ CaSO_4(s) + 6\ H_3PO_4(aq) + 2\ HF(g)$$

Because of their economic importance, calcium carbonate and calcium oxide are of special interest. The thermal decomposition of $CaCO_3$ to lime, CaO, is one of the oldest chemical transformations known. Among industrial chemicals produced today, lime is fifth with 16.7 billion kg produced in 1993. Most limestone and lime are used in the chemicals industry, and one third of the lime is used to make steel by the basic oxygen process (Chapter 23).

Limestone has been used in agriculture for centuries. It is spread on fields to neutralize acidic compounds in the soil and supply the essential nutrient Ca^{2+}. Because magnesium carbonate also is often present in limestone, "liming" a field also supplies Mg^{2+}, another important plant nutrient.

For several thousand years, lime has been used as mortar (a lime, sand, and water paste) to secure stones to one another in building houses, walls, and roads. The Chinese used it in setting stones in the Great Wall. The Romans perfected its use, and the fact that many of their constructions still stand is testament both to their skill and the usefulness of lime. In 312 BC, the famous Appian Way, a Roman highway stretching from Rome to Brindisi (a distance of about 350 miles) was begun, and lime mortar was used between several layers of its stones.

Mining phosphate near Ft. Meade, Florida. Phosphate ore, called matrix, is a loose conglomerate of phosphate rock, clay, and sand. It is used to prepare phosphoric acid and fertilizer. (Runk/Schoenberger from Grant Heilman)

The utility of mortar depends on some simple chemistry. Mortar consists of one part lime to three parts sand, with water added to make a thick paste. The first reaction that occurs is the formation of $Ca(OH)_2$, the process referred to as "slaking," the product being called slaked lime. When the mortar is placed between bricks or stone blocks, it slowly absorbs CO_2 from the air, and the slaked lime reverts to calcium carbonate.

$$Ca(OH)_2(s) + CO_2(g) \rightleftharpoons CaCO_3(s) + H_2O(\ell)$$

Although sand mixed into mortar is chemically inert, the grains are bound together by the particles of calcium carbonate, and a hard material results.

"Hard water" contains dissolved ions, chiefly Ca^{2+} and Mg^{2+}. Water containing dissolved CO_2 reacts with limestone.

$$CaCO_3(s) + H_2O(\ell) + CO_2(g) \rightleftharpoons Ca^{2+}(aq) + 2\ HCO_3^-(aq)$$

This reaction is reversible, however. When hard water is heated, the solubility of CO_2 drops, and the equilibrium shifts to the left. If this happens in a heating system or steam-generating plant, the walls of the hot water pipes can become coated or even blocked with solid $CaCO_3$. In your house, you may notice a coating of calcium carbonate on the inside of cooking pots.

The last equation describes the chemistry of caves as well. The acidic oxide CO_2 reacts with $Ca(OH)_2$ to produce white, solid $CaCO_3$. When further CO_2 is available, however, the $CaCO_3$ can redissolve because of the formation of aqueous Ca^{2+} and HCO_3^- ions. (See Section 19.10.)

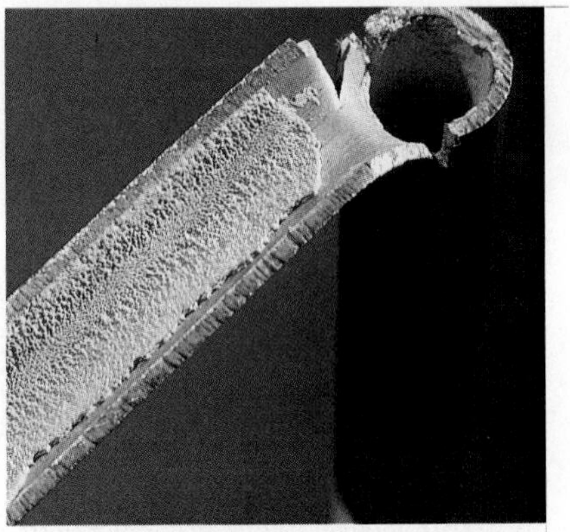

A section of water pipe that has been coated on the inside with $CaCO_3$ deposited from hard water. (Courtesy of Betz Laboratories)

22.5 ALUMINUM

Aluminum is the third most abundant element in earth's crust (7.4%). You are familiar with the metal because it is widely used in packaging (aluminum foil and as aluminum cans) and as a structural material.

Aluminum melts at 660 °C and has a density of 2.70 g/cm^3.

Pure aluminum is soft and weak; moreover, it loses strength rapidly above 300 °C. What you recognize as aluminum is actually aluminum alloyed with small amounts of other elements, which strengthens the metal and improves its properties. A large passenger plane may use more than 50 tons of aluminum alloy. A typical alloy may contain about 4% copper with smaller amounts of silicon, magnesium, and manganese. To make a softer, more corrosion-resistant alloy for window frames, furniture, highway signs, and cooking utensils, however, only manganese may be included.

Aluminum is readily oxidized, as seen from the value of the

$$Al^{3+}(aq) + 3\ e^- \longrightarrow Al(s) \qquad E° = -1.66\ V$$

redox couple in Table 21.1. Aluminum's resistance to corrosion is therefore unexpected. We know, however, that this corrosion resistance is due to the formation of a thin, tough, and transparent skin of oxide, Al_2O_3, that adheres to the metal surface.

$$4\ Al(s) + 3\ O_2(g) \longrightarrow 2\ Al_2O_3(s) \qquad \Delta H° = -3351.4\ kJ$$

An important feature of the protective oxide layer is that it rapidly self-repairs. If you scratch the surface coating, a new layer of oxide immediately forms over the damaged area.

The corrosion-resistance of aluminum can be defeated in the presence of chloride ion. See Figure 21.17.

Metallurgy of Aluminum

Aluminum is found in varying amounts in nature as aluminosilicates, minerals such as clay that are based on aluminum, silicon, and oxygen. As these minerals are weathered, they gradually break down to various forms of hydrated aluminum oxide, $Al_2O_3 \cdot n\ H_2O$, called *bauxite.*

Aluminum (*right*) does not react with HNO_3 because of the protective oxide coating on the surface of the metal. In contrast, copper reacts vigorously with nitric acid (*left*) to give $Cu(NO_3)_2$ and NO_2 gas. (C. D. Winters)

Aluminum is obtained by electrolysis of bauxite (Figure 21.22). For this process it is first necessary to purify the ore, separating iron and silicon oxides. Purification is done by the *Bayer process,* which uses the amphoteric, basic, or acidic nature of the various oxides. Silica, SiO_2, is an acidic oxide, Al_2O_3 is amphoteric, and Fe_2O_3 is a basic oxide. The first two oxides therefore dissolve in a hot concentrated solution of caustic soda (NaOH), leaving insoluble Fe_2O_3 to be filtered out.

$$Al_2O_3(s) + 2\ NaOH(aq) + 3\ H_2O(\ell) \longrightarrow 2\ Na[Al(OH)_4](aq)$$
$$SiO_2(s) + 2\ NaOH(aq) + 2\ H_2O(\ell) \longrightarrow Na_2[Si(OH)_6](aq)$$

By treating the solution containing aluminum and silicon anions with CO_2, Al_2O_3 is precipitated, and the silicate ion remains in solution. Recall that CO_2 forms the weak acid H_2CO_3 in water, so Al_2O_3 precipitation is an acid-base reaction

$$H_2CO_3(aq) + 2\ Na[Al(OH)_4](aq) \longrightarrow Na_2CO_3(aq) + Al_2O_3(s) + 5\ H_2O(\ell)$$

Properties of Aluminum and Its Compounds

Aluminum dissolves in HCl(aq) but not in nitric acid. The latter is a powerful oxidizing agent and a source of oxygen atoms, so it oxidizes the surface of aluminum rapidly, and the film of Al_2O_3 protects the metal from further attack. In fact, nitric acid is often shipped in aluminum tanker trucks.

Various salts of aluminum dissolve in water, giving the hydrated $Al^{3+}(aq)$ ion. As described in Chapter 17, however, these solutions are acidic, due to the following equilibrium:

The two aluminum compounds in this equation are representatives of a class of compounds known as *coordination compounds.* We say more about this area of chemistry in the next chapter.

$$[Al(H_2O)_6]^{3+} + H_2O(\ell) \rightleftharpoons [Al(H_2O)_5(OH)]^{2+} + H_3O^+(aq)$$

The addition of acid shifts the equilibrium to the left, whereas base causes the equilibrium to shift to the right. Eventually, with further hydroxide, the hydrated oxide $Al_2O_3 \cdot 3\ H_2O$ (=2 $Al(OH)_3$) precipitates.

Aluminum oxide, Al_2O_3, which can be formed by dehydrating $Al(OH)_3$, is quite insoluble in water and generally resistant to chemical attack. In the crystalline form, aluminum oxide is known as *corundum.* This material is extraordinarily hard, a property that leads to its use as the abrasive in grinding wheels, "sandpaper," and toothpaste.

Some gems are impure aluminum oxide. Rubies, beautiful red crystals prized for jewelry and used in some lasers, are Al_2O_3 contaminated with a small amount of Cr^{3+} (Figure 3.12). The Cr^{3+} ions replace some of the Al^{3+} ions in the crystal lattice. Blue sapphires occur when Fe^{2+} and Ti^{4+} impurities are present in Al_2O_3. Synthetic rubies were first made in 1902, and the worldwide capacity is now about 200,000 kg per year; much of this production is used for jewel bearings in watches and instruments (Figure 22.5).

Figure 22.5 Synthetic rubies, crystals of Al_2O_3 containing a small amount of Cr^{3+} in place of Al^{3+} in the crystal lattice. (Kurt Nassau)

22.6 SILICON

Silicon is the second most abundant element in the earth's crust. It is almost inevitable, therefore, that silicon compounds would be important in the development of society. Pottery, made of silicon-based natural materials, was made at least 6000 years ago in the Middle East, and sophisticated techniques were developed by the Chinese 5000 years ago. Silicon-based semiconductors have fueled the computer revolution of the past decade.

The name "silicon" is derived from the Latin word *silex,* meaning flint, a silicate mineral often used by prehistoric people to make knives and other tools. Today we are surrounded by silicon-containing materials: bricks, pottery, porcelain, lubricants, sealants, computer chips, and solar cells.

Reasonably pure silicon is made in large quantities by heating pure silica sand with purified coke to approximately 3000 °C in an electric furnace.

$$SiO_2(s) + 2\ C(s) \longrightarrow Si(\ell) + 2\ CO(g)$$

The molten silicon is drawn off the bottom of the furnace and allowed to cool to a shiny blue-gray solid (Figure 2.14 and page 487). For applications in the electronics industry, extremely high-purity silicon is needed. The crude silicon is first chlorinated to form silicon tetrachloride, a liquid with a boiling point of only 57.6 °C.

Elemental silicon has the diamond structure (Figure 3.5) with tetrahedral silicon atoms linked together in an infinite lattice.

$$Si(s) + 2\ Cl_2(g) \longrightarrow SiCl_4(\ell)$$

The volatile tetrachloride is purified by distillation and then reduced to silicon using very pure magnesium or zinc.

$$SiCl_4(g) + 2\ Mg(s) \longrightarrow 2\ MgCl_2(s) + Si(s)$$

The magnesium chloride is washed out with water. The silicon is then remelted and cast into bars. A final purification is carried out by *zone refining,* a process in which a special heating device is used to melt a narrow segment of the silicon rod. The heater is then moved slowly down the rod. Impurities contained in the silicon tend to remain in the liquid phase because the melting point of a mixture is lower than that of the pure "solvent" (Chapter 14). Therefore, the silicon that crystallizes above the heated zone is of a higher purity (Figure 22.6).

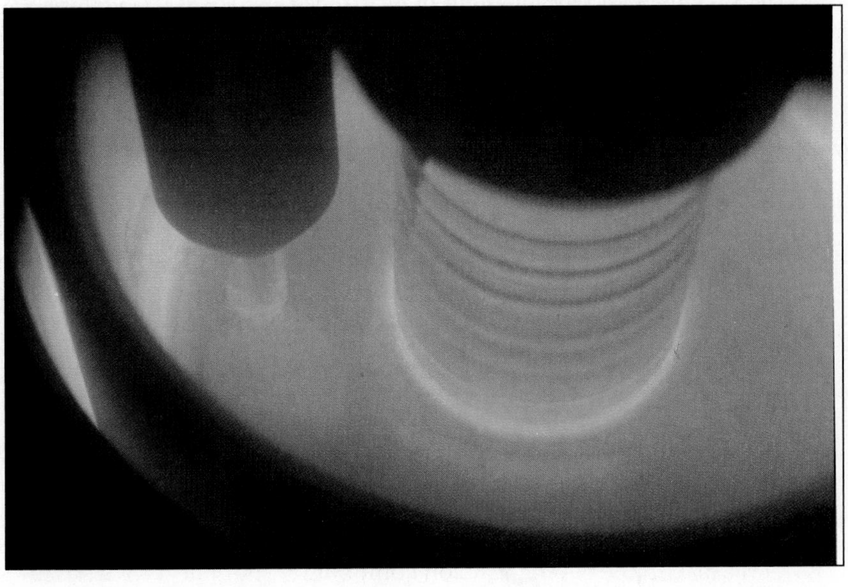

(a)

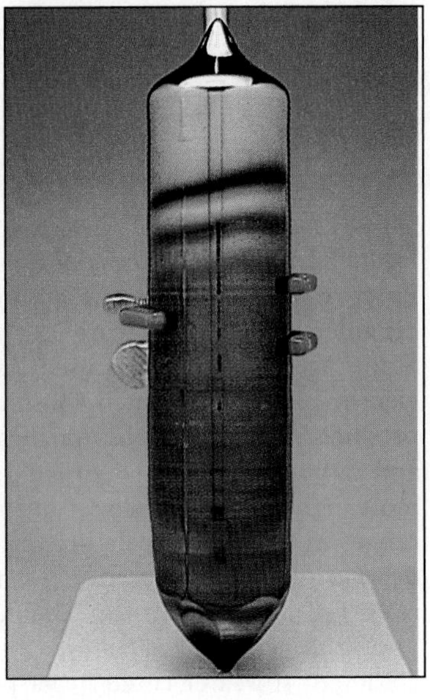

(b)

Figure 22.6 Making pure silicon. (a) A mechanically rotated seed crystal of pure silicon is slowly withdrawn from the molten silicon. Because the temperature of the crystal is lower than the melt, the liquid freezes on the seed crystal as it is withdrawn. (Courtesy of Great Western Silicone) (b) A solid cylinder of nearly pure silicon made by this technique. The crystal shown in this picture is approximately 11 in. long. (C. D. Winters)

Silicon Dioxide

The simplest oxide of silicon is SiO_2, commonly called silica. Silica is a major constituent of many rocks, such as granite and sandstone. Quartz is a pure crystalline form of silica. The introduction of impurities into quartz produces gemstones such as amethyst (Figure 22.7).

The fact that SiO_2 is a high-melting solid (quartz melts at 1610 °C) should catch your attention because CO_2, the oxide of the element above silicon in the periodic table, is a gas at room temperature and one atmosphere. This great disparity in properties arises from the different structures of CO_2 and SiO_2. Carbon dioxide is a molecular species with the carbon atom linked to each oxygen by a double bond. In contrast, SiO_2 has silicon and oxygen atoms bonded together in a giant network (Figure 13.36). This structure is preferred over a simple molecular structure because the energy of two Si═O double bonds is much less than that of four Si—O single bonds. In fact, the contrast between SiO_2 and CO_2 exemplifies a more general phenomenon. Multiple bonds, often encountered between second-period elements, are rare among elements in the lower periods.

Other examples of the preference of third-period elements to form single bonds rather than multiple bonds are seen in the structures of S_8 vs. O_2 and P_4 vs. N_2.

Crystalline quartz is used to control the frequency of almost all radio and television transmissions. These and related applications use so much quartz that there is not enough natural quartz to fulfill demand, so quartz is synthesized. Noncrystalline, or vitreous, quartz, made by melting pure silica sand, is placed in a steel "bomb" and dilute aqueous NaOH is added. A "seed" crystal is placed in the mixture, just as you might place a seed crystal in a hot sugar solution to grow rock candy. When the mixture is heated above the critical temperature of water (above 400 °C and 1700 atm) over a period of days, pure quartz crystallizes.

Silica is resistant to attack by all acids except HF, with which it reacts to give SiF_4 and H_2O. It does dissolve slowly in hot, molten NaOH or Na_2CO_3 to give Na_4SiO_4.

Figure 22.7 Various forms of quartz. Pure, colorless quartz was used as an ornamental material as early as the Stone Age; now, quartz is used for its electric properties in such consumer products as phonographs, watches, and radios. Purple amethyst is the most highly prized variety of quartz. It ranges in color from pale lilac to a deep, royal purple. The name comes from the Greek *amethustos,* meaning "not drunken"; it was believed that an amethyst wearer could never become intoxicated. (C. D. Winters)

The structure of a repeating unit of quartz, SiO_2.

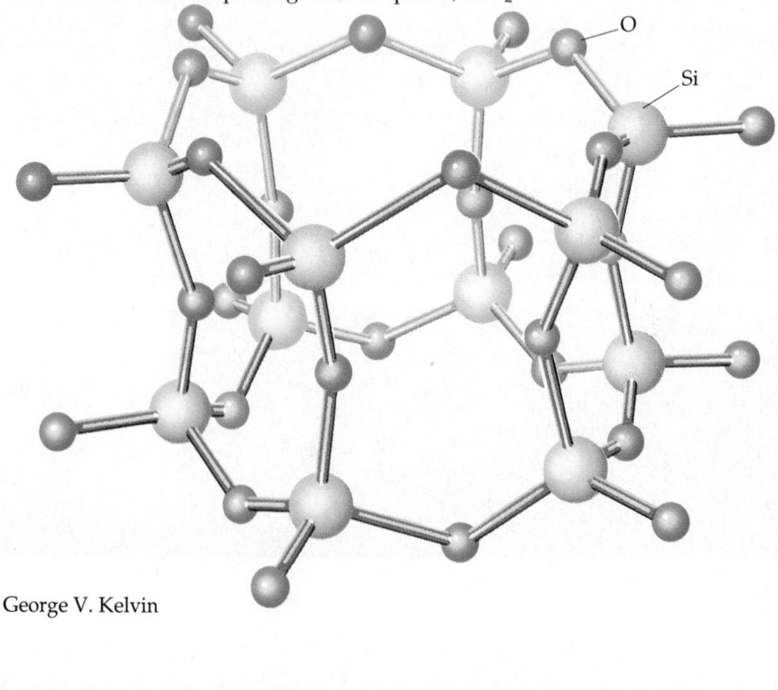

© George V. Kelvin

$$SiO_2(s) + 4\ HF(\ell) \longrightarrow SiF_4(g) + 2\ H_2O(\ell)$$
$$2\ Na_2CO_3 + SiO_2 \longrightarrow Na_4SiO_4 + 2\ CO_2$$

When the molten mixture has cooled, hot water under pressure is added. This partially dissolves the material to give a solution of sodium silicate. After filtering off insoluble sand or glass, the solvent is evaporated, leaving sodium silicate, called *water glass.* The biggest single use of this material is in household and industrial detergents. A sodium silicate solution maintains pH by its buffering ability and can degrade animal and vegetable fats and oils. Sodium silicate is also used in various adhesives and binders, especially for gluing corrugated cardboard boxes.

If sodium silicate is treated with acid, a gelatinous precipitate of noncrystalline, or amorphous, SiO_2, called *silica gel,* is obtained. After being washed and dried, silica gel is a very porous material with dozens of uses. Because it can absorb up to 40% of its own weight of water, you may know it as a drying agent. (Small packets of silica gel are often placed in packing boxes of merchandise during storage.) When stained with $(NH_4)_2CoCl_4$, it is a humidity detector, turning pink when hydrated, but remaining blue when dry. Finally, silicates are used to clarify beer; passage through a bed of silica gel removes minute particles that make the brew cloudy.

Quartz crystals can be grown from silica at high temperatures under pressure in an autoclave. Here they are being removed from the autoclave. (Kurt Nassau)

The Silicate Minerals

The **silicate minerals** are a world in themselves. All silicates are built from tetrahedral SiO_4 units. The greatly differing properties of silicate materials are created when many of these tetrahedral SiO_4 units link together.

The simplest silicates, *orthosilicates,* contain SiO_4^{4-} anions. The 4− charge of the anion can be balanced by four M^+ ions, two M^{2+} ions, or a combination of ions. For example, calcium orthosilicate, Ca_2SiO_4, is one of the components of *Portland cement. Olivine,* one of the most important minerals in earth's mantle, contains Mg^{2+}, Fe^{2+}, and Mn^{2+}; the Fe^{2+} ion gives olivine its characteristic olive color.

A group of minerals called *pyroxenes* have as their basic structural unit an extended chain of linked SiO_4 tetrahedra (Figure 22.8). If two such chains link by sharing oxygen atoms, the result is an *amphibole* (Figure 22.9), of which the *asbestos* minerals are one example. As a result of its chain structure, asbestos is a

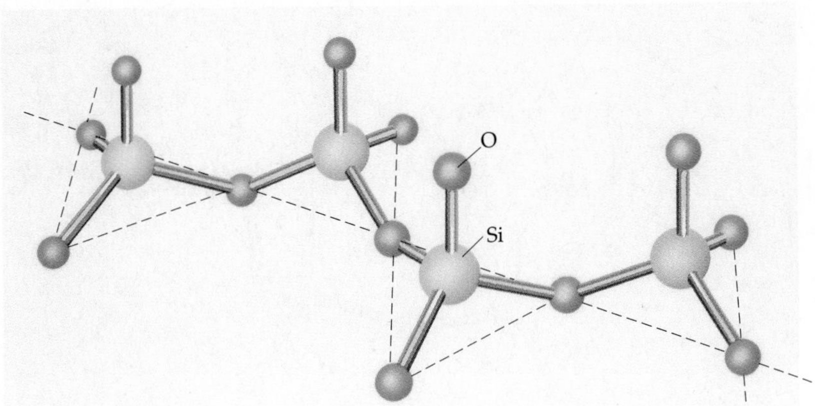

© George V. Kelvin

Figure 22.8 Pyroxene. The SiO_4 units share a common O atom.

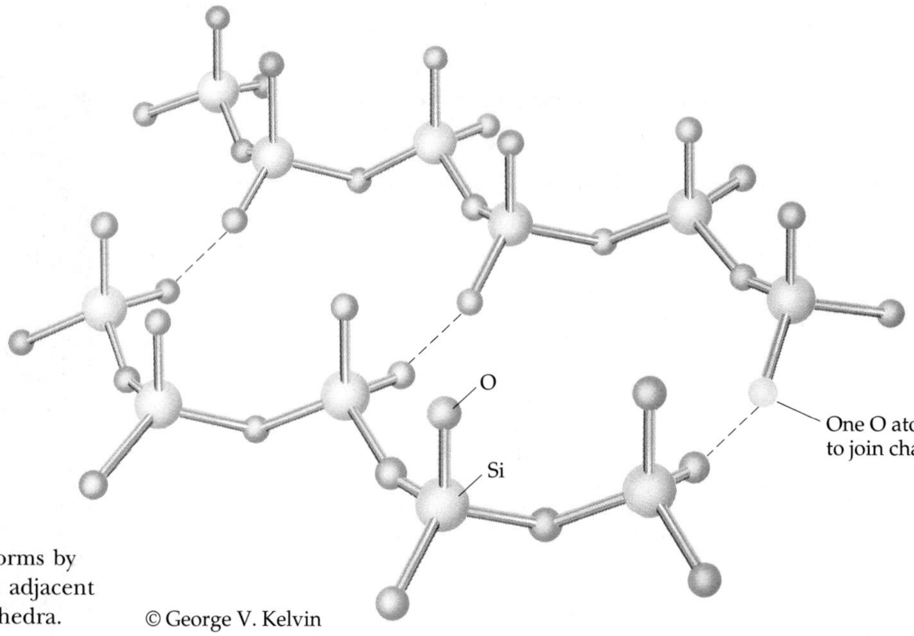

Figure 22.9 An amphibole forms by eliminating O atoms between adjacent chains of silicon-oxygen tetrahedra.

fibrous material. The asbestos minerals are best known for their very low thermal conductivity, which led to their use in insulation and fireproofing.

Linking many silicate chains together produces a sheet of SiO_4 tetrahedra (Figure 22.10). Substances in this category include the mineral mica and clays. The molecular sheet of SiO_4 tetrahedra leads to the characteristic appearance of mica, which is often found as ''books'' of thin, silicate sheets. Mica is used in furnace windows and as insulation, and flecks of mica give the glitter to ''metallic'' paints.

Clay minerals are essential components of soils, and are the raw material for pottery, bricks, and tiles. Clays come from the weathering and decomposition of

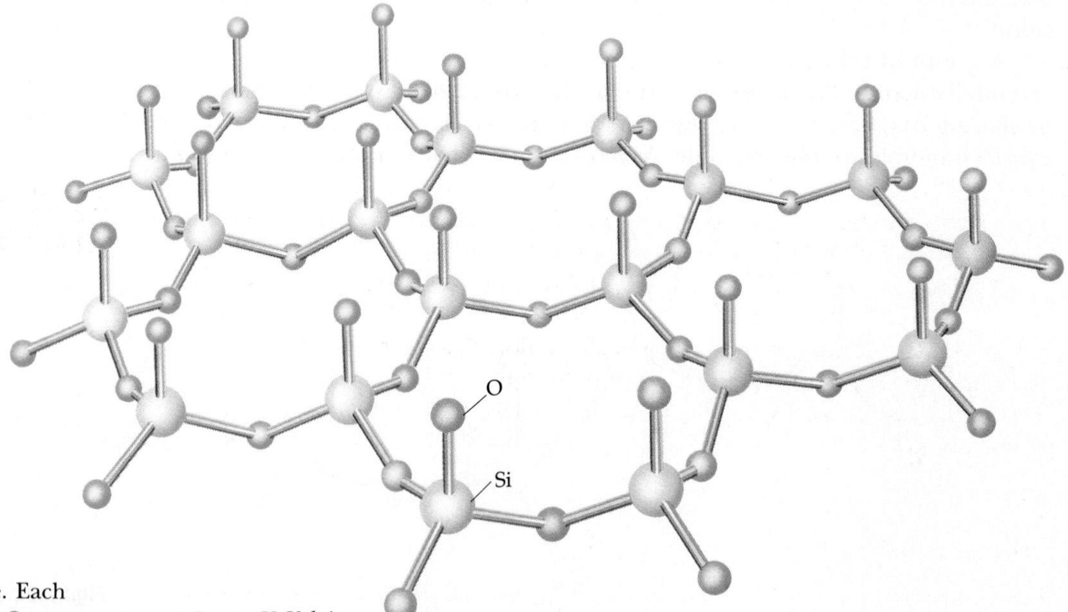

Figure 22.10 Mica structure. Each SiO_4 unit shares three of its O atoms.

igneous rocks. Specifically, the aluminosilicate *kaolinite* comes from the weathering of feldspar, an idealized version of the reaction being

$$\underset{\text{feldspar}}{2\ KAlSi_3O_8(s)} + CO_2(g) + 2\ H_2O(\ell) \longrightarrow \underset{\text{kaolinite}}{Al_2(OH)_4Si_2O_5(s)} + 4\ SiO_2(s) + K_2CO_3(aq)$$

Its structure consists of layers of SiO_4 tetrahedra linked into sheets, as in mica, but these sheets are interleaved with six-coordinate Al^{3+} ions. The aluminum atoms are positioned in the lattice to be surrounded octahedrally by O atoms of the silicon-oxygen sheets and OH^- ions (Figure 22.11).

China clay, or kaolin, is primarily kaolinite. It is practically free of iron (a common impurity), and so it is colorless, making it particularly valuable. The predominant use for kaolin in the United States is for paper filling and coating, but some is also used for china, crockery, and earthenware.

All clays are aluminosilicates in that they contain both aluminum and silicon. In other clay minerals, however, some Si^{4+} ions are also replaced by Al^{3+} ions. To make up for the "missing" positive charge, 1+ for every Si^{4+} replaced by an Al^{3+}, nature adds positive ions such as Na^+ and Mg^{2+}, which are located between the aluminosilicate sheets. This gives these materials interesting properties, one of which is their use in medicines. Several remedies for the relief of upset stomach contain highly purified clays, which absorb excess stomach acid as well as potentially harmful bacteria and their toxins by exchanging the intersheet cations in the clays for the toxins, which are often organic cations. Indeed, this is a remedy learned from other societies, where clay has long been eaten for medicinal purposes.

Mica, a layered silicate. (C. D. Winters)

Other aluminosilicates are *feldspars* (among the most common minerals; they make up about 60% of earth's crust) and *zeolites*. Both materials are again composed of SiO_4 tetrahedra with some of the silicon atoms replaced by Al atoms. Because the silicon atoms formally bear a 4+ valence and are replaced by Al^{3+} ions, other positive ions are present for charge balance. Typically, alkali and alkaline earth ions serve this purpose. For example, the synthetic zeolite "Linde A" has the formula $Na_{12}(Al_{12}Si_{12}O_{48}) \cdot 27\ H_2O$.

The structure of a zeolite is illustrated on page 433. The main feature of zeolite structures is their regularly shaped tunnels and cavities. Hole diameters are typically 300 to 1000 pm, and small molecules such as water can fit into the cavities of the zeolite. As a result, zeolites can be used as drying agents to selectively absorb water from air or a solvent. Small amounts are sealed into multipane windows to keep the air dry between the panes.

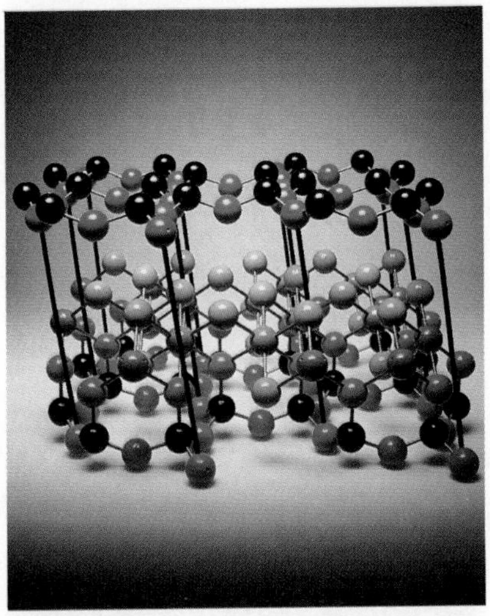

Figure 22.11 A model of kaolinite, an aluminosilicate. Each Si atom (black) is surrounded tetrahedrally by O atoms (red) to give rings consisting of six Si atoms and six O atoms. The layer of Al ions (light blue) is attached through O atoms to the Si—O rings. Hydroxide ions (light green) act as bridges between Al ions. The net result is a layered structure that gives clays their slipperiness and workability when wet. (C. D. Winters)

Zeolites are also used as catalysts. Mobil Oil Corporation, for example, has patented a process in which the one-carbon compound methyl alcohol, CH_3OH, forms gasoline in the presence of specially tailored zeolites. Finally, zeolites are used as water-softening agents in detergents, because the sodium ions of the zeolite can be exchanged for Ca^{2+} ions in hard water, effectively removing Ca^{2+} from the water.

Silicone Polymers

Just as silicon reacts readily with chlorine to produce $SiCl_4$, a similar reaction occurs between silicon and methyl chloride, CH_3Cl. This reaction forms

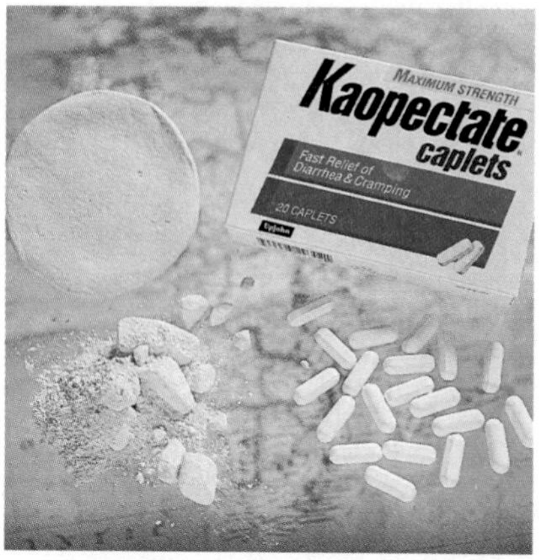

A commercial remedy for diarrhea contains one type of clay. The off-white objects in the photo are pieces of clay purchased in a market in Ghana, West Africa. This clay is made to be eaten as a remedy for stomach ailments. The practice of eating clay is widespread among many cultures of the world. (C. D. Winters)

$(CH_3)_2SiCl_2$; the structure of this compound has two methyl groups (CH_3) and two Cl atoms bound to tetrahedral silicon.

$$\mathrm{Si(s) + 2\ CH_3Cl(g) \xrightarrow{\text{Cu powder catalyst/300 °C}} (CH_3)_2SiCl_2(\ell)}$$

Unlike CCl_4 and other compounds with C—Cl bonds, halides based on other Group 4A elements hydrolyze readily. For $(CH_3)_2SiCl_2$, this reaction with water initially produces $(CH_3)_2Si(OH)_2$; however, the molecules condense together, eliminating water.

$$\mathrm{(CH_3)_2SiCl_2 + 2\ H_2O \longrightarrow (CH_3)_2Si(OH)_2 + 2\ HCl}$$
$$\mathrm{n(CH_3)_2Si(OH)_2 \longrightarrow -\!\!\left[(CH_3)_2SiO\right]\!\!-_n + n\ H_2O}$$

The product of this reaction again illustrates the reluctance of third-period elements to form multiple bonds. Instead of forming a molecular compound like the carbon analogue, acetone, $(CH_3)_2C{=}O$, the product contains a chain of alternating silicon and oxygen atoms.

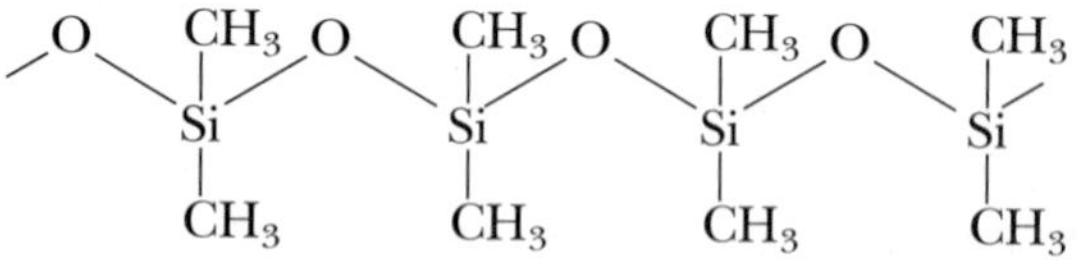

This is a *polymer* called polydimethylsiloxane, a member of the *silicone* polymer family (Figure 22.12). Silicones are nontoxic and have good stability to heat, light, and oxygen; they are chemically inert and have valuable antistick and antifoam properties. They can be made in the form of oils, greases, and resins, or with rubber-like properties (''Silly Putty,'' for example). Approximately 300,000 tons are made worldwide annually and are used in a wide variety of products: as lubricants, as the antistick material for peel-off labels, in lipstick, suntan lotion, and car polish, and as the antifoam substance in stomach remedies.

22.7 NITROGEN AND PHOSPHORUS

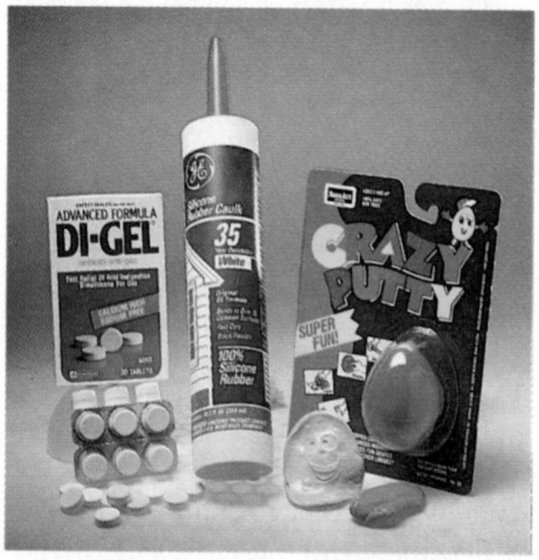

Figure 22.12 Some examples of silicones, polymers with repeating —Si—O— units. (C. D. Winters)

Neither nitrogen nor phosphorus is among the 10 most common elements in earth's crust, but that fact understates the importance of these elements, which are essential to life on this planet. Nitrogen and its compounds, such as ammonia, nitric acid, ammonium nitrate, and urea, play a key role in our economy, and phosphoric acid is an important commodity chemical. The major use of all of these chemicals is in fertilizers.

Over 200 different phosphorus-containing minerals are known; all are *orthophosphates,* that is, they contain the tetrahedral PO_4^{3-} ion or a derivative of this ion. By far the largest source of this element is the *apatite* mineral family, members of which have the general formula $3\ Ca_3(PO_4)_2 \cdot CaX_2$ (X = F, Cl, OH).

Nitrogen is found primarily as N_2 in the atmosphere where it constitutes 78.1% by volume (or 75.5% by weight). Both phosphorus and nitrogen, however, are part of every living organism. The element phosphorus is contained in biochemicals called nucleic acids and phospholipids, and nitrogen occurs in proteins and nucleic acids. Nitrogen and phosphorus constitute about 3% and 1.2% by weight, respectively, of the human body.

CURRENT ISSUES IN CHEMISTRY

The Asbestos Controversy

Exposure to asbestos is known to cause various lung diseases, so the Environmental Protection Agency (EPA) requires that schools be inspected for asbestos and that it be removed if asbestos fibers are found in the air. If the testing and removal program is extended to all public and commercial buildings, the cost could be as much as $150 billion dollars. Does the available scientific evidence support this concern? There has been a stream of articles in the scientific literature regarding this question. One of the most complete and definitive was by B. Mossman and coworkers in the journal *Science.**

The term "asbestos" refers to naturally occurring hydrated silicates that crystallize in a fibrous manner. These minerals are generally subdivided into two forms: serpentine and amphibole fibers. Approximately 5 million tons of the serpentine form of asbestos, chrysotile, are mined each year, chiefly in Canada and the Soviet Union; this is essentially the only form used commercially in the United States. Another form, an amphibole called crocidolite, is mined in small quantities, mainly in South Africa. The two minerals differ greatly in composition, color, shape, solubility, and persistence in human tissue. Crocidolite is blue, relatively insoluble, and persists in tissue. Its fibers are long, thin, and straight, and they penetrate narrow lung passages. In contrast, chrysotile is white, tends to be soluble, and disappears in tissue. Its fibers are curly; they ball up like yarn and are more easily rejected by the body.

Asbestos minerals have been used in a variety of ways because of their heat resistance and high tensile strength. For many years they were sprayed onto surfaces as fireproofing, but this use is now banned in the United States. Asbestos minerals continue to be used, however, in cement construction materials (roofing and cement pipes), friction materials (brake linings and clutch pads), asphalt coatings and sealants, and similar products.

Occupational exposure to asbestos can cause four types of medical problems: asbestosis, lung cancer, mesothelioma of lung tissue, and benign changes in lung tissue. Asbestosis is a nonmalignant scarring of lung tissue that led to disability and death in many asbestos workers exposed before the enforcement of occupational standards. Lung cancers are rare except among smokers. Mesothelioma is an extremely rare but fatal tumor in lung tissue; only 53 documented cases of chrysotile-induced mesothelioma have been reported, and 41 of these occurred in individuals exposed to mine dust contaminated with tremolite, an amphibole. Approximately 20% to 30% of the mesotheliomas occur in the general adult population with no known occupational exposure to asbestos.

Increasingly, scientists have questioned the extent of risk posed by asbestos and the extensive and costly measures that have been required to deal with it. Mossman and her coworkers stated that "chrysotile asbestos, the type of fiber found predominantly in U.S. schools and buildings, is not a health risk in the nonoccupational environment. Clearly, the asbestos panic in the United States must be curtailed. . . . Prevention (especially in adolescents) of tobacco smoking, the principal cause of lung cancer in the general population, is both a more promising and more rational approach to eliminating lung tumors than asbestos abatement."

A sample of chrysotile, one of the minerals in the asbestos family. (C. D. Winters)

*B. T. Mossman, J. Bignon, M. Corn, et al.: *Science,* Vol. 247, p. 294, 1990.

The Elements: Nitrogen and Phosphorus

Nitrogen (N_2) is a colorless gas that liquifies at 77 K (−196 °C) (Figure 2.14). Its most notable feature is its reluctance to react with other elements or compounds. This comes about because the N≡N triple bond has a large dissociation energy (945.4 kJ/mol) and because the molecule is nonpolar. Nitrogen does react, however, with hydrogen to give ammonia (page 778) and with a few metals to give nitrides, compounds with the N^{3-} ion such as Mg_3N_2

$$3\ Mg(s) + N_2(g) \longrightarrow \underset{\text{magnesium nitride}}{Mg_3N_2(s)}$$

Elemental nitrogen, N_2, is a very useful material. The largest quantity of this gas is used to provide a nonoxidizing atmosphere for packaged foods and wine, for example, and to pressurize electric cables and telephone wires. Liquid nitrogen is used as a coolant in a variety of ways, including freezing soft materials such as rubber so they can be ground to a powder and preserving biological samples (e.g., blood and semen).

Phosphorus is unique in that it was first isolated from an animal source rather than from a mineral. In 1669, an alchemist obtained phosphorus from the distillation of putrefied urine.

Elemental phosphorus is produced in large quantities by the reduction of phosphate minerals such as apatite.

$$2\ Ca_3(PO_4)_2(s) + 10\ C(s) + 6\ SiO_2(s) \longrightarrow P_4(g) + 6\ CaSiO_3(s) + 10\ CO(g)$$

The allotrope of phosphorus that is white or yellowish (Figure 3.3) has the molecular formula P_4, and consists of a tetrahedron of phosphorus atoms.

Oxidation States of Nitrogen

−3	NH_3 ammonia
−2	N_2H_4 hydrazine
−1	N_2H_2 diimine
0	N_2 nitrogen
+1	N_2O nitrous oxide
+2	NO nitric oxide
+3	NO_2^- nitrite ion
+4	NO_2 nitrogen dioxide
	N_2O_4 dinitrogen tetraoxide
+5	HNO_3 nitric acid
	NO_3^- nitrate ion

Nitrogen Compounds

One of the most interesting features of nitrogen is the wide diversity of its compounds. In its compounds, nitrogen is known to exist in all oxidation states between −3 and +5. This is the maximum range available for a second-period element. Several of these compounds are especially significant and are discussed later.

Ammonia and Nitrogen Fixation

Nitrogen gas, N_2, cannot be used by plants until it is "fixed," that is, converted into a form that can be used by living systems. Nitrogen fixation is done naturally by organisms such as blue-green algae. A few field crops, such as alfalfa and soybeans, fix nitrogen; this is actually done by nitrogen-fixing bacteria that have a symbiotic relationship with the plant. Because most plants cannot fix N_2, however, it is necessary that the nitrogen be provided by an external source. This is especially true of the new varieties of wheat, corn, and rice that grow fast or that have been bred to provide higher levels of protein. Thus, the production of nitrogen-containing fertilizers is a huge part of the chemical industry today.

A feasible process for fixing nitrogen in the form of NH_3 was devised by Fritz Haber. The Haber process, as it is called, produces ammonia by direct combination of nitrogen and hydrogen (page 778).

Prior to World War I, the majority of commercial nitrogen fertilizer came from deposits of Chilean saltpeter ($NaNO_3$) and the excrement of bats and sea birds (guano).

The Haber Process

$$\tfrac{1}{2}N_2(g) + \tfrac{3}{2}\ H_2(g) \rightleftharpoons NH_3(g)$$

Nitrogen from the air is free, but H_2 must be made, the most general route being from natural gas by steam re-forming (page 1008). The Haber process is so efficient that the cost of ammonia is now almost entirely the cost of the hydrogen consumed in making the NH_3.

Ammonia is a gas at room temperature and pressure with a very penetrating odor. It condenses to a liquid at −33 °C under 1 atm pressure. Solutions in water, often referred to as ammonium hydroxide, are basic, due to the reaction of ammonia with water (Section 17.5).

$$NH_3(aq) + H_2O(\ell) \rightleftharpoons NH_4^+(aq) + OH^-(aq) \qquad K_b = 1.8 \times 10^{-5} \text{ at } 25\ °C$$

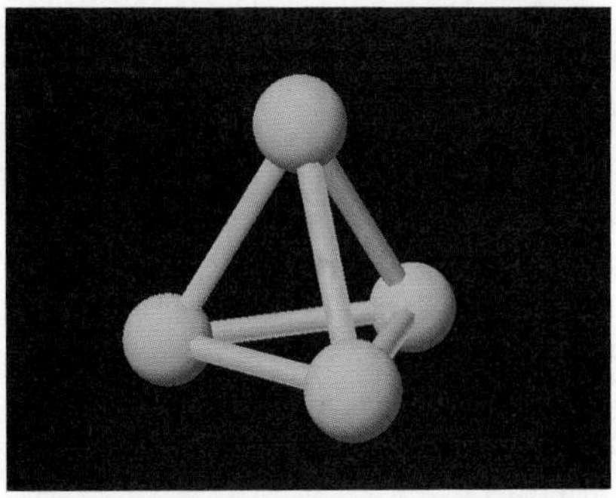

Yellow phosphorus is a tetrahedron of phosphorus atoms.

Hydrazine

Another nitrogen-hydrogen compound is hydrazine, N_2H_4, a colorless fuming liquid with an ammonia-like odor (mp, 2.0 °C; bp, 113.5 °C). About 9.1 million kilograms of hydrazine is produced annually by the *Raschig process*—the oxidation of ammonia with alkaline sodium hypochlorite in the presence of gelatin.

$$2\ NH_3(aq) + NaOCl(aq) \longrightarrow N_2H_4(aq) + NaCl(aq) + H_2O(\ell)$$

Not surprisingly, hydrazine is a base; in water it has an equilibrium constant for the first ionization of 8.5×10^{-7}.

$$N_2H_4(aq) + H_2O(\ell) \rightleftharpoons N_2H_5^+(aq) + OH^-(aq) \qquad K_b = 8.5 \times 10^{-7}$$

It is also a strong reducing agent, as reflected in the $E°$ value in basic solution.

$$N_2(g) + 4\ H_2O(\ell) + 4\ e^- \longrightarrow N_2H_4(aq) + 4\ OH^-(aq) \qquad E° = -1.16\ V$$

This ability is exploited in the use of hydrazine to treat waste water from chemical plants. It removes ions such as CrO_4^{2-} by reducing them and thus preventing them from entering the environment. A related use is the treatment of water boilers in large electric-generating plants. Oxygen dissolved in the water is a serious problem in these plants because the dissolved gas can oxidize the metal of the boiler and pipes and lead to corrosion. Hydrazine is added to reduce the dissolved oxygen to water.

$$N_2H_4(aq) + O_2(g) \longrightarrow N_2(g) + 2\ H_2O(\ell)$$

Oxides of Nitrogen

Nitrogen is unique among elements in the number of binary oxides it forms. The most common of these are listed in Table 22.4. It is interesting to note that all of these oxides are thermodynamically unstable with respect to decomposition to N_2 and O_2; all have positive ΔG_f° values but most are slow to decompose and are said to be kinetically stable.

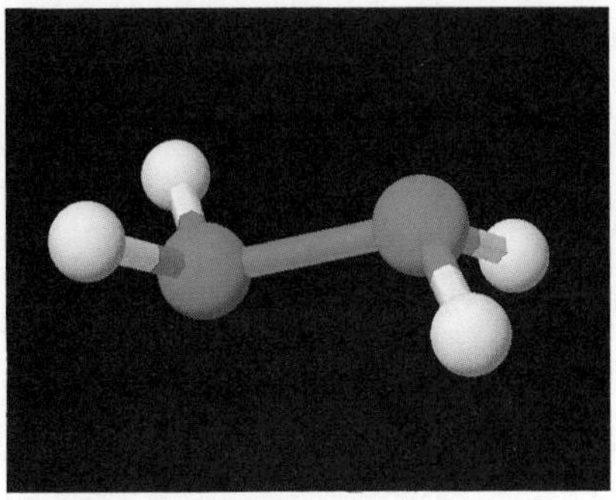

Hydrazine, N_2H_4.

Dinitrogen oxide N_2O is a nontoxic, odorless, and tasteless gas having nitrogen with the lowest oxidation number (+1) in the series of nitrogen oxides. It can be made by the careful decomposition of ammonium nitrate at 250 °C.

$$NH_4NO_3(s) \longrightarrow N_2O(g) + 2\ H_2O(g)$$

As a gas it is used as an anesthetic in minor surgery and has come to be called "laughing gas" because of its effects. Because it is soluble in vegetable fats, the

TABLE 22.4 Some Oxides of Nitrogen

Formula	Name	Structure	Nitrogen Oxidation Number	Description
N_2O	Dinitrogen oxide (nitrous oxide)	:N≡N—Ö: linear	+1	Colorless gas (laughing gas)
NO	Nitrogen monoxide (nitric oxide)	:Ṅ=Ö:	+2	Colorless gas, odd-electron molecule (paramagnetic)
N_2O_3	Dinitrogen trioxide	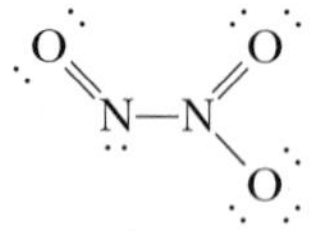 planar	+3	Blue solid (mp, −100.7 °C), reversibly dissociates to NO and NO_2
NO_2	Nitrogen dioxide	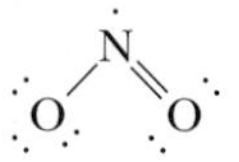	+4	Brown, paramagnetic gas
N_2O_4	Dinitrogen tetraoxide	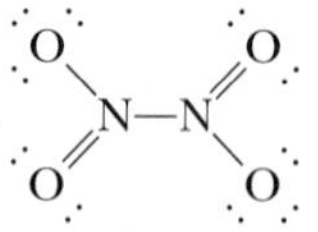 planar	+4	Colorless liquid/gas, dissociates to NO_2 (Fig. 16.7)
N_2O_5	Dinitrogen pentaoxide	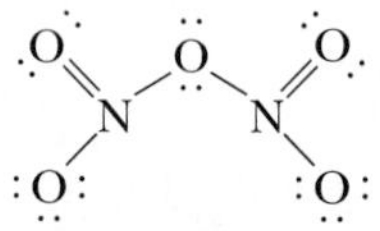	+5	Colorless solid

largest commercial use of N_2O is as a propellent and aerating agent in cans of whipped cream.

See "NO is no Dud!", Chapter 9, page 414.

Nitrogen monoxide NO, is a simple odd-electron molecule. As you will see later on, NO is an intermediate in the synthesis of nitric acid by the oxidation of ammonia. On a laboratory scale, however, it can be synthesized conveniently by reaction of a mild reducing agent with another nitrogen oxide in which N has a higher oxidation number.

$$KNO_2(aq) + KI(aq) + H_2SO_4(aq) \longrightarrow NO(g) + K_2SO_4(aq) + H_2O(\ell) + \tfrac{1}{2}\, I_2(aq)$$

Nitrogen monoxide, NO, has recently been the subject of intense research because it has been found to be important in a number of biochemical processes, and it may be beneficial to newborns with breathing problems.

Nitrogen dioxide The brown gas you see when a bottle of nitric acid is allowed to stand in the sunlight is NO_2, nitrogen dioxide. This gas is also a culprit in air pollution.

$$2\ HNO_3(aq) \longrightarrow 2\ NO_2(g) + H_2O(\ell) + \tfrac{1}{2}O_2(g)$$

Nitrogen dioxide is also formed when NO reacts with oxygen (page 414)

$$2\ NO(g) + O_2(g) \longrightarrow 2\ NO_2(g)$$

Dinitrogen tetraoxide Like NO, the dioxide NO_2 is an odd-electron molecule, but unlike NO, NO_2 molecules can dimerize to give a species that obeys the octet rule. Two molecules of NO_2 combine to form N_2O_4, dinitrogen tetraoxide, a molecule with an N—N single bond.

$$\underset{\text{colorless}}{2\ NO(g) + O_2(g)} \longrightarrow \underset{\text{deep brown gas}}{2\ NO_2(g)} \rightleftharpoons \underset{\text{colorless (mp, } -11.2\ ^\circ\text{C)}}{N_2O_4(g)}$$

When N_2O_4 is frozen (mp −11.2 °C), the solid consists entirely of N_2O_4 molecules; as the solid melts and the temperature increases to the boiling point, dissociation to NO_2 begins to occur. At the boiling point (21.5 °C) and 1 atm pressure, the distinctly brown gas phase consists of 15.9% NO_2 and 84.1% N_2O_4 (page 413).

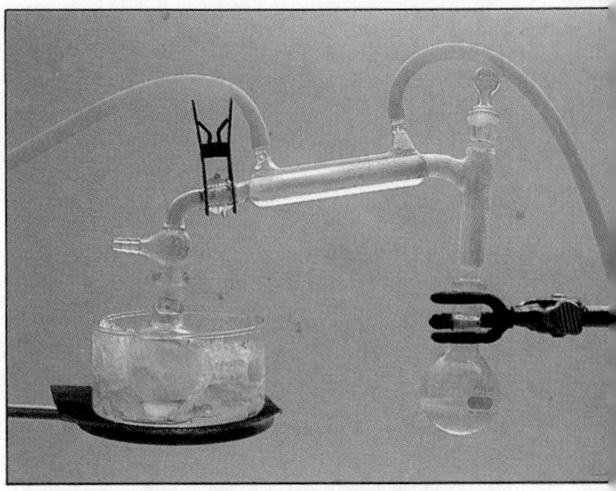

The preparation of nitric acid by the reaction of sulfuric acid and sodium nitrate. Pure HNO_3 is colorless, but some acid decomposes to give brown NO_2, and it is this gas that fills the apparatus and colors the liquid in the distillation flask. (C. D. Winters)

Nitric acid Nitrogen dioxide and N_2O_4 react with water to form nitric acid, HNO_3, so the moist gases are not only toxic but highly corrosive as well.

$$N_2O_4(g) + H_2O(\ell) \longrightarrow HNO_3(aq) + HNO_2(aq)$$

Nitric acid has been known for centuries and is still one of the most important compounds in our economy. The oldest way to make the acid is to treat Chilean saltpeter, $NaNO_3$, with sulfuric acid.

$$2\ NaNO_3(s) + H_2SO_4(aq) \longrightarrow 2\ HNO_3(aq) + Na_2SO_4(s)$$

The anhydride of nitric acid is *dinitrogen pentaoxide,* N_2O_5, which is made by chemically dehydrating nitric acid.

$$4\ HNO_3 + P_4O_{10} \rightarrow 2\ N_2O_5(s) + 4\ (HPO_3)$$

Enormous quantities of nitric acid are now produced, however, from ammonia in the multistep *Ostwald process,* which was described in Section 15.7 as an example of industrial catalysis. Roughly 20% of the ammonia produced every year is converted to nitric acid. The acid has many uses, but by far the greatest amount is turned into ammonium nitrate by neutralization of nitric acid with ammonia.

Nitric acid is a powerful oxidizing agent, as the large, positive $E°$ values for the following half-reactions illustrate.

$$NO_3^-(aq) + 4\ H_3O^+(aq) + 3\ e^- \longrightarrow NO(g) + 6\ H_2O(\ell) \qquad E° = +0.96\ V$$
$$NO_3^-(aq) + 2\ H_3O^+(aq) + e^- \longrightarrow NO_2(g) + 3\ H_2O(\ell) \qquad E° = +0.80\ V$$

Concentrated nitric acid attacks and oxidizes almost all metals. In this process, the nitrate ion is usually reduced to one of the nitrogen oxides. Which oxide is formed depends on the metal and on reaction conditions. With copper, for example, either NO_2 (page 1018) or NO is produced, depending on the concentration of the acid.

In concentrated acid:

$$Cu(s) + 4\ H_3O^+(aq) + 2\ NO_3^-(aq) \longrightarrow Cu^{2+}(aq) + 6\ H_2O(\ell) + 2\ NO_2(g)$$

In dilute acid:

$$3\ Cu(s) + 8\ H_3O^+(aq) + 2\ NO_3^-(aq) \longrightarrow 3\ Cu^{2+}(aq) + 12\ H_2O(\ell) + 2\ NO(g)$$

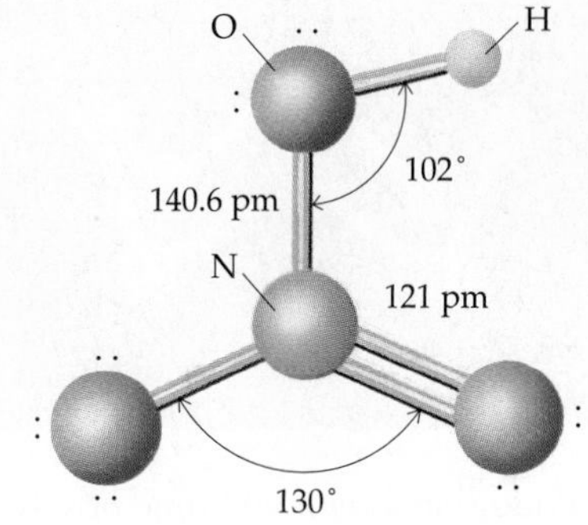

The structure of nitric acid, HNO_3.

At least four metals are not attacked by nitric acid: Au, Pt, Rh, and Ir. These came to be known as the "noble metals." The alchemists of the 14th century,

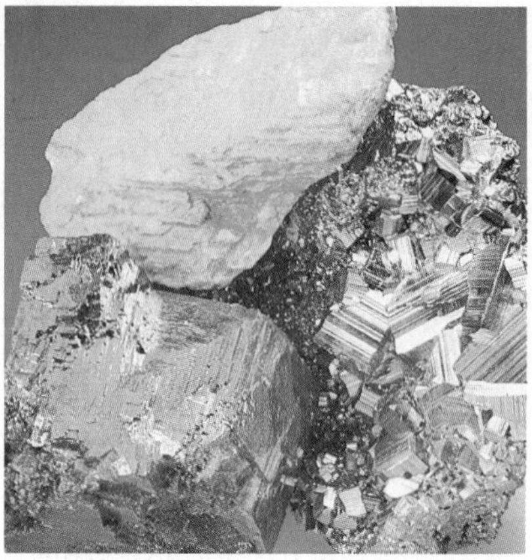

Figure 22.13 Virtually all of the metals in Groups 8B, 1B, and 2B, and the metals and metalloids of Groups 3A through 6A are found in nature as sulfides. Some common sulfur-containing minerals are golden iron pyrite, FeS_2; black galena, PbS, on quartz; and yellow orpiment, As_2S_3. (C. D. Winters)

however, knew that if you mixed HNO_3 with HCl in a ratio of about 1:3, this *aqua regia,* or "kingly water," would attack even the noblest of metals. The reaction of platinum in aqua regia is

$$5\ Pt(s) + 4\ NO_3^-(aq) + 30\ Cl^-(aq) + 24\ H_3O^+(aq) \longrightarrow 5\ [PtCl_6]^{2-}(aq) + 2\ N_2(g) + 36\ H_2O(\ell)$$

22.8 OXYGEN AND SULFUR

Oxygen is by far the most abundant element in the earth's crust, representing just under 50% by weight. It appears widely as elemental oxygen in the atmosphere, in water, and in a great many minerals. Scientists believe that elemental oxygen did not appear on this planet until about 2 billion years ago. The theory is that oxygen was formed on the planet by plants in the process of photosynthesis.

Sulfur is 15th in abundance in earth's crust. It too is found in its elemental form in nature, but only in certain concentrated deposits (page 805). Generally, it exists in the form of sulfur-containing compounds in natural gas and oil and as metal sulfide minerals. There are many common sulfide-containing minerals (Figure 22.13), including cinnabar, HgS; galena, PbS; and iron pyrite, FeS_2, or "fool's gold," a salt of the disulfide ion S_2^{2-}. Sulfur also occurs as sulfate ion (for example, in gypsum, $CaSO_4 \cdot 2\ H_2O$, Figure 4.16). Sulfur oxides (SO_2 and SO_3) are also found in nature, primarily as products of volcanic activity.

In the United States, most sulfur, about 10 million tons per year, is obtained from deposits along the Gulf of Mexico. These occur in the caprock over subterranean salt domes, typically at a depth of 150 to 750 m below the surface in layers perhaps 30 m thick. The theory explaining their existence is that the sulfur was formed by anaerobic ("without air") bacteria feeding on sedimentary sulfate deposits such as gypsum ($CaSO_4 \cdot 2\ H_2O$).

Recall from Chapter 19 and the discussion of "black smokers" on the sea floor (page 182) that many metal sulfides are poorly soluble in water, so these compounds can accumulate in geologic zones and not be dispersed by ground water.

Preparation and Properties of the Elements

Pure oxygen is obtained by fractionation of air and is third in industrial production in the United States. Very pure oxygen can be made in the laboratory by electrolysis of water and by the catalyzed decomposition of metal chlorates such as $KClO_3$.

$$2\ KClO_3(s) \longrightarrow 2\ KCl(s) + 3\ O_2(g)$$

At room temperature and pressure oxygen is a colorless gas, but it is pale blue when condensed to the liquid at −183 °C (Figure 10.15). As described in Section 10.2, diatomic oxygen is paramagnetic because it has two unpaired electrons.

Ozone, O_3, is a second, less stable, allotrope of oxygen. It is a blue, diamagnetic gas with an odor so strong that it can be detected in concentrations as low as 0.05 ppm. Ozone is conveniently synthesized by passing O_2 through an electric discharge, or by irradiation of O_2 with ultraviolet light. The gas is in the news constantly because of the realization that the earth's protective layer of ozone is being disrupted by chemicals such as the chlorofluorocarbons (see page 499).

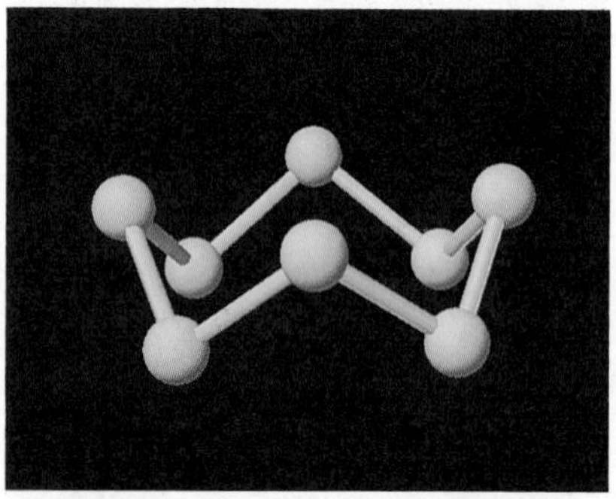

The most important allotrope of sulfur consists of a puckered, eight-member ring of sulfur atoms. (Susan Young)

Sulfur has perhaps more allotropes than any other element. The most common and most stable allotrope is the yellow, orthorhombic form, which consists of S_8 molecules with the sulfur atoms arranged in a crown-shaped ring. Less stable allotropes have rings of 6 to 20 sulfur atoms. Another form of sulfur has a molecular structure with chains of sulfur atoms (Figure 3.4).

Sulfur is obtained from underground deposits by a process developed by Herman Frasch about 1900. Superheated water (at 165 °C) and then air are forced into the deposit; the sulfur is melted (mp, 113 °C) and is forced to the surface as a frothy, yellow stream where it solidifies (Figure 22.14).

The largest use of sulfur by far is the production of *sulfuric acid,* H_2SO_4, the compound produced in largest quantity by the chemical industry (page 805). In the United States, roughly 70% of the sulfuric acid is used to manufacture "super-phosphate" fertilizer, and smaller amounts are used, for example, in the conversion of ilmenite, a titanium-bearing ore, to TiO_2, which is then used as a white pigment in paint, plastics, and paper. The acid is also used to make iron and steel, petroleum products, synthetic polymers, and paper.

Figure 22.14 About 10% of the sulfur used in the United States is obtained from underground deposits using the Frasch process. In this photo the sulfur has been forced to the surface by volcanic activity and has deposited on the rim of a steam vent. (David Cavagnaro)

The Chemistry of Sulfur

Hydrogen sulfide, H_2S, has a structure resembling that of water. Unlike water, however, hydrogen sulfide is a gas under standard conditions (mp, −85.6 °C; bp, −60.3 °C) because only very weak hydrogen bonding occurs between molecules compared with the bonding in water (Figure 13.8). Hydrogen sulfide is a deadly poison, comparable to hydrogen cyanide, but the sulfide fortunately has a terrible odor and is detected in concentrations as low as 0.02 ppm. You must be careful with H_2S, however, because it has an anesthetic effect, and your nose rapidly loses its ability to detect H_2S. Death occurs at H_2S concentrations of 100 ppm.

As illustrated by Figure 22.13, sulfur is often found as the sulfide ion in conjunction with a number of metals. The recovery of metals from their sulfide ores is usually done by heating, or *roasting,* the ore in air. The outcome of this process is the conversion of the metal sulfide to either a metal oxide or the metal itself, the sulfur appearing as SO_2.

$$2\ ZnS(s) + 3\ O_2(g) \longrightarrow 2\ ZnO(s) + 2\ SO_2(g)$$
$$2\ PbO(s) + PbS(s) \longrightarrow 3\ Pb(s) + SO_2(g)$$

Sulfur dioxide (SO_2) and trioxide (SO_3) are the most important oxides of the element. The former is produced on an enormous scale by the combustion of sulfur and by roasting sulfide ores (especially iron pyrites, FeS_2) in air. The combustion of sulfur in sulfur-containing coal and fuel oil is a particularly large environmental problem. It has been estimated that about 200 million tons of sulfur are released into the atmosphere each year by human activities, primarily in the form of SO_2; this is more than half of the total emitted by all other natural sources of sulfur in the environment.

Sulfur dioxide is a colorless, toxic gas with a choking odor. It readily dissolves in water. The most important reaction of SO_2 is oxidation to the trioxide.

$$SO_2(g) + \tfrac{1}{2}\ O_2(g) \longrightarrow SO_3(g) \qquad \Delta H° = -98.9\ kJ/mol$$

Sulfur trioxide is extremely reactive and is very difficult to handle; it is almost always deliberately converted to sulfuric acid by reaction with water (page 805).

Some common products containing sulfur or sulfur-based compounds. (C. D. Winters)

22.9 CHLORINE

Yellow-green chlorine gas was first made by the Swedish chemist Karl Wilhelm Scheele (1742–1786) in 1774 by a reaction still used today as a laboratory source of Cl_2 (Figure 22.15).

$$2\ NaCl(s) + 2\ H_2SO_4(aq) + MnO_2(s) \longrightarrow Na_2SO_4(aq) + MnSO_4(aq) + 2\ H_2O(\ell) + Cl_2(g)$$

Chlorine exists in the earth's crust as chloride ion in sea water and in brine wells. It is the 11th ranked element in abundance, and the most abundant halogen. Today, chlorine is made in enormous quantities by electrolysis of brine (Chapter 21) and ranks eighth among industrial chemicals. Almost 70% of the chlorine manufactured is used for the production of organic chemicals including vinyl chloride, which is then converted to polyvinyl chloride (PVC), a plastic used in many consumer items (Section 11.8).

Environmental groups have been outspoken in their belief that chlorine use should be severely curtailed. The subject has generated considerable controversy. See page 88 for an essay on this matter.

One of the first properties that Scheele recognized about Cl_2 was its ability to bleach textiles and paper. Soon thereafter, chlorine's ability to act as a disinfectant in water was also recognized. Today, these two uses consume about 20% of the Cl_2 produced.

Chlorine Compounds

Hydrogen Chloride

Hydrogen chloride, in the form of hydrochloric acid, ranks 26th among industrial chemicals. The gas can be prepared by the reaction of hydrogen and chlorine, but the rapid, exothermic reaction is difficult to control. The classical method of making laboratory quantities of HCl uses the reaction of NaCl and sulfuric acid. This procedure takes advantage of the fact that HCl is a gas.

$$2\ NaCl(s) + H_2SO_4(\ell) \longrightarrow Na_2SO_4(s) + 2\ HCl(g)$$

Gaseous hydrogen chloride has a sharp, irritating odor. It dissolves in water to give hydrochloric acid, a strong acid. Gaseous HCl reacts with metals, metal oxides, and metal hydrides to give metal chlorides and, depending on the reactant, water or hydrogen.

$$Mg(s) + 2\ HCl(g) \longrightarrow MgCl_2(s) + H_2(g)$$
$$ZnO(s) + 2\ HCl(g) \longrightarrow ZnCl_2(s) + H_2O(g)$$
$$NaH(s) + HCl(g) \longrightarrow NaCl(s) + H_2(g)$$

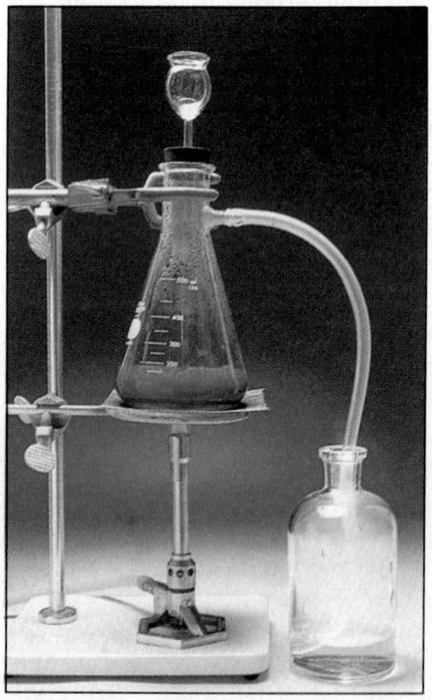

Figure 22.15 Chlorine is prepared by oxidation of chloride ion using a strong oxidizing agent. Here, oxidation of NaCl is accomplished using $K_2Cr_2O_7$ in H_2SO_4. The Cl_2 gas is bubbled into water.

Oxoacids of Chlorine

Oxoacids of chlorine range from HOCl, in which chlorine has an oxidation number of +1, to $HClO_4$, in which the chlorine oxidation number is equal to the group number, +7. All are strong oxidizing agents.

Acid	Name	Anion	Name
HOCl	Hypochlorous	OCl^-	Hypochlorite
HOClO	Chlorous	ClO_2^-	Chlorite
$HOClO_2$	Chloric	ClO_3^-	Chlorate
$HOClO_3$	Perchloric	ClO_4^-	Perchlorate

A CLOSER LOOK — *The Chemistry of Fireworks*

Black powder, a mixture of potassium nitrate, charcoal, and sulfur, was one of the more important developments in the history of the human race. The discovery, apparently well before AD 1000, is usually attributed to the Chinese, who also began the development of fireworks. Black powder made its way to Europe by the 1300s. By the American Revolution, fireworks formulations and manufacturing methods had been worked out that are still in use today.

There are several important components to typical fireworks. First, there must be an oxidizer, and today this is usually potassium perchlorate ($KClO_4$) or potassium chlorate ($KClO_3$) as well as potassium nitrate. Potassium salts are used instead of sodium salts because the latter have two important drawbacks. They are hygroscopic, meaning they absorb water from the air and so do not remain dry on storage. In addition, sodium salts give off an intense, yellow light when heated, and the light is so bright it hides other colors.

Potassium perchlorate is generally safer to use than the chlorate salt. The problem is that it may become difficult to obtain. The only supplier of perchlorates in the United States makes ammonium perchlorate, the oxidizer used in the Space Shuttle solid-fuel booster rockets. Each Shuttle launch requires about 1.5 million lb of NH_4ClO_4, about twice the annual consumption of $KClO_4$ in the United States. Thus, the fireworks industry may have trouble getting enough $KClO_4$ if Shuttle launchings become very frequent (or if the ammonium perchlorate production facilities are disabled, as they were a few years ago by an explosion).

Fireworks. (C. D. Winters)

The parts of any fireworks display we remember best are the vivid colors and brilliant flashes. White light can be produced by oxidizing magnesium or aluminum metal at high temperatures, and the flashes you see at rock concerts or similar events are typically mixtures of Mg and $KClO_4$.

Yellow light is easiest to produce because sodium salts give an intense light. Fireworks mixtures usually contain sodium in the form of such nonhygroscopic compounds as cryolite, Na_3AlF_6. Strontium salts are most often used to produce a red light, and green is produced by barium salts such as $Ba(NO_3)_2$. But the next time you see a fireworks display, watch for the ones that are blue. You will undoubtedly see very few because this color is by far the most difficult to achieve. The best way to produce the blue color is by the decomposition of copper(I) chloride at low temperatures, but pyrotechnic chemists keep looking for better, more brilliant blues.

(Telegraph Colour Library/FPG International)

Hypochlorous acid, HOCl, which forms when chlorine dissolves in water, was discovered over two centuries ago in the original work on chlorine. In this reaction, part of the chlorine is oxidized to hypochlorite ion and part is reduced to chloride ion.

$$Cl_2(g) + 2\ H_2O(\ell) \rightleftharpoons H_3O^+(aq) + HOCl(aq) + Cl^-(aq)$$

Chlorine, chloride ion, and hypochlorous acid exist in equilibrium; a low pH favors Cl_2, whereas a high pH favors the products.

If, instead of dissolving Cl_2 in pure water, the element is added to cold aqueous NaOH, hypochlorite ion and chloride ion form.

$$Cl_2(g) + 2\ OH^-(aq) \rightleftharpoons OCl^-(aq) + Cl^-(aq) + H_2O(\ell)$$

A reaction in which an element or compound is simultaneously oxidized and reduced is called a disproportionation reaction. Here Cl_2 is oxidized to OCl^- and reduced to Cl^-.

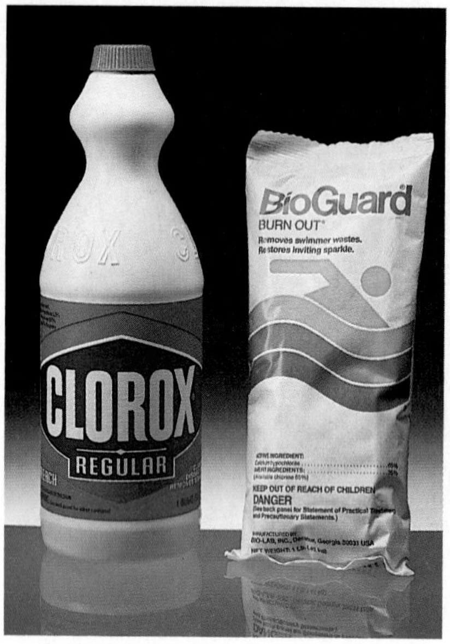

Common household bleach contains hypochlorite ion, OCl^-, formed when Cl_2 dissolves in base. "Swimming pool chlorine" is calcium hypochlorite.
(C. D. Winters)

The resulting alkaline solution is the "liquid bleach" used in home laundries. Note that the reaction equation bears a close resemblance to the equation for the reaction of Cl_2 with water. With basic conditions, however, the equilibrium has shifted far to the right. The bleaching action of this solution is a result of the oxidizing ability of OCl^-. Most dyes are colored organic compounds, and hypochlorite ion can oxidize a dye to colorless products.

When calcium hydroxide is used for this reaction in place of sodium hydroxide, solid $Ca(OCl)_2$ is the product. This compound is easily handled and is the "chlorine" that is sold for swimming pool sterilization.

When a basic solution of hypochlorite ion is heated, a further disproportionation reaction ensues, forming chlorate ion and chloride ion.

$$3\ OCl^-(aq) \longrightarrow ClO_3^-(aq) + 2\ Cl^-(aq)$$

Sodium and potassium chlorates are made by this reaction in large quantities. The sodium salt, for example, is reduced to ClO_2, which is used for bleaching paper pulp. Some $NaClO_3$ is also converted to potassium chlorate, $KClO_3$, the preferred oxidizer in fireworks and a component of safety matches.

Perchlorates

Perchlorates, salts containing ClO_4^-, are the most stable oxochlorine compounds, although they remain powerful oxidants. Pure perchloric acid, $HClO_4$, is a colorless liquid that explodes if shocked. It explosively oxidizes organic materials and rapidly oxidizes silver and gold. Dilute aqueous solutions of the acid are safer to handle, however, because they have less oxidizing power.

Perchlorate salts of most metals exist. Although many are relatively stable, some are dangerously unpredictable. *Great care should be used when handling any perchlorate salt.* Ammonium perchlorate, for example, bursts into flame if heated above 200 °C.

$$2\ NH_4ClO_4(s) \longrightarrow N_2(g) + Cl_2(g) + 2\ O_2(g) + 4\ H_2O(g)$$

This property of the ammonium salt accounts for its use as the oxidizer in the solid booster rockets for the Space Shuttle. The solid propellent in these rockets is largely NH_4ClO_4, the remainder being the reducing agent, powdered aluminum. Each Shuttle launch requires about 750 tons of ammonium perchlorate, and more than half of the sodium perchlorate currently manufactured is con-

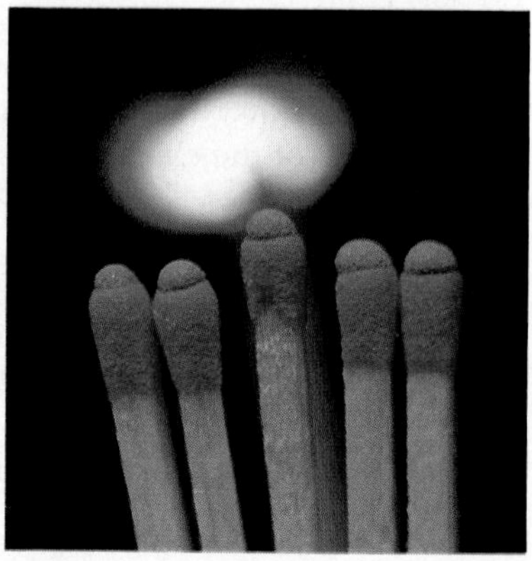

The head of a "strike anywhere" match contains, among other things, P_4S_3, and an oxidizing agent, potassium chlorate. Safety matches have sulfur (3%–5%) and $KClO_3$ (45%–55%) in the head and red phosphorus (page 103) in the striking strip.

verted to the ammonium salt. The process for doing this is an exchange reaction that takes advantage of the fact that ammonium perchlorate is less soluble in water than sodium perchlorate.

$$NaClO_4(aq) + NH_4Cl(aq) \longrightarrow NaCl(aq) + NH_4ClO_4(s)$$

The booster rockets of the Space Shuttle are fueled with a mixture of NH_4ClO_4 and Al powder. (NASA)

CHAPTER HIGHLIGHTS

When you have finished studying this chapter you should be able to

- Predict general types of chemical behavior of the A Group elements (Section 22.1).
- Predict similarities and differences among the elements in a given group, based on periodic properties (Section 22.1).
- Know which reactions produce ionic compounds, and predict formulas for common ions and common ionic compounds based on electron configurations (Section 22.1).
- Recognize when a formula is incorrectly written, based on general principles governing electron configurations (Sections 22.2–22.9).
- Be able to summarize briefly a series of facts about the most common compounds of main group elements (ionic or covalent structure, color, solubility, simple reaction chemistry) (Sections 22.2–22.9).
- Identify uses of common elements and compounds and understand the chemistry that relates to the usage (Sections 22.2–22.9).

STUDY QUESTIONS

Review Questions

1. The periodic table is one of the most useful sources of information available to a chemist. List at least three types of information that can be obtained from the periodic table.
2. Give formulas for the following common acids: nitric acid, sulfuric acid, hydrobromic acid, perchloric acid, carbonic acid. What is the oxidation number of the central atom in each of these compounds? How are these related to the periodic group number of the element?
3. Define the term "amphoteric." Write chemical equations to illustrate the amphoteric character of $Al(OH)_3$.
4. Give examples of two basic oxides. Write equations illustrating the formation of each oxide from its component elements. Write another chemical equation that illustrates the basic character of each oxide.
5. Give examples of two acidic oxides. Write equations illustrating the formation of each oxide from its component elements. Write another chemical equation that illustrates the acidic character of each oxide.
6. Write complete electron configurations using the spectroscopic notation for each of the elements in the second period. How many valence electrons are present on each atom?
7. Write complete electron configurations using the spectroscopic notation for the main group elements in the third period. How many valence electrons are present on each atom?
8. What is the general electron configuration of the elements in Group 4A in the periodic table? Relate this to the formulas of some compounds formed by these elements.
9. Give the name and symbol of each element with the valence configuration [noble gas]ns^2np^1.
10. Give symbols and names for four monatomic ions that have the same electron configuration as argon.
11. Name one cation and one anion that is isoelectronic with Br^-.
12. Write a balanced chemical equation for the reaction of an alkali metal with a halogen. Use M to represent the metal and X to represent the halogen. Is the reaction likely to be exothermic or endothermic? Is the product ionic or covalent?
13. Write a balanced chemical equation for the general reaction of an alkaline earth metal with oxygen. Use the symbol M to represent the alkaline earth element. Is the reaction likely to

be exothermic or endothermic? Is the product ionic or covalent?

14. List, in order, the 10 most abundant elements in the earth's crust. Identify one or more chemical species that occur in the earth's crust containing each of the main group elements on this list.

15. Sum the percentage abundances of the 10 most abundant elements in the earth's crust. How much of the earth's crust is made up of the remaining elements?

16. Would you expect to find calcium naturally occurring in the earth's crust as a free element? Why or why not?

17. Place the following oxides in order of increasing basicity: CO_2, SiO_2, and SnO_2.

18. Place the following oxides in order of increasing basicity: Na_2O, Al_2O_3, SiO_2, and SO_3.

19. Complete and balance equations for the following reactions. (Assume an excess of oxygen for (d).)
(a) $Li(s) + Cl_2(g)$
(b) $Ca(s) + O_2(g)$
(c) $B(s) + F_2(g)$
(d) $C(s) + O_2(g)$

20. Complete and balance equations for the following reactions:
(a) $Na(s) + I_2(s)$
(b) $Ca(s) + S_8(s)$
(c) $Al(s) + O_2(g)$
(d) $Si(s) + Cl_2(g)$

Hydrogen

21. Write balanced chemical equations for the reaction of hydrogen gas with oxygen, chlorine, nitrogen.

22. Write an equation for the reaction of sodium and hydrogen. Name the product. Is it ionic or covalent? Predict one physical property and one chemical property of this compound.

23. One of the pieces of evidence for the hydride ion in metal hydrides comes from electrochemistry. Predict the reactions that occur at each electrode when molten LiH is electrolyzed.

24. To store 2.88 kg of gasoline with an energy equivalence of 1.43×10^8 J requires a volume of 4.1 L. In comparison, 1.0 kg of H_2 has the same energy equivalence. What volume is required if this quantity of H_2 gas is to be stored at 25 °C and 1.0 atm of pressure?

25. A method recently suggested for the preparation of hydrogen (and oxygen) proceeds as follows:
(a) Sulfuric acid and hydrogen iodide are formed from sulfur dioxide, water, and iodine.
(b) The sulfuric acid from the first step is decomposed by heat to water, sulfur dioxide, and oxygen.
(c) The hydrogen iodide from the first step is decomposed with heat to hydrogen and iodine.
Write a balanced equation for each of these steps and show that their sum is the decomposition of water to hydrogen and oxygen.

26. Write a balanced chemical equation for the preparation of H_2 by the reaction of CH_4 and water. Using data in Appendix K, calculate $\Delta H°$, $\Delta G°$, and $\Delta S°$ for this reaction.

27. Calculate and compare the masses of H_2 expected for 100% reaction of steam (H_2O) with CH_4 and with coal (assume this to be pure carbon).

Alkali Metals

28. Write equations for the reaction of sodium with each of the halogens. Predict several physical properties that are common to all of the alkali metal halides.

29. Sodium peroxide is the primary product when sodium metal is burned in oxygen. Write a balanced equation for this reaction.

30. Write balanced equations for the reaction of lithium, sodium, and potassium with O_2. Specify which metal largely forms oxides, which one forms peroxides, and which one forms superoxides.

31. A piece of sodium catches on fire in the laboratory! How do you extinguish the fire? What is the worst thing you could do?

32. The electrolysis of aqueous NaCl gives NaOH, Cl_2, and H_2.
(a) Write a balanced equation for the process.
(b) In 1993, in the United States, 25.71 billion lb of NaOH and 24.06 billion lb of Cl_2 were produced. Does the ratio of masses of NaOH and Cl_2 produced agree with the ratio of masses expected from the balanced equation? If not, what does this tell you about the way in which NaOH or Cl_2 are actually produced? Is the electrolysis of aqueous NaCl the only source of these chemicals?

33. What is oxidized, and what is reduced, when an aqueous solution of KCl is electrolyzed under the same conditions as the electrolysis of aqueous NaCl? Predict the products that are formed if CsI is electrolyzed under these conditions.

Alkaline Earths

34. When magnesium burns in air, it forms both an oxide and a nitride. Write balanced equations for the formation of both compounds.

35. Calcium reacts with hydrogen gas at elevated temperatures to form a hydride. This compound reacts readily with water, so it is an excellent drying agent for organic solvents.
(a) Write a balanced equation showing the formation of calcium hydride from Ca and H_2.
(b) Write a balanced equation for the reaction of calcium hydride with water.

36. Name three uses of limestone. Write a balanced equation for the reaction of limestone with CO_2 in water.

37. Explain what is meant by "hard water." What causes hard water and what problems are associated with it?

38. Calcium oxide, CaO, is used to remove SO_2 from power plant exhaust because the two react to give solid $CaSO_3$. How many grams of SO_2 can be removed using 1000. kg of CaO?

39. Calcium hydroxide, $Ca(OH)_2$, has a K_{sp} of 7.9×10^{-6}, whereas that for $Mg(OH)_2$ is 1.5×10^{-11}. Calculate the equilibrium constant for the reaction

$$Ca(OH)_2(s) + Mg^{2+}(aq) \rightleftharpoons Ca^{2+}(aq) + Mg(OH)_2(s)$$

and explain why this reaction can be used in the commercial isolation of magnesium from sea water.

Aluminum

40. Write an equation for the reactions of aluminum with: HCl(aq), Cl_2, and O_2.

41. Even though the table of $E°$ values in Chapter 21 shows that metallic aluminum is a very good reducing agent, a typical sample of the metal resists attack by air and water. Explain this observation.

42. Aluminum dissolves readily in hot aqueous base (NaOH) to give the aluminate ion, $Al(OH)_4^-$, and H_2. Write a balanced equation for this reaction. If you begin with 13.2 g of Al, what volume of H_2 gas (in milliliters) is produced when the gas is measured at 735 torr and 22.5 °C?

43. Alumina, Al_2O_3, is amphoteric, so it dissolves when heated strongly or "fused" with an acidic oxide or basic oxide.
(a) Write a balanced equation for the reaction of alumina with silica, an acidic oxide, to give aluminum metasilicate, $Al_2(SiO_3)_3$.
(b) Write a balanced equation for the reaction of alumina with the basic oxide CaO to give calcium aluminate, $Ca(AlO_2)_2$.

44. Aluminum sulfate, with a worldwide production of about 3 million tons, is the most important commercial aluminum compound after aluminum oxide and aluminum hydroxide. Write a balanced equation for the reaction of aluminum oxide with sulfuric acid to give aluminum sulfate. If you want to manufacture 1.00 kg of aluminum sulfate, how many kilograms of aluminum oxide and sulfuric acid must you use?

45. Gallium hydroxide, like aluminum hydroxide, is amphoteric. Write balanced equations showing how $Ga(OH)_3$ can dissolve in both HCl and NaOH. What volume of 0.0112 M HCl do you need to react completely with 1.25 g of $Ga(OH)_3$?

46. Halides of the Group 3A elements are excellent Lewis acids. When a Lewis base such as Cl^- interacts with $AlCl_3$, the ion $AlCl_4^-$ is formed. What is the structure of this ion? What is the hybridization of the aluminum atom in $AlCl_4^-$?

47. "Aerated" concrete bricks are widely used building materials. They are obtained by mixing gas-forming additives with a moist mixture of lime, sand, and possibly cement. Industrially, the following reaction is important:

$$2\ Al(s) + 3\ Ca(OH)_2(s) + 6\ H_2O(\ell) \longrightarrow [3\ CaO \cdot Al_2O_3 \cdot 6\ H_2O](s) + 3\ H_2(g)$$

Assume that the mixture of reactants contains 0.56 g of Al in each brick. What volume of hydrogen gas do you expect at 26 °C and atmospheric pressure (745 mm Hg)?

Silicon

48. What is the structure of SiO_2?
(a) Why is there a difference in the structures of SiO_2 and CO_2?
(b) Explain why SiO_2 has a very high melting point.

49. Describe how ultrapure silicon can be produced from sand.

50. One starting material to make silicones (see Figure 22.12) is dichlorodimethylsilane, $(CH_3)_2SiCl_2$. It is made by treating silicon powder at about 300 °C with CH_3Cl in the presence of a copper-containing catalyst.
(a) Write a balanced equation for the reaction.
(b) Assume you carry out the reaction on a small scale with 2.65 g of silicon. To measure the CH_3Cl gas, you fill a 5.60-L flask at 24.5 °C. What pressure of CH_3Cl gas must you have in the flask in order to have the stoichiometrically correct amount of the compound?
(c) What mass of $(CH_3)_2SiCl_2$ is produced? (Assume 100% yield.)

51. Silicates are enormously important in our lithosphere.
(a) Name one mineral that is a simple orthosilicate.
(b) Name one mineral that is an amphibole.
(c) Name one mineral that is a sheet silicate.
(d) Give one example of an aluminosilicate.

Nitrogen and Phosphorus

52. Use data in Appendix K to calculate the enthalpy change for the reaction

$$2\ NO(g) + O_2(g) \longrightarrow 2\ NO_2(g)$$

(a) Are the nitrogen oxides in this reaction stable with respect to their elements?
(b) Is the reaction of NO with oxygen exothermic or endothermic?
(c) Are the compounds in this reaction diamagnetic or paramagnetic?

53. The overall reaction involved in the industrial synthesis of nitric acid is

$$NH_3(g) + 2\ O_2(g) \longrightarrow HNO_3(aq) + H_2O(\ell)$$

Calculate $\Delta G°$ for the reaction and then the equilibrium constant, at 25 °C.

54. A major use of hydrazine, N_2H_4, is in steam boilers in power plants.
(a) The reaction of hydrazine with O_2 dissolved in water gives N_2 and water. Write a balanced equation for this reaction.
(b) Oxygen gas, O_2, dissolves in water to the extent of 3.08 cm^3 (gas at STP) in 100. mL of water at 20 °C. To consume all of the dissolved O_2 in 3.00×10^4 L of water (enough to fill a small swimming pool), how many grams of N_2H_4 are needed?

55. Before hydrazine came into use to remove dissolved oxygen in the water in steam boilers, Na_2SO_3 was commonly used.

$$Na_2SO_3(aq) + \tfrac{1}{2}\ O_2(aq) \longrightarrow Na_2SO_4(aq)$$

How many grams of Na_2SO_3 is required to remove O_2 from 30,000 L of water as outlined in Study Question 54?

56. A common analytical method for hydrazine involves its oxidation with iodate ion, IO_3^-. In the process, hydrazine behaves as a four-electron reducing agent.

$$N_2(g) + 5\ H_3O^+(aq) + 4\ e^- \longrightarrow N_2H_5^+(aq) + 5\ H_2O(\ell) \qquad E° = -0.23\ V$$

Write the balanced equation for the reaction of hydrazine in acid solution ($N_2H_5^+$) with $IO_3^-(aq)$ to give N_2 and I_2. Calculate $E°$ for this reaction.

57. The steering rockets in the Space Shuttle use N_2O_4 and a derivative of hydrazine, dimethylhydrazine. This mixture is called a *hypergolic fuel* because it ignites when the reactants come into contact.

$$H_2NN(CH_3)_2(\ell) + 2\ N_2O_4(\ell) \longrightarrow 3\ N_2(g) + 4\ H_2O(g) + 2\ CO_2(g)$$

(a) Which is the oxidizing agent, and which is the reducing agent in this reaction?

(b) The same propulsion system was used by the Lunar Lander on moon missions in the 1970s. If the Lander used 4100 kg of $H_2NN(CH_3)_2$, how many kilograms of N_2O_4 were required to react with it? How many kilograms of each of the reaction products was generated?

58. Unlike carbon, which can form extended chains of atoms, nitrogen can form chains of very limited length. Draw the Lewis electron dot structure of the azide ion, N_3^-.

59. Calcium hydrogen phosphate, $CaHPO_4$, is used as an abrasive in toothpaste. Write a balanced equation showing a possible preparation for this compound.

Oxygen and Sulfur

60. In the "contact process" for making sulfuric acid, sulfur is first burned to SO_2. Environmental restrictions allow no more than 0.30% of this SO_2 to be vented to the atmosphere.

(a) If enough sulfur is burned in a plant to produce 1.80×10^6 kg of pure, anhydrous H_2SO_4 per day, how much SO_2 is allowed to be exhausted to the atmosphere?

(b) One way to prevent even this much SO_2 from reaching the atmosphere is to "scrub" the exhaust gases with hydrated lime.

$$Ca(OH)_2(s) + SO_2(g) \longrightarrow CaSO_3(s) + H_2O(\ell)$$
$$2\ CaSO_3(s) + O_2(g) \longrightarrow 2\ CaSO_4(s)$$

What mass of $Ca(OH)_2$ (in kilograms) is needed to remove the SO_2 calculated in part (a)?

61. A sulfuric acid plant produces an enormous amount of heat. To keep costs as low as possible, much of this heat is used to make steam to generate electricity. Some of the electricity is used to run the plant, and the excess is sold to the local electrical utility. Three reactions are important in sulfuric acid production: (1) burning S to SO_2; (2) oxidation of SO_2 to SO_3; and (3) reaction of SO_3 with H_2O.

$$SO_3(g) + H_2O\ (\text{in 98\% } H_2SO_4) \longrightarrow H_2SO_4(\ell)$$

If the enthalpy change of the third reaction is −130 kJ/mol, estimate the total heat produced per mole of H_2SO_4 produced. How much heat is produced per ton (907 kg) of H_2SO_4?

62. In addition to the simple sulfide ion, S^{2-}, there are polysulfides, S_n^{2-}. (Such ions are chains of sulfur atoms, not rings.) Draw a Lewis electron dot structure of the S_3^{2-} ion. What geometry would you predict for this ion?

63. Sulfur forms a range of compounds with fluorine. Draw Lewis electron dot structures for S_2F_2 (connectivity is FSSF), SF_2, SF_4, and SF_6. What is the oxidation number of sulfur in each of these compounds? Describe the structure of each compound.

Chlorine

64. The halogen oxides and oxoanions are generally good oxidizing agents. For example, the bromate ion oxidizing half-reaction has an $E°$ value of 1.44 V in acid solution.

$$BrO_3^-(aq) + 6\ H_3O^+(aq) + 6\ e^- \longrightarrow Br^-(aq) + 9\ H_2O(\ell)$$

Can you oxidize aqueous 1.0 M Mn^{2+} to aqueous MnO_4^- with 1.0 M bromate ion?

65. The hypohalite ions, OX^-, are the salts of weak acids. Calculate the pH of a 0.10 M solution of NaOCl. What is the concentration of HOCl in this solution?

66. Halogens combine with one another to form *interhalogens* such as BrF_3. Sketch a possible structure for this molecule and decide whether the Br—F angles are less than or greater than ideal.

67. Halogens form polyhalide ions. Sketch Lewis structures and molecular structures for (a) I_3^-, (b) $BrCl_2^-$, and (c) ClF_2^+. Are there any similarities or differences?

General Questions

68. Using data in Appendix K and those given below, calculate $\Delta G°$ values for the decomposition of MCO_3 to MO and CO_2 where M = Mg, Ca, and Ba. What is the relative tendency of these carbonates to decompose?

Compound	**$\Delta G_f°$ (kJ/mol)**
$MgCO_3$	−1012.1
$BaCO_3$	−1137.6
BaO	−525.1

69. Ammonium perchlorate is used as the oxidizer in the solid-fuel booster rockets of the Space Shuttle. If one launch requires 750 tons (6.8×10^5 kg) of the salt, and the salt decomposes according to the equation on page 1034, what mass of water is produced? What mass of O_2 is produced? If the O_2 produced is assumed to react with the powdered aluminum present in the rocket engine, how much aluminum is neces-

sary to use up all of the O_2, and how much Al_2O_3 is produced?

70. Metals react with hydrogen halides [such as HCl] to give the metal halide and hydrogen.

$$M(s) + n\ HX(g) \longrightarrow MX_n(s) + \frac{n}{2}H_2(g)$$

The free energy change for the reaction is

$$\Delta G^\circ_{rxn} = \Delta G^\circ_f(MX_n) - n\ \Delta G^\circ_f\ [HX(g)]$$

(a) ΔG°_f for HCl(g) is −95.3 kJ/mol. What must be the value for $\Delta G^\circ_f(MX_n)$ for the reaction to be spontaneous?
(b) Which of the following metals is predicted to react spontaneously with HCl(g): Ba, Pb, Hg, or Ti?

71. The boron in boric acid, $B(OH)_3$, is bonded to three —OH groups. (In the solid state, the —OH groups are in turn hydrogen-bonded to —OH groups in neighboring molecules.)
(a) Draw the Lewis structure for boric acid.
(b) What is the hybridization of the boron in the acid?
(c) Sketch a picture showing how hydrogen bonding can occur between neighboring molecules.
(d) Boric acid is an acid because it reacts with water to give the borate and hydronium ions.

$$B(OH)_3(aq) + 2\ H_2O(\ell) \rightleftharpoons B(OH)_4^-(aq) + H_3O^+(aq)$$

Given the nature of bonding in boric acid, why do you believe this reaction is possible? Is boric acid acting as a Lewis acid or Lewis base in this reaction?

72. The structure of nitric acid is illustrated on page 1029.
(a) Why are the N—O bonds the same length, and why are both shorter than the N—OH bond length?
(b) Rationalize the bond angles in the molecule.
(c) What is the hybridization of the central N atom? Describe how this atom can form an N—O pi bond.

73. Like many metals, aluminum reacts readily with halogens. Aluminum bromide, the product of the reaction illustrated in Figure 4.1, has the structure illustrated here in the solid and gaseous phases. That is, it is a dimer of the monomer units $AlBr_3$ with a Br atom bridging from one Al atom to another. (This form of halogen bridging is fairly common in chemistry.) The Br—Al—Br angle is 115° and the Al—Br—Al angle is 87°.

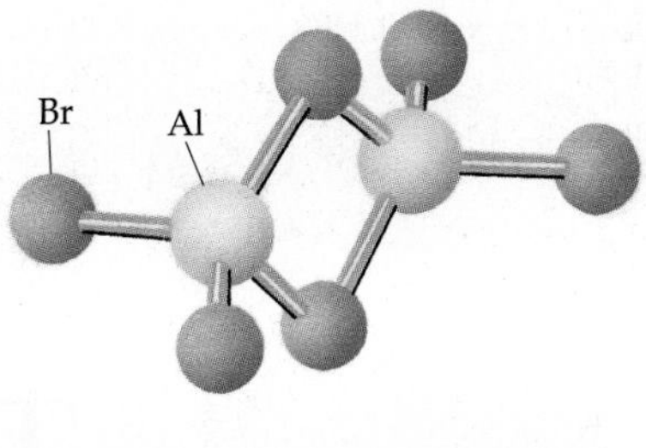

(a) What is the hybridization of the Al atoms?
(b) Draw the Lewis structure for an $AlBr_3$ monomer.
(c) How might a Br atom form a bridge from one Al atom to another? Describe the bonding in the Al_2Br_6 molecule.

Conceptual Questions

74. Identify the lettered compounds in the following reaction scheme. When 1.00 g of a white solid A is strongly heated, you obtain another white solid, B, and a gas. (An experiment is carried out on the gas, showing that it exerts a pressure of 209 mm Hg in a 450-mL flask at 25 °C.) Bubbling the gas into a solution of $Ca(OH)_2$ gives another white solid, C. If the white solid B is added to water, the resulting solution turns red litmus paper blue. To the solution of B, you add dilute, aqueous HCl and evaporate to dryness to yield a white solid D. When D is placed in a Bunsen flame, it colors the flame green. Finally if the aqueous solution of B is treated with sulfuric acid, a white precipitate, E, forms.

75. Use ΔH°_f data in Appendix K to calculate the enthalpy of the reaction

$$2\ N_2(g) + 5\ O_2(g) + 2\ H_2O(\ell) \longrightarrow 4\ HNO_3(aq)$$

Speculate on whether such a reaction can be used to fix nitrogen. Would research to find ways to accomplish this reaction be a good endeavor?

76. You are given air and water for starting materials, along with whatever laboratory equipment you need. Describe how you could synthesize ammonium nitrate.

77. On page 1014 it is noted that many alkali metal compounds melt in fire, whereas alkaline earth compounds do not melt. What might be the underlying reason for this difference in behavior?

78. In Problem-Solving Tips and Ideas 22.1, it is noted that Group 1A elements never form 2+ or 3+ ions. Why?

Summary Question

79. Magnesium chemistry.
(a) Magnesium is obtained from sea water. If sea water is 0.050 M in magnesium ion, what volume of sea water (in liters) must be treated to obtain 1.00 kg of magnesium metal? What mass of lime (CaO, in kilograms) must be used to precipitate the magnesium in this volume of sea water?
(b) When 1000. kg of molten $MgCl_2$ is electrolyzed to produce magnesium, how many kilograms of metal are produced at the cathode? What is produced at the anode? How many kilograms of the other product are produced? What is the total number of Faradays of electricity used in the process?
(c) One industrial process has an energy consumption of 8.4 kwh/lb of Mg. How many joules is therefore required per mole? How does this energy compare with the energy of the process

$$MgCl_2(s) \longrightarrow Mg(s) + Cl_2(g)$$

CHAPTER 23

The Transition Elements

A Chemical Puzzler

This photo shows crystals of rhodochrosite, a crystalline form of manganese(II) carbonate. They are red due to the presence of the Mn(II) ion. Many other gem stones are colored because of transition metal ions. Rubies are red due to a trace of Cr^{3+} in a lattice of aluminum oxide, and citrine is yellow because there is a trace of Fe^{3+} in a lattice of SiO_2. Why does the presence of transition metal ions result in color? (Brian Parker/ Tom Stack & Associates)

The transition elements, the large block of elements in the central portion of the periodic table, form a bridge between the *s*-block elements at the left and the *p*-block metals, metalloids, and nonmetals on the right (Figure 23.1). The first three rows of these elements (Sc to Zn, Y to Cd, and La to Hg) are called the **transition elements,** or **transition metals.** These elements are also referred to as the ***d*-block elements** because their occurrence in the periodic table coincides with the filling of the *d* orbitals. Contained within this group of elements are two subgroups sometimes called inner transition elements; these are the **lanthanide elements** that occur between La and Hf, and the **actinide elements** that occur between Ac and element 104. Because these subgroups arise as the *f* orbitals are filled, they are also called ***f*-block elements.**

In this chapter, the primary focus is on the *d*-block elements, and within this group, we concentrate mainly on the elements in the fourth period, that is, the elements of the first transition series, the group from scandium to zinc.

23.1 PROPERTIES OF THE TRANSITION ELEMENTS

The *d*-block elements include the most common metal used in construction and manufacturing (iron), metals that are valued for their beauty (gold, silver, and platinum), metals used in coins (nickel, copper), and metals used in modern technology (titanium). Copper, silver, gold, and iron were known and used in early civilization. This group of elements contains the densest elements known (osmium, $d = 22.49\ g/cm^3$, and iridium, $d = 22.41\ g/cm^3$), the metals with the

For historical reasons, the lanthanide elements are sometimes called the "rare earths," though many of these elements are actually not rare. For example, cerium (Ce, element 58) is half as abundant as chlorine and about five times as abundant as lead in earth's crust.

1A																	8
H	2A											3A	4A	5A	6A	7A	He
Li	Be											B	C	N	O	F	Ne
Na	Mg	3B	4B	5B	6B	7B		8B		1B	2B	Al	Si	P	S	Cl	Ar
K	Ca	Sc	Ti	V	Cr	Mn	Fe	Co	Ni	Cu	Zn	Ga	Ge	As	Se	Br	Kr
Rb	Sr	Y	Zr	Nb	Mo	Tc	Ru	Rh	Pd	Ag	Cd	In	Sn	Sb	Te	I	Xe
Cs	Ba	*La	Hf	Ta	W	Re	Os	Ir	Pt	Au	Hg	Tl	Pb	Bi	Po	At	Rn
Fr	Ra	†Ac															

* Lanthanides	Ce	Pr	Nd	Pm	Sm	Eu	Gd	Tb	Dy	Ho	Er	Tm	Yb	Lu
† Actinides	Th	Pa	U	Np	Pu	Am	Cm	Bk	Cf	Es	Fm	Md	No	Lr

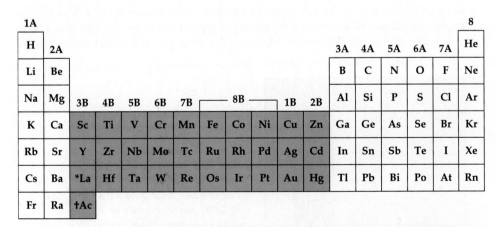

Figure 23.1 Periodic table showing the *d*-block, or transition elements (red), and the *f*-block elements, the lanthanides (blue) and the actinides (yellow).

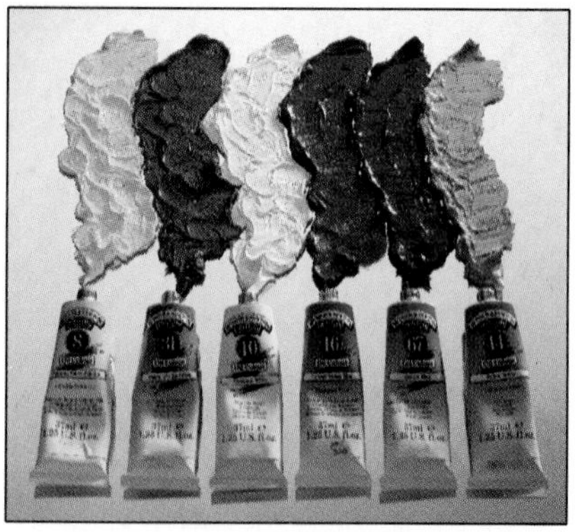

Figure 23.2 Paint pigments often contain transition metal compounds: Green, Cr_2O_3; white, TiO_2 and ZnO; purple, $Mn_3(PO_4)_2$; blue, cobalt and aluminum oxides; and ochre, Fe_2O_3. The yellow pigment is a cadmium compound. (C. D. Winters)

highest and lowest melting points (tungsten, mp = 3410 °C, and mercury, mp = −38.9 °C), and one of two radioactive elements with atomic number less than 83 (technetium (Tc), atomic number 43; the other is promethium (Pm), atomic number 61, in the *f*-block).

Certain *d*-block elements are particularly important in living organisms. For example, cobalt is the crucial element in vitamin B_{12}, a compound that acts as a catalyst in the metabolism of carbohydrates, fats, and proteins. Hemoglobin and myoglobin, compounds in biochemical oxidation-reduction processes, contain iron. Molybdenum and iron, together with sulfur, form the reactive portion of nitrogenase, a biological catalyst used by nitrogen-fixing organisms to convert atmospheric nitrogen into ammonia. Copper and zinc are important in other biological catalysts.

Some "bad actors" also appear in this group of elements. Mercury, for example, is toxic and is a threat in the environment. Quite a number of other metals are toxic as well, so that disposal of "heavy-metal" wastes is generally a significant problem.

Many transition metal compounds are highly colored, which makes them useful as pigments in paints and dyes. Prussian blue, $Fe_4[Fe(CN)_6]_3$, is the "bluing agent" in laundry bleach and in engineering blueprints. A common pigment (artist's cadmium-yellow) contains cadmium sulfide (CdS), and the pigment in most white paints is titanium(IV) oxide (Figure 23.2).

The presence of transition metal ions in crystalline silicates or alumina causes a common material to be transformed into a gemstone. Iron(II) is the cause of the yellow color in citrine, whereas chromium(III) causes the red color of a ruby (Figure 3.12). Transition metal complexes in small quantities add color to glass. For example, blue glass is made by adding a small amount of a cobalt (III) oxide, and green glass is made by adding Cr_2O_3. Old window panes sometimes take on a purple color over time as a consequence of oxidation of traces of manganese(II) ion present in the glass to purple permanganate (MnO_4^-) (Figure 23.3).

Knowing that the transition elements are metals, we can predict typical metallic properties. We expect most transition elements to be solids with relatively high melting and boiling points, to have a metallic sheen, and to be conductors of electricity and heat. Metals typically undergo oxidation reactions, and we can expect formation of ionic compounds to be typical behavior. We can also anticipate some differences, however. Mercury, for example, is a liquid rather than a

Figure 23.3 Colored glass can be made by adding small amounts of metal oxides to clear glass. Blue glass often contains cobalt(II) oxide, copper or chromium oxides give green glass, nickel or cobalt oxides give a purple color, copper or selenium oxide gives red, and an iridescent green color is due to uranium oxide. (C. D. Winters)

solid. Whereas the oxidation of iron is well known, and a problem that we go to some lengths to prevent, silver and gold are used in coins and jewelry because they are relatively less likely to oxidize, and the oxidation products of copper are often considered to have aesthetic value.

Let us look more closely at the properties of the transition elements, concentrating especially on the underlying principles that govern the characteristics of these elements.

The black coating of tarnish that develops on silver is the result of oxidation. Similarly, oxidation is the cause of the blackening of the surface of copper and the attractive green "patina" that copper surfaces sometimes attain.

Electron Configurations

Because the chemical behavior of an element is related to its electronic structure, it is important to know the electron configurations of the *d*-block elements and their common ions (Section 8.4). Recall that the configuration of these metals has the general form [noble gas core] $ns^a(n-1)d^b$, that is, valence electrons for the transition elements reside in the ns and $(n-1)d$ subshells (see Tables 8.2 and 8.4).

Oxidation of these elements results in positive ions. In this process, all the *s* electrons are lost, and, in some instances, one or more *d* electrons are lost as well. The resulting ions have the electron configuration [noble gas core] $(n-1)d^x$. In contrast to ions formed by main group elements, transition metal ions do not have rare gas configurations, and their compounds often possess unpaired electrons, leading to paramagnetic behavior as discussed later in this section.

Oxidation Numbers

Oxidation numbers of +2 and +3 are commonly observed in compounds of the first transition series. Examples of oxidation reactions of transition metals include those with oxygen to form metal oxides, with halogens to form metal halides, and with aqueous acid to form the hydrated metal ion (Table 23.1 and Figure 23.4). With iron, for example, oxidation processes usually convert Fe

TABLE 23.1 Common Oxidation Products of Elements in the First Transition Series*

Element	Reaction with O_2	Reaction with Cl_2	Reaction with Aqueous HCl
Scandium	Sc_2O_3	$ScCl_3$	$Sc^{3+}(aq)$
Titanium	TiO_2	$TiCl_4$	$Ti^{3+}(aq)$
Vanadium	V_2O_5	NR	NR
Chromium	Cr_2O_3	$CrCl_3$	$Cr^{2+}(aq)$
Manganese	Mn_3O_4	$MnCl_2$	$Mn^{2+}(aq)$
Iron	Fe_2O_3	$FeCl_3$	$Fe^{2+}(aq)$
Cobalt	Co_3O_4	$CoCl_2$	$Co^{2+}(aq)$
Nickel	NiO	$NiCl_2$	$Ni^{2+}(aq)$
Copper	CuO	$CuCl_2$	NR
Zinc	ZnO	$ZnCl_2$	$Zn^{2+}(aq)$

*NR = no reaction.

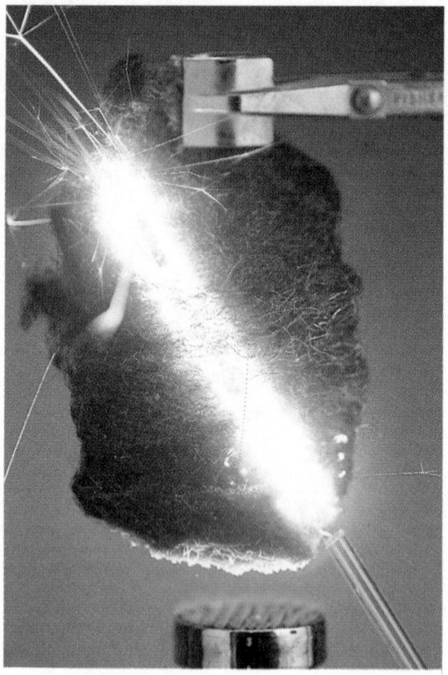

(a)

(b)

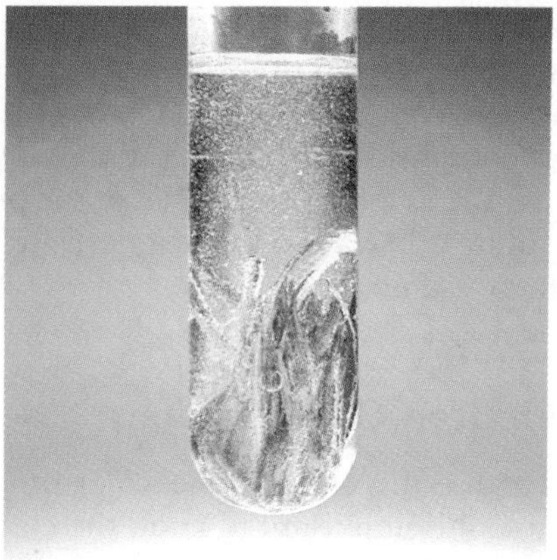

(c)

Figure 23.4 Typical reactions of transition metals. Most metals react with oxygen, with halogens, and with acids under appropriate conditions. Here steel wool reacts with O_2 (a), with chlorine gas, Cl_2 (b), and iron filings react with aqueous HCl (c). (C. D. Winters)

([Ar] $3d^6 4s^2$) to Fe^{2+} ([Ar] $3d^6$) or to Fe^{3+} ([Ar] $3d^5$). The rusting of iron is a well-known example of oxidation. Rusting is actually a complicated process that requires both oxygen and water to proceed, and the product is a hydrated iron(III) oxide. Iron reacts with chlorine to given $FeCl_3$, and it reacts with H_3O^+(aq) to produce Fe^{2+}(aq) and H_2.

Recall from Chapter 21 that tables of electrochemical potentials are a source of useful information on oxidations and reductions. Standard reduction potentials for the elements of the first transition series are shown in Table 23.2. All of these metals except copper can be oxidized by H_3O^+(aq).

Despite the preponderance of +2 and +3 compounds of these elements, the known range of identified oxidation states is considerably broader (Figures

TABLE 23.2 Standard Aqueous Reduction Potentials of Elements of the First Transition Series

Element	Half-Reaction	$E°$ (V)
Scandium	$Sc^{3+} + 3\ e^- \rightarrow Sc$	−2.08
Titanium	$Ti^{2+} + 2\ e^- \rightarrow Ti$	−1.63
Vanadium	$V^{2+} + 2\ e^- \rightarrow V$	−1.2
Chromium	$Cr^{2+} + 2\ e^- \rightarrow Cr$	−1.18
Manganese	$Mn^{2+} + 2\ e^- \rightarrow Mn$	−0.91
Iron	$Fe^{2+} + 2\ e^- \rightarrow Fe$	−0.44
Cobalt	$Co^{2+} + 2\ e^- \rightarrow Co$	−0.28
Nickel	$Ni^{2+} + 2\ e^- \rightarrow Ni$	−0.23
Copper	$Cu^{2+} + 2\ e^- \rightarrow Cu$	+0.34
Zinc	$Zn^{2+} + 2\ e^- \rightarrow Zn$	−0.76

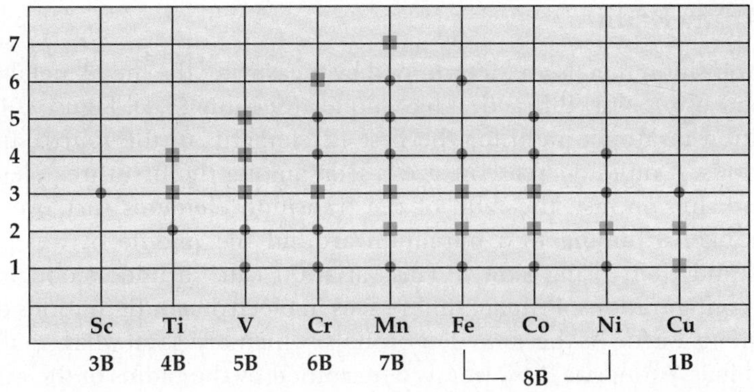

Figure 23.5 Oxidation numbers for the fourth-period transition elements, except zinc. More common oxidations states are in red.

23.5 and 23.6). In examples earlier in this text, we encountered chromium in the +6 oxidation state (CrO_4^{2-}, $Cr_2O_7^{2-}$), manganese in the +7 oxidation state (MnO_4^-) and silver and copper as +1 ions. The most common oxidation state of titanium is +4, seen for example in the common minerals of this element: TiO_2 (rutile) and ilmenite ($FeTiO_3$).

The most common oxidation state for the lanthanide elements is +3, but compounds with metal oxidation states of +2 and +4 are also encountered for some metals.

Higher oxidation states are more common in compounds of the elements in the second and third transition series, whereas +2 and +3 ions are less often encountered. For example, the naturally occurring sources of molybdenum and tungsten are the ores molybdenite (MoS_2) and wolframite (WO_3). In contrast, the principal ore of chromium is chromite, $FeO \cdot Cr_2O_3$. This general trend is carried over in the *f*-block. The lanthanides generally form 3+ ions, whereas actinide elements usually also have higher oxidation numbers, such as +4 and even +6. For example, UO_3 is a common oxide of uranium, and UF_6 is a compound important in reprocessing uranium fuel for nuclear reactors (Section 24.6).

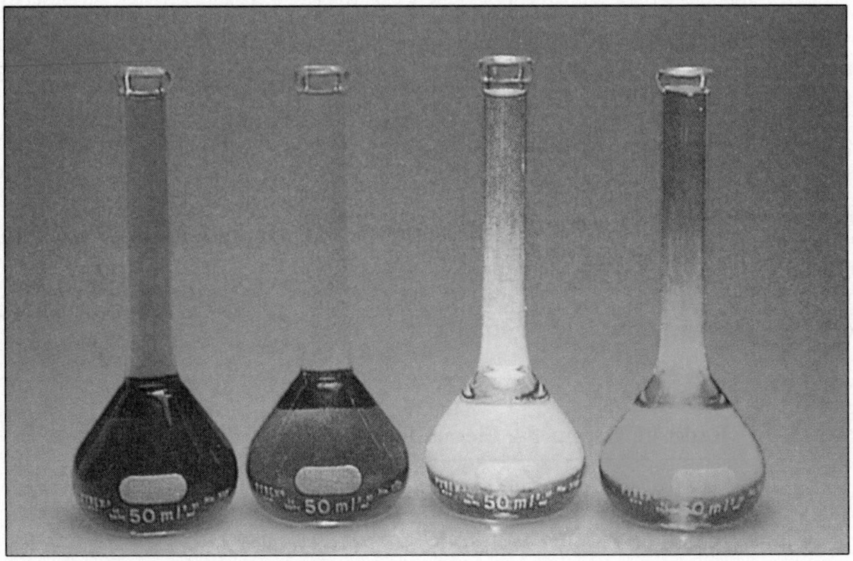

Figure 23.6 Some common compounds of chromium illustrating oxidation numbers +3 [in $Cr(NO_3)_3$ (violet) and $CrCl_3$(green)] and +6 [in K_2CrO_4 (yellow) and $K_2Cr_2O_7$ (orange)]. (C. D. Winters)

Metal Atom Radii

Radii of transition metals are determined by measuring the metal-metal distance in the solid using x-ray diffraction techniques (Section 13.4). Figure 8.11 shows the variation in atomic radii for the first 12 elements in the fourth, fifth, and sixth periods. A rapid decrease in size is seen among the first three elements in the period. In contrast, the radii of the transition elements vary over a fairly narrow range, dropping to a minimum around the middle of this group of elements and then rising slowly. This variation can be understood based on electron configurations. Primary differences in electron configurations of transition elements occur in the $(n - 1)d$ orbitals, which are somewhat smaller than the *ns* orbitals. Atom size, however, is determined by the radius of the outermost orbital, which for these elements is the *ns* orbital ($n = 4$, 5, or 6). As we progress from left to right in the periodic table, the effect caused by the increasing nuclear charge is mostly canceled by the increasing number of electrons in the $(n - 1)d$ orbitals. In the first half of the transition series the *ns* electrons experience a small increase in net nuclear attraction across the series and decrease slightly in size. The small rise in radius in the second half of this group of elements is due to the continually increasing electron-electron repulsions as the *d* subshell is completed.

The overall decrease in metal radii across the fourth-period transition metals has a noticeable effect on the radii of main group elements that follow (Table 23.3). Instead of the normal increase in size down a periodic group, we see that gallium is actually smaller than aluminum.

An interesting observation to be made from Figure 8.11 is that radii of the transition elements in the fifth and sixth periods are almost identical. The reason for this is that the lanthanide elements are inserted into the table just before the third series of transition elements. The filling of 4*f* orbitals through the lanthanide elements is accompanied by a steady contraction in size (not unlike the overall size decrease across each series of transition elements). At the point where the 5*d* orbitals begin to fill again, the radii have decreased to a size similar to that of elements in the previous period. Because this is such a noticeable effect with significant consequences, it is given a specific name, the **lanthanide contraction.**

The effect of the lanthanide contraction carries over into the *p* block as well. Thus, the increase in radius from indium to thallium is smaller than expected (see Table 23.3).

The similar radii of the transition metals and their ions affect their chemistry. For example, the "platinum group metals" (Ru, Os, Rh, Ir, Pd, and Pt) form similar compounds. Thus, it is not surprising that minerals containing these metals are found in the same geologic zones.

TABLE 23.3 Radii of Group 3A Elements

Element	Period	Radius (pm)
Boron	2	85
Aluminum	3	143
Gallium	4	135
Indium	5	167
Thallium	6	170

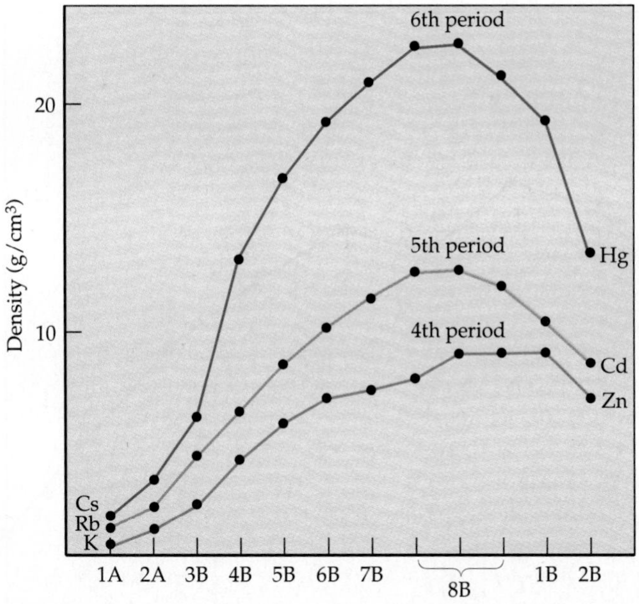

Figure 23.7 Densities of the transition metals (and the preceding *s*-block elements) as a function of periodic group.

Density

A consequence of the variation in metal radii is that the densities of the metals first increase and then decrease across a period (Figure 23.7). This variation reflects the slow increase in atomic mass and the change in volume. Although the overall change in radii among these elements is small, the effect is magnified because the volume is actually changing with the cube of the radius ($V = \frac{4}{3}\pi r^3$).

The lanthanide contraction is the reason that elements in the sixth period have the highest density. The relatively small radii of sixth-period transition metals, combined with the fact that their atomic mass is considerably larger than their counterparts in the fifth period, causes sixth-period metal densities to be very large.

Melting Point

The periodic variation of melting points of the transition elements is illustrated in Figure 23.8. The metals with the highest melting points occur in the middle of each series.

The melting point of any substance reflects the forces of attraction between elementary particles. With metals, we are dealing with forces of attraction between atoms of metals, which we could also call metallic bonding (Section 10.3). Again, electron configurations provide us with an explanation. The variation in melting point indicates the strongest metallic bonds occur when the *d* subshell is about half-filled. This is also the point at which the largest number of unpaired electrons occurs in the isolated atoms. We can conclude that the *d* electrons are playing an important role in metallic bonding in the solid metal.

Magnetism

Between zero and five unpaired electrons occur in the *d* orbitals of transition metal ions and up to 7 unpaired electrons are in *f* orbitals of the lanthanide and actinide metal ions in their compounds. If a species contains unpaired electrons,

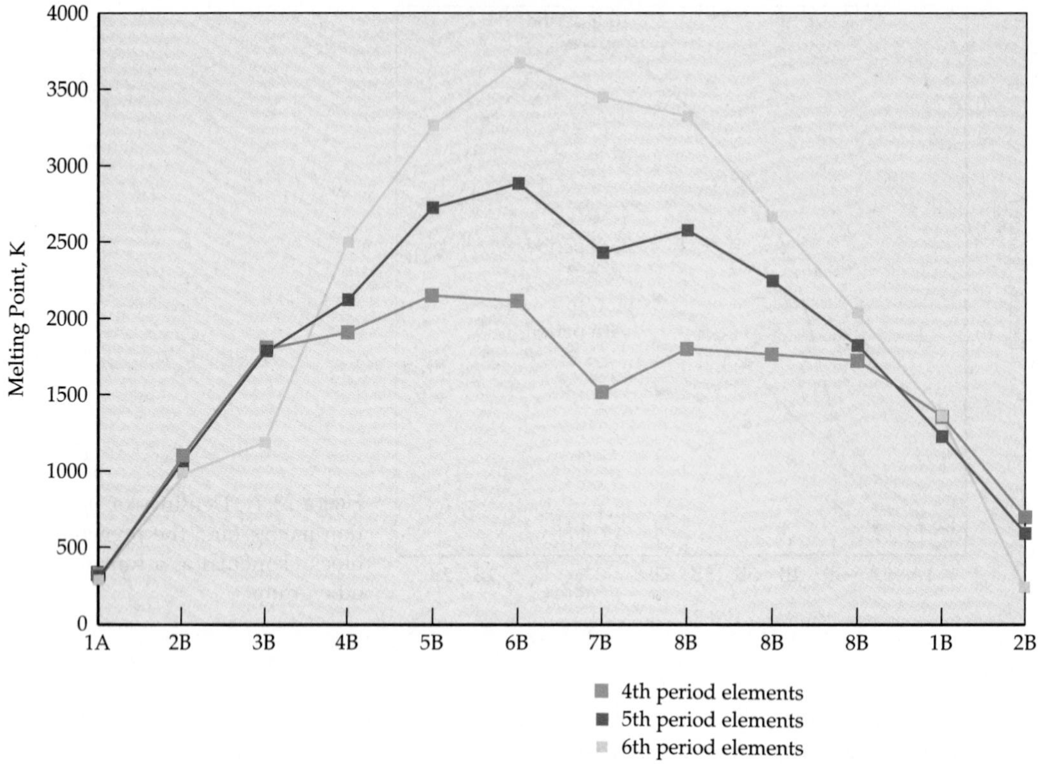

Figure 23.8 Melting points of the transition metals as a function of periodic group. ■ fourth-period elements; □ fifth-period elements; ◆ sixth-period elements.

it is **paramagnetic** and attracted by a magnet (Section 8.1). As we will discover in Section 23.5, the paramagnetism of transition metal compounds is an important feature that must be accounted for in a description of the species.

Recall that unpaired electrons can act as subatomic bar magnets. If, in a solid metal, the atomic magnets of a group of atoms (called a domain) interact so that their magnets are all oriented in the same direction, the magnitude of the magnetic effect is much larger than paramagnetism. We call this effect **ferromagnetism** (see page 361). Iron, cobalt, and nickel metals are ferromagnetic. Ferromagnetic materials are also unique in that, once the electron magnets are aligned by an external magnetic field, the metal is permanently magnetized. In such a case, the magnetism can be eliminated only by heating or vibrating the metal to rearrange the electron spin domains.

Antiferromagnetism occurs when electron spins on adjacent atoms in a domain are in opposite directions.

23.2 COMMERCIAL PRODUCTION OF TRANSITION METALS

Most metals are found in nature as oxides, sulfides, halides, carbonates, or other salts. Many of the metal-containing mineral deposits are of little value, either because the deposit has too low a concentration of the desired metal or because the metal is difficult to separate from impurities. The relatively few minerals from which elements can be obtained profitably are called ores, which are listed on page 169.

Very few ores are chemically pure substances. They are usually mixtures of the desired mineral and large quantities of impurities such as sand and clay, called **gangue** (pronounced "gang"). A major step in obtaining the desired metal must therefore separate its mineral from the gangue. The second major step involves converting the ore to the metal. Pyrometallurgy and hydrometallurgy are two methods of recovering metals from their ores. **Pyrometallurgy,** a high-temperature method, is illustrated by iron production. **Hydrometallurgy** uses aqueous solutions at relatively low temperatures for the extraction of metals, such as copper, zinc, tungsten, and gold.

Iron Production

The production of iron from its ores involves oxidation-reduction reactions carried out in a blast furnace (Figure 23.9). The furnace is charged at the top with a mixture of ore (usually hematite, Fe_2O_3), coke (which is primarily carbon), and limestone ($CaCO_3$), and a blast of hot air is forced in at the bottom. The coke burns with such an intense heat that the temperature at the bottom is almost

In 1993, production of steel in the United States was 98 million tons of raw steel and 53 million tons of iron.

Coke is made by heating coal in a tall narrow oven that is sealed to keep out oxygen. Heating drives off a number of volatile byproducts, such as benzene and ammonia, leaving nearly pure carbon.

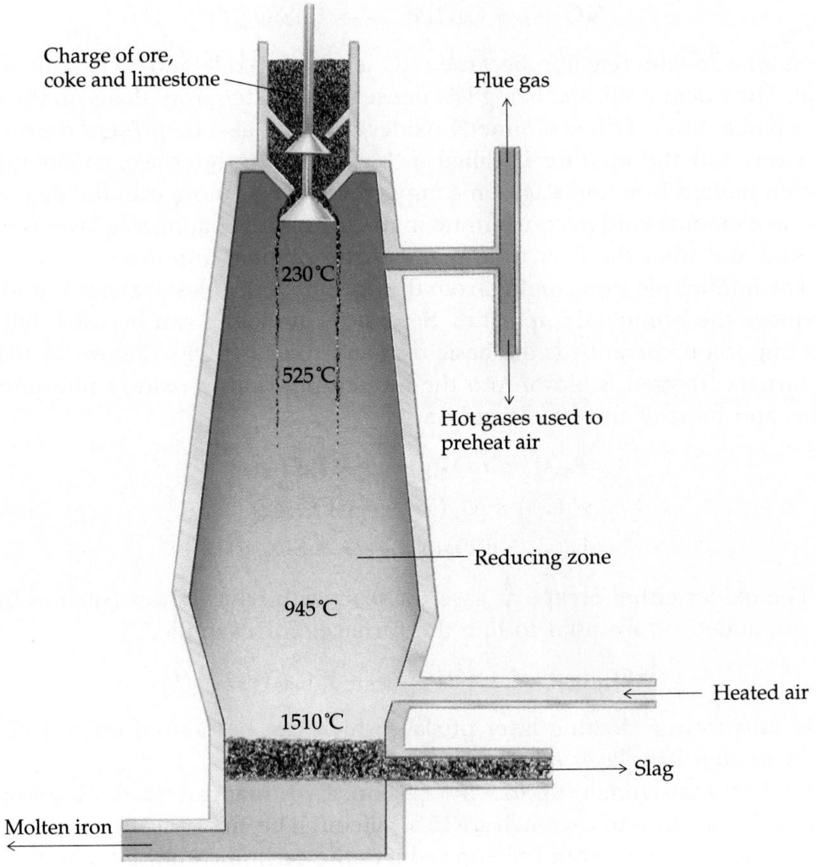

Figure 23.9 A blast furnace used for the reduction of iron ore to iron. The largest modern furnaces have hearths 14 m in diameter and can produce up to 10,000 tons of iron per day.

2000 °C, and a temperature of about 200 °C is attained at the top of the furnace. The quantity of oxygen is controlled so that carbon monoxide is the primary product.

$$2\ C(s, coke) + O_2(g) \longrightarrow 2\ CO(g) + heat$$

Both carbon and carbon monoxide participate in the reduction of iron(III) oxide to give impure metal.

$$Fe_2O_3(s) + 3\ CO(g) \longrightarrow 2\ Fe(\ell) + 3\ CO_2(g)$$
$$Fe_2O_3(s) + 3\ C(s) \longrightarrow 2\ Fe(\ell) + 3\ CO(g)$$

Much of the carbon dioxide formed in the reduction process (and from heating limestone) is itself reduced on contact with unburned coke and produces more reducing agent.

$$CO_2(g) + C(s) \longrightarrow 2\ CO(g)$$

Production of 1 ton of pig iron requires about 1.7 tons of iron ore, 0.5 ton of coke, 0.25 ton of limestone, and 2 tons of air.

The molten iron flows down through the furnace and collects at the bottom, where it can be tapped off through an opening in the side. When cooled, the impure iron is called "cast iron" or "pig iron." The material is brittle and soft (rather undesirable properties for most uses), due to the presence of small amounts of impurities, such as elemental carbon, phosphorus, and sulfur.

Iron ores generally contain silicate minerals and silicon dioxide. Lime, formed from added limestone, reacts with these materials to give calcium silicate, which is molten at the temperature of the blast furnace.

$$SiO_2(s) + CaO(s) \longrightarrow CaSiO_3(\ell)$$

This is an acid-base reaction because CaO is a basic oxide and SiO_2 is an acidic oxide. The calcium silicate, being less dense than molten iron, floats on the iron as a separate layer. Other nonmetal oxides that may also be present dissolve in this layer, and the mixture is called a "slag." At the interface, or boundary, between molten iron and slag, ionic impurities tend to move into the slag layer, whereas elements tend to concentrate in the iron. The floating slag layer is easily removed and frees the iron from many of the original impurities.

The impure pig iron coming from the bottom of the blast furnace is purified to remove the nonmetal impurities. Several technologies can be used, but the most important currently is the **basic oxygen furnace** (BOF) (Figure 23.10). In this furnace, oxygen is blown into the molten pig iron to oxidize phosphorus, sulfur, and most of the excess carbon.

$$P_4(s) + 5\ O_2(g) \longrightarrow P_4O_{10}(g)$$
$$C(s) + O_2(g) \longrightarrow CO_2(g)$$
$$S_8(s) + 8\ O_2(g) \longrightarrow 8\ SO_2(g)$$

The oxides either escape as gases or react with basic oxides (such as CaO) that are added or are used to line the furnace. For example,

$$P_4O_{10}(g) + 6\ CaO(s) \longrightarrow 2\ Ca_3(PO_4)_2(\ell)$$

These salts form a floating layer of slag, which can be poured off to free the denser molten iron layer of impurities.

Pig iron may contain up to 4.5% carbon, 1.7% manganese, 0.3% phosphorus, 0.04% sulfur, and as much as 15% silicon. The final purification step removes much of the P, S, and Si and reduces the carbon content to about 1.3%.

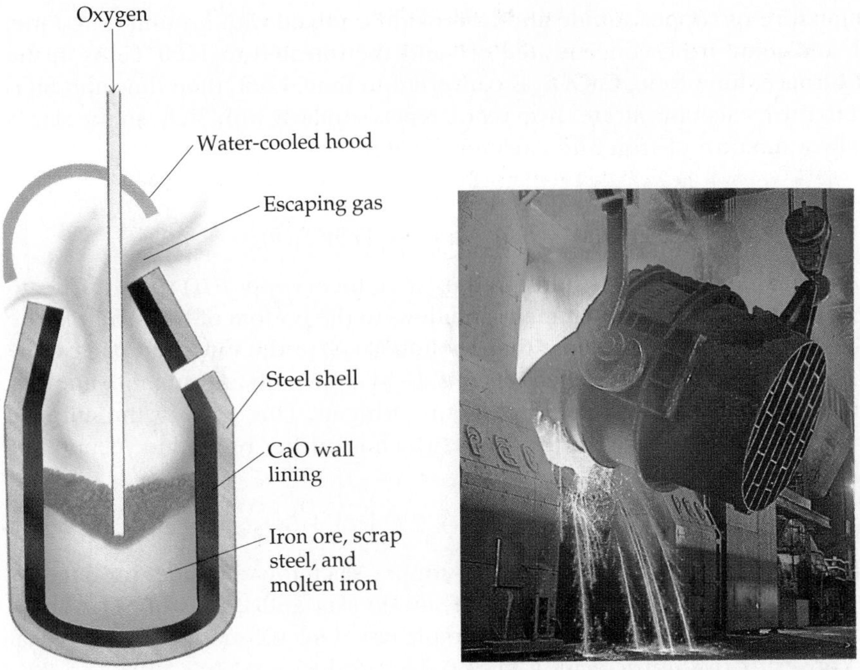

(a) (b)

Figure 23.10 A "basic oxygen furnace." (a) Much of the steel produced today is made by blowing oxygen through a furnace charged with scrap and molten iron from a blast furnace. Measured amounts of alloying elements determine the particular steel produced. (b) Molten steel being poured from a basic oxygen furnace. (b, Courtesy Bethlehem Steel Corporation)

The result is ordinary *carbon steel.* Almost any degree of flexibility, hardness, strength, and malleability can be achieved in carbon steel by proper cooling, reheating, and tempering. The drawbacks to this material, however, are that it corrodes easily and loses its properties when heated strongly.

During the processing of steel, other transition metals, such as chromium, manganese, and nickel, can be added to produce alloys that have specific physical, chemical, and mechanical properties. An important type of steel is *stainless steel,* which typically contains 18% to 20% Cr and 8% to 12% Ni. Stainless steel's usefulness results from its resistance to corrosion. Another alloy of iron of interest is *alnico V.* This alloy, used in loudspeaker magnets because of its permanent magnetism, contains five elements: Al (8%), Ni (14%), Co (24%), Cu (3%), and Fe (51%).

Native copper and two minerals of copper. Azurite [2 $CuCO_3 \cdot Cu(OH)_2$] is blue, and malachite [$CuCO_3 \cdot Cu(OH)_2$] is green. (C. D. Winters)

Copper Production

Copper-bearing ores include chalcopyrite ($CuFeS_2$), chalcocite (Cu_2S), and covellite (CuS). Because the ores generally have a low percentage of copper, enrichment is necessary. This is carried out by a process known as flotation. First, the ore is finely powdered. Oil is then added and the mixture agitated with soapy water in a large tank (Figure 1.18). At the same time, compressed air is forced through the mixture, and the lightweight, oil-covered copper sulfide particles are carried to the top as a frothy mixture. The heavier gangue settles to the bottom of the tank, and the copper-laden froth is skimmed off.

In the pyrometallurgy of copper, the enriched ore is **roasted** with enough air to convert any iron to its oxide while leaving copper sulfide.

$$2\ CuFeS_2(s) + 3\ O_2(g) \longrightarrow 2\ CuS(s) + 2\ FeO(s) + 2\ SO_2(g)$$

This mixture of copper sulfide and iron oxide is mixed with ground limestone, sand, and some fresh concentrated ore and then heated to 1100 °C. As in the blast furnace, limestone, $CaCO_3$, is converted to lime, CaO; then lime and SiO_2 react to form calcium silicate. Iron oxide reacts similarly with SiO_2, so the slag is actually a mixture of iron and calcium silicates.

$$CaO(s) + SiO_2(s) \longrightarrow CaSiO_3(\ell)$$
$$FeO(s) + SiO_2(s) \longrightarrow FeSiO_3(\ell)$$

At the same time, excess sulfur in the ore reduces copper(II) sulfide, CuS, to copper(I) sulfide, Cu_2S, which melts and flows to the bottom of the furnace. The iron-containing slag is less dense than molten Cu_2S, so the Cu_2S and slag can be separated easily. The Cu_2S, called *copper matte,* is tapped off and run into another furnace (the converter), where it is "blown" with air. This converts the sulfur to SO_2 and produces impure copper metal, which is further refined in yet another furnace.

$$Cu_2S(\ell) + O_2(g) \longrightarrow 2\ Cu(\ell) + SO_2(g)$$

In the pyrometallurgical recovery of copper, each ton of copper produced is accompanied by about 1.5 tons of iron silicate slag and 2 tons of SO_2. These byproducts must be disposed of, not a simple task. One solution for SO_2 disposal is to convert it to sulfuric acid, a marketable product.

Hydrometallurgy avoids some of the energy costs and pollution problems of pyrometallurgy. One method of copper recovery now in use in Arizona leaches, or dissolves, the impurities from the ore by treating the ore with a solution of copper(II) chloride and iron(III) chloride.

$$CuFeS_2(s) + 3\ CuCl_2(aq) \longrightarrow 4\ CuCl(s) + FeCl_2(aq) + 2\ S(s)$$
$$CuFeS_2(s) + 3\ FeCl_3(aq) \longrightarrow CuCl(s) + 4\ FeCl_2(aq) + 2\ S(s)$$

Figure 23.11 An open pit copper mine near Bagdad, Arizona. Additional copper can be extracted from the waste rock by bacteria. (James Cowlin/Image Enterprises)

Copper is obtained in the form of copper(I) chloride by this reaction. To return copper to solution, sodium chloride is added because the soluble complex ion $[CuCl_2]^-$ is formed in the presence of excess chloride ion.

$$CuCl(s) + Cl^-(aq) \longrightarrow [CuCl_2]^-(aq)$$

Copper(I) compounds are unstable with respect to Cu(0) and Cu(II), so the $[CuCl_2]^-$ ion disproportionates to the metal and $CuCl_2$; the latter is used to continue the leaching process.

$$2\ [CuCl_2]^-(aq) \longrightarrow Cu(s) + CuCl_2(aq) + 2\ Cl^-(aq)$$

Approximately 10% of the copper produced in the United States is actually obtained by using bacteria. Acidified water is sprayed onto copper-mining wastes that contain low levels of copper (Figure 23.11). As the water trickles down through the crushed rock, the bacterium *Thiobacillus ferrooxidans,* which thrives in the presence of acid and sulfur, breaks down the iron sulfides in the rock and converts iron(II) to iron(III). The iron(III) ion in turn oxidizes the sulfide ion of copper sulfide, leaving copper(II) ion in the water. Then the copper(II) ion is reduced to metallic copper by reaction with iron.

$$Cu^{2+}(aq) + Fe(s) \longrightarrow Cu(s) + Fe^{2+}(aq)$$

Whatever method is used to recover copper from its ores, the final step in the refining process is purification by electrolysis. Thin sheets of pure copper metal and slabs of impure copper are immersed in a solution containing $CuSO_4$ and H_2SO_4 (Figure 23.12). The pure copper sheets are the cathode of an electrolysis cell, and the impure slabs are the anode. Copper is oxidized to copper(II) ions at the anode; at the cathode, copper(II) ions in solution are reduced to pure copper.

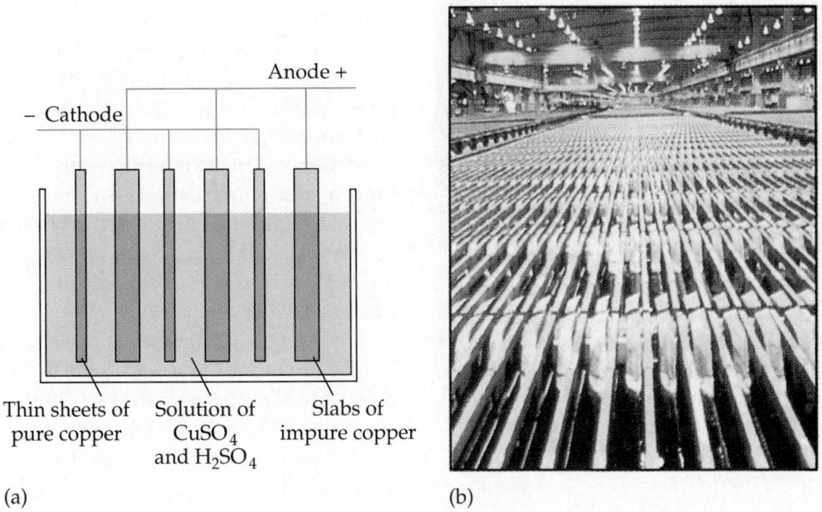

Figure 23.12 Electrolytic refining of copper. (a) Slabs of impure copper, called "blister copper," form the anode and pure copper is deposited at the cathode. (b) Electrolysis cells for refining copper.

CURRENT ISSUES IN CHEMISTRY

Metal with a Memory

In the early 1960s, William J. Buehler, a metallurgical engineer at the Naval Ordinance Laboratory in White Oak, Maryland, was experimenting with binary alloys, that is, alloys made up of two metals. He was looking for a material that was resistant to impact and heat because it was to be used in the nose cone of a Navy missile. It was also important that the material be fatigue-resistant, that is, it should not lose its desirable properties when heated or handled. An alloy of nickel and titanium appeared to have some desirable properties, so Buehler prepared long, thin strips of this alloy to demonstrate that it could be folded and unfolded many times without breaking. At a meeting to discuss this material, a colleague wanted to see what happened when the strip was heated. He held a pipe lighter to a folded-up piece of metal and was amazed to observe that the metal strip immediately unfolded to its original shape. Thus, memory metal was discovered. This unusual alloy is now called nitinol, a name constructed out of nickel, titanium, Naval Ordinance Lab.

The shape that NiTi "remembers" is established by heating the alloy to between 500 and 550 °C for about an hour and then allowing it to cool. At the low temperature, the alloy is fairly soft and may be bent and twisted out of shape. When warmed, the metal returns to its original shape. The temperature at which the change in shape occurs varies along with small differences in the Ni-Ti ratio. Depending on composition, materials that change shape at temperatures ranging from −125 °C to about 70 °C are possible, greatly increasing the number of possible uses for this intriguing material.

Memory metal never made it into missile nose cones, but it has found a wide variety of other uses, and some of the most interesting are in medicine. For example, bone anchors can be made of nitinol; on warming to body temperature the alloy expands and locks the anchor in place. It is possible to thread a filament of nitinol into a vein and have the filament rearrange itself to form a fine-mesh screen to filter out blood clots. Nitinol can also be used in orthodontics; braces made of nitinol remember their shape and apply a steady constant pressure to move teeth into position. Eyeglass frames of nitinol can be twisted into odd shapes only to return to their original shape when warmed to body temperature.

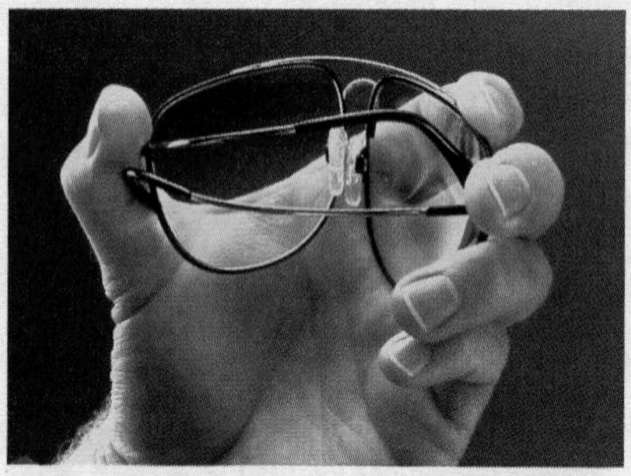

These sunglass frames are made of nitinol, so they snap back to the proper fit even after being twisted like a pretzel. The nitinol used in these frames has a critical temperature below room temperature, so the metal readily returns to its "memorized" shape. Similar alloys are used for wires in dental braces and surgical anchors, which cannot be heated after insertion. (NASA/Science Source/Photo Researchers, Inc.)

What is this remarkable alloy and how does it work? When heated above 500 °C the alloy crystallizes in a structure with eight atoms of nickel at the corners of a cube and a titanium atom in the center. Extended in three dimensions, the crystal creates interpenetrating cubic lattices of nickel and titanium atoms.

The key to understanding the behavior of this material is knowing that solid nitinol undergoes a transition between two solid phases. Solid phases differ in important ways at the atomic level. For example, a solid-solid phase transition you have heard about is the change between the diamond and graphite phases of carbon.

At higher temperatures, nitinol exists in what is called the austenite phase, but at temperatures lower than the transition temperature the atomic arrangement distorts slightly as the material shifts to the martensite phase. There is no visible change in the shape of the metal when the phase change occurs; the change occurs entirely at the atomic level. The distortion of the atomic structure, however, imposes a strain on the system. Therefore, when heated, the process reverses and the metal atoms move back into their original positions to relieve the strain.

For more information, see G. B. Kauffman and I. Mayo: "Memory metal." *Chem Matters,* October, 1993, p. 4; and "The metal with a memory." *Invention and Technology,* Fall, 1993, p. 18.

23.3 COORDINATION COMPOUNDS

When a metal salt dissolves in water, water molecules cluster around the ions. We commonly designate this by adding "aq" to the formula of these ions. The negative end of the polar water molecule is attracted to the positively charged metal ion (Figure 17.12), and the positive end of the water molecule is attracted to the anion. Indeed, the energy of the ion-solvent interaction is what drives the solution process.

When salts crystallize from aqueous solution, the compound often retains some number of water molecules. Chemists indicate this by appending the appropriate number of water molecules to the formula for the salt. For example, hydrated iron(II) chloride is written $FeCl_2 \cdot 6\ H_2O$. This formula indicates that six water molecules per $FeCl_2$ unit are somehow involved in the crystalline structure of solid iron(II) chloride.

Species in which a metal cation is associated with a number of anions were encountered earlier in the book. One example is the ferrocyanide ion, $[Fe(CN)_6]^{4-}$, that exists in solutions containing Fe^{2+} and CN^-, in salts such as potassium ferrocyanide, $K_4Fe(CN)_6$, and in the pigment Prussian blue (see Figure 21.16). In the metallurgy of copper, the solubility of copper(I) chloride in aqueous NaCl was due to formation of a complex ion $[CuCl_2]^-$.

Hydrates and complex ions are examples of a very large group of substances known as **coordination compounds,** which are species in which a metal ion is associated with a group of neutral molecules or anions. They may be neutral molecules, but species with an overall positive and negative charge are also included in this category. Coordination compounds are the subject of this section.

Figure 23.13 Coordination compounds. Top: blue $Cu(NH_3)_4SO_4 \cdot H_2O$ and green $Cu[(CH_3)_2SO]_2Cl_2$. Bottom: green $K_3[Fe(C_2O_4)_3]$, red $[Co(NH_3)_4CO_3]NO_3$, and yellow $[Co(NH_3)_5(NO_2)]Cl_2$. (C. D. Winters)

All metals form coordination compounds. Because there are more transition metals than main group metals, and because coordination compounds of these elements are particularly interesting, this subject is generally considered as part of transition metal chemistry.

Complexes and Ligands

A formula like $FeCl_2 \cdot 6\ H_2O$ identifies the compound's stoichiometry but fails to give information about its structure. The preferred method of writing the formula for a coordination compound places the metal atom or ion and the molecules or anions directly bonded to the metal within brackets. Thus, it is actually preferable to write the formula for this compound as $[Fe(H_2O)_6]Cl_2$, to show that the iron(II) ion and six water molecules are a single structural unit.

The part of the formula of $[Fe(H_2O)_6]Cl_2$ within the brackets is a cation, $[Fe(H_2O)_6]^{2+}$. The two chloride ions are not part of this structural unit but are present as counterions to balance the charge in this ionic compound. The species $[Fe(H_2O)_6]^{2+}$ is called a **coordination complex,** or **complex ion.** Coordination complexes can be cations as in this example, anions (like $[Fe(CN)_6]^{4-}$), or neutral species (Figure 23.13).

The molecules or ions attached to the metal ion are called **ligands,** from the Latin verb *ligare,* meaning "to bind." Ligands have at least one atom that has a lone pair of electrons, and this lone pair gives a ligand the ability to bond to the metal. The classical description of bonding in coordination complexes is that the lone pair of electrons on the ligand is shared with the metal ion. Therefore, a ligand is a *Lewis base* because it furnishes the electron pair, and the metal ion is a *Lewis acid* (see Section 17.11). The bond between ligand and metal is a Lewis acid-Lewis base interaction, as in the bond between a nickel(II) ion and an ammonia molecule.

$$Ni^{2+} + :NH_3 \longrightarrow [Ni \leftarrow NH_3]^{2+}$$

The classical name given to a bond in which one of the atoms formally contributes the pair of electrons is "coordinate covalent bond." The term "coordination complex" derives from this.

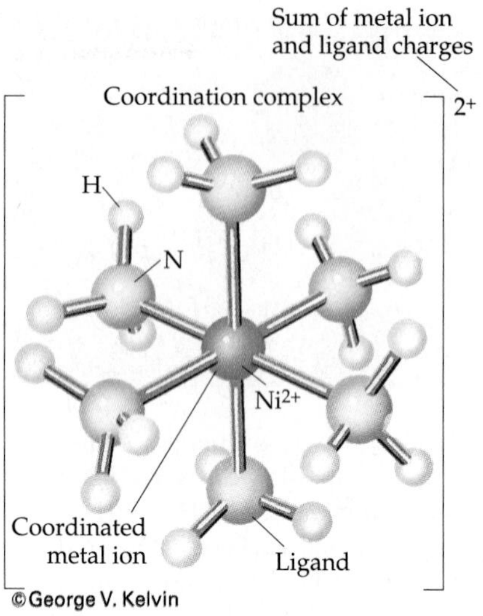

$[Ni(NH_3)_6]^{2+}$

The number of ligand atoms attached to a metal is called the **coordination number.** Coordination complexes also have a definite geometry or structure. The nickel ion in the complex $[Ni(NH_3)_6]^{2+}$, drawn here, has a coordination number of six, and the six ligands are in a regular octahedral geometry around the central metal ion.

Ligands like H_2O and NH_3, with only a single Lewis base atom, are termed monodentate. The word "dentate" stems from the Latin word *dentis* meaning "tooth," so NH_3 is a "one-toothed" ligand. Other ligands, however, have more than one donor atom. When several atoms separate the Lewis base sites, it is possible for two or more atoms in the same ligand to bind to one metal atom. Ligands that contain two or more atoms attached to the metal are called **polydentate ligands.** Ethylenediamine (1,2-diaminoethane, $H_2NCH_2CH_2NH_2$, often abbreviated as *en*) and oxalate ion, $C_2O_4^{2-}$, are examples of common **bidentate** ligands (Figure 23.14). Ethylenediaminetetraacetate ion ($EDTA^{4-}$) is an exam-

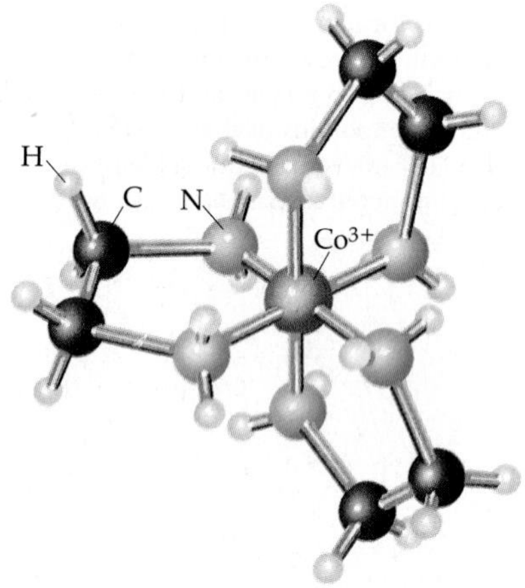

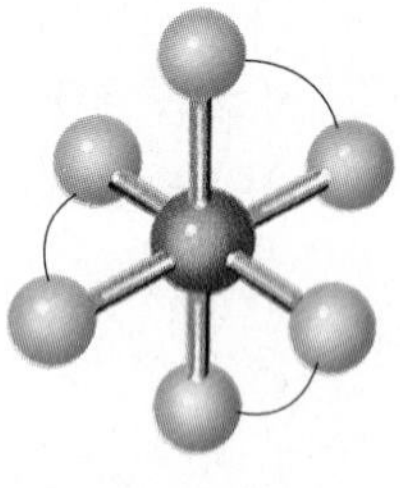

Figure 23.14 The structure of $[Co(en)_3]^{3+}$, a complex formed from three bidentate ethylenediamine ligands and the Co^{3+} ion. Structures of complexes with bidentate ligands are often drawn schematically, as in the drawing at the bottom.

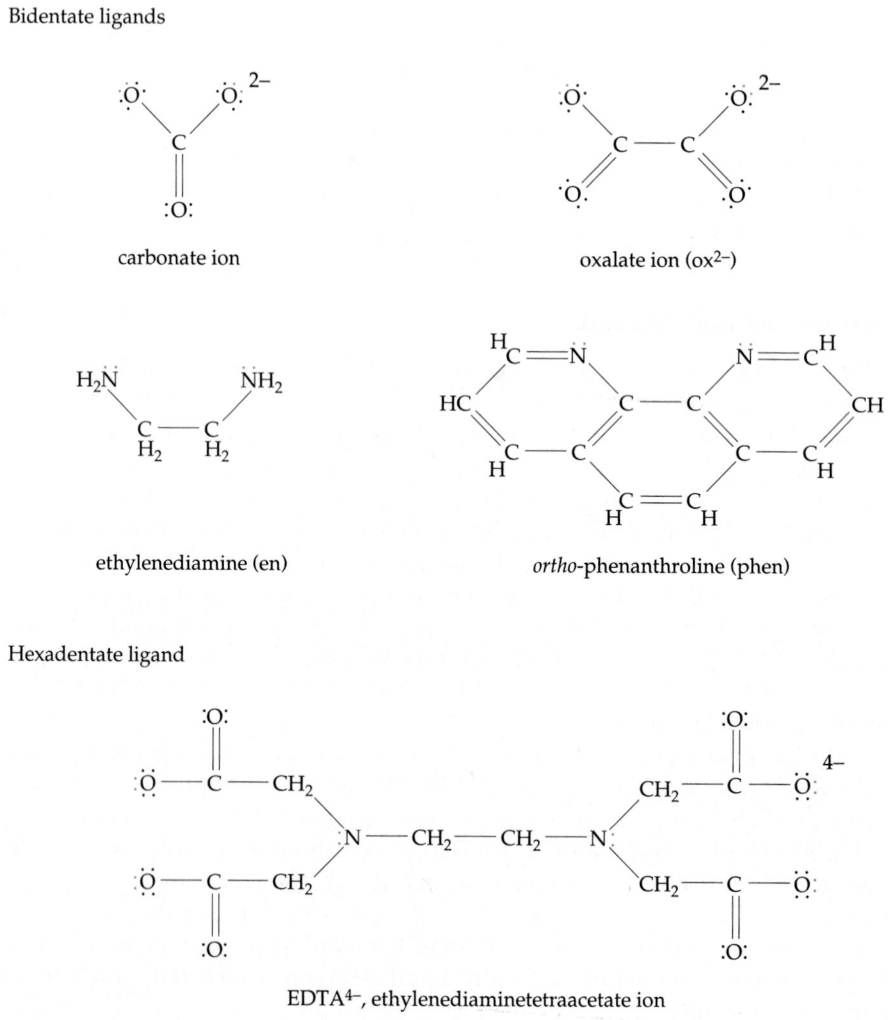

Figure 23.15 Some common polydentate ligands.

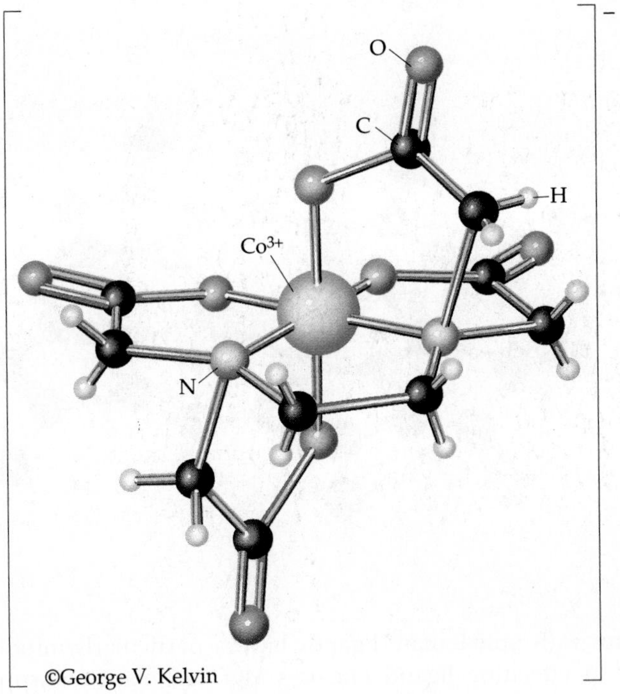

Figure 23.16 The structure of $[Co(EDTA)]^-$.

ple of a **hexadentate** ligand (Figure 23.16). Several other polydentate ligands are illustrated in Figure 23.15.

An important structural characteristic of a molecule or ion that allows it to act as a polydentate ligand is its ability to form five- and six-membered rings. Ethylenediamine bonds to a metal to form a five-membered ring with two carbon atoms, two nitrogen atoms, and the metal ion. Structures containing five-membered and six-membered rings allow the atoms to achieve their normal bond angles and distances.

Polydentate ligands are also called **chelating ligands,** or chelates (pronounced ''key-late''). The name derives from the Greek word *chele* meaning ''claw.'' Because two bonds must be broken to separate a ligand from the metal, complexes having chelating ligands are extra stable. Chelated complexes are important in everyday life. One way to clean the rust out of water-cooled automobile engines and steam boilers, for example, is to add a solution of oxalic acid. Iron oxide dissolves in the presence of the acid to give the water-soluble iron oxalate complex.

$$Fe_2O_3(s) + 6\ H_2C_2O_4(aq) + 3\ H_2O(\ell) \longrightarrow 2\ [Fe(C_2O_4)_3]^{3-}(aq) + 6\ H_3O^+(aq)$$

$EDTA^{4-}$ is an excellent chelating ligand; it encapsulates and firmly binds metal ions. It is often added, for example, to commercial salad dressing to remove traces of metal ions from solution because these metal ions can otherwise act as catalysts for the oxidation of the oils in the product. Without $EDTA^{4-}$, the dressing would quickly become rancid. Another use is in bathroom cleansers, in which $EDTA^{4-}$ removes deposits of $CaCO_3$ and $MgCO_3$ left by hard water. The $EDTA^{4-}$ coordinates to Ca^{2+} or Mg^{2+}, creating a soluble complex ion. Figure 23.16(a) shows the structure of $[Co(EDTA)]^-$. Note how the ligand is oriented to allow carbon, nitrogen, oxygen, and cobalt atoms to achieve the desired bond angles.

Ligands that are capable of being bidentate (or polydentate) ligands can act as monodentate ligands, coordinating through just one of the donor atoms. These ligands can also ''bridge'' two or more metal atoms.

A regular pentagon (five-sided figure) has 110° angles. This value is very close to the tetrahedral angles of 109.5° preferred by carbon and nitrogen.

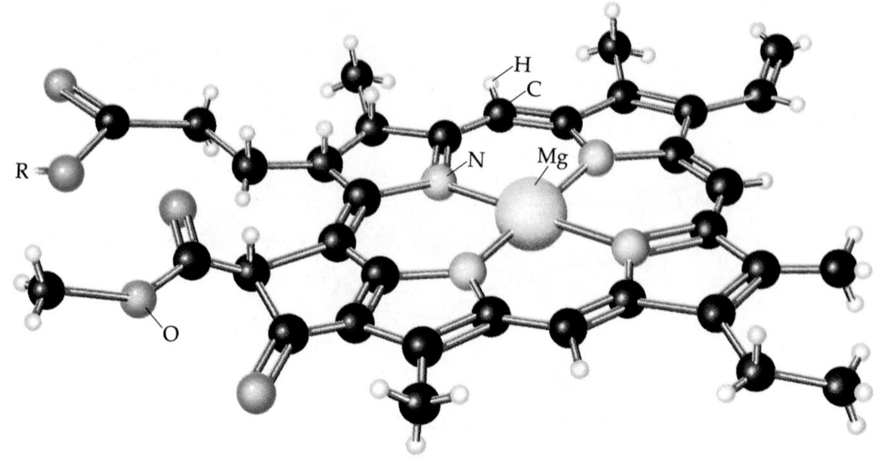

Figure 23.17 The structure of chlorophyll.

Complexes with polydentate ligands have a particularly important role in biochemistry. A chelating ligand encloses Mg^{2+} in plant chlorophyll (Figure 23.17). There are similar ligands in hemoglobin and myoglobin.

It is useful to be able to predict the formula of a coordination complex, given the metal ion and ligands, and to predict the oxidation number of the coordinate metal ion. The following examples explore these questions.

EXAMPLE 23.1 *Coordination Complexes*

Write the formula of the coordination complex in which the metal ion is coordinated to six Lewis base sites.

1. One Ni^{2+} ion is bound to two water molecules and two bidentate oxalate ions.
2. A Co^{3+} ion is bound to one Cl^- ion, one ammonia molecule, and two bidentate ethylenediamine molecules.

Solution

1. In the nickel(II) complex, the ligands consist of two neutral molecules and two oxalate with 2− charges. When these are combined with Ni^{2+}, the net charge is 2−.

$$Ni^{2+} + 2\ H_2O + 2\ C_2O_4^{2-} \longrightarrow [Ni(C_2O_4)_2(H_2O)_2]^{2-}$$

2. In the cobalt(III) complex there are two neutral en molecules, one neutral NH_3 molecule, and one Cl^- ion. When these are combined with Co^{3+}, the net charge is 2+.

$$Co^{3+} + 2\ H_2NCH_2CH_2NH_2 + NH_3 + Cl^- \longrightarrow [Co(H_2NCH_2CH_2NH_2)_2(NH_3)Cl]^{2+}$$

EXAMPLE 23.2 *Formulas of Coordination Compounds*

Give the oxidation number of the metal ion in each of the following complexes:

1. $[Co(en)_2(NO_2)_2]^+$
2. $Pt(NH_3)_2(C_2O_4)$

Solution

1. In this cobalt complex there are two neutral, bidentate ethylenediamine molecules and two nitrite ions, NO_2^-. Because the overall charge on the ion is 1+, the cobalt ion must be 3+.

$$Co^{3+} + 2\ H_2NCH_2CH_2NH_2 + 2\ NO_2^- \longrightarrow [Co(H_2NCH_2CH_2NH_2)_2(NO_2)_2]^+$$

2. Platinum is coordinated to two neutral ammonia molecules and one bidentate oxalate ion. Thus, platinum is in the form of the Pt^{2+} ion.

$$Pt^{2+} + 2\ NH_3 + C_2O_4^{2-} \longrightarrow Pt(NH_3)_2(C_2O_4)$$

EXERCISE 23.1 ***Formulas of Coordination Complexes***

1. Give the oxidation number of platinum in $Pt(NH_3)_2Cl_2$.
2. What is the formula of a complex assembled from one Co^{3+} ion, five ammonia molecules, and one monodentate carbonate ion? ■

Naming Coordination Compounds

Just as there are rules for naming simple inorganic and organic compounds, coordination compounds are named according to an established system. For example, the following compounds are named according to the rules outlined below.

Compound	Systematic Name
$[Ni(H_2O)_6]SO_4$	Hexaaquanickel(II) sulfate
$[Cr(en)_2(CN)_2]Cl$	Dicyanobis(ethylenediamine)chromium(III) chloride
$K[Pt(NH_3)Cl_3]$	Potassium amminetrichloroplatinate(II)

As you read through the rules, notice how they apply to the examples above:

1. In naming a coordination compound that is a salt, name the cation first and then the anion. (This is how all salts are commonly named).
2. When giving the name of the complex ion or molecule, name the ligands first, in alphabetical order, followed by the name of the metal.
   a. If a ligand is an anion whose name ends in *-ite* or *-ate,* the final *e* is changed to *o* (as in sulfate → sulfato or nitrite → nitrito).
   b. If the ligand is an anion whose name ends in *-ide,* the ending is changed to *o* (as in chloride → chloro or cyanide → cyano).
   c. If the ligand is a neutral molecule, its common name is usually used. The important exceptions to this rule are water, which is called *aqua,* ammonia, which is called *ammine,* and CO, called *carbonyl.*
   d. When there is more than one of a particular monodentate ligand with a simple name, the number of ligands is designated by the appropriate prefix: *di, tri, tetra, penta,* or *hexa.* If the ligand name is complicated (whether monodentate or bidentate), the prefix changes to *bis, tris, tetrakis, pentakis,* or *hexakis,* followed by the ligand name in parentheses.

3. If the complex ion is an anion, the suffix *-ate* is added to the metal name.
4. Following the name of the metal, the oxidation number of the metal is given in Roman numerals.

Complexes can be considerably more complicated than those described in this chapter; then, even more rules of nomenclature must be applied. The brief rules just outlined, however, are sufficient for the vast majority of complexes.

EXAMPLE 23.3 ***Naming Coordination Compounds***

Name the following compounds:

1. $[Cu(NH_3)_4]SO_4$
2. $K_2[CoCl_4]$
3. $Co(phen)_2Cl_2$
4. $[Co(en)_2(H_2O)Cl]Cl_2$

Solution

1. The sulfate ion has a 2− charge, so the complex ion has a 2+ charge (that is, $[Cu(NH_3)_4]^{2+}$). Because NH_3 is a neutral molecule, the copper ion is Cu^{2+}. The compound's name is therefore tetraamminecopper(II) sulfate.
2. Two K^+ ions occur in this compound, so the complex ion has a 2− charge ($[CoCl_4]^{2-}$). Because four Cl^- ions occur in the complex ion, the cobalt center is Co^{2+}. Thus, the name of the compound is potassium tetrachlorocobaltate(II).
3. This is a neutral compound. Because two Cl^- ions and two neutral phen (phenanthroline) ligands are bonded to a cobalt ion, the metal ion must be Co^{2+}. This means the compound name is dichlorobis(phenanthroline) cobalt(II).
4. Here the complex ion has a 2+ charge because it is associated with two uncoordinated Cl^- ions. The cobalt ion must be Co^{3+} because it is bonded to two neutral en (ethylenediamine) ligands, one neutral water, and one Cl^-. The name is aquachlorobis(ethylenediamine)cobalt(III) chloride.

EXERCISE 23.2 ***Naming Coordination Compounds***

Name the compounds $[Ru(phen)_2(H_2O)CN]Cl$ and $Pt(NH_3)_2Cl_2$. ■

23.4 STRUCTURES OF COORDINATION COMPOUNDS AND ISOMERS

Common Geometries

The geometry of a coordination complex is defined by the arrangement of donor atoms of the ligands around the central metal ion. The metal ion in a coordination compound may have a coordination number between 2 and 12. Only complexes with coordination numbers of two, four, and six are very common, however, so we concentrate on species such as $[ML_2]^{n\pm}$, $[ML_4]^{n\pm}$, and $[ML_6]^{n\pm}$, where M is the metal ion and L is a monodentate ligand.

Complexes with the General Formula $[ML_2]^{n\pm}$

This stoichiometry is often encountered with metal ions with a 1+ charge. One example, the copper complex $[CuCl_2]^-$, was mentioned earlier in this chapter. Another example is the complex ion that forms when AgCl(s) dissolves in aqueous ammonia (Section 19.9).

$$AgCl(s) + 2\ NH_3(aq) \longrightarrow [Ag(NH_3)_2]^+(aq) + Cl^-(aq)$$

In all cases, complexes of this stoichiometry have a linear geometry, that is, the two ligands are on opposite sides of the metal, with an L—M—L bond angle of 180°.

Complexes with the General Formula $[ML_4]^{n\pm}$

The VSEPR theory of Chapter 9 might lead us to expect tetrahedral structures for $[ML_4]^{n\pm}$ complexes. Indeed, this is observed for such complexes as $TiCl_4$, $[CoCl_4]^{2-}$, $[NiCl_4]^{2-}$, and $[Zn(NH_3)_4]^{2+}$. For a tetrahedral complex, L—M—L bond angles are 109.5°. A large number of $[ML_4]^{n\pm}$ complexes, however, are *square-planar;* the four ligands lie in a plane surrounding the metal atom in the center with L—M—L bond angles of 90°. Square-planar complexes are particularly common with metal ions that have the electron configuration [noble gas] $(n - 1)d^8$. This includes many complexes of Pt^{2+} and Pd^{2+} and some of Ni^{2+}, such as $[Ni(CN)_4]^{2-}$.

Complexes with the General Formula $[ML_6]^{n\pm}$

With *very* rare exceptions, the six ligands in an $[ML_6]^{n\pm}$ complex are arranged at the corners of an octahedron, with the metal at the center, as in $[Ni(NH_3)_6]^{2+}$ (page 1056). The iron(II) complexes $[Fe(H_2O)_6]^{2+}$ and $[Fe(CN)_6]^{4-}$ mentioned earlier in this chapter also have octahedral geometry.

The examples given above were illustrated using complexes with monodentate ligands. The structural principles apply equally well to complexes with polydentate ligands, however. The structural geometry is defined by the metal and the donor atoms attached to it.

Isomerism

Isomerism is one of the most interesting aspects of molecular structure. Molecules that have the same molecular formula but different bonding arrangements of atoms are called **structural isomers.** We already encountered structural isomers in organic chemistry (Section 11.1). Compounds like butane and 2-methylpropane are structural isomers (page 499). In a second type of isomerism, **stereoisomerism,** the atom-to-atom bonding sequence is the same, but the atoms differ in their arrangement in space. Two types of stereoisomerism occur. One is **geometric isomerism,** in which the atoms making up a molecule are arranged in different geometrical relationships (Section 11.2). *Cis*-2-butene and *trans*-2-butene are geometric isomers (page 504). The second type of stereoisomerism is **optical isomerism,** which arises when a molecule and its mirror image do not superimpose. Both geometric and optical isomerism are encountered in coordination chemistry.

Geometric Isomerism

Geometric isomers result when the atoms bonded *directly* to the metal have a different spatial arrangement (different bond angles). The simplest example of geometric isomerism is *cis-trans* isomerism, which occurs in both square-planar and octahedral complexes. We can illustrate *cis-trans* isomerism with the square-planar complex $Pt(NH_3)_2Cl_2$. This complex is formed from Pt^{2+}, two NH_3 molecules, and two Cl^- ions. The two Cl^- ions, for example, can be either adjacent to each other (*cis*) or on opposite sides of the complex (*trans*). The *cis* isomer is effective in the treatment of testicular, ovarian, bladder, and osteogenic sarcoma cancers, but the *trans* isomer has no effect on these diseases.

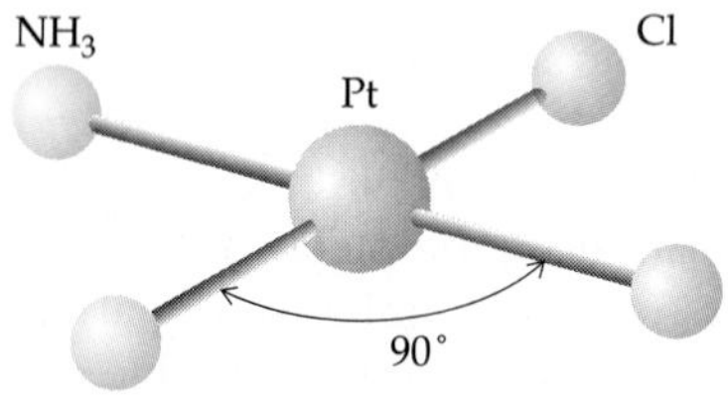

cis

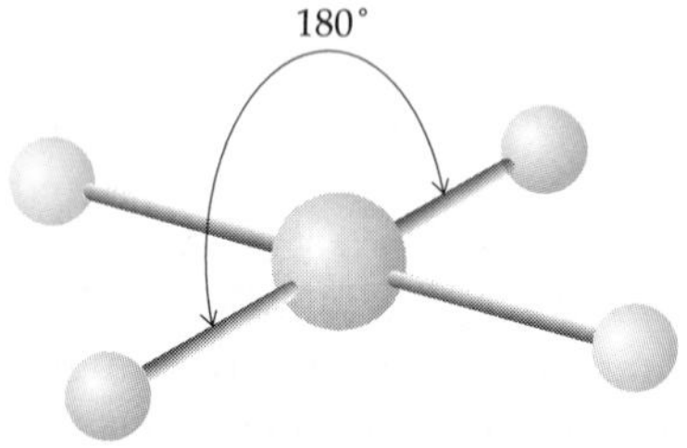

trans

Cis and *trans* isomers also arise with $Pt(NH_3)_2(Cl)(NO_2)$; the two ammonia molecules can either be in adjacent positions (at 90°) or at opposite positions (180°).

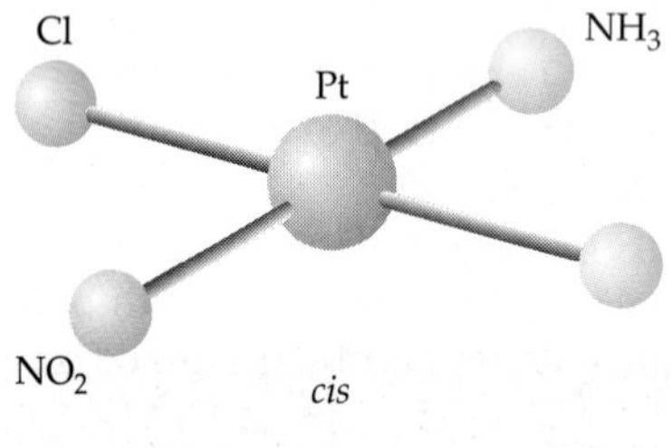

cis

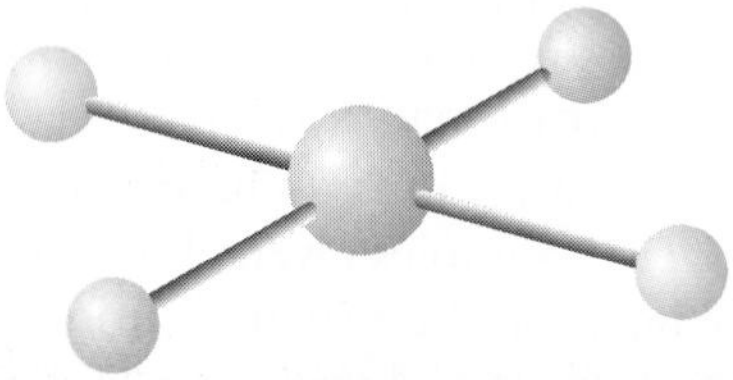

trans

Cis-trans isomerism in an octahedral complex can be illustrated with $[Co(H_2NCH_2CH_2NH_2)_2Cl_2]^+$, an octahedral complex with two bidentate ethylenediamine ligands and two chloride ligands. In this complex, the two Cl^- ions occupy positions that are either adjacent (*cis* isomer) or opposite (*trans* isomer) (Figure 23.18). It is interesting that these isomers have different colors: the *cis* isomer is green, and the *trans* isomer is purple.

Cis-trans isomerism is *not possible* for tetrahedral complexes. All L—M—L angles in tetrahedron geometry are 109.5°, and all ligands are adjacent in this three-dimensional structure.

Another type of geometrical isomerism, called *mer-fac* isomerism, occurs for octahedral complexes with the general formula MX_3Y_3. In a *fac* isomer, three similar ligands lie at the corners of a triangular face of the octahedron (*fac* =

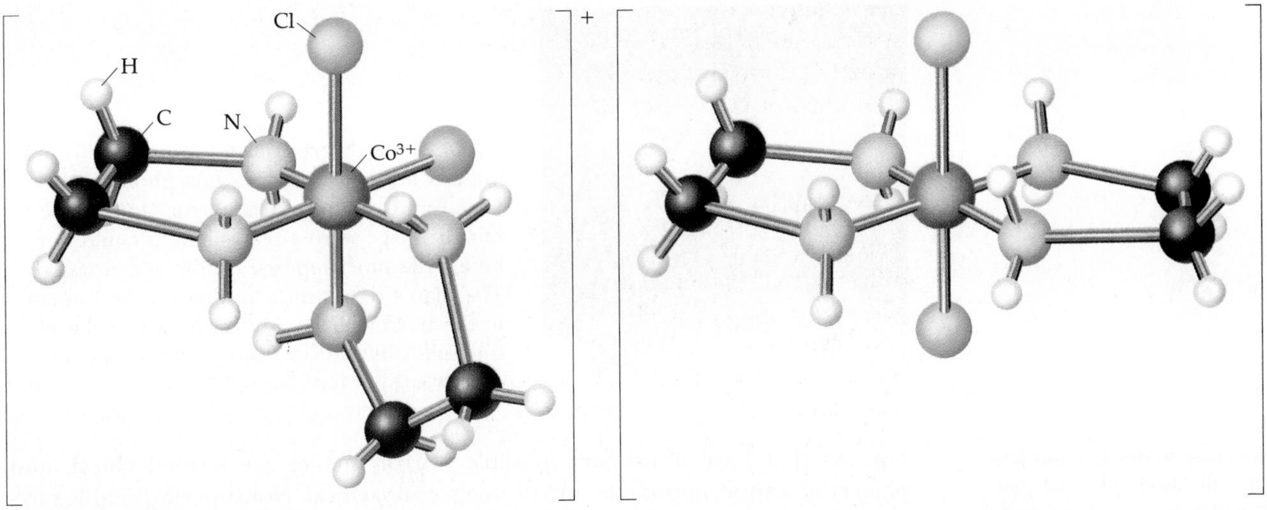

Figure 23.18 *Cis* and *trans* isomers of $[Co(en)_2Cl_2]^+$.

facial), whereas in the *mer* isomer, the ligands follow a meridian (*mer* = *meridional*). *Fac* and *mer* isomers of $Cr(NH_3)_3Cl_3$ are shown in Figure 23.19.

Optical Isomerism

Everyone has, at one time or another, tried to put a left shoe on a right foot, or a left-handed glove on a right hand. It doesn't work very well. Even though our two hands and two feet appear generally similar, there is a very important distinction between them. Left hands and feet are mirror images of right hands and feet, and, most importantly, these mirror images cannot be superimposed (Figure 23.20).

Certain molecules have the same characteristic as gloves and hands: A given structure and its mirror image cannot be superimposed. Molecules (and other

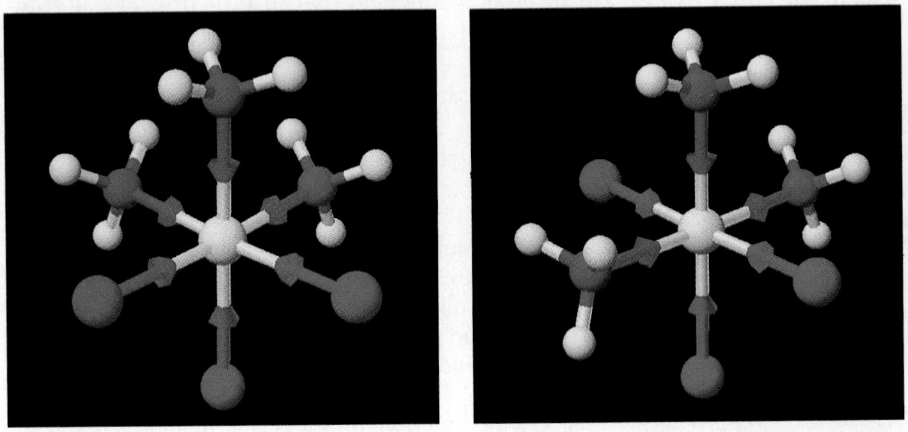

(a) (b)

Figure 23.19 *Fac* and *mer* isomers of $Cr(NH_3)_3Cl_3$. (a) In the *fac* isomer the three NH_3 ligands (or the three Cl^- ligands) are arranged on one octahedral face. (b) In the *mer* isomer, the three NH_3 ligands (or the three Cl^- ligands) are arranged around the meridian on the molecule. (Susan Young)

Figure 23.20 Mirror images of two wood carvings. Consider the carving of a photographer (*right*) and its image in a mirror (*left*). If the mirror image were a real statue, it could not be exactly superimposed on the actual statue. The man's right arm is resting on the camera in the mirror image, but in the actual statue the man's left arm is resting on the camera. (C. D. Winters)

Although we chose to describe this phenomenon in a discussion of metal complexes, it is important to realize that it is also a particularly important structural feature in organic and biological chemistry.

Enantiomers are possible when a molecule is **asymmetrical,** that is, lacking symmetry.

objects) that have nonsuperimposable mirror images are termed **chiral,** and objects with superimposable mirror images are **achiral.** Nonsuperimposable molecules are known as **enantiomers.** Many common objects have a similar property. For example, some seashells are chiral, and wood screws and machine bolts are also chiral, distinguished by left-handed or right-handed threads.

Enantiomers have the same stoichiometry *and* the same atom-to-atom bonding sequences, but they differ in the details of the arrangement of atoms in space. The most common type of chirality that occurs in chemistry involves carbon atoms bonded to four different groups. An example of such a compound is lactic acid, $CH_3CH(OH)COOH$. The enantiomers or mirror images of lactic acid are shown in Figure 23.21.

To see that the enantiomers of lactic acid differ, imagine that you are looking down the H—C bond from the "top" of the molecule. In one of the enantiomers the three other groups (CH_3, OH, and COOH) are arranged in a clockwise order, whereas in the second enantiomer these groups appear in counterclockwise order. It is impossible to superimpose the two mirror images; however you rotate the molecule, two groups of the four groups never line up.

The two enantiomers of a compound have the same physical properties, such as melting point, boiling point, density, and solubility in common solvents. They differ in a very significant way, however: When a beam of plane-polarized light is passed through a solution of a pure enantiomer, the plane of polarization is twisted in one direction (Figure 23.22). The two enantiomers rotate polarized light to an equal extent, *but in opposite directions.* Because of this, chiral com-

A mixture of equal quantities of two enantiomers is called a **racemic** mixture. Racemic mixtures do not rotate polarized light because the effects of the two optical isomers cancel.

Many sea shells are chiral, virtually all that are are right-handed. If you cup a shell in your right hand, with thumb extended and pointing from the narrow end to the wide end, your fingers follow the curvature of the shell as it curls from the outside toward the center. Here is a person holding left and right handed shells. (C. D. Winters)

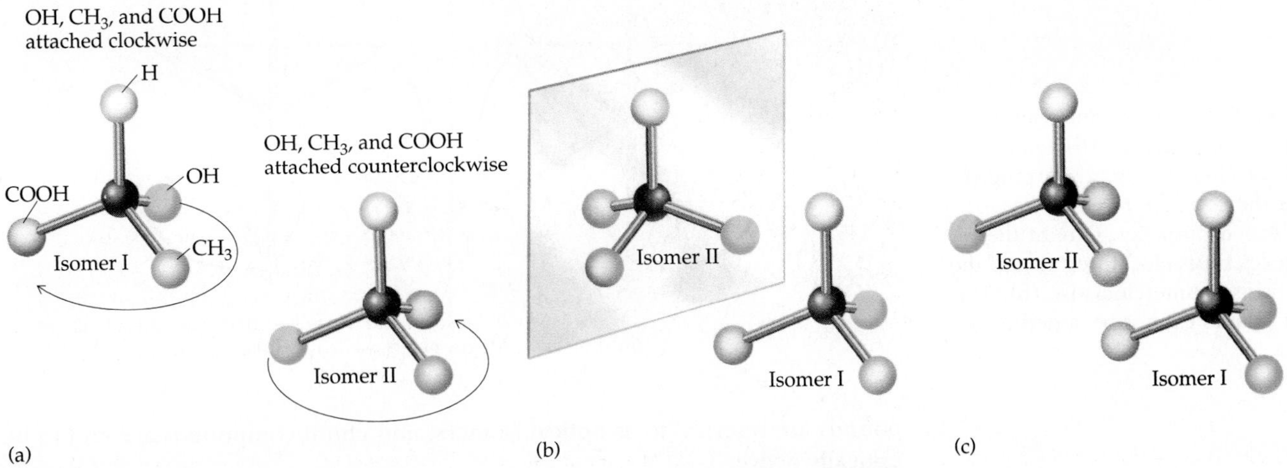

Figure 23.21 The two enantiomeric forms of lactic acid, $CH_3CH(OH)(CO_2H)$. (a) In Isomer I the groups —OH, —CH_3, and —COOH are attached in a clockwise manner. In Isomer II the groups —OH, —CH_3, and —COOH are attached in a counterclockwise manner. (b) Isomer I is placed in front of a mirror, and its mirror image is Isomer II. (c) The isomers are nonsuperimposable.

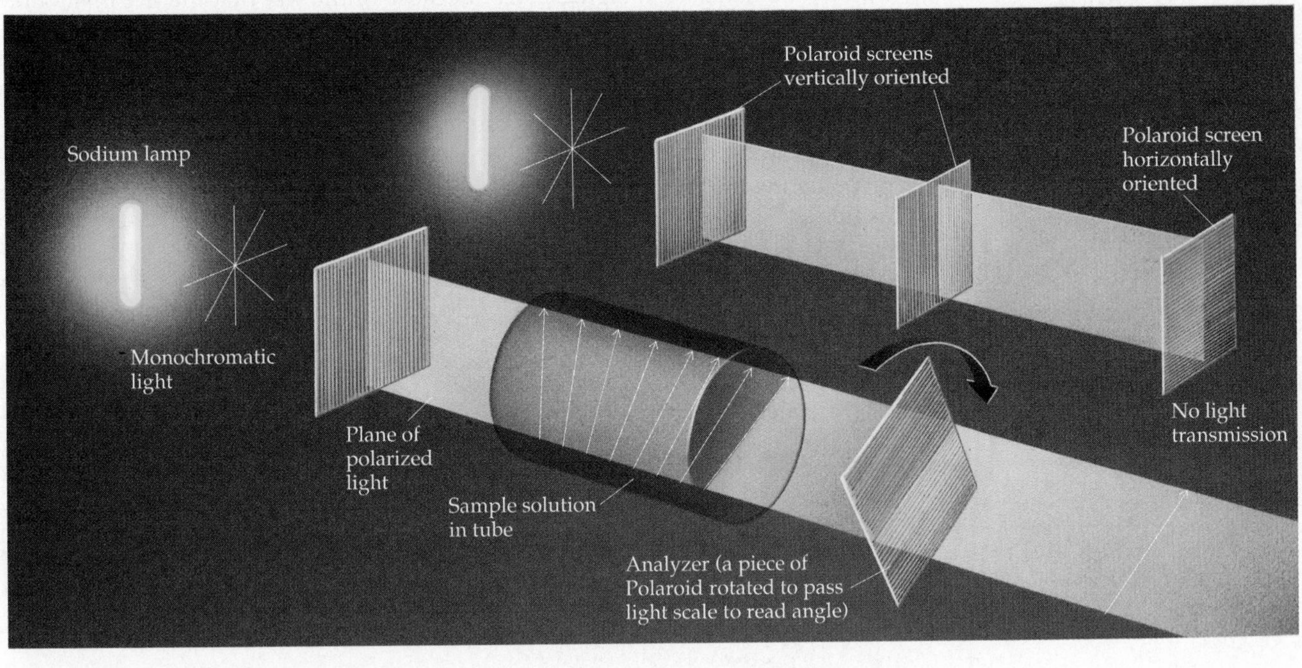

Figure 23.22 Schematic drawing of the rotation of plane-polarized light by an optical isomer. (*Top*) Monochromatic light (light of only one wavelength) is produced by a sodium lamp. After it passes through a Polaroid filter, the light is vibrating in only one direction—it is polarized. Polarized light passes through a second Polaroid filter if the filters are parallel to the first filter, but not if the second filter is perpendicular. (*Bottom*) A solution of an optical isomer placed between the first and second Polaroid filters causes rotation of the plane of polarized light. The angle of rotation can be determined by rotating the second filter until maximum light transmission occurs. The magnitude and direction of rotation are unique physical properties of the optical isomer being tested.

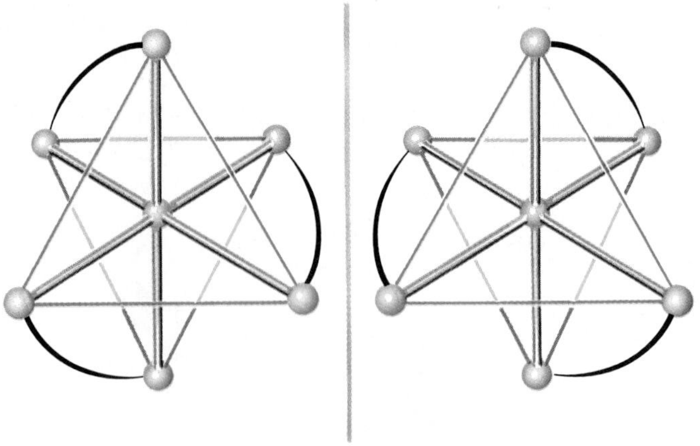

Figure 23.23 Optical isomerism in complexes of the type M(bidentate)$_3$. The three chelate rings are arranged so that the complex resembles a three-bladed propeller. One of the complexes twists clockwise (a) and the other twists counterclockwise (b). The mirror images cannot be superimposed.

pounds are referred to as **optical isomers,** and chiral compounds are said to be **optically active.**

The possibility for chirality arises for a number of coordination compounds based on octahedral geometry, only one of which is described here. In this situation a metal atom coordinates to three bidentate ligands, as for instance in the complex ion $[Co(en)_3]^{3+}$, which has three bidentate ethylenediamine ligands coordinated to a cobalt(III) ion. The structure of this complex is portrayed schematically in Figure 23.23. Because of the way that the five-membered chelate rings are arranged, mirror images of this molecule do not superimpose. Solutions of each optical isomer rotate polarized light in opposite directions.

Other stoichiometries in which optical activity occurs include *cis*-$[M(bidentate)_2X_2]^{n+}$ and *fac*-$[ML_3XYZ]^{n+}$.

Square-planar complexes are incapable of optical isomerism that is based at the metal; mirror images are always superimposable. Although optical isomers of tetrahedral complexes are possible, no examples of stable complexes with a metal bonded tetrahedrally to four different types of ligands are known to exist.

EXAMPLE 23.4 *Isomerism*

Which of the following complexes exhibits geometric or optical isomerism or both?

1. $[Co(NH_3)_4Cl_2]^+$
2. $[Ru(phen)_3]Cl_2$
3. $[Pt(CN)_2Cl_2]^{2-}$

Solution

1. This Co^{3+} complex has an octahedral structure, and two geometric isomers are possible. One isomer has two Cl^- ions in *cis* positions; in the other isomer the Cl^- ligands are *trans.*

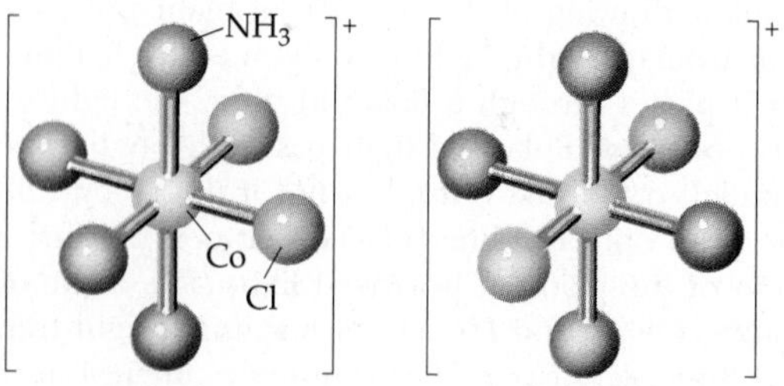

2. In $[Ru(phen)_3]^{2+}$, the Ru^{2+} ion is surrounded by three bidentate phen ligands. Two optical isomers are possible for this complex.

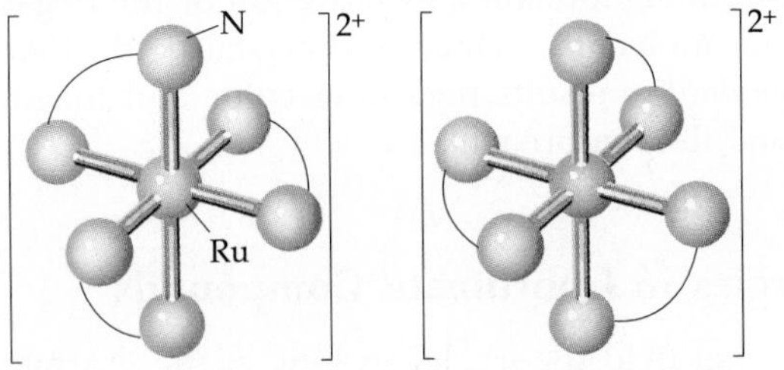

3. This platinum(II) complex has a square-planar geometry. *Cis* and *trans* isomers are possible.

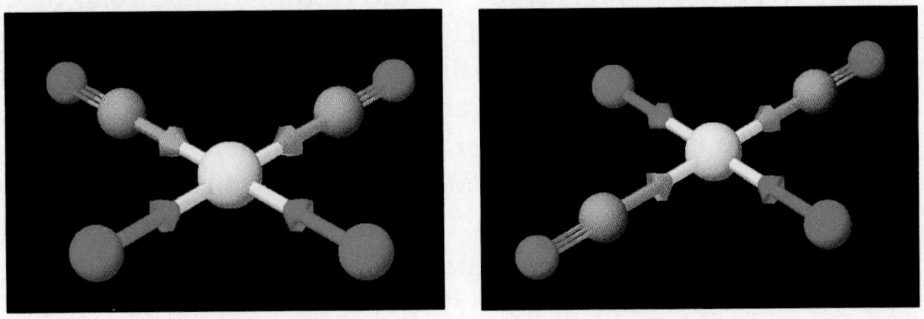

EXERCISE 23.3 ***Identifying Isomers***

What type of isomers are possible for $[Co(en)_2(CN)_2]Br$? ■

23.5 BONDING IN COORDINATION COMPOUNDS

Metal-ligand bonding in a coordination complex was described earlier as the interaction between a Lewis acid (the metal ion) and a Lewis base (the ligand). This valence bond picture represents the ligand-metal bond as covalent, with an electron pair shared between the metal and the ligand donor atom. This model is frequently used, but it is not adequate to explain such properties of complexes as their color (see Figures 23.6 and 23.13) and magnetism. As a result, other bonding models have largely superseded the valence bond model. Currently, the bonding in coordination complexes is usually described by either **molecular orbital theory** or by **crystal field theory.**

Molecular orbital theory and crystal field theory approach metal-ligand bonding using metal *d* orbitals and ligand lone pair orbitals. As ligands approach the metal to form bonds, two effects occur: (a) the metal and ligand orbitals overlap, and (b) the electrons of the metal repulse the electrons of the ligand. Molecular orbital theory takes both effects into account, whereas the crystal field model focuses on metal-ligand electron repulsion. The molecular orbital model assumes that metal and ligand bond through the molecular orbitals formed by

Crystal field theory was first developed to explain the properties of metal ions in crystalline ionic solids. Solid ionic compounds contain metal ions surrounded by a group of negative ions (see the structure of NaCl, page 123). The negative ions in the crystal created a "field" of charges in a specific geometric arrangement around the metal.

atomic orbital overlap between metal and ligand. In contrast, the crystal field model assumes that the positive metal ion and negative ligand lone pair are attracted *electrostatically,* that is, the bond arises from the attractive force between a positively charged metal ion and a negative ion or the negative end of a polar molecule. Both the molecular orbital and crystal field models ultimately produce the same *qualitative* results regarding color and magnetic behavior. We focus on the crystal field approach.

d-Orbital Energies in Coordinate Compounds

To understand crystal field theory, let us look at the d orbitals in more detail. Our particular interest is in the orientation of the d orbitals relative to the positions of ligands in a metal complex. In Figure 23.24 the five d orbitals are grouped into two sets: the $d_{x^2-y^2}$ and d_{z^2} orbitals in one set and the d_{xy}, d_{xz}, and d_{yz} orbitals in another. The $d_{x^2-y^2}$ and d_{z^2} orbitals have their greatest probability *along the x-, y-, and z-axes,* whereas the orbitals of the second group have their greatest probability *between these axes.* The d orbitals are divided into these two sets because we have chosen to assign the ligands in square-planar and octahedral complexes to lie along the x-, y-, and z-axes.

In an isolated atom or ion, the d orbitals are degenerate, that is, they have the same energy. For a metal atom or ion in a coordination complex, however, the d orbitals have different energies. According to the crystal field model, repulsion between d electrons and the electron pairs of the ligands destabilizes the d orbitals (causes their energy to become higher). Electrons in the various d orbitals are not affected equally, however, because of their placement in space relative to the position of the ligand lone pairs. In an octahedral complex, the d orbitals have different energies (Figure 23.25). Electrons in the $d_{x^2-y^2}$ and d_{z^2} orbitals experience a larger repulsion because these orbitals point directly at the incoming ligand electron pairs. A smaller repulsive effect is experienced by electrons in the d_{xy}, d_{xz}, and d_{yz} orbitals. The difference in degree of repulsion means an energy difference exists between the two sets of orbitals. This difference, called the *crystal field splitting* and given the symbol Δ_o, is a function of the metal and the ligands and varies predictably from one complex to another.

A different splitting pattern is encountered with square-planar complexes (Figure 23.26). If the four ligands are in the xy-plane, the $d_{x^2-y^2}$ orbital is at highest energy. The d_{z^2} energy is shifted upward to a lesser extent, whereas the d_{xy} orbital (which lies in the xy-plane) is found at higher energy than the d_{xz} and d_{yz}, both of which are partially pointing in the z direction.

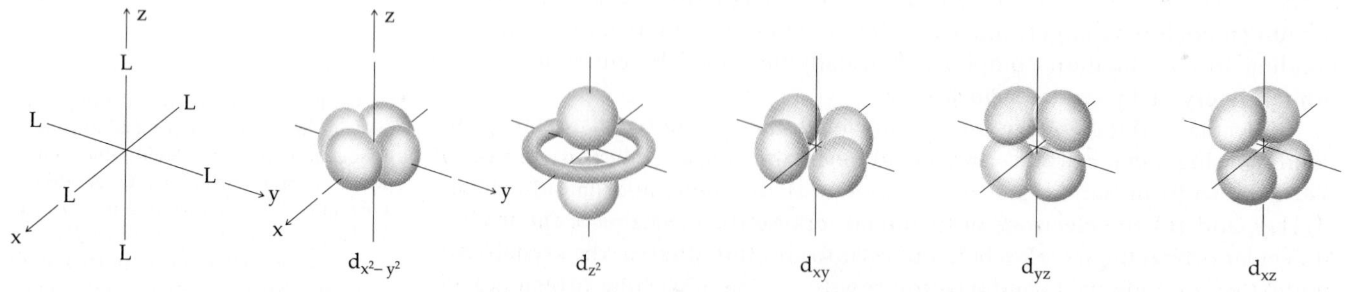

Figure 23.24 The five d orbitals and their spatial relation to ligands on the x-, y-, and z-axes.

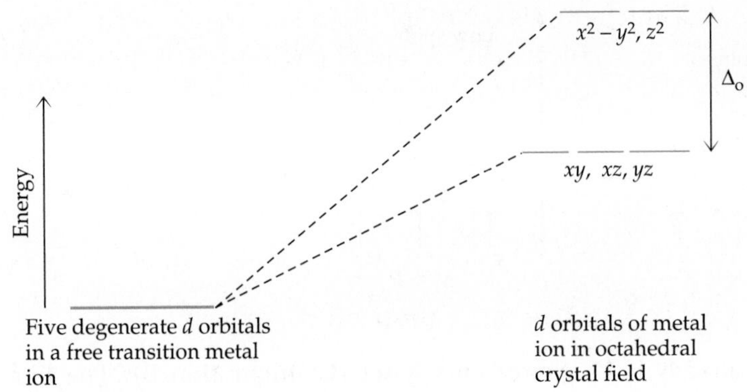

Figure 23.25 An octahedral complex. The *d*-orbital energy changes as six ligands approach the metal ion along the *x*-, *y*-, and *z*-axes. The splitting of the *d* orbitals is labeled Δ_o.

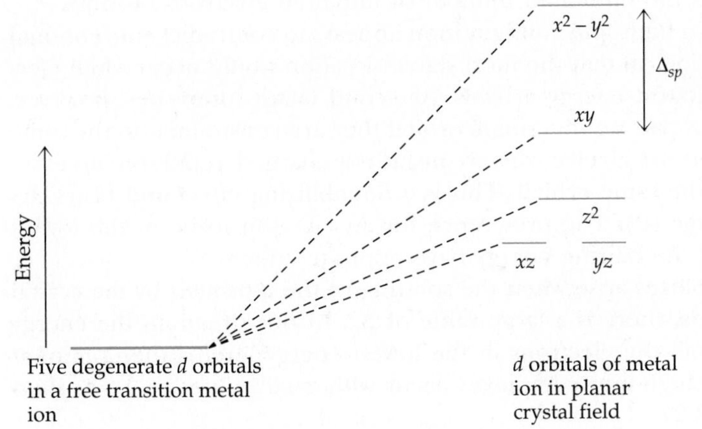

Figure 23.26 Splitting of the *d* orbitals in a square-planar complex.

Magnetic Properties of Coordination Compounds

The *d*-orbital splitting in coordination complexes is the cause of both the magnetic behavior and color for these species. To understand these properties, we must first understand how to assign electrons to the various orbitals in square-planar and octahedral complexes.

A gaseous Cr^{2+} ion has the electron configuration [Ar] $3d^4$. (The term "gaseous" in this context is used by scientists to denote that an atom or ion is isolated, that all other particles are an infinite distance away.) In such an ion, the five $3d$ orbitals have the same energy. The four electrons reside singly in different *d* orbitals, according to Hund's rule, and the Cr^{2+} ion has four unpaired electrons.

Cr^{2+} ion [Ar] [↑|↑|↑|↑|] [ ]
$3d$ $4s$

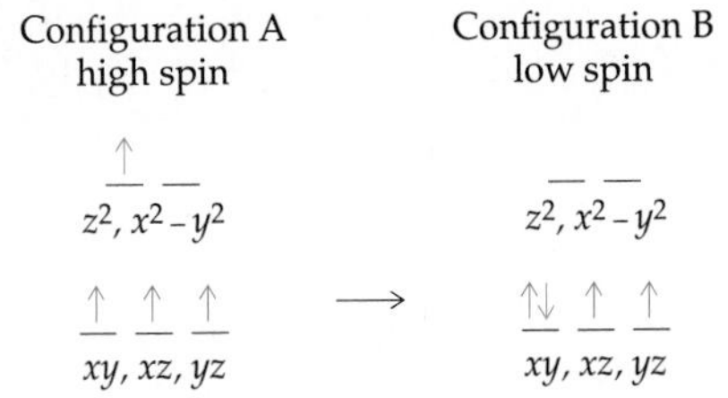

When the Cr^{2+} ion is part of an octahedral complex, however, the five *d* orbitals are no longer degenerate. These orbitals divide into two sets with the d_{xy}, d_{xz}, and d_{yz} orbitals, which are at a lower energy than the $d_{x^2-y^2}$ and d_{z^2} orbitals. Having two sets of orbitals means that two different electron configurations are possible. Three of the four *d* electrons in Cr^{2+} are assigned to the lower energy d_{xy}, d_{xz}, and d_{yz} orbitals. The fourth electron, however, can either be assigned to an orbital in the higher energy $d_{x^2-y^2}$ and d_{z^2} set (Configuration A) or pair up

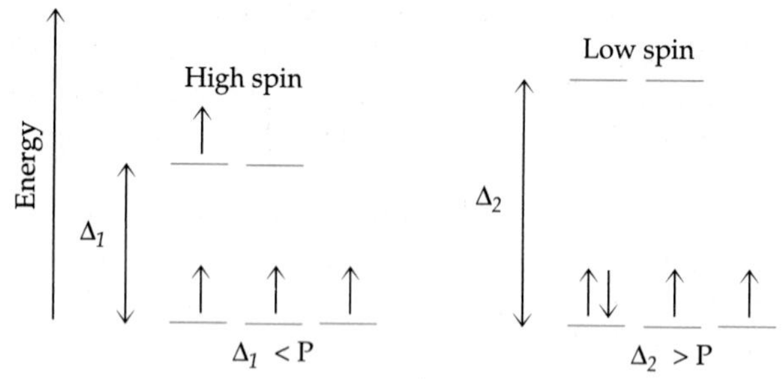

Figure 23.27 High- and low-spin cases for $[CrL_6]^{2+}$ complexes. When the crystal field splitting is smaller than the pairing energy (P), the electrons prefer to remain unpaired, and the complex has four unpaired electrons. When the crystal field splitting is larger than the pairing energy, then the four electrons are in the lowest energy orbitals, and the complex has two unpaired electrons.

with an electron already in the lower energy set (Configuration B). The first arrangement is commonly called **high spin** because it has the maximum number of unpaired electrons, four. In contrast, the second arrangement is called **low spin** because it has the minimum number of unpaired electrons possible.

At first glance, a high-spin configuration appears to contradict conventional thinking. It seems logical that the most stable situation would occur when electrons occupy the lowest energy orbitals. A second factor intervenes, however. When two electrons pair up in a single orbital they are constrained to the same region of space. Because all electrons are negatively charged, repulsion increases when they are in the same orbital. This is a destabilizing effect and bears the name **pairing energy (P).** The preference for an electron to be in the lowest energy orbital and the pairing energy have opposite effects.

Low-spin complexes arise when the splitting of the *d* orbitals by the crystal field is large, that is, there is a large value of Δ_o. In this situation, the energy gained by putting all the electrons in the lowest energy level is the dominant effect. Conversely, high-spin complexes occur with small values of Δ_o, as illustrated in Figure 23.27.

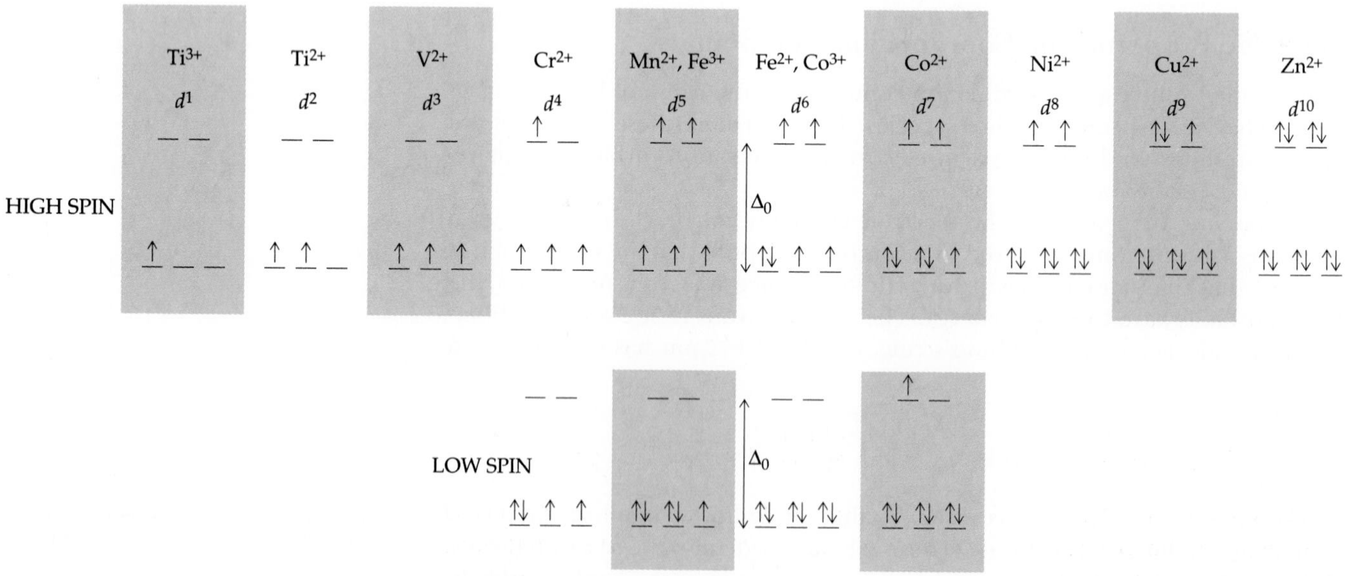

Figure 23.28 Electron configurations for octahedral complexes of metal ions having from d^1 to d^{10} configurations. Only the d^4 through d^7 cases have both high-spin and low-spin configurations.

Actually, for octahedral complexes, there is a choice between high and low spin only for configurations d^4 through d^7 (Figure 23.28). Complexes of the d^6 metal ion, Fe^{2+}, for example, can have either high spin or low spin. The complex formed when the ion is placed in water, $[Fe(H_2O)_6]^{2+}$, is high spin, whereas $[Fe(CN)_6]^{4-}$ is low spin.

It is possible to tell the difference between low- and high-spin complexes by determining the magnetic behavior of the substance. The high-spin complex $[Fe(H_2O)_6]^{2+}$ has four unpaired electrons and is *paramagnetic* (attracted by a magnet), whereas the low-spin $[Fe(CN)_6]^{4-}$ complex has no unpaired electrons and is *diamagnetic* (repelled by a magnet) (see Section 8.1).

Electron Configuration for Fe^{2+} in an Octahedral Complex

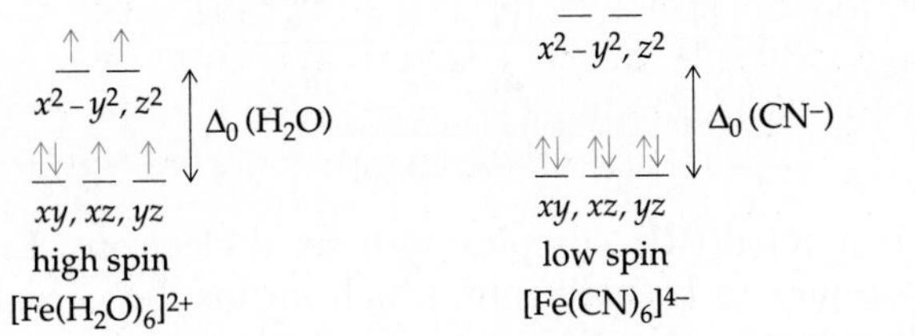

The Ni^{2+}, Pd^{2+}, and Pt^{2+} ions have the electron configuration [noble gas]$(n - 1)d^8$ and are especially prone to form square-planar complexes. In a square-planar complex there are four sets of orbitals (Figure 23.26). Although high- and low-spin configurations would appear to be possible, only low-spin complexes are known.

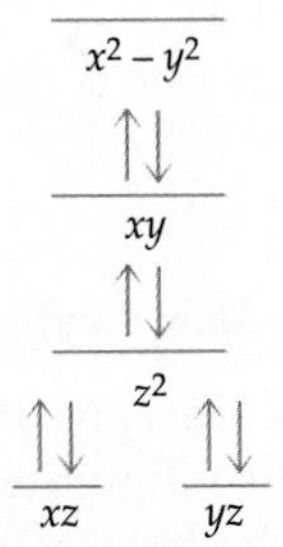

Electron configuration in d^8 square-planar complexes.

EXAMPLE 23.5 *High- and Low-Spin Complexes and Magnetism*

Depict the electron configurations for each of the following complexes, and tell how many unpaired electrons are present in each. Describe each complex as paramagnetic or diamagnetic.

1. Low-spin $[Co(NH_3)_6]^{3+}$
2. High-spin $[CoF_6]^{3-}$

Solution

1. We assume this is an octahedral complex because there are six ligands surrounding cobalt. Furthermore, because the NH_3 ligands are neutral molecules and because the overall charge on the complex is 3+, this complex is based on the Co^{3+} ion. The cobalt(III) ion has an electron configuration of [Ar] $3d^6$. Experiment shows that the complex ion $[Co(NH_3)_6]^{3+}$ is low spin. To obtain the low-spin configuration for the complex ion, the lower energy set of orbitals is filled entirely before adding electrons to the higher energy orbitals. This d^6 complex ion has no unpaired electrons and so is diamagnetic.

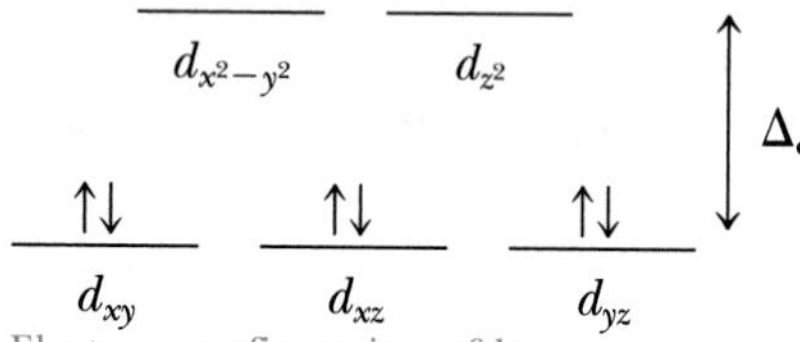

Electron configuration of low-spin, octahedral $[Co(NH_3)_6]^{3+}$

2. Again, this is a cobalt(III) complex with six *d* electrons. Experiment now shows this complex to be high spin, which means the crystal field splitting (Δ_o) is smaller than in $[Co(NH_3)_6]^{3+}$. To obtain the electron configuration for the d^6 Co^{3+} metal ion in $[CoF_6]^{3-}$, place one electron in each of the five *d* orbitals, and then place the sixth electron in one of the lower energy orbitals. The complex has four unpaired electrons and is paramagnetic.

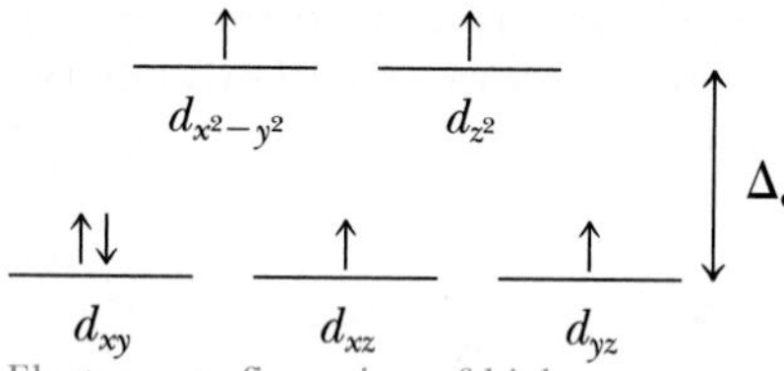

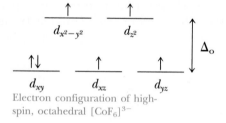

Electron configuration of high-spin, octahedral $[CoF_6]^{3-}$

EXERCISE 23.4 ***High- and Low-Spin Configurations and Magnetism***

For each of the following complex ions, give the oxidation number of the metal ion, depict the low- and high-spin configurations, give the number of unpaired electrons in each state, and tell whether each is paramagnetic or diamagnetic.

(a) $[Ru(H_2O)_6]^{2+}$

(b) $[Ni(NH_3)_6]^{2+}$ ■

23.6 THE COLORS OF COORDINATION COMPOUNDS

One of the most interesting properties of the transition elements is that their compounds are usually colored, whereas compounds of main group metals are usually colorless (see Figures 23.6, 23.13, and 23.29). With an understanding of *d*-orbital splitting, we can now explain the origin of the colors for complexes. First, however, let us look more closely at what we mean by color.

Figure 23.29 Compounds of the transition elements are often colored; those of main group elements are usually colorless. Pictured, from left to right, are aqueous solutions of nitrate salts of Fe^{3+}, Co^{2+}, Ni^{2+}, Cu^{2+}, and Zn^{2+}. (C. D. Winters)

Color

Visible light radiation with wavelengths from 400 nm to 700 nm (Section 7.1) represents a very small portion of the electromagnetic spectrum. Within this region of the spectrum are all the colors you see when white light is passed through a prism: red, orange, yellow, green, blue, indigo, and violet (ROYGBIV). Each color is identified with a narrow wavelength range (Figure 7.3).

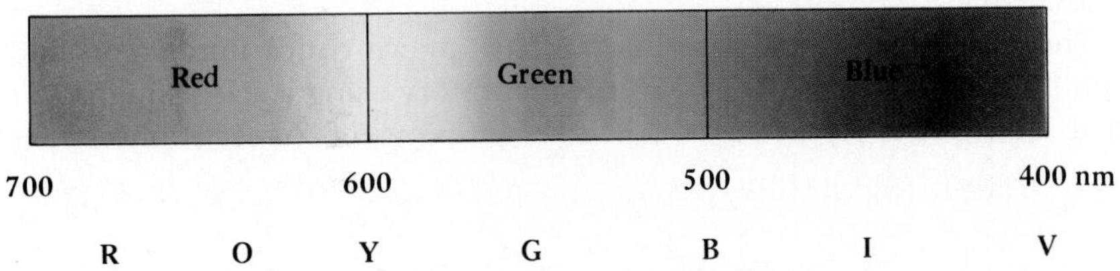

The blue in ROYGBIV is actually cyan, C, according to now accepted color industry standards. The highlights in this book are printed in cyan, for example. Note also that magenta doesn't have its own wavelength region, it is a superposition of B and R, and the orange is simply a hue of yellow with some red content.

Isaac Newton did experiments with light and established that the mind's *perception* of color requires only three colors! When we see white light, we are seeing a mixture of all of the colors as the superposition of red, green, and blue. If one or more of these colors is absorbed, the light of the other colors can then pass through to your eyes. Your mind then perceives and interprets the color.

To understand the perceived colors of compounds, we divide the visible spectral range into three broad regions: red, green, and blue (Figure 23.30). The *primary* colors—red, green, and blue—appear at the corners of the triangle superimposed on the color discs, and the *secondary* colors—yellow, cyan, and magenta—appear at the edges of the triangle.

Figure 23.31 presents this information in another way. A color arises in two ways: by *addition* of two other colors or by the *subtraction* of light of a particular color from white light. For example, adding blue and green light leads to a color called cyan. Alternatively, if red light is subtracted from white light, cyan light remains.

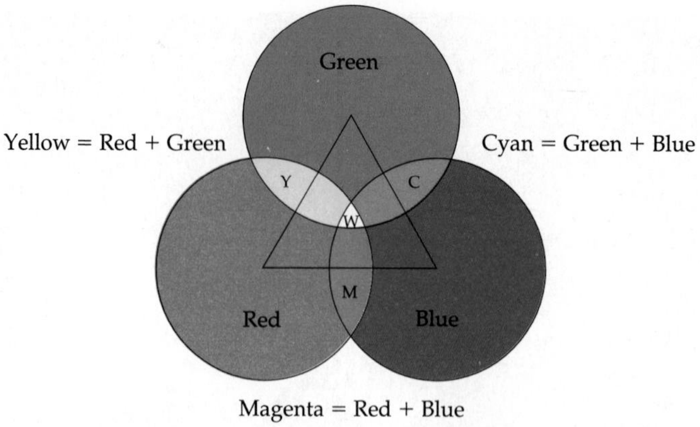

Figure 23.30 Color discs can be used to explain the perception of color. The three primary colors are red, green, and blue. Light of two primary colors can add together to give a third color. For example, cyan arises when green and blue light are added, and yellow arises from the addition of green and red. Alternatively, we can think of color as arising from the subtraction of a light from white light. For example if red light is removed from white light, the color cyan results.

Now let us apply these ideas to transition metal complexes. A solution of $[Ni(H_2O)_6]^{2+}$ is green. We know that this color is the result of removing red (R) and blue (B) light from white light. As white light passes through an aqueous solution of Ni^{2+} (see Figure 23.29), blue and red light are absorbed but green light is allowed to pass, and so this is the color we perceive. Similarly, the $[Co(NH_3)_6]^{3+}$ ion (see Figure 23.29) is yellow-orange because blue (B) light has been selectively absorbed; that is, the solution allows red (R) and green (G) light to pass.

The qualitative conclusions that we have drawn concerning colors and absorption of light are confirmed in the laboratory using a scientific instrument called a spectrophotometer. A schematic drawing of a spectrophotometer is

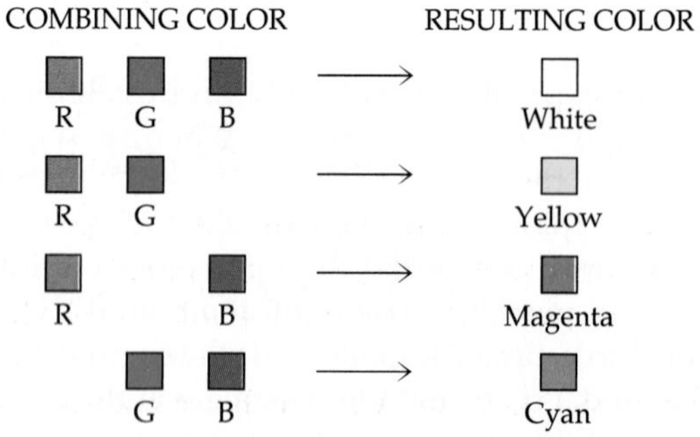

Figure 23.31 Another way to view colors: The three secondary colors shown at the right—cyan, yellow, and magenta—are derived from addition of the indicated primary colors. Alternatively, a secondary color results if one of the three primary colors is subtracted. For example, yellow is viewed as arising *either* from the addition of red and green *or* by subtraction (or absence) of blue light from white light.

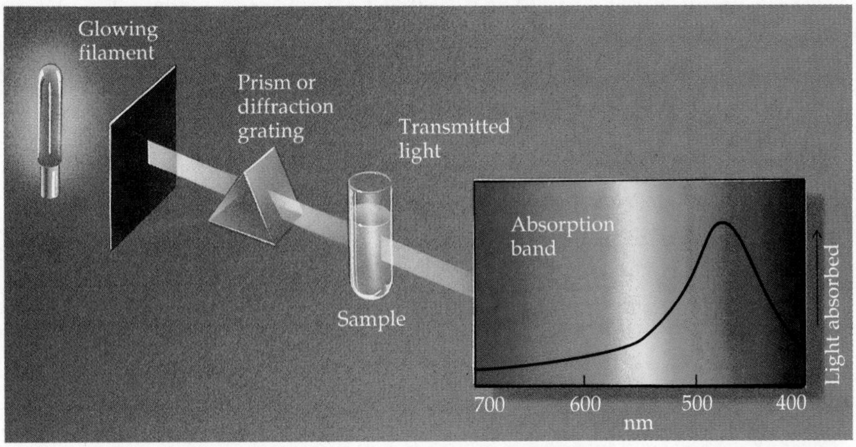

Figure 23.32 A spectrophotometer, and the absorption spectrum of $[Co(NH_3)_6]^{3+}$.

shown in Figure 23.32. White light from a glowing filament is first passed through a device (a prism or diffraction grating) that divides the light according to its respective frequencies. The instrument selects a specific frequency to pass through a solution of the compound to be studied. If light of a given frequency is not absorbed, its intensity is unchanged when it emerges from the sample. On the other hand, if light of this frequency is absorbed, the light emerging from the sample has a lower intensity.

Modern spectrometers are designed to make absorption measurements over a range of frequencies. This range, or **absorption spectrum** for the sample, is a graph of the frequency or wavelength of the light against the intensity of light absorbed at that frequency or wavelength. When light in a certain frequency range is absorbed the plot shows an *absorption band.* For example, $[Co(NH_3)_6]^{3+}$ absorbs in the blue region, leaving its complementary color yellow (Y = R + G) to pass through to the eye.

Scientists are not limited to the visible region of the spectrum when making measurements on absorption of electromagnetic radiation. Ultraviolet and infrared spectrometers are common pieces of scientific apparatus in a chemistry laboratory. They differ from the visible spectrometer depicted in Figure 23.32 only in that the light source must generate radiation in the correct region of the electromagnetic spectrum and the detector must be able to detect this radiation.

The Absorption of Light by Coordination Complexes

The color of a transition metal complex results from the absorption of light in the visible region of the spectrum. Although the details of the process by which light is absorbed are beyond the scope of this book, we can describe this process in qualitative terms.

Chapter 7 described atomic spectra, which are obtained when electrons are excited from one energy level to another. The absorptions correspond to specific wavelengths, and the energy of the light absorbed or emitted is related to the energy levels of the atom or ion under study. Actually, the concept that light is absorbed when electrons move between energy levels applies to all substances, and this is the basis for the spectrum (and hence the colors) for transition metal coordination complexes. Because most transition metal complexes are colored, we conclude that the energy levels in these complexes are spaced so that visible light is absorbed.

In coordination complexes, the splitting between *d* orbitals often corresponds to the energy of visible light, so light in the visible region of the spectrum is absorbed when electrons move from a lower energy *d* orbital to a higher energy *d* orbital. This transition of an electron between two orbitals in a complex is labeled a *d-to-d transition.* Qualitatively, such a transition for $[Co(NH_3)_6]^{3+}$ might be represented, using an energy level diagram such as that shown here.

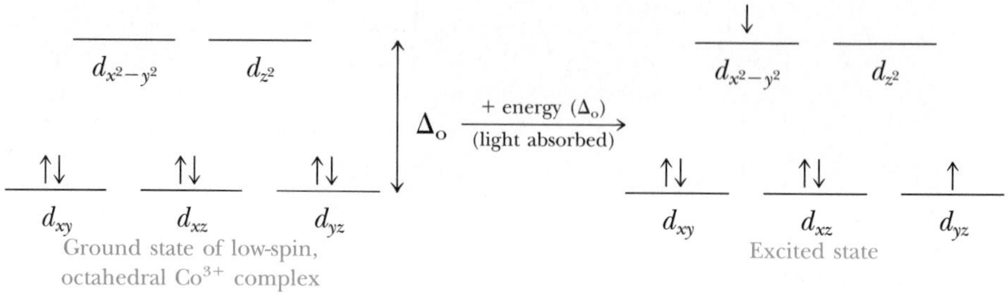

The Spectrochemical Series of Ligands

Experiments with coordination complexes have revealed that, for a given metal ion, some ligands cause a small energy separation of the *d* orbitals, whereas others cause a large separation. In other words, some ligands create a small crystal field, and others create a large one.

Spectroscopic data for several cobalt(III) complexes are presented in Table 23.4. We can see from this information why different complexes have the indicated colors.

- Both $[Co(NH_3)_6]^{3+}$ and $[Co(en)_3]^{3+}$ are yellow-orange, because they absorb light in the blue portion of the visible spectrum. Note, by the way, that these compounds have very similar spectra, which is to be expected because both have six amine donor atoms.
- Although $[Co(CN)_6]^{3-}$ does not have an absorption band in the visible region, it is pale yellow. This is due to the fact that the absorption in the ultraviolet is broad and extends at least minimally into the visible region, resulting in the absorption of a small amount of blue light.

In Table 23.4 notice the connection between the wavelength of the absorbed light, its relative energy, and the color of the complex.

TABLE 23.4 The Colors of Some Complexes of the Co^{3+} Ion*

Complex Ion	Wavelength of Light Absorbed (nm)	Color of Light Absorbed	Color of Complex
$[CoF_6]^{3-}$	700	Red	Green
$[Co(C_2O_4)_3]^{3-}$	600, 420	Yellow, violet	Dark green
$[Co(H_2O)_6]^{3+}$	600, 400	Yellow, violet	Blue-green
$[Co(NH_3)_6]^{3+}$	475, 340	Blue, ultraviolet	Yellow-orange
$[Co(en)_3]^{3+}$	470, 340	Blue, ultraviolet	Yellow-orange
$[Co(CN)_6]^{3-}$	310	Ultraviolet	Pale yellow

*The complex with fluoride ion, $[CoF_6]^{3-}$, is high spin and has one absorption band. The other complexes are low spin and have two absorption bands. In all but one case, one of these absorptions is in the visible region of the spectrum. The wavelengths refer to the center of that absorption band.

- $[Co(C_2O_4)_3]^{3-}$ and $[Co(H_2O)_6]^{3+}$ have fairly similar absorptions, in the yellow and violet regions. Their colors are shades of green with a small difference due to the relative amount of light of each color being absorbed.

There is a progression of absorption maxima among the complexes listed, ranging from 700 nm for $[CoF_6]^{3-}$ to 310 nm for $[Co(CN)_6]^{3-}$. The ligand is the primary factor that changes from member to member in this series, so we can conclude that the energy of the light absorbed by the complex is related to different crystal field splitting Δ_o caused by the different ligands. Fluoride ion has caused the smallest splitting of the *d* orbitals among the complexes listed, whereas cyanide caused the largest splitting.

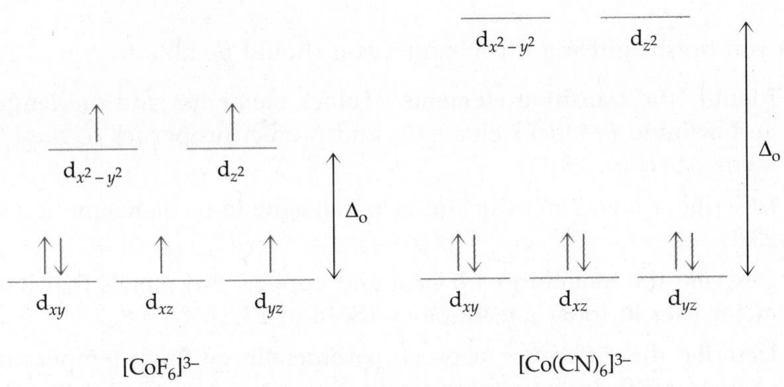

Measurements on spectra of many other complexes provide similar results. Based on this information, it is possible to list ligands in order of their ability to split the *d* orbitals. This list, called the **spectrochemical series** because it was determined by spectroscopy, is shown here. (The series contains many more ligands than are listed here.)

Spectrochemical Series

$$\text{Halides} < C_2O_4^{2-} < H_2O < NH_3 = \text{en} < \text{phen} < CN^-$$

small orbital splitting	large orbital splitting
small Δ_o	large Δ_o
weak field ligands	strong field ligands

The spectrochemical series is applicable to a wide range of metal complexes and ligands. The ability of crystal field theory to explain differences in the color of transition metal complexes is one of the strengths of this theory.

From the relative position of the ligand in the series, you can now also make predictions about a compound's magnetic behavior. Recall that d^4, d^5, d^6, and d^7 complexes can be high or low spin, depending on the crystal field splitting Δ_o. Complexes formed with ligands near the left end of the spectrochemical series are expected to have small Δ_o values and thus are likely to be high spin. In contrast, complexes with ligands near the right end are expected to have large Δ_o values and have low-spin configurations. The complex $[CoF_6]^{3-}$ is high spin, whereas $[Co(NH_3)_6]^{3+}$ and the others shown in Table 23.4 are low spin.

EXAMPLE 23.6 ***Spectrochemical Series***

An aqueous solution of $[Fe(H_2O)_6]^{2+}$ is light blue-green. Do you expect the d^6 Fe^{2+} ion in this complex to have a high- or low-spin configuration?

Solution Using the color wheel in Figure 23.30, we see that the blue-green color arises when the complex absorbs red light. The *d*-orbital splitting must be small, meaning that $[Fe(H_2O)_6]^{2+}$ has a good chance of being a high-spin complex. The presence of four unpaired electrons can be verified experimentally.

CHAPTER HIGHLIGHTS

When you finish studying this chapter, you should be able to

- Identify the **transition elements** (*d*-block elements) and the **lanthanide** and **actinide** (*f*-block) elements, and predict properties of these elements (Section 23.1).
- Describe a metal atom or ion as **paramagnetic** or diamagnetic (Section 23.1).
- Describe the metallurgy for iron and copper, two metals that have major uses in today's technology (Section 23.2).
- Describe the difference between **pyrometallurgy** (high temperature processes) and **hydrometallurgy** (techniques using water) (Section 23.2).
- Identify and describe important features of coordination complexes: ligands (monodentate and polydentate), the charge on the complex ion, and the oxidation number of the central metal ion (Section 23.3).
- Provide the systematic name for a coordination complex (Section 23.3).
- Recognize examples of common coordination numbers and their geometries (2, linear; 4, tetrahedral and square-planar; 6, octahedral) (Section 23.4).
- Recognize and draw isomers of coordination complexes (Section 23.4).
- Know that there are two common types of isomers of coordination compounds: **geometric** and **optical isomers** (Section 23.4).
- Recognize that square-planar complexes may exist as geometric isomers but not optical isomers. Octahedral complexes can exhibit both forms of isomerism (Section 23.4).
- Explain the **bonding** in coordination compounds using **crystal field theory** (Section 23.5).
- Rationalize the existence of **high-** and **low-spin** complexes (Section 23.5).
- Understand why complexes are colored and how the color of a complex can be explained using the crystal field model of bonding (Section 23.6).

STUDY QUESTIONS

Review Questions

1. What feature of electronic structure distinguishes transition elements, lanthanide elements, and actinide elements from the other elements and from one another?
2. Write out electron configurations for each element in the first transition series.
3. Write out the electron configuration for the 2+ ion of each metal in the first transition series.
4. Identify two physical properties and two chemical properties for each element in the first transition series.
5. Identify a common use for the following elements: Ti, Cr, Mn, Fe, Ni, Cu.
6. What are the two major types of processes for the recovery of metals from their ores? Give an example of the use of each type of process.
7. What transition elements are commonly used in the manufacture of permanent magnets?
8. Define the following terms: (a) transition element, (b) coordination compound, (c) complex ion, (d) ligand, (e) chelate, (f) bidentate. Give an example to illustrate each word or phrase.
9. Define the terms "diamagnetic" and "paramagnetic." What feature of electronic structure distinguishes these properties?
10. What are three common metal coordination numbers encountered in coordination chemistry, and what structure or structures are possible for each?
11. Give an example of (a) geometric isomerism and (b) optical isomerism.
12. According to the crystal field model, what is the origin of the splitting of metal *d* orbitals into two sets in an octahedral complex?
13. What factors determine whether a complex is high or low spin?

Numerical and Other Questions

Configurations and Physical Properties

14. The metal atoms in the chromium compounds pictured in Figure 23.6 have two different oxidation numbers. Using an orbital box diagram, show the electron configuration for chromium in each oxidation state. In which of the two oxidation states is chromium expected to be paramagnetic?

15. Give the electron configuration for each ion listed here, and tell whether it is paramagnetic or diamagnetic.
(a) Y^{3+}
(b) Pt^{2+}
(c) Rh^{3+}
(d) V^{2+}
(e) Ce^{4+}
(f) U^{4+}

16. Identify two transition metal ions with the following electron configurations:
(a) [Ar] $3d^6$
(b) [Ar] $3d^{10}$
(c) [Ar] $3d^5$
(d) [Ar] $3d^8$

17. Identify another ion of a first series transition metal that is isoelectronic with each of the following:
(a) Fe^{3+}
(b) Fe^{2+}
(c) Zn^{2+}
(d) Cr^{3+}

18. Which element in each of the following pairs should be denser? Explain each answer briefly.
(a) Ti or Fe
(b) Ti or Os
(c) Ti or Zr
(d) Zr or Hf

Metallurgy

19. The following equations represent various ways of obtaining transition metals from their compounds. Balance each equation.
(a) $Cr_2O_3(s) + Al(s) \rightarrow Al_2O_3(s) + Cr(s)$
(b) $TiCl_4(\ell) + Mg(s) \rightarrow Ti(s) + MgCl_2(s)$
(c) $[Ag(CN)_2]^-(aq) + Zn(s) \rightarrow Ag(s) + [Zn(CN)_4]^{2-}(aq)$

20. In the first step in the recovery of copper, an ore such as chalcopyrite, $CuFeS_2$, is roasted in air to give CuS, FeO, and SO_2. If you begin with 1 ton (908 kg) of chalcopyrite, how many tons of SO_2 is produced?

21. Titanium is the seventh most abundant metal in earth's crust. It is strong, lightweight, and resistant to corrosion, properties which lead to its use in aircraft engines. To obtain metallic titanium, ilmenite ($FeTiO_3$), an ore of titanium, is first treated with sulfuric acid to give $Ti(SO_4)_2$ and $FeSO_4$. After separating these compounds, the latter substance is converted to TiO_2 in basic solution.

$$FeTiO_3(s) + 3\ H_2SO_4(\ell) \longrightarrow FeSO_4(aq) + Ti(SO_4)_2(aq) + 3\ H_2O(\ell)$$

$$Ti^{4+}(aq) + 4\ OH^-(aq) \longrightarrow TiO_2(s) + 2\ H_2O(\ell)$$

How many liters of 18.0 M H_2SO_4 is required to react completely with 1.00 kg of ilmenite? How many kilograms of TiO_2 can theoretically be produced by this sequence of reactions?

22. In the process described in Study Question 21, ilmenite ore is leached with sulfuric acid. This leads to the significant environmental problem of disposal of the iron(II) sulfate (which, in its hydrated form, is commonly called "copperas"). To avoid this, it has been suggested that HCl be used to leach ilmenite so that the iron-containing product is

$FeCl_2$. This can be treated with water and air to give commercially useful iron(III) oxide and regenerate HCl by the reaction

$$2\ FeCl_2(aq) + 2\ H_2O(\ell) + \tfrac{1}{2}\ O_2(g) \longrightarrow Fe_2O_3(s) + 4\ HCl(aq)$$

(a) Write a balanced equation for the treatment of ilmenite with aqueous HCl to give iron(II) chloride, titanium(IV) oxide, and water.
(b) If the equation written in part (a) is combined with the preceding equation for oxidation of $FeCl_2$ to Fe_2O_3, is the HCl used in the first step recovered in the second step?
(c) How many grams of iron(III) oxide can be obtained from 1 ton (908 kg) of ilmenite in this process?

23. Worldwide production of nickel is 750,000 tons per year, most of which comes from Canada. The common mineral containing nickel is pentlandite, NiS. Roasting of Ni in air produces NiO, which is reduced to the metal using H_2.

$$NiO(s) + H_2(g) \longrightarrow Ni(s) + H_2O(g)$$

The *Mond process* was, at one time, the method used to purify the metal. At a temperature of about 50 °C and at atmospheric pressure, the impure metal reacts with CO to give a volatile compound tetracarbonylnickel(0), $Ni(CO)_4$.

$$Ni(s) + 4\ CO(g) \rightleftarrows Ni(CO)_4(\ell)$$

If this compound is passed into another part of the reactor and heated to 250 °C, it reverts to pure Ni and CO. You wish to produce 1 ton (908 kg) of pure nickel. How much NiS, H_2, and CO are required?

Ligands and Formulas of Complexes

24. Which of the following ligands are expected to be monodentate and which are polydentate?
(a) CH_3NH_2 (e) en
(b) $C_2O_4^{2-}$ (f) phen
(c) Br^- (g) N^{3-}
(d) CH_3CN

25. Only one of the following nitrogen compounds or ions, NH_4^+, NH_3, or NH_2^-, is incapable of serving as a ligand. Identify this species and explain your answer.

26. Give the oxidation number of the metal ion in each of the following compounds:
(a) $[Mn(NH_3)_6]SO_4$
(b) $K_3[Co(CN)_6]$
(c) $[Co(NH_3)_4Cl_2]Cl$
(d) $Mn(en)_2Cl_2$

27. Give the formula of a complex constructed from one Ni^{2+} ion, one ethylenediamine ligand, three ammonia molecules, and one water molecule. Is the complex neutral or is it charged? If charged, give the charge.

28. Give the formula of a complex formed from one Co^{3+} ion, two ethylenediamine molecules, one water molecule, and one chloride ion. Is the complex neutral or charged? If charged, give the net charge on the ion.

Naming

29. Write formulas for the following ions or compounds:
(a) dichlorobis(ethylenediamine)nickel(II)
(b) potassium tetrachloroplatinate(II)
(c) potassium dicyanocuprate(I)
(d) diaquatetraammineiron(II)

30. Write formulas for the following ions or compounds:
(a) diamminetriaquahydroxochromium(III) nitrate
(b) hexaammineiron(III) nitrate
(c) pentacarbonyliron(0)
(d) ammonium tetrachlorocuprate(II)

31. Name the following ligands:
(a) OH^- (c) I^-
(b) O^{2-} (d) $C_2O_4^{2-}$

32. Name the following compounds pictured in Figure 23.13:
(a) $Cu(NH_3)_4SO_4 \cdot H_2O$
(b) $K_3[Fe(C_2O_4)_3]$
(c) $[Co(NH_3)_4CO_3]NO_3$
(d) $[Co(NH_3)_5(NO_2)]Cl_2$

33. Name the following ions or compounds:
(a) $[Ni(C_2O_4)_2(H_2O)_2]^{2-}$
(b) $[Co(en)_2(NO_2)_2]^+$
(c) $[Co(en)_2(NH_3)Cl]^{2+}$
(d) $Pt(NH_3)_2(C_2O_4)$

34. Give the name or formula for each ion or compound, as appropriate.
(a) pentaaquahydroxoiron(III) ion
(b) $K_2[Ni(CN)_4]$
(c) $K[Cr(C_2O_4)_2(H_2O)_2]$
(d) Ammonium hexachloroplatinate(IV)

35. Give the name or formula for each ion or compound, as appropriate.
(a) dichlorotetraaquachromium(III) chloride
(b) $[Cr(NH_3)_5SO_4]Cl$
(c) sodium tetrachlorocobaltate(II)
(d) $[Co(C_2O_4)_3]^{3-}$

Isomerism

36. Draw all possible geometric isomers of
(a) $Fe(NH_3)_4Cl_2$
(b) $Pt(NH_3)_2(SCN)(Br)$ (SCN is bonded to Pt^{2+} through S)
(c) $Co(NH_3)_3(NO_2)_3$ (NO_2 is bonded to Co^{3+} through N)
(d) $[Co(en)Cl_4]^-$

37. Which of the following complexes can have geometrical isomers? (If isomers are possible, draw the structures of the isomers and label them as *cis* or *trans,* or as *fac* or *mer* if appropriate).
(a) $[Co(H_2O)_4Cl_2]^+$
(b) $Co(H_2O)_3F_3$
(c) $[Pt(NH_3)Br_3]^-$
(d) $[Co(en)_2(NH_3)Cl]^{2+}$

38. Determine whether the following molecules possess a chiral center:
(a) CH_2Cl_2
(b) $H_2NCH(CH_3)CO_2H$
(c) $ClCH(OH)CH_2Cl$
(d) $CH_3CH_2CH{=}CHC_6H_5$

39. Decide whether each of the following molecules has an enantiomer:

(a) 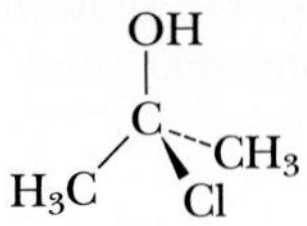 (b) 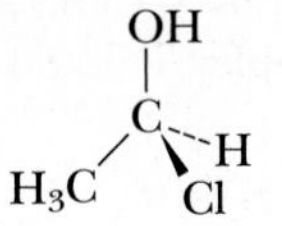 (c) 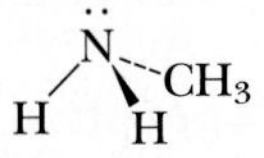

40. Four isomers are possible for $[Co(en)(NH_3)_2(H_2O)Cl]^{2+}$. (Two of the four have optical isomers and so each has a nonsuperimposable mirror image.) Draw the structures of the four isomers.

41. Draw all possible isomers (geometric and optical) for the ion $[Cr(C_2O_4)_2(H_2O)_2]^-$.

42. Determine whether the following complexes have a chiral metal center:
(a) $[Fe(en)_3]^{2+}$
(b) *fac*-$[Co(en)(H_2O)Cl_3]^+$
(c) *cis*-$[Co(en)_2Br_2]^+$
(d) $Pt(NH_3)(H_2O)(Cl)(NO_2)$ (square-planar Pt)

Magnetism of Coordination Complexes

43. What *d* electron configurations exhibit both high and low spin in octahedral complexes?

44. Can $[Ni(NH_3)_6]^{2+}$ form both low-spin and high-spin complexes? Explain your answer.

45. Depict high- and low-spin configurations for each of the complexes below. Tell whether each is diamagnetic or paramagnetic. Give the number of unpaired electrons for the paramagnetic cases.
(a) $[Fe(CN)_6]^{4-}$ (c) $[Fe(H_2O)_6]^{3+}$
(b) $[Co(NH_3)_6]^{3+}$ (d) $[CrF_6]^{4-}$

46. From experiment we know that $[CoF_6]^{3-}$ is paramagnetic and $[Co(NH_3)_6]^{3+}$ is diamagnetic. Using the crystal field model, depict the electron configuration for each ion. What can you conclude about the effect of the ligand on the magnitude of Δ_o?

47. From experiment we know that $[Mn(H_2O)_6]^{2+}$ has five unpaired electrons, whereas $[Mn(CN)_6]^{4-}$ has one unpaired electron. Using the crystal field model, depict the electron configuration for each ion. What can you conclude about the effect of the ligand on the magnitude of Δ_o?

Color

48. Arrange the following ligands in order of increasing crystal field splitting:
(a) CN^- (c) F^-
(b) NH_3 (d) H_2O

49. In water, the titanium(III) ion, $[Ti(H_2O)_6]^{3+}$, is violet. (Its broad absorption band occurs at about 500 nm.) What color light is absorbed by the ion?

50. The chromium(II) ion in water, $[Cr(H_2O)_6]^{2+}$, absorbs light with a wavelength of about 700 nm. What color is the solution?

General Questions

51. For the complex $[Fe(H_2O)_6]^{2+}$ identify
(a) The coordination number of iron
(b) The coordination geometry for iron
(c) The oxidation state of iron
(d) The number of unpaired electrons, assuming high spin
(e) Whether the complex is diamagnetic or paramagnetic

52. For the complex $[Co(en)(NH_3)_2Cl_2]^+$ identify
(a) The coordination number of cobalt
(b) The coordination geometry for cobalt
(c) The oxidation state of cobalt
(d) The number of unpaired electrons, assuming low spin
(e) Whether the complex is diamagnetic or paramagnetic

53. Predict whether each complex below is high or low spin and whether each is paramagnetic or diamagnetic, based on the position of the ligands in the spectrochemical series. If paramagnetic, give the number of unpaired electrons. Use the crystal field model to find the electron configuration of each ion.
(a) $[Fe(CN)_6]^{4-}$
(b) $[MnF_6]^{4-}$
(c) $[Cr(en)_3]^{3+}$
(d) $[Cu(phen)_3]^{2+}$

54. In $Pt(NH_3)_2(C_2O_4)$ the metal ion is surrounded by a square plane of coordinating atoms. Draw a structure for this molecule. Give the oxidation number of the platinum and the name of the compound.

55. A complex formed from a cobalt(III) ion, five ammonia molecules, a bromide ion, and a sulfate ion exists in two forms: one dark violet (A) and the other violet-red (B). The dark violet form (A) gives a precipitate with $BaCl_2$ but none with $AgNO_3$. Form B behaves in the opposite manner. This tells you that one form is $[Co(NH_3)_5Br]SO_4$ and the other is $[Co(NH_3)_5(SO_4)]Br$. Which compound is A, and which is B? (*Note:* Only when ions such as Br^- or SO_4^{2-} are *not* directly coordinated to the metal ion can they form free ions in aqueous solution.)

56. A platinum-containing compound, known as Magnus's green salt, has the formula $[Pt(NH_3)_4][PtCl_4]$ (where both platinum ions are Pt^{2+}). Name the compound.

57. Give the formula of a complex ion formed from one Pt^{2+} ion, one nitrite ion (NO_2^-, which binds to Pt^{2+} through N), one chloride ion, and two ammonia molecules. Are isomers possible? If so, draw the structure of each isomer and tell what type of isomerism is observed. Name the ion.

58. A 0.213-g sample of uranyl(VI) nitrate, $UO_2(NO_3)_2$, is dissolved in 20.0 mL of 1.0 M H_2SO_4 and shaken with Zn. The

zinc reduces the uranyl ion, UO_2^{2+}, to an ion with a lower oxidation number, U^{n+} ($n < 6$). The uranium-containing solution, after reduction, is titrated with 0.0173 M $KMnO_4$. The potassium permanganate oxidizes the uranium back to the +6 oxidation state. Given that 12.47 mL of the potassium permanganate is required for titration to a permanent pink color at the equivalence point, calculate the oxidation number of the uranium after the uranyl(VI) nitrate is reduced with zinc. Write a balanced, net ionic equation for the oxidation of U^{n+} (in which you now know n) by MnO_4^- in acid solution to give UO_2^{2+} and Mn^{2+}.

59. Comment on the fact that, although an aqueous solution of cobalt(III) sulfate is diamagnetic, the solution becomes paramagnetic when a large excess of fluoride ion is added.

60. Experiments show that $K_4[Cr(CN)_6]$ is paramagnetic and has two unpaired electrons. In contrast, the related complex $K_4[Cr(NCS)_6]$ is paramagnetic with four unpaired electrons. Account for these differences using the crystal field model. Predict where the SCN^- ion occurs in the spectrochemical series relative to CN^-.

61. In this question, we wish to explore the differences between metal coordination by monodentate and bidentate ligands. Formation constants, K_f, for $[Ni(NH_3)_6]^{2+}(aq)$ and $[Ni(en)_3]^{2+}(aq)$ follow.

$$Ni^{2+}(aq) + 6\ NH_3(aq) \rightleftharpoons [Ni(NH_3)_6]^{2+}(aq) \qquad K_f = 10^8$$

$$Ni^{2+}(aq) + 3\ en(aq) \rightleftharpoons [Ni(en)_3]^{2+}(aq) \qquad K_f = 10^{18}$$

The difference in K_f between these complexes, reflecting a large increase in stability of the chelated complex, is caused by the *chelate effect*. Recall that K is related to the standard free energy of the reaction by $\Delta G° = -RT \ln K$ and $\Delta G° = \Delta H° - T\Delta S°$. Here we know from experiment that $\Delta H°$ for the NH_3 reaction is −109 kJ/mol, and $\Delta H°$ for the en reaction is −117 kJ/mol. Is the difference in $\Delta H°$ sufficient to account for the 10^{10} difference in K_f? Comment on the role of entropy in the second reaction compared with that in NH_3 reaction.

62. The glycinate ion, $H_2NCH_2CO_2^-$ (formed by deprotonation of the amino acid glycine), can function as a bidentate ligand, as pictured here.

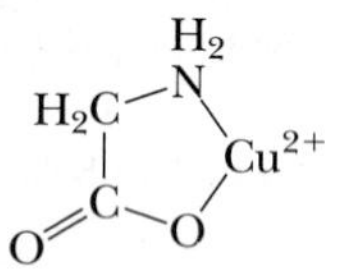

Draw all of the isomers that can be formed by the complex ion $Cu(H_2NCH_2CO_2)_2(H_2O)_2$.

63. You have a sample of an alloy of copper and aluminum and wish to determine the weight percentage of each element in the mixture. A 2.1309-g sample was first dissolved in a mixture of HCl and HNO_3. The resulting solution was made basic with excess ammonia, and the $Al(OH)_3$ that precipitated was collected and dried in a furnace to give 3.8249 g of Al_2O_3. What is the weight percentage of Al and Cu in the alloy?

64. Three different compounds of chromium(III) with water and chloride ion have the same composition: 19.51% Cr, 39.92% Cl, and 40.57% H_2O. One of the compounds is violet and dissolves in water to give a complex ion with a 3+ charge and three Cl^- ions. All three chloride ions precipitate immediately as AgCl on adding $AgNO_3$. Draw the structure of the complex ion and name the compound.

Conceptual Questions

65. It is usually observed that stability of analogous complexes $[ML_6]^{n+}$ is in the order $Mn^{2+} < Fe^{2+} < Co^{2+} < Ni^{2+} < Cu^{2+} > Zn^{2+}$. (This order of ions is called the *Irving-Williams series*.) Look up the values of formation constants for ammonia complexes of Co^{2+}, Ni^{2+}, Cu^{2+}, and Zn^{2+} in Appendix I and verify this statement. (See also Section 19.9.)

66. Describe an experiment to determine the following:
   (a) Whether the cation in $[Fe(H_2O)_6]Cl_2$ is a low-spin or a high-spin complex
   (b) Whether nickel in $K_2[NiCl_4]$ is square-planar or tetrahedral
   (c) Whether a complex has the structure $[Co(NH_3)_5Br]SO_4$ or $[Co(NH_3)_5(SO_4)]Br$

67. How many geometric isomers of the complex $[Cr(dmen)_3]^{3+}$ can exist? dmen is the bidentate ligand 1,1-dimethylethylenediamine, $(CH_3)_2NCH_2CH_2NH_2$.

68. Diethylenetriamine (dien), $H_2NCH_2CH_2NHCH_2CH_2NH_2$, is capable of serving as a tridentate ligand.
   (a) Draw the structure of *fac*-Cr(dien)Cl_3 and *mer*-Cr(dien)Cl_3.
   (b) Two different geometric isomers of *mer*-Cr(dien)Cl_2Br are possible. Draw the structure for each.
   (c) Three different geometric isomers are possible for $[Cr(dien)_2]^{3+}$. Two have the dien ligand in a *fac* configuration, and one has the ligand in the *mer* orientation. Draw the structure of each isomer.

69. The square-planar complex Pt(en)Cl_2 has chloride ligands in *cis* configuration. No *trans* isomer is known. Based on the bond lengths and bond angles of carbon and nitrogen in the ethylenediamine ligand, explain why the *trans* compound is not possible.

70. The complex ion $[Co(CO_3)_3]^{3-}$, an octahedral complex with bidentate carbonate ions as ligands, has one absorption in the visible region of the spectrum at 640 nm. From this information
   (a) Predict the color of this complex, and explain your reasoning.
   (b) Place the carbonate ion in the proper place in the spectrochemical series.
   (c) Predict whether $[Co(CO_3)_3]^{3-}$ will be paramagnetic or diamagnetic.

Darleane C. Hoffman

(Lawrence Berkeley Laboratory)

One of the world's foremost authorities on the heaviest elements, Dr. Darleane Christian Hoffman, says she was once "a fairly shy girl from a small town." Born Nov. 8, 1926, in Terril, Iowa, she spent her childhood in a series of different communities where her father was mathematics teacher and school superintendent. Her mother, a homemaker, nurtured Darleane's interest in music and art.

Though she entered Iowa State College as an applied art major, Hoffman graduated in 1948 with a B.S. degree in chemistry. After completing her Ph.D. in physical chemistry (nuclear) from Iowa State in 1951, she joined the staff of Oak Ridge National Laboratory for one year.

Hoffman moved to the Los Alamos National Laboratory in 1953, where she was ultimately named leader of the Isotope and Nuclear Chemistry Division. In 1984 she became professor of nuclear chemistry at the University of California, Berkeley, and leader of the Heavy Element Nuclear and Radiochemistry Group at Lawrence Berkeley Laboratory. In addition, she is currently director of the Glenn T. Seaborg Institute for Transactinium Science at Lawrence Livermore National Laboratory.

The book, Women in Chemistry and Science, *describes Hoffman's discovery of naturally occurring plutonium-244 as "an experimental tour de force." Among other fundamental discoveries, her research team is credited with confirming the discovery of element 106, and with conducting the first aqueous chemistry studies of element 105, hahnium.*

Hoffman's many honors include the 1990 Garvan Medal, presented by the American Chemical Society (ACS) to an outstanding woman chemist, and the ACS Award in Nuclear Chemistry in 1983. She has chaired the National Academy of Sciences' Committee on Nuclear and Radiochemistry and served on the International Union of Pure and Applied Chemistry's Commission on Radiochemistry and Nuclear Techniques. She and her husband, Dr. Marvin Hoffman, a nuclear physicist, have a daughter, Maureane, and a son, Daryl, both of whom are M.D.'s. Her daughter has a Ph.D. as well.

Role models and mentors

When I first went to Iowa State College, I wasn't sure whether I wanted to study art or mathematics. Ultimately, I decided to give art a try. During my first semester, however, I had to take chemistry, English, history, and physical education in addition to applied art, and I quickly realized that I liked chemistry better than anything else. This was fortunate, because I decided my applied art skills were marginal!

I had a marvelous chemistry teacher named Professor Nellie Nayor who, because of her excellent teaching, inspired me to change majors. When I went to talk to my advisor, an applied art teacher, she asked, "Do you think chemistry is a suitable profession for a woman?" I said, "Well, I don't know, but that's what I want to do. I have this marvelous woman teacher, so obviously it can be suitable for a woman," and I became a chemistry major in the spring of 1945. Although there were very few women attending chemistry classes, this was not a problem for me. I never felt that I suffered any discrimination because I was a young woman.

During my junior year, the Institute of Atomic Research at Iowa State University advertised positions for two research assistants. I was a fairly shy girl from a small town and I didn't know whether I dared to try or not! But I applied and was accepted for the job, which led to my work with Prof. Donald S. Martin, Jr., an inorganic and nuclear chemist.

We started out making Geiger counters to detect radioactivity. My first job was a simple one: I had to split mica into very thin sheets to cover the counter windows. We then used these detectors in our studies of radioactivity, and it was this research that prompted me to become a nuclear and radiochemist.

At the edge of a new frontier

My current research focuses on the chemical and nuclear properties of the heaviest elements, which I define as those with proton numbers greater than 100. Fermium has proton number 100. The next heaviest element, 101, is mendelevium. Up to fermium, you can make the elements by neutron capture reactions in reactors with a very high flux (or flow) of neutrons. Beginning with mendelevium, however, you have to produce these elements one atom at a time, in a cyclotron or linear accelerator.

My research team has been investigating the aqueous chemistry of elements 104, 105, 106. Recently, for example, we studied the properties of hahnium, element 105, to find out whether it would behave like its lighter homologs in Group 5B of the periodic table. The research was conducted using the Lawrence Berkeley Laboratory's 88-inch cyclotron. Oxygen-18 atoms are ionized and then accelerated in the cyclotron so that they impinge on the back of a berkelium-249 target. In this way, oxygen-18 and berkelium fuse to make element 105.

The longest-lived known isotope of element 105 has a half-life of 30 seconds. Consequently, my research team of graduate students had to work very quickly, using either manual or automated techniques. Using manual techniques, the students performed 800 one-minute chemical separations. It sounds tedious, but the goal of exploring the chemistry of element 105 for the first time is exciting. We were able to show that element 105's most stable oxidation state in aqueous solution is +5—just like its homologs, niobium and tantalum.

I'm often asked why I choose to focus on these very heaviest elements, which last only a few seconds. The answer is that I like the idea of working at the edge of stability. It's like wanting to climb the highest mountain, or to do or see something that nobody has ever seen before.

The thrill of discovery

When my colleagues and I isolated plutonium-244 from some natural samples, it was a real thrill because we had worked very hard to process the samples from Precambrian bastnasite, a rare earth ore. We had done the plutonium separations and the blank separations, and then we sent the samples away to be analyzed at a state-of-the-art mass spectroscopy laboratory.

I remember going to the opera that night, thinking, "I know we've got it!" Of course, I didn't know for sure, but the following week, my colleagues at the mass spectrometry laboratory called and said, "Yes, we have seen the atoms of plutonium-244." Previously, plutonium-244 had been identified only in non-natural samples.

Another big moment for me was the discovery that the spontaneous fission of fermium tends to be symmetrical. That is, instead of fissioning into two unequal parts, which is what happens in conventional fission, fermium separates into two nearly equal mass fragments. This was very exciting because it had not been predicted. In fact, when I described the phenomenon at a meeting, I was told by the physicists, "Well, you chemists just don't understand these things. You'd better go back and do it over." But my findings proved to be correct, and it created a real renaissance in the study of spontaneous fission.

The future of nuclear power

Nuclear energy now generates about 20% of the electrical power consumed in the United States. In France, it amounts to nearly 80% of the country's total electricity consumption. Unlike coal-fired electrical generating plants, nuclear power is clean; it doesn't generate greenhouse gases. So we definitely need nuclear power in our energy future, and I believe that it can be produced safely.

Many people point to the Chernobyl disaster in Russia as proof that nuclear power isn't safe. But that reactor had a completely different design than U.S. reactors.

A perceived barrier to the use of nuclear power in this country is the question of where to store spent reactor fuel. I personally don't see this as a big problem. You can store the spent fuel in cooling pools at the reactor site, or you can store it elsewhere when the pools become full. Spent fuel rods are insoluble. They can therefore be encapsulated and transported easily to an underground repository such as the one proposed at Yucca Mountain near Las Vegas, Nevada. Multiple barriers at the site, including the natural sorptive properties of the geologic medium, would prevent migration of the fuel fission products.

I've been told that people who live near nuclear power plants in France receive free electricity. Perhaps we need similar incentive programs in the United States for communities willing to safely store spent nuclear fuel.

What's in a name? The heaviest elements

Uranium was the heaviest element known for more than 150 years. Since 1940, however, researchers have produced a host of additional elements. Glenn T. Seaborg, a professor of chemistry at the University of California, Berkeley, is credited with co-discovering ten transuranium elements—most notably, plutonium in 1941. The very heaviest elements, from nobelium to meitnerium (and now elements 110 and 111), have been identified in the past 40 years.

The names for elements 104 through 109 are the subject of some controversy. Although new elements have traditionally been named by the researchers who discovered them, the International Union of Pure and Applied Chemistry (IUPAC) has recommended different names for elements 104–109 than those suggested by the American Chemical Society. The controversy should become even more interesting now that the discovery of elements 110 and 111 has just been reported.

Now we are all wondering why the IUPAC's Commission on Inorganic Nomenclature chose to disregard the names proposed by the discoverers. What they've said is that the discoverers have no right to name the elements, that they can only suggest names to the IUPAC. This goes against all historical precedent.

The IUPAC Commission also decided elements should not be named after living people. The discoverers recently named element 106 seaborgium in honor of Glenn Seaborg, but IUPAC has decided not to use this name. This is another decision with which we strongly disagree. A person is penalized for longevity, even though his scientific accomplishments have stood the test of time.

CHAPTER 24

Nuclear Chemistry

Chemical Puzzler

What radioactive element is contained in many household smoke detectors? How do smoke detectors work?

A recent report issued by the National Research Council stated that "The future vigor and prosperity of American medicine, science, technology, and national defense clearly depend on continued use and development of nuclear techniques and use of radioactive isotopes." Nuclear chemistry, a subject that bridges chemistry and physics, has a significant influence on our society. Radioactive isotopes are now widely used in medicine, and some of the latest diagnostic techniques such as positron emission tomography (PET) scans depend on radioactivity. Similarly, your home may be protected with a smoke detector that contains a radioactive element, and research in all fields of science uses radioactive elements and their compounds. The national security of the United States since World War II has depended on nuclear weapons, and a number of nations around the world depend on nuclear reactors as a source of electricity. No matter what your reason for taking a college course in chemistry—to prepare for a career in one of the sciences or simply to gain knowledge as a concerned citizen—you should know something about nuclear chemistry. This chapter, therefore, considers changes in the atomic nucleus and the effects of those changes, the fissioning and fusion of nuclei and the energy that can be derived from such changes, the units used to measure radioactivity, and the uses of radioactive isotopes.

On August 2, 1939, as the world was on the brink of World War II, Albert Einstein sent a letter to President Franklin D. Roosevelt. In this letter, which profoundly changed the course of history, Einstein called attention to work being done on the physics of the atomic nucleus. He said he and others believed this work suggested the possibility that "uranium may be turned into a new and important source of energy . . . and [that it was] conceivable . . . that extremely powerful bombs of a new type may thus be constructed. . . . "

Powerful indeed! Einstein's letter was the beginning of the Manhattan Project, the project that led to the detonation of the first atomic bomb at 5:30 AM on July 16, 1945, in the desert of New Mexico. The rest of the world would learn the truth of the power locked in the atomic nucleus a few weeks later, on August 6 and August 9, when the United States used atomic weapons against Japan. J. Robert Oppenheimer, the director of the atomic bomb project, is said to have recalled the following words from the sacred Hindu epic, Bhagavad-Gita, at the moment of the explosion of the first atomic bomb.

The Making of the Atomic Bomb by Richard Rhodes (Simon and Schuster, 1986) is a comprehensive history of the exploration of atomic physics in this century and of the events leading up to the development of atomic weapons. This very readable book is highly recommended.

If the radiance of a thousand suns
Were to burst at once upon the sky,
That would be like the splendor of the Mighty One . . .
I am become Death,
The shatterer of worlds.

In the almost 50 years since the first—and thankfully only—use of an atomic weapon in war, more powerful weapons have been developed and stockpiled by a number of nations. With the end of the Cold War, fears of a nuclear holocaust are fading, but they are being replaced to some extent by the concern that Third World nations have developed nuclear weapons. The respected magazine *The Bulletin of Atomic Scientists* has used for many years the symbol of a clock with its hands near midnight, illustrating the danger faced by the world from atomic weapons. Even with the end of the Cold War, the hands have moved back only a little.

Although nuclear reactions have been associated in the public consciousness with weapons for more than 40 years, nuclear chemistry has been continuously developed for the generation of electric power, for the diagnosis and treatment of disease, and for food preservation.

24.1 THE NATURE OF RADIOACTIVITY

In Chapter 2 (Section 2.2) we described the work of Henri Becquerel and Marie and Pierre Curie with radioactive elements. Their discovery of radioactivity was fundamental to opening the way to our current understanding of atomic structure.

By the early part of this century it was known that there are three basic kinds of radiation: alpha (α), beta (β), and gamma rays (γ). In the late 19th century Ernest Rutherford and J. J. Thomson were studying the radiation from uranium and thorium. Rutherford found that "There are present at least two distinct types of radiation—one that is readily absorbed, which will be termed for convenience α [alpha] radiation, and the other of a more penetrative character, which will be termed β [beta] radiation." **Alpha radiation,** he discovered, was composed of particles, which, when passed through an electric field, were attracted to the negative side of the field (Figure 2.3); indeed, his later studies showed these particles to be helium nuclei, ${}^{4}_{2}He^{2+}$, which were ejected at high speeds from a radioactive element (Table 24.1). As might be expected, such massive particles have limited penetrating power and can be stopped by several sheets of ordinary paper or clothing (Figure 24.1).

See the biographies of Marie Curie (page 63) and Ernest Rutherford (page 70).

In the same experiment, Rutherford also found that because the beam of radiation was attracted to the electrically positive plate, β radiation must be composed of negatively charged particles. Becquerel's work showed that these

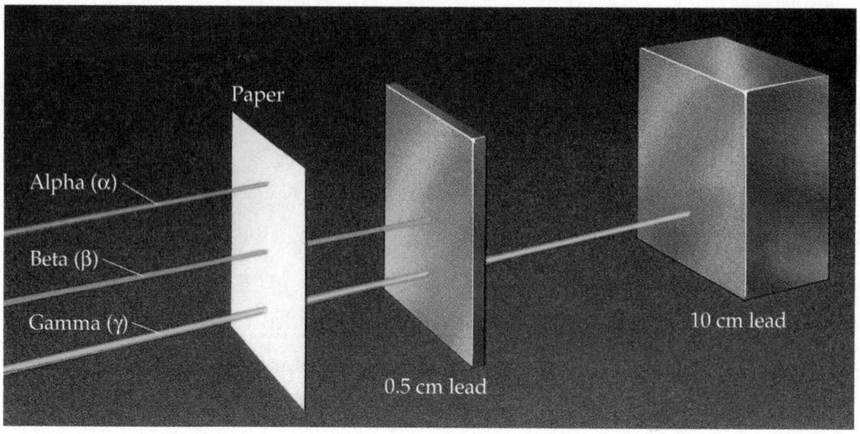

Figure 24.1 The relative penetrating ability of the three major types of nuclear radiation. Heavy, highly charged α particles interact with matter most strongly and so are stopped by a piece of paper or a layer of skin. Beta (β) particles and positrons are lighter, have a lower charge, and so interact to a smaller extent with matter; they are stopped by about 0.5 cm of lead. Gamma (γ) rays are uncharged, massless particles and are the most penetrating.

TABLE 24.1 Characteristics of α, β, and γ Emissions

Name	Symbols	Charge	Mass (g/particle)
Alpha	${}^{4}_{2}He^{2+}$, ${}^{4}_{2}\alpha$	+2	6.65×10^{-24}
Beta	${}^{0}_{-1}e$, ${}^{0}_{-1}\beta$	−1	9.11×10^{-28}
Gamma	${}^{0}_{0}\gamma$, γ	0	0

particles have an electric charge and mass equal to those of an electron. Thus, **beta (β) particles** are electrons ejected at high speeds from some radioactive nuclei. They are more penetrating than α particles, because at least a $\frac{1}{8}$ in. piece of aluminum is necessary to stop β particles, and they penetrate several millimeters of living bone or tissue.

Rutherford hedged his bets when he said there were *at least* two types of radiation. Indeed, a third type was later discovered by P. Villard, a Frenchman, who named it **γ (gamma) radiation,** using the third letter in the Greek alphabet in keeping with Rutherford's scheme. Unlike α and β radiation, which are particulate in nature, γ radiation is a form of electromagnetic radiation like x-radiation, although γ rays are even more energetic than x-rays (Figure 7.3). Furthermore, γ rays have no electrical charge and so are not affected by an electrical field (Figure 2.3). Finally, γ radiation is the most penetrating; it can pass completely through the human body. Thick layers of lead or concrete are required to minimize penetration.

24.2 NUCLEAR REACTIONS

Equations for Nuclear Reactions

Ernest Rutherford found that radium not only emits α particles but that it also produces the radioactive gas radon in the process. Such observations led Rutherford and Frederick Soddy, in 1903, to propose the revolutionary theory that radioactivity is the result of a natural change of the isotope of one element into the isotope of a *different* element. In such changes, called **nuclear reactions,** or *transmutations,* an unstable nucleus emits radiation and is converted into a more stable nucleus of a different element. Thus, a nuclear reaction results in a change in atomic number and often a change in mass number as well. For example, the reaction studied by Rutherford can be written as

$$^{226}_{88}\text{Ra} \longrightarrow {}^{4}_{2}\text{He} + {}^{222}_{86}\text{Rn}$$

Notice that when a radioactive atom decays, the emission of a charged particle leaves a charged atom. Thus, when ^{226}Ra decays it gives a helium-4 cation (He^{2+}) and a ^{222}Rn anion (Rn^{2-}). By convention, the ion charges are not shown in balanced equations for nuclear reactions.

In this balanced equation the subscripts are the atomic numbers and the superscripts are the mass numbers.

The atoms in molecules and ions are rearranged in a chemical change; they are not created or destroyed. The number of atoms remains the same. Similarly, in nuclear reactions the total number of nuclear particles, or **nucleons** (protons plus neutrons), remains the same. The essence of nuclear reactions, however, is that one nucleon can change into a different nucleon. A proton can change to a neutron or a neutron can change to a proton, but the total number of nucleons remains the same. Therefore, *the sum of the mass numbers of reacting nuclei must equal the sum of the mass numbers of the nuclei produced.* Furthermore, to maintain charge balance, *the sum of the atomic numbers of the products must equal the sum of the reactants.* These principles may be verified for the preceding nuclear equation.

	$^{226}_{88}$**Ra**	$\longrightarrow$	$^{4}_{2}$**He**	+	$^{222}_{86}$**Rn**
	radium-226		α particle		radon-222
mass number: (protons + neutrons)	226	$\longrightarrow$	4	+	222
atomic number: (protons)	88	$\longrightarrow$	2	+	86

Reactions Involving α and β Particles

One way a radioactive isotope can disintegrate or decay is to eject an α particle from the nucleus as illustrated by the conversion of radium to radon and by the following reaction:

	$^{234}_{92}$**U**	$\longrightarrow$	$^{4}_{2}$**He**	+	$^{230}_{90}$**Th**
	uranium-234		α particle		thorium-230
mass number:	234	$\longrightarrow$	4	+	230
atomic number:	92	$\longrightarrow$	2	+	90

Notice that in α emission the *atomic number decreases by two* units and the *mass number decreases by four* units for each α particle emitted.

Emission of a β particle is another way for an isotope to decay. For example, loss of a β particle by uranium-235 is represented by

	$^{235}_{92}$**U**	$\longrightarrow$	$^{0}_{-1}\beta$	+	$^{235}_{93}$**Np**
	uranium-235		β particle		neptunium-235
mass number:	235	$\longrightarrow$	0	+	235
atomic number:	92	$\longrightarrow$	−1	+	93

Because a β particle has a charge of -1, electric balance makes the atomic number of the product *greater* by one than that of the reacting nucleus. The mass number does not change, however. The mass number of 0 for the electron is due to the small mass of the particle (only 1/1836 the mass of a proton).

How does a nucleus, composed only of protons and neutrons, eject an electron? It is generally accepted that a series of steps is involved, but the net process is

$$\underset{\text{neutron}}{{}^{1}_{0}\text{n}} \longrightarrow \underset{\text{electron}}{{}^{0}_{-1}\beta} + \underset{\text{proton}}{{}^{1}_{1}\text{p}}$$

where we use the symbol p for a proton. The ejection of a β particle always means that a new element is formed with an atomic number *one unit greater* than the decaying nucleus.

A nucleus formed as a result of an α or β emission is generally in an excited state and so also emits a γ ray.

In many cases, the emission of an α or β particle results in the formation of an isotope that is also unstable and therefore radioactive. The new radioactive isotope may therefore undergo a number of successive transformations until a stable, nonradioactive isotope is finally produced. Such a series of reactions is called a **radioactive series.** One such series begins with uranium-238 and ends with lead-206, as illustrated in Figure 24.2. The first step in the series is

$${}^{238}_{92}\text{U} \longrightarrow {}^{4}_{2}\text{He} + {}^{234}_{90}\text{Th}$$

and the equation for the final step, the conversion of polonium-210 to lead-206, is

$${}^{210}_{84}\text{Po} \longrightarrow {}^{4}_{2}\text{He} + {}^{206}_{82}\text{Pb}$$

EXAMPLE 24.1 *Radioactive Series*

The second, third, and fourth steps in the uranium-238 series in Figure 24.2 involve emission of first a β particle, then another β particle, and finally an α particle. Write equations to show the products of these steps.

Solution The product of the first step, thorium-234, is our starting point. Figure 24.2 shows that the mass remains the same during the second step but that the atomic number increases by 1 to 91, a result shown by the balanced equation

$$\underset{\text{thorium-234}}{{}^{234}_{90}\text{Th}} \longrightarrow {}^{0}_{-1}\beta + \underset{\text{protactinium-234}}{{}^{234}_{91}\text{Pa}}$$

In the third step Figure 24.2 shows that the mass again stays constant, and the atomic number increases once again by 1.

$$\underset{\text{protactinium-234}}{{}^{234}_{91}\text{Pa}} \longrightarrow {}^{0}_{-1}\beta + \underset{\text{uranium-234}}{{}^{234}_{92}\text{U}}$$

Finally, the fourth step involves α particle emission, so both the mass number and atomic number decline. This is again confirmed in Figure 24.2.

$$\underset{\text{uranium-234}}{{}^{234}_{92}\text{U}} \longrightarrow {}^{4}_{2}\text{He} + \underset{\text{thorium-230}}{{}^{230}_{90}\text{Th}}$$

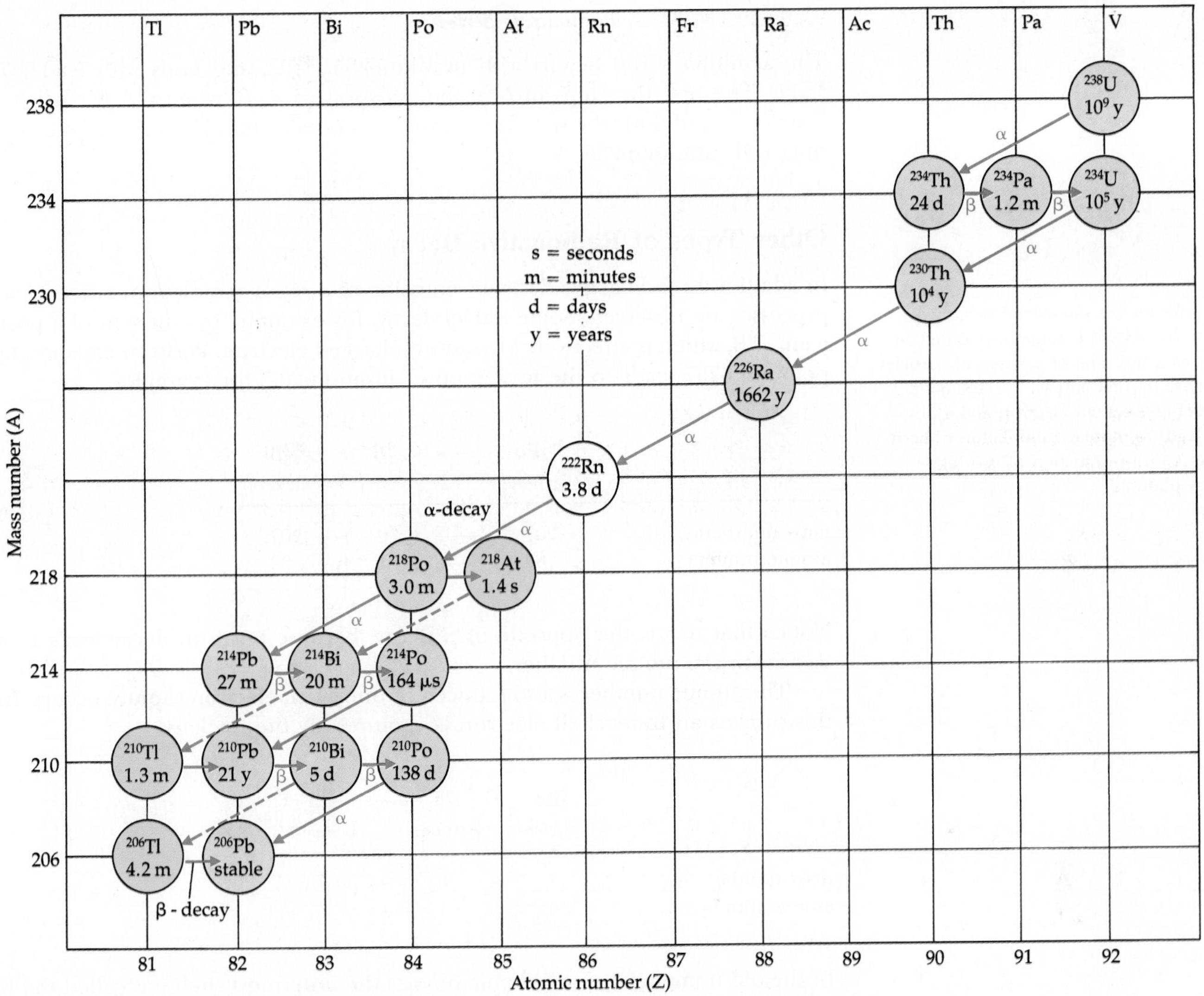

Figure 24.2 A radioactive series beginning with uranium-238 and ending with lead-206. In the first step, for example, ^{238}U emits an α particle to give thorium-234, $^{234}_{90}Th$. This radioactive isotope then emits a β particle to give protactinium-234, $^{234}_{91}Pa$. The protactinium-234 then emits another β particle to continue the series, which finally ends at lead-206, $^{206}_{82}Pb$. The half-life is given for each isotope (see Section 24.4).

EXERCISE 24.1 ***Nuclear Reactions: α and β Emission***

1. Write an equation showing the emission of an α particle by an isotope of neptunium, $^{237}_{93}Np$, to produce an isotope of protactinium.
2. Write an equation showing the emission of a β particle by an isotope of sulfur, $^{35}_{16}S$, to produce an isotope of chlorine. ■

EXERCISE 24.2 ***Radioactive Series***

The actinium series begins with uranium-235, $^{235}_{92}U$, and ends with lead-207, $^{207}_{82}Pb$. The first five steps involve the emission of α, β, α, α, and β particles, respectively. Identify the radioactive isotope produced in each of the steps beginning with uranium-235. ■

Other Types of Radioactive Decay

The positron was discovered by Carl Anderson in 1932. It is sometimes called an "antielectron," one of a group of particles that have become known as "antimatter." Contact between an electron and a positron leads to mutual annihilation of both particles with production of two high-energy photons.

In addition to radioactive decay by emission of α, β, or γ radiation, other decay processes are observed. Some nuclei decay, for example, by emission of a **positron,** $^{0}_{+1}\beta$, which is effectively a positively charged electron. Positron emission by polonium-207 leads to the formation of bismuth-207, for example.

	$^{207}_{84}Po$	$\longrightarrow$	$^{0}_{+1}\beta$	+	$^{207}_{83}Bi$
	polonium-207		positron		bismuth-207
mass number:	207	$\longrightarrow$	0	+	207
atomic number:	84	$\longrightarrow$	+1	+	83

Notice that this is the opposite of β decay, because positron decay leads to a *decrease* in the atomic number.

The atomic number is also reduced by one when **electron capture** occurs. In this process an inner-shell electron is captured by the nucleus.

	$^{7}_{4}Be$	+	$^{0}_{-1}e$	$\longrightarrow$	$^{7}_{3}Li$
	beryllium-7		electron		lithium-7
mass number:	7	+	0	$\longrightarrow$	7
atomic number:	4	+	−1	$\longrightarrow$	3

In the old nomenclature of atomic physics the innermost shell was called the K shell, so the electron capture decay mechanism is sometimes called *K capture.*

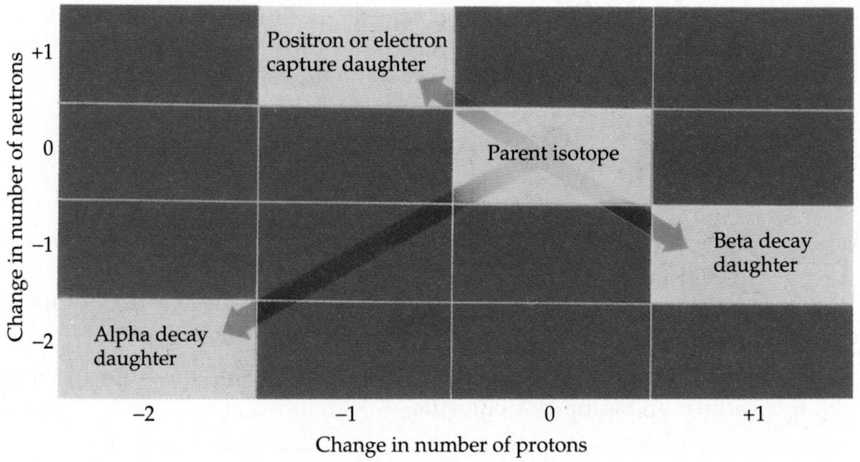

A memory aid for predicting the product of α, β, and γ emission.

In summary, a radioactive nucleus can decay in four common ways, as summarized in the figure on the previous page. In nuclear chemistry, the radioactive isotope that begins a process is called the "parent" and the product is called a "daughter" isotope.

EXERCISE 24.3 ***Nuclear Reactions***

Complete the following nuclear equations. Indicate the symbol, the mass number, and the atomic number of "?".

1. $^{13}_{7}N \rightarrow {}^{13}_{6}C + ?$
2. $^{41}_{20}Ca + {}^{0}_{-1}e \rightarrow ?$
3. $^{90}_{38}Sr \rightarrow {}^{90}_{39}Y + ?$
4. $^{11}_{6}C \rightarrow {}^{11}_{5}B + ?$
5. $^{22}_{11}Na \rightarrow ? + {}^{0}_{+1}\beta$ ■

24.3 STABILITY OF ATOMIC NUCLEI

The fact that some nuclei are unstable (radioactive), and others are stable (nonradioactive), leads us to consider the reasons for stability. Figure 24.3 shows the naturally occurring isotopes of the elements from hydrogen to bismuth. It is quite astonishing that there are so few. Why not hundreds more?

As illustrated by Figure 24.3, there are very few stable combinations of protons and neutrons. Examining those combinations can give us some insight into what factors affect nuclear stability.

In its simplest and most abundant form, hydrogen has only one nuclear particle, the proton. In addition, the element has two other well-known isotopes:

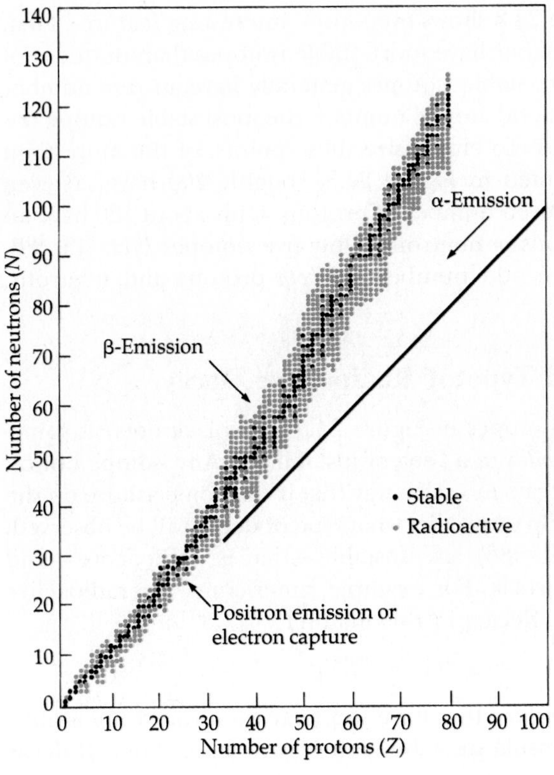

Figure 24.3 A plot of the number of neutrons (N) versus the number of protons (Z) for stable and radioactive isotopes from hydrogen ($Z = 1$) through bismuth ($Z = 83$). The effects of α, β, and positron emission or electron capture are indicated by arrows. For example, radioactive isotopes (indicated by red dots) that lie above the band of stable isotopes (indicated by black dots) decay by β emission. The arrow indicates that this raises the value of Z (by one unit per β particle) and lowers the value of N by one unit.

nonradioactive deuterium, with one proton and one neutron (^2_1H = D), and radioactive tritium, with one proton and two neutrons (^3_1H = T). Helium, the next element, has two protons and two neutrons in its most stable isotope. At the end of the actinide series is element 103, lawrencium, one isotope of which has a mass number of 257 and 154 neutrons. From hydrogen to lawrencium, except for ^1_1H and ^3_2He, the mass numbers of stable isotopes are always *at least twice as large* as the atomic number. In other words, except for ^1_1H and ^3_2He, every isotope of every element has a nucleus containing *at least* one neutron for every proton. Apparently the tremendous *repulsive* forces between the positively charged protons in the nucleus are moderated by the presence of neutrons with no electric charge.

1. For light elements up to Ca ($Z = 20$), the stable isotopes usually have equal numbers of protons and neutrons, or perhaps one more neutron than protons. Examples include ^7_3Li, $^{12}_6C$, $^{16}_8O$, and $^{32}_{16}S$.
2. Beyond calcium the neutron-proton ratio becomes increasingly greater than 1. The band of stable isotopes deviates more and more from the line $N = Z$. It is evident that more neutrons are needed for nuclear stability in the heavier elements. For example, whereas one stable isotope of Fe has 26 protons and 30 neutrons, one of the stable isotopes of platinum has 78 protons and 117 neutrons.
3. Beyond bismuth (83 protons and 126 neutrons) all isotopes are unstable and radioactive. Beyond this point there is apparently no nuclear "super glue" strong enough to hold heavy nuclei together. Furthermore, the rate of disintegration becomes greater the heavier the nucleus. For example, half of a sample of $^{238}_{92}U$ disintegrates in a billion years, whereas half of a sample of $^{257}_{103}Lr$ is gone in only 8 s.
4. A very careful look at Figure 24.3 shows even more interesting features. First, elements of even atomic number have more stable isotopes than do those of odd atomic number. Second, stable isotopes generally have an *even* number of neutrons. For elements of odd atomic number, the most stable isotope has an even number of neutrons. To emphasize these points, of the more than 300 stable isotopes represented in Figure 24.3, roughly 200 have an even number of neutrons *and* an even number of protons. Only about 120 have an odd number of either protons *or* neutrons. Only five isotopes (^2_1H, ^6_3Li, $^{10}_5B$, and $^{14}_7N$, for example) have odd numbers of *both* protons and neutrons.

The Band of Stability and Type of Radioactive Decay

The narrow "band" of stable isotopes in Figure 24.3 (the black dots) is sometimes called the *peninsula of stability* in a "sea of instability." Any isotope not on this peninsula (the red dots) decays in such a way that it can come ashore on the peninsula, and the chart can help us predict what type of decay will be observed.

All elements beyond Bi ($Z = 83$) are unstable—that is, radioactive—and most decay by ejecting an α particle. For example, americium, the radioactive element used in smoke alarms, decays in this manner.

$$^{243}_{95}Am \longrightarrow {}^4_2He + {}^{239}_{93}Np$$

Beta emission occurs in isotopes that have too many neutrons to be stable, that is, isotopes *above* the peninsula of stability in Figure 24.3. When β decay converts a neutron to a proton and an electron, which is then ejected, the mass number remains constant, but the number of neutrons drops.

$$^{60}_{27}\text{Co} \longrightarrow {}^{0}_{-1}\beta + {}^{60}_{28}\text{Ni}$$

Conversely, lighter isotopes that have too few neutrons—isotopes *below* the peninsula of stability—attain stability by positron emission (because this converts a proton to a neutron in one step) or by electron capture.

$$^{13}_{7}\text{N} \longrightarrow {}^{0}_{+1}\beta + {}^{13}_{6}\text{C}$$

$$^{41}_{20}\text{Ca} + {}^{0}_{-1}\text{e} \longrightarrow {}^{41}_{19}\text{K}$$

EXERCISE 24.4 ***Nuclear Stability***

For each of the following unstable isotopes, write an equation for its probable mode of decay.

a. silicon-32, $^{32}_{14}\text{Si}$

b. titanium-45, $^{45}_{22}\text{Ti}$

c. plutonium-239, $^{239}_{94}\text{Pu}$

Binding Energy

As proved by Ernest Rutherford's experiments (Chapter 2), the nucleus of the atom is extremely small. Yet the nucleus can contain up to 83 protons before becoming unstable. This is evidence that a very strong short-range binding force must be able to overcome the electrostatic repulsive force of a number of protons packed into such a tiny volume. A measure of the force holding the nucleus together is the nuclear **binding energy.** This energy (E_b) is defined as the negative of the energy change (ΔE) that would occur if a nucleus were formed directly from its component protons and neutrons. For example, if a mole of protons and a mole of neutrons directly formed a mole of deuterium nuclei, the energy change would be more than 200 million kJ!

$$^{1}_{1}\text{H} + {}^{1}_{0}\text{n} \longrightarrow {}^{2}_{1}\text{H} \qquad \Delta E = -2.15 \times 10^{8}\ \text{kJ}$$

$$\text{Binding energy} = -\Delta E = E_b = +2.15 \times 10^{8}\ \text{kJ}$$

This nuclear synthesis reaction is highly exothermic (and so E_b is very positive), an indication of the strong attractive forces holding the nucleus together. The deuterium nucleus is more stable than an isolated proton and an isolated neutron, just as the H_2 molecule is more stable than two isolated H atoms. Recall, however, that the energy released when a mole of H—H covalent bonds form is only 436 kJ, a tiny fraction of the energies released when protons and neutrons coalesce to form a nucleus.

To understand the enormous energy released during the formation of an atomic nucleus, we turn to an experimental observation and a theory. The experimental observation is that the mass of a nucleus is always less than the sum of the masses of its constituent protons and neutrons.

$$\underset{1.007825\ \text{g/mol}}{^{1}_{1}\text{H}} + \underset{1.008665\ \text{g/mol}}{^{1}_{0}\text{n}} \longrightarrow \underset{2.01410\ \text{g/mol}}{^{2}_{1}\text{H}}$$

$$\begin{aligned}\text{Change in mass} = \Delta m &= \text{mass of product} - \text{sum of masses of reactants}\\ &= 2.01410 \text{ g/mol} - 2.016490 \text{ g/mol}\\ &= -0.00239 \text{ g/mol}\end{aligned}$$

The quantity Δm is sometimes called the mass defect.

The theory is that the "missing mass," Δm, has been converted to energy, and this is the energy we described as the binding energy.

The relation between mass and energy is contained in Albert Einstein's 1905 theory of special relativity, which holds that mass and energy are simply different manifestations of the same quantity. Einstein stated that the energy of a body is equivalent to its mass times the square of the speed of light, $E = mc^2$. So, to calculate the energy change in a process in which the mass has changed, the equation becomes

$$\Delta E = (\Delta m)\, c^2$$

We can calculate ΔE in joules if the change in mass is given in kilograms and the velocity of light is in meters per second (because $1 \text{ J} = 1 \text{ kg} \cdot \text{m}^2/\text{s}^2$). For the formation of deuterium nuclei from protons and neutrons, we have

$$\begin{aligned}\Delta E &= (-2.39 \times 10^{-6} \text{ kg})(3.00 \times 10^{8} \text{ m/s})^2 = -2.15 \times 10^{11} \text{ J}\\ &= -2.15 \times 10^{8} \text{ kJ}\end{aligned}$$

This is the value of ΔE given at the beginning of this section for the change in energy when a mole of protons and a mole of neutrons form a mole of deuterium nuclei.

A helium nucleus is composed of two protons and two neutrons. As expected, the binding energy, E_b, is very large, even larger than for deuterium.

$$2\ {}^{1}_{1}\text{H} + 2\ {}^{1}_{0}\text{n} \longrightarrow {}^{4}_{2}\text{He} \qquad E_b = +2.73 \times 10^{9} \text{ kJ/mol of helium nuclei}$$

To compare nuclear stabilities more directly, however, nuclear scientists generally calculate the **binding energy per nucleon.** For helium-4 this is

$$E_b/\text{mol nucleons} = \frac{2.73 \times 10^{9} \text{ kJ}}{4 \text{ mol nucleons}} = 6.83 \times 10^{8} \text{ kJ/mol nucleons}$$

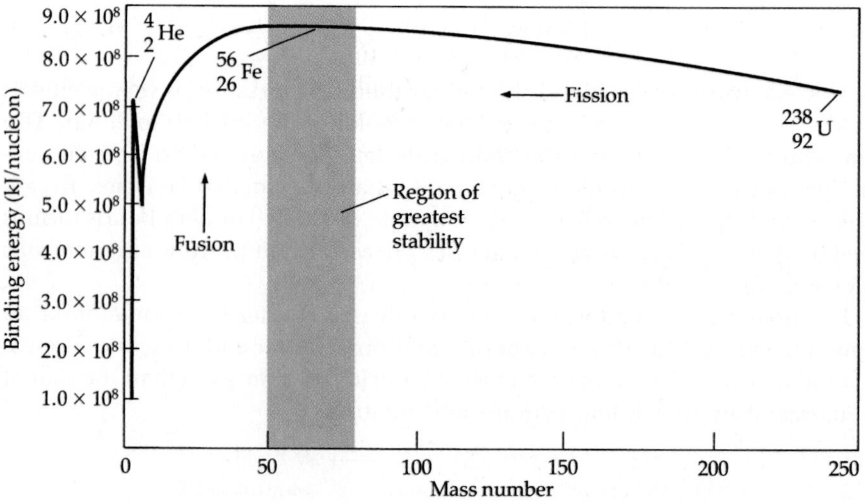

Figure 24.4 The relative stability of nuclei. This "curve of binding energy" was derived by calculating the binding energy per nucleon (in kilojoules per mole) for the most abundant isotope of the elements from hydrogen to uranium.

The greater the binding energy per nucleon, the greater is the stability of the nucleus. Scientists have calculated the binding energies of a great number of nuclei and have plotted them as a function of mass number (Figure 24.4). It is very interesting—and important—that the point of maximum stability occurs in the vicinity of iron-56, $^{56}_{26}Fe$. This means that *all elements are thermodynamically unstable with respect to iron.* That is, very heavy nuclei may split, or undergo **fission,** with the release of enormous quantities of energy, to give more stable nuclei with atomic numbers nearer iron. In contrast, two very light nuclei may come together and undergo **fusion** exothermically to form heavier nuclei.

EXERCISE 24.5 ***Binding Energy***

Calculate the binding energy, in kilojoules per mole, for the formation of lithium-6.

$$3\ ^{1}_{1}H + 3\ ^{1}_{0}n \longrightarrow\ ^{6}_{3}Li$$

The necessary masses are $^{1}_{1}H$ = 1.00783 g/mol, $^{1}_{0}n$ = 1.00867 g/mol, and $^{6}_{3}Li$ = 6.015125 g/mol. Is the binding energy greater than or less than that for helium-4? Finally, compare the binding energy per nucleon of $^{6}_{3}Li$ and helium-4. Which nucleus is the more stable? ■

24.4 RATES OF DISINTEGRATION REACTIONS

Cobalt-60 is used as a source of β particles and γ rays to treat malignancies in the human body. Although the isotope is radioactive, it is nonetheless reasonably stable because only half of a sample of cobalt-60 decays in a little over 5 years. On the other hand, copper-64, which is used in the form of copper acetate to detect brain tumors, decays much more rapidly; half of the radioactive copper decays in slightly less than 13 h. These two radioactive isotopes are clearly different in their rates of decay.

Half-Life

The relative stabilities of radioactive isotopes are often expressed just as we have done: in terms of the time required for half of the sample to decay. This is called the **half-life, $t_{1/2}$,** of a radioactive isotope. As illustrated by Table 24.2, isotopes have widely varying half-lives; some take years for half of the sample to decay, and others decay to half the original number of atoms in a fraction of a second.

TABLE 24.2 Half-Lives of Some Common Radioactive Isotopes

Isotope	Decay Process	Half-Life
$^{238}_{92}U$	$^{238}_{92}U \longrightarrow\ ^{234}_{90}Th + ^{4}_{2}He$	4.51×10^{9} years
$^{3}_{1}H$ (tritium)	$^{3}_{1}H \longrightarrow\ ^{3}_{2}He + ^{0}_{-1}\beta$	12.26 years
$^{14}_{6}C$ (carbon-14)	$^{14}_{6}C \longrightarrow\ ^{14}_{7}N + ^{0}_{-1}\beta$	5730 years
$^{131}_{53}I$	$^{131}_{53}I \longrightarrow\ ^{131}_{54}Xe + ^{0}_{-1}\beta$	8.05 days

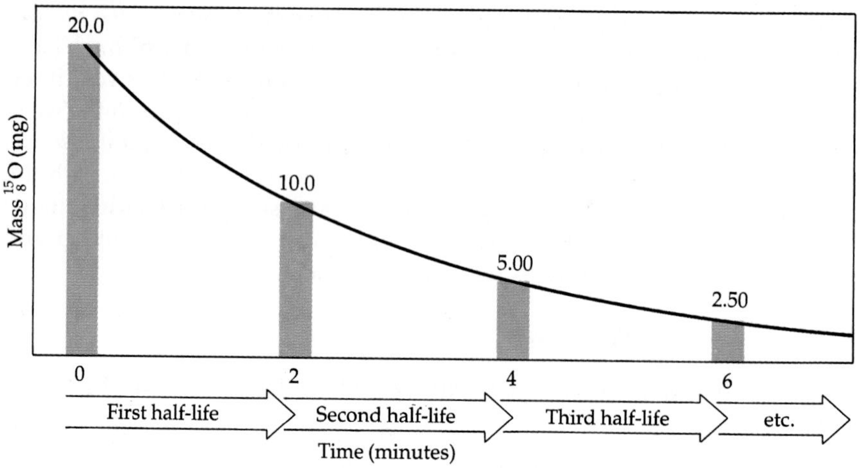

Figure 24.5 Decay of 20 mg of oxygen-15. After each half-life period of 2.0 min, the quantity present at the beginning of the period is reduced by half. The plot is based on the following data:

Number of Half-Lives	Fraction of Initial Quantity Remaining	Quantity Remaining (mg)
0	1	20.0 (initial)
1	$\frac{1}{2}$	10.00
2	$\frac{1}{4}$	5.00
3	$\frac{1}{8}$	2.50
4	$\frac{1}{16}$	1.25
5	$\frac{1}{32}$	0.625

For nuclear disintegration reactions, the half-life is a constant, independent of temperature and of the number of radioactive nuclei present.

As an example of the concept of half-life, consider the decay of oxygen-15, $^{15}_{8}O$, by positron emission.

$$^{15}_{8}O \longrightarrow {}^{15}_{7}N + {}^{0}_{+1}\beta$$

The half-life of oxygen-15 is 2.0 min. This means that half of the quantity of $^{15}_{8}O$ present at any given time disintegrates every 2.0 min. Thus, if we begin with 20 mg of $^{15}_{8}O$, 10 mg of the isotope remains after 2.0 min. After 4.0 min (two half-lives), only half of the remainder, 5.0 mg, is still there. After 6.0 min (three half-lives), only half of the 5.0 mg is still present, or 2.5 mg. The amounts of $^{15}_{8}O$ present at various times are illustrated in Figure 24.5.

EXAMPLE 24.2 *Half-Life*

Tritium ($^{3}_{1}H$), a radioactive isotope of hydrogen, has a half-life of 12.3 years.

$$^{3}_{1}H \longrightarrow {}^{0}_{-1}\beta + {}^{3}_{2}He$$

If you begin with 1.5 mg of the isotope, how many milligrams remain after 49.2 years?

Solution First, we find the number of half-lives in the given period of 49.2 years.

Because the half-life is 12.3 years, the number of half-lives is

$$49.2 \text{ years} \times \frac{1 \text{ half-life}}{12.3 \text{ years}} = 4.00 \text{ half-lives}$$

This means that the initial quantity of 1.5 mg is reduced four times by $\frac{1}{2}$.

$$1.5 \text{ mg} \times \frac{1}{2} \times \frac{1}{2} \times \frac{1}{2} \times \frac{1}{2} = 1.5 \times \left(\frac{1}{2}\right)^4 = 1.5 \text{ mg} \times \frac{1}{16} = 0.094 \text{ mg}$$

After 49.2 years, only 0.094 mg of the original 1.5 mg remains.

EXERCISE 24.6 ***Radioactivity and Half Life***

Strontium-90, $^{90}_{38}Sr$, is a radioisotope ($t_{1/2}$ = 28 years) produced in atomic bomb explosions. Its long life and tendency to concentrate in bone marrow make it particularly dangerous to people and animals.

1. The isotope decays with loss of a β particle; write a balanced equation showing the other product of decay.
2. A sample of the isotope emits 2000 β particles per minute. How many half-lives and how many years are necessary to reduce the emission to 125 β particles per minute? ■

Rate of Radioactive Decay

To determine the half-life of a radioactive element, the *rate of decay* must be measured. That is, we must measure the number of atoms that disintegrate per second or per hour or per year.

The equations used in describing the rate of radioactive decay are those of kinetics (see Chapter 15).

The rate of nuclear decay is often described in terms of the **activity** (A) of the sample, the number of disintegrations observed per unit time. The activity is *proportional* to the number of radioactive atoms present (N).

$$\text{Rate of radioactive decay} \equiv \text{activity } (A) \propto \text{number of radioactive atoms present } (N)$$

This proportionality can also be expressed in the form

$$A = kN \tag{24.1}$$

$$\frac{\text{Disintegrations}}{\text{Time}} = \frac{\text{disintegrations}}{(\text{number of atoms})(\text{time})} \times \text{number of atoms}$$

where k is the proportionality constant or *decay constant.* In the language of kinetics, Equation 24.1 is simply a rate law that is first order in the number of atoms in the sample, and k is the rate constant.

The activity of a sample can be measured with a device such as a Geiger counter (Figure 24.6). Let us say the activity is measured at some time t_0 and then measured again after a few minutes, hours, or days. If the initial activity is A_0 at t_0, then a second measurement gives a smaller activity A at a later time t. From Equation 24.1, you can see that the ratio of the activity A at some time t to the activity at the beginning of the experiment (A_0) must be equal to the ratio

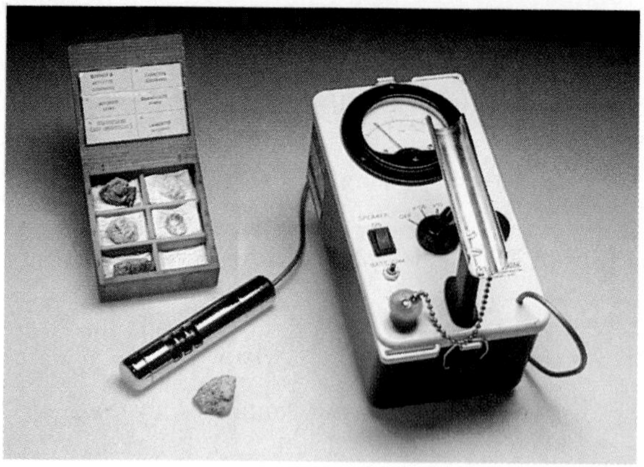

Figure 24.6 A Geiger counter with a sample of carnotite, a mineral containing uranium oxide. The Geiger counter was invented by Hans Geiger and Ernest Rutherford in 1908. A charged particle (such as an α and β particle), when entering a gas-filled tube, ionizes the gas. These gaseous ions are attracted to electrically charged plates and thereby give rise to a "pulse" or momentary flow of electric current. The current is amplified and used to operate a counter. (C. D. Winters)

of the number of radioactive atoms N that are present at time t to the number present at the beginning of the experiment (N_0).

$$\frac{A}{A_0} = \frac{kN}{kN_0}$$

or

$$\frac{A}{A_0} = \frac{N}{N_0} = \text{fraction of radioactive atoms still present in a sample after some time has elapsed}$$

This means that experimental information is related directly to the fraction of radioactive atoms remaining in a sample after some time has passed.

An extraordinarily useful equation relates the time period over which a sample is observed (t) to the fraction of radioactive atoms present after that amount of time has passed.

Equation 24.2 is an integrated first order rate equation and is derived in the same way as Equation 15.1, page 710.

$$\ln \frac{N}{N_0} = -kt \qquad \textbf{(24.2)}$$

Equation 24.2 is useful in three ways:

- If A/A_0 (and thus N/N_0) is measured in the laboratory over some time period t, then k can be calculated. The decay constant k can then be used to determine the half-life of the sample, as illustrated in Example 24.3.
- If k is known, the fraction of a radioactive sample still present after some time t has elapsed can be calculated.
- If k is known for a particular radioactive isotope, you can calculate the time required for that isotope to decay to a fraction of the original activity.

Now we are in a position to see how the half-life of a radioactive isotope $t_{1/2}$ is determined. The half-life is the time needed for half of the material present at the beginning of the experiment (N_0) to disappear. Thus, when time = $t_{1/2}$, then $N = \frac{1}{2}N_0$. This means that

$$\ln = \frac{\frac{1}{2}N_0}{N_0} = -kt_{1/2}$$

or

$$\ln \frac{1}{2} = -kt_{1/2}$$

$$-0.693 = -kt_{1/2}$$

and we arrive at a simple equation that connects the half-life and decay constant.

$$t_{1/2} = \frac{0.693}{k} \qquad \textbf{(24.3)}$$

Equation 24.3 is identical to Equation 15.4. See page 717 for a further discussion of the half-life of reactions.

The half-life $t_{1/2}$ is found by calculating k from Equation 24.2, where N and N_0 in turn come from laboratory measurements over the period t.

EXAMPLE 24.3 *Determination of Half-Life*

A sample of radon initially undergoes 7.0×10^4 α particle disintegrations per second (dps). After 6.6 days, it undergoes only 2.1×10^4 α particle dps. What is the half-life of this isotope of radon?

Solution Experiment has provided us with both A and A_0.

$$A = 2.1 \times 10^4 \text{ dps} \qquad A_0 = 7.0 \times 10^4 \text{ dps}$$

and the time ($t = 6.6$ days). We can therefore find the value of k. Because $N/N_0 = A/A_0$,

$$\ln\left(\frac{2.1 \times 10^4}{7.0 \times 10^4}\right) = -k(6.6 \text{ days})$$

$$\ln (0.30) = -k(6.6 \text{ days})$$

$$k = -\frac{\ln (0.30)}{6.6 \text{ days}} = -\frac{(-1.20)}{6.6 \text{ days}} = 0.18 \text{ day}^{-1}$$

and from k we can obtain $t_{1/2}$.

$$t_{1/2} = \frac{0.693}{k} = \frac{0.693}{0.18 \text{ day}^{-1}} = 3.8 \text{ days}$$

EXAMPLE 24.4 *Time and Radioactivity*

Some high-level radioactive waste with a half-life $t_{1/2}$ of 200. years is stored in underground tanks. What time is required to reduce an activity of 6.50×10^{12} disintegrations per minute (dpm) to a fairly harmless activity of 3.00×10^{-3} dpm?

Solution The data give you the initial activity ($A_0 = 6.50 \times 10^{12}$ dpm) and the activity after some elapsed time ($A = 3.00 \times 10^{-3}$ dpm). To find the elapsed time t, you must first find k from the half-life.

$$k = \frac{0.693}{t_{1/2}} = \frac{0.693}{200. \text{ years}} = 0.00347 \text{ year}^{-1}$$

With k known, the time t can be calculated.

$$\ln\left(\frac{3.00 \times 10^{-3}}{6.50 \times 10^{12}}\right) = -[0.00347\ \text{year}^{-1}]t$$

$$-35.312 = -[0.00347\ \text{year}^{-1}]t$$

$$t = \frac{-35.312}{-(0.00347)\ \text{year}^{-1}}$$

$$= 1.02 \times 10^4\ \text{years}$$

EXERCISE 24.7 ***Rate of Radioactive Decay***

Gallium citrate, containing the radioactive isotope gallium-67, is used medically as a tumor-seeking agent. It has a half-life of 77.9 h. How much time is needed for a sample of gallium citrate to decay to 10% of its original activity? ■

Radiochemical Dating

For excellent articles on the Shroud of Turin and carbon-14 dating, see *Chem Matters* magazine, February, 1989, pages 8–15.

Scientists have used radiochemical dating to determine the ages of rocks, fossils, and artifacts that date back many years. For example, radiochemical methods were used recently to show that the Shroud of Turin was created somewhere around 1300 AD, and not at the time of Christ as has been alleged for many centuries (Figure 24.7).

In 1946, Willard Libby developed the technique of determining age using radioactive carbon-14 ($^{14}_{6}C$). Carbon is an important building block of all living systems, and so all organisms contain the three isotopes of carbon: ^{12}C, ^{13}C, and ^{14}C. The first two are stable and have been around since the universe was created. In contrast, carbon-14 is being created continuously by cosmic radiation (as outlined below). It is also radioactive and decays to nitrogen-14 by β emission.

$$^{14}_{6}C \longrightarrow {}^{0}_{-1}\beta + {}^{14}_{7}N$$

Willard Libby and his apparatus for carbon-14 dating. (Oesper Collection in the History of Chemistry/University of Cincinnati)

Because the half-life of ^{14}C is known to be 5.73×10^3 years, the amount of the isotope present (N) can be measured from the activity of a sample. If the amount of ^{14}C originally in the sample (N_0) is known, then the age of the sample can be found from Equation 24.2.

This method of age determination clearly depends on knowing how much ^{14}C was originally in the sample. The answer to this question comes from work by physicist Serge Korff who discovered, in 1929, that ^{14}C is continually generated in the upper atmosphere. High-energy cosmic rays smash into gases in the upper atmosphere and force them to eject neutrons. These free neutrons collide with nitrogen atoms in the atmosphere and produce carbon-14.

$$^{14}_{7}N + {}^{1}_{0}n \longrightarrow {}^{14}_{6}C + {}^{1}_{1}H$$

Throughout the *entire* atmosphere, only about 7.5 kg of ^{14}C is produced per year. This tiny amount of radioactive carbon is incorporated into CO_2, however, and then becomes part of the carbon cycle and is distributed worldwide. The continual formation of ^{14}C, exchange of the isotope within the oceans, atmosphere, and biosphere, and decay of living matter keep the supply of ^{14}C constant.

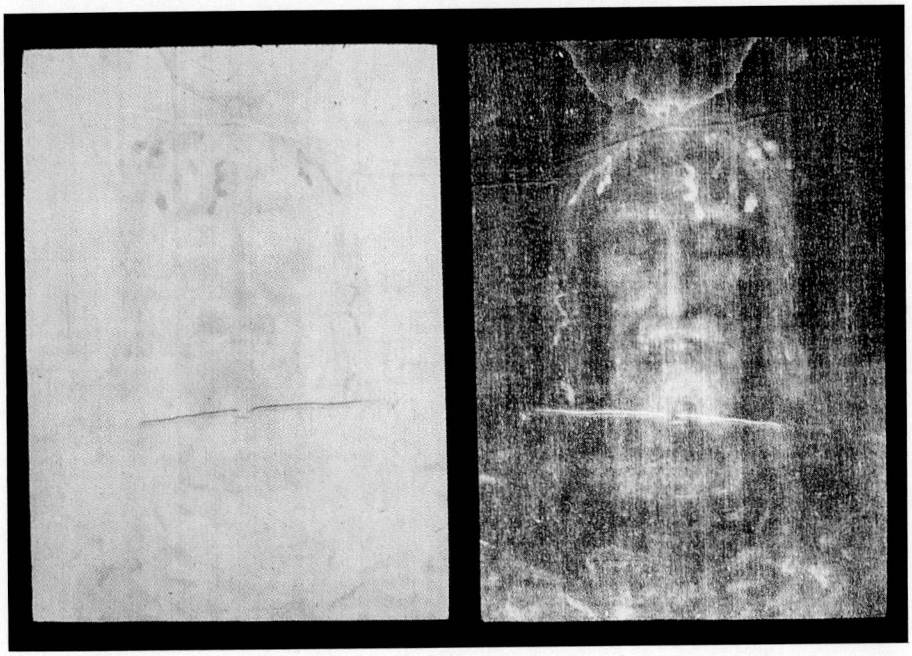

Figure 24.7 The Shroud of Turin is a linen cloth over 4 m long. It bears a faint, straw-colored image of an adult male of average build who had apparently been crucified. Reliable records of the shroud date to about 1350, but for these past 600 years it has been alleged to be the burial shroud of Jesus Christ. Numerous chemical and other tests have been done on tiny fragments of the shroud in recent years. The general conclusion has been that the image was not painted on the cloth by any traditional method, but no one could say exactly how the image had been created. Recent advances in radiochemical dating methods, however, led to a new effort in 1987–1988 to estimate the age of the cloth. Using radioactive ^{14}C, the flax from which the linen was made was shown to have been grown between 1260 and 1390 A.D. There is no chance that the cloth was made at the time of Christ. (Santi Visalli/The Image Bank)

Plants absorb carbon dioxide from the atmosphere, convert it into food, and so incorporate the carbon-14 into living tissue. It has been established that the β activity of carbon-14 in *living* plants and in the air is constant at about 14 disintegrations per minute per gram of carbon. When the plant dies or is ingested by an animal, however, carbon-14 disintegration continues *without the ^{14}C being replaced;* consequently, the activity decreases with passage of time. The smaller the activity of carbon-14, the longer the period between the death of the plant and the present time. Assuming that ^{14}C activity was about the same hundreds of years ago as it is now, measurement of the ^{14}C β activity of an artifact can be used to date the article.

EXAMPLE 24.5 *Radiochemical Dating*

The so-called Dead Sea Scrolls, Hebrew manuscripts of the books of the Old Testament, were found in 1947. The activity of carbon-14 in the linen wrappings of the book of Isaiah is about 11 disintegrations per minute per gram (d/min · g). Calculate the approximate age of the linen.

Solution We use Equation 24.2

$$\ln\left(\frac{N}{N_0}\right) = -kt$$

where N is proportional to the activity at the present time (11 d/min · g) and N_0 is proportional to the activity of carbon-14 in the living material (14 d/min · g). To calculate the time elapsed since the linen wrappings were part of a living plant, we first need k, the rate constant. From the text you know that $t_{1/2}$ is 5.73×10^3 years, so

$$k = \frac{0.693}{t_{1/2}} = \frac{0.693}{5.73 \times 10^3 \text{ years}} = 1.21 \times 10^{-4} \text{ year}^{-1}$$

Now everything is in place to calculate t.

$$\ln\left(\frac{11\ \text{d/min}\cdot\text{g}}{14\ \text{d/min}\cdot\text{g}}\right) = -[1.21\times10^{-4}\ \text{year}^{-1}]t$$

$$t = \frac{\ln 0.79}{-[1.21\times10^{-4}\ \text{year}^{-1}]}$$

$$= \frac{-0.24}{-[1.21\times10^{-4}\ \text{year}^{-1}]}$$

$$= 2.0\times10^{3}\ \text{years}$$

The linen is therefore about 2000 years old.

EXERCISE 24.8 ***Radiochemical Dating***

A wooden Japanese temple guardian statue of the Kamakura period (AD 1185–1334) had a carbon-14 activity of 12.9 d/min · g in 1990. What is the age of the statue? In what year was the statue made? The initial activity of carbon-14 was 14 d/min · g, and $t_{1/2} = 5.73\times10^{3}$ years. ■

24.5 ARTIFICIAL TRANSMUTATIONS

In the course of his experiments, Rutherford found in 1919 that α particles ionize atomic hydrogen, knocking off an electron from each atom. If atomic nitrogen was used instead, he found that bombardment with α particles *also produced protons.* Quite correctly he concluded that the α particles had knocked a proton out of the nitrogen nucleus and that an isotope of another element had been produced. Nitrogen had undergone a *transmutation* to oxygen.

$$^{4}_{2}\text{He} + ^{14}_{7}\text{N} \longrightarrow ^{17}_{8}\text{O} + ^{1}_{1}\text{H}$$

Enrico Fermi (1901–1954), the Italian-born American physicist who first experimentally observed nuclear fission and who demonstrated a nuclear chain reaction. (AIP/Niels Bohr Library)

Rutherford had proposed that protons and neutrons are the fundamental building blocks of nuclei. Although Rutherford's search for the neutron was not successful, it was found by James Chadwick in 1932 as a product of the α-particle bombardment of beryllium.

$$^{4}_{2}\text{He} + ^{9}_{4}\text{Be} \longrightarrow ^{12}_{6}\text{C} + ^{1}_{0}\text{n}$$

Changing one element into another by α-particle bombardment has its limitations. Before a positively charged particle (such as the α particle) can be captured by a positively charged nucleus, the particle must have sufficient kinetic energy to overcome the repulsive forces developed as the particle approaches the nucleus. But the neutron is electrically neutral, so Enrico Fermi (1934) reasoned that a nucleus would not oppose its entry. By this approach, practically all elements have since been transmuted, and a number of *transuranium elements* (elements beyond uranium) have been prepared. For example, uranium-238 forms neptunium-239 on neutron bombardment,

$$^{1}_{0}\text{n} + ^{238}_{92}\text{U} \longrightarrow ^{239}_{92}\text{U} \longrightarrow ^{239}_{93}\text{Np} + ^{0}_{-1}\beta$$

and the isotope of neptunium decays to plutonium-239

$$^{239}_{93}\text{Np} \longrightarrow ^{0}_{-1}\beta + ^{239}_{94}\text{Pu}$$

CURRENT ISSUES IN CHEMISTRY

Two New Elements Found in 1994!

Most periodic tables show that element 109, which was discovered in 1982, is the heaviest known element. Although more than 10 years have passed since this discovery, attempts to synthesize even heavier elements have continued in laboratories in Berkeley, California, Darmstadt, Germany, and Dubna, Russia. There have been tantalizing leads, but no success. Then, within the space of a month, the periodic table was extended by two more elements! Element 110 was discovered on November 9, 1994, by a team of European scientists at the Gesellschaft für Schwerionenforschung (Society for Heavy Ion Research) in Darmstadt, Germany, and the same team found element 111 on December 8, 1994.

Three atoms of element 110 were formed by bombarding a lead target with nickel atoms for two days. On average, about 3 trillion particles per second struck the lead target, but most went through without reaction. In at least three instances, however, atoms of nickel-62 fused with atoms of lead-208 and produced atoms of element 110 plus a neutron.

$$^{62}_{28}\text{Ni} + ^{208}_{82}\text{Pb} \longrightarrow {}^{269}110 + ^{1}_{0}\text{n}$$

The atoms quickly decayed by alpha emission to element 108 and finally to nobelium (element 102).

Almost exactly one month later the team in Darmstadt produced element 111, an element in the same group as copper, silver, and gold. They again used nickel atoms as the projectiles, but this time the target was bismuth.

$$^{64}_{28}\text{Ni} + ^{209}_{83}\text{Bi} \longrightarrow {}^{272}111 + ^{1}_{0}\text{n}$$

Again three atoms were formed after three days, and again they decay rapidly ($t_{1/2}$ = 1.5 milliseconds) by alpha emission as expected for a very heavy element (see Figure 24.3).

Are more elements possible? The most recent work suggests that there may be more, and scientists are looking for a predicted "island of stability" in the region of 114 protons and 184 neutrons.

For more information see Darleane C. Hoffman, "The Heaviest Elements," *Chemical and Engineering News,* May 2, 1994, page 24. See also *ibid,* March 13, 1995, pages 35–40.

Of the 111 elements known at present, only elements up to americium exist in nature (except for Tc, Pm, At, and Fr). The transuranium elements are all synthetic. Up to element 101, mendelevium, all of the elements can be made by bombarding the nucleus of a lighter element with small particles such as $^{4}_{2}\text{He}$ or $^{1}_{0}\text{n}$. Beyond 101, though, special techniques using heavier particles are required and are still being developed. For example, lawrencium is made by bombarding californium-252 with boron nuclei,

$$^{10}_{5}\text{B} + ^{252}_{98}\text{Cf} \longrightarrow ^{257}_{103}\text{Lr} + 5\ ^{1}_{0}\text{n}$$

and the latest element to be discovered, 111, was made by firing nickel atoms at bismuth atoms.

EXERCISE 24.9 ***Nuclear Transmutations***

Complete the following nuclear equations, indicating the symbol, the mass number, and the atomic number of the remaining product.

1. $^{13}_{6}\text{C} + ^{1}_{0}\text{n} \rightarrow ^{4}_{2}\text{He} + ?$
2. $^{14}_{7}\text{N} + ^{4}_{2}\text{He} \rightarrow ^{1}_{0}\text{n} + ?$
3. $^{253}_{99}\text{Es} + ^{4}_{2}\text{He} \rightarrow ^{1}_{0}\text{n} + ?$ ■

Glenn Theodore Seaborg (1912–) began his college education as a literature major but changed to science in his junior year at the University of California. He shared the Nobel Prize in 1951 with E. M. McMillan (1907–), who started Seaborg in this area of research. (Lawrence Berkeley Laboratory)

PORTRAIT OF A SCIENTIST

Glenn Seaborg and the Transuranium Elements

Among the most significant contributions to the modern periodic chart is that made by Nobel laureate Glenn Seaborg (born 1912). Among other things he demonstrated the importance of maintaining the courage of one's convictions. Thanks to his insights it is now very well established that the transuranium elements (atomic numbers greater than 92), a number of which he either discovered or helped to discover in the course of the Manhattan Project during World War II, are members of the actinide series.

Until Seaborg offered his version of the periodic table, chemists were convinced that Th, Pa, and U belonged in the main body of the table, Th under Hf, Pa under Ta, and U under W. When Seaborg proposed that Th was the beginning of the actinides and that the transuranium elements belonged as a group under the lanthanides. Some prominent inorganic chemists, many of them Seaborg's friends, tried to discourage his publication of this finding in the open literature. One very prominent inorganic chemist felt that Seaborg would ruin his scientific reputation. Nevertheless Seaborg, strongly convinced, persisted and, as a result, on Seaborg's expansion of the periodic table it was possible to predict accurately the properties of many of the as yet undiscovered transuranium elements. Subsequent preparation of these elements in atomic accelerators proved him right and it was fitting that he was awarded the Nobel Prize in 1951 for his work. The name seaborgium for synthetic element number 106 was recently approved by the American Chemical Society. The name has not yet been officially accepted internationally, however, as no element has ever been named for a living person. Indeed, his daughter was said to have remarked that her father must surely be dead when she heard of the naming of the element.

24.6 NUCLEAR FISSION

See the biographies of Lise Meitner (page 1110) and Niels Bohr (page 327). Element 109 is named meitnerium and 107 is named nielsbohrium (but unofficially as yet).

In 1938, the radiochemists Otto Hahn and Fritz Strassman found some barium in a sample of uranium that had been bombarded with neutrons. Further work by Lise Meitner, Otto Frisch, Niels Bohr, and Leo Szilard confirmed that a uranium-235 nucleus had captured a neutron to form uranium-236 and that this heavier isotope had undergone **nuclear fission;** that is, the nucleus had split in two (Figure 24.8)

$$^{235}_{92}\text{U} + ^{1}_{0}\text{n} \longrightarrow ^{236}_{92}\text{U} \longrightarrow ^{141}_{56}\text{Ba} + ^{92}_{36}\text{Kr} + 3\ ^{1}_{0}\text{n} \qquad \Delta E = -2 \times 10^{10}\ \text{kJ/mol}$$

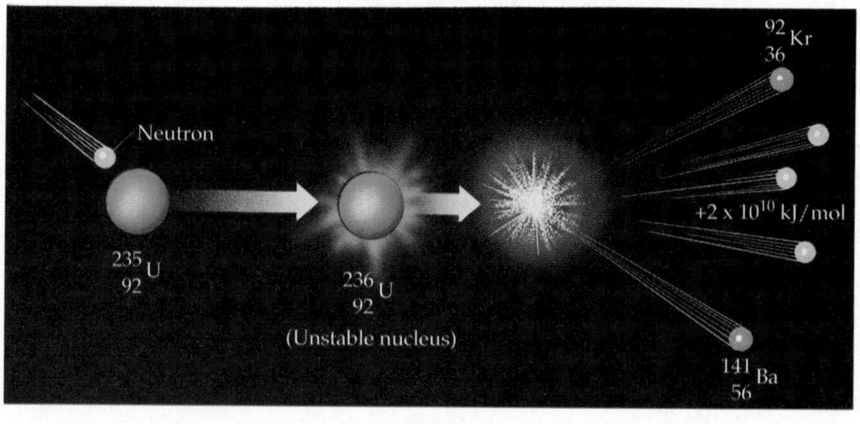

Figure 24.8 The fission of a $^{236}_{92}\text{U}$ nucleus that arises from the bombardment of $^{235}_{92}\text{U}$ with a neutron. The electric repulsion between protons rips the nucleus apart, producing 2×10^{10} kJ/mol of energy.

CURRENT ISSUES IN CHEMISTRY

Protect Your Home with a Smoke Detector

Radioactive isotopes are used in many ways, one of which may already be familiar to you. That is, some home smoke alarms use the disintegration of a radioactive element, americium, as a way to detect smoke particles in the air. As shown here, a weak radioactive source in the smoke alarm ionizes the air, thus setting up a small current in an electric circuit. If smoke is present, the ions become attached to the smoke particles. The slower movement of the heavier, charged smoke particles reduces the current in the circuit and sets off the alarm.

Americium-241 was prepared in 1944 by G. T. Seaborg, R. A. James, L. O. Morgan, and A. Ghiorso at what is now the Argonne National Laboratory. It arises from successive neutron capture by ^{239}Pu, followed by β decay. The isotope is now obtained in kilogram quantities as a byproduct of processing plutonium produced in nuclear reactors. The isotope ^{241}Am has a half-life of 432 years, whereas ^{243}Am has a half-life of 7370 years. Both decay by α emission.

$$^{241}_{95}Am \longrightarrow ^{4}_{2}He + ^{237}_{93}Np$$

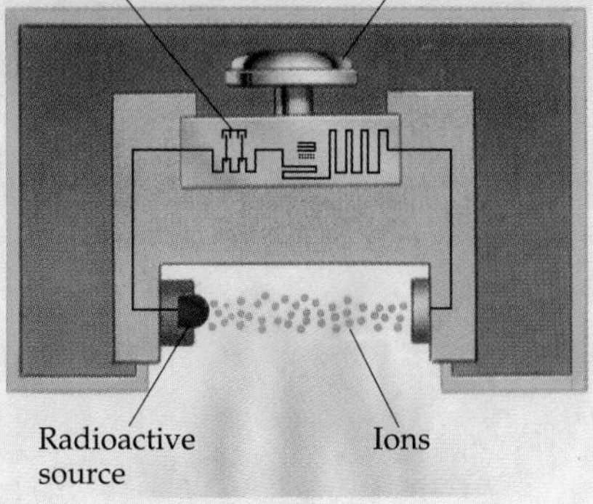

Diagram of a smoke alarm.

A household smoke alarm. (C. D. Winters)

The fact that the fission reaction produces more neutrons than are required to begin the process is important. In the previous nuclear reaction, bombardment with a single neutron produces 3 neutrons capable of inducing 3 more fission reactions, which release 9 neutrons to induce 9 more fissions, from which 27 neutrons are obtained, and so on. Because the fission of uranium-236 is extremely rapid, this sequence of reactions can be an explosive chain reaction as illustrated in Figure 24.9. If the amount of uranium-235 is small, so few neutrons are captured by ^{235}U nuclei that the chain reaction cannot be sustained. In an atomic bomb, two small pieces of uranium, neither capable of sustaining a chain reaction, are brought together to form one larger piece capable of supporting a chain reaction, and an explosion results.

Rather than allow a fission reaction to run away explosively, engineers can slow it by limiting the number of neutrons available, and energy can be derived safely and used as a heat source in a power plant (Figure 24.10). In a **nuclear,** or **atomic reactor,** the rate of fission is controlled by inserting cadmium rods or other "neutron absorbers" into the reactor. The rods absorb the neutrons that cause fission reactions; by withdrawing or inserting the rods, the rate of the fission reaction can be increased or decreased.

Not all nuclei can be made to fission on colliding with a neutron, but ^{235}U and ^{239}Pu are two isotopes for which fission is possible. Natural uranium contains an average of only 0.72% of the fissionable 235 isotope; more than 99% of

If the uranium-235 content of a sample is over 90%, it is considered of weapons quality.

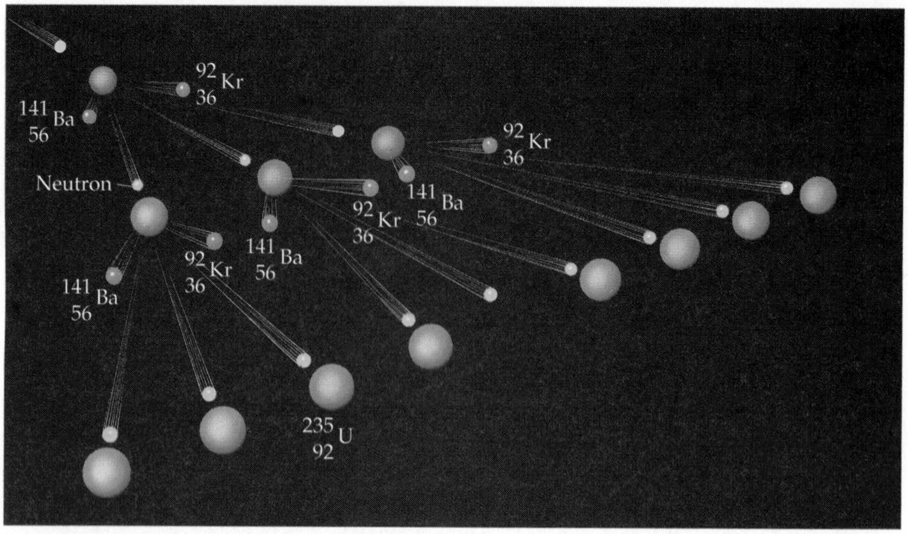

Figure 24.9 Illustration of a chain reaction initiated by capture of a stray neutron. (Many pairs of different isotopes are produced, but only one kind of pair is shown.)

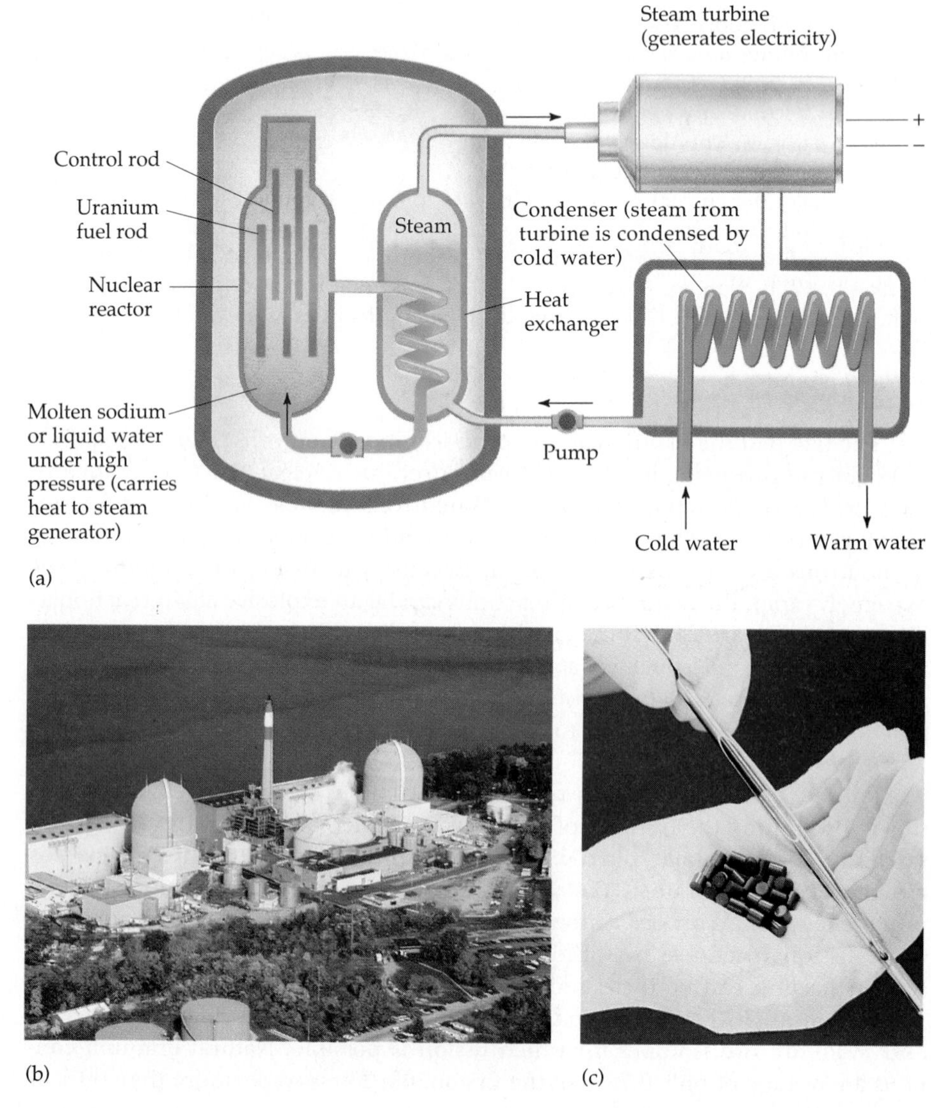

Figure 24.10 Nuclear power plants. (a) Liquid water (or liquid sodium) is circulated through the reactor, where the liquid is heated to about 325 °C. When this hot liquid is circulated through a steam generator, water in the generator is turned to steam, which in turn drives a steam turbine. After passing through the turbine, the steam is converted back to liquid water and is recirculated through the steam generator. Enormous quantities of outside cooling water from rivers or lakes are necessary to condense the steam. (This basic system is the same as in any power plant, except that the water of circulating liquid is heated initially by coal, gas, or oil-fired burners.) (b) A nuclear power plant at Indian Pointe, New York. (c) Uranium pellets used in the reactor fuel rods. (b, Joe Azzara/The Image Bank; c, D.O.E./Science Source/Photo Researchers, Inc.)

the natural element is uranium-238, which is fissionable only with high energy neutrons. Because the percentage of natural ^{235}U is too small to sustain a chain reaction, uranium for nuclear power fuel must be enriched. To accomplish this, some of the ^{238}U isotope in a sample is separated, thereby raising the concentration of ^{235}U. One way to do this is by gaseous diffusion (as described in Section 12.7).

There is of course some controversy surrounding the use of nuclear power plants, particularly in the United States. Their proponents regard nuclear power to be an essential part of an advancing, technologically dependent society. The health of our economy and our standard of living are dependent on inexpensive, reliable, and safe sources of energy. Just within the past few years the demand for electric power has once again begun to exceed the supply, so many believe nuclear power plants should be built to meet the demand. Nuclear power plants are capable of supplying these demands, and they can be the source of "clean" energy in that they do not pollute the atmosphere with ash, smoke, or oxides of sulfur, nitrogen, or carbon. In addition, they help to ensure that our supplies of fossil fuels will not be depleted in the near future, and they free us of dependence on such fuels from other countries. There are currently more than 100 operating plants in the United States, and more than 350 worldwide. The nuclear plants in the United States supply about 20% of the nation's electric energy; only coal-fired plants contribute a greater share (57%) (Figure 24.11).

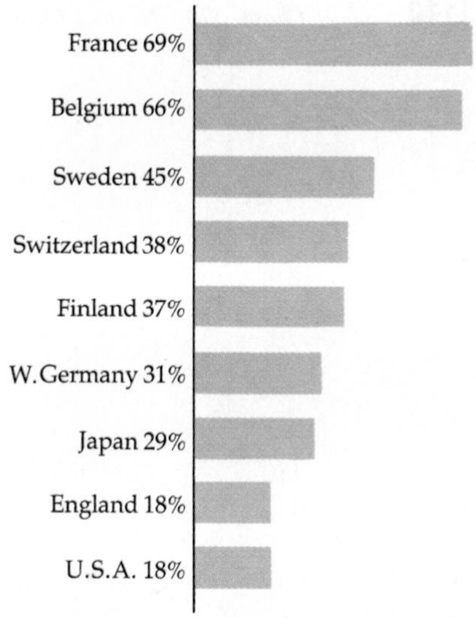

Figure 24.11 The approximate share of electricity generated by nuclear power in various countries.

There are *no* new nuclear power plants now under construction in the United States because these plants do have disadvantages. One problem is presented by the reactor fission products. Although some are put to various uses (Section 24.9), many are not suitable as a fuel or for other purposes. Because these products are often highly radioactive, their disposal poses an enormous problem. Perhaps the most reasonable suggestion is that radioactive wastes can be converted to a glassy material having a volume of about 2 m^3 per reactor per year; this relatively small volume of material can then be stored underground in geological formations, such as salt deposits, that are known to be stable for hundreds of millions of years.

Nuclear power plants generate large amounts of electricity worldwide, but they also generate highly radioactive wastes. Some of these "high-level" wastes have very long half-lives, up to tens of thousands of years. In addition, they can be very dangerous in high concentrations. Therefore, the storage of wastes is a formidable problem. They are now stored in large double-walled tanks buried in the ground, but better long-term solutions must be found. (U. S. Department of Energy)

24.7 NUCLEAR FUSION

Tremendous amounts of energy are generated when comparatively light nuclei combine to form heavier nuclei. Such a reaction is called **nuclear fusion,** and one of the best examples is the fusion of hydrogen nuclei (protons) to give helium nuclei.

$$4\ {}^{1}_{1}H \longrightarrow {}^{4}_{2}He + 2\ {}^{\ 0}_{+1}\beta \qquad \Delta E = -2.5 \times 10^9 \text{ kJ}$$

This reaction is the source of the energy from our sun and other stars, and it is the beginning of the synthesis of the elements in the universe. Temperatures of 10^6 to 10^7 K, found in the core and radiative zone of the sun, are required to bring the positively charged nuclei together with enough kinetic energy to overcome nuclear repulsions.

Deuterium—heavy hydrogen—can also be fused to give helium-3,

$${}^{2}_{1}H + {}^{2}_{1}H \longrightarrow {}^{3}_{2}He + {}^{1}_{0}n \qquad \Delta E = -3.2 \times 10^8 \text{ kJ}$$

or deuterium can be fused with tritium, a radioactive isotope of hydrogen, to give helium-4.

$${}^{2}_{1}H + {}^{3}_{1}H \longrightarrow {}^{4}_{2}He + {}^{1}_{0}n \qquad \Delta E = -1.7 \times 10^9 \text{ kJ}$$

Both of these reactions evolve an enormous quantity of energy, so it has been the dream of nuclear physicists to try to harness them to provide power for the nations of the world.

At the very high temperatures that allow fusion reactions to occur rapidly, atoms do not exist as such; instead, there is a **plasma** consisting of unbound nuclei and electrons. To achieve the high temperatures required for the fusion reaction of the hydrogen bomb, a fission bomb (atomic bomb) is first set off. One type of hydrogen bomb depends on the production of tritium ($^{3}_{1}H$) in the bomb. In this type, lithium-6 deuteride (LiD, a solid salt) is placed around an ordinary $^{235}_{92}U$ or $^{239}_{94}Pu$ fission bomb, and the fission is set off in the usual way. A $^{6}_{3}Li$ nucleus absorbs one of the neutrons produced and splits into tritium and helium.

$$^{6}_{3}Li + ^{1}_{0}n \longrightarrow ^{3}_{1}H + ^{4}_{2}H$$

The temperature reached by the fission of uranium or plutonium is high enough to bring about the fusion of tritium and deuterium and the release of 1.7×10^{9} kJ/mol of ^{3}He. A 20-megaton bomb usually contains about 300 lb of lithium deuteride, as well as a considerable amount of plutonium and uranium.

PORTRAIT OF A SCIENTIST

Lise Meitner (1878–1968)

Lise Meitner. (AIP-Emilio Segré Visual Archives, Herzfeld Collection)

Element number 109 is named meitnerium to honor the contributions of Lise Meitner. She was born in November, 1878, in Vienna, Austria, the third of eight children. Fortunately, she had the support crucial to the success of women in science: her father demonstrated a strong interest in her education and accomplishments, and her mother supported him in this. At the turn of the century young women were not prepared academically to attend a university. Nonetheless, she had private tutors, and was finally allowed to take the university entrance examinations when she was almost 23 years old. She began as a student at the University of Vienna shortly thereafter to study science. According to one biographer, "women university students were widely regarded as freaks," but Lise Meitner found a mentor, Ludwig Boltzmann, one of the giants of science in this century. He was an enthusiastic and emotional lecturer who drew her to the study of physics. She earned her Ph.D. in physics in 1905, only the second woman to be awarded a doctorate in physics in the university's 500-year history.

Boltzmann took his own life in 1906, so Meitner moved to Berlin. There she began working with a young chemist, Otto Hahn, in the new field of radiochemistry. The problem was that Hahn worked in an institute directed by Emil Fischer (Nobel laureate, 1902), and Fischer absolutely forbade women in his laboratory. They reached a compromise, however: Meitner could set up a laboratory in a damp basement carpentry shop as long as she did not come upstairs. And she could use the restroom in a neighboring hotel! Later, Meitner was taken on as an assistant by Max Planck (Nobel laureate, 1918) who became one of her greatest supporters.

Working with Hahn, Meitner discovered protactinium (Pa, element 91) in 1918 and carried out many significant studies on radioactive elements. Her greatest contribution to 20th-century science, however, was to explain the process of nuclear fission. She and her nephew, Otto Frisch, also a physicist, published a paper in 1939 that first used the term "nuclear fission." When she did this, in 1938–1939, she was living in exile in Sweden. She had fled Germany because of her Jewish ancestry.

Meitner was suggested for a Nobel prize for many years, and her coworker Hahn received the chemistry prize in 1944 for his work on nuclear fission. Lise Meitner never did receive that accolade. Nonetheless, the leader of the team that discovered element 109 in Germany recently said that "She should be honored as the most significant woman scientist of this century."

Controlling a nuclear fusion reaction in order to harness it for peaceful uses has been extraordinarily difficult and has not yet been achieved. Three critical requirements must be met for controlled fusion. First, the temperature must be high enough for fusion to occur. The fusion of deuterium and tritium, for example, requires a temperature of 100 million degrees or more. Second, the plasma must be confined long enough to release a net output of energy. Third, the energy must be recovered in some usable form.

In spite of the problems in controlling fusion, a number of attractive features encourage research on nuclear fusion. For example, the hydrogen fuel (in water) is cheap and abundant. Furthermore, most radioisotopes produced by fusion have short half-lives and so are a serious radiation hazard for only a short time.

Unfortunately, fusion reactions have not yet been "controlled." No physical device can contain the plasma without cooling it below the critical fusion temperature. Magnetic "bottles" (enclosures in space bounded by magnetic fields) have confined the plasma, but not for long enough periods.

The possibility of "cold fusion," the room-temperature fusion of deuterium atoms to provide energy, was announced in early 1989. Although this "discovery" is now largely discredited, work continues. The reaction to the announcement by the international scientific community illustrates well how the process of scientific investigation works. See *Journal of Chemical Education.* Vol. 66, p. 449, 1989.

24.8 RADIATION EFFECTS AND UNITS OF RADIATION

All three types of radiation (α, β, and γ) disrupt normal cell processes in living organisms, and the potential for serious radiation damage to humans is well known. The biological effects of the atomic bombs exploded at Hiroshima and Nagasaki, Japan, at the close of World War II in 1945 have been well documented. Controlled exposure however, can be beneficial in destroying unwanted tissue, as in the radiation therapy used in treating some types of cancer.

To quantify radiation and its effects, particularly on humans, several units have been developed. The **röntgen,** or roentgen (R), is a measure of radiation exposure and is proportional to the amount of ionization produced in air by x-rays and γ-rays. A normal chest x-ray exposes you to about 0.1 R.

The **rad** (for "radiation absorbed dose") measures the radiation dose to tissue rather than to air. The röntgen and rad are similar in size. One rad represents a dose of 1.00×10^{-5} J absorbed per gram of material. In more meaningful terms, a whole-body dose of 450 rad would be fatal to about 50 percent of the population.

Different types of radiation have different biological effects. A rad of alpha particles can produce 10 to 20 times as much of an effect as a rad of x-rays, for example. To take these differences into account, a unit called the **rem** (standing for *r*öntgen *e*quivalent *m*an) is used. The dose in rems is the product of the absorbed dose in rads times a "quality factor." The "quality factor" is 1 for gamma and beta radiation, 5 for low energy neutrons and protons, and 10 to 20 for alpha particles and high energy neutrons and protons. Because most radiation doses are fairly small, the millirem or mrem is commonly used, where 1 mrem = 10^{-6} rem.

Finally, the **curie** (Ci) is commonly used as a unit of activity. One curie represents the quantity of any radioactive isotope that undergoes 3.7×10^{10} dps.

Another unit applied to radioactive substances is the becquerel (Bq), where 1 Bq = 1 dps.

Humans are constantly exposed to natural and artificial **background radiation,** estimated to be about 200 mrem per year (Table 24.3). More than half of this is from natural background radiation sources: cosmic radiation and radioactive elements and minerals found naturally in the earth and air.

Cosmic radiation, emitted by the sun and other stars, continually bombards the earth and accounts for about 40% of background radiation. The remainder

TABLE 24.3 Radiation Exposure for One Year from Natural and Artificial Sources*

	Millirem/Year	Percentage
Natural Sources		
Cosmic radiation	50.0	25.8
The earth	47.0	24.2
Building materials	3.0	1.5
Inhaled from the air	5.0	2.6
Elements found naturally in human tissues	21.0	10.8
Subtotal	126.0	64.9
Medical Sources		
Diagnostic x-rays	50.0	25.8
Radiotherapy	10.0	5.2
Internal diagnosis	1.0	0.5
Subtotal	61.0	31.5
Other Artificial Sources		
Nuclear power industry	0.85	0.4
Luminous watch dials, TV tubes, industrial wastes	2.0	1.0
Fallout from nuclear testing	4.0	2.1
Subtotal	6.9	3.5
Total	193.9	99.9

*From J. R. Amend, B. P. Mundy, and M. T. Arnold: *General, Organic, and Biochemistry,* 2nd ed., p. 356. Philadelphia, Saunders College Publishing, 1993.

comes from elements such as ^{40}K. Because potassium (which is present to the extent of about 0.3 g/kg of soil) is essential to all living organisms, we all carry some radioactive potassium. Other radioactive elements found in some abundance on the earth are thorium-232, uranium-238, and radium-226. Thorium, for example, is found to the extent of 12 g/1000 kg of soil. Its oxide, ThO_2, glows very brightly when heated, so it was used until very recently in the mantles of lanterns that are used by campers.

Roughly 17% of our annual exposure comes from medical procedures such as diagnostic x-rays and the use of radioactive compounds to trace the body's functions. Finally, another 17% comes from such sources as the radioactive products from testing nuclear explosives in the atmosphere, x-ray generators, televisions, nuclear power plants and their wastes, nuclear weapons manufacture, and nuclear fuel processing.

Burning fossil fuels (coal and oil) releases naturally occurring radioactive isotopes into the atmosphere. This has added significantly to the background radiation in recent years.

Radon

Radon is a chemically inert gas, in the same periodic group as helium, neon, argon, and krypton. The trouble with radon is that it is radioactive. As Figure 24.2 shows, $^{222}_{86}Rn$ is part of the chain of events beginning with the decay of uranium-238. (Other isotopes of Rn are products of other decay series.) In February 1989, *Chemical and Engineering News* said that "In three short years, the radioactive gas radon has progressed from relative obscurity to a cause of high anxiety as an indoor air pollutant.''*

Radon occurs naturally in our environment. Because it comes from natural uranium deposits, the amount depends on local geology, but it is believed to account for less than half of normal background radioactivity. Furthermore, the gas is chemically inert and has a relatively short half-life (3.82 days). It is not trapped by chemical processes in the soil or water and is free to seep up from the ground and into underground mines or into homes through pores in block walls, cracks in the basement floor or walls, or around pipes. When breathed by humans occupying that space, the radon-222 isotope can decay inside the lungs to give polonium, a radioactive element that is not a gas and is not chemically inert.

$$^{222}_{86}\text{Rn} \longrightarrow {}^{4}_{2}\text{He} + {}^{218}_{84}\text{Po} \qquad t_{1/2} = 3.82 \text{ days}$$

$$^{218}_{84}\text{Po} \longrightarrow {}^{4}_{2}\text{He} + {}^{214}_{82}\text{Pb} \qquad t_{1/2} = 3.05 \text{ months}$$

Polonium-218 can therefore lodge in body tissues where it undergoes α decay to give lead-214, itself a radioactive isotope. The range of an α particle is quite small, perhaps 0.7 mm (about the thickness of a sheet of paper). This is, however, approximately the thickness of the epithelial cells of the lungs, so the radiation can damage these tissues and induce lung cancer.

Virtually every home in the United States is believed to have some level of radon gas. To test for the presence of the gas, you can purchase testing kits of various kinds (Figure 24.12). There is currently a great deal of controversy over the level of radon that is considered ''safe.'' The U. S. Environmental Protection Agency has set a standard of 4 pCi/L of air as an ''action level.'' There are some who believe 1.5 pCi is close to the average level, and that only about 2% of homes contain over 8 pCi/L. If your home shows higher levels of radon gas than this, you should probably have it tested further and perhaps take corrective actions, such as sealing cracks around the foundation and in the basement. Keep in mind the relative risks involved, however (see page 13). A 1.5 pCi/L level of radon leads to a lung cancer risk about the same as the risk of your dying in an accident in your home.

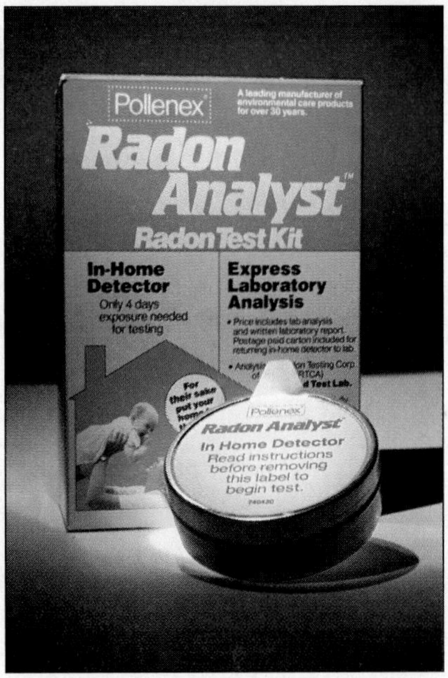

Figure 24.12 A commercially available kit to test for radon gas in the home. (C. D. Winters)

The abbreviation pCi stands for a picocurie, that is, 1×10^{-12} of a curie.

24.9 APPLICATIONS OF RADIOACTIVITY

Food Irradiation

Although uncontrolled radioactivity may be harmful, the radiation from radioisotopes can be put to beneficial use. For example, consider the importance of killing pests that would destroy food during storage. In some parts of the world stored-food spoilage may claim up to 50% of the food crop. In our society, refrigeration, canning, and chemical additives lower this figure considerably. Still, there are problems with food spoilage, and food protection costs amount to a sizable fraction of the final cost of food. Food irradiation with γ rays from sources such as ^{60}Co and ^{137}Cs is commonly used in European countries, Canada, and Mexico. Some irradiated foods are sold in the United States as well. Foods may be pasteurized by irradiation to retard the growth of organisms, such as bacteria, molds, and yeasts. This irradiation prolongs shelf life under refriger-

*D. J. Hanson, *Chemical and Engineering News,* February 6, 1989, p. 7.

Assessing Your Exposure to Radiation

The Committee on Biological Effects of Ionizing Radiation of the National Academy of Sciences issued a report in 1980 that contained a survey for individual evaluation of exposure to ionizing radiation. The following table is adapted from this report. By adding up your exposure, you can compare your annual dose to the United States annual average of 180 to 200 mrem.

(Adapted from A. R. Hinrichs, *Energy,* pp. 335–336. Philadelphia, Saunders College Publishing, 1992.)

	Common Sources of Radiation	Your Annual Dose (mrem)
Where You Live	**Location:** Cosmic radiation at sea level	26
	For your elevation (in feet), add this number of mrem	____
	Elevation 1000, *mrem* 2; *Elevation* 4000, *mrem* 15; *Elevation* 7000, *mrem* 40	
	Elevation 2000, *mrem* 5; *Elevation* 5000, *mrem* 21; *Elevation* 8000, *mrem* 53	
	Elevation 3000, *mrem* 9; *Elevation* 6000, *mrem* 29; *Elevation* 9000, *mrem* 70	
	Ground: U. S. average	26
	House construction: For stone, concrete, or masonry building, add 7	____
What You Eat, Drink, and Breathe	**Food, water, air:** U. S. average	24
	Weapons test fallout	4
How You Live	**X-ray and radiopharmaceutical diagnosis**	
	Number of chest x-rays ____ × 10	____
	Number of lower gastrointestinal tract x-rays ____ × 500	____
	Number of radiopharmaceutical examinations ____ × 300	____
	(Average dose to total U. S. population = 92 mrem)	
	Jet plane travel: For each 2500 miles add 1 mrem	____
	TV viewing: Number of hours per day ____ × 0.15	____
How Close You Live to a Nuclear Plant	**At site boundary:** average number of hours per day ____ × 0.2	____
	One mile away: average number of hours per day ____ × 0.02	____
	Five miles away: average number of hours per day ____ × 0.002	____
	Over 5 miles away: .... none	____
	Note: Maximum allowable dose determined by "as low as reasonably achievable" (ALARA) criteria established by the U.S. Nuclear Regulatory Commission. Experience shows that your actual dose is substantially less than these limits.	
	Your total annual dose in mrem	

Compare your annual dose to the U.S. annual average of 180 mrem.
One mrem per year is a risk equal to increasing your diet by 4%, or taking a 5-day vacation in the Sierra Nevada (CA) mountains.

*Based on the "BEIR Report III"—National Academy of Sciences, Committee on Biological Effects of Ionizing Radiation, *The Effects on Populations of Exposure to Low Levels of Ionizing Radiation,* National Academy of Sciences, Washington, D.C., 1980.

TABLE 24.4 Examples of Irradiated Foodstuffs

Food	Purpose	Status
Potatoes	Retardation of sprouts	FDA approved
Wheat	Insect disinfection	FDA approved
Wheat flour	Insect disinfection	FDA approved
Spices	Retardation of microbe growth	FDA approved
Grapefruit	Mold control	Approved for export
Strawberries	Mold control	Approved for export
Fish	Microbe control	Approved for export
Shrimp	Microbe control	Approved for export

Figure 24.13 Irradiated strawberries.

ation in much the same way that heat pasteurization protects milk. Chicken normally has a three-day refrigerated shelf life; after irradiation, it may have a three-week refrigerated shelf life.

The FDA may soon permit irradiation up to 100 kilorads for the pasteurization of foods. Radiation levels in the 1- to 5-Mrad range sterilize; that is, every living organism is killed. Foods irradiated at these levels will keep indefinitely when sealed in plastic or aluminum-foil packages. Indeed, radiation-sterilized foods can last three to seven years without refrigeration. Ham, beef, turkey, and corned beef sterilized by radiation have been used on many space shuttle flights. An astronaut said that "The beautiful thing was that it didn't disturb the taste, which made the meals much better than the freeze-dried and other types of foods we had."

1 Mrad = 1×10^6 rad

More than 40 classes of foods are already irradiated in 24 countries. In the United States, only a small number of foods may be irradiated (Figure 24.13 and Table 24.4).

Recent studies indicate that there may be harmful health effects from several common agricultural fumigants; irradiation of fruits and vegetables could be an effective alternative to some chemical fumigants. The agricultural products may be picked, packed, and readied for shipment. After that, the entire shipping container can be passed through a building containing a strong source of radiation. This type of sterilization offers greater worker safety because it lessens chances of exposure to harmful chemicals, and it protects the environment by avoiding contamination of water supplies with these toxic chemicals.

Radioactive Tracers

The chemical behavior of a radioisotope is almost identical to that of the nonradioactive isotopes of the same element, because the energies of the valence electrons are nearly the same in both atoms. Chemists can therefore use radioactive isotopes as **tracers** in chemical reactions and biological processes. To use a tracer, a chemist prepares a reactant compound in which one of the elements consists of both radioactive and stable isotopes, and introduces it into the reaction (or feeds it to an organism). After the reaction, the chemist measures the radioactivity of the products (or determines which parts of the organism contain the radioisotope) by using a Geiger-Müller counter or similar instrument. Several radioisotopes commonly used as tracers are listed in Table 24.5.

TABLE 24.5 Radioisotopes Used as Tracers

Isotope	Half-Life	Use
^{14}C	5730 years	CO_2 for photosynthesis research
^{3}H	12.26 years	Tag hydrocarbons
^{35}S	87.9 days	Tag pesticides, measure air flow
^{32}P	14.3 days	Measure phosphorus uptake by plants

For example, plants are known to take up phosphorus-containing compounds from the soil through their roots. The use of the radioactive phosphorus isotope ^{32}P, a β-emitter, presents a way not only of detecting the uptake of phosphorus by a plant but also of measuring the speed of uptake under various conditions. Plant biologists can grow hybrid strains of plants that can absorb phosphorus quickly, and then they can test this ability with the radioactive phosphorus tracer. This type of research leads to faster maturing crops, better yields per acre, and more food or fiber at less expense.

Important characteristics of a pesticide can be measured by tagging the pesticide with radioisotopes that have short half-lives and then applying it to a test field. Following the tagged pesticide can provide information on its tendency to accumulate in the soil, to be taken up by the plant, and to accumulate in run-off surface water. This is done with a high degree of accuracy by counting the disintegrations of the radioactive tracer. After these tests are completed, the radioactive isotopes in the tagged pesticides decay to a harmless level in a few days or a few weeks because of the short half-lives of the species used. This type of research leads to safer, more effective pesticides.

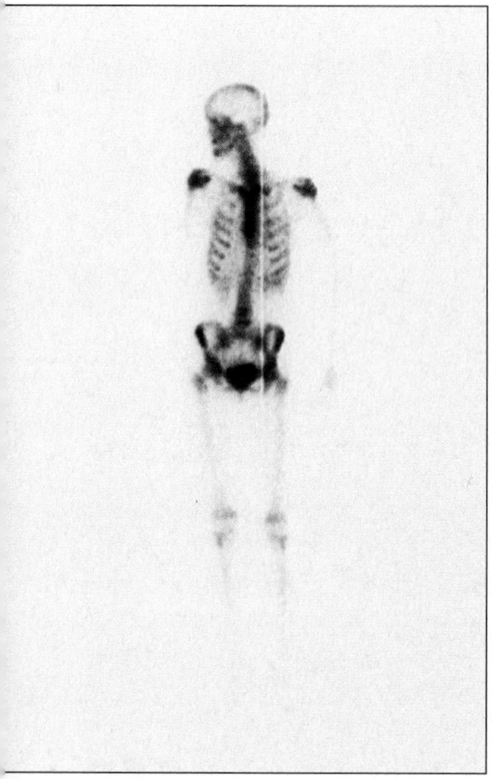

Figure 24.14 A whole-body scan. Phosphate with technetium-99m was injected into the blood and then absorbed by the bones and kidneys. This picture was taken three hours after injection. (SUNY Upstate Medical Center)

Medical Imaging

Radioactive isotopes are also used in **nuclear medicine** in two different ways: diagnosis and therapy. In the diagnosis of internal disorders such as tumors, physicians need information on the locations of abnormal tissue. This is done by **imaging,** a technique in which the radioisotope, either alone or combined with some other chemical, accumulates at the site of the disorder. There, acting like a homing device, the radioisotope disintegrates and emits its characteristic radiation, which is detected. Modern medical diagnostic instruments not only determine where the radioisotope is located in the patient's body but also construct an image of the area within the body where the radioisotope is concentrated.

Four of the most common diagnostic radioisotopes are given in Table 24.6. All are made in a particle accelerator in which heavy, charged nuclear particles are made to react with other atoms. Each of these radioisotopes produces γ radiation, which in low doses is less harmful to the tissue than ionizing radiations, such as β or α particles. By the use of special carrier compounds, these radioisotopes can be made to accumulate in specific areas of the body. For example, the pyrophosphate ion, $P_4O_7^{4-}$, can bond to the technetium-99m radioisotope; together they accumulate in the skeletal structure where abnormal bone metabolism is occurring (Figure 24.14). The technetium-99m radioisotope is metastable, as denoted by the letter *m*; this term means that the nucleus loses energy by disintegrating to a more stable version of the same isotope,

$$^{99m}Tc \longrightarrow {}^{99}Tc + \gamma$$

TABLE 24.6 Diagnostic Radioisotopes

Radioisotope	Name	Half-Life (Hours)	Uses
^{99m}Tc*	Technetium-99m	6.0	To the thyroid, brain, kidneys
^{201}Tl	Thallium-201	74	To the heart
^{123}I	Iodine-123	13.3	To the thyroid
^{67}Ga	Gallium-67	77.9	To various tumors and abscesses

*The technetium-99m isotope is the one most commonly used for diagnostic purposes. The *m* stands for "metastable," a term explained in the text.

and the γ rays are detected. Such investigations often pinpoint bone tumors.

Positron emission tomography (PET) is a form of nuclear imaging that uses **positron emitters,** such as carbon-11, fluorine-18, nitrogen-13, or oxygen-15. All these radioisotopes are neutron-deficient, have short half-lives, and therefore must be prepared in a cyclotron immediately before use. When these radioisotopes decay, a proton is converted into a neutron, a positron, and a neutrino (ν),

$$^{1}_{1}p \longrightarrow {}^{1}_{0}n + {}^{0}_{1}\beta + \nu$$

Because matter is virtually transparent to neutrinos, they escape undetected, but the positron travels less than a few millimeters before it encounters an electron and undergoes antimatter-matter annihilation.

$$^{0}_{1}\beta + {}^{0}_{-1}\beta \longrightarrow 2\ \gamma$$

The annihilation event produces two γ rays that radiate in opposite directions and are detected by two scintillation detectors located 180° apart in the PET scanner. By detecting several million annihilation γ rays within a circular slice around the subject over approximately 10 min, the region of tissue containing the radioisotope can be imaged with computer signal-averaging techniques (Figure 24.15).

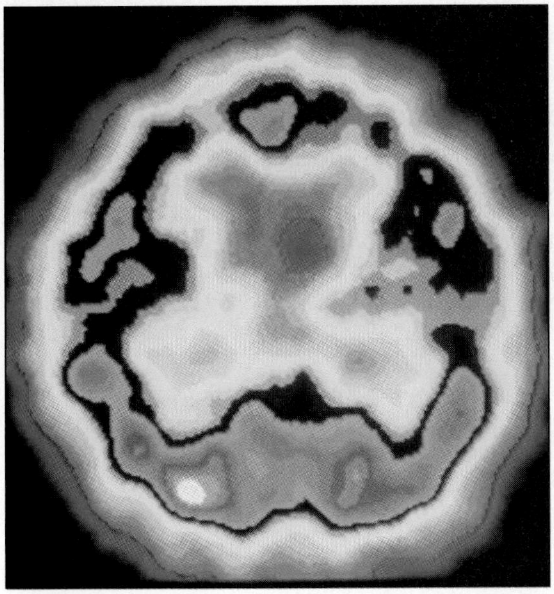

Figure 24.15 A PET scan (positron emission tomography) of a normal human brain. (CEA-ORSAY/CNRI/Science Photo Library)

CHAPTER HIGHLIGHTS

Having studied this chapter, you should be able to

- Characterize the three major types of radiation observed in radioactive decay: α, β, and γ (Section 24.1).
- Write a balanced equation for a nuclear reaction or transmutation (Section 24.2).
- Decide whether a particular radioactive isotope will decay by α, β, or positron emission or by electron capture (Sections 24.2 and 24.3).
- Calculate the binding energy for a particular isotope and understand what this energy means in terms of nuclear stability (Section 24.3).
- Use the equation $\ln (N/N_0) = -kt$ (Equation 24.2), which relates (through the decay constant k) the time period over which a sample is observed (t) to the number of radioactive atoms present at the beginning (N_0) and end (N) of the period (Section 24.4).
- Calculate the half-life of a radioactive isotope ($t_{1/2}$) from the activity of a sample, or use the half-life to find the time required for an isotope to decay to a particular activity (Section 24.4).
- Describe nuclear chain reactions, nuclear fission, and nuclear fusion (Sections 24.6 and 24.7).
- Describe some sources of background radiation and the units used to measure radiation (Section 24.8).
- Relate some uses of radioisotopes (Section 24.9).

STUDY QUESTIONS

Review Questions

1. Name three people who made important contributions to nuclear chemistry and indicate their contributions.
2. Tell the similarities and differences in α particles, β particles, positrons, and γ rays.
3. What is the binding energy of a nucleus?
4. If the mass number of an isotope is *much greater than* twice the atomic number, what type of radioactive decay might you expect?
5. If the number of neutrons in an isotope is *much less than* the number of protons, what type of radioactive decay might you expect?
6. What is the difference between fission and fusion? Illustrate your answer with an example of each.
7. Is there a nuclear reactor producing electric power in your state? If so, where is it located?
8. One reaction occurring in the hydrogen bomb is that between tritium and deuterium. What type of reaction is this, fusion or fission?

$$^{3}_{1}H + ^{2}_{1}H \longrightarrow ^{1}_{0}n + ^{4}_{2}He$$

9. What are some of the advantages and disadvantages of nuclear power plants?
10. Name at least two uses of radioactive isotopes (outside of their use in power reactors and weapons).

Nuclear Reactions

11. Complete the following nuclear equations. Write the mass number and atomic number for the remaining particle, as well as the element symbol where possible.
(a) $^{54}_{26}Fe + ^{4}_{2}He \rightarrow 2\ ^{1}_{1}H + ?$
(b) $^{27}_{13}Al + ^{4}_{2}He \rightarrow ^{30}_{15}P + ?$
(c) $^{32}_{16}S + ^{1}_{0}n \rightarrow ^{1}_{1}H + ?$
(d) $^{96}_{42}Mo + ^{2}_{1}H \rightarrow ^{1}_{0}n + ?$
(e) $^{98}_{42}Mo + ^{1}_{0}n \rightarrow ^{99}_{43}Tc + ?$

12. Complete the following nuclear equations. Write the mass number and atomic number for the remaining particle, as well as the element symbol where possible.
(a) $^{9}_{4}Be + ? \rightarrow ^{6}_{3}Li + ^{4}_{2}He$
(b) $? + ^{1}_{0}n \rightarrow ^{24}_{11}Na + ^{4}_{2}He$
(c) $^{40}_{20}Ca + ? \rightarrow ^{40}_{19}K + ^{1}_{1}H$

(d) $^{241}_{95}Am + ^{4}_{2}He \rightarrow ^{243}_{97}Bk + ?$
(e) $^{246}_{96}Cm + ^{12}_{6}C \rightarrow 4\ ^{1}_{0}n + ?$
(f) $^{238}_{92}U + ? \rightarrow ^{249}_{100}Fm + 5\ ^{1}_{0}n$

13. Complete the following nuclear equations. Write the mass number and atomic number for the remaining particle, as well as the element symbol where possible.
(a) $^{111}_{47}Ag \rightarrow ^{104}_{48}Cd + ?$
(b) $^{87}_{36}Kr \rightarrow ^{0}_{-1}\beta + ?$
(c) $^{231}_{91}Pa \rightarrow ^{227}_{89}Ac + ?$
(d) $^{230}_{90}Th \rightarrow ^{4}_{2}He + ?$
(e) $^{82}_{35}Br \rightarrow ^{82}_{36}Kr + ?$
(f) $? \rightarrow ^{24}_{12}Mg + ^{0}_{-1}\beta$

14. Complete the following nuclear equations. Write the mass number and atomic number for the remaining particle, as well as the element symbol where possible.
(a) $^{19}_{10}Ne \rightarrow ^{0}_{+1}\beta + ?$
(b) $^{59}_{26}Fe \rightarrow ^{0}_{-1}\beta + ?$
(c) $^{40}_{19}K \rightarrow ^{0}_{-1}\beta + ?$
(d) $^{37}_{18}Ar + ^{0}_{-1}e$ (electron capture) $\rightarrow ?$
(e) $^{55}_{26}Fe + ^{0}_{-1}e$ (electron capture) $\rightarrow ?$
(f) $^{26}_{13}Al \rightarrow ^{25}_{12}Mg + ?$

15. One radioactive series that begins with uranium-235 and ends with lead-207 undergoes the sequence of emission reactions: $\alpha, \beta, \alpha, \beta, \alpha, \alpha, \alpha, \alpha, \beta, \beta, \alpha$. Identify the radioisotope produced in each of the *first five steps.*

16. One radioactive series that begins with uranium-235 and ends with lead-207 undergoes the sequence of emission reactions: $\alpha, \beta, \alpha, \beta, \alpha, \alpha, \alpha, \alpha, \beta, \beta, \alpha$. Identify the radioisotope produced in each of the *last six steps.* (See Study Question 15.)

Nuclear Stability

17. Boron has two stable isotopes, ^{10}B (abundance = 19.78%) and ^{11}B (abundance = 80.22%). Calculate the binding energies per nucleon of these two nuclei and compare their stabilities.

$$5\ ^{1}_{1}H + 5\ ^{1}_{0}n \rightarrow ^{10}_{5}B$$
$$5\ ^{1}_{1}H + 6\ ^{1}_{0}n \rightarrow ^{11}_{5}B$$

The required masses (in grams per mole) are: $^{1}_{1}H = 1.00783$; $^{1}_{0}n = 1.00867$; $^{10}_{5}B = 10.01294$; and $^{11}_{5}B = 11.00931$.

18. Calculate the binding energy in kilojoules per mole of P for the formation of $^{30}_{15}P$

$$15\ ^{1}_{1}H + 15\ ^{1}_{0}n \rightarrow ^{30}_{15}P$$

and for the formation of $^{31}_{15}P$.

$$15\ ^{1}_{1}H + 16\ ^{1}_{0}n \rightarrow ^{31}_{15}P$$

Which is the more stable isotope? The required masses (in grams per mole) are $^{1}_{1}H = 1.00783$; $^{1}_{0}n = 1.00867$; $^{30}_{15}P = 29.97832$; and $^{31}_{15}P = 30.97376$.

Rates of Disintegration Reactions

19. Copper-64 is used in the form of copper acetate to study brain tumors. It has a half-life of 12.8 h. If you begin with 15.0 mg of ^{64}Cu-labeled copper acetate, what mass in milligrams remains after 2 days and 16 h?

20. Gold-198 is used as the metal in the diagnosis of liver problems. The half-life of ^{198}Au is 2.7 days. If you begin with 5.6 mg of this gold isotope, what mass remains after 10.8 days?

21. Iodine-131 is used in the form of sodium iodide to treat cancer of the thyroid.
(a) The isotope decays by ejecting a β particle. Write a balanced equation to show this process.
(b) The isotope has a half-life of 8.05 days. If you begin with 25.0 mg of radioactive Na ^{131}I, what mass remains after 32.2 days (about a month)?

22. Phosphorus-32 is used in the form of Na_2HPO_4 in the treatment of chronic myeloid leukemia, among other things.
(a) The isotope decays by emitting a β particle. Write a balanced equation to show this process.
(b) The half-life of ^{32}P is 14.3 days. If you begin with 9.6 mg of radioactive Na_2HPO_4, what mass remains after 28.6 days (about one month)?

23. In 1984, cobalt-60 was involved in the worst accident with radioactive isotopes in North America (*Science 84,* December 1984, p. 28). The half-life of the isotope is 5.3 years. Starting with 10.0 mg of ^{60}Co, how much will remain after 21.2 years? Approximately how much will remain after a century?

24. Gallium-67 ($t_{1/2} = 77.9$ h) is used in the medical diagnosis of certain kinds of tumors. If you ingest a compound containing 0.15 mg of this isotope, what mass in milligrams remains in your body after 13 days? (Assume none is excreted.)

25. Radioisotopes of iodine are widely used in medicine. For example, iodine-131 ($t_{1/2} = 8.05$ days) is used to treat thyroid cancer. If you ingest a sample of NaI containing ^{131}I, how much time is required for the isotope to fall to 5.0% of its original activity?

26. The noble gas radon has been the focus of much attention recently because it can be found in homes. Radon-222 emits α particles and has a half-life of 3.82 days.
(a) Write a balanced equation to show this process.
(b) How long does it take for a sample of radon to decrease to 10.0% of its original activity?

27. A sample of wood from a Thracian chariot found in an excavation in Bulgaria has a ^{14}C activity of 11.2 disintegrations per minute per gram. Estimate the age of the chariot and the year it was made. ($t_{1/2}$ for ^{14}C is 5.73×10^3 years and the activity of ^{14}C in living material is 14.0 disintegrations per minute per gram.)

28. A piece of charred bone found in the ruins of an American Indian village has a ^{14}C to ^{12}C ratio of 0.72 times that found in living organisms. Calculate the age of the bone fragment. (See Study Question 27 for required data on carbon-14.)

Nuclear Transmutations

29. There are two isotopes of americium, both with half-lives sufficiently long to allow the handling of massive quantities. Americium-241, for example, has a half-life of 248 years as an α emitter, and it is used in gauging the thickness of materials and in smoke detectors. The isotope is formed from ${}^{239}Pu$ by absorption of two neutrons followed by emission of a β particle. Write a balanced equation for this process.

30. Americium-240 is made by bombarding a plutonium-239 atom with an α particle. In addition to ${}^{240}Am$, the products are a proton and two neutrons. Write a balanced equation for this process.

31. To synthesize the heavier transuranium elements, a lighter nucleus must be bombarded with a relatively large particle. If you know the products are californium-246 and 4 neutrons, with what particle would you bombard uranium-238 atoms?

32. The element with the highest known atomic number is 111. It is thought that still heavier elements are possible, especially with $Z = 114$ and $N = 184$. To this end, serious attempts have been made to force calcium-40 and curium-248 to merge. What would be the atomic number of the element formed?

Nuclear Fission and Power

33. The average energy output of a good grade of coal is 2.6×10^7 kJ/ton. Fission of 1 mol of ${}^{235}U$ releases 2.1×10^{10} kJ. Find the number of tons of coal needed to produce the same energy as 1 lb of ${}^{235}U$. (See Appendix C for conversion factors.)

34. A concern in the nuclear power industry is that, if nuclear power becomes more widely used, there may be serious shortages in worldwide supplies of fissionable uranium. One solution is to build "breeder" reactors that manufacture more fuel than they consume. One such cycle works as follows:
(a) A "fertile" ${}^{238}U$ nucleus collides with a neutron to produce ${}^{239}U$.
(b) ${}^{239}U$ decays by β emission ($t_{1/2} = 24$ min) to give an isotope of neptunium.
(c) This neptunium isotope decays by β emission to give a plutonium isotope.
(d) The plutonium isotope is fissionable. On collision of one of these plutonium isotopes with a neutron, fission occurs with energy, at least two neutrons, and other nuclei as products.

Write an equation for each of the steps, and explain how this process can be used to breed more fuel than the reactor originally contained and still produce energy.

Uses of Radioisotopes

35. In order to measure the volume of the blood system of an animal, the following experiment was done. A 1.0-mL sample of an aqueous solution containing tritium with an activity of 2.0×10^6 disintegrations per second (dps) was injected into the bloodstream. After time was allowed for complete circulatory mixing, a 1.0-mL blood sample was withdrawn and found to have an activity of 1.5×10^4 dps. What was the volume of the circulatory system? (The half-life of tritium is 12.3 years, so this experiment assumes that only a negligible amount of tritium has decayed in the time of the experiment.)

36. Radioactive isotopes are often used as "tracers" to follow an atom through a chemical reaction, and the following is an example. Acetic acid reacts with methanol, CH_3OH, by eliminating a molecule of H_2O to form methyl acetate, $CH_3CO_2CH_3$. Explain how you would use the radioactive isotope ${}^{18}O$ to show whether the oxygen atom in the water product comes from the —OH of the acid or the —OH of the alcohol.

$$\underset{\text{acetic acid}}{H_3C\overset{\overset{O}{\|}}{C}OH} + \underset{\text{methanol}}{HOCH_3} \longrightarrow \underset{\text{methyl acetate}}{H_3C\overset{\overset{O}{\|}}{C}OCH_3} + H_2O$$

General Questions

37. Complete the following nuclear equations. Write the mass number and atomic number for the remaining particle, as well as the element symbol where possible.
(a) ${}^{13}_{6}C + ? \rightarrow {}^{14}_{6}C$
(b) ${}^{40}_{18}Ar + ? \rightarrow {}^{43}_{19}K + {}^{1}_{1}H$
(c) ${}^{250}_{98}Cf + {}^{11}_{5}B \rightarrow 4\ {}^{1}_{0}n + ?$
(d) ${}^{53}_{24}Cr + {}^{4}_{2}He \rightarrow ? + {}^{56}_{26}Fe$
(e) ${}^{212}_{84}Po \rightarrow {}^{208}_{82}Pb + ?$
(f) ${}^{122}_{53}I \rightarrow {}^{0}_{+1}\beta + ?$
(g) $? \rightarrow {}^{23}_{11}Na + {}^{0}_{-1}\beta$
(h) ${}^{137}_{53}I \rightarrow {}^{1}_{0}n + ?$

38. The following reaction sequence involves four unknown elements, Q, Δ, Σ, and Π. Based on the reactions in the sequence, identify the unknown elements.

$$Pb + Cr \longrightarrow Q$$
$$Cf + O \longrightarrow Q$$
$$Q \longrightarrow \alpha + \Delta$$
$$\Delta \longrightarrow \alpha + \Sigma$$
$$\Sigma \longrightarrow \alpha + \Pi$$

39. The oldest known fossil cells form a biological cluster found in South Africa. The fossil has been dated by the reaction

$${}^{87}Rb \rightarrow {}^{87}Sr + {}^{0}_{-1}\beta \qquad t_{1/2} = 4.9 \times 10^{10}\ y$$

If the ratio of the present quantity of ${}^{87}Rb$ to the original quantity is 0.951, calculate the age of the fossil cells.

40. Balance the following reactions used for the synthesis of transuranium elements.
(a) ${}^{238}_{92}U + {}^{14}_{7}N \rightarrow ? + 5\ {}^{1}_{0}n$
(b) ${}^{238}_{92}U + ? \rightarrow {}^{249}_{100}Fm + 5\ {}^{1}_{0}n$

(c) ${}^{253}_{99}Es + ? \rightarrow {}^{256}_{101}Md + {}^{1}_{0}n$
(d) ${}^{246}_{96}Cm + ? \rightarrow {}^{254}_{102}No + 4\ {}^{1}_{0}n$
(e) ${}^{252}_{98}Cf + ? \rightarrow {}^{257}_{103}Lr + 5\ {}^{1}_{0}n$

41. On December 2, 1942, the first man-made self-sustaining nuclear fission chain reactor was operated by Enrico Fermi and others under the University of Chicago Stadium. In June, 1972, *natural* fission reactors, which operated billions of years ago, were discovered in Oklo, Gabon. At present, natural uranium contains 0.72% ^{235}U. How many years ago did natural uranium contain 3.0% ^{235}U, sufficient to sustain a natural reactor? ($t_{1/2}$ for ^{235}U is 7.04×10^8 y.)

42. Predict the probable mode of decay for each of the following radioactive isotopes and write an equation to show the products of decay:
(a) radon-218
(b) californium-240
(c) cobalt-61
(d) carbon-11

List of Appendices

APPENDIX A

Some Mathematical Operations

The mathematical skills required in this introductory course are basic skills in algebra and a knowledge of (a) exponential (or scientific) notation, (b) logarithms, and (c) quadratic equations. This appendix reviews each of the final three topics.

A.1 ELECTRONIC CALCULATORS

The directions for calculator use in this section are given for calculators using "algebraic" logic. Such calculators are the most common type used by students in introductory courses. The procedures differ slightly for calculators using RPN logic (such as those made by Hewlett-Packard).

The advent of inexpensive electronic calculators a few years ago has made calculations in introductory chemistry much more straightforward. You are well advised to purchase a calculator that has the capability of performing calculations in scientific notation, has both base-10 and natural logarithms, and is capable of raising any number to any power and of finding any root of any number. In the following discussion, we point out in general how these functions of your calculator can be used.

Although electronic calculators have greatly simplified calculations, they have also forced us to focus again on significant figures. A calculator easily handles eight or more significant figures, but real laboratory data are never known to this accuracy. You are therefore urged to review Section 1.7 on handling numbers.

A.2 EXPONENTIAL (SCIENTIFIC) NOTATION

In exponential, or scientific, notation, a number is expressed as a product of two numbers: $N \times 10^n$. The first number, N, is the so-called *digit term* and is a number between 1 and 10. The second number, 10^n, the *exponential term,* is some integer power of 10. For example, 1234 is written in scientific notation as 1.234×10^3, or 1.234 multiplied by 10 three times.

$$1234 = 1.234 \times 10^1 \times 10^1 \times 10^1 = 1.234 \times 10^3$$

Conversely, a number less than 1, such as 0.01234, is written as 1.234×10^{-2}. This notation tells us that 1.234 should be divided twice by 10 to obtain 0.01234.

$$0.01234 = \frac{1.234}{10^1 \times 10^1} = 1.234 \times 10^{-1} \times 10^{-1} = 1.234 \times 10^{-2}$$

Some other examples of scientific notation are

$10000 = 1 \times 10^4$	$12345 = 1.2345 \times 10^4$
$1000 = 1 \times 10^3$	$1234 = 1.234 \times 10^3$
$100 = 1 \times 10^2$	$123 = 1.23 \times 10^2$
$10 = 1 \times 10^1$	$12 = 1.2 \times 10^1$
$1 = 1 \times 10^0$	(any number to the zero power = 1)
$1/10 = 1 \times 10^{-1}$	$0.12 = 1.2 \times 10^{-1}$
$1/100 = 1 \times 10^{-2}$	$0.012 = 1.2 \times 10^{-2}$
$1/1000 = 1 \times 10^{-3}$	$0.0012 = 1.2 \times 10^{-3}$
$1/10000 = 1 \times 10^{-4}$	$0.00012 = 1.2 \times 10^{-4}$

When converting a number to scientific notation, notice that the exponent n is positive if the number is greater than one and negative if the number is less than 1. The value of n is the number of places by which the decimal is shifted to obtain the number in scientific notation.

$$1\ 2\ 3\ 4\ 5. = 1.2345 \times 10^4$$

Decimal shifted 4 places to the left. Therefore, n is positive and equal to 4.

$$0.0\ 0\ 1\ 2 = 1.2 \times 10^{-3}$$

Decimal shifted 3 places to the right. Therefore, n is negative and equal to 3.

If you wish to convert a number in scientific notation to the usual form, the procedure is simply reversed.

$$6\ 2\ 7\ 3 \times 10^2 = 627.3$$

Decimal point moved 2 places to the right, since n is positive and equal to 2.

$$0\ 0\ 6.273 \times 10^{-3} = 0.006273$$

Decimal point shifted 3 places to the left, since n is negative and equal to 3.

Two final points must be made concerning scientific notation. First, if you are used to working on a computer, you may be in the habit of writing a number such as 1.23×10^3 as 1.23E3 or 6.45×10^{-5} as 6.45E-5. Second, some electronic calculators allow you to convert numbers readily to the scientific notation. If you have such a calculator, you can change a number shown in the usual form to scientific notation simply by pressing the EE or EXP key and then the "=" key.

1. Adding and Subtracting Numbers

When adding or subtracting two numbers, first convert them to the same powers of 10. The digit terms are then added or subtracted as appropriate.

$$(1.234 \times 10^{-3}) + (5.623 \times 10^{-2}) = (0.1234 \times 10^{-2}) + (5.623 \times 10^{-2}) = 5.746 \times 10^{-2}$$

$$(6.52 \times 10^2) - (1.56 \times 10^3) = (6.52 \times 10^2) - (15.6 \times 10^2) = -9.1 \times 10^2$$

2. Multiplication

The digit terms are multiplied in the usual manner, and the exponents are added algebraically. The result is expressed with a digit term with only one nonzero digit to the left of the decimal.

$$\begin{aligned}(1.23 \times 10^3)(7.60 \times 10^2) &= (1.23)(7.60) \times 10^{3+2} \\ &= 9.35 \times 10^5\end{aligned}$$

$$\begin{aligned}(6.02 \times 10^{23})(2.32 \times 10^{-2}) &= (6.02)(2.32) \times 10^{23-2} \\ &= 13.966 \times 10^{21} \\ &= 1.40 \times 10^{22} \text{ (answer in three significant figures)}\end{aligned}$$

3. Division

The digit terms are divided in the usual manner, and the exponents are subtracted algebraically. The quotient is written with one nonzero digit to the left of the decimal in the digit term.

$$\frac{7.60 \times 10^3}{1.23 \times 10^2} = \frac{7.60}{1.23} \times 10^{3-2} = 6.18 \times 10^1$$

$$\frac{6.02 \times 10^{23}}{9.10 \times 10^{-2}} = \frac{6.02}{9.10} \times 10^{(23)-(-2)} = 0.662 \times 10^{25} = 6.62 \times 10^{24}$$

4. Powers of Exponentials

When raising a number in exponential notation to a power, treat the digit term in the usual manner. The exponent is then multiplied by the number indicating the power.

$$\begin{aligned}(1.25 \times 10^3)^2 &= (1.25)^2 \times 10^{3\times 2} \\ &= 1.5625 \times 10^6 = 1.56 \times 10^6\end{aligned}$$

$$\begin{aligned}(5.6 \times 10^{-10})^3 &= (5.6)^3 \times 10^{(-10)\times 3} \\ &= 175.6 \times 10^{-30} = 1.8 \times 10^{-28}\end{aligned}$$

Electronic calculators usually have two methods of raising a number to a power. To square a number, enter the number and then press the "x^2" key. To raise a number to any power, use the "y^x" key. For example, to raise 1.42×10^2 to the fourth power,

1. Enter 1.42×10^2.
2. Press "y^x".
3. Enter 4 (this should appear on the display).
4. Press "=" and 4.0659×10^8 appears on the display.

As a final step, express the number in the correct number of significant figures (4.07×10^8) in this case.

5. Roots Of Exponentials

Unless you use an electronic calculator, the number must first be put into a form in which the exponential is exactly divisible by the root. The root of the digit term is found in the usual way, and the exponent is divided by the desired root.

$$\sqrt{3.6 \times 10^7} = \sqrt{36 \times 10^6} = \sqrt{36} \times \sqrt{10^6} = 6.0 \times 10^3$$

$$\sqrt[3]{2.1 \times 10^{-7}} = \sqrt[3]{210 \times 10^{-9}} = \sqrt[3]{210} \times \sqrt[3]{10^{-9}} = 5.9 \times 10^{-3}$$

To take a square root on an electronic calculator, enter the number and then press the "$\sqrt{x}$" key. To find a higher root of a number, such as the fourth root of 5.6×10^{-10},

1. Enter the number.
2. Press the "$\sqrt[x]{y}$" key. (On most calculators, the sequence you actually use is to press "2ndF" and then "$\sqrt[x]{y}$." Alternatively, you press "INV" and then "y^x.")
3. Enter the desired root, 4 in this case.
4. Press "=". The answer here is 4.8646×10^{-3}, or 4.9×10^{-3}.

A general procedure for finding any root is to use the "y^x" key. For a square root, x is 0.5 (or $\frac{1}{2}$), whereas it is 0.33 (or $\frac{1}{3}$) for a cube root, 0.25 (or $\frac{1}{4}$) for a fourth root, and so on.

A.3 LOGARITHMS

Two types of logarithms are used in this text: (a) common logarithms (abbreviated log) whose base is 10 and (b) natural logarithms (abbreviated ln) whose base is e (=2.71828).

$$\log x = n, \text{ where } x = 10^n$$

$$\ln x = m, \text{ where } x = e^m$$

Most equations in chemistry and physics were developed in natural, or base e, logarithms, and we follow this practice in this text. The relation between log and ln is

$$\ln x = 2.303 \log x$$

Despite the different bases of the two logarithms, they are used in the same manner. What follows is largely a description of the use of common logarithms.

A common logarithm is the power to which you must raise 10 to obtain the number. For example, the log of 100 is 2, since you must raise 10 to the second power to obtain 100. Other examples are

$$\begin{aligned}
\log 1000 &= \log (10^3) = 3 \\
\log 10 &= \log (10^1) = 1 \\
\log 1 &= \log (10^0) = 0 \\
\log \frac{1}{10} &= \log (10^{-1}) = -1 \\
\log \frac{1}{10000} &= \log (10^{-4}) = -4
\end{aligned}$$

To obtain the common logarithm of a number other than a simple power of 10, you must resort to a log table or an electronic calculator. For example,

$$\begin{aligned}
\log 2.10 &= 0.3222, \text{ which means that } 10^{0.3222} = 2.10 \\
\log 5.16 &= 0.7126, \text{ which means that } 10^{0.7126} = 5.16 \\
\log 3.125 &= 0.49485, \text{ which means that } 10^{0.49485} = 3.125
\end{aligned}$$

To check this on your calculator, enter the number, and then press the "log" key. When using a log table, the logs of the first two numbers can be read directly from the table. The log of the third number (3.125), however, must be interpolated. That is, 3.125 is midway between 3.12 and 3.13, so the log is midway between 0.4942 and 0.4955.

To obtain the natural logarithm ln of the numbers shown here, use a calculator having this function. Enter each number and press "ln."

$$\ln 2.10 = 0.7419, \text{ which means that } e^{0.7419} = 2.10$$
$$\ln 5.15 = 1.6409, \text{ which means that } e^{1.6409} = 5.16$$

To find the common logarithm of a number greater than 10 or less than 1 with a log table, first express the number in scientific notation. Then find the log of each part of the number and add the logs. For example,

$$\log 241 = \log (2.41 \times 10^2) = \log 2.41 + \log 10^2$$
$$= 0.382 + 2 = 2.382$$

$$\log 0.00573 = \log (5.73 \times 10^{-3}) = \log 5.73 + \log 10^{-3}$$
$$= 0.758 + (-3) = -2.242$$

Significant Figures and Logarithms

LOGARITHMS AND NOMENCLATURE: The number to the left of the decimal in a logarithm is called the **characteristic,** and the number to the right of the decimal is the **mantissa.**

Notice that the mantissa has as many significant figures as the number whose log was found. (So that you could more clearly see the result obtained with a calculator or a table, this rule was not strictly followed until the last two examples.)

Obtaining Antilogarithms

If you are given the logarithm of a number, and find the number from it, you have obtained the "antilogarithm," or "antilog," of the number. Two common procedures used by electronic calculators to do this are:

Procedure A	**Procedure B**
1. Enter the log or ln.	1. Enter the log or ln.
2. Press 2ndF.	2. Press INV.
3. Press 10^x or e^x.	3. Press log or ln x.

Test one or the other of these procedures with the following examples:

1. Find the number whose log is 5.234.

Recall that $\log x = n$, where $x = 10^n$. In this case $n = 5.234$. Enter that number in your calculator, and find the value of 10^n, the antilog. In this case,

$$10^{5.234} = 10^{0.234} \times 10^5 = 1.71 \times 10^5$$

Notice that the characteristic (5) sets the decimal point; it is the power of 10 in the exponential form. The mantissa (0.234) gives the value of the number x. Thus, if you use a log table to find x, you need only look up 0.234 in the table and see that it corresponds to 1.71.

2. Find the number whose log is −3.456.

$$10^{-3.456} = 10^{0.544} \times 10^{-4} = 3.50 \times 10^{-4}$$

Notice here that −3.456 must be expressed as the sum of −4 and +0.544.

Mathematical Operations Using Logarithms

Because logarithms are exponents, operations involving them follow the same rules as the use of exponents. Thus, multiplying two numbers can be done by adding logarithms.

$$\log xy = \log x + \log y$$

For example, we multiply 563 by 125 by adding their logarithms and finding the antilogarithm of the result.

$$\begin{aligned} \log 563 &= 2.751 \\ \log 125 &= \underline{2.097} \\ \log xy &= 4.848 \\ xy &= 10^{4.848} = 10^4 \times 10^{0.848} = 7.04 \times 10^4 \end{aligned}$$

One number (x) can be divided by another (y) by subtraction of their logarithms.

$$\log \frac{x}{y} = \log x - \log y$$

For example, to divide 125 by 742,

$$\begin{aligned} \log 125 &= \quad 2.097 \\ -\log 742 &= \quad \underline{2.870} \\ \log \frac{x}{y} &= -0.773 \\ \frac{x}{y} &= 10^{-0.773} = 10^{0.227} \times 10^{-1} = 1.68 \times 10^{-1} \end{aligned}$$

Similarly, powers and roots of numbers can be found using logarithms.

$$\log x^y = y(\log x)$$

$$\log \sqrt[y]{x} = \log x^{1/y} = \frac{1}{y} \log x$$

As an example, find the fourth power of 5.23. We first find the log of 5.23 and then multiply it by 4. The result, 2.874, is the log of the answer. Therefore, we find the antilog of 2.874.

$$\begin{aligned} (5.23)^4 &= ? \\ \log (5.23)^4 &= 4 \log 5.23 = 4\,(0.719) = 2.874 \\ (5.23)^4 &= 10^{2.874} = 748 \end{aligned}$$

As another example, find the fifth root of 1.89×10^{-9}.

$$\sqrt[5]{1.89 \times 10^{-9}} = (1.89 \times 10^{-9})^{1/5} = ?$$

$$\log (1.89 \times 10^{-9})^{1/5} = \frac{1}{5} \log (1.89 \times 10^{-9}) = \frac{1}{5} (-8.724) = -1.745$$

The answer is the antilog of -1.745.

$$(1.89 \times 10^{-9})^{1/5} = 10^{-1.745} = 1.80 \times 10^{-2}$$

A.4 QUADRATIC EQUATIONS

Algebraic equations of the form $ax^2 + bx + c = 0$ are called **quadratic equations.** The coefficients a, b, and c may be either positive or negative. The two roots of the equation may be found using the *quadratic formula.*

$$x = \frac{-b \pm \sqrt{b^2 - 4ac}}{2a}$$

As an example, solve the equation $5x^2 - 3x - 2 = 0$. Here $a = 5$, $b = -3$, and $c = -2$. Therefore,

$$x = \frac{3 \pm \sqrt{(-3)^2 - 4(5)(-2)}}{2(5)}$$

$$= \frac{3 \pm \sqrt{9 - (-40)}}{10} = \frac{3 \pm \sqrt{49}}{10} = \frac{3 \pm 7}{10}$$

$$x = 1 \text{ and } -0.4$$

How do you know which of the two roots is the correct answer? You have to decide in each case which root has physical significance. It is *usually* true in this course, however, that negative values are not significant.

When you have solved a quadratic expression, you should always check your values by substitution into the original equation. In the previous example, we find that $5(1)^2 - 3(1) - 2 = 0$ and that $5(-0.4)^2 - 3(-0.4) - 2 = 0$.

The most likely place you will encounter quadratic equations is in the chapters on chemical equilibria, particularly in Chapters 16 through 18. Here you will often be faced with solving an equation such as

$$1.8 \times 10^{-4} = \frac{x^2}{0.0010 - x}$$

This equation can certainly be solved using the quadratic equation (to give $x = 3.4 \times 10^{-4}$). You may find the *method of successive approximations* to be especially convenient, however. Here we begin by making a reasonable approximation of x. This approximate value is substituted into the original equation, and this is solved to give what is hoped to be a more correct value of x. This process is repeated until the answer converges on a particular value of x, that is, until the value of x derived from two successive approximations is the same.

Step 1: First assume that x is so small that $(0.0010 - x) \approx 0.0010$. This means that

$$x^2 = 1.8 \times 10^{-4}(0.0010)$$

$$x = 4.2 \times 10^{-4} \text{ (to 2 significant figures)}$$

Step 2: Substitute the value of x from Step 1 into the denominator of the original equation, and again solve for x.

$$x^2 = 1.8 \times 10^{-4}(0.0010 - 0.00042)$$

$$x = 3.2 \times 10^{-4}$$

Step 3: Repeat Step 2 using the value of x found in that step.

$$x = \sqrt{1.8 \times 10^{-4}(0.0010 - 0.00032)} = 3.5 \times 10^{-4}$$

Step 4: Continue repeating the calculation, using the value of x found in the previous step.

$$x = \sqrt{1.8 \times 10^{-4}(0.0010 - 0.00035)} = 3.4 \times 10^{-4}$$

Step 5: $$x = \sqrt{1.8 \times 10^{-4}(0.0010 - 0.00034)} = 3.4 \times 10^{-4}$$

Here we find that iterations after the fourth step give the same value for x, indicating that we have arrived at a valid answer (and the same one obtained from the quadratic formula).

Here are several final thoughts on using the method of successive approximations. First, in some cases the method does not work. Successive steps may give answers that are random or that diverge from the correct value. In Chapters 16 through 18, you confront quadratic equations of the form $K = x^2/(C - x)$. The method of approximations works as long as $K < 4C$ (assuming one begins with $x = 0$ as the first guess, that is, $K \approx x^2/C$). This is always going to be true for weak acids and bases (the topic of Chapters 17 and 18), but it may *not* be the case for problems involving gas phase equilibria (Chapter 16), where K can be quite large.

Second, values of K in the equation $K = x^2/(C - x)$ are usually known only to two significant figures. We are therefore justified in carrying out successive steps until two answers are the same to two significant figures.

Finally, we highly recommend this method of solving quadratic equations, especially those in Chapters 17 and 18. If your calculator has a memory function, successive approximations can be carried out easily and rapidly.

APPENDIX B*

Some Important Physical Concepts

MATTER

The tendency to maintain a constant velocity is called inertia. Thus, unless acted on by an unbalanced force, a body at rest remains at rest, and a body in motion remains in motion with uniform velocity. Matter is anything that exhibits inertia; the quantity of matter is its mass.

MOTION

Motion is the change of position or location in space. Objects can have the following classes of motion:

- Translation occurs when the center of mass of an object changes its location. Example: a car moving on the highway.
- Rotation occurs when each point of a moving object moves in a circle about an axis through the center of mass. Examples: a spinning top, a rotating molecule.
- Vibration is a periodic distortion of and then recovery of original shape. Examples: a struck tuning fork, a vibrating molecule.

FORCE AND WEIGHT

Force is that which changes the velocity of a body; it is defined as

$$\text{Force} = \text{mass} \times \text{acceleration}$$

The SI unit of force is the **newton,** N, whose dimensions are kilograms per meter per second squared ($kg \cdot m/s^2$). A newton is therefore the force needed to

*Adapted from F. Brescia, J. Arents, H. Meislich, et al.: *General Chemistry,* 5th ed. Philadelphia, Harcourt Brace, 1988.

change the velocity of a mass of 1 kilogram by 1 meter per second in a time of 1 second.

Because the earth's gravity is not the same everywhere, the weight corresponding to a given mass is not a constant. At any given spot on earth gravity is constant, however, and therefore weight is proportional to mass. When a balance tells us that a given sample (the "unknown") has the same weight as another sample (the "weights," as given by a scale reading or by a total of counterweights), it also tells us that the two masses are equal. The balance is therefore a valid instrument for measuring the mass of an object independently of slight variations in the force of gravity.

PRESSURE†

Pressure is force per unit area. The SI unit, called the pascal, Pa, is

$$1 \text{ pascal} = \frac{1 \text{ newton}}{\text{m}^2} = \frac{1 \text{ kg} \cdot \text{m/s}^2}{\text{m}^2} = \frac{1 \text{ kg}}{\text{m} \cdot \text{s}^2}$$

10^5 Pa = 1 bar

The International System of Units also recognizes the bar, which is 10^5 Pa and which is close to standard atmospheric pressure (Table 1).

Chemists also express pressure in terms of the heights of liquid columns, especially water and mercury. This usage is not completely satisfactory, because the pressure exerted by a given column of a given liquid is not a constant but depends on the temperature (which influences the density of the liquid) and the location (which influences gravity). Such units are therefore not part of the SI, and their use is now discouraged. The older units have not been removed from books and journals, however, and chemists must still be familiar with them.

The pressure of a liquid or a gas depends only on the depth (or height) and is exerted equally in all directions. At sea level, the pressure exerted by the earth's atmosphere supports a column of mercury about 0.76 m (76 cm, or 760 mm) high.

One **standard atmosphere** (atm) is the pressure exerted by exactly 76 cm of mercury at 0 °C (density 13.5951 g/cm^3) and at standard gravity, 9.80665 m/s^2. The **bar** is equivalent to 1.01325 atm. One **torr** is the pressure exerted by exactly 1 mm of mercury at 0 °C and standard gravity.

$1 \text{ atm} \times \frac{760 \text{ mmHg}}{\text{atm}} \times \frac{1 \text{ torr}}{\text{mmHg}} = 760 \text{ torr}$

TABLE 1 Pressure Conversions

From	To	Multiply By
atmosphere	mm Hg	760 mm Hg/atm (exactly)
atmosphere	lb/in.^{2}	14.6960 lb/(in.^{2} atm)
atmosphere	kPa	101.325 kPa/atm
bar	Pa	10^5 Pa/bar (exactly)
bar	lb/in.^{2}	14.5038 lb/(in.^{2} bar)
mm mercury	torr	1 torr/mm Hg (exactly)

†See Section 12.1

ENERGY AND POWER

The SI unit of energy is the product of the units of force and distance, or kilograms per meter per second squared ($kg \cdot m/s^2$) times meters ($\times$ m), which is $kg \cdot m^2/s^2$; this unit is called the **joule,** J. The joule is thus the work done when a force of 1 newton acts through a distance of 1 meter.

Work may also be done by moving an electric charge in an electric field. When the charge being moved is 1 coulomb (C), and the potential difference between its initial and final positions is 1 volt (V), the work is 1 joule. Thus,

$$1 \text{ joule} = 1 \text{ coulomb volt (CV)}$$

Another unit of electric work that is not part of the International System of Units but is still in use is the **electron volt,** eV, which is the work required to move an electron against a potential difference of 1 volt. (It is also the kinetic energy acquired by an electron when it is accelerated by a potential difference of 1 volt.) Because the charge on an electron is 1.602×10^{-19} C, we have

$$1 \text{ eV} = 1.602 \times 10^{-19} \text{ CV} \cdot \frac{1 \text{ J}}{1 \text{ CV}} = 1.602 \times 10^{-19} \text{ J}$$

If this value is multiplied by Avogadro's number, we obtain the energy involved in moving 1 mole of electronic charges (1 faraday) in a field produced by a potential difference of 1 volt:

$$1 \frac{\text{eV}}{\text{particle}} = \frac{1.602 \times 10^{-19} \text{ J}}{\text{particle}} \cdot \frac{6.022 \times 10^{23} \text{ particles}}{\text{mol}} \cdot \frac{1 \text{ kJ}}{1000 \text{ J}} = 96.49 \text{ kJ/mol}$$

Power is the amount of energy delivered per unit time. The SI unit is the watt, W, which is a joule per second. One kilowatt, kW, is 1000 W. Watt hours and kilowatt hours are therefore units of energy (Table 2). For example, 1000 watts, or 1 kilowatt, is

$$1.0 \times 10^3 \text{ W} \cdot \frac{1 \text{ J}}{1 \text{ W} \cdot \text{s}} \cdot \frac{3.6 \times 10^3 \text{ s}}{1 \text{ h}} = 3.6 \times 10^6 \text{ J}$$

TABLE 2 Energy Conversions

From	To	Multiply By
calorie (cal)	joule	4.184 J/cal (exactly)
kilocalorie (kcal)	cal	10^3 cal/kcal (exactly)
kilocalorie	joule	4.184×10^3 J/kcal (exactly)
liter atmosphere (L · atm)	joule	101.325 J/L · atm
electron volt (eV)	joule	1.60218×10^{-19} J/eV
electron volt per particle	kilojoules per mole	96.485 kJ · particle/eV · mol
coulomb volt (CV)	joule	1 CV/J (exactly)
kilowatt hour (kWh)	kcal	860.4 kcal/kWh
kilowatt hour	joule	3.6×10^6 J/kWh (exactly)
British thermal unit (Btu)	calorie	252 cal/Btu

APPENDIX C

Abbreviations and Useful Conversion Factors

TABLE 3 Some Common Abbreviations and Standard Letter Symbols

Term	Abbreviation	Term	Abbreviation
Activation energy	E_a	Faraday constant	F
Ampere	A	Gas constant	R
Aqueous solution	aq	Gibbs free energy	G
Atmosphere, unit of pressure	atm	Standard free energy	G°
Atomic mass unit	amu	Standard free energy of formation	ΔG_f°
Avogadro constant	N_A	Free energy change for reaction	ΔG_{rxn}°
Bar, unit of pressure	bar	Half-life	$t_{1/2}$
Body-centered cubic	bcc	Heat	q
Bohr radius	a_0	Hertz	Hz
Boiling point	bp	Hour	h
Celsius temperature, °C	t	Joule	J
Charge number of an ion	z	Kelvin	K
Coulomb, electric charge	C	Kilocalorie	kcal
Curie, radioactivity	Ci	Liquid	ℓ
Cycles per second, hertz	Hz	Logarithm, base 10	log
Debye, unit of electric dipole	D	Logarithm, base e	ln
Electron	e^-	Minute	min
Electron volt	eV	Molar	M
Electronegativity	χ	Molar mass	M
Energy	E	Mole	mol
Enthalpy	H	Osmotic pressure	Π
Standard enthalpy	H°	Pascal, unit of pressure	Pa
Standard enthalpy of formation	ΔH_f°	Planck's constant	h
Standard enthalpy of reaction	ΔH_{rxn}°	Pound	lb
Entropy	S	Pressure	
Standard entropy	S°	In atmospheres	atm
Entropy change for reaction	ΔS_{rxn}°	In millimeters of mercury	mm Hg
Equilibrium constant	K	Proton number	Z
Concentration basis	K_c	Rate constant	k
Pressure basis	K_p	Simple cubic (unit cell)	sc
Ionization weak acid	K_a	Standard temperature and pressure	STP
Ionization weak base	K_b	Volt	V
Solubility product	K_{sp}	Watt	W
Formation constant	K_{form}	Wavelength	λ
Ethylenediamine	en		
Face-centered cubic	fcc		

FUNDAMENTAL UNITS OF THE SI SYSTEM

The metric system was begun by the French National Assembly in 1790 and has undergone many modifications. The International System of Units or *Système International* (SI), which represents an extension of the metric system, was adopted by the 11th General Conference of Weights and Measures in 1960. It is constructed from seven base units, each of which represents a particular physical quantity (Table 4).

TABLE 4 SI Fundamental Units

Physical Quantity	Name of Unit	Symbol
Length	meter	m
Mass	kilogram	kg
Time	second	s
Temperature	kelvin	K
Amount of substance	mole	mol
Electric current	ampere	A
Luminous intensity	candela	cd

The first five units listed in Table 4 are particularly useful in general chemistry and are defined as follows:

1. The *meter* was redefined in 1960 to be equal to 1,650,763.73 wavelengths of a certain line in the emission spectrum of krypton-86.
2. The *kilogram* represents the mass of a platinum-iridium block kept at the International Bureau of Weights and Measures at Sèvres, France.
3. The *second* was redefined in 1967 as the duration of 9,192,631,770 periods of a certain line in the microwave spectrum of cesium-133.
4. The *kelvin* is 1/273.15 of the temperature interval between absolute zero and the triple point of water.
5. The *mole* is the amount of substance that contains as many entities as there are atoms in exactly 0.012 kg of carbon-12 (12 g of ^{12}C atoms).

PREFIXES USED WITH TRADITIONAL METRIC UNITS AND SI UNITS

Decimal fractions and multiples of metric and SI units are designated by using the prefixes listed in Table 5. Those most commonly used in general chemistry appear in italics.

TABLE 5 Traditional Metric and SI Prefixes

Factor	Prefix	Symbol	Factor	Prefix	Symbol
10^{12}	tera	T	10^{-1}	*deci*	d
10^{9}	giga	G	10^{-2}	*centi*	c
10^{6}	mega	M	10^{-3}	*milli*	m
10^{3}	*kilo*	k	10^{-6}	micro	μ
10^{2}	hecto	h	10^{-9}	*nano*	n
10^{1}	deka	da	10^{-12}	*pico*	p
			10^{-15}	femto	f
			10^{-18}	atto	a

DERIVED SI UNITS

In the International System of Units, all physical quantities are represented by appropriate combinations of the base units listed in Table 4. A list of the derived units frequently used in general chemistry is given in Table 6.

TABLE 6 Derived SI Units

Physical Quantity	Name of Unit	Symbol	Definition
Area	square meter	m^2	
Volume	cubic meter	m^3	
Density	kilogram per cubic meter	kg/m^3	
Force	newton	N	$kg \cdot m/s^2$
Pressure	pascal	Pa	N/m^2
Energy	joule	J	$kg \cdot m^2/s^2$
Electric charge	coulomb	C	$A \cdot s$
Electric potential difference	volt	V	$J/(A \cdot s)$

TABLE 7 Common Units of Mass and Weight

1 Pound = 453.59 Grams
1 pound = 453.59 grams = 0.45359 kilogram
1 kilogram = 1000 grams = 2.205 pounds
1 gram = 10 decigrams = 100 centigrams = 1000 milligrams
1 gram = 6.022×10^{23} atomic mass units
1 atomic mass unit = 1.6605×10^{-24} gram
1 short ton = 2000 pounds = 907.2 kilograms
1 long ton = 2240 pounds
1 metric tonne = 1000 kilograms = 2205 pounds

TABLE 8 Common Units of Length

1 Inch = 2.54 Centimeters (Exactly)

1 mile = 5280 feet = 1.609 kilometers
1 yard = 36 inches = 0.9144 meter
1 meter = 100 centimeters = 39.37 inches = 3.281 feet = 1.094 yards
1 kilometer = 1000 meters = 1094 yards = 0.6215 mile
1 Ångstrom = 1.0×10^{-8} centimeter = 0.10 nanometer = 100 picometers
= 1.0×10^{-10} meter = 3.937×10^{-9} inch

TABLE 9 Common Units of Volume

1 quart = 0.9463 liter
1 liter = 1.0567 quarts

1 liter = 1 cubic decimeter = 1000 cubic centimeters = 0.001 cubic meter
1 milliliter = 1 cubic centimeter = 0.001 liter = 1.056×10^{-3} quart
1 cubic foot = 28.316 liters = 29.924 quarts = 7.481 gallons

APPENDIX D

Physical Constants

Quantity	Symbol	Traditional Units	SI Units
Acceleration of gravity	g	980.6 cm/s	9.806 m/s
Atomic mass unit (1/12 the mass of ^{12}C atom)	amu or u	1.6605×10^{-24} g	1.6605×10^{-27} kg
Avogadro's number	N	6.0221367×10^{23} particles/mol	6.0221367×10^{23} particles/mol
Bohr radius	a_0	0.52918 Å 5.2918×10^{-9} cm	5.2918×10^{-11} m
Boltzmann constant	k	1.3807×10^{-16} erg/K	1.3807×10^{-23} J/K
Charge-to-mass ratio of electron	e/m	1.7588×10^{8} C/g	1.7588×10^{11} C/kg
Electronic charge	e	1.6022×10^{-19} C 4.8033×10^{-10} esu	1.6022×10^{-19} C
Electron rest mass	m_e	9.1094×10^{-28} g 0.00054858 amu	9.1094×10^{-31} kg
Faraday constant	F	96,485 C/mol e^- 23.06 kcal/V · mol e^-	96,485 C/mol e^- 96,485 J/V · mol e^-
Gas constant	R	$0.08206 \dfrac{L \cdot atm}{mol \cdot K}$ $1.987 \dfrac{cal}{mol \cdot K}$	$8.3145 \dfrac{Pa \cdot dm^3}{mol \cdot K}$ 8.3145 J/mol · K
Molar volume (STP)	V_m	22.414 L/mol	22.414×10^{-3} m^3/mol 22.414 dm^3/mol
Neutron rest mass	m_n	1.67495×10^{-24} g 1.008665 amu	1.67493×10^{-27} kg
Planck's constant	h	6.6261×10^{-27} erg · s	$6.6260755 \times 10^{-34}$ J · s
Proton rest mass	m_p	1.6726×10^{-24} g 1.007276 amu	1.6726×10^{-27} kg
Rydberg constant	R_α	3.289×10^{15} cycles/s 2.1799×10^{-11} erg	1.0974×10^{7} m^{-1} 2.1799×10^{-18} J
Velocity of light (in a vacuum)	c	2.9979×10^{10} cm/s (186,282 miles/s)	2.9979×10^{8} m/s

$\pi = 3.1416$
$e = 2.7183$
$\ln X = 2.303 \log X$
$2.303\ R = 4.576$ cal/mol · K = 19.15 J/mol · K
$2.303\ RT$ (at 25 °C) = 1364 cal/mol = 5709 J/mol

TABLE 10 Specific Heats and Heat Capacities for Some Common Substances

Substance	Specific Heat J/g · K	Molar Heat Capacity J/mol · K
Al (s)	0.902	24.3
Ca (s)	0.653	26.2
Cu (s)	0.385	24.5
Fe (s)	0.451	24.8
Hg (ℓ)	0.138	27.7
H_2O (s), ice	2.06	37.7
H_2O (ℓ), water	4.18	75.3
H_2O (g), steam	2.03	36.4
C_6H_6 (ℓ), benzene	1.74	136
C_6H_6 (g), benzene	1.04	81.6
C_2H_5OH (ℓ), ethanol	2.46	113
C_2H_5OH (g), ethanol	0.954	420
$(C_2H_5)_2O$ (ℓ), diethyl ether	3.74	172
$(C_2H_5)_2O$ (g), diethyl ether	2.35	108

TABLE 11 Heats of Transformation and Transformation Temperatures of Several Substances

Substance	MP (°C)	Heat of Fusion		BP (°C)	Heat of Vaporization	
		J/g	*kJ/mol*		*J/g*	*kJ/mol*
Elements*						
Al	660	395	10.7	2467	12083	326
Ca	839	230	9.3	1493	3768	151
Cu	1083	205	13.0	2570	4799	305
Fe	1535	267	14.9	2750	6285	351
Hg	−38.8	11	2.3	357	294	59.0
Compounds						
H_2O	0.00	333	6.02	100.0	2260	40.7
CH_4	−182	58.6	0.92	−164	—	—
C_2H_5OH	−117	109	5.02	78.0	855	39.3
C_6H_6	5.48	127	9.92	80.1	395	30.8
$(C_2H_5)_2O$	−116	97.9	7.66	35	351	26.0

*Data for the elements are taken from "The Periodic Table Stack," the Macintosh version of KC? Discoverer from *JCE: Software.*

A P P E N D I X **E**

Vapor Pressure of Water at Various Temperatures

TABLE 12 Vapor Pressure of Water at Various Temperatures

Temperature °C	Vapor Pressure torr	Temperature °C	Vapor Pressure torr	Temperature °C	Vapor Pressure torr	Temperature °C	Vapor Pressure torr
−10	2.1	21	18.7	51	97.2	81	369.7
−9	2.3	22	19.8	52	102.1	82	384.9
−8	2.5	23	21.1	53	107.2	83	400.6
−7	2.7	24	22.4	54	112.5	84	416.8
−6	2.9	25	23.8	55	118.0	85	433.6
−5	3.2	26	25.2	56	123.8	86	450.9
−4	3.4	27	26.7	57	129.8	87	468.7
−3	3.7	28	28.3	58	136.1	88	487.1
−2	4.0	29	30.0	59	142.6	89	506.1
−1	4.3	30	31.8	60	149.4	90	525.8
0	4.6	31	33.7	61	156.4	91	546.1
1	4.9	32	35.7	62	163.8	92	567.0
2	5.3	33	37.7	63	171.4	93	588.6
3	5.7	34	39.9	64	179.3	94	610.9
4	6.1	35	42.2	65	187.5	95	633.9
5	6.5	36	44.6	66	196.1	96	657.6
6	7.0	37	47.1	67	205.0	97	682.1
7	7.5	38	49.7	68	214.2	98	707.3
8	8.0	39	52.4	69	223.7	99	733.2
9	8.6	40	55.3	70	233.7	100	760.0
10	9.2	41	58.3	71	243.9	101	787.6
11	9.8	42	61.5	72	254.6	102	815.9
12	10.5	43	64.8	73	265.7	103	845.1
13	11.2	44	68.3	74	277.2	104	875.1
14	12.0	45	71.9	75	289.1	105	906.1
15	12.8	46	75.7	76	301.4	106	937.9
16	13.6	47	79.6	77	314.1	107	970.6
17	14.5	48	83.7	78	327.3	108	1004.4
18	15.5	49	88.0	79	341.0	109	1038.9
19	16.5	50	92.5	80	355.1	110	1074.6
20	17.5						

APPENDIX **F**

Ionization Constants for Weak Acids at 25 °C

Ionization Constants for Weak Acids at 25 °C

Acid	Formula and Ionization Equation	K_a
Acetic	$CH_3CO_2H \rightleftharpoons H^+ + CH_3CO_2^-$	1.8×10^{-5}
Arsenic	$H_3AsO_4 \rightleftharpoons H^+ + H_2AsO_4^-$	$K_1 = 2.5 \times 10^{-4}$
	$H_2AsO_4^- \rightleftharpoons H^+ + HAsO_4^{2-}$	$K_2 = 5.6 \times 10^{-8}$
	$HAsO_4^{2-} \rightleftharpoons H^+ + AsO_4^{3-}$	$K_3 = 3.0 \times 10^{-13}$
Arsenous	$H_3AsO_3 \rightleftharpoons H^+ + H_2AsO_3^-$	$K_1 = 6.0 \times 10^{-10}$
	$H_2AsO_3^- \rightleftharpoons H^+ + HAsO_3^{2-}$	$K_2 = 3.0 \times 10^{-14}$
Benzoic	$C_6HgCO_2H \rightleftharpoons H^+ + C_6H_5CO_2^-$	6.3×10^{-5}
Boric	$H_3BO_3 \rightleftharpoons H^+ + H_2BO_3^-$	$K_1 = 7.3 \times 10^{-10}$
	$H_2BO_3 \rightleftharpoons H^+ + HBO_3^{2-}$	$K_2 = 1.8 \times 10^{-13}$
	$HBO_3^{2-} \rightleftharpoons H^+ + BO_3^{3-}$	$K_3 = 1.6 \times 10^{-14}$
Carbonic	$H_2CO_3 \rightleftharpoons H^+ + HCO_3^-$	$K_1 = 4.2 \times 10^{-7}$
	$HCO_3^- \rightleftharpoons H^+ + CO_3^{2-}$	$K_2 = 4.8 \times 10^{-11}$
Citric	$H_3C_6H_5O_7 \rightleftharpoons H^+ + H_2C_6H_5O_7^-$	$K_1 = 7.4 \times 10^{-3}$
	$H_2C_6H_5O_7^- \rightleftharpoons H^+ + HC_6H_5O_7^{2-}$	$K_2 = 1.7 \times 10^{-5}$
	$HC_6H_5O_7^{2-} \rightleftharpoons H^+ + C_6H_5O_7^{3-}$	$K_3 = 4.0 \times 10^{-7}$
Cyanic	$HOCN \rightleftharpoons H^+ + OCN^-$	3.5×10^{-4}
Formic	$HCO_2H \rightleftharpoons H^+ + HCO_2^-$	1.8×10^{-4}
Hydrazoic	$HN_3 \rightleftharpoons H^+ + N_3^-$	1.9×10^{-5}
Hydrocyanic	$HCN \rightleftharpoons H^+ + CN^-$	4.0×10^{-10}
Hydrofluoric	$HF \rightleftharpoons H^+ + F^-$	7.2×10^{-4}
Hydrogen peroxide	$H_2O_2 \rightleftharpoons H^+ + HO_2^-$	2.4×10^{-12}
Hydrosulfuric	$H_2S \rightleftharpoons H^+ + HS^-$	$K_1 = 1 \times 10^{-7}$
	$HS^- \rightleftharpoons H^+ + S^{2-}$	$K_2 = 1 \times 10^{-19}$
Hypobromous	$HOBr \rightleftharpoons H^+ + OBr^-$	2.5×10^{-9}
Hypochlorous	$HOCl \rightleftharpoons H^+ + OCl^-$	3.5×10^{-8}
Nitrous	$HNO_2 \rightleftharpoons H^+ + NO_2^-$	4.5×10^{-4}
Oxalic	$H_2C_2O_4 \rightleftharpoons H^+ + HC_2O_4^-$	$K_1 = 5.9 \times 10^{-2}$
	$HC_2O_4^- \rightleftharpoons H^+ + C_2O_4^{2-}$	$K_2 = 6.4 \times 10^{-5}$

Acid	Formula and Ionization Equation	K_a
Phenol	$C_6H_5OH \rightleftharpoons H^+ + C_6H_5O^-$	1.3×10^{-10}
Phosphoric	$H_3PO_4 \rightleftharpoons H^+ + H_2PO_4^-$	$K_1 = 7.5 \times 10^{-3}$
	$H_2PO_4^- \rightleftharpoons H^+ + HPO_4^{2-}$	$K_2 = 6.2 \times 10^{-8}$
	$HPO_4^{2-} \rightleftharpoons H^+ + PO_4^{3-}$	$K_3 = 3.6 \times 10^{-13}$
Phosphorous	$H_3PO_3 \rightleftharpoons H^+ + H_2PO_3^-$	$K_1 = 1.6 \times 10^{-2}$
	$H_2PO_3 \rightleftharpoons H^+ + HPO_3^{2-}$	$K_2 = 7.0 \times 10^{-7}$
Selenic	$H_2SeO_4 \rightleftharpoons H^+ + HSeO_4^-$	K_1 = very large
	$HSeO_4^- \rightleftharpoons H^+ + SeO_4^{2-}$	$K_2 = 1.2 \times 10^{-2}$
Selenous	$H_2SeO_3 \rightleftharpoons H^+ + HSeO_3^-$	$K_1 = 2.7 \times 10^{-3}$
	$HSeO_3^- \rightleftharpoons H^+ + SeO_3^{2-}$	$K_2 = 2.5 \times 10^{-7}$
Sulfuric	$H_2SO_4 \rightleftharpoons H^+ + HSO_4^-$	K_1 = very large
	$HSO_4^- \rightleftharpoons H^+ + SO_4^{2-}$	$K_2 = 1.2 \times 10^{-2}$
Sulfurous	$H_2SO_3 \rightleftharpoons H^+ + HSO_3^-$	$K_1 = 1.7 \times 10^{-2}$
	$HSO_3^- \rightleftharpoons H^+ + SO_3^{2-}$	$K_2 = 6.4 \times 10^{-8}$
Tellurous	$H_2TeO_3 \rightleftharpoons H^+ + HTeO_3^-$	$K_1 = 2 \times 10^{-3}$
	$HTeO_3^- \rightleftharpoons H^+ + TeO_3^{2-}$	$K_2 = 1 \times 10^{-8}$

APPENDIX G

Ionization Constants for Weak Bases at 25 °C

Ionization Constants for Weak Bases at 25 °C

Base	**Formula and Ionization Equation**	K_b
Ammonia	$NH_3 + H_2O \rightleftharpoons NH_4^+ + OH^-$	1.8×10^{-5}
Aniline	$C_6H_5NH_2 + H_2O \rightleftharpoons C_6H_5NH_3^+ + OH^-$	4.0×10^{-10}
Dimethylamine	$(CH_3)_2NH + H_2O \rightleftharpoons (CH_3)_2NH_2^+ + OH^-$	7.4×10^{-4}
Ethylenediamine	$(CH_2)_2(NH_2)_2 + H_2O \rightleftharpoons (CH_2)_2(NH_2)_2H^+ + OH^-$	$K_1 = 8.5 \times 10^{-5}$
	$(CH_2)_2(NH_2)_2H^+ + H_2O \rightleftharpoons (CH_2)_2(NH_2)_2H_2^{2+} + OH^-$	$K_2 = 2.7 \times 10^{-8}$
Hydrazine	$N_2H_4 + H_2O \rightleftharpoons N_2H_5^+ + OH^-$	$K_1 = 8.5 \times 10^{-7}$
	$N_2H_5^+ + H_2O \rightleftharpoons N_2H_6^{2+} + OH^-$	$K_2 = 8.9 \times 10^{-16}$
Hydroxylamine	$NH_2OH + H_2O \rightleftharpoons NH_3OH^+ + OH^-$	6.6×10^{-9}
Methylamine	$CH_3NH_2 + H_2O \rightleftharpoons CH_3NH_3^+ + OH^-$	5.0×10^{-4}
Pyridine	$C_5H_5N + H_2O \rightleftharpoons C_5H_5NH^+ + OH^-$	1.5×10^{-9}
Trimethylamine	$(CH_3)_3N + H_2O \rightleftharpoons (CH_3)_3NH^+ + OH^-$	7.4×10^{-5}

APPENDIX **H**

Solubility Product Constants for Some Inorganic Compounds at 25 °C

Solubility Product Constants for Some Inorganic Compounds at 25 °C

Substance	K_{sp}	Substance	K_{sp}
Aluminum compounds		**Calcium compounds**	
$AlAsO_4$	1.6×10^{-16}	$Ca_3(AsO_4)_2$	6.8×10^{-19}
$Al(OH)_3$	1.9×10^{-33}	$CaCO_3$	3.8×10^{-9}
$AlPO_4$	1.3×10^{-20}	$CaCrO_4$	7.1×10^{-4}
Antimony compounds		$CaC_2O_4 \cdot H_2O$*	2.3×10^{-9}
Sb_2S_3	1.6×10^{-93}	CaF_2	3.9×10^{-11}
Barium compounds		$Ca(OH)_2$	7.9×10^{-6}
$Ba_3(AsO_4)_2$	1.1×10^{-13}	$CaHPO_4$	2.7×10^{-7}
$BaCO_3$	8.1×10^{-9}	$Ca(H_2PO_4)_2$	1.0×10^{-3}
$BaC_2O_4 \cdot 2H_2O$*	1.1×10^{-7}	$Ca_3(PO_4)_2$	1.0×10^{-25}
$BaCrO_4$	2.0×10^{-10}	$CaSO_3 \cdot 2H_2O$*	1.3×10^{-8}
BaF_2	1.7×10^{-6}	$CaSO_4 \cdot 2H_2O$*	2.4×10^{-5}
$Ba(OH)_2 \cdot 8H_2O$*	5.0×10^{-3}	**Chromium compounds**	
$Ba_3(PO_4)_2$	1.3×10^{-29}	$CrAsO_4$	7.8×10^{-21}
$BaSeO_4$	2.8×10^{-11}	$Cr(OH)_3$	6.7×10^{-31}
$BaSO_3$	8.0×10^{-7}	$CrPO_4$	2.4×10^{-23}
$BaSO_4$	1.1×10^{-10}	**Cobalt compounds**	
Bismuth compounds		$Co_3(AsO_4)_2$	7.6×10^{-29}
$BiOCl$	7.0×10^{-9}	$CoCO_3$	8.0×10^{-13}
$BiO(OH)$	1.0×10^{-12}	$Co(OH)_2$	2.5×10^{-16}
$Bi(OH)_3$	3.2×10^{-40}	CoS (α)	5.9×10^{-21}
BiI_3	8.1×10^{-19}	$Co(OH)_3$	4.0×10^{-45}
$BiPO_4$	1.3×10^{-23}	**Copper compounds**	
Bi_2S_3	1.6×10^{-72}	$CuBr$	5.3×10^{-9}
Cadmium compounds		$CuCl$	1.9×10^{-7}
$Cd_3(AsO_4)_2$	2.2×10^{-32}	$CuCN$	3.2×10^{-20}
$CdCO_3$	2.5×10^{-14}	Cu_2O ($Cu^+ + OH^-$)†	1.0×10^{-14}
$Cd(CN)_2$	1.0×10^{-8}	CuI	5.1×10^{-12}
$Cd_2[Fe(CN)_6]$	3.2×10^{-17}	Cu_2S	1.6×10^{-48}
$Cd(OH)_2$	1.2×10^{-14}	$CuSCN$	1.6×10^{-11}
CdS	3.6×10^{-29}	$Cu_3(AsO_4)_2$	7.6×10^{-36}

Solubility Product Constants for Some Inorganic Compounds at 25 °C (*Continued*)

Substance	K_{sp}
$CuCO_3$	2.5×10^{-10}
$Cu_2[Fe(CN)_6]$	1.3×10^{-16}
$Cu(OH)_2$	1.6×10^{-19}
CuS	8.7×10^{-36}
Gold compounds	
AuBr	5.0×10^{-17}
AuCl	2.0×10^{-13}
AuI	1.6×10^{-23}
$AuBr_3$	4.0×10^{-36}
$AuCl_3$	3.2×10^{-25}
$Au(OH)_3$	1×10^{-53}
AuI_3	1.0×10^{-46}
Iron compounds	
$FeCO_3$	3.5×10^{-11}
$Fe(OH)_2$	7.9×10^{-15}
FeS	4.9×10^{-18}
$Fe_4[Fe(CN)_6]_3$	3.0×10^{-41}
$Fe(OH)_3$	6.3×10^{-38}
Fe_2S_3	1.4×10^{-88}
Lead compounds	
$Pb_3(AsO_4)_2$	4.1×10^{-36}
$PbBr_2$	6.3×10^{-6}
$PbCO_3$	1.5×10^{-13}
$PbCl_2$	1.7×10^{-5}
$PbCrO_4$	1.8×10^{-14}
PbF_2	3.7×10^{-8}
$Pb(OH)_2$	2.8×10^{-16}
PbI_2	8.7×10^{-9}
$Pb_3(PO_4)_2$	3.0×10^{-44}
$PbSeO_4$	1.5×10^{-7}
$PbSO_4$	1.8×10^{-8}
PbS	8.4×10^{-28}
Magnesium compounds	
$Mg_3(AsO_4)_2$	2.1×10^{-20}
$MgCO_3 \cdot 3H_2O$*	4.0×10^{-5}
MgC_2O_4	8.6×10^{-5}
MgF_2	6.4×10^{-9}
$Mg(OH)_2$	1.5×10^{-11}
$MgNH_4PO_4$	2.5×10^{-12}
Manganese compounds	
$Mn_3(AsO_4)_2$	1.9×10^{-11}
$MnCO_3$	1.8×10^{-11}
$Mn(OH)_2$	4.6×10^{-14}
MnS	5.1×10^{-15}
$Mn(OH)_3$	$\approx 1 \times 10^{-36}$
Mercury compounds	
Hg_2Br_2	1.3×10^{-22}
Hg_2CO_3	8.9×10^{-17}
Hg_2Cl_2	1.1×10^{-18}
Hg_2CrO_4	5.0×10^{-9}
Hg_2I_2	4.5×10^{-29}
$Hg_2O \cdot H_2O$ ($Hg_2^{2+} + 2OH^-$)*†	1.6×10^{-23}

Substance	K_{sp}
Hg_2SO_4	6.8×10^{-7}
Hg_2S	5.8×10^{-44}
$Hg(CN)_2$	3.0×10^{-23}
$Hg(OH)_2$	2.5×10^{-26}
HgI_2	4.0×10^{-29}
HgS	3.0×10^{-53}
Nickel compounds	
$Ni_3(AsO_4)_2$	1.9×10^{-26}
$NiCO_3$	6.6×10^{-9}
$Ni(CN)_2$	3.0×10^{-23}
$Ni(OH)_2$	2.8×10^{-16}
NiS (α)	3.0×10^{-21}
NiS (β)	1.0×10^{-26}
NiS (γ)	2.0×10^{-28}
Silver compounds	
Ag_3AsO_4	1.1×10^{-20}
AgBr	3.3×10^{-13}
Ag_2CO_3	8.1×10^{-12}
AgCl	1.8×10^{-10}
Ag_2CrO_4	9.0×10^{-12}
AgCN	1.2×10^{-16}
$Ag_4[Fe(CN)_6]$	1.6×10^{-41}
Ag_2O ($Ag^+ + OH^-$)†	2.0×10^{-8}
AgI	1.5×10^{-16}
Ag_3PO_4	1.3×10^{-20}
Ag_2SO_3	1.5×10^{-14}
Ag_2SO_4	1.7×10^{-5}
Ag_2S	1.0×10^{-49}
AgSCN	1.0×10^{-12}
Strontium compounds	
$Sr_3(AsO_4)_2$	1.3×10^{-18}
$SrCO_3$	9.4×10^{-10}
$SrC_2O_4 \cdot 2H_2O$*	5.6×10^{-8}
$SrCrO_4$	3.6×10^{-5}
$Sr(OH)_2 \cdot 8H_2O$*	3.2×10^{-4}
$Sr_3(PO_4)_2$	1.0×10^{-31}
$SrSO_3$	4.0×10^{-8}
$SrSO_4$	2.8×10^{-7}
Tin compounds	
$Sn(OH)_2$	2.0×10^{-26}
SnI_2	1.0×10^{-4}
SnS	1.0×10^{-28}
$Sn(OH)_4$	1.0×10^{-57}
SnS_2	1.0×10^{-70}
Zinc compounds	
$Zn_3(AsO_4)_2$	1.1×10^{-27}
$ZnCO_3$	1.5×10^{-11}
$Zn(CN)_2$	8.0×10^{-12}
$Zn_3[Fe(CN)_6]$	4.1×10^{-16}
$Zn(OH)_2$	4.5×10^{-17}
$Zn_3(PO_4)_2$	9.1×10^{-33}
ZnS	1.1×10^{-21}

*Since $[H_2O]$ does not appear in equilibrium constants for equilibria in aqueous solution in general, it does *not* appear in the K_{sp} expressions for hydrated solids.

†Very small amounts of oxides dissolve in water to give the ions indicated in parentheses. Solid hydroxides are unstable and decompose to oxides as rapidly as they are formed.

APPENDIX I

Formation Constants for Some Complex Ions in Aqueous Solution

Formation Constants for Some Complex Ions in Aqueous Solution

Formation Equilibrium	*K*
$Ag^+ + 2Br^- \rightleftharpoons [AgBr_2]^-$	1.3×10^7
$Ag^+ + 2Cl^- \rightleftharpoons [AgCl_2]^-$	2.5×10^5
$Ag^+ + 2CN^- \rightleftharpoons [Ag(CN)_2]^-$	5.6×10^{18}
$Ag^+ + 2S_2O_3^{2-} \rightleftharpoons [Ag(S_2O_3)_2]^{3-}$	2.0×10^{13}
$Ag^+ + 2NH_3 \rightleftharpoons [Ag(NH_3)_2]^+$	1.6×10^7
$Al^{3+} + 6F^- \rightleftharpoons [AlF_6]^{3-}$	5.0×10^{23}
$Al^{3+} + 4OH^- \rightleftharpoons [Al(OH)_4]^-$	7.7×10^{33}
$Au^+ + 2CN^- \rightleftharpoons [Au(CN)_2]^-$	2.0×10^{38}
$Cd^{2+} + 4CN^- \rightleftharpoons [Cd(CN)_4]^{2-}$	1.3×10^{17}
$Cd^{2+} + 4Cl^- \rightleftharpoons [CdCl_4]^{2-}$	1.0×10^4
$Cd^{2+} + 4NH_3 \rightleftharpoons [Cd(NH_3)_4]^{2+}$	1.0×10^7
$Co^{2+} + 6NH_3 \rightleftharpoons [Co(NH_3)_6]^{2+}$	7.7×10^4
$Cu^+ + 2CN^- \rightleftharpoons [Cu(CN)_2]^-$	1.0×10^{16}
$Cu^+ + 2Cl^- \rightleftharpoons [CuCl_2]^-$	1.0×10^5
$Cu^{2+} + 4NH_3 \rightleftharpoons [Cu(NH_3)_4]^{2+}$	6.8×10^{12}
$Fe^{2+} + 6CN^- \rightleftharpoons [Fe(CN)_6]^{4-}$	7.7×10^{36}
$Hg^{2+} + 4Cl^- \rightleftharpoons [HgCl_4]^{2-}$	1.2×10^{15}
$Ni^{2+} + 4CN^- \rightleftharpoons [Ni(CN)_4]^{2-}$	1.0×10^{31}
$Ni^{2+} + 6NH_3 \rightleftharpoons [Ni(NH_3)_6]^{2+}$	5.6×10^8
$Zn^{2+} + 4OH^- \rightleftharpoons [Zn(OH)_4]^{2-}$	2.9×10^{15}
$Zn^{2+} + 4NH_3 \rightleftharpoons [Zn(NH_3)_4]^{2+}$	2.9×10^9

APPENDIX J

Standard Reduction Potentials in Aqueous Solution at 25 °C

Standard Reduction Potentials in Aqueous Solution at 25 °C

Acidic Solution	**Standard Reduction Potential, E^0 (volts)**
F_2 (g) + $2e^-$ $\longrightarrow$ $2F^-$ (aq)	2.87
Co^{3+} (aq) + e^- $\longrightarrow$ Co^{2+} (aq)	1.82
Pb^{4+} (aq) + $2e^-$ $\longrightarrow$ Pb^{2+} (aq)	1.8
H_2O_2 (aq) + $2H^+$ (aq) + $2e^-$ $\longrightarrow$ $2H_2O$	1.77
NiO_2 (s) + $4H^+$ (aq) + $2e^-$ $\longrightarrow$ Ni^{2+} (aq) + $2H_2O$	1.7
PbO_2 (s) + SO_4^{2-} (aq) + $4H^+$ (aq) + $2e^-$ $\longrightarrow$ $PbSO_4$ (s) + $2H_2O$	1.685
Au^+ (aq) + e^- $\longrightarrow$ Au (s)	1.68
2HClO (aq) + $2H^+$ (aq) + $2e^-$ $\longrightarrow$ Cl_2 (g) + $2H_2O$	1.63
Ce^{4+} (aq) + e^- $\longrightarrow$ Ce^{3+} (aq)	1.61
$NaBiO_3$ (s) + $6H^+$ (aq) + $2e^-$ $\longrightarrow$ Bi^{3+} (aq) + Na^+ (aq) + $3H_2O$	≈1.6
MnO_4^- (aq) + $8H^+$ (aq) + $5e^-$ $\longrightarrow$ Mn^{2+} (aq) + $4H_2O$	1.51
Au^{3+} (aq) + $3e^-$ $\longrightarrow$ Au (s)	1.50
ClO_3^- (aq) + $6H^+$ (aq) + $5e^-$ $\longrightarrow$ $\frac{1}{2}Cl_2$ (g) + $3H_2O$	1.47
BrO_3^- (aq) + $6H^+$ (aq) + $6e^-$ $\longrightarrow$ Br^- (aq) + $3H_2O$	1.44
Cl_2 (g) + $2e^-$ $\longrightarrow$ $2Cl^-$ (aq)	1.358
$Cr_2O_7^{2-}$ (aq) + $14H^+$ (aq) + $6e^-$ $\longrightarrow$ $2Cr^{3+}$ (aq) + $7H_2O$	1.33
$N_2H_5^+$ (aq) + $3H^+$ (aq) + $2e^-$ $\longrightarrow$ $2NH_4^+$ (aq)	1.24
MnO_2 (s) + $4H^+$ (aq) + $2e^-$ $\longrightarrow$ Mn^{2+} (aq) + $2H_2O$	1.23
O_2 (g) + $4H^+$ (aq) + $4e^-$ $\longrightarrow$ $2H_2O$	1.229
Pt^{2+} (aq) + $2e^-$ $\longrightarrow$ Pt (s)	1.2
IO_3^-(aq) + $6H^+$ (aq) + $5e^-$ $\longrightarrow$ $\frac{1}{2}I_2$ (aq) + $3H_2O$	1.195
ClO_4^- (aq) + $2H^+$ (aq) + $2e^-$ $\longrightarrow$ ClO_3^- (aq) + H_2O	1.19
Br_2 (ℓ) + $2e^-$ $\longrightarrow$ $2Br^-$ (aq)	1.066
$AuCl_4^-$ (aq) + $3e^-$ $\longrightarrow$ Au (s) + $4Cl^-$ (aq)	1.00
Pd^{2+} (aq) + $2e^-$ $\longrightarrow$ Pd (s)	0.987
NO_3^- (aq) + $4H^+$ (aq) + $3e^-$ $\longrightarrow$ NO (g) + $2H_2O$	0.96

Standard Reduction Potentials in Aqueous Solution at 25 °C

Acidic Solution	Standard Reduction Potential, E^0 (volts)
NO_3^- (aq) + $3H^+$ (aq) + $2e^-$ ⟶ HNO_2 (aq) + H_2O	0.94
$2Hg^{2+}$ (aq) + $2e^-$ ⟶ Hg_2^{2+} (aq)	0.920
Hg^{2+} (aq) + $2e^-$ ⟶ Hg (ℓ)	0.855
Ag^+ (aq) + e^- ⟶ Ag (s)	0.7994
Hg_2^{2+} (aq) + $2e^-$ ⟶ 2Hg (ℓ)	0.789
Fe^{3+} (aq) + e^- ⟶ Fe^{2+} (aq)	0.771
$SbCl_6^-$ (aq) + $2e^-$ ⟶ $SbCl_4^-$ (aq) + $2Cl^-$ (aq)	0.75
$[PtCl_4]^{2+}$ (aq) + $2e^-$ ⟶ Pt (s) + $4Cl^-$ (aq)	0.73
O_2 (g) + $2H^+$ (aq) + $2e^-$ ⟶ H_2O_2 (aq)	0.682
$[PtCl_6]^{2-}$ (aq) + $2e^-$ ⟶ $[PtCl_4]^{2-}$ (aq) + $2Cl^-$ (aq)	0.68
H_3AsO_4 (aq) + $2H^+$ (aq) + $2e^-$ ⟶ H_3AsO_3 (aq) + H_2O	0.58
I_2 (s) + $2e^-$ ⟶ $2I^-$ (aq)	0.535
TeO_2 (s) + $4H^+$ (aq) + $4e^-$ ⟶ Te (s) + $2H_2O$	0.529
Cu^+ (aq) + e^- ⟶ Cu (s)	0.521
$[RhCl_6]^{3-}$ (aq) + $3e^-$ ⟶ Rh (s) + $6Cl^-$ (aq)	0.44
Cu^{2+} (aq) + $2e^-$ ⟶ Cu (s)	0.337
$HgCl_2$ (s) + $2e^-$ ⟶ 2Hg (ℓ) + $2Cl^-$ (aq)	0.27
AgCl (s) + e^- ⟶ Ag (s) + Cl^- (aq)	0.222
SO_4^{2-} (aq) + $4H^+$ (aq) + $2e^-$ ⟶ SO_2 (g) + $2H_2O$	0.20
SO_4^{2-} (aq) + $4H^+$ (aq) + $2e^-$ ⟶ H_2SO_3 (aq) + H_2O	0.17
Cu^{2+} (aq) + e^- ⟶ Cu^+ (aq)	0.153
Sn^{4+} (aq) + $2e^-$ ⟶ Sn^{2+} (aq)	0.15
S (s) + $2H^+$ (aq) + $2e^-$ ⟶ H_2S (aq)	0.14
AgBr (s) + e^- ⟶ Ag (s) + Br^- (aq)	0.0713
$2H^+$ (aq) + $2e^-$ ⟶ H_2 (g) (reference electrode)	0.0000
N_2O (g) + $6H^+$ (aq) + H_2O + $4e^-$ ⟶ $2NH_3OH^+$ (aq)	−0.05
Pb^{2+} (aq) + $2e^-$ ⟶ Pb (s)	−0.126
Sn^{2+} (aq) + $2e^-$ ⟶ Sn (s)	−0.14
AgI (s) + e^- ⟶ Ag (s) + I^- (aq)	−0.15
$[SnF_6]^{2-}$ (aq) + $4e^-$ ⟶ Sn (s) + $6F^-$ (aq)	−0.25
Ni^{2+} (aq) + $2e^-$ ⟶ Ni (s)	−0.25
Co^{2+} (aq) + $2e^-$ ⟶ Co (s)	−0.28
Tl^+ (aq) + e^- ⟶ Tl (s)	−0.34
$PbSO_4$ (s) + $2e^-$ ⟶ Pb (s) + SO_4^{2-} (aq)	−0.356
Se (s) + $2H^+$ (aq) + $2e^-$ ⟶ H_2Se (aq)	−0.40
Cd^{2+} (aq) + $2e^-$ ⟶ Cd (s)	−0.403
Cr^{3+} (aq) + e^- ⟶ Cr^{2+} (aq)	−0.41
Fe^{2+} (aq) + $2e^-$ ⟶ Fe (s)	−0.44
$2CO_2$ (g) + $2H^+$ (aq) + $2e^-$ ⟶ $(COOH)_2$ (aq)	−0.49
Ga^{3+} (aq) + $3e^-$ ⟶ Ga (s)	−0.53
HgS (s) + $2H^+$ (aq) + $2e^-$ ⟶ Hg (ℓ) + H_2S (g)	−0.72
Cr^{3+} (aq) + $3e^-$ ⟶ Cr (s)	−0.74
Zn^{2+} (aq) + $2e^-$ ⟶ Zn (s)	−0.763
Cr^{2+} (aq) + $2e^-$ ⟶ Cr (s)	−0.91
FeS (s) + $2e^-$ ⟶ Fe (s) + S^{2-} (aq)	−1.01
Mn^{2+} (aq) + $2e^-$ ⟶ Mn (s)	−1.18
V^{2+} (aq) + $2e^-$ ⟶ V (s)	−1.18
CdS (s) + $2e^-$ ⟶ Cd (s) + S^{2-} (aq)	−1.21
ZnS (s) + $2e^-$ ⟶ Zn (s) + S^{2-} (aq)	−1.44
Zr^{4+} (aq) + $4e^-$ ⟶ Zr (s)	−1.53
Al^{3+} (aq) + $3e^-$ ⟶ Al (s)	−1.66

Standard Reduction Potentials in Aqueous Solution at 25 °C (*Continued*)

Acidic Solution	Standard Reduction Potential, E^0 (volts)
Mg^{2+} (aq) + $2e^-$ $\longrightarrow$ Mg (s)	−2.37
Na^+ (aq) + e^- $\longrightarrow$ Na (s)	−2.714
Ca^{2+} (aq) + $2e^-$ $\longrightarrow$ Ca (s)	−2.87
Sr^{2+} (aq) + $2e^-$ $\longrightarrow$ Sr (s)	−2.89
Ba^{2+} (aq) + $2e^-$ $\longrightarrow$ Ba (s)	−2.90
Rb^+ (aq) + e^- $\longrightarrow$ Rb (s)	−2.925
K^+ (aq) + e^- $\longrightarrow$ K (s)	−2.925
Li^+ (aq) + e^- $\longrightarrow$ Li (s)	−3.045

Basic Solution	Standard Reduction Potential, E^0 (volts)
ClO^- (aq) + H_2O + $2e^-$ $\longrightarrow$ Cl^- (aq) + $2OH^-$ (aq)	0.89
OOH^- (aq) + H_2O + $2e^-$ $\longrightarrow$ $3OH^-$ (aq)	0.88
$2NH_2OH$ (aq) + $2e^-$ $\longrightarrow$ N_2H_4 (aq) + $2OH^-$ (aq)	0.74
ClO_3^- (aq) + $3H_2O$ + $6e^-$ $\longrightarrow$ Cl^- (aq) + $6OH^-$ (aq)	0.62
MnO_4^- (aq) + $2H_2O$ + $3e^-$ $\longrightarrow$ MnO_2 (s) + $4OH^-$ (aq)	0.588
MnO_4^- (aq) + e^- $\longrightarrow$ MnO_4^{2-} (aq)	0.564
NiO_2 (s) + $2H_2O$ + $2e^-$ $\longrightarrow$ $Ni(OH)_2$ (s) + $2OH^-$ (aq)	0.49
Ag_2CrO_4 (s) + $2e^-$ $\longrightarrow$ 2Ag (s) + CrO_4^{2-} (aq)	0.446
O_2 (g) + $2H_2O$ + $4e^-$ $\longrightarrow$ $4OH^-$ (aq)	0.40
ClO_4^- (aq) + H_2O + $2e^-$ $\longrightarrow$ ClO_3^- (aq) + $2OH^-$ (aq)	0.36
Ag_2O (s) + H_2O + $2e^-$ $\longrightarrow$ 2Ag (s) + $2OH^-$ (aq)	0.34
$2NO_2^-$ (aq) + $3H_2O$ + $4e^-$ $\longrightarrow$ N_2O (g) + $6OH^-$ (aq)	0.15
N_2H_4 (aq) + $2H_2O$ + $2e^-$ $\longrightarrow$ $2NH_3$ (aq) + $2OH^-$ (aq)	0.10
$[Co(NH_3)_6]^{3+}$ (aq) + e^- $\longrightarrow$ $[Co(NH_3)_6]^{2+}$ (aq)	0.10
HgO (s) + H_2O + $2e^-$ $\longrightarrow$ Hg (ℓ) + $2OH^-$ (aq)	0.0984
O_2 (g) + H_2O + $2e^-$ $\longrightarrow$ OOH^- (aq) + OH^- (aq)	0.076
NO_3^- (aq) + H_2O + $2e^-$ $\longrightarrow$ NO_2^- (aq) + $2OH^-$ (aq)	0.01
MnO_2 (s) + $2H_2O$ + $2e^-$ $\longrightarrow$ $Mn(OH)_2$ (s) + $2OH^-$ (aq)	−0.05
CrO_4^{2-} (aq) + $4H_2O$ + $3e^-$ $\longrightarrow$ $Cr(OH)_3$ (s) + $5OH^-$ (aq)	−0.12
$Cu(OH)_2$ (s) + $2e^-$ $\longrightarrow$ Cu (s) + $2OH^-$ (aq)	−0.36
S (s) + $2e^-$ $\longrightarrow$ S^{2-} (aq)	−0.48
$Fe(OH)_3$ (s) + e^- $\longrightarrow$ $Fe(OH)_2$ (s) + OH^- (aq)	−0.56
$2H_2O$ + $2e^-$ $\longrightarrow$ H_2 (g) + $2OH^-$ (aq)	−0.8277
$2NO_3^-$ (aq) + $2H_2O$ + $2e^-$ $\longrightarrow$ N_2O_4 (g) + $4OH^-$ (aq)	−0.85
$Fe(OH)_2$ (s) + $2e^-$ $\longrightarrow$ Fe (s) + $2OH^-$ (aq)	−0.877
SO_4^{2-} (aq) + H_2O + $2e^-$ $\longrightarrow$ SO_3^{2-} (aq) + $2OH^-$ (aq)	−0.93
N_2 (g) + $4H_2O$ + $4e^-$ $\longrightarrow$ N_2H_4 (aq) + $4OH^-$ (aq)	−1.15
$[Zn(OH)_4]^{2-}$ (aq) + $2e^-$ $\longrightarrow$ Zn (s) + $4OH^-$ (aq)	−1.22
$Zn(OH)_2$ (s) + $2e^-$ $\longrightarrow$ Zn (s) + $2OH^-$ (aq)	−1.245
$[Zn(CN)_4]^{2-}$ (aq) + $2e^-$ $\longrightarrow$ Zn (s) + $4CN^-$ (aq)	−1.26
$Cr(OH)_3$ (s) + $3e^-$ $\longrightarrow$ Cr (s) + $3OH^-$ (aq)	−1.30
SiO_3^{2-} (aq) + $3H_2O$ + $4e^-$ $\longrightarrow$ Si (s) + $6OH^-$ (aq)	−1.70

APPENDIX **K**

Selected Thermodynamic Values*

Selected Thermodynamic Values*

Species	ΔH_f°(298.15K) kJ/mol	S°(298.15K) J/K · mol	ΔG_f°(298.15K) kJ/mol
Aluminum			
Al(s)	0	28.3	0
$AlCl_3$(s)	−704.2	110.67	−628.8
Al_2O_3(s)	−1675.7	50.92	−1582.3
Barium			
$BaCl_2$(s)	−858.6	123.68	−810.4
BaO(s)	−553.5	70.42	−525.1
$BaSO_4$(s)	−1473.2	132.2	−1362.2
Beryllium			
Be(s)	0	9.5	0
$Be(OH)_2$	−902.5	51.9	−815.0
Bromine			
Br(g)	111.884	175.022	82.396
Br_2(ℓ)	0	152.2	0
Br_2(g)	30.907	245.463	3.110
BrF_3(g)	−255.60	292.53	−229.43
HBr(g)	−36.40	198.695	−53.45
Calcium			
Ca(s)	0	41.42	0
Ca(g)	178.2	158.884	144.3
Ca^{2+}(g)	1925.90	—	—
CaC_2(s)	−59.8	69.96	−64.9
$CaCO_3$(s; calcite)	−1206.92	92.9	−1128.79
$CaCl_2$(s)	−795.8	104.6	−748.1
CaF_2(s)	−1219.6	68.87	−1167.3
CaH_2(s)	−186.2	42	−147.2
CaO(s)	−635.09	39.75	−604.03

Selected Thermodynamic Values* (*Continued*)

Species	ΔH_f°(298.15K) kJ/mol	S°(298.15K) J/K · mol	ΔG_f°(298.15K) kJ/mol
$CaS(s)$	−482.4	56.5	−477.4
$Ca(OH)_2(s)$	−986.09	83.39	−898.49
$Ca(OH)_2(aq)$	−1002.82	−74.5	−868.07
$CaSO_4(s)$	−1434.11	106.7	−1321.79
Carbon			
C(s, graphite)	0	5.740	0
C(s, diamond)	1.895	2.377	2.900
C(g)	716.682	158.096	671.257
$CCl_4(\ell)$	−135.44	216.40	−65.21
$CCl_4(g)$	−102.9	309.85	−60.59
$CHCl_3$(liq)	−134.47	201.7	−73.66
$CHCl_3(g)$	−103.14	295.71	−70.34
CH_4(g, methane)	−74.81	186.264	−50.72
C_2H_2(g, acetylene)	226.73	200.94	209.20
C_2H_4(g, ethylene)	52.26	219.56	68.15
C_2H_6(g, ethane)	−84.68	229.60	−32.82
C_3H_8(g, propane)	−103.8	269.9	−23.49
C_6H_6(ℓ, benzene)	49.03	172.8	124.5
CH_3OH(ℓ, methanol)	−238.66	126.8	−166.27
CH_3OH(g, methanol)	−200.66	239.81	−161.96
C_2H_5OH(ℓ, ethanol)	−277.69	160.7	−174.78
C_2H_5OH(g, ethanol)	−235.10	282.70	−168.49
CO(g)	−110.525	197.674	−137.168
$CO_2(g)$	−393.509	213.74	−394.359
$CS_2(g)$	117.36	237.84	67.12
$COCl_2(g)$	−218.8	283.53	−204.6
Cesium			
Cs(s)	0	85.23	0
$Cs^+(g)$	457.964	—	—
CsCl(s)	−443.04	101.17	−414.53
Chlorine			
Cl(g)	121.679	165.198	105.680
$Cl^-(g)$	−233.13	—	—
$Cl_2(g)$	0	223.066	0
HCl(g)	−92.307	186.908	−95.299
HCl(aq)	−167.159	56.5	−131.228
Chromium			
Cr(s)	0	23.77	0
$Cr_2O_3(s)$	−1139.7	81.2	−1058.1
$CrCl_3(s)$	−556.5	123.0	−486.1
Copper			
Cu(s)	0	33.150	0
CuO(s)	−157.3	42.63	−129.7
$CuCl_2(s)$	−220.1	108.07	−175.7
Fluorine			
$F_2(g)$	0	202.78	0
F(g)	78.99	158.754	61.91
$F^-(g)$	−255.39	—	—
$F^-(aq)$	−332.63	−13.8	−278.79
HF(g)	−271.1	173.779	−273.2
HF(aq)	−332.63	−13.8	−278.79

Species	ΔH_f°(298.15K) kJ/mol	S°(298.15K) J/K · mol	ΔG_f°(298.15K) kJ/mol
Hydrogen			
$H_2(g)$	0	130.684	0
$H(g)$	217.965	114.713	203.247
$H^+(g)$	1536.202	—	—
$H_2O(\ell)$	−285.830	69.91	−237.129
$H_2O(g)$	−241.818	188.825	−228.572
$H_2O_2(\ell)$	−187.78	109.6	−120.35
Iodine			
$I_2(s)$	0	116.135	0
$I_2(g)$	62.438	260.69	19.327
$I(g)$	106.838	180.791	70.250
$I^-(g)$	−197.	—	—
$ICl(g)$	17.78	247.551	−5.46
Iron			
$Fe(s)$	0	27.78	0
$FeO(s)$	−272	—	—
Fe_2O_3(s, hematite)	−824.2	87.40	−742.2
Fe_3O_4(s, magnetite)	−1118.4	146.4	−1015.4
$FeCl_2(s)$	−341.79	117.95	−302.30
$FeCl_3(s)$	−399.49	142.3	−344.00
FeS_2(s, pyrite)	−178.2	52.93	−166.9
$Fe(CO)_5(\ell)$	−774.0	338.1	−705.3
Lead			
$Pb(s)$	0	64.81	0
$PbCl_2(s)$	−359.41	136.0	−314.10
PbO(s, yellow)	−217.32	68.70	−187.89
$PbS(s)$	−100.4	91.2	−98.7
Lithium			
$Li(s)$	0	29.12	0
$Li^+(g)$	685.783	—	—
$LiOH(s)$	−484.93	42.80	−438.95
$LiOH(aq)$	−508.48	2.80	−450.58
$LiCl(s)$	−408.701	59.33	−384.37
Magnesium			
$Mg(s)$	0	32.68	0
$MgCl_2(s)$	−641.32	89.62	−591.79
$MgCO_3(s)$	−1095.8	65.7	−1012.1
$MgO(s)$	−601.70	26.94	−569.43
$Mg(OH)_2(s)$	−924.54	63.18	−833.51
$MgS(s)$	−346.0	50.33	−341.8
Mercury			
$Hg(\ell)$	0	76.02	0
$HgCl_2(s)$	−224.3	146.0	−178.6
HgO(s, red)	−90.83	70.29	−58.539
HgS(s, red)	−58.2	82.4	−50.6
Nickel			
$Ni(s)$	0	29.87	0
$NiO(s)$	−239.7	37.99	−211.7
$NiCl_2(s)$	−305.332	97.65	−259.032
Nitrogen			
$N_2(g)$	0	191.61	0
$N(g)$	472.704	153.298	455.563

Selected Thermodynamic Values* (*Continued*)

Species	ΔH_f°(298.15K) kJ/mol	S°(298.15K) J/K · mol	ΔG_f°(298.15K) kJ/mol
NH_3(g)	−46.11	192.45	−16.45
N_2H_4(ℓ)	50.63	121.21	149.34
NH_4Cl(s)	−314.43	94.6	−202.87
NH_4Cl(aq)	−299.66	169.9	−210.52
NH_4NO_3(s)	−365.56	151.08	−183.87
NH_4NO_3(aq)	−339.87	259.8	−190.56
NO(g)	90.25	210.76	86.55
NO_2(g)	33.18	240.06	51.31
N_2O(g)	82.05	219.85	104.20
N_2O_4(g)	9.16	304.29	97.89
NOCl(g)	51.71	261.69	66.08
HNO_3(ℓ)	−174.10	155.60	−80.71
HNO_3(g)	−135.06	266.38	−74.72
HNO_3(aq)	−207.36	146.4	−111.25
Oxygen			
O_2(g)	0	205.138	0
O(g)	249.170	161.055	231.731
O_3(g)	142.7	238.93	163.2
Phosphorus			
P_4(s, white)	0	164.36	0
P_4(s, red)	−70.4	91.2	−48.4
P(g)	314.64	163.193	278.25
PH_3(g)	5.4	310.23	13.4
PCl_3(g)	−287.0	311.78	−267.8
P_4O_{10}(s)	−2984.0	228.86	−2697.7
H_3PO_4(s)	−1279.0	110.5	−1119.1
Potassium			
K(s)	0	64.18	0
KCl(s)	−436.747	82.59	−409.14
$KClO_3$(s)	−397.73	143.1	−296.25
KI(s)	−327.90	106.32	−324.892
KOH(s)	−424.764	78.9	−379.08
KOH(aq)	−482.37	91.6	−440.50
Silicon			
Si(s)	0	18.83	0
$SiBr_4$(ℓ)	−457.3	277.8	−443.9
SiC(s)	−65.3	16.61	−62.8
$SiCl_4$(g)	−657.01	330.73	−616.98
SiH_4(g)	34.3	204.62	56.9
SiF_4(g)	−1614.94	282.49	−1572.65
SiO_2(s, quartz)	−910.94	41.84	−856.64
Silver			
Ag(s)	0	42.55	0
Ag_2O(s)	−31.05	121.3	−11.20
AgCl(s)	−127.068	96.2	−109.789
$AgNO_3$(s)	−124.39	140.92	−33.41
Sodium			
Na(s)	0	51.21	0
Na(g)	107.32	153.712	76.761
Na^+(g)	609.358	—	—
NaBr(s)	−361.062	86.82	−348.983

Species	ΔH_f°(298.15K) kJ/mol	S°(298.15K) J/K · mol	ΔG_f°(298.15K) kJ/mol
NaCl(s)	−411.153	72.13	−384.138
NaCl(g)	−176.65	229.81	−196.66
NaCl(aq)	−407.27	115.5	−393.133
NaOH(s)	−425.609	64.455	−379.494
NaOH(aq)	−470.114	48.1	−419.150
Na_2CO_3(s)	−1130.68	134.98	−1044.44
Sulfur			
S(s, rhombic)	0	31.80	0
S(g)	278.805	167.821	238.250
S_2Cl_2(g)	−18.4	331.5	−31.8
SF_6(g)	−1209	291.82	−1105.3
H_2S(g)	−20.63	205.79	−33.56
SO_2(g)	−296.830	248.22	−300.194
SO_3(g)	−395.72	256.76	−371.06
$SOCl_2$(g)	−212.5	309.77	−198.3
H_2SO_4(ℓ)	−813.989	156.904	−690.003
H_2SO_4(aq)	−909.27	20.1	−744.53
Tin			
Sn(s, white)	0	51.55	0
Sn(s, gray)	−2.09	44.14	0.13
$SnCl_4$(ℓ)	−511.3	258.6	−440.1
$SnCl_4$(g)	−471.5	365.8	−432.2
SnO_2(s)	−580.7	52.3	−519.6
Titanium			
Ti(s)	0	30.63	0
$TiCl_4$(ℓ)	−804.2	252.34	−737.2
$TiCl_4$(g)	−763.2	354.9	−726.7
TiO_2	−939.7	49.92	−884.5
Zinc			
Zn(s)	0	41.63	0
$ZnCl_2$(s)	−415.05	111.46	−369.398
ZnO(s)	−348.28	43.64	−318.30
ZnS(s, sphalerite)	−205.98	57.7	−201.29

*Taken from *The NBS Tables of Chemical Thermodynamic Properties,* 1982.

APPENDIX L

Answers to Exercises

CHAPTER 1

Chemical Puzzler: The iron in some breakfast cereals is present as metallic iron, and it can be removed using a magnet. (See Figure 1.16. Try this yourself with some breakfast cereal.)

1.1. Iron: Lustrous solid, metallic, conducts heat and electricity, malleable, ductile
Water: Colorless liquid at room temperature, melting point is 0 °C and boiling point is 100 °C
Salt: White solid, high melting point
Oxygen: Colorless gas at room temperature.

1.2. $100.\ \text{g}\left(\dfrac{1\ \text{cm}^3}{1.12 \times 10^{-3}\ \text{g}}\right) = 8.93 \times 10^4\ \text{cm}^3$

1.3. $77\ \text{K} - 273.15 = -196\ °\text{C}$

1.4. **(a)** sodium, chlorine, chromium
(b) Zn, Ni, K

1.5. Chemical changes: propane burns in air (combustion).
Physical changes: water boils.
Energy is evolved in the combustion and transferred to the water and to the glass container holding the water.

1.6. The beaker on the left holds a homogeneous solution, whereas the beaker on the right contains a heterogeneous mixture of sand and iron chips.

1.7.

$$10.0\ \text{in.}\left(\frac{2.54\ \text{cm}}{\text{in.}}\right) = 25.4\ \text{cm}$$

$$25.4\ \text{cm}\left(\frac{1\ \text{m}}{100\ \text{cm}}\right) = 0.254\ \text{cm}$$

$$25.4\ \text{cm}\left(\frac{10\ \text{mm}}{\text{cm}}\right) = 254\ \text{mm}$$

$$8.00\ \text{in.}\left(\frac{2.54\ \text{cm}}{\text{in.}}\right) = 20.3\ \text{cm (or 203 mm)}$$

$$25.4\ \text{cm}\ (20.3\ \text{cm}) = 516\ \text{cm}^2$$

1.8.

$$\text{Area of sheet} = (2.50\ \text{cm})^2 = 6.25\ \text{cm}^2$$

$$\text{Volume} = 1.656\ \text{g}\left(\frac{1\ \text{cm}^3}{21.45\ \text{g}}\right) = 0.07720\ \text{cm}^3$$

$$\text{Thickness} = \text{volume/area} = 0.0124\ \text{cm}$$

$$0.0124\ \text{cm}\ (10\ \text{mm/cm}) = 0.124\ \text{mm}$$

1.9. **(a)** $750\ \text{mL}\ (1\ \text{L}/1000\ \text{mL}) = 0.750\ \text{L}$
(b) $2.0\ \text{qt} = 0.50\ \text{gal}$
$0.50\ \text{gal}\ (3.786\ \text{L/gal}) = 1.9\ \text{L} = 1.9\ \text{dm}^3$

1.10. **(a)** $500.\ \text{mg}\ (1\ \text{g}/1000\ \text{mg}) = 0.500\ \text{g}$
$0.500\ \text{g}\ (1\ \text{kg}/1000\ \text{g}) = 5.00 \times 10^{-4}\ \text{kg}$
(b)

$$\text{Length} = 4.5\ \text{m}\ (100\ \text{cm/m}) = 450\ \text{cm}$$

$$\text{Width} = 1.5\ \text{m}\ (100/\text{m}) = 150\ \text{cm}$$

$$\text{Volume} = (450\ \text{cm})(150\ \text{cm})(25\ \text{cm}) = 1.7 \times 10^6\ \text{cm}^3$$

$$\text{Mass} = 1.7 \times 10^6\ \text{cm}^3\left(\frac{7.874\ \text{g}}{\text{cm}^3}\right)\left(\frac{1\ \text{kg}}{1000\ \text{g}}\right) = 1.3 \times 10^4\ \text{kg}$$

1.11. **(a)** 11.24, has two places to the right of the decimal as in 11.19. Product is 0.60; two significant figures.
(b) 1900, has two significant figures.

1.12. $15.0\ \text{g earring}\left(\dfrac{0.58\ \text{g gold}}{\text{g earring}}\right) = 8.7\ \text{g gold}$

CHAPTER 2

Chemical Puzzler: Counting jelly beans is similar to the problem of counting atoms. You can find the mass of a single jelly bean. Then, assuming that all jelly beans are exactly alike, you could weigh the jelly beans in a jar and calculate the number of beans from the mass. This is what we do when relating the mass of an element to the number of atoms present (or, in Chapter 3, relating the mass of a compound to the number of molecules present.) In the case of atoms, we know that 1 *m* of an element always contains Avogadro's number of atoms.

2.1. **(a)** Molecules of water leave the clothes and enter the atmosphere.
(b) Molecules of water from the air condense on a cold glass.
(c) Molecules of sugar break away from one another in the solid and mix with molecules of water to form a solution.

(d) Molecules move faster as the temperature increases. Therefore, sugar molecules more rapidly break away from the solid and mix with water molecules.

2.2. An atom is about 10^5 times larger than the nucleus. Therefore, the nucleus has a radius of $100\ m(1/10^5) = 0.001$ m, or about 1 mm. A thumb tack has a diameter of about 1 cm.

2.3. (a) A (for Cu) = 34 n + 29 p = 63
(b) Nickel-59 has 28 protons, 28 electrons, and (59 − 28) = 31 neutrons.

2.4. ${}^{28}_{14}Si$, ${}^{29}_{14}Si$, ${}^{30}_{14}Si$

2.5. (0.7577)(34.96885 amu) + (0.2423)(36.96590 amu) = 35.45 amu

2.6. Eight elements in the third period: sodium (Na), magnesium (Mg), and aluminum (Al) are metals. Silicon (Si) is a metalloid. Phosphorus (P), sulfur (S), chlorine (Cl), and argon (Ar) are nonmetals.

2.7. (a) InF_3
(b) Na_2O and NaCl: O requires 2 Na, and Cl requires 1.
(c) $MgCl_2$

2.8. (a) $2.5 \text{ mol Al}\left(\frac{26.98 \text{ g}}{\text{mol}}\right) = 67 \text{ g}$

(b) $454 \text{ g S}\left(\frac{1 \text{ mol}}{32.07 \text{ g}}\right) = 14.2 \text{ mol}$

2.9. $1.0 \times 10^{24} \text{ atoms}\left(\frac{1 \text{ mol}}{6.02 \times 10^{23} \text{ atoms}}\right)\left(\frac{195 \text{ g}}{\text{mol}}\right) = 320 \text{ g Pt}$

$$\text{Volume} = 320 \text{ g}\left(\frac{1 \text{ cm}^3}{21.45 \text{ g}}\right) = 15 \text{ cm}^3$$

$$\text{Area} = \frac{15 \text{ cm}^3}{0.10 \text{ cm}} = 150 \text{ cm}^2$$

$$\text{Length of a side} = \text{Length of a side} = \sqrt{\text{area}} = \sqrt{150 \text{ cm}^2} = 12 \text{ cm}$$

CHAPTER 3

Chemical Puzzler: The compound used to make "disappearing ink" is hydrated cobalt(II) chloride. When painted onto a sheet of paper, you cannot see the faint pink solution. If the paper is heated, however, the compound is dehydrated to form blue $CoCl_2$, and this is visible. See Section 3.8 for a discussion of hydrated compounds.

3.1. The layers of graphite are bound to one another only weakly. They slide over one another.

3.2. Styrene = C_8H_8

3.3. Glycine formula = $C_2H_5NO_2$

3.4. (1) K^+; (2) Se^{2-}; (3) Be^{2+}; (4) V^{2+}; (5) Co^{2+} or Co^{3+}; (6) Cs^+

3.5. Part 1: (a) 1 Na^+ and 1 F^- ion. (b) 1 Cu^{2+} and 2 NO_3^- ion. (c) 1 Na^+ and 1 $CH_3CO_2^-$ ion
Part 2: $FeCl_2$ and $FeCl_3$
Part 3: Na_2S, Na_3PO_4, BaS, $Ba_3(PO_4)_2$

3.6. Br_2 and K are elements, and KBr is an ionic compound. Br_2 is an orange-brown liquid, whereas K is a shiny metallic solid. KBr is also a solid (white).

3.7. Part 1:
(a) NH_4NO_3 (d) V_2O_3
(b) $CoSO_4$ (e) $Ba(CH_3CO_2)_2$
(c) $Ni(CN)_2$ (f) $Ca(OCl)_2$
Part 2:
(a) Magnesium bromide
(b) Lithium carbonate
(c) Potassium hydrogen sulfite
(d) Potassium permanganate
(e) Ammonium sulfide
(f) Copper(I) chloride and copper(II) chloride

3.8. (1) $C_{12}H_{26}$ and $C_{24}H_{50}$
(2) Hexadecane = $C_{16}H_{34}$

3.9. Part 1:
(a) CO_2 (e) BF_3
(b) PI_3 (f) O_2F_2
(c) SCl_2 (g) C_9H_{20}
(d) XeO_3
Part 2:
(a) Dinitrogen tetrafluoride
(b) Hydrogen bromide
(c) Sulfur tetrafluoride
(d) Chlorine trifluoride
(e) Boron trichloride
(f) Tetraphosphorus decaoxide
(g) Heptane

3.10. Part 1:
(a) $CaCO_3$ = 100.1 g/mol
(b) Caffeine = 194.2 g/mol

Part 2: $454 \text{ g}\left(\frac{1 \text{ mol}}{100.1 \text{ g}}\right) = 4.54 \text{ mol}$

Part 3: $2.50 \times 10^{-3} \text{ mol}\left(\frac{194.2 \text{ g}}{\text{mol}}\right) = 0.486 \text{ g}$

3.11. 1. NaCl has 23.0 g of Na (39.3%) and 35.5 g of Cl (60.7%) in 1.00 mol.
2. C_8H_{18} has 96.1 g of C (84.1%) and 18.1 g of H (15.9%) in 1.00 mol.
3. $(NH_4)_2SO_4$ has a molar mass of 132.15 g/mol. It has 28.0 g of N (21.2%), 8.06 g of H (6.10%), 32.1 g of S (24.3%), and 64.0 g of O (48.4%) in 1.00 mol.

3.12.

$$78.14 \text{ g B}\left(\frac{1 \text{ mol}}{10.88 \text{ g}}\right) = 7.228 \text{ mol B}$$

$$21.86 \text{ g H}\left(\frac{1 \text{ mol}}{1.008 \text{ g}}\right) = 21.69 \text{ mol H}$$

Ratio of H to B = 21.69 mol H/7.228 mol B = 3.000 H to 1.000 B
Empirical formula = BH_3
The molecular weight (27.7 g/mol) is twice the empirical formula weight (13.8 g/formula unit), so the molecular formula is $(BH_3)_2$ or B_2H_6.

3.13. Mass of chlorine (Cl) = 2.108 g − 0.532 g = 1.576 g Cl

$$0.532 \text{ g Ti} \left(\frac{1 \text{ mol}}{47.88 \text{ g}}\right) = 0.0111 \text{ mol Ti}$$

$$1.576 \text{ g Cl} \left(\frac{1 \text{ mol Cl}}{35.453 \text{ g}}\right) = 0.04445 \text{ mol Cl}$$

Ratio of Cl to Ti = 0.04445 mol Cl/0.0111 mol Ti = 4 Cl to 1 Ti
Empirical formula = $TiCl_4$

3.14. Iron(III) oxide contains 2 mol of Fe per mol of Fe_2O_3, or 111.7 g Fe per 159.7 g Fe_2O_3.

$$50.0 \text{ g Fe} \left(\frac{159.7 \text{ g Fe}_2\text{O}_3}{111.7 \text{ g Fe}}\right) = 71.5 \text{ g Fe}_2\text{O}_3$$

3.15. Mass of water = 1.056 g compound − 0.838 g $RuCl_3$ = 0.218 g H_2O

$$0.218 \text{ g H}_2\text{O} \left(\frac{1 \text{ mol}}{18.02 \text{ g}}\right) = 0.0121 \text{ mol H}_2\text{O}$$

$$0.838 \text{ g RuCl}_3 \left(\frac{1 \text{ mol}}{207.4 \text{ g}}\right) = 0.00404 \text{ mol RuCl}_3$$

Ratio of H_2O to $RuCl_3$ = 0.0121 mol H_2O/0.00404 mol $RuCl_3$ = 3 H_2O to 1 $RuCl_3$. Therefore, the value of x is 3 (for $RuCl_3 \cdot 3\,H_2O$).

CHAPTER 4

Chemical Puzzler: Both of the observations here fall into the category of acid-base reactions. An Alka-Seltzer forms bubbles of CO_2 by the reaction of citric acid with sodium hydrogen carbonate.

$$C_3H_5O(CO_2H)_3(aq) + NaHCO_3(aq) \longrightarrow NaC_3H_5O(CO_2)(CO_2H)_2(aq) + H_2O(\ell) + CO_2(g)$$

When lemon juice (which contains citric acid) is added to tea, the acid reacts with the compound that gives tea its color and changes it to a new compound with a different color. This is an acid-base indicator, many of which exist in nature (see page 170 and Figure 5.12).

4.1. 1. Stoichiometric coefficients: 4 for Fe, 3 for O_2, and 2 for Fe_2O_3
2. 8000 atoms of Fe require $(\frac{3}{4}) \times 8000$ = 6000 molecules of O_2

4.2. 1. $C_5H_{12}(g) + 8\,O_2(g) \rightarrow 5\,CO_2(g) + 6\,H_2O(\ell)$
2. $2\,Pb(C_2H_5)_4(\ell) + 27\,O_2(g) \rightarrow 2\,PbO(s) + 16\,CO_2(g) + 20\,H_2O(\ell)$

4.3. $MgSO_4 \cdot 7\,H_2O$ is an electrolyte, and methanol is a nonelectrolyte.

4.4. 1. KNO_3 is soluble and gives K^+ and NO_3^- ions.
2. $CaCl_2$ is soluble and gives Ca^{2+} and Cl^- ions.
3. CuO is not water-soluble.
4. $NaCH_3CO_2$ is soluble and gives Na^+ and $CH_3CO_2^-$ ions.

4.5. 1. H^+ (or H_3O^+) and ClO_4^-
2. Ca^{2+} and OH^-

4.6. 1. SeO_2 is an acidic oxide (like SO_2).
2. MgO is a basic oxide.
3. P_4O_{10} is an acidic oxide.

4.7. 1. $BaCl_2(aq) + Na_2SO_4(aq) \longrightarrow BaSO_4(s) + 2\,NaCl(aq)$

$$Ba^{2+}(aq) + SO_4^{2-}(aq) \longrightarrow BaSO_4(s)$$

2. $Pb(NO_3)_2(aq) + 2\,KCl(aq) \longrightarrow PbCl_2(s) + 2\,KNO_3(aq)$

$$Pb^{2+}(aq) + 2\,Cl^-(aq) \longrightarrow PbCl_2(s)$$

4.8. $2\,AgNO_3(aq) + K_2CrO_4(aq) \longrightarrow Ag_2CrO_4(s) + 2\,KNO_3(aq)$

$$2\,Ag^+(aq) + CrO_4^{2-}(aq) \longrightarrow Ag_2CrO_4(s)$$

4.9. $Mg(OH)_2(s) + 2\,HCl(aq) \longrightarrow MgCl_2(aq) + 2\,H_2O(\ell)$

$$Mg(OH)_2(s) + 2\,H^+(aq) \longrightarrow Mg^{2+}(aq) + 2\,H_2O(\ell)$$

4.10. $PbCO_3(s) + HNO_3(aq) \longrightarrow Pb(NO_3)_2(aq) + H_2O(\ell) + CO_2(g)$

Lead(II) carbonate + nitric acid ⟶ lead(II) nitrate + water + carbon dioxide

4.11. 1. Gas-forming reaction

$$CuCO_3(s) + H_2SO_4(aq) \longrightarrow CuSO_4(aq) + H_2O(\ell) + CO_2(g)$$

2. Acid-base reaction

$$Ba(OH)_2(s) + 2\,HNO_3(aq) \longrightarrow Ba(NO_3)_2(aq) + 2\,H_2O(\ell)$$

3. Precipitation reaction

$$ZnCl_2(aq) + (NH_4)_2S(aq) \longrightarrow ZnS(s) + 2\,NH_4Cl(aq)$$

4.12. 1. Fe = +3
2. S = +6
3. C = +4
4. C = −1

4.13. $Cr_2O_7^{2-}$ is reduced and so is the oxidizing agent because the oxidation number of Cr changes from +6 in this ion to +3 in Cr^{3+}. C_2H_5OH is oxidized and so is the reducing agent because the oxidation number of C is changed from −2 in C_2H_5OH to 0 in CH_3CO_2H.

4.14. 1. This is an acid-base reaction.
2. This is an oxidation-reduction because Cu is oxidized (oxidation number changes from 0 to +2 in $CuCl_2$), and Cl_2 is reduced (oxidation number changes from 0 to −1).
3. This is a gas-forming reaction.
4. This is an oxidation-reduction because S in $S_2O_3^{2-}$ is oxidized (oxidation number changes from +2 to $+2\frac{1}{2}$ in $S_4O_6^{2-}$), and I_2 is reduced (oxidation number changes from 0 to −1).

CHAPTER 5

Chemical Puzzler: The reaction ceases when the acetic acid has been consumed. This is a situation that chemists called a reaction with a "limiting reactant." One of the reactants is in short supply (here it is the vinegar) and limits the quantity of product (here the CO_2).

5.1. $454\ \text{g}\ O_2\ (1\ \text{mol}/32.00\ \text{g}) = 14.2\ \text{mol}\ O_2$

$$14.2\ \text{mol}\ O_2 \left(\frac{2\ \text{mol C}}{1\ \text{mol}\ O_2}\right)(12.01\ \text{g C/mol}) = 341\ \text{g C}$$

$$14.2\ \text{mol}\ O_2 \left(\frac{2\ \text{mol CO}}{1\ \text{mol}\ O_2}\right)(28.01\ \text{g CO/mol}) = 795\ \text{g CO}$$

5.2. $$20.0\ \text{g}\ S_8 \left(\frac{1\ \text{mol}}{256.5\ \text{g}}\right) = 0.0780\ \text{mol}\ S_8$$

$$160\ \text{g}\ O_2 \left(\frac{1\ \text{mol}}{32.00\ \text{g}}\right) = 5.0\ \text{mol}\ O_2$$

$$S_8(s) + 8\ O_2(g) \longrightarrow 8\ SO_2(g)$$

$$0.0780\ \text{mol}\ S_8 \left(\frac{8\ \text{mol}\ O_2}{1\ \text{mol}\ S_8}\right) =$$
$0.624\ \text{mol}\ O_2$ required by $0.0780\ \text{mol}\ S_8$

$$5.0\ \text{mol}\ O_2 \left(\frac{1\ \text{mol}\ S_8}{8\ \text{mol}\ O_2}\right) =$$
$0.625\ \text{mol}\ S_8$ required by $5.0\ \text{mol}\ O_2$

Sulfur is the limiting reagent, so some O_2 remains after reaction.

$$\text{Moles of } O_2 \text{ in excess} = 5.0\ \text{mol} - 0.624\ \text{mol} = 4.4\ \text{mol}$$

$$0.0780\ \text{mol}\ S_8 \left(\frac{8\ \text{mol}\ SO_2}{1\ \text{mol}\ S_8}\right) =$$
$0.624\ \text{mol}\ SO_2$ produced

$$0.624\ \text{mol}\ SO_2\ (64.07\ \text{g/mol}) = 40.0\ \text{g}\ SO_2$$

5.3. $$18.9\ \text{g}\ NaBH_4 \left(\frac{1\ \text{mol}}{37.83\ \text{g}}\right) = 0.500\ \text{mol}$$

$$0.500\ \text{mol}\ NaBH_4 \left(\frac{2\ \text{mol}\ B_2H_6}{3\ \text{mol}\ NaBH_4}\right) =$$
$0.333\ \text{mol}\ B_2H_6$ expected

$$0.333\ \text{mol}\ B_2H_6\ (27.67\ \text{g/mol}) =$$
$9.22\ \text{g}\ B_2H_6$ expected

$$\left(\frac{7.50\ \text{g}}{9.22\ \text{g}\ B_2H_6}\right)100\% = 81.4\%\ \text{yield}$$

5.4. Mass of water $= 2.357\ \text{g} - 2.108\ \text{g} = 0.249\ \text{g}$

$$\left(\frac{0.249\ \text{g}\ H_2O}{2.357\ \text{g sample}}\right)100\% = 10.6\%\ H_2O;\ 71.6\%\ BaCl_2 \cdot 2\ H_2O.$$

5.5. $$1.612\ \text{g} \left(\frac{1\ \text{mol}\ CO_2}{44.01\ \text{g}}\right)\left(\frac{1\ \text{mol C}}{1\ \text{mol}\ CO_2}\right) = 0.03663\ \text{mol C}$$

$$0.7425\ \text{g}\ H_2O \left(\frac{1\ \text{mol}\ H_2O}{18.01\ \text{g}}\right)\left(\frac{2\ \text{mol H}}{1\ \text{mol}\ H_2O}\right) = 0.08245\ \text{mol H}$$

$$\frac{0.08245\ \text{mol H}}{0.03663\ \text{mol C}} = \frac{2.25\ \text{H}}{1\ \text{C}} = \frac{9\ \text{H}}{4\ \text{C}}$$

The empirical formula is C_4H_9, which has a molar mass of 57 g/mol. The molecular formula is $(C_4H_9)_2$, or C_8H_{18}.

5.6. $$0.600\ \text{g} \left(\frac{1\ \text{mol}\ CO_2}{44.01\ \text{g}}\right)\left(\frac{1\ \text{mol C}}{1\ \text{mol}\ CO_2}\right) = 0.0136\ \text{mol C}$$

$$0.0136\ \text{mol C}\ (12.01\ \text{g/mol}) = 0.163\ \text{g C}$$

$$0.163\ \text{g}\ H_2O \left(\frac{1\ \text{mol}\ H_2O}{18.01\ \text{g}}\right)\left(\frac{2\ \text{mol H}}{1\ \text{mol}\ H_2O}\right) = 0.0181\ \text{mol H}$$

$$0.0181\ \text{mol H}\ (1.008\ \text{g/mol}) = 0.0182\ \text{g H}$$

Mass of O in unknown =
$$0.400\ \text{g} - \text{mass of C} - \text{mass of H} = 0.219\ \text{g O}$$

$$0.219\ \text{g O} \left(\frac{1\ \text{mol}}{16.00\ \text{g}}\right) = 0.0137\ \text{mol O}$$

The ratio of C to O is 1 C/1 O. The ratio of H to C is

$$\frac{0.0181\ \text{mol H}}{0.0136\ \text{C}} = \frac{1.33\ \text{H}}{1\ \text{C}} = \frac{4\ \text{H}}{3\ \text{C}}$$

The empirical formula is $C_3H_4O_3$.

5.7. $$0.452\ \text{g} \left(\frac{1\ \text{mol}\ CO_2}{44.01\ \text{g}}\right)\left(\frac{1\ \text{mol C}}{1\ \text{mol}\ CO_2}\right) = 0.0103\ \text{mol C}$$

$$0.0103\ \text{mol C}\ (12.01\ \text{g/mol}) = 0.123\ \text{g C}$$

$$0.0924\ \text{g}\ H_2O \left(\frac{1\ \text{mol}\ H_2O}{18.01\ \text{g}}\right)\left(\frac{2\ \text{mol H}}{1\ \text{mol}\ H_2O}\right) = 0.0103\ \text{mol H}$$

$$0.0103\ \text{mol H}\ (1.008\ \text{g/mol}) = 0.0103\ \text{g H}$$

Mass of Cr in unknown =
$$0.178\ \text{g} - \text{mass of C} - \text{mass of H} = 0.0447\ \text{g Cr}$$

$$0.0447\ \text{g Cr} \left(\frac{1\ \text{mol}}{52.00\ \text{g}}\right) = 8.60 \times 10^{-4}\ \text{mol Cr}$$

The ratio of C to H is 1 C/1 H. The ratio of C to Cr is 12 to 1. Therefore, the empirical formula is $C_{12}H_{12}\ Cr$.

5.8. $$26.3\ \text{g} \left(\frac{1\ \text{mol}\ NaHCO_3}{84.01\ \text{g}}\right) = 0.313\ \text{mol}\ NaHCO_3$$

$$\frac{0.313\ \text{mol}\ NaHCO_3}{0.200\ \text{L}} = 1.57\ \text{M}$$

5.9. HCl solution: $[H^+] = 1.0$ M, $[Cl^-] = 1.0$ M, and total ion concentration is 2.0 M. Na_2SO_4 solution: $[Na^+] =$

1.00 M, $[SO_4^{2-}]$ = 0.500 M, and total ion concentration is 1.50 M.

5.10. 0.500 L (0.0200 mol/L) = 0.0100 mol $KMnO_4$ required.

$$0.0100 \text{ mol } KMnO_4\ (158.0 \text{ g/mol}) = 1.58 \text{ g } KMnO_4$$

Place 1.58 g $KMnO_4$ in a 500-mL flask, and add water to the mark on the flask.

5.11. Add 125 mL of water to 125 mL of 2.00 M NaOH.

5.12. $$\frac{(0.15 \text{ M})(0.0060 \text{ L})}{0.010 \text{ L}} = 0.090 \text{ M}$$

5.13. $$(0.0500 \text{ L})(0.450 \text{ M}) = 0.0225 \text{ mol HCl}$$

$$(0.0225 \text{ mol HCl})\left(\frac{2 \text{ mol NaCl}}{2 \text{ mol HCl}}\right) = 0.0225 \text{ mol NaCl}$$

$$(0.0225 \text{ mol NaCl})(58.44 \text{ g/mol}) = 1.31 \text{ g NaCl}$$

5.14. $$(0.02833 \text{ L})(0.953 \text{ M}) = 0.0270 \text{ mol NaOH}$$

$$(0.0270 \text{ mol NaOH})\left(\frac{1 \text{ mol } CH_3CO_2H}{1 \text{ mol NaOH}}\right) = 0.0270 \text{ mol } CH_3CO_2H$$

$$0.0270 \text{ mol } CH_3CO_2H\ (60.05 \text{ g/mol}) = 1.62 \text{ g } CH_3CO_2H$$

$$0.0270 \text{ mol } CH_3CO_2H/0.0250 \text{ L} = 1.08 \text{ M}$$

5.15. $$(0.02967 \text{ L})(0.100 \text{ M}) = 0.00297 \text{ mol HCl}$$

$$0.00297 \text{ mol HCl}\left(\frac{1 \text{ mol NaOH}}{1 \text{ mol HCl}}\right) = 0.00297 \text{ mol NaOH}$$

$$\frac{0.00297 \text{ mol NaOH}}{0.0250 \text{ L}} = 0.119 \text{ M NaOH}$$

5.16. $$(0.02030 \text{ L})(0.196 \text{ M } Na_2S_2O_3) = 0.00398 \text{ mol } Na_2S_2O_3$$

$$0.00398 \text{ mol } Na_2S_2O_3\left(\frac{1 \text{ mol } I_2}{2 \text{ mol } Na_2S_2O_3}\right) = 0.00199 \text{ mol } I_2$$

0.00199 mol I_2 was not used in reaction with ascorbic acid.

$$I_2 \text{ originally added} = (0.05000 \text{ L})(0.0520 \text{ M}) = 0.00260 \text{ mol}$$

I_2 used in reaction with ascorbic acid =

$$0.00260 \text{ mol} - 0.00199 \text{ mol} = 6.1 \times 10^{-4} \text{ mol } I_2$$

$$(6.10 \times 10^{-4} \text{ mol } I_2)\left(\frac{1 \text{ mol } C_6H_8O_6}{1 \text{ mol } I_2}\right)(176.1 \text{ g/mol}) = 0.11 \text{ g } C_6H_8O_6$$

CHAPTER 6

Chemical Puzzler: This question is answered thoroughly on the CD-ROM version of the textbook (where the answer is shown to be 39 peanuts). The quantity of thermal energy that can be derived from a peanut can be determined with a calorimeter (Section 6.9). We also know from experiment the quantity of heat required to heat water to its boiling point and then to evaporate the water (see Sections 6.2 and 6.3). Knowing the quantity of heat required to heat and then evaporate a cup of water, and the quantity of heat available from one burning peanut, we can then calculate the number of peanuts required.

6.1. The egg is dropped into the pan, and potential energy is converted into kinetic energy as it falls. Heat energy is transferred to the egg from burning gas or from a hot electric coil. If gas is used, for example, chemical potential energy is converted to heat energy.

6.2. 1. $$160 \text{ Cal}\left(\frac{1000 \text{ cal}}{\text{Calorie}}\right)\left(\frac{4.184 \text{ J}}{\text{calorie}}\right) = 6.7 \times 10^5 \text{ J}$$

2. $$(75 \text{ W})(3.0 \text{ h})\left(\frac{3600 \text{ s}}{\text{h}}\right) = 8.1 \times 10^5 \text{ Ws} = 8.1 \times 10^5 \text{ J}$$

3. $$(16 \text{ kJ})\left(\frac{1 \text{ kcal}}{4.184 \text{ kJ}}\right) = 3.8 \text{ kcal}$$

6.3. $$q = 24.1 \times 10^3 \text{ J} = (250.\text{ g})(0.902 \text{ J/g} \cdot \text{K})(T_{final} - 5.0\ °\text{C})$$
$$T_{final} = 111.9°\text{C}$$

6.4. $$(15.5 \text{ g})(C_{metal})(18.9\ °\text{C} - 100.0\ °\text{C}) = -(55.5 \text{ g})(4.184 \text{ J/g} \cdot \text{K})(18.9\ °\text{C} - 16.5\ °\text{C})$$
$$C_{metal} = 0.44 \text{ J/g} \cdot \text{K}$$

6.5. $$(400.\text{ g iron})(0.451 \text{ J/g} \cdot \text{K})(32.8\ °\text{C} - T_{initial}) = -(1000.\text{ g})(4.184 \text{ J/g} \cdot \text{K})(32.8\ °\text{C} - 20.0\ °\text{C})$$
$$T_{initial} = 330.\ °\text{C}$$

6.6. $$(25.0 \text{ g } CH_3OH)(2.53 \text{ J/g} \cdot \text{K})(64.6\ °\text{C} - 25.0\ °\text{C}) = 2.50 \times 10^3 \text{ J}$$

$$(25.0 \text{ g } CH_3OH)(2.00 \times 10^3 \text{ J/g}) = 5.00 \times 10^4 \text{ J}$$
$$\text{Total heat energy} = 5.25 \times 10^4 \text{ J}$$

6.7. $$(1.32 \text{ g ice})(333 \text{ J/g ice}) = 440.\text{ J}$$

$$440.\text{ J} = -(9.85 \text{ g})(C_{metal})(0.0\ °\text{C} - 100.0\ °\text{C})$$
$$C_{metal} = 0.446 \text{ J/g} \cdot \text{K}$$

6.8. 1. $(10.0 \text{ g})(1 \text{ mol } I_2/253.8 \text{ g})(62.4 \text{ kJ/mol}) = 2.46 \text{ kJ}$
2. The process is exothermic.

$$(3.45 \text{ g})\left(\frac{1 \text{ mol } I_2}{253.8 \text{ g}}\right)(62.4 \text{ kJ/mol}) = 0.848 \text{ kJ}$$

6.9. $$(12.6 \text{ g } H_2O)\left(\frac{1 \text{ mol}}{18.02 \text{ g}}\right)(285.8 \text{ kJ/mol}) = 2.00 \times 10^2 \text{kJ}$$

6.10.
$$2\ CO_2(g) + 3\ H_2O(\ell) \longrightarrow C_2H_6(g) + \tfrac{7}{2}\ O_2(g) \qquad \Delta H = +1559.7 \text{ kJ}$$
$$2\ C(s) + 2\ O_2(g) \longrightarrow 2\ CO_2(g) \qquad \Delta H = 2(-393.5 \text{ kJ})$$
$$3\ H_2(g) + \tfrac{3}{2}\ O_2(g) \longrightarrow 3\ H_2O(\ell) \qquad \Delta H = 3(-285.8 \text{ kJ})$$
$$2\ C(s) + 3\ H_2(g) \longrightarrow C_2H_6(g) \qquad \Delta H = -84.7 \text{ kJ}$$

6.11. $$\Delta H_{overall} = -413.7 \text{ kJ} + 106.8 \text{ kJ} = -306.9 \text{ kJ}$$

$$(454 \text{ g})\left(\frac{1 \text{ mol PbS}}{239.3 \text{ g}}\right)(-306.9 \text{ kJ}) = 582 \text{ kJ}$$

6.12. $2\ C(s) + 3\ H_2(g) + \frac{1}{2}\ O_2(g) \longrightarrow C_2H_5OH(\ell)$

$\Delta H_f^\circ = -277.7$ kJ/mol

$Cu(s) + \frac{1}{8}\ S_8(s) + 2\ O_2(g) \longrightarrow CuSO_4(s)$

$\Delta H_f^\circ = -771.4$ kJ/mol

$\Delta H^\circ_{rxn} = (-277.7\ kJ/mol)(1.5\ mol) = -4.2 \times 10^2$ kJ

6.13. $\Delta H^\circ_{rxn} = 6\ \Delta H_f^\circ[CO_2(g)] + 3\ \Delta H_f^\circ[H_2O\ell] - \{\Delta H_f^\circ[C_6H_6(\ell)] + \frac{15}{2}\ \Delta H_f^\circ[O_2(g)]\}$

$= 6\ mol(-393.5\ kJ/mol) + 3\ mol\ (-285.8\ kJ/mol) - 1\ mol\ (+49.0\ kJ/mol) - 0$

$= -3267.4$ kJ

6.14. Heat transferred to calorimeter water =

$(1.50 \times 10^3\ g)(4.184\ J/g \cdot K)(2.32\ K) = 14.6 \times 10^3$ J

Heat transferred to calorimeter bomb =

$(837\ J/K)(2.32\ K) = 1.94 \times 10^3$ J

Total heat transferred by 1.00 g sucrose = −16.5 kJ

Heat transferred per mole =

$(-16.5\ kJ/g)(342.2\ g/mol) = -5650$ kJ

6.15. Mass of final solution = 400. g

$q = (4.2\ J/g \cdot K)(400.\ g)(26.60 - 25.10)\ °C = 2.5 \times 10^3$ J for 0.0800 mol HCl

2.5 kJ/0.0800 mol HCl = 32 kJ/mol of HCl

ΔH of neutralization = −32 kJ

6.16. $Ba(s) + \frac{1}{2}\ O_2(g) \longrightarrow BaO(s)$

$\Delta H_f^\circ[BaO(s)] = -553.5$ kJ

$H_2(g) + \frac{1}{2}\ O_2(g) \longrightarrow H_2O(\ell)$

$\Delta H_f^\circ[H_2O(\ell)] = -285.8$ kJ

$BaO(s) + H_2O(\ell) \longrightarrow Ba(OH)_2(s)$

$\Delta H^\circ_{rxn} = -105.4$ kJ

$Ba(s) + O_2(g) + H_2(g) \longrightarrow Ba(OH)_2(s)$

$\Delta H^\circ_{rxn} = -944.7$ kJ

CHAPTER 7

Chemical Puzzler: When heated, metal ions absorb energy and move from the ground state to an excited state. (You can picture what happens by imagining an electron moving from an orbital in its highest, filled subshell to an orbital in a still higher energy subshell.) When the metal ion returns to the ground state, it loses energy, and this energy appears as light. See Section 7.3 for a discussion of this in the case of the H atom.

7.1. 1. Highest frequency = violet; lowest frequency = red
2. FM radio has a lower frequency than a microwave oven.
3. The wavelength of x-rays is shorter than that of ultraviolet light.

7.2. 1. 10 cm
2. 6.67 cm
3. 2 waves; five nodes (one at each end and three in the middle)

7.3. Blue light: 4.00×10^2 nm = 4.00×10^{-7} m

$\nu = (2.998 \times 10^8\ m/s)/(4.00 \times 10^{-7}\ m) = 7.50 \times 10^{14}/s$

$E = (6.626 \times 10^{-34}\ J \cdot s)(7.50 \times 10^{14}/s) = 4.97 \times 10^{-19}$ J/photon

Microwaves:

$E = (6.626 \times 10^{-34}\ J \cdot s)(2.45 \times 10^9\ /s) = 1.62 \times 10^{-24}$ J/photon

$$\frac{E(\text{blue light})}{E\ (\text{microwaves})} = 3.1 \times 10^5$$

Blue light is almost half a million times more energetic than microwaves.

7.4. Energy for $n = 3 = \frac{-Rhc}{n^2} = \frac{-Rhc}{9}$

If $-Rhc = -2.179 \times 10^{-18}$ J/atom (see Example 7.3), then

$$\frac{-Rhc}{9} = -2.421 \times 10^{-19}\ \text{J/atom}$$

$$(-2.421 \times 10^{-19}\ \text{J/atom})(6.022 \times 10^{23}\ \text{atm/mol})\left(\frac{1\ \text{kJ}}{1000\ \text{J}}\right) = 145.8\ \text{kJ/mol}$$

7.5. Use Equation 7.5, where $n_{final} = 1$ and $n_{initial} = 2$.

$$\Delta E = -Rhc\left[\left(\frac{1}{1^2}\right) - \left(\frac{1}{2^2}\right)\right]$$

$$= -(2.179 \times 10^{-18}\ \text{J/atom})\left(\frac{3}{4}\right)$$

$$= 1.634 \times 10^{-18}\ \text{J/atom}$$

$$\lambda = hc/\Delta E$$

$$= (6.626 \times 10^{-34}\ \text{J} \cdot \text{s})\frac{2.998 \times 10^8\ \text{m/s}}{1.634 \times 10^{-18}\ \text{J/atom}}$$

$\lambda = 1.216 \times 10^{-7}$ m, or 121.6 nm; $\nu = 2.466 \times 10^{15}/s$

7.6. First calculate the velocity of the neutron:

$$v = [2E/m]^{1/2}$$

$$= [2(6.21 \times 10^{-21}\ \text{kg} \cdot \text{m}^2/\text{s}^2)]/(1.675 \times 10^{-27}\ \text{kg}]^{1/2}$$

$$= 2723\ \text{m/s}$$

With the velocity known, the wavelength can be calculated:

$$\lambda = \frac{h}{mv} = \frac{6.626 \times 10^{-34}(\text{kg} \cdot \text{m}^2/\text{s}^2)\text{s}}{(1.675 \times 10^{-31}\ \text{kg})(2723\ \text{m/s})}$$

$$= 1.453 \times 10^{-10}\ \text{m}$$

7.7. 1. $\ell = 0$ and 1
2. $m_\ell = +1$, 0, and −1; subshell label = *p*
3. *d* subshell
4. $\ell = 0$ and $m_\ell = 0$ for an *s* subshell.
5. three orbitals in a *p* subshell
6. *f* subshell has seven values of m_ℓ and seven orbitals.

7.8. 1.

Orbital	n	ℓ
$6s$	6	0
$4p$	4	1
$5d$	5	2
$4f$	4	3

2. A $4p$ orbital has one nodal plane, and a $6d$ orbital has two nodal planes.

CHAPTER 8

Chemical Puzzler: Iron is a good reducing agent because, like all metals, its ionization energy is relatively low (Section 8.6). Chlorine is a good oxidizing agent because, like many nonmetals, it has a relatively large affinity for electrons (Section 8.6). The iron ion in $FeCl_3$ is Fe^{3+}, which has unpaired electrons, making it paramagnetic (Sections 8.1 and 8.5).

8.1. 1. $4s (n + \ell = 4)$ filled before $4p\ (n + \ell = 5)$
2. $6s (n + \ell = 6)$ filled before $5d\ (n + \ell = 7)$
3. $5s (n + \ell = 5)$ filled before $4f (n + \ell = 7)$

8.2. 1. Cl
2. Cl has the spectroscopic notation given in part 1.

[Ne] $3s$ [↑↓] $3p$ [↑↓][↑↓][↑]

3. Ca has two valence electrons, the $4s$ electrons.

[Ar] $4s$ [↑↓]

For ↑, $n = 4$, $\ell = 0$, $m_\ell = 0$, and $m_s = +\frac{1}{2}$
For ↓, $n = 4$, $\ell = 0$, $m_\ell = 0$, and $m_s = -\frac{1}{2}$

8.3. See Table 8.2

8.4.
V^{2+}: [Ar] $4s$ [ ] $3d$ [↑][↑][↑][ ][ ]
V^{3+}: [Ar] $4s$ [ ] $3d$ [↑][↑][ ][ ][ ]
Co^{3+}: [Ar] $4s$ [ ] $3d$ [↑↓][↑][↑][↑][↑]

All three ions are paramagnetic.

8.5. Radii are in the order C < Si < Al. See Figure 8.10.

8.6. 1. H—O distance = 37 pm + 73 pm = 110 pm
H—S distance = 37 pm + 103 pm = 140 pm
2. Br has a radius of 228 pm/2 = 114 pm. The Br—Cl distance is 114 pm + 100 pm = 214 pm

8.7. 1. Radii are in the order C < Si < Al. See Figure 8.10.
2. Ionization energies: Al < Si < C
3. Al should have a less negative *EA* then Si

8.8. The ion radii are in the order $N^{3-} > O^{2-} > F^-$. All three ions have 10 electrons, but N has only 7 protons, O has 8, and F has 9.

8.9.
Enthalpy of formation of Na(g) = +107.3 kJ/mol
Enthalpy of formation of I(g) = +106.8 kJ/mol
Ionization energy of Na(g) = +496 kJ/mol
Electron affinity of I(g) = −295.2 kJ/mol
Formation of NaI(s) from ions = −702 kJ/mol
Sum = calculated ΔH of formation = −287 kJ/mol

(The enthalpy of formation of NaI(s) is −287.78 kJ/mol.)

CHAPTER 9

Chemical Puzzler: Water is affected by a static electric charge because water molecules are polar (Section 9.6).

9.1. Ca, two valence electrons
As, five valence electrons
Br, seven valence electrons
Br dot symbol: $\cdot\ddot{\underset{..}{Br}}:$

9.2. 3, 4, 5, and 6

9.3. 1. C_2H_4 has six bond pairs and no lone pairs.
SCl_2 has two bond pairs and eight lone pairs.
2. The N_2 molecule has three bond pairs and two lone pairs.

9.4. Sulfate and phosphate ions.

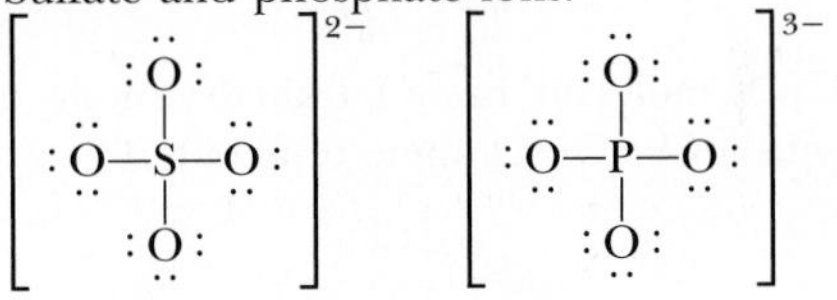

9.5.

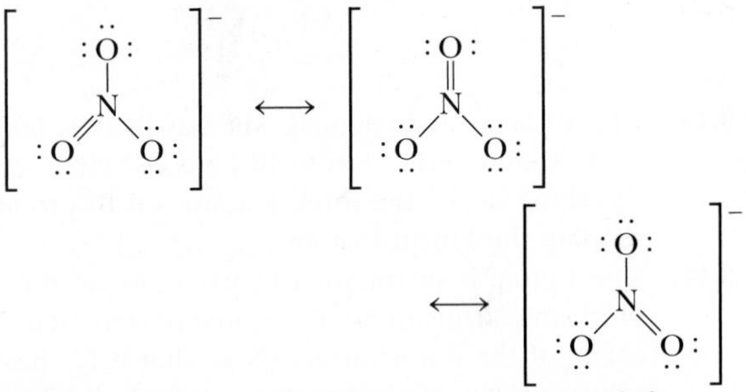

9.6. Note that there are five electron pairs around the central atom.

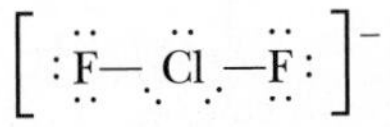

9.7. 1. C—N (order = 1) > C=N (order = 2) < C≡N (order = 3)
2.

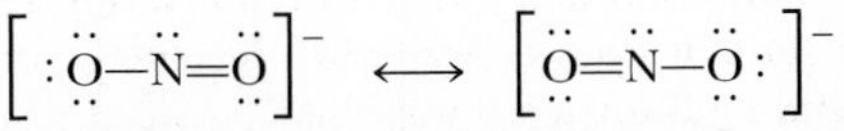

The bond order in NO_2^- is 1.5. Therefore, the NO bond length (124 pm) should be between the length of a N—O single bond (136 pm) and a N=O double bond (115 pm).

9.8.
$$CH_4(g) + 2\ O_2(g) \longrightarrow CO_2(g) + 2\ H_2O(g)$$

Break 4 C—H bonds and 2 O=O bond =
4 mol (414 kJ/mol) + 2 mol (498 kJ/mol) = 2652 kJ

Make 2 C=O bonds and 4 H—O bonds =
2 mol (803 kJ/mol) + 4 mol (464 kJ/mol) = 3462 kJ

$$\Delta H^\circ_{rxn} = 2652\ kJ - 3462\ kJ = -810\ kJ$$

(Calculated from ΔH_f°, the value of ΔH_{rxn}° is -802.3 kJ)

9.9. 1. The H atom is the positive atom in each case. H—F ($\Delta\chi = 1.9$) is more polar than H—I ($\Delta\chi = 0.4$).
2. B—F ($\Delta\chi = 2.0$) is much more polar than B—C ($\Delta\chi = 0.5$).
3. C—Si is more polar ($\Delta\chi = 0.7$) than C—S (which is not polar at all, $\Delta\chi = 0$)

9.10. 1. SF_4: Oxidation number of S = +4 and oxidation number of F = −1.
2. CO_3^{2-}: Oxidation number of C = +4 and oxidation number of O = −2.
3. SO_3: Oxidation number of S = +6 and oxidation number of O = −2.

9.11. The formal charge on the F atom is 1+, a highly unlikely charge for the very electronegative F atom.

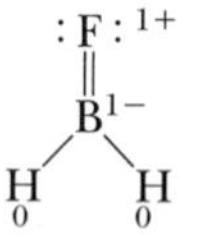

9.12. The CH_2Cl_2 molecule has a tetrahedral molecular shape with a Cl—C—Cl angle of about 109°.

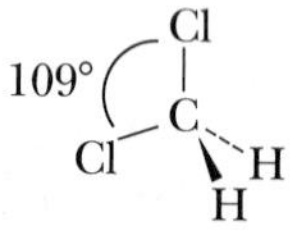

9.13. BF_3 is planar and trigonal, whereas BF_4^- is tetrahedral. Adding an F^- ion to BF_3 has the effect of "pushing back" the three F atoms on BF_3 to accommodate the fourth F atom.

9.14. The I atom is surrounded by five pairs, so the electron-pair geometry is trigonal-bipyramidal. The shape of the ion is linear. (Note that ICl_2^- has the same number of electron pairs as XeF_2 in Figure 9.13.)

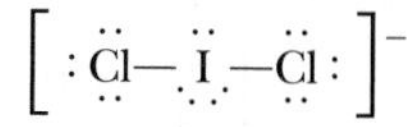

9.15. **(a)** Phosphate ion has a tetrahedral electron pair geometry, and the shape of the ion is likewise tetrahedral.

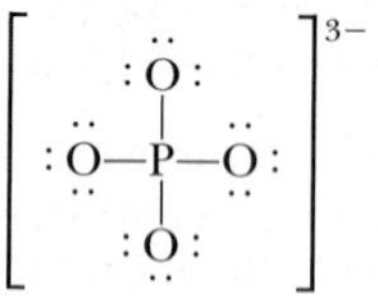

(b) The sulfite ion has a tetrahedral electron pair geometry, and the shape of the ion is pyramidal.

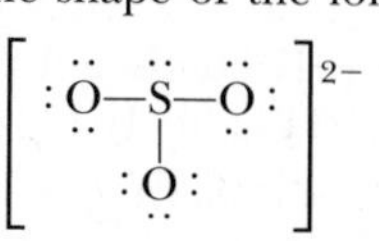

(c) The IF_5 ion has an octahedral electron pair geometry and a square-pyramidal molecular shape. (Lone pairs on the F atoms are not shown.)

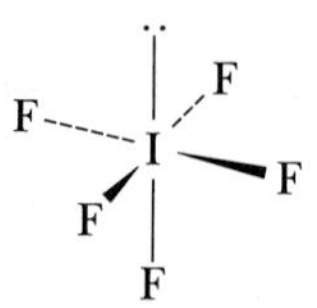

9.16. 1. $BFCl_2$, polar, the negative side is the F atom because F is the most electronegative atom in the molecule.

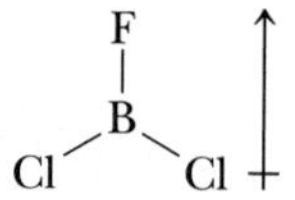

2. NH_2Cl, polar, the negative side is the Cl atom.

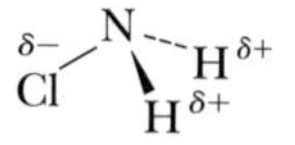

3. SCl_2, polar, the Cl atoms are on the negative side.

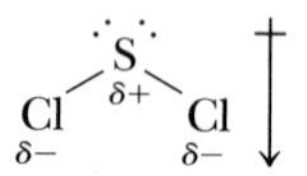

CHAPTER 10

Chemical Puzzler: The properties of the O_2 molecule are not satisfactorily explained by the simple electron dot picture. Instead, a more sophisticated bonding model—molecular orbital theory (Section 10.2)—provides a better understanding of the O_2 molecule and many others as well.

10.1. SCl_2 has an S atom surrounded by two lone pairs and two bond pairs. The electron-pair geometry is tetrahedral, so the hybridization of the S atom is sp^3.

10.2. The Xe atom in XeF_4 is surrounded by two lone pairs and four bond pairs. The electron-pair geometry is therefore octahedral. This means the hybridization of the Xe atom is sp^3d^2.

10.3. The bonding in N_2 is identical to that in CO (one sigma bond, two pi bonds, and one lone pair on each atom), except that both atoms are of course N. Each N atom is *sp*-hybridized. The assignments of the five valence electrons of each N and their roles in bonding are

↑↓	↑	↑ ↑
sp	*sp*	2*p* 2*p*
lone pair	N—N sigma bond	for the two N—N pi bonds

10.4.

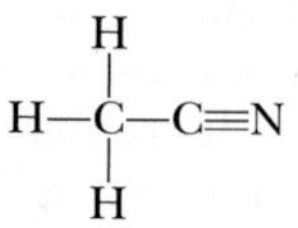

(a) Bond angles: H—C—H = 109°; H—C—C = 109°; C—C—N = 180°

(b) Atom hybridizations: The CH_3 carbon has a tetrahedral electron-pair geometry and is *sp*^{3}-hybridized. The CN carbon has a linear electron-pair geometry and is *sp*-hybridized. The N atom has a linear electron-pair geometry (the triple bond and lone pair are 180° apart) and can be considered *sp*-hybridized.

10.5. MO configuration of H_2^+: $(\sigma_{1s})^1$. The ion therefore has a bond order of $\frac{1}{2}$. This is the same bond order as in the ions He_2^+ and H_2^-.

10.6. The anion in $NaLi_2$ is Li_2^-. Its electron configuration is $(\sigma_{1s})^2(\sigma^*_{1s})^2(\sigma_{2s})^2(\sigma^*_{2s})^1$. This gives a bond order of $\frac{1}{2}$, implying that the anion might be stable enough to prepare a salt such as $NaLi_2$.

10.7. O_2^+: [core electrons]$(\sigma_{2s})^2(\sigma^*_{2s})^2(\pi_{2p})^4(\sigma_{2p})^2(\pi^*_{2p})^1$
The net bond order is 2.5, a higher bond order than in O_2 and thus a stronger bond. The ion is paramagnetic to the extent of one electron.

CHAPTER 11

Chemical Puzzler: The stretchiness of bubble gum is due to SBR (styrene butadiene rubber), a polymer. See page 534 for more information.

11.1. Structural isomers and names of C_6H_{14} isomers:

$CH_3CH_2CH_2CH_2CH_2CH_3$

hexane

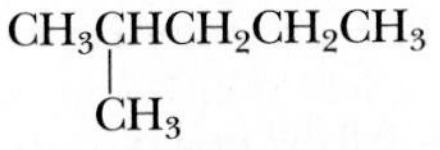

2-methylpentane

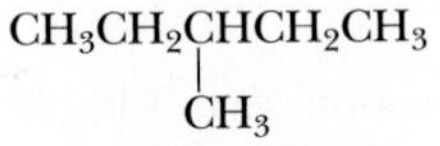

3-methylpentane

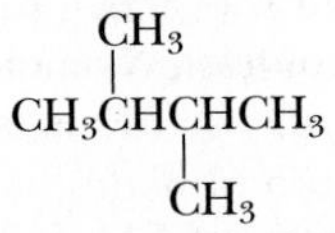

2,3-dimethylbutane

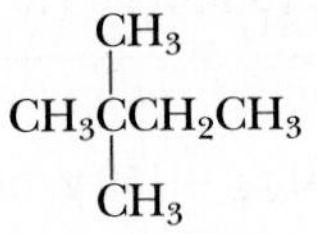

2,2-dimethylbutane

11.2. C_7H_{16} with an unbranched chain is heptane.

11.3.

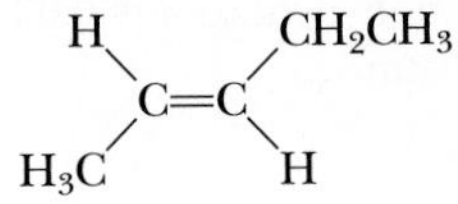

trans-2-pentane

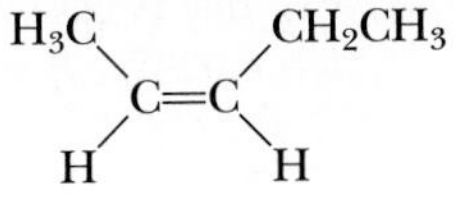

cis-2-pentane

11.4. 1.

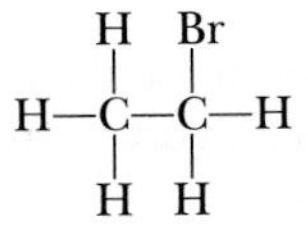

2.

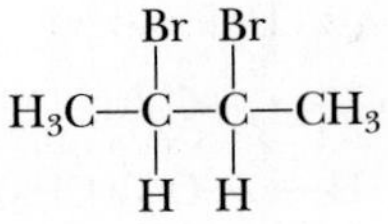

2,3-dibromobutane

11.5. 1. *p*-Bromonitrobenzene; *o*-chlorohydroxobenzene (or *o*-chlorophenol); 1-chloro-2,4-dimethylbenzene.

2.

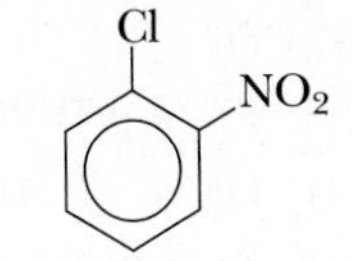

11.6. 1. 1-Propanol and 2-propanol

2.

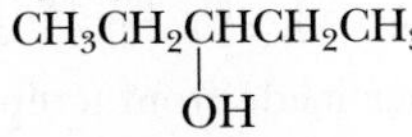

3-pentanol

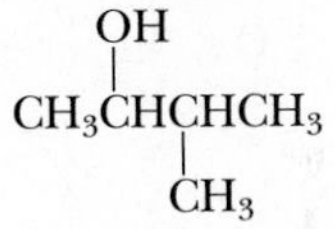

3-methyl-2-butanol

3. Both alcohols in part 2 are secondary.

11.7. 1. $CH_3CH_2CH_2OH \longrightarrow CH_3CH{=}CH_2 + H_2O$

2. $CH_3CH_2CH_2OH + HI \longrightarrow CH_3CH_2CH_2I + H_2O$

11.8.

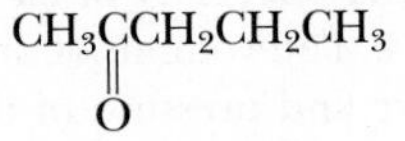

2-pentanone

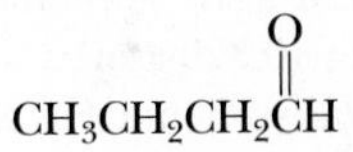

butanal

11.9. 1. 3-Hexanone

2. Pentanal

11.10. 1. Product is 1-propanol

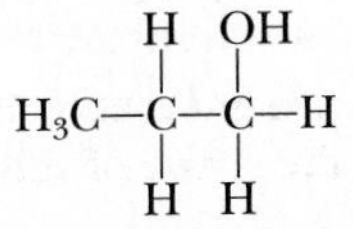

2. $Na_2Cr_2O_7$ is an oxidizing agent (see Table 4.2). The product is a carboxylic acid (see page 517).

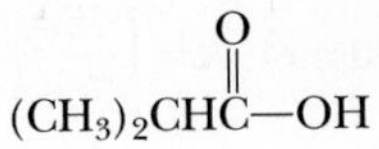

11.11. 1. Hexanoic acid

2. Propyl acetate

11.12. 1.

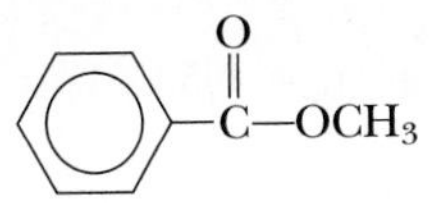

2.

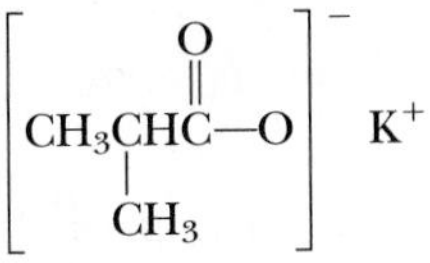

11.13. **(a)** Glyceryl tripalmitate,

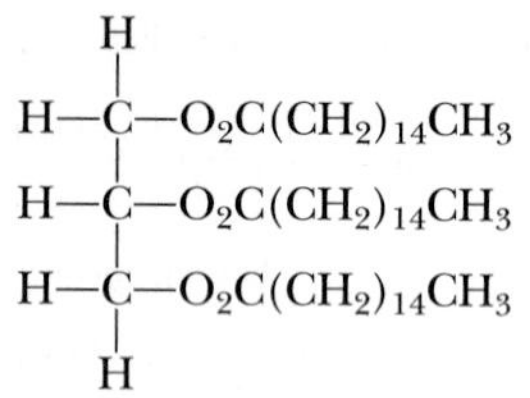

(b) Products are glycerol and sodium palmitate, $NaCH_3(CH_2)_{14}CO_2$

11.14. Three units of polymer of acrylonitrile.

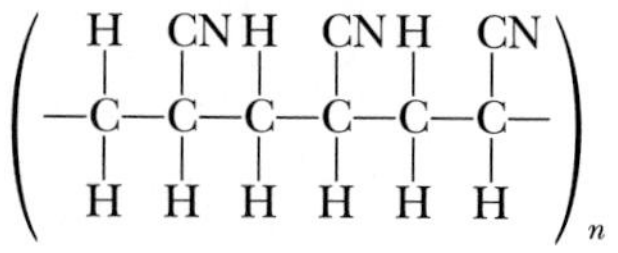

11.15. A polymer made from terephthalic acid and ethylene glycol is a polyester.

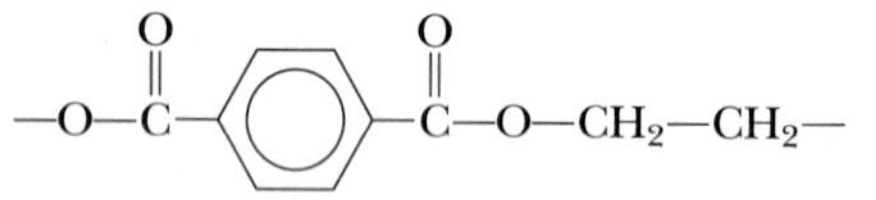

CHAPTER 12

Chemical Puzzler: The chemistry of air bags is described on pages 547 and 564. The volume of an air bag can be affected by the temperature and pressure of the surrounding air. The effect of T and P on gas volume is described in this chapter.

12.1. 0.83 bar (0.82 atm) > 75 kPa (0.74 atm) > 0.63 atm > 250. mm Hg (0.329 atm)

12.2. $P_1 = 55$ mm Hg and $V_1 = 125$ mL

$$P_2 = 78 \text{ mm Hg and } V_2 = ?$$

$$V_2 = P_1V_1/P_2 = 88 \text{ mL}$$

12.3. $T_1 = 298$ K and $V_1 = 45$ L

$$T_2 = 263 \text{ K and } V_2 = ?$$

$$V_2 = V_1\left(\frac{T_2}{T_1}\right) = 40. \text{ L}$$

12.4. $22.4 \text{ L } CH_4\left(\frac{2 \text{ L } O_2}{1 \text{ L } CH_4}\right) = 44.8 \text{ L } O_2$ required

44.8 L of H_2O and 22.4 L of CO_2 are produced.

12.5. $n = 1300$ mol, $P = \left(\frac{750}{760}\right)$ atm, $T = 293$ K

$$V = \frac{nRT}{P} = 3.2 \times 10^4 \text{ L}$$

12.6. $P_1 = 150$ atm, $T_1 = 303$ K, $V_1 = 20.$ L

$$P_2 = \left(\frac{755}{760}\right) \text{ atm}, \ T_2 = 295 \text{ K}, \ V_2 = ?$$

$$V_2 = V_1\left(\frac{P_1}{P_2}\right)\left(\frac{T_2}{T_1}\right) = 2.9 \times 10^3 \text{ L; fill 590 balloons}$$

12.7. $M = 28.96$ g/mol, $P = 1.00$ atm, $T = 288$ K

$$d = \frac{PM}{RT} = 1.23 \text{ g/L}$$

12.8. $P = 0.737$ atm, $V = 0.125$ L, $T = 296.2$ K

$$n = \frac{PV}{RT} = 3.79 \times 10^{-3} \text{ mol}$$

$$\text{Molar mass} = \frac{0.105 \text{ g}}{3.79 \times 10^{-3} \text{ mol}} = 27.7 \text{ g/mol}$$

12.9.

$$180 \text{ g}\left(\frac{1 \text{ mol } N_2H_4}{32.0 \text{ g}}\right) = 5.6 \text{ mol } N_2H_4$$

$$5.6 \text{ mol } N_2H_4\left(\frac{1 \text{ mol } O_2}{1 \text{ mol } N_2H_4}\right) = 5.6 \text{ mol } O_2$$

$$V(O_2) = \frac{nRT}{P} = 140 \text{ L when } n = 5.6 \text{ mol}, \ T = 294 \text{ K},$$

and $P = 0.99$ atm.

12.10. $P_{\text{halothane}} =$

$$\frac{(0.0760 \text{ mol})(0.082057 \text{ L} \cdot \text{atom/K} \cdot \text{mol})(298.2 \text{ K})}{5.00 \text{ L}}$$

$$P_{\text{halothane}} = 0.372 \text{ atm (or 283 mm Hg)}$$

$$P(O_2) = 3.59 \text{ atm (or 2730 mm Hg)}$$

$$P_{\text{total}} = 3.96 \text{ atm (or 3010 mm Hg)}$$

12.11. $P = 742$ mm Hg − vapor pressure H_2O = (742 − 22.4)mm Hg = 720. mm Hg = 0.947 atm

$$n = \frac{PV}{RT} = 0.0137 \text{ mol } N_2 = 0.383 \text{ g } N_2$$

(when $V = 0.352$ L and $T = 295$ K)

12.12. Using Equation 12.15 with $M = 4.00$ g/mol (or 4.00×10^{-3} kg/mol), $T = 298$ K, and $R = 8.314$ J/K · mol, one obtains a root mean square speed of 1360 m/s for He. In contrast, N_2 molecules have a much smaller rms speed (515 m/s) owing to their greater mass.

12.13. The molar mass of CH_4 is 16.0 g/mol. Therefore,

$$\frac{\text{Rate for } CH_4}{\text{Rate for unk}} = \frac{n \text{ molecules}/1.50 \text{ min}}{n \text{ molecules}/4.73 \text{ min}} = \sqrt{\frac{M_{\text{unk}}}{16.0}}$$

$$M_{\text{unk}} = 159 \text{ g/mol}$$

12.14. For $n = 10.0$ mol, $V = 1.00$ L, $T = 298$ K:

(a) Ideal gas law: $P = nRT/V = 245$ atm

(b) Van der Waals's equation (where $a = 0.034$ and $b = 0.0237$): $P = 320$ atm

CHAPTER 13

Chemical Puzzler: Ice is less dense than liquid water because of the open structure of ice (Figure 13.10), a result of hydrogen bonding. This means that ice cubes float and that a given mass of water occupies a larger volume when frozen than in the liquid state. This latter phenomenon means that expansion occurs when water freezes, allowing it to expand out of its container.

13.1. Ca^{2+} should have a more negative hydration energy because its radius is much less than that of Ba^{2+} (see Figure 8.15).

13.2. (a) O_2 interactions occur by induced dipole/induced dipole forces, the weakest of all intermolecular forces. (b) $MgSO_4$ consists of the ions Mg^{2+} and SO_4^{2-}, so ion-dipole forces are involved when the salt dissolves in water. The common, hydrated salt $MgSO_4 \cdot 7\ H_2O$ (epsomite) is widely used in agriculture and medicine. (c) Dipole-induced dipole forces exist between H_2O and O_2. Order of strength is $O_2—O_2 < O_2—H_2O < MgSO_4—H_2O$.

13.3. $$(1.00 \times 10^3\ \text{g})\left(\frac{1\ \text{mol}}{32.04\ \text{g}}\right)(35.21\ \text{kJ/mol}) = 1.10 \times 10^3\ \text{kJ}$$

13.4. 1. Vapor pressure of ethanol at 40 °C is about 120 mm Hg.
2. The equilibrium vapor pressure at 60 °C is about 370 mm Hg. The only way a pressure of 600 mm Hg can be observed is if the temperature of the alcohol is 75 °C.

13.5. $$P = \frac{nRT}{V}$$
$$= (0.028\ \text{mol})(0.0821\ \text{L} \cdot \text{atm/K} \cdot \text{mol})(333\ \text{K})/5.0\ \text{L}$$
$$= 0.15\ \text{atm} = 120\ \text{mm Hg}$$

Appendix E gives a vapor pressure of H_2O at 60 °C of 149 mm Hg. The calculated pressure of water in the flask is smaller than this, so all the water (0.50 g) evaporates. With 2.0 g, however, the calculated pressure (460 mm Hg) is much larger than the vapor pressure of water at 60 °C, so only enough water can evaporate to give an equilibrium pressure of 149 mm Hg.

13.6. Glycerol has three —OH groups per molecule that can be used in hydrogen bonding, as compared with only one for ethanol. The viscosity of glycerol should be greater than that of ethanol.

13.7. $$\text{Mass of unit cell} = \left(\frac{4\ \text{Au atoms}}{\text{unit cell}}\right)(\text{mass of one atom})$$
$$= \left(\frac{4\ \text{Au atoms}}{1\ \text{unit cell}}\right)\left(\frac{196.97\ \text{g}}{\text{mol}}\right)\left(\frac{1\ \text{mol}}{6.022 \times 10^{23}\ \text{atoms}}\right)$$
$$= 1.308 \times 10^{-21}\ \text{g/unit cell}$$

$$\text{Volume of unit cell} = (\text{mass of unit cell})(\text{density})$$
$$= \left(\frac{1.308 \times 10^{-21}\ \text{g}}{\text{unit cell}}\right)\left(\frac{1\ \text{cm}^3}{19.32\ \text{g}}\right)$$
$$= 6.772 \times 10^{-23}\ \text{cm}^3/\text{unit cell}$$

$$\text{Unit cell edge} = (\text{volume})^{1/3} = 4.076 \times 10^{-8}\ \text{cm}$$

$$\text{Atom radius} = \frac{1}{4}(\text{diagonal distance}) = \frac{1}{4}(\sqrt{2})(\text{edge})$$

$$\text{Atom radius} = 1.441 \times 10^{-8}\ \text{cm, or } 144.1\ \text{pm}$$

(Literature value for the gold radius is 144.2 pm.)

13.8. (8 corner Cl^- ions)(1/8 ion per corner) = 1 net Cl^- ion in the unit cell. Because there is 1 Cs^+ ion in the center of the unit cell, the formula of the salt must be CsCl.

13.9. $$\text{Cube edge} = 2(\text{radius of K}^+) + (\text{radius of Cl}^-)$$
$$= 2(152.0\ \text{pm}) + 2(167.0\ \text{pm})$$
$$= 638.0\ \text{pm}\ (=6.38 \times 10^{-8}\ \text{cm})$$

$$\text{Volume} = (\text{edge})^3 = (6.38 \times 10^{-8}\ \text{cm})^3 = 2.60 \times 10^{-22}\ \text{cm}^3$$

Mass of KCl unit cell =
$$(4\ \text{KCl/cell})(74.55\ \text{g/mol})(1\ \text{mol}/6.022 \times 10^{23}\ \text{KCl}) = 4.95 \times 10^{-22}\ \text{g/cell}$$

$$\text{Density} = (4.95 \times 10^{-22}\ \text{g/cell})/(2.60 \times 10^{-22}\ \text{cm}^3/\text{cell}) = 1.91\ \text{g/cm}^3$$

(Literature density of KCl = 1.99 g/cm³.)

13.10. **(a)** $$100.0\ \text{g H}_2\text{O}\left(\frac{1\ \text{mol}}{18.02\ \text{g}}\right)(6.02\ \text{kJ/mol}) = 33.4\ \text{kJ}$$

(b) $$100.0\ \text{g C}_8\text{H}_{18}\left(\frac{1\ \text{mol}}{114.23\ \text{g}}\right)(20.65\ \text{kJ/mol}) = 18.08\ \text{kJ}$$

The intermolecular forces are weaker in the nonpolar hydrocarbon octane, so solid octane requires less energy to melt than an equal mass of water.

CHAPTER 14

Chemical Puzzler: This is an illustration of osmosis (page 680). When an egg is placed in pure water, water enters the egg through the membrane, and the egg swells. When the egg is in a concentrated salt solution, water leaves the egg, and it shrivels.

14.1. 10.0 g sugar = 0.0292 mol and 250. g water = 13.9 mol

$$X_{\text{sugar}} = \frac{0.0292\ \text{mol sugar}}{0.0292\ \text{mol sugar} + 13.9\ \text{mol water}} = 0.00210$$

$$\frac{0.0292\ \text{mol sugar}}{0.250\ \text{kg}} = 0.117\ \text{molal}$$

$$\text{Percentage sugar} = \frac{10.0\ \text{g sugar}}{260.\ \text{g solution}} \times 100\%$$
$$= 3.85\%\ \text{sugar}$$

14.2. $$1.08 \times 10^4\ \text{ppm} = 1.08 \times 10^4\ \text{mg/L}$$
$$1.08 \times 10^4\ \text{mg/L}\ (1\ \text{g}/1000\ \text{mg}) = 10.8\ \text{g/L}$$

14.3. 1. For AgCl: Enthalpy of solution = 912 kJ/mol + (−851 kJ/mol) = +61 kJ/mol

For RbF: Enthalpy of solution = 789 kJ/mol + (−792 kJ/mol) = −3 kJ/mol
AgCl, an insoluble salt, has a large, positive enthalpy of solution, whereas the soluble salt RbF has a negative enthalpy of solution.

2. $$\Delta H^\circ_{solution}(NH_4NO_3) = \Delta H^\circ_f[NH_4NO_3(aq)] - \Delta H^\circ_f[NH_4NO_3(s)] = -339.9 \text{ kJ/mol} - (-365.6 \text{ kJ/mol}) = 25.7 \text{ kJ/mol}$$

14.4. $C(CO_2) = (4.48 \times 10^{-5}$ M/mm Hg)(253 mm Hg) $= 1.13 \times 10^{-2}$ M

14.5. Solution consists of sucrose (0.0292 mol) and water (12.5 mol).

$$X_{water} = \frac{12.5 \text{ mol } H_2O}{12.5 \text{ mol } H_2O + 0.0292 \text{ mol sucrose}} = 0.998$$

$$P_{water} = (0.998)(149.4 \text{ mm Hg}) = 149 \text{ mm Hg}$$

Even with 10.0 g of sugar, the vapor pressure of water has changed very little.

14.6. $$P_{solution} = X_{benzene}P^\circ_{benzene}$$

$$94.85 \text{ mm Hg} = X_{benzene}(95.00 \text{ mm Hg})$$

$$X_{benzene} = 0.9984$$

$$X_{benzene} = \frac{1.28 \text{ mol benzene}}{? \text{ mol nitro} + 1.28 \text{ mol benzene}}$$

$$? \text{ mol nitro} = 0.00202$$

$$\text{Molar mass nitroglycerin} = (0.454 \text{ g}/0.00202 \text{ mol}) = 225 \text{ g/mol}$$

(Nitroglycerin is $C_3H_5N_3O_9$ with a molar mass of 227 g/mol)

14.7. $$\text{Concentration } (m) = \Delta t/K = 1.0\ ^\circ\text{C}/(+0.512\ ^\circ\text{C}/m) = 2.0\ m$$

$$(2.0 \text{ mol/kg}) \cdot (0.100 \text{ kg}) = 0.20 \text{ mol}$$

$$0.20 \text{ kg } (62.07 \text{ g/mol}) = 12 \text{ g glycol}$$

14.8. $$\Delta t_{bp} = 80.23\ ^\circ\text{C} - 80.10\ ^\circ\text{C} = 0.13\ ^\circ\text{C}$$

$$\text{Concentration } (m) = \Delta t_{bp}/K_{bp} = 0.13\ ^\circ\text{C}/(2.53\ ^\circ\text{C/molal}) = 0.051\ m$$

$$0.051 \text{ mol/kg } (0.100 \text{ kg}) = 0.0051 \text{ mol}$$

$$0.640 \text{ g}/0.0051 \text{ mol} = 130 \text{ g/mol}$$

(Azulene has the formula $C_{10}H_8$ with a molar mass of 128.2 g/mol.)

14.9. Molality of HOC_2H_4OH (ethylene glycol) = 8.06 mol/3.00 kg = 2.69 m

$$\Delta t_{fp} = K_{fp}m = (-1.86 \text{ degree/molal})(2.69 \text{ molal}) = -4.99 \text{ degree}$$

500. g of glycol is not sufficient to keep the plumbing from freezing at −25 °C.

14.10. $$25.0 \text{ g of NaCl} = 0.428 \text{ mol}$$

$$\text{Concentration } (m) = \frac{0.428 \text{ mol}}{0.500 \text{ kg}} = 0.856\ m$$

$$\Delta t_{fp} = K_{fp} \cdot m \cdot i = (-1.86\ ^\circ\text{C}/m)(0.856\ m)(1.85) = -2.94\ ^\circ\text{C}$$

14.11. $$M = \frac{\Pi}{RT} = \frac{(364 \text{ mm Hg}/760. \text{ mm Hg atm}^{-1})}{(0.0821 \text{ L} \cdot \text{atm/K} \cdot \text{mol})(298 \text{ k})} = 1.96 \times 10^{-2} \text{ mol/L}$$

$$(1.96 \times 10^{-2} \text{ mol/L})(0.0250 \text{ L}) = 4.89 \times 10^{-4} \text{ mol}$$

$$\text{Molar mass} = \frac{(144 \times 10^{-3})}{(4.89 \times 10^{-4} \text{ mol})} = 294 \text{ g/mol}$$

The actual molar mass of aspartame ($C_{14}H_{18}N_2O_5$) is 294.3 g/mol.

CHAPTER 15

Chemical Puzzler: The catalyst in the converter is involved in the chemical reactions in the converter, but it is not consumed by those reactions. This is a property of all catalysts, as described in Section 15.7. See page 737 in particular.

15.1. $$-\left(\frac{1}{2}\frac{\Delta[NOCl]}{\Delta T}\right) = \frac{1}{2}\left(\frac{\Delta[NO]}{\Delta t}\right) = \frac{\Delta[Cl_2]}{\Delta t}$$

15.2. For the first 2 hours:

$$-\frac{\Delta\text{Sucrose}}{\Delta t} = \frac{(0.034 - 0.050)\text{mol/L}}{2 \text{ h}} = \frac{-0.0080 \text{ mol/L}}{\text{h}}$$

For the last 2 hours:

$$-\frac{\Delta\text{Sucrose}}{\Delta t} = \frac{(0.010 - 0.015)\text{mol/L}}{2 \text{ h}} = \frac{-0.0025 \text{ mol/L}}{\text{h}}$$

Notice that the rate is much slower over the last 2 h; less reactant is available, so the rate of reaction is smaller.

$$\text{Instantaneous rate at 4 h} = 0.0045 \text{ mol/L} \cdot \text{h}$$

Notice that this rate is intermediate between the rate over the first 2 h and that over the last 2 h.

15.3. 1. Second order with respect to NO and first order with respect to H_2.
2. Increases by a factor of 4.
3. The reaction rate is halved.

15.4. Rate of reaction = (0.090/h)(0.020 mol/L) = 0.0018 mol/L · h
Cl^- appears at a rate of 0.0018 mol/L · h.

15.5. **(a)** $m = 1$. The initial rates and initial concentrations are directly proportional. For example, as the concentration is doubled, the rate is doubled.
(b) Taking the data from Experiment 1, we have

$$k = \frac{\text{Rate}}{[\text{Reactant}]} = \frac{1.3 \times 10^{-7}\ \text{mol/(L} \cdot \text{min)}}{1.0 \times 10^{-3}\ \text{mol/L}}$$
$$= 1.3 \times 10^{-4}/\text{min}$$

15.6.
$$\ln\left(\frac{[\text{sucrose}]}{[\text{sucrose}]_0}\right) = -kt$$

$$\ln\left(\frac{[\text{sucrose}]}{(0.010)}\right) = -(0.21/\text{h})(5.00\ \text{h}) = -1.1\ \text{h}$$

$$\ln[\text{sucrose}] - \ln(0.010) = -1.1\ \text{h}$$
$$\ln[\text{sucrose}] - (-4.61) = -1.1$$
$$\ln[\text{sucrose}] = -5.7$$
$$[\text{sucrose}]\ \text{after}\ 5.00\ \text{h} = 0.003$$

15.7.
$$\left(\frac{1}{[\text{HI}]}\right) - \left(\frac{1}{[\text{HI}]_0}\right) = kt$$
When $[\text{HI}]_0 = 0.010$ M, $k = 30.$L/mol · min), and $t =$ 10. min, [HI] = 0.0025 M

15.8. Concentration versus time

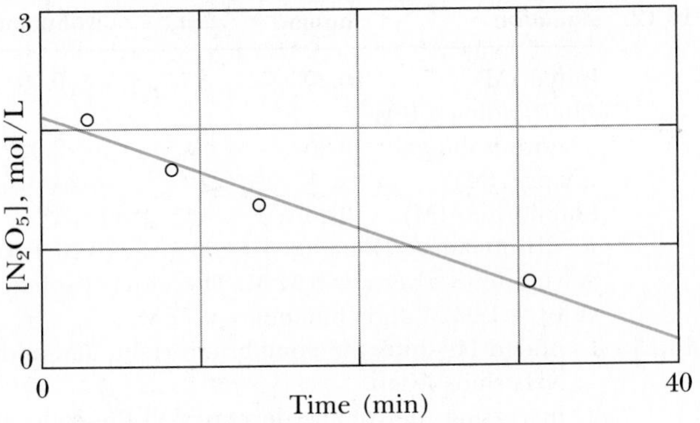

ln $[N_2O_5]$ versus time

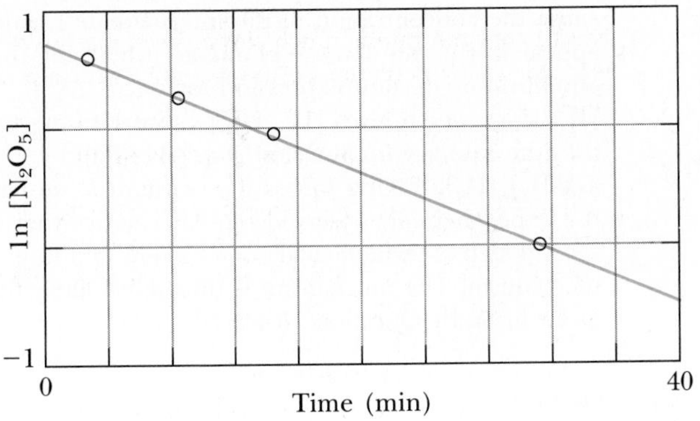

$1/[N_2O_5]$ versus time

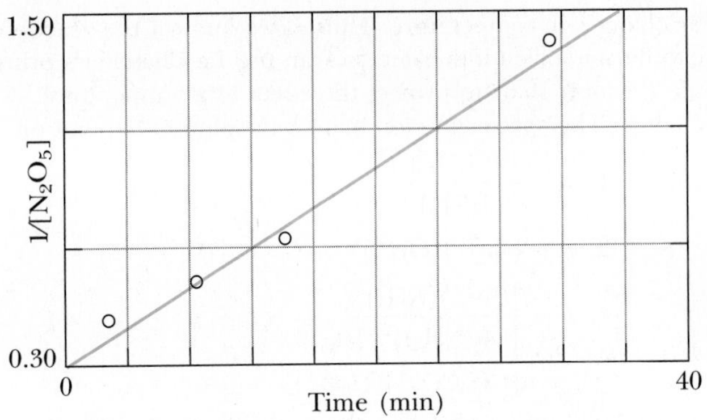

Only the plot of $\ln[N_2O_5]$ versus time is linear; the reaction is first order in N_2O_5. The slope of the line is −0.038 for $k = 0.038$/min.

15.9. **(a)** $t_{1/2} = \dfrac{0.693}{5.40 \times 10^{-2}/\text{h}} = 12.8\ \text{h}$
(b) 51.2 h = 4.00 half-lives. After 4.00 half-lives the fraction remaining is 1/16.
(c) $\ln(\text{Fraction remaining}) = -kt = -(5.40 \times 10^{-2}/\text{h})(18\ \text{h}) = -0.97.$

$$\text{Fraction remaining} = \frac{[\text{A}]}{[\text{A}]_0} = 0.38$$

15.10. Substituting the given values into Equation 15.7, we have

$$\ln\frac{1.00 \times 10^4}{4.5 \times 10^3} = -\frac{E_a}{8.31 \times 10^{-3}\ \text{kJ/K} \cdot \text{mol}}\left[\frac{1}{283} - \frac{1}{274}\right]$$

$$E_a = 57\ \text{kJ}$$

15.11. All three steps are bimolecular. N_2O_2, the product of the first step, is used in the second step, and N_2O, a product of the second step, is consumed in the third step. Therefore, adding the three reactions gives the equation for the overall process.

15.12. 1. $2\ NH_3(aq) + OCl^-(aq) \longrightarrow N_2H_4(aq) + Cl^-(aq) + H_2O(\ell)$
2. Step 2 is rate-determining.
3. Rate = $k[NH_2Cl][NH_3]$
4. NH_2Cl, $N_2H_5^+$, OH^-

15.13. 1. True
2. A catalyst is never consumed in the overall reaction.
3. A catalyst may be in the same phase or in a different phase than the reactants.

CHAPTER 16

Chemical Puzzler: This is an example of a chemical equilibrium, a reversible chemical process. The reactants are converted to products at the same time that products are converted back to reactants. All chemical equilibria are affected by changes in temperature. Thus, if we know if the reaction is exothermic or endothermic, we can use Le Chatelier's principle (Section 16.6) to predict the effect of a temperature change. The species predominating in solution depend on T.

16.1. 1. $K = \frac{[PCl_3][Cl_2]}{[PCl_5]}$

2. $K = [Cu^{2+}][OH^-]^2$

3. $K = \frac{[Cu^{2+}][NH_3]^4}{[Cu(NH_3)_4{}^{2+}]}$

4. $K = \frac{[CH_3CO_2{}^-][H_3O^+]}{[CH_3CO_2H]}$

16.2. 1. $K_{new} = [K_{old}]^2 = 6.3 \times 10^{-58}$

2. $K = \frac{1}{(6.3 \times 10^{-58})} = 1.6 \times 10^{57}$

16.3. $K_p = \left[(7.9 \times 10^{11})\left(\frac{1}{4.8 \times 10^{-41}}\right)(1/2.2 \times 10^{-15})\right]^{1/2}$

$= \sqrt{7.5 \times 10^{66}} = 2.7 \times 10^{33}$

16.4. Concentration of Ag^+ in the AgCl beaker (1.3×10^{-5} M) is greater than in the AgI beaker (1.2×10^{-8}). Both are reactant-favored.

16.5. **(a)** $Q = \frac{[\text{isobutane}]}{[\text{butane}]} = \frac{2.18}{0.97} = 2.3$

$Q < K$ (=2.5), so the reaction is not at equilibrium; the reaction produces more isobutane on proceeding to equilibrium.

(b) $Q = \frac{[\text{isobutane}]}{[\text{butane}]} = \frac{2.60}{0.75} = 3.5$

$Q > K$ (=2.5), so the reaction is not at equilibrium; the reaction consumes isobutane on proceeding to equilibrium.

16.6. $Q = \frac{[NO]^2}{[N_2][O_2]} = \frac{(4.2 \times 10^{-3})^2}{(0.50)(0.25)} = 1.4 \times 10^{-4}$

$Q < K$, so the reaction is not at equilibrium. The reaction consumes N_2 and O_2 and produces NO on proceeding to equilibrium.

16.7.

Equation	H_2	+ I_2	$\rightleftharpoons$ 2 HI
Initial (mol)	9.838×10^{-4}	13.77×10^{-4}	0
Change (mol)	-9.05×10^{-4}	-9.05×10^{-4}	$+18.1 \times 10^{-4}$
Equilibrium (mol)	7.93×10^{-5}	4.725×10^{-4}	18.1×10^{-4}
Equilibrium (M)	0.00793	0.04725	0.181

K(for volume of 10.0 mL) = 87.4

K(for volume of 20.0 mL) = 87.4

Notice that the volume of the container does not affect the value of K in this case.

16.8.

Equation	$C_6H_{10}I_2$	$\rightleftharpoons$ C_6H_{10}	+ I_2
Initial (M)	0.050	0	0
Change (M)	−0.035	+0.035	+0.035
Equilibrium (M)	0.015	0.035	0.035

$K = 0.082$

16.9.

Equation	H_2	+ I_2	$\rightleftharpoons$ 2 HI
Initial (M)	6.00×10^{-3}	6.00×10^{-3}	0
Change (M)	$-x$	$-x$	$+2x$
Equilibrium (M)	$0.00600 - x$	$0.00600 - x$	$+2x$

$K_c = 33 = \frac{(2x)^2}{(0.00600 - x)^2}$

$x = 0.0045$ M, so $[H_2] = [I_2] = 0.0015$ M and [HI] = 0.0090 M

16.10.

Equation	**C(s) +**	CO_2	$\rightleftharpoons$ **2 CO**
Initial (M)		0.012	0
Change (M)		$-x$	$+x$
Equilibrium (M)		$0.012 - x$	$2x$

$K_c = 0.012 = \frac{(2x)^2}{(0.012 - x)}$

$x = [CO_2] = 0.0057$ M and $2x = [CO] = 0.011$ M

16.11. 1. The reaction is endothermic. As the temperature increases, the equilibrium shifts to the right, and [NOCl] decreases.

2. The reaction is exothermic. As the temperature increases, the equilibrium shifts to the left, and $[SO_3]$ decreases.

16.12.

Equation	**butane**	$\rightleftharpoons$ **isobutane**
Initial (M)	0.20	0.50
After adding 2.0 M more isobutane	0.20	2.0 + 0.50
Change (M)	$+x$	$-x$
Equilibrium (M)	$0.20 + x$	$2.50 - x$

$K = [\text{isobutane}]/[\text{butane}] = (2.50 - x)/(0.20 + x)$. Solving for x gives $x = 0.57$ M. Therefore, [isobutane] = 1.93 M and [butane] = 0.77 M.

16.13. 1. Added H_2 shifts the equilibrium right, and added NH_3 shifts it left.

2. Increasing the volume decreases all the concentrations. The equilibrium shifts to the left, toward the side with the greater number of molecules.

16.14. Rate law for the slow step is Rate $= k_2[I]^2[H_2]$. Because the concentration of the intermediate I cannot appear in the rate law, we eliminate it by using the equilibrium constant expression for step 1 ($K = [I]^2/[I_2]$), which gives $[I]^2 = K[I_2]$. Substituting this into the rate law for the slow step gives Rate $= k_2\{K[I_2]\}[H_2]$. Taking k_2K as the constant k, we have the experimentally observed rate law. Notice that the second step is termolecular, not a likely step in a mechanism. The mechanism is thus more likely that given in Study Question 16.44.

CHAPTER 17

Chemical Puzzler: Soft drinks are usually acidic, aspirin is a weak acid and so is the citric acid found in citrus fruits such as lemons. Dish detergents are basic, and oven cleaner is often quite basic. Fertilizers contain basic salts, such as phosphates and acidic metal cations.

17.1. 1. $H_3PO_4(aq) + H_2O(\ell) \longrightarrow H_3O^+(aq) + H_2PO_4^-(aq)$
2. $H_2O(\ell) + CN^-(aq) \longrightarrow OH^-(aq) + HCN(aq)$; CN^- is a Brønsted base.
3. $H_2C_2O_4(aq) + H_2O(\ell) \longrightarrow H_3O^+(aq) + HC_2O_4^-(aq)$
$HC_2O_4^-(aq) + H_2O(\ell) \longrightarrow H_3O^+(aq) + C_2O_4^{2-}(aq)$

17.2. 1. HBr is an acid, and Br^- is its conjugate base. NH_3 is a base, and its conjugate acid is NH_4^+.
2. The conjugate base of H_2S is HS^-, and that for NH_4^+ is NH_3.
3. The conjugate acid for NO_3^- is HNO_3, and that for HPO_4^{2-} is $H_2PO_4^-$.

17.3. 1. The NH_4^+ ion is a stronger Brønsted acid than HCO_3^-. This means that HCO_3^- has the stronger conjugate base.
2. The CN^- ion is a stronger Brønsted base than SO_4^{2-}.

17.4. 1. HSO_4^- is a stronger acid than NH_4^+, so the equilibrium lies predominantly to the right. (We arrive at the same answer by knowing that NH_3 is a stronger base than SO_4^{2-}.
2. H_2S is a stronger acid than HCO_3^-, so the equilibrium lies predominantly to the left. (We arrive at the same answer by knowing that CO_3^{2-} is a stronger base than HS^-.)

17.5. The NH_4^+ ion is an acid, and NH_3 is its conjugate base. The SO_4^{2-} ion is a base, and HSO_4^- is its conjugate acid. Thus, the net ionic net equation is

$$NH_4^+(aq) + SO_4^{2-}(aq) \rightleftharpoons NH_3(aq) + HSO_4^-(aq)$$

The HSO_4^- ion is a stronger acid than NH_4^+, and NH_3 is a stronger base than SO_4^{2-}. Therefore, the equilibrium lies predominantly to the left.

17.6. 1. A K_a of 1.4×10^{-4} places lactic acid between formic acid and benzoic acid in strength.
2. $CH_3CHOHCO_2^-(aq) + H_2O(\ell) \rightleftharpoons CH_3CHOHCO_2H(aq) + OH^-(aq)$

17.7. $[H_3O^+] = 0.0020 \text{ mol}/0.50 \text{ L} = 0.0040 \text{ M}$

$$[OH^-] = \frac{1.0 \times 10^{-14}}{[H_3O^+]} = 2.5 \times 10^{-12} \text{ M}$$

17.8.
$$K_b = \frac{K_w}{K_a} = \frac{1.0 \times 10^{-14}}{1.4 \times 10^{-4}} = 7.1 \times 10^{-11}$$

This places the lactate ion between formate ion and benzoate in base strength.

17.9. $\text{pH} = 4.32$: $[H_3O^+] = 10^{-\text{pH}} = 10^{-4.32} = 4.8 \times 10^{-5} \text{ M}$

$\text{pOH} = 14.00 - 4.32 = 9.68$; $[OH^-] = 10^{-\text{pOH}} = 10^{-9.68} = 2.1 \times 10^{-10} \text{ M}$

17.10. $[H_3O^+] = 10^{-\text{pH}} = 10^{-2.94} = 1.1 \times 10^{-3} \text{ M}$

$$K_a = \frac{[H_3O^+][CH_3CH_2CO_2^-]}{[CH_3CH_2CO_2H]} = \frac{(1.1 \times 10^{-3})(1.1 \times 10^{-3})}{0.10 - 1.1 \times 10^{-3}}$$

$K_a = 1.3 \times 10^{-5}$

17.11. $K_a = 1.8 \times 10^{-5} = \dfrac{[H_3O^+][CH_3CO_2^-]}{[CH_3CO_2H]} = \dfrac{(x)(x)}{0.10 - x}$

$x = [H_3O^+] = 1.3 \times 10^{-3} \text{ M}$; $\text{pH} = 2.87$

17.12. $K_a = 7.2 \times 10^{-4} = \dfrac{[H_3O^+][F^-]}{[HF]} = \dfrac{(x)(x)}{0.015 - x}$

x is found by solving the quadratic equation or by using the method of successive approximations (Appendix A).

$x = [H_3O^+] = 2.9 \times 10^{-3} \text{ M}$; $\text{pH} = 2.53$

$[F^-] = 2.9 \times 10^{-3} \text{ M}$; $[HF] = 0.12 \text{ M}$

17.13. $NH_3(aq) + H_2O(\ell) \rightleftharpoons NH_4^+(aq) + OH^-(aq)$

$$K_b = 1.8 \times 10^{-5} = \frac{[NH_4^+][OH^-]}{[NH_3]} = \frac{(x)(x)}{0.025 - x}$$

$x = [OH^-] = 6.7 \times 10^{-4} \text{ M}$;
$\text{pOH} = 3.17$ and $\text{pH} = 10.83$

17.14. $K_a = 5.6 \times 10^{-10} = \dfrac{[NH_3][H_3O^+]}{[NH_4^+]} = \dfrac{(x)(x)}{0.50 - x}$

$x = [H_3O^+] = 1.7 \times 10^{-5} \text{ M}$; $\text{pH} = 4.78$

17.15. **(a)** NaCl is neutral, and the pH is equal to 7.
(b) $FeCl_3$ is acidic, and the pH is less than 7.
(c) NH_4NO_3 is acidic, and the pH is less than 7.
(d) Na_2HPO_4 is basic, and the pH is greater than 7.

17.16. $H_2C_2O_4(aq) + H_2O(\ell) \longrightarrow H_3O^+(aq) + HC_2O_4^-(aq)$
(The K_a values are found in Appendix F)

$$K_{a1} = 5.9 \times 10^{-2} = \frac{[H_3O^+][HC_2O_4^-]}{[H_2C_2O_4]} = \frac{(x)(x)}{0.10 - x}$$

x is found by solving the quadratic equation or by using the method of successive approximations (Appendix A).

$x = [H_3O^+] = [HC_2O_4^-] = 5.3 \times 10^{-2} \text{ M}$; $\text{pH} = 1.28$

$[C_2O_4^{2-}] = K_{a2} = 6.4 \times 10^{-5} \text{ M}$

17.17. 1. H_2SO_4(strong acid) $> H_2SO_3(K_{a1} = 1.7 \times 10^{-2})$
2. $H_3AsO_4(K_{a1} = 2.5 \times 10^{-4}) > H_3AsO_3(K_{a1} = 6.0 \times 10^{-10})$
In both cases the stronger acid has more O atoms attached to the central atom (S or As).

17.18. 1. PH_3 is a Lewis base. There is a lone pair of electrons on the P atom.
2. BCl_3 is a Lewis acid. The central B atom is surrounded by only three bond pairs of electrons (and no lone pairs). Therefore, it can accept a pair of electrons from an electron-pair donor.

3. H_2S is a Lewis base as there are two lone pairs of electrons on the S atom.
4. HS^- is analogous with OH^-, an excellent Lewis base.

CHAPTER 18

Chemical Puzzler: If your blood pH is too high, it can be lowered by increasing the blood CO_2 concentration. This leads to additional hydronium ion in the blood, thus lowering the pH.

$$CO_2(g) + H_2O(\ell) \rightleftharpoons H_2CO_3(aq)$$

$$H_2CO_3(aq) + H_2O(\ell) \rightleftharpoons H_3O^+(aq) + HCO_3^-(aq)$$

By breathing into a paper bag, the CO_2 is your exhaled breath is captured and you breathe air that is higher in CO_2 concentration.

18.1. Moles of HCl = (0.050 L)(0.20 M) = 0.010 mol

$$\text{Moles of aniline} = (0.93\text{ g})\left(\frac{1\text{ mol}}{93.1\text{ g}}\right) = 0.010\text{ mol}$$

Because 1 mol of aniline requires 1 mol of HCl, and because they were present initially in equal molar quantities, the reaction completely consumes the acid and base. The solution after reaction contains 0.010 mol of $C_6H_5NH_3^+$ in 0.050 L of solution, so its concentration is 0.20 M. It is the conjugate acid of the weak base aniline, so

$$C_6H_5NH_3^+(aq) + H_2O(\ell) \rightleftharpoons C_6H_5NH_2(aq) + H_3O^+(aq)$$

$$K_a = 2.4 \times 10^{-5} = \frac{[H_3O^+][C_6H_5NH_2]}{[C_6H_5NH_3^+]} = \frac{x^2}{0.20}$$

$x = [H_3O^+] = 2.2 \times 10^{-3}$ M, which gives a pH of 2.66.

18.2. $$(0.976\text{ g})\left(\frac{1\text{ mol}}{122.1\text{ g}}\right) = 0.00799\text{ mol benzoic acid}$$

Moles of NaOH required = 0.00799 mol

$$(0.00799\text{ mol NaOH})\left(\frac{1\text{ L}}{0.100\text{ mol}}\right) = 0.0799\text{ L or }79.9\text{ mL NaOH}$$

Reaction gives 0.00799 mol of benzoate ion, $C_6H_5CO_2^-$.

$$\text{Concentration benzoate ion} = \frac{0.00799\text{ mol}}{0.0799\text{ L}} = 0.100\text{ M}$$

Hydrolysis of benzoate ion:

$$C_6H_5CO_2^-(aq) + H_2O(\ell) \rightleftharpoons C_6H_5CO_2H(aq) + OH^-(aq)$$

$$K_b = 1.6 \times 10^{-10} = \frac{[C_6H_5CO_2H][OH^-]}{[C_6H_5CO_2^-]}$$

$[OH^-] = 4.0 \times 10^{-6}$ M;

pOH = 5.40 and pH = 8.60

18.3. Equal numbers of moles of the acid and base were mixed, so reaction was complete to produce $CH_3CO_2^-$ (a weak base) and $C_5H_5NH^+$ (a weak acid).

K_b for $CH_3CO_2^- = 5.6 \times 10^{-10}$ and

K_a for $C_5H_5NH^+ = 6.7 \times 10^{-6}$

The acid is stronger than the base here, so the solution containing the two ions is slightly acidic.

18.4. **(a)** pH of 0.30 M formic acid

$$K_a = 1.8 \times 10^{-4} = \frac{[H_3O^+][HCO_2^-]}{[HCO_2H]} = \frac{x^2}{0.30}$$

$x = [H_3O^+] = 7.3 \times 10^{-3}$ M, which gives a pH of 2.14.

(b) pH of 0.30 M formic acid + 0.010 M $NaHCO_2$

Equation	HCO_2H	$\rightleftharpoons$ H_3O^+	+ HCO_2^-
Initial concentrations (M)	0.30	0	0.10
Change (M)	$-x$	$+x$	$+x$
At equilibrium (M)	$0.30 - x$	x	$0.10 + x$

$$K_a = 1.8 \times 10^{-4} = \frac{[H_3O^+][HCO_2^-]}{[HCO_2H]} = \frac{(x)(0.10 + x)}{0.30 - x}$$

Assuming that x is small compared with 0.10 or 0.30, we find that $x = [H_3O^+] = 5.4 \times 10^{-4}$ M, which gives a pH of 3.27.

18.5. **(a)** Find pH of buffer before adding HCl.

$$[H_3O^+]\text{ before adding HCl} = \frac{[\text{acid}]}{[\text{conjugate base}]}K_a = \frac{0.50}{0.70}(1.8 \times 10^{-4})$$

$[H_3O^+] = 1.3 \times 10^{-4}$, which gives a pH of 3.89.

(b) Add 10. mL of HCl(=0.010 mol). The reaction occurring is

$$H_3O^+(aq) + HCO_2^-(aq) \rightleftharpoons HCO_2H(aq)$$

The molar quantities and concentrations at this stage are as follows:

Equation	$[H_3O^+]$ from HCl	$[HCO_2^-]$ from buffer	$[HCO_2H]$ from buffer
Before reaction (mol)	0.010	0.35	0.25
Change when HCl reaction occurs (mol)	−0.010	−0.010	+0.010
After HCl reaction (mol)	0	0.34	0.26
After reaction (mol/L in 0.510 L solution)	0	0.67	0.51

To find $[H_3O^+]$ at this stage, we use the chemical equation for the ionization of formic acid.

$$HCO_2\ (aq) + H_2O(\ell) \rightleftharpoons HCO_2^-(aq) + H_3O^+(aq)$$

$[H_3O^+]$ after adding excess

$$\text{HCl} = \frac{[\text{acid}]}{[\text{conjugate base}]}K_a = \frac{0.51}{0.67}(1.8 \times 10^{-4}) = 1.4 \times 10^{-4}\text{ M}$$

From $[H_3O^+] = 1.4 \times 10^{-4}$ M the pH is 3.86. Only a small change in pH (3.89 → 3.86) occurred on adding a concentrated acid.

18.6. K_a for $HCO_3^- = 4.8 \times 10^{-11}$, so $pK_a = 10.32$

$$15.0 \text{ g } NaHCO_3 = 0.179 \text{ mol and } 18.0 \text{ g } Na_2CO_3 = 0.170 \text{ mol}$$

$$pH = 10.32 + \log\left(\frac{0.170 \text{ mol}}{0.179 \text{ mol}}\right) = 10.30$$

18.7. A pH of 5.00 corresponds to $[H_3O^+] = 1.0 \times 10^{-5}$ M. This means that

$$[H_3O^+] = 1.0 \times 10^{-5} \text{ M} = \frac{[CH_3CO_2H]}{[CH_3CO_2^-]}(1.8 \times 10^{-5})$$

The ratio $[CH_3CO_2H]/[CH_3CO_2^-]$ must be 1/1.8 to achieve the correct hydronium ion concentration. Therefore, 1.0 mol of CH_3CO_2H is mixed with 1.8 mol of a salt of $CH_3CO_2^-$ (say $NaCH_3CO_2$) in some amount of water. (The volume of water is not critical; only the *relative* amounts of acid and conjugate base are important.)

18.8. **(a)** Using the expression in Example 18.7 to find $[H_3O^+]$ before the equivalence point, we have

$$[H_3O^+] = \frac{0.00500 \text{ mol HCl} - (0.0400 \text{ L NaOH})(0.100 \text{ M})}{0.0500 \text{ L HCl} + 0.0400 \text{ L NaOH}} = 0.0111 \text{ M}$$

$$pH = 1.954$$

(b) After 60.0 mL of base has been added, the HCl has been completely consumed, and we have added 10.0 mL of base in excess of the equivalence point.

$$[OH^-] = \frac{\text{moles excess base}}{\text{total volume}} = \frac{(0.0100 \text{ L})(0.100 \text{ M})}{0.050 \text{ L acid} + 0.060 \text{ L NaOH}} = 9.1 \times 10^{-3} \text{ M}$$

pOH = 2.04, which gives a pH of 11.96

(c) When 49.9 mL of NaOH has been added, we are still 0.1 mL short of the equivalence point. Therefore, $[H_3O^+]$ is calculated as in part (a). This gives $[H_3O^+] = 1.00 \times 10^{-4}$ M, or a pH of 4.000. Even this close to the equivalence point the solution is still relatively acidic.

18.9. The equation for the reaction occurring here is

$$NaOH(aq) + CH_3CO_2H(aq) \rightleftharpoons NaCH_3CO_2(aq) + H_2O(\ell)$$

$$(0.0350 \text{ L NaOH})(0.100 \text{ M}) = 0.00350 \text{ mol NaOH}$$

This leads to 0.00350 mol of $NaCH_3CO_2$.

Acid remaining = 0.0100 mol acid − 0.00350 mol NaOH = 0.00650 mol

$$[H_3O^+] = \left(\frac{\text{acid remaining}}{\text{conjugate base formed}}\right) K_a$$

$$[H_3O^+] = \left(\frac{0.0065 \text{ mol}}{0.00350}\right)(1.8 \times 10^{-5}) = 3.3 \times 10^{-5} \text{ M}$$

$$pH = 4.48$$

18.10. $(0.100 \text{ L})(0.100 \text{ M } NH_3) = 0.0100 \text{ mol } NH_3$
At the equivalence point, the reaction has produced 0.0100 mol NH_4^+. The titration required 100.0 mL of 0.100 M HCl, so the total solution volume at the equivalence point is 200. mL. Thus, $[NH_4^+] = 0.0500$ M.

$$K_a \text{ for } NH_4^+ = 5.6 \times 10^{-10} = \frac{[H_3O^+]^2}{[NH_4^+]}$$

When $[NH_4^+] = 0.0500$ M, $[H_3O^+] = 5.3 \times 10^{-6}$ M and pH = 5.28.

18.11. The equivalence point occurs at a pH of about 5.2 to 5.3. A suitable indicator might be methyl orange, which is yellow in a basic solution but red to red-orange in an acidic solution (Figure 17.7). The color change occurs around pH 4.5–5.0. Other choices of indicator include bromcresol green, bromphenol blue, and methyl red.

CHAPTER 19

Chemical Puzzler: The chemistry of caves is described on pages 750–751 and 906–908 as well as in Figures 16.1, 16.2, and 19.1.

19.1. 1. $BaSO_4(s) \rightleftharpoons Ba^{2+}(aq) + SO_4^{2-}(aq)$
$K_{sp} = [Ba^{2+}][SO_4^{2-}] = 1.1 \times 10^{-10}$
2. $BI_3(s) \rightleftharpoons Bi^{3+}(aq) + 3\ I^-(aq)$
$K_{sp} = [Bi^{3+}][I^-]^3 = 8.1 \times 10^{-19}$
3. $Ag_2CO_3(s) \rightleftharpoons 2\ Ag^+(aq) + CO_3^{2-}(aq)$
$K_{sp} = [Ag^+]^2[CO_3^{2-}] = 8.1 \times 10^{-12}$

19.2. $[Ba^{2+}] = 7.5 \times 10^{-3}$ M and $[F^-] = 2 \times [Ba^{2+}] = 1.5 \times 10^{-2}$ M

$$K_{sp} = [Ba^{2+}][F^-]^2 = (7.5 \times 10^{-3})(1.5 \times 10^{-2})^2 = 1.7 \times 10^{-6}$$

19.3. K_{sp} for CuI $= 5.1 \times 10^{-12} = [Cu^+][I^-] = (x)(x)$

$$x = (5.1 \times 10^{-12})^{1/2} = 2.3 \times 10^{-6} \text{ M}$$

$$K_{sp} \text{ for } Mg(OH)_2 = 1.5 \times 10^{-11} = [Mg^{2+}][OH^-]^2 = (x)(2x)^2$$

$$x = 1.6 \times 10^{-4} \text{ M}$$

19.4. 1. Solubility of AgCl ($K_{sp} = 1.8 \times 10^{-10}$) is greater than the solubility of AgCN ($K_{sp} = 1.2 \times 10^{-16}$).
2. Solubility of $Ca(OH)_2$ ($K_{sp} = 7.9 \times 10^{-6}$) is greater than the solubility of $Mg(OH)_2$ ($K_{sp} = 1.5 \times 10^{-11}$).
3. Solubility of $MgCO_3$ ($K_{sp} = 4.0 \times 10^{-5}$) is greater than the solubility of $CaCO_3$ ($K_{sp} = 3.8 \times 10^{-9}$).

19.5. If $[Pb^{2+}] = 1.1 \times 10^{-3}$ M, then $[I^-]$

$$= 2\ (1.1 \times 10^{-3}\ M) = 2.2 \times 10^{-3}\ M$$

$$Q = [Pb^{2+}][I^-]^2 = (1.1 \times 10^{-3}\ M)(2.2 \times 10^{-3}\ M)^2 = 5.3 \times 10^{-9}$$

The value of Q is less than K_{sp} (8.7×10^{-9}), so PbI_2 can dissolve to a greater extent.

19.6. $Q = [Sr^{2+}][SO_4^{2-}] = (2.5 \times 10^{-4}\ M)(2.5 \times 10^{-4}\ M) = 6.3 \times 10^{-8}$

Q is less than K_{sp}, so no precipitation occurs.

19.7.

$$[I^-] = \left\{\frac{K_{sp}}{[Pb^{2+}]}\right\}^{1/2} = \left[\frac{(8.7 \times 10^{-9})}{(0.050\ M)}\right]^{1/2} = 4.2 \times 10^{-4}\ M$$

The I^- concentration needed to precipitate the lead(II) ion is 4.2×10^{-4} M. The lead(II) ion concentration remaining in solution when I^- reaches 0.0015 M.

$$[Pb^{2+}] = \frac{K_{sp}}{[I^-]^2} = \frac{(8.7 \times 10^{-9})}{(0.0015)^2} = 3.9 \times 10^{-3}\ M$$

19.8.

$$[Ag^+] \text{ after mixing} = \frac{(0.0010\ M)(0.1000\ L)}{(0.1050\ L)} = 9.5 \times 10^{-4}\ M$$

$$[Cl^-] \text{ after mixing} = \frac{(0.025\ M)(0.0050\ L)}{(0.1050\ L)} = 1.2 \times 10^{-3}\ M$$

$$Q = [Ag^+][Cl^-] = (9.5 \times 10^{-4}\ M)(1.2 \times 10^{-3}\ M) = 1.1 \times 10^{-6}$$

Q is greater than K_{sp} (1.8×10^{-10}), so precipitation occurs.

19.9. Solubility in pure water: $[Ba^{2+}] = [SO_4^{2-}] = (K_{sp})^{1/2} = 1.0 \times 10^{-5}$ M
Solubility with added Ba^{2+} ion: $[SO_4^{2-}] =$

$$\frac{K_{sp}}{(x + 0.010\ M)} \approx \frac{K_{sp}}{(0.010\ M)}$$

Solubility $= 1.1 \times 10^{-8}$ M

19.10. Solubility in pure water: $[Zn^{2+}] = x$. Because $[CN^-] = 2x$, we have $K_{sp} = 8.0 \times 10^{-12} = (x)(2x)^2$ and $x = 1.3 \times 10^{-4}$ M.

Solubility in the presence of 0.10 M CN^-: $[Zn^{2+}] = x$. Because $[CN^-] = 2x + 0.10$, we have $K_{sp} = 8.0 \times 10^{-12} = (x)(2x + 0.10)^2 \approx (x)(0.10)^2$. Solving, $x = 8 \times 10^{-10}$ M.

19.11. 1. $[Cl^-]$ needed to precipitate AgCl

$$= \frac{K_{sp}}{[Ag^+]} = \frac{(1.8 \times 10^{-10})}{(0.0010\ M)} = 1.8 \times 10^{-7}\ M$$

$$[Cl^-] \text{ needed to precipitate } PbCl_2 = \left\{\frac{K_{sp}}{[Pb^{2+}]}\right\}^{1/2} = \left[\frac{(1.7 \times 10^{-5})}{(0.0010\ M)}\right]^{1/2} = 0.13\ M$$

Silver chloride precipitates before $PbCl_2$.

2. When $[Cl^-] = 0.13$ M, $[Ag^+] = K_{sp}/(0.13) = 1.4 \times 10^{-9}$ M

19.12. 1. Can be separated by adding Cl^- to give AgCl(s) and leave Bi^{3+} in solution.
2. Can be separated by adding S^{2-} or OH^- to give FeS(s) or $Fe(OH)_2$(s) and leave K^+ in solution because both K_2S and KOH are water-soluble.

19.13. K for the overall reaction

$$= K_{sp} \text{ for AgCl} \cdot \left(\frac{1}{K_{sp}} \text{ for AgBr}\right) = \frac{1.8 \times 10^{-10}}{3.3 \times 10^{-13}} = 550$$

AgCl can be converted to AgBr.

19.14. 0.010 mol AgCl gives 0.010 M $Ag(NH_3)_2^+$ and 0.010 M Cl^- when dissolved completely in NH_3. Solving for $[NH_3]$ as in Example 19.12 gives $[NH_3] = 0.19$ M. The quantity of NH_3 available $= (0.100\ L)(4.0\ M) = 0.40$ mol. Only 0.020 mol of NH_3 is required to form 0.010 M $Ag(NH_3)_2^+$ and to achieve a concentration of 0.19 M. Therefore, there is sufficient NH_3 to dissolve the AgCl completely.

CHAPTER 20

Chemical Puzzler: Dissolving substances is often driven by a positive entropy change. See Figure 20.5 and page 933.

20.1. Entropy change for liquid to vapor:

$$\Delta S° = \frac{30{,}900\ J/mol}{353.3\ K} = +87.5\ J/K \cdot mol$$

$$\Delta S° \text{ for vapor} \longrightarrow \text{liquid} = -87.5\ J/K \cdot mol$$

20.2. 1. $S°$ for solid CO_2 is less than for CO_2 vapor. For a given compound, molecules in the vapor state always have higher entropy because the vapor state is more disordered than the solid state.
2. KCl dissolved in water forms ions separated by water molecules (see Figure 20.5), a more highly disordered state than solid KCl. Thus, the entropy of the dissolved KCl ($S° = 159.0\ J/K \cdot mol$) is larger than that of solid KCl ($S° = 82.6\ J/K \cdot mol$).
3. In this reaction a mole of solid has given rise to a mole of solid compound and a mole of gas. The entropy has increased.

20.3.

$$Ca(s) + C(s) + \tfrac{3}{2}\ O_2(g) \longrightarrow CaCO_3(s)$$

$$\Delta S° = S°[CaCO_3(s)] - \{S°[Ca(s)] + S°[C(graphite)] + \tfrac{3}{2}\ S°[O_2(g)]\}$$

$$= (1\ mol)(92.9\ J/K \cdot mol) - [(1\ mol)(41.4\ J/K \cdot mol)$$

$+ (1 \text{ mol})(5.7 \text{ J/K}\cdot\text{mol})$
$+ (\frac{3}{2} \text{ mol})(205.1 \text{ J/K}\cdot\text{mol})]$
$= -261.9 \text{ J/K}$

Here two solid and one gaseous reactant have been converted to a solid product. The entropy of the system has declined.

20.4. (a) Both $\Delta H°$ and $\Delta S°$ are negative, so the outcome depends on T.
(b) Both $\Delta H°$ and $\Delta S°$ are positive, so the outcome depends on T.
(c) Reaction is exothermic ($\Delta H°$ is negative), and $\Delta S°$ is positive. The reaction is product-favored.
(d) Both $\Delta H°$ and $\Delta S°$ are negative, so the outcome depends on T.

20.5. $\Delta H°_{rxn} = 2\,\Delta H_f°[HCl(g)] = -184.6 \text{ kJ}$

$$\Delta S°_{surroundings} = -[-184.6 \times 10^3 \text{ J/298 K}] = +620.\text{ J/K}$$
$$\Delta S°_{rxn} = 2\,S°[HCl(g)] - \{S°[H_2(g)] + S°[Cl_2(g)]\}$$
$$= 2 \text{ mol } (186.9 \text{ J/K}\cdot\text{mol}) - [1 \text{ mol } (130.7 \text{ J/K}\cdot\text{mol}) + 1 \text{ mol } (223.1 \text{ J/K}\cdot\text{mol}]$$
$$= +20.0 \text{ J/K}$$
$$\Delta S°_{universe} = 640.\text{ J/K}$$

20.6. $\Delta H°_{rxn} = \Delta H_f°[NH_3(g)] - \{\frac{1}{2}\Delta H_f°[N_2(g)] + \frac{3}{2}\Delta H_f°[H_2(g)]\}$
$= (1 \text{ mol})(-46.11 \text{ kJ/mol}) - [\frac{1}{2}\text{ mol}(0) + \frac{3}{2}\text{ mol}(0)]$
$= -46.11 \text{ kJ}$

$$\Delta S°_{rxn} = S°[NH_3(g)] - \{\tfrac{1}{2}\,S°[N_2(g)] + \tfrac{3}{2}\,S°[H_2(g)]\}$$
$$= -99.38 \text{ J/K}$$
$$\Delta G°_{rxn} = \Delta H°_{rxn} - T\Delta S°_{rxn}$$
$$= -46.11 \text{ kJ} - (298 \text{ K})(-99.38 \text{ J/K})\left(\frac{1 \text{ kJ}}{1000 \text{ J}}\right)$$
$$= -16.5 \text{ kJ}$$

20.7. 1. $C(graphite) + O_2(g) \longrightarrow CO_2(g)$
2. -394.4 kJ/mol
3. $(2.5 \text{ mol})(-394.4 \text{ kJ/mol}) = -990 \text{ kJ}$

20.8. $C_6H_6(\ell) + \frac{15}{2}\,O_2(g) \longrightarrow 6\,CO_2(g) + 3\,H_2O(\ell)$

$$\Delta G°_{rxn} = 6\,\Delta G_f°[CO_2(g)] + 3\,\Delta G_f°[H_2O(\ell)] - \{\Delta G_f°[C_6H_6(\ell)] + \tfrac{15}{2}\,\Delta G_f°[O_2(g)]\}$$
$$= (6 \text{ mol})(-394.359 \text{ kJ/mol}) + (3 \text{ mol})(-237.129 \text{ kJ/mol}) - [(1 \text{ mol})(124.5 \text{ kJ/mol}) + (\tfrac{15}{2}\text{ mol})(O)]$$
$$= -3202.0 \text{ kJ}$$

20.9. (1) $CaO(s) + CO_2(g) \longrightarrow CaCO_3(s)$

$$\Delta G°_{rxn} = \Delta G_f°[CaCO_3(s)] - \{\Delta G_f°[CaCO(s)] + \Delta G_f°[CO_2(g)]\}$$
$$= (1 \text{ mol})(-1128.79 \text{ kJ/mol}) - [(1 \text{ mol})(-604.03 \text{ kJ/mol}) + (1 \text{ mol})(-394.359 \text{ kJ/mol})]$$
$$= -130.40 \text{ kJ}$$

(2) $CaCO_3(s) \longrightarrow CaO(s) + CO_2(g)$

$\Delta G°$ for the decomposition $= -\Delta G°$ for the formation $= +130.40$ kJ. The formation reaction is product-favored under standard conditions.

20.10. $MgO(s) + C(graphite) \longrightarrow Mg(s) + CO(g)$

$\Delta H°_{rxn} = +491.18 \text{ kJ}$ and $\Delta S°_{rxn} = 197.67 \text{ J/K}$

$$T = \frac{\Delta H°_{rxn}}{\Delta S°_{rxn}} = 491.18 \text{ kJ}/(0.198 \text{ kJ/K})$$
$$= 2480 \text{ K (or about 2200 °C)}$$

20.11. (a) $\Delta G°_{rxn} = \Delta G_f°[SO_2(s)] = -300.194 \text{ kJ/mol}$
$(-300.194 \text{ kJ})(1000 \text{ J/kJ}) = -(8.314510 \text{ J/K}\cdot\text{mol})(298 \text{ K}) \ln K_p$
$\ln K_p = 1.21 \times 10^2$ and so $K_p = 4.15 \times 10^{52}$
(b) $\Delta G°_{rxn} = +130.4$ kJ from Exercise 20.9.

$\ln K_p = -52.6$ and so $K_p = 1.39 \times 10^{-23}$

20.12.

$SnO_2(s) \longrightarrow Sn(s) + O_2(g)$	$\Delta G_f° = +519.6$ kJ
$C(s) + O_2(g) \longrightarrow CO_2(g)$	$\Delta G_f° = -394.4$ kJ
$SnO_2(s) + C(s) \longrightarrow Sn(s) + CO_2(g)$	$\Delta G_f° = +125.2$ kJ

The decomposition of SnO_2 is more product-favored in the presence of carbon. Indeed, in practice tin(IV) oxide is readily reduced by carbon.

CHAPTER 21

Chemical Puzzler: Batteries and their chemistry are described throughout this chapter, but especially in Section 21.6

21.1.

Oxidizing agent:	$I_2(s) + 2e^- \longrightarrow 2\,I^-(aq)$
Reducing agent:	$2\,[Cr^{2+}(aq) \longrightarrow Cr^{3+}(aq) + e^-]$
	$2\,Cr^{2+}(aq) + I_2(s) \longrightarrow Cr^{3+}(aq) + 2\,I^-(aq)$

Cr^{2+} is oxidized by I_2, and I_2 is reduced by Cr^{2+}.

21.2. Oxidizing agent:
$3\,[NO_3^-(aq) + 2\,H^+(aq) + e^- \longrightarrow NO_2(g) + H_2O(\ell)]$
Reducing agent:
$Co(s) \longrightarrow Co^{3+}(aq) + 3e^-$

$CO(s) + 3\,NO_3^-(aq) + 6\,H^+(aq) \longrightarrow Co^{3+}(aq) + 3\,NO_2(g) + 3\,H_2O(\ell)$

Cobalt metal is oxidized by nitric acid, and nitric acid is reduced by cobalt metal.

21.3. Oxidizing agent:
$3\,[S(s) + H_2O(\ell) + 2e^- \longrightarrow HS^-(aq) + OH^-(aq)]$
Reducing agent:
$2\,[Al(s) + 3\,OH^-(aq) \longrightarrow Al(OH)_3(s) + 3e^-]$

$2\,Al(s) + 3\,S(s) + 3\,OH^-(aq) + 3\,H_2O(\ell) \longrightarrow 2\,Al(OH)_3(s) + 3\,HS^-(aq)$

Aluminum metal is oxidized by sulfur, which is reduced by aluminum.

21.4. Oxidation at the anode: $Ni(S) \longrightarrow Ni^{2+}(aq) + 2e^-$

Reduction at the cathode: $e^- + Ag^+(aq) \longrightarrow Ag(s)$

Electrons flow from the anode (Ni) to the cathode (Ag), and NO_3^- ions flow from the cathode compartment (containing a declining concentration of Ag^+) to the anode compartment (where there is an increasing concentration of Ni^{2+}).

21.5. $\Delta G^\circ_{rxn} = -(2.00 \text{ mol e}^-)(9.65 \times 10^4 \text{ J/V}\cdot\text{mol})(-0.76 \text{ V})(1.0 \text{ kJ}/1000 \text{ J}) = +150 \text{ kJ}$

The reaction is not product-favored as written.

21.6.

$$Fe(s) \longrightarrow Fe^{2+}(aq) + 2e^- \quad E^\circ = +0.44 \text{ V}$$
$$Cu^{2+}(aq) + 2e^- \longrightarrow Cu(s) \quad E^\circ = +0.34 \text{ V}$$
$$Fe(s) + Cu^{2+}(aq) \longrightarrow Fe^{2+}(aq) + Cu(s) \quad E^\circ_{net} = +0.78 \text{ V}$$

21.7.

$$H_2O_2(aq) + 2\,H_3O^+(aq) + 2e^- \longrightarrow 4\,H_2O(\ell) \quad E^\circ = +1.77 \text{ V}$$
$$2\,Br^-(aq) \longrightarrow Br_2(\ell) + 2e^- \quad E^\circ = -1.08 \text{ V}$$
$$H_2O_2(aq) + 2\,H_3O^+(aq) + 2\,Br^-(aq) \longrightarrow 4\,H_2O(\ell) + Br_2(\ell) \quad E^\circ_{net} = +0.69 \text{ V}$$

The reaction is product-favored.

21.8.

$$2\,Al(s) \longrightarrow 2\,Al^{3+}(aq) + 6\,e^- \quad E^\circ = +1.66 \text{ V}$$
$$3\,Zn^{2+}(aq) + 6e^- \longrightarrow 3\,Zn(s) \quad E^\circ = -0.76 \text{ V}$$
$$2\,Al(s) + 3\,Zn^{2+}(aq) \longrightarrow 2\,Al^{3+}(aq) + 3\,Zn(s) \quad E^\circ_{net} = +0.90 \text{ V}$$

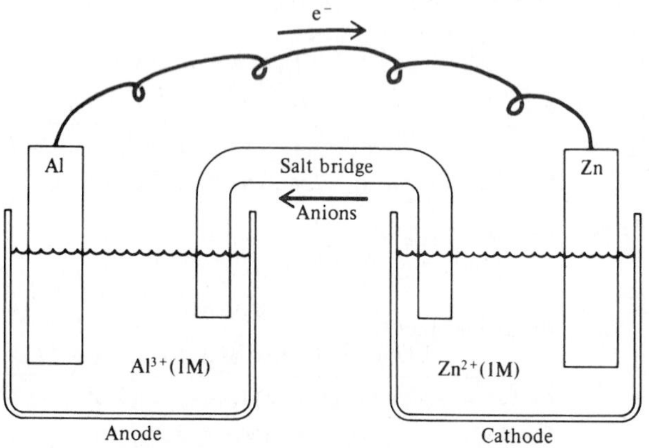

Electrons flow through the external wire from Al (anode) to Zn (cathode). Anions flow through the salt bridge from the Zn^{2+}-containing compartment to the Al^{3+}-containing compartment.

21.9.

$$2\,[Ag^+(aq) + e^- \longrightarrow Ag(s)] \quad E^\circ = +0.800 \text{ V}$$
$$Hg(\ell) \longrightarrow Hg^{2+}(aq) + 2e^- \quad E^\circ = -0.855 \text{ V}$$
$$2\,Ag^+(aq) + Hg(\ell) \longrightarrow 2\,Ag(s) + Hg^{2+}(aq) \quad E^\circ_{net} = -0.055 \text{ V}$$

The reaction is not product-favored under standard conditions.

$$E_{net} = E^\circ_{net} - \frac{0.0257}{n}\ln\frac{[Hg^{2+}]}{[Ag^+]^2}$$
$$= -0.055 - \frac{0.0257}{2}\ln\frac{[0.0010]}{[0.80]^2}$$
$$= -0.055 \text{ V} + 0.083 \text{ V} = 0.028 \text{ V}$$

The reaction is product-favored under the new conditions.

21.10.

$$\ln K = \frac{nE^\circ}{0.0257 \text{ V}} = \frac{(2)(-0.055 \text{ V})}{0.0257 \text{ V}}$$
$$= -4.3, \quad \text{and so} \quad K = 0.014.$$

Notice that K is less than 1, as expected for a reaction with a negative E°.

21.11. **(a)** $NaBr(\text{molten}) + \text{electricity} \longrightarrow Na(s) + Br_2(\ell)$
(b) $NaBr(aq) + \text{electricity} \longrightarrow OH^-(aq)$ and $H_2(g)$ from water + $Br_2(\ell)$ from $Br^-(aq)$
(c) $SnCl_2(aq) + \text{electricity} \longrightarrow Sn(s) + Cl_2(g)$

21.12. $(25 \times 10^3 \text{ A})(3600 \text{ s/h}) = 9.0 \times 10^7 \text{ C/h}$

$(9.0 \times 10^7 \text{ C/h})(1 \text{ mol e}^-/96{,}500 \text{ C}) = 9.3 \times 10^2 \text{ mol e}^-/\text{h}$

$(9.3 \times 10^2 \text{ mol e}^-/\text{h})(1 \text{ mol Na}/1 \text{ mol e}^-)(23.0 \text{ g/mol}) = 2.1 \times 10^4 \text{ g Na/h}$

21.13. (See Exercise 21.12.)

$(1.00 \times 10^3 \text{ g Na})(1 \text{ mol Na}/22.99 \text{ g})(96485 \text{ C/mol Na}) = 4.20 \times 10^6 \text{ C}$

$(4.20 \times 10^6 \text{ C})(7.0 \text{ V})(1 \text{ J/V}\cdot\text{C})(1 \text{ kwh}/3.60 \times 10^6 \text{ J}) = 8.2 \text{ kwh}$

CHAPTER 22

Chemical Puzzler: Aluminum does not react with nitric acid because it has an unreactive coating of aluminum oxide (see Section 22.5). Copper has no such inert coating.

22.1.
1. $2\,Na(s) + Br_2(\ell) \longrightarrow 2\,NaBr(s)$
2. $Ca(s) + Se(s) \longrightarrow CaSe(s)$
3. $2\,Al(s) + 3\,Cl_2(g) \longrightarrow 2\,AlCl_3(s)$
4. $4\,K(s) + O_2(g) \longrightarrow 2\,K_2O(s)$

22.2. 1. H_2Te 3. $SeCl_6$
2. Na_3AsO_4 4. $HBrO_4$

22.3. 1. NH_4^+ (ammonium ion) 3. N_2H_4 (hydrazine)
2. O_2^{2-} (peroxide ion) 4. NF_3(nitrogen trifluoride)

22.4. **(a)** ClO is an odd-electron molecule, with Cl having the unlikely oxidation number of +1.
(b) Na_2Cl has chlorine with the unlikely charge of 2+. The ionization energy of Cl (Chapter 8) is high, so loss of two electrons is unlikely.
(c) This compound has calcium as Ca^+. Calcium generally loses two electrons to form the Ca^{2+} ion.
(d) C_3H_7 is unlikely. No structure can be drawn that has seven H atoms and four bonds to each C atom (see Chapter 11).

CHAPTER 23

Chemical Puzzler: Compounds of the transition metals are generally not white. The reason for this is explained by the mode of bonding in such compounds, as outlined in Section 23.5. The important aspect of this is that an electron can be promoted from a filled or partially filled *d*-type orbital to a higher energy *d*-type orbital. The energy required to do this is supplied by light in the visible region of the spectrum, thus removing some wavelengths of light from the incident light. You see the wavelengths of light that remain, and you observe this as a color.

23.1. 1. The platinum in $Pt(NH_3)_2Cl_2$ is Pt^{2+}.
2. $[Co(NH_3)_5CO_3]^+$

23.2. 1. Aquacyanobis(phenanthroline)ruthenium(II) chloride
2. Diaminedichloroplatinum(II)

23.3. **(a)** *cis-trans* isomers (see figure of $[Co(en)_2Cl_2]^+$ on page 1063.

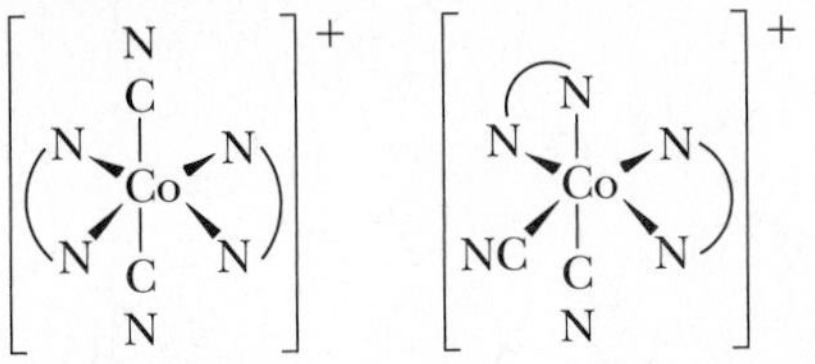

trans *cis*

(b) optical isomers

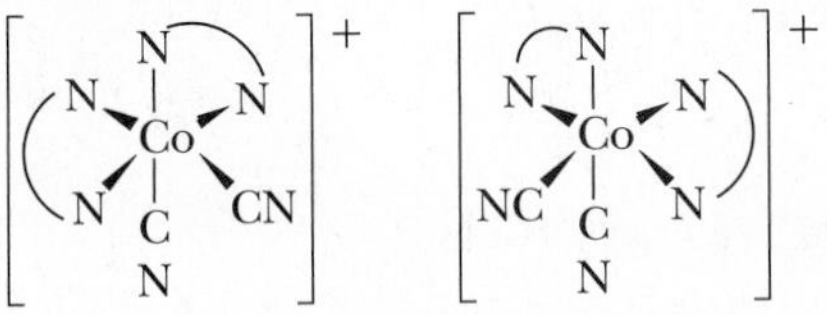

23.4. **(a)** Ru^{2+} has a d^6 configuration and so has four unpaired electrons in the high-spin state and none in the low-spin state. High-spin Ru^{2+} is paramagnetic.

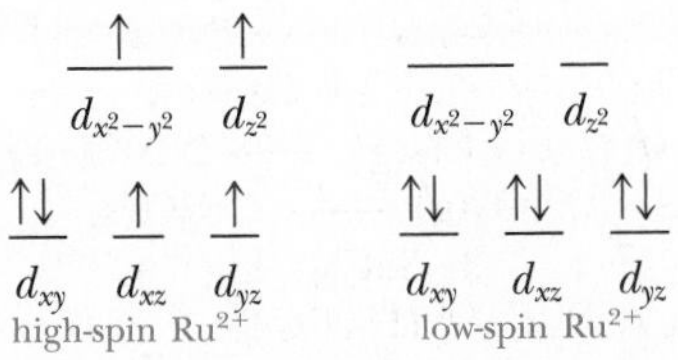

(b) Ni^{2+} has a d^8 configuration. Thus, there is only one way to arrange electrons in the two sets of *d* orbitals, and there are two unpaired electrons. The Ni^{2+} ion is paramagnetic.

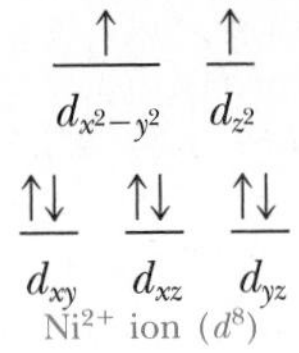

CHAPTER 24

Chemical Puzzler: The radioactive element in smoke detectors is often americium, element 95 (Am). The operation of a smoke detector is explained on page 1107.

24.1. 1. $^{237}_{93}Np \longrightarrow {}^{4}_{2}He + {}^{233}_{91}Pa$
2. $^{35}_{16}S \longrightarrow {}^{0}_{-1}e + {}^{35}_{17}Cl$

24.2. $^{235}_{92}U \longrightarrow {}^{4}_{2}He + {}^{231}_{90}Th$

$^{231}_{90}Th \longrightarrow {}^{0}_{-1}e + {}^{231}_{91}Pa$

$^{231}_{91}Pa \longrightarrow {}^{4}_{2}He + {}^{227}_{89}Ac$

$^{227}_{89}Ac \longrightarrow {}^{4}_{2}He + {}^{223}_{87}Fr$

$^{223}_{87}Fr \longrightarrow {}^{0}_{-1}e + {}^{223}_{88}Ra$

24.3. Positron emission: $^{13}_{7}N \longrightarrow {}^{13}_{6}C + {}^{0}_{+1}e$
K-capture: $^{41}_{20}Ca + {}^{0}_{-1}e \longrightarrow {}^{41}_{19}K$
β-emission: $^{90}_{38}Sr \longrightarrow {}^{0}_{-1}e + {}^{90}_{39}Y$
Positron emission: $^{11}_{6}C \longrightarrow {}^{11}_{5}B + {}^{0}_{+1}e$
Positron emission: $^{22}_{11}Na \longrightarrow {}^{22}_{10}Ne + {}^{0}_{+1}e$

24.4. β-decay: $^{32}_{14}Si \longrightarrow {}^{0}_{-1}e + {}^{32}_{15}P$
Positron emission: $^{45}_{22}Ti \longrightarrow {}^{0}_{+1}e + {}^{45}_{21}Sc$
α-emission: $^{239}_{94}Pu \longrightarrow {}^{4}_{2}He + {}^{235}_{92}U$

24.5. Mass defect $= \Delta m = -0.03438$ g/mol

$$\Delta E = (-34.38 \times 10^{-6}\ \text{kg/mol})(2.998 \times 10^{8}\ \text{m/s})^2 = -3.090 \times 10^{12}\ \text{J/mol}$$

E_b per mole of nucleons $= 5.150 \times 10^8$ kJ/mol of nucleons

E_b for ^6Li is smaller than for ^4He.

24.6. 1. $^{90}_{38}Sr \longrightarrow {}^{0}_{-1}e + {}^{90}_{39}Y$
2. Disintegrations are reduced to 1000 after 1 half-life, to 500 after 2 half-lives, to 250 after 3 half-lives, and to 125 after 4 half-lives. Thus, a total of $4 \times 28 = 112$ years is required.

24.7. $k = 0.693/t_{1/2} = 8.90 \times 10^{-3}\ \text{h}^{-1}$

$$\ln(0.10/1.0) = -kt = -(8.90 \times 10^{-3}\ \text{h}^{-1})t$$

$$t = 259\ \text{h}$$

24.8.

$$\ln\left(\frac{12.9}{14.0}\right) = -(1.21 \times 10^{-4}/\text{yr})t$$

$$t = 680\ \text{years}$$

$$1990 - 680 = 1310\ \text{AD}$$

24.9. 1. $^{13}_{6}C + {}^{1}_{0}n \longrightarrow {}^{4}_{2}He + {}^{10}_{4}Be$
2. $^{14}_{7}N + {}^{4}_{2}He \longrightarrow {}^{1}_{0}n + {}^{17}_{9}F$
3. $^{253}_{99}Es + {}^{4}_{2}He \longrightarrow {}^{1}_{0}n + {}^{256}_{101}Md$

APPENDIX M

Answers to Selected Study Questions

CHAPTER 1

1.14. 557 g

1.16. 498 g; 0.498 kg

1.18. 0.911 g/cm^3

1.20. The most likely metal is aluminum, with a density of 2.68 g/cm^3.

1.22. 298 K

1.24. **(a)** 289 K **(b)** 1.0×10^2 °C **(c)** 230 K

1.26. Liquid; body temperature (37 °C) is above the melting point of gallium.

1.28. **(a)** carbon
(b) sodium
(c) chlorine
(d) phosphorus
(e) magnesium
(f) calcium

1.30. **(a)** Li
(b) Ti
(c) Fe
(d) Si
(e) Co
(f) Zn

1.32. 190 mm; 0.19 m

1.34. 5.3 cm^3; 5.3×10^{-4} m^2

1.36. 800. cm^3; 0.800 L; 8.00×10^{-4} m^3

1.38. 0.00563 kg; 5630 mg

1.40.

Milligrams	Grams	Kilograms
693	0.693	6.93×10^{-4}
156	0.156	1.56×10^{-4}
2.23×10^6	2.23×10^3	2.23

1.42. 22 cm × 28 cm; 6.0×10^2 cm^2

1.44. 9.29 g

1.46. 3.62×10^{-3}

1.48. 4.72×10^{-4}

1.50. 0.146

1.52. 80.1% silver, 19.9% copper

1.54. 245 g of sulfuric acid

1.56. 0.197 nm; 197 pm

1.58. 0.178 nm^3; 1.78×10^{-22} cm^3

1.60. The troy ounce (31.103 g) is larger.

1.62. 0.995 g of platinum

1.64. 15% lost; 3630 kernels/lb

1.66. The water level rises 18.0 mL.

1.68. 90 tons/year

1.70. 294 ft

1.72. Nickel, with a density of 8.91 g/cm^3

1.74. 5×10^{17} g; diameter = 8×10^5 cm, or about 5 miles

1.76. 14.3 nm

1.78. 2.50 mm; a simpler method involves using a measuring device to measure a representative sample of lead shot, then calculating the average diameter.

1.81. The sample's density and melting point can be compared with that of silver to prove whether or not the sample is silver.

1.83. The mass of the object is determined and then the volume is determined by submersion in a known volume of liquid. The increase in volume equals the volume of the irregularly shaped object. The density can be calculated by dividing its mass by its volume.

1.85. A clean, dry 10.0-mL graduated cylinder can be weighed (empty) and then filled (to the 10.0-mL mark) with the unknown liquid and reweighed. The difference in mass equals the mass of 10.0 mL of the liquid. Depending on the balance used in the experiment, the accuracy is ±0.01 g in mass and ±0.1 mL in volume The calculated density has no more than three significant figures, limited by the volume (10.0 mL).

1.87. **(a)** The water can be evaporated by heating the solution, leaving the salt behind.

(b) A magnet attracts the iron filings away from the lead.

(c) Mixing the solids with water dissolves only the sugar. Filtration separates the solid sulfur from the solution. Finally, the sugar can be separated from the water by boiling the solution.

CHAPTER 2

2.21. **(a)** 9 **(b)** 48 **(c)** 70

2.23. **(a)** $^{23}_{11}Na$ **(b)** $^{39}_{18}Ar$ **(c)** $^{70}_{31}Ga$

2.25.

	Electrons	Protons	Neutrons
(a)	20	20	20
(b)	50	50	69
(c)	94	94	150

2.27.

Symbol	^{45}Sc	^{33}S	^{17}O	^{56}Mn
Number of protons	21	16	8	25
Number of neutrons	24	17	9	31
Number of electrons in the neutral atom	21	16	8	25

2.29. 95 protons, 95 electrons, 146 neutrons

2.31. $^{19}_{9}X$, $^{20}_{9}X$, and $^{21}_{9}X$; not $^{9}_{18}X$

2.33. (^{6}Li mass)(% abundance) + (^{7}Li mass)(% abundance) = atomic weight of Li

(6.015121 amu)(0.0750) + (7.016003 amu)(0.9250) = 6.94 amu

2.35. ^{69}Ga abundance is 60.12%, ^{71}Ga abundance is 39.88%.

2.37. Group 4A has five elements in it:

C	carbon	nonmetal
Si	silicon	metalloid
Ge	germanium	metalloid
Sn	tin	metal
Pb	lead	metal

2.39. The seventh period is incomplete. The majority are called transition elements and are metals. All of these elements are unstable and radioactive.

2.41. **(a)** 27 g B **(b)** 0.48 g O_2 **(c)** 6.98×10^{-2} g Fe **(d)** 2.61×10^{3} g He

2.43. **(a)** 1.9998 mol Cu **(b)** 0.499 mol Ca **(c)** 0.6208 mol Al **(d)** 3.1×10^{-4} mol K **(e)** 2.1×10^{-5} mol Am

2.45. 2.19 mol Na; 1.32×10^{24} atoms Na

2.47. 2.39 mol Pt; 21.7 cm^3

2.49. 4.131×10^{23} atoms Cr

2.51. 1.0552×10^{-22} g/Cu atom

2.53. ^{39}K must be more abundant because the average atomic weight of K is 39.0983, which is closest to the mass number of ^{39}K.

2.55. ^{121}Sb and ^{123}Sb; ^{121}Sb is more abundant; ^{121}Sb has 51 protons, 51 electrons, and 70 neutrons; the atomic weight of Sb is approximately 122 amu.

2.57. **(a)** Ti; atomic number 22; atomic mass 47.88

(b) Ti is in Period 4 and Group 4B; other elements in Group 4B are Zr, Hf, and Rf.

(c) Low reactivity, noncorrosive, low density, easily fabricated

(d) Melting point 1660 °C, boiling point 3287 °C, low density, high strength, resistant to dilute sulfuric and hydrochloric acids. Natural titanium exists as five stable isotopes. Titanium metal is considered to be physiologically inert.

2.59. **(a)** Co, Ni, and Cu, with densities of approximately 9 g/cm^3; all are metals.

(b) B in the second period; Al in the third period; both are in Group 3A.

(c) H, He, N, O, F, Ne, Cl, Ar, and Kr do not appear in the plot; all are gases.

2.61. **(a)** 0.159 mol Pt

(b) 370 g Pt; 1.90 mol Pt

2.63. 3.40 mol Cu; 2.0×10^{24} atoms Cu

2.65.

S	N
B	I

2.68. We expect all Group 2A elements to form compounds with oxygen in the same 1 to 1 ratio: BeO, MgO, CaO, SrO, BaO, and RaO.

2.70. With the exception of the Li/Be/B area, the elements of lowest atomic number are present in greatest abundance; the higher the atomic number, the lower the abundance. In general, the elements with even atomic numbers are more abundant than those with odd atomic numbers.

2.72. **(a)** a, b, f

(b) h, i

(c) c, d, g

(d) a, d, f

(e) c

(f) b, g, i

(g) e, h

2.74. **(a)** 15.873 amu; 5.9745×10^{23} atoms/mol

(b) 1.00801 amu; 6.02236×10^{23} atoms/mol

2.76. mass P/mass O = 1.94; atomic mass of P = 31.0 amu

CHAPTER 3

3.13. Sucrose ($C_{12}O_{22}H_{11}$) contains more O atoms and more total atoms.

3.15. **(a)** 1 Ca atom, 2 C atoms, 4 O atoms

(b) 8 C atoms, 8 H atoms

(c) 2 Cu atoms, 1 C atom, 5 O atoms, 2 H atoms

(d) 1 Pt atom, 2 N atoms, 6 H atoms, 2 Cl atoms

(e) 4 K atoms, 1 Fe atom, 6 C atoms, 6 N atoms

3.17. **(a)** $C_3H_6O_3$ **(b)** $C_6H_8O_7$

3.19. Aluminum: Al^{3+}; selenium: Se^{2-}

3.21. **(a)** Mg^{2+} **(b)** Zn^{2+} **(c)** Fe^{2+} and Fe^{3+} **(d)** Ga^{3+}

3.23. (a) Sr^{2+} (b) Al^{3+} (c) S^{2-} (d) Co^{2+} (e) Ti^{4+} (f) HCO_3^- (g) ClO_4^- (h) NH_4^+

3.25. (a) 2 K^+, 1 S^{2-}
(b) 1 Ni^{2+}, 1 SO_4^{2-}
(c) 3 NH_4^+, 1 PO_4^{3-}
(d) 1 Ca^{2+}, 2 ClO^-
(e) 1 K^+, 1 MnO_4^-

3.27. CoO and Co_2O_3

3.29. (a) $AlCl_3$ (b) NaF (c) correct (d) correct

3.31. Magnesium oxide, MgO; charges of the ions are greater and ionic radii are smaller; this results in stronger attractions among the ions and a higher melting temperature.

3.33. (a) potassium sulfide (b) nickel(II) sulfate
(c) ammonium phosphate
(d) calcium hypochlorite

3.35. (a) $(NH_4)_2CO_3$ (b) CaI_2 (c) $CuBr_2$
(d) $AlPO_4$ (e) $AgCH_3CO_2$

3.37.

K_2CO_3	potassium carbonate
KBr	potassium bromide
KNO_3	potassium nitrate
$BaCO_3$	barium carbonate
$BaBr_2$	barium bromide
$Ba(NO_3)_2$	barium nitrate
$(NH_4)_2CO_3$	ammonium carbonate
NH_4Br	ammonium bromide
NH_4NO_3	ammonium nitrate

3.39. (a) nitrogen trifluoride (b) hydrogen iodide
(c) boron tribromide (d) hexane

3.41. (a) C_4H_{10} (b) N_2O_5 (c) C_9H_{20} (d) $SiCl_4$
(e) B_2O_3

3.43. (a) 159.7 g/mol
(b) 67.81 g/mol
(c) 44.02 g/mol
(d) 197.9 g/mol
(e) 176.1 g/mol

3.45. (a) 0.0312 mol CH_3OH (b) 0.0101 mol Cl_2CO
(c) 0.0125 mol NH_4NO_3
(d) 4.06×10^{-3} mol $MgSO_4 \cdot 7\,H_2O$

3.47. 47.1 mol C_2H_3CN

3.49. (a) 1.80×10^{-3} mol $C_9H_8O_4$; 2.266×10^{-2} mol $NaHCO_3$; 5.205×10^{-3} mol $C_6H_8O_7$
(b) 1.08×10^{21} molecules $C_9H_8O_4$

3.51. 12.5 mol SO_3; 7.52×10^{24} molecules SO_3; 7.52×10^{24} S atoms; 2.26×10^{25} O atoms

3.53. (a) 86.59% Pb, 13.41% S
(b) 81.71% C, 18.29% H
(c) 24.77% Co, 29.80% Cl, 5.08% H, 40.35% O
(d) 35.00% N, 5.04% H, 59.96% O

3.55. (a) 62.50 g/mol
(b) 38.44% C, 4.84% H, 56.72% Cl
(c) 174 g C

3.57. The molar mass of the empirical formula is 59.04 g/mol, therefore the molecular formula is $C_4H_6O_4$.

3.59. The empirical formula is CH; the molecular formula is C_2H_2.

3.61. The empirical formula is N_2O_3.

3.63. The empirical formula is $C_8H_8O_3$; the molecular formula is also $C_8H_8O_3$.

3.65. The empirical formula is C_2H_6As; the molecular formula is $C_4H_{12}As_2$.

3.67. There are seven molecules of H_2O per $MgSO_4$ formula unit.

3.69. There are six F atoms for every S atom; the formula is SF_6.

3.71. 2.31×10^3 g PbS

3.73. 75.1 kg $Ca_3(PO_4)_2$

3.75. 0.787 metric tons Na and 1.21 metric tons Cl

3.77. (a) $C_7H_5N_3O_6$; 227.1 g/mol; 18.50% N, 37.02% C
(b) $C_3H_7NO_3$; 105.1 g/mol; 13.33% N, 34.28% C
Trinitrotoluene has a larger percentage of nitrogen and of carbon.

3.79. 1.495×10^{-22} g/molecule

3.81. C_2H_6SO; 78.13 g/mol; 30.75% C, 7.74% H, 41.03% S; 0.128 mol C_2H_6SO; 4.10 g S

3.83. (a) $C_{10}H_{10}Fe$
(b) 4.50×10^{-2} g Fe; 4.86×10^{20} atoms Fe
(c) Carbon accounts for 64.56% of the mass of the molecule.

3.85. (a) Not likely ionic
(b) Li_2Te; lithium telluride
(c) Not likely ionic
(d) MgF_2; magnesium fluoride
(e) Not likely ionic
(f) In_2S_3; indium sulfide
(g) Not likely ionic

3.87. (a) NaClO; ionic
(b) $Al(ClO_4)_3$; ionic
(c) $KMnO_4$; ionic
(d) KH_2PO_4; ionic
(e) ClF_3
(f) BBr_3
(g) $Ca(CH_3CO_2)_2$; ionic
(h) $(NH_4)_2SO_3$; ionic
(i) S_2Cl_2
(j) PF_3

3.89. The empirical formula is C_5H_4; the molecular formula is $C_{10}H_8$.

3.91. The empirical formula is ICl_3; the molecular formula is I_2Cl_6.

3.93. 7.36 kg Fe

3.95. 148 g Sb_2S_3

3.97. There are four CO molecules for every Ni atom; $Ni(CO)_4$.

3.99. Molar mass of E = 12 g/mol; the element is carbon.

3.101. 5.5 g of compound B is mixed with 1.00 g of A.

3.103. Element A is 26 g/mol; element Z is 18 g/mol.

3.105. Molar mass is 211 g/mol; element M (179 g/mol) could be Hf or Ta.

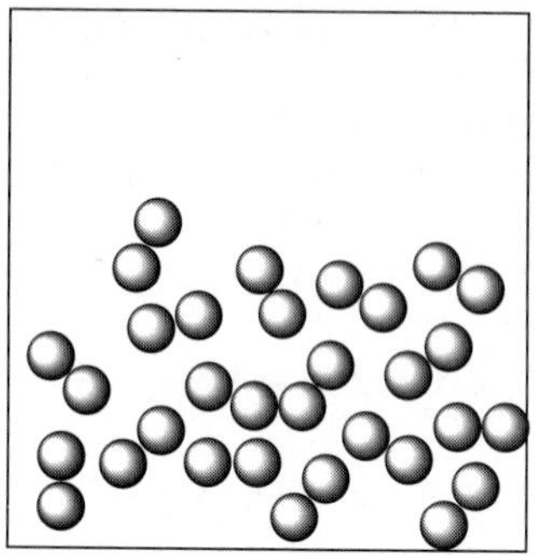

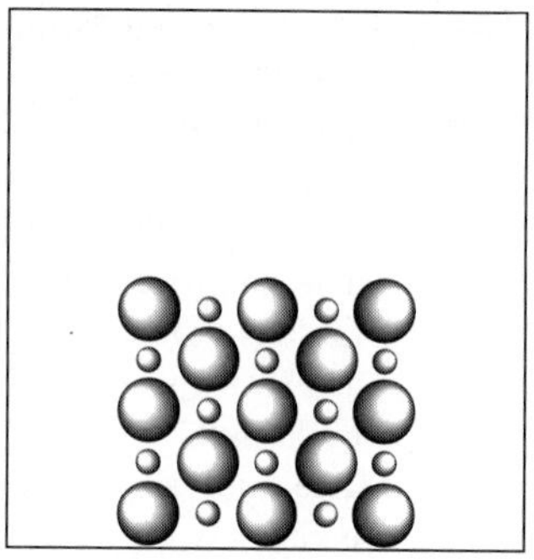

3.107. $Br_2(\ell)$ $LiF(s)$

3.110. The Al^{3+} ion should have the strongest attraction for nearby water molecules primarily because of the greater charge on the ion compared with the other ions listed.

3.112. Answer (d) $C_9H_{12}O_5$ is correct. The other students apparently did not correctly calculate the number of moles of material in 100. g or they improperly calculated the ratio of those moles in determining their empirical formula.

3.115. **(a)** 0.0130 mol Ni **(b)** NiF_2
(c) nickel(II) fluoride

CHAPTER 4

4.15. **(a)** $4\,Cr(s) + 3\,O_2(g) \rightarrow 2\,Cr_2O_3(s)$
(b) $Cu_2S(s) + O_2(g) \rightarrow 2\,Cu(s) + SO_2(g)$
(c) $C_6H_5CH_3(\ell) + 9\,O_2(g) \rightarrow 4\,H_2O(\ell) + 7\,CO_2(g)$

4.17. **(a)** $3\,MgO(s) + 2\,Fe(s) \rightarrow Fe_2O_3(s) + 3\,Mg(s)$
iron(III) oxide magnesium
(b) $AlCl_3(s) + 3\,H_2O(\ell) \rightarrow Al(OH)_3(s) + 3\,HCl(aq)$
aluminum hydroxide hydrogen chloride
(c) $2\,NaNO_3(s) + H_2SO_4(\ell) \rightarrow Na_2SO_4(s) + 2\,HNO_3(g)$
sodium sulfate nitric acid

4.19. **(a)** $CO_2(g) + 2\,NH_3(g) \rightarrow CO(NH_2)_2(s) + H_2O(\ell)$
(b) $UO_2(s) + 4\,HF(aq) \rightarrow UF_4(s) + 2\,H_2O(aq)$
$UF_4(s) + F_2(g) \rightarrow UF_6(s)$
(c) $TiO_2(s) + 2\,Cl_2(g) + 2\,C(s) \rightarrow TiCl_4(\ell) + 2\,CO(g)$
$TiCl_4(\ell) + 2\,Mg(s) \rightarrow Ti(s) + 2\,MgCl_2(s)$

4.21. **(a)** $FeCl_2$ is soluble.
(b) $AgNO_3$ is soluble.
(c) NaCl and $KMnO_4$ are soluble.

4.23. **(a)** $NaCH_3CO_2$ **(b)** Fe_2S_3
(c) KOH **(d)** $PbCl_2$

4.25. **(a)** K^+ and I^- ions **(b)** K^+ and SO_4^{2-} ions
(c) K^+ and HSO_4^- ions **(d)** K^+ and CN^- ions

4.27. **(a)** soluble; Ba^{2+} and Cl^- ions
(b) soluble; Cr^{2+} and NO_3^- ions
(c) soluble; Pb^{2+} and NO_3^- ions
(d) insoluble

4.29. soluble: $CuCl_2$ and $Cu(NO_3)_2$
insoluble: $Cu(OH)_2$ and CuS

4.31. $HNO_3(aq) \rightarrow H^+(aq) + NO_3^-(aq)$

4.33. $H_2C_2O_4(aq) \longrightarrow H^+(aq) + HC_2O_4^-(aq)$
$HC_2O_4^-(aq) \longrightarrow H^+(aq) + C_2O_4^{2-}(aq)$

4.35. $MgO(s) + H_2O(\ell) \rightarrow Mg(OH)_2(s)$

4.37. **(a)** $Zn(s) + 2\,HCl(aq) \longrightarrow H_2(g) + ZnCl_2(aq)$
$Zn(s) + 2\,H^+(aq) \longrightarrow H_2(g) + Zn^{2+}(aq)$
(b) $Mg(OH)_2(s) + 2\,HCl(aq) \longrightarrow MgCl_2(aq) + 2\,H_2O(\ell)$
$Mg(OH)_2(s) + 2\,H^+(aq) \longrightarrow Mg^{2+}(aq) + 2\,H_2O(\ell)$
(c) $2\,HNO_3(aq) + CaCO_3(s) \longrightarrow Ca(NO_3)_2(aq) + H_2O(\ell) + CO_2(g)$
$2\,H^+(aq) + CaCO_3(s) \longrightarrow Ca^{2+}(aq) + H_2O(\ell) + CO_2(g)$

4.39. **(a)** $Ba(OH)_2(s) + 2\,HNO_3(aq) \longrightarrow Ba(NO_3)_2(aq) + 2\,H_2O(\ell)$
$Ba(OH)_2(s) + 2\,H^+(aq) \longrightarrow Ba^{2+}(aq) + 2\,H_2O(\ell)$
(b) $BaCl_2(aq) + Na_2CO_3(aq) \longrightarrow BaCO_3(s) + 2\,NaCl(aq)$
$Ba^{2+}(aq) + CO_3^{2-}(aq) \longrightarrow BaCO_3(s)$
(c) $2\,Na_3PO_4(aq) + 3\,Ni(NO_3)_2(aq) \longrightarrow Ni_3(PO_4)_2(s) + 6\,NaNO_3(aq)$
$2\,PO_4^{3-}(aq) + 3\,Ni^{2+}(aq) \longrightarrow Ni_3(PO_4)_2(s)$

4.41. **(a)** $K_2CO_3(aq) + Cu(NO_3)_2(aq) \longrightarrow CuCO_3(s) + 2\,KNO_3(aq)$
precipitation reaction
(b) $Pb(NO_3)_2(aq) + 2\,HCl(aq) \longrightarrow PbCl_2(s) + 2\,HNO_3(aq)$
precipitation reaction
(c) $MgCO_3(s) + 2\,HCl(aq) \longrightarrow MgCl_2(aq) + CO_2(g) + H_2O(\ell)$
gas-forming reaction

4.43. **(a)** $MnCl_2(aq) + Na_2S(aq) \longrightarrow MnS(s) + 2\,NaCl(aq)$
$Mn^{2+}(aq) + S^{2-}(aq) \longrightarrow MnS(s)$
precipitation reaction
(b) $K_2CO_3(aq) + ZnCl_2(aq) \longrightarrow ZnCO_3(s) + 2\,KCl(aq)$
$CO_3^{2-}(aq) + Zn^{2+}(aq) \longrightarrow ZnCO_3(s)$
precipitation reaction
(c) $K_2CO_3(aq) + 2\,HClO_4(aq) \longrightarrow 2\,KClO_4(aq) + CO_2(g) + H_2O(\ell)$
$CO_3^{2-}(aq) + 2\,H^+(aq) \longrightarrow CO_2(g) + H_2O(\ell)$
gas-forming reaction

4.45. $CdCl_2(aq) + 2\,NaOH(aq) \longrightarrow Cd(OH)_2(s) + 2\,NaCl(aq)$
$Cd^{2+}(aq) + 2\,OH^-(aq) \longrightarrow Cd(OH)_2(s)$

4.47. **(a)** $NiCl_2(aq) + (NH_4)_2S(aq) \rightarrow NiS(s) + 2\,NH_4Cl(aq)$
(b) $3\,Mn(NO_3)_2(aq) + 2\,Na_3PO_4(aq) \rightarrow Mn_3(PO_4)_2(s) + 6\,NaNO_3(aq)$

4.49. $Pb(NO_3)_2(aq) + 2\,KOH(aq) \longrightarrow Pb(OH)_2(s) + 2\,KNO_3(aq)$
lead(II) nitrate; potassium hydroxide; lead(II) hydroxide; potassium nitrate

4.51. **(a)** $2\ CH_3CO_2H(aq)$ (acetic acid) + $Mg(OH)_2(s)$ (magnesium hydroxide) ⟶ $Mg(CH_3CO_2)_2(aq)$ (magnesium acetate) + $2\ H_2O(\ell)$ (water)

(b) $HClO_4(aq)$ (perchloric acid) + $NH_3(aq)$ (ammonia) ⟶ $NH_4ClO_4(aq)$ (ammonium perchlorate)

4.53. $Ba(OH)_2(s) + 2\ HNO_3(aq) \rightarrow Ba(NO_3)_2(aq) + 2\ H_2O(\ell)$

4.55. $MnCO_3(s)$ (manganese(II) carbonate) + $2\ HCl(aq)$ (hydrochloric acid) ⟶ $MnCl_2(aq)$ (manganese(II) chloride) + $CO_2(g)$ (carbon dioxide) + $H_2O(\ell)$ (water)

4.57. **(a)** Br is +5 and O is −2
(b) C is +3 and O is −2
(c) F is 0
(d) Ca is +2 and H is −1
(e) H is +1, Si is +4, and O is −2
(f) S is +6 and O is −2

4.59. Reactants: Na is +1, I is −1, H is +1, S is +6, O is −2, Mn is +4
Products: Na is +1, I is 0, H is +1, S is +6, O is −2, Mn is +2

4.61. **(a)** precipitation reaction
(b) oxidation-reduction reaction
The oxidation number of Ca changes from 0 to +2, whereas that of O changes from 0 to −2.
(c) acid-base reaction

4.63. **(a)** Mg is oxidized and is the reducing agent. O_2 is reduced and is the oxidizing agent.
(b) C is oxidized, and C_2H_4 is the reducing agent. O_2 is reduced and is the oxidizing agent.
(c) Si is oxidized and is the reducing agent. Cl_2 is reduced and is the oxidizing agent.

4.65. Cl^- is a spectator ion.
$MgCO_3(s) + 2\ H^+(aq) \longrightarrow Mg^{2+}(aq) + CO_2(g) + H_2O(\ell)$
This is a gas-forming reaction,

4.67. **(a)** $(NH_4)_2S(aq) + Hg(NO_3)_2(aq) \rightarrow HgS(s) + 2\ NH_4NO_3(aq)$
(b) ammonium sulfide, mercury(II) nitrate, mercury(II) sulfide, ammonium nitrate
(c) precipitation reaction

4.69. $Cu_3(CO_3)_2(OH)_2(s) + 6\ HCl(aq) \rightarrow 3\ CuCl_2(aq) + 2\ CO_2(g) + 4\ H_2O(\ell)$

4.71. **(a)** $MnCl_2(aq) + Na_2S(aq) \longrightarrow MnS(s) + 2\ NaCl(aq)$
$Mn^{2+}(aq) + S^{2-}(aq) \longrightarrow MnS(s)$
precipitation reaction
(b) $K_2CO_3(aq) + ZnCl_2(aq) \longrightarrow ZnCO_3(s) + 2\ KCl(aq)$
$CO_3^{2-}(aq) + Zn^{2+}(aq) \longrightarrow ZnCO_3(s)$
precipitation reaction
(c) $K_2CO_3(aq) + 2\ HClO_4(aq) \longrightarrow 2\ KClO_4(aq) + CO_2(g) + H_2O(\ell)$
$CO_3^{2-}(aq) + 2\ H^+(aq) \longrightarrow CO_2(g) + H_2O(\ell)$
gas-forming reaction

4.73. C is oxidized, and $C_6H_8O_6$ is the reducing agent. Br is reduced, and Br_2 is the oxidizing agent.

4.75. Precipitation reaction: $BaCl_2(aq) + Na_2SO_4(aq) \rightarrow BaSO_4(s) + 2\ NaCl(aq)$
Gas-forming reaction: $BaCO_3(s) + H_2SO_4(aq) \rightarrow BaSO_4(s) + CO_2(g) + H_2O(\ell)$

4.78. **(a)** $CaF_2(s)$ (calcium fluoride) + $H_2SO_4(aq)$ (sulfuric acid) ⟶ $2\ HF(g)$ (hydrogen fluoride) + $CaSO_4(s)$ (calcium sulfate)
(b) It is both a precipitation reaction and a gas-forming reaction.
(c) CCl_4: carbon tetrachloride; HCl: hydrogen chloride
(d) The empirical formula is CCl_3F.

CHAPTER 5

5.1. 36.6 g CO_2
5.3. 0.699 g Ga and 0.75 g As needed.
5.5. 5.64 g $CoCl_2$; 0.0876 g H_2
5.7. 1.1 mol O_2; 35 g O_2
5.9. **(a)** 318 g Fe
(b) 239 g CO
5.11. **(a)** sulfur dioxide, calcium carbonate, calcium sulfate, carbon dioxide
(b) 234 g $CaCO_3$
(c) 319 g $CaSO_4$
5.13. **(a)** $NH_4NO_3(s) \rightarrow N_2O(g) + 2\ H_2O(\ell)$
(b) 5.50 g N_2O; 4.50 g H_2O
5.15. **(a)** Cl_2 is the limiting reactant.
(b) 5.08 g $AlCl_3$
(c) 1.67 g Al left unreacted
5.17. CO is the limiting reactant. 1.3 g of H_2 are left unreacted. Theoretical yield of CH_3OH is 85.2 g,
5.19. CaO is the limiting reactant; the maximum possible yield of NH_3 is 68.0 g; 10.3 g of NH_4Cl are left unreacted.
5.21. 73.5%
5.23. Theoretical yield is 74.0 g; 88.1%.
5.25. 1.14 g $CuSO_4 \cdot 5\ H_2O$; 91.6%
5.27. 83.9% $Al(C_6H_5)_3$
5.29. Empirical formula is CH.
5.31. Empirical formula is $C_3H_6O_2$.
5.33. Empirical formula is SiH_4.
5.35. 0.254 M Na_2CO_3; 0.508 M Na^+; 0.254 M CO_3^{2-}
5.37. 0.494 g $KMnO_4$
5.39. 5.08×10^3 mL
5.41. 0.0100 M
5.43. **(a)** 0.0450 M H_2SO_4
(b) 0.125 M H_2SO_4
(c) 0.150 M H_2SO_4
(d) 0.250 M H_2SO_4
5.45. **(a)** 0.12 M Ba^{2+}; 0.24 M Cl^-
(b) 0.0125 M Cu^{2+}; 0.0125 M SO_4^{2-}
(c) 0.146 M Al^{3+}; 0.438 M Cl^-
(d) 1.000 M K^+; 0.500 M $Cr_2O_7^{2-}$

5.47. 0.205 g Na_2CO_3
5.49. 60. g NaOH; 53 g Cl_2
5.51. 193 mL
5.53. 1.5×10^3 mL
5.55. 0.18 g AgCl; NaCl is in excess; 8.3×10^{-3} M NaCL remains.
5.57. 40.9 mL
5.59. 42.5 mL
5.61. 0.0219 g citric acid/100. mL
5.63. The unknown acid is citric acid.
5.65. 0.500 g vitamin C
5.67. **(a)** $2\,Fe(s) + 3\,Cl_2(g) \rightarrow 2\,FeCl_3(s)$
(b) 19.0 g Cl_2 required; 0.179 mol $FeCl_3$; 29.0 g $FeCl_3$
(c) 63.8%
5.69. The metal is Cu.
5.71. **(a)** $(NH_4)_2PtCl_4$ is limiting and NH_3 is in excess.
(b) 12.5 g $Pt(NH_3)_2Cl_2$
(c) 1.46 g NH_3
5.73. 252 g CCl_4; 85.8 g Cl_2 remains
5.75. B_5H_7
5.77. $[H^+] = 0.102$ M, $[Cl^-] = 0.102$ M
5.79. 0.567 g B_2H_6
5.81. 31.7% Pb
5.83. (b)
5.85. $K_3[Fe(C_2O_4)_3]$
5.87. 2.0 kg Cl_2
5.89. Ti_2O_3
5.91. **(a)** $\frac{x}{y} = \frac{2}{1}$
(b) Element A is carbon.
5.93. Empirical formula is $FeBr_3$.

$$2\,Fe(s) + 3\,Br_2(\ell) \rightarrow 2\,FeBr_3(s)$$

iron(III) bromide
5.96. **(e)** The same as the first student's. Both students started with the same volume of acid (the same number of moles), so both require the same amount of base to titrate the sample.
5.98. **(a)** $MgBr_2$, magnesium bromide; $CaBr_2$, calcium bromide; $SrBr_2$, strontium bromide
(b)
$$Mg(s) + Br_2(\ell) \longrightarrow MgBr_2(s)$$
$$Ca(s) + Br_2(\ell) \longrightarrow CaBr_2(s)$$
$$Sr(s) + Br_2(\ell) \longrightarrow SrBr_2(s)$$
(c) oxidation-reduction reactions
(d) For $MgBr_2$: 1.5 g Mg produces 11.5 g of compound when Mg and Br_2 are mixed in a stoichiometric ratio.

$$1.50\text{ g Mg} \cdot \left(\frac{1\text{ mol Mg}}{24.31\text{ g}}\right) = 0.0617\text{ mol Mg}$$

$$(11.5 - 1.50)\text{ g Br} \cdot \left(\frac{1\text{ mol Br}}{79.90\text{ g}}\right) = 0.125\text{ mol Br}$$

$$\frac{0.125\text{ mol Br}}{0.0617\text{ mol Mg}} = \frac{2\text{ mol Br}}{1\text{ mol Mg}};$$

the empirical formula is $MgBr_2$.

The atomic mass differences of Ca, Mg, and Sr account for the differences in masses of metal and masses of product.

CHAPTER 6

6.11. 399 Calories
6.13. 3.9×10^{11} J/day
6.15. 0.235 J/g · K
6.17. Ethylene glycol requires more heat (1.51×10^4 J).
6.19. 104 kJ
6.21. Water requires more heat.
6.23. 6.2 °C
6.25. 37 °C
6.27. 0.24 J/g · K
6.29. 330 kJ
6.31. 905 kJ
6.33. 48.2 kJ
6.35. exothermic; 2.38 kJ evolved
6.37. exothermic; 3.29×10^4 kJ evolved
6.39. $\Delta H^\circ_{rxn} = 619$ kJ; molar heat of combustion = 310 kJ
6.41. $\Delta H^\circ_f = 50.5$ kJ
6.43. −434.6 kJ; 260 kJ of heat evolved from 250 g of lead
6.45. $2\,Cr(s) + \frac{3}{2}\,O_2(g) \longrightarrow Cr_2O_3(s)$
6.47. **(a)** exothermic
(b) $6\,C(\text{graphite}) + 6\,H_2(g) + 3\,O_2(g) \longrightarrow C_6H_{12}O_6(s)$
6.49. $\Delta H^\circ_{rxn} = 30.1$ kJ
6.55. **(a)** −14.5 kJ
(b) 0.0626 kJ is evolved.
6.57. $\Delta H^\circ_f(C_{10}H_8) = 77.9$ kJ
6.59. 6.62 kJ evolved
6.61. 394 kJ evolved/mol C
6.63. $\Delta H_{rxn} = -56$ kJ/mol
6.65. 0.489 J/g · K
6.67. The ethanol gives up more heat.
6.69. $\Delta H^\circ_{rxn} = -3.529 \times 10^5$ J; 0.43 g Mg required.
6.71. Al 24.3 J/mol · K; Fe 25.2 J/mol · K; Cu 25.5 J/mol · K; Au 25.2 J/mol · K The molar heat capacities are very similar (≈25 J/mol · K). If Ag also has a molar heat capacity of ≈25 J/mol · K, its specific heat is ≈0.23 J/g · K.
6.73. $\Delta H^\circ_f = -56$ kJ
6.75. $\Delta H^\circ_f = 50.$ kJ
6.77. **(a)** Step 1: −2511 kJ (exothermic); Step 2: −566 kJ (exothermic); Step 3: −2446.8 kJ (exothermic); Step 4: −746.6 kJ (exothermic)
(b) In Step 3, 705 kJ evolved for 15.0 g C_2H_2. In Step 1, 723 kJ evolved for 15.0 g C_2H_2.
6.79. Step 1: −137.3 kJ; Step 2: 275.4 kJ; Step 3: 103.7 kJ

Overall process: $H_2O(g) \longrightarrow H_2(g) + \frac{1}{2}\,O_2(g) \qquad \Delta H_{rxn} = 241.8$

The overall process is endothermic.
6.81. CH_3OH −22.67 kJ/g
C_8H_{18} −47.81 kJ/g

6.83. N_2H_4 -16.67 kJ/g
$N_2H_4(CH_3)_2$ -30.00 kJ/g

6.85. 2.36×10^7 J/m^2

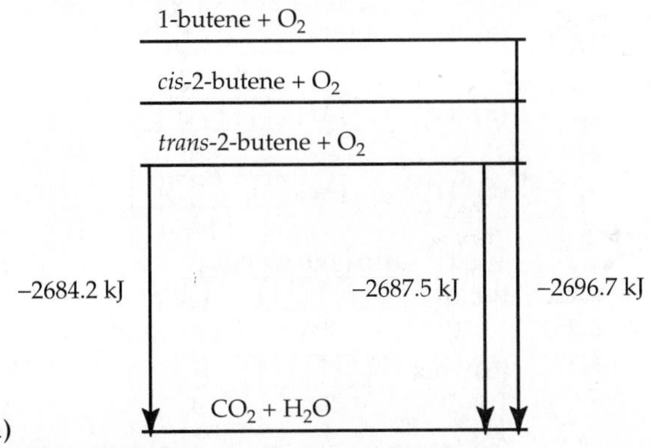

6.87. **(a)**
(b) -3.3 kJ/mol
(c) ΔH_f° (*cis*-2-butene) $= -29.7$ kJ/mol
ΔH_f° (*trans*-2-butene) $= -33.0$ kJ/mol

6.90. The extra heat is the result of the action of the compressor to compress the refrigerant gas into liquid, which then releases heat through the coils on the back of the refrigerator.

6.92. **(a)** 1.7×10^8 g SO_2; 2.7×10^6 mol SO_2
(b) 1.1×10^{13} g MgO; 3.4×10^{13} g $MgSO_4$
(c) 1919.0 kJ evolved
(d) 3.65×10^9 kJ produced per day

CHAPTER 7

7.19. **(a)** red, orange, yellow
(b) blue
(c) blue

7.21. 3.9×10^6 m; 2.5×10^3 mi

7.23. 6.0×10^{14} s^{-1}; 4.0×10^{-19} J/photon; 2.4×10^5 J/mol

7.25. 7.5677×10^{14} s^{-1}; 5.0143×10^{-19} J/photon; 3.0197×10^5 J/mol

7.27. **(a)** 253.652 nm is the most energetic line.
(b) 1.18190×10^{15} s^{-1}; 7.83139×10^{-19} J/photon
(c) yes; 404.656 nm (violet) and 435.833 nm (blue)

7.29. FM radio station < microwaves < yellow light < x-rays

7.31. 600 nm; found in the visible portion of the spectrum

7.33. **(a)** six possible emission lines
(b) highest energy from $n = 4$ to $n = 1$
(c) longest wavelength from $n = 4$ to $n = 3$

7.35. (a) and (d) require longer wavelengths.

7.37. 102.5 nm; ultraviolet portion of the spectrum

7.39. The atom must absorb 2.093×10^{-18} J.

7.41. 2.9×10^{-10} m

7.43. 2.2×10^{-25} nm; 1.2×10^{-21} m/s

7.45.

Quantum number	Atomic property
n	Orbital size
m_ℓ	Relative orbital orientation
ℓ	Orbital shape

7.47. **(a)** $\ell = 0, 1, 2, 3$
(b) $m_\ell = -2, -1, 0, 1, 2$
(c) $n = 4$, $\ell = 0$, $m_\ell = 0$
(d) $n = 4$, $\ell = 3$, $m_\ell = -3, -2, -1, 0, 1, 2, 3$

7.49.

n	ℓ	m_ℓ
4	1	−1
4	1	0
4	1	1

7.51. 4 subshells; 4*s*, 4*p*, 4*d*, 4*f*

7.53. **(a)** The value of ℓ is limited to a maximum of $(n - 1)$.
(b) If $\ell = 0$, m_ℓ can only have a value of 0.
(c) If $\ell = 0$, m_ℓ can only have a value of 0.

7.55. **(a)** 7
(b) 25
(c) None; ℓ cannot have a value equal to n.
(d) 1

7.57. **(a)** no planar nodes
(b) two planar nodes
(c) three planar nodes

7.59. 2*d* cannot exist because ℓ values are limited to $(n - 1)$. Here, $n = \ell = 2$.
3*f* cannot exist because ℓ values are limited to $(n - 1)$. Here $n = \ell = 3$.

7.61. **(a)**

	n	ℓ	m_ℓ
2*p*:	2	1	−1
	2	1	0
	2	1	0

(b)

3*d*:	3	2	−2
	3	2	−1
	3	2	0
	3	2	1
	3	2	2

(c)

4*f*:	4	3	−3
	4	3	−2
	4	3	−1
	4	3	0
	4	3	1
	4	3	2
	4	3	3

7.63. **(e)** *s* orbital

7.65. 500. m; 3.98×10^{-28} J/photon; red light has an energy of 2.90×10^{-19} J/photon.

7.67. 7.2×10^{25} photons

7.69. 4.0 h

7.71. 8.720×10^{-18} J/ion, or 5251 kJ/mol; the ionization energy for He^+ is four times larger than the H ionization engery.

7.73. The shortest wavelength is the transition from $n = \infty$ to $n = (\infty - 1)$, a wavelength of 91.1 nm.

7.75. 0.1058 nm

7.77. **(a)** 3 **(b)** 3 **(c)** 1 **(d)** 25 **(e)** 5 **(f)** 5
(g) 7 **(h)** 1

7.79. **(a)** size and energy; shape
(b) 0, 1, 2
(c) f
(d) 4; 2; −2

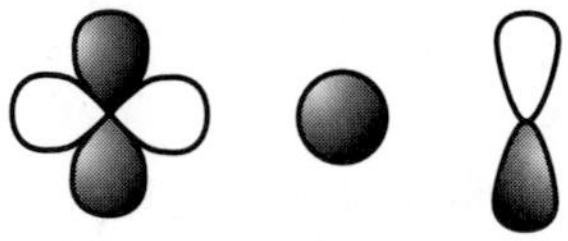

(e)

letter	d	s	p
ℓ value	2	0	1
nodal planes	2	0	1

(f) f
(g) $2d$, $4g$
(h) $n = 2$, $\ell = 1$, $m_\ell = 2$ is not valid
(i) (i) 3
(ii) 9
(iii) none
(iv) 1

7.81. lower

7.86. N = 1, L = 1, M = −1, 0, 1
N = 2, L = 2, M = −1, 0, 1
N = 3, L = 3, M = −1, 0, 1
a total of nine orbitals

7.90. At $y = d$, probability is also 1.0×10^{-4}. At $z = \frac{1}{2}d$, probability is greater than 1.0×10^{-4}.

7.91. **(a)** Group VII B, Period 5
(b) $n = 5$, $\ell = 0$, $m_\ell = 0$
(c) $3.41 \times 10^{19}\ s^{-1}$; 8.79×10^{-12} m
(d) (i) $HTcO_4(aq) + NaOH(aq) \longrightarrow NaTcO_4(aq) + H_2O(\ell)$
(ii) 8.5×10^{-3} g $NaTcO_4$; 1.8×10^{-3} g NaOH

CHAPTER 8

8.13. Mg: $1s^22s^22p^63s^2$ [↑↓] [↑↓] [↑↓|↑↓|↑↓] [↑↓] (1s 2s 2p 3s)

Cl: $1s^22s^22p^63s^23p^5$ [↑↓] [↑↓] [↑↓|↑↓|↑↓] [↑↓] [↑↓|↑↓|↑] (1s 2s 2p 3s 3p)

8.15. V: $1s^22s^22p^63s^23p^63d^34s^2$

8.17. Ge: $1s^22s^22p^63s^23p^63d^{10}4s^24p^2$
[Ar] $3d^{10}4s^24p^2$

8.19. **(a)** Sr: [Kr] $5s^2$
(b) Zr: [Kr] $4d^25s^2$
(c) Rh: [Kr] $4d^85s^1$
(d) Sn: [Kr] $4d^{10}5s^25p^2$

8.21. **(a)** Eu: [Xe] $4f^76s^2$
(b) Yb: [Xe] $4f^{14}6s^2$

8.23. **(a)** Pu: [Rn] $5f^67s^2$
(b) Es: [Rn] $5f^{11}7s^2$

8.25. **(a)** Na^+: [↑↓] [↑↓] [↑↓|↑↓|↑↓] (1s 2s 2p)
(b) Al^{3+}: [↑↓] [↑↓] [↑↓|↑↓|↑↓] (1s 2s 2p)
(c) Cl^-: [↑↓] [↑↓] [↑↓|↑↓|↑↓] [↑↓] [↑↓|↑↓|↑↓] (1s 2s 2p 3s 3p)

8.27. **(a)** Ti: [Ar] [↑|↑| | |] (3d) [↑↓] (4s)
(b) Ti^{2+}: [Ar] [↑|↑| | |] (3d) [ ] (4s)
(c) Ti^{4+}: [Ar] [| | | |] (3d) [ ] (4s)
Ti^{2+} is paramagnetic.

8.29. **(a)** [Ar] [↑|↑|↑|↑|↑] (3d) [↑↓] (4s)
(b) [Ar] [↑|↑|↑|↑|↑] (3d) [ ] (4s)
(c) yes; there are five unpaired electrons.

8.31. **(a)** [Kr] $4d^65s^2$. Table 8.2 shows [Kr] $4d^75s^1$
There is a difference in the $4d$ and $5s$ electron distribution.
[Kr] [↑|↑|↑|↑|↑] (4d) [ ] (5s)
(b) Yes.

8.33. **(a)** Sm: [Xe] [↑|↑|↑|↑|↑|↑|] (4f) [↑↓] (6s)
Sm^{3+}: [Xe] [↑|↑|↑|↑|↑| |] (4f) [ ] (6s)
(b) Ho: [Xe] [↑↓|↑↓|↑↓|↑↓|↑|↑|↑] (4f) [↑↓] (6s)
Ho^{3+}: [Xe] [↑↓|↑↓|↑↓|↑|↑|↑|↑] (4f) [ ] (6s)

8.35. Co^{3+}; four unpaired electrons
Ti^{2+}; two unpaired electrons

8.37. Only Zn^{2+} is diamagnetic; Mn^{2+} has five unpaired electrons.

8.39. Mg[Ne] [↑↓] (3s)

$n = 3$, $\ell = 0$, $m_\ell = 0$, $m_s = +\frac{1}{2}$
$n = 3$, $\ell = 0$, $m_\ell = 0$, $m_s = -\frac{1}{2}$

8.41. Ti[Ar] [↑|↑| | |] (3d) [↑↓] (4s)

$n = 3$, $\ell = 2$, $m_\ell = -2$, $m_s = +\frac{1}{2}$
$n = 3$, $\ell = 2$, $m_\ell = -1$, $m_s = +\frac{1}{2}$
$n = 4$, $\ell = 0$, $m_\ell = 0$, $m_s = +\frac{1}{2}$
$n = 4$, $\ell = 0$, $m_l = 0$, $m_s = -\frac{1}{2}$

8.43. **(a)** 6
(b) 18
(c) None; ℓ cannot have a value equal to n.
(d) 1
(d) None; if $\ell = 0$, $m_\ell = 0$.

8.45. **(a)** ℓ cannot have a value equal to n; $n = 2$, $\ell = 1$, $m_\ell = 0$, $m_s = +\frac{1}{2}$
(b) m_s cannot equal 0; $n = 2$, $\ell = 1$, $m_\ell = -1$, $m_s = +\frac{1}{2}$

(c) The oxygen atom (in its ground state) cannot have a value of n that is greater than 2; $n = 2$, $\ell = 1$, $m_\ell = +1$, $m_s = +\frac{1}{2}$

8.47.
N—Cl 70 + 100 = 170 pm
P—Cl 110 + 100 = 210 pm
As—Cl 120 + 100 = 220 pm
Sb—Cl 140 + 100 = 240 pm
Bi—Cl 150 + 100 = 250 pm

8.49. C < B < Al < Na < K

8.51. (a) Cl^-
(b) Al
(c) In

8.53. (c) Li < Si < C < Ne

8.55. K < Li < C < N

8.57. (a) K (b) C (c) K < Li < C < N

8.59. (a) S < O < F IE increases, moving up a group and to the right across a period.
(b) O IE increases, moving up a group.
(c) Cl The affinity for an electron increases across a period and up a group.
(d) O^{2-} Anions are larger than their neutral atoms; F^- has 10 electrons and 9 protons, whereas O^{2-} has 10 electrons and only 8 protons.

8.61. ΔH_f° (LiCl) = −400. kJ (calculated)

8.63. The lattice energy decreases as the halide gets heavier (larger radius). RbBr lattice energy ≈ 670 kJ.

8.65. element 109 [Rn] $5f^{14}6d^{7}7s^{2}$ Co, Rh, Ir

8.67. (b) is not allowable since the value of ℓ cannot equal the value of n.

8.69. Element 71, Lu, has four complete shells.

8.71. $n = 4$, $\ell = 1$, $m_\ell = -1$, $m_s = +\frac{1}{2}$
$n = 4$, $\ell = 1$, $m_\ell = -1$, $m_s = -\frac{1}{2}$
$n = 4$, $\ell = 1$, $m_\ell = 0$, $m_s = +\frac{1}{2}$
$n = 4$, $\ell = 1$, $m_\ell = 0$, $m_s = -\frac{1}{2}$
$n = 4$, $\ell = 1$, $m_\ell = +1$, $m_s = +\frac{1}{2}$
$n = 4$, $\ell = 1$, $m_\ell = +1$, $m_s = -\frac{1}{2}$

8.73. Tc and Rh

8.75. (a) alkaline earth metal
(b) nonmetal (halogen)
(c) element B
(d) element B

8.77. $S^{2-} > Cl^- > K^+ > Ca^{2+} > Ar$

8.79. In the group 4A element C, an added electron is assigned to a previously unoccupied orbital. In contrast, in N the added electron is assigned an orbital already occupied by an electron. Electron-electron repulsions in N cause the EA to be less negative than expected.

8.81. Since IE's increase across a period, the first IE of Ca is greater than that of K. However, after each has lost an electron, K has the noble gas configuration of Ar, whereas Ca^+ has [Ar] $4s^1$. To remove an inner electron from K^+ requires more energy than removing a second $4s$ electron from Ca^+.

8.83. (a) Na (b) C (c) Na < Al < B < C

8.85. (a) Ti (b) Group 4B, Period 4 (c) transition element
(d) paramagnetic, two unpaired electrons
(e)

	n	ℓ	m_ℓ	m_s
1	3	2	−2	$+\frac{1}{2}$
2	3	2	−1	$+\frac{1}{2}$
4	4	0	0	$-\frac{1}{2}$

(f) Electrons 3 and 4 are removed, leaving a paramagnetic Ti^{2+} ion with two unpaired electrons.

8.87. The increasing nuclear charge attracts the electrons with increasing force, causing a shrinkage of all orbitals and hence a decrease in size.

8.89. As the transition metals add electrons to the $(n-1)d$ level, they repel the ns electrons, partially offsetting the increase in nuclear charge. Consequently, the ns electrons experience only a slightly increasing nuclear attraction, and there is only a slight decline in radii.

8.91. Na has a very large difference between the first and second ionization energies because the first electron is removed from the $3s$ orbital and the second from the $2p$ orbital.

8.93. Cs^+ (noble gas configuration), Se^{2-} (noble gas configuration)

8.95. The size of the atoms decreases across a period, and the mass increases. Therefore, the ratio of mass to volume (density) increases.

8.97. Co^{2+}: [Ar] $3d^7$ three unpaired electrons
Co^{3+}: [Ar] $3d^6$ four unpaired electrons
The two ions have different numbers of unpaired electrons, so $Co(NO_3)_2$ is more paramagnetic than $CoCl_2$. The magnetic moment of the reaction product can be measured to determine the identity of the product.

8.100. (a)

$1s$	$2s$	$2p$	$3s$	$3p$
[↑↓]	[↑↓]	[↑↓][↑↓][↑↓]	[↑↓]	[↑↓][↑][↑]

(b) $n = 3$, $\ell = 1$, $m_\ell = -1$ (or 0 or +1), $m_s = +\frac{1}{2}$ (or $-\frac{1}{2}$)
(c) Smallest IE = S; smallest radius = O
(d) S is smaller than S^{2-}
(e) 804 g Cl_2
(f) Cl_2 is the limiting reactant; 16.8 g $SOCl_2$ is produced.
(g) −212.5 kJ

CHAPTER 9

9.26.

Atom	Group number	Number of valence electrons
N	5A	5
B	3A	3
Na	1A	1
Mg	2A	2
F	7A	7
S	6A	6

9.28. Group 1A: one bond Group 2A: two bonds
Group 3A: three bonds Group 4A: four bonds
Group 5A: three bonds Group 6A: two bonds
Group 7A: one bond

9.30.
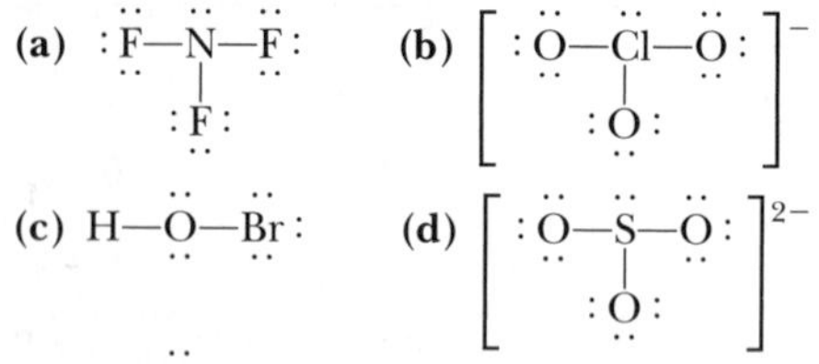

9.32.
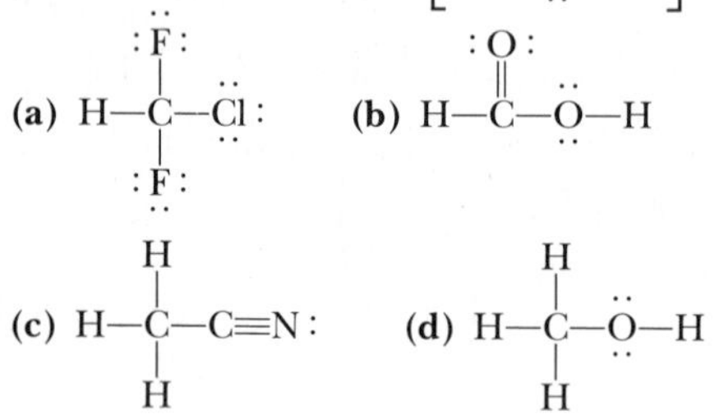

9.34.
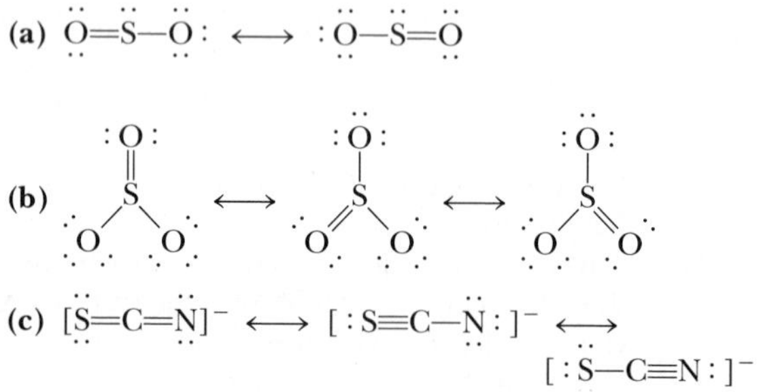

9.36.
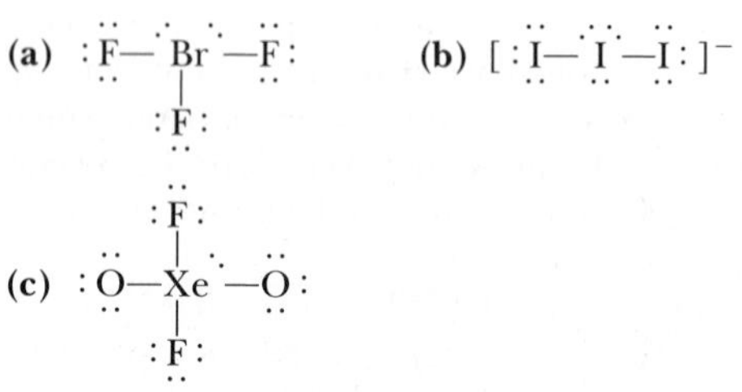

9.38. (a) H_2CO One C═O bond, bond order = 2
Two C—H bonds, bond order = 1
(b) SO_3^{2-} Three S—O bonds, bond order = 1
(c) NO_2^+ Two N═O bonds, bond order = 2

9.40. (a) B—Cl is shorter (b) C—O is shorter
(c) P—O is shorter (d) C═O is shorter

9.42. The carbon monoxide C≡O bond is shorter and stronger than the formaldehyde C═O bond.

9.44. The average bond order in NO_2^+ is 2, whereas it is only 1.33 in NO_3^-. Therefore, the NO bonds in NO_3^- are longer than in NO_2^+.

9.46. enthalpy of hydrogenation = −128 kJ

9.48. reaction enthalpy = −87 kJ

9.50. D_{O-F} = 195 kJ/mol

9.52. (a) C—O > C—N
+⟶ +⟶
(b) P—O > P—S
+⟶ +⟶
(c) P—N > P—H
+⟶ not polar
(d) B—I > B—H
+⟶ +⟶

9.54. (a) The C—H and C═O bonds are polar, whereas the C═C and C—C bonds are nonpolar.
(b) The most polar bond is C═O. The O atom is the negative end of the bond dipole.

9.56. (a) H = +1 and O = −2
(b) H = +1 and O = −1
(c) S = +4 and O = −2
(d) N = +1 and O = −2
(e) Cl = +1 and O = −2

9.58. (a) H = 0 and O = 0 (b) C = 0 and H = 0
(c) N = +1 and O = 0 (d) H = 0, O = 0, F = 0

9.60.
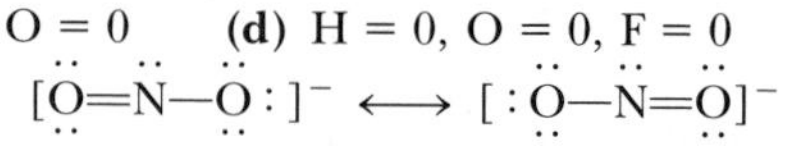

Formal charges: 0 0 −1 −1 0 0

9.62. (a)
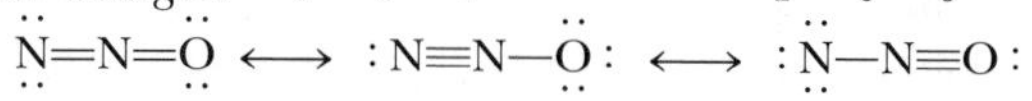

(b) −1 +1 0 0 +1 −1 −2 +1 +1
(c) The center resonance structure is most reasonable since the oxygen atom bears the negative formal charge.

9.64. (a)
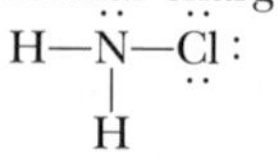
The electron-pair geometry is tetrahedral; the molecular geometry is trigonal-pyramidal.

(b)
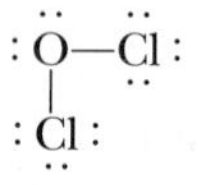
The electron-pair geometry is tetrahedral; the molecular geometry is bent.

(c)
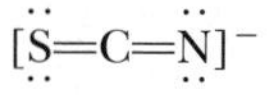
The electron-pair geometry is linear; the molecular geometry is linear.

(d)
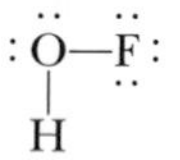
The electron-pair geometry is tetrahedral; the molecular geometry is bent.

9.66. As seen below, 16-electron molecules or ions are linear, 18-electron molecules or ions have a trigonal-planar electron-pair geometry, and 20-electron molecules or ions have a tetrahedral electron-pair geometry.

(a)
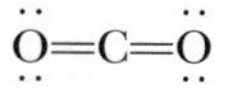
The electron-pair geometry is linear; the molecular geometry is linear.

(b)
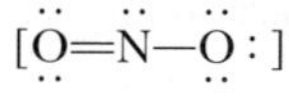
The electron-pair geometry is trigonal-planar; the molecular geometry is bent.

(c)
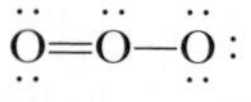
The electron-pair geometry is trigonal-planar; the molecular geometry is bent.

(d)
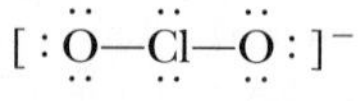
The electron-pair geometry is tetrahedral; the molecular geometry is bent.

(e)
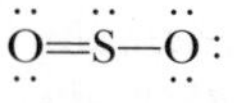
The electron-pair geometry is trigonal-planar; the molecular geometry is bent.

9.68. (a)
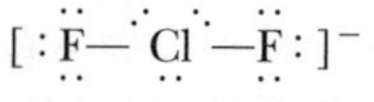
The electron-pair geometry is trigonal-bipyramidal; the molecular geometry is linear.

(b) 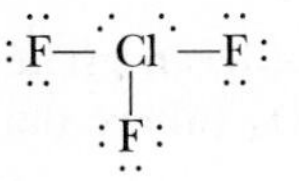 The electron-pair geometry is trigonal-bipyramidal; the molecular geometry is T-shaped.

(c) 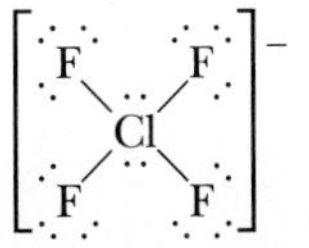 The electron-pair geometry is octahedral; the molecular geometry is square-planar.

(d) 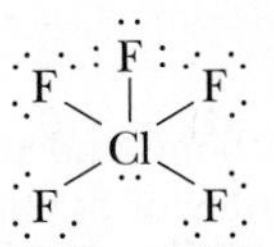 The electron-pair geometry is octahedral; the molecular geometry is square-pyramidal.

9.70. **(a)** 120° **(b)** 120° **(c)** 1 = 109°; 2 = 180°

9.72. 1 = 120°; 2 = 109°; 3 = 120°

9.74. **(a)** SeF_4 is "see-saw"-shaped. One F—Se—F angle is 120°, and the other two are 90°.

(b) 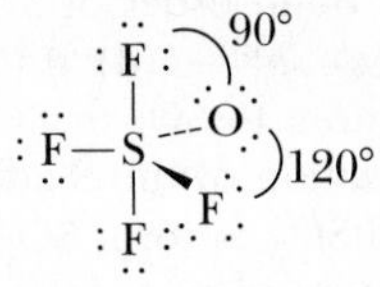

(c) All F—Br—F angles are 90°.

9.76. NO_2^+ has a linear geometry (O—N—O angle = 180°), whereas NO_2^- has a trigonal-planar electron-pair geometry (O—N—O angle = 120°).

9.78. (i) H_2O has the most polar bonds.
(ii) CO_2 and CCl_4 are not polar.
(iii) In ClF the F atom is more negatively charged.

9.80. 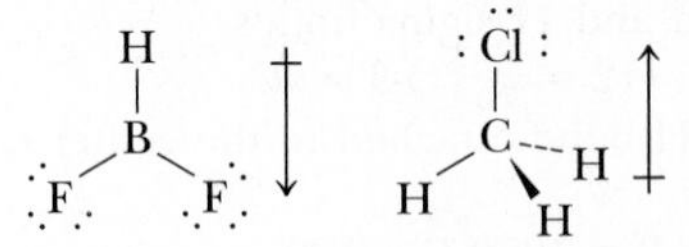

9.82. Group 1A = 1; Group 3A = 3; Group 4A = 4

9.84. The N—O bonds in NO_2^- have a bond order of 1.5, whereas in NO_2^+ the bond order is 2. The shorter bonds (110 pm) are the N—O bonds with the higher bond order (2), whereas the N—O bonds with a bond order of 1.5 are longer (124 pm).

9.86. **(a)** 1 = 120°, 2 = 180°
(b) The C=C bond is shorter than the C—C bonds.
(c) The C=C bond is shorter than the C—C bonds.
(d) The C≡N bond is most polar, and N is the negative end of the bond dipole.

9.88. All molecules except BF_3 are based on a tetrahedron of electron pairs.

(a) 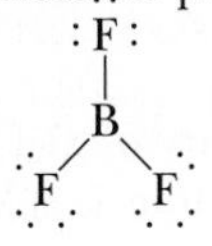 The electron-pair geometry is trigonal-planar; the molecular geometry is trigonal-planar.

(b) 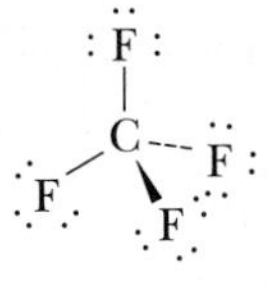 The electron-pair geometry is tetrahedral; the molecular geometry is tetrahedral.

(c) 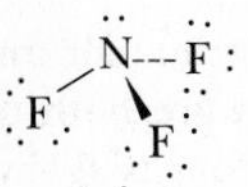 The electron-pair geometry is tetrahedral; the molecular geometry is trigonal-pyramidal.

(d) 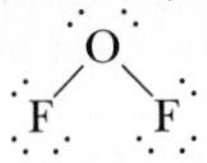 The electron-pair geometry is tetrahedral; the molecular geometry is bent.

(e) H—F: The electron-pair geometry is tetrahedral; the molecular geometry is linear.

9.90. **(a)** In XeF_2 the fluorine atoms are axial becuase this results in a 120° angle between the lone pair electrons, minimizing electron-electron repulsion.
(b) If ClF_3 were trigonal-planar, the lone pair electrons would be in the axial positions of the trigonal bipyramid. It is usually preferable to place these pairs in the equatorial plane, so the trigonal-planar structure of ClF_3 is not observed.

9.92. **(a)** 1 = 120°; 2 = 120°; 3 = 120°; 4 = 120°; and 5 = 109°
(b) The C=O bond is the most polar bond.

9.94. **(a)** [C=N=O]⁻ ⟷ [:C≡N—O:]⁻ ⟷ [:C—N≡O:]⁻

Formal charges: −2 +1 0; −1 +1 −1; −3 +1 +1

(b) The center resonance structure is the most reasonable since it is the only one with a negative formal charge on oxygen.
(c) The most reasonable resonance structure has a negative formal charge on the less electronegative carbon and a positive formal charge on the more electronegative nitrogen. Therefore, it is likely to be an unstable ion.

9.96. $D_{B—F}$ = 646 kJ/mol

9.98. **(a)** $Cl_2(g) + 3\,F_2(g) \longrightarrow 2\,ClF_3(g)$
(b) F_2 is limiting; 1.62 g ClF_3 is produced.

(c, d) 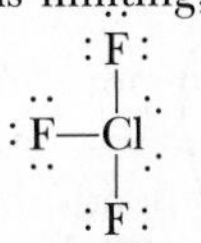 The electron-pair geometry is trigonal-bipyramidal.

(e) If the geometry of ClF_3 were planar and trigonal, the molecule would be nonpolar. Therefore, this geometry is ruled out. Another structural arrangement is also polar, however. Other methods would be needed to prove the molecule is actually T-shaped.
(f) $\Delta H_f^\circ = -405$ kJ/mol

CHAPTER 10

10.15. 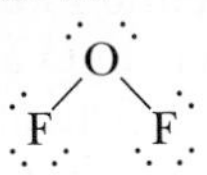 The electron-pair geometry is tetrahedral; the molecular geometry is bent. The O atom is sp^3-hybridized and the O—F bonds form by overlap of an O sp^3 orbital with an F p orbital.

10.17. 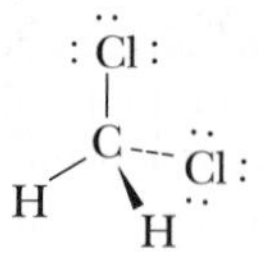

The electron-pair and molecular geometry are both tetrahedral. The C atom is sp^3-hybridized. C—H bonds form by overlap of a C sp^3 orbital with an H *s* orbital. The C—Cl bonds form by overlap of a C sp^3 orbital with a Cl *p* orbital.

10.19. **(a)** sp^2 **(b)** *sp* **(c)** sp^3 **(d)** sp^2

10.21. **(a)** Both C and O are sp^3-hybridized.
(b) C of CH_3 is s^3-hybridized; other C atoms are sp^2-hybridized.
(c) The C of $—CH_2—$ is sp^3-hybridized, the C=O carbon is sp^2-hybridized, and the N atom is sp^3-hybridized.

10.23. :F—Xe—F: The electron-pair geometry is trigonal-bipyramidal, and the molecular geometry is linear. The Xe atom is sp^3d-hybridized.

10.25. **(a)** sp^3d^2 **(b)** sp^3d **(c)** sp^3d

10.27. O=C=O The central atom is *sp*-hybridized. This leaves two unhybridized *p* orbitals on the C atom to be used in pi bonding. The C atom *sp* hybrid orbitals are used to form sigma bonds with the O atoms.

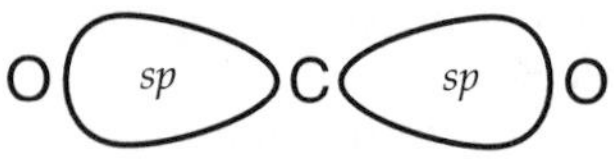

The unhybridized 2*p* orbitals are used to form pi bonds to the O atoms. One 2*p* C atom orbital is illustrated below, forming one of the C=O pi bonds.

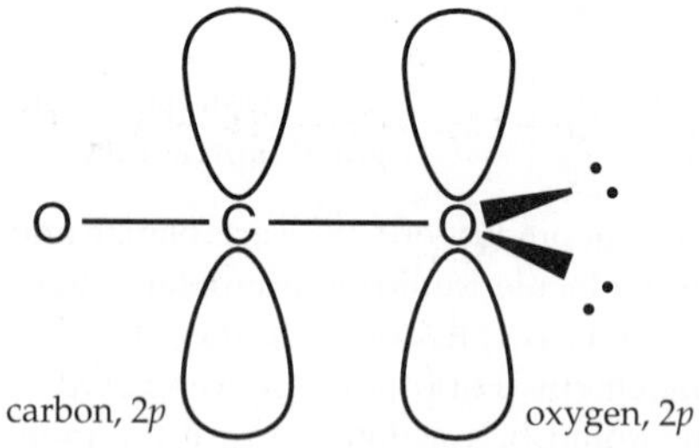

10.29. 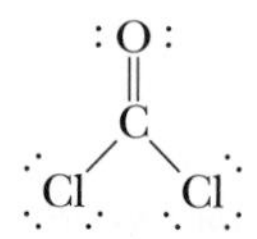

The C atom is sp^2-hybridized, each of these hybrid atomic orbitals being used to form a sigma bond (one to O and two to the Cl atoms). The unhybridized 2*p* orbital is used to form a CO pi bond as in CO_2 in Study Question 10.27.

10.31. H_2^+, $(\sigma_{1s})^1(\sigma^*_{1s})^0$. Bond order = $\frac{1}{2}$. The H—H bond in H_2^+ is weaker than in H_2 (where the bond order is 1).

10.33. 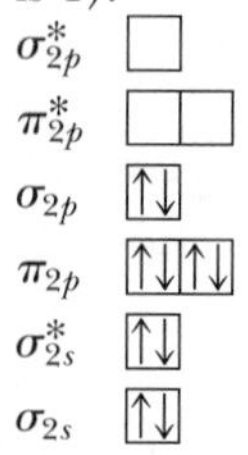

The C_2^{2-} ion has one sigma bond and two pi bonds, for a carbon-carbon bond order of 3. On adding two electrons to C_2 the bond order increases by 1. The C_2^{2-} ion is diamagnetic.

10.35. CO: [core electrons] $(\sigma_{2s})^2(\sigma_{2s})^2(\pi_{2p})^4(\sigma_{2p})^2$. The molecule is diamagnetic with one sigma bond and two pi bonds, for a net bond order of 3.

10.37. **(a)** sp^2 **(b)** sp^2 **(c)** sp^3 **(d)** sp^3
The addition of an O atom to SO_2 or SO_3^{2-} does not change the hybridization of the S atom. The addition of electrons to SO_3 to form SO_3^{2-} causes the S atom hybridization to change from sp^2 to sp^3.

10.39. The N atom is sp^2-hybridized. The N=O bond consists of a sigma bond resulting from overlap of an N sp^2 hybrid orbital with an O sp^2 hybrid orbital, and a pi bond formed from overlap of unhybridized 2*p* orbitals on N and O.

10.41. **(a)** Carbon 1: sp^2 Carbon 2: sp^2
(b) A = 120°; B = 120°; and C = 120°

10.43. **(a)** 1 pi bond and 11 sigma bonds
(b) C-1 = sp^3; C-2 = sp^2; O-3 = sp^3
(c) The O—H bond attached to the center carbon is shorter.
(d) A = 109°; B = 109°; C = 120°

10.45. **(a)** Trigonal-planar in BF_3, tetrahedral in $H_3N—BF_3$
(b) Boron is sp^2-hybridized in BF_3, sp^3-hybridized in $H_3N—BF_3$.
(c) Yes

10.47.

	N_2	N_2^+	N_2^-
(a)	diamagnetic	paramagnetic	paramagnetic
(b)	2 pi bonds	2 pi bonds	$1\frac{1}{2}$ pi bonds
(c)	bond order = 3	$2\frac{1}{2}$	$2\frac{1}{2}$

(d) $N_2 < N_2^+ \approx N_2^-$
—increasing bond length →

(e) $N_2^+ \approx N_2^- < N_2$
—increasing bond strength →

10.49. **(a)** sp^3d in SbF_5, sp^3d^2 in SbF_6^-

(b) 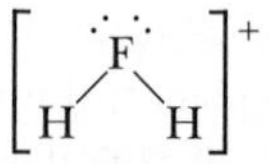

The geometry of H_2F^+ is bent, and the F atom is sp^3-hybridized.

10.51. 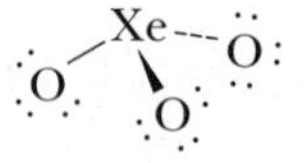

The electron-pair geometry is tetrahedral; the molecular geometry is trigonal-pyramidal, Xe is sp^3-hybridized.

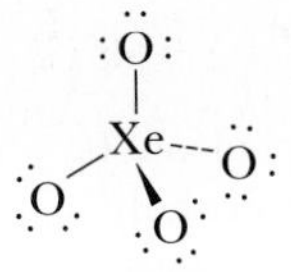

The electron-pair and molecular geometry are tetrahedral; Xe is sp^3-hybridized.

10.53. NO, Ne_2^+, and CN are paramagnetic.

10.55. **(a)** both C atoms: $sp^2 \rightarrow sp^3$
(b) P: $sp^3 \rightarrow sp^3d$
(c) Xe: $sp^3d \rightarrow sp^3d^2$
(d) Sn: $sp^3 \rightarrow sp^3d^2$

10.57. **(a)** C-1 = sp^3; C-2 = sp; N = sp^2
(b) C—N═C angle = 120°; N═C═S angle = 180°
(c) The carbon 2s and 2p_x orbitals are used to form two sp hybrid orbitals. The carbon 2p_y and 2p_z orbitals are used to form pi bonds with N and S.

10.58. 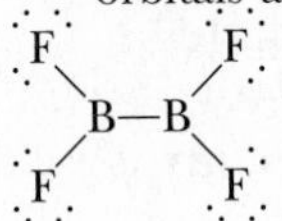

B is sp^2-hybridized; F—B—F = 120°.

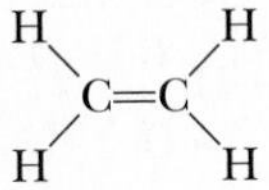

C is sp^2-hybridized; H—C—H = 120°.

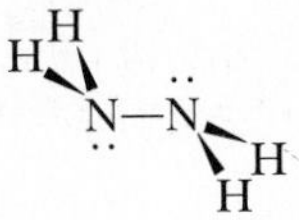

N is sp^3-hybridized; H—N—H = 109°.

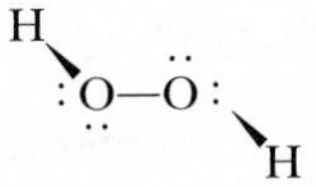

O is sp^3-hybridized; H—O—O = 109°. The major difference is that the carbon compound has a double bond.

10.60. **(a)** −80 kJ, exothermic

(b) 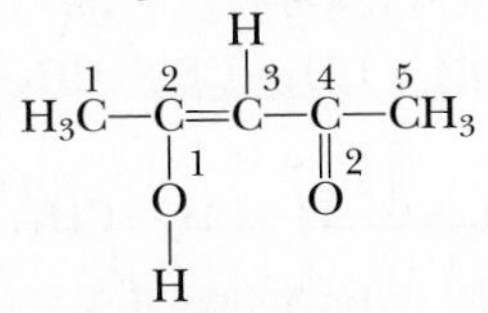

enol form

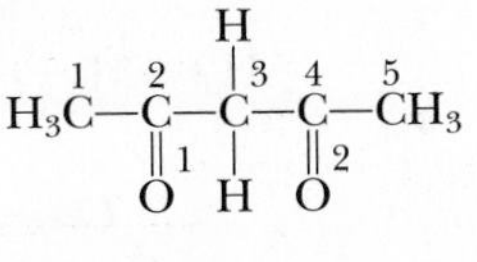

keto form

Enol form: C-1 = sp^3, C-2 = sp^2, C-3 = sp^2, C-4 = sp^2, C-5 = sp^3, O-1 = sp^3, O-2 = sp^2. In the keto form, O-1 changes from sp^3 to sp^2, and C-3 changes from sp^2 to sp^3.

(c)

Carbon Number	Shape	Enol	Keto
C-1	electron-pair geometry	tetrahedral	tetrahedral
	molecular geometry	tetrahedral	tetrahedral
C-2	electron-pair geometry	trigonal planar	trigonal planar
	molecular geometry	trigonal planar	trigonal planar
C-3	electron-pair geometry	trigonal planar	tetrahedral
	molecular geometry	trigonal planar	tetrahedral
C-4	electron-pair geometry	trigonal planar	trigonal planar
	molecular geometry	trigonal planar	trigonal planar
C-5	electron-pair geometry	tetrahedral	tetrahedral
	molecular geometry	tetrahedral	tetrahedral

(d) 6.80 g $CrCl_3$; 12.9 g acetylacetone; 5.15 g NaOH

CHAPTER 11

11.1. octane

11.3. **(c)** $C_{14}H_{30}$ is an alkane.
(b) C_5H_{10} could be a cycloalkane.

11.5. 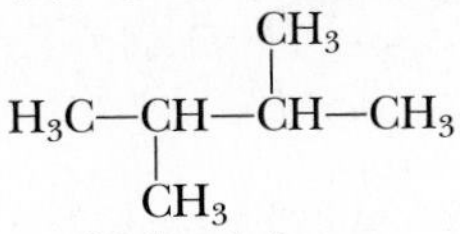

2,3-dimethylbutane

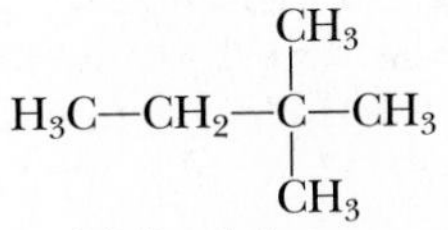

2,2-dimethylbutane

11.7. **(a)** 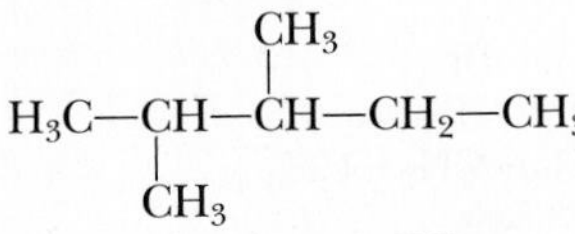

(b) 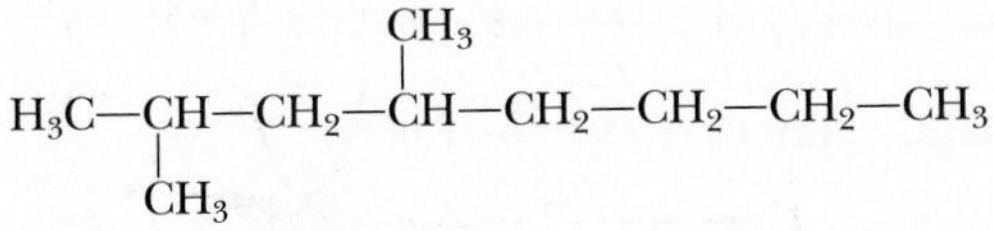

(c) 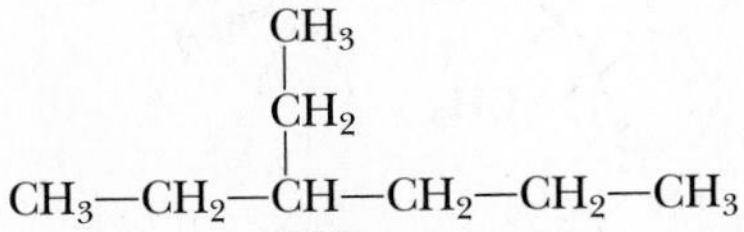

(d) 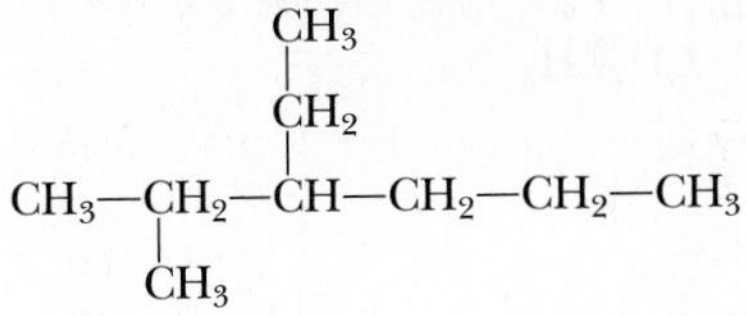

11.9. **(a)** 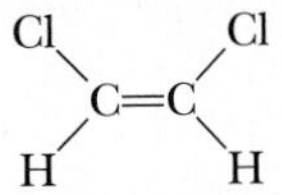

cis-1,2-dichloroethene

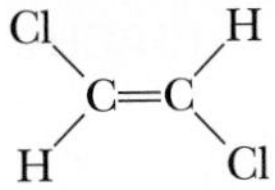

trans-1,2-dichloroethene

(b) 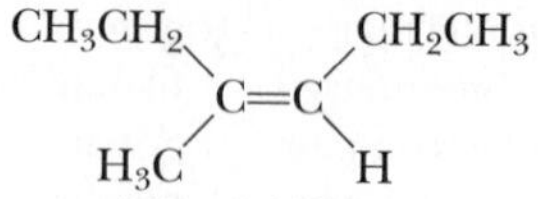

cis-3-methyl-3-hexene

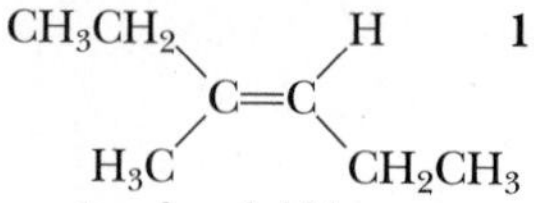

trans-3-methyl-3-hexene

11.11. (a) $H_2C{=}CH{-}CH_2{-}CH_2{-}CH_3$

1-pentene

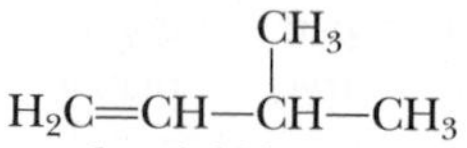

3-methyl-1-butene

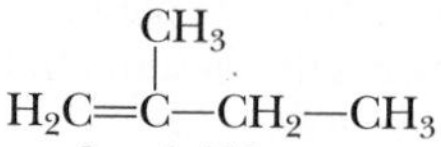

2-methyl-1-butene

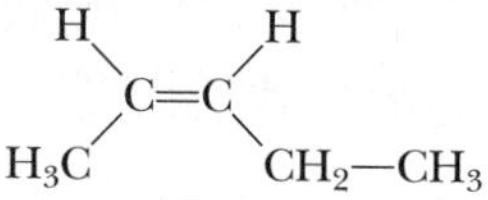

cis-2-pentene

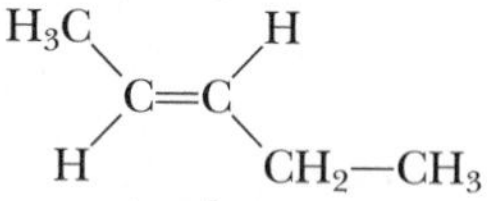

trans-2-pentene

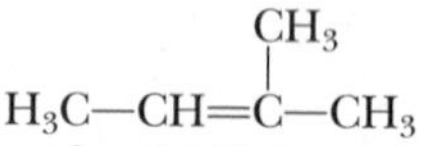

2-methyl-2-butene

11.12. (b) 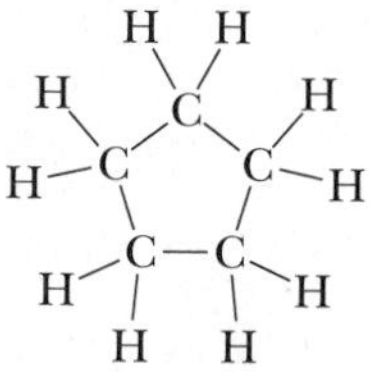

cyclopentane

11.13. (a) 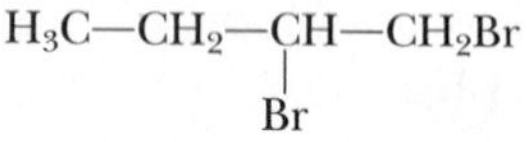

1,2-dibromobutane

(b) $H_3C{-}CH_2{-}CH_2{-}CH_3$

butane

11.15. (a) 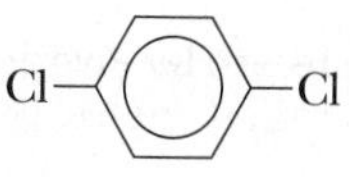

(b) 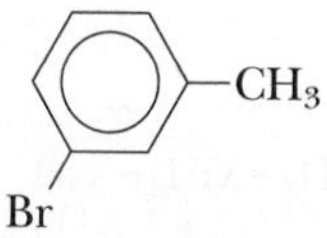

(c) 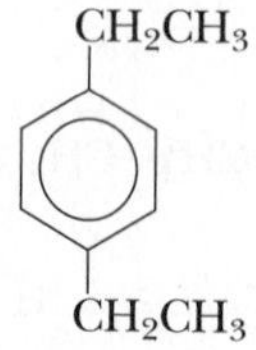

11.17. 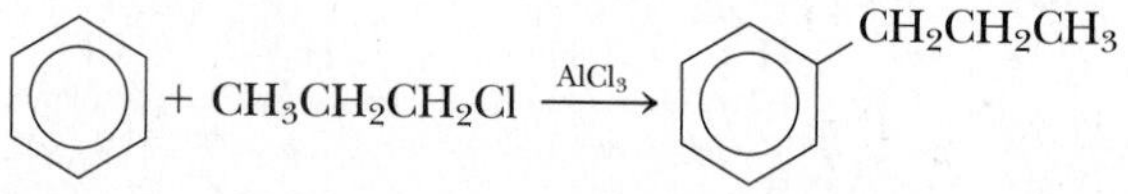

11.19. (a) 1-propanol, primary alcohol
(b) 1-butanol, primary alcohol
(c) 2-methyl-2-propanol, tertiary alcohol
(d) 2-methyl-2-butanol, tertiary alcohol

11.21. $CH_3{-}CH_2{-}CH_2{-}CH_2{-}OH$ 1-butanol

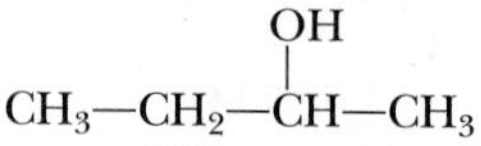 2-butanol

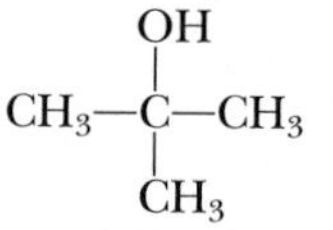 2-methyl-2-propanol

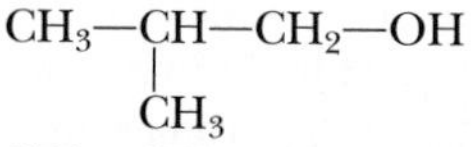 2-methyl-1-propanol

11.23. (a) $CH_3{-}CH_2{-}CH{=}CH_2$
(b) $CH_3{-}CH_2{-}CH_2{-}ONa + \frac{1}{2}H_2$
(c) $CH_3{-}CH_2{-}CH_2{-}CH_2Br + H_2O$

11.25. (a) 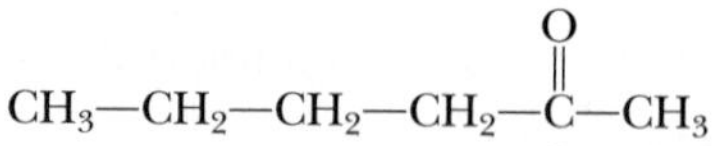

(b) 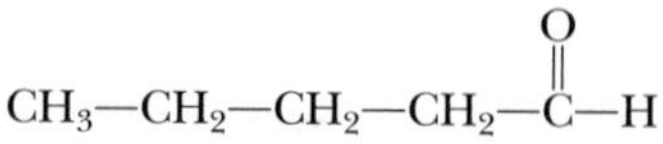

(c) 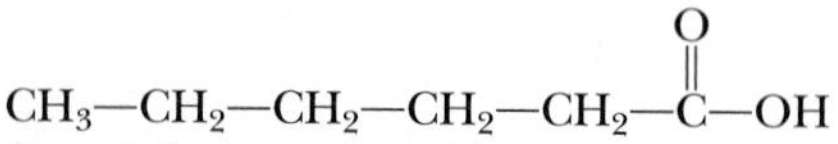

11.27. (a) 3-methylpentanoic acid
(b) methyl propanoate
(c) butyl ethanoate
(d) *p*-bromobenzoic acid

11.29. (a) 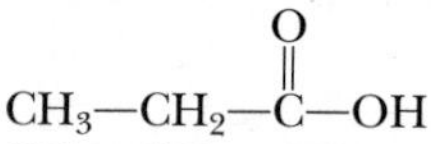 propanoic acid

(b) $CH_3{-}CH_2{-}CH_2{-}OH$ 1-propanol

(c) 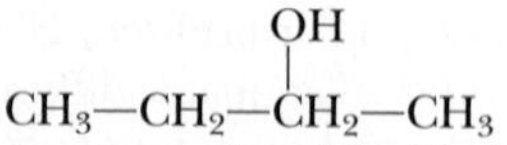 2-butanol

11.31. React propanoic acid with 1-propanol in the presence of a strong acid.

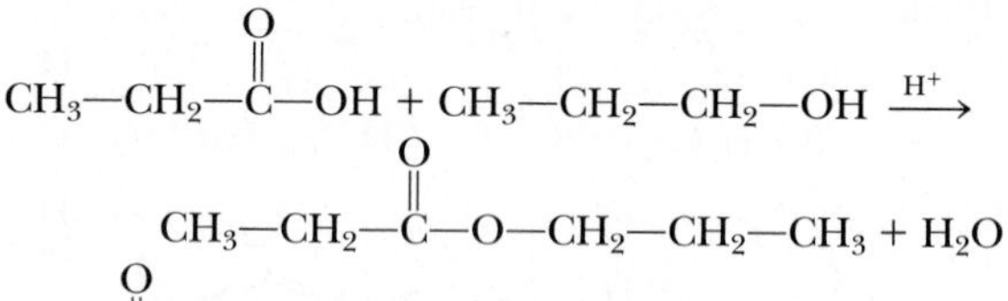

11.33. 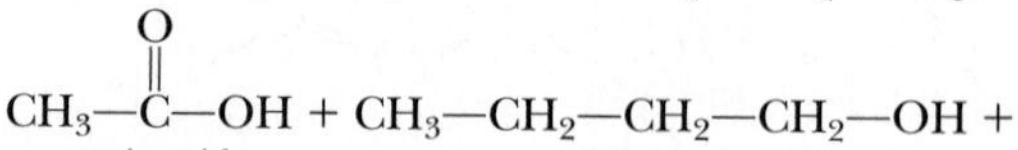

acetic acid — 1-butanol — Na^+ sodium ion

11.35. (a) 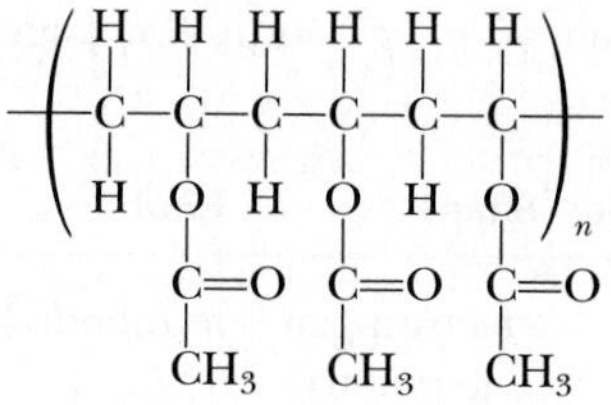

(b) Polyvinyl alcohol can be formed by the hydrolysis of polyvinyl acetate, followed by treatment with a strong acid such as HCl.

11.37. 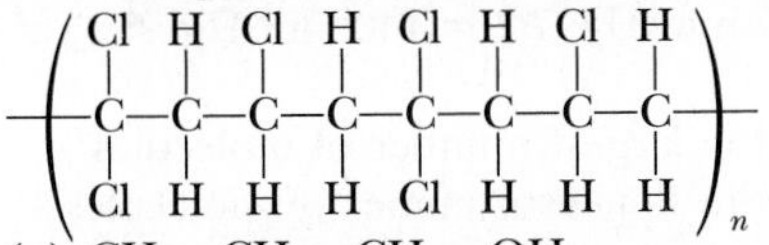

11.39. **(a)** $CH_3-CH_2-CH_2-OH$
1-propanol

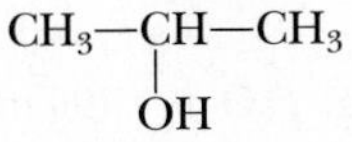
2-propanol

$CH_3-O-CH_2-CH_3$
methylethyl ether

(b) 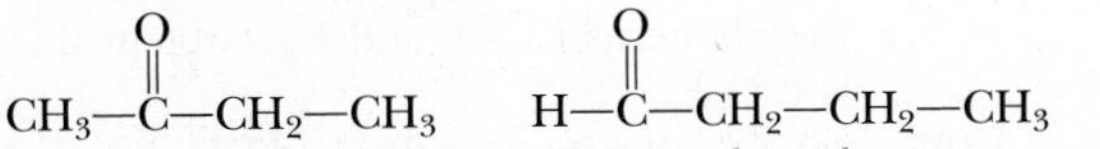

11.41. 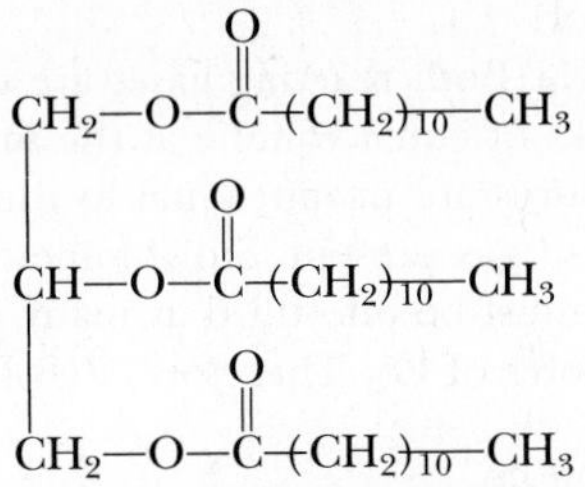
glyceryl trilaurate

Upon saponification, glyceryl trilaurate forms glycerol and 3 mol of sodium laurate, a soap.

11.43. **(a)** 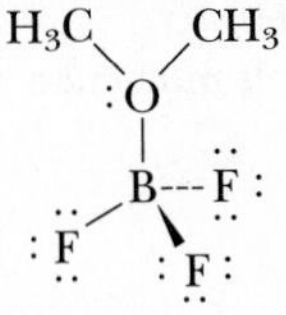

(b) The O atom is sp^3-hybridized before and after bond formation. The B atom is sp^2-hybridized before bond formation and sp^3-hybridized after.

(c) Before reaction: 120° After reaction: 109°

11.45. 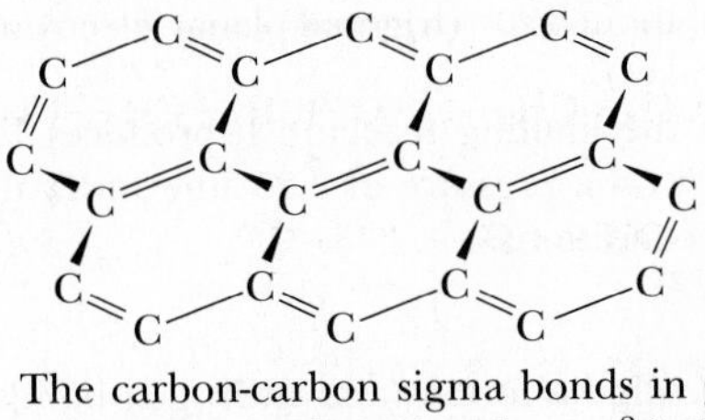

The carbon-carbon sigma bonds in graphite are formed by overlap of carbon sp^2 hybrid orbitals. The pi bonds are formed from overlap of unhybridized carbon $2p$ orbitals.

11.47. **(a)** $H_3C-CH_2-CH=CH_2 + Br_2 \longrightarrow$
1-butene

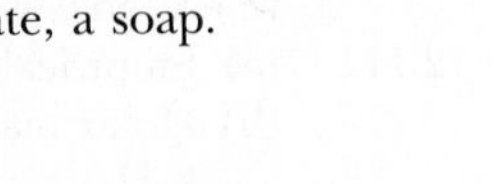
1,2-dibromobutane

(b) $CH\equiv C-CH_3 + H_2 \xrightarrow{\text{catalyst}} CH_2=CH-CH_3$
propyne → propene

(c) $CH\equiv C-CH_3 + 2\,H_2 \xrightarrow{\text{catalyst}} CH_3-CH_2-CH_3$
propyne → propane

(d) $H_3C-CH=CH-CH_3 + H_2O \longrightarrow$
2-butene

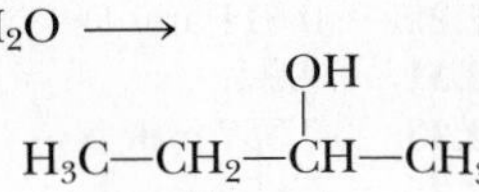
2-butanol

11.49. 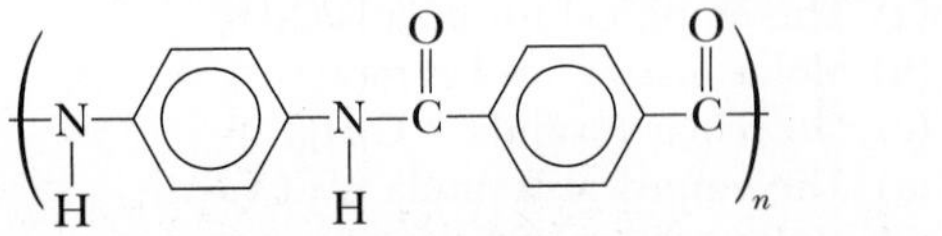

11.51. 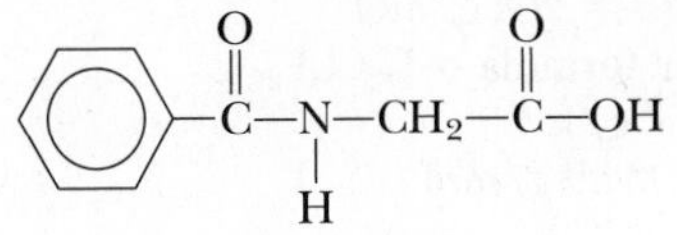
hippuric acid

Hippuric acid is an acid because it contains a carboxylic acid group.

11.53. $CH_3-CH_3 + \frac{7}{2}\,O_2 \longrightarrow 2\,CO_2 + 3\,H_2O$

$\Delta H^\circ_{rxn} = -1559.7$ kJ

$CH_3-CH_2-OH + 3\,O_2 \longrightarrow 2\,CO_2 + 3\,H_2O$

$\Delta H^\circ_{rxn} = -1366.7$ kJ

Ethane has a more negative enthalpy of combustion. The heat realized by combustion of ethanol is less negative, so partially oxidizing ethane to form ethanol decreases the amount of energy per mole available from combustion of the substance.

11.55. The alkene should react with elemental bromine to give colorless products. The cyclopentane does not react with elemental bromine.

11.57. **(a)** Rotation about the carbon-carbon double bond in 2-butene requires considerable energy because it involves breaking a carbon-carbon pi bond.

(b) Although there is free rotation about single bonds, there are some energy barriers due to electron-pair repulsions between adjacent bonding pairs and steric repulsions between atoms.

11.59. **(a)** The empirical formula is CHO.

(b) The molecular weight is 116 g/mol; the molecular formula is $C_4H_4O_4$.

(c) 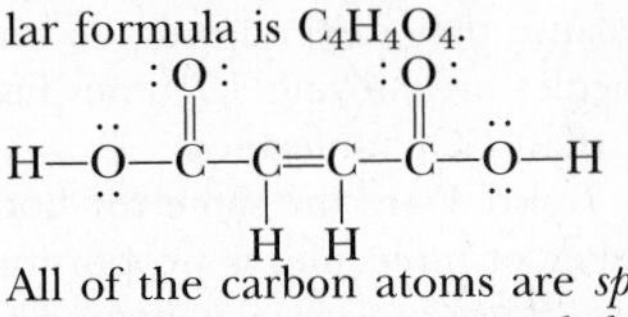

(d) All of the carbon atoms are sp^2-hybridized.

(e) All the bond angles around the carbon atoms are 120°.

CHAPTER 12

12.11. **(a)** 0.954 atm
(b) 5.1×10^2 mm Hg
(c) 99 kPa
(d) 76 kPa
(e) 0.993 bar
(f) 125 torr

12.13. 56.3 mm Hg; 0.0741 atm; 56.3 torr; 7.51 kPa

12.15. 16.9 mm Hg

12.17. 3.7 L

12.19. 0.50 L O_2 gas

12.21. 123 mm Hg

12.23. 82.3 mm Hg

12.25. 224 mL

12.27. 9.72 atm

12.29. 0.811 atm
12.31. 2.9 L
12.33. 0.337 mol N_2
12.35. 57.5 g/mol
12.37. **(a)** The empirical formula is C_2H_5.
(b) Molar mass = 58.2 g/mol
(c) Molecular formula = C_4H_{10}
12.39. **(a)** The empirical formula is CCl_2F.
(b) Molar mass = 204 g/mol
(c) Molecular formula = $C_2Cl_4F_2$
12.41. $d = 3.7 \times 10^{-4}$ g/L
12.43. Molar mass = 34.0 g/mol
12.45. 10.2 g O_2
12.47. 58°C
12.49. 22.2 mm Hg
12.51. 58 g NaN_3
12.53. 1.7 atm of O_2 gas
12.55. 2 atm F_2 gas remaining
12.57. $Fe(CO)_5$
12.59. **(a)** KO_2 is consumed completely.
(b) 1.65 atm O_2 gas
12.61. 0.047 g H_2
12.63. 2.1 atm H_2S; 1.4 atm CO_2; and 43 atm N_2
12.65. **(a)** 2.0 g He
(b) 0.46 atm He
(c) 0.5 atm O_2
(d) X(He) = 0.46 and X(O_2) = 0.54
12.67. P(Cl_2) = 250. mm Hg
12.69. P(N_2) = 732 mm Hg
12.71. **(a)** true
(b) False. Because the masses of N_2 and O_2 are the same, there are more moles of the lighter compound (N_2) than of the heavier compound (O_2). Therefore, there are more molecules of N_2 than of O_2.
12.73. **(a)** The average kinetic energies of the gases are the same because they depend only on temperature.
(b) H_2 molecules are moving 4.7 times faster, on average, than CO_2 molecules.
(c) Because *T* and *V* are the same for both gases, the number of molecules is proportional to *P*. Therefore, there are twice as many CO_2 molecules as H_2 molecules.
(d) Because there are more molecules present of CO_2 than of H_2, the mass of CO_2 is greater than the mass of H_2.
12.75. 3.65×10^4 cm/s
12.77. $CH_2F_2 < Ar < N_2 < CH_4$
12.79. He effuses 3.159 times faster than Ar.
12.81. 36 g/mol
12.83. 29.5 atm from van der Waals's equation; 49.3 atm from the ideal gas law
12.85.

	atm	mm Hg	kPa	bar
(a)	1	760.	101.325	1.013
(b)	0.780	593	79.0	0.790
(c)	131	9.98×10^4	1.33×10^4	133
(d)	0.333	2.53×10^2	33.7	0.337

12.87. 44.0 g/mol
12.89. 306 K or 33 °C
12.91. 17 °C
12.93. P_{He} = 87 mm Hg; P_{Ar} = 140 mm Hg; P_{total} = 227 mm Hg
12.95. **(d)** has the largest number of molecules.
(c) has the smallest number of molecules.
(b) has the greatest mass of gas.
12.97. 64.6% $NaNO_2$
12.99. **(a)** 3.8 g NiO
(b) $P(Cl_2)$ = 120 mm Hg; $P(O_2)$ = 190 mm Hg; P_{total} = 310 mm Hg
12.101. 7.71×10^{17} molecules
12.103. Molar mass = 86.4 g/mol. The only way to combine the elements Cl, O, and F to obtain this molar mass is ClO_2F.
12.105. **(a)** O_2(32 g/mol) < B_2H_6(27.7 g/mol) < H_2O(18 g/mol)
(b) 15 atm of B_2H_6. Both reacting gases are contained in tanks of equal volume at the same *T*, so their pressures are proportional to the number of moles of gas present. Stoichiometry demands there must be one third as many moles of B_2H_6 as moles of O_2. Therefore, $P(B_2H_6) = \frac{1}{3}P(O_2)$.
12.107. 0.0125 mol He
12.109. Gases meet approximately 30 cm from the end where NH_3 is injected.
12.111. **(a)** Empirical formula = B_2H_3
(b) Molar mass = 98.4 g/mol; molecular formula = B_8H_{12}
12.119. **(a)** 19 valence electrons
(b) 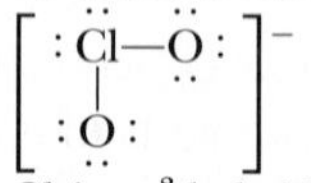
(c) Cl is sp^3-hybridized, and the ion shape is bent.
(d) The angle in ClO_2^- is about 109° (electron-pair geometry is tetrahedral), whereas the angle in O_3 is about 120° (trigonal-planar electron-pair geometry).
(e) Cl_2 is the limiting reactant. It produces 11.2 g ClO_2 with a pressure of 3.25 atm under the specified conditions.

CHAPTER 13

13.16. **(a)** Dipole-dipole forces (and hydrogen bonds) in water; **(b)** Ion-ion forces in solid KCl; (c) Ion-dipole forces assist in dissolving the KCl in water.
13.18. Ion-dipole forces
13.20. **(a)** Induced dipole/induced dipole; **(b)** induced dipole/induced dipole; **(c)** dipole-dipole; and **(d)** dipole-dipole (and H-bonds)
13.22. The order that might be expected is Ne < CH_4 < CO < CCl_4. In fact, however, the boiling points are in the order Ne (−246 °C) < CO(−191 °C) < CH_4(−161 °C) < CCl_4(77 °C). The position of the middle two compounds is difficult to predict because

of the importance of induced dipole forces and the very weak dipole in CO.

13.24. No there is no significant hydrogen bonding in H_2S.

13.26. CH_3I. Forces involved are dipole-dipole attractions and induced dipole/induced dipole attractions. CH_3Cl has a larger dipole (1.892 D) than CH_3I (1.62 D), but the larger I atom in CH_3I is more polarizable, yielding greater induced dipole attractive forces and a higher boiling point.

13.28. **(c)** HF; **(d)** CH_3CO_2H; and **(f)** CH_3OH.

13.30. **(a)** LiCl because Li^+ is smaller than Cs^+ and exerts a greater attractive force on water.
(b) $Mg(NO_3)_2$ because magnesium has a 2+ charge (Mg^{2+}), whereas sodium is only 1+ (Na^+).
(c) $NiCl_2$ because nickel has a 2+ charge (Ni^{2+}), whereas rubidium is only 1+ (Rb^+).

13.32. 2.00 kJ

13.34. **(a)** Slightly greater than 400 mm Hg
(b) Diethyl ether < ethanol < water
(c) Diethyl ether is a vapor, whereas ethanol and water are predominantly liquids.

13.36. **(a)** Approximate vapor pressure is 600 mm Hg. About 0.2 g of the ether evaporates to create this pressure.
(b) At ice temperature the vapor pressure is less than 200 mm Hg, so some ether vapor condenses to liquid.

13.38. **(a)** CS_2, 360 mm Hg; CH_3NO_2, 80 mm Hg
(b) CS_2, induced dipole attractions; CH_3NO_2, dipole-dipole
(c) CS_2, 46 °C; CH_3NO_2, 101 °C.
(d) about 39 °C
(e) about 35 °C

13.40. Propane can be liquefied as long as T does not exceed T_c, 96.7 °C.

13.42. Assume that black = A and white = B. The following pattern leads to a formula of AB because there is a net of $2A$ ($= 1 + 4 \cdot \frac{1}{4}$) and $2B$ ($= 4 \cdot \frac{1}{2}$) inside the dashed line.

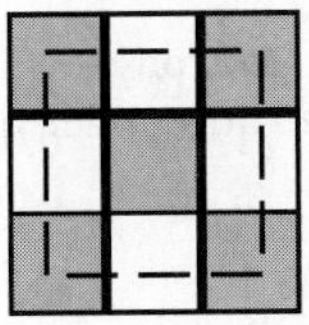

13.44. Mass of unit cell = 2.66×10^{-22} g. Volume of unit cell = 1.73×10^{-22} cm³. Length of cell edge = 5.57×10^{-8} cm. Radius of Ca atom = 197 pm

13.46. body-centered cubic

13.48. face-centered cubic

13.50. Unit cell contains one Ca^{2+} ion, one Ti^{4+} ion (= 8 corners × $\frac{1}{8}$ per corner), and three O^{2-} ions (= 12 edges × $\frac{1}{4}$ per edge).

13.52. The unit cell consists of a face-centered cubic unit cell of Zn^{2+} ions (= 4 Zn^{2+} ions) and S^{2-} ions in holes (in the four so-called "tetrahedral holes" in the lattice).

13.54. This is a face-centered cubic unit cell of C atoms (= 4 C atoms) with 4 more C atoms in holes inside the lattice. Therefore, there is a total of 8 C atoms in the unit cell.
(a) 4.55×10^{-23} cm³
(b) 357 pm

13.56. **(a)** The solid phase is denser than the liquid phase.
(b) vapor
(c) T_c = 31 °C

13.58. Heat evolved on cooling to boiling point = 1350 J
Heat evolved on converting from vapor to liquid at boiling point = 3.33 kJ.
Heat evolved on cooling liquid from boiling point to −40 °C = 122 J
Total heat evolved = 4.80 kJ

13.60. $MgCl_2$, because the lattice energy depends inversely on the size of the ions involved, the smaller the ions the larger the lattice energy. Here Mg^{2+} is smaller than Ba^{2+}. See pages 386–389 and Figure 13.39.

13.62. As explained in the answer to Question 13.60, lattice energy depends on ion size (and charge). CsF has a higher melting point (682 °C) than CsI (621 °C). The cation-anion separation in CsI is greater than in CsF, leading to a weaker ion-ion attraction and to a smaller lattice energy.

13.64. Heat evolved on freezing, $q = -1.97$ kJ. Heat required to melt the solid, $q = +1.97$ kJ

13.66. $Ar < CO_2 < CH_3OH < CaO$

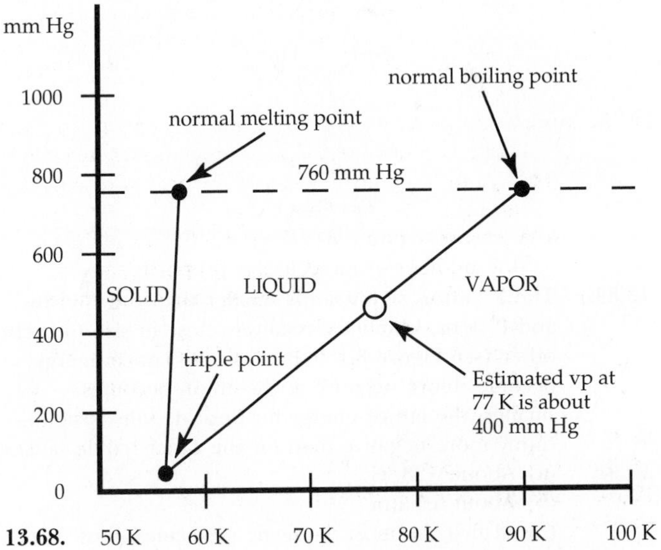

13.68. The density of solid O_2 is greater than that of the liquid. This is indicated by the fact that solid-liquid equilibrium line has a positive slope.

13.70. Acetone is a polar molecule. It can interact with the dipole of water. In addition, the electronegative O atom can be involved in hydrogen bonding with water.

13.72. Be^{2+} ion has a smaller radius than the Ba^{2+} ion. Thus, the former should have the higher enthalpy of solution because it can interact more strongly with water.

13.74. All evaporate.

13.76. $HOCH_2CH_2OH$ are more viscous than C_2H_5OH because the glycol has more sites for hydrogen bonding.

13.78. **(a)** ICl
(b) Kr
(c) C_2H_5OH
(d) piperidine (H-bonding possible)

13.80. 47.6% empty

13.82. **(a)** Ethanol does not have the capacity for hydrogen bonding that water does. Ethanol has only one O—H bond, whereas H_2O has two.
(b) A possible structure for the ion involves hydrogen bonding between HF and an F^- ion.

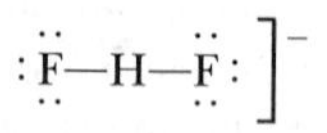

(c) A mixture of ethanol and water involves strong hydrogen bonding between the molecules. It is stronger than in ethanol alone and causes the volume to decrease.

13.85. A unit cell of the NaCl can only have a 1:1 cation to anion ratio.

13.87. CO_2 has polar C═O bonds. This makes the C atom positive and the O atoms negative. Thus, water molecules can interact with the partially charged atoms and this assists in dissolving the gas in water.

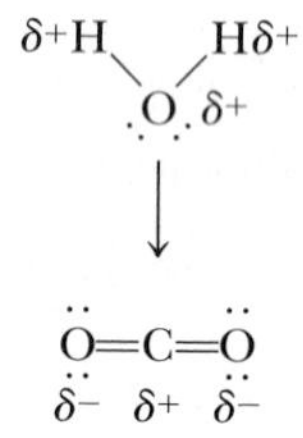

13.89. The F^- ion is significantly smaller than the Cl^-, Br^-, and I^- ions, which are relatively close in size to each other (see Figure 8.15). Because the lattice energy becomes more negative as the anion becomes smaller, the lattice energy for fluoride salts is significantly more negative than for the other halide salts.

13.91. **(a)** About −28 °C
(b) About 6.5 atm
(c) At first the most energetic molecules, those moving with the greatest velocity, escape from the cylinder. This leaves less energetic molecules with a lower velocity, and they escape less rapidly. When the most energetic molecules have escaped, the less energetic ones remain, and, because temperature is proportional to energy, their average temperature is lower. In fact, it is so low that water vapor condenses on the flask.
(d) If the flask is cooled to −78 °C, most of the vapor condenses to liquid with a very low vapor pressure. The valve can be opened, and the liquid poured out.

13.92. Each of the possible unit cells has one net cation (A) and one net anion (B). One possible cell includes one large sphere (A) and one fourth of each of four small spheres (B). Another possible cell includes one small sphere (B) and one fourth of each of four large spheres (A). Still another one includes $2[= 1 + 4(\frac{1}{4})]$ small spheres and $2[= 2(\frac{1}{2})]$ large spheres. See Question 13.42.

13.93. **(a)** See page 428 for SO_2 resonance structures. The O—S—O angle is approximately 120 °, the structural pair geometry is trigonal-planar, and the best description of S-atom hybridization is sp^2.
(b) dipole-dipole
(c) $CH_4 < NH_3 < SO_2 < H_2O$
(d) $SO_2(g) + O_2(g) \longrightarrow SO_3(g)$
$\Delta H° = -98.89$ kJ/mol
$H_2O(\ell) + SO_3(g) \longrightarrow H_2SO_4(aq)$
$\Delta H° = -227.72$ kJ/mol

CHAPTER 14

14.13. **(a)** Assume total volume is 0.500 L, so molarity is 0.0382 M.
(b) 0.0382 molal
(c) Mole fraction malic acid = 6.88×10^{-4}
(d) 0.509%

14.15.

Compound	Molality	%	Mole Fraction
KI	0.15	2.4	2.7×10^{-3}
C_2H_5OH	0.67	3.0	0.012
$C_{12}H_{22}O_{11}$	0.10	3.3	0.0018

14.17. **(a)** 4.25 g $NaNO_3$; mole fraction $NaNO_3$ = 0.00359

14.19. 468 g glycol; 7.93 molal

14.21.

Compound	Grams	g H_2O	m	X
K_2CO_3	0.432	250.	0.0125	2.25×10^{-4}
C_2H_5OH	13.5	150.	1.95	0.0340
$NaNO_3$	270.	555	5.72	0.0934

14.23. 16.2 molal; 37.1%

14.25. **(a)** Molal fraction NaOH = 0.162
(b) 30.0% NaOH
(c) 9.97 M

14.27. Molality of $Ca(NO_3)_2$ = 0.0163 m; total molality is three times the compound molality or 0.0489 m.

14.29. 2.6×10^{-5} m

14.31. option (c)

14.33. More soluble. Because the enthalpy of solution is positive (an endothermic process), raising the temperature should increase the solubility. The enthalpy of formation of $NH_4Cl(aq)$ is −299.6 kJ/mol.

14.35. Choices (b), (c), and (d) are all nonpolar solvents, which should solvate the sodium ion very poorly. Therefore, only with water should Na^+ have a negative enthalpy of solvation.

14.37. 6.6×10^{-5} M

14.39. 1.13×10^3 mm Hg

14.41. Vapor pressure of water over the solution = 35.0 mm Hg

14.43. 1.04×10^3 g glycol

14.45. 100. g/mol

14.47. 100.26 °C

14.49. 62.51 °C

14.51. 580 g glycerol; mole fraction glycerol = 0.13

14.53. Increasing boiling point: 0.10 *m* sugar < 0.10 *m* KCl < 0.080 *m* $MgCl_2$

14.55. 180 g/mol

14.57. Molar mass = 180 g/mol; molecular formula = $C_{14}H_{10}$

14.59. **(a)** 8.60 mol/kg
(b) 28.4%

14.61. −25.1 °C

14.63. 170 g/mol

14.65. $[(C_2H_5)_2AlF]_4$

14.67. Increasing melting point: 0.08 m $CaCl_2$ < 0.1 m NaCl < 0.04 m Na_2SO_4 < 0.1 m sugar

14.69. **(a)** −0.348 °C
(b) 100.10 °C
(c) Π = 4.58 atm
The osmotic pressure is the easiest to measure in the laboratory.

14.71. 6.0×10^3 g/mol

14.73. **(a)** $BaCl_2(aq) + Na_2SO_4(aq) \longrightarrow BaSO_4(s) + 2\ NaCl(aq)$
(b) Small particles of insoluble $BaSO_4$ are formed rapidly. As described for AgCl particles on page 686, the $BaSO_4$ particles can accumulate a charge and resist forming larger particles that can precipitate.
(c) As the $BaSO_4$ particles "age," they lose their charge, form larger particles, and precipitate.

14.75. **(a)** 0.10 *m* Na_2SO_4
(b) less soluble
(c) 0.15 *m* Na_2SO_4

14.77. **(a)** Mole fraction = 0.0404
(b) 0.913 m
(c) 729 mm Hg
(d) 79.5 °C

14.79. 55.3 M; 55.5 molal

14.81. 472 g/mol

14.83. **(a)** Because the solubility increases with increasing temperature, the enthalpy of solution is expected to be positive.
(b) KBr, because the effective concentration of the ionic solute is about 0.40 *m*, whereas it is only 0.30 *m* in sugar.
(c) Na_2CO_3, because the effective concentration of this ionic solute is about 0.30 *m*, whereas that of the other ionic solution, NH_4NO_3, is about 0.24 *m*.

14.85. 404 g/mol

14.87. 106.34 °C

14.89. 0.45 g Cl_2

14.91. **(a)** $x(C_2H_5OH) = 0.235$ and $X(H_2O) = 0.765$
(b) P_{total} = 23.6 mm Hg

14.93. **(a)** 338 g/mol
(b) Π = 2.93 atm at 25 °C

14.95. 888 g NH_4CHO_2

14.97. Desired water pressure = 8.5 mm Hg
Mole fraction water = 0.55
Quantity of glycerol = 8.2 mol
Weight % glycerol = 81%

14.99. KBr

14.101. **(a)** 0.040 mol/L
(b) 1.4%

14.103. **(a)** Egg shell is composed of $CaCO_3$ (in a protein binder) so the carbonate reacts with the acetic acid in vinegar to give calcium acetate and carbonic acid (which gives CO_2 gas and water).
(b) Concentration of material inside the egg membrane is higher than that of vinegar. Water flows into the egg through the semipermeable membrane, and the egg swells.
(c) Concentration of water in the corn syrup is higher than in the egg. Therefore, water flows out of the egg to a greater extent than it flows in. The result is that egg shrivels.

14.105. **(a)** In benzene, molar mass = 120 g/mol
(b) In water, molar mass = 62.4 g/mol
In benzene acetic acid is a so-called dimer. That is, two molecules are bound together, presumably through hydrogen bonding.

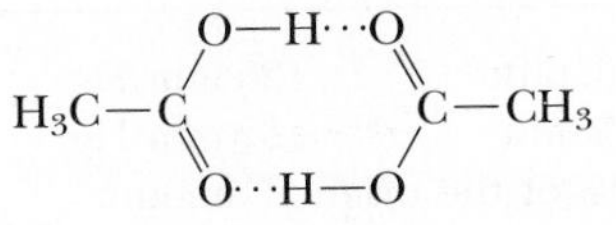

14.107. The —OH group in both CH_3OH and C_2H_5OH forms hydrogen bonds with water, which overcomes the lack of interaction of the CH-containing group. In the case of alcohols with long carbon chains, however, the hydrocarbon chains are hydrophobic and dominate the interaction with water, making such alcohols only poorly soluble.

14.109. Starch colloids are hydrophilic, whereas hydrocarbons are hydrophobic.

14.111. Empirical and molecular formulas are $C_{18}H_{24}Cr$.

CHAPTER 15

15.18. **(a)** Rate $= -\frac{1}{2}\left(\frac{\Delta[O_3]}{\Delta t}\right) = +\frac{1}{3}\left(\frac{\Delta[O_2]}{\Delta t}\right)$

(b) $\text{Rate} = -\frac{1}{2}\left(\frac{\Delta[\text{HOF}]}{\Delta t}\right) = +\frac{1}{2}\left(\frac{\Delta[\text{HF}]}{\Delta t}\right) = +\left(\frac{\Delta[\text{O}_2]}{\Delta t}\right)$

(c) $\text{Rate} = -\frac{1}{2}\left(\frac{\Delta[\text{NO}]}{\Delta t}\right) = -\left(\frac{\Delta[\text{Br}_2]}{\Delta t}\right) = +\frac{1}{2}\left(\frac{\Delta[\text{BrNO}]}{\Delta t}\right)$

15.20. **(a)** Rate for 0 to 10 s = 0.167 mol/L · s
Rate for 10 to 20 s = 0.119 mol/L · s
Rate for 20 to 30 s = 0.0089 mol/L · s
Rate for 30 to 40 s = 0.0070 mol/L · s
The rate decreases because there is less A available in each successive time period.
(b) Rate of appearance of B is twice the rate of disappearance of A.
(c) Rate when [A] = 0.75 mol/L is 0.011 mol/L · s

15.22. **(a)** Rate = $k[NO_2][O_3]$
(b) Tripling $[NO_2]$ triples the reaction rate.
(c) Halving $[O_3]$ halves the reaction rate.

15.24. **(a)** Rate = $k[NO]^2[O_2]$
(b) $k = 25\ L^2/mol^2 \cdot s$
(c) 1.3×10^{-3} mol/L · s
(d) NO is disappearing at a rate of 10.0×10^{-4} mol/L · s, and NO_2 is appearing at the same rate.

15.26. $5.57 \times 10^{-3}\ min^{-1}$ or $9.28 \times 10^{-5}\ s^{-1}$

15.28. 13 h

15.30. 2.94 s

15.32. **(a)** $t_{1/2} = 2.08 \times 10^5$ h
(b) 4.8×10^5 h

15.34. 75.8 s

15.36. 4.48×10^{-3} mol azomethane remains and 0.0300 mol N_2 form.

15.38.

	P(total)	*P*(HOF)
After 30. min	125 mm Hg	50. mm Hg
After 45 min	132 mm Hg	35 mm Hg

15.40. **(a)** 3.0% of the original amount
(b) 0.14% of the original amount

15.42. **(a)** A graph of ln [sucrose] versus time is a straight line, so the reaction is first order in sucrose.
(b) Rate = k [sucrose], where $k = 3.7 \times 10^{-3}\ min^{-1}$ (as calculated from the slope of the ln [sucrose] versus time plot).
(c) [sucrose] = 0.166 mol/L

15.44. **(a)** A graph of ln $[N_2O]$ versus time is a straight line, so the reaction is first order in N_2O.
(b) $k = 1.28 \times 10^{-2}\ min^{-1}$, so rate = 4.5×10^{-4} mol/L · min

15.46. Reaction is exothermic, releasing 26 kJ/mol

15.48. 102 kJ/mol

15.50. **(a)** E_a = 53.0 kJ/mol
(b) A = 8.0×10^7, and k at 311 K = 0.0985 min^{-1}

15.52. **(a)** Rate = k [NO][NO_3]; bimolecular
(b) Rate = k [Cl][H_2]; bimolecular
(c) Rate = k [$(CH_3)_3CBr$]; unimolecular

15.54. **(a)** Second step (the slow step)
(b) Rate = k [O_3][O]
(c) Step 1 is unimolecular, and Step 2 is bimolecular.

15.56. **(a)** true
(b) False. A catalyst is never consumed by the overall reaction.
(c) False. A catalyst may be in a different phase from the reactants.
(d) True. However, usually a catalyst is used to alter the mechanism of a given reaction to produce products with a lower activation energy.
(e) true

15.58. 8×10^2 kg/L · h

15.60. **(a)** Rate = k [CO][NO_2]
(b) First order with respect to both CO and NO_2
(c) k = 1.9 L/mol · h

15.62. **(a)** 0.59 mg of NO_x
(b) 75 h

15.64. $2\ O_3(g) \longrightarrow 3\ O_2(g)$
Ozone decomposes to oxygen molecules, resulting in a net loss of ozone in the stratosphere. Cl is a catalyst, and ClO is an intermediate.

15.66. About 7.5 times faster

15.68. $k = 0.037\ hr^{-1}$ and $t_{1/2}$ = 19 hr.

15.70. k at 338.0 K = 0.275 min^{-1}

15.72. **(a)** 0.250 g after 125 min and 0.144 g after 145 min
(b) 43.9 min
(c) $\frac{1}{64}$

15.74. **(a1)** −37 kJ
(a2) −214 kJ
(a3) −251 kJ
(b) E_a = 20 kJ

15.76. The danger is the potential for an explosion caused by a spark igniting very fine dust or airborne powder. The particles are so fine that the rate of the ignition is extremely rapid. See Figure 15.6 for the effect of particle size on reaction rate.

15.78. Very finely divided metal powder provides a much larger surface on which reaction can occur.

15.80. **(a)** 3.3×10^{13} atoms/L
(b) 1.4×10^{11} atoms/L

CHAPTER 16

16.7. **(a)** $K_c = \frac{[H_2O]^2[O_2]}{[H_2O_2]^2}$
(b) $K_c = \frac{[PCl_5]}{[PCl_3][Cl_2]}$
(c) $K_c = \frac{[CO_2]}{[CO][O_2]^{1/2}}$
(d) $K_c = \frac{[CO]^2}{[CO_2]}$

(e) $K_c = \dfrac{[CO_2]}{[CO]}$

16.9. (e), $K_2 = \dfrac{1}{K_1^2}$

16.11. $K = (1.6)(1.5) = 2.4$

16.13. $K_c = 2.25 \times 10^{22}$

16.15. $Q = 1.0 \times 10^3$, so $Q > K_c$, and the reaction shifts to the left.

16.17. $Q = 5.0 \times 10^{-4}$, so $Q < K_c$, and the reaction shifts toward products.

16.19. $K_c = 279$

16.21. (a) $K_c = 1.6$
(b) Mol CO = mol H_2O = 0.064 mol

16.23. $K_c = 0.035$

16.25. $K_c = 0.029$

16.27. $K_p = 2.31 \times 10^{-4}$

16.29. [isobutane] = 0.024 M and [butane] = 0.010 M

16.31. [CO] = [Br_2] = 0.034 M

16.33. 0.0730 mol NO_2; 16.7% dissociated

16.35. (a) [CO] = [Br_2] = 7.0×10^{-4} M and [$COBr_2$] = 3×10^{-6} M
(b) $P_{total} = 0.040$ atm

16.37. [I] = 0.0424 M and [I_2] = 0.479 M

16.39. (a) left
(b) left
(c) left
(d) right

16.41. (a) and (b) [isobutane] = 2.9 M and [butane] = 1.1 M

16.43. Rate = $k[H_2]^{1/2}[CO]$

16.45. Mechanism 2. Rate law for Mechanism 1 is Rate = $k[H_2][NO]$, which does not match the experimental rate law. However, the observed rate law can be derived from Mechanism 2.

16.47. (a) $K_p = K_c$ because $\Delta n = 0$
(b) 4.1×10^{-2}
(c) 5.9×10^{-2}

16.49. Reaction is not at equilibrium. $Q (= 1.4 \times 10^{-4})$ is less than K_p, so the reaction proceeds further toward products.

16.51. $K_c = 0.080$

16.53. $P(NO_2) = 0.40$ atm and $P(N_2O_4) = 1.1$ atm

16.55. (a) The percentage of acid forming dimer is 84%.
(b) As T increases, the equilibrium shifts left.

16.57. $K_p = 3.40$

16.59. [Zn^{2+}] = [CO_3^{2-}] = 3.9×10^{-6} M

16.61. 1.7×10^{18} O atoms

16.63. (a) Fraction dissociated = 0.15
(b) The fraction dissociated increases.

16.65. (a) [SO_2] = [Cl_2] = 0.030 M and [SO_2Cl_2] = 0.020 M
Fraction dissociated = 0.60
(b) [SO_2] = 0.025 M, [Cl_2] = 0.044 M, and [SO_2Cl_2] = 0.025 M
Fraction dissociated = 0.50

(c) Agrees with result expected from Le Chatelier's principle

16.67. (a) Second step
(b) From Step 1: $[O] = \dfrac{K[O_3]}{[O_2]}$
Substituting into the rate equation:
$$\text{Rate} = k[O][O_3] = k\left\{\frac{K[O_3]}{[O_2]}\right\} = \frac{k'[O_3]^2}{[O_2]}$$

16.69. About 1×10^{-3} atm

CHAPTER 17

17.12. (a) CN^-, cyanide ion
(b) SO_4^{2-}, sulfate ion
(c) F^-, fluoride ion
(d) NO_2^-, nitrite ion
(e) CO_3^{2-}, carbonate ion

17.14. (a) Products: H_3O^+ and NO_3^-
Acid is HNO_3, and conjugate base is NO_3^-.
Base is H_2O, and conjugate acid is H_3O^+.
(b) Products: H_3O^+ and SO_4^{2-}
Acid is HSO_4^-, and conjugate base is SO_4^{2-}.
Base is H_2O, and conjugate acid is H_3O^+.
(c) Products: H_2O and HF
Acid is H_3O^+, and conjugate base is H_2O.
Base is F^-, and conjugate acid is HF.

17.16. $CO_3^{2-}(aq) + H_2O(\ell) \rightleftharpoons HCO_3^-(aq) + OH^-(aq)$

17.18. Acid: $HPO_4^{2-}(aq) + H_2O(\ell) \rightleftharpoons PO_4^{3-}(aq) + H_3O^+(aq)$
Base: $HPO_4^{2-}(aq) + H_2O(\ell) \rightleftharpoons H_2PO_4^-(aq) + OH^-(aq)$

17.20.

	Acid (A)	Base (B)	Conjugate of A	Conjugate of B
(a)	HCO_2H	H_2O	HCO_2^-	H_3O^+
(b)	H_2S	NH_3	HS^-	NH_4^+
(c)	HSO_4^-	OH^-	SO_4^{2-}	H_2O

17.22. (a) HF is the strongest acid, and NH_4^+ is the weakest.
(b) F^-
(c) HF
(d) NH_4^+

17.24. (a) NH_3 is the strongest base, and C_5H_5N is the weakest.
(b) $C_5H_5NH^+$
(c) C_5H_5N has the strongest conjugate acid (because it is the weakest base), and NH_3 has the weakest conjugate acid (because it is the strongest base).

17.26. HClO has the strongest conjugate base because HClO is the weakest acid of the group. (See Table 17.4).

17.28. $H_2PO_4^-(aq) + NH_4^+(aq) \rightleftharpoons H_3PO_4(aq) + NH_3(aq)$
Equilibrium lies predominantly to the left because H_3PO_4 is a stronger acid than NH_4^+.

17.30. (a) Right; CO_3^{2-} is a stronger base than HS^-, and H_2S is a stronger acid than HCO_3^-.

(b) Left; HSO_4^- is a stronger acid than HCN, and CN^- is a stronger base than SO_4^{2-}.
(c) Left; NH_2^- is a very strong base in water.
(d) Left; HSO_4^- is a stronger acid than CH_3CO_2H.

17.32. Wine is acidic; $[H_3O^+] = 4.0 \times 10^{-4}$ M

17.34. pH = 2.89; $[OH^-] = 7.7 \times 10^{-12}$ M

17.36. pH = 11.48

17.38. (a) 1.00; 0.10; 1.0×10^{-13}; acidic
(b) 10.50; 3.2×10^{-11}; 3.2×10^{-4}; basic
(c) 4.89; 1.3×10^{-5}; 7.7×10^{-10}; acidic
(d) 10.36; 4.3×10^{-11}; 2.3×10^{-4}; basic

17.40. (a) 1.6×10^{-4} M; (b) moderately weak

17.42. (a) 2.1×10^{-3} M; (b) $K_a = 3.6 \times 10^{-4}$

17.44. $K_b = 6.6 \times 10^{-9}$

17.46. $[H_3O^+] = [A^-] = 1.3 \times 10^{-5}$ M; [HA] = 0.040 M

17.48. $[H_3O^+] = [CN^-] = 3.2 \times 10^{-6}$ M; [HCN] = 0.025 M; pH = 5.50

17.50. $[H_3O^+] = 6.4 \times 10^{-4}$ M; pH = 3.19

17.52. (a) 4-chlorobenzoic acid is the stronger acid.
(b) Benzoic acid, the weaker acid, has the higher pH.

17.54. $[OH^-] = [M^+] = 8.4 \times 10^{-3}$ M; [MOH] = 0.14 M.

17.56. $[OH^-] = 1.1 \times 10^{-2}$ M; pOH = 1.96 and pH = 12.04

17.58. pH = 3.25

17.60. Highest pH, Na_2S. Lowest pH, $AlCl_3$. $AlCl_3$ gives the $Al(H_2O)_6^{3+}$ ion in solution, and this is a stronger acid than the $H_2PO_4^-$ ion.

17.62. $[H_3O^+] = 1.1 \times 10^{-5}$ M; pH = 4.98

17.64. $[OH^-] = [HCN] = 3.3 \times 10^{-3}$ M; $[Na^+] = 0.441$ M; $[H_3O^+] = 3.0 \times 10^{-12}$ M; pH = 11.52.

17.66. (a) $K_a = 2.5 \times 10^{-5}$
(b) pH = 2.85

17.68. pH = 12.29

17.70 (a) pH = 1.17
(b) 6.2×10^{-8} M

17.72. (a) $[OH^-] = [N_2H_5^+] = 9.2 \times 10^{-5}$ M; $[N_2H_6^{2+}] = 8.9 \times 10^{-16}$ M
(b) pH = 9.96

17.74. (a) Lewis acid
(b) Lewis base
(c) Lewis base
(d) Lewis base
(e) Lewis acid

17.76. Lewis acid

17.78. ICl_3 is a Lewis acid. I in ICl_3 is sp^3d-hybridized, whereas it is sp^3d^2 hybridized in ICl_4^-.

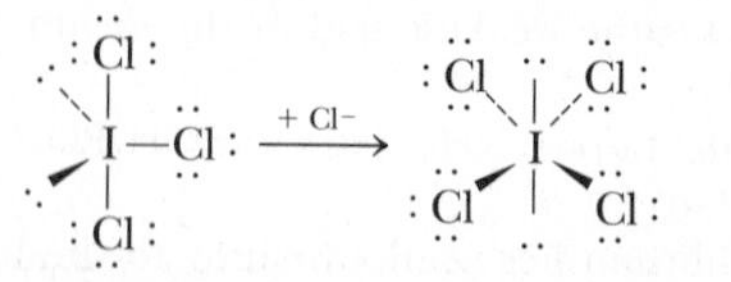

ICl_3 is T-shaped ICl_4^- is square-planar

17.80. H^- is a Lewis base, and H_2O is a Lewis acid. The resulting solution is basic.

17.82. pH = 2.644

17.84. $H_2S(aq) + CH_3CO_2^-(aq) \rightleftharpoons HS^-(aq) + CH_3CO_2H(aq)$
Because CH_3CO_2H is a stronger acid than H_2S, the equilibrium lies predominantly to the left.

17.86. pH = 12.13

17.88. $K_a = 1.4 \times 10^{-5}$

17.90. (a) $K_b = 4.3 \times 10^{-4}$
(b) $C_4H_9NH_3^+$ is stronger than $Ni(H_2O)_5OH^+$ but weaker than PO_4^{3-}.

17.92. (a) less than 7
(b) less than 7
(c) equals 7
(d) greater than 7
(e) greater than 7
(f) equals 7
(g) greater than 7
(h) equals 7
(i) less than 7

17.94. pH = 9.26 and pOH = 4.74

17.96. (a) CH_3CO_2H, NH_4Cl
(b) NH_3; Na_2CO_3; $NaCH_3CO_2$
(c) CH_3CO_2H

17.98. $HCl < CH_3CO_2H < NaCl < NH_3 < NaCN < NaOH$

17.100. pH = 1.02; $[HC_2O_4^-] = 0.095$ M; $[C_2O_4^{2-}] = 6.4 \times 10^{-5}$ M

17.102. $I_2(aq) + I^-(aq) \rightleftharpoons I_3^-(aq)$
I_2 is the Lewis acid and I^- is the Lewis base.

17.104. (a) Weakest acid ($X = CH_3$) has the highest pH, whereas the strongest acid ($X = NO_2$) has the lowest pH.
(b) The strongest base has $X = CH_3$, whereas the weakest base has $X = NO_2$.

17.106. (a) Acid strength increases as more Cl atoms replace H atoms in the CH_3 group. The greater the number of Cl atoms, the greater the attraction of bonding electrons toward the electronegative Cl atoms. This decreases the O—H bond strength, and the H atom of the —OH group is easier to ionize.
(b) Highest pH (weakest acid) = CH_3CO_2H
Lowest pH (strongest acid) = Cl_3CCO_2H

17.108. (a) $HClO_4 + H_2SO_4 \rightleftharpoons ClO_4^- + H_3SO_4^+$

(b) and (c) 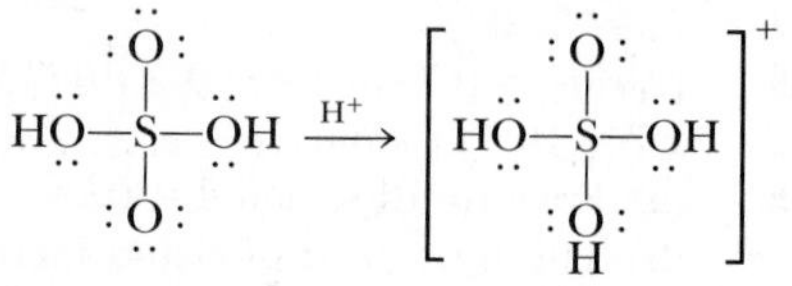

The H_2SO_4 molecule functions as a Lewis base in donating a pair of electrons on an O atom to H^+.

17.111. (a) 93 mL
(b) All the indicated angles are about 109°.
(c) pH = 8.02

CHAPTER 18

18.8. $K = 6.3 \times 10^9$; equilibrium lies to the right.

18.10. $[H_3O^+] = 1.89 \times 10^{-9}$ M and pH = 8.72

18.12. **(a)** Greater than 7; **(b)** less than 7; **(c)** equal to 7

18.14. $[OH^-] = 1.7 \times 10^{-3}$ M; $[H_3O^+] = 5.9 \times 10^{-12}$ M; $[Na^+] = [C_6H_5O^-] = 3.79 \times 10^{-2}$ M; pH = 11.23

18.16. At equivalence point: $[NH_4^+] = 6.25 \times 10^{-3}$ M and $[H_3O^+] = 1.9 \times 10^{-6}$ M; $[OH^-] = 5.3 \times 10^{-9}$ M; pH = 5.72. Original concentration of $NH_3 = 1.54 \times 10^{-2}$ M

18.18. **(a)** decrease pH; **(b)** increase pH; **(c)** pH stays the same.

18.20. **(a)** decreases pH
(b) increases pH

18.22. pH = 9.25

18.24. pH = 9.11; pH of buffer solution is lower than that of the original NH_3 solution.

18.26. 4.8 g

18.28. **(a)** pH = 3.90; **(b)** Diluting the solution with 500 mL of pure water does not change the pH.

18.30. NH_3 and NH_4Cl

18.32. **(a)** pH = 4.95; **(b)** pH = 5.05

18.34. **(a)** pH = 9.55; **(b)** pH = 9.50

18.36. pK_a of acetic acid = 4.74; pH = 4.92.

18.38. **(a)** pH = 3.59; **(b)** 0.45

18.40. The curve begins at a pH of 13.00. Although the pH drops slowly as HCl is added at first (it is 12.52 after 15.0 mL of HCl has been added), it drops sharply (vertically) when 30.0 mL of 0.10 M HCl has been added. The midpoint of the vertical portion of the curve is at a pH of 7. When slightly more than 30.0 mL of HCl has been added, the curve levels out at a pH of about 1.5.

18.42. **(a)** pH of original NH_3 solution = 11.15
(b) pH at equivalence point (after adding 27.5 mL of HCl) = 5.27
(c) pH at midpoint = 9.26
(d) Bromcresol green or methyl red
(e)

Acid Added (mL)	pH
5.00	9.91
15.0	9.18
20.0	8.83
22.0	8.65
30.0	2.34

18.44. **(a)** thymol blue; **(b)** phenolphthalein; **(c)** methyl orange or bromcresol green

18.46. **(a)** $CH_3CH_2CO_2H(aq) + OH^-(aq) \longrightarrow CH_3CH_2CO_2^-(aq) + H_2O(\ell)$
(b) 0.60 mol acid and 0.40 mol conjugate base
(c) pH = 4.71
(d) pH does not change

18.48. $[NH_4^+] = 9.0 \times 10^{-5}$ M; $[OH^-] = [Na^+] = 0.040$ M; $[NH_3] = 0.20$ M

18.50. **(a)** pH of acetic acid solution = 2.77; **(b)** pH after adding sodium acetate = 5.20

18.52. 28.6 mL of NaOH is required to reach the equivalence point. The concentration of HCO_2^- at this point is 0.0560 M. The pH is 8.25.

18.54. **(a)** pH = 4.99; **(b)** pH = 5.13

18.56. 0.08 mol

18.58. **(a)** Original pH = 8.52
(b) At equivalence point $[H_3O^+] = 5.1 \times 10^{-4}$ M and pH = 3.29.
(c) pH at midpoint = 4.62
(d) bromphenol blue
(e)

Volume Base Added	Solution pH
5.00	5.37
10.0	4.98
15.0	4.70
20.0	4.43
24.0	4.19
30.0	3.60

18.60. $\dfrac{[H_2PO_4^-]}{[HPO_4^{2-}]} = 0.65$

18.62. pH = 6.73

18.64. **(a)** pH of original solution = 3.31
(b) pH at equivalence point = 9.31
(c) pH at midpoint = 5.82

18.66. $[C_2O_4^{2-}]$ at equivalence point = 0.0333 M
$[OH^-]$ at equivalence point = 2.3×10^{-6} M
pH = 8.36

18.68. **(a)** HB is the stronger acid. The pH at the equivalence point is controlled by the conjugate base of the acid. The stronger the acid, the weaker the conjugate base, and the lower the pH.
(b) A^- is the stronger conjugate base.

18.70. **(a)** The fraction of acetic acid declines because the proton combines with the hydroxide ion to form water, leaving an equivalent amount of acetate ion in solution. The quantity of acetic acid present decreases and the pH rises.
(b) At pH of 4 the fraction of acid is greater than that of the acetate ion. The situation is reversed at a pH of 6; only about 5% of the acetic acid remains.
(c) At the cross-over point, $[CH_3CO_2H] = [CH_3CO_2^-]$, so $[H_3O^+] = K_a = 1.8 \times 10^{-5}$. This leads to a pH of 4.74.

18.73. **(a)**

Bond	Angle
C—C—C in ring	120°
C—C═O	120°
C—O—H	109°
C—C—H	120°

(b) C atoms of the ring are sp^2-hybridized as in the C atom of the CO_2H group.
(c) pH of salicylic acid solution = 2.44
(d) About 11% present as salicylate ion in gastric juice
(e) pH at the midpoint is 2.96 and at the equivalence point is 7.36.

CHAPTER 19

19.8. **(a)** AgCl and $PbCl_2$
(b) $ZnCO_3$; ZnS; $Zn_3(PO_4)_2$
(c) FeS; $FeCO_3$; $Fe(OH)_2$

19.10. **(a)** soluble; **(b)** insoluble; **(c)** insoluble; **(d)** insoluble

19.12. **(a)** AgI **(b)** $PbCl_2$ **(c)** no precipitate

19.14. **(a)** $AgCN(s) \rightleftarrows Ag^+(aq) + CN^-(aq)$
$K_{sp} = [Ag^+][CN^-] = 1.2 \times 10^{-16}$
(b) $PbCO_3(s) \rightleftarrows Pb^{2+}(aq) + CO_3^{2-}(aq)$
$K_{sp} = [Pb^{2+}][CO_3^{2-}] = 1.5 \times 10^{-13}$
(c) $AuI_3(s) \rightleftarrows Au^{3+}(aq) + 3\ I^-(aq)$
$K_{sp} = [Au^{3+}][I^-]^3 = 1.0 \times 10^{-46}$

19.16. $K_{sp} = 3.2 \times 10^{-6}$

19.18. $K_{sp} = 7.9 \times 10^{-6}$

19.20. $K_{sp} = 8.1 \times 10^{-12}$

19.22. $K_{sp} = 7.9 \times 10^{-6}$

19.24. **(a)** 1.1×10^{-8} M; **(b)** 1.5×10^{-6} g/L

19.26. 0.21 mg $RaSO_4$

19.28. 1.2×10^{-3} mol/L; 0.073 g/L

19.30. **(a)** AgSCN; **(b)** $SrSO_4$; **(c)** MgF_2; **(d)** AgI

19.32. $BaCO_3$ (solubility = 9.0×10^{-5} M) < Ag_2CO_3 (solubility = 1.3×10^{-4} M) < BaF_2 (solubility = 7.5×10^{-3} M)

19.34. All $NiCO_3$ dissolves.

19.36. **(a)** $Q < K_{sp}$, no precipitate;
(b) $Q > K_{sp}$, $NiCO_3$ precipitates.

19.38. $Q(=1.6 \times 10^{-8}) > K_{sp}$; therefore, $Zn(OH)_2$ precipitates.

19.40. $[OH^-] = 1.6 \times 10^{-5}$ M

19.42. $Q(=2.7 \times 10^{-10}) > K_{sp}$, so a precipitate of $BaSO_4$ forms.

19.44. $Q(=1.7 \times 10^{-8}) < K_{sp}$, so no precipitate of $PbCl_2$ forms.

19.46. $BaSO_4$ ($K_{sp} = 1.1 \times 10^{-10}$) is precipitated first, then $PbSO_4$ ($K_{sp} = 1.8 \times 10^{-8}$), and finally $SrSO_4$ ($K_{sp} = 2.8 \times 10^{-7}$).

19.48. $Fe(OH)_3$ precipitates first, followed by $Al(OH)_3$ and then $Pb(OH)_2$.

19.50. 1.0×10^{-6} M in pure water, but only 1.0×10^{-10} M in 0.010 M in NaSCN

19.52. **(a)** 2.0×10^{-3} mg/mL; **(b)** 6.8×10^{-13} mg/mL

19.54. **(a)** Oxalate concentration should be slightly less than 4.3×10^{-3} M to precipitate CaC_2O_4 and leave Mg^{2+} in solution.
(b) $[Ca^{2+}] = 5.3 \times 10^{-7}$ M

19.56. **(a)** $PbCO_3$ precipitates first.
(b) $[CO_3^{2-}] = 1.7 \times 10^{-7}$ M

19.58. **(a)** Na^+ salts are generally water-soluble, whereas many barium salts are not soluble. Therefore, we only need to add a reagent such as SO_4^{2-} to precipitate Ba^{2+} as $BaSO_4$ and leave Na^+ in solution.
(b) Separation as hydroxides should be feasible. The concentration of OH^- needed to precipitate $Bi(OH)_3$ is about 10^6 less than that needed for $Cd(OH)_2$.

19.60.

$$\begin{array}{ll} AgBr(s) \rightleftarrows Ag^+(aq) + Br^-(aq) & K = K_{sp} \text{ for AgBr} = 3.3 \times 10^{-13} \\ Ag^+(aq) + 2\ NH_3(aq) \rightleftarrows [Ag(NH_3)_2]^+ & K = K_{formation} = 1.6 \times 10^{7} \\ \hline AgBr(s) + 2\ NH_3(aq) \rightleftarrows Br^-(aq) + [Ag(NH_3)_2]^+(aq) & K_{net} = 5.3 \times 10^{-6} \end{array}$$

19.62.

$$\begin{array}{ll} AgCl(aq) \rightleftarrows Ag^+(aq) + Cl^-(aq) & K = K_{sp} \text{ for AgCl} = 1.8 \times 10^{-10} \\ Ag^+(aq) + I^-(aq) \rightleftarrows AgI(s) & K = 1/K_{sp} \text{ for AgI} = 1/1.5 \times 10^{-16} \\ \hline AgCl(s) + I^-(aq) \rightleftarrows AgI(s) + Cl^-(aq) & K_{net} = 1.2 \times 10^{6} \end{array}$$

Yes, it is possible to precipitate AgI by adding I^- ion to a precipitate of AgCl since K_{net} is much greater than 1.

19.64. $K_{net} = 5.3 \times 10^{-6}$ for $AgBr(s) + 2\ NH_3(aq) \rightleftarrows Ag(NH_3)_2^+(aq) + Br^-(aq)$
All the AgBr does not dissolve. If all the AgBr dissolved, the concentrations of $Ag(NH_3)_2^+$ and Br^- would be 1.0×10^{-4} M. To have these concentrations requires $[NH_3]$ at equilibrium to be 0.043 M. After 5.0 mL of 2.5 M NH_3 to 1.0 L of water, $[NH_3]$ is only 0.012 M.

19.66. $Ba(OH)_2$ and $BaCO_3$; both have a basic anion.

19.68. $BiCO_3$ and $BiPO_4$; both have a basic anion.

19.70. Aragonite is more soluble than calcite.

19.72. $Q(=2.0 \times 10^{-9}) > K_{sp}$; therefore, AgCl forms.

19.74. $BaSO_4$ precipitates first followed by $CaSO_4$. $MgSO_4$ is water-soluble.

19.76. No $Ca(OH)_2$ precipitates; $Q(=1.5 \times 10^{-9}) < K_{sp}$.

19.78. $K_{sp} = 4.5 \times 10^{-17}$

19.80. $K_{sp} = 1.5 \times 10^{-11}$

19.82. $[Ag^+] = 6.9 \times 10^{-6}$ M; $[NO_3^-] = 0.235$ M; $[K^+] = 0.618$ M; $[CrO_4^{2-}] = 0.191$ M

19.84. **(a)** $K_{net} = 6.6$ for $AgBr(s) + 2\ S_2O_3^{2-}(aq) \rightleftarrows Ag(S_2O_3)_2^{3-}(aq) + Br^-(aq)$
(b) 2.0 g of $Na_2S_2O_3$ is required.

19.86. $[Zn^{2+}] = 2 \times 10^{-6}$ M

19.88. If the pH is decreased (by adding acid), the OH^- ions that are produced by any $Ni(OH)_2$ that dissolves are consumed. This removes a product of the equilibrium

$$Ni(OH)_2(s) \rightleftharpoons Ni^{2+}(aq) + 2\ OH^-(aq)$$

causing the equilibrium to shift to the right, dissolving more $Ni(OH)_2$.

19.90. $AgCN(s) + CN^-(aq) \rightleftharpoons Ag(CN)_2^-(aq)$
$K_{net} = 670$

Adding CN^- ion does not affect $[Ag^+]$, but the CN^- ion concentration increases.

19.92. **(a)** Increasing the CO_2 pressure should increase the $BaCO_3$ solubility because the equilibrium is shifted to the right.
(b) Decreasing the pH (by adding acid) should increase the $BaCO_3$ solubility.

19.94. **(a)** $AlCl_3(aq) + H_3PO_4(aq) \longrightarrow AlPO_4(s) + 3\ HCl(aq)$
(b) $AlCl_3$ is the limiting reagent; the theoretical yield of $AlPO_4$ is 139 g.
(c) $[Al^{3+}] = [PO_4^{3-}] = 1.1 \times 10^{-10}$
(d) Solubility increases, owing to the formation of the weak acid HPO_4^{2-}
$AlPO_4(s) + 3\ HCl(aq) \longrightarrow Al^{3+}(aq) + HPO_4^{2-}(aq)$
(e) Some $AlPO_4$ (0.46 g) forms because $Q > K_{sp}$.

CHAPTER 20

20.6. **(a)** CO_2 vapor; **(b)** dissolved sugar; **(c)** alcohol/water mixture

20.8. **(a)** $AlCl_3$; **(b)** CH_3CH_2I; **(c)** $NH_4Cl(aq)$

20.10. **(a)** $+3.363$ J/K; graphite is less ordered than diamond (page 105), so the entropy increases.
(b) -102.50 J/K; a highly disorganized gas is condensed to an organized solid, so the entropy declines.
(c) $+93.3$ J/K; a liquid is converted to a vapor, so the entropy increases.

20.12. **(a)** $\Delta S° = +84.4$ J/K for the vaporization of liquid diethyl ether.
(b) $\Delta S° = -84.4$ J/K for the condensation of the ether.

20.14. $\Delta S° = -270.1$ J/K

20.16. **(a)** $\Delta S° = -163.34$ J/K;
(b) $\Delta S° = -305.32$ J/K;
(c) $\Delta S° = -151.9$ J/K

20.18. **(a)** $\Delta S° = -252.2$ J/K; **(b)** $\Delta S° = +217.8$ J/K

20.20. $\Delta H°_{rxn}$ is positive. $\Delta S°_{rxn}$ is positive because 1 molecule (H_2O) is being split into two, gaseous molecules (H_2 and O_2). Cannot decide if reaction is product- or reactant-favored; it depends on T.

20.22. **(a)** Both $\Delta H°_{rxn}$ and $\Delta S°_{rxn}$ are negative, so no prediction can be made. It is likely, though, that the reaction is product-favored at relatively low temperatures.
(b) $\Delta H°_{rxn}$ is positive and $\Delta S°_{rxn}$ is negative so the reaction is reactant-favored.

20.24. $\Delta S°_{rxn} = 46.37$ J/K; $\Delta H°_{rxn} = -1427.8$ kJ; and
$\Delta S°_{surr} = 4790$ J/K
$\Delta S°_{universe} = 46.37\ J/K + 4790\ J/K = 4840\ J/K$
The reaction is product-favored, as expected for an exothermic reaction that produces 5 moles of gaseous products from 4.5 moles of gaseous reactants.

20.26. **(a)** $\Delta H°_{rxn} = \Delta H_f°[SnCl_4(\ell)] = -511.3$ kJ; $\Delta S°_{rxn} = -239.1$ J/K
$\Delta G°_{rxn} = -440.1$ kJ/mol at 298 K
The reaction is product-favored and is enthalpy-driven. In this case $\Delta G°_{rxn} = \Delta G_f°$
(b) $\Delta H°_{rxn} = -176.01$ kJ; $\Delta S°_{rxn} = -284.8$ J/K
$\Delta G°_{rxn} = -91.10$ kJ/mol at 298 K
The reaction is product-favored and is enthalpy-driven.

20.28. **(a)** Calculated $\Delta G_f°[CS_2(g)] = +67.12$ kJ/mol; formation is reactant-favored.
(b) Calculated $\Delta G_f°[N_2H_4(\ell)] = +149.55$ kJ/mol; formation is reactant-favored.
(c) Calculated $\Delta G_f°[COCl_2(g)] = -204.5$ kJ/mol; formation is product-favored.

20.30. $2\ Fe(s) + 3/2\ O_2(g) \longrightarrow Fe_2O_3(s)$
$\Delta G_f°[Fe_2O_3(s)] = -742.2$ kJ/mol
$\Delta G°_{rxn}$ for 454 g $= -2110$ kJ

20.32. **(a)** $\Delta G°_{rxn} = \Delta G_f°[CaCl_2(s)] = -748.1$ kJ; product-favored.
(b) $\Delta G°_{rxn} = 2 \cdot -\Delta G_f°[HgO(s)] = -2(-58.539\ kJ) = +117.078$ kJ; reactant-favored.
(c) $\Delta G°_{rxn} = -301.39$ kJ; product-favored.

20.34. $\Delta G_f°[BaCO_3(s)] = -1137.6$ kJ/mol

20.36. $\Delta H°_{rxn} = -125.52$ kJ; $\Delta S°_{rxn} = -129.9$ J/K; $\Delta G°_{rxn} = -86.82$ kJ at 298 K; Reaction is product-favored.

20.38. $K_p = 7 \times 10^{-16}$
A positive $\Delta G°_{rxn}$ indicates that K is less than 1. Both are consistent with a reactant-favored reaction.

20.40. **(a)** $\Delta G°_{rxn} = -100.97$ kJ; product-favored at 298 K
(b) $K_p = 5 \times 10^{17}$; a negative free energy change indicates an equilibrium constant greater than 1. Both are consistent with product-favored.

20.42. **(a)** Step 1: $\Delta G°_{rxn} = +490.1$ kJ and $K_p = 1 \times 10^{-86}$
Step 2: $\Delta G°_{rxn} = -737.2$ kJ and $K_p = 2 \times 10^{129}$
(b) $\Delta G°_{rxn}$ overall $= \Delta G°_{rxn}$ (step 1) $+ \Delta G°_{rxn}$ (step 2) $= -247.1$ kJ
K_p (overall) $= 2 \times 10^{43}$
The overall free energy change is the sum of the free energy changes for each step, and K_p (overall) is the product of the K_p values of the two steps.
(c) The overall reaction is enthalpy-driven as indicated by the overall enthalpy change for the reaction.
$\Delta H°_{rxn}$ overall $= \Delta H°_{rxn}$ (step 1) $+ \Delta H°_{rxn}$ (step 2) $= +546.2\ kJ - 804.2\ kJ = -258.0$ kJ

20.44. The reaction is expected to be exothermic, and so the value of $\Delta H°_{rxn}$ is expected to be negative (-296.83 kJ). The reaction produces 1 mole of a gas from a solid and 1 mole of a gas. The gaseous product has more atoms than the gaseous reactant, so one might expect the value of $\Delta S°_{rxn}$ to be positive (11.28 J/K). Both point to a product-favored reaction, which we know this to be (see Figure 16.4).

20.46. **(a)** Splitting of water: A liquid produces 1.5 moles of gas. Therefore, $\Delta S°_{rxn}$ is expected to be positive

(+163 J/K). Energy is required to split water (to break the O—H bonds), so ΔH°_{rxn} is positive (+285.83 kJ). Thus, ΔG°_{rxn} is positive, and the reaction is reactant-favored.
(b) One mole of liquid nitroglycerin gives a number of moles of gas, and the reaction evolves heat. Therefore, ΔS°_{rxn} is positive and ΔH°_{rxn} is negative. Both predict a negative ΔG°_{rxn}.
(c) Seventeen moles of gas combine to produce 34 moles of gaseous products, so ΔS°_{rxn} is expected to be positive. From experience you know the reaction is exothermic, so ΔH°_{rxn} is expected to be negative. This should lead to a product-favored reaction.

20.48. ΔH°_{rxn} = +126.9 kJ; ΔS°_{rxn} = +81 J/K; ΔG°_{rxn} = +103 kJ at 298 K. Reaction is reactant-favored.

20.50. ΔG°_{rxn} = +6.98 kJ at 700 K

20.52. ΔG°_{rxn} = −97.9 kJ at 298 K; product-favored and enthalpy-driven.

20.54. **(a)** ΔG°_{rxn} = +91.4 kJ at 298 K
(b) $K_p = 9.71 \times 10^{-17}$
(c) Reaction is not product-favored at 298 K. T at which ΔG°_{rxn} is zero is 981.3 K.

20.56. **(a)** The value of ΔS°_{rxn} (+162.0 J/K) is positive as expected for a reaction that converts 1 mole of liquid to 1.5 moles of gas.
(b) The reaction is not product-favored at 298 K (ΔG°_{rxn} = +115.6 kJ).
(c) It becomes product-favored at temperatures over 1012 K (739 °C).

20.58. ΔG°_{rxn} = +22.52 kJ at 298 K. The reaction is not product-favored, but it can become so at temperatures above 468 K or 195 °C.

20.60. ΔS°_{rxn} = −200.3 J/K. The entropy is expected to decline as a gas dissolves in water. The system is becoming more orderly.

20.62. From experience with seashells and limestone, which are largely $CaCO_3$, it is expected that the reaction is not product-favored at ordinary temperatures. However, it has been mentioned (page 806) that lime, CaO, is produced by heating $CaCO_3$. Therefore, the reaction becomes product-favored at higher temperatures. Calculations show that ΔG°_{rxn} = +130.5 kJ at 298 K and that ΔG°_{rxn} = 0 kJ at 1110 K.

20.64. **(a)** ΔG°_{rxn} = +173.10 kJ at 298 K.
(b) ΔG°_{rxn} = +156.4 kJ at 700. K and $K_p = 4 \times 10^{-9}$.
(c) $P(O_2) = P(N_2)$ = 1.00 atm and P(NO) = 6×10^{-5} atm

20.66. **(a)** ΔH°_{rxn} = −62.10 kJ; ΔS°_{rxn} = −132.74 J/K; ΔG°_{rxn} = −22.52 kJ at 298 K. Reaction is product-favored. (This is the reverse of the reaction in Study Question 58.)
(b) $K_p = 8.9 \times 10^3 = 1/P(O_2)$; $P(O_2) = 1.1 \times 10^{-4}$ atm
(c) If $P(O_2)$ = 1.00 atm, then K_p = 1.00 and ΔG°_{rxn} = 0. T at this point is 467.8 K or 194.7 °C.

CHAPTER 21

21.12. **(a)** Oxidation; reducing agent

$$Cr(s) \longrightarrow Cr^{3+}(aq) + 3\,e^-$$

(b) Oxidation; reducing agent

$$AsH_3(g) \longrightarrow As(s) + 3\,H^+(aq) + 3\,e^-$$

(c) Reduction; oxidizing agent

$$VO_3^-(aq) + 6\,H^+(aq) + 3\,e^- \longrightarrow V^{2+}(aq) + 3\,H_2O(\ell)$$

21.14. **(a)** Reduction; oxidizing agent

$$Cr_2O_7^{2-}(aq) + 14\,H^+(aq) + 6\,e^- \longrightarrow 2\,Cr^{3+}(aq) + 7\,H_2O(\ell)$$

(b) Oxidation; reducing agent

$$CH_3CHO(aq) + H_2O(\ell) \longrightarrow CH_3CO_2H(aq) + 2\,H^+(aq) + 2\,e^-$$

(c) Oxidation; reducing agent

$$Bi^{3+}(aq) + 3\,H_2O(\ell) \longrightarrow HBiO_3(aq) + 5\,H^+(aq) + 2\,e^-$$

21.16. **(a)** Oxidation; reducing agent

$$Sn(s) + 4\,OH^-(aq) \longrightarrow Sn(OH)_4^{2-}(aq) + 2\,e^-$$

(b) Reduction; oxidizing agent

$$MnO_4^-(aq) + 2\,H_2O(\ell) + 3\,e^- \longrightarrow MnO_2(s) + 4\,OH^-(aq)$$

(c) Reduction; oxidizing agent

$$ClO^-(aq) + H_2O(\ell) + 2\,e^- \longrightarrow Cl^-(aq) + 2\,OH^-(aq)$$

21.18. **(a)**

$$Cl_2(aq) + 2\,e^- \longrightarrow 2\,Cl^-(aq)$$
$$2\,Br^-(aq) \longrightarrow Br_2(aq) + 2\,e^-$$
$$\overline{Cl_2(aq) + 2\,Br^-(aq) \longrightarrow 2\,Cl^-(aq) + Br_2(aq)}$$

(b)

$$Sn(s) \longrightarrow Sn^{2+}(aq) + 2\,e^-$$
$$2\,H^+(aq) + 2\,e^- \longrightarrow H_2(g)$$
$$\overline{Sn(s) + 2\,H^+(aq) \longrightarrow Sn^{2+}(aq) + H_2(g)}$$

(c)

$$Zn(s) \longrightarrow Zn^{2+}(aq) + 2\,e^-$$
$$2[VO^{2+}(aq) + 2\,H^+(aq) + e^- \longrightarrow V^{3+}(aq) + H_2O(\ell)]$$
$$\overline{Zn(s) + 2\,VO^{2+}(aq) + 4\,H^+(aq) \longrightarrow Zn^{2+}(aq) + 2\,V^{3+}(aq) + 2\,H_2O(\ell)}$$

21.20. **(a)**

$$2[Ag^+(aq) + e^- \longrightarrow Ag(s)]$$
$$HCHO(aq) + H_2O(\ell) \longrightarrow HCO_2H(aq) + 2\,H^+(aq) + 2\,e^-$$
$$\overline{2\,Ag^+(aq) + HCHO(aq) + H_2O(\ell) \longrightarrow 2\,Ag(s) + HCO_2H(aq) + 2\,H^+(aq)}$$

(b)
$$3[H_2S(aq) \longrightarrow S(s) + 2\ H^+(aq) + 2\ e^-]$$
$$Cr_2O_7^{2-}(aq) + 14\ H^+(aq) + 6\ e^- \longrightarrow 2\ Cr^{3+}(aq) + 7\ H_2O(\ell)$$
$$3\ H_2S(aq) + Cr_2O_7^{2-}(aq) + 8\ H^+(aq) \longrightarrow 2\ Cr^{3+}(aq) + 3\ S(s) + 7\ H_2O(\ell)$$

(c)
$$3[Zn(s) \longrightarrow Zn^{2+}(aq) + 2\ e^-]$$
$$2[VO_3^-(aq) + 6\ H^+(aq) + 3\ e^- \longrightarrow V^{2+}(aq) + 3\ H_2O(\ell)]$$
$$3\ Zn(s) + 2\ VO_3^-(aq) + 12\ H^+(aq) \longrightarrow 3\ Zn^{2+}(aq) + 2\ V^{2+}(aq) + 6\ H_2O(\ell)$$

21.22. **(a)**
$$Zn(s) + 2\ OH^-(aq) \longrightarrow Zn(OH)_2(s) + 2\ e^-$$
$$ClO^-(aq) + H_2O(\ell) + 2\ e^- \longrightarrow Cl^-(aq) + 2\ OH^-(aq)$$
$$Zn(s) + ClO^-(aq) + H_2O(\ell) \longrightarrow Zn(OH)_2(s) + Cl^-(aq)$$

(b)
$$3[ClO^-(aq) + H_2O(\ell) + 2\ e^- \longrightarrow Cl^-(aq) + 2\ OH^-(aq)]$$
$$2[CrO_2^-(aq) + 4\ OH^-(aq) \longrightarrow CrO_4^{2-}(aq) + 2\ H_2O(\ell) + 3\ e^-]$$
$$3\ ClO^-(aq) + 2\ CrO_2^-(aq) + 2\ OH^-(aq) \longrightarrow 3\ Cl^-(aq) + 2\ CrO_4^{2-}(aq) + H_2O(\ell)$$

(c)
$$5[Br_2(aq) + 2\ e^- \longrightarrow 2\ Br^-(aq)]$$
$$Br_2(aq) + 12\ OH^-(aq) \longrightarrow 2\ BrO_3^-(aq) + 6\ H_2O(\ell) + 10\ e^-$$
$$3\ Br_2(aq) + 6\ OH^-(aq) \longrightarrow 5\ Br^-(aq) + BrO_3^-(aq) + 3\ H_2O(\ell)$$

21.24. Oxidation; anode compartment:
$Cr(s) \longrightarrow Cr^{3+}(aq) + 3\ e^-$
Reduction; cathode compartment:
$Fe^{2+}(aq) + 2\ e^- \longrightarrow Fe(s)$

21.26. If the tin electrode is negative, it is furnishing electrons to the external circuit. The source of the electrons is the oxidation reaction $Sn(s) \rightarrow Sn^{2+}(aq) + 2e^-$; this tells us the electrode is the anode. The copper half-cell, with a positive charge on the electrode, reflects a deficiency in negative charge due to the reduction of Cu^{2+} to copper metal: $Cu^{2+}(aq) + 2e^- \rightarrow Cu(s)$. Therefore, this is the cathode.

21.28. $\Delta G^\circ_{rxn} = -562$ kJ

21.30. **(a)** $E^\circ_{net} = -1.298$ V; not product-favored as written
(b) $E^\circ_{net} = -0.51$ V; not product-favored as written
(c) $E^\circ_{net} = -1.021$ V; not product-favored as written

21.32. **(a)** $Sn^{2+}(aq) + 2\ Ag(s) \longrightarrow Sn(s) + 2\ Ag^+(aq)$
$E^\circ_{net} = -0.94$ V; not product-favored as written
(b) Balanced as written. $E^\circ_{net} = +0.91$ V; product-favored
(c) $I_2(aq) + 2\ Br^-(aq) \longrightarrow 2\ I^-(aq) + Br_2(aq)$
$E^\circ_{net} = -0.531$ V; not product-favored as written

21.34. **(a)** Weakest oxidizing agent, V^{2+};
(b) strongest oxidizing agent, Cl_2;
(c) strongest reducing agent, V;
(d) weakest reducing agent, Cl^-;
(e) Pb does not reduce V^{2+};
(f) I_2 does not oxidize Cl^-;
(g) Cl_2 and I_2 can be reduced by Pb.

31.36. **(a)** $V(s) + Cl_2(g) \longrightarrow V^{2+}(aq) + 2\ Cl^-(aq)$
$E^\circ_{net} = 2.54$ V
(b) $V(s) + I_2(g) \longrightarrow V^{2+}(aq) + 2\ I^-(aq)$
$E^\circ_{net} = 1.72$ V

21.38. **(a)** $Zn(s) + 2\ Ag^+(aq) \longrightarrow Zn^{2+}(aq) + 2\ Ag(s)$
$E^\circ = +1.56$ V
(b) Zn = anode and Ag = cathode
(c–f)

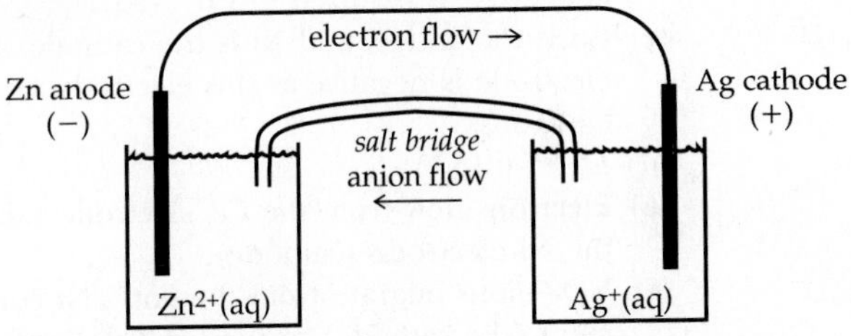

21.40. $E = 0.177$ V; the cell potential (E) is much less positive when the dissolved reagents are 0.10 M ($E^\circ_{net} = 0.236$ V). (Note that I_2 is a solid as given in Appendix J and is not included when solving the Nernst equation.)

21.42. **(a)** $Fe^{2+}(aq) + Ag^+(aq) \longrightarrow Ag(s) + Fe^{3+}(aq)$
(b) $E^\circ_{net} = +0.028$ V
(c) $E_{net} = -0.031$ V. The net cell reaction now is the reverse of (a).

21.44. **(a)** 9.5×10^7
(b) 1.1×10^{-18}

21.46. 735 s or 12.2 min

21.48. 5.47 g Ni

21.50. 0.10 A

21.52. 270 kg Al

21.54. 190 g Pb

21.56. 12 W

21.58. 2.97×10^{16} C; 3.79×10^{10} kwh

21.60. **(a)** Oxidation occurs at the positive anode:
$$2\ I^-(aq) \longrightarrow I_2(s) + 2\ e^-$$
(b) Reduction occurs at the negative cathode:
$$2\ H_2O(\ell) + 2\ e^- \longrightarrow 2\ OH^-(aq) + H_2(g)$$
(c) 1.2 g I_2(s), 0.52 g KOH, and 0.0094 g H_2 is expected.

21.62. **(a)** Anode (oxidation of Br^-):
$2\ Br^-(aq) \longrightarrow Br_2(\ell) + 2e^-$
Cathode (reduction of water):
$2\ H_2O(\ell) + 2e^- \longrightarrow 2\ OH^-(aq) + H_2(g)$

(b) Anode (oxidation of F^-):
$2\ F^- \longrightarrow F_2(g) + 2e^-$
Cathode (reduction of Na^+):
$Na^+ + e^- \longrightarrow Na(\ell)$

(c) Anode (oxidation of water):
$6\ H_2O(\ell) \longrightarrow O_2(g) + 4\ H_3O^+(aq) + 4e^-$
Cathode (reduction of water):
$2\ H_2O(\ell) + 2e^- \longrightarrow 2\ OH^-(aq) + H_2(g)$

21.64. (a)
$$I^-(aq) + 3\ H_2O(\ell) \longrightarrow IO_3^-(aq) + 6\ H^+(aq) + 6\ e^-$$
$$3[Br_2(\ell) + 2\ e^- \longrightarrow 2\ Br^-(aq)]$$
$$I^-(aq) + 3\ Br_2(\ell) + 3\ H_2O(\ell) \longrightarrow IO_3^-(aq) + 6\ Br^-(aq) + 6\ H^+(aq)$$

(b)
$$5[U^{4+}(aq) + 2\ H_2O(\ell) \longrightarrow UO_2^+(aq) + 4\ H^+(aq) + e^-]$$
$$MnO_4^-(aq) + 8\ H^+(aq) + 5\ e^- \longrightarrow Mn^{2+}(aq) + 4\ H_2O(\ell)$$
$$5\ U^{4+}(aq) + MnO_4^-(aq) + 6\ H_2O(\ell) \longrightarrow 5\ UO_2^+(aq) + Mn^{2+}(aq) + 12\ H^+(aq)$$

(c)
$$2\ I^-(aq) \longrightarrow I_2(s) + 2\ e^-$$
$$MnO_2(s) + 4\ H^+(aq) + 2\ e^- \longrightarrow Mn^{2+}(aq) + 2\ H_2O(\ell)$$
$$2\ I^-(aq) + MnO_2(s) + 4\ H^+(aq) \longrightarrow I_2(s) + Mn^{2+}(aq) + 2\ H_2O(\ell)$$

21.66. (a) Balanced equation: $Ni^{2+}(aq) + Cd(s) \longrightarrow Ni(s) + Cd^{2+}(aq)$

(b) Cd is the reducing agent, and $Ni^{2+}(aq)$ is the oxidizing agent. $Ni^{2+}(aq)$ is reduced to Ni(s), and Cd(s) is oxidized to $Cd^{2+}(aq)$.

(c) Cd is the anode, and Ni is the cathode. The Cd electrode is negative as this electrode supplies electrons.

(d) $E^\circ_{net} = 0.15$ V

(e) Electrons flow from the Cd electrode (anode) to the Ni electrode (cathode).

(f) NO_3^- ions migrate from the Ni^{2+}/Ni compartment (the cathode) to the Cd/Cd^{2+} compartment (the anode).

(g) $\ln K = +nE^\circ/0.0257 = +(2.00\text{ mol})(0.15\text{ V})/(0.0257\text{ V}\cdot\text{mol})$
$\ln K = 11.67$, which gives $K = 1.2 \times 10^5$

(h) For $[Cd^{2+}] = 0.010$ M and $[Ni^{2+}] = 1.0$ M, and $E^\circ = 0.15$ V, we have

$$E = E^\circ - \frac{0.0592}{2}\log\frac{[Cd^{2+}]}{[Ni^{2+}]} = 0.21\text{ V}$$

The reaction is still product-favored in the direction written.

(i) The battery has, in the beginning, 1.00 mol each of Ni^{2+} and Cd^{2+}. In addition,

$$50.0\text{ g Ni}\cdot\frac{1\text{ mol}}{58.69\text{ g}} = 0.852\text{ mol Ni}$$

$$50.0\text{ g Cd}\cdot\frac{1\text{ mol}}{112.4\text{ g}} = 0.445\text{ mol Cd}$$

The limiting reagent here is 0.445 mol Cd. Therefore, we use this to find the lifetime of the battery.

$$0.445\text{ mol Cd}\cdot\frac{2\text{ mol e}^-}{1\text{ mol Cd}}\cdot\frac{96500\text{ C}}{\text{mol e}^-} = 8.59\times10^4\text{ C}$$

$$8.59\times10^4\text{ C}\cdot\frac{1\text{ A}\cdot\text{s}}{\text{C}}\cdot\frac{1}{0.050\text{ A}} = 1.7\times10^6\text{ s}$$

(about 20 days)

21.68. $E^\circ = -0.185$ V for $Md^{3+}(aq) + e^- \longrightarrow Md^{2+}(aq)$

21.70. $E^\circ = 2.087$ V

21.72. Concentrations after 48.0 h: $[Cu^{2+}] = 0.64$ M and $[Sn^{2+}] = 1.36$ M.
Cell potential after 48.0 h: 0.47 V

21.74. (a) E° for magnesium oxidation is 2.37 V, whereas that for iron oxidation is 0.44 V. Magnesium is the better oxidizing agent, and the metal oxidizes more readily than iron.
(b) 42 years

21.76. (a) $Al(s) + 3\ Cl_2(g) \longrightarrow 2\ Al^{3+}(aq) + 6\ Cl^-(aq)$
(b) Al is the anode and the Cl_2/Cl^- compartment is the cathode.
(c) $E^\circ_{net} = 3.02$ V
(d) If the pressure of Cl_2 decreases, the cell potential should also decrease. Now E_{net} is 3.01 V.
(e) 119 h

21.78. (a) 5.0×10^{-4} W/g
(b) 3.1×10^{-4} W/g
(c) The Ag/Zn battery

21.80. (a) A and C are stronger reducing agents than H_2.
(b) C is the strongest reducing agent of the series.
(c) D is a stronger reducing agent than B.
Order of strength as reducing agents: $B < D < H_2 < A < C$

21.82. (a) 82.96% efficient
(b) 94.54% efficient
(c) If energy is used to order the products (such as creating liquid water instead of water vapor), less energy is available to do work, and the efficiency is lower.

21.84. 290 h

CHAPTER 22

22.8. ns^2np^2. Common oxidation numbers are +2, +4, and −4.

22.10. S^{2-}, sulfide ion; Cl^-, chloride ion; K^+, potassium ion; Ca^{2+}, calcium ion.

22.12. $2\ M(s) + X_2(g) \longrightarrow 2\ MX(s)$. Reaction is likely to be exothermic, and the product is ionic. See Figure 2.21.

22.14. See page 1001. Some representative compounds are SiO_2, Fe_2O_3, $CaCO_3$, NaCl, KCl, $MgCl_2$, TiO_2.

22.16. Ca is very reactive with water (to give $Ca(OH)_2$) and air (to give CaO).

22.18. $SO_3 < SiO_2 < Al_2O_3 < Na_2O$

22.20. **(a)** $2\ Na(s) + I_2(s) \longrightarrow 2\ NaI(s)$
(b) $8\ Ca(s) + S_8(s) \longrightarrow 8\ CaS(s)$
(c) $4\ Al(s) + 3\ O_2(g) \longrightarrow 2\ Al_2O_3(s)$
(d) $Si(s) + 2\ Cl_2(g) \longrightarrow SiCl_4(\ell)$

22.22. $2\ Na(s) + H_2(g) \longrightarrow 2\ NaH(s)$
The solid, ionic product is sodium hydride. It should react with water to produce NaOH and H_2 gas. (See Figure 17.2.)

22.24. 1.2×10^4 L

22.26. $CH_4(g) + H_2O(g) \longrightarrow CO(g) + 3\ H_2(g)$
$\Delta H^\circ_{rxn} = +206.1$ kJ; $\Delta S^\circ_{rxn} = +214.7$ J/K; and $\Delta G^\circ_{rxn} = +142.1$ kJ

22.28.

$$2\ Na(s) + F_2(g) \longrightarrow 2\ NaF(s)$$
$$2\ Na(s) + Cl_2(g) \longrightarrow 2\ NaCl(s)$$
$$2\ Na(s) + Br_2(\ell) \longrightarrow 2\ NaBr(s)$$
$$2\ Na(s) + I_2(s) \longrightarrow 2\ NaI(s)$$

All sodium halides are crystalline solids with high melting points. All are readily soluble in water.

22.30.

$$4\ Li(s) + O_2(g) \longrightarrow 2\ Li_2O(s)$$
lithium oxide
$$2\ Na(s) + O_2(g) \longrightarrow Na_2O_2(s)$$
sodium peroxide
$$K(s) + O_2(g) \longrightarrow KO_2(s)$$
potassium superoxide

22.32. $2\ NaCl(aq) + 2\ H_2O(\ell) \longrightarrow 2\ NaOH(aq) + Cl_2(g) + H_2(g)$
The theoretical weight ratio of products from the reaction (80.0 g NaOH/70.9 g Cl_2 = 1.12) is larger than the weight ratio of products actually produced (=1.07). This indicates that some Cl_2 is produced by some other route in addition to the electrolysis of NaCl.

22.34.

$$2\ Mg(s) + O_2(g) \longrightarrow 2\ MgO(s)$$
$$3\ Mg(s) + N_2(g) \longrightarrow Mg_3N_2(s)$$

22.36. **(a)** Produce CaO for the steel industry.
(b) To add to agricultural land to rise soil pH.
(c) Produce CaO for use as an inexpensive industrial base.
$CaCO_3(s) + CO_2(g) + H_2O(\ell) \longrightarrow Ca^{2+}(aq) + HCO_3^-(aq)$

22.38. 1.142×10^6 g SO_2

22.40. $2\ Al(s) + 6\ HCl(aq) \longrightarrow 2\ Al^{3+}(aq) + 6\ Cl^-(aq) + 3\ H_2(g)$

$$2\ Al(s) + 3\ Cl_2(g) \longrightarrow 2\ AlCl_3(s)$$
$$4\ Al(s) + 3\ O_2(g) \longrightarrow 2\ Al_2O_3(s)$$

22.42. $2\ Al(s) + 2\ OH^-(aq) + 6\ H_2O(\ell) \longrightarrow 2\ Al(OH)_4^-(aq) + 3\ H_2(g)$
1.84×10^4 mL of H_2 gas is produced.

22.44. $Al_2O_3(s) + 3\ H_2SO_4(aq) \longrightarrow Al_2(SO_4)_3(aq) + 3\ H_2O(\ell)$
0.298 kg Al_2O_3 and 0.860 kg H_2SO_4

22.46. $AlCl_4^-$ is tetrahedral with a sp^3-hybridized Al atom.

22.48. SiO_2 is a network solid. See page 1020.
(a) The energy of four Si—O bonds is greater than the energy of two Si=O bonds (as SiO_2 would have if its structure were analogous to the structure of CO_2).
(b) SiO_2 is a network solid with extended bonding, making it a very stable solid with a high melting point.

22.50. **(a)** $Si(s) + 2\ CH_3Cl(g) \longrightarrow (CH_3)_2SiCl_2(\ell)$
(b) $P(CH_3Cl) = 0.823$ atm
(c) 12.2 g product

22.52. The enthalpy change for the reaction is $\Delta H^\circ_{rxn} = -114.14$ kJ.
(a) The values for the free energy of formation of both NO(g) and $NO_2(g)$ are positive, indicating that neither is thermodynamically stable relative to the elements.
(b) The reaction of NO with O_2 is exothermic, as indicated by the enthalpy change for the reaction.
(c) Both NO and NO_2 are paramagnetic as both are odd electron molecules.

22.54. **(a)** $N_2H_4(aq) + O_2(aq) \longrightarrow N_2(g) + 2\ H_2O(\ell)$
(b) 1.32×10^3 g N_2H_4

22.56. $5\ N_2H_5^+(aq) + 4\ IO_3^-(aq) \rightarrow 5\ N_2(g) + 2\ I_2(s) + H_3O^+(aq) + 11\ H_2O(\ell)$

22.58. Azide ion

[:N̈=N=N̈:]$^-$

22.60. **(a)** 3.5×10^3 kg SO_2
(b) 4.0×10^3 kg $Ca(OH)_2$

22.62. The S_3^{2-} ion has tetrahedral geometry.

[:S̈—S̈—S̈:]$^{2-}$

22.64. E°_{net} is −0.07 V, so bromate ion is not a sufficiently good oxidizing agent under standard conditions to oxidize Mn^{2+} to MnO_4^- ($E^\circ = -1.51$ V).

22.66. BrF_3 has a trigonal-bipyramidal electron-pair geometry, and its molecular structure is T-shaped. The equatorial lone pairs force the axial Br—F bonds toward the equatorial F atoms, and the F—Br—F bond angles are slightly less than 90°.

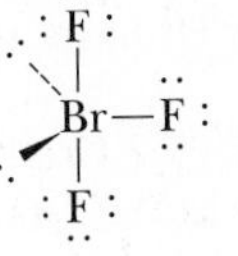

22.68.

Compound	ΔG°_{rxn}
$MgCO_3$	+48.3 kJ
$CaCO_3$	+130.40 kJ
$BaCO_3$	+218.1 kJ

The stability of the metal carbonates increases on moving down the periodic group.

22.70. **(a)** ΔG_f° should be more negative than n (−95.3 kJ).
(b) Ba, Pb, and Ti should react spontaneously with HCl(aq).

22.72. **(a)** If resonance structures are written, we see that the two N—O bonds have the same bond order (1.5) and so are expected to have the same bond length. On the other hand, the N—OH bond is a single bond, and so is longer than the other N—O bonds.
(b) The O—N—O bond angles are in the range expected for a planar-trigonal N atom. The H—O—N angle is closer to the tetrahedral angle expected for an O atom surrounded by two lone electron pairs and two bond pairs.
(c) The central N atom is sp^2-hybridized. This means a p orbital remains unhybridized, and this orbital can overlap with a p-type orbital on an O atom to form a N—O pi bond. (See the discussion of O_3 bonding on page 485.

22.74. A = $BaCO_3$; B = BaO; C = $CaCO_3$; D = $BaCl_2$; and E = $BaSO_4$

22.79. **(a)** 820 L of sea water requires 2.31 kg of CaO.
(b) 255.3 kg Mg is produced at the cathode. Cl_2 (744.8 kg) is produced at the anode. 2.101×10^4F is used.
(c) The commercial process uses 1600 kJ/mol of Mg. This is greater than the enthalpy change for the decomposition of $MgCl_2$ (+641 kJ/mol).

CHAPTER 23

23.14. Cr^{3+} has three unpaired electrons in the $3d$ orbitals, and it is paramagnetic. In contrast, Cr^{6+} would have no electrons remaining in either the $4s$ or $3d$ subshell.

23.16. **(a)** Fe^{2+} and Co^{3+}
(b) Zn^{2+} and Cu^{+}
(c) Mn^{2+} and Fe^{3+}
(d) Ni^{2+} and Cu^{3+}

23.18. **(a)** Fe, smaller atom than Ti but with a greater mass
(b) Os, smaller atom than Ti but with a greater mass
(c) Zr, radius is slightly larger than that of Ti but it has a larger mass.
(d) Hf, radius is similar to Zr but Hf has a significantly larger mass.

23.19. **(a)** $Cr_2O_3(s) + 2\ Al(s) \rightarrow Al_2O_3(s) + 2\ Cr(s)$
(b) $TiCl_4(\ell) + 2\ Mg(s) \rightarrow Ti(s) + 2\ MgCl_2(s)$
(c) $2[Ag(CN)_2]^-(aq) + Zn(s) \rightarrow 2\ Ag(s) + [Zn(CN)_4]^{2-}(aq)$

23.21. 1.10 L H_2SO_4; 0.527 kg TiO_2

23.23. 1.41×10^3 kg NiS; 1.16×10^3 kg NiO; 31.2 kg H_2; and 1.73×10^3 kg CO

23.24. Monodentate ligands: a, c, d, and g
Bidentate ligands: b, e, and f

23.26. **(a)** Mn^{2+}
(b) Co^{3+}
(c) Co^{3+}
(d) Mn^{2+}

23.28. $[Co(en)_2(H_2O)Cl]^{2+}$

23.29. **(a)** $Ni(en)_2Cl_2$
(b) K_2PtCl_4
(c) $KCu(CN)_2$
(d) $[Fe(H_2O)_2(NH_3)_4]^{2+}$

23.31. **(a)** hydroxo
(b) oxo
(c) iodo
(d) oxalato

23.33. **(a)** diaquabis(oxalato)nickelate(II) ion
(b) bis(ethylenediamine)dinitritocobalt(III) ion
(c) amminechlorobis(ethylenediamine)cobalt(III) ion
(d) diammineoxalatoplatinum(II)

23.35. **(a)** $[Cr(H_2O)_4Cl_2]Cl$
(b) pentaamminesulfatochromium(III) chloride
(c) $Na_2[CoCl_4]$
(d) tris(oxalato)cobaltate(III) ion

23.36. **(a)**

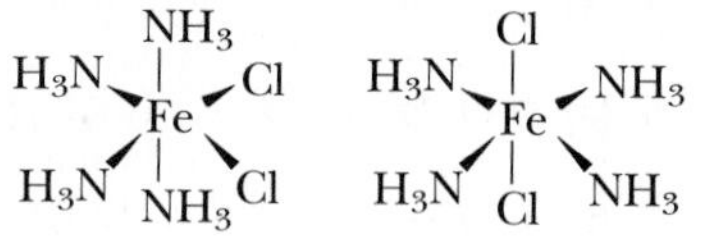

cis *trans*

(b)

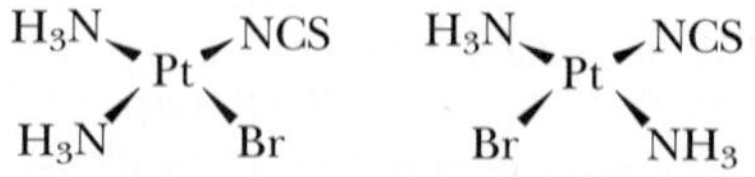

cis *trans*

(c)

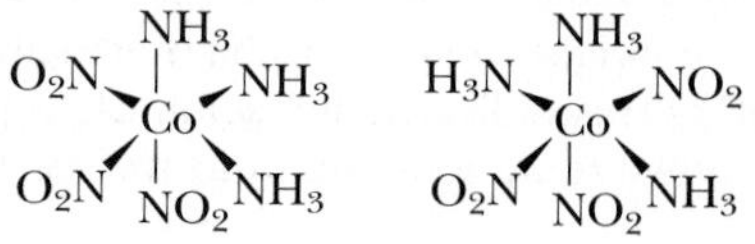

fac *mer*

(d)

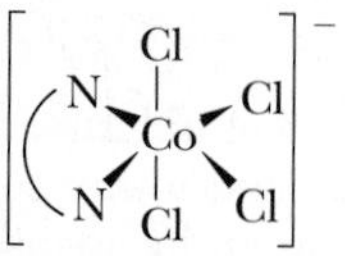

Only one isomer is possible for this ion. (N—N is the ethylendiamine ligand.)

23.38. **(a)** No chiral center

(b)

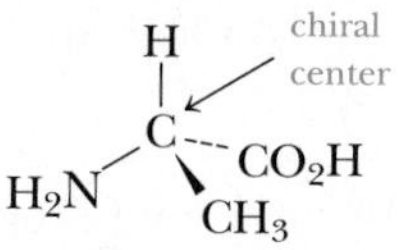

(c)

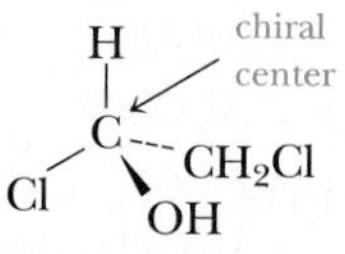

(d) No chiral centers

23.40. Isomers of $[Co(en)(NH_3)_2(H_2O)Cl]^{2+}$, ethylenediamine (en) is depicted by N N

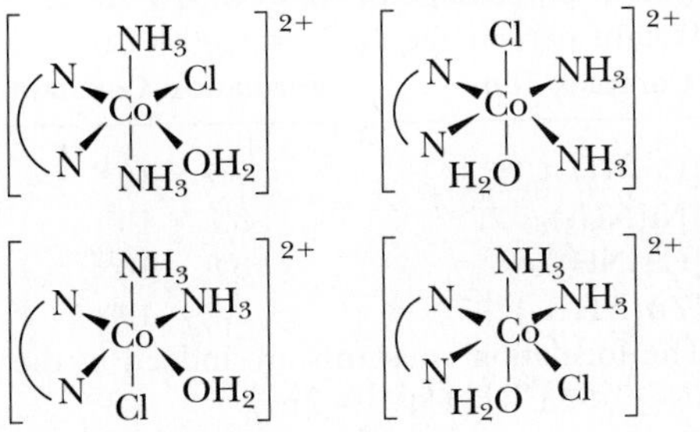

23.42. **(a)** The iron is a chiral center

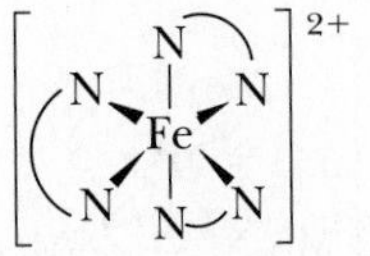

(b) The cobalt ion is not a chiral center in the *fac* isomer.

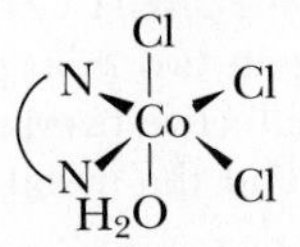

(c) The cobalt ion is a chiral center.

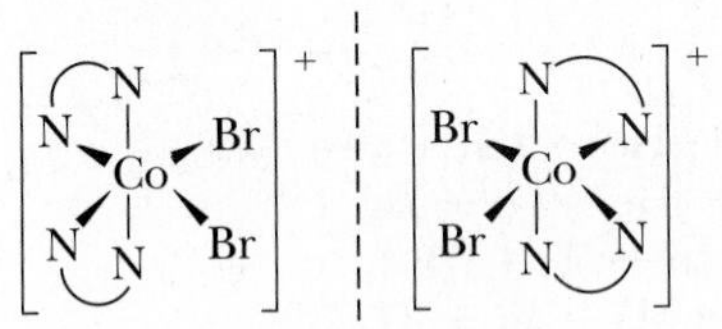

mirror images

(d) Chirality is not possible for square-planar complexes.

23.43. d^4, d^5, d^6, and d^7

23.45. **(a)** d^6 Fe^{2+} in $[Fe(CN)_6]^{4-}$. Note that only the low-spin complex actually exists. See Question 23.53.

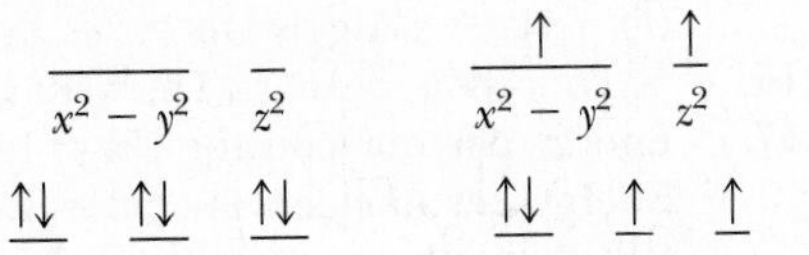

Low-spin, diamagnetic — High-spin, paramagnetic, 4 unpaired electrons

(b) d^6 Co^{3+} in $[Co(NH_3)_6]^{3+}$. See the configurations in part (a).

(c) d^5 Fe^{3+} in $[Fe(H_2O)_6]^{3+}$

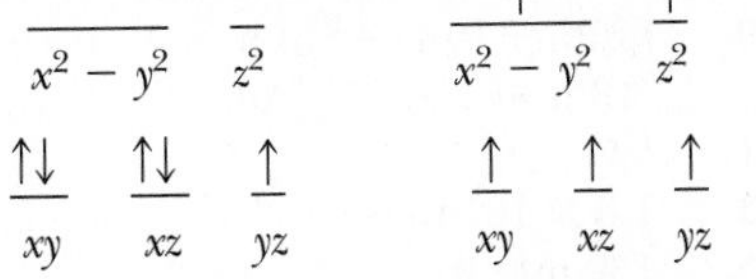

Low-spin, paramagnetic, 1 unpaired electron — High-spin, paramagnetic, 5 unpaired electrons

(d) d^3 Cr^{3+} in $[CrF_6]^{3-}$. There is no difference in high and low spin for a d^3 complex. It is paramagnetic to the extent of three unpaired electrons. See Question 23.43.

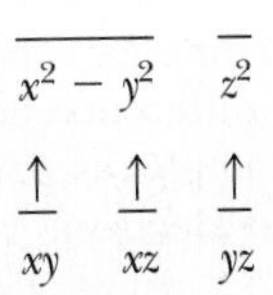

23.47. The Mn^{2+} ion has a d^5 configuration. In the high-spin state it has five unpaired electrons, whereas it has one unpaired electron in the low-spin state. (See Question 23.45, part c.) This means the CN^- ligand leads a to an increased value of Δ_O.

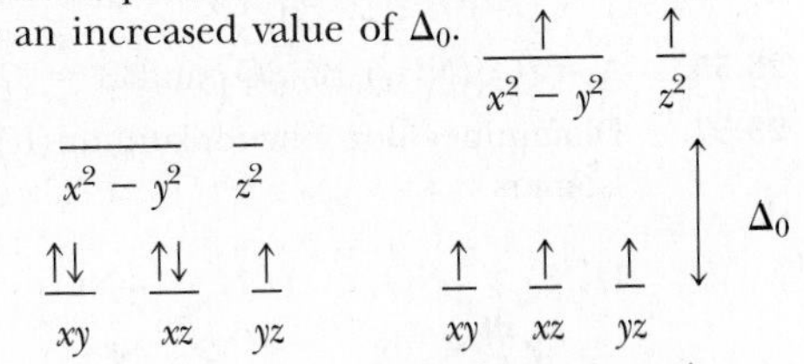

Low-spin, paramagnetic, 1 unpaired electron — High-spin, paramagnetic, 5 unpaired electrons

23.48. $F^- < H_2O < NH_3 < CN^-$

23.50. The ion absorbs in the red region of the spectrum, so it passes blue and green light to give a blue or cyan color.

23.51. **(a)** 6
(b) octahedral
(c) Fe^{2+}
(d) 4 unpaired electrons; high spin (see Question 23.45c).
(e) paramagnetic (4 unpaired electrons).

23.53. **(a)** The CN^- ligand is a strong field ligand and should lead to low-spin complexes. In the case of $[Fe(CN)_6]^{4-}$, where Fe^{2+} is a d^6 ion, the complex ion is diamagnetic.

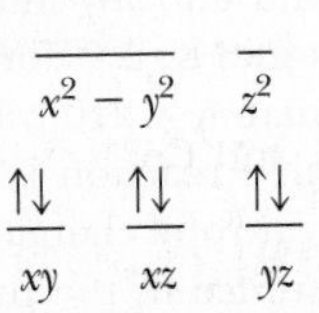

(b) The F^- ligand is a weak field ligand, so the complex based on Mn^{2+} (d^5) is expected to be high spin. The complex is paramagnetic with five unpaired electrons.

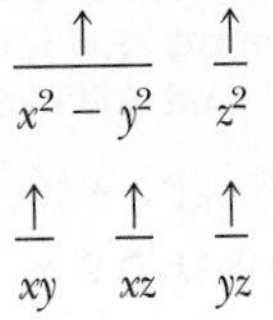

(c) Only high-spin complexes are possible for Cr^{3+} (d^3), so the complex is paramagnetic with three unpaired electrons.

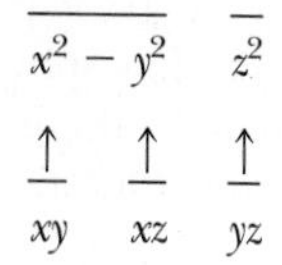

(d) This complex is based on Cu^{2+}, a d^9 ion. There is no difference between high and low spin, so the complex is paramagnetic and has one unpaired electron.

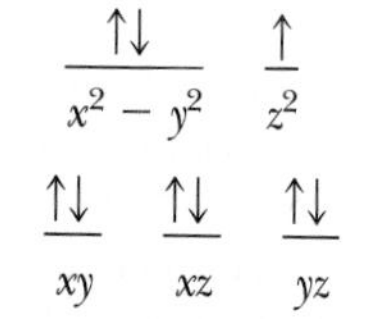

23.55. A = $[Co(NH_3)_5Br]SO_4$ and B = $[Co(NH_3)_5SO_4]Br$

23.57. Diamminechloronitritoplatinum(II) has geometric isomers.

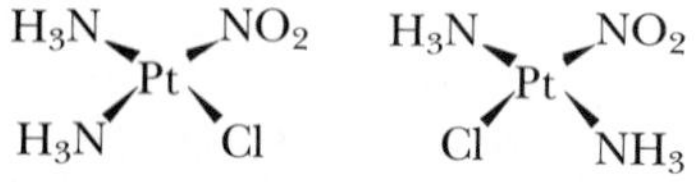

cis *trans*

23.59. Excess F^- ion displaces H_2O from the coordination sphere of $[Co(H_2O)_6]^{3+}$ (which does not exist because Co^{3+} (d^6) is a powerful oxidizing agent in water), a low-spin ion. The F^- ion is a weaker field ligand, so $[CoF_6]^{3-}$ is a high-spin complex and is paramagnetic by four unpaired electrons.

23.61. K_{form} for the ammine complex = 10^8, so $\Delta G = -45.6$ kJ

K_{form} for the en complex = 10^{18}, so $\Delta G = -102.7$ kJ

The difference in ΔG values for the two complexes is about 57 kJ, much greater than the difference in the enthalpies of the reactions. Thus, enthalpy changes alone cannot account for the great difference in ΔG values, and entropy must play a role in these reactions. At 298 K, $\Delta S°$ for the hexammine complex is approximately −210 J/K, whereas that for the ethylenediamine reaction is about −48 J/K. This difference in entropy changes seems reasonable in view of the difference in the number of molecules reacting. For the ammine reaction, seven ions or molecules form one complex ion, whereas in the ethylenediamine reaction only four molecules or ions form one complex ion.

Another view, and one that provides more insight into chelation reactions, is obtained if we calculate $\Delta H°$, $\Delta S°$, and $\Delta G°$ for the *exchange* of ligands.

$$[Ni(NH_3)_6]^{2+} + 3\ en \longrightarrow [Ni(en)_3]^{2+} + 6\ NH_3$$

$$\Delta H° = -8\ kJ,\ \Delta S° = +165\ J/K,\ and\ \Delta G° = -57.1\ kJ$$

This shows that the chelated complex is much more stable than the complex ion with monodentate ligands and that the reaction is primarily *entropy-driven.*

23.63. Weight percentage of Al = 95.00% Al
Weight percentage of Cu = 5.00% Cu

23.65.

Complex Ion	Formation Constant
$[Co(NH_3)_6]^{2+}$	7.7×10^4
$[Ni(NH_3)_6]^{2+}$	5.6×10^8
$[Cu(NH_3)_4]^{2+}$	6.8×10^{12}
$[Zn(NH_3)_4]^{2+}$	2.9×10^9

The formation constants are indeed in the order $Co^{2+} < Ni^{2+} < Cu^{2+} > Zn^{2+}$

23.67. Two geometric isomers are possible (but each is chiral).

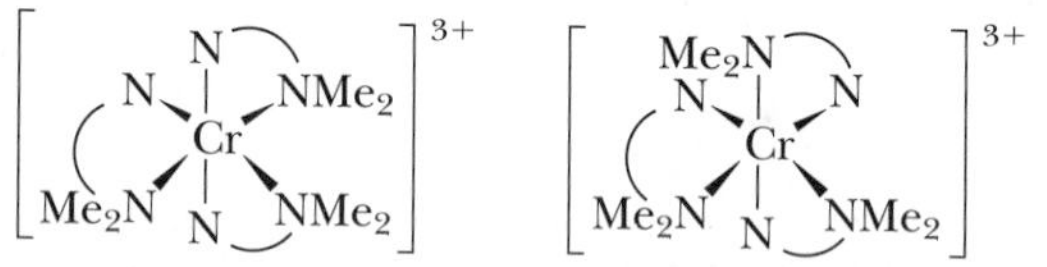

mer *fac*

23.69. The N—C—C—N chain of the $H_2NCH_2CH_2NH_2$ ligand is not long enough to span two *trans* positions about a metal ion and allow effective overlap between the N atom lone pairs and the metal atom hybrid orbitals.

CHAPTER 24

24.11. **(a)** $^{54}_{26}Fe + ^{4}_{2}He \rightarrow 2^{1}_{1}H + ^{56}_{26}Fe$
(b) $^{27}_{13}Al + ^{4}_{2}He \rightarrow ^{30}_{15}P + ^{1}_{0}n$
(c) $^{32}_{16}S + ^{1}_{0}n \rightarrow ^{1}_{1}H + ^{32}_{15}P$
(d) $^{96}_{42}Mo + ^{2}_{1}H \rightarrow ^{1}_{0}n + ^{97}_{43}Tc$
(e) $^{98}_{42}Mo + ^{1}_{0}n \rightarrow ^{99}_{43}Tc + ^{0}_{-1}e$

24.13. **(a)** $^{111}_{47}Ag \rightarrow ^{111}_{48}Cd + ^{0}_{-1}\beta$
(b) $^{87}_{36}Kr \rightarrow ^{0}_{-1}e + ^{87}_{37}Rb$
(c) $^{231}_{91}Pa \rightarrow ^{227}_{89}Ac + ^{4}_{2}He$
(d) $^{230}_{90}Th \rightarrow ^{4}_{2}He + ^{226}_{88}Ra$
(e) $^{82}_{35}Br \rightarrow ^{82}_{36}Kr + ^{0}_{-1}e$
(f) $^{24}_{11}Na \rightarrow ^{24}_{12}Mg + ^{0}_{-1}e$

24.15. $^{231}_{90}Th$; $^{231}_{91}Pa$; $^{227}_{89}Ac$; $^{227}_{90}Th$; $^{223}_{88}Ra$

24.17. Energy per nucleon for $^{10}B = 1.038 \times 10^{-12}$ J
Energy per nucleon for $^{11}B = 1.111 \times 10^{-12}$ J
^{11}B is slightly more stable and is the more abundant isotope.

24.19. 0.469 mg

24.21. **(a)** $^{131}_{53}I \rightarrow ^{0}_{-1}e + ^{131}_{54}Xe$
(b) 1.56 mg

24.23. 0.625 mg; 2×10^{-5} mg

24.25. 35 days

24.27. 1850 years old; made about 145 AD

24.29. $^{239}_{94}Pu + 2^{1}_{0}n \rightarrow ^{241}_{94}Pu$
$^{241}_{94}Pu \rightarrow ^{0}_{-1}e + ^{241}_{95}Am$

24.31. $^{12}_{6}C$

24.33. 1.6×10^3 tons

24.35. 130 mL

24.37. **(a)** ${}^{13}_{6}C + {}^{1}_{0}n \rightarrow {}^{14}_{6}C$
(b) ${}^{40}_{18}Ar + {}^{4}_{2}He \rightarrow {}^{43}_{19}K + {}^{1}_{1}H$
(c) ${}^{250}_{98}Cf + {}^{11}_{5}B \rightarrow 4{}^{1}_{0}n + {}^{257}_{103}Lr$
(d) ${}^{53}_{24}Cr + {}^{4}_{2}He \rightarrow {}^{1}_{0}n + {}^{56}_{26}Fe$
(e) ${}^{212}_{84}Po \rightarrow {}^{208}_{82}Pb + {}^{4}_{2}He$
(f) ${}^{122}_{53}I \rightarrow {}^{0}_{+1}\beta + {}^{122}_{52}Te$
(g) ${}^{23}_{10}Ne \rightarrow {}^{23}_{11}Na + {}^{0}_{-1}\beta$
(h) ${}^{137}_{53}I \rightarrow {}^{1}_{0}n + {}^{136}_{53}I$

24.39. 3.6×10^9 years

24.41. 1.5×10^9 years

Index/Glossary

absolute temperature scale. *See* Kelvin temperature scale
absolute zero, 23, 553
absorption spectrum A plot of the intensity of light absorbed by a sample as a function of the wavelength of the light, 1075
abundance, isotopic, 75
 of elements, *98,* 1001
accuracy The agreement between a measured quantity and the accepted value, 43
acetaldehyde, 518t
 Lewis dot structure, 407
acetamide, Lewis dot structure, 407
acetaminophen, 147, 527
acetanilide, 527
acetate ion, resonance structures, 409
acetic acid, 519–520
 as weak Brønsted acid, 800–801
 buffer solution, 860–862
 dimerization, 791
 electron distribution in, *836*
 equilibrium in water, 752
 Lewis dot structure, 407
 orbital hybridization in, 472
 production, 305, 520
  catalyst in, 736–737
 reaction with ammonia, 850
 reaction with sodium cyanide, 803
 structure, 166, *807*
acetic anhydride, 219, 247
acetone, 518t
 hydrogenation of, 421
acetylacetone, 493
acetylene, bond order in, 415
 pi bonds in, 471
 standard molar enthalpy of formation, 289
acetylsalicylic acid, 491, 826, 882

Italicized page numbers indicate pages containing illustrations. Glossary terms, printed in boldface, are defined here as well as in the text. Some terms used generally in the text are defined here without giving specific page references.

achiral compound, 1064
acid(s) A compound that ionizes in water to form hydronium ion and an anion. *See also* Brønsted acid(s), Lewis acid(s)
 Arrhenius definition of, 796
 as catalyst, 792
 bases and, 170–175, 795. *See also* acid-base reaction(s)
 Brønsted definition of, 797
 carboxylic. *See* carboxylic acid(s)
 common, 172t
 ionization constants for, 808t
 Lewis definition of, 837
 molecular structure, 833–836
 monoprotic, 797
 polyprotic, 797, 798t, 830–833
 properties of, 170, 794–848
 strength of, 800–804, 802t
  electron affinity and, 834
 strong, 800, 804–806
 weak, 800, 806–810
  equilibria involving, 819–827
  equilibrium constant expression for, 756
acid ionization constant (K_a) The equilibrium constant for the ionization of an acid in aqueous solution, 806–807, 808t
 determination by titration, 870
 relation to base ionization constant, 812
acid rain, 175
acid solution, balancing redox equations for, 952–956
acid-base pairs, conjugate, 799, 801
acid-base reaction(s) An exchange reaction between an acid and a base that produces a salt and water, 178, 181–185, 849–882
 direction of, 802–804, 850
 strong acid–strong base, 851
 strong acid–weak base, 851
 summary of, 856
 titration using, 238–242, 867–874
 weak acid–strong base, 853
 weak acid–weak base, 855
acrolein, 491
acrylonitrile, 111
actinide(s) The series of elements between actinium and element 104 in the periodic table, 83, 373, 1041
activation energy (E_a) The minimum amount of energy that must be absorbed by a system to cause it to react, 720
 calculating, 726
 reduction by catalyst, 734
activity (*A*) A measure of the rate of nuclear decay, the number of disintegrations observed in a sample per unit time, 1099
actual yield The measured amount of product obtained from a chemical reaction, 219
addition polymer(s) A synthetic organic polymer formed by directly joining monomer units, 528–534
addition reaction(s), of alkanes and alkenes, 507
adduct, acid-base, 837
adhesive force A force of attraction between molecules of two different substances, 620
adipic acid, 536
aerosol, 684
air, components of, 567t
 density, 561
 pollution, 413
 radioactive carbon in, 1102
 separation, 33
air bag, reaction in, 547
alcohol(s) Any of a class of organic compounds characterized by the presence of a hydroxyl group bonded to a saturated carbon atom, 512–516
 common, 512t
 naming, 514
 reactions of, 516, 517
aldehyde(s) Any of a class of organic compounds characterized by the presence of a carbonyl group, in which the carbon atom is bonded to at least one hydrogen atom, 517
 common, 518t

first-order reaction, 710–712
half-life and, 717–719
radioactive decay as, 1099
Fischer, Emil, 1110
fission The highly exothermic process by which very heavy nuclei split to form lighter nuclei, 1097, 1106–1109
flotation, *33,* 1051
fluorapatite, 1016
fluorine, electron affinity, 382
electronegativity , 424
hydrogen bonding and, 601
ionic radius, 385
fluorite, *49,* 651
fluorspar, 1014
foam, 684
fool's gold, *901,* 1030
force The physical quantity that causes a change in the motion of a mass if it is free to move, 549
intermolecular. *See* intermolecular forces
formal charge The charge on an atom in a molecule or ion calculated by assuming equal sharing of the bonding electrons, 427
formaldehyde, 518t
Lewis dot structure, 407
pi bonds in, 470
synthesis, 789
formation, standard molar enthalpy of, 288–293
standard molar free energy of, 930
formation constant An equilibrium constant for the formation of a complex ion, 905
formic acid, 520t, *807,* 823
in water, equilibrium constant expression for, 756, 757
reaction with sodium hydroxide, 853
formula, condensed, 498
empirical, 137, 223–228
molecular. *See* molecular formula
structural, 110, 498
formula unit The smallest collection of atoms that represents the ratio of elements in an ionic or molecular compound, 122
formula weight, 132
fossil fuels, 298
Franklin, Benjamin, 61
Franklin, Rosalind, 3
Frasch process, for sulfur mining, 1031
free energy. *See* Gibbs free energy
free energy change (ΔG), 928
equilibrium constant and, 935–940
temperature and, 933–935
free radical(s) A neutral atom or molecule containing an unpaired electron, 413
in addition polymerization, 529–530
freezing point depression, 675–677
constant for (K_{fp}), 673t, 675
frequency (ν) The number of complete waves passing a point in a given amount of time, 313
FRH (Flameless Ration Heater), 283
Frisch, Otto, 1106, 1110
froth flotation, *33,* 1051
fuel cell A voltaic cell in which reactants are continuously added, 979–980
efficiency of, 999
Fuller, R. Buckminster, 106
fullerenes, 108
fulminate ion, 456
functional group A structural fragment found in all members of a class of compounds, 110
fusion, heat of, 271
fusion, nuclear The highly exothermic process by which comparatively light nuclei combine to form heavier nuclei, 1097, 1109–1111

galena, *901,* 1030
Galilei, Galileo, 59
gallium arsenide, *245*
Galvani, Luigi, 950
galvanic cell(s), 950
galvanized iron, 984
gamma ray(s) Electromagnetic radiation having energies greater than those of x-rays, emitted by certain radioactive substances, 62, 1088
gangue A mixture of sand and clay in which a desired mineral is usually found, 1049
gas(es) The phase of matter in which a substance has no definite shape and a volume determined only by the size of its container, 17
behavior of, 546–592
laws governing, 550–563
collection over water, 570–572
compressibility of, 550–552
concentration, effect on equilibrium position, 777
density, calculation from ideal gas law, 560
ideal, 557
kinetic-molecular theory of, 572–576
mixtures of, partial pressures in, 567–572
noble. See noble gas(es)
nonideal, 581–583
pressure of, 547–550
in equilibrium constant expression, 755
properties of, 547–550
solubility in water, 609t
speeds of molecules in, 574–576
supercritical, 619
gas constant (R) The proportionality constant in the ideal gas law, 0.082057 L · atm/mol · K or 8.314510 J/mol · K, 557
in Maxwell's equation, 576
in osmotic pressure equation, 681
gas laws, 550–563
applications of, 578–580
gas-forming reaction(s), 178, 186
Gay-Lussac, Joseph, 555
Gay-Lussac's law Volumes of gases (at the same temperature and pressure) combine with each other in the ratio of small whole numbers, 555
Geiger, Hans, 68, 1100
Geiger counter, 1099, *1100*
gel, 684
gems, aluminum oxide in, 1018
colors of, transition elements and, 1042
general gas law An equation derived from the ideal gas law that allows calculation of pressure, temperature, and volume when a given amount of gas undergoes a change in conditions, 558
units in, 559
geometric isomers Isomers in which the atoms of the molecule are arranged in different geometric relationships, 1061–1063
geometry, molecular, 434
geothermal energy, 298
Germer, L. H., 333
Gibbs, J. Willard, *927*
Gibbs free energy (G) A thermodynamic state function relating enthalpy, temperature, and entropy, 927
gibbsite, 202
Gillespie, Ronald J., 430
glass, as solution, 655
energy use in production, 299
fibers in composite plastics, 538
glassware, laboratory, *40*
Glauber's salt, 143t
glucose, enzymatic degradation, 783
hydrogen sulfide as energy source for formation, 946
oxidation of, 999
structure, *28*
glycerol, 514–515
glycinate ion, as ligand, 1082
glycine, 113, 542
gold, accumulation by bacteria, 27
Goodyear, Charles, 533
Graf Zeppelin, 57
Graham, Thomas, 576, 684
Graham's law The rates of effusion of two gases at the same temperature and pressure are inversely proportional to the square roots of their molar masses, 577
gram (g), 41
graphite, fibers in composite plastics, 538
structure, 104, *105,* 495
Grasselli Brown, Jeanette, 544–545
greenhouse effect, 908
ground state The state of an atom in which all electrons are in the lowest possible energy levels, 328
group(s) The vertical columns in the periodic table, 80
functional, 110
ion charge and, 115
Group 1A elements, 82, 1003–1005, 1011. *See also* alkali metal(s)

sulfurous acid, 845
sulfuryl chloride, 791
supercritical gas, 619
superoxides, 141, 483, 1013
superphosphate fertilizer, 805
supersaturated solution(s) A solution that temporarily contains more than the saturation amount of solute, 659
reaction quotient in, 890
surface area, of colloid, 685
reaction rates and, 704
surface tension The energy required to disrupt the surface of a liquid, 620
of water, 604t
reduction by detergent, 689
surfactants, 688
surroundings Everything outside the system in a thermodynamic process, 274
sweat, evaporation of, 615
system The substance(s) of primary concern in a thermodynamic process, 274
systematic name, alkanes, 501
alkenes and alkynes, 503
Système International d'Unités, 35
Szilard, Leo, 1106

talc, 1014
tartaric acid, 186, 521t
technetium-99m, medical imaging with, 1116
temperature A physical property that determines the direction of spontaneous heat transfer between objects, 22–24, 261
calorimetry and, 293–297, 919
critical, 618
density of water and, 606–607
direction of spontaneous reaction and, 926t, 928
equilibrium constant and, 772–774
free energy and, 933–935
gas, volume and, 553–554
heat and, 261
kinetic energy of gas molecules and, 572
kinetic-molecular theory and, 20
reaction direction and, 926t
reaction rate and, 724–727
scales, 22–24
solubility and, 666–668
vapor pressure and, 616
termolecular process, 728
tertiary alcohols, 514
tertiary amines, 526
Terylene, 536
tetrafluoroethylene, 577
tetragonal unit cell, *624*
tetrahedral molecular geometry, *431,* 432, *468,* 1061
orbital hybridization and, 460
thallium(I) chloride, density, 649
theoretical yield The amount of product theoretically obtainable from the given amounts of reactants, 219
theory A unifying principle that explains a body of facts and the laws based on them, 7
atomic. *See* atomic theory of matter
kinetic-molecular, 18, 595
molecular orbital, 476–486, 1067
quantum. *See* wave mechanics
special relativity, binding energy and, 1096
valence bond, 458–475
thermal conductivity, 267
thermal energy. *See* heat
thermite reaction The production of molten iron by reduction of iron(III) oxide with aluminum, *196,* 934
thermodynamic equilibrium constant, 936
thermodynamics The science of heat or energy flow in chemical reactions, 258
first law of, 277, 940
second law of, 923, 940
third law of, 919, 940
time and, 940–941
versus kinetics, 915
thermometer, 261
thermoplastic polymer(s) A polymer that softens but is unaltered on heating, 527
thermosetting polymer(s) A polymer that degrades or decomposes on heating, 527
third law of thermodynamics The entropy of a pure, perfectly formed crystal at 0 K is zero, 919, 940
Thomson, Joseph John, 65, 67, 1087
Thomson, William (Lord Kelvin), 23, 553
thorium, 1089, 1112
thunderstorm, heat released by, 280
thymine, 113
thymol blue, 875
time, thermodynamics and, 940–941
tin iodide, *139*
titanium, production of, 1079
titanium(IV) oxide, 1042
titanium tetrachloride, preparation, 217, 945
reaction with water, *246*
titration(s) A procedure for the quantitative analysis of a substance by means of an essentially complete reaction in solution with a reagent of known concentration
acid-base, 238–242
curves for, 867–874
polyprotic acid or base, 872–874
strong acid–strong base, 867–869
weak acid–strong base, 869–872
oxidation-reduction, 242–244
toluene, 508
tonicity, 683
torr A unit of pressure equivalent to one millimeter of mercury, 548
Torricelli, Evangelista, 548
total reaction order The sum of the exponents of all the concentration terms in the reaction's rate equation, 706
tourmaline, *52*
transition element(s), 80, 83, 1040–1082
biological importance, 1042
commercial production, 1048–1053
densities of, 1047
electron configurations of, 373, 1043
properties of, 1041–1048
transmutation, 1089
artificial, 1104
transuranium elements, names of, 1084
preparation of, 1104, 1106
trichloroacetic acid, 151
triclinic unit cell, *624*
trigonal planar molecular geometry, *431,* 432, *468*
orbital hybridization and, 463
trigonal unit cell, *624*
trimethylamine, 526, 809, 845, 879
trimethylborane, 791
trinitrotoluene (TNT), 150
triple bond A bond formed by sharing three pairs of electrons, one pair in a sigma bond and the other two in pi bonds, 470
triple point The temperature and pressure at which the solid, liquid, and vapor phases of a substance are in equilibrium, 642
tritium, 73
half-life of, 1098
trona, 1013
T-shaped molecular geometry, *438*

ultraviolet catastrophe, 319
ultraviolet radiation, absorption by ozone, 738
ultraviolet spectroscopy, 1075
uncertainty principle, 336–337
unimolecular process, 728
unit cell(s) The smallest repeating unit in a crystal lattice, 622
shapes of, *624*
unsaturated compound(s) A hydrocarbon containing double or triple carbon–carbon bonds, 502
unsaturated solution(s) A solution in which the concentration of solute is less than the saturation amount, 659
uranium, decay to neptunium, 1089
decay to thorium, 1089
fission of, 1106
isotopes, enrichment of, 1109
separation by gas effusion, 580
radioactive series from, *1091*
uranium(IV) oxide, formula, 154
urea, Lewis dot structure, 407

valence band, 486
valence bond theory A model of bonding in which a bond arises from the overlap of atomic orbitals on two atoms to give a bonding orbital with electrons localized between the atoms, 458–475
valence electron(s) The outermost and most